Standard Atomic Weights of the Elements 2009, IUPAC

Based on Relative Atomic Mass of $^{12}C = 12$, where ^{12}C is a neutral atom in its nuclear and electronic ground state.[1]

Name	Symbol	Atomic Number	Atomic Weight
Actinium[2]	Ac	89	(227)
Aluminum	Al	13	26.981 5385(7)
Americium[2]	Am	95	(243)
Antimony	Sb	51	121.760(1)
Argon	Ar	18	39.948(1)
Arsenic	As	33	74.921 595(6)
Astatine[2]	At	85	(210)
Barium	Ba	56	137.327(7)
Berkelium[2]	Bk	97	(247)
Beryllium	Be	4	9.012 1831(5)
Bismuth	Bi	83	208.980 40(1)
Bohrium[2]	Bh	107	(270)
Boron	B	5	10.811(7)
Bromine	Br	35	79.904(1)
Cadmium	Cd	48	112.414(4)
Calcium	Ca	20	40.078(4)
Californium[2]	Cf	98	(251)
Carbon	C	6	12.0107(8)
Cerium	Ce	58	140.116(1)
Cesium	Cs	55	132.905 451 96(6)
Chlorine	Cl	17	35.453(2)
Chromium	Cr	24	51.9961(6)
Cobalt	Co	27	58.933 194(4)
Copernicium[2]	Cn	112	(285)
Copper	Cu	29	63.546(3)
Curium[2]	Cm	96	(247)
Darmstadtium[2]	Ds	110	(281)
Dubnium[2]	Db	105	(268)
Dysprosium	Dy	66	162.500(1)
Einsteinium[2]	Es	99	(252)
Erbium	Er	68	167.259(3)
Europium	Eu	63	151.964(1)
Fermium[2]	Fm	100	(257)
Flerovium[2]	Fl	114	(287)
Fluorine	F	9	18.998 403 163(6)
Francium[2]	Fr	87	(223)
Gadolinium	Gd	64	157.25(3)
Gallium	Ga	31	69.723(1)
Germanium	Ge	32	72.64(1)
Gold	Au	79	196.966 569(5)
Hafnium	Hf	72	178.49(2)
Hassium[2]	Hs	108	(277)
Helium	He	2	4.002 602(2)
Holmium	Ho	67	164.930 33(2)
Hydrogen	H	1	1.00794(7)
Indium	In	49	114.818(3)
Iodine	I	53	126.904 47(3)
Iridium	Ir	77	192.217(3)
Iron	Fe	26	55.845(2)
Krypton	Kr	36	83.798(2)
Lanthanum	La	57	138.905 47(7)
Lawrencium[2]	Lr	103	(262)
Lead	Pb	82	207.2(1)
Lithium	Li	3	[6.941(2)][†]
Livermorium[2]	Lv	116	(293)
Lutetium	Lu	71	174.9668(1)
Magnesium	Mg	12	24.3050(6)
Manganese	Mn	25	54.938 044(3)
Meitnerium[2]	Mt	109	(278)
Mendelevium[2]	Md	101	(258)
Mercury	Hg	80	200.59(2)
Molybdenum	Mo	42	95.95(1)
Neodymium	Nd	60	144.242(3)
Neon	Ne	10	20.1797(6)
Neptunium[2]	Np	93	(237)
Nickel	Ni	28	58.6934(4)
Niobium	Nb	41	92.906 37(2)
Nitrogen	N	7	14.0067(2)
Nobelium[2]	No	102	(259)
Osmium	Os	76	190.23(3)
Oxygen	O	8	15.9994(3)
Palladium	Pd	46	106.42(1)
Phosphorus	P	15	30.973 761 998(5)
Platinum	Pt	78	195.084(9)
Plutonium[2]	Pu	94	(244)
Polonium[2]	Po	84	(209)
Potassium	K	19	39.0983(1)
Praseodymium	Pr	59	140.907 66(2)
Promethium[2]	Pm	61	(145)
Protactinium[2]	Pa	91	231.035 88(2)
Radium[2]	Ra	88	(226)
Radon[2]	Rn	86	(222)
Rhenium	Re	75	186.207(1)
Rhodium	Rh	45	102.905 50(2)
Roentgenium[2]	Rg	111	(281)
Rubidium	Rb	37	85.4678(3)
Ruthenium	Ru	44	101.07(2)
Rutherfordium[2]	Rf	104	(267)
Samarium	Sm	62	150.36(2)
Scandium	Sc	21	44.955 908(5)
Seaborgium[2]	Sg	106	(271)
Selenium	Se	34	78.971(8)
Silicon	Si	14	28.0855(3)
Silver	Ag	47	107.8682(2)
Sodium	Na	11	22.989 769 28(2)
Strontium	Sr	38	87.62(1)
Sulfur	S	16	32.065(5)
Tantalum	Ta	73	180.947 88(2)
Technetium[2]	Tc	43	(98)
Tellurium	Te	52	127.60(3)
Terbium	Tb	65	158.925 35(2)
Thallium	Tl	81	204.3833(2)
Thorium[2]	Th	90	232.0377(4)
Thulium	Tm	69	168.934 22(2)
Tin	Sn	50	118.710(7)
Titanium	Ti	22	47.867(1)
Tungsten	W	74	183.84(1)
Uranium[2]	U	92	238.028 91(3)
Vanadium	V	23	50.9415(1)
Xenon	Xe	54	131.293(6)
Ytterbium	Yb	70	173.054(5)
Yttrium	Y	39	88.905 84(2)
Zinc	Zn	30	65.38(2)
Zirconium	Zr	40	91.224(2)
—[2,3]		113	(286)
—[2,3]		115	(289)
—[2,3]		117	(293)
—[2,3]		118	(294)

1. The atomic weights of many elements vary depending on the origin and treatment of the sample. This is particularly true for Li; commercially available lithium-containing materials have Li atomic weights in the range of 6.939 and 6.996. Uncertainties are given in parentheses following the last significant figure to which they are attributed.
2. Elements with no stable nuclide; the value given in parentheses is the atomic mass number of the isotope of longest known half-life. However, three such elements (Th, Pa, and U) have a characteristic terrestrial isotopic composition, and the atomic weight is tabulated for these.
3. Not yet named.

FIFTH EDITION

Chemistry

THE MOLECULAR SCIENCE

John W. Moore
University of Wisconsin–Madison

Conrad L. Stanitski
Franklin and Marshall College

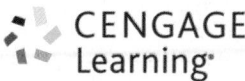

CENGAGE Learning

Australia • Brazil • Mexico • Singapore • United Kingdom • United States

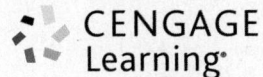

CENGAGE Learning

Chemistry: The Molecular Science,
Fifth Edition
John W. Moore, Conrad L. Stanitski

Product Director: Mary Finch
Product Manager: Maureen Rosener
Managing Developer: Peter McGahey
Associate Content Developer: Brendan Killion
Product Assistant: Karolina Kiwak
Media Developer: Rebecca Berardy Schwartz
Marketing Director: Nicole Hamm
Content Project Manager: Teresa L. Trego
Art Director: Maria Epes
Manufacturing Planner: Judy Inouye
Rights Acquisitions Specialist: Dean Dauphinais
Production and Composition: Graphic World Inc.
Photo Researcher: PreMedia Global
Text Researcher: Pablo D'Stair
Copy Editor: Graphic World Inc.
Illustrator: Graphic World Inc.
Text Designer: Delgado and Company
Cover Designer: Lee Friedman Studios
Cover Image: John P Kelly/The Image Bank/ Getty Images

Library of Congress Control Number: 2013952142

Student Edition:
ISBN-13: 978-1-285-19904-7
ISBN-10: 1-285-19904-9

Loose-leaf Edition:
ISBN-13: 978-1-305-25668-2
ISBN-10: 1-305-25668-9

Cengage Learning
200 First Stamford Place, 4th Floor
Stamford, CT 06902
USA

Cengage Learning is a leading provider of customized learning solutions with office locations around the globe, including Singapore, the United Kingdom, Australia, Mexico, Brazil, and Japan. Locate your local office at **www.cengage.com/global.**

Cengage Learning products are represented in Canada by Nelson Education, Ltd.

To learn more about Cengage Learning Solutions, visit **www.cengage.com.**

Purchase any of our products at your local college store or at our preferred online store **www.cengagebrain.com.**

Printed in the United States of America
2 3 4 5 18 17 16 15

To All Students of Chemistry

We intend that this book will help you to discover that chemistry is relevant to your lives and careers, full of beautiful ideas and phenomena, and of great benefit to society. May your study of this fascinating subject be exciting, successful, and fun!

We thank our wives—Betty (JWM) and Barbara (CLS)—for their patience, support, understanding, and love.

It does not do harm to the mystery
to know a little more about it.
—Richard Feynman

About the Authors

© Dr. Donal R. Neu

© Dr. Donal R. Neu

John W. Moore received an A.B. magna cum laude from Franklin and Marshall College and a Ph.D. from Northwestern University. He held a National Science Foundation (NSF) postdoctoral fellowship at the University of Copenhagen and taught at Indiana University and Eastern Michigan University before joining the faculty of the University of Wisconsin–Madison in 1989. At the University of Wisconsin, Dr. Moore is W. T. Lippincott Professor of Chemistry and Director of the Institute for Chemical Education. He was Editor of the *Journal of Chemical Education* from 1996 to 2009. Among his many awards are the American Chemical Society (ACS) George C. Pimentel Award in Chemical Education, the James Flack Norris Award for Excellence in Teaching Chemistry, and the CMA CATALYST National Award for Excellence in Chemistry Teaching. He is a Fellow of the ACS and of the American Association for the Advancement of Science (AAAS). He has won two major awards from the University of Wisconsin: the Wisconsin Power and Light Underkofler Award for Excellence in Teaching (1995) and the Benjamin Smith Reynolds Award for excellence in teaching chemistry to engineering students (2003). Dr. Moore has received a series of major grants from the NSF to support development of online chemistry learning materials for the ChemEd DL and the National Science Distributed Learning (NSDL) initiative.

Conrad L. Stanitski is currently a Visiting Scholar at Franklin and Marshall College and is Distinguished Emeritus Professor of Chemistry at the University of Central Arkansas. He received his B.S. in Science Education from Bloomsburg State College, M.A. in Chemical Education from the University of Northern Iowa, and Ph.D. in Inorganic Chemistry from the University of Connecticut. He has co-authored chemistry textbooks for science majors, allied health science students, nonscience majors, and high school chemistry students. Among Dr. Stanitski's many awards are the American Chemical Society George C. Pimentel Award in Chemical Education, the CMA CATALYST National Award for Excellence in Chemistry Teaching, the Gustav Ohaus–National Science Teachers Association Award for Creative Innovations in College Science Teaching, the Thomas R. Branch Award for Teaching Excellence, the Samuel Nelson Gray Distinguished Professor Award from Randolph-Macon College, and the 2002 Western Connecticut American Chemical Society Section Visiting Scientist Award. He was Chair of the American Chemical Society Division of Chemical Education (2001) and has been an elected Councilor for that division. He is a Fellow of the American Association for the Advancement of Science (AAAS). An instrumental and vocal performer, he also enjoys jogging, tennis, rowing, and reading.

Brief Contents

Contents

1 The Nature of Chemistry 1

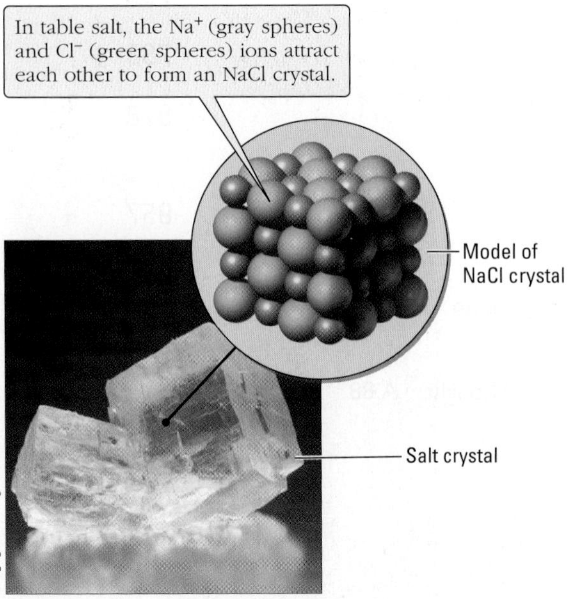

In table salt, the Na^+ (gray spheres) and Cl^- (green spheres) ions attract each other to form an NaCl crystal.
Model of NaCl crystal
Salt crystal
Cengage Learning/Charles D. Winters

2 Chemical Compounds 42

3 Chemical Reactions 91

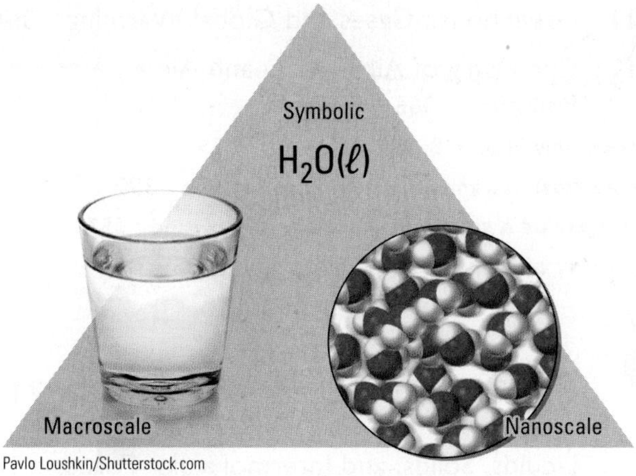
Symbolic
$H_2O(\ell)$
Macroscale
Nanoscale
Pavlo Loushkin/Shutterstock.com

4 Energy and Chemical Reactions 149

5 Electron Configurations and the Periodic Table 190

6 Covalent Bonding 241

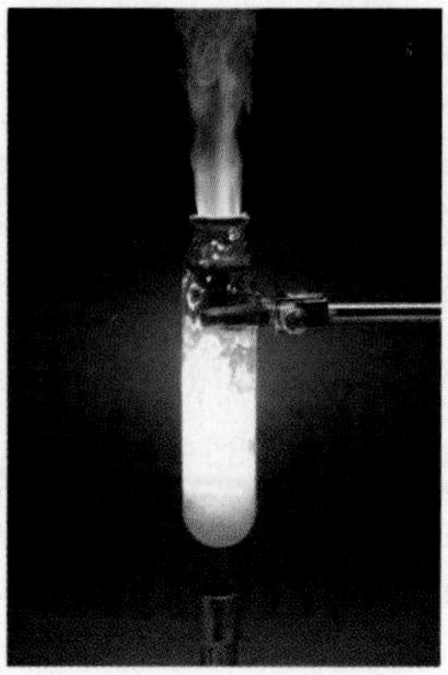
Cengage Learning/Charles D. Winters

I_2 vapor
Mixture of Al + I_2
© Cengage Learning/ James Maynard

7 Molecular Structures 284

8 Properties of Gases 326

9 Liquids, Solids, and Materials 371

10 Fuels, Organic Chemicals, and Polymers 422

11 Chemical Kinetics: Rates of Reactions 475

12 Chemical Equilibrium 524

CO2
coral, CaCO3
HCl(aq)

Cengage Learning/Charles D. Winters

© Cengage Learning/James Maynard

13 The Chemistry of Solutes and Solutions 564

14 Acids and Bases 606

15 Additional Aqueous Equilibria 653

16 Thermodynamics: Directionality of Chemical Reactions 694

17 Electrochemistry and Its Applications 738

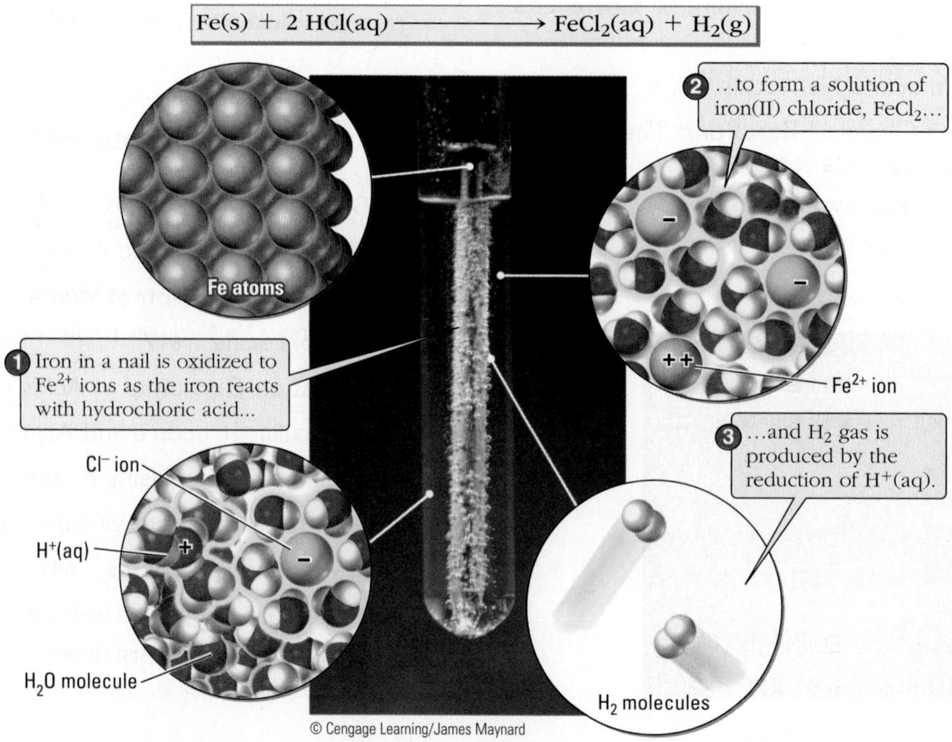

Fe(s) + 2 HCl(aq) → FeCl2(aq) + H2(g)
Fe atoms
1 Iron in a nail is oxidized to Fe2+ ions as the iron reacts with hydrochloric acid...
2 ...to form a solution of iron(II) chloride, FeCl2...
3 ...and H2 gas is produced by the reduction of H+(aq).
Fe2+ ion
Cl− ion
H+(aq)
H2O molecule
H2 molecules

© Cengage Learning/James Maynard

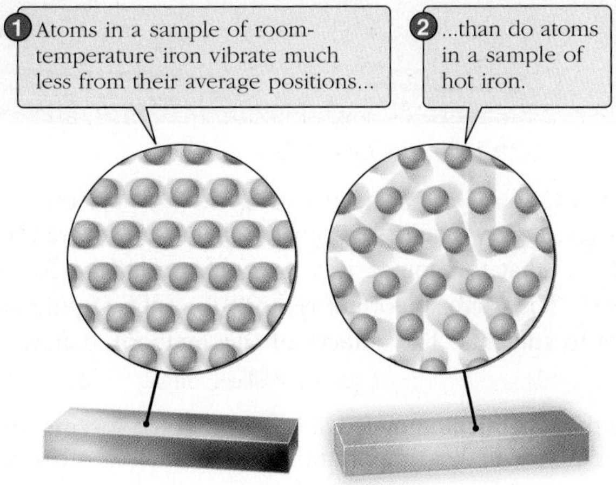
1 Atoms in a sample of room-temperature iron vibrate much less from their average positions...
2 ...than do atoms in a sample of hot iron.

Preface

Students have many reasons for taking a two-semester general chemistry course for science majors, but the most likely is that the course is a pre- or co-requisite for other science-related courses or careers. There are important reasons for such requirements, but they are not always obvious to students. The authors of this textbook believe very strongly that

- students need to recognize that chemical knowledge is essential for solving important problems and that chemistry makes important contributions to other disciplines; and
- it is essential that students gain a working knowledge of how chemistry principles are applied to solve problems in a broad spectrum of applications.

Examples of such applications are creating new and improving existing chemical pathways that lead to the more efficient synthesis of new pharmaceuticals; developing a deeper understanding of alternative energy sources to mitigate global warming; and understanding how new, more efficient catalysts could help to decrease air pollution and to minimize production of chemical waste from industrial processes. Knowledge of chemistry provides a way of interpreting macroscale phenomena at the molecular level that can be applied to many critical 21st-century problems, including those just given. This fifth edition of *Chemistry: The Molecular Science* continues our tradition of integrating other sciences with chemistry and has been updated to include a broad range of recent chemical innovations that illustrate the importance of multidisciplinary science.

Goals

Our overarching goal is to involve science and engineering students in active study of what modern chemistry is, how it applies to a broad range of disciplines, and what effects it has on their own lives. We maintain a high level of rigor so that students in mainstream general chemistry courses for science majors and engineers will learn the concepts and develop the problem-solving skills essential to their future ability to use chemical ideas effectively. We have selected and carefully refined the book's many unique features in support of this goal.

More specifically, we intend that this textbook will help students develop

- a broad overview of chemistry and chemical reactions;
- an understanding of the most important concepts and models used by chemists and scientists in chemistry-related fields;
- the ability to apply the facts, concepts, and models of chemistry appropriately to new situations in chemistry, to other sciences and engineering, and to other disciplines;
- knowledge of the many practical applications of chemistry in other sciences, in engineering, and in other fields;
- an appreciation of the many ways that chemistry affects the daily lives of all people, students included; and
- motivation to study in ways that help all students achieve real learning that results in long-term retention of facts and concepts, and how to apply them.

Because modern chemistry is inextricably entwined with so many other disciplines, we have integrated organic chemistry, biochemistry, environmental chemistry, industrial chemistry, and materials chemistry into the discussions of chemical principles and facts.

Applications in these areas are discussed together with the principles on which they are based. This approach serves to motivate students whose interests lie in related disciplines and also gives a more accurate picture of the multidisciplinary collaborations so prevalent in contemporary chemical research and modern industrial chemistry.

Audience

Chemistry: The Molecular Science, Fifth Edition, is intended for mainstream general chemistry courses for students who expect to pursue further study in science, engineering, or science-related disciplines. Those planning to major in chemistry, biochemistry, biological sciences, engineering, geological sciences, environmental science, agricultural sciences, materials science, physics, and many related areas will benefit from this book and its approach. The book has an extensive glossary and an excellent index, making it especially useful as a reference for study or review for standardized examinations, such as the MCAT.

We assume that the students who use this book have a basic foundation in mathematics (algebra and geometry) and in general science. Almost all will also have had a chemistry course before coming to college. The book is suitable for the typical two-semester sequence of general chemistry, and it has also been used quite successfully in a one-semester accelerated course that presumes students have a strong background in chemistry and mathematics.

New in This Edition

This fifth edition of *Chemistry: The Molecular Science* has undergone major revisions—far more extensive than any of the previous editions, even though each of them was significantly revised. Specifically, we have made these major changes from the fourth edition:

- Evaluated all chapter-end Questions for Review and Thought and in-chapter Problem-Solving Examples and Exercises with regard to conceptual level, using Bloom's taxonomy of educational objectives and aiming for comprehensive coverage of topics and strong development of students' conceptual abilities;
- Based on the assessment of conceptual level, culled lower-level and redundant end-of-chapter questions and added 328 new questions, many of which require higher-level thinking;
- Based on the conceptual-level evaluation and modifications of chapter content, replaced or revised 17 Problem-Solving Examples, 29 Problem-Solving Practice problems, and 60 Exercises;
- Revised all Problem-Solving Examples to include explicitly our problem-solving strategy (analyze, plan, execute, and check) so that students have clear guidance in how to approach and solve problems;
- Combined Chapters 1 to 5 from the fourth edition into Chapters 1 to 3 in this edition, re-ordering and modernizing the content to improve clarity and aid learning;
- Significantly revised all other chapters; in particular, Chapters 8, 9, and 17 (fifth edition) had sections rewritten, removed, added, or re-ordered;
- Revised our award-winning art program to better suit today's visually oriented students, greatly increasing the use of text and pointers to draw students' attention to the important information in each figure;
- Enhanced our emphasis on applications of chemistry in other sciences and in daily life by incorporating suggestions of an Applications Advisory Board into the text of most chapters;
- Began each chapter with an engaging photo accompanied by a series of related questions that are designed to pique students' interest and that are answered in the chapter;

- Continued to use published research as a guide, for example, in changing notation for the solvated proton from H_3O^+(aq) to H^+(aq) (see *Journal of Chemical Education*, Vol. 88, No. 7, p. 875, 2011);
- Removed from the printed text two features, *Chemistry in the News* and *Chemistry You Can Do,* making them available online;
- Developed many new *Chemistry in the News* items;
- Updated all *Tools of Chemistry* features and added new *Portrait of a Scientist* items to include more that feature members of groups underrepresented in science;
- Updated thermodynamic notation to reflect IUPAC conventions (such as $\Delta_r H°$ for reaction enthalpy change) and revised unit conventions for reaction enthalpies and Gibbs free energies to include units per mole.

A hallmark of this book is its emphasis on conceptual understanding as opposed to memorization or rote answering of questions. To enhance this approach, Dr. Kristin Briney evaluated the conceptual level of every chapter's content and every set of chapter-end problems. She categorized all problems using Bloom's Taxonomy of educational objectives, identifying areas where new conceptual problems were needed and determining whether all concepts included in each chapter were assessed by chapter-end problems. Based on this analysis, we have written new chapter-end questions, exercises, and problem-solving practice problems. Where Dr. Briney identified redundancies in chapter-end questions, we deleted questions to make room for the newly written items. This analysis has provided excellent guidance and greatly enhanced an already strong feature of the book.

Our revision of Chapters 1 to 5 (which are now Chapters 1 to 3) involved far more than simply combining the content of five chapters and reapportioning it into three. For example, material about the periodic table from Chapters 2 and 3 is now in Chapter 1, where it helps students to see how the field of chemistry is organized. We also modernized our approach to chemical reactions, categorizing them as precipitation, acid-base, gas forming, and redox with less emphasis on combination, exchange, etc. Combining Chapters 1 to 5 also reduced their length from 210 pages to 169 pages—a welcome change in this era of long, expensive textbooks.

The art in this textbook has been a strong point from the first edition, which won an award for excellence. In this fifth edition every figure and photograph has been re-examined to enhance its pedagogical impact. Related figures have been combined; diagrams that were too complex have been broken into two separate figures. We continue to have many figures that emphasize the macroscopic, symbolic, and nanoscale views of chemical processes, and we have increased our use of balloon text and pointers to emphasize what students should be looking at (see p. xxv). Often this has allowed us to shorten the written text, bearing out the adage that a picture is worth 1000 words. These enhanced figures are also very important in the online version of this text, consolidating text and images so that students have less reading on a computer screen.

To support our emphasis on developing students' ability to approach problems systematically and logically, we have revised all Problem-Solving Examples with subheadings to remind students that they should analyze the problem, plan a solution, execute the plan, and check that the result is reasonable. This strategy is also emphasized in the solutions manuals developed by Dr. Judy Ozment. We have added 328 new questions at the ends of the chapters and have paid special attention to the section headed More Challenging Questions, to which we have added a significant number of new questions in nearly every chapter. To help students develop their conceptual problem-solving skills, we have created 29 new Problem-Solving Practice problems and 60 new Exercises.

Our emphasis on applications of chemistry in other disciplines and in daily life has been enhanced through many suggestions from an Applications Advisory Board. We have incorporated the board's recommendations in the text. We have also carefully selected chapter-opening photographs that relate to each chapter's concepts; in the captions of these photos we ask questions about what is shown that are designed to pique students' interest in the chapter; all such questions are answered later in the chapter.

In addition to these global changes, revisions specific to each chapter include

Chapter 1

- Replaced Section 1-2 with an entirely new example of how chemistry is applied to societal problems;
- Added five new questions about real-world situations that are answered later in the book;
- Revised or replaced three Exercises and one Problem-Solving Practice problem;
- Added, revised, or replaced 12 figures;
- Added a new section, Measurements, Units, and Calculations, that includes SI units and significant digits;
- Emphasized a general approach to solving problems and demonstrated how to apply it to a specific problem;
- Added two new sections, The Periodic Table and The Biological Periodic Table; and
- Added 15 new end-of-chapter questions, 10 of which are More Challenging Questions.

Chapter 2

- Combined content from fourth-edition Chapters 2 and 3;
- Deleted sections on units and significant figures, which are now in Chapter 1;
- Changed order of presentation so that ions and ionic compounds precede molecular compounds;
- Enhanced discussion of identifying ionic and molecular compounds and relating compound type to properties;
- Juxtaposed discussion of amount of substance, molar mass, composition and formulas, and determining formulas;
- Added, revised, or replaced 12 figures;
- Revised or replaced seven Exercises and eight Problem-Solving Practice problems; and
- Added 40 new end-of-chapter questions and deleted 137 that were in the fourth edition Chapters 2 and 3.

Chapter 3

- Combined content from fourth-edition Chapters 4 and 5 in this chapter;
- Reorganized so the order is now chemical equations and balancing; precipitation, acid-base, and redox reactions; stoichiometry, limiting reactant, percent yield, and formula from mass composition; and solution stoichiometry;
- De-emphasized nomenclature of combination, decomposition, displacement, and exchange reactions;
- Revised or replaced 11 Exercises and six Problem-Solving Practice problems;
- Added, revised, or replaced 20 figures; and
- Added 13 new end-of-chapter questions; deleted 131 questions that were in the fourth edition Chapters 4 and 5.

Chapter 4

- Revised or replaced eight figures; deleted five figures;
- Revised or replaced seven Exercises and two Problem-Solving Practice problems;
- Added subsection on power and unit W;
- Added new Section 4-3, Keeping Track of Energy Transfers;

- Deleted Section 6.5, Thermochemical Expressions, and incorporated material into Section 4-5;
- Consolidated Sections 6.11 and 6.12 into new Section 4-11; reduced coverage of food and fuel slightly;
- Replaced an *Estimation* feature; and
- Added seven new end-of-chapter questions; deleted 58 questions.

Chapter 5

- Revised or replaced 25 figures; deleted 3 figures;
- Revised material on waves and moved it before electromagnetic radiation;
- Revised or replaced four Exercises;
- Added Section 5-5e to separate discussion of one-electron atoms/ions from discussion of atoms/ions with more than a single electron;
- Revised discussion of quantum numbers to emphasize electron shells and placed summary at end;
- Moved discussion of paramagnetism before section on ion electron configurations so that paramagnetic experimental results could be used in both atoms and ions; and
- Expanded discussion of effective nuclear charge and improved figures showing periodic trends;
- Explicitly introduced the idea that radii are based on the assumption that atoms and ions are spherical;
- Added one end-of-chapter learning goal and significantly modified several others; and
- Added 30 new end-of-chapter questions; deleted 43 questions.

Chapter 6

- Revised or replaced 11 figures;
- Revised or replaced two Problem-Solving Practice problems;
- Revised Section 6-1 extensively, explicitly defining valence bond theory;
- Added a new Problem-Solving Example;
- Revised tables showing formal charges to better relate formal charge to structure;
- Revised material on molecular orbital (MO) theory to better show formation of MOs, to better define bond order, and to more clearly show formation of sigma and pi orbitals;
- Revised discussion of electron delocalization by deleting nitrate and expanding ozone discussion;
- Revised several learning goals and added seven key terms; and
- Added 20 new end-of-chapter questions; deleted 31 questions.

Chapter 7

- Revised or replaced 17 figures;
- Major revision of Section 7-1 involved deleting some text and replacing it with balloon text in figures;
- Revised discussion of VSEPR to explain more explicitly how electron-region geometry defines molecular geometry;
- Moved discussion of hydrocarbons with many C atoms to section on molecules with more than a single central atom;
- Revised Sections 7-3a, 7-3b, and 7-3c to reduce text length and use text in figures to explain hybrid orbital formation;
- Introduced figures with balloon text to explain sideways overlap of *p* orbitals and to show sigma framework and pi framework in multiple bonds;

- Revised text and figure to better explain alignment of polar molecules in electric field; added figure showing why CO_2 is nonpolar and SO_2 is polar;
- Revised definition of hydrogen bond to agree with new IUPAC definition and to introduce the idea that there is a covalent component in a hydrogen bond;
- Included more questions about molecular geometry in Summary Problem; and
- Added 15 new end-of-chapter questions; deleted 17 questions.

Chapter 8

- Revised or added 14 figures;
- Revised or replaced seven Exercises and six Problem-Solving Practice problems;
- Revised or added three Problem-Solving Examples;
- Revised the introduction, which now notes similarity of physical behavior of gases, contrasting with the lack of similarity in solids and liquids;
- Revised the section on gas pressure;
- Used kinetic-molecular theory early but moved more detailed treatment to Section 8-7;
- Revised treatment of combined gas law to include n;
- Moved gas density and molar mass to directly follow the ideal gas law;
- Shifted Law of Combining Volumes into a revised section on gases in chemical reactions;
- Applied gas theory to new context: self-contained self-rescue breathing devices;
- Moved discussion of the atmosphere later in the chapter and used it to introduce an updated and enhanced discussion of stratospheric ozone depletion, greenhouse gases and climate change, and tropospheric air pollution; and
- Added 23 new end-of-chapter questions; deleted 35 questions.

Chapter 9

- Revised or replaced 24 figures;
- Revised or replaced four Exercises and six Problem-Solving Practice problems;
- Added three new Problem-Solving Examples;
- Wrote a new introduction that compares two important differences between properties of gases and properties of liquids and solids: molar volumes and intermolecular forces;
- Reorganized to increase coverage of intermolecular forces and to move material on surface tension, capillary action, and meniscuses to the section on unusual properties of water;
- Incorporated a new vapor pressure section that includes polar and nonpolar compounds; revised the discussion of the Clausius-Clapeyron equation;
- Separated vaporization/condensation from melting/freezing and sublimation/deposition; discussed critical T and P as properties of liquids, not just in phase diagrams;
- Moved discussion of types of solids earlier so that phase diagrams make more sense; moved phase diagram of water to section on special properties of water;
- Revised table showing types of solids to include images of example solids and diagrams of nanoscale structures;
- Reorganized the discussion of crystalline solids and calculations related to unit cells;
- Introduced Nobel Prize work of Daniel Shechtman on quasicrystals;
- Added new Estimation box;
- Revised the Summary Problem: Parts I and II are new;
- Added two new learning goals; and
- Added 22 new end-of-chapter questions; deleted 20 questions.

Chapter 10

- Revised or replaced 17 figures;
- Revised or replaced four Exercises and three Problem-Solving Practice problems;
- Updated one table;
- Updated information on reformulated gasolines, energy sources, recycling of plastics;
- Added a subsection on hydraulic fracturing (fracking);
- Reduced text in Section 10-1, letting figures tell the story of distillation, cracking, and reforming in petroleum refining;
- More clearly defined each organic functional group at the beginning of its section;
- Added discussion of planarity of peptide bond;
- Added one learning goal and deleted one that was low in Bloom's taxonomy;
- Added nine new Key Terms; and
- Added 11 new end-of-chapter questions (six more challenging); rewrote 25 questions.

Chapter 11

- Revised or replaced 16 figures;
- Revised or replaced one Exercise;
- Revised one Problem-Solving Example;
- Updated one table;
- Updated text description of reason for activation energy in Section 11-4; and
- Revised presentation of Step 3 in mechanism in Section 11-7 based on student questions;
- Updated information on catalyst usage in industry;
- Added one learning goal and revised three others; and
- Reduced the number of end-of-chapter questions by 37 (to 122).

Chapter 12

- Revised or replaced 12 figures;
- Revised one Problem-Solving Example;
- Updated one table;
- Revised example in Section 12-6a involving dissolution of carbonates; related it to ocean acidification and coral reefs;
- Revised Section 12-6b to provide a different example of changing solution volume;
- Added a completely new Summary Problem; and
- Reduced the number of end-of-chapter questions to 125 (from 158); reordered questions to better fit topic headings.

Chapter 13

- Revised or replaced 24 figures;
- Revised one Problem-Solving Example;
- Added table of freezing point and boiling point constants in Section 13-7;
- Reorganized Sections 1 to 3 to make better use of entropy as a means of explaining solubility and better delineate enthalpy and entropy of solution based on prior discussion in Chapter 12;
- Expanded discussion of solubility of ionic solids and repositioned it to precede solubility of gases;

- Revised discussion of softening water using ion exchange;
- Modified Summary Problem; and
- Added 16 new end-of-chapter questions; deleted nine questions.

Chapter 14

- Revised or replaced 25 figures; added five new figures;
- Revised or replaced three Exercises and one Problem-Solving Practice problem;
- Added new Section 14-6e on periodic variation of acid-base properties of oxides;
- To reinforce pedagogy, added color coding to section teaching how to solve equilibrium problems;
- Included method of successive approximations for solving equilibrium problems; emphasized 5% rule more explicitly;
- Added one learning goal;
- Added three new Key Terms; and
- Replaced 20 end-of-chapter questions with new ones more relevant to chapter topics and higher in Bloom's taxonomy.

Chapter 15

- Revised or replaced 25 figures; added one new figure to better explain buffers;
- Revised one Exercise;
- Revised two Problem-Solving Examples;
- Revised one table;
- Consolidated Sections 15-1e and 15-1f;
- Updated data on sulfur and nitrogen oxide emissions and distribution of acid rain;
- Added a second part to the Summary Problem; and
- Added 15 new end-of-chapter questions; modified 15 questions; deleted four questions.

Chapter 16

- Revised or replaced 17 figures;
- Revised Estimation box;
- Revised one Problem-Solving Example;
- Rewrote introduction to better accommodate new chapter-opening photo and caption;
- Defined third law of thermodynamics more explicitly;
- Shortened text in Section 16-3b because Figure 16.3 tells the story;
- Rewrote Section 16-8a to use electric cars as example of charge/discharge battery cycle;
- Updated correlations of end-of-chapter questions with learning goals;
- Added three new Key Terms; and
- Added one new end-of-chapter question; deleted 16 questions.

Chapter 17

- Revised or replaced 20 figures; added 11 new ones;
- Added two new Problem-Solving Examples; modified two others;
- Revised or replaced three Exercises and added five new Exercises; added two new Problem-Solving Practice problems;

- Added one new table;
- Added a new Estimation box and a new Portrait of a Scientist;
- Moved material on balancing redox equations to Appendix F;
- Moved section on neuron cells to online only;
- Added subsection and figure to explain shorthand cell notation;
- Wrote new introduction to using standard half-cell potentials;
- Revised pH meter section;
- Deleted discussion of mercury cells and non-alkaline dry cells; explained why NiCads need to be recycled;
- Added new material on lithium-ion batteries and plug-in hybrid electric vehicles;
- Replaced Summary Problem with a new one; and
- Added 12 new end-of-chapter questions; deleted five questions.

Chapter 18

- Revised or replaced eight figures; added three new figures;
- Revised or replaced four Exercises;
- Revised two tables;
- Revised Section 18-3b to define nuclear binding energy analogously to bond energy;
- Revised discussion of $E = mc^2$ to make equivalence of mass and energy clearer;
- Shortened discussion of half-life and related it better to kinetics chapter;
- Added information on newly synthesized elements;
- Added new sections on nuclear power plant accidents including Fukushima, and nuclear power pros and cons;
- Deleted one Estimation box;
- Replaced Summary Problem; and
- Added 18 new end-of-chapter questions, one of them more challenging.

Chapter 19

- Added two new figures; revised 19 figures;
- Added two new Exercises; modified one Exercise; deleted one Exercise;
- De-emphasized Frasch process because it currently produces very little sulfur;
- Replaced Summary Problem; and
- Added 22 new end-of-chapter questions; revised 10 questions.

Chapter 20

- Revised or replaced one figure;
- Added one new learning goal;
- Revised Summary Problem; and
- Added 22 new end-of-chapter questions; revised 10 questions.

Appendices

- Updated Appendix C to include latest values of physical constants and references to sources of data;
- Updated Appendix D with most recent references on electron configurations of the elements;
- Created new Appendix F: Balancing Redox Equations; and
- Combined former Appendix F with Appendix G so that all acid/base ionization constants are in a single appendix.

Features

We strongly encourage students to understand concepts and to learn to apply those concepts to problem solving. We believe that such understanding is essential if students are to be able to use what they learn from this book in subsequent courses and in their future careers. All too often we hear professors in courses for which general chemistry is a prerequisite complain that students have not retained what they were taught in general chemistry. This book is unique in its thoughtful choice of features that address this issue and help students achieve long-term retention of the material.

Problem Solving

This book places major emphasis on helping students learn to approach and solve real problems. Problem solving is introduced in Chapter 1, and a framework is built there that is followed throughout the book. Four important components of our strategy for teaching problem solving are

- *Problem-Solving Example/Problem-Solving Practice* problems that outline how to approach and solve a specific problem, check the answer, and practice a similar problem;
- *Estimation* boxes that help students learn how to do back-of-the-envelope calculations and apply concepts to new situations;
- *Exercises,* many of which deal with conceptual learning and are identified as *Conceptual Exercises,* that follow introduction of new material and for which answers are not immediately available, forcing students to work out the Exercise before seeing the answer; and
- *General Questions, Applying Concepts, More Challenging Questions,* and *Conceptual Challenge Problems* at the end of each chapter that are not keyed to specific textual material and require integration of concepts and out-of-the-box thinking to solve.

Problem-Solving Example/Problem-Solving Practice Each chapter contains many worked-out *Problem-Solving Examples*—a total of 242 in the book as a whole. Most consist of seven parts:

- a *Question* (problem);
- a *Result,* stated briefly;
- an *Analyze* section that outlines one approach to analyzing the problem;
- a *Plan* section that illustrates how to plan a solution;
- an *Execute* section that shows how the plan can be carried out;

  (The preceding three sections are designed to provide pedagogically sound help for students whose answer did not agree with ours.)
- a *Reasonable Result Check* section marked with a ☑ that indicates how a student could check whether a result is reasonable; and
- a companion *Problem-Solving Practice* that directly provides a similar question or questions, with answers appearing only in Appendix K.

We encourage students to first work out an answer without looking at either the Result or the Analyze, Plan, and Execute sections, and only then to compare their answer with ours. If their answer did not agree with ours, students are asked to repeat their work. Only then do we suggest that they look at the Analyze, Plan, and Execute sections, which are couched in conceptual as well as numeric terms to improve students' understanding, not just their ability to answer an identical question on an exam. The Reasonable Result Check section helps students learn how to use estimated results and other criteria to decide whether a result is reasonable, an ability that will serve them well in the future. By providing related *Problem-Solving Practice* problems that are answered only in the back of the book, we encourage students to immediately consolidate their thinking and improve their ability to apply their new understanding to other problems based on the same concept.

The first Problem-Solving Example and Problem-Solving Practice in Chapter 1 is shown below. It explicitly describes the strategy of analyzing the problem, planning a solution, executing the plan, checking that the answer is reasonable, and solving another similar problem. In a section of the text immediately following this example, we explicitly point out this problem-solving strategy and how the structure of the problem-solving examples supports it.

PROBLEM-SOLVING EXAMPLE 1.1

Density

In an old movie, thieves are shown running off with pieces of gold bullion that are about a foot long and have a square cross section of about six inches. The volume of each piece of gold is 7000 mL. Is what the movie shows physically possible? [Hint: calculate the mass of gold and express the result in pounds (lb). 1 lb = 454 g.]

Result Probably not; 1.4×10^5 g; 300 lb

Strategy and Explanation A good approach to problem solving is to (1) analyze the problem, (2) plan a solution, (3) execute the plan, and (4) check your result to see whether it is reasonable. (These four steps are described in more detail in Appendix A-1.)

Analyze the problem.

Step 1: *Analyze the problem.* You are asked to calculate the mass of the gold, and you know the volume is 7000 mL (one significant figure).

Plan a solution.

Step 2: *Plan a solution.* Density relates mass and volume and is the appropriate proportionality factor, so look up the density in a table. Mass is proportional to volume, so the volume either has to be multiplied by the density or divided by the density. Use the units to decide which. Use the information that 1 lb = 454 g to obtain a conversion factor for the units.

Execute the plan.

Step 3: *Execute the plan.* According to Table 1.1, the density of gold is 19.32 g/mL. Setting up the calculation so that the unit (milliliter) cancels gives

$$7000 \text{ mL} \times \frac{19.32 \text{ g}}{1 \text{ mL}} = 1.35 \times 10^5 \text{ g}$$

This can be converted to pounds

$$1.35 \times 10^5 \text{ g} \times \frac{1 \text{ lb}}{454 \text{ g}} = 297 \text{ lb which rounds to 300 lb}$$

Notice that the result is expressed to one significant figure, because the volume was given to only one significant figure and only multiplications and divisions were done. The intermediate result, 1.35×10^5 g, was expressed to more significant figures because rounding should only be done at the end of a calculation.

Check that the result is reasonable.

☑ **Reasonable Result Check** The units are mass units, so they are reasonable. Gold is nearly 20 times denser than water. A liter (1000 mL) of water is about a quart and a quart of water (two pints) weighs about two pounds. Seven liters (7000 mL) of water should weigh 14 lb, and 20 times 14 gives 280 lb, which rounds to 300 lb, so 300 lb is reasonable. The movie is not—few people could run while carrying a 300-lb object!

Solve another related problem.

PROBLEM-SOLVING PRACTICE 1.1

Calculate the volume occupied by a 4.33-g sample of benzene.

Estimation Enhancing students' abilities to estimate results is the goal of the *Estimation* boxes found in most chapters. These are a unique feature of this book. Each Estimation poses a problem that relates to the content of the chapter in which it appears and for which an approximate solution suffices. Students gain knowledge of various means of approximation, such as back-of-the-envelope calculations and graphing, and are encouraged to use diverse sources of information, such as encyclopedias, handbooks, and the Internet.

Exercises To further ensure that students do not merely memorize algorithmic solutions to specific problems, we provide 353 *Exercises,* which immediately follow introduction of new concepts within each chapter. Often the results that students obtain from a numeric Exercise provide insights into the concepts. Most Exercises are thought provoking and require that students apply conceptual thinking. Exercises that are conceptual rather than mathematical are clearly designated as shown below.

CONCEPTUAL EXERCISE 5.13

***h* Atomic Orbitals**

Using the same reasoning developed for *s, p, d,* and *f* atomic orbitals, determine the *n* value of the first shell that could contain *h* atomic orbitals. How many *h* atomic orbitals are in that shell?

Exercises that are designed to test understanding of a concept are identified as conceptual.

Problem-Solving Examples, Problem-Solving Practice problems, Estimation boxes, and Exercises are all designed to stimulate active thinking and participation by students as they read the text and to help them hone their understanding of concepts. The grand total of more than 600 of these **active-learning items** exceeds the number found in any similar textbook.

End-of-Chapter Questions At the end of each chapter we provide *Review Questions, Topical Questions, General Questions, Applying Concepts, More Challenging Questions,* and *Conceptual Challenge Problems.* Topical Questions are keyed to the sections in the chapter and to the learning goals in the "Having studied this chapter you should be able to" section at the end of each chapter. General Questions typically involve only one concept or topic, but students are required to think about which concept is needed to answer the question; no immediate indication is given regarding where to look in the chapter for the concept. Applying Concepts questions explicitly require conceptual thinking instead of numerical calculations and are designed to test students' understanding of concepts. It has been clearly established by research on cognition in both chemistry and physics that many students can correctly answer numerical-calculation questions, yet not understand concepts well enough to answer simple conceptual questions. Applying Concepts questions have been designed to address this issue. More Challenging Questions are provided so that students' minds can be stretched to link two or more concepts and apply them to a problem. Conceptual Challenge Problems require out-of-the-box thinking and are suitable for group work by students.

Conceptual Understanding

We believe that a sound conceptual foundation is the best means by which students can approach and solve a wide variety of real-world problems. This approach is supported by considerable evidence in the literature: Students learn better and retain what they learn longer when they have mastered fundamental concepts. Chemistry requires familiarity with at least three conceptual levels:

- **Macroscale** (laboratory and real-world phenomena)
- **Nanoscale** (models involving particles: atoms, molecules, and ions)
- **Symbolic** (chemical formulas and equations, as well as mathematical equations)

These three conceptual levels are explicitly defined in Chapter 1 (see the figure below). This chapter emphasizes the value of the chemist's unique perspective on science and the world with a specific example of how chemical thinking can help solve a real-world problem—how tube wells in Bangladesh that were designed to prevent disease

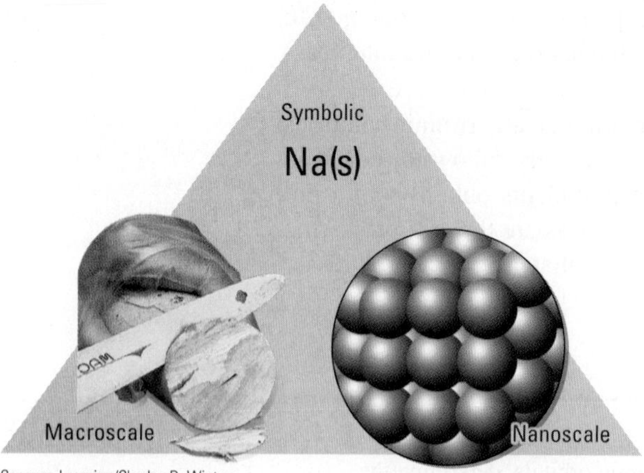
Symbolic
Na(s)
Macroscale
Nanoscale

Cengage Learning/Charles D. Winters

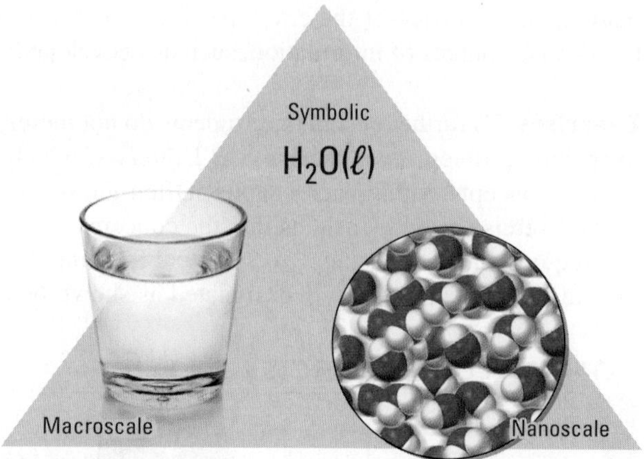
Symbolic
H2O(ℓ)
Macroscale
Nanoscale

Pavlo Loushkin/Shutterstock.com

resulted in arsenic contamination and how that contamination is being mitigated. This theme of conceptual understanding and its application to problems continues throughout the book. Many of the problem-solving features already mentioned have been specifically designed to support conceptual understanding.

The entire book has been assessed with respect to conceptual level by Dr. Kristin Briney. Using Bloom's Taxonomy as a guide, she analyzed Problem-Solving Examples, Exercises, learning goals at the end of each chapter, and the organization and content of end-of-chapter questions. All of these have been revised in response to this analysis. We believe that the conceptual level and consistency of this textbook make it significantly better than others.

Units are introduced on a need-to-know basis at the first point in the book where they contribute to the discussion. Units for length, mass, volume, and density are defined in Chapter 1, in the discussion of the international system of units. Energy units are defined in Chapter 4, where they are first needed to deal with kinetic and potential energy, work, and heat. In each case, defining units at the time when the need for them can be made clear allows definitions that would otherwise appear pointless and arbitrary to support the development of closely related concepts.

We use real chemical systems in examples and problems whenever possible, both in the text and in the end-of-chapter questions. In the kinetics chapter, for example, the text and problems utilize real reactions and real data from which to determine reaction rates or orders. Instead of $A + B \rightarrow C + D$, students will find $I^- + CH_3Br \rightarrow CH_3I + Br^-$. Data have been taken from the recent research literature. The same approach is employed in many other chapters, where real chemical systems are used as examples.

Most important, we provide **clear, direct, thorough, and understandable explanations** of all topics, including those such as stoichiometry, chemical kinetics, chemical thermodynamics, chemical equilibrium, and electrochemistry that many students find daunting. The methods of science and concepts such as chemical and physical properties; purification and separation; the relation of macroscale, nanoscale, and symbolic representations; elements and compounds; and kinetic-molecular theory are introduced in Chapter 1 so that they can be used throughout the later discussion. Rather than being bogged down with discussions of units and nomenclature, students begin this book with an overview of what real chemistry is about—together with fundamental ideas that they will need to understand it.

Visualization for Understanding

The **illustrations** in *Chemistry: The Molecular Science* have been designed to engage today's visually oriented students. The success of the illustration program is exemplified by the fact that the first edition was awarded a national prize for visual excellence.

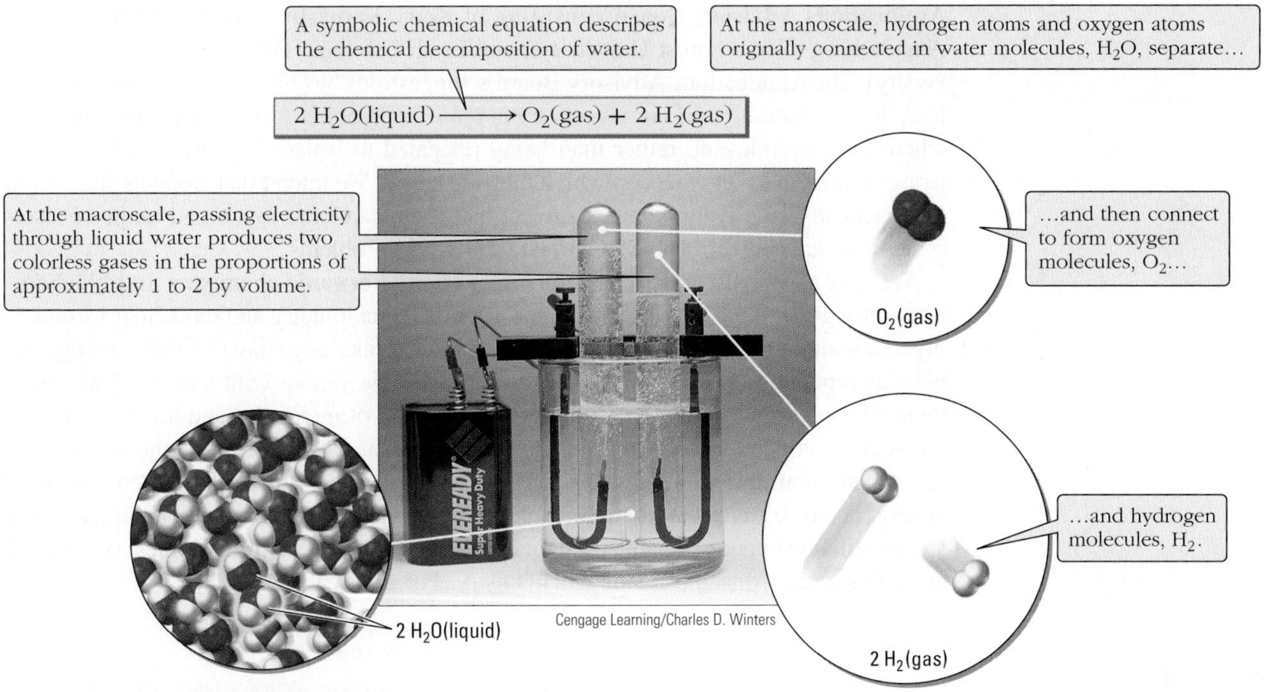
A symbolic chemical equation describes the chemical decomposition of water.
At the nanoscale, hydrogen atoms and oxygen atoms originally connected in water molecules, H_2O, separate...
$2\ H_2O(liquid) \longrightarrow O_2(gas) + 2\ H_2(gas)$
At the macroscale, passing electricity through liquid water produces two colorless gases in the proportions of approximately 1 to 2 by volume.
...and then connect to form oxygen molecules, O_2...
$O_2(gas)$
...and hydrogen molecules, H_2.
$2\ H_2O(liquid)$
Cengage Learning/Charles D. Winters
$2\ H_2(gas)$

Nevertheless, for this edition we have re-examined carefully each piece of art and revised more than 300 figures. In most cases these revisions have expanded the use of macroscale/nanoscale illustrations of the type shown here and there has been much more use of text and pointers to call students' attention to the important ideas or observations in each figure. Illustrations help students to visualize atoms and molecules and to make connections among macroscale observations, nanoscale models, and symbolic representations of chemistry. Excellent color photographs of substances and reactions, many by Charles D. Winters and James L. Maynard, are presented together with greatly magnified illustrations of the atoms, molecules, and/or ions involved. Often these are accompanied by the symbolic formula for a substance or equation for a reaction, as in the example shown. These **nanoscale views of atoms, molecules, and ions** have been generated with molecular modeling software and then combined by a skilled artist with the photographs and formulas or equations. Similar illustrations appear in exercises, examples, and end-of-chapter problems, thereby ensuring that students are tested on the ideas the illustrations represent. Consistent color-coding is used throughout, as illustrated in the Style Key. This provides an exceptionally effective way for students to learn how chemists think about the nanoscale world of atoms, molecules, and ions.

Often the story is carried solely by an illustration that includes text and pointers to indicate the most important parts of the figure. This too is illustrated in the example figure. Text and pointers are also used to explain the operation of instruments, apparatus, and experiments; to clarify the development of a mathematical derivation; or to point out salient features of graphs or nanoscale pictures. Throughout the book visual interest is high, and visualizations of many kinds are used to support conceptual development. This more effective use of illustrations has enabled us to reduce the length of textual descriptions and explanations.

Interdisciplinary Applications

Whenever possible we include **practical applications**, especially those applications that students will revisit when they study other natural science and engineering disciplines. To enhance and improve this aspect of the book, we have asked for advice from an

Applications Advisory Board consisting of Kerry Karukstis (Harvey Mudd College), Angela King (Wake Forest University), and Erich Uffelman (Washington and Lee University). The Applications Advisory Board's suggestions have been incorporated into the text, the illustrations, and our *Chemistry in the News* feature. Applications are integrated where they are relevant, rather than being relegated to isolated chapters and separated from the principles and facts on which they are based. We intend that students should see that chemistry is a lively, relevant subject that is fundamental to a broad range of disciplines and that can help solve important, real-world problems.

We have especially emphasized the **integration of organic chemistry and biochemistry** throughout the book. In many areas, such as stoichiometry and molecular formulas, organic compounds provide excellent examples. To take advantage of this synergy, we have incorporated basic organic topics into the text beginning with Chapter 2 and used them wherever they are appropriate. In the discussion of molecules and the properties of molecular compounds, for example, the concepts of structural formulas and isomers are developed naturally and effectively. Many of the principles that students encounter in general chemistry are directly applicable to biochemistry, and a large percentage of the students in most general chemistry courses are planning careers in biological or medical areas that make constant use of biochemistry. For this reason, we have chosen to deal

STYLE KEY

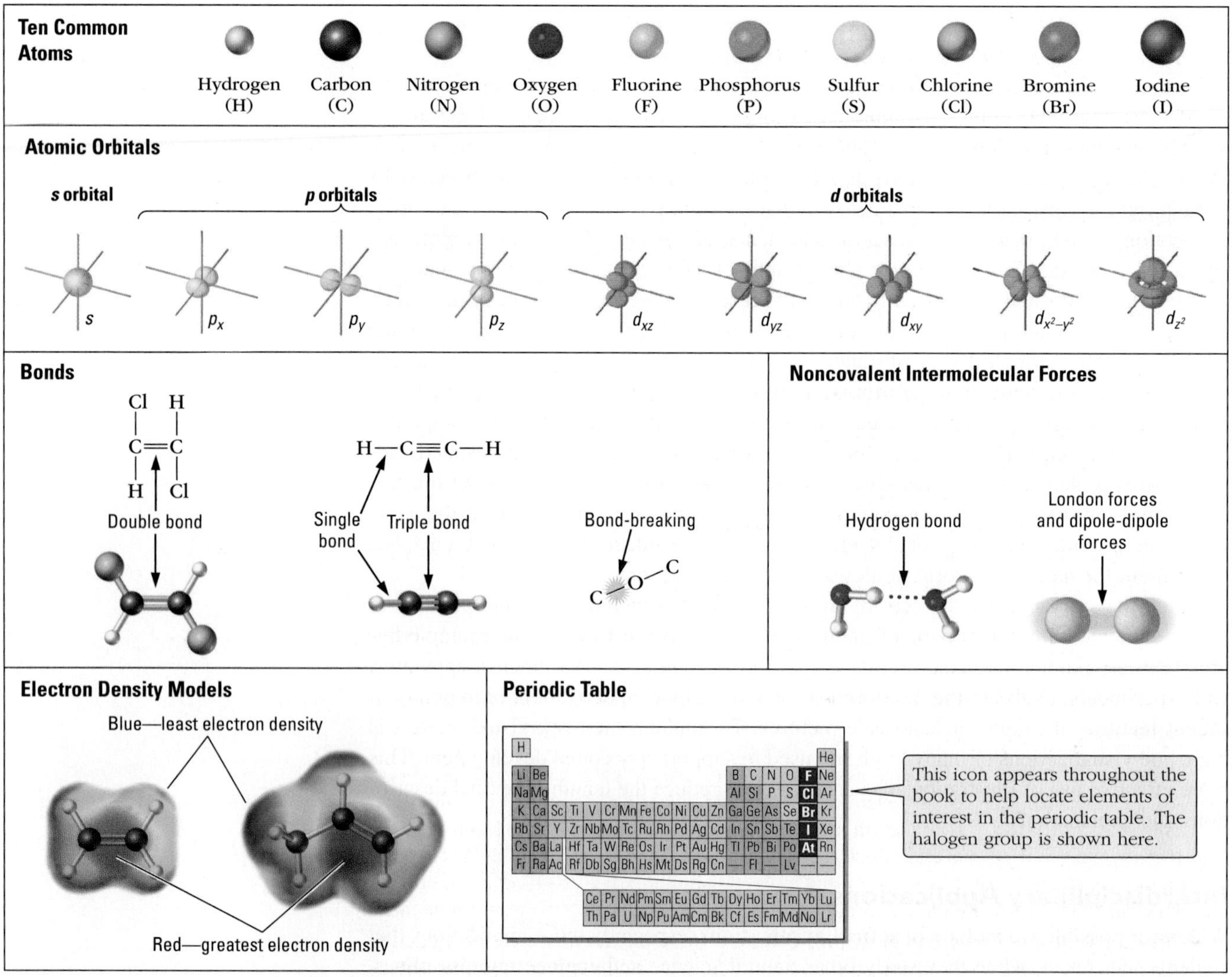
Ten Common Atoms
Hydrogen (H)
Carbon (C)
Nitrogen (N)
Oxygen (O)
Fluorine (F)
Phosphorus (P)
Sulfur (S)
Chlorine (Cl)
Bromine (Br)
Iodine (I)
Atomic Orbitals
s orbital
p orbitals
d orbitals
s
p_x
p_y
p_z
d_{xz}
d_{yz}
d_{xy}
$d_{x^2-y^2}$
d_{z^2}
Bonds
Double bond
Single bond
Triple bond
Bond-breaking
Noncovalent Intermolecular Forces
Hydrogen bond
London forces and dipole-dipole forces
Electron Density Models
Blue—least electron density
Red—greatest electron density
Periodic Table
This icon appears throughout the book to help locate elements of interest in the periodic table. The halogen group is shown here.

with fundamental biochemical topics in juxtaposition with the general chemistry principles that underlie them.

Here are some examples of integration of organic and biochemistry; the book contains many more:

- Section 1-14, *The Biological Periodic Table,* describes the many elements that are essential to living systems and why they are important.
- Section 2-9, *Organic Molecular Compounds,* introduces simple hydrocarbons and the concept of isomerism as a natural part of the discussion of molecular compounds.
- Section 4-11, *Fuels for Society and Our Bodies,* applies thermochemical and calorimetric principles learned earlier in the chapter to the caloric values of proteins, fats, and carbohydrates in food.
- Section 7-7, *Biomolecules: DNA and the Importance of Molecular Structure,* extends the concepts of molecular shapes and intermolecular forces developed earlier in Chapter 7 to the structure and function of DNA, explaining how chemical principles can be applied to the storage and transmission of genetic information.
- Chapter 10, *Fuels, Organic Chemicals, and Polymers,* builds on principles and facts introduced earlier, applying them to organic molecules and functional groups selected for their relevance to synthetic and natural polymers. Proteins and polysaccharides illustrate the importance of biopolymers.
- Section 11-9, *Enzymes: Biological Catalysts,* applies kinetics principles developed earlier in the chapter and ideas about molecular structure from earlier chapters to enzyme catalysis and the way in which it is influenced by protein structure.
- Section 15-1a, *Buffer Action,* is introduced using blood buffer systems.
- Section 16-9, *Gibbs Free Energy and Biological Systems,* discusses the role of Gibbs free energy and coupling of thermodynamic systems in metabolism, making clear the fact that metabolic pathways are governed by the rules of thermodynamics.

Environmental and industrial chemistry are also integrated. In Chapter 4, *Energy and Chemical Reactions,* thermochemical principles are used to evaluate the energy densities of fuels. In Chapter 8, *Properties of Gases,* we apply what students are learning about gases to stratospheric ozone depletion and the consequences of combustion on global warming. Chapter 10, *Fuels, Organic Chemicals, and Polymers,* discusses energy resources, hydraulic fracking, and recent developments in recycling plastics. In Chapter 11, *Chemical Kinetics: Rates of Reactions,* the importance of catalysts is illustrated by several industrial processes and exhaust-emission control on automobiles. In Chapter 14, *Acids and Bases,* practical acid-base chemistry illustrates many of the principles students learn. In Chapter 19, *The Chemistry of the Main-Group Elements,* and Chapter 20, *Chemistry of Selected Transition Elements and Coordination Compounds,* principles developed in earlier chapters are applied to uses of the elements and to extraction of elements from their ores. Through these many applications students in a variety of disciplines will discover that chemistry is fundamental to their other studies.

Other Features

Additional features of the book that we have designed specifically to address the needs of students are:

- ***Chemistry You Can Do*** features associated with most chapters are available online. A *Chemistry You Can Do* experiment requires only simple equipment and familiar chemicals available at home or on a college campus, can be performed in a kitchen or residence hall room, and illustrates a topic included in the chapter. Including these experiments reflects our goal that students should be involved in doing chemistry, and they ought to learn that common household materials are also chemicals.
- ***Chemistry in the News*** boxes available online bring to the attention of students the latest discoveries in chemistry and applications of chemistry, making clear

that chemistry is continually changing and developing—it is not merely a static compendium of items to memorize. These boxes have been updated, and 18 are new to this edition.

- ***Tools of Chemistry*** boxes provide examples of how chemists use modern instrumentation to solve challenging problems. They introduce to students the excitement and broad range of chemical measurements. They are placed in the text strategically to support students' understanding of a topic being discussed; for example, the Tools of Chemistry on mass spectrometry accompanies the discussion of isotopes.
- ***Portrait of a Scientist*** items show that, like any other human pursuit, chemistry depends on people. These biographical sketches of men and women who have advanced our understanding or applied chemistry imaginatively to important problems bring the human side of chemistry to students using this book; we have selected many of them to illustrate the diversity of people who do science.

End-of-Chapter Study Aids

At the end of each chapter, students will find many ways to test and consolidate their learning.

- A **Summary Problem** brings together concepts and problem-solving skills from throughout the chapter. Students are challenged to answer a multifaceted question that builds on and integrates the chapter's content. Because these problems are an important aid to student understanding, we have revised or created new summary problems in 14 chapters.
- **Having studied this chapter, you should be able to** highlights the learning goals for the chapter, provides references to the sections in the chapter that address each goal, and identifies end-of-chapter questions appropriate to test each goal. Because of their importance in defining conceptual goals for students, all learning goals were examined for conceptual level and their correlation with end-of-chapter questions has been updated.
- **Key Terms** are listed, with references to the sections where they are defined. New key terms have been added to the already extensive glossary.

A broad range of chapter-end **Questions for Review and Thought** are provided to serve as a basis for homework or in-class problem solving.

- **Review Questions**, which are not answered in the back of the book, test vocabulary and simple concepts.
- **Topical Questions** are keyed to the major topics in the chapter and are listed under headings that correspond with each section in the chapter. Questions are often accompanied by photographs, graphs, and diagrams that make the situations described more concrete and realistic. Usually a question that is answered at the end of the book is paired with a similar one that is not.
- **General Questions** are not explicitly keyed to chapter topics. These questions are designed to help students analyze problems and learn to apply appropriate ideas to solving them.
- **Applying Concepts** includes questions specifically designed to test conceptual learning. Many of these questions include diagrams of atoms, molecules, or ions and require students to relate macroscopic observations, atomic-scale models, and symbolic formulas and equations.
- **More Challenging Questions** require students to think more deeply and conceptually. They integrate multiple concepts and are higher in Bloom's taxonomy than typical end-of chapter questions.
- **Conceptual Challenge Problems**, most of which were written by H. Graden Kirksey, emeritus faculty member of the University of Memphis, are especially important in helping students assess and improve their conceptual thinking ability. Designed for group work, the Conceptual Challenge Problems are rigorous and thought provoking. Much effective learning can be induced by dividing a class into groups of three or four students and then assigning these groups to work collaboratively on these problems.

Organization

The order of chapters reflects the most common division of content between the first and second semesters of a typical general chemistry course. The first two chapters briefly review basic material that most students should have encountered in high school. Next, the book develops the ideas of chemical reactions, stoichiometry, and energy transfers during reactions. Throughout these early chapters, organic chemistry, biochemistry, and applications of chemistry are integrated. We then deal with the electronic structure of atoms, bonding and molecular structures, and the way in which structure affects properties. To finish up a first-semester course, there are adjacent chapters on gases and on liquids and solids.

Next, we extend our integration of organic chemistry in a chapter that describes the role of organic chemicals in fuels, polymers, and biopolymers. Chapters on kinetics and equilibrium establish fundamental understanding of how fast reactions will go and what concentrations of reactants and products will remain when equilibrium is reached. These ideas are then applied to solutions, as well as to acid-base and solubility equilibria in aqueous solutions. A chapter on thermodynamics and Gibbs free energy is followed by one on electrochemistry, which makes use of thermodynamic ideas. Finally, the book focuses on nuclear chemistry and the descriptive chemistry of main-group and transition elements.

To help students connect chemical ideas that are closely related but are presented in different chapters, we have included **numerous cross references (indicated by the ← symbol)**. These cross references will help students link a concept being developed in the chapter they are currently reading with an earlier, related principle or fact. They also provide many opportunities for students to review material encountered earlier.

Varying Chapter Order

A number of variations in the order of presentation are possible. For example, in the classes of one of the authors, the first six sections of Chapter 16 on thermodynamics follow Chapter 11 on chemical kinetics and precede Chapter 12 on equilibrium. Section 12-7 is omitted, and the last five sections of Chapter 16 follow Chapter 12. The material on thermochemistry in Chapter 4 could be postponed and combined with Chapter 16 on thermodynamics with only minor adjustments in the teaching of other chapters, so long as the treatment of thermochemistry precedes the material in Chapter 11, which uses thermochemical concepts in the discussion of activation energy. Many other reorderings of chapters or sections within chapters are possible. The numerous cross references will aid students in picking up concepts that they would be assumed to know, had the chapters been taught consecutively.

At the University of Wisconsin–Madison, this textbook is used in a one-semester accelerated course that is required for most engineering students. We assume substantial high-school background in both chemistry and mathematics, and the syllabus includes Chapters 1, 5, 6, 7, part of 9, 10, 11, 12, 14, 15, 16, and 17. This presentation strategy works quite well, and some engineering students have commented favorably on the inclusion of practical applications of chemistry, such as octane rating and catalysis, in which they were interested.

Chemistry: The Molecular Science, Fifth Edition, can be divided into a number of sections, each of which treats an important aspect of chemistry:

Fundamental Ideas of Chemistry

Chapter 1, **The Nature of Chemistry**, is designed to capture students' interest from the start by concentrating on chemistry (not on math, units, and significant figures, which are outlined in Chapter 1 but treated comprehensively in Appendix A). It asks, "Why Care

About Chemistry?" and then tells a story of the application of chemical principles to problems of water quality in Bangladesh that illustrates how chemistry is important in the real world and describes interdisciplinary chemical research. Chapter 1 also introduces major concepts that bear on all of chemistry, including the periodic table, the kinetic-molecular theory, and the three conceptual levels with which students must be familiar—macroscale, nanoscale, and symbolic.

Chapter 2, Chemical Compounds, concentrates on thorough, understandable treatment of the concepts of atomic structure, ionic and molecular compounds, molar mass, and determining formulas of compounds. The important theme of structure is introduced and used to explain properties of ionic and molecular compounds; it is reinforced by showing several ways that organic structures can be written.

Chemical Reactions

Chapter 3, Chemical Reactions, begins a two-chapter sequence that treats chemical reactions qualitatively and quantitatively. Students first learn how to interpret, construct, and balance equations. We next introduce precipitation reactions, the dissociation of ions when ionic compounds dissolve, and net ionic equations. Acid-base and redox reactions further illustrate net ionic equations. Students learn how to recognize a redox reaction from the chemical nature of the reactants (not just by using oxidation numbers). The fundamental principle of stoichiometry—that amounts of reactants and products are proportional and that the proportionality involves mole ratios derived from coefficients in balanced equations—is introduced. A clear conceptual foundation is provided that enhances students' understanding of stoichiometry. The chapter concludes with solution stoichiometry and titration.

Chapter 4, Energy and Chemical Reactions, begins with a thorough and straightforward introduction to forms of energy, conservation of energy, heat and work, system and surroundings, and exothermic and endothermic processes. Carefully designed figures help students to understand thermodynamic principles. Heat capacity, heats of changes of state, and heats of reactions are clearly explained, as are calorimetry and standard enthalpy changes. These ideas are then applied to fossil fuel combustion and to metabolism of biochemical fuels (proteins, carbohydrates, and fats).

Electrons, Bonding, and Structure

Chapter 5, Electron Configurations and the Periodic Table, introduces spectra, quantum theory, and quantum numbers, using color-coded illustrations to visualize the different energy levels of *s, p, d,* and *f* orbitals. The *s*-, *p*-, *d*-, and *f*-block locations in the periodic table are used to predict electron configurations.

Chapter 6, Covalent Bonding, provides simple, stepwise guidelines for writing Lewis structures, with many examples of how to use them. The role of single and multiple bonds in hydrocarbons is smoothly integrated with the introduction to covalent bonding. The discussion of polar bonds is enhanced by molecular models that show variations in electron density. Molecular orbital theory is introduced at the end of the chapter.

Chapter 7, Molecular Structures, provides a thorough presentation of valence-shell electron-pair repulsion (VSEPR) theory and orbital hybridization. Molecular geometry and polarity are extensively illustrated with computer-generated models, and the relation of structure, polarity, and hydrogen bonding to attractions among molecules is clearly developed and illustrated in Problem-Solving Examples. The importance of intermolecular forces is emphasized early and then reinforced by describing how noncovalent attractions determine the structure of DNA.

States of Matter; Materials; Important Molecular Substances

Chapter 8, **Properties of Gases**, briefly introduces kinetic-molecular theory early, using it to interpret each of the gas laws. Many Conceptual Exercises throughout the chapter emphasize qualitative understanding of gas properties. Mathematical problem solving focuses on the ideal gas law or the combined gas law. Gas stoichiometry is applied to the practical problem of self-contained self-rescue breathing devices. Then, the properties of gases are applied to chemical reactions in the atmosphere, specifically the role of ozone in the stratosphere and the role of CO_2 in global warming.

Chapter 9, **Liquids, Solids, and Materials**, begins with a discussion of intermolecular forces and their effect on properties. Properties of liquids are discussed next: vaporization and condensation, vapor pressure and boiling points, and critical temperature and pressure. Solids, including types of solids and their associated properties, phase changes of solids, and heating curves are discussed before phase diagrams are introduced. The unique and vitally important properties of water are covered thoroughly. The principles of crystal structure are introduced using cubic unit cells only. The fact that much current chemical research involves materials is illustrated by the discussions of metals, *n*- and *p*-type semiconductors, insulators, superconductors, network solids, carbon nanotubes, cement, ceramics and ceramic composites, and glasses, including shatter-resistant glass for smartphone screens.

Chapter 10, **Fuels, Organic Chemicals, and Polymers**, offers a distinctive combination of topics of major relevance to industrial, energy, and environmental concerns. Petroleum, natural gas, and coal are discussed as resources for energy and chemical materials. Enough organic functional groups are introduced so that students can understand polymer formation, and the idea of condensation polymerization is extended to carbohydrates and proteins, which are compared with synthetic polymers.

Reactions: How Fast and How Far?

Chapter 11, **Chemical Kinetics: Rates of Reactions**, presents one of the most difficult topics in the course with extraordinary clarity. Defining reaction rate, finding rate laws from initial rates and integrated rate laws, and using the Arrhenius equation are thoroughly developed. How molecular changes during unimolecular and bimolecular elementary reactions relate to activation energy initiates the treatment of reaction mechanisms (including those with an initial fast equilibrium). Catalysis is shown to involve changing a reaction mechanism. Both enzymes and industrial catalysts are described using concepts developed earlier in the chapter.

Chapter 12, **Chemical Equilibrium**, emphasizes both a qualitative understanding of the nature of equilibrium and solving mathematical problems. That equilibrium results from equal but opposite reaction rates is fully explained. Both Le Chatelier's principle and the reaction quotient, *Q*, are used to predict shifts in equilibria. A unique section on equilibrium at the nanoscale introduces briefly and qualitatively how enthalpy changes and entropy changes affect equilibria. Optimizing the yield of the Haber-Bosch ammonia synthesis elegantly illustrates how kinetics, equilibrium, and enthalpy and entropy changes control the outcome of a chemical reaction.

Reactions in Aqueous Solution

Chapter 13, **The Chemistry of Solutes and Solutions**, builds on principles previously introduced, showing the influence of enthalpy and entropy on solution properties. Understanding of solubility, Henry's law, concentration units (including ppm and ppb), and colligative properties (including osmosis) is reinforced by applying these ideas to water as a resource, hard water, and municipal water treatment.

Chapter 14, **Acids and Bases**, concentrates initially on the Brønsted-Lowry acid-base concept, clearly delineating proton transfers using color coding and molecular models. In addition to a full exploration of pH and the meaning and use of K_a and K_b, acid strength is related to molecular structure, and the acid-base properties of carboxylic acids, amines, and amino acids are introduced. Lewis acids and bases are defined and illustrated using examples. Student interest is enhanced by a discussion of everyday uses of acids and bases.

Chapter 15, **Additional Aqueous Equilibria**, extends the treatment of acid-base and solubility equilibria to buffers, titration, and precipitation. The Henderson-Hasselbalch equation, which is widely used in biochemistry, is applied to buffer pH. Calculations of points on titration curves are shown, and the interpretation of several types of titration curves provides conceptual understanding. Acid-base concepts are applied to the formation of acid rain. The final section deals with the various factors that affect solubility (pH, common ions, complex ions, and amphoterism) and with selective precipitation.

Thermodynamics and Electrochemistry

Chapter 16, **Thermodynamics: Directionality of Chemical Reactions**, explores the nature and significance of entropy, both qualitatively and quantitatively. The signs of Gibbs free energy changes are related to the easily understood classification of reactions as reactant- or product-favored, with the discussion deliberately avoiding the often-misinterpreted term "spontaneous." The thermodynamic significance of coupling one reaction with another is illustrated using industrial, metabolic, and photosynthetic examples. Energy conservation is defined thermodynamically. A closing section reinforces the important distinction between thermodynamic and kinetic stability.

Chapter 17, **Electrochemistry and Its Applications**, uses the nitrogen cycle to review oxidation numbers and defines half-reactions; balancing redox equations is treated in detail in Appendix F. Electrochemical cells, cell potentials, standard half-cell potentials, the relation of cell potential to Gibbs free energy, and the effect of concentrations on cell potential are all explored. Batteries, including Li-ion batteries, and fuel cells provide important examples. Electrolysis is presented as a means of causing reactant-favored reactions to occur. Electroplating provides examples of calculations involving number of electrons transferred. The chapter closes with a discussion of corrosion.

Nuclear Chemistry

Chapter 18, **Nuclear Chemistry**, deals with radioactivity, nuclear reactions, nuclear stability, and rates of disintegration reactions. Also provided are thorough descriptions of nuclear fission, fission power plants and their pros and cons, and nuclear fusion. There is extensive discussion of nuclear radiation, background radiation, and applications of radioisotopes.

Descriptive Chemistry

Chapter 19, **The Chemistry of the Main-Group Elements**, consists of two main parts. The first part tells the interesting story of how the elements were formed and which are most important on Earth. The physical separation of nitrogen, oxygen, and sulfur from natural sources, and the extraction of sodium, chlorine, magnesium, and aluminum by electrolysis, provide important industrial examples as well as an opportunity for students to apply principles learned earlier in the book. The second part (Section 19-6) discusses the properties, chemistry, and uses of the elements of Groups 1A to 7A and their compounds in a systematic way, based on groups of the periodic table. Trends in atomic and ionic radii, melting points and boiling points, and densities of each group's elements are summarized. Group 8A is covered briefly.

Chapter 20, Chemistry of Selected Transition Elements and Coordination Compounds, treats a few important elements in depth and integrates the review of principles learned earlier. Iron, copper, chromium, silver, and gold provide an interesting, motivating collection of elements from which students can learn the principles of transition metal chemistry. In addition to the treatment of complex ions and coordination compounds, this chapter includes an extensive section on crystal-field theory, electron configurations, color, and magnetism in coordination complexes.

Alternate Editions

Chemistry: The Molecular Science, Fifth Edition
Hybrid Version with Access (24 months) to OWLv2 with MindTap Reader
ISBN: 978-1-285-46184-7
This briefer, paperbound version of *Chemistry: The Molecular Science, Fifth Edition* does not contain the end-of-chapter problems, which can be assigned in OWL, the online homework and learning system for this book. Access to OWLv2 and the MindTap Reader eBook is included with the Hybrid version. The MindTap Reader is the full version of the text, with all end-of-chapter questions and problem sets.

Supporting Materials

Please visit **http://www.cengage.com/chemistry/moore/CTMS5e** for information about student and instructor resources for this text.

Reviewers

Reviewers have played a critical role in the preparation of this textbook. The individuals listed below helped to shape this text into one that is not merely accurate and up to date, but a valuable practical resource for teachers and learners of chemistry.

Fifth Edition

Special thanks to these reviewers

Rachel Bain, *University of Wisconsin–Madison,* who reviewed the entire text, all figures, and all tables for accuracy. Her thorough review resulted in many improvements and rooted out numerous inaccuracies.

Elizabeth A. Moore, *University of Wisconsin–Madison,* who updated and corrected all InDesign files from the fourth edition; more importantly, she improved clarity of text and figures—many of the new figures in this edition are the result of her sharp eye and cogent, constructive, critical comments.

Judy Ozment, *Penn State University,* has solved all of the chapter-end questions and made many helpful suggestions that have improved the presentation and pedagogy of the questions. Judy was aided in her work by Karen Pesis, *American River College*, and Arya Kermanshah, *Penn State University*.

The Applications Advisory Board, which provided many suggestions for new applications of chemistry in real-world situations:
Kerry Karukstis, *Harvey Mudd College*
Angela King, *Wake Forest University*
Erich Uffelman, *Washington and Lee University*

Pre-Revision Reviewers

Kevin Crawford, *University of Wisconsin–Oshkosh*
Kevin Davies, *Florida Gulf Coast University*

Edward Delafuente, *Kennesaw State University*
William Deutschman, *Westminster College*
Michael Garlick, *Delta College*
Paul Hooker, *Westminster College*
Michael Hurst, *Georgia Southern University*
Gerald Korenowski, *Rensselaer Polytechnic Institute*
Roy McClean, *United States Naval Academy*
Dillip Mohanty, *Central Michigan University*
Douglas Mulford, *Emory University*
Ruth Robinson, *University of Minnesota–Twin Cities*
Mark Schraf, *West Virginia University*
Robert Snipp, *Creighton University*
Laura Starkey, *California State Polytechnic University–Pomona*
Christy Vogel, *Cabrillo College*

Manuscript Reviewers
Margaret Czerw, *Raritan Valley Community College*
Bradley Fahlman, *Central Michigan University*
David Hanson, *Stony Brook University*
Michael Hull, *Northwest Missouri State University*
Daniel King, *Drexel University*
Joe March, *University of Alabama–Birmingham*
Raymond Sadeghi, *University of Texas–San Antonio*

Fourth Edition

Reviewers
Margaret Czerw, *Raritan Valley Community College*
Michelle Driessen, *University of Minnesota*
Harold Goldwhite, *California State University–Los Angeles*
Steven C. Haefner, *Bridgewater State College*
David M. Hanson, *Stony Brook University*
Andy Jorgensen, *University of Toledo*
Roy Kennedy, *Massachusetts Bay Community College*
Mahesh Mahanthappa, *University of Wisconsin–Madison*
Joe L. March, *University of Alabama–Birmingham*
Wyatt R. Murphy, Jr., *Seton Hall University*
Jeff R. Schoonover, *St. Mary's University*
Clarissa Sorensen-Unruh, *Central New Mexico Community College*
Anton Wallner, *Barry University*
Kathy Thrush Shaginaw meticulously evaluated all art in this fourth edition. She provided a detailed review of each figure with suggestions for improving an already excellent illustration program. Her work in this regard was outstanding and has resulted in figures that will help students learn more effectively.

Accuracy Reviewers
Patrick J. Desrochers, *University of Central Arkansas*
Paul T. Kaiser, *United States Naval Academy*
Karen Pesis, *American River College*

Acknowledgments

No project on the scale of a textbook revision is accomplished solely by the authors. We have had assistance of the very highest quality in all aspects of production of this book, and we extend hearty thanks to everyone who contributed to the project.

Lisa Lockwood and Maureen Rosener, product managers, have overseen the entire project and have collaborated with the author team on decisions and initiatives that have greatly improved what was already an excellent, rigorous, mainstream general chemistry textbook.

Peter McGahey, content developer, provided expert advice and active support throughout the revision. He assembled an excellent group of expert reviewers, obtained reviews from them in timely fashion, and provided feedback based on their comments that was invaluable. He has also served as an interface between the authors and the many other members of the production staff.

Teresa Trego, content project manager, oversaw production of the book. We thank her for her invaluable contribution. Lisa Weber served as media developer, handling OWL and online materials for the book. Elizabeth Woods, media developer, ably handled all of the ancillary print materials. We also thank Karolina Kiwak, chemistry product assistant, for handling many tedious tasks.

The success of a book such as this one depends also on its being adopted and read. Nicole Hamm, marketing director, and Janet del Mundo, marketing manager, direct the marketing and sales programs. We thank them and the many local representatives throughout the country who will help get this book to students who can benefit from it.

This book is beautiful to look at, and its beauty is more than skin deep. The illustration program has been carefully designed to support student learning in every possible way. The many photographs of Charles D. Winters of Oneonta, New York, and James L. Maynard, *University of Wisconsin–Madison*, provide students with close-up views of chemistry in action. We thank Jim for doing many new shoots for this new edition. Cheryl DuBois, PreMedia Global, carried out photo research; we thank her for her part in the illustration program.

Dan Fitzgerald, project manager, together with the staff at Graphic World Publishing Services, handled copy editing, layout, and production of the book. Dan worked calmly and effectively with the authors to make certain that this book is of the highest possible quality. Dan's constant concern for high quality while keeping the project on track extended well beyond the call of duty. We thank all of the staff at Graphic World who contributed to this edition.

Judy Ozment has solved all of the end-of-chapter questions in this book for all five editions. She has produced excellent student solution manuals and answers to selected questions at the end of the book. Judy is diligent in finding ways in which questions can be stated more clearly, pointing out cases where data used in a question are inconsistent with other material in the book, and situations where authors may not have asked what they wanted to ask. For all of her work and help we thank her profusely. Karen Pesis and Arya Kermanshah provided a second set of eyes for Judy and we thank them for their excellent work.

Elizabeth Moore, *University of Wisconsin–Madison*, provided especially useful comments regarding graphics and how they could best be used to enhance student learning. She also contributed to improving the layout of the book and to enhancing the quality of the text. Rachel Bain, *University of Wisconsin–Madison*, has provided by far the most thorough and helpful accuracy review we have encountered in many years of preparing textbooks. We thank her for the long hours she put into this project and for finding errors that had previously eluded many pairs of critical eyes.

Many of the take-home *Chemistry You Can Do* experiments that accompany this book online were adapted from activities published by the Institute for Chemical Education as *Fun with Chemistry: Volumes I and II,* by Mickey and Jerry Sarquis of Miami University (Ohio). Some were adapted from Classroom Activities published in the *Journal of Chemical Education.* Conceptual Challenge Problems at the end of most chapters were written by H. Graden Kirksey, emeritus faculty of the University of Memphis, and we very much appreciate his contribution.

We also thank the many teachers, colleagues, students, and others who have contributed to our knowledge of chemistry and helped us devise better ways to help others learn it. Collectively, the authors of this book have many years of experience teaching and learning, and we have tried to incorporate as much of that as possible into our presentation of chemistry.

Finally, we thank our families and friends who have supported all of our efforts—and who can reasonably expect more of our time and attention now that this new edition is complete.

We hope that using this book results in a lively and productive experience for both faculty and students.

John W. Moore
Madison, Wisconsin

Conrad L. Stanitski
Lancaster, Pennsylvania

Special Features

ESTIMATION

TOOLS OF CHEMISTRY

PORTRAIT OF A SCIENTIST

The Nature of Chemistry | 1

Rafiqur Rahman/Reuters/Landov

In many parts of the world, water that is safe to drink can be hard to find. This boy is collecting drinking water from a well, signified as safe by the green pipe, at a village 275 kilometers (172 miles) from Dhaka, the capital of Bangladesh. In that country, the water from many wells is contaminated by unsafe levels of arsenic. Why is arsenic so widespread in Bangladeshi wells? How is the level of arsenic in well water tested accurately? Below what arsenic concentration is the well water safe to drink? How has chemistry been applied to remove arsenic, making water reasonably safe to drink by this boy and other people in his village? How is chemistry communicated? You will find answers to these questions in this chapter. You will also learn how chemistry can help to identify and solve many problems faced by people throughout the world. Safe drinking water is only one of thousands of examples of chemistry's crucial contributions to modern society.

Welcome to the world of chemical science! This chapter describes how modern chemical research is done and how it can be applied to questions and problems that affect our daily lives. It also provides an overview of the methods of science and the fundamental ideas of chemistry. These ideas are extremely important and very powerful. They will be applied over and over throughout your study of chemistry and of many other sciences.

1-1 Why Care About Chemistry?

Why study chemistry? There are many good reasons. **Chemistry** is *the science of matter and its transformations from one form to another*. **Matter** is *anything that has mass and occupies space*. Consequently, chemistry has enormous impact on our daily lives, on other sciences, and even on areas as diverse as art, music, cooking, and recreation. Chemical transformations happen all the time, everywhere. Chemistry is intimately involved in the air we breathe and the reasons we need to breathe it; in purifying the water we drink; in growing, cooking, and digesting the food we eat; and in the discovery and production of medicines to help maintain health. Chemists continually provide new ways of transforming matter into different forms with useful properties. Examples include plastic DVD discs; the microchips and batteries in cell phones, tablets, or laptop computers; the carbon-fiber composites in high-end bicycles; and the materials in hip replacements, surgical stents, and other medical devices.

Very human accounts of how fascinating—even romantic—chemistry can be are provided by Primo Levi in his autobiography, *The Periodic Table* (New York: Schocken Books, 1984), and by Oliver Sacks in *Uncle Tungsten: Memories of a Chemical Boyhood* (New York: Knopf, 2001). Levi was sentenced to a death camp during World War II but survived because the Nazis found his chemistry skills useful; those same skills made him a special kind of writer. Sacks describes how his mother and other relatives encouraged his interest in metals, diamonds, magnets, medicines, and other chemicals, and how he learned that "science is a territory of freedom and friendship in the midst of tyranny and hatred."

Chemists are people who are fascinated by matter and its transformations—as you are likely to be after seeing and experiencing chemistry in action. Chemists have a unique and spectacularly successful way of thinking about and interpreting the material world around them—an atomic and molecular perspective. Knowledge and understanding of chemistry are crucial in biology, pharmacology, medicine, geology, materials science, many branches of engineering, and other sciences. Modern research is often done by teams of scientists whose members represent several of these different disciplines. In such teams, ability to communicate and collaborate is just as important as knowledge in a single field. Studying chemistry can help you learn how chemists think about the world and solve problems, which in turn can lead to effective collaborations. Such knowledge will be useful in many career paths and will help you become a better-informed citizen in a world that is becoming technologically more and more complex—and interesting.

Atoms are the extremely small particles that are the building blocks of all matter. In molecules, atoms combine to give the smallest particles with the properties of a particular substance (Section 1-10).

Chemistry, and the chemist's way of thinking, can help answer a broad range of questions—questions that might arise in your mind as you carefully observe the world around you. Here are some that have occurred to us and are answered later in this book; the section number following each question indicates where you can find the answer. There are probably many more questions like these that have occurred to you. We encourage you to add them to the list and think about them as you study chemistry.

- How many chemical elements are known and how are they formed? (Sections 1-12 and 19-1)
- Why do some substances dissolve in water but others do not? (Section 13-1)
- Why does salt help to clear snow and ice from roads? (Section 13-7c)
- What is the difference between a saturated fat, an unsaturated fat, and a polyunsaturated fat? Why are trans fats harmful? (Section 10-5c)
- Where does the energy come from to make my muscles work? (Sections 4-11 and 16-9a)
- How is carbon dioxide involved in global warming? (Section 8-11)
- Can a nuclear power plant explode like an atomic bomb? (Section 18-6a)
- Why is iron strongly attracted to a magnet but most substances are not? (Section 5-8a)
- How can a disease be caused or cured by a tiny change in a molecule? (Section 10-7e)
- How does my cell phone battery work? (Section 17-8c)
- Why is only the tip of an iceberg (about 10%) above water? (Section 9-5a)

© Cengage Learning/James Maynard

Chemistry in daily life. Chemists transform matter to make the materials in smartphones, medicines, LED flashlights, and many other items we use every day.

1-2 Cleaning Drinking Water

How modern science works and why the chemist's unique perspective is so valuable can be seen through an example. (At this point you need not fully understand the science, so don't worry if some words or ideas are unfamiliar.) From many possibilities, we have chosen to discuss an issue that directly affects more than a hundred million people worldwide—the toxic effects of arsenic in drinking water. The story of what has been called the largest mass poisoning in history has unfolded in the delta of the Ganges River in Bangladesh and northeastern India during the past 30 years.

"The whole of science is nothing more than a refinement of everyday thinking." —Albert Einstein

The element arsenic has been known for about 800 years, and its compounds have been known much longer than that. Orpiment, As_2S_3, a bright yellow mineral known to the ancient Greeks as arsenikon, probably gave the element its name. Ancient Chinese writings describe the toxicity of arsenic compounds and their use as pesticides. They also indicate that those who worked with the arsenical pesticides for two years or more began to develop symptoms of poisoning. Arsenic is a popular poison in detective stories because arsenic was often used by real-life killers. Arsenic compounds were readily available as weed or insect killers and could easily be added surreptitiously to the victim's food or drink. How then could it be that arsenic would accidentally be present in the drinking water supply of large numbers of people?

In the 1960s the United Nations Children's Fund (UNICEF) began a campaign to reduce the number of children (and adults) in Bangladesh dying of diseases such as dysentery and cholera caused by drinking water contaminated by bacteria. Most drinking water came from rivers, shallow hand-dug wells, and other surface sources. Ineffective sewage treatment and annual flooding caused by monsoon rains almost guaranteed that this surface water would be contaminated. Therefore, UNICEF and other aid agencies funded and encouraged installation of wells from which people could obtain uncontaminated water. Called tube wells because a steel tube was driven 20 to 100 feet into the ground and a hand pump installed, these wells halved the mortality rate for children under five, saving 125,000 children every year.

At the time the wells were installed, nobody checked the water for contaminants other than bacteria, apparently on the assumption that groundwater would be pure enough to drink. In many parts of the Ganges delta there are high concentrations of arsenic compounds in sediments just below the surface. Groundwater can dissolve these compounds from the sediments, producing dangerous concentrations of arsenic (above 50 ppb, parts per billion; 1 ppb is one gram of arsenic per billion grams of water). One in every five wells was found to be contaminated. Wells with more than 50 ppb arsenic serve about 20 million people. It is estimated that more than 100 million people in Bangladesh, India, and surrounding countries are drinking water that contains a greater concentration of arsenic than 10 ppb, the maximum safe level recommended by the World Health Organization.

In the early 1980s scientists A. K. Chakraborty and K. C. Saha at the School of Tropical Medicine in Kolkata (Calcutta) began to see many patients who had spots on their skin that changed from white to brown to black. There was thickening and eventual cracking of the skin, as well as other clinical symptoms of arsenic poisoning. Chakraborty and Saha were aware of water analyses carried out as early as 1978 that revealed high concentrations of arsenic. In 1987 they published in the *Indian Journal of Medical Research* their conclusions that people in both India and Bangladesh were being poisoned by arsenic in the water they were drinking. It was not until the latter part of the 1990s, however, that the full scale of the problem was recognized and a major program of testing water from the tube wells began.

According to UNICEF there are nearly nine million tube wells in Bangladesh. More than five million wells have been tested for arsenic between 2000 and the

STRDEL/AFP/Getty Images

Figure 1.1 A SONO filter in a village hut in Bangladesh. The red container holds the composite iron matrix. The aqua bucket holds the sand and charcoal filters.

present. If the concentration of arsenic was below 50 ppb, then the well's pump was painted green to indicate that the water was reasonably safe to drink. If a well tested above 50 ppb, then it was painted red. About 20% of the pumps were painted red. In more than 8300 villages 80% or more of the pumps were red, leaving the inhabitants with very few sources of clean water to drink. Once they knew which wells were contaminated, nearly a third of the people switched to a clean well, but more than half of the people with contaminated water have done nothing and are still exposed to unacceptable levels of arsenic.

This problem was so severe that in 2005 the U. S. National Academy of Engineering announced the Grainger Challenge Prize for Sustainable Development, which would award $1 million for development of an inexpensive system to reduce arsenic to safe levels in drinking water. In February 2007 the National Academy announced a winner: Abul Hussam, a professor of chemistry and biochemistry at George Mason University. Hussam invented the SONO filter, which effectively removes arsenic from drinking water, lasts for many years, and binds the arsenic so tightly that there is little or no hazard associated with disposing of used filters. A photograph of the SONO filter is shown in Figure 1.1. Although it looks crude, a simple device is exactly what is needed in a country where the average annual income is $850 per person.

The chemistry of the SONO filter is interesting, but most of the details are proprietary (that is, the inventor is keeping them secret). In groundwater the pH is between 6.5 and 7.5 and arsenic is present in both the +3 oxidation state (as H_3AsO_3) and the +5 oxidation state (as $H_2AsO_4^-$ or $HAsO_4^{2-}$). The filter must remove all three chemical species. Its primary active material is called a composite iron matrix (CIM). The CIM is made from cast-iron chips by a special process. The upper bucket of the filter contains three layers: 10 kg coarse river sand on top, then 5 to 10 kg CIM, and 10 kg brick chips and coarse river sand at the bottom. Water passes slowly through this mixture and drains into the lower bucket, which contains 10 kg coarse river sand on top, wood charcoal to adsorb organic material, 9 kg fine river sand, and 3.5 kg brick chips. Water is tapped from the bottom of the lower bucket. The taps are designed to keep the flow of water slow to ensure that the water remains in contact with the contents of both buckets for long enough to remove the dissolved arsenic species.

One kilogram (kg) weighs 2.2 pounds.

In the upper bucket, arsenic in the +3 oxidation state is oxidized to the +5 state by oxygen in the air. Manganese, which constitutes 1% to 2% by weight of the CIM, catalyzes the oxidation. The negative ions containing arsenic ($H_2AsO_4^-$ or $HAsO_4^{2-}$) become tightly attached to iron atoms at the surfaces of iron particles in the CIM, replacing hydroxide ions (OH^-) that are present on the iron surfaces. The chemical reactions are

$$=\text{FeOH} + \text{H}_2\text{AsO}_4^- \longrightarrow =\text{FeHAsO}_4^- + \text{H}_2\text{O}$$

and

$$=\text{FeOH} + \text{HAsO}_4^{2-} \longrightarrow =\text{FeAsO}_4^{2-} + \text{H}_2\text{O}$$

where =Fe represents an iron atom on the surface of an iron turning. The arsenic species are so tightly attached to the iron atoms that they do not come off, no matter how much water passes through the filter.

Chemists have contributed in many other ways to alleviating the problem of contaminated drinking water in Bangladesh. Hauke Harms of the Helmholtz Center for Environmental Research in Leipzig, Germany, has devised an inexpensive, portable kit that can detect four grams of arsenic in a billion grams of water (4 ppb)—a quantity smaller than a needle in a haystack. Ashok Gadgil of the University of California, Berkeley, has developed a method for removing arsenic from water that uses a simple battery with iron and copper electrodes to precipitate iron oxide (rust) that traps arsenic. These and many other approaches depend on applying the methods of science to improve everyday life.

1-3 How Science Is Done

The story of arsenic in drinking water illustrates many aspects of how people do science and how scientific knowledge changes and improves over time. In antiquity it was known that compounds of arsenic were poisons, even though the element itself had not been discovered. For example, the Chinese used As_2O_3 to rid their fields of pests and they knew that workers who produced the poison usually lost their hair and became ill with less than two years of exposure. This led to the reasonable hypothesis that "pee-song" (as the poison was known in Chinese) is harmful to humans. A **hypothesis** is *an idea that is tentatively proposed as an explanation for some observation and provides a basis for experimentation.* The hypothesis that pee-song or "white arsenic" (as it was known in Western countries) could be used as a poison was amply verified by the Borgias and others who poisoned their rivals to gain power or money.

Testing a hypothesis may involve collecting qualitative data, quantitative data, or both. Qualitative data, the observation that people drinking water from tube wells in Bangladesh developed spots on their skin that changed color and the skin eventually became hard and cracked, provided a clue that arsenic poisoning might be occurring. **Qualitative** means *something that does not involve numbers.* The qualitative data led to quantitative studies showing that one in five of the wells tested had arsenic levels greater than 50 ppb. **Quantitative** means *measuring something and reporting results involving numeric quantities.* Other quantitative clinical studies had already demonstrated that levels of arsenic above 50 ppb would lead to poisoning and possibly cancer.

A scientific **law** is *a statement that summarizes and explains a wide range of experimental results and has not been contradicted by experiments.* A law can predict unknown results and also can be disproved or falsified by new experiments. When the results of a new experiment contradict a law, that's exciting to a scientist. If enough scientists repeat the experiment and get the same contradictory result, then the law must be modified to account for the new results—or even discarded altogether.

A well-tested hypothesis is designated as a **theory**—*a unifying principle that explains a body of facts and the laws based on them.* (Notice that in everyday speech the word "theory" is often used to designate what a scientist means by "hypothesis".) A theory usually suggests new hypotheses and experiments, and, like a law, it may have to be modified or even discarded if contradicted by new experimental results. A **model** *makes a theory more concrete, often in a tangible or a mathematical form.* Models of molecules and ions, for example, tell us how the atoms are connected in arsenic species (such as H_3AsO_3) and give clues to why arsenic could be tightly bound to iron atoms at the surface of a cast-iron chip. Molecular models can be constructed by using spheres to represent atoms and sticks to represent the connections between the atoms. Or a computer can be used to calculate the locations of the atoms and display model molecular structures on a screen (as was done to create the models of arsenic-containing species shown nearby). The theories that matter is made of atoms and molecules, that atoms are arranged in specific molecular structures, and that the properties of matter depend on those structures are fundamental to chemists' unique atomic/molecular perspective on the world and to nearly everything modern chemists do. Clearly it is important that you become as familiar as you can with these theories and with models based on them.

Another important aspect of the way science is done involves communication. Science is based on experiments and on hypotheses, laws, and theories that can be contradicted by experiments. Therefore, it is essential that experimental results be communicated

How science is done is dealt with in *Oxygen*, a play written by chemists Carl Djerassi and Roald Hoffmann that premiered in 2001. By revisiting the discovery of oxygen, the play provides many insights regarding the process of science and the people who make science their life's work.

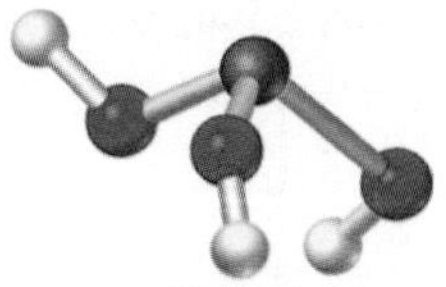

H_3AsO_3

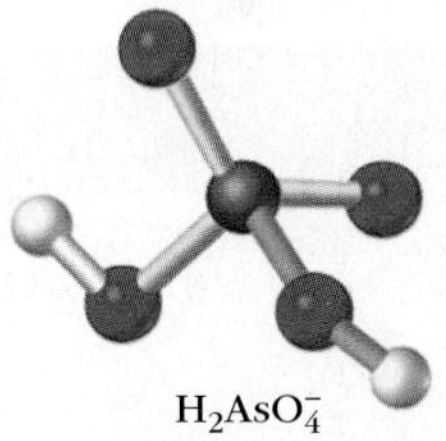

$H_2AsO_4^-$

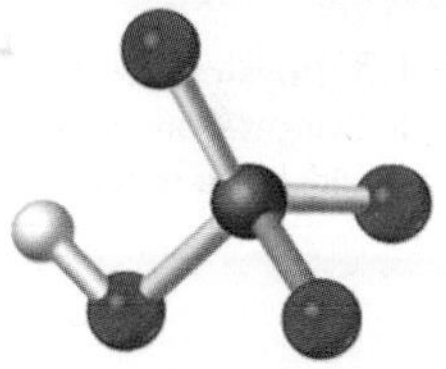

$HAsO_4^{2-}$

Models of structures of arsenic-containing species found in drinking water in Bangladesh.

In this book, models of molecular structures are used to help you visualize chemistry at the atomic and molecular levels. Atoms are color coded: H, light gray; C, dark gray; O, red. A chart showing the full color key is inside the back cover.

Hydrogen (H) | Carbon (C) | Oxygen (O)

Figure 1.2 Substances have characteristic physical properties. The diamond and gold in the ring are discernibly different from each other and from the copper pipe, iron nails, liquid water, and plastic cup.

to all scientists working in any specific area of research as quickly and accurately as possible. Scientific communication allows contributions to be made by scientists in different parts of the world and greatly enhances the rapidity with which science can develop. In addition, communication among members of scientific research teams is crucial to their success. Poor communication (and failure of some experts to read and act on published information) kept the full enormity of the arsenic problem from becoming known for almost 10 years. Had communication been better, many fewer residents of Bangladesh and India would be ill today. The importance of scientific communication is emphasized by the fact that the Internet was created not by commercial interests, but by scientists who saw its great potential for communicating scientific information.

In the remainder of this chapter we discuss fundamental concepts of chemistry that have been revealed by applying the processes of science to the study of matter. We begin by considering how matter can be classified according to characteristic properties.

1-4 Identifying Matter: Physical Properties

One type of matter can be distinguished from another by observing its properties and classifying the matter according to those properties. A **substance** is *a type of matter that has the same properties and the same composition throughout* a sample. Each substance has characteristic properties that are different from the properties of any other substance (Figure 1.2). In addition, one sample of a substance has the same composition as every other sample of that substance—it consists of the same stuff in the same proportions.

You can distinguish sugar from water because you know that sugar consists of small, white particles of solid, while water is a colorless liquid. Metals can be recognized as a class of substances because they usually are solids, have high densities, feel cold to the touch, and have shiny surfaces. *Properties that can be observed and measured without changing the composition of a substance* are called **physical properties**.

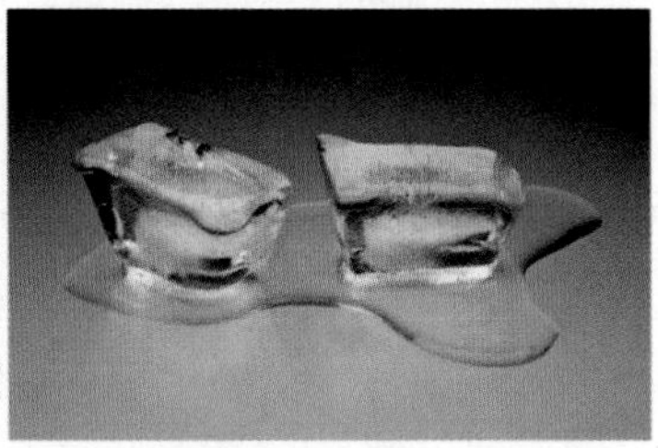

Figure 1.3 Physical change. When ice melts it changes—physically—from a solid to a liquid, but it is still water.

1-4a Physical Change

As a substance's temperature or pressure changes, or if it is mechanically manipulated, some of its physical properties may change. *Changes in the physical properties of a substance* are called **physical changes**. The same substance is present before and after a physical change, but the substance's physical state or the gross size and shape of its pieces may have changed. Examples are melting a solid (Figure 1.3), boiling a liquid, hammering a copper wire into a flat shape, and grinding sugar into a fine powder.

Some Physical Properties
Temperature
Pressure
Mass
Volume
State (solid, liquid, gas)
Melting point
Boiling point
Density
Color
Shape of solid crystals
Hardness, brittleness
Heat capacity
Thermal conductivity
Electrical conductivity

1-4b States of Matter

An easily observed and very useful property of matter is its **physical state**. Is it a *solid, liquid, or gas?* A **solid** can be recognized because it *has a rigid shape and a fixed volume* that changes very little as temperature and pressure change. Like a solid, a **liquid** *has a fixed volume, but* a liquid *is fluid—it takes the shape of its container* and has no definite form of its own. A **gas** *is* also *fluid but expands to fill whatever container it occupies*; gas volume varies considerably with temperature and pressure. For most substances, when compared at the same conditions, the volume of the solid is slightly less than the volume of the same mass of liquid, but the volume of the same mass of gas is much, much larger. As the temperature is raised, most solids melt to form liquids; eventually, if the temperature is raised enough, most liquids boil to form gases. Whether a substance is solid, liquid, or gas under typical conditions is an important physical property that can help to identify the substance.

1-4c Melting and Boiling Point

Another property that helps to identify a substance is the substance's **melting point**, *the temperature at which the solid melts* (or *the temperature at which the liquid freezes*, the **freezing point**, which is the same temperature). Also characteristic is the substance's **boiling point**, *the temperature at which the liquid boils*. If two or more substances are in a mixture, the melting point depends on how much of each is present, but for a single substance the melting point is always the same. This is also true of the boiling point (as long as the pressure on the boiling liquid is the same). In addition, the melting point of a pure crystalline sample of a substance is sharp—there is almost no change in temperature as the sample melts. When a mixture of two or more substances melts, the temperature when liquid first appears can be quite different from the temperature when the last of the solid is gone.

Temperature is *the property of matter that determines whether one object can transfer energy to another object by heating it*. It is represented by the symbol *T*. Energy transfers of its own accord from an object at a higher temperature to a cooler object. In the United States, everyday temperatures are reported using the Fahrenheit temperature scale. On this scale the freezing point of water is by definition 32 °F and the boiling point is 212 °F. The **Celsius temperature scale** is used in most countries of the world and in science. On this scale, *0 °C is the freezing point and 100 °C is the boiling point of pure water at a pressure of one atmosphere*. The number of units between the freezing and boiling points of water is 180 Fahrenheit degrees and 100 Celsius degrees. This means that the Celsius degree is almost twice as large as the Fahrenheit degree. It takes only 5 Celsius degrees to cover the same temperature range as 9 Fahrenheit degrees, and this relationship can be used to calculate a temperature on one scale from a temperature on the other (see Appendix B-2).

If you need to convert from the Fahrenheit to the Celsius scale or from Celsius to Fahrenheit, an explanation of how to do so is in Appendix B-2.

Because temperatures in scientific studies are usually measured in Celsius units, there is little need to make conversions to and from the Fahrenheit scale, but it is quite useful to be familiar with how large various Celsius temperatures are. For example, it is useful to know that water freezes at 0 °C and boils at 100 °C, a comfortable room temperature is about 22 °C, your body temperature is 37 °C, and the hottest water you could put your hand into without serious burns is about 60 °C.

CONCEPTUAL EXERCISE 1.1

Temperature

(a) Determine which temperature is higher, 110 °C or 180 °F.
(b) Determine which temperature is lower, 36 °C or 100 °F.
(c) The melting point of gallium is 29.8 °C. If you hold a sample of gallium in your hand, will it melt? Explain your answer.

Answers to **EXERCISES** are provided at the back of this book in Appendix L.

EXERCISES that are labeled **CONCEPTUAL** are designed to test your understanding of one or more concepts; they usually involve qualitative rather than quantitative thinking.

1-4d Density

Another physical property that is often used to help identify a substance is **density**, *the ratio of the mass of a sample to its volume*. If you have ten pounds of sugar, it occupies ten times the volume that one pound of sugar does. In mathematical terms, a substance's volume is directly proportional to its mass. This means that a substance's density has the same value regardless of how big the sample is.

$$\text{Density} = \frac{\text{mass}}{\text{volume}} \qquad d = \frac{m}{V}$$

The density of a substance varies depending on the temperature and the pressure. Densities of liquids and solids change very little as pressure changes, and they change less with temperature than do densities of gases. Because the volume of a gas varies significantly with temperature and pressure, the density of a gas can help identify the gas only if the temperature and pressure are specified.

Even if they look similar, you can tell a sample of aluminum from a sample of lead by picking each up. Your brain will automatically estimate which sample has greater

Table 1.1 Densities at 20 °C

Substance	Density (g/mL)
Butane	0.579
Ethanol	0.789
Benzene	0.880
Water	0.998
Bromobenzene	1.49
Magnesium	1.74
Sodium chloride	2.16
Aluminum	2.70
Titanium	4.50
Zinc	7.14
Iron	7.86
Nickel	8.90
Copper	8.96
Lead	11.34
Mercury	13.55
Gold	19.32

Useful sources of data are listed in Appendix C at the back of this book. A good print source is the *CRC Handbook of Chemistry and Physics,* published by the CRC Press. Information is also available via the Internet—for example, the National Institute for Standards and Technology's Webbook at http://webbook.nist.gov.

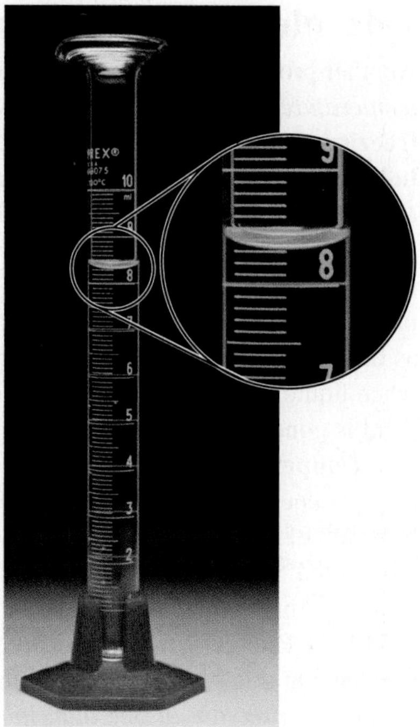

Cengage Learning/Charles D. Winters

Figure 1.4 Graduated cylinder containing 8.30 mL of liquid.

mass for the same volume, telling you which is the lead. Data in Table 1.1 indicate that aluminum has a density of 2.70 g/mL, placing it among the least dense metals. Lead's density is 11.34 g/mL, so a sample of lead is much heavier than a sample of aluminum of the same size.

Suppose that you are trying to identify a liquid that you think might be ethanol (ethyl alcohol), and you want to determine its density. You could weigh a clean, dry graduated cylinder and then add some of the liquid to it (Figure 1.4). Suppose that, from the markings on the cylinder, you read the volume of liquid to be 8.30 mL (at 20 °C). You could then weigh the cylinder with the liquid and subtract the mass of the empty cylinder to obtain the mass of liquid. Suppose the liquid mass is 6.544 g. The density can then be calculated as

$$d = \frac{m}{V} = \frac{6.544\ \text{g}}{8.30\ \text{mL}} = 0.788\ \text{g/mL}$$

From a table of physical properties of various substances you find that the density of ethanol is 0.789 g/mL, which helps confirm your suspicion that the substance is ethanol.

CONCEPTUAL EXERCISE 1.2

Density of Liquids

When 5.0 mL each of linseed oil, water, and kerosene are put into a large test tube, they form three layers, as shown in the liquid densities photo nearby.

(a) List the three liquids in order of increasing density (smallest density first, largest density last).

(b) If an additional 2.0 mL of linseed oil is poured into the test tube, what will happen? Draw the tube and describe its appearance.

(c) If 5.0 mL of kerosene is added to the test tube as well as the 2.0 mL of linseed oil in part (b), will there be a permanent change in the order of liquids from top to bottom of the tube? Why or why not?

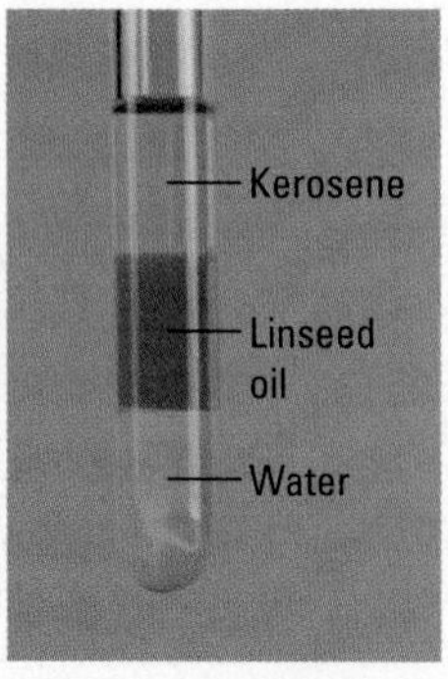
Kerosene
Linseed oil
Water

© Cengage Learning/James Maynard

Liquid densities. Kerosene, linseed oil, and water have different densities.

EXERCISE 1.3

Physical Properties and Changes

Identify each physical property and physical change mentioned in each of these statements. Also identify the qualitative and the quantitative information given in each statement.

(a) The blue chemical compound azulene melts at 99 °C.
(b) The white crystals of table salt are cubic.
(c) A sample of lead has a mass of 0.123 g and melts at 327 °C.
(d) Ethanol is a colorless liquid that vaporizes easily; it boils at 78 °C and its density at 20 °C is 0.789 g/mL.

1-5 Measurements, Units, and Calculations

Determining a quantitative property such as density requires measurements and calculations. A **measurement** involves *comparing the quantity to be measured*, such as the volume of ethanol, *with a unit,* such as mL. We count how many of the units match the quantity being measured. In Figure 1.4, the volume of ethanol was measured in a graduated cylinder calibrated in milliliters and tenths of a milliliter. From the graduations you can count 8 mL and 3 tenths of a mL. The last digit, 0, is estimated between the graduations. This gives a volume of 8.30 mL. The calibrations helped us to count 8.30 of the units, mL.

A **physical quantity** is *the result of a measurement and consists of a number and a unit*. The number and the unit behave mathematically as if they were multiplied; that is, 8.30 mL is 8.30 times the unit mL. Both number and unit should be included in calculations. For example, the densities in Table 1.1 have units of grams per milliliter, g/mL, because density is defined as the mass of a sample divided by its volume. When a mass is divided by a volume, the units (g for the mass and mL for the volume) are also divided. The result is grams divided by milliliters, g/mL. Both numbers *and units* follow the rules of algebra. This is an example of **dimensional analysis**, *a method of using units in calculations to check for correctness*. More detailed descriptions of dimensional analysis are given in Appendix A-2. We will use this technique for problem solving throughout the book.

In this book, units and dimensional analysis techniques are introduced at the first point where you need to know them. Appendices A and B provide all of this information in one place.

1-5a SI Units

Because nearly every scientific measurement is reported as a number times a unit, units are extremely important. Consequently, scientists have agreed on a single system of units that are carefully and reproducibly defined and are used universally. This has the advantage that when a scientist anywhere in the world reports the result of a measurement, other scientists, no matter what their nationality, can immediately interpret that result correctly.

The International System of Units, officially named in French as *Système International d'Unités* (**SI units**), is *the measurement system of science*. Originally known as the *metric system*, SI units have two main features: different sized units are indicated by prefixes based on powers of 10, and all units are derived from seven fundamental units, called base units. Here we briefly describe each of these two features; complete information is in Appendix B.

SI base units are *a set of units from which all other SI units are derived*. The seven SI base units are listed in Table 1.2. For example, the unit of volume, liter (L), is derived from the SI unit of length, meter (m). The volume of a cube equals the length of an edge cubed: if l is the edge length,

Table 1.2 SI Base Units

Physical Quantity	Name of Unit	Symbol
Length	Meter	m
Mass	Kilogram	kg
Time	Second	s
Temperature	Kelvin	K
Amount of substance	Mole	mol
Electric current	Ampere	A
Luminous intensity	Candela	cd

Other SI units are derived from these fundamental units.

Table 1.3 Some Prefixes Used in the SI and Metric Systems

Prefix	Abbreviation	Meaning	Example
mega	M	10^6	1 megaton = 1×10^6 tons
kilo	k	10^3	1 kilometer (km) = 1×10^3 meter (m) 1 kilogram (kg) = 1×10^3 gram (g)
deci	d	10^{-1}	1 decimeter (dm) = 1×10^{-1} m 1 deciliter (dL) = 1×10^{-1} liter (L)
centi	c	10^{-2}	1 centimeter (cm) = 1×10^{-2} m
milli	m	10^{-3}	1 milligram (mg) = 1×10^{-3} g
micro	μ	10^{-6}	1 micrometer (μm) = 1×10^{-6} m
nano	n	10^{-9}	1 nanometer (nm) = 1×10^{-9} m 1 nanogram (ng) = 1×10^{-9} g
pico	p	10^{-12}	1 picometer (pm) = 1×10^{-12} m
femto	f	10^{-15}	1 femtogram (fg) = 1×10^{-15} g

then the volume is l^3. Thus units of volume are units of length cubed and the SI unit of volume is m^3 (cubic meter or meter cubed). A meter is about the distance from your nose to the tip of the fingers on your outstretched arm, much bigger than most chemistry laboratory apparatus, so a cubic meter is quite a large volume for laboratory work. Instead we use a cubic decimeter. (The prefix deci means 1/10, so a cubic decimeter is the volume of a cube 1/10 of a meter on a side, which is about the same volume as a quart.) The liter is defined as one cubic decimeter (1 L = 1 dm^3). [Notice that a cubic decimeter is 1/10 meter times 1/10 meter times 1/10 meter and is therefore $(1/10)^3 = 1/1000$ of a cubic meter. The exponent 3 applies to the prefix as well as to the unit: 1 dm^3 = 1 $(dm)^3$, not 1 $d(m^3)$.].

Using 1×10^{-3} to represent 1/1000 or one one-thousandth is called scientific notation. For example, 0.000001 is 1×10^{-6} (the decimal point moves six places to the right to give the −6 exponent) and 2,000,000 is 2×10^6 (the decimal point moves six places to the left to give the +6 exponent). If you are not familiar with scientific notation, study Appendix A-5.

An **SI prefix** *represents a number that is a power of 10; that number is multiplied times a base unit to get a new unit that is larger or smaller than the base unit.* SI prefixes commonly encountered in chemistry are listed in Table 1.3. For example, the unit of volume used in densities in Table 1.1 is the milliliter (mL). According to Table 1.3, the prefix milli and its abbreviation m mean multiply the unit by 1/1000 (1×10^{-3}). Thus a milliliter is 1/1000 the volume of a liter. It is important to know the SI prefixes, because they are used throughout science to indicate the sizes of units.

EXERCISE 1.4

SI Units and Prefixes

Show mathematically that 1 mL = 1 cm^3.

1-5b Significant Digits and Rounding Numbers

To determine the number of significant figures in a quantity, read the number from left to right and count all digits, starting with the first digit that is not zero. All the digits counted are significant except any zeros that are used only to position the decimal point, such as the zeros in 4500, which has two significant figures.

In the experiment where ethanol was identified based on its density, you probably noticed that the measured density, 0.788 g/mL, was not identical to the density reported in the Table 1.1, 0.789 g/mL. This is because no scientific measurement can be completely accurate. When the volume of ethanol was measured using a graduated cylinder, the value recorded was 8.30 mL, but as you can see from the picture of the graduated cylinder (Figure 1.4) another observer might have recorded 8.31 mL. Because the measurement involves estimating the last digit (on the right), we would not record 8.300 mL—that would be too exact. We say that the volume of ethanol, 8.30 mL, has three significant digits (significant figures) because there is uncertainty in the rightmost digit.

When we use this result in a calculation, the result of the calculation will also have some uncertainty. For example, if the volume had been recorded as 8.31 mL, then the density would be calculated as 6.544 g/8.31 mL = 0.787 g/mL, which is 0.001 g/mL less than calculated before. To avoid reporting a result more accurately than the experimental measurements warrant, we report all digits that are certain and one digit that is uncertain. In this case, 0.78 gives the certain digits, and the last digit (0.007 or 0.008) is uncertain, so the result is reported to three significant digits, 0.788 g/mL. The next two paragraphs briefly describe rules for handling significant digits in calculations.

If you divide 6.544 by 8.30 on a scientific calculator, the calculator might report 0.78843375, but only the first three digits to the right of the decimal point are meaningful. Therefore the result should be rounded to three significant digits, giving 0.788.

In *addition or subtraction*, the number of decimal places in the result equals the number of decimal places in the number with the fewest decimal places. Suppose you add these three numbers:

0.12	2 significant figures	2 decimal places
1.6	2 significant figures	1 decimal place
10.976	5 significant figures	3 decimal places
12.696 should be rounded to 12.7		1 decimal place

This sum should be reported as 12.7, a number with one decimal place, because 1.6 has only one decimal place; 12.7 has three significant figures.

In *multiplication or division*, the number of significant figures in the result is the same as that in the quantity with the fewest significant figures.

$$\frac{0.7608}{0.0546} = 13.9 \quad \text{or, in scientific notation,} \quad 1.39 \times 10^{1}$$

The numerator, 0.7608, has four significant figures, but the denominator, 0.0546, has only three, so the result must be reported with three significant figures.

When you *round a number* to reduce the number of digits,

- The last digit retained is left unchanged if the following digit is less than 5 (4.327 rounded to two significant digits is 4.3)
- The last digit retained is increased by 1 if the following digit is greater than 5 or is a 5 followed by other non-zero digits (4.573 rounded to two significant figures is 4.6)
- If the last digit retained is followed by a single digit 5 only or by a 5 followed by zeroes, then the last digit retained is increased by 1 if it is odd and remains the same if it is even (4.75 rounded to two significant digits is 4.8; 4.850 rounded to two significant digits is 4.8)

Rounding Numbers to Three Significant Digits

Full Number	Rounded
12.696	12.7
16.249	16.2
18.35	18.4
18.45	18.4
24.752	24.7
18.351	18.4

When using a calculator, include all digits allowed by the calculator and round only at the end of the calculation. Rounding more than once introduces small errors that can accumulate later in the calculation. If your results do not quite agree with those in the appendices of this book, rounding too early may be the source of the disagreement.

More examples of deciding how many significant digits should be reported in the result of a calculation and of rounding numbers are given in Appendix A-3. If you are uncertain about determining the number of significant figures or about rounding, you should study Appendix A-3.

EXERCISE 1.5

Significant Digits and Rounding Numbers

To test your understanding of significant digits and rounding, answer these questions and check your results with the results in the back of the book. If you don't know how to answer or if you get more than one or two wrong, study Appendix A-3 carefully.

(a) How many significant digits are in each number?

(i) 1.25 g (ii) 0.00125 g (iii) 0.020 g (iv) 100 g
(v) 100 cm/m (vi) 4280. m (vii) π = 3.1415926 . . .

In this book we use the convention that if a decimal point is given, all digits to the left of it are significant. 7000 mL has one significant figure. 7000. mL has four significant figures.

(b) Round each number to three significant digits.
   (i) 12.696 (ii) 12.249 (iii) 18.35 (iv) 18.45
   (v) 24.752 (vi) 18.351

(c) Do these calculations and round each result to the proper number of significant digits.
   (i) $15.80 + 0.0060 \times 2.0 + 0.081$ (ii) $\dfrac{55.0}{12.34}$
   (iii) $\dfrac{12.7732 - 2.3317}{5.007}$ (iv) $2.16 \times 103 + 4.01 \times 102$

1-5c Calculations Involving Density

Because mercury and mercury vapor are poisonous, carrying a gallon of it around is not a good idea unless the mercury is in a sealed container.

Suppose that you want to know whether you could lift a gallon (3784 mL) of the liquid metal mercury. To answer the question, calculate the mass of the mercury using the density, 13.55 g/mL, obtained from Table 1.1. One way to do this is to use the equation that defines density, $d = m/V$. Then solve algebraically for m, and calculate the result:

$$m = V \times d = 3784\ \cancel{\text{mL}} \times \frac{13.55\ \text{g}}{1\ \cancel{\text{mL}}} = 51{,}270\ \text{g} \qquad [1.1]$$

This equation emphasizes the fact that mass is proportional to volume, because the volume is multiplied by a proportionality constant, the density. Notice also that the units of volume (mL) appeared once in the denominator of a fraction and once in the numerator, thereby dividing out (canceling) and leaving only mass units (g). The result, 51,270 g, is more than 100 pounds, so you could probably lift the mercury, but not easily.

In Equation 1.1, a known quantity (the volume) was multiplied by a proportionality factor (the density), and the units canceled, giving a result (the mass) with appropriate units. A general approach to this kind of problem is to recognize that the quantity you want to calculate (the mass) is proportional to a quantity whose value you know (the volume). Then use a proportionality factor that relates the two quantities, setting things up so that the units cancel.

$$\cancel{\text{known quantity}}\ \text{units} \times \frac{\text{desired quantity units}}{\cancel{\text{known quantity}}\ \text{units}} = \text{desired quantity units}$$

proportionality (conversion) factor

$$3784\ \text{mL} \times \frac{13.55\ \text{g}}{1\ \text{mL}} = 51{,}270\ \text{g}$$

A **proportionality factor** is *a ratio (fraction) whose numerator and denominator have different units but refer to the same thing*. In the preceding example, the proportionality factor is the density, which relates the mass and volume of the same sample of mercury. A proportionality factor is often called a **conversion factor** because it enables us to convert from one kind of unit to a different kind of unit. When you do this kind of calculation, it is important to *check that the units of the result are appropriate*. For example, if you are calculating a volume, the result should be in L, mL, cm^3, or some other volume unit.

Because a conversion factor is a fraction, every conversion factor can be expressed in two ways. The conversion factor in the example just given could be expressed either as the density or as its reciprocal:

$$\frac{13.55\ \text{g}}{1\ \text{mL}} \quad \text{or} \quad \frac{1\ \text{mL}}{13.55\ \text{g}}$$

The first fraction enables conversion from volume units (mL) to mass units (g). The second allows mass units to be converted to volume units. Which conversion factor to use depends on which units are in the known quantity and which units are in the quantity that we want to calculate. Setting up the calculation so that the units cancel ensures that we are using the appropriate conversion factor. (See Appendix A-2 for more examples.)

PROBLEM-SOLVING EXAMPLE 1.1

Density

In an old movie, thieves are shown running off with pieces of gold bullion that are about a foot long and have a square cross section of about six inches. The volume of each piece of gold is 7000 mL. Is what the movie shows physically possible? [Hint: calculate the mass of gold and express the result in pounds (lb). 1 lb = 454 g.]

Result Probably not; 1.4×10^5 g; 300 lb

Strategy and Explanation A good approach to problem solving is to (1) analyze the problem, (2) plan a solution, (3) execute the plan, and (4) check your result to see whether it is reasonable. (These four steps are described in more detail in Appendix A-1.)

Step 1: *Analyze the problem.* You are asked to calculate the mass of the gold, and you know the volume is 7000 mL (one significant figure).

Step 2: *Plan a solution.* Density relates mass and volume and is the appropriate proportionality factor, so look up the density in a table. Mass is proportional to volume, so the volume either has to be multiplied by the density or divided by the density. Use the units to decide which. Use the information that 1 lb = 454 g to obtain a conversion factor for the units.

Step 3: *Execute the plan.* According to Table 1.1, the density of gold is 19.32 g/mL. Setting up the calculation so that the unit (milliliter) cancels gives

$$7000 \text{ mL} \times \frac{19.32 \text{ g}}{1 \text{ mL}} = 1.35 \times 10^5 \text{ g}$$

This can be converted to pounds

$$1.35 \times 10^5 \text{ g} \times \frac{1 \text{ lb}}{454 \text{ g}} = 297 \text{ lb which rounds to 300 lb}$$

Notice that the result is expressed to one significant figure, because the volume was given to only one significant figure and only multiplications and divisions were done. The intermediate result, 1.35×10^5 g, was expressed to more significant figures because rounding should only be done at the end of a calculation.

☑ **Reasonable Result Check** The units are mass units, so they are reasonable. Gold is nearly 20 times denser than water. A liter (1000 mL) of water is about a quart and a quart of water (two pints) weighs about two pounds. Seven liters (7000 mL) of water should weigh 14 lb, and 20 times 14 gives 280 lb, which rounds to 300 lb, so 300 lb is reasonable. The movie is not—few people could run while carrying a 300-lb object!

Rounding the intermediate result, 1.35×10^5, to 1×10^5 would have given a quite different result of 200 lb.

The checkmark ☑ symbol accompanied by the words "Reasonable Result Check" will be used throughout this book to indicate how to check the answer to a problem to make certain a reasonable result has been obtained.

PROBLEM-SOLVING PRACTICE 1.1

Calculate the volume occupied by a 4.33-g sample of benzene.

PROBLEM-SOLVING PRACTICE answers are provided at the back of this book in Appendix K.

1-5d Problem-Solving Strategy

This book includes many examples, like Problem-Solving Example 1.1, that illustrate general problem-solving techniques and ways to approach specific types of problems. Usually, each of these examples states a problem; gives the result; explains one way to analyze the problem, plan a solution, and execute the plan; and describes a way to check that the result is reasonable. We urge you to first try to solve the problem on your own. Then check to see whether your result matches the one given. If it does not match, try again before reading the explanation. After you have tried twice, read the explanation to find out why your reasoning differs from that given. If your result is correct, but your reasoning differs from the explanation, you may have discovered an alternative solution to the problem. Finally, work out the Problem-Solving Practice that accompanies the example. It relates to the same concept and allows you to improve your problem-solving skills.

1-6 Chemical Changes and Chemical Properties

In addition to its physical properties, another way to identify a substance is to observe how it reacts chemically. For example, if you heat a white, granular solid carefully and it caramelizes (turns brown and becomes a syrupy liquid—see Figure 1.5), it is a good bet that the white solid is ordinary table sugar (sucrose). When heated gently, sucrose decomposes to give water and other new substances. If you heat sucrose very hot, it will char, leaving behind a black residue that is mainly carbon (and is hard to clean up). If you drip some water onto a sample of sodium metal, the sodium will react violently with the water, producing an aqueous solution of lye (sodium hydroxide) and a flammable gas, hydrogen (Figure 1.6). These are examples of a **chemical change** or **chemical reaction**, *processes in which one or more substances* (the **reactants**) *are transformed into one or more different substances* (the **products**). Reactant substances are replaced by product substances as the reaction occurs. This process is indicated by writing the reactants, an arrow, and then the products:

Sucrose	$\longrightarrow$	Carbon + Water
Reactant	changes to	Products

Chemical reactions make chemistry interesting, exciting, and valuable. If you know how, you can make pure drinking water, clothing from crude petroleum, or even a silk purse from a sow's ear (it has been done). This is a very empowering idea, and human society has gained a great deal from it. Our way of life is greatly enhanced by our ability to use and

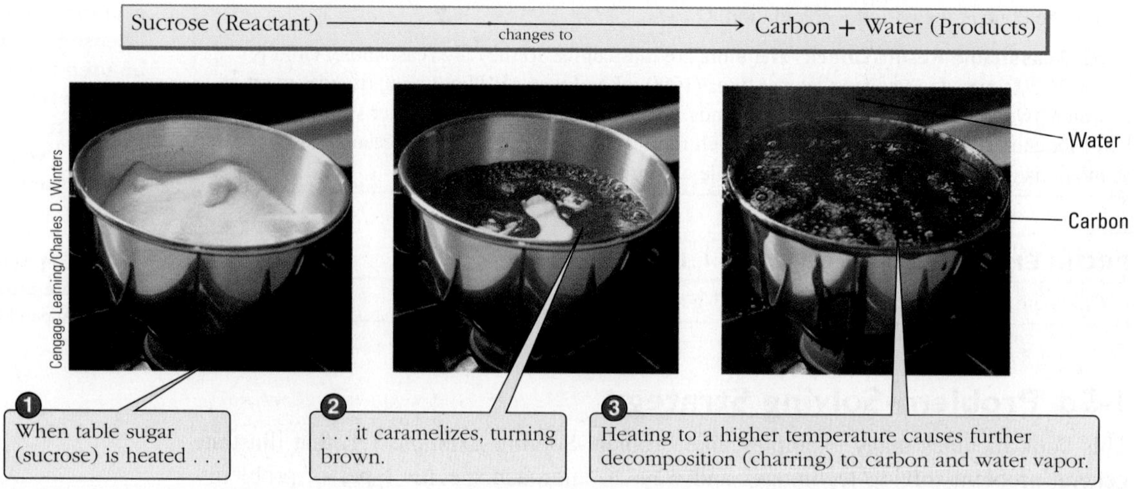
Sucrose (Reactant) changes to Carbon + Water (Products)
Water
Carbon
Cengage Learning/Charles D. Winters
1 When table sugar (sucrose) is heated . . .
2 . . . it caramelizes, turning brown.
3 Heating to a higher temperature causes further decomposition (charring) to carbon and water vapor.

Figure 1.5 Chemical change. Heat can caramelize or char sugar.

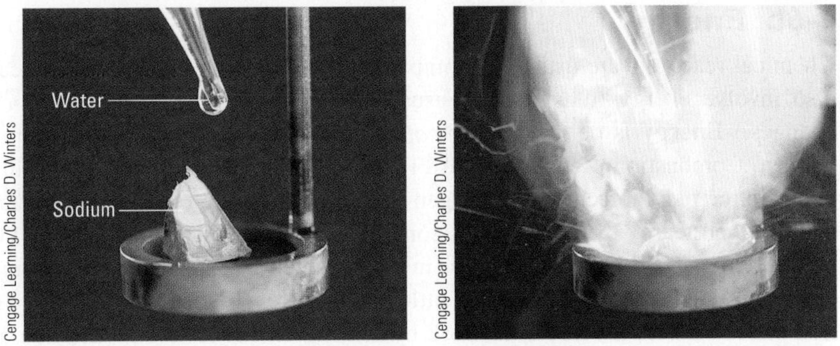
Water
Sodium
Cengage Learning/Charles D. Winters
Cengage Learning/Charles D. Winters

Figure 1.6 Chemical change. When a drop of water contacts sodium, a violent reaction produces flammable hydrogen gas and a solution of sodium hydroxide (lye). The solution is alkaline (basic). Motion, large temperature change, and emission of light when substances are mixed are all evidence that a chemical reaction is occurring.

control chemical reactions. And life itself is based on chemical reactions. Biological cells are filled with water-based solutions in which thousands of chemical reactions are happening all the time.

1-6a Chemical Properties

A substance's **chemical properties** *describe the kinds of chemical reactions the substance can undergo*. One chemical property of metallic sodium is that it reacts rapidly with water to produce hydrogen and a solution of sodium hydroxide (Figure 1.6). Because it also reacts rapidly with air and a number of other substances, sodium is said to have a more general chemical property: It is highly reactive. A chemical property of substances known as metal carbonates is that they produce carbon dioxide when treated with an acid (Figure 1.7). Fuels are substances that have the chemical property of reacting with oxygen or air and at the same time transferring large quantities of energy to their surroundings. An example is natural gas (mainly methane), which is shown reacting with oxygen from the air in a gas stove in Figure 1.8. A substance's chemical properties tell us how it will behave when it contacts air or water, when it is heated or cooled, when it is exposed to sunlight, or when it is mixed with another substance. Such knowledge is very useful to chemists, biochemists, geologists, chemical engineers, and many other kinds of scientists.

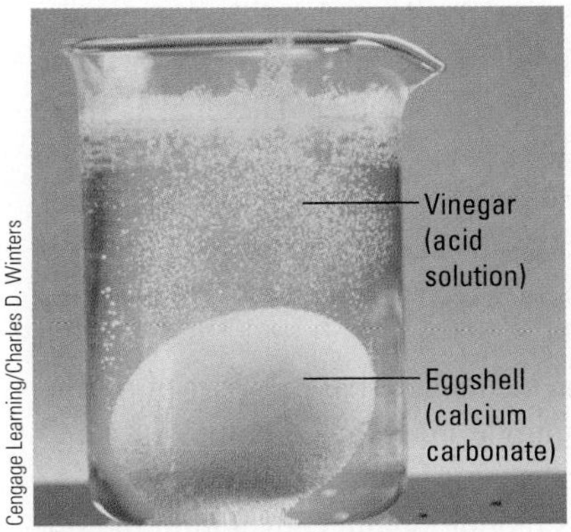
Vinegar (acid solution)
Eggshell (calcium carbonate)
Cengage Learning/Charles D. Winters

Figure 1.7 Chemical change. Vinegar, an acid, reacts with eggshell, mainly calcium carbonate, and colorless carbon dioxide gas bubbles away. The gas bubbles are one kind of evidence of a chemical reaction.

Cengage Learning/George Semple

Figure 1.8 Combustion of natural gas. Natural gas, which in the United States consists mostly of methane, burns in air, transferring energy that raises the temperature of its surroundings.

Cengage Learning/Charles D. Winters

Figure 1.9 Transforming energy. In each of these light sticks a chemical reaction transforms energy stored in molecules into light.

1-6b Energy

Chemical reactions are usually accompanied by transfers of energy. (Physical changes also involve energy transfers, but usually they are smaller than those for chemical changes.) **Energy** is defined as *the capacity to do work*—that is, to make something happen. Combustion of a fuel, as in Figure 1.8, transforms energy stored in chemical bonds in the fuel molecules and oxygen molecules into motion of the product molecules and of other nearby molecules. This corresponds to a higher temperature in the vicinity of the flame. The chemical reaction in a light stick transforms energy stored in molecules into light energy, with only a little heat transfer (Figure 1.9). A chemical reaction in a battery makes a cell phone work by forcing electrons to flow through an electric circuit.

Energy supplied from somewhere else can cause chemical reactions to occur. For example, photosynthesis takes place when sunlight illuminates green plants. Some of the sunlight's energy is stored in carbohydrate molecules and oxygen molecules that are produced from carbon dioxide and water by photosynthesis. Aluminum, which you may have used as foil to wrap and store food, is produced by passing electricity through a molten, aluminum-containing ore (Section 19-4d). You consume and metabolize food, using the energy stored in food molecules to cause chemical reactions to occur in the cells of your body. The relation between chemical changes and energy is an important theme of chemistry.

CONCEPTUAL EXERCISE 1.6

Chemical and Physical Changes

Identify the chemical and physical changes that are described in this statement: Propane gas burns, and the high temperature produced by the combustion reaction hard-boils an egg.

1-7 Classifying Matter: Substances and Mixtures

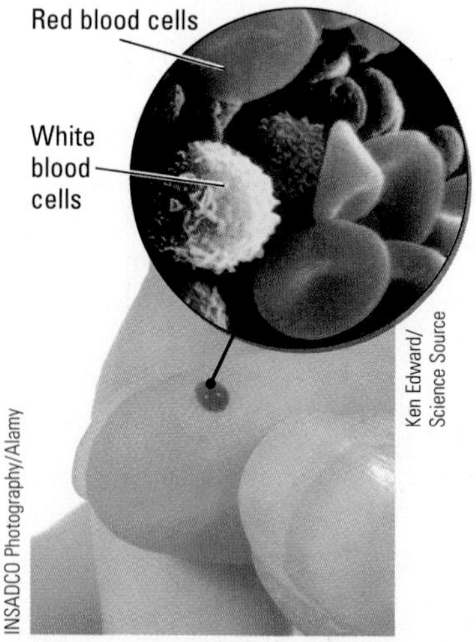
Red blood cells
White blood cells
Ken Edward/ Science Source
INSADCO Photography/Alamy

Figure 1.10 A heterogeneous mixture. Blood appears to be uniform to the unaided eye, but a microscope reveals that it is not homogeneous. The properties of red blood cells differ from the properties of the surrounding blood plasma, for example.

Once its chemical and physical properties are known, a sample of matter can be classified on the basis of those properties. Most of the matter we encounter every day is like the composite iron matrix of a SONO filter, like concrete, or like the carbon fiber composite frame of a high-tech bicycle—not uniform throughout. There are variations in color, hardness, and other properties from one part of a sample to another. This makes these materials complicated, but also interesting. A major advance in chemistry occurred when it was realized that it was possible to separate several component substances from such nonuniform samples. For example, the fact that groundwater can separate (dissolve) arsenic-containing substances from soil is the means by which arsenic gets into drinking water in Bangladesh.

Often, as in the case of soil or concrete, we can easily see that one part of a sample is different from another part. In other cases a sample may appear completely uniform to the unaided eye, but a microscope can reveal that it is not. For example, blood appears smooth in texture, but magnification reveals red and white cells within the liquid (Figure 1.10). The same is true of milk. *A mixture in which the uneven texture of the material can be seen with the naked eye or with a microscope* is classified as a **heterogeneous mixture.** Properties in one region are different from the properties in another region.

A **homogeneous mixture**, or **solution**, is *a completely uniform mixture that consists of two or more substances in the same phase*—solid, liquid, or gas (Figure 1.11). No amount of optical magnification will reveal different properties in one region of a solution compared with those in another. Heterogeneity exists in a solution only at the scale of atoms and molecules, which are too small to be seen with visible light. Examples of solutions are clear air (mostly a mixture of nitrogen and oxygen gases), sugar water, and some brass alloys (which are homogeneous mixtures of copper and zinc). The properties of a homogeneous mixture are the same everywhere in any particular sample, but they can vary from one sample to another depending on how much of one component is present relative to another component.

Cengage Learning/Charles D. Winters

Figure 1.11 A solution. When solid salt (sodium chloride) is stirred into liquid water, the salt dissolves to form a homogeneous liquid mixture. Each portion of the solution has exactly the same saltiness as every other portion, and other properties are also the same throughout the solution.

1-7a Separation and Purification

Earlier in this chapter we stated that a substance has characteristic properties that distinguish it from all other substances. However, for those characteristic properties to be observed, the substance must be **purified**; that is, it must be *separated from all other substances*. The melting point of an impure substance is different from that of the purified substance. The color and appearance of a mixture may also differ from those of a pure substance. Therefore, when we talk about the properties of a substance, it is assumed that we are referring to a pure substance—one from which all other substances have been separated.

Purification usually has to be done in several repeated steps and monitored by observing some property of the substance being purified. For example, iron can be separated from a heterogeneous mixture of iron and sulfur with a magnet, as shown in Figure 1.12. In this example, color, which depends on the relative quantities of iron and sulfur, indicates purity. The bright yellow color of pure sulfur is assumed to indicate that all the iron has been removed.

Concluding that a substance is pure on the basis of a single property of the mixture could be misleading because other methods of purification might change some other properties of the sample. It is reasonable to call sulfur pure only after a variety of methods of purification fail to change its physical and chemical properties. Purification is important because it allows us to attribute properties (such as suitability of water for humans to drink) to specific substances and then to study systematically which kinds of substances have properties that we find useful. In some cases, insufficient purification of

Cengage Learning/Charles D. Winters
1 Iron and sulfur can be separated by stirring with a magnet.
2 The first time that the magnet is removed, much of the iron is removed with it.
3 The sulfur still looks dirty because a small quantity of iron remains.
4 Repeated extractions of iron with the magnet leave a bright yellow sample of sulfur that cannot be purified further by this technique.

Figure 1.12 Separating a mixture: iron and sulfur.

Cengage Learning/Charles D. Winters
Ramona Heim/Shutterstock.com

(a) (b)

Figure 1.13 Purity and purification. (a) A quartz crystal is pure silicon dioxide, one of the few pure substances that occurs in nature. (b) Silicon used in manufacture of solar cells must have less than one gram impurity per 1000 tons of silicon.

a substance has led scientists to attribute to that substance properties that were actually due to a tiny trace of another substance (an impurity).

Only a few substances occur in nature in pure form. Gold, diamonds, and silicon dioxide (quartz; see Figure 1.13a) are examples. We live in a world of mixtures; all living things, the air and food on which we depend, and many products of technology are mixtures. Much of what we know about chemistry, however, is based on separating and purifying the components of those mixtures and then determining their properties. To date, more than 70 million substances have been reported, and many more are being discovered or synthesized by chemists every year. When pure, each of these substances has its own particular composition and its own characteristic properties.

A good example of the importance of purification is the high-purity silicon needed to produce computer chips and solar cells (Figure 1.13b). In a billion grams (about 1000 tons) of silicon there has to be less than one gram of impurity. Once the silicon has been purified, small but accurately known quantities of specific substances, such as boron or arsenic, can be introduced to give the electronic chip the desired properties. (See Section 9-10.)

1-7b Detection and Analysis

The example of arsenic in drinking water shows that it is very important to know whether a substance is present in a sample and to be able to find out how much of it is there. Does an ore contain enough of a valuable metal to make it worthwhile to mine the ore? Is there enough mercury in a sample of fish to make it unsafe for humans to eat the fish?

One part per million (ppm) means we can find one gram of a substance in one million grams of total sample. That corresponds to one-tenth of a drop of water in a bucket of water. One part per billion corresponds to a drop in a swimming pool, and one part per trillion corresponds to a drop in a large supermarket.

Answering questions like these is the job of *analytical chemists,* and they improve their methods every year. For example, in 1960 mercury could be detected at a concentration of one part per million, in 1970 the detection limit was one part per billion, and by 1980 the limit had dropped to one part per trillion. Thus, in 20 years the ability to detect small concentrations of mercury had increased by a factor of one million. This improvement has an important effect. Because we can detect smaller and smaller concentrations of contaminants, such contaminants can be found in many more samples. A few decades ago, toxic substances were usually not found when food, air, or water was tested, but that did not mean they were not there. It just meant that our analytical methods were unable to detect them. Today, with much better methods, toxic substances can be detected in most samples, which prompts demands that concentrations of such substances should be reduced to zero.

Although we expect that chemistry will push detection limits lower and lower, there will always be a limit below which an impurity will be undetectable. Proving that there are no contaminants in a sample will never be possible. This is a specific instance of the general rule that it is impossible to prove a negative. To put this idea another way, it will never be possible to prove that we have produced a completely pure sample of a substance, and therefore it is unproductive to legislate that there should be zero contamination in food or other substances. It is more important to use chemical analysis to determine a safe level of a toxin than to try to prove that the toxin is completely absent. In some cases, very small concentrations of a substance are beneficial but larger concentrations are toxic. (An example of this is selenium in the human diet.) Analytical chemistry can help us to determine the ranges of concentration that best serve humankind.

Absence of evidence is not evidence of absence.

1-8 Classifying Matter: Elements and Compounds

Most of the substances separated from mixtures can be converted to two or more simpler substances by chemical reactions—a process called *decomposition*. Substances are often decomposed by heating them, illuminating them with high-intensity light, or passing electricity through them. For example, table sugar (sucrose) can be separated from sugar cane and purified. When heated, sucrose decomposes via a complex series of chemical changes (caramelization—shown earlier in Figure 1.5) that produces the brown color and flavor of caramel candy. If heated for a longer time at a high enough temperature, sucrose is converted completely to two other substances, carbon and water. Furthermore, if the water is collected, it can be decomposed still further to hydrogen and oxygen by passing a direct electric current through it. However, nobody has found a way to decompose carbon, hydrogen, or oxygen.

A *substance* like carbon, hydrogen, or oxygen *that cannot be changed by chemical reactions into two or more new substances* is called a **chemical element** (or just element). A *substance that can be decomposed*, like sucrose or water, is a **chemical compound** (or just compound). When elements are chemically combined in a compound, their original characteristic properties—such as color, hardness, and melting point—are replaced by the characteristic properties of the compound. For example, in sucrose these three elements are chemically combined:

In 1661 Robert Boyle was the first to propose that elements could be defined by the fact that they could not be decomposed into two or more simpler substances.

- Carbon, which is usually a black powder, but is also commonly encountered in the form of diamonds
- Hydrogen, a colorless, flammable gas with the lowest density known
- Oxygen, a colorless gas necessary for human respiration

As you know from experience, sucrose is a white, crystalline powder that is completely unlike any of these three elements (Figure 1.14).

If a compound consists of two or more different elements, how is it different from a mixture? There are two ways: (1) A compound has specific composition; and (2) a compound has specific properties. Both the composition and the properties of a mixture can vary. A solution of sugar in water can be very sweet or only a little sweet, depending on how much sugar has been dissolved. There is no particular composition of a sugar solution that is favored over any other, and each different composition has its own set of properties. On the other hand, 100.0 g pure water always contains 11.2 g hydrogen and 88.8 g oxygen. Pure water always melts at 0.0 °C and boils at 100 °C (at one atmosphere pressure), and it is always a colorless liquid at room temperature.

Oxygen
Hydrogen
Carbon
Sucrose
Cengage Learning/Charles D. Winters

Figure 1.14 A compound and its elements. Table sugar, sucrose, is composed of the elements carbon, oxygen, and hydrogen. When elements are combined in a compound, the properties of the elements are no longer evident. Only the properties of the compound can be observed.

PROBLEM-SOLVING EXAMPLE 1.2

Elements and Compounds

A shiny, hard solid (substance A) is heated in the presence of carbon dioxide gas (which contains only carbon and oxygen). After a few minutes, a white solid (substance B) and a black solid (substance C) are formed. No other substances are found. When the black solid is heated in the presence of pure oxygen, carbon dioxide is formed. Decide whether each substance (A, B, and C) is an element or a compound, and give a reason for your choice in each case. If there is insufficient evidence to decide, say so. *(Remember: Try to answer the question on your own. Then look at the result to check your result. If your result does not agree, try again. Only look at the explanation after that.)*

Result A, insufficient evidence; B, compound; C, element

Analyze You know the properties of three substances and need to classify the substances.

Plan Compare the properties with properties of elements and compounds.

Execute Substance C must be an element, because it combines with oxygen to form a compound, carbon dioxide, that contains only two elements; in fact, substance C must be carbon. Substance B must be a compound, because it must contain oxygen (from the carbon dioxide) and at least one other element (from substance A). There is not enough evidence to decide whether substance A is an element or a compound. If substance A is an element, then substance B must be an oxygen compound of that element. However, there could be two or more elements in substance A (that is, it could be a compound), and the compound could still combine with oxygen from carbon dioxide to form a new compound.

☑ **Reasonable Result Check** Substance C is black, and carbon (graphite) is black. You could test experimentally to see whether substance A could be decomposed by heating it in a vacuum; if two or more new substances were formed, then substance A would have to be a compound. If there was no change, substance A could be an element, but it could also be a compound that is hard to decompose.

PROBLEM-SOLVING PRACTICE 1.2

A student grinds an unknown sample (A) to a fine powder and attempts to dissolve the sample in 100 mL pure water. Some solid (B) remains undissolved. When the water is separated from the solid and allowed to evaporate, a white powder (C) forms. The dry white powder (C) is found to weigh 0.034 g. All of sample C can be dissolved in 25 mL pure water. Can you say whether each sample A, B, and C is an element, a compound, or a mixture? Explain briefly.

1-8a Types of Matter

Desalinization of water (removing salt) could provide drinking water for large numbers of people who live in dry climates near the ocean. However, distillation requires a lot of energy resources and, therefore, is expensive. When solar energy can be used to evaporate the water, desalinization is less costly. For more information about desalinization of water, see Section 13-7f.

What we have just said about separating mixtures to obtain elements or compounds and decomposing compounds to obtain elements leads to a useful way to classify matter (Figure 1.15). Heterogeneous mixtures such as iron with sulfur can be separated using simple manipulation—such as a magnet. Homogeneous mixtures are somewhat more difficult to separate, but physical processes will serve. For example, salt water can be purified for drinking by distilling: heating to evaporate the water and cooling to condense the water vapor back to liquid. When enough water has evaporated, salt crystals form and they can be separated from the solution. Most difficult of all is separating elements that are combined in a compound. Such a separation requires a chemical change, which may involve reactions with other substances or sizable inputs of energy.

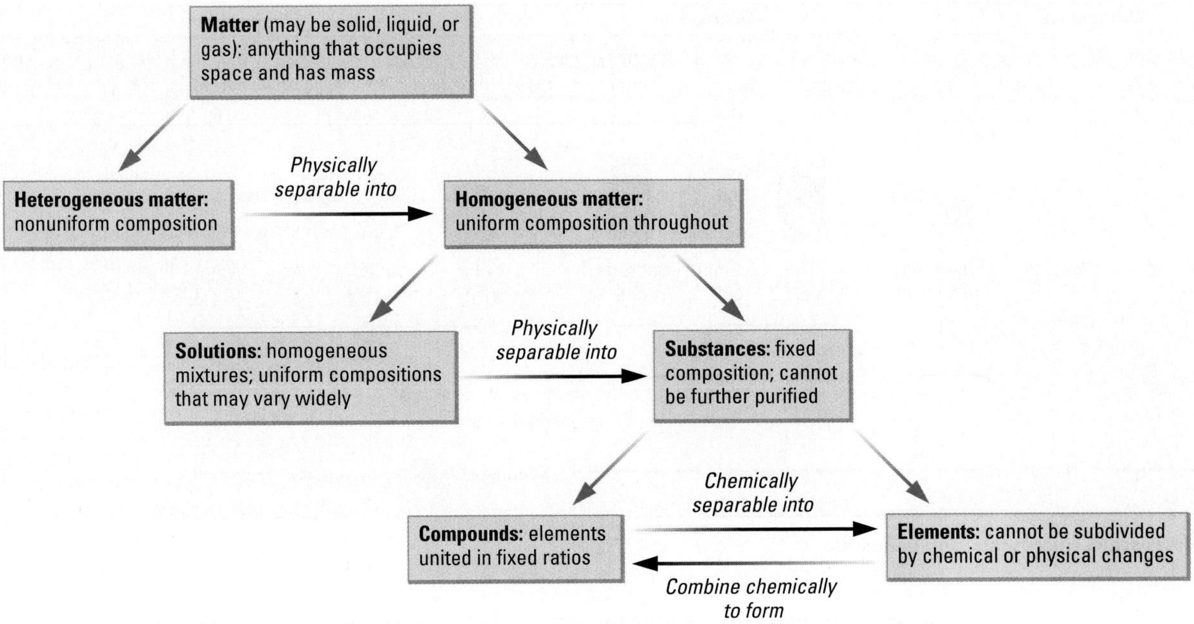
Matter (may be solid, liquid, or gas): anything that occupies space and has mass
Heterogeneous matter: nonuniform composition
Physically separable into
Homogeneous matter: uniform composition throughout
Solutions: homogeneous mixtures; uniform compositions that may vary widely
Physically separable into
Substances: fixed composition; cannot be further purified
Compounds: elements united in fixed ratios
Chemically separable into
Combine chemically to form
Elements: cannot be subdivided by chemical or physical changes

Figure 1.15 A scheme for classifying matter. The simplest matter (element) is on the right and the most complex matter (heterogeneous matter) is on the left.

EXERCISE 1.7

Classifying Matter

Classify each of these with regard to the type of matter described:

(a) Sugar dissolved in water
(b) The drink in a glass of carbonated beverage
(c) Used motor oil freshly drained from a car
(d) The diamond in a piece of jewelry
(e) A 25-cent coin minted in 2000
(f) A single crystal of sugar

1-9 Nanoscale Theories and Models

To further illustrate how the methods of science are applied to matter, we now consider how a theory based on atoms and molecules can account for the physical properties, chemical properties, and classification scheme that we have just described. Physical and chemical properties can be observed by the unaided human senses and refer to *samples of matter large enough to be seen, measured, and handled.* Such samples are said to be **macroscopic** or at the **macroscale**. By contrast, *samples of matter so small that they have to be viewed with a microscope* are **microscopic** or **microscale** samples. Blood cells and bacteria, for example, are matter at the microscale. The matter that really interests chemists is at the **nanoscale**. The term is based on the SI units prefix nano, which indicates something one billion times smaller than something else (see Table 1.3, Sec. 1-5a). One nanometer (nm) is 1×10^{-9} m; **nanoscale** involves *anything with at least one dimension smaller than about 100 nm.* The sizes of atoms and molecules are at the nanoscale. An average-sized atom such as a sulfur atom has a diameter of two tenths of a nanometer ($0.2 \text{ nm} = 2 \times 10^{-10}$ m). A water molecule is about the same size and a sugar

Jacob Bronowski, in a television series and book titled *The Ascent of Man*, had this to say about the importance of imagination: "There are many gifts that are unique in man; but at the centre of them all, the root from which all knowledge grows, lies the ability to draw conclusions from what we see to what we do not see . . ."

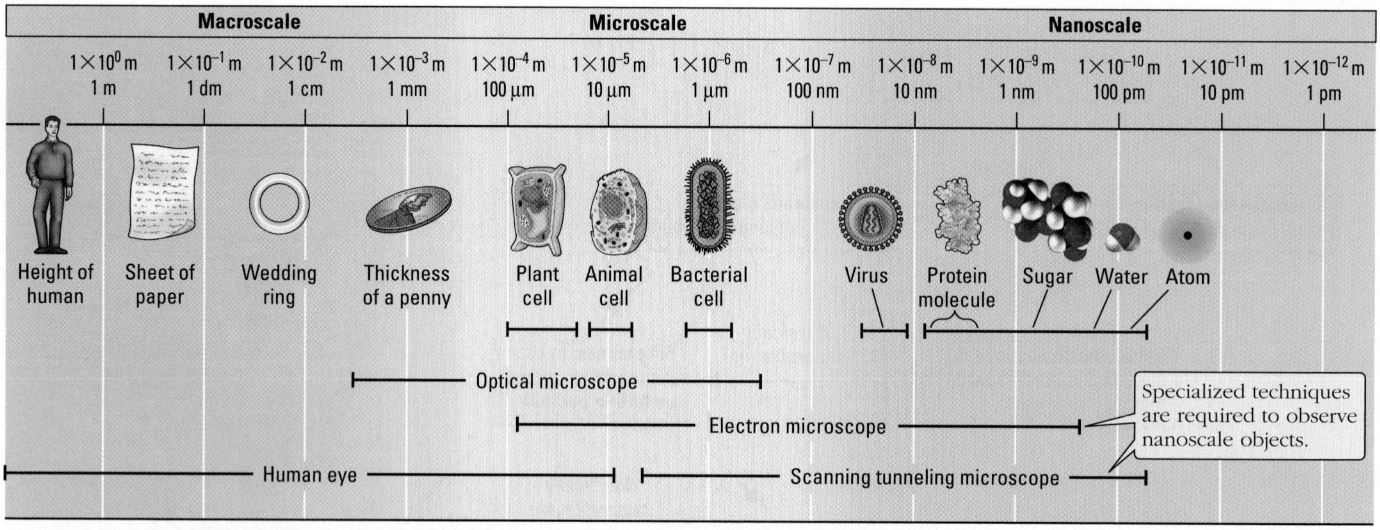
Macroscale
Microscale
Nanoscale
1×10^0 m 1 m
1×10^-1 m 1 dm
1×10^-2 m 1 cm
1×10^-3 m 1 mm
1×10^-4 m 100 μm
1×10^-5 m 10 μm
1×10^-6 m 1 μm
1×10^-7 m 100 nm
1×10^-8 m 10 nm
1×10^-9 m 1 nm
1×10^-10 m 100 pm
1×10^-11 m 10 pm
1×10^-12 m 1 pm
Height of human
Sheet of paper
Wedding ring
Thickness of a penny
Plant cell
Animal cell
Bacterial cell
Virus
Protein molecule
Sugar
Water
Atom
Optical microscope
Electron microscope
Human eye
Scanning tunneling microscope
Specialized techniques are required to observe nanoscale objects.

Figure 1.16 Macroscale, microscale, and nanoscale.

molecule is about one nanometer across. Figure 1.16 indicates the relative sizes of various objects at the macroscale, microscale, and nanoscale.

Earlier, we described the chemist's unique atomic and molecular perspective. It is a fundamental idea of chemistry that matter is the way it is because of the nature of its constituent atoms and molecules. Those atoms and molecules are very, very tiny. Therefore,

ESTIMATION | How Tiny Are Atoms and Molecules?

It is often useful to estimate an approximate value for something. Usually this estimation can be done quickly, and often it can be done without a calculator. The idea is to pick round numbers that you can work with in your head, or to use some other method that allows a quick estimate. If you really need an accurate value, an estimate is still useful to check whether the accurate value is in the right ballpark. Often an estimate is referred to as a "back-of-the-envelope" calculation, because estimates might be done over lunch on any piece of paper that is at hand. Some estimates are referred to as "order-of-magnitude calculations" because only the power of ten (the order of magnitude) in the answer is obtained. To help you develop estimation skills, most chapters in this book will provide you with an example of estimating something.

To get a more intuitive feeling for how small atoms and molecules are, estimate how many hydrogen atoms could fit inside a 12-oz (355-mL) soft-drink can. Make the same estimate for protein molecules. Use the approximate sizes given in Figure 1.16.

Because 1 mL is the same volume as a cube 1 cm on each side (1 cm^3), the volume of the can is the same as the volume of 355 cubes 1 cm on each side. Therefore we can first estimate how many atoms would fit into a 1-cm cube and then multiply that number by 355.

According to Figure 1.16, a typical atom has a diameter slightly less than 100 pm. Because this is an estimate, and to make the numbers easy to handle, assume that we are dealing with an atom that is 100 pm in diameter. Then the atom's diameter is 100×10^{-12} m $= 1 \times 10^{-10}$ m, and it will require 10^{10} of these atoms lined up in a row to make a length of 1 m. Since 1 cm is $\frac{1}{100}$, that is, 10^{-2}, of a meter, only $10^{-2} \times 10^{10} = 10^8$ atoms would fit in 1 cm.

In three dimensions, there could be 10^8 atoms along each of the three perpendicular edges of a 1-cm cube (the *x*, *y*, and *z* directions). The one row along the *x*-axis could be repeated 10^8 times along the *y*-axis; then that layer of atoms could be repeated 10^8 times along the *z*-axis. Therefore, the number of atoms that we estimate would fit inside the cube is $10^8 \times 10^8 \times 10^8 = 10^{24}$ atoms. Multiplying this by 355 gives $355 \times 10^{24} = 3.6 \times 10^{26}$ atoms in the soft-drink can.

This estimate is a bit low. A hydrogen atom's diameter is less than 100 pm, so more hydrogen atoms would fit inside the can. Also, atoms are usually thought of as spheres, so they could pack together more closely than they would if just lined up in rows. Therefore, an even larger number of atoms than 3.6×10^{26} could fit inside the can.

For a typical protein molecule, Figure 1.16 indicates a diameter on the order of 5 nm = 5000 pm. That is 50 times bigger than the 100-pm diameter we used for the hydrogen atom. Thus, there would be 50 times fewer protein molecules in the *x* direction, 50 times fewer in the *y* direction, and 50 times fewer in the *z* direction. Therefore, the number of protein molecules would be fewer by $50 \times 50 \times 50 = 125{,}000$. The number of protein molecules can thus be estimated as $(3.6 \times 10^{26}) / (1.25 \times 10^5)$. Because 3.6 is roughly three times 1.25, and because we are estimating, not calculating accurately, we can take the result to be 3×10^{21} protein molecules. That's still a whole lot of molecules!

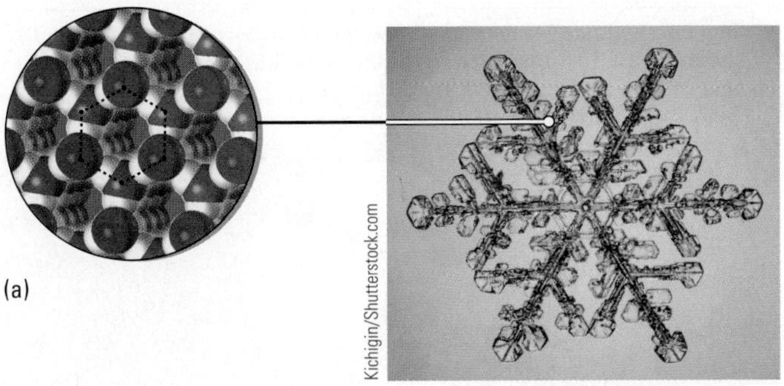

(a)

Kichigin/Shutterstock.com

(b)

Figure 1.17 Structure and form. (a) In the nanoscale structure of ice, each water molecule occupies a position in a regular array or lattice that includes hexagonal units (outlined). (b) The form of a snowflake reflects the hexagonal symmetry of the nanoscale structure of ice.

we need to use imagination creatively to discover useful theories that connect the behavior of tiny nanoscale constituents to the observed behavior of chemical substances at the macroscale. Learning chemistry enables you to "see" in the things all around you nanoscale structure that cannot be seen with your eyes.

1-9a Kinetic-Molecular Theory

A theory that deals with matter at the nanoscale, the **kinetic-molecular theory**, states that *all matter consists of extremely tiny particles (atoms or molecules) that are in constant motion*. In a solid these particles are packed closely together in a regular array, as shown in Figure 1.17. The particles vibrate back and forth about their average positions, but seldom does a particle in a solid squeeze past its immediate neighbors to come into contact with a new set of particles. Because the particles are packed so tightly and in such a regular arrangement, a solid is rigid, its volume is fixed, and the volume of a given mass is small. The external shape of a solid often reflects the internal arrangement of its particles. This relation between the observable structure of the solid and the arrangement of the particles from which it is made is one reason that scientists have long been fascinated by the shapes of crystals and minerals such as quartz (Figure 1.13a) and by snowflakes (Figure 1.17).

The late Richard Feynman, a Nobel laureate in physics, said, "If in some cataclysm all of scientific knowledge were to be destroyed, and only one sentence passed on to the next generation of creatures, what statement would contain the most information in the fewest words? I believe it is the atomic hypothesis, that all things are made of atoms, little particles that move around in perpetual motion . . ."

The kinetic-molecular theory of matter can also be used to interpret the properties of liquids, as shown in Figure 1.18. Liquids are fluid because the atoms or molecules are arranged more haphazardly than in solids. Particles are not confined to specific locations but rather can move past one another. No particle goes very far without bumping into another—the particles in a liquid interact with their neighbors continually. Because the particles are usually a little farther apart in a liquid than in the corresponding solid, the volume is usually a little bigger. (Ice and liquid water, which are shown in Figure 1.18, are an important exception to this last generality. As you can see from the figure, the water molecules in ice are arranged so that there are empty hexagonal channels. When ice melts, these channels become partially filled by water molecules, accounting for the slightly smaller volume of the same mass of liquid water.)

Because the nature of the particles is relatively unimportant in determining the behavior of gases, all gases can be described fairly accurately by the ideal gas law, which is introduced in Chapter 8.

Like liquids, gases are fluid because their nanoscale particles can easily move past one another. As shown in Figure 1.18, the particles fly about to fill any container they are in; hence, a gas has no fixed shape or volume. In a gas, the particles are much farther apart than in a solid or a liquid. They move significant distances before hitting other particles or the walls of the container. The particles also move quite rapidly. In air at room temperature, for example, the average molecule is going faster than 1000 miles per hour. A particle hits another particle every so often, but most of the time each is quite far away

In gaseous water (water vapor), the molecules are much farther apart than in liquid or solid, and they move relatively long distances before colliding with other molecules.
In solid water (ice), each water molecule is close to its neighbors and restricted to vibrating back and forth around a specific location.
In liquid water, the molecules are close together, but they can move past each other; each molecule can move only a short distance before bumping into one of its neighbors.
Volodymyr Goinyk/Shutterstock.com

Figure 1.18 Nanoscale representations of water in three states of matter: solid (ice), liquid, and gas (water vapor).

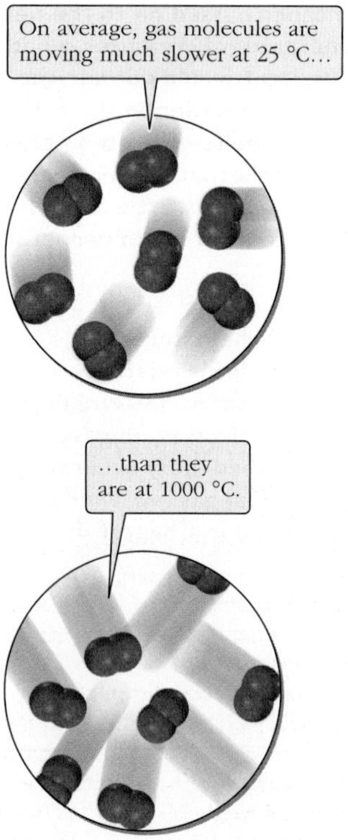
On average, gas molecules are moving much slower at 25 °C...
...than they are at 1000 °C.

Figure 1.19 Molecular speed and temperature.

from all the others. Consequently, the nature of the particles is much less important in determining the properties of a gas.

Temperature can also be interpreted using the kinetic-molecular theory. The higher the temperature is, the more active the nanoscale particles are. A solid melts when its temperature is raised to the point where the particles vibrate fast enough and far enough to push each other out of the way and move out of their regularly spaced positions. The substance becomes a liquid because the particles are now behaving as they do in a liquid, bumping into one another and pushing past their neighbors. As the temperature increases, the particles move even faster, until finally they can escape the clutches of their comrades and become independent; the substance becomes a gas. *Increasing temperature corresponds to faster and faster motions of atoms and molecules.* This is a general rule that you will find useful in many future discussions of chemistry (Figure 1.19).

Using the kinetic-molecular theory to interpret the properties of solids, liquids, and gases and the effect of changing temperature provides a very simple example of how chemists use nanoscale theories and models to interpret and explain macroscale observations. In the remainder of this chapter and throughout your study of chemistry, you should try to imagine how the atoms and molecules are arranged and what they are doing whenever you consider a macroscale sample of matter. That is, you should try to develop the chemist's special perspective on the relation of nanoscale structure to macroscale behavior. Such a perspective will help you learn chemistry more effectively.

CONCEPTUAL EXERCISE 1.8

Kinetic-Molecular Theory

Use the idea that matter consists of tiny particles in motion to interpret each observation.

(a) An ice cube in sunlight melts, and the liquid water eventually evaporates.
(b) Wet clothes hung on a line eventually dry.
(c) Moisture appears on the outside of a glass of ice water.
(d) Evaporation of a solution of sugar in water forms crystals.

1-10 The Atomic Theory

The existence of elements can be explained by a nanoscale theory involving particles, just as the properties of solids, liquids, and gases can be. This theory, which is closely related to the kinetic-molecular theory, is called the atomic theory. It was proposed in 1803 by John Dalton, and says that an element cannot be decomposed into two or more new substances because at the nanoscale it consists of one and only one kind of atom and atoms are indivisible under the conditions of chemical reactions. An **atom** is the *smallest particle of an element that embodies the chemical properties of that element.* An element, such as the sample of copper in Figure 1.20, is made up entirely of atoms of the same kind.

Our bodies are made up of atoms from the distant past—atoms from other people and other things. Some of the carbon, hydrogen, and oxygen atoms in our carbohydrates have come from the breaths (first and last) of both famous and ordinary persons of the past.

The fact that a compound can be decomposed into two or more different substances can be explained by saying that each compound must contain two or more different kinds of atoms. Dalton proposed that those atoms were somehow connected or bonded to each other to form molecules. A **molecule** is *the smallest particle of a substance that exists independently and embodies the chemical properties of the substance; it is a collection of atoms connected by bonds.* Decomposing a compound involves separating at least one type of atom from atoms of other kinds. For example, charring of sugar corresponds to separating atoms that were bonded together in sugar molecules: carbon atoms are separated from oxygen atoms and hydrogen atoms. Dalton used symbols to represent atoms. A white circle represented a hydrogen atom and a black circle represented a carbon atom, for example. Today a **chemical symbol** is *a one- or two-letter abbreviation for the name of an element*; examples are H for hydrogen, C for carbon, and Ca for calcium. Dalton introduced the idea of a chemical formula for a compound. In a **chemical formula**, *symbols for different kinds of atoms are written together with subscripts to show how many atoms of each kind are in a molecule*. For example, in the formula for water, H_2O, the symbol H for hydrogen is written with a subscript 2 and O for oxygen has no subscript, indicating two hydrogen atoms and one oxygen atom in the molecule. (If the subscript is omitted, it is understood to be 1.)

Chemical bonds are strong attractions that hold atoms together. Chemical bonding is discussed in detail in Chapter 6.

hydrogen atom

oxygen atom

water, H_2O

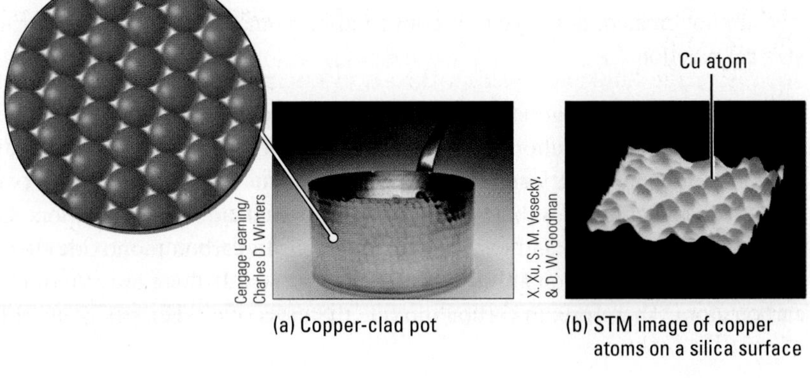
Cengage Learning/
Charles D. Winters
X. Xu, S. M. Vesecky,
& D. W. Goodman
Cu atom

(a) Copper-clad pot

(b) STM image of copper atoms on a silica surface

Figure 1.20 Elements, atoms, and the nanoscale world of chemistry. (a) A macroscopic sample of copper metal on a copper-clad pot with a nanoscale, magnified representation of a tiny portion of its surface. It is clear that all the atoms in the sample of copper are the same kind of atoms. (b) A scanning tunneling microscopy (STM) image, enhanced by a computer, of a layer of copper atoms on the surface of silica (a compound of silicon and oxygen). The section of the layer shown is 1.7 nm by 1.7 nm and the rows of atoms are separated by about 0.44 nm.

According to the modern atomic theory, atoms of the same element have the same chemical properties but are not necessarily identical in all respects. The discussion of isotopes in Chapter 2 shows how atoms of the same element can differ in mass.

Dalton also said that each kind of atom must have its own properties—in particular, a characteristic mass. An atom's characteristic mass is called its atomic weight; the mass of a molecule is called its molecular weight. This idea allowed Dalton's theory to account for the masses of different elements that combine via chemical reactions to form compounds. We will explore this idea more thoroughly in Chapter 2.

An important success of Dalton's ideas was that they could be used to interpret known chemical facts quantitatively. Two laws known in Dalton's time could be explained by the atomic theory.

In the 18th century, the great French chemist, Antoine Lavoisier performed experiments in which he carefully weighed the reactants before a chemical reaction, and carefully collected and weighed all reaction products afterward. Lavoisier found that if he began with 5.000 g reactants they always formed 5.000 g products. These results led to the **law of conservation of matter** (also called the law of conservation of mass): *There is no detectable change in mass during a chemical reaction.* The atomic theory says that mass is conserved because the same number of atoms of each kind is present before and after a reaction, and each of those kinds of atoms has its same characteristic mass before and after the reaction.

The other law was based on the observation that in a chemical compound the proportions of the elements by mass are always the same. Water always contains 1 g hydrogen for every 8 g oxygen, and carbon monoxide always contains 4 g oxygen for every 3 g carbon. The **law of constant composition** summarizes such observations: *A chemical compound always contains the same elements in the same proportions by mass.* The atomic theory explains this observation by saying that atoms of different elements always combine in the same ratio in a compound. For example, in carbon monoxide there is always one carbon atom for each oxygen atom. If the mass of an oxygen atom is $\frac{4}{3}$ times the mass of a carbon atom, then the ratio of mass of oxygen to mass of carbon in carbon monoxide will always be 4:3.

Dalton's theory has been modified to account for discoveries since his time. The *modern* **atomic theory** is based on these assumptions:

All matter is composed of atoms, which are extremely tiny. Interactions among atoms account for the properties of matter.

All atoms of a given element have the same chemical properties. Atoms of different elements have different chemical properties.

Compounds are formed by the chemical combination of two or more different kinds of atoms. Atoms usually combine in the ratio of small whole numbers. For example, in a carbon monoxide molecule there is one carbon atom and one oxygen atom; a carbon dioxide molecule consists of one carbon atom and two oxygen atoms.

A chemical reaction involves joining, separating, or rearranging atoms. Atoms in the reactant substances form new combinations in the product substances. Atoms are not created, destroyed, or converted into other kinds of atoms during a chemical reaction.

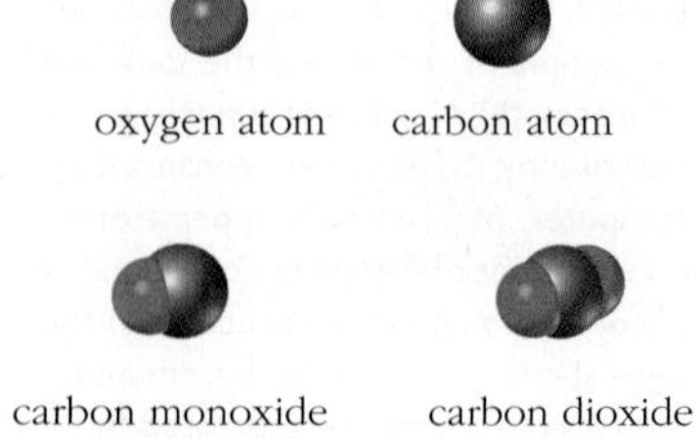
oxygen atom
carbon atom
carbon monoxide
carbon dioxide

The hallmark of a good theory is that it suggests new experiments, and this was true of the atomic theory. Dalton realized that it predicted a law that had not yet been discovered. If compounds are formed by combining atoms of different elements on the nanoscale, then in some cases there might be more than a single combination. An example is carbon monoxide, CO, and carbon dioxide, CO_2. In carbon monoxide there is one oxygen atom for each carbon atom, while in carbon dioxide there are two oxygen atoms per carbon atom. Therefore, in carbon dioxide the mass of oxygen per gram of carbon ought to be twice as great as it is in carbon monoxide (because twice as many oxygen atoms will weigh twice as much). Dalton called this the **law of multiple proportions**, and he carried out quantitative experiments seeking data to confirm or deny it. Dalton and others obtained data consistent with the law of multiple proportions, thereby enhancing acceptance of the atomic theory.

1-11 Communicating Chemistry: Symbolism

Chemical symbols—such as H, O, C, Na, Cl, or Mt—are a shorthand way of indicating what kind of atoms we are talking about. Chemical formulas tell us how many atoms of an element are combined in a molecule and in what ratios atoms are combined in compounds. For example, the formula Cl_2 tells us that there are two chlorine atoms in a chlorine molecule. The formulas CO and CO_2 tell us that carbon and oxygen form two different compounds—one that has equal numbers of C and O atoms, and one that has twice as many O atoms as C atoms. In other words, chemical symbols and formulas symbolize the nanoscale composition of each substance.

Model of Cl_2

Chemical symbols and formulas also represent the macroscale properties of elements and compounds. That is, the symbol Na brings to mind a soft, highly reactive metal, and the formula H_2O represents a colorless liquid that freezes at 0 °C, boils at 100 °C, and reacts violently with Na. Because chemists are familiar with both the nanoscale and macroscale characteristics of substances, they usually use symbols to abbreviate their representations of both. Symbols are also useful for representing chemical reactions. For example, the charring of sucrose mentioned earlier is represented by

Sucrose	$\longrightarrow$	Carbon	+	Water
$C_{12}H_{22}O_{11}$	$\longrightarrow$	12 C	+	11 H_2O
Reactant	changes to		Products	

As shown in Figure 1.21, the symbolic aspect of chemistry is the third part of the chemist's special view of the natural world. It is important that you become familiar and comfortable with using chemical symbols and formulas to represent chemical substances and their reactions. Figure 1.22 shows how chemical symbolism can be applied to the process of decomposing water with electricity (electrolysis of water).

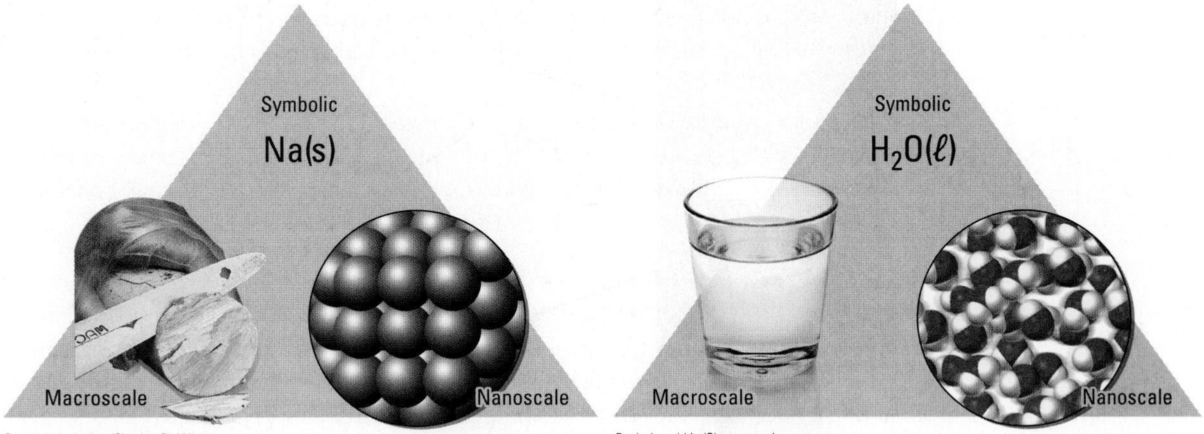
Symbolic
Na(s)
Macroscale
Nanoscale
Symbolic
$H_2O(\ell)$
Macroscale
Nanoscale

Cengage Learning/Charles D. Winters Pavlo Loushkin/Shutterstock.com

Figure 1.21 Three ways of representing an element, sodium, and a compound, water.

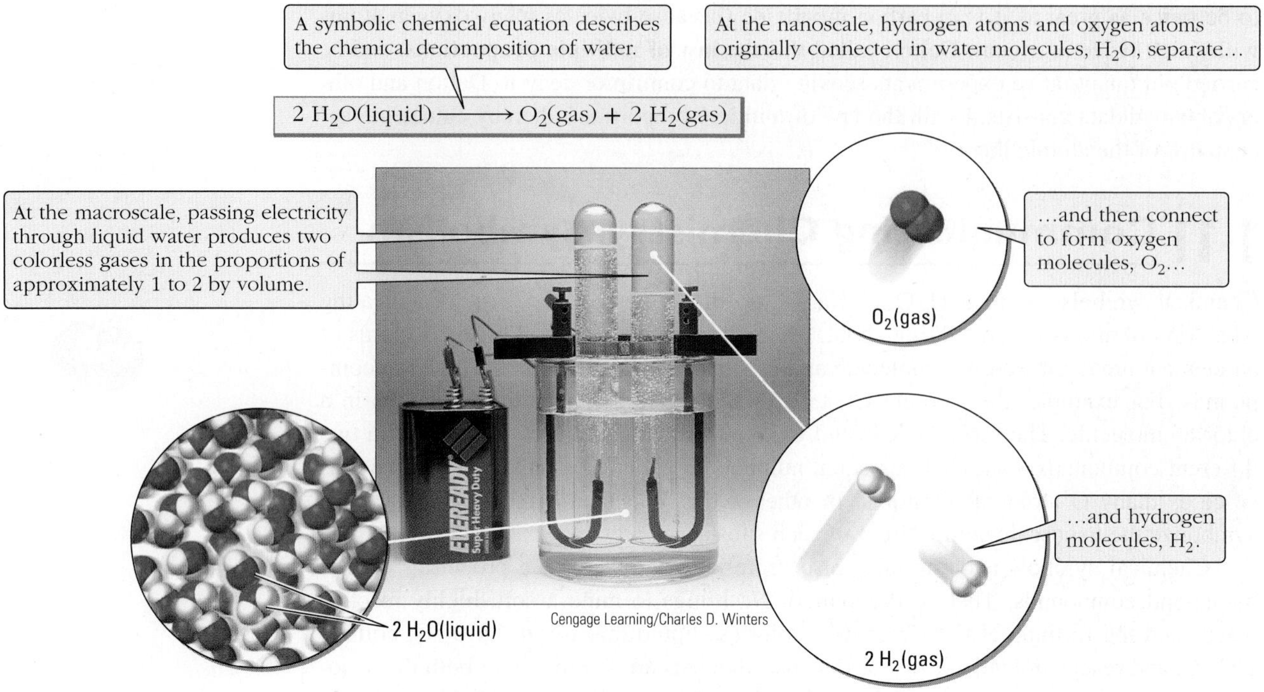
A symbolic chemical equation describes the chemical decomposition of water.
At the nanoscale, hydrogen atoms and oxygen atoms originally connected in water molecules, H_2O, separate...
2 H_2O(liquid) → O_2(gas) + 2 H_2(gas)
At the macroscale, passing electricity through liquid water produces two colorless gases in the proportions of approximately 1 to 2 by volume.
...and then connect to form oxygen molecules, O_2...
O_2(gas)
EVEREADY
Super Heavy Duty
...and hydrogen molecules, H_2.
2 H_2O(liquid)
Cengage Learning/Charles D. Winters
2 H_2(gas)

Figure 1.22 Symbolic, macroscale, and nanoscale representations of a chemical reaction: electrolysis of water.

PROBLEM-SOLVING EXAMPLE 1.3

Macroscale, Nanoscale, and Symbolic Representations

The figure shows a sample of water boiling. In spaces labeled C, indicate whether the macroscale or the nanoscale is represented. In spaces labeled B, draw the molecules that would be present with appropriate distances between them. One of the circles represents a bubble of gas within the liquid. The other represents the liquid. In space A, write a symbolic representation of the boiling process.

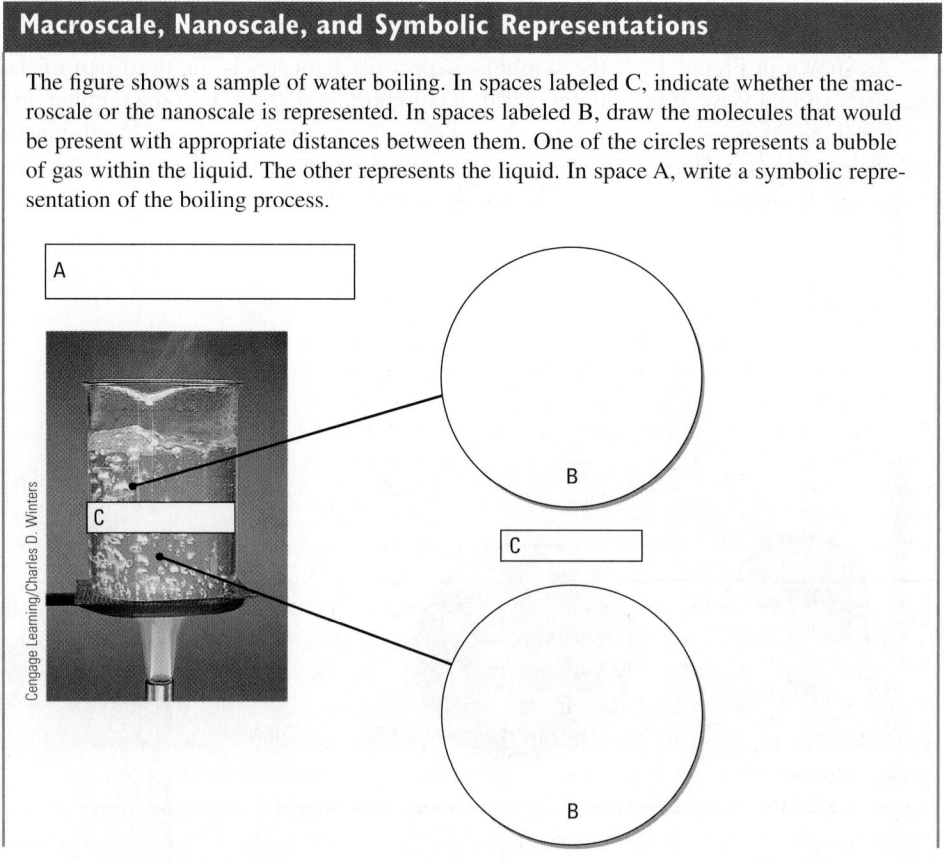

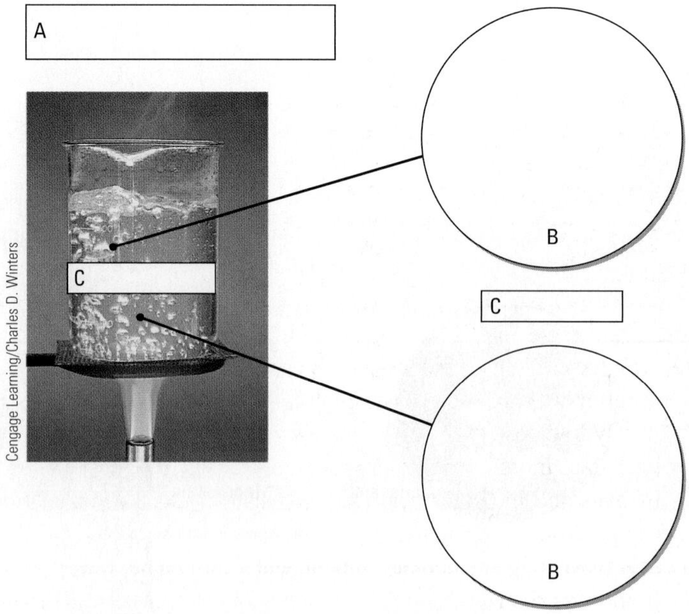
A
C
B
C
B
Cengage Learning/Charles D. Winters

Result

A H_2O(liquid) $\longrightarrow$ H_2O(gas)

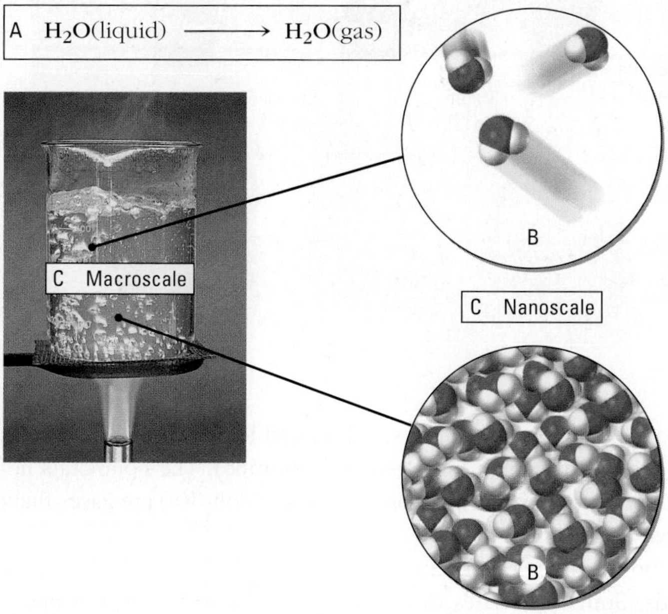
B
C Macroscale
C Nanoscale
B
Cengage Learning/Charles D. Winters

Analyze From a macroscale picture you are asked to make nanoscale and symbolic representations.

Plan Apply your knowledge of nanoscale representations of liquids and gases and your knowledge of chemical symbols and formulas.

Execute Each water molecule consists of two hydrogen atoms and one oxygen atom. In liquid water the molecules are close together and oriented in various directions. In a bubble of gaseous water the molecules are much farther apart—there are fewer of them per unit volume. The symbolic representation involves H_2O(liquid) changing to H_2O(gas).

PROBLEM-SOLVING PRACTICE 1.3

Draw a nanoscale representation and a symbolic representation for melting copper. Describe the macroscale properties of solid copper and liquid copper.

1-12 The Chemical Elements

Because the chemical elements are the building blocks of all matter, it is important to know their properties. The vast majority of the elements are **metals**—only 25 are not. You are probably familiar with many properties of metals. At room temperature they are solids (except for mercury, which is a liquid), they conduct electricity (and conduct better as the temperature decreases), they are ductile (can be drawn into wires), they are malleable (can be rolled into sheets), and they can form alloys (solutions of one or more metals in another metal). In a solid metal, individual metal atoms are packed close to each other, so metals usually have fairly high densities. Figure 1.23 shows some common metals. Iron (Fe) and aluminum (Al) are used in automobile parts because of their ductility, malleability, and relatively low cost. Copper (Cu) is used in electrical wiring because it conducts electricity better than most metals. Gold (Au) is used for the vital electrical contacts in automobile air bags and in some computers because it does not corrode and is an excellent electrical conductor.

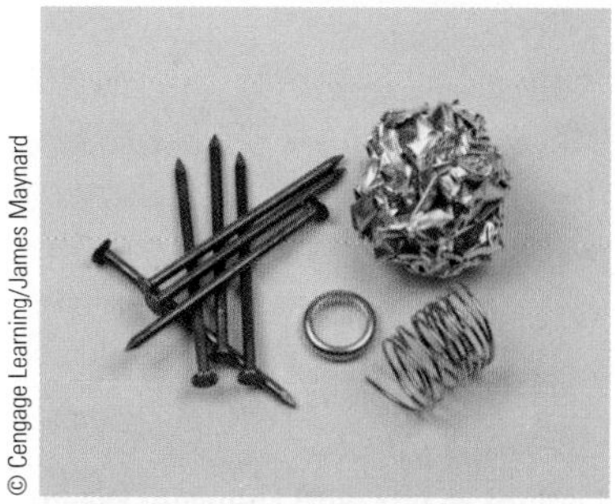
© Cengage Learning/James Maynard

Figure 1.23 Some metallic elements—iron, gold, aluminum, and copper. Metals are malleable, ductile, and conduct electricity.

In contrast, **nonmetals** usually do not conduct electricity. Nonmetals are more diverse in their physical properties than are metals (Figure 1.24). At room temperature

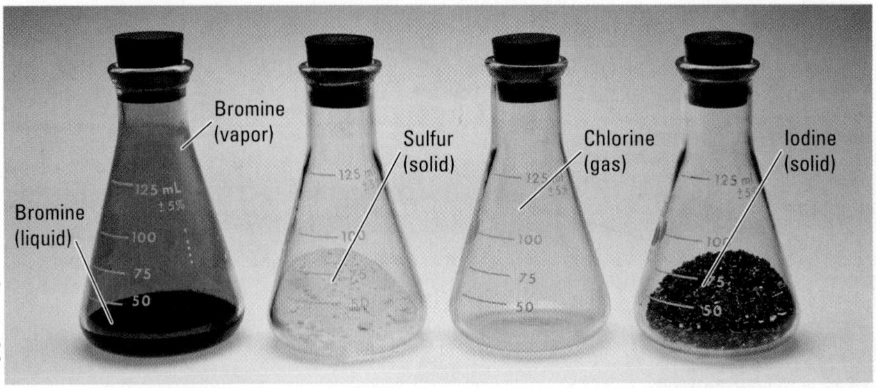
Bromine
(vapor)
Sulfur
(solid)
Chlorine
(gas)
Iodine
(solid)
Bromine
(liquid)
125 mL
±5%
100
75
50

Cengage Learning/Charles D. Winters

Figure 1.24 Some nonmetallic elements—bromine, sulfur, chlorine, and iodine. Nonmetals occur as solids, liquids, and gases and have very low electrical conductivities. Bromine is the only nonmetal that is a liquid at room temperature.

some nonmetals are solids (such as phosphorus, sulfur, and iodine), bromine is a liquid, and others are gases (such as hydrogen, nitrogen, and chlorine). The nonmetals helium (He), neon (Ne), argon (Ar), krypton (Kr), xenon (Xe), and radon (Rn) are gases that consist of individual atoms.

A few elements—boron, silicon, germanium, arsenic, antimony, and tellurium—are classified as **metalloids**. Some properties of metalloids are typical of metals and other properties are characteristic of nonmetals. For example, some metalloids are shiny like metals, but they do not conduct electricity as well as metals. Many of them are semiconductors and are essential for the electronics industry. (See Sections 9-9 and 9-10.)

1-12a Groups of Elements

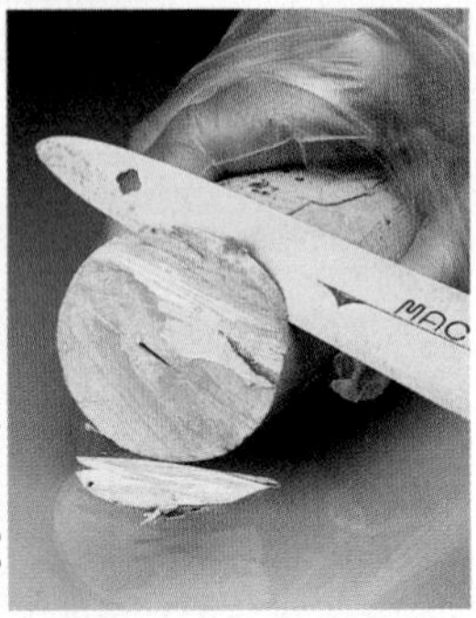

Cengage Learning/Charles D. Winters

An alkali metal, sodium, being cut with a knife.

In addition to these broad classifications, there are groups of elements whose properties are very similar. For example, all of the **alkali metals**, lithium (Li), sodium (Na), potassium (K), rubidium (Rb), cesium (Cs), and francium (Fr), are soft solids (they can easily be cut with a knife). They all have low melting points and react violently with water. (See the reaction of sodium with water in Figure 1.6.) The reaction products are a flammable gas, hydrogen, and a water solution that feels soapy and burns the skin—an *alkal*ine solution. All of the alkali metals react with the element chlorine to form compounds in which chlorine atoms and alkali metal atoms are combined in a 1 : 1 ratio; that is, the formulas are all MCl, where M represents Li, Na, K, Rb, Cs, or Fr. For example, sodium reacts with chlorine to form sodium chloride, NaCl, which is ordinary table salt. Sodium chloride is a fundamental component of the human diet, so throughout history NaCl has been sought as a dietary necessity and a commodity in commerce. The word "salary" is from Latin *salarium*, money given Roman soldiers to purchase salt, or salt used as a means of payment of the soldiers.

Another group of very similar elements is the **alkaline-earth metals**: beryllium (Be); magnesium (Mg); calcium (Ca); strontium (Sr); barium (Ba); and radium (Ra).

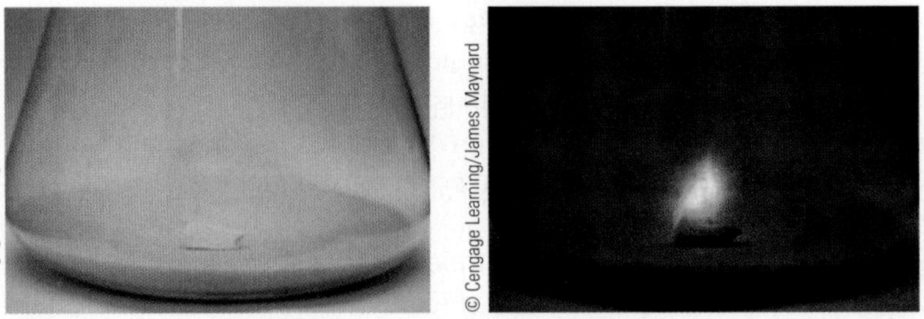

© Cengage Learning/James Maynard

© Cengage Learning/James Maynard

Sodium reacts with chlorine. When sodium metal is placed in yellow-green chlorine gas (left), the two elements react vigorously (right) to form sodium chloride.

These metals are all harder than the alkali metals, they all combine with oxygen from the air to form oxides, and they all react with chlorine to form compounds with the general formula MCl_2. Two of the alkaline-earth elements, calcium and magnesium, are the fifth and sixth most abundant elements in Earth's crust; they are present in a vast array of chemical compounds.

A third group of closely related elements is the **halogens**: fluorine (F); chlorine (Cl); bromine (Br); iodine (I); and astatine (At). These elements all react readily with metals, they all consist of **diatomic molecules** *(molecules that contain two atoms, such as* Cl_2*)*, and they all react with alkali metals to form salts, MX (where M represents Li, Na, K, Rb, Cs, or Fr, and X represents F, Cl, Br, I, or At). In these salts, the ratio of metal atoms to halogen atoms is 1:1. The name of this group, halogen, is derived from the Greek word *hals*, which means salt.

Another group of elements is the **noble gases**: helium (He); neon (Ne); argon (Ar); krypton (Kr); xenon (Xe); and radon (Rn). None of these elements reacts readily with any of the other elements, and some of them form no compounds at all. They are all colorless, odorless gases. Because they are so unreactive they are mainly found in Earth's atmosphere, where they make up about 1% of the gases. The noble gases were all discovered between 1894 and 1900, although helium (Greek *helios*, sun) was detected in the sun as early as 1868.

1-13 The Periodic Table

Dmitri Mendeleev (1834–1907), while looking for better ways to teach chemistry, realized that listing the elements in order of increasing atomic weight revealed a periodic repetition of their properties. For example, if you list the first dozen elements in order of atomic weight, they are

H He Li Be B C N O F Ne Na Mg

Notice that He, Li, and Be are a noble gas, an alkali metal, and an alkaline-earth metal. The same sequence, noble gas, alkali metal, alkaline-earth metal, occurs eight elements later with Ne, Na, and Mg. Like keys on a piano, the elements' properties repeat at regular intervals—they are periodic.

Cengage Learning/George Semple

Periodicity of piano keys.

Mendeleev summarized his findings in the table that has come to be called the **periodic table**. By lining up the elements in horizontal rows in order of increasing atomic weight and starting a new row when he came to an element with properties similar to one already in the previous row, he saw that the resulting columns contained elements with similar properties. (Each of the groups of similar elements discussed earlier is in a column in the periodic table.) Mendeleev found that some positions in his table were not filled. He predicted that new elements would be found that filled the gaps, and he predicted properties of the undiscovered elements. Three of the missing elements—scandium (Sc), gallium (Ga), and germanium (Ge)—were soon discovered. They had properties very close to those Mendeleev had predicted, which was strong support for his ideas.

In the late 1860s, when Mendeleev was developing his ideas about the periodic table, the noble gases had not yet been discovered. Mendeleev's list of elements did not include He and Ne.

The symbols for all known elements are listed in the periodic table in Figure 1.25. (Names as well as symbols are listed in the periodic table inside the front cover of this book.) The first letter of each symbol is capitalized; the second letter, if there is one, is lowercase, as in He, the symbol for helium. Elements discovered a long time ago have symbols with Latin, Greek, or other origins, such as Au for gold (from *aurum,* meaning "bright dawn") and Fe for iron (from *ferrum*). The names of more recently discovered elements are derived from their place of discovery or from a person or place of significance (Table 1.4, page 33).

Main-group metals
Transition metals
Metalloids
Nonmetals, noble gases

A label of A denotes main-group elements…

…and a label of B denotes transition elements, the system used most commonly at present in the United States.

The international system is to number the groups from 1 to 18.

Period	1A (1)	2A (2)	3B (3)	4B (4)	5B (5)	6B (6)	7B (7)	8B (8)	8B (9)	8B (10)	1B (11)	2B (12)	3A (13)	4A (14)	5A (15)	6A (16)	7A (17)	8A (18)	Period
1	1 H																	2 He	1
2	3 Li	4 Be											5 B	6 C	7 N	8 O	9 F	10 Ne	2
3	11 Na	12 Mg											13 Al	14 Si	15 P	16 S	17 Cl	18 Ar	3
4	19 K	20 Ca	21 Sc	22 Ti	23 V	24 Cr	25 Mn	26 Fe	27 Co	28 Ni	29 Cu	30 Zn	31 Ga	32 Ge	33 As	34 Se	35 Br	36 Kr	4
5	37 Rb	38 Sr	39 Y	40 Zr	41 Nb	42 Mo	43 Tc	44 Ru	45 Rh	46 Pd	47 Ag	48 Cd	49 In	50 Sn	51 Sb	52 Te	53 I	54 Xe	5
6	55 Cs	56 Ba	57 La	72 Hf	73 Ta	74 W	75 Re	76 Os	77 Ir	78 Pt	79 Au	80 Hg	81 Tl	82 Pb	83 Bi	84 Po	85 At	86 Rn	6
7	87 Fr	88 Ra	89 Ac	104 Rf	105 Db	106 Sg	107 Bh	108 Hs	109 Mt	110 Ds	111 Rg	112 Cn	113 —	114 Fl	115 —	116 Lv	117 —	118 —	7

	Period															Period
Lanthanides	6	58 Ce	59 Pr	60 Nd	61 Pm	62 Sm	63 Eu	64 Gd	65 Tb	66 Dy	67 Ho	68 Er	69 Tm	70 Yb	71 Lu	6
Actinides	7	90 Th	91 Pa	92 U	93 Np	94 Pu	95 Am	96 Cm	97 Bk	98 Cf	99 Es	100 Fm	101 Md	102 No	103 Lr	7

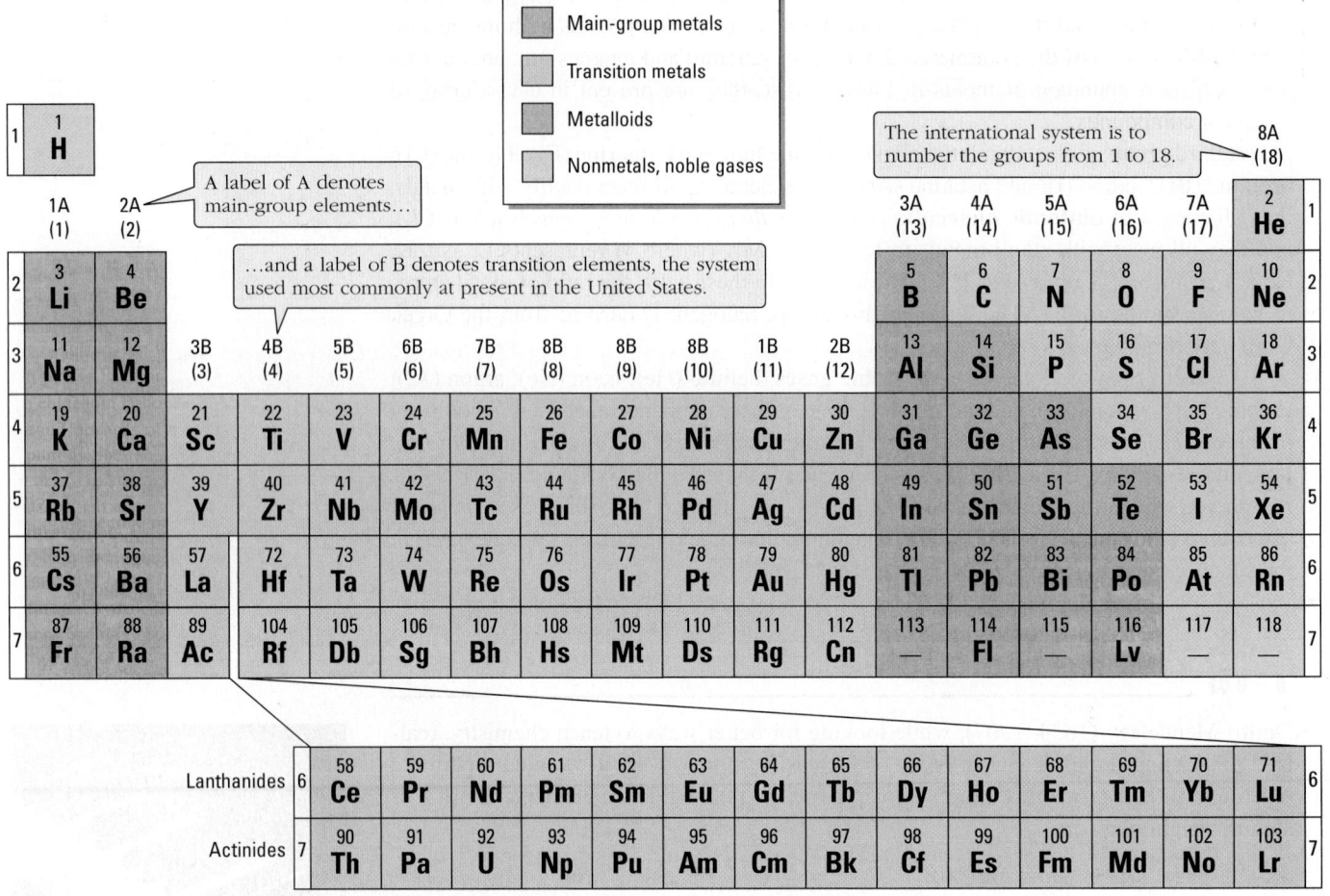

Figure 1.25 Modern periodic table of the elements. Elements are listed in order of increasing atomic number across horizontal rows called periods. Groups are vertical columns of elements. Some groups have common names: Group 1A, alkali metals; Group 2A, alkaline-earth metals; Group 7A, halogens; Group 8A, noble gases.

Elements are being synthesized even now. Element 117 was made in 2010, but only a few atoms were produced.

Ancient people knew of nine elements—gold (Au), silver (Ag), copper (Cu), tin (Sn), lead (Pb), mercury (Hg), iron (Fe), sulfur (S), and carbon (C). Most of the other naturally occurring elements were discovered during the 1800s, as one by one they were separated from minerals in Earth's crust or from Earth's oceans or atmosphere. Currently, 118 elements are known, but only 90 occur in nature. Elements such as technetium (Tc), neptunium (Np), mendelevium (Md), seaborgium (Sg), meitnerium (Mt), copernicium (Cn), and flerovium (Fl) have been made using nuclear reactions (see Chapter 18), beginning in the 1930s.

EXERCISE 1.9

Elements

Use Table 1.4, the periodic table inside the front cover, and/or the list of elements inside the front cover to answer these questions.

(a) Four elements are named for planets in our solar system (including the ex-planet Pluto). Give their names and symbols.

(b) One element is named for a state in the United States. Name the element and give its symbol.

(c) Two elements are named in honor of women. What are their names and symbols?

(d) Several elements are named for countries or regions of the world. Find at least four of these and give names and symbols.

(e) List the symbols of all elements that are nonmetals.

Table 1.4 Discovery and Name of Some Chemical Elements

Element	Symbol	Date Discovered	Discoverer	Derivation of Name/Symbol
Copernicium	Cn	1996	Sigurd Hofmann, et al.	Honoring Nicolaus Copernicus
Curium	Cm	1944	G. Seaborg, et al.	Honoring Marie and Pierre Curie, Nobel Prize winners for discovery of radioactive elements
Hydrogen	H	1766	H. Cavendish	Greek, *hydro* (water) and *genes* (generator)
Meitnerium	Mt	1982	P. Armbruster, et al.	Honoring Lise Meitner, codiscoverer of nuclear fission
Mendelevium	Md	1955	G. Seaborg, et al.	Honoring Dmitri Mendeleev, who devised the periodic table
Mercury	Hg	Ancient	Ancient	For Mercury, messenger of the gods, because it flows quickly; symbol from Greek *hydrargyrum,* liquid silver
Polonium	Po	1898	M. Curie and P. Curie	In honor of Poland, Marie Curie's native country
Seaborgium	Sg	1974	A. Ghiorso, et al.	Honoring Glenn Seaborg, Nobel Prize winner for synthesis of new elements
Sodium	Na	1807	H. Davy	Latin, *soda* (sodium carbonate); symbol from Latin *natrium*
Tin	Sn	Ancient	Ancient	German, *Zinn;* symbol from Latin, *stannum*

EXERCISE 1.10

Element Names

On June 14, 2000, a major daily American newspaper published this paragraph:

> "ABC's *Who Wants to Be a Millionaire* crowned its fourth million-dollar winner Tuesday night. Bob House . . . [answered] the final question: Which of these men does not have a chemical compound named after him? (a) Enrico Fermi, (b) Albert Einstein, (c) Niels Bohr, (d) Isaac Newton"

What is wrong with the question? What is the correct answer to the question after it is properly posed? (The question was properly posed and correctly answered on the TV show.)

1-13a Periodic Table Features

The periodic table is arranged so that *elements with similar chemical properties occur in vertical columns* called **groups**. For example, in Figures 1.25 and 1.26, the alkali metals are found in group 1A. The periodic table commonly used in the United States has groups numbered 1 through 8 (Figures 1.25 and 1.26), with each number followed by either an A or a B. *Elements in the A groups (Groups 1A and 2A on the left and Groups 3A through 8A at the right)* are collectively known as **main-group elements** or **representative elements**. The *B groups (in the middle of the table)* are called **transition elements**. The International Union of Pure and Applied Chemistry numbers the groups differently: from 1 to 18 in numeric order. Figures 1.25 and 1.26 and the table inside the front cover include this numbering system as well.

The *horizontal rows of the periodic table* are called **periods**, and they are numbered from top to bottom of the table, beginning with 1 for the period containing H and He. Sodium (Na) is, for example, in Group 1A and is the first element in the third period. Silver (Ag) is in Group 1B and is in the fifth period.

The table in Figures 1.25 and 1.26 and inside the front cover helps us to recognize that most elements are metals (gray and blue), far fewer elements are nonmetals (lavender), and only six are metalloids (orange). Elements become less metallic from left to right across a period, and eventually one or more nonmetals are found in each period.

Dmitri Mendeleev
1834–1907

Baldwin H. Ward & Kathryn C. Ward/CORBIS

Originally from Siberia, Mendeleev (1834–1907) spent most of his life in St. Petersburg. He taught at the University of St. Petersburg, where he wrote books and published his concept of chemical periodicity, which helped systematize inorganic chemistry. Later in life he moved on to other interests, including studying the natural resources of Russia and their commercial applications.

Many interactive periodic tables are available on the Internet. Two to explore are at http://www.chemeddl.org/resources/ptl and http://periodictable.com.

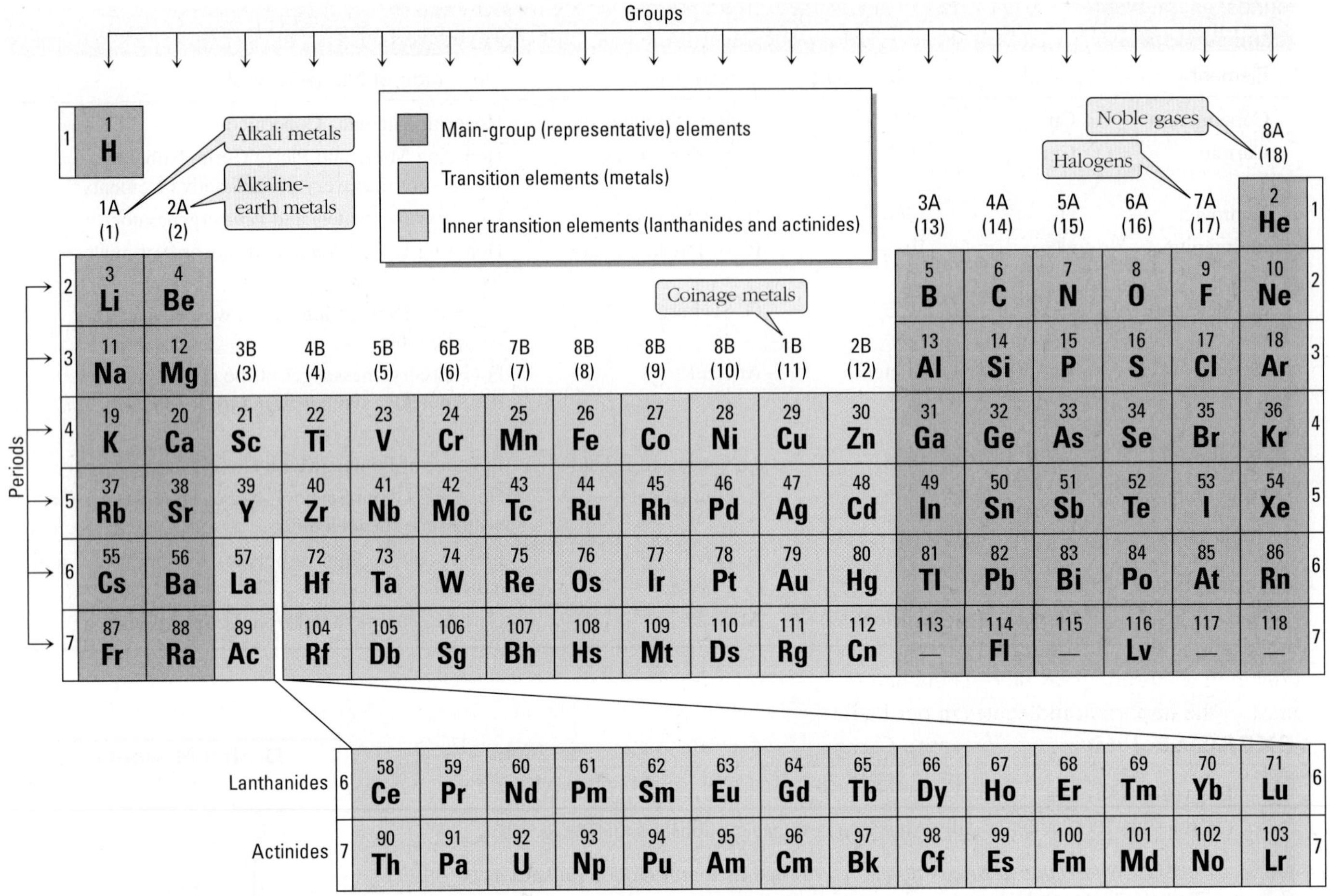
Groups
Main-group (representative) elements
Transition elements (metals)
Inner transition elements (lanthanides and actinides)
Alkali metals
Alkaline-earth metals
Coinage metals
Halogens
Noble gases
Periods
1A (1)
2A (2)
3B (3)
4B (4)
5B (5)
6B (6)
7B (7)
8B (8)
8B (9)
8B (10)
1B (11)
2B (12)
3A (13)
4A (14)
5A (15)
6A (16)
7A (17)
8A (18)
1 1 H 2 He 1
2 3 Li 4 Be 5 B 6 C 7 N 8 O 9 F 10 Ne 2
3 11 Na 12 Mg 13 Al 14 Si 15 P 16 S 17 Cl 18 Ar 3
4 19 K 20 Ca 21 Sc 22 Ti 23 V 24 Cr 25 Mn 26 Fe 27 Co 28 Ni 29 Cu 30 Zn 31 Ga 32 Ge 33 As 34 Se 35 Br 36 Kr 4
5 37 Rb 38 Sr 39 Y 40 Zr 41 Nb 42 Mo 43 Tc 44 Ru 45 Rh 46 Pd 47 Ag 48 Cd 49 In 50 Sn 51 Sb 52 Te 53 I 54 Xe 5
6 55 Cs 56 Ba 57 La 72 Hf 73 Ta 74 W 75 Re 76 Os 77 Ir 78 Pt 79 Au 80 Hg 81 Tl 82 Pb 83 Bi 84 Po 85 At 86 Rn 6
7 87 Fr 88 Ra 89 Ac 104 Rf 105 Db 106 Sg 107 Bh 108 Hs 109 Mt 110 Ds 111 Rg 112 Cn 113 — 114 Fl 115 — 116 Lv 117 — 118 — 7
Lanthanides 6 58 Ce 59 Pr 60 Nd 61 Pm 62 Sm 63 Eu 64 Gd 65 Tb 66 Dy 67 Ho 68 Er 69 Tm 70 Yb 71 Lu 6
Actinides 7 90 Th 91 Pa 92 U 93 Np 94 Pu 95 Am 96 Cm 97 Bk 98 Cf 99 Es 100 Fm 101 Md 102 No 103 Lr 7

Figure 1.26 Organization of the periodic table. It is important to be able to recognize regions of the periodic table and groups of related elements.

EXERCISE 1.11

Exploring the Periodic Table

1. What are the symbols for the elements in periodic Group 4A (Group 14)?
2. How many elements are in the fifth period?
3. Locate each of these elements in the periodic table. What group and which period is each element in?
   (a) Ti (b) Si (c) Rf (d) Se (e) Fe
4. Locate each of these groups of elements and give their group numbers.
   (a) halogens (b) alkaline-earth metals (c) noble gases
5. Elements that consist of diatomic molecules are hydrogen, nitrogen, oxygen, fluorine, chlorine, bromine, iodine, astatine, and (probably) element 117.
   (a) On a copy of the periodic table, circle the symbols of these elements.
   (b) Devise a rule related to the periodic table that will help you to remember which elements consist of diatomic molecules.

1-13b The Transition Elements, Lanthanides, and Actinides

The transition elements (also known as the transition metals) fill the middle of the periodic table in Periods 4 through 7, and most are found in nature only in compounds. The notable exceptions are gold, silver, platinum, copper, and liquid mercury, which can be found in elemental form. Iron, zinc, copper, and chromium are among the most important commercial metals. Because of their vivid colors, transition-metal compounds are used for pigments (Section 20-7).

The **lanthanides** and **actinides** are *metallic elements listed separately in two rows at the bottom of the periodic table*. They are also referred to as inner-transition elements. The lanthanides are part of the sixth period and the actinides are part of the seventh period; they are listed below the other elements to keep the periodic table from becoming too wide and cumbersome. The lanthanide elements are widely dispersed throughout Earth's surface and therefore difficult to obtain. Some of them have specific uses that make them commercially important. For example, magnets used in computer disc drives, crucial components of hybrid automobiles, and generators for wind-energy turbines all contain the lanthanide neodymium, Nd. The actinide elements that follow uranium, U, as well as the elements to the right of actinium, Ac, are synthesized by special nuclear reaction techniques (see Chapter 18).

1-13c Other Representative Elements: Groups 3A to 6A (13 to 16)

More information about the representative elements is available in Chapter 19.

We have already discussed many of the representative elements: Group 1A (the alkali metals); Group 2A (the alkaline-earth metals); Group 7A (the halogens); and Group 8A (the noble gases). Groups 3A through 6A contain the most abundant elements in Earth's crust and atmosphere, as well as elements—carbon, nitrogen, and oxygen—present in most of the important molecules in our bodies.

Aluminum, the second element in *Group 3A*, is the most abundant metal on Earth's surface; seven percent of the crust is aluminum. Aluminum is always found combined with other elements—most often with oxygen and silicon.

In *Group 4A*, both carbon and silicon are very important. Silicon is the second most abundant element in Earth's crust. Like aluminum, it is always combined with other elements—most often with oxygen in silicate minerals such as sand and quartz (which are mainly silicon dioxide).

Because carbon atoms bond strongly with each other, and because each carbon atom can form four bonds, huge numbers of carbon compounds exist. Organic chemistry is the branch of chemistry devoted to the study of carbon compounds. Carbon atoms also provide the framework for the molecules essential to living things, which are the subject of the branch of chemistry known as biochemistry.

In *Group 5A*, nitrogen and phosphorus are very important. Nitrogen is the most abundant element in Earth's atmosphere. N_2 molecules constitute 78% of all molecules in the air we breathe. Because nitrogen is much less reactive than oxygen, nitrogen is not found extensively in Earth's solid crust. Phosphorus is one of the elements that can limit the growth of plants, so phosphorus compounds are important components of fertilizers.

Oxygen is the first element in *Group 6A*. It is the most abundant of all elements at the surface of the earth (47%), and it is second most abundant in the atmosphere (21%). Oxygen is highly reactive and combines with most other elements (except most noble gases) to form compounds. It is a good thing that the atmosphere contains mostly nitrogen, because pure oxygen gas would make fires uncontrollable.

Groups 4A to 6A each begin with one or more nonmetals, include one or more metalloids, and end with a metal. Group 4A, for example, contains carbon, a nonmetal, two metalloids (Si and Ge), and two metals (Sn and Pb). Group 3A starts with boron (B), a metalloid, and contains four metals (Al, Ga, In, and Tl; not enough atoms of the element numbered 113 have been synthesized so far to be able to tell whether it is a metal).

EXERCISE 1.12

The Periodic Table

1. How many (a) metals, (b) nonmetals, and (c) metalloids are in the fourth period of the periodic table? Give the name and symbol for each element.
2. Which groups of the periodic table contain (a) only metals, (b) only nonmetals, (c) only metalloids?
3. Which period of the periodic table contains the most metals?

1-14 The Biological Periodic Table

Most of the more than 100 known elements are not directly involved with our personal health and well-being. However, more than 25 of the elements, listed in Figure 1.27, are absolutely essential to human life. Among these essential elements are metals, nonmetals, and metalloids from across the periodic table. All are necessary as part of a well-balanced diet.

Table 1.5 lists the major elements in the human body in order of their relative abundances per million atoms in the body, showing the preeminence of four of the nonmetals—*hydrogen, oxygen, carbon, and nitrogen*—the **building-block elements**. These four elements contribute most of the atoms in the *biologically significant chemicals*—the **biochemicals**—that make up all plants and animals. With few exceptions, a major one being water, the biochemicals that these nonmetals form are organic compounds.

Organic compounds are discussed in Chapter 2 and Chapter 10.

Ions and their reactions are discussed in Chapter 2 and Chapter 3.

Nonmetals are also present in the body as anions, negatively charged atoms or groups of atoms. Phosphorus-containing anions are found in bones and teeth; phosphorus is also found in cell membranes and DNA. Sulfur is a component of anions in body fluids and is also found in proteins. Chlorine, in the form of chloride anions, is essential for digestion of food by stomach acid. Many metals are present as cations, positively charged atoms or groups of atoms. Sodium and potassium cations are found in the bloodstream and in cellular fluids. Calcium cations are found in bones and teeth. Metal cations are also incorporated into large biochemical molecules (for example, iron in hemoglobin and cobalt in vitamin B-12).

If you weigh 150 lb, about 90 lb (60%) is water, 30 lb is fat, and the remaining 30 lb is a combination of proteins, carbohydrates, and calcium, phosphorus, and other dietary minerals.

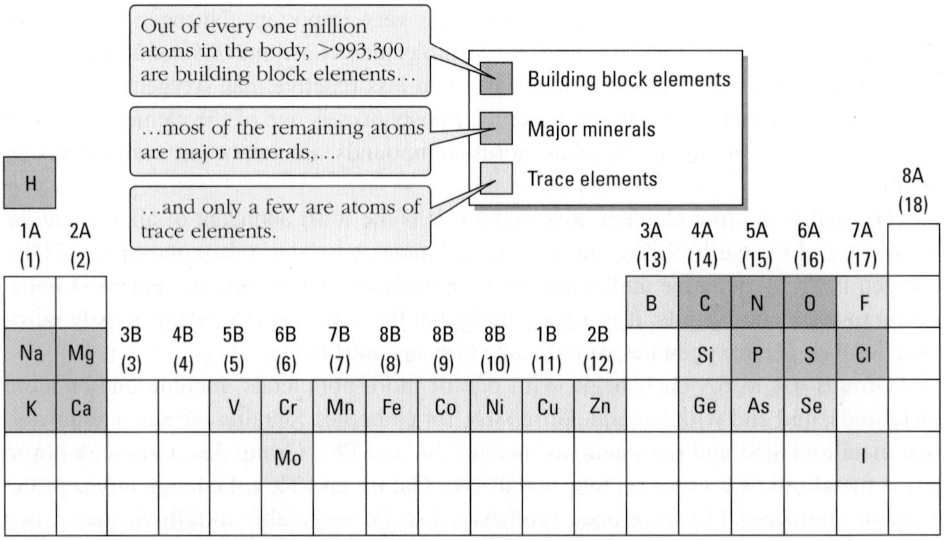

Out of every one million atoms in the body, >993,300 are building block elements...
...most of the remaining atoms are major minerals,...
...and only a few are atoms of trace elements.
Building block elements
Major minerals
Trace elements
H
1A (1)
2A (2)
3B (3)
4B (4)
5B (5)
6B (6)
7B (7)
8B (8)
8B (9)
8B (10)
1B (11)
2B (12)
3A (13)
4A (14)
5A (15)
6A (16)
7A (17)
8A (18)
B C N O F
Na Mg
Si P S Cl
K Ca V Cr Mn Fe Co Ni Cu Zn Ge As Se
Mo I

Figure 1.27 Elements essential to human health. Four elements—C, H, N, and O—form the many organic compounds that make up living organisms. The major minerals are required in relatively large amounts; trace elements are required in lesser amounts.

Table 1.5 Major Elements of the Human Body

Element	Symbol	Relative Abundance*	Element	Symbol	Relative Abundance*
Hydrogen	H	628,000	Phosphorus	P	2100
Oxygen	O	257,000	Sulfur	S	500
Carbon	C	95,000	Sodium	Na	420
Nitrogen	N	13,600	Potassium	K	330
Calcium	Ca	2400	Chlorine	Cl	270
			Magnesium	Mg	130

*Relative abundance in atoms per million atoms in the body.

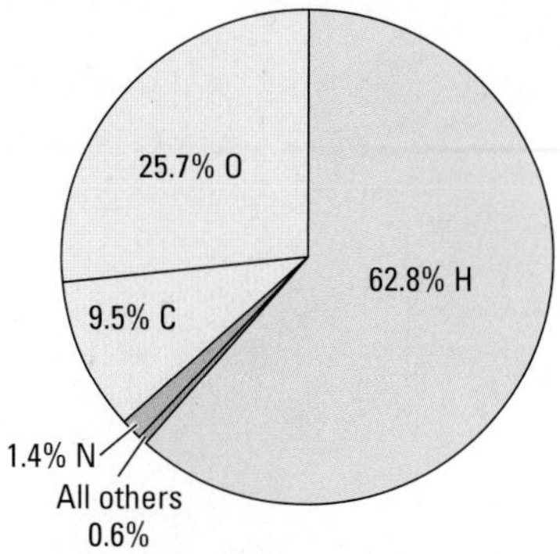
25.7% O
62.8% H
9.5% C
1.4% N
All others
0.6%

The four building-block elements are more than 99% of the atoms in the human body.

1-14a The Dietary Minerals

The general term **dietary minerals** refers to *the essential elements other than carbon, hydrogen, oxygen, or nitrogen*. The dietary necessity and effects of these elements go far beyond those implied by their collective presence as only about 4% of our body weight. They exemplify the old saying, "Good things come in small packages." Because the body uses them efficiently, recycling them through many reactions, dietary minerals are required in only small quantities, but their absence from your diet can cause significant health problems.

The dietary minerals indicated in Figure 1.27 are further classified as major minerals (more abundant) and trace elements (less plentiful). **Major minerals** are *present in quantities greater than 0.01% of body mass* (100 mg per kg)—more than 6 g for a 60-kg (132-lb) individual. **Trace elements** are *present in smaller quantities than 0.01% of body mass* (sometimes far smaller). The necessary daily total intake of iodine is only 150 μg, for example.

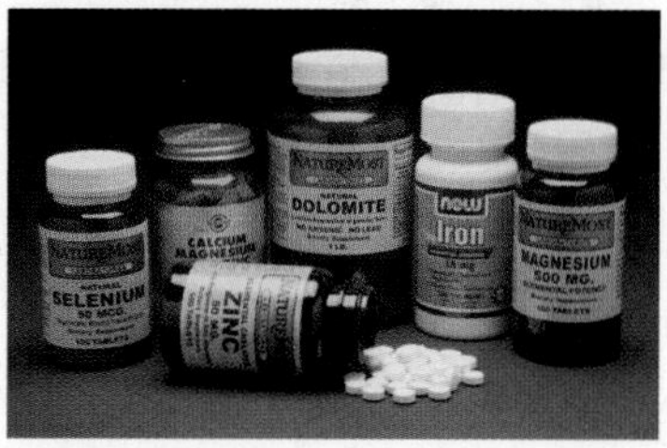
DOLOMITE
CALCIUM MAGNESIUM
SELENIUM
ZINC
Iron
MAGNESIUM

Cengage Learning/Charles D. Winters

Vitamin and mineral supplements.

EXERCISE 1.13

Essential Elements

Using Figure 1.27, identify (a) the essential nonmetals, (b) the essential alkaline-earth metals, (c) the essential halide ions, and (d) four essential transition metals.

1-15 Modern Chemical Sciences

Our goal in this chapter has been to make clear many of the reasons you should care about chemistry. Chemistry is fundamental to understanding many other sciences and to understanding how the material world around us works. Chemistry provides a unique, nanoscale perspective that has been highly successful in stimulating scientific inquiry and in the development of high-tech materials, modern medicines, and many other advances that benefit us every day. Chemistry is happening all around us and within us all the time. Knowledge of chemistry is a key to understanding and making the most of our internal and external environments and to making better decisions about how we live our lives and structure our economy and society. Finally, chemistry—the properties of elements and compounds, the nanoscale theories and models that interpret those properties, and the changes of one kind of substance into another—is just plain interesting and fun. Chemistry presents an intellectual challenge and provides ways to satisfy intellectual curiosity while helping us to better understand the world in which we live.

Modern chemistry overlaps more and more with biology, medicine, engineering, and other sciences. Because it is central to understanding matter and its transformations, chemistry becomes continually more important in a world that relies on chemical knowledge to produce the materials and energy required for a comfortable and productive way of life. The breadth of chemistry is recognized by the term **chemical sciences**, which *includes chemistry, chemical engineering, chemical biology, materials chemistry, industrial chemistry, environmental chemistry, medicinal chemistry, and many other fields.* Practitioners of the chemical sciences produce new plastics, medicines, superconductors, composite materials, and electronic devices. The chemical sciences enable better understanding of the mechanisms of biological processes, how proteins function, and how to imitate the action of organisms that carry out important functions. The chemical sciences enable us to measure tiny concentrations of substances, separate one substance from another, and determine whether a given substance is helpful or harmful to humans. Practitioners of the chemical sciences are involved in "green chemistry"—creating new industrial processes and products that are less hazardous, produce less pollution, and generate smaller quantities of wastes. The enthusiasm of chemists for research in all of these areas and the many discoveries that are made every day offer ample evidence that chemistry is an energetic and exciting science. We hope that this excitement is evident in this chapter and in the rest of the book.

The report "Beyond the Molecular Frontier: Challenges for Chemistry and Chemical Engineering" is available from the National Academies Press, http://www.nap.edu/catalog/10633.html.

Near the beginning of this chapter we listed questions you may have wondered about that will be answered in this book. More important by far, however, are questions that chemists or other scientists cannot yet answer. Here are some big challenges for chemists of the future, as envisioned by a blue-ribbon panel of experts convened by the U. S. National Academies of Science, Engineering, and Medicine.

- How can chemists design and synthesize new substances with well-defined properties that can be predicted ahead of time? Example substances are medicines, electronic devices, composite materials, and polymeric plastics.
- How can chemists learn from nature to produce new substances efficiently and use biological processes as models for industrial production? An example is extracting a compound from the English yew and then processing it chemically to produce paclitaxel, an anti-cancer drug.
- How can chemists design mixtures of molecules that will assemble themselves into useful, more complex structures? Such self-assembly processes, in which each different molecule falls of its own accord into the right place, could be used to create a variety of new nanostructures.
- How can chemists learn to measure more accurately how much of a substance is present, determine the substance's properties, and predict how long it will last? Sensors are now being invented that combine chemical and biological processes to allow rapid, accurate determination of composition and structure.
- How can chemists devise better theories that will predict more accurately the behavior of molecules and of large collections of molecules?
- How can chemists devise new ways of making large quantities of materials without significant negative consequences for Earth's environment and the many species that inhabit our planet?
- How can chemists use improved knowledge of the molecular structures of genes and proteins to create new medicines and therapies to deal with viral diseases such as AIDS; major killers such as cancer, stroke, and heart disease; and psychological problems?
- How can chemists find alternatives to fossil fuels, avoid global warming, enable mobility for all without polluting the planet, and make use of the huge quantities of solar energy that are available?
- How can chemists contribute to national and global security?

- How can chemists improve the effectiveness of education, conveying chemical knowledge to the many students and others who can make use of it in their chosen fields?

A major goal of the authors of this book is to help you along the pathway to becoming a scientist. We encourage you to choose one of the preceding problems or another problem of similar importance and devote your life's work to finding new approaches and useful solutions to it. The future of our society depends on it!

SUMMARY PROBLEM

What kinds of fuels will power transportation systems in the future? This is a major issue facing our society because there is clear scientific evidence that carbon dioxide emissions from combustion of petroleum-based fuels contribute to global climate change. The scientific method and scientific data can help policymakers develop approaches to this problem that will be most beneficial to the world as a whole.

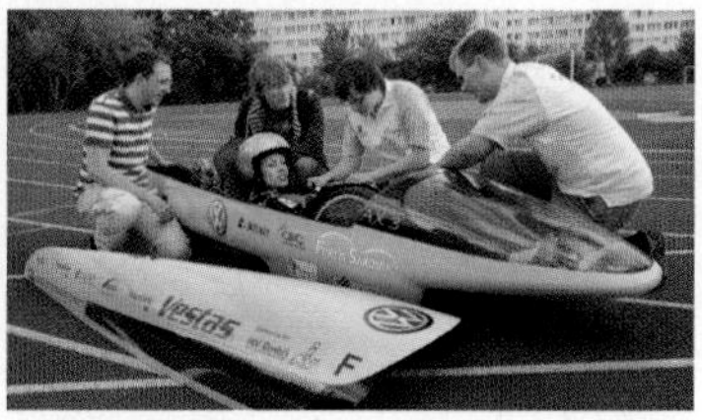
WOLFGANG THIEME/dpa/Landov

Students and their entry in the Shell Eco-marathon.

Efforts are under way to maximize the fuel economy of our transportation system and many students are involved in such efforts. The photograph shows a group of students and their entry in the Shell Eco-marathon, a worldwide program in which students are challenged to design and produce a vehicle that can travel the farthest on the least quantity of energy. As of 2012, the record was 2903 km/L. That's more than the distance from Chicago to Los Angeles on a single liter of fuel!

A variety of units are used to describe quantities of energy and it is important in any scientific analysis of fuels and energy that units be carefully defined. For national and global energy use, typical SI units are petajoule (1 PJ = 10^{15} J) and exajoule (1 EJ = 10^{18} J). Other often-used units are million tonnes of oil equivalent for petroleum (1 Mtoe = 41.9 PJ) and terawatt-hour electric for electric power (1 $TW_eh = 10^{12}\ W_eh$ = 3.6 PJ). Joule is the SI unit of energy and watt is the SI unit of power (energy per unit time). $1\ J = 1\ W \times 1\ s$.

An all-electric vehicle (such as the Nissan Leaf) does not have a typical internal-combustion engine. It is powered by batteries that are recharged by plugging the car into a wall outlet connected to the electric power grid. Suppose that there are 300,000 all-electric vehicles on the road and, on average, each of them travels 10,000 miles per year (1 mi = 1.6 km) and has an efficiency of 6 km/kW_eh. Calculate the total annual energy use of the 300,000 all-electric vehicles.

A potential fuel of the future is hydrogen. A gas at room temperature, hydrogen burns in air to form a single reaction product, water. One way to generate hydrogen for use as a fuel is to electrolyze water, a process in which a direct electric current generates hydrogen and oxygen in separate containers. Based only on information given in this paragraph, what can you conclude about whether hydrogen, oxygen, and water are elements, compounds, or mixtures? Explain your reasoning.

Biogas, that is, methane, CH_4, can be produced by bacterial action on municipal solid waste. Methane can be compressed, stored in strong tanks, and burned in engines to power vehicles such as buses. When methane burns it combines with oxygen, O_2, from the air and produces carbon dioxide, CO_2, and water, H_2O. Write a nanoscale and a symbolic representation of the combustion of methane. Make certain that the representation is consistent with the modern atomic theory (← **Sec. 1-10**).

HAVING STUDIED THIS CHAPTER . . .

. . . you should be able to:

- Appreciate the power of chemistry to answer intriguing questions (Section 1-1).
- Describe the approach used by scientists in solving problems (Sections 1-2, 1-3).
- Define quantitative and qualitative observations (Section 1-3). End-of-chapter Question 9
- Identify the physical properties of matter or physical changes occurring in a sample of matter (Section 1-4). Question 11

- Describe characteristic properties of the three states of matter—gases, liquids, and solids (Section 1-4).
- Estimate Celsius temperatures for commonly encountered situations (Section 1-4). Question 13
- Apply the ideas of measurement, physical quantities, and the International System of Units (Section 1-5). Questions 15, 17
- Be able to determine the number of significant digits in a number and round numbers to an appropriate number of significant digits (Section 1-5). Questions 19, 21
- Calculate density and use it to identify substances or explain their behavior (Section 1-5). Questions 23, 27, 100, 102, 104
- Solve problems using proportionality (conversion) factors, dimensional analysis, and the Analyze, Plan, Execute, Check Result problem-solving strategy (Section 1-5). Question 25
- Identify the chemical properties of matter or chemical changes occurring in a sample of matter (Section 1-6). Questions 29, 31, 33
- Explain the difference between homogeneous and heterogeneous mixtures (Section 1-7). Question 35
- Describe the importance of separation, purification, and analysis (Section 1-7). Question 37
- Understand the difference between a chemical element and a chemical compound (Sections 1-8, 1-10). Questions 39, 43
- Classify matter (Section 1-8, Figure 1.15). Question 41
- Identify relative sizes at the macroscale, microscale, and nanoscale levels (Section 1-9). Question 45
- Describe the kinetic-molecular theory at the nanoscale level (Section 1-9). Questions 47, 49, 106
- Use the postulates of modern atomic theory to explain macroscale observations about elements, compounds, conservation of mass, constant composition, and multiple proportions (Section 1-10). Questions 51, 53
- Apply and interpret macroscale, nanoscale, and symbolic representations of substances and chemical processes (Section 1-11). Questions 55, 57, 59
- Distinguish metals, nonmetals, and metalloids according to their properties (Section 1-12). Question 61
- Identify and name groups of elements that have similar properties (Section 1-12). Question 63
- Define periods, groups, main-group (representative) elements, and transition elements in the periodic table (Section 1-13). Questions 65, 67
- Locate in the periodic table alkali metals, alkaline-earth metals, halogens, noble gases, lanthanides, and actinides (Section 1-13, Figure 1.26). Question 69
- Identify essential elements, building-block elements, and other features of the biological periodic table (Section 1-14).
- Identify areas in which chemistry can contribute to societal needs (Section 1-15).

KEY TERMS

These terms were defined and given in boldface type in this chapter. Be sure to understand each of these terms and the concepts with which they are associated.

actinides (Section 1-13b)
alkali metals (1-12a)
alkaline-earth metals (1-12a)
atom (1-10)
atomic theory (1-10)
biochemicals (1-14)
boiling point (1-4c)
building-block elements (1-14)
Celsius temperature scale (1-4c)
chemical change (1-6)
chemical compound (1-8)
chemical element (1-8)
chemical formula (1-10)
chemical properties (1-6a)
chemical reaction (1-6)
chemical sciences (1-15)
chemical symbol (1-10)
chemistry (1-1)
conservation of matter, law of (1-10)
constant composition, law of (1-10)
conversion factor (1-5c)
density (1-4d)
diatomic molecule (1-12a)
dietary minerals (1-14a)
dimensional analysis (1-5)
energy (1-6b)
freezing point (1-4c)
gas (1-4b)
groups (in periodic table) (1-13a)
halogens (1-12a)
heterogeneous mixture (1-7)
homogeneous mixture (1-7)
hypothesis (1-3)
kinetic-molecular theory (1-9a)
lanthanides (1-13b)
law (1-3)
liquid (1-4b)
macroscale (1-9)
macroscopic (1-9)
main-group elements (1-13a)
major minerals (1-14a)
matter (1-1)
measurement (1-5)
melting point (1-4c)
metals (1-12)
metalloids (1-12)
microscale (1-9)
microscopic (1-9)
model (1-3)
molecule (1-10)
multiple proportions, law of (1-10)
nanoscale (1-9)
noble gases (1-12a)
nonmetals (1-12)
periodic table (1-13)
periods (in periodic table) (1-13a)
physical changes (1-4a)
physical properties (1-4)
physical quantity (1-5)
physical state (1-4b)
products (1-6)
proportionality factor (1-5c)
purified (1-7a)
qualitative (1-3)
quantitative (1-3)
reactants (1-6)
representative elements (1-13a)
SI base units (1-5a)
SI prefix (1-5a)
SI units (1-5a)
solid (1-4b)
solution (1-7)
substance (1-4)
temperature (1-4c)
theory (1-3)
trace elements (1-14a)
transition elements (1-13a)

QUESTIONS FOR REVIEW AND THOUGHT

Interactive versions of these problems are assignable in OWL.

Red-numbered questions have short answers at the back of this book in Appendix M and full solutions in the *Student Solutions Manual.*

Review Questions

These questions test vocabulary and simple concepts.

1. What is meant by nanoscale? Why is structure at the nanoscale important?
2. Choose an object in your room, such as a cell phone or television set. Write down five qualitative observations and five quantitative observations regarding the object you chose.
3. What are three important characteristics of a scientific law? Name two laws that were mentioned in this chapter. State each of the laws that you named.
4. How does a scientific theory differ from a law? How are theories and models related?
5. What is the unique perspective that chemists use to make sense out of the material world? Give at least one example of how that perspective can be applied to a significant problem.
6. Give two examples of situations in which purity of a chemical substance is important.

Topical Questions

These questions are keyed to the major topics in the chapter. Usually a question that is answered at the back of this book is paired with a similar one that is not.

Why Care About Chemistry? (Section 1-1)

7. Make a list of at least four issues faced by our society that require scientific studies and scientific data before a democratic society can make informed, rational decisions. Exchange lists with another student and evaluate the quality of each other's choices.
8. Make a list of at least four questions you have wondered about that may involve chemistry. Compare your list with a list from another student taking the same chemistry course. Evaluate the quality of each other's questions and decide how "chemical" they are.

How Science Is Done (Section 1-3)

9. Which of these statements are qualitative? Which are quantitative? Explain your choice in each case.
   (a) Sodium is a silvery-white metal.
   (b) Aluminum melts at 660 °C.
   (c) Carbon makes up about 23% of the human body by mass.
   (d) Pure carbon occurs in different forms: graphite, diamond, and fullerenes.
10. Which of the these statements are qualitative? Which are quantitative? Explain your choice in each case.
    (a) The atomic mass of carbon is 12.011 (12.011 atomic mass units).
    (b) Pure aluminum is a silvery-white metal that is nonmagnetic, has a low density, and does not produce sparks when struck.
    (c) Sodium has a density of 0.968 g/mL.
    (d) In animals the sodium cation, Na^+, is the main extracellular cation and is important for nerve function.

Identifying Matter: Physical Properties (Section 1-4)

11. The elements sulfur and bromine are shown in the photograph. Based on the photograph, describe as many properties of each sample as you can. Are any properties the same? Which properties are different?

Cengage Learning/Charles D. Winters

Sulfur and bromine. The sulfur is on the dish; the bromine is in a closed flask.

12. In the accompanying photo, you see a crystal of the mineral calcite surrounded by piles of calcium and carbon, two of the elements that combine to make the mineral. (The other element combined in calcite is oxygen.) Based on the photo, describe some of the physical properties of the elements and the mineral. Are any properties the same? Are any properties different?

Cengage Learning/Charles D. Winters

Calcite (the transparent, cube-like crystal) and two of its constituent elements, calcium (chips) and carbon (black grains). The calcium chips are covered with a thin film of calcium oxide.

13. The melting point of gallium is 29.76 °C. If you hold a sample of gallium in your hand, will it melt? Explain briefly.

14. These temperatures are measured at various locations during the same day in the winter in North America: −10 °C at Montreal, 28 °F at Chicago, 20 °C at Charlotte, and 40 °F at Philadelphia. Which city is the warmest? Which city is the coldest?

Measurements, Units, and Calculations (Section 1-5)

15. A crystal of fluorite (a mineral that contains calcium and fluorine) has a mass of 2.83 g. What is this mass in kilograms? Give the symbols for the elements in this crystal.
16. Suppose a room is 18 m long, 15 m wide, and the distance from floor to ceiling is 2.9 m. What is the room's volume in cubic meters? In cubic centimeters? In liters?
17. The current world record for the 100-m dash is 9.58 seconds held by Usain Bolt of Jamaica. Calculate whether he could have been arrested for exceeding a 25 mi/h speed limit while setting that record.
18. Body mass index (BMI) is the ratio of a person's mass (kg) divided by the square of the height (m). In general, a BMI >25 indicates that the person is overweight.
    (a) Calculate the BMI of someone who is 76.0 in tall and weighs 176 lb.
    (b) Should the person be considered as being overweight? Explain.
19. How many significant figures are present in these measured quantities?
    (a) 1374 kg
    (b) 0.00348 s
    (c) 5.619 mm
    (d) 2.475×10^{-3} cm
    (e) 33.1 mL
    (f) 2300. m
20. How many significant figures are present in these measured quantities?
    (a) 1.022×10^2 km
    (b) 34 m²
    (c) 0.042 L
    (d) 28.2 °C
    (e) 323. mg
    (f) 420 g
21. Perform these calculations and express the result with the proper number of significant figures.
    (a) $\dfrac{4.850\text{ g} - 2.34\text{ g}}{1.3\text{ mL}}$
    (b) $V = \frac{4}{3}\pi r^3$ where $r = 4.112$ cm
    (c) $(4.66 \times 10^{-3}) \times 4.666$
    (d) $\dfrac{0.003400}{65.2}$
22. Perform these calculations and express the result with the proper number of significant figures.
    (a) $2221.05 - \dfrac{3256.5}{3.20}$
    (b) $343.2 \times (2.01 \times 10^{-3})$
    (c) $S = 4\pi r^2$ where $r = 2.55$ cm
    (d) $\dfrac{2802}{15} - (0.0025 \times 10{,}000.)$
23. A 105.5-g sample of a metal was placed into water in a graduated cylinder, and it completely submerged. The water level rose from 25.4 mL to 37.2 mL. Use data in Table 1.1 to identify the metal.
24. An irregularly shaped piece of lead weighs 10.0 g. It is carefully lowered into a graduated cylinder containing 30.0 mL ethanol, and it sinks to the bottom of the cylinder. To what volume reading does the ethanol rise?
25. An unknown sample of a metal is 1.0 cm thick, 2.0 cm wide, and 10.0 cm long. Its mass is 54.0 g. Use data in Table 1.1 to identify the metal. (Remember that 1 cm³ = 1 mL.)
26. Calculate the volume of a 23.4-g sample of bromobenzene, density = 1.49 g/mL.
27. Calculate the mass of a sodium chloride crystal if the dimensions of the crystal are 10 cm thick by 12 cm long by 15 cm wide. (Remember that 1 cm³ = 1 mL.)
28. Calculate the volume occupied by a 4.33-g sample of iron.

Chemical Changes and Chemical Properties (Section 1-6)

29. In each case, identify the *italicized* property as a physical or chemical property. Give a reason for your choice.
    (a) The normal *color* of the element bromine is red-orange.
    (b) Iron is *transformed into rust* in the presence of air and water.
    (c) Dynamite can *explode*.
    (d) Aluminum metal, the *shiny* "foil" you use in the kitchen, *melts* at 660 °C.
30. In each case, identify the *italicized* property as a physical or a chemical property. Give a reason for your choice.
    (a) Dry ice *sublimes* (changes directly from a solid to a gas) at −78.6 °C.
    (b) Methanol (methyl alcohol) *burns in air* with a colorless flame.
    (c) Sugar is *soluble in water*.
    (d) Hydrogen peroxide, H_2O_2, *decomposes to form oxygen*, O_2, *and water*, H_2O.
31. In each case, describe the change as a chemical or physical change. Give a reason for your choice.
    (a) A cup of household bleach changes the color of your favorite T-shirt from purple to pink.
    (b) The fuels in the space shuttle (hydrogen and oxygen) combine to give water and provide the energy to lift the shuttle into space.
    (c) An ice cube in your glass of lemonade melts.
32. In each case, describe the change as a chemical or physical change. Give a reason for your choice.
    (a) Salt dissolves when you add it to water.
    (b) Food is digested and metabolized in your body.
    (c) Crystalline sugar is ground into a fine powder.
    (d) When potassium is added to water there is a purplish-pink flame and the water becomes basic (alkaline).
33. In each situation, decide whether a chemical reaction is releasing energy and causing work to be done, or whether an outside source of energy is forcing a chemical reaction to occur.
    (a) Your body converts excess intake of food into fat molecules.
    (b) Sodium reacts with water as shown in Figure 1.5.
    (c) Sodium azide in an automobile air bag decomposes, causing the bag to inflate.
    (d) An egg is hard-boiled in a pan on your kitchen stove.

34. While camping in the mountains you build a small fire out of tree limbs you find on the ground near your campsite. The dry wood crackles and burns brightly and warms you. Before slipping into your sleeping bag for the night, you put the fire out by dousing it with cold water from a nearby stream. Steam rises when the water hits the hot coals. Describe the physical and chemical changes in this scene.

Classifying Matter: Substances and Mixtures (Section 1-7)

35. Small chips of iron are mixed with sand (see photo). Is this a homogeneous or heterogeneous mixture? Suggest a way to separate the iron and sand from each other.

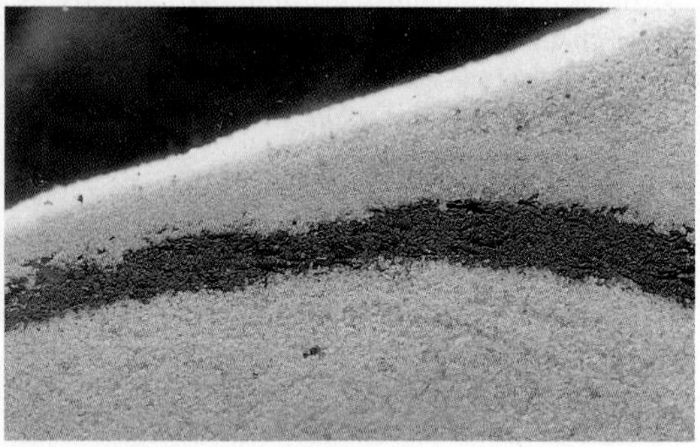

Cengage Learning/Charles D. Winters

Layers of sand, iron, and sand.

36. Identify each of these as a homogeneous or a heterogeneous mixture.
    (a) An asphalt (blacktop) road (b) Clear ocean water
    (c) Iced tea with ice cubes (d) Filtered apple cider
37. Devise and describe an experiment to
    (a) Separate table salt (sodium chloride) from water.
    (b) Separate iron filings from small pieces of magnesium.
    (c) Separate the element zinc from sugar (sucrose).
38. Devise and describe an experiment to
    (a) Separate sucrose (table sugar) from water.
    (b) Separate the element sulfur from table salt (sodium chloride).
    (c) Separate iron filings from granular zinc.

Classifying Matter: Elements and Compounds (Section 1-8)

39. For each of the changes described, decide whether two or more elements formed a compound or if a compound decomposed (to form elements or other compounds). Explain your reasoning in each case.
    (a) Upon heating, a blue powder turned white and lost mass.
    (b) A white solid forms three different gases when heated. The total mass of the gases is the same as that of the solid.
40. For each of the changes described, decide whether two or more elements formed a compound or if a compound decomposed (to form elements or other compounds). Explain your reasoning in each case.
    (a) After a reddish-colored metal is placed in a flame, it turns black and has a higher mass.
    (b) A white solid is heated in oxygen and forms two gases. The mass of the gases is the same as the masses of the solid and the oxygen.
41. Classify each of these as an element, a compound, a heterogeneous mixture, or a homogeneous mixture. Explain your choice in each case.
    (a) Chunky peanut butter (b) Distilled water
    (c) Platinum (d) Air
42. Classify each of these as an element, a compound, a heterogeneous mixture, or a homogeneous mixture. Explain your choice in each case.
    (a) Table salt (sodium chloride)
    (b) Methane (which burns in pure oxygen to form only carbon dioxide and water)
    (c) Chocolate chip cookie
    (d) Silicon
43. A black powder is placed in a long glass tube. Hydrogen gas is passed into the tube so that the hydrogen sweeps out all other gases. The powder is then heated with a Bunsen burner. The powder turns red-orange, and water vapor can be seen condensing at the unheated far end of the tube. The red-orange color remains after the tube cools.
    (a) Was the original black substance an element? Explain briefly.
    (b) Is the new red-orange substance an element? Explain briefly.
44. A finely divided black substance is placed in a glass tube filled with air. When the tube is heated with a Bunsen burner, the black substance turns red-orange. The total mass of the red-orange substance is greater than that of the black substance.
    (a) Can you conclude that the black substance is an element? Explain briefly.
    (b) Can you conclude that the red-orange substance is a compound? Explain briefly.

Nanoscale Theories and Models (Section 1-9)

45. The image of part of a DNA molecule below was obtained with a scanning tunneling microscope. The scales are given in hundreds of Angstrom units. Is this image at the macroscale, the microscale, or the nanoscale?

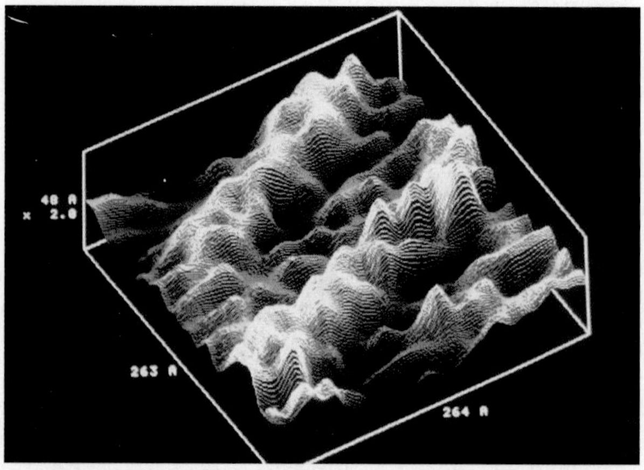
263 Å
264 Å

Science Source

46. The image below shows several examples of tobacco mosaic virus. Is this virus at the macroscale, the microscale, or the nanoscale? (The scale bar at the bottom of the image is 100 nm long.)

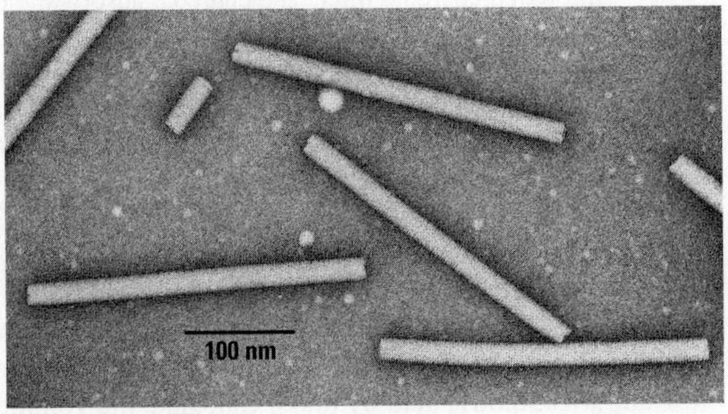
100 nm

Robert G. Milne, Plant Virus Institute, National Research Council, Turin, Italy

47. When you open a can of a carbonated drink, the carbon dioxide gas inside expands rapidly as it rushes from the can. Describe this process in terms of the kinetic-molecular theory.
48. Sometimes, after clothes are washed, they are hung in the sun to dry. Describe the change or changes that occur in terms of the kinetic-molecular theory. Are the changes that occur physical or chemical changes?
49. Sucrose has to be heated to a high temperature before it caramelizes. Use the kinetic-molecular theory to explain why sugar caramelizes only at high temperatures.
50. Give a nanoscale interpretation of the fact that at the melting point the density of solid mercury is greater than the density of liquid mercury, and at the boiling point the density of liquid mercury is greater than the density of gaseous mercury.

The Atomic Theory (Section 1-10)

51. Explain in your own words, by writing a short paragraph, how the atomic theory explains conservation of mass during a chemical reaction and during a physical change.
52. Explain in your own words, by writing a short paragraph, how the atomic theory explains constant composition of chemical compounds.
53. Explain in your own words, by writing a short paragraph, how the atomic theory predicts the law of multiple proportions.
54. The element chromium forms three different oxides (that contain only chromium and oxygen). The percentage of chromium (number of grams of chromium in 100 g oxide) in these compounds is 52.0%, 68.4%, and 76.5%. Do these data conform to the law of multiple proportions? Explain why or why not.

Communicating Chemistry: Symbolism (Section 1-11)

55. Write a chemical formula for each substance, and draw a picture of how the nanoscale particles are arranged at room temperature.
    (a) Water, a liquid whose molecules contain two hydrogen atoms and one oxygen atom each
    (b) Nitrogen, a gas that consists of diatomic molecules
    (c) Neon
    (d) Chlorine
56. Write a chemical formula for each substance and draw a picture of how the nanoscale particles are arranged at room temperature.
    (a) Iodine, a solid that consists of diatomic molecules
    (b) Ozone, a gas that consists of triatomic molecules
    (c) Helium
    (d) Carbon dioxide
57. Write a nanoscale representation and a symbolic representation and describe what happens at the macroscale when sulfur reacts chemically with oxygen to form sulfur dioxide gas.
58. Write a nanoscale representation and a symbolic representation and describe what happens at the macroscale when carbon monoxide reacts with oxygen to form carbon dioxide.
59. Write a nanoscale representation and a symbolic representation and describe what happens at the macroscale when iodine sublimes (passes directly from solid to gas with no liquid formation) to form iodine vapor.
60. Write a nanoscale representation and a symbolic representation and describe what happens at the macroscale when bromine evaporates to form bromine vapor.

The Chemical Elements (Section 1-12)

61. Name and give the symbols for two elements that
    (a) Are metals.
    (b) Are nonmetals.
    (c) Are metalloids.
    (d) Are alkaline-earth elements.
62. Name and give the symbols for two elements that
    (a) Are gases at room temperature.
    (b) Are solids at room temperature.
    (c) Do not consist of molecules.
    (d) Consist of diatomic molecules.

The Periodic Table (Section 1-13)

63. Name and give symbols for three transition metals in the fourth period. Look up each of your choices in a dictionary, a book such as *The Handbook of Chemistry and Physics*, or on the Internet, and make a list of their properties. Also list the uses of each element.
64. Name two halogens. Look up each of your choices in a dictionary, in a book such as *The Handbook of Chemistry and Physics*, or on the Internet, and make a list of their properties. Also list any uses of each element that are given by the source.
65. How many elements are there in Group 4A of the periodic table? Give the name and symbol of each of these elements. Tell whether each is a metal, nonmetal, or metalloid.
66. How many elements are there in the fourth period of the periodic table? Give the name and symbol of each of these elements. Tell whether each is a metal, metalloid, or nonmetal.

67. The symbols for the four elements whose names begin with the letter I are In, I, Ir, and Fe. Match each symbol with one of these phrases below.
   (a) A halogen
   (b) A main-group metal
   (c) A transition metal in Period 6
   (d) A transition metal in Period 4
68. The symbols for four of the eight elements whose names begin with the letter S are Si, Ag, Na, and S. Match each symbol with one of these phrases below.
   (a) A solid nonmetal (b) An alkali metal
   (c) A transition metal (d) A metalloid
69. Use the periodic table to identify these elements:
   (a) Name an element in Group 2A.
   (b) Name an element in the third period.
   (c) What element is in the second period in Group 4A?
   (d) What element is in the third period in Group 6A?
   (e) What halogen is in the fifth period?
   (f) What alkaline-earth element is in the third period?
   (g) What noble gas element is in the fourth period?
   (h) What nonmetal is in Group 6A and the second period?
   (i) Name a metalloid in the fourth period.
70. Use the periodic table to identify these elements:
   (a) Name an element in Group 2B.
   (b) Name an element in the fifth period.
   (c) What element is in the sixth period in Group 4A?
   (d) What element is in the third period in Group 5A?
   (e) What alkali metal is in the third period?
   (f) What noble gas is in the fifth period?
   (g) Name the element in Group 6A and the fourth period. Is it a metal, nonmetal, or metalloid?
   (h) Name a metalloid in Group 5A.

The Biological Periodic Table (Section 1-14)

71. Which types of compounds contain the majority of the oxygen found in the human body?
72. (a) In what form are metals found in the body, as atoms or as ions?
   (b) What are two uses for metals in the human body?
73. Distinguish between major minerals and trace elements.
74. Name a mineral that is essential at smaller concentrations but toxic at higher concentrations.

General Questions

These questions are not explicitly keyed to chapter topics; many require integration of several concepts.

75. Classify the information in each of these statements as quantitative or qualitative and as relating to a physical or chemical property.
   (a) A white chemical compound has a mass of 1.456 g. When placed in water containing a dye, it causes the red color of the dye to fade to colorless.
   (b) A sample of lithium metal, with a mass of 0.6 g, was placed in water. The metal reacted with the water to produce the compound lithium hydroxide and the element hydrogen.
76. Classify the information in each of these statements as quantitative or qualitative and as relating to a physical or chemical property.
   (a) A liter of water, colored with a purple dye, was passed through a charcoal filter. The charcoal adsorbed the dye, and colorless water came through. Later, the purple dye was removed from the charcoal and retained its color.
   (b) When a white powder dissolved in a test tube of water, the test tube felt cold. Hydrochloric acid was then added, and a white solid formed.
77. The label on a bale of mulch indicates a volume of 1.45 ft^3. The label also states that the mulch in the bale will cover an area of a garden 6 ft × 6 ft to a depth of 1 in. Account for the discrepancy in the given volumes.
78. The density of a solution of sulfuric acid is 1.285 g/cm^3, and it is 38.08% acid by mass. Calculate the volume of solution (in mL) that contains 125 g of sulfuric acid.
79. In addition to the metric units of nm and pm, a commonly used unit is the angstrom, where $1 \text{ Å} = 1 \times 10^{-10}$ m. If the distance between a Pt atom and an N atom in a compound is 1.97 Å, what is the distance in nm? In pm?
80. The cancer drug cisplatin contains 65.0% platinum. You have a 1.53-g sample of the compound; calculate the mass of platinum it contains.
81. The density of solid potassium is 0.86 g/mL. The density of solid calcium is 1.55 g/mL, almost twice as great. However, the mass of a potassium atom is only slightly less than the mass of a calcium atom. Provide a nanoscale explanation of these facts.
82. The density of gaseous helium at 25 °C and normal atmospheric pressure is 1.64×10^{-4} g/mL. At the same temperature and pressure the density of argon gas is 1.63×10^{-3} g/mL. The mass of an atom of argon is almost exactly ten times the mass of an atom of helium. Provide a nanoscale explanation of why the densities differ as they do.
83. The gauge number of a wire is related to the diameter of the wire. Suppose you have a 10.0-lb spool of 12-gauge aluminum wire (diameter 0.0808 in). Calculate the length of wire (in meters) on the spool. Assume that the wire is a cylinder ($V = \pi r^2 \ell$, where V is the volume, r is the radius, and ℓ is the length) and obtain the density of aluminum from Table 1.1. (1 in = 2.54 cm; 1 lb = 453.59 g)
84. The dimensions of aluminum foil in a box for sale at a supermarket are $66\frac{2}{3}$ yards by 12 inches. The mass of the foil is 0.83 kg. Calculate the thickness of the foil (in cm). (1 in = 2.54 cm)
85. Hexane (density = 0.66 g/cm^3), perfluorohexane (density = 1.669 g/cm^3), and water are immiscible liquids; that is, they do not dissolve in one another. You place 10 mL of each in a graduated cylinder, along with pieces of high-density polyethylene (HDPE, density 0.97 g/mL), polyvinyl chloride (PVC, density = 1.36 g/cm^3), and Teflon (density = 2.3 g/cm^3). None of these common plastics dissolves in these liquids. Describe what you expect to see.

86. You can figure out whether a substance floats or sinks if you know its density and the density of the liquid. In which of the liquids listed below will high-density polyethylene (HDPE) float? HDPE, a common plastic, has a density of 0.97 g/mL. It does not dissolve in any of these liquids.

Substance	Density (g/mL)	Properties, Uses
Ethylene glycol	1.113	Toxic; the major component of automobile antifreeze
Water	0.9982	
Ethanol	0.7893	The alcohol in alcoholic beverages
Methanol	0.7917	Toxic; gasoline additive to prevent gas line freezing
Acetic acid	1.0498	Component of vinegar
Glycerol	1.2611	Solvent used in home care products

87. Eleven of the elements in the periodic table are found in nature as gases at room temperature. List them. Where are they located in the periodic table?
88. Ten of the elements are O, H, Ar, Al, Ca, Br, Ge, K, Cu, and P. Pick the one that best fits each description: (a) an alkali metal; (b) a noble gas; (c) a transition metal; (d) a metalloid; (e) a Group 1 nonmetal; (f) an alkaline-earth metal; (g) a halogen; (h) a nonmetal that is a solid.
89. The periodic table shown below is color coded gray, blue, orange, and lavender. Identify the color of the area (or colors of the areas) in which you would expect to find each type of element.
    (a) A colorless gas
    (b) A solid that is ductile and malleable
    (c) A solid that has poor electrical conductivity

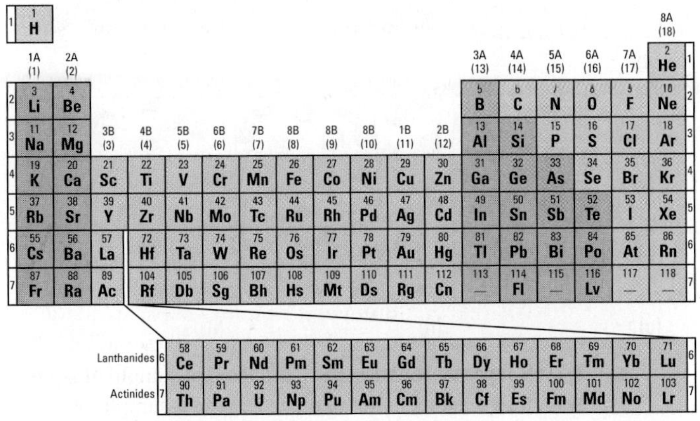

1A (1) 2A (2) 3B (3) 4B (4) 5B (5) 6B (6) 7B (7) 8B (8) 8B (9) 8B (10) 1B (11) 2B (12) 3A (13) 4A (14) 5A (15) 6A (16) 7A (17) 8A (18)
1 H 2 He
3 Li 4 Be 5 B 6 C 7 N 8 O 9 F 10 Ne
11 Na 12 Mg 13 Al 14 Si 15 P 16 S 17 Cl 18 Ar
19 K 20 Ca 21 Sc 22 Ti 23 V 24 Cr 25 Mn 26 Fe 27 Co 28 Ni 29 Cu 30 Zn 31 Ga 32 Ge 33 As 34 Se 35 Br 36 Kr
37 Rb 38 Sr 39 Y 40 Zr 41 Nb 42 Mo 43 Tc 44 Ru 45 Rh 46 Pd 47 Ag 48 Cd 49 In 50 Sn 51 Sb 52 Te 53 I 54 Xe
55 Cs 56 Ba 57 La 72 Hf 73 Ta 74 W 75 Re 76 Os 77 Ir 78 Pt 79 Au 80 Hg 81 Tl 82 Pb 83 Bi 84 Po 85 At 86 Rn
87 Fr 88 Ra 89 Ac 104 Rf 105 Db 106 Sg 107 Bh 108 Hs 109 Mt 110 Ds 111 Rg 112 Cn 113 — 114 Fl 115 — 116 Lv 117 — 118 —
Lanthanides 58 Ce 59 Pr 60 Nd 61 Pm 62 Sm 63 Eu 64 Gd 65 Tb 66 Dy 67 Ho 68 Er 69 Tm 70 Yb 71 Lu
Actinides 90 Th 91 Pa 92 U 93 Np 94 Pu 95 Am 96 Cm 97 Bk 98 Cf 99 Es 100 Fm 101 Md 102 No 103 Lr

90. The periodic table shown here is color coded gray, blue, orange, and lavender. Identify the color of the area (or colors of the areas) in which you would expect to find each type of element.
    (a) A shiny solid that conducts electricity
    (b) A gas whose molecules consist of single atoms
    (c) An element that is a semiconductor
    (d) A yellow solid that has very low electrical conductivity
91. Which two elements from this list exhibit the greatest similarity in physical and chemical properties? Explain your choice. S, Ga, Se, Ti.
92. Which two elements from this list exhibit the greatest similarity in physical and chemical properties? Explain your choice. Mg, Br, Si, Sr.
93. In the past, when someone discovered a new substance, it was relatively easy to show that the substance was not an element, but quite difficult to prove that the substance was an element. Explain why this is so, and relate your explanation to the discussion of scientific laws and theories in Section 1-3.
94. Soap can be made by mixing animal or vegetable fat with a concentrated solution of lye and heating it in a large vat. Suppose that 3.24 kg vegetable fat is placed in a large iron vat and then 50.0 L water and 5.0 kg lye (sodium hydroxide, NaOH) are added. The vat is placed over a fire and heated for two hours, and soap forms.
    (a) Classify each of the materials identified in the soapmaking process as a substance or a mixture. For each substance, indicate whether it is an element or a compound. For each mixture, indicate whether it is homogeneous or heterogeneous.
    (b) Assuming that the fat and lye are completely converted into soap, what mass of soap is produced?
    (c) What physical and chemical processes occur as the soap is made?
95. The densities of several elements are given in Table 1.1.
    (a) Of the elements nickel, gold, lead, and magnesium, which will float on liquid mercury at 20 °C?
    (b) Of the elements titanium, copper, iron, and gold, which will float highest on the mercury? That is, which element will have the smallest fraction of its volume below the surface of the liquid?
96. The element platinum has a solid-state structure in which platinum atoms are arranged in a cubic shape that repeats throughout the solid. The length of an edge of the cube is 392 pm (1 pm $= 1 \times 10^{-12}$ m). Calculate the volume of the cube in cubic meters.
97. The compound sodium chloride has a solid-state structure in which there is a repeating cubic arrangement of sodium ions and chloride ions. The volume of the cube is 1.81×10^{-22} cm^3. Calculate the length of an edge of the cube in pm (1 pm $= 1 \times 10^{-12}$ m).

Applying Concepts

These questions test conceptual learning.

98. You are given a mixture of sand, sugar, and sulfur. Write a detailed description of a procedure that would completely separate each component from this mixture.
99. You are given a mixture of sand, salt (sodium chloride), and naphthalene, a white solid that is not soluble in water. Write a detailed description of a procedure that would completely separate each component from this mixture.
100. Using Table 1.1, but without using your calculator, decide which has the larger mass:
    (a) 20. mL butane or 20. mL bromobenzene
    (b) 10. mL benzene or 1.0 mL gold
    (c) 0.732 mL copper or 0.732 mL lead

101. Using Table 1.1, but without using your calculator, decide which has the larger volume:
   (a) 1.0 g ethanol or 1.0 g bromobenzene
   (b) 10. g aluminum or 12 g water
   (c) 20 g gold or 40 g magnesium
102. At 25 °C the density of water is 0.997 g/mL, whereas the density of ice at −10 °C is 0.917 g/mL.
   (a) If a plastic soft-drink bottle (volume = 250 mL) is completely filled with pure water, capped, and then frozen at −10 °C, what volume will the solid occupy?
   (b) What will the bottle look like when you take it out of the freezer?
103. When water alone (instead of engine coolant, which contains water and other substances) was used in automobile radiators to cool cast-iron engine blocks, it sometimes happened in winter that the engine block would crack, ruining the engine. Cast iron is not pure iron and is relatively hard and brittle. Explain in your own words how the engine block in a car might crack in cold weather.
104. Water does not dissolve in bromobenzene.
   (a) If you pour 2 mL water into a test tube that contains 2 mL bromobenzene, which liquid will be on top?
   (b) If you pour 2 mL ethanol carefully into the test tube containing bromobenzene and water described in part (a) without shaking or mixing the liquids, what will happen?
   (c) What will happen if you thoroughly stir the mixture in part (b)?
105. Water does not mix with either benzene or bromobenzene when it is stirred together with either of them, but benzene and bromobenzene do mix.
   (a) If you pour 2 mL bromobenzene into a test tube, then add 2 mL water and stir, what would the test tube look like a few minutes later?
   (b) Suppose you add 2 mL benzene to the test tube in part (a), pouring the benzene carefully down the side of the tube so that the liquids do not mix. Describe the appearance of the test tube now.
   (c) If the test tube containing all three liquids is thoroughly shaken and then allowed to stand for five minutes, what will the tube look like?
106. The figure shows a nanoscale view of the atoms of mercury in a thermometer registering 10 °C.

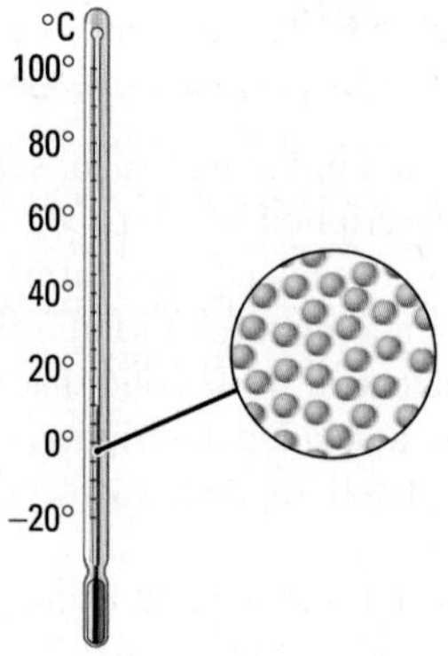
°C
100°
80°
60°
40°
20°
0°
−20°

Which nanoscale drawing best represents the atoms in the liquid in this same thermometer at 90 °C? (Assume that the same volume of liquid is shown in each nanoscale drawing.)

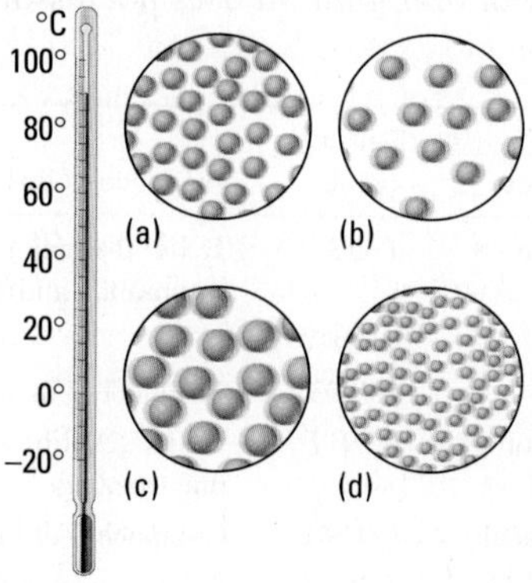
°C
100°
80°
60°
40°
20°
0°
−20°
(a)
(b)
(c)
(d)

107. Answer these questions using figures (a) through (i). (Each question may have more than one answer.)

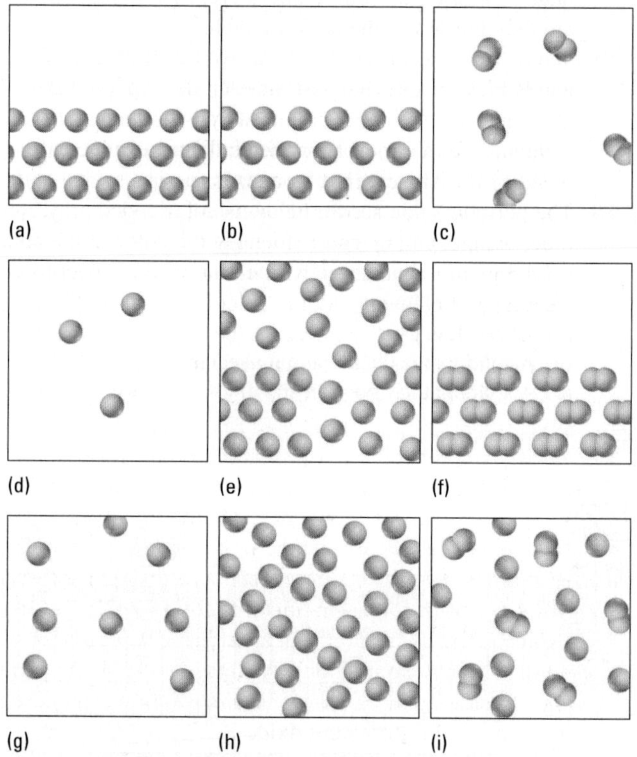
(a)
(b)
(c)
(d)
(e)
(f)
(g)
(h)
(i)

   (a) Which represents nanoscale particles in a sample of solid?
   (b) Which represents nanoscale particles in a sample of liquid?
   (c) Which represents nanoscale particles in a sample of gas?
   (d) Which represents nanoscale particles in a sample of an element?
   (e) Which represents nanoscale particles in a sample of a compound?

(f) Which represents nanoscale particles in a sample of a pure substance?

(g) Which represents nanoscale particles in a sample of a mixture?

108. Air mostly consists of diatomic molecules of nitrogen (about 80%) and oxygen (about 20%). Draw a nanoscale picture of a sample of air that contains a total of 10 molecules.

109. Most pure samples of metals are malleable, which means that if you try to grind up a sample of a metal by pounding on it with a hard object, the pieces of metal change shape but do not break apart. Solid samples of nonmetallic elements, such as sulfur or graphite, are often brittle and break into smaller particles when hit by a hard object. Devise a nanoscale theory about the structures of metals and nonmetals that can account for this difference in macroscale properties.

More Challenging Questions

These questions require more thought and integrate several concepts.

110. Rivers add salt (sodium chloride, NaCl) to the oceans of the world at a rate of approximately 2×10^{16} g/yr. Assume that Earth is a sphere with a diameter of 8,000 mi, 67% of which is covered by oceans to a depth of 1 mi. The average sodium chloride concentration of the oceans is 3% NaCl by mass and the average density of seawater is 1.03 g/cm^3. If the rate at which the NaCl addition to the oceans has been constant, calculate the approximate age of the oceans. Comment on the reasonableness of your answer given that the age of Earth is 4.5×10^9 yr.

111. When 1.0000 g silver chloride (AgCl) is all converted to silver iodide (AgI), 1.6381 g AgI is formed. You know that an atom of iodine has 3.580 times the mass of an atom of chlorine; calculate the mass of silver in each sample. Also calculate the mass of chlorine in the AgCl sample and the mass of iodine in the AgI sample.

112. Rust (an oxide of iron) can be converted to iron metal by reacting it with aluminum metal to form an aluminum oxide. You know that an iron atom has 2.069 times the mass of an aluminum atom. The mass of rust is 35.48 g and, when all of the iron has been replaced by aluminum, the mass of the aluminum oxide is 22.65 g. Calculate the mass of oxygen in both samples. Also calculate the mass of iron in the rust and the mass of aluminum in the final sample.

113. Normal respiration rate of humans is 12-20 breaths/min. When inhaling, you take in approximately 500 mL of air, each milliliter of which contains about 2.5×10^{19} molecules. Suppose you delivered a 10 minute speech in a class and, due to the stress you feel during that time, your respiration rate was 20 breaths/min.

(a) Calculate the number of air molecules you inhaled during your speech.

(b) There are approximately 1.1×10^{44} air molecules in the entire atmosphere. Calculate the fraction of all air molecules in the atmosphere you inhaled during your speech.

(c) Now, well after your speech is over, take a breath. Estimate the number of molecules in that breath that also were in the air you inhaled and exhaled during your speech.

114. In April 2010 the main pipe of the Deepwater Horizon oil-drilling platform burst, spewing crude oil into the Gulf of Mexico. On April 25, 2010, the spilled crude oil that reached the surface of the Gulf formed an oil slick 120 nm thick covering an area of 1.5×10^3 km^2. Calculate the number of barrels of crude oil in this slick; 1 barrel = 42 gal.

115. You and your lab partner are each given metal cubes that look similar. Your assignment is to make length and mass measurements and use only these data to determine whether the metal is the same in each cube. Your cube is 1.32 cm on each edge and has a mass of 16.23 g. Your partner's cube has a mass of 24.64 g and each edge measures 1.46 cm. Your partner says that the metal is the same in each cube; you don't agree. Refute your partner's conclusion.

116. Fritz Haber, a German chemist, proposed extracting gold from seawater as a way to pay off Germany's debt, $\$28.8 \times 10^6$, after World War I. The value of gold at the time was \$21.25/troy oz (1 troy oz = 31.103 g). The gold concentration in seawater is 0.15 mg gold/ton seawater (1 ton = 2000 lb). Assume the density of seawater is 1.03 g/cm^3.

(a) Calculate the volume (in cubic kilometers) of seawater that would have had to be processed to obtain the required mass of gold.

(b) By comparison, an Olympic-sized swimming pool is 50 m $\times$ 25 m $\times$ 2.0 m. Calculate the number of Olympic-sized swimming pools required to hold the volume of seawater needed in part (a).

117. Prior to 1734, only 14 elements were known. The rest of the known elements have been discovered across the next nearly three centuries. From 1805–1825, 12 new elements were discovered. A cluster of six new elements was isolated and identified from 1895–1905. Beginning in 1940 to the present, 19 new elements have been characterized. During each of these periods of discovery, a single scientific discovery or breakthrough led to the discovery of the new elements. Use the Internet to determine what each of the discoveries was and how it led to the discovery of new elements.

118. A group of astronauts in a spaceship accidentally encounters a space warp that traps them in an alternative universe where the chemical elements are quite different from the ones they are used to. For the elements they discovered, the astronauts found these properties:

Atomic Symbol	Atomic Weight	State	Color	Electrical Conductivity	Chemical Reactivity
A	3.2	Solid	Silvery	High	Medium
D	13.5	Gas	Colorless	Very low	Very high
E	5.31	Solid	Golden	Very high	Medium
G	15.43	Solid	Silvery	High	Medium
J	27.89	Solid	Silvery	High	Medium
L	21.57	Liquid	Colorless	Very low	Medium
M	11.23	Gas	Colorless	Very low	Very low
Q	8.97	Liquid	Colorless	Very low	Medium
R	1.02	Gas	Colorless	Very low	Very high
T	33.85	Solid	Colorless	Very low	Medium
X	23.68	Gas	Colorless	Very low	Very low
Z	36.2	Gas	Colorless	Very low	Medium
Ab	29.85	Solid	Golden	Very high	Medium

(a) Arrange these elements into a periodic table.
(b) If a new element, X, with atomic weight 25.84 is discovered, what would its properties be? Where would it fit in the periodic table you constructed?
(c) Are there any elements that have not yet been discovered? If so, what would their properties be?

119. You have some metal shot (small spheres like BBs), and you want to identify the metal. You have a flask that is known to contain exactly 100.0 mL when filled with liquid to a mark in the flask's neck. When the flask is filled with water at 20 °C, the mass of flask and water is 122.3 g. The water is emptied from the flask and 20 of the small spheres of metal are carefully placed in the flask. The 20 small spheres had a mass of 42.3 g. The flask is again filled to the mark with water at 20 °C and weighed. This time the mass is 159.9 g.
(a) What metal is in the spheres? (Assume that the spheres are all the same and consist of pure metal.)
(b) Calculate the volume occupied by 500 spheres.

120. When volumes of liquids are mixed, the resulting volume is not always equal to the sum of the volume of each liquid. For example, when 50.0 mL of ethanol (d = 0.789 g/mL) is mixed with 50.0 mL of water (0.998 g/mL) at 25 °C, the resulting volume is only 95.6 mL. Calculate the density of the solution.

121. Consider a 1.5 in. diameter solid sphere of sterling silver, which is a uniform mixture (an alloy) of 7.5% copper and 92.5% silver, by mass. The densities of the metals are: copper = 8.96 g/cm^3; silver = 10.5 g/cm^3. Calculate the mass of the sphere.

122. The element zinc reacts with the element sulfur to form a white solid compound, zinc sulfide. When a sample of zinc that weighs 65.4 g reacts with sulfur, it is found that the zinc sulfide produced weighs exactly 97.5 g.
(a) Calculate the mass of sulfur in the zinc sulfide.
(b) Calculate the mass of zinc sulfide that could be produced from 20.0 g zinc.

123. The element magnesium reacts with the element oxygen to form a white solid compound, magnesium oxide. When a sample of magnesium that weighs 24.30 g reacts with oxygen, it is found that the magnesium oxide produced weighs exactly 40.30 g.
(a) Calculate the mass of oxygen in the magnesium oxide.
(b) Calculate the mass of magnesium oxide that could be produced from 40.0 g magnesium.

124. A chemist analyzed several portions taken from different parts of a sample that contained only iron, Fe, and sulfur, S. She reported the results in the table. Could this sample be a compound of iron and sulfur? Explain why or why not.

Portion	Total Mass (g)	Mass of Fe (g)
1	1.518	0.964
2	2.056	1.203
3	1.873	1.290

125. A chemist analyzed several portions taken from different parts of a sample that contained only selenium, Se, and oxygen, O. She reported the results in the table. Could this sample be a compound of selenium and oxygen? Explain why or why not.

Portion	Total Mass (g)	Mass of Se (g)
1	1.518	1.08
2	2.056	1.46
3	1.873	1.33

126. When 12.6 g calcium carbonate (the principal component of chalk) is treated with 63.0 mL hydrochloric acid (muriatic acid sold in hardware stores; density = 1.096 g/mL), the calcium carbonate reacts with the acid, goes into solution, and carbon dioxide gas bubbles out of the solution. After all of the carbon dioxide has escaped, the solution weighs 76.1 g. Calculate the volume (in liters) of carbon dioxide gas that was produced. (The density of the carbon dioxide gas is 1.798 g/L.)

127. When 15.6 g sodium carbonate (washing soda used in tie dyes) is treated with 63.0 mL hydrochloric acid (muriatic acid sold in hardware stores; density = 1.096 g/mL), the sodium carbonate reacts with the acid, goes into solution, and carbon dioxide gas bubbles out of the solution. All of the carbon dioxide is collected and its volume is found to be 3.29 L. Calculate the mass of solution that remains. (The density of the carbon dioxide gas is 1.798 g/L.)

Conceptual Challenge Problems

These rigorous, thought-provoking problems integrate conceptual learning with problem solving and are suitable for group work.

CP1.A (Section 1-3) Suppose you are trying to get lemon juice and you have no juicer. Some people say that you can get more juice from a lemon if you roll it on a hard surface, applying pressure with the palm of your hand before you cut it and squeeze out the juice. Others claim that you will get more juice if you first heat the lemon in a microwave and then cut and squeeze it. Apply the methods of science to arrive at a technique that will give the most juice from a lemon. Carry out experiments and draw conclusions based on them. Try to generate a hypothesis to explain your results.

CP1.B (Section 1-3) If you drink orange juice soon after you brush your teeth, the orange juice tastes quite different. Apply the methods of science to find what causes this effect. Carry out experiments and draw conclusions based on them.

CP1.C (Section 1-3) Some people use expressions such as "a rolling stone gathers no moss" and "where there is no light there is no life." Why do you believe these are "laws of nature"?

CP1.D (Section 1-3) Parents teach their children to wash their hands before eating. (a) Do all parents accept the germ theory of disease? (b) Are all diseases caused by germs?

CP1.E (Section 1-9) In Section 1-9 you read that, on an atomic scale, all matter is in constant motion. (For example, the average speed of a molecule of nitrogen or oxygen in the air is greater than 1000 miles per hour at room temperature.) (a) What evidence can you put forward that supports the kinetic-molecular theory? (b) Suppose you accept the notion that molecules of air are moving at speeds near 1000 miles per hour. What can you then reason about the paths that these molecules take when moving at this speed?

CP1.F (Section 1-9) Some scientists think there are living things smaller than bacteria (*New York Times,* January 18, 2000, p. D1). Called "nanobes," they are roughly cylindrical and range from 20 to 150 nm long and about 10 nm in diameter. One approach to determining whether nanobes are living is to estimate how many atoms and molecules could make up a nanobe. If the number is too small, then there would not be enough DNA, protein, and other biological molecules to carry out life processes. To test this method, estimate an upper limit for the number of atoms that could be in a nanobe. (Use a small atom, such as hydrogen.) Also estimate how many protein molecules could fit inside a nanobe. Do your estimates rule out the possibility that a nanobe could be living? Explain why or why not.

CP1.G (Section 1-9) Helium-filled balloons rise and will fly away unless tethered by a string. Use the kinetic-molecular theory to explain why a helium-filled balloon is "lighter than air."

CP1.H (Section 1-15) The life expectancy of U. S. citizens in 1992 was 76 years. In 1916 the life expectancy was only 52 years. This is an increase of 46% in a lifetime. (a) Could this astonishing increase occur again? (b) To what single source would you attribute this noteworthy increase in life expectancy? Why did you identify this one source as being most influential?

2 | Chemical Compounds

Coke
PEPSI
Diet

Left: egorr/iStockphoto.com. Center: Lee Sutterby/iStockphoto.com. Right: Dennis MacDonald/PhotoEdit.

What do hair, salmon, and soft drinks have in common? All can be analyzed by determining the ratios of naturally occurring isotopes they contain. For example, hair growth depends on diet, especially drinking water. Water in different locales contains slightly different ratios of the isotopes oxygen-18 and oxygen-16. Such ratios in strands of hair from a crime scene or in drinking water or soft drinks can determine whether a victim had been in a particular location. Differences in the ratio of carbon isotopes can indicate whether a salmon (or another fish) was wild or farm raised. Thus, it is worthwhile to know: What are isotopes? How do they differ? Do elements other than oxygen and carbon have isotopes? How are isotope ratios measured experimentally? These questions will be answered in this chapter (see Section 2-3).

Among the most important things that chemists do are synthesize chemical compounds, determine properties of compounds, and find ways in which those compounds can enhance our lives. Often chemists can custom-design a new compound to have desirable properties. More than 70 million compounds are known and the number is growing rapidly—by about 15,000 every day. By definition a chemical compound contains two or more elements. Those elements consist of atoms—the basic building blocks of matter. Consequently we begin our study of compounds by delving more deeply into the nature of atoms.

2-1 Atomic Structure: Subatomic Particles

According to Dalton's atomic theory (← **Sec. 1-10**) elements differ from one another because their atoms are different. But *how* do hydrogen atoms differ from oxygen atoms? The difference involves smaller particles within atoms—known as *subatomic particles*. The term **atomic structure** refers to *the identity and arrangement of these subatomic particles in each atom*. An understanding of the details of atomic structure aids in the understanding of how elements combine to form compounds and how atoms are rearranged in chemical reactions. Atomic structure also accounts for the properties of materials.

Electrical charges played an important role in many experiments from which the theory of atomic structure was derived. Two types of electrical charge exist: positive and negative. ***Electrical charges of the same type repel one another, and charges of the opposite type attract one another.***

CONCEPTUAL EXERCISE 2.1

Electrical Charge

When you comb your hair on a dry day, your hair sticks to the comb. How could you explain this in terms of a nanoscale model in which atoms are made up of even smaller particles, some positively charged and some negatively charged?

Answers to **EXERCISES** are provided at the back of this book in Appendix L.

EXERCISES that are labeled **CONCEPTUAL** are designed to test your understanding of one or more concepts; they usually involve qualitative rather than quantitative thinking.

2-1a Radioactivity

In 1896 Henri Becquerel discovered that a sample of a uranium ore emitted invisible radiation that darkened a photographic plate, even though the plate was covered by a protective black paper. In 1898 Marie and Pierre Curie isolated two new elements, polonium and radium, that emitted the same kind of radiation. Marie suggested that *atoms of such elements spontaneously emit radiation* and named the phenomenon **radioactivity**.

Like the sensor in a cell phone or digital camera, a photographic plate senses light (radiation). Where radiation is absorbed, the photographic plate darkens.

Atoms of radioactive elements can emit three types of radiation: alpha (α), beta (β), and gamma (γ). When passed between electrically charged plates (Figure 2.1), alpha and beta radiation are deflected, but gamma radiation is not. This result was explained by saying that alpha and beta radiation consist of a stream of charged particles—they have mass

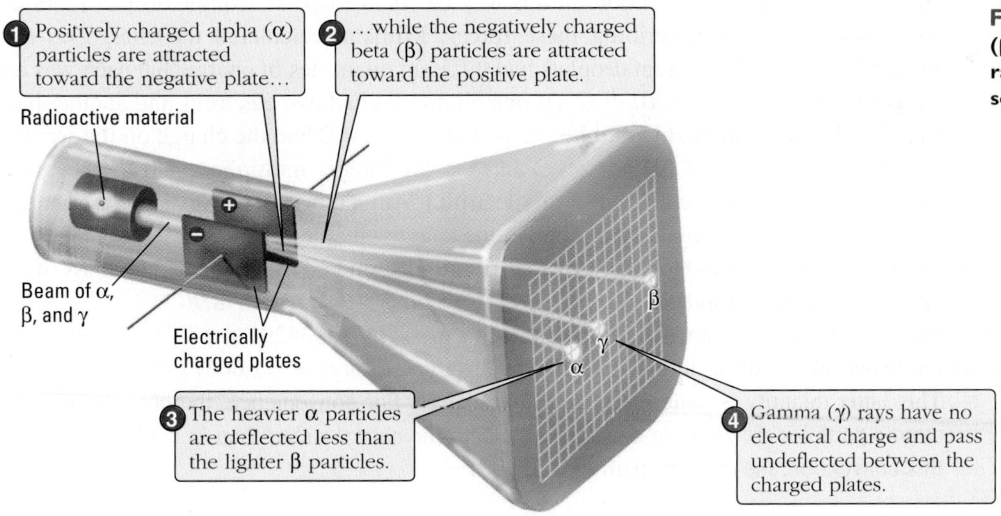
1 Positively charged alpha (α) particles are attracted toward the negative plate...
2 ...while the negatively charged beta (β) particles are attracted toward the positive plate.
Radioactive material
+
−
Beam of α, β, and γ
Electrically charged plates
β
γ
α
3 The heavier α particles are deflected less than the lighter β particles.
4 Gamma (γ) rays have no electrical charge and pass undeflected between the charged plates.

Figure 2.1 The alpha (α), beta (β), and gamma (γ) rays from a radioactive sample can be separated by an electrical field.

and are a form of matter. Gamma radiation is like visible light—it has no mass or charge and is not deflected by electrically charged plates. Alpha particles were found to have a 2+ charge, and beta particles a 1− charge. Despite their greater charge, in the experiment shown in Figure 2.1, alpha particles are deflected less, which means that they must be heavier than beta particles. Marie Curie's idea that the alpha and beta particles come from atoms implies that there must be smaller particles (subatomic particles) inside an atom. Thus, Dalton's idea that atoms are indivisible (← **Sec. 1-10**) is not entirely true—but it is true during a chemical reaction.

The electrical charges are written 2+ or 1− because the same convention is used in writing the formulas for ions. For example, calcium ion, which has a 2+ charge, is written Ca^{2+}.

2-1b Electrons

Further evidence that atoms are composed of subatomic particles came from experiments with cathode-ray tubes, glass tubes from which most of the air was removed and that had a metal electrode sealed into each end. When a sufficiently high voltage was applied to the electrodes, a beam of cathode rays flowed from the metal atoms of the negatively charged electrode (the *cathode*) to the positively charged electrode (the *anode*). Cathode rays travel in straight lines, are attracted toward positively charged plates, can be deflected by a magnetic field, and can cause gases and fluorescent materials to glow. These properties of cathode rays can be explained as those of a stream of negatively charged particles, each of which produces a light flash when it hits a fluorescent screen. Sir Joseph John Thomson suggested that cathode rays consist of the same particles that had earlier been named electrons and had been suggested to be the carriers of electricity. He also observed that cathode rays were produced from cathodes made of different metals, which implied that electrons are constituents of the atoms of each of those different metallic elements. Thus, the **electron** was identified as *a subatomic, negatively charged particle.*

In 1897 Thomson used a specially designed cathode-ray tube to simultaneously apply electric and magnetic fields to a beam of cathode rays. By balancing the electric field against the magnetic field and using the basic laws of electricity and magnetism, Thomson calculated the *ratio* of mass to charge for the electrons in the cathode-ray beam: 5.60×10^{-9} grams per coulomb (g/C). [The coulomb, C, is the SI base unit of electrical charge (← **Sec. 1-5a**).]

See Appendix A-2 for a review of scientific notation, which is used to represent very small or very large numbers as powers of 10. For example, 0.000001 is 1×10^{-6} (the decimal point moves six places to the right to give the −6 exponent) and 2,000,000 is 2×10^{6} (the decimal point moves six places to the left to give the +6 exponent).

Twelve years later, Robert A. Millikan used a cleverly devised experiment to measure the charge of an electron (Figure 2.2). A mist of tiny oil droplets was sprayed into a chamber. As the droplets settled slowly through the air, they were exposed to X-rays, which caused electrons to be transferred from gas molecules in the air to the droplets. Using a small microscope to observe individual droplets, Millikan adjusted the electrical charge of plates above and below the droplets so that the electrostatic attraction just balanced the gravitational attraction. In this way he could suspend a single droplet motionless. From equations describing these forces, Millikan calculated the charge on the suspended droplets. Different droplets had different quantities of charge, but each was an integer multiple of 1.60×10^{-19} C. (If one oil droplet had two electrons and another had five, the charge of the first would be $-2 \times 1.60 \times 10^{-19}$ C and the charge on the second would be $-5 \times 1.60 \times 10^{-19}$ C; −2 and −5 are integer multipliers and the negative signs reflect the negative charge of an electron.) Millikan assumed this smallest charge to be the fundamental quantity of charge, the charge on an electron.

Given this value and the mass-to-charge ratio determined by Thomson, the mass of an electron could be computed: $(1.60 \times 10^{-19}\ \text{C})(5.60 \times 10^{-9}\ \text{g/C}) = 8.96 \times 10^{-28}$ g. Currently the most accurate value for the electron's mass is $9.10938291 \times 10^{-28}$ g, and the currently accepted most accurate value for the electron's charge is $-1.602176565 \times 10^{-19}$ C. This latter quantity is called the *electronic charge.* For convenience, the charges on subatomic particles are given in multiples of electronic charge rather than in coulombs. Using this convention, the charge on an electron is 1−.

Electron
Charge = 1−
Mass = $9.10938291 \times 10^{-28}$ g

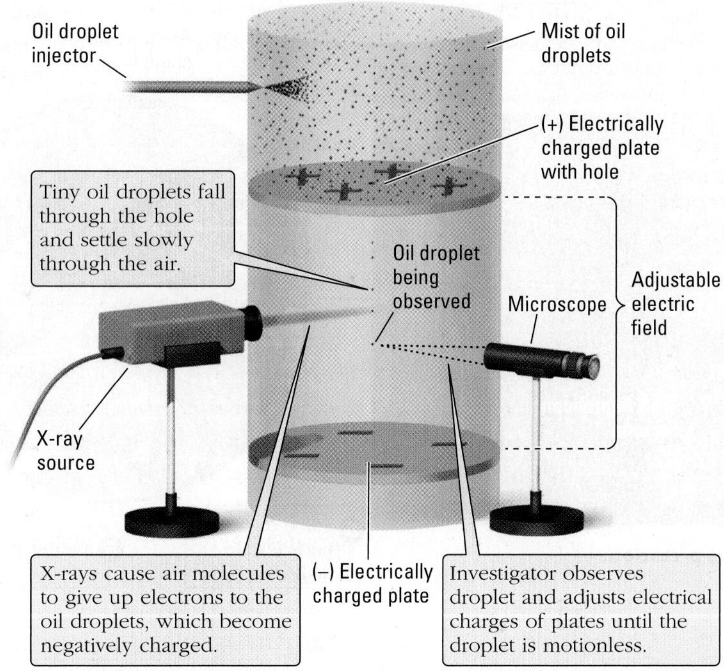
Oil droplet injector
Mist of oil droplets
(+) Electrically charged plate with hole
Tiny oil droplets fall through the hole and settle slowly through the air.
Oil droplet being observed
Microscope
Adjustable electric field
X-ray source
X-rays cause air molecules to give up electrons to the oil droplets, which become negatively charged.
(–) Electrically charged plate
Investigator observes droplet and adjusts electrical charges of plates until the droplet is motionless.

Figure 2.2 Millikan oil-drop experiment. From the known mass of the droplets and the applied voltage at which the charged droplets were held stationary, Millikan could calculate the charges on the droplets.

Other experiments provided further evidence that the electron is a *fundamental* particle of matter; that is, it is present in *all* matter. For example, the beta radiation emitted by radioactive elements was found to have the same properties as cathode rays, so beta particles must be electrons.

2-1c Protons

When atoms *lose* electrons, the atoms become positively charged. When atoms *gain* electrons, the atoms become negatively charged. Such *charged atoms, or charged groups of atoms*, are known as **ions**. From experiments with positive ions, formed by removing electrons from atoms, the existence of a positively charged fundamental particle was deduced. Positively charged particles with different mass-to-charge ratios were formed by atoms of different elements. The variation in masses showed that atoms of different elements must contain different numbers of positive particles. Those from hydrogen atoms had the smallest mass-to-charge ratio, indicating that they must be *the fundamental positively charged particles of atomic structure.* Such a particle is called a **proton**. The proton mass is known from experiment to be $1.672621777 \times 10^{-24}$ g, which is about 1800 times the mass of an electron. The charge on a proton is $+1.602176565 \times 10^{-19}$ C, equal in size, but opposite in sign, to the charge on an electron. The proton's charge is represented by 1+. Thus, an atom that has lost two electrons has a charge of 2+.

As mass increases, mass-to-charge ratio increases for a given amount of charge. For a fixed charge, doubling the mass will double the mass-to-charge ratio. For a fixed mass, doubling the charge will halve the mass-to-charge ratio.

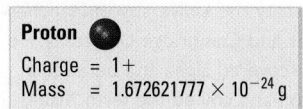

Proton
Charge = 1+
Mass = $1.672621777 \times 10^{-24}$ g

2-2 The Nuclear Atom

Once it was known that there were subatomic particles, the next question scientists wanted to answer was, How are these particles arranged in an atom? During 1910 and 1911 Ernest Rutherford reported experiments (Figure 2.3) that led to a better understanding of atomic structure. Alpha particles (which have the same mass as helium atoms and

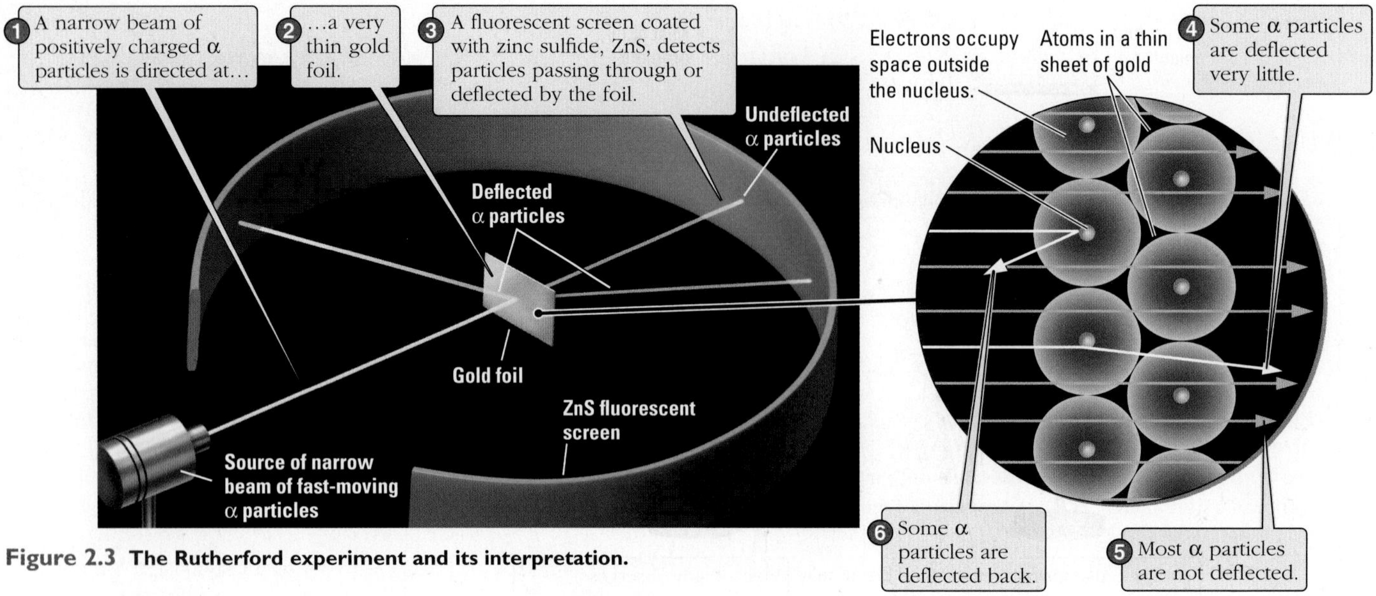
1 A narrow beam of positively charged α particles is directed at...
2 ...a very thin gold foil.
3 A fluorescent screen coated with zinc sulfide, ZnS, detects particles passing through or deflected by the foil.
Undeflected α particles
Deflected α particles
Gold foil
ZnS fluorescent screen
Source of narrow beam of fast-moving α particles
Electrons occupy space outside the nucleus.
Atoms in a thin sheet of gold
4 Some α particles are deflected very little.
Nucleus
6 Some α particles are deflected back.
5 Most α particles are not deflected.

Figure 2.3 The Rutherford experiment and its interpretation.

Alpha particles are four times heavier than the lightest atoms, which are hydrogen atoms.

a 2+ charge) were allowed to hit a very thin sheet of gold foil. Almost all of the alpha particles passed through undeflected. However, a very few alpha particles were deflected through large angles, and some came almost straight back toward the source. Rutherford described this unexpected result by saying, "It was about as credible as if you had fired a 15-inch [artillery] shell at a piece of tissue paper and it came back and hit you."

Ernest Rutherford
1871–1937

Bettmann/Corbis

Born on a farm in New Zealand, Rutherford worked at McGill University, the University of Manchester, and Cambridge University. He discovered alpha and beta radiation and coined the term "half-life." For demonstrating that alpha radiation is composed of helium nuclei and that beta radiation consists of electrons, Rutherford received the Nobel Prize in Chemistry in 1908. At Cambridge University, he guided the work of no fewer than ten future Nobel Prize recipients. Element 104 is named in Rutherford's honor.

2-2a The Nucleus

The only way to account for the observations was to conclude that all of the positive charge and most of the mass of the atom are concentrated in a very small region (Figure 2.3). Rutherford called this *tiny atomic core* the **nucleus**. Only such a region could be sufficiently dense and highly charged to repel an alpha particle back toward its origin. From their results, Rutherford and his associates calculated values for the charge and radius of the gold nucleus. The currently accepted values are a charge of 79+ and a radius of 5.1×10^{-15} m. This makes the nucleus about 10,000 times *smaller* than the atom so most of the volume of the atom is occupied by the electrons. How the electrons could fill all of the space outside the nucleus was a puzzle to Rutherford and other scientists of the time. The arrangement of electrons in atoms is now well understood and is described in Chapter 5. Figure 2.4 indicates the relative sizes of nuclei and atoms.

PROBLEM-SOLVING EXAMPLE 2.1

Alpha Particles and Gold Sheets

Gold was used in Rutherford's experiments because it is more malleable than most metals and could be formed into sheets as little as 1 μm thick. If an alpha particle went straight through a 1.3-μm gold sheet, how many gold atoms would it pass? (Assume that the gold atoms are arranged in a row from one side of the sheet of gold to the other and that the diameter of a gold atom is 288 pm.)

Result 4500 atoms

Analyze The thickness of the sheet and the diameter of a gold atom are given in different units; convert to the same units and take a ratio.

Plan Convert units to the SI base unit, m, and then divide the thickness of the foil by the diameter of an atom. Use the table of SI prefixes in Table 1.3 (← **Sec. 1-5a**) or Appendix B, Table B.2 to obtain conversion factors. Make certain that units cancel.

Execute According to Table 1.3, 1 μm = 1×10^{-6} m, so

$$\text{thickness of Au sheet} = 1.3\ \mu\text{m} \times \frac{1 \times 10^{-6}\ \text{m}}{1\ \mu\text{m}} = 1.3 \times 10^{-6}\ \text{m}$$

From Table 1.2, 1 pm = 1×10^{-12} m, from which we obtain

$$\text{thickness of Au atom} = 288\ \text{pm} \times \frac{1 \times 10^{-12}\ \text{m}}{1\ \text{pm}} = 2.88 \times 10^{-10}\ \text{m}$$

Notice that in each case the conversion factor was arranged so that units cancelled and we obtained the unit m. The ratio of thickness of sheet to diameter of atom is

$$\frac{\text{thickness of Au sheet}}{\text{thickness of Au atom}} = \frac{1.3 \times 10^{-6}\ \text{m}}{2.88 \times 10^{-10}\ \text{m}} = 4.5 \times 10^{3}$$

Thus the sheet is 4500 atoms thick. Notice that because the gold sheet is so thin and because there was no need for precision regarding its thickness, the thickness, 1.3 μm, was reported to only two significant figures. Because only multiplication and division are involved, the number of significant figures in the result is the number of significant figures in the least precise number—two significant figures in the number 1.3. Both 4.5×10^{-5} and 4500 contain two significant figures.

☑ **Reasonable Result Check** Atoms are very tiny, so we expect a large number in the result. Atoms are on the nanoscale, below 1 nm in diameter. The ratio of μm to nm is 10^{-6} m/10^{-9} m = 10^3 or 1000, so the number should be at least 1000.

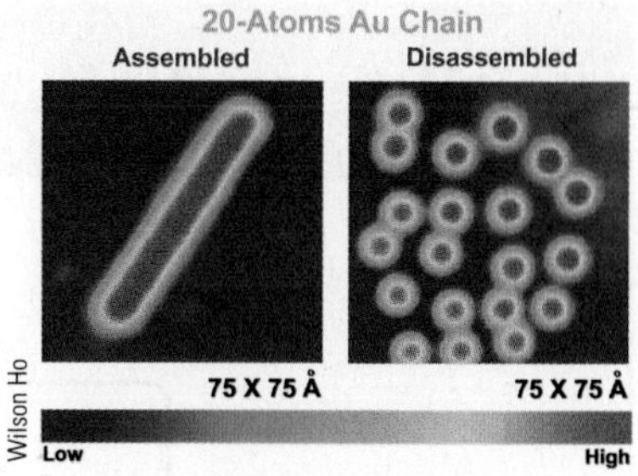
20-Atoms Au Chain
Assembled
Disassembled
75 X 75 Å
75 X 75 Å
Low
High
Wilson Ho

Atomic wires: chains of gold atoms on a germanium surface are shown in this image from a scanning tunneling microscope (STM).

In this calculation, the number displayed by a calculator was 4.513888889e03. This was rounded to 4.5×10^3 because there are only two significant figures in the least precise number, 1.3. The number 4500 has only two significant figures because there is no decimal point following the second zero.

PROBLEM-SOLVING PRACTICE 2.1

Prior to 1983 U. S. one-cent coins were made of copper. If copper atoms are lined up in a row from front side of a penny to the back, it takes 3.97×10^6 atoms. The thickness of a one-cent coin is 1.0 mm. What is the radius of a copper atom?

PROBLEM-SOLVING PRACTICE answers are provided at the back of this book in Appendix K.

2-2b Neutrons

Atoms are electrically neutral (no net charge), so they must contain equal numbers of protons and electrons. However, most neutral atoms have masses greater than the sum of the masses of their protons and electrons. The additional mass indicates that subatomic

In 1920 Ernest Rutherford proposed that the nucleus might contain an uncharged particle whose mass approximated that of a proton.

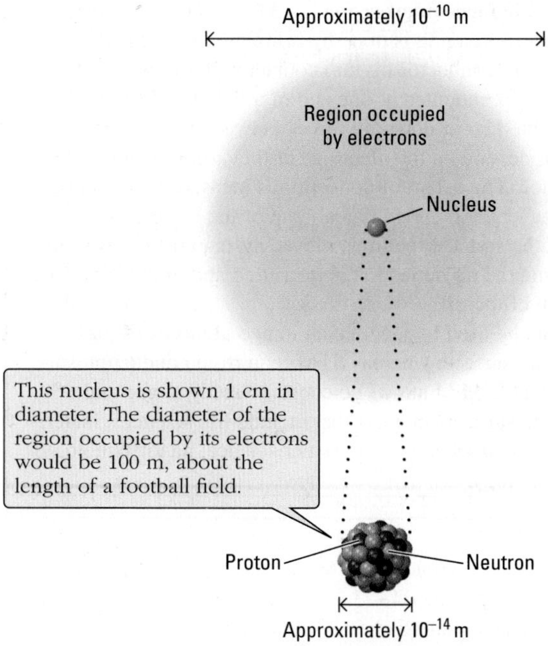
Approximately 10^{-10} m
Region occupied by electrons
Nucleus
This nucleus is shown 1 cm in diameter. The diameter of the region occupied by its electrons would be 100 m, about the length of a football field.
Proton
Neutron
Approximately 10^{-14} m

Figure 2.4 Relative sizes of an atom and its nucleus (not to scale).

TOOLS OF CHEMISTRY | "Seeing" Atoms: Scanning Tunneling Microscopy and Atomic Force Microscopy

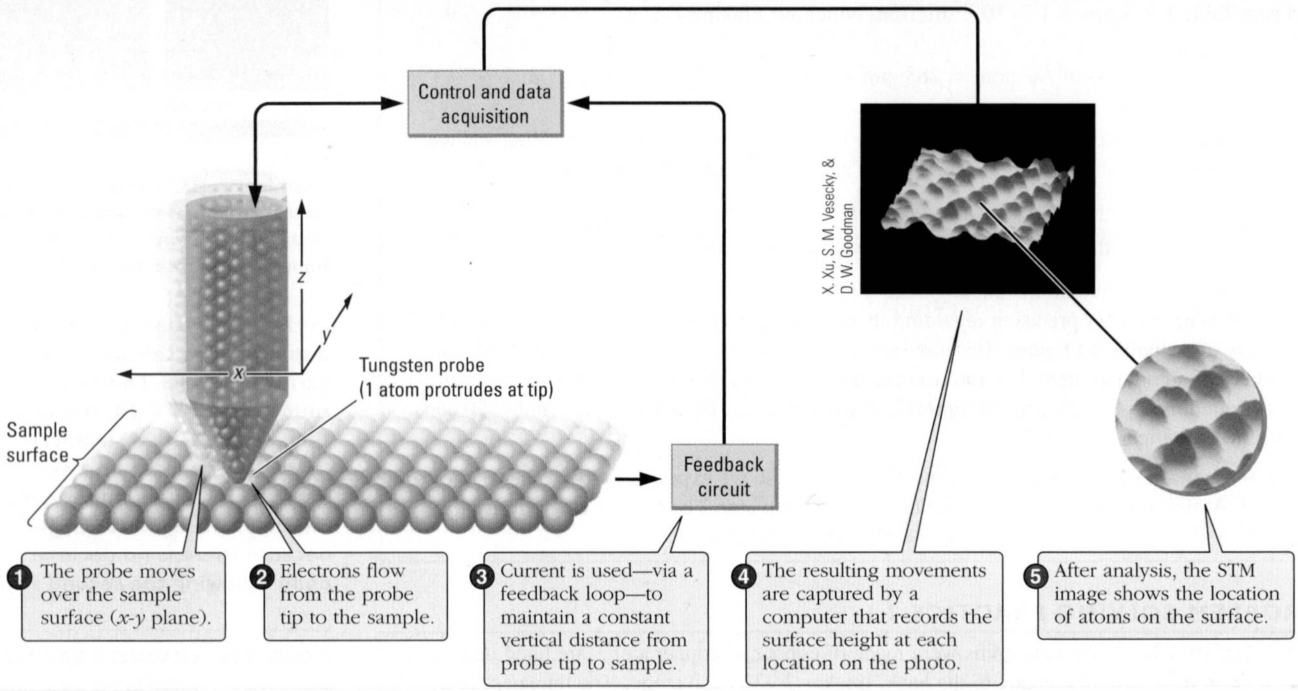

Control and data acquisition
z
y
x
Tungsten probe
(1 atom protrudes at tip)
Sample surface
Feedback circuit
X. Xu, S. M. Vesecky, & D. W. Goodman
1 The probe moves over the sample surface (x-y plane).
2 Electrons flow from the probe tip to the sample.
3 Current is used—via a feedback loop—to maintain a constant vertical distance from probe tip to sample.
4 The resulting movements are captured by a computer that records the surface height at each location on the photo.
5 After analysis, the STM image shows the location of atoms on the surface.

Schematic diagram of scanning tunneling microscopy.

The **scanning tunneling microscope** (STM) is *an analytical instrument that provides images of individual atoms or molecules on a surface*. To do this, a metal probe in the shape of a needle with an extremely fine point (a nanoscale tip) is brought extremely close (within one or two atomic diameters, a few tenths of a nanometer) to the sample surface being examined. When the tip is close enough to the sample, electrons jump between the probe and the sample. The size and direction of this electron flow (the current) depend on the applied voltage, the distance between probe tip and sample, and the identity and location of the nearest sample atom on the surface and its closest neighboring atoms.

The probe tip is attached to a control mechanism that maintains a constant distance between the tip and the sample. The current provides an extremely sensitive measure of the interatomic separation between probe tip and sample. The probe tip is scanned across the surface to form a topographic map of that part of the surface. The STM image, which appears much like a photographic image, shows the locations of atoms on the surface being investigated. An STM image of gold atoms on a germanium surface appeared earlier in this chapter.

The STM has been applied to a wide variety of problems throughout science and engineering. The greatest number of studies have focused on the properties of clean surfaces that have been modified. The STM can also be used to study electrode surfaces in a liquid. Applications of STM to biological molecules on surfaces represent another growing area of scientific research. The STM can be used to move atoms on a surface, and researchers have taken advantage of this capability to assemble atomic wires (shown earlier) and other nanostructures.

The STM was invented by Gerd Binnig and Heinrich Rohrer at IBM, Zurich, in 1981, and they shared a Nobel Prize in Physics in 1986 for this work.

The **atomic force microscope** (AFM) is a close relative of the STM. AFM imaging is done by bringing a very small ceramic or semiconductor tip into contact with the surface being studied. The tip, mounted at one end of a tiny, flexible cantilever, is deflected up or down by forces between the tip and the surface that depend on the identities of the atoms or molecules at the surface. The extent of deflection is measured with a laser beam. To generate a topographical map of the bumps and grooves of the surface, the tip is moved systematically over the surface while the tip deflection is measured and graphed versus the position of the tip.

AFM can be used to generate an image of atoms or molecules on any surface, whereas STM requires a conducting surface. Like STM, AFM allows scientists to study atoms and molecules on surfaces in areas that include life sciences, materials science, electrochemistry, polymer science, nanotechnology, and biotechnology.

particles with mass but no charge must also be present. Because they have no charge, these particles are more difficult to detect experimentally. In 1932 James Chadwick devised a clever experiment that detected the neutral particles by having them knock protons out of atoms and then detecting the protons. This *neutral subatomic particle* is called the **neutron**; it has no electrical charge and a mass of $1.674927351 \times 10^{-24}$ g, nearly the same as the mass of a proton.

Neutron
Charge = 0
Mass = $1.674927351 \times 10^{-24}$ g

In summary, there are three primary subatomic particles: protons, neutrons, and electrons.

- Protons and neutrons make up the nucleus, providing most of the atom's mass; the protons provide all of its positive charge.
- The nuclear radius is approximately 10,000 times smaller than the radius of the atom.
- Negatively charged electrons outside the nucleus occupy most of the volume of the atom, but contribute very little mass.
- A neutral atom has no net electrical charge because the number of electrons outside the nucleus equals the number of protons inside the nucleus.

To chemists, the electrons are the most important subatomic particles because they are the first part of the atom to contact another atom. The electrons in atoms largely determine how elements combine to form chemical compounds.

PROBLEM-SOLVING EXAMPLE 2.2

Density of an Atom

The diameter of a gold atom was given earlier as 288 pm. The mass of a gold atom is 3.27×10^{-22} g. Calculate the density of a gold atom. Express the result in g/cm^3 and g/mL. (Assume that atoms are spherical. 1 mL = 1 cm^3.)

Result 26.1 g/cm^3; 26.1 g/mL

Analyze Density is mass/volume; mass is given and volume can be calculated from the diameter.

Plan Use the formula for volume of a sphere, calculate the volume, and divide the mass by the volume. If necessary, look up the formula for volume of a sphere.

Execute The formula involves the radius, which is half the diameter. The volume is

$$V = \frac{4}{3}\pi r^3 = \frac{4}{3}\pi\left(\frac{288 \text{ pm}}{2}\right)^3 = 1.251 \times 10^7 \text{ pm}^3$$

To express this volume in cm^3, use appropriate conversion factors based on SI prefixes. A picometer is 1×10^{-12} m and a centimeter is 1×10^{-2} m so the volume is

$$V = 1.251 \times 10^7 \cancel{\text{pm}}^3 \times \left(\frac{1 \times 10^{-12} \cancel{\text{m}}}{1 \cancel{\text{pm}}}\right)^3 \times \left(\frac{1 \text{ cm}}{1 \times 10^{-2} \cancel{\text{m}}}\right)^3 = 1.251 \times 10^{-23} \text{ cm}^3$$

Now calculate the density.

$$d = \frac{m}{V} = \frac{3.27 \times 10^{-22} \text{ g}}{1.251 \times 10^{-23} \text{ cm}^3} = 2.61 \times 10^1 \text{ g/cm}^3 = 2.61 \times 10^1 \text{ g/mL}$$

☑ **Reasonable Result Check** Densities of substances reported in Table 1.1 (← **Sec. 1-4d**) range from less than 1 g/mL to about 20 g/mL; this value is a bit above the range but still reasonable. The density of gold in the table is 19.32 g/mL and one might expect the density of a gold atom to be the same as the density of gold itself; however, in gold there is empty space between spherical atoms (see figure nearby), so the density of the gold is less than the density of a gold atom.

In the conversion of pm^3 to cm^3, each conversion factor is cubed because we are converting units cubed.

Because the mass had three significant digits and only multiplication and division were done, the result should be rounded to three significant digits.

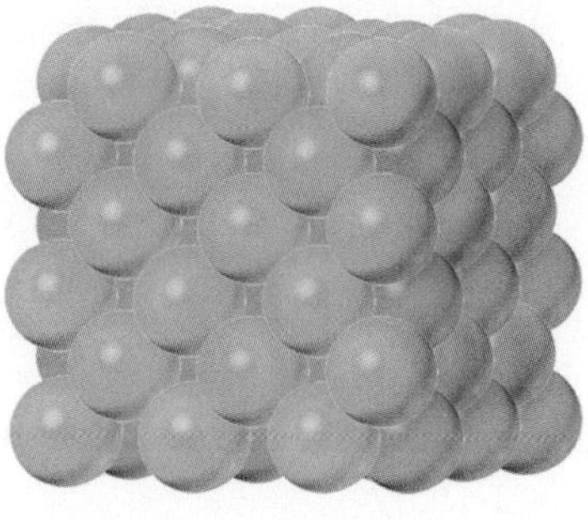

The arrangement of atoms in gold.

PROBLEM-SOLVING PRACTICE 2.2

Given the radius of 5.15×10^{-15} m that Rutherford calculated for a gold nucleus, calculate the density of a gold nucleus. Express your result in g/cm^3. Assume that all of the mass of a gold atom is in the spherical nucleus.

CONCEPTUAL EXERCISE 2.2

Describing Atoms

If an atom had a radius of 100 m, it would approximately fill a football stadium.

(a) What would the approximate radius of the nucleus of such an atom be?

(b) What common spherical object is about that size?

2-2c Atomic Number and Mass Number

Experiments done early in the 20th century found that all atoms of the same element have the same number of protons in their nuclei. The *number of protons in the nucleus of an atom is called* the **atomic number** of the element and is given the symbol *Z*. In the periodic table on the inside front cover of this book, the atomic number for each element is written above the element's symbol. For example, a copper atom has a nucleus containing 29 protons, so its atomic number is 29 ($Z = 29$). A lead atom (Pb) has 82 protons in its nucleus, so the atomic number for lead is 82.

In Mendeleev's periodic table (← Sec. 1-13) the elements were placed in order of atomic weights. In a few cases (such as Ar and K or Co and Ni) the order of atomic weights differs from the order of atomic numbers. Because atomic numbers reflect the nuclear structure of the atom, atomic numbers are more fundamental and are used to determine the order of elements in the modern periodic table.

The **mass number**, **A**, of an atom is *the sum of the number of protons and the number of neutrons in the atom.* For example, a copper atom with 29 protons and 34 neutrons in its nucleus has a mass number, *A*, of 63. A lead atom with 82 protons and 126 neutrons has $A = 208$. The name mass number comes from the fact that most of the mass of an atom is due to the masses of the protons and neutrons in its nucleus. The total number of protons and neutrons gives an approximation of the mass of the atom.

An atom of known composition, such as a lead atom with 126 neutrons, can be represented this way:

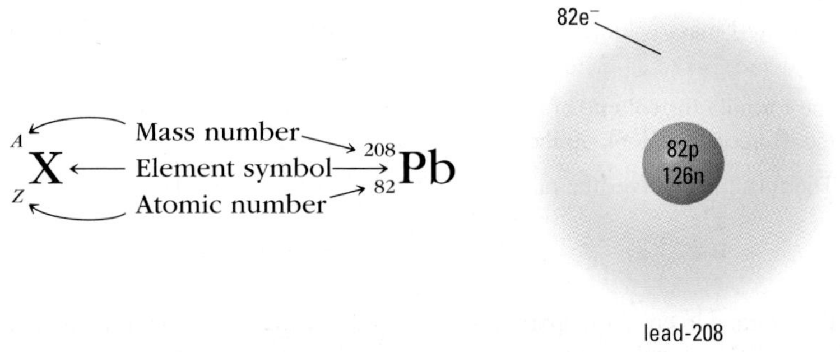
A
X
Z
Mass number
Element symbol
Atomic number
208
82
Pb
82e⁻
82p
126n
lead-208

The mass number, a superscript, and the atomic number, a subscript, precede the element's unique one- or two-letter chemical symbol (← **Sec. 1-10**). You have probably noticed that the *Z* part of the notation is redundant. For example, the representation ^{208}Pb is sufficient, because the symbol Pb tells us that the atom is lead and lead atoms by definition always contain 82 protons. Whether we use $^{208}_{82}Pb$, ^{208}Pb, or the alternative notation lead-208, we simply say "lead-208."

PROBLEM-SOLVING EXAMPLE 2.3

Atomic Nuclei

Iodine-131 is used in medicine to assess thyroid gland function. Write the symbol representation for iodine-131. How many protons and neutrons are there?

Result $^{131}_{53}I$; 53 protons and 78 neutrons

Analyze The mass number is given; the atomic number is needed.

Plan Obtain the atomic number from the name of the element using the periodic table inside the front cover. Use it to write the symbol and calculate the number of neutrons.

Execute The atomic number of iodine (I) is 53. Therefore, the symbol representation is

$$^{131}_{53}\text{I}$$

Because the mass number is the sum of the number of protons and neutrons,

$$\text{Mass number} = \text{number of protons} + \text{number of neutrons}$$
$$131 = 53 + \text{number of neutrons}$$
$$\text{Number of neutrons} = 131 - 53 = 78$$

PROBLEM-SOLVING PRACTICE 2.3

(a) Determine the mass number of a phosphorus atom with 16 neutrons.
(b) How many protons, neutrons, and electrons are there in a neon-22 atom?
(c) Write the symbol for the atom with 82 protons and 125 neutrons.

2-3 Isotopes and Average Atomic Mass

Atoms are extremely small. One teaspoon of water contains about three times as many atoms as the Atlantic Ocean contains teaspoons of water. It is not surprising that determining the mass of an atom is not an easy task. Chemists initially determined atomic masses by measuring the mass of one element that combined with another. For example, when hydrogen combined with fluorine to form HF, it was known that for every 1.0 g hydrogen there would be 19.0 g fluorine. Thus it was reasonable to conclude that the mass of a fluorine atom was 19.0 times the mass of a hydrogen atom. That is, atomic masses were obtained by comparing the mass of each atom with the mass of a standard atom. Even if the exact mass of the standard atom was not known, the masses of all other atoms relative to the standard could be determined.

Today the masses of atoms of every other element are compared with the mass of a carbon atom that has six protons and six neutrons in its nucleus. *Carbon-12 is defined as having a mass of exactly 12* **unified atomic mass units (u)**. The symbol amu is also used and in biochemistry, the name dalton (Da) is used for the unified atomic mass unit. In terms of macroscale mass units, $1 \text{ u} = 1.66054 \times 10^{-24}$ g. For example, when an experiment shows that a gold atom is 16.4 times as massive as the standard carbon atom, we then can calculate the mass of the gold atom in unified atomic mass units and in grams.

$$\text{Mass of gold atom} = 16.4 \times (\text{mass of carbon-12}) = 16.4 \times 12 \text{ u} = 197 \text{ u}$$

$$197 \text{ u} \times \frac{1.66054 \times 10^{-24} \text{ g}}{1 \text{ u}} = 3.27 \times 10^{-22} \text{ g}$$

The masses of the electron, proton, and neutron, expressed in unified atomic mass units, are given in Table 2.1. As you can see, the masses of the proton and the neutron are very close to 1 and the mass of the electron is very small. This is why the mass number approximates the actual atomic mass.

The relative masses of atoms can be determined very precisely by using an instrument called a mass spectrometer. (See the *Tools of Chemistry* description later in this section.) Consider the mass spectrum of the element boron (B), a relatively rare element present in compounds used in laundry detergents, mild antiseptics, and Pyrex cookware. Boron's mass spectrum is shown in Figure 2.5. Notice that there are two peaks, one at mass about 10 u and one at mass about 11 u. The two peaks mean that there must be two different charge/mass ratios. Because of the way the ions are produced, essentially all of

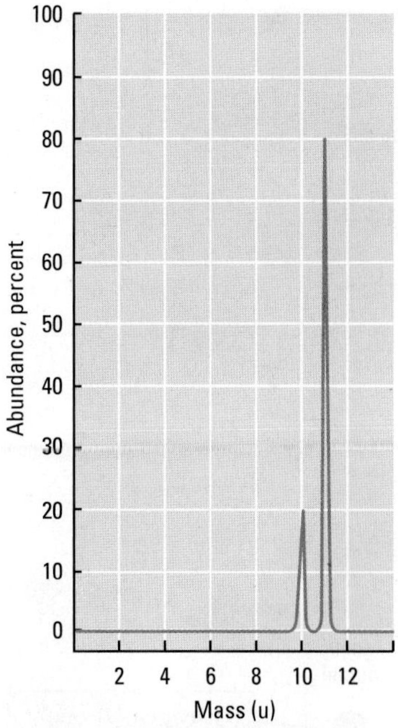
Abundance, percent
100
90
80
70
60
50
40
30
20
10
0
2
4
6
8
10
12
Mass (u)

Figure 2.5 Mass spectrum of boron. The two peaks in the spectrum indicate that some boron atoms have relative mass of about 10 and others have relative mass of about 11.

Table 2.1 Subatomic Particles

Particle	Mass (u)
Electron	0.000548579
Proton	1.00728
Neutron	1.00866

Two elements cannot have the same atomic number. If two atoms differ in their number of protons, they are atoms of different elements. If only their number of neutrons differs, they are isotopes of a single element, such as neon-20, neon-21, and neon-22.

them have 1+ charges, so the different peaks correspond with different masses. That is, boron atoms can have two different masses, one about 10 u and another about 11 u. The peak at 11 u is much stronger—roughly 80% of the boron atoms have relative mass of about 11 u; only 20% of the atoms have mass of about 10 u.

Mass spectrometric analysis of most naturally occurring elements reveals that not all atoms of the same element have the same mass. For example, all silicon atoms have 14 protons, but some silicon nuclei have 14 neutrons, others have 15, and others have 16. Thus, naturally occurring silicon (atomic number 14) is always a mixture of silicon-28, silicon-29, and silicon-30 atoms. *Atoms of the same element that have different numbers of neutrons are called* **isotopes**. Isotopes have the *same* atomic number (Z) but *different* mass numbers (A). Naturally occurring silicon has three isotopes:

The International Union of Pure and Applied Chemistry (IUPAC) has a useful periodic table of the isotopes at http://www.ciaaw.org/pubs/Periodic_Table_Isotopes.pdf.

Isotope	^{28}Si	^{29}Si	^{30}Si
Atomic number (Z), protons	14	14	14
Neutrons	14	15	16
Mass number (A), protons + neutrons	28	29	30

PROBLEM-SOLVING EXAMPLE 2.4

Isotopes

Chlorine has two isotopes, one with 18 neutrons and the other with 20 neutrons. Determine the mass numbers and write the symbols of these isotopes.

Result The mass numbers are 35 and 37. The symbols are $^{35}_{17}Cl$ and $^{37}_{17}Cl$.

Analyze We know the number of neutrons but need to find the number of protons.

Plan Use the entry for chlorine in the periodic table to find the atomic number (number of protons); add that to the number of neutrons to get the mass number.

Execute Chlorine is in Group 7A, Period 3; it has atomic number 17,

Isotope 1: 17 protons + 18 neutrons = mass number 35 (chlorine-35)

Isotope 2: 17 protons + 20 neutrons = mass number 37 (chlorine-37)

Place the atomic number at the bottom left and the mass number at the top left.

$$^{35}_{17}Cl \text{ and } ^{37}_{17}Cl.$$

PROBLEM-SOLVING PRACTICE 2.4

Naturally occurring magnesium has three isotopes with 12, 13, and 14 neutrons. Determine the mass numbers and write the symbols of these three isotopes.

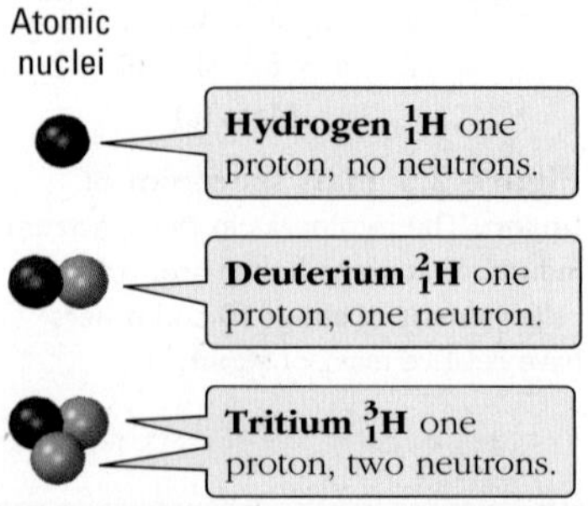
Atomic nuclei
Hydrogen ^1_1H one proton, no neutrons.
Deuterium ^2_1H one proton, one neutron.
Tritium ^3_1H one proton, two neutrons.

Hydrogen isotopes.

We usually refer to a particular isotope by giving its mass number. For example, $^{238}_{92}U$ is referred to as uranium-238. But a few isotopes have distinctive names and symbols because of their importance, such as the isotopes of hydrogen. All hydrogen isotopes have just one proton. When the single proton is the only nuclear particle, the element is called hydrogen (or sometimes protium). With one neutron as well as one proton present, the isotope ^2_1H is called either deuterium (symbol D) or heavy hydrogen. When two neutrons are present, the isotope ^3_1H is called tritium (symbol T).

CONCEPTUAL EXERCISE 2.3

Isotopes

A student in your chemistry class tells you that nitrogen-14 and nitrogen-15 are not isotopes because they have the same number of protons. How would you refute this statement?

TOOLS OF CHEMISTRY | Mass Spectrometer

A **mass spectrometer** is *an instrument that can measure the mass-to-charge ratio of atoms or molecules* that have lost or gained electrons and therefore are positive or negative ions. Atoms or molecules in a gaseous sample of the substance being analyzed pass through a stream of high-energy electrons. Collisions between the electrons and the sample's atoms (or molecules) produce positive ions, mostly with 1+ charge. A negatively charged grid attracts the positive ions and they form a beam. The beam of ions passes through a mass analyzer in which electric or magnetic fields change the paths of the ions. (This is similar to the deflection of charged alpha and beta particles depicted in Figure 2.1.) If the ions are moving at the same speed and all have the same charge, ions with larger mass are deflected less and ions with smaller mass are deflected more. This happens because greater force is needed to change the path of a heavier particle. The deflected ions hit a detector, which produces a signal proportional to the number of ions that strike it in a given time. The angle of deflection determines the mass/charge ratio, which determines the atomic mass relative to the standard, ${}^{12}_{6}C$ ions.

In a typical mass spectrometer, electric or magnetic fields are varied to focus ions of different masses on the stationary detector at different times. The mass spectrometer records the signal (proportional to the number of ions) as the electric or magnetic fields are varied systematically. After processing by a computer, the data are plotted as *a graph of the ion abundance versus the relative mass of the ions*, which is a **mass spectrum**.

The mass spectrum of neon is shown in the figure. The beam of Ne^+ ions produces three signals because three isotopes are present: ${}^{20}Ne$, with atomic mass of 19.9924 u and abundance of 90.48%; ${}^{21}Ne$, with atomic mass of 20.9938 u and abundance of only 0.27%; and ${}^{22}Ne$, with atomic mass of 21.9914 u and abundance of 9.25%.

Contemporary chemical research uses mass spectrometers to investigate details of molecular structure of substances ranging from simple organic and inorganic molecules to complex biomolecules such as proteins.

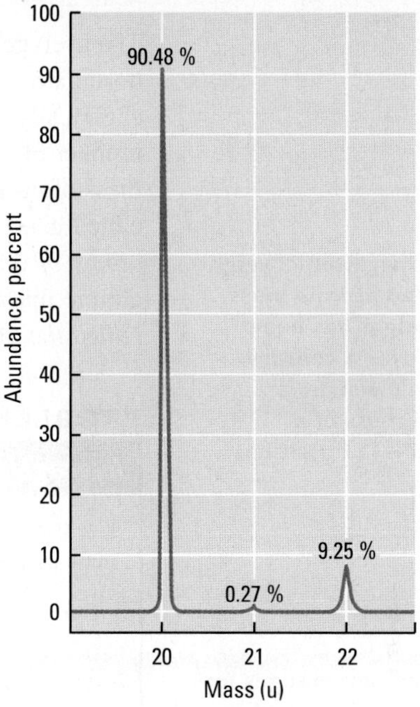
100
90
80
70
60
50
40
30
20
10
0
Abundance, percent
90.48 %
0.27 %
9.25 %
20
21
22
Mass (u)

Mass spectrum of neon. The principal peak corresponds to the most abundant isotope, neon-20. The height of each peak indicates the percent relative abundance of each isotope.

2-3a Average Atomic Mass: Atomic Weight

The mass spectrometer provides highly accurate measurements of relative masses of atoms. For example, in the mass spectrum of boron shown in Figure 2.5, quantitative data from the mass spectrometer indicate that the relative mass of an atom of one isotope, boron-10, is 10.0129 u. For boron-11, the mass of one atom is 11.0093 u. As you can see, although an atom's mass approximately equals its mass number, the actual mass is not an integer.

The mass spectrum also indicates the abundance of atoms having each mass. For example, mass spectrum data show that 19.91% of all boron atoms are boron-10 and 80.09% of all boron atoms are boron-11. These percentages are examples of **percent abundance**, *the fraction of atoms of each isotope in a natural sample of an element.*

$$\text{Percent abundance} = \frac{\text{number of atoms of a given isotope}}{\text{total number of atoms of all isotopes of an element}} \times 100\%$$

In a naturally occurring sample of boron, almost exactly 80 out of every 100 atoms are boron-11; 20 out of every 100 atoms are boron-10. Thus, in a sample of 100 boron atoms, the total mass is

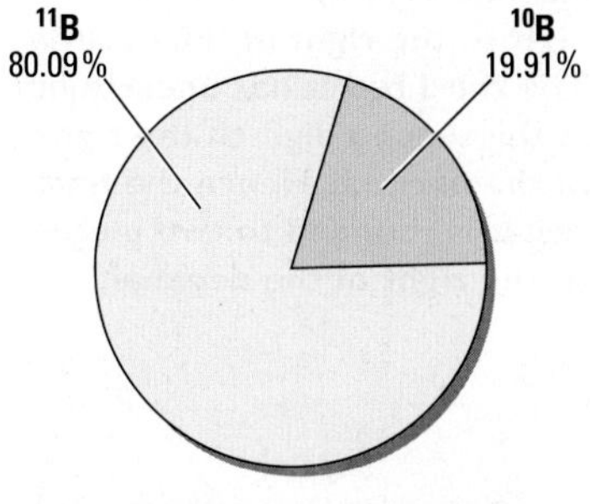
¹¹B
80.09%
¹⁰B
19.91%

Percent abundance of boron-10 and boron-11.

Percent means "per 100". Thus a percentage tells how many atoms of one isotope there are per hundred atoms of all isotopes.

$$\begin{aligned} \text{Mass of 20 boron-10 atoms} &= 20 \times 10.0129 \text{ u} = 200.259 \text{ u} \\ \text{Mass of 80 boron-11 atoms} &= 80 \times 11.0093 \text{ u} = \underline{880.744 \text{ u}} \\ \text{Total mass of 100 boron atoms} &= 1081.003 \text{ u} \end{aligned}$$

The average mass of a boron atom is $\frac{1}{100}$ of 1081.003 u, that is, 10.810 u. Because not all boron atoms have the same mass, when we determine the mass of boron that combines chemically with a given mass of another element, it is the *average* mass of a very large number of boron atoms that is important. The average mass takes into account the exact mass of each isotope and the percent abundance of that isotope. This type of average is called a *weighted average. The weighted average mass of a representative sample of atoms of an element is called the* **average atomic mass** and is expressed in unified atomic mass units. *The weighted average mass relative to the standard* $^{12}_{6}C = 12$ *u, is called the* **atomic weight** of that element.

Strictly speaking, atomic weight should be called atomic mass, but atomic weight has been used for well over a century. Because atomic weight is a ratio with the mass of a carbon-12 atom (12 u), it has no units.

PROBLEM-SOLVING EXAMPLE 2.5

Average Atomic Mass

Use these data from the mass spectrum of neon to calculate the weighted average mass of neon atoms.

Isotope	Mass (u)	Abundance
neon-20	19.9924	90.48%
neon-21	20.9938	0.27%
neon-22	21.9914	9.25%

The sum of the percentages for the composition of a sample must be 100%.

Result 20.18 u

Analyze Mass and abundance data are given; calculate a weighted average.

Plan Multiply each mass by its abundance and sum the masses. Remember that percent means per 100.

Execute

$$\text{neon-20:} \quad 19.9924 \text{ u} \times 90.48\% = 19.9924 \text{ u} \times \frac{90.48}{100} = 18.0891 \text{ u}$$

$$\text{neon-21:} \quad 20.9938 \text{ u} \times 0.27\% = 20.9938 \text{ u} \times \frac{0.27}{100} = 0.0567 \text{ u}$$

$$\text{neon-22:} \quad 21.9914 \text{ u} \times 9.25\% = 21.9914 \text{ u} \times \frac{9.25}{100} = 2.0342 \text{ u}$$

Average atomic mass = 18.0891 u + 0.0567 u + 2.0342 u = 20.1800 u, which, rounded to two places to the right of the decimal point, is 20.18 u.

The first intermediate result has four significant figures and so there is uncertainty in the second digit to the right of the decimal point. The second result has uncertainty in the third digit to the right of the decimal. The third result has uncertainty in the second digit to the right of the decimal. Hence the final result is rounded to two places to the right of the decimal.

☑ **Reasonable Result Check** The result should be between 20 u and 22 u, and it is. A quick estimate can be made by ignoring neon-21 and approximating neon-20 as 90% and neon-22 as 10%. Because the neon atoms are mostly neon-20, the average should be closer to 20 u than to 22 u. The 10% neon-22 should increase the mass by 10% of the difference between neon-20 and neon-22, which is 10% of 2 or 0.2. The result, 20.18, is about 0.2 larger than 20, so the result is reasonable.

PROBLEM-SOLVING PRACTICE 2.5

Calculate the average atomic mass for lithium, given that the mass of lithium-6 is 6.015121 u, its abundance is 7.500%, the mass of lithium-7 is 7.016003 u, and its abundance is 92.50%. Compare your result with the atomic weight of lithium given in the periodic table inside the front cover of this book.

The atomic weight of each stable (nonradioactive) element has been determined; these values appear in the periodic table in the inside front cover of this book. In the periodic table, each element's box contains the atomic number, the symbol, and the weighted average atomic weight. For example, the periodic table entry for titanium is

22	← Atomic number
Ti	← Symbol
47.867	← Atomic weight

A.B.G./Shutterstock.com

The exterior of the Guggenheim Museum in Bilbao, Spain, is covered with titanium metal, which provides a special sheen.

For many elements, there is only a single isotope or the abundances of the isotopes are the same no matter where a sample is collected. However, in some cases this is not true. For example, the atomic weight of lithium varies depending on where the lithium is mined. Average atomic masses for different samples of lithium can vary from as low as 6.938 u to as high as 6.997 u.

CONCEPTUAL **EXERCISE 2.4**

Isotopic Abundance

Naturally occurring magnesium contains three isotopes: ^{24}Mg (78.99%), ^{25}Mg (10.00%), and ^{26}Mg (11.01%). Estimate the atomic weight of Mg to two significant figures. Compare your estimate with the atomic weight in the periodic table inside the front cover. Also calculate the unweighted arithmetic average (mean) of the mass numbers. Is the arithmetic average smaller or larger than your estimate? Explain the difference.

CONCEPTUAL **EXERCISE 2.5**

Percent Abundance

Gallium has two abundant isotopes that differ by two in their mass numbers. Gallium's atomic weight is 69.72. Based on this information, your friend says that the percent abundance of each of the two gallium isotopes cannot be 50%. Is your friend correct? Explain why or why not.

2-4 Ions and Ionic Compounds

We mentioned earlier that it is the electrons in an atom that interest chemists, because the electrons occupy most of the space in an atom. When one atom comes into contact with another it is the electrons of both atoms that interact. One type of interaction is for one or more electrons to be removed from one atom, forming a positive ion. Those electrons can be transferred to another atom, forming a negative ion. Because opposite charges attract, the positive ion attracts the negative ion. This bonds the two atoms (ions) together. *A compound whose nanoscale composition consists of positive and negative ions is classified as an* **ionic compound**. Many common substances, such as table salt, NaCl; lime, CaO; and lye, NaOH, are ionic compounds. Because table salt is a typical example, ionic compounds are often called *salts*.

Cengage Learning/Charles D. Winters

Ionic compounds. Red iron(III) oxide, black copper(II) bromide, CaF_2 (front crystal), and NaCl (rear crystal).

Ionic compounds form when metals react with nonmetals:

- **Metal atoms can have electrons removed relatively easily to form positive ions, cations** (pronounced CAT-ions).
- **The quantity of positive charge on the positive ion equals the number of electrons removed from the neutral metal atom.**

In a *cation* there are always *fewer* electrons than protons. For example, Figure 2.6 shows that an electrically neutral sodium atom, which has 11 protons and 11 electrons, forms a sodium cation by removal of one electron. The sodium cation, symbol Na^+, has 11 protons but only 10 electrons, and thus a *net* 1+ charge. When two electrons are removed from a neutral magnesium atom, it forms a 2+ magnesium ion, Mg^{2+}, which has 12 protons and 10 electrons.

- **Nonmetal atoms can gain electrons to form negatively charged ions, anions** (pronounced ANN-ions).
- **The quantity of negative charge on the negative ion equals the number of electrons gained by the neutral nonmetal atom.**

The terms "cation" and "anion" are derived from the Greek words *ion* (traveling), *cat* (down), and *an* (up).

In an *anion* there are always *more* electrons than protons. Figure 2.6 shows that a neutral chlorine atom (17 protons, 17 electrons) gains an electron to form a chlor*ide* ion, Cl^-. With 17 protons and 18 electrons, the chloride ion has a *net* 1− charge. A sulfur atom that gains two electrons forms a sulf*ide* ion, S^{2-}.

2-4a Monoatomic Ions

A **monoatomic ion** *is a single atom that has lost or gained electrons*. The charges of the common monoatomic ions are given in Figure 2.7. Study this figure carefully, because it provides rules by which you can predict the charges of many ions. There are also some cases that do not follow the rules, so make certain that you can remember them. Then, without looking at Figure 2.7, test yourself by assigning charges to the ions listed in Problem-Solving Example 2.6.

PROBLEM-SOLVING EXAMPLE 2.6

Predicting Charges of Monoatomic Ions

Using a periodic table, but not Figure 2.7, predict the charges on ions of strontium, phosphorus, tellurium, and iron. Write a symbol for each ion.

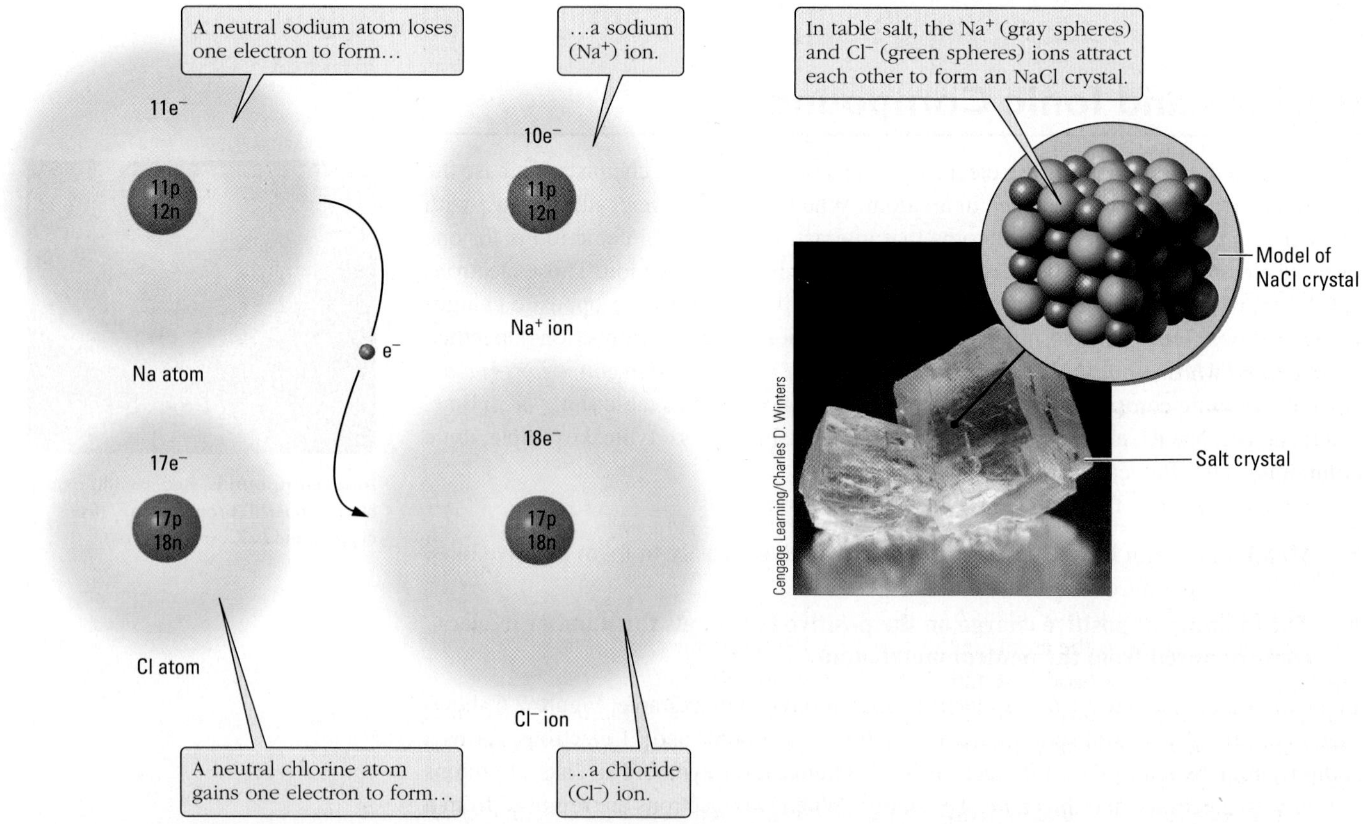
A neutral sodium atom loses one electron to form...
...a sodium (Na+) ion.
In table salt, the Na+ (gray spheres) and Cl− (green spheres) ions attract each other to form an NaCl crystal.
11e−
11p 12n
10e−
11p 12n
Na+ ion
e−
Na atom
Model of NaCl crystal
17e−
17p 18n
18e−
17p 18n
Salt crystal
Cengage Learning/Charles D. Winters
Cl atom
Cl− ion
A neutral chlorine atom gains one electron to form...
...a chloride (Cl−) ion.

Figure 2.6 Formation of the ionic compound NaCl. A chlorine atom removes an electron from a sodium atom, forming a sodium ion; the electron is transferred to the chlorine atom, forming a chloride ion.

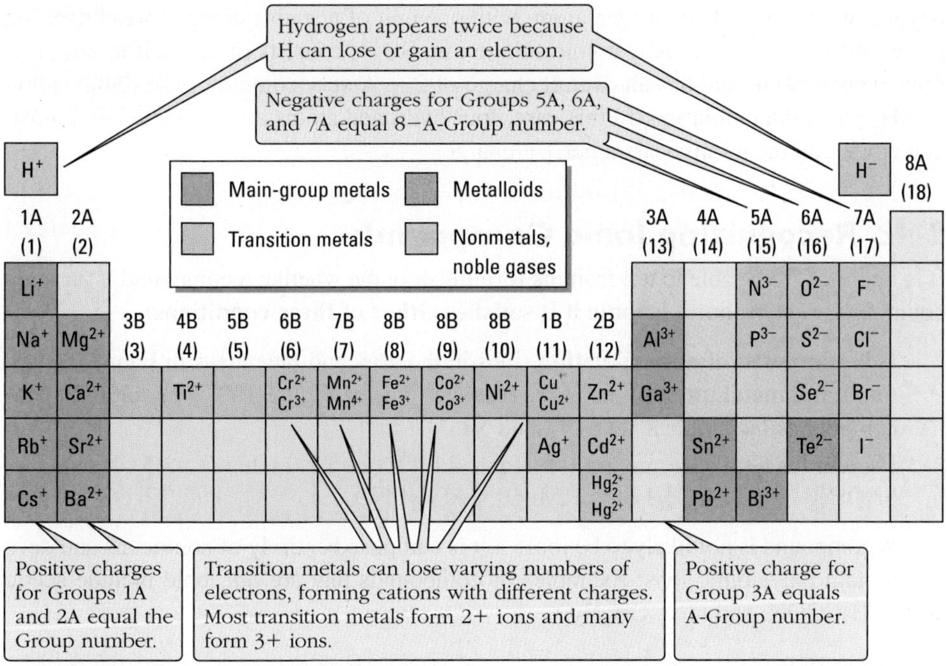

Hydrogen appears twice because H can lose or gain an electron.
Negative charges for Groups 5A, 6A, and 7A equal 8−A-Group number.
Main-group metals
Metalloids
Transition metals
Nonmetals, noble gases
H+
H−
1A (1)
2A (2)
3B (3)
4B (4)
5B (5)
6B (6)
7B (7)
8B (8)
8B (9)
8B (10)
1B (11)
2B (12)
3A (13)
4A (14)
5A (15)
6A (16)
7A (17)
8A (18)
Li+
Na+ Mg2+
K+ Ca2+ Ti2+ Cr2+ Cr3+ Mn2+ Mn4+ Fe2+ Fe3+ Co2+ Co3+ Ni2+ Cu+ Cu2+ Zn2+ Ga3+ Se2− Br−
Rb+ Sr2+ Ag+ Cd2+ Sn2+ Te2− I−
Cs+ Ba2+ Hg2 2+ Hg2+ Pb2+ Bi3+
Al3+ P3− S2− Cl−
N3− O2− F−
Positive charges for Groups 1A and 2A equal the Group number.
Transition metals can lose varying numbers of electrons, forming cations with different charges. Most transition metals form 2+ ions and many form 3+ ions.
Positive charge for Group 3A equals A-Group number.

Figure 2.7 Charges on some common monoatomic cations and anions. Note that metals usually form cations. The cation charge is given by the group number in the case of the main-group elements of Groups 1A, 2A, and 3A *(gray)*. For transition elements *(blue)*, the positive charge is variable, and other ions in addition to those illustrated are possible. Nonmetals *(lavender)* usually form anions that have a negative charge equal to 8 minus the A-group number.

Result Sr^{2+}, P^{3-}, Te^{2-}, and Fe^{2+} and Fe^{3+}

Analyze Apply rules learned from Figure 2.7.

Plan Find each element in the periodic table and use its position to answer the question.

Execute Strontium is a Group 2A metal, so it loses two electrons to give the Sr^{2+} cation.

$$Sr \rightarrow Sr^{2+} + 2\,e^-$$

Phosphorus is a Group 5A nonmetal, so it gains $8 - 5 = 3$ electrons to give the P^{3-} anion.

$$P + 3\,e^- \rightarrow P^{3-}$$

Tellurium is a Group 8A nonmetal, so it gains two electrons to give Te^{2-}.

$$Te + 2\,e^- \rightarrow Te^{2-}$$

Iron is a transition metal, so it has more than one ion. Most transition metals form 2+ ions and many form 3+ ions. Iron is one of those; the ions are Fe^{2+} and Fe^{3+}.

It is extremely important that you know the ions commonly formed by the elements shown in Figure 2.7 so that you can recognize ionic compounds from their formulas and write their formulas as reaction products (Section 2-4d).

PROBLEM-SOLVING PRACTICE 2.6

Decide whether each ion listed below is likely to be found in a common ionic compound. Explain your choice in each case.

(a) Ca^{4+} (b) Cr^{2+} (c) Sr^- (d) Se^{2+}

2-4b Polyatomic Ions

A **polyatomic ion** *is a unit of two or more atoms that bears a net electrical charge.* In many chemical reactions the atoms in the polyatomic ion remain exactly the same, so the polyatomic ion can be treated as a unit. Table 2.2 lists some common polyatomic ions. A compound that contains any of these polyatomic ions *must* be an ionic compound. *It is important to know the names, formulas, and charges of the common polyatomic ions listed in Table 2.2.*

Polyatomic ions are found in many places—oceans, minerals, living cells, and foods. For example, hydrogen carbonate (bicarbonate) ion, HCO_3^-, is present in rain water, seawater, blood, and baking soda. This polyatomic ion consists of one carbon atom, three oxygen atoms, and one hydrogen atom, with one unit of negative charge spread over the group of five atoms. The polyatomic sulfate ion, SO_4^{2-}, consists of one sulfur atom and four oxygen atoms and has an overall charge of 2−. A very common polyatomic cation is NH_4^+, the ammonium ion. In this case, four hydrogen atoms are connected to a nitrogen atom, and the group bears a net 1+ charge.

Cengage Learning/Marna Clarke

Potassium dichromate, $K_2Cr_2O_7$. This beautiful orange-red compound contains potassium ions, K^+, and dichromate ions, $Cr_2O_7^{2-}$.

2-4c Recognizing Ionic Compounds

It is important to be able to tell from its formula or name whether a compound is an ionic compound. **A compound is ionic if it satisfies either of these conditions:**

- **It is composed of a metal cation** (ions in the gray and blue areas in Figure 2.7) **and a nonmetal anion** (ions in the lavender area of Figure 2.7). Examples of such compounds include NaCl, $CaCl_2$, and KI.
- **It includes a polyatomic ion** (see Table 2.2). Examples include $CaSO_4$, $NaNO_3$, $Sr(OH)_2$, NH_4Cl, $(NH_4)_2SO_4$, $KMnO_4$, and $K_2Cr_2O_7$.

A compound is not likely to be ionic if it is composed entirely of nonmetals and does not contain polyatomic ions. Examples of compounds that are *not* ionic include acetic acid, CH_3COOH, and urea, CH_4N_2O.

CONCEPTUAL EXERCISE 2.6

Recognizing Ionic Compounds

Predict whether each compound is likely to be ionic. Explain your predictions.

(a) Li_2CO_3	(b) $(NH_4)_2SO_3$	(c) $C_{10}H_{22}$
(d) N_2H_4	(e) Na_2S	(f) P_4S_3

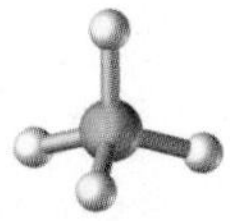

NH_4^+
ammonium ion

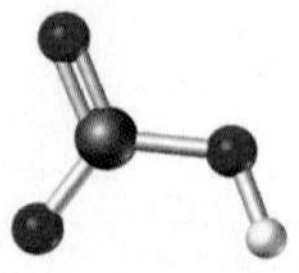

HCO_3^-
hydrogen carbonate ion

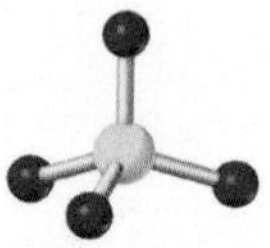

SO_4^{2-}
sulfate ion

Table 2.2 Common Polyatomic Ions

Cations			
NH_4^+	Ammonium	Hg_2^{2+}	Mercury(I)
Anions (1−)			
OH^-	Hydroxide	NO_2^-	Nitrite
HSO_4^-	Hydrogen sulfate	NO_3^-	Nitrate
CH_3COO^-	Acetate	MnO_4^-	Permanganate
ClO^-	Hypochlorite	$H_2PO_4^-$	Dihydrogen phosphate
ClO_2^-	Chlorite	CN^-	Cyanide
ClO_3^-	Chlorate	HCO_3^-	Hydrogen carbonate (bicarbonate)
ClO_4^-	Perchlorate		
Anions (2−)			
SO_3^{2-}	Sulfite	CO_3^{2-}	Carbonate
SO_4^{2-}	Sulfate	HPO_4^{2-}	Monohydrogen phosphate
$S_2O_3^{2-}$	Thiosulfate	$C_2O_4^{2-}$	Oxalate
$Cr_2O_7^{2-}$	Dichromate	CrO_4^{2-}	Chromate
Anion (3−)			
PO_4^{3-}	Phosphate		

2-4d Writing Formulas for Ionic Compounds

All compounds are electrically neutral; that is, there is no net electrical charge. Therefore, when cations and anions combine to form an ionic compound, the *total* positive charge of all the cations must equal the *total* negative charge of all the anions. For example, consider the ionic compound formed when potassium reacts with sulfur. Potassium is a Group 1A metal, so a potassium atom loses one electron to become a K^+ ion. Sulfur is a Group 6A nonmetal, so a sulfur atom gains two electrons to become an S^{2-} ion. To make the compound electrically neutral, two K^+ ions (total charge 2+) are needed for each S^{2-} ion. Consequently, the compound has the formula K_2S. The subscripts in an ionic compound formula show the numbers of ions included in the simplest formula unit. In this case, the subscript 2 indicates two K^+ ions for every S^{2-} ion.

Similarly, aluminum oxide, a combination of Al^{3+} and O^{2-} ions, has the formula Al_2O_3: 2 Al^{3+} gives 6+ charge; 3 O^{2-} gives 6− charge; total charge = 0.

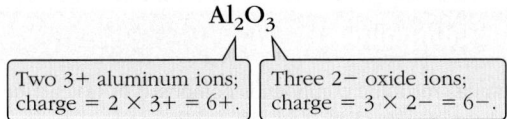

Al_2O_3

Two 3+ aluminum ions; charge = 2 × 3+ = 6+.

Three 2− oxide ions; charge = 3 × 2− = 6−.

Notice that in writing the formula for an ionic compound, *the cation symbol is written first, followed by the anion symbol.* The charges of the ions are *not* included in the formulas of ionic compounds.

Let's now consider several ionic compounds of magnesium, a Group 2A metal that forms Mg^{2+} ions.

Combining Ions	Overall Charge	Formula
Mg^{2+} and Br^-	(2+) + 2 (1−) = 0	$MgBr_2$
Mg^{2+} and SO_4^{2-}	(2+) + (2−) = 0	$MgSO_4$
Mg^{2+} and OH^-	(2+) + 2 (1−) = 0	$Mg(OH)_2$
Mg^{2+} and PO_4^{3-}	3 (2+) + 2 (3−) = 0	$Mg_3(PO_4)_2$

Notice in the latter two cases that when a polyatomic ion occurs more than once in a formula, the polyatomic ion's formula is put in parentheses followed by the necessary subscript.

$Mg_3(PO_4)_2$

Three 2+ magnesium ions; charge = 3 × 2+ = 6+.

Two 3− phosphate ions; charge = 2 × 3− = 6−.

PROBLEM-SOLVING EXAMPLE 2.7

Ions in Ionic Compounds

For each of these compounds, give the symbol or formula of each ion present and indicate how many of each ion are represented in the formula.

(a) K_2SO_4 (b) Na_2S (c) $Mg(CH_3COO)_2$ (d) $(NH_4)_2CO_3$

Result

(a) Two K^+, one SO_4^{2-} (b) Two Na^+, one S^{2-}
(c) One Mg^{2+}, two CH_3COO^- (d) Two NH_4^+, one CO_3^{2-}

Analyze Do what the question asks.

Plan Recognize ions and apply rules for formulas of ionic compounds.

Execute

(a) Potassium is a Group 1A element and therefore forms a 1+ ion. Sulfate is a polyatomic ion with 2− charge. Two K^+ ions balance the 2− charge of SO_4^{2-}.

(b) Sodium is a Group 1A element and therefore forms Na^+. The S^{2-} ion is formed from sulfur, which is a Group 6A element, by gaining two electrons (8 − 6 = 2). Two Na^+ ions and one S^{2-} ion maintain electrical neutrality.
(c) Magnesium is a Group 2A element and therefore forms Mg^{2+} ions. To form an electrically neutral compound, two acetate ions, CH_3COO^-, each with a 1− charge, are necessary to offset the 2+ charge on the Mg^{2+}.
(d) Ammonium ions each have a 1+ charge. Each carbonate ion has a 2− charge. Two ammonium ions are needed to balance the carbonate ion's 2− charge.

PROBLEM-SOLVING PRACTICE 2.7

Determine which ions and how many of each are present in each formula.
(a) $CaSO_3$ (b) $Mg_3(PO_4)_2$

PROBLEM-SOLVING EXAMPLE 2.8

Formulas of Ionic Compounds

Write the correct formulas for ionic compounds composed of (a) calcium and fluoride ions, (b) barium and phosphate ions, (c) Fe^{3+} and nitrate ions, and (d) sodium and carbonate ions.

Result
(a) CaF_2 (b) $Ba_3(PO_4)_2$ (c) $Fe(NO_3)_3$ (d) Na_2CO_3

Analyze Determine correct formulas for ionic compounds.

Plan Locate each element in the periodic table. Use these locations (groups) to determine the cation (metal) and anion (nonmetal) charges. Metal atoms form cations, and nonmetal atoms form anions. Identify any polyatomic anions and their charges. The sum of the positive charges of the cations must equal the sum of the negative charges of the anions.

Execute
(a) Calcium is a Group 2A metal, so it forms 2+ ions. Fluorine is a Group 7A nonmetal, so it forms 1− ions. We need two F^- ions for every Ca^{2+} ion to make CaF_2 electrically neutral.
(b) Barium is a Group 2A metal, so it forms 2+ ions. Phosphate is a 3− polyatomic ion. For electrical neutrality, we need three Ba^{2+} ions and two PO_4^{3-} ions to form $Ba_3(PO_4)_2$. Because there is more than one polyatomic phosphate ion, its formula is enclosed in parentheses followed by the proper subscript.
(c) Iron is in its Fe^{3+} state. Nitrate is a 1− polyatomic ion. Therefore, we need three nitrate ions for each Fe^{3+} ion to form $Fe(NO_3)_3$. The polyatomic nitrate ion is enclosed in parentheses followed by the proper subscript.
(d) Carbonate is a 2− polyatomic ion that combines with two Na^+ ions to form Na_2CO_3. The polyatomic carbonate ion is not enclosed in parentheses because the formula contains only one carbonate.

PROBLEM-SOLVING PRACTICE 2.8

Write formulas for ionic compounds composed of
(a) magnesium and bromine (b) lithium ions and carbonate ions
(c) copper and iodine (d) ammonium ions and chloride ions

2-5 Naming Ions and Ionic Compounds

The name of a compound is often given on bottles of reagents in a laboratory. If you don't know what the name of a compound means in terms of the formula, you are likely to choose the wrong compound. In some cases this could be disastrous. For example, sodium chloride is ordinary table salt; sodium cyanide is a poison. Mixing them up is not a good idea. It is important to be able to write the formula of a compound given its name and to write the name of a compound given its formula. The rules in this section explain how to do this and should be learned thoroughly.

2-5a Naming Positive Ions

Most cations used in this book are metal ions that can be named by these rules. The ammonium ion, NH_4^+, is the major exception; it is a polyatomic ion composed of nonmetal atoms.

1a. ***For metals that form only one kind of cation, the name is simply the name of the metal plus the word "ion."*** For example, Mg^{2+} is the magnesium ion. For these ions you can predict the charge from the periodic table.

1b. ***For metals (mostly transition metals) that can form more than one kind of cation, the name of each ion must indicate its charge. The charge is indicated by a Roman numeral in parentheses immediately following the ion's name. (This is called the* Stock system.*)*** For example, Cu^{2+} is the copper(II) ion and Cu^+ is the copper(I) ion.

Bill Crumbleholme

Serp/Shutterstock.com

The red ceramic glaze includes copper(I) oxide, Cu_2O; the black machine parts are coated with copper(II) oxide, CuO. The different copper-ion charges give different colors.

The Stock system is named after Alfred Stock (1876–1946), a German chemist famous for his work on the hydrogen compounds of boron and silicon.

2-5b Naming Negative Ions

These rules apply to naming anions.

2a. ***A monoatomic anion is named by adding* -ide *to the stem of the name of the nonmetal element from which the ion is derived.*** For example, a *phosph*orus atom gives a *phosph*ide ion, and a *chlor*ine atom forms a *chlor*ide ion. Anions of Group 7A elements, the halogens, are collectively called **halide ions**.

2b. ***The names of the most common polyatomic ions are given in Table 2.2*** (← **Sec. 2-4c**). Most must be memorized. However, some guidelines can help, especially for **oxoanions**, which are *polyatomic ions containing oxygen.*

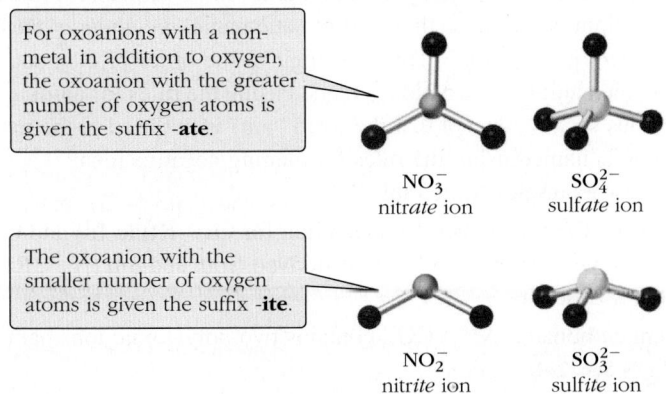

For oxoanions with a nonmetal in addition to oxygen, the oxoanion with the greater number of oxygen atoms is given the suffix **-ate**.

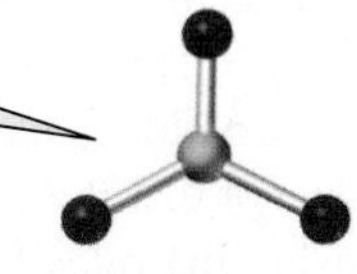

NO_3^-
nitr*ate* ion

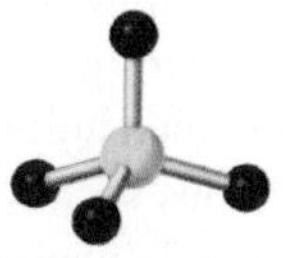

SO_4^{2-}
sulf*ate* ion

The oxoanion with the smaller number of oxygen atoms is given the suffix **-ite**.

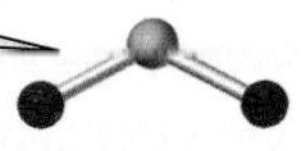

NO_2^-
nitr*ite* ion

SO_3^{2-}
sulf*ite* ion

When more than two different oxoanions of a given nonmetal exist, a more extended naming scheme is used. When there are four different oxoanions, the two middle ones are named according to the *-ate* and *-ite* endings; the oxoanion containing the largest number of oxygen atoms is given the prefix *per-* and the suffix *-ate,* and the oxoanion containing the smallest number is given the prefix *hypo-* and the suffix *-ite.* The oxoanions of chlorine are good examples:

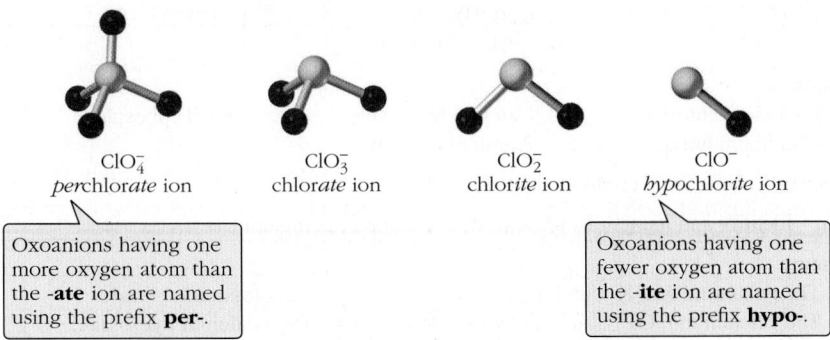

ClO_4^-
*per*chlor*ate* ion

ClO_3^-
chlor*ate* ion

ClO_2^-
chlor*ite* ion

ClO^-
*hypo*chlor*ite* ion

Oxoanions having one more oxygen atom than the **-ate** ion are named using the prefix **per-**.

Oxoanions having one fewer oxygen atom than the **-ite** ion are named using the prefix **hypo-**.

The same naming rules also apply to the oxoanions of bromine.

Table 2.3 Names of Some Useful Ionic Compounds

Common Name	Systematic Name	Formula
Baking soda	Sodium hydrogen carbonate	$NaHCO_3$
Lime	Calcium oxide	CaO
Milk of magnesia	Magnesium hydroxide	$Mg(OH)_2$
Table salt	Sodium chloride	NaCl
Smelling salts	Ammonium carbonate	$(NH_4)_2CO_3$
Lye	Sodium hydroxide	NaOH

Note that the negative charge of the polyatomic ion decreases by one for each hydrogen added.

Oxoanions containing hydrogen are named simply by adding the word "hydrogen" before the name of the oxoanion, for example, hydrogen sulfate ion, HSO_4^-. When an oxoanion of a given nonmetal can combine with different numbers of hydrogen atoms, we must use prefixes to indicate which ion we are talking about: *di*hydrogen phosphate for $H_2PO_4^-$ and *mono*hydrogen phosphate for HPO_4^{2-}. Because some hydrogen-containing oxoanions have common names that are used often, you should know them. For example, the hydrogen carbonate ion, HCO_3^-, is often called the bicarbonate ion.

2-5c Naming Ionic Compounds

Table 2.3 lists common names and systematic names of several useful ionic compounds with which you might be familiar. In the systematic name, *the name of the cation comes first, then the name of the anion.* (This is similar to the order of element symbols in a formula: cation, then anion.) The cation is named using the rules for naming positive ions given in a previous section, except that the word "ion" is not used in the name of a compound. The anion is named using the rules for naming negative ions.

Consider these examples from Table 2.3:

- Calcium oxide, CaO, is named from calcium for Ca^{2+} (Rule 1a) and oxide for O^{2-} (Rule 2a). Likewise, sodium chloride is derived from sodium (Na^+, Rule 1a) and chloride (Cl^-, Rule 2a).
- Ammonium carbonate, $(NH_4)_2CO_3$, contains two polyatomic ions named in Table 2.2 (← **Sec. 2-4c**).
- In the name copper(II) sulfate, the (II) indicates that Cu^{2+} is present, not Cu^+, the other possibility.

PROBLEM-SOLVING EXAMPLE 2.9

Ionic Compounds: Names from Formulas

Name each of these ionic compounds.

(a) KCl (b) $Ca(OH)_2$ (c) $Fe_3(PO_4)_2$
(d) $Al(NO_3)_3$ (e) $(NH_4)_2SO_4$

Result

(a) Potassium chloride (b) Calcium hydroxide (c) Iron(II) phosphate
(d) Aluminum nitrate (e) Ammonium sulfate

Analyze Name ionic compounds given formulas.

Plan Identify and name ions present; derive name of compound.

Execute

(a) The potassium ion, K^+, and the chloride ion, Cl^-, form potassium chloride.
(b) The calcium ion, Ca^{2+}, and the hydroxide ion, OH^-, form calcium hydroxide.
(c) Two PO_4^{3-} ions give 6− charge; because there are three iron ions they must have 2+ charge. The iron(II) ion, Fe^{2+}, and the phosphate ion, PO_4^{3-}, form iron(II) phosphate.

(d) The aluminum ion, Al^{3+}, and the nitrate ion, NO_3^-, form aluminum nitrate.
(e) The ammonium ion, NH_4^+, and the sulfate ion, SO_4^{2-}, combine to form ammonium sulfate.

PROBLEM-SOLVING PRACTICE 2.9

Name each of these ionic compounds:

(a) KNO_2	(b) $NaHSO_3$	(c) $Mn(OH)_2$
(d) $Mn_2(SO_4)_3$	(e) Ba_3N_2	(f) LiH

PROBLEM-SOLVING EXAMPLE 2.10

Ionic Compounds: Formulas from Names

Write the correct formula for each of these ionic compounds:

(a) Ammonium sulfide	(b) Potassium sulfate
(c) Copper(II) nitrate	(d) Iron(II) chloride

Result (a) $(NH_4)_2S$ (b) K_2SO_4 (c) $Cu(NO_3)_2$ (d) $FeCl_2$

Analyze Write the formula given the name of an ionic compound.

Plan Determine the charge of each ion and then make certain that the total charge for the formula is zero.

Execute
(a) The ammonium cation is NH_4^+ and the sulfide ion is S^{2-}, so two NH_4^+ (total 2+) are needed for one S^{2-} (2−) to make the electrically neutral $(NH_4)_2S$.
(b) The potassium cation is K^+ and the sulfate anion is SO_4^{2-}, so two K^+ (total 2+) are needed for each SO_4^{2-} (2−) to make the electrically neutral K_2SO_4.
(c) The copper(II) cation is Cu^{2+} (2+ charge) and the anion is NO_3^-, so two NO_3^- (total 2−) are needed to give electrically neutral $Cu(NO_3)_2$.
(d) The iron(II) ion is Fe^{2+} and the chloride ion is Cl^-, so two chloride ions are needed for each iron(II) ion.

PROBLEM-SOLVING PRACTICE 2.10

Write the correct formula for each of these ionic compounds:

(a) Potassium dihydrogen phosphate	(b) Copper(I) hydroxide
(c) Sodium hypochlorite	(d) Ammonium perchlorate
(e) Chromium(III) chloride	(f) Iron(II) sulfite

2-6 Ionic Compounds: Bonding and Properties

The book *Salt: A World History*, by Mark Kurlansky, is a compelling account of table salt's importance through the ages.

The properties of an ionic compound differ significantly from those of its component elements. Consider the familiar ionic compound, table salt (sodium chloride, NaCl), composed of Na^+ and Cl^- ions. Sodium chloride is a white, crystalline, water-soluble solid. It is very different from its component elements, metallic sodium and gaseous chlorine. Sodium is an extremely reactive metal that reacts violently with water. Chlorine is a diatomic, toxic gas that reacts with water. Sodium *ions* and chloride *ions* do not undergo such reactions, and NaCl dissolves uneventfully in water.

Ionic compounds usually are solids at room temperature, and their crystals are hard, brittle, and can be cleaved easily. They have high melting and boiling points and are poor conductors of heat. They are good conductors of electricity when molten, but poor conductors when solid. Any substance that you would classify as an ionic compound based on the rules in Section 2-4c has these properties.

STERLING
SALT

Cengage Learning/Charles D. Winters

Sodium chloride, NaCl. This common ionic compound contains sodium ions (Na^+) and chloride ions (Cl^-).

2-6a Coulomb's Law

In ionic compounds, *electrostatic forces attract cations and anions to each other. This attraction is called* **ionic bonding**. The strength of the electrostatic force dictates many of the properties of ionic compounds. *The attraction between oppositely charged ions increases with charge and decreases with the square of the distance between the ions.* This is known as **Coulomb's law**.

Quantitatively, Coulomb's law is given by the equation

$$\text{Force} = F = k\,\frac{Q_1Q_2}{d^2}$$

where Q_1 and Q_2 are the magnitudes of the charges on the two interacting particles, d is the distance between the two particles, and k is a proportionality constant. For ions separated by the same distance, the attractive force between 2+ and 2− ions is four times greater than that between 1+ and 1− ions. The attractive force also increases as the distance between the centers of the ions decreases. Thus, a small cation and a small anion will attract each other more strongly than will larger ions. (We discuss the sizes of ions in Section 5-10.)

2-6b The Crystal Lattice

In solid ionic compounds, cations and anions are held by ionic bonding in *an orderly array* called a **crystal lattice**, *in which each cation is surrounded by anions and each anion is surrounded by cations.* Such an arrangement maximizes the attraction between cations and anions and minimizes the repulsion between ions of like charge. In sodium chloride, as shown in Figure 2.8, six chloride ions surround each sodium ion, and six sodium ions surround each chloride ion. As indicated in the formula, there is one sodium ion for each chloride ion.

The formula of an ionic compound indicates only the smallest whole-number ratio of the number of cations to the number of anions in the compound. In NaCl that ratio is 1 : 1. An Na^+Cl^- pair is referred to as a formula unit of sodium chloride. A **formula unit** is *a grouping of chemical symbols that indicates the ratio of number of atoms of each kind that make up a substance.* Note that the formula unit of an ionic compound has no independent existence outside of the crystal—it is different from a molecule.

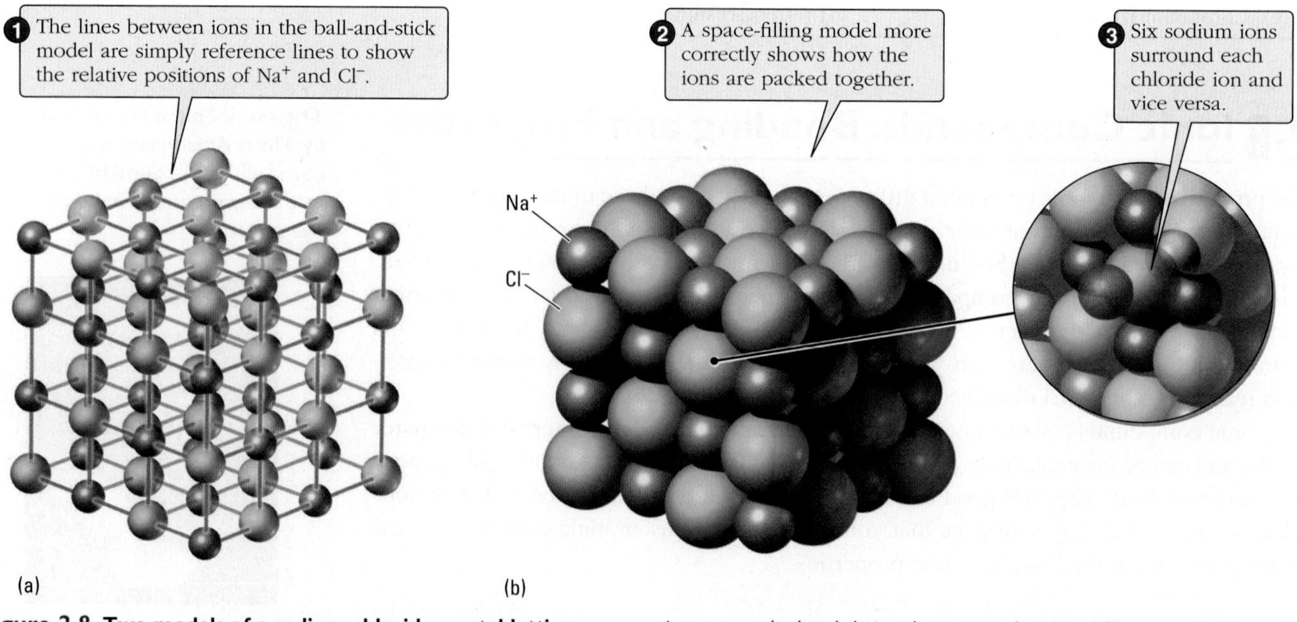

1 The lines between ions in the ball-and-stick model are simply reference lines to show the relative positions of Na⁺ and Cl⁻.
2 A space-filling model more correctly shows how the ions are packed together.
3 Six sodium ions surround each chloride ion and vice versa.
Na⁺
Cl⁻
(a)
(b)

Figure 2.8 Two models of a sodium chloride crystal lattice. (a) This ball-and-stick model illustrates how the ions are arranged but shows the ions too far apart. (b) A space-filling model shows how the ions are packed and their relative sizes, but it is difficult to see the locations of ions other than those on the faces of the crystal lattice.

2-6c Characteristic Properties of Ionic Compounds

The regular array of cations and anions in a crystal lattice and the strong electrostatic attractions that hold the ions rather rigidly in position help us understand characteristic properties of ionic compounds that were listed earlier.

Ionic compounds have distinctive crystalline shapes and are easily cleaved. Crystalline shapes are distinctive because the cations and anions are held rather rigidly in position. Such alignment creates planes of ions within the crystals and the angles between those planes determine the angles between sides of macroscopic crystals. For example, if you look closely (perhaps with a magnifying glass) at table salt, you will see that many of the salt crystals have 90° angles between sides. This is consistent with the 90° angles between layers of ions shown in Figure 2.8. Ionic crystals also can be cleaved—split parallel to the planes of ions (Figure 2.9). When an outside force causes one plane to shift slightly relative to the next, ions of like charge are brought close together and repel strongly. The repulsion causes the layers on opposite sides of the cleavage plane to separate, and the crystal splits.

Ionic compounds have high melting points and are solids at room temperature. Melting points are high because, according to the kinetic-molecular theory (← **Sec. 1-9a**), melting requires that ions break out of the rigid array in the solid and move about independently in the liquid. Because of the strong attractions among the cations and anions, a high temperature is required before the motion of the ions is great enough to overcome these attractions. Melting points are also related to the charges and sizes of the ions. For ions of similar size, such as the cations Na^+ and Ca^{2+} and the anions O^{2-} and F^-, Coulomb's law predicts that *the larger the charges, the greater the attraction and the higher the melting point.* For example, CaO (composed of doubly charged Ca^{2+} and O^{2-} ions) melts at 2613 °C, whereas NaF (composed of singly charged Na^+ and F^- ions) melts at 996 °C. For ions of similar charge but different sizes, such as F^- and the much larger I^-, Coulomb's law predicts that *the smaller the ion, the higher the melting point.* For example, NaF melts at 996 °C, whereas NaI melts at 661 °C.

Ionic compounds do not conduct electricity when solid but do conduct when molten. Electric current involves movement of charged nanoscale particles such as electrons or ions. Because the ions in a crystal can only vibrate about fixed positions, ionic solids do not conduct electricity. However, when an ionic solid melts, the liquid conducts

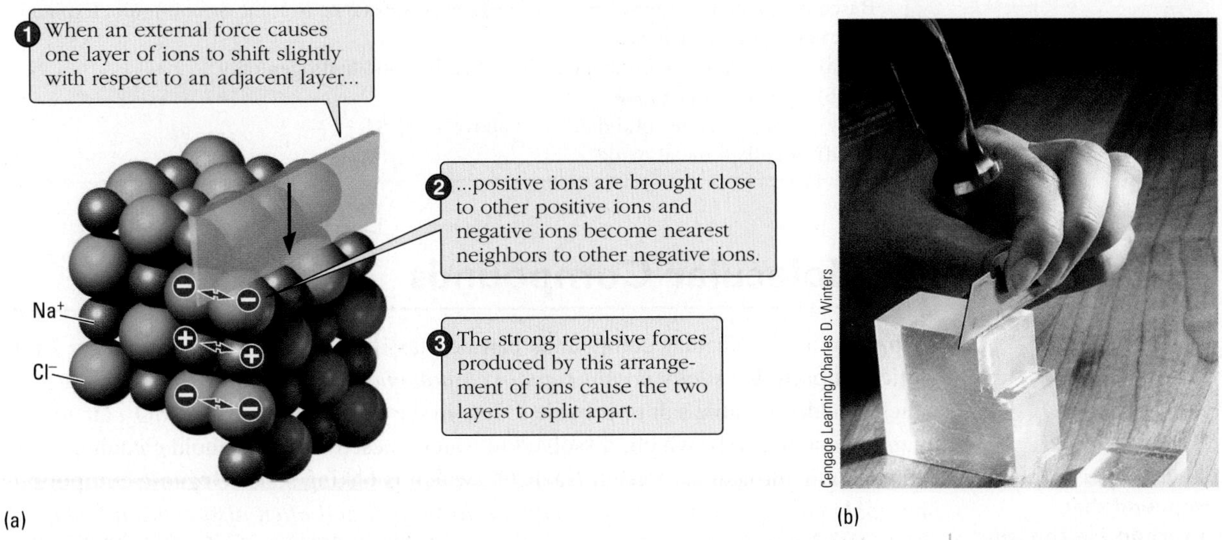
1 When an external force causes one layer of ions to shift slightly with respect to an adjacent layer...
2 ...positive ions are brought close to other positive ions and negative ions become nearest neighbors to other negative ions.
3 The strong repulsive forces produced by this arrangement of ions cause the two layers to split apart.
Na+
Cl−
Cengage Learning/Charles D. Winters
(a)
(b)

Figure 2.9 Cleavage of an ionic crystal. (a) Diagram of the forces involved in cleaving an ionic crystal. (b) A sharp blow on a knife edge lying along a plane of a salt crystal causes the crystal to split.

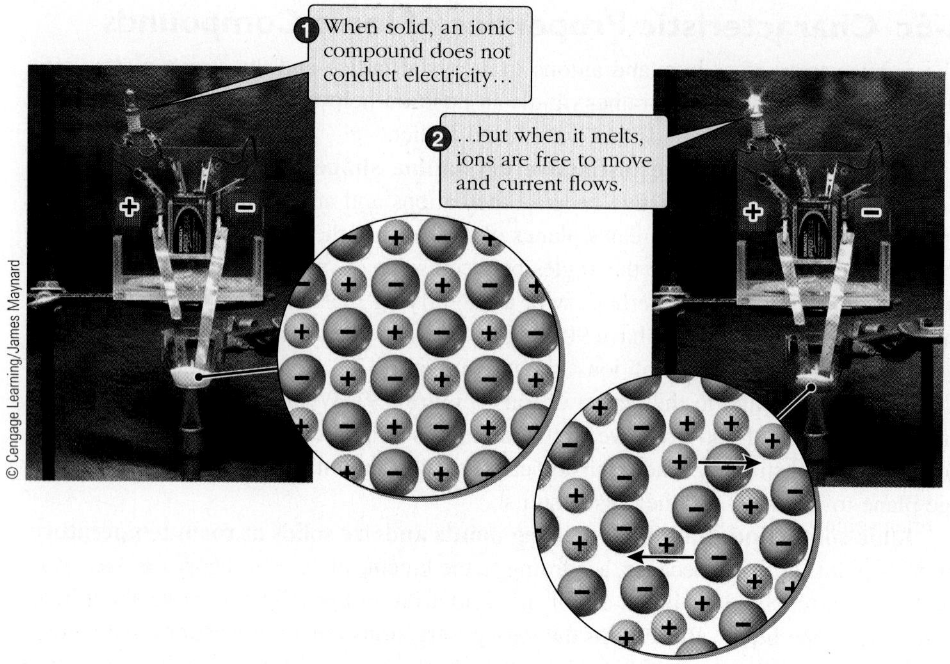

1 When solid, an ionic compound does not conduct electricity…
2 …but when it melts, ions are free to move and current flows.
© Cengage Learning/James Maynard
© Cengage Learning/James Maynard

Figure 2.10 When it melts, an ionic compound conducts electric current, lighting the bulb.

electricity as shown in Figure 2.10. This is because the ions are now free to move relative to each other. Cations moving in one direction and anions moving in the opposite direction carry an electric current in the molten ionic compound, just as electrons carry current in a copper wire.

Recognizing from its name or formula that a compound is ionic is important because once you know a compound is ionic you can predict that it will have these characteristic properties. On the macroscale, if a compound has high melting and boiling points, if its crystals can be cleaved, and if it conducts electricity when molten but not when solid, the compound is probably ionic. The properties of ionic compounds are quite different from those of another major class of compounds, molecular compounds.

CONCEPTUAL EXERCISE 2.7

Identifying Ionic Compounds

Based on the name or description, classify each substance as ionic or non-ionic. Explain your reasoning in each case.

(a) A soft, waxy solid that consists of carbon and hydrogen only
(b) Ammonium oxalate
(c) A hard, brittle solid that melts above 1000 °C
(d) An alkali metal oxide

2-7 Molecular Compounds

Both ionic and molecular compounds can be classified as organic or inorganic. **Inorganic compounds** *usually do not contain carbon and are often ionic.* Examples are KI, which is added to table salt to provide iodine in the diet; sulfur dioxide, SO_2, an air pollutant; ammonia, NH_3, which, dissolved in water, is used as a household cleaning agent; and sodium hydrogen carbonate, $NaHCO_3$, which is baking soda. **Organic compounds** *invariably contain carbon, usually contain hydrogen, and often also contain other elements.* Organic compounds are of great interest because they are the basis for the clothes we wear, the food we eat, the fuels we burn, and the living organisms in our environment. For example, ethanol, which has the formula C_2H_6O, is familiar as a component of

Notice that $NaHCO_3$ is an inorganic compound that does contain carbon—in the HCO_3^- ion.

Table 2.4 Properties of Molecular and Ionic Compounds

Molecular Compounds	Ionic Compounds
Usually formed by combination of nonmetals with other nonmetals; some involve metalloids or certain metals	Typically formed by combination of reactive metals with reactive nonmetals
Consists of atoms combined into molecules	Consists of ions packed in a crystal lattice
Gases, liquids, solids	Crystalline solids
Brittle and weak or soft and waxy solids	Hard and brittle solids
Low melting points	High melting points
Low boiling points (−250 to 600 °C)	High boiling points (700 to 3500 °C)
Poor conductors of electricity both as solid and liquid	Good conductors of electricity when molten; poor conductors of electricity when solid
Poor conductors of heat	Poor conductors of heat
Solubility depends on molecular structure; often soluble in organic solvents	Often soluble in water
Examples: hydrocarbons, H_2O, CO_2, sugar	Examples: NaCl, CaF_2, NH_4NO_3

"alcoholic" beverages and of motor fuel; methane, CH_4, is the major component of natural gas; and glycine, $C_2H_5NO_2$, is a component of proteins in our bodies.

The majority of organic compounds, as well as many other compounds, are molecular compounds. *In a* **molecular compound**, *at the nanoscale, atoms of two or more different elements are combined into the independent units known as molecules.* A molecule is a collection of atoms connected by chemical bonds (← **Sec. 1-10**). Every day we inhale, exhale, metabolize, and in other ways use thousands of molecular compounds. Water, carbon dioxide, sucrose (table sugar), and caffeine, as well as carbohydrates, proteins, and fats, are among the many common molecular compounds in our bodies.

Some elements are composed of molecules and have molecular formulas. In oxygen, for example, two oxygen atoms are joined in an O_2 molecule (← Sec. 1-12a).

2-7a Characteristic Properties of Molecular Compounds

Most compounds that are not ionic compounds are molecular compounds. Molecular compounds are usually formed from nonmetals or sometimes metalloids. Only a small fraction of them involve metals. That is, in most molecular compounds, all of the elements are nonmetals or perhaps metalloids.

The properties of molecular compounds, whether inorganic or organic, are quite different from the properties of ionic compounds. Molecular compounds usually have low melting and boiling points, do not conduct electricity, and, if they are solids, are soft and waxy. Table 2.4 contrasts the properties of molecular compounds with those of ionic compounds. If you can recognize from its composition that a compound is molecular, then it will have these properties.

CONCEPTUAL EXERCISE 2.8

Ionic and Molecular Compounds

Predict whether each substance is ionic or molecular.

(a) A soft, waxy solid (b) $(NH_4)_2Cr_2O_7$ (c) $C_2H_5NO_2$
(d) A hard, brittle solid (e) Rb_2O (f) A gas at room temperature

2-7b Molecular Formulas

The composition of a molecular compound is represented symbolically by its molecular formula. In a **molecular formula**, *the kinds of atoms combined to make one molecule are indicated by element symbols, and the number of each kind of atom is indicated by a*

H_2O	H—O—H

Space-filling model　　Ball-and-stick model

In this figure and throughout the book, atoms in molecular models are color-coded: H, light gray; C, dark gray; O, red. A chart showing the full color key is inside the back cover.

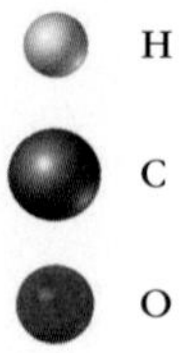
H
C
O

subscript. For example, the molecular formula for water, H_2O, shows that there are three atoms per molecule—two hydrogen atoms and one oxygen atom. The subscript to the right of each element's symbol indicates the number of atoms of that element present in the molecule. If the subscript is omitted, it is understood to be 1, as for the O in H_2O. Usually the element symbols are written so that the element farthest from the upper right corner of the periodic table is written first (H is written before O).

The formula of a molecular compound, especially an organic compound, can be written in several different ways. The molecular formula given previously for ethanol, C_2H_6O, is one example. For an organic compound, the symbols of the elements other than carbon are frequently written in alphabetical order, and each has a subscript indicating the total number of atoms of that type in the molecule, as illustrated by C_2H_6O. Because of the huge number of organic compounds, this formula may not give sufficient information to indicate what compound is represented. Such identification requires more information about how the atoms are connected to each other. A **structural formula** *shows exactly how atoms are connected by chemical bonds.* A **chemical bond** *is an attractive force between two atoms holding them together in a molecule.* In ethanol, for example, the first carbon atom is bonded to three hydrogen atoms, and the second carbon atom is bonded to the first carbon atom, to two hydrogen atoms, and to an —OH group.

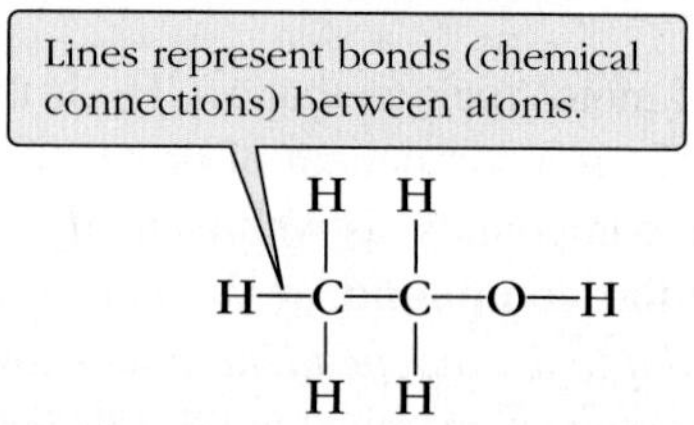
Lines represent bonds (chemical connections) between atoms.
H H
H—C—C—O—H
H H

A structural formula can also be written as a **condensed formula** to *show how the atoms are grouped together in the molecule.* Each carbon atom and its hydrogen atoms

Table 2.5 Examples of Simple Molecular Compounds

Name	Molecular Formula	Number and Kind of Atoms in Molecule	Molecular Model
Carbon dioxide	CO_2	3 total: 1 carbon, 2 oxygen	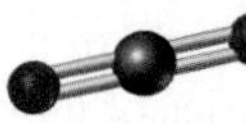
Ammonia	NH_3	4 total: 1 nitrogen, 3 hydrogen	
Nitrogen dioxide	NO_2	3 total: 1 nitrogen, 2 oxygen	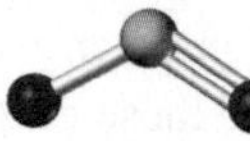
Carbon tetrachloride	CCl_4	5 total: 1 carbon, 4 chlorine	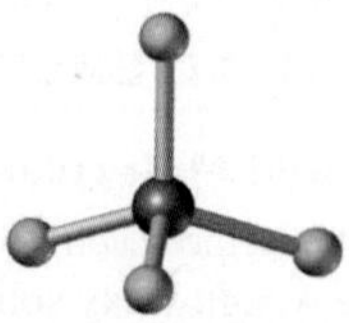
Octane	C_8H_{18}	26 total: 8 carbon, 18 hydrogen	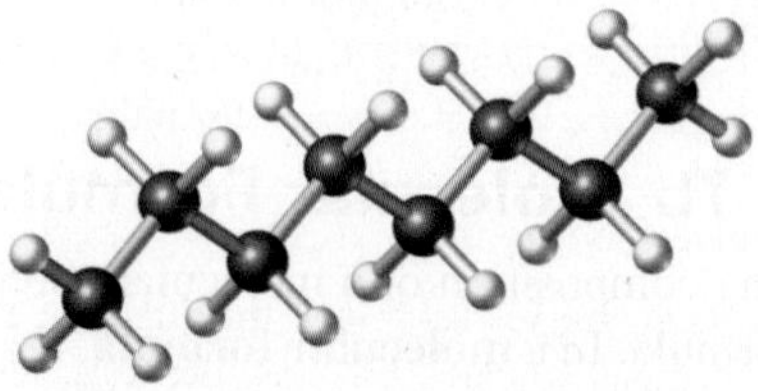

are written without connecting lines (CH_3, CH_2, or CH). Other groups are usually written on the same line with the carbon and hydrogen atoms if the groups are at the beginning or end of the molecule. Otherwise, they are connected above or below the line by straight lines to the respective carbon atoms. Condensed formulas emphasize the atoms or groups of atoms connected to each carbon atom. For ethanol, the condensed formula is CH_3CH_2OH. If you compare this with the structural formula for ethanol, you should be able to see that they represent the same structure: there is a C atom connected to three H atoms, then a C atom connected to two H atoms, and then an O atom connected to an H atom.

To summarize, three ways of writing formulas are shown here for ethanol. In addition, two types of molecular models are shown. Study these representations and make certain you can see the relationships among them.

Molecular formula	Condensed formula	Structural formula	Ball-and-stick model	Space-filling model
C_2H_6O	CH_3CH_2OH	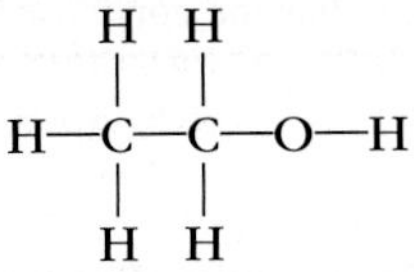		

Additional examples of molecular formulas with ball-and-stick molecular models are in Table 2.5.

PROBLEM-SOLVING EXAMPLE 2.11

Formulas for Molecular Compounds

(a) Write the molecular formulas for these molecules:

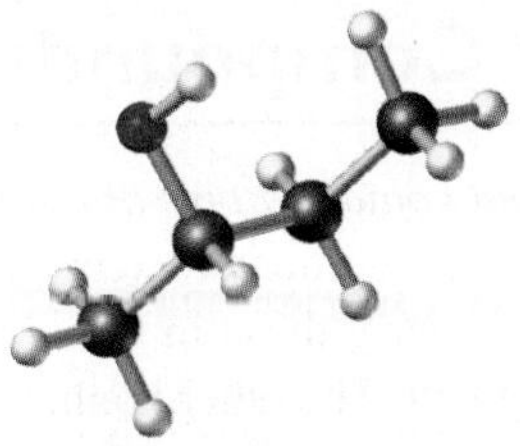

2-butanol

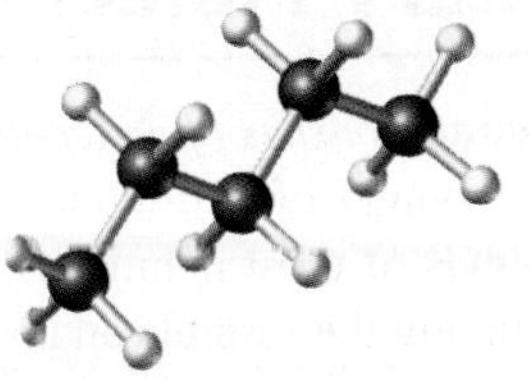

pentane

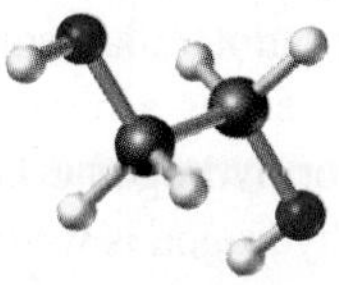

ethylene glycol

(b) Write the condensed formulas for 2-butanol, pentane, and ethylene glycol.
(c) Write the structural formula for the molecule with molecular formula CH_4.

Result

(a) 2-butanol, $C_4H_{10}O$; pentane, C_5H_{12}; ethylene glycol, $C_2H_6O_2$

(b)

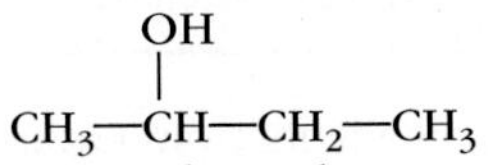

2-butanol

$CH_3CH_2CH_2CH_2CH_3$
pentane

$HOCH_2CH_2OH$
ethylene glycol

(c) 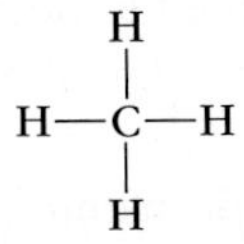

Analyze Models of molecular structures are given; from them you are asked to write molecular and condensed formulas. In part (c) a molecular formula is given and you are asked to draw a structural formula; recognize that carbon forms four bonds in each structure given so far.

Plan In part (a), count the atoms of each type in each model to obtain the molecular formula. In part (b) follow the rules for writing condensed formulas. In part (c) draw four bonds to H atoms around the C atom.

Execute

(a) 2-butanol: 4 C atoms, 10 H atoms, and 1 O atom; formula is $C_4H_{10}O$.
pentane: 5 C atoms and 12 H atoms; formula is C_5H_{12}.
ethylene glycol: 2 C atoms, 6 H atoms, and 2 O atoms; formula is $C_2H_6O_2$.

(b) For 2-butanol, starting at the left there is CH_3 group, CHOH, CH_2, and CH_3; write the groups in this order with the —OH group above its C (see Result for formula). Pentane and ethylene glycol are done similarly.

(c) Drawing four bonds to H around C gives the structural formula in the Result.

PROBLEM-SOLVING PRACTICE 2.11

Write the molecular formulas for these compounds. In part (c) write a structural formula.

(a) Adenosine triphosphate (ATP), an energy source in biochemical reactions, contains 10 carbon, 11 hydrogen, 13 oxygen, 5 nitrogen, and 3 phosphorus atoms per molecule.

(b) Capsaicin, the active ingredient in chili peppers, has 18 carbon atoms, 27 hydrogen atoms, 3 oxygen atoms, and 1 nitrogen atom per molecule.

(c) Diethyl ether, which is an anesthetic, has the condensed formula $CH_3CH_2OCH_2CH_3$.

INGREDIENTS: Water, Propylene Glycol, Corrosion Inhibitor Packag
This product is considered GRAS (Generally Regarded As Safe) by
Antifreeze

Cengage Learning/Charles D. Winters

This automotive coolant contains propylene glycol.

EXERCISE 2.9

Structural, Condensed, and Molecular Formulas

A molecular model of propylene glycol, used in some "environmentally friendly" antifreezes, is shown here. Write the molecular formula, the structural formula, and the condensed formula for propylene glycol.

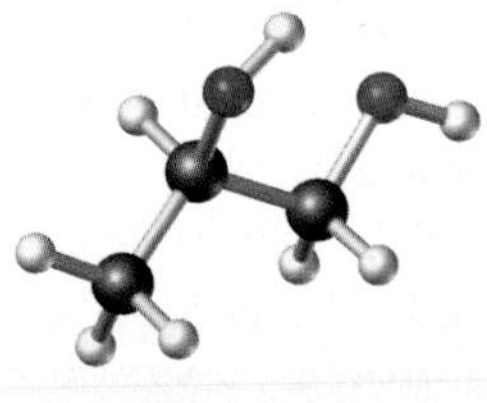

propylene glycol

2-8 Naming Binary Molecular Compounds

Names of binary compounds containing hydrogen and carbon are discussed in the next section.

A **binary molecular compound** *consists of molecules that contain atoms of only two elements.* There is a binary compound of hydrogen with every nonmetal except the noble gases. For hydrogen compounds of most nonmetals, particularly those in Groups 6A and 7A, the hydrogen is written first in the formula and named first. The other nonmetal is then named, with the nonmetal's name changed to end in *-ide.* For example, HCl is named hydrogen chloride.

Formula	Name	Formula	Name
HCl	Hydrogen chloride	HI	Hydrogen iodide
HBr	Hydrogen bromide	H_2Se	Hydrogen selenide

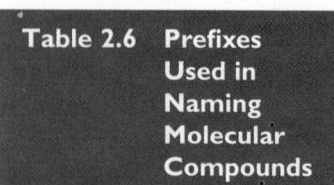

Table 2.6 Prefixes Used in Naming Molecular Compounds

Prefix	Number
Mono-	1
Di-	2
Tri-	3
Tetra-	4
Penta-	5
Hexa-	6
Hepta-	7
Octa-	8
Nona-	9
Deca-	10

Many binary molecular compounds contain nonmetallic elements from Groups 4A, 5A, 6A, and 7A of the periodic table. In these compounds the elements are listed in formulas and names in the order of the group numbers, and prefixes are used to designate the number of a particular kind of atom. The prefixes are listed in Table 2.6. Table 2.7 illustrates how these prefixes are applied.

Some binary molecular compounds were named before systematic naming rules were developed. Their *common names* must simply be learned.

Formula	Common Name	Formula	Common Name
H_2O	Water	NO	Nitric oxide
NH_3	Ammonia	N_2O	Nitrous oxide ("laughing gas")
N_2H_4	Hydrazine	PH_3	Phosphine

Table 2.7 Examples of Binary Molecular Compounds

Molecular Formula	Name	Use
CO	Carbon monoxide	Steel manufacturing
NO_2	Nitrogen dioxide	Preparation of nitric acid
N_2O	Dinitrogen oxide (nitrous oxide)	Anesthetic; spray can propellant
N_2O_5	Dinitrogen pentaoxide	Forms nitric acid
PBr_3	Phosphorus tribromide	Forms phosphorous acid
PBr_5	Phosphorus pentabromide	Forms phosphoric acid
SF_6	Sulfur hexafluoride	Transformer insulator
P_4O_{10}	Tetraphosphorus decaoxide	Drying agent

PROBLEM-SOLVING EXAMPLE 2.12

Naming Binary Molecular Compounds

Name these compounds: (a) CO_2, (b) IF_7, (c) SO_3, (d) N_2O_4, (e) PCl_5.

Result

(a) Carbon dioxide (b) Iodine heptafluoride (c) Sulfur trioxide
(d) Dinitrogen tetraoxide (e) Phosphorus pentachloride

Analyze You are asked to name compounds given formulas.

Plan Follow rules for naming binary molecular compounds; use prefixes in Table 2.6.

Execute

(a) Use *di-* to represent the two oxygen atoms; name is carbon dioxide.
(b) Use *hepta-* for the seven fluorine atoms; name is iodine heptafluoride.
(c) Use *tri-* for the three oxygen atoms; name is sulfur trioxide.
(d) Use *di-* for the two nitrogen atoms and *tetra-* for the four oxygen atoms; name is dinitrogen tetraoxide.
(e) Use *penta-* for the five chlorine atoms; name is phosphorus pentachloride.

PROBLEM-SOLVING PRACTICE 2.12

Name these compounds: (a) SO_2, (b) P_4O_6, (c) CCl_4.

EXERCISE 2.10

Names and Formulas of Binary Molecular Compounds

Give the formula for each of these binary compounds involving nonmetals:

(a) Carbon disulfide (b) Phosphorus trichloride (c) Sulfur dibromide
(d) Selenium dioxide (e) Oxygen difluoride (f) Xenon trioxide

2-9 Organic Molecular Compounds

Millions of organic compounds are known. They vary enormously in structure and function, ranging from the simple molecule methane (CH_4, the major constituent of natural gas) to large, complex biochemical molecules such as proteins, which often contain hundreds or thousands of atoms. One reason for the enormous variety of organic compounds is that each carbon atom can form strong, stable chemical bonds with up to four other carbon atoms. Through their carbon-carbon bonds, carbon atoms can form chains, branched chains, rings, and other more complicated structures.

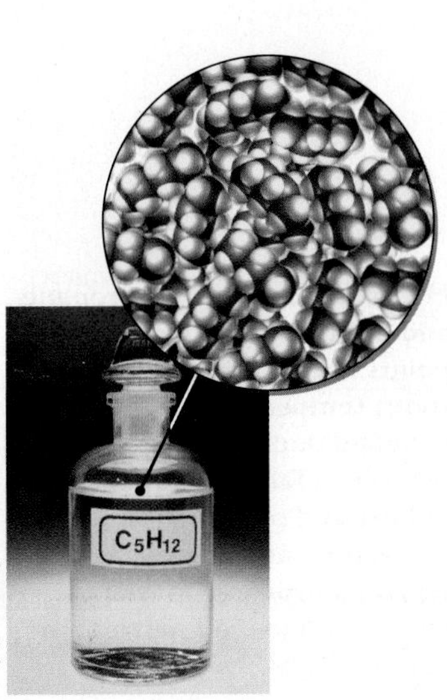
C_5H_{12}

Cengage Learning/Charles D. Winters

Pentane is a molecular compound used as a solvent.

A **hydrocarbon** is *a binary compound of carbon and hydrogen;* that is, hydrocarbons are composed of only carbon and hydrogen. An **alkane** is *a hydrocarbon that contains the greatest possible ratio of hydrogen to carbon.* Alkanes are the main components of automobile fuels and motor oils. Table 2.8 provides some information about the first ten alkanes. The first four (methane, ethane, propane, butane) have common names that must be memorized. For $n = 5$ or greater, the names are systematic: the prefixes of Table 2.6 indicate the number of carbon atoms in the molecule; the ending *-ane* indicates that the compound is an alk*ane.* For example, the five-carbon alkane is *pentane.*

Straight chain means a chain of carbon atoms with no branches to other carbon atoms; each carbon atom connects with no more than two other carbon atoms. As you can see from the molecular model of butane, the chain is not actually straight, but rather a zigzag.

Methane, the simplest alkane, makes up about 85% of natural gas in the United States. Methane is also known to be one of the greenhouse gases (Section 8-11), meaning that it is one of the chemicals implicated in the problem of global warming. Ethane, propane, and butane are used as heating fuel for homes and in industry. In these simple alkanes, the carbon atoms are connected in unbranched (straight) chains, and each carbon atom is connected to either two or three hydrogen atoms. The general formula for noncyclic alkanes is C_nH_{2n+2}, where n is an integer.

In an unbranched alkane molecule, each of the *n* carbon atoms has two hydrogen atoms bonded to it. Two carbon atoms (one at each end of the molecule) have an extra hydrogen atom each. Thus, if *n* is the number of carbon atoms, the number of hydrogen atoms is 2*n* + 2.

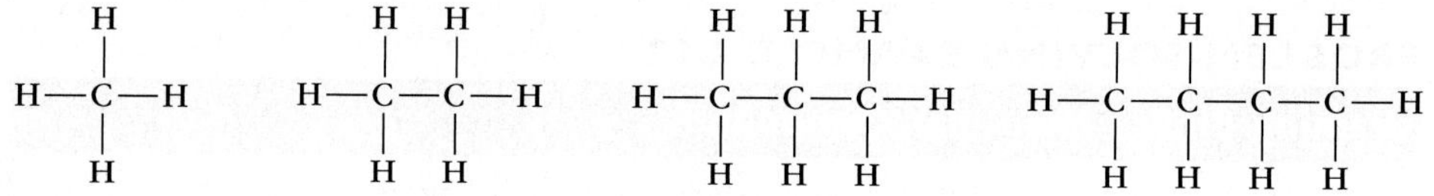

methane ethane propane butane

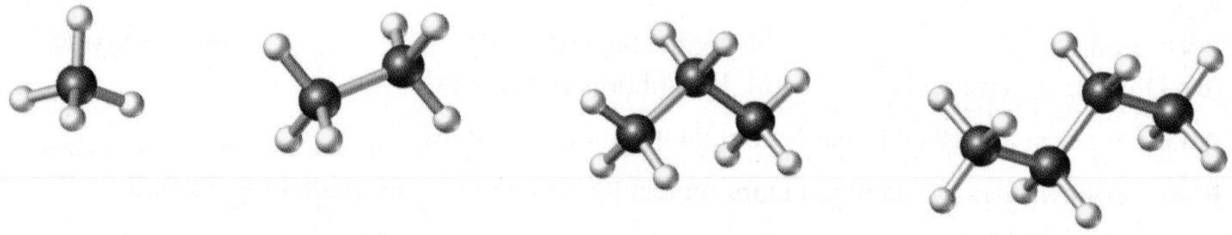

Larger alkanes have longer chains of carbon atoms with hydrogens attached to each carbon. For example, heptane, C_7H_{16}, is found in gasoline,

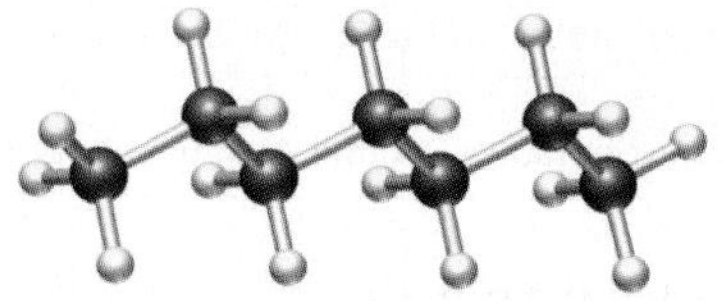

$CH_3(CH_2)_5CH_3$
heptane

Hydrocarbons such as propane and butane, which have boiling points not too far below room temperature, can be liquefied under higher pressures. Tanks of liquefied butane and propane are used for heating and cooking. The liquefied hydrocarbons are known as liquefied petroleum gas or LP gas.

Table 2.8 The First Ten Alkane Hydrocarbons, C_nH_{2n+2}

Molecular Formula	Name	Boiling Point (°C)	Melting Point (°C)	Physical State at Room Temperature
CH_4	Methane	−161.5	−184	Gas
C_2H_6	Ethane	−88.6	−172	Gas
C_3H_8	Propane	−42.17	−189.9	Gas
C_4H_{10}	Butane	−0.5	−138	Gas
C_5H_{12}	Pentane	36.2	−131.5	Liquid
C_6H_{14}	Hexane	69.0	−94.3	Liquid
C_7H_{16}	Heptane	98.427	−90.5	Liquid
C_8H_{18}	Octane	125.8	−56.5	Liquid
C_9H_{20}	Nonane	150.798	−53.7	Liquid
$C_{10}H_{22}$	Decane	174	−31	Liquid

and eicosane, $C_{20}H_{42}$, is found in paraffin wax.

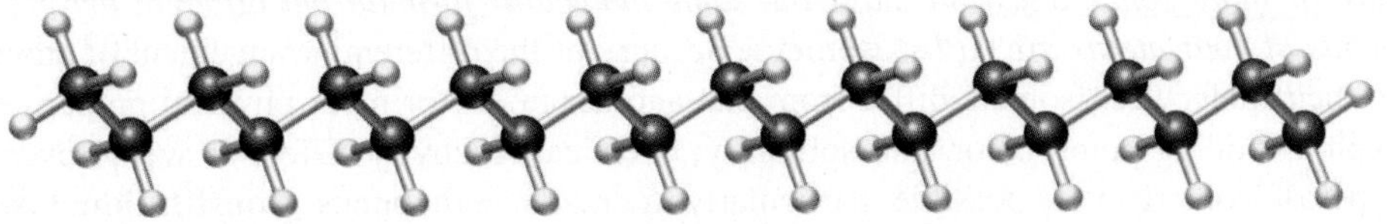

$CH_3(CH_2)_{18}CH_3$
eicosane

There are also cyclic hydrocarbons in which the carbon atoms are connected in rings, for example, cyclopentane. These cyclic alkanes have the general formula C_nH_{2n}, because each carbon atom has two hydrogen atoms.

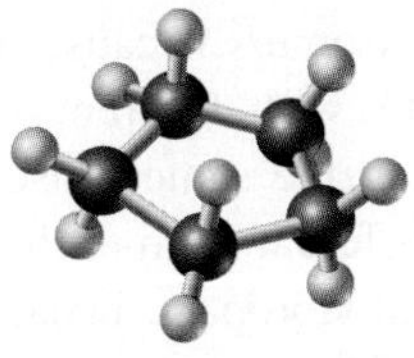

cyclopentane, C_5H_{10}

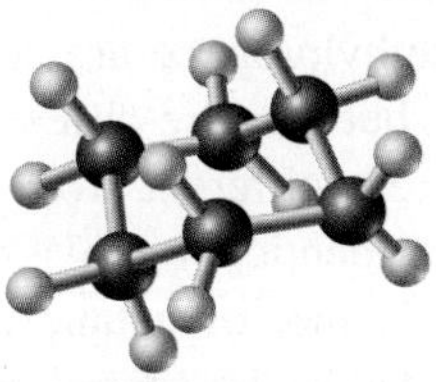

cyclohexane, C_6H_{12}

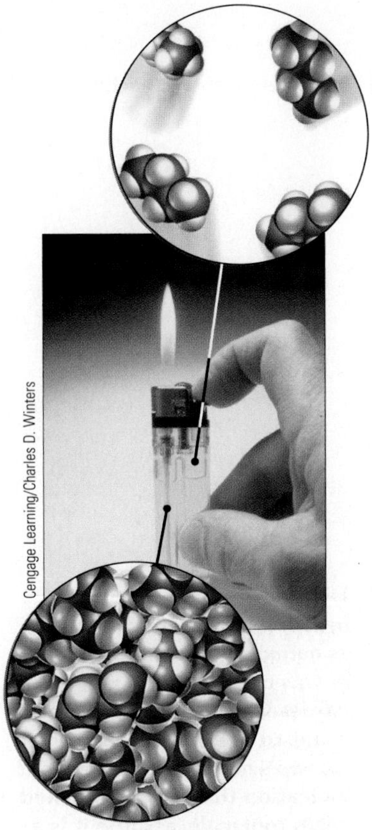
Cengage Learning/Charles D. Winters

Butane, $CH_3CH_2CH_2CH_3$, is the fuel in this lighter. Butane is present in the liquid and gaseous states in the lighter because gases can be liquefied at higher pressures and the inside of the lighter is pressurized.

EXERCISE 2.11

Alkane Molecular Formulas

(a) Write the structural formulas for the alkanes containing 16 and 28 carbon atoms.
(b) How many hydrogen atoms are present in tetradecane, which has 14 carbon atoms?
(c) Verify that each of these formulas corresponds to the general formula for noncyclic alkanes.

The molecular structures of hydrocarbons provide the framework for the discussion of the structures of all other organic compounds. If a different atom or combination of atoms replaces one or more of the hydrogen atoms in the molecular structure of an alkane, a compound with different properties results. A hydrogen atom in an alkane structure can be replaced by a single atom such as a halogen, for example. In this way the structure of ethane, CH_3CH_3, becomes the structure of chloroethane, CH_3CH_2Cl. The replacement can also be a group of atoms such as an oxygen bonded to a hydrogen, —OH, so the structure of ethane, CH_3CH_3, can be changed to the structure of ethanol, CH_3CH_2OH.

As is true of other types of compounds, molecular structures determine the properties of organic compounds. For example, the boiling point of ethane, CH_3CH_3, is −88.6 °C, but the boiling point of ethanol, CH_3CH_2OH, where an —OH group substitutes for a hydrogen atom, is 78.5 °C—quite a difference.

EXERCISE 2.12

Alkane Boiling Points

Table 2.8 gives the boiling points for the first ten noncyclic alkane hydrocarbons.
(a) Describe the general trend in boiling point as the number of carbon atoms increases.
(b) Is the change in boiling point constant from one noncyclic alkane to the next?
(c) Propose an explanation for the manner in which the boiling point changes and for the manner in which the change in boiling point behaves.

2-9a Isomers

Two or more compounds that have the same molecular formula but different arrangements of their atoms are called **isomers**. Because of the different arrangement of atoms in their molecules, isomers differ from one another in one or more physical properties, such as boiling point, color, and solubility; chemical reactivity differs as well. Several types of isomerism are possible, particularly in organic compounds. **Constitutional isomers** (also called **structural isomers**) *are compounds with the same molecular formula that differ in the order in which their atoms are bonded.*

The first three alkanes—methane, ethane, and propane—have only one possible structural arrangement. When we come to an alkane with four carbon atoms, C_4H_{10}, there are two possible arrangements—a *straight* chain of four carbons (butane) or a *branched* chain of three carbons with the fourth carbon attached to the central atom of the chain of three (methylpropane), as shown in Table 2.9.

Butane and methylpropane are constitutional isomers because they have the same molecular formula, but they are different compounds with different properties. Two constitutional isomers are different from each other in the same sense that two different structures built with identical Lego™ blocks are different from each other.

Historically, straight-chain hydrocarbons were referred to as *normal* hydrocarbons, and *n-* was used as a prefix in their names. The current practice is not to use *n-*. If a hydrocarbon's name is given without indication that it is a branched-chain molecule, assume it is a straight-chain hydrocarbon.

The branched isomer of butane is called methylpropane because it has a *methyl* group, —CH_3, bonded to the central carbon atom. The name methyl comes from methane, CH_4. A methyl group is the fragment of a methane molecule that remains when a hydrogen atom is removed:

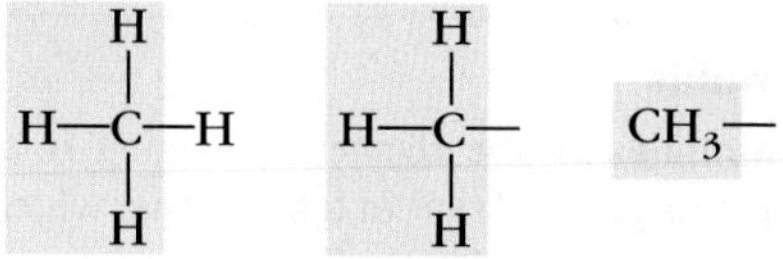

methyl group

Table 2.9 Isomers of C_4H_{10}

	Butane	Methylpropane
Molecular Formula	C_4H_{10}	C_4H_{10}
Condensed Formula	$CH_3CH_2CH_2CH_3$	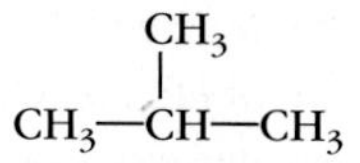
Structural Formula	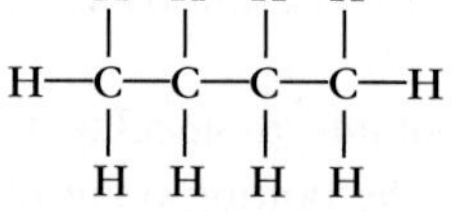	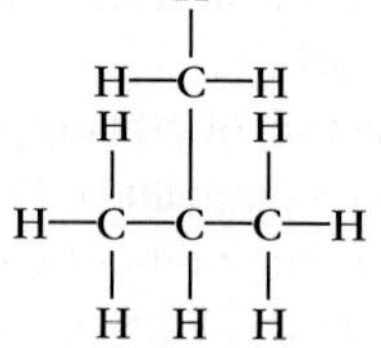
Molecular Model	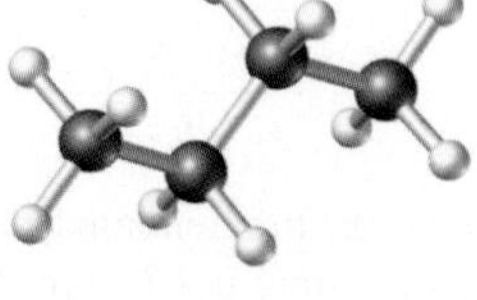	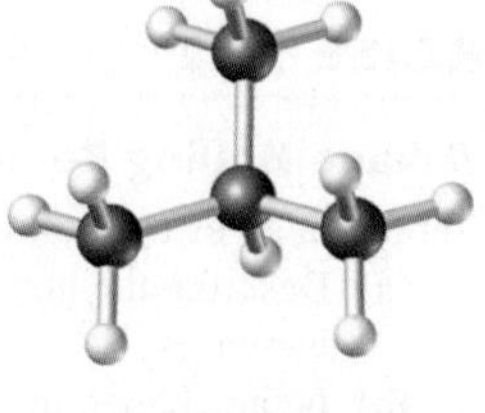
Melting Point (°C)	−138	−159.6
Boiling Point (°C)	−0.5	−11.6

For noncyclic alkanes with five carbon atoms there are three structures. One is a straight chain. The second is a branched chain with four carbon atoms in a row and a methyl group on the second carbon atom down the chain. The third isomer has three carbon atoms in a row and two methyl groups, both on the second carbon atom.

Table 2.10 Alkane Isomers	
Molecular Formula	Number of Isomers
C_4H_{10}	2
C_5H_{12}	3
C_6H_{14}	5
C_7H_{16}	9
C_8H_{18}	18
C_9H_{20}	35
$C_{10}H_{22}$	75
$C_{12}H_{26}$	355
$C_{15}H_{32}$	4347
$C_{20}H_{42}$	366,319
$C_{30}H_{62}$	4,111,846,763
$C_{40}H_{82}$	62,491,178,805,831

CONCEPTUAL EXERCISE 2.13

Pentane Isomers

Draw structural formulas and condensed formulas for the three constitutional isomers of an alkane with five carbon atoms.

Table 2.10 shows the number of isomers for some alkanes. The number of alkane constitutional isomers grows rapidly as the number of carbon atoms increases because of the possibility of chain branching (Estimation box). Chain branching is another reason for the enormous number of organic compounds.

ESTIMATION | Number of Alkane Isomers

The number of possible carbon compounds is truly enormous. Table 2.10 shows how the number of isomers of the simplest hydrocarbon compounds, alkanes, increases as the number of carbon atoms increases. How could we use these data to estimate the number of alkane isomers for a much larger number of carbon atoms? More specifically, let's estimate the number of alkane isomers for C_{40} and check the result against the last entry in the table.

To picture the growth rate, we could plot the number of alkane isomers versus the number of carbon atoms. If we made such a linear plot—that is, with the *x*-axis as the number of carbon atoms and the *y*-axis as the number of alkane isomers—we would see very little, because the plot would be rising so fast. To keep the final point on the plot, the *y*-axis would be so expanded that all the other points would be squashed toward the bottom of the plot. Therefore, to make these data easier to view, we plot the logarithm of the number of isomers, $\log(N_i)$, versus the number of carbon atoms (see the figure). The points lie on a slightly concave-upward curve, but a line fitted through them would be reasonably close to a straight line.

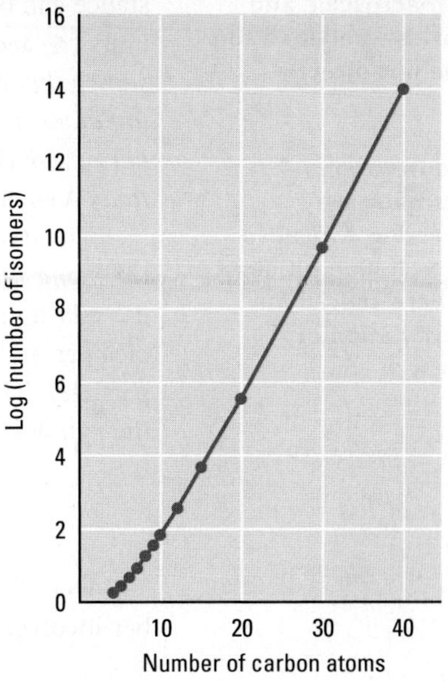

Log (number of isomers)
0
2
4
6
8
10
12
14
16
10
20
30
40
Number of carbon atoms

Semilog plot of the number of isomers versus the number of carbon atoms for alkanes.

Now we are ready to make our estimate. To estimate how many isomers there are for C_{40}, we will extrapolate from the C_{20} and C_{30} points. The $\log(N_i)$ for C_{20} is 5.56, and the $\log(N_i)$ for C_{30} is 9.61; the difference is $9.61 - 5.56 = 4.05$. Our estimate of the $\log(N_i)$ at C_{40} will be this increment added to the value of the $\log(N_i)$ for C_{30}:

$$\begin{aligned}\log(N_i) \text{ for } C_{40} &= [\log(N_i) \text{ for } C_{30}] + \text{increment}\\ &= 9.61 + 4.05 = 13.66\end{aligned}$$

To calculate the number of isomers at C_{40}, we take the antilog(13.66) $= 4.57 \times 10^{13}$. The actual number of alkane isomers for C_{40} is given in Table 2.10 as 6.25×10^{13}. We assumed the graph is straight but actually it is curving upward. Consequently the estimate should be lower than the actual value. Our estimate is well within what is called an *order of magnitude*—it gives the correct power of ten.

CONCEPTUAL **EXERCISE 2.14**

Hexane Isomers

According to Table 2.10, five constitutional isomers are possible for alkanes with six carbon atoms. Write structural and condensed formulas for these isomers.

2-10 Amount of Substance: The Mole

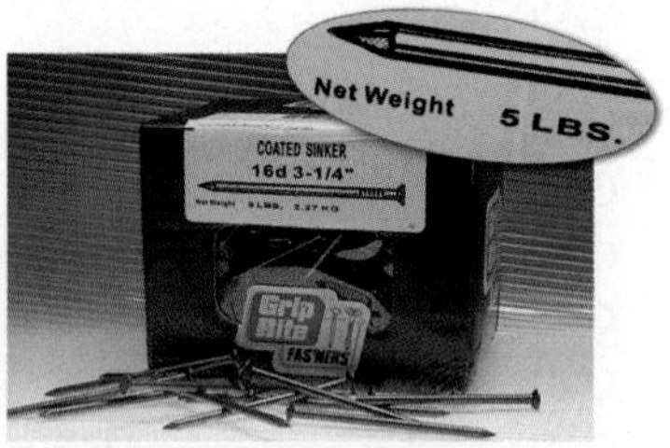
Net Weight 5 LBS.
COATED SINKER
16d 3-1/4"
Grip Rite
FASTENERS

Cengage Learning/Charles D. Winters

Items can be counted by weighing. Knowing how many nails there are in a pound tells us whether this 5-lb box contains enough nails for a project.

At least one website is available to tell how many nails of different kinds there are in a pound. See http://diyhousetips.com/206/nails-per-pound/.

The mole is the connection between the macroscale and nanoscale worlds—between the visible and the not directly visible.

Avogadro's constant and Avogadro's number are named for Amadeo Avogadro (1776–1856), an Italian physicist whose ideas were crucial for correct determination of atomic weights.

The notation mol^{-1} is equivalent to $\frac{1}{mol}$.

Although Avogadro's constant is known to nine significant figures, we will most often use it rounded to four significant figures, $6.022 \times 10^{23}\ mol^{-1}$.

The essence of chemistry is that it interprets macroscale events, such as chemical reactions, in nanoscale terms involving atoms and molecules. Consequently it is important to be able to determine the amounts of atoms or molecules that we are dealing with. As noted earlier, atoms are much too small to be seen directly or weighed individually—even on the most sensitive laboratory balance. However, it is possible to weigh a very large number of atoms and use the mass to figure out how many atoms or molecules we have.

Counting atoms by weighing is similar to buying nails at a hardware store. Nails are packaged by the pound. If we know how many nails of a given kind there are in a pound, then we can figure out how many pounds to purchase to complete a construction project. That is, the number of individual nails is proportional to the mass of nails, and if we need 1000 nails, we can use the proportionality factor to figure out how many pounds of nails to purchase. We can count nails by weighing them, and we can count atoms the same way.

Chemists have defined *a quantity called* **amount of substance** that *is proportional to the number of atoms, molecules, or other particles* and a unit by which amount of substance can be measured. This chemical counting unit is one of the seven fundamental SI units (← **Sec. 1-5a**); it is called the **mole** (symbol **mol**). *One mole is the amount of substance that contains as many atoms, molecules, ions, or other nanoscale entities as there are atoms in* exactly *12 g of carbon-12.* The best experimental determination of the *number of atoms in exactly 12 g of carbon-12 is $6.02214129 \times 10^{23}$. This number is referred to as* **Avogadro's number**.

The essential point to understand about the mole is that ***one mole always contains the same number of particles***. No matter what substance or what kind of particles we are talking about, there are always $6.02214129 \times 10^{23}$ per mole. The number of particles per mole is a proportionality factor that relates amount of substance and number of particles. The *number of particles per mole* is called the **Avogadro constant** and given the symbol N_A.

$$\text{Avogadro constant} = N_A = \frac{6.022141 \times 10^{23}}{\text{mol}} = 6.022141 \times 10^{23}\ \text{mol}^{-1}$$

One difficulty in comprehending Avogadro's constant is the sheer size of the number involved. Writing it out fully yields

$$6.02214129 \times 10^{23} = 602{,}214{,}129{,}000{,}000{,}000{,}000{,}000$$

or $602{,}214.129 \times 1$ million $\times$ 1 million $\times$ 1 million. If you poured Avogadro's number of marshmallows over the continental United States, the marshmallows would cover the country to a depth of approximately 650 miles. Or, if one mole of pennies were divided evenly among every man, woman, and child in the United States, your share alone would more than pay off the national debt (about \$17 trillion, or $\$17 \times 10^{12}$). It is helpful to think of the mole as a counting unit, analogous to the counting units we use for ordinary items such as eggs by the dozen, pencils by the gross (144), or sheets of paper by the ream (500 sheets). Atoms, molecules, and other particles in chemistry are counted by the mole.

2-11 Molar Mass

The mass of 1000 large roofing nails is greater than the mass of 1000 small finishing nails. The same applies to atoms: different kinds of atoms have different masses, so the mass of a mole of hydrogen atoms is different from the mass of a mole of carbon atoms. *The* **molar mass** *of any substance is the mass of one mole of that substance.* The relative mass of each kind of atom, its atomic weight, tells us what mass is needed to make a mole of that kind of atom. The atomic weights listed in a periodic table are numerically the same as the molar mass expressed in units of g/mol. For example, the atomic weight of copper in the periodic table inside the front cover of this book is 63.546; the molar mass of copper is therefore 63.546 g/mol.

Each of the six samples of elements shown in Figure 2.11 contains one mole of atoms. For each element in the figure, the mass in grams (the macroscale) is numerically equal to the atomic weight in atomic mass units (the nanoscale). For example,

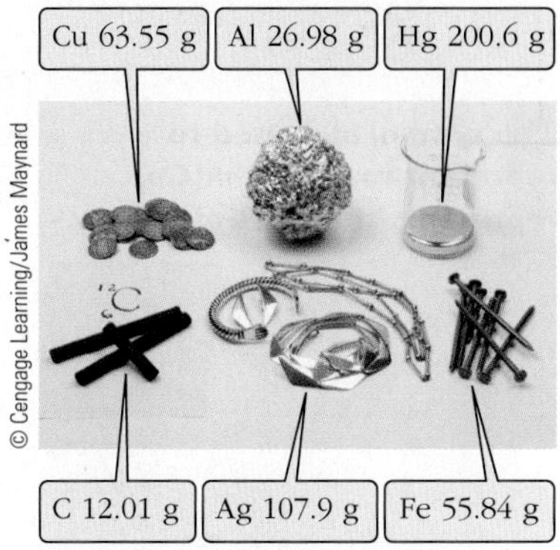
Cu 63.55 g
Al 26.98 g
Hg 200.6 g
C 12.01 g
Ag 107.9 g
Fe 55.84 g
© Cengage Learning/James Maynard

Figure 2.11 Each one-mole sample of an element has a different mass.

Quantity	Cu	Al
Atomic weight in periodic table	63.546	26.9815
Mass of 1 atom	63.546 u	26.9815 u
Mass of 1 mol atoms	63.546 g	26.9815 g
Molar mass	63.546 g/mol	26.9815 g/mol

Each molar mass of copper or aluminum contains Avogadro's number of atoms. Think of a mole as analogous to a dozen. We could have a dozen golf balls, a dozen baseballs, or a dozen bowling balls, 12 items in each case. Each dozen does not weigh the same, however, because the individual items do not weigh the same: 45 g per golf ball, 146 g per baseball, and 7200 g per bowling ball. *Molar mass differs from one element to the next because the atoms of different elements have different masses.*

Understanding the idea of a mole and applying it properly are *essential* to doing quantitative work in chemistry and in many other sciences. In particular, it is very important to be able to calculate the amount of substance if the mass is known or to calculate the mass if the amount of substance is known. Each of these quantities is proportional to the other, and the proportionality factor is the molar mass or its reciprocal. In effect there are two conversions of units:

$$\text{grams} \longrightarrow \text{moles} \qquad \text{and} \qquad \text{moles} \longrightarrow \text{grams}.$$

To do these and many other calculations in chemistry, it is helpful to use dimensional analysis (← **Sec. 1-5c**). Include the units for each quantity in a calculation and make certain that the units cancel. This ensures that the result will have appropriate units.

Let's see how these concepts apply to calculating mass from amount or amount from mass. *The proportionality factor is provided by the molar mass of the substance, units g/mol, or by the reciprocal of the molar mass, units mol/g.*

Mass ⇄ Amount conversions for substance A

Mass A ⟶ amount A

$$\text{Grams A} \times \underbrace{\frac{1 \text{ mol A}}{\text{grams A}}}_{\frac{1}{\text{molar mass}}} = \text{moles A}$$

Amount A ⟶ mass A

$$\text{Moles A} \times \underbrace{\frac{\text{grams A}}{1 \text{ mol A}}}_{\text{molar mass}} = \text{grams A}$$

Suppose you need 0.250 mol Cu for an experiment. How many grams of Cu should you use? The atomic weight of Cu in the periodic table is 63.546, so the molar mass of

Cu is 63.546 g/mol. To calculate the mass of 0.250 mol Cu, you need the proportionality factor 63.546 g Cu/1 mol Cu.

The symbol *m* is used to represent mass and *m*(Cu) represents the mass of copper.

$$\text{Mass of Cu} = m(\text{Cu}) = 0.250 \text{ mol Cu} \times \frac{63.55 \text{ g Cu}}{1 \text{ mol Cu}} = 15.9 \text{ g Cu}$$

In this book we will, when possible, *use one more significant figure in the molar mass than in any of the other data in the problem.* In the problem just completed, note that we used four significant figures in the molar mass of Cu when three were given in the amount of Cu (0.250 mol Cu). Using one more significant figure in the molar mass guarantees that its precision is greater than that of the other numbers so the molar mass does not limit the precision of the result.

In the laboratory we often know the mass of a substance and want to know how many moles we have. An example is calculating the amount of bromine in 10.00 g bromine. Bromine is one of the elements that consists of diatomic molecules, Br_2 molecules. Therefore, there are 2 mol Br atoms in 1 mol Br_2 molecules. We need to double the atomic weight of Br to obtain the molar mass of Br_2: 2×79.904 g/mol = 159.81 g/mol. To calculate the amount of bromine molecules in 10.00 g Br_2, use the reciprocal of the molar mass of Br_2 as the proportionality factor, 1 mol Br_2/159.81 g Br_2.

The symbol *n* is used to represent amount of substance and *n*(Br_2) represents the amount of bromine.

$$\text{Amount of Br}_2 = n(\text{Br}_2) = 10.00 \text{ g Br}_2 \times \frac{1 \text{ mol Br}_2}{159.81 \text{ g Br}_2} = 6.257 \times 10^{-2} \text{ mol Br}_2$$

PROBLEM-SOLVING EXAMPLE 2.13

Mass and Amount

Iodine reacts with aluminum as shown here. You have a 10.00-g sample of aluminum and you want to react it with iodine so that you have the same number of iodine atoms as aluminum atoms in the reactants. Calculate the mass of iodine that you would need for the reaction.

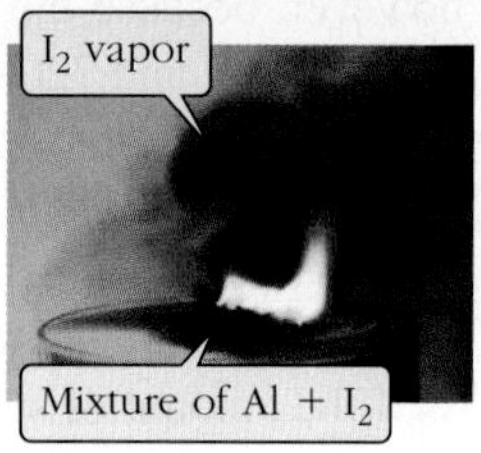
I_2 vapor
Mixture of Al + I_2

© Cengage Learning/James Maynard

Result 4.703×10^1 g

Analyze You know the mass of aluminum and you want to calculate a mass of iodine. The sample of iodine must contain the same number of I atoms as the aluminum sample contains Al atoms. A mole of atoms of any kind contains the same number of atoms.

Plan Calculate the amount of Al atoms. Then calculate the mass of the same amount of I atoms. Use molar masses as proportionality factors.

Execute

$$n(\text{Al}) = 10.00 \text{ g Al} \times \frac{1 \text{ mol Al}}{26.981 \text{ g Al}} = 3.7063 \times 10^{-1} \text{ mol Al}$$

The amount of I atoms is equal to the amount of Al atoms, so

$$m(\text{I}) = 3.7063 \times 10^{-1} \text{ mol I} \times \frac{126.90 \text{ g I}}{1 \text{ mol I}} = 4.703 \times 10^{1} \text{ g I}$$

☑ **Reasonable Result Check** The atomic weight of I rounds to 127 (about 125); the atomic weight of Al rounds to 27 (about 25). 125/25 = 5, so an I atom is about five times as massive as an Al atom. The mass of I should be about five times the mass of Al and 47 is about 5×10, so the result is reasonable.

PROBLEM-SOLVING PRACTICE 2.13

Calculate (a) the amount of Ti atoms in 4.00 g titanium and (b) the mass of 3.00×10^{-2} mol silver (Ag).

CONCEPTUAL EXERCISE 2.15

Molar Mass

In Problem-Solving Example 2.13 you calculated the mass of iodine from the amount of I atoms. Iodine is an element that consists of diatomic molecules, I_2. Why is this fact not used in the solution of Problem-Solving Example 2.13?

EXERCISE 2.16

Grams, Moles, and the Avogadro Constant

You have a 10.00-g sample of lithium and a 10.00-g sample of iridium. How many atoms are in each sample, and how many more atoms are in the lithium sample than in the iridium sample?

CONCEPTUAL EXERCISE 2.17

Mass and Number of Particles

Samples of Al, P_4, and Cl_2 have the same mass. Which sample contains the largest number of atoms? Which contains the smallest number of atoms? Explain.

2-11a Molar Masses of Molecular Compounds

The familiar molecular formula, H_2O, shows us that there are two H atoms for every O atom in a water molecule (← **Sec. 2-7b**). In two water molecules, therefore, there are four H atoms and two O atoms; in a dozen water molecules, there are two dozen H atoms and one dozen O atoms. We can extend this until we have one mole of water molecules (Avogadro's number of molecules, 6.022×10^{23}), which contains two moles of hydrogen atoms and one mole of oxygen atoms. We can also say that in 1.000 mol water molecules there are 2.000 mol H atoms and 1.000 mol O atoms:

H_2O	H	O
6.022×10^{23} H_2O molecules	2 (6.022×10^{23} H atoms)	6.022×10^{23} O atoms
1.000 mol H_2O molecules	2.000 mol H atoms	1.000 mol O atoms
18.0152 g H_2O	2 (1.0079 g H) = 2.0158 g H	15.9994 g O

The mass of one mole of water molecules—the *molar mass*—is the mass of two moles of H atoms plus the mass of one mole of O atoms: 2.0158 g + 15.9994 g = 18.0152 g per mole of water. For chemical compounds, the *molar mass,* in grams per mole, is *numerically the same* as the **molecular weight**, *the sum of the atomic weights of all the atoms in the compound's formula*. Molar masses of several molecular compounds are shown in Table 2.11.

Table 2.11 Molar Masses of Some Molecular Compounds.

Compound	Structural Formula	Atomic Masses, Molecular Mass	Molar Mass
Ammonia NH_3	H—N(—H)—H	N: 14.01 u; H: 1.01 u 14.01 u + 3 (1.01 u) = 17.04 u	17.04 g/mol
Sulfur dioxide SO_2	O=S—O	S: 32.07 u; O: 16.00 u 32.07 u + 2 (16.00 u) = 64.07 u	64.07 g/mol
Glycerol $C_3H_8O_3$	CH_2OH—CHOH—CH_2OH	C: 12.01 u; H: 1.01 u; O: 16.00 u 3 (12.01 u) + 8 (1.01 u) + 3 (16.00 u) = 92.11 u	92.11 g/mol

Table 2.12 Molar Masses of Some Ionic Compounds.

Compound	Atomic Masses, Formula Mass	Molar Mass
NaCl, Sodium chloride	Na: 23.00 u; Cl: 35.45 u 23.00 u + 35.45 u = 58.45 u	58.45 g/mol
K_2S, Potassium sulfide	K: 39.10 u; S: 32.06 u 2 (39.10 u) + 32.06 u = 110.26 u	110.26 g/mol
$MgBr_2$, Magnesium bromide	Mg: 24.30 u; Br: 79.90 u 24.30 u + 2 (79.90 u) = 184.10 u	184.10 g/mol
$Ca_3(PO_4)_2$, Calcium phosphate	Ca: 40.08 u; P: 30.97 u; O: 16.00 u 3 (40.08) + 2 (30.97) + 8 (16.00) = 310.18 u	310.18 g/mol

Notice that $Ca_3(PO_4)_2$ has 2 P atoms and 2 × 4 = 8 O atoms because there are two PO_4^{3-} ions in the formula.

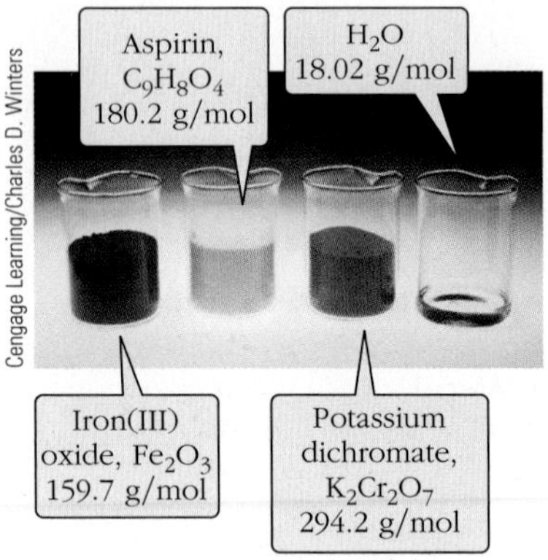
Aspirin, $C_9H_8O_4$ 180.2 g/mol
H_2O 18.02 g/mol
Iron(III) oxide, Fe_2O_3 159.7 g/mol
Potassium dichromate, $K_2Cr_2O_7$ 294.2 g/mol
Cengage Learning/Charles D. Winters

One-mole quantities of four compounds; two compounds are ionic and two are molecular.

2-11b Molar Masses of Ionic Compounds

Because ionic compounds do not contain individual molecules, the term "formula weight" is often used instead of "molecular weight," but the molar mass is calculated the same way. An ionic compound's **formula weight** *is the sum of the atomic weights of all the atoms in the compound's formula*. The molar mass of an ionic compound, expressed in grams per mole (g/mol), is numerically equivalent to its formula weight (Table 2.12). The term "molar mass" is used for both molecular and ionic compounds and we will use molar mass exclusively from now on.

EXERCISE 2.18

Molar Masses

Calculate the molar mass of each compound:

(a) K_2HPO_4
(b) $C_{27}H_{46}O$ (cholesterol)
(c) $Mn_2(SO_4)_3$
(d) $C_8H_{10}N_4O_2$ (caffeine)

2-11c Ionic Hydrates

An ionic compound that has water molecules trapped among the ions in its crystal lattice is known as an **ionic hydrate** or just a hydrate. *The associated water is called the* **water of hydration**. For example, the beautiful deep-blue compound named copper(II) sulfate pentahydrate, $CuSO_4 \cdot 5H_2O$, has five moles of water trapped in its crystal lattice per mole of copper(II) sulfate. The "$\cdot 5H_2O$" and the term "pentahydrate" indicate five moles of water associated with every mole of copper(II) sulfate. *The molar mass of a hydrate includes the mass of the water of hydration.* Thus, the molar mass of $CuSO_4 \cdot 5H_2O$ is 249.7 g/mol: the molar mass of $CuSO_4$ plus five times the molar mass of water; that is,

White $CuSO_4$
Blue $CuSO_4 \cdot 5H_2O$
Cengage Learning/Charles D. Winters

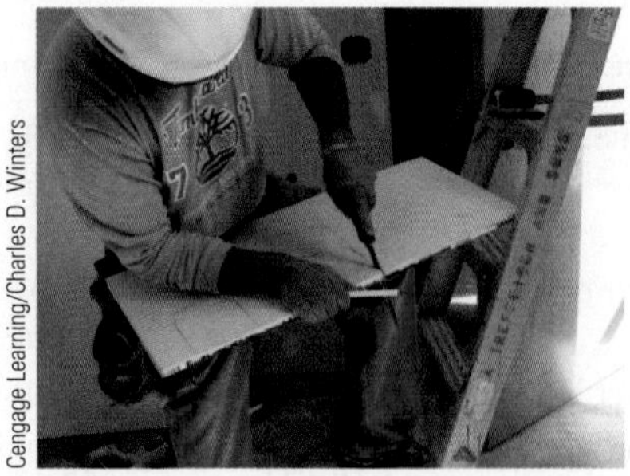
Cengage Learning/Charles D. Winters

Ionic hydrates. Heating $CuSO_4 \cdot 5H_2O$ drives off water of hydration *(left)*. Gypsum in wallboard is hydrated calcium sulfate, $CaSO_4 \cdot 2H_2O$ *(right)*.

Table 2.13 Some Common Hydrated Ionic Compounds

Formula	Systematic Name	Common Name	Uses
$Na_2CO_3 \cdot 10H_2O$	Sodium carbonate decahydrate	Washing soda	Water softener
$MgSO_4 \cdot 7H_2O$	Magnesium sulfate heptahydrate	Epsom salt	Dyeing and tanning
$CaSO_4 \cdot 2H_2O$	Calcium sulfate dihydrate	Gypsum	Wallboard
$CaSO_4 \cdot \frac{1}{2}H_2O$	Calcium sulfate hemihydrate	Plaster of Paris	Casts, molds

159.6 g/mol + 90.1 g/mol = 249.7 g/mol. There are many ionic hydrates, including the frequently encountered ones listed in Table 2.13.

One commonly used hydrate may well be in the walls of your room. Plasterboard (sometimes called wallboard, sheetrock, or gypsum board) contains hydrated calcium sulfate, or gypsum, $CaSO_4 \cdot 2H_2O$, sandwiched between two thicknesses of paper. Gypsum is a natural mineral that can be mined. It is also formed when sulfur dioxide is removed from electric power plant exhaust gases by reacting the SO_2 with an aqueous slurry of lime, calcium oxide.

Heating gypsum to 180 °C drives off some of the water of hydration to form calcium sulfate hemihydrate, $CaSO_4 \cdot \frac{1}{2}H_2O$, commonly called *Plaster of Paris.* This compound is widely used in casts for broken limbs. When water is added to it, it forms a thick slurry that can be poured into a mold or spread over a part of the body. As the slurry hardens, it takes on additional water of hydration and its volume increases, forming a rigid protective cast.

Calcium sulfate *hemi*hydrate contains one water molecule per two $CaSO_4$ units. The prefix *hemi-* refers to $\frac{1}{2}$ as in the familiar word "hemisphere."

EXERCISE 2.19

Amount of an Ionic Hydrate

A home remedy calls for 2 teaspoons (20. g) Epsom salt (formula is in Table 2.13).

(a) Calculate the molar mass of Epsom salt.
(b) Calculate the amount of the hydrate in 20. g Epsom salt.
(c) Calculate the number of O atoms in 20. g Epsom salt.

2-11d Gram-Mole Calculations

As was true for elements (← **Sec. 2-11**), the molar mass of a compound (or its reciprocal) is a proportionality factor that relates amount of substance to mass (or mass to amount of substance). Because we know that mass is proportional to amount and amount is proportional to mass, we can use molar mass to calculate one from the other. The next two examples show how this is done.

PROBLEM-SOLVING EXAMPLE 2.14

Amount from Mass

Ammonium carbonate, $(NH_4)_2CO_3$, produces ammonia gas when heated. Calculate the amount of ammonium carbonate in 20.0 g of this ionic compound. Express the result in moles.

Result 0.208 mol $(NH_4)_2CO_3$

Analyze We know the mass and are asked to calculate the amount. These two quantities are proportional and the proportionality factor is molar mass.

Plan Calculate the molar mass. Then use it to calculate amount. Make certain the units cancel.

Execute Molar mass = (2 × 14.0067 g N/mol N) + (2 × 4 × 1.0079 g H/mol H) + (1 × 12.0107 g C/mol C) + (3 × 15.9994 g O/mol O) = 96.09 g $(NH_4)_2CO_3$/mol $(NH_4)_2CO_3$

$$\text{Amount} = 20.0 \text{ g } (NH_4)_2CO_3 \times \frac{1 \text{ mol } (NH_4)_2CO_3}{96.09 \text{ g } (NH_4)_2CO_3} = 0.208 \text{ mol } (NH_4)_2CO_3$$

☑ **Reasonable Result Check** The molar mass of $(NH_4)_2CO_3$ is about 100 g/mol, so 20 g is about 0.2 mol, which is close to the result of our exact calculation.

PROBLEM-SOLVING PRACTICE 2.14

Calculate the amount of substance in 12.0 g of each of these ionic compounds:
(a) $Ca(NO_3)_2$ (b) $KMnO_4$ (c) $NiCl_2 \cdot 6H_2O$

PROBLEM-SOLVING EXAMPLE 2.15

Mass from Amount

Cortisone, $C_{21}H_{28}O_5$, is an anti-inflammatory steroid. Calculate the mass of 5.00×10^{-3} mol cortisone.

Result 1.80 g

Analyze You are given an amount and asked to calculate a mass. Molar mass relates mass to amount.

Plan Calculate the molar mass and then use it to calculate the mass from the amount.

Execute Molar mass = (21 × 12.0107 g C/mol C) + (28 × 1.0079 g H/mol H) + (5 mol O × 15.9994 g O/mol O) = 360.4 g cortisone/mol cortisone

$$\text{Mass} = 5.00 \times 10^{-3} \text{ mol cortisone} \times \frac{360.4 \text{ g cortisone}}{1 \text{ mol cortisone}} = 1.80 \text{ g cortisone}$$

☑ **Reasonable Result Check** The amount of cortisone, 5.00×10^{-3} mol, is $\frac{1}{200}$ mol, and this fraction of the molar mass is about (360 g/mol)/200 mol = 1.8 g. The result is reasonable.

PROBLEM-SOLVING PRACTICE 2.15

(a) Calculate the mass of 5.00×10^{-3} mol sucrose, $C_{12}H_{22}O_{11}$.
(b) Calculate the mass of 3.00×10^{-6} mol adrenocorticotropic hormone (ACTH), which has a molar mass of approximately 4600 g/mol.

Once we have calculated how many moles of an element or compound we have, it is also possible to calculate the number of atoms or molecules that are present. The proportionality factor relating amount of substance and number of atoms or molecules is the Avogadro constant, 6.022×10^{23}/mol.

PROBLEM-SOLVING EXAMPLE 2.16

Mass and Number of Atoms

Sucralose, an artificial sweetener (Splenda®), $C_{12}H_{19}Cl_3O_8$, is about 600 times sweeter than sucrose. You have a sample of sucralose with a mass of 10.0 g. Calculate the number of carbon atoms in the sample.

Result 1.82×10^{23} C atoms

Analyze You are given the mass of the sample. You know the molecular formula of sucralose, which gives the number of carbon atoms per molecule. The number of molecules is related to the amount by the Avogadro constant. The amount is related to the mass by the molar mass.

Plan Calculate the molar mass of sucralose. Use the molar mass to calculate the amount of sucralose from the mass. Use the Avogadro constant to calculate the number of sucralose molecules from the amount. Multiply that number by 12 because the formula indicates 12 C atoms per molecule of sucralose. Include units in each calculation and make certain the units cancel.

Execute Molar mass = $(12 \times 12.011 + 19 \times 1.008 + 3 \times 35.453 + 8 \times 15.999)$ g/mol = 397.64 g/mol

$$\text{Amount of sucralose} = 10.0 \text{ g sucralose} \times \frac{1 \text{ mol sucralose}}{397.64 \text{ g sucralose}} = 0.02515 \text{ mol sucralose}$$

$$\text{Number of molecules} = 0.02515 \text{ mol} \times \frac{6.022 \times 10^{23} \text{ molecules}}{1 \text{ mol}} = 1.514 \times 10^{22} \text{ molecules}$$

$$\text{Number of C atoms} = 1.514 \times 10^{22} \text{ molecules} \times \frac{12 \text{ C atoms}}{1 \text{ molecule}} = 1.82 \times 10^{23} \text{ C atoms}$$

Notice that the three calculations that involve proportionalities could all have been done in a single step:

$$N_{\text{C atoms}} = 10.0 \text{ g} \times \frac{1 \text{ mol sucralose}}{397.64 \text{ g sucralose}} \times \frac{6.022 \times 10^{23} \text{ molecules}}{1 \text{ mol}} \times \frac{12 \text{ C atoms}}{1 \text{ molecule}}$$

$$= 1.82 \times 10^{23} \text{ C atoms}$$

Once you have planned your approach to a problem, you can set up the entire calculation first and then use your calculator to do the arithmetic all at once. This saves time and usually is more accurate, so it is recommended.

☑ **Reasonable Result Check** Because the molar mass of sucralose is nearly 400 g/mol, 10.0 g sucralose is about 10/400 = 1/40 or 0.025 mol, so the calculation of amount of sucralose is reasonable. The number of molecules should be about 1/40 times the Avogadro constant, and it is. The number of carbon atoms should be 12 times about 1.5×10^{22}, which is 18×10^{22}, which is 1.8×10^{23}, so the number of carbon atoms is reasonable.

PROBLEM-SOLVING PRACTICE 2.16

(a) Calculate the number of O atoms in a 12.0-g sample of each compound: aspirin, $C_9H_8O_4$; $Ca_3(PO_4)_2$.
(b) Calculate the mass of each compound that contains 6.022×10^{23} O atoms: $Mn_2(SO_4)_3$; caffeine, $C_8H_{10}N_4O_2$.

CONCEPTUAL EXERCISE 2.20

Amount of Substance and Formulas

Is this statement true? "Two different compounds have the same formula. Therefore, 100 g of each compound contains the same amount of substance." Justify your answer.

2-12 Composition and Chemical Formulas

A chemical formula gives two kinds of information: *(1) the number of atoms of each type per molecule or formula unit;* and *(2) the amount of each element in a mole of the compound.* The latter relationship provides the information needed to find the percent composition of the compound by mass. The **percent composition by mass** *is the mass of each element divided by the total mass of compound and expressed as a percent.*

It is important to recognize that the percent composition of a compound by mass is independent of the quantity of the compound. The percent composition by mass remains the same whether a sample contains 1 mg, 1 g, or 1 kg of the compound.

PROBLEM-SOLVING EXAMPLE 2.17

Percent Composition by Mass

Propane, C_3H_8, is the fuel used in gas grills. Calculate the percentages by mass of carbon and hydrogen in propane.

Result 81.72% carbon and 18.28% hydrogen

Analyze We need to find the mass of carbon and the mass of hydrogen in a given mass of compound and then calculate percentages. The formula says there is 3 mol C and 8 mol H per 1 mol propane.

Plan Calculate the mass of 1 mol propane. Also calculate the mass of 3 mol C and the mass of 8 mol H. Then calculate percent for each element.

Execute Molar mass $(C_3H_8) = (3 \times 12.011 + 8 \times 1.0079)$ g/mol $= 44.096$ g/mol

In 44.096 g (1 mol) propane there is 3 mol C and 8 mol H.

$$\text{Mass of C} = 3 \text{ mol C} \times \frac{12.011 \text{ g C}}{1 \text{ mol C}} = 36.033 \text{ g C}$$

$$\text{Mass of H} = 8 \text{ mol H} \times \frac{1.0079 \text{ g H}}{1 \text{ mol H}} = 8.0632 \text{ g H}$$

$$\text{Percent C} = \frac{36.033 \text{ g}}{44.096 \text{ g}} \times 100\% = 81.71\%$$

$$\text{Percent H} = \frac{8.0632 \text{ g}}{44.096 \text{ g}} \times 100\% = 18.29\%$$

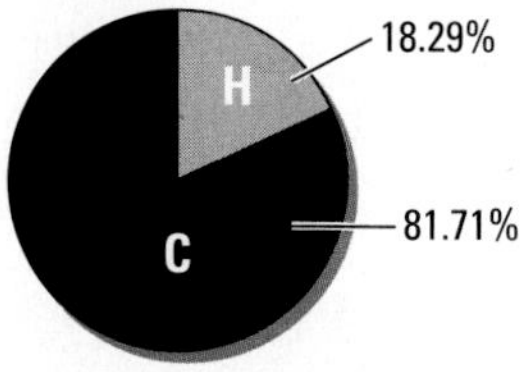
H
18.29%
C
81.71%

Mass percent carbon and hydrogen in propane, C_3H_8.

These results can also be expressed as 81.71 g C per 100.0 g C_3H_8 and 18.29 g H per 100.0 g C_3H_8.

☑ **Reasonable Result Check** Each carbon atom has 12 times the mass of a hydrogen atom. Propane has approximately $12 \times 3 = 36$ u carbon and approximately $1 \times 8 = 8$ u hydrogen. So the percentage of carbon should be about $36/8 = 4.5$ times larger than the percentage of hydrogen. This agrees with our more carefully calculated result.

PROBLEM-SOLVING PRACTICE 2.17

Calculate the percentage of each element in silicon dioxide, SiO_2.

Notice that the percentages calculated in Problem-Solving Example 2.17 add up to 100%. Therefore, once we calculated the percentage of carbon, we also could have determined the percentage of hydrogen simply by subtracting: 100% − 81.71% C = 18.29% H. Calculating all percentages and adding them to confirm that they give 100% is a good way to check for errors.

Epsom
SALT
RESEALABLE POUCH

Cengage Learning/Charles D. Winters

Epsom salt, $MgSO_4 \cdot 7H_2O$.

PROBLEM-SOLVING EXAMPLE 2.18

Percent Composition of Hydrated Salt

Calculate (a) the percent composition for Epsom salt, $MgSO_4 \cdot 7H_2O$, and (b) the percent by mass of water in Epsom salt.

Result (a) 9.86% Mg; 13.00% S; 71.40% O; 5.72% H (b) 51.16% water

Analyze The formula indicates that there is 1 mol Mg, 1 mol S, (4 + 7) mol O, (7 × 2) mol H, and 7 mol H_2O in 1 mol $MgSO_4 \cdot 7H_2O$. The percentages are the mass of each amount divided by the mass of 1 mol $MgSO_4 \cdot 7H_2O$.

Plan Calculate the molar mass of Epsom salt to obtain the mass of one mole. Calculate the mass of each element and of water in 1 mol Epsom salt. Calculate percentages.

Execute Molar mass $= 24.31 + 32.07 + (4 \times 15.9994 + 7 \times 15.9994) + 14 \times 1.0079 =$ 246.49 g/mol Epsom salt

$$\text{Mass Mg} = 1 \text{ mol Mg} \times \frac{24.3050 \text{ g}}{1 \text{ mol Mg}} = 24.3050 \text{ g}$$

$$\text{Mass S} = 1 \text{ mol S} \times \frac{32.065 \text{ g}}{1 \text{ mol S}} = 32.065 \text{ g}$$

$$\text{Mass O} = 11 \text{ mol O} \times \frac{15.9994 \text{ g}}{1 \text{ mol O}} = 175.993 \text{ g}$$

$$\text{Mass H} = 14 \text{ mol H} \times \frac{1.0079 \text{ g}}{1 \text{ mol H}} = 14.111 \text{ g}$$

$$\text{Mass H}_2\text{O} = 7 \text{ mol H}_2\text{O} \times \frac{(2 \times 1.0079 + 15.9994) \text{ g}}{1 \text{ mol H}_2\text{O}} = 126.106 \text{ g}$$

$$\text{Percent Mg} = \frac{24.3050 \text{ g}}{246.49 \text{ g}} \times 100\% = 9.8604\%$$

$$\text{Percent S} = \frac{32.065 \text{ g}}{246.49 \text{ g}} \times 100\% = 13.001\%$$

$$\text{Percent O} = \frac{175.993 \text{ g}}{246.49 \text{ g}} \times 100\% = 71.400\%$$

$$\text{Percent H} = \frac{14.111 \text{ g}}{246.49 \text{ g}} \times 100\% = 5.7248\%$$

$$\text{Percent H}_2\text{O} = \frac{126.106 \text{ g}}{246.49 \text{ g}} \times 100\% = 51.161\%$$

☑ **Reasonable Result Check** In the formula of the hydrated salt, there are seven waters with a combined mass of $7 \times 18 = 126$ g, and there are six other atoms with molar masses ranging between 16 and 32 that total to 120 g. Thus, the hydrated salt should be about 50% water by weight, and it is. There are 11 oxygen atoms in the formula, so oxygen should have the largest percent by weight, and it does. The percentages sum to 99.986% which is within rounding error of 100%.

PROBLEM-SOLVING PRACTICE 2.18

Calculate the mass percent of each element in hydrated sodium sulfate, $Na_2SO_4 \cdot 10H_2O$.

EXERCISE 2.21

Composition

Express the composition of each compound first as the mass of each element in 1.000 mol of the compound and then as the mass percent of each element:

(a) SF_6 (b) $C_{12}H_{22}O_{11}$ (c) $Al_2(SO_4)_3$ (d) $U(OTeF_5)_6$

2-12a Empirical and Molecular Formulas

Historically, chemical formulas were obtained by experimentally determining the masses of two or more elements that combined to form a compound. Such data give the percent composition by mass of a compound and we can determine the formula of a compound from mass percent data. This depends on the fact that the subscripts in a formula indicate how many moles of each element there are per mole of the compound.

Consider diborane, a compound consisting of boron and hydrogen that ignites explosively when exposed to air. Experiments show that diborane is 78.13% B and 21.87% H. Based on these percentages, a 100.0-g diborane sample contains 78.13 g B and 21.87 g H. From this information we can calculate the amount of each element in the sample:

$$\text{Amount B} = 78.13 \text{ g B} \times \frac{1 \text{ mol B}}{10.811 \text{ g B}} = 7.227 \text{ mol B}$$

$$\text{Amount H} = 21.87 \text{ g H} \times \frac{1 \text{ mol H}}{1.0079 \text{ g H}} = 21.70 \text{ mol H}$$

To determine the formula from these data, we next apply the idea that the subscripts in a formula indicate the relative number of moles of each element. For example, the formula H_2O indicates the ratio 2 mol H : 1 mol O. Thus we find the amount of each element *relative to the amount of the other element or elements*—in this case, the ratio of amount of hydrogen to amount of boron. There are about three times as many moles of H atoms as there are moles of B atoms. To calculate the ratio exactly, we divide the larger amount by the smaller amount. For diborane that ratio is

$$\frac{\text{Amount H}}{\text{Amount B}} = \frac{21.70 \text{ mol H}}{7.227 \text{ mol B}} = \frac{3.003 \text{ mol H}}{1.000 \text{ mol B}}$$

This ratio confirms that there are three moles of H atoms for every one mole of B atoms and that there are three hydrogen atoms for each boron atom. This information gives the formula BH_3.

For a molecular compound such as diborane, the molecular formula must also accurately reflect the *total number of atoms in a molecule of the compound.* The calculation we have done gives the *simplest possible ratio of atoms in the molecule,* and BH_3 is the simplest formula for diborane. *A formula that reports the simplest possible whole-number ratio of atoms in the molecule is called an* **empirical formula**. Multiples of the simplest formula that maintain the ratio 3 H : 1 B are possible, such as B_2H_6, B_3H_9, and so on.

One way to determine which of these formulas is the actual molecular formula involves *experimentally determining* the molar mass of the compound. The experimental molar mass can be compared with the molar mass predicted by the empirical formula. If the two molar masses are the same, the empirical and molecular formulas are the same. However, if the experimentally determined molar mass is some multiple of the empirical formula value, the molecular formula is that multiple of the empirical formula. In the case of diborane, experiments indicate that the molar mass is 27.67 g/mol. The empirical formula, BH_3, predicts a molar mass of 13.83 g/mol, so the molecular formula must be a multiple of the empirical formula. That multiple is 27.67/13.83 = 2.00. Thus, the molecular formula of diborane is $B_{(2\times1)}H_{(2\times3)}$, or B_2H_6.

$\frac{\text{Molecular formula molar mass}}{\text{Empirical formula molar mass}} = n$, where n is an integer. If $n = 1$, the molecular formula and the empirical formula are the same. When $n > 1$, the subscripts in the molecular formula are all n times the subscripts in the empirical formula.

PROBLEM-SOLVING EXAMPLE 2.19

Determining Formulas

An oxide of phosphorus contains 56.34% P and 43.66% O. Its experimentally determined molar mass is 219.90 g/mol. Determine the formula of this compound.

Result Empirical formula: P_2O_3; molecular formula: P_4O_6

Analyze The compound contains only nonmetals, so it is probably molecular; hence a molecular formula must be determined. You know the mass of each element per 100 g compound, so you can calculate the relative amounts of each element and determine an empirical formula. The molar mass tells what multiple of the empirical formula is needed for the molecular formula.

Plan Calculate the amount of each element. Use mole ratios to determine the empirical formula. Compare the experimental molar mass with the molar mass from the empirical formula.

Execute A 100-g sample of the compound contains 56.34 g P and 43.66 g O; thus

$$\text{Amount P} = 56.34 \text{ g P} \times \frac{1 \text{ mol P}}{30.9738 \text{ g P}} = 1.819 \text{ mol P}$$

$$\text{Amount O} = 43.66 \text{ g O} \times \frac{1 \text{ mol O}}{15.9994 \text{ g O}} = 2.729 \text{ mol O}$$

The ratio of the larger amount to the smaller amount is

$$\frac{\text{Amount O}}{\text{Amount P}} = \frac{2.729 \text{ mol O}}{1.819 \text{ mol P}} = \frac{1.500 \text{ mol O}}{1.000 \text{ mol P}}$$

The ratio is 1.5 O atoms per 1.0 P atom, but we cannot have half an atom. If both numbers in the ratio are doubled, we have 3.0 O atoms per 2.0 P atoms, which gives the empirical formula P_2O_3. The corresponding molar mass is (2 × 30.9738 + 3 × 15.9994) g/mol = 109.95 g/mol, about half the experimental molar mass:

$$\frac{\text{Experimental molar mass}}{\text{Molar mass from formula}} = \frac{219.90 \text{ g/mol}}{109.95 \text{ g/mol}} = 2.000$$

Thus the subscripts in the empirical formula P_2O_3 need to be doubled to obtain the molecular formula, which is P_4O_6.

☑ **Reasonable Result Check** The molar mass of P is nearly twice the molar mass of O. If the subscripts in the formula were equal, then we should have twice the mass of P as mass of O. If the subscript of O was twice that of P, then the masses should be equal. The actual percentages are 56% P and 44% O, so the ratio of O to P must be between 2:1 and 1:1. Our result of 1.5:1 is reasonable.

PROBLEM-SOLVING PRACTICE 2.19

Hydrazine is composed of 87.42% nitrogen and 12.58% hydrogen by mass. Its experimentally determined molar mass is 32.05 g/mol. Determine the empirical and molecular formulas of hydrazine.

It is not necessary to know percent composition to determine a chemical formula. All that is needed is a way to determine the masses of the elements that combine to form a compound. If you know the mass of each element in a sample of a compound, then you can calculate the amount of each element and from the ratios of amounts you can determine the formula.

PROBLEM-SOLVING EXAMPLE 2.20

Formula from Mass Data

A 63.47-g sample of an ionic compound contains 17.17 g sodium and 10.46 g nitrogen; the rest is oxygen. The compound's molar mass is 84.9947 g/mol. Determine its formula.

Result $NaNO_3$.

Analyze The compound is ionic, so it does not consist of molecules and does not have a molecular formula; determine the empirical formula. The mass of oxygen in the sample is not given, but it can be calculated by subtracting the other masses from the mass of the sample. Because all masses refer to the same sample, the masses can be used to calculate the amounts of each element in the sample. The empirical formula can be determined from the relative amounts of the elements compound. The molar mass is not needed to determine the formula, but it can be used to check the result.

Plan Calculate the mass of oxygen by subtraction. Then calculate the amount of each element and determine the mole ratios.

Execute

$m(\text{O}) = m(\text{sample}) - m(\text{Na}) - m(\text{N}) = 63.47 \text{ g} - 17.17 \text{ g} - 10.46 \text{ g} = 35.84 \text{ g}$

$$\text{Amount Na} = 17.17 \text{ g Na} \times \frac{1 \text{ mol Na}}{22.9898 \text{ g Na}} = 0.7469 \text{ mol Na}$$

$$\text{Amount N} = 10.46 \text{ g N} \times \frac{1 \text{ mol N}}{14.0067 \text{ g N}} = 0.7468 \text{ mol N}$$

$$\text{Amount O} = 35.84 \text{ g O} \times \frac{1 \text{ mol O}}{15.9994 \text{ g O}} = 2.240 \text{ mol O}$$

When there are three or more amounts, divide each of the others by the smallest to obtain mole ratios.

$$\frac{\text{Amount Na}}{\text{Amount N}} = \frac{0.7469 \text{ mol Na}}{0.7468 \text{ mol N}} = \frac{1.000 \text{ mol Na}}{1.000 \text{ mol N}}$$

$$\frac{\text{Amount O}}{\text{Amount N}} = \frac{2.240 \text{ mol O}}{0.7468 \text{ mol N}} = \frac{2.999 \text{ mol O}}{1.000 \text{ mol N}}$$

Na and N are in a 1 : 1 mole ratio, but the ratio O : N is 2.999. There is experimental error in the mass data, so it is reasonable to round 2.999 to 3. The formula is $NaNO_3$.

☑ **Reasonable Result Check** The molar mass is given, so calculate the molar mass from the formula $NaNO_3$.

$$\text{Molar mass} = (22.9898 + 14.0067 + 3 \times 15.9994) \text{ g/mol} = 84.9947 \text{ g/mol}$$

The result is reasonable.

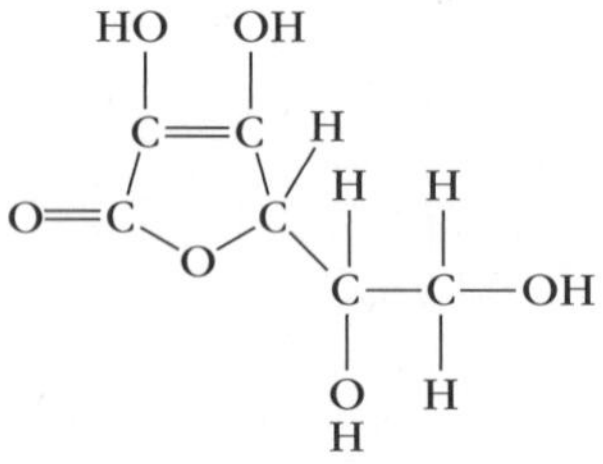

vitamin C

PROBLEM-SOLVING PRACTICE 2.20

Vitamin C (ascorbic acid) contains 40.9% C, 4.58% H, and 54.5% O and has an experimentally determined molar mass of 176.13 g/mol. Determine its empirical and molecular formulas.

EXERCISE 2.22

Formula of a Hydrate

A sample of a hydrate was heated to drive off water. After the water was gone the sample was found to contain 11.013 g Ca, 8.811 g S, and 17.59 g O. The water was collected and weighed; its mass was 9.903 g. Use this information to determine the formula of the hydrate.

SUMMARY PROBLEM

Part I

An isotope of an element contains 63 protons and 91 neutrons.

(a) Identify the element and give its symbol.

(b) Give the element's atomic number.

(c) Give the mass number of the isotope.

(d) This element has two naturally occurring isotopes. Given the information in the table, calculate the atomic weight of the element.

Isotope	Mass Number	Percent Abundance	Isotopic Mass (u)
1	151	47.80	150.920
2	153	52.20	152.921

(e) In which region of the periodic table is the element found? Explain your answer.

(f) Is the element a metal, metalloid, or nonmetal? Explain your answer.

(g) This element, used in compact fluorescent light bulbs and computer screens, has an atomic radius of 180 pm. Calculate how long the chain of atoms would be if all the atoms in a 1.25-mg sample of this element were put into a row.

Part II

Consider these three nitrogen-containing compounds: ammonium dichromate, ammonium nitrate, and trinitrotoluene. A molecule of trinitrotoluene contains seven carbon atoms, five hydrogen atoms, three nitrogen atoms, and six oxygen atoms. All three compounds are solids at room temperature and all three decompose at high temperatures.

(a) Write the correct formula for each compound.

(b) Determine which compound has the highest mass percent nitrogen.

(c) Which compound has the lowest melting point? Explain your choice.

(d) Which compounds, if any, conduct electric current at room temperature? At a much higher temperature? Explain.

(e) If they could be melted without decomposing, which compound (if any) would conduct electricity?

Part III

The mineral fluorapatite has the formula $Ca_5(PO_4)_3F$. Hypo, a substance used when photographs were developed chemically, has the formula $Na_2S_2O_3 \cdot 5H_2O$. Weddellite has the formula $CaC_2O_4 \cdot 2H_2O$ and is involved in formation of kidney stones.

(a) Identify the ions and their charges in these three compounds.

(b) Calculate the molar mass of each compound.

Part IV

Dioxathion, a pesticide, contains carbon, hydrogen, oxygen, phosphorus, and sulfur. The compound has this mass percent composition: 31.57% C; 5.74% H; 21.03% O. The mass percent sulfur is 2.07 times that of phosphorus. Dioxathion has a molar mass of 456.64 g/mol.

(a) Determine the empirical formula of dioxathion.

(b) Determine the molecular formula of dioxathion.

(c) A 57.50 mg dose of this pesticide is administered in a laboratory test. Calculate the amount of dioxathion in this dose.

(d) Calculate the number of molecules of dioxathion in the dose.

HAVING STUDIED THIS CHAPTER . . .

. . . you should be able to:

- Describe radioactivity, electrons, protons, and neutrons, and the general structure of the atom (Sections 2-1, 2-2). End-of-chapter Questions 7, 9, 11
- Define isotope and determine the atomic number, mass number, and number of neutrons for a specified isotope (Section 2-3). Questions 18, 20
- Find the atomic number and atomic weight for any element and explain how these two terms differ (Section 2-3). Questions 24, 119
- Understand the kinds of information that can be obtained from scanning tunneling microscopy and mass spectrometry. (Sections 2-2, 2-3). Questions 14, 16
- Calculate the average atomic mass or atomic weight of an element from isotopic abundances and isotopic masses (Section 2-3). Questions 22, 26, 28
- Know the charges on monoatomic ions of metals and nonmetals (Section 2-4; Figure 2.7). Questions 30, 32, 34
- Know the names and formulas of polyatomic ions (Section 2-4; Table 2.2).

- Be able to recognize ionic compounds and describe their properties (Sections 2-4, 2-6). Questions 38, 48, 117
- Given their names, write the formulas of ionic compounds; given formulas, write names (Section 2-5). Questions 36, 40, 42, 46, 127
- Interpret the meaning of molecular formulas, condensed formulas, and structural formulas (Section 2-7). Questions 52, 54, 56
- Be able to recognize molecular compounds, describe their properties, and compare them with the properties of ionic compounds (Section 2-7). Questions 50, 113
- Name binary molecular compounds, including straight-chain alkanes (Sections 2-8, 2-9). Questions 58, 60
- Define the term "isomer"; write structural formulas for and identify straight- and branched-chain alkane constitutional isomers (Section 2-9). Question 64, 126
- Define the unit "mole" and explain its importance as a means of counting nanoscale particles (Section 2-10). Questions 68, 123
- Given a chemical formula, calculate molar mass; use the molar mass to relate mass and amount of substance (Section 2-11). Questions 72, 76, 82
- Given the amount of substance, use the Avogadro constant to calculate the number of particles (atoms, molecules, ions) (Section 2-11). Questions 74, 78, 80
- Based on its formula, describe the nanoscale makeup of a hydrated ionic compound and calculate its molar mass (Section 2-11). Question 89
- Given a chemical formula, calculate mass percent composition (Section 2-12). Question 87
- Use composition based on mass and molar mass to determine the empirical and molecular formulas of a compound (Section 2-12). Questions 91, 96, 98

KEY TERMS

alkane (Section 2-9)
amount of substance (2-10)
atomic force microscope (2-2b)
atomic number (Z) (2-2c)
atomic structure (2-1)
atomic weight (2-3a)
average atomic mass (2-3a)
Avogadro constant (N_A) (2-10)
Avogadro's number (2-10)
binary molecular compound (2-8)
chemical bond (2-7b)
condensed formula (2-7b)
constitutional isomers (2-9a)
Coulomb's law (2-6a)
crystal lattice (2-6b)
electron (2-1b)
empirical formula (2-12a)
formula unit (2-6b)
formula weight (2-11b)
hydrocarbon (2-9)
inorganic compounds (2-7)
ions (2-1c)
ionic bonding (2-6a)
ionic compound (2-4)
ionic hydrate (2-11c)
isomers (2-9a)
isotopes (2-3)
mass number (A) (2-2c)
mass spectrometer (2-3)
mass spectrum (2-3)
molar mass (2-11)
mole (mol) (2-10)
molecular compound (2-7)
molecular formula (2-7b)
molecular weight (2-11a)
monoatomic ion (2-4a)
neutron (2-2b)
nucleus (2-2a)
organic compounds (2-7)
oxoanions (2-5b)
percent abundance (2-3a)
percent composition by mass (2-12)
polyatomic ion (2-4b; Table 2.2)
proton (2-1c)
radioactivity (2-1a)
scanning tunneling microscope (2-2b)
structural formula (2-7b)
structural isomers (2-9a)
unified atomic mass unit (u) (2-3)
water of hydration (2-11c)

QUESTIONS FOR REVIEW AND THOUGHT

Red-numbered questions have short answers at the back of this book in Appendix M and fully worked solutions in the *Student Solutions Manual.*

Review Questions

These questions test vocabulary and simple concepts.

1. Identify the fundamental unit of electrical charge.
2. The positively charged particle in an atom is called the proton.
   (a) Calculate how much heavier a proton is than an electron.
   (b) What is the difference in the charge on a proton and an electron?
3. In any given *neutral* atom, how many protons are there compared with the number of electrons?
4. Atoms of elements can have varying numbers of neutrons in their nuclei.
   (a) What are species called that have varying numbers of neutrons for the same element?
   (b) How do the mass numbers vary for these species?
   (c) What are two common elements that exemplify this property?
5. Define these terms: (a) unified atomic mass unit; (b) mass number; (c) molar mass; (d) isotope.
6. A dictionary defines the word "compound" as a "combination of two or more parts." What are the "parts" of a chemical compound? Identify three pure (or nearly pure) compounds you have encountered today. What is the difference between a compound and a mixture?

Topical Questions

These questions are keyed to the major topics in the chapter. Usually a question that is answered at the back of the book is paired with a similar one that is not.

Atomic Structure: Subatomic Particles (Section 2-1)

7. Complete the table below.

Name	Electric Charge (C)	Mass (g)	Deflected by Electric Field?
______	1.6022×10^{-19}	1.6726×10^{-24}	______
alpha particle	______	______	______
______	-1.6022×10^{-19}	______	______

8. Complete the table below.

Name	Electric Charge (C)	Mass (g)	Deflected by Electric Field?
______	______	1.6726×10^{-24}	______
______	______	0	no
beta ray	______	______	______

The Nuclear Atom (Section 2-2)

9. If the nucleus of an atom were the size of a golf ball (4-cm diameter), what would be the diameter of the atom?
10. If the nucleus of an atom were the size of Earth, would the moon be within the atom? Would the sun be within the atom? (Use the Internet to find information about the distances from Earth to the moon and sun.)
11. Match these by placing the correct notation in the appropriate blank.

    $^{67}_{34}Se$ $^{67}_{33}As$ $^{67}_{35}Br$ $^{72}_{36}Kr$

    ___ a. Contains 33 neutrons
    ___ b. Contains greatest number of neutrons
    ___ c. Contains equal number of protons and neutrons
    ___ d. Contains the same number of *neutrons* as there are protons in As-67
12. Match these by placing the correct notation in the appropriate blank.

    $^{112}_{50}Sn$ $^{115}_{50}Sn$ $^{112}_{51}Sb$ $^{115}_{49}In$

    ___ a. Contains 65 neutrons
    ___ b. Contains fewest number of neutrons
    ___ c. Contains greatest number of neutrons
    ___ d. Contains the same number of *neutrons* as there are protons in Sm-142

Tools of Chemistry (Sections 2-2 and 2-3)

13. Both scanning tunneling microscopy and mass spectrometry use electrons to help measure different properties of atoms and molecules. Describe the role that electrons play in each technique.
14. What nanoscale species are moving through a mass spectrometer during its operation?
15. What is plotted on the *x*-axis and on the *y*-axis in a mass spectrum? What information does a mass spectrum convey?
16. Bromine has two isotopes, bromine-79 (50.69% abundance) and bromine-81 (49.31% abundance). Draw a graph of the mass spectrum obtained when a sample of Br_2 gas is run through a mass spectrometer.
17. Chlorine consists of two isotopes, chlorine-35 (75.77% abundance) and chlorine-37 (24.23% abundance). Draw a graph of the mass spectrum obtained from a sample of Cl_2 gas.

Isotopes and Average Atomic Mass (Section 2-3)

18. Uranium-235 and uranium-238 differ in terms of the number of subatomic particles. For which subatomic particle is the number different and by how much?
19. Strontium 90 is a product formed when atomic bombs exploded. How do strontium-90 and strontium-88 differ in terms of the number of subatomic particles?
20. How many electrons, protons, and neutrons are present in an atom of cobalt-60?

21. The artificial radioactive element technetium is used in many medical studies. Give the number of electrons, protons, and neutrons in an atom of technetium-99.
22. The atomic weight of bromine is 79.904. The natural abundance of ^{81}Br, atomic weight 80.916289 u, is 49.31%. What is the atomic weight of the only other natural isotope of bromine?
23. The atomic weight of boron is 10.811. The natural abundance of ^{10}B is 19.91%. Determine the atomic weight of the only other natural isotope of boron.
24. Give the complete symbol $^{A}_{Z}X$ for each of these atoms: (a) sodium with 12 neutrons, (b) argon with 21 neutrons, and (c) gallium with 38 neutrons.
25. Give the complete symbol $^{A}_{Z}X$ for each of these atoms: (a) nitrogen with 8 neutrons, (b) zinc with 34 neutrons, and (c) xenon with 75 neutrons.
26. Verify that the average atomic mass of lithium is 6.941, given this information:

    ^{6}Li, exact mass = 6.015121 u
    percent abundance = 7.500%
    ^{7}Li, exact mass = 7.016003 u
    percent abundance = 92.50%

27. Verify that the average atomic mass of magnesium is 24.3050, given this information:

    ^{24}Mg, exact mass = 23.985042 u
    percent abundance = 78.99%
    ^{25}Mg, exact mass = 24.985837 u
    percent abundance = 10.00%
    ^{26}Mg, exact mass = 25.982593 u
    percent abundance = 11.01%

28. Gallium has two naturally occurring isotopes, ^{69}Ga and ^{71}Ga, with masses of 68.9257 u and 70.9249 u, respectively. Calculate the abundances of these isotopes of gallium.
29. Argon has three naturally occurring isotopes: 0.3336% ^{36}Ar, 0.063% ^{38}Ar, and 99.60% ^{40}Ar. Estimate the average atomic mass of argon. If the masses of the isotopes are 35.968 u, 37.963 u, and 39.962 u, respectively, calculate the average atomic mass of natural argon.

Ions and Ionic Compounds (Section 2-4)

30. For each of these metals, write the chemical symbol for the corresponding monoatomic ion (with charge).
    (a) Lithium (b) Strontium
    (c) Aluminum (d) Zinc
31. For each of these nonmetals, write the chemical symbol for the corresponding monoatomic ion (with charge).
    (a) Nitrogen (b) Sulfur
    (c) Chlorine (d) Iodine
32. Predict the charges for monoatomic ions of these elements.
    (a) Magnesium (b) Phosphorus
    (c) Iron (d) Selenium
33. Predict the charges for monoatomic ions of these elements.
    (a) Gallium (b) Fluorine
    (c) Silver (d) Nitrogen
34. Cobalt is a transition metal and thus can form ions with at least two different charges. Write the formulas for the compounds formed between cobalt ions and the oxide ion.
35. Although not a transition element, lead can form two cations: Pb^{2+} and Pb^{4+}. Write the formulas for the compounds of these ions with the chloride ion.
36. Which of these are the correct formulas of compounds? For those that are not, give the correct formula.
    (a) AlCl (b) NaF_2
    (c) Ga_2O_3 (d) MgS
37. Which of these are the correct formulas of compounds? For those that are not, give the correct formula.
    (a) Ca_2O (b) $SrCl_2$
    (c) Fe_2O_5 (d) K_2O
38. Predict which compounds are ionic. Explain your answers.
    (a) CF_4 (b) $SrBr_2$
    (c) $Co(NO_3)_3$ (d) SiO_2
    (e) KCN (f) SCl_2
39. Predict which compounds are ionic. Explain your answers.
    (a) NaH (b) HCl
    (c) NH_3 (d) CH_4
    (e) HI

Naming Ions and Ionic Compounds (Section 2-5)

40. Determine the chemical formulas for barium sulfate, magnesium nitrate, and sodium acetate. Each compound contains a monoatomic cation and a polyatomic anion. What are the names and electrical charges of these ions?
41. Write the chemical formula for calcium nitrate, barium chloride, and ammonium phosphate. What are the names and charges of all the ions in these three compounds?
42. Write the chemical formulas for these compounds.
    (a) Nickel(II) nitrate (b) Sodium bicarbonate
    (c) Lithium hypochlorite (d) Magnesium chlorate
    (e) Calcium sulfite
43. Write the chemical formulas for these compounds.
    (a) Iron(III) nitrate (b) Potassium carbonate
    (c) Sodium phosphate (d) Calcium chlorite
    (e) Sodium sulfate
44. Give the correct formula for each of these ionic compounds.
    (a) Ammonium carbonate (b) Calcium iodide
    (c) Copper(II) bromide (d) Aluminum phosphate
45. Give the correct formula for each of these ionic compounds.
    (a) Calcium hydrogen carbonate
    (b) Potassium permanganate
    (c) Magnesium perchlorate
    (d) Ammonium monohydrogen phosphate
46. Correctly name each of these ionic compounds.
    (a) K_2S (b) $NiSO_4$
    (c) $(NH_4)_3PO_4$ (d) $Al(OH)_3$
    (e) $Co_2(SO_4)_3$
47. Correctly name each of these ionic compounds.
    (a) KH_2PO_4 (b) $CuSO_4$
    (c) $CrCl_3$ (d) $Ca(CH_3COO)_2$
    (e) $Fe_2(SO_4)_3$

Ionic Compounds: Bonding and Properties (Section 2-6)

48. Solid magnesium oxide melts at 2800 °C. This property, combined with the fact that magnesium oxide is not an electrical conductor, makes it an ideal heat insulator for electric wires in cooking ovens and toasters. In contrast, solid NaCl melts at the relatively low temperature of 801 °C. What is the formula of magnesium oxide? Suggest a reason that it has a melting temperature so much higher than that of NaCl.
49. Assume you have an unlabeled bottle containing a white, crystalline powder. The powder melts at 310 °C. You are told that it could be NH_3, NO_2, or $NaNO_3$. What do you think it is and why?

Molecular Compounds (Section 2-7)

50. Identify each compound as ionic or molecular based on its formula or properties.
    (a) Rb_2O
    (b) C_6H_{12}
    (c) Liquid at room temperature
    (d) Conducts electricity when molten
51. Identify each compound as ionic or molecular based on its formula or properties.
    (a) Can be cleaved with a sharp wedge
    (b) Melts at −22.3 °C
    (c) $MgBr_2$
    (d) $C_5H_{10}O_2N$
52. Given these condensed formulas, write the structural and molecular formulas.
    (a) CH_3OH (b) $CH_3CH_2NH_2$
    (c) $CH_3CH_2SCH_2CH_3$ (d) CH_3CH_2SH
53. Write molecular and structural formulas for these compounds.
    (a) $CH_3CH_2NHCH_2CH_3$ (b) CH_3NH_2
    (c) $CH_3CHClCH_3$ (d) $CH_2OHCHOHCH_2OH$
54. Write the molecular formula for each substance.
    (a) The hydrocarbon heptane, which has seven carbon atoms and 16 hydrogen atoms
    (b) Acrylonitrile (the basis of Orlon and Acrilan fibers), which has three carbon atoms, three hydrogen atoms, and one nitrogen atom
55. Write the molecular formula for each substance.
    (a) Fenclorac, an anti-inflammatory drug, which has 14 carbon atoms, 16 hydrogen atoms, two chlorine atoms, and two oxygen atoms.
    (b) Vitamin B-12, which has 63 carbon atoms, 88 hydrogen atoms, one cobalt atom, 14 nitrogen atoms, 14 oxygen atoms, and one phosphorus atom.
56. Give the total number of atoms of each element in one formula unit of each of these compounds.
    (a) CaC_2O_4 (b) $C_6H_5CHCH_2$
    (c) $(NH_4)_2SO_4$ (d) $Pt(NH_3)_2Cl_2$
    (e) $K_4Fe(CN)_6$
57. Give the total number of atoms of each element in each of these molecules.
    (a) $C_6H_5COOC_2H_5$ (b) $HOOCCH_2CH_2COOH$
    (c) $NH_2CH_2CH_2COOH$ (d) $C_{10}H_9NH_2Fe$
    (e) $C_6H_2CH_3(NO_2)_3$

Naming Binary Molecular Compounds (Section 2-8)

58. Give the correct name for each compound.
    (a) SO_2 (b) CCl_4
    (c) P_4S_{10} (d) SF_4
59. Give the correct name for each compound.
    (a) HBr (b) ClF_3
    (c) Cl_2O_7 (d) BI_3
60. Write the correct formula for each compound.
    (a) Nitrogen triiodide (b) Carbon disulfide
    (c) Dinitrogen tetraoxide (d) Selenium hexafluoride
61. Write the correct formula for each compound.
    (a) Bromine trichloride
    (b) Xenon trioxide
    (c) Diphosphorus tetrafluoride
    (d) Oxygen difluoride

Organic Molecular Compounds (Section 2-9)

62. In a noncyclic alkane other than methane, what is the maximum number of hydrogen atoms that can be bonded to one carbon atom?
63. In a noncyclic alkane, what is the maximum number of carbon atoms that can be bonded to one carbon atom?
64. Consider two molecules that are constitutional isomers.
    (a) What is the same on the molecular level between these two molecules?
    (b) What is different on the molecular level between these two molecules?
65. Draw structural formulas for the five constitutional isomers of C_6H_{14}.
66. The noncyclic hydrocarbon eicosane has 20 carbon atoms in each molecule. How many hydrogen atoms are in each molecule?
67. A cyclic hydrocarbon has 16 hydrogen atoms in each molecule. How many carbon atoms are there per molecule? What is the name of the compound?

Amount of Substance: The Mole (Section 2-10)

68. If you divide Avogadro's number of pennies among the nearly 300 million people in the United States, and if each person could count one penny each second every day of the year for eight hours per day, calculate how long it would take to count the pennies.
69. Why do you think it is more convenient to use some chemical counting unit when doing calculations (chemists have adopted the unit of the mole, but it could have been something different) rather than using individual molecules?

Molar Mass (Section 2-11)

70. Calculate the mass of
    (a) 2.5 mol boron (b) 0.015 mol O_2
    (c) 1.25×10^{-3} mol iron (d) 653 mol helium
71. Calculate the mass of
    (a) 6.03 mol gold (b) 0.045 mol uranium
    (c) 15.6 mol Ne (d) 3.63×10^{-4} mol plutonium
72. Calculate the amount of substance for each sample:
    (a) 127.08 g Cu (b) 20.0 g calcium
    (c) 16.75 g Al (d) 0.012 g potassium
    (e) 5.0 mg americium

73. Calculate the amount of substance for each sample:
(a) 16.0 g Na (b) 0.0034 g platinum
(c) 1.54 g P (d) 0.876 g arsenic
(e) 0.983 g Xe

74. If you have a 35.67-g piece of chromium metal on your car, calculate how many chromium atoms you have.

75. If you have a ring that contains 1.94 g gold, calculate how many gold atoms are in the ring.

76. You have a pure sample of the antiseptic aminacrine, $C_{13}H_{10}N_2$.
(a) Calculate the mass in grams of 0.06500 mol aminacrine.
(b) Calculate the number of aminacrine molecules in a 0.2480-g sample.
(c) Calculate the number of nitrogen atoms in this 0.2480-g sample.
(d) Calculate the mass of N in 100. g aminacrine.

77. You have a pure sample of apholate, $C_{12}H_{24}N_9P_3$, a highly effective commercial insecticide.
(a) Calculate the molar mass of apholate.
(b) Calculate the mass of N in 100. g apholate.
(c) A sample containing 250.0 mg apholate is sprayed on an agricultural field. Calculate the mass of phosphorus in this sample of apholate; express your result in grams.
(d) Calculate the number of phosphorus atoms in this sample of apholate.

78. You have a U. S. penny that weighs 2.458 g and contains 2.40% copper by mass.
(a) Calculate the number of pennies needed to contain 2.458 g Cu.
(b) Calculate the amount of copper in the penny.
(c) Calculate the number of copper atoms in the penny.

79. You have a sterling silver fork that contains 92.5% silver and weighs 43.2 g.
(a) Calculate the mass of silver in the fork.
(b) Calculate the amount of silver in the fork.
(c) Calculate the number of silver atoms in the fork.

80. Fill in this table for 1 mol methanol, CH_3OH.

	CH_3OH	Carbon	Hydrogen	Oxygen
Amount of substance	______	______	______	______
Number of molecules or atoms	______	______	______	______
Molar mass	______	______	______	______

81. Fill in this table for 1 mol glucose, $C_6H_{12}O_6$.

	$C_6H_{12}O_6$	Carbon	Hydrogen	Oxygen
Amount of substance	______	______	______	______
Number of molecules or atoms	______	______	______	______
Molar mass	______	______	______	______

82. Calculate the amount of substance in 1.00 g of each compound.
(a) CH_3OH, methanol
(b) Cl_2CO, phosgene, a poisonous gas
(c) Ammonium nitrate
(d) Magnesium sulfate heptahydrate (Epsom salt)
(e) Silver acetate

83. Calculate the amount of substance in 0.250 g of each compound.
(a) $C_7H_5NO_3S$, saccharin, an artificial sweetener
(b) $C_{13}H_{20}N_2O_2$, procaine, a painkiller used by dentists
(c) $C_{20}H_{14}O_4$, phenolphthalein, a dye

84. Acetaminophen, an analgesic, has the molecular formula $C_8H_9O_2N$.
(a) Calculate the molar mass of acetaminophen.
(b) Calculate the amount of acetaminophen in 5.32 g acetaminophen.
(c) Calculate the mass of acetaminophen in 0.166 mol acetaminophen.

85. An Alka-Seltzer tablet contains 324 mg aspirin, $C_9H_8O_4$; 1904 mg $NaHCO_3$; and 1000 mg citric acid, $C_6H_8O_7$. (The last two compounds react with each other to provide the "fizz," bubbles of CO_2, when the tablet is put into water.)
(a) Calculate the amount of each substance in the tablet.
(b) If you take one tablet, calculate how many aspirin molecules you are consuming.

86. Calculate how many water molecules are in one drop of water. (One drop of water is $\frac{1}{20}$ mL, and the density of water is 1.0 g/mL.)

Composition and Chemical Formulas (Section 2-12)

87. Calculate the molar mass of each of these compounds and the mass percent of each element.
(a) PbS, lead(II) sulfide, galena
(b) C_2H_6, ethane, a hydrocarbon fuel
(c) CH_3COOH, acetic acid, an important ingredient in vinegar
(d) NH_4NO_3, ammonium nitrate, a fertilizer

88. Three oxygen-containing compounds of iron are $FeCO_3$, Fe_2O_3, and Fe_3O_4. Calculate the percent iron by mass in each iron compound.

89. The copper-containing compound $Cu(NH_3)_4SO_4 \cdot H_2O$ is a beautiful blue solid. Calculate the molar mass of the compound and the mass percent of each element.

90. The compound $Co(NO_3)_2 \cdot 6H_2O$ is a red solid that absorbs water from the atmosphere. Calculate its molar mass and the mass percent of each element. Also calculate the mass percent water.

91. Quinine (molar mass = 324.41 g/mol) is used as a cardiac depressant. It has this percent composition by mass: 74.04% C; 7.46% H; 8.64% N; and 9.86% O. Use these data to determine (a) the empirical formula and (b) the molecular formula of quinine.

92. About a century ago, Paul Ehrlich discovered Salvarsan, the first arsenical antibiotic that cured a targeted disease, syphilis. The compound was the six hundred and sixth compound he tried against the disease, but the first one found to be effective against it. The compound has molar mass 549.102 g/mol and this mass percent composition: 39.37% C; 3.304% H; 7.653% N; 8.741% O; 40.93% As.
(a) Calculate the empirical formula of Salvarsan.
(b) Determine the molecular formula of this compound.

93. The mineral uraninite is a uranium oxide that is 84.80% uranium by mass. Show calculations to determine the correct empirical formula of uraninite.
94. Carbonic anhydrase, an important enzyme in mammalian respiration, is a large zinc-containing protein with a molar mass of 3.00×10^4 g/mol. The zinc is 0.218% by mass of the protein. Determine how many zinc atoms each carbonic anhydrase molecule contains.
95. Nitrogen fixation in the root nodules of peas and other legumes occurs with a reaction involving a molybdenum-containing enzyme named *nitrogenase.* This enzyme contains two Mo atoms per molecule and is 0.0872% Mo by mass. Calculate the molar mass of the enzyme.
96. Disilane, Si_2H_x, contains 90.28% silicon by mass. Calculate the value of x in this compound.
97. Chalky, white crystals in mineral collections are often labeled borax, which has the molecular formula $Na_2B_4O_7 \cdot 10H_2O$, when actually they are partially dehydrated samples with the molecular formula $Na_2B_4O_7 \cdot 5H_2O$, which is more stable under the storage conditions. Real crystals of borax are colorless and transparent.
    (a) Calculate the percent mass that the mineral has lost when it partially dehydrates.
    (b) Is the percent boron by mass the same in both compounds?
98. A well-known reagent in analytical chemistry, dimethylglyoxime, has the empirical formula C_2H_4NO. If its molar mass is 116.1 g/mol, determine the molecular formula of the compound.
99. The molecular formula of ascorbic acid (vitamin C) is $C_6H_8O_6$. What is its empirical formula?
100. The alum used in cooking is potassium aluminum sulfate hydrate, $KAl(SO_4)_2 \cdot xH_2O$. To find the value of x, you can heat a sample of the compound to drive off all the water and leave only $KAl(SO_4)_2$. Assume that you heat 4.74 g of the hydrated compound and that it loses 2.16 g water. Calculate the value of x.

General Questions

These questions are not explicitly keyed to chapter topics; many require integration of several concepts.

101. The density of a solution of sulfuric acid is 1.285 g/cm^3, and it is 38.08% acid by mass. Calculate the volume of the acid solution (in mL) you need to supply 125 g of sulfuric acid.
102. The cancer drug cisplatin contains 65.0% platinum. If you have 1.53 g of the compound, calculate how many grams of platinum the sample contains.
103. Ethyl alcohol, C_2H_5OH, has a density of 0.789 g/mL at 25 °C. Water weighs 1.00 kg/L at 25 °C. Calculate the volume of ethanol that contains the same number of molecules as 1.00 L water.
104. A common fertilizer used on lawns is designated as "16-4-8." These numbers mean that the fertilizer contains 16% nitrogen-containing compounds, 4.0% phosphorus-containing compounds, and 8.0% potassium-containing compounds. You buy a 40.0-lb bag of this fertilizer and use all of it on your lawn. Calculate the mass (g) of the phosphorus-containing compound you are putting on your lawn. If the phosphorus-containing compound consists of 43.64% phosphorus (the rest is oxygen), calculate the mass (g) of phosphorus that are in 40.0 lb of this fertilizer.
105. The fluoridation of city water supplies has been practiced in the United States for several decades because there is scientific evidence that fluoride prevents tooth decay, especially in young children. Fluoridation is done by continuously adding sodium fluoride to water as it comes from a reservoir. Assume you live in a medium-sized city of 150,000 people and that each person uses 175 gal water per day. Calculate how many tons of sodium fluoride you must add to the water supply each year (365 days) to have the required fluoride concentration of 1 part per million (that is, 1 ton of fluoride per million tons of water). (Sodium fluoride is 45.0% fluoride, and 1 U. S. gallon of water has a mass of 8.34 lb.)
106. Which one of these symbols conveys more information about the atom: ^{37}Cl or $_{17}Cl$? Explain.
107. Gems and precious stones are measured in carats, a mass unit equivalent to 200. mg. If you have a 2.3-carat diamond in a ring, calculate the amount of carbon you have.
108. The Statue of Liberty in New York harbor is made of 2.00×10^5 lb copper sheets bolted to an iron framework. Calculate the mass and the amount of copper in the statue (1 lb = 454 g).
109. Two different halogens are present in a series known as interhalogen compounds. Name these interhalogen compounds: (a) IBr; (b) BrF_3; (c) I_2Cl_6; (d) ClF_5; (e) IF_7.
110. A helium atom consists of two protons, two electrons, and two neutrons.
    (a) Explain how these particles are arranged to make the atom.
    (b) Calculate the mass of the atom in unified atomic mass units. (Assume no change in mass of the particles when they are in the atom.)
    (c) Calculate the mass of the atom in grams. (Assume no change in mass of the particles when they are in the atom.)
    (d) Calculate the mass (in grams) of a mole of these helium atoms. Compare your result with the atomic weight given in the periodic table inside the front cover of the book.
111. It's final exam time and a student drinks a 1.93-oz bottle of 5-Hour Energy® to stay awake. The drink contains, among other substances, 212 mg caffeine, $C_8H_{10}N_4O_2$.
    (a) Calculate the mass percent nitrogen in caffeine.
    (b) Calculate the number of caffeine molecules that the student ingested.
    (c) Calculate the number of carbon atoms in this mass of caffeine.
    (d) An 8-oz cup of regular coffee contains approximately 100 mg caffeine. Calculate how many times greater the caffeine concentration (mg/oz) is in the 5-Hour Energy® drink than in the regular coffee.
112. (a) Calculate the mass of one molecule of nitrogen.
    (b) Calculate the mass of one molecule of oxygen.
    (c) Calculate the ratio of masses of these two molecules. How does it compare with the ratio of the atomic weights of N and O?

113. For each pair of elements, (i) through (vii),
   (a) Determine whether an ionic compound, a molecular compound, or no compound would form.
   (b) Write an appropriate formula for each compound you expect to form and name the compound.
      (i) Chlorine and bromine
      (ii) Lithium and tellurium
      (iii) Sodium and argon
      (iv) Magnesium and fluorine
      (v) Nitrogen and bromine
      (vi) Indium and sulfur
      (vii) Selenium and bromine

114. For each substance, (i) through (viii),
   (a) Write the correct formula.
   (b) Decide whether the substance is ionic or molecular.
      (i) Sodium hypochlorite
      (ii) Tetraphosphorus decaoxide
      (iii) Potassium permanganate
      (iv) Potassium dihydrogen phosphate
      (v) Chlorine trifluoride
      (vi) Boron tribromide
      (vii) Calcium acetate
      (viii) Sodium sulfite

115. Pepto-Bismol, which helps provide relief for an upset stomach, contains 300 mg bismuth subsalicylate, $C_7H_5BiO_4$, per tablet.
   (a) You take two tablets for your stomach distress. Calculate the amount of the "active ingredient" you are taking.
   (b) What mass of Bi did the two tablets contain?

Applying Concepts

These questions test conceptual learning.

116. Draw a diagram to indicate the arrangement of nanoscale particles of each substance. Consider each drawing to hold a very tiny portion of each substance. Each drawing should contain at least 16 particles, and it need not be three-dimensional.

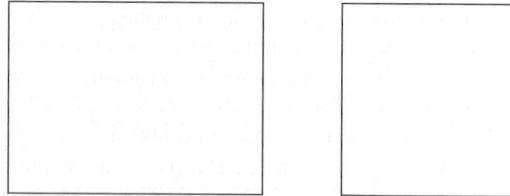

$Br_2(\ell)$ $\quad$ LiF(s)

117. Draw diagrams of each nanoscale situation given in parts (a) and (b). Represent atoms or monoatomic ions as circles; represent molecules or polyatomic ions by overlapping circles for the atoms that make up the molecule or ion; and distinguish among different kinds of atoms by labeling or shading the circles. In each case draw representations of at least five nanoscale particles. Your diagrams can be two-dimensional.
   (a) A crystal of solid sodium chloride
   (b) The sodium chloride from part (a) after it has been melted

118. Draw diagrams of each nanoscale situation given in parts (a)–(c). Represent atoms or monoatomic ions as circles; represent molecules or polyatomic ions by overlapping circles for the atoms that make up the molecule or ion; and distinguish among different kinds of atoms by labeling or shading the circles. In each case draw representations of at least five nanoscale particles. Your diagrams can be two-dimensional.
   (a) A sample of solid lithium nitrate, $LiNO_3$
   (b) A sample of molten lithium nitrate
   (c) A molten sample of lithium nitrate after electrodes have been placed into it and a direct current applied to the electrodes

119. Which sets of values are possible? Why are the others not possible? Explain your reasoning.

	Mass Number	Atomic Number	Number of Protons	Number of Neutrons
(a)	19	42	19	23
(b)	235	92	92	143
(c)	53	131	131	79
(d)	32	15	15	15
(e)	14	7	7	7
(f)	40	18	18	40

120. Which sets of values are possible? Why are the others not possible? Explain your reasoning.

	Mass Number	Atomic Number	Number of Protons	Number of Neutrons
(a)	53	25	25	29
(b)	195	78	195	117
(c)	33	16	16	16
(d)	52	24	24	28
(e)	35	17	18	17

121. Potassium has three stable isotopes, ^{39}K, ^{40}K, and ^{41}K, but ^{40}K has a very low natural abundance. Which of the other two is the more abundant? (No calculation should be necessary.)

122. Lithium has two stable isotopes, ^{6}Li and ^{7}Li. The atomic weight of lithium is 6.941. Without doing a calculation, explain which is the more abundant isotope.

123. Which member of each pair has the greater number of atoms? Explain why.
   (a) 1 mol Cl or 1 mol Cl_2
   (b) 1 molecule O_2 or 1 mol O_2
   (c) 1 nitrogen atom or 1 nitrogen molecule
   (d) 6.032×10^{23} fluorine molecules or 1 mol fluorine molecules
   (e) 20.3 g Ne or 1 mol Ne
   (f) 1 molecule Br_2 or 159.8 g Br_2
   (g) 107.9 g Ag or 9.6 g Li
   (h) 58.9 g Co or 58.9 g Cu
   (i) 1 g calcium or 6.022×10^{23} calcium atoms
   (j) 1 g chlorine atoms or 1 g chlorine molecules

124. Which member of each pair has the greater mass? Explain why.

(a) 1 mol iron or 1 mol aluminum
(b) 6.022×10^{24} lead atoms or 1 mol lead
(c) 1 copper atom or 1 mol copper
(d) 1 mol Cl or 1 mol Cl_2
(e) 1 g oxygen atoms or 1 g oxygen molecules
(f) 23.4 g Mg or 1 mol Mg
(g) 1 mol Na or 1 g Na
(h) 4.1 g He or 6.022×10^{23} He atoms
(i) 1 molecule I_2 or 1 mol I_2
(j) 1 oxygen molecule or 1 oxygen atom

125. One way to solve problems is to find an analogy between what you know and what you need to determine. For example, if you know that the formula of magnesium oxide is MgO, and that sulfur is in Group 6A along with oxygen, by analogy the formula of magnesium sulfide is MgS. By analogy with sulfur or phosphorus compounds, name these compounds:
(a) Na_2SeO_3 (b) $AlSbO_4$
(c) K_3AsO_4 (d) Ag_2TeO_4

126. When asked to draw all the possible constitutional isomers for C_3H_8O, a student drew these structures. The student's instructor said some of the structures were identical.
(a) How many actual isomers are there?
(b) Which structures are identical?
(i) $CH_3—CH_2—CH_2—OH$
(ii) $CH_3—CH_2—O—CH_3$
(iii) $CH_3—O—CH_2—CH_3$
(iv) $HO—CH_2—CH_2$
|
CH_3
(v) $CH_3—CH—CH_3$
|
OH
(vi) $HO—CH—CH_3$
|
CH_3

127. The formula for thallium nitrate is $TlNO_3$. Based on this information, what would be the formulas for thallium carbonate and thallium sulfate?

128. The name given with each of these formulas is incorrect. What are the correct names?
(a) CaF_2, calcium difluoride
(b) CuO, copper oxide
(c) $NaNO_3$, sodium nitroxide
(d) NI_3, nitrogen iodide
(e) $FeCl_3$, iron(I) chloride
(f) Li_2SO_4, dilithium sulfate

More Challenging Questions

These questions require more thought and integrate several concepts.

129. A high-quality analytical balance can weigh accurately to the nearest 1.0×10^{-4} g. A sample of carbon weighed on this balance has a mass of 1.000 mg. Calculate the number of carbon atoms in the sample. Given the precision of the balance, determine the maximum and minimum number of carbon atoms that could be in the sample.

130. The element bromine is Br_2, so the mass of a Br_2 molecule is the sum of the mass of its two atoms. Bromine has two isotopes. The mass spectrum of Br_2 produces three peaks with relative masses of 157.836, 159.834, and 161.832, and relative heights of 6.337, 12.499, and 6.164, respectively.
(a) What isotopes of bromine are present in each of the three peaks?
(b) What is the mass of each bromine isotope?
(c) What is the average atomic mass of bromine?
(d) What is the abundance of each of the two bromine isotopes?

131. Uranium is used as a fuel, primarily in the form of uranium(IV) oxide, in nuclear power plants. This question considers some uranium chemistry.
(a) A small sample of uranium metal (0.169 g) is heated to 900 °C in air to give 0.199 g of a dark green oxide, U_xO_y. How many moles of uranium metal were used? What is the empirical formula of the oxide U_xO_y? What is the name of the oxide? How many moles of U_xO_y must have been obtained?
(b) The oxide U_xO_y is obtained if $UO_2NO_3 \cdot n\,H_2O$ is heated to temperatures greater than 800 °C in air. However, if you heat it gently, only the water of hydration is lost. If you have 0.865 g $UO_2NO_3 \cdot n\,H_2O$ and obtain 0.679 g UO_2NO_3 on heating, how many molecules of water of hydration were there in each formula unit of the original compound?

132. A mixture contains only $MgSO_4$ and $(NH_4)_2SO_4$. If the mass percent of $MgSO_4$ in the mixture is 32.0%, what is the mass percent of sulfate in the mixture?

133. Hemoglobin is an iron-containing protein (molar mass 64,458 g/mol) that is responsible for oxygen transport in our blood. Hemoglobin is 0.35% iron by mass. Calculate how many iron atoms are in each hemoglobin molecule.

134. There are three naturally occurring isotopes of potassium: ^{39}K 38.963707 u; ^{40}K 39.963999 u; and ^{41}K 40.961825 u. The average atomic mass of potassium is 39.0983 u and the natural abundance of the lightest isotope is 93.2581%. Calculate the natural abundances of the other two isotopes.

135. The diatomic compound BrCl is a reddish-brown gas. Consider the naturally occurring isotopes of each element:

Isotope	Natural Abundance, %
Br-79	50.69
Br-81	49.31
Cl-35	75.77
Cl-37	24.23

(a) Name the compound.
(b) Determine how many different types of BrCl molecules are possible by using the sum of mass numbers as the criterion for type.
(c) Determine which is the most abundant type in (b).
(d) Determine which is the second most abundant type.

136. The stainless steel used in the Gateway Arch in St. Louis contains 72.0% Fe, 19.0% Cr, and the remainder is

nickel. A 10.0-g sample of this stainless steel is treated to convert the metals to their oxides: 10.3 g Fe_2O_3, 2.71 g Cr_2O_3, and 1.14 g NiO. Calculate the mass percent of each metal in the sample.

137. Galinstan, a gallium-indium-tin alloy, is a liquid at room temperature and is used as a nontoxic replacement for mercury in thermometers. Its mass ratio of gallium-to-indium is 3.186. The mole ratio of indium-to-tin is 2.223. Calculate the mass percent composition of galinstan.

138. An adult human body contains 6.0 L blood, which contains about 15.5 g hemoglobin per 100.0 mL blood. The molar mass of hemoglobin is approximately 64,500 g/mol and there is 4 mol iron per 1 mol hemoglobin. A news item claims that there is sufficient iron in the hemoglobin of the body that this iron, if it were in the form of metallic iron, could make a 3-in. iron nail that weighs approximately 3.7 g. Show sufficient calculations to either support or refute the claim.

139. A 1.546-g sample of magnesium metal is heated in sufficient air at a high temperature so that all of the magnesium reacts. The reaction forms 2.512 g MgO and a small quantity of another magnesium-containing compound that is 72.24% magnesium by mass.
(a) Determine the formula of the other magnesium-containing compound.
(b) Name this compound.
(c) Calculate what fraction of the original Mg is in this second compound.

140. There are four binary potassium compounds of oxygen. They contain these mass percents of potassium: Compound I, 83.0%; Compound II, 55.0%; Compound III, 44.9%; and Compound IV, 71.0%. One compound has molar mass equal to 110.2 g/mol. Use this information to determine the chemical formula of each compound.

141. Direct reaction of fluorine with xenon produces three different xenon fluorides. One of the compounds, call it compound "I", contains twice the mass of fluorine as another xenon fluoride. Let's call the latter one compound "II". The third compound, compound "III", contains 1.5 times the mass of fluorine contained in compound "I". Compound II contains 77.5% Xe.
(a) Determine the formula of each compound.
(b) Name each compound.

142. A 20.00 g mixture of PCl_3 and PCl_5 contains 79.50 mass percent chlorine.
(a) Name each compound.
(b) Calculate the individual masses of PCl_3 and PCl_5 in the mixture.

143. The present average concentration (mass percent) of magnesium ions in seawater is 0.13%. A chemistry textbook estimates that if 1.00×10^8 tons Mg were taken out of the sea each year, it would take one million years for the Mg concentration to drop to 0.12%. Do sufficient calculations to either verify or refute this statement. Assume that Earth is a sphere with a diameter of 8000 mi, 67% of which is covered by oceans to a depth of 1 mi, and that no Mg is washed back into the oceans at any time.

144. Through a series of reactions, a 12.3-g sample of potassium carbonate was chemically reacted so that all of its carbon was found in $K_2Zn_3[Fe(CN)_6]_2$. Calculate the mass of $K_2Zn_3[Fe(CN)_6]_2$ formed.

145. A 4.22-g mixture of calcium chloride and sodium chloride was treated so that all of the calcium was converted to calcium carbonate. This product was then heated, converting it to 0.959 g pure calcium oxide. Calculate the mass percent of calcium chloride in the original mixture.

146. A certain metal, M, forms two oxides, M_2O and MO. If the percent by mass of M in M_2O is 73.4%, calculate the percent by mass in MO.

147. If you heat Al with an element from Group 6A, an ionic compound is formed that contains 18.55% Al by mass.
(a) What is the likely charge on the nonmetal in the compound formed?
(b) Using X to represent the nonmetal, what is the empirical formula for this ionic compound?
(c) Which element in Group 6A has been combined with Al?

Conceptual Challenge Problems

These rigorous, thought-provoking problems integrate conceptual learning with problem solving and are suitable for group work.

CP2.A (Section 2-1) Suppose you are faced with a problem similar to the one faced by Robert Millikan when he analyzed data from his oil-drop experiment. The masses of three stacks of dimes are given. What do you conclude to be the mass of a dime, and what is your argument?

Stack 1 = 9.12 g Stack 2 = 15.96 g Stack 3 = 27.36 g

CP2.B (Section 2-12) The age of the universe is unknown, but some conclude from measuring Hubble's constant that the age is about 18 billion years old, which is about four times the age of Earth. If so, calculate the age of the universe in seconds. If you had a sample of carbon with the same number of carbon atoms as there have been seconds since the universe began, determine whether you could measure this sample on a laboratory balance that can detect masses as small as 0.1 mg.

CP2.C (Section 2-12) A chemist analyzes three compounds and reports these data for the percent by mass of the elements Ex, Ey, and Ez in each compound.

Compound	% Ex	% Ey	% Ez
A	37.485	12.583	49.931
B	40.002	6.7142	53.284
C	40.685	5.1216	54.193

Assume that you accept the notion that the numbers of atoms of the elements in compounds are in small whole-number ratios and that the number of atoms in a sample of any element is directly proportional to that sample's mass. What is possible for you to know about the empirical formulas for these three compounds?

CP2.D (Section 2-12) The table nearby displays on each horizontal row an empirical formula for one of the three compounds noted in CP3.A.

Compound A	Compound B	Compound C
________	$ExEy_2Ez$	________
$Ex_6Ey_8Ez_3$	________	________
________	________	Ex_3Ey_2Ez
________	$Ex_9Ey_2Ez_6$	________
________	________	$ExEy_2Ez_3$
$Ex_3Ey_8Ez_3$	________	________

Based only on what was learned in that problem, determine the empirical formula for each of the other two compounds in that row.

CP2.E (Section 2-12)

(a) Suppose that a chemist now determines that the ratio of the masses of equal numbers of atoms of Ez and Ex atoms is 1.3320 g Ez/1 g Ex. With this added information, what can now be known about the formulas for compounds A, B, and C in Problem CP2.C?

(b) Suppose that this chemist further determines that the ratio of the masses of equal numbers of atoms of Ex and Ey is 11.916 g Ex/1 g Ey. Determine the ratio of the masses of equal numbers of Ez and Ey atoms.

(c) If the mass ratios of equal numbers of atoms of Ex, Ey, and Ez are known, what can be known about the formulas of the three compounds A, B, and C?

Chemical Reactions | 3

pmarage/Fotolia LLC

High in the Andes mountains of southwestern Bolivia huge piles of salt rest on top of water saturated with salt at Salar de Uyuni salt flat—the world's largest at nearly 4,000 square miles. Approximately 25,000 tons of salt are harvested annually. The very dry air at 12,000 feet above sea level allows water to evaporate readily from the saturated solution, forming a crust of solid salt containing lithium, sodium, potassium, and magnesium chlorides. The saturated salt solution conducts an electric current; why? (Section 3-3) Lithium chloride could be formed by reaction of an acid and a base; what are acids and bases, and how do chemists symbolize reactions by which they product salts? (Section 3-4) How can we report how much salt there is in a solution such as the water at Salar de Uyuni? (Section 3-10) These and many similar questions are answered in this chapter.

Chemical reactions occur everywhere on Earth, from the tiniest biological cells to geological processes that form mineral deposits and continents. During a chemical reaction, reactants are converted to products that have properties different from those of the reactants. The *chemical properties* (← **Sec. 1-6**) of a substance consist of the chemical reactions that involve the substance as either reactant or product. A major emphasis of chemistry is to understand chemical reactions and answer questions such as these: When two or more substances are mixed, will a reaction occur? If a reaction occurs, what products form? How much reactant is needed to produce a desired quantity of product?

Many chemical reactions fall into three general categories, each of which will be considered subsequently in greater detail.

Ba^{2+} and Na^+ do not react with each other, and neither do Cl^- and SO_4^{2-}.

1. ***Precipitation reactions*** occur in aqueous solutions when at least one *insoluble product,* a precipitate, is formed. For example, when an aqueous solution of barium chloride, $BaCl_2$, is added to an aqueous solution of sodium sulfate, Na_2SO_4, insoluble barium sulfate, $BaSO_4$, precipitates from solution (Figure 3.1).
2. ***Acid-base reactions*** in aqueous solutions involve transfer of a hydrogen ion, H^+, from one reactant to another. For example, when hydrochloric acid, HCl, reacts with aqueous sodium hydroxide, NaOH, an H^+ from the acid is transferred to a hydroxide ion, OH^-, from NaOH to form water, HOH (H_2O). Sodium ions, Na^+, and chloride ions, Cl^-, remain in the solution as sodium chloride, NaCl (Figure 3.2).
3. ***Oxidation-reduction (redox) reactions*** involve transfer of one or more electrons from one reactant to another reactant. The reaction of magnesium metal with oxygen is an example of a redox reaction (Figure 3.3). Two electrons are transferred from each magnesium metal atom to an oxygen atom to form magnesium oxide, MgO, an ionic compound containing magnesium ions, Mg^{2+}, and oxide ions, O^{2-}.

Although we observe chemical reactions at the macroscale in the everyday world, understanding reactions requires that we imagine and interpret them at the atomic and molecular level: the nanoscale (← **Sec. 1-9**). Such nanoscale interpretations are summarized in *chemical equations*—symbolic representations (← **Sec. 1-11**) of the conversion of reactants to products. A correctly written chemical equation specifies reactants and products by their chemical formulas and must be *balanced* so that the same number of atoms of each kind is present before and after the reaction.

There must be an accounting for *all* atoms in a chemical reaction.

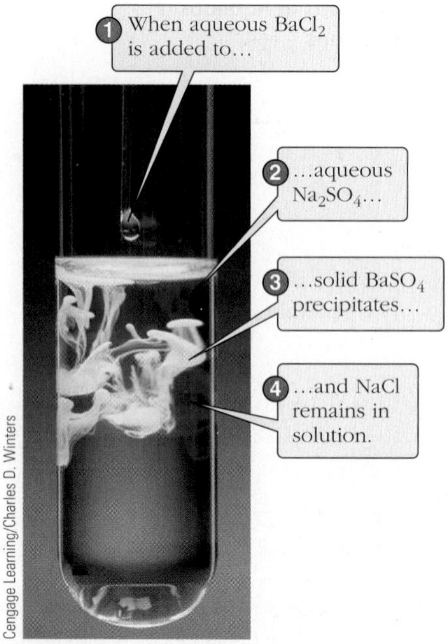
1 When aqueous BaCl2 is added to...
2 ...aqueous Na2SO4...
3 ...solid BaSO4 precipitates...
4 ...and NaCl remains in solution.
Cengage Learning/Charles D. Winters

Figure 3.1 Precipitation of barium sulfate.

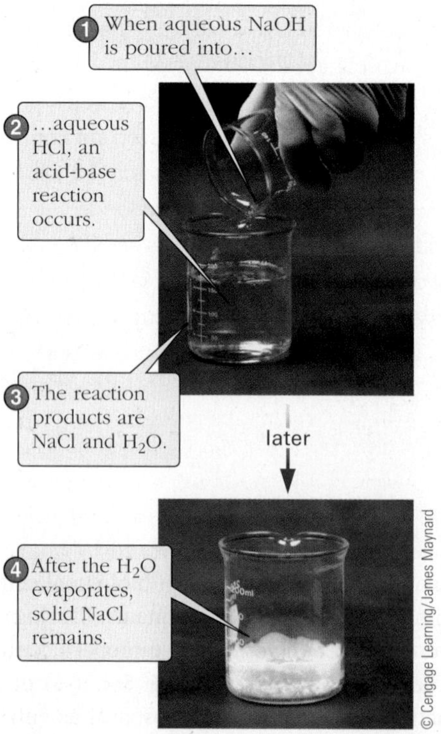
1 When aqueous NaOH is poured into...
2 ...aqueous HCl, an acid-base reaction occurs.
3 The reaction products are NaCl and H2O.
later
4 After the H2O evaporates, solid NaCl remains.
© Cengage Learning/James Maynard

Figure 3.2 Reaction of sodium hydroxide with hydrochloric acid.

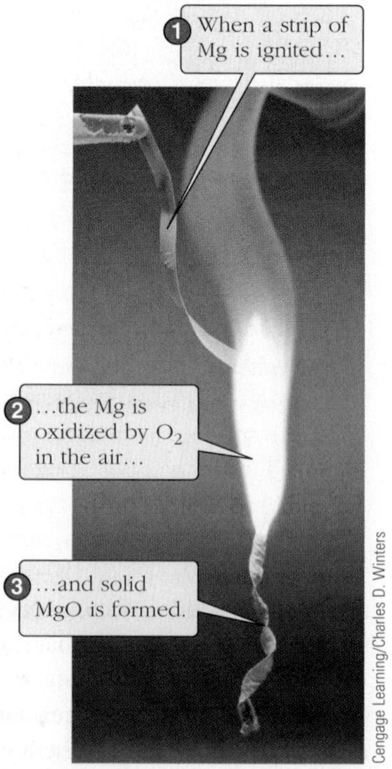
1 When a strip of Mg is ignited...
2 ...the Mg is oxidized by O2 in the air...
3 ...and solid MgO is formed.
Cengage Learning/Charles D. Winters

Figure 3.3 Oxidation of magnesium in air.

3-1 Chemical Equations

For any chemical reaction (← **Sec. 1-6**) we can write the general equation

$$\text{Reactant(s)} \longrightarrow \text{Products(s)}$$

which we would read as "reactants yield products" or "reactants form products" or "reactants change to products." The reactants and products can be elements, compounds, or ions. To write a chemical equation for a specific reaction, the first step is to figure out what the reactants are and what the products are.

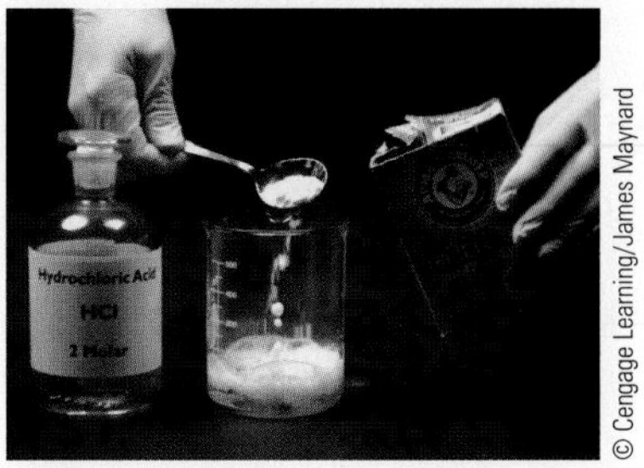
Hydrochloric Acid
HCl
2 Molar
© Cengage Learning/James Maynard

Figure 3.4 The reaction of sodium hydrogen carbonate (white solid) with hydrochloric acid produces carbon dioxide gas and aqueous sodium chloride.

For example, consider the reaction of an antacid, baking soda (sodium hydrogen carbonate), with excess stomach acid (hydrochloric acid). Figure 3.4 shows the reaction of sodium hydrogen carbonate with hydrochloric acid, a solution of hydrogen chloride in water. Based on their chemical properties, the gas bubbles escaping from the liquid can be identified as carbon dioxide. It can also be shown that some water is produced (in addition to the water already present in the solution of hydrogen chloride). If all of the water is evaporated after the reaction ends, sodium chloride remains. Therefore, we can write this equation:

$$\underset{\text{hydrogen carbonate}}{\text{sodium}} + \underset{\text{chloride}}{\text{hydrogen}} \longrightarrow \underset{\text{dioxide}}{\text{carbon}} + \text{water} + \underset{\text{chloride}}{\text{sodium}}$$

Usually we write chemical formulas rather than names for the reactants and products because the formulas specify clearly what elements are in each compound and how many atoms are in each molecule or formula unit. In addition, the physical states of reactants and products are usually indicated by placing a symbol after the formula: (s) for solid; (ℓ) for liquid; and (g) for gas. The symbol (aq) is used for substances dissolved in water—in *aq*ueous solution.

Based on what you learned about ionic and molecular compounds in Sections 2-4 through 2-9, formulas can be written for each reactant and each product. Sodium hydrogen carbonate is an ionic compound that consists of sodium ions and (polyatomic) hydrogen carbonate ions (← **Sec. 2-4d**); its formula is $NaHCO_3$ and it is the white solid in Figure 3.4. Hydrogen chloride is a binary molecular compound (← **Sec. 2-7**); its formula is HCl and it is in aqueous solution. The name carbon dioxide implies a formula CO_2 (← **Sec. 2-8**); it is a gas. The aqueous solution of sodium chloride contains Na^+ ions and Cl^- ions; because the charges on the ions are equal in magnitude, the formula is NaCl (← **Sec. 2-5c**). The formula for water is H_2O and it is a liquid. Thus the equation becomes

$$NaHCO_3(s) + HCl(aq) \longrightarrow CO_2(g) + H_2O(\ell) + NaCl(aq)$$

The formulas in the equation correspond with our experimental determination of what the reactants are and what the products are. Once established, the formulas cannot be changed because different formulas correspond with different substances. For example, if you changed the subscript 2 in CO_2 to a 1, the formula would be CO, the formula of carbon monoxide. Carbon monoxide is highly poisonous and CO is *not* the gas that is produced in the reaction shown in Figure 3.1—baking soda could not be used as an antacid if it produced a highly poisonous gas when it reacted with excess stomach acid!

The previous equation can be interpreted on the nanoscale by saying that one formula unit of sodium hydrogen carbonate reacts with one molecule of hydrogen chloride to form one molecule of carbon dioxide, one molecule of water, and one formula unit of sodium chloride (dissolved in water). It follows logically that if we had four formula units of sodium hydrogen carbonate we would need four molecules of hydrogen chloride to react with them and we would get four molecules (or formula units) of each of the products. Scaling up further, if we had Avogadro's number ($N_A = 6.022 \times 10^{23}$) of formula units of sodium hydrogen carbonate (1 mol $NaHCO_3$), we would need Avogadro's number of molecules of hydrogen chloride (1 mol HCl) to react with it and we would get

Recall that there are 6.022×10^{23} molecules in a mole of any molecular compound (← Sec. 2-11a) and 6.022×10^{23} formula units in a mole of any ionic compound (← Sec. 2-11b).

Avogadro's number of molecules of each product. That is, at the macroscale, this equation can be interpreted by substituting "mole" for "molecule" or "formula unit."

$NaHCO_3(s)$ +	$HCl(aq)$	$\longrightarrow$	$CO_2(g)$ +	$H_2O(\ell)$ +	$NaCl(aq)$
1 formula unit	1 molecule		1 molecule	1 molecule	1 formula unit
4 formula units	4 molecules		4 molecules	4 molecules	4 formula units
N_A formula units	N_A molecules		N_A molecules	N_A molecules	N_A formula units
1 mol	1 mol		1 mol	1 mol	1 mol

$N_A = 6.022 \times 10^{23}$

Once formulas for reactants and products have been written it is important to make certain that a chemical equation conforms to the law of conservation of matter and Dalton's atomic theory (← **Sec. 1-10**). **Because matter is neither created nor destroyed when an ordinary chemical reaction occurs, the total number of each kind of atom in the products must equal the total number of the same kind of atom in the reactants.**

By "atom" we mean not only an atom in a molecule, such as H in H_2O, but also an atom that has become an ion, such as Na^+ or Cl^- in NaCl.

We can verify the second statement by counting atoms of each kind in the formulas for reactants and products for the equation we have been discussing:

	$NaHCO_3(s)$ +	$HCl(aq)$	$\longrightarrow$	$CO_2(g)$ +	$H_2O(\ell)$ +	$NaCl(aq)$	Reactants	Products
Na	1	0	$\longrightarrow$	0	0	1	1	1
H	1	1	$\longrightarrow$	0	2	0	2	2
C	1	0	$\longrightarrow$	1	0	0	1	1
O	3	0	$\longrightarrow$	2	1	0	3	3
Cl	0	1	$\longrightarrow$	0	0	1	1	1

Because the mass of each atom is the same regardless of whether it is in a reactant or a product, verifying that the same number of atoms of each kind is present before and after a reaction also guarantees that the total mass of products equals the total mass of reactants. *When a chemical equation conforms to the law of conservation of matter it is a* **balanced chemical equation**.

In a chemical reaction there is no *detectable* change in mass, but there is a very tiny change because of the equivalence of mass and energy expressed in the equation $E = mc^2$. If the energy of reaction products is less than the energy of reactants, the mass of the products is also less. The difference is very small because c^2 is very big ($c^2 = 8.99 \times 10^{16}$ m^2 s^{-2}).

When a chemical reaction is over, the same atoms are present that were there when the reaction started. Because reactants change into products when a chemical reaction takes place, we often say that the reactants are "consumed" or "used up" and that the products are "formed." This does not mean that the atoms of which the reactants were composed were consumed, nor that atoms were created when the products formed. The same atoms are still there, but they are arranged differently. For example, in the reaction of $NaHCO_3$ with HCl, the H atoms in the product H_2O are H atoms that were originally in $NaHCO_3$ and HCl. Reactants disappear and products appear, but atoms are neither created nor destroyed in a chemical reaction.

CONCEPTUAL EXERCISE 3.1

Answers to **EXERCISES** are provided at the back of this book in Appendix L.

EXERCISES that are labeled **CONCEPTUAL** are designed to test your understanding of one or more concepts; they usually involve qualitative rather than quantitative thinking.

Chemical Equations

When washing soda, Na_2CO_3, reacts with sulfuric acid, H_2SO_4, the equation is

$$Na_2CO_3(aq) + H_2SO_4(aq) \longrightarrow CO_2(g) + H_2O(\ell) + Na_2SO_4(aq)$$

(a) Describe in words the meaning of this chemical equation.
(b) Verify that the equation conforms to the law of conservation of matter.

3-2 Balancing Chemical Equations

Most chemical equations differ from the reaction of $NaHCO_3$ with HCl in that they are not balanced when correct formulas are first written for reactants and products. Such unbalanced equations need to be balanced. This is done by adjusting **coefficients**, *numbers that precede each formula and indicate how many of each formula must be involved in the reaction*. For example, consider the reaction of hydrogen gas with chlorine gas to form hydrogen chloride gas shown in Figure 3.5. Hydrogen and chlorine are elements

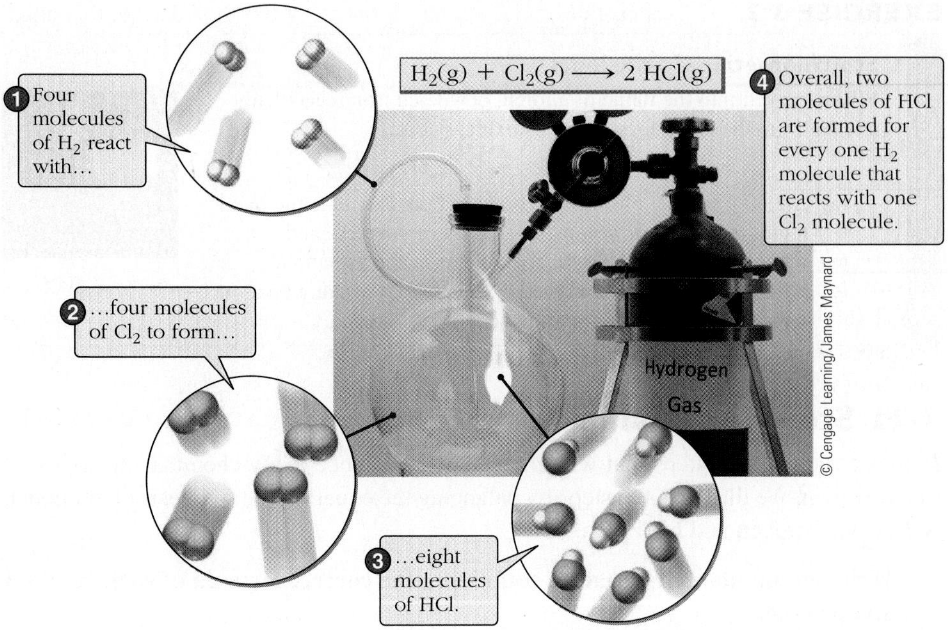

$H_2(g) + Cl_2(g) \longrightarrow 2\ HCl(g)$
1 Four molecules of H_2 react with...
2 ...four molecules of Cl_2 to form...
3 ...eight molecules of HCl.
4 Overall, two molecules of HCl are formed for every one H_2 molecule that reacts with one Cl_2 molecule.
Hydrogen Gas
© Cengage Learning/James Maynard

Figure 3.5 Colorless hydrogen gas, H_2, and pale yellow chlorine gas, Cl_2, react to form hydrogen chloride, HCl, in a bright flame.

that consist of diatomic molecules; hydrogen chloride has the formula HCl. Thus we can write

$$H_2(g) + Cl_2(g) \longrightarrow 2\ HCl(g)$$

The coefficient 2 is needed so that the total number of H atoms on the right equals the 2 H atoms in the H_2 molecule on the left. The 2 also balances the number of Cl atoms. (When a formula is written with no coefficient, this means that the coefficient is 1; this is analogous to the convention that in a chemical formula such as $CaCl_2$ the subscript 1 is not written with Ca.) This equation can be interpreted as

$H_2(g)$	+	$Cl_2(g)$	$\longrightarrow$	$2\ HCl(g)$
1 molecule		1 molecule		2 molecules
1 mol		1 mol		2 mol

The elements that consist of diatomic molecules are highlighted in red in this periodic table.

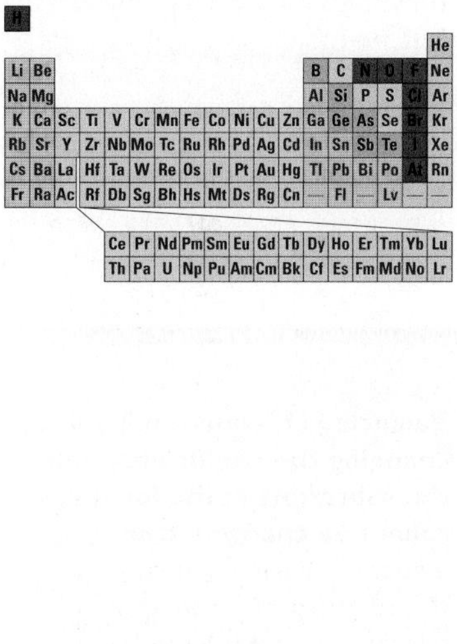

We can use this interpretation to verify that balancing the atoms also balances the masses. The mass of 1 mol H_2 is 2.016 g and the mass of 1 mol Cl_2 is 70.906 g, giving a total mass for the reactants of 72.922 g. The mass of 2 mol HCl should be the same, and it is:

$$\text{Mass HCl} = 2.0000 \text{ mol HCl} \times \frac{36.461 \text{ g HCl}}{1 \text{ mol HCl}} = 72.922 \text{ g HCl}$$

Relationships among the masses of reactants and products in chemical reactions are part of **stoichiometry** (stow-eee-key-AHM-uh-tree). The *coefficients in a balanced chemical equation are referred to as* **stoichiometric coefficients**.

One more point about balanced chemical equations: It is conventional, but not always essential, to make the coefficients the smallest possible whole numbers. For example, in Figure 3.5 the enlarged circles showing the nanoscale correspond with this equation:

$$4\ H_2(g) + 4\ Cl_2(g) \longrightarrow 8\ HCl(g)$$

This equation is balanced (there are eight H atoms and eight Cl atoms on each side), but the coefficients are four times the smallest whole numbers. We would write the balanced equation with coefficients of 1 for each reactant and 2 for HCl, the product, as we did previously. It is important to recognize, however, that the balanced equation does represent this and many other cases where all of the coefficients have been multiplied by the same factor.

EXERCISE 3.2

Stoichiometric Coefficients

When sprayed into the flame of a torch, powdered iron reacts with oxygen from the air to form iron(III) oxide, Fe_2O_3:

© Cengage Learning/James Maynard

$$4\ Fe(s) + 3\ O_2(g) \longrightarrow 2\ Fe_2O_3(s)$$

(a) If 2.50 g Fe_2O_3 is formed by this reaction, calculate the maximum *total* mass of iron metal and oxygen that reacted.
(b) Identify the stoichiometric coefficients in this equation.
(c) If 10,000 oxygen atoms reacted, calculate how many Fe atoms were needed to react with this quantity of oxygen.

3-2a Steps for Balancing an Equation

Here is a sequence of steps that will enable you to balance many chemical equations by trial and error. We illustrate the steps by balancing the equation for synthesis of ammonia, NH_3, from nitrogen and hydrogen.

Millions of tons of ammonia are manufactured annually by this reaction using nitrogen extracted from air and hydrogen obtained from petroleum refining.

- **Write an unbalanced equation containing the correct formula of each reactant and product.**

$$\text{(unbalanced equation)}\ N_2 + H_2 \longrightarrow NH_3$$

Elemental nitrogen and hydrogen consist of diatomic molecules, so each reactant formula needs a subscript 2; the product formula was given.

Elemental nitrogen is nitrogen as the element—not combined with any other element.

Clearly, both nitrogen and hydrogen atoms are unbalanced. There are two nitrogen atoms on the left and only one on the right, and two hydrogen atoms on the left and three on the right.

- **Choose an atom that appears in the fewest formulas and insert coefficients to balance that atom.** Both N and H appear in one formula on the left and one on the right, so either could be chosen. Start by using a coefficient of 2 on the right to balance the nitrogen atoms: 2 NH_3 indicates two ammonia molecules, each containing a nitrogen atom and three hydrogen atoms. On the right we now have two nitrogen atoms and six hydrogen atoms.

$$\text{(unbalanced equation)}\ N_2 + H_2 \longrightarrow 2\ NH_3$$

Balancing an equation involves changing the coefficients, but the subscripts in the formulas *cannot* be changed. Changing a subscript means changing the identity of a reactant or product, and we have already determined the correct formulas for the reactants and products.

- **Insert coefficients to balance atoms of the remaining elements, starting with those that appear in the fewest formulas.** In this case only one element is left. Balance the six hydrogen atoms on the right using a coefficient of 3 for the H_2 to furnish six hydrogen atoms on the left.

$$\text{(balanced equation)}\ N_2 + 3\ H_2 \longrightarrow 2\ NH_3$$

- **Verify that the number of atoms of each element is balanced.** Do an atom count to check that the numbers of nitrogen and hydrogen atoms are the same on each side of the equation.

(balanced equation)	N_2	+	$3\ H_2$	$\longrightarrow$	$2\ NH_3$
atom count:	2 N	+	(3×2) H	=	2 N + (2×3) H
	2 N	+	6 H	=	2 N + 6 H

The physical states of the reactants and products are frequently included in the balanced equation. Thus, the final equation for ammonia formation is

$$N_2(g) + 3\ H_2(g) \longrightarrow 2\ NH_3(g)$$

PROBLEM-SOLVING EXAMPLE 3.1

Balancing a Chemical Equation

Ammonia gas reacts with oxygen gas to form nitrogen monoxide gas and water vapor at 1000 °C. Write a balanced equation for this reaction.

Result $4\ NH_3(g) + 5\ O_2(g) \longrightarrow 4\ NO(g) + 6\ H_2O(g)$

Analyze Balancing an equation means finding the smallest whole number coefficients that result in the same number of each kind of atom on each side of the equation.

Plan Use the stepwise approach to balancing chemical equations.

Execute

- **Write an unbalanced equation containing the correct formulas of all reactants and products.** The formula for ammonia is NH_3. Oxygen consists of diatomic molecules. The name nitrogen monoxide means the formula is NO (← **Sec. 2-8**). The formula for water is H_2O. The unbalanced equation is

(unbalanced equation) $NH_3(g) + O_2(g) \longrightarrow NO(g) + H_2O(g)$

- **Choose an atom that appears in the fewest formulas and insert coefficients to balance that atom.** Nitrogen and hydrogen each appear in two formulas; oxygen is in three. Nitrogen is balanced, so choose hydrogen. There are three hydrogen atoms on the left and two on the right. Whenever three and two atoms must be balanced, use coefficients to give six atoms on each side of the equation. A coefficient of 2 on the left and 3 on the right gives six hydrogen atoms on each side.

(unbalanced equation) $2\ NH_3(g) + O_2(g) \longrightarrow NO(g) + 3\ H_2O(g)$

- **Insert coefficients to balance atoms of the remaining elements, starting with those that appear in the fewest formulas.** Nitrogen is in two formulas; oxygen is in three. There are now two nitrogen atoms on the left and one on the right, so we balance nitrogen by using the coefficient 2 for the NO molecule.

(unbalanced equation) $2\ NH_3(g) + O_2(g) \longrightarrow 2\ NO(g) + 3\ H_2O(g)$

Now there are two oxygen atoms on the left and five on the right. We use a coefficient of $\frac{5}{2}$ with O_2 to balance the atoms of O.

(balanced equation) $2\ NH_3(g) + \frac{5}{2}\ O_2(g) \longrightarrow 2\ NO(g) + 3\ H_2O(g)$

The equation is now balanced, but it is customary to adjust the coefficients to the smallest possible whole-number values. Therefore, we multiply each coefficient by 2 to get the final balanced equation.

(final balanced equation) $4\ NH_3(g) + 5\ O_2(g) \longrightarrow 4\ NO(g) + 6\ H_2O(g)$

☑ **Reasonable Result Check** Verify that the number of atoms of each element is balanced.

(final balanced equation) $4\ NH_3(g) + 5\ O_2(g) \longrightarrow 4\ NO(g) + 6\ H_2O(g)$

4 N		=	4 N		
12 H		=			12 H
	10 O	=	4 O	+	6 O

$\frac{5}{2}\ O_2$ is equivalent to 5 O.

PROBLEM-SOLVING PRACTICE 3.1

Balance these equations:

(a) $Cr(s) + Cl_2(g) \longrightarrow CrCl_3(s)$
(b) $As_2O_3(s) + H_2(g) \longrightarrow As(s) + H_2O(\ell)$
(c) $C_3H_8 + O_2 \longrightarrow CO_2 + H_2O$
(d) $C_2H_5OH + O_2 \longrightarrow CO_2 + H_2O$

PROBLEM-SOLVING PRACTICE answers are provided at the back of this book in Appendix K.

When one or more polyatomic ions (← **Sec. 2-4a**) appear on both sides of a chemical equation, each one is treated as a unit during the balancing steps. When such an ion must have a subscript in the chemical formula, the polyatomic ion is enclosed in parentheses. For example, in the equation for the reaction between sodium phosphate and barium nitrate to produce barium phosphate and sodium nitrate,

$$2\ Na_3PO_4(aq) + 3\ Ba(NO_3)_2(aq) \longrightarrow Ba_3(PO_4)_2(s) + 6\ NaNO_3(aq)$$

the nitrate ions, NO_3^-, and phosphate ions, PO_4^{3-}, are present before and after the reaction. They can be balanced as units, similar to atoms. The polyatomic ions are enclosed in parentheses when they occur more than once in a formula.

Fe(NO3)3(aq)
Na2CrO4(aq)

Reaction of iron(III) nitrate with sodium chromate.

PROBLEM-SOLVING EXAMPLE 3.2

Balancing Polyatomic Ions

Sodium chromate and iron(III) nitrate react in aqueous solution to form a precipitate of solid iron(III) chromate; sodium nitrate remains in solution after the reaction. Write the balanced equation for this reaction.

Result $3\ Na_2CrO_4(aq) + 2\ Fe(NO_3)_3(aq) \longrightarrow Fe_2(CrO_4)_3(s) + 6\ NaNO_3(aq)$

Analyze Balance the equation, which involves polyatomic ions.

Plan Use the stepwise approach to balancing chemical equations. Treat each polyatomic ion the same way you would treat an atom.

Execute

- **Write an unbalanced equation containing the correct formulas of all reactants and products.** Use rules for writing formulas in Chapter 2 (← **Sec. 2-4d**):

$$\text{(unbalanced)}\quad Na_2CrO_4(aq) + Fe(NO_3)_3(aq) \longrightarrow Fe_2(CrO_4)_3(s) + NaNO_3(aq)$$

- **Choose an atom that appears in the fewest formulas and insert coefficients to balance that atom.** Treat the four species Na^+, CrO_4^{2-}, Fe^{3+}, and NO_3^- as if they were atoms. Each appears in one formula on each side of the equation. Begin with any of them; CrO_4^{2-} appears once on the left and three times on the right so insert a coefficient 3 on the left.

$$\text{(unbalanced)}\quad 3\ Na_2CrO_4(aq) + Fe(NO_3)_3(aq) \longrightarrow Fe_2(CrO_4)_3(s) + NaNO_3(aq)$$

- **Insert coefficients to balance atoms of the remaining elements, starting with those that appear in the fewest formulas.** There are six sodium ions on the left and only one on the right so insert a coefficient of 6 on the right.

$$\text{(unbalanced)}\quad 3\ Na_2CrO_4(aq) + Fe(NO_3)_3(aq) \longrightarrow Fe_2(CrO_4)_3(s) + 6\ NaNO_3(aq)$$

  There are six nitrate ions on the right and three on the left. Use a coefficient of 2 on the left to balance the nitrate ions. This also balances the iron(III) ions.

$$\text{(balanced)}\quad 3\ Na_2CrO_4(aq) + 2\ Fe(NO_3)_3(aq) \longrightarrow Fe_2(CrO_4)_3(s) + 6\ NaNO_3(aq)$$

  The equation is now balanced with the smallest whole-number coefficients.

☑ **Reasonable Result Check** Verify that the number of atoms of each element is balanced.

$$3\ Na_2CrO_4(aq) + 2\ Fe(NO_3)_3(aq) \longrightarrow Fe_2(CrO_4)_3(s) + 6\ NaNO_3(aq)$$

Na_2CrO_4		$Fe(NO_3)_3$		$Fe_2(CrO_4)_3$		$NaNO_3$
6 Na			=			6 Na
3 Cr			=	3 Cr		
		2 Fe	=	2 Fe		
		6 N	=			6 N
12 O	+	18 O	=	12 O	+	18 O

PROBLEM-SOLVING PRACTICE 3.2

Balance these equations.

(a) $Na_2Cr_2O_7(aq) + AgNO_3(aq) \longrightarrow Ag_2Cr_2O_7(s) + NaNO_3(aq)$

(b) $NaHCO_3(aq) + CaCl_2(aq) \longrightarrow CaCO_3(s) + H_2CO_3(aq) + NaCl(aq)$

3-3 Precipitation Reactions

Because atoms, molecules, or ions can move about freely in a liquid and can easily interact with other atoms, molecules, or ions, many reactions occur in solutions. This is especially true of ionic compounds, many of which are soluble in water and form aqueous solutions. As a result, oceans, rivers, lakes, and even the tap water in our residences contain many kinds of ions in solution. This makes the solubilities of ionic compounds and the properties of ions in aqueous solution of great practical importance.

3-3a Ionic Compounds in Aqueous Solution

When an ionic compound dissolves in water, the ions in the compound separate and become surrounded by water molecules. *The process by which cations and anions separate when an ionic compound dissolves is called* **dissociation**; the compound is said to *dissociate*. For example, when solid sodium chloride dissolves in water the sodium chloride dissociates into Na^+ and Cl^- ions that become surrounded by water molecules, as illustrated by Figure 3.6. The dissociation of NaCl in water can be represented by the balanced chemical equation:

$$NaCl(aq) \longrightarrow Na^+(aq) + Cl^-(aq)$$

Polyatomic ions in an ionic compound remain intact during the dissociation of the compound in water. For example, solid magnesium acetate, $Mg(CH_3COO)_2$, dissociates in water into magnesium ions and acetate ions:

$$Mg(CH_3COO)_2(aq) \longrightarrow Mg^{2+}(aq) + 2\ CH_3COO^-(aq)$$

Dissociation means that when we write NaCl(aq) it is understood that the solution contains aqueous sodium ions, $Na^+(aq)$, and aqueous chloride ions, $Cl^-(aq)$. These ions are free to move about the solution where they may encounter ions of other kinds (if another ionic compound is in the same solution).

PROBLEM-SOLVING EXAMPLE 3.3

Dissociation of Ionic Compounds

Write balanced equations for the dissociation of each of these ionic compounds when it dissolves in water.

(a) $CaBr_2$ (b) $Fe(NO_3)_3$ (c) $V_2(SO_4)_3$ (d) Na_3PO_4 (e) $Sr(ClO_4)_2$

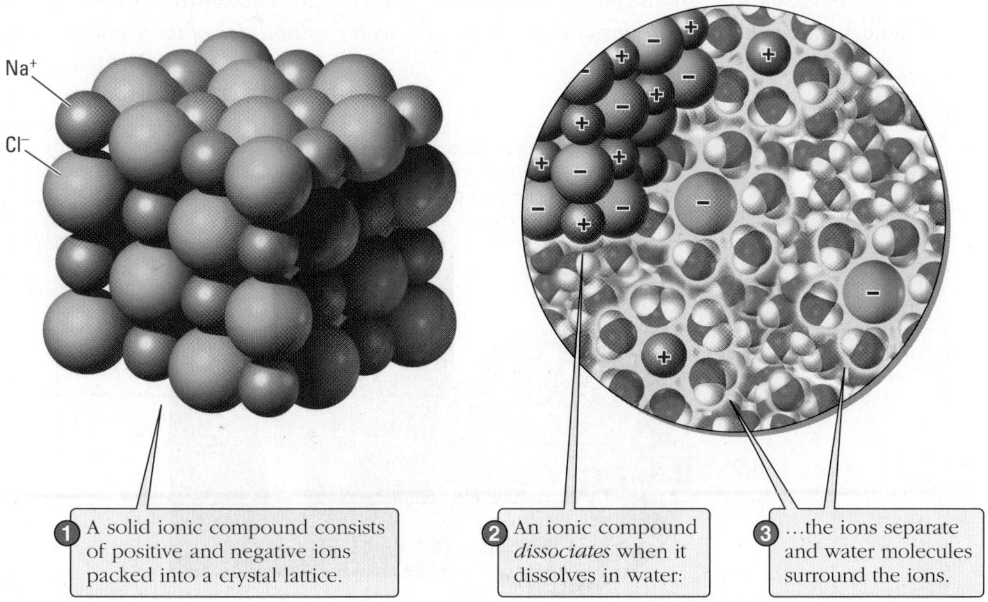
Na+
Cl−
1 A solid ionic compound consists of positive and negative ions packed into a crystal lattice.
2 An ionic compound *dissociates* when it dissolves in water:
3 ...the ions separate and water molecules surround the ions.

Figure 3.6 Dissociation of ions in an ionic compound, NaCl. Dissociation is separation of the ions present in a solid ionic compound when the compound dissolves in water.

Result

(a) $CaBr_2(aq) \longrightarrow Ca^{2+}(aq) + 2\ Br^-(aq)$
(b) $Fe(NO_3)_3(aq) \longrightarrow Fe^{3+}(aq) + 3\ NO_3^-(aq)$
(c) $V_2(SO_4)_3(aq) \longrightarrow 2\ V^{3+}(aq) + 3\ SO_4^{2-}(aq)$
(d) $Na_3PO_4(aq) \longrightarrow 3\ Na^+(aq) + PO_4^{3-}(aq)$
(e) $Sr(ClO_4)_2(aq) \longrightarrow Sr^{2+}(aq) + 2\ ClO_4^-(aq)$

Analyze The ions that dissociate are the ions present in the ionic solid. Polyatomic ions remain intact during dissociation.

Plan Identify ions in each compound and determine proper coefficients based on subscripts in the formulas. Use (aq) to indicate ions in aqueous solution.

Execute Part (c) is given as an example; other parts are similar.

(c) $V_2(SO_4)_3$ consists of V^{3+} ions and SO_4^{2-} ions. (V^{3+} is 3+ because two 3+ ions are needed to balance three 2− sulfate ions). The subscripts indicate 2 V^{3+} ions and 3 SO_4^{2-} ions so the products are $2\ V^{3+}(aq) + 3\ SO_4^{2-}(aq)$.

1 mol $V_2(SO_4)_3$ contains 2 mol V^{3+} ions along with 3 mol SO_4^{2-} ions.

PROBLEM-SOLVING PRACTICE 3.3

Write a balanced equation for dissociation of each of these ionic compounds when it dissolves in water.

(a) $(NH_4)_2S$ (b) $KMnO_4$ (c) $K_2C_2O_4$ (d) Li_2CO_3 (e) $Co(SCN)_2$

3-3b Electrolytes

Cations and anions also move about freely in molten ionic compounds (← Sec. 2-6c), which is the reason they conduct electricity.

Ionic compounds are **electrolytes**, *substances that conduct an electric current when dissolved in water.* Aqueous solutions of ionic compounds conduct electricity because the compound dissociates and its cations and anions can move freely in the solution (Figure 3.7). There are two categories of electrolytes: A **strong electrolyte** is *a compound that completely dissociates into its ions when it forms an aqueous solution;* a **weak electrolyte** is *a compound that dissociates only partially when it forms an aqueous solution*. Aqueous solutions of weak electrolytes conduct less electricity; we will discuss them later.

3-3c Solubility Rules: Aqueous Solubility of Ionic Compounds

Ionic compounds that you encounter frequently, such as table salt, NaCl, and baking soda, $NaHCO_3$, are usually soluble in water. Although many ionic compounds are water-soluble, some are only slightly soluble, and others dissolve hardly at all. The **solubility rules** given in Table 3.1 are *general guidelines for predicting the water solubilities of ionic compounds*

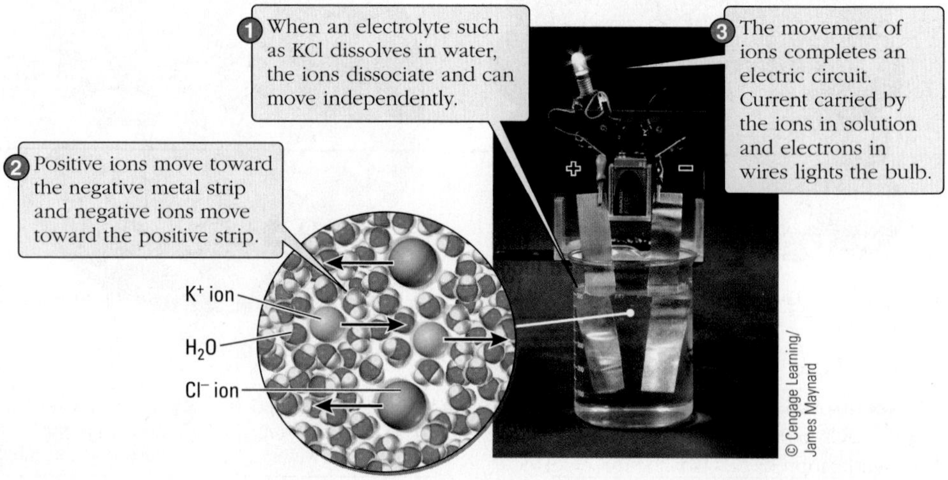
1 When an electrolyte such as KCl dissolves in water, the ions dissociate and can move independently.
2 Positive ions move toward the negative metal strip and negative ions move toward the positive strip.
3 The movement of ions completes an electric circuit. Current carried by the ions in solution and electrons in wires lights the bulb.
K+ ion
H2O
Cl− ion
© Cengage Learning/ James Maynard

Figure 3.7 An aqueous solution of an ionic compound conducts electricity because the ions move.

Table 3.1 Solubility Rules for Ionic Compounds

Usually Soluble	
Group 1A (Li^+, Na^+, K^+, Rb^+, Cs^+), NH_4^+	All Group 1A (alkali metal) and ammonium (NH_4^+) compounds are soluble.
Nitrates: NO_3^-	All nitrates are soluble.
Chlorides, bromides, iodides: Cl^-, Br^-, I^-	All common chlorides, bromides, and iodides are soluble except those of Ag^+, Hg_2^{2+}, and Pb^{2+}. $AgCl$, Hg_2Cl_2, $PbCl_2$, $AgBr$, Hg_2Br_2, $PbBr_2$, AgI, Hg_2I_2, and PbI_2 are insoluble.
Sulfates: SO_4^{2-}	Most sulfates are soluble; exceptions include $CaSO_4$, $SrSO_4$, $BaSO_4$, and $PbSO_4$.
Chlorates: ClO_3^-	All chlorates are soluble.
Perchlorates: ClO_4^-	All perchlorates are soluble.
Acetates: CH_3COO^-	All acetates are soluble.
Usually Insoluble	
Phosphates: PO_4^{3-}	All phosphates are insoluble except those of NH_4^+ and Group 1A ions (alkali metal cations).
Carbonates: CO_3^{2-}	All carbonates are insoluble except those of NH_4^+ and Group 1A ions (alkali metal cations).
Chromates: CrO_4^{2-}	All chromates are insoluble except those of NH_4^+, Mg^{2+}, and Group 1A ions (alkali metal cations).
Hydroxides: OH^-	All hydroxides are insoluble except those of NH_4^+ and Group 1A (alkali metal cations). $Sr(OH)_2$, $Ba(OH)_2$, and $Ca(OH)_2$ are slightly soluble.
Oxalates: $C_2O_4^{2-}$	All oxalates are insoluble except those of NH_4^+ and Group 1A (alkali metal cations).
Sulfides: S^{2-}	All sulfides are insoluble except those of NH_4^+, Group 1A (alkali metal cations) and Group 2A (MgS, CaS, and BaS are sparingly soluble).

based on the ions they contain. ***If a compound contains at least one of the ions indicated for soluble compounds in Table 3.1, then the compound is at least moderately soluble.***

Suppose you want to know whether nickel(II) sulfate, $NiSO_4$, is soluble in water. Nickel(II) sulfate contains Ni^{2+} and SO_4^{2-} ions. Although Ni^{2+} is not mentioned in Table 3.1, substances containing SO_4^{2-} are described as soluble (except for $CaSO_4$, $SrSO_4$, $BaSO_4$, and $PbSO_4$). Because $NiSO_4$ contains an ion, SO_4^{2-}, that indicates solubility and $NiSO_4$ is not one of the sulfate exceptions, it is predicted to be soluble.

PROBLEM-SOLVING EXAMPLE 3.4

Using Solubility Rules

Predict whether each compound is water-soluble.

(a) $CaCl_2$ (b) $Fe(OH)_3$ (c) NH_4NO_3
(d) $CuCO_3$ (e) $Ni(ClO_3)_2$

Result (a) $CaCl_2$, soluble. (b) $Fe(OH)_3$, insoluble. (c) NH_4NO_3, soluble. (d) $CuCO_3$, insoluble. (e) $NiClO_3$, soluble.

Analyze To apply the solubility rules, identify the ions present and check their aqueous solubility.

Plan Figure out from each formula what ions are present; then use Table 3.1.

Execute

(a) $CaCl_2$ contains Ca^{2+} and Cl^- ions. All chlorides are soluble, with a few exceptions for transition metals, so calcium chloride is soluble.

(b) $Fe(OH)_3$ contains Fe^{3+} and OH^- ions. All hydroxides are insoluble except those of alkali metals and a few others, so iron(III) hydroxide is insoluble.
(c) NH_4NO_3 contains NH_4^+ and NO_3^- ions. All ammonium salts are soluble, and all nitrates are soluble, so NH_4NO_3 is soluble.
(d) $CuCO_3$ contains Cu^{2+} and CO_3^{2-} ions. All carbonates are insoluble except those of ammonium and alkali metal ions; copper(II) ion is not one of the exceptions, so $CuCO_3$ is insoluble.
(e) $Ni(ClO_3)_2$ contains Ni^{2+} and ClO_3^- ions. All chlorates are soluble, so $Ni(ClO_3)_2$ is soluble.

PROBLEM-SOLVING PRACTICE 3.4

Predict whether each of these compounds is likely to be water-soluble.

(a) NaF (b) $Ca(CH_3COO)_2$ (c) $SrCl_2$
(d) MgO (e) $PbCl_2$ (f) HgS

3-3d Precipitation Reactions

A **precipitation reaction** *is a reaction in which one or more insoluble products form from soluble reactants.* Formation of a precipitate removes ions from the solution and causes the reaction to occur. Consider a mixture of aqueous solutions of lead(II) nitrate, $Pb(NO_3)_2$, and potassium chromate, K_2CrO_4. Data in Table 3.1 indicate that these two compounds are water soluble; they dissociate into their component ions.

$$Pb(NO_3)_2(aq) \longrightarrow Pb^{2+}(aq) + 2\ NO_3^-(aq)$$
$$K_2CrO_4(aq) \longrightarrow 2\ K^+(aq) + CrO_4^{2-}(aq)$$

Thus, the aqueous mixture contains Pb^{2+}, NO_3^-, K^+, and CrO_4^{2-} ions. Because of their like charges, the Pb^{2+} and K^+ cations repel each other; the NO_3^- and CrO_4^{2-} anions also repel. Therefore, cations do not precipitate with other cations and anions do not precipitate with anions. Two new cation-anion combinations are possibilities: Pb^{2+} with CrO_4^{2-} and K^+ with NO_3^-.

Notice that Pb^{2+} ends up with CrO_4^{2-} and K^+ ends up with NO_3^-. Reactions in which ions exchange partners like this are called *exchange reactions*.

$$Pb(NO_3)_2(aq) + K_2CrO_4(aq) \longrightarrow 2\ KNO_3 + PbCrO_4$$

A precipitate will form if either of the two products is insoluble (or if both are). Checking Table 3.1 we see that KNO_3 is soluble, but $PbCrO_4$ is not soluble. Therefore, Pb^{2+} and CrO_4^{2-} ions combine to form solid lead(II) chromate, $PbCrO_4(s)$, which precipitates from the solution (Figure 3.8). Potassium nitrate is in aqueous solution so the other product in the equation is $KNO_3(aq)$.

$$Pb(NO_3)_2(aq) + K_2CrO_4(aq) \longrightarrow 2\ KNO_3(aq) + PbCrO_4(s)$$

Could we mix two ion-containing aqueous solutions and not have a reaction? Consider mixing strontium nitrate and sodium bromide solutions. Both compounds are water-soluble and dissociate, so the mixture contains Sr^{2+}, NO_3^-, Na^+, and Br^- ions. If any cation-anion pair forms an insoluble compound, precipitation will occur. Possible cation-anion pairs in this case are Sr^{2+} and Br^- or Na^+ and NO_3^-. Based on the solubility rules (Table 3.1), most bromides are soluble and strontium bromide is not an exception; all nitrates and all sodium compounds are soluble. Therefore, there is no precipitate to remove ions from solution and no chemical reaction (abbreviated N.R.) occurs.

Pb(NO3)2(aq)
K2CrO4(aq)

Figure 3.8 Precipitation of lead(II) chromate.

PROBLEM-SOLVING EXAMPLE 3.5

Ion Combinations and Precipitation

Predict whether mixing aqueous solutions of these pairs of ionic compounds produces a precipitate. Write a balanced chemical equation for each reaction that produces a precipitate. If no reaction occurs, write N.R.

(a) $(NH_4)_2S$ and $Cu(NO_3)_2$ (b) $ZnCl_2$ and Na_2CO_3 (c) $CaCl_2$ and KNO_3

Result

(a) CuS precipitates. $(NH_4)_2S(aq) + Cu(NO_3)_2(aq) \longrightarrow CuS(s) + 2\ NH_4NO_3(aq)$
(b) $ZnCO_3$ precipitates. $ZnCl_2(aq) + Na_2CO_3(aq) \longrightarrow ZnCO_3(s) + 2\ NaCl(aq)$
(c) No reaction occurs (N.R.). Both possible products, $Ca(NO_3)_2$ and KCl, are soluble.

Analyze Given formulas of reactants, some containing polyatomic ions, predict products that are insoluble or soluble.

Plan Identify ions in each formula, consider which new cation-anion combinations can form, and decide whether any are insoluble. Treat each polyatomic ion as a unit.

Execute

(a) Possible products are CuS(s) and NH_4NO_3. All nitrates are soluble, so NH_4NO_3 remains in solution. Most sulfides are not soluble so CuS(s) precipitates. (See Result for the equation.)
(b) The exchange reaction between $ZnCl_2$ and Na_2CO_3 forms the insoluble product $ZnCO_3(s)$ and leaves soluble NaCl in solution.
(c) No precipitate forms when $CaCl_2$ and KNO_3 are mixed because each product, $Ca(NO_3)_2$ and KCl, is soluble. No reaction occurs because all four of the ions (Ca^{2+}, Cl^-, K^+, NO_3^-) that were in the initial solution remain in solution in the products.

PROBLEM-SOLVING PRACTICE 3.5

Predict the products and write a balanced chemical equation for the reaction in aqueous solution between each pair of ionic compounds.
(a) $NiCl_2$ and NaOH (b) K_2CO_3 and $CaBr_2$

3-3e Net Ionic Equations

In the preceding section we wrote this equation for precipitation of $PbCrO_4(s)$:

$$Pb(NO_3)_2(aq) + K_2CrO_4(aq) \longrightarrow 2\ KNO_3(aq) + PbCrO_4(s)$$

This is called an *overall equation.* But remember that three of the formulas represent soluble ionic compounds; for example, the first formula, $Pb(NO_3)_2(aq)$, represents an aqueous solution that consists of water molecules, $Pb^{2+}(aq)$ ions, and $NO_3^-(aq)$ ions. It would be more accurate to write the equation this way:

$$Pb^{2+}(aq) + 2\ NO_3^-(aq) + 2\ K^+(aq) + CrO_4^{2-}(aq) \longrightarrow 2\ K^+(aq) + 2\ NO_3^-(aq) + PbCrO_4(s)$$

Notice that NO_3^- and K^+ ions were in solution at the beginning of the reaction and are still in solution at the end of the reaction. *Any ion that is present but is not involved directly in the reaction* is called a **spectator ion**. Like the spectators at a play or game, such ions are not part of the real action (in this case, the precipitation) and can be left out of the equation that represents the chemical change. This gives

$$Pb^{2+}(aq) + CrO_4^{2-}(aq) \longrightarrow PbCrO_4(s)$$

An equation that includes only the symbols or formulas of ions or molecules in solution that undergo change is called a **net ionic equation**.

We use the reaction of NaCl(aq) with $AgNO_3(aq)$ (Figure 3.9) to illustrate how to decide whether a reaction occurs and how to write a net ionic equation.

- ***Determine what chemical species (ions in this case) are present in the reactants.*** Soluble ionic compounds dissociate into their component ions. Nitrates are soluble so $AgNO_3$ is soluble; sodium compounds are soluble so NaCl is soluble.

$AgNO_3(aq)$ consists of $Ag^+(aq) + NO_3^-(aq)$.

NaCl(aq) consists of $Na^+(aq) + Cl^-(aq)$.

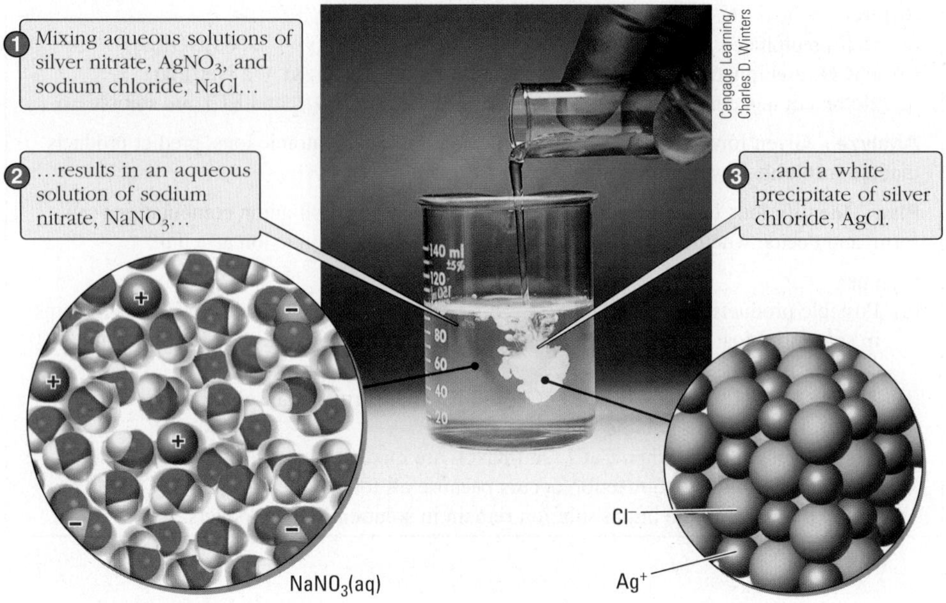

1 Mixing aqueous solutions of silver nitrate, AgNO3, and sodium chloride, NaCl...
2 ...results in an aqueous solution of sodium nitrate, NaNO3...
3 ...and a white precipitate of silver chloride, AgCl.
Cengage Learning/ Charles D. Winters
NaNO3(aq)
Cl−
Ag+

Figure 3.9 **Precipitation of silver chloride.**

- ***Decide whether a reaction occurs based on possible products and their solubilities.*** Two new cation-anion combinations are possible: Na^+ with NO_3^- and Ag^+ with Cl^-. Nitrates are soluble, so $NaNO_3$ is soluble. Most chlorides are soluble, but AgCl is one of the insoluble chlorides (the others are Hg_2Cl_2 and $PbCl_2$). Thus, a reaction occurs: AgCl(s) precipitates (Figure 3.9).

- ***Write the overall equation and balance it.*** This equation is balanced:

$$AgNO_3(aq) + NaCl(aq) \longrightarrow AgCl(s) + NaNO_3(aq)$$

A complete ionic equation shows all ions, including spectator ions.

- ***Write a complete ionic equation with ions from each soluble compound shown separately.*** Remember that $NaNO_3$ consists of $Na^+(aq) + NO_3^-(aq)$. The precipitate is represented by its complete formula because its ions are in a crystal lattice (← **Sec. 2-6b**), not free to move about in the solution.

$$Ag^+(aq) + NO_3^-(aq) + Na^+(aq) + Cl^-(aq) \longrightarrow AgCl(s) + Na^+(aq) + NO_3^-(aq)$$

Spectator ions are those that are in solution in the products as well as the reactants.

- ***To obtain the net ionic equation, cancel the spectator ions from each side of the complete ionic equation.*** Sodium ions and nitrate ions are the spectator ions in this example, and we cancel them from the complete ionic equation to give the net ionic equation.

  Complete ionic equation:

$$Ag^+(aq) + \cancel{NO_3^-(aq)} + \cancel{Na^+(aq)} + Cl^-(aq) \longrightarrow AgCl(s) + \cancel{Na^+(aq)} + \cancel{NO_3^-(aq)}$$

  Net ionic equation:

$$Ag^+(aq) + Cl^-(aq) \longrightarrow AgCl(s)$$

- ***Check that the sum of the charges is the same on each side of the net ionic equation.*** The sum of charges is zero on each side: On the left (1+) + (1−) = 0; on the right AgCl is a solid ionic compound with zero net charge.

The *charge must be the same on both sides of a balanced equation because electrons are neither created nor destroyed in a chemical reaction*; this is known as the **law of conservation of electric charge**.

PROBLEM-SOLVING EXAMPLE 3.6

Net Ionic Equations

Determine whether a reaction occurs when aqueous solutions of potassium iodide, KI(aq), and lead nitrate, $Pb(NO_3)_2$(aq), are mixed. If so, write the net ionic equation. If not, write N.R.

Result A precipitation occurs: $Pb^{2+}(aq) + 2\ I^-(aq) \longrightarrow PbI_2(s)$.

Analyze Given reactants, decide if a reaction occurs; write a net ionic equation.

Plan Determine which ions are present in the reactants and whether a reaction occurs. Write an overall balanced equation. Write a complete ionic equation and cancel aqueous ions that appear on both sides to get the net ionic equation.

Execute Species present in reactants are Pb^{2+}(aq), NO_3^-(aq), K^+(aq), and I^-(aq).

$Pb(NO_3)_2$(aq) consists of Pb^{2+}(aq) and 2 NO_3^-(aq).

KI(aq) consists of K^+(aq) and I^-(aq).

The formula $Pb(NO_3)_2$ contains 1 Pb^{2+} ion and 2 NO_3^- ions.

Does a reaction occur? Possible products are PbI_2 and KNO_3. PbI_2 is insoluble, an exception to the rule that iodides are soluble. KNO_3 is soluble because potassium compounds and nitrates are soluble. A reaction occurs: PbI_2(s) precipitates.

The balanced overall reaction is

$$Pb(NO_3)_2(aq) + 2\ KI(aq) \longrightarrow PbI_2(s) + 2\ KNO_3(aq)$$

The complete ionic equation is

$$Pb^{2+}(aq) + 2\ NO_3^-(aq) + 2\ K^+(aq) + 2\ I^-(aq) \longrightarrow PbI_2(s) + 2\ K^+(aq) + 2\ NO_3^-(aq)$$

The net ionic equation is

$$Pb^{2+}(aq) + 2\ I^-(aq) \longrightarrow PbI_2(s)$$

☑ **Reasonable Result Check** Atoms balance. Charges balance. (Reactants have net charge $(2+) + 2 \times (1-) = 0$. Product has zero charge.)

PROBLEM-SOLVING PRACTICE 3.6

In each case, decide whether a reaction will occur. If so, write a balanced overall equation and a balanced net ionic equation. If no reaction occurs, write N.R.
(a) $BaCl_2$ and Na_2SO_4 (b) $(NH_4)_2S$ and $FeCl_2$ (c) $(NH_4)_2SO_4$ and KCl

CONCEPTUAL EXERCISE 3.3

Net Ionic Equations

There are some reactions where both of the possible products are insoluble and precipitate from aqueous solution. Use information in Table 3.1 to devise an example of such a reaction. Write a balanced net ionic equation for the reaction.

3-4 Acid-Base Reactions

Acids and bases are two extremely important classes of compounds. You encounter them often in everyday life (vinegar is an acid; lye is a base), and you will encounter them often in the laboratory. Acids have many properties in common, as do bases. Some properties of acids are related to properties of bases.

- Acidic solutions change the color of litmus from blue to red, and basic solutions change the color of litmus from red to blue.
- Acidic solutions cause the dye phenolphthalein to be colorless, but basic solutions make phenolphthalein pink.

Litmus is a dye derived from lichens.

Phenolphthalein is a synthetic dye.

© Cengage Learning/James Maynard

Adding lye to vinegar causes phenolphthalein to turn pink.

- If an acid has made litmus red, adding a base reverses the effect, making the litmus blue. If a base has made litmus blue, adding an acid reverses the effect, making the litmus red.
- A base can *neutralize* the effect of an acid, and an acid can *neutralize* the effect of a base. Acids and bases are chemical opposites.

Acids have other characteristic properties. They taste sour, they produce bubbles of gas when reacting with limestone, and they react with many metals to produce a flammable gas. Although you should never taste substances in a chemistry laboratory, you have probably experienced the sour taste of at least one acid—vinegar, which is a dilute solution of acetic acid in water. Bases, in contrast, have a bitter taste. Soap, for example, contains a base. Rather than dissolving metals, bases often cause metal ions to form insoluble compounds that precipitate from solution as metal hydroxides. Such precipitates can be made to dissolve by adding an acid, another case in which an acid counteracts a property of a base.

3-4a Acids

The properties of acids in aqueous solutions can be explained by defining an **acid** as *any substance that increases the concentration of aqueous hydrogen ions, $H^+(aq)$, when dissolved in pure water*. ***The properties acidic solutions have in common are the properties of $H^+(aq)$.*** The "(aq)" is important here, because a hydrogen ion, H^+, is a hydrogen atom that has lost its single electron; that is, H^+ is a proton—the nucleus of a hydrogen atom. Remember that the diameter of the nucleus of an atom is about 1/10,000 the diameter of the atom (← **Sec. 2-2a**). Consequently, a proton, H^+, is much smaller than a hydrogen atom and has a very high ratio of positive charge to size. This very concentrated positive charge interacts strongly with electrons in oxygen atoms of water molecules and H^+ combines with H_2O to form H_3O^+, known as the *hydronium ion*. Other water molecules are attracted to each H_3O^+ ion, forming even larger clusters—some with more than 20 water molecules. $H^+(aq)$ is used to represent H_3O^+ and these larger clusters of water molecules.

One cluster of water molecules around H^+ in aqueous solution is $H_9O_4^+$.

Most acids are molecular compounds, which do not consist of ions (← **Sec. 2-7**). When a molecule of an acid dissolves in water, the molecule ionizes; **ionization** is a process in which a molecule is *transformed into positive and negative ions*. One of the ions formed by an acid is always $H^+(aq)$. An *acid that is entirely converted to ions (completely*

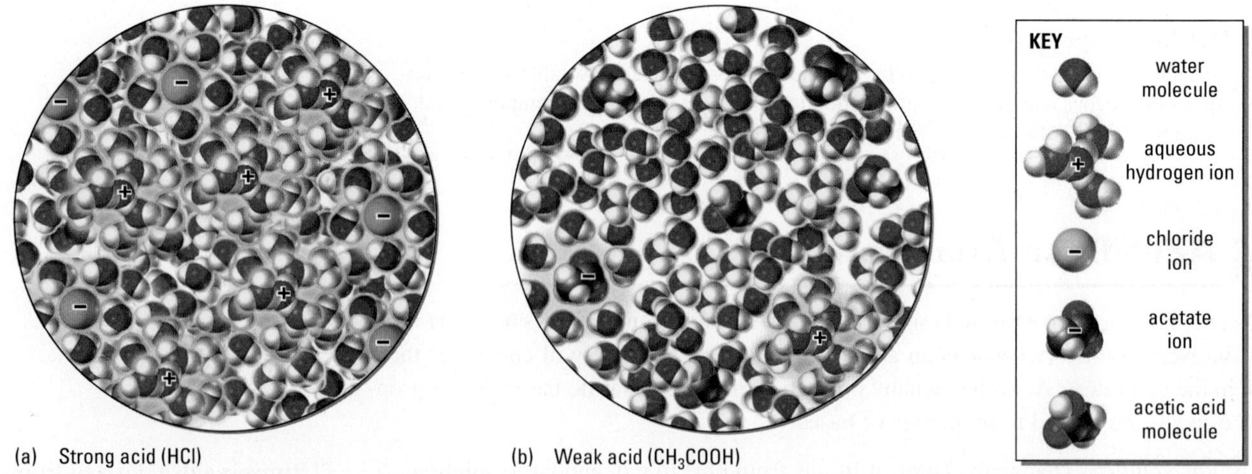
KEY
water molecule
aqueous hydrogen ion
chloride ion
acetate ion
acetic acid molecule
(a) Strong acid (HCl)
(b) Weak acid (CH₃COOH)

Figure 3.10 The ionization of acids in water. (a) A strong acid such as hydrochloric acid, HCl, is completely ionized in water; all the HCl molecules ionize to form $H^+(aq)$ and $Cl^-(aq)$ ions. (b) Weak acids such as acetic acid, CH_3COOH, are only slightly ionized in water. Nonionized acetic acid molecules far outnumber $H^+(aq)$ and $CH_3COO^-(aq)$ ions formed by the ionization of acetic acid molecules. Ions are highlighted in yellow to make them easier to distinguish from water molecules.

ionized) when dissolved in water is a **strong electrolyte** *and is called a* **strong acid**. For example, a very familiar strong acid is hydrochloric acid, which ionizes completely in aqueous solution to form H^+(aq) and chloride ions (Figure 3.10a).

The terms strong electrolyte and weak electrolyte also apply to substances other than acids: for example, bases.

$$HCl(aq) \longrightarrow H^+(aq) + Cl^-(aq)$$

In contrast, *an acid that ionizes only slightly is a* **weak electrolyte** *and is called a* **weak acid**. (Table 3.2 lists common strong and week acids.) For example, when acetic acid, CH_3COOH, dissolves in water, usually fewer than 5% of the acetic acid molecules are ionized at any time. The remainder of the acetic acid exists as nonionized molecules. Because acetic acid is only slightly ionized in aqueous solution, it is a weak electrolyte and classified as a weak acid (Figure 3.10b).

$$\underset{\text{Mostly nonionized (molecular) form}}{CH_3COOH(aq)} \rightleftharpoons \underset{\text{a few positive and negative aqueous ions}}{H^+(aq) + CH_3COO^-(aq)}$$

The double arrow in this equation for the ionization of acetic acid signifies a characteristic property of the reaction of a weak electrolyte with water: There is a *dynamic equilibrium* in which the nonionized, molecular form, CH_3COOH, is ionizing at exactly the same rate that the ions, H^+(aq) and CH_3COO^-(aq), are reacting to form nonionized acetic acid molecules. Dynamic equilibrium will be described in more detail in Chapter 12.

Some common acids, such as sulfuric acid, can provide more than 1 mol H^+ ions per mole of acid:

$$\underset{\text{sulfuric acid}}{H_2SO_4(aq)} \longrightarrow H^+(aq) + \underset{\text{hydrogen sulfate ion}}{HSO_4^-(aq)}$$

$$\underset{\text{hydrogen sulfate ion}}{HSO_4^-(aq)} \rightleftharpoons H^+(aq) + \underset{\text{sulfate ion}}{SO_4^{2-}(aq)}$$

Table 3.2 Common Acids and Bases

Strong Acids (Strong Electrolytes)		Strong Bases (Strong Electrolytes)	
HCl	Hydrochloric acid	LiOH	Lithium hydroxide
HNO_3	Nitric acid	NaOH	Sodium hydroxide
H_2SO_4	Sulfuric acid	KOH	Potassium hydroxide
$HClO_4$	Perchloric acid	$Ca(OH)_2$	Calcium hydroxide‡
HBr	Hydrobromic acid	$Sr(OH)_2$	Strontium hydroxide‡
HI	Hydroiodic acid	$Ba(OH)_2$	Barium hydroxide‡
Weak Acids* (Weak Electrolytes)		**Weak Bases† (Weak Electrolytes)**	
H_3PO_4	Phosphoric acid	NH_3	Ammonia
CH_3COOH	Acetic acid	CH_3NH_2	Methylamine
H_2CO_3	Carbonic acid		
HCN	Hydrocyanic acid		
HCOOH	Formic acid		
C_6H_5COOH	Benzoic acid		

*Many weak acids, such as acetic and benzoic, are organic compounds that contain the —COOH group.

†Organic amines (related to ammonia), such as methylamine, contain the —NH_2 group and are weak bases.

‡The hydroxides of calcium, barium, and strontium are only slightly soluble, but all of the solute that dissolves is dissociated into ions.

The first ionization reaction is essentially complete, so sulfuric acid is considered a strong acid (and a strong electrolyte as well). However, the hydrogen sulfate ion, like acetic acid, is only partially ionized, so it is a weak acid.

CONCEPTUAL EXERCISE 3.4

Dissociation of Acids

Phosphoric acid, H_3PO_4, has three protons that can ionize. Write the equations for its three ionization reactions, each of which is a dynamic equilibrium.

3-4b Bases

A **base** is *a substance that increases the concentration of aqueous hydroxide ions, $OH^-(aq)$, when dissolved in pure water.* ***The properties that basic solutions have in common are properties attributable to the aqueous hydroxide ion, $OH^-(aq)$.*** A *soluble ionic compound that contains hydroxide ions*, such as sodium hydroxide or potassium hydroxide, is a strong electrolyte and a **strong base** (Table 3.2).

$$NaOH(s) \xrightarrow{H_2O} Na^+(aq) + OH^-(aq)$$

A base that dissolves slightly in water, such as $Ba(OH)_2$, can still be a strong electrolyte and a strong base if the amount of the compound that dissolves dissociates completely into ions.

Ammonia, NH_3, is another very common base. Although ammonia does not have an OH^- ion as part of its formula, it produces OH^- ions by reacting with water.

$$\underset{\text{Larger concentration of } NH_3}{NH_3(aq) + H_2O(\ell)} \rightleftharpoons \underset{\text{much smaller concentration of ions}}{NH_4^+(aq) + OH^-(aq)}$$

In the equilibrium between NH_3 and the NH_4^+ and OH^- ions, only a small concentration of the ions is present, so ammonia is a weak electrolyte (<5% ionized). *A weak electrolyte that produces OH^- ions* is a **weak base** (Table 3.2).

To summarize:

Acids and bases that are strong electrolytes are strong acids and bases. Acids and bases that are weak electrolytes are weak acids and bases.

- *Strong electrolytes* are compounds that exist *completely* as ions in aqueous solutions. They can be ionic compounds (salts or strong bases) or molecular compounds that are strong acids and ionize completely.
- *Weak electrolytes* are molecular compounds that are weak acids or bases.
- *Nonelectrolytes* are molecular compounds that do not ionize in aqueous solution. Most molecular compounds are nonelectrolytes.

CONCEPTUAL EXERCISE 3.5

Acids and Bases

(a) What ions are produced when perchloric acid, $HClO_4$, dissolves in water?

(b) Calcium hydroxide is only slightly soluble in water. What little does dissolve, however, is dissociated. What ions are produced? Write an equation for the dissociation of calcium hydroxide.

PROBLEM-SOLVING EXAMPLE 3.7

Strong Electrolytes, Weak Electrolytes, and Nonelectrolytes

Identify whether each of these substances in an aqueous solution is a strong electrolyte, a weak electrolyte, or a nonelectrolyte: HBr, hydrogen bromide; LiOH, lithium hydroxide; HCOOH, formic acid; CH_3CH_2OH, ethanol.

Result HBr strong; LiOH strong; HCOOH weak; CH_3CH_2OH nonelectrolyte.

Analyze Based on formulas and names, identify strengths of electrolytes. Strong electrolytes consist entirely of ions in aqueous solution. Weak electrolytes are weak acids or bases. Nonelectrolytes are substances that are neither strong nor weak electrolytes.

Plan Soluble ionic compounds that are not acids or bases are strong electrolytes. For acids and bases, refer to Table 3.2.

Execute Hydrogen bromide is a common strong acid and, therefore, is a strong electrolyte. Lithium hydroxide is a strong base that is an ionic compound, so it is a strong electrolyte. Formic acid contains a —COOH group and is a weak acid, so it is a weak electrolyte. Ethanol is a molecular compound that does not contain a —COOH group or a —NH_2 group, so it is probably a nonelectrolyte.

PROBLEM-SOLVING PRACTICE 3.7

Look back through the discussion of electrolytes and Table 3.2 and identify at least one strong electrolyte, one weak electrolyte, and one nonelectrolyte different from those discussed in Problem-Solving Example 3.7.

3-4c Neutralization Reactions

In an **acid-base neutralization reaction**, *an acid reacts with a base and each neutralizes the properties of the other*. When aqueous solutions of a strong acid (such as HCl) and a strong base (such as NaOH) are mixed, the ions in solution are $H^+(aq)$ and the anion from the acid, the metal cation, and the $OH^-(aq)$ from the base:

From hydrochloric acid: $H^+(aq)$, $Cl^-(aq)$

From sodium hydroxide: $Na^+(aq)$, $OH^-(aq)$

As in precipitation reactions (← **Sec. 3-3d**), a reaction will occur *whenever a compound is formed that removes ions from solution*. In an acid-base reaction, that compound is water, HOH, formed by combination of $H^+(aq)$ with $OH^-(aq)$. In the case of HCl plus NaOH, water and a *salt*, sodium chloride (← **Sec. 2-4**), form.

$$\underset{\text{acid}}{HCl(aq)} + \underset{\text{base}}{NaOH(aq)} \longrightarrow \underset{\text{water}}{HOH(\ell)} + \underset{\text{salt}}{NaCl(aq)}$$

Notice that HOH is equivalent to H_2O. We wrote HOH here to emphasize that water is formed from $H^+(aq)$ and $OH^-(aq)$.

When a strong acid and a strong base react, they neutralize each other. This happens because the hydrogen ions from the acid react with hydroxide ions from the base to form water, a molecular compound. The other ions form a **salt**, *an ionic compound whose cation comes from the base and whose anion comes from the acid*. If the water is evaporated, the solid salt remains.

Note that the cation of the salt comes from the base, and the anion of the salt comes from the acid.

The neutralization reaction above can be written more generally as

$$\underset{\text{acid}}{HX(aq)} + \underset{\text{base}}{MOH(aq)} \longrightarrow \underset{\text{water}}{HOH(\ell)} + \underset{\text{salt}}{MX(aq)}$$

This is an *exchange reaction* (← **Sec. 3-3d**) in which the $H^+(aq)$ ions from the aqueous acid and the $M^+(aq)$ ions from the metal hydroxide exchange partners.

The salt that forms in a neutralization reaction depends on the acid and base that react. The salt magnesium chloride is formed when a commercial antacid containing magnesium hydroxide is swallowed to neutralize excess hydrochloric acid in the stomach.

$$\underset{\text{hydrochloric acid}}{2\ HCl(aq)} + \underset{\text{magnesium hydroxide}}{Mg(OH)_2(s)} \longrightarrow 2\ H_2O(\ell) + \underset{\text{magnesium chloride}}{MgCl_2(aq)}$$

Milk of magnesia consists of a suspension of finely divided particles of $Mg(OH)_2(s)$ in water.

Organic acids, such as acetic acid, which contain —COOH, a grouping of atoms called the acid group, also neutralize bases to form salts. The H in the —COOH group is the acidic proton. Its removal generates the —COO^- anion. The reaction of propanoic acid, CH_3CH_2COOH,

The —COOH structure, called the *acid group*, is present in all organic acids and imparts acidic properties to compounds containing it.

and sodium hydroxide produces the salt sodium propanoate, $NaCH_3CH_2COO$, containing sodium ions, Na^+, and propanoate ions, $CH_3CH_2COO^-$. Sodium propanoate is commonly used as a food preservative.

This specific reaction is an example of the general equation for a neutralization reaction.

$$\underset{\text{propanoic acid}}{CH_3CH_2COOH(aq)} + NaOH(aq) \longrightarrow HOH(\ell) + \underset{\text{sodium propanoate}}{NaCH_3CH_2COO(aq)}$$

Although a propanoic acid molecule contains six H atoms, it is only the H atom in the acid group —COOH that is involved in this neutralization reaction.

PROBLEM-SOLVING EXAMPLE 3.8

Balancing Neutralization Equations

Write a balanced chemical equation for the reaction of nitric acid, HNO_3, with calcium hydroxide, $Ca(OH)_2$, in aqueous solution.

Result $2\ HNO_3(aq) + Ca(OH)_2(aq) \longrightarrow Ca(NO_3)_2(aq) + 2\ H_2O(\ell)$

Analyze Reactants are an acid and a base. This is a neutralization reaction producing a salt and water. Hydrogen ions and hydroxide ions exchange partners.

Plan Follow the usual procedure for balancing chemical equations; for neutralization it is usually best to balance ions first and H and O last.

Execute

(unbalanced equation) $HNO_3(aq) + Ca(OH)_2(aq) \longrightarrow Ca(NO_3)_2(aq) + H_2O(\ell)$

The calcium ions are in balance, but we need to add a coefficient of 2 to the nitric acid because two nitrate ions appear in the products.

(unbalanced equation) $2\ HNO_3(aq) + Ca(OH)_2(aq) \longrightarrow Ca(NO_3)_2(aq) + H_2O(\ell)$

To balance H and O, count four hydrogen atoms in the reactants (two from nitric acid and two from calcium hydroxide); this requires a coefficient of 2 for water. This balances the equation.

(balanced equation) $2\ HNO_3(aq) + Ca(OH)_2(aq) \longrightarrow Ca(NO_3)_2(aq) + 2\ H_2O(\ell)$

☑ **Reasonable Result Check** There are eight oxygen atoms in the reactants (six from nitric acid and two from calcium hydroxide), and there are eight oxygen atoms in the products (six from calcium nitrate and two from water).

PROBLEM-SOLVING PRACTICE 3.8

Write a balanced equation for the reaction of phosphoric acid, H_3PO_4, with sodium hydroxide, NaOH.

PROBLEM-SOLVING EXAMPLE 3.9

Acids, Bases, and Salts

Name and write the formula for the acid and base that form each salt. Write a balanced equation for the reaction that forms each salt: (a) $CaSO_4$; (b) $KClO_4$.

Result (a) Calcium hydroxide, $Ca(OH)_2$, and sulfuric acid, H_2SO_4.

$$Ca(OH)_2(aq) + H_2SO_4(aq) \longrightarrow CaSO_4(s) + 2\ H_2O(\ell)$$

(b) Potassium hydroxide, KOH, and perchloric acid, $HClO_4$.

$$KOH(aq) + HClO_4(aq) \longrightarrow KClO_4(aq) + H_2O(\ell)$$

Analyze A salt is formed from the cation of a base and the anion of an acid.

Plan Find an acid with the appropriate anion and a base with the appropriate cation. Write a formula and name the compound. Then, write a balanced equation with the acid and base as reactants and the salt and water as products.

Execute

(a) $CaSO_4$ contains calcium ions and sulfate ions. Ca^{2+} requires 2 OH^- to balance charge giving $Ca(OH)_2$, calcium hydroxide. SO_4^{2-} requires 2 H^+ to balance charge giving H_2SO_4, sulfuric acid. Neutralization of $Ca(OH)_2$ with H_2SO_4 produces $CaSO_4$ and water.

$$Ca(OH)_2(aq) + H_2SO_4(aq) \longrightarrow CaSO_4(s) + 2\ H_2O(\ell)$$

Table 3.1 indicates that $CaSO_4$ is insoluble so it is written $CaSO_4(s)$.

(b) $KClO_4$ contains potassium ions and perchlorate ions. K^+ requires 1 OH^- to balance charge giving KOH, potassium hydroxide. ClO_4^- requires 1 H^+ to balance charge giving $HClO_4$, perchloric acid. Neutralization of KOH with $HClO_4$ produces $KClO_4$ and water.

$$KOH(aq) + HClO_4(aq) \longrightarrow KClO_4(aq) + H_2O(\ell)$$

☑ **Reasonable Result Check** The final neutralization equations each have the same types and numbers of atoms on each side. Water is formed in each case.

PROBLEM-SOLVING PRACTICE 3.9

Identify the acid and the base that react to form (a) $MgSO_4$ and (b) $SrCO_3$.

3-4d Net Ionic Equations for Acid-Base Reactions

Acid-base reactions can be written as net ionic equations just as precipitation reactions were. Consider the reaction of calcium hydroxide, $Ca(OH)_2$, with hydrochloric acid, HCl. The product salt, $CaCl_2$, is soluble (Table 3.1) so the overall balanced equation is

$$2\ HCl(aq) + Ca(OH)_2(aq) \longrightarrow 2\ H_2O(\ell) + CaCl_2(aq)$$

From the overall equation we can write this complete ionic equation:

$$2\ H^+(aq) + \cancel{2\ Cl^-(aq)} + \cancel{Ca^{2+}(aq)} + 2\ OH^-(aq) \longrightarrow 2\ H_2O(\ell) + \cancel{Ca^{2+}(aq)} + \cancel{2\ Cl^-(aq)}$$

Canceling spectator ions, in this case calcium ions and chloride ions, from each side of the complete ionic equation yields this net ionic equation:

$$2\ H^+(aq) + 2\ OH^-(aq) \longrightarrow 2\ H_2O(\ell)$$

or, using the smallest whole-number coefficients,

$$H^+(aq) + OH^-(aq) \longrightarrow H_2O(\ell)$$

There is *conservation of charge* in the net ionic equation. On the left, $(1+) + (1-) = 0$; on the right, water has zero net charge.

Notice that because the ions in the salt are spectator ions, the equation

$$H^+(aq) + OH^-(aq) \longrightarrow H_2O(\ell)$$

*is **always** the net ionic equation for a neutralization reaction between a strong acid and a strong base that yields a soluble salt.*

Because weak acids and weak bases are only partially ionized, net ionic equations for neutralizations involving a weak acid, a weak base, or both a weak acid and a weak base should include molecular formulas. Consider neutralization of a weak acid, HCN, by a strong base, KOH. The overall equation is

$$HCN(aq) + KOH(aq) \longrightarrow KCN(aq) + H_2O(\ell)$$

The weak acid HCN is only partially ionized, so we leave it in the molecular form, but KOH and KCN are strong electrolytes. The complete ionic equation is

$$HCN(aq) + \cancel{K^+(aq)} + OH^-(aq) \longrightarrow \cancel{K^+(aq)} + CN^-(aq) + H_2O(\ell)$$

Canceling spectator ions yields

$$HCN(aq) + OH^-(aq) \longrightarrow CN^-(aq) + H_2O(\ell)$$

There is a 1− charge on each side of the net ionic equation so charge is conserved.

PROBLEM-SOLVING EXAMPLE 3.10

Net Ionic Equation: Neutralization

Write a balanced overall equation and a net ionic equation for the reaction of acetic acid, CH_3COOH, with calcium hydroxide, $Ca(OH)_2$.

Result $2\ CH_3COOH(aq) + Ca(OH)_2(aq) \longrightarrow Ca(CH_3COO)_2(aq) + 2\ H_2O(\ell)$

Net ionic equation: $CH_3COOH(aq) + OH^-(aq) \longrightarrow CH_3COO^-(aq) + H_2O(\ell)$

Analyze Formulas for the acid and base are given. Acetic acid is a weak acid; calcium hydroxide is a strong base. Products of the neutralization reaction are water and the salt calcium acetate, $Ca(CH_3COO)_2$, formed from the base's cation, Ca^{2+}, and the acid's anion, CH_3COO^-. All acetates are soluble (Table 3.1).

Plan Write an overall equation and balance it; balance ions and then H and O. Write the weak acid in molecular form, write the strong base as ions, write the soluble salt as ions, and write water as a molecule in the complete ionic equation. Cancel spectator ions to get the net ionic equation.

Execute

(unbalanced) $CH_3COOH(aq) + Ca(OH)_2(aq) \longrightarrow Ca(CH_3COO)_2(aq) + H_2O(\ell)$

Balance ions first. Add a coefficient of 2 to the acetic acid because two acetate ions appear in the product, $Ca(CH_3COO)_2$.

(unbalanced) $2\ CH_3COOH(aq) + Ca(OH)_2(aq) \longrightarrow Ca(CH_3COO)_2(aq) + H_2O(\ell)$

Two $H^+(aq)$ from the acetic acid react with two $OH^-(aq)$ from the calcium hydroxide producing two water molecules so put a coefficient of 2 with H_2O.

(balanced) $2\ CH_3COOH(aq) + Ca(OH)_2(aq) \longrightarrow Ca(CH_3COO)_2(aq) + 2\ H_2O(\ell)$

The complete ionic equation is

$$2\ CH_3COOH(aq) + \cancel{Ca^{2+}(aq)} + 2\ OH^-(aq) \longrightarrow \cancel{Ca^{2+}(aq)} + 2\ CH_3COO^-(aq) + 2\ H_2O(\ell)$$

The calcium ions are spectator ions; cancel them to give the net ionic equation.

$$CH_3COOH(aq) + OH^-(aq) \longrightarrow CH_3COO^-(aq) + H_2O(\ell)$$

☑ **Reasonable Result Check** Each side of the net ionic equation has the same charge (1−) and the same number and types of atoms. The result is reasonable.

PROBLEM-SOLVING PRACTICE 3.10

Write a balanced equation for the reaction of hydroiodic acid, HI, with calcium hydroxide, $Ca(OH)_2$. Then, write the balanced complete ionic equation and the net ionic equation for this neutralization reaction.

EXERCISE 3.6

Neutralizations and Net Ionic Equations

Write balanced complete ionic equations and net ionic equations for the neutralization reactions of these acids and bases:

(a) HCl and KOH

(b) H_2SO_4 and $Ba(OH)_2$ (Remember that sulfuric acid can provide 2 mol $H^+(aq)$ per 1 mol sulfuric acid.)

(c) CH_3COOH and NaOH

EXERCISE 3.7

Net Ionic Equations and Antacids

The commercial antacids Maalox, Di-Gel tablets, and Mylanta contain aluminum hydroxide or magnesium hydroxide that reacts with excess hydrochloric acid in the stomach. Write the balanced complete ionic equation and net ionic equation for the soothing neutralization reaction of aluminum hydroxide with HCl. Assume that dissolved aluminum hydroxide is completely dissociated.

Cengage Learning/Charles D. Winters
CO_2
coral, $CaCO_3$
HCl(aq)

Figure 3.11 A gas-forming reaction: coral, calcium carbonate, reacts with hydrochloric acid.

3-4e Reactions in Which Gases Are Formed

One of the properties of acids mentioned early in our discussion of acids and bases is that acids react with limestone ($CaCO_3$), generating bubbles of gas. This property is illustrated in Figure 3.11, which shows the reaction of hydrochloric acid with coral, which is mainly calcium carbonate.

$$CaCO_3(s) + 2\,HCl(aq) \longrightarrow CaCl_2(aq) + H_2O(\ell) + CO_2(g)$$

Carbon dioxide always forms when acids react with metal carbonates. Carbonic acid, H_2CO_3, is produced initially, but it is unstable and decomposes to water and carbon dioxide, the products shown in the overall equation. As Earth's oceans become more acidic due to industrial pollutants and increasing levels of dissolved CO_2, dissolution of coral is becoming a significant environmental issue. *A reaction in which a gas is generated* is called a **gas-forming reaction**.

Like formation of a precipitate or formation of water, formation of a gas removes ions from solution and causes a reaction to occur. Another example of a gas-forming reaction is the reaction of metal sulfites with acids, which produces the foul-smelling gas SO_2. When sulfites are acidified, H_2SO_3 is formed initially. It is unstable and, like carbonic acid, quickly decomposes to sulfur dioxide and water.

$$CaSO_3(s) + 2\,HCl(aq) \longrightarrow CaCl_2(aq) + H_2O(\ell) + SO_2(g)$$

A third example is reaction of a metal sulfide with acid. This produces another foul-smelling gas, H_2S.

$$Na_2S(aq) + 2\,HCl(aq) \longrightarrow 2\,NaCl(aq) + H_2S(g)$$

EXERCISE 3.8

A Gas-Forming Reaction

Write the net ionic equation for sodium sulfide reacting with hydrochloric acid.

CONCEPTUAL EXERCISE 3.9

Gas-Forming Reactions

Predict the products and write the balanced overall equation and the net ionic equation for each of these gas-generating reactions.

(a) $Na_2CO_3(aq) + H_2SO_4(aq) \longrightarrow$
(b) $FeS(s) + HCl(aq) \longrightarrow$
(c) $K_2SO_3(aq) + HCl(aq) \longrightarrow$

CONCEPTUAL EXERCISE 3.10

Reactions and Equations

Predict the products of each reaction. Write an overall balanced chemical equation and a net ionic equation for each reaction.

(a) $BaCl_2(aq) + Na_2C_2O_4(aq) \longrightarrow$ (b) $Sr(OH)_2(s) + HNO_3(aq) \longrightarrow$
(c) $(NH_4)_3PO_4(aq) + NiCl_2(aq) \longrightarrow$ (d) $KOH(aq) + HClO_4(aq) \longrightarrow$

Cengage Learning/Charles D. Winters

$2\,Mg(s) + O_2(g) \rightarrow 2\,MgO(s)$. Magnesium is oxidized and oxygen is reduced as the white solid magnesium oxide, MgO, forms.

3-5 Oxidation-Reduction and Electron Transfer

As mentioned at the beginning of this chapter, shown in Figure 3.3, and reproduced here, the reaction of magnesium metal with oxygen produces magnesium oxide. This is a dramatic change in that neither reactant consists of ions, yet the product is an ionic compound that contains Mg^{2+} and O^{2-} ions. This reaction is an example of an **oxidation-reduction reaction**, *one in which electrons are transferred from one reactant atom, molecule, or ion to another.* Oxidation-reduction is, somewhat strangely, abbreviated **redox** (for *red*uction-*ox*idation).

- ***Oxidation is a loss of electrons from an atom, molecule, or ion.* After the loss of electrons, the atom, molecule, or ion has been oxidized.**

In the reaction with oxygen, each magnesium metal atom is oxidized; it loses two electrons, its charge goes from zero to 2+, and a magnesium ion forms: $Mg \rightarrow 2\ e^- + Mg^{2+}$. Magnesium metal atoms have been oxidized to Mg^{2+} ions.

- ***Reduction is a gain of electrons by an atom, molecule, or ion.* After the gain of electrons, the atom, molecule, or ion has been reduced.**

When oxygen reacts with magnesium metal, each oxygen atom is reduced; it gains two electrons, its charge goes from zero to 2−, and an oxide ion is formed: $O + 2\ e^- \rightarrow O^{2-}$. Oxygen atoms have been reduced to O^{2-} ions.

We can summarize these redox processes in two fundamental conclusions:

- **In every redox reaction, the transfer of electrons occurs *simultaneously*. Electrons are lost by one reactant while being gained by another reactant. When one reactant is oxidized, another reactant must be reduced.**
- **Oxidation (electron loss) is the opposite of reduction (electron gain).**

3-5a Oxidizing Agents and Reducing Agents

Note that oxidizing and reducing agents are *always* reactants, not products.

An **oxidizing agent** *is a substance that causes something else to be oxidized.* (It is the agent that causes oxidation.) A **reducing agent** *is a substance that causes something else to be reduced.* Oxidizing agents and reducing agents are *reactants* that transfer electrons between them. As a redox reaction proceeds:

- The **oxidizing agent** is *reduced* by *taking electrons from the reducing agent.* In this way, the oxidizing agent *causes the oxidation* (electron loss) *of the reducing agent.*
- The **reducing agent** is *oxidized* by *donating electrons to the oxidizing agent.* By doing so, the reducing agent *causes the reduction* (electron gain) *of the oxidizing agent.*
- As a result, **oxidizing agents are reduced; reducing agents are oxidized.**

For example, in the formation of magnesium oxide, magnesium metal atoms are oxidized (donate electrons), so magnesium metal is the reducing agent; oxygen atoms are reduced (gain electrons), making oxygen the oxidizing agent. Magnesium metal, the reducing agent, *causes oxygen atoms to be reduced*; oxygen, the oxidizing agent, *causes magnesium metal atoms to be oxidized.*

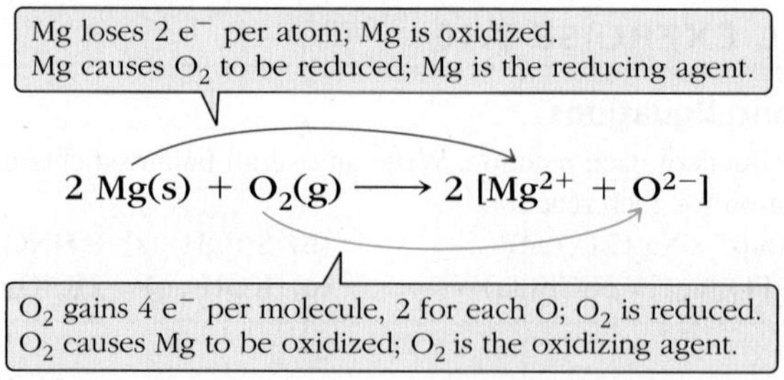
Mg loses 2 e⁻ per atom; Mg is oxidized.
Mg causes O₂ to be reduced; Mg is the reducing agent.
2 Mg(s) + O₂(g) ⟶ 2 [Mg²⁺ + O²⁻]
O₂ gains 4 e⁻ per molecule, 2 for each O; O₂ is reduced.
O₂ causes Mg to be oxidized; O₂ is the oxidizing agent.

3-5b Identifying Oxidizing and Reducing Agents

Most metals are reducing agents and those in periodic Groups 1A and 2A are strong reducing agents. A few nonmetallic elements, such as oxygen and the halogens (F_2, Cl_2, Br_2, and I_2) are strong oxidizing agents; they oxidize most metals and many nonmetals.

In the periodic table, strong reducing agents are on the left and toward the bottom; strong oxidizing agents are at the right and near the top (except for the noble gases, which are unreactive).

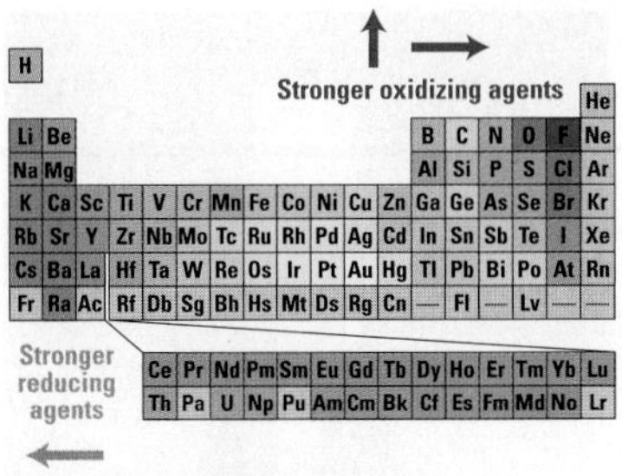
Stronger oxidizing agents
Stronger reducing agents

Elements that are reducing agents are blue. Elements that are oxidizing agents are red. The darker the color the stronger the reducing or oxidizing agent. Gray indicates no data.

A halogen can oxidize a reactive metal to form an ionic metal halide whose formula can be predicted from the charges on the metal cation and halide anion. For example, consider the reaction of sodium metal with chlorine gas:

Na metal loses 1 e^- per atom to form a Na^+ ion. Na metal is oxidized. Na is the reducing agent because it reduces Cl_2.

$$2\ Na(s) + Cl_2(g) \longrightarrow 2\ [Na^+ + Cl^-]$$

Cl_2 gains 2 e^- per molecule to form 2 Cl^-. Cl_2 is reduced. Cl_2 is the oxidizing agent because it oxidizes Na.

Here, sodium metal atoms donate electrons to chlorine atoms (in diatomic chlorine molecules). Therefore, sodium metal atoms are oxidized to Na^+ ions; by gaining the donated electrons, chlorine atoms are reduced to chloride ions, Cl^-. Metallic sodium atoms are the reducing agent; chlorine atoms are the oxidizing agent. Two moles of sodium atoms donate two moles of electrons to one mole of diatomic chlorine molecules.

Fluorine is a particularly strong oxidizing agent. Chlorine is not as strong as fluorine, bromine is less strong than chlorine, and iodine is least strong, though still an oxidizing agent. A consequence of these differences is that a more reactive *diatomic halogen* will oxidize the *halide ion* of another halogen of lesser oxidizing ability. For example, $Cl_2(aq)$ oxidizes $Br^-(aq)$ to form Br_2 and $Cl^-(aq)$ (Figure 3.12).

$$Cl_2(aq) + 2\ Br^-(aq) \longrightarrow Br_2(aq) + 2\ Cl^-(aq)$$

CONCEPTUAL **EXERCISE 3.11**

Oxidation and Reduction

Write the balanced chemical equation, identify what is oxidized and what is reduced, and identify the oxidizing and reducing agents in these redox reactions:

(a) $Br_2(aq) + I^-(aq) \longrightarrow$
(b) $Br_2(\ell) + K(s) \longrightarrow$

Oxidation and reduction occur readily when a strong oxidizing agent comes in contact with a strong reducing agent. It can be dangerous to mix a strong oxidizing agent with a strong reducing agent. A violent reaction, even an explosion, may take place. Reagents should not be stored on laboratory shelves in alphabetical order, because such an

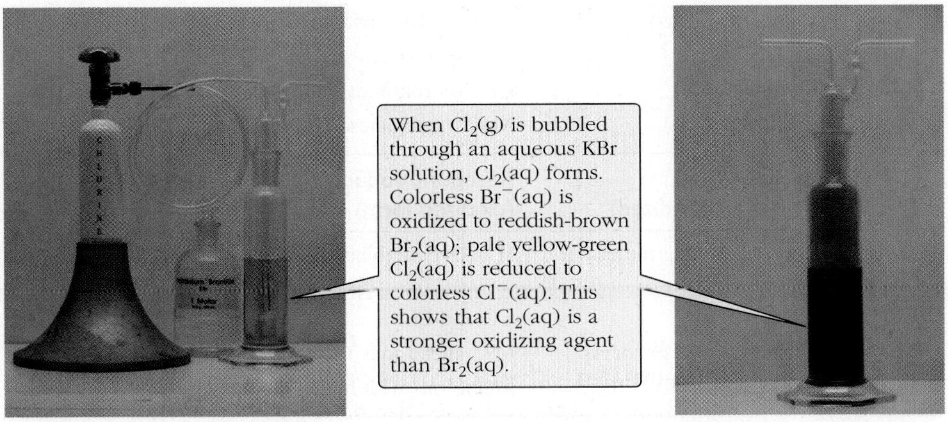
CHLORINE
When $Cl_2(g)$ is bubbled through an aqueous KBr solution, $Cl_2(aq)$ forms. Colorless $Br^-(aq)$ is oxidized to reddish-brown $Br_2(aq)$; pale yellow-green $Cl_2(aq)$ is reduced to colorless $Cl^-(aq)$. This shows that $Cl_2(aq)$ is a stronger oxidizing agent than $Br_2(aq)$.

Figure 3.12 Oxidation of Br^- by Cl_2.

Cengage Learning/Charles D. Winters

Figure 3.13 Reaction of copper metal with concentrated nitric acid. The reddish-brown gas is NO_2.

ordering may place a strong oxidizing agent next to a strong reducing agent. In particular, swimming pool chemicals that contain chlorine and are strong oxidizing agents should not be stored in the hardware store or the garage next to easily oxidized materials such as ammonia. Knowing the common oxidizing and reducing agents (Table 3.3) enables you to predict whether a redox reaction will take place and, in some cases, can allow you to predict what products will form.

The reaction of concentrated nitric acid, HNO_3, with copper metal (Figure 3.13) illustrates the use of Table 3.3. Nitrate ion, NO_3^-, in concentrated nitric acid is a powerful oxidizing agent in acidic solution.

$$Cu(s) + 4\,H^+(aq) + 2\,NO_3^-(aq) \longrightarrow Cu^{2+}(aq) + 2\,NO_2(g) + 2\,H_2O(\ell)$$

In this reaction, nitrate ions oxidize metallic copper atoms to Cu^{2+} ions; thus nitrate ion is the oxidizing agent. Simultaneously, copper metal atoms, the reducing agent, reduce nitrate ions to NO_2.

Some metal ions such as Fe^{2+} can also be reducing agents because they can be oxidized to ions of higher charge. Aqueous Fe^{2+} ion reacts readily with the strong oxidizing agent MnO_4^-, the permanganate ion. The Fe^{2+} ion is oxidized to Fe^{3+}, and the MnO_4^- ion is reduced to the Mn^{2+} ion.

$$5\,Fe^{2+}(aq) + MnO_4^-(aq) + 8\,H^+(aq) \longrightarrow 5\,Fe^{3+}(aq) + Mn^{2+}(aq) + 4\,H_2O(\ell)$$

Carbon can reduce many metal oxides to metals, and it is widely used in the metals industry to obtain metals from their compounds in ores. For example, iron is manufactured by heating an iron ore, such as hematite, Fe_2O_3, in a blast furnace with coke, which is mainly carbon.

$$2\,Fe_2O_3(s) + 3\,C(s) \longrightarrow 4\,Fe(\ell) + 3\,CO_2(g)$$

Finally, H_2 gas is a common reducing agent, widely used in the laboratory and in industry. For example, H_2 readily reduces copper(II) oxide to copper metal (Figure 3.14).

$$\underset{\text{reducing agent}}{H_2(g)} + \underset{\text{oxidizing agent}}{CuO(s)} \longrightarrow Cu(s) + H_2O(g)$$

Table 3.3 Common Oxidizing and Reducing Agents

Oxidizing Agent (oxidizing agents are reduced)	Reaction Product (reduced form)
O_2	O^{2-} or an oxygen-containing molecular compound such as H_2O
H_2O_2 (hydrogen peroxide)	$H_2O(\ell)$
F_2, Cl_2, Br_2, or I_2 (halogens)	F^-, Cl^-, Br^-, or I^- (halide ions)
HNO_3 (nitric acid)	Nitrogen oxides such as NO and NO_2
$Cr_2O_7^{2-}$ (dichromate ion)	Cr^{3+} (chromium(III) ion), in acid solution
MnO_4^- (permanganate ion)	Mn^{2+} (manganese(II) ion), in acid solution
Reducing Agent (reducing agents are oxidized)	**Reaction Product (oxidized form)**
H_2 or hydrogen-containing molecular compound	H^+ or H combined in H_2O
C (coke or charcoal)	CO and CO_2
M, metals such as Na, K, Fe, or Al	M^{n+}, metal ions such as Na^+, K^+, Fe^{3+}, or Al^{3+}
Some metal ions such as Cr^{2+}, Fe^{2+}	Metal ions with higher charge: Cr^{3+}, Fe^{3+}

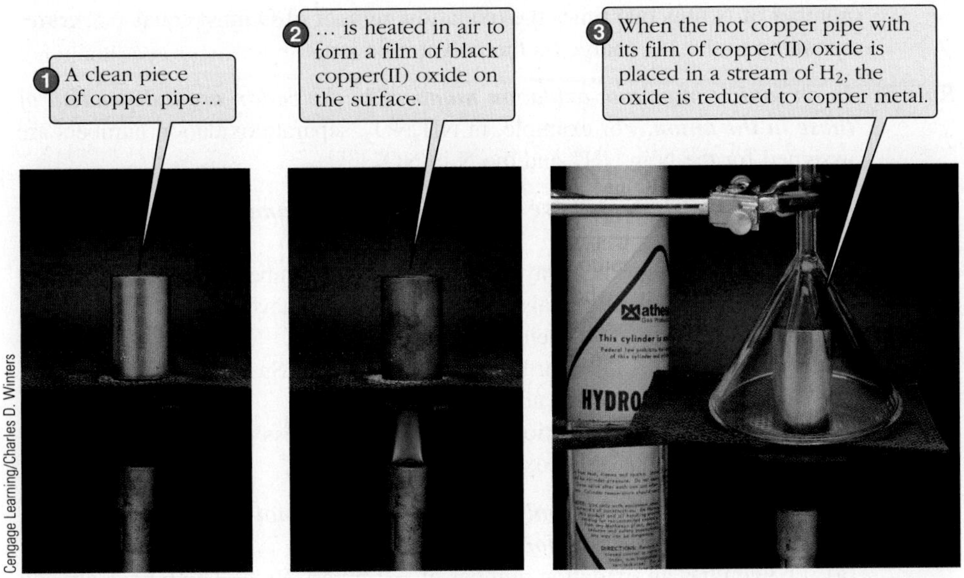
1 A clean piece of copper pipe...
2 ... is heated in air to form a film of black copper(II) oxide on the surface.
3 When the hot copper pipe with its film of copper(II) oxide is placed in a stream of H₂, the oxide is reduced to copper metal.
HYDRO

Cengage Learning/Charles D. Winters

Figure 3.14 Reduction of copper(II) oxide, CuO, with hydrogen, H_2.

CONCEPTUAL EXERCISE 3.12

Oxidation-Reduction Reactions

Decide which of these reactions are oxidation-reduction reactions. In each case explain your choice. Identify the oxidizing and reducing agents in the redox reactions.

(a) $NaOH(aq) + HNO_3(aq) \longrightarrow NaNO_3(aq) + H_2O(\ell)$
(b) $4\ Cr(s) + 3\ O_2(g) \longrightarrow 2\ Cr_2O_3(s)$
(c) $NiCO_3(s) + 2\ HCl(aq) \longrightarrow NiCl_2(aq) + H_2O(\ell) + CO_2(g)$
(d) $Cu(s) + Cl_2(g) \longrightarrow CuCl_2(s)$

3-5c Oxidation Numbers and Electron Transfer

When magnesium reacted with oxygen to form magnesium oxide, the transfer of electrons caused the charge on a magnesium atom to increase from 0 to 2+ and the charge on an oxygen atom to decrease from 0 to 2−. That observation is the basis for a system used to keep track of electron transfers during redox reactions. The **oxidation number (oxidation state)** *of an atom in a chemical formula is a measure of the apparent positive charge on the atom.* An atom by itself has an equal number of protons and electrons and thus has no net charge. If the atom forms an ion its positive charge can increase, as in Mg^{2+}, or it can decrease (the atom's charge can become negative) as in O^{2-}.

How electrons participate in bonding atoms in molecules is the subject of Chapter 6.

When an atom is in a molecule or a polyatomic ion its oxidation number is not as obvious. These rules have been devised to assign oxidation numbers in such cases, as well as for the simple cases just described.

The rules given here are a modified version of those devised by Holder, Johnson, and Karol, *Journal of Chemical Education,* Volume 79, Number 4, April 2002, p 465.

Rule 1: ***The oxidation number of an atom of an element is 0.*** Examples are Fe in metallic iron, or Cl in Cl_2; each has oxidation number 0.

Rule 2: ***The oxidation number of a monoatomic ion equals its charge.*** Thus, the oxidation number of Cu^{2+} is +2; and that of S^{2-} is −2.

In this book, oxidation numbers are written as +1, +2, etc., whereas charges on ions are written as 1+, 2+, etc.

Rule 3: ***The sum of the oxidation numbers over all atoms in a complete formula for a compound is 0.*** For example, in H_2O, two times the oxidation number of each of the two H atoms plus one times the oxidation number of the one O atom must equal zero.

Rule 4: ***The sum of the oxidation numbers over all atoms in a polyatomic ion equals the charge on the ion.*** For example, in NO_2^- ion, one times the oxidation number of N plus two times the oxidation number of O must equal −1, corresponding to the 1− charge on the polyatomic ion.

Rule 5: ***In an ionic compound, oxidation numbers in the cation are independent of those in the anion.*** For example, in NH_4NO_2, separate oxidation numbers are assigned for the N in NH_4^+ and the N in NO_2^-.

Rule 6: ***Atoms of some elements have the same oxidation number in almost all their compounds.***

(a) An atom of fluorine always has an oxidation number of −1.
(b) Atoms of the alkali metals (Li, Na, K, Rb, Cs) have an oxidation number of +1 in nearly all of their compounds.
(c) Atoms of the alkaline-earth metals (Be, Mg, Ca, Sr, Ba) have an oxidation number of +2 in nearly all of their compounds.
(d) Hydrogen has an oxidation number of +1 unless it is combined with a metal, in which case its oxidation number is −1.

Rule 7: ***Apply Rules 7(a), and 7(b) only if Rules 1 to 6 have not determined oxidation numbers for all atoms in a formula.***

(a) Oxygen has an oxidation number of −2 unless Rules 1 to 6 have already given oxygen a different oxidation number.
(b) In binary compounds of nonmetals, the element closer to fluorine in the periodic table is given a negative oxidation number equal to the charge on its common monoatomic ion (← **Figure 2.7, Sec. 2-4a**).

As examples of applying these rules consider these compounds: Hg_2SO_3, H_2O_2, SF_4, and K_3N.

Hg_2SO_3 is an ionic compound, consisting of Hg_2^{2+} ions and SO_3^{2-} ions, so we can consider each ion separately (Rule 5). In Hg_2^{2+} the sum of the oxidation numbers for the two Hg atoms must be 2+, the charge on the ion (Rule 4), so the oxidation number of each Hg is +1. In the sulfite ion, SO_3^{2-}, the net charge is 2−. Rules 1 to 6 do not assign oxidation numbers for S or for O, so we apply Rule 7. Because each oxygen is −2 (Rule 7a), the oxidation number of sulfur in sulfite ion must be +4: (+4) + 3(−2) = 2− (Rule 4).

Oxidation numbers are written above the chemical symbols for the atoms to which they apply.

$$\overset{+4}{S}\overset{-2}{O}{}_3^{2-}$$

These rules for assigning oxidation number will work for most of the formulas you will encounter in this book, but there are some formulas for which they do not work.

In H_2O_2, Rule 6d assigns oxidation number +1 to H; therefore the oxidation number of O must be −1: 2(+1) +2(−1) = 0 (Rule 3). (Notice that because the oxidation number of O was assigned using Rules 1 to 6, Rule 7 was not used.)

In SF_4, the oxidation number of F is −1 (Rule 6a), so the oxidation number of S must be +4: (+4) + 4(−1) = 0 (Rule 3).

In K_3N, the oxidation number of K must be +1 (Rule 6b), so the oxidation number of N is −3: 3(+1) + (−3) = 0 (Rule 3).

EXERCISE 3.13

Assigning Oxidation Numbers

Assign oxidation numbers to each atom in each formula.

(a) Mg_3N_2 (b) ClF_3 (c) NH_4NO_3

Because oxidation numbers indicate transfers of electrons, **oxidation numbers always change during a redox reaction**. If there is no change in oxidation numbers in a reaction, the reaction is not a redox reaction. Also, because there must be the same number of electrons before and after a reaction, any increase in oxidation number for an atom from a reactant to a product must be balanced by an equal decrease in oxidation number somewhere else.

- **Oxidation is defined as an increase in oxidation number from reactant to product; the number becomes more positive.**
- **Reduction is defined as a reduction in oxidation number from reactant to product; the number becomes less positive (more negative).**

Every reaction in which a free (uncombined) element reacts to form a compound is a redox reaction. The oxidation number of the free element must increase or decrease from its original value of zero. Let's apply oxidation numbers to a redox reaction involving a free element—one we discussed earlier (← **Sec. 3-5b**), the reaction of copper metal with concentrated nitric acid.

$$Cu(s) + 4\ H^+(aq) + 2\ NO_3^-(aq) \longrightarrow Cu^{2+}(aq) + 2\ NO_2(g) + 2\ H_2O(\ell)$$

- Copper metal is in its elemental, uncombined state, so its oxidation number is 0 (Rule 1).
- H^+ has an oxidation number of +1, the charge on the ion (Rule 2).
- The overall charge on nitrate is 1−, so the sum of the oxidation numbers of oxygen and nitrogen must total −1 (Rule 4). Neither oxygen nor nitrogen can be assigned an oxidation number based on Rules 1 to 6, so we apply Rule 7. The oxidation number of each oxygen is −2 (Rule 7a) so the three oxygen atoms total −6. To achieve the overall 1− charge, the oxidation number of nitrogen in nitrate ion must be +5: $-1 = 3(-2) + (+5)$.
- Cu^{2+} has an oxidation number of +2, the charge on the ion (Rule 2).
- Rule 7a assigns an oxidation number of −2 to *each* oxygen and Rule 4 requires that the sum of oxidation numbers must be 0 for a neutral molecule such as NO_2. Thus, the oxidation number of nitrogen here is +4: $0 = 2(-2) + (+4)$.
- Water is a neutral molecule so its net charge is 0 (Rule 4) due to +1 for each hydrogen (Rule 6d) and −2 for oxygen: $0 = 2(+1) + (-2)$.
- The oxidation number of copper changes from 0 in copper metal to +2 in Cu^{2+}. This is an *increase* in oxidation number so copper metal is *oxidized.*
- The oxidation number of N changes from +5 in nitrate ion to +4 in NO_2. This *decrease* in oxidation number indicates that N in nitrate ion is *reduced.*

PROBLEM-SOLVING EXAMPLE 3.11

Applying Oxidation Numbers

The U. S. space program used the reaction of gaseous dinitrogen tetraoxide, N_2O_4, with hydrazine, N_2H_4, as a rocket fuel:

$$N_2O_4(g) + 2\ N_2H_4(g) \longrightarrow 3\ N_2(g) + 4\ H_2O(g)$$

Assign oxidation numbers to each atom in the equation. Identify which element has been oxidized, which has been reduced, and the oxidizing and reducing agents.

Result

$$\overset{+4\ -2}{N_2O_4}(g) + 2\ \overset{-2\ +1}{N_2H_4}(g) \longrightarrow 3\ \overset{0}{N_2}(g) + 4\ \overset{+1\ -2}{H_2O}(g)$$

Analyze Assign oxidation numbers. If the oxidation number of an atom is higher in the products, the reactant atom has been oxidized; if it is lower, the reactant atom has been reduced. The oxidizing agent *causes* something to be oxidized; the reducing agent *causes* something to be reduced.

Plan Use the seven rules to assign oxidation numbers. Note the differences in oxidation numbers from reactants to products for each atom. Determine what is oxidized and what is reduced. The substance that is reduced is the oxidizing agent. The substance that is oxidized is the reducing agent.

NASA Images

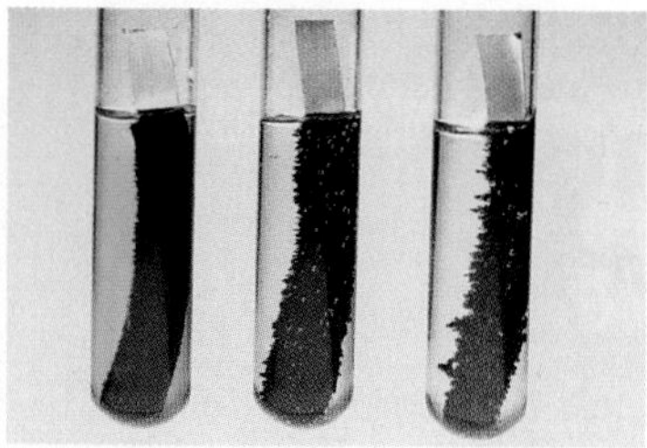

Cengage Learning/Charles D. Winters

Figure 3.15 Zinc reacts with copper(II) sulfate. When a strip of zinc metal is placed in a solution of copper(II) sulfate, copper metal forms on the zinc surface, the color of copper(II) ions disappears, and colorless zinc ions are formed in the solution.

Execute Consider each reactant and product separately.

- Dinitrogen tetraoxide is a neutral molecule. Neither N nor O can be assigned using Rules 1 to 6, so use Rule 7a: O has an oxidation number of -2. The four oxygen atoms total -8; the two nitrogen atoms must total $+8$ to balance the oxygen total (Rule 3), thus each nitrogen atom is $+4$: $0 = 2(+4) + 4(-2)$.
- Hydrazine is a neutral molecule so its charge is 0, as is the sum of the oxidation numbers of N and H (Rule 3). Hydrogen has an oxidation number of $+1$ when combined with a nonmetal, such as N, so H is $+1$. The four hydrogen atoms total $+4$. To balance this, the sum of the oxidation numbers of nitrogen atoms must be -4 and *each* nitrogen atom is -2: $0 = 2(-2) + 4(+1)$.
- N_2 is an element, so its oxidation number is 0 (Rule 1).
- The oxidation numbers of hydrogen and oxygen in water are $+1$ and -2.
- Note that the oxidation numbers of hydrogen and oxygen do not change as reactants change to products.
- Hydrazine was oxidized; the oxidation number of its nitrogen *increased* from -2 to 0. Therefore, hydrazine is the reducing agent.
- In dinitrogen tetraoxide, nitrogen was reduced; its oxidation number *decreased* from $+4$ to 0. Dinitrogen tetraoxide is the oxidizing agent.
- Nitrogen was present in each reactant but played a different role. Nitrogen in N_2O_4 was reduced; N_2O_4 was the oxidizing agent. Nitrogen in hydrazine was oxidized; N_2H_4 was the reducing agent. In NASA parlance, N_2H_4 was the fuel and N_2O_4 was the oxidizer.

PROBLEM-SOLVING PRACTICE 3.11

Assign oxidation numbers to each atom in each formula in the equation. Determine which element has been oxidized and which has been reduced. Identify the oxidizing agent and the reducing agent.

$$Sb_2S_3(s) + 3\ Fe(s) \longrightarrow 3\ FeS(s) + 2\ Sb(s)$$

Table 3.4 Activity Series of Metals

Displace H_2 from $H_2O(\ell)$, steam, or acid	Li K Ba Sr Ca Na	Ease of oxidation decreases; strength as reducing agent decreases. ↓
Displace H_2 from steam or acid	Mg Al Mn Zn Cr	
Displace H_2 from acid	Fe Ni Sn Pb	
	H_2	
Do not displace H_2 from $H_2O(\ell)$, steam, or acid	Sb Cu Hg Ag Pd Pt Au	

In rare cases, *the same reactant can serve as both the oxidizing and the reducing agent.* This is called a **disproportionation reaction**. Disproportionation occurs when chlorine reacts with water to form a mixture of hydrochloric acid and hypochlorous acid, HOCl.

$$Cl_2(g) + H_2O(\ell) \longrightarrow H^+(aq) + Cl^-(aq) + HOCl(aq)$$

Chlorine, Cl_2, with oxidation number 0 for each Cl, disproportionates into chloride ions, Cl^-, with -1 oxidation number for Cl, and hypochlorous acid, HOCl, with $+1$ oxidation number for Cl. Half of the chlorine atoms in Cl_2 are reduced from 0 to -1; the other half are oxidized from 0 to $+1$.

3-5d Metals as Reducing Agents: The Activity Series of Metals

Most metals react with solutions containing ions of other metals. For example, Figure 3.15 shows what happens when a strip of zinc is placed in a solution of $CuSO_4(aq)$. Zinc is oxidized to zinc ions and copper(II) ions are reduced to copper metal. The copper forms on the surface of the zinc strip. The disappearance of blue color from the solution indicates that copper(II) ions have reacted.

$$Zn(s) + CuSO_4(aq) \longrightarrow ZnSO_4(aq) + Cu(s)$$

Extensive studies with many metals have generated the activity series of metals shown in Table 3.4. This **activity series of metals** *ranks metals in order of decreasing ability to donate electrons during redox reactions*. Metals at the top of the series can donate electrons readily and therefore are powerful reducing agents. Metals at the lower end of the series are very poor reducing agents. ***A metal higher in the series will reduce ions of any metal lower in the series.***

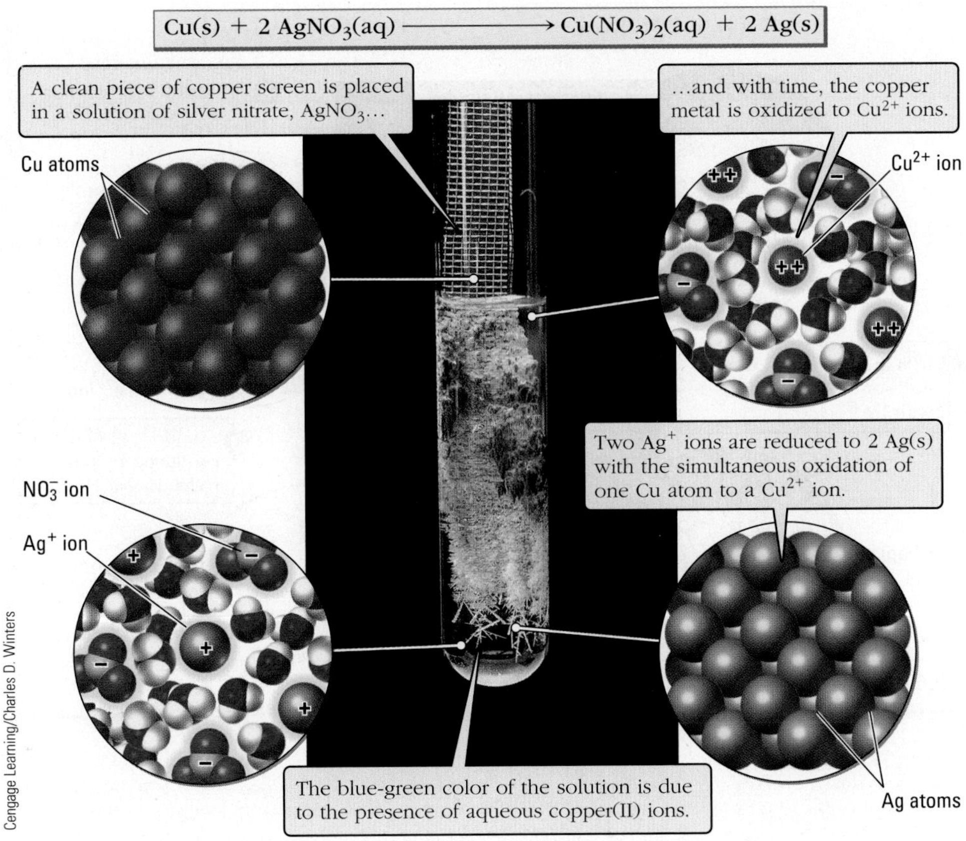
Cu(s) + 2 AgNO3(aq) ⟶ Cu(NO3)2(aq) + 2 Ag(s)
A clean piece of copper screen is placed in a solution of silver nitrate, AgNO3...
...and with time, the copper metal is oxidized to Cu2+ ions.
Cu atoms
Cu2+ ion
Two Ag+ ions are reduced to 2 Ag(s) with the simultaneous oxidation of one Cu atom to a Cu2+ ion.
NO3− ion
Ag+ ion
The blue-green color of the solution is due to the presence of aqueous copper(II) ions.
Ag atoms
Cengage Learning/Charles D. Winters

Figure 3.16 Oxidation of copper metal to Cu^{2+}; reduction of Ag^+ to metallic silver.

For example, zinc is higher in the series than copper. As we have seen, metallic zinc reduces Cu^{2+}(aq) ions in copper(II) sulfate solution to copper metal. At the same time the zinc is oxidized and the product is $ZnSO_4$, a salt containing Zn^{2+} ions. Copper is higher in the series than silver; copper metal reduces Ag^+ ions from silver nitrate solution to metallic silver (Figure 3.16).

$$Cu(s) + 2\,AgNO_3(aq) \longrightarrow Cu(NO_3)_2(aq) + 2\,Ag(s)$$

Notice that nitrate ions, NO_3^-, are spectator ions in this reaction.

Metals above *hydrogen* in the series react with acids whose anions are not oxidizing agents, such as HCl, to form hydrogen, H_2, plus a salt containing the cation of the metal and the anion of the acid (Figure 3.17).

$$Fe(s) + 2\,HCl(aq) \longrightarrow H_2(g) + FeCl_2(aq)$$

It is no accident that gold and silver have been used extensively for coinage since antiquity. These metals do not react with air, water, or even common acids, thus maintaining their luster (and value) for many years. Their low reactivity explains why they occur naturally as free metals and have been known as elements since antiquity. These metals are discussed in Chapter 20.

PROBLEM-SOLVING EXAMPLE 3.12

Activity Series of Metals

Predict whether each reaction occurs. Complete and balance overall and net ionic equations for the reactions that occur. If there is no reaction, write N.R.

(a) $Cr(s) + MgCl_2(aq) \longrightarrow$ (b) $Al(s) + Pb(NO_3)_2(aq) \longrightarrow$

(c) Mg(s) + hydrochloric acid $\longrightarrow$

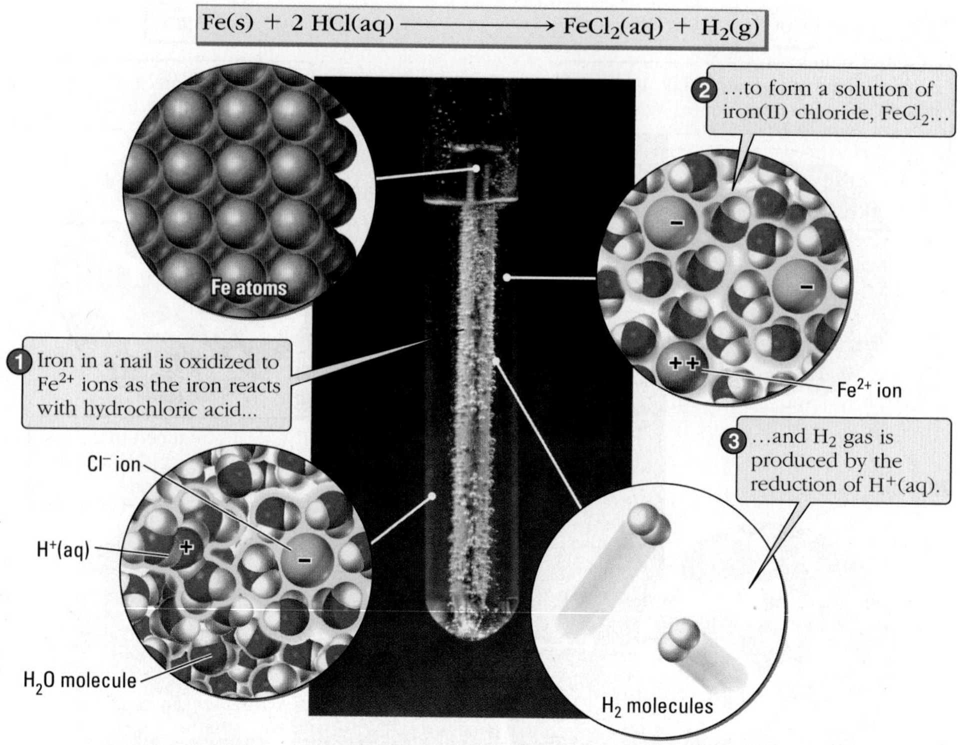

Fe(s) + 2 HCl(aq) ⟶ FeCl2(aq) + H2(g)
Fe atoms
1 Iron in a nail is oxidized to Fe2+ ions as the iron reacts with hydrochloric acid...
2 ...to form a solution of iron(II) chloride, FeCl2...
Fe2+ ion
3 ...and H2 gas is produced by the reduction of H+(aq).
Cl− ion
H+(aq)
H2O molecule
H2 molecules
© Cengage Learning/James Maynard

Figure 3.17 Reaction of a metal, iron, with an acid, HCl, to produce hydrogen.

Result

(a) No reaction (N.R.)

(b) $2\,Al(s) + 3\,Pb(NO_3)_2(aq) \longrightarrow 2\,Al(NO_3)_3(aq) + 3\,Pb(s)$

$2\,Al(s) + 3\,Pb^{2+}(aq) \longrightarrow 2\,Al^{3+}(aq) + 3\,Pb(s)$

(c) $Mg(s) + 2\,HCl(aq) \longrightarrow MgCl_2(aq) + H_2(g)$

$Mg(s) + 2\,H^+(aq) \longrightarrow Mg^{2+}(aq) + H_2(g)$

Analyze A metal and an aqueous solution are the reactants in each case. The solutions contain either a salt of a metal cation or an acid. The activity series of metals in Table 3.4 can be used to predict whether metals react with either metal cations or aqueous hydrogen ions (acids). If a reaction occurs, the products will be a cation of the metal reactant and either a solid metal (from the metal cation in solution) or hydrogen gas (from $H^+(aq)$).

Plan If metal A is higher in the activity series than metal B whose cation is in solution, then metal A will reduce the cation to metal B. If metal A is higher in the activity series than H_2, then metal A will reduce $H^+(aq)$, forming $H_2(g)$. In either case, if a reaction occurs, the cation of metal A will appear in solution. Apply this reasoning to each case. If a reaction occurs, write overall and net ionic equations.

Execute

(a) Chromium is below magnesium in the activity series, so chromium will not reduce magnesium ions from magnesium chloride. No reaction occurs.

(b) Aluminum is above lead in the activity series, so aluminum will displace lead ions from a lead(II) nitrate solution to form metallic lead and Al^{3+} ions.

$$2\,Al(s) + 3\,Pb(NO_3)_2(aq) \longrightarrow 2\,Al(NO_3)_3(aq) + 3\,Pb(s)$$
$$2\,Al(s) + 3\,Pb^{2+}(aq) \longrightarrow 2\,Al^{3+}(aq) + 3\,Pb(s)$$

(c) Magnesium is above hydrogen in the activity series, so magnesium will displace $H^+(aq)$ from HCl to form the metal salt, $MgCl_2$, plus $H_2(g)$.

$$Mg(s) + 2\,HCl(aq) \longrightarrow MgCl_2(aq) + H_2(g)$$
$$Mg(s) + 2\,H^+(aq) \longrightarrow Mg^{2+}(aq) + H_2(g)$$

PROBLEM-SOLVING PRACTICE 3.12

Predict whether each reaction occurs. If a reaction occurs, identify what has been oxidized, what has been reduced, the oxidizing agent, and the reducing agent. If not, write N.R.

(a) $2\ Al(s) + 3\ CuSO_4(aq) \longrightarrow Al_2(SO_4)_3(aq) + 3\ Cu(s)$

(b) $2\ Al(s) + Cr_2O_3(s) \longrightarrow Al_2O_3(s) + 2\ Cr(s)$

(c) $Pt(s) + 4\ HCl(aq) \longrightarrow PtCl_4(aq) + 2\ H_2(g)$

(d) $Au(s) + 3\ AgNO_3(aq) \longrightarrow Au(NO_3)_3(aq) + 3\ Ag(s)$

CONCEPTUAL EXERCISE 3.14

Reaction Product Prediction

For each pair of reactants, predict whether a reaction occurs. If a reaction occurs, identify the type of reaction and predict what products form. If not, write N.R. Which reactions are redox reactions?

(a) Combustion of ethanol: $CH_3CH_2OH(\ell) + O_2(g) \longrightarrow ?$

(b) $Fe(s) + HNO_3(aq) \longrightarrow ?$

(c) $AgNO_3(aq) + KBr(aq) \longrightarrow ?$

3-6 The Mole and Chemical Reactions

Our discussion of chemical reactions has been *qualitative* so far. The focus has been on three major types of reactions and the products they produce. We have not considered *quantitative* information, such as the amounts (moles) or masses (grams) of reactants or products that participate in a reaction. Such information involves stoichiometry (← **Sec. 3-2**). ***The fundamental principle of chemical stoichiometry is that the amount (moles) of a reaction product that forms is proportional to the amount of a reactant that reacts.*** That is, if we double the amount of a reactant that reacts, we double the amount of product that forms. The converse is also true: If we want to get twice the amount of product we will need to use twice the amount of reactant.

Stoichiometric relationships are exemplified by a redox reaction we described earlier, the reaction of dinitrogen tetraoxide with hydrazine. In the balanced chemical equation that follows, stoichiometric coefficients indicate the number of molecules of each reactant that reacts and the number of molecules of each product that forms. The stoichiometric coefficients also indicate the amount (moles) of each reactant that reacts and the amount of each product that forms (← **Sec. 3-2a**).

$$N_2O_4(g) \quad + \quad 2\ N_2H_4(g) \longrightarrow 3\ N_2(g) \quad + \quad 4\ H_2O(g)$$

The stoichiometric coefficients in a balanced chemical equation can be used to obtain mole ratios. A **mole ratio (stoichiometric factor)** *is a proportionality factor that relates the amount of one substance in the equation to the amount of another.* The amount of a reactant that reacts is proportional to the amount of a product that forms; it is also proportional to the amount of another reactant that reacts. Because of these proportionalities, mole ratios are essential for quantitative calculations involving chemical reactions. Here are some mole ratios the example equation provides:

The mole ratio is also known as the stoichiometric factor or stoichiometric ratio.

$$\frac{1\ \text{mol}\ N_2O_4}{2\ \text{mol}\ N_2H_4} \qquad \frac{1\ \text{mol}\ N_2O_4}{4\ \text{mol}\ H_2O} \qquad \frac{3\ \text{mol}\ N_2}{2\ \text{mol}\ N_2H_4}$$

These mole ratios tell us that 1 mol N_2O_4 must react for every 2 mol N_2H_4 that reacts; or 1 mol N_2O_4 must react for every 4 mol H_2O that forms; or 3 mol N_2 must be produced for each 2 mol N_2H_4 that reacts.

CONCEPTUAL EXERCISE 3.15

Mole Ratios

Write four more mole ratios that can be obtained from the example balanced equation. Explain in words what each mole ratio means in terms of the amounts of reactants or products.

Mole ratios are significant because we can use them to calculate the amount of one reactant or product from the amount of another reactant or product. For example, we can calculate the amount of N_2, $n(N_2)$, produced when 0.400 mol N_2H_4 reacts fully with N_2O_4. Because the units of each amount are moles, this is often called a *moles-to-moles conversion.*

Amount of N_2H_4 that reacts → Amount of N_2 that forms

$$n(N_2) = 0.400 \text{ mol } N_2H_4 \times \frac{3 \text{ mol } N_2}{2 \text{ mol } N_2H_4} = 0.600 \text{ mol } N_2$$

3-6a Masses of Reactants and Products

Molar mass links the amount (moles) of atoms, molecules, or formula units with the mass (grams) of atoms, molecules, or ionic compounds. When molar mass is combined with a balanced chemical equation, the masses of the reactants and products can be calculated. *In this way the nanoscale of chemical reactions is linked with the macroscale, at which we can measure masses of reactants and products by weighing.* For the reaction of N_2O_4 with N_2H_4, we can use the molar mass of each substance and its coefficient to calculate the mass of each reactant that reacts and the mass of each product that forms.

$N_2O_4(g)$	+	$2\ N_2H_4(g)$	$\longrightarrow$	$3\ N_2(g)$	+	$4\ H_2O(g)$
1 mol N_2O_4		2 mol N_2H_4		3 mol N_2		4 mol H_2O
92.0 g N_2O_4		64.0 g N_2H_4		84.0 g N_2		72.0 g H_2O
Reactants: 156.0 g total				Products: 156.0 g total		

The total mass of the products formed equals the total mass of the reactants that reacted, as must always be the case for a balanced chemical equation.

Combined with mole ratios, molar masses enable us to calculate what mass of product forms when a given amount of reactant reacts, or to calculate what mass of reactant must react to form a given mass of a product. For example, suppose we had started with 0.500 mol N_2O_4 and it reacted completely with hydrazine. What mass of water would be produced? We could use a mole ratio from the equation to calculate the amount of water from the amount of N_2O_4 and then use the molar mass of H_2O to calculate the mass of water.

In the second factor in the calculation, once the units mol N_2O_4 cancel, we have units mol H_2O, which means we have calculated the amount of H_2O.

Amount N_2O_4 → Amount H_2O → Mass H_2O

$$m(H_2O) = 0.500 \text{ mol } N_2O_4 \times \left[\frac{4 \text{ mol } H_2O}{1 \text{ mol } N_2O_4}\right] \times \left[\frac{18.0 \text{ g } H_2O}{1 \text{ mol } H_2O}\right] = 36.0 \text{ g } H_2O$$

The mole ratio (4 mol H_2O):(1 mol N_2O_4) provides the “bridge” between N_2O_4 and H_2O. This is a very important point. *Coefficients in the chemical equation, together with the* *mole ratios they provide, enable us to calculate a property of one substance, the mass of water, based on a property of another substance, the amount of N_2O_4.* This calculation began with an amount (moles) and calculated a mass (grams), so it is often referred to as a *moles-to-grams* calculation.

EXERCISE 3.16

Amount and Mass in Chemical Reactions

Verify that 10.8 g water is produced by the reaction of sufficient O_2 with 0.300 mol CH_4.

One more step can be added to the calculations enabled by a balanced chemical equation. If we know the mass of one substance, we can use molar masses and a mole ratio to calculate the mass of another. For example, calculate the mass (grams) of hydrazine needed to react completely with 0.460 g N_2O_4.

Notice that these steps reflect this book's problem-solving strategy: Analyze; Plan; Execute; and Check for a reasonable result.

Step 1: ***Begin with a balanced chemical equation that provides the correct stoichiometric coefficients.***

$$N_2O_4(g) + 2\,N_2H_4(g) \longrightarrow 3\,N_2(g) + 4\,H_2O(g)$$

Step 2: ***Analyze the problem to determine what is known and what is unknown; plan a solution.*** In this case you know the mass of N_2O_4. From that mass you can use the molar mass to calculate the amount of N_2O_4. The balanced equation gives a mole ratio from which you can calculate the amount of N_2H_4, and from its molar mass you can calculate the mass of N_2H_4. A diagram may help to remember the plan:

Mass N_2O_4 → Amount N_2O_4 → Amount N_2H_4 → Mass N_2H_4

Step 3: ***Execute: Use mole ratios and molar masses as required.***

$$m(N_2H_4) = 0.460\text{ g } N_2O_4 \times \left[\frac{1\text{ mol } N_2O_4}{92.0\text{ g } N_2O_4}\right] \times \left[\frac{2\text{ mol } N_2H_4}{1\text{ mol } N_2O_4}\right] \times \left[\frac{32.0\text{ g } N_2H_4}{1\text{ mol } N_2H_4}\right]$$

$$= 0.320\text{ g } N_2H_4$$

Step 4: ***Check your result to see whether it is reasonable.*** The starting mass of N_2O_4, 0.460 g, is 0.00500 mol N_2O_4, which requires twice that amount of hydrazine because the mole ratio is 1 : 2. The 0.0100 mol hydrazine would be about 0.32 g because the molar mass of hydrazine is 32.0 g. Therefore, the result is reasonable.

A variety of different questions about the amounts and masses of reactants and products of chemical reactions can be answered using molar masses and mole ratios. Figure 3.18 illustrates schematically the steps involved.

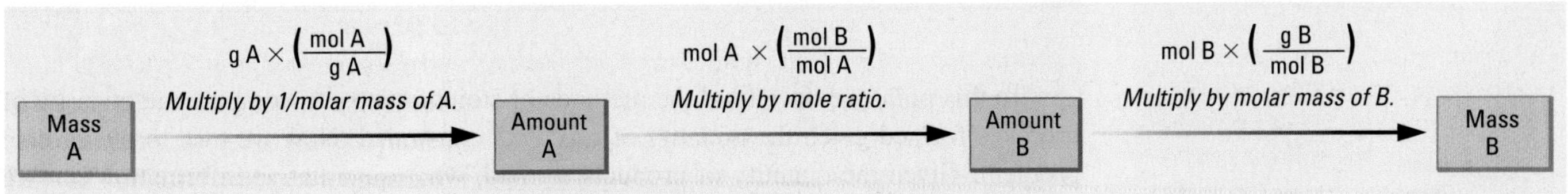
Mass A
g A × (mol A / g A)
Multiply by 1/molar mass of A.
Amount A
mol A × (mol B / mol A)
Multiply by mole ratio.
Amount B
mol B × (g B / mol B)
Multiply by molar mass of B.
Mass B

Figure 3.18 Stoichiometric relationships in a chemical reaction. The mass or amount of one reactant or product (A) is related to the mass or amount of another reactant or product (B) by the series of calculations shown.

PROBLEM-SOLVING EXAMPLE 3.13

Moles and Grams in Chemical Reactions

An iron ore named hematite, Fe_2O_3, reacts with carbon monoxide, CO, to form iron and carbon dioxide. Calculate the amount (mol) and the mass (g) of iron that are produced when 45.0 g hematite reacts with sufficient CO.

Result 0.564 mol and 31.5 g Fe

Analyze You are given the mass of one substance, Fe_2O_3, and asked to calculate the amount and mass of another substance, Fe, involved in a chemical reaction. The relationships illustrated in Figure 3.18 connect masses and amounts of reactants or products.

Plan Write a balanced chemical equation for the reaction. Calculate the molar mass of Fe_2O_3 and use it to calculate the amount of Fe_2O_3. Use the mole ratio $Fe:Fe_2O_3$ to calculate the amount of Fe. Use the molar mass of Fe to calculate the mass of Fe.

Execute The balanced equation is

$$Fe_2O_3(s) + 3\ CO(g) \longrightarrow 2\ Fe(s) + 3\ CO_2(g)$$

Molar mass (hematite) $= (2 \times 55.845 + 3 \times 15.9994)$ g/mol $= 159.69$ g/mol.

$$n(\text{hematite reacted}) = 45.0 \text{ g hematite} \times \frac{1 \text{ mol hematite}}{159.69 \text{ g hematite}} = 0.282 \text{ mol hematite}$$

$$n(\text{Fe formed}) = 0.282 \text{ mol hematite} \times \frac{2 \text{ mol Fe}}{1 \text{ mol hematite}} = 0.564 \text{ mol Fe}$$

Molar mass (Fe) = 55.845 g/mol

$$m(\text{Fe formed}) = 0.564 \text{ mol Fe} \times \frac{55.845 \text{ g Fe}}{1 \text{ mol Fe}} = 31.5 \text{ g Fe}$$

☑ **Reasonable Result Check** The mass of hematite, 45 g, is a little less than a third the molar mass, 160 g, so the amount of hematite should be less than 0.33 mol; it is. The balanced equation shows that for every one mole of hematite reacted, two moles of iron are produced. Therefore, approximately 0.33 mol hematite should produce twice that amount, approximately 0.6 mol iron. Six tenths of the molar mass of iron is about 33 g, so the final result is reasonable.

PROBLEM-SOLVING PRACTICE 3.13

Calculate the mass of carbon monoxide required to react completely with 0.433 mol hematite.

EXERCISE 3.17

Metabolism of Sugar

A 12-fluid ounce can of a soft drink (355 mL) contains 35.0 g sugar, which can be considered to be sucrose (table sugar), $C_{12}H_{22}O_{11}$. When you drink the soda, the sucrose is metabolized. Metabolism involves reaction with oxygen to produce carbon dioxide and water. (a) Write a balanced chemical equation for this reaction. (b) Calculate the mass of O_2 consumed and the masses of CO_2 and H_2O produced.

To this point we have used the methods of stoichiometry to compute the quantity of products formed given the quantity of reactants consumed. Now we turn to the reverse problem: Given the quantity of products formed, what quantitative information can we deduce about the reactants? Questions such as these are often solved by *analytical chemistry,* a field in which chemists creatively identify pure substances and measure the quantities of components of mixtures (← **Sec. 1-7b**). The analysis of mixtures is often challenging. It can take a great deal of imagination to figure out how to use chemistry to determine what, and how much, is present in a mixture such as an environmental sample containing air or water pollutants.

PROBLEM-SOLVING EXAMPLE 3.14

Analyzing a Metal-Containing Sample

Chromium metal is obtained from chromium(III) oxide, Cr_2O_3, by reacting the oxide with aluminum metal.

$$Cr_2O_3(s) + 2\ Al(s) \longrightarrow 2\ Cr(s) + Al_2O_3(s)$$

An analyst found that 10.0 g of an impure sample containing Cr_2O_3 yielded 0.821 g chromium metal. Calculate the mass percent of Cr_2O_3 in the sample.

Result 12.0%

Analyze The mass percent of Cr_2O_3 is

$$\text{Mass percent } Cr_2O_3 = \frac{\text{mass of } Cr_2O_3}{\text{mass of sample}} \times 100\%$$

so to calculate it we need to know the mass of Cr_2O_3 and the mass of the sample. The mass of the sample is given. The mass of Cr_2O_3 can be obtained from the mass of Cr and the balanced equation.

Plan Look up the molar mass of Cr; calculate the molar mass of Cr_2O_3. Calculate the amount of Cr and use the mole ratio 1 Cr_2O_3 : 2 Cr to calculate the amount of Cr_2O_3. Use the molar mass of Cr_2O_3 to calculate the mass of Cr_2O_3. Divide by the mass of the sample and multiply by 100%.

Execute The molar mass of Cr_2O_3 is 151.9904 g/mol, and the molar mass of Cr is 51.9961 g/mol.

$$m(Cr_2O_3) = 0.821 \text{ g Cr} \times \frac{1 \text{ mol Cr}}{51.9961 \text{ g Cr}} \times \frac{1 \text{ mol } Cr_2O_3}{2 \text{ mol Cr}} \times \frac{151.9904 \text{ g } Cr_2O_3}{1 \text{ mol } Cr_2O_3} = 1.20 \text{ g } Cr_2O_3$$

$$\text{Mass percent } Cr_2O_3 = \frac{1.20 \text{ g } Cr_2O_3}{10.0 \text{ g sample}} \times 100\% = 12.0\%$$

☑ **Reasonable Result Check** If the sample had been 100% Cr_2O_3, then it would have contained 10.0 g Cr_2O_3 and would have produced 6.8 g Cr:

$$m(Cr) = 10.0 \text{ g } Cr_2O_3 \times \frac{1 \text{ mol } Cr_2O_3}{152 \text{ g } Cr_2O_3} \times \frac{2 \text{ mol Cr}}{1 \text{ mol } Cr_2O_3} \times \frac{52 \text{ g Cr}}{1 \text{ mol Cr}} = 6.8 \text{ g Cr}$$

This reaction actually produced only 0.82 g Cr. Because 0.82 is a bit more than 10% of 6.8, the more precisely calculated result is reasonable.

Notice that the calculations outlined in the plan have all been set up so that they can be done in a single step. This usually is less prone to error and saves time.

PROBLEM-SOLVING PRACTICE 3.14

The purity of magnesium metal can be determined by reacting the metal with sufficient hydrochloric acid to form $MgCl_2$, evaporating the water from the resulting solution, and weighing the solid $MgCl_2$ formed.

$$Mg(s) + 2\ HCl(aq) \longrightarrow MgCl_2(aq) + H_2(g)$$

Calculate the percentage of magnesium in a 1.72-g sample that produced 6.46 g $MgCl_2$ when reacted with sufficient HCl.

3-7 Limiting Reactant

In the previous section, we used phrases such as "When 0.25 mol N_2H_4 *reacts fully* with N_2O_4" and "Calculate the mass of carbon monoxide required to *react completely* with 0.433 mol hematite." What do "reacts fully" and "react completely" mean in this context? In each case, we assumed that there was exactly the right amount of each reactant and that all reactants were entirely converted to products when the reaction was over. This is rarely the case when chemists carry out an actual reaction, whether using small quantities in a laboratory or on a much larger scale to synthesize a pharmaceutical or other industrial chemical product. The cheaper reactant is often used in excess (beyond the required stoichiometric amount) to ensure that as much as possible of the more expensive reactant is converted to the desired product.

A **limiting reactant** *is a reactant that is completely converted to products.* Once the limiting reactant has all been converted to products there is none of it left to react. Thus the limiting reactant limits the amounts of products that form. Consequently, ***the amount(s) of product(s) formed is(are) always determined (limited) by the initial amount of the limiting reactant.***

Consider this analogy to a chemistry limiting reactant. You are assembling gift boxes of chocolates. Every gift box (the "product") consists of one carton with 24 sections, each containing a piece of chocolate (the "reactants"). You could write this "equation," analogous to a balanced chemical equation:

$$1 \text{ carton} + 24 \text{ chocolates} \longrightarrow 1 \text{ gift box of chocolates}$$

The candy store has on hand 12 cartons and 120 chocolates. How many complete gift boxes can be made?

If all 12 cartons are used (assuming sufficient chocolates):

$$\text{Number of gift boxes} = 12 \text{ cartons} \times \frac{1 \text{ gift box}}{1 \text{ carton}} = 12 \text{ gift boxes}$$

If all 120 chocolates are used (assuming sufficient cartons):

$$\text{Number of gift boxes} = 120 \text{ chocolates} \times \frac{1 \text{ gift box}}{24 \text{ chocolates}} = 5 \text{ gift boxes}$$

Therefore, only five *complete* gift boxes can be made. Once the 120 chocolates are put into five cartons, there are no more chocolates to fill the remaining seven cartons. The chocolates are the limiting reactant; you ran out of them before using all the available cartons. The chocolates limited the production to only five complete gift boxes. Seven unfilled cartons remain; they are in excess.

Before we can calculate how much product will be produced by a chemical reaction, the limiting reactant must be identified. *The amount of the limiting reactant that is converted to product determines the amount of product that can be formed.*

PROBLEM-SOLVING EXAMPLE 3.15

Limiting Reactant: Amounts

Calculate the amount (moles) of nitrogen that is produced when 1.00 mol dinitrogen tetraoxide and 1.00 mol hydrazine are mixed and this reaction occurs.

$$N_2O_4(g) + 2\,N_2H_4(g) \longrightarrow 3\,N_2(g) + 4\,H_2O(g)$$

Result 1.50 mol N_2

Analyze The amount of nitrogen produced will be determined by the limiting reactant. The limiting reactant reacts completely; some of the other reactant is left. Mole ratios for reactants can be obtained from coefficients in the balanced equation.

Plan Compare the actual amounts of reactants with the mole ratio given by the coefficients. Figure out which reactant is entirely used up and which is left over.

Execute Two moles of N_2H_4 are required to react with 1 mol N_2O_4, but only 1 mol N_2H_4 is available. Thus, N_2H_4 is the limiting reactant. N_2O_4 is in excess by 0.5000 mol. Two moles of hydrazine reacted produces 3 mol N_2.

$$n(N_2) = \text{amount of } N_2 = 1.00 \text{ mol } N_2H_4 \times [3 \text{ mol } N_2/2 \text{ mol } N_2H_4] = 1.50 \text{ mol } N_2$$

☑ **Reasonable Result Check** The balanced chemical equation shows that if 2 mol N_2H_4 reacts, 3 mol N_2 will be produced; only 1 mol N_2H_4 reacted so the amount of N_2 should be half of 3 mol or 1.5 mol. The result is reasonable.

PROBLEM-SOLVING PRACTICE 3.15

Calculate the amount (mol) of water produced if 3.00 mol hydrazine is mixed with 1.25 mol dinitrogen tetraoxide and a reaction takes place.

We now consider the case where the quantities of reactants are given in grams, not moles, and the limiting reactant must be determined. Two approaches can be used: (1) the mole ratio method; and (2) the mass method.

3-7a Mole Ratio Method

Use the molar mass of each reactant to calculate the amount of each reactant available; use the results to calculate the *actual* mole ratio. Compare this ratio with the theoretical mole ratio from coefficients in the balanced chemical equation. *If the actual mole ratio is less than the theoretical mole ratio, the substance in the numerator is the limiting reactant.* If the actual mole ratio is larger than the theoretical mole ratio, the substance in the numerator is in excess.

3-7b Mass Method

Calculate the mass of product that would be produced from the available quantity of each reactant, assuming that an excess of the other reactant is available. *The limiting reactant is the one that produces the smaller mass of product.*

PROBLEM-SOLVING EXAMPLE 3.16

Limiting Reactant

Once again we use the reaction between dinitrogen tetraoxide and hydrazine.

$$N_2O_4(g) + 2\ N_2H_4(g) \longrightarrow 3\ N_2(g) + 4\ H_2O(g)$$

If the reaction starts with 500. g of each reactant: (a) Determine the limiting reactant. (b) Calculate the mass (g) of water produced. Assume that all of the limiting reactant is converted to products.

Result (a) N_2O_4 (b) 391 g H_2O

Analyze You are asked to determine a limiting reactant and the mass of a product, given masses of reactants. Use either the mole ratio method or the mass method.

Mole Ratio Method

Plan Calculate molar masses of reactants and use them to calculate amounts of reactants. Compare actual mole ratios with the theoretical mole ratio given by the coefficients in the balanced equation. Identify the limiting reactant and use its amount to calculate the amount and mass of product.

Execute The molar masses of the reactants were calculated earlier (← **Sec. 3-6a**): N_2O_4 = 92.0 g/mol; N_2H_4 = 32.0 g/mol. The amounts of each reactant are:

$$\text{Amount}(N_2O_4) = 500.\ \text{g}\ N_2O_4 \times \frac{1\ \text{mol}\ N_2O_4}{92.0\ \text{g}\ N_2O_4} = 5.43\ \text{mol}\ N_2O_4$$

$$\text{Amount}(N_2H_4) = 500.\ \text{g}\ N_2H_4 \times \frac{1\ \text{mol}\ N_2H_4}{32.0\ \text{g}\ N_2H_4} = 15.6\ \text{mol}\ N_2H_4$$

$$\text{Actual mole ratio} = \frac{5.43\ \text{mol}\ N_2O_4}{15.6\ \text{mol}\ N_2H_4} = \frac{0.348\ \text{mol}\ N_2O_4}{1\ \text{mol}\ N_2H_4}$$

$$\text{Theoretical mole ratio} = \frac{1 \text{ mol } N_2O_4}{2 \text{ mol } N_2H_4} = \frac{0.500 \text{ mol } N_2O_4}{1 \text{ mol } N_2H_4}$$

The actual mole ratio is less than the theoretical mole ratio, which means that the actual amount of N_2O_4 is not enough to react with all of the N_2H_4; N_2O_4 is the limiting reactant and it must be used to calculate the mass of water produced.

$$\text{Mass}(H_2O) = 5.43 \text{ mol } N_2O_4 \times \left[\frac{4 \text{ mol } H_2O}{1 \text{ mol } N_2O_4}\right] \times \left[\frac{18.0 \text{ g } H_2O}{1 \text{ mol } H_2O}\right] = 391 \text{ g } H_2O$$

Note: The same limiting reactant would be obtained had we calculated the actual mole ratio the other way, with hydrazine in the numerator.

$$\text{Actual mole ratio} = \frac{15.6 \text{ mol } N_2H_4}{5.43 \text{ mol } N_2O_4} = \frac{2.87 \text{ mol } N_2H_4}{1 \text{ mol } N_2O_4}$$

$$\text{Theoretical mole ratio} = \frac{2 \text{ mol } N_2H_4}{1 \text{ mol } N_2O_4}$$

In this case, the actual mole ratio is *greater* than the theoretical mole ratio. This means that hydrazine is in excess and cannot be the limiting reactant. Therefore N_2O_4 must be the limiting reactant (as we found previously).

Mass Method

Plan Calculate the mass of water that would be produced if 500. g N_2O_4 reacted, assuming that sufficient N_2H_4 is present. Then, calculate the mass of water that would be produced if all 500. g hydrazine reacted, assuming that sufficient N_2O_4 is present. Determine which reactant produces the smaller mass of product. That reactant is the limiting reactant.

Execute Calculate the mass of H_2O based on the initial mass of each reactant:

$$m(H_2O, \text{ if } N_2O_4 \text{ limiting}) =$$
$$500. \text{ g } N_2O_4 \times \left[\frac{1 \text{ mol } N_2O_4}{92.0 \text{ g } N_2O_4}\right] \times \left[\frac{4 \text{ mol } H_2O}{1 \text{ mol } N_2O_4}\right] \times \left[\frac{18.0 \text{ g } H_2O}{1 \text{ mol } H_2O}\right] = 391 \text{ g } H_2O$$

$$m(H_2O, \text{ if } N_2H_4 \text{ limiting}) =$$
$$500. \text{ g } N_2H_4 \times \left[\frac{1 \text{ mol } N_2H_4}{32.0 \text{ g } N_2H_4}\right] \times \left[\frac{4 \text{ mol } H_2O}{2 \text{ mol } N_2H_4}\right] \times \left[\frac{18.0 \text{ g } H_2O}{1 \text{ mol } H_2O}\right] = 563 \text{ g } H_2O$$

The limiting reactant produces the smaller mass of product so N_2O_4 is the limiting reactant.

☑ **Reasonable Result Check** The molar mass of N_2O_4 is about three times larger than the molar mass of N_2H_4; this means that for the same mass the amount of N_2O_4 should be about 1/3 the amount of N_2H_4, and it is. If the amount of N_2O_4 were 1/2 the amount of N_2H_4, then they would be in the theoretical mole ratio, but the amount of N_2O_4 is only 1/3, so N_2O_4 is the limiting reactant. The molar mass of N_2O_4 is a little less than 100 g/mol, so 500 g should be a bit more than 5 mol; 5.43 mol N_2O_4 is reasonable. 5.43 mol N_2O_4 produces four times that amount of water, a bit more than 20 mol, which is about 400 g water. The result is reasonable.

PROBLEM-SOLVING PRACTICE 3.16

Using the same reaction, calculate the mass of nitrogen produced when 200. g N_2O_4 and 100. g hydrazine react.

Once the limiting reactant has been determined, it is possible to calculate the mass of excess reactant remaining after the reaction is complete.

PROBLEM-SOLVING EXAMPLE 3.17

Limiting Reactant and Excess Reactant

Powdered aluminum reacts with iron(III) oxide in the *thermite reaction* to form molten iron and aluminum oxide:

$$2\ Al(s) + Fe_2O_3(s) \longrightarrow 2\ Fe(\ell) + Al_2O_3(s)$$

Liquid iron is produced because the reaction releases enough energy to raise the temperature above the melting point of iron. This liquid iron can be used to weld steel railroad rails. Suppose that a mixture of 100. g Al and 100. g Fe_2O_3 reacts: (a) Calculate the mass (g) of liquid iron formed; (b) Calculate the mass of excess reactant that remains after the reaction is complete.

Result (a) 69.9 g Fe (b) 66.4 g Al

Analyze To determine the mass of product, it is first necessary to determine whether there is a limiting reactant. The limiting reactant must be used to calculate the mass of product. The mass of the excess reactant that remains is the difference between the initial mass and the mass that reacted.

Plan Determine the amount of each reactant available. Use the mass method to determine the limiting reactant. Calculate the mass of $Fe(\ell)$ that forms from the amount of the limiting reactant. Calculate the mass of excess reactant that reacted from the amount of the limiting reactant. Subtract that mass from the initial mass of the excess reactant.

Execute Calculate the amount of each reactant:

$$n(\text{Al}) = 100.\ \text{g Al} \times \frac{1\ \text{mol Al}}{26.98\ \text{g Al}} = 3.71\ \text{mol Al}$$

$$n(\text{Fe}_2\text{O}_3) = 100.\ \text{g Fe}_2\text{O}_3 \times \frac{1\ \text{mol Fe}_2\text{O}_3}{159.69\ \text{g Fe}_2\text{O}_3} = 0.626\ \text{mol Fe}_2\text{O}_3$$

Use the mass method to determine the limiting reactant:

$$m(\text{Fe, if Al limiting}) = 3.71\ \text{mol Al} \times \frac{2\ \text{mol Fe}}{2\ \text{mol Al}} \times \frac{55.845\ \text{g Fe}}{1\ \text{mol Fe}} = 207\ \text{g Fe}$$

$$m(\text{Fe, if Fe}_2\text{O}_3\ \text{limiting}) = 0.626\ \text{mol Fe}_2\text{O}_3 \times \frac{2\ \text{mol Fe}}{1\ \text{mol Fe}_2\text{O}_3} \times \frac{55.845\ \text{g Fe}}{1\ \text{mol Fe}} = 69.9\ \text{g Fe}$$

Iron(III) oxide is the limiting reactant because it produces less iron. Aluminum is in excess. The mass of iron formed is 69.9 g Fe, based on the limiting reactant. The amount of Al that reacted can also be calculated from the amount of Fe_2O_3:

$$n(\text{Al, reacted}) = 0.626\ \text{mol Fe}_2\text{O}_3 \times \frac{2\ \text{mol Al}}{1\ \text{mol Fe}_2\text{O}_3} = 1.25\ \text{mol Al}$$

Subtracting the amount of Al that reacted from the initial amount gives the amount of Al that did not react (3.71 mol Al − 1.25 mol Al). The mass of excess Al is

$$m(\text{Al, in excess}) = (3.71 - 1.25)\ \text{mol Al} \times \frac{26.98\ \text{g Al}}{1\ \text{mol Al}} = 66.4\ \text{g Al}$$

☑ **Reasonable Result Check** The molar masses of Fe_2O_3 (160 g/mol) and Fe (56 g/mol) are in the ratio of approximately 3 : 1. One mole of Fe_2O_3 produces 2 mol Fe according to the balanced equation. So a given mass of Fe_2O_3 should produce about two thirds as much Fe, and this agrees with our more exact calculation.

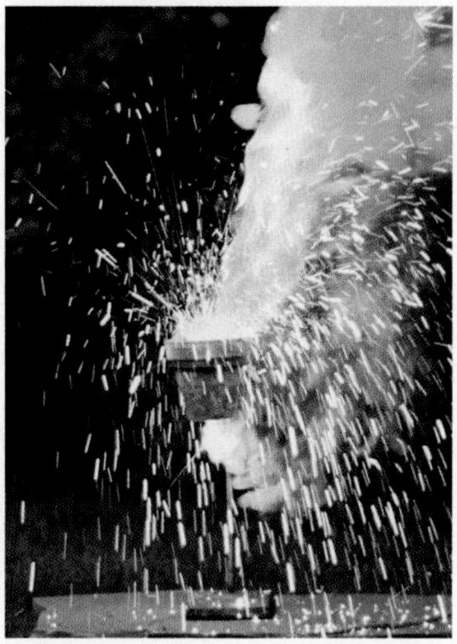

Cengage Learning/Charles D. Winters

Powdered aluminum reacts with iron(III) oxide in the thermite reaction.

Notice that we could have used the mole ratio method to determine which reactant was limiting.

It is useful to calculate the quantity of excess reactant remaining to verify that the reactant in excess is not the limiting reactant.

PROBLEM-SOLVING PRACTICE 3.17

At high temperatures, silicon dioxide reacts with carbon to produce silicon carbide, SiC, also known as carborundum, an important industrial abrasive.

$$SiO_2(g) + 3\ C(s) \longrightarrow SiC(s) + 2\ CO(g)$$

If the reaction is run with 100. g of each reactant: (a) Determine the limiting reactant; and (b) Calculate the mass of excess reactant remaining after the reaction is over.

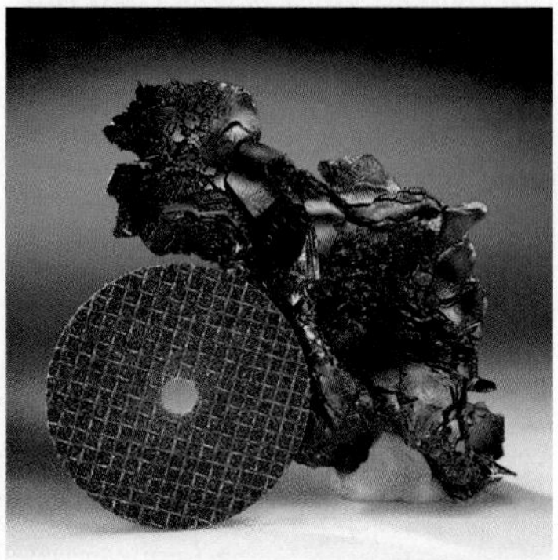

Cengage Learning/Charles D. Winters

Silicon carbide, SiC, is the abrasive on the grinding wheel *(left)*. Naturally occurring silicon carbide *(right)* is also known as carborundum.

EXERCISE 3.18

Limiting Reactant

Urea is used as a fertilizer because it can react with water to release ammonia, which provides nitrogen to plants.

$$(NH_2)_2CO(s) + H_2O(\ell) \longrightarrow 2\ NH_3(aq) + CO_2(g)$$

(a) When 300. g urea and 100. g water are combined, calculate the mass of ammonia and the mass of carbon dioxide that form.

(b) Calculate the mass of the excess reactant that remains after reaction.

3-8 Evaluating Chemical Synthesis: Percent Yield

A reaction that converts the limiting reactant into the maximum possible quantity of product is said to have a 100% yield. This *maximum possible quantity of product is called the* **theoretical yield**. Often the **actual yield**, *the quantity of desired product actually obtained from a synthesis* in a laboratory or industrial chemical plant, is less than the theoretical yield.

The efficiency of a particular synthesis method is evaluated by calculating the **percent yield**, which is defined as *the ratio of actual to theoretical yield*:

$$\text{Percent yield} = \frac{\text{actual yield}}{\text{theoretical yield}} \times 100\%$$

Percent yield can be applied, for example, to the synthesis of aspirin. Suppose a student carried out the synthesis and obtained 2.2 g aspirin rather than the calculated theoretical yield of 2.6 g. The percent yield is

$$\text{Percent yield} = \frac{\text{actual yield of product}}{\text{theoretical yield of product}} \times 100\% = \frac{2.2\ \text{g}}{2.6\ \text{g}} \times 100\% = 85\%$$

Because the amount of substance is proportional to mass, percent yield can also be calculated from the actual amount produced in a synthesis divided by the theoretical amount (and multiplied by 100%).

Percent yield is especially important in industrial syntheses, such as the synthesis of drugs, where 10 or 20 reactions might be involved, one after the other. If the yield is only 50% in each of 20 successive reactions, the overall yield is less than one millionth. The synthesis would be worthless because it would not produce enough of the final product.

Although we hope to obtain as close to the theoretical yield as possible when carrying out a reaction, few reactions or experimental manipulations are 100% efficient, despite controlled conditions and careful laboratory techniques. Side reactions can occur that form products other than the desired one, and during the isolation and purification of the desired product, some of it may be lost. When chemists report the synthesis of a new compound or the development of a new synthesis, they also report the percent yield of the reaction or the overall series of reactions. Other chemists who wish to repeat the synthesis then have an idea of how much product can be expected from a certain amount of reactants.

PROBLEM-SOLVING EXAMPLE 3.18

Calculating Percent Yield

Methanol, CH_3OH, is an excellent fuel, and it can be produced from carbon monoxide and hydrogen.

$$CO(g) + 2\ H_2(g) \longrightarrow CH_3OH(\ell)$$

If 500. g CO reacts with sufficient H_2 and 485 g CH_3OH is produced, calculate the percent yield of the reaction.

Result 85.0%

Analyze "Sufficient H_2" means that CO is limiting. Calculate the theoretical yield of CH_3OH and compare it with the mass actually produced.

Plan Calculate the amount of CO. Use a mole ratio from the balanced equation to calculate the amount of CH_3OH and the mass of CH_3OH that could be produced if the synthesis were perfect. Calculate percent yield.

Execute The molar masses are CO, 28.01 g/mol and CH_3OH, 32.0 g/mol. The calculation of theoretical yield can be done in a single step. The first factor calculates the amount of CO. The second factor is the mole ratio. The third factor calculates the mass of CH_3OH.

$$\text{Theoretical yield} = 500.\text{ g CO} \times \frac{1 \text{ mol CO}}{28.0 \text{ g CO}} \times \frac{1 \text{ mol CH}_3\text{OH}}{1 \text{ mol CO}} \times \frac{32.0 \text{ g CH}_3\text{OH}}{1 \text{ mol CH}_3\text{OH}}$$

$$= 571 \text{ g CH}_3\text{OH}$$

The problem states that 485 g CH_3OH was produced, so

$$\text{Percent yield} = \frac{\text{actual yield}}{\text{theoretical yield}} \times 100\% = \frac{485 \text{ g CH}_3\text{OH}}{571 \text{ g CH}_3\text{OH}} \times 100\% = 85.0\%$$

☑ **Reasonable Result Check** The molar mass of CH_3OH is slightly greater than that of CO, and an equal number of moles of CO and CH_3OH are in the balanced equation. So *x* g CO should produce somewhat more than *x* g CH_3OH. Slightly less is actually produced, however, so a percent yield slightly less than 100% is about right.

PROBLEM-SOLVING PRACTICE 3.18

Assume the methanol synthesis has an 85.0% yield and you want to make 1.00 kg CH_3OH. Calculate the mass (g) of H_2 you should use if you have sufficient CO.

PROBLEM-SOLVING EXAMPLE 3.19

Percent Yield

Ammonia can be produced from the reaction of a metal oxide such as calcium oxide with ammonium chloride:

$$\text{CaO(s)} + 2\ \text{NH}_4\text{Cl(s)} \longrightarrow 2\ \text{NH}_3\text{(g)} + \text{H}_2\text{O(g)} + \text{CaCl}_2\text{(s)}$$

Calculate the mass of calcium oxide needed to react with excess ammonium chloride to produce 1.00 g ammonia; the expected percent yield is 25%.

Result 6.6 g CaO

Analyze The theoretical yield of product must be used to calculate the amount of a reactant. An actual yield of 1.00 g ammonia is desired and this is only 25% of the theoretical yield. The definition of percent yield relates actual yield and theoretical yield. From the theoretical yield the starting amount and mass of the limiting reactant can be calculated. Ammonium chloride is in excess so CaO is limiting.

Plan Rearrange the definition of percent yield algebraically to solve for theoretical yield. Remember that 25% is 25/100 = 0.25. Use molar masses and a mole ratio to calculate the amount and mass of CaO from the theoretical yield of ammonia.

Execute

$$\text{Theoretical yield} = \frac{\text{actual yield}}{\text{percent yield}} = \frac{1.00 \text{ g NH}_3}{0.25} = 4.00 \text{ g NH}_3$$

$$m(\text{CaO}) = 4.00 \text{ g NH}_3 \times \frac{1 \text{ mol NH}_3}{17.03 \text{ g NH}_3} \times \frac{1 \text{ mol CaO}}{2 \text{ mol NH}_3} \times \frac{56.077 \text{ g CaO}}{1 \text{ mol CaO}} = 6.6 \text{ g CaO}$$

☑ **Reasonable Result Check** To check the result, start with 6.6 g CaO and estimate how much NH_3 would form. The molar mass of CaO is approximately 56 g/mol, so 6.6 g CaO is approximately 0.12 mol CaO. The coefficients in the balanced equation tell us that this would produce twice as many moles of NH_3 or 0.24 mol NH_3, which is 0.24 × 17 g/mol = 4 g NH_3. This is the quantity of NH_3 we would get if the yield were 100%. The actual yield is 25%, so the amount of ammonia expected is reduced to one fourth, that is, approximately 1 g NH_3, which is what was desired. The result, 6.6 g CaO, is reasonable.

PROBLEM-SOLVING PRACTICE 3.19

You heat 2.50 g copper with an excess of sulfur and synthesize 2.53 g copper(I) sulfide, Cu_2S:

$$16\ Cu(s) + S_8(s) \longrightarrow 8\ Cu_2S(s)$$

Your laboratory instructor expects students to have at least a 60% yield for this reaction. Did your synthesis meet this standard?

EXERCISE 3.19

Percent Yield

Percent yield can be reduced by side reactions that produce undesired product(s) and by poor laboratory technique in isolating and purifying the desired product. Identify two other factors that could lead to a low percent yield.

3-8a Atom Economy—Another Approach to Tracing Starting Materials

When chemical reactions are carried out on an industrial scale, huge quantities of reactants and products are involved. The field of green chemistry has arisen in response to the environmental and health problems sometimes associated with industrial chemistry. **Green chemistry** is the *design, development, and implementation of chemical products and processes to reduce or eliminate the use and generation of substances hazardous to human health and the environment.* Maximizing percent yield is one way to reduce generation of hazardous substances, but it is not the only one. Another measure of the success of a chemical reaction is **atom economy**, *the fraction of the mass of starting materials that ends up in the desired final product or products.* The greater the fraction of starting atoms incorporated into the desired final product, the fewer waste by-products created. Atom economy is defined by this equation:

$$\text{Atom economy} = \frac{\text{sum of atomic weights of all atoms in the useful product}}{\text{sum of atomic weights of all atoms in reactants}}$$

Atom economy is usually expressed as a percentage, that is,

$$\text{Percent atom economy} = \text{atom economy} \times 100\%$$

Reactions for which all atoms in the reactants are found in the desired product have a percent atom economy of 100%. As an example, consider this reaction, which was used previously in Problem-Solving Example 3.18:

$$CO(g) + 2\ H_2(g) \longrightarrow CH_3OH(\ell)$$

The sum of atomic weights of all atoms in the reactants is 12.011 u + 15.9994 u + {2 × (2 × 1.0079 u)} = 32.042 u. The atomic weight of all the atoms in the product is 12.011 u + {3 × (1.0079 u)} + 15.9994 u + 1.0079 u = 32.042 u. The percent atom economy for this reaction is (32.042/32.042) × 100% = 100%.

Many other synthesis reactions, however, generate other products in addition to the desired product. In such cases, the percent atom economy can be far less than 100%. Devising strategies for synthesis of desired compounds with the least waste is a major goal of the current push toward green chemistry. Green chemistry also aims to prevent pollution at its source rather than having to clean up problems after they occur. Each year since 1996 the U. S. Environmental Protection Agency has given Presidential Green Chemistry Challenge Awards for noteworthy green chemistry advances.

One of the 2012 Green Chemistry Challenge Awards was given for developing a family of catalysts that convert carbon dioxide and carbon monoxide into biodegradable polymers. This discovery is being commercialized to produce high-performance products, such as paints, adhesives, foams, and plastics. (See http://www2.epa.gov/green-chemistry/2012-academic-award-coates.)

EXERCISE 3.20

Atom Economy

Calculate the percent atom economy for the reaction of coke (mainly carbon) with steam to generate hydrogen gas and carbon dioxide gas. The desired product is hydrogen, which is used in other industrial processes.

3-9 Composition and Empirical Formulas

In Chapter 2 (← **Sec. 2-12a**) we showed how empirical and molecular formulas can be derived from percent composition and molar mass data. One way to obtain composition data is *combustion analysis*. In combustion analysis, a known, small mass of compound, especially organic compounds of carbon and hydrogen, is burned in excess oxygen converting all the carbon in the compound to carbon dioxide and all its hydrogen into water. The combustion products are collected and weighed (Figure 3.19). The mass of each product and the balanced combustion equation are used to calculate the mass of carbon and of hydrogen in the original compound. If the compound also contains oxygen, the mass of oxygen in the original sample is calculated by difference.

$$\text{Mass of oxygen} = \text{mass of sample} - (\text{mass of C} + \text{mass of H})$$

Consider this application of combustion analysis. Butyric acid, an organic compound with an extremely unpleasant odor, contains only carbon, hydrogen, and oxygen. When 1.200 g butyric acid was burned, 2.41 g CO_2 and 0.982 g H_2O were produced. Calculate the empirical formula of butyric acid. In a separate experiment, the molar mass of butyric acid was determined to be 88.1 g/mol. Determine butyric acid's molecular formula.

All of the carbon and hydrogen in the butyric acid are burned to form CO_2 and H_2O, respectively. Use the mass of CO_2 to calculate the mass of C in the original butyric acid sample; use the mass of H_2O formed to calculate the mass of H. Calculate the mass of O in the sample by difference. Then, find the amount of each element in the sample, the mole ratios, and the empirical formula. To determine the molecular formula, compare the formula mass of the empirical formula with the known molar mass.

- *Calculate the mass of each element in the sample*:

$$m(\text{C, in sample}) = 2.41 \text{ g } CO_2 \times \frac{1 \text{ mol } CO_2}{44.01 \text{ g } CO_2} \times \frac{1 \text{ mol C}}{1 \text{ mol } CO_2} \times \frac{12.01 \text{ g C}}{1 \text{ mol C}}$$
$$= 0.658 \text{ g C}$$

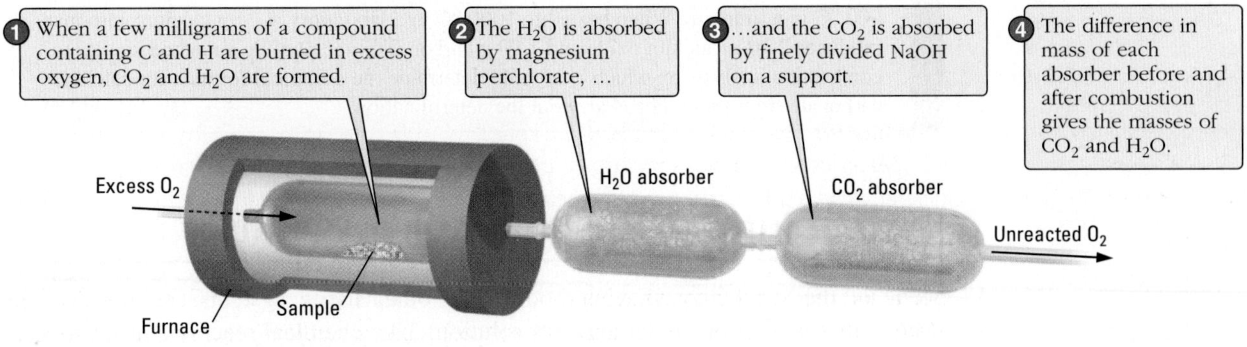

1 When a few milligrams of a compound containing C and H are burned in excess oxygen, CO₂ and H₂O are formed.
2 The H₂O is absorbed by magnesium perchlorate, ...
3 ...and the CO₂ is absorbed by finely divided NaOH on a support.
4 The difference in mass of each absorber before and after combustion gives the masses of CO₂ and H₂O.
Excess O₂
H₂O absorber
CO₂ absorber
Unreacted O₂
Sample
Furnace

Figure 3.19 Combustion analysis.

$$m(\text{H, in sample}) = 0.982 \text{ g } H_2O \times \frac{1 \text{ mol } H_2O}{18.02 \text{ g } H_2O} \times \frac{2 \text{ mol H}}{1 \text{ mol } H_2O} \times \frac{1.0079 \text{ g H}}{1 \text{ mol H}}$$
$$= 0.110 \text{ g H}$$

$$m(\text{O, in sample}) = 1.200 \text{ g} - 0.658 \text{ g C} - 0.110 \text{ g H} = 0.432 \text{ g O}$$

- *Calculate the amount of each element in the sample*:

$$n(\text{C, in sample}) = 0.658 \text{ g C} \times \frac{1 \text{ mol C}}{12.01 \text{ g C}} = 0.0548 \text{ mol C}$$

$$n(\text{H, in sample}) = 0.110 \text{ g H} \times \frac{1 \text{ mol H}}{1.0079 \text{ g H}} = 0.109 \text{ mol H}$$

$$n(\text{O, in sample}) = 0.432 \text{ g O} \times \frac{1 \text{ mol O}}{16.00 \text{ g O}} = 0.0270 \text{ mol O}$$

- *Obtain mole ratios by dividing the amount of each element by the smallest amount*:

$$\frac{0.0548 \text{ mol C}}{0.0270 \text{ mol O}} = \frac{2.03 \text{ mol C}}{1.00 \text{ mol O}} \text{ and } \frac{0.109 \text{ mol H}}{0.0270 \text{ mol O}} = \frac{4.03 \text{ mol H}}{1.00 \text{ mol O}}$$

The mole ratios show that for every oxygen atom in the molecule, there are two carbon atoms and four hydrogen atoms. Therefore, the empirical formula of butyric acid is C_2H_4O, which has an empirical formula mass of 44.05 g/mol.

- *Compare the experimental molar mass to the empirical formula mass to determine the molecular formula.*

The experimental molar mass is known to be 88.10 g/mol, twice the empirical formula mass. Therefore, the molecular formula of butyric acid is $C_4H_8O_2$, twice the empirical formula.

The molar mass of butyric acid, $C_4H_8O_2$, is 4(12.01) + 8(1.0079) + 2(16.00) = 88.10 g/mol, so the result is reasonable.

CONCEPTUAL EXERCISE 3.21

Formula from Combustion Analysis

Phenol is a compound of carbon, hydrogen, and oxygen used commonly as a disinfectant. Combustion of a 175-mg sample of phenol yielded 491 mg CO_2 and 100. mg H_2O.

(a) Calculate the empirical formula of phenol.
(b) Identify what other information is needed to determine whether the empirical formula is the actual molecular formula.

CONCEPTUAL EXERCISE 3.22

Formula from Masses of Products

Calcium carbonate forms carbon dioxide and calcium oxide when it is heated above 900 °C in a limekiln. When heated to 1000 °C in a laboratory, 4.31 g calcium carbonate produces 2.40 g calcium oxide and 1.90 g carbon dioxide. Outline a method similar to combustion analysis by which you could determine the empirical formula for calcium carbonate from these data. Carry out the determination.

3-10 Solution Concentration: Molarity

Many of the substances in your body or in other living systems are dissolved in water—that is, they are in an aqueous solution. Like chemical reactions in living systems, many reactions studied in the chemical laboratory are carried out in solution.

Frequently, this chemistry must be done quantitatively. For example, intravenous fluids administered to patients contain many compounds (salts, nutrients, drugs, and so on), and the quantity of each must be known accurately. To accomplish this task, we make solutions in which we know the amount or mass of each component per unit volume of solution. Then, we can measure volumes of solution rather than masses of solids, liquids, and gases.

A solution is a homogeneous mixture (← **Sec. 1-7**) of a **solute**, *a substance that has been dissolved*, and the **solvent**, *the substance in which one or more solutes has been dissolved.* To know the quantity of solute in a given volume of a liquid solution requires knowing the *composition* of the solution—the relative quantities of solute and solvent. There are many ways of expressing solution composition. The most useful of these for studying chemical reactions in solution is molarity.

3-10a Molarity

The **concentration** or **molarity** of a solution is *the amount of solute per unit volume of solution.* It is usually expressed in moles per liter (mol/L, abbreviated M).

$$\text{Molarity} = \frac{\text{amount of solute (moles)}}{\text{volume of solution (liters)}}$$

Note that the volume term in the denominator is volume of *solution,* not volume of solvent. The International Union of Pure and Applied Chemistry uses the term concentration and the abbreviation c to represent what we call molarity in this book. We will use the abbreviation c in equations.

If, for example, 40.0 g (1.00 mol) NaOH is dissolved in sufficient water to produce a solution with a total volume of 1.00 L, the solution has a concentration of 1.00 mol NaOH per 1.00 L of solution, which is a 1.00 molar solution. The molarity of this solution is reported as 1.00 M, where the capital M stands for the units mol/L. Molarity is also represented by the letter c, sometimes followed by the formula of a compound or ion, such as $c(\text{NaOH})$ or $c(\text{OH}^-)$. Writing $c(\text{NaOH}) = 2.00$ M indicates that 2 mol NaOH is dissolved per liter of solution.

molarity of NaOH solution → $c(\text{NaOH}) = \dfrac{2 \text{ mol NaOH}}{1 \text{ L solution}} = 2 \text{ M}$ ← 2 mol NaOH per liter of solution

A solution of known molarity can be made by adding the required amount of solute to a volumetric flask, adding some solvent to dissolve all the solute, and then adding sufficient solvent with continual mixing to fill the flask to the mark etched on the flask's neck, as shown in Figure 3.20 for a $KMnO_4$ solution. The etched mark indicates the liquid level equal to the specified volume of the flask.

PROBLEM-SOLVING EXAMPLE 3.20

Molarity

In aqueous solution, potassium permanganate, $KMnO_4$, is a strong oxidizing agent that is often used in laboratory experiments. Calculate the molarity of a solution made by placing 0.433 g $KMnO_4$ in a 500.0-mL volumetric flask and adding water until the solution volume is exactly 500.0 mL.

Result 0.00548 M

Analyze Molarity is amount per unit volume. The amount of solute can be calculated from the mass. The solution volume is given.

Plan Use the mass and the molar mass to calculate the amount of solute. Change units for the volume from mL to L. Then, divide the amount by the volume.

1 Combine ~240 mL pure (deionized) H_2O with 0.198 g (0.00125 mol) $KMnO_4$ in a 250.0-mL volumetric flask.
2 Shake the flask to dissolve the $KMnO_4$. After the solid dissolves…
3 …add pure water to the mark etched in the neck.
4 The flask now contains 250.0 mL of 0.00500-M $KMnO_4$ solution.
Deionized H2O
KMnO4
© Cengage Learning/James Maynard

Figure 3.20 Preparing 250.0 mL of a 0.00500-M aqueous solution from solid $KMnO_4$ and pure water.

Execute The molar mass of $KMnO_4$ is 158.03 g/mol.

$$n(\text{KMnO}_4) = 0.433 \text{ g KMnO}_4 \times \frac{1 \text{ mol KMnO}_4}{158.03 \text{ g KMnO}_4} = 2.74 \times 10^{-3} \text{ mol KMnO}_4$$

$$\text{Molarity of KMnO}_4 = c(\text{KMnO}_4) = \frac{2.74 \times 10^{-3} \text{ mol KMnO}_4}{500.0 \text{ mL solution}} \times \frac{1000 \text{ mL}}{1 \text{ L}}$$
$$= 5.48 \times 10^{-3} \text{ mol/L}$$

Notice that the last factor in the calculation converts volume units from mL to L.

☑ **Reasonable Result Check** We have about half a gram of solute in a half liter of solution. This is equivalent to having a gram of solute in a liter of solution. The molar mass of solute is about 160 g/mol. The molarity should be about 1/160 = 0.00625, which is close to our more exact result.

PROBLEM-SOLVING PRACTICE 3.20

Calculate the molarity of a solution containing 36.0 g Na_2SO_4 in 750. mL solution.

EXERCISE 3.23

Cholesterol Molarity

A blood serum cholesterol level greater than 240 mg cholesterol per deciliter (0.100 L) of blood usually indicates the need for medical intervention. Calculate this serum cholesterol level in mol/L. Cholesterol's molecular formula is $C_{27}H_{46}O$.

Sometimes the molarity of a particular ion in a solution is required, a value that depends on the nature of the solute. For example, potassium chromate is a soluble ionic compound and a strong electrolyte that completely dissociates in solution to form 2 mol K^+ ions and 1 mol CrO_4^{2-} ions for each mole of K_2CrO_4 that dissolves:

$$\underset{1 \text{ mol}}{K_2CrO_4(aq)} \longrightarrow \underset{2 \text{ mol}}{2\,K^+(aq)} + \underset{1 \text{ mol}}{CrO_4^{2-}(aq)} \qquad \text{100\% dissociated}$$

The K^+ concentration is twice the K_2CrO_4 concentration because each mole of K_2CrO_4 contains 2 mol K^+. Therefore, a 0.00283-M K_2CrO_4 solution has a K^+ concentration of 2×0.00283 M = 0.00566 M and a CrO_4^{2-} concentration of 0.00283 M.

EXERCISE 3.24

Molarity of Ions in Solution

A student dissolves 6.37 g aluminum nitrate in sufficient water to make 250.0 mL of solution. Calculate (a) the molarity of aluminum nitrate in this solution and (b) the molarity of aluminum ions and of nitrate ions in this solution.

CONCEPTUAL EXERCISE 3.25

Preparing a Solution

When solutions are prepared, the final volume of solution can be different from the sum of the volumes of the solute and solvent because some expansion or contraction can occur. Why is it always better to describe solution preparation as "adding enough solvent" to make a certain volume of solution?

3-10b Preparing a Solution of Known Molarity from a Pure Solute

In Problem-Solving Example 3.20, we described finding the molarity of a $KMnO_4$ solution that was prepared from known quantities of solute and solution. More frequently, a solid or liquid solute (sometimes even a gas) must be used to make up a solution of known molarity. The problem becomes one of calculating what mass of solute to use to provide the proper amount of solute.

Consider a laboratory experiment that requires 2.00 L of 0.750-M NH_4Cl solution. What mass of NH_4Cl must be dissolved in water to make 2.00 L of solution? The amount of NH_4Cl required can be calculated from the molarity.

$$n(\text{NH}_4\text{Cl}) = \left(\frac{0.750 \text{ mol NH}_4\text{Cl}}{\text{L solution}}\right) \times 2.00 \text{ L solution} = 1.500 \text{ mol NH}_4\text{Cl}$$

M × L = mol/L × L = mol

Then the molar mass can be used to calculate the mass of NH_4Cl needed.

$$m(\text{NH}_4\text{Cl}) = 1.500 \text{ mol NH}_4\text{Cl} \times \left(\frac{53.49 \text{ g NH}_4\text{Cl}}{\text{mol NH}_4\text{Cl}}\right) = 80.2 \text{ g NH}_4\text{Cl}$$

The solution can be prepared by placing 80.2 g NH_4Cl into a beaker, dissolving it in pure water, rinsing all of the solution into a volumetric flask, and adding distilled water with thorough mixing until the solution volume is 2.00 L.

PROBLEM-SOLVING EXAMPLE 3.21

Solute Mass and Molarity

Describe how to prepare 500.0 mL of 0.0250-M $K_2Cr_2O_7$ solution starting with solid potassium dichromate.

Result Dissolve 3.68 g $K_2Cr_2O_7$ in water and add enough water to make 500.0 mL solution.

Analyze Solution concentration and volume of solution are known. The amount of solute is proportional to the volume and the proportionality factor is molarity. From the amount of solute, its mass can be calculated.

Plan Use the concentration to calculate how many moles of $K_2Cr_2O_7$ are needed and use the molar mass to calculate the mass.

Execute $$n(\text{K}_2\text{Cr}_2\text{O}_7) = 500.0 \text{ mL solution} \times \frac{1 \text{ L}}{1000 \text{ mL}} \times \frac{0.0250 \text{ mol K}_2\text{Cr}_2\text{O}_7}{1 \text{ L solution}}$$
$$= 1.25 \times 10^{-2} \text{ mol K}_2\text{Cr}_2\text{O}_7$$

The molar mass of potassium dichromate is 294.2 g/mol.

$$m(K_2Cr_2O_7) = 1.25 \times 10^{-2} \text{ mol } K_2Cr_2O_7 \times \frac{294.2 \text{ g } K_2Cr_2O_7}{1 \text{ mol } K_2Cr_2O_7} = 3.68 \text{ g } K_2Cr_2O_7$$

To prepare the solution, place 3.68 g $K_2Cr_2O_7$ into a 500-mL volumetric flask, add enough pure water to dissolve the solute, and then add water, shaking thoroughly to mix the solution and water, to bring the solution volume up to the mark on the flask. This results in 500.0 mL of 0.0250-M $K_2Cr_2O_7$ solution.

☑ **Reasonable Result Check** The molar mass of $K_2Cr_2O_7$ is about 300 g/mol, and we want a 0.025 M solution, so we need about 300 g/mol × 0.025 mol/L = 7.5 g/L. But only one-half liter is required, so 0.5 L × 7.5 g/L = 3.75 g is needed, which agrees with our more accurate result.

PROBLEM-SOLVING PRACTICE 3.21

Describe how you would prepare each solution.
(a) 1.00 L of 0.125-M Na_2CO_3 from solid Na_2CO_3
(b) 500. mL of 0.0215-M $KMnO_4$ from solid $KMnO_4$

3-10c Preparing a Solution by Dilution

Consider two cases: A teaspoonful of sugar, $C_{12}H_{22}O_{11}$, is dissolved in a glass of water and a teaspoonful of sugar is dissolved in a swimming pool full of water. The swimming pool and the glass contain the same amount (mol) of sugar, but the concentration of sugar in the swimming pool is far less because the volume of solution in the pool is much greater than that in the glass.

Frequently, solutions of the same solute need to be available at several different molarities. For example, hydrochloric acid is often used at concentrations of 6.0 M, 1.0 M, and 0.050 M. To make these solutions, chemists often use a concentrated solution of known molarity and dilute it. **Dilution** *means adding solvent* (in this case water) *to make a solution of lower concentration* (see Figure 3.21). Because the process of dilution involves adding solvent only, *the amount of solute in the diluted solution is exactly the same as it was in the concentrated solution.* Diluting a solution increases the volume, so the molarity is lowered, even though the amount of solute remains unchanged.

The amount of solute always equals the molarity times the volume, so

$$c(\text{conc.}) \times V(\text{conc.}) = \text{amount of solute} = c(\text{dil.}) \times V(\text{dil.})$$

where c(conc.) and V(conc.) represent the molarity and the volume of the concentrated solution and c(dil.) and V(dil.) represent the molarity and volume of the dilute solution.

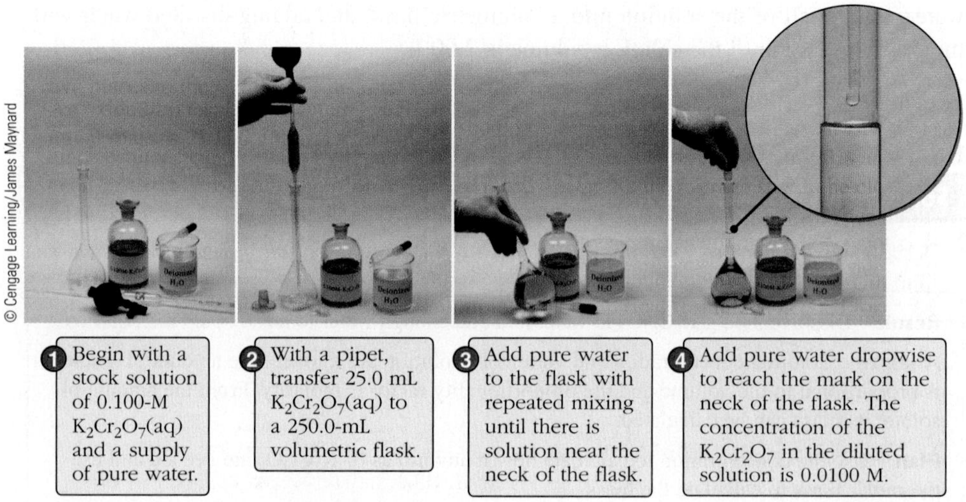
© Cengage Learning/James Maynard
1 Begin with a stock solution of 0.100-M $K_2Cr_2O_7$(aq) and a supply of pure water.
2 With a pipet, transfer 25.0 mL $K_2Cr_2O_7$(aq) to a 250.0-mL volumetric flask.
3 Add pure water to the flask with repeated mixing until there is solution near the neck of the flask.
4 Add pure water dropwise to reach the mark on the neck of the flask. The concentration of the $K_2Cr_2O_7$ in the diluted solution is 0.0100 M.

Figure 3.21 Solution preparation by dilution.

We can calculate, for example, the concentration of a hydrochloric acid solution made by diluting 25.0 mL of 6.0-M HCl to 500. mL. In this case, we want to determine c(dil.) when c(conc.) = 6.0 M, V(conc.) = 0.0250 L, and V(dil.) = 0.500 L. We algebraically rearrange the relationship to get the concentration of the diluted HCl.

$$c(\text{dil.}) = \frac{c(\text{conc.}) \times V(\text{conc.})}{V(\text{dil.})} = \frac{\left(6.0\frac{\text{mol}}{\text{L}}\right) \times 0.0250\text{ L}}{0.500\text{ L}} = 0.30\text{ mol/L}$$

A quick and useful check on a dilution calculation is to make certain that the molarity of the diluted solution is lower than that of the concentrated solution.

EXERCISE 3.26

Amount of Solute in Solutions

Consider 100. mL of 6.0-M HCl solution, which is diluted with water to yield 500. mL of 1.20-M HCl. Show that 100. mL of the more concentrated solution contains the same amount of HCl as 500. mL of the more dilute solution.

PROBLEM-SOLVING EXAMPLE 3.22

Solution Concentration and Dilution

Describe how to prepare 500.0 mL of 1.00-M H_2SO_4 solution from a concentrated sulfuric acid solution that is 18.0 M.

Result Add 27.8 mL of the concentrated sulfuric acid slowly, carefully, and with stirring to about 450 mL water. After the sulfuric acid is thoroughly mixed with the water, add water to make up a total volume of 500.0 mL solution.

Analyze The concentrations of both solutions are given. The volume of the diluted solution is given.

Plan Rearrange the dilution equation to calculate the volume:

$$c(\text{conc.}) \times V(\text{conc.}) = c(\text{dil.}) \times V(\text{dil.})$$

Execute

$$V(\text{conc.}) = \frac{c(\text{dil.}) \times V(\text{dil.})}{c(\text{conc.})} = \frac{1.00\text{ mol/L} \times 0.5000\text{ L}}{18.0\text{ mol/L}} \times \frac{1000\text{ mL}}{1\text{ L}} = 27.8\text{ mL}$$

Add 27.8 mL of concentrated sulfuric acid slowly, with stirring, to about 450 mL distilled water. When the solution has cooled to room temperature, add sufficient water with thorough mixing to bring the final volume to 500.0 mL. This gives a 1.00-M sulfuric acid solution.

☑ **Reasonable Result Check** The ratio of molarities is 18 : 1, so the ratio of volumes should be 1 : 18, and it is.

Use caution when diluting a concentrated acid or a concentrated base. The more concentrated solution should be added slowly to the solvent (water) so that the heat released during the dilution is rapidly dissipated into a large volume of water. If water is added to the acid or base, the heat released could be sufficient to vaporize the solution, spraying the acid or base over you and anyone nearby.

PROBLEM-SOLVING PRACTICE 3.22

A laboratory procedure calls for 50.0 mL of 0.150-M NaOH. You have available 100. mL of 0.500-M NaOH. What volume of the more concentrated solution should be diluted to make the desired solution?

CONCEPTUAL EXERCISE 3.27

Solution Concentration

The molarity of a solution can be decreased by dilution. How could the molarity of a solution be increased without adding additional solute?

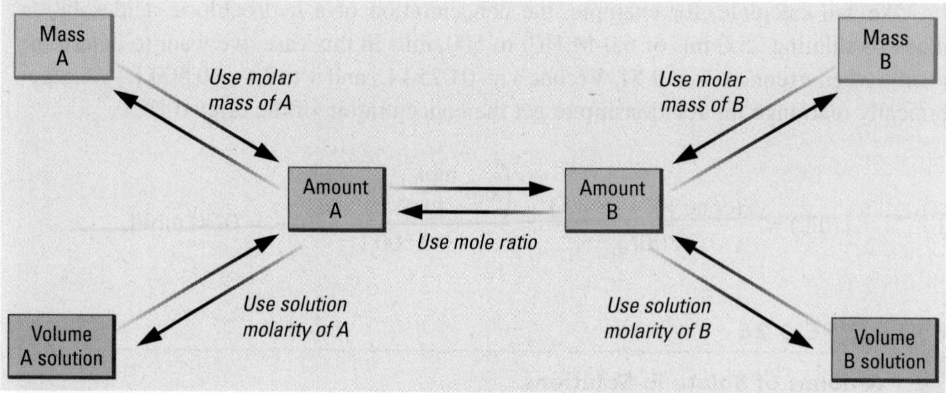

Mass A
Use molar mass of A
Mass B
Use molar mass of B
Amount A
Amount B
Use mole ratio
Volume A solution
Use solution molarity of A
Use solution molarity of B
Volume B solution

Figure 3.22 Stoichiometric relationships for a chemical reaction in aqueous solution.

3-11 Stoichiometry in Aqueous Solutions

Many kinds of reactions—acid-base neutralization (← **Sec. 3-4c**), precipitation (← **Sec. 3-3**), and redox (← **Sec. 3-5**)—occur in aqueous solutions. In such reactions, molarity allows us to make conversions between volumes of solutions and moles of reactants and products as given by the stoichiometric coefficients. Molarity is used to link mass, amount (moles), and volume of solution (Figure 3.22).

PROBLEM-SOLVING EXAMPLE 3.23

Solution Reaction Stoichiometry

Limestone, $CaCO_3$, reacts with HCl to produce the salt $CaCl_2$, carbon dioxide, and water:

$$CaCO_3(s) + 2\ HCl(aq) \longrightarrow CaCl_2(aq) + CO_2(g) + H_2O(\ell)$$

Calculate the mass of $CaCO_3$ that reacts completely with 10.0 mL of 3.00-M HCl.

Result 1.50 g $CaCO_3$

Analyze You know the volume and concentration of HCl and are asked to calculate the mass of $CaCO_3$ that reacts with it. Amount is proportional to volume and concentration is the proportionality factor. The balanced chemical equation gives a mole ratio between HCl and $CaCO_3$. Molar mass relates amount and mass. These relationships are diagrammed in Figure 3.22.

Plan Use the solution volume and concentration to calculate the amount of HCl. Use a mole ratio to calculate the amount of $CaCO_3$. Use the molar mass of $CaCO_3$ to calculate its mass.

Execute

$$n(\text{HCl}) = 10.0\ \text{mL HCl} \times \frac{1\ \text{L}}{1000\ \text{mL}} \times \frac{3.00\ \text{mol HCl}}{1\ \text{L HCl}} = 0.0300\ \text{mol HCl}$$

$$n(\text{CaCO}_3) = 0.0300\ \text{mol HCl} \times \frac{1\ \text{mol CaCO}_3}{2\ \text{mol HCl}} = 0.0150\ \text{mol CaCO}_3$$

$$m(\text{CaCO}_3) = 0.0150\ \text{mol CaCO}_3 \times \frac{100.09\ \text{g CaCO}_3}{1\ \text{mol CaCO}_3} = 1.50\ \text{g CaCO}_3$$

Notice that the calculation could have been done all at once. The relevant part of Figure 3.22 is provided to show what each factor does.

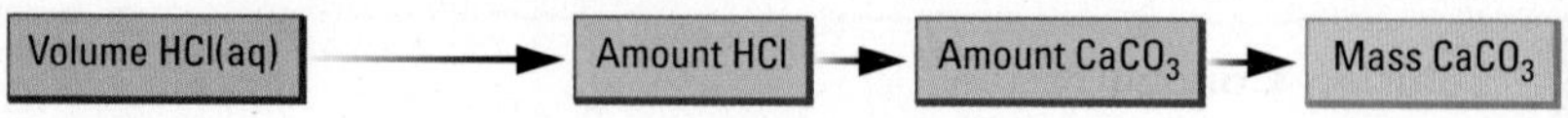

Volume HCl(aq)
Amount HCl
Amount CaCO3
Mass CaCO3

$$m(\text{CaCO}_3) =$$

$$10.0\ \text{mL HCl} \times \frac{1\ \text{L}}{1000\ \text{mL}} \times \frac{3.00\ \text{mol HCl}}{1\ \text{L HCl}} \times \frac{1\ \text{mol CaCO}_3}{2\ \text{mol HCl}} \times \frac{100.09\ \text{g CaCO}_3}{1\ \text{mol CaCO}_3}$$

$$= 1.50\ \text{g CaCO}_3$$

☑ **Reasonable Result Check** The HCl concentration is 3 M; we have 0.0100 L solution, so we have 0.03 mol HCl. Because of the 1 : 2 stoichiometry, HCl will react with half as many moles of $CaCO_3$, or 0.015 mol $CaCO_3$. The molar mass of $CaCO_3$ is about 100 g/mol, so 0.15 mol is about 0.15 mol × 100 g/mol = 1.5 g, which agrees with the more exact result.

PROBLEM-SOLVING PRACTICE 3.23

In a recent year, 1.2×10^{10} kg sodium hydroxide (NaOH) was produced in the United States by passing an electric current through brine, an aqueous solution of sodium chloride.

$$2\,NaCl(aq) + 2\,H_2O(\ell) \longrightarrow 2\,NaOH(aq) + Cl_2(g) + H_2(g)$$

Calculate the volume of brine needed to produce this mass of NaOH. (*Note*: 1.0 L brine contains 360 g dissolved NaCl.)

PROBLEM-SOLVING EXAMPLE 3.24

Solution Reaction Stoichiometry

When aqueous potassium iodide, KI, is added to aqueous lead(II) nitrate, $Pb(NO_3)_2$, a brilliant yellow precipitate of PbI_2 is produced.

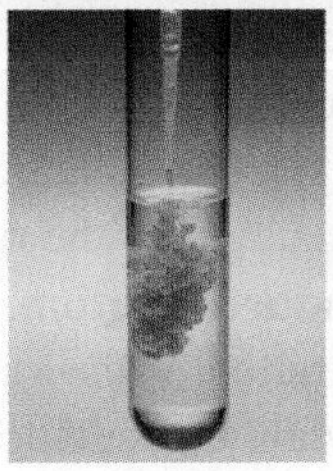
Cengage Learning/Charles D. Winters

$$Pb(NO_3)_2(aq) + 2\,KI(aq) \longrightarrow PbI_2(s) + 2\,KNO_3(aq)$$

Calculate the minimum volume (mL) of 3.00-M KI required to react completely with 55.0 mL of 0.740-M $Pb(NO_3)_2$.

Result 27.1 mL of 3.00 M KI solution

Analyze The volume of potassium iodide solution is desired. The volume and molarity of the lead(II) nitrate solution are known. The concentration of potassium iodide is known. The balanced equation provides a mole ratio. The diagram in Figure 3.22 summarizes the relationships involved.

Plan Calculate the amount of $Pb(NO_3)_2$ from the volume and concentration of its solution. Use the 2 : 1 mole ratio to calculate the amount of KI. Calculate the volume of KI from the molarity of the KI solution. Do the calculations all at once.

Execute

$$V(\text{KI solution}) = 55.0\text{ mL} \times \frac{1\text{ L}}{1000\text{ mL}} \times \frac{0.740\text{ mol Pb(NO}_3)_2}{1\text{ L}} \times$$
$$\frac{2\text{ mol KI}}{1\text{ mol Pb(NO}_3)_2} \times \frac{1\text{ L}}{3.00\text{ mol KI}} \times \frac{1000\text{ mL}}{1\text{ L}} = 27.1\text{ mL}$$

☑ **Reasonable Result Check** We have 0.055 L × 0.74 mol/L ≅ 0.041 mol $Pb(NO_3)_2$. We have 0.0271 L × 3 mol/L ≅ 0.081 mol KI. This is the 2 : 1 ratio of reactants needed according to the balanced equation. The result is reasonable.

PROBLEM-SOLVING PRACTICE 3.24

Insoluble solid silver bromide, AgBr, can be dissolved by aqueous sodium thiosulfate, $Na_2S_2O_3$, as described by this net ionic equation:

$$AgBr(s) + 2\,S_2O_3^{2-}(aq) \longrightarrow Ag(S_2O_3)_2^{3-}(aq) + Br^-(aq)$$

If you want to dissolve 50.0 mg AgBr, calculate the minimum volume of 0.0150-M $Na_2S_2O_3$ you must use. Express your result in milliliters.

EXERCISE 3.28

Molarity in Medicine

Sodium chloride is used in intravenous solutions for medical applications. The NaCl concentration in such solutions must be accurately known and can be assessed by reacting the solution with an experimentally determined volume of $AgNO_3$ solution of known concentration. The net ionic equation is

$$Ag^+(aq) + Cl^-(aq) \longrightarrow AgCl(s)$$

Suppose that a chemical technician uses 19.3 mL of 0.200-M $AgNO_3$ to convert all the NaCl in a 25.0-mL sample of an intravenous solution to AgCl. Calculate the molarity of NaCl in the solution.

3-12 Titrations in Aqueous Solutions

Acid-base titrations are described more extensively in Chapter 15.

One important quantitative use of aqueous solution reactions is to determine the unknown concentration of a reactant in a solution, such as the concentration of HCl in a solution of HCl. This is done with a titration. In a **titration**, *one solution is added to another solution until just enough of the first solution has been added to react with all of the second solution.* Apparatus is used so that the volume of each solution can be determined. *The point at which the amount of reactant in the solution being added is just enough to react completely with the reactant in the other solution is called the* **equivalence point**. At the equivalence point, the amounts of the two reactants are in exactly the stoichiometric ratio (mole ratio) given by the balanced chemical equation for the reaction that occurs. This means that the mole ratio can be used to calculate the amount of one reactant from the amount of the other. Usually one of the solutions in a titration is a **standard solution**, *a solution whose concentration is known accurately*. From the concentration of the standard solution, the concentration of the other solution can be determined by calculations such as those in the preceding section.

An example titration is shown in Figure 3.23. It involves the titration of hydrogen peroxide solution, which is colorless, with potassium permanganate solution, the dark purple solution shown in Figure 3.20 (← **Sec. 3-10a**):

$$2\ KMnO_4(aq) + 5\ H_2O_2(aq) + 3\ H_2SO_4(aq) \longrightarrow$$
$$2\ MnSO_4(aq) + 5\ O_2(aq) + 8\ H_2O(aq) + K_2SO_4(aq)$$

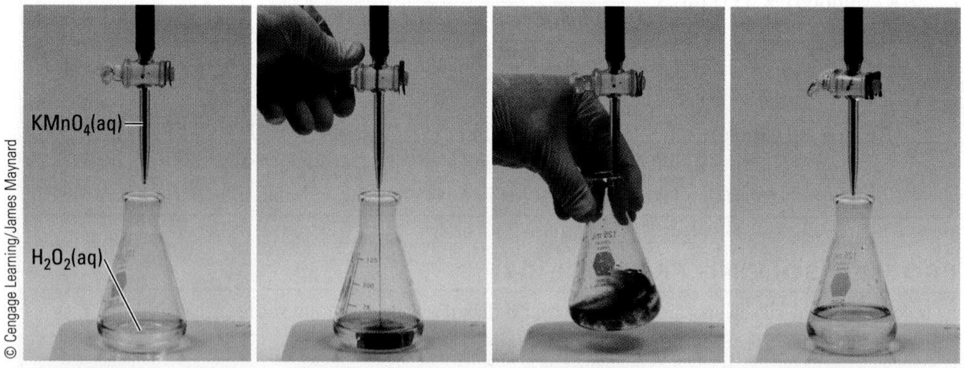

KMnO4(aq)
H2O2(aq)
© Cengage Learning/James Maynard

1 $KMnO_4$(aq) of known concentration is in a buret, a measuring device calibrated in divisions of 0.10 mL. A 25.0-mL sample of H_2O_2(aq) of unknown concentration is in the flask.

2 When $KMnO_4$(aq) is added to the flask, the purple color is visible until it mixes with the H_2O_2(aq) and reacts.

3 As long as there is enough H_2O_2(aq) to react with all of the $KMnO_4$(aq) added, the purple color disappears.

4 When all of the H_2O_2(aq) has reacted, the next drop of $KMnO_4$(aq) remains unreacted and a trace of purple color persists. This signals the end point of the titration.

Figure 3.23 Titration of H_2O_2(aq) with a standard solution of $KMnO_4$(aq).

$KMnO_4$ solution is added from a buret, which is marked in units of milliliters and tenths of a milliliter. This enables the volume of $KMnO_4$ added to be measured. A measured volume of H_2O_2 solution is placed in the flask. As $KMnO_4$ is added its color disappears because it reacts with the H_2O_2. When all of the H_2O_2 has been used up, the next drop of $KMnO_4$ has nothing to react with so its color remains. The first appearance of the color of permanganate indicates that the titration is finished—the amount of $KMnO_4$ added is equivalent to the amount of H_2O_2 that had been in the flask. By equivalent, we mean that 2 mol $KMnO_4$ has been added for every 5 mol H_2O_2 that was originally in the flask. The reactants are in the mole ratio given by the balanced equation.

EXERCISE 3.29

Redox Reaction?

Show that the reaction of $KMnO_4$ with H_2O_2 is a redox reaction.

Titrations may involve redox reactions like the permanganate-peroxide example. They often involve acid-base reactions as well. In most titrations none of the reactants is colored, so some other method of finding the equivalence point is needed. An **indicator** is *a substance that can be added in a titration to show when the equivalence point has been reached.* Phenolphthalein, for example, is commonly used as the indicator in strong acid-strong base titrations because it is colorless in acidic solutions and pink in basic solutions. *The point at which an indicator is seen to change color is called the* **end point**.

PROBLEM-SOLVING EXAMPLE 3.25

Acid-Base Titration

A student has an aqueous solution of calcium hydroxide that is approximately 0.10 M. She titrated a 50.0-mL sample of the calcium hydroxide solution with a standardized solution of 0.300-M $HNO_3(aq)$. To reach the end point, 41.4 mL of the HNO_3 solution was needed. Determine the molarity of the calcium hydroxide solution.

Result 0.124 M $Ca(OH)_2$

Analyze At the end point the reactants are in the mole ratio given by the balanced equation for the reaction. We know the volume and exact concentration of the HNO_3 solution, and we know the volume of the $Ca(OH)_2$ solution.

Plan From the volume and molarity of HNO_3, calculate the amount of titrant reacted. From the mole ratio in the equation, calculate the amount of $Ca(OH)_2$ reacted. Use the volume of $Ca(OH)_2$ solution to calculate its concentration.

Execute The balanced equation for the reaction is

$$2\ HNO_3(aq) + Ca(OH)_2(aq) \longrightarrow Ca(NO_3)_2(aq) + 2\ H_2O(\ell)$$

$$n(HNO_3) = 0.0414\ \text{L}\ HNO_3 \times \frac{0.300\ \text{mol}\ HNO_3}{1.00\ \text{L}\ HNO_3\ \text{solution}} = 0.0124\ \text{mol}\ HNO_3$$

$$n(Ca(OH)_2) = 1.24 \times 10^{-2}\ \text{mol}\ HNO_3 \times \frac{1\ \text{mol}\ Ca(OH)_2}{2\ \text{mol}\ HNO_3} = 6.21 \times 10^{-3}\ \text{mol}\ Ca(OH)_2$$

$$c(Ca(OH)_2) = \frac{6.21 \times 10^{-3}\ \text{mol}\ Ca(OH)_2}{0.0500\ \text{L}\ Ca(OH)_2\ \text{solution}} = 0.124\ \text{M}\ Ca(OH)_2$$

☑ **Reasonable Result Check** The net ionic equation for the titration reaction is

$$H^+(aq) + OH^-(aq) \longrightarrow H_2O(\ell)$$

At the equivalence point, the amount of $H^+(aq)$ added and the amount of $OH^-(aq)$ originally in the sample must be equal. The amount of each reactant is its volume multiplied by its molarity. For the HNO_3 we have 0.0414 L × 0.300 M = 0.0124 mol HNO_3. Because of the subscript 2 in the formula $Ca(OH)_2$ we have 0.050 L × 0.124 M × 2 = 0.0124 mol OH^-. The result is reasonable.

PROBLEM-SOLVING PRACTICE 3.25

In a titration, a 20.0-mL sample of sulfuric acid, H_2SO_4, was titrated to the end point with 41.3 mL of 0.100 M NaOH. Calculate the molarity of the H_2SO_4 solution.

SUMMARY PROBLEM

Part I

Aqueous solutions of ammonium sulfide and mercury(II) nitrate react and a precipitate forms.

(a) Write the overall balanced chemical equation and indicate the state (aq) or (s) for each compound.

(b) Name each product.

(c) Write the complete ionic equation.

(d) Write the net ionic equation.

Part II

In a blast furnace at high temperature, iron(III) oxide in ore reacts with carbon monoxide to produce metallic iron and carbon dioxide. The liquid iron produced is cooled and weighed. The reaction is run repeatedly with the same initial mass of iron(III) oxide, 19.0 g, but differing initial masses of carbon monoxide. The masses of iron obtained are shown in this graph.

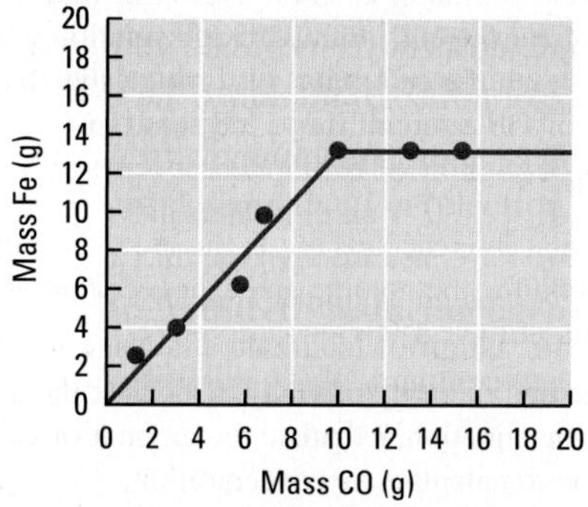
Mass Fe (g)
0
2
4
6
8
10
12
14
16
18
20
0 2 4 6 8 10 12 14 16 18 20
Mass CO (g)

(a) Write the balanced chemical equation for this reaction.

(b) Calculate the mass of CO required to react completely with 19.0 g iron(III) oxide.

(c) Calculate the mass of carbon dioxide produced when the reaction converts 10.0 g iron(III) oxide completely to products.

(d) From the graph, determine which reactant is limiting when less than 10.0 g carbon monoxide is available to react with 19.0 g iron(III) oxide.

(e) From the graph, determine which reactant is limiting when more than 10.0 g carbon monoxide is available to react with 19.0 g iron(III) oxide.

(f) Calculate the percent yield if 24.0 g iron(III) oxide reacted with 20.0 g carbon monoxide to produce 15.9 g metallic iron.

(g) Calculate the minimum mass of additional limiting reactant required to react with all of the excess of nonlimiting reactant from part (f).

Part III

During coal mining, FeS_2, which contains the S_2^{2-} disulfide ion, reaches the surface where it reacts with oxygen and water to form iron(II) sulfate and sulfuric acid.

(a) Write the balanced chemical equation for this reaction.

(b) In the presence of oxygen, water, and a sufficiently high hydroxide ion concentration, aqueous Fe^{2+} precipitates as iron(III) hydroxide. Write the balanced chemical equation for this reaction. Make certain that charge is conserved.

Part IV

Gold metal can be separated from gold-bearing ore by treating the ore with aqueous NaCN in the presence of oxygen. The reaction has the *unbalanced* net ionic equation given here; the spectator ion, Na^+, is left out of the equation.

$$Au(s) + CN^-(aq) + O_2(g) + H_2O(\ell) \longrightarrow [Au(CN)_2]^-(aq) + OH^-(aq)$$

(a) Balance the net ionic equation.

(b) Which reactant is oxidized? What are the oxidation numbers of this species as a reactant and as a product?

(c) Which reactant is reduced? What are the oxidation numbers of this species as a reactant and as a product?

(d) What is the oxidizing agent? The reducing agent?

(e) Calculate the mass of NaCN needed to prepare 1.0 L of 0.75-M NaCN.

(f) Calculate the mass (g) of NaCN required to extract all of the gold from one metric ton (1000 kg) of the ore.

(g) Calculate the volume (L) of 0.75-M NaCN required to extract all of the gold from one metric ton of the ore.

(h) The gold content of the solution can be determined by titrating with aqueous silver nitrate. The net ionic equation for the titration reaction is

$$[Au(CN)_2]^-(aq) + Ag^+(aq) \longrightarrow Ag[Au(CN)_2](s)$$

If 32.45 mL of 0.100-M $AgNO_3$(aq) is required to titrate 25.00 mL of the solution obtained by treating a sample of ore with aqueous sodium cyanide, determine the mass of gold that was present in the ore sample.

HAVING STUDIED THIS CHAPTER . . .

. . . you should be able to:

- Balance chemical equations and interpret the information they convey (Sections 3-1, 3-2). End-of-chapter Questions 12, 18, 20, 118, 123
- Use solubility rules to predict solubility (Section 3-3). Questions 29, 133
- Predict products of common types of exchange reactions: precipitation, acid-base, and gas-forming (Sections 3-3, 3-4). Questions 33, 38, 44
- Write a net ionic equation for a given reaction in aqueous solution (Sections 3-3, 3-4). Questions 35, 42, 120
- Recognize common acids and bases as strong or weak acids and write balanced equations for neutralization reactions (Section 3-4). Questions 40, 44
- Identify the acid and base used to form a specific salt (Section 3-4). Questions 42, 131

- Recognize oxidation-reduction reactions and common oxidizing and reducing agents (Section 3-5). Questions 52, 55
- Assign oxidation numbers to reactants and products in a redox reaction; identify what has been oxidized and reduced; identify the oxidizing and reducing agents (Section 3-5). Question 49
- Use the activity series of metals to predict products of redox reactions (Section 3-5). Question 57
- Use a balanced equation and its mole ratios to calculate the amount or mass of one reactant or product from the amount or mass of another reactant or product (Section 3-6). Questions 59, 61, 67, 126
- Determine which reactant is the limiting reactant and calculate the quantity of other reactants consumed or of products obtained (Section 3-7). Questions 69, 71
- Do calculations involving actual yield, theoretical yield, and percent yield; differentiate among them (Section 3-8). Questions 73, 76
- Determine the empirical formula of an unknown compound using composition by mass or combustion analysis data (Section 3-9). Questions 78, 80
- Calculate molarity given amount or mass of solute and volume of solution (Section 3-10). Questions 82, 84
- Determine how to prepare a solution of a given molarity from the solute and water, or by dilution of a more concentrated solution (Section 3-10). Questions 86, 88, 135
- Solve stoichiometry problems by using solution molarities (Section 3-11). Question 90
- Apply titration principles to determine the concentration of an unknown aqueous solution (Section 3-12). Questions 93, 97

KEY TERMS

acid (Section 3-4a)
acid-base neutralization reaction (3-4c)
activity series of metals (3-5d)
actual yield (3-8)
atom economy (3-8a)
balanced chemical equation (3-1)
base (3-4b)
coefficients (3-2)
concentration (3-10a)
conservation of electric charge, law of (3-3e)
dilution (3-10c)
disproportionation reaction (3-5c)
dissociation (3-3a)
electrolytes (3-3b)
end point (3-12)
equivalence point (3-12)
gas-forming reaction (3-4e)
green chemistry (3-8a)
indicator (3-12)
ionization (3-4a)
limiting reactant (3-7)
molarity (3-10a)
mole ratio (stoichiometric factor) (3-6)
net ionic equation (3-3e, 3-4)
neutralization reaction (3-4c)
oxidation (3-5)
oxidation number (oxidation state) (3-5c)
oxidation-reduction reaction (3-5)
oxidizing agent (3-5a)
percent yield (3-8)
precipitation reaction (3-3d)
redox (3-5)
reducing agent (3-5a)
reduction (3-5)
salt (3-4c)
solubility rules (3-3c)
solute (3-10)
solvent (3-10)
spectator ion (3-3e)
standard solution (3-12)
stoichiometric coefficients (3-2)
stoichiometry (3-2, 3-6, 3-11)
strong acid (3-4a)
strong base (3-4b)
strong electrolyte (3-3b, 3-4a)
theoretical yield (3-8)
titration (3-12)
weak acid (3-4a)
weak base (3-4b)
weak electrolyte (3-3b, 3-4a)

QUESTIONS FOR REVIEW AND THOUGHT

Red-numbered questions have short answers at the back of this book in Appendix M and fully worked solutions in the *Student Solutions Manual.*

Review Questions

These questions test vocabulary and simple concepts.

1. What information does a balanced chemical equation provide?
2. Find two examples of acid-base reactions in this chapter. Write balanced equations for these reactions, and name the reactants and products.
3. For each of the following, does the oxidation number increase or decrease in the course of a redox reaction?
   (a) An oxidizing agent
   (b) A reducing agent
   (c) A substance undergoing oxidation
   (d) A substance undergoing reduction
4. What is meant by the statement, "The reactants were present in stoichiometric amounts"?
5. Write all the possible mole ratios for the reaction

$$3\ MgO(s) + 2\ Fe(s) \longrightarrow Fe_2O_3(s) + 3\ Mg(s)$$

6. Given the reaction

$$2\ Fe(s) + 3\ Cl_2(g) \longrightarrow 2\ FeCl_3(s)$$

fill in the missing conversion factors for the scheme

$$\begin{array}{ccc} g\ Cl_2 & \xrightarrow{?} & g\ FeCl_3 \\ \downarrow ? & & \downarrow ? \\ mol\ Cl_2 & \xrightarrow{?} & mol\ FeCl_3 \end{array}$$

7. When asked, "What is the limiting reactant?" you might be tempted to choose the reactant with the smallest mass. Why is this not a good strategy?
8. Why can't the product of a reaction ever be the limiting reactant?
9. Does the limiting reactant determine the theoretical yield, actual yield, or both? Explain.

Topical Questions

These questions are keyed to the major topics in the chapter. Usually a question that is answered at the back of this book is paired with a similar one that is not.

Chemical Equations (Section 3-1)

10. For this reaction, fill in the table with the indicated quantities for the balanced equation.

$$4\ NH_3(g) + 5\ O_2(g) \longrightarrow 4\ NO(g) + 6\ H_2O(g)$$

	NH_3	O_2	NO	H_2O
No. of molecules				
No. of atoms				
Amount of molecules				
Mass				
Total mass of reactants				
Total mass of products				

11. For this reaction, fill in the table with the indicated quantities for the balanced equation.

$$2\ C_2H_6(g) + 7\ O_2(g) \longrightarrow 4\ CO_2(g) + 6\ H_2O(g)$$

	C_2H_6	O_2	CO_2	H_2O
No. of molecules				
No. of atoms				
Amount of molecules				
Mass				
Total mass of reactants				
Total mass of products				

12. This diagram shows A (blue spheres) reacting with B (tan spheres). Write a balanced equation that describes the reaction depicted in this diagram.

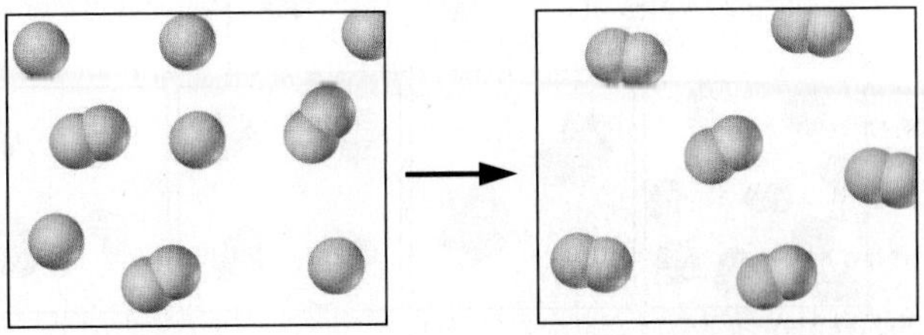

13. This diagram shows A (blue spheres) reacting with B (tan spheres). Write a balanced equation that describes the reaction shown in the diagram.

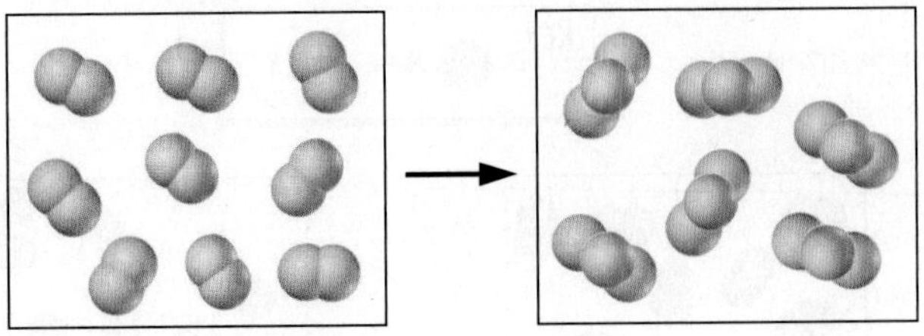

14. Given this equation,

$$4\ A_2 + 3\ B \longrightarrow B_3A_8$$

use a diagram to illustrate the number of molecules of reactant A and product B_3A_8 that would be needed/produced from the reaction of six atoms of B.

15. The elements X (blue) and Y (pink) react by this equation

$$X_2 + 2\,Y_2 \longrightarrow 2\,XY_2$$

(a) Which drawing represents the reactants?
(b) Which drawing represents the products?

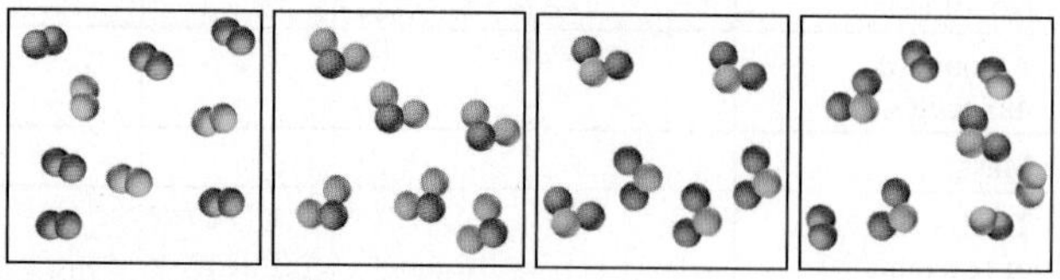

(a) (b) (c) (d)

16. The elements X (blue) and Y (pink) react by this equation

$$2\,X_2 + Y_2 \longrightarrow 2\,X_2Y$$

(a) Which drawing represents the reactants?
(b) Which drawing represents the products?

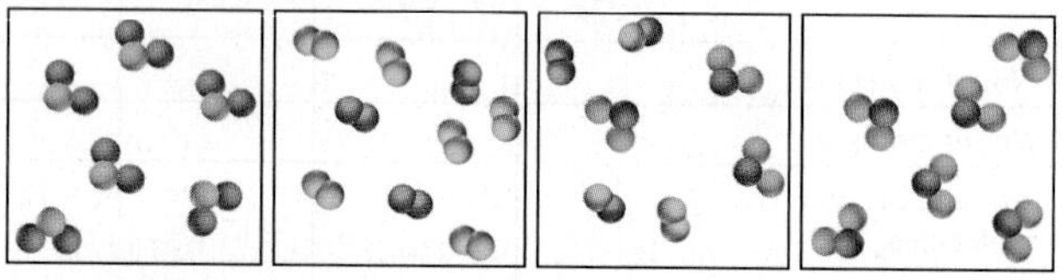

(a) (b) (c) (d)

Balancing Chemical Equations (Section 3-2)

17. Balance this equation and determine which box represents reactants and which box represents products.

$$Sb(g) + Cl_2(g) \longrightarrow SbCl_3(g)$$

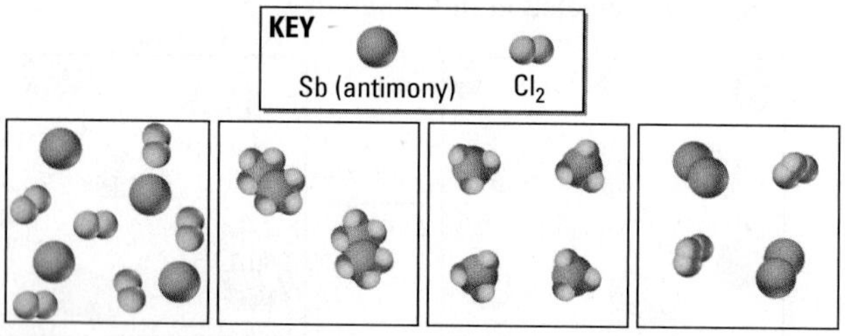
KEY
Sb (antimony)
Cl2

(a) (b) (c) (d)

18. Write a balanced chemical equation that represents the reaction shown in the two drawings.

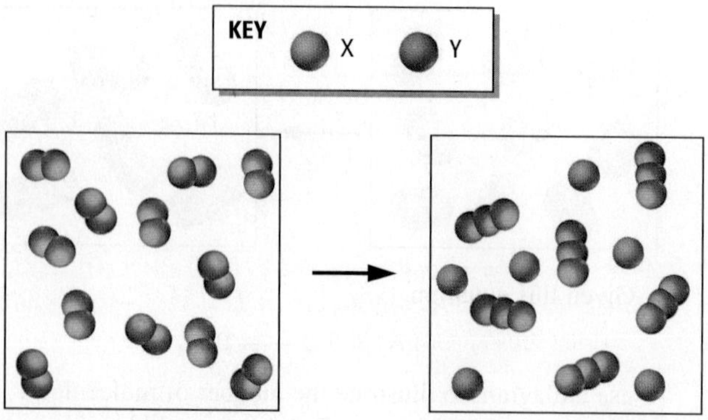
KEY
X
Y

19. Write a balanced chemical equation that represents the reaction shown in the two drawings.

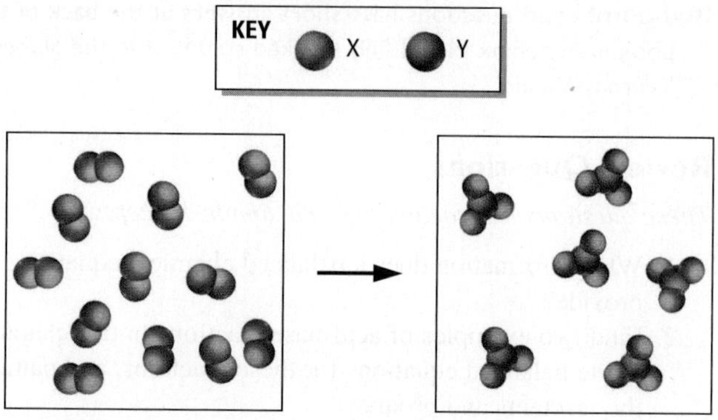
KEY
X
Y

20. Balance these equations.
(a) $UO_2(s) + HF(\ell) \longrightarrow UF_4(s) + H_2O(\ell)$
(b) $B_2O_3(s) + HF(\ell) \longrightarrow BF_3(g) + H_2O(\ell)$
(c) $BF_3(g) + H_2O(\ell) \longrightarrow HF(\ell) + H_3BO_3(s)$

21. Balance these equations.
(a) $MgO(s) + Fe(s) \longrightarrow Fe_2O_3(s) + Mg(s)$
(b) $H_3BO_3(s) \longrightarrow B_2O_3(s) + H_2O(\ell)$
(c) $NaNO_3(s) + H_2SO_4(aq) \longrightarrow Na_2SO_4(aq) + HNO_3(g)$

22. Balance these equations.
(a) Reaction to produce hydrazine, N_2H_4:

$$H_2NCl(aq) + NH_3(g) \longrightarrow NH_4Cl(aq) + N_2H_4(aq)$$

(b) Reaction of the fuels (dimethylhydrazine and dinitrogen tetraoxide) used in the Space Shuttle:

$$(CH_3)_2N_2H_2(\ell) + N_2O_4(g) \longrightarrow N_2(g) + H_2O(g) + CO_2(g)$$

(c) Reaction of calcium carbide with water to produce acetylene, C_2H_2:

$$CaC_2(s) + H_2O(\ell) \longrightarrow Ca(OH)_2(s) + C_2H_2(g)$$

23. Balance these equations.
(a) Reaction of calcium cyanamide to produce ammonia:

$$CaNCN(s) + H_2O(\ell) \longrightarrow CaCO_3(s) + NH_3(g)$$

(b) Reaction to produce diborane, B_2H_6:

$$NaBH_4(s) + H_2SO_4(aq) \longrightarrow B_2H_6(g) + H_2(g) + Na_2SO_4(aq)$$

(c) Reaction to rid water of hydrogen sulfide, H_2S, a foul-smelling compound:

$$H_2S(aq) + Cl_2(aq) \longrightarrow S_8(s) + HCl(aq)$$

24. Balance these combustion reactions.
(a) $C_6H_{12}O_6 + O_2 \longrightarrow CO_2 + H_2O$
(b) $C_5H_{12} + O_2 \longrightarrow CO_2 + H_2O$
(c) $C_7H_{14}O_2 + O_2 \longrightarrow CO_2 + H_2O$
(d) $C_2H_4O_2 + O_2 \longrightarrow CO_2 + H_2O$

Precipitation Reactions (Section 3-3)

25. For each substance, what ions are present in an aqueous solution?
(a) KOH
(b) K_2SO_4
(c) $NaNO_3$
(d) NH_4Cl

26. For each substance, what ions are present in an aqueous solution?
    (a) CaI_2 (b) $Mg_3(PO_4)_2$
    (c) NiS (d) $MgBr_2$
27. Which substance conducts electricity when dissolved in water?
    (a) NaCl (b) $CH_3CH_2CH_3$ (propane)
    (c) CH_3OH (methanol) (d) $Ca(NO_3)_2$
28. Which substance conducts electricity when dissolved in water?
    (a) NH_4Cl (b) $CH_3CH_2CH_2CH_3$ (butane)
    (c) $C_{12}H_{22}O_{11}$ (table sugar) (d) $Ba(NO_3)_2$
29. Predict whether each compound is soluble in water. Indicate which ions are present in solution for the water-soluble compounds.
    (a) Potassium monohydrogen phosphate
    (b) Sodium hypochlorite (c) Magnesium chloride
    (d) Calcium hydroxide (e) Aluminum bromide
30. Predict whether each compound is soluble in water. Indicate which ions are present in solution for the water-soluble compounds.
    (a) Ammonium nitrate (b) Barium sulfate
    (c) Potassium acetate (d) Calcium carbonate
    (e) Sodium perchlorate
31. Which drawing is the best nanoscale representation of an aqueous solution of calcium chloride? (Water molecules are not shown for simplicity.)

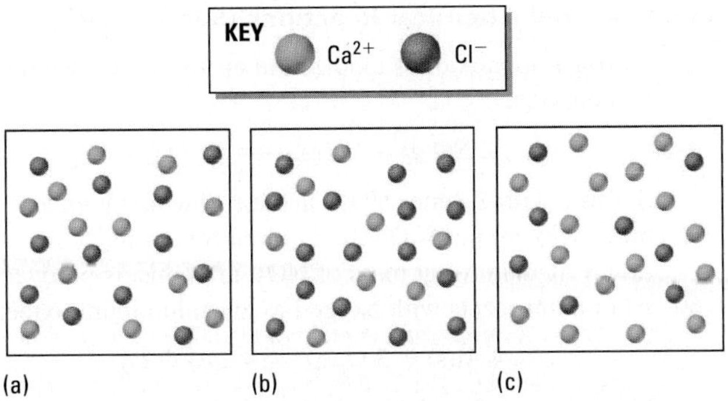
KEY
Ca^{2+}
Cl^-
(a)
(b)
(c)

32. Which drawing is the best nanoscale representation of an aqueous solution of magnesium nitrate? (Water molecules are not shown for simplicity.)

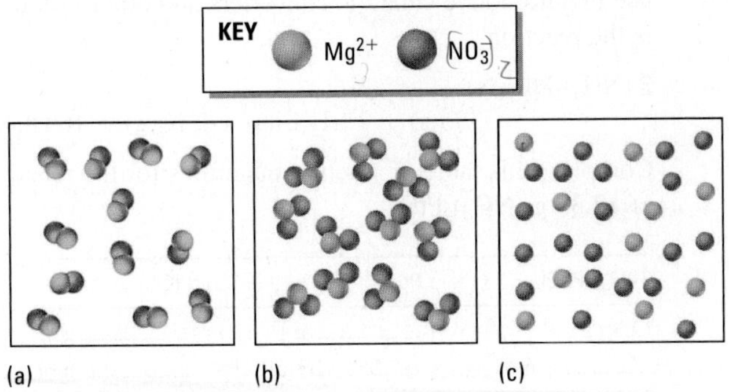
KEY
Mg^{2+}
NO_3^-
(a)
(b)
(c)

33. If aqueous solutions of potassium carbonate and copper(II) nitrate are mixed, a precipitate is formed. Write the complete and net ionic equations for this reaction, and name the precipitate.
34. If aqueous solutions of potassium sulfide and iron(III) chloride are mixed, a precipitate is formed. Write the complete and net ionic equations for this reaction, and name the precipitate.
35. Balance each equation; then write the complete ionic and net ionic equations. Refer to Table 3.1 for information on solubility. Show states (s, ℓ, g, aq) for all reactants and products.
    (a) $Ca(OH)_2 + CoCl_2 \longrightarrow Co(OH)_2 + CaCl_2$
    (b) $BaCl_2 + Na_2CO_3 \longrightarrow BaCO_3 + NaCl$
    (c) $Na_3PO_4 + Ni(NO_3)_2 \longrightarrow Ni_3(PO_4)_2 + NaNO_3$
36. Balance each equation; then write the complete ionic and net ionic equations. Refer to Table 3.1 for information on solubility. Show states (s, ℓ, g, aq) for all reactants and products.
    (a) $ZnCl_2 + KOH \longrightarrow KCl + Zn(OH)_2$
    (b) $AgNO_3 + KI \longrightarrow AgI + KNO_3$
    (c) $NaOH + FeCl_2 \longrightarrow Fe(OH)_2 + NaCl$
37. Aluminum is obtained from bauxite, which is not a specific mineral but a name applied to a mixture of minerals. One of those minerals, which can dissolve in acids, is gibbsite, $Al(OH)_3$. Write balanced overall and net ionic equations for the reaction of gibbsite with sulfuric acid.
38. Write an overall balanced equation for the precipitation reaction that occurs when aqueous lead(II) nitrate is mixed with an aqueous solution of potassium chloride. Name each reactant and product. Indicate the state of each substance (s, ℓ, g, or aq).
39. Write an overall balanced equation for the precipitation reaction that occurs when aqueous copper(II) nitrate is mixed with an aqueous solution of sodium carbonate. Name each reactant and product. Indicate the state of each substance (s, ℓ, g, or aq).

Acid-Base Reactions (Section 3-4)

40. Classify each of these as an acid or a base. Which are strong and which are weak? What ions are produced when each is dissolved in water?
    (a) KOH (b) $Mg(OH)_2$
    (c) HClO (d) HBr
    (e) LiOH (f) H_2SO_3
41. Classify each of these as an acid or a base. Which are strong and which are weak? What ions are produced when each is dissolved in water?
    (a) HNO_3 (b) $Ca(OH)_2$
    (c) NH_3 (d) H_3PO_4
    (e) KOH (f) CH_3COOH
42. Identify the acid and base used to form these salts, and write the overall neutralization reaction in both complete and net ionic form.
    (a) $NaNO_2$ (b) $CaSO_4$
    (c) NaI (d) $Mg_3(PO_4)_2$
43. Identify the acid and base used to form these salts, and write the overall neutralization reaction in both complete and net ionic form.
    (a) $NaCH_3COO$ (b) $CaCl_2$
    (c) LiBr (d) $Ba(NO_3)_2$

44. Classify each of these exchange reactions as an acid-base reaction, a precipitation reaction, or a gas-forming reaction. Predict the products of the reaction and then balance the completed equation.
(a) $MnCl_2(aq) + Na_2S(aq) \longrightarrow$
(b) $Na_2CO_3(aq) + ZnCl_2(aq) \longrightarrow$
(c) $K_2CO_3(aq) + HClO_4(aq) \longrightarrow$

45. Classify each of these exchange reactions as an acid-base reaction, a precipitation reaction, or a gas-forming reaction. Predict the products of the reaction and then balance the completed equation.
(a) $Fe(OH)_3(s) + HNO_3(aq) \longrightarrow$
(b) $FeCO_3(s) + H_2SO_4(aq) \longrightarrow$
(c) $FeCl_2(aq) + (NH_4)_2S(aq) \longrightarrow$
(d) $Fe(NO_3)_2(aq) + Na_2CO_3(aq) \longrightarrow$

46. The beautiful mineral rhodochrosite is manganese(II) carbonate. Write an overall balanced equation for the reaction of the mineral with hydrochloric acid. Name each reactant and product.

47. Identify each substance as a strong electrolyte, weak electrolyte, or nonelectrolyte.
(a) Na_2CO_3 (b) H_2CO_3
(c) HNO_3 (d) KOH

48. Identify each substance as a strong electrolyte, weak electrolyte, or nonelectrolyte.
(a) $K_2Cr_2O_7$ (b) HCOOH
(c) NH_3 (d) HI

Oxidation-Reduction Reactions (Section 3-5)

49. Determine the oxidation number of Cl in each formula.
(a) HCl (b) HClO
(c) $HClO_2$ (d) $HClO_3$
(e) $HClO_4$

50. Determine the oxidation number of Mn in each of these species.
(a) $(MnF_6)^{3-}$ (b) Mn_2O_7
(c) MnO_4^- (d) $Mn(CN)_6^-$
(e) MnO_2

51. Sulfur can exist in many oxidation states. Determine the oxidation state of S in each species.
(a) H_2S (b) S_8
(c) SCl_2 (d) SO_3^{2-}
(e) K_2SO_4

52. Classify each reaction as oxidation-reduction, acid-base, or precipitation.
(a) $CdCl_2(aq) + Na_2S(aq) \longrightarrow CdS(s) + 2\ NaCl(aq)$
(b) $2\ Ca(s) + O_2(g) \longrightarrow 2\ CaO(s)$
(c) $Ca(OH)_2(s) + 2\ HCl(aq) \longrightarrow CaCl_2(aq) + 2\ H_2O(\ell)$

53. Classify each reaction as oxidation-reduction, acid-base, or precipitation.
(a) $Zn(s) + 2\ NO_3^-(aq) + 4\ H^+(aq) \longrightarrow Zn^{2+}(aq) + 2\ NO_2(g) + 2\ H_2O(\ell)$
(b) $Zn(OH)_2(s) + H_2SO_4(aq) \longrightarrow ZnSO_4(aq) + 2\ H_2O(\ell)$
(c) $Ca(s) + 2\ H_2O(\ell) \longrightarrow Ca(OH)_2(s) + H_2(g)$

54. Identify the region of the periodic table where the elements are good reducing agents. Identify the region where the elements are good oxidizing agents.

55. Which substances are oxidizing agents?
(a) Zn (b) O_2 (c) HNO_3
(d) MnO_4^- (e) H_2 (f) H^+

56. Which substances are reducing agents?
(a) Ca (b) Ca^{2+} (c) $Cr_2O_7^{2-}$
(d) Al (e) Br_2 (f) H_2

57. (a) In what groups of the periodic table are the most reactive metals found? Where do we find the least reactive metals?
(b) Silver (Ag) does not react with 1-M HCl solution. Will Ag react with a solution of aluminum nitrate, $Al(NO_3)_3$? If so, write a chemical equation for the reaction.
(c) Lead (Pb) will react very slowly with 1-M HCl solution. Aluminum will react with lead(II) sulfate solution, $PbSO_4$. Will Pb react with an $AgNO_3$ solution? If so, write a chemical equation for the reaction.
(d) On the basis of the information obtained in answering parts (a), (b), and (c), arrange Ag, Al, and Pb in decreasing order of reactivity.

58. Using the activity series of metals (Table 3.4), predict whether these reactions will occur in aqueous solution.
(a) $Mg(s) + Ca(s) \longrightarrow Mg^{2+}(aq) + Ca^{2+}(aq)$
(b) $2\ Al^{3+}(aq) + 3\ Pb^{2+}(aq) \longrightarrow 2\ Al(s) + 3\ Pb(s)$
(c) $H_2(g) + Zn^{2+}(aq) \longrightarrow 2\ H^+(aq) + Zn(s)$
(d) $Mg(s) + Cu^{2+}(aq) \longrightarrow Mg^{2+}(aq) + Cu(s)$
(e) $Pb(s) + 2\ H^+(aq) \longrightarrow H_2(g) + Pb^{2+}(aq)$
(f) $2\ Ag^+(aq) + Cu(s) \longrightarrow 2\ Ag(s) + Cu^{2+}(aq)$
(g) $2\ Al^{3+}(aq) + 3\ Zn(s) \longrightarrow 3\ Zn^{2+}(aq) + 2\ Al(s)$

The Mole and Chemical Reactions (Section 3-6)

59. Nitrogen monoxide is oxidized in air to give brown nitrogen dioxide.

$$2\ NO(g) + O_2(g) \longrightarrow 2\ NO_2(g)$$

Starting with 2.2 mol NO, calculate how many moles and how many grams of O_2 are required for complete reaction. Calculate what mass of NO_2, in grams, is produced.

60. Aluminum reacts with oxygen to give aluminum oxide.

$$4\ Al(s) + 3\ O_2(g) \longrightarrow 2\ Al_2O_3(s)$$

If you have 6.0 mol Al, calculate the amount (mol) and mass (g) of O_2 needed for complete reaction. Calculate the mass of Al_2O_3, in grams, that is produced.

61. The final step in the manufacture of platinum metal (for use in automotive catalytic converters and other products) is the reaction

$$3\ (NH_4)_2PtCl_6(s) \longrightarrow 3\ Pt(s) + 2\ NH_4Cl(s) + 2\ N_2(g) + 16\ HCl(g)$$

Complete this table of reaction quantities for the reaction of 12.35 g $(NH_4)_2PtCl_6$.

$(NH_4)_2PtCl_6$	Pt	HCl
12.35 g	______ g	______ g
______ mol	______ mol	______ mol

62. Disulfur dichloride, S_2Cl_2, is used to vulcanize rubber. It can be made by treating molten sulfur with gaseous chlorine.

$$S_8(\ell) + 4\ Cl_2(g) \longrightarrow 4\ S_2Cl_2(g)$$

Complete this table of reaction quantities for the production of 103.5 g S_2Cl_2.

S_8	Cl_2	S_2Cl_2
______ g	______ g	103.5 g
______ mol	______ mol	______ mol

63. Iron reacts with oxygen to give iron(III) oxide, Fe_2O_3.
   (a) Write a balanced equation for this reaction.
   (b) An ordinary iron nail (assumed to be pure iron) has a mass of 5.58 g; calculate the mass (in grams) of Fe_2O_3 that is produced when the nail is converted completely to Fe_2O_3.
   (c) Calculate the mass of O_2 (in grams) required for the reaction.
64. Nitroglycerin decomposes violently according to the equation

$$4\ C_3H_5(NO_3)_3(\ell) \longrightarrow 12\ CO_2(g) + 10\ H_2O(\ell) + 6\ N_2(g) + O_2(g)$$

   Calculate the mass (in grams) of each gaseous product produced from 1.00 g nitroglycerin.
65. Chlorinated fluorocarbons, such as CCl_2F_2, have been banned from use in automobile air conditioners because the compounds are destructive to the stratospheric ozone layer. Researchers at MIT have found an environmentally safe way to decompose these compounds by treating them with sodium oxalate. The products of the reaction are carbon, carbon dioxide, sodium chloride, and sodium fluoride.
   (a) Write a balanced equation for this reaction of CCl_2F_2.
   (b) Calculate the mass of sodium oxalate needed to remove 76.8 g CCl_2F_2.
   (c) Calculate the mass of CO_2 produced.
66. Careful decomposition of ammonium nitrate gives laughing gas (dinitrogen monoxide) and water.
   (a) Write a balanced equation for this reaction.
   (b) Beginning with 10.0 g ammonium nitrate, calculate the masses of N_2O and water that are produced.
67. In making iron from iron ore, this reaction occurs.

$$Fe_2O_3(s) + 3\ CO(g) \longrightarrow 2\ Fe(s) + 3\ CO_2(g)$$

   (a) Calculate the mass of iron (in grams) that can be obtained from 1.00 kg iron(III) oxide.
   (b) Calculate the mass of CO required.
68. Cisplatin, $Pt(NH_3)_2Cl_2$, a drug used in the treatment of cancer, can be made by the reaction of K_2PtCl_4 with ammonia, NH_3. Besides cisplatin, the other product is KCl.
   (a) Write a balanced equation for this reaction.
   (b) To obtain 2.50 g cisplatin, calculate what masses (in grams) of K_2PtCl_4 and ammonia you need.

Limiting Reactant (Section 3-7)

69. Aluminum chloride, Al_2Cl_6, is an inexpensive reagent used in many industrial processes. It is made by treating scrap aluminum with chlorine according to the balanced equation

$$2\ Al(s) + 3\ Cl_2(g) \longrightarrow Al_2Cl_6(s)$$

   (a) Determine which reactant is limiting if 2.70 g Al and 4.05 g Cl_2 are mixed.
   (b) Calculate what mass of Al_2Cl_6 can be produced.
   (c) Calculate what mass of the excess reactant remains when the reaction is complete.
70. Methanol, CH_3OH, is a clean-burning, easily handled fuel. It can be made by the direct reaction of CO and H_2.

$$CO(g) + 2\ H_2(g) \longrightarrow CH_3OH(\ell)$$

   (a) For a mixture of 12.0 g H_2 and 74.5 g CO, determine the limiting reactant.
   (b) Calculate the mass (g) of the excess reactant left after reaction is complete.
   (c) Calculate the theoretical mass of methanol that can be obtained.
71. This reaction can be used to generate hydrogen gas from methane:

$$CH_4(g) + H_2O(g) \longrightarrow CO(g) + 3\ H_2(g)$$

   If you use 500. g CH_4 and 1300. g water:
   (a) Determine the limiting reactant.
   (b) Calculate the mass (g) of H_2 that can be produced.
   (c) Calculate the mass (g) of the excess reactant remaining when the reaction is complete.
72. Aspirin is produced by the reaction of salicylic acid and acetic anhydride.

$$\underset{\text{salicylic acid}}{2\ C_7H_6O_3(s)} + \underset{\text{acetic anhydride}}{C_4H_6O_3(\ell)} \longrightarrow \underset{\text{aspirin}}{2\ C_9H_8O_4(s)} + H_2O(\ell)$$

   If you mix 100. g of each of the reactants, calculate the maximum mass of aspirin that can be obtained.

Percent Yield (Section 3-8)

73. Iron oxide can be reduced to the metal as follows:

$$Fe_2O_3(s) + 3\ CO(g) \longrightarrow 2\ Fe(s) + 3\ CO_2(g)$$

   Calculate the mass of iron that can be obtained from 1.00 kg of the iron oxide. If 654 g Fe was obtained from the reaction, calculate the percent yield.
74. Quicklime, CaO, is formed when calcium hydroxide is heated.

$$Ca(OH)_2(s) \longrightarrow CaO(s) + H_2O(\ell)$$

   The theoretical yield is 65.5 g but only 36.7 g quicklime is produced. Calculate the percent yield.
75. Diborane, B_2H_6, is valuable for the synthesis of new organic compounds. The boron compound can be made by the reaction

$$2\ NaBH_4(s) + I_2(s) \longrightarrow B_2H_6(g) + 2\ NaI(s) + H_2(g)$$

   Suppose you use 1.203 g $NaBH_4$ and excess iodine, and you isolate 0.295 g B_2H_6. Calculate the percent yield of B_2H_6.
76. Disulfur dichloride, which has a revolting smell, can be prepared by directly combining S_8 and Cl_2, but it can also be made by this reaction:

$$3\ SCl_2(\ell) + 4\ NaF(s) \longrightarrow SF_4(g) + S_2Cl_2(\ell) + 4\ NaCl(s)$$

   Calculate the mass of SCl_2 needed to react with excess NaF to prepare 1.19 g S_2Cl_2 if the expected yield is 51%.

77. The ceramic silicon nitride, Si_3N_4, is made by heating silicon and nitrogen at an elevated temperature.

$$3\ Si(s) + 2\ N_2(g) \longrightarrow Si_3N_4(s)$$

Calculate the mass of silicon that must combine with excess N_2 to produce 1.0 kg Si_3N_4 if this process is 92% efficient.

Composition and Empirical Formulas (Section 3-9)

78. Propionic acid, an organic acid, contains only C, H, and O. When 0.236 g of the acid burns completely in O_2 it gives 0.421 g CO_2 and 0.172 g H_2O. Determine the empirical formula of the acid.
79. Quinone, which is used in the dye industry and in chemical photography, is an organic compound containing only C, H, and O. Determine the empirical formula if 0.105 g of the compound gives 0.257 g CO_2 and 0.0350 g H_2O when burned completely.
80. L-Dopa is a drug used for the treatment of Parkinson's disease. Elemental analysis shows it to be 54.82% carbon, 7.10% nitrogen, 32.46% oxygen, and the remainder hydrogen.
    (a) Determine L-dopa's empirical formula.
    (b) The molar mass of L-dopa is 197.19 g/mol. Determine its molecular formula.
81. Write the balanced chemical equation for the complete combustion of adipic acid, an organic acid containing 49.31% C, 6.90% H, and the remainder O, by mass.

Solution Concentration: Molarity (Section 3-10)

82. Assume that 6.73 g Na_2CO_3 is dissolved in enough water to make 250. mL solution.
    (a) Calculate the molarity of the sodium carbonate.
    (b) Calculate the concentrations of the Na^+ and CO_3^{2-} ions.
83. Some $K_2Cr_2O_7$, with a mass of 2.335 g, is dissolved in enough water to make 500. mL solution.
    (a) Calculate the molarity of the potassium dichromate.
    (b) Calculate the concentrations of the K^+ and $Cr_2O_7^{2-}$ ions.
84. Calculate the volume of 0.123-M NaOH that contains 25.0 g NaOH. Express your result in milliliters.
85. Calculate the volume of 2.06-M $KMnO_4$ that contains 322 g solute.
86. You need 1.00 L of 0.125-M H_2SO_4. Which method is best to prepare this solution? Explain your choice.
    (a) Dilute 36.0 mL of 1.25-M H_2SO_4 to a volume of 1.00 L.
    (b) Dilute 20.8 mL of 6.00-M H_2SO_4 to a volume of 1.00 L.
    (c) Add 50.0 mL of 3.00-M H_2SO_4 to 950. mL water.
    (d) Add 500. mL of 0.500-M H_2SO_4 to 500. mL water.
87. You need 300. mL of 0.500-M $K_2Cr_2O_7$. Which method is best to prepare this solution? Explain your choice.
    (a) Dilute 250. mL of 0.600-M $K_2Cr_2O_7$ to 300. mL.
    (b) Add 50.0 mL water to 250. mL of 0.250-M $K_2Cr_2O_7$.
    (c) Dilute 125 mL of 1.00-M $K_2Cr_2O_7$ to 300. mL.
    (d) Add 30.0 mL of 1.50-M $K_2Cr_2O_7$ to 270. mL of water.
88. You need to make a 0.300-M solution of $NiSO_4$(aq). Calculate the mass of $NiSO_4 \cdot 6H_2O$ you should put into a 0.500-L volumetric flask.
89. You wish to make a 0.200-M solution of $CuSO_4$(aq). Calculate the mass of $CuSO_4 \cdot 5H_2O$ required to make 0.500 L of solution.

Stoichiometry in Aqueous Solutions (Section 3-11)

90. Calculate the volume, in milliliters, of 0.125-M HNO_3 required to react completely with 1.30 g $Ba(OH)_2$.

$$2\ HNO_3(aq) + Ba(OH)_2(s) \longrightarrow Ba(NO_3)_2(aq) + 2\ H_2O(\ell)$$

91. Diborane, B_2H_6, can be produced by this reaction:

$$2\ NaBH_4(s) + H_2SO_4(aq) \longrightarrow 2\ H_2(g) + Na_2SO_4(aq) + B_2H_6(g)$$

Calculate the volume, in milliliters, of 0.0875-M H_2SO_4 needed to completely react with 1.35 g $NaBH_4$.

92. You mix 25.0 mL of 0.234-M $FeCl_3$ solution with 42.5 mL of 0.453-M NaOH.
    (a) Calculate the maximum mass, in grams, of $Fe(OH)_3$ that will precipitate.
    (b) Determine which reactant is in excess.
    (c) Calculate the concentration of the excess reactant remaining in solution after the maximum mass of $Fe(OH)_3$ has precipitated.

Titrations (Section 3-12)

93. A soft drink contains an unknown mass of citric acid, $C_3H_5O(COOH)_3$. It requires 6.42 mL of 9.580×10^{-2}-M NaOH to neutralize the citric acid in 10.0 mL of the soft drink.

$$C_3H_5O(COOH)_3(aq) + 3\ NaOH(aq) \longrightarrow Na_3C_3H_5O(COO)_3(aq) + 3\ H_2O(\ell)$$

(a) Determine which step in these calculations for the mass of citric acid in 1 mL soft drink is incorrect? Why?
    (i) n (NaOH) = (6.42 mL)(1L/1000 mL)(9.580×10^{-2} mol/L)
    (ii) n (citric acid) = (6.15×10^{-4} mol NaOH) × (3 mol citric acid/1 mol NaOH)
    (iii) m (citric acid in sample) = (1.85×10^{-3} mol citric acid) × (192.12 g/mol citric acid)
    (iv) m (citric acid in 1 mL soft drink) = (0.354 g citric acid)/(10 mL soft drink)
(b) Determine the correct result.

94. Vitamin C is the compound $C_6H_8O_6$. Besides being an acid, it is a reducing agent that reacts readily with bromine, Br_2, a good oxidizing agent.

$$C_6H_8O_6(aq) + Br_2(aq) \longrightarrow 2\ HBr(aq) + C_6H_6O_6(aq)$$

Suppose a 1.00-g chewable vitamin C tablet requires 27.85 mL of 0.102-M Br_2 to react completely.
(a) Determine which step in these calculations for the mass, in grams, of vitamin C in the tablet is incorrect. Why?
    (i) n (Br_2) = (27.85 mL)(0.102 mol/L)
    (ii) n ($C_6H_8O_6$) = (2.84 mol Br_2)(1 mol $C_6H_8O_6$/1 mol Br_2)

(iii) m ($C_6H_8O_6$) = (2.84 mol $C_6H_8O_6$)(176 g/mol $C_6H_8O_6$)

(iv) m ($C_6H_8O_6$) = (500. g $C_6H_8O_6$)/(1 g tablet)

(b) Determine which is the correct result.

95. A sample of a mixture of oxalic acid, $H_2C_2O_4$, and sodium chloride, NaCl, has a mass of 4.554 g. If a volume of 29.58 mL of 0.550-M NaOH is required to neutralize all the $H_2C_2O_4$, calculate the mass percent of oxalic acid in the mixture. Oxalic acid and NaOH react according to this equation:

$$H_2C_2O_4(aq) + 2\ NaOH(aq) \longrightarrow Na_2C_2O_4(aq) + 2\ H_2O(\ell)$$

96. Potassium hydrogen phthalate, $KHC_8H_4O_4$, is used to standardize solutions of bases. The acidic anion reacts with bases according to this net ionic equation:

$$HC_8H_4O_4^-(aq) + OH^-(aq) \longrightarrow H_2O(\ell) + C_8H_4O_4^{2-}(aq)$$

A 0.902-g sample of potassium hydrogen phthalate requires 26.45 mL NaOH to react; determine the molarity of the NaOH.

97. You are given 0.954 g of an unknown acid, H_2A, which reacts with NaOH according to the balanced equation

$$H_2A(aq) + 2\ NaOH(aq) \longrightarrow Na_2A(aq) + 2\ H_2O(\ell)$$

If 36.04 mL of 0.509-M NaOH is required to react with all of the acid, calculate the molar mass of the acid.

General Questions

These questions are not explicitly keyed to chapter topics; many require integration of several concepts.

98. In an experiment, 1.056 g of a metal carbonate containing an unknown metal M was heated to give the metal oxide and 0.376 g CO_2.

$$MCO_3(s) \longrightarrow MO(s) + CO_2(g)$$

Determine the identity of the metal M.

(a) Ni (b) Cu
(c) Co (d) Ba

99. Uranium(VI) oxide reacts with bromine trifluoride to give uranium(IV) fluoride, an important step in the purification of uranium ore.

$$6\ UO_3(s) + 8\ BrF_3(\ell) \longrightarrow 6\ UF_4(s) + 4\ Br_2(\ell) + 9\ O_2(g)$$

You begin with 365 g each of UO_3 and BrF_3; determine the maximum yield, in grams, of UF_4.

100. Silicon and hydrogen form a series of interesting compounds, Si_xH_y. To find the formula of one of them, a 6.22-g sample of the compound is burned in oxygen. All of the Si is converted to 11.64 g SiO_2 and all of the H to 6.980 g H_2O. Determine the empirical formula of the silicon compound.

101. Boron forms an extensive series of compounds with hydrogen, all with the general formula B_xH_y. To analyze one of these compounds, you burn it in air and isolate the boron in the form of B_2O_3 and the hydrogen in the form of water. You find that 0.1482 g B_xH_y gives 0.4221 g B_2O_3 when burned in excess O_2. Determine the empirical formula of B_xH_y.

102. Determine the limiting reactant for the reaction

$$4\ KOH + 2\ MnO_2 + O_2 + Cl_2 \longrightarrow 2\ KMnO_4 + 2\ KCl + 2\ H_2O$$

(a) if 5 mol of each reactant is present. (b) if 5 g of each reactant is present.

103. The Hargreaves process is an industrial method for making sodium sulfate for use in papermaking.

$$4\ NaCl + 2\ SO_2 + 2\ H_2O + O_2 \longrightarrow 2\ Na_2SO_4 + 4\ HCl$$

(a) If you start with 10. mol of each reactant, determine which one limits the amount of Na_2SO_4 produced.
(b) Determine the amount of Na_2SO_4 produced if you start with 100. g of each reactant.

104. Hydrogen sulfide gas is bubbled into an acidified solution of potassium permanganate; elemental sulfur precipitates. The *unbalanced* equation is

$$H_2S(g) + MnO_4^-(aq) + H^+(aq) \longrightarrow Mn^{2+}(aq) + S(s) + H_2O(\ell)$$

(a) Potassium ions are not given in the equation. Explain why.
(b) Explain why this is a redox equation.
(c) Identify the oxidizing and reducing agents.

105. An aqueous hydrochloric acid solution is 37.0% HCl by mass. The density of the solution is 1.185 g/mL. Calculate the molarity of HCl in this solution.

106. Azurite is a copper-containing mineral that often forms beautiful crystals. Its formula is $Cu_3(CO_3)_2(OH)_2$. Write a balanced equation for the reaction of this mineral with hydrochloric acid.

107. Determine which of these are redox reactions, which are acid-base reactions, and which are gas-forming reactions. Identify the oxidizing and reducing agents in each of the redox reactions. Identify the acid and base in each acid-base reaction.

(a) $NaOH(aq) + H_3PO_4(aq) \longrightarrow NaH_2PO_4(aq) + H_2O(\ell)$
(b) $NH_3(g) + CO_2(g) + H_2O(\ell) \longrightarrow NH_4HCO_3(aq)$
(c) $TiCl_4(g) + 2\ Mg(\ell) \longrightarrow Ti(s) + 2\ MgCl_2(\ell)$
(d) $NaCl(s) + NaHSO_4(aq) \longrightarrow HCl(g) + Na_2SO_4(aq)$

108. Chlorofluorocarbons are involved in creating the hole in the ozone layer. Their manufacture begins with the preparation of HF from the mineral fluorspar, CaF_2, according to this unbalanced equation:

$$CaF_2(s) + H_2SO_4(aq) \longrightarrow HF(g) + CaSO_4(s)$$

HF is combined with, for example, CCl_4 in the presence of $SbCl_5$ to make CCl_2F_2, called dichlorodifluoromethane or CFC-12, and other chlorofluorocarbons.

$$2\ HF(g) + CCl_4(\ell) \longrightarrow CCl_2F_2(g) + 2\ HCl(g)$$

(a) Balance the first equation and name each substance.
(b) Is the first reaction best classified as an acid-base reaction, an oxidation-reduction reaction, or a precipitation reaction?
(c) Give the names of the compounds CCl_4, $SbCl_5$, and HCl.
(d) Another chlorofluorocarbon produced in the reaction is composed of 8.74% C, 77.43% Cl, and 13.83% F. Determine the empirical formula of the compound.

109. Vitamin C is ascorbic acid, $HC_6H_7O_6$, which can be titrated with a strong base.

$$HC_6H_7O_6(aq) + NaOH(aq) \longrightarrow NaC_6H_7O_6(aq) + H_2O(\ell)$$

A student dissolved a 500.0-mg vitamin C tablet in 200.0 mL water and then titrated it with 0.1250-M NaOH. It required 21.30 mL of the base to reach the equivalence point. Calculate the mass percentage of the tablet that is impurity.

110. Write the balanced chemical equation for the complete combustion of malonic acid, an organic acid containing 34.62% C, 3.88% H, and the remainder O, by mass.

111. You have an organic liquid that contains either ethyl alcohol, C_2H_5OH, or methyl alcohol, CH_3OH, or both. You burned a sample of the liquid weighing 0.280 g to form 0.385 g $CO_2(g)$. Determine the composition of the liquid.

112. You are given an acid and told only that it could be citric acid (molar mass = 192.1 g/mol) or tartaric acid (molar mass = 150.1 g/mol). To determine which acid you have, you react it with NaOH. The appropriate reactions are

Citric acid: $C_6H_8O_7(aq) + 3\ NaOH(aq) \longrightarrow Na_3C_6H_5O_7(aq) + 3\ H_2O(\ell)$

Tartaric acid: $C_4H_6O_6(aq) + 2\ NaOH(aq) \longrightarrow Na_2C_4H_4O_6(aq) + 2\ H_2O(\ell)$

You find that a 0.956-g sample requires 29.1 mL of 0.513-M NaOH to reach the equivalence point. Determine which is the unknown acid.

113. In the past, devices for testing a driver's breath for alcohol depended on this reaction:

$$3\ C_2H_5OH(aq) + 2\ K_2Cr_2O_7(aq) + 8\ H_2SO_4(aq) \longrightarrow 3\ CH_3COOH(aq) + 2\ Cr_2(SO_4)_3(aq) + 2\ K_2SO_4(aq) + 11\ H_2O(\ell)$$

(a) Write the net ionic equation for this reaction. (b) What oxidation numbers are changing in the course of this reaction? (c) Which substances are being oxidized and reduced? (d) Which substance is the oxidizing agent and which is the reducing agent?

114. The salt calcium sulfate is sparingly soluble in water with a solubility of 0.209 g/100 mL water at 30° C. If you stirred 0.550 g $CaSO_4$ into 100.0 mL water at 30° C, calculate the molarity of the resulting solution. Calculate the mass of $CaSO_4$ that would remain undissolved.

115. Determine the molarity of water in pure water.

Applying Concepts

These questions test conceptual learning.

116. When these pairs of reactants are combined in a beaker, (a) describe in words what the contents of the beaker would look like before and after any reaction occurs; (b) use different circles for atoms, molecules, and ions to draw a nanoscale (particulate-level) diagram of what the contents would look like; and (c) write a chemical equation to represent symbolically what the contents would look like.

LiCl(aq) and $AgNO_3(aq)$

NaOH(aq) and HCl(aq)

117. When these pairs of reactants are combined in a beaker, (a) describe in words what the contents of the beaker would look like before and after any reaction occurs; (b) use different circles for atoms, molecules, and ions to draw a particulate-level diagram of what the contents would look like; and (c) write a chemical equation to represent symbolically what the contents would look like.

$CaCO_3(s)$ and HCl(aq)

$NH_4NO_3(aq)$ and KOH(aq)

118. Chemical equations can be interpreted on either a nanoscale level (atoms, molecules, ions) or a mole level (moles of reactants and products). Write word statements to describe the combustion of butane on a nanoscale level and a mole level.

$$2\ C_4H_{10}(g) + 13\ O_2(g) \longrightarrow 8\ CO_2(g) + 10\ H_2O(\ell)$$

119. What is the single product of this hypothetical reaction? Explain why you gave the formula you wrote.

$$3\ A_2B_3 + B_3 \longrightarrow 6\ \_\_\_\_\_\_$$

120. If 1.5 mol Cu reacts with a solution containing 4.0 mol $AgNO_3$, what ions will be present in the solution at the end of the reaction?

$$Cu(s) + 2\ AgNO_3(aq) \longrightarrow Cu(NO_3)_2(aq) + 2\ Ag(s)$$

121. Ammonia can be formed by a direct reaction of nitrogen and hydrogen.

$$N_2(g) + 3\ H_2(g) \longrightarrow 2\ NH_3(g)$$

A tiny portion of the starting mixture is represented by the diagram, where the blue circles represent N and the white circles represent H.

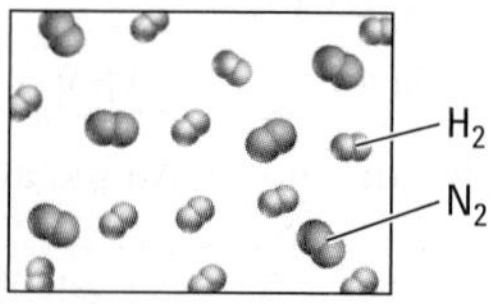
H2
N2

Which of these represents the product mixture?

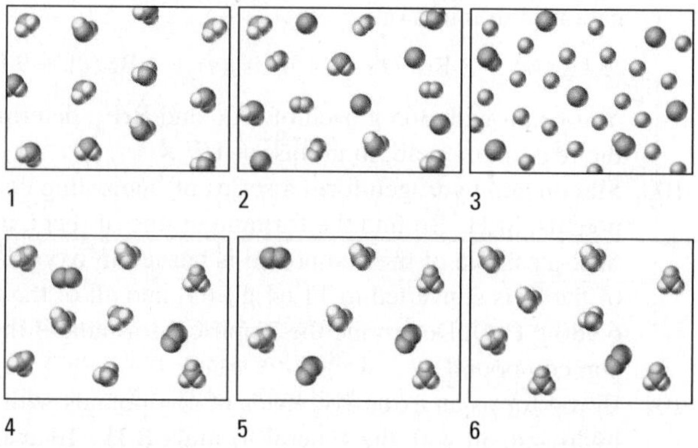
1
2
3
4
5
6

For the reaction of the given sample, which of these statements is true?

(a) N_2 is the limiting reactant.

(b) H_2 is the limiting reactant.

(c) NH_3 is the limiting reactant.
(d) No reactant is limiting; they are present in the correct stoichiometric ratio.

122. Carbon monoxide burns readily in oxygen to form carbon dioxide.

$$2\,CO(g) + O_2(g) \longrightarrow 2\,CO_2(g)$$

The box on the left represents a tiny portion of a mixture of CO and O_2. If these molecules react to form CO_2, what should the contents of the box on the right look like?

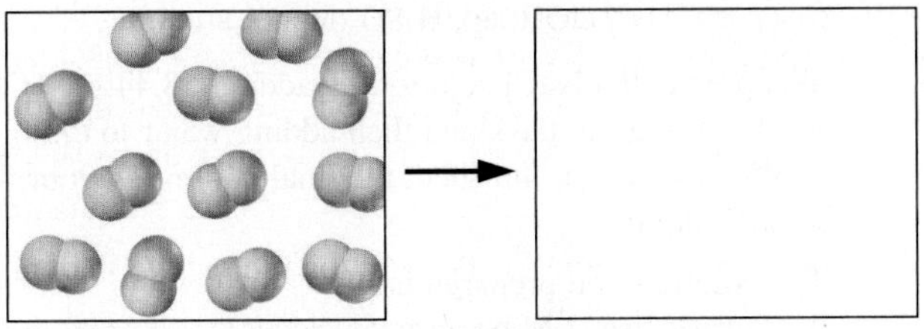

123. Which chemical equation best represents the reaction taking place in the illustration?
(a) $X_2 + Y_2 \longrightarrow n\,XY_3$
(b) $X_2 + 3\,Y_2 \longrightarrow 2\,XY_3$
(c) $6\,X_2 + 6\,Y_2 \longrightarrow 4\,XY_3 + 4\,X_2$
(d) $6\,X_2 + 6\,Y_2 \longrightarrow 4\,X_3Y + 4\,Y_2$

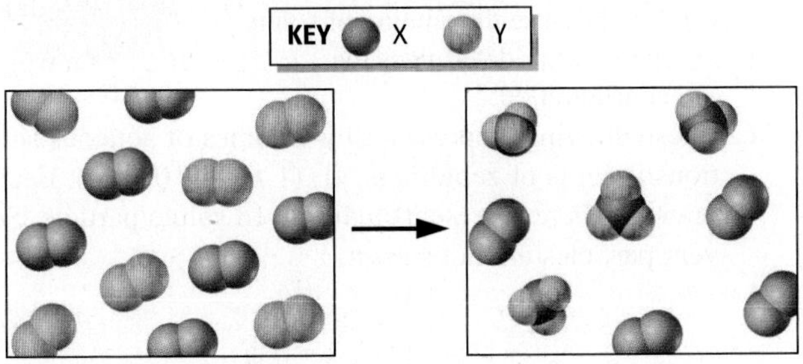
KEY
X
Y

124. Write a balanced chemical equation that represents the reaction shown in the two drawings.

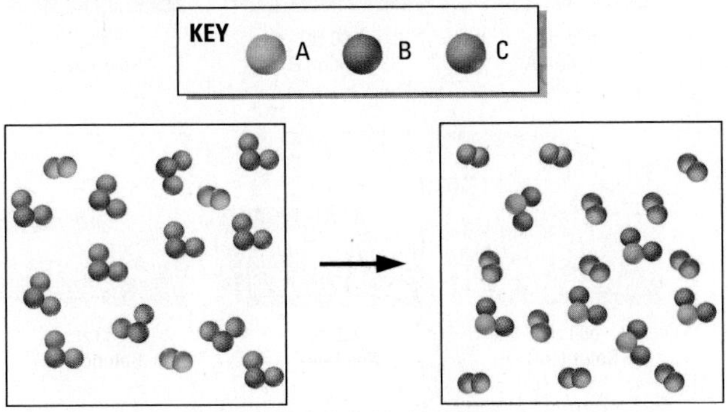
KEY
A
B
C

125. A student set up an experiment for six different trials of the reaction between 1.00-M aqueous acetic acid, CH_3COOH, and solid sodium hydrogen carbonate, $NaHCO_3$.

$$CH_3COOH(aq) + NaHCO_3(s) \longrightarrow NaCH_3CO_2(aq) + CO_2(g) + H_2O(\ell)$$

The volume of acetic acid was kept constant, but the mass of sodium bicarbonate increased with each trial. The results of the tests are shown in the figure.

(a) In which trial(s) is the acetic acid the limiting reactant?
(b) In which trial(s) is sodium bicarbonate the limiting reactant?
(c) Explain your reasoning in parts (a) and (b).

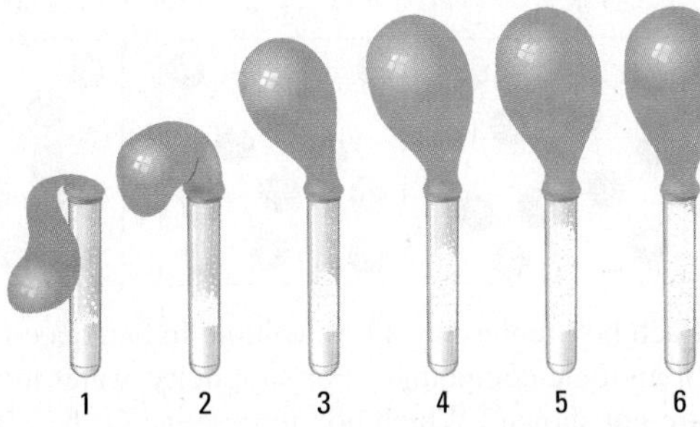
1
2
3
4
5
6

126. A weighed sample of a metal is added to liquid bromine and allowed to react completely. The product substance is then separated from any leftover reactants and weighed. This experiment is repeated with several masses of the metal but with the same volume of bromine. This graph indicates the results. Explain why the graph has the shape that it does.

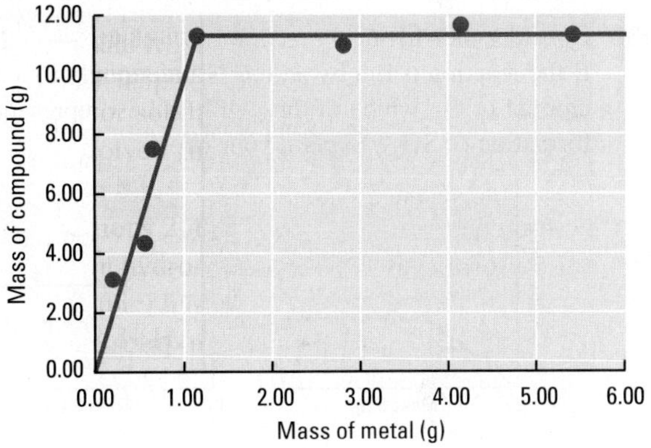
12.00
10.00
8.00
6.00
4.00
2.00
0.00
Mass of compound (g)
0.00
1.00
2.00
3.00
4.00
5.00
6.00
Mass of metal (g)

127. A series of experimental measurements like the ones described in Question 126 is carried out for iron reacting with bromine. This graph is obtained. What is the empirical formula of the compound formed by iron and bromine? Write a balanced equation for the reaction between iron and bromine. Name the product.

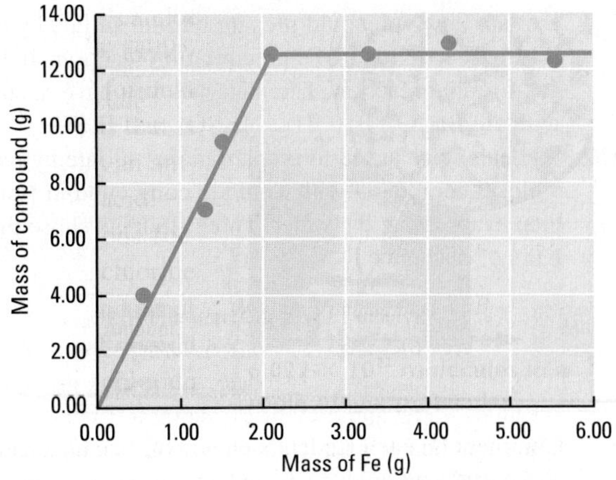
14.00
12.00
10.00
8.00
6.00
4.00
2.00
0.00
Mass of compound (g)
0.00
1.00
2.00
3.00
4.00
5.00
6.00
Mass of Fe (g)

128. Each box represents a tiny volume in an aqueous solution of an ionic compound. (For simplicity, water molecules are not shown.) Which box represents: (a) $MgCl_2$, (b) K_2SO_4, (c) NH_4Cl? Explain your reasoning.

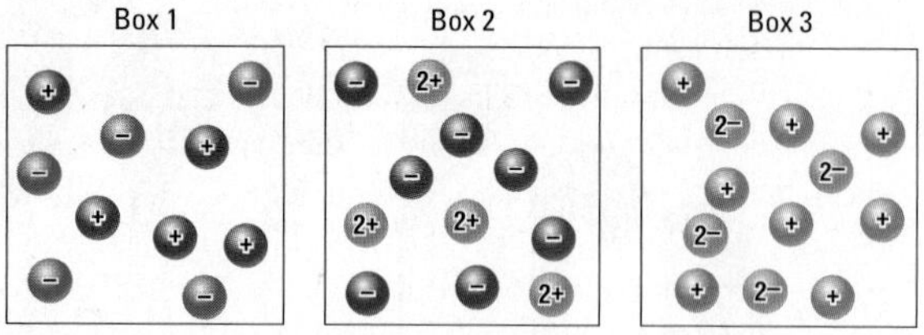
Box 1
Box 2
Box 3

129. Each box represents a tiny volume in an aqueous solution of an ionic compound. (For simplicity, water molecules are not shown.) Which box represents: (a) Na_2CO_3, (b) NH_4NO_3, (c) $CaBr_2$? Explain your reasoning.

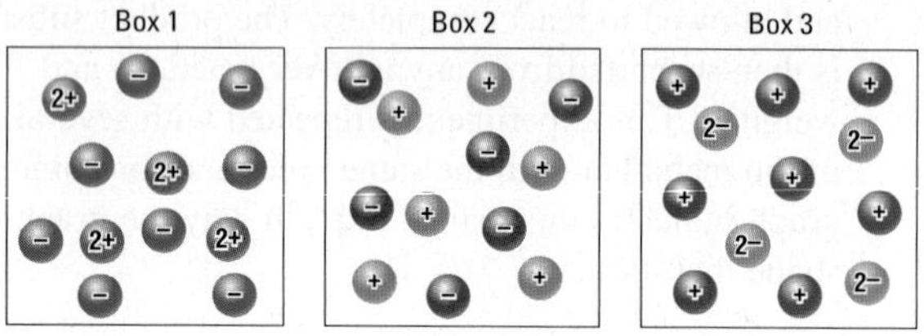
Box 1
Box 2
Box 3

130. Consider the chemical reaction $2\ S + 3\ O_2 \longrightarrow 2\ SO_3$. If the reaction is run by adding S indefinitely to a fixed amount of O_2, which of these graphs best represents the formation of SO_3? Explain your choice.

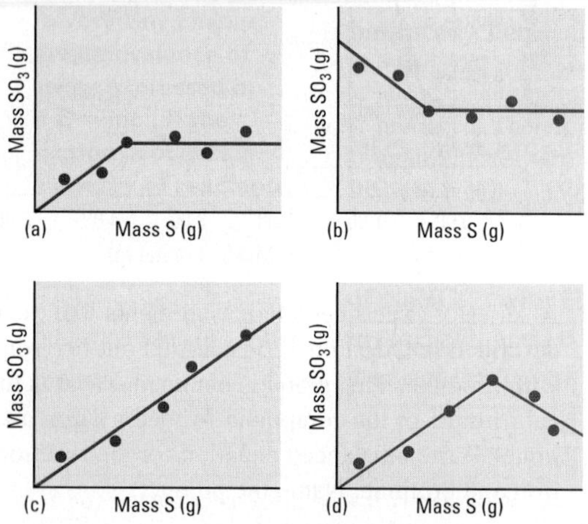
Mass SO3 (g)
(a) Mass S (g)
Mass SO3 (g)
(b) Mass S (g)
Mass SO3 (g)
(c) Mass S (g)
Mass SO3 (g)
(d) Mass S (g)

131. Explain how you could prepare barium sulfate by (a) an acid-base reaction, (b) a precipitation reaction, and (c) a gas-forming reaction. The materials you have to start with are $BaCO_3$, $Ba(OH)_2$, Na_2SO_4, and H_2SO_4.

132. Students were asked to prepare nickel sulfate by reacting a nickel compound with a sulfate compound in water and then evaporating the water. Three students chose these pairs of reactants:

Student 1	$Ni(OH)_2$ and H_2SO_4
Student 2	$Ni(NO_3)_2$ and Na_2SO_4
Student 3	$NiCO_3$ and H_2SO_4

Comment on each student's choice of reactants and how successful you think each student will be at preparing nickel sulfate by the procedure indicated.

133. An unknown solution contains either lead ions or barium ions, but not both. Which one of these solutions could you use to tell whether the ions present are Pb^{2+} or Ba^{2+}? Explain the reasoning behind your choice.

$HCl(aq)$, $H_2SO_4(aq)$, $H_3PO_4(aq)$

134. An unknown solution contains either calcium ions or strontium ions, but not both. Which one of these solutions could you use to tell whether the ions present are Ca^{2+} or Sr^{2+}? Explain the reasoning behind your choice.

$NaOH(aq)$, $H_2SO_4(aq)$, $H_2S(aq)$

135. You prepared a NaCl solution by adding 58.44 g NaCl to a 1-L volumetric flask and then adding water to dissolve it. When you were finished, the final volume in your flask looked like this:

Fill mark
1.00-L flask

The solution you prepared is
(a) Greater than 1 M because you added more solvent than necessary.
(b) Less than 1 M because you added less solvent than necessary.
(c) Greater than 1 M because you added less solvent than necessary.
(d) Less than 1 M because you added more solvent than necessary.
(e) 1 M because the amount of solute, not solvent, determines the concentration.

136. These drawings represent tiny volumes of aqueous solutions in units of zeptoliters, zL (1 zL $= 10^{-21}$ L). Each orange sphere represents a dissolved solute particle. Solvent particles are not shown.

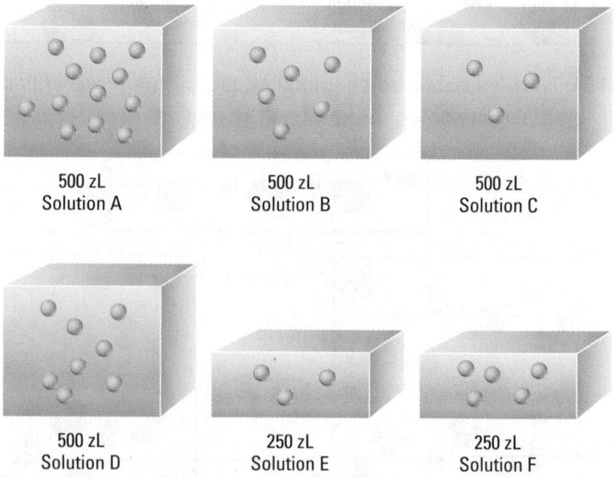
500 zL
Solution A
500 zL
Solution B
500 zL
Solution C
500 zL
Solution D
250 zL
Solution E
250 zL
Solution F

(a) Which solution is most concentrated?
(b) Which solution is least concentrated?
(c) Which two solutions have the same concentration?
(d) When solutions E and F are combined, the resulting solution has the same concentration as solution ______.
(e) When solutions B and E are combined, the resulting solution has the same concentration as solution ______.
(f) If you evaporate half of the water from solution B, the resulting solution will have the same concentration as solution ______.

(g) If you take half of solution A and add 250. zL water, the resulting solution will have the same concentration as solution ______.

137. Ten milliliters of a solution of an acid is mixed with 10 mL of a solution of a base. When the mixture was tested with litmus paper, the blue litmus turned red, and the red litmus remained red. Which of these interpretations is (are) correct?
(a) The mixture contains more hydrogen ions than hydroxide ions.
(b) The mixture contains more hydroxide ions than hydrogen ions.
(c) When an acid and a base react, water is formed, so the mixture cannot be acidic or basic.
(d) If the acid was HCl and the base was NaOH, the concentration of HCl in the initial acidic solution must have been greater than the concentration of NaOH in the initial basic solution.
(e) If the acid was H_2SO_4 and the base was NaOH, the concentration of H_2SO_4 in the initial acidic solution must have been greater than the concentration of NaOH in the initial basic solution.

138. A chemical company was interested in characterizing a competitor's organic acid (it consists of C, H, and O). After determining that it was a diacid, H_2X, a 0.1235-g sample was neutralized with 15.55 mL of 0.1087-M NaOH. Next, a 0.3469-g sample was burned completely in pure oxygen, producing 0.6268 g CO_2 and 0.2138 g H_2O.
(a) Calculate the molar mass of H_2X.
(b) Calculate the empirical formula for the diacid.
(c) Calculate the molecular formula for the diacid.

More Challenging Questions

These questions require more thought and integrate several concepts.

139. Various masses of the three Group 2A elements magnesium, calcium, and strontium were allowed to react with the same mass of liquid bromine, Br_2. After the reaction was complete, the reaction product was freed of excess reactant(s) and weighed. In each case, the mass of product was plotted against the mass of metal used in the reaction (as shown here).

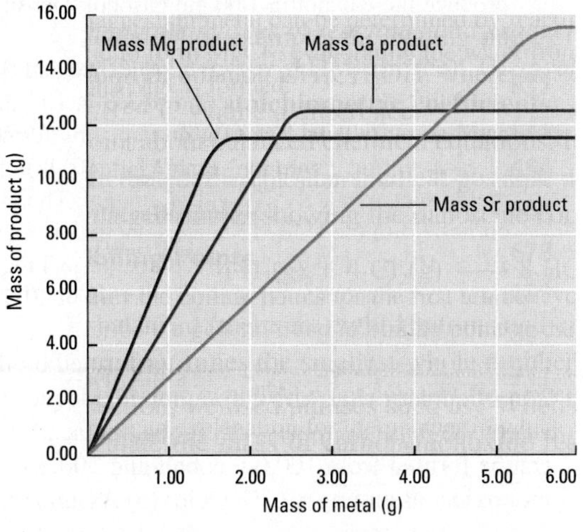
Mass Mg product
Mass Ca product
Mass Sr product
Mass of product (g)
Mass of metal (g)
16.00
14.00
12.00
10.00
8.00
6.00
4.00
2.00
0.00
1.00
2.00
3.00
4.00
5.00
6.00

(a) Based on your knowledge of the reactions of metals with halogens, what product is predicted for each reaction? What are the name and formula for the reaction product in each case?
(b) Write a balanced equation for the reaction occurring in each case.
(c) What kind of reaction occurs between the metals and bromine—that is, is the reaction an acid-base reaction, a precipitation reaction, or an oxidation-reduction reaction?
(d) Each plot shows that the mass of product increases with increasing mass of metal used, but the plot levels out at some point. Use these plots to verify your prediction of the formula of each product, and explain why the plots become level at different masses of metal and different masses of product.

140. Four groups of students from an introductory chemistry laboratory are studying the reactions of solutions of alkali metal halides with aqueous silver nitrate, $AgNO_3$. They use these salts.

Group A: NaCl
Group B: KCl
Group C: NaBr
Group D: KBr

Each of the four groups dissolves 0.004 mol of their salt in some water. Each then adds various masses of silver nitrate, $AgNO_3$, to their solutions. After each group collects the precipitated silver halide, the mass of this product is plotted versus the mass of $AgNO_3$ added. The results are given on this graph.

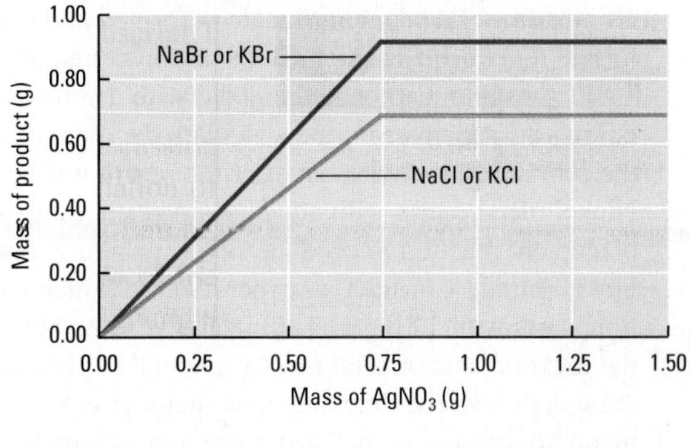
NaBr or KBr
NaCl or KCl
Mass of product (g)
Mass of AgNO3 (g)
1.00
0.80
0.60
0.40
0.20
0.00
0.00
0.25
0.50
0.75
1.00
1.25
1.50

(a) Write the balanced net ionic equation for the reaction observed by each group.
(b) Explain why the data for groups A and B lie on the same line, whereas those for groups C and D lie on a different line.
(c) Explain the shape of the plot observed by each group. Why do the plots level off at the same mass of added $AgNO_3$ (0.75 g) but give different masses of product?

141. One way to determine the stoichiometric relationships among reactants is continuous variations. In this process, a series of reactions is carried out in which the reactants are varied systematically, while keeping the total volume of each reaction mixture constant. When the reactants combine stoichiometrically, they react completely; none is in excess. These data were collected to determine the stoichiometric relationship for the reaction.

$$mX^{n+} + nY^{m-} \longrightarrow X_mY_n$$

Trial	A	B	C	D	E
0.10 M X^{n+}	7 mL	6 mL	5 mL	4 mL	3 mL
0.20 M Y^{m-}	3 mL	4 mL	5 mL	6 mL	7 mL
Excess X^{n+} present?	Yes	Yes	Yes	No	No
Excess Y^{m-} present?	No	No	No	No	Yes

(a) Determine in which trial the reactants are present in stoichiometric amounts.
(b) Calculate how many moles of X^{n+} reacted in that trial.
(c) Calculate how many moles of Y^{m-} reacted in that trial.
(d) Calculate the whole-number ratio of X^{n+} to Y^{m-}.
(e) Determine what the chemical formula is for the product X_mY_n in terms of x and y.

142. Hydrogen gas $H_2(g)$ is reacted with a sample of $Fe_2O_3(s)$ at 400 °C. Two products are formed: water vapor and a black solid compound that is 72.3% Fe and 27.7% O by mass. Write the balanced chemical equation for the reaction.

143. In a reaction, 1.2 g element A reacts with exactly 3.2 g oxygen to form an oxide, AO_x; 2.4 g element A reacts with exactly 3.2 g oxygen to form a second oxide, AO_y.
(a) Determine the ratio $x:y$.
(b) If $x = 2$, determine what the identity of element A might be.

144. A sample of a compound with the formula X_2S_3 has a mass of 10.00 g. It is then roasted (reacted with oxygen) to convert it to X_2O_3. After roasting, it weighs 7.410 g. Calculate the atomic mass of X.

145. When solutions of silver nitrate and sodium carbonate are mixed, solid silver carbonate is formed and sodium nitrate remains in solution. If a solution containing 12.43 g sodium carbonate is mixed with a solution containing 8.37 g silver nitrate, calculate the mass of each of the four substances present after the reaction is complete.

146. Nickel metal reacts with aqueous silver nitrate in a displacement reaction to produce silver metal and aqueous nickel nitrate. Consider an experiment in which the reaction starts with 12.0 g nickel metal and stops before all the nickel reacts. A total of 24.0 g metal is present when the reaction stops. Calculate how many grams of each metal are present in the 24.0-g mixture of metals.

147. Dolomite, found in soil, is $CaMg(CO_3)_2$. If a 20.0-g sample of soil is titrated with 65.25 mL of 0.2500-M HCl, calculate the mass percent of dolomite in the soil sample.

148. A 60.0-mL sample of 2.00-M NaCl and a 40.0-mL sample of 0.500-M NaCl are mixed. Then, additional distilled water is added until the total volume is 500. mL. Calculate the molarity of the NaCl in the final solution.

149. Chlorine has several oxidation states. (a) Determine the oxidation state of chlorine in Cl_2, HOCl, ClO_2, NH_4Cl, and NaCl. (b) Chloride ion is the most stable form. Predict which of the other substances given in part (a) is an oxidizing or a reducing agent and rank them in decreasing order of their oxidizing strengths.

150. In the laboratory, you are given four unlabeled bottles. Each bottle contains a *different* aqueous solution: HCl, $AgNO_3$, Na_2CO_3, or NaOH. Devise an experimental procedure to identify the solute in each of the four bottles using *only* the aqueous solutions provided in the four bottles.

151. A mountain lake that is 4.0 km × 6.0 km with an average depth of 75 m has an $H^+(aq)$ concentration of 1.3×10^{-6} M. Calculate the mass of calcium carbonate that would have to be added to the lake to change the $H^+(aq)$ concentration to 6.3×10^{-8} M. Assume that all the carbonate is converted to carbon dioxide, which bubbles out of the solution.

152. A 7.290-mg mixture containing only cyclohexane and acetaldehyde, C_2H_4O, is analyzed by combustion analysis, which yielded 21.999 mg CO_2. Calculate the mass percent of acetaldehyde in the mixture.

153. Arsenic is often present in polluted water as sodium dihydrogen arsenate. It can be removed from water by reaction with iron(III) chloride, which precipitates iron(III) arsenate. (a) Write the balanced chemical equation for this reaction. (b) Write the net ionic equation. (c) Calculate the minimum volume of 0.150-M iron(III) chloride required to precipitate all of the arsenate from 25.0 mL of a solution that contains 0.025-M arsenate. (d) Calculate the mass of iron(III) arsenate that precipitates.

154. The mass percent of ammonia in a commercial window cleaner was determined by titration with HCl. A 9.360-g sample of the window cleaner was diluted with 35.778 g water and 4.188 g of this diluted solution was titrated using 13.58 mL of 0.1093-M HCl to reach the equivalence point. Calculate the mass percent of ammonia in the window cleaner.

155. A commercial antacid tablet contains $NaAl(OH)_2CO_3$ plus inert materials. A 0.500-g tablet is titrated and requires 27.60 mL of 0.425-M HCl to reach the equivalence point. Calculate the mass percent of $NaAl(OH)_2CO_3$ in the tablet.

156. Ethanol, C_2H_5OH, is a gasoline additive that can be produced by fermentation of glucose.

$$C_6H_{12}O_6 \longrightarrow 2\ C_2H_5OH + 2\ CO_2$$

(a) Calculate the mass (g) of ethanol produced by the fermentation of 1.000 lb glucose.
(b) Gasohol is a mixture of 10.00 mL ethanol per 90.00 mL gasoline. Calculate the mass (in g) of glucose required to produce the ethanol in 1.00 gal gasohol. Density of ethanol = 0.785 g/mL.
(c) By 2022, the U. S. Energy Independence and Security Act calls for annual production of 3.6×10^{10} gal of ethanol, no more than 40% of it produced by fermentation of corn. Fermentation of 1 ton (2.2×10^3 lb) of corn yields approximately 106 gal of ethanol. The average corn yield in the United States is about 2.1×10^5 lb per 1.0×10^5 m^2. Calculate the acreage (in m^2) required to raise corn solely for ethanol production in 2022 in the United States.

157. Dicobalt octacarbonyl, $Co_2(CO)_8$, is formed by the reaction of cobalt metal with carbon monoxide gas. (a) Calculate the maximum mass of dicobalt octacarbonyl that can be formed from 100.0 g cobalt and 200.0 g carbon monoxide, assuming 89.7% yield. (b) Assuming that the

deviation from 100% yield was entirely due to loss of product during purification, calculate the additional mass of the limiting reactant that would be needed to react completely with the residual excess reactant. (c) Calculate the total mass of dicobalt octacarbonyl produced in the reactions in parts (a) and (b).

Conceptual Challenge Problems

These rigorous, thought-provoking problems integrate conceptual learning with problem solving and are suitable for group work.

CP3.A (Section 3-1) How could you show that when baking soda reacts with the acetic acid, CH_3COOH, in vinegar, all of the carbon and oxygen atoms in the carbon dioxide produced come from the baking soda alone and none comes from the acetic acid in vinegar?

CP3.B (Section 3-5) There is a conservation of the number of electrons exchanged during redox reactions, which is tantamount to stating that electrical charge is conserved during chemical reactions. The assignment of oxidation numbers is an arbitrary yet clever way to do the bookkeeping for these electrons. What makes it possible to assign the same oxidation number to all elements that are not bound to other elements not in chemical compounds?

CP3.C (Section 3-5) Consider these redox reactions:

$$HIO_3 + FeI_2 + HCl \longrightarrow FeCl_3 + ICl + H_2O$$

$$CuSCN + KIO_3 + HCl \longrightarrow CuSO_4 + KCl + HCN + ICl + H_2O$$

(a) Identify the species that have been oxidized or reduced in each of the reactions.

(b) After you have correctly identified the species that have been oxidized or reduced in each equation, you might like to try using oxidation numbers to balance each equation. This will be a challenge because, as you have discovered, more than one kind of atom is oxidized or reduced, although in all cases the product of the oxidation and reduction is unambiguous. Record the initial and final oxidation states of each kind of atom that is oxidized or reduced in each equation. Then decide on the coefficients that will equalize the oxidation number changes and satisfy any other atom balancing needed. Finally, balance the equation by adding the correct coefficients to it.

CP3.D (Section 3-5) A student was given four metals (A, B, C, and D) and solutions of their corresponding salts (AZ, BZ, CZ, and DZ). The student was asked to determine the relative reactivity of the four metals by reacting the metals with the solutions. The student's laboratory observations are indicated in the table. Arrange the four metals in order of decreasing activity.

Metal	AZ(aq)	BZ(aq)	CZ(aq)	DZ(aq)
A	No reaction	No reaction	No reaction	No reaction
B	Reaction	No reaction	Reaction	No reaction
C	Reaction	No reaction	No reaction	No reaction
D	Reaction	Reaction	Reaction	No reaction

CP3.E (Section 3-6) In Exercise 3.17 it was not possible to find the mass of O_2 directly from a knowledge of the mass of sucrose. Are there chemical reactions in which the mass of a product or another reactant can be known directly if you know the mass of a reactant? Cite a few of these reactions.

CP3.F (Section 3-6) Glucose, $C_6H_{12}O_6$, a monosaccharide, and sucrose, $C_{12}H_{22}O_{11}$, a disaccharide, undergo complete combustion with O_2 (metabolic conversion) to produce H_2O and CO_2.

(a) Determine how many moles of O_2 are needed per mole of each sugar for the reaction to proceed.

(b) Determine how many grams of O_2 are needed per mole of each sugar for the reaction to proceed.

(c) Determine which combustion reaction produces more H_2O per gram of sugar. Determine how many grams of H_2O are produced per gram of each sugar.

CP3.G (Section 3-10) Describe how you would prepare 1 L of 1.00×10^{-6} M NaCl (molar mass = 58.44 g/mol) solution by using a balance that can measure mass only to 0.01 g.

Energy and Chemical Reactions

4

Cengage Learning/Charles D. Winters

Burning a fuel, such as charcoal, transfers a lot of energy to anything in contact with the reactant and product molecules. In this case a metal grill, four hamburgers, and the air above the fire are all heated. The energy released when a fuel such as charcoal burns can be transformed to provide many of the benefits of our technology-intensive society. Why is energy transferred to the surroundings by some chemical reactions? (Section 4-4) What makes a good fuel? (Section 4-11) How do chemists measure the quantity of energy transferred when a fuel burns? (Section 4-8) How do foods fuel our bodies? (Section 4-10) These and many related questions are answered in this chapter.

In our industrialized, high-technology, appliance-oriented society, the average use of energy per person is at nearly its highest point in history. The United States, with less than 5% of the world's population, consumes about 20% of the world's energy resources. In every year since 1958 we have consumed more energy resources than have been produced within our borders. Most of the energy we use comes from chemical reactions: combustion of the fossil fuels coal, petroleum, and natural gas. The rest comes from hydroelectric power plants,

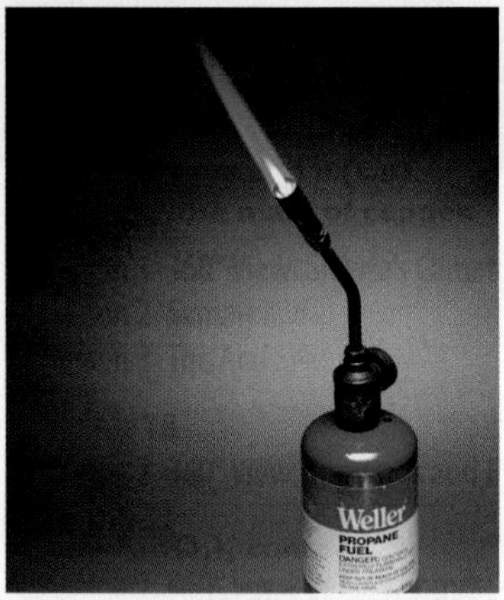
Weller
PROPANE
FUEL
Charles D. Winters/Science Source

Propane torch for soldering.

"A theory is the more impressive the greater the simplicity of its premises is, the more different kinds of things it relates, and the more extended is its area of applicability. Therefore, the deep impression which classical thermodynamics made upon me. It is the only physical theory of universal content concerning which I am convinced that, within the framework of the applicability of its basic concepts, it will never be overthrown." (Albert Einstein, quoted in Schlipp, P. A. [ed.] "Albert Einstein: Philosopher-Scientist." In *The Library of Living Philosophers*, Vol. VII. Autobiographical notes, 3rd ed. LaSalle, IL: Open Court Publishing, 1969; p. 33.)

nuclear power plants, solar energy and wind collectors, and burning wood and other plant material. Both U. S. and world energy use are growing rapidly.

Chemical reactions involve transfers of energy. When a fuel burns, the energy of the products is less than the energy of the reactants. The leftover energy shows up in anything that is in contact with the reactants and products. For example, propane is an important fuel used in soldering torches. When propane burns in air, the carbon, hydrogen, and oxygen atoms in the C_3H_8 and O_2 reactant molecules rearrange to form CO_2 and H_2O product molecules.

$$C_3H_8(g) + 5\,O_2(g) \longrightarrow 3\,CO_2(g) + 4\,H_2O(g)$$

Because of the way their atoms are connected, the CO_2 and H_2O molecules have less total energy than the C_3H_8 and O_2 reactant molecules did. After the reaction, some energy that was in the reactants is not contained in the product molecules. That energy heats everything that is close to where the reaction takes place. The reaction transfers energy to its surroundings.

For the past hundred years or so, most of the energy society has used has come from combustion of fossil fuels, and this will continue to be true well into the future. Consequently, it is very important to understand how energy and chemical reactions are related and how chemistry might be used to alter our dependence on fossil fuels. This requires knowledge of **thermodynamics**, *the science of heat, work, and transformations of one to the other.* The fastest-growing new industries in the twenty-first century may well be those that capitalize on such knowledge and the new chemistry and chemical industries it spawns.

4-1 The Nature of Energy

What is energy? Where does the energy we use come from? And how can chemical reactions result in the transfer of energy to or from their surroundings? **Energy**, represented by E, was defined in Chapter 1 (← **Sec. 1-6b**) as *the capacity to do work.* If you climb a mountain or a staircase, you work against the force of gravity as you move upward, and your gravitational energy increases. The energy you use to do this work is released when food you have eaten is metabolized (undergoes chemical reactions) within your body. Energy from food enables you to work against the force of gravity as you climb, and it warms your body (climbing makes you hotter as well as higher). Therefore our study of the relations between energy and chemistry also needs to consider processes that involve work and processes that involve heat.

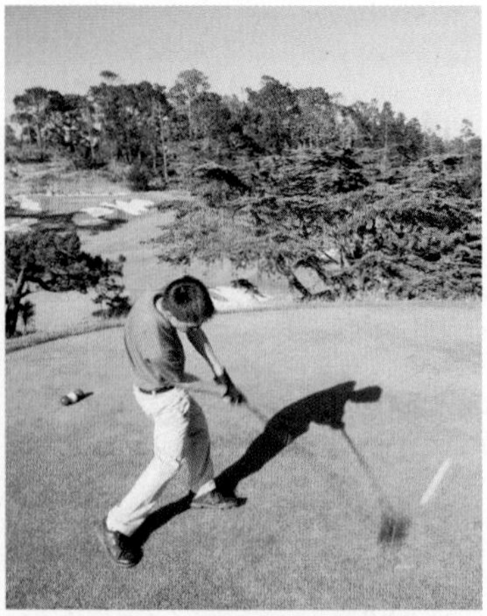
Riddle Photography/Shutterstock.com

Figure 4.1 Kinetic energy. As it speeds away from the club, the golf ball has kinetic energy that depends on its mass and velocity.

Energy can be classified as kinetic or potential. **Kinetic energy** is *energy that something has because it is moving* (Figure 4.1). Examples of kinetic energy are

- Energy of motion of a macroscale object, such as a moving baseball or automobile; this is often called *mechanical energy.*
- Energy of motion of nanoscale objects such as atoms, molecules, or ions; this is often called *thermal energy.*
- Energy of motion of electrons through an electrical conductor; this is often called *electrical energy.*

- Energy of periodic motion of nanoscale particles when a macroscale sample is alternately compressed and expanded (as when a sound wave passes through air).

Kinetic energy, E_k, can be calculated as $E_k = \frac{1}{2}mv^2$, where m represents the mass and v represents the velocity of a moving object.

Potential energy is *energy that something has as a result of its position and some force that is capable of changing that position.* Examples include

- Energy that a ball held in your hand has because the force of gravity attracts it toward the floor; this is often called *gravitational energy.*
- Energy that charged particles have because they attract or repel each other; this is often called *electrostatic energy.* An example is the potential energy of positive and negative ions close together.
- Energy resulting from attractions and repulsions among electrons and atomic nuclei in molecules; this is often called *chemical potential energy* and is the kind of energy stored in foods and fuels.

Potential energy can be calculated in different ways depending on the type of force that is involved. For example, near the surface of Earth, gravitational potential energy, E_p, can be calculated as $E_p = mgh$, where m is mass, g is the gravitational constant ($g = 9.8 \text{ m/s}^2$), and h is the height above the surface.

Potential energy can be converted to kinetic energy and vice versa. As droplets of water fall over a waterfall (Figure 4.2), the potential energy they had at the top is converted to kinetic energy—they move faster and faster. Conversely, the kinetic energy of falling water could drive a water wheel to pump water to an elevated reservoir, where its potential energy would be higher.

Gary Muth/iStockphoto.com

Figure 4.2 Gravitational potential energy. Because of its position relative to Earth, water on the brink of a waterfall has potential energy that could be used to do work; that energy could be used to generate electricity, for example, as in a hydroelectric power plant.

4-1a Energy Units

The SI unit of energy is the *joule* (rhymes with rule), symbol J. The joule is a derived unit, which means that it can be expressed as a combination of other more fundamental units: $1 \text{ J} = 1 \text{ kg m}^2/\text{s}^2$. If a 2.0-kg object (which weighs about $4\frac{1}{2}$ pounds) is moving with a velocity of 1.0 meter per second (roughly 2 miles per hour), its kinetic energy is

$$E_k = \frac{1}{2}mv^2 = \frac{1}{2} \times (2.0 \text{ kg})(1.0 \text{ m/s})^2 = 1.0 \text{ kg m}^2/\text{s}^2 = 1.0 \text{ J}$$

This is a relatively small quantity of energy. Because the joule is so small, we often use the kilojoule (1 kilojoule = 1 kJ = 1000 J) as a unit of energy.

The joule is the unit of energy in the International System of units (SI units). SI units are described in Chapter 1 and Appendix B. A joule is approximately the quantity of energy required for two human heartbeats.

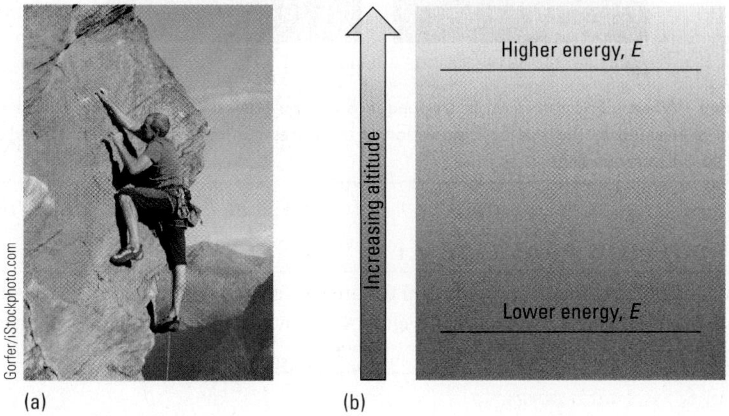
Gorfer/iStockphoto.com
Increasing altitude
Higher energy, E
Lower energy, E
(a)
(b)

Rock climbing. (a) Climbing requires energy. (b) The higher the altitude, the greater the climber's gravitational energy.

1 SACHET (16kJ) IS
EQUAL

Cengage Learning/Charles D. Winters

Figure 4.3 Food energy. As its label shows, the sweetener in this packet supplies 16 kJ of nutritional energy.

Another energy unit is the *calorie*, symbol cal. By definition 1 cal = 4.184 J exactly. A calorie is very close to the quantity of energy required to raise the temperature of one gram of water by one degree Celsius. (The calorie was originally defined as the quantity of energy required to raise the temperature of 1 g $H_2O(\ell)$ from 14.5 °C to 15.5 °C.) The "calorie" that you hear about in connection with nutrition and dieting is actually a kilocalorie (kcal) and is usually represented with a capital C. Thus, a breakfast cereal that gives you 100 Calories of nutritional energy actually provides 100 kcal = 100×10^3 cal. In many countries food energy is reported in kilojoules rather than in Calories. For example, the label on the packet of nonsugar sweetener shown in Figure 4.3 indicates that it provides 16 kJ of nutritional energy.

The food Calorie measures how much energy is released when a given quantity of food undergoes combustion with oxygen.

PROBLEM-SOLVING EXAMPLE 4.1

Energy Units

A single Fritos snack chip has a food energy of 5.0 Cal. Calculate this energy in joules.

Result 2.1×10^4 J

Analyze We are asked to express an energy in different units.

Plan Use the definitions 1 Cal = 1 kcal and kilo = 1000 to generate proportionality (conversion) factors.

Execute

$$E = 5.0 \text{ Cal} \times \frac{1 \text{ kcal}}{1 \text{ Cal}} \times \frac{1000 \text{ cal}}{1 \text{ kcal}} \times \frac{4.184 \text{ J}}{1 \text{ cal}} = 2.1 \times 10^4 \text{ J}$$

☑ **Reasonable Result Check** 2.1×10^4 J is 21 kJ. Because 1 Cal = 1 kcal = 4.184 kJ, the result in kJ should be about four times the original 5 Cal (that is, about 20 kJ); the result is reasonable.

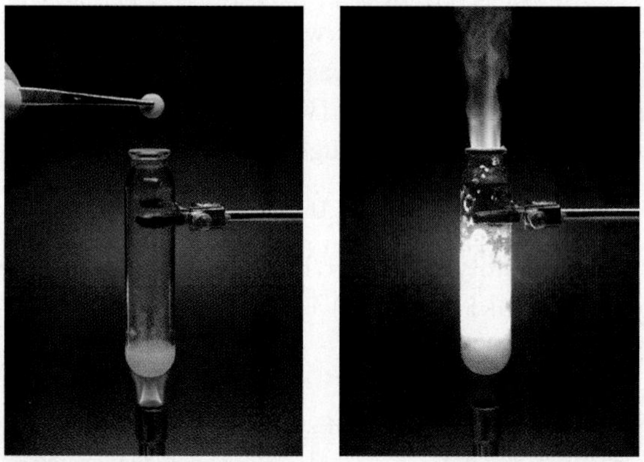

Photos: Cengage Learning/Charles D. Winters

(a) (b)

Food energy. When a Fritos chip (a) is dropped into molten potassium chlorate, the chip burns (b) in oxygen generated by thermal decomposition of the potassium chlorate, and about 20 kJ is transferred to the surroundings.

PROBLEM-SOLVING PRACTICE 4.1

(a) If you eat a hot dog, it will provide 160 Calories of energy. Express this energy in joules.
(b) The packet of nonsugar sweetener in Figure 4.3 provides 16 kJ of nutritional energy. Express this energy in kilocalories.

PROBLEM-SOLVING PRACTICE answers are provided at the back of this book in Appendix K.

Another useful energy-related term is **power**: *energy per unit time*. The unit of power is the watt, abbreviated W. One watt is one joule per second, 1 J/s. A 60-W incandescent light bulb requires 60 J every second; replacing it with an 8-W LED bulb would save 52 J every second the bulb is lighted!

ESTIMATION | Counting Calories

Many sweetened soft drinks contain high-fructose corn syrup—a significant source of Calories in the diet. Estimate the weight loss that would occur in a year if a person who typically drank one 12-oz sweetened soft drink per day switched to water.

From an encyclopedia or the Internet, you can learn that a typical sweetened soft drink contains 39 g sugar; when metabolized fructose generates 3.73 kcal/g, excess caloric intake is usually stored in fat molecules, and fat provides 8.8 kcal/g when metabolized. Assume that the sugar is all fructose. Using round numbers to make the calculation easier, the decreased caloric intake is about 40 g/d $\times$ $3\frac{3}{4}$ kcal/g $=$ 150 kcal/d. In a year, that is about 150 kcal/d $\times$ 360 d/y $=$ 54,000 kcal/y.

Assume further that the weight loss is all due to metabolism of fat, which is where the body stored excess caloric intake from the soft drink. To find the mass of fat equivalent to the 54,000 kcal, divide by the approximately 9 kcal/g for metabolism of fat. This gives about 6000 g or 6 kg. A kilogram is about 2 lb, so the weight loss is about 12 lb.

This estimate assumes that there was no change in diet other than replacing the soft drink with water and that fat, which produces more energy than carbohydrate or protein when metabolized, was consumed to replace the calories formerly supplied by the soft drink. The calculated weight loss would have been greater for carbohydrate or protein. A can of sugary beverage a day makes a difference!

4-2 Conservation of Energy

When you dive from a diving board into a pool of water, several transformations of energy occur (Figure 4.4). Eventually, you float on the surface and the water becomes still. However, on average, the water molecules are moving a little faster in the vicinity of your point of impact; that is, the temperature of the water is now a little higher. Energy has been transformed from potential to kinetic and from macroscale kinetic to nanoscale kinetic (that is, thermal). Nevertheless, the total quantity of energy, kinetic plus potential, is the same before and after the dive. In many, many experiments, the total energy has always been found to be the same before and after an event. These experiments are summarized by the **law of conservation of energy**, which states that *energy can neither be created nor destroyed—the total energy of the universe is constant.* This is also called the **first law of thermodynamics**.

In Section 1-9a the kinetic-molecular theory was described qualitatively. A corollary to this theory is that *molecules move faster, on average, as the temperature increases.*

The nature of scientific laws is discussed in Chapter 1 (← Sec. 1-3).

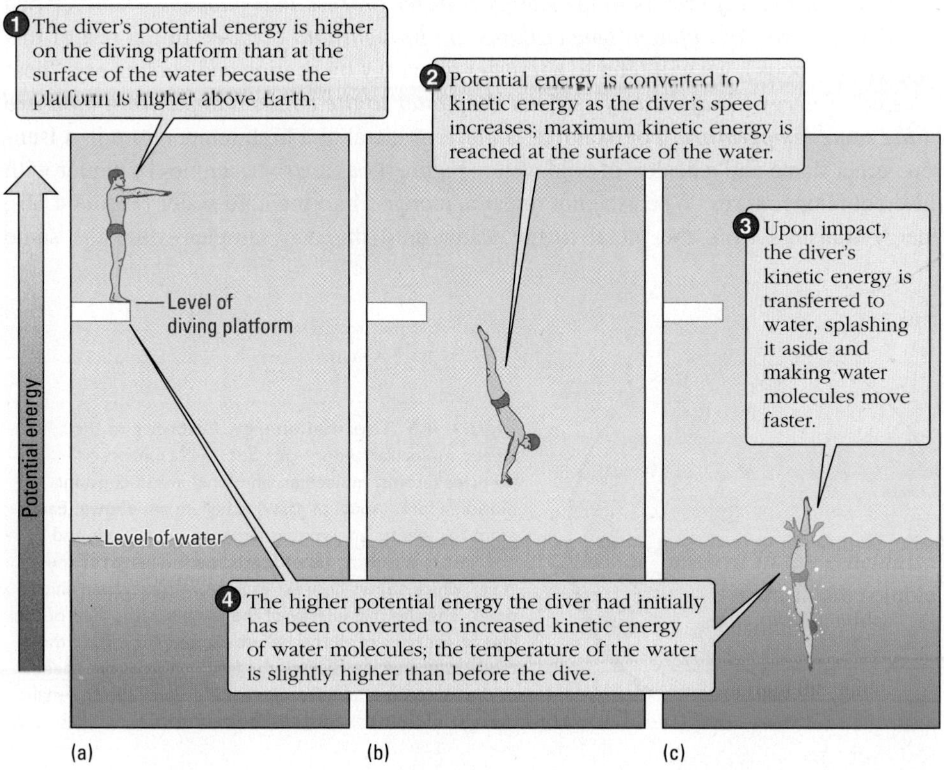
1 The diver's potential energy is higher on the diving platform than at the surface of the water because the platform is higher above Earth.
2 Potential energy is converted to kinetic energy as the diver's speed increases; maximum kinetic energy is reached at the surface of the water.
3 Upon impact, the diver's kinetic energy is transferred to water, splashing it aside and making water molecules move faster.
4 The higher potential energy the diver had initially has been converted to increased kinetic energy of water molecules; the temperature of the water is slightly higher than before the dive.
Level of diving platform
Level of water
Potential energy
(a)
(b)
(c)

Figure 4.4 Energy transformations. Potential and kinetic energy are interconverted when someone dives into water but the total energy remains the same.

Answers to **EXERCISES** are provided at the back of this book in Appendix L.

EXERCISES that are labeled **CONCEPTUAL** are designed to test your understanding of one or more concepts; they usually involve qualitative rather than quantitative thinking.

CONCEPTUAL EXERCISE 4.1

Energy Transfers

You toss a rubber ball up into the air. It falls to the floor, bounces for a while, and eventually comes to rest. Several energy transfers are involved. Describe them and the changes they cause.

4-2a Energy and Working

Work and _heat_ refer to the quantity of energy transferred from one object or sample to another by working or heating _processes_. However, we often talk about work and heat as if they were forms of energy; rather, they are processes that transfer energy from one form or one place to another.

Work is required to cause some chemical and biochemical processes to occur. Examples are moving ions across a cell membrane and synthesizing adenosine triphosphate (ATP) from adenosine diphosphate (ADP).

When a force acts on an object and moves the object, the change in the object's kinetic energy is equal to the work done on the object. Work has to be done, for example, to accelerate a car from 0 to 60 miles per hour or to hit a baseball out of a stadium. Work is also required to increase the potential energy of an object. Thus, work has to be done to raise an object against the force of gravity (as in an elevator), to separate a sodium ion, Na^+, from a chloride ion, Cl^-, or to move an electron away from an atomic nucleus. The work done on an object corresponds to the quantity of energy transferred to that object; that is, doing **work** (or **working**) on an object is *a process that transfers energy to an object.* Conversely, if an object does work on something else, the quantity of energy associated with the object must decrease. In the rest of this chapter (and book), we will refer to a transfer of energy by doing work as a "work transfer."

4-2b Energy, Temperature, and Heating

Transferring energy by heating is a process, but it is common to talk about that process as if heat were a form of energy. It is often said that one sample transfers heat to another, when what is meant is that one sample transfers energy by heating the other.

According to the kinetic-molecular theory (← **Sec. 1-9a**), all matter consists of nanoscale particles that are in constant motion (Figure 4.5). Therefore, all matter has thermal energy. For a given sample, the quantity of thermal energy is greater the higher the temperature is. Temperature can be measured with a thermometer, a device in which a liquid is confined in a small-diameter tube. When a thermometer is heated, energy transfers to the thermometer, the atoms or molecules in the liquid move about more rapidly, and the volume of the spaces between the particles increases slightly. Consequently, the liquid expands and rises higher in the thermometer tube.

Heat (or **heating**) refers to *the energy transfer process that happens whenever two samples of matter at different temperatures are brought into contact.* In the rest of this chapter (and book), we will refer to a transfer of energy by heating and cooling as a "heat transfer". ***Energy always transfers from the hotter to the cooler sample until both are at the same temperature.*** For example, a piece of metal at a high temperature in a Bunsen burner flame and a beaker of cold water (Figure 4.6a) are two samples of matter with different temperatures. When the hot metal is plunged into the cold water (Figure 4.6b), energy transfers from the metal to the water until the two samples reach the same

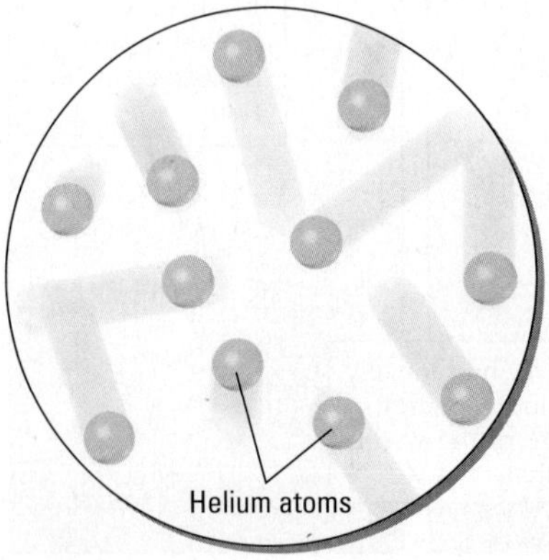
Helium atoms

Figure 4.5 Thermal energy. According to the kinetic-molecular theory (← **Sec. 1-9a**), nanoscale particles (atoms, molecules, and ions) are in constant motion. Here, atoms of gaseous helium are shown. Each atom has kinetic energy that depends on its mass and how fast it is moving (as indicated by the length of the "tail," which shows how far each atom travels per unit time). The thermal energy of the sample is the sum of the kinetic energies of all the helium atoms. The higher the temperature of the helium, the faster the average speed of the molecules, and therefore the greater the thermal energy.

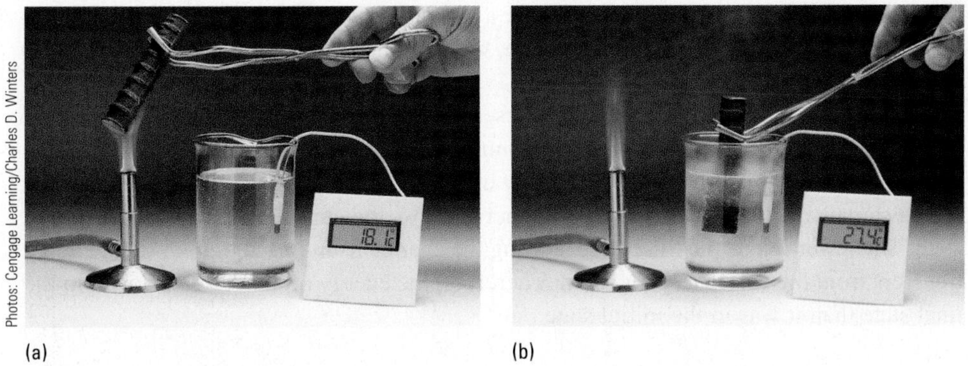
Photos: Cengage Learning/Charles D. Winters
18.1
27.4
(a)
(b)

Figure 4.6 Energy transfer by heating. (a) Water in a beaker with a probe that measures temperature in °C is heated when a hotter sample (a steel bar) is plunged into the water. There is a transfer of energy from the hotter metal bar to the cooler water. (b) Eventually, enough energy is transferred so that the bar and the water reach the same temperature—that is, thermal equilibrium is achieved.

temperature. Once that happens, the metal and water are said to be in **thermal equilibrium**, *a state in which samples of matter initially at different temperatures have transferred energy to reach a common temperature*.

It is important to understand that ***the process of heating involves transfer of kinetic energy at the nanoscale***. Atoms in the hot iron (Figure 4.7) are vibrating much farther from their average positions than in cold iron. This means that atoms at the surface of the hot iron collide more vigorously with water molecules in the surrounding liquid. On average, these collisions transfer kinetic energy from the iron atoms to the water molecules: Iron atoms move less and water molecules move more. When thermal equilibrium is reached, the average kinetic energy of the particles in each sample is the same.

Usually most objects in a given region, such as your room, are at about the same temperature—at thermal equilibrium. A fresh cup of coffee, which is hotter than room temperature, transfers energy by heating the rest of the room until the coffee cools off (and the rest of the room warms up a tiny bit). A can of cold soft drink, which is much cooler than its surroundings, receives energy from everything else until it warms up (and your room cools off a little). Because the total quantity of material in your room is very

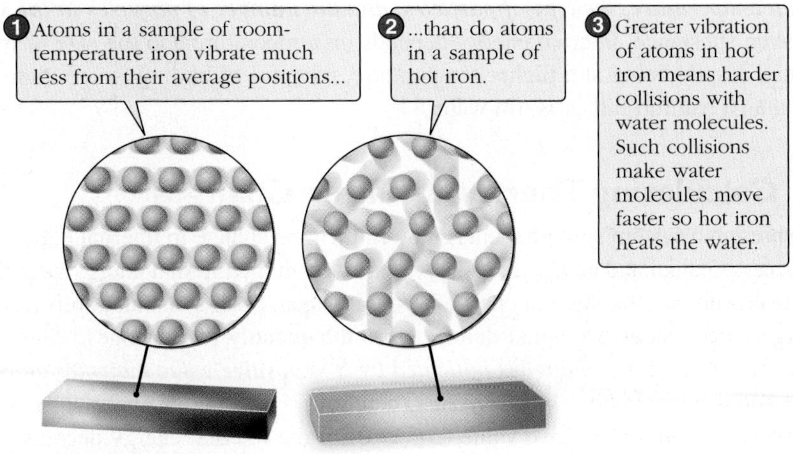
1 Atoms in a sample of room-temperature iron vibrate much less from their average positions...
2 ...than do atoms in a sample of hot iron.
3 Greater vibration of atoms in hot iron means harder collisions with water molecules. Such collisions make water molecules move faster so hot iron heats the water.

Figure 4.7 Cold and hot iron.

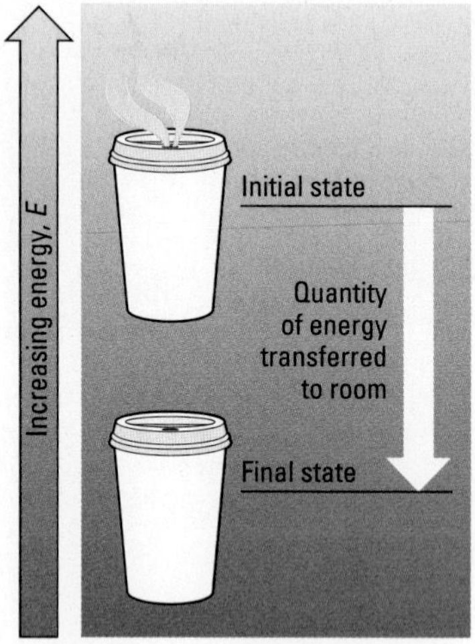
Increasing energy, E
Initial state
Quantity of energy transferred to room
Final state

Figure 4.8 Energy diagram. A cup of hot coffee (initial state) has more thermal energy than a cup of coffee at room temperature (final state). Therefore the hot coffee is higher in the diagram. The quantity of energy transferred to the room as the coffee cools is represented by the arrow from the initial to the final state.

much greater than that in a cup of coffee or a can of soda, the room temperature changes only a tiny bit to reach thermal equilibrium, whereas the temperature of the coffee or the soda changes a lot.

A diagram such as Figure 4.8 can be used to show the energy transfer from a cup of hot coffee to your room. The upper horizontal line represents the energy of the hot coffee and the lower line represents the energy of the room-temperature coffee. Because the coffee started at a higher temperature (higher energy), the upper line is labeled the initial state. The lower line is the final state. During the change from initial to final state, energy transfers from the coffee to your room. Therefore, the energy of the coffee is lower in the final state than it was in the initial state.

CONCEPTUAL EXERCISE 4.2

Energy Diagrams

(a) Draw an energy diagram like the one in Figure 4.8 for warming a can of cold soft drink to room temperature. Label the initial and final states and use an arrow to represent the change in energy of the can of soda.

(b) Draw a second energy diagram, to the same scale, to show the change in energy of the room as the can of cold drink warms to room temperature.

4-3 Keeping Track of Energy Transfers

In thermodynamics it is useful to define a *region of primary concern* as the **system**. Then we can decide whether energy transfers into or out of the system and keep track of how much energy transfers in each direction. *Everything that can exchange energy with the system* is defined as the **surroundings**. A system may be delineated by an actual physical boundary, such as the inside surface of a flask or the membrane of a cell in your body. Or the boundary may be indistinct, as in the case of the solar system within its surroundings, the rest of the galaxy. In the case of a hot cup of coffee in your room, the cup and the coffee might be the system, and your room would be the surroundings. For a chemical reaction, the system is usually defined to be all of the atoms that make up the reactants. These same atoms will be bonded in a different way in the products after the reaction, and it is their energy before and after reaction that interests us most.

The **internal energy** of a system is the *sum of the individual energies (kinetic and potential) of all nanoscale particles (atoms, molecules, or ions)* in that system. Increasing the temperature increases the internal energy because it increases the average speed of motion of nanoscale particles. ***The total internal energy of a sample of matter depends on temperature, the type of particles, and the number of particles in the sample.*** For a given substance, internal energy depends on temperature and the size of the sample. Thus, despite being at a higher temperature, a cupful of boiling water contains less energy than a bathtub full of warm water.

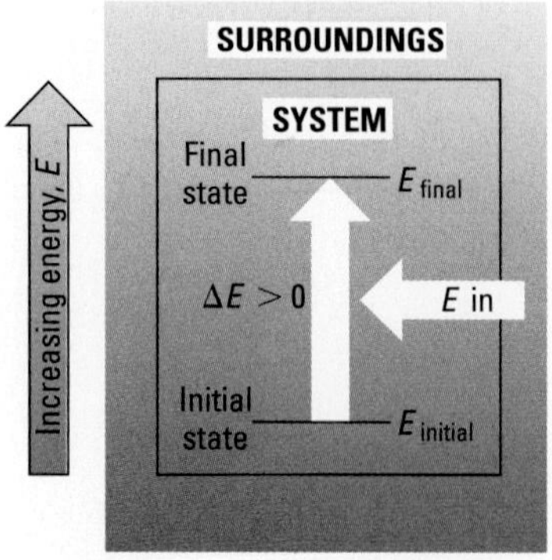
SURROUNDINGS
SYSTEM
Increasing energy, E
Final state
E final
ΔE > 0
E in
Initial state
E initial

ΔE positive: Internal energy increases.

4-3a Calculating Thermodynamic Changes

If we represent a system's internal energy by E, then the change in internal energy during any process is calculated as $E_{final} - E_{initial}$. That is, from the internal energy after the process is over, subtract the internal energy before it began. Such a calculation is designated by using a Greek letter Δ (capital delta) before the quantity that changes. Thus, $E_{final} - E_{initial} = \Delta E$. *Whenever a change is indicated by* Δ, *a positive value indicates an increase and a negative value indicates a decrease.* Therefore, if the internal energy increases during a process, ΔE has a positive value ($\Delta E > 0$); if the internal energy decreases, ΔE is negative ($\Delta E < 0$).

A good analogy to this thermodynamic calculation is your bank account. Assume that in your account (the system) you have a balance B of \$260 (B_{initial}), and you withdraw \$60 in spending money. After the withdrawal the balance is \$200 (B_{final}). The change in your balance is

$$\text{Change in balance} = \Delta B = B_{\text{final}} - B_{\text{initial}} = \$200 - \$260 = -\$60$$

The negative sign on the \$60 indicates that money has been withdrawn from the account (system) and transferred to you (the surroundings). The cash itself is not negative, but during the process of withdrawing your money the balance in the bank went down, so ΔB was negative. Similarly, the *magnitude* of change of a thermodynamic quantity is a number with no sign. To indicate the *direction* of a change, we attach a negative sign (transferred out of the system) or a positive sign (transferred into the system).

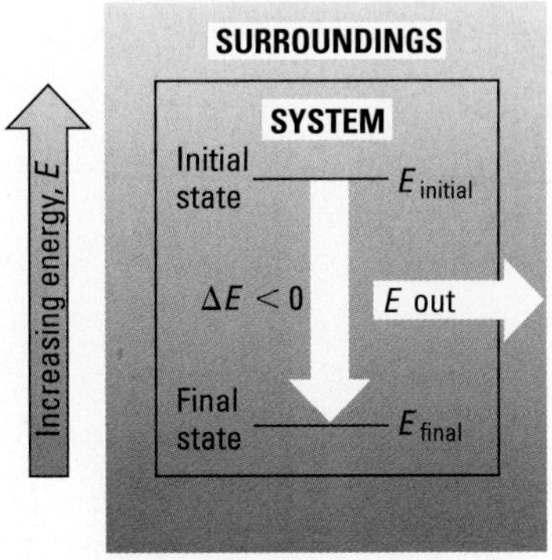
SURROUNDINGS
SYSTEM
Increasing energy, E
Initial state
E initial
ΔE < 0
E out
Final state
E final

ΔE negative: Internal energy decreases.

CONCEPTUAL EXERCISE 4.3

Direction of Energy Transfer

It takes about 1.5 kJ to raise the temperature of a can of beverage from 25.0 °C to 26.0 °C. You put the can of beverage into a refrigerator to cool it from room temperature (25.0 °C) to 1.0 °C.

(a) Calculate what quantity of heat transfer is required. Express your result in kilojoules.
(b) What is a reasonable choice of system for this situation?
(c) What constitutes the surroundings?
(d) What is the sign of ΔE for this situation? What is the calculated value of ΔE?
(e) Draw an energy diagram showing the system, the surroundings, the change in energy of the system, and the energy transfer between the system and the surroundings.

4-3b Conservation of Energy and Chemical Reactions

For many chemical reactions the only energy transfer processes are heating and doing work. If no other energy transfers (such as emitting light) take place, the law of conservation of energy for any system can be written as

$$\Delta E = q + w \qquad [4.1]$$

where q represents the quantity of energy transferred by heating the system, and w represents the quantity of energy transferred by doing work on the system. The relationships among ΔE, q, and w for a system are shown in Figure 4.9. If energy is transferred *into* the system from the surroundings by heating, then q is *positive*; if energy is transferred *into* the system because the surroundings do *work on* the system, then w is *positive*. If energy is transferred *out of* the system by *heating* the surroundings, then q has a *negative* value; if energy is transferred *out of* the system because *work* is done on the surroundings, then

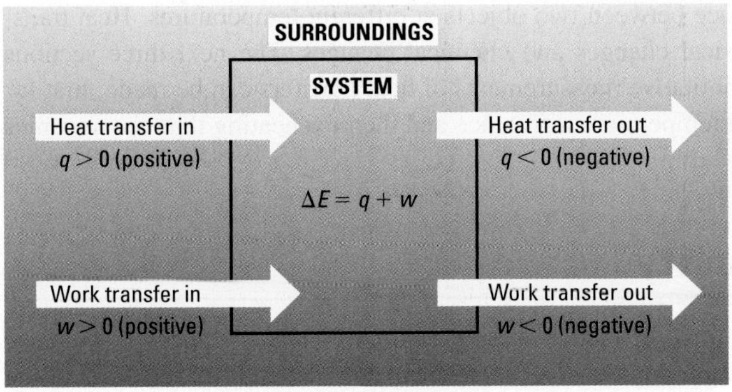
SURROUNDINGS
SYSTEM
Heat transfer in
q > 0 (positive)
Heat transfer out
q < 0 (negative)
ΔE = q + w
Work transfer in
w > 0 (positive)
Work transfer out
w < 0 (negative)

Figure 4.9 Internal energy, heat, and work. Energy transfers between a thermodynamic system and its surroundings are shown schematically.

w has a *negative* value. The *magnitudes* of q and w indicate the *quantities* of energy transferred, and the *signs* of q and w indicate the *direction* in which the energy is transferred.

PROBLEM-SOLVING EXAMPLE 4.2

Internal Energy, Heat, and Work

A fuel cell based on the reaction of hydrogen with oxygen powers a small automobile by running an electric motor. The motor draws 75.0 kilowatts (75.0 kJ/s) and runs for 2 minutes and 20 seconds. During this period, 5.0×10^3 kJ must be carried away from the fuel cell to prevent it from overheating. The system is defined to be the hydrogen and oxygen that react. Calculate the change in the system's internal energy.

Result -1.55×10^4 kJ

Analyze Change in internal energy depends on work done on the system and heat transfer into the system. The system is the chemicals that react. The motor powers a car so the system does work on the surroundings and w must be negative. The system heats the surroundings so q must be negative.

Plan Calculate the work done based on the power and how long the motor runs. The heat transfer is $q = -5.0 \times 10^3$ kJ. Use the equation $\Delta E = q + w$.

Execute

$$w = -75.0\frac{\text{kJ}}{\text{s}} \times 140.\ \text{s} = -1.05 \times 10^4\ \text{kJ} = -10.5 \times 10^3\ \text{kJ}$$

$$\Delta E = q + w = (-5.0 \times 10^3\ \text{kJ}) + (-10.5 \times 10^3\ \text{kJ}) = -1.55 \times 10^4\ \text{kJ}$$

Thus the internal energy of the reaction product (water) is 1.55×10^4 kJ lower than the internal energy of the reactants (hydrogen and oxygen).

☑ **Reasonable Result Check** ΔE is negative, which is reasonable because energy transferred from the reaction heats and does work on the surroundings.

PROBLEM-SOLVING PRACTICE 4.2

Suppose that the internal energy decreases by 2400 J when a mixture of natural gas (methane) and oxygen is ignited and burns and the surroundings are heated by 1.89 kJ. Calculate the work done by this system on the surroundings.

So far we have seen that

- energy transfers can occur either by heating or by working;
- it is convenient to define a system so that energy transfers into a system (positive) and out of a system (negative) can be accounted for; and
- the internal energy of a system can change as a result of heating or doing work on the system.

Our primary interest in this chapter is heat transfers (the "thermo" in thermodynamics). Heat transfers can take place between two objects at different temperatures. Heat transfers also accompany physical changes and chemical changes. The next three sections (4-4 to 4-6) show how quantitative measurements of heat transfers can be made, first for heating that results from a temperature difference and then for heating that accompanies a physical change.

4-4 Heat Capacity

The **heat capacity** of a sample of matter is *the quantity of energy required to increase the temperature of that sample by one degree.* Heat capacity depends on the mass of the sample and the substance of which it is made (or substances, if it is not pure). To determine the quantity of energy transferred by heating, we usually measure the change in temperature of a substance whose heat capacity is known. Often that substance is water.

4-4a Specific Heat Capacity

To make useful comparisons among samples of different substances with different masses, the **specific heat capacity** (which is sometimes just called *specific heat*) is defined as *the quantity of energy needed to increase the temperature of one gram of a substance by one degree Celsius*. For water at 15 °C, the specific heat capacity is 1.00 cal g^{-1} $°C^{-1}$ or 4.184 J g^{-1} $°C^{-1}$; for common window glass, it is only 0.8 J g^{-1} $°C^{-1}$. That is, it takes about five times as much heat transfer of energy to raise the temperature of a gram of water by 1 °C as it does for a gram of glass. Like density (← **Sec. 1-4d**), specific heat capacity is a property that can be used to distinguish one substance from another. It can also be used to distinguish a pure substance from a solution or mixture, because the specific heat capacity of a mixture will vary with the proportions of its components.

The notation J g^{-1} $°C^{-1}$ means that the units are joules divided by grams and divided by degrees Celsius; that is, J/(g °C). We will use negative exponents to show unambiguously which units are in the denominator whenever the denominator includes two or more units.

The specific heat capacity, *c*, of a substance can be determined experimentally by measuring the quantity of energy transferred to or from a known mass of the substance as its temperature rises or falls. We assume that there is no work transfer of energy to or from the sample and we treat the sample as a thermodynamic system, so $\Delta E = q$.

$$\text{Specific heat capacity} = \frac{\text{quantity of heat energy transfer}}{\text{sample mass} \times \text{temperature change}} = \frac{q}{m \times \Delta T} \qquad [4.2]$$

Suppose that for a 25.0-g sample of ethylene glycol, 90.7 J is required to change the temperature from 22.4 °C to 23.9 °C. (Ethylene glycol is used as a coolant in automobile engines.) Thus,

Ethylene glycol makes a good automobile-engine coolant because it has a relatively high heat capacity and is soluble in water.

$$\Delta T = (23.9\ °C - 22.4\ °C) = 1.5\ °C$$

From Equation 4.2, the specific heat capacity of ethylene glycol is

$$c = \frac{q}{m \times \Delta T} = \frac{90.7\ J}{25.0\ g \times 1.5\ °C} = 2.4\ J\ g^{-1}\ °C^{-1}$$

The specific heat capacities of many substances have been determined. A few values are listed in Table 4.1. Notice that water has one of the highest values. This is important

The high specific heat capacity of water helps to keep your body temperature constant. Water accounts for a large fraction of your body mass, and warming or cooling that water requires a lot of energy transfer.

Table 4.1 Representative Specific Heat Capacities

Substance	Specific Heat Capacity (J g^{-1} $°C^{-1}$)	Substance	Specific Heat Capacity (J g^{-1} $°C^{-1}$)
Elements		***Compounds***	
Aluminum, Al	0.902	Ammonia, $NH_3(\ell)$	4.70
Carbon (graphite), C	0.720	Water, liquid, $H_2O(\ell)$	4.184
Iron, Fe	0.451	Water, solid, $H_2O(s)$	2.06
Copper, Cu	0.385	Ethanol, $C_2H_5OH(\ell)$	2.46
Gold, Au	0.128	Ethylene glycol (antifreeze) $HOCH_2CH_2OH(\ell)$	2.42
Common solids		Propylene glycol (antifreeze) $HOCH_2CHOHCH_3(\ell)$	2.5
Wood	1.76		
Concrete	0.88	Carbon tetrachloride, $CCl_4(\ell)$	0.861
Glass	0.84	A chlorofluorocarbon, $CCl_2F_2(\ell)$	0.598
Granite	0.79		

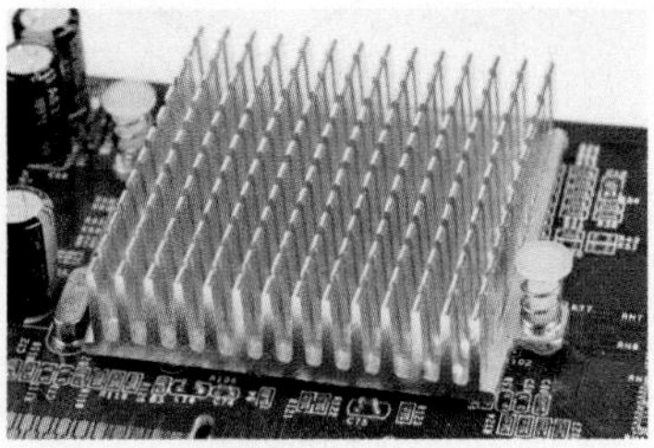
Popovici Ioan/Shutterstock.com

Because of its high specific heat capacity compared with other metals, aluminum is used in heat sinks for electronic devices.

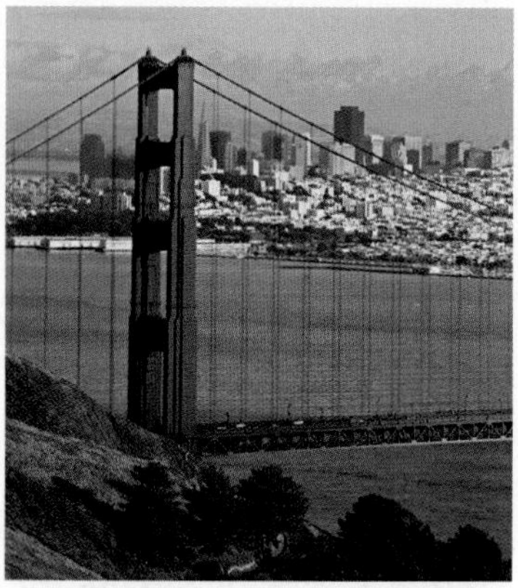
lucato/iStockphoto.com

Moderation of microclimate by water. In cities near bodies of water (such as San Francisco, shown here), summertime temperatures are lower near the waterfront than they are a few kilometers away. Wintertime temperatures are higher, so the climate is better in both seasons.

because a high specific heat capacity means that a great deal of energy must be transferred to a large body of water to raise its temperature by just one degree. Conversely, a lot of energy must be transferred away from the water before its temperature falls by one degree. Thus, a lake or ocean can store an enormous quantity of energy and thereby moderate local temperatures. This has a profound influence on weather near lakes or oceans.

When the specific heat capacity of a substance is known, you can calculate the temperature change that should occur when a given quantity of energy is transferred to or from a sample of known mass. More important, by measuring the temperature change and the mass of a substance, you can calculate q, the quantity of energy transferred to or from it by heating. For these calculations it is convenient to rearrange Equation 4.2 algebraically as

$$\Delta T = \frac{q}{c \times m} \quad \text{or} \quad q = c \times m \times \Delta T \qquad [4.2']$$

PROBLEM-SOLVING EXAMPLE 4.3

Using Specific Heat Capacity

If 100.0 g water is cooled from 25.3 °C to 16.9 °C, calculate the quantity of energy transferred from the water.

Result 3.5 kJ transferred from the water

Analyze Treat the water as a system. The quantity of energy is proportional to the specific heat capacity of water, the mass of water, and the change in temperature. The temperature change and mass of water are given.

Plan Look up the specific heat capacity of water in Table 4.1; use Equation 4.2′.

Execute

$$\Delta E = q = c \times m \times \Delta T$$
$$= 4.184 \text{ J g}^{-1}\,°\text{C}^{-1} \times 100.0 \text{ g} \times (16.9\ °\text{C} - 25.3\ °\text{C}) = -3.5 \times 10^3 \text{ J} = -3.5 \text{ kJ}$$

☑ **Reasonable Result Check** It requires about 4 J to heat 1 g water by 1 °C. In this case the temperature change is not quite 10 °C and we have 100 g water, so q should be about $4 \text{ J g}^{-1}\,°\text{C}^{-1} \times (-10\ °\text{C}) \times 100 \text{ g} = -4000 \text{ J} = -4 \text{ kJ}$, which it is. The sign is negative because energy is transferred out of the water as it cools.

PROBLEM-SOLVING PRACTICE 4.3

A piece of aluminum with a mass of 250. g is at an initial temperature of 5.0 °C. If 24.1 kJ is supplied to warm the Al, calculate its final temperature. Obtain the specific heat capacity of Al from Table 4.1.

CONCEPTUAL EXERCISE 4.4

Specific Heat Capacity and Temperature Change

Suppose you put two 50-mL beakers in a refrigerator so energy is transferred out of each sample at the same constant rate. If one beaker contains 10. g pulverized glass and one contains 10. g carbon (graphite), determine which beaker has the lower temperature after 3 min in the refrigerator.

4-4b Molar Heat Capacity

It is often useful to know the heat capacity of a sample in terms of the same number of particles instead of the same mass. For this purpose we use the **molar heat capacity**, symbol c_m. This is *the quantity of energy that must be transferred to increase the temperature of one mole of a substance by 1 °C*. The molar heat capacity is easily calculated from the specific heat capacity by using the molar mass of the substance. For example, the specific heat capacity of liquid ethanol is given in Table 4.1 as 2.46 J g^{-1} °C^{-1}. The molecular formula of ethanol is CH_3CH_2OH so its molar mass is 46.07 g/mol. The molar heat capacity is

$$c_m = \frac{2.46 \text{ J}}{\text{g °C}} \times \frac{46.07 \text{ g}}{\text{mol}} = 113 \text{ J mol}^{-1} \text{ °C}^{-1}$$

CONCEPTUAL EXERCISE 4.5

Molar Heat Capacity

Calculate the molar heat capacities of all the metals listed in Table 4.1. Compare these with the value just calculated for ethanol. Based on your results, suggest a way to predict the molar heat capacity of a metal. Can this same rule be applied to other kinds of substances?

As you should have found in Conceptual Exercise 4.5, molar heat capacities of metals are very similar. This can be explained if we consider what happens on the nanoscale when a metal is heated. The energy transferred by heating a solid makes the atoms vibrate more extensively about their average positions in the solid crystal lattice. Every metal consists of many, many atoms, all of the same kind and packed closely together; that is, the structures of all metals are very similar. As a consequence, the ways that the metal atoms can vibrate (and therefore the ways that their energies can be increased) are very similar. No matter what the metal, nearly the same quantity of energy must be transferred per metal atom to increase the temperature by one degree. The quantity of energy per mole is therefore very similar for all metals.

PROBLEM-SOLVING EXAMPLE 4.4

Direction of Energy Transfer

People sometimes drink hot tea to keep warm. Suppose that you drink a 250.-mL cup of tea that is at 65.0 °C. Estimate the quantity of energy transferred to your body and the surrounding air when the temperature of the tea drops to 37.0 °C (normal body temperature).

Result 29.3 kJ transferred out of the tea

Analyze The volume of tea and the initial and final temperatures are given. Reasonable assumptions can be made to obtain the mass and specific heat capacity.

Plan Treat the tea as the system. Assume the density of tea, which is mostly water, is 1.00 g/mL, so the tea has a mass of 250. g; also assume the specific heat capacity of tea is the same as that of water, 4.184 J g^{-1} °C^{-1}. Use Equation 4.2′.

Execute

$$\Delta E = q = c \times m \times \Delta T = 4.184 \text{ J g}^{-1} \text{ °C}^{-1} \times 250. \text{ g} \times (37.0 - 65.0) \text{ °C}$$
$$= -2.93 \times 10^4 \text{ J} = -29.3 \times 10^3 \text{ J} = -29.3 \text{ kJ}$$

The negative sign of the result indicates that 29.3 kJ is transferred *from* the tea (system) *to* the surroundings (you) as the temperature of the tea decreases.

☑ **Reasonable Result Check** Estimate the heat transfer as a bit more than (4 × 25 × 250) J = (100 × 250) J = 25,000 J = 25 kJ, which it is. The transfer is from the tea so q should be negative, which it is.

Usually the surroundings contain a great deal more matter than the system and hence have a much greater heat capacity. Consequently the change in temperature of the surroundings is often so small that it cannot be measured.

PROBLEM-SOLVING PRACTICE 4.4

Assume that the same cup of tea described in Problem-Solving Example 4.4 is warmed from 37.0 °C to 65.0 °C and there is no work done by the heating process. Calculate ΔE for this process.

PROBLEM-SOLVING EXAMPLE 4.5

Transfer of Energy by Heating

Suppose that you have 100. mL H_2O at 20.0 °C and you add to the water 55.0 g iron pellets that had been heated to 425 °C. Calculate the temperature of both the water and the iron when thermal equilibrium is reached.

Result $T_{final} = 42.7$ °C

Analyze Thermal equilibrium means that the water and the iron pellets will have the same final temperature, which is what we want to calculate. Consider the iron to be the system and the water to be the surroundings. The mass of iron and volume of water are given. Some assumptions are useful: Assume that there is no energy transfer to the glass beaker, the air, or anything else but the water; that no work is done; that no liquid water vaporizes; and that the density of water is 1.00 g/mL. With these assumptions, the quantity of energy transferred to the water and the quantity transferred from the iron are equal. They are opposite in algebraic sign because energy was transferred *from* the iron as its temperature dropped, and energy was transferred *to* the water to raise its temperature.

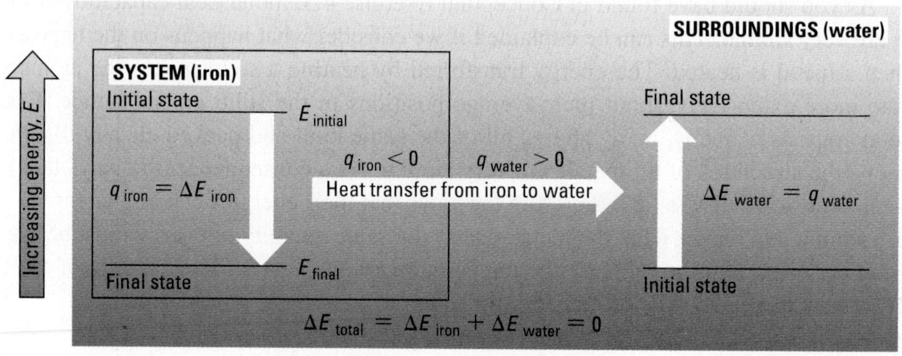
SURROUNDINGS (water)
SYSTEM (iron)
Increasing energy, E
Initial state
E initial
q iron < 0
q water > 0
q iron = ΔE iron
Heat transfer from iron to water
ΔE water = q water
Final state
E final
Final state
Initial state
ΔE total = ΔE iron + ΔE water = 0

The quantities of energy transferred have opposite signs because they take place from the iron (negative) to the water (positive).

Plan According to our analysis, $\Delta E_{water} = -\Delta E_{iron}$ and $q_{water} = -q_{iron}$. Use Equation 4.2′ for iron and for water. Look up specific heat capacities in Table 4.1.

Execute The mass of water is 100. mL × 1.00 g/mL = 100. g. $T_{initial}$ for the iron is 425 °C and $T_{initial}$ for the water is 20.0 °C.

$$q_{water} = -q_{iron}$$

$$c_{water} \times m_{water} \times \Delta T_{water} = -c_{iron} \times m_{iron} \times \Delta T_{iron}$$

$$(4.184 \text{ J g}^{-1}\text{ °C}^{-1})(100.\text{ g})(T_{final} - 20\text{ °C}) = -(0.451 \text{ J g}^{-1}\text{ °C}^{-1})(55.0\text{ g})(T_{final} - 425\text{ °C})$$

$$(418.4 \text{ J °C}^{-1})T_{final} - (8.368 \times 10^3 \text{ J}) = -(24.80 \text{ J °C}^{-1})T_{final} + (1.054 \times 10^4 \text{ J})$$

$$(443.2 \text{ J °C}^{-1})\ T_{final} = 1.891 \times 10^4 \text{ J}$$

Solving, we find $T_{final} = 42.7$ °C. The iron has cooled a lot ($\Delta T_{iron} = -382$ °C) and the water has warmed a little ($\Delta T_{water} = 22.7$ °C).

☑ **Reasonable Result Check** As a check, note that the final temperature must be between the two initial values, which it is. Also, don't be concerned by the fact that transferring the same quantity of energy resulted in two very different values of ΔT; this difference arises because the specific heat capacities and masses of iron and water are different. There is much less iron and its specific heat capacity is smaller, so its temperature changes much more than the temperature of the water.

PROBLEM-SOLVING PRACTICE 4.5

A 400.-g iron bar is heated in a flame and then immersed in 1000. g water in a beaker. The initial temperature of the water was 20.0 °C, and both the iron and the water are at 32.8 °C at the end of the experiment. Calculate the original temperature of the hot iron bar. (Make reasonable assumptions.)

4-5 Energy and Enthalpy

Using heat capacity we can account for transfers of energy between samples of matter as a result of temperature differences. But energy transfers also accompany physical or chemical changes, *even though there may be no change in temperature.* We will first consider the simpler case of physical change and then apply the same ideas to chemical changes.

4-5a Conservation of Energy and Changes of State

As shown in Figure 4.10, when ice is heated at a slow, constant rate from −50 °C to +50 °C the temperature does not increase at a constant rate. At 0 °C, the temperature remains constant, despite the fact that energy is still being transferred to the sample. As long as ice is melting, thermal energy must be continually supplied to overcome forces that hold the water molecules in their regularly spaced positions in the nanoscale structure of solid ice. Overcoming these forces raises the potential energy of the water molecules and therefore requires a transfer of energy into the system.

Melting a solid is an example of a **change of state** or **phase change**, *a physical process in which one state of matter is transformed into another.* During any phase change, the temperature remains constant, but energy must be continually transferred into the system or out of the system because the nanoscale particles have higher potential energy in one phase than in the other. As shown in Figure 4.10, the quantity of energy transferred during a phase change is significant.

Changes of state (between solid and liquid, liquid and gas, or solid and gas) are described in more detail in Sections 9-2, 9-3, and 9-4. Because the temperature remains constant during a change of state, melting points and boiling points can be measured relatively easily and used to identify substances (← Sec. 1-4c).

Another example of a phase change is boiling a liquid or condensing a gas. Consider a system that consists of water at its boiling temperature in a container with a balloon attached (Figure 4.11). The system is under a constant atmospheric pressure. If the water is heated, it will boil, the temperature will remain at 100 °C, and the steam produced by boiling the water will inflate the balloon (Figure 4.11b). If the heating stops, then the water will stop boiling, some of the steam will condense to liquid, and the volume of steam will decrease (Figure 4.11c). There will be heat transfer of energy to the

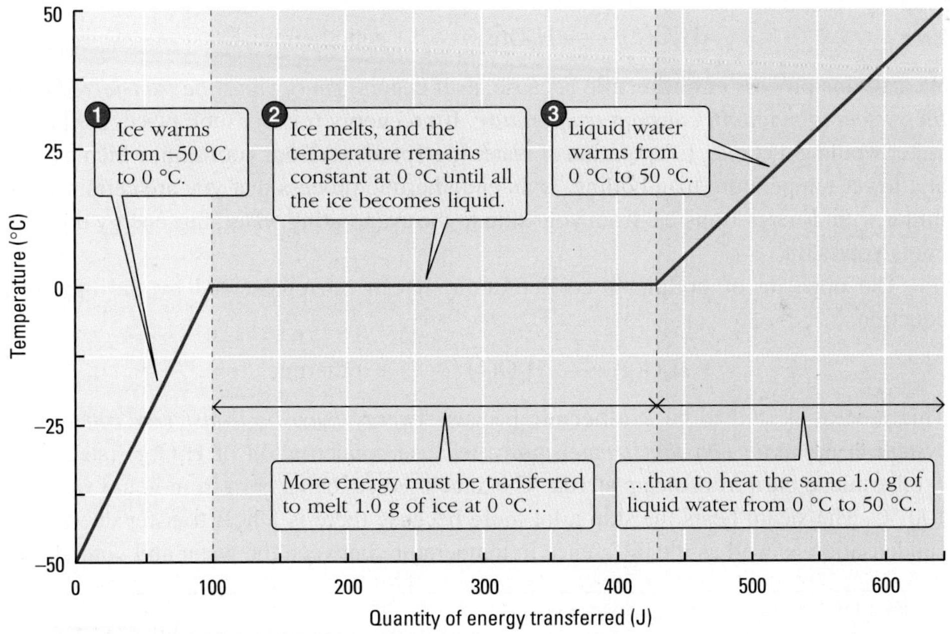

50
25
0
−25
−50
Temperature (°C)
1 Ice warms from −50 °C to 0 °C.
2 Ice melts, and the temperature remains constant at 0 °C until all the ice becomes liquid.
3 Liquid water warms from 0 °C to 50 °C.
More energy must be transferred to melt 1.0 g of ice at 0 °C...
...than to heat the same 1.0 g of liquid water from 0 °C to 50 °C.
0
100
200
300
400
500
600
Quantity of energy transferred (J)

Figure 4.10 Heating graph. When a 1.0-g sample of ice is heated at a constant rate, the temperature does not always increase at a constant rate.

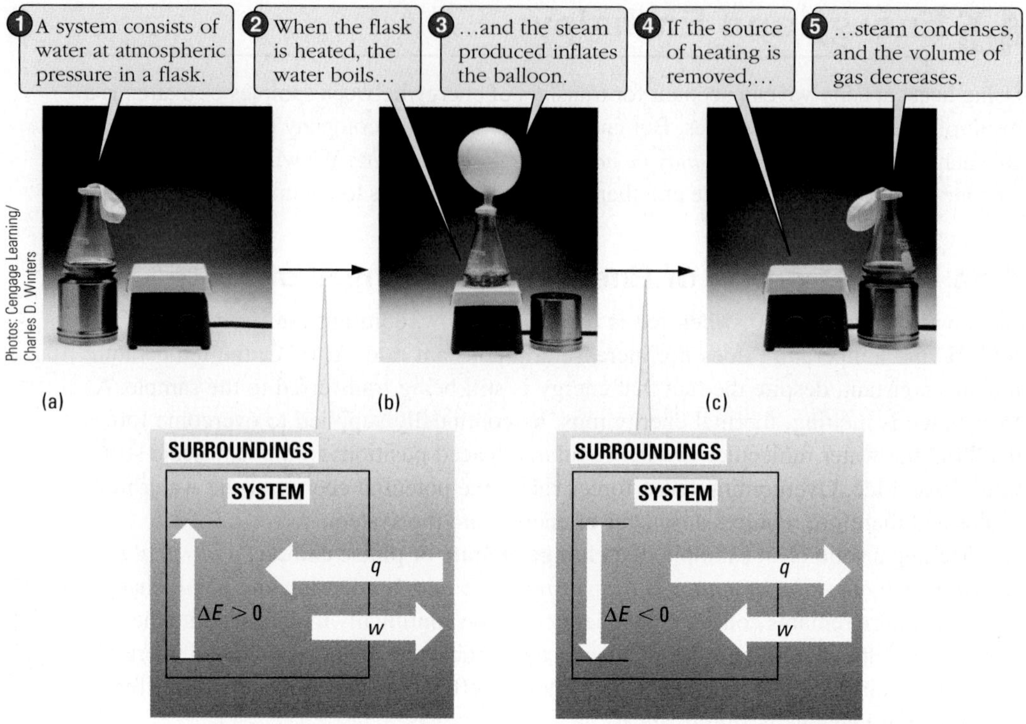

1 A system consists of water at atmospheric pressure in a flask.
2 When the flask is heated, the water boils...
3 ...and the steam produced inflates the balloon.
4 If the source of heating is removed,...
5 ...steam condenses, and the volume of gas decreases.
Photos: Cengage Learning/ Charles D. Winters
(a)
(b)
(c)
SURROUNDINGS
SYSTEM
q
w
ΔE > 0
SURROUNDINGS
SYSTEM
q
w
ΔE < 0

Figure 4.11 Boiling water at constant pressure. When water boils, the steam pushes against atmospheric pressure and does work on the surrounding atmosphere. (The balloon allows the expansion of the steam to be seen.) For any constant-pressure process, if a change in volume occurs, work is done.

surroundings. However, as long as steam is condensing to liquid water, the temperature will remain at 100 °C. In summary, transferring energy *into* the system produces more steam; transferring energy *out of* the system results in less steam. Both boiling and condensing occur at the same temperature—the boiling point.

The boiling process can be represented by the equation

$$H_2O(\ell) \longrightarrow H_2O(g) \qquad \text{endothermic}$$

Thermic **or** ***thermo*** **comes from the Greek word** ***thermé,*** **meaning "heat."** ***Endo*** **comes from the Greek word** ***endon,*** **meaning "within or inside."** ***Endothermic*** **therefore indicates transfer of energy** ***into*** **the system.**

We call this process **endothermic** because, as it occurs, *energy must be transferred* ***into*** *the system to maintain constant temperature*. If no energy transfer took place, the liquid water would get cooler. Evaporation of water (perspiration) from your skin, which occurs at a lower temperature than boiling, is an endothermic process that you are certainly familiar with. Energy transfers from your skin to the evaporating water; this energy transfer cools your skin.

The opposite of boiling is condensation. It can be represented by the opposite equation,

$$H_2O(g) \longrightarrow H_2O(\ell) \qquad \text{exothermic}$$

Exo **comes from the Greek word exō, meaning "out of."** ***Exothermic*** **indicates transfer of energy** ***out of*** **the system.**

This process is said to be **exothermic** because *energy must be transferred* ***out of*** *the system to maintain constant temperature*. Because condensation of $H_2O(g)$ (steam) is exothermic, a burn from steam at 100 °C is much worse than a burn from liquid water at 100 °C. The steam heats the skin a lot more because there is a heat transfer due to the condensation as well as the difference in temperature between the water and your skin.

Phase Change	Direction of Heat Energy Transfer	Sign of q	Type of Change
$H_2O(\ell) \rightarrow H_2O(g)$	Surroundings → system	Positive ($q > 0$)	*Endo*thermic
$H_2O(g) \rightarrow H_2O(\ell)$	System → surroundings	Negative ($q < 0$)	*Exo*thermic

The system in Figure 4.11 can be analyzed by using the law of conservation of energy, Equation 4.1, $\Delta E = q + w$. Vaporizing 1.0 g water requires heat transfer of 2260 J, so $q = 2260$ J (a positive value because the transfer is from the surroundings to the system). At the same time, the expansion of the steam pushes back the atmosphere, doing work. The quantity of work is more difficult to calculate, but it is clear that w must be negative, because the system does work on the surroundings. Therefore, the internal energy of the system is increased by the quantity of heating and decreased by the quantity of work done.

Now suppose that the heating is stopped and the direction of heat transfer is reversed. The water stops boiling and some of the steam condenses to liquid water. The balloon deflates and the atmosphere pushes back the steam. If 1.0 g steam condenses, then $q = -2260$ J. Because the surrounding atmosphere pushes on the system, the surroundings have done work on the system, which makes w positive. As long as steam is condensing to liquid, the temperature remains at 100 °C. The internal energy of the system is increased by the work done on it and decreased by the heat transfer of energy to the surroundings.

The device described here is a crude example of a steam engine. Burning fuel boils water, and the steam does work. In a real steam engine the steam would drive a piston and then be allowed to escape, providing a means for the system to continually do work on its surroundings. Systems that convert heat into work are called heat engines. Another example is an internal combustion engine in an automobile, which converts heat from the combustion of fuel into work to move the car.

4-5b Enthalpy: Heat Transfer at Constant Pressure

In the previous section it was clear that work was done. When the balloon's volume increased, the balloon pushed aside air that had occupied the space the gas in the balloon expanded to fill. Work is done when a force moves something through some distance. If the flask containing the boiling water had been sealed with a solid stopper, nothing would have moved and no work would have been done. Therefore, in a closed container where the system's volume is constant, $w = 0$, and

$$\Delta E = q + w = q + 0 = q_V$$

The subscript V indicates constant volume; that is, q_V is the heat transfer into a constant-volume system. This means that ***if a process is carried out in a closed container, the measured heat transfer is*** **ΔE**.

In plants, animals, laboratories, and the environment, physical processes and chemical reactions seldom take place in closed containers. Instead they are carried out in contact with the atmosphere. For example, the vaporization of water shown in Figure 4.11 took place under conditions of constant atmospheric pressure, and the expanding steam had to push back the atmosphere. In such a case,

$$\Delta E = q_P + w_{\text{atm}}$$

That is, ΔE differs from the heat transfer at constant pressure, q_P, by the work done to push back the atmosphere, w_{atm}.

To see how much work is required, consider the idealized system for vaporization of water shown nearby. There is a cylinder with a weightless piston. (The purpose of the piston is to distinguish the water system from the surrounding air.) When some of the water in the bottom of the cylinder boils, the volume of the system increases. The piston and the atmosphere are forced upward. The system (water and vapor) does work to raise the surrounding air. The work can be calculated as $w_{\text{atm}} = -(\text{force} \times \text{distance}) = -(F \times d)$. (The negative sign indicates that the system is doing work on the surroundings.) The distance the piston moves is $d = h_2 - h_1$. The force can be calculated from the pressure, P, which is defined as force per unit area, A. Because $P = F/A$, the force is $F = P \times A$, and the work is $w_{\text{atm}} = -F \times d = -P \times A \times d$. The change in the volume of the system, ΔV, is the volume of the cylinder through which the piston moves. This is calculated as the area of the base times the height, or $\Delta V = A \times d$. Thus, the work is $w_{\text{atm}} = -P \times A \times d = -P\Delta V$. The work of pushing back the atmosphere is always equal to the atmospheric pressure times the change in volume of the system.

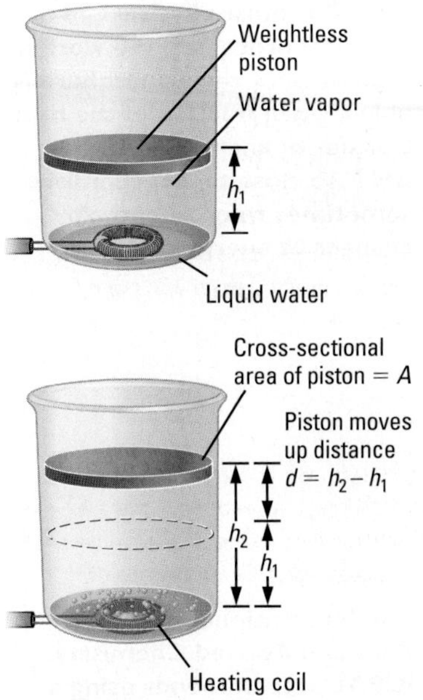
Weightless piston
Water vapor
h_1
Liquid water
Cross-sectional area of piston = A
Piston moves up distance $d = h_2 - h_1$
h_2
h_1
Heating coil

Vaporization of water. When a sample of water boils at constant pressure, energy must be supplied to expand the steam against atmospheric pressure.

Thus, the law of conservation of energy for a constant-pressure process is

$$\Delta E = q_P + w_{atm} \qquad \text{or} \qquad q_P = \Delta E - w_{atm} = \Delta E + P\Delta V$$

This equation says that when we carry out reactions in beakers or other containers open to the atmosphere, the heat transfer differs from the energy change by an easily calculated term, $P\Delta V$. Therefore it is convenient to use q_P to characterize energy transfers in typical chemical and physical processes. *The quantity of thermal energy transferred into a system at constant pressure,* q_P, is called the **enthalpy change (ΔH)** of the system. Thus,

Because it is equal to the heat transfer of energy at constant pressure, and because most chemical reactions are carried out at atmospheric (constant) pressure, the enthalpy change for a process is often called the *heat of that process.* For example, the enthalpy change for vaporization or boiling is also called the heat of vaporization.

$$\Delta H = q_P = \Delta E + P\Delta V$$

ΔH accounts for all the energy transferred except the quantity that does the work of pushing back the atmosphere. For processes that do not involve gases, w_{atm} is very small. Even when gases are involved, w_{atm} is usually much smaller than q_P. That is, ΔH is closely related to the change in the internal energy of the system but is slightly different in magnitude. ***Whenever heat transfer is measured at constant pressure, it is ΔH that is determined.***

PROBLEM-SOLVING EXAMPLE 4.6

Changes of State, ΔH, and ΔE

The pressure unit bar is very close to the standard atmosphere: 1 bar = 0.98692 atm = 1 × 10^5 kg m^{-1} s^{-2}. 1 bar × 1 L = 1 × 10^5 kg m^{-1} s^{-2} × 1 × 10^{-3} m^3 = 100 J. (Pressure units are discussed further in Section 8-1.)

Methanol, CH_3OH, boils at 65.0 °C. When 5.0 g methanol boils at 1 bar, the volume of $CH_3OH(g)$ is 4.32 L greater than the volume of the liquid. The heat transfer is 5865 J, and the process is endothermic. Calculate ΔH and ΔE, given that the units 1 L × 1 bar = 100.0 J.

Result ΔH = 5865 J; ΔE = 5433 J

Analyze The volume change and heat transfer are given. The process takes place at constant pressure. Because energy is transferred to the system, ΔH is positive.

Plan At constant pressure, $\Delta H = q_P$. Use $\Delta H = \Delta E + P\Delta V$ to calculate ΔE. Use the unit conversion 1 L × 1 bar = 100.0 J to obtain a result in joules.

Execute $\Delta H = q_P = 5865$ J.

$$P\Delta V = 1 \text{ bar} \times 4.32 \text{ L} = 4.32 \text{ L bar} = 4.32 \times 100.0 \text{ J} = 432 \text{ J.}$$

Rearrange the equation $\Delta H = \Delta E + P\Delta V$ and calculate ΔE.

$$\Delta E = \Delta H - P\Delta V = 5865 \text{ J} - 432 \text{ J} = 5433 \text{ J}$$

☑ **Reasonable Result Check** Boiling is an endothermic process, so ΔH must be positive. Because the system did work on the surroundings, the change in internal energy must be less than the enthalpy change. The result is reasonable.

The results of this example show that ΔE differs by less than 10% from ΔH—that is, by 432 J out of 5865 J, which is 7.4%. For most physical and chemical processes, the work of pushing back the atmosphere is only a small fraction of the heat transfer of energy. Because ΔH is so close to ΔE, chemists sometimes refer to enthalpy changes as energy changes.

PROBLEM-SOLVING PRACTICE 4.6

When potassium melts at atmospheric pressure, the heat transfer is 14.6 cal/g. The density of liquid potassium at its melting point is 0.82 g/mL, and that of solid potassium is 0.86 g/mL. Given that a volume change of 1.00 mL at atmospheric pressure corresponds to 0.10 J, calculate ΔH and ΔE for melting 1.00 g potassium.

4-5c Melting (Fusion) and Freezing

The *quantity of heat energy transfer to a solid as it melts at constant pressure* is called the **fusion enthalpy ($\Delta_{fus}H$)**. The subscript "fus" indicates the process (in this case fusion or melting) to which the enthalpy change applies. The symbol $\Delta_{fus}H$ can be read as "the change upon fusion (melting) in enthalpy." For ice the fusion enthalpy is 333 J/g at 0 °C. This quantity of energy could raise the temperature of 1.00 g iron from 0 °C to 738 °C (red hot), or it could melt 0.50 g ice and heat the liquid water from 0 °C to 80 °C.

The International Union of Pure and Applied Chemistry, IUPAC, recommends using a subscript on Δ to indicate the process to which Δ applies. You will also see the notation ΔH_{fusion}, or ΔH_{fus}, which means the same thing.

The opposite of melting is freezing. When water freezes, the quantity of energy transferred is the same as when the same mass of water melts, but energy transfers in the opposite direction—from the system to the surroundings. Thus, under the same conditions of temperature and pressure, $\Delta_{fus}H = -\Delta_{freezing}H$.

TONY RANZE/AFP/Getty Images

Spraying with water prevents crops from freezing. As water freezes heat transfer to the oranges (surroundings) prevents them from freezing.

EXERCISE 4.6

Melting Ice and Heating Iron

Do a calculation to verify the statement that the quantity of energy required to melt 1.00 g ice is sufficient to heat a 1.00-g sample of iron from 0 °C to 738 °C.

4-5d Vaporization and Condensation

The *quantity of heat energy transfer at constant pressure to convert a liquid to vapor (gas)* is called the **vaporization enthalpy ($\Delta_{vap}H$)**. For water it is 2260. J/g at 100 °C. This is considerably larger than the fusion enthalpy, because the water molecules become completely separated during the transition from liquid to vapor. As they separate, a great deal of energy is required to overcome attractions among the water molecules. Therefore, the potential energy of the vapor is considerably higher than that of the liquid. Although 333 J can melt 1.00 g ice at 0 °C, it will boil only 0.147 g water at 100 °C.

$$m(H_2O) = 333\ J \times \frac{1.00\ g\ H_2O}{2260\ J} = 0.147\ g\ H_2O\ \text{vaporized}$$

The opposite of vaporization is condensation. Therefore, under the same conditions of temperature and pressure, $\Delta_{vap}H = -\Delta_{condensation}H$.

CONCEPTUAL EXERCISE 4.7

Heating and Cooling Graphs

(a) Assume that a 1.0-g sample of ice at −5 °C is heated at a uniform rate until the temperature is 105 °C. Draw a graph like the one in Figure 4.10 to show how temperature varies with energy transferred. Your graph should be drawn to approximately the correct scale.

(b) Assume that a 0.50-g sample of water is cooled [at the same uniform rate as the heating in part (a)] from 105 °C to −5 °C. Draw a cooling curve to show how temperature varies with energy transferred. Your graph should be drawn to the same scale as in part (a).

Evaporation of water cools you when you perspire. If you work up a real sweat, then lots of water evaporates, producing a large cooling effect. People who exercise vigorously in cool weather need to carry a warm sweatshirt or jacket to prevent this effect from causing a chill when they stop.

EXERCISE 4.8

Changes of State

Assume you have 1 cup of ice (237 g) at 0.0 °C. Calculate how much heating is required to melt the ice, warm the resulting water to 100.0 °C, and then boil the water to vapor at 100.0 °C. (*Hint:* Do three separate calculations and then add the results.)

4-5e State Functions and Path Independence

Both energy and enthalpy are *state functions.* A **state function** is *a property whose value is invariably the same if a system is in the same state.* A system's state is defined by its temperature, pressure, volume, mass, and composition. For the same initial and final states, ***a change in a state function does not depend on the path by which the system changes from one state to another.*** Returning to the bank-account analogy (← **Sec. 4-3a**), your bank balance is independent of the path by which you change it. If you have \$1000 in the bank (initial state) and withdraw \$100, your balance will go down to \$900 (final state) and $\Delta B = -\$100$. If instead you had deposited \$500 and withdrawn \$600 you would have achieved the same change of $\Delta B = -\$100$ by a different pathway, and your final balance would still be \$900. Figure 4.12 shows three ways the temperature of the same sample of water could change from 25 °C to 37 °C. Because energy is a state function, the energy change is identical in all three cases.

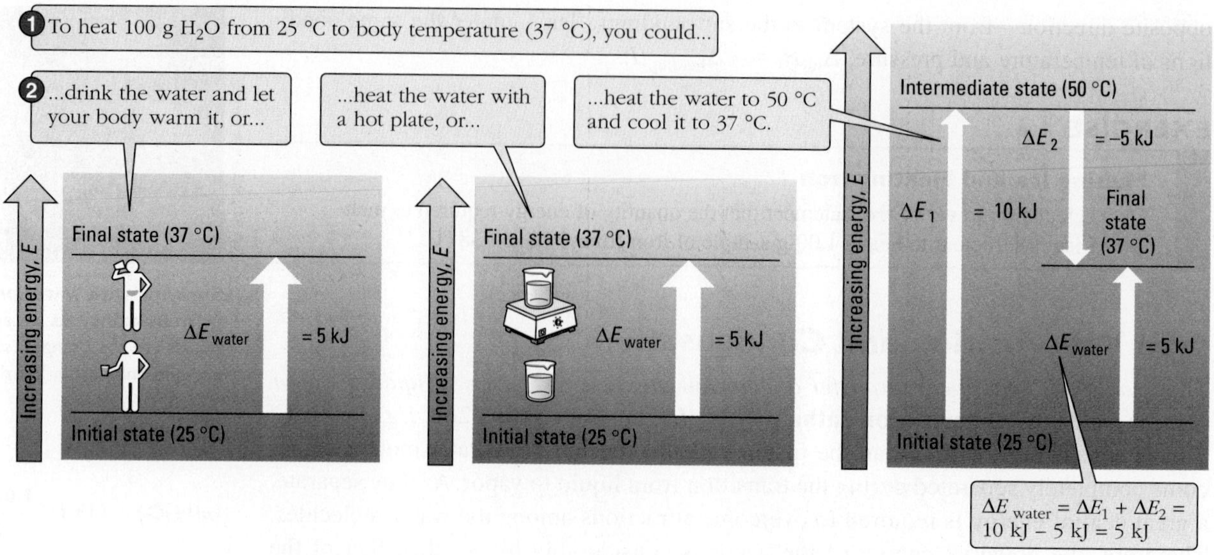

1 To heat 100 g H2O from 25 °C to body temperature (37 °C), you could...
2 ...drink the water and let your body warm it, or...
...heat the water with a hot plate, or...
...heat the water to 50 °C and cool it to 37 °C.
Increasing energy, E
Final state (37 °C)
ΔE water = 5 kJ
Initial state (25 °C)
Increasing energy, E
Final state (37 °C)
ΔE water = 5 kJ
Initial state (25 °C)
Increasing energy, E
Intermediate state (50 °C)
ΔE2 = −5 kJ
ΔE1 = 10 kJ
Final state (37 °C)
ΔE water = 5 kJ
Initial state (25 °C)
ΔE water = ΔE1 + ΔE2 = 10 kJ − 5 kJ = 5 kJ

Figure 4.12 Energy change is independent of path. In all three cases, the energy change of the water is the same.

The fact that changes in a state function are independent of the sequence of events by which change occurs is important, because it allows us to apply laboratory measurements to real-life situations. For example, if you measure in the laboratory the heat transfer when 1.0 g glucose (dextrose sugar) burns in exactly the amount of oxygen required to convert it to carbon dioxide and water, you will find that 15.5 kJ is transferred to the surroundings. When you eat something that contains 1.0 g glucose and your body metabolizes the glucose (producing the same products at the same temperature and pressure), there is the same transfer of energy. Thus, laboratory measurements can be used to determine how much energy you can get from a given quantity of food, which is the basis for the caloric values listed on labels.

4-6 Reaction Enthalpies for Chemical Reactions

Like phase changes, chemical reactions can be exothermic or endothermic, but chemical reactions usually involve much larger energy transfers than do phase changes. Indeed, a significant temperature change is one piece of evidence that a chemical reaction has taken place. The large energy transfers that occur during chemical reactions result from breaking and forming chemical bonds as reactants are converted into products. These energy transfers have important applications in living systems, in industrial processes, in heating or cooling your home, and in many other situations.

Hydrogen is an excellent fuel. It produces very little pollution when it burns in air, and its reaction with oxygen to form water is highly exothermic. It was used as a fuel in the Space Shuttle, for example. To indicate the heat transfer when hydrogen burns, we write a **thermochemical expression**, *a balanced chemical equation (including states of matter) together with the corresponding value of the reaction enthalpy change.* A thermochemical expression for reaction of hydrogen and oxygen to form water is

$$2\,H_2(g) + O_2(g) \longrightarrow 2\,H_2O(g) \qquad \Delta_r H^\circ = -483.6 \text{ kJ/mol} \quad (25\ ^\circ\text{C}) \qquad [4.3]$$

The **standard reaction enthalpy, ($\Delta_r H^\circ$)**, is the *enthalpy of pure, unmixed reaction products minus the enthalpy of pure, unmixed reactants at the standard pressure of 1 bar*

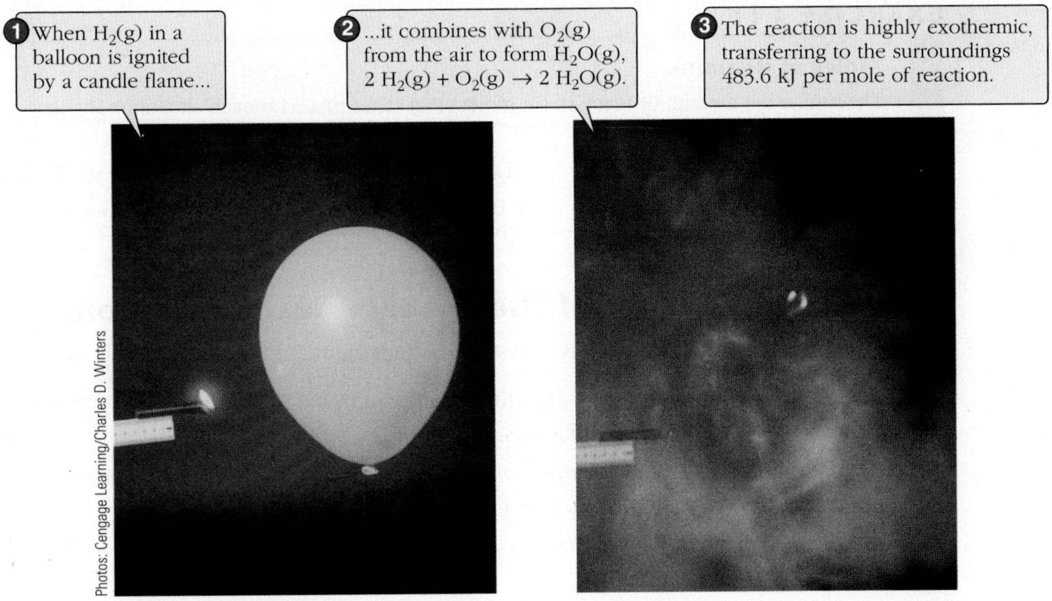
1 When $H_2(g)$ in a balloon is ignited by a candle flame...
2 ...it combines with $O_2(g)$ from the air to form $H_2O(g)$, $2\ H_2(g) + O_2(g) \rightarrow 2\ H_2O(g)$.
3 The reaction is highly exothermic, transferring to the surroundings 483.6 kJ per mole of reaction.
Photos: Cengage Learning/Charles D. Winters

Combustion of hydrogen is highly exothermic.

and a specified temperature. In this case the reactants are two molecules of hydrogen gas and one molecule of oxygen gas; the product is two molecules of water in the gas phase. In $\Delta_r H°$ the subscript "r" stands for "reaction" and the superscript "°" means "standard" so the symbol $\Delta_r H°$ is pronounced "delta-reaction-aitch-standard."

In 1982 the International Union of Pure and Applied Chemistry chose a pressure of 1 bar as the standard for tabulating thermochemical data.

Because the value of the reaction enthalpy depends on the pressure at which the process is carried out, standard reaction enthalpies are reported at the same standard pressure, 1 bar. (The bar is a unit of pressure that is very close to the pressure of Earth's atmosphere at sea level; you may have heard this unit used in a weather report.) The value of the reaction enthalpy also varies slightly with temperature, so temperature should be specified. For thermochemical expressions in this book, the temperature can be assumed to be 25 °C, unless some other temperature is given.

In Thermochemical Expression 4.3 the units for $\Delta_r H°$ are kilojoules per mole. The "per mole" means per mole of reaction; a **mole of reaction** *means that the process specified occurs 6.022×10^{23} times—that is, a mole of reaction processes (specified by the equation) takes place.* For this equation, a mole of reaction occurs when 2 mol $H_2(g)$ reacts with 1 mol $O_2(g)$ to form 2 mol $H_2O(g)$. After 1 mol reaction has occurred, 483.6 kJ is transferred to the surroundings. When the product water molecules are first formed, their temperature is much higher than 25 °C. For them to cool to the specified temperature of 25 °C, energy must be transferred to the surroundings. This energy transfer from the hotter reaction products to the cooler surroundings follows the same principles regarding heating and cooling that we discussed earlier (← **Sec. 4-2b**).

Talking about a mole of reaction is similar to talking about a dozen trips to the store. Counting units such as mole or dozen can be applied to processes (reactions or trips) as well as to physical entities (atoms or eggs).

CONCEPTUAL **EXERCISE 4.9**

Interpreting Thermochemical Expressions

What part of Thermochemical Expression 4.3 indicates that energy is transferred to the surroundings from the system when the combustion of hydrogen occurs?

CONCEPTUAL **EXERCISE 4.10**

Thermochemical Expressions

Explain why it is essential to specify the state (s, ℓ, or g) of each reactant and each product in a thermochemical expression.

EXERCISE 4.11

Moles of Reaction

Determine the amount of reaction (in moles) that takes place for each situation. The thermochemical expression is

$$N_2(g) + 3\ H_2(g) \longrightarrow 2\ NH_3(g) \qquad \Delta_r H° = -92.22\ kJ$$

(a) 1 mol NH_3 is formed. (b) 0.375 mol H_2 reacts. (c) 4.42 g N_2 reacts.

4-6a Characteristics of Thermochemical Expressions

All thermochemical expressions have these important characteristics:

- The *sign of* $\Delta_r H°$ indicates the direction of energy transfer; if the reaction process is reversed, the sign of $\Delta_r H°$ changes.
- The magnitude of $\Delta_r H°$ is the *difference in enthalpy when reactants change to products*; it depends on the reactant *state* and product *state*.
- The *quantity of energy transferred is proportional to the amount of reaction* that occurs.

This is similar to water falling from top to bottom of a waterfall. The decrease in potential energy from the top to the bottom of the waterfall is exactly equal to the increase in potential energy from the bottom to the top. The signs are opposite because in one case potential energy is transferred *from* the water and in the other case it is transferred to the water.

Sign of $\Delta_r H°$ The process in Thermochemical Expression 4.3 is exothermic, because $\Delta_r H°$ is negative. A negative $\Delta_r H°$ means that the enthalpy of the products is lower than the enthalpy of the reactants. Less enthalpy for the products means the extra enthalpy must go somewhere: It shows up in the surroundings. Formation of 2 mol water vapor involves heat transfer 483.6 kJ from the reacting chemicals to the surroundings. If 2 mol water vapor decomposes to hydrogen and oxygen (the reverse process), the magnitude of $\Delta_r H°$ is the same, but the sign is opposite, indicating heat transfer of energy from the surroundings to the system. These relationships are diagrammed in Figure 4.13.

$$2\ H_2O(g) \longrightarrow 2\ H_2(g) + O_2(g) \qquad \Delta_r H° = +483.6\ kJ/mol \quad (25\ °C) \qquad [4.4]$$

The reverse of an exothermic process is endothermic. The magnitude of the energy transfer is the same, but the direction of transfer is opposite.

States If the product of the reaction in Thermochemical Expression 4.3 is liquid water instead of water vapor, the product state is different and the magnitude of $\Delta_r H°$ is also different:

$$2\ H_2(g) + O_2(g) \longrightarrow 2\ H_2O(\ell) \qquad \Delta_r H° = -571.7\ kJ/mol \quad (25\ °C) \qquad [4.5]$$

Our discussion of phase changes (← **Secs. 4-5c and 4-5d**) showed that an enthalpy change occurs when a substance changes state. Vaporizing 2 mol $H_2O(\ell)$ requires 88.0 kJ.

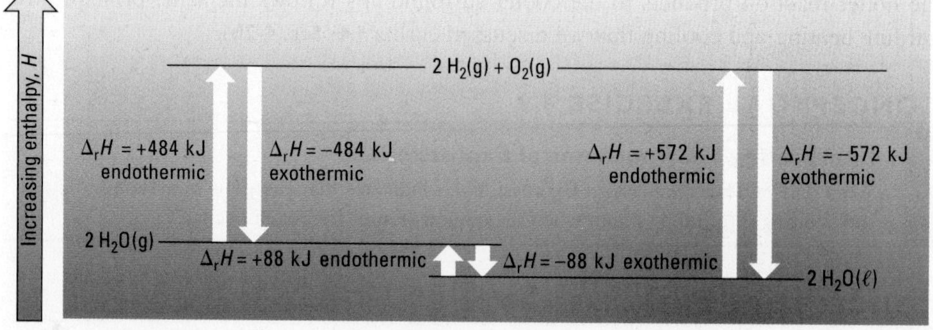
Increasing enthalpy, H
2 H2(g) + O2(g)
ΔrH = +484 kJ endothermic
ΔrH = −484 kJ exothermic
ΔrH = +572 kJ endothermic
ΔrH = −572 kJ exothermic
2 H2O(g)
ΔrH = +88 kJ endothermic
ΔrH = −88 kJ exothermic
2 H2O(ℓ)

Figure 4.13 Enthalpy diagram. Water vapor [2 mol $H_2O(g)$], liquid water [2 mol $H_2O(\ell)$], and an equivalent quantity of hydrogen and oxygen gases [2 mol $H_2(g)$ and 1 mol $O_2(g)$] all have different enthalpy values. Because 2 $H_2O(\ell)$ has less enthalpy than 2 $H_2O(g)$, $\Delta_r H°$ is larger (more negative) when $H_2O(\ell)$ forms than when $H_2O(g)$ forms.

Forming 2 mol $H_2O(\ell)$ from 2 $H_2(g)$ and $O_2(g)$ is 571.6 kJ − 483.6 kJ = 88.0 kJ more exothermic than is forming 2 mol $H_2O(g)$. Figure 4.13 shows the relationships among these quantities. The enthalpy of the reactants [2 $H_2(g)$ and $O_2(g)$] is greater than that of the product [2 $H_2O(\ell)$]. Because the system has less enthalpy after the reaction than before, the law of conservation of energy requires that 483.6 kJ must be transferred *to* the surroundings as the reaction takes place. $H_2O(\ell)$ has even less enthalpy than $H_2O(g)$, so when $H_2O(\ell)$ is formed, even more energy, 571.6 kJ, must be transferred to the surroundings.

Quantity of Energy Is Proportional to Amount of Reaction Thermochemical expressions obey the rules of stoichiometry (← **Sec. 3-6**). The more reaction there is, the more energy is transferred. We can calculate how much heat transfer occurs from the amount of a reactant that is consumed or the amount of a product that is formed because either of these quantities determines the amount (in moles) of reaction that occurs.

The direct proportionality between quantity of reaction and quantity of heat transfer is in line with your everyday experience. Burning twice as much charcoal in a grill produces twice as much heating.

PROBLEM-SOLVING EXAMPLE 4.7

Calculating Heat Transfer from $\Delta_r H°$

Powerful storms often depend on energy transfers involving condensation of water vapor in the air. Calculate the energy transferred to the surroundings when water vapor condenses at 25 °C to give rain in a thunderstorm. Suppose that one inch of rain falls over one square mile of ground, so that 6.6×10^{10} mL has fallen. The thermochemical expression for condensation of water at 25 °C is

$$H_2O(g) \longrightarrow H_2O(\ell) \qquad \Delta_r H° = -44.0 \text{ kJ/mol}$$

Result 1.6×10^{11} kJ

Analyze The thermochemical expression indicates that 44.0 kJ is transferred to the surroundings per mole of reaction. Because the coefficient of $H_2O(\ell)$ in the equation is 1, 1 mol reaction corresponds with condensation of 1 mol H_2O. The amount of water that condenses can be calculated from the mass, which can be calculated from the given volume using the density.

Plan Look up the density of water, $d_{H_2O(\ell)} = 1.0$ g/mL. Set up a series of proportionality factors to obtain the result.

Execute Energy transferred =

$$6.6 \times 10^{10} \text{ mL} \times \frac{1.0 \text{ g}}{1 \text{ mL}} \times \frac{1 \text{ mol } H_2O}{18 \text{ g}} \times \frac{1 \text{ mol reaction}}{1 \text{ mol } H_2O} \times \frac{44.0 \text{ kJ}}{\text{mol reaction}}$$
$$= 1.6 \times 10^{11} \text{ kJ}$$

☑ **Reasonable Result Check** The quantity of water is about 10^{11} g. The energy transfer is 44 kJ for 1 mol (18 g) water. Since 44 is about twice 18, this is about 2 kJ/g. Therefore, the number of kJ transferred should be about twice the number of grams, or about 2×10^{11} kJ, and it is.

Agronomists and meteorologists measure quantities of rainwater in units of acre-feet; an acre-foot is enough water to cover an acre of land to a depth of one foot.

The explosion of 1000 tons of dynamite is equivalent to 4.2×10^9 kJ. Thus, the energy transferred by our hypothetical thunderstorm is about the same as that released when 38,000 tons of dynamite explodes! No wonder storms can be so destructive.

PROBLEM-SOLVING PRACTICE 4.7

The reaction enthalpy for sublimation of 1 mol solid iodine at 25 °C and 1 bar is 62.4 kJ. (Sublimation means changing directly from solid to gas.)

$$I_2(s) \longrightarrow I_2(g) \qquad \Delta_r H° = 62.4 \text{ kJ}$$

(a) Calculate the heat transfer required to vaporize 10.0 g solid iodine.
(b) If 3.42 g iodine vapor changes to solid iodine, calculate the heat transfer.
(c) Is the process in part (b) exothermic or endothermic?

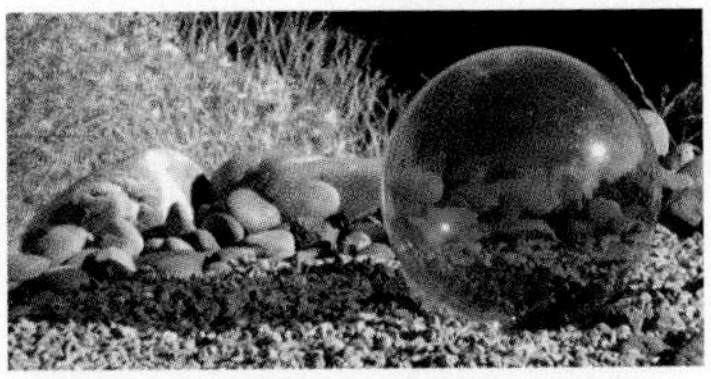
Richard Ramette

Iodine "thermometer." A glass sphere containing a few iodine crystals rests on the ground in desert sunshine. The higher the temperature, the more iodine sublimes and the darker the beautiful violet color becomes.

Like the quantities of products that can be produced by a reaction, the energy transfer associated with a reaction is determined by the amount of the *limiting reactant* that reacts (← **Sec. 3-7**). If there is a limiting reactant, its amount must be used to determine the amount of reaction and the heat energy transfer. When reactions take place in solution, the concentration (molarity) and volume of the solution can be used to calculate the amount of a reactant (← **Sec. 3-11**) and hence the amount of reaction that occurs.

EXERCISE 4.12

Reaction Enthalpy and Stoichiometry

Calculate the heat transfer when 0.5000 mol H_2(g) is mixed with 0.5000 mol O_2(g) and the mixture is ignited to form water vapor at 25 °C.

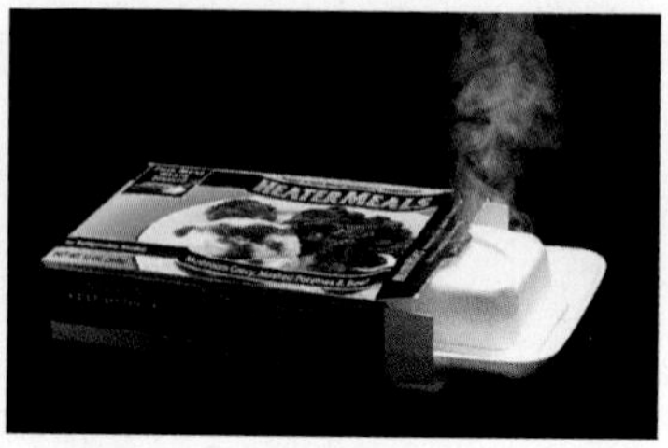
HEATER MEALS
Cengage Learning/Charles D. Winters

PROBLEM-SOLVING EXAMPLE 4.8

Calculating Energy Transferred

A "Heater Meal" has a flameless heater; you just pour water into it and it heats your dinner (see photo nearby). "Heater Meals" use the reaction

$$\text{Mg(s)} + 2\,\text{H}_2\text{O}(\ell) \longrightarrow \text{Mg(OH)}_2\text{(s)} + 2\,\text{H}_2\text{(g)} \qquad \Delta_r H^\circ = -352.88 \text{ kJ/mol}$$

Suppose the "Heater Meal" contains 1.50 g Mg and you pour in 5.00 mL H_2O (5.00 g H_2O). Calculate the heat transfer due to the reaction.

Result −21.8 kJ

Analyze The masses of water and magnesium are given. From them we can calculate the amount of each reactant and the amount of reaction when each reactant is consumed. Whichever reactant produces the smaller amount of reaction is limiting and determines the heat transfer.

Plan Use molar masses of Mg and H_2O to calculate the amount of each. Use coefficients in the balanced equation to determine the amount of reaction each reactant could provide. Use the smaller amount of reaction to calculate the heat transfer.

Execute

$$n(\text{rxn, Mg}) = 1.50\text{ g} \times \frac{1\text{ mol Mg}}{24.30\text{ g}} \times \frac{1\text{ mol reaction}}{1\text{ mol Mg}} = 0.06173\text{ mol reaction}$$

$$n(\text{rxn, H}_2\text{O}) = 5.00\text{ g} \times \frac{1\text{ mol H}_2\text{O}}{18.01\text{ g}} \times \frac{1\text{ mol reaction}}{2\text{ mol H}_2\text{O}} = 0.1388\text{ mol reaction}$$

Because it produces a smaller amount of reaction, Mg is limiting.

$$\text{Energy transferred} = 0.06173\text{ mol reaction} \times \frac{-352.88\text{ kJ}}{1\text{ mol reaction}} = -21.8\text{ kJ}$$

☑ **Reasonable Result Check** The amount of Mg is less than 0.1 mol (1.5/24 is less than 0.1). The amount of water is more than 0.25 mol (5/18 is greater than 5/20, which is 0.25). Even though 2 mol water is required for 1 mol Mg, there is more than twice the amount of water, so Mg is reasonable as the limiting reactant. It takes about 4 J to heat a gram of water by 1 °C. About 20 kJ should be able to heat 1000 g water by 5 °C. Given that a meal might contain only about a pound (454 g) and the heat capacity would be less than for water, the meal could be heated well above room temperature, so the energy transfer is reasonable.

PROBLEM-SOLVING PRACTICE 4.8

When aqueous solutions of silver nitrate and sodium chloride are mixed, silver chloride precipitates. The reaction enthalpy for this reaction is −69 kJ/mol. Calculate the energy transfer when 25.0 mL of 0.873-M silver nitrate solution reacts with 15.0 mL of 1.00-M sodium chloride solution.

4-7 Where Does the Energy Come From?

During melting or boiling, nanoscale particles (atoms, molecules, or ions) that attract each other are separated, which increases their potential energy. This requires transfer of energy from the surroundings to enable the particles to overcome their mutual attractions. During a chemical reaction, chemical compounds are created or broken down; that is, reactant molecules are converted into product molecules. Atoms in molecules are held together by chemical bonds. When existing chemical bonds are broken and new chemical bonds are formed, atomic nuclei and electrons move farther apart or closer together, and their energy increases or decreases. These energy differences are usually much greater than those for phase changes.

Consider the reaction of hydrogen gas with chlorine gas to form hydrogen chloride gas.

$$H_2(g) + Cl_2(g) \longrightarrow 2\ HCl(g) \qquad [4.6]$$

When this reaction occurs, the two hydrogen atoms in an H_2 molecule separate, as do the two chlorine atoms in a Cl_2 molecule. In the product the atoms are combined in a different way—as two HCl molecules. We can think of this change as involving two steps:

$$H_2(g) + Cl_2(g) \longrightarrow 2\ H(g) + 2\ Cl(g) \longrightarrow 2\ HCl(g)$$

The first step is to break all bonds in the reactant H_2 and Cl_2 molecules. The second step is to form the bonds in the two product HCl molecules. The net effect of these two steps is the same as for Equation 4.6: One hydrogen molecule and one chlorine molecule change into two hydrogen chloride molecules. The enthalpy changes for these two processes are shown in Figure 4.14.

The reaction of hydrogen with chlorine actually occurs by a complicated series of steps, but the details of how the atoms rearrange do not matter, because enthalpy is a state function and the initial and final states are the same. This means that we can concentrate on products and reactants and not worry about exactly what happens in between.

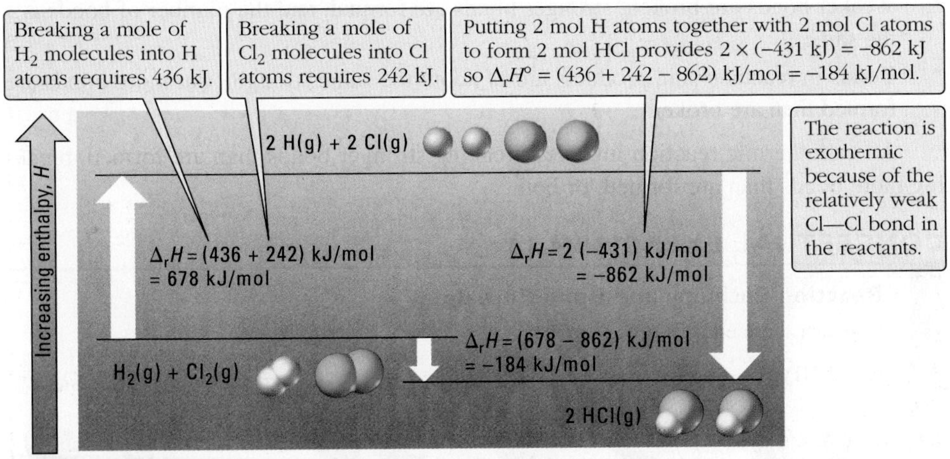
Breaking a mole of H₂ molecules into H atoms requires 436 kJ.
Breaking a mole of Cl₂ molecules into Cl atoms requires 242 kJ.
Putting 2 mol H atoms together with 2 mol Cl atoms to form 2 mol HCl provides 2 × (−431 kJ) = −862 kJ so ΔrH° = (436 + 242 − 862) kJ/mol = −184 kJ/mol.
The reaction is exothermic because of the relatively weak Cl—Cl bond in the reactants.
Increasing enthalpy, H
2 H(g) + 2 Cl(g)
ΔrH = (436 + 242) kJ/mol = 678 kJ/mol
ΔrH = 2 (−431) kJ/mol = −862 kJ/mol
ΔrH = (678 − 862) kJ/mol = −184 kJ/mol
H₂(g) + Cl₂(g)
2 HCl(g)

Figure 4.14 **Stepwise energy changes in the reaction $H_2(g) + Cl_2(g) \longrightarrow 2\ HCl(g)$.**

Another analogy for the enthalpy change for a reaction is the change in altitude when you climb a mountain. No matter which route you take to the summit (which atoms you separate or combine first), the difference in altitude between the summit and where you started to climb (the enthalpy difference between products and reactants) is the same.

CONCEPTUAL **EXERCISE 4.13**

Reaction Pathways and Reaction Enthalpy

Suppose that the reaction enthalpy differed depending on the pathway a reaction took from reactants to products. For example, suppose that 190. kJ was released when a mole of $H_2(g)$ and a mole of $Cl_2(g)$ combined to form two moles of HCl(g) (Equation 4.6), but that only 185 kJ was released when the same reactant molecules were broken into atoms that then recombined to form HCl(g) (Equation 4.7). Would this violate the first law of thermodynamics? Explain why or why not.

4-7a Bond Enthalpies

Bond enthalpy and bond energy differ because a volume change occurs when one molecule changes to two atoms at constant pressure. Therefore work is done on the surroundings (← Sec. 4-5b) and $\Delta_r E \neq \Delta_r H$. For a more detailed discussion, see Treptow, R. S., *Journal of Chemical Education*, Vol. 72, 1995; p. 497.

Separating two atoms that are bonded together requires a transfer of energy into the system, because work must be done against the force holding the pair of atoms together. *The reaction enthalpy when two bonded atoms in a gas-phase molecule are separated completely at constant pressure* is called the **bond enthalpy** (or the **bond energy**—the two terms are often used interchangeably). The bond enthalpy is usually expressed per mole of bonds. For example, the bond enthalpy for a Cl_2 molecule is 242 kJ/mol, so we can write

$$Cl_2(g) \longrightarrow 2\ Cl(g) \qquad \Delta_r H^\circ = 242\ \text{kJ/mol} \qquad [4.7]$$

Bond *breaking* is *endo*thermic.

Bond enthalpies are always positive, and they range in magnitude from about 150 kJ/mol to a little more than 1000 kJ/mol. ***Bond breaking is always endothermic***, because there is always a transfer of energy into the system (in this case, the mole of Cl_2 molecules). The transfer is needed because the separated atoms have higher energy than when they were bonded.

Bond *making* is *exo*thermic.

Conversely, when atoms form a bond, energy is transferred to the surroundings because the potential energy of the atoms is lower when they are bonded together. Conservation of energy requires that if the system's energy goes down, the energy of the surroundings must go up. Thus, ***formation of bonds from separated atoms is always exothermic***. Figure 4.14 shows how these generalizations apply to the reaction of $H_2(g)$ with $Cl_2(g)$ to form HCl(g).

Bond enthalpies provide a way to see what makes a process exothermic or endothermic. If, as in Figure 4.14, the total energy transferred out of the system when new bonds form is greater than the total energy transferred in to break all of the bonds in the reactants, then the reaction is exothermic. In terms of bond enthalpies *either or both of these can make a reaction exothermic:*

- Weaker bonds are broken, stronger bonds are formed, and the number of bonds is the same.
- Bonds in reactants and products are of about the same strength, but more bonds are formed than are broken.

An endothermic reaction involves breaking stronger bonds than are formed, breaking more bonds than are formed, or both.

CONCEPTUAL **EXERCISE 4.14**

Reaction Enthalpy and Bond Enthalpies

Consider these endothermic reactions:

(a) $2\ HF(g) \longrightarrow H_2(g) + F_2(g)$ (b) $2\ H_2O(g) \longrightarrow 2\ H_2(g) + O_2(g)$

In which case is formation of weaker bonds more important in making the reaction endothermic? In which case is formation of fewer bonds more important?

4-8 Measuring Reaction Enthalpies: Calorimetry

A thermochemical expression tells us how much energy is transferred as a chemical process occurs. This knowledge enables us to calculate useful results such as the heat transfer when a fuel is burned. Also, when reactions are carried out on a larger scale—say, in a chemical plant that manufactures sulfuric acid—the surroundings must have enough cooling capacity to prevent an exothermic reaction from overheating, speeding up, running out of control, and possibly damaging the plant. For these and many other reasons it is useful to know as many $\Delta_r H°$ values as possible. For many reactions, direct experimental measurements can be made by using a **calorimeter**, *a device that measures heat transfers*. Calorimetric measurements can be made at constant volume or at constant pressure.

4-8a Constant-Volume Calorimetry

Often, in finding heats of combustion or the caloric value of foods, where at least one of the reactants is a gas, the measurement is done at constant volume in a *bomb calorimeter* (Figure 4.15). The "bomb" is a capped cylinder about the size of a large fruit juice can with heavy steel walls so that it can contain high pressures. A weighed sample of a combustible solid or liquid is placed in a dish inside the bomb. The bomb is then filled with pure $O_2(g)$ and placed in a water-filled container with well-insulated walls. The sample is ignited, usually by an electrical spark. When the sample burns, it warms the bomb and the water around it to the same temperature.

In this configuration, the oxygen and the compound represent the *system* and the bomb and the water around it are the *surroundings*. For the system, $\Delta_r E = q + w$. Because there is no change in volume of the sealed, rigid bomb, $w = 0$. Therefore, $\Delta_r E = q_V$. To calculate q_V and $\Delta_r E$, we can sum the energy transfers from the reaction to the bomb and to the water. Because each of these is a transfer out of the system, each will be negative. For example, the energy transfer to heat the water can be calculated as $c_{water} \times m_{water} \times \Delta T_{water}$, where c_{water} is the specific heat capacity of water, m_{water} is the mass of water, and ΔT_{water} is the change in temperature of the water. Because this energy transfer is out of the system, the energy transfer from the system to the water is negative, that is, $-(c_{water} \times m_{water} \times \Delta T_{water})$. Problem-Solving Example 4.9 illustrates how this works.

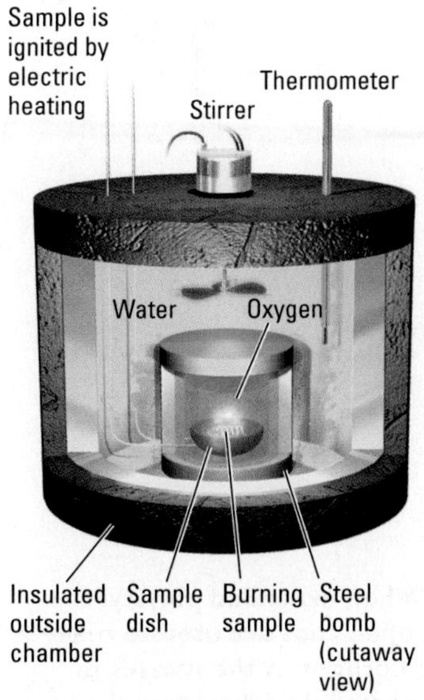
Sample is ignited by electric heating
Stirrer
Thermometer
Water
Oxygen
Insulated outside chamber
Sample dish
Burning sample
Steel bomb (cutaway view)

Figure 4.15 Combustion (bomb) calorimeter. A sample in a strong container (bomb) full of oxygen is ignited by electric heating. The heat transfer from combustion of the sample raises the temperature of the bomb and the surrounding water.

PROBLEM-SOLVING EXAMPLE 4.9

Measuring Reaction Energy with a Bomb Calorimeter

A 3.30-g sample of the sugar glucose, $C_6H_{12}O_6(s)$, was placed in a bomb calorimeter, ignited, and burned to form carbon dioxide and water. The temperature of the water and the bomb changed from 22.4 °C to 34.1 °C. If the calorimeter contained 850. g water and had a heat capacity of 847 J/°C, calculate $\Delta_r E$ for combustion of 1 mol glucose. (The heat capacity of the bomb is the energy transfer required to raise the bomb's temperature by 1 °C.)

Result −2810 kJ/mol

Analyze The energy transfer for 1 mol glucose is asked for. The energy transfer for the given mass (3.30 g) of glucose can be calculated from the change in temperature of the bomb and its heat capacity (given) and the temperature change of water and its heat capacity. The molar mass relates energy transfer per gram to energy transfer per mole of glucose.

Plan Sum the energy transfer to the bomb and to the water to obtain the total energy transfer. Use the molar mass of glucose to obtain energy transfer per mole.

Execute $\Delta T = (34.1 - 22.4)\ °C = 11.7\ °C$

Energy transferred from system to bomb = −(heat capacity of bomb × ΔT)

$$= -\left(\frac{847\ \text{J}}{°\text{C}} \times 11.7\ °\text{C}\right) = -9910\ \text{J} = -9.910\ \text{kJ}$$

Energy transferred from system to water $= -(c \times m \times \Delta T)$

$$= -\left(\frac{4.184\ \text{J}}{\text{g}\ °\text{C}} \times 850.\ \text{g} \times 11.7\ °\text{C}\right) = -41{,}610\ \text{J} = -41.61\ \text{kJ}$$

$$\Delta_r E = q_V = -9.910\ \text{kJ} - 41.61\ \text{kJ} = -51.52\ \text{kJ}$$

$$\text{Energy transfer per mole } C_6H_{12}O_6 = \frac{-51.52\ \text{kJ}}{3.30\ \text{g}} \times \frac{180.16\ \text{g}}{1\ \text{mol}} = -2.81 \times 10^3\ \text{kJ/mol}$$

☑ **Reasonable Result Check** The result is negative, which correctly reflects the fact that burning sugar is exothermic. A mole of glucose (180 g) is more than a third of a pound, and a third of a pound of sugar contains quite a bit of energy (many Calories), so a result in thousands of kilojoules is reasonable.

PROBLEM-SOLVING PRACTICE 4.9

In Problem-Solving Example 4.1, a single Fritos chip was oxidized by potassium chlorate. Suppose that a single chip weighing 1.0 g is placed in a bomb calorimeter that has a heat capacity of 877 J/°C. The calorimeter contains 832 g water. When the bomb is filled with excess oxygen and the chip is ignited, the temperature rises from 20.64 °C to 25.43 °C. Use these data to verify the statement that the chip provides 5 Cal when metabolized.

CONCEPTUAL EXERCISE 4.15

Comparing Enthalpy Change and Energy Change

Write a balanced equation for the combustion of glucose to form CO_2(g) and $H_2O(\ell)$. Use what you already know about the volume of a mole of any gas at a given temperature and pressure (or look in Section 8-3c) to predict whether $\Delta_r H$ would differ significantly from $\Delta_r E$ for the reaction in Problem-Solving Example 4.9.

4-8b Constant-Pressure Calorimetry

When reactions take place in solution, it is much easier to use a calorimeter that is open to the atmosphere. An example, often encountered in introductory chemistry courses, is the *coffee cup calorimeter* shown in Figure 4.16. The nested coffee cups (which are made of expanded polystyrene) provide good thermal insulation; reactions can occur when solutions are poured together in the inner cup. Because a coffee cup calorimeter is a constant-pressure device, the measured heat transfer is q_P, which can be used to calculate $\Delta_r H°$ as shown in Problem-Solving Example 4.10.

When expanded polystyrene coffee cups are used to make a calorimeter, the masses of substances other than the solvent water are often so small that their heat capacities can be ignored; all of the energy of a reaction can be assumed to be transferred to the water.

PROBLEM-SOLVING EXAMPLE 4.10

Measuring Reaction Enthalpy with a Coffee Cup Calorimeter

A coffee cup calorimeter is used to determine $\Delta_r H°$ for the reaction

$$NaOH(aq) + HCl(aq) \longrightarrow H_2O(\ell) + NaCl(aq) \qquad \Delta_r H° = ?$$

When 250. mL of 1.00-M NaOH was added to 250. mL of 1.00-M HCl at 1 bar, the temperature of the solution increased from 23.4 °C to 30.4 °C. Use this information to determine $\Delta_r H°$ and complete the thermochemical expression. Assume that the heat capacities of the coffee cups, the temperature probe, and the stirrer are negligible; that the solution has the same density and the same specific heat capacity as water; and that there is no change in volume of the solutions upon mixing.

Result $\Delta_r H° = -58$ kJ/mol

Analyze Calculating $\Delta_r H°$ is the goal. The temperature change and volume of the product solution are known; from them the heat transfer can be calculated. Because the reaction system heats the solution, q_P is negative. In the chemical equation, 1 mol of each reactant or product corresponds with 1 mol reaction. The amount of each reactant can be obtained from the solution volume and molarity.

Plan Use the definition of specific heat capacity, Equation 4.2′ (← **Sec. 4-4a**), to calculate q_P. Look up the density and specific heat capacity for water. Use the concentration and volume of each reactant solution to calculate the amount of reactant and amount of reaction; if there is a limiting reactant, use the smaller amount of reaction. Calculate the heat transfer per mole of reaction.

Execute
$$q_P = -c \times m \times \Delta T$$
$$= -(4.184 \text{ J g}^{-1}\ °\text{C}^{-1})(500.\text{ g})(30.4\ °\text{C} - 23.4\ °\text{C})$$
$$= -1.46 \times 10^4 \text{ J} = -14.6 \text{ kJ}$$

$$n(\text{rxn, HCl}) = 250.\text{ mL} \times \frac{1.00 \text{ mol HCl}}{1000 \text{ mL}} \times \frac{1.00 \text{ mol rxn}}{1.00 \text{ mol HCl}} = 0.250 \text{ mol rxn}$$

$$n(\text{rxn, NaOH}) = 250.\text{ mL} \times \frac{1.00 \text{ mol NaOH}}{1000 \text{ mL}} \times \frac{1.00 \text{ mol rxn}}{1.00 \text{ mol NaOH}} = 0.250 \text{ mol rxn}$$

The reactants are in the stoichiometric ratio; neither reactant is limiting.

$$\Delta_r H° = \frac{-14.6 \text{ kJ}}{0.250 \text{ mol rxn}} = -58 \text{ kJ/mol}$$

☑ **Reasonable Result Check** The temperature of the surroundings increased, so the reaction is exothermic and $\Delta_r H°$ must be negative. The temperature of 500. g solution went up 7.0 °C, so the heat transfer was about $(500 \times 7 \times 4)$ J = 14,000 J = 14 kJ. This corresponded to one-quarter mole of each reactant, so the heat transfer per mole must be about 4×14 kJ = 56 kJ. $\Delta_r H°$ should be about −56 kJ, so the result is reasonable. There are only two significant figures in ΔT (7.0 °C) so only two significant figures are given in the answer.

(a)

Photos: Cengage Learning/Charles D. Winters

(b)

Figure 4.16 Coffee-cup calorimeter. (a) Components: two coffee cups, a cork lid, a temperature probe, and a stirrer. (b) Assembled calorimeter. There is essentially no energy transfer into or out of the calorimeter, so the heat capacity of a solution and its change in temperature can be used to calculate q_P and $\Delta_r H$.

PROBLEM-SOLVING PRACTICE 4.10

Suppose that 100. mL of 1.00-M HCl and 100. mL of 0.50-M NaOH, both at 20.4 °C, are mixed in a coffee cup calorimeter. Use the result from Problem-Solving Example 4.10 to predict what will be the highest temperature reached in the calorimeter after mixing the solutions. Make assumptions similar to those made in Problem-Solving Example 4.10.

CONCEPTUAL EXERCISE 4.16

Calorimetry

In Problem-Solving Example 4.10, ΔT was observed to be 7.0 °C for mixing 250. mL of 1.00-M HCl and 250. mL of 1.00-M NaOH in a coffee cup calorimeter. Predict ΔT for mixing

(a) 200. mL of 1.0-M HCl and 200. mL 1.0-M NaOH.
(b) 100. mL of 1.0-M H_2SO_4 and 100. mL 1.0-M NaOH.

EXERCISE 4.17

Reaction Enthalpy for Dissolving

The process of dissolving solid ammonium nitrate is often used in cold packs. When 7.07 g NH_4NO_3 is added to 150. mL H_2O in a coffee cup calorimeter and the mixture is stirred to dissolve all of the NH_4NO_3, the temperature falls from 22.3 °C to 19.2 °C. Write a balanced equation for dissolving 1 mol NH_4NO_3 in water and calculate the reaction enthalpy for dissolving NH_4NO_3. Assume that the solution has the same specific heat capacity as water and the density of the solution is 1.00 g/mL. Express your result in kJ/mol.

4-9 Hess's Law

Hess's law is based on a fact we mentioned earlier (← Sec. 4-5e). A system's enthalpy will be the same no matter how the system is prepared. Therefore, at 25 °C and 1 bar, the initial system, $H_2(g) + \frac{1}{2} O_2(g)$, has a particular enthalpy value. The final system, $H_2O(\ell)$, also has a characteristic (but different) enthalpy. Whether initial system changes to final system by a single step or by the two-step process of chemical equations (a) and (b), the reaction enthalpy will be the same.

It requires 1 mol $H_2O(g)$ to cancel 1 mol $H_2O(g)$. If the coefficient of $H_2O(g)$ on one side of the chemical equation had been different from the other side, $H_2O(g)$ could not have been completely canceled.

Calorimetry works well for some reactions, but for many others it is difficult to use. Besides, it would be very time-consuming to measure values for every conceivable reaction, and it would take a great deal of space to tabulate so many values. Fortunately, there is a better way. It is based on **Hess's law**, which states that *if the equation for a reaction is the sum of the equations for two or more other reactions, then $\Delta_r H°$ for the first reaction must be the sum of $\Delta_r H°$ values of the other reactions.* Hess's law is a corollary of the law of conservation of energy. It works even if the overall reaction does not actually occur by way of the separate equations that are summed.

For example, in Figure 4.13 (← **Sec. 4-6a**) we noted that the formation of liquid water from its elements $H_2(g)$ and $O_2(g)$ could be thought of as two successive changes: (a) formation of water vapor from the elements and (b) condensation of water vapor to liquid water. As shown below, the equation for formation of liquid water can be obtained by adding algebraically the chemical equations for these two steps. Therefore, according to Hess's law, the $\Delta_r H°$ value can be found by adding the $\Delta_r H°$ values for the two steps.

$$\text{(a)} \quad H_2(g) + \tfrac{1}{2} O_2(g) \longrightarrow \cancel{H_2O(g)} \quad \Delta_r H_a^\circ = -241.8 \text{ kJ/mol}$$

$$\text{(b)} \quad \cancel{H_2O(g)} \longrightarrow H_2O(\ell) \quad \Delta_r H_b^\circ = -44.0 \text{ kJ/mol}$$

$$\text{(a)} + \text{(b)} \quad H_2(g) + \tfrac{1}{2} O_2(g) \longrightarrow H_2O(\ell) \quad \Delta_r H^\circ = \Delta_r H_a^\circ + \Delta_r H_b^\circ = -285.8 \text{ kJ/mol}$$

Here, 1 mol $H_2O(g)$ is a product of the first reaction and a reactant in the second. Thus, $H_2O(g)$ can be canceled. This is similar to adding two algebraic equations: If the same quantity or term appears on both sides of the equation, it cancels. The net result is an equation for the overall reaction and its associated reaction enthalpy. This overall enthalpy change applies even if the liquid water is formed directly from hydrogen and oxygen.

A useful approach to Hess's law is

- **begin with the thermochemical expression for which you want to calculate $\Delta_r H°$; call this the target expression;**
- **in the target expression, identify which reactants and which products are desired; determine the appropriate coefficients;**
- **look at the known thermochemical expressions and decide how each needs to be changed to give reactants and products with the coefficients that are in the target expression.**

For example, suppose you want the thermochemical expression for the reaction

$$\tfrac{1}{2} CH_4(g) + O_2(g) \longrightarrow \tfrac{1}{2} CO_2(g) + H_2O(\ell) \qquad \Delta_r H^\circ = ?$$

and you already know these thermochemical expressions

$$\text{(a)} \quad CH_4(g) + 2\, O_2(g) \longrightarrow CO_2(g) + 2\, H_2O(g) \qquad \Delta_r H_a^\circ = -802.34 \text{ kJ/mol}$$

$$\text{(b)} \quad H_2O(\ell) \longrightarrow H_2O(g) \qquad \Delta_r H_b^\circ = 44.01 \text{ kJ/mol}$$

The target expression has $\frac{1}{2}$ $CH_4(g)$ as a reactant; it also has $\frac{1}{2}$ $CO_2(g)$ and 1 $H_2O(\ell)$ as products. Expression (a) has the same reactants and products, but each coefficient is twice as big; also, water is in the gaseous state in expression (a). If we change each coefficient and the $\Delta_r H°$ value of expression (a) to one half their original values, we have this thermochemical expression:

$$\text{(a}'\text{)} \quad \tfrac{1}{2} CH_4(g) + O_2(g) \longrightarrow \tfrac{1}{2} CO_2(g) + H_2O(g) \qquad \Delta_r H_{a'}^\circ = -401.17 \text{ kJ/mol}$$

which differs from the target expression only in the phase of water. Expression (b) has liquid water on the left and gaseous water on the right, but our target expression has liquid water on the right. If the equation in (b) is reversed (which changes the sign of $\Delta_r H°$), the thermochemical expression becomes (b′) $H_2O(g) \longrightarrow H_2O(\ell)$ $\Delta_r H^\circ_{b'} = -44.01$ kJ/mol

Summing the expressions (a′) and (b′) gives the target expression, from which $H_2O(g)$ has been canceled on each side.

(a′ + b′) $\frac{1}{2} CH_4(g) + O_2(g) \longrightarrow \frac{1}{2} CO_2(g) + H_2O(\ell)$ $\Delta_r H^\circ = \Delta_r H^\circ_{a'} + \Delta_r H^\circ_{b'}$

$$\Delta_r H^\circ = (-401.17 \text{ kJ}) + (-44.01 \text{ kJ}) = -445.18 \text{ kJ/mol}$$

PROBLEM-SOLVING EXAMPLE 4.11

Using Hess's Law

In designing a chemical plant for manufacturing the plastic polyethylene, you need to know the reaction enthalpy for the removal of H_2 from C_2H_6 (ethane) to give C_2H_4 (ethylene), a key step in the process.

$$C_2H_6(g) \longrightarrow C_2H_4(g) + H_2(g) \qquad \Delta_r H^\circ = ?$$

From experiments you know these thermochemical expressions:

(a) $2\,C_2H_6(g) + 7\,O_2(g) \longrightarrow 4\,CO_2(g) + 6\,H_2O(\ell)$ $\Delta_r H^\circ_a = -3119.4$ kJ/mol

(b) $C_2H_4(g) + 3\,O_2(g) \longrightarrow 2\,CO_2(g) + 2\,H_2O(\ell)$ $\Delta_r H^\circ_b = -1410.9$ kJ/mol

(c) $2\,H_2(g) + O_2(g) \longrightarrow 2\,H_2O(\ell)$ $\Delta_r H^\circ_c = -571.7$ kJ/mol

Use this information to find the value of $\Delta_r H^\circ$ for the formation of ethylene from ethane.

Result $\Delta_r H^\circ = 137.0$ kJ/mol

Analyze Compare reactions (a), (b), and (c) with the target expression and decide how to change each to match the target expression.

- Reaction (a) involves 2 mol ethane on the reactant side, but the target expression requires only 1 mol ethane.
- Reaction (b) has 1 mol C_2H_4 as a reactant, but 1 mol C_2H_4 is a product in the target expression.
- Reaction (c) has 2 mol H_2 as a reactant, but 1 mol H_2 is a product in the target expression.

Plan Multiply expression (a) by $\frac{1}{2}$ to give an expression (a′) that has 1 mol ethane on the reactant side. This halves the reaction enthalpy. $\Delta_r H^\circ_{a'} = \frac{1}{2}\,\Delta_r H^\circ_a$

Reverse expression (b) so that C_2H_4 is on the product side, giving expression (b′). This reverses the sign of the reaction enthalpy. $\Delta_r H^\circ_{b'} = -\Delta_r H^\circ_b$

Reverse expression (c) and multiply all coefficients by $\frac{1}{2}$. This changes the sign and halves the reaction enthalpy. $\Delta_r H^\circ_{c'} = -\frac{1}{2}\,\Delta_r H^\circ_c$

Then, sum expressions (a′), (b′), and (c′) to give the desired expression.

Execute

(a′) = $\frac{1}{2}$ (a) $C_2H_6(g) + \cancel{\frac{7}{2}\,O_2(g)} \longrightarrow \cancel{2\,CO_2(g)} + \cancel{3\,H_2O(\ell)}$ $\Delta_r H^\circ_{a'} = -1559.7$ kJ/mol

(b′) = −(b) $\cancel{2\,CO_2(g)} + \cancel{2\,H_2O(\ell)} \longrightarrow C_2H_4(g) + \cancel{3\,O_2(g)}$ $\Delta_r H^\circ_{b'} = 1410.9$ kJ/mol

(c′) = $-\frac{1}{2}$ (c) $\cancel{H_2O(\ell)} \longrightarrow H_2(g) + \cancel{\frac{1}{2}\,O_2(g)}$ $\Delta_r H^\circ_{c'} = 285.8$ kJ/mol

The sum of expressions (a′), (b′), and (c′) is the desired expression.

$$C_2H_6(g) \longrightarrow C_2H_4(g) + H_2(g) \qquad \Delta_r H^\circ = 137.0 \text{ kJ/mol}$$

There was $\frac{7}{2}$ mol $O_2(g)$ on the reactant side and $(3 + \frac{1}{2}) = \frac{7}{2}$ mol $O_2(g)$ on the product side. There was 3 mol $H_2O(\ell)$ on each side and 2 mol $CO_2(g)$ on each side. Therefore, $O_2(g)$, $CO_2(g)$, and $H_2O(\ell)$ all canceled.

☑ **Reasonable Result Check** The overall process involves breaking a molecule apart into simpler molecules, which is likely to involve breaking bonds. Therefore it should be endothermic, and $\Delta_r H^\circ$ should be positive.

Soft
OFF!
Comet
Rolaids
Cengage Learning/Charles D. Winters

Polyethylene is a common plastic. Many products are packaged in polyethylene bottles.

PROBLEM-SOLVING PRACTICE 4.11

When iron is obtained from iron ore, an important reaction is conversion of $Fe_3O_4(s)$ to FeO(s). Write a balanced equation for this reaction. Then use these thermochemical expressions to calculate $\Delta_r H°$ for the reaction.

$$3\ Fe(s) + 2\ O_2(g) \longrightarrow Fe_3O_4(s) \qquad \Delta_r H_1° = -1118.4\ kJ$$
$$Fe(s) + \tfrac{1}{2}\ O_2(g) \longrightarrow FeO(s) \qquad \Delta_r H_2° = -272.0\ kJ$$

Both "heat of formation" and "enthalpy of formation" were formerly used to describe standard formation enthalpies. You should be aware that these two terms mean the same thing as standard formation enthalpy.

At 25 °C and 1 bar, carbon can exist as graphite, diamond, buckminsterfullerene, and other forms. Such forms are called allotropes. Oxygen has two allotropes, O_2 and O_3.

4-10 Standard Formation Enthalpies

Hess's law makes it possible to tabulate $\Delta_r H°$ values for a relatively few reactions and, by suitable combinations of these few reactions, to calculate $\Delta_r H°$ values for a great many other reactions. To make such a tabulation we use standard formation enthalpies. The **standard formation enthalpy ($\Delta_f H°$)** is the *standard reaction enthalpy for formation of one mole of a compound from its elements in their standard states.* The subscript f in $\Delta_f H°$ indicates *formation* of the compound. The **standard state** is *the physical state in which an element or compound exists at 1 bar and a specified temperature.* At 25 °C the standard state for hydrogen is $H_2(g)$ and for sodium chloride is NaCl(s). For an element that can exist in several different forms at 1 bar and 25 °C, the most stable form is usually selected as the standard state. For example, graphite, not diamond or buckminsterfullerene, is the standard state for carbon because it is more stable; $O_2(g)$, which is more stable than $O_3(g)$, is the standard state for oxygen.

Some examples of thermochemical expressions involving standard formation enthalpies are

$$H_2(g) + \tfrac{1}{2}\ O_2(g) \longrightarrow H_2O(\ell) \qquad \Delta_r H° = \Delta_f H°\{H_2O(\ell)\} = -285.8\ kJ/mol$$
$$2\ C(graphite) + 2\ H_2(g) \longrightarrow C_2H_4(g) \qquad \Delta_r H° = \Delta_f H°\{C_2H_4(g)\} = 52.26\ kJ/mol$$
$$2\ C(graphite) + 3\ H_2(g) + \tfrac{1}{2}\ O_2 \longrightarrow C_2H_5OH(\ell)$$
$$\Delta_r H° = \Delta_f H°\{C_2H_5OH(\ell)\} = -227.69\ kJ/mol$$

Notice that in each case 1 mol of a compound in its standard state is *formed directly from appropriate amounts of elements in their standard states.*

PROBLEM-SOLVING EXAMPLE 4.12

Thermochemical Expressions for Standard Formation Enthalpies

Two examples of thermochemical expressions at 25 °C and 1 bar where $\Delta_r H°$ is *not* a standard formation enthalpy are

$$MgO(s) + SO_3(g) \longrightarrow MgSO_4(s) \qquad \Delta_r H° = -287.5\ kJ/mol \quad [4.8]$$
$$P_4(s) + 6\ Cl_2(g) \longrightarrow 4\ PCl_3(\ell) \qquad \Delta_r H° = -1278.8\ kJ/mol \quad [4.9]$$

Rewrite Thermochemical Expressions 4.8 and 4.9 so that they represent standard formation enthalpies for the products given. (The standard formation enthalpy for $MgSO_4(s)$ has the value −1284.9 kJ/mol at 25 °C.)

Result

$$Mg(s) + \tfrac{1}{8}\ S_8(s) + 2\ O_2(g) \longrightarrow MgSO_4(s) \qquad \Delta_f H°\{MgSO_4(s)\} = -1284.9\ kJ/mol$$
$$\tfrac{1}{4}\ P_4(s) + \tfrac{3}{2}\ Cl_2(g) \longrightarrow PCl_3(\ell) \qquad \Delta_f H°\{PCl_3(\ell)\} = -319.7\ kJ/mol$$

Analyze To represent a standard formation enthalpy, a thermochemical expression must have an equation where reactants are all elements and the product is 1 mol of a compound. Knowledge of the properties of elements is needed to decide in which form each element should appear.

Plan For Thermochemical Expression 4.8 there is 1 mol $MgSO_4(s)$ on the right side, but the reactants are not elements. The left side should contain the most stable forms of magnesium, sulfur, and oxygen, which are Mg(s), S_8(s), and O_2(g). Use the standard formation enthalpy value given.

Rewrite Thermochemical Expression 4.9 so the right side involves only 1 mol $PCl_3(\ell)$, and reduce the coefficients of the elements on the left side in proportion—that is, divide all coefficients by 4. Then, $\Delta_r H°$ must also be divided by 4.

Execute See the Result section.

☑ **Reasonable Result Check** Check each thermochemical expression carefully to make certain the substance whose standard enthalpy you want is on the right side and has a coefficient of 1. For $PCl_3(\ell)$, $\Delta_f H°$ should be one fourth of about −1300 kJ, and it is.

PROBLEM-SOLVING PRACTICE 4.12

Write an appropriate thermochemical expression in each case. (You may need to use fractional coefficients.)
(a) The standard formation enthalpy of NH_3(g) at 25 °C is −46.11 kJ/mol.
(b) The standard formation enthalpy of CO(g) at 25 °C is −110.525 kJ/mol.

CONCEPTUAL **EXERCISE 4.18**

Standard Formation Enthalpies of Elements

Write the thermochemical expression that corresponds to the standard formation enthalpy of N_2(g).

(a) What process, if any, takes place in the chemical equation?
(b) What does this imply about the reaction enthalpy?

Table 4.2 and Appendix J list values of $\Delta_f H°$, obtained from the National Institute for Standards and Technology (NIST), for many compounds. Notice that no values are listed in these tables for elements in their standard-state forms, such as C(graphite) or O_2(g). As you probably realized from Conceptual Exercise 4.18, *standard formation enthalpies for the elements in their standard states are zero,* because forming an element in its standard state from the same element in its standard state involves no chemical or physical change.

Hess's law can be used to find the standard reaction enthalpy for any reaction if there is a set of reactions whose reaction enthalpies are known and whose chemical equations, when added together, will give the equation for the desired reaction. For example, suppose you are a chemical engineer and want to know how much heating is required to decompose limestone (calcium carbonate) to lime (calcium oxide) and carbon dioxide. The target expression for which you want to know $\Delta_r H°$ is

$$CaCO_3(s) \longrightarrow CaO(s) + CO_2(g) \qquad \Delta_r H° = ?$$

As a first approximation you can assume that all substances are in their standard states at 25 °C and look up the standard formation enthalpy of each substance in Table 4.2 or Appendix J. This gives three thermochemical expressions:

$$\text{(a)} \quad Ca(s) + C(\text{graphite}) + \tfrac{3}{2}\,O_2(g) \longrightarrow CaCO_3(s) \qquad \Delta_r H°_a = -1206.9 \text{ kJ/mol}$$

$$\text{(b)} \quad Ca(s) + \tfrac{1}{2}\,O_2(g) \longrightarrow CaO(s) \qquad \Delta_r H°_b = -635.1 \text{ kJ/mol}$$

$$\text{(c)} \quad C(\text{graphite})\ O_2(g) \longrightarrow CO_2(g) \qquad \Delta_r H°_c = -393.5 \text{ kJ/mol}$$

Now combine the three expressions to get the target expression for decomposition of limestone.

In expression (a), $CaCO_3$(s) is a product, but it appears in the target expression as a reactant. Therefore, the equation in (a) must be reversed, and the sign of $\Delta_r H°_a$ must also be reversed. On the other hand, CaO(s) and CO_2(g) are products in the desired

Photos: Courtesy of João Paiva

Lime production. At high temperature in a lime kiln, calcium carbonate (limestone, $CaCO_3$) decomposes to calcium oxide (lime, CaO) and carbon dioxide (CO_2).

expression, so expressions (b) and (c) can be added with the same direction and sign of $\Delta_r H°$ as they have above:

(a′) = −(a)	$CaCO_3(s) \longrightarrow \cancel{Ca(s)} + \cancel{C(graphite)} + \cancel{\frac{3}{2} O_2(g)}$	
		$\Delta_r H°_{a'} = 1206.9$ kJ/mol
(b)	$\cancel{Ca(s)} + \cancel{\frac{1}{2} O_2(g)} \longrightarrow CaO(s)$	$\Delta_r H°_{b'} = -635.1$ kJ/mol
(c)	$\cancel{C(graphite)} + \cancel{O_2(g)} \longrightarrow CO_2(g)$	$\Delta_r H°_{c'} = -393.5$ kJ/mol
	$CaCO_3(s) \longrightarrow CaO(s) + CO_2(g)$	$\Delta_r H° = 178.3$ kJ/mol

When the expressions are added, 1 C(graphite) and 1 Ca(s) and $\frac{3}{2}$ $O_2(g)$ appear on both sides and cancel. Thus, the sum of these chemical equations is the desired one for the decomposition of calcium carbonate, and the sum of the reaction enthalpies gives the desired enthalpy change.

Another very useful conclusion can be drawn from this example. The calculation can be written mathematically as

$$\Delta_r H° = \Delta_f H°\{CaO(s)\} + \Delta_f H°\{CO_2(g)\} - \Delta_f H°\{CaCO_3(s)\}$$
$$= (-635.1 \text{ kJ/mol}) + (-393.5 \text{ kJ/mol}) - (-1206.9 \text{ kJ/mol}) = 178.3 \text{ kJ/mol}$$

Table 4.2 Selected Standard Formation Enthalpies, $\Delta_f H°$, at 25 °C

Formula	Name	$\Delta_f H°$ (kJ/mol)	Formula	Name	$\Delta_f H°$ (kJ/mol)
$Al_2O_3(s)$	Aluminum oxide	−1675.7	HI(g)	Hydrogen iodide	26.48
$BaCO_3(s)$	Barium carbonate	−1216.3	KF(s)	Potassium fluoride	−567.27
$CaCO_3(s)$	Calcium carbonate	−1206.92	KCl(s)	Potassium chloride	−436.747
CaO(s)	Calcium oxide	−635.09	KBr(s)	Potassium bromide	−393.8
C (s, diamond)	Diamond	1.895	MgO(s)	Magnesium oxide	−601.70
$CCl_4(\ell)$	Carbon tetrachloride	−135.44	$MgSO_4(s)$	Magnesium sulfate	−1284.9
$CH_4(g)$	Methane	−74.81	$Mg(OH)_2(s)$	Magnesium hydroxide	−924.54
$C_2H_5OH(\ell)$	Ethyl alcohol	−277.69	NaF(s)	Sodium fluoride	−573.647
CO(g)	Carbon monoxide	−110.525	NaCl(s)	Sodium chloride	−411.153
$CO_2(g)$	Carbon dioxide	−393.509	NaBr(s)	Sodium bromide	−361.062
$C_2H_2(g)$	Acetylene (ethyne)	226.73	NaI(s)	Sodium iodide	−287.78
$C_2H_4(g)$	Ethylene (ethene)	52.26	$NH_3(g)$	Ammonia	−46.11
$C_2H_6(g)$	Ethane	−84.68	NO(g)	Nitrogen monoxide	90.25
$C_3H_8(g)$	Propane	−103.8	$NO_2(g)$	Nitrogen dioxide	33.18
$C_4H_{10}(g)$	Butane	−126.148	$O_3(g)$	Ozone	142.7
$C_6H_{12}O_6(s)$	α-D-Glucose	−1274.4	$PCl_3(\ell)$	Phosphorus trichloride	−319.7
$CuSO_4(s)$	Copper(II) sulfate	−771.36	$PCl_5(s)$	Phosphorus pentachloride	−443.5
$H_2O(g)$	Water vapor	−241.818	$SiO_2(s)$	Silicon dioxide (quartz)	−910.94
$H_2O(\ell)$	Liquid water	−285.830	$SnCl_2(s)$	Tin(II) chloride	−325.1
HF(g)	Hydrogen fluoride	−271.1	$SnCl_4(\ell)$	Tin(IV) chloride	−511.3
HCl(g)	Hydrogen chloride	−92.307	$SO_2(g)$	Sulfur dioxide	−296.830
HBr(g)	Hydrogen bromide	−36.40	$SO_3(g)$	Sulfur trioxide	−395.72

From Wagman, D. D., Evans, W. H., Parker, V. B., Schumm, R. H., Halow, I., Bailey, S. M., Churney, K. L., and Nuttall, R. The NBS Tables of Chemical Thermodynamic Properties. *Journal of Physical and Chemical Reference Data,* Vol. 11, Suppl. 2, 1982. (NBS, the National Bureau of Standards, is now NIST, the National Institute for Standards and Technology.)

which involves adding the $\Delta_f H°$ values for the products of the reaction, CaO(s) and CO_2(g), and subtracting the $\Delta_f H°$ value for the reactant, $CaCO_3$(s). The mathematics can be summarized by this equation, where Σ means summation:

$$\Delta_r H° = \Sigma\{(\text{coefficient of product}) \times \Delta_f H°(\text{product})\} - \Sigma\{(\text{coefficient of reactant}) \times \Delta_f H°(\text{reactant})\} \quad [4.10]$$

According to this equation you should **follow these steps to calculate the standard reaction enthalpy from standard formation enthalpies:**

- **Multiply the standard formation enthalpy of each product by the coefficient of that product in the equation and then sum over all products.**
- **Multiply the standard formation enthalpy of each reactant by the coefficient of that reactant in the equation and then sum over all reactants.**
- **Subtract the sum for the reactants from the sum for the products.**

This is a useful shortcut for writing the thermochemical expressions for all appropriate formation reactions and applying Hess's law.

PROBLEM-SOLVING EXAMPLE 4.13

Using Standard Formation Enthalpies

Benzene, C_6H_6, is a commercially important hydrocarbon that is present in gasoline, where it enhances the octane rating. Calculate the enthalpy of combustion per mole of benzene. For benzene, $\Delta_f H°\{C_6H_6(\ell)\} = 49.0$ kJ/mol. Use Table 4.2 for any other values you may need.

Result $\Delta_r H° = -3267.5$ kJ

Analyze To calculate $\Delta_r H°$ you need a thermochemical expression with a balanced equation for the reaction. You also need standard formation enthalpies for all compounds (and any elements not in their standard states) involved in the reaction.

Plan Write a balanced equation for the reaction. Look up $\Delta_f H°$ values for CO_2(g) and $H_2O(\ell)$. O_2(g) is in its standard state so its value is zero. Multiply each $\Delta_f H°$ by the corresponding coefficient in the equation. Sum the values for the products; subtract the sum of the values for the reactants.

Execute

$$C_6H_6(\ell) + \tfrac{15}{2}\,O_2(g) \longrightarrow 6\,CO_2(g) + 3\,H_2O(\ell)$$

$$\Delta_f H°\{CO_2(g)\} = -393.509 \text{ kJ/mol} \qquad \Delta_f H°\{H_2O(\ell)\} = -285.830 \text{ kJ/mol}$$

$$\begin{aligned}\Delta_r H° &= [6 \times \Delta_f H°\{CO_2(g)\} + 3 \times \Delta_f H°\{H_2O(\ell)\}] - [1 \times \Delta_f H°\{C_6H_6(\ell)\}] \\ &= [6 \times (-393.509 \text{ kJ/mol}) + 3 \times (-285.830 \text{ kJ/mol})] - [1 \times (49.0 \text{ kJ/mol})] \\ &= -3267.5 \text{ kJ/mol}\end{aligned}$$

☑ **Reasonable Result Check** As expected, the reaction enthalpy for combustion of a fuel is negative and large.

PROBLEM-SOLVING PRACTICE 4.13

Nitroglycerin is a powerful explosive because it decomposes exothermically and four different gases are formed.

$$2\,C_3H_5(NO_3)_3(\ell) \longrightarrow 3\,N_2(g) + \tfrac{1}{2}\,O_2(g) + 6\,CO_2(g) + 5\,H_2O(g)$$

For nitroglycerin, $\Delta_f H°\{C_3H_5(NO_3)_3(\ell)\} = -364$ kJ/mol. Using data from Table 4.2, calculate the energy transfer when 10.0 g nitroglycerin explodes.

When the reaction enthalpy is known, it is possible to use that information to calculate $\Delta_f H°$ for one substance in the reaction provided that $\Delta_f H°$ values are known for all of the rest of the substances. Problem-Solving Example 4.14 indicates how to do this.

Reatha Clark King
1938–

Reatha Clark King

Reatha Clark King was born in Georgia. She obtained degrees from Clark Atlanta University and the University of Chicago and began her career with the National Bureau of Standards (now the National Institute for Standards and Technology, NIST), where she determined standard formation enthalpies of fluorine compounds that were important to the U. S. space program and NASA. She became a dean at York College, president of Metropolitan State University (Minneapolis), and served as president of the General Mills Foundation.

PROBLEM-SOLVING EXAMPLE 4.14

Standard Formation Enthalpy from Combustion Enthalpy

Glyceryl tristearate, $C_{57}H_{110}O_6$, is a solid saturated fat that stores energy in the human body. The combustion enthalpy is $\Delta_c H° = -37{,}760$ kJ/mol for glyceryl tristearate at 25 °C. Use data from Table 4.2 to calculate the standard formation enthalpy of glyceryl tristearate. (Assume that liquid water is a combustion product.)

Result $\Delta_f H° \{C_{57}H_{110}O_6\} = -390$ kJ/mol

Analyze $\Delta_c H°$, which equals $\Delta_r H°$, is known for the combustion reaction; it can also be calculated from $\Delta_f H°$ data for the substances in the chemical equation for combustion. The unknown $\Delta_f H°(C_{57}H_{110}O_6)$ can be calculated if all other $\Delta_f H°$ values are substituted into Equation 4.10 for the combustion reaction.

Plan Write a balanced chemical equation for combustion of $C_{57}H_{110}O_6(s)$ to form $CO_2(g)$ and $H_2O(\ell)$. Look up $\Delta_f H°$ values for all substances (except $O_2(g)$). Write Equation 4.10 including the known value of $\Delta_c H°$. Solve for $\Delta_f H°(C_{57}H_{110}O_6)$.

Execute

$$C_{57}H_{110}O_6(s) + \tfrac{163}{2}\, O_2(g) \longrightarrow 57\, CO_2(g) + 55\, H_2O(\ell) \qquad \Delta_c H° = -37{,}760 \text{ kJ/mol}$$

$$\Delta_f H°\{CO_2(g)\} = -393.509 \text{ kJ/mol} \qquad \Delta_f H°\{H_2O(\ell)\} = -285.830 \text{ kJ/mol}$$

$$\begin{aligned} \Delta_r H° &= -37{,}760 \text{ kJ/mol} \\ &= [57 \times \Delta_f H°\{CO_2(g)\} + 55 \times \Delta_f H°\{H_2O(\ell)\}] - [1 \times \Delta_f H°\{C_{57}H_{110}O_6\}] \\ &= [57 \times (-393.509 \text{ kJ/mol}) + 55 \times (-285.830 \text{ kJ/mol})] - \Delta_f H°\{C_{57}H_{110}O_6\} \end{aligned}$$

$$\begin{aligned} \Delta_f H°\{C_{57}H_{110}O_6\} &= -22{,}430 \text{ kJ/mol} - 15{,}720 \text{ kJ/mol} + 37{,}760 \text{ kJ/mol} \\ &= -390 \text{ kJ/mol} \end{aligned}$$

PROBLEM-SOLVING PRACTICE 4.14

Use data from Table 4.2 to calculate the standard combustion enthalpy for conversion of sulfur dioxide, $SO_2(g)$, to sulfur trioxide, $SO_3(g)$.

4-11 Fuels for Society and Our Bodies

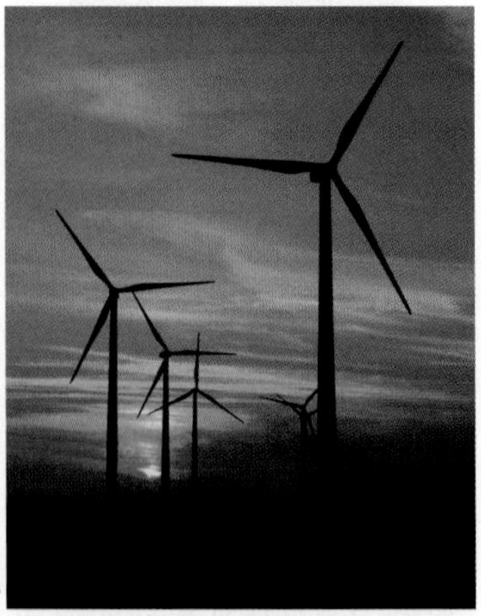
LightShaper/Shutterstock.com

Windmills in a wind farm.

Geothermal energy is derived from hot springs or other sources of heat within Earth.

Energy used by our society and by living systems comes from many sources. When you drive a car, the gasoline or diesel fuel that provides the energy may have come from the United States or another country. Some vehicles are powered by natural gas (buses), electricity (electric cars), or a combination of electricity and gasoline (hybrid cars). Electricity used in your home might have come from a power plant that burns coal or natural gas, from a nuclear power plant, or from a wind farm. Foods contain carbohydrate, fat, protein, and other nutrients that provide energy. The ideas developed in this chapter apply to all of these energy sources, which, in principle, are interchangeable: The work a given quantity of energy can do is the same, regardless of the energy's source.

A **chemical fuel** is *any substance that reacts rapidly and exothermically with atmospheric oxygen and is available at reasonable cost and in reasonable quantity.* When a fuel burns, the products should create as little environmental damage as possible. As indicated in Figure 4.17, most of the fuels that supply us with thermal energy are fossil fuels: coal, petroleum, and natural gas. Biofuels—wood, peat, other plant matter, or ethanol derived from plants—are a distant second among chemical fuels. A significant quantity of energy comes from nuclear reactors and hydroelectric power plants, and a much smaller (see Figure 10.4) quantity comes from solar, wind, and geothermal sources.

Foods consist mainly of carbohydrate, fat, and protein. *Carbohydrates* have the general formula $C_x(H_2O)_y$. Carbohydrates are metabolized quickly and are not usually stored in the body. They are converted in the intestines to glucose, $C_6H_{12}O_6$, which is soluble in

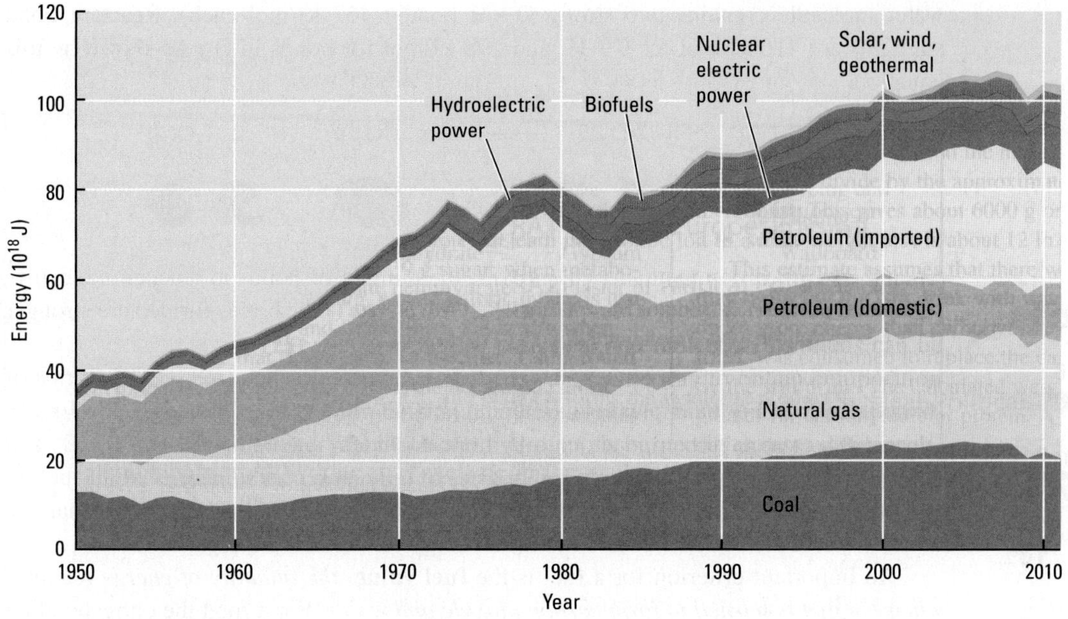

120
100
80
60
40
20
0
Energy (10^{18} J)
1950
1960
1970
1980
1990
2000
2010
Year
Hydroelectric power
Biofuels
Nuclear electric power
Solar, wind, geothermal
Petroleum (imported)
Petroleum (domestic)
Natural gas
Coal

Figure 4.17 Energy resources. Use of energy resources in the United States is plotted from 1950 to 2011. (An energy resource is a naturally occurring fuel, such as petroleum, or a continuous supply, such as sunlight.) At the midpoint of the twentieth century, coal and petroleum were almost equally important, with natural gas coming in third. Today, petroleum and natural gas are used in greater quantities than coal, and more than half of U. S. petroleum is imported. Nuclear electric power did not exist in 1950 but contributes significantly to energy resources today, whereas hydroelectric electricity generation has grown only slightly since 1950.

blood and thereby can be transported throughout the body. Glucose is metabolized to $CO_2(g)$ and $H_2O(\ell)$,

$$C_6H_{12}O_6(s) + 6\ O_2(g) \longrightarrow 6\ CO_2(g) + 6\ H_2O(\ell) \qquad \Delta_r H° = -2801.6 \text{ kJ/mol}$$

This overall reaction transfers energy that maintains body temperature, powers muscles, transmits nerve impulses, and causes chemical reactions that construct and repair tissues to occur. Energy from carbohydrates that is not needed for these purposes is stored in fats.

This discussion answers the question "Where does the energy come from to make my muscles work?" that was posed in Chapter 1 (← Sec. 1-1).

Fat molecules consist largely of long chains of carbon atoms with hydrogen atoms attached. They are excellent storehouses for energy because they are insoluble in water and therefore are not excreted easily. Like carbohydrates, fats are metabolized to $CO_2(g)$ and $H_2O(\ell)$. For example, the fat glyceryl tristearate is oxidized according to the thermochemical expression

$$2\ C_{57}H_{110}O_6(s) + 163\ O_2(g) \longrightarrow 114\ CO_2(g) + 110\ H_2O(\ell) \qquad \Delta_r H° = -75{,}520 \text{ kJ}$$

Protein contains nitrogen in addition to carbon, hydrogen, and oxygen. The nitrogen is metabolized in the body to produce new proteins or urea, $(NH_2)_2CO(aq)$, which is excreted. The carbon, hydrogen, and oxygen are metabolized in pathways that release energy. On average, protein metabolism releases 4 Cal/g (17 kJ/g).

From an energy perspective, what nanoscale characteristics make a good fuel or food? If some or all of the bonds in reactant molecules are weak, or if bonds in products of combustion or metabolism are strong, then the reaction is exothermic. Or, if more bonds are formed the reaction is exothermic.

Coal, petroleum, and natural gas consist mainly of hydrocarbon molecules. When these fuels burn completely in air, they produce water and carbon dioxide. A carbon dioxide molecule contains two very strong carbon-oxygen bonds (803 kJ/mol each), and a water molecule contains two strong O—H bonds (467 kJ/mol each). Reactant bond strengths are 416 kJ/mol for C—H and 498 kJ/mol for bonds in O_2. As shown by this equation,

$$CH_4(g) + 2\ O_2(g) \longrightarrow CO_2(g) + 2\ H_2O(g)$$

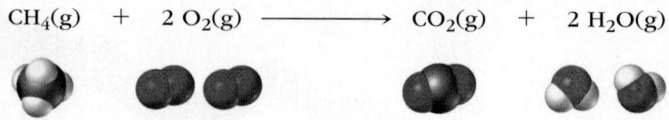

when the hydrocarbon methane, CH_4, burns, the number of bonds in reactant molecules equals the number of bonds in product molecules. Because the bonds formed are stronger than the bonds broken, the reaction is exothermic.

Metabolism of carbohydrates, fats, and proteins can also be understood in terms of bond enthalpies. For example, fats contain mostly C—H and C—C bonds (356 kJ/mol). When fats are metabolized, each carbon atom becomes bonded to oxygen in CO_2 and each hydrogen becomes bonded to oxygen in H_2O. Because the number of bonds before and after reaction is nearly the same, and because the bonds formed are stronger than the bonds broken, metabolism of fats transfers energy to the surroundings.

An important criterion for a fuel is the **fuel value**, *the quantity of energy released when 1 g fuel is burned to form carbon dioxide and water*. For a food the corresponding criterion is **caloric value**, *the energy transferred to the body when 1 g food is metabolized*. Caloric values for some foods are in Table 4.3.

Another important characteristic of a fuel is its **energy density**, *the quantity of energy released per unit volume of fuel*. For gaseous fuels the fuel value may be high, but the energy density is low because the density of a gas is low. Fuels with low energy density require a large volume for storage and therefore are not as convenient to use in cars or airplanes, where portability is important. Gaseous fuels, such as natural gas, can be supplied to fixed locations via pipes (mains) or, like propane and butane, can be condensed and stored as liquids under pressure (liquefied petroleum gas or LP gas).

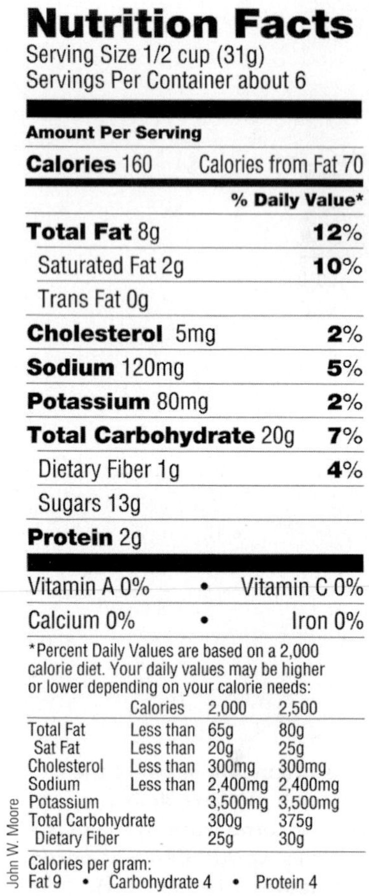
Nutrition Facts
Serving Size 1/2 cup (31g)
Servings Per Container about 6

Amount Per Serving	
Calories 160	Calories from Fat 70
	% Daily Value*
Total Fat 8g	**12%**
Saturated Fat 2g	**10%**
Trans Fat 0g	
Cholesterol 5mg	**2%**
Sodium 120mg	**5%**
Potassium 80mg	**2%**
Total Carbohydrate 20g	**7%**
Dietary Fiber 1g	**4%**
Sugars 13g	
Protein 2g	
Vitamin A 0% •	Vitamin C 0%
Calcium 0% •	Iron 0%

*Percent Daily Values are based on a 2,000 calorie diet. Your daily values may be higher or lower depending on your calorie needs:

		Calories 2,000	2,500
Total Fat	Less than	65g	80g
Sat Fat	Less than	20g	25g
Cholesterol	Less than	300mg	300mg
Sodium	Less than	2,400mg	2,400mg
Potassium		3,500mg	3,500mg
Total Carbohydrate		300g	375g
Dietary Fiber		25g	30g

Calories per gram:
Fat 9 • Carbohydrate 4 • Protein 4

John W. Moore

Nutrition label from a package of caramelized popcorn.

Table 4.3 Composition and Caloric Values of Some Foods

Food	Approximate Composition per 100. g			Caloric Value	
	Fat	Carbohydrate	Protein	Cal/g	kJ/g
Apple	0.5	13.0	0.4	0.59	2.47
Brownie with nuts	16.0	64.0	4.0	4.04	16.9
Cheese pizza	10.2	25.8	11.2	2.41	10.1
Green beans	0.0	7.0	1.9	0.38	0.00
Hamburger	30.0	0.0	22.0	3.60	15.06
Microwave popcorn (popped)	7.1	11.4	2.9	1.00	4.18
Peanuts (unsalted)	50.0	21.4	28.6	5.71	23.91

PROBLEM-SOLVING EXAMPLE 4.15

Comparing Fuels and Foods

Evaluate each fuel listed on the basis of fuel value and each food on the basis of caloric value. Assume that $H_2O(g)$ is formed for fuels and $H_2O(\ell)$ is formed for foods; use data from thermochemical expressions in this section, Table 4.2, or Appendix J. Which provides the largest fuel value or caloric value?

(a) Methane, $CH_4(g)$
(b) Glucose, $C_6H_{12}O_6(s)$
(c) Glyceryl tristearate, $C_{57}H_{110}O_6(s)$
(d) Ethanol (fuel), $C_2H_5OH(\ell)$

Result

(a) 50.013 kJ/g CH_4
(b) 15.551 kJ/g $C_6H_{12}O_6$
(c) 42.36 kJ/g $C_{57}H_{110}O_6$
(d) 26.807 kJ/g C_2H_5OH

Methane has the highest fuel value; fat has the highest caloric value.

Analyze From a balanced chemical equation and thermodynamic data, reaction enthalpies can be calculated; from them energy transfer per gram can be obtained.

Plan Write balanced equations and calculate the reaction enthalpy or find thermochemical expressions in the preceding text. Then, use molar mass to calculate the energy transfer per gram of each substance.

Execute For methane:

$$CH_4(g) + 2\ O_2(g) \longrightarrow CO_2(g) + 2\ H_2O(g)$$

$$\Delta_r H^\circ = [(-393.509\ \text{kJ/mol}) + 2(-241.818\ \text{kJ/mol})] - (-74.81\ \text{kJ/mol})$$
$$= -802.34\ \text{kJ/mol}$$

$$\text{Fuel value} = \frac{802.34\ \text{kJ}}{1\ \text{mol rxn}} \times \frac{1\ \text{mol rxn}}{1\ \text{mol CH}_4} \times \frac{1\ \text{mol CH}_4}{16.0426\ \text{g CH}_4} = 50.013\ \text{kJ/g CH}_4$$

Calculate the fuel/caloric values for the other substances in a similar manner.

PROBLEM-SOLVING PRACTICE 4.15

Evaluate each fuel listed on the basis of fuel value and energy density. Use data from Appendix J. Which fuel provides the largest fuel value? Which provides the greatest energy density? Assume that when the fuels burn, carbon is converted to gaseous CO_2, hydrogen to water vapor, and nitrogen to N_2 gas. The densities are given with the substances.

(a) Octane, $C_8H_{18}(\ell)$ (0.703 g/mL)
(b) Hydrazine, $N_2H_4(\ell)$ (1.004 g/mL)
(c) Methanol, $CH_3OH(\ell)$ (0.792 g/mL)

CONCEPTUAL EXERCISE 4.19

Fuel/Caloric Values and Oxygen

Correlate the fuel values and caloric values calculated in Problem-Solving Example 4.15 with the relative number of oxygen atoms per molecule. Provide a nanoscale explanation for any correlation you find.

There is a balance between the quantity of food energy that is taken into our bodies and the quantity that is used for body functions. If food intake exceeds consumption, the body stores energy in fat molecules. If consumption exceeds intake, some fat is burned (metabolized) to provide the needed energy. The **basal metabolic rate (BMR)** is the minimum energy intake required to maintain a body that is awake and at rest. The BMR varies considerably depending on age, gender, and body mass. For a 70-kg (155-lb) human between 18 and 30 years old, the average BMR is 1750 Cal/day for a male and 1525 Cal/day for a female. The basal metabolic rate is approximately 1 Cal $kg^{-1}\ h^{-1}$; that is, about 1 Calorie is expended per hour for each kilogram of body mass. Thus, the average 60-kg (132–lb) person has a daily BMR of

$$\frac{1\ \text{Cal}}{\text{kg} \times \text{h}} \times 60\ \text{kg} \times \frac{24\ \text{h}}{\text{day}} = 1440\ \text{Cal/day}$$

This value depends on the level of muscular activity: Walking or other light work requires 2.5 times the BMR; heavy work, such as playing basketball or soccer, requires 7 times the BMR. Therefore, the appropriate food energy intake varies greatly from one individual to another.

EXERCISE 4.20

Power of a Person

Use the average BMR values in the text to calculate the power required to sustain a 70-kg male (a) at rest and (b) playing basketball. Express your result in watts. (1 W = 1 J/s.) Repeat the calculation for yourself (using your body mass). Compare your results with the power of a typical incandescent light bulb.

EXERCISE 4.21

Calories and Exercise

A 70-kg, 22-year-old female eats 50. g unsalted peanuts and then exercises by playing basketball for 30 min.

(a) Determine what fraction of the food energy of the peanuts comes from fat, carbohydrate, and protein.

(b) Is the exercise sufficient to use up the energy provided by the food?

SUMMARY PROBLEM

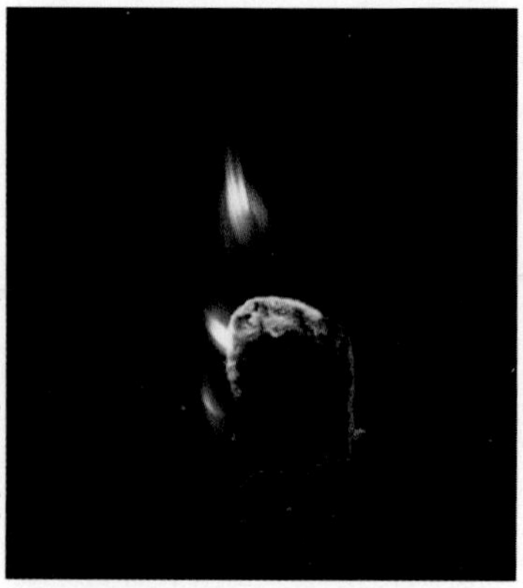

Burning sugar cube.

Sugar cubes are sometimes used to sweeten hot coffee or tea. A sugar cube is 1 cm on a side, weighs 4.5 g, and is mostly sucrose, $C_{12}H_{22}O_{11}$. It will ignite and burn if its surface is coated with a fine powder such as talcum powder or powdered cinnamon. You have 250 mL coffee in a 150-g glass mug; it has gotten cold and you want to reheat it from 25 °C to 60. °C. Your microwave is not working so you decide to burn sugar under the coffee mug to heat it.

(a) Choose an appropriate system and describe the energy transfers that occurred as the hot coffee cooled to room temperature. Make a diagram similar to Figure 4.9 and assign values to each of the four energy transfers in the diagram: heat transfer in, heat transfer out, work transfer in, and work transfer out. (Assume that the coffee has the same specific heat capacity and density as water and was at 60. °C initially.)

(b) Based on its formula, explain why sucrose cannot be a protein.

(c) The standard formation enthalpy of sucrose is −2221.2 kJ/mol. Calculate the standard combustion enthalpy for sucrose. Assume that the combustion products are carbon dioxide gas and water vapor.

(d) Sucrose is composed of equal amounts of two other sugars, glucose, $C_6H_{12}O_6$, and fructose, $C_6H_{12}O_6$. The two components can be obtained from sucrose by reacting it with a common substance. Examine the formulas of sucrose, glucose, and fructose and decide what the other substance is. Write a balanced equation for the reaction.

(e) Calculate the caloric value of sucrose. Compare your result with the caloric values of glucose (15.555 kJ/g) and fructose (15.600 kJ/g).

(f) In part (e) you should have found the caloric value of sucrose is about the same as that of glucose and fructose; explain why this is.

(g) Notice that the caloric values of glucose and fructose are different but their chemical formulas are the same. What can you conclude about the molecular structures of these two compounds from this information?

(h) Determine the minimum number of sugar cubes you would have to burn to reheat your coffee and mug. (Assume that the heating is 50% efficient; that is, only half the heat transfer from combustion of the sugar is to the coffee and mug.)

HAVING STUDIED THIS CHAPTER . . .

. . . you should be able to:

- Understand the difference between kinetic energy and potential energy (Section 4-1). End-of-chapter Question 17
- Be familiar with typical units for energy and power; be able to convert from one unit to another (Sections 4-1 and 4-11). Questions 9, 11, 13
- Understand conservation of energy and energy transfer by heating and working (Section 4-2). Questions 15, 17
- Recognize and use thermodynamic terms: system, surroundings, heat, work, temperature, thermal equilibrium, exothermic, endothermic, and state function (Sections 4-2, 4-3, and 4-4). Questions 19, 21, 35
- Use specific heat capacity, molar heat capacity, and the sign conventions for transfer of energy (Section 4-4). Questions 23, 25, 27, 31, 33
- Distinguish between the change in internal energy and the change in enthalpy for a system (Section 4-5). Question 41
- Use thermochemical expressions and calculate heat transfers given mass or amount of reactant or product (Sections 4-5 and 4-6). Questions 39, 43, 45, 55
- Understand the origin of the reaction enthalpy for a chemical reaction in terms of bond enthalpies (Section 4-7). Questions 59, 61
- Describe how calorimeters can measure the quantity of thermal energy transferred during a reaction (Section 4-8). Questions 64, 68, 70
- Apply Hess's law to find the reaction enthalpy for a reaction (Section 4-9). Questions 72, 74
- Use standard formation enthalpies to calculate the heat energy transfer when a reaction takes place (Section 4-10). Questions 76, 77, 79, 83
- Define and give examples of some chemical fuels and foods and evaluate their abilities to provide heating (Section 4-11). Questions 87, 89, 91

KEY TERMS

basal metabolic rate (BMR) (Section 4-11)

bond energy (4-7a)

bond enthalpy (4-7a)

caloric value (4-11)

calorimeter (4-8)

change of state (4-5a)

chemical fuel (4-11)

conservation of energy, law of (4-2)

endothermic (4-5a)

energy (4-1)

energy density (4-11)

enthalpy change (ΔH) (4-5b)

exothermic (4-5a)

first law of thermodynamics (4-2)

fuel value (4-11)

fusion enthalpy ($\Delta_{fus}H$) (4-5c)

heat/heating (4-2b)

heat capacity (4-4)

Hess's law (4-9)

internal energy (4-3)

kinetic energy (4-1)

molar heat capacity (4-4b)

mole of reaction (4-6)

phase change (4-5a)

potential energy (4-1)

power (4-1a)

specific heat capacity (4-4a)

standard formation enthalpy ($\Delta_f H°$) (4-10)

standard reaction enthalpy ($\Delta_r H°$) (4-6)

standard state (4-10)

state function (4-5e)

surroundings (4-3)

system (4-3)

thermal equilibrium (4-2b)

thermochemical expression (4-6)

thermodynamics (Introduction)

vaporization enthalpy ($\Delta_{vap}H$) (4-5d)

work/working (4-2a)

QUESTIONS FOR REVIEW AND THOUGHT

Red-numbered questions have short answers at the back of this book in Appendix M and fully worked solutions in the *Student Solutions Manual.*

Review Questions

These questions test vocabulary and simple concepts.

1. Name two laws stated in this chapter and explain each in your own words.
2. For each situation, define a system and its surroundings, and give the direction of heat transfer:
   (a) Propane is burning in a Bunsen burner in the laboratory.
   (b) After you have a swim, water droplets on your skin evaporate.
   (c) Water, originally at 25 °C, is placed in the freezing compartment of a refrigerator.
   (d) Two chemicals are mixed in a flask on a laboratory bench. A reaction occurs and heat is evolved.
3. What is the value of the standard formation enthalpy for any element under standard conditions?
4. Criticize each of these statements:
   (a) Formation enthalpy refers to a reaction in which 1 mol of one or more reactants produces some quantity of product.
   (b) The standard formation enthalpy of O_2 as a gas at 25 °C and a pressure of 1 atm is 15.0 kJ/mol.
5. What is required for heat transfer of energy from one sample of matter to another to occur?
6. Name two exothermic processes and two endothermic processes that you encountered recently and that were not associated with your chemistry course.
7. Explain what is meant by (a) energy density of a fuel and (b) caloric value of a food. Why is each of these terms important?
8. Explain in your own words why it is useful in thermodynamics to distinguish a system from its surroundings.

Topical Questions

These questions are keyed to the major topics in this chapter. Usually a question that is answered at the back of this book is paired with a similar one that is not.

The Nature of Energy (Section 4-1)

9. (a) A 2-inch piece of two-layer chocolate cake with frosting provides 1670 kJ of energy. Calculate this in Cal.
   (b) If you were on a diet that calls for eating no more than 1200 Cal per day, calculate how many joules you could consume per day.
10. Sulfur dioxide, SO_2, is found in wines and in polluted air. If a 32.1-g sample of sulfur is burned in the air to get 64.1 g SO_2, 297 kJ of energy is released. Express this energy in (a) joules, (b) calories, and (c) kilocalories.
11. Melting lead requires 5.50 cal/g. Calculate how many joules are required to melt 1.00 lb (454 g) lead.
12. On a sunny day, solar energy reaches Earth at a rate of 4.0 J min^{-1} cm^{-2}. Suppose a house has a square, flat roof of dimensions 12 m by 12 m. Calculate how much solar energy reaches this roof in 1.0 h. (*Note:* This is why roofs painted with light-reflecting paint keep buildings cooler than black, unpainted roofs. The painted roofs reflect most of this energy rather than absorb it.)
13. The energy unit used by electrical utilities in their monthly bills is the *kilowatt hour* (kWh; 1 kilowatt used for 1 hour). Calculate how many joules there are in a kilowatt hour. If electricity costs $.09 per kilowatt hour, calculate how much it costs per megajoule.
14. A 100-W light bulb is left on for 14 h. Calculate how many joules are used. With electricity at $0.09 per kWh, calculate how much it costs to leave the light on for 14 h.

Conservation of Energy (Section 4-2)

15. Describe how energy is changed from one form to another in these processes:
   (a) At a July 4th celebration, a match is lit and ignites the fuse of a rocket firecracker, which fires off and explodes at an altitude of 1000 ft.
   (b) A gallon of gasoline is pumped from an underground storage tank into the fuel tank of your car, and you use it up by driving 25 mi.
16. Analyze transfer of energy from one form to another in each situation.
   (a) In a Space Shuttle, hydrogen and oxygen combine to form water, boosting the shuttle into orbit above Earth.
   (b) You eat a package of Fritos, go to class and listen to a lecture, walk back to your dorm, and climb the stairs to the fourth floor.
17. Analyze this situation in terms of potential and kinetic energy of water molecules: Water flows over a waterfall; the temperature of water at the bottom is higher than at the top.
18. Suppose that you are studying kinetic energy of helium molecules: A helium weather balloon rises to an altitude of 40,000 ft; the temperature of the gas drops to −70 °F.
   (a) Make an appropriate choice of system and surroundings and describe it unambiguously.
   (b) Explain why you chose the system and surroundings you did.
   (c) Identify transfers of energy and material into and out of the system that would be important for you to monitor in your study.

Energy Transfers (Section 4-3)

19. Solid ammonium chloride is added to water in a beaker and dissolves. The beaker becomes cold to the touch.
    (a) Make an appropriate choice of system and surroundings and describe it unambiguously.
    (b) Explain why you chose the system and surroundings you did.
    (c) Identify transfers of energy and material into and out of the system that would be important for you to monitor in your study.
    (d) Is the process of dissolving $NH_4Cl(s)$ in water exothermic or endothermic? Explain your answer.
20. A bar of Monel (an alloy of nickel, copper, iron, and manganese) is heated until it melts, poured into a mold, and solidifies.
    (a) Make an appropriate choice of system and surroundings and describe it unambiguously.
    (b) Explain why you chose the system and surroundings you did.
    (c) Identify transfers of energy and material into and out of the system that would be important for you to monitor in your study.
21. Make a diagram like the one in Figure 4.9 for a system in which 127.6 kJ of work is done on the surroundings and there is 843.2 kJ of heat transfer into the system. Use the diagram to help you determine ΔE_{system}.
22. Make a diagram like the one in Figure 4.9 for a system in which 876.3 J of work is done on the surroundings and there is 37.4 J of heat transfer into the system. Use the diagram to help you determine ΔE_{system}.

Heat Capacity (Section 4-4)

23. Determine which requires greater transfer of energy: (a) cooling 10.0 g water from 50. °C to 20. °C or (b) cooling 20.0 g Cu from 37 °C to 25 °C.
24. Determine which requires more energy transfer: (a) warming 15.0 g water from 25 °C to 37 °C or (b) warming 60.0 g aluminum from 25 °C to 37 °C.
25. You hold a gram of copper in one hand and a gram of aluminum in the other. Each metal was originally at 0 °C. (Both metals are in the shape of a little ball that fits into your hand.) If energy is transferred to both at the same rate, which will warm to your body temperature first? Explain your answer.
26. Ethylene glycol, $HOCH_2CH_2OH$, is often used as a coolant in cars. Determine which requires greater transfer of thermal energy to warm from 25.0 °C to 100.0 °C, pure water or an equal mass of pure ethylene glycol.
27. The specific heat capacity of benzene, C_6H_6, is 1.74 J g^{-1} K^{-1}. Calculate its molar heat capacity.
28. The specific heat capacity of carbon tetrachloride, CCl_4, is 0.861 J g^{-1} K^{-1}. Calculate its molar heat capacity.
29. A 237-g piece of molybdenum, initially at 100.0 °C, is dropped into 244 g water at 10.0 °C. When the system comes to thermal equilibrium, the temperature is 14.9 °C. Calculate the specific heat capacity of molybdenum.
30. A sample of glass beads weighs 34.5 g. The beads are heated to 100.0 °C in a boiling water bath and poured into a beaker containing 100.0 g H_2O at 25.0 °C. When thermal equilibrium is reached, the temperature of the glass and the water is 29.9 °C. Calculate the specific heat capacity of the glass.
31. A piece of iron (400. g) is heated in a flame and then plunged into a beaker containing 1.00 kg water. The original temperature of the water was 20.0 °C, but it is 32.8 °C after the iron bar is put in and thermal equilibrium is reached. Calculate the original temperature of the hot iron bar.
32. A 192-g piece of copper was heated to 100.0 °C in a boiling water bath and then put into a beaker containing 750. mL water (density = 1.00 g/cm^3) at 4.0 °C. Calculate the final temperature of the copper and water after they come to thermal equilibrium.
33. An unknown metal requires 34.7 J to heat a 23.4-g sample of it from 17.3 °C to 28.9 °C. Determine which of the metals in Table 4.1 is most likely to be the unknown.
34. An unknown metal requires 336.9 J to heat a 46.3-g sample of it from 24.3 °C to 43.2 °C. Determine which of the metals in Table 4.1 is most likely to be the unknown.

Energy and Enthalpy (Section 4-5)

35. A chemical reaction occurs, and 20.7 J is transferred from the chemical system to its surroundings. (Assume that no work is done.)
    (a) What is the algebraic sign of $\Delta T_{surroundings}$?
    (b) What is the algebraic sign of $\Delta_r E_{system}$?
36. A phase transition occurs in a sample of an alloy, and 437 kJ transfers from the surroundings to the alloy. (Assume that no work is done.)
    (a) What is the algebraic sign of ΔT_{alloy}?
    (b) What is the algebraic sign of $\Delta_r E_{alloy}$?
37. The heat of fusion of mercury is 2.72 cal/g. Calculate the quantity of energy transferred when 4.37 mol Hg freezes at a temperature of −39 °C.
38. Chloromethane, CH_3Cl, arises from microbial fermentation and is found throughout the environment. It is also produced industrially and is used in the manufacture of various chemicals and has been used as a topical anesthetic. Calculate how much energy is required to convert 92.5 g liquid to a vapor at its boiling point, −24.09 °C. (The heat of vaporization of CH_3Cl is 21.40 kJ/mol.)
39. The freezing point of mercury is −38.8 °C. Calculate what quantity of energy, in joules, is released to the surroundings if 1.00 mL mercury is cooled from 23.0 °C to −38.8 °C and then frozen to a solid. (The density of liquid mercury is 13.6 g/cm^3. Its specific heat capacity is 0.140 J g^{-1} K^{-1} and its heat of fusion is 11.4 J g^{-1}.)
40. Calculate the quantity of heating required to convert the water in four ice cubes (60.1 g each) from $H_2O(s)$ at 0 °C to $H_2O(g)$ at 100. °C. The enthalpy of fusion of ice is 333 J/g and the enthalpy of vaporization of liquid water is 2260 J/g.

41. What is the sign of w for these processes if they occur at constant pressure? Consider only $P\Delta V$ work from gases.
(a) $Fe_2S_3(s) + 6\ HNO_3(aq) \longrightarrow 2\ Fe(NO_3)_3(aq) + 3\ H_2S(g)$
(b) $C_3H_8(g) + 5\ O_2(g) \longrightarrow 3\ CO_2(g) + 4\ H_2O(\ell)$

42. Assume that these reactions occur under constant atmospheric pressure. What is the sign of w for each?
(a) $CaO(s) + 3\ C(s) \longrightarrow CaC_2(s) + CO(g)$
(b) $2\ C_6H_6(\ell) + 15\ O_2(g) \longrightarrow 12\ CO_2(g) + 6\ H_2O(\ell)$

43. Calculate how much energy must be transferred to vaporize 125 g benzene, C_6H_6, at its boiling point, 80.1 °C. (The enthalpy of vaporization of benzene is 30.8 kJ/mol.)

44. The enthalpy of fusion (melting) of water is 6.0 kJ/mol. Calculate the quantity of energy that must be transferred to melt 25.0 g H_2O at 0 °C.

Reaction Enthalpies for Chemical Reactions (Section 4-6)

45. Energy is stored in the body in adenosine triphosphate, ATP, which is formed by the reaction between adenosine diphosphate, ADP, and dihydrogen phosphate ions.

$$ADP^{3-}(aq) + H_2PO_4^{2-}(aq) \longrightarrow ATP^{4-}(aq) + H_2O\ (\ell) \qquad \Delta_r H^\circ = 20.5\ \text{kJ/mol}$$

Is the reaction endothermic or exothermic?

46. Calcium carbide, CaC_2, is manufactured by reducing lime with carbon at high temperature. (The carbide is used in turn to make acetylene, an industrially important organic chemical.)

$$CaO(s) + 3\ C(s) \longrightarrow CaC_2(s) + CO(g) \qquad \Delta_r H^\circ = 464.8\ \text{kJ/mol}$$

Is the reaction endothermic or exothermic?

47. When calcium oxide, CaO(s), dissolves in water the water becomes hot. Write a chemical equation for this process and indicate whether it is exothermic or endothermic.

48. When table salt is dissolved in water, the temperature drops slightly. Write a chemical equation for this process and indicate if it is exothermic or endothermic.

49. Given the thermochemical expression

$$H_2O(s) \longrightarrow H_2O(\ell) \qquad \Delta_r H^\circ = 6.0\ \text{kJ/mol}$$

calculate what quantity of energy is transferred to the surroundings when
(a) 34.2 mol liquid water freezes.
(b) 100.0 g liquid water freezes.

50. Given the thermochemical expression

$$CaO(s) + 3\ C(s) \longrightarrow CaC_2(s) + CO(g) \qquad \Delta_r H^\circ = 464.8\ \text{kJ/mol}$$

calculate the quantity of energy transferred when
(a) 34.8 mol CO(g) is formed by this reaction.
(b) A metric ton (1000 kg) of $CaC_2(s)$ is manufactured.
(c) 0.432 mol carbon reacts with CaO(s).

51. Determine the amount of reaction (in moles) that takes place for each process

$$2\ NO(g) + \tfrac{1}{2}\ O_2(g) \longrightarrow N_2O_3(g)$$

(a) 2 mol O_2 reacts
(b) 0.115 mol N_2O_3 forms
(c) 4.73 g NO reacts

52. Determine the amount of reaction (in moles) that takes place for each process

$$\tfrac{1}{2}\ Fe_2O_3(s) \longrightarrow FeO(s) + \tfrac{1}{4}\ O_2(g)$$

(a) 2 mol O_2 forms
(b) 0.824 mol Fe_2O_3 reacts
(c) 1.34 g FeO forms

53. Isooctane (2,2,4-trimethylpentane), one of the many hydrocarbons that make up gasoline, burns in air to give water and carbon dioxide.

$$2\ C_8H_{18}(\ell) + 25\ O_2(g) \longrightarrow 16\ CO_2(g) + 18\ H_2O(\ell) \qquad \Delta_r H^\circ = -10{,}922\ \text{kJ/mol}$$

Calculate the enthalpy change if you burn 1.00 L isooctane (density = 0.69 g/mL).

54. When $KClO_3(s)$, potassium chlorate, is heated, it melts and decomposes to form oxygen gas. [Molten $KClO_3$ was shown reacting with a Fritos chip earlier in this chapter (← **Sec. 4-1a**).] The thermochemical expression for decomposition of potassium chlorate is

$$2\ KClO_3(s) \longrightarrow 2\ KCl(s) + 3\ O_2(g) \qquad \Delta_r H^\circ = -89.4\ \text{kJ/mol}$$

Calculate q at constant pressure for
(a) Formation of 97.8 g KCl(s).
(b) Production of 24.8 mol $O_2(g)$.
(c) Decomposition of 35.2 g $KClO_3(s)$.

55. "Gasohol," a mixture of gasoline and ethanol, C_2H_5OH, is used as automobile fuel. The alcohol releases energy in a combustion reaction with O_2.

$$C_2H_5OH(\ell) + 3\ O_2(g) \longrightarrow 2\ CO_2(g) + 3\ H_2O(\ell)$$

If 0.115 g ethanol evolves 3.62 kJ when burned at constant pressure, calculate the combustion enthalpy for ethanol.

56. White phosphorus, P_4, ignites in air to produce P_4O_{10}.

$$P_4(s) + 5\ O_2(g) \longrightarrow P_4O_{10}(s)$$

When 3.56 g P_4 is burned, 85.8 kJ of thermal energy is evolved at constant pressure. Calculate the combustion enthalpy of P_4.

57. Acetic acid, CH_3CO_2H, is made industrially by the reaction of methanol and carbon monoxide.

$$CH_3OH(\ell) + CO(g) \longrightarrow CH_3COOH(\ell) \qquad \Delta_r H^\circ = -135.3\ \text{kJ/mol}$$

If you produce 1.00 L acetic acid (d = 1.044 g/mL) by this reaction, calculate how much energy is transferred out of the system.

58. When wood is burned we may assume that the reaction is the combustion of cellulose (empirical formula, CH_2O).

$$CH_2O(s) + O_2(g) \longrightarrow CO_2(g) + H_2O(g) \qquad \Delta_r H^\circ = -458\ \text{kJ/mol}$$

Calculate how much energy is released when a 10.0-lb wood log burns completely. (Assume the wood is 100% dry and burns via the reaction given.)

Where Does the Energy Come From? (Section 4-7)

Use these bond enthalpy values to answer Questions 59–62.

Bond	Bond Enthalpy (kJ/mol)	Bond	Bond Enthalpy (kJ/mol)
H—F	566	F—F	158
H—Cl	431	Cl—Cl	242
H—Br	366	Br—Br	193
H—I	299	I—I	151
H—H	436		

59. Which molecule, HF, HCl, HBr, or HI, has the strongest chemical bond?
60. Which molecule, F_2, Cl_2, Br_2, or I_2, has the weakest chemical bond?
61. For the reactions of molecular hydrogen with fluorine and with chlorine:
    (a) Calculate the enthalpy change for breaking all the bonds in the reactants.
    (b) Calculate the enthalpy change for forming all the bonds in the products.
    (c) From the results in parts (a) and (b), calculate the enthalpy change for the reaction.
    (d) Which reaction is most exothermic?
62. For the reactions of molecular hydrogen with bromine and with iodine:
    (a) Calculate the enthalpy change for breaking all the bonds in the reactants.
    (b) Calculate the enthalpy change for forming all the bonds in the products.
    (c) From the results in parts (a) and (b), calculate the enthalpy change for the reaction.
    (d) Which reaction is most exothermic?
63. A diamond can be considered a giant all-carbon supermolecule in which almost every carbon atom is bonded to four other carbons. When a diamond cutter cleaves (splits) a diamond, carbon-carbon bonds must be broken. Is the cleavage (splitting) of a diamond endothermic or exothermic? Explain.

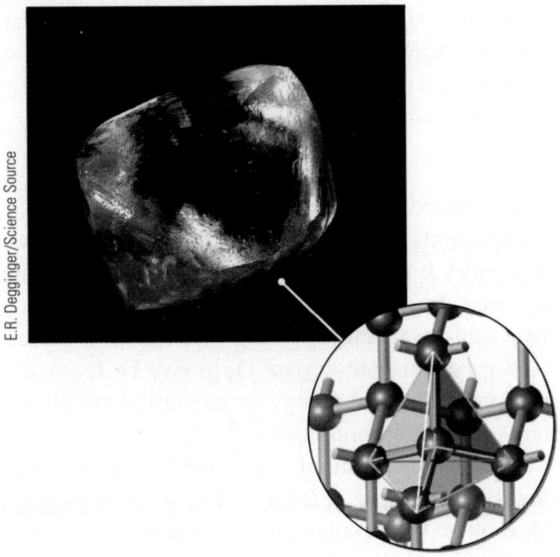
E.R. Degginger/Science Source

Measuring Reaction Enthalpies: Calorimetry (Section 4-8)

64. When 0.100 g CaO(s) is added to 125 g H_2O at 23.6 °C in a coffee cup calorimeter, this reaction occurs.

$$CaO(s) + H_2O(\ell) \longrightarrow Ca(OH)_2(aq) \quad \Delta_r H^\circ = -81.9 \text{ kJ/mol}$$

Calculate the final temperature of the solution.

65. A coffee cup calorimeter can be used to investigate the "cold pack reaction," the process that occurs when solid ammonium nitrate dissolves in water:

$$NH_4NO_3(s) \longrightarrow NH_4^+(aq) + NO_3^-(aq)$$

Suppose 25.0 g solid NH_4NO_3 at 23.0 °C is added to 250. mL H_2O at the same temperature. After all of the solid dissolves, the temperature is measured to be 15.6 °C. Calculate the reaction enthalpy for the cold pack reaction. (Assume that the specific heat capacity of the solution is the same as for water.) Is the reaction endothermic or exothermic?

66. When a 13.0-g sample of NaOH(s) dissolves in 400.0 mL water in a coffee cup calorimeter, the temperature of the water changes from 22.6 °C to 30.7 °C. Assuming that the specific heat capacity of the solution is the same as for water, calculate
    (a) The heat transfer from system to surroundings.
    (b) $\Delta_r H$ for the reaction.

$$NaOH(s) \longrightarrow Na^+(aq) + OH^-(aq)$$

67. Suppose that you mix 200.0 mL of 0.200-M RbOH(aq) with 100. mL of 0.400-M HBr(aq) in a coffee cup calorimeter. If the temperature of each of the two solutions was 24.40 °C before mixing, and the temperature rises to 26.18 °C,
    (a) Calculate the heat transfer as a result of the reaction.
    (b) Write the thermochemical expression for the reaction.
68. A 0.692-g sample of glucose, $C_6H_{12}O_6$, is burned in a constant-volume calorimeter. The temperature rises from 21.70 °C to 25.22 °C. The calorimeter contains 575 g water, and the bomb has a heat capacity of 650 J/K. Determine $\Delta_r E$ per mole of glucose.
69. Benzoic acid, $C_7H_6O_2$, occurs naturally in many berries. Suppose you burn 1.500 g of the compound in a combustion calorimeter and find that the temperature of the calorimeter increases from 22.50 °C to 31.69 °C. The calorimeter contains 775 g water, and the bomb has a heat capacity of 893 J $°C^{-1}$. Calculate $\Delta_r E$ per mole of benzoic acid.
70. Design an experiment to directly measure the reaction enthalpy for this reaction

$$2\ C_8H_{18}(\ell) + 25\ O_2(g) \longrightarrow 16\ CO_2(g) + 18\ H_2O(g)$$

Describe the apparatus and how the experiment would be carried out.

71. Design an experiment to directly measure the reaction enthalpy for this reaction

$$2\ NaOH(aq) + H_2SO_4(aq) \longrightarrow Na_2SO_4(aq) + 2\ H_2O(\ell)$$

Describe the apparatus and how the experiment would be carried out.

Hess's Law (Section 4-9)

72. These reaction enthalpies can be measured:

$$C_2H_4(g) + 3\,O_2(g) \longrightarrow 2\,CO_2(g) + 2\,H_2O(\ell) \quad \Delta_rH^\circ = -1411.1 \text{ kJ/mol}$$

$$C_2H_5OH(\ell) + 3\,O_2(g) \longrightarrow 2\,CO_2(g) + 3\,H_2O(\ell) \quad \Delta_rH^\circ = -1367.5 \text{ kJ/mol}$$

Use these values and Hess's law to determine the reaction enthalpy for

$$C_2H_4(g) + H_2O(\ell) \longrightarrow C_2H_5OH(\ell)$$

73. Three reactions very important to the semiconductor industry are
   (a) The reduction of silicon dioxide to crude silicon,

$$SiO_2(s) + 2\,C(s) \longrightarrow Si(s) + 2\,CO(g) \quad \Delta_rH^\circ = 689.9 \text{ kJ/mol}$$

   (b) The formation of silicon tetrachloride from crude silicon,

$$Si(s) + 2\,Cl_2(g) \longrightarrow SiCl_4(g) \quad \Delta_rH^\circ = -657.01 \text{ kJ/mol}$$

   (c) The reduction of silicon tetrachloride to pure silicon with magnesium,

$$SiCl_4(g) + 2\,Mg(s) \longrightarrow 2\,MgCl_2(s) + Si(s) \quad \Delta_rH^\circ = -625.6 \text{ kJ/mol}$$

Calculate the overall enthalpy change when 1.00 mol sand, SiO_2, changes into very pure silicon by this series of reactions.

74. You wish to know the standard formation enthalpy of liquid PCl_3.

$$P_4(s) + 6\,Cl_2(g) \longrightarrow 4\,PCl_3(\ell)$$

These reaction enthalpies have been determined experimentally:

$$P_4(s) + 10\,Cl_2(g) \longrightarrow 4\,PCl_5(s) \quad \Delta_rH^\circ = -1774.0 \text{ kJ/mol}$$

$$PCl_3(\ell) + Cl_2(g) \longrightarrow PCl_5(s) \quad \Delta_rH^\circ = -123.8 \text{ kJ/mol}$$

Calculate the formation enthalpy for $PCl_3(\ell)$.

75. Calculate the standard reaction enthalpy, Δ_rH°, for formation of 1 mol strontium carbonate (the material that gives the red color in fireworks) from its elements.

$$Sr(s) + C(\text{graphite}) + \tfrac{3}{2}\,O_2(g) \longrightarrow SrCO_3(s)$$

The information available is

$$Sr(s) + \tfrac{1}{2}\,O_2(g) \longrightarrow SrO(s) \quad \Delta_rH^\circ = -592 \text{ kJ/mol}$$

$$SrO(s) + CO_2(g) \longrightarrow SrCO_3(s) \quad \Delta_rH^\circ = -234 \text{ kJ/mol}$$

$$C(\text{graphite}) + O_2(g) \longrightarrow CO_2(g) \quad \Delta_rH^\circ = -394 \text{ kJ/mol}$$

Standard Formation Enthalpies (Section 4-10)

76. Write a balanced thermochemical expression depicting the formation of 1 mol of each compound. Standard formation enthalpies are found in Appendix J.
   (a) $Al_2O_3(s)$ (b) $TiCl_4(\ell)$
   (c) $NH_4NO_3(s)$ (d) $CH_3OH(\ell)$

77. Given

$$2\,Al_2O_3(s) \longrightarrow 4\,Al(s) + 3\,O_2(g) \quad \Delta_rH^\circ = 3351.4 \text{ kJ/mol}$$

   (a) Calculate the standard formation enthalpy of aluminum oxide.
   (b) Calculate the energy transfer for formation of 12.50 g aluminum oxide.

78. Calorimetric measurements show that the reaction of magnesium with chlorine releases 26.4 kJ per gram of magnesium reacted.
   (a) Write a balanced chemical equation for the reaction.
   (b) Calculate the standard formation enthalpy of magnesium chloride.

79. We burn 3.47 g lithium in excess oxygen at constant atmospheric pressure to form Li_2O. Then, we bring the reaction mixture back to 25 °C. In this process 146 kJ of heat is given off. Calculate the standard formation enthalpy of Li_2O.

80. When 43.2 g Mg reacts with sufficient sulfur, 615 kJ is transferred to the surroundings. Calculate the standard formation enthalpy for MgS.

81. Use data in Appendix J to find the reaction enthalpy for
   (a) $CaCO_3(s) \longrightarrow CaO(s) + CO_2(g)$
   (b) $2\,HI(g) + F_2(g) \longrightarrow 2\,HF(g) + I_2(s)$
   (c) $SF_6(g) + 3\,H_2O(\ell) \longrightarrow 6\,HF(g) + SO_3(g)$

82. Use data in Appendix J to calculate reaction enthalpy when
   (a) 0.054 g sulfur burns, forming $SO_2(g)$
   (b) 0.20 mol $HgO(s)$ decomposes to $Hg(\ell)$ and $O_2(g)$
   (c) 2.40 g $NH_3(g)$ is formed from $N_2(g)$ and excess $H_2(g)$
   (d) 1.05×10^{-2} mol carbon is oxidized to $CO_2(g)$

83. The reaction enthalpy for oxidation of styrene, C_8H_8, has been measured by calorimetry.

$$C_8H_8(\ell) + 10\,O_2(g) \longrightarrow 8\,CO_2(g) + 4\,H_2O(\ell) \quad \Delta_rH^\circ = -4395.0 \text{ kJ/mol}$$

Use this value, along with the data from Table 4.2, to calculate the standard formation enthalpy of styrene, in kJ/mol.

84. Oxygen is not normally found in positive oxidation states, but when it is combined with fluorine in the compound oxygen difluoride, oxygen is in a +2 oxidation state. Oxygen difluoride is a strong oxidizing agent that oxidizes water to produce O_2 and HF.

$$OF_2(g) + H_2O(g) \longrightarrow 2\,HF(g) + O_2(g)$$

The standard formation enthalpy of oxygen difluoride is 24.5 kJ/mol. Using this information and data from Appendix J, calculate the standard reaction enthalpy for the reaction of oxygen difluoride with water.

85. Iron can react with oxygen to give iron(III) oxide. If 5.58 g Fe is heated in pure O_2 to give $Fe_2O_3(s)$, calculate how much thermal energy is transferred out of this system (at constant pressure).

86. The formation of aluminum oxide from its elements is highly exothermic. If 2.70 g Al metal is burned in pure O_2 to give Al_2O_3, calculate how much thermal energy is evolved in the process (at constant pressure).

Fuels (Section 4-11)

87. You want to heat the air in your house with natural gas, CH_4. Assume your house has 275 m^2 (about 2800 ft^2) of floor area and that the ceilings are 2.50 m (about 8 ft) from the floors. The air in the house has a molar heat capacity of 29.1 J mol^{-1} K^{-1}. (The number of moles of air in the house can be found by assuming that the average molar mass of air is 28.9 g/mol and that the density of air at these temperatures is 1.22 g/L.) Calculate what mass of methane you have to burn to heat the air from 15.0 °C to 22.0 °C.
88. If you want to convert 56.0 g ice (at 0 °C) to water at 75.0 °C, calculate how many grams of propane, C_3H_8, you would have to burn to supply the energy to melt the ice and then warm it to the final temperature (at 1 bar).
89. Companies around the world are constantly searching for compounds that can be used as substitutes for gasoline in automobiles. One possibility is methanol, CH_3OH, a compound that can be made relatively inexpensively from coal. The alcohol has a smaller energy per gram than gasoline, but with its higher octane rating it burns more efficiently than gasoline in internal combustion engines. (It also contributes smaller quantities of some air pollutants.) Compare the quantity of thermal energy produced per gram of CH_3OH and C_8H_{18} (octane), the latter being representative of the compounds in gasoline.
90. Both hydrazine and 1,1-dimethylhydrazine react with O_2 and can be used as rocket fuels.

$$\underset{\text{hydrazine}}{N_2H_4(\ell)} + O_2(g) \longrightarrow N_2(g) + 2\ H_2O(g)$$

$$\underset{\text{1,1-dimethylhydrazine}}{N_2H_2(CH_3)_2(\ell)} + 4\ O_2(g) \longrightarrow 2\ CO_2(g) + 4\ H_2O(g) + N_2(g)$$

The standard formation enthalpy of liquid hydrazine is 50.6 kJ/mol, and that of liquid 1,1-dimethylhydrazine is 49.2 kJ/mol. By doing appropriate calculations, decide whether the reaction of hydrazine or 1,1-dimethylhydrazine with oxygen gives more heat per gram (at constant pressure). (Other formation enthalpy data can be obtained from Table 4.2.)
91. Suppose you eat a quarter-pound hamburger (no bread, cheese, or other items—just meat) and then take a walk. Determine how long you will need to walk to use up the caloric value of the hamburger.
92. If you eat a quarter-pound of cheese pizza and then go out and play soccer vigorously, determine how long it will take before you use up the calories in the pizza you ate.

General Questions

These questions are not explicitly keyed to chapter topics; many require integration of several concepts.

93. Calculate the quantity of energy, in joules, required to raise the temperature of 454 g tin from room temperature, 25.0 °C, to its melting point, 231.9 °C, and then melt the tin at that temperature. (The specific heat capacity of tin is 0.227 J g^{-1} K^{-1}, and the enthalpy of fusion of this metal is 59.2 J/g.)
94. A 25.0-mL sample of benzene at 19.9 °C was cooled to its melting point, 5.5 °C, and then frozen. Calculate how much heat transfer to the surroundings occurred in this process. (The density of benzene is 0.880 g/mL; its specific heat capacity is 1.74 J g^{-1} K^{-1}, and its enthalpy of fusion is 127 J/g.)
95. You add 100.0 g water at 60.0 °C to 100.0 g ice at 0.00 °C. Some of the ice melts and cools the water to 0.00 °C. Calculate what mass of ice has melted when the ice and water mixture reaches a uniform temperature of 0 °C.
96. The combustion of diborane, B_2H_6, proceeds according to the equation

$$B_2H_6(g) + 3\ O_2(g) \longrightarrow B_2O_3(s) + 3\ H_2O(\ell)$$

and 2166 kJ is liberated per mole of reaction (at constant pressure). Calculate the standard formation enthalpy of $B_2H_6(g)$ using this information, the data in Table 4.2, and the fact that the standard formation enthalpy for $B_2O_3(s)$ is −1273 kJ/mol.
97. From these enthalpies of reaction,

$$CaCO_3(s) \longrightarrow CaO(s) + CO_2(g) \qquad \Delta_r H° = 178.3\ \text{kJ/mol}$$

$$CaO(s) + H_2O(\ell) \longrightarrow Ca(OH)_2(s) \qquad \Delta_r H° = -65.2\ \text{kJ/mol}$$

$$Ca(OH)_2(s) \longrightarrow Ca^{2+}(aq) + 2\ OH^-(aq) \qquad \Delta_r H° = -16.7\ \text{kJ/mol}$$

calculate $\Delta_r H°$ for

$$Ca^{2+}(aq) + 2\ OH^-(aq) + CO_2(g) \longrightarrow CaCO_3(s) + H_2O(\ell)$$

98. The Romans used calcium oxide, CaO, to produce a strong mortar to build stone structures. The CaO was mixed with water to give $Ca(OH)_2$, which reacted slowly with CO_2 in the air to give $CaCO_3$.

$$Ca(OH)_2(s) + CO_2(g) \longrightarrow CaCO_3(s) + H_2O(g)$$

(a) Calculate the standard reaction enthalpy.
(b) Calculate the heat transfer when 1.00 kg $Ca(OH)_2$ reacts with a stoichiometric amount of CO_2.
99. Given this thermochemical expression and data in Table 4.2, calculate the standard formation enthalpy for liquid hydrazine, N_2H_4.

$$N_2H_4(\ell) + O_2(g) \longrightarrow N_2(g) + 2\ H_2O(g) \qquad \Delta_r H° = -534\ \text{kJ/mol}$$

100. How fast (in meters per second) must an iron ball with a mass of 56.6 g be traveling to have a kinetic energy of 15.75 J?
101. In 1947, Texas City, Texas, was devastated by the explosion of a shipload of ammonium nitrate, a compound intended to be used as a fertilizer. When heated, ammonium nitrate can decompose exothermically to N_2O and water.

$$NH_4NO_3(s) \longrightarrow N_2O(g) + 2\ H_2O(g)$$

If the heat from this exothermic reaction is contained, higher temperatures are generated, at which point ammonium nitrate can decompose explosively to N_2, H_2O, and O_2.

$$2\ NH_4NO_3(s) \longrightarrow 2\ N_2(g) + 4\ H_2O(g) + O_2(g)$$

If oxidizable materials are present, fires can break out, as was the case at Texas City. Using the information in Appendix J, answer the following questions.
(a) Calculate the standard reaction enthalpy for the first reaction.
(b) If 8.00 kg ammonium nitrate explodes (the second reaction), calculate how much thermal energy is evolved (under standard conditions).

102. One method of producing H_2 on a large scale is this chemical cycle.

Step 1: $SO_2(g) + 2\ H_2O(g) + Br_2(g) \longrightarrow H_2SO_4(\ell) + 2\ HBr(g)$
Step 2: $H_2SO_4(\ell) \longrightarrow H_2O(g) + SO_2(g) + \frac{1}{2}\ O_2(g)$
Step 3: $2\ HBr(g) \longrightarrow H_2(g) + Br_2(g)$

Using the table of standard formation enthalpies in Appendix J, calculate $\Delta_rH°$ for each step. Write the equation for the overall process and calculate the reaction enthalpy. Is the overall process exothermic?

Applying Concepts

These questions test conceptual learning.

103. Based on your experience, when ice melts to liquid water, is the process exothermic or endothermic? When liquid water freezes to ice at 0 °C, is this exothermic or endothermic? (Assume that the ice/water is the system in each case.) Explain your answers.

104. You pick up a six-pack of soft drinks from the floor, but it slips from your hand and smashes onto your foot. Comment on the work and energy involved in this sequence. What forms of energy are involved at what stages of the process?

105. In the figure, a heating graph is shown for a substance, *X*, at constant pressure.
(a) Which has the largest specific heat capacity, *X*(s), *X*(ℓ), or *X*(g)?
(b) Which is smaller, the heat of fusion or the heat of vaporization?
(c) What is the algebraic sign of the enthalpy of vaporization at the boiling point?

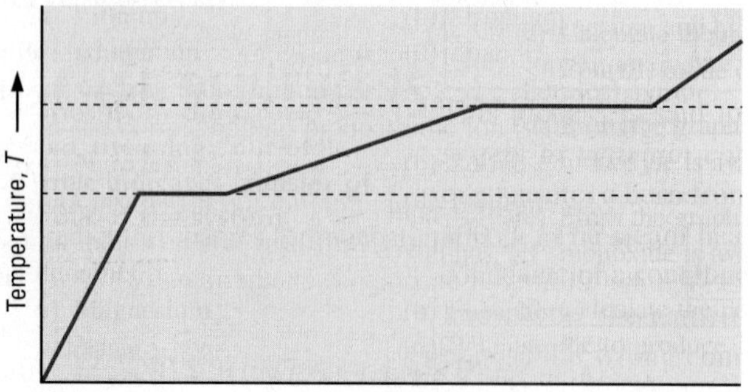
Temperature, T
Energy transferred

106. In the figure, a cooling graph is shown for a substance, *Y*, at constant pressure.
(a) Which has the smallest specific heat capacity, *Y*(s), *Y*(ℓ), or *Y*(g)?
(b) Which is larger, the heat of fusion or the heat of vaporization?
(c) What is the algebraic sign of the enthalpy of fusion at the melting point?

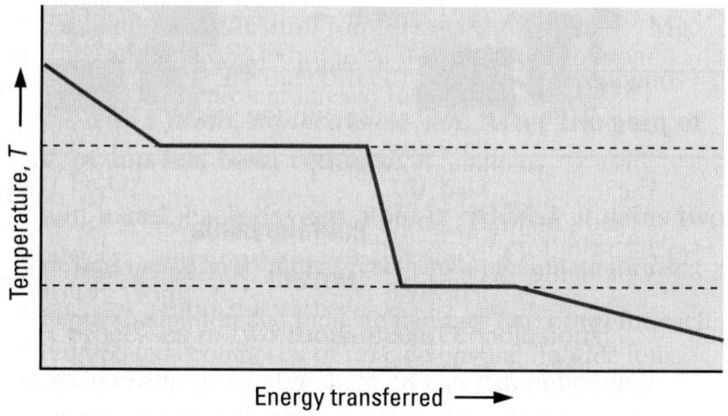
Temperature, T
Energy transferred

107. The specific heat capacity of copper is 0.385 J g^{-1} °C^{-1}, whereas it is 0.128 J g^{-1} °C^{-1} for gold. Assume you place 100. g of each metal, originally at 25 °C, in a boiling water bath at 100 °C. If energy is transferred to each metal at the same rate, determine which piece of metal will reach 100 °C first.

108. Consider this graph, which presents data for a 1.0-g mass of each of four substances, A, B, C, and D. Which substance has the highest specific heat capacity? Explain your answer.

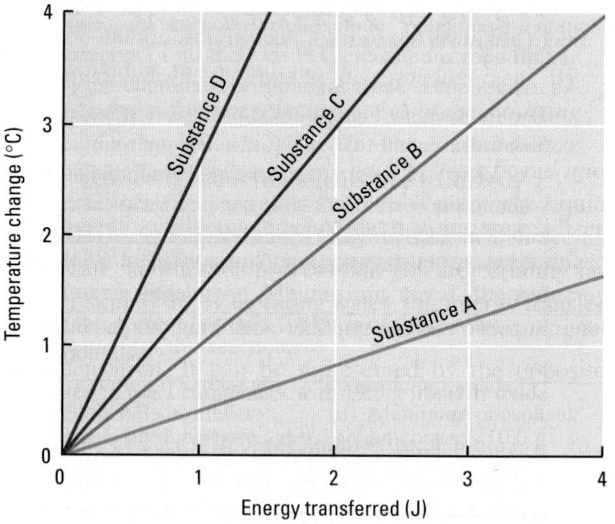
Temperature change (°C)
4
3
2
1
0
Substance D
Substance C
Substance B
Substance A
0
1
2
3
4
Energy transferred (J)

109. Based on the graph that accompanies Question 108, determine how much thermal energy would need to be transferred to 10 g substance B to raise its temperature from 35 °C to 38 °C.

110. The sketch shows two identical beakers with different volumes of water at the same temperature. Is the thermal energy content of beaker 1 greater than, less than, or equal to that of beaker 2? Explain your reasoning.

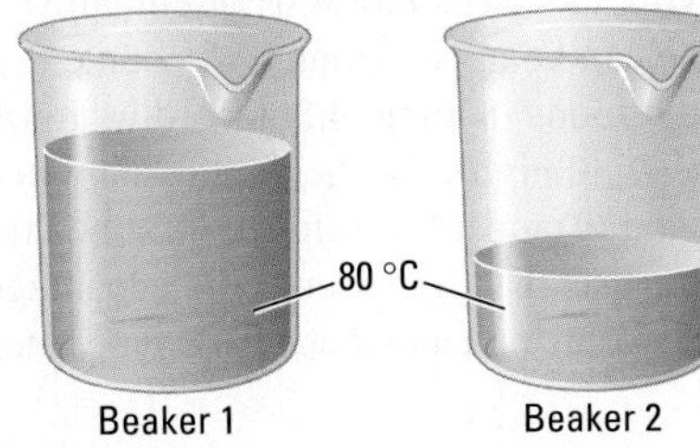

80 °C
Beaker 1
Beaker 2

111. If the same quantity of thermal energy were transferred to each beaker in Question 110, would the temperature of beaker 1 be greater than, less than, or equal to that of beaker 2? Explain your reasoning.
112. Consider this thermochemical expression

$$2\ S(s) + 3\ O_2(g) \longrightarrow 2\ SO_3(g) \qquad \Delta_r H° = -791\ kJ/mol$$

and the standard formation enthalpy for $SO_3(g)$ listed in Table 4.2. Why are the enthalpy values different?
113. In this chapter, the symbols $\Delta_f H°$ and $\Delta_r H°$ were used to denote a change in enthalpy. What is similar and what is different about the enthalpy changes they represent?
114. A fully inflated Mylar party balloon (the kind that does not expand once it is fully inflated) is heated, and 310 J transfers to the gas in the balloon. Calculate $\Delta_r E$ and w for the gas.
115. A sample of Dry Ice, $CO_2(s)$, is placed into a flask that is then tightly stoppered. Because the Dry Ice is very cold, 350 J transfers from the surroundings to the $CO_2(s)$. Calculate $\Delta_r E$ and w for the contents of the flask.

More Challenging Questions

These questions require more thought and integrate several concepts.

116. Oxygen difluoride, OF_2, is a colorless, very poisonous gas that reacts rapidly and exothermically with water vapor to produce O_2 and HF.

$$OF_2(g) + H_2O(g) \longrightarrow 2\ HF(g) + O_2(g)$$

When 4.32 g OF_2 reacts with 3.37 g H_2O the heat transfer at constant pressure is 26.0 kJ. Using this information and Table 4.2 or Appendix J, calculate the standard formation enthalpy of $OF_2(g)$.
117. Suppose that 28.89 g $ClF_3(g)$ is reacted with 57.3 g Na(s) to form NaCl(s) and NaF(s). The reaction occurs at 1 bar and 25.0 °C. Calculate the quantity of energy transferred between the system and surroundings and describe in which direction the energy is transferred.
118. You want to determine the value for the standard formation enthalpy of $CaSO_4(s)$.

$$Ca(s) + S(s) + 2\ O_2(g) \longrightarrow CaSO_4(s)$$

This synthesis reaction cannot be done directly. You know, however, that both calcium and sulfur react with oxygen to produce oxides in reactions that can be studied calorimetrically. You also know that the basic oxide CaO reacts with the acidic oxide $SO_3(g)$ to produce $CaSO_4(s)$ with $\Delta_r H° = -403.3$ kJ/mol. Outline a method for determining $\Delta_f H°$ for $CaSO_4(s)$, and identify the information that must be collected by experiment. Using information in Appendix J, confirm that $\Delta_f H°$ for $CaSO_4(s) = -1434.11$ kJ/mol.
119. When Serena Williams serves a tennis ball it is estimated that the ball is traveling about 130 miles per hour when it comes off the racquet. Find the mass of a tennis ball on the Internet and calculate the kinetic energy of the served ball. Calculate the minimum quantity of peanuts that Williams would have to eat to provide the energy she imparted to the ball.
120. A student had five beakers, each containing 100. mL of 0.500-M NaOH(aq) and all at room temperature (20.0 °C). The student planned to add a carefully weighed quantity of solid ascorbic acid, $C_6H_8O_6$, to each beaker, stir until it dissolved, and measure the increase in temperature. After the fourth experiment, the student was interrupted and called away. The data table looked like this:

Experiment	Mass of Ascorbic Acid (g)	Final Temperature (°C)
1	2.20	21.7
2	4.40	23.3
3	8.81	26.7
4	13.22	26.6
5	17.62	—

(a) Predict the temperature the student would have observed in experiment 5. Explain why you predicted this temperature.
(b) For each experiment indicate which is the limiting reactant, sodium hydroxide or ascorbic acid.
(c) When ascorbic acid reacts with NaOH, how many hydrogen ions are involved? One, as in the case of HCl? Two, as in the case of H_2SO_4? Or three, as in the case of phosphoric acid, H_3PO_4? Explain clearly how you can tell, based on the student's calorimeter data.

121. In their home laboratory, two students do an experiment (a rather dangerous one—don't try it without proper safety precautions!) with drain cleaner (Drāno, a solid) and toilet bowl cleaner (The Works, a liquid solution). The students measure 1 teaspoon (tsp) of Drāno into each of four Styrofoam coffee cups and dissolve the solid in half a cup of water. Then they wash their hands and go have lunch. When they return, they measure the temperature of the solution in each of the four cups and find it to be 22.3 °C. Next they measure into separate small empty cups 1, 2, 3, and 4 tablespoons (Tbsp) of The Works. In each cup they add enough water to make the total volume

4 Tbsp. After a few minutes they measure the temperature of each cup and find it to be 22.3 °C. Finally the two students take each cup of The Works, pour it into a cup of Drāno solution, and measure the temperature over a period of a few minutes. Their results are reported in the table.

Experiment	Volume of The Works (Tbsp)	Highest Temperature (°C)
1	1	28.0
2	2	33.6
3	3	39.3
4	4	39.4

Discuss these results and interpret them in terms of the thermochemistry and stoichiometry of the reaction. Is the reaction exothermic or endothermic? Why is more energy transferred in some cases than others? For each experiment, which reactant, Drāno or The Works, is limiting? Why are the final temperatures nearly the same in experiments 3 and 4? What can you conclude about the stoichiometric ratio between the two reactants?

122. On March 4, 1996, five railroad tank cars carrying liquid propane derailed in Weyauwega, Wisconsin, forcing evacuation of the town for more than a week. Residents who lived within the square mile centered on the accident were unable to return to their homes for more than two weeks. Evaluate whether this evacuation was reasonable and necessary by considering these questions.
    (a) Estimate the volume of a railroad tank car. Obtain the density of liquid propane, C_3H_8, at or near its boiling point and calculate the mass of propane in the five tank cars. Obtain the data you need to calculate the energy transfer if all the propane burned at once. (Assume that the reaction takes place at room temperature.)
    (b) The enthalpy of decomposition of TNT, $C_7H_5N_3O_6$, to water, nitrogen, carbon monoxide, and carbon is -1066.1 kJ/mol. Calculate how many metric kilotons (1 tonne = 1 metric ton = 1 Mg = 1×10^6 g) of TNT would provide energy transfer equivalent to that produced by combustion of propane in the five tank cars.
    (c) Find the energy transfer (in kilotons of TNT) resulting from the nuclear fission bombs dropped on Hiroshima and Nagasaki, Japan, in 1945, and the energy transfer for modern fission weapons. What is the largest nuclear weapon thought to have been detonated to date? Compare the energy of the Hiroshima and Nagasaki bombs with the Weyauwega propane spill. What can you conclude about the wisdom of evacuating the town?
    (d) Compare the energy that would have been released by burning the propane with the energy of a hurricane.
123. In some cities, taxicabs run on liquefied propane fuel instead of gasoline. This practice extends the lifetime of the vehicle and produces less pollution. Given that it costs about \$2000 to modify the engine of a taxicab to run on propane and that the cost of gasoline and liquid propane are \$3.50 per gallon and \$2.50 per gallon, respectively, make reasonable assumptions and figure out how many miles a taxi would have to go so that the decreased fuel cost would balance the added cost of modifying the taxi's motor. [For enthalpy calculations, gasoline can be approximated as octane, $C_8H_{18}(\ell)$.]
124. Rank the foods in Table 4.3 according to (a) their caloric value (low to high); (b) their fat content per 100 g (low to high); and (c) most healthful to least healthful food. Do the three lists match? Discuss the advantages and disadvantages of using caloric value or fat content to evaluate a food.

Conceptual Challenge Problems

These rigorous, thought-provoking problems integrate conceptual learning with problem solving and are suitable for group work.

CP4.A (Section 4-2) Suppose a scientist discovered that energy was not conserved, but rather that 1×10^{-7}% of the energy transferred from one system vanishes before it enters another system. How would this affect electric utilities, thermochemical experiments in scientific laboratories, and scientific thinking?

CP4.B (Section 4-2) Suppose that someone were to tell your instructor during class that energy is not always conserved. This person states that he or she had previously learned that in the case of nuclear reactions, mass is converted into energy according to Einstein's equation $E = mc^2$. Hence, energy is continuously produced as mass is changed into energy. Your instructor quickly responds by giving this assignment to the class: "Please write a paragraph or two to refute or clarify this student's thesis." What would you write?

CP4.C (Section 4-3) The specific heat capacities at 25 °C for three metals with widely differing molar masses are 3.6 J g^{-1} °C^{-1} for Li, 0.24 J g^{-1} °C^{-1} for Ag, and 0.13 J g^{-1} °C^{-1} for Th. Suppose that you have three samples, one of each metal and each containing the same number of atoms.
(a) Is the energy transfer required to increase the temperature of each sample by 1 °C significantly different from one sample to the next?
(b) What interpretation can you make about temperature based on the result you found in part (a)?

CP4.D (Section 4-4) During one of your chemistry classes a student asks the instructor, "Why does hot water freeze more quickly than cold water?"
(a) What do you expect the instructor to say to correctly answer the student's question?
(b) In one experiment, two 100.-g samples of water were placed in identical containers on the same surface 10. cm apart in a room at −25 °C. One sample had an initial temperature of 78 °C, while the second was at 24 °C. The second sample took 151 min to freeze, and the first took 166 min (only 10% longer) to freeze. Clearly the cooler sample froze more quickly, but not nearly as quickly as one might have expected. Explain how this can be so.

CP4.E (Section 4-10) Assume that glass has the same properties as pure SiO_2. The thermal conductivity (the rate at which heat transfer occurs through a substance) for aluminum is 200 times that for SiO_2.

(a) Is it more efficient in time and energy to bake brownies in an aluminum pan or a glass pan? Explain your answer.

(b) It is said that things cook more evenly in a glass pan than in an aluminum pan. Are there scientific data that indicate that this statement is reasonable? What are the data, if available?

CP4.F (Section 4-11) Suppose that you are an athlete and exercise a lot. Consequently you need a large caloric intake each day. Choose at least ten foods that you normally eat and evaluate each one to find which provides the most calories per dollar.

5 | Electron Configurations and the Periodic Table

Smileus/Shutterstock.com

The spectacular colors given off by exploding fireworks are due to changes in the energy levels of electrons in various metal ions: strontium–crimson; calcium–red/orange; copper–blue/green; barium–green; and strontium/copper mixtures–purple. What is an energy level? (Section 5-3a) How is the color of light related to electron energies? (Sections 5-1 and 5-3b) What do electron energy levels tell us about the behavior of electrons in atoms? (Section 5-5) In this chapter we discuss the energies of electrons in atoms, the electron configurations of atoms, and how these are related to atomic properties such as sizes of atoms, formation of ions, and stability of ionic compounds.

The periodic table was created by Dmitri Mendeleev to summarize experimental observations (← **Sec. 1-13**). He had no theory or model to explain why, for example, all alkaline-earth metals combine with oxygen in a 1 : 1 atom ratio—he just knew they did. In the early years of the twentieth century, however, it became evident that atoms contain electrons. As a result of these findings, explanations of periodic trends in physical and chemical properties began to be based on an understanding of the arrangement of electrons within atoms—on what we now call *electron configurations*. Studies of the interaction of light with atoms and molecules revealed that electrons in atoms are arranged in energy levels, also called shells. Electrons in the outermost shell are the *valence electrons; the number of valence electrons and the shell in which they occur are the chief factors that determine an atom's chemical reactivity.* This chapter describes the relationship of atomic electron configurations to atomic properties and the periodic table.

5-1 Electromagnetic Radiation and Matter

Theories about the energy and arrangement of electrons in atoms are based on experimental studies of the interaction of matter with electromagnetic radiation. A good example is the interaction of matter with visible light (one form of electromagnetic radiation). Interestingly, matter in some form is always associated with any color of light our eyes can see. The red glow of a neon sign is produced when electrons in neon atoms lose energy they gained when electricity was passed through the neon; fireworks displays are visible because of light emitted by metal atoms as their electrons lose extra energy they gained from heating when the firework exploded. When an electron in an atom has gained extra energy, from electricity, heating, or some other way, the atom is said to be "excited." The extra energy can be released as visible light, decreasing the atom's energy. We can account for energy transfer into or out of an atom using the same ideas we developed for thermodynamic systems in Chapter 4 (← **Sec. 4-3**). Studying the quantity of energy gained or released when an atom is excited can help us understand the arrangement of electrons within the atom.

Café

Steven Taylor/Photodisc/Getty Images

The various colors of neon-type signs are due to emissions of visible light from atoms of different gases.

All **electromagnetic radiation** consists of *oscillating, perpendicular electric and magnetic fields that travel through space at the same rate.* (In a perfect vacuum, that rate is the *speed of light*: 186,000 mi/s, or 2.998×10^8 m/s.) One way to describe oscillations is in terms of frequency and wavelength, the same description that is used for waves in water. As illustrated in Figure 5.1, the **wavelength**, symbol λ, is the *distance between adjacent crests (or troughs) in a wave*, and the **frequency**, symbol ν, is the *number of complete waves passing a point in a given period of time*—that is, waves per second or simply per second, 1/s or s^{-1}. A frequency of $4.80 \times 10^{14}\ s^{-1}$ means that 4.80×10^{14} waves pass a fixed point every second. The unit s^{-1} is given the name *hertz* (Hz). The frequency of electromagnetic radiation is related to its wavelength by the equation

$$\nu\lambda = c$$

The speed of light through matter depends on the substance and the wavelength of the light. This is the reason a glass prism disperses light and is the explanation for rainbows.

Reciprocal units such as 1/s are often represented in the negative exponent form, s^{-1}. The hertz unit (s^{-1}) was named in honor of Heinrich Hertz (1857–1894), a German physicist.

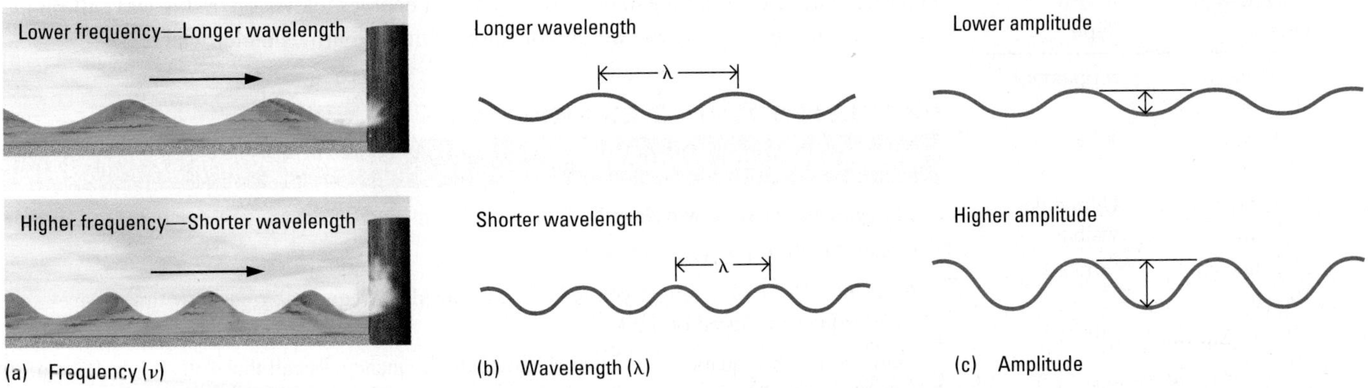
Lower frequency—Longer wavelength
Higher frequency—Shorter wavelength
(a) Frequency (ν)
Longer wavelength
λ
Shorter wavelength
λ
(b) Wavelength (λ)
Lower amplitude
Higher amplitude
(c) Amplitude

Figure 5.1 Wavelength, frequency, and amplitude of water waves. (a) The waves are moving from left to right toward the post at the same speed. The upper wave has a long wavelength (large λ) and low frequency (the number of times per second the peak of the wave hits the post). The lower wave has a shorter wavelength and a higher frequency (its peaks hit the post more often in a given time). (b) Variation in wavelength. (c) Variation in amplitude.

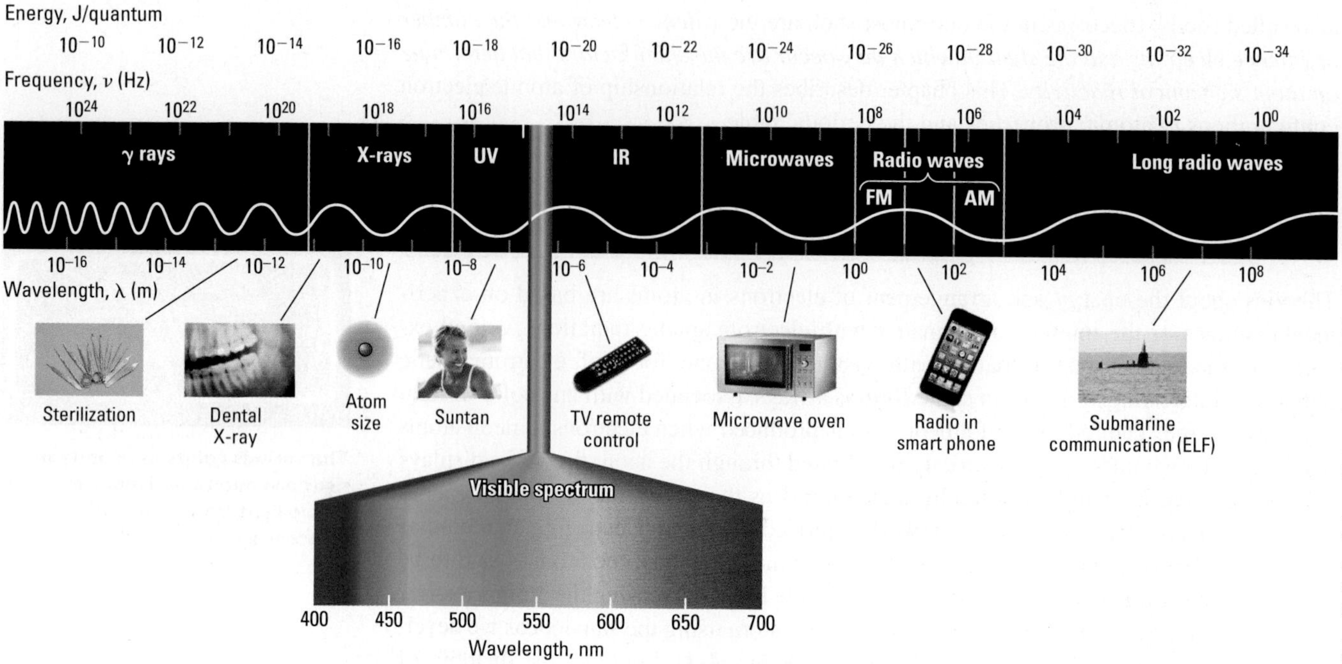
Energy, J/quantum
10^{-10} 10^{-12} 10^{-14} 10^{-16} 10^{-18} 10^{-20} 10^{-22} 10^{-24} 10^{-26} 10^{-28} 10^{-30} 10^{-32} 10^{-34}
Frequency, ν (Hz)
10^{24} 10^{22} 10^{20} 10^{18} 10^{16} 10^{14} 10^{12} 10^{10} 10^{8} 10^{6} 10^{4} 10^{2} 10^{0}
γ rays
X-rays
UV
IR
Microwaves
Radio waves
FM
AM
Long radio waves
10^{-16} 10^{-14} 10^{-12} 10^{-10} 10^{-8} 10^{-6} 10^{-4} 10^{-2} 10^{0} 10^{2} 10^{4} 10^{6} 10^{8}
Wavelength, λ (m)
Sterilization
Dental X-ray
Atom size
Suntan
TV remote control
Microwave oven
Radio in smart phone
Submarine communication (ELF)
Visible spectrum
400 450 500 550 600 650 700
Wavelength, nm

Figure 5.2 Electromagnetic radiation. Wavelengths of electromagnetic radiation are associated with common ways it impacts our lives. Electromagnetic radiation that humans can see (colored section) is only a small part of the entire spectrum.

where c is the speed of light, 2.998×10^8 m/s. This equation says that the larger the frequency is, the smaller the wavelength must be (and *vice versa*).

For water waves, the intensity (strength) is related to the *amplitude* (height of the wave crest) (Figure 5.1). The higher the amplitude, the more energy the wave can transfer. For example, on a given day the frequency and wavelength of waves coming ashore in an ocean or lake stay pretty much the same, but when a wave of higher amplitude comes in, it may knock you over.

As Figure 5.2 shows, visible light is only a small fraction of all wavelengths in the electromagnetic spectrum. Radiation with wavelengths too short for us to see includes ultraviolet (UV) radiation (the type that leads to sunburn), X-rays (such as those used for dental diagnosis), and gamma (γ) rays (emitted by radioactive disintegration of some atoms and used to sterilize food). Infrared radiation (the type sensed by our skin as heat from a fire) has wavelengths longer than those of visible light. Longer still is the wavelength of the radiation that heats food in a microwave oven, that enables television, radio, and cell-phone transmissions, and that allows communication with submarines under water.

Useful Wavelength Units for Electromagnetic Radiation

Wavelength Unit	Radiation Type
Picometer, 10^{-12} m	γ (gamma)
Ångström, 10^{-10} m	X-ray
Nanometer, 10^{-9} m	Ultraviolet, visible
Micrometer, 10^{-6} m	Infrared
Millimeter, 10^{-3} m	Infrared
Centimeter, 10^{-2} m	Microwave
Meter, m	TV, radio, cell phone

PROBLEM-SOLVING EXAMPLE 5.1

Wavelength and Frequency

Blue-green light has a wavelength of 495 nm. Calculate the frequency of this light.

Result 6.06×10^{14} Hz ($6.06 \times 10^{14}\ s^{-1}$)

Analyze Given wavelength you are asked to calculate frequency; frequency times wavelength equals the speed of light.

Plan Solve the equation $\nu\lambda = c$ to calculate the frequency. Recall that 1 m = 1×10^9 nm. Substitute into the equation and convert nanometers to meters.

Execute

$$\nu = \frac{c}{\lambda} = \frac{2.998 \times 10^8 \text{m/s}}{495 \text{ nm}} \times \frac{1 \text{ nm}}{1 \times 10^{-9} \text{m}} = 6.06 \times 10^{14}\ s^{-1} = 6.06 \times 10^{14} \text{ Hz}$$

☑ **Reasonable Result Check** Compare the calculated frequency ($6.06 \times 10^{14}\ s^{-1}$) with the frequency of the visible region of the electromagnetic spectrum illustrated in Figure 5.2. The wavelength and frequency fall into the visible region and thus the result is reasonable.

PROBLEM-SOLVING PRACTICE 5.1

In the upper atmosphere there is solar radiation with a frequency of 9.99×10^{14} Hz. Calculate its wavelength in (a) meters; (b) nanometers.

PROBLEM-SOLVING PRACTICE answers are provided at the back of this book in Appendix K.

CONCEPTUAL EXERCISE 5.1

Frequency and Wavelength

A fellow chemistry student says that low-frequency radiation is short-wavelength radiation. You disagree. Explain why the other student is wrong.

Answers to **EXERCISES** are provided at the back of this book in Appendix L.

EXERCISES that are labeled **CONCEPTUAL** are designed to test your understanding of one or more concepts; they usually involve qualitative rather than quantitative thinking.

CONCEPTUAL EXERCISE 5.2

In the Atmosphere

One type of solar radiation in the upper atmosphere has a frequency of 7.898×10^{14} Hz; another type has a frequency of 1.20×10^{15} Hz. (a) In what region of the electromagnetic spectrum does this solar radiation occur? (b) Which of the two types of radiation has the shorter wavelength? Explain your answer.

5-2 Planck's Quantum Theory

Have you ever looked at a wire in a toaster when electricity heats it? As shown in Figure 4.7 (← **Sec. 4-2b**) at higher temperatures metal atoms vibrate farther from their average positions. The atoms gain energy, which can later be emitted as radiation. At first, the wire emits infrared radiation that you can feel but not see. As the wire gets hotter, it begins to glow red, then orange, and at a high enough temperature it can become white hot (Figure 5.3).

© Cengage Learning/James Maynard

Figure 5.3 Electric toaster wire at three temperatures.

Figure 5.4 shows the experimentally determined **spectrum**, *a graph of intensity of radiation versus wavelength*, for a typical heated object. Notice how the wavelength of maximum emission becomes shorter (moves from infrared to yellow to green) as the temperature increases from 3500 K to 5000 K to 5780 K. In trying to explain the nature of these emissions from hot objects, late nineteenth-century scientists assumed that vibrating atoms in a hot wire caused electromagnetic vibrations, that is, light waves. Using the laws of physics known at the time, however, the scientists were unable to predict the experimentally observed spectrum.

In 1900, Max Planck (1858–1947) offered an explanation for the spectrum of a heated body. His ideas contained the seeds of a revolution in scientific thought. Planck made what was at that time an incredible assumption: When an atom in a hot object vibrates, the energies of the vibrations all have to be integer multiples of some minimum energy. When the atom emits radiation, it does so only in packets having that minimum quantity of energy. That is, just as an atom is the smallest packet of an element, there must be a *small packet of energy such that no smaller quantity can be emitted.* Planck called this packet of energy a **quantum**. He further asserted that the energy of a quantum is proportional to the frequency of the radiation according to the equation

A quantum is the smallest possible unit of a distinct quantity—for example, the smallest possible unit of energy for electromagnetic radiation of a given frequency.

$$E_{\text{quantum}} = h\nu_{\text{radiation}}$$

The *proportionality constant h* is called **Planck's constant**; it has the value 6.626×10^{-34} J s and relates the frequency of radiation to the energy *per quantum.*

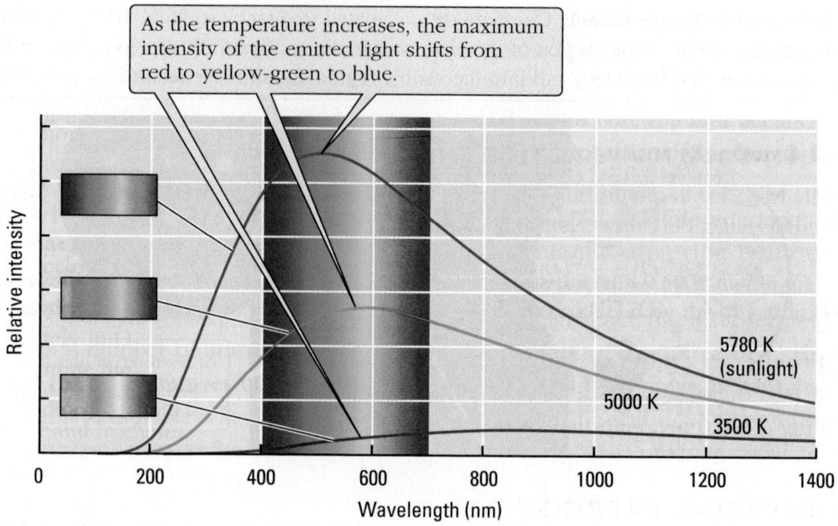
As the temperature increases, the maximum intensity of the emitted light shifts from red to yellow-green to blue.
Relative intensity
5780 K (sunlight)
5000 K
3500 K
0
200
400
600
800
1000
1200
1400
Wavelength (nm)

Figure 5.4 Heated objects and temperature. The spectrum of radiation given off by a heated object. At very high temperatures, the heated object becomes "white hot" when all wavelengths of visible light are being emitted.

Orange light has a measured frequency of $4.80 \times 10^{14}\ s^{-1}$. The energy of *one* quantum of orange light is therefore

$$E = h\nu = (6.626 \times 10^{-34}\ J\ s)(4.80 \times 10^{14}\ s^{-1}) = 3.18 \times 10^{-19}\ J$$

Max Planck won the 1918 Nobel Prize in Physics for his quantum theory.

The theory based on Planck's work is called the **quantum theory**. By using his quantum theory, Planck was able to calculate results that agreed very well with the experimentally measured spectra of heated objects, spectra for which the laws of classical physics had no explanation.

PROBLEM-SOLVING EXAMPLE 5.2

Calculating Quantum Energies

In the stratosphere, ultraviolet radiation with a frequency of $1.36 \times 10^{15}\ s^{-1}$ can break C—Cl bonds in chlorofluorocarbons (CFCs), which can lead to stratospheric ozone depletion. Calculate the energy per quantum of this radiation.

Result 9.01×10^{-19} J

Analyze According to Planck's quantum theory, the energy per quantum of radiation is proportional to frequency: $E = h\nu$.

Plan Substitute values for h and ν into the equation and cancel units appropriately.

Execute

$$E = h\nu = (6.626 \times 10^{-34}\ J\ s)(1.36 \times 10^{15}\ s^{-1}) = 9.01 \times 10^{-19}\ J$$

☑ **Reasonable Result Check** The calculated energy, 9.01×10^{-19} J/quantum = 0.901×10^{-18} J/quantum, is within the energy of the ultraviolet region of the electromagnetic spectrum as illustrated in Figure 5.2; the answer is reasonable.

PROBLEM-SOLVING PRACTICE 5.2

Which has more energy,

(a) one quantum of microwave radiation or one quantum of ultraviolet radiation?

(b) one quantum of blue light or one quantum of green light?

A very important relationship links the energy and wavelength of a quantum of radiation. Because $E = h\nu$ and $\nu = c/\lambda$, then

$$E = \frac{hc}{\lambda}$$

where h is Planck's constant, c is the velocity of light, and λ is the wavelength of the radiation. Note that energy and wavelength are inversely proportional: ***The energy per quantum of radiation increases as the wavelength gets shorter.*** This relationship can be seen in Figure 5.2 where both wavelength and energy per quantum are shown.

For example, red light and blue light have wavelengths of 656.3 nm and 434.1 nm, respectively. Because of its shorter wavelength, blue light has a higher energy per quantum than red light. We can calculate their different energies by applying the previous equation.

$$E_{red} = \frac{(6.626 \times 10^{-34}\ \text{J s})(2.998 \times 10^{8}\ \text{m/s})}{(656.3\ \text{nm})(1\ \text{m}/10^{9}\ \text{nm})} = 3.027 \times 10^{-19}\ \text{J}$$

$$E_{blue} = \frac{(6.626 \times 10^{-34}\ \text{J s})(2.998 \times 10^{8}\ \text{m/s})}{(434.1\ \text{nm})(1\ \text{m}/10^{9}\ \text{nm})} = 4.576 \times 10^{-19}\ \text{J}$$

EXERCISE 5.3

Carbon Dioxide, Energy, and Global Warming

Atmospheric carbon dioxide is a greenhouse gas associated with global warming. The carbon-to-oxygen bonds in CO_2 absorb electromagnetic radiation at 4.257 μm and at 15.00 μm. The 4.257-μm absorption increases the energy of stretching the bonds; the 15.00-μm absorption increases the energy of bending them. Which absorption requires greater energy? Calculate the energies to verify your prediction.

5-2a The Photoelectric Effect

When a theory can accurately predict experimental results, the theory is usually regarded as useful. Planck's quantum theory was not widely accepted at first because of its radical assertion that energy is quantized. The quantum theory of electromagnetic energy was firmly accepted only after quanta were used by Albert Einstein (1879–1955) to explain another phenomenon, the photoelectric effect.

In the early 1900s it was known that *certain metals emit electrons when illuminated by light of certain wavelengths* (Figure 5.5). This is known as the **photoelectric effect**. For each photosensitive metal there is a threshold wavelength long enough so that no photoelectric effect is observed, no matter how intense the light shining on the metal is. For example, for calcium, ultraviolet light causes emission of electrons but visible light does not; the threshold wavelength is 427 nm. For cesium, Cs, yellow light causes the metal to emit electrons, but infrared does not; the threshold is 590 nm. Figure 5.5 illustrates schematically how the photoelectric effect can be measured experimentally.

Einstein's explanation of these experimental observations is described in Figure 5.5. He assumed that Planck's quantum of radiation was a *massless particle of electromagnetic radiation*, which he called a **photon**. That is, he described light as a stream of photons that have particle-like properties as well as wave-like properties. To remove one electron from a photosensitive metal surface requires a certain minimum quantity of energy; we call it E_{min}. Only a photon whose E is greater than E_{min} has enough energy to knock an electron loose. Because each photon has an energy given by $E = h\nu$, an electron cannot be ejected unless a photon with a high enough frequency (and short enough wavelength) strikes the photosensitive surface. If a metal requires photons of green light to eject electrons from its surface, then yellow light, red light, or light of any other lower frequency (longer wavelength) will not have sufficient energy per photon to cause the photoelectric effect with that metal. This brilliant deduction about the quantized nature of light and how it relates to light's interaction with matter won Einstein the Nobel Prize for Physics in 1921.

Many people think that Einstein won the Nobel Prize for his theory of relativity, but he didn't.

The energies of photons are important for practical reasons. Because of its higher energy, a photon of ultraviolet light can damage skin, while a photon of visible light cannot. We use sunblocks containing molecules that selectively absorb ultraviolet photons to protect our skin from solar UV radiation. X-ray photons are even more energetic than ultraviolet photons and can disrupt molecules at the cellular level, causing genetic damage, among other effects. For this reason we try to limit our exposure to X-rays even more than we limit our exposure to ultraviolet light.

In the quantum theory, the intensity of radiation is proportional to the number of photons. The total energy is proportional to the number of photons times the energy per photon.

Look again at the graph in Figure 5.5. Notice how, at a short enough wavelength, a higher-intensity light source causes a higher photoelectric current. Because there are more photons, higher intensities of ultraviolet light (or longer exposure) can cause greater damage to the skin than lower intensities (or less exposure time). The same holds true for other high-energy forms of electromagnetic radiation, such as X-rays.

Developments such as Einstein's explanation of the photoelectric effect led eventually to acceptance of what is referred to as the *dual nature* of light. Depending on the experimental circumstances, visible light and all other forms of electromagnetic radiation appear to have either "wave" or "particle" characteristics. However, both ideas are needed to fully explain light's behavior. Classical wave theory fails to explain the photoelectric effect. But it does explain quite well the *refraction,* or bending, of light by a prism and the diffraction of light by a diffraction grating, a device with a series of parallel, closely spaced small slits. When waves of light pass through such adjacent narrow slits, the waves are scattered so that the emerging light waves spread out, a phenomenon called *diffraction.* The semicircular waves emerging from the narrow slits can either

Refraction is the bending of light as it crosses the boundary from one medium to another—for example, from air to water. Diffraction is the bending of light around the edges of objects, such as slits in a diffraction grating.

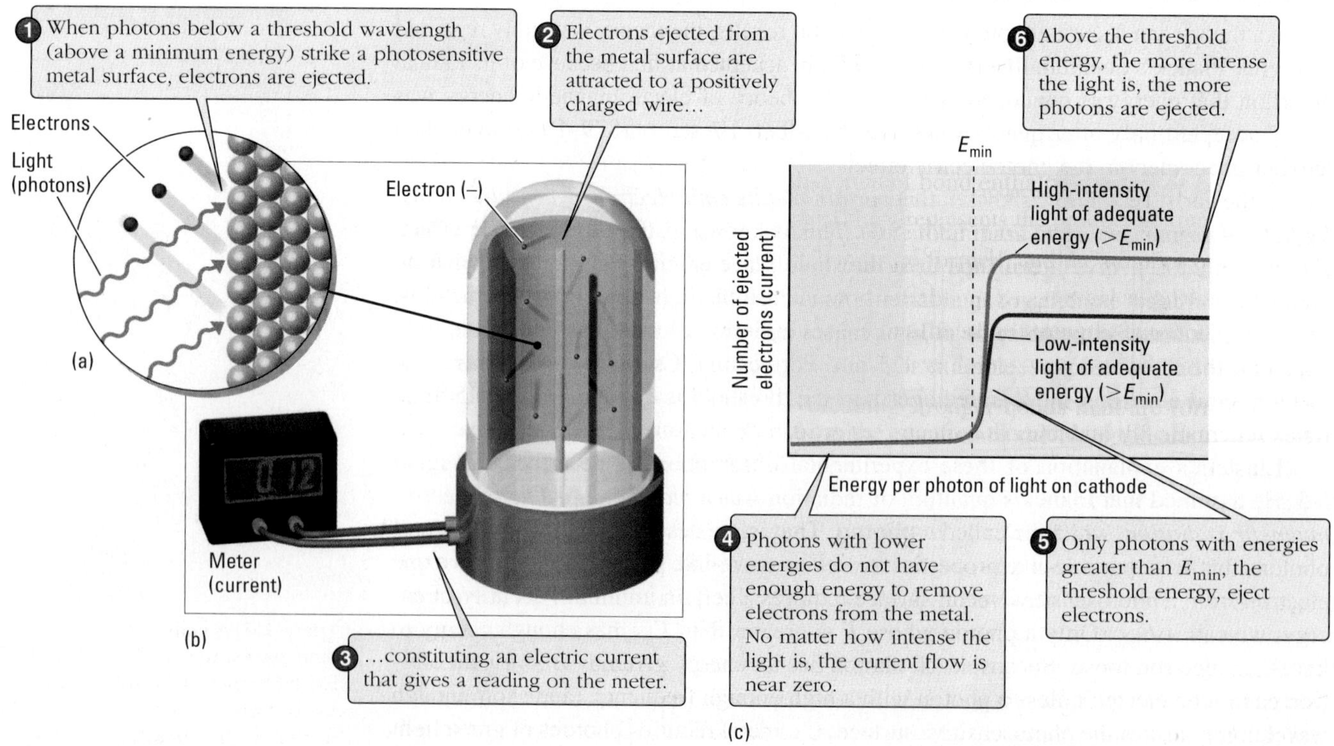

1 When photons below a threshold wavelength (above a minimum energy) strike a photosensitive metal surface, electrons are ejected.
Electrons
Light (photons)
(a)
2 Electrons ejected from the metal surface are attracted to a positively charged wire…
Electron (–)
0.12
Meter (current)
(b)
3 …constituting an electric current that gives a reading on the meter.
6 Above the threshold energy, the more intense the light is, the more photons are ejected.
E_{min}
High-intensity light of adequate energy ($>E_{min}$)
Low-intensity light of adequate energy ($>E_{min}$)
Number of ejected electrons (current)
Energy per photon of light on cathode
4 Photons with lower energies do not have enough energy to remove electrons from the metal. No matter how intense the light is, the current flow is near zero.
5 Only photons with energies greater than E_{min}, the threshold energy, eject electrons.
(c)

Figure 5.5 The photoelectric effect.

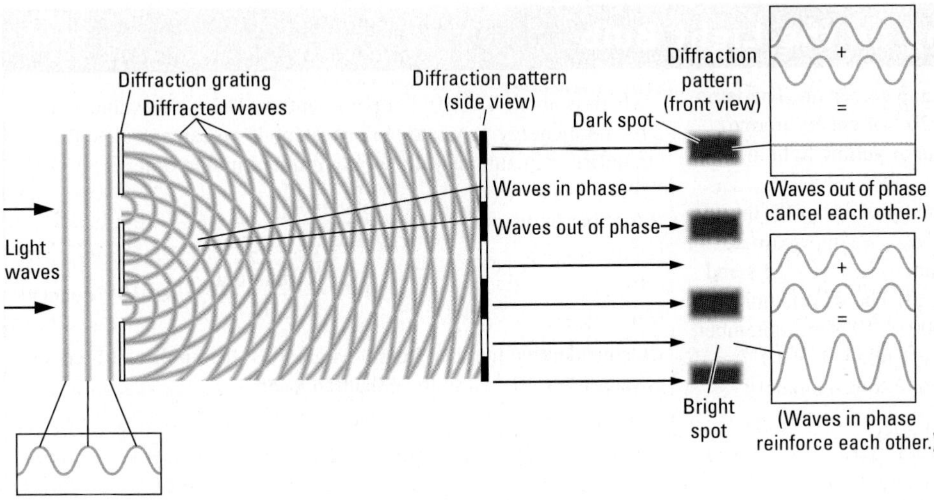
Diffraction grating
Diffracted waves
Diffraction pattern
(side view)
Diffraction
pattern
(front view)
Dark spot
Waves in phase
Waves out of phase
(Waves out of phase
cancel each other.)
Light
waves
Bright
spot
(Waves in phase
reinforce each other.)
Light waves, crests

Figure 5.6 Diffraction of light produces a pattern of dark and bright spots.

amplify or cancel each other. Such behavior creates a diffraction pattern of dark and bright spots, as seen in Figure 5.6.

It is important to realize, however, that this "dual nature" description arises because of our attempts to explain observations by using inadequate models. Light is not changing back and forth from being a wave to being a particle, but has a single, consistent nature that can be described by modern quantum theory. The dual-nature description arises when we try to explain our observations by using our familiarity with classical models for "wave" or "particle" behavior.

Prior to Einstein's explanation of the photoelectric effect, classical physics considered light as being only wave-like.

5-3 The Bohr Model of the Hydrogen Atom

Within about a decade (1900–1911), three major discoveries were made about the atom and the nature of electromagnetic radiation. The first two, discussed in the previous section, were Max Planck's suggestion that energy was quantized and, five years later, Albert Einstein's application of the quantum idea to explain the photoelectric effect. The third came in 1911 when Ernest Rutherford demonstrated experimentally that atoms contain a dense positive core, the nucleus, surrounded by electrons (← **Sec. 2-2a**). Niels Bohr linked these three powerful ideas when, in 1913, he used quantum theory to explain the behavior of the electron in a hydrogen atom. In doing so, Bohr developed a mathematical model to explain how excited hydrogen atoms emit or absorb only certain wavelengths of light, a phenomenon that had been unexplained for nearly a half century. To understand what Bohr did, we turn first to the visible spectrum. The spectrum of white light, such as that from the sun or an incandescent light bulb, consists of a rainbow of colors, as shown in Figure 5.7. This **continuous spectrum** *includes light of all wavelengths* in the visible region.

Earlier we described neon-type advertising signs in which a high voltage is applied to a gaseous element at low pressure (← **Sec. 5-1**). Under these circumstances, energy transfers to the atoms and they become excited. The excited atoms then emit, as electromagnetic radiation, energy that was previously absorbed. When light from such a source passes through a prism onto a white surface a line spectrum appears. A **line emission spectrum** *consists of a few colored lines at specific wavelengths characteristic of the element that emits the radiation.* The line spectra of the visible light emitted by excited atoms of hydrogen, mercury, and neon are shown in Figure 5.8.

Roger Antrobus/Corbis

Figure 5.7 A continuous spectrum from white light. When white light is passed through slits to produce a narrow beam and then refracted in a glass prism, the various colors blend smoothly into one another.

ESTIMATION | Turning on the Light Bulb

You turn the switch and a living room lamp comes on. The lamp emits yellow light (585 nm), producing 20 J of energy in one second. Approximately how many quanta of yellow light are given off by the lamp in that time?

To find out, we first determine the energy of each 585-nm quantum by using the relationship $E = hc/\lambda$. We approximate by rounding the value of Planck's constant to 7×10^{-34} J s and that of c, the speed of light, to 3×10^{8} m/s. The wavelength of the quanta we round to approximately 600×10^{-9} m; remember we have to convert nanometers to meters: $1 \text{ nm} = 1 \times 10^{-9}$ m. Therefore, the energy, E, of each quantum is approximately

$$\frac{(7 \times 10^{-34} \text{ J s})(3 \times 10^{8} \text{ m/s})}{600 \times 10^{-9} \text{ m}}$$

which is about 4×10^{-19} J per quantum. Using this value and the total energy generated in one second, we can calculate the number of quanta of yellow light emitted in that time.

$$\text{Number of quanta} = \frac{\text{total energy}}{\text{energy of each quantum}} =$$

$$20 \text{ J} \times \frac{1 \text{ quantum}}{4 \times 10^{-19} \text{ J}} = 5 \times 10^{19} \text{ quanta}$$

a considerable number, nearly as great as the number of liters of water (3×10^{20} L) in the Atlantic Ocean.

Hydrogen has the simplest line emission spectrum. In the 1880s, wavelengths of lines in the visible region of the hydrogen spectrum (Figure 5.8) were accurately measured. Subsequently other series of lines were discovered in the ultraviolet region ($\lambda < 400$ nm) and in the infrared region ($\lambda > 700$ nm). The fact that only specific wavelengths (lines) are found in the hydrogen emission spectrum indicates that only photons with specific energies are emitted. This differs from the continuous spectrum where a broad range of photon energies is emitted.

EXERCISE 5.4

Hydrogen Emission Spectrum

The lines in the visible emission spectrum of hydrogen (Figure 5.8) are at 656.4 nm, 486.2 nm, 434.1 nm, and 410.2 nm. (a) What color is each line? (b) Calculate the energy of a photon corresponding with each line.

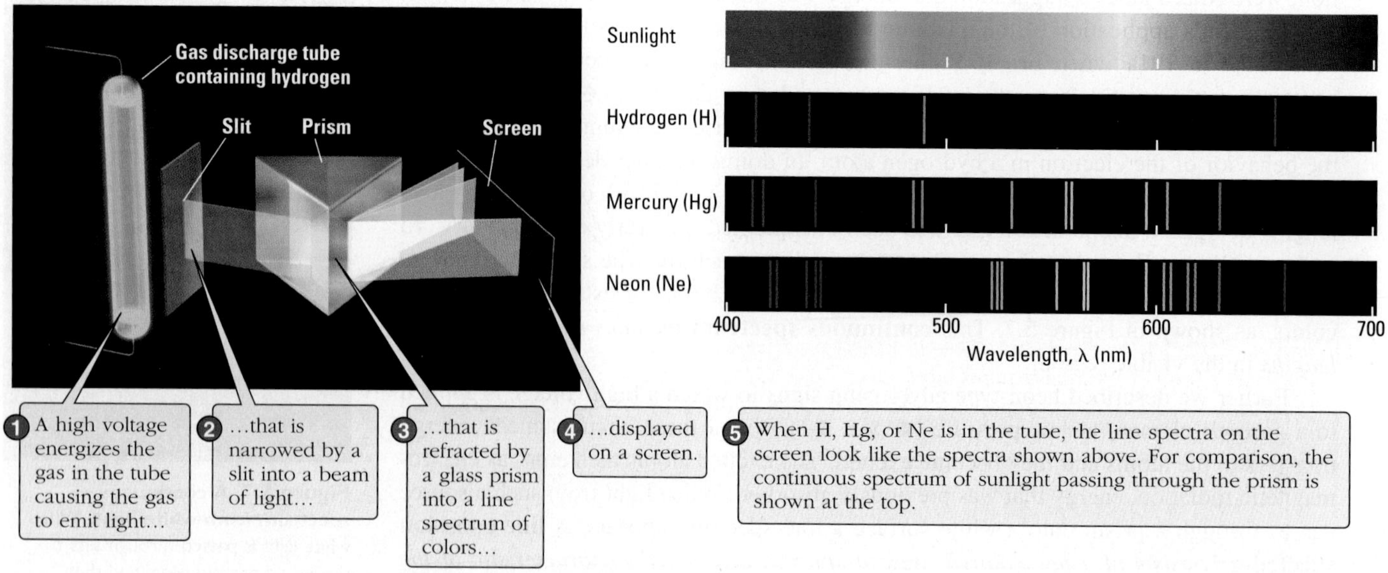
Gas discharge tube containing hydrogen
Slit
Prism
Screen
Sunlight
Hydrogen (H)
Mercury (Hg)
Neon (Ne)
400
500
600
700
Wavelength, λ (nm)
1 A high voltage energizes the gas in the tube causing the gas to emit light...
2 ...that is narrowed by a slit into a beam of light...
3 ...that is refracted by a glass prism into a line spectrum of colors...
4 ...displayed on a screen.
5 When H, Hg, or Ne is in the tube, the line spectra on the screen look like the spectra shown above. For comparison, the continuous spectrum of sunlight passing through the prism is shown at the top.

Figure 5.8 Line emission spectra of hydrogen, mercury, and neon. Excited gaseous elements produce characteristic line spectra that can be used to identify the elements as well as to determine how much of an element is present in a sample.

5-3a Energy Levels

Niels Bohr provided the first theoretical explanation of a line emission spectrum by audaciously assuming that the single electron of a hydrogen atom moved in a circular orbit around the nucleus. He used Coulomb's law (← **Sec. 2-6a**) to relate the energy of the electron to the radius of its orbit (distance from the nucleus) and introduced quantum theory into his atomic model by invoking Planck's idea that energies are quantized. In the Bohr model, the electron could circle the nucleus only in orbits of certain radii. Each radius corresponded to *a specific energy of the atom*, that is, to an **energy level**. The radius of the electron's orbit could only change if the atom gained or lost a certain quantity of energy. The *allowed orbits were numbered using a positive integer,* n, known as the **principal quantum number**.

In the Bohr model, the value of n for the possible orbits can be any integer from 1 to infinity (∞). *The energy of the electron and the size of its orbit increase as the value of* n *increases.* The orbit of lowest energy, $n = 1$, is closest to the nucleus, and the electron of a hydrogen atom is normally in this energy level. *Any atom with its electrons arranged to give the lowest total energy* is said to be in its **ground state**. Coulomb's law says that there is an attractive force between the positive nucleus and the negative electron, so energy must be transferred into an atom to move the electron farther from the nucleus (increase its potential energy by doing work against the attractive force). When the electron is in an orbit of larger radius (larger n), the hydrogen atom is in a higher-energy state. *Any state that has greater energy than the lowest-energy (ground) state* is called an **excited state**. Figure 5.9 shows the relationship between n and the energy levels for a hydrogen atom. When an atom changes from one energy level to another, the electron is said to undergo a *transition*.

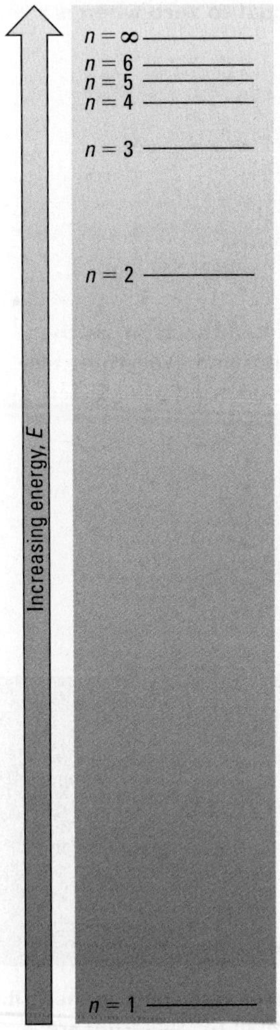
n = ∞
n = 6
n = 5
n = 4
n = 3
n = 2
n = 1
Increasing energy, E

Figure 5.9 Energy levels for hydrogen atom.

PROBLEM-SOLVING EXAMPLE 5.3

Electron Transitions

Consider these four possible electron transitions in a hydrogen atom:

(a) Which transition(s) represent a loss of energy?

(b) For which transition does the atom gain the greatest quantity of energy?

(c) Which transition corresponds to emission of the greatest quantity of energy?

	$n_{initial}$	n_{final}
(1)	2	5
(2)	5	3
(3)	7	2
(4)	4	6

Result

(a) (2) and (3) (b) (1) (c) (3)

Analyze A transition from a higher to a lower energy level involves transfer of energy out of the atom. A transition from a lower to a higher energy level requires energy transfer into the atom. The quantity of energy gained or lost is the difference in energy between the two energy levels involved in the transition.

Plan Compare initial and final energy levels (Figure 5.9) for each transition.

Execute (a) The $n_7 \longrightarrow n_2$ and $n_5 \longrightarrow n_3$ transitions release energy.

(b) The $n_2 \longrightarrow n_5$ transition requires more energy than the $n_4 \longrightarrow n_6$ one.

(c) The $n_7 \longrightarrow n_2$ transition represents the greatest energy difference between the energy levels specified by the values of n given.

PROBLEM-SOLVING PRACTICE 5.3

Identify two electron transitions in a hydrogen atom that would be of greater energy than any of those listed in Problem-Solving Example 5.3.

CONCEPTUAL EXERCISE 5.5

Many Spectral Lines, but Only One Kind of Atom

The hydrogen atom contains only one electron, but there are several lines in its line emission spectrum (Figure 5.8). How does the Bohr theory explain this?

5-3b Photon Energies and Energy-Level Differences

Based on his theory, Bohr was able to calculate the allowed energies of the hydrogen atom. They are restricted by *n*, the principal quantum number:

$$E_{\text{atom}} = -\left(\frac{1}{n^2}\right)2.179 \times 10^{-18}\ \text{J} \quad (n = 1,2,3,\ldots;\ \text{that is, a positive integer})$$

Bohr was able to calculate the constant, 2.179×10^{-18} J, called the Rydberg constant, from his theory and it agreed very well with the experimentally determined value. According to this equation, the energy of a hydrogen atom in its ground state ($n = 1$) is

$$E_{\text{atom}} = -\left(\frac{1}{n^2}\right)2.179 \times 10^{-18}\ \text{J} = -\left(\frac{1}{1^2}\right)2.179 \times 10^{-18}\ \text{J} = -2.179 \times 10^{-18}\ \text{J}$$

Division by infinity makes E equal to zero when $n = \infty$.

The negative sign in the equation arises due to a choice Bohr made. He selected zero as the energy of a completely separated electron and proton ($n = \infty$). This is the highest potential energy the electron can have and still be part of the atom. As a consequence of this choice, all other energies for the electron in the atom are negative. Consider this analogy: A book resting on a table is arbitrarily designated as having zero potential energy. As the book falls from the table to a chair, to a stool, and then to the floor, its potential energy decreases. When it is on the floor, its potential energy is negative with respect to when it was on the table or the chair or the stool. Such is also the case for an electron going from a higher energy level to a lower energy level. The electron must lose energy. Correspondingly, the electron's energy in all its allowed energy states within the atom must be less than zero; that is, the energy must be negative.

Think of the atom as the system and everything else as the surroundings (← Sec. 4-3).

The difference in energy, ΔE, when the hydrogen atom moves from an initial state, E_i, to a final state, E_f, can be calculated as $\Delta E = E_f - E_i$. When an atom moves from a higher energy level to a lower energy level (one with a lower n value) energy is emitted and ΔE is negative. When an atom moves from a lower energy state to a higher one, energy must be absorbed (ΔE is positive). If an atom undergoes a transition from an initial state with principal quantum number n_i to a state with principal quantum number n_f the energy change is

$$\Delta E_{\text{atom}} = E_f - E_i = -2.179 \times 10^{-18}\ \text{J}\left(\frac{1}{n_f^2} - \frac{1}{n_i^2}\right)$$

If $n_f > n_i$, energy is absorbed; if $n_f < n_i$, energy is emitted.

The Bohr model reproduces the experimental observation that only certain frequencies of light can be absorbed or emitted by an atom. This is because the energy of the photon absorbed or emitted must be the same as the difference in energy, ΔE, between the two energy levels involved. If ΔE is positive, then a photon must have transferred energy *to* the atom (the photon must have been *absorbed*). If ΔE is negative, then the atom must have transferred energy to a photon (the photon must have been *emitted*). According to Planck's quantum theory, the energy of the photon equals Planck's constant times the frequency,

$$E = h\nu = \frac{hc}{\lambda}$$

Frank Leather/Eye Ubiquitous/Corbis

Electron transitions in the visible region of the spectrum are responsible for the impressive displays of color in fireworks.

This allows us to calculate the frequency and wavelength of the photon absorbed or emitted. Notice that the energy of the photon is always positive, but the sign of the energy change of the atom indicates whether the photon is absorbed (positive change in E_{atom}) or emitted (negative change in E_{atom}). Figure 5.10 shows the energy levels of a hydrogen atom together with transitions that give rise to emission lines in the ultraviolet, the visible, and the infrared.

Absorption of radiation involves transition from a lower to higher energy level. For example, the frequency of light absorbed in the $n = 2$ to $n = 3$ transition for a hydrogen atom electron can be determined by calculating the change in energy of the atom

This is analogous to the bank account analogy for energy transfer described in Chapter 4 (← Sec. 4-3a). The quantity of energy transferred (the energy of the photon) is always positive. The sign of ΔE indicates the direction of transfer.

$$\Delta E_{atom} = E_f - E_i = -2.179 \times 10^{-18}\text{ J}\left(\frac{1}{n_f^2} - \frac{1}{n_i^2}\right)$$

$$= -2.179 \times 10^{-18}\text{ J}\left(\frac{1}{3^2} - \frac{1}{2^2}\right) = -2.179 \times 10^{-18}\text{ J}\left(\frac{1}{9} - \frac{1}{4}\right)$$

$$= -2.179 \times 10^{-18}\text{ J}(0.1111 - 0.2500) = 3.026 \times 10^{-19}\text{ J}$$

The positive energy change indicates that the atom absorbed a photon whose energy is 3.026×10^{-19} J. Therefore, the frequency can be calculated as

$$\nu = \frac{E_{photon}}{h} = \frac{3.026 \times 10^{-19}\text{ J}}{6.626 \times 10^{-34}\text{ J s}} = 4.567 \times 10^{14}\text{ s}^{-1}$$

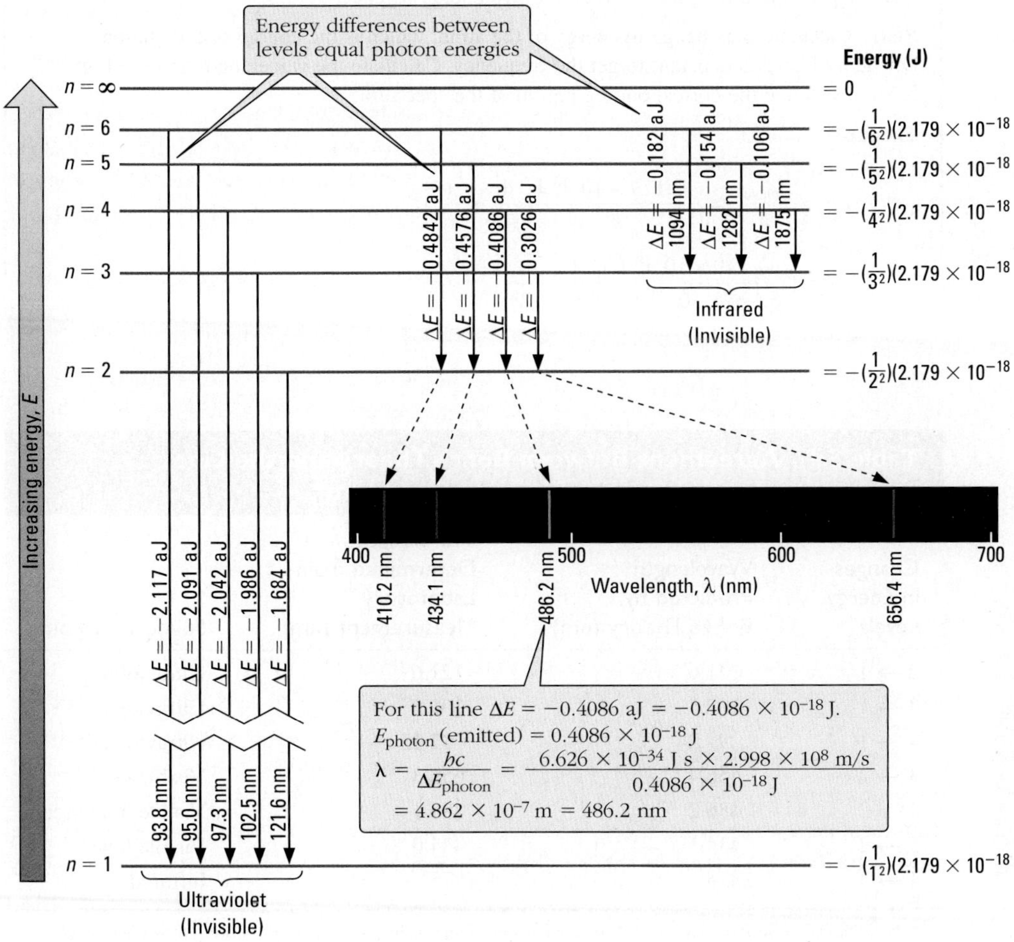

Energy differences between levels equal photon energies
Energy (J)
n = ∞
= 0
n = 6
= −(1/6²)(2.179 × 10⁻¹⁸
n = 5
= −(1/5²)(2.179 × 10⁻¹⁸
n = 4
= −(1/4²)(2.179 × 10⁻¹⁸
n = 3
= −(1/3²)(2.179 × 10⁻¹⁸
n = 2
= −(1/2²)(2.179 × 10⁻¹⁸
n = 1
= −(1/1²)(2.179 × 10⁻¹⁸
Increasing energy, E
ΔE = −0.182 aJ 1094 nm
ΔE = −0.154 aJ 1282 nm
ΔE = −0.106 aJ 1875 nm
Infrared (Invisible)
ΔE = −0.4842 aJ
ΔE = −0.4576 aJ
ΔE = −0.4086 aJ
ΔE = −0.3026 aJ
400
500
600
700
Wavelength, λ (nm)
410.2 nm
434.1 nm
486.2 nm
656.4 nm
ΔE = −2.117 aJ 93.8 nm
ΔE = −2.091 aJ 95.0 nm
ΔE = −2.042 aJ 97.3 nm
ΔE = −1.986 aJ 102.5 nm
ΔE = −1.684 aJ 121.6 nm
Ultraviolet (Invisible)
For this line ΔE = −0.4086 aJ = −0.4086 × 10⁻¹⁸ J.
E_photon (emitted) = 0.4086 × 10⁻¹⁸ J
λ = hc/ΔE_photon = (6.626 × 10⁻³⁴ J s × 2.998 × 10⁸ m/s)/(0.4086 × 10⁻¹⁸ J)
= 4.862 × 10⁻⁷ m = 486.2 nm

Figure 5.10 Electron transitions for hydrogen atom emission spectrum. Energy differences in attojoules, aJ; 1 aJ = 1×10^{-18} J.

The wavelength (λ) of the light absorbed can be obtained from its frequency by using the relationship $\lambda = c/\nu$, where c is the velocity of light (2.998×10^8 m/s). For the $n = 2$ to $n = 3$ transition,

$$\lambda = \frac{2.998 \times 10^8 \text{ m/s}}{4.567 \times 10^{14} \text{ s}^{-1}} = 6.565 \times 10^{-7} \text{ m} = 656.4 \text{ nm}$$

As shown in Figure 5.10, this absorption is in the red region of visible light.

Data in Table 5.1 indicate that there is exceptionally good agreement between the experimentally measured wavelengths and those calculated by the Bohr theory. Thus, Niels Bohr had tied the unseen (the atom) to the seen (the observable lines of the hydrogen emission spectrum)—a fantastic achievement!

Niels Bohr
1885–1962

Library of Congress Prints and Photographs Division[LC-B2-5894-8]

Born in Copenhagen, Denmark, Bohr worked with J. J. Thomson and Ernest Rutherford in England, where he began to develop the ideas that led to the publication of his explanation of atomic spectra. He received the Nobel Prize in Physics in 1922 for this work.

As the director of the Institute of Theoretical Physics in Copenhagen, Bohr was a mentor to many young physicists, seven of whom later received Nobel Prizes for their studies in physics or chemistry, including Werner Heisenberg, Wolfgang Pauli, and Linus Pauling.

PROBLEM-SOLVING EXAMPLE 5.4

Electron Transitions

(a) Calculate the frequency and wavelength (nm) corresponding to the $n = 2$ to $n = 5$ transition in a hydrogen atom.
(b) In what region of the spectrum does this transition occur?

Result

(a) 6.906×10^{14} s^{-1}; 434.1 nm (b) Visible region

Analyze The goal is to calculate frequency and wavelength and from them determine the region of the spectrum. The frequency can be calculated from the energy of the photon, which equals the difference between the two energy levels.

Plan Calculate the change in energy of the atom, which is the energy of the photon. Divide by Planck's constant to get the frequency. Calculate the wavelength and use Figure 5.2 to determine the corresponding region of the spectrum.

Execute

$$\nu = \frac{E_{\text{photon}}}{h} = \frac{-2.179 \times 10^{-18} \text{ J}\left(\frac{1}{n_f^2} - \frac{1}{n_i^2}\right)}{h}$$

$$= \left(\frac{-2.179 \times 10^{-18} \text{ J}}{6.626 \times 10^{-34} \text{ J s}}\right)\left(\frac{1}{5^2} - \frac{1}{2^2}\right)$$

Table 5.1 Agreement Between Bohr's Theory and the Lines of the Hydrogen Emission Spectrum*

Changes in Energy Levels	Wavelength Predicted by Bohr's Theory (nm)	Wavelength Determined from Laboratory Measurement (nm)	Spectral Region
$2 \rightarrow 1$	121.6	121.6	Ultraviolet
$3 \rightarrow 1$	102.5	102.6	Ultraviolet
$4 \rightarrow 1$	97.24	97.25	Ultraviolet
$3 \rightarrow 2$	656.4	656.3	Visible red
$4 \rightarrow 2$	486.2	486.1	Visible blue-green
$5 \rightarrow 2$	434.1	434.0	Visible blue
$4 \rightarrow 3$	1875	1875	Infrared

*These lines are typical; other lines could be cited as well, with equally good agreement between theory and experiment. The unit of wavelength is the nanometer (nm), 10^{-9} m.

$$= \left(\frac{-2.179 \times 10^{-18}\ \text{J}}{6.626 \times 10^{-34}\ \text{J s}}\right)\left(\frac{1}{25} - \frac{1}{4}\right) = (3.289 \times 10^{15}\ \text{s}^{-1})(0.2500 - 0.04000)$$

$$= (3.289 \times 10^{15}\ \text{s}^{-1})(0.2100) = 6.906 \times 10^{14}\ \text{s}^{-1}$$

Calculate the wavelength from the frequency.

$$\lambda = \frac{c}{\nu} = \frac{2.998 \times 10^{8}\ \text{m/s}}{6.906 \times 10^{14}\ \text{s}^{-1}} = 4.341 \times 10^{-7}\ \text{m} = 434.1\ \text{nm}$$

According to Figure 5.2, 434.1 nm is in the visible region of the spectrum.

☑ **Reasonable Result Check** According to Figure 5.10, transitions from higher levels to the $n = 2$ level are in the visible region of the spectrum, so the answer is reasonable.

PROBLEM-SOLVING PRACTICE 5.4

(a) Calculate the frequency and the wavelength of the line for the $n = 6$ to $n = 4$ transition.
(b) Is this wavelength longer or shorter than that of the $n = 7$ to $n = 4$ transition?

CONCEPTUAL EXERCISE 5.6

Conversions

Show that the value of the Rydberg constant per photon, 2.179×10^{-18} J, is equivalent to 1312 kJ/mol photons.

5-4 Beyond the Bohr Model: The Quantum Mechanical Model of the Atom

The Bohr atomic model was accepted almost immediately. Bohr's success with the hydrogen atom soon led to attempts by him and others to extend the same model to more complex atoms. Before long it became apparent, however, that line spectra for elements other than hydrogen had more lines than could be explained by the simple Bohr model. A totally different approach was needed to explain electron behavior in atoms or ions with more than one electron. The new approach again was radically different from previous ideas.

Spectroscopy is the science of measuring spectra. Many kinds of spectroscopy have emerged since the first studies of simple line spectra. Some spectral measurements are done for quantitative analytical purposes; others are done to determine molecular structure.

In 1924, the young physicist Louis de Broglie (1892–1987) posed the question: *If light can be viewed in terms of both wave and particle properties, why can't particles of matter, such as electrons, be treated the same way?* And so de Broglie proposed the revolutionary idea that electrons could have wave-like properties. De Broglie proposed that the wavelength (λ) of an electron (or any other particle) depends on its mass (m), its velocity (v), and Planck's constant (h):

$$\lambda = \frac{h}{mv}$$

The *product of* m × v *for any object* is the **momentum** of the object. Note that the smaller the mass of the particle is, or the smaller the particle's velocity is, the larger the wavelength is. An electron has a very small mass (9.11×10^{-28} g) and can move fast, so its wavelength can be large enough that we observe wave-like behavior. A macroscopic object, such as a tennis ball, has very much smaller wavelengths and does not move as fast, so we do not observe wave-like behavior.

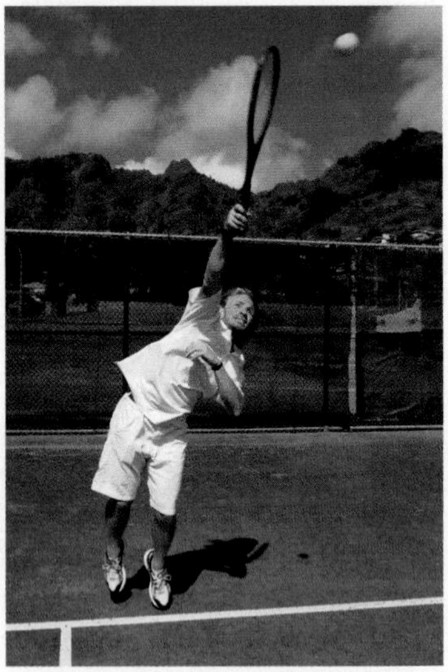

Michael Mitchell/Shutterstock.com

When a tennis ball is served at 125 mi/h its wavelength is far too short to be observed.

PROBLEM-SOLVING EXAMPLE 5.5

Tennis Balls and Electrons

At the U. S. Open tennis tournament, serves routinely reach more than 100 mi/h. Calculate the de Broglie wavelength (nm) of a tennis ball traveling at 56.0 m/s (125 mi/h) with that of an electron traveling at 5.0×10^6 m/s. Masses: $m_{electron} = 9.11 \times 10^{-31}$ kg; $m_{tennis\ ball} = 0.0567$ kg.

Result The wavelength of the electron is *much* longer than that of the tennis ball: $\lambda_{electron} = 0.15$ nm; $\lambda_{tennis\ ball} = 2.09 \times 10^{-25}$ nm.

Analyze The de Broglie equation relates wavelength to mass and velocity.

Plan Substitute the mass and velocity given for an electron and a tennis ball into the de Broglie equation and use proportionality factors to obtain units of nm.

Execute Planck's constant is

$$6.626 \times 10^{-34}\ \text{J s, and } 1\ \text{J} = \frac{1\ \text{kg m}^2}{\text{s}^2}\ \text{so that } h = 6.626 \times 10^{-34}\ \text{kg m}^2\ \text{s}^{-1}$$

For the electron:

$$\lambda = \frac{6.626 \times 10^{-34}\ \text{kg m}^2\ \text{s}^{-1}}{(9.11 \times 10^{-31}\ \text{kg})(5.0 \times 10^6\ \text{m/s})} = 1.5 \times 10^{-10}\ \text{m} \times \frac{1\ \text{nm}}{10^{-9}\ \text{m}} = 0.15\ \text{nm}$$

For the tennis ball:

$$\lambda = \frac{6.626 \times 10^{-34}\ \text{kg m}^2\text{s}^{-1}}{(0.0567\ \text{kg})(56.0\ \text{m/s})} = 2.09 \times 10^{-34}\ \text{m} \times \frac{1\ \text{nm}}{10^{-9}\ \text{m}} = 2.09 \times 10^{-25}\ \text{nm}$$

The wavelength of the electron is in the X-ray region of the electromagnetic spectrum (Figure 5.2, ← **Sec. 5-1**). The wavelength of the tennis ball is far too short to observe.

PROBLEM-SOLVING PRACTICE 5.5

Calculate the de Broglie wavelength of a neutron moving at 10% the velocity of light. The mass of a neutron is 1.67×10^{-24} g.

Many scientists found de Broglie's ideas hard to accept, but in 1927 C. Davisson and L. H. Germer, researchers at the Bell Telephone Laboratories, found that a beam of electrons is diffracted by planes of atoms in a thin sheet of metal foil (Figure 5.11) in the same way that light waves are diffracted by a diffraction grating. Because diffraction is readily explained by the wave properties of light (Figure 5.6), it followed that electrons also can be described by the equations of waves under some circumstances.

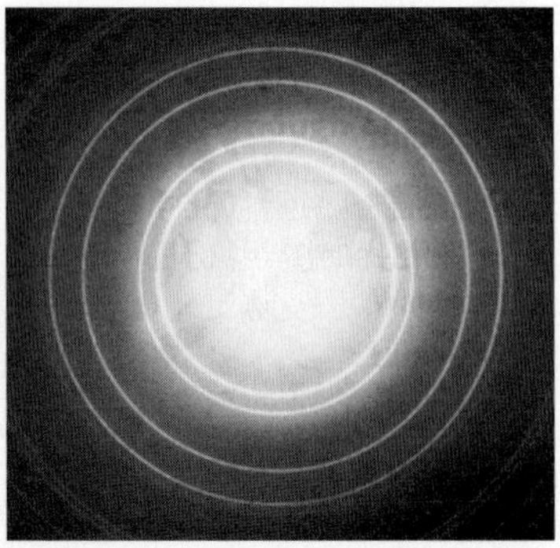

Hagley Museum and Library

Figure 5.11 Electron diffraction pattern obtained from aluminum foil.

A few years after de Broglie's hypothesis about the wave nature of the electron, Werner Heisenberg (1901–1976) proposed the **uncertainty principle**, which states that *it is impossible to simultaneously determine the exact position and the exact momentum of an electron.* This limitation is not a problem for a macroscopic object because when we see an object the energy of photons used to locate the object does not cause a measurable change in the object's position or momentum. However, the very act of measurement affects the position and momentum of an electron because of its very small size and mass. The wavelength of light used to locate an object has to be smaller than the size of the object, so photons with short wavelengths (and high energies) are required to locate an electron; when such photons collide with the electron, its momentum is changed. If lower-energy photons were used to avoid affecting the momentum, little information would be obtained about the location of the electron because the wavelength would be much larger than the size of the electron. Consider an analogy in photography. If you take a picture of a car race with a high shutter speed, you get a clear picture of the cars but you can't tell how fast they are going or even whether they are moving. With a slow shutter speed, you can tell from the blur of the car images something about the speed and direction, but you have less information about where the car is.

The Heisenberg uncertainty principle illustrated another inadequacy in the Bohr model—its representation of the electron in the hydrogen atom in terms of well-defined orbits about the nucleus. In practical terms, the best we can do is to represent the *probability* of finding an electron of a given energy and momentum within a given space. This probability-based model of the atom is what chemists now use.

In 1926, Erwin Schrödinger (1877–1961) combined de Broglie's hypothesis with mathematical equations suitable for wave motion to describe the hydrogen atom with a *wave equation*. The wave equation involves mathematical functions (based on exponential, sine, or cosine functions) called *wave functions* and represented by the Greek letter psi, ψ. Wave functions can be used to calculate the energies of the allowed energy states of a hydrogen atom. Each energy state corresponds to one or more wave functions. *Each of the wave functions that result from solving the Schrödinger wave equation* is called an **orbital**.

A wave function is a complicated mathematical equation that has no direct physical meaning. However, the square of the wave function, ψ^2, is related to the **electron density**, *the probability of finding the electron in a given region of space*. Electron density is also called probability density. Figure 5.12 shows two ways to represent the electron density for a ground-state hydrogen atom. On the left is a *dot-density* diagram—the density of dots is proportional to the electron probability; this is sometimes called an electron-cloud picture. On the right is a *boundary-surface* diagram; this shows the shape of an orbital by drawing a surface within which a large percentage (in this case, 90%) of the electron density is found. A 100% probability is not chosen because such a surface would have an infinite radius. An analogy is a dart board, which has a finite size; normally, dart players hit the board at least 90% of the time. But if you wanted to be certain that any player, no matter how far away or how inept, would be able to hit the board on 100% of his or her throws, then the board would have to be infinitely large.

Note that an atomic *orbital* (quantum mechanical model) is not the same as an *orbit* (Bohr model). In the quantum mechanical model, the principal quantum number, n, is related to the most probable distance of the electron from the nucleus, not to the radius of a well-defined orbit.

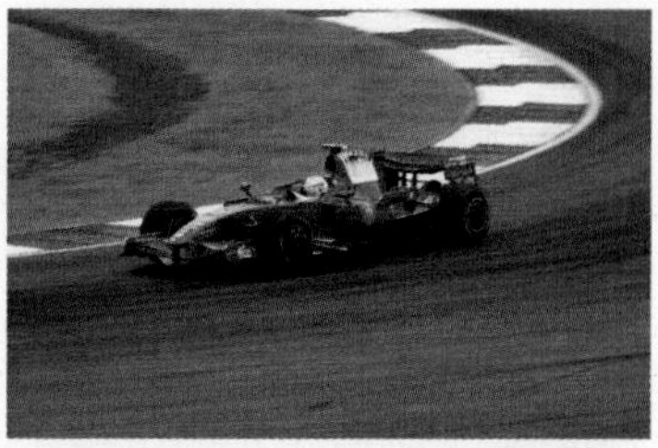
Chen WS/Shutterstock.com

PHOENIX INTERNATIONAL RACEWAY
Al Bello/Getty Images Sport/Getty Images

Speeding race cars. Top photo taken at high shutter speed. Bottom photo taken at low shutter speed.

The density of dots means the number of dots within a specified volume.

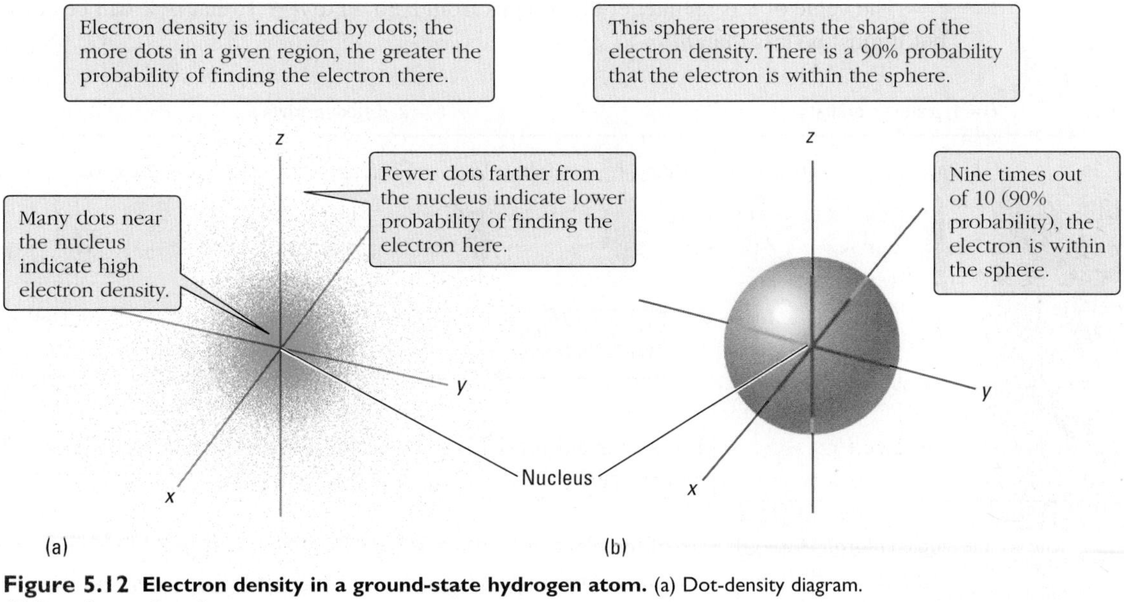
Electron density is indicated by dots; the more dots in a given region, the greater the probability of finding the electron there.
This sphere represents the shape of the electron density. There is a 90% probability that the electron is within the sphere.
z
Fewer dots farther from the nucleus indicate lower probability of finding the electron here.
Many dots near the nucleus indicate high electron density.
y
x
Nucleus
z
Nine times out of 10 (90% probability), the electron is within the sphere.
y
x
(a)
(b)

Figure 5.12 Electron density in a ground-state hydrogen atom. (a) Dot-density diagram. (b) Boundary-surface diagram.

5-5 Quantum Numbers, Energy Levels, and Atomic Orbitals

A hydrogen atom has many energy levels and it also has many atomic orbitals. Boundary-surface diagrams for some of them are shown in Figure 5.13. As they developed the quantum mechanical model of the atom, Schrödinger and others found *four related sets of numbers that define the shapes of orbitals and the allowed energies of a hydrogen atom*; these are called **quantum numbers**. The four quantum numbers are designated n, ℓ, m_ℓ, and m_s.

5-5a First Quantum Number, *n*: Principal Energy Levels (Shells)

The first quantum number, ***n***, the **principal quantum number**, *designates the main electron energy levels in an atom and the size of the electron-density distribution.* This is the same quantum number that Bohr used to predict the hydrogen-atom spectrum and the energy of an electron in a hydrogen atom. It has only integer values, starting with 1 and going to ∞:

$$n = 1, 2, 3, 4, \ldots \infty$$

Each value of n designates a principal energy level, also called an electron shell. A **principal energy level**, or **shell**, is *a collection of atomic orbitals that has the same principal quantum number,* n. An electron is in the first shell when $n = 1$, in the second shell when $n = 2$, and so on. As n increases, the energy of the electron increases as well, and the electron, on average, is farther away from the nucleus and is less tightly bound to it. The larger size of orbitals with larger n values and the number of different kinds of orbitals for each n value can be seen in Figure 5.13.

5-5b Second Quantum Number, ℓ: Atomic Orbital Shapes

Within each principal energy level (shell) is one or more *subshells*. The second quantum number, ℓ, the **orbital quantum number**, (sometimes called the *azimuthal quantum number*) *designates different subshells within a particular shell.* A **subshell** is *one or more atomic orbitals with the same* n *and* ℓ *quantum numbers*.

The value of ℓ is an integer that ranges from zero up to $n - 1$; that is, ℓ can be zero, but it must be less than n.

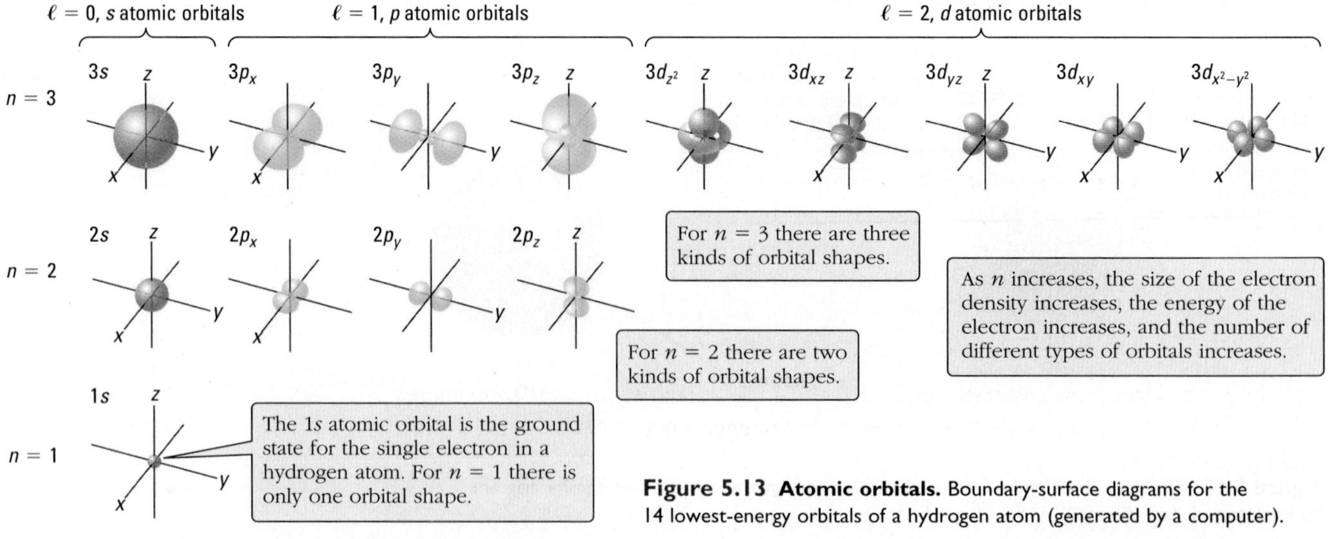

ℓ = 0, s atomic orbitals
ℓ = 1, p atomic orbitals
ℓ = 2, d atomic orbitals
n = 3
3s
3px
3py
3pz
3dz²
3dxz
3dyz
3dxy
3dx²−y²
n = 2
2s
2px
2py
2pz
n = 1
1s
x
y
z
For n = 3 there are three kinds of orbital shapes.
As n increases, the size of the electron density increases, the energy of the electron increases, and the number of different types of orbitals increases.
For n = 2 there are two kinds of orbital shapes.
The 1s atomic orbital is the ground state for the single electron in a hydrogen atom. For n = 1 there is only one orbital shape.

Figure 5.13 Atomic orbitals. Boundary-surface diagrams for the 14 lowest-energy orbitals of a hydrogen atom (generated by a computer).

$$\ell = 0, 1, 2, 3, \ldots, (n - 1)$$

According to this relationship, if $n = 1$, then ℓ must be zero. Thus, in the first shell ($n = 1$) there is only one subshell. The second shell ($n = 2$) has two subshells—one with an ℓ value of 0 and a second with an ℓ value of 1. Similarly,

for $n = 3$;	$\ell = 0, 1,$ or 2	(three subshells)
for $n = 4$;	$\ell = 0, 1, 2$ or 3	(four subshells)

Rather than using ℓ values, subshells are more commonly designated by letters: *s, p, d,* or *f.* The first four types of subshells are known as the *s* subshell, *p* subshell, *d* subshell, and *f* subshell.

The letters *s, p, d,* and *f* derived historically from spectral lines called sharp, principal, diffuse, and fundamental.

ℓ value	0	1	2	3
Subshell	*s*	*p*	*d*	*f*

A number (the *n* value) and a letter (*s, p, d,* or *f*) are used to designate a specific subshell.

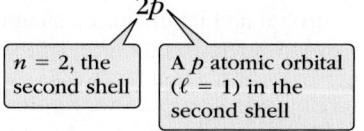

2*p*
- $n = 2$, the second shell
- A *p* atomic orbital ($\ell = 1$) in the second shell

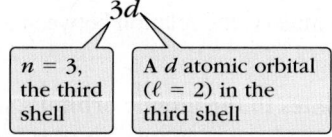

3*d*
- $n = 3$, the third shell
- A *d* atomic orbital ($\ell = 2$) in the third shell

For the first four principal energy levels, these designations are as follows.

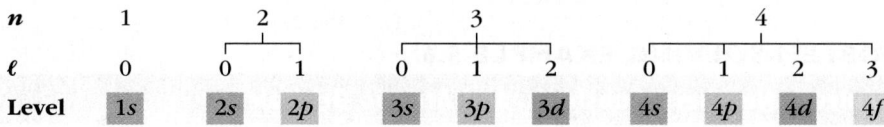

n	1	2		3			4			
ℓ	0	0	1	0	1	2	0	1	2	3
Level	1*s*	2*s*	2*p*	3*s*	3*p*	3*d*	4*s*	4*p*	4*d*	4*f*

In Figure 5.13 the subshells for each *n* value are colored the same as in the designations above. Notice that within a subshell all of the orbitals have similar shapes. For example, each *p* orbital has two parts at an angle of 180° to each other on opposite sides of the nucleus. ***The shape of the electron cloud corresponding to an atomic orbital is determined by the value of ℓ.***

EXERCISE 5.7

Subshells and Quantum Numbers

Give the subshell designation for an electron with these quantum numbers.

(a) $n = 5, \ell = 2$ (b) $n = 4, \ell = 3$ (c) $n = 6, \ell = 1$

EXERCISE 5.8

Subshell Designations

Explain why each statement is incorrect.

(a) 3*f* is an appropriate subshell designation.

(b) $n = 2, \ell = 2$ correctly designates a subshell.

5-5c Third Quantum Number, m_ℓ: Orientation of Atomic Orbitals

The **magnetic quantum number, m_ℓ,** *designates the orientation of different atomic orbitals within the same subshell.* It can have any integer value between ℓ and $-\ell$, including zero. Thus,

$$m_\ell = \ell, (\ell - 1), \ldots, +1, 0, -1, \ldots, -\ell$$

For an *s* subshell, $\ell = 0$ so m_ℓ can have only one value—zero. An *s* subshell, regardless of its *n* value—1*s*, 2*s*, and so on—contains only one orbital. For a *p* subshell, $\ell = 1$, and so m_ℓ can be +1, 0, or −1. Within each *p* subshell there are three *different* atomic orbitals: one with $m_\ell = +1$, another with $m_\ell = 0$, and a third with $m_\ell = -1$. *In general, there is a total of* $2\ell + 1$ *atomic orbitals within a subshell of quantum number* ℓ.

The m_ℓ value, in conjunction with the ℓ value, is related to the shape and orientation of an atomic orbital in space. The *s* atomic orbitals are spherical and there is only one possible value of m_ℓ so only a single orientation is possible. (Figure 5.13, ← **Sec. 5-5b**). The three *p* atomic orbitals are dumbbell-shaped and oriented at right angles to each other, with maximum electron density directed along either the *x*-, *y*-, or *z*-axis. They are usually designated as p_x, p_y, and p_z. The five *d* atomic orbitals in the 3*d* sublevel ($\ell = 2$; $2\ell + 1 = 5$) are also illustrated in Figure 5.13. They are designated d_{z^2}, d_{xz}, d_{yz}, d_{xy}, and $d_{x^2-y^2}$.

We will discuss *d* atomic orbitals further in Chapter 20 in connection with their role in the chemistry of transition-metal ions.

Table 5.2 summarizes the relationships among *n*, ℓ, and m_ℓ. Notice from both Figure 5.13 and Table 5.2 that ***the total number of atomic orbitals in a shell equals*** $\mathbf{n^2}$. For example, an $n = 3$ shell has a total of $3^2 = 9$ atomic orbitals: one 3*s* + three 3*p* + five 3*d*.

We leave the discussion of *f* atomic orbitals, which are very complex, to subsequent courses.

In summary, the relation between an atomic orbital and its first three quantum numbers is

- ***n* relates to the atomic orbital's size.**
- **ℓ relates to the atomic orbital's shape.**
- **m_ℓ relates to the atomic orbital's orientation.**

PROBLEM-SOLVING EXAMPLE 5.6

Quantum Numbers, Subshells, and Atomic Orbitals

Consider the $n = 4$ principal energy level.

(a) Without referring to Table 5.2, predict the number of subshells in this level.

(b) Identify each of the subshells by its number and letter designation (as in 1*s*) and give its ℓ values.

(c) Determine how many atomic orbitals each subshell has and identify the m_ℓ value for each orbital.

(d) What is the total number of atomic orbitals in the $n = 4$ level?

Table 5.2 Relationships Among *n*, ℓ, and m_ℓ for the First Four Shells

n	ℓ	Subshell Label	Possible m_ℓ Values	Number of Atomic Orbitals in Subshell, $2\ell + 1$	Total Number of Atomic Orbitals in Shell, n^2
1	0	1*s*	0	1	1
2	0	2*s*	0	1	
2	1	2*p*	1, 0, −1	3	4
3	0	3*s*	0	1	
3	1	3*p*	1, 0, −1	3	
3	2	3*d*	2, 1, 0, −1, −2	5	9
4	0	4*s*	0	1	
4	1	4*p*	1, 0, −1	3	
4	2	4*d*	2, 1, 0, −1, −2	5	
4	3	4*f*	3, 2, 1, 0, −1, −2, −3	7	16

Result

(a) Four subshells
(b) 4*s*, 4*p*, 4*d*, and 4*f*; $\ell = 0, 1, 2$, and 3, respectively
(c) One 4*s* atomic orbital, three 4*p* atomic orbitals, five 4*d* atomic orbitals, and seven 4*f* atomic orbitals
(d) 16 orbitals

Analyze The goal is to characterize all subshells and orbitals in the $n = 4$ shell.

Plan Apply the rules for quantum numbers given in the preceding sections.

Execute

(a) There are *n* subshells in the *n*th shell. Thus, the $n = 4$ level contains four subshells.
(b) The number refers to the principal quantum number, *n*; the letter is associated with the ℓ quantum number. The four sublevels correspond to the four possible ℓ values:

Sublevels	4*s*	4*p*	4*d*	4*f*
ℓ value	0	1	2	3

(c) There are $2\ell + 1$ atomic orbitals within a sublevel. Only one 4*s* atomic orbital is possible ($\ell = 0$, so m_ℓ must be zero). There are three 4*p* atomic orbitals ($\ell = 1$) with m_ℓ values of 1, 0, or -1. There are five 4*d* atomic orbitals ($\ell = 2$) corresponding to the five allowed values for m_ℓ: 2, 1, 0, -1, and -2. There are seven 4*f* atomic orbitals ($\ell = 3$), each with one of the seven permitted values of m_ℓ: 3, 2, 1, 0, -1, -2, and -3.
(d) The total number of atomic orbitals in a level is n^2. Therefore, the $n = 4$ level has a total of 16 atomic orbitals.

☑ **Reasonable Result Check** Counting the orbitals identified in part (c) gives a total of $1 + 3 + 5 + 7 = 16$, verifying the result in part (d).

PROBLEM-SOLVING PRACTICE 5.6

(a) Identify the subshell with $n = 6$ and $\ell = 2$.
(b) How many atomic orbitals are in this subshell?
(c) What are the m_ℓ values for these atomic orbitals?

5-5d Fourth Quantum Number, m_s: Electron Spin

When spectroscopists more closely studied emission spectra of hydrogen and sodium atoms, they discovered that what were originally thought to be single lines were actually very closely spaced pairs of lines. In 1925 the Dutch physicists George Uhlenbeck and Samuel Goudsmit proposed that the line splitting could be explained by assuming that each electron in an atom can exist in one of two possible spin states. To visualize these states, consider an electron as a charged sphere rotating about an axis through its center (Figure 5.14). Such a spinning charge generates a magnetic field, so that each electron acts like a tiny bar magnet with north and south magnetic poles. Only two directions of spin are possible—clockwise or counterclockwise. Spins in opposite directions produce oppositely directed magnetic fields, which result in two slightly different energies. This slight difference in energy splits the spectral lines into closely spaced pairs.

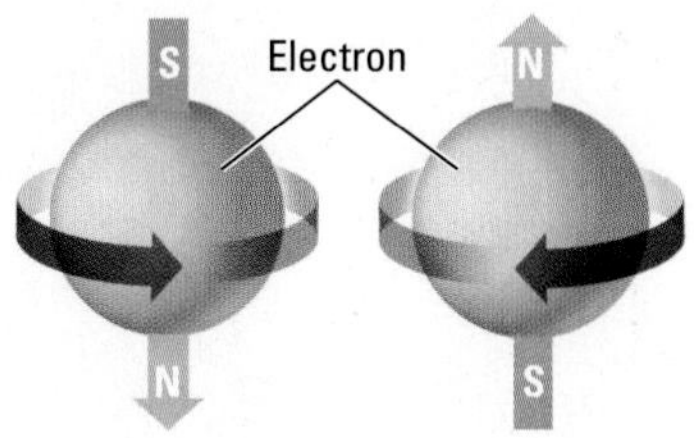
Electron
S
N
N
S
Figure 5.14 Electron spin can have only two directions.

Thus, to describe an electron in an atom completely, a fourth quantum number, m_s, called the **spin quantum number**, *designates the direction of electron spin*. This spin quantum number can have either of only two values, $+\frac{1}{2}$ or $-\frac{1}{2}$. That is, either $m_s = +\frac{1}{2}$ or $m_s = -\frac{1}{2}$.

EXERCISE 5.9

Quantum Numbers

Give all possible sets of four quantum numbers for an electron in a hydrogen atom that is in

(a) the 3*s* atomic orbital.
(b) the 2*p* subshell.
(c) the $4d_{xy}$ atomic orbital.

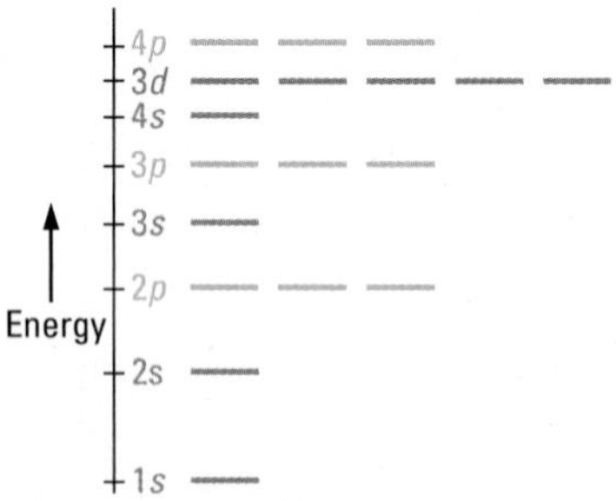
4p
3d
4s
3p
3s
2p
Energy
2s
1s

Figure 5.15 The relative ordering of atomic orbital energy levels in a many-electron atom from the 1*s* level through the 4*p* sublevels. Atomic orbitals in different subshells of the same shell have different energy. The energy differences between subshells decrease as *n* increases.

5-5e Atoms with More Than a Single Electron

So far we have limited our discussion to the orbitals and quantum numbers of a hydrogen atom. It turns out that the same orbital types and quantum numbers can be applied to any atom. The situation is more complicated, however, because with two or more electrons we have to take account of the fact that the electrons repel each other as well as being attracted by the nucleus.

Repulsions between two electrons in different orbitals depend on the shapes of the orbitals. Therefore, atomic orbitals in different subshells, but with the same *n* value, have different energies. The energies of subshells within a given shell always increase in the order $ns < np < nd < nf$. Consequently, orbitals in a 3*p* subshell have higher energy than those in a 3*s* subshell, but less energy than those in a 3*d* subshell. The 3*d* orbitals have the highest energy in the $n = 3$ level. Figure 5.15 shows the relative energies of atomic orbitals in many-electron atoms through the 4*p* subshell.

To make quantum theory consistent with experiment, Wolfgang Pauli stated in 1925 what is now known as the **Pauli exclusion principle**: *No more than two electrons can occupy the same atomic orbital in an atom, and these electrons must have opposite spins.* This is equivalent to saying that ***no two electrons in the same atom can have the same set of four quantum numbers*** n, ℓ, m_ℓ, and m_s. For example, two electrons can occupy the same 3*p* atomic orbital only if one of them has $m_s = +\frac{1}{2}$ and the other has $m_s = -\frac{1}{2}$. This is because both electrons have the same first three quantum numbers, for example, $n = 3$, $\ell = 1$, and $m_\ell = +1$. If the two electrons are to occupy the same orbital the Pauli principle says they must differ in their m_s values.

Electrons are said to have *parallel* spins if they have the same m_s quantum number (both $+\frac{1}{2}$ or both $-\frac{1}{2}$). Electrons are said to be *paired* when they are in the same atomic orbital and have opposite spins—one $+\frac{1}{2}$ and the other $-\frac{1}{2}$. In most multielectron atoms most of the electrons are paired.

CONCEPTUAL EXERCISE 5.10

Orbitals and Quantum Numbers

Give sets of four quantum numbers for two electrons that are (a) in the same shell and subshell, but in different atomic orbitals; and (b) in the same shell, but in different subshells and in different atomic orbitals.

EXERCISE 5.11

Quantum Number Comparisons

Two electrons in the same atom have these sets of quantum numbers: electron (a) has 3, 1, 0, $+\frac{1}{2}$; electron (b) has 3, 1, -1, $+\frac{1}{2}$. Show that these two electrons are not in the same atomic orbital. Which subshell are these electrons in?

The restriction that only two electrons can occupy a single atomic orbital establishes a maximum number of electrons for each shell. Each shell has n^2 orbitals, and each orbital can accommodate two electrons, so ***the* n*th shell can accommodate a maximum of* $2n^2$ *electrons.*** This is summarized in Table 5.3. For example, the $n = 2$ shell has

Table 5.3 Number of Electrons Accommodated in Electron Shells and Subshells

Electron Shell (n)	Subshells Available (number=n)	Number of Atomic Orbitals Available ($=2\ell + 1$)	Number of Electrons Possible in Subshell	Maximum Electrons for nth Shell ($=2n^2$)
1	s	1	2	2
2	s	1	2	
	p	3	6	8
3	s	1	2	
	p	3	6	
	d	5	10	18
4	s	1	2	
	p	3	6	
	d	5	10	
	f	7	14	32
5*	—	—	—	—
6*	—	—	—	—
7†	s	1	2	
	p	3	6	

*Atomic orbitals in the fifth and sixth shells are not listed. Completing Exercises 5.12 and 5.13 will provide much of the information for the $n = 5$ and $n = 6$ sections of this table.

†Subshells for $n = 7$ are listed only if they are occupied by electrons in ground-state atoms of known elements.

The results expressed in this table were predicted by the Schrödinger theory and have been confirmed by experiment.

one s and three p atomic orbitals, each of which can accommodate two paired electrons. Therefore, this shell can accommodate a total of $2 \times (2)^2$ = eight electrons (two in the $2s$ orbital and three pairs in the three $2p$ orbitals). In summary,

- ***Each principal energy level (shell) has a quantum number* n *= 1, 2, 3, . . .***
- ***Within each principal energy level there are* n *subshells designated as* s, p, d, *and* f *subshells.***
- ***The number of atomic orbitals in each subshell is $2\ell + 1$: one* s *atomic orbital ($\ell = 0$), three* p *atomic orbitals ($\ell = 1$), five* d *atomic orbitals ($\ell = 2$), and seven* f *atomic orbitals ($\ell = 3$).***
- ***Within a principal energy level* n*, there are* n^2 *atomic orbitals.***
- ***Within a principal energy level* n*, there are at most* $2n^2$ *electrons.***

A quantum number analogy: A ticket for a reserved seat on a train is analogous to a set of four quantum numbers, n, ℓ, m_ℓ, and m_s. The ticket specifies a particular train, a certain car on the train, a pair of seats in that car, and a specific seat in that pair. The pair of seats is analogous to an atomic orbital, and the occupants of the row are like two electrons in the same orbital (same train, car, and row), but with opposite spins (different seats).

EXERCISE 5.12

Maximum Number of Electrons

(a) Without looking at Table 5.3, determine the maximum number of electrons in the $n = 3$ level. Identify the atomic orbital of each electron.

(b) What is the maximum number of electrons in the $n = 5$ level? Identify the atomic orbital of each electron. (*Hint:* g and h atomic orbitals follow f atomic orbitals.)

CONCEPTUAL EXERCISE 5.13

h Atomic Orbitals

Using the same reasoning developed for s, p, d, and f atomic orbitals, determine the n value of the first shell that could contain h atomic orbitals. How many h atomic orbitals are in that shell?

5-6 Shapes of Atomic Orbitals

As noted in the previous section, the ℓ and m_ℓ quantum numbers determine the shapes and spatial orientations of atomic orbitals. It is important to be familiar with **orbital shapes**, *the sizes and 3-D orientations of atomic orbitals.*

5-6a *s* Atomic Orbitals ($\ell = 0$)

Figure 5.16a is the same electron density distribution shown in Figure 5.12a.

Figure 5.16a shows a dot-density diagram of the 1*s* atomic orbital of a hydrogen atom with its electron density decreasing rapidly as the distance from the nucleus increases. Note from Figure 5.16a that the probability of finding a 1*s* electron is the same *in any direction* at the same distance from the nucleus. Thus, the 1*s* atomic orbital is spherical. The value r_{90} in the figure is the radius of the sphere within which 90% of the electron density is found.

Figure 5.16b shows another way to represent the electron density of a 1*s* electron in a hydrogen atom. This *graph of probability of finding the electron at a distance,* r, *from the nucleus versus the distance,* r, is known as a **radial distribution plot**. In this case, the *y*-axis represents the probability of finding the electron on a sphere of radius, *r*, and the *x*-axis represents the radius of the sphere. The surface area of a sphere gets bigger the larger the radius is, so near the nucleus the sphere's area is small. Even though the electron density is large, the small area makes the probability of finding the electron on the sphere very small. As *r* increases the area of the sphere increases and, even though the electron density is somewhat less, the probability of finding the electron at a given *r* value is greater. As *r* increases still further the electron density decreases so rapidly that the probability of finding the electron at a given *r* goes down. The greatest probability of finding the 1*s* electron in a ground-state hydrogen atom is at 0.0529 nm (52.9 pm) from the nucleus. Note from the figure that at very large distances the probability drops very close to zero, but does not quite become zero. This indicates that there is a very small, but finite, probability of the electron being very far away from the nucleus.

For *n* values other than one, the shape of *s* atomic orbitals remains the same; they are all spherical. Their sizes, however, increase as *n* increases, as shown in Figure 5.16c. For example, the boundary surface of a 3*s* orbital has a greater volume (and radius) than that of a 2*s* atomic orbital, which is greater than that of a 1*s* atomic orbital.

In Figure 5.16b, the y-axis probability values equal the square of the wave function, ψ^2, times the area of the surface of a sphere with radius *r*: The area of the surface of a sphere is $4\pi r^2$, so the y-axis value is $4\pi r^2\psi^2$.

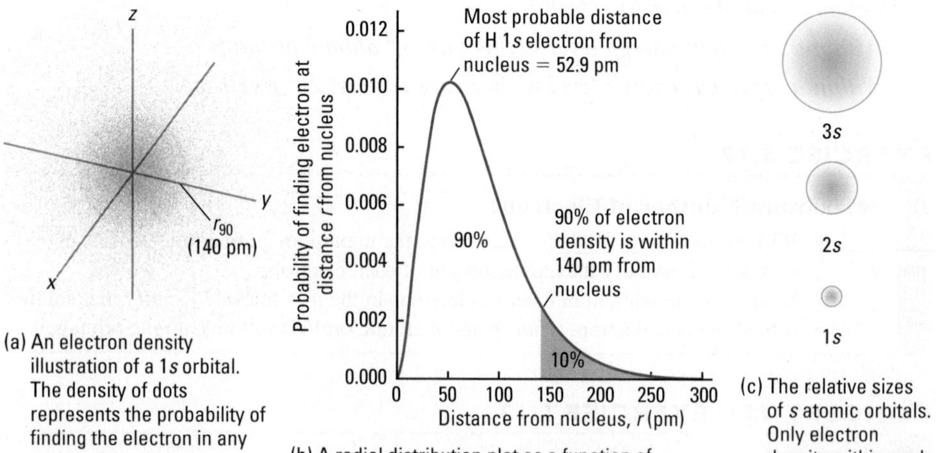
z
y
x
r_{90} (140 pm)
Probability of finding electron at distance r from nucleus
0.012
0.010
0.008
0.006
0.004
0.002
0.000
Most probable distance of H 1s electron from nucleus = 52.9 pm
90%
90% of electron density is within 140 pm from nucleus
10%
0 50 100 150 200 250 300
Distance from nucleus, r (pm)
3s
2s
1s

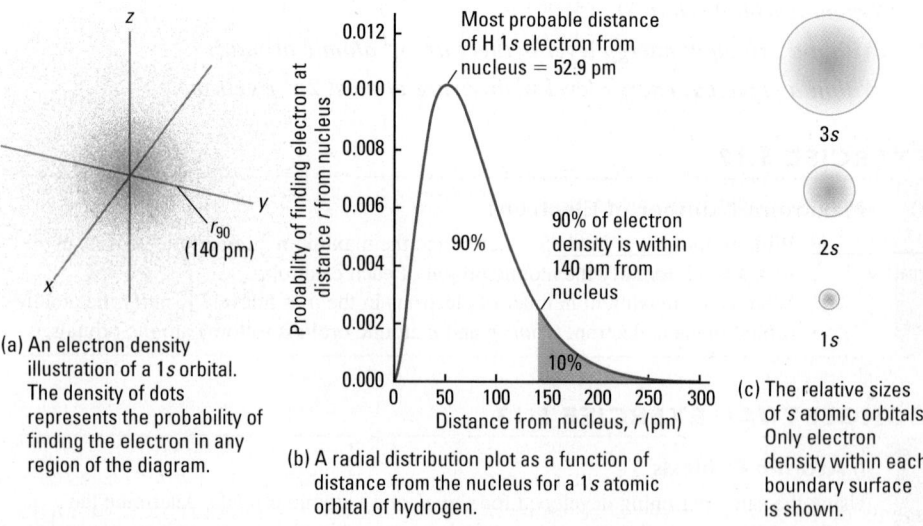

(a) An electron density illustration of a 1*s* orbital. The density of dots represents the probability of finding the electron in any region of the diagram.

(b) A radial distribution plot as a function of distance from the nucleus for a 1*s* atomic orbital of hydrogen.

(c) The relative sizes of *s* atomic orbitals. Only electron density within each boundary surface is shown.

Figure 5.16 Depicting *s* atomic orbitals.

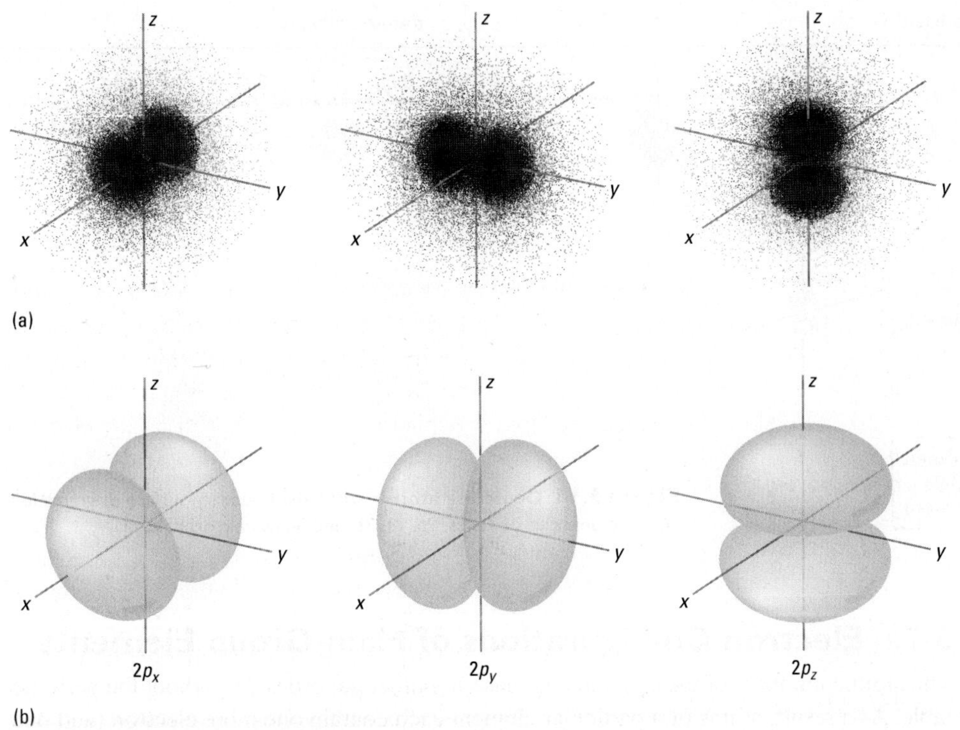
z
y
x
(a)
2p_x
2p_y
2p_z
(b)

Figure 5.17 Depicting *p* atomic orbitals. (a) Electron density distributions for the three 2*p* atomic orbitals. (b) The corresponding boundary surfaces for the set of three 2*p* orbitals.

5-6b *p* Atomic Orbitals ($\ell = 1$)

Unlike *s* atomic orbitals, which are all spherical, the *p* atomic orbitals, those for which $\ell = 1$, are all dumbbell-shaped (Figure 5.17). In a *p* atomic orbital, there are two lobes with electron density on either side of the nucleus. The *p* atomic orbitals within a given subshell (same *n* and ℓ value) differ from each other in their orientation in space; the three *p* atomic orbitals within the same subshell are mutually perpendicular and lie along the *x*-, *y*-, and *z*-axes. These orientations correspond to the three allowed m_ℓ values of +1, 0, and −1.

5-6c *d* Atomic Orbitals ($\ell = 2$)

Each energy level with $n > 2$ contains a subshell for which $\ell = 2$, consisting of five *d* atomic orbitals with corresponding m_ℓ values of +2, +1, 0, −1, and −2. The *d* atomic orbitals consist of two different types of shapes and spatial orientations. Three of them (d_{xz}, d_{yz}, and d_{xy}) each have four lobes that lie in the plane of but between the designated *x*-, *y*-, or *z*-axes; the other pair—the $d_{x^2-y^2}$ and d_{z^2} orbitals—have their principal electron density along the designated axes. For example, the d_{xz} orbital is in the *xz* plane and the regions of high electron density are between the *x*- and *z*-axes; the $d_{x^2-y^2}$ orbital lies in the *xy* plane and the regions of high electron density are along the *x*- and *y*-axes. The shapes of five *d* atomic orbitals, along with those of *s* and *p* atomic orbitals, are illustrated in Figure 5.18. The *d* atomic orbitals are important in the chemistry of transition-metal ions and we will discuss them further in Chapter 20.

5-7 Atom Electron Configurations

The complete description of the atomic orbitals occupied by all the electrons in an atom or monoatomic ion is called its **electron configuration**. The chemical similarities of elements in the same periodic table group are explained by the similar electron configurations of their atoms.

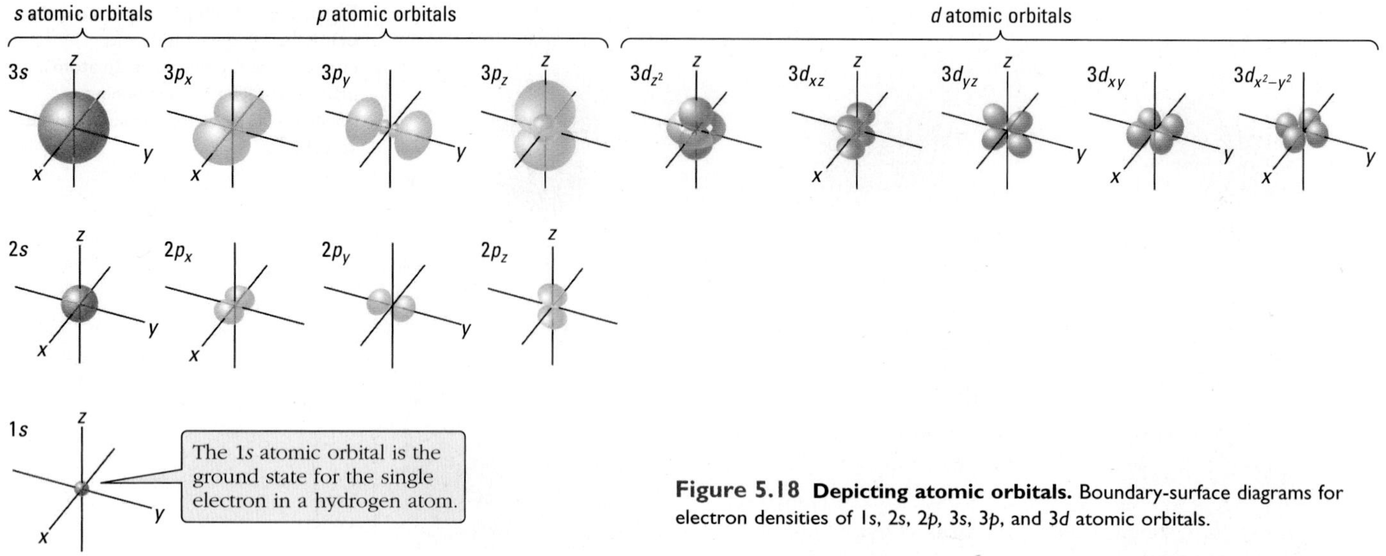
s atomic orbitals
p atomic orbitals
d atomic orbitals
3s
3px
3py
3pz
3dz2
3dxz
3dyz
3dxy
3dx2−y2
2s
2px
2py
2pz
1s
x
y
z
The 1s atomic orbital is the ground state for the single electron in a hydrogen atom.

Figure 5.18 Depicting atomic orbitals. Boundary-surface diagrams for electron densities of 1s, 2s, 2p, 3s, 3p, and 3d atomic orbitals.

5-7a Electron Configurations of Main-Group Elements

The atomic numbers of the elements increase in numerical order throughout the periodic table. As a result, atoms of a particular element each contain one more electron (and one more proton) than atoms of the preceding element. Each additional proton is in the nucleus, but how do we know which shell and atomic orbital each new electron occupies? An important principle for answering this question is this: *For an atom in its ground state, electrons are found in the energy shells, subshells, and atomic orbitals that produce the lowest energy for the atom.* Electrons fill atomic orbitals starting with the 1*s* atomic orbital and work upward in the subshell energy order shown in Figure 5.15 (← **Sec. 5-5e**).

To better understand how this filling of atomic orbitals works, consider the experimentally determined electron configurations of the first ten elements, which are written in three different ways in Table 5.4—condensed, expanded, and atomic orbital box diagram. Because electrons assigned to the $n = 1$ shell are closest to the nucleus and therefore lowest in energy, electrons are assigned to it first (H and He). At the left in Table 5.4,

Table 5.4 Electron Configurations of the First Ten Elements

	Electron Configurations		Atomic Orbital Box Diagrams				
	Condensed	*Expanded*	1s	2s	2p		
H	$1s^1$		↑				
He	$1s^2$		↑↓				
Li	$1s^22s^1$		↑↓	↑			
Be	$1s^22s^2$		↑↓	↑↓			
B	$1s^22s^22p^1$		↑↓	↑↓	↑		
C	$1s^22s^22p^2$	$1s^22s^22p^12p^1$	↑↓	↑↓	↑	↑	
N	$1s^22s^22p^3$	$1s^22s^22p^12p^12p^1$	↑↓	↑↓	↑	↑	↑
O	$1s^22s^22p^4$	$1s^22s^22p^22p^12p^1$	↑↓	↑↓	↑↓	↑	↑
F	$1s^22s^22p^5$	$1s^22s^22p^22p^22p^1$	↑↓	↑↓	↑↓	↑↓	↑
Ne	$1s^22s^22p^6$	$1s^22s^22p^22p^22p^2$	↑↓	↑↓	↑↓	↑↓	↑↓

Hund's rule says that electrons do not pair up in the 2*p* orbitals until they have to.

Each box represents an orbital.

Up and down arrows represent electrons and their spins.

the occupied atomic orbitals and the number of electrons in each atomic orbital are represented by this notation:

$$\text{H} \qquad 1s^1$$

one electron in *s* subshell

subshell (*s*), $\ell = 0$

principal quantum number, *n*

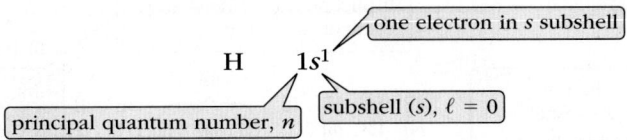

At the right in the table, each occupied atomic orbital is represented by a box in which electrons are shown as arrows: ↑ for a single electron in an orbital and ↑↓ for paired electrons in an atomic orbital.

In helium the two electrons are paired in the 1*s* atomic orbital so that the lowest energy shell ($n = 1$) is filled. The electron configuration is $1s^2$. Beginning with lithium, after the $n = 1$ shell is filled, electrons are assigned to the next lowest unoccupied energy level (Table 5.4), the $n = 2$ shell. This second shell can hold eight electrons ($2n^2$), and its atomic orbitals are occupied sequentially in the eight elements from lithium to neon. Notice in the periodic table inside the front cover that these are the eight elements of the second period.

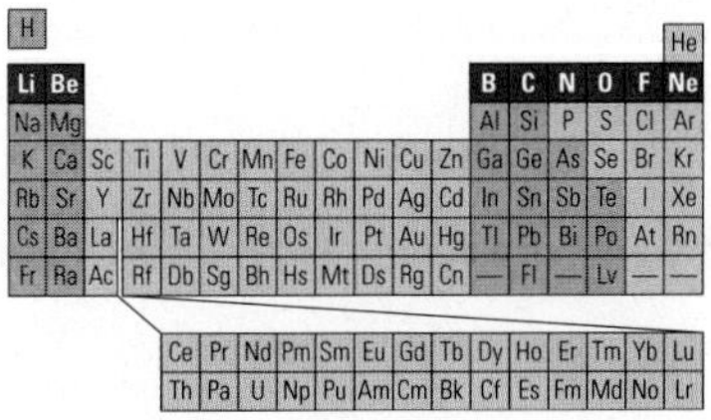

Second-period elements.

As happens in each principal energy level (and each period), the first two electrons fill the *s* atomic orbital. In the second period this occurs in Li ($1s^22s^1$) and in Be ($1s^22s^2$), at which point the 2*s* atomic orbital is completely filled. The next element is boron, B ($1s^22s^22p^1$); the fifth (and last) electron goes into a 2*p* atomic orbital. The three 2*p* atomic orbitals are of equal energy, and it does not matter which 2*p* atomic orbital is occupied first. Adding a second *p* electron for the next element, carbon ($1s^22s^22p^2$), presents a choice. Does the second 2*p* electron in the carbon atom pair with the existing electron in a 2*p* atomic orbital, or does it occupy a 2*p* atomic orbital by itself? It has been shown experimentally that both 2*p* electrons have the same spin. (The 2*p* electrons have *parallel* spins.) Hence, they must occupy different 2*p* atomic orbitals; otherwise, they would violate the Pauli exclusion principle. The expanded electron configurations in the middle of Table 5.4 show that, in boron, carbon, and nitrogen atoms, electrons occupy the 2*p* atomic orbitals individually.

Electron configurations where electrons with parallel spins occupy different orbitals minimize electron–electron repulsions, making the total energy of the set of electrons as low as possible. **Hund's rule** summarizes how subshells are filled: *The most stable arrangement of electrons in the same subshell has the maximum number of unpaired electrons, all with the same spin:* ***Electrons pair only after each atomic orbital in a subshell is occupied by a single electron.*** The general result of Hund's rule is that in *p, d,* or *f* subshells, each successive electron enters a different atomic orbital of the subshell until the subshell is half-full. After that, electrons pair in the atomic orbitals one by one. We can see this in Table 5.4 for the elements that follow boron—carbon, nitrogen, oxygen, fluorine, and neon.

Electron configurations in atomic-number order are in Appendix D.

Suppose you need to predict the electron configuration of a ground-state phosphorus atom. You can use a periodic table (inside the front cover) to do so. The expanded periodic table in Figure 5.19 shows how. First, locate phosphorus in the periodic table; it is in the third period, $n = 3$ and Group 5A. Then, start with H at the top of the periodic table and work your way along each row (in order of increasing atomic number) until you reach phosphorus, writing electron configurations as you go. In the first row (first period), Figure 5.19 indicates that the 1*s* atomic orbital is being filled, so write $1s^2$. By the end of the second row (second period) Figure 5.19 indicates that the 2*s* orbital and then the three 2*p* atomic orbitals have been filled, so the electron configuration at that point is $1s^22s^22p^6$. In the third period, the 3*s* atomic orbital fills and you are three squares into the 3*p* block when you reach phosphorus (Group 5A). Thus, the 3*p* atomic orbitals are only partly filled and the complete electron configuration is

$$\text{P} \qquad 1s^22s^22p^63s^23p^3$$

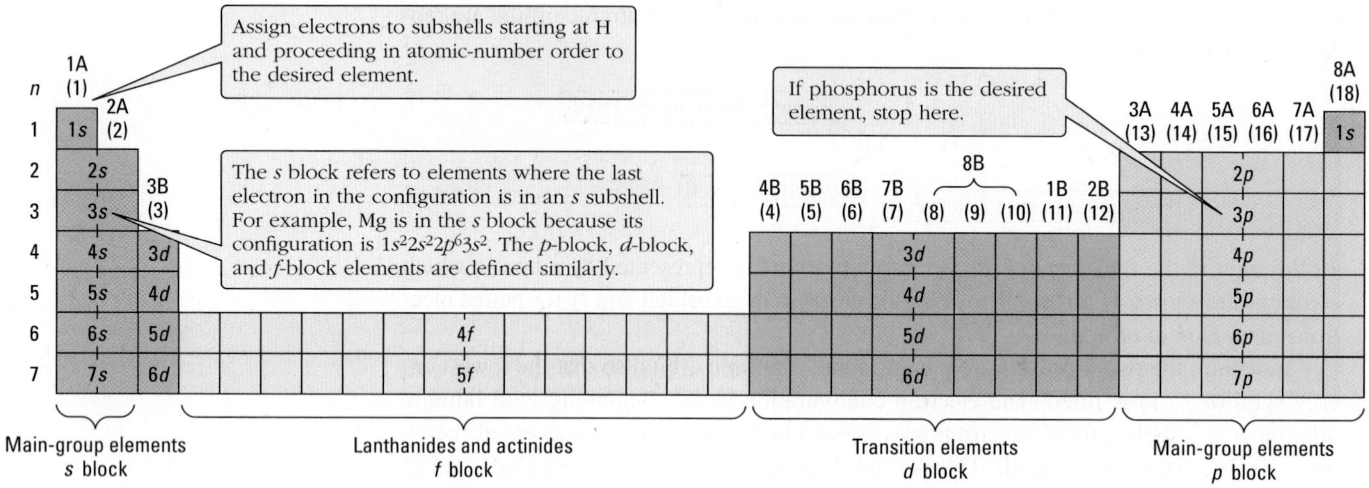

Assign electrons to subshells starting at H and proceeding in atomic-number order to the desired element.
If phosphorus is the desired element, stop here.
The s block refers to elements where the last electron in the configuration is in an s subshell. For example, Mg is in the s block because its configuration is 1s²2s²2p⁶3s². The p-block, d-block, and f-block elements are defined similarly.
n
1A (1)
2A (2)
3B (3)
4B (4) 5B (5) 6B (6) 7B (7) 8B (8) (9) (10) 1B (11) 2B (12)
3A (13) 4A (14) 5A (15) 6A (16) 7A (17)
8A (18)
1 2 3 4 5 6 7
1s 2s 3s 4s 5s 6s 7s
3d 4d 5d 6d
4f 5f
2p 3p 4p 5p 6p 7p
Main-group elements s block
Lanthanides and actinides f block
Transition elements d block
Main-group elements p block

Figure 5.19 Electron configuration and the periodic table.

You can check this by noting that phosphorus has atomic number 15 and therefore must have 15 electrons. Adding the superscripts in the electron configuration gives $2 + 2 + 6 + 2 + 3 = 15$.

There is another way to write this electron configuration. The **noble-gas notation** *uses brackets around the symbol of the noble gas at the end of the period* preceding *the element whose configuration is being written*. In the case of phosphorus, the preceding noble gas is Ne, at the end of the second period, so [Ne] is used to represent the electron configuration up to the end of that period, $1s^2 2s^2 2p^6$. The electron configuration for phosphorus in noble-gas notation is [Ne] $3s^2 3p^3$. Noble-gas notation is used to present the ground-state electron configurations for all elements in Appendix D.

According to Hund's rule, the three electrons in the $3p$ atomic orbitals of the phosphorus atom must be unpaired. To show this, you can write the *expanded* electron configuration

$$\text{P} \qquad 1s^2 2s^2 2p^2 2p^2 2p^2 3s^2 3p^1 3p^1 3p^1$$

or the atomic orbital box diagram

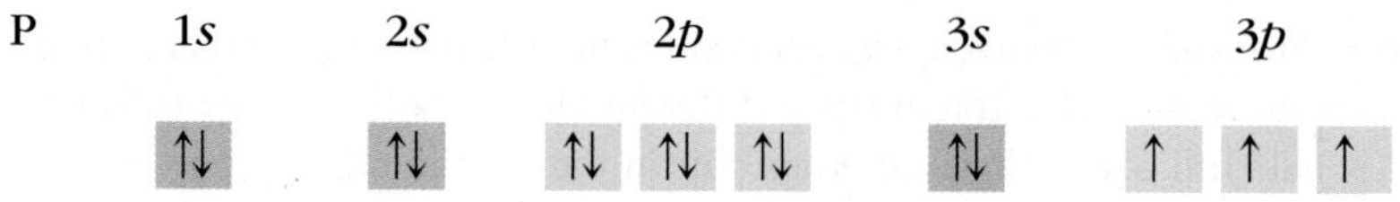

P
1s
2s
2p
3s
3p

Thus, all the electrons are paired except for the three electrons in $3p$ atomic orbitals. These three electrons each occupy different $3p$ atomic orbitals, and they have parallel spins. Partial atomic orbital box diagrams, which show only the $n = 3$ orbitals, are given in Figure 5.20 for all elements in Period 3.

At this point, you should be able to write electron configurations and atomic orbital box diagrams for main-group elements through Ca, atomic number 20, using only the periodic table to assist you.

5-7b Electron Configurations of Transition Elements

The **transition elements** are *the elements in B groups in Periods 4 through 7* in the middle of the periodic table. For these elements the last electron filled into the electron configuration is in a *d* subshell. For example, the electron configuration for the first transition element, scandium, is Sc [Ar] $4s^2 3d^1$. Notice that in each period in which they occur, the transition elements are preceded by two *s*-block elements (in the case of Sc,

Atomic number/ element	Partial atomic orbital box diagram (3s and 3p sublevels only) 3s	3p	Electron configuration	Noble-gas notation
$_{11}Na$	[↑]	[ ] [ ] [ ]	$[1s^22s^22p^6]\,3s^1$	$[Ne]\,3s^1$
$_{12}Mg$	[↑↓]	[ ] [ ] [ ]	$[1s^22s^22p^6]\,3s^2$	$[Ne]\,3s^2$
$_{13}Al$	[↑↓]	[↑] [ ] [ ]	$[1s^22s^22p^6]\,3s^23p^1$	$[Ne]\,3s^23p^1$
$_{14}Si$	[↑↓]	[↑] [↑] [ ]	$[1s^22s^22p^6]\,3s^23p^2$	$[Ne]\,3s^23p^2$
$_{15}P$	[↑↓]	[↑] [↑] [↑]	$[1s^22s^22p^6]\,3s^23p^3$	$[Ne]\,3s^23p^3$
$_{16}S$	[↑↓]	[↑↓] [↑] [↑]	$[1s^22s^22p^6]\,3s^23p^4$	$[Ne]\,3s^23p^4$
$_{17}Cl$	[↑↓]	[↑↓] [↑↓] [↑]	$[1s^22s^22p^6]\,3s^23p^5$	$[Ne]\,3s^23p^5$
$_{18}Ar$	[↑↓]	[↑↓] [↑↓] [↑↓]	$[1s^22s^22p^6]\,3s^23p^6$	$[Ne]\,3s^23p^6$

Figure 5.20 Partial atomic orbital diagrams for Period 3 elements.

these are K and Ca). This might be a surprise, because $n = 3$ ($3d$) usually indicates lower energy than $n = 4$ ($4s$), but Figure 5.19 and the periodic table clearly predict that the electron configuration for Sc has only a single d electron; the experimental data in Appendix D confirm this prediction.

In general, (n − 1)d atomic orbitals are filled after ns atomic orbitals and before filling of np atomic orbitals begins. For example, comparing Ca [Ar] $4s^2$, Mn [Ar] $3d^54s^2$, and Ge [Ar] $3d^{10}4s^24p^2$ illustrates that the $4s$ atomic orbitals are filled before $3d$, and $3d$ before $4p$. Figure 5.21 summarizes the order of filling of subshells as the electron configuration of an atom is built up.

Occupancy of d atomic orbitals begins with the $n = 3$ level and therefore with the first transition element, scandium, which has the configuration [Ar] $3d^14s^2$. After scandium comes titanium with [Ar] $3d^24s^2$ and vanadium with [Ar] $3d^34s^2$. The next element, chromium, is predicted to be [Ar] $3d^44s^2$, but that is incorrect; based on spectroscopic and magnetic measurements, the correct configuration is [Ar] $3d^54s^1$. This illustrates one of several cases where predicted and experimental configurations differ. When half-filled

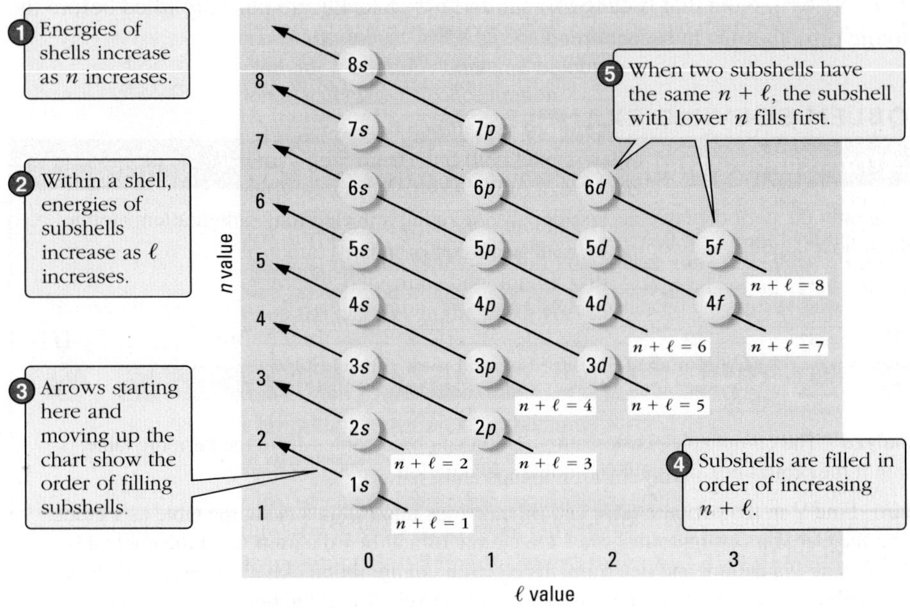
1 Energies of shells increase as n increases.
2 Within a shell, energies of subshells increase as ℓ increases.
3 Arrows starting here and moving up the chart show the order of filling subshells.
4 Subshells are filled in order of increasing n + ℓ.
5 When two subshells have the same n + ℓ, the subshell with lower n fills first.
n value
8
7
6
5
4
3
2
1
8s
7s
7p
6s
6p
6d
5s
5p
5d
5f
4s
4p
4d
4f
3s
3p
3d
2s
2p
1s
n + ℓ = 1
n + ℓ = 2
n + ℓ = 3
n + ℓ = 4
n + ℓ = 5
n + ℓ = 6
n + ℓ = 7
n + ℓ = 8
0
1
2
3
ℓ value

Figure 5.21 Order of subshell filling.

Cr is [Ar] $4s^1 3d^5$ not [Ar] $4s^2 3d^4$; Cu is [Ar] $4s^1 3d^{10}$ not [Ar] $4s^2 3d^9$.

Unpredicted electron configurations also occur with silver and gold, elements that are in the same group as copper.

or filled *d* subshells are possible, they sometimes occur, even though this forces an *s* subshell to also be half-filled. (Hund's rule indicates that half-filled subshells minimize electron repulsions.) As a result, *transition elements that have a half-filled* s *subshell and half-filled or filled* d *subshells have relatively stable electron configurations*. Chromium ([Ar] $4s^1 3d^5$) and copper ([Ar] $4s^1 3d^{10}$) are examples.

The number of unpaired electrons for atoms of most transition elements can be predicted according to Hund's rule by placing valence electrons in atomic orbital diagrams. For example, the electron configuration of Co ($Z = 27$) is [Ar] $3d^7 4s^2$, and the number of unpaired electrons is three, corroborated by experimental measurements, and seen from this atomic orbital box diagram.

		3*d*					4*s*
Co	[Ar]	↑↓	↑↓	↑	↑	↑	↑↓

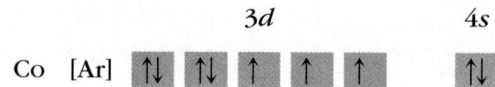

EXERCISE 5.14

Unpaired Electrons

Use atomic orbital box diagrams to determine which chromium ground-state configuration has the greater number of unpaired electrons: [Ar] $3d^4 4s^2$ or [Ar] $3d^5 4s^1$.

5-7c Elements with Incompletely Filled *f* Atomic Orbitals

In the elements of the sixth and seventh periods, *f* subshell atomic orbitals exist and can be filled (Figure 5.19). The elements (all metals) for which *f* subshells are filling are sometimes called the *inner transition* elements or, more usually, *lanthanides* (for lanthanum, the element just before those filling the 4*f* subshell) and *actinides* (for actinium, the element just before those filling the 5*f* subshell). The lanthanides start with lanthanum (La), which has the electron configuration [Xe] $5d^1 6s^2$. The next element, cerium (Ce), begins a separate row at the bottom of the periodic table, and it is with these elements that *f* atomic orbitals are filled (Appendix D). The electron configuration of Ce is [Xe] $4f^1 5d^1 6s^2$. Each of the lanthanide elements, from Ce to Lu, continues to add 4*f* electrons until the seven 4*f* atomic orbitals are filled by 14 electrons in lutetium (Lu, [Xe] $4f^{14} 5d^1 6s^2$). Note that both the $n = 5$ and $n = 6$ levels are partially filled before the 4*f* atomic orbitals start to be occupied.

PROBLEM-SOLVING EXAMPLE 5.7

Electron Configurations

Using only the periodic table as a guide, give the complete electron configuration, atomic orbital box diagram, and noble-gas notation for vanadium, V.

Result $1s^2 2s^2 2p^6 3s^2 3p^6 3d^3 4s^2$; [Ar] $3d^3 4s^2$

1*s*	2*s*	2*p*			3*s*	3*p*			3*d*					4*s*
↑↓	↑↓	↑↓	↑↓	↑↓	↑↓	↑↓	↑↓	↑↓	↑	↑	↑			↑↓

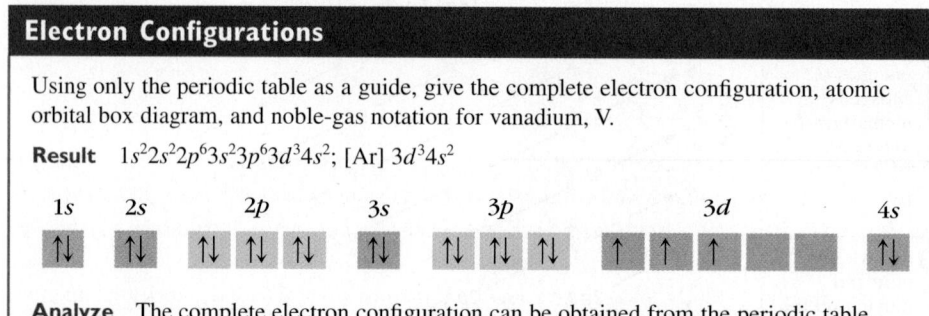

Analyze The complete electron configuration can be obtained from the periodic table. From it the orbital box diagram and noble-gas notation can be obtained.

Plan Find V in the periodic table and fill electrons into orbitals using the table as a guide. From the electron configuration, draw the atomic orbital box diagram. Find the noble gas that precedes vanadium and determine its electron configuration. All electrons not included in the noble gas's configuration need to be written explicitly in the noble-gas notation.

Execute The complete electron configuration and atomic orbital box diagram are shown in the Result. The first 18 electrons are represented by $1s^2 2s^2 2p^6 3s^2 3p^6$, the electron configuration of argon (Ar). The noble-gas notation is [Ar] $3d^3 4s^2$.

☑ **Reasonable Result Check** Vanadium has atomic number 23; there are 23 electrons in the electron configuration and the atomic orbital box diagram.

PROBLEM-SOLVING PRACTICE 5.7

(a) Write the electron configuration of silicon in the noble-gas notation.
(b) Determine how many unpaired electrons a silicon atom has by drawing the atomic orbital box diagram.

EXERCISE 5.15

Highest Energy Electrons in Ground-State Atoms

Give the electron configuration of electrons in the highest occupied principal energy level (highest *n*) in a ground-state chlorine atom. Do the same for a selenium atom.

5-7d Valence Electrons

Elements in the same group in the periodic table have similar chemical properties (← **Sec. 1-13a**). For example, the formulas of the oxides of elements in Group 4A are similar: CO_2, SiO_2, GeO_2, SnO_2, and PbO_2. Elements in different periodic groups display different chemical behaviors. This can be explained if we assume that elements in the same group all have the same number and arrangement of electrons in the outermost region of their atoms. These are the electrons that can easily interact with electrons from another atom. This idea originated with Gilbert N. Lewis (1875–1946), who suggested as early as 1902 that electrons in atoms might be arranged in shells, starting close to the nucleus and building outward. The *electrons in the outermost parts of the atom that determine chemical behavior* are called **valence electrons**. *Electrons in filled inner shells and subshells of an atom* are referred to as **core electrons**.

For main-group elements (those in groups designated by a number and the letter A), the only valence electrons are those in the shell with the highest principal quantum number *n*. All of the other electrons, those with smaller values of *n*, are core electrons. ***For main-group elements, the number of valence electrons is given by the A-group number.*** Some examples of core and valence electrons for main-group elements are

Element	Electron Configuration	Core Electrons	Valence Electrons	Periodic Group
Na	[Ne] $3s^1$	[Ne] ($1s^2 2s^2 2p^6$)	$3s^1$	1A
Si	[Ne] $3s^2 3p^2$	[Ne] ($1s^2 2s^2 2p^6$)	$3s^2 3p^2$	4A
As	[Ar] $3d^{10} 4s^2 4p^3$	[Ar] $3d^{10}$	$4s^2 4p^3$	5A

Notice that for a given main-group atom the valence electrons all have the same principal quantum number and all are in the outermost shell. In the case of As, the filled *d* subshell ($3d^{10}$) is part of the core even though it filled after the noble-gas configuration of Ar was complete.

For transition metals and inner transition metals (*d*-block and *f*-block elements), electrons in incompletely filled *d* and *f* subshells that are not part of the outermost shell may also be valence electrons. Examples are scandium, Sc: [Ar] $3d^1 4s^2$, which has three valence electrons, and cerium, Ce: [Xe] $4f^1 5d^1 6s^2$, which has four valence electrons. These numbers of valence electrons account for the formation of compounds that contain Sc^{3+} ions and Ce^{4+} ions in which three and four electrons have been lost, respectively, from the neutral atom. Examples of core and valence electrons for transition and inner transition elements are

Element	Electron Configuration	Core Electrons	Valence Electrons	Periodic Group
Mn	[Ar] $3d^5 4s^2$	[Ar]	$3d^5 4s^2$	7B
Ta	[Xe] $4f^{14} 5d^3 6s^2$	[Xe] $4f^{14}$	$5d^3 6s^2$	5B
U	[Rn] $5f^3 6d^1 7s^2$	[Rn]	$5f^3 6d^1 7s^2$	Not numbered; 6 valence electrons

While teaching his students about atomic structure, Lewis introduced the practice of representing the valence electrons as dots. In a **Lewis dot symbol** an *element symbol is surrounded by a number of dots equal to the number of valence electrons.* The element symbol represents the atomic nucleus together with the core electrons. The dots are placed to the right of the symbol, to the left, above, and below. Dots are added one at a time until all valence electrons are represented. If there are more than four valence electrons, dots are paired with ones already there. Table 5.5 shows Lewis dot symbols for the atoms of the elements in Periods 2 and 3.

Table 5.5 Lewis Dot Symbols for Atoms

	1A ns^1	2A ns^2	3A ns^2np^1	4A ns^2np^2	5A ns^2np^3	6A ns^2np^4	7A ns^2np^5	8A ns^2np^6
Period 2	Li·	·Be·	·Ḃ·	·Ċ·	·N̈·	:Ö·	:F̈·	:N̈e:
Period 3	Na·	·Mg·	·Ȧl·	·Ṡi·	·P̈·	:S̈·	:C̈l·	:Är:

Figure 5.19 (← Sec. 5-7a) defines and illustrates *s*-block, *p*-block, *d*-block, and *f*-block elements.

Main-group elements in Groups 1A and 2A are known as ***s*-block elements**, and their valence electrons are *s* electrons (ns^1 for Group 1A, ns^2 for Group 2A). *Elements in the main groups at the right in the periodic table, Groups 3A through 8A,* are known as ***p*-block elements**. Their valence electrons include the outermost *s* and *p* electrons. Notice in Table 5.5 how the Lewis dot symbols show that ***in each A group the number of valence electrons is equal to the group number.***

PROBLEM-SOLVING EXAMPLE 5.8

Valence Electrons

(a) Using the noble-gas notation, write the electron configuration for bromine. Identify its core and valence electrons.
(b) Write the Lewis dot symbol for bromine.

Result

(a) [Ar] $3d^{10}4s^24p^5$ Core electrons are [Ar] $3d^{10}$; valence electrons are $4s^24p^5$.
(b) :B̤̈r·

Analyze The electron configuration can be written using the periodic table. Valence electrons are all electrons in the outermost shell plus electrons in incompletely filled *d* and *f* subshells that are not part of the outermost shell. Core electrons are all electrons that are not valence electrons.

Plan Locate bromine, Br, in the periodic table and build up its electron configuration. Identify valence electrons. The rest of the electrons are core electrons.

Execute Bromine is element 35 and so has 35 electrons. Its full electron configuration is $1s^22s^22p^63s^23p^63d^{10}4s^24p^5$. The first 18 electrons can be represented by [Ar] to the noble-gas notation, [Ar] $3d^{10}4s^24p^5$. The ten $3d$ electrons make up a complete subshell so they are not valence electrons. The core electrons are [Ar] $3d^{10}$. The valence electrons are the electrons in the outermost shell, $4s^24p^5$. The Lewis structure has seven dots and is shown in the Result.

☑ **Reasonable Result Check** Bromine, atomic number 35, is in Group 7A, so it should have seven valence electrons (seven dots in the Lewis structure), and it does; 35 − 7 = 28, which is the correct number of core electrons.

PROBLEM-SOLVING PRACTICE 5.8

Use the noble-gas notation to write electron configurations and Lewis dot symbols for Se and Te. What do these configurations illustrate about elements in the same main group?

It is hard to overemphasize how useful the periodic table is as a guide to electron configurations. As another example, determine the electron configuration and the number of unpaired electrons for tellurium, Te. Because Te is in Group 6A it has *six* valence electrons with an outer electron configuration of ns^2np^4. And, because Te is in the fifth period, $n = 5$. The complete electron configuration is determined by starting with the electron configuration of krypton [$1s^22s^22p^63s^23p^63d^{10}4s^24p^6$], the noble gas at the end of the fourth period. Then, add the filled $4d^{10}$ subshell and the six valence electrons of Te ($5s^25p^4$) to give $1s^22s^22p^63s^23p^63d^{10}4s^24p^64d^{10}5s^25p^4$, or [Kr] $4d^{10}5s^25p^4$. To predict the number of unpaired electrons, look at the outermost subshell's electron configuration, because the inner shells (represented by [Kr] $4d^{10}$ in this case) are completely filled with paired electrons. For Te, [Kr] $4d^{10}5s^25p^25p^15p^1$ indicates two p atomic orbitals with unpaired electrons for a total of two unpaired electrons.

5-8 Ion Electron Configurations

In studying ionic compounds (← **Sec. 2-4a**), you learned that atoms from Groups 1A through 3A form monoatomic positive ions (cations) with charges equal to their group numbers—for example, Li^+, Mg^{2+}, and Al^{3+}. Nonmetals in Groups 5A through 7A that form monoatomic ions do so by adding electrons to form negative ions (anions) with *charges equal to eight minus the A group number.* Examples of such anions are N^{3-}, O^{2-}, and F^-. Here's the explanation.

The electron configuration of an ion is derived from the electron configuration of the atom from which the ion was formed. ***When atoms from* s- *and* p-*block elements form ions, electrons are removed from or added to an atom's valence shell so that a noble-gas configuration is achieved.***

- Atoms from Groups 1A, 2A, and 3A lose 1, 2, or 3 electrons to form 1+, 2+, or 3+ ions, respectively; the positive ions have the same electron configuration as the *preceding* noble gas.
- Atoms from Groups 7A, 6A, and some in 5A gain 1, 2, or 3 electrons to form 1−, 2−, or 3− ions, respectively; the negative ions have the same electron configuration as the *next* noble gas.
- Metal atoms *lose* electrons, forming cations with a positive charge equal to the A group number; nonmetals *gain* electrons, forming anions with a negative charge equal to the A group number minus eight.

These relationships apply to Groups 1A to 3A and 5A to 7A.

Atoms and ions that have the same electron configuration are said to be **isoelectronic**. Each row in Figure 5.22 shows isoelectronic ions and the noble-gas atom that has the same electron configuration, emphasizing that ***metal ions are isoelectronic with the preceding noble-gas atom, while nonmetal ions have the electron configuration of the next noble-gas atom.***

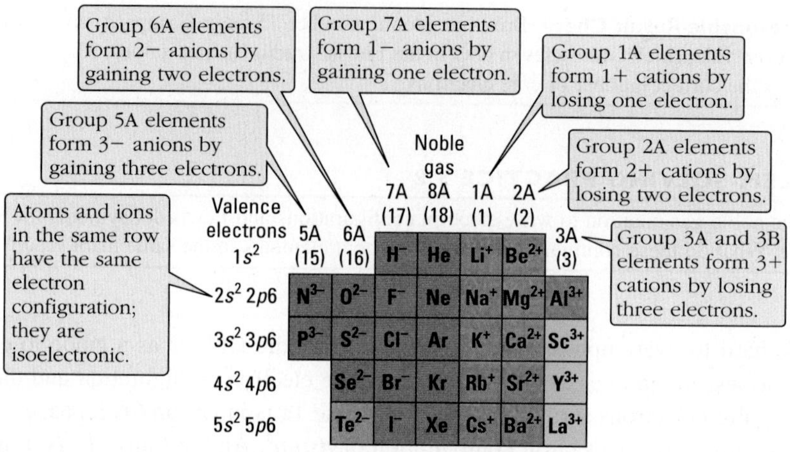
Group 6A elements form 2− anions by gaining two electrons.
Group 7A elements form 1− anions by gaining one electron.
Group 1A elements form 1+ cations by losing one electron.
Group 5A elements form 3− anions by gaining three electrons.
Noble gas
Group 2A elements form 2+ cations by losing two electrons.
Atoms and ions in the same row have the same electron configuration; they are isoelectronic.
Valence electrons
5A (15)
6A (16)
7A (17)
8A (18)
1A (1)
2A (2)
3A (3)
Group 3A and 3B elements form 3+ cations by losing three electrons.
1s2
2s2 2p6
3s2 3p6
4s2 4p6
5s2 5p6
H− He Li+ Be2+
N3− O2− F− Ne Na+ Mg2+ Al3+
P3− S2− Cl− Ar K+ Ca2+ Sc3+
Se2− Br− Kr Rb+ Sr2+ Y3+
Te2− I− Xe Cs+ Ba2+ La3+

Figure 5.22 Anions, atoms, and cations with isoelectronic ground-state noble-gas configurations.

PROBLEM-SOLVING EXAMPLE 5.9

Atom and Ion Electron Configurations

Complete this table.

Neutral Atom	Neutral Atom Electron Configuration	Ion	Ion Electron Configuration
Se	________	________	[Kr]
Ba	________	Ba^{2+}	________
Br	________	Br^-	________
________	[Kr] $5s^1$	Rb^+	________
________	[Ne] $3s^23p^3$	________	[Ar]

Result

Neutral Atom	Neutral Atom Electron Configuration	Ion	Ion Electron Configuration
Se	[Ar] $3d^{10}4s^24p^4$	Se^{2-}	[Kr]
Ba	[Xe] $6s^2$	Ba^{2+}	[Xe]
Br	[Ar] $3d^{10}4s^24p^5$	Br^-	[Kr]
Rb	[Kr] $5s^1$	Rb^+	[Kr]
P	[Ne] $3s^23p^3$	P^{3-}	[Ar]

Analyze In each row either an atom or ion symbol is given or an atom or ion electron configuration is given. Electron configurations for atoms can be determined from the periodic table. Nonmetal atoms form anions by gaining electrons to attain a noble-gas electron configuration. Metal atoms form cations by losing electrons to attain a noble-gas electron configuration. From an atom electron configuration the total number of electrons (the atomic number) can be determined.

Plan Analyze each row of the table using the rules just described. Use the periodic table to determine electron configurations and to identify elements.

Execute Se is a nonmetal; a neutral Se atom has the electron configuration [Ar] $3d^{10}4s^24p^4$, so it gains two electrons in the 4*p* subshell to form Se^{2-} and achieve the noble-gas configuration of krypton, [Kr] (36 electrons).

Barium (Group 2A) has the electron configuration [Xe] $6s^2$; to form Ba^{2+}, it loses two 6*s* electrons to acquire the electron configuration of xenon, [Xe].

Bromine (Group 7A) has the electron configuration [Ar] $3d^{10}4s^24p^4$; to form Br^-, it gains one electron to achieve the electron configuration of krypton, [Kr].

The electron configuration of [Kr] $5s^1$ indicates 36 + 1 = 37 electrons in the neutral atom; rubidium, Rb, has atomic number 37. To form Rb^+, a neutral Rb atom loses the $5s^1$ electron leaving a [Kr] configuration.

The [Ne] $3s^23p^3$ configuration is for a phosphorus atom, P. By gaining three electrons, a neutral phosphorus atom becomes a P^{3-} ion with 18 electrons and the [Ar] configuration.

PROBLEM-SOLVING PRACTICE 5.9

(a) What Period 3 anion with a 2− charge has the [Ar] electron configuration?
(b) What Period 4 cation with a 2+ charge has the electron configuration of argon?

5-8a Paramagnetism and Unpaired Electrons

The magnetic properties of a spinning electron were described in Section 5-5d. In atoms and ions that have filled subshells, all the electrons are paired (have opposite spins) and their magnetic fields effectively cancel each other. *Substances in which all electrons are paired* are **diamagnetic**; they are very weakly repelled by magnetic fields. Atoms or ions with unpaired electrons are attracted to a magnetic field; the more unpaired electrons they have, the greater is the attraction. *Substances with unpaired electrons* are **paramagnetic**.

Permanent magnets, such as those holding notes on a refrigerator, exhibit a third kind of magnetism that is much stronger than paramagnetism. In a **ferromagnetic** substance, the *spins of unpaired electrons in a cluster of atoms (called a domain) in a solid are all aligned in the same direction*. The difference between ferromagnetism and paramagnetism is explained in Figure 5.23. Elements of the iron, cobalt, and nickel subgroups in the periodic table exhibit ferromagnetism, as do many alloys, such as alnico (an alloy of aluminum, nickel, and cobalt), and some metal oxides, such as CrO_2 and Fe_3O_4. Computer hard-drive storage media use very small magnetic domains of a ferromagnetic cobalt alloy to store bits of binary data; a magnetized region represents a one and an unmagnetized region that represents a zero.

This answers the question posed in Chapter 1 (← Sec. 1-1), "Why is iron strongly attracted to a magnet, but most substances are not?"

5-8b Transition-Metal Ions

The closeness in energy of the 4*s* and 3*d* subshells was mentioned earlier in connection with the electron configurations of transition-metal atoms. The 5*s* and 4*d* subshells are also close to each other in energy, as are the 6*s*, 4*f*, and 5*d* subshells and the 7*s*, 5*f*, and 6*d* subshells. The *ns* and $(n - 1)d$ subshells are so close in energy that once *d* electrons

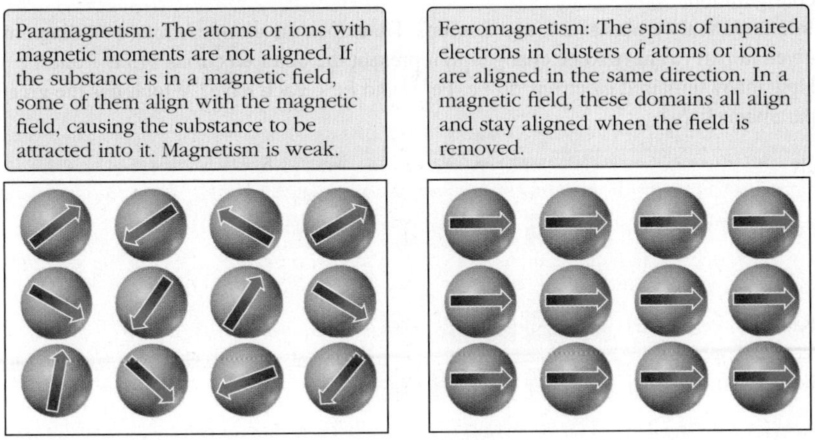
Paramagnetism: The atoms or ions with magnetic moments are not aligned. If the substance is in a magnetic field, some of them align with the magnetic field, causing the substance to be attracted into it. Magnetism is weak.
Ferromagnetism: The spins of unpaired electrons in clusters of atoms or ions are aligned in the same direction. In a magnetic field, these domains all align and stay aligned when the field is removed.

Figure 5.23 Paramagnetism and ferromagnetism.

are added, the $(n - 1)d$ subshell becomes slightly lower in energy than the ns subshell. As a result, *the* ns *electrons are at higher energy and are always removed before* (n − 1)d *electrons when transition metals form cations.* A simple way to remember this is that ***when a transition-metal atom loses electrons to form a cation, the electrons in the outermost shell (highest* n*) are removed first.***

A neutral nickel atom, Ni^0, with an equal number of protons (28) and electrons (28), has no net charge. An Ni^{2+} ion has a net charge of 2+ because it has 28 protons, but only 26 electrons.

For example, when a nickel atom, Ni, [Ar] $3d^84s^2$, forms a Ni^{2+} ion, it loses its two $4s$ electrons ($n = 4$ is the outermost shell):

$$Ni^0 \longrightarrow Ni^{2+} + 2\,e^-$$

This is shown by these atomic orbital box diagrams.

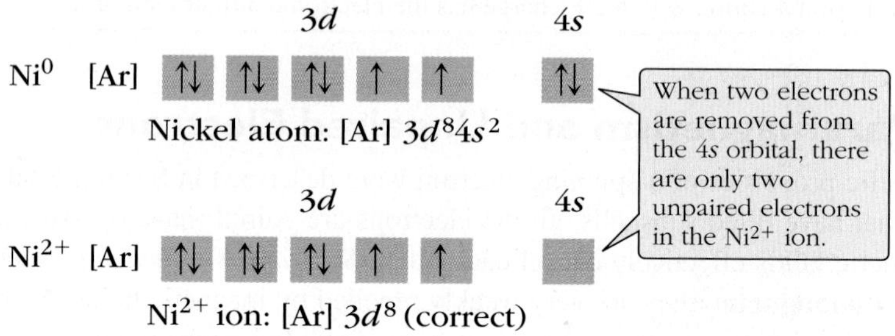
3d
4s
Ni0 [Ar]
Nickel atom: [Ar] 3d84s2
3d
4s
Ni2+ [Ar]
Ni2+ ion: [Ar] 3d8 (correct)
When two electrons are removed from the 4s orbital, there are only two unpaired electrons in the Ni2+ ion.

That two $4s$ electrons are lost is corroborated by the experimental evidence that Ni^{2+} has the quantity of paramagnetism expected for two unpaired electrons. If the two electrons had been removed from the $3d$ subshell rather than the $4s$ subshell, a Ni^{2+} ion would have had four unpaired electrons, resulting in greater paramagnetism.

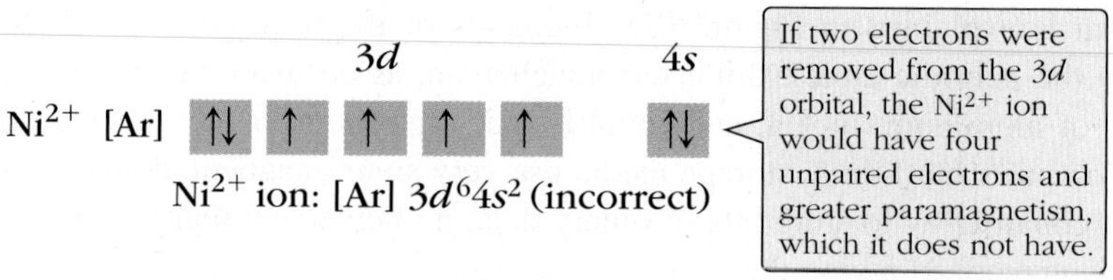
3d
4s
Ni2+ [Ar]
Ni2+ ion: [Ar] 3d64s2 (incorrect)
If two electrons were removed from the 3d orbital, the Ni2+ ion would have four unpaired electrons and greater paramagnetism, which it does not have.

Atoms or ions of inner transition elements (lanthanides and actinides) can have as many as seven unpaired electrons in the f subshell, as occurs in Eu^{2+} and Gd^{3+} ions.

PROBLEM-SOLVING EXAMPLE 5.10

Electron Configurations for Transition Elements and Ions

(a) Write the electron configuration for the Fe atom using the noble-gas notation. Then draw the atomic orbital box diagram for the electrons beyond the preceding noble-gas configuration.

(b) Iron commonly exists as 2+ and 3+ ions. How does the atomic orbital box diagram given in part (a) have to be changed to represent the outer electrons of Fe^{2+} and Fe^{3+}?

(c) How many unpaired electrons do Fe, Fe^{2+}, and Fe^{3+} each have? Which has the greatest paramagnetism?

Result

(a)

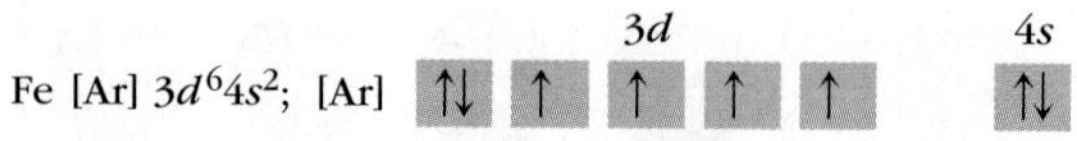
3d
4s
Fe [Ar] 3d64s2; [Ar]

(b)

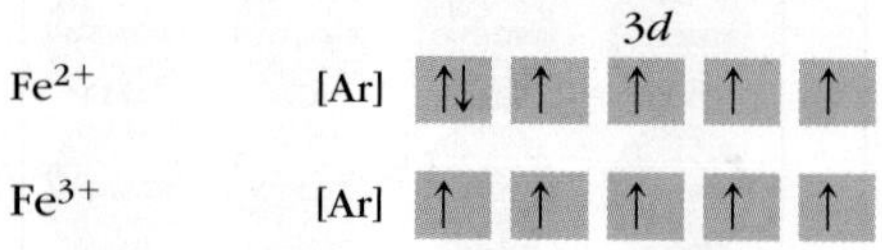
3d
Fe2+
[Ar]
Fe3+
[Ar]

(c) Fe has four unpaired electrons; Fe^{2+} has four unpaired electrons; Fe^{3+} has five unpaired electrons. Fe^{3+} has the greatest paramagnetism.

Analyze Atom electron configurations can be derived using the periodic table. Ion electron configurations are derived by removing or adding electrons to an atom electron configuration. When positive ions form, electrons are removed first from the outermost shell. Atomic orbital box diagrams are derived from electron configurations and Hund's rule. Strength of paramagnetism depends on number of unpaired electrons.

Plan In the periodic table iron is eight spaces beyond Ar so a neutral iron atom has eight more electrons than Ar; its electron configuration is [Ar] $3d^6 4s^2$. For the atomic orbital box diagram of a neutral iron atom, each *d* atomic orbital gets one electron before pairing occurs (Hund's rule). To form Fe^{2+}, an Fe atom loses its two 4*s* electrons. To form Fe^{3+}, an Fe atom loses two 4*s* electrons and one of its paired 3*d* electrons. Count the unpaired electrons in the orbital box diagram for each species; whichever has the most unpaired electrons is the most paramagnetic.

Execute See Result.

The fact that the two 4*s* electrons are removed to form Fe^{2+} shows that the 4*s* electrons are valence electrons. The fact that a 3*d* electron is removed when Fe^{3+} forms shows that the 3*d* electrons are also valence electrons.

PROBLEM-SOLVING PRACTICE 5.10

Look up the electron configuration of a ground-state copper atom in Appendix D. Use it to explain why copper readily forms the Cu^+ ion.

CONCEPTUAL EXERCISE 5.16

Unpaired Electrons

Fluoride ion, F^-, has no unpaired electrons. Vanadium forms four binary fluorides—VF_2, VF_3, VF_4, and VF_5. Assume that all four are ionic compounds. (a) Which fluoride is diamagnetic? (b) Which fluoride has the greatest attraction to a magnetic field? (c) Which fluoride has two unpaired electrons per vanadium?

5-9 Periodic Trends: Atomic Radii

Using knowledge of electron configurations, we can now answer fundamental questions about why atoms of different elements fit as they do in the periodic table, as well as explain trends in the properties of the elements in the table.

It is often useful to assume that atoms are spherical, even when they are in molecules. The **atomic radius**, then, is *the radius of a sphere that represents an atom.* Figure 5.24 shows that if two atoms are bonded together in a molecule, the sum of their radii

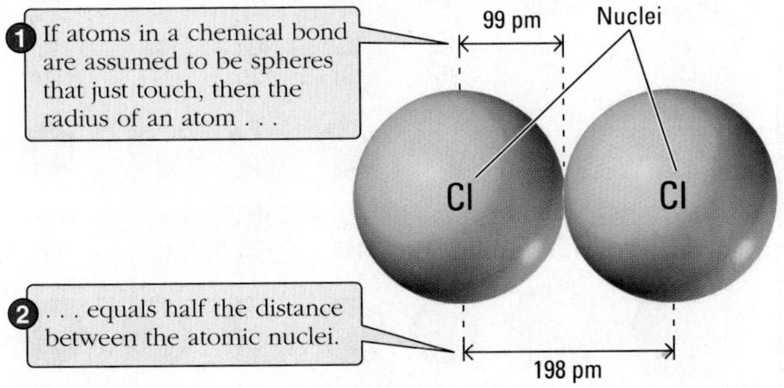
1 If atoms in a chemical bond are assumed to be spheres that just touch, then the radius of an atom . . .
99 pm
Nuclei
Cl
Cl
2 . . . equals half the distance between the atomic nuclei.
198 pm

Figure 5.24 Atomic radius of chlorine. The atomic radius is taken to be one half of the internuclear distance in the Cl_2 molecule.

should equal the distance between their centers (where their nuclei are). For atoms that form simple diatomic molecules, such as Cl_2, the atomic radius can be defined experimentally by determining the distance between the nuclei. One half of this distance is a good estimate of the atom's radius. In the Cl_2 molecule, the internuclear distance is 198 pm. Dividing by 2 gives a radius of 99 pm for Cl. The C—C distance in diamond is 154 pm, so the radius of the carbon atom is 77 pm. To test these estimates, we can add them together to estimate the distance between Cl and C in CCl_4. The estimated distance, 99 pm + 77 pm = 176 pm, is in good agreement with the experimentally measured C—Cl distance of 176 pm.

This approach can be extended to other atomic radii. The radii of O, Si, and S atoms can be estimated by measuring the O—H, Si—Cl, and H—S distances in H_2O, $SiCl_4$, and H_2S, and then subtracting the H and Cl radii found from H_2 and Cl_2. By this and other techniques, a reasonable set of atomic radii for main-group elements has been assembled (Figure 5.25).

Main-group elements are in the A groups in the periodic table (Figure 5.19, ← Sec. 5-7a).

5-9a Atomic Radii of the Main-Group Elements

The radius of an atom is determined by the size of the electron density distribution of electrons in the atom's outermost shell. This in turn depends on the attraction between the nucleus and those outermost electrons and on repulsions between other electrons and the outermost electrons. Such attractions and repulsions follow Coulomb's law (← **Sec. 2-6a**). Figure 5.25 shows atomic radii for main-group elements arranged in a periodic table. ***For main-group elements, atomic radii increase going down a group in the***

According to Coulomb's law, opposite charges attract and like charges repel. The larger the charges are, the greater the force of attraction or repulsion is. The greater the distance between charges is, the weaker the force is.

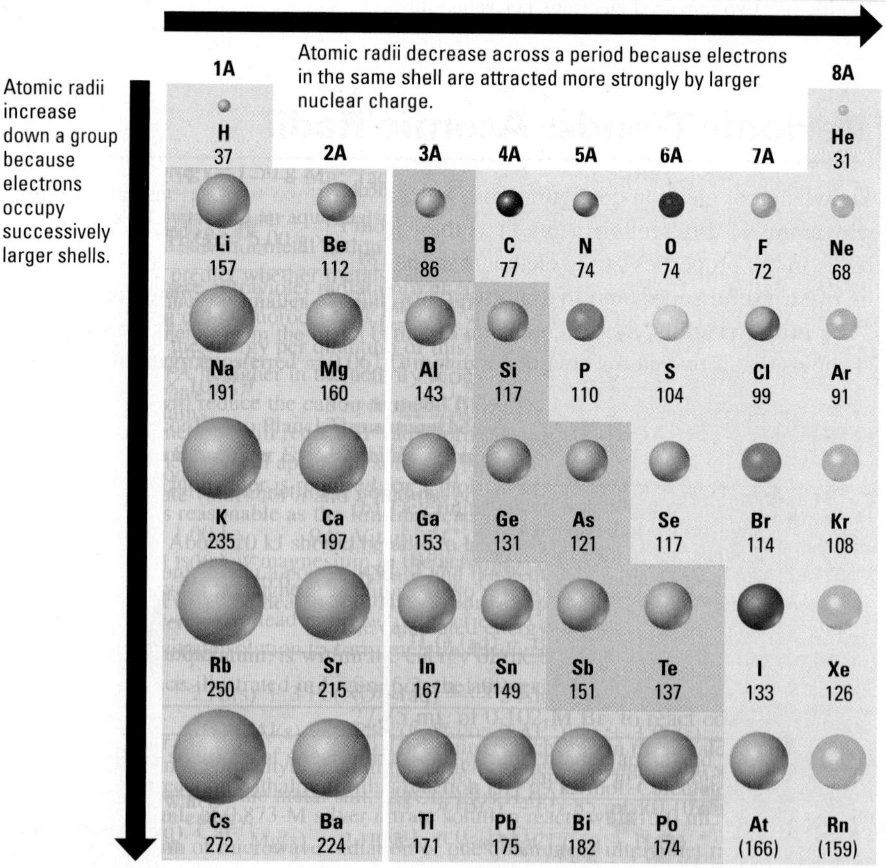
Atomic radii decrease across a period because electrons in the same shell are attracted more strongly by larger nuclear charge.
Atomic radii increase down a group because electrons occupy successively larger shells.
1A
2A
3A
4A
5A
6A
7A
8A
H 37
He 31
Li 157
Be 112
B 86
C 77
N 74
O 74
F 72
Ne 68
Na 191
Mg 160
Al 143
Si 117
P 110
S 104
Cl 99
Ar 91
K 235
Ca 197
Ga 153
Ge 131
As 121
Se 117
Br 114
Kr 108
Rb 250
Sr 215
In 167
Sn 149
Sb 151
Te 137
I 133
Xe 126
Cs 272
Ba 224
Tl 171
Pb 175
Bi 182
Po 174
At (166)
Rn (159)

Figure 5.25 Atomic radii of the main-group elements (in picometers; 1 pm = 10^{-12} m).

periodic table and decrease going across a period. These trends reflect three important effects:

- *Atomic radii increase from the top to the bottom of a group in the periodic table,* because electrons occupy atomic orbitals that are successively larger as the value of *n,* the principal quantum number, increases.
- There is a large increase in atomic radius going from any noble-gas atom to the following Group 1A atom, where the outermost electron is assigned to the next larger shell (higher *n* value). For example, compare the atomic radii of Ne (68 pm) and Na (191 pm) or Ar (91 pm) and K (235 pm).
- Atomic radii decrease from left to right across a period. The *n* value of the outermost atomic orbitals stays the same, so we might expect the radii of the occupied atomic orbitals to remain approximately constant. However, in crossing a period, as each successive electron is added, the nuclear charge also increases by the addition of one proton. According to Coulomb's law, increased charge causes increased attraction between the nucleus and electrons. The increased attraction is somewhat stronger than the increasing repulsion between electrons, causing atomic radii to decrease (Figure 5.25).

PROBLEM-SOLVING EXAMPLE 5.11

Atomic Radii and Periodic Trends

Using only a periodic table, list these atoms in order of decreasing radius: Se, Cl, Ge, K, S.

Result K > Ge > Se > S > Cl

Analyze Atomic radius increases down a group and decreases across a period.

Plan Locate the elements in the periodic table; apply the periodic trends.

Execute Se, Ge, and K are in the fourth period; Cl and S are in the third period. Because size decreases across a period, K is larger than Ge, which is larger than Se; also, S is larger than Cl. Because size increases down a group, S is smaller than Se. The relative sizes of the radii are (largest) K > Ge > Se > S > Cl (smallest).

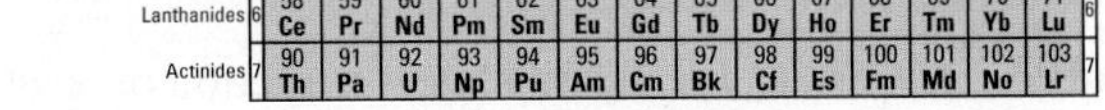
1A (1)
2A (2)
3B (3) 4B (4) 5B (5) 6B (6) 7B (7) 8B (8) 8B (9) 8B (10) 1B (11) 2B (12)
3A (13) 4A (14) 5A (15) 6A (16) 7A (17)
8A (18)
Lanthanides
Actinides

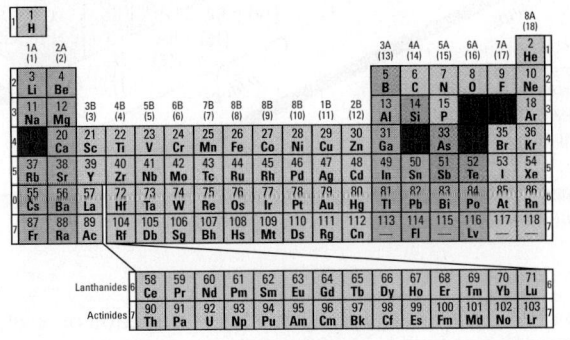

PROBLEM-SOLVING PRACTICE 5.11

Using just a periodic table, arrange these atoms in order of increasing atomic radius: B, Mg, K, Na.

5-9b Atomic Radii of Transition Metals

Periodic trends in the atomic radii of main-group and transition-metal atoms are illustrated in Figure 5.26. The sizes of transition-metal atoms change very little across a period, especially from Group 5B (V, Nb, or Ta), because the sizes are determined by the radius of the outermost shell—an *ns* atomic orbital ($n = 4$, 5, or 6) occupied by at least one electron. As electrons and protons are added to transition-metal atoms, the electrons are added to $(n - 1)d$ atomic orbitals, which are in a smaller shell and therefore between the *ns* electrons and the nucleus. As the number of electrons in these $(n - 1)d$ atomic orbitals increases, they increasingly repel the *ns* electrons away from the nucleus. This

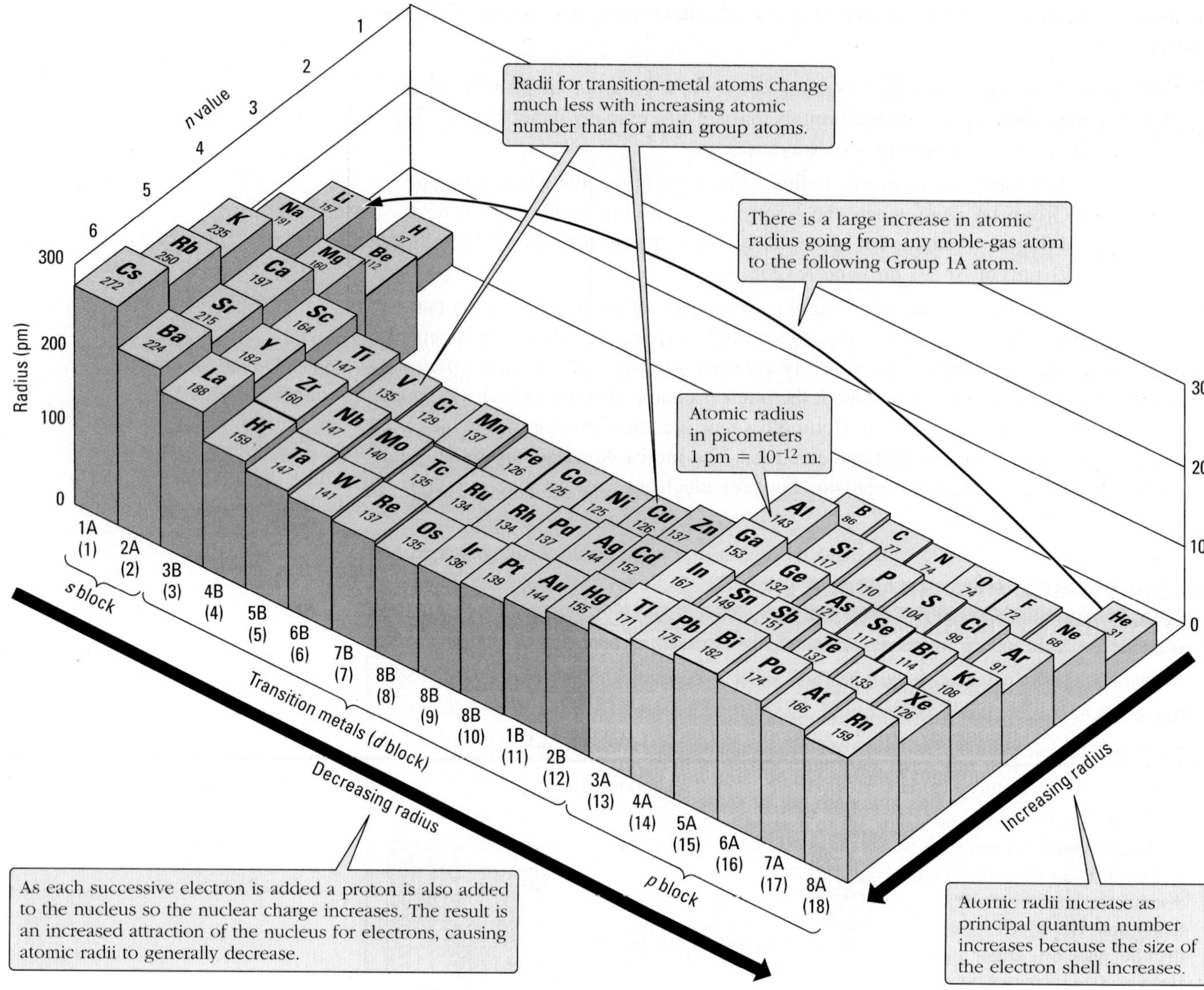

Radii for transition-metal atoms change much less with increasing atomic number than for main group atoms.
There is a large increase in atomic radius going from any noble-gas atom to the following Group 1A atom.
Atomic radius in picometers 1 pm = 10^{-12} m.
As each successive electron is added a proton is also added to the nucleus so the nuclear charge increases. The result is an increased attraction of the nucleus for electrons, causing atomic radii to generally decrease.
Atomic radii increase as principal quantum number increases because the size of the electron shell increases.
n value
Radius (pm)
Decreasing radius
Increasing radius
s block
Transition metals (d block)
p block

Figure 5.26 Atomic radii of *s*-, *p*-, and *d*-block elements in the first six periods.

partly compensates for the increased attraction of *ns* electrons as nuclear charge increases across the periods. Consequently, the *ns* electrons experience only a slightly increasing nuclear attraction, and the radii remain nearly constant until the slight rise at the copper and zinc groups due to the continually increasing electron-to-electron repulsions as the *d* subshell is filled.

The similar radii of the transition metals and the similar sizes of their ions have an important effect on their chemistry—transition elements in neighboring groups tend to be more alike in their properties than main-group elements in neighboring groups. The nearly identical radii of the fifth- and sixth-period transition elements lead to difficult problems in separating them from one another. The metals Ru, Os, Rh, Ir, Pd, and Pt are called the "platinum-group metals" because they occur together in nature. Their radii and chemistry are so similar that their minerals are similar and are found in the same geologic zones.

5-9c Effective Nuclear Charge

The repulsion of electrons in the outermost shells of an atom by electrons in inner shells can be described in terms of an *effective* nuclear charge. In a many-electron atom, such as a phosphorus atom, $Z = 15$, the outermost electrons are attracted by the 15 protons in the nucleus, but they are also repelled by the other 14 electrons. The electrons in the first and second shells ($n = 1$ and $n = 2$) are closer to the nucleus most of the time than are the valence electrons, so those core electrons repel the valence electrons away from the nucleus. This repulsion is described by saying that there is a **screening effect**: the *inner electrons screen the valence electrons from the charge of the protons in the nucleus*. This qualitative discussion can be made quantitative by defining a screening constant, σ, for each electron. The screening constant measures how much the other electrons screen the electron in question from the nuclear charge. The effective nuclear charge, Z^*, is then calculated as $Z^* = Z - \sigma$. The **effective nuclear charge** is *the positive charge on the nucleus, Z, corrected for the repulsions, σ, all other electrons exert on a given electron.* Within a period, electrons are being added to the same shell and each added electron is not as effective in screening the nuclear charge as are the inner electrons. Thus, the effective nuclear charge increases across a period. This increase in effective nuclear charge predicts that the atomic radii should decrease across a period.

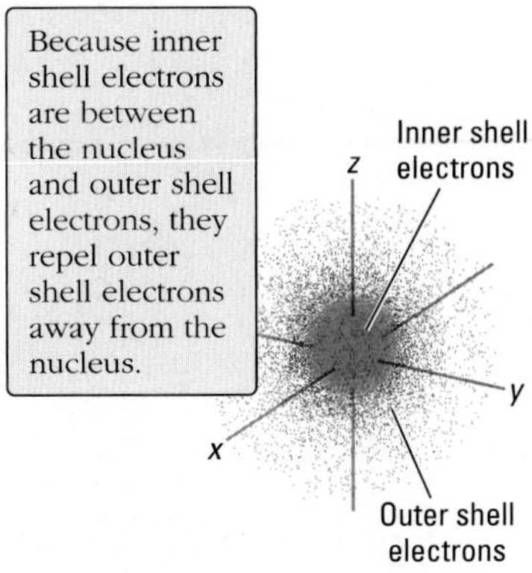
Because inner shell electrons are between the nucleus and outer shell electrons, they repel outer shell electrons away from the nucleus.
Inner shell electrons
z
y
x
Outer shell electrons

Effective nuclear charge and screening.

5-10 Periodic Trends: Ionic Radii

The **ionic radius** is *the radius of a sphere representing a positive or negative ion.* A consistent set of ionic radii can be determined from the distances between adjacent positive and negative ions in crystalline ionic solids. For example, the distance between centers of an Na^+ ion and a Cl^- ion in a crystal of NaCl (← **Sec. 2-6b**) can be measured and compared with distances between Na^+ ions and other anions to obtain a radius for Na^+. Ionic radii allow predictions of distances between ions in crystals for which no measurements have been made. Figure 5.27 shows both atomic and ionic radii for the periodic groups highlighted at the left of the diagram. Important trends with respect to cations are:

- **In the same periodic group, cations increase in size down the group.**
- **The radius of the cation is always smaller than the radius of the neutral atom from which the cation is formed.**
- **When an atom can form more than one cation, the larger the charge is, the smaller the ion is.**

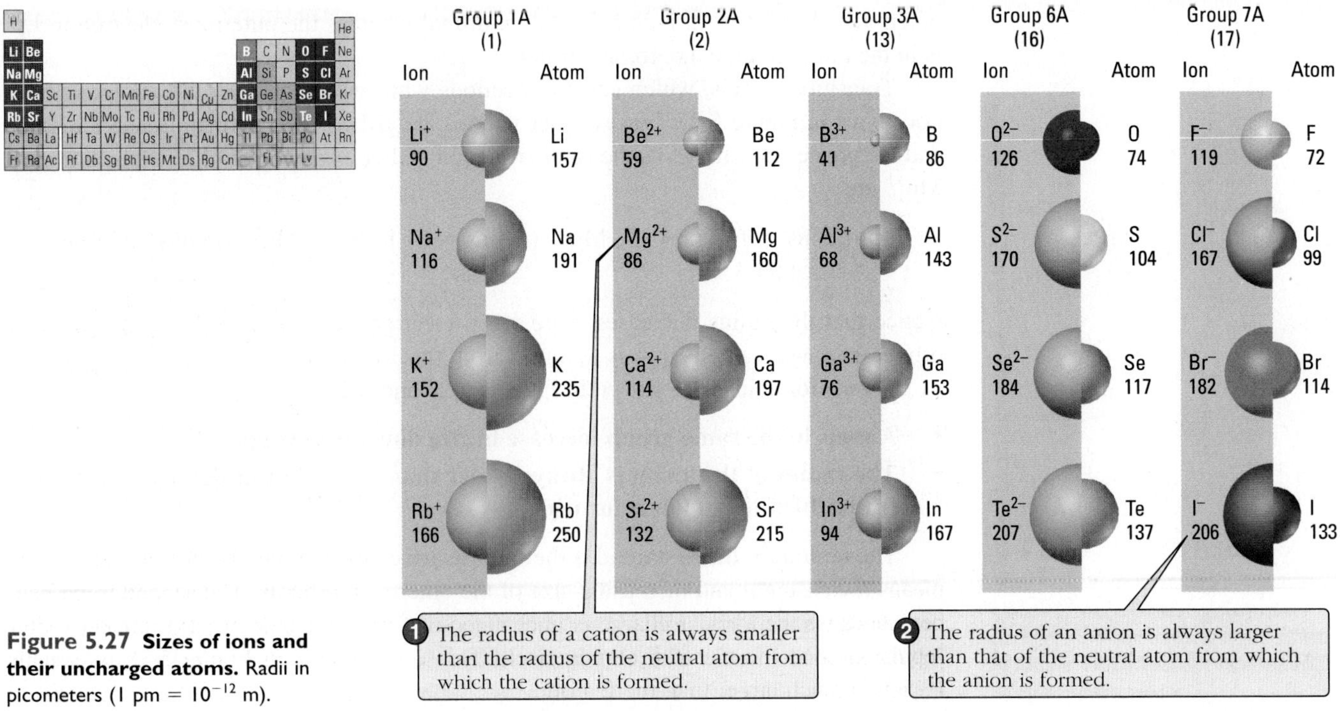
Group 1A (1)
Ion Atom
Li+ 90 Li 157
Na+ 116 Na 191
K+ 152 K 235
Rb+ 166 Rb 250
Group 2A (2)
Ion Atom
Be2+ 59 Be 112
Mg2+ 86 Mg 160
Ca2+ 114 Ca 197
Sr2+ 132 Sr 215
Group 3A (13)
Ion Atom
B3+ 41 B 86
Al3+ 68 Al 143
Ga3+ 76 Ga 153
In3+ 94 In 167
Group 6A (16)
Ion Atom
O2– 126 O 74
S2– 170 S 104
Se2– 184 Se 117
Te2– 207 Te 137
Group 7A (17)
Ion Atom
F– 119 F 72
Cl– 167 Cl 99
Br– 182 Br 114
I– 206 I 133
1 The radius of a cation is always smaller than the radius of the neutral atom from which the cation is formed.
2 The radius of an anion is always larger than that of the neutral atom from which the anion is formed.

Figure 5.27 Sizes of ions and their uncharged atoms. Radii in picometers (1 pm = 10^{-12} m).

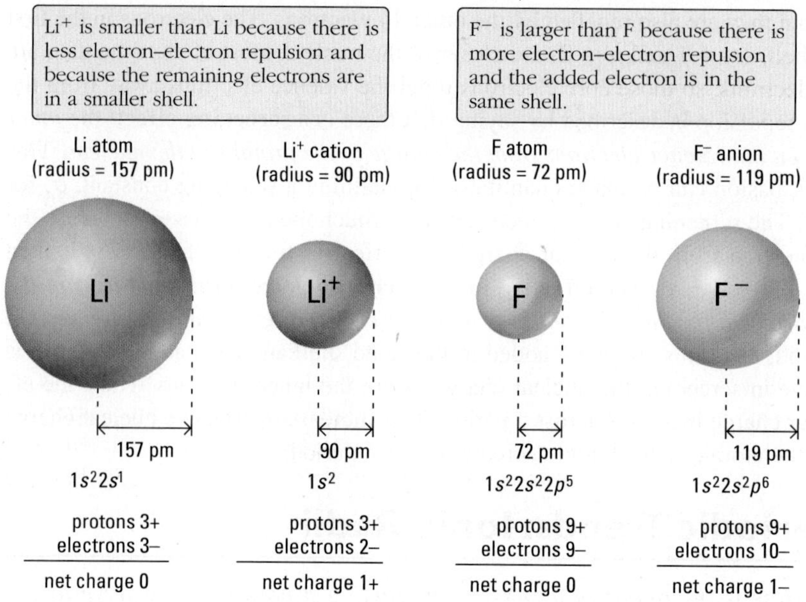
Li+ is smaller than Li because there is less electron–electron repulsion and because the remaining electrons are in a smaller shell.
F− is larger than F because there is more electron–electron repulsion and the added electron is in the same shell.
Li atom (radius = 157 pm)
Li+ cation (radius = 90 pm)
F atom (radius = 72 pm)
F− anion (radius = 119 pm)
Li
Li+
F
F−
157 pm
90 pm
72 pm
119 pm
1s2 2s1
1s2
1s2 2s2 2p5
1s2 2s2 2p6
protons 3+
electrons 3−
net charge 0
protons 3+
electrons 2−
net charge 1+
protons 9+
electrons 9−
net charge 0
protons 9+
electrons 10−
net charge 1−

Figure 5.28 Comparison of sizes of atoms and positive and negative ions.

An example of the first trend is that Rb^+ is larger than Na^+, which is larger than Li^+. This is because the outermost electrons in Rb^+ are in the fourth shell, which is much larger than the second and first shells where the outermost electrons of Na^+ and Li^+ are. The second trend is illustrated by lithium as shown in Figure 5.28. The radius of an uncharged Li atom is 157 pm, whereas that for an Li^+ ion is only 90 pm. When an electron is removed from a lithium atom, the nuclear charge remains the same (3+), but there are fewer electrons repelling each other. The two remaining electrons are more strongly attracted toward the nucleus, decreasing the radius. The radius also decreases because the outermost electron in Li is in a 2*s* atomic orbital whereas the outermost electron in Li^+ is in the much smaller 1*s* atomic orbital.

Transition metals, which can form cations with two or more different charges, are not shown in Figure 5.27; however, the cations are always smaller than the parent atom, and the greater the charge is, the smaller the ion is. For example, radii for Mn, Mn^{2+} and Mn^{3+} are

Mn (radius = 137 pm)	Mn^{2+} (radius = 81 pm)	Mn^{3+} (radius = 72 pm)
$[Ar]3d^54s^2$	$[Ar]3d^5$	$[Ar]3d^4$

Notice that the radius decreases more when electrons are removed from the larger 4*s* orbital than when an electron is removed from the 3*d* orbital.

Trends for anions are shown in Figures 5.27 and 5.28.

- **Anions in the same group increase in size down the group.**
- **The radius of the anion is always larger than the radius of the neutral atom from which the anion is formed.**

The reason for the first trend is the same as for cations: going down a periodic group means increasing *n* and increasing size of the outermost orbitals. The second trend happens because the nuclear charge is unchanged and the added electron(s) introduce new repulsions and the electron clouds swell. This can be seen in Figure 5.28 where the F^- ion is much larger than the F atom.

The effect of changing nuclear charge is evident when the sizes of isoelectronic ions are compared. Consider the isoelectronic ions O^{2-}, F^-, Na^+, and Mg^{2+}, each of which contains ten electrons.

Isoelectronic Ions	O^{2-}	F^-	Na^+	Mg^{2+}
Ionic radius (pm)	126	119	116	86
Number of protons	8	9	11	12
Number of electrons	10	10	10	10

The O^{2-} ion has only eight protons in its nucleus to attract the ten electrons, while F^- has nine, Na^+ has eleven, and Mg^{2+} has twelve. Thus, there is an increasing number of protons to attract the same number of electrons with approximately the same electron–electron repulsions. As the proton-to-electron ratio increases in an isoelectronic series of ions, the overall electron cloud is attracted more tightly to the nucleus and the ion size shrinks.

PROBLEM-SOLVING EXAMPLE 5.12

Trends in Ion Sizes

Using only a periodic table, choose the smaller atom or ion for each pair: (a) an Fe atom or an Fe^{2+} ion; (b) a P atom or a P^{3-} ion; (c) an Fe^{2+} or an Fe^{3+} ion.

Result

(a) Fe^{2+} ion (b) P atom (c) Fe^{3+} ion

Analyze Radii are being compared for ions with uncharged atoms or with other ions. The larger the nuclear charge and the smaller the electron repulsions are, the smaller an ion is. If the outermost electrons in an ion are in a smaller shell than in the atom, the ion is smaller.

Plan Apply the rules just stated to each pair.

Execute An Fe^{2+} ion is smaller than a Fe atom because the nuclear charge is the same and there are fewer electrons (less electron–electron repulsion). A P atom is smaller than a P^{3-} ion for the same reason. Both iron ions contain 26 protons. Because Fe^{3+} has one fewer electron than Fe^{2+}, there is less electron–electron repulsion and the radius of Fe^{3+} is smaller than that of Fe^{2+}.

PROBLEM-SOLVING PRACTICE 5.12

Which of these isoelectronic ions, Ba^{2+}, Cs^+, or La^{3+}, is (a) the largest? (b) the smallest? Explain your reasoning.

5-11 Periodic Trends: Ionization Energies

An element's chemical reactivity is determined, in part, by how easily valence electrons are removed from its atoms, a process that requires energy transfer into the atom. The **ionization energy (IE)** of an atom is *the energy needed to remove an electron from that atom in the gas phase.* For a gaseous sodium atom, the ionization process is

Recall that Coulomb's law says that the greater the distance between two opposite charges is, the smaller the force of attraction is.

$$Na(g) \longrightarrow Na^+(g) + e^- \qquad \Delta E = \text{ionization energy (IE)}$$

For an atom with more than a single electron there is more than one ionization energy. The first ionization energy involves removing an electron from an uncharged atom, the second ionization energy involves removing an electron from a 1+ ion, and so on.

Because energy is always required to remove an electron, the process is always endothermic and the sign of the ionization energy is positive. The more difficult an electron is to remove, the greater the ionization energy is. As illustrated in Figures 5.29 and 5.30,

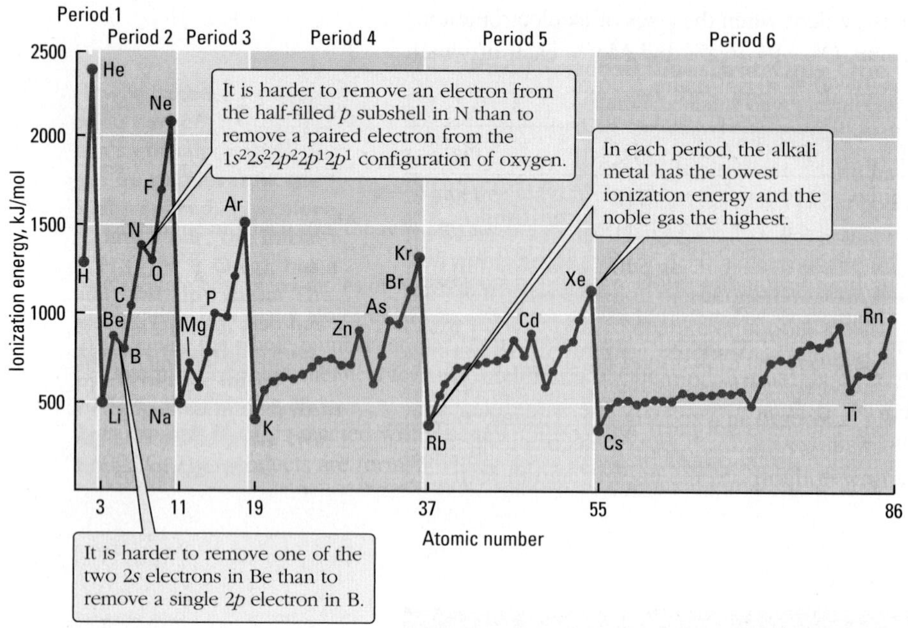
Period 1
Period 2
Period 3
Period 4
Period 5
Period 6
Ionization energy, kJ/mol
2500
2000
1500
1000
500
Atomic number
3
11
19
37
55
86
H
He
Li
Be
B
C
N
O
F
Ne
Na
Mg
P
Ar
K
Zn
As
Br
Kr
Rb
Cd
Xe
Cs
Tl
Rn
It is harder to remove an electron from the half-filled p subshell in N than to remove a paired electron from the 1s²2s²2p²2p¹2p¹ configuration of oxygen.
In each period, the alkali metal has the lowest ionization energy and the noble gas the highest.
It is harder to remove one of the two 2s electrons in Be than to remove a single 2p electron in B.

Figure 5.29 First ionization energies for elements in the first six periods plotted against atomic number.

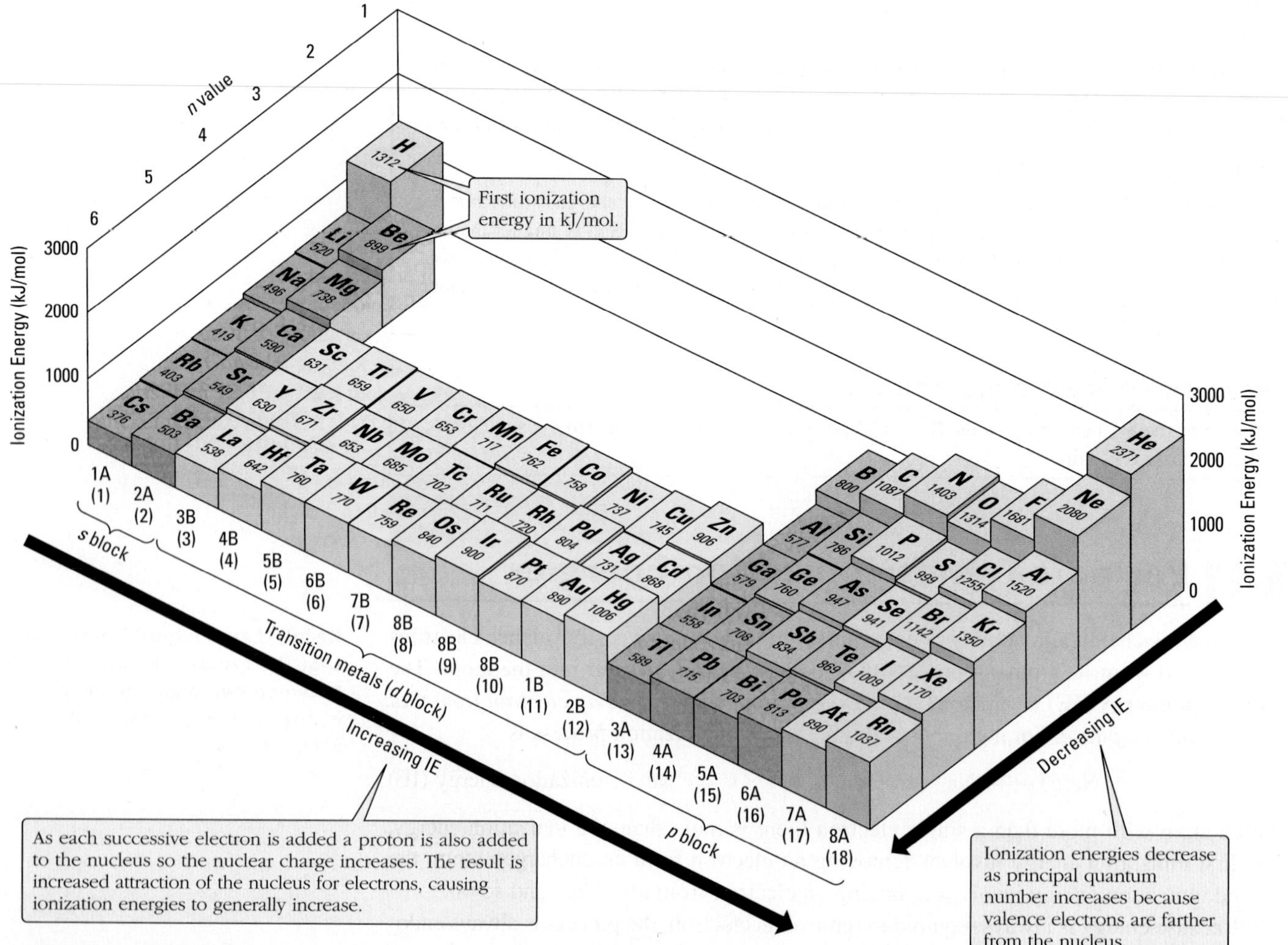
n value
1
2
3
4
5
6
Ionization Energy (kJ/mol)
3000
2000
1000
0
H 1312
Li 520
Be 899
Na 496
Mg 738
K 419
Ca 590
Sc 631
Ti 659
V 650
Cr 653
Mn 717
Fe 762
Co 758
Ni 737
Cu 745
Zn 906
Rb 403
Sr 549
Y 630
Zr 671
Nb 653
Mo 685
Tc 702
Ru 711
Rh 720
Pd 804
Ag 731
Cd 868
Cs 376
Ba 503
La 538
Hf 642
Ta 760
W 770
Re 759
Os 840
Ir 900
Pt 870
Au 890
Hg 1006
B 800
C 1087
N 1403
O 1314
F 1681
Ne 2080
He 2371
Al 577
Si 786
P 1012
S 999
Cl 1255
Ar 1520
Ga 579
Ge 760
As 947
Se 941
Br 1142
Kr 1350
In 558
Sn 708
Sb 834
Te 869
I 1009
Xe 1170
Tl 589
Pb 715
Bi 703
Po 813
At 890
Rn 1037
First ionization energy in kJ/mol.
1A (1)
2A (2)
3B (3)
4B (4)
5B (5)
6B (6)
7B (7)
8B (8)
8B (9)
8B (10)
1B (11)
2B (12)
3A (13)
4A (14)
5A (15)
6A (16)
7A (17)
8A (18)
s block
Transition metals (d block)
p block
Increasing IE
Decreasing IE
As each successive electron is added a proton is also added to the nucleus so the nuclear charge increases. The result is increased attraction of the nucleus for electrons, causing ionization energies to generally increase.
Ionization energies decrease as principal quantum number increases because valence electrons are farther from the nucleus.

Figure 5.30 First ionization energies (kJ/mol) for *s*-, *p*-, and *d*-block elements in the first six periods.

the first ionization energy generally decreases down a group and increases across a period. The decrease down a group reflects the increasing radii of the atoms—it is easier to remove a valence electron from a larger atom because the greater distance between charges makes the force of attraction between the electron and the nucleus smaller. For example, the ionization energy of Rb (403 kJ/mol) is less than that of K (419 kJ/mol) or Na (496 kJ/mol). Ionization energies increase across a period for the same reason that the radii decrease. Increasing nuclear charge attracts electrons in the same shell more tightly. However, as explained in Figure 5.29, the trend across a given period is not always smooth. The single *np* electron of a Group 3A element is more easily removed than one of the two *ns* electrons in the preceding Group 2A element. Also, in the second, third, and fourth periods, Group 6A elements (ns^2np^4) have smaller ionization energies than the Group 5A elements (ns^2np^3) that precede them. In Group 6A, two electrons must be assigned to the same *p* atomic orbital. Thus, greater electron repulsion is experienced by the fourth *p* electron, making it easier to remove.

Ionization energies increase more gradually across the period for the transition and inner transition elements than for the main-group elements (Figure 5.29). Just as the atomic radii of transition elements change very little across a period, the ionization energy for the removal of an *ns* electron also shows small changes, until the *d* orbitals are filled (Zn, Cd, Hg).

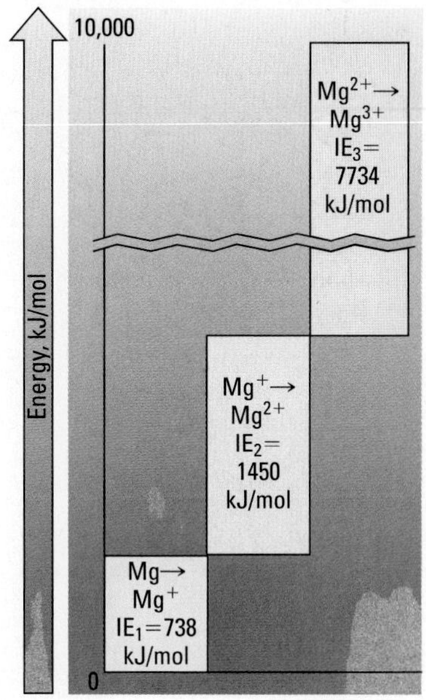
10,000
Energy, kJ/mol
0
Mg→ Mg+ IE1=738 kJ/mol
Mg+→ Mg2+ IE2= 1450 kJ/mol
Mg2+→ Mg3+ IE3= 7734 kJ/mol

First three ionization energies for Mg.

Every atom with more than a single electron has a series of ionization energies. For example, the first three ionization energies (IEs) of Mg are

$$Mg(g) \longrightarrow Mg^+(g) + e^- \qquad IE_1 = 738 \text{ kJ/mol}$$
$$1s^22s^22p^63s^2 \qquad 1s^22s^22p^63s^1$$

$$Mg^+(g) \longrightarrow Mg^{2+}(g) + e^- \qquad IE_2 = 1450 \text{ kJ/mol}$$
$$1s^22s^22p^63s^1 \qquad 1s^22s^22p^6$$

This ionization energy is bigger because the electron is being removed from a positive ion.

$$Mg^{2+}(g) \longrightarrow Mg^{3+}(g) + e^- \qquad IE_3 = 7734 \text{ kJ/mol}$$
$$1s^22s^22p^6 \qquad 1s^22s^22p^5$$

This ionization energy is much bigger because a core electron is being removed from the second shell, which is much closer to the nucleus, rather than from the third (larger) shell.

Notice that removing each subsequent electron requires more energy, and the jump from the second (IE_2) to the third (IE_3) ionization energy of Mg is particularly great. The great difference between the second and third ionization energies for Mg is excellent experimental evidence for the existence of electron shells in atoms. Removal of the first core electron requires much more energy than removal of a valence electron. Table 5.6 gives the successive ionization energies of the second-period elements.

PROBLEM-SOLVING EXAMPLE 5.13

Ionization Energies

Using only a periodic table, arrange these atoms in order of increasing first ionization energy: Al, Ar, Cl, Na, K, Si.

Result K < Na < Al < Si < Cl < Ar

Analyze Ionization energy increases across a period and decreases down a group in the periodic table.

Plan Locate the period and group of each of the six elements; apply the rule.

Execute Ar, Al, Cl, Na, and Si are all in Period 3, so their order is Na < Al < Si < Cl < Ar. K is below Na in Group 1A; thus K has a lower first ionization energy than that of Na. Therefore, the final order is K < Na < Al < Si < Cl < Ar.

PROBLEM-SOLVING PRACTICE 5.13

Use only a periodic table to arrange these elements in decreasing order of their first ionization energy: Na, F, N, and P. Explain your reasoning.

Table 5.6 Successive Ionization Energies for Second-Period Atoms (MJ/mol)

	Li $1s^22s^1$	Be $1s^22s^2$	B $1s^22s^22p^1$	C $1s^22s^22p^2$	N $1s^22s^22p^3$	O $1s^22s^22p^4$	F $1s^22s^22p^5$	Ne $1s^22s^22p^6$
IE_1	0.52	0.90	0.80	1.09	1.40	1.31	1.68	2.08
IE_2	7.30	1.76	2.43	2.35	2.86	3.39	3.37	3.95
IE_3	11.81	14.85	3.66	4.62	4.58	5.30	6.05	6.12
IE_4		21.01	25.02	6.22	7.48	7.47	8.41	9.37
IE_5			32.82	37.83	9.44	10.98	11.02	12.18
IE_6				47.28	53.27	13.33	15.16	15.24
IE_7					64.37	71.33	17.87	20.00
IE_8		Core electrons				84.08	92.04	23.07
IE_9							106.43	115.38
IE_{10}								131.43

Note: In each of these elements the core electrons are those in the 1*s* orbital.

5-12 Periodic Trends: Electron Affinities

Two sign conventions are used for electron affinity—the one used here and an alternative one. The latter uses the opposite sign—for example, +328 kJ/mol for the electron affinity of fluorine.

The **electron affinity (EA)** of an element is *the energy change when an electron is added to a gaseous atom to form a 1− ion.* As the term implies, the electron affinity is a measure of the attraction an atom has for an additional electron. For example, the electron affinity of fluorine is −328 kJ/mol.

$$\underset{[He]\,2s^22p^5}{F(g)} + e^- \longrightarrow \underset{[He]\,2s^22p^6}{F^-(g)} \qquad \Delta E = EA = -328 \text{ kJ/mol}$$

This large negative value indicates that a fluorine atom readily accepts an electron. As expected, fluorine and the rest of the halogens have large negative electron affinities because, by acquiring an electron, the halogen atoms achieve a stable octet of valence electrons, ns^2np^6 (Table 5.7). Notice from Table 5.7 that electron affinities generally become more negative across a period toward the halogen, which is only one electron away from a noble-gas configuration.

Some elements have a positive electron affinity; this means that the negative ion is less stable than the neutral atom.

Table 5.7 Electron Affinity (EA) by Periodic Table Group (kJ/mol)

1A (1)	2A (2)	3A (13)	4A (14)	5A (15)	6A (16)	7A (17)	8A (18)
H −73							He +50
Li −60	Be +50	B −27	C −122	N +70	O −141	F −328	Ne +120
Na −53	Mg +40	Al −43	Si −134	P −72	S −200	Cl −349	Ar +96
K −48	Ca −2.4	Ga −30	Ge −120	As −78	Se −195	Br −325	Kr +96
Rb −47	Sr −5	In −30	Sn −120	Sb −103	Te −190	I −295	Xe +80

First electron affinity (kJ/mol)

Electron affinities tend to become more negative across a period, but they depend a lot on electron configurations.

Halogens have the most negative electron affinities.

$$\underset{[He]\,2s^22p^6}{Ne} + e^- \longrightarrow \underset{[He]\,2s^22p^63s^1}{Ne^-} \qquad EA = +120 \text{ kJ/mol}$$

The Ne^- ion is unstable and would revert back to the neutral neon atom and an electron. From their electron configurations we can understand why the Ne^- ion would be less stable than the neutral neon atom; the anion would exceed the stable octet of valence electrons of an Ne atom.

5-13 Energy, Ions, and Ionic Compounds

Energy is usually released when a metal and a nonmetal react to form a salt. For example, the vigorous reaction of sodium metal and chlorine gas forms solid sodium chloride, NaCl, and transfers energy to the surroundings (Figure 5.31).

$$Na(s) + \tfrac{1}{2}\,Cl_2(g) \longrightarrow NaCl(s) \qquad \Delta_r H° = \Delta_f H° = -410.9 \text{ kJ/mol}$$

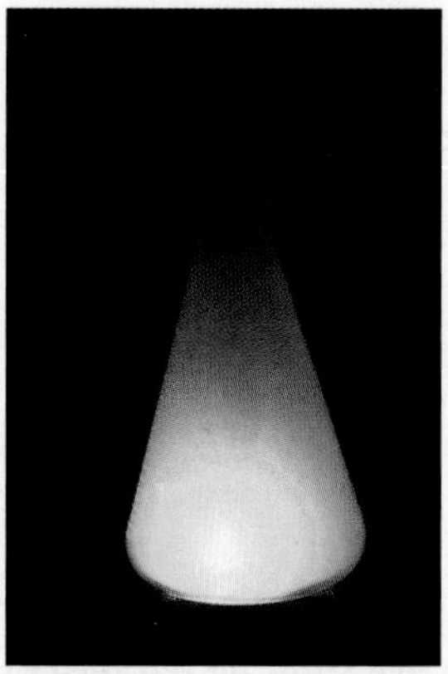

Cengage Learning/Charles D. Winters

Figure 5.31 The reaction of sodium with chlorine. The reaction forms Na^+ and Cl^- ions as well as liberating heat and light. Ionic compounds such as NaCl were first discussed in Section 2-4.

Let's consider the energy changes in the formation of another ionic compound, solid potassium fluoride. It is formed by the reaction of potassium, a very reactive alkali metal, and fluorine, a highly reactive halogen. Potassium fluoride contains K^+ and F^- ions grouped in a crystalline lattice of alternating cations and anions (Figure 5.32). Potassium atoms have low ionization energy and readily lose their $4s^1$ valence electrons; fluorine atoms have a high electron affinity and accept the electrons into a $2p$ orbital. Thus, electron transfer occurs from potassium metal atoms to fluorine atoms during KF formation.

$$K\ 1s^22s^22p^63s^23p^64s^1 \longrightarrow K^+\ 1s^22s^22p^63s^23p^6 + e^- \qquad \text{Ionization energy} = 419 \text{ kJ}$$

$$F\ 1s^22s^22p^5 + e^- \longrightarrow F^-\ 1s^22s^22p^6 \qquad \text{Electron affinity} = -328 \text{ kJ}$$

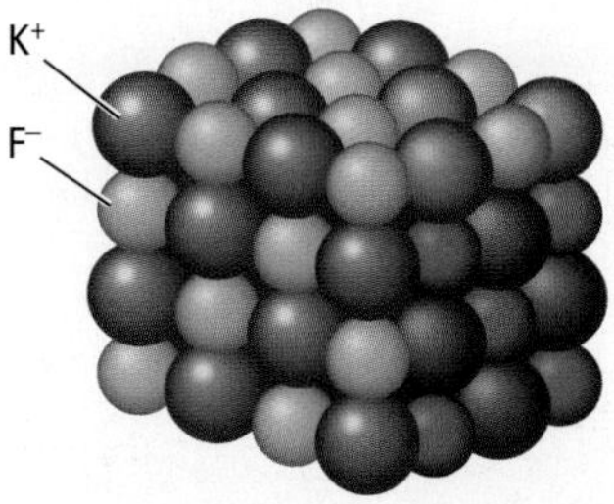
K+
F−

Figure 5.32 A potassium fluoride crystal illustrating the arrangement of K^+ and F^- ions in a lattice structure.

The transfer of electrons converts potassium and fluorine atoms into K^+ and F^- ions, each of which has an octet of outer electrons, a very stable noble-gas configuration.

Electron configurations: K: $[Ar]4s^1$; F: $[He]2s^22p^5$; K^+: [Ar]; F^-: [Ne] ($[He]2s^22p^6$)

The standard formation enthalpy, $\Delta_f H°$ of KF is -567.27 kJ/mol.

$$K(s) + \tfrac{1}{2}\,F_2(g) \longrightarrow KF(s) \qquad \Delta_f H° = -567.27 \text{ kJ}$$

It is possible to write *a series of other thermochemical expressions that sum to the standard formation enthalpy expression for any ionic compound.* Such a series of steps is called a **Born-Haber cycle**. A Born-Haber cycle for potassium fluoride is described below and illustrated in Figure 5.33. Hess's law (← **Sec. 4-9**) can be applied to such a multistep sequence and the standard formation enthalpy to calculate the **lattice energy** of the KF crystal, *the reaction enthalpy when 1 mol of an ionic solid forms from its separated, gaseous ions.*

This cycle is named for Max Born and Fritz Haber, two Nobel Prize-winning German scientists.

Step 1: *Convert solid potassium to gaseous potassium atoms.* This is the sublimation enthalpy, $\Delta_r H_1°$, which is known to be 89 kJ for 1 mol potassium.

$$K(s) \longrightarrow K(g) \qquad \Delta_r H_1° = 89 \text{ kJ/mol}$$

Step 2: *Dissociate F_2 molecules into F atoms.* One mol of KF contains 1 mol fluoride ions, so $\frac{1}{2}$ mol F_2 is required. Breaking the bonds in diatomic fluorine molecules requires energy. This energy, $\Delta_r H_2°$, is 79 kJ/mol, one-half the bond energy of F_2, that is, $\frac{1}{2}$ (158 kJ/mol).

$$\tfrac{1}{2}\,F_2(g) \longrightarrow F(g) \qquad \Delta_r H_2° = \tfrac{1}{2}\text{ BE of } F_2 = \tfrac{1}{2}\,(158 \text{ kJ/mol}) = 79 \text{ kJ/mol}$$

Step 3: *Ionize gaseous potassium atoms to form gaseous potassium ions.* The energy, $\Delta_r H_3°$, required to remove the $4s^1$ electron is the first ionization energy, IE_1, of potassium, 419 kJ/mol, which can be obtained from a table of ionization energies (Figure 5.30).

$$K(g) \longrightarrow K^+(g) + e^- \qquad \Delta_r H_3° = IE_1 = 419 \text{ kJ/mol}$$

Step 4: *Form F^- ions from F atoms.* The addition of a sixth electron to the $2p$ subshell completes the octet of valence electrons for F^-. For the formation of 1 mol F^- ions, this energy, $\Delta_r H_4^\circ$, is the electron affinity (EA) of F, −328 kJ/mol, obtained from Table 5.7.

$$F(g) + e^- \longrightarrow F^-(g) \qquad \Delta_r H_4^\circ = EA = -328 \text{ kJ/mol}$$

Step 5: *Combine gaseous K^+ and F^- ions into a crystal lattice.* The cations and anions assemble into a crystal, releasing a significant quantity of energy due to the electrostatic attraction between oppositely charged ions. This energy, $\Delta_r H_5^\circ$, is the lattice energy.

$$K^+(g) + F^-(g) \longrightarrow KF(s) \qquad \Delta_r H_5^\circ = \text{lattice energy of KF}$$

The five steps do not describe what actually happens when K(s) reacts with $F_2(g)$. They do allow us to apply known thermochemical data to the problem.

The sum of the enthalpy changes in the five steps of the cycle represents the enthalpy change for the overall reaction of solid potassium metal with gaseous fluorine to produce solid potassium fluoride (Figure 5.33). According to Hess's law, the sum of these enthalpy changes must equal the standard formation enthalpy ($\Delta_f H^\circ$) of KF, which is experimentally known to be −567.27 kJ.

$$\Delta_f H^\circ = -567.27 \text{ kJ} = \Delta_r H_1^\circ + \Delta_r H_2^\circ + \Delta_r H_3^\circ + \Delta_r H_4^\circ + \Delta_r H_5^\circ$$

Although data for Steps 1 to 4 can be determined experimentally, lattice energies cannot be measured directly. Instead, we can apply Hess's law in a Born-Haber cycle to calculate the lattice energy for KF, $\Delta_r H_5^\circ$, using data from Steps 1 to 4 and the enthalpy of formation, −567.27 kJ.

$$\Delta_r H_5^\circ = \text{lattice energy of KF} = (-567.27 \text{ kJ/mol}) - (89 \text{ kJ/mol}) - (79 \text{ kJ/mol}) - (419 \text{ kJ/mol}) - (-328 \text{ kJ/mol}) = -826 \text{ kJ/mol}$$

Notice that the formation of the gaseous cations and anions from the respective elements (Steps 1 to 4) is not energetically favorable (Figure 5.33). The net enthalpy change for these four steps in the cycle is positive, +259 kJ. The solid compound forms because of the very large lattice energy released during the assembling of the gaseous cations and anions into the solid crystal lattice. It is the lattice energy, not the formation of K^+ and F^- ions, that causes solid KF to form (Figure 5.33).

The strength of the ionic bonds in a crystal lattice—its lattice energy—depends on the strength of the electrostatic interactions among the crystal's cations and anions. In a

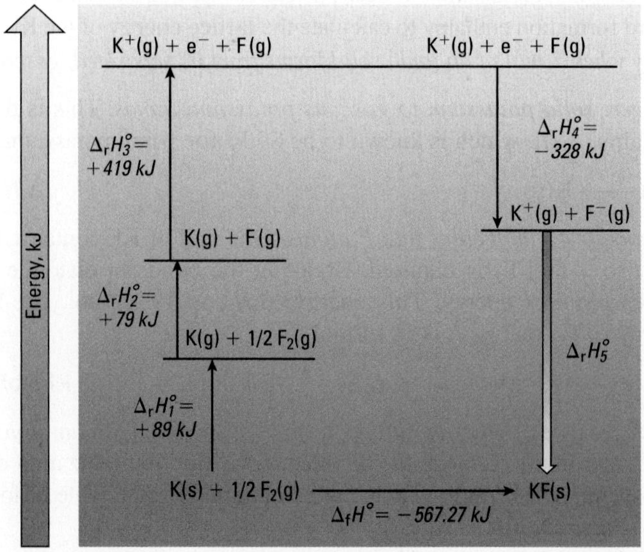

K⁺(g) + e⁻ + F(g)
K⁺(g) + e⁻ + F(g)
ΔrH3° = +419 kJ
ΔrH4° = −328 kJ
K⁺(g) + F⁻(g)
K(g) + F(g)
Energy, kJ
ΔrH2° = +79 kJ
K(g) + 1/2 F2(g)
ΔrH5°
ΔrH1° = +89 kJ
K(s) + 1/2 F2(g)
KF(s)
ΔfH° = −567.27 kJ

Figure 5.33 **Born-Haber cycle for KF.**

crystal lattice there are attractive forces between oppositely charged ions and repulsive forces between ions of like charge. All these forces obey Coulomb's law (← **Sec. 2-6a**). Both attractive and repulsive forces are affected in the same proportion by two factors:

- *The sizes of the charges of the cations and anions.* As the charges get larger (from 1+ and 1−, to 2+ and 2−, etc.), both attractive and repulsive forces among the ions increase.
- *The distances among the ions.* The forces increase as the distances decrease. The shortest distance is $r_+ + r_-$, where r_+ is the cation radius and r_- is the anion radius. The distances among ions of like charge increase or decrease in proportion to this shortest distance. The smaller the ionic radii, the shorter the distances among the ions.

Ions are arranged in a crystal so that the attractive forces predominate over the repulsive forces. This makes the energy lower when the ions are in a crystal lattice than when they are completely separated from one another. Because the energy of the crystal lattice is lower than that for the separated ions (that is, a mole of KF(s) is lower in energy than a mole of separated K^+ and F^- ions), the lattice energy is negative. *The stronger the net ionic attractions are, the more negative the lattice energy is, and the more stable the crystal lattice is.* Thus, crystal lattices are more stable for small ions of large charge.

This is reflected in the fact that compounds with high lattice energies have high melting points. The entries in Table 5.8 exemplify this trend. Consider NaCl and BaO in Table 5.8. In both compounds the sum of the ionic radii of the cation and anion is about the same; it is 283 pm for NaCl and 275 pm for BaO. However, the product of the charges differs by a factor of four: [(1+) × (1−)] for NaCl versus [(2+) × (2−)] for BaO. This charge factor of four is reflected in a lattice energy of (−3054 kJ/mol) and a melting point (1973 °C) for BaO that are much larger than the lattice energy (−786 kJ/mol) and the melting point (800 °C) of NaCl.

The effect of cation size can be seen by comparing MgO and BaO as shown in Table 5.8. Because Mg^{2+} (r_+ = 86 pm) is much smaller than Ba^{2+} (r_+ = 149 pm), the 2+ charge on magnesium ion is more concentrated (higher charge density) than the 2+ charge on the barium ion. Consequently, the net attractive force between Mg^{2+} and O^{2-} is greater than the net attractive force between Ba^{2+} and O^{2-}, and the lattice energy of MgO is more negative. This results in a higher melting point for MgO (2825 °C) than for BaO (1973 °C).

CONCEPTUAL EXERCISE 5.17

Melting Points of Ionic Solids

Consider these ionic compounds: KCl, CaS, CaO, SrSe, and LiF. Without looking up any values, predict the order of melting points from lowest to highest. Explain the reasons for your predictions.

Table 5.8 Effect of Ion Size and Charge on Lattice Energy and Melting Point

Compound	Charges of Ions	$r_+ + r_-$	Lattice Energy, kJ/mol	Melting Point, °C
NaCl	1+, 1−	116 pm + 167 pm = 283 pm	790	800
BaO	2+, 2−	149 pm + 126 pm = 275 pm	3054	1973
MgO	2+, 2−	86 pm + 126 pm = 212 pm	3791	2825

SUMMARY PROBLEM

(a) Without looking back in the chapter, draw and label the first five energy levels of a hydrogen atom. Next, indicate the $2 \longrightarrow 1$, $3 \longrightarrow 1$, $5 \longrightarrow 2$, and $4 \longrightarrow 3$ transitions in a hydrogen atom. Look in Table 5.1 (← **Sec. 5-3b**) to get the measured wavelengths and spectral regions for these transitions. Calculate the frequency (ν) for each transition. Calculate the energy of the photon emitted by each transition.

(b) The photoelectric threshold is the minimum wavelength (energy) a photon must have to produce a photoelectric effect for a metal. These three metals exhibit photoelectric effects when photons of sufficient energies strike their surfaces.

Element	Photoelectric Threshold (nm)
Lithium	430
Potassium	540
Cesium	580

Which photon energies calculated in part (a) are sufficient to cause a photoelectric effect in lithium, in potassium, and in cesium?

(c) Niobium (Nb) and tantalum (Ta) are transition elements named after the mythical king Tantalus, son of Zeus, and his daughter Niobe. Niobium metal is a component of stainless steels fabricated for high temperature use. Tantalum metal is extremely corrosion resistant and thus is used in the manufacturing of surgical implants.

(i) Write the full electron configuration for niobium and for tantalum.

(ii) Using the noble-gas notation, write the electron configuration for an Nb^{4+} ion.

(iii) Write the atomic orbital box diagram for a Ta^{5+} ion.

(iv) How many unpaired electrons are there in an Nb^{5+} ion? Explain your answer.

(v) Are either an Nb^{4+} ion or an Nb^{5+} ion paramagnetic? Explain your answer.

(iv) Explain why niobium and tantalum occur together in natural deposits of their compounds.

(d) An element is brittle with a steel-gray appearance. It is a relatively poor electrical conductor and forms a molecular chloride with a low boiling temperature and a molecular hydride that decomposes at room temperature. Is the element an alkaline-earth metal, a transition metal, a metalloid, or a halogen? Explain your answer.

(e) Describe two ways in which *p* atomic orbitals and *d* atomic orbitals are (i) alike; (ii) different.

HAVING STUDIED THIS CHAPTER . . .

. . . you should be able to:

- Use the relationships among frequency, wavelength, and the speed of light for electromagnetic radiation; based on wavelength or frequency, determine the region of the electromagnetic spectrum (Section 5-1). End-of-chapter Questions 14, 16, 26
- Given any two of the three, calculate frequency, wavelength, or energy per photon for electromagnetic radiation (Section 5-2). Questions 15, 17
- Explain the photoelectric effect based on Planck's quantum theory (Section 5-2). Questions 20, 22
- Use the Bohr model of the atom to interpret line emission spectra and the energy absorbed or emitted when electrons in atoms change energy levels (Section 5-3). Questions 28, 31

- Calculate the frequency, energy, and wavelength of a photon involved in an electron transition in a hydrogen atom; determine in what region of the electromagnetic spectrum the emission occurs (Section 5-3). Question 32
- Explain how the quantum mechanical model of the atom represents energy and probable location of an electron (Section 5-4). Question 39
- Apply quantum numbers (Section 5-5). Questions 43, 45
- Understand electron spin and how it affects electron configurations (Section 5-5). Question 41
- Describe and explain the relationships among shells, subshells, and atomic orbitals (Section 5-5). Questions 49, 52
- Sketch the shapes of boundary surfaces for *s*, *p*, and *d* atomic orbitals (Section 5-6). Question 50
- Using the periodic table, write electron configurations of atoms and ions of main-group and transition elements (Sections 5-7, 5-8). Questions 56, 58, 62
- Explain variations in number of valence electrons, electron configurations, ion formation, and paramagnetism of transition metals (Section 5-8). Questions 64, 68
- Describe trends in atomic radii, based on electron configurations (Section 5-9). Question 83
- Describe the role of effective nuclear charge in the trend of atomic radii across a period (Section 5-9). Question 81
- Describe trends in ionic radii and explain why ions differ in size from their atoms (Section 5-10). Question 85
- Use electron configurations to explain trends in the ionization energies of the elements (Section 5-11). Questions 79, 87
- Define electron affinity and describe periodic trends (Section 5-12). Questions 79, 92
- Discuss the energy changes that occur during the formation of an ionic compound from its elements and use them to predict properties of ionic compounds (Section 5-13). Question 97
- Use a Born-Haber cycle to determine the lattice energy of an ionic compound (Section 5-13). Question 95

KEY TERMS

atomic radius (Section 5-9)
Born-Haber cycle (5-13)
continuous spectrum (5-3)
core electrons (5-7d)
diamagnetic (5-8a)
effective nuclear charge (5-9c)
electromagnetic radiation (5-1)
electron affinity (5-12)
electron configuration (5-7)
electron density (5-4)
energy level (5-3a; 5-5a)
excited state (5-3a)
ferromagnetic (5-8a)
frequency (ν) (5-1)
ground state (5-3a)
Hund's rule (5-7a)
ionic radius (5-10)
ionization energy (IE) (5-11)
isoelectronic (5-8)
ℓ (5-5b)
lattice energy (5-13)
Lewis dot symbol (5-7d)
line emission spectrum (5-3)
m_ℓ (5-5c)
m_s (5-5d)
magnetic quantum number (m_ℓ) (5-5c)
momentum (5-4)
n (5-5a)
noble-gas notation (5-7a)
orbital (5-4)
orbital shape (5-6)
***p*-block elements** (5-7d)
paramagnetic (5-8a)
Pauli exclusion principle (5-5e)
photoelectric effect (5-2a)
photon (5-2a)
Planck's constant (h) (5-2)
principal energy level (5-5a)
principal quantum number (n) (5-3a)

quantum (5-2)
quantum number (5-5)
quantum theory (5-2)
radial distribution plot (5-6a)
***s*-block elements** (5-7d)
screening effect (5-9c)
shell (5-5a)
spectrum (5-2)
subshell (5-5b)
transition elements (5-7b)
uncertainty principle (5-4)
valence electrons (5-7d)
wavelength (λ) (5-1)

QUESTIONS FOR REVIEW AND THOUGHT

Red-numbered questions have short answers at the back of this book in Appendix M and fully worked solutions in the *Student Solutions Manual.*

Review Questions

These questions test vocabulary and simple concepts.

1. Describe how the frequency of electromagnetic radiation is related to its wavelength.
2. Light is given off by a sodium- or mercury-containing streetlight when the atoms are excited in some way. The light you see arises for which of these reasons?
   (a) Electrons moving from a given quantum level to one of higher *n*.
   (b) Electrons being removed from the atom, thereby creating a metal cation.
   (c) Electrons moving from a given quantum level to one of lower *n*.
   (d) Electrons whizzing about the nucleus in an absolute frenzy.
3. What is Hund's rule? Give an example of its use.
4. Explain what it means when someone says, "An electron occupies the $3p_x$ orbital".
5. Describe the changes in atomic size and ionization energy across a period and down a group.
6. Why is the radius of Na^+ much smaller than the radius of Na? Why is the radius of Cl^- much larger than the radius of Cl?
7. Write electron configurations to show the first two ionization steps for potassium. Explain why the second ionization energy is much larger than the first.
8. Explain how and why the sizes of atoms change across a period of the periodic table.
9. Write the electron configurations for the valence electrons of elements in the first three periods in Groups 1A through 8A.

Topical Questions

These questions are keyed to the major topics in the chapter. Usually a question that is answered at the back of this book is paired with a similar one that is not.

Electromagnetic Radiation (Section 5-1)

10. NASA operates a jetliner fitted with a special telescope that looks into outer space. The telescope has a camera that detects wavelengths between 5 and 40 μm. During eight-hour flights at 12,000 m above Earth, the plane will gather data of the nearby universe.
    (a) In what region of the electromagnetic spectrum does the camera operate?
    (b) Calculate the range of frequencies the camera detects.
11. Many marine organisms exhibit bioluminescence, which occurs when excited molecules return to their lowest energy (ground) state by releasing photons in the visible (and other) region of the electromagnetic spectrum. The starfish, *Plutonaster bifrons*, bioluminesces at 525 nm.
    (a) What color is the bioluminescence?
    (b) Calculate the frequency of this bioluminescence.
12. Many marine organisms exhibit bioluminescence, which occurs when excited singlet-state molecules return to their lowest energy (ground) state by releasing photons in the visible region (and beyond) of the electromagnetic spectrum. *Photostomias guernei*, a spiny-rayed oceanic fish, bioluminesces at 470. nm.
    (a) What color is the bioluminescence?
    (b) Calculate the frequency of this bioluminescence.
13. Based on Figure 5.2, which shows the regions of the electromagnetic spectrum, answer these questions.
    (a) Which radiation involves less energy per photon, radio waves or infrared radiation?
    (b) Which radiation has higher frequency, radio waves or microwaves?
14. The colors of the visible spectrum and the wavelengths corresponding to the colors are given in Figure 5.2.
    (a) Which colors of light involve photons with less energy per photon than yellow light?
    (b) Which color of visible light has photons of greater energy, green or violet?
    (c) Which color of light has the greater frequency, blue or green?

Planck's Quantum Theory (Section 5-2)

15. Calculate the energy of a photon of blue light that has a wavelength of 450 nm.
16. Calculate the wavelength and energy per photon associated with one quantum of laser light that has a frequency of $4.57 \times 10^{14}\ s^{-1}$.

17. Stratospheric ozone absorbs damaging UV-C radiation from the sun, preventing the radiation from reaching Earth's surface. Calculate the frequency and energy per photon of UV-C radiation that has a wavelength of 270. nm.
18. When someone uses a sunscreen, which kind of radiation is blocked? How does the sunscreen protect your skin from this type of radiation?
19. Describe the role Einstein's explanation of the photoelectric effect played in the development of the quantum theory.
20. Light of very long wavelength strikes a photosensitive metallic surface and no electrons are ejected. Explain why increasing the intensity of this light on the metal still will not cause the photoelectric effect.
21. A bright red light strikes a photosensitive surface and no electrons are ejected, even though dim blue light ejects electrons from the surface. Explain.
22. To eject electrons from the surface of potassium metal requires a minimum energy of 3.68×10^{-19} J. When 600.-nm photons shine on a potassium surface, will they cause the photoelectric effect? Explain.
23. To eject electrons from a gold surface requires photons with a frequency equal to or greater than 1.29×10^{15} Hz. Will photons in the visible region of the spectrum dislodge electrons from a gold surface? Explain.
24. A photoemissive material has a threshold energy, $E_{min} = 5 \times 10^{-19}$ J. Will 300. nm radiation eject electrons from the material? Explain.

The Bohr Model of the Hydrogen Atom (Section 5-3)

25. Flame tests depend on emissions in the visible region of the spectrum to identify elements in a sample. Barium-containing compounds emit at 493 nm; strontium-containing compounds emit at 642 nm.
    (a) Determine the identifying color of the flame test in each case.
    (b) Calculate the energy (J/photon) associated with each of these emissions.
    (c) Explain why the barium emission occurs at lower wavelength than the strontium does.
26. In Problem-Solving Example 5.4, the wavelength of an $n = 2$ to $n = 5$ transition in a hydrogen atom was calculated to be 434.1 nm. In Table 4.2, the standard formation enthalpy for liquid water is given as −285.830 kJ/mol. Use this information to calculate which process involves more energy per gram of hydrogen—the electronic transition from $n = 2$ to $n = 5$ in a hydrogen atom or the combustion of hydrogen and oxygen gas to produce liquid water.
27. Energy is emitted from an atom when an electron moves from a(n) ________ state to the ________ state. The energy of the emitted radiation corresponds to the ________ between the two energy levels.
28. For which of these transitions in a hydrogen atom is energy absorbed? Emitted?
    (a) $n = 1$ to $n = 3$ (b) $n = 5$ to $n = 1$
    (c) $n = 2$ to $n = 4$ (d) $n = 5$ to $n = 4$
29. Which transition involves the emission of less energy in the H atom, an electron moving from $n = 4$ to $n = 3$ or an electron moving from $n = 3$ to $n = 1$? (See Figure 5.10.)
30. For the transitions in Question 28:
    (a) which involve the ground state?
    (b) which involves the greatest energy change?
    (c) which absorbs the most energy?
31. If energy is absorbed by a hydrogen atom in its ground state, the atom is excited to a higher energy state. For example, the excitation of an electron from the energy level with $n = 1$ to a level with $n = 4$ requires radiation with a wavelength of 97.3 nm. Which of these transitions requires radiation of a wavelength longer than this? (See Figure 5.10.)
    (a) $n = 2$ to $n = 4$ (b) $n = 1$ to $n = 3$
    (c) $n = 1$ to $n = 5$ (d) $n = 3$ to $n = 5$
32. Calculate the energy and wavelength of the photon associated with the electron transition from $n = 2$ to $n = 5$ in the hydrogen atom.
33. Calculate the energy and the wavelength of the photon associated with an electron transition from $n = 1$ to $n = 4$ in the hydrogen atom.
34. Spectroscopists have observed He^+ in outer space. This ion is a one-electron species like a neutral hydrogen atom. Calculate the energy of the photon emitted for the transition from the $n = 5$ to the $n = 3$ state in this ion using the equation: $E_n = -Z^2/n^2\ (2.179 \times 10^{-18}\ \text{J})$. Z is the positive charge of the nucleus and n is the principal quantum number. In what part of the electromagnetic spectrum does this radiation lie?
35. The Bohr equation for hydrogen can be modified to apply to one-electron species other than uncharged hydrogen atoms, for example Li^{2+}, to calculate the energy of electron transitions in the ion. The modified equation is $E_n = -Z^2/n^2\ (2.179 \times 10^{-18}\ \text{J})$. Z is the positive charge of the nucleus and n is the principal quantum number. Calculate the energy of the photon emitted for the transition from the $n = 4$ to the $n = 1$ state in this ion. In what region of the electromagnetic spectrum does it lie?
36. The Brackett series of emissions has $n_f = 4$.
    (a) Calculate the wavelength, in nanometers, of the photon emitted by the $n = 7$ to $n = 4$ transition.
    (b) In what region of the electromagnetic spectrum is the emitted radiation?

Quantum Mechanical Model of the Atom (Section 5-4)

37. Arrange these items from smallest to largest de Broglie wavelength: baseball; bowling ball; electron moving at the velocity of light; the Moon; neon atom.

38. Arrange these items from largest to smallest de Broglie wavelength: basketball; proton; potassium atom; the planet Venus; soccer ball.
39. Based on the uncertainty principle, explain why the dot-density diagram in Figure 5.12 is a good way to represent the electron density in an atom.
40. Compare and contrast the representations of electron density in an atom provided by dot-density diagrams and boundary-surface diagrams.

Quantum Numbers (Section 5-5)

41. Consider these sets of four quantum numbers for electrons in an atom.

    (i) $n = 2, \ell = 1, m_\ell = 1, m_s = +\frac{1}{2}$

    (ii) $n = 2, \ell = 0, m_\ell = 1, m_s = 0$

    (iii) $n = 2, \ell = 1, m_\ell = 1, m_s = -\frac{1}{2}$

    (iv) $n = 2, \ell = 0, m_\ell = 1, m_s = -\frac{1}{2}$

    (a) Identify the two electrons that have the same spin.
    (b) Identify the two electrons that are in the same atomic orbital.
42. Give possible values for all four quantum numbers for a 7*g* electron.
43. Assign a correct set of four quantum numbers for
    (a) *Each* electron in a boron atom.
    (b) The 3*s* electrons in a magnesium atom.
    (c) A 3*d* electron in an iron atom.
44. Assign a correct set of four quantum numbers for
    (a) *Each* electron in a nitrogen atom.
    (b) The valence electron in a sodium atom.
    (c) A 3*d* electron in a nickel atom.
45. Some of these sets of quantum numbers (n, ℓ, m_ℓ, m_s) could not occur. Explain why.
    (a) $2, 1, 2, +\frac{1}{2}$ (b) $3, 2, 0, -\frac{1}{2}$
    (c) $1, 0, 0, 1$ (d) $3, 3, 2, -\frac{1}{2}$
    (e) $2, 0, 0, +\frac{1}{2}$
46. One electron has the set of quantum numbers $n = 3$, $\ell = 1$, $m_\ell = -1$, and $m_s = +\frac{1}{2}$; another electron has the set $n = 3$, $\ell = 1$, $m_\ell = 1$, and $m_s = +\frac{1}{2}$.
    (a) Could the electrons be in the same atom? Explain.
    (b) Could they be in the same atomic orbital? Explain.
47. Assign a correct set of four quantum numbers for the circled electrons in these orbital diagrams.

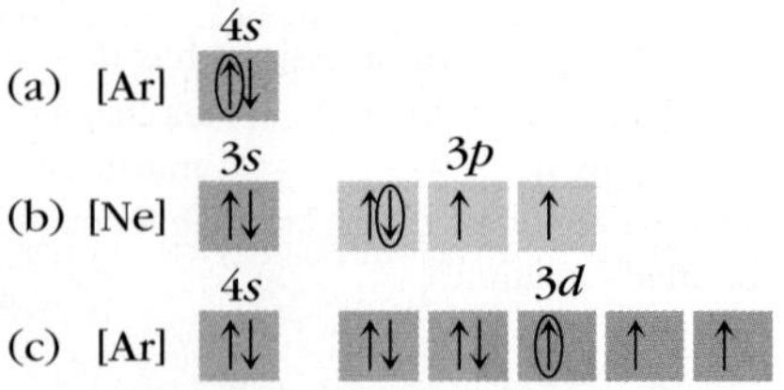

(a) [Ar]
4s
(b) [Ne]
3s
3p
(c) [Ar]
4s
3d

48. Give the n, ℓ, and m_ℓ values for
    (a) Each atomic orbital in the 6*f* sublevel.
    (b) Each atomic orbital in the $n = 5$ level.
49. How many elements are there in the fourth period of the periodic table? Based on quantum theory, explain why it is not possible for there to be another element in this period.

Shapes of Atomic Orbitals (Section 5-6)

50. From memory, sketch the shape of the boundary surface for each of these atomic orbitals:
    (a) $4d_{x^2-y^2}$ (b) 2*s* (c) $3p_y$
51. From memory, sketch the shape of the boundary surface for each of these atomic orbitals:
    (a) $2p_z$ (b) 4*s* (c) $3d_{xy}$
52. How many subshells are there in the electron shell with the principal quantum number $n = 4$?
53. How many subshells are there in the electron shell with the principal quantum number $n = 5$?

Electron Configurations (Sections 5-7 and 5-8)

54. Titanium metal and Cr^{2+} have the same number of electrons. However, the electron configuration of Ti is [Ar] $4s^23d^2$, but that of Cr^{2+} is [Ar] $3d^4$. Explain.
55. Consider a 2+ ion that has six 3*d* electrons; which ion is it? Which 2+ ion would have only three 3*d* electrons?
56. Germanium had not been discovered in the 1870s when Mendeleev formulated his ideas of chemical periodicity. He predicted its existence, however, and germanium was found in 1886 by Winkler. Write the electron configuration of germanium.
57. Write electron configurations for these atoms.
    (a) Strontium (Sr), named for a town in Scotland.
    (b) Tin (Sn), a metal used in the ancient world. Alloys of tin (solder, bronze, and pewter) are important.
58. Name an element of Group 6A. What does the group designation tell you about the electron configuration of the element?
59. Name an element of Group 3A. What does the group designation tell you about the electron configuration of the element?
60. (a) Which ions in this list are likely to be found in ionic compounds: K^{2+}, Cs^+, Al^{4+}, F^{2-}, Se^{2-}?
    (b) Which, if any, of these ions have a noble-gas configuration?
61. These ground-state atomic orbital diagrams are incorrect. Explain why they are incorrect and how they should be corrected.

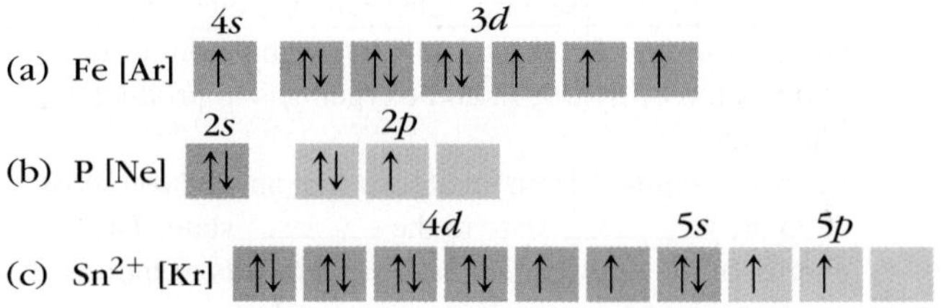

(a) Fe [Ar]
4s
3d
(b) P [Ne]
2s
2p
(c) Sn2+ [Kr]
4d
5s
5p

62. Write the atomic orbital diagram for the 4*s* and 3*d* electrons in a
   (a) vanadium atom. (b) V^{2+} ion.
   (c) V^{4+} ion.
63. Write the atomic orbital diagram for the 4*s* and 3*d* electrons in a
   (a) manganese atom. (b) Mn^{+} ion.
   (c) Mn^{5+} ion.
64. Give the electron configurations of Mn, Mn^{2+}, and Mn^{3+}. Use atomic orbital box diagrams to determine the number of unpaired electrons for each species.
65. Write the electron configurations of chromium: Cr, Cr^{2+}, and Cr^{3+}. Use atomic orbital box diagrams to determine the number of unpaired electrons for each species.
66. Write electron configurations for these elements.
   (a) Zirconium (Zr). This metal is exceptionally resistant to corrosion and so has important industrial applications. Moon rocks show a surprisingly high zirconium content compared with rocks on Earth.
   (b) Rhodium (Rh), which is used in jewelry and in industrial catalysts.
67. The lanthanides, or rare earths, are only "medium rare," because all can be purchased for a reasonable price. Give electron configurations for atoms of these lanthanides.
   (a) Europium (Eu), the most expensive of the rare earth elements; 1 g can be purchased for about $1.40.
   (b) Ytterbium (Yb). Less expensive than Eu, Yb costs only about $0.35 per gram. It was named for the village of Ytterby in Sweden, where a mineral source of the element was found.
68. Locate these elements in the periodic table, and then draw a Lewis dot symbol that represents the number of valence electrons for an atom of each.
   (a) Sr (b) Br
   (c) Ga (d) Sb
69. Locate these elements in the periodic table, and then draw a Lewis dot symbol that represents the number of valence electrons for an atom of each.
   (a) F (b) In
   (c) Te (d) Cs
70. Give the electron configurations of these ions, and indicate which ones are isoelectronic.
   (a) Ca^{2+} (b) K^{+} (c) O^{2-}
71. Give the electron configurations of these ions, and indicate which ones are isoelectronic.
   (a) Na^{+} (b) Al^{3+} (c) Cl^{-}
72. (a) What is the electron configuration for an atom of tin?
   (b) What are the electron configurations for Sn^{2+} and Sn^{4+} ions?
73. What is the electron configuration for
   (a) a bromine atom? (b) a bromide ion?
74. (a) In the first transition series (in row four of the periodic table), which elements would you predict to be diamagnetic?
   (b) Which element in this series has the greatest number of unpaired electrons?
75. Nearly all first-row transition elements form 2+ ions.
   (a) For which of these elements are the 2+ ions paramagnetic?
   (b) For which element do compounds containing 2+ ions and chloride ions have the greatest paramagnetism? (Chloride ions have no unpaired electrons.)
76. How do the spins of unpaired electrons from paramagnetic and ferromagnetic materials differ in their behavior in a magnetic field?
77. Consider titanium metal and its two oxides, TiO and TiO_2. The oxide ion has no unpaired electrons.
   (a) Which of these titanium species is diamagnetic? Explain your answer.
   (b) Which titanium species will be most attracted to a magnetic field? Explain your answer.
78. The acetylacetonate ion, $(acac)^{-}$, has no unpaired electrons. It forms compounds with Fe^{2+} and with Fe^{3+} ions whose formulas are $Fe(acac)_2$ and $Fe(acac)_3$, respectively. Explain which compound has the greater attraction to a magnetic field.

Periodic Trends (Sections 5-9 to 5-12)

79. Use electron configurations to explain why
   (a) sulfur has a lower electron affinity than chlorine.
   (b) boron has a lower first ionization energy than beryllium.
   (c) chlorine has a lower first ionization energy than fluorine.
   (d) oxygen has a lower first ionization energy than nitrogen.
   (e) iodine has a lower electron affinity than bromine.
80. A fellow chemistry student states that periodic trends cover the main-group elements but do not apply to the transition metals. Is the student correct? Explain.
81. Arrange these elements in order of increasing effective nuclear charge: N, F, B, O, and C.
82. Arrange these ions in order of decreasing size: Be^{2+}, Rb^{+}, Ca^{2+}, Mg^{2+}.
83. Arrange these elements in order of increasing atomic size: Ca, Rb, P, Ge, Sr. (Try arranging these without looking at Figure 5.25 and then check yourself by looking up the necessary atomic radii.)
84. Arrange these elements in order of increasing atomic size: Al, B, C, K, Na. (Try arranging these without looking at Figure 5.25 and then check yourself by looking up the necessary atomic radii.)
85. Select the atom or ion in each pair that has the smaller radius.
   (a) Cs or Rb (b) O^{2-} or O
   (c) Br or As (d) Ba or Ba^{2+}
   (e) Cl^{-} or Ca^{2+}

86. Select the atom or ion in each pair that has the larger radius.
(a) Cl or Cl^- (b) Ca or Ca^{2+}
(c) Al or N (d) Cl^- or K^+
(e) In or Sn
87. Which of these groups of elements is arranged correctly in order of increasing ionization energy?
(a) C, Si, Li, Ne (b) Ne, Si, C, Li
(c) Li, Si, C, Ne (d) Ne, C, Si, Li
88. Rank these ionization energies (IE) from the smallest to the largest value. Briefly explain your answer.
(a) First IE of Be (b) First IE of Li
(c) Second IE of Be (d) Second IE of Na
(e) First IE of K
89. Predict which of these elements would have the greatest difference between the first and second ionization energies: Si, Na, P, Mg. Briefly explain your answer.
90. Compare the elements B, Al, C, Si.
(a) Which has the most metallic character?
(b) Which has the largest atomic radius?
(c) Arrange the three elements B, Al, and C in order of increasing first ionization energy.
91. Compare the elements Li, K, C, N.
(a) Which has the largest atomic radius?
(b) Arrange the elements in order of increasing first ionization energy.
92. The first electron affinity of oxygen is negative, the second is positive. Explain why this change in sign occurs.
93. Explain why nitrogen has a higher first ionization energy than does carbon, the element that precedes it in the periodic table.
94. Which group of the periodic table has elements with high first ionization energies *and* very negative electron affinities? Explain this behavior.

Energy, Ions, and Ionic Compounds (Section 5-13)

95. Determine the lattice energy for LiCl(s) given these data: Sublimation enthalpy of Li, 161 kJ/mol; IE_1 for Li, 520 kJ/mol; BE of Cl_2(g), 242 kJ/mol; electron affinity of Cl, −349 kJ/mol; formation enthalpy of LiCl(s), −408.7 kJ/mol.
96. The lattice energy of KCl(s) is −719 kJ/mol. Use data from the text to calculate the formation enthalpy of KCl.
97. Which ionic compound has the lowest melting point? Explain your choice.
LiCl NaBr KCl
98. Which ionic compound has the largest lattice energy? Explain your choice.
MgS RbI Li_2S

General Questions

These questions are not explicitly keyed to chapter topics; many require integration of several concepts.

99. Give the symbol of the ground-state atom that
(a) is in Group 8A but has no *p* electrons.
(b) has a single electron in the 3*d* subshell.
(c) forms a 1+ ion with a $1s^22s^22p^6$ electron configuration.
100. Give the symbol of all the ground-state atoms that have
(a) no *p* electrons.
(b) from two to four *d* electrons.
(c) from two to four *s* electrons.
101. Answer these questions about the elements X and Z, which have the electron configurations shown.

$$X = [Kr]\ 4d^{10}5s^1 \qquad Z = [Ar]\ 3d^{10}4s^24p^4$$

(a) Is element X a metal or a nonmetal?
(b) Which element has the larger atomic radius?
(c) Which element would have the greater first ionization energy?
102. (a) Rank these elements in order of increasing atomic radius: O, S, F. Briefly explain your reasoning.
(b) Which element has the largest first ionization energy: P, Si, S, Se? Briefly explain your reasoning.
103. (a) Rank these in order of increasing radius: Ne, O^{2-}, N^{3-}, F^-. Briefly explain your reasoning.
(b) Place these elements in order of increasing first ionization energy: Cs, Sr, Ba. Briefly explain your reasoning.
104. Name the element corresponding to each of these characteristics.
(a) The element whose atoms have the electron configuration $1s^22s^22p^63s^23p^4$
(b) The element in the alkaline-earth group that has the largest atomic radius
(c) The element in Group 5A whose atoms have the largest first ionization energy
(d) The element whose 2+ ion has the configuration $[Kr]\ 4d^6$
(e) The element whose neutral atoms have the electron configuration $[Ar]\ 3d^{10}4s^1$
105. The ionization energies for the removal of the first electron from atoms of Si, P, S, and Cl are listed below. Briefly rationalize this trend.

Element	First Ionization Energy (kJ/mol)
Si	786
P	1060
S	999
Cl	1255

106. Answer these questions about the elements with the electron configurations shown.

$$X = [Ar]\ 3d^84s^2 \qquad Z = [Ar]\ 3d^{10}4s^24p^5$$

(a) An atom of which element is expected to have the larger first ionization energy?
(b) An atom of which element would be the smaller of the two?
107. Place these atoms and ions in order of decreasing size: Ar, K^+, Cl^-, S^{2-}, Ca^{2+}. Briefly explain your reasoning.
108. Which of these ions are unlikely, and why: Cs^+, In^{4+}, V^{6+}, Te^{2-}, Sn^{5+}, I^-? Briefly explain your reasoning.
109. Rank these in order of increasing first ionization energy: Zn, Ca, Ca^{2+}, Cl^-. Briefly explain your answer.

110. Below is a grid with nine lettered boxes, each of which contains an item that is used to answer the questions that follow. Items may be used more than once and there may be more than one correct item in response to a question. The n_m are principal quantum numbers for H-atom energy levels.

A $n_4 \longrightarrow n_2$	**B** $n_3 \longrightarrow n_1$	**C** > 700 nm
D $n_4 \longrightarrow n_1$	**E** $n_2 \longrightarrow n_7$	**F** $n_3 \longrightarrow n_5$
G 486 nm	**H** $n_6 \longrightarrow n_3$	**I** < 400 nm

Place the letter(s) of the correct selection(s) on the appropriate line.
(a) Ultraviolet emission(s) ________
(b) Highest energy absorption ________
(c) Highest energy emission ________
(d) Visible region emission(s) ________
(e) Infrared region emission(s) ________

111. Below is a grid with nine lettered boxes, each of which contains an item that is used to answer the questions that follow. Items may be used more than once and there may be more that one correct item in response to a question.

A Fe	**B** Br^-	**C** K^+
D Ne	**E** Fe^{3+}	**F** F
G K	**H** Sc	**I** Ar

Place the letter(s) of the correct selection(s) on the appropriate line.
(a) Neutral atom with a radius greater than the radius of Cl.
____
(b) Radius is less than the radius of Si. ________
(c) Isoelectronic species. ________
(d) Radius is greater than that of Br. ________
(e) Neutral atom with highest first ionization energy.
____
(f) Iron-containing species with radius less than the radius of Fe^{2+}. ________

112. Criticize these statements. If a statement is incorrect, rewrite it so that it is correct.
(a) The energy of a photon is inversely proportional to its frequency.
(b) The energy of the hydrogen electron is inversely proportional to its principal quantum number *n*.
(c) Electrons start to enter the fourth energy level as soon as the third level is full.
(d) Light emitted by an $n = 4$ to $n = 2$ transition has a lower frequency than that from an $n = 5$ to $n = 2$ transition.

113. A general chemistry student tells a chemistry classmate that when an electron goes from a 2*d* atomic orbital to a 1*s* atomic orbital, it emits more energy than that for a 2*p* to 1*s* transition. The other student is skeptical and says that such an energy change is not possible and explains why. What explanation was given?

114. A certain minimum energy, E_{min}, is required to eject an electron from a photosensitive surface. Any energy absorbed beyond this minimum gives kinetic energy to the ejected electron. When 540.-nm light falls on a cesium surface, an electron is ejected with a kinetic energy of 6.69×10^{-20} J. When the wavelength is 400 nm, the kinetic energy is 1.96×10^{-19} J.
(a) Calculate E_{min} for cesium, in joules.
(b) Calculate the longest wavelength, in nanometers, that will eject an electron from cesium.

115. Suppose a new element, unbinilium, has recently been discovered. Its atomic number is 120.
(a) Write the electron configuration of the element.
(b) Name another element you would expect to find in the same periodic table group as unbinilium.
(c) Write the formulas for the compounds of unbinilium with O and Cl. Use X for the chemical symbol.

116. When sulfur dioxide reacts with chlorine, the products are thionyl chloride, $SOCl_2$, and dichlorine monoxide, OCl_2.

$$SO_2(g) + 2\ Cl_2(g) \longrightarrow SOCl_2(g) + OCl_2(g)$$

(a) In what period of the periodic table is S located?
(b) Give the complete electron configuration of S. Do *not* use the noble-gas notation.
(c) An atom of which element involved in this reaction (O, S, or Cl) should have the smallest first ionization energy? The smallest radius?
(d) You want to make 675 g $SOCl_2$. Calculate what mass, in grams, of Cl_2 is required.
(e) If you use 10.0 g SO_2 and 20.0 g Cl_2, determine the theoretical yield of $SOCl_2$.

Applying Concepts

These questions test conceptual learning.

117. The ionization energy of H(g) is 1312 kJ/mol. The second ionization energy of He(g) is almost exactly four times this value, and the third ionization energy of Li(g) is almost exactly nine times that of hydrogen. What relationship, if any, do these data suggest between ionization energy and a fundamental characteristic of elements?

118. The so-called northern lights (*aurora borealis*) are caused by ionization and dissociation of O_2 and N_2 molecules through collisions with high-energy electrons ejected from the Sun's surface. Light in the visible region of the electromagnetic spectrum is emitted when the excited ions and atoms return to their ground states.

Species	Wavelength Emitted, nm
O_2^+	~630
N_2^+	391.4; 470.0
O	557.7; 630.0

(a) List the emissions in order of increasing energy.
(b) Calculate the energy of the photons emitted at 391.4 nm.
(c) Give the approximate color of (i) the 557.7 nm emission; (ii) the 470.0 emission.
(d) List the species, if any, that are: (i) paramagnetic; (ii) diamagnetic. Explain your choices.

119. In the upper atmosphere, H—O bonds in water vapor are broken by high-energy photons. To break an H—O bond in water vapor requires an average of 467 kJ/mol.
(a) Calculate the minimum energy of a photon that can break an H—O bond.
(b) From what region of the electromagnetic spectrum does this photon come?

120. Microwave ovens, commonly used to heat water in beverages and foods, emit radiation with a wavelength of 12.2 cm.
(a) Calculate the amount (moles) of photons of this microwave radiation required to raise the temperature of 230.0 g water (such as in a cup of coffee, which is mainly water) from 24.0 °C to 55.0 °C.
(b) As noted in Chapter 4, the watt, W, is a unit of power: 1 W = 1 J/s. If the microwave oven is rated at 800 W, calculate the time needed to heat the water in part (a). Assume that all the energy is delivered to the water.

121. Use electron configurations to explain why
(a) the electron affinity of selenium is lower than that of bromine.
(b) the first ionization energy of aluminum is lower than that of magnesium.
(c) sulfur has a lower first ionization energy than phosphorus.
(d) bromine has a lower first ionization energy than chlorine.

122. In a hypothetical universe, atoms have two *s* orbitals, four *p* orbitals, six *d* orbitals, and eight *f* orbitals.
(a) Draw a diagram of the periodic table for this universe, a table that retains the general positions of each block relative to the others.
(b) Write the electron configuration for the first *f*-block element in this universe.

123. Write the formula for a compound formed by chlorine and element X, if element X has the electronic configuration $1s^22s^22p^63s^1$.

124. Write the formula for a compound formed by potassium and element Z, if element Z has the electronic configuration $1s^22s^22p^63s^23p^4$.

125. Which of these electron configurations are for atoms in the ground state? In excited states? Which are impossible?
(a) $1s^22s^1$
(b) $1s^22s^22p^3$
(c) [Ne] $3s^23p^34s^1$
(d) [Ne] $3s^23p^64s^33d^2$
(e) [Ne] $3s^23p^64f^4$
(f) $1s^22s^22p^43s^2$

126. Which of these electron configurations are for atoms in the ground state? In excited states? Which are impossible?
(a) $1s^22s^2$
(b) $1s^22s^23s^1$
(c) [Ne] $3s^23p^84s^1$
(d) [He] $2s^22p^62d^2$
(e) [Ar] $4s^23d^3$
(f) [Ne] $3s^22p^54s^1$

127. Using the information in Appendix D and Figure 5.19, write the electron configuration for the undiscovered element with an atomic number of 164. Where would this element be located in the periodic table?

128. Refer to the graph below to answer these questions.

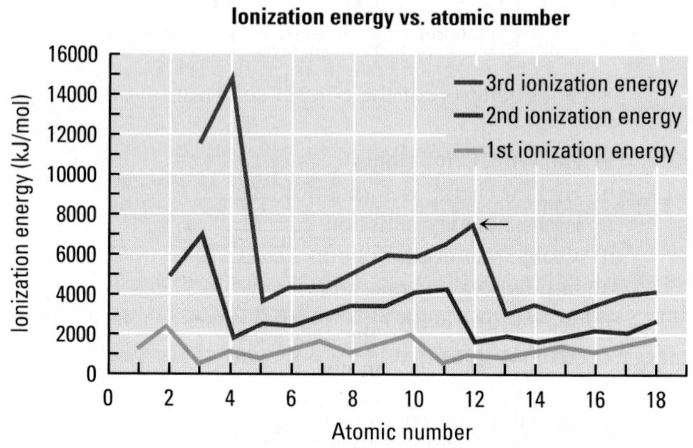
Ionization energy vs. atomic number
Ionization energy (kJ/mol)
16000
14000
12000
10000
8000
6000
4000
2000
0
0 2 4 6 8 10 12 14 16 18
Atomic number
3rd ionization energy
2nd ionization energy
1st ionization energy

(a) Based on the graphic data, ionization energies __________ (decrease, increase) left to right and __________ (decrease, increase) top to bottom on the periodic table.
(b) Which element has the largest first ionization energy?
(c) A plot of the fourth ionization energy versus atomic number for elements 1 through 18 would have peaks at which atomic numbers?
(d) Why is there no third ionization energy for helium?
(e) What is the reason for the large second ionization energy for lithium?
(f) Find the arrow pointing to the third ionization energy curve. Write the equation for the process corresponding to this data point.

129. Use Coulomb's law to predict which substance in each of these pairs has the larger lattice energy.
(a) CaO or KI
(b) CaF_2 or BaF_2
(c) KCl or LiBr

More Challenging Questions

These questions require more thought and integrate several concepts.

The next two questions require use of this set of guidelines. To determine σ in the effective nuclear charge equation $Z^* = Z - \sigma$, write the electron configuration for an atom. Group the subshells this way:

$$(1s)\ (2s, 2p)\ (3s, 3p)\ (3d)\ (4s, 4p)\ (4d)\ (4f)\ (5s, 5p)\ \ldots$$

Choose the electron for which you want to calculate Z^*. Sum these contributions to σ for all other electrons: Electrons in groups to the right of the selected electron contribute zero. Electrons in the same group as the selected electron contribute 0.35 per electron (except for (1*s*) where the contribution is 0.30). If the selected electron is in an (*ns*, *np*) group, electrons in the $n - 1$ shell contribute 0.85 and electrons in the shells with lower *n* contribute 1.0. If the selected electron is in an (*nd*) or (*nf*) group, all electrons in groups to the left count 1.0.

130. Calculate the effective nuclear charge, Z^*, on these electrons in a tin atom (a) 5*s*; (b) 5*p*; and (c) 4*d*.
131. (a) Calculate the effective nuclear charge, Z^*, on a 2*p* electron in O^{2-}, F^-, Na^+, and Mg^{2+}.
    (b) Is the calculated effective nuclear charge consistent with the relative sizes of these ions? Explain your answer.
132. An element has an ionization energy of 1.66×10^{-18} J. The three longest wavelengths in its absorption spectrum are 253.7 nm, 185.0 nm, and 158.5 nm.
    (a) Construct an energy-level diagram similar to Figure 5.9 for this element.
    (b) Indicate all possible emission lines. Start from the highest energy level on the diagram.
133. The energy of a photon needed to cause ejection of an electron from a photoemissive metal is expressed as the sum of the binding energy of the photon plus the kinetic energy of the emitted electron. When photons of 4.00×10^{-7} m light strike a calcium metal surface, electrons are ejected with a kinetic energy of 6.3×10^{-20} J.
    (a) Calculate the binding energy of the calcium electrons.
    (b) The minimum frequency at which the photoelectric effect occurs for a metal is that for which the emitted electron has no kinetic energy. Calculate the minimum frequency required for the photoelectric effect to occur with calcium metal.
134. The energy of a photon needed to cause ejection of an electron from a photoemissive metal is expressed as the sum of the binding energy of the photon plus the kinetic energy of the emitted electron. Calculate the kinetic energy of an electron that is emitted from a strontium metal surface irradiated with photons of 4.20×10^{-7} m light. The binding energy of strontium is 4.39×10^{-19} J.
135. According to a relationship developed by Niels Bohr, for an atom or ion that has a single electron, the total energy of an electron in a stable orbit of quantum number *n* is $E_n = -[Z^2/n^2](2.179 \times 10^{-18}\text{ J})$ where *Z* is the atomic number. Calculate the ionization energy for the electron in a ground-state hydrogen atom.
136. According to a relationship developed by Niels Bohr, for an atom or ion that has a single electron, the total energy, E_n, of an electron in a stable orbit of quantum number *n* is $E_n = -[Z^2/n^2](2.179 \times 10^{-18}\text{ J})$ where *Z* is the atomic number. Calculate the ionization energy for the electron in a ground-state He^+ ion.
137. Oxygen atoms are smaller than nitrogen atoms, yet oxygen has a lower first ionization energy than nitrogen. Explain.
138. Beryllium atoms are larger than boron atoms, yet boron has a lower first ionization energy than beryllium. Explain.
139. The element meitnerium (Mt) honors Lise Meitner for her role in the discovery of nuclear fission.
    (a) Meitnerium atoms have the same outer electron configuration as what transition metal?
    (b) Using the noble-gas notation, write the electron configuration for a ground-state Mt atom.
140. Suppose two electrons in the same system each have $n = 3$, $\ell = 0$.
    (a) How many different electron arrangements would be possible if the Pauli exclusion principle did *not* apply in this case?
    (b) How many would apply if it *is* operative?
141. Only a few atoms of element 112, copernicium, have ever been synthesized so its chemical properties are difficult to determine experimentally.
    (a) The chemical properties of the element can be expected to be those of what kind of element—main group, transition metal, lanthanide, or actinide?
    (b) Using the noble-gas notation, write the electron configuration for a ground-state atom of element 112 to corroborate your answer to part (a).
142. You are given the atomic radii of 110. pm, 117 pm, 121 pm, 132 pm, and 153 pm, but do not know to which element (As, Ga, Ge, P, Si) these values correspond. Which must be the value for Ge?
143. (a) Use a Born-Haber cycle to calculate $\Delta_f H°$ CaCl(s). Use these data (kJ/mol): $\Delta_{sublimation}H$ Ca(s) = 178; lattice energy CaCl(s) = −744; obtain other necessary data from tables and figures in this book.
    (b) Use the answer from Part (a) and other $\Delta_f H°$ to calculate $\Delta_r H°$ for the reaction
    $$2\,CaCl(s) \longrightarrow CaCl_2(s) + Ca(s)$$
    (c) Based on the answers to (a) and (b), is $CaCl_2$(s) more stable than CaCl(s)? Explain your answer.
144. Solid silver(I) chloride and sodium chloride have the same type of crystalline lattice, but differ in their $\Delta_f H°$.
    (a) Use a Born-Haber cycle to calculate $\Delta_f H°$ AgCl(s). Use these data (kJ/mol): $\Delta_{sublimation}H$ Ag(s) = 284; lattice energy AgCl(s) = −910.; obtain other necessary data from tables and figures in this book.
    (b) Use a Born-Haber cycle to calculate $\Delta_f H°$ NaCl(s). Use these data (kJ/mol): $\Delta_{sublimation}H$ Na(s) = 108; lattice energy NaCl(s) = −787; obtain other necessary data from tables and figures in this book.
    (c) Comment on the main factors for the differences in $\Delta_f H°$ of these two compounds.

Conceptual Challenge Problems

These rigorous, thought-provoking problems integrate conceptual learning with problem solving and are suitable for group work.

CP5.A (Section 5-2) Planck stated in 1900 that the energy of a single photon of electromagnetic radiation was directly proportional to the frequency of the radiation ($E = h\nu$). The constant h is known as Planck's constant and has a value of 6.626×10^{-34} J s. Soon after Planck's statement, Einstein proposed his famous equation ($E = mc^2$), which states that the total energy in any system is equal to its mass times the speed of light squared.

According to the de Broglie relation, what is the apparent mass of a photon emitted by an electron undergoing a change from the second to the first energy level in a hydrogen atom? How does the photon mass compare with the mass of the electron (9.109×10^{-31} kg)?

CP5.B (Section 5-7) When D. I. Mendeleev proposed a periodic law around 1870, he asserted that the properties of the elements are a periodic function of their atomic weights. Later, after H. G. J. Moseley measured the charge on the nuclei of atoms, the periodic law could be revised to state that the properties of the elements are a periodic function of their atomic numbers. What is another way to define the periodic function that relates the properties of the elements?

CP5.C (Sections 5-3 and 5-11) Figure 5.10 shows a diagram of the energy states that an electron can occupy in a hydrogen atom. Use this diagram to show that the first ionization energy for hydrogen, given in Figure 5.30, is correct.

Covalent Bonding | 6

gasprl3/iStockphoto.com

Human body cells, such as those of these hikers, contain thousands of compounds, most of which consist of atoms covalently bonded to each other. What is a covalent bond? (Section 6-1) How do we represent such bonding? Is there a general guideline for doing so? (Sections 6-2 to 6-5) How do the lengths and other characteristics of covalent bonds affect the properties of molecular substances? (Sections 6-6 to 6-7) What energy changes occur when covalent bonds are broken and formed? (Section 6-6) This chapter explains the various aspects of covalent bonding.

Atoms of elements are rarely found uncombined in nature. Only the noble gases consist of individual uncombined atoms. Most nonmetallic elements consist of molecules, and in a solid metallic element each atom is closely surrounded by eight or twelve neighbors. What makes atoms stick to one another? Interactions among valence electrons of bonded atoms form the glue, but how? In ionic compounds, the transfer of one or more valence electrons from an atom of a metal to the valence shell of an atom of a nonmetal produces positively charged ions (cations) and negatively charged ions (anions). The electrostatic attractions among these oppositely charged ions hold them in a regular arrangement in a crystal lattice, such as in the ionic compounds NaCl (Na^+ and Cl^-) and CaO (Ca^{2+} and O^{2-}). Ionic compounds are solids at room temperature and conduct an electric current when melted due to movement of the mobile cations and anions (← **Sec. 2-6c**). Many other substances, however, are composed of molecules, not ions. Examples are CO_2, H_2O, and $C_6H_{12}O_6$ (glucose). Molecular compounds have properties very different from those of ionic compounds. The melting points of molecular compounds are lower, and, when melted, these compounds do not conduct an electric current because they do not contain ions (← **Sec. 2-7a**).

In molecular compounds, atoms are held together by bonds consisting of one or more pairs of electrons *shared* between the bonded atoms. The attraction of positively charged nuclei for electrons between them pulls the nuclei together. This simple idea can account for the bonding in nearly all molecular compounds, allowing us to correlate their structures with their physical and chemical properties. This chapter describes the bonding found in molecules ranging from simple diatomic gases to hydrocarbons and more complex molecules.

6-1 Covalent Bonding

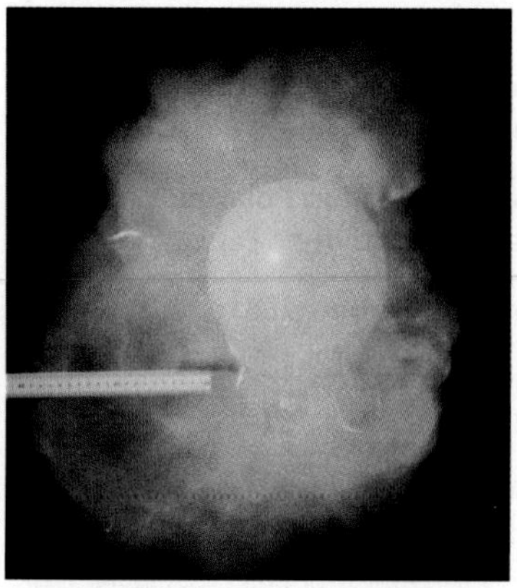

Water is formed by reaction of hydrogen with oxygen. Water is a molecular compound in which the atoms are connected by covalent bonds—shared electron pairs.

An explanation of bonding in the simplest stable molecule H_2 was proposed in 1916 by G. N. Lewis, who suggested that valence electrons rearrange to give noble-gas electron configurations when atoms join together chemically. Lewis regarded the lack of reactivity of the noble gases as due to their stable electron configuration, a $1s^2$ configuration for He or an ns^2np^6 configuration for Ne, Ar, Kr, and Xe. To achieve these stable configurations, atoms share electrons. Lewis proposed that a **covalent bond** is *a force attracting two atoms together when one or more pairs of electrons are shared between the bonded atoms*. Covalent bonds connect the atoms in molecular (covalent) compounds. For example, in H_2O, the two H atoms are held to the O atom by covalent bonds. This *description of covalent bonds involving pairs of valence electrons localized in bonds between atoms* is called the **valence bond theory** (valence bond model).

But why does sharing electrons provide an attractive force between bonded atoms? Consider the formation of the simplest molecule, H_2. Figure 6.1 shows how overlap of the $1s$ orbitals on two H atoms concentrates electron density between the nuclei, attracting the atoms together. If the atoms get too close, repulsions between the negatively charged electrons and repulsions between the positively charged nuclei force the atoms apart.

Experimental data and calculations indicate that an H_2 molecule has its lowest potential energy and is therefore most stable when the nuclei are 74 pm apart (Figure 6.2). At that distance—*the bond length*—the attractive and repulsive electrostatic forces are balanced. If the nuclei get closer than 74 pm, repulsion of each nucleus by the other raises the potential energy. When the H nuclei are 74 pm apart, it takes 436 kJ of energy to

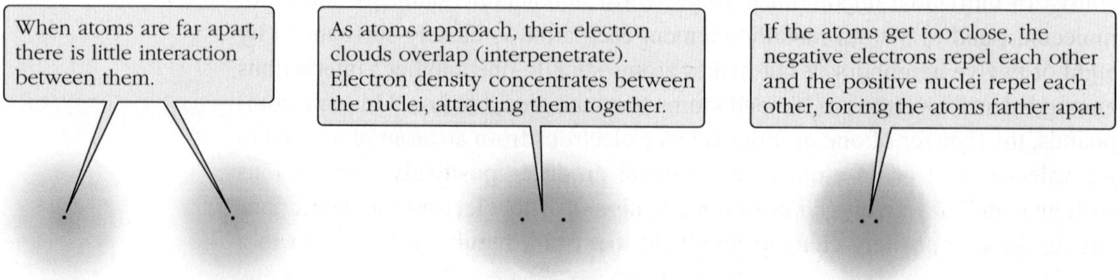
When atoms are far apart, there is little interaction between them.
As atoms approach, their electron clouds overlap (interpenetrate). Electron density concentrates between the nuclei, attracting them together.
If the atoms get too close, the negative electrons repel each other and the positive nuclei repel each other, forcing the atoms farther apart.

Figure 6.1 Overlap of atomic orbitals attracts atoms together forming a chemical bond.

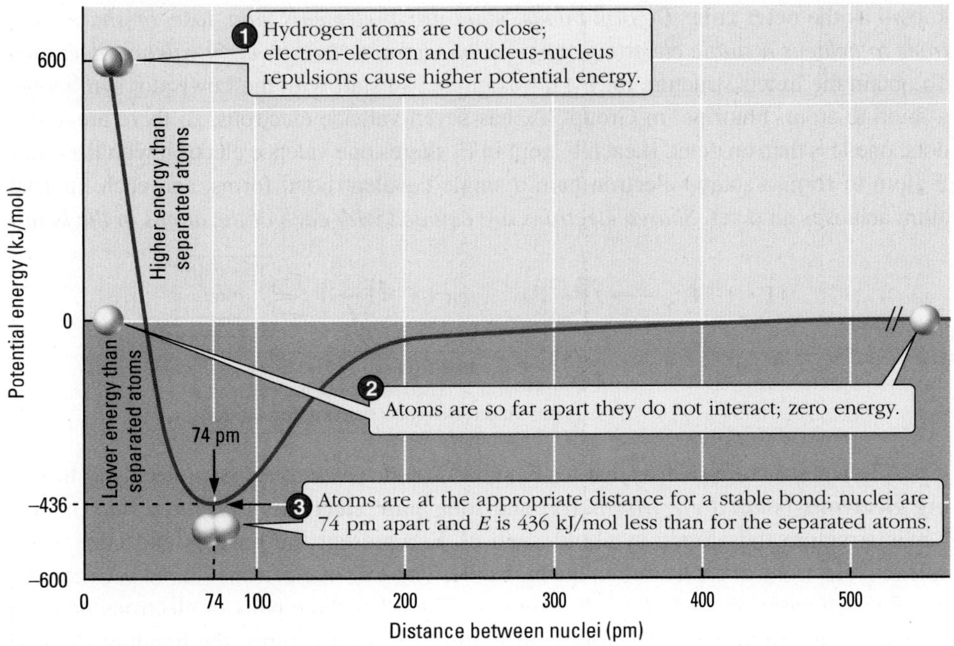

600
0
–436
–600
Potential energy (kJ/mol)
Higher energy than separated atoms
Lower energy than separated atoms
1 Hydrogen atoms are too close; electron-electron and nucleus-nucleus repulsions cause higher potential energy.
2 Atoms are so far apart they do not interact; zero energy.
3 Atoms are at the appropriate distance for a stable bond; nuclei are 74 pm apart and E is 436 kJ/mol less than for the separated atoms.
74 pm
0
74 100
200
300
400
500
Distance between nuclei (pm)

Figure 6.2 H—H bond formation from isolated H atoms. Energy is at a minimum at an internuclear distance of 74 pm.

Lowering potential energy is favorable to bond formation.

break the hydrogen-to-hydrogen covalent bonds when a mole of gaseous H_2 molecules is converted into isolated H atoms. This energy is the *bond energy* of H_2 (← **Sec. 4-7**).

6-2 Single Covalent Bonds and Lewis Structures

The shared electron pairs of covalent bonds occupy the same shell as the valence electrons of each atom. For main-group elements, the electrons commonly contribute to a noble-gas configuration on each atom. Lewis further proposed that by counting the valence electrons of an atom, it would be possible to predict how many bonds that atom can form. *The number of covalent bonds an atom can form is determined by the number of electrons that the atom must share to achieve a noble-gas configuration.*

A **single covalent bond** *is formed when two atoms share one pair of electrons.* The simplest examples are the bonds in diatomic molecules such as H_2, F_2, and Cl_2, which, like other simple molecules, can be represented by Lewis structures. A **Lewis structure** for a molecule *shows all valence electrons as dots or some as dots and others as lines that represent covalent bonds.*

Lewis structures are drawn by starting with the Lewis dot symbols for the atoms (see Table 5.5, ← **Sec. 5-7d**) and arranging the valence electrons until each atom in the molecule has a noble-gas configuration. For example, the Lewis structure for H_2 shows two bonding electrons shared between two hydrogen atoms. The shared electron pair of a *single* covalent bond is often represented by a line instead of a pair of dots.

$$\mathrm{H\!:\!H} \qquad \text{or} \qquad \mathrm{H{-}H}$$

Note that *each* hydrogen atom shares the pair of electrons, thereby achieving the same two-electron configuration as helium, the simplest noble gas.

Atoms with more than two valence electrons achieve a noble-gas structure by sharing enough electrons to attain an octet of valence electrons (an ns^2np^6 configuration.) This is

Gilbert Newton Lewis
1875–1946

Oseper Collection in the History of Chemistry. University of Cincinnati

In 1916, G. N. Lewis introduced the theory of the shared electron pair chemical bond in a paper published in the *Journal of the American Chemical Society.* His theory revolutionized chemistry, and it is in honor of this contribution that we refer to "electron dot" structures as Lewis structures. Of particular interest in this text is the extension of his theory of bonding to a generalized theory of acids and bases (Section 14-9).

Achieving a noble-gas configuration is also referred to as "obeying the octet rule" because all noble gases except helium have eight valence electrons.

known as the **octet rule**: *To form bonds, main-group elements gain, lose, or share electrons to achieve a stable electron configuration characterized by eight valence electrons.* To obtain the Lewis structure for F_2, for example, we start with the Lewis dot symbol for a fluorine atom. Fluorine, in Group 7A, has seven valence electrons, so there are seven dots, one less than an octet. If each F atom in F_2 shares one valence electron with the other F atom to form a shared electron pair, a single covalent bond forms, and each fluorine atom achieves an octet. *Shared electrons are counted with each of the atoms in the bond.*

The term "lone pairs" will be used in this text to refer to unshared electron pairs.

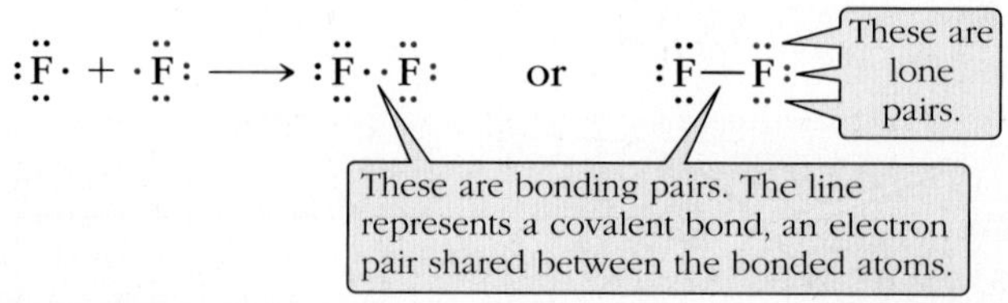
or
These are lone pairs.
These are bonding pairs. The line represents a covalent bond, an electron pair shared between the bonded atoms.

Diamagnetism (← Sec. 5-8a) of substances like F_2 indicates that each electron in a bonding or lone pair of electrons has its spin opposite that of the other electron.

A Lewis structure such as that for F_2 shows valence electrons in a molecule as **bonding electrons** (*shared electron pairs*) and **lone pair electrons** (*unshared pairs*). In a Lewis structure the atomic symbols, such as F, represent the nucleus and core (nonvalence) electrons of each atom in the molecule. The pair of electrons shared between the two fluorine atoms is a *bonding electron pair.* The other three pairs of electrons on each fluorine atom are *lone pair electrons.* In writing Lewis structures, the bonding pairs of electrons are usually indicated by lines connecting the atoms they hold together; lone pairs are usually represented by pairs of dots.

What about Lewis structures for molecules such as H_2O or NH_3? Oxygen (Group 6A) has six valence electrons and must share two electrons to satisfy the octet rule. This can be accomplished by forming covalent bonds with two hydrogen atoms.

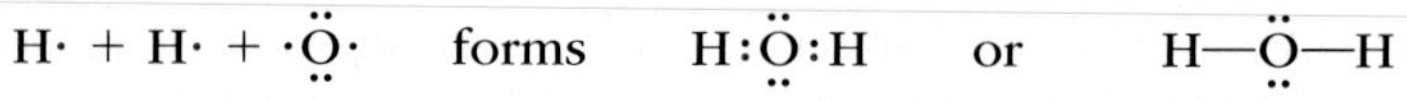
forms
or

Nitrogen (Group 5A) in NH_3 must share three electrons to achieve a noble-gas configuration, which can be done by forming covalent bonds with three hydrogen atoms.

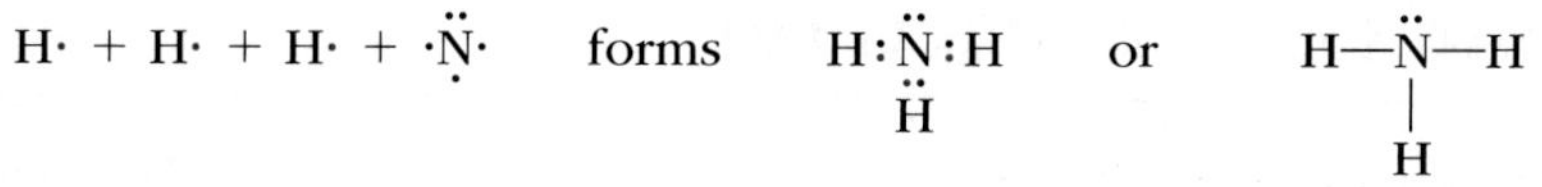
forms
or

Main-group elements are those in the groups labeled "A" in the periodic table inside the front cover of this book.

From the Lewis structures of F_2, H_2O, and NH_3, we can make an important generalization: *The number of electrons that an atom of a main-group element must share to achieve an octet equals eight minus its A-group number.* Carbon, for example, which is in Group 4A, needs to share four electrons to reach an octet.

Group Number	Number of Valence Electrons	Number of Electrons Shared to Complete an Octet	Example	Lewis Structure
4A	4	(8 − A-group number) = 8 − 4 = 4	C in CH_4	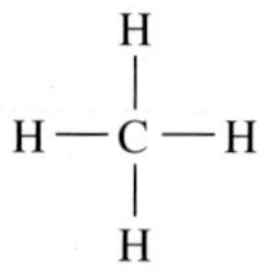
5A	5	8 − 5 = 3	N in NF_3	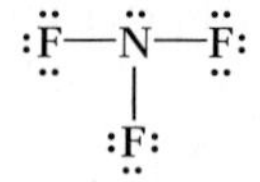
6A	6	8 − 6 = 2	O in H_2O	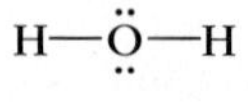
7A	7	8 − 7 = 1	F in HF	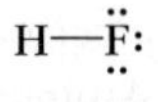

Many essential biochemical molecules contain carbon, hydrogen, oxygen, and nitrogen atoms. The structures of these molecules are dictated by the number of bonds C, O, and N need to complete their octets, as pointed out in the table. For example, glycerol, $C_3H_8O_3$,

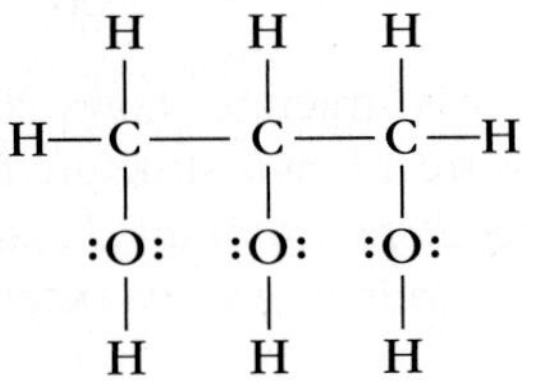

glycerol

is vital to the formation and metabolism of fats. In glycerol, each carbon atom has four shared pairs to complete an octet; each oxygen atom completes an octet by sharing two pairs and having two lone pairs; and each hydrogen atom achieves a $1s^2$ configuration by sharing a single pair of electrons. The same type of electron sharing occurs in deoxyribose, $C_5H_{10}O_4$, an essential component of DNA (deoxyribonucleic acid). Four of the carbon atoms are joined in a ring with an oxygen atom.

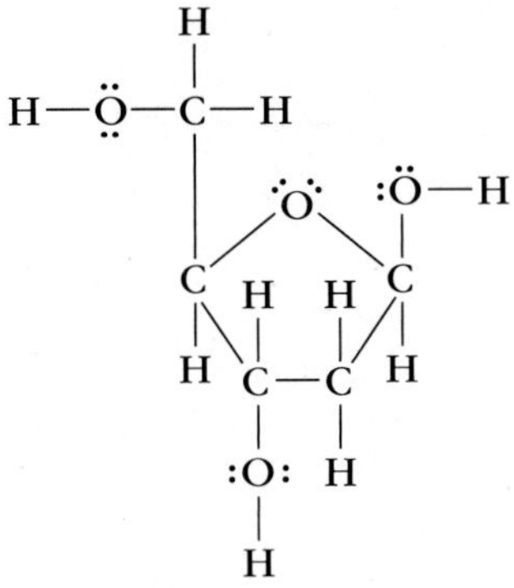

deoxyribose

6-2a Guidelines for Writing Lewis Structures

Guidelines have been developed for writing Lewis structures correctly, and we will illustrate using these guidelines to write the Lewis structure for PCl_3.

1. *Count the total number of valence electrons in the molecule or ion.* Use the A-group number in the periodic table as a guide to indicate the number of valence electrons in each atom. The total of the A-group numbers equals the total number of valence electrons of the atoms in a neutral molecule. For a negative ion, add electrons equal to the ion's charge. For a positive ion, subtract the number of electrons equal to the charge. For example, add one electron for the negative charge of OH^-, total of eight valence electrons: $6 + 1 + 1 = 8$; subtract one electron for the positive charge of NH_4^+, total of eight valence electrons: $5 + (4 \times 1) - 1 = 8$.

   Because PCl_3 is a neutral molecule, its number of valence electrons is five for P (it is in Group 5A) and seven for *each* Cl (chlorine is in Group 7A): Total number of valence electrons $= 5 + (3 \times 7) = 26$.

2. *Use atomic symbols to draw a skeleton structure by joining the atoms with shared pairs of electrons (a single line for each shared pair).* A skeleton structure indicates the attachment of terminal atoms to a central atom. The central atom is usually the one written first in the molecular formula and is the one that can form the most bonds, such as Si in $SiCl_4$ and P in PO_4^{3-}. Hydrogen, oxygen, and the halogens are often terminal atoms.

   In PCl_3, the central atom is phosphorus, so we draw a skeleton structure with P as the central atom and the three terminal chlorine atoms arranged around it. The three bonding pairs account for six of the total of 26 valence electrons counted in Step 1.

Although H is given first in the formulas of H_2O and H_2O_2, for example, it is not the central atom. H is never the central atom in a molecule or ion because a hydrogen atom forms only one covalent bond.

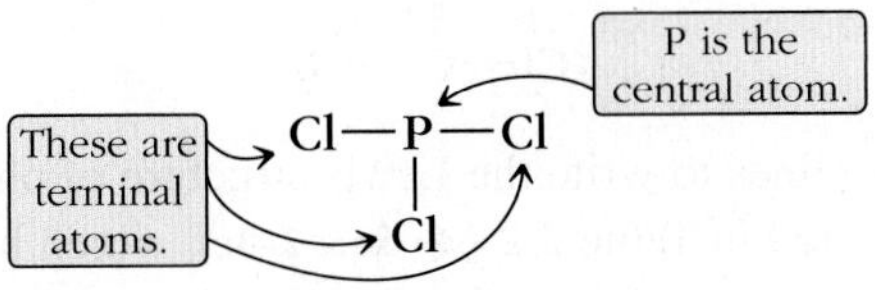
P is the central atom.
These are terminal atoms.

3. *Place lone pairs of electrons around each atom (except H) to satisfy the octet rule, starting with the terminal atoms.* Using lone pairs in this way on the Cl atoms accounts for 18 of the remaining 20 valence electrons, leaving two electrons for a lone pair on the P atom. When you check for octets, remember that shared electrons are counted as "belonging" to *each* of the atoms bonded by the shared pair. Thus, each

P—Cl bond has two shared electrons that count for phosphorus and also count for chlorine.

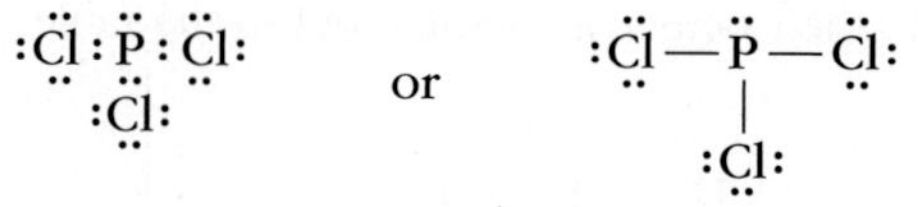

Counting dots and lines in this Lewis structure shows 26 electrons, accounting for all valence electrons. This is the correct Lewis structure for PCl_3.

The next two steps apply when Steps 1 through 3 result in a structure that does not use all the valence electrons or fails to give an octet of electrons to each atom that should have an octet.

4. *Place any remaining electrons on the central atom, even if it will give the central atom more than an octet.* If the central atom is from the third period, or further down in the periodic table, it is large enough to accommodate more than an octet of electrons (see Section 6-10). Sulfur tetrafluoride, SF_4, is such a molecule. It has a total of 34 valence electrons—six for S (Group 6A) plus seven for each F (Group 7A). In the Lewis structure, each F atom has an octet of electrons giving a total of 32. The remaining two valence electrons are a lone pair on the central S atom, which, as a third-period element, can accommodate more than an octet of electrons.

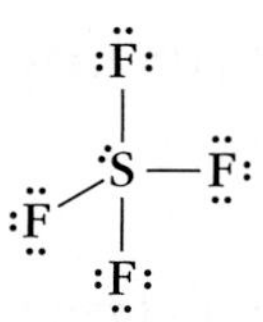

5. *If the number of electrons around the central atom is less than eight, change single bonds to the central atom to multiple bonds.* Some atoms can share more than one pair of electrons, resulting in a double covalent bond (two shared pairs) or a triple covalent bond (three shared pairs), known as *multiple bonds.* Where multiple bonds are needed to complete an octet, use one or more lone pairs of electrons from the terminal atoms to form *double* (two shared pairs) or *triple* (three shared pairs) covalent bonds until the central atom and all terminal atoms have octets. This guideline is not needed for PCl_3, but is illustrated here with $COCl_2$.

Double bond: two shared pairs of electrons, as in C═C.

Triple bond: three shared electron pairs, as in C≡N.

There are 24 valence electrons in $COCl_2$—four for carbon (Group 4A), six for oxygen (Group 6A), plus seven for each chlorine (Group 7A). However, these 24 valence electrons are not enough to give each atom an octet of electrons using just single bonds and lone pairs (Guideline 3) as seen from the skeleton structure for $COCl_2$.

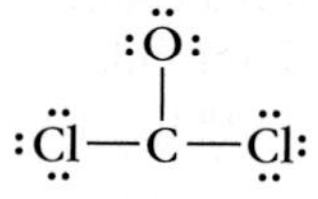

In this structure, carbon does not have an octet. We use one of the lone pairs on oxygen as a shared pair with carbon, which changes the C—O bond to a C═O double bond giving carbon an octet (Guideline 5).

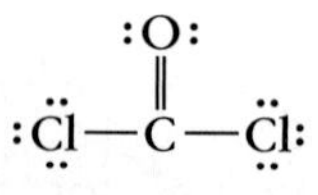

We apply these guidelines to write the Lewis structure of phosphate ion, PO_4^{3-}, one of the polyatomic ions listed in Table 2.2 (← **Sec. 2-4c**), which has 32 valence electrons (five from P, six from each O, and three for the 3− charge). Phosphorus is the central atom and oxygens are the terminal atoms, giving a skeleton structure of

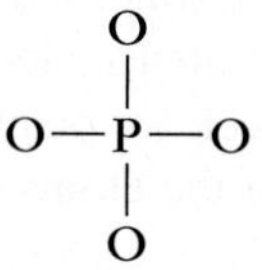

The four single P—O bonds account for eight of the valence electrons and provide phosphorus with an octet of electrons. The remaining 24 valence electrons are distributed as three lone pairs around each oxygen to complete their octets.

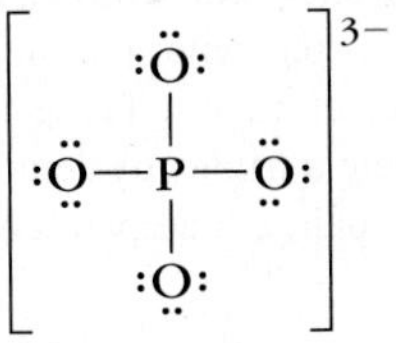

The brackets indicate that the polyatomic phosphate ion, consisting of one phosphorus atom and four oxygen atoms, collectively has a net charge of 3−. Such brackets are used with other polyatomic ions as well.

This is the correct Lewis structure for phosphate ion.

PROBLEM-SOLVING EXAMPLE 6.1

Lewis Structures

Write Lewis structures for these molecules or ions.

(a) HOCl (b) ClO_3^- (c) SF_6

(d) ClF_4^+ (e) HCN

Result

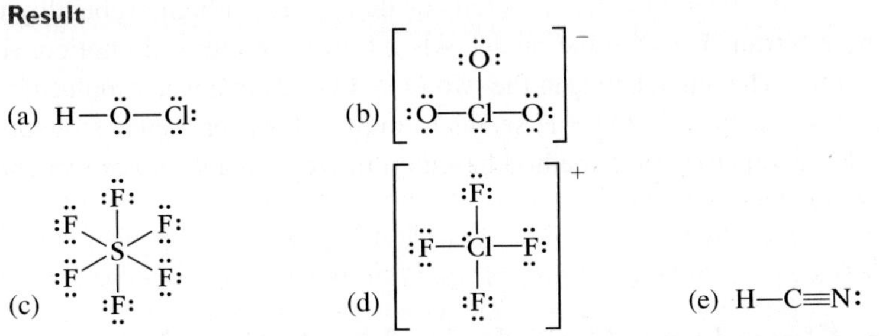

Analyze A Lewis structure involves sharing pairs of electrons to achieve a noble-gas electron configuration for all atoms, if possible.

Plan Apply the guidelines given previously. Total the valence electrons for all atoms and distribute them so that each atom has a noble-gas configuration. The central atom is usually the one written first in the formula and is the one that can form the most bonds; H is never the central atom.

Execute

(a) HOCl: There are 14 valence electrons—one from hydrogen (Group 1A), six from oxygen (Group 6A), and seven from chlorine (Group 7A). The central atom is O and the skeleton structure is H—O—Cl. This arrangement uses four of the 14 valence electrons. Use the remaining ten valence electrons by placing three lone pairs on the chlorine and two lone pairs on the oxygen. Oxygen and chlorine have octets (Ne configuration) and hydrogen has two electrons.

(b) ClO_3^-: There are 26 valence electrons—seven from chlorine and six from each oxygen, plus one for the 1− charge of the ion. In the skeleton structure, chlorine is the central atom with the three oxygen atoms each single bonded to it. This uses six of the 26 valence electrons, leaving 20 for lone pairs. Placing three lone pairs on each oxygen and a lone pair on chlorine satisfies the octet rule for each atom and uses the remaining valence electrons.

(c) SF_6: Sulfur forms more bonds than fluorine does, so sulfur is the central atom. There are 48 valence electrons: six from sulfur (Group 6A) and seven from each fluorine (Group 7A) for a total of 42 from fluorine. Sulfur is in Period 3 and can accommodate more than an octet, allowing for six bond pairs with fluorine atoms. The remaining 36 valence electrons are distributed as three lone pairs around each of the six fluorine atoms.

(d) ClF_4^+: Chlorine is first in the formula and is the central atom. There are 34 valence electrons: seven from chlorine, seven for each fluorine, and one subtracted for the

1+ charge. There are 8 electrons (4 pairs) in the four Cl—F bonds. The remaining 34 − 8 = 26 valence electrons are distributed as three lone pairs on each fluorine and one lone pair on the chlorine, a Period 3 element that can accommodate more than an octet of electrons.

(e) HCN: There are 10 valence electrons—one from H, four from C, and five from N. H cannot be a central atom, so C is the central atom in the skeleton structure H—C—N, which uses four of the ten valence electrons. Three electron pairs remain. Share two pairs between N and C to complete a triple covalent bond, C≡N, giving C an octet (Guideline 5). The other electron pair, a lone pair, completes an octet on N.

H—C—N̈: gives H—C≡N:

PROBLEM-SOLVING PRACTICE answers are provided at the back of this book in Appendix K.

PROBLEM-SOLVING PRACTICE 6.1

Write Lewis structures for (a) NF_3, (b) N_2H_4, and (c) ClO_4^-.

Although Lewis structures are useful for predicting the number of covalent bonds an atom will form, they do not give an accurate representation of where electrons are located in a molecule. Bonding electrons do not stay in fixed positions between nuclei, as Lewis's dots might imply. Instead, quantum mechanics tells us that there is a high probability of finding the bonding electrons between the nuclei. Also, Lewis structures do not convey the shapes of molecules. The angle between the two O—H bonds in a water molecule is not 180°, as the Lewis structure H—Ö—H seems to imply. However, Lewis structures can be used to predict geometries by a method based on the repulsions between valence shell electron pairs (see Section 7-1).

6-3 Single Covalent Bonds in Hydrocarbons

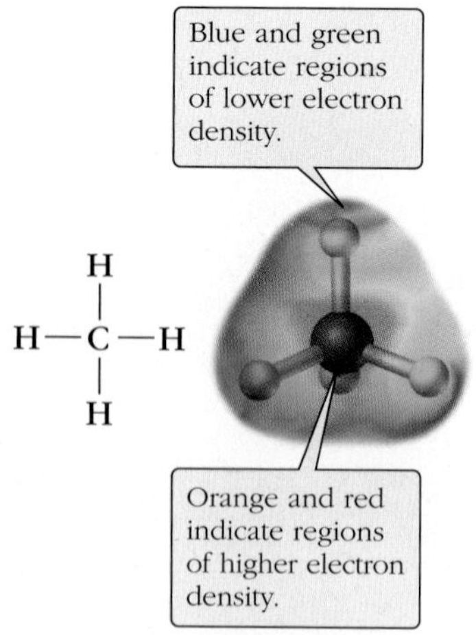

Blue and green indicate regions of lower electron density.
Orange and red indicate regions of higher electron density.

Lewis structure and electron density model of methane.

In hydrocarbons (← **Sec. 2-9**), carbon's four valence electrons are shared with hydrogen atoms or other carbon atoms. In methane, CH_4, the simplest hydrocarbon, the four valence electrons are shared with electrons from four hydrogen atoms, forming four single covalent bonds. The bonding in methane can be represented by a Lewis structure and an *electron density model.* In an electron density model, a ball-and-stick model is surrounded by a space-filling model that represents the distribution of electron density on the surface of the molecule.

Carbon is unique among the elements because of the ability of its atoms to form strong bonds with one another as well as with atoms of hydrogen, oxygen, nitrogen, sulfur, and the halogens. The strength of the carbon-carbon bond permits long chains to form:

—C—C—C—C—C—C—C—C—C—C—C—C—

Review the discussion on alkanes in Chapter 2 (← Sec. 2-9). See Table 2.8 for a list of selected alkanes.

Because each carbon atom can form four covalent bonds, such chains contain numerous sites to which other atoms (including more carbon atoms) can bond, leading to isomeric structures (← **Sec. 2-9a**) and the great variety of carbon compounds.

Hydrocarbons contain only carbon and hydrogen atoms. An alkane contains only C—H and C—C single covalent bonds and is often referred to as a **saturated hydrocarbon** because *each carbon is bonded to a maximum number of hydrogen atoms*

(← **Sec. 2-9**). The carbon atoms in alkanes with four or more carbon atoms per molecule can be arranged in either a straight chain or a branched chain (← **Sec. 2-9a**).

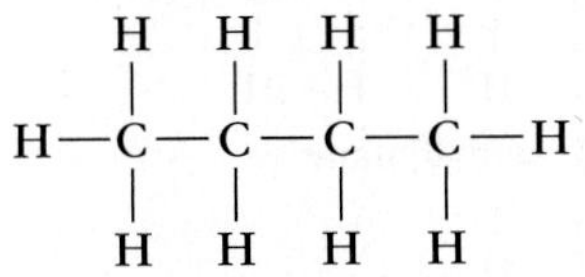

butane

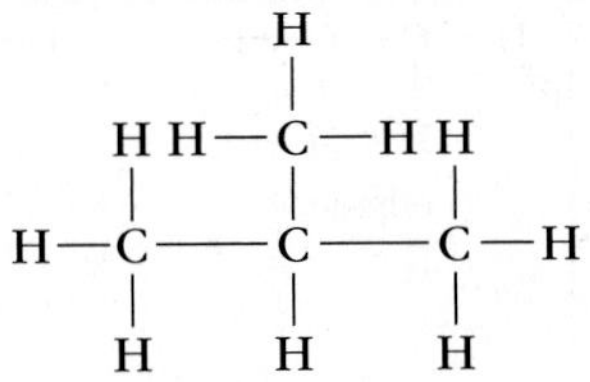

2-methylpropane

Rules for naming organic compounds are given in Appendix E.

In addition to straight-chain and branched-chain alkanes, there are *cycloalkanes,* saturated hydrocarbon compounds consisting of carbon atoms joined in rings of $—CH_2—$ units. The simplest cycloalkane is cyclopropane; other common cycloalkanes include cyclobutane, cyclopentane, and cyclohexane.

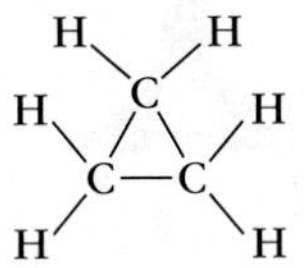

cyclopropane

or

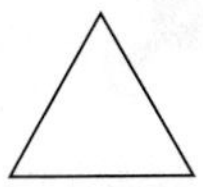

cyclopropane
C_3H_6

cyclobutane
C_4H_8

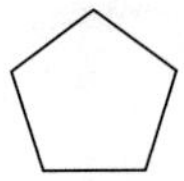

cyclopentane
C_5H_{10}

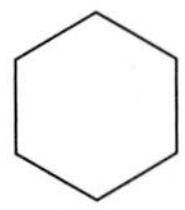

cyclohexane
C_6H_{12}

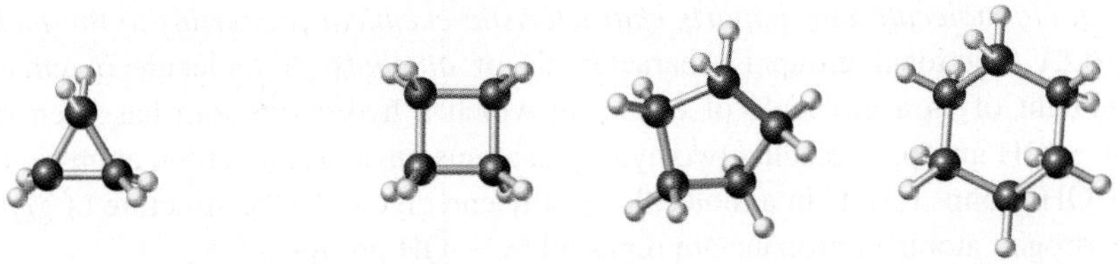

The cycloalkanes shown above are drawn using a common convention, called a **line drawing** or **line structure**, in which *only bonds between carbon atoms are shown.* For example, each line in the drawing of cyclohexane represents a C—C bond. It is assumed that you understand that at the intersection of two lines there is a carbon atom and each carbon atom forms four bonds. Therefore, neither hydrogen atoms nor bonds from carbon to hydrogen atoms are shown. You have to fill in hydrogen atoms and C—H bonds for yourself. To show how this works, cyclopropane is represented both as a drawing with all bonds and atoms showing and as a line drawing in which only C—C bonds are shown.

EXERCISE 6.1

Representing Structures of Saturated Hydrocarbons

Write the Lewis structure, the line structure, and the molecular formula for each compound: (a) cyclooctane; (b) octane.

Answers to **EXERCISES** are provided at the back of this book in Appendix L.

The great variety of organic compounds can also be accounted for by the fact that one or many carbon-hydrogen bonds in hydrocarbons can be replaced by bonds between carbon and other atoms. For example, the new bonds to carbon can connect to individual halogen atoms, thus creating entirely different compounds. Consider the new substances that result when a chlorine atom replaces a hydrogen atom in ethane, 2-methylbutane, and cyclopropane.

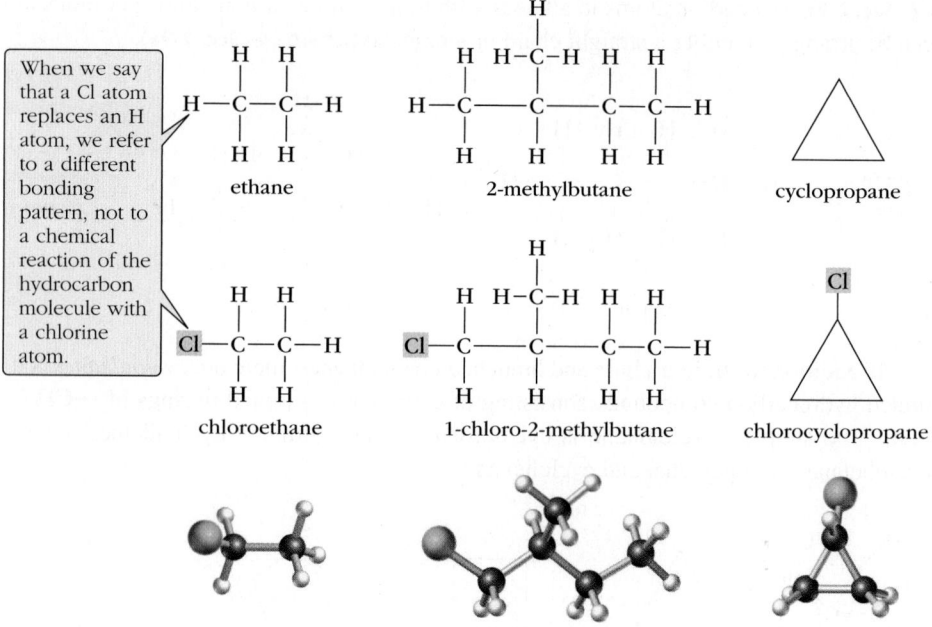

When we say that a Cl atom replaces an H atom, we refer to a different bonding pattern, not to a chemical reaction of the hydrocarbon molecule with a chlorine atom.

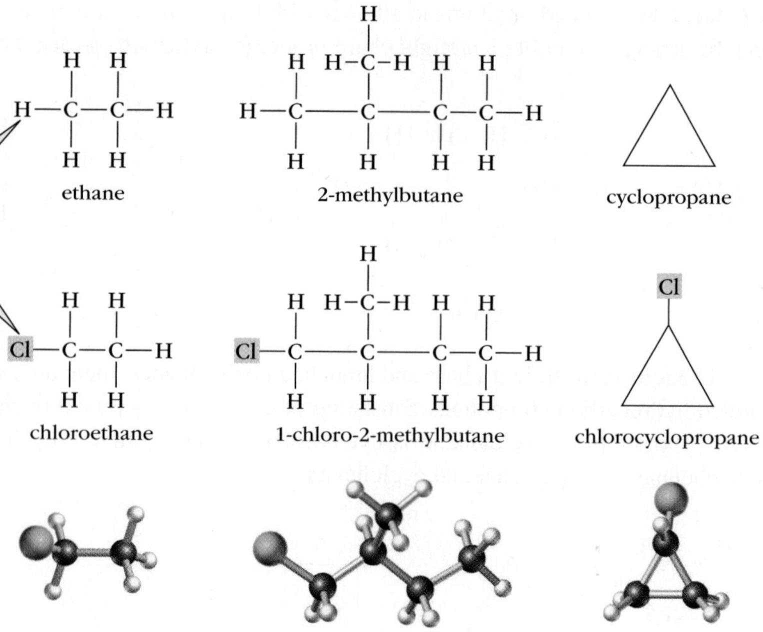
ethane
2-methylbutane
cyclopropane
chloroethane
1-chloro-2-methylbutane
chlorocyclopropane

The —OH group in an alcohol is not an ion; it is different from the OH^- ion in a base.

Chapter 10 describes additional functional groups. Appendix E-2 includes a list of functional groups.

In another case, consider how an —OH functional group can replace one or more hydrogen atoms in an alkane. A **functional group** is *a distinctive group of atoms in an organic molecule that imparts characteristic chemical properties to the molecule*. The —OH functional group is characteristic of *alcohols*. A molecule of ethanol can be thought of as a molecule of ethane in which a hydrogen atom has been replaced by an —OH group. Replacing two hydrogen atoms on adjacent carbon atoms in ethane with —OH groups results in a molecule of ethylene glycol. In the structure of glycerol, three hydrogen atoms in propane are replaced by —OH groups.

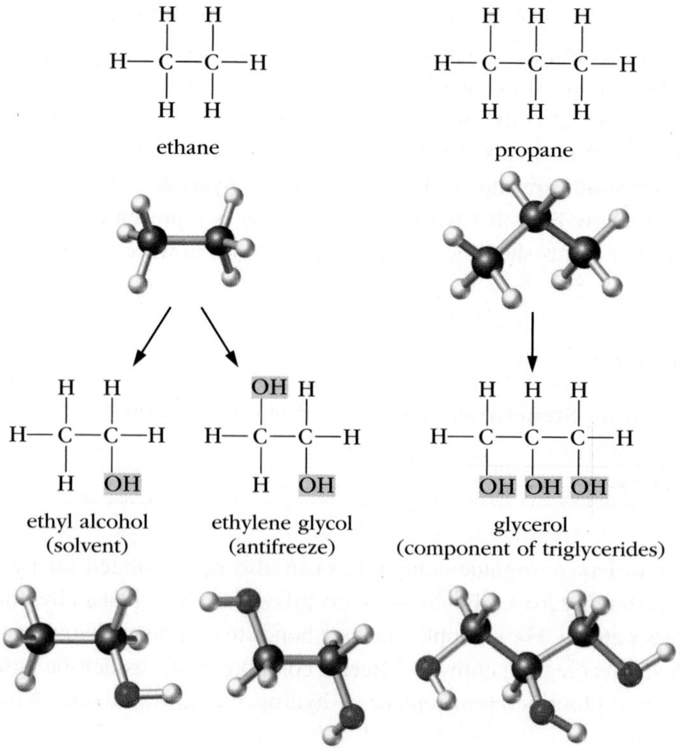
ethane
propane
ethyl alcohol
(solvent)
ethylene glycol
(antifreeze)
glycerol
(component of triglycerides)

PROBLEM-SOLVING EXAMPLE 6.2

Structural Formulas and Cl Substitution

Three hydrogens in propane can be replaced with three Cl atoms to form five different structures. Draw structural formulas for two of five possible compounds.

Result

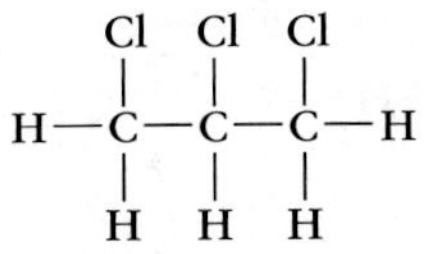

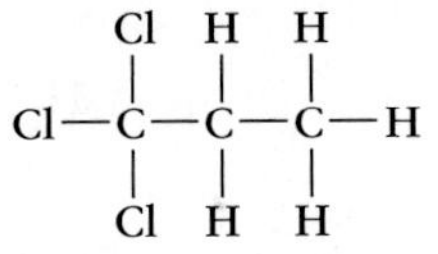

Analyze The three chlorine atoms can be put individually, one on each carbon atom. Alternatively, there can be one, two, or three chlorine atoms on a terminal carbon atom; the middle carbon can have one or two chlorine atoms on it.

Plan Place the chlorine atoms so that the resulting structures are different.

Execute Two of the possibilities are to place one chlorine atom on each carbon atom and to place all three chlorine atoms on a terminal carbon atom.

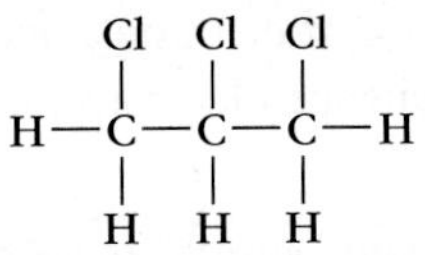

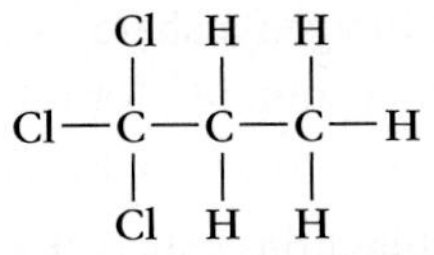

When structural formulas are drawn lone pairs of electrons that would have been included in a Lewis structure are often omitted. This has been done in this Problem-Solving Example.

PROBLEM-SOLVING PRACTICE 6.2

There are three other possible structures for three Cl atoms on a three-carbon chain. Draw these structures.

6-4 Multiple Covalent Bonds

A nonmetal atom with fewer than seven valence electrons can form covalent bonds in more than one way. The atom can share a single electron with another atom that also contributes a single electron; this process forms a shared electron pair—a *single* covalent bond. But some atoms can also share two or three pairs of electrons with another atom, in which case there will be two or three bonds, respectively, between the two atoms. *A bond where* two *shared pairs of electrons join the same two atoms* is called a **double bond**. *A bond where* three *shared pairs join the same two atoms* is called a **triple bond**. *Any bond with more than a single pair of electrons* is referred to as a **multiple covalent bond**. For example, nitrous acid, HNO_2, contains an N=O double bond, and hydrogen cyanide, HCN, contains a C≡N triple bond.

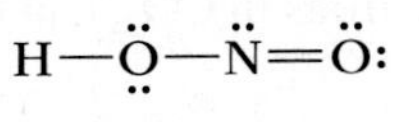

nitrous acid

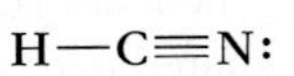

hydrogen cyanide

Guideline 5 (← **Sec. 6-2a**) says that, in molecules where there are not enough electrons to complete all octets using only single bonds, one or more lone pairs of electrons from the terminal atoms can be shared with the central atom to form double or triple bonds, so that all atoms have octets of electrons. Let's apply this guideline to the Lewis structure for formaldehyde, H_2CO. There are 12 valence electrons (Guideline 1): two from two H atoms (Group 1A), four from the C atom (Group 4A), and six from the O atom (Group 6A). To complete noble-gas configurations, H should form one bond, C four bonds, and O two bonds. Because C forms the most bonds, it is the central atom, and we can write this skeleton structure (Guideline 2).

You might have written the skeleton structure O—C—H—H, but remember that H forms only one bond. Another possible skeleton is H—O—C—H, but with this skeleton it is impossible to achieve an octet around carbon without having more than two bonds to oxygen.

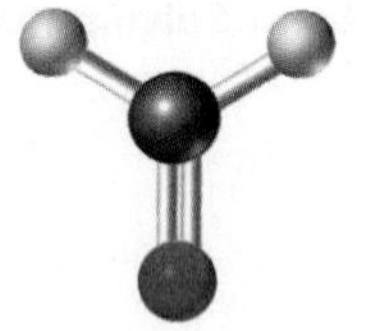

Formaldehyde

Formaldehyde. A gas at room temperature, formaldehyde is the simplest compound with an aldehyde functional group.

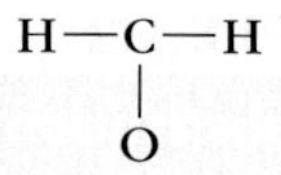

Putting bonding pairs and lone pairs in the skeleton structure according to Guideline 3 yields a structure using all 12 valence electrons in which oxygen has an octet, but carbon does not.

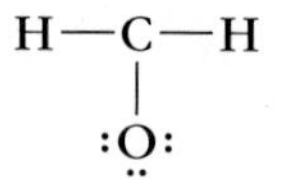

We use one of the lone pairs on oxygen as a shared pair with carbon to change the C—O single bond to a C=O double bond (Guideline 5).

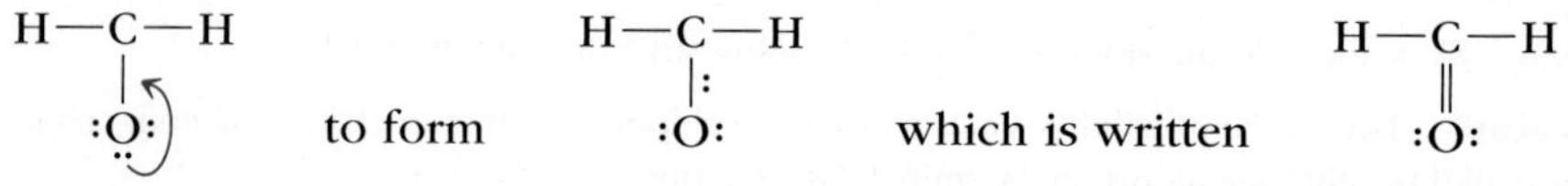

This gives carbon and oxygen a share in an octet of electrons, and each hydrogen has a share of two electrons, accounting for all 12 valence electrons and verifying that this is the correct Lewis structure for formaldehyde.

The >C=O combination, called the *carbonyl group,* is part of several functional groups that are very important in organic and biochemical molecules (see Section 10-4). The carbonyl-containing 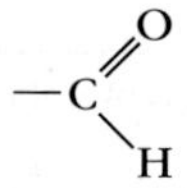 group that appears in formaldehyde is known as the *aldehyde functional group*.

As another example of multiple bonds, let's write the Lewis structure for molecular nitrogen, N_2. There are ten valence electrons (five from each N). If two lone pairs of electrons (one pair from each N) become bonding pairs to give a triple bond, the octet rule is satisfied. This is the correct Lewis structure of N_2.

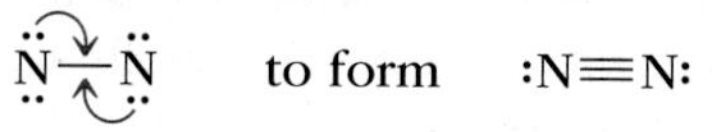

CONCEPTUAL EXERCISE 6.2

Lewis Structures

Why is :N̈—N̈: an incorrect Lewis structure for N_2?

EXERCISES that are labeled **CONCEPTUAL** are designed to test your understanding of one or more concepts; they usually involve qualitative rather than quantitative thinking.

A molecule can have more than one multiple bond, as in carbon dioxide, where carbon is the central atom. There are 16 valence electrons in CO_2, and the skeleton structure uses four of them (two shared pairs):

O—C—O

Adding lone pairs to give each O an octet of electrons uses up the remaining 12 electrons, but leaves C needing four more valence electrons to complete an octet.

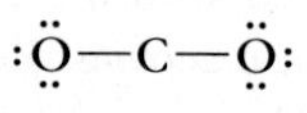

With no more valence electrons available, the only way that carbon can have four more valence electrons is to use one lone pair of electrons on each oxygen to form a covalent bond to carbon to give it an octet. In this way the 16 valence electrons are accounted for, and each atom has an electron octet.

PROBLEM-SOLVING EXAMPLE 6.3

Lewis Structures

Write Lewis structures for (a) carbon monoxide, CO, an air pollutant; (b) nitrosyl chloride, ClNO, an unstable solid; (c) N_3^-, a polyatomic ion; and (d) acetylene, C_2H_2, a fuel.

Result

(a) :C≡O: (b) :C̤̈l—N̈=Ö: (c) $[\ddot{N}=N=\ddot{N}]^-$ (d) H—C≡C—H

Analyze A Lewis structure involves sharing pairs of electrons to achieve a noble-gas electron configuration for all atoms.

Plan Follow the guidelines given previously to write proper Lewis structures (← **Sec. 6-2a**). Assign the valence electrons to atoms in each molecule or polyatomic ion. Where appropriate, use lone pairs for multiple bonds to achieve an octet.

Execute

(a) The molecule CO contains ten valence electrons (four from C and six from O). Start with this skeleton structure, which uses one electron pair.

C—O

Distributing the remaining eight electrons as lone pairs around the C and O atoms satisfies the octet rule for one of the atoms, but not both. Change lone pairs to bond pairs to achieve an appropriate Lewis structure.

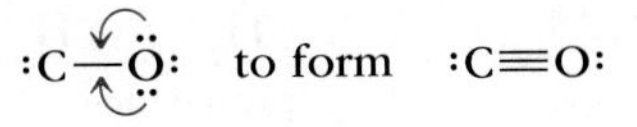

(b) In ClNO, nitrogen is the central atom (it can form more bonds than can oxygen or chlorine) and there are 18 valence electrons (five from N, six from O, and seven from Cl). The skeleton structure is

Cl—N—O

Placing lone pairs on the terminal atoms and a lone pair on the N atom uses the remaining valence electrons, but leaves the N atom without an octet.

:C̤̈l—N̈—Ö̤:

Converting a lone pair on the oxygen to a bonding pair results in the correct Lewis structure, with each atom having an octet of electrons.

:C̤̈l—N̈=Ö:

(c) There are 16 valence electrons in N_3^- (15 from three N atoms and one for the ion's 1− charge). The skeleton structure with lone pairs on the outer two nitrogen atoms uses all 16 valence electrons.

a plausible Lewis structure.

(d) Acetylene has ten valence electrons (four from each C and one from each H).

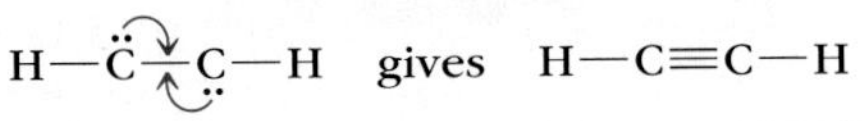

PROBLEM-SOLVING PRACTICE 6.3

Write Lewis structures for (a) nitrosyl ion, NO^+; (b) silicon sulfide dichloride, $SiSCl_2$.

CONCEPTUAL EXERCISE 6.3

Lewis Structures

Which of these are appropriate Lewis structures and which are not? Explain what is wrong with the incorrect ones.

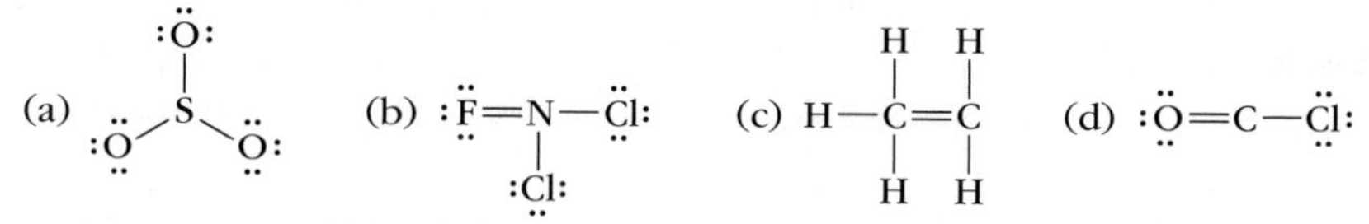
(a)
(b)
(c)
(d)

6-5 Multiple Covalent Bonds in Hydrocarbons

In many compounds, carbon atoms are connected by double bonds or triple bonds. An **alkene** is *a hydrocarbon that has one or more carbon-carbon double bonds,* C═C. The general formula for alkenes with one double bond is C_nH_{2n}, where n = 2, 3, 4, and so on. The first two members of the alkene series are ethene, CH_2═CH_2, and propene, CH_3CH═CH_2; they are commonly called ethylene and propylene.

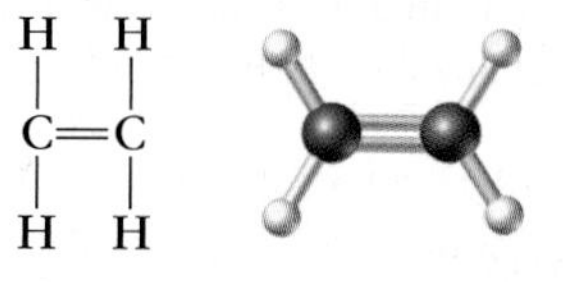
ethylene

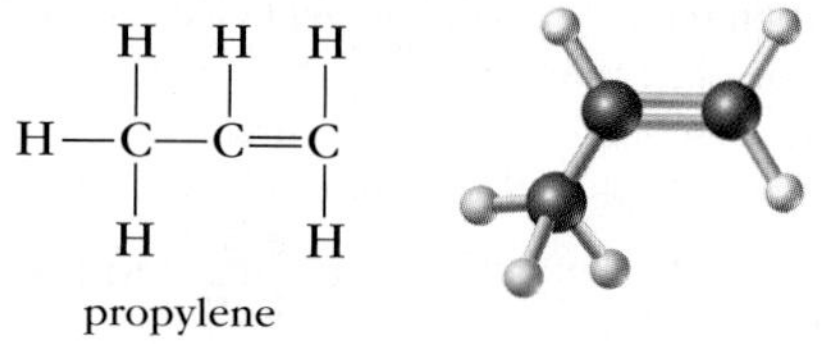
propylene

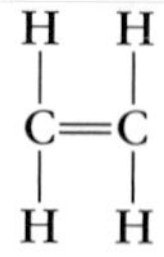
unsaturated

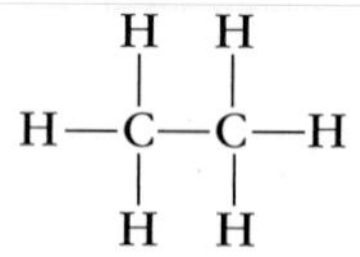
saturated

Every alkene is an **unsaturated hydrocarbon**—*it contains fewer hydrogen atoms than the corresponding alkane* (ethene, CH_2═CH_2; ethane, CH_3—CH_3). Carbon atoms connected by double bonds are the *unsaturated sites.*

Alkenes are named by using the name of the corresponding alkane (← **Sec. 2-9**) to indicate the number of carbons and the suffix *-ene* to indicate one or more double bonds. The first member, ethene (ethylene), is the most important raw material in the organic chemical industry, where it is used in making polyethylene, antifreeze (ethylene glycol), ethanol, and other chemicals.

CONCEPTUAL EXERCISE 6.4

Alkenes

(a) Write the molecular formula and structural formula of an alkene with five carbon atoms and one C═C double bond.

(b) How many different alkenes have five carbon atoms and one C═C double bond?

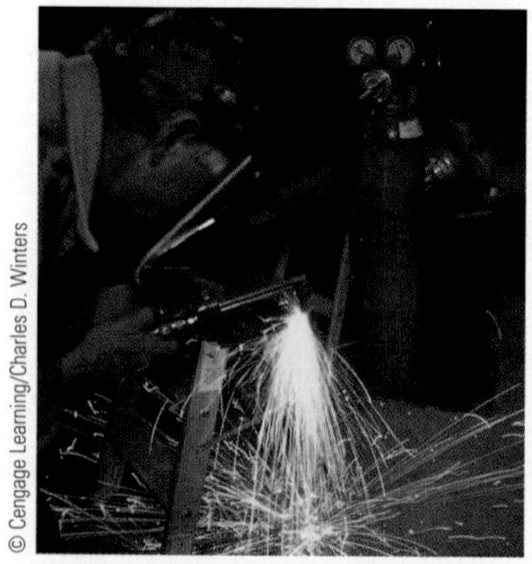
© Cengage Learning/Charles D. Winters

An oxyacetylene torch cutting steel. A mixture of acetylene and oxygen burns with a flame hot enough (3100 °C) to cut steel.

A hydrocarbon with one or more triple bonds, —C≡C—, *per molecule* is an **alkyne**. The general formula for alkynes with one triple bond is C_nH_{2n-2}, where n = 2, 3, 4, and so on. The simplest alkyne is ethyne, commonly called acetylene, C_2H_2.

H—C≡C—H

acetylene

6-5a Double Bonds and Isomerism

The C═C double bond creates an important difference between alkanes and alkenes—the degree to which one end of a carbon-carbon bond can rotate with respect to the other end. The C—C single bonds in alkanes allow the carbon atoms to rotate freely around

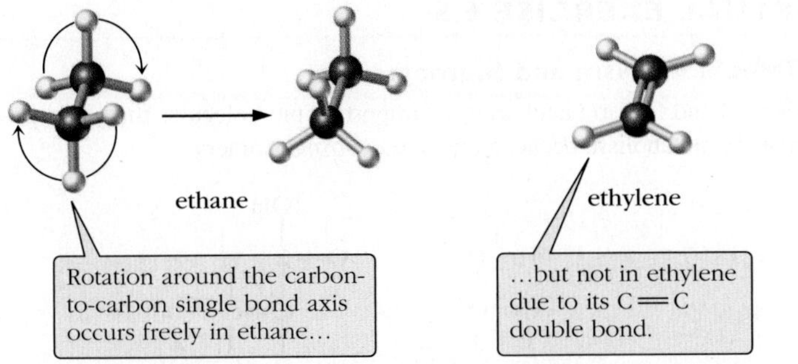
ethane
ethylene
Rotation around the carbon-to-carbon single bond axis occurs freely in ethane...
...but not in ethylene due to its C=C double bond.

Figure 6.3 Rotation is restricted around a carbon-carbon double bond.

the C—C bond axis (Figure 6.3). But in alkenes, the C=C double bond prevents such free rotation. This limitation is responsible for the *cis-trans* isomerism of alkenes.

Two or more compounds with the same molecular formula but different arrangements of atoms are known as *isomers* (← **Sec. 2-9a**). ***Cis-trans*** **isomerism** (also called **geometric isomerism**) *occurs when molecules differ in their arrangement of atoms on either side of a C=C double bond because there is no rotation around the C=C double bond.* As shown in Figure 6.4, when *two atoms or groups of atoms are attached on the same side of the C=C bond*, the compound is the ***cis*** **isomer**; the groups or atoms are said to be *cis* to each other. *When two atoms or groups of atoms are on opposite sides*, the compound is the ***trans*** **isomer**; the groups are *trans* to each other. An example is the *cis* and *trans* isomers of ClHC=CHCl, 1,2-dichloroethene. Because of their different geometries, the two isomers have different physical properties, including melting point, boiling point, and density as shown in Figure 6.4.

Cis-trans isomerism in alkenes is possible *only when each of the carbon atoms connected by the double bond has two different groups attached.* (For the sake of brevity, the word "groups" refers to both atoms and groups of atoms.) For example, two chlorine atoms can also bond to the *first* carbon to give 1,1-dichloroethene, which does not have *cis* and *trans* isomers because each carbon atom is attached to two identical atoms (one carbon to two chlorines, the other carbon to two hydrogens).

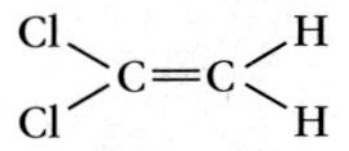

1,1-dichloroethene

The 1 and 2 indicate that the two chlorine atoms are attached to the first and second carbon atoms, respectively.

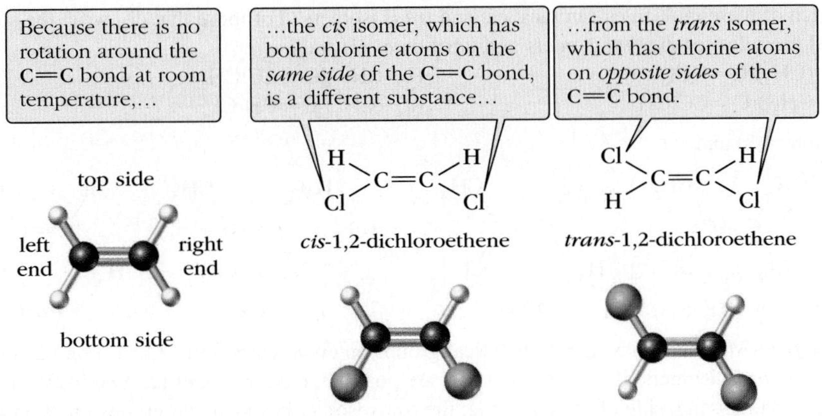
Because there is no rotation around the C=C bond at room temperature,...
...the cis isomer, which has both chlorine atoms on the same side of the C=C bond, is a different substance...
...from the trans isomer, which has chlorine atoms on opposite sides of the C=C bond.
top side
left end
right end
bottom side
cis-1,2-dichloroethene
trans-1,2-dichloroethene

Physical Property	*cis*-1,2-dichloroethene	*trans*-1,2-dichloroethene
Melting point	−80.0 °C	−49.8 °C
Boiling point (at 1 atm)	60.1 °C	48.7 °C
Density (at 25 °C)	1.284 g/mL	1.2565 g/mL

Figure 6.4 ***Cis*** **and** ***trans*** **isomers are different substances because of restricted rotation around C=C bonds.**

CONCEPTUAL EXERCISE 6.5

Cis-Trans Isomerism and Biomolecules

Maleic acid and fumaric acid are very important biomolecules that undergo different reactions in metabolism because they are *cis-trans* isomers.

maleic acid   fumaric acid

Identify the *cis* isomer and the *trans* isomer.

When there are four or more carbon atoms in an alkene, the possibility exists for *cis* and *trans* isomers even when only carbon and hydrogen atoms are present. For example, 2-butene has both *cis* and *trans* isomers. (The 2 indicates that the double bond is at the second carbon atom, with the straight carbon chain beginning with carbon 1.)

Physical Property	*cis*-2-butene	*trans*-2-butene
Melting point	−138.9 °C	−105.5 °C
Boiling point (at 1 atm)	3.7 °C	0.88 °C

PROBLEM-SOLVING EXAMPLE 6.4

Cis and *Trans* Isomers

Which of these molecules can have *cis* and *trans* isomers? For those that do, write the structural formulas for the two isomers and label them *cis* and *trans*.

(a) $(CH_3)_2C{=}CCl_2$
(b) $CH_3ClC{=}CClCH_3$
(c) $CH_3BrC{=}CClCH_3$
(d) $(CH_3)_2C{=}CBrCl$

Result (b) and (c)

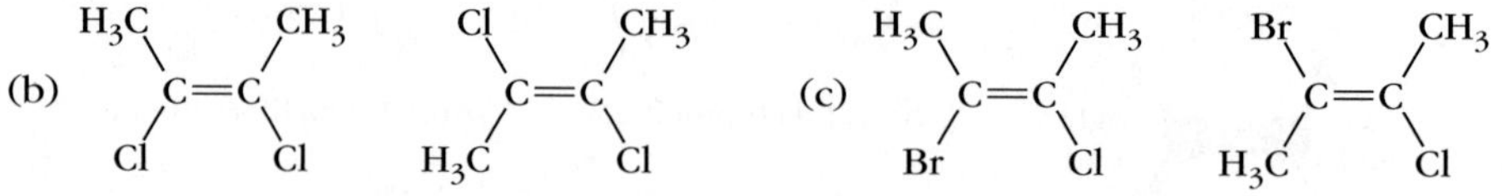

cis   *trans*   *cis*   *trans*

Analyze Molecules having two identical groups on either carbon in a C═C bond cannot have *cis-trans* isomers. If *cis-trans* isomers are possible, the *cis* isomer has two identical groups on the same side of the molecule; the *trans* isomer has identical groups on opposite sides of the molecule.

Plan Carefully consider each structure and apply the analysis just stated.

Execute

(a) Both carbons have two identical groups so *cis-trans* isomers are not possible.
(b) Each carbon has two different groups, CH_3 and Cl; the two CH_3 groups can both be on the same side of the C═C bond (*cis*) or on opposite sides (*trans*).
(c) The same argument as in (b) holds true for the CH_3 groups.
(d) There are two CH_3 groups on the same carbon, so *cis-trans* isomers are not possible.

PROBLEM-SOLVING PRACTICE 6.4

Which of these molecules can have *cis* and *trans* isomers? For those that do, write the structural formulas for the two isomers and label them *cis* and *trans*.

(a) 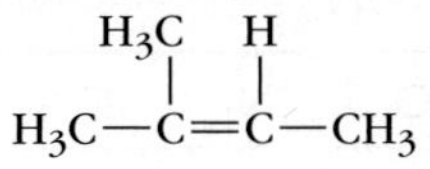

(b) 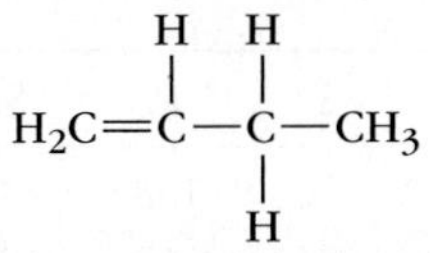

(c) 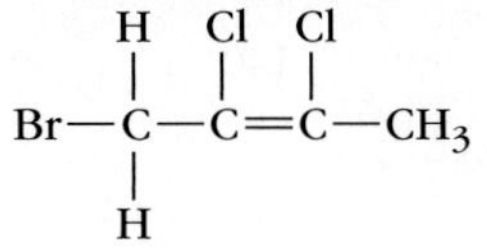

6-6 Bond Properties: Bond Length; Bond Energy

6-6a Bond Length

Bond length, *the distance between nuclei of two bonded atoms*, is determined primarily by the sizes of the atoms themselves (← **Sec. 5-9**). Assuming the atoms are spheres, the bond length is the sum of the atomic radii of the two bonded atoms (Figure 6.5). Bond lengths are given in Table 6.1. As expected, the bond length is greater for larger atoms. Figure 6.5 illustrates the change in atomic size across Periods 2 and 3, and down Groups 4A–6A. Thus, single bonds with carbon increase in length along the series:

Increase in bond length ⟶

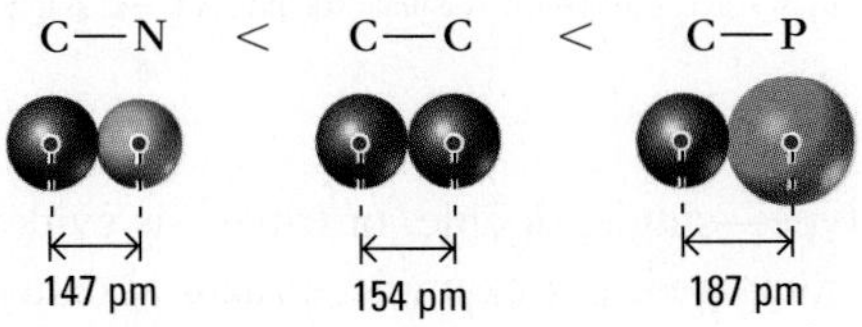
C—N < C—C < C—P
147 pm
154 pm
187 pm

In the case of multiple bonds, a C═O bond is shorter than a C═S bond because O is a smaller atom than S. Likewise, a C≡N bond is shorter than a C≡C bond because N is a smaller atom than C (Figure 6.5). Each of these trends can be predicted from the relative sizes shown in Figure 6.5 and is confirmed by the average bond lengths given in Table 6.1.

Because atomic radii are averages over a large number of bonds, the sum of atomic radii is not necessarily exactly equal to a particular bond length.

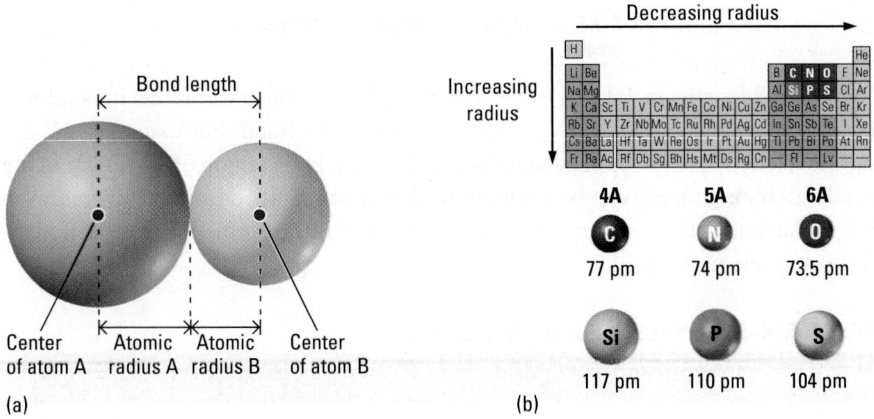
Bond length
Center of atom A
Atomic radius A
Atomic radius B
Center of atom B
(a)
Decreasing radius
Increasing radius
4A
5A
6A
C
N
O
77 pm
74 pm
73.5 pm
Si
P
S
117 pm
110 pm
104 pm
(b)

Figure 6.5 Bond length. (a) Bond length is the sum of atomic radii. (b) Radii depend on position in the periodic table: relative radii for some elements in Groups 4A, 5A, and 6A.

Table 6.1 Average Single and Multiple Bond Lengths (in picometers, pm)*

Single Bonds

	I	Br	Cl	S	P	Si	F	O	N	C	H
H	161	142	127	132	138	145	92	94	98	110	74
C	210	191	176	181	187	194	141	143	147	154	
N	203	184	169	174	180	187	134	136	140		
O	199	180	165	170	176	183	130	148			
F	197	178	163	168	174	181	128				
Si	250	231	216	221	227	234					
P	243	224	209	214	220						
S	237	218	203	208							
Cl	232	213	200								
Br	247	228									
I	266										

Multiple Bonds

N═N	120	C═C	134
N≡N	110	C≡C	121
C═N	127	C═O	122
C≡N	115	C≡O	113
O═O (in O_2)	121	N≡O	108
N═O	115		

*1 pm = 10^{-12} m. Bond lengths are given in picometers (pm) in this table, but many scientists use nanometers (1 nm = 10^3 pm) or the older unit of Ångströms (Å). 1 Å equals 100 pm. A C—C single bond is 0.154 nm, 1.54 Å, or 154 pm in length.

The effect of bond type—single, double, or triple—is evident when bonds between the same two atoms are compared. For example, structural data show that the bonds become shorter in the series C—O > C═O > C≡O. As the electron density between the atoms increases (number of bonds increases), the bond lengths decrease because the atoms are pulled together more strongly.

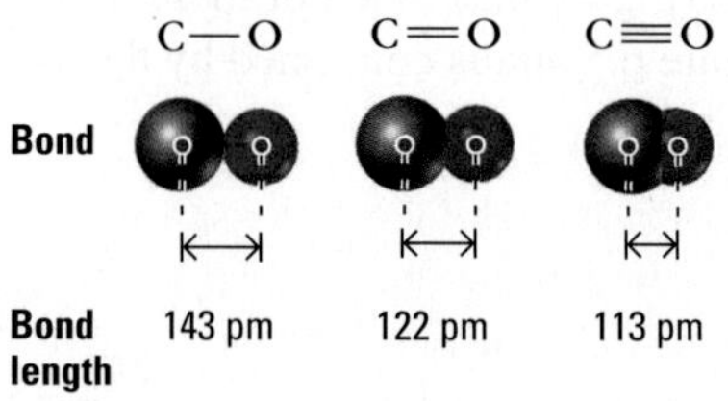
C—O
C═O
C≡O
Bond
Bond length
143 pm
122 pm
113 pm

The bond lengths in Table 6.1 are average values, because variations in neighboring parts of a molecule can affect the length of a particular bond. For example, the C—H bond has a length of 105.9 pm in acetylene, HC≡CH, but a length of 109.3 pm in methane, CH_4. Although there can be a variation of as much as 10% from the average values listed in Table 6.1, the average bond lengths are useful for estimating bond lengths and building models of molecules.

PROBLEM-SOLVING EXAMPLE 6.5

Bond Lengths

In each pair of bonds, predict which is shorter. Explain why.

(a) P—O or S—O (b) C≡C or C═C (c) C—S or C—Cl

Result The shorter bonds are (a) S—O, (b) C≡C, (c) C—Cl

Analyze Bond lengths depend first on atomic radii; when radii are similar, triple bonds are shorter than double bonds, which are shorter than single bonds.

Plan Locate each atom in the periodic table and apply periodic trends in radii.

Execute

(a) S—O is shorter than P—O because a P atom is larger than an S atom.
(b) C≡C is shorter than C=C because the more electrons that are shared by atoms, the more closely the atoms are pulled together.
(c) C—Cl is shorter than C—S because a Cl atom is smaller than an S atom.

PROBLEM-SOLVING PRACTICE 6.5

Explain the increasing order of bond lengths in these pairs of bonds.

(a) C—S is shorter than C—Si. (b) C—Cl is shorter than C—Br.
(c) N≡O is shorter than N=O.

6-6b Bond Enthalpies

In any chemical reaction, bonds are broken and new bonds are formed.

- ***Bond breaking* is an *endothermic* process that requires transfer of energy into a system.**
- ***Bond making* is an *exothermic* process in which energy is transferred out of a system.**

These changes contribute to the enthalpy change for the overall reaction (← **Sec. 4-7**). **Bond enthalpy (bond energy)** *is the enthalpy change that occurs when the bond between two bonded atoms in the gas phase is broken and the atoms are separated completely at constant pressure.* Data on the strengths of bonds between atoms in gas-phase molecules are summarized in Table 6.2. The quantity of energy released when 1 mol of a particular bond is made equals that needed when 1 mol of that bond is broken. For example, the H—Cl bond energy is 431 kJ/mol, indicating that 431 kJ must be supplied to break 1 mol H—Cl bonds ($\Delta H° = +431$ kJ). Conversely, when 1 mol H—Cl bonds is formed, 431 kJ is released ($\Delta H° = -431$ kJ).

The two terms "bond enthalpy" and "bond energy" are often used interchangeably, although they are not quite equal.

You have seen that as the number of bonding electrons between a pair of atoms increases (single to double to triple bonds), the bond length decreases. It is therefore reasonable to expect that multiple bonds are stronger than single bonds. *As the electron density between two atoms increases, the bond gets shorter and stronger.* For example, the bond energy of C=O in CO_2 is 803 kJ/mol and that of C≡O is 1073 kJ/mol. In fact, the C≡O triple bond in carbon monoxide is the strongest known covalent bond.

The data in Table 6.2 can help us understand why an element such as nitrogen, which forms many compounds with oxygen, is unreactive enough to remain in Earth's atmosphere as N_2 molecules even though there is plenty of O_2 to react with. Reactions in which N_2 combines with other elements are less likely to occur, because they require breaking a very strong N≡N bond (946 kJ/mol). This allows us to inhale and exhale N_2 without its undergoing any chemical change. If this were not the case and N_2 reacted readily at body temperature (37 °C) to form oxides and other compounds, there would be severe consequences for us. In fact, life on Earth as we know it would not be possible.

We can use data from Table 6.2 to estimate $\Delta_r H°$ for reactions such as the reaction of hydrogen gas with chlorine gas:

$$H_2(g) + Cl_2(g) \rightarrow 2\ HCl(g).$$

- Breaking the covalent bond of H_2 requires an *input* to the system of 436 kJ/mol.
- Breaking the covalent bond of Cl_2 requires an *input* to the system of 242 kJ/mol. In each of these cases the sign of the enthalpy change is *positive*.

The bond enthalpies in Table 6.2 are for gas-phase reactions. If liquids or solids are involved, there are additional energy transfers for the phase changes needed to convert the liquids or solids to the gas phase. We shall restrict our use of bond enthalpies to gas-phase reactions for that reason.

Table 6.2 Average Bond Enthalpies (in kJ/mol)*

Single Bonds

	I	Br	Cl	S	P	Si	F	O	N	C	H
H	299	366	431	347	322	323	566	467	391	416	436
C	213	285	327	272	264	301	485	336	285	356	
N	—	—	193	—	~200	335	272	201	160		
O	201	—	205	—	~340	368	190	146			
F	—	238	255	326	490	582	158				
Si	234	310	391	226	—	226					
P	184	264	319	—	209						
S	—	213	255	226							
Cl	209	217	242								
Br	180	193									
I	151										

Multiple Bonds

N=N	418	C=C	598
N≡N	946	C≡C	813
C=N	616	C=O (as in CO_2, O=C=O)	803
C≡N	866	C=O (as in H_2C=O)	695
O=O (in O_2)	498	C≡O	1073

*Data from Cotton, F. A., Wilkinson, G., and Gaus, P. L. *Basic Inorganic Chemistry,* 3rd ed. New York: Wiley, 1995; p. 12.

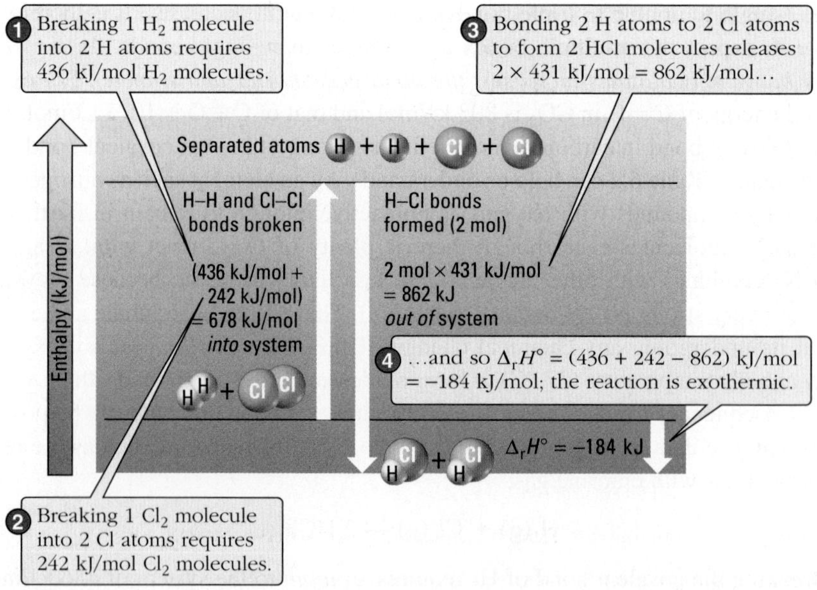
1 Breaking 1 H_2 molecule into 2 H atoms requires 436 kJ/mol H_2 molecules.
3 Bonding 2 H atoms to 2 Cl atoms to form 2 HCl molecules releases 2 × 431 kJ/mol = 862 kJ/mol...
Separated atoms H + H + Cl + Cl
Enthalpy (kJ/mol)
H–H and Cl–Cl bonds broken
(436 kJ/mol + 242 kJ/mol) = 678 kJ/mol *into* system
H–Cl bonds formed (2 mol)
2 mol × 431 kJ/mol = 862 kJ *out of* system
4 ...and so $\Delta_r H°$ = (436 + 242 − 862) kJ/mol = −184 kJ/mol; the reaction is exothermic.
$\Delta_r H°$ = −184 kJ
2 Breaking 1 Cl_2 molecule into 2 Cl atoms requires 242 kJ/mol Cl_2 molecules.

Figure 6.6 Stepwise energy (enthalpy) changes in the reaction of hydrogen with chlorine.

- Forming an H—Cl covalent bond in HCl transfers 431 kJ/mol *out of the system.* Because 2 HCl bonds are formed, there will be 2 × 431 kJ/mol = 862 kJ transferred *out,* making the sign of this enthalpy change *negative.*

If we represent bond enthalpy by the letter *D* (the *D* refers to *d*issociation of the bond), with a subscript to show which bond it refers to, then for the reaction $H_2(g) + Cl_2(g) \rightarrow 2\ HCl(g)$ the net energy transfer is

$$\begin{aligned}\Delta_r H^\circ &= \{[(1\text{ H—H bond}) \times D_{H—H}] + [(1\text{ Cl—Cl bond}) \times D_{Cl—Cl}]\}\\ &\quad -[(2\text{ H—Cl bonds}) \times D_{H—Cl}]\\ &= \{[(1)(436\text{ kJ/mol})] + [(1)(242\text{ kJ/mol})]\} - [(2)(431\text{ kJ/mol})]\\ &= -184\text{ kJ}\end{aligned}$$

This is nearly the same as the experimentally determined value of −184.614 kJ.

As illustrated in Figure 6.6, we can think of the process in the calculation in terms of breaking all the bonds in each reactant molecule and then forming all the bonds in the product molecules. Each bond enthalpy was multiplied by the number of bonds that were broken or formed. For bonds in reactant molecules, we added the bond enthalpies because *breaking bonds is endothermic.* For products, we subtracted the bond enthalpies because *bond formation is exothermic.* This can be summarized in this equation:

$$\begin{aligned}\Delta_r H^\circ = &\sum[(\text{number of bonds}) \times D(\text{bonds broken})]\\ &-\sum[(\text{number of bonds}) \times D(\text{bonds formed})]\end{aligned} \quad [6.1]$$

The Σ (Greek capital letter *sigma*) represents summation. We add the bond enthalpies for all bonds broken, and we subtract the bond enthalpies for all bonds formed.

This equation and the values in Table 6.2 allow us to estimate enthalpy changes for a wide variety of gas-phase reactions. There are several important points to keep in mind when doing such calculations.

- **The enthalpies listed are often average bond enthalpies and may vary depending on the molecular structure.** For example, the enthalpy of a C—H bond is given as 416 kJ/mol, but C—H bond strengths are affected by other atoms and bonds in the same molecule. Depending on the structure of the molecule, the enthalpy required to break a C—H bond may vary by 30 to 40 kJ/mol, so the values in Table 6.2 can be used only to *estimate* an enthalpy change, not to calculate it exactly.
- **The enthalpies in Table 6.2 are for breaking bonds in molecules in the gas phase.** If a reactant or a product is in the liquid or solid state, the energy required to convert it to or from the gas phase will also contribute to the enthalpy change of a reaction and must be accounted for.
- **Multiple bonds, shown as double and triple lines between atoms, are listed at the bottom of Table 6.2.** In some cases, different enthalpies are given for multiple bonds in specific molecules such as $O_2(g)$ or $CO_2(g)$.

© Cengage Learning/Charles D. Winters

Burning methane in a Bunsen burner. Combustion of methane is highly exothermic, releasing energy to the surroundings for every mole of CH_4 that burns. (See Problem-Solving Practice 6.6.)

PROBLEM-SOLVING EXAMPLE 6.6

Estimating $\Delta_r H^\circ$ from Bond Enthalpies

The conversion of diazomethane to ethene and nitrogen is given by this equation:

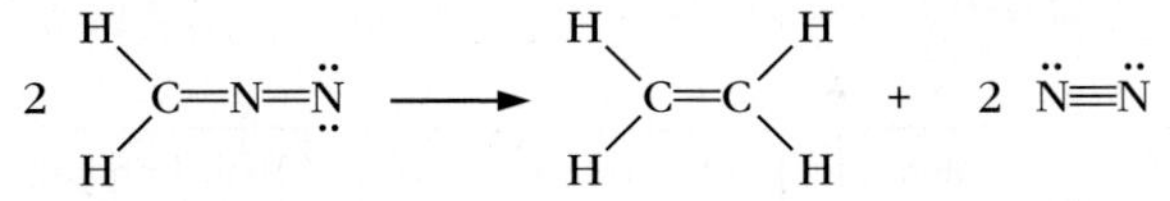

diazomethane ethene

Using the bond enthalpies in Table 6.2, calculate the $\Delta_r H^\circ$ for this reaction.

Result $\Delta_r H^\circ = -422$ kJ

Analyze A chemical equation and Lewis structures are given; this allows us to compare bonds broken with bonds formed. Bond breaking is endothermic; bond making is exothermic.

Plan Determine the number of each kind of bond that is broken or formed. Use bond enthalpies in Table 6.2 and Equation 6.1 to calculate $\Delta_r H°$ for the reaction.

Execute

$$\begin{aligned}\Delta_r H° &= \Sigma[(\text{number of bonds}) \times D(\text{bonds broken})] \\ &\quad -\Sigma[(\text{number of bonds}) \times D(\text{bonds formed})] \\ &= \{[(4\ \text{C—H}) \times D_{\text{C—H}}] + [(2\ \text{C=N}) \times D_{\text{C=N}}] + [(2\ \text{N=N}) \times D_{\text{N=N}}]\} \\ &\quad -\{[(4\ \text{C—H}) \times D_{\text{C—H}}] + [(1\ \text{C=C}) \times D_{\text{C=C}}] + [(2\ \text{N}{\equiv}\text{N}) \times D_{\text{N}\equiv\text{N}}]\} \\ &= \{(4)(416\ \text{kJ/mol}) + (2)(616\ \text{kJ/mol}) + (2)(418\ \text{kJ/mol})\} \\ &\quad - \{(4)(416\ \text{kJ/mol}) + [(1)(598\ \text{kJ/mol}) + (2)(946\ \text{kJ/mol})\} \\ &= (3732\ \text{kJ/mol} - 4154\ \text{kJ/mol}) = -422\ \text{kJ/mol}\end{aligned}$$

☑ **Reasonable Result Check** Note that an N≡N triple bond, 946 kJ/mol, is more than 500 kJ/mol stronger than an N=N double bond, 418 kJ/mol. The C=C double bond, 598 kJ/mol, is about the same strength as a C=N double bond, 616 kJ/mol. Now analyze the bonds broken and bonds formed. There are 4 C—H bonds broken in the reactants and 4 C—H bonds formed in the products; this results in no overall energy transfer. There are 2 C=N bonds broken in the reactants and only 1 C=C bond formed in the products; thus about 600 kJ/mol must be transferred into the system to break the extra double bond in the reactants. There are 2 N=N double bonds in the reactants and 2 N≡N triple bonds in the products; thus about 2 × 500 kJ/mol is released when the stronger bonds form. Overall there should be about 1000 kJ/mol − 600 kJ/mol = 400 kJ/mol released, which agrees with the calculated $\Delta_r H° = -422$ kJ/mol.

Linus Pauling
1901–1994

Oesper Collection in the History of Chemistry. University of Cincinnati

The son of a druggist, Linus Pauling completed his Ph.D. in chemistry and then traveled to Europe, where he worked briefly with Erwin Schrödinger and Niels Bohr (see Chapter 5). For his bonding theories and his work with structural aspects of proteins, Pauling was awarded the Nobel Prize in Chemistry in 1954. Shortly after World War II, Pauling and his wife began a crusade to limit nuclear weapons, a crusade culminating in the limited test ban treaty of 1963. For this effort, Pauling was awarded the 1962 Nobel Peace Prize, the first time any person received two unshared Nobel Prizes.

Linus Pauling died in August 1994, leaving a remarkable legacy of breakthrough scientific research and a social consciousness to be remembered in guarding against the possible misapplications of technology.

PROBLEM-SOLVING PRACTICE 6.6

Use Equation 6.1 and values from Table 6.2 to estimate the enthalpy change when methane, CH_4, and oxygen combine according to the equation:

$$CH_4(g) + 2\,O_2(g) \longrightarrow CO_2(g) + 2\,H_2O(g)$$

CONCEPTUAL EXERCISE 6.6

Bond Length

Arrange C=N, C≡N, and C—N in order of decreasing bond length. Is the order for decreasing bond energy the same or the reverse order? Explain.

6-7 Bond Properties: Polarity; Electronegativity

In a molecule such as H_2 or F_2, where both atoms are the same, there is *equal sharing of the bonding electron pair* and the bond is a **nonpolar covalent bond** as shown in Figure 6.7. When two different atoms are bonded, however, the sharing of the bonding electrons is usually *unequal* and results in a displacement of the bonding electrons toward one of the atoms. If the displacement is complete, electron transfer from one atom to the other

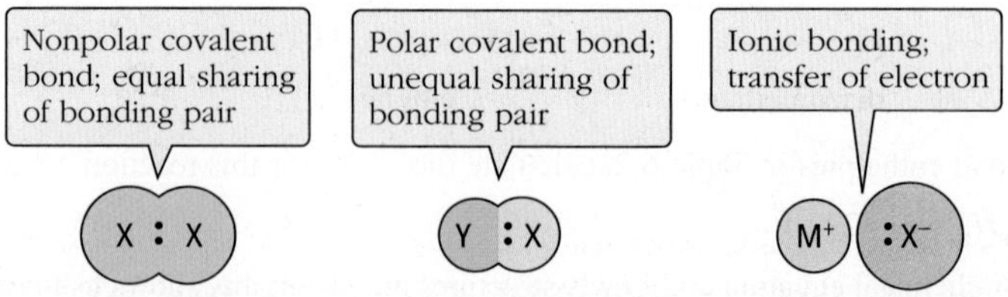
Nonpolar covalent bond; equal sharing of bonding pair
Polar covalent bond; unequal sharing of bonding pair
Ionic bonding; transfer of electron
X : X
Y :X
M+
:X−

Figure 6.7 Nonpolar covalent, polar covalent, and ionic bonding.

creates positive and negative ions and the bond is ionic. If the displacement is less than complete, the *bonding electrons are shared unequally,* and the bond is a **polar covalent bond** (Figure 6.7). As you will see in Chapters 7 and 10, properties of molecules are dramatically affected by bond polarity.

Based on an analysis of bond energies, Linus Pauling first proposed the concept of electronegativity in 1932. **Electronegativity** represents *the ability of an atom in a covalent bond to attract* shared *electrons to itself*—creating a displacement of those bonding electrons toward the more electronegative atom.

Pauling's electronegativity values are relative numbers with an arbitrary value of 4.0 assigned to fluorine, the most electronegative element. More recent experiments and theoretical calculations have modified many of his original values. The current values are given in Figure 6.8. The nonmetal with the next highest electronegativity is oxygen, with a value of 3.4, followed by nitrogen (3.0) and chlorine (2.7). Elements with electronegativities of 2.5 or more are all nonmetals in the upper-right corner of the periodic table. By contrast, elements with electronegativities of 1.1 or less are all metals on the left side of the periodic table. These elements are often referred to as the most electro*positive* elements; *they are the metals that invariably form ionic compounds.* Between these two extremes are most of the remaining metals (largely transition metals) with electronegativities between 1.1 and 1.7, the metalloids with electronegativities between 1.6 and 2.1, and some nonmetals (P, Se) with electronegativities between 2.0 and 2.4.

These periodic trends in electronegativities are evident in Figure 6.8:

- **Electronegativity increases across a period.**
- **Electronegativity decreases down a group.**
- **In general, electronegativity increases diagonally upward and to the right in the periodic table.**

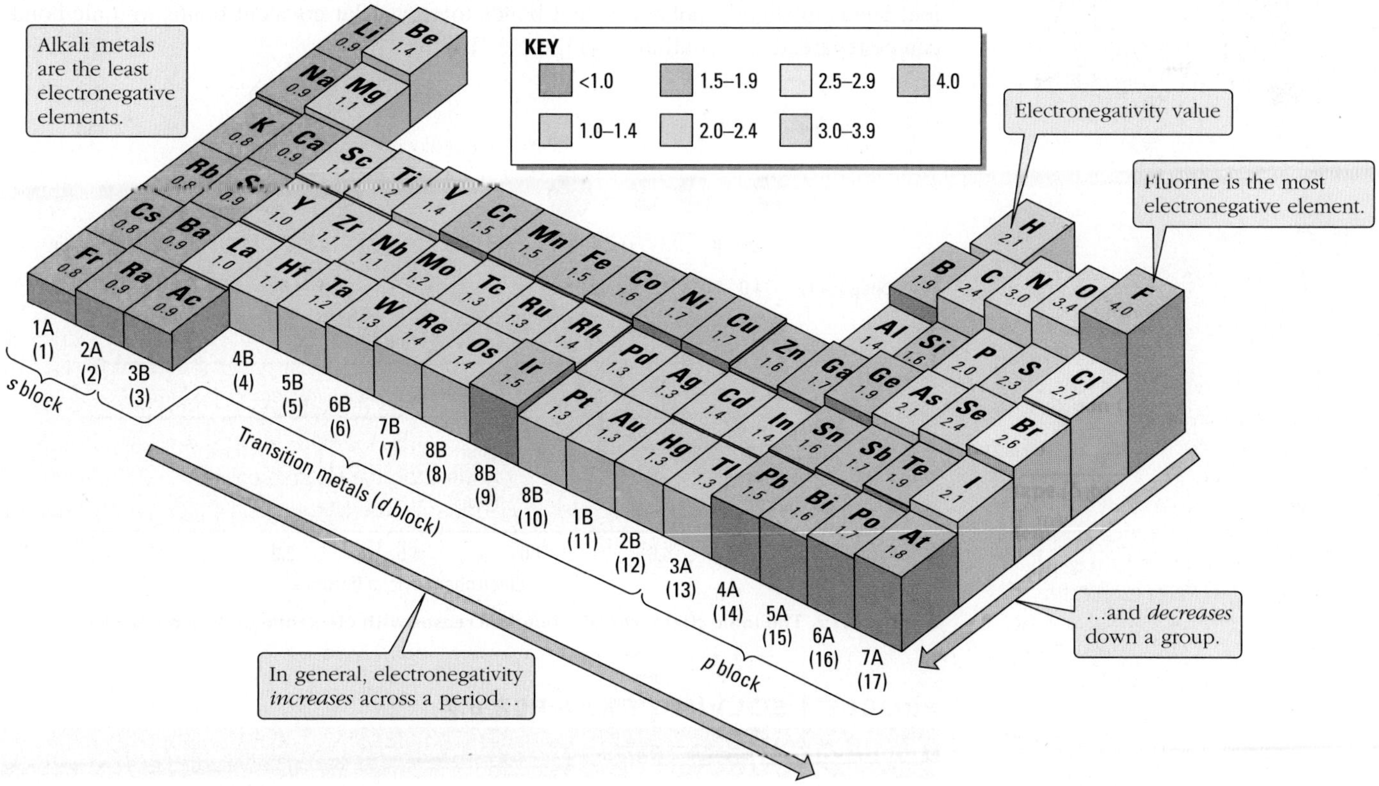
Alkali metals are the least electronegative elements.
KEY
<1.0
1.0–1.4
1.5–1.9
2.0–2.4
2.5–2.9
3.0–3.9
4.0
Electronegativity value
Fluorine is the most electronegative element.
Li 0.9
Be 1.4
Na 0.9
Mg 1.1
K 0.8
Ca 0.9
Sc 1.1
Ti 1.2
V 1.4
Cr 1.5
Mn 1.5
Fe 1.5
Co 1.6
Ni 1.7
Cu 1.7
Zn 1.6
Ga 1.7
Ge 1.9
As 2.1
Se 2.4
Br 2.6
Rb 0.8
Sr 0.9
Y 1.0
Zr 1.1
Nb 1.1
Mo 1.2
Tc 1.3
Ru 1.3
Rh 1.4
Pd 1.3
Ag 1.3
Cd 1.4
In 1.4
Sn 1.6
Sb 1.7
Te 1.9
I 2.1
Cs 0.8
Ba 0.9
La 1.0
Hf 1.1
Ta 1.2
W 1.3
Re 1.4
Os 1.4
Ir 1.5
Pt 1.3
Au 1.3
Hg 1.3
Tl 1.3
Pb 1.5
Bi 1.6
Po 1.7
At 1.8
Fr 0.8
Ra 0.9
Ac 0.9
H 2.1
B 1.9
C 2.4
N 3.0
O 3.4
F 4.0
Al 1.4
Si 1.6
P 2.0
S 2.3
Cl 2.7
1A (1)
2A (2)
3B (3)
4B (4)
5B (5)
6B (6)
7B (7)
8B (8)
8B (9)
8B (10)
1B (11)
2B (12)
3A (13)
4A (14)
5A (15)
6A (16)
7A (17)
s block
Transition metals (d block)
p block
In general, electronegativity increases across a period...
...and decreases down a group.

Figure 6.8 Periodic trends in electronegativities. Data from Allred, A. L., and Rochow, G. E. *Journal of Inorganic and Nuclear Chemistry,* Vol. 5, 1958; pp. 264, 269; Little, E. J., and Jones, M. M. *Journal of Chemical Education,* Vol. 37, 1960; p. 261; and Meek, T. L. *Journal of Chemical Education,* Vol. 83, 2005; p. 325.

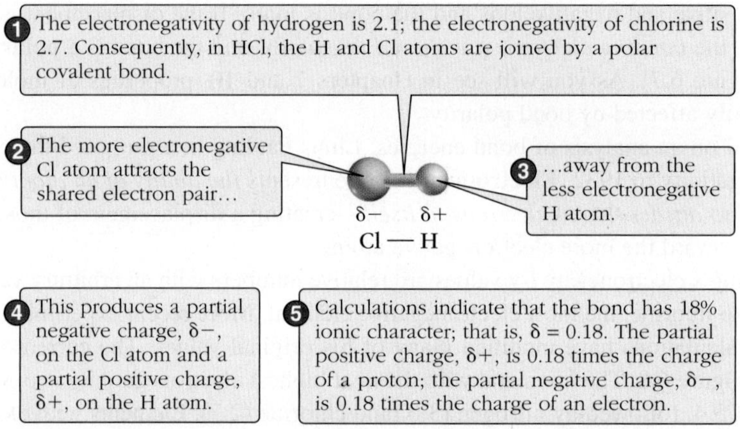
1 The electronegativity of hydrogen is 2.1; the electronegativity of chlorine is 2.7. Consequently, in HCl, the H and Cl atoms are joined by a polar covalent bond.
2 The more electronegative Cl atom attracts the shared electron pair...
3 ...away from the less electronegative H atom.
δ− δ+
Cl — H
4 This produces a partial negative charge, δ−, on the Cl atom and a partial positive charge, δ+, on the H atom.
5 Calculations indicate that the bond has 18% ionic character; that is, δ = 0.18. The partial positive charge, δ+, is 0.18 times the charge of a proton; the partial negative charge, δ−, is 0.18 times the charge of an electron.

Figure 6.9 The polar covalent bond in hydrogen chloride, HCl.

Metals, which are the least electronegative elements, typically lose electrons when they form ionic compounds. Nonmetals, which are the most electronegative elements, typically gain electrons.

Electronegativity values are approximate and are primarily used to predict the polarity of covalent bonds. Bond polarity is indicated by writing δ+ by the *less* electronegative atom and δ− by the *more* electronegative atom, where δ stands for partial charge. For example, the polar H—Cl bond in hydrogen chloride can be represented as shown in Figure 6.9.

Except for bonds between identical atoms (equal electronegativity), all bonds are polar to some extent, and the difference in electronegativity values is a qualitative measure of the degree of polarity. *The greater the difference in electronegativity between two atoms, the more polar will be the bond between them.* The change from nonpolar covalent bonds to slightly polar covalent bonds to very polar covalent bonds to ionic bonds can be regarded as a continuum (Figure 6.10).

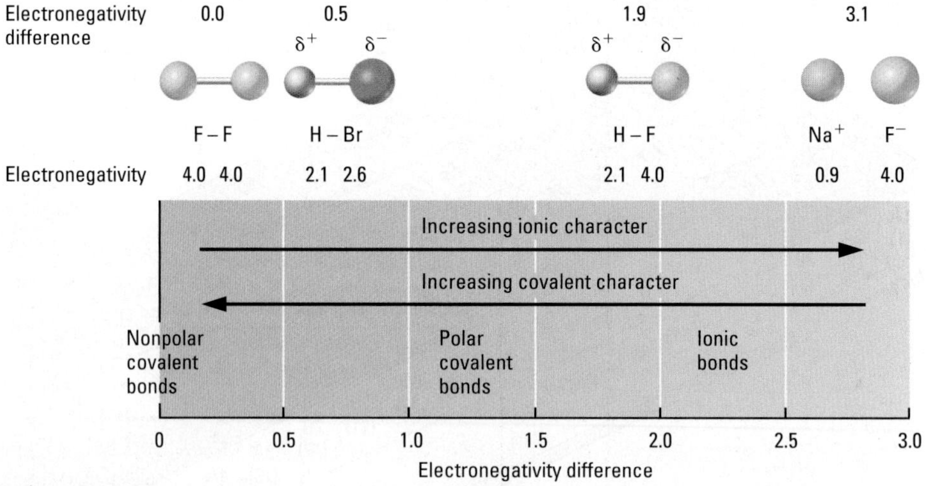
Electronegativity difference
0.0
0.5
1.9
3.1
δ+ δ−
δ+ δ−
F – F
H – Br
H – F
Na+ F−
Electronegativity
4.0 4.0
2.1 2.6
2.1 4.0
0.9 4.0
Increasing ionic character
Increasing covalent character
Nonpolar covalent bonds
Polar covalent bonds
Ionic bonds
0
0.5
1.0
1.5
2.0
2.5
3.0
Electronegativity difference

Figure 6.10 The ionic character of a bond increases with electronegativity difference.

PROBLEM-SOLVING EXAMPLE 6.7

Bond Polarity

For each of these bond pairs, indicate the partial positive and negative atoms and tell which is the more polar bond.

(a) Cl—F and Br—F (b) O—N and P—O

Result

(a) $\overset{\delta+}{Br}—\overset{\delta-}{F}$ (more polar); $\overset{\delta-}{F}—\overset{\delta+}{Cl}$ (b) $\overset{\delta+}{P}—\overset{\delta-}{O}$ (more polar); $\overset{\delta-}{O}—\overset{\delta+}{N}$

Analyze Signs of partial charges and polarities of bonds are to be found. The more electronegative atom has a negative partial charge; the less electronegative atom has a positive partial charge. Partial charges are larger for larger electronegativity differences.

Plan Use periodic trends to predict relative electronegativities. Recall that electronegativities increase from lower left to upper right and increase very rapidly at the upper right.

Execute Apply the analysis and plan.

(a) F is above Cl, which is above Br in the periodic table. Therefore F is the most electronegative and Br the least. In both bonds F is partially negative. Because Br is the least electronegative, the difference in electronegativity is greater between Br and F than between Cl and F.

(b) P is below N in the periodic table and N is to the left of O; therefore the electronegativity of P is less than N, which is less than O. Therefore O is partially negative in each bond. P—O is more polar than N—O because P is less electronegative than N.

☑ **Reasonable Result Check** Consult Figure 6.8 to check electronegativity values: F, 4.0; Cl, 2.7; Br, 2.6; O, 3.4; N, 3.0; and P, 2.0. This verifies both signs of partial charges and electronegativity differences predicted from the periodic table.

PROBLEM-SOLVING PRACTICE 6.7

For each of these pairs of bonds, decide which is the more polar. For the more polar bond in each case, indicate the partial positive and partial negative atoms.

(a) B—C and B—Cl (b) N—H and C—H

CONCEPTUAL EXERCISE 6.7

Bond Types

Use Figures 6.8 and 6.10 to explain why

(a) NaCl is considered to be an ionic compound rather than a polar covalent compound.

(b) BrF is considered to be a polar covalent compound rather than an ionic compound.

The effect of electronegativity difference on properties of substances is nicely illustrated by properties of chlorine compounds with other third-period elements. The third period begins with the ionic compounds NaCl and $MgCl_2$; both are crystalline solids at room temperature. The electronegativity differences between the metal and chlorine are 1.8 and 1.6, respectively. Aluminum chloride is less ionic; with an aluminum-chlorine electronegativity difference of 1.3, its bonding is polar covalent. As electronegativity increases across the period, the other Period 3 chlorides—$SiCl_4$, PCl_3, and SCl_2—are molecular compounds, with decreasing electronegativity differences between the nonmetal and chlorine from Si to S. This results in a decrease in bond polarity from Si—Cl to P—Cl to S—Cl bonds, culminating in no electronegativity difference in Cl—Cl bonds.

Compound	$SiCl_4$	PCl_3	SCl_2	Cl_2
Bond	Si—Cl	P—Cl	S—Cl	Cl—Cl
Electronegativity difference	1.1	0.7	0.4	0.0

Electronegativities: Si = 1.6; P = 2.0; S = 2.3; Cl = 2.7.

6-8 Formal Charge

Lewis structures depict how valence electrons are distributed in a molecule or ion. For some molecules or ions, more than one Lewis structure can be written, each of which obeys the octet rule. Which of the structures is more correct? How do you decide? Using formal charge is one way to do so. **Formal charge** *is the charge a bonded atom would have if its bonding electrons were shared equally.* In calculating formal charges, these assignments are made:

- *All of the lone pair electrons are assigned to the atom on which they are found.*
- *Half of the bonding electrons are assigned to each atom in the bond.*
- *The sum of the formal charges must equal the actual charge: zero for molecules and the ionic charge for an ion.*

Thus, in assigning formal charge to an atom in a Lewis structure,

$$\text{Formal charge} = [\textit{number valence electrons in an atom}] - [(\textit{number lone pair electrons}) + \tfrac{1}{2}\,(\textit{number bonding electrons})]$$

We can, for example, apply these rules to calculate the formal charges of the atoms in a cyanate ion, [:N≡C—Ö:]⁻

	[:N≡C—Ö:]⁻		
Valence electrons	5	4	6
Lone pair electrons	2	0	6
$\frac{1}{2}$ shared electrons	3	4	1
Formal charge	0	0	−1

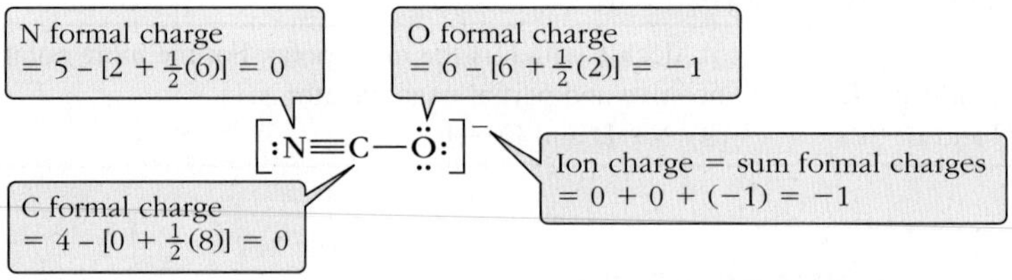
N formal charge
= 5 – [2 + ½(6)] = 0
O formal charge
= 6 – [6 + ½(2)] = −1
[:N≡C—Ö:]⁻
Ion charge = sum formal charges
= 0 + 0 + (−1) = −1
C formal charge
= 4 – [0 + ½(8)] = 0

Note that the sum of the formal charges equals −1, the charge on the ion. The −1 formal charge is on oxygen, the most electronegative atom in the ion.

It is important to recognize that formal charges do *not* indicate actual charges on atoms. Formal charge is a useful way to determine the most likely structure from among several Lewis structures. In evaluating possible structures with different formal charge distributions, these principles apply:

- *Smaller formal charges are more favorable than larger ones.*
- *Negative formal charges should reside on the more electronegative atoms and positive formal charges on less electronegative atoms.*
- *Like charges should not be on adjacent atoms.*

PROBLEM-SOLVING EXAMPLE 6.8

Formal Charges

(a) Two possible Lewis structures for N_2O are :Ö=N=N̈: and :Ö—N≡N:. Determine the formal charges on each atom and the preferred structure.

(b) Determine the formal charges on each atom of these two Lewis structures of dichlorine monoxide, Cl_2O. Which structure is preferred?

Structure 1 :Ö—C̈l—C̈l: Structure 2 :C̈l—Ö—C̈l:

Result

(a) 0, +1, −1: :Ö=N=N̈:	−1, +1, 0: :Ö—N≡N:	(preferred)
(b) −1, +1, 0: :Ö—C̈l—C̈l:	0, 0, 0: :C̈l—Ö—C̈l:	(preferred)

Analyze Formal charges and preferred structures are desired. Formal charge is the charge an atom in a molecule would have if all electrons were shared equally. Preferred structures have minimum formal charge; if there are formal charges, negative charges should be associated with the most electronegative atom.

Plan Count valence electrons for each atom. Assign both electrons in each lone pair to the corresponding atom and assign half the bonding electrons to each atom in the bond. Formal charge is the number of valence electrons minus the number of assigned electrons. From Figure 6.8, O is more electronegative than either N or Cl, so preferred structures should have negative formal charge on O.

Execute The steps in the Plan are summarized in a table for each Lewis structure.

(a)

	:Ö=N=N̈:			:Ö̤—N≡N:		
Valence electrons	6	5	5	6	5	5
Lone pair electrons	4	0	4	6	0	2
$\frac{1}{2}$ shared electrons	2	4	2	1	4	3
Formal charge	0	+1	−1	−1	+1	0

Formal charges are low in both structures; the second structure is preferred because the negative formal charge is on the more electronegative atom (O).

(b)

	:Ö̤—C̤̈l—C̤̈l:			:C̤̈l—Ö̤—C̤̈l:		
Valence electrons	6	7	7	7	6	7
Lone pair electrons	6	4	6	6	4	6
$\frac{1}{2}$ shared electrons	1	2	1	1	2	1
Formal charge	−1	+1	0	0	0	0

The second structure is preferred because it has all zero formal charges.

☑ **Reasonable Result Check** In all cases the formal charges sum to zero as they should for the neutral molecules N_2O and Cl_2O; the results are reasonable.

PROBLEM-SOLVING PRACTICE 6.8

A third Lewis structure can be written for N_2O, which also obeys the octet rule. Write this other Lewis structure and determine the formal charges on its atoms. Is this Lewis structure preferred to either or both of the structures already drawn?

CONCEPTUAL EXERCISE 6.8

Formal Charge

Determine the formal charge of each atom in hydrazine, H_2NNH_2.

6-9 Lewis Structures and Resonance

Ozone, O_3, is an unstable, pale blue, diamagnetic, gaseous form of oxygen with a pungent odor. Ozone and oxygen, O_2, are allotropes—different forms of the same element that exist in the same physical state at the same temperature and pressure. Depending on its location, ozone can be either beneficial or harmful. A very important layer of ozone in the stratosphere (the region of the atmosphere between 10 and 50 km above Earth's surface) protects Earth and its inhabitants from intense ultraviolet solar radiation (Section 8-10), but ozone pollution in the lower atmosphere (troposphere) causes respiratory problems (Section 8-12b).

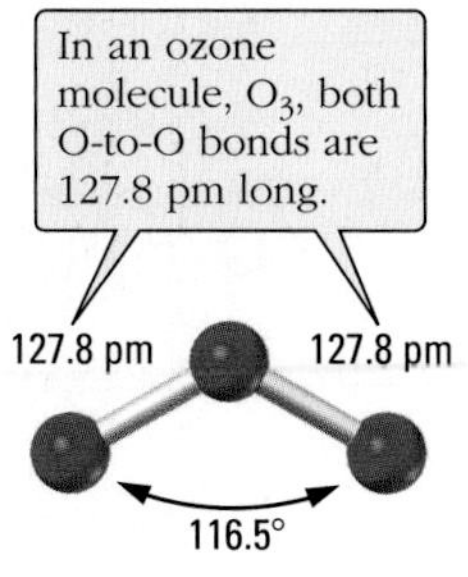
In an ozone molecule, O_3, both O-to-O bonds are 127.8 pm long.
127.8 pm
127.8 pm
116.5°
ozone

Experimental determinations of the structure of ozone show that the lengths for the two oxygen-oxygen bonds are the same, 127.8 pm. Equal bond lengths imply that both bonds contain the same number of bond pairs. However, using the guidelines for writing Lewis structures, you might come to a different conclusion. Two possible Lewis structures are

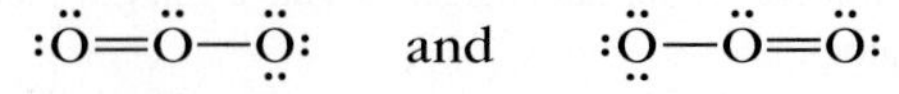

Each Lewis structure shows a double bond on one side of the central O atom and a single bond on the other side. Because double bonds between the same atoms are shorter than single bonds, if either Lewis structure were the actual structure of O_3, then one bond, O=O, should be shorter than the other, O—O. That the oxygen-to-oxygen bonds in ozone are neither double bonds nor single bonds is supported by the fact that the experimentally determined bond length, 127.8 pm, is longer than the O=O bond length, 121 pm, but shorter than the O—O bond length, 148 pm. No *single* Lewis structure can be written that is consistent with all of the experimental data for ozone. The actual structure of O_3 is a composite of the two Lewis structures: both contribute to its properties.

When no single Lewis structure is adequate, the actual structure, which cannot be drawn, is thought of as *a composite of two or more Lewis structures* and is called a **resonance hybrid**. The *two or more Lewis structures that can be drawn for the same molecule* are called **resonance structures**. It is conventional to connect the resonance structures with a double-headed arrow, ⟷, to indicate that the actual bonding is a composite of these structures.

The double-headed arrow does not imply that one resonance structure converts to the other.

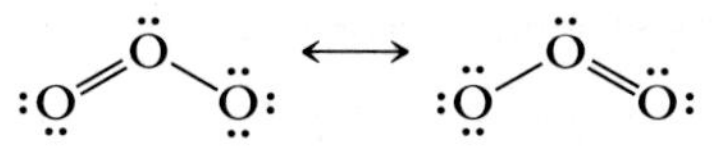

Resonance structures of ozone

When drawing resonance structures, keep several important things in mind:

"Resonance" is an unfortunate term, because it implies that the molecule somehow "resonates," moving in some way to form different kinds of molecules, which is not true. There is only one kind of ozone molecule.

- Lewis structures contributing to the resonance hybrid structure differ only in the assignment of electron pair positions; *the positions of the atoms don't change.*
- Contributing Lewis structures differ in the number of bond pairs between one or more specific pairs of atoms.
- The resonance hybrid represents a single composite structure, not different structures that are continually changing back and forth.

Resonance is useful whenever there is a choice about which of two or three atoms contribute lone pairs to achieve an octet of electrons about a central atom by multiple bond formation. For example, consider the Lewis structure of the carbonate ion, CO_3^{2-}, which has 24 valence electrons (four from C, 18 from three O atoms, and two for the 2− charge). Writing the skeleton structure and putting in lone pairs so that each O has an octet uses all 24 electrons but leaves carbon without an octet:

Writing the Lewis structures of oxygen-containing anions often requires using resonance.

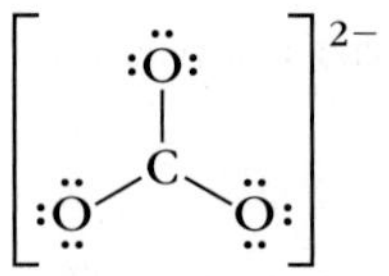

To give carbon an octet requires moving a lone pair from one of the three oxygen atoms to form a double bond. This can be done in three equivalent ways:

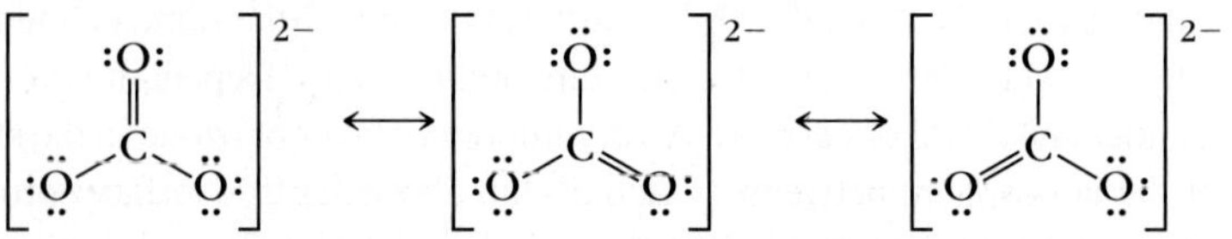

The three resonance structures contribute to the resonance hybrid. As with ozone, experiments confirm that there are no single and double bonds in the carbonate ion. All three carbon-oxygen bond distances are 129 pm, intermediate between the C—O single bond (143 pm) and the C=O double bond (122 pm) distances.

6-9a Resonance and the Structure of Benzene

To nineteenth-century chemists, the C_6H_6 molecular formula of benzene implied that it was an unsaturated compound because it lacked the ratio of carbon to hydrogen found in saturated noncyclic hydrocarbons, C_nH_{2n+2}. A six-membered ring structure with alternating double bonds uses all the available valence electrons and gives each carbon atom an octet of valence electrons.

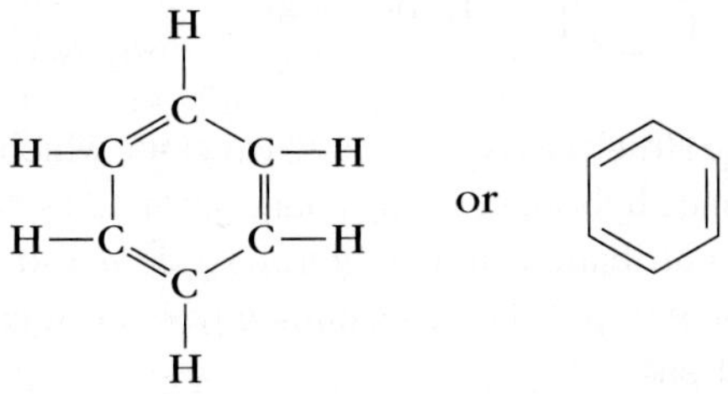
H
H—C
C—H
H—C
C—H
C
H
or

Bonding in benzene can be described better using the molecular orbital theory (see Section 6-12).

This structure implies that alternating C—C single bonds and C=C double bonds are present in benzene. But, carbon compounds that contain C=C double bonds react with bromine, Br_2, by adding bromine atoms directly to the double bond.

Addition reaction (occurs only with C=C double bonds):

H
H
C=C
H
H
+ Br2
Br—C—C—Br
H H
H H

Benzene, however, does *not* react this way under the same conditions. Rather, benzene reacts with bromine by a *substitution* reaction in which a bromine atom replaces (substitutes for) a hydrogen atom of benzene to produce bromobenzene. The displaced hydrogen atom combines with the remaining bromine atom to form hydrogen bromide, HBr. The fact that benzene reacts by substitution rather than addition is strong experimental evidence that benzene does not contain C=C bonds.

Substitution reaction:

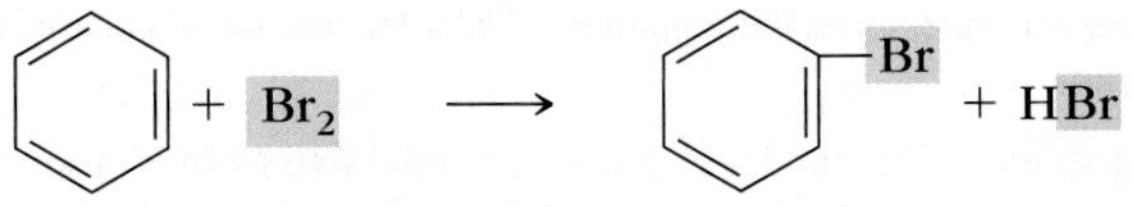
+ Br2
Br
+ HBr

In 1872 Friedrich August Kekulé proposed that benzene could be represented by a combination of two structures, now called resonance structures, indicated by the double-headed arrow.

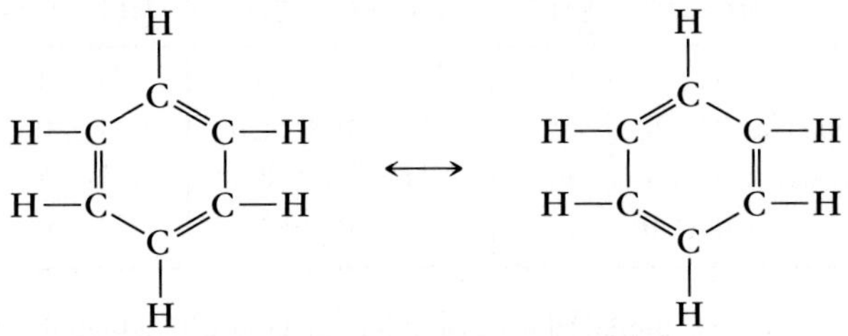
H
H—C
C—H
H—C
C—H
C
H
H
H—C
C—H
H—C
C—H
C
H

Neither of these alternating single- and double-bond resonance structures, however, accurately represents benzene. Experimental structural data for benzene indicate a planar, symmetric molecule in which all six carbon-carbon bonds are equivalent. Each carbon-carbon bond is 139 pm long, intermediate between the length of a C—C single bond

(154 pm) and a C=C double bond (134 pm). Benzene is a resonance hybrid of these resonance structures—it is a molecule in which the six electrons of the suggested three double bonds are spread (delocalized) uniformly around the ring. The six *delocalized electrons* are shared equally by all six carbon atoms. When hydrogen and carbon atoms are not shown, the benzene ring is sometimes written as a hexagon with a circle in the middle. The circle represents the six delocalized electrons spread evenly over all of the carbon atoms. Each corner in the hexagon represents one carbon atom and one hydrogen atom, and each line represents a single C—C bond.

Sometimes a dotted circle rather than a solid circle is used to represent the delocalized electrons.

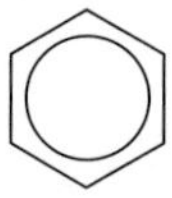

rather than

Unlike the resonance structures of CO_3^{2-} and benzene, which are similar and contribute equally to the resonance hybrid, some resonance structures are significantly different and cannot make equal contributions to the hybrid. As demonstrated in Problem-Solving Example 6.9, formal charge is useful in determining which resonance structures make the most important contributions.

PROBLEM-SOLVING EXAMPLE 6.9

Writing Resonance Structures

Draw three Lewis structures for thiocyanate ion, NCS^-. Determine which resonance structure makes the greatest contribution to the resonance hybrid. Explain.

Result

Structure 1: $[:\text{N}{\equiv}\text{C}{-}\underset{\cdot\cdot}{\overset{\cdot\cdot}{\text{S}}}:]^-$

Structure 2: $[:\underset{\cdot\cdot}{\overset{\cdot\cdot}{\text{N}}}{-}\text{C}{\equiv}\text{S}:]^-$

Structure 3: $[:\overset{\cdot\cdot}{\text{N}}{=}\text{C}{=}\overset{\cdot\cdot}{\text{S}}:]^-$

Structure 3, $[:\overset{\cdot\cdot}{\text{N}}{=}\text{C}{=}\overset{\cdot\cdot}{\text{S}}:]^-$, is preferred.

Analyze Use the guidelines given previously (← **Sec. 6-2a**) to write correct Lewis structures for all resonance structures. Formal charges should be small and negative formal charges should be on more electronegative atoms.

Plan For each Lewis structure, first count the valence electrons, then assign the number of lone pair electrons to each atom, and give half the bonding electrons to each atom in the bond. Substitute these numbers into the equation to calculate the formal charge around each atom in the Lewis structure.

Execute The thiocyanate ion has 16 valence electrons: five from N, four from C, six from S, and one to account for the 1− charge. All structures have 16 valence electrons and differ only in the placement of the multiple bonds, showing that they are correct resonance structures. The formal charges are shown in the table.

	$[:\text{N}{\equiv}\text{C}{-}\underset{\cdot\cdot}{\overset{\cdot\cdot}{\text{S}}}:]^-$			$[:\underset{\cdot\cdot}{\overset{\cdot\cdot}{\text{N}}}{-}\text{C}{\equiv}\text{S}:]^-$			$[:\overset{\cdot\cdot}{\text{N}}{=}\text{C}{=}\overset{\cdot\cdot}{\text{S}}:]^-$		
Valence e^-	5	4	6	5	4	6	5	4	6
Lone pair e^-	2	0	6	6	0	2	4	0	4
$\frac{1}{2}$ shared e^-	3	4	1	1	4	3	2	4	2
Formal charge	0	0	−1	−2	0	+1	−1	0	0

The second structure is least preferred because it has an atom with a high formal charge (−2 on N). The third structure makes the greatest contribution to the resonance hybrid because it has low formal charge and the negative charge is on the most electronegative atom.

☑ **Reasonable Result Check** Sum the formal charges on each Lewis structure. The formal charges should add up to be 1−, the net charge on the thiocyanate ion. They do; the results are reasonable.

PROBLEM-SOLVING PRACTICE 6.9

The nitrogen-oxygen bond lengths in NO_2^- are both 124 pm. Compare this with the bond distances given in Table 6.1 for N—O and N=O bond lengths. Account for any difference.

CONCEPTUAL EXERCISE 6.9

Resonance Structures

Why is $[:\ddot{N}—O\equiv C:]^-$ not a resonance structure for cyanate ion?

6-10 Exceptions to the Octet Rule

Many molecules and polyatomic ions have structures that are not consistent with the octet rule (← **Sec. 6-2**). Three kinds of exceptions occur: (1) molecules or ions with central atoms having fewer than eight electrons; (2) molecules or ions with an odd number of valence electrons; and (3) molecules or ions with central atoms having more than an octet of electrons.

6-10a Fewer Than Eight Valence Electrons

Boron trifluoride, BF_3, is a molecule with less than an octet of valence electrons around the central boron atom. Boron, a Group 3A element, has only three valence electrons; each fluorine contributes seven, for a total of 24 valence electrons. Although the Lewis structure has an octet around each fluorine atom, there are only six electrons around the B atom, an exception to the octet rule. Because it lacks an octet around boron, BF_3 is very reactive—for example, it readily combines with NH_3 to form a compound with the formula H_3NBF_3. The bonding between BF_3 and NH_3 can be explained by using the lone pair of electrons on N to form a covalent bond with B in BF_3. In this case, the nitrogen lone pair provides *both* of the shared electrons, resulting in an octet of electrons for both B and N. This type of bond is known as a coordinate covalent bond.

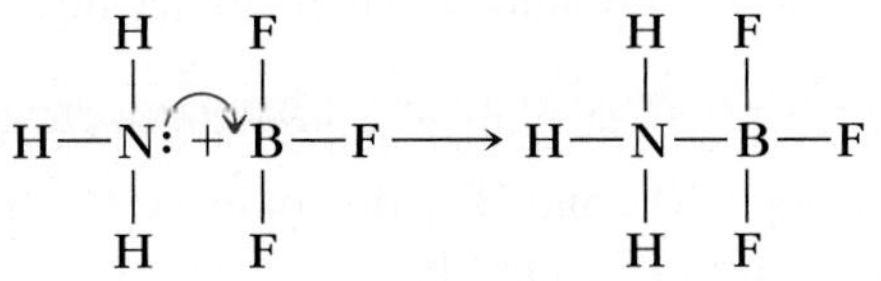

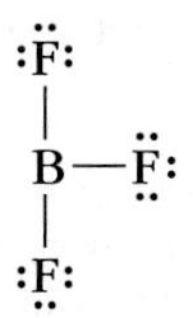

Lewis structure of boron trifluoride. This structure has zero formal charge on every atom. Moving a lone pair from F to form a B=F bond would place a positive formal charge on F. Because F has the highest electronegativity of any element, this is inappropriate. The structure shown is preferred and boron is electron deficient.

6-10b Odd Number of Valence Electrons

All the molecules we have discussed up to this point have contained only *pairs* of valence electrons. However, a few stable molecules have an odd number of valence electrons. For example, NO has 11 valence electrons, and NO_2 has 17 valence electrons. The most plausible Lewis structures for these molecules are

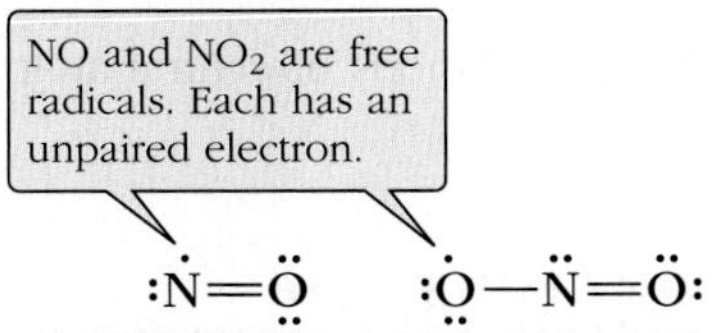
NO and NO₂ are free radicals. Each has an unpaired electron.

An atom or molecule that has an unpaired electron is known as a **free radical**. How do unpaired electrons affect reactivity? Simple free radicals such as atoms of H· and Cl· are very reactive and readily combine with other atoms to give molecules such as H_2, Cl_2, and HCl. Therefore, we would expect free radical molecules to be more reactive than molecules that have all electrons paired, and they are. A free radical either combines with another free radical to form a more stable molecule in which the electrons are paired, or it reacts with other molecules to produce new free radicals. These kinds of reactions are central to the formation of addition polymers (Section 10-6a) and air pollutants (Section 8-12). For example, when gaseous NO and NO_2 are released in vehicle exhaust, the colorless NO reacts with O_2 in the air to form brown NO_2. The NO_2 decomposes in the presence of sunlight to give NO and O, both of which are free radicals.

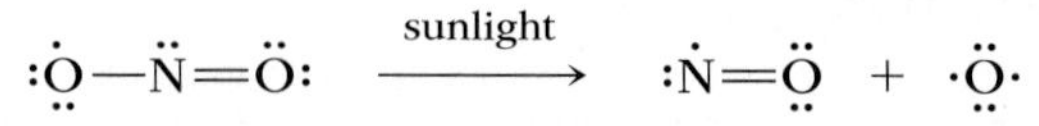
sunlight

Figure 6.11 Paramagnetism of liquid oxygen. Paramagnetic substances are attracted into a magnetic field. Because O_2 is paramagnetic, liquid oxygen, the boiling liquid in the photo, can be suspended between the poles of a magnet.

NO_2 is a free radical; N_2O_4 is not because it has no unpaired electrons.

The free O atom reacts with O_2 in the air to give ozone, O_3, a tropospheric air pollutant that affects the respiratory system (Section 8-12). Free radicals also have a tendency to combine with themselves to form dimers, substances made from two smaller units. For example, when NO_2 gas is cooled it dimerizes to N_2O_4.

As expected, NO and NO_2 are paramagnetic (← **Sec. 5-8a**) because of their odd numbers of electrons. Experimental evidence indicates that O_2 is also paramagnetic (Figure 6.11) with two unpaired electrons and a double bond. The predicted Lewis structure for O_2 shows a double bond, but in that case all the electrons would be paired. It is impossible to write a conventional Lewis structure of O_2 that is in agreement with the experimental results. The molecular orbital theory (Section 6-12) explains bonding in a way that accounts for the paramagnetism of O_2.

6-10c More Than Eight Valence Electrons

See Suidan, L., *et al.* *Journal of Chemical Education*, Vol. 72, 1995; p. 583 for a discussion of bonding in exceptions to the octet rule.

Exceptions to the octet rule are most common among molecules or ions with an "expanded octet"—that is, more than eight electrons in the valence shell around a central atom. For example, sulfur and phosphorus commonly form stable molecules and ions in which the Lewis structure has S or P surrounded by more than an octet of valence electrons. Expanded octets are almost never found with elements in the first or second period, where central atoms are too small to accommodate more than four terminal atoms. The octet rule reliably predicts formulas for all stable molecules in which the central atom is C, N, O, or F; when the central atom is from the third period or below, the octet rule predicts some, but not all, of the formulas for stable molecules.

Guideline 4 (← Sec. 6-2a) says to place any remaining electrons on the central atom once the terminal atoms all have an octet.

Compounds with expanded octets occur mainly with elements in the third period and beyond. For example, the Period 3 elements P and S form the known compounds PF_5 and SF_4, but their Period 2 analogs, NF_5 and OF_4, do not exist. Phosphorus as a central atom can violate the octet rule to form PF_5 and PF_6^-.

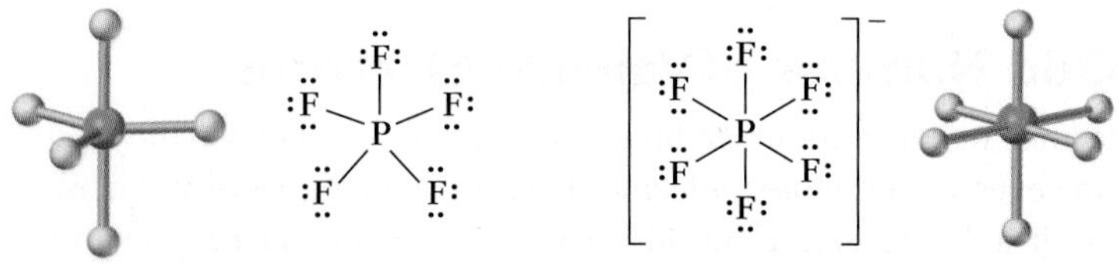

phosphorus pentafluoride

phosphorus hexafluoride ion

Table 6.3 illustrates molecules and ions of central atoms beyond the second period that violate the octet rule.

PROBLEM-SOLVING EXAMPLE 6.10

Exceptions to the Octet Rule

Write the Lewis structure for (a) tellurium tetrabromide, $TeBr_4$; (b) triiodide ion, I_3^-; and (c) boric acid, $B(OH)_3$.

Result

(a) 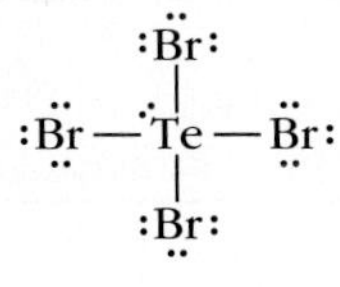 (b) 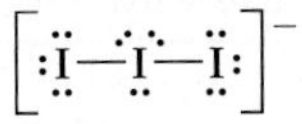 (c) 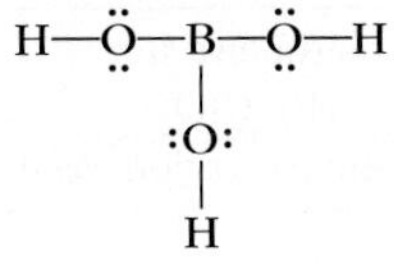

Analyze In each case, the central atom will be an exception to the octet rule. Large central atoms such as Te and I can have greater than an octet of electrons around them. Boron will be an exception to the octet rule by having fewer than eight electrons around the boron atom.

Plan Use the general guidelines for writing Lewis structures (← **Sec. 6-2a**) paying particular attention to Guideline 4 about having more than an octet around the central atom. In cases where the octet is exceeded, place the "extra" electrons as lone pairs on the central atom.

Execute

(a) $TeBr_4$ has 34 valence electrons, eight of which are distributed among four Te—Br bonds. Of the remaining 26 lone pair electrons, 24 complete octets for the Br atoms. The other two electrons form a lone pair on Te, which has a total of ten electrons (five pairs: four shared, one unshared) around it, acceptable for a Period 5 element.

(b) There are 22 valence electrons: seven from each iodine atom and one for the 1− charge on the ion. Forming two I—I single bonds with the central I atom uses two of the 11 electron pairs. Distributing six of the remaining nine electron pairs as lone pairs on terminal I atoms to satisfy the octet rule uses a total of eight electron pairs. The remaining three electron pairs are placed on the central iodine, which can accommodate more than eight electrons because it is from the fifth period.

(c) $B(OH)_3$ uses six of its 24 valence electrons to form three B—O bonds and an additional six electrons for three O—H bonds. The remaining 12 electrons complete two lone pairs on each oxygen atom, giving each oxygen atom an octet. This leaves boron with only three electron pairs, an exception to the octet rule.

Table 6.3 Examples of Lewis Structures with More Than Eight Electrons Around the Central Atom*

	Periodic Group				
	1A	5A	6A	7A	8A
Central atoms with five valence pairs		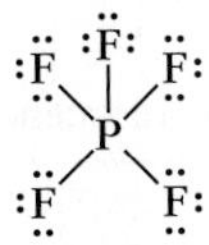	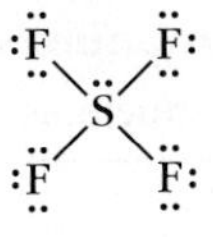	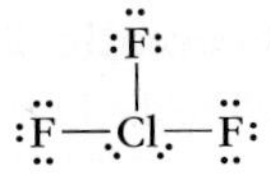	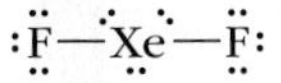
Formula	—	PF_5	SF_4	ClF_3	XeF_2
Bonding pairs	—	5	4	3	2
Lone pairs	—	0	1	2	3
Central atoms with six valence pairs	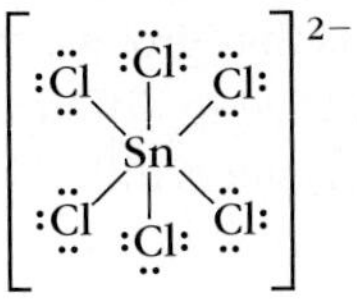	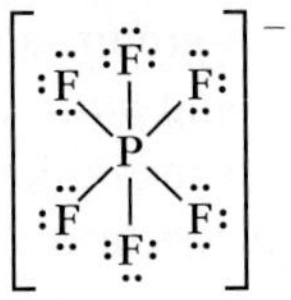	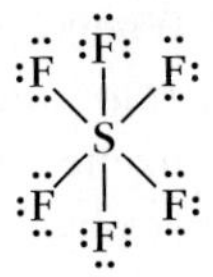	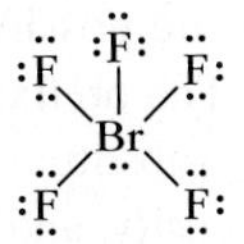	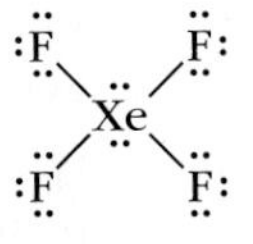
Formula	$SnCl_6^{2-}$	PF_6^-	SF_6	BrF_5	XeF_4
Bonding pairs	6	6	6	5	4
Lone pairs	0	0	0	1	2

*In each case, the numbers of bond pairs and lone pairs about the central atom are given.

PROBLEM-SOLVING PRACTICE 6.10

Write the Lewis structure for:
(a) BeF_2 (b) ClO_2 (c) PCl_5 (d) BH_2^+ (e) IF_7
Indicate which are exceptions to the octet rule and why.

6-11 Aromatic Compounds

Benzene, C_6H_6, is an important industrial chemical that is used, along with its derivatives, to manufacture plastics, detergents, pesticides, drugs, and other organic chemicals. It is the simplest member of a very large family of compounds known as aromatic compounds. An **aromatic compound** is *a compound containing one or more benzene or benzene-like rings*. The word "aromatic" is derived from "aroma," which describes the rather strong and often pleasant odors of these compounds.

As noted earlier (← **Sec. 6-9a**), the benzene ring can be represented by a hexagon with a circle in the middle. Whenever you see a formula with one or more carbon rings, each with a central circle or alternating double and single bonds throughout, the compound is aromatic, like benzene. Benzaldehyde and toluene are examples of the many aromatic compounds with functional groups or alkyl groups bonded to the aromatic ring. Naphthalene is representative of a large group of aromatic compounds with more than one ring joined by common carbon-carbon bonds. Benzaldehyde is used in synthetic almond and cherry food flavoring, toluene and benzene boost the octane rating of gasoline, and naphthalene is a moth repellent in one kind of moth balls.

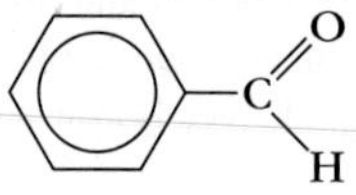
O
C
H

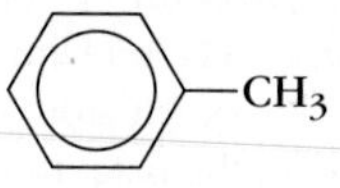
CH3

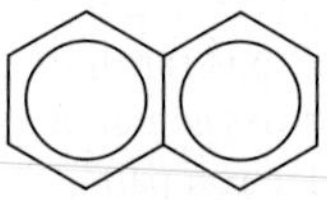

benzaldehyde toluene naphthalene

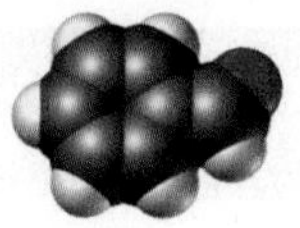

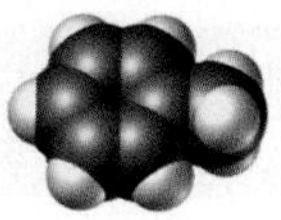

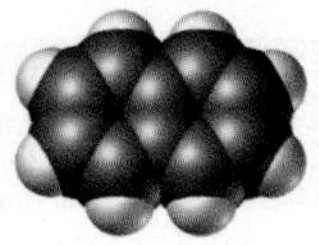

EXERCISE 6.10

Aromatic Compounds

Write the Lewis structures for the resonance structures of toluene.

6-11a Constitutional Isomers of Aromatic Compounds

Because benzene is a planar molecule, constitutional isomers are possible when two or more groups (or atoms) are substituted for hydrogen atoms on the benzene ring. If two groups are substituted for two hydrogen atoms on the benzene ring, three constitutional isomers are possible. When the two groups are methyl groups, the compound is called xylene. The prefixes *ortho-*, *meta-*, and *para-* are used to differentiate the three isomers of any disubstituted benzene. These constitutional isomers differ in melting point, boiling point, density, and other physical properties.

Ortho—a prefix indicating that two substituents are on _adjacent_ carbon atoms on a benzene ring. _Meta_—two substituents attached to carbon atoms _separated by one carbon atom_ on a benzene ring. _Para_—two substituents attached to carbon atoms _separated by two carbon atoms_ on a benzene ring.

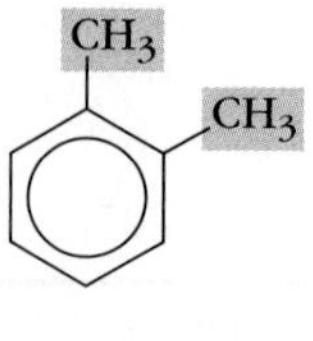
CH3
CH3

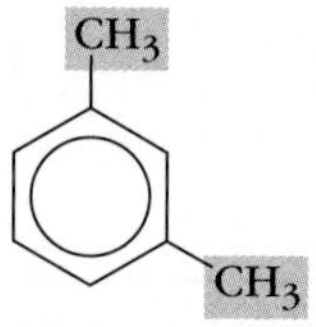
CH3
CH3

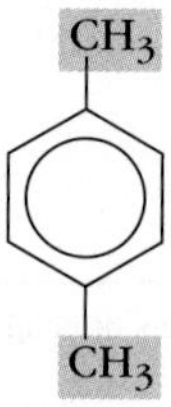
CH3
CH3

ortho-xylene *meta*-xylene *para*-xylene

Physical Property	*ortho*-xylene	*meta*-xylene	*para*-xylene
Melting point, °C	−25.2	−47.8	13.2
Boiling point, °C	144.5	139.1	138.4
Density, g/mL (at 25 °C)	0.876	0.860	0.857

If more than two groups are attached to the benzene ring, numbers must be used to identify them and their positions, as for the three trichlorobenzenes:

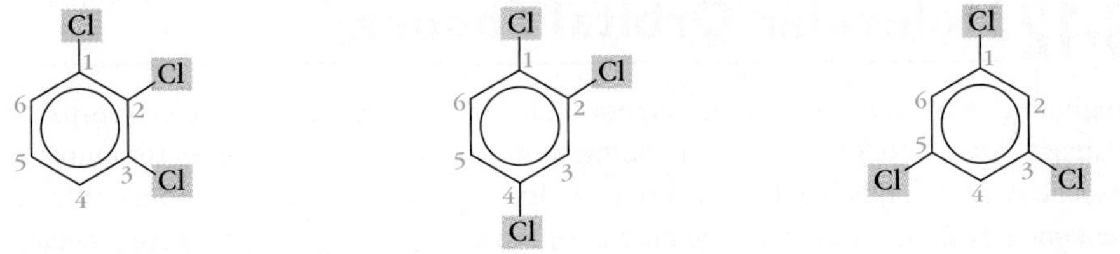

1,2,3-trichlorobenzene 1,2,4-trichlorobenzene 1,3,5-trichlorobenzene

Because there is no other way to attach three chlorine atoms to a benzene ring, only three trichlorobenzenes exist.

EXERCISE 6.11

Constitutional Isomers of Aromatic Compounds

Write the structural formula of 1,2,4-trimethylbenzene. Explain why 1,5,6-trimethylbenzene is not different from 1,2,3-trimethylbenzene.

Throughout this chapter, you have seen simple as well as complex structural formulas. The structural formulas of many biochemically active molecules are large and seemingly quite complicated. But, by closely examining the structural formulas, you can recognize the individual parts of the molecule. For example, by looking at the structural formulas of vitamins A and E that follow, you should recognize that the molecules are almost entirely assembled from structural parts that you now understand—rings and chains of carbon atoms, including an aromatic ring, C=C double bonds, and an alcohol group. There is also a six-membered ring containing one oxygen atom in the place of one of the carbon atoms.

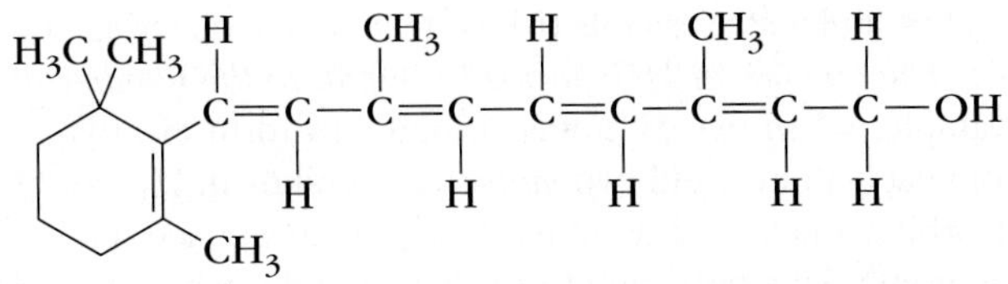

vitamin A

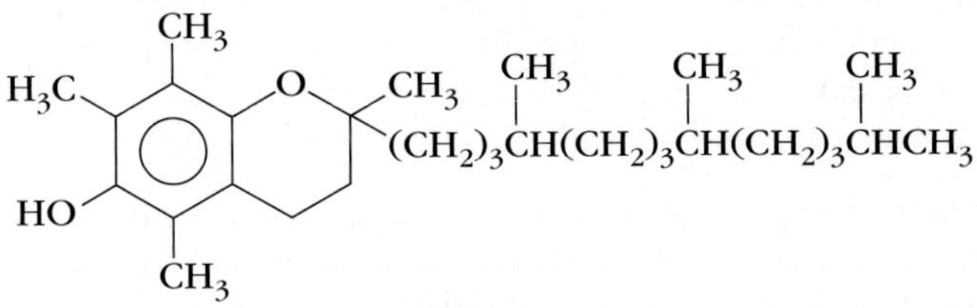

vitamin E

CONCEPTUAL EXERCISE 6.12

Knowing Your Vitamins

1. Use the structural formula for vitamin A to answer these questions.
   (a) How many carbon atoms and how many hydrogen atoms does it contain?

(b) Locate the carbon atom that is bonded directly to four carbon atoms.
(c) How many C=C double bonds are there in this molecule?

2. Consider the structural formula for vitamin E.
   (a) Determine its molecular formula.
   (b) How many C=C double bonds does it have?
   (c) Which is likely to be the most polar bond in the molecule?

6-12 Molecular Orbital Theory

Beginning in Section 6-1, and throughout this chapter, we have used electron-pair bonds (valence bond theory) to explain bonding in molecules, drawing Lewis structures to represent covalent bonds involving overlap of atomic orbitals on pairs of atoms. Where more than one Lewis structure could be drawn for the same arrangement of atomic nuclei, as in O_3, resonance (← **Sec. 6-9**) was used to account for observed properties of molecules.

A major weakness of the valence bond theory is that it does not always correctly predict the magnetic properties of substances. An important example is O_2, which is paramagnetic (← **Sec. 6-10b**), meaning that O_2 molecules must have unpaired electrons. Diatomic oxygen has an even number of valence electrons (12), and the octet rule predicts that all of these electrons should be paired. According to valence bond theory, O_2 should be diamagnetic, but experimental measurements show that it is not.

6-12a Molecular Orbitals

This discrepancy between experiment and theory for O_2 (and other molecules) can be resolved by using an alternative model of covalent bonding, the molecular orbital (MO) approach. **Molecular orbital (MO) theory** *treats bonding in terms of molecular orbitals.* A **molecular orbital (MO)** is *an orbital that can extend over all atoms in an entire molecule.* Unlike electron-pair bonds, molecular orbitals are not confined to two atoms at a time. The MO approach results in a set of energy levels for the molecule that is analogous to the atomic energy levels from which atomic electron configurations were obtained (Figure 5.15, ← **Sec. 5-5e**). It involves three basic operations.

Step 1: Combine the valence atomic orbitals (AOs) of all atoms in the molecule to give a new set of molecular orbitals (MOs) that are characteristic of the molecule as a whole. *The number of MOs formed is equal to the number of AOs combined.* For example, when two H atoms combine to form H_2, two *s atomic* orbitals, one from each atom, yield two *molecular* orbitals in H_2. For O_2 there are one *s* atomic orbital and three *p* atomic orbitals in the valence shell of each of the two oxygen atoms. This gives eight AOs, which combine to give eight MOs in O_2.

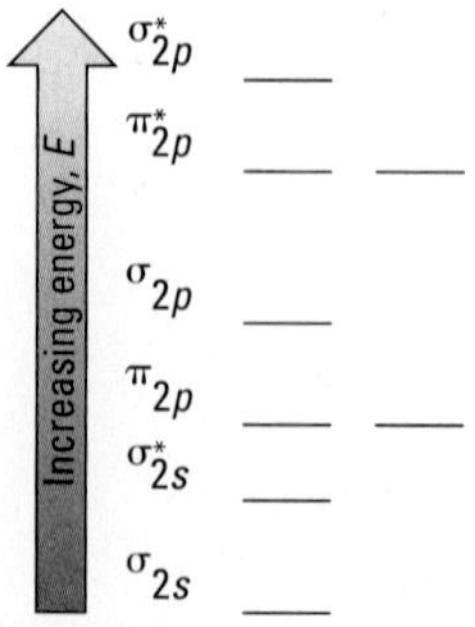

σ^*_{2p}
π^*_{2p}
σ_{2p}
π_{2p}
σ^*_{2s}
σ_{2s}
Increasing energy, E

Figure 6.12 Relative order of filling molecular orbitals.

Step 2: Arrange the MOs in order of increasing energy, in a molecular-orbital energy-level diagram. An example, for diatomic molecules of elements in the second period, is shown in Figure 6.12. The relative energies of MOs are usually deduced from experiments involving spectral and magnetic properties.

Step 3: Distribute the *valence electrons* of the molecule among the available MOs, filling the lowest-energy MO first and continuing to build up the electron configuration of the molecule in the same way that electron configurations of atoms are built up (← **Sec. 5-7a**). As electrons are added,

(a) *Each MO can hold a maximum of two electrons.* In a filled molecular orbital the two electrons have opposed spins, in accordance with the Pauli exclusion principle (← **Sec. 5-5e**).

(b) *Electrons go into the lowest-energy MO available.* A higher-energy molecular orbital starts to fill only when every molecular orbital below it has its quota of two electrons.
(c) *Hund's rule is obeyed.* When two molecular orbitals of equal energy are available to two electrons, one electron goes into each, giving two half-filled molecular orbitals.

6-12b Molecular Orbitals for Diatomic Molecules

To illustrate molecular orbital theory, we apply it to the diatomic molecules of the elements in the first two periods of the periodic table.

When two atomic orbitals of equal energy combine, the resulting two MOs lie above and below the energy of the atomic orbitals.

Hydrogen and Helium (Combining Two 1s Atomic Orbitals) When two hydrogen atoms (or two helium atoms) come close together, two 1*s* atomic orbitals overlap and combine to give two MOs. One MO has an energy lower than that of the atomic orbitals from which it was formed; the other MO is of higher energy (Figure 6.13). A molecule that has two electrons in the lower-energy MO is lower in energy (more stable) than the isolated atoms. That lowering of energy corresponds to the bond energy. *An MO that is lower in energy than the AOs from which it is derived*, such as the lower-energy MO in Figure 6.13, is called a **bonding molecular orbital**. If electrons are placed in the higher-energy MO, the molecule's energy is higher than the energy of the isolated atoms. This unstable situation is the opposite of bonding; *a MO that is higher in energy than the AOs from which it is derived* is called an **antibonding molecular orbital**.

The electron density in these MOs is shown as boundary surface diagrams in Figure 6.13. The bonding MO has higher electron density between the nuclei than the sum of the individual atomic orbitals. This attracts the nuclei together and forms the bond. In the antibonding orbital, the electron density between the nuclei is smaller than it would have been in the atoms alone. The positively charged nuclei have less electron "glue" to hold them together, they repel each other, and the molecule flies apart—the opposite of a bond.

The electron density in both MOs has circular symmetry around a line connecting the two nuclei, which means that both are sigma, σ, orbitals. In MO notation, the 1*s* bonding orbital is designated as σ_{1s}. The antibonding orbital is given the symbol σ^*_{1s}. An asterisk designates an antibonding molecular orbital.

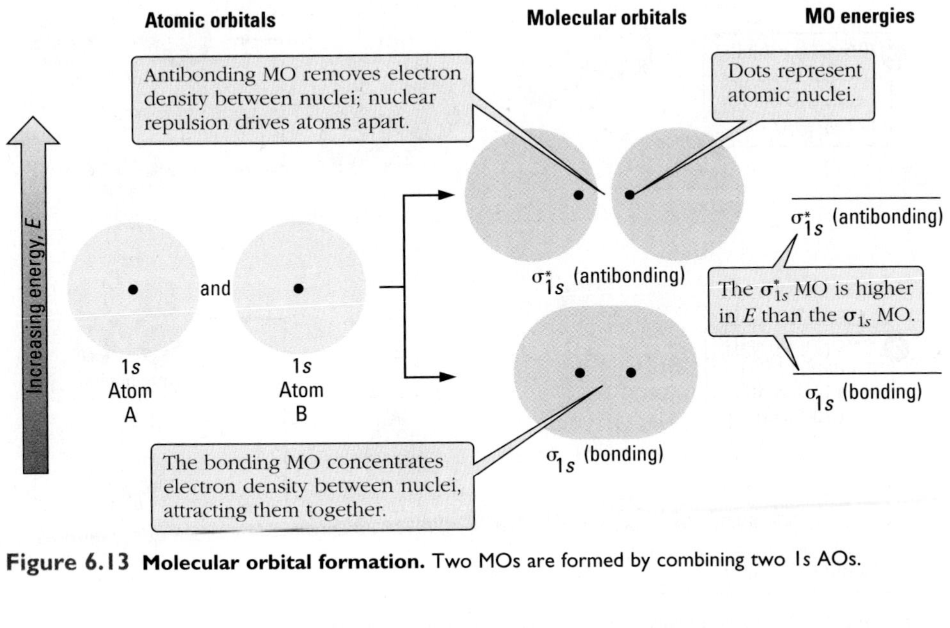
Atomic orbitals
Molecular orbitals
MO energies
Antibonding MO removes electron density between nuclei; nuclear repulsion drives atoms apart.
Dots represent atomic nuclei.
Increasing energy, E
and
1s Atom A
1s Atom B
σ^*_{1s} (antibonding)
σ^*_{1s} (antibonding)
The σ^*_{1s} MO is higher in E than the σ_{1s} MO.
σ_{1s} (bonding)
σ_{1s} (bonding)
The bonding MO concentrates electron density between nuclei, attracting them together.

Figure 6.13 Molecular orbital formation. Two MOs are formed by combining two 1s AOs.

In the H_2 molecule, there are two 1*s* electrons. They fill the σ_{1s} orbital, giving a single bond. The molecular electron configuration is $(\sigma_{1s})^2$. The **bond order**, *the number of electron-pair bonds,* is one. If an He_2 molecule could form, there would be four electrons—two from each atom. These would fill the bonding and antibonding orbitals. One bond and one antibond give a bond order of zero in He_2. The bond order is calculated as

$$\text{Bond order} = \text{number of bonds} = \frac{n_B - n_A}{2}$$

where n_B is the number of electrons in bonding molecular orbitals and n_A is the number of electrons in antibonding molecular orbitals; dividing by 2 accounts for the fact that two electrons are needed for a bond. In H_2, $n_B = 2$ and $n_A = 0$, so we have one bond. For He_2, $n_B = n_A = 2$, so the number of bonds is zero. MO theory predicts that He_2 should not exist, and it does not.

Second-Period Elements (Combining 2*s* and 2*p* Atomic Orbitals) Three of the elements in the second period form familiar diatomic molecules: N_2, O_2, and F_2. Less common, but also known, are Li_2, B_2, and C_2, which have been observed as gases. The molecules Be_2 and Ne_2 are either highly unstable or nonexistent. To see what MO theory predicts about the stability of diatomic molecules from the second period, consider the valence atomic orbitals, 2*s* and 2*p*.

Combining two 2*s* atomic orbitals, one from each atom, gives two MOs. These are very similar to the ones shown in Figure 6.13. They are designated as σ_{2s} (sigma, bonding, 2*s*) and σ^*_{2s} (sigma, antibonding, 2*s*).

In an isolated atom, there are three 2*p* atomic orbitals, oriented at right angles to each other. We call these atomic orbitals p_x, p_y, and p_z. Figure 6.14 shows these *p* orbitals and explains what MOs they form when two atoms approach along the *x*-axis. The two p_x

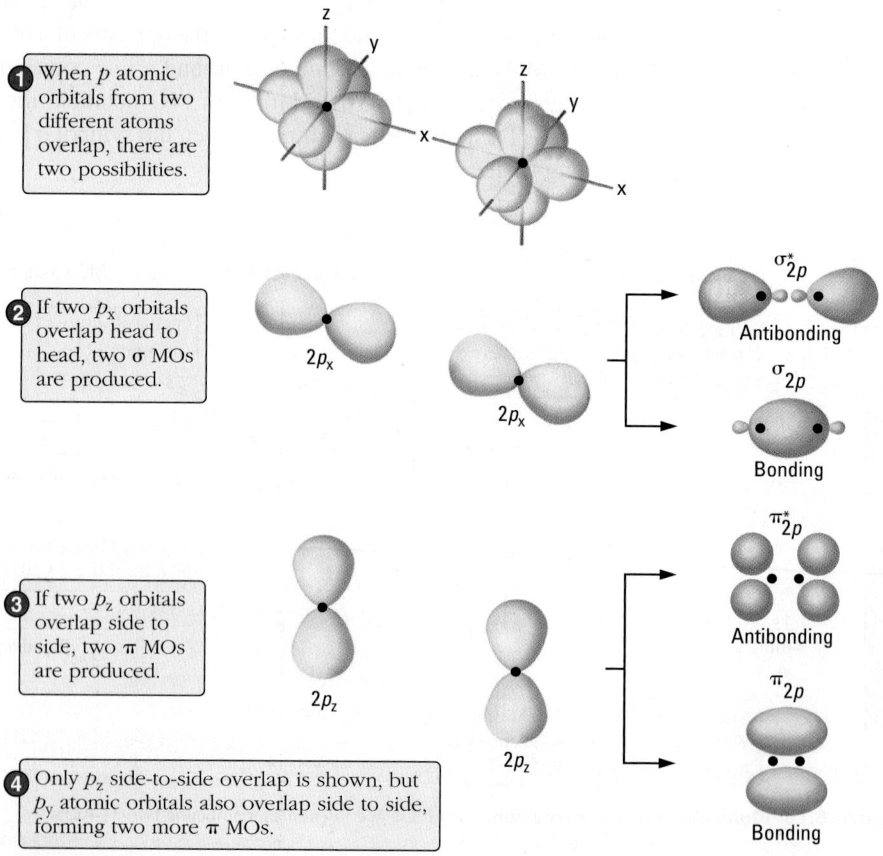

1 When p atomic orbitals from two different atoms overlap, there are two possibilities.
z
y
x
z
y
x
2 If two p_x orbitals overlap head to head, two σ MOs are produced.
$2p_x$
$2p_x$
σ^*_{2p}
Antibonding
σ_{2p}
Bonding
3 If two p_z orbitals overlap side to side, two π MOs are produced.
$2p_z$
$2p_z$
π^*_{2p}
Antibonding
π_{2p}
Bonding
4 Only p_z side-to-side overlap is shown, but p_y atomic orbitals also overlap side to side, forming two more π MOs.

Figure 6.14 Forming molecular orbitals from *p* atomic orbitals.

Table 6.4 MO Electron Configurations, Predicted Properties, and Observed Properties of Diatomic Molecules of Second-Period Elements

	Occupancy of Molecular Orbitals							
	σ_{2s}	σ^*_{2s}	π_{2p}	π_{2p}	σ_{2p}	π^*_{2p}	π^*_{2p}	σ^*_{2p}
Li_2	(↑↓)	( )	( )	( )	( )	( )	( )	( )
Be_2	(↑↓)	(↑↓)	( )	( )	( )	( )	( )	( )
B_2	(↑↓)	(↑↓)	(↑)	(↑)	( )	( )	( )	( )
C_2	(↑↓)	(↑↓)	(↑↓)	(↑↓)	( )	( )	( )	( )
N_2	(↑↓)	(↑↓)	(↑↓)	(↑↓)	(↑↓)	( )	( )	( )
O_2	(↑↓)	(↑↓)	(↑↓)	(↑↓)	(↑↓)	(↑)	(↑)	( )
F_2	(↑↓)	(↑↓)	(↑↓)	(↑↓)	(↑↓)	(↑↓)	(↑↓)	( )
Ne_2	(↑↓)	(↑↓)	(↑↓)	(↑↓)	(↑↓)	(↑↓)	(↑↓)	(↑↓)

	Predicted Properties		Observed Properties	
	Number of Unpaired e$^-$	Bond Order	Number of Unpaired e$^-$	Bond Energy (kJ/mol)
Li_2	0	1	0	105
Be_2	0	0	0	Unstable
B_2	2	1	2	289
C_2	0	2	0	598
N_2	0	3	0	946
O_2	2	2	2	498
F_2	0	1	0	158
Ne_2	0	0	0	Nonexistent

atomic orbitals overlap head to head to form two orbitals that are symmetric around the line connecting the two nuclei. That is, they form a sigma bonding orbital, σ_{2p}, and a sigma antibonding orbital, σ^*_{2p}.

The situation is quite different when the p_z atomic orbitals overlap. Because they are oriented parallel to one another, *they overlap side to side* (Figure 6.14). The two MOs formed in this case are pi, π, orbitals; one is a bonding MO, π_{2p}, and the other is an antibonding MO, π^*_{2p}. Similarly, the p_y atomic orbitals of the two atoms interact to form another pair of pi MOs, π_{2p} and π^*_{2p}, which are not shown in Figure 6.14. This is the reason that in Figure 6.12 there are two π_{2p} and two π^*_{2p} MO energy levels.

If you look down the internuclear axis, a π molecular orbital looks like a p atomic orbital.

The relative energies of the MOs available for occupancy by the valence electrons of diatomic molecules formed from second-period atoms were shown in Figure 6.12. To obtain the MO electron configuration of each diatomic molecule formed by second-period elements, we fill the available MOs in order of increasing energy. This order of energies applies to Li_2 through N_2, but for O_2 and F_2 the energy of the σ_{2p} orbital drops below the energy of the two π_{2p} orbitals. This does not affect the MO configurations, because in O_2 and F_2 the σ_{2p} and π_{2p} orbitals are completely filled. The results are shown in Table 6.4. MO theory correctly predicts the number of unpaired electrons in each molecule and the bonding in the molecule.

PROBLEM-SOLVING EXAMPLE 6.11

Filling Molecular Orbitals

Without consulting Table 6.4, apply the energy diagram in Figure 6.12 to C_2 and fill the molecular orbitals appropriately. Determine the molecular electron configuration, number of unpaired electrons, and the bond order.

Result $(\sigma_{2s})^2 (\sigma^*_{2s})^2 (\pi_{2p})^4$; no unpaired electrons; bond order = 2

Analyze First, determine the number of valence electrons. Then, use the MO diagram in Figure 6.12 to fill valence electrons into the MOs in proper order of increasing energy.

Plan There are eight valence electrons, four from each carbon atom, that need to be placed into the MO diagram.

Execute Place the eight valence electrons correctly into the MO energy-level diagram to obtain the electron configuration in the Result. The bond order is (6 − 2)/2 = 2. There are no unpaired electrons.

☑ **Reasonable Result Check** Verify your result using Table 6.4.

PROBLEM-SOLVING PRACTICE 6.11

Repeat Problem-Solving Example 6.11, but use N_2 instead of C_2.

There is also a general correlation between the predicted bond order, $(n_B - n_A)/2$, and the experimental bond energy. The bond order of two for C_2 and O_2 implies a double bond, which is expected to be stronger than the single bonds in Li_2, B_2, and F_2 (bond order = 1). The bond order of three for N_2 (triple bond) implies a still stronger bond. A major triumph of MO theory is its ability to explain the properties of O_2. The bond order of two and the presence of two unpaired electrons in the two π^*_{2p} MOs explain how the molecule can have a double bond and at the same time be paramagnetic.

Molecular orbital theory can also be applied to predict the properties of diatomic molecules containing different kinds of atoms, such as NO. There are 11 valence electrons in this molecule (five from nitrogen and six from oxygen). These 11 valence electrons are placed into molecular orbitals of increasing energy in the molecular orbital diagram:

	σ_{2s}	σ^*_{2s}	π_{2p}	π_{2p}	σ_{2p}	π^*_{2p}	π^*_{2p}	σ^*_{2p}
NO	(↑↓)	(↑↓)	(↑↓)	(↑↓)	(↑↓)	(↑)	( )	( )

There are eight electrons in bonding orbitals (two in σ_{2s}, four in π_{2p}, two in σ_{2p}) and three electrons in antibonding orbitals (two in σ^*_{2s}, one in π^*_{2p}). Therefore, there is one unpaired electron (in the π^*_{2p} molecular orbital). The bond order is

$$\text{Bond order} = \text{number of bonds} = \frac{8-3}{2} = \frac{5}{2} = 2.5$$

PROBLEM-SOLVING EXAMPLE 6.12

Electron Structure of Nitrosyl Ion

Using MO theory, predict the bond order and number of unpaired electrons in the nitrosyl ion, NO^+. Which has the stronger bond, NO^+ or NO?

Result Bond order = 3; there are no unpaired electrons. NO^+ has the stronger bond.

Analyze First, determine the number of valence electrons. Then, construct a molecular orbital diagram, filling the electrons into available MOs (Figure 6.12) in order of increasing energy. The stronger bond has larger bond order.

Plan Oxygen is in Group 6A of the periodic table, so an oxygen atom has six valence electrons. Nitrogen is in Group 5A so it has 5 valence electrons. The 1+ charge indicates loss of 1 electron, so there are 6 + 5 − 1 = 10 valence electrons.

Execute Complete the molecular orbital diagram for nitrosyl ion.

	σ_{2s}	σ^*_{2s}	π_{2p}	π_{2p}	σ_{2p}	π^*_{2p}	π^*_{2p}	σ^*_{2p}
NO^+	(↑↓)	(↑↓)	(↑↓)	(↑↓)	(↑↓)	( )	( )	( )

The bond order is (8 − 2)/2 = 3. There are no unpaired electrons. These conclusions are in agreement with the Lewis structure (valence bond theory) for the nitrosyl ion: $:N{\equiv}O:^+$. The bond order calculated previously for NO is 2.5; therefore, NO^+ has the stronger bond.

PROBLEM-SOLVING PRACTICE 6.12

Use MO theory to predict the bond order and the number of unpaired electrons in the superoxide ion, O_2^-, and the peroxide ion, O_2^{2-}.

6-12c Polyatomic Molecules; Delocalized π Electrons

The bonding in molecules containing more than two atoms can also be described in terms of MOs. We will not attempt to do this; the energy-level structure is considerably more complex than that for diatomic molecules. However, one point is worth mentioning: In polyatomic species, *the MOs can be spread over the entire molecule* rather than being localized between two atoms.

No single Lewis structure can represent the experimental fact that ozone, O_3, has two O-to-O bonds of equal length (← **Sec. 6-9**). Valence bond theory considers ozone as a resonance hybrid of two structures, each of which contains one double bond and one single bond. MO theory considers the electron pair in the single bond and one of the two electron pairs in the double bond to involve sigma MOs. The second electron pair in the double bond is in a pi molecular orbital that encompasses all three O atoms. The pi electrons are said to be delocalized over the entire three-oxygen chain of the molecule. Because the pi molecular orbital extends over two bonding regions between O atoms, it contributes $\frac{1}{2}$ to the bond order of each. Thus each O-to-O bond in O_3 has a single sigma bond and half a pi bond. This is intermediate between a single and a double bond. This explains the fact that both O-to-O bonds are equal in length and intermediate between O—O and O═O.

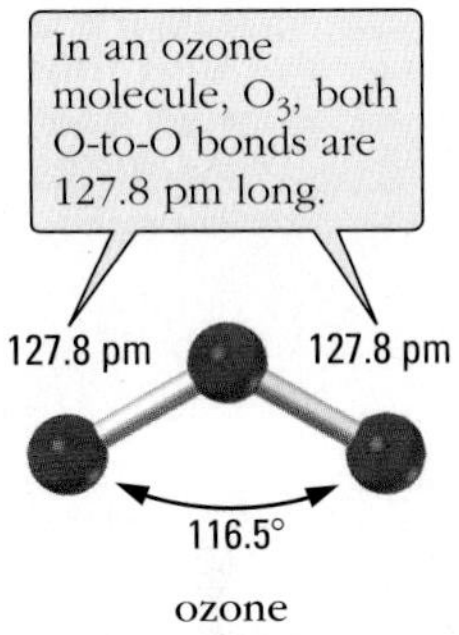

In an ozone molecule, O₃, both O-to-O bonds are 127.8 pm long.
127.8 pm
127.8 pm
116.5°
ozone

Benzene, C_6H_6, is another species in which delocalized pi molecular orbitals play an important role. There are 30 valence electrons—15 electron pairs—in this molecule; 12 of the electron pairs form the sigma-bond framework:

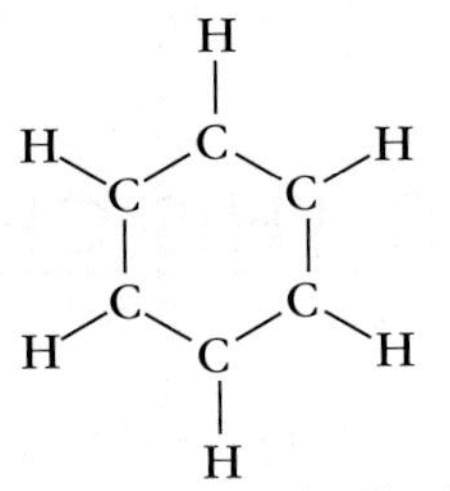

Figure 6.15 shows how the remaining six electrons—three electron pairs—are located in three π orbitals, which according to MO theory extend over more than two atoms. These delocalized electrons are spread over all six C atoms and are represented simply as

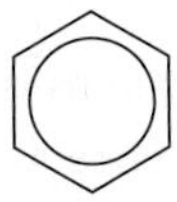

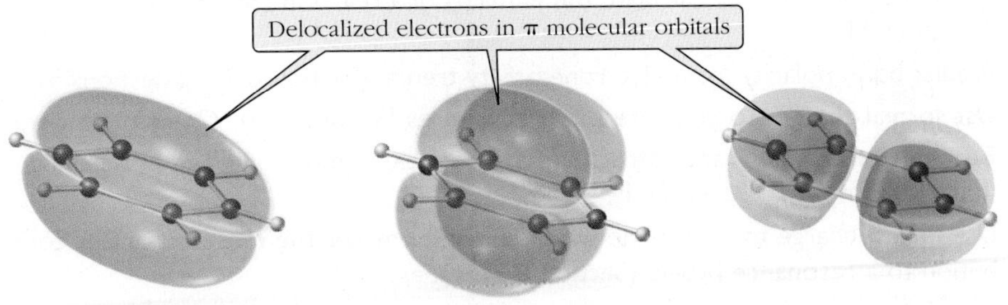

Delocalized electrons in π molecular orbitals

Figure 6.15 Electron delocalization in benzene. Three of the 15 valence electron pairs in the benzene molecule occupy three delocalized π MOs. Electron pairs in π MOs add 0.5 to the bond order between each pair of carbon atoms, giving a total of $6 \times 0.5 = 3$ additional bonds.

SUMMARY PROBLEM

Glycine, $C_2H_5NO_2$, is a naturally occurring amino acid with this Lewis structure,

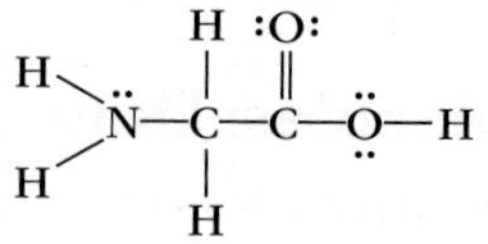

Glycine contains the amine functional group, $—NH_2$, and a carboxylic acid functional group, —COOH.

(a) List the C—H, O—H, C—N, N—H, C—C, C—O, and C=O bonds in glycine in order of increasing bond length.

(b) List the C—H, O—H, C—N, C—C, C—O, and C=O bonds in glycine in order of increasing bond strength.

(c) Arrange these bonds in glycine in order of increasing bond polarity: C—N, C—O, C—H, N—H, O—H, and C—C.

(d) The ionization of glycine in solution forms $H^+(aq)$ and aqueous glycinate ions. Write the structural formula for glycinate ion.

(e) Glycinate ion is an example of a resonance hybrid. Write the Lewis structures that contribute to this resonance hybrid.

(f) Glycine also forms a zwitterion. In the zwitterion an H^+ is released by the carboxylic acid group and attaches to the nitrogen of the amine group. Write the Lewis structure of the zwitterion form of glycine.

(g) Using the Lewis structure drawn for part (f), calculate the formal charges on each of its atoms. Determine the net charge of the zwitterion.

HAVING STUDIED THIS CHAPTER . . .

. . . you should be able to:

- Describe changes in electron distribution and energy that occur when a covalent bond forms (Section 6-1). End-of-chapter Question 15
- Use Lewis structures to represent covalent bonds in molecules and polyatomic ions (Sections 6-1 to 6-4). Questions 15, 17, 21
- Describe bonding in saturated and unsaturated hydrocarbons (Sections 6-3 and 6-5). Questions 23, 25
- Recognize molecules that can have *cis-trans* isomerism (Section 6-5). Question 27
- Predict bond lengths from periodic trends in atomic radii (Section 6-6). Question 29
- Relate bond energy to bond length (Section 6-6). Question 31
- Use bond enthalpies to calculate the enthalpy of a reaction (Section 6-6). Question 34
- Predict bond polarity from electronegativity trends (Section 6-7). Questions 37, 39
- Use formal charges to compare Lewis structures (Section 6-8). Questions 41, 45
- Recognize when resonance structures are possible in molecules and polyatomic ions. (Section 6-9). Question 49
- Use formal charge to determine which structure makes the most important contribution to a resonance hybrid (Section 6-9). Question 51
- Recognize three types of exceptions to the octet rule (Section 6-10). Question 60

- Describe bonding and constitutional isomerism in aromatic compounds (Section 6-11). Questions 63, 65
- Use molecular orbital theory to explain bonding in diatomic molecules and ions (Section 6-12). Questions 67, 69

KEY TERMS

addition reaction (Section 6-9a)

alkene (6-5)

alkyne (6-5)

antibonding molecular orbital (6-12b)

aromatic compound (6-11)

bond energy (6-6b)

bond enthalpy (6-6b)

bond length (6-6a)

bond order (6-12b)

bonding electrons (6-2)

bonding molecular orbital (6-12b)

***cis* isomer** (6-5a)

***cis-trans* isomerism** (6-5a)

covalent bond (6-1)

double bond (6-4)

electronegativity (6-7)

formal charge (6-8)

free radical (6-10b)

functional group (6-3)

geometric isomerism (6-5a)

Lewis structure (6-2)

line drawing (6-3)

line structure (6-3)

lone pair electrons (6-2)

molecular orbital (MO) (6-12a)

molecular orbital (MO) theory (6-12a)

multiple covalent bond (6-4)

nonpolar covalent bond (6-7)

octet rule (6-2)

polar covalent bond (6-7)

resonance hybrid (6-9)

resonance structures (6-9)

saturated hydrocarbon (6-3)

single covalent bond (6-2)

substitution reaction (6-9a)

***trans* isomer** (6-5a)

triple bond (6-4)

unsaturated hydrocarbon (6-5)

valence bond theory (6-1)

QUESTIONS FOR REVIEW AND THOUGHT

Red-numbered questions have short answers at the back of this book in Appendix M and fully worked solutions in the *Student Solutions Manual.*

Review Questions

These questions test vocabulary and simple concepts.

1. Explain the difference between an ionic bond and a covalent bond.
2. What kind of bonding (ionic or covalent) would you predict for the products resulting from the following combinations of elements?
   (a) $Na + I_2$ (b) $C + S_8$
   (c) $Mg + Br_2$ (d) $P_4 + Cl_2$
3. What characteristics must atoms A and X have if they are able to form a covalent bond A—X with each other? A polar covalent bond with each other?
4. Indicate how alkanes, alkenes, and alkynes differ by giving the structural formula of a compound in each class that contains three carbon atoms.
5. Refer to Table 6.3 and answer these questions:
   (a) Do any molecules exist that have more than eight valence electrons around a central atom that is a second-period element?
   (b) What is the maximum number of bond pairs and lone pairs that surround the central atom in any of the molecules in Table 6.3?
6. While sulfur forms the compounds SF_4 and SF_6, no equivalent compounds of oxygen, OF_4 and OF_6, are known. Explain.
7. Which of these molecules have an odd number of valence electrons: NO_2, SCl_2, NH_3, NO_3?
8. Write resonance structures for NO_2^-. Predict a value for the N—O bond length based on bond lengths given in Table 6.1, and explain your answer.
9. Consider these structures for the formate ion, HCO_2^-. Designate which two are resonance structures and which is equivalent to one of the resonance structures.

(a) 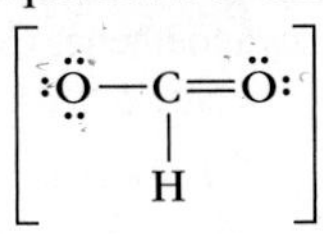

(b) 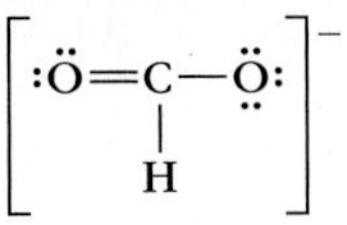

(c) 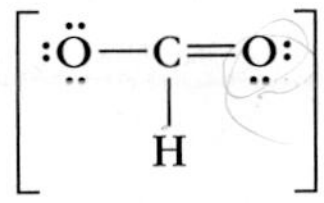

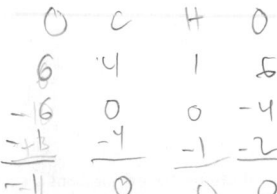

10. Consider a series of molecules in which the C atom is bonded to atoms of second-period elements: C—O, C—F, C—N, C—C, and C—B. Place these bonds in order of increasing bond length.
11. Describe the trends in bond length and bond energy for single, double, and triple carbon-to-oxygen bonds.
12. Why is *cis-trans* isomerism not possible for alkynes?

Topical Questions

These questions are keyed to the major topics in the chapter. Usually a question that is answered at the back of this book is paired with a similar one that is not.

Covalent Bonding (Section 6-1)

13. Explain in your own words why the energy of two H atoms is lower when the atoms are 74 pm apart than when the atoms are (a) 25 pm apart; (b) 100 pm apart.
14. As two F atoms approach, the energy decreases until they are 128 pm apart; then it increases. At 128 pm, E = −158 kJ/mol. Draw a diagram like Figure 6.2 for the F_2 molecule.

Lewis Structures (Sections 6-2, 6-4)

15. Write Lewis structures for these molecules or ions.
    (a) ClF (b) H_2Se
    (c) BF_4^- (d) PO_4^{3-}
16. Write the Lewis structures of (a) dichlorine monoxide, Cl_2O; (b) hydrogen peroxide, H_2O_2; (c) borohydride ion, BH_4^-; (d) phosphonium ion, PH_4^+; and (e) PCl_5.
17. Write Lewis structures for these molecules or ions.
    (a) CH_3Cl (b) SiO_4^{4-}
    (c) ClF_4^+ (d) C_2H_6
18. Write Lewis structures for these molecules.
    (a) Formic acid, HCOOH, in which atomic arrangement is

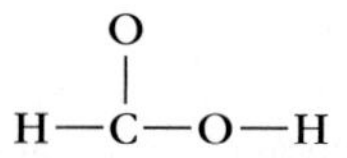

    (b) Acetonitrile, CH_3CN
    (c) Vinyl chloride, CH_2CHCl, the molecule from which PVC plastics are made
19. Write Lewis structures for these molecules.
    (a) Tetrafluoroethylene, C_2F_4, the molecule from which Teflon is made
    (b) Acrylonitrile, CH_2CHCN, the molecule from which Orlon is made
20. Write Lewis structures for
    (a) N_2^+ (b) XeF_7^- (c) tetracyanoethene, C_6N_4

21. Which of these are correct Lewis structures and which are incorrect? Explain what is wrong with the incorrect ones.

(a) 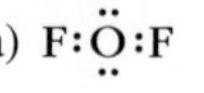
OF_2

(b) 
O_2

(c) 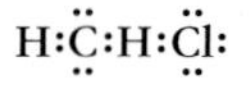
CH_3Cl

(d) 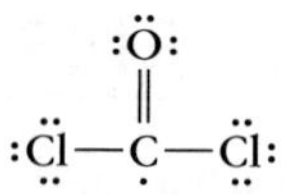
CCl_2O

(e) 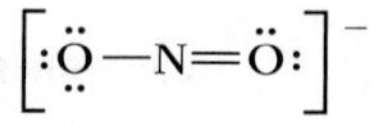
NO_2^-

22. Which of these are correct Lewis structures and which are incorrect? Explain what is wrong with the incorrect ones.

(a) 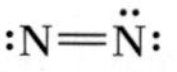
N_2

(b) 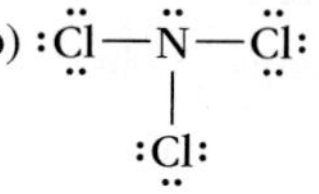
NCl_3

(c) 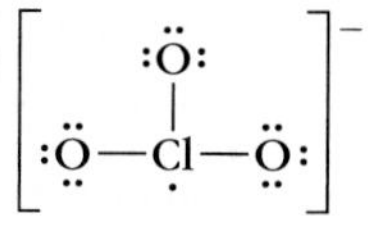
ClO_3^-

(d) 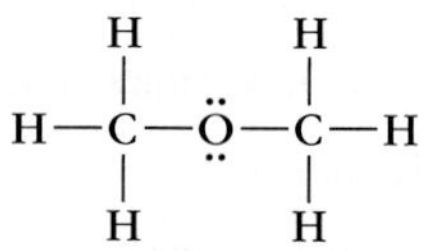
$(CH_3)_2O$

(e) 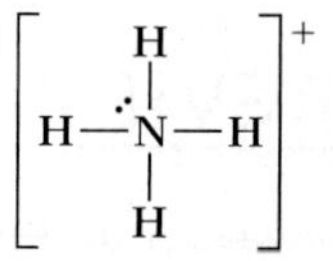
NH_4^+

Bonding in Hydrocarbons (Sections 6-3, 6-5)

23. Write the structural formulas for all the branched-chain compounds with the molecular formula C_6H_{14}.
24. Write structural formulas for two straight-chain alkenes with the formula C_5H_{10}. Are these the only two structures that meet these specifications?
25. From their molecular formulas, classify each of these straight-chain hydrocarbons as an alkane, an alkene, or an alkyne.
    (a) C_5H_8 (b) $C_{24}H_{50}$ (c) C_7H_{14}
26. From their molecular formulas, classify each of these straight-chain hydrocarbons as an alkane, an alkene, or an alkyne.
    (a) $C_{21}H_{44}$ (b) C_4H_6 (c) C_8H_{16}

27. In each case, tell whether *cis* and *trans* isomers exist. If they do, write structural formulas for the two isomers and label each *cis* or *trans*. For those that cannot have *cis-trans* isomers, explain why.
(a) Br_2CH_2
(b) $CH_3CH_2CH{=}CHCH_2CH_3$
(c) $CH_3CH{=}CHCH_3$
(d) $CH_2{=}CHCH_2CH_3$

28. Which of these molecules can have *cis* and *trans* isomers? For those that do, write the structural formulas of the two isomers and label each *cis* or *trans*. For those that cannot have these isomers, explain why.
(a) $CH_3CH_2BrC{=}CBrCH_3$
(b) $(CH_3)_2C{=}C(CH_3)_2$
(c) $CH_3CH_2IC{=}CICH_2CH_3$
(d) $CH_3ClC{=}CHCH_3$
(e) $(CH_3)_2C{=}CHCH_3$

Bond Length and Bond Energy (Section 6-6)

29. For each pair of bonds, predict which is the shorter.
(a) B—Cl or Ga—Cl (b) C—O or Sn—O
(c) P—S or P—O
(d) The C=C or the C=O bond in acrolein

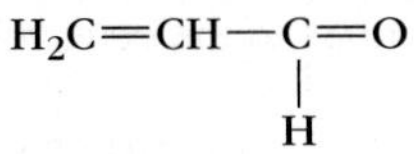

H₂C=CH—C=O
H

30. For each pair of bonds, predict which is the shorter.
(a) Si—N or P—O (b) Si—O or C—O
(c) C—F or C—Br
(d) The C=C or the C≡N bond in acrylonitrile, $H_2C{=}CH{-}C{\equiv}N$

31. Using only a periodic table (not a table of bond energies), predict which is the strongest bond.
(a) Si—F (b) P—S (c) Si—O

32. When sulfur is heated to above 720 °C, the major component is S_2 in which the bond distance is 189 pm. This is significantly shorter than the sulfur-to-sulfur distance in S_8, which has a ring structure. Propose a reason for this difference.

33. Which bond requires more energy to break: the carbon-oxygen bond in formaldehyde, H_2CO, or the carbon-oxygen bond in carbon monoxide, CO?

34. Estimate $\Delta_r H°$ for forming 2 mol ammonia from molecular nitrogen and molecular hydrogen. Is this reaction exothermic or endothermic?

35. Estimate $\Delta_r H°$ for the conversion of 1 mol carbon monoxide to carbon dioxide by combination with molecular oxygen. Is this reaction exothermic or endothermic?

36. Light of appropriate wavelength can break chemical bonds. Light having $\lambda < 240$ nm can dissociate gaseous O_2. It requires light with $\lambda < 819$ nm to dissociate gaseous H_2O_2 to 2 OH. Assume that all of the photon energy is used solely for these dissociations.
(a) Calculate the energy required to dissociate (i) O_2 and (ii) H_2O_2.
(b) Consider the results of part (a). How well do they correlate with the Lewis structures of O_2 and H_2O_2? Explain your answer.

Electronegativity and Bond Polarity (Section 6-7)

37. For each pair of bonds, indicate the more polar bond and use δ+ or δ− to show the partial charge on each atom.
(a) C—O and C—N (b) B—O and P—S
(c) P—H and P—N (d) B—H and B—Cl

38. For each pair of bonds, identify the more polar one and use δ+ or δ− to indicate the partial charge on each atom.
(a) B—Cl and B—O (b) O—F and O—Se
(c) S—Cl and B—F (d) N—H and N—F

39. Urea is used in plastics and fertilizers.

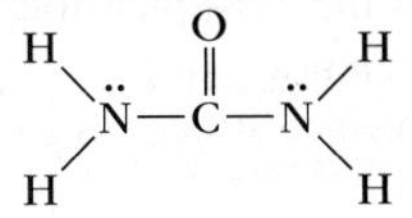

O
H
H
N—C—N
H
H

urea

(a) Which bonds in the molecule are polar and which are nonpolar?
(b) Which is the most polar bond in the molecule? Which atom is the partial negative end of this bond?

40. Acrolein is the starting material for certain plastics.

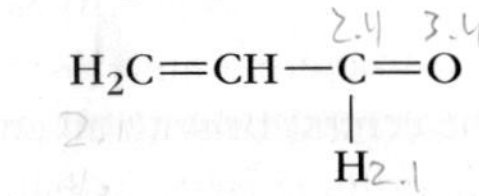

H₂C=CH—C=O
H

acrolein

(a) Which bonds in the molecule are polar and which are nonpolar?
(b) Which is the most polar bond in the molecule? Which atom is the partial negative end of this bond?

Formal Charge (Section 6-8)

41. Write correct Lewis structures and assign a formal charge to each atom.
(a) CH_3CHO (b) N_3^- (c) CH_3CN

42. Write correct Lewis structures and assign a formal charge to each atom.
(a) KrF_4 (b) ClO_3^- (c) SO_2Cl_2

43. Write the correct Lewis structure and assign a formal charge to each atom in fulminate ion, CNO^-.
44. Peroxydisulfate ion, $S_2O_8^{2-}$, contains an —O—O— linkage between two sulfate ions, SO_4^{2-}. Write the correct Lewis structures and assign a formal charge to each atom in the peroxydisulfate and sulfate ions.
45. Two Lewis structures can be written for nitrosyl chloride, which contains one nitrogen, one oxygen, and one chlorine atom per molecule. Write the two Lewis structures and assign a formal charge to each atom.
46. Two Lewis structures can be written for nitrosyl fluoride, which contains one nitrogen, one oxygen, and one fluorine atom per molecule. Write the two Lewis structures and assign a formal charge to each atom.
47. Use Lewis structures and formal charges to determine the bond type (single, double, or triple) for each bond in
(a) POF_3 (b) SOF_4

Resonance (Section 6-9)

48. Consider the SCO_2^{2-} ion in which each S and O atom is bonded to a central C atom. Use Lewis structures and formal charges to write resonance structures and to determine which is the most plausible resonance form. Explain your choice.
49. Write all resonance structures for
(a) nitric acid

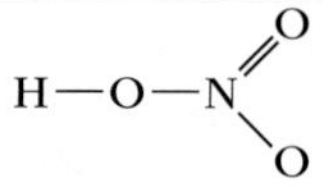
H—O—N
O
O

(b) nitrate ion, NO_3^-
50. Write all the resonance structures for
(a) SO_3 (b) SCN^-
51. Several Lewis structures can be written for perbromate ion, BrO_4^-, the central Br with all single Br—O bonds, or with one, two, or three Br=O double bonds. Draw the Lewis structures of these possible resonance structures, and use formal charges to predict which makes the greatest contribution to the resonance hybrid.
52. Use formal charges to predict which of the resonance structures makes the greatest contribution to the resonance hybrid for (a) HNO_3 and (b) SO_3. (Resonance structures were drawn in Questions 49 and 50.)
53. Several Lewis structures can be written for thiosulfate ion, $S_2O_3^{2-}$. Write the Lewis structures of these possible resonance structures. Predict which structure makes the most important contribution to the resonance hybrid.
54. Compare the carbon-oxygen bond lengths in the formate ion, HCO_2^-, and in the carbonate ion, CO_3^{2-}. In which ion is the bond longer? Explain briefly.
55. Compare the nitrogen-oxygen bond lengths in NO_2^+ and in NO_3^-. In which ion are the bonds longer? Explain briefly.
56. Draw resonance structures for each of these ions: NSO^- and SNO^-. (The atoms are bonded in the order given in each case, that is, S is the central atom in NSO^-.)
(a) Use formal charges to determine which ion is likely to be more stable.
(b) Explain why the two ions cannot be considered resonance structures of each other.
57. Three known isomers exist of N_2CO, with the atoms in these sequences: NOCN; ONNC; and ONCN. Write resonance structures for each isomer and use formal charge to predict which isomer is the most stable.

Exceptions to the Octet Rule (Section 6-10)

58. Write the Lewis structure for
(a) BrF_5 (b) IF_5 (c) IBr_2^-
59. Write the Lewis structure for
(a) BrF_3 (b) I_3^- (c) XeF_4
60. Which of these molecules have Lewis structures that involve exceptions to the octet rule? Classify each exception.
(a) PCl_3 (b) SnF_4 (c) BCl_3
(d) NO
61. Which of these molecules have Lewis structures that involve exceptions to the octet rule? Classify each exception.
(a) SF_6 (b) NH_4NO_3 (c) SeF_2
(d) ClO_2
62. The NO^{2-} radical dianion of nitrogen oxide has been formed, as well as N_2^{3-}, the radical trianion of diatomic nitrogen, electronically analogous to superoxide ion, O_2^-. Write Lewis structures for these two nitrogen radical anions.

Aromatic Compounds (Section 6-11)

63. All carbon-to-carbon bond lengths are identical in benzene. Does this argue for or against the presence of C=C bonds in benzene? Explain.
64. Carbon-to-carbon double bonds, C=C, react by addition. Cite experimental evidence that benzene does not have C=C bonds.
65. Three dibromobenzenes are known. Write the Lewis structure and name for each compound.
66. Write the structural formula for 1,2-diiodobenzene (also known as *ortho*-diiodobenzene). Write the structural formulas for the *meta* and *para* isomers as well.

Molecular Orbital Theory (Section 6-12)

67. Use MO theory to predict the MO diagram, the number of bonds, and the number of unpaired electrons in
(a) peroxide ion, O_2^{2-} (b) B_2^+
(c) Li_2^+ (d) O_2^+

68. Use MO theory to predict the number of electrons in each of the molecular orbitals, the number of bonds, and the number of unpaired electrons in
(a) CO (b) F_2^- (c) NO^-
69. Use molecular orbital theory to predict the arrangement of electrons in MOs, the bond order, and the number of unpaired electrons in
(a) BN (b) cyanide ion, CN^-
70. Both polyatomic ions and uncharged molecules can be detected using spectroscopic measurements. Two examples of polyatomic ions are He_2^{2+} and HHe^+. Predict the arrangement of electrons in MOs and the bond order for each ion.

General Questions

These questions are not explicitly keyed to chapter topics; many require integration of several concepts.

71. Using just a periodic table (not a table of electronegativities), decide which of these is likely to be the most polar bond. Explain your answer.
(a) C—F (b) S—F
(c) Si—F (d) O—F
72. The C—Br bond length in CBr_4 is 191 pm; the Br—Br distance in Br_2 is 228 pm. Estimate the radius of a C atom in CBr_4. Use this value to estimate the C—C distance in ethane, $H_3C—CH_3$. How does your calculated bond length agree with the measured value of 154 pm? Are radii of atoms exactly the same in every molecule?
73. Your friend says, "Elements that are close together in the periodic table form covalent bonds, whereas elements that are far apart form ionic bonds." Is your friend correct? Why or why not?
74. Acrylonitrile is the building block of the synthetic fiber Orlon.

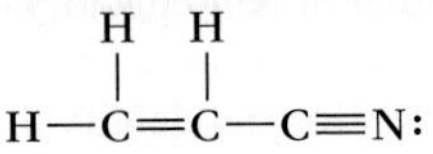

acrylonitrile

(a) Which carbon-carbon bond is shorter?
(b) Which carbon-carbon bond is stronger?
(c) Which bond is the most polar? What is the partial negative end of that bond?
75. In nitryl chloride, NO_2Cl, there is no oxygen-oxygen bond. Write a Lewis structure for the molecule. Write any resonance structures for this molecule.
76. Write Lewis structures for
(a) SCl_2 (b) Cl_3^+ (c) $SOCl_2$
(d) $ClOClO_3$ (contains Cl—O—Cl bond)
77. Write a Lewis structure for each molecule. Which formulas represent alkanes? Which are probably aromatic? Which fall into neither category?
(a) C_8H_{10} (b) $C_{10}H_8$ (c) C_6H_{12}
(d) C_6H_{14} (e) C_8H_{18} (f) C_6H_{10}

Applying Concepts

These questions test conceptual learning.

78. Based on the structural formulas given for these *cis* or *trans* alkenes,

(a) 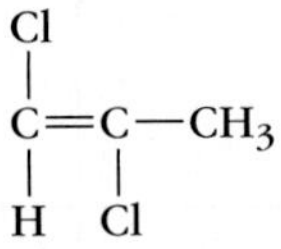 *trans*-1,2-dichloropropene

(b) 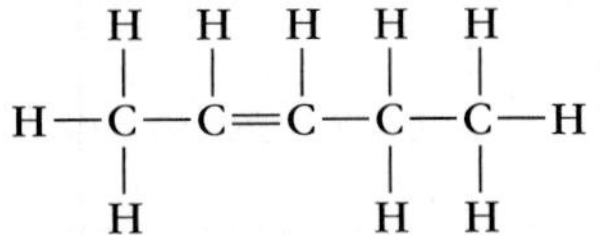 *cis*-2-pentene

(c) 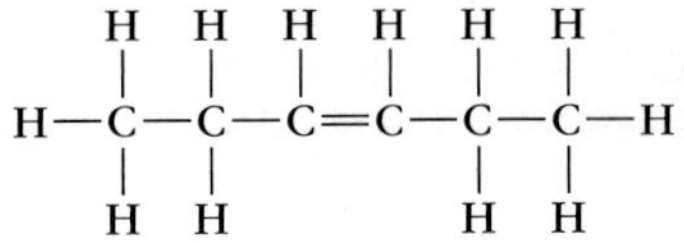 *cis*-3-hexene

(d) 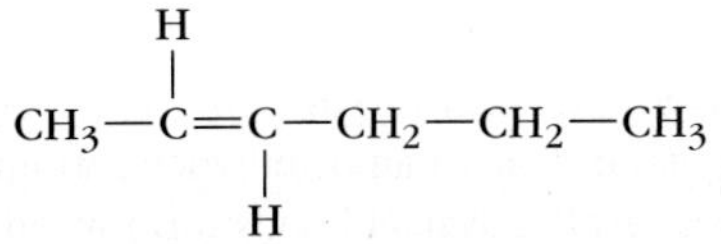 *trans*-2-hexene

write the structural formulas for
(a) *cis*-1,2-dichloropropene (b) *trans*-2-pentene
(c) *trans*-3-hexene (d) *cis*-2-hexene
79. A student drew this incorrect Lewis structure for ClO_3^-. What errors were made when determining the number of valence electrons?

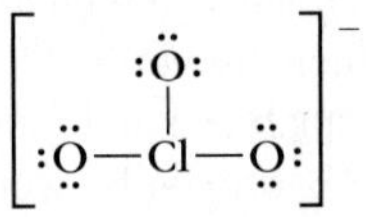

80. This Lewis structure for SF_5^+ is drawn incorrectly. What error was made when determining the number of valence electrons?

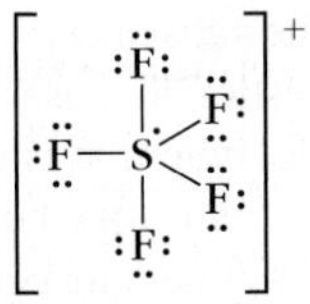

81. Tribromide, Br_3^-, and triiodide, I_3^-, ions are often found in aqueous solutions, but trifluoride ion, F_3^-, is so rare that its bond strength was only measured in 2000. Explain.
82. Explain why nonmetal atoms in Period 3 and beyond can accommodate greater than an octet of electrons and those in Period 2 cannot do so.
83. Why is this not an example of resonance structures?

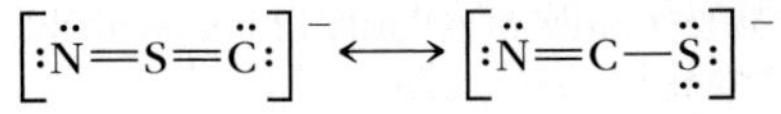

Grid for Question 87

1 Si—Br > P—Br > S—Br > Cl—Br	2 Si—F	3 S—Cl < Se—Cl < O—Cl
4 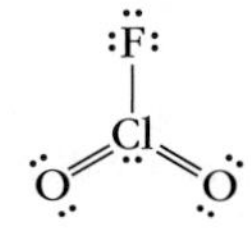	5 Cl—Br > S—Br > P—Br > Si—Br	6 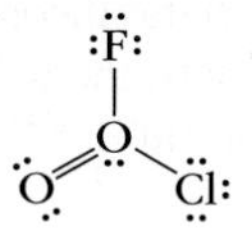
7 Si—H	8 Si—Si	9 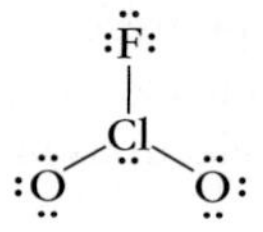
10 O—Cl > S—Cl > Se—Cl	11 C—F > C=N > C—P	12 C—P < C—F < C=N

84. In another universe, elements try to achieve a nonet (nine valence electrons) instead of an octet when forming chemical bonds. As a result, covalent bonds form when a trio of electrons is shared between two atoms. Draw Lewis structures for the compounds that would form between (a) hydrogen and oxygen, and (b) hydrogen and fluorine.
85. The elements As, Br, Cl, S, and Se have electronegativity values of 2.6, 2.3, 2.1, 2.7, and 2.4, but not in that order. Using the periodic trend for electronegativity, assign the values to the elements. Which assignments are you certain about? Which are you not?
86. A substance is analyzed and found to contain 85.7% carbon and 14.3% hydrogen by weight. A gaseous sample of the substance is found to have a density of 1.87 g/L, and 1 mol of it occupies a volume of 22.4 L. What are two possible Lewis structures for molecules of the compound? (*Hint:* First determine the empirical formula and molar mass of the substance.)
87. The grid for Question 87 has 12 numbered boxes, each of which contains an item that is used to answer the questions that follow. Items may be used more than once and there may be more than one correct item in response to a question.
Place the number(s) of the correct selection(s) on the appropriate line.
(a) Shortest Si-containing bond. ________
(b) Correct sequence of bond lengths for Period 3 bromides. ________
(c) Resonance hybrid structure(s) of chloryl fluoride, $FClO_2$. ________
(d) Correct order of polarity of bonds in chlorides of Group 6A. ________
(e) Correct sequence of bond enthalpies of carbon-containing bonds. ________
(f) Weakest silicon-containing bond (of those given). ________
88. The grid for Question 88 has 12 numbered boxes, each of which contains an item that is used to answer the questions that follow. Items may be used more than once and there may be more than one correct item in response to a question.
Place the number(s) of the correct selection(s) on the appropriate line.
(a) Correct sequence of bond enthalpies for multiple bonds. ________
(b) Correct sequence of bond lengths for multiple bonds. ________
(c) Resonance hybrid structure(s) of $[ClO_2]^+$. ________
(d) Correct order of bond polarity of hydrides of Period 2. ________
(e) *cis* and *trans* forms (identify the *cis* then the *trans* form). ________
89. Which of these molecules is least likely to exist: NF_5, PF_5, SbF_5, or IF_5? Explain why.
90. Write the Lewis structure for nitrosyl fluoride, FNO. Using only a periodic table, identify
(a) which is the longer bond.
(b) which is the stronger bond.
(c) which is the more polar bond.
91. Write the Lewis structure for nitrosyl chloride, ClNO. Using only a periodic table, identify
(a) which is the longer bond.
(b) which is the stronger bond.
(c) which is the more polar bond.

Grid for Question 88

1 C≡C < N≡N < C≡O	**2** 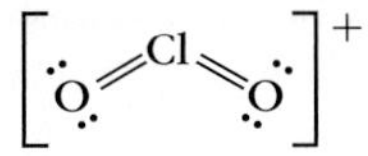	**3** 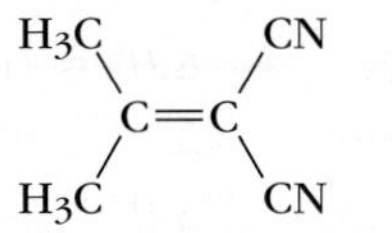
4 C—H < N—H < O—H < F—H	**5** 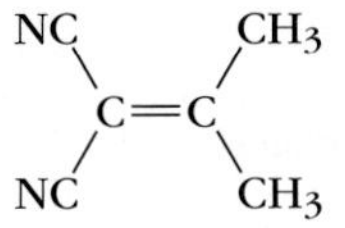	**6** 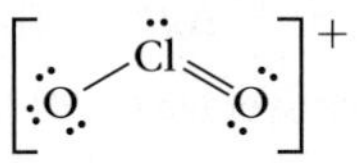
7 N≡N < C≡O < C≡C	**8** C—H > N—H > O—H > F—H	**9** 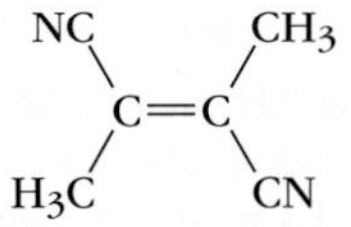
10 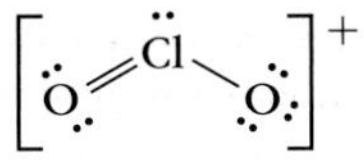	**11** 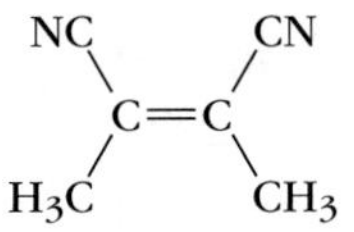	**12** C≡C > N≡N > C≡O

92. Methylcyanoacrylate is the active ingredient in "super" glues. Its Lewis structure is

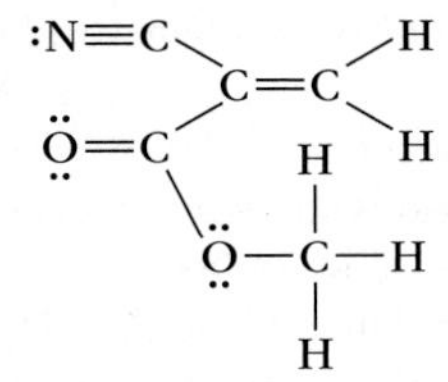

In this molecule, which is the
(a) weakest carbon-containing bond?
(b) strongest carbon-containing bond?
(c) most polar bond?

93. Aspirin is made from salicylic acid, which has this Lewis structure:

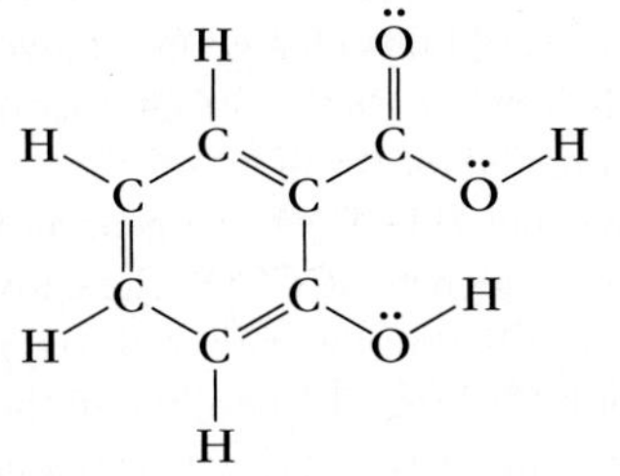

(a) Which is the longest carbon-carbon bond?
(b) Which is the strongest carbon-oxygen bond?
(c) Draw resonance structures for this molecule.

94. For many people, tears flow when they cut raw onions. The tear-inducing agent is thiopropionaldehyde-S-oxide, C_3H_6OS. Write the Lewis structure for this molecule.

95. Phosphorus and sulfur form a series of compounds, one of which is tetraphosphorus trisulfide. Write the Lewis structure for this molecule.

96. Tetrasulfur tetranitride reacts with disulfur dichloride to form S_4N_3Cl, a salt.

$$3\ S_4N_4 + 2\ S_2Cl_2 \longrightarrow 4\ S_4N_3Cl$$

Write a plausible Lewis structure for the two reactants and the cation of the salt.

More Challenging Questions

These questions require more thought and integrate several concepts.

97. Elemental phosphorus has the formula P_4. Propose a Lewis structure for this molecule. [*Hints:* (1) Each phosphorus atom is bonded to three other phosphorus atoms. (2) Visualize the structure three-dimensionally, not flat on a page.]

98. When we estimate $\Delta_r H°$ from bond enthalpies we assume that all bonds of the same type (single, double, triple) between the same two atoms have the same energy, regardless of the molecule in which they occur. The purpose of this problem is to show you that this is only an

approximation. You will need these standard enthalpies of formation:

C(g)	$\Delta_f H° = 716.7$ kJ/mol
CH(g)	$\Delta_f H° = 596.3$ kJ/mol
CH_2(g)	$\Delta_f H° = 392.5$ kJ/mol
CH_3(g)	$\Delta_f H° = 146.0$ kJ/mol
H(g)	$\Delta_f H° = 218.0$ kJ/mol

(a) What is the average C—H bond energy in methane, CH_4?

(b) Using bond enthalpies, estimate $\Delta_r H°$ for the reaction

$$CH_4(g) \longrightarrow C(g) + 2\ H_2(g)$$

(c) By heating CH_4 in a flame it is possible to produce the reactive gaseous species CH_3, CH_2, CH, and even carbon atoms, C. Experiments give these values of $\Delta_r H°$ for the reactions shown:

$$CH_3(g) \longrightarrow C(g) + H_2(g) + H(g) \qquad \Delta_r H° = 788.7 \text{ kJ}$$

$$CH_2(g) \longrightarrow C(g) + H_2(g) \qquad \Delta_r H° = 324.2 \text{ kJ}$$

$$CH(g) \longrightarrow C(g) + H(g) \qquad \Delta_r H° = 338.3 \text{ kJ}$$

For each of the reactions in part (c), draw a diagram similar to Figure 6.6. Then, calculate the average C—H bond energy in CH_3, CH_2, and CH. Comment on any trends you see.

99. Nitrosyl azide, N_4O, is a pale yellow solid first synthesized in 1993. Write the Lewis structure for nitrosyl azide.

100. Write the Lewis structures for
(a) $(Cl_2PN)_3$ (b) $(Cl_2PN)_4$

101. Nitrous oxide, N_2O, is a linear molecule that has the two nitrogen atoms adjacent to each other. Using concepts from this chapter, explain why the structure is not NON with the nitrogen atoms each bonded to the oxygen.

102. The azide ion, N_3^-, has three resonance hybrid structures.
(a) Write the Lewis structure of each.
(b) Use formal charges to determine which is the most favorable resonance structure.

103. Hydrazoic acid, HN_3, has three resonance hybrid structures.
(a) Write the Lewis structure of each.
(b) Use formal charges to determine which is the least favorable resonance structure.

104. Write the Lewis structures for
(a) $(SO_3)_3$ (b) $FXeN(SO_2F)_2$

105. Experimental evidence indicates the existence of HC_3N molecules in interstellar clouds. Write a plausible Lewis structure for this molecule.

106. Molecules of SiC_3 have been discovered in interstellar clouds. Write a plausible Lewis structure for this molecule.

107. One of the structural isomers of C_3H_6OS is the compound that makes you cry when you slice onions. Write Lewis structures for two isomers of this molecule.

108. Piperine, the active ingredient in black pepper, has this structural formula:

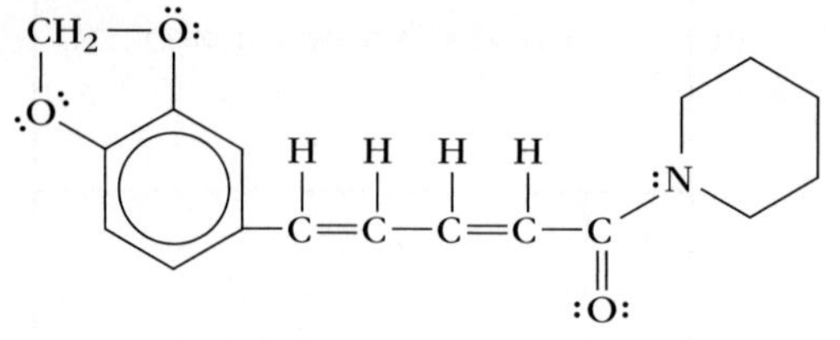

(a) Write the molecular formula of piperine.
(b) Identify the shortest carbon-to-carbon bond in piperine.
(c) Identify the shortest carbon-to-oxygen bond in piperine.
(d) Identify the strongest carbon-to-carbon bond in piperine.
(e) Identify the most polar bond in piperine.

109. Sulfur and oxygen form a series of 2− anions including sulfite, SO_3^{2-}, and sulfate, SO_4^{2-}. In addition to these, there are three other more complex anions—dithionite, $S_2O_4^{2-}$, dithionate, $S_2O_6^{2-}$, and tetrathionate, $S_4O_6^{2-}$. Write correct Lewis structures for dithionite, dithionate, and tetrathionate ions.

110. Gaseous molecules in the ground vibrational state have a vibrational energy given by the equation $E = \frac{1}{2}h\nu$ where h is Planck's constant and ν is the frequency of the vibration. Consider a water vapor molecule, HOH, in which one of the lighter isotope hydrogen atoms, H, is replaced by the heavier isotope deuterium atom, D, to form HOD.
(a) Will an HOD molecule have a higher or lower ground vibrational state energy than an HOH molecule? Explain your answer.
(b) Which bond is stronger: O—H or O—D? Explain your answer.

111. Gamma-aminobutanoic acid, $C_4H_9O_2N$, is a neurotransmitter. The molecule contains a four-carbon chain.
(a) Write the Lewis structure for this molecule.
(b) Which bond is the longest?
(c) Which bond is the most polar?
(d) Which bond is the strongest?

112. Suppose in building up molecular orbitals, the π_{2p} were placed above the σ_{2p}. Prepare a diagram similar to Figure 6.11 based on these changes. For which species in Table 6.4 would this change in relative energies of the MOs affect the prediction of number of bonds and number of unpaired electrons?

113. In carbon suboxide, C_3O_2, a linear molecule, the atoms are in the sequence OCCCO. The carbon-to-carbon bond distance is 128 pm; the carbon-to-oxygen bond distance is 119 pm. Describe the bonding in this molecule.

114. Compare and contrast the valence-bond and the molecular-orbital descriptions of bonding in the gaseous, diatomic S_2 molecule. The sulfur-to-sulfur bond distance is 189 pm. A chemistry classmate says, "The valence bond description of this molecule is not correct." Explain whether the classmate's comment is accurate.

115. Write Lewis structures for these three anions: C_2^{2-} (acetylide); N_3^- (azide); and O_3^- (ozonide). Write resonance structures if there are any. Use formal charges to determine which is the more likely resonance structure.

116. The bond length can be considered as approximately the sum of the atomic radii of the two bonded atoms. The atomic radii for single-bonded carbon and oxygen are 77 pm and 74 pm, respectively. For double-bonded C and O the values are 67 pm and 60 pm, respectively. Use these data to estimate the carbon-to-oxygen bond length in
(a) methanol, CH_3OH;
(b) dimethyl ether, CH_3OCH_3;
(c) formaldehyde, H_2CO.
Explain any difference among the bond lengths (see Table 6.1 as a reference).

117. Consider the $SC[N(CH_3)_2]_2$ molecule in which S is bonded to C and each CH_3 group is bonded to N. Use Lewis structures and formal charges to write resonance structures and to determine which is the most plausible resonance structure.

118. Consider the reaction: $P_4(g) \longrightarrow 2\ P_2(g)$ for which $\Delta_r H = 229$ kJ/mol. The bond energy of a P—P single bond is 209 kJ/mol.
(a) Calculate the bond energy of a phosphorus-to-phosphorus triple bond.
(b) Compare this calculated value with the bond energy of N_2 and propose an explanation for the difference in bond energies between P_2 and N_2.

Conceptual Challenge Problems

These rigorous, thought-provoking problems integrate conceptual learning with problem solving and are suitable for group work.

CP6.A (Section 6-2) What deficiency is acknowledged by chemists when they write the formula for silicon dioxide as $(SiO_2)_n$ but write CO_2 for carbon dioxide?

CP6.B (Section 6-7) Without referring to the periodic table, write an argument to predict how the electronegativities of elements change based on the composition of their atoms.

CP6.C (Section 6-7) How would you rebut the statement, "There are no ionic bonds, only polar covalent bonds"?

7 | Molecular Structures

matuchaki/Shutterstock.com

Water is one of the most common yet important compounds on Earth. A water molecule's two hydrogen atoms are covalently bonded to a central oxygen atom. But, water is more complex than its simple molecular formula, H_2O, might suggest. What would a water molecule look like, if we could see it? (Section 7-1) How do chemists predict the shape of water and other molecules? (Section 7-2) How does bonding and shape of water molecules affect the properties of water? (Section 7-6) Why are shapes of molecules and forces between molecules important in living organisms? (Section 7-7) These and many other questions about the interactions and forces between water molecules and many other kinds of molecules are described in this chapter.

Although the composition, empirical formula, molecular formula, and Lewis structure of a substance provide important information, they are not sufficient to predict or explain the properties of most molecular compounds. Also important are the arrangement of the atoms and how they occupy three-dimensional space—the shape of a molecule. Because of differences in the sequence in which their atoms are joined, molecules can have the same numbers of the same kinds of atoms, yet have different properties. For example, ethanol (in alcoholic beverages) and dimethyl ether (a refrigerant) have the same molecular formula,

C_2H_6O, but the C, H, and O atoms of the two molecules are arranged differently. The melting points (m.p.) of the two compounds differ by 27 °C and their boiling points (b.p.) by 103 °C.

The ideas about molecular shape presented in this chapter are crucial to understanding the relationships between the structure and the properties of molecules, properties related to the behavior of molecules in living organisms, the design of molecules that are effective drugs, and many other aspects of modern chemistry.

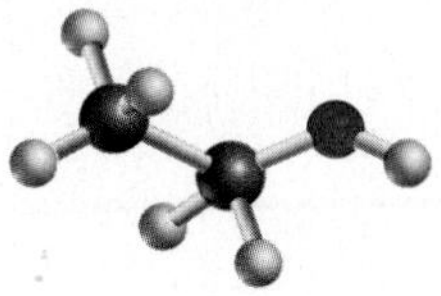

ethanol
m.p. = −114.1 °C;
b.p. = 78.3 °C

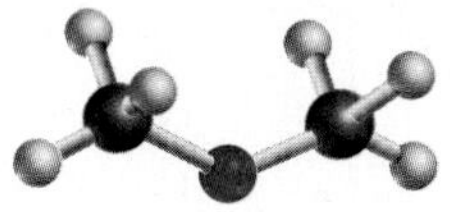

dimethyl ether
m.p. = −141.5 °C;
b.p. = −24.8 °C

7-1 Molecular Models

Molecules are three-dimensional aggregates of atoms, much too small to see. Molecular sizes depend on the sizes of electron clouds of the valence electrons (outermost electron shells) of atoms in the molecules. Molecular shapes depend also on the angles between bonds. Our ability to understand three-dimensional molecular structures is aided by the use of molecular models, which make atomic sizes and bond angles visible. Molecular models for a water molecule are shown in Figure 7.1.

Probably the best example of the impact molecular models can have on the advancement of science is the double helix model of DNA used by James Watson and Francis Crick, which revolutionized the understanding of human heredity and genetic disease. We will discuss DNA structure at the end of this chapter. Watson and Crick's DNA model was a physical model, assembled atom by atom from wooden balls and metal rods, like the ball-and-stick model of water shown in Figure 7.1. Today, research in chemistry, biochemistry, and other biomedical fields depends on computer-generated pictures of molecular models that can be manipulated using a mouse (Figure 7.1). The computer programs that generate these pictures rely on the most accurate experimentally derived data on atomic radii, bond lengths, and bond angles.

Figure 7.2 is a computer-generated model of hemoglobin, the remarkable and essential protein in the blood that takes up and releases oxygen. This vital protein is made up of four long chains of atoms. Each chain is indicated by a different colored ribbon. In a hemoglobin molecule, oxygen binds to Fe^{2+} ions at four sites where individual atoms are shown in gray, red, and blue in Figure 7.2. Computer graphics can rotate such computer-generated models to provide different views of the molecule. In the right side of Figure 7.2 the structure has been rotated to better show the oxygen-binding sites.

Ball-and-stick model kits are available in many campus bookstores. The models are easy to assemble and will help you to visualize the molecular geometries described in this chapter. Ball-and-stick models are relatively inexpensive compared with space-filling models.

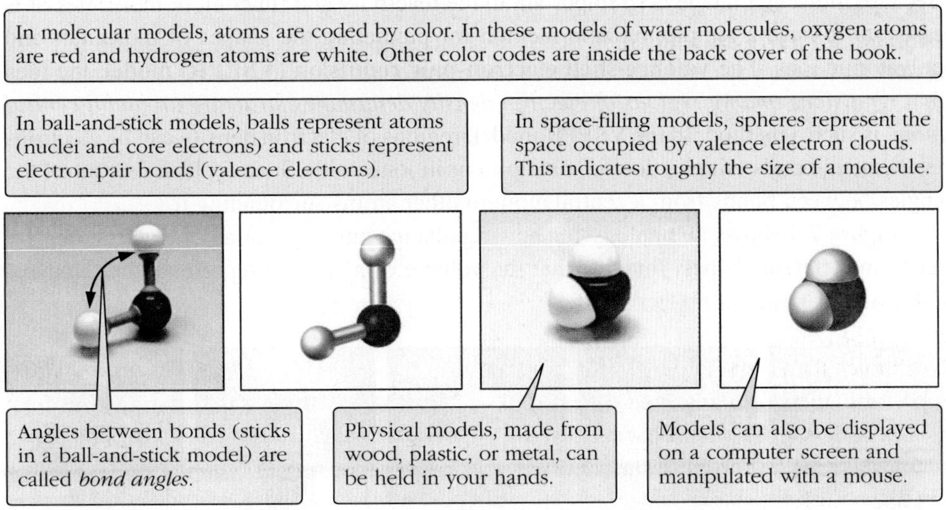
In molecular models, atoms are coded by color. In these models of water molecules, oxygen atoms are red and hydrogen atoms are white. Other color codes are inside the back cover of the book.
In ball-and-stick models, balls represent atoms (nuclei and core electrons) and sticks represent electron-pair bonds (valence electrons).
In space-filling models, spheres represent the space occupied by valence electron clouds. This indicates roughly the size of a molecule.
Angles between bonds (sticks in a ball-and-stick model) are called *bond angles*.
Physical models, made from wood, plastic, or metal, can be held in your hands.
Models can also be displayed on a computer screen and manipulated with a mouse.

Photos: © Cengage Learning/ Charles D. Winters

Figure 7.1 Ball-and-stick and space-filling models of a water molecule.

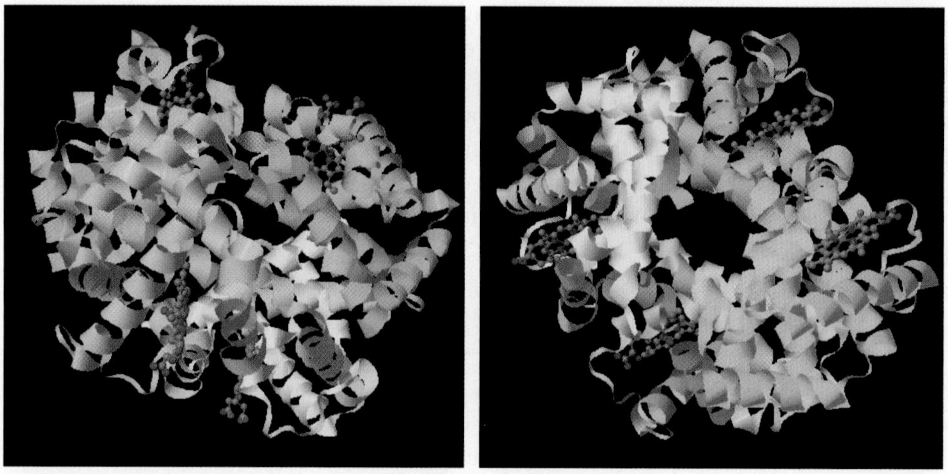

Figure 7.2 Two views of hemoglobin, a protein essential to human life. Because the molecule is so large, ribbons are used to represent many atoms. The structure at the left was rotated using a computer mouse to obtain the structure at the right.

Both physical and computer-generated models show that molecules are three-dimensional. To convey a three-dimensional perspective for a molecule drawn on a flat surface, such as the page of a book, chemists use solid wedges (▬) to represent bonds extending in front of the page and dashed lines (---) or dashed wedges (⋯) to represent bonds behind the page. Bonds that lie in the plane of the page are indicated by a line (—), as illustrated in this perspective drawing for methane, a tetrahedral molecule:

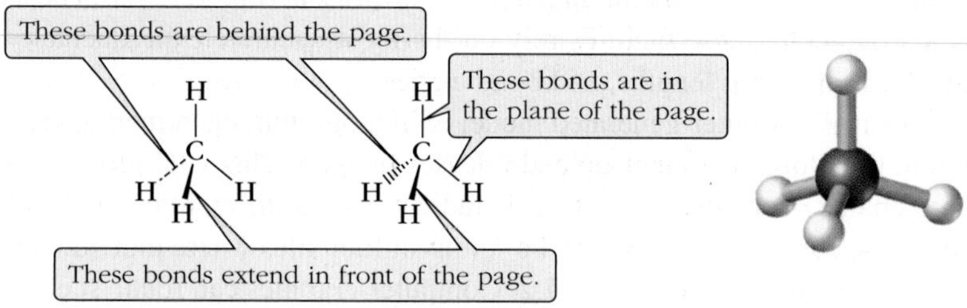

These bonds are behind the page.
These bonds are in the plane of the page.
These bonds extend in front of the page.
H
C
H
H
H

7-2 Predicting Molecular Shapes: VSEPR

Because molecular shape is so important to the reactivity and function of molecules, it is essential to have a simple, reliable method for predicting the shapes of molecules and polyatomic ions. The **valence-shell electron-pair repulsion (VSEPR) model**, the idea that *repulsions among regions of electron density determines structure around a central atom*, is such a method. In the VSEPR model, regions of electron density can be electrons in chemical bonds (single, double, or triple) or in lone pairs. Such repulsions control the angles between bonds from a central atom to other atoms surrounding it.

Figure 7.3 shows an analogy for how repulsions among regions of electron density result in different shapes. Imagine that the volume of a balloon represents the repulsive

Photos: © Cengage Learning/Charles D. Winters

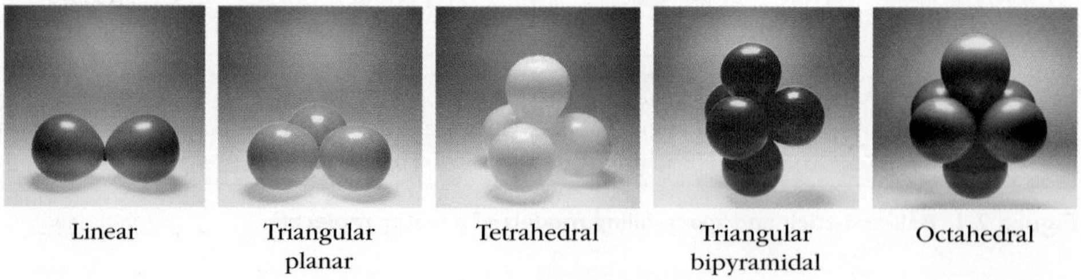

Linear
Triangular planar
Tetrahedral
Triangular bipyramidal
Octahedral

Figure 7.3 Balloon models of geometries predicted by the VSEPR model. The balloons are as far apart as possible. (Corresponding molecular structures are in Table 7.1.)

force of a region of electron density that prevents other regions of electron density from occupying the same space. When two or more balloons are tied together at a central point, the cluster of balloons assumes the shapes shown in Figure 7.3. The central point represents the nucleus and the core electrons of a central atom. Each balloon represents a region of electron density that can bind another atom to the central atom by electron pair bonding. To minimize repulsions among like electrical charges, regions of electron density are oriented as far apart as possible. Because electrons are constrained by electrostatic attraction to be near the nucleus, the shapes illustrated in the balloon analogy are those from which VSEPR theory predicts molecular shape.

The term "molecular geometry" is often used rather than "molecular shape."

To determine the three-dimensional structure of a molecule, we first determine the orientation of regions of electron density around each atom that bonds to two or more other atoms. Then, other atoms are bonded to each central atom using some or all of the regions of electron density (which can be single, double, or triple bonds). The atoms surrounding a central atom are called *terminal atoms*, and it is these atoms that determine the molecular shape around the central atom. Thus, we can distinguish two categories of geometries:

- **Electron-region geometry** is *determined by the number of regions of electron density around a central atom.*
- **Molecular geometry** (molecular shape) is *the arrangement of terminal atoms in space around a central atom.*

For central atoms with no lone pair electrons, electron-region geometry and molecular geometry are the same, but when lone electron pairs are present on the central atom, electron-region and molecular geometries are *not* the same. We can summarize a broad range of molecular structures using an AX_nE_m shorthand notation: A (a central atom) is bonded to X_n (*n* terminal atoms, X); E_m represents *m* lone pair electrons, E, on the central atom. When there are no lone pairs, the symbolism is E_0. For example, an AX_3E_0 designation denotes a central atom (A) with no lone pairs (E_0) and whose shared pairs are bonded to three terminal atoms (X_3), such as in BF_3.

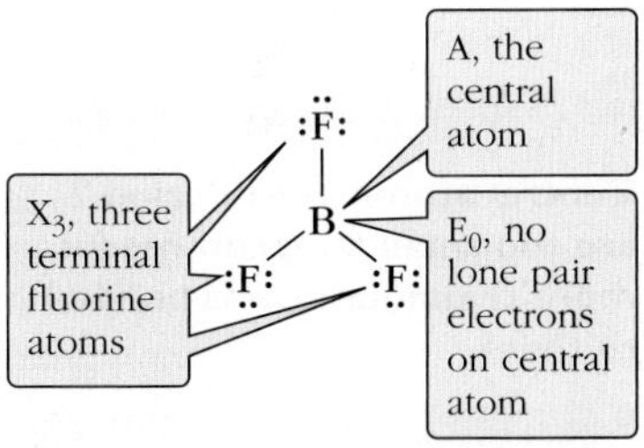

A, the central atom
X_3, three terminal fluorine atoms
E_0, no lone pair electrons on central atom
:F:
B
:F:
:F:

Using the AX_3E_0 designation for boron trifluoride.

7-2a Central Atoms with Only Bonding Pairs, AX_nE_0

The simplest application of VSEPR is to molecules and polyatomic ions in which every region of electron density surrounding a central atom bonds to another atom and the central atom has no lone pairs, only bonding pairs. Table 7.1 illustrates geometries predicted by the VSEPR model for molecules of types AX_2E_0 to AX_6E_0.

The predicted bond angles given in the examples in Table 7.1 are in agreement with experimental values obtained from structural studies.

Table 7.1 Geometries Predicted by VSEPR for Molecules with No Lone Pairs on Central Atom

Molecular model	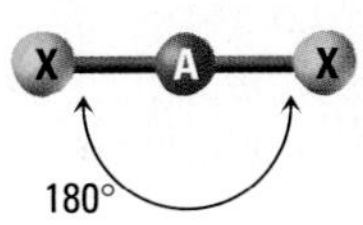 X A X 180°	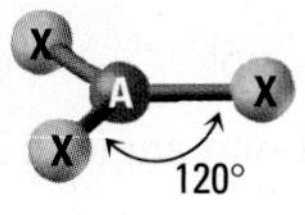 X A X X 120°	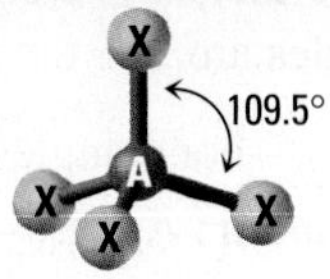 X A X X X 109.5°	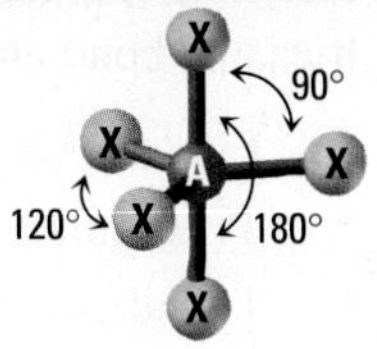 X A X X X X 90° 120° 180°	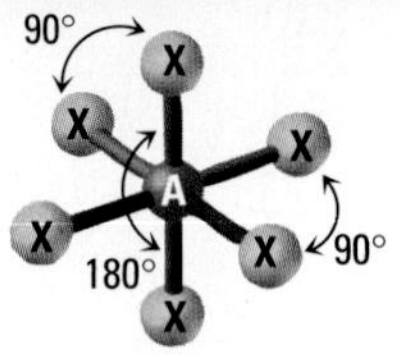 X A X X X X X X 90° 180° 90°
Type	AX_2E_0	AX_3E_0	AX_4E_0	AX_5E_0	AX_6E_0
Electron-region geometry	Linear	Triangular planar	Tetrahedral	Triangular bipyramidal	Octahedral
Molecular geometry	Linear	Triangular planar	Tetrahedral	Triangular bipyramidal	Octahedral
Examples	BeF_2, CO_2	BF_3, CO_3^{2-}	CH_4, $SiCl_4$	PCl_5, PF_5	SF_6

As shown in Figure 7.1 and Table 7.1, the **bond angle** is *the angle between the bonds of two atoms that are bonded to the same third atom.* In a methane molecule, CH_4, for example, the four C—H bonds point toward the corners of a tetrahedron and all the H—C—H bond angles are 109.5°.

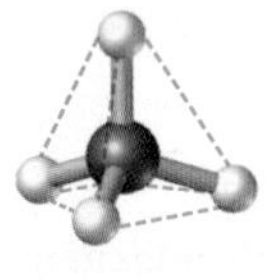

The four bonds in methane, CH_4, point toward the corners of a tetrahedron (blue dashed lines); the bond angles are all 109.5°.

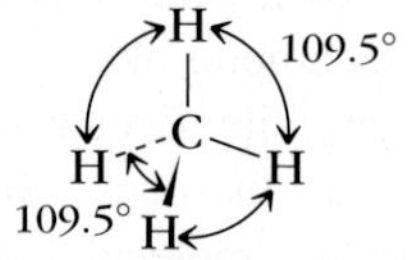
H
109.5°
C
H
H
109.5°
H

methane, CH_4, a tetrahedral molecule

Notice that except for the triangular bipyramidal structure (fifth column in Table 7.1), all bond angles in each structure are equal. The triangular bipyramidal structure has 90° angles between the top and bottom atoms and the three around the middle, but 120° angles between pairs of bonds around the middle.

The geometries illustrated in Table 7.1 are the ones found experimentally for typical molecules and polyatomic ions in which there are no lone pairs on the central atom. The tetrahedral geometry (fourth column in Table 7.1) is by far the most common, because, with four bonding electron pairs, the central atom obeys the octet rule. The central atom has less than an octet in the linear and triangular planar geometries (second and third columns); in these structures there are two bonding pairs and three bonding pairs, respectively. The central atom has more than an octet in the trigonal bipyramidal and octahedral geometries in the last two columns in Table 7.1; these two geometries have five bonding pairs and six bonding pairs, respectively. The trigonal bipyramidal and octahedral structures are expected only when the central atom is an element in Period 3 or higher (← **Sec. 6-10c**).

Central atoms from Period 2 are too small to accommodate more than four atoms bonded to them.

Suppose you want to use VSEPR concepts to predict the shape of $SiCl_4$. First, draw the Lewis structure, with Si as the central atom. From the Lewis structure, you can see that around Si there are four regions of electron density (four single bonds to Cl) and no lone pairs. This is classified as AX_4E_0, where A is Si, X is Cl, and X_4 indicates four Cl atoms bonded to A; E_0 denotes no lone pairs on the central atom. Based on the balloon analogy (Figure 7.3), you would predict a tetrahedral electron-region geometry and a tetrahedral molecular geometry (Table 7.1). This is in agreement with structural studies of $SiCl_4$, which indicate a tetrahedral molecule with all Cl—Si—Cl bond angles 109.5°.

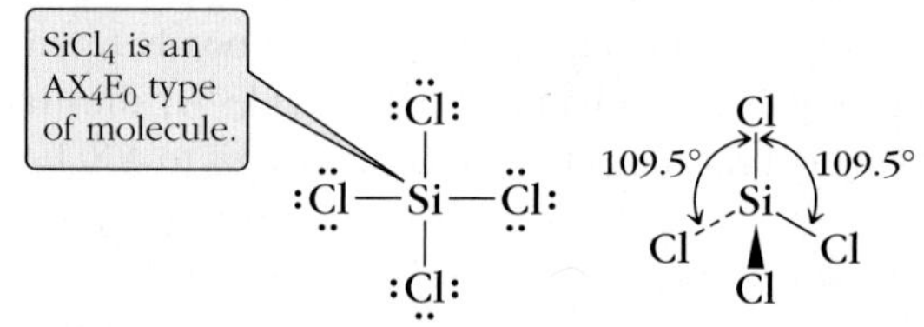
SiCl4 is an AX4E0 type of molecule.
:Cl:
:Cl—Si—Cl:
:Cl:
Cl
109.5°
109.5°
Si
Cl
Cl
Cl

In summary, molecules and polyatomic ions whose central atoms have *no* lone pair electrons (E_0) have the *same* electron-region *and* molecular geometries. These geometries and bond angles are

- **Linear:** AX_2E_0—bond angles 180°.
- **Triangular planar:** AX_3E_0—bond angles 120°.
- **Tetrahedral:** AX_4E_0—bond angles 109.5°.
- **Triangular bipyramidal:** AX_5E_0—bond angles 90°, 120°, and 180°.
- **Octahedral:** AX_6E_0—bond angles 90° and 180°.

PROBLEM-SOLVING EXAMPLE 7.1

Molecular Geometry

Use the VSEPR model to predict the electron-region geometry, molecular geometry, and bond angles of (a) BF_3 and (b) CBr_4.

Result

(a) Triangular planar electron-region and molecular geometries with 120° F—B—F bond angles

(b) Tetrahedral electron-region and molecular geometries with 109.5° Br—C—Br bond angles

Analyze Electron-region geometries, molecular geometries, and bond angles are desired. VSEPR concepts enable such predictions.

Plan Apply the VSEPR guidelines: Write the correct Lewis structure. Determine the number of electron regions around the central atom and classify the molecule according to its AX_nE_m type. Use the balloon analogy (Figure 7.3) to choose the correct electron-region geometry around the central atom. Use electron regions to bond terminal atoms to the central atom and determine the molecular shape.

Execute

(a) The Lewis structure of BF_3 is

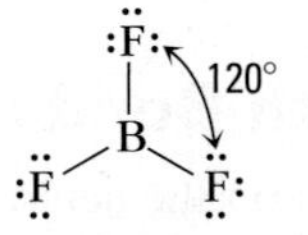
120°

BF_3 is an AX_3E_0-type molecule with three B—F single bonds and no lone pairs around the central B atom. Three regions of electron density are at 120° angles and each bonds an F atom to B. Therefore, BF_3 has triangular planar electron-region geometry and molecular geometry with all atoms in the same plane.

(b) CBr_4, with four regions of electron density around C, is an AX_4E_0-type molecule. Four regions of electron density gives both tetrahedral electron-region and molecular geometries with 109.5° Br—C—Br bond angles.

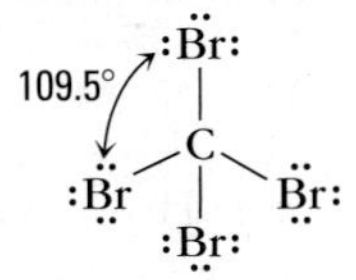
109.5°

PROBLEM-SOLVING PRACTICE 7.1

Identify the electron-region geometry, the molecular geometry, and the bond angles for BeF_2.

PROBLEM-SOLVING PRACTICE answers are provided at the back of this book in Appendix K.

7-2b Multiple Bonds and Molecular Geometry

Double bonds and triple bonds affect molecular geometry because they are shorter and stronger than single bonds, but multiple bonds have only a minor effect on bond angles. Why? All electron pairs in a multiple bond are shared between the same two nuclei and therefore occupy the same region of space. Because they must remain in that region, two electron pairs in a double bond or three electron pairs in a triple bond are like a single, slightly fatter balloon, rather than two or three balloons. Hence, ***for the purpose of determining molecular geometry, the electron pairs in a multiple bond constitute a single region of electron density.*** For example, use Table 7.1 to compare BeF_2 with CO_2. BeF_2 is a linear molecule with the two Be—F single bonds 180° apart. In CO_2, each C=O double bond is a single region of electron density, just as each Be—F single bond is, so the structure of CO_2 is also linear.

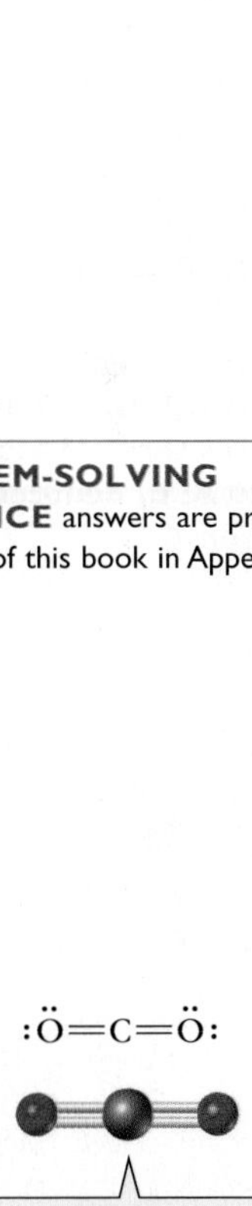

Carbon dioxide is a linear molecule because it has two regions of electron density (two double bonds) around the central carbon atom.

When resonance structures are possible, the geometry can be predicted from the individual resonance structures or from the resonance hybrid. For example, the geometry of the CO_3^{2-} ion is predicted to be triangular planar because in each resonance structure there are three regions of electron density, one double bond and two single bonds. In the resonance hybrid the carbon atom also has three regions of electron density, three equivalent bonds and no lone pairs. Each of the three bonds is intermediate between a C—O single bond and a C═O double bond.

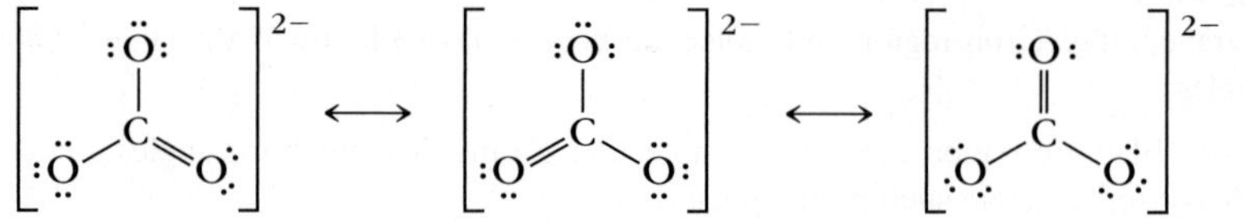

Answers to **EXERCISES** are provided at the back of this book in Appendix L.

EXERCISES that are labeled **CONCEPTUAL** are designed to test your understanding of one or more concepts; they usually involve qualitative rather than quantitative thinking.

CONCEPTUAL EXERCISE 7.1

Geometries

Based on the discussion so far, identify a characteristic that is common to all situations where electron-region geometry and molecular geometry are the same for a molecule or a polyatomic ion.

7-2c Central Atoms with Bonding Pairs and Lone Pairs

How does the presence of lone pairs on the central atom affect the geometry of a molecule or polyatomic ion? The easiest way to visualize this situation is to return to the balloon model (Figure 7.3) and realize that the electron regions on the central atom do not all have to be bonds. To predict the electron-region geometry, molecular geometry, and bond angles, we apply the VSEPR model to the total number of electron regions—that is, to lone pairs as well as bonds around the central atom. Ammonia, NH_3, illustrates guidelines for doing so.

1. ***Draw the Lewis structure.*** The Lewis structure for NH_3 is

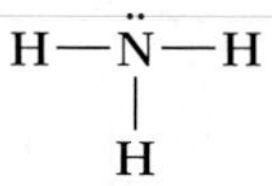

In ammonia, A is N, X is H, and X_3 indicates three H atoms bonded to A; E_1 denotes one lone pair on A.

2. ***Determine the electron-region geometry from the number of bonds and the number of lone pairs around the central atom.*** In NH_3, there are three bonds and one lone pair—four regions of electron density. Thus, NH_3 is an AX_3E_1-type molecule and the *electron-region geometry* is tetrahedral. (The electron-region geometry includes the spatial positions of all bonds and lone pairs.) To represent this geometry, draw a tetrahedral structure with N as the central atom and the three bonds represented by lines, wedges, or dashed lines, since they are single covalent bonds. Draw the lone pair as a balloon shape to indicate its spatial position in the tetrahedron.

Tetrahedral electron-region geometry for NH_3 (arrangement of electron regions, including lone pairs).

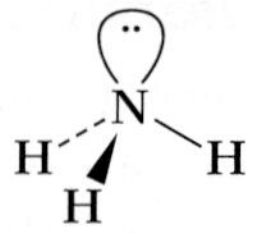

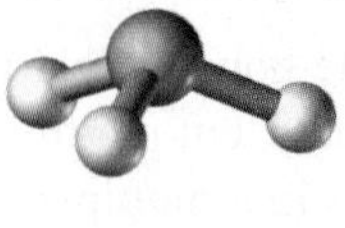

Triangular pyramid molecular geometry for NH_3 (arrangement of atomic nuclei and chemical bonds).

3. ***Determine the molecular geometry.*** Positions of atoms are specified in molecular geometry, but *not* positions of lone pairs. The *molecular geometry* of ammonia is described as a triangular pyramid because the three hydrogen atoms form a triangular base with the nitrogen atom at the apex of the pyramid. That is, the molecular geometry specifies only the locations of the three H atoms and the N atom.
4. ***Predict bond angles, using the idea that lone pairs occupy more space around the central atom than do bonding pairs.*** Bonding pairs are concentrated in the bonding region between two atoms by the strong attractive forces of two positive nuclei and are, therefore, relatively compact, or "skinny." For a lone pair, there is only one nucleus attracting the electron pair. As a result, lone pairs are less compact. Using the balloon analogy, a lone pair is like a fatter balloon that takes up more room and squeezes the thinner balloons closer together. Thus, *bond angles adjacent to lone*

The success of the VSEPR model in predicting molecular shapes indicates that it is appropriate to account for the effects of lone pairs in this way.

pairs are smaller than those predicted for perfect geometric shapes (those with no lone pairs). In ammonia, the electron-region geometry is tetrahedral, so the expected H—N—H bond angles are 109.5°. However, the experimentally determined bond angles in NH_3 are 107.5°. This is attributed to the bulkier lone pair forcing the bonding pairs closer together, thereby reducing the bond angle from 109.5° to 107.5°.

Predicting electron-region geometry and molecular geometry for NH_3 is summarized by this sequence:

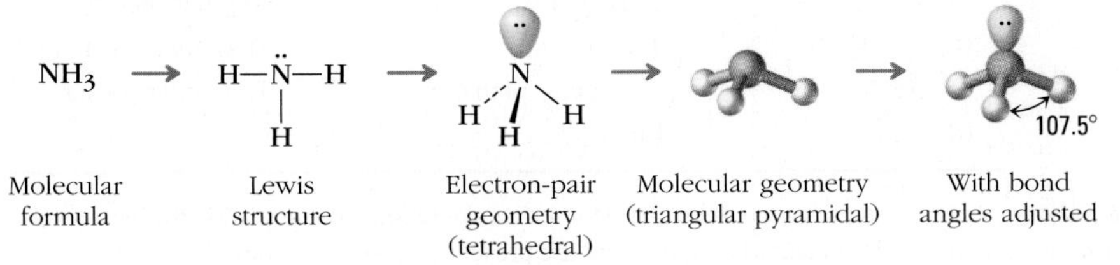
NH3
H—N—H
H
N
H
H
H
107.5°
Molecular formula
Lewis structure
Electron-pair geometry (tetrahedral)
Molecular geometry (triangular pyramidal)
With bond angles adjusted

The effect that increasing numbers of lone pairs on the central atom have on bond angles is illustrated by the structures of CH_4, NH_3, and H_2O.

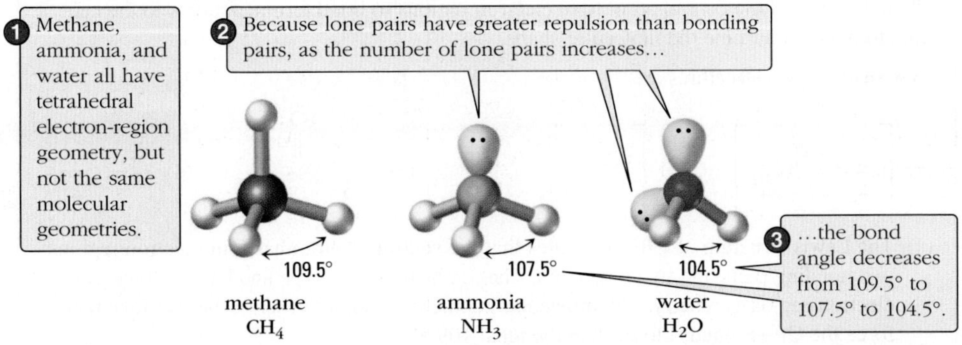
1 Methane, ammonia, and water all have tetrahedral electron-region geometry, but not the same molecular geometries.
2 Because lone pairs have greater repulsion than bonding pairs, as the number of lone pairs increases...
3 ...the bond angle decreases from 109.5° to 107.5° to 104.5°.
109.5°
107.5°
104.5°
methane
CH4
ammonia
NH3
water
H2O

Tables 7.2 and 7.3 summarize the electron-region geometry and molecular geometry for cases of one, two, or three lone pairs on a central atom. (Refer back to Table 7.1 for examples having no lone pairs.) To check your understanding of the VSEPR model, try to explain the electron-region geometry, the molecular geometry, and bond angles given in Table 7.2.

Table 7.2 Central Atom with Lone Pairs: Three and Four Electron Regions

Molecular model	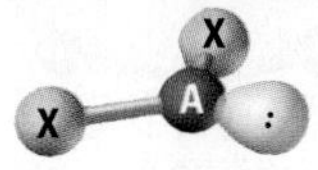	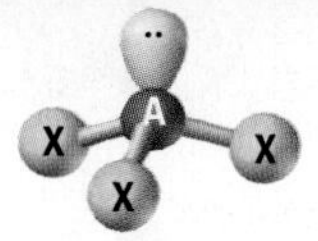	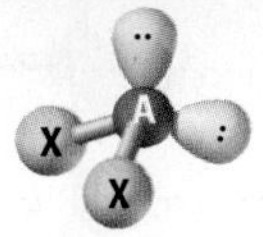
	Three electron regions; *one* lone pair	*Four* electron regions; *one* lone pair	*Four* electron regions; *two* lone pairs
Type	AX_2E_1	AX_3E_1	AX_2E_2
Electron-region geometry	Triangular planar	Tetrahedral	Tetrahedral
Molecular geometry	Angular (bent)	Triangular pyramidal	Angular (bent)
Examples	$SnCl_2$, $GeCl_2$	NH_3, NCl_3	H_2O, OF_2, OCl_2

PROBLEM-SOLVING EXAMPLE 7.2

Molecular Structure

Predict the electron-region geometry, molecular geometry, and bond angles of (a) ClF_2^+, (b) $SnCl_3^-$, (c) SeO_3, and (d) CS_2.

Result

Species	Electron-region Geometry	Molecular Geometry
(a) ClF_2^+	Tetrahedral	Angular (bent)
(b) $SnCl_3^-$	Tetrahedral	Triangular pyramidal
(c) SeO_3	Triangular planar	Triangular planar
(d) CS_2	Linear	Linear

Analyze Electron-region and molecular geometries are predicted using VSEPR. Deviations from ideal bond angles occur when lone pairs are present on a central atom.

Plan Apply the VSEPR guidelines: Write the correct Lewis structure. Determine the number of electron regions around the central atom and classify the molecule according to its AX_nE_m type. Use the balloon analogy (Figure 7.3) to choose the correct electron-region geometry around the central atom. Use electron regions to bond terminal atoms to the central atom and determine the molecular shape.

Execute Lewis structures are

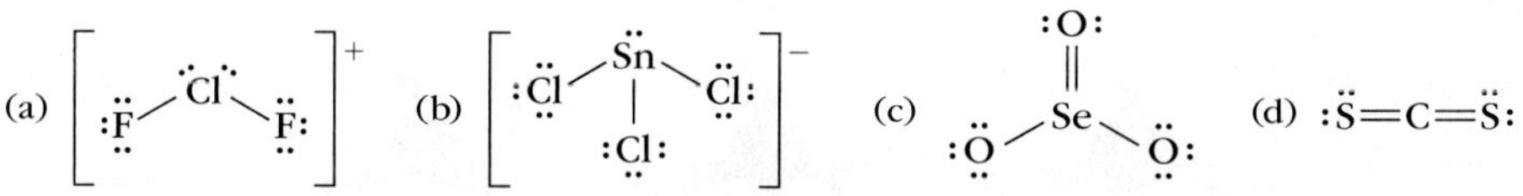

(a) The Lewis structure of ClF_2^+ indicates that the central Cl atom has four electron regions: two bonding pairs and two lone pairs. The ion is an AX_2E_2 type and has a tetrahedral electron-region geometry. The molecular geometry is angular (bent); the two lone pairs force the Cl—F bonds closer than the ideal 109.5°.

(b) In $SnCl_3^-$ the central tin atom is surrounded by three bonding pairs and one lone pair (AX_3E_1 type). These four electron pairs give a tetrahedral electron-region geometry. The molecular geometry is triangular pyramidal. The lone pair pushes the Sn—Cl bonds closer together than the purely 109.5° tetrahedral angle expected for four bonding-pair–bonding-pair repulsions.

Table 7.3 Central Atom with Lone Pairs: Three, Four, Five, and Six Electron Regions

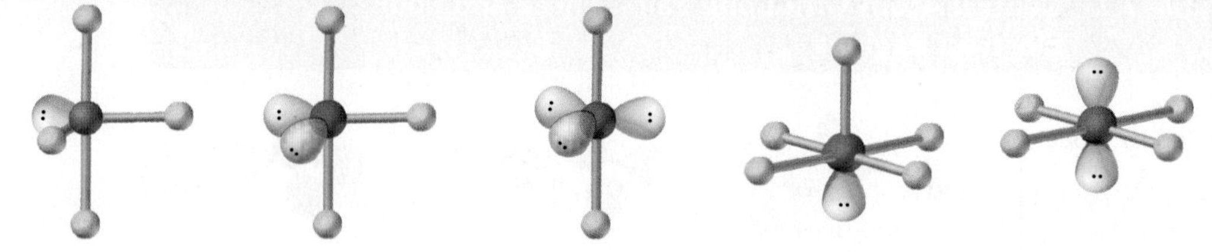

Molecular model	*Five* electron pairs; *one* lone pair	*Five* electron pairs; *two* lone pairs	*Five* electron pairs; *three* lone pairs	*Six* electron pairs; *one* lone pair	*Six* electron pairs; *two* lone pairs
Type	AX_4E_1	AX_3E_2	AX_2E_3	AX_5E_1	AX_4E_2
Electron-pair geometry	Triangular bipyramidal	Triangular bipyramidal	Triangular bipyramidal	Octahedral	Octahedral
Molecular geometry	Seesaw	T-shaped	Linear	Square pyramidal	Square planar
Examples	SF_4	ClF_3	XeF_2	BrF_5	$[ICl_4]^-$

(c) The two bonding pairs in the Se=O double bond count as one region of electron density for determining the geometry of the molecule, giving three bonding regions around Se (AX_3E_0 type). Only one of the three resonance structures is shown; the resonance hybrid has three equivalent regions of electron density around the Se atom. With no lone pairs around Se, the electron-region and molecular geometries are the same, triangular planar, with 120° angles.

(d) Each C=S double bond is treated as one region of electron density. There are no lone pairs on the central C atom (AX_2E_0 type). Therefore, the electron-region geometry and molecular geometry are both linear, with 180° angles.

PROBLEM-SOLVING PRACTICE 7.2

Determine electron-region geometries, molecular geometries, and bond angles for (a) BrO_3^-, (b) SeF_2, and (c) NO_2^-.

7-2d Central Atoms with Five or Six Electron Regions

When a central atom has five or six electron pairs, some of which are lone pairs, the electron-region geometries are triangular bipyramidal or octahedral, respectively.

Triangular Bipyramidal Structure When there are five electron regions around a central atom, the electron regions point toward the corners of a **triangular bipyramid,** *a geometric solid with six triangular faces and five corners.*

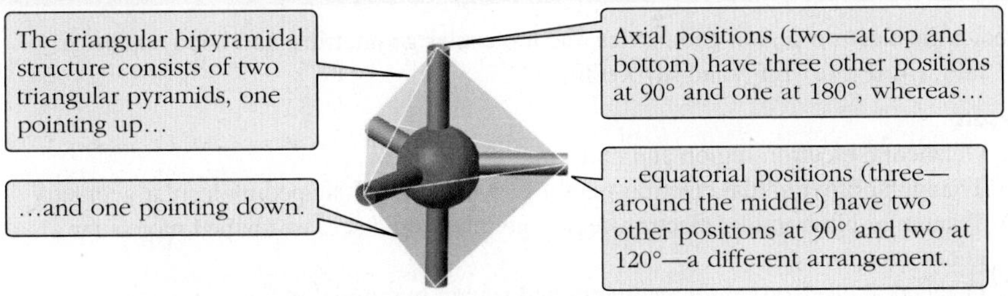
The triangular bipyramidal structure consists of two triangular pyramids, one pointing up...
...and one pointing down.
Axial positions (two—at top and bottom) have three other positions at 90° and one at 180°, whereas...
...equatorial positions (three—around the middle) have two other positions at 90° and two at 120°—a different arrangement.

Triangular bipyramidal structure

In the *triangular bipyramidal structure, positions in the equator of an imaginary globe around the central atom* are called **equatorial positions**; the *north and south poles* are called **axial positions**. A general rule for VSEPR theory is that *electron regions with the strongest repulsions should be placed where there are the fewest other electron regions at the smallest angle in the structure.* In a triangular bipyramidal structure, the smallest angle is 90°. Equatorial positions have only two 90° neighbors, while axial positions have three. Therefore, *lone pairs occupy equatorial positions rather than axial positions.* An example is ClF_3 (Table 7.3), which has two lone pairs—both in equatorial positions.

Octahedral Structure When there are six electron regions around a central atom, the electron regions point toward the corners of an **octahedron,** *a geometric solid with eight faces that are equilateral triangles and six corners.*

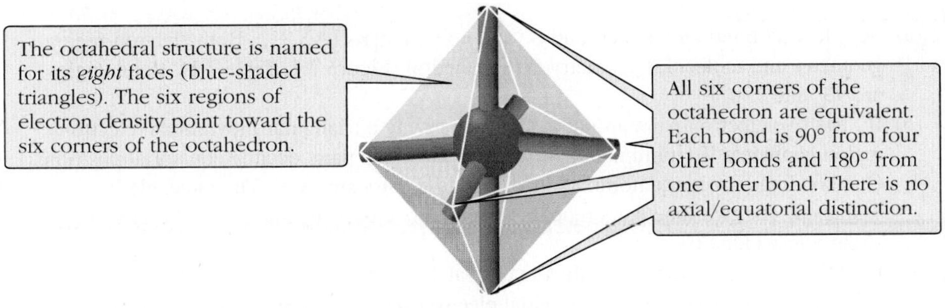
The octahedral structure is named for its *eight* faces (blue-shaded triangles). The six regions of electron density point toward the six corners of the octahedron.
All six corners of the octahedron are equivalent. Each bond is 90° from four other bonds and 180° from one other bond. There is no axial/equatorial distinction.

Octahedral structure

In Table 7.3, the lone pair in BrF_5 could have been located in any one of the six positions around Br. We arbitrarily chose the "down" position.

Because all bonds in an octahedron are equivalent, if there is one lone pair on the central atom, it makes no difference which apex the lone pair occupies. An example is BrF_5, whose molecular geometry is *square pyramidal* (Table 7.3). There is a *square plane* containing the central atom and four of the atoms bonded to it; the fifth bonded atom is directly above the central atom and equidistant from the first four.

The synthesis of XeF_4 was a surprise to chemists because all noble gases have stable electron configurations (← **Sec. 6-2**) and were not expected to form compounds. Nevertheless, you can use the VSEPR model to predict the correct geometry. XeF_4 has 36 valence electrons (eight from Xe and seven from each of four F atoms). There are eight electrons in four bonding pairs around Xe, and a total of 24 electrons in the lone pairs on the four F atoms. That leaves four electrons in two lone pairs on the Xe atom. Xe is in Period 5, so it can accommodate more than an octet (← **Sec. 6-10c**). The total of six electron regions on Xe leads to a prediction of an octahedral electron-region geometry (AX_4E_2). The two lone pairs are placed at opposite corners of the octahedron to keep them as far apart as possible. The result is a square planar molecular geometry for the XeF_4 molecule (cover the lone pairs in the drawing for XeF_4 in Table 7.3 to see this). This shape agrees with experimental structural results.

When XeF_4 was first synthesized many chemists thought its structure was tetrahedral; VSEPR theory correctly predicted and experiments confirmed that it is square planar.

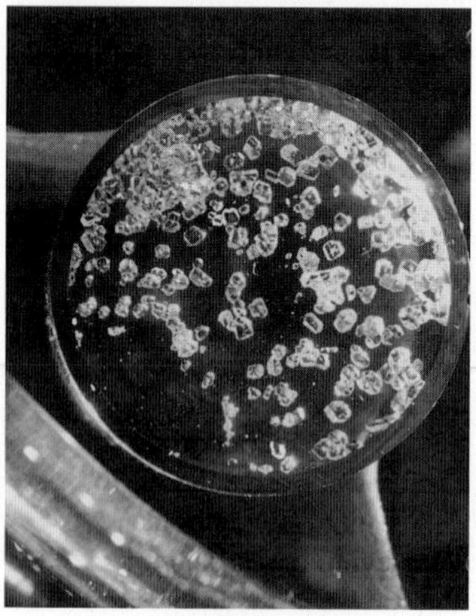
Science Source

Xenon tetrafluoride crystals.

PROBLEM-SOLVING EXAMPLE 7.3

VSEPR and Molecular Shape

Predict the electron-region geometries, the molecular geometries, and bond angles of (a) SeF_6, (b) IF_3, (c) TeF_4, and (d) $XeOF_4$.

Result

(a) Octahedral electron-region and molecular geometries
(b) Triangular bipyramidal electron-region geometry and T-shaped molecular geometry
(c) Triangular bipyramidal electron-region geometry and seesaw-shaped molecular geometry
(d) Octahedral electron-region geometry and square pyramidal molecular geometry

Analyze Electron-region and molecular geometries are predicted using VSEPR. Deviations from ideal bond angles occur when lone pairs are present. All central atoms are from Period 4 or beyond so more than an octet can be accommodated.

Plan Apply the VSEPR guidelines: Write the correct Lewis structure. Determine the number of electron regions around the central atom and classify the molecule according to its AX_nE_m type. Use the balloon analogy (Figure 7.3) to choose the correct electron-region geometry around the central atom. Use electron regions to bond terminal atoms to the central atom and determine the molecular shape. Minimize repulsions between lone pairs at the smallest angle in each structure.

Execute The Lewis structures are

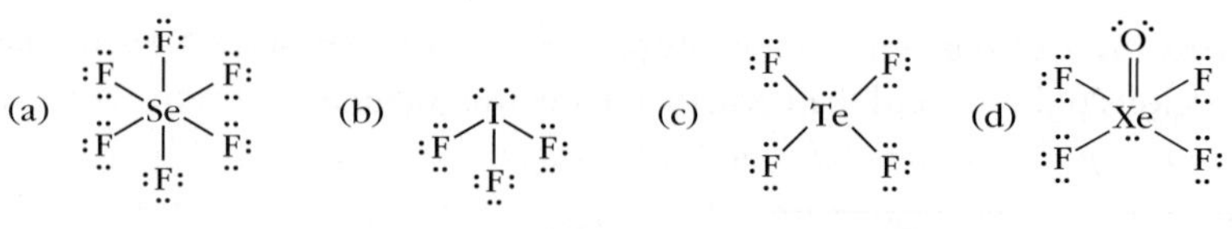
(a)
Se
F
(b)
I
F
(c)
Te
F
(d)
Xe
O
F

(a) SeF_6 has six bonding pairs around Se and no lone pairs (AX_6E_0). Both electron-region geometry and molecular geometry are octahedral (Figure 7.1, Table 7.1). Bond angles are 90° and 180°.
(b) IF_3 has five electron pairs around I (AX_3E_2), giving a triangular bipyramidal electron-region geometry. To minimize repulsion, the two lone pairs occupy equatorial positions. One bonding pair is equatorial and two bonding pairs are axial. The molecule is T-shaped. The two lone pairs repel the bond pairs closer together, so the angles are slightly less than 90°.
(c) In TeF_4, an AX_4E_1-type molecule, the central Te atom is surrounded by five electron pairs, giving it a triangular bipyramidal electron-region geometry. The lone pair is in an equatorial position, creating a seesaw-shaped molecule.

(d) In $XeOF_4$, an AX_5E_1-type molecule, the central Xe atom has a lone pair, four bonding pairs to terminal fluorine atoms, and a double bond to O (which counts as a single region of electron density). The electron-region geometry is octahedral. Because double bonds and lone pairs are "fatter balloons", the double bond and lone pair occupy opposite corners of the octahedron in a square-pyramidal molecular geometry (Table 7.3). The O-Xe-F angles are approximately 90° because the greater repulsion of the "fatter" lone pair on one side approximately cancels that of the double bond on the other side.

PROBLEM-SOLVING PRACTICE 7.3

Determine the electron-region geometry and the molecular geometry of (a) ClF_2^- and (b) XeO_3.

CONCEPTUAL EXERCISE 7.2

Two Dissimilar Shapes

Triangular bipyramidal and square pyramidal molecular geometries each have a central atom with five terminal atoms around it, but the molecular shapes are different. Explain how and why these two shapes differ.

7-2e Molecules with More than a Single Central Atom

The VSEPR model also can be used to predict the geometry around atoms in molecules with more than one central atom. Each atom in such a molecule (except, of course, for hydrogen, which can form only one bond) is treated as a central atom. Methane, which has a tetrahedral shape, is the smallest member of the large family of saturated hydrocarbons called alkanes (← **Sec. 6-3**). It is important to recognize that every carbon atom in an alkane has four bonds and hence a tetrahedral environment. For example, the carbon atoms in propane and in the much longer carbon chain of hexadecane do not lie in a straight line because of the tetrahedral geometry about each carbon atom.

propane
C_3H_8

hexadecane
$C_{16}H_{34}$

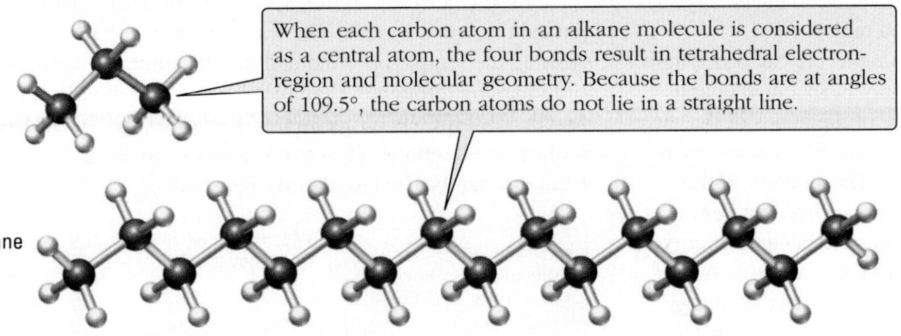
When each carbon atom in an alkane molecule is considered as a central atom, the four bonds result in tetrahedral electron-region and molecular geometry. Because the bonds are at angles of 109.5°, the carbon atoms do not lie in a straight line.

The tetrahedron is arguably the most important shape in chemistry because of its predominance in carbon chemistry and carbon-based biomolecules.

Another example is lactic acid, a compound important in carbohydrate metabolism. The structure around each atom in lactic acid is shown here:

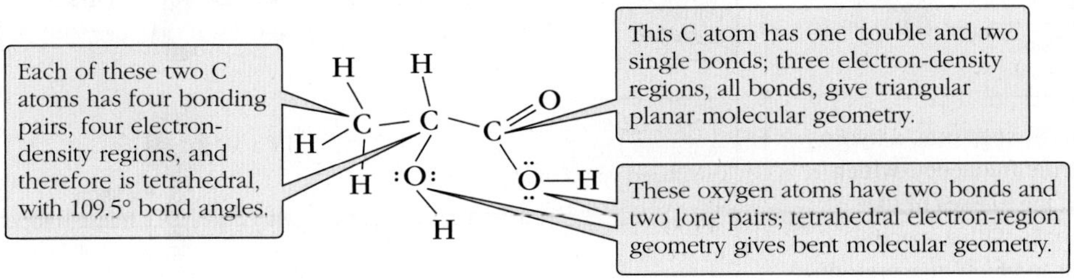
Each of these two C atoms has four bonding pairs, four electron-density regions, and therefore is tetrahedral, with 109.5° bond angles.
This C atom has one double and two single bonds; three electron-density regions, all bonds, give triangular planar molecular geometry.
These oxygen atoms have two bonds and two lone pairs; tetrahedral electron-region geometry gives bent molecular geometry.

TOOLS OF CHEMISTRY | Infrared Spectroscopy

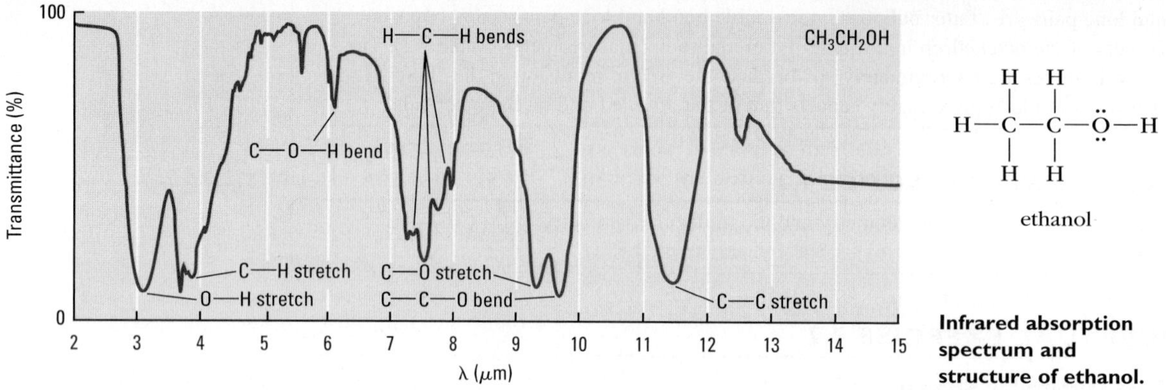
100
0
Transmittance (%)
H—C—H bends
CH₃CH₂OH
C—O—H bend
C—H stretch
O—H stretch
C—O stretch
C—C—O bend
C—C stretch
2 3 4 5 6 7 8 9 10 11 12 13 14 15
λ (μm)
ethanol

Infrared absorption spectrum and structure of ethanol.

Molecular structures are often determined by **spectroscopy**, *measuring how absorption or emission of electromagnetic radiation by a sample of matter varies with wavelength.* Each region of the electromagnetic spectrum (← **Sec. 5-1**) can be used as the basis for a particular spectroscopic method. Recall that electromagnetic radiation is emitted or absorbed in quantized packets of energy called photons and that the energy of a photon is represented by $E = h\nu = hc/\lambda$, where ν is the frequency and λ is the wavelength of the light. Molecules may absorb several different electromagnetic radiation frequencies depending on the energy differences between their allowed energy levels. Each frequency absorbed must provide the exact packet of energy needed to lift a molecule from one energy level to another. Infrared (IR) spectroscopy uses the interaction of infrared radiation with matter to study molecular structure. It is particularly useful for learning about molecular structure because the energy of the internal motions of molecules is similar to the energy of photons with frequencies in the infrared region.

Covalent bonds between atoms in a molecule can be thought of as springs that can bend or stretch only in specified quantities. For example, bending or stretching of the bonds of the water molecule, as shown, occurs at specific frequencies, each corresponding to a specific energy level. The strength of the covalent bonds determines what frequency of infrared light is necessary for changing from one stretching or bending energy level to another. The molecule can only be excited from one of these allowed energy states to another by a quantity of energy that exactly matches the energy difference between the states.

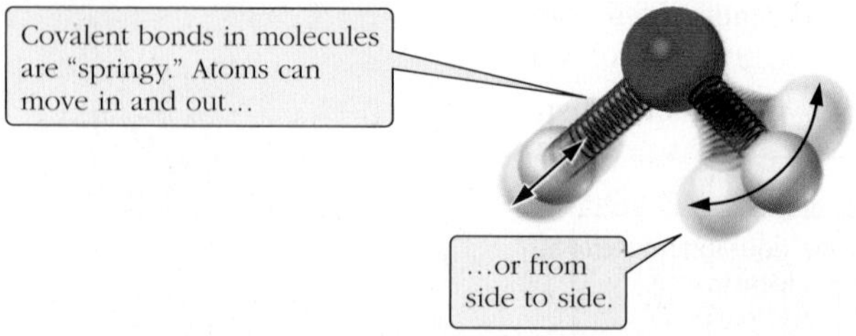
Covalent bonds in molecules are "springy." Atoms can move in and out...
...or from side to side.

As an example of how this affects an IR spectrum, a hydrogen chloride molecule vibrates at a specific frequency, which depends on the strength of the bond and the masses of the vibrating atoms. The frequency corresponds with a specific energy level. Photons that are too low or too high in energy do not cause vibration at the next higher energy level, and the radiation passes through the molecule without being absorbed. In contrast, photons of the proper energy are absorbed and increase the molecule's vibrational energy.

Because the covalent bonds in molecules differ in strength and number, and the atoms have different masses, the molecular motions and the number of vibrational energy levels vary; hence, the wavelengths of infrared radiation absorbed by different molecules vary. Thus, infrared spectroscopy can be used to learn about the structures of molecules and to analyze an unknown material by matching its infrared spectrum with that of a known compound. In fact, the infrared frequencies absorbed by a molecule are so characteristic of the bending and stretching of various bonds that the infrared spectrum of a molecule is regarded as its *fingerprint.*

How infrared spectroscopy aids in the identification of ethanol is shown by its infrared absorption spectrum. Each absorption is labeled with the type of bond vibration that causes it. To obtain such an IR absorption spectrum, infrared radiation is passed through a sample. The wavelength of the IR radiation striking the sample is varied and the transmittance (percent of light passing through the sample) is measured at each wavelength. Absorptions (drops in transmittance) are characteristic of molecular vibrations. The overall result is an absorption spectrum that can be used to identify the sample.

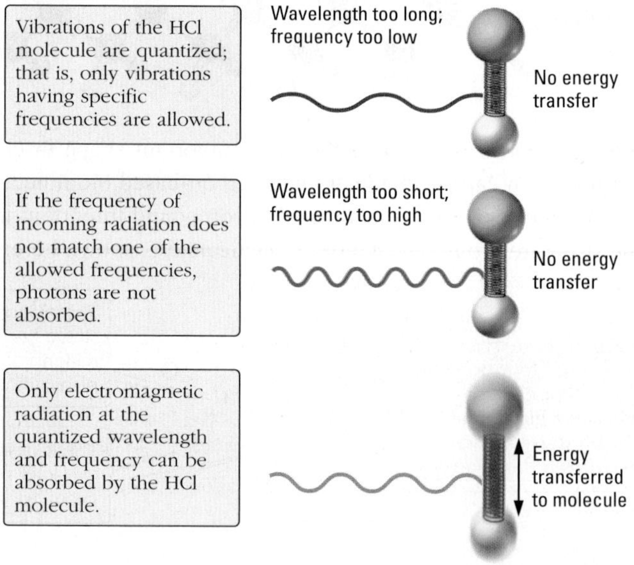
Vibrations of the HCl molecule are quantized; that is, only vibrations having specific frequencies are allowed.
Wavelength too long; frequency too low
No energy transfer
If the frequency of incoming radiation does not match one of the allowed frequencies, photons are not absorbed.
Wavelength too short; frequency too high
No energy transfer
Only electromagnetic radiation at the quantized wavelength and frequency can be absorbed by the HCl molecule.
Energy transferred to molecule

The combined effect of these geometries is shown in these molecular models.

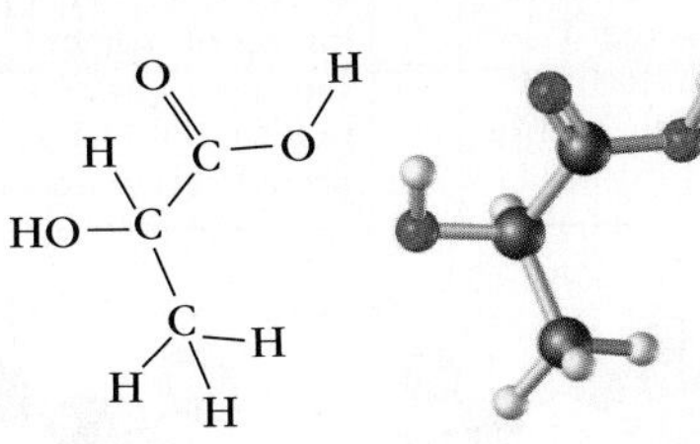
O
H
H
C—O
HO—C
C—H
H
H

lactic acid

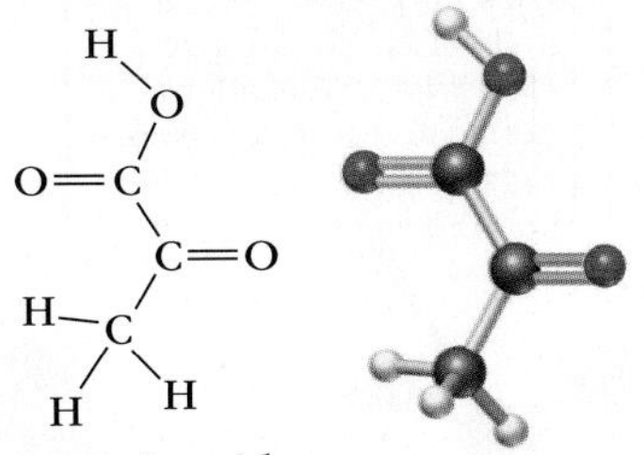
H
O
O=C
C=O
H—C
H
H

pyruvic acid

Denis Kuvaev/Shutterstock.com

Lactic acid has built up in the cells of this sprinter.

Lactic acid, a waste product of glucose metabolism, builds up rapidly in muscles during strenuous short-term exercise such as sprinting (100 to 800 m) and swimming (100 or 200 m). During such exercise, the accumulated lactic acid irritates the muscle tissue, causing muscle fatigue and soreness. But after the exercise stops, the lactic acid is converted to pyruvic acid, which is removed from the muscles, and the soreness and fatigue dissipate.

EXERCISE 7.3

Molecules with Several Central Atoms

Draw a Lewis diagram for each molecule; predict the geometry around each atom.

(a) ethene, C_2H_4 (b) C_2H_6O (there are two isomeric Lewis structures)

7-2f Chiral Molecules

Are you right-handed or left-handed? Regardless of the preference, we learn at a very early age that a right-handed glove doesn't fit the left hand, and vice versa. Our hands are mirror images of one another and are not superimposable (Figure 7.4). *An object that cannot be superimposed on its mirror image* is called **chiral**. *Objects that are superimposable on their mirror images* are **achiral**. *A chiral molecule and its* nonsuperimposable *mirror image* are called **enantiomers**; they are two different *isomeric structures* (← **Sec. 2-9a**) that differ just as your left and right hands differ. The simplest case of chirality is a tetrahedral carbon atom that is bonded to four *different* atoms or groups of atoms. Such a *chiral carbon atom* is a *chiral center* or *asymmetric carbon atom*. A molecule with just one chiral center is always a chiral molecule.

Lactic acid has one C atom bonded to four different groups: $—CH_3$, —OH, —H, and —COOH. Therefore, lactic acid is chiral. Figure 7.5 shows how the tetrahedral geometry around the central carbon atom results in two different arrangements of the four groups—two different *isomers*. When Isomer II is placed in front of a mirror (Figure 7.5) you can see that Isomer I is its mirror image. These two mirror-image, chiral molecules *cannot be superimposed on one another* and therefore *are* enantiomers. *Enantiomers* are also referred to as **optical isomers** because when plane polarized light passes through one isomer, the plane of polarization is rotated in a different direction from the other isomer.

Chirality is a part of our everyday life; chiral objects include hands (and feet), screws and corkscrews, helical seashells (most spiral like a right-handed screw), and creeping vines that wind around a tree or post. What is not as well known is that a large number of the molecules in plants and animals are chiral. Lactic acid is found in nature in both enantiomeric forms. During the contraction of muscles, the body produces one enantiomer. The other enantiomer is produced when milk sours.

More typically, only one form of the chiral molecule (left-handed or right-handed) is found in nature. In Figure 7.5, R and S represent the "handedness" of enantiomers. This "handedness" is also sometimes represented by a different system: D for right-handed

In *A Midsummer Night's Dream*, Shakespeare wrote of the chirality of honeysuckle and woodbine plants. Queen Titania says to Bottom, ". . . Sleep thou, and I will wind thee in my arms . . . So doth the woodbine the sweet honeysuckle gently entwist. . . ." Woodbine spirals clockwise and honeysuckle twists in a counterclockwise direction.

In *Through the Looking Glass* (the companion volume to *Alice in Wonderland*), Alice speculates to her cat that "Perhaps looking glass milk isn't good to drink. . ."

Cengage Learning/Charles D. Winters

Figure 7.4 Nonsuperimposable mirror images. Right- and left-handed seashells are mirror images, as are the hands holding them.

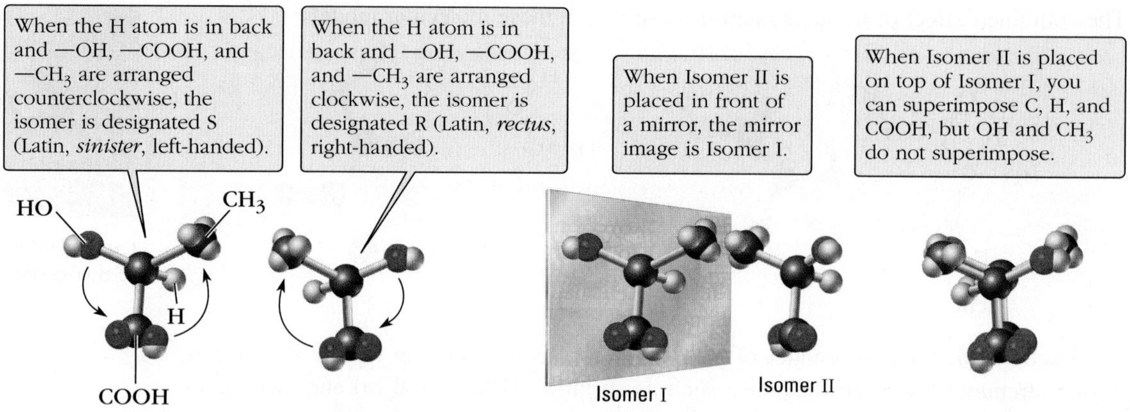
When the H atom is in back and —OH, —COOH, and —CH3 are arranged counterclockwise, the isomer is designated S (Latin, *sinister*, left-handed).
When the H atom is in back and —OH, —COOH, and —CH3 are arranged clockwise, the isomer is designated R (Latin, *rectus*, right-handed).
When Isomer II is placed in front of a mirror, the mirror image is Isomer I.
When Isomer II is placed on top of Isomer I, you can superimpose C, H, and COOH, but OH and CH3 do not superimpose.
HO
CH3
H
COOH
Isomer I
Isomer II

Figure 7.5 The enantiomers of lactic acid. You could superimpose C, H, and COOH, but then OH and CH_3 would not superimpose.

(D for "dextro" from the Latin *dexter,* "right") and L for left-handed (L for "levo" from the Latin *laevus,* "left"). Most natural sugars, including glucose, sucrose, and deoxyribose, the sugar found in DNA, are right-handed (D form). All but one of the 20 amino acids from which proteins are made are chiral, and they are left-handed amino acids (L form). Nature's preference for L-amino acids has provoked much discussion and speculation among scientists, but there is still no widely accepted explanation of this preference.

Large organic molecules may have two or more chiral carbon atoms within the same molecule. At each such carbon atom (a chiral center), two arrangements of the molecule are possible. The maximum number of possible isomers increases exponentially with the number of different chiral carbon atoms; with n different chiral carbon atoms there are up to 2^n possible isomers. (The words "up to" are used because in some cases a mirror image may not be different from the mirrored structure.) The widely used artificial sweetener aspartame (NutraSweet®) has two chiral centers and two pairs of enantiomers. One enantiomer has a sweet taste, while another enantiomer is bitter, indicating that the receptor sites on our taste buds respond differently to the "handedness" of aspartame enantiomers and therefore must be chiral! In another example, D-glucose is sweet and nutritious, while L-glucose is also sweet but cannot be metabolized by the body.

PROBLEM-SOLVING EXAMPLE 7.4

Handedness in Asparagine

Asparagine is a naturally occurring amino acid first isolated as a white powder from asparagus juice. Later, a second form of asparagine was isolated from a sprouting vetch plant. D-Asparagine is sweet; the L isomer is tasteless. The Lewis structure of asparagine is

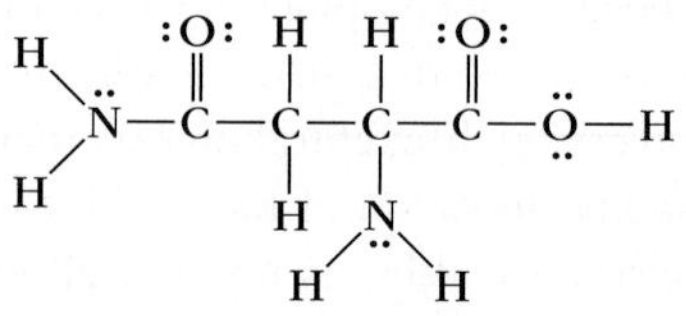

Identify which carbon atom in asparagine is the chiral center (chiral carbon).

Result

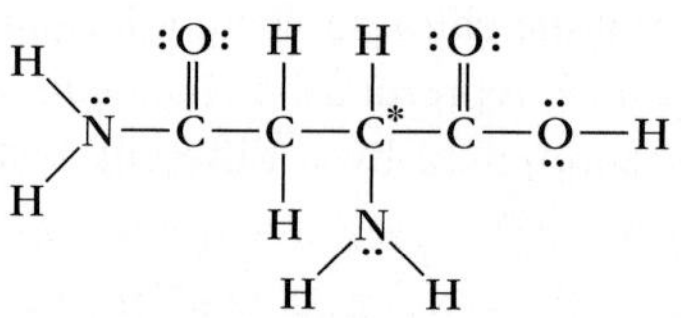

*Chiral center

Analyze A carbon atom with four different atoms or groups attached to it is a chiral center. A double-bonded carbon, such as C=O, has only three groups, so it cannot be a chiral center. The $—CH_2—$ carbon cannot be a chiral center because it has two of the same atoms (hydrogen) bonded to it.

Plan Examine the Lewis structure to find a C atom with four different groups.

Execute The carbon next to the —COOH group is the chiral center. It has four different atoms or groups of atoms attached to it: (1) a hydrogen atom; (2) a —COOH group; (3) an $—NH_2$ group; and (4) the rest of the molecule, starting with the $—CH_2—$ group.

PROBLEM-SOLVING PRACTICE 7.4

Aspartame, a widely used sugar substitute, is a chiral molecule with the structural formula

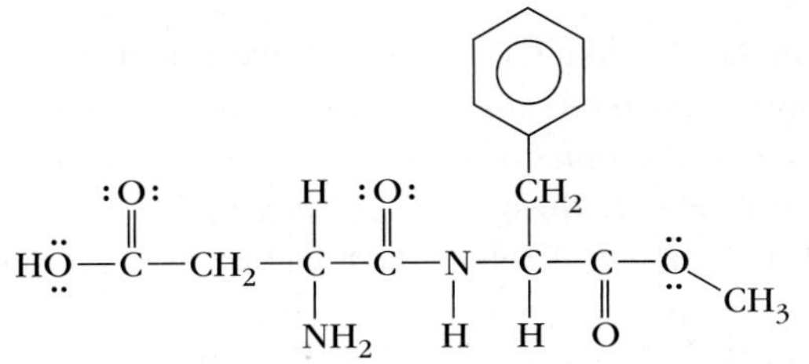

Identify all asymmetric (chiral) carbon atoms.

7-3 Hybridization: Atomic Orbitals Consistent with Molecular Shapes

Although Lewis structures are helpful in assigning molecular geometries, they indicate nothing about the atomic orbitals occupied by bonding and lone pair electrons. The *valence bond model* (← **Sec. 6-1**) describes a covalent bond as the result of an overlap of orbitals on each of two bonded atoms. For H_2, for example, the bond is a shared electron pair located in the overlapping *s* atomic orbitals.

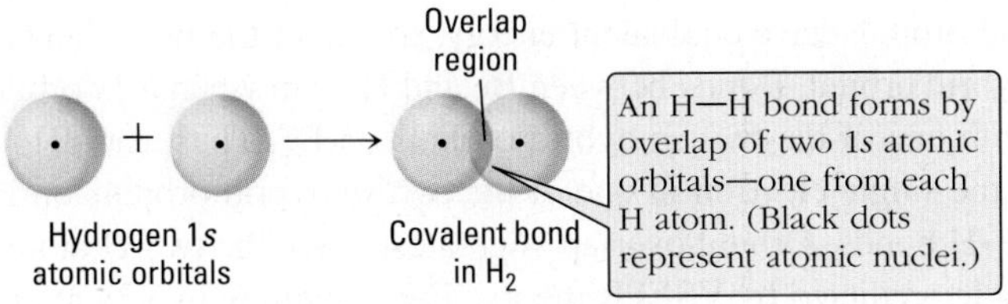
Overlap region
Hydrogen 1s atomic orbitals
Covalent bond in H2
An H—H bond forms by overlap of two 1s atomic orbitals—one from each H atom. (Black dots represent atomic nuclei.)

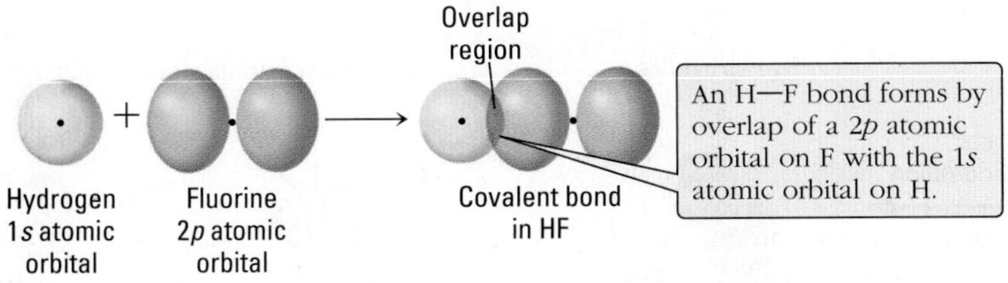
Overlap region
Hydrogen 1s atomic orbital
Fluorine 2p atomic orbital
Covalent bond in HF
An H—F bond forms by overlap of a 2p atomic orbital on F with the 1s atomic orbital on H.

In hydrogen fluoride, HF, the overlap occurs between a fluorine 2*p* atomic orbital with a single electron and the 1*s* atomic orbital of a hydrogen atom.

To account for bonding in molecules with central atoms such as Be, B, and C, however, the simple valence bond model of overlapping *s* and *p* atomic orbitals must be modified. Consider the ground-state electron configurations for Be, B, and C atoms:

			1s	2s	2p		
Be	$1s^2 2s^2$		↑↓	↑↓			
B	$1s^2 2s^2 2p^1$		↑↓	↑↓	↑		
C	$1s^2 2s^2 2p^2$	$1s^2 2s^2 2p_x^1 2p_y^1$	↑↓	↑↓	↑	↑	

The simple valence bond model predicts that Be, with an outer s^2 configuration like He, should form no covalent bonds because there are no unpaired electrons to pair with electrons from another atom. B, with one unpaired electron ($2p^1$), should form only one bond and C ($2p^2$), should form only two bonds with its two unpaired $2p$ electrons. But, BeH_2, BF_3, and CH_4 all exist, as well as other compounds in which these atoms have two, three, and four bonds.

Hybridization is described as a combination of the mathematical functions (wave functions) that represent *s* and *p* atomic orbitals to give new functions that predict shapes of the hybrid orbitals.

Another theory, the *molecular orbital theory*, is also used to describe chemical bonding, as well as the magnetic properties of molecules. Molecular orbital theory was discussed in Section 6-12.

To account for molecules like CH_4, in which the actual molecular geometry is incompatible with simple overlap of *s* and *p* atomic orbitals, *atomic orbitals of the proper energy and orientation in the same atom are combined* to form **hybrid orbitals**. ***The total number of hybrid orbitals formed always equals the number of atomic orbitals that were combined (hybridized).*** Each of the resulting hybrid orbitals can form a bond with another atom or contain a lone pair of electrons. The hybrid orbitals have the same shape and are oriented in the same directions as the balloons in Figure 7.3 (← **Sec. 7-2**). Compared with unhybridized atomic orbitals, hybrid orbitals are better able to overlap with bonding orbitals on other atoms, resulting in stronger bonds, and provide more locations where bonds can be formed, resulting in more bonds.

7-3a *sp* Hybrid Orbitals

The simplest hybrid orbitals are ***sp* hybrid orbitals** formed by the *combination of one* s *atomic orbital and one* p *atomic orbital in the valence shell (same* n *value) in the same atom*; these *two* atomic orbitals combine to form *two sp* hybrid orbitals (Figure 7.6). The two *sp* hybrid orbitals are 180° from each other, the same orientation predicted earlier (Figure 7.3) for two regions of electron density.

$$\text{One } s \text{ atomic orbital} + \text{one } p \text{ atomic orbital} \longrightarrow \text{two } sp \text{ hybrid orbitals}$$

beryllium hydride, BeH_2

In the gas phase, molecules of BeH_2 exist. The central Be atom has two valence electrons. Two *sp* hybrid orbitals have equivalent energy, so one of the two valence electrons occupies each *sp* hybrid orbital. Bonds between Be and H form when a 1*s* orbital on each H atom overlaps with one of the two *sp* hybrid orbitals on Be. Thus, each H 1*s* electron can be paired with the single electron in one of the two *sp* hybrid orbitals on Be to form two equivalent Be—H bonds. Orbital overlap is greatest when the two H atoms are 180° apart—the same angle predicted by VSEPR theory. The remaining two 2*p* atomic orbitals in Be are *unhybridized* and are at 90° to each other and to the two *sp* hybrid orbitals.

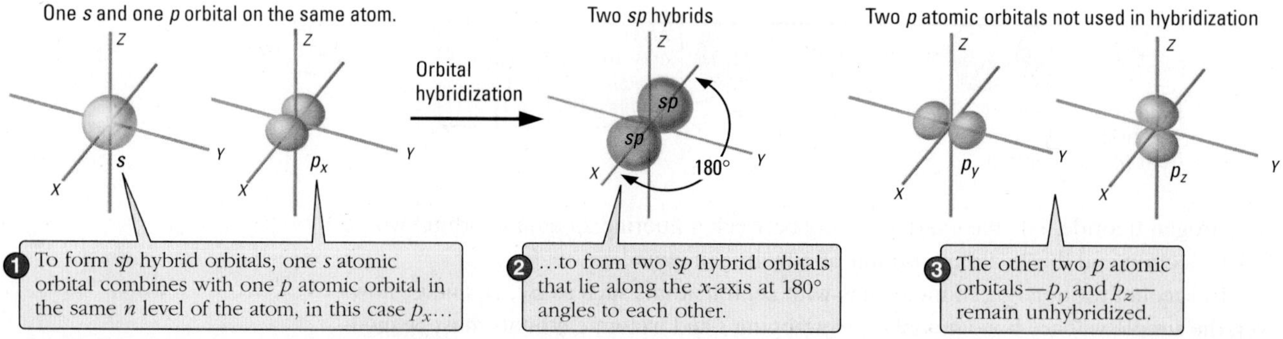
One s and one p orbital on the same atom.
Orbital hybridization
Two sp hybrids
Two p atomic orbitals not used in hybridization
Z
Y
X
s
px
sp
sp
180°
py
pz
1 To form sp hybrid orbitals, one s atomic orbital combines with one p atomic orbital in the same n level of the atom, in this case px...
2 ...to form two sp hybrid orbitals that lie along the x-axis at 180° angles to each other.
3 The other two p atomic orbitals—py and pz—remain unhybridized.

Figure 7.6 Formation of two *sp* hybrid orbitals from one *s* orbital and one p_x orbital in the same atom.

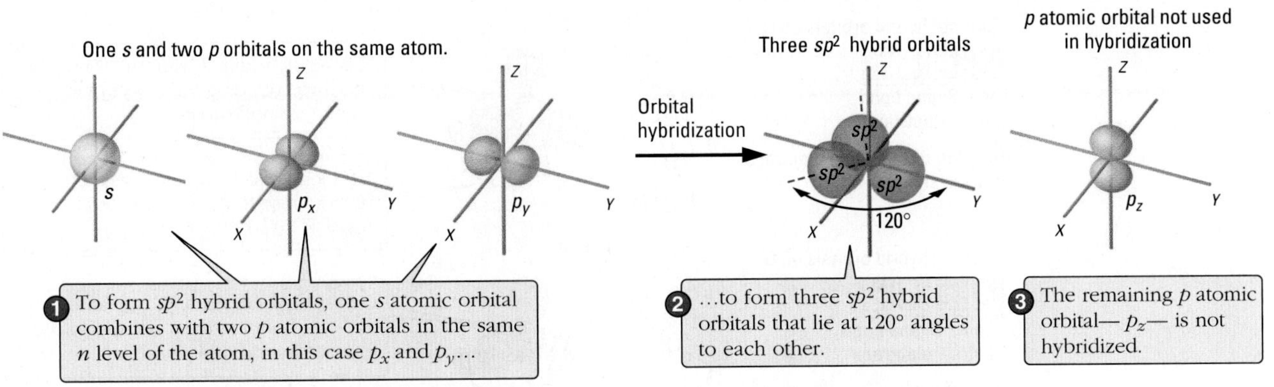
One s and two p orbitals on the same atom.
Z
s
p_x
p_y
X
Y
Orbital hybridization
Three sp^2 hybrid orbitals
sp^2
120°
p atomic orbital not used in hybridization
p_z
1 To form sp^2 hybrid orbitals, one s atomic orbital combines with two p atomic orbitals in the same n level of the atom, in this case p_x and p_y...
2 ...to form three sp^2 hybrid orbitals that lie at 120° angles to each other.
3 The remaining p atomic orbital— p_z— is not hybridized.

Figure 7.7 Three sp^2 hybrid orbitals form from one s and two p atomic orbitals in the same atom.

7-3b sp^2 Hybrid Orbitals

When *three atomic orbitals in the valence shell (same* n *value) in the same atom—for example, a* 2s *and two* 2p *atomic orbitals—are hybridized they form three* **sp^2 hybrid orbitals** (Figure 7.7). The sum of the superscripts in sp^2 (1 + 2 = 3) indicates the number of atomic orbitals that have hybridized—three in this case.

One *s* atomic orbital + two *p* atomic orbitals ⟶ three sp^2 hybrid orbitals

In BF_3, there are three sp^2 hybridized orbitals on boron; they are 120° apart in a plane. Each sp^2 hybrid orbital contains one electron from B and one from one of the F atoms. This forms three equivalent B—F bonds. One of the boron 2*p* atomic orbitals remains unhybridized and contains no electrons.

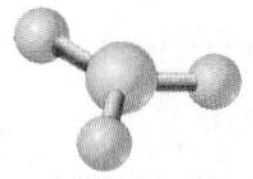

boron trifluoride, BF_3

7-3c sp^3 Hybrid Orbitals

When one s *and the three* p *atomic orbitals in the valence shell (same* n *value) in a central atom are hybridized, four* hybrid orbitals, called **sp^3 hybrid orbitals**, are formed (Figure 7.8).

One *s* atomic orbital + three *p* atomic orbitals ⟶ four sp^3 hybrid orbitals

The four sp^3 hybrid orbitals are equivalent and directed to the corners of a tetrahedron. Because all of the *s* and *p* atomic orbitals are hybridized, there are no unhybridized *p* orbitals on the central atom.

The sp^3 hybridization is consistent with the fact that carbon forms four tetrahedral bonds in CH_4 and in all other single-bonded carbon compounds. Overlap of each half-filled sp^3 hybrid orbital on carbon with half-filled orbitals from four hydrogen atoms forms four equivalent C—H bonds. The central atoms of Period 2 and 3 elements that obey the

methane, CH_4

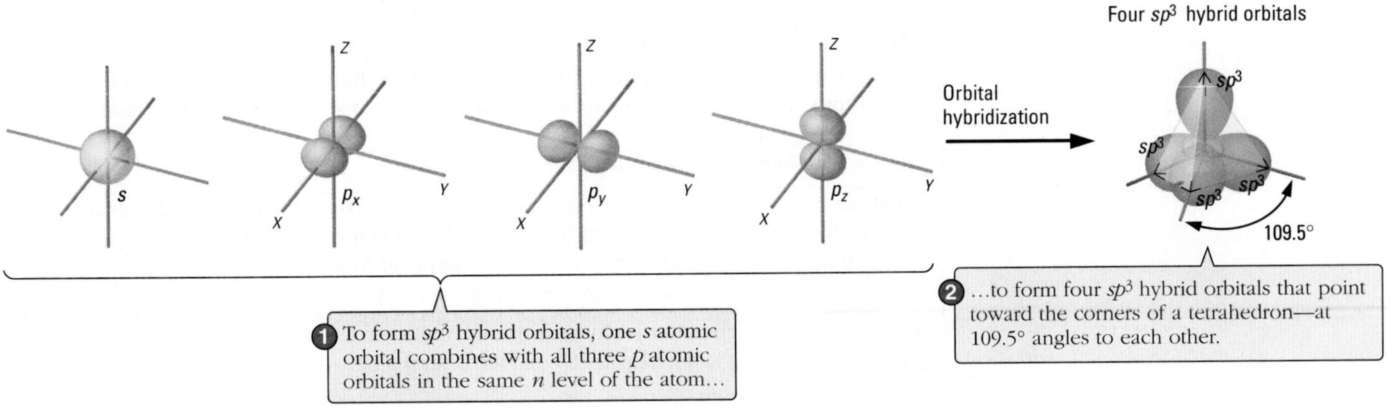
Z
s
p_x
p_y
p_z
X
Y
Orbital hybridization
Four sp^3 hybrid orbitals
sp^3
109.5°
1 To form sp^3 hybrid orbitals, one s atomic orbital combines with all three p atomic orbitals in the same n level of the atom...
2 ...to form four sp^3 hybrid orbitals that point toward the corners of a tetrahedron—at 109.5° angles to each other.

Figure 7.8 Formation of four sp^3 hybrid orbitals from one s and three p atomic orbitals in the same atom.

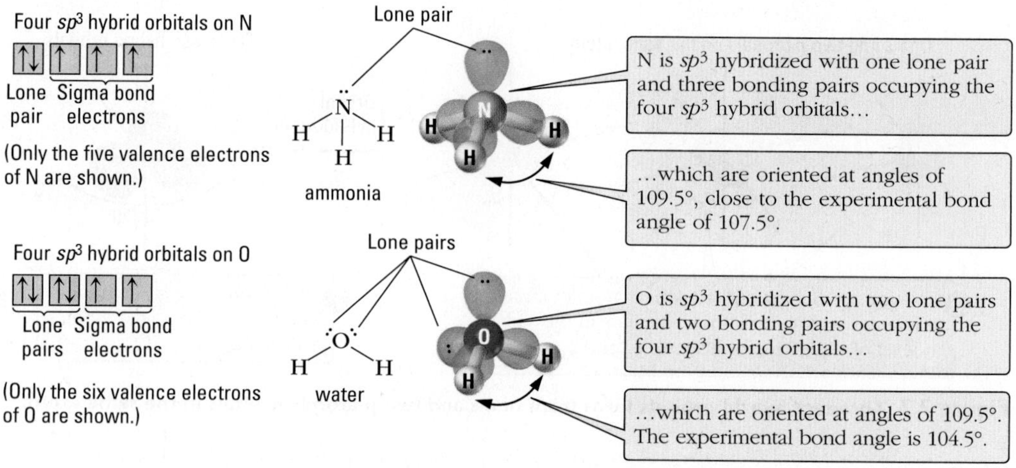
Four sp^3 hybrid orbitals on N
Lone pair
Sigma bond electrons
(Only the five valence electrons of N are shown.)
Lone pair
ammonia
N is sp^3 hybridized with one lone pair and three bonding pairs occupying the four sp^3 hybrid orbitals…
…which are oriented at angles of 109.5°, close to the experimental bond angle of 107.5°.
Four sp^3 hybrid orbitals on O
Lone pairs
Sigma bond electrons
(Only the six valence electrons of O are shown.)
Lone pairs
water
O is sp^3 hybridized with two lone pairs and two bonding pairs occupying the four sp^3 hybrid orbitals…
…which are oriented at angles of 109.5°. The experimental bond angle is 104.5°.

Figure 7.9 Hybridization of nitrogen in ammonia and oxygen in water.

Electron-region Geometry (VSEPR)	Hybrid Orbitals
Linear	Two *sp*
Triangular planar	Three sp^2
Tetrahedral	Four sp^3

octet rule commonly have sp^3 hybridization. Figure 7.9 illustrates sp^3 hybridization for NH_3 and H_2O, showing that lone pairs as well as bonding pairs can occupy hybrid orbitals. Notice that the hybrid orbitals are consistent with the molecular geometries predicted using VSEPR theory: triangular pyramid for NH_3 and bent (or angular) for H_2O.

Table 7.4 summarizes information about hybrid orbitals formed from *s* and *p* atomic orbitals.

A bond in which there is head-to-head orbital overlap so that the electron density of the bond is symmetric around the bonding axis is called a **sigma bond (σ bond)**. The single bonds in diatomic molecules such as H_2 and Cl_2 and the single bonds in BeH_2, BF_3, and CH_4 are sigma bonds, as are the N—H and O—H bonds in ammonia and water. In general, bonds involving hybrid orbitals on one of the bonded atoms or both bonded atoms are sigma bonds.

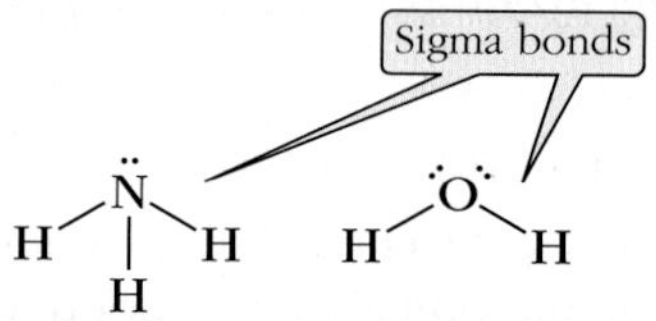
Sigma bonds

PROBLEM-SOLVING EXAMPLE 7.5

Hybridization of Atomic Orbitals

Describe the hybridization around the central atom and the bonding in
(a) NCl_3. (b) BCl_3.
(c) CH_3OH (hybridization of C and O). (d) $AlCl_4^-$.

Result

(a) sp^3 with three N—Cl sigma bonds (b) sp^2 with three B—Cl sigma bonds
(c) sp^3 on C as well as O with three C—H sigma bonds, one C—O sigma bond, and one O—H sigma bond
(d) sp^3 with four sigma bonds between Al and Cl atoms

Analyze Atoms that form more than one bond use one hybrid orbital for each sigma bond and one for each lone pair of electrons. The number of sigma bonds and lone pairs, which equals the number of hybrid orbitals, can be determined from a Lewis structure. Each hybrid orbital bonds a terminal atom or contains a lone pair.

Plan Write the correct Lewis structure of each molecule or ion. Determine the number of sigma bonding pairs and lone pairs around the central atom.

Execute Lewis structures are

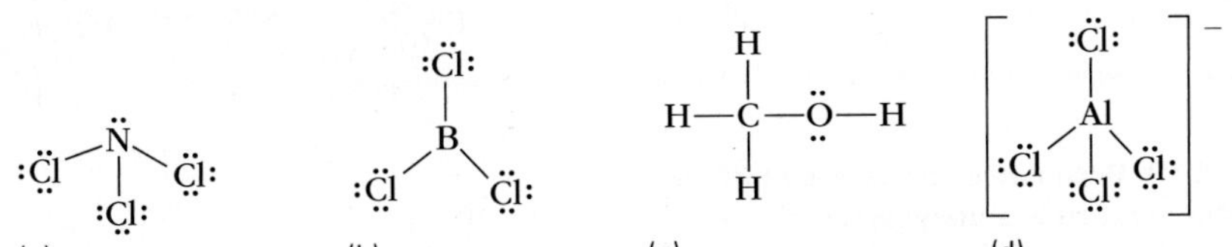

(a) (b) (c) (d)

(a) There are four electron pairs around nitrogen in NCl_3, a compound analogous to NH_3. Each electron pair is in an sp^3 hybrid orbital. The three N—Cl bonds are sigma bonds formed by the head-to-head overlap of three half-filled sp^3 hybrid orbitals on nitrogen with those of chlorine. The lone pair on nitrogen is in a filled sp^3 hybrid orbital.

(b) BCl_3 is analogous to BF_3; therefore, boron has sp^2 orbital hybridization with boron forming three sigma bonds to chlorine.

(c) Carbon forms four sigma bonds in CH_3OH (three to hydrogen atoms and one to oxygen) and so has four bonding pairs, all in sp^3 hybrid orbitals. Oxygen has two sigma bonds formed by the head-to-head overlap of two half-filled sp^3 hybrid orbitals on oxygen with one on carbon and one on hydrogen. The two lone pairs on oxygen are in filled sp^3 hybrid orbitals.

(d) The four sigma bonds in $AlCl_4^-$ involve four bonding pairs in sp^3 hybrid orbitals on aluminum that overlap head-to-head with sp^3 hybrid orbitals on chlorine atoms.

PROBLEM-SOLVING PRACTICE 7.5

Describe the hybridization around the central atom and the bonding in

(a) $BeCl_2$. (b) NH_4^+.
(c) CH_3OCH_3 (hybridization of C and O). (d) BF_4^-.

Table 7.4 Hybrid Orbitals and Their Geometries

	Linear	Triangular Planar	Tetrahedral
Atomic orbitals mixed	One *s* and one *p*	One *s* and two *p*	One *s* and three *p*
Hybrid orbitals formed	Two *sp*	Three sp^2	Four sp^3
Unhybridized orbitals remaining	Two *p*	One *p*	None

7-3d Expanded Octets and Hybridization

Molecules with expanded octets (← **Sec. 7-2d**) are sometimes described as being *hypervalent* because in a Lewis diagram the central atom has more than an octet of electron pairs in its valence shell. This situation occurs with elements from the third period and beyond, such as in PF_5 and SF_6. The structures of hypervalent molecules can be predicted based on the number of electron regions around the central atom using VSEPR theory; however, using quantum mechanics to explain the structures is more complicated.

Until recently, the favored explanation was that *d* atomic orbitals, as well as *s* and *p* atomic orbitals, are involved in hybrid orbitals on the central atom in a hypervalent molecule. Two types of hybrid orbitals are dsp^3 (the hybridized combination of one *d* atomic orbital, one *s* atomic orbital, and three *p* atomic orbitals, which allows for 10 valence electrons) and d^2sp^3 (the hybridized combination of two *d* atomic orbitals, one *s* atomic orbital, and three *p* atomic orbitals, which allows for 12 valence electrons). However, a growing number of publications based on quantum mechanical calculations report that there is *little to no* hybridization of *d* atomic orbitals in hypervalent molecules or polyatomic ions where a *nonmetal* atom is the central atom. Such molecules or ions

See, for example:

Gillespie, R. J., and Popelier, P. L. A. *Chemical Bonding and Molecular Geometry*. London: Oxford University Press, 2001.

Mitchell, T. A., et al. *Journal of Chemical Education*, Vol. 84, 2007; p. 629.

Galbraith, J. M. *Journal of Chemical Education*, Vol. 84, 2007; p. 783.

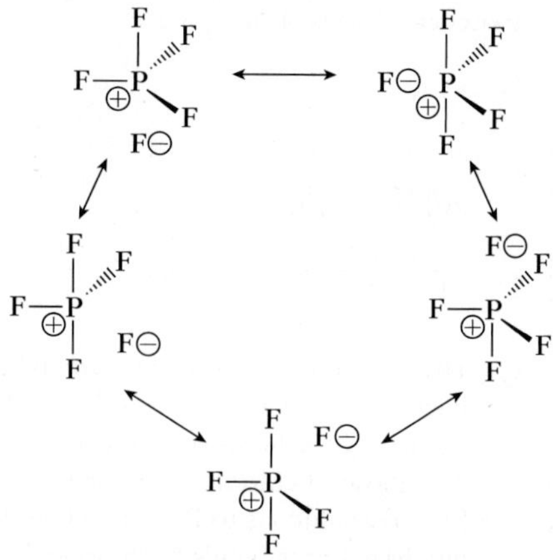

Figure 7.10 Resonance structures of PF_5 in which the octet rule is obeyed. (From T. A. Mitchell, et al., *Journal of Chemical Education,* Vol. 84, 2007; p. 629. Copyright © 2007, Division of Chemical Education, Inc. Reprinted by permission.)

include triangular bipyramidal or octahedral ones, such as PF_5, PF_6^-, and SF_6. Quantum chemistry calculations using both valence bond (← **Sec. 6-1**) and molecular orbital (← **Sec. 6-12**) theories indicate that for hypervalent nonmetal central atoms, the energy of the *d* atomic orbitals is too high for them to hybridize with lower-energy *s* and *p* atomic orbitals. Therefore, participation of *d* atomic orbitals is very small and designations such as dsp^3 and d^2sp^3 are not appropriate. The stability of species such as PF_5, PF_6^-, and SF_6 is more directly related to the size of the central atom and the high electronegativity of the terminal atoms than to *d* atomic orbital involvement.

There is ample experimental evidence that *d* atomic orbitals do participate in bonding when the central atom is a transition metal ion or atom. Such *d*-orbital involvement is described in detail in Chapter 20.

These findings raise the point of whether the Lewis structure and bonding in a molecule such as PF_5 can be described in such a way that the octet rule is not exceeded, that is, without an expanded octet. An alternate approach that does so involves five equivalent resonance structures, each of which has four covalent bonds and one ionic bond as shown in Figure 7.10.

This discussion of hypervalent molecules nicely illustrates the continuing development of the theory of chemical bonding and molecular structure. How we explain bonding in hypervalent molecules has changed over time, and may change again.

The highly electronegative terminal fluorine atoms draw sufficient electron density away from the much less electronegative phosphorus atom to leave a net of eight electrons around the phosphorus. These eight electrons are shared in four covalent bonds between phosphorus and four fluorine atoms, conforming to the octet rule. In this model, the P atom has a formal charge of 1+ (← **Sec. 6-8**); each resonance form also has a countervailing 1− fluoride ion to establish the ionic bond. (Notice that the formal charges are not the minimum possible values; this is because of the very strong electron-attracting ability of the terminal fluorine atoms.) This method can be extended to SF_6 as well as to polyatomic ions such as SO_4^{2-} and ClO_4^-, where Lewis structures are frequently written that have more than eight electrons in the valence shell of the central nonmetal atom.

For example, the Lewis structure for SO_4^{2-} is often written

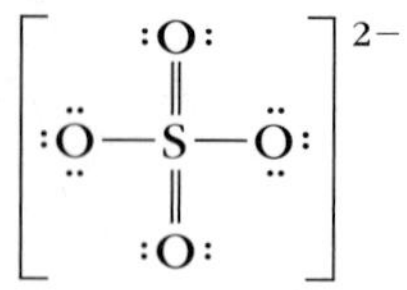

which indicates 12 valence electrons around S.

7-4 Hybridization: Molecules with Multiple Bonds

In a multiple bond, only one of the shared electron pairs occupies a hybrid orbital. On a central atom with a multiple bond, the hybrid orbitals contain

- All electron pairs that form single bonds to terminal atoms.
- All lone pairs.
- *One* of the shared electron pairs in a double or triple bond.

The electrons in *unhybridized* atomic orbitals are used to form the second bond of a double bond and the second and third bonds in a triple bond. Such bonds, called **pi bonds (π bonds)**, are *formed by the* sideways *(edgewise) overlap of parallel* p *atomic orbitals.*

It is also possible for *d* atomic orbitals to overlap above and below a bond axis, forming a pi bond.

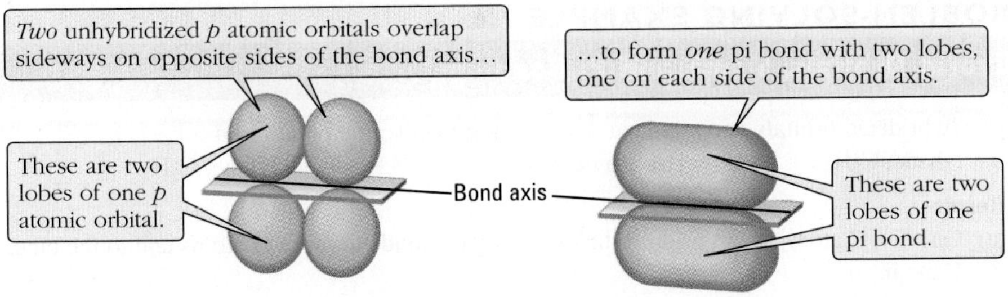
Two unhybridized *p* atomic orbitals overlap sideways on opposite sides of the bond axis...
...to form *one* pi bond with two lobes, one on each side of the bond axis.
These are two lobes of one *p* atomic orbital.
Bond axis
These are two lobes of one pi bond.

Figure 7.11 Pi bonding involves sideways overlap of *p* atomic orbitals.

In contrast to sigma bonds, in which orbital overlap is symmetric around the bond axis, pi bonds involve overlap above and below the bond axis, as shown in Figure 7.11.

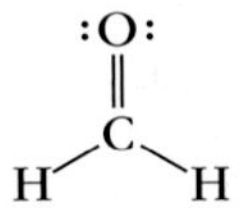
formaldehyde

The bonding in formaldehyde, H_2CO, exemplifies sigma and pi bonding. Formaldehyde has two C—H single bonds and one C=O double bond. The triangular planar electron-region geometry suggests sp^2 hybridization, which provides three sp^2 hybrid orbitals for three sigma bonds.

As seen in Figure 7.12, the C and O atoms in formaldehyde are sp^2 hybridized. With its three sp^2 hybrid orbitals the C atom forms two sigma bonds to the two H atoms and one sigma bond to the O atom. The three sp^2 hybrid orbitals on the O atom are involved in a sigma bond to the C atom and two lone pairs. This constitutes the sigma-bonding framework. On both the C atom and the O atom there is one unhybridized *p* orbital. The two *p* orbitals overlap side by side to form a pi bond, which is the second half of the C=O double bond. This is the pi bonding framework. The C=O double bond consists of one sigma plus one pi bond.

Because pi bonds have less orbital overlap than sigma bonds, pi bonds are usually weaker than sigma bonds. Thus, a C=C double bond is stronger (and shorter) than a C—C single bond, but not twice as strong; correspondingly, a C≡C triple bond, although stronger (and shorter) than a C=C double bond, is not three times stronger than a C—C single bond.

Bond enthalpies are given in Table 6.2 (← Sec. 6-6b).

Consider what would happen if one end of a pi bond were to rotate around the sigma bond axis relative to the other end. Because the *p* atomic orbitals need to be parallel to overlap, the pi bond would be broken and the molecule containing the pi bond would be less stable by a quantity of energy equal to the strength of the pi bond. The molecule must gain this quantity of energy for rotation around the pi bond to happen. At room temperature there is insufficient energy and there is no rotation around a double bond. A consequence of this barrier to rotation is *cis-trans isomerism* (← **Sec. 6-5a**).

At temperatures well above room temperature, molecules have greater kinetic energy and some molecules may have enough energy for rotation around a double bond to occur.

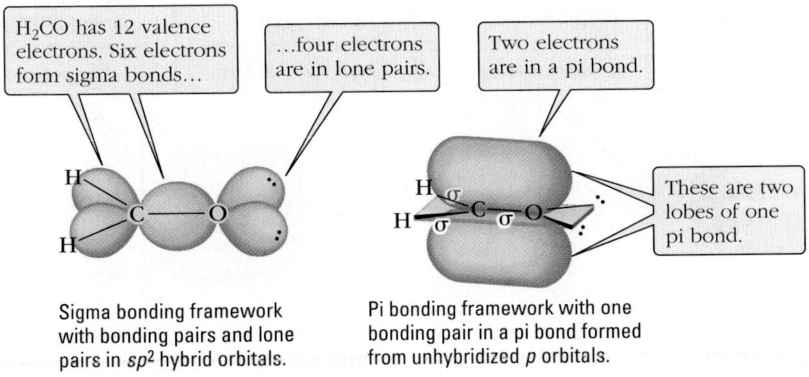
H_2CO has 12 valence electrons. Six electrons form sigma bonds...
...four electrons are in lone pairs.
Two electrons are in a pi bond.
H
H
C
O
H
H
C
O
σ
σ
σ
These are two lobes of one pi bond.
Sigma bonding framework with bonding pairs and lone pairs in sp^2 hybrid orbitals.
Pi bonding framework with one bonding pair in a pi bond formed from unhybridized *p* orbitals.

Figure 7.12 Sigma and pi bonding in formaldehyde.

PROBLEM-SOLVING EXAMPLE 7.6

Hybrid Orbitals; Sigma and Pi Bonding

Use hybridized orbitals and sigma and pi bonding to describe bonding in
(a) ethane, C_2H_6. (b) ethylene, C_2H_4. (c) acetylene, C_2H_2.

Result

(a) On each C atom, sp^3 hybrid orbitals form sigma bonds to three H atoms and to the other C atom.
(b) On each C atom, sp^2 hybrid orbitals form sigma bonds to two H atoms and to the other C atom; unhybridized 2*p* atomic orbitals form one pi bond between C atoms.
(c) On each C atom, *sp* hybrid orbitals form sigma bonds to one H and the other C and two unhybridized 2*p* atomic orbitals form two pi bonds between C atoms. (See Figure 7.13.)

Analyze Electron-region geometry around each atom (other than H) can be obtained from a Lewis structure and VSEPR theory. Hybridization can be determined from electron-region geometry. Sigma bonds and lone pairs involve hybrid orbitals. Pi bonds involve sideways overlap of unhybridized *p* orbitals. Multiple bonds involve sigma and pi bonding.

Plan Write a Lewis structure for each molecule. Determine electron-region geometry and hence hybridization on each atom. Form sigma bonds and lone pairs using hybrid orbitals. Form multiple bonds from one sigma and one or two pi bonds.

Execute The correct Lewis structures are

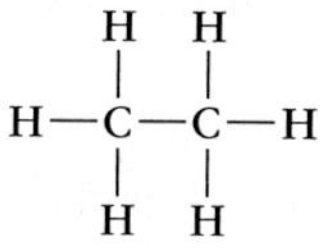
ethane

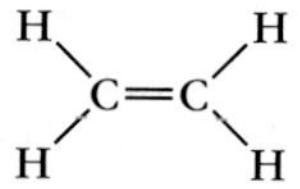
ethylene

acetylene

(a) Each carbon atom in ethane has four regions of electron density and therefore is sp^3 hybridized. One sp^3 hybrid orbital on each C atom overlaps head-to-head with an sp^3 hybrid orbital on the other C atom, forming a sigma bond. The other three sp^3 hybrid orbitals form sigma bonds to H atoms.
(b) In ethylene, each C atom has three regions of electron density, which requires sp^2 hybridization. Two of the three sp^2 hybrid orbitals on each carbon form sigma bonds with hydrogens; the third sp^2 hybrid orbital forms a C—C sigma bond to the other C atom. Sideways overlap of parallel unhybridized *p* atomic orbitals from each carbon forms a pi bond, completing the C=C double bond.

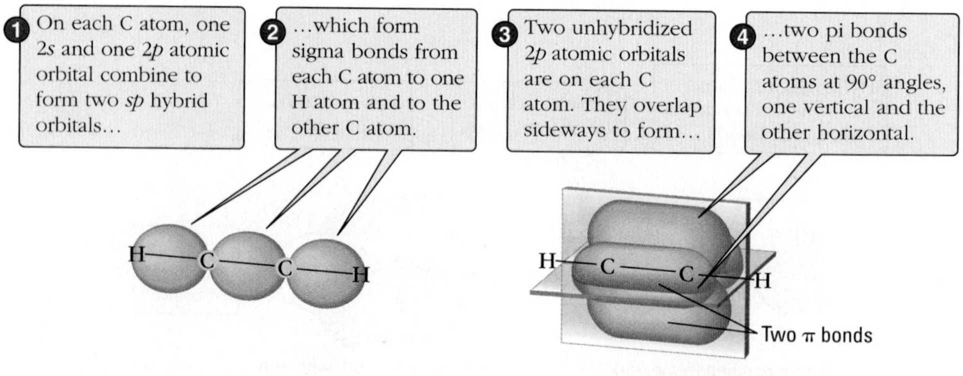
1 On each C atom, one 2s and one 2p atomic orbital combine to form two sp hybrid orbitals…
2 …which form sigma bonds from each C atom to one H atom and to the other C atom.
3 Two unhybridized 2p atomic orbitals are on each C atom. They overlap sideways to form…
4 …two pi bonds between the C atoms at 90° angles, one vertical and the other horizontal.
H
C
C
H
H
C
C
H
Two π bonds

Sigma bonding framework with bonding pairs in *sp* hybrid orbitals at 180° angles.

Pi bonding framework with two bonding pairs in two pi bonds at 90° to each other.

Figure 7.13 Sigma and pi bonding in acetylene.

(c) In acetylene, each C atom has two regions of electron density, implying *sp* hybridization. On each C atom one *sp* hybrid orbital forms a sigma bond to H; the other *sp* hybrid orbital joins the C atoms by a sigma bond. The two unhybridized *p* atomic orbitals on each carbon overlap sideways to form two pi bonds, completing the triple bond, which consists of one sigma and two pi bonds at right angles (90°) to each other (Figure 7.13).

PROBLEM-SOLVING PRACTICE 7.6

Using hybridization and sigma and pi bonding, explain the bonding in (a) HCN and (b) H_2CNH.

CONCEPTUAL EXERCISE 7.4

Pi Bonding

Explain why pi bonding is not possible for an sp^3 hybridized carbon atom.

7-5 Molecular Polarity

Polar bonds occur when two bonded atoms have different electronegativities (← **Sec. 6-7**). Figure 7.14 shows example diatomic molecules with nonpolar (Cl_2) and polar (HCl) bonds. In HCl, *electron density is more concentrated at one end of the molecule than the other*, so HCl is a **polar molecule**. A polar molecule is a *permanent* dipole; that is, it always has a partial positive electric charge at one end and a partial negative charge at the other. A **nonpolar molecule** *has no overall separation of electric charge.*

In an electric field created by a pair of oppositely charged plates (Figure 7.15), the partially positive end of a polar molecule is attracted toward the negative plate, and the partially negative end is attracted toward the positive plate. This tends to turn a polar molecule so it aligns with the field. The extent to which a molecule lines up with the field depends on its dipole moment. The **dipole moment**, μ, is *the product of the magnitude of the partial charges ($\delta-$) and ($\delta+$) times the distance of separation between them.* The derived unit of the dipole moment is the coulomb-meter (C m); a more convenient derived unit is the debye (D), defined as $1\ D = 3.34 \times 10^{-30}$ C m. Some typical experimental values are listed in Table 7.5. *Nonpolar molecules have zero dipole moment ($\mu = 0$);* ***polar molecules have positive dipole moments that increase with greater molecular polarity.***

Table 7.5 Dipole Moments of Selected Molecules

Molecule	Dipole Moment, μ (D)
H_2	0
HF	1.78
HCl	1.07
HBr	0.79
HI	0.38
ClF	0.88
BrF	1.29
BrCl	0.52
H_2O	1.85
H_2S	0.95
CO_2	0
NH_3	1.47
NF_3	0.23
NCl_3	0.39
CH_4	0
CH_3Cl	1.92
CH_2Cl_2	1.60
$CHCl_3$	1.04
CCl_4	0

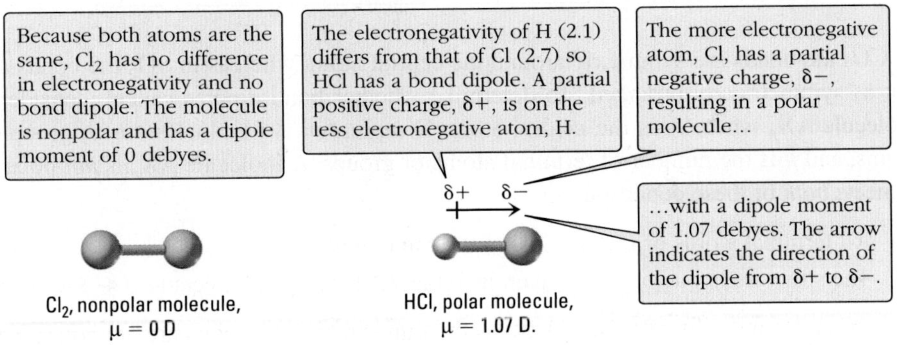
Because both atoms are the same, Cl2 has no difference in electronegativity and no bond dipole. The molecule is nonpolar and has a dipole moment of 0 debyes.
The electronegativity of H (2.1) differs from that of Cl (2.7) so HCl has a bond dipole. A partial positive charge, δ+, is on the less electronegative atom, H.
The more electronegative atom, Cl, has a partial negative charge, δ−, resulting in a polar molecule...
δ+ δ−
...with a dipole moment of 1.07 debyes. The arrow indicates the direction of the dipole from δ+ to δ−.
Cl2, nonpolar molecule, μ = 0 D
HCl, polar molecule, μ = 1.07 D.

Figure 7.14 Examples of nonpolar (Cl_2) and polar (HCl) molecules.

In physics a dipole is defined as pointing from negative to positive electric charge, but in chemistry it is electron distribution in a bond or molecule that causes a dipole. Thus, chemists draw dipoles pointing from positive charge to negative charge.

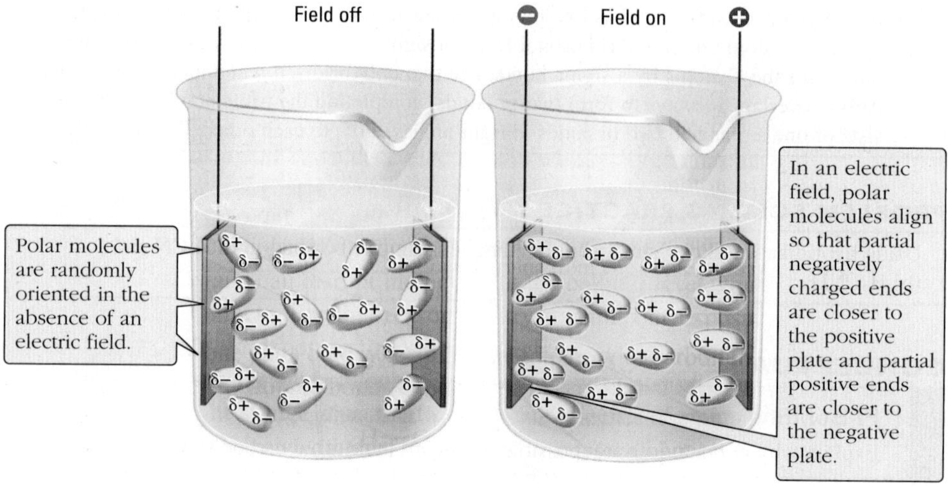
Field off
Field on
Polar molecules are randomly oriented in the absence of an electric field.
In an electric field, polar molecules align so that partial negatively charged ends are closer to the positive plate and partial positive ends are closer to the negative plate.

Figure 7.15 Polar molecules align with an electric field.

Combining bond dipoles to obtain a molecular dipole involves addition of vectors (the bond dipoles) to obtain a resultant vector (the molecular dipole).

The net dipole moment of a molecule is the sum of its bond dipoles. To predict whether a molecule is polar, we need to answer two questions:

- **Does the molecule have at least one polar bond?**
- **How are polar bonds positioned relative to one another?**

A diatomic molecule has only one bond; if that bond is polar, the molecule must be polar, with the partial negative charge at the more electronegative atom, as in HCl (Figure 7.14). In a polyatomic molecule with polar bonds, there are two possibilities. For certain molecular shapes, there is no net dipole ($\mu = 0$) because the bond dipoles sum to zero; the *molecule* is nonpolar even though the *bonds* are polar. In other polyatomic molecules, the molecular shape is such that the bond dipoles sum to a net dipole ($\mu > 0$) and the molecule is polar. CO_2 is an example of the first type and SO_2 is an example of the second type. Because O (3.4) is more electronegative than either C (2.4) or S (2.3), bonds in each molecule are polar.

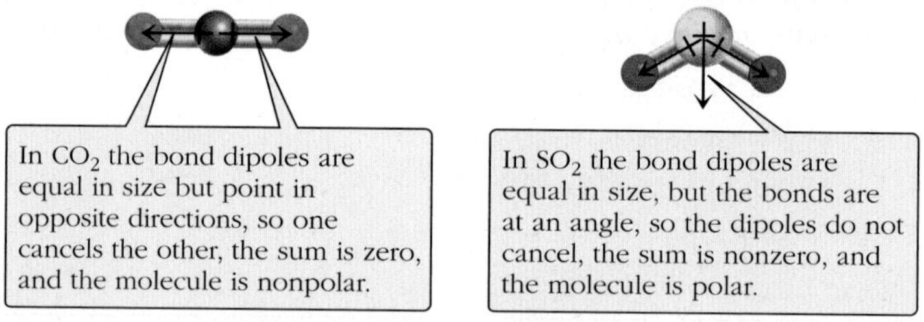
In CO_2 the bond dipoles are equal in size but point in opposite directions, so one cancels the other, the sum is zero, and the molecule is nonpolar.
In SO_2 the bond dipoles are equal in size, but the bonds are at an angle, so the dipoles do not cancel, the sum is nonzero, and the molecule is polar.

In CO_2 the bonds are symmetric around the central atom, whereas in SO_2 the bonds are not as symmetrically arranged. To determine whether bond dipoles cancel, consider a molecule AX_n where A is the central atom, X represents a terminal atom or group of atoms, and n is the number of terminal atoms or groups. A molecule AX_n is *not* polar if it meets *both* of these conditions:

- All terminal atoms or groups, X, are the same, *and*
- The structure is one of those given in Table 7.1 for AX_nE_0 molecules (← **Sec. 7-2a**).

That is, molecules with all terminal atoms the same and no lone pairs are nonpolar; individual bond dipoles cancel due to the symmetry of the structure. Examples of the application of these rules are given in Figure 7.16, where molecular structures are color coded: red indicates high electron density and blue indicates low electron density. From the

Peter Debye
1884–1966

Imagno/Hulton Archive/Getty Images

Peter Debye was one of the leading figures in physical chemistry during the early- to mid-twentieth century. His many contributions included the first major X-ray diffraction studies with randomly oriented microcrystals, the seminal theoretical work in conjunction with Hückel on interactions of ionic solutes in solution, and the defining research on molecular polarities. Debye received the 1936 Nobel Prize in Chemistry. He left Germany in 1940 because he was being required to give up his Dutch citizenship and become a German citizen. He accepted an invitation to join the chemistry faculty at Cornell University where he remained until the end of his career.

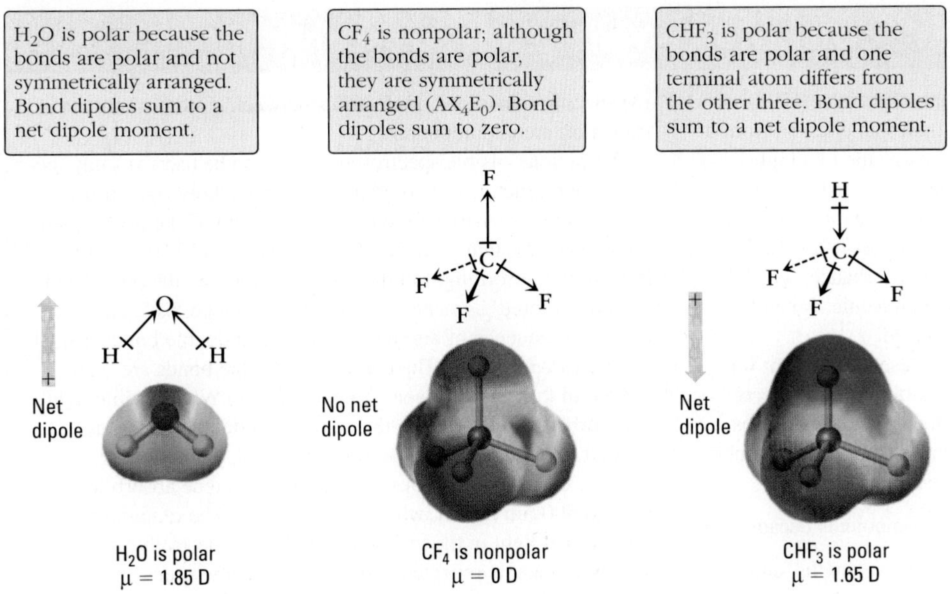
H_2O is polar because the bonds are polar and not symmetrically arranged. Bond dipoles sum to a net dipole moment.
CF_4 is nonpolar; although the bonds are polar, they are symmetrically arranged (AX_4E_0). Bond dipoles sum to zero.
CHF_3 is polar because the bonds are polar and one terminal atom differs from the other three. Bond dipoles sum to a net dipole moment.
Net dipole
No net dipole
Net dipole
H_2O is polar
$\mu = 1.85$ D
CF_4 is nonpolar
$\mu = 0$ D
CHF_3 is polar
$\mu = 1.65$ D

Figure 7.16 Determining whether a molecule is polar or nonpolar from its structure.

colors it is clear that CF_4 has symmetrical electron density whereas H_2O and CHF_3 are unsymmetrical.

Using a microwave oven to make popcorn or to heat dinner is an everyday application of the fact that water is a polar molecule. Most foods have a high water content. When water molecules in foods absorb microwave radiation, they rotate, turning so that the dipoles align with the crests and troughs of the oscillating electric field of the microwave radiation. Microwave ovens create microwave radiation that oscillates at 2.45 GHz (2.45 gigahertz, $2.45 \times 10^9\ s^{-1}$), very nearly the optimum frequency to rotate water molecules. So the leftover pizza warms up in a hurry for a late-night snack.

PROBLEM-SOLVING EXAMPLE 7.7

Molecular Polarity

Are (a) boron trifluoride, BF_3, and (b) dichloromethane, CH_2Cl_2, polar or nonpolar? If a molecule is polar, indicate the direction of the net dipole.

Result (a) BF_3 is nonpolar; (b) CH_2Cl_2 is polar, with the partial negative end at the Cl atoms.

Analyze A molecule is polar if it has polar bonds that are not symmetrical. Bond polarity depends on electronegativity difference, which can be predicted from the periodic table. A net dipole points from the positive to the negative side of a molecule.

Plan Write the correct Lewis structure for each molecule and use it to assess whether the bonds are arranged symmetrically or unsymmetrically. Use the periodic table to predict the polarity of the bonds. If the bonds are polar and arranged unsymmetrically around the central atom, the molecule is polar. The net dipole points from the less electronegative atoms to the more electronegative atoms.

Execute

(a) F is more electronegative than B so the B—F bonds are polar. BF_3 has three regions of electron density around B and is triangular planar, a symmetrical AX_3E_0 arrangement. The bond dipole of each B—F bond is canceled by the sum of the bond dipoles of the other two B—F bonds. Therefore, BF_3 is nonpolar.

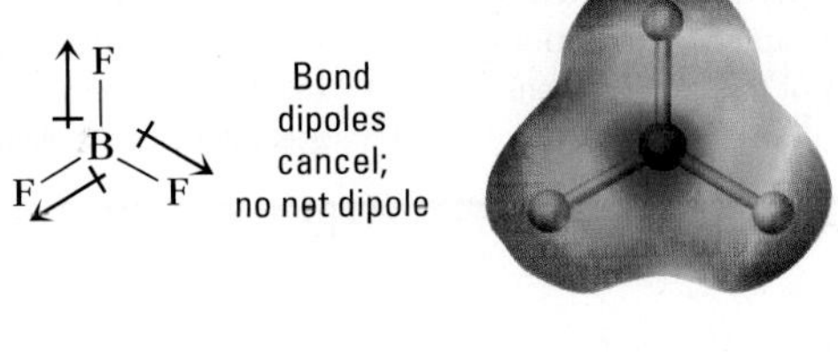
Bond dipoles cancel; no net dipole

TOOLS OF CHEMISTRY | Ultraviolet-Visible Spectroscopy

When a molecule absorbs radiation in the ultraviolet (UV) or visible region, the absorbed energy promotes an electron from a lower energy orbital in the ground-state molecule to a higher energy orbital in the excited state molecule. The maximum UV-visible absorption occurs at a wavelength characteristic of the molecule's structure. This absorption can be determined from a UV-visible spectrum, a plot of the absorbance, which is the intensity of radiation absorbed by the molecule, versus the wavelength at which the absorbance occurred.

Ultraviolet-visible spectral data can be used to account for several structural features, including (1) differentiation of *cis* from *trans* isomers, because a *trans* isomer generally absorbs UV-visible radiation at a longer wavelength than the *cis* isomer; (2) confirmation of the presence (or absence) of carbonyl $\rangle C{=}O$ and aromatic groups in organic compounds because they absorb UV radiation at characteristic wavelengths; and (3) specific transitions of electrons between *d* atomic orbitals in transition metal compounds.

Ultraviolet-visible spectroscopy can also be used to study the molecular structures of colored compounds, those that absorb radiation in the visible region of the spectrum. Generally, organic compounds containing only sigma bonds are colorless; that is, they do not absorb light in the visible region. On the other hand, brightly colored pigments, such as beta-carotene, have an extended sequence of alternating single and double bonds, called a *conjugated* system. (The conjugated double bonds are highlighted in the beta-carotene structure below.) When visible light is absorbed, pi electrons in the conjugated double bonds are excited to higher energy levels. Carrots look yellow-orange because electron transitions in their beta-carotene absorb in the 400- to 500-nm region, which is blue-green. The remaining, unabsorbed light in the visible region of the spectrum, which is yellow-orange, is reflected and we see a yellow-orange color.

beta-carotene

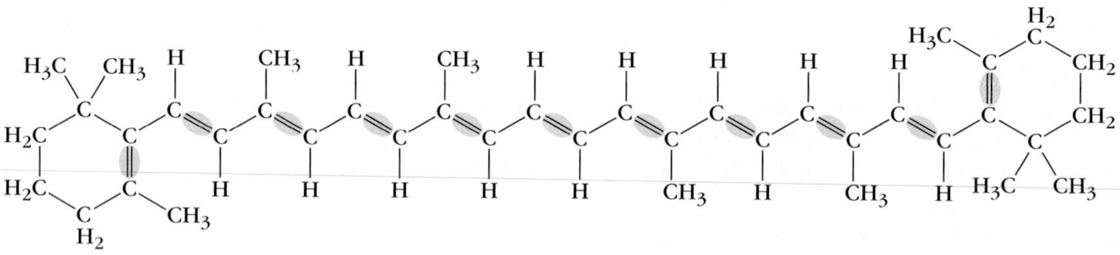

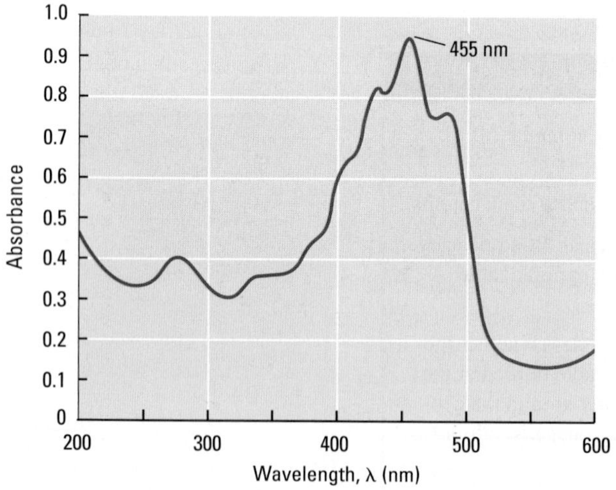
455 nm
Absorbance
1.0
0.9
0.8
0.7
0.6
0.5
0.4
0.3
0.2
0.1
0
200
300
400
500
600
Wavelength, λ (nm)

The UV-visible spectrum of beta-carotene. Beta-carotene absorbs in the blue-green region of the visible spectrum, indicated by a maximum absorbance at 455 nm. This peak is caused by transitions of pi electrons in the conjugated double bonds of the beta-carotene molecule.

© Cengage Learning/James Maynard

Some vegetables that are rich in beta-carotene. The best sources of beta-carotene have intensely yellow-orange color—on the outside, on the inside, or both. Sweet potatoes, carrots, butternut squash, red-leaf lettuce, and many green vegetables are rich in beta-carotene. In general, the more intensely colored a vegetable is the greater its concentration of beta-carotene is.

(b) From the Lewis structure for CH_2Cl_2, you might be tempted to say that CH_2Cl_2 is nonpolar because the hydrogens appear across from each other, as do the chlorines. In fact, CH_2Cl_2 is a tetrahedral molecule in which no atom is directly across from another atom; all angles are near 109.5°.

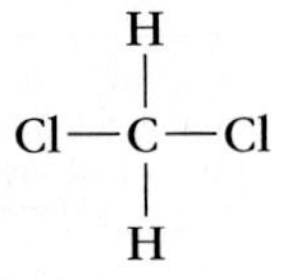

CH_2Cl_2,

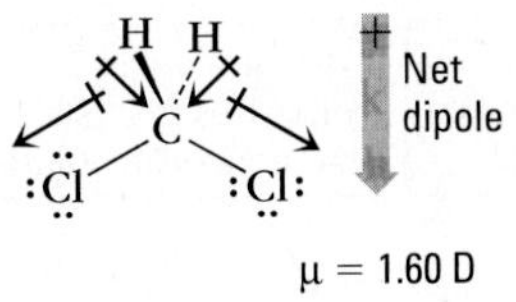
Net
dipole
μ = 1.60 D

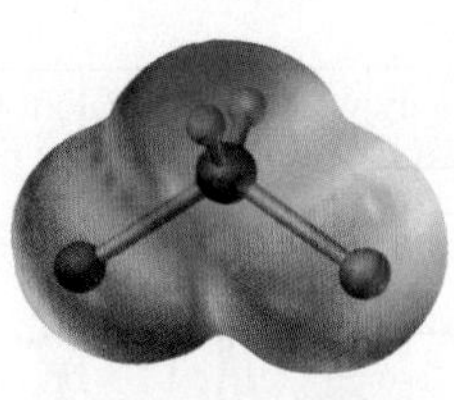

Because Cl is more electronegative than C, while H is less electronegative, negative charge is drawn away from H atoms and toward Cl atoms. As a result, CH_2Cl_2 has a net dipole (μ = 1.60 D), with the partial negative end at the Cl atoms and the partial positive end between the H atoms.

PROBLEM-SOLVING PRACTICE 7.7

Decide whether each molecule is polar and, if so, which region is partially positive and which is partially negative.

(a) $BFCl_2$ (b) NH_2Cl (c) SCl_2

EXERCISE 7.5

Dipole Moments

Explain the differences in the dipole moments of

(a) HI (0.38 D) and HBr (0.79 D).

(b) CH_3Cl (1.92 D), CH_3Br (1.81 D), and CH_3I (1.62 D).

7-6 Noncovalent Interactions and Forces Between Molecules

Molecules are attracted to one another. If there were no attractions, then there would be no liquid or solid molecular compounds, only gaseous ones. The attraction between molecules is electrostatic. It results from attraction between opposite electric charges, regardless of whether the charges are permanent or temporary. For example, polar molecules attract each other due to the partial positive and negative charges of their permanent dipoles.

It is important to distinguish noncovalent interactions—forces between molecules—from covalent bonds, which are much stronger. Atoms within the same molecule are held together by covalent chemical bonds with strengths ranging from 150 to 1000 kJ/mol. Covalent bonds are quite strong forces. For example, it takes 1664 kJ to break 4 mol C—H covalent bonds and separate the one C atom and four H atoms in all the molecules in 1 mol methane molecules:

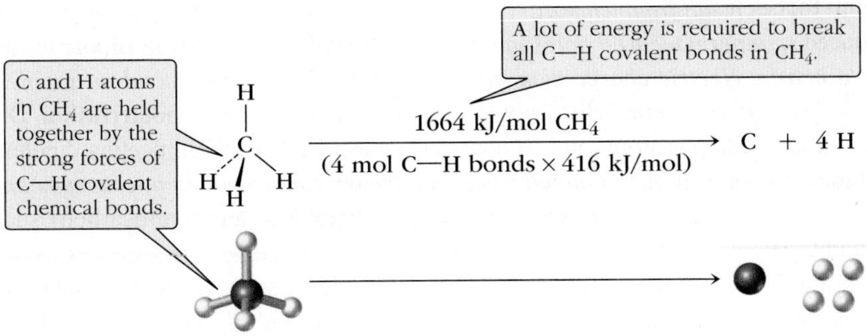
A lot of energy is required to break all C—H covalent bonds in CH4.
C and H atoms in CH4 are held together by the strong forces of C—H covalent chemical bonds.
1664 kJ/mol CH4
(4 mol C—H bonds × 416 kJ/mol)
C + 4 H

$CH_4(g) \rightarrow C(g) + 4\ H(g)$

$\Delta_r H° = 1664$ kJ/mol $CH_4(g)$

Weaker forces attract one molecule to another molecule. In contrast to the 1664 kJ/mol it takes to atomize methane, only 8.2 kJ is required to pull 1 mol of methane molecules that are close together in liquid methane away from each other to evaporate liquid methane to a gas.

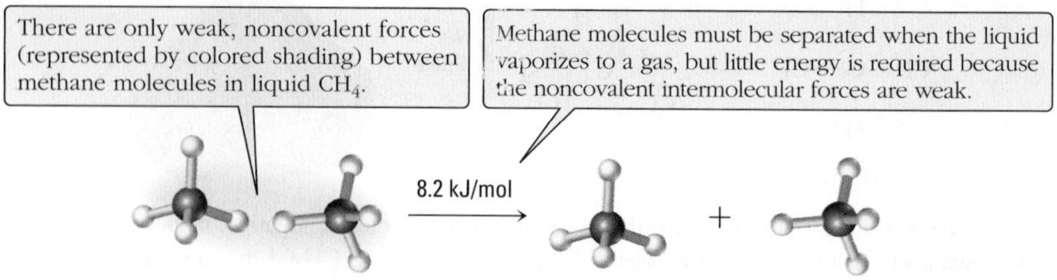
There are only weak, noncovalent forces (represented by colored shading) between methane molecules in liquid CH_4.
Methane molecules must be separated when the liquid vaporizes to a gas, but little energy is required because the noncovalent intermolecular forces are weak.
8.2 kJ/mol
+

$CH_4(\ell) \rightarrow CH_4(g)$

$\Delta_{vap}H° = 8.2$ kJ/mol $CH_4(\ell)$

It requires about 200 times more energy to decompose 1 mol methane gas into its atoms than to evaporate that amount of liquid methane.

***Intra*molecular means within a single molecule; *inter*molecular means *between* two or more separate molecules.**

We use the term **noncovalent interactions** to refer to *all forces of attraction* other than *covalent bonding* (← **Sec. 6-1**), *ionic bonding* (← **Sec. 2-6a**), *or metallic bonding* (*metallic bonding* is discussed in Section 9-9). *Noncovalent interactions acting between one molecule and another* are referred to as **intermolecular forces**. Intermolecular forces are weaker than covalent bonds because they do not result from the sharing of electron pairs between atoms. However, intermolecular forces affect melting points, boiling points, and other physical properties of molecular substances. If a single molecule is large enough, noncovalent interactions can occur between different parts of the same molecule (*intra*molecular forces). Such intermolecular and intramolecular noncovalent interactions maintain biologically important molecules in the exact shapes required to carry out their functions. Chymotrypsin, an enzyme with a molar mass of more than 25,000 g/mol, is folded into its biochemically active shape with the assistance of noncovalent interactions between strategically placed atoms. Figure 7.17 shows a ribbon model of chymotrypsin. Intramolecular noncovalent interactions help maintain the shapes of thousands of different protein molecules in our bodies.

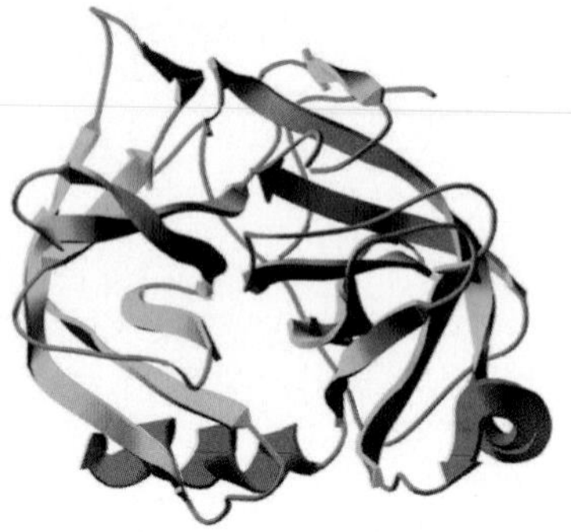

Figure 7.17 A ribbon model of a chymotrypsin molecule. Noncovalent interactions help fold chymotrypsin into its biochemically active shape. Red and green ribbons show portions where the protein chain is folded into helical or sheet structures, respectively; the narrow purple ribbons represent simpler, unfolded protein regions.

The next few sections explore three types of noncovalent interactions: *London forces, dipole-dipole attractions, and hydrogen bonding.*

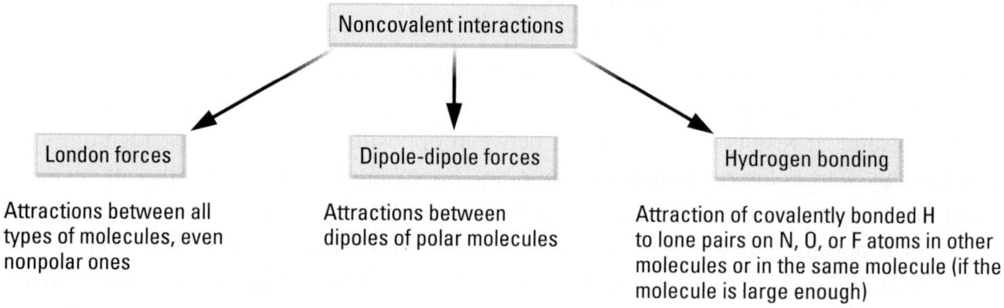
Noncovalent interactions
London forces
Attractions between all types of molecules, even nonpolar ones
Dipole-dipole forces
Attractions between dipoles of polar molecules
Hydrogen bonding
Attraction of covalently bonded H to lone pairs on N, O, or F atoms in other molecules or in the same molecule (if the molecule is large enough)

7-6a London (Dispersion) Forces

London forces are named to recognize the work of Fritz London, who extensively studied the origins and nature of such forces.

London forces, also known as *dispersion forces,* occur in *all* molecular substances. **London forces** *result from temporary attractions between the positive and negative ends of* induced *(nonpermanent) dipoles in adjacent molecules*. An **induced dipole** is *caused when a nearby electric charge momentarily distorts the electron cloud in a molecule*. This kind of *shift in electron distribution in a molecule* is known as **polarization**. Due to polarization, noble-gas atoms and all nonpolar molecules have these fleeting dipoles and are attracted to one another. *London forces are the* only *noncovalent interactions among nonpolar molecules*, including oxygen, nitrogen, halogens, and hydrocarbons such as CH_4 and C_2H_6.

Figure 7.18 illustrates how one Cl_2 molecule with a momentary unevenness in its electron distribution can induce a dipole in a neighboring Cl_2 molecule and how the dipoles align to produce a net attraction. London forces range in energy from

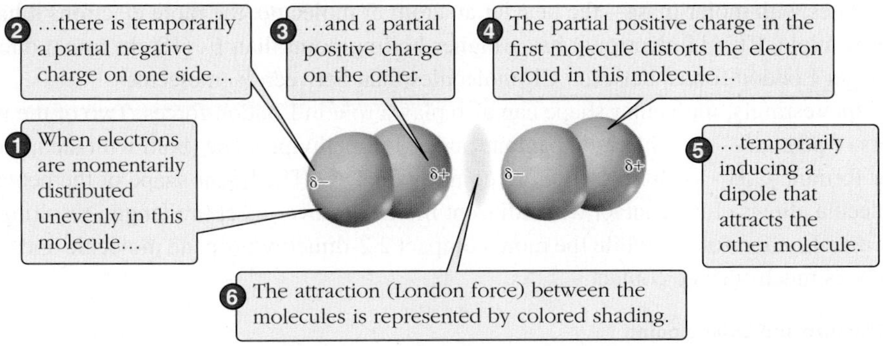
1 When electrons are momentarily distributed unevenly in this molecule…
2 …there is temporarily a partial negative charge on one side…
3 …and a partial positive charge on the other.
4 The partial positive charge in the first molecule distorts the electron cloud in this molecule…
5 …temporarily inducing a dipole that attracts the other molecule.
6 The attraction (London force) between the molecules is represented by colored shading.
δ−
δ+
δ−
δ+

Figure 7.18 Origin of London forces between two adjacent chlorine molecules.

approximately 0.05 to 40 kJ/mol. Their strength depends on how readily electrons in a molecule can be polarized. In general, polarization is greater when a molecule contains more electrons and the electrons are less tightly attracted to the nucleus; *London forces increase with increasing number of electrons in a molecule.* Thus, large molecules with many electrons, such as Br_2 and I_2, are more polarizable and have stronger London forces. Smaller molecules (F_2, N_2, O_2) have fewer electrons and weaker London forces.

For a liquid to boil, its molecules must have enough energy to overcome noncovalent intermolecular attractive forces among the molecules in the liquid. The higher the temperature is, the greater the average energy of molecules is (← **Sec. 1-9a**). When intermolecular forces are stronger, more energy is required to overcome them and a higher temperature is needed to boil a liquid. Thus, *the boiling point of a liquid depends on the nature and strength of intermolecular forces.*

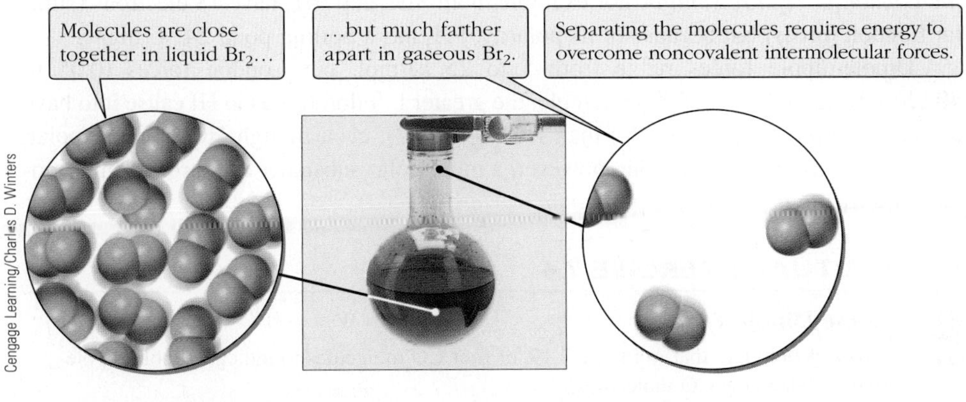
Molecules are close together in liquid Br₂…
…but much farther apart in gaseous Br₂.
Separating the molecules requires energy to overcome noncovalent intermolecular forces.
Cengage Learning/Charles D. Winters

Table 7.6 lists boiling points of several groups of nonpolar molecules. It is clear that *boiling points increase as the total number of electrons increases.* (This effect also

Table 7.6 Effect of Number of Electrons on Boiling Points of Nonpolar Molecular Substances

Noble Gases			Halogens			Hydrocarbons		
	No. e's	bp (°C)		No. e's	bp (°C)		No. e's	bp (°C)
He	2	−269	F_2	18	−188	CH_4	10	−161
Ne	10	−246	Cl_2	34	−34	C_2H_6	18	−88
Ar	18	−186	Br_2	70	59	C_3H_8	26	−42
Kr	36	−152	I_2	106	184	C_4H_{10}	34	0

correlates with molar mass—the heavier an atom or molecule, the more electrons it has.) For example, Cl_2 (34 electrons) has a higher boiling point than F_2 (18 electrons) due to stronger London forces between Cl_2 molecules than between F_2 molecules.

Interestingly, molecular shape can also play a role in London forces. Two of the isomers of pentane—straight-chain pentane and 2,2-dimethylpropane (both with the molecular formula C_5H_{12})—differ in boiling point by 26.5 °C. The linear shape of the pentane molecule allows close contact with adjacent molecules over its entire length, resulting in stronger London forces, while the more compact 2,2-dimethylpropane molecule does not allow as much surface contact.

Structure and boiling point. The boiling points of pentane and 2,2-dimethylpropane differ because of differences in their molecular structures even though the total number of electrons in each molecule is the same.

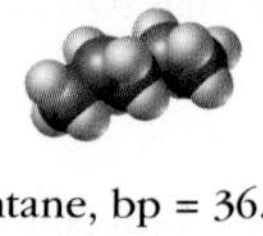

pentane, bp = 36.0 °C

2,2-dimethylpropane, bp = 9.5 °C

7-6b Dipole-Dipole Attractions

A noncovalent interaction called **dipole-dipole attraction** *occurs between permanent dipoles in adjacent* polar *molecules or between two polar groups in the same large molecule*. Molecules that are permanent dipoles attract each other when the partial positive region of one is close to the partial negative region of another (Figure 7.19).

The boiling points of several nonpolar and polar substances with comparable numbers of electrons, and therefore comparable London forces, are given in Table 7.7. In general, *the more polar its molecules, the higher the boiling point of a substance, provided the London forces are similar*. For example, both SiH_4 and PH_3 have 18 electrons (similar London forces) but because of its polarity, PH_3 has a boiling point 24 °C higher.

Dipole-dipole forces range from 5 to 25 kJ/mol, but London forces (0.05 to 40 kJ/mol) can be stronger. For example, the greater London forces in HI cause it to have a higher boiling point (−36 °C) than HCl (−85 °C), even though HCl is more polar. When London forces are similar, however, a more polar substance has stronger intermolecular attractions than a less polar one.

Table 7.7 Numbers of Electrons and Boiling Points of Nonpolar and Polar Substances

Nonpolar Molecules		
	No. e's	bp (°C)
SiH_4	18	−112
GeH_4	36	−90
Br_2	70	59
Polar Molecules		
	No. e's	bp (°C)
PH_3	18	−88
AsH_3	36	−62
ICl	70	97

CONCEPTUAL EXERCISE 7.6

Dipole-Dipole Forces

Draw a sketch, like that in Figure 7.19, of four CO molecules to indicate dipole-dipole forces between the CO molecules.

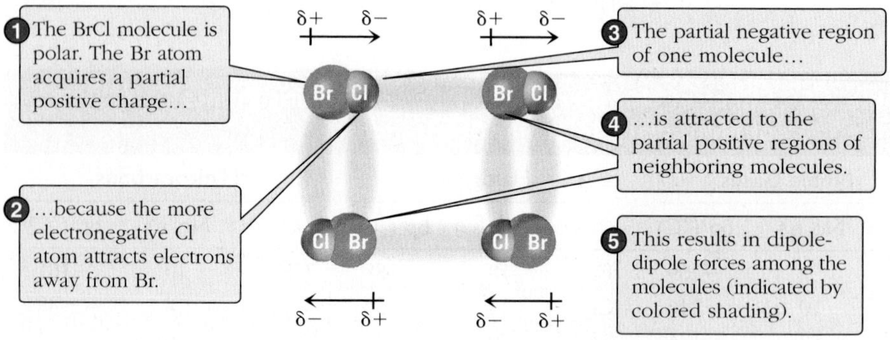

1 The BrCl molecule is polar. The Br atom acquires a partial positive charge...
2 ...because the more electronegative Cl atom attracts electrons away from Br.
δ+ δ−
δ+ δ−
Br Cl
Br Cl
Cl Br
Cl Br
δ− δ+
δ− δ+
3 The partial negative region of one molecule...
4 ...is attracted to the partial positive regions of neighboring molecules.
5 This results in dipole-dipole forces among the molecules (indicated by colored shading).

Figure 7.19 Dipole-dipole attractions between polar BrCl molecules.

PROBLEM-SOLVING EXAMPLE 7.8

London and Dipole-Dipole Forces

What kind of forces must be overcome to
(a) evaporate gasoline, a mixture of hydrocarbons?
(b) convert solid carbon dioxide (Dry Ice) into a vapor?
(c) decompose ammonium nitrite, NH_4NO_2, into N_2 and H_2O?
(d) boil ClF?
(e) convert P_4 to P_2?

Result
(a) and (b) London forces
(c) Covalent bonds between N and H atoms, and O and N atoms; ionic attractions between NH_4^+ and NO_2^- ions
(d) London forces and dipole-dipole forces
(e) Covalent bonds between phosphorus atoms

Analyze Noncovalent intermolecular attractions between molecules must be broken when a molecular substance changes physical state. In such a change:

- dipole-dipole forces are involved if the molecules are polar.
- London forces are present between all molecules, polar as well as nonpolar.

A chemical reaction requires breaking covalent bonds *within* a molecule or polyatomic ion, or separation or combination of charged ions.

Plan Consider whether each change given is a change in physical state or a chemical reaction.

Execute
(a) Gasoline is a mixture of hydrocarbons, which are nonpolar molecules. To evaporate, the hydrocarbon molecules must overcome London forces among them in the liquid to escape from each other and enter the gas phase.
(b) Solid carbon dioxide converts to a vapor when the thermal energy becomes great enough to overcome some of the noncovalent forces of attraction (London forces) among the nonpolar CO_2 molecules. Noncovalent forces still exist in the gaseous CO_2.
(c) Decomposing ammonium nitrite into N_2 and H_2O requires breaking the N—H and N—O covalent bonds in the NH_4^+ and NO_2^- polyatomic ions and rearranging the atoms to form nitrogen and water. These bond-breaking changes involve much greater energy (the bond enthalpy) than that needed to melt the compound, which involves overcoming cation-anion attractions.
(d) Dipole-dipole forces, like those between BrCl molecules in Figure 7.19, attract the polar ClF molecules to each other in the liquid. London forces are also present. These collective forces must be overcome when ClF molecules separate from their neighbors at the liquid surface and enter the gaseous state.
(e) Covalent bonds must be broken between phosphorus atoms in P_4.

PROBLEM-SOLVING PRACTICE 7.8

Explain the principal type of forces that must be overcome for
(a) Kr to melt.
(b) propane to release C and H_2.

7-6c Hydrogen Bonds

A **hydrogen bond** is *the attraction of a hydrogen atom in a molecule or part of a molecule X—H for another atom or group of atoms*. A hydrogen bond is differentiated because it is stronger than would be predicted by London forces and dipole forces alone. For strong hydrogen bonding there must be a hydrogen atom covalently bonded to F, O, or N (designated X) and a lone electron pair on a small, very electronegative atom,

Hydrogen bonds can form between molecules or within large molecules. These are known as *inter*molecular and *intra*molecular hydrogen bonds, respectively.

usually F, O, or N (designated :Z). The X—H · · · :Z angle is usually 180°. The H atom and the lone pair may be in two different molecules or in different parts of the same large molecule.

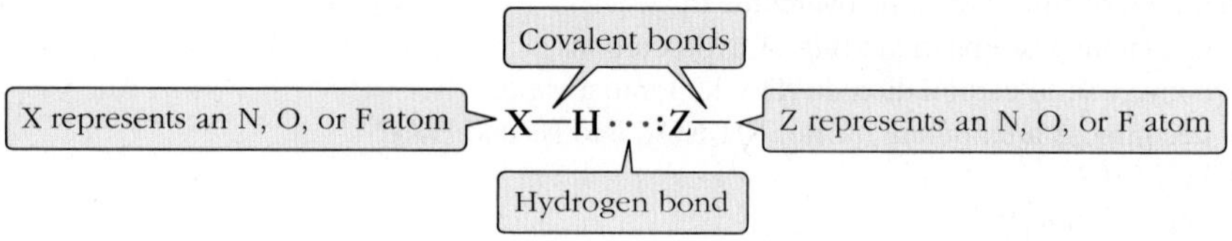
Covalent bonds
X represents an N, O, or F atom
X—H···:Z—
Z represents an N, O, or F atom
Hydrogen bond

A hydrogen bond can be thought of as a resonance hybrid of these two structures

X—H :Z— $\overset{-1}{\text{X}}$: H—$\overset{+1}{\text{Z}}$—

This resonance provides greater stability than would a dipole-dipole attraction alone. (Formal charges are shown in the Lewis structures.)

Type of Force	Magnitude of Force (kJ/mol)
Covalent bond	150–1000
London force	0.05–40
Dipole-dipole	5–25
Hydrogen bond	10–40

The X—H group is referred to as a *hydrogen-bond donor*; the :Z group is a *hydrogen-bond acceptor*. The greater the electronegativity of the atom connected to H is, the stronger the hydrogen bond is. Hydrogen bond "bridges" from a hydrogen atom to a lone pair on an N or O atom play an essential role in determining the folding (three-dimensional structure) of protein molecules.

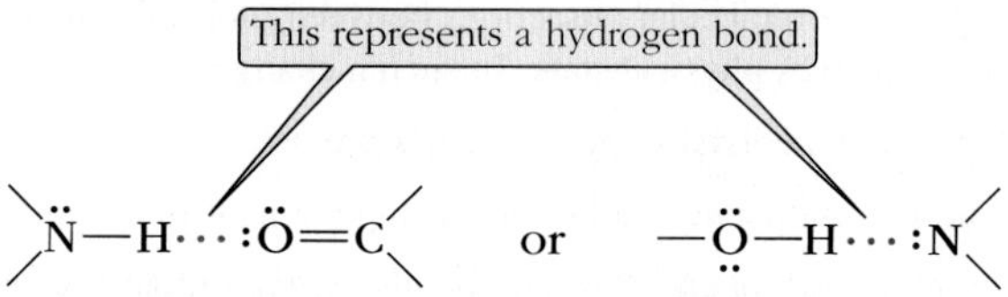
This represents a hydrogen bond.
or

Hydrogen-bond strengths range from 10 to 40 kJ/mol—less than for covalent bonds, but greater than dipole-dipole attractions and greater than London forces in small molecules. However, there are often a great many hydrogen bonds in a sample of matter so the overall effect can be very dramatic. An example is the very different melting and boiling points of ethanol and dimethyl ether, which were mentioned at the beginning of this chapter. Figure 7.20 explains the different physical properties of these two compounds.

The hydrogen halides also illustrate the significant effects of hydrogen bonding. The boiling point of hydrogen fluoride, HF, is 19.5 °C, much higher than the boiling points

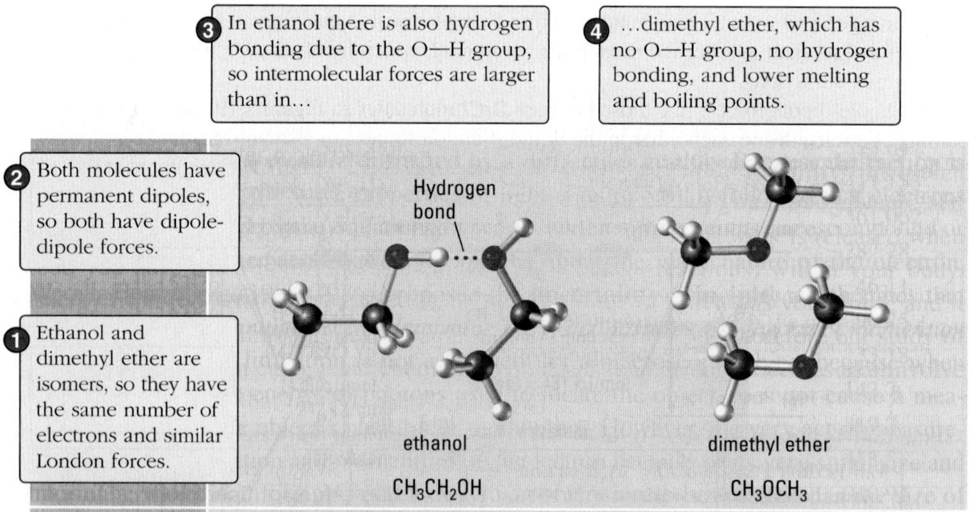
3 In ethanol there is also hydrogen bonding due to the O—H group, so intermolecular forces are larger than in…
4 …dimethyl ether, which has no O—H group, no hydrogen bonding, and lower melting and boiling points.
2 Both molecules have permanent dipoles, so both have dipole-dipole forces.
1 Ethanol and dimethyl ether are isomers, so they have the same number of electrons and similar London forces.
Hydrogen bond
ethanol
dimethyl ether

	ethanol CH_3CH_2OH	dimethyl ether CH_3OCH_3
Dipole moment, D	1.69	1.30
Melting point, °C	−114.1	−141.5
Boiling point, °C	78.29	−24.8

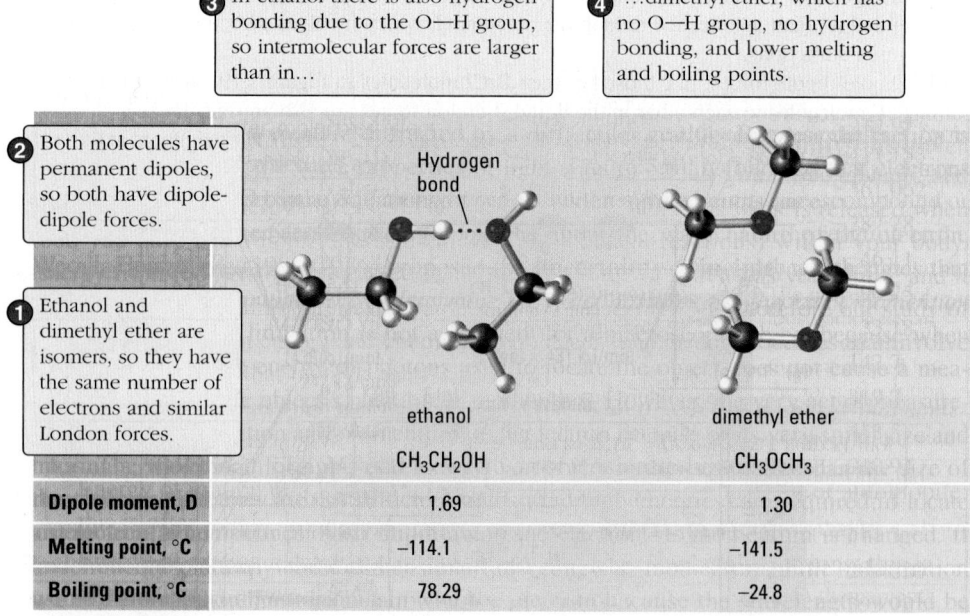

Figure 7.20 Noncovalent interactions, melting points, and boiling points in ethanol and dimethyl ether.

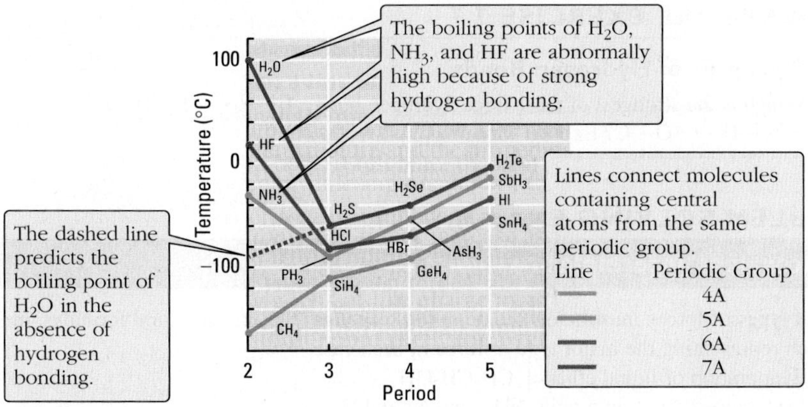
The boiling points of H2O, NH3, and HF are abnormally high because of strong hydrogen bonding.
Lines connect molecules containing central atoms from the same periodic group.
Line Periodic Group
4A
5A
6A
7A
The dashed line predicts the boiling point of H2O in the absence of hydrogen bonding.
Temperature (°C)
100
0
−100
H2O
HF
NH3
PH3
CH4
H2S
HCl
SiH4
H2Se
HBr
GeH4
AsH3
H2Te
SbH3
HI
SnH4
2
3
4
5
Period

Figure 7.21 Boiling points of some simple hydrogen-containing binary compounds.

of the other hydrogen halides (HCl, −85 °C, HBr, −67 °C, and HI, −35 °C). All of the hydrogen halides are polar and all have London forces (with HI the strongest because it has the most electrons). The high boiling point of HF is attributed to strong hydrogen bonds. In solid HF, hydrogen bonds hold the molecules in the zig-zag structure shown.

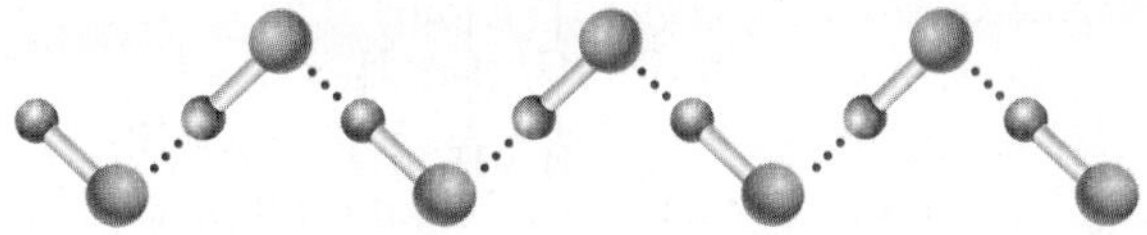

Hydrogen bonding in solid HF

The especially strong hydrogen bonding when H is bonded to N, O, or F is illustrated in Figure 7.21, where H_2O, HF, and NH_3 are seen to have unusually high boiling points.

Hydrogen bonding among water molecules is responsible for many of the unique properties of water (Section 9-5), including the hexagonal shapes of snowflakes. Except for water, all hydrogen compounds in Figure 7.21 are gases at room temperature. That H_2O is a liquid indicates very strong intermolecular attraction. Also notice that the boiling point of H_2O is about 200 °C higher than would be predicted if hydrogen bonding were not present. Figure 7.22 shows that in ice each water molecule participates in four hydrogen bonds and there are two hydrogen bonds per water molecule.

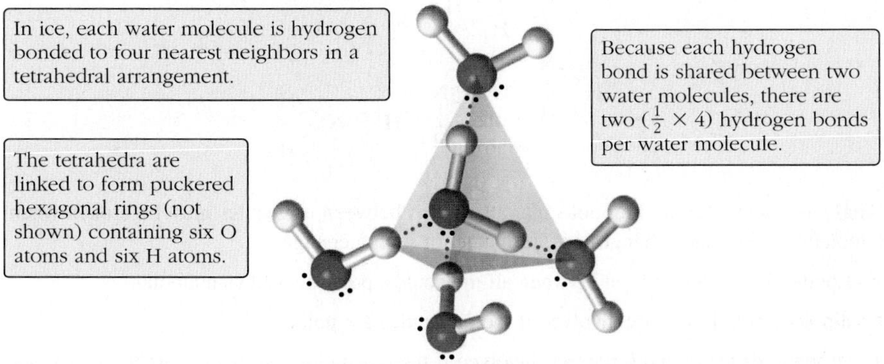
In ice, each water molecule is hydrogen bonded to four nearest neighbors in a tetrahedral arrangement.
The tetrahedra are linked to form puckered hexagonal rings (not shown) containing six O atoms and six H atoms.
Because each hydrogen bond is shared between two water molecules, there are two ($\frac{1}{2} \times 4$) hydrogen bonds per water molecule.

Figure 7.22 Hydrogen bonding between one water molecule and four nearest neighbors.

Scott Camazine/Science Source

Ice crystals. The crystal shapes reflect the hexagonal orientation of water molecules in ice that results from hydrogen bonding.

CONCEPTUAL EXERCISE 7.7

Strengths of Hydrogen Bonds

Which is the strongest of these three hydrogen bonds: F—H · · · F—H, O—H · · · N—C, or N—H · · · O—C? Explain why.

PROBLEM-SOLVING EXAMPLE 7.9

Noncovalent Intermolecular Forces

What types of forces must be overcome in these changes? Using structural formulas, make a sketch representing the major type of force in each case.

(a) Evaporation of liquid ethanol, CH_3CH_2OH
(b) Decomposition of ammonia, NH_3, into N_2 and H_2
(c) Sublimation of iodine, I_2, that is, its conversion from a solid directly into a vapor
(d) Boiling of liquid NH_3

Result

(a) Hydrogen bonding, dipole forces, and London forces:

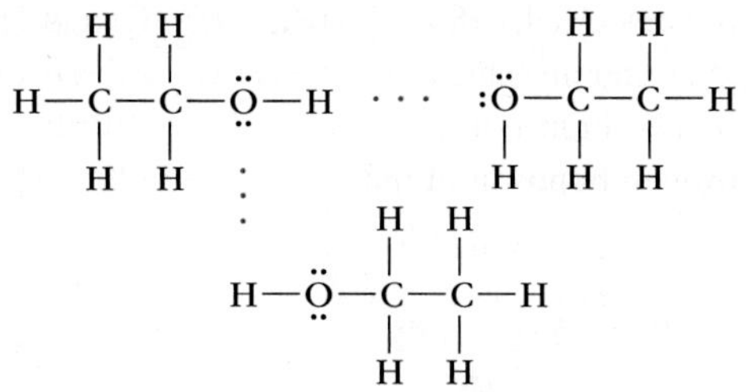

(b) Covalent N—H bonds:

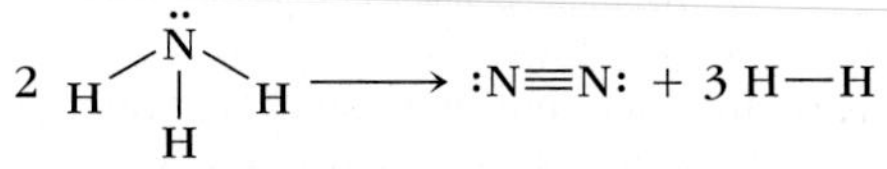

(c) London forces between I_2 molecules:

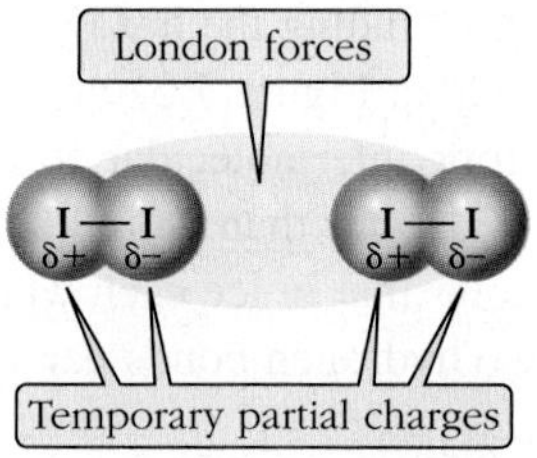
London forces
I—I
δ+ δ−
I—I
δ+ δ−
Temporary partial charges

(d) Hydrogen bonding, dipole forces, and London forces:

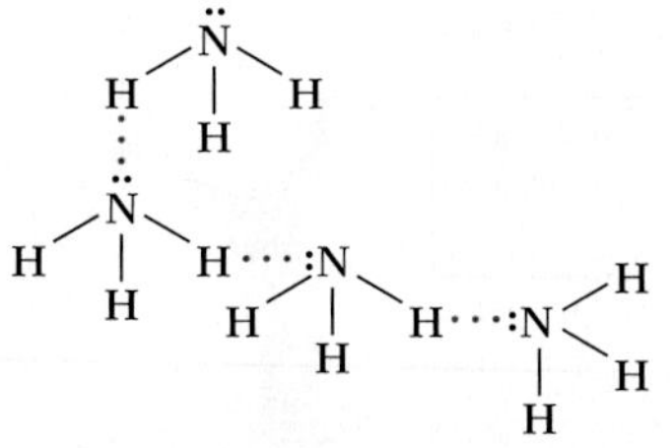

Analyze Noncovalent intermolecular attractions between molecules must be broken when a molecular substance changes physical state. In such a change:

- London forces are present between all molecules, polar as well as nonpolar.
- dipole-dipole forces are involved if the molecules are polar.
- strong hydrogen bond forces are involved if there is an H atom bonded to N, O, or F and an N, O, or F with a lone pair of electrons.

A chemical reaction requires breaking covalent bonds *within* a molecule or polyatomic ion, or separation or combination of charged ions.

Of the four cases given, three of them—(a), (c), and (d)—involve a change in the physical state of a molecular compound and one, (b), involves a chemical reaction.

Plan For each physical change consider the molecular structure to identify the types of noncovalent intermolecular forces. For each chemical reaction consider the type of chemical bonding (covalent or ionic).

Execute

(a) Ethanol molecules have an H atom covalently bonded to an O atom; the O atom has a lone pair of electrons. Therefore, ethanol molecules can hydrogen bond to each other and hydrogen bonds must be overcome. The bent —O—H group is polar, so there are dipole-dipole forces; London forces are always present.

(b) This is not a case of overcoming an intermolecular force, but rather an *intra*molecular one, covalent bonds. The decomposition of ammonia involves breaking the N—H covalent bonds in the NH_3 molecule.

(c) Iodine is composed of nonpolar I_2 molecules, which are held to each other in the solid by London forces. Therefore, these forces must be overcome for the iodine molecules in the solid to escape from one another and become gaseous.

(d) Ammonia molecules have an H atom covalently bonded to an N atom; the N atom has a lone pair of electrons. Therefore, hydrogen bonds must be overcome for the molecules at the surface of the liquid to escape into the vapor. Because NH_3 is polar, dipole-dipole forces must also be overcome; London forces are always present.

250 ml
PYREX

Cengage Learning/Charles D. Winters

Sublimation of iodine. When heated, solid iodine in the bottom of the beaker sublimes directly to violet-colored iodine vapor. When it contacts the cold surface of the ice-filled, round flask at the top of the beaker, the iodine vapor deposits as dark-colored iodine crystals.

PROBLEM-SOLVING PRACTICE 7.9

Decide what types of intermolecular forces are involved in the attraction between molecules of (a) N_2 and N_2, (b) CO_2 and H_2O, and (c) CH_3OH and NH_3.

7-6d Noncovalent Forces in Living Cells

Solubility depends on intermolecular forces. For example, hydrocarbons, such as octane, C_8H_{18}, do not readily dissolve in water—gasoline and water do not mix. In hydrocarbons, the principal intermolecular forces are London forces. In water, strong hydrogen bonding and dipole-dipole forces attract the molecules to each other. This leads to a rule that often predicts solubility: *like dissolves like,* where the word "like" refers to the main type of intermolecular forces. Polar, hydrogen bonding substances dissolve in other polar, hydrogen bonding substances; nonpolar substances are soluble in other nonpolar substances.

The like-dissolves-like rule applies to many biological structures, including cell membranes. The basic structure of a cell membrane is a **lipid bilayer**, two aligned layers of *phospholipids*. Phospholipids are glycerol derivatives having two long, nonpolar fatty acid chains (alkane-like) with a polar phosphate derivative at one end, the polar head, of the molecule (Figure 7.23). The polar end of phospholipid molecules is **hydrophilic**, that is, "water-loving"; it is *attracted to water molecules.* The polar end is oriented toward the aqueous environments inside and outside of the cell. The long, nonpolar chains of the fatty

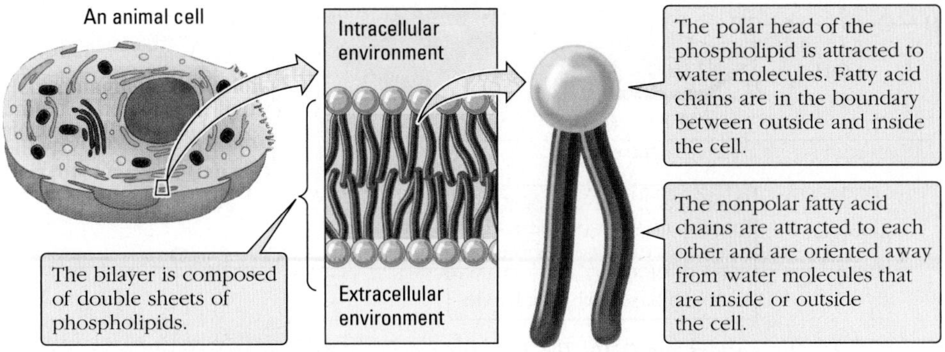
An animal cell
Intracellular environment
Extracellular environment
The polar head of the phospholipid is attracted to water molecules. Fatty acid chains are in the boundary between outside and inside the cell.
The nonpolar fatty acid chains are attracted to each other and are oriented away from water molecules that are inside or outside the cell.
The bilayer is composed of double sheets of phospholipids.

Figure 7.23 Lipid bilayer in cell membranes.

acids are **hydrophobic**, or "water-fearing"; they are *held together by London forces* and point away from the aqueous environments. The result is separation of the interior of a cell from the surrounding fluids by the lipid bilayer, a separation that allows control over which ions or molecules enter a cell and which chemical reactions can occur within the cell.

In our bodies, noncovalent interactions bring molecules of metabolic products, drugs, hormones, and neurotransmitters (nervous system messengers) into contact with cellular receptors. The receptors are often proteins embedded in the cell membrane's lipid bilayer. Hydrogen bonding, dipole-dipole forces, and London forces help to align the target molecule with its particular site on the receptor surface. Once a chemical messenger connects with its receptor, the messenger may enter the cell to initiate a chemical reaction, or a reaction may occur as a result of changes initiated by the receptor.

7-7 Biomolecules: DNA and the Importance of Molecular Structure

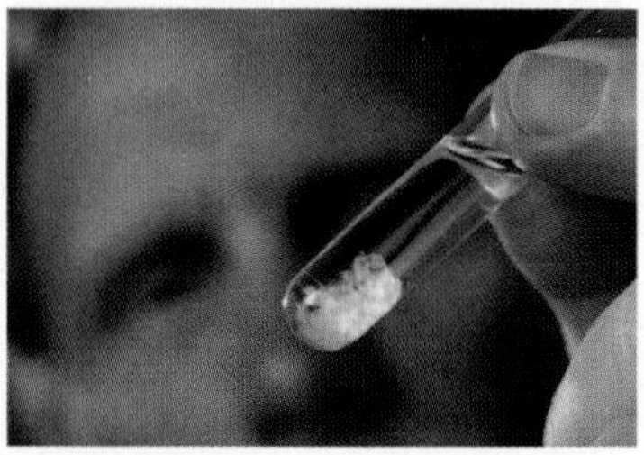

Volker Steger/Science Source

DNA. Solid DNA is precipitated from solution.

Nowhere do the shape of a molecule and noncovalent forces play a more intriguing and important role than in the structure and function of **deoxyribonucleic acid (DNA)**, *the molecule that stores the genetic code*. Whether you are male or female, have blue or brown eyes, or have curly or straight hair depends on your genetic makeup. These and all your other physical traits are determined by the composition of the approximately 1 m of tightly coiled DNA that makes up the 23 double-stranded chromosomes in the nucleus of each of your cells.

7-7a Nucleotides of DNA

DNA is a **polymer**, *a molecule composed of many small, repeating units bonded together*. Each *repeating unit in DNA*, called a **nucleotide**, has the three connected parts shown in Figure 7.24—one sugar unit, one phosphate group (phosphoric acid unit), and

The deoxyribose units in the backbone chain are joined through phosphate diester linkages. Phosphoric acid is H_3PO_4, $O{=}P(OH)_3$.

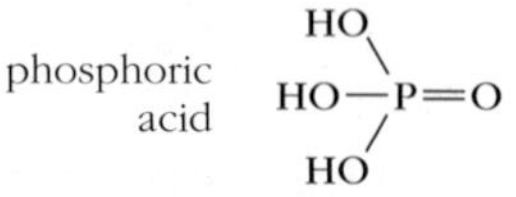
phosphoric
acid

Replacing two H atoms with two CH_3 groups gives a phosphate diester, $O{=}P(OH)(OCH_3)_2$.

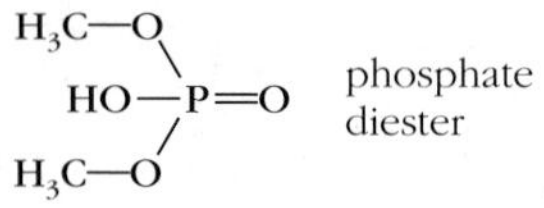
phosphate
diester

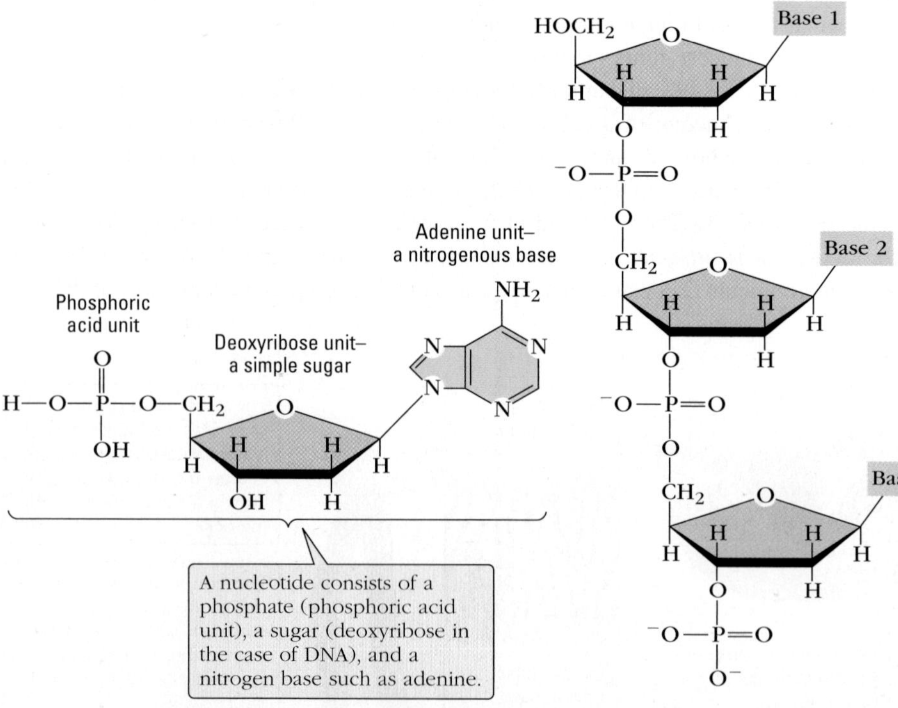
Base 1
Base 2
Base 3
Adenine unit–
a nitrogenous base
Phosphoric
acid unit
Deoxyribose unit–
a simple sugar
A nucleotide consists of a phosphate (phosphoric acid unit), a sugar (deoxyribose in the case of DNA), and a nitrogen base such as adenine.

Figure 7.24 Nucleotides and DNA. The phosphate groups and the deoxyribose sugar groups form the *backbone* of DNA. The genetic code is carried by the nitrogenous bases.

a cyclic nitrogen compound known as a *nitrogen base.* In DNA the sugar is always deoxyribose and the base is one of four bases—*adenine (A), thymine (T), guanine (G),* or *cytosine (C).* The bases are often referred to by their single-letter abbreviations.

A DNA segment with nucleotides bonded together is shown in Figure 7.24. The phosphate units join nucleotides into a polynucleotide chain that has a backbone of alternating deoxyribose and phosphate units in a long strand with the various nitrogenous bases extending out from the sugar-phosphate backbone. The order of the nucleotides (and thus the particular sequence of bases) along the DNA strand carries the genetic code from one generation to the next.

DOE/Science Source

Francis H. C. Crick *(right)* and James D. Watson. Working in the Cavendish Laboratory at Cambridge, England, Watson and Crick built a scale model of the double-helical structure, based on X-ray data. Knowing distances and angles between atoms, they compared the task to working on a three-dimensional jigsaw puzzle. Watson, Crick, and Maurice Wilkins received the 1962 Nobel Prize in Physiology or Medicine for their work relating to the structure of DNA.

7-7b The Double Helix: The Watson-Crick Model

In the early 1950s, Erwin Chargaff measured the quantities of nitrogen bases in DNA samples from a broad range of organisms. Based on his analyses, it was apparent that in any given organism, from a human genius to a bacterium,

- the base composition is the same in all cells of the organism and is characteristic of that organism.
- the quantities of adenine and thymine are equal, as are the quantities of guanine and cytosine.
- the sum of the quantities of adenine plus guanine equals the sum of the quantities of cytosine plus thymine.

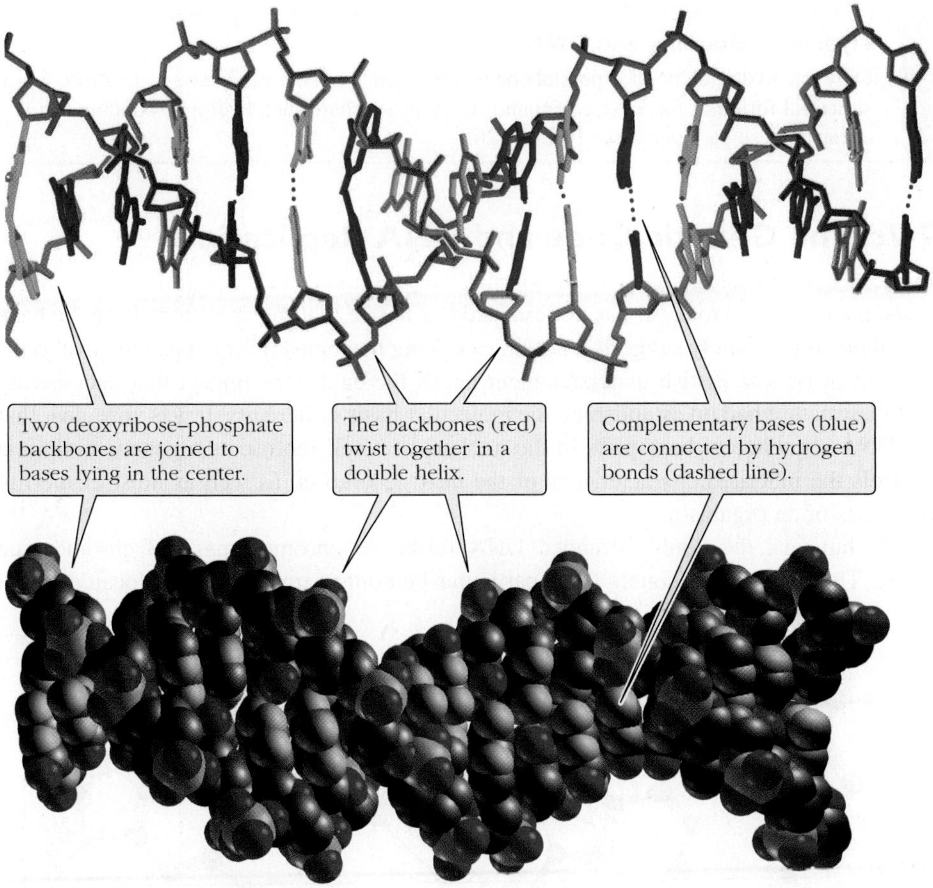
Two deoxyribose–phosphate backbones are joined to bases lying in the center.
The backbones (red) twist together in a double helix.
Complementary bases (blue) are connected by hydrogen bonds (dashed line).

Figure 7.25 Double-stranded DNA. The illustration shows the deoxyribose-phosphate backbone of each strand as phosphate groups (orange and red) linking five-membered rings of deoxyribose joined to bases. The complementary bases (*shown in blue*) are connected by hydrogen bonds: two between adenine and thymine, and three for guanine and cytosine.

Rosalind Franklin
1920–1958

Science Source

Rosalind Franklin received her degrees from Cambridge University. Her most famous work was the first X-ray photographs of DNA, done in collaboration with Maurice Wilkins at King's College in London, showing two forms of DNA, one of them a helix. By interpreting these X-ray data, James Watson and Francis Crick derived their now-famous double helix model of DNA for which they, along with Wilkins, received the 1962 Nobel Prize in Physiology or Medicine. Rosalind Franklin's untimely death of ovarian cancer at the age of 37 resulted in her not sharing that award. Nobel Prizes are not awarded posthumously.

This information implied that the bases occurred in pairs: adenine with thymine and guanine with cytosine.

Using X-ray crystallography data gathered by Rosalind Franklin and Maurice Wilkins on the relative positions of atoms in DNA, plus the concept of A-T and G-C base pairs, the American biologist James D. Watson and the British physicist Francis H. C. Crick proposed a double helix structure for DNA in 1953.

In the three-dimensional Watson-Crick structure, two polynucleotide DNA strands wind around each other to form a double helix. Remarkable insight into how hydrogen bonding could stabilize DNA ultimately led Watson and Crick to propose the double helix structure. Hydrogen bonds form between specific base pairs lying opposite each other in the two polynucleotide strands. Adenine is hydrogen-bonded to thymine and guanine is hydrogen-bonded to cytosine to form **complementary base pairs**; that is, *each base in one strand hydrogen bonds to its complementary base in the other strand* (Figure 7.25). The result is that the base pairs (A on one strand with T on the other strand, or C with G) are stacked one above the other in the interior of the double helix (*blue* in Figure 7.25).

Two hydrogen bonds occur between every adenine and thymine pair; three occur between each guanine and cytosine pair. These hydrogen bonds hold the double helix together. Note in Figure 7.26 the similar structures of thymine and cytosine, and of adenine and guanine. If adenine and guanine try to pair, there is insufficient space between the strands to accommodate the bulky pair; thymine and cytosine are too small to pair and to align properly.

CONCEPTUAL EXERCISE 7.8

Hydrogen Bonding and DNA

Only one hydrogen bond is possible between G and T or between C and A. Use the structural formulas for these compounds to indicate where such hydrogen bonding can occur in these base pairs (see Figure 7.26).

7-7c The Genetic Code and DNA Replication

Watson, J. D., and Crick, F. H. C., *Nature*, Vol. 171, 1953, p. 737.

In the historic article in a scientific journal that described their revolutionary findings, Watson and Crick wrote, "It has not escaped our notice that the specific pairing we have postulated immediately suggests a possible copying mechanism for the genetic material." This bit of typical British understatement belies the enormous impact that deciphering DNA's structure had on establishing the molecular basis of heredity. It was now clear that the DNA sequence of base pairs in the nucleus of a cell represents a genetic code that controls the inherited characteristics of the next generation, as well as most of the life processes of an organism.

The total sequence of base pairs in a plant or animal cell is called its genome.

In humans, the double-stranded DNA forms 46 chromosomes (23 chromosome pairs). The *gene* that accounts for a particular hereditary trait is usually located in the

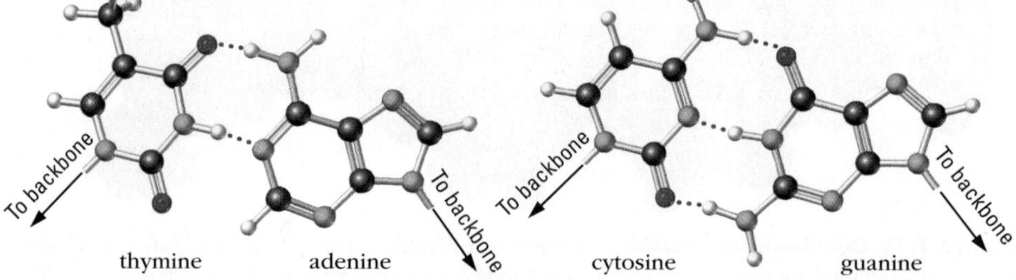
To backbone
thymine
adenine
To backbone
To backbone
cytosine
guanine
To backbone

Figure 7.26 **Hydrogen bonding (· · ·) between complementary base pairs, T-A and C-G, in DNA.**

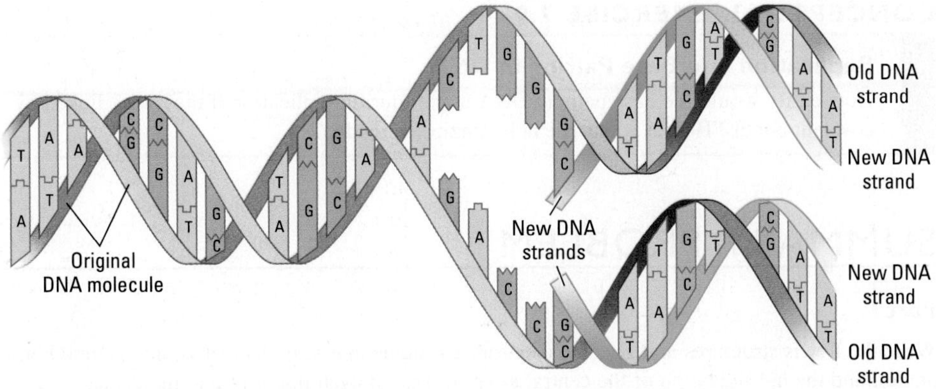
Old DNA strand
New DNA strand
New DNA strands
New DNA strand
Old DNA strand
Original DNA molecule

Figure 7.27 DNA replication. When the original DNA helix (*orange*) unwinds, each half is a template on which nucleotides from the cell environment assemble to produce a complementary strand (*green*). Each new double strand is an exact replica of the original.

same position on the same chromosome. Each **gene** is *a unique sequence of nitrogen bases that codes for the synthesis of a single protein within the body*. The gene can be "read" by chemical processes involving RNA (ribonucleic acid, which has a structure closely related to DNA's structure) and used to control synthesis of a protein molecule. The protein then goes on to play its role in the growth and functioning of the individual. (Proteins are discussed in Section 10-7d.)

Each organism (except viruses) begins life as a single cell. The double-stranded DNA in that cell consists of a single strand from each parent. During regular cell division, both DNA strands are accurately copied, with a remarkably low incidence of error. The *copying of a DNA molecule*, called **replication**, takes place as enzymes aid in breaking hydrogen bonds between strands, unzipping the DNA helix. New nucleotides are sequentially brought into the proper places on each of two new strands so that they pair with bases on the old strands. Thus, each strand of the original DNA serves as a template from which a complementary strand is produced (Figure 7.27). The process is termed *semiconservative* because in each of two new cells, each chromosome consists of one DNA strand from the parent cell and one newly made strand.

ESTIMATION | Base Pairs and DNA

Human DNA contains about three billion base pairs, which are an average distance of 0.34 nm apart in the DNA molecule. Only about 2% of this DNA consists of unique genes. The number of genes is estimated to be 2×10^4.

(a) Calculate the average number of base pairs per gene.

(b) Calculate approximately how long (m) a DNA molecule is.

(a) We begin by calculating the average number of base pairs/gene, starting with approximately 3×10^9 base pairs, of which there are 2 base pairs in genes per 100 base pairs. We get

$$\frac{3 \times 10^9 \text{ base pairs}}{\text{DNA molecule}} \times \frac{2 \text{ base pairs in genes}}{100 \text{ base pairs}} = \frac{6 \times 10^7 \text{ base pairs in genes}}{\text{DNA molecule}}$$

Thus, there are about 6×10^7 base pairs in genes/DNA. Using this information and the fact that DNA contains about 2×10^4 genes, there are roughly 3×10^3 base pairs/gene.

$$\frac{6 \times 10^7 \text{ base pairs in genes}}{\text{DNA molecule}} \times \frac{1 \text{ DNA molecule}}{2 \times 10^4 \text{ genes}} \cong \frac{3 \times 10^3 \text{ base pairs}}{\text{gene}}$$

(b) To approximate how many meters long a DNA molecule is, we use the information that the distance between base pairs in DNA is 0.34 nm and 1 m = 10^9 nm. Therefore,

$$\frac{3 \times 10^9 \text{ base pairs}}{\text{DNA molecule}} \times \frac{0.34 \text{ nm}}{\text{base pair}} \times \frac{1 \text{ m}}{10^9 \text{ nm}} \cong \frac{1 \text{ m}}{\text{DNA molecule}}$$

This distance, approximately 1 m (about three feet), would be much too large to fit into a cell. Therefore, the DNA molecule must be tightly coiled, which it is.

CONCEPTUAL EXERCISE 7.9

Replication and Base Pairing in DNA

How easily would the base pairs in DNA unpair during replication if they were linked by covalent bonds? How is it that the helix unzips so readily?

SUMMARY PROBLEM

Part I

Write the Lewis structures and give the electron-region geometry, molecular geometry, and bond angles, and the hybridization of the central atom of these polyatomic ions and molecules.

(a) BrF_2^+ (b) $OCCl_2$ (c) CH_3^+ (d) SeCS (e) CH_3^-

Part II

Aspartame is a commonly used artificial sweetener (NutraSweet®) that was discovered accidentally by a chemist in the laboratory carelessly licking his fingers after synthesizing a new compound (aspartame). The sweetener's popularity stems from the fact that aspartame tastes over 100 times sweeter than sugar. Aspartame has this structural formula.

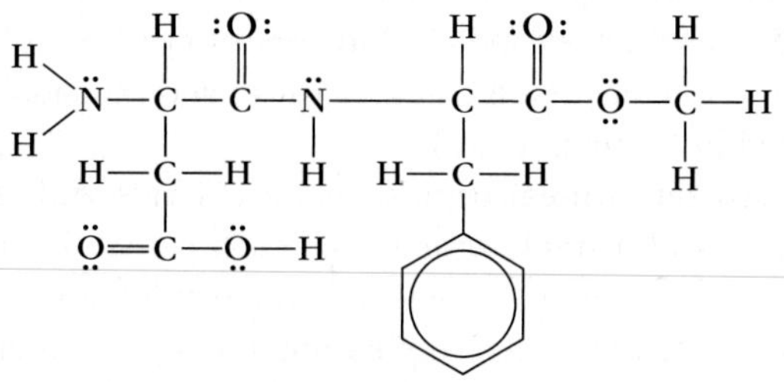

(a) Write the molecular formula of aspartame.

(b) Identify three atoms that have triangular planar molecular geometry.

(c) Identify three atoms that have tetrahedral molecular geometry.

(d) Identify two atoms that have bent (angular) molecular geometry.

(e) Identify the hybridization of each nitrogen atom.

(f) Identify the hybridization of carbon in the CH_2 groups.

(g) Identify the hybridization of the carbon atom in the C═O group.

(h) Identify the hybridization of the oxygen atom to which a methyl (—CH_3) group is attached.

(i) Label each of the sigma and pi bonds involving carbon and those involving nitrogen.

Part III

The structural formula for the open-chain form of glucose is

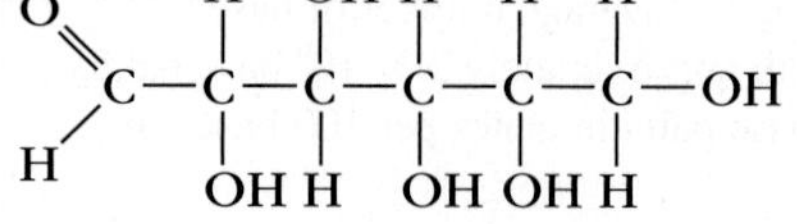

Glucose dissolves readily in water. Use molecular structure principles to explain why glucose is so water-soluble.

HAVING STUDIED THIS CHAPTER . . .

. . . you should be able to:

- Recognize and use the various ways that the shapes of molecules are represented by models and on the printed page (Section 7-1).
- Predict shapes of molecules and polyatomic ions by using the VSEPR model (Section 7-2). End-of-chapter Questions 13, 15, 19, 22
- Identify chiral centers and predict the existence of enantiomers (Section 7-2). Questions 28, 30
- Determine the orbital hybridization of a central atom and the associated electron-region and molecular geometries (Section 7-3). Questions 32, 34, 37
- Describe in terms of sigma and pi bonding the orbital hybridization of a central atom and the associated molecular geometry for molecules containing single and multiple bonds (Sections 7-3 and 7-4). Questions 35, 41
- Describe covalent bonding between two atoms in terms of sigma or pi bonds, or both (Section 7-4). Questions 39, 41
- Use molecular structure and electronegativities to predict the polarities of molecules (Section 7-5). Questions 43, 45, 46
- Describe the different types of noncovalent interactions and use them to explain melting points, boiling points, and solubility (Section 7-6). Questions 48, 49, 55, 60
- Explain how different types of chemical bonding affect the structure of DNA (Section 7-7). Question 61
- Compare infrared and UV-visible spectroscopy and interpret spectra from each technique *(Tools of Chemistry)*. Questions 63, 65

KEY TERMS

achiral (Section 7-2f)
axial positions (7-2d)
bond angle (7-2a)
chiral (7-2f)
complementary base pairs (7-7b)
deoxyribonucleic acid (DNA) (7-7)
dipole-dipole attraction (7-6b)
dipole moment (7-5)
electron-region geometry (7-2)
enantiomers (7-2f)
equatorial positions (7-2d)
gene (7-7c)
hybrid orbitals (7-3)
hydrogen bond (7-6c)
hydrophilic (7-6d)
hydrophobic (7-6d)
induced dipole (7-6a)
intermolecular forces (7-6)
lipid bilayer (7-6d)
London forces (7-6a)
molecular geometry (7-2)
noncovalent interactions (7-6)
nonpolar molecule (7-5)
nucleotide (7-7a)
octahedron (7-2d)
optical isomers (7-2f)
pi bond (π bond) (7-4)
polar molecule (7-5)
polarization (7-6a)
polymer (7-7a)
replication (7-7c)
sigma bond (σ bond) (7-3c)
***sp* hybrid orbitals** (7-3a)
sp^2 hybrid orbitals (7-3b)
sp^3 hybrid orbitals (7-3c)
spectroscopy (*Tools of Chemistry*, 7-2e, 7-5)
triangular bipyramid (7-2d)
valence-shell electron-pair repulsion (VSEPR) model (7-2)

QUESTIONS FOR REVIEW AND THOUGHT

Red-numbered questions have short answers at the back of this book in Appendix M and fully worked solutions in the *Student Solutions Manual.*

Review Questions

These questions test vocabulary and simple concepts.

1. Describe the VSEPR model. How is the model used to predict molecular structure?
2. What is the difference between the electron-region geometry and the molecular geometry of a molecule? Use the water molecule as an example in your discussion.
3. Designate the electron-region geometry for each case from two to six electron pairs around a central atom.
4. What are the molecular geometries for each of these?

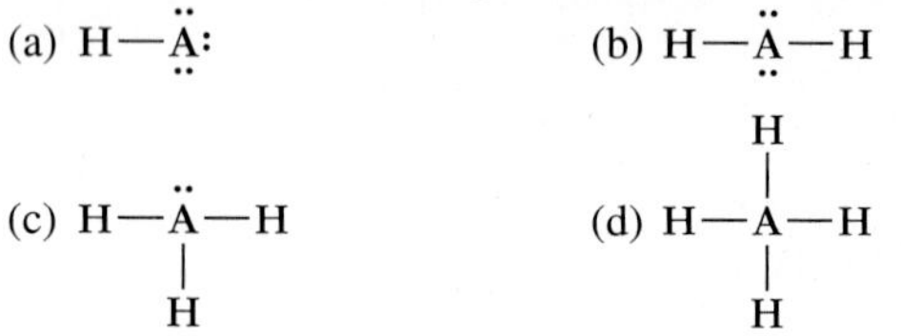

Estimate the H—A—H bond angle for each of the last three.

5. If you have three electron regions around a central atom, how can you have a triangular planar molecule? An angular molecule? What bond angles are predicted in each case?
6. Use VSEPR to explain why ethylene is a planar molecule.
7. How can a molecule with polar bonds be nonpolar? Give an example.
8. Explain the significance of Erwin Chargaff's research to the understanding of the structure of DNA.
9. (a) Describe how nucleotides are joined into a polynucleotide chain in DNA.
   (b) Explain the role of nitrogen bases in the DNA double helix.
10. Explain why the infrared spectrum of a molecule is referred to as its "fingerprint."
11. How is the frequency of infrared radiation absorbed by a molecule related to the motion of atoms in a molecule?

Topical Questions

These questions are keyed to the major topics in the chapter. Usually a question that is answered at the back of this book is paired with a similar one that is not.

Molecular Models/Shapes (Sections 7-1, 7-2)

12. Use the various molecular modeling techniques (ball-and-stick, space-filling, two-dimensional pictures using wedges and dashed lines) to illustrate these simple molecules:
    (a) NH_3 (b) H_2O
    (c) CO_2
13. Draw the Lewis structure and identify the molecular shape of each molecule.
    (a) BeH_2 (b) CH_2Cl_2
    (c) BH_3 (d) $SeCl_6$
    (e) PF_3
14. Draw the Lewis structure for each molecule or ion. Describe the electron-region geometry and the molecular geometry.
    (a) NH_2Cl (b) OF_2
    (c) SCN^- (d) HOF
15. In each of these ions, three oxygen atoms are attached to a central atom. Draw the Lewis structure for each ion, and then describe the electron-region geometry and the molecular geometry. Comment on similarities and differences in the series.
    (a) BO_3^{3-} (b) CO_3^{2-}
    (c) SO_3^{2-} (d) ClO_3^-
16. In each of these molecules or ions, two oxygen atoms are attached to a central atom. Draw the Lewis structure for each, and then describe the electron-region geometry and the molecular geometry. Comment on similarities and differences in the series.
    (a) CO_2 (b) NO_2^-
    (c) SO_2 (d) O_3
    (e) ClO_2^-
17. Write Lewis structures for $XeOF_2$ and $ClOF_3$. Use VSEPR theory to predict the electron-region and molecular geometries of these molecules, and note any differences between these geometries.
18. Write Lewis structures for HCP and $[IOF_4]^-$. Use VSEPR theory to predict the electron-region and molecular geometries of these species, and note any differences between these geometries.
19. These are examples of molecules and an ion that do not obey the octet rule. After drawing the Lewis structure, describe the electron-region geometry and the molecular geometry for each.
    (a) SiF_6^{2-} (b) SF_4
    (c) PF_5 (d) XeF_4
20. These are examples of molecules and ions that do not obey the octet rule. After drawing the Lewis structure, describe the electron-region geometry and the molecular geometry for each.
    (a) ClF_2^- (b) ClF_3
    (c) ClF_4^- (d) ClF_5
21. Explain why $(I_3)^+$ is bent, but $(I_3)^-$ is linear.
22. Give the approximate values for the indicated bond angles. The figures given are not the actual molecular shapes; they are only to show the bonds being considered.
    (a) O—S—O angle in SO_2 (b) F—B—F angle in BF_3

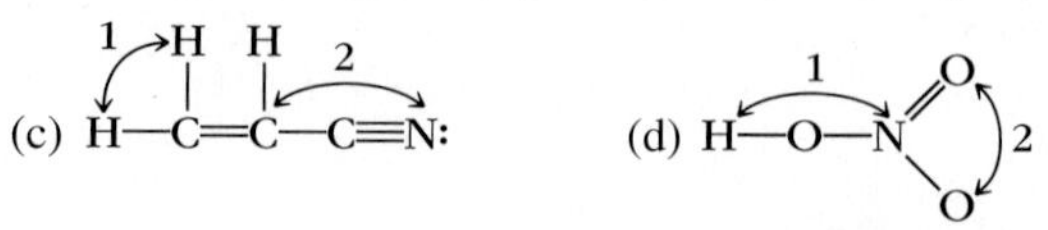

23. Give approximate values for the indicated bond angles. The figures given are not the actual molecular shapes, but are used only to show the bonds being considered.
(a) angle in SCl_2 (b) angle in N_2O
(c) 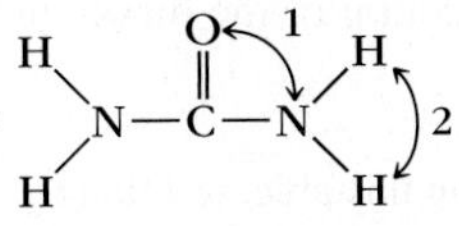 (d) 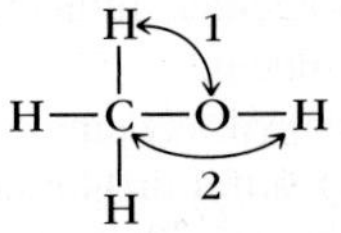

24. Give approximate values for the indicated bond angles.
(a) F—Se—F angles in SeF_4
(b) angles in SOF_4 (The O atom is in an equatorial position.)
(c) angles in BrF_5

25. Give approximate values for the indicated bond angles.
(a) angles in SF_6
(b) angle in XeF_2
(c) angle in ClF_2^-

26. Which has the greater O—N—O bond angle, NO_2 or NO_2^+? Explain your answer.

27. Compare the F—Cl—F angles in ClF_2^+ and ClF_2^-. From Lewis structures, determine the approximate bond angle in each ion. Explain which ion has the greater angle and why.

Chiral Molecules (Section 7-2f)

28. Circle the chiral centers, if any, in these molecules.
(a) 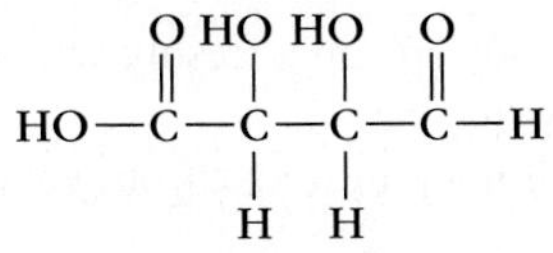
(b) 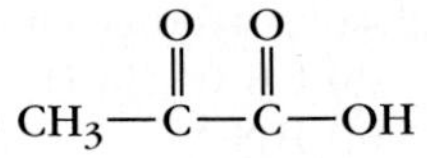
(c) 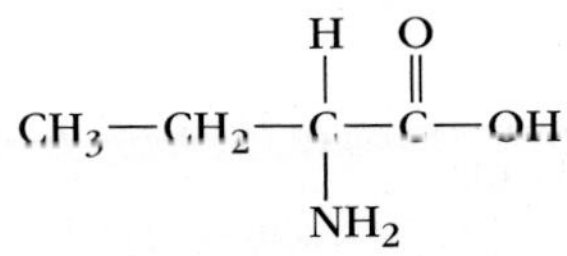

29. Circle the chiral centers, if any, in these molecules.
(a) 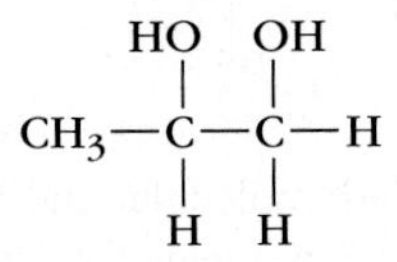 (b) 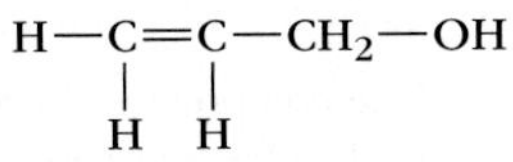
(c) 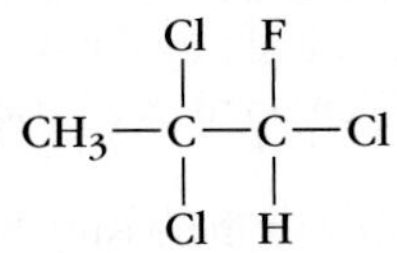

30. Circle the chiral centers, if any, in these compounds.
(a) 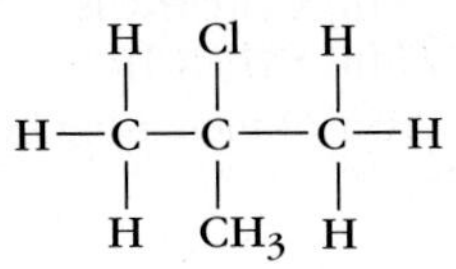
(b) 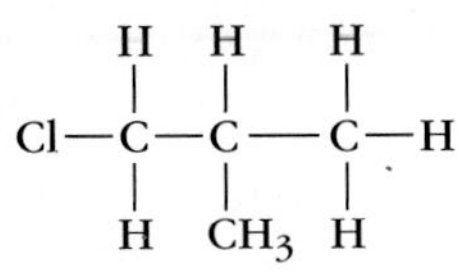
(c) 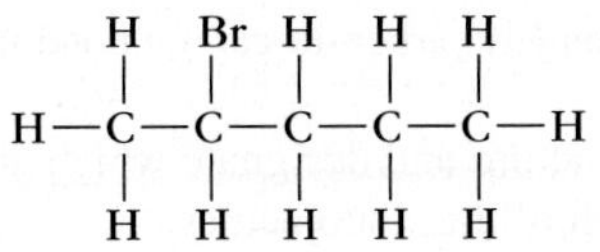
(d) 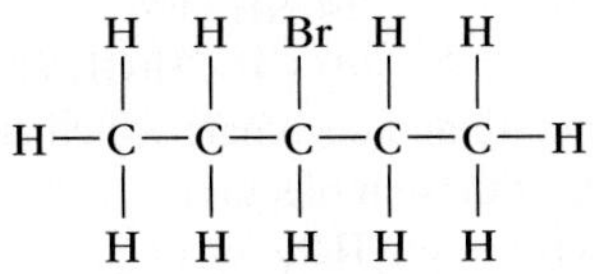

31. Circle the chiral centers, if any, in these compounds.
(a) 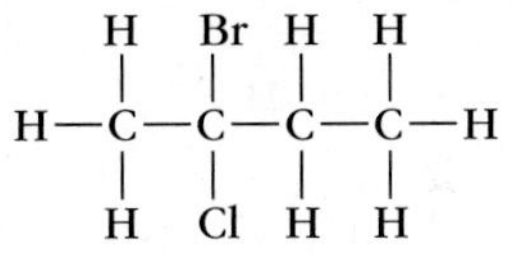
(b) 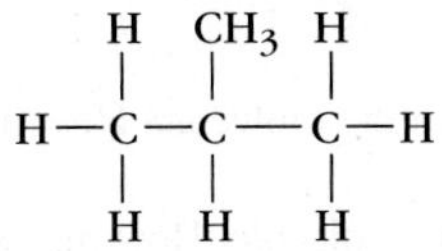
(c) 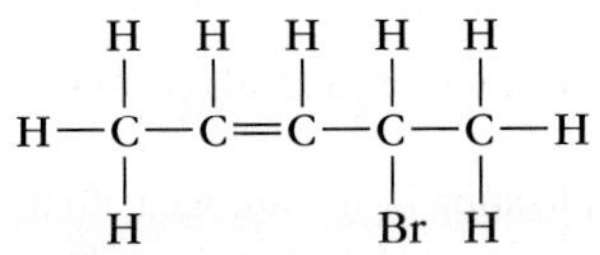

Hybridization (Sections 7-3, 7-4)

32. Describe the geometry and hybridization of carbon in chloroform, $CHCl_3$.

33. Describe the geometry and hybridization for each C and O atom in ethylene glycol, $HOCH_2CH_2OH$, the main component in antifreeze.

34. Describe the hybridization around the central atom and the bonding in SCl_2 and OCS.

35. The hybridization of the two carbon atoms differs in an acetic acid, CH_3COOH, molecule.
(a) Designate the correct hybridization for each carbon atom in this molecule.
(b) What is the approximate bond angle around each carbon?

36. The hybridization of the two nitrogen atoms differs in NH_4NO_3.
(a) Designate the correct hybridization for each nitrogen atom.
(b) What is the approximate bond angle around each nitrogen?

37. Identify the type of hybridization, approximate bond angles for the N, C, and O atoms, and shortest carbon-to-oxygen bond length in alanine, an amino acid, whose Lewis structure is

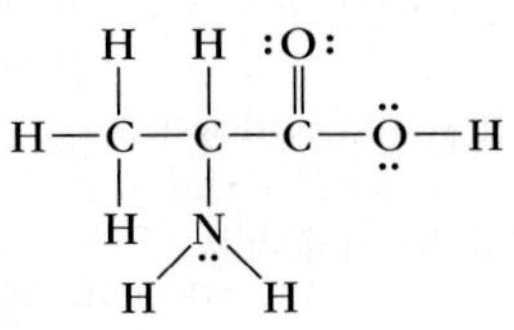

38. (a) Identify the type of hybridization and approximate bond angle for each carbon atom in CH_3CH_2CCH.
(b) Which is the shortest carbon-to-carbon bond length in this molecule?

(c) Which is the strongest carbon-to-carbon bond in this molecule?

39. Write the Lewis structure and designate which are sigma and pi bonds in each of these molecules.
(a) OCS (b) NH_2OH
(c) CH_2CHCHO (d) $CH_3CH(OH)COOH$

40. Write the Lewis structure and designate which are sigma and pi bonds in each of these molecules.
(a) HCN (b) N_2H_2 (c) HN_3

41. Acrylonitrile is polymerized to manufacture carpets and wigs. Its Lewis structure is

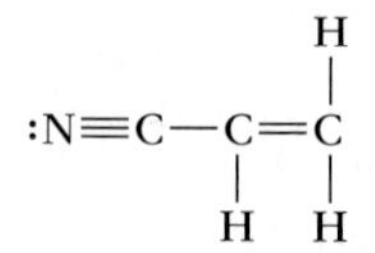

(a) Locate the sigma bonds in the molecule. How many are there?
(b) Locate the pi bonds in the molecule. How many are there?
(c) What is the hybridization of the carbon atom that is bonded to nitrogen?
(d) What is the hybridization of the nitrogen atom?
(e) What is the hybridization of the hydrogen-bearing carbon atoms?

42. Methylcyanoacrylate is the active ingredient in "super glues." Its Lewis structure is

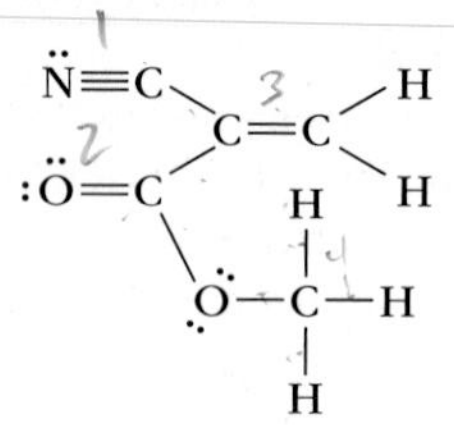

(a) How many sigma bonds are in the molecule?
(b) How many pi bonds are in the molecule?
(c) What is the hybridization of the carbon atom bonded to nitrogen?
(d) What is the hybridization of the carbon atom bonded to oxygen?
(e) What is the hybridization of the double-bonded oxygen?

Molecular Polarity (Section 7-5)

43. Which of these molecules is (are) polar? For each polar molecule, what is the direction of polarity; that is, which is the partial negative end and which is the partial positive end of the molecule?
(a) CO_2 (b) HBF_2
(c) CH_3Cl (d) SO_3

44. Which of these molecules has a net dipole moment? For each of these polar molecules, indicate the direction of the dipole in the molecule.
(a) XeF_2 (b) H_2S
(c) CH_2Cl_2 (d) HCN

45. Explain the differences in the dipole moments of
(a) BrF (1.29 D) and BrCl (0.52 D).
(b) H_2O (1.86 D) and H_2S (0.95 D).

46. Which of these molecules is polar? For each of the polar molecules, indicate the direction of the dipole in the molecule.
(a) hydroxylamine, NH_2OH
(b) sulfur dichloride, SCl_2, an unstable, red liquid

47. Which of these molecules has a net dipole moment? For each of the polar molecules, indicate the direction of the dipole in the molecule.
(a) nitrosyl fluoride, FNO
(b) disulfur difluoride, S_2F_2

Noncovalent Interactions (Section 7-6)

48. Construct a table that includes all the types of noncovalent interactions and comment about the strength of each. Also include an example of a substance that exhibits each type of noncovalent interaction in the table.

49. Use molecular structures and noncovalent interactions to explain why dimethyl ether, $(CH_3)_2O$, is completely miscible in water, but dimethylsulfide, $(CH_3)_2S$, is only slightly water soluble.

50. Explain in terms of noncovalent interactions why water and ethanol are miscible, but water and cyclohexane are not.

51. Explain why water "beads up" on a freshly waxed car, but not on a dirty, unwaxed car.

52. Explain why water will not remove tar from your shoe, but kerosene will.

53. Which of these form intermolecular hydrogen bonds?
(a) CH_2Br_2 (b) $CH_3OCH_2CH_3$
(c) H_2NCH_2COOH (d) H_2SO_3
(e) CH_3CH_2OH

54. Consider this structure.

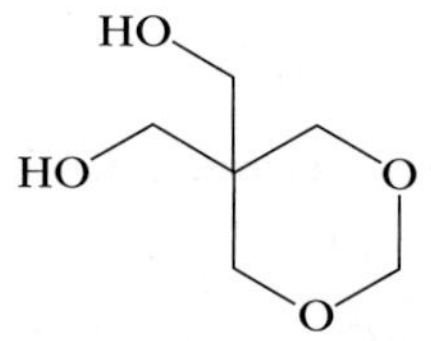

The compound exhibits both intramolecular and intermolecular hydrogen bonding. Draw Lewis structures to represent both types of hydrogen bonding. Draw all lone pairs and represent each hydrogen bond as a series of three dots · · ·.

55. Arrange these substances in order of increasing boiling point. Explain your reasoning.
(a) CH_3CH_2OH (b) $CH_3CH_2CH_3$
(c) $CBr_3CBr_2CBr_3$ (d) CH_3OCH_3

56. Arrange the noble gases in order of increasing boiling point. Explain your reasoning.

57. The structural formula for vitamin C is

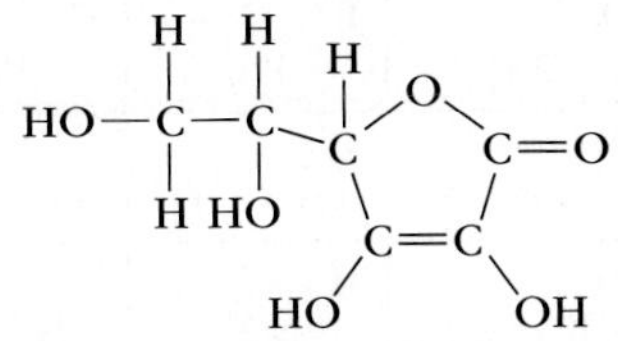

Give a molecular-level explanation why vitamin C is a water-soluble rather than a fat-soluble vitamin.

58. The structural formula of alpha-tocopherol, a form of vitamin E, is

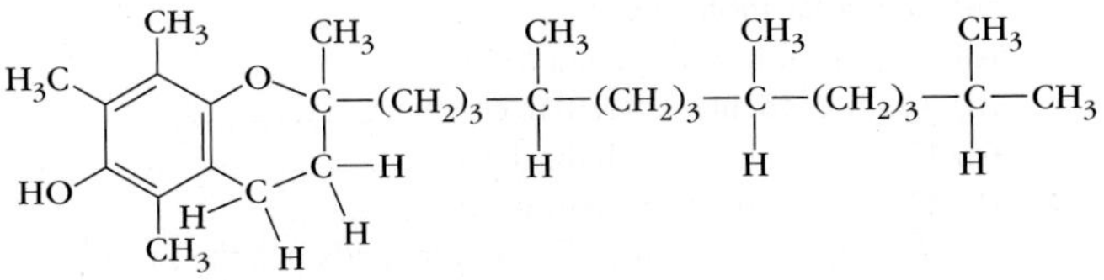

Give a molecular-level explanation why alpha-tocopherol dissolves in fat, but not in water.

59. Which is most soluble in cyclohexane, C_6H_{12}? Least soluble? Explain your reasoning.
(a) NaCl (b) CH_3CH_2OH (c) C_3H_8

60. What types of forces must be overcome in each change?
(a) sublimation of solid $C_{10}H_8$
(b) melting of propane, C_3H_8
(c) decomposition of water into H_2 and O_2
(d) evaporation of liquid PCl_3
(e) unzipping the DNA double helix during replication

Biomolecules (Section 7-7)

61. One strand of DNA contains the base sequence T-C-G. Draw a structure of this section of DNA that shows the hydrogen bonding between the base pairs of this strand and its complementary strand.

62. Discuss the differences between the extent of hydrogen bonding for the pairs G-C and A-T in nucleic acids. If a strand of DNA has more G-C pairs than A-T pairs in the double helix, the melting point (unwinding point) increases. The melting point for strands with more A-T pairing will decrease in comparison. Explain.

Tools of Chemistry: Molecular Structure Determination by Spectroscopy (Sections 7-2e, 7-5)

63. Compare IR and UV spectroscopy in these ways:
(a) energy of radiation
(b) wavelength of radiation
(c) frequency of radiation

64. Ultraviolet and infrared spectroscopies probe different aspects of molecules.
(a) Ultraviolet radiation is energetic enough to cause transitions in the energies of which electrons in an atom?
(b) What does infrared spectroscopy tell us about a given molecule?

65. The infrared spectrum of ethanol is given in *Tools of Chemistry: Infrared Spectroscopy* (← **Sec. 7-2e**).
(a) Which stretching requires the lowest quantity of energy?
(b) The highest quantity of energy?

66. The UV-visible spectrum of beta-carotene given in *Tools of Chemistry: Ultraviolet-Visible Spectroscopy* (← **Sec. 7-5**), shows a maximum absorbance at 455 nm. Estimate the wavelengths of two other peaks in the spectrum and indicate which peak identifies the more energetic electron transition.

General Questions

These questions are not explicitly keyed to chapter topics; many require integration of several concepts.

67. Methylcyanoacrylate is the active ingredient in "super glues." Its Lewis structure is

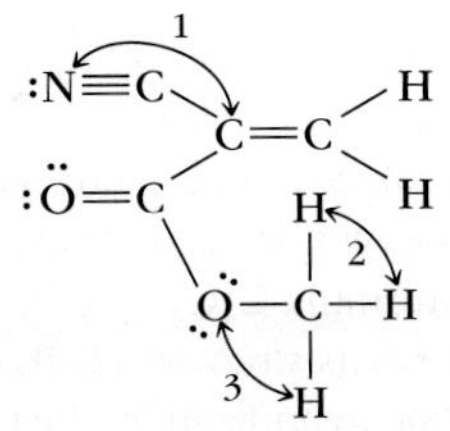

(a) Give values for the three bond angles indicated.
(b) Indicate the most polar bond in the molecule.
(c) Circle the shortest carbon-oxygen bond.
(d) Circle the shortest carbon-carbon bond.

68. Acrylonitrile is polymerized to manufacture carpets and wigs. Its Lewis structure is

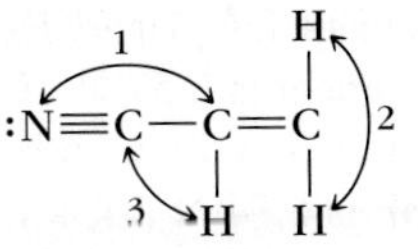

(a) Give approximate values for the indicated bond angles.
(b) Which is the most polar bond in the molecule?

69. In addition to CO and CO_2, there are other carbon oxides. One is tricarbon dioxide, C_3O_2, also called carbon suboxide, a foul-smelling gas.
(a) Write the Lewis structure of this compound.
(b) What is the value of the C-to-C-to-O bond angle in carbon suboxide?
(c) What is the value of the C-to-C-to-C bond angle in tricarbon dioxide?

70. Use Lewis structures and VSEPR theory to predict the electron-region and molecular geometries of
(a) $PSCl_3$. (b) SOF_6.
(c) $[S_2O_4]^{2-}$. (d) $[TeF_4]^{2-}$.
Note any differences between these geometries.

71. In addition to CO, CO_2, and C_3O_2, there is another molecular oxide of carbon, pentacarbon dioxide, C_5O_2, a yellow solid.
(a) What is the approximate C-to-C-to-O bond angle in pentacarbon dioxide?
(b) What is the approximate C-to-C-to-C bond angle in this compound?

Grid for Question 79

A HCN	B PO_4^{3-}	C PH_3 or PF_3
D SiH_4	E Cl_2O	F NH_2Cl
G HF or F_2	H CH_4	I OF_2

72. Cyanidin chloride is an anthocyanin found in strawberries, apples, and cranberries.

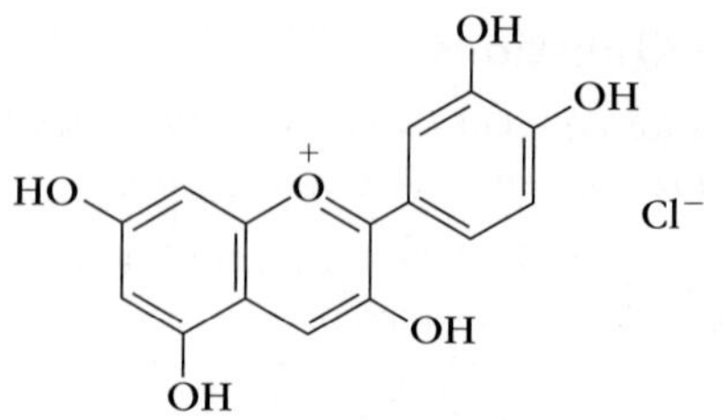

Would you expect this compound to absorb in the visible region of the spectrum? Explain your answer.

73. Carbon and oxygen form the squarate ion, a polyatomic ion with the formula $C_4O_4^{2-}$.
    (a) Write the Lewis structure for this ion.
    (b) Describe the hybridization of each carbon atom.
    (c) Based on your answers to parts (a) and (b), is there any inconsistency between the molecular shape and the predicted bond angles?
74. In addition to carbonate ion, carbon and oxygen form the croconate ion, a polyatomic ion with the formula $C_5O_5^{2-}$.
    (a) Write the Lewis structure for this ion.
    (b) Describe the hybridization of each carbon atom.
    (c) Is the croconate ion planar? Explain your reasoning.
75. The dipole moment is 3.57×10^{-30} C m for the HCl molecule, and the bond length is 127.4 pm; the dipole moment of HF is 5.94×10^{-30} C m, with bond length of 91.68 pm. Use the definition of dipole moment as a product of partial charge on each atom times the distance of separation (Section 7-5) to calculate the quantity of charge in coulombs that is separated by the bond length in each dipolar molecule. Use your result to show that fluorine is more electronegative than chlorine.
76. In the gas phase, positive and negative ions form ion pairs that are like molecules. An example is KF, which is found to have a dipole moment of 28.7×10^{-30} C m and a distance of separation between the two ions of 217.2 pm. Use this information and the definition of dipole moment to calculate the partial charge on each atom. Compare your result with the expected charge, which is the charge on an electron, -1.602×10^{-19} C. Based on your result, is KF really completely ionic?
77. How can a diatomic molecule be nonpolar? Polar?
78. Explain clearly in your own words how a molecule can have polar bonds yet have a dipole moment of zero.
79. The grid for Question 79 has nine lettered boxes, each of which contains an item that is used to answer the questions that follow. Items may be used more than once and there may be more than one correct item in response to a question.

Grid for Question 80

A PH_4^+	B PF_3	C PF_5
D O=C=S	E O=C=O	F PF_3 or PF_5
G ClF or ClF_3	H NO_3^-	I H_2NOH

Place the letter(s) of the correct selection(s) on the appropriate line.
(a) Electron-region geometry is the same as the molecular geometry ________
(b) Nonpolar molecule ________
(c) Linear molecular geometry ________
(d) Angular (bent) molecular geometry ________
(e) Central atom is sp^3 hybridized ________
(f) Central atom is *sp* hybridized ________
(g) Which one in each pair of compounds has the lower boiling point? ________
(h) Which one in each pair of compounds has the higher vapor pressure? ________
(i) Which one in each pair of compounds has the higher dipole moment? ________
(j) Has dipole-dipole and hydrogen bonding intermolecular forces ________

80. The grid for Question 80 has nine lettered boxes, each of which contains an item that is used to answer the questions that follow. Items may be used more than once and there may be more than one correct item in response to a question.
Place the letter(s) of the correct selection(s) on the appropriate line.
    (a) Electron-region geometry is the same as the molecular geometry ________
    (b) Nonpolar molecule ________
    (c) Linear molecular geometry ________
    (d) Triangular planar molecular geometry ________
    (e) Central atom is sp^3 hybridized ________
    (f) Central atom is sp^2 hybridized ________
    (g) Has dipole-dipole and hydrogen bonding intermolecular forces ________

Applying Concepts

These questions test conceptual learning.

81. Explain why the boiling point (5.9 °C) of methanethiol, CH_3SH, is much lower than the boiling point (64.7 °C) of methanol, CH_3OH.
82. Explain why, even though CO and N_2 each have a total of 14 electrons, the melting and boiling points of N_2 are slightly lower than those of CO.
83. Azidotrifluoromethane, CF_3N_3, is a colorless gas that is stable at room temperature.
    (a) Write a plausible Lewis structure for this compound and estimate the N—N—C bond angle.
    (b) Estimate the N—N—N bond angle.
    (c) Identify the hybridization of the central nitrogen atom and the hybridization of the carbon atom.

(d) An intermediate compound with the molecular formula $CF_3N_3H_2$ forms during the synthesis of azidotrifluoromethane. Draw the correct Lewis structures for two plausible resonance hybrids of this intermediate. Identify the hybridization of each nitrogen atom in each of the two structures.

84. The nitride ion, N_3^-, has the same number of electrons as (is isoelectronic with) carbon dioxide.
(a) Write correct Lewis structures for two resonance hybrids of nitride ion.
(b) Identify the hybridization of the central N atom in each case.

85. Complete this table.

Molecule or Ion	Electron-region Geometry	Molecular Geometry	Bond Angle
ICl_2^+			
I_3^-			
ICl_3			
ICl_4^-			
IO_4^-			
IF_4^+			
IF_5			
IF_6^+			

86. Complete this table.

Molecule or Ion	Electron-region Geometry	Molecular Geometry	Bond Angle
SO_2			
SCl_2			
SO_3			
SO_3^{2-}			
SF_4			
SO_4^{2-}			
SF_5^+			
SF_6			

87. What are the types of forces, in addition to London forces, that are overcome in these changes? Using structural formulas, make a sketch representing the major type of force in each case.
(a) the evaporation of liquid methanol, CH_3OH
(b) the decomposition of hydrogen peroxide, H_2O_2, into water and oxygen
(c) the melting of urea, H_2NCONH_2
(d) the boiling of liquid HCl

88. ACE solution was a mixture used as an anesthetic in the mid- to late-1800s. It contained ethanol (CH_3CH_2OH), chloroform ($CHCl_3$), and diethyl ether, $CH_3CH_2—O—CH_2CH_3$. Explain, on a molecular basis, why chloroform and diethyl ether are miscible, that is, they dissolve in each other.

89. Benzene, C_6H_6, and pyridine, C_5H_5N, molecules are each in a ring structure, hydrogen atoms bonded to the carbon atoms in the ring. Use molecular structures and noncovalent interactions to explain why pyridine is completely miscible in water, but benzene is only very slightly water soluble.

90. Name a Group 1A to 8A element that could be the central atom (X) in these compounds.
(a) XH_3 with one lone pair of electrons
(b) XCl_3 (c) XF_5
(d) XCl_3 with two lone pairs of electrons

91. Name a Group 1A to 8A element that could be the central atom (X) in these compounds.
(a) XCl_2
(b) XH_2 with two lone pairs of electrons
(c) XF_4 with one lone pair of electrons
(d) XF_4

92. What is the maximum number of water molecules that could hydrogen-bond directly to an acetic acid molecule? Draw in the water molecules and use dotted lines to show the hydrogen bonds.

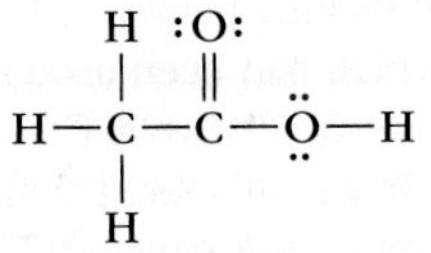

93. What is the maximum number of water molecules that could hydrogen-bond directly to an ethylamine molecule? Draw in the water molecules and use dotted lines to show the hydrogen bonds.

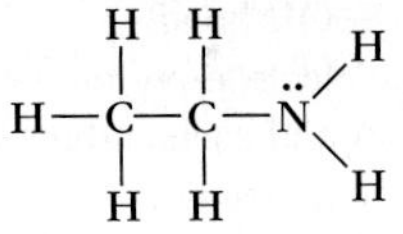

94. These are responses students wrote when asked to give an example of hydrogen bonding. Which is/are correct?
(a) H—H· · ·H—H (c) H—F
(b) 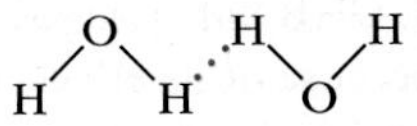 (d) 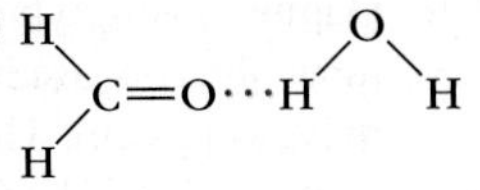

95. Which of these are examples of hydrogen bonding?
(a) 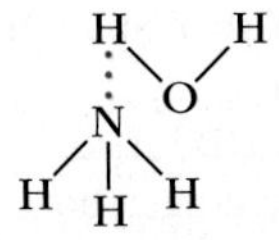
(b) H—H
(c) 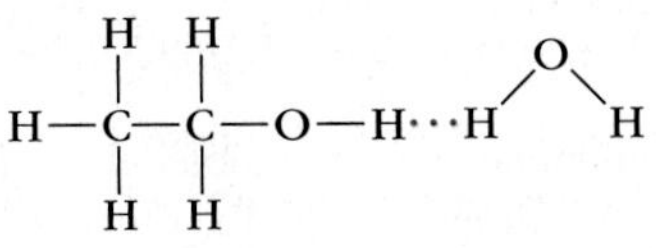
(d) 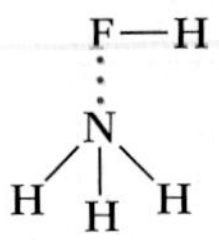

96. In another universe, elements try to achieve a nonet (nine valence electrons) instead of an octet when forming chemical bonds. As a result, covalent bonds form when a trio of electrons is shared between two atoms. Two compounds in this other universe are H_3O and H_2F. Draw their Lewis structures, then determine their electron-trio geometry and molecular geometry.

More Challenging Questions

These questions require more thought and integrate several concepts.

97. One of the three isomers of dichlorobenzene, $C_6H_4Cl_2$, has a dipole moment of zero. Draw the structural formula of this isomer and explain your choice.
98. There are three isomers with the formula $C_6H_6O_2$. Each isomer contains a benzene ring to which two —OH groups are attached.
    (a) Write the Lewis structures for the three isomers.
    (b) Taking their molecular structure and the likelihood of hydrogen bonding into account, list them in order of increasing melting point.
99. Halothane, which had been used as an anesthetic, has the molecular formula $CHBrClCF_3$.
    (a) Write the Lewis structure for halothane.
    (b) Is halothane a polar molecule? Explain your answer.
    (c) Can hydrogen bonding occur in halothane? Explain.
100. Ketene, C_2H_2O, is a reactant for synthesizing cellulose acetate, which is used to make films, fibers, and fashionable clothing.
    (a) Write the Lewis structure of ketene. Ketene does not contain an —OH bond.
    (b) Identify the electron-region geometry and the molecular geometry around each carbon atom and all the bond angles in the molecule.
    (c) Identify the hybridization of each carbon and oxygen atom.
    (d) Is the molecule polar or nonpolar? Use appropriate data to support your answer.
101. Gamma hydroxybutyric acid, GHB, infamous as a "date rape" drug, is used illicitly because of its effects on the nervous system. The condensed molecular formula for GHB is $HO(CH_2)_3COOH$.
    (a) Write the Lewis structure for GHB.
    (b) Identify the hybridization of the carbon atom in the CH_2 groups and of the terminal carbon.
    (c) Is hydrogen bonding possible in GHB? If so, write Lewis structures to illustrate the hydrogen bonding.
    (d) Which carbon atoms are involved in sigma bonds? In pi bonds?
    (e) Which oxygen atom is involved in sigma bonds? In pi bonds?
102. There are two compounds with the molecular formula HN_3. One is called hydrogen azide; the other is cyclotriazene.
    (a) Write the Lewis structure for each compound.
    (b) Designate the hybridization of each nitrogen in hydrogen azide.
    (c) What is the hybridization of each nitrogen in cyclotriazene?
    (d) How many sigma bonds are in hydrogen azide? In cyclotriazene?
    (e) How many pi bonds are in hydrogen azide? In cyclotriazene?
    (f) Give approximate values for the N-to-N-to-N bond angles in each molecule.
103. Nitrosyl azide, a yellow solid first synthesized in 1993, has the molecular formula N_4O.
    (a) Write its Lewis structure.
    (b) What is the hybridization on the terminal nitrogen?
    (c) What is the hybridization on the "central" nitrogen?
    (d) Which is the shortest nitrogen-nitrogen bond?
    (e) Give the approximate bond angle between the three nitrogens, beginning with the nitrogen that is bonded to oxygen.
    (f) Give the approximate bond angle between the last three nitrogens, those not involved in bonding to oxygen.
    (g) How many sigma bonds are there? How many pi bonds?
104. Piperine, the active ingredient in black pepper, has this Lewis structure.

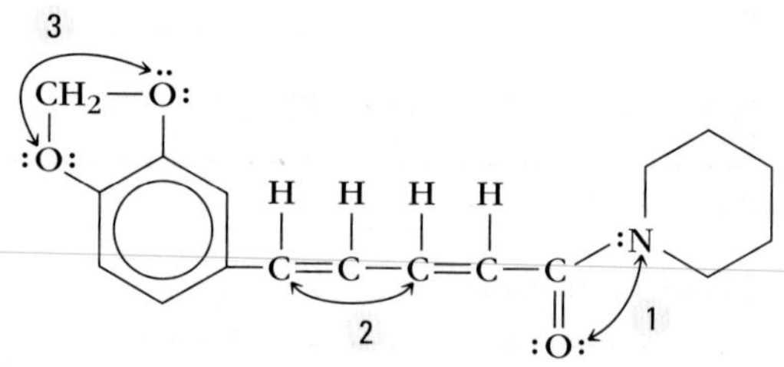
3
CH2—O:
:O:
H H H H
C=C—C=C—C
:N
2
1
:O:

    (a) Give the values for the indicated bond angles.
    (b) What is the hybridization of the nitrogen?
    (c) What is the hybridization of the oxygens?
105. Three compounds have the molecular formula N_2H_2.
    (a) Write the correct Lewis structure for each compound.
    (b) If there are geometric isomers, properly label each.
106. Two compounds have the molecular formula N_3H_3. One of the compounds, triazene, contains an N=N bond; the other compound, triaziridene, does not.
    (a) Write the correct Lewis structures for each compound.
    (b) Approximate the bond angle between the three nitrogen atoms in each compound.
107. This compound is commonly called acetylacetone. As shown, it exists in two forms, one called the *enol* form and the other called the *keto* form.

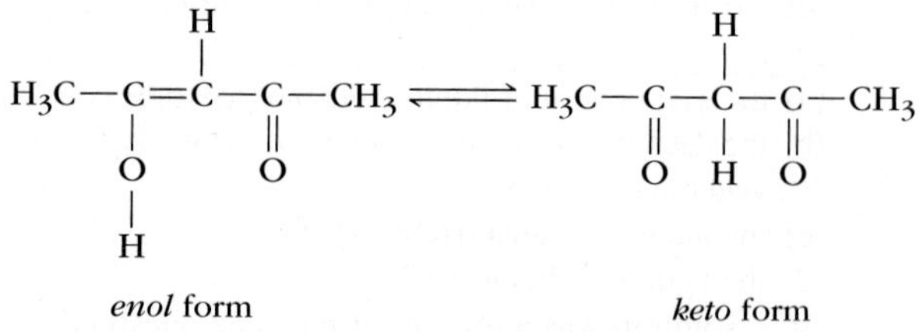
H
H3C—C=C—C—CH3
O
O
H
H
H3C—C—C—C—CH3
O H O
enol form
keto form

While in the *enol* form, the molecule can lose H^+ from the —OH group to form an anion. One of the most interesting aspects of this anion (sometimes called the *acac* ion) is that one or more of them can react with a transition metal cation to give very stable, highly colored compounds.

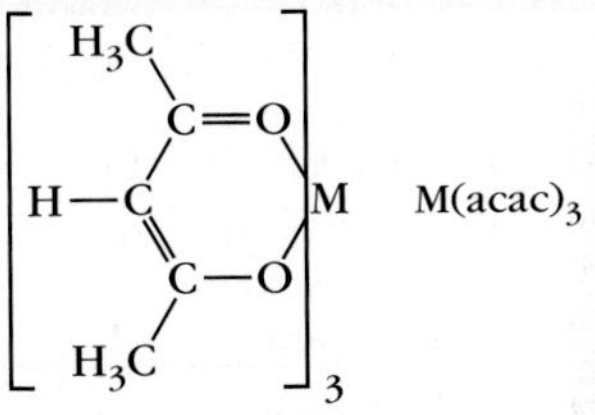
H₃C
C=O
H—C
M
C—O
H₃C
3
M(acac)₃

(a) Using bond enthalpies, calculate the reaction enthalpy for the *enol* to *keto* change. Is the reaction exothermic or endothermic?
(b) What are the electron-region and "molecular" geometries around each C atom in the *keto* and *enol* forms? What changes (if any) occur when the *keto* form changes to the *enol* form?
(c) If you wanted to prepare 15.0 g deep red $Cr(acac)_3$ using this reaction,

$$CrCl_3 + 3\ H_3C{-}C(OH){=}CH{-}C(OH){-}CH_3 + 3\ NaOH \rightarrow Cr(acac)_3 + 3\ H_2O + 3\ NaCl$$

calculate the mass (g) of each reactant you would need.

108. The compound S_4N_4 is a bright orange solid that is insoluble in water. Its eight-membered ring has several resonance structures, one of which is

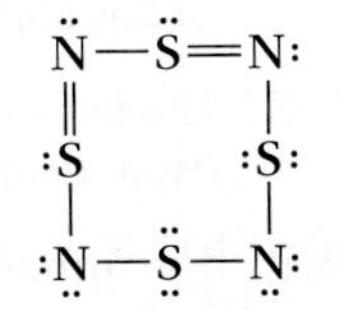

(a) Identify the hybridization of single-bonded N atoms.
(b) Identify the hybridization of double-bonded S atoms.
(c) Identify the hybridization of double-bonded N atoms.
(d) How many sigma bonds and how many pi bonds are in this molecule?
(e) Identify where the sigma and the pi bonds are in this molecule.
(f) Write the Lewis structure for at least one other resonance hybrid structure.

109. The bond angles around the central N in this series—NH_3, $N(CH_3)_3$, $N(SiH_3)_3$, and $N(GeH_3)_3$—are 107.5°, 110.9°, 120°, and 120°, respectively. Explain the trend in these bond angles.
110. There are two crystalline forms of oxalic acid, HOOC—COOH, both involving hydrogen bonding. The *alpha* form consists of three-dimensional zig-zag chains of oxalic acid molecules; the *beta* form has long, linear chains of oxalic acid molecules. Draw Lewis structures that illustrate hydrogen bonding in each form.
111. Consider a hydrogen bond X—H · · · :Z—.
(a) What kinds of vibrations can the H atom in the hydrogen bond undergo? Draw a diagram showing how the H atom would move in each vibration.
(b) Suppose that a D atom participated in the hydrogen bond (D is the isotope of hydrogen with one neutron and one proton in the nucleus.). How would substitution of D for H affect the vibrations you described in part (a)?

Conceptual Challenge Problems

These rigorous, thought-provoking problems integrate conceptual learning with problem solving and are suitable for group work.

CP7.A (Section 7-2) What advantages does the VSEPR model of chemical bonding have compared with the Lewis dot formulas predicted by the octet rule?

CP7.B (Section 7-2) The VSEPR model does not differentiate between single bonds and double bonds for predicting molecular shapes. What experimental evidence supports this?

CP7.C (Section 7-3) What evidence could you present to show that two carbon atoms joined by a single sigma bond are able to rotate about an axis that coincides with the bond, but two carbon atoms bonded by a double bond cannot rotate about an axis along the double bond?

CP7.D (Sections 7-2 and 7-3) Compare and contrast VSEPR theory and hybrid orbital theory with regard to which theory correctly predicts molecular shapes and bond angles, and also with regard to each theory's basis in quantum theory.

8 | Properties of Gases

Michael Ventura/Alamy

Emergency personnel are called on to enter and remain in situations where the air is toxic. To avoid being poisoned, these personnel wear a self-contained self-rescue breathing apparatus that generates oxygen for them to breathe, keeping them safe. What constitutes clean, breathable air? (Section 8-9) How is oxygen generated in the apparatus, and how can the volume of oxygen that is generated be calculated? (Section 8-5) Gas pressure is important in the apparatus; what is pressure and how does it affect gas behavior? (Sections 8-1 to 8-3) These questions and many others, all related to the properties of gases, will be answered in this chapter.

Early chemists were attracted to the study of gases. The atmosphere was an ample source from which they could sample and study gaseous behavior. Until about 1750, air was the only identified gas. Many fundamental concepts about gases were derived from the study of air. Careful scientific observations led to natural laws that explain the macroscale and nanoscale behavior of gases.

All pure gases exhibit remarkably similar physical behavior, much more so than liquids or solids. For example:

- 1 mol gaseous N_2, O_2, Ar, CH_4, CO_2, and almost all other gases has the *same* volume—*22.4 L*—at 0 °C and 1 atm pressure. This volume can be

22.4 L is about the total volume of four fully inflated soccer balls or three basketballs.

changed by altering the temperature, pressure, or the amount of gas, or by altering any combination of these variables.

- The dependency of a gas's volume on temperature, pressure, or amount of the gas is essentially the same for any gas, a behavior that led to a mathematical expression—*the ideal gas law*—that quantitatively accounts for such dependencies.
- Gases are mobile, expanding to uniformly fill any sealed container that holds them. If the seal is removed, the gas diffuses from the container into its surroundings to fill all space available.
- In a mixture of gases, such as air, the different gases mix uniformly in all proportions.
- Gases can be compressed readily, much more than liquids or solids.

An important property of any gas is its pressure. Measurements of gas pressure were fundamental to early studies of air and other gases.

8-1 Gas Pressure

We all have experienced a balloon filled with air. The firmness of the balloon reflects that the gas inside the balloon exerts pressure on the "skin" of the balloon; the greater the internal pressure, the firmer the balloon. As gas molecules strike the balloon's inner surface, they exert a force on that surface. The **pressure** of the gas is the *force exerted on a unit area*

$$\text{Pressure} = \frac{\text{force}}{\text{area}}$$

The SI unit of force is the **newton** (N), which *equals a force of 1 kg m/s^2. A pressure of one newton per square meter (N/m^2) is defined as one* **pascal (Pa)**. Several units, in addition to Pa, are used to express pressure (Table 8.1).

One newton is approximately the gravitational force exerted by Earth on an apple.

- **Kilopascal**: *one kilopascal, 1 kPa = 10^3 Pa,* is approximately equal to the pressure exerted by a mass of 10. g resting on a 1 cm^2 area at Earth's surface. Automobile tires are typically inflated to a pressure of 200–240 kPa.
- **Bar**: *one bar, equal to 100,000 Pa,* is approximately the pressure exerted by the atmosphere at Earth's surface. Weather maps show pressure contours in millibars. Standard thermodynamic properties (← **Secs. 4-5b, 4-10**) are at 1 bar.

Table 8.1 Pressure Units

SI Unit: Pascal (Pa)		
1 Pa	=	1 N m^{-2} = 1 kg m^{-1} s^{-2}; 1 kPa = 10^3 Pa
Other Common Units		
1 bar	=	10^5 Pa = 100 kPa; 1 millibar (mbar) = 10^2 Pa
1 atm	=	1.01325 × 10^5 Pa = 101.325 kPa (this conversion is exact)*
1 atm	=	760 Torr = 760 mmHg (this conversion is exact)*
1 atm	=	14.7 lb/in^2 (psi) = 1.01325 bar

*Exact conversion factors do not limit the number of significant figures in calculations.

Karsten Schneider/Science Source

Atmospheric pressure. The pressure is low in the center of a tropical depression over the Gulf of Mexico. Lines represent constant atmospheric pressure in millibars (mbar).

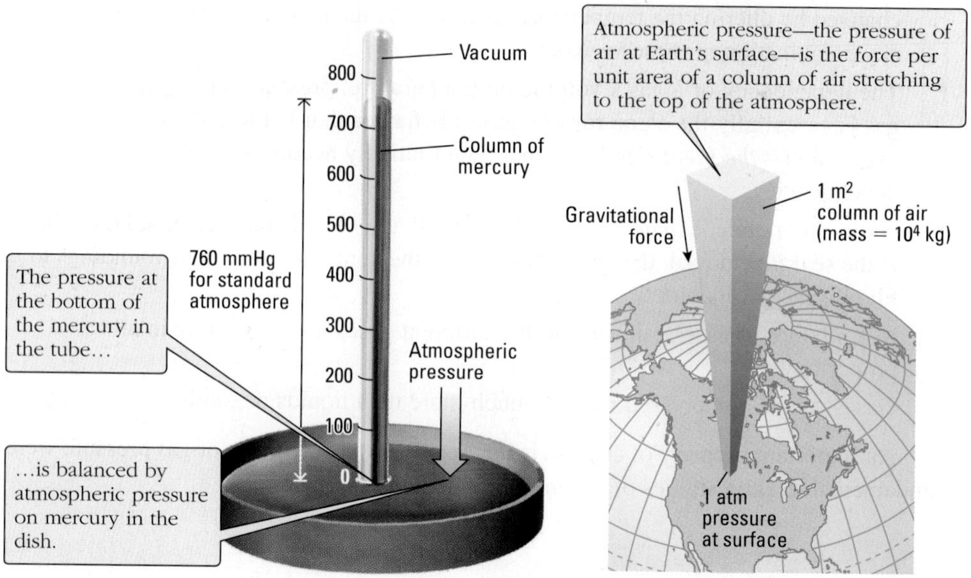
Vacuum
800
700
600
500
400
300
200
100
0
Column of mercury
760 mmHg for standard atmosphere
The pressure at the bottom of the mercury in the tube...
...is balanced by atmospheric pressure on mercury in the dish.
Atmospheric pressure
Atmospheric pressure—the pressure of air at Earth's surface—is the force per unit area of a column of air stretching to the top of the atmosphere.
Gravitational force
1 m²
column of air
(mass = 10⁴ kg)
1 atm
pressure
at surface

Figure 8.1 A Torricellian barometer measures atmospheric pressure.

- **Pound per square inch** (psi, lb/in^2): *a force of 1 lb on a 1 in^2 area at Earth's surface exerts a pressure of 1 lb/in^2 = 1 psi*. For example, bicycle tires for road bikes are usually inflated to 90–110 psi (620–760 kPa).
- **Atmosphere**: one atmosphere, 1 atm, is *defined as 101.325 kPa*. This is approximately the pressure of the atmosphere at Earth's surface.
- **Millimeter of mercury** (mmHg): also known as the **torr** (Torr), in honor of Evangelista Torricelli, who invented the mercury barometer in 1643; *760 mmHg = 760 Torr = 1 atm*.

A **barometer** is *a device for measuring atmospheric pressure*. A mercury-filled barometer (Figure 8.1) is made by taking a glass tube closed at one end, filling it with mercury, covering the open end of the tube, and inverting the open end into a pan containing a pool of mercury. When the cover is removed, some mercury from the tube flows into the pool creating an almost perfect vacuum above the mercury surface in the tube. Mercury flows out of the tube until the pressure at the bottom of the mercury in the tube exactly equals the pressure of the atmosphere on the surface of the pool of mercury. Thus, the height of the mercury column indicates the atmospheric pressure. The height of the mercury in a barometer is conveniently measured in millimeters; hence the unit mmHg.

The height of mercury in a barometer varies with weather conditions and with altitude above sea level. A "high-pressure" weather system causes the mercury column to rise; "low pressure", usually associated with stormy weather, causes the mercury level in a barometer to fall. Because there is less air above a mountaintop than at sea level, the air pressure is less and can support only a shorter column of mercury. Typical atmospheric pressure in the mile-high city of Denver, CO is 625 mmHg, or about 0.82 atm.

Because pressures can be expressed in so many different ways, it is important to learn to make conversions among the various pressure units.

PROBLEM-SOLVING EXAMPLE 8.1

Converting Pressure Units

Convert these pressures:

(a) an automobile tire pressure of 303 kPa to psi.

(b) an eye's internal pressure of 15.5 mmHg to psi.

(c) air pressure of 0.350 atm at the top of Mount Everest to torr and to kPa.

Result (a) 44.0 lb/in^2 (b) 0.300 lb/in^2 (c) 266 Torr; 35.5 kPa

Analyze These are all unit conversions.

Plan Use the unit equivalencies in Table 8.1 as conversion factors.

Execute

$$\text{(a) } P = 303 \text{ kPa} \times \frac{1 \text{ atm}}{101.3125 \text{ kPa}} \times \frac{14.7 \text{ lb/in}^2}{1 \text{ atm}} = 44.0 \text{ lb/in}^2$$

$$\text{(b) } P = 15.5 \text{ mmHg} \times \frac{1 \text{ atm}}{760 \text{ mmHg}} \times \frac{14.7 \text{ lb/in}^2}{1 \text{ atm}} = 0.300 \text{ lb/in}^2$$

$$\text{(c) } P = 0.350 \text{ atm} \times \frac{760 \text{ Torr}}{1 \text{ atm}} = 266 \text{ Torr}; P = 0.350 \text{ atm} \times \frac{101.325 \text{ kPa}}{1 \text{ atm}} = 35.5 \text{ kPa}$$

PROBLEM-SOLVING PRACTICE 8.1

Convert these pressures:
(a) a bicycle tire pressure of 100. psi to kPa.
(b) a systolic blood pressure of 135 mmHg to Pa.
(c) a pressure of 690 Torr inside an airplane cabin at a high altitude to psi and to bar.

PROBLEM-SOLVING PRACTICE answers are provided at the back of this book in Appendix K.

EXERCISE 8.1

Atmospheric Pressure and Weather

On June 24, 2003, a barometric pressure of 25.17 in of mercury occurred in a tornado near Manchester, South Dakota, likely the lowest pressure reading ever recorded in a tornado. For comparison, on August 28, 2005 Hurricane Katrina, a Category 5 hurricane, had a minimum central pressure of 26.64 inches. Convert these two pressures into: (a) torr; (b) mbar; (c) kPa; and (d) atm.

Answers to **EXERCISES** are provided at the back of this book in Appendix L.

National Oceanic and Atmospheric Acministration NOAA

Hurricane Katrina.

EXERCISE 8.2

Atmospheric Pressure and Water Wells

Some water wells use atmospheric pressure to force water to the surface. Such a well is similar to a barometer in that the atmospheric pressure at the bottom of the well must be equal to or greater than the pressure of water in the well pipe. If the atmospheric pressure is 0.973 atm, determine the maximum depth in meters and in feet for such a well.

8-2 Kinetic-Molecular Theory

The fact that all gases have similar physical properties can be interpreted at the nanoscale level using the *kinetic-molecular theory* (← **Sec. 1-9a**). According to the theory, a gas consists of tiny molecules in *constant, rapid, random motion.* Four fundamental postulates, developed from experimental studies of gases, are the foundation of the kinetic-molecular theory; a fifth postulate is closely associated with it.

In the kinetic-molecular theory, the term "molecule" is taken to include atoms of the monoatomic noble gases He, Ne, Ar, Kr, and Xe.

1. ***A gas is composed of molecules whose size is much smaller than the distances between them.*** This postulate accounts for the ease with which gases can be compressed and for the fact that gases at ordinary temperature and pressure mix completely with each other. These facts imply that there must be much unoccupied space in gases that provides substantial room for additional molecules in a sample of gas.
2. ***Gas molecules move randomly at various speeds and in every possible direction.*** This postulate is consistent with the fact that gases quickly and completely fill any container in which they are placed.

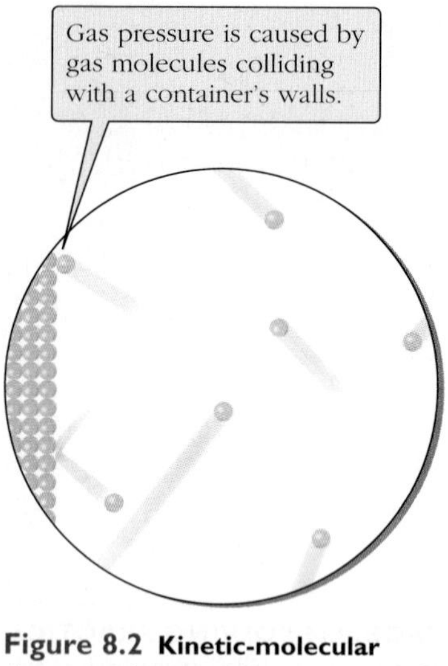
Gas pressure is caused by gas molecules colliding with a container's walls.

Figure 8.2 Kinetic-molecular theory and pressure.

3. ***Except when gas molecules collide, forces of attraction and repulsion between them are negligible; collisions with molecules in the walls of a solid container account for gas pressure (Figure 8.2).*** This postulate is consistent with the fact that all gases behave in the same way, regardless of the types of noncovalent interactions among their molecules.
4. ***When collisions between molecules occur, the collisions are elastic.*** In an elastic collision, the speeds of colliding molecules may change, but the total kinetic energy of two colliding molecules is the same after a collision as before the collision. This postulate is consistent with the fact that a gas sample at constant temperature never "runs down," with all molecules falling to the bottom of the container.
5. ***The average kinetic energy of gas molecules is proportional to the absolute temperature.*** Though not part of the kinetic-molecular theory, this useful postulate is consistent with the fact that gas molecules escape through a tiny hole faster as the temperature increases, and with the fact that rates of chemical reactions are faster at higher temperatures.

CONCEPTUAL EXERCISE 8.3

Seeing Through Gases

Explain why gases are transparent to light. (*Hint:* You may need to refer to Chapter 5 in addition to using some of the concepts in this section.)

EXERCISES that are labeled **CONCEPTUAL** are designed to test your understanding of one or more concepts; they usually involve qualitative rather than quantitative thinking.

CONCEPTUAL EXERCISE 8.4

The Kinetic-Molecular Theory

Use the kinetic-molecular theory to explain why the pressure goes up when gas molecules are added to a sample of gas in a fixed-volume container at constant temperature.

8-3 The Behavior of Ideal Gases: Gas Laws

The kinetic-molecular theory explains gas behavior on the nanoscale. On the macroscale, gases have been studied for hundreds of years. How gas volume changes when temperature, pressure, or amount of gas are changed has been summarized in mathematical equations called *gas laws* that are named for their discoverers. *A gas that behaves exactly as described by the gas laws* is called an **ideal gas**. At room temperature and atmospheric pressure, most gases behave nearly ideally. However, at pressures much higher than 1 atm or at temperatures just above the boiling point, gases deviate substantially from ideal behavior (Section 8-8). Each of the gas laws can be explained by the kinetic-molecular theory.

8-3a The Pressure-Volume Relationship: Boyle's Law

Boyle's law states that *the volume* (V) *of an ideal gas is inversely proportional to the pressure* (P) *when temperature* (T) *and amount* (n, *moles) are constant.*

$\propto$ means "is proportional to."

$$V \propto \frac{1}{P} \qquad V = \text{constant} \times \frac{1}{P} \qquad PV = \text{constant} \qquad (\text{unchanging } T \text{ and } n)$$

Boyle's law explains how expanding and contracting your chest cavity (changing its volume) leads to pressure changes that in turn lead to inhalation and exhalation of air.

The value of the constant depends on the temperature, T, and the amount of the gas, n. The inverse relationship between V and P is shown graphically in Figure 8.3. For a gas sample under two sets of pressure and volume conditions, but the same T and n, Boyle's law can be written as

$$P_1V_1 = \text{constant} = P_2V_2 \qquad (\text{unchanging } T \text{ and } n)$$

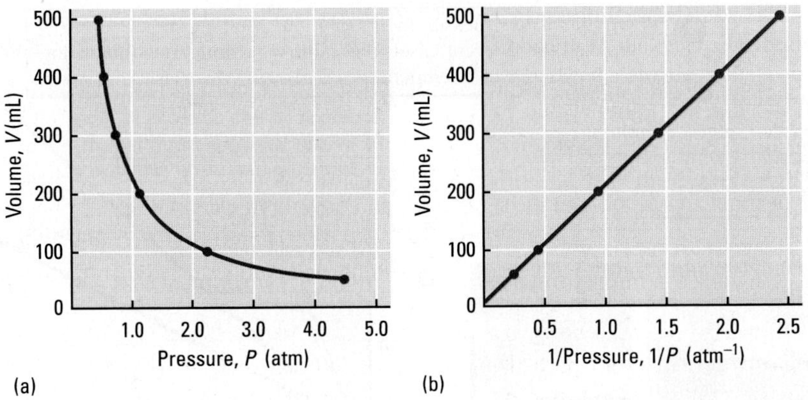
500
400
300
200
100
0
Volume, V (mL)
1.0 2.0 3.0 4.0 5.0
Pressure, P (atm)
(a)
500
400
300
200
100
0
Volume, V (mL)
0.5 1.0 1.5 2.0 2.5
1/Pressure, 1/P (atm^{-1})
(b)

Figure 8.3 Boyle's law. (a) This curve shows that as pressure increases, volume decreases. (b) A plot of *V* versus 1/*P* is a straight line, showing that *V* is inversely proportional to *P*.

In terms of the kinetic-molecular theory, a decrease in volume of a gas increases its pressure because there is less room for the gas molecules to move around in before they collide with the walls of the container. Thus, there are more frequent collisions with the walls. Each collision contributes to the pressure on the container walls, and more collisions mean a higher pressure. When you pump up a tire with a bicycle pump, the gas in the pump is squeezed into a smaller volume by application of pressure. This property is called *compressibility.* In contrast to gases, liquids and solids are only slightly compressible.

Robert Boyle studied the compressibility of gases in 1661 by pouring mercury into a J-shaped tube containing a sample of trapped gas. Each time he added more mercury, at constant temperature, the pressure of the trapped gas increased and its volume decreased. Data from Figure 8.4 illustrate Boyle's law.

$$V_2 = V_1 \times \frac{P_1}{P_2} = 30.0 \text{ cm}^3 \times \frac{760 \text{ mmHg}}{1065 \text{ mmHg}} = 21.4 \text{ cm}^3$$

CONCEPTUAL EXERCISE 8.5

Visualizing Boyle's Law

Many cars have gas-filled shock absorbers to give the car and its occupants a smooth ride. Suppose that four NFL linemen get into a four passenger car with gas-filled shock absorbers. Write a macroscale and a nanoscale description of the gas inside the shock absorbers before and after the linemen got aboard.

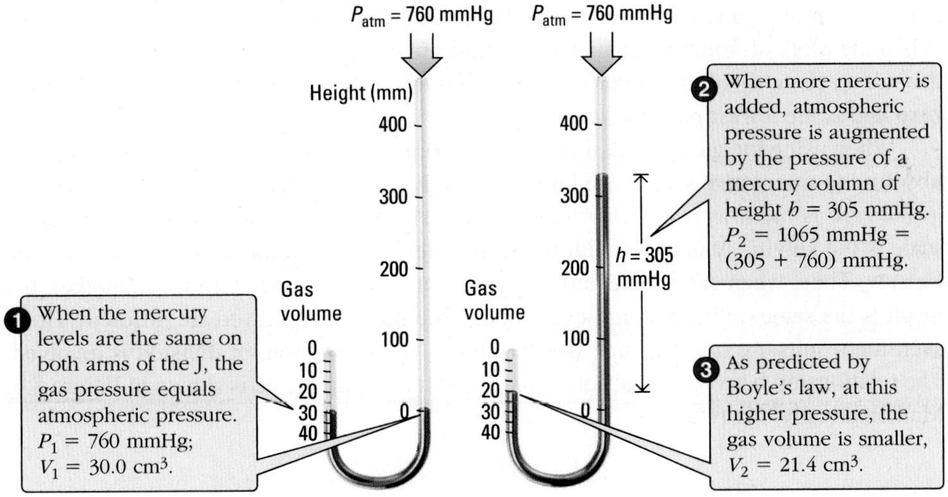
P_{atm} = 760 mmHg
P_{atm} = 760 mmHg
Height (mm)
400
300
200
100
0
400
300
200
100
0
Gas volume
0
10
20
30
40
Gas volume
0
10
20
30
40
h = 305 mmHg
1 When the mercury levels are the same on both arms of the J, the gas pressure equals atmospheric pressure. P_1 = 760 mmHg; V_1 = 30.0 cm^3.
2 When more mercury is added, atmospheric pressure is augmented by the pressure of a mercury column of height h = 305 mmHg. P_2 = 1065 mmHg = (305 + 760) mmHg.
3 As predicted by Boyle's law, at this higher pressure, the gas volume is smaller, V_2 = 21.4 cm^3.

Figure 8.4 Boyle's law experiment: volume is inversely proportional to pressure.

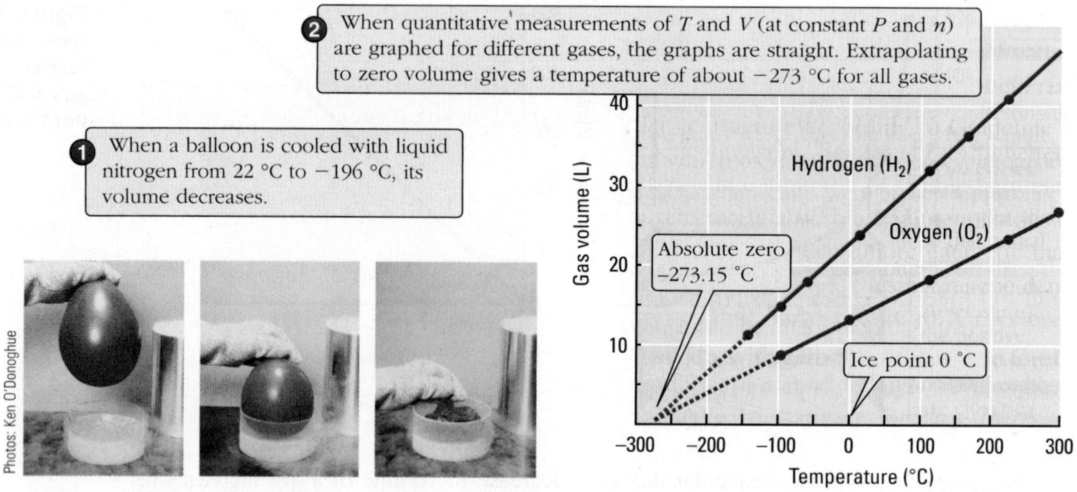

2 When quantitative measurements of T and V (at constant P and n) are graphed for different gases, the graphs are straight. Extrapolating to zero volume gives a temperature of about −273 °C for all gases.
1 When a balloon is cooled with liquid nitrogen from 22 °C to −196 °C, its volume decreases.
Photos: Ken O'Donoghue
Gas volume (L)
40
30
20
10
Hydrogen (H2)
Oxygen (O2)
Absolute zero −273.15 °C
Ice point 0 °C
−300 −200 −100 0 100 200 300
Temperature (°C)

Figure 8.5 Charles's law: gas volume is proportional to absolute temperature.

8-3b The Temperature-Volume Relationship: Charles's Law

Charles's law states that *the volume* (V) *of an ideal gas is directly proportional to absolute temperature* (T) *at constant pressure* (P) *and amount* (n, *moles).*

$$V \propto T \qquad V = \text{constant} \times T \qquad \frac{V}{T} = \text{constant} \qquad (\text{unchanging } P \text{ and } n)$$

The value of the constant depends on pressure and the amount of the gas. If the volume, V_1, and temperature, T_1, of a sample of gas are known, then the volume, V_2, at some other temperature, T_2, at the same pressure is given by

$$\frac{V_1}{T_1} = \frac{V_2}{T_2} \qquad (\text{unchanging } P \text{ and } n)$$

In terms of the kinetic-molecular theory, higher temperature means faster molecular motion and a higher average kinetic energy. The more rapidly moving molecules therefore strike the walls of a container more often, and each collision exerts greater force. For the pressure to remain constant, the volume of the container must expand.

Charles studied samples of carbon dioxide, hydrogen, oxygen, and nitrogen.

In 1787 Jacques Charles discovered that the volume of a fixed quantity of a gas at constant pressure increases with increasing temperature. Figure 8.5 shows how the volume of 1.0 mol H_2 and the volume of 0.55 mol O_2 change with temperature at $P = 1$ atm. When the plots of volume versus temperature are extended toward lower temperatures, they all reach zero volume at the same temperature, −273.15 °C. This is called **absolute zero** and is *the lowest possible temperature.*

The absolute temperature scale is also called the thermodynamic temperature scale.

When using the gas law relationships, temperature must be expressed in terms of the **absolute temperature scale**, also known as the **Kelvin temperature scale**, *the scale on which zero is the lowest possible temperature.* The SI unit of the absolute temperature scale is the kelvin, symbol K (with no degree sign), which is the same size as a degree Celsius. Thus, when ΔT is calculated by subtracting one temperature from another, the result is the same on both scales, even though the numbers involved are different. However, the results are *not* the same when addition, multiplication, or division is involved. The relationship between the absolute scale and the Celsius scale is shown in Figure 8.6. Mathematically this relationship is

$$T(\text{K}) = t(°\text{C}) + 273.15$$

Thus, 25.00 °C (the temperature of a warm room) is the same as 298.15 K.

Suppose you want to calculate the new volume, V_2, when 450.0 mL of a gas, V_1, is cooled from 60.0 °C, T_1, to 20.0 °C, T_2, at constant pressure. First, convert the temperatures to kelvins by adding 273.15 to the Celsius values: 60.0 °C becomes 333.2 K and 20.0 °C becomes 293.2 K. Solving Charles's law algebraically for the two sets of conditions at constant pressure gives

$$V_2 = \frac{V_1T_2}{T_1} = \frac{450.0 \text{ mL} \times 293.2 \text{ K}}{333.2 \text{ K}} = 396.0 \text{ mL}$$

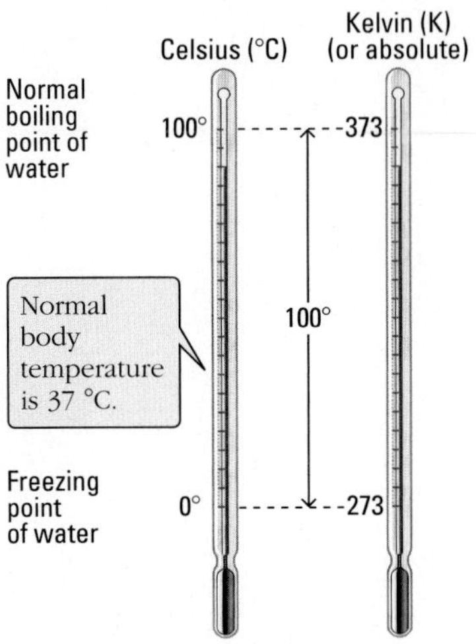
Celsius (°C)
Kelvin (K) (or absolute)
Normal boiling point of water
100°
373
Normal body temperature is 37 °C.
100°
Freezing point of water
0°
273

Figure 8.6 Temperature scales. Zero on the absolute scale is the lowest possible temperature (0 K or −273.15 °C).

CONCEPTUAL **EXERCISE 8.6**

Visualizing Charles's Law

Consider a collection of gas molecules at temperature T_1. The temperature increases to T_2. Use the ideas of the kinetic-molecular theory and explain why the volume has to be larger if the pressure remains constant. What would have to be done to maintain a constant volume if the temperature increased?

8-3c The Amount-Volume Relationship: Avogadro's Law

Avogadro's law *states that the volume* (V) *of an ideal gas varies directly with amount* (n, *moles) when temperature* (T) *and pressure* (P) *are constant.*

$$V \propto n \qquad V = \text{constant} \times n \qquad \frac{V}{n} = \text{constant} \qquad (\text{unchanging } T \text{ and } P)$$

The value of the constant depends on T and P. Avogadro's law means, for example, that at constant temperature and pressure, if the amount of gas doubles, the volume doubles. At the same temperature and pressure, the volumes of two different amounts of ideal gases are related as follows:

$$\frac{V_1}{n_1} = \frac{V_2}{n_2} \qquad (\text{unchanging } T \text{ and } P)$$

Another way to state the relationship between the amount of gas and its volume is illustrated in Table 8.2. *Equal volumes of gases contain equal numbers of molecules at*

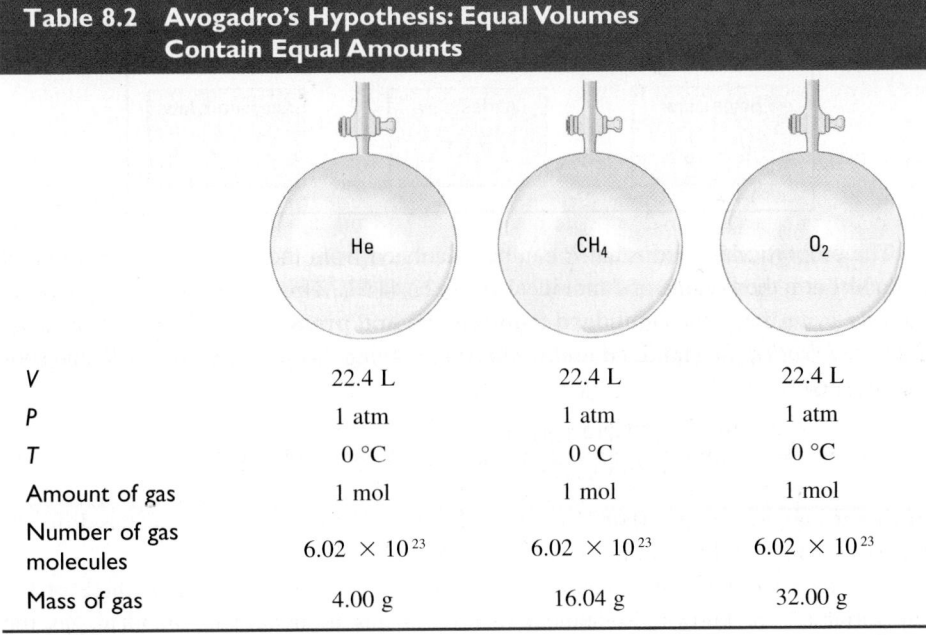

Table 8.2 Avogadro's Hypothesis: Equal Volumes Contain Equal Amounts

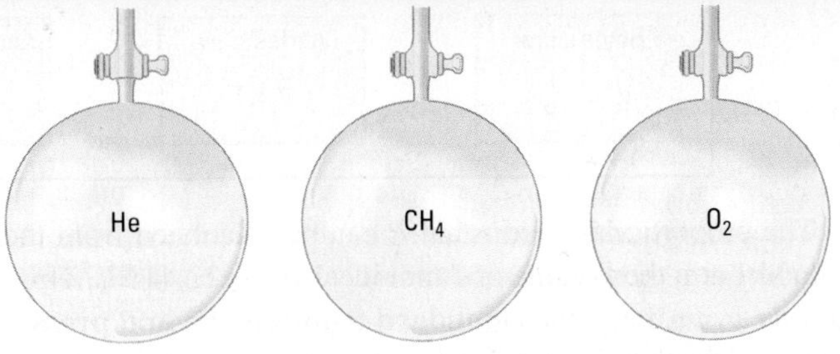

	He	CH_4	O_2
V	22.4 L	22.4 L	22.4 L
P	1 atm	1 atm	1 atm
T	0 °C	0 °C	0 °C
Amount of gas	1 mol	1 mol	1 mol
Number of gas molecules	6.02×10^{23}	6.02×10^{23}	6.02×10^{23}
Mass of gas	4.00 g	16.04 g	32.00 g

In Table 8.2 each gas sample has the same volume, temperature, and pressure, so each container holds the same number of molecules. Because the molar masses of the three gases are different, the mass of gas in each container is different.

the same temperature and pressure. Experiments show that at 0 °C and 1 atm pressure, 22.4 L of any ideal gas contains 1 mol of the gas (6.02×10^{23} gas molecules).

In terms of the kinetic-molecular theory, increasing the number of gas molecules increases the number of collisions with the container walls in proportion to the number of molecules. Constant temperature means that the added molecules have the same average molecular kinetic energy as the molecules to which they were added. Consequently each collision makes the same contribution to the pressure. More collisions would increase the pressure if the volume were held constant. To maintain constant pressure, the volume must increase.

8-3d The Ideal Gas Law

The three gas laws just discussed focus on the effects of changes in P, T, or n on gas volume:

- ***Boyle's law and pressure*** $V \propto 1/P$
- ***Charles's law and temperature*** $V \propto T$
- ***Avogadro's law and amount (mol)*** $V \propto n$

These three gas laws can be combined to give the ideal gas law, which summarizes the relationships among volume, temperature, pressure, and amount.

$$V \propto \frac{nT}{P}$$

To make this proportionality into an equation, a proportionality constant, R, named the **ideal gas constant**, is used. The equation becomes

$$V = R\frac{nT}{P}$$

and, on rearranging, gives the equation called the **ideal gas law**.

$$PV = nRT$$

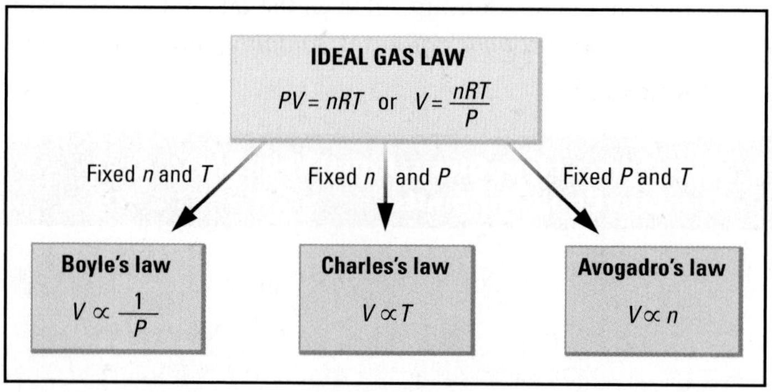
IDEAL GAS LAW
PV = nRT or V = nRT/P
Fixed n and T
Fixed n and P
Fixed P and T
Boyle's law
V ∝ 1/P
Charles's law
V ∝ T
Avogadro's law
V ∝ n

The proportionality constant R can be calculated from the experimental fact that at 0 °C and 1 atm the volume of 1 mol ideal gas is 22.414 L. This temperature and pressure, *0 °C and 1 atm, are called* **standard temperature and pressure (STP)**, and *the volume, 22.414 L, is called the* **standard molar volume**. Solving the ideal gas law for R, and substituting, gives

$$R = \frac{PV}{nT} = \frac{(22.414\ \text{L})(1\ \text{atm})}{(1\ \text{mol})\ (273.15\ \text{K})} = 0.082057\ \text{L atm mol}^{-1}\ \text{K}^{-1}$$

which is usually rounded to 0.0821 L atm $\text{mol}^{-1}\ \text{K}^{-1}$. The ideal gas constant has different numerical values in different units, as shown in Table 8.3.

The ideal gas law can be used to calculate P, V, n, or T whenever three of the four variables are known. Remember that for a gas to behave as an ideal gas the

Table 8.3 Values of R, in Different Units

$$R = 0.08206\frac{\text{L atm}}{\text{mol K}}$$

$$R = 62.36\frac{\text{mmHg L}}{\text{mol K}}$$

$$R = 8.314\frac{\text{kPa dm}^3}{\text{mol K}}$$

$$R = 8.314\frac{\text{J}}{\text{mol K}}$$

temperature must be well above its boiling point and the pressure should be no more than a few atmospheres.

PROBLEM-SOLVING EXAMPLE 8.2

Using the Ideal Gas Law

Calculate the volume that 0.40 g methane, CH_4, occupies at 25 °C and 1.0 atm.

Result 0.61 L or 6.1×10^2 mL CH_4

Analyze The mass, temperature, and pressure of methane are given. The amount (n) is proportional to the mass of methane. Volume can be calculated from $PV = nRT$.

Plan Use $PV = nRT$ and make certain that units cancel those in the gas constant, $R = 0.0821$ L atm mol^{-1} K^{-1}; that is, T in kelvins, P in atmospheres, n in moles.

Execute

$$n(CH_4) = (0.40 \text{ g } CH_4) \times \frac{1 \text{ mol } CH_4}{16.04 \text{ g } CH_4} = 0.0249 \text{ mol } CH_4$$

$$T = (25 + 273.15) \text{ K} = 298 \text{ K}$$

Solve for V in the ideal gas equation. Cancellation of units gives a result in liters.

$$V = \frac{nRT}{P} = \frac{(0.0249 \text{ mol})(0.0821 \text{ L atm mol}^{-1} \text{ K}^{-1})(298 \text{ K})}{1.0 \text{ atm}}$$

$$= 0.61 \text{ L or } 6.1 \times 10^2 \text{ mL } CH_4$$

☑ **Reasonable Result Check** The temperature and pressure are close to STP. The mass of methane corresponds to about 0.025 mol, so the volume should also be about 0.025 times the volume of a mole of gas (22.4 L × 0.025 = 0.56 L). This approximate result is close to our more exact calculation.

Notice that the result is given with two significant digits because the mass of methane and the pressure had only two significant digits.

PROBLEM-SOLVING PRACTICE 8.2

Calculate the volume of 2.64 mol N_2 at 0.640 atm and 31 °C.

When considering a gas sample under two sets of conditions (n_1, P_1, V_1, T_1; and n_2, P_2, V_2, and T_2), the ideal gas law can be written as

$$R = \frac{P_1V_1}{n_1T_1} \quad \text{and} \quad R = \frac{P_2V_2}{n_2T_2}$$

Each quotient is equal to R, so we can set the two quotients equal to each other.

$$\frac{P_1V_1}{n_1T_1} = \frac{P_2V_2}{n_2T_2}$$

When we are dealing with the same sample of gas, the amount of gas, n, is constant, so that $n_1 = n_2$. Then this equation simplifies to what is known as the **combined gas law**:

$$\frac{P_1V_1}{T_1} = \frac{P_2V_2}{T_2}$$

PROBLEM-SOLVING EXAMPLE 8.3

The Combined Gas Law

Helium-filled balloons are used to carry scientific instruments high into the atmosphere. Suppose that such a balloon is launched on a summer day when the temperature at ground level is 22.5 °C and the barometer reading is 754 mmHg. If the balloon's volume is 1.00×10^6 L at launch, what will its volume be at a height of 37 km, where the pressure is 76.0 mmHg and the temperature is 240. K? (Assume that no helium escapes.)

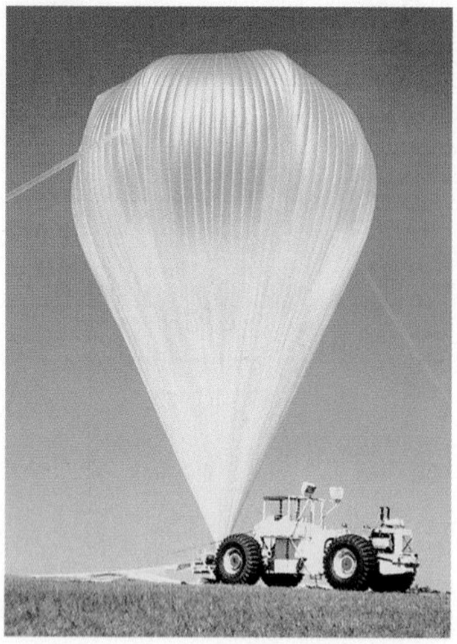
NASA

A weather balloon filled with helium. As it ascends into the atmosphere, does the volume increase or decrease?

Result 8.05×10^6 L

Analyze The amount of helium is the same. T, P, and V are known at ground level and T and P are known at 37 km, so the combined gas law can be used.

Plan Use subscript 1 to indicate initial conditions (ground level) and subscript 2 to indicate final conditions (high altitude); solve the combined gas law for V_2.

Execute

Initial: $P_1 = 754$ mmHg $\quad T_1 = (22.5 + 273.15)$ K $= 295.6$ K $\quad V_1 = 1.00 \times 10^6$ L

Final: $P_2 = 76.0$ mmHg $\quad T_2 = 240.$ K

$$V_2 = \frac{P_1V_1T_2}{P_2T_1} = \frac{(754 \text{ mmHg})(1.00 \times 10^6 \text{ L})(240. \text{ K})}{(76.0 \text{ mmHg})(295.6 \text{ K})} = 8.05 \times 10^6 \text{ L}$$

Thus, the final volume is about eight times larger. For this reason, weather balloons are never fully inflated at launch. A great deal of room has to be left so that the helium can expand at high altitudes.

☑ **Reasonable Result Check** The pressure dropped by a factor of about 10; the temperature decreased only about 20%, so the volume should increase by a factor of about 20% less than 10, that is, by a factor of 8, and it does.

PROBLEM-SOLVING PRACTICE 8.3

A small sample of a gas is prepared in the laboratory and found to occupy 21 mL at a pressure of 710. mmHg and a temperature of 22.3 °C. The next morning the temperature has changed to 26.5 °C, and the pressure is found to be 740. mmHg. No gas has escaped from the container.

(a) Calculate the volume that the gas sample now occupies.

(b) Assume that the pressure does not change. Calculate the volume the gas would occupy at the new temperature.

It is important to remember that Boyle's, Charles's, and Avogadro's laws do not depend on the identity of the gas being studied. These laws reflect properties of all gases and therefore describe the behavior of any gaseous substance.

In summary, for problems involving gases, there are two useful equations:

- ***When three of the variables* P, V, n, *and* T *are given, and the value of the fourth variable is needed, use the ideal gas law.***

$$PV = nRT \qquad [8.1]$$

- ***When one set of conditions is given and one of the variables under a new set of conditions is needed, then use this equation and cancel any of the four variables that do not change.***

$$\frac{P_1V_1}{n_1T_1} = \frac{P_2V_2}{n_2T_2} \qquad [8.2]$$

PROBLEM-SOLVING EXAMPLE 8.4

Pressure and Volume

Consider a gas sample with a volume of 1.12 L at a pressure of 0.523 atm. Calculate the volume this sample occupies when the pressure is increased to 2.00 atm. Assume constant T and n and ideal gas behavior.

Result 0.293 L or 293 mL

Analyze T and n are constant and the gas is ideal.

Plan Cancel T and n in Equation 8.2 and solve for V_2.

Execute $P_1 = 0.523$ atm, $P_2 = 2.00$ atm, and $V_1 = 1.12$ L.

$$V_2 = \frac{P_1V_1}{P_2} = \frac{(0.523 \text{ atm})(1.12 \text{ L})}{2.00 \text{ atm}} = 0.293 \text{ L} = 293 \text{ mL}$$

☑ **Reasonable Result Check** V is inversely proportional to P. P increases by a bit less than 4 times so V_2 should be a bit more than $\frac{1}{4}V_1$. The result is reasonable.

PROBLEM-SOLVING PRACTICE 8.4

At a pressure of exactly 1 atm and some temperature, a gas sample occupies 400. mL. Calculate the volume of the gas at the same temperature if the pressure is decreased to 0.750 atm.

CONCEPTUAL EXERCISE 8.7

Predicting Gas Behavior

Name three ways the volume occupied by a sample of gas can be decreased.

8-4 Gas Density, Molar Mass, and the Ideal Gas Law

The ideal gas law can be used to relate gas density to molar mass. The definition of density is mass per unit volume (← **Sec. 1-4d**). The densities of gases are extremely variable because the volume of a sample of gas (but not its mass) varies with temperature and pressure. However, once T and P are specified, the density of a gas can be calculated from the ideal gas law. Additionally, because equal volumes of gas at the same T and P contain equal numbers of molecules, the densities of different gases are directly proportional to their molar masses. As a result, experimental gas densities can be used to determine molar masses.

To derive the relationship between gas density and molar mass, start with the ideal gas law, $PV = nRT$. For any substance, the amount (n) equals the mass (m) divided by the molar mass (M), so we will substitute m/M for n in the ideal gas equation:

$$PV = \frac{m}{M}RT$$

Density (d) is mass divided by volume (m/V), and we can rearrange the equation so that m/V is on one side:

$$d = \frac{m}{V} = \frac{PM}{RT}$$

Thus, the density of a gas is directly proportional to its molar mass, M.

Consider the densities of the three pure gases He, O_2, and SF_6. If we consider 1 mol of each of these gases at 25 °C and 0.750 atm, we can see from the table below that density (at the same conditions) increases with molar mass.

	He	O_2	SF_6
Molar mass, g/mol	4.003	31.999	146.06
Density (25 °C), g/L	0.123	0.981	4.48

Because the densities of gases are quite small they are usually expressed in units of g/L.

PROBLEM-SOLVING EXAMPLE 8.5

Gas Density

Calculate the density of Cl_2 at 25 °C and 0.75 atm.

Result 2.2 g/L

Analyze Gas density is proportional to molar mass; assume ideal gas behavior.

Plan Calculate the molar mass of Cl_2. Convert the temperature to kelvins. Use these values, the pressure, and the gas constant to calculate the density of Cl_2.

Execute $$\text{Molar mass of } Cl_2 = M = 2\text{ Cl} \times \frac{35.5\text{ g Cl}}{1\text{ mol Cl}} = 71.0\text{ g/mol}$$

$$T = (25 + 273.15)\text{ K} = 298\text{ K}$$

$$d = \frac{PM}{RT} = \frac{(0.75\text{ atm})(71.0\text{ g/mol})}{(0.0821\text{ L atm mol}^{-1}\text{ K}^{-1})(298\text{ K})} = 2.2\text{ g/L}$$

☑ **Reasonable Result Check** Density is proportional to molar mass, so the density of 2.2 g/L for Cl_2 (M = 71.0 g/mol) should be greater than the density of O_2 (d = 0.981 g/L; M = 32.0 g/mol), but about half that of SF_6 (d = 4.48 g/L; M = 146 g/mol), which it is. The result is reasonable.

PROBLEM-SOLVING PRACTICE 8.5

Calculate the density of SO_2 at 25.0 °C and 2.60 atm.

CONCEPTUAL EXERCISE 8.8

Comparing Densities

Express the gas density of He (0.166 g/L) in grams per milliliter (g/mL) and compare that value with the density of metallic lithium (0.53 g/mL). The mass of an Li atom (7 u) is less than twice that of an He atom (4 u). What do these densities tell you about how closely Li atoms are packed compared with He atoms when each element is in its standard state? To which of the concepts of the kinetic-molecular theory does this comparison apply?

CONCEPTUAL EXERCISE 8.9

Densities of Gas Mixtures

Assume a mixture of equal volumes of the two gases nitrogen and oxygen. Would the density of this gas mixture be higher, lower, or the same as the density of air at the same temperature and pressure? (Air consists of 21% oxygen, 78% nitrogen, and 1% argon.) Explain your answer.

PROBLEM-SOLVING EXAMPLE 8.6

Calculating Molar Mass

You have a 0.555-g sample of gas, a compound of phosphorus and fluorine, in a sealed 100.-mL flask at 20.0 °C. The pressure in the flask is 0.959 atm. From these data:
(a) Calculate the molar mass of the gas; (b) propose a molecular formula consistent with the molar mass.

Result (a) Molar mass = 139 g/mol; (b) P_2F_4

Analyze The molar mass can be calculated from the modified ideal gas equation using the mass, volume, temperature, and pressure of the sample. We want to determine the formula P_aF_b where a and b are the relative amounts of phosphorus and fluorine, respectively, expressed as small whole numbers (← **Sec. 2-12a**). The sum of the molar masses of the atoms in the formula must equal the experimental molar mass. Trial and error can find the formula.

Plan Calculate the molar mass. Assume different formulas and determine whether any of them correspond with the calculated molar mass.

Execute

(a) $M = \dfrac{mRT}{PV} = \dfrac{(0.555 g)(0.0821 \text{ L atm mol}^{-1}\text{ K}^{-1})(273.15 + 20.0)\text{ K}}{(0.959 \text{ atm})(0.100 \text{ L})} = 139 \text{ g/mol}$

(b) The molar masses of phosphorus and fluorine are 30.97 and 19.00 g/mol, respectively (one more significant figure than other data given). The simplest formula is PF; if this were correct, M = 49.97 g/mol, but we know that M = 139 g/mol, so PF cannot be correct. Doubling this formula to P_2F_2 gives M = 99.94 g/mol; tripling the simplest formula gives a molar mass greater than 139 g/mol; PF cannot be the simplest formula. Next, try a formula with two phosphorus atoms and different numbers of fluorine atoms. Two moles of phosphorus atoms account for 61.94 g, requiring (139 − 62.0) g = 77 g F.

$$\text{Amount of F} = 77 \text{ g F} \times \frac{1 \text{ mol F}}{19.00 \text{ g F}} = 4.05 \text{ mol F}$$

Thus, a = 2 and b = 4; the correct formula is P_2F_4.

The empirical (simplest) formula is PF_2 (← Sec. 2-12a).

☑ **Reasonable Result Check** A gas volume of 0.100 L is about 0.100 L/22.4 L mol^{-1} = 0.0045 mol gas in the 0.555-g sample. Therefore, 0.555 g/0.0045 mol = 123 g/mol, a bit less than 139 g/mol. This is understandable because 22.4 L is the volume at STP (1 atm and 0 °C). At the higher temperature and slightly lower pressure in the experiment the volume would be larger. The result is still reasonable; P_2F_4 is a known compound.

PROBLEM-SOLVING PRACTICE 8.6

The pressure of a 20.0-g sample of a pure gas in a 2.50-L sealed flask at 10.0 °C is 1.41 atm. Calculate the molar mass of the gas and from the molar mass, determine the identity of the gas.

ESTIMATION | Helium Balloon Buoyancy

How large a helium balloon is needed to lift a weather-instruments package weighing 500 lb? The buoyancy of He is the difference in density between air and He at the given temperature and pressure.

First, estimate the density of helium. Assume a temperature of 21 °C and a pressure of 1 atm. Then,

$$d_{\text{He}} = \frac{PM}{RT} = \frac{(1 \text{ atm})(4.00 \text{ g mol}^{-1})}{(0.0821 \text{ L atm mol}^{-1}\text{ K}^{-1})[(273 + 21)\text{K}]}$$

$$= 0.166 \text{ g/L}$$

Next, estimate the density of air at the same temperature and pressure. The same calculation can be done using a weighted average molar mass. Air is 21% O_2 molecules, 78% N_2 molecules, and 1% Ar molecules. Thus a rough molar mass is 0.21 × 32.0 g/mol + 0.78 × 28.0 g/mol + 0.01 × 39.9 g/mol = 29.0 g/mol. The density of air is

$$d_{\text{air}} = \frac{PM}{RT} = \frac{(1 \text{ atm})(29.0 \text{ g mol}^{-1})}{(0.0821 \text{ L atm mol}^{-1}\text{ K}^{-1})[(273 + 21)\text{K}]}$$

$$= 1.20 \text{ g/L}$$

The mass difference between 1 L He and 1 L air is 1.20 g/L − 0.166 g/L = 1.03 g/L. Thus, in air, a liter of He can lift 1.03 g.

We can now estimate the size of the balloon. Ignoring the mass of the balloon material, the total mass to be lifted is 500 lb × 454 g/lb = 2.27 × 10^5 g. Because He can lift 1 g/L, the volume of He is 2.27 × 10^5 L.

Let's change this to cubic meters.

$$V = 2.27 \times 10^5 \text{ L} \times \frac{1 \text{ dm}^3}{1 \text{ L}} \times \left(\frac{1 \text{ m}}{10 \text{ dm}}\right)^3 = 2.27 \times 10^2 \text{ m}^3$$

Assume that the balloon is spherical; the volume and radius are related by

$$V = \frac{4}{3}\pi r^3 \quad \text{or} \quad r = \left(\frac{3V}{4\pi}\right)^{1/3}$$

The radius of the balloon is

$$r = \left(\frac{3V}{4\pi}\right)^{1/3} = \left(\frac{3 \times 227 \text{ m}^3}{4 \times 3.14}\right)^{1/3} = 54.2^{1/3} \text{ m} = 3.78 \text{ m}$$

The diameter of a spherical He balloon that can lift 500 lb is approximately 7.5 m (24 ft).

CONCEPTUAL EXERCISE 8.10

Helium-Filled Balloon in Car

If you have to slam on the brakes when driving a car, you will be thrown forward against the seat belt. So will a book or bag of groceries on the seat next to you. This happens because everything in the car is moving and keeps moving unless there is a force to slow its motion. A helium-filled balloon suspended in midair inside a car moves toward the back of the car when the brakes are slammed on. Explain why the helium-filled balloon behaves exactly the opposite of everything else.

8-5 Quantities of Gases in Chemical Reactions

Volumes, masses, or amounts can be related in calculations based on reaction stoichiometry (← **Sec. 3-6**) through the application of the ideal gas law and the law of combining volumes. In 1808, the French scientist Joseph Gay-Lussac experimentally determined that *the volumes of reacting gases were always in the ratios of small whole numbers, at the same temperature and pressure*. This is known as the **law of combining volumes**. Shortly after Gay-Lussac's discovery, Amadeo Avogadro proposed that Gay-Lussac's observations indicated that, *at the same temperature and pressure, equal volumes of gases contain the same number of molecules* (Table 8.2, ← **Sec. 8-3c**). How this applies to the volumes of gases before and after a chemical reaction is illustrated in Figure 8.7.

PROBLEM-SOLVING EXAMPLE 8.7

Using Avogadro's Law and the Law of Combining Volumes

Nitrogen and hydrogen gases react to form ammonia gas in the reaction

$$N_2(g) + 3\ H_2(g) \longrightarrow 2\ NH_3(g)$$

If 500. mL N_2 at 1 atm and 25 °C were available for reaction, calculate the volume of H_2, at the same temperature and pressure, required in the reaction.

Result 1.50 L H_2

Analyze All gases are at the same T and P, so volumes of gases involved in the reaction are proportional to the coefficients in the balanced chemical equation.

Plan Use the ratio of coefficients for $N_2(g)$ and $H_2(g)$ to relate $V(H_2)$ to $V(N_2)$.

Execute

$$\text{Volume of } H_2 = 500.\ \text{mL } N_2 \times \frac{3\ \text{mL } H_2}{1\ \text{mL } N_2} = 1.50\ \text{L } H_2$$

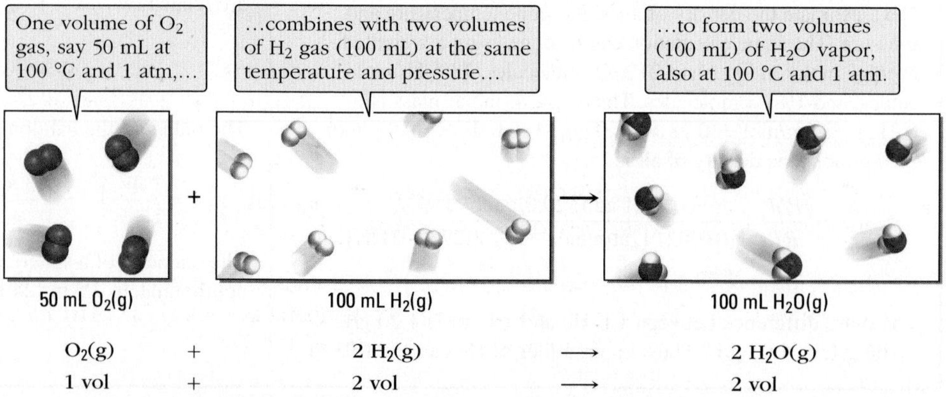
One volume of O_2 gas, say 50 mL at 100 °C and 1 atm,...
...combines with two volumes of H_2 gas (100 mL) at the same temperature and pressure...
...to form two volumes (100 mL) of H_2O vapor, also at 100 °C and 1 atm.
+
50 mL $O_2(g)$
100 mL $H_2(g)$
100 mL $H_2O(g)$
$O_2(g)$ + $2\ H_2(g)$ ⟶ $2\ H_2O(g)$
1 vol + 2 vol ⟶ 2 vol

Figure 8.7 Law of combining volumes. When gases react to form other gases, all at the same T and P, the gas volumes are proportional to the coefficients in the chemical equation.

PROBLEM-SOLVING PRACTICE 8.7

Nitrogen monoxide, NO, combines with oxygen to form nitrogen dioxide.

$$2\ NO(g) + O_2(g) \longrightarrow 2\ NO_2(g)$$

If 1.0 L oxygen gas at 30.25 °C and 0.975 atm is used and there is an excess of NO, calculate the volume of NO gas, at the same temperature and pressure, that will be converted to NO_2.

Michael Ventura/Alamy

Firefighters wearing self-contained self-rescue breathing devices.

Consider the application of the law of combining volumes and the ideal gas law to the generation of oxygen in self-contained self-rescue breathing devices such as those worn by firefighters and others who must enter environments that contain potentially toxic fumes. In such breathing devices, oxygen is generated by the reaction of potassium superoxide, KO_2, with exhaled carbon dioxide.

$$4\ KO_2(s) + 2\ CO_2(g) \longrightarrow 2\ K_2CO_3(s) + 3\ O_2(g)$$

Here are some questions to ask about this reaction. For each question, recognize what information is known and what is needed to answer the question. Then, decide what relationship(s) connects the two.

1. *How many liters of O_2 are produced from 0.50 L CO_2 and sufficient KO_2?* We know the volume of CO_2 and that there is sufficient KO_2. We need to determine the volume (L) of O_2 produced. Because both CO_2 and O_2 are gases, the coefficients of the balanced chemical equation represent volumes of these gases; we can use the law of combining volumes to calculate the result.

$$V(O_2) = 0.50\ \text{L}\ CO_2 \times \frac{3\ \text{L}\ O_2}{2\ \text{L}\ CO_2} = 0.75\ \text{L}\ O_2$$

2. *How many moles of KO_2 are required to react completely with 100. g CO_2?* We know the mass of one reactant (CO_2) and we need to calculate the amount of the other reactant, KO_2. To do so, we use the stoichiometric factors in the balanced chemical equation (← **Sec. 3-6**); no gas law calculations are necessary.

$$n(KO_2) = 100.\ \text{g}\ CO_2 \times \frac{1\ \text{mol}\ CO_2}{44.0\ \text{g}\ CO_2} \times \frac{4\ \text{mol}\ KO_2}{2\ \text{mol}\ CO_2} = 4.55\ \text{mol}\ KO_2$$

3. *How many grams of CO_2 are required to react with sufficient KO_2 to produce 5.00 L O_2 at STP?* The needed information is the mass of CO_2, which can be calculated once the amounts of O_2 and CO_2 are determined taking into account that 1 mol O_2 at STP has $V = 22.4$ L (the standard molar volume of any gas).

$$m(CO_2) = 5.00\ \text{L}\ O_2 \times \frac{1\ \text{mol}\ O_2}{22.4\ \text{L}\ O_2} \times \frac{2\ \text{mol}\ CO_2}{3\ \text{mol}\ O_2} \times \frac{44.0\ \text{g}\ CO_2}{1\ \text{mol}\ CO_2} = 6.55\ \text{g}\ CO_2$$

4. *What volume (L) of oxygen can be generated at 37 °C and 768 Torr by the complete reaction of 100. g CO_2?* The amount of O_2 can be calculated using stoichiometric factors as done in Part 2 above.

$$n(O_2) = 100.\ \text{g}\ CO_2 \times \frac{1\ \text{mol}\ CO_2}{44.0\ \text{g}\ CO_2} \times \frac{3\ \text{mol}\ O_2}{2\ \text{mol}\ CO_2} = 3.41\ \text{mol}\ O_2$$

The ideal gas law can be used to calculate the volume of O_2 generated, once the pressure is converted to atm (1.01 atm) and the temperature to kelvins (310 K) to have units consistent with those of R, 0.0812 L atm $\text{mol}^{-1}\ \text{K}^{-1}$.

$$V(O_2) = \frac{nRT}{P} = \frac{(3.41\ \text{mol})(0.0821\ \text{L atm mol}^{-1}\ \text{K}^{-1})(310\ \text{K})}{(1.01\ \text{atm})} = 85.9\ \text{L}\ O_2$$

5. *What is the minimum mass (g) of KO_2 required in the breathing apparatus if it is to provide sufficient oxygen for at least 3.00 h to the wearer who exhales at an average rate of 16.0 breaths a minute and exhales 20.0 mL CO_2 per breath at 37 °C and 1.02 atm?* First, determine the volume of CO_2 exhaled during the 3.00 h. From the volume use the ideal gas law to calculate the amount of CO_2 at the given temperature (310 K) and pressure (1.02 atm). Then, use that amount and stoichiometric factors to find the minimum amount and mass of KO_2.

$$V(\text{CO}_2\text{ exhaled}) = 3.00\text{ h} \times \frac{60.0\text{ min}}{1.00\text{ h}} \times \frac{16.0\text{ breaths}}{1.00\text{ min}} \times \frac{20.0\text{ mL CO}_2}{1\text{ breath}}$$

$$= 5.76 \times 10^4\text{ mL CO}_2 = 57.6\text{ L CO}_2$$

$$n(\text{CO}_2) = \frac{PV}{RT} = \frac{(1.02\text{ atm})(57.6\text{ L})}{(0.0821\text{ L atm mol}^{-1}\text{ K}^{-1})(310\text{ K})} = 2.31\text{ mol CO}_2$$

$$m(\text{KO}_2) = 2.31\text{ mol CO}_2 \times \frac{4\text{ mol KO}_2}{2\text{ mol CO}_2} \times \frac{71.1\text{ g KO}_2}{1\text{ mol KO}_2} = 327\text{ g KO}_2$$

The 327 g KO_2 corresponds to about 12 oz—three quarters of a pound.

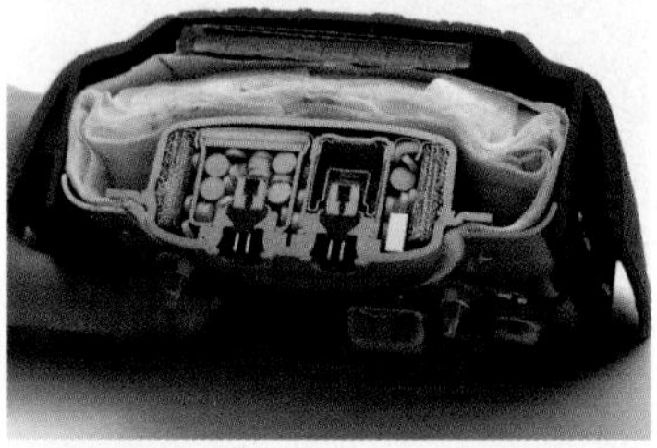

When a car decelerates in a collision, an electrical contact is made in the sensor unit. The propellant (green solid) detonates, releasing nitrogen gas, and the folded nylon bag explodes out of the plastic housing.

Driver's-side air bags inflate with 35–70 L of N_2 gas, whereas passenger air bags hold about 60–160 L.

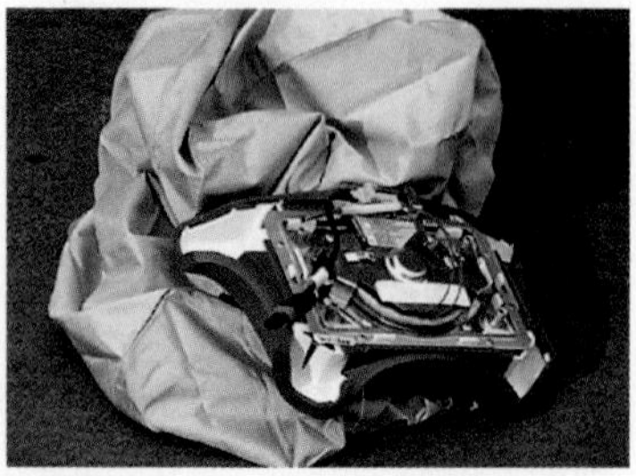

The bag deflates within 0.2 s, the gas escaping through holes in the bottom of the bag.

Photos: Autoliv/ASP - Kotz

PROBLEM-SOLVING EXAMPLE 8.8

Gas Stoichiometry and Automobile Air Bags

Within 40 milliseconds after a collision, automobile air bags deploy and inflate with nitrogen gas. The nitrogen is generated initially by the decomposition of sodium azide, NaN_3:

$$2\,\text{NaN}_3(s) \longrightarrow 2\,\text{Na}(\ell) + 3\,\text{N}_2(g) \qquad [8.3]$$

The residual sodium metal, however, is a hazard because it reacts with water, such as rain, to release flammable hydrogen gas. To prevent such an occurrence, potassium nitrate, KNO_3, in the air bag system reacts rapidly with the sodium metal to release additional nitrogen gas and form stable potassium and sodium oxides:

$$10\,\text{Na}(\ell) + 2\,\text{KNO}_3(s) \longrightarrow 5\,\text{Na}_2\text{O}(s) + \text{K}_2\text{O}(s) + \text{N}_2(g) \qquad [8.4]$$

The volumes of inflated air bags differ: 35–70 L on the driver's side; 60–160 L on the passenger's side. Based on Equation 8.3, calculate the mass of sodium azide required to generate sufficient nitrogen to inflate a driver's-side air bag to 55.0 L at 26.0 °C and 1.10 atm.

Result 107 g NaN_3

Analyze The volume, temperature, and pressure are specified, so the ideal gas law can be used to calculate the amount of N_2. A mole ratio from the balanced overall equation and the molar mass of sodium azide can be used to calculate the mass.

Plan First, use the ideal gas law to calculate the amount of N_2. Then, use the mole ratio 10 mol NaN_3 : 16 mol N_2 to calculate the amount of NaN_3 and use the molar mass of NaN_3 to calculate its mass.

Execute

$$n(\text{N}_2) = \frac{PV}{RT} = \frac{(1.10\text{ atm})(55.0\text{ L})}{(0.0821\text{ L atm mol}^{-1}\text{ K}^{-1})(299.2\text{ K})} = 2.463\text{ mol N}_2$$

$$m(\text{NaN}_3) = 2.463\text{ mol N}_2 \times \frac{2\text{ mol NaN}_3}{3\text{ mol N}_2} \times \frac{65.01\text{ g NaN}_3}{1\text{ mol NaN}_3} = 107\text{ g NaN}_3$$

☑ **Reasonable Result Check** The 107 g NaN_3 is about 1.7 mol, which would generate 1.5 times that amount of nitrogen, approximately 2.6 mol. This is close to the calculated value of 2.46 mol, so the result is reasonable.

PROBLEM-SOLVING PRACTICE 8.8

Based on the conditions described in Problem-Solving Example 8.8, calculate the minimum mass of potassium nitrate needed to react with all of the sodium formed by Equation 8.3. Also, determine the additional volume of nitrogen generated at 26 °C and 1.10 atm by the reaction in Equation 8.4.

EXERCISE 8.11

Air Bags

In Problem-Solving Example 8.8, two chemical equations are given for reactions that occur when an air bag inflates. Combine the two equations to form an overall balanced chemical equation. {The process is similar to combining equations when using Hess's law (← **Sec. 4-9**).} Use calculations based on the overall equation to calculate the total volume of N_2 produced by both reactions. Compare your result with the sum of the volumes calculated from Problem-Solving Example 8.8 and Problem-Solving Practice 8.8.

8-6 Gas Mixtures and Partial Pressures

Our atmosphere is a mixture of nitrogen, oxygen, argon, carbon dioxide, varying concentrations of water vapor, and small concentrations of several other gases. What we call atmospheric pressure is the sum of the pressures exerted by all these individual gases. The same is true of every gas mixture. Consider the mixture of nitrogen and oxygen illustrated on the right side in Figure 8.8. The pressure exerted by the mixture is equal to the sum of the pressures that the nitrogen alone and the oxygen alone exert in the same volume at the same temperature and pressure (left side of Figure 8.8). *The pressure that one gas in a mixture of gases would exert if it occupied the same volume by itself (at the same* T *and* n*) is called the* **partial pressure** *of that gas.*

John Dalton was the first to observe that ***the total pressure exerted by a mixture of gases is the sum of the partial pressures of the individual gases in the mixture.*** This

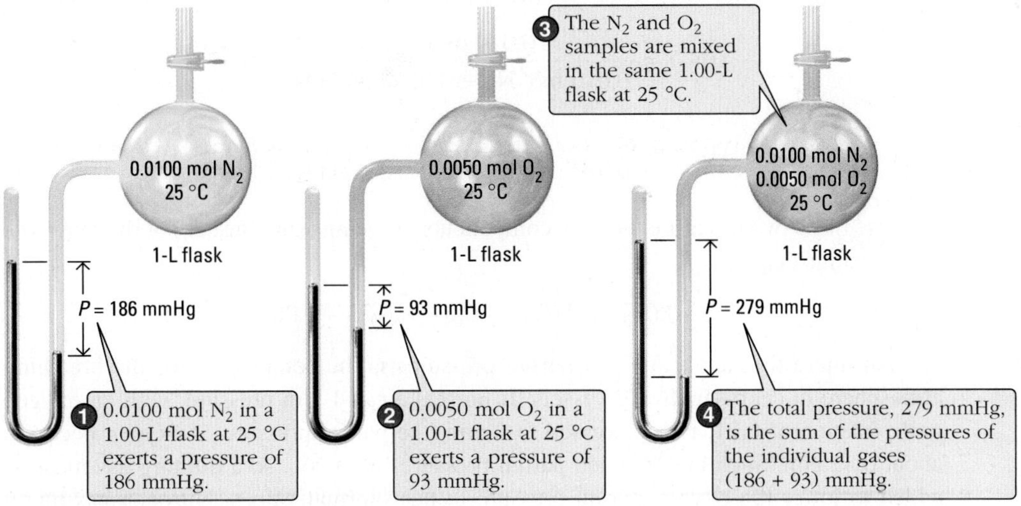

3 The N_2 and O_2 samples are mixed in the same 1.00-L flask at 25 °C.
0.0100 mol N_2 25 °C
1-L flask
P = 186 mmHg
0.0050 mol O_2 25 °C
1-L flask
P = 93 mmHg
0.0100 mol N_2 0.0050 mol O_2 25 °C
1-L flask
P = 279 mmHg
1 0.0100 mol N_2 in a 1.00-L flask at 25 °C exerts a pressure of 186 mmHg.
2 0.0050 mol O_2 in a 1.00-L flask at 25 °C exerts a pressure of 93 mmHg.
4 The total pressure, 279 mmHg, is the sum of the pressures of the individual gases (186 + 93) mmHg.

Figure 8.8 Dalton's law of partial pressures.

statement, known as **Dalton's law of partial pressures**, is a consequence of the fact that gas molecules behave independently of one another. As a demonstration of Dalton's law, consider the three main components of dry air, for which the total amount of gas is

$$n_{total} = n(N_2) + n(O_2) + n(Ar)$$

If we replace n in the ideal gas law with n_{total}, the sum of the individual amounts of gases, the equation becomes

$$P_{total}V = n_{total}RT$$

$$P_{total} = \frac{n_{total}RT}{V} = \frac{[n(N_2) + n(O_2) + n(Ar)]\,RT}{V}$$

Expanding the right side of this equation and rearranging gives

$$P_{total} = \frac{n(N_2)RT}{V} + \frac{n(O_2)RT}{V} + \frac{n(Ar)RT}{V} = P(N_2) + P(O_2) + P(Ar)$$

The quantities $P(N_2)$, $P(O_2)$, and $P(Ar)$ are the partial pressures of the three major components of the atmosphere. Dalton's law means that the pressure exerted by the atmosphere is the sum of the pressures due to nitrogen, oxygen, argon, and the other much less abundant components.

We can write a ratio of the partial pressure of one of the components, A, of a gas mixture over the total pressure,

$$\frac{P_A}{P_{total}} = \frac{n_A(RT/V)}{n_{total}(RT/V)}$$

After canceling R, T, and V terms on the right-hand side of this equation, we get

$$\frac{P_A}{P_{total}} = \frac{n_A}{n_{total}}$$

The ratio of the pressures is the same as the ratio of the amount of gas A, n_A, to the total amount of gas, n_{total}. *This ratio* (n_A/n_{total}) *is called the* **mole fraction** *of A and is given the symbol* X_A. Hence, rearranging the equation gives

$$P_A = X_A P_{total}$$

In the gas mixture in Figure 8.8, the mole fractions of nitrogen and oxygen are

$$X(N_2) = \frac{0.010 \text{ mol } N_2}{0.010 \text{ mol } N_2 + 0.0050 \text{ mol } O_2} = 0.67$$

$$X(O_2) = \frac{0.0050 \text{ mol } O_2}{0.010 \text{ mol } N_2 + 0.0050 \text{ mol } O_2} = 0.33$$

Because these two gases are the only components of the mixture, the sum of the two mole fractions must equal 1:

$$X(N_2) + X(O_2) = 0.67 + 0.33 = 1.00$$

An interesting application of partial pressures is the composition of the breathing atmosphere in deep-sea-diving vessels. If normal air at 1 atm pressure, with an oxygen mole fraction of 0.21, is compressed to 2 atm, the partial pressure of oxygen becomes about 0.42 atm. Such high oxygen partial pressures are toxic, so a diluting gas must be added to lower the oxygen partial pressure to near-normal values. Nitrogen gas might seem the logical choice because it is the diluting gas in the atmosphere. The problem is

The solubility of nitrogen (or any gas) in the blood increases in proportion to the partial pressure of nitrogen.

that nitrogen is fairly soluble in the blood at increased pressures, and at high concentrations causes *nitrogen narcosis,* a condition similar to alcohol intoxication.

Helium is much less soluble in the blood and is therefore a good substitute for nitrogen in a deep-sea-diving atmosphere. However, using helium leads to interesting side effects. As we will see in Section 8-7, He atoms, on average, move faster than N_2 molecules at the same temperature. Because He atoms strike a diver's skin more often than would the N_2 molecules He atoms are more efficient at transferring energy away. This effect causes divers to complain of feeling chilled while breathing a helium/oxygen mixture.

PROBLEM-SOLVING EXAMPLE 8.9

Calculating Partial Pressures

Mixtures of oxygen and helium are used in scuba diving tanks to provide the proper oxygen pressure for divers. Suppose that you want to fill a 10.0-L scuba tank with He and O_2 so that the partial pressure of the oxygen is 21.0% of the total pressure (the same as in air at 1.00 atm). You start with 56.0 L He at 25 °C and 1.00 atm. Calculate the volume (L) of O_2 at the same *T* and *P* that you should add to get the desired mixture. Calculate the partial pressure of each gas in the scuba tank.

Result $V(O_2) = 14.9$ L; $P(O_2) = 1.49$ atm; $P(\text{He}) = 5.60$ atm

Analyze At the same *T* and *V* (in the tank), the amount of each gas is proportional to the partial pressure of each gas. If the partial pressure of O_2 is 21.0% of the total pressure, then the amount of O_2 must be 21% of the total. Thus the mixture must contain 21.0% O_2 and 79.0% He on a mole basis. At the same *T* and *P* (before compressing the gases), *V* is proportional to *n*, so 56.0 L He is 79.0% of the total volume. This allows us to calculate the total volume of gas needed and the volume of O_2 by difference. The partial pressure of each gas in the tank can be calculated from the amount of each gas using the ideal gas law.

KevinPanizza/iStockphoto.com

Scuba diver under water.

Plan Calculate the total volume from the fact that 79.0% of the volume is 56.0 L. Calculate the volume of O_2 at 1.00 atm and 25 °C by subtracting the volume of He from the total volume. From each volume, calculate the amount of gas added to the tank; from each amount, calculate the partial pressure of each gas.

Execute

$$V_{\text{total}} = \frac{56.0\text{ L}}{0.790} = 70.9\text{ L}$$

$$V(O_2) = V_{\text{total}} - V(\text{He}) = 70.9\text{ L} - 56.0\text{ L} = 14.9\text{ L}$$

$$n(\text{He}) = \frac{PV}{RT} = \frac{(1.00\text{ atm})(56.0\text{ L})}{0.0821\text{ L atm mol}^{-1}\text{ K}^{-1})(298\text{ K})} = 2.29\text{ mol He}$$

$$P(\text{He}) = \frac{nRT}{V} = \frac{(2.29\text{ mol})(0.0821\text{ L atm mol}^{-1}\text{ K}^{-1})(298\text{ K})}{10.0\text{ L}} = 5.60\text{ atm}$$

$$n(O_2) = \frac{PV}{RT} = \frac{(1.00\text{ atm})(14.9\text{ L})}{0.0821\text{ L atm mol}^{-1}\text{ K}^{-1})(298\text{ K})} = 0.609\text{ mol } O_2$$

$$P(O_2) = \frac{nRT}{V} = \frac{(0.609\text{ mol})(0.0821\text{ L atm mol}^{-1}\text{ K}^{-1})(298\text{ K})}{10.0\text{ L}} = 1.49\text{ atm}$$

The partial pressure of oxygen in a scuba tank is the same under water as it is in air at atmospheric pressure.

☑ **Reasonable Result Check** The amount of O_2 needed is 21% of the initial mixture—about one fifth. Thus, four fifths is He and the volume of He should be about four times the volume of O_2, that is, $4 \times 15\text{ L} = 60\text{ L}$, which is close to our more exact calculation. Another check is that the final partial pressure of He should be about four times the final partial pressure of O_2, and it is.

PROBLEM-SOLVING PRACTICE 8.9

A mixture of 7.0 g N_2 and 6.0 g H_2 is confined in a 5.0-L reaction vessel at 500. °C. Assume that no reaction occurs and calculate the total pressure. Then, calculate the mole fraction and partial pressure of each gas.

CONCEPTUAL EXERCISE 8.12

Pondering Partial Pressures

What happens to the partial pressure of each gas in a mixture when the volume is decreased by (a) lowering the temperature or (b) increasing the total pressure?

EXERCISE 8.13

Partial Pressures

A sealed 355-mL flask contains 0.146 g neon gas, Ne, and an unknown amount of argon gas, Ar, at 35 °C and a total pressure of 626 mmHg. Calculate the mass, in grams, of Ar in the flask.

PROBLEM-SOLVING EXAMPLE 8.10

Partial Pressure of Air Pollutants

Nitrogen oxides will be discussed in Section 8-12b.

The reaction between nitrogen monoxide, NO, and oxygen gas to form NO_2 occurs in the atmosphere and can contribute to air pollution.

$$2\ NO(g) + O_2(g) \longrightarrow 2\ NO_2(g)$$

Suppose that the two reactant gases are kept in separate containers, as shown here. Then suppose the valve is opened and the reaction proceeds to completion. Assume a constant temperature of 25 °C.

(a) Determine what gases remain at the end of the reaction, and calculate the amount of each that is present.

(b) Calculate their partial pressures, and determine the total pressure in the system.

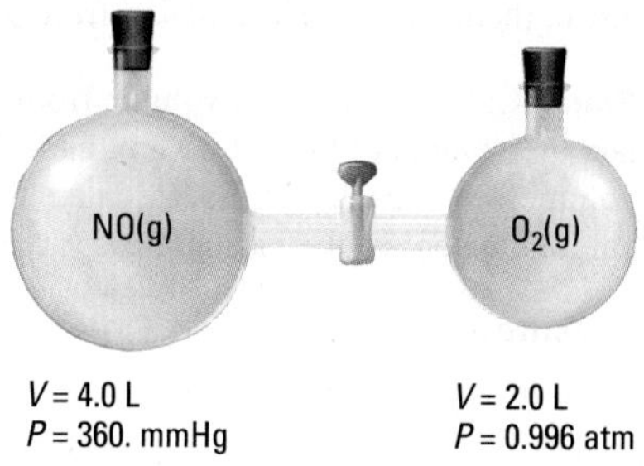

NO(g)
O₂(g)
V = 4.0 L
P = 360. mmHg
V = 2.0 L
P = 0.996 atm

Result

(a) No NO remains; 0.0427 mol O_2 and 0.0774 mol NO_2 remain

(b) $P_{NO_2} = 0.316$ atm; $P_{O_2} = 0.174$ atm; $P_{total} = 0.490$ atm

Analyze

(a) The product gas is certainly present after the reaction; if one reactant is limiting, then the other (excess) reactant gas will also be present. The mole ratio method (← **Sec. 3-7a**) can tell whether there is a limiting reactant so we need to know the amount of each reactant. The amounts of excess reactant and product can be calculated from the amount of limiting reactant and a mole ratio.

(b) Partial pressures are proportional to amounts and total pressure is the sum of partial pressures.

Plan

(a) From the initial *V* and *P* of each gas given in the figure, calculate the amount of each gas. Find the limiting reactant by comparing the ratio of actual amounts to the theoretical ratio from coefficients in the equation: 2 mol NO : 1 mol O_2. From the amount of limiting reactant, calculate the amount of the other reactant that reacted and subtract from the total amount to get the amount remaining. From the amount of limiting reactant, calculate the amount of product formed.

(b) Based on the amount of excess reactant remaining and the amount of product, use the ideal gas law to calculate the partial pressures. Sum the partial pressures.

Execute

(a) $$n(\text{NO}) = \frac{[(360./760)\ \text{atm}](4.0\ \text{L})}{(0.0821\ \text{L atm mol}^{-1}\ \text{K}^{-1})(298\ \text{K})} = 0.0774\ \text{mol NO}$$

$$n(\text{O}_2) = \frac{(0.996\ \text{atm})(2.0\ \text{L})}{(0.0821\ \text{L atm mol}^{-1}\ \text{K}^{-1})(298\ \text{K})} = 0.0814\ \text{mol O}_2$$

$$\text{Theoretical ratio: } \frac{2\ \text{mol NO}}{1\ \text{mol O}_2} \quad \text{Actual ratio: } \frac{0.0774\ \text{mol NO}}{0.0814\ \text{mol O}_2} = \frac{0.951\ \text{mol NO}}{1\ \text{mol O}_2}$$

NO is the limiting reactant because 0.951 mol is less than 2 mol.

$$n(\text{O}_2\ \text{reacted}) = 0.0774\ \text{mol NO} \times \frac{1\ \text{mol O}_2}{2\ \text{mol NO}} = 0.0387\ \text{mol O}_2$$

$$n(\text{O}_2\ \text{remaining}) = (0.0814 - 0.0387)\ \text{mol} = 0.0427\ \text{mol}$$

$$n(\text{NO}_2\ \text{formed}) = 0.0774\ \text{mol NO} \times \frac{2\ \text{mol NO}_2}{2\ \text{mol NO}} = 0.0774\ \text{mol NO}_2\ \text{formed}$$

(b) $$P(\text{NO}_2) = \frac{(0.0774\ \text{mol})(0.0821\ \text{L atm mol}^{-1}\ \text{K}^{-1})(298\ \text{K})}{6.0\ \text{L}} = 0.316\ \text{atm}$$

$$P(\text{O}_2) = \frac{(0.0427\ \text{mol})(0.0821\ \text{L atm mol}^{-1}\ \text{K}^{-1})(298\ \text{K})}{6.0\ \text{L}} = 0.174\ \text{atm}$$

$$P_{\text{total}} = 0.316\ \text{atm} + 0.174\ \text{atm} = 0.490\ \text{atm}$$

PROBLEM-SOLVING PRACTICE 8.10

Consider the case with the same reactants in the same vessels as in Problem-Solving Example 8.10, but with an NO pressure of 1.00 atm and O_2 pressure of 0.400 atm. Calculate the final pressure in the system after the reaction is complete.

Table 8.4 Vapor Pressure of Water at Different Temperatures

T (°C)	P_{water} (mmHg)
0	4.6
10	9.2
20	17.5
25	23.8
30	31.8
40	55.3
50	92.5
60	149.4
70	233.7
80	355.1
90	525.8
100	760.0

8-6a Collecting a Gas over Water

While studying chemical reactions that produce a gas as a product, it is often necessary to determine the amount of the product gas. One convenient way to do this, for gases that are insoluble in water, involves collecting the gas over water. Figure 8.9 shows how such an experiment can measure the volume of gas collected. The *total pressure of the mixture of gases in the tube equals the partial pressure of the gas being collected plus the partial pressure of the water vapor*. The partial pressure of water in the gaseous mixture is the vapor pressure of water, and it depends on the temperature of the liquid water as shown in Table 8.4.

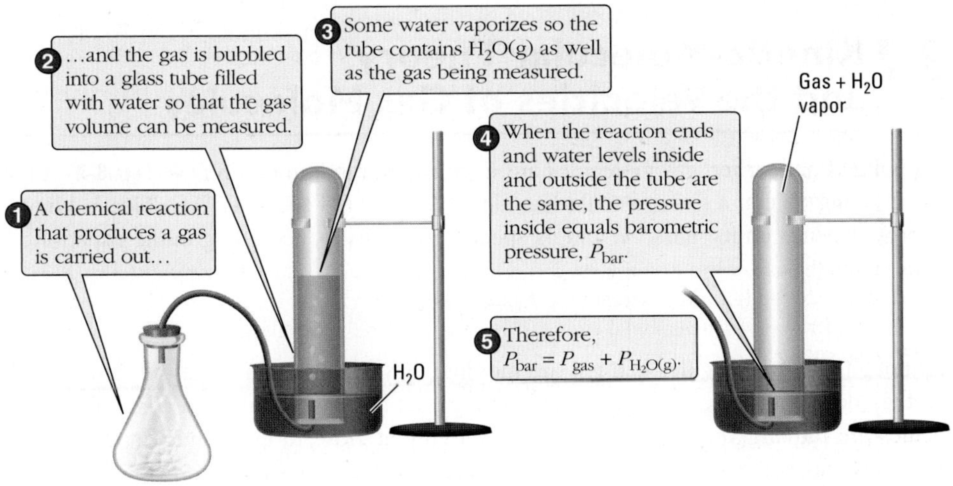

1 A chemical reaction that produces a gas is carried out…
2 …and the gas is bubbled into a glass tube filled with water so that the gas volume can be measured.
3 Some water vaporizes so the tube contains $H_2O(g)$ as well as the gas being measured.
4 When the reaction ends and water levels inside and outside the tube are the same, the pressure inside equals barometric pressure, P_{bar}.
5 Therefore, $P_{bar} = P_{gas} + P_{H_2O(g)}$.
H_2O
Gas + H_2O vapor

Figure 8.9 Collecting a gas over water.

To calculate the amount of gaseous product, we first subtract the vapor pressure of the water from the total gas pressure (the barometric pressure), to give the partial pressure of the gaseous product. Substituting this partial pressure and the known volume and temperature into the ideal gas law allows the calculation of the number of moles of the gaseous product.

H—C≡C—H

acetylene

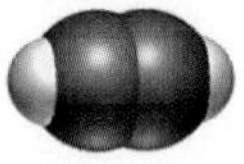

PROBLEM-SOLVING EXAMPLE 8.11

Collecting a Gas over Water

Before electric lamps were available, the reaction between calcium carbide, CaC_2, and water was used to produce acetylene, C_2H_2, to generate a bright flame in miners' lamps.

$$CaC_2(s) + 2\ H_2O(\ell) \longrightarrow C_2H_2(g) + Ca(OH)_2(aq)$$

At a barometric pressure of 745 mmHg and a temperature of 24.0 °C, 625 mL gas was collected over water. Calculate the number of milligrams of acetylene produced. The vapor pressure of water at 24.0 °C is 22.4 mmHg.

Result 635 mg acetylene

Analyze The gas is collected over water, so the measured pressure is the partial pressure of acetylene plus the vapor pressure of water. The amount of acetylene can be calculated from its partial pressure using the ideal gas law. The mass of acetylene can be obtained from the amount.

Plan Subtract the vapor pressure of water from the total measured pressure. Calculate the amount of acetylene using the ideal gas law and the mass using the molar mass. Change mass units to milligrams.

Execute $P_{\text{acetylene}} = P_{\text{total}} - P_{\text{water}} = 745\ \text{mmHg} - 22.4\ \text{mmHg} = 723\ \text{mmHg}$

$$n(C_2H_2) = \frac{PV}{RT} = \frac{[(723/760)\ \text{atm}](0.625\ \text{L})}{(0.0821\ \text{L atm mol}^{-1}\ \text{K}^{-1})(273.15 + 24.0)\ \text{K}} = 0.0244\ \text{mol}\ C_2H_2$$

$$m(C_2H_2) = 0.0244\ \text{mol} \times 26.04\ \text{g/mol} = 0.635\ \text{g} = 635\ \text{mg acetylene}$$

☑ **Reasonable Result Check** The quantities entered into the ideal gas equation can be approximated as $n = \dfrac{(1)(0.6)}{(0.082)(300)} = 0.02\ \text{mol}$, which is close to our more exact result.

PROBLEM-SOLVING PRACTICE 8.11

Zinc metal reacts with HCl to produce hydrogen gas, H_2.

$$2\ HCl(aq) + Zn(s) \longrightarrow ZnCl_2(aq) + H_2(g)$$

The H_2 can be collected over water. If you collected 260. mL H_2 at 25 °C, and the total pressure was 740. mmHg, calculate how many milligrams of H_2 were collected.

8-7 Kinetic-Molecular Theory and the Velocities of Gas Molecules

As pointed out earlier, gas molecules are in constant, random motion (← **Sec. 8-2**). Like any moving object, a gas molecule has kinetic energy (← **Sec. 4-1**). An object's kinetic energy depends on its mass, m, and its speed or velocity, v, according to the equation

$$E_k = \frac{1}{2}mv^2$$

Although all the molecules in a gas are moving, they do not all move at the same speed, so they do not all have the same kinetic energy. At any given time in a gas, a few molecules are moving very quickly, most are moving at close to the average speed, and a few may be in the process of colliding with a surface, in which case their speed is

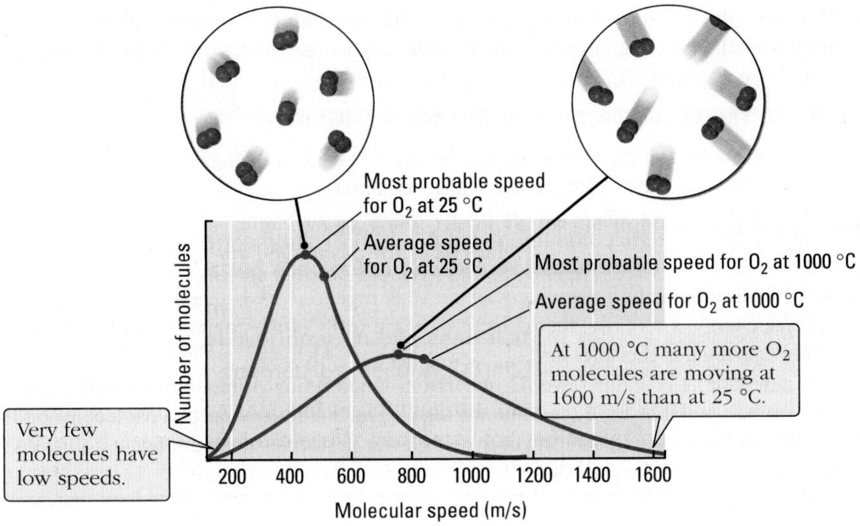
Most probable speed for O_2 at 25 °C
Average speed for O_2 at 25 °C
Most probable speed for O_2 at 1000 °C
Average speed for O_2 at 1000 °C
At 1000 °C many more O_2 molecules are moving at 1600 m/s than at 25 °C.
Very few molecules have low speeds.
Number of molecules
200 400 600 800 1000 1200 1400 1600
Molecular speed (m/s)

Figure 8.10 Distribution of molecular speeds. The number of gas molecules with a given speed is plotted versus that speed.

momentarily zero. The speed of any individual molecule changes from time to time as it collides with and exchanges energy with other molecules.

8-7a Velocities of Gas Molecules

The relative number of molecules that have a given speed can be measured experimentally. Figure 8.10 is a graph of the number of gas molecules that have a particular speed plotted versus speed. The higher a point is on the curve, the greater the number of molecules moving at that speed. Notice in the plot that some oxygen molecules are moving quickly (have high kinetic energy) while others are moving slowly (have low kinetic energy). The *maximum* (peak) in the distribution curve is the *most probable* speed. For oxygen gas at 25 °C, for example, the maximum in the curve occurs at a speed of about 450 m/s (1000 mph); most of the oxygen molecules have speeds in the range from 200 m/s to 700 m/s. The curves are not symmetric due to the fact that molecules with speeds higher than the most probable speed outnumber those with speeds lower than the most probable speed. As a consequence, the average speed is shifted toward a somewhat higher value and the average speed (shown in Figure 8.10 for O_2) is a little faster than the most probable speed.

The areas are the same under the two curves in Figure 8.10 representing the two oxygen gas samples at different temperatures because the total number of oxygen molecules is the same in both samples. However, the distributions of the speeds of the oxygen molecules differ at the different temperatures. Note in Figure 8.10 that as *temperature increases,* the *most probable speed increases,* and the *number of molecules moving very quickly increases.*

At a given temperature, the average kinetic energy ($\frac{1}{2}$ mv^2) *of the molecules of one gas is the same as for any other gas.* For this to be so, the larger *m* is, the smaller the average *v* must be. That is, *at the same temperature, heavier gas molecules have slower average speed than lighter ones.* Figure 8.11 illustrates this relationship for several gases of differing molar mass. From the graph you can also see that average speeds range from a few hundred to a few thousand meters per second. However, the peaks in the curves (most probable speed) for the heavier molecules, O_2 and N_2 in this case, occur at a much lower speed than that for helium, the lightest gas present. As usual, the average speed for each type of molecule is a little faster than its most probable speed because the curves are not symmetric.

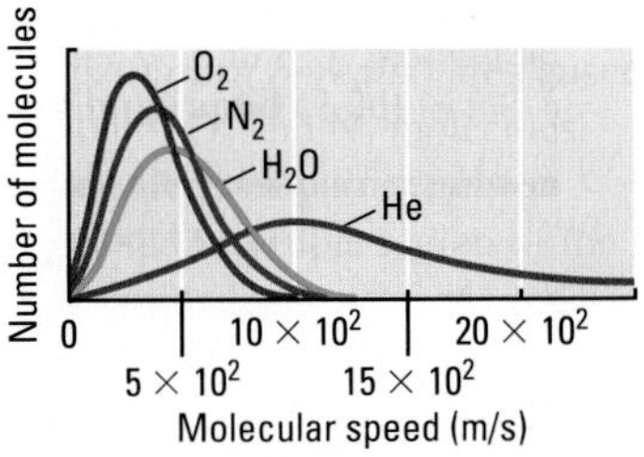
O_2
N_2
H_2O
He
Number of molecules
0
5×10^2
10×10^2
15×10^2
20×10^2
Molecular speed (m/s)

Figure 8.11 The effect of molar mass on the distribution of molecular speeds at a given temperature. On average, heavier molecules move slower than lighter molecules.

PROBLEM-SOLVING EXAMPLE 8.12

Kinetic-Molecular Theory

Consider a sample of O_2 gas at 25 °C in a rigid container. Use the kinetic-molecular theory to answer these questions.

(a) How does the pressure of the gas change if the temperature is raised to 40 °C?
(b) How does the pressure change if some of the oxygen molecules are removed from the sample (at constant T)?

Result (a) The pressure increases. (b) The pressure decreases.

Analyze According to the kinetic-molecular theory (← **Sec. 8-2**) gas pressure results from collisions of gas molecules with the walls of a container.

Plan Apply the kinetic-molecular theory to each situation.

Execute
(a) At a higher temperature the molecules move at higher average speed and have higher average energy. Therefore, the molecules hit the walls of the container harder and more often, causing a higher pressure.
(b) The new sample contains fewer O_2 molecules than originally, but they are at the same temperature and thus have the same average speed and kinetic energy. Molecules hit the walls with the same impact; however, with fewer molecules there are fewer collisions with the walls and the pressure is lower.

PROBLEM-SOLVING PRACTICE 8.12

(a) How does kinetic-molecular theory explain the change in pressure when the volume of a sample of gas decreases while the temperature remains constant?
(b) How does kinetic-molecular theory explain the change in pressure when gas molecules are added to a sample of gas in a rigid container at constant temperature?

EXERCISE 8.14

Average Kinetic Energies

Arrange these gaseous substances in order of increasing average kinetic energy of their molecules at 25 °C: Cl_2, H_2, NH_3, and SF_6.

A molecule is the smallest particle of a substance that exists independently; for noble gases like He and Ar, the smallest particle is an atom.

CONCEPTUAL EXERCISE 8.15

Molecular Kinetic Energies

Using Figure 8.11 as a guide, plot the number of molecules versus molecular speed for a sample of helium at 25 °C. Now assume that an equal number of molecules of argon, also at 25 °C, are added to the helium. What would the distribution curve for the mixture of gases look like (a single distribution curve for all molecules in the mixture)?

CONCEPTUAL EXERCISE 8.16

Molecular Motion

Suppose you have two helium-filled balloons of about equal size. You put one of them in the freezer compartment of your refrigerator and leave the other one out in your room. After a few hours you take the balloon from the freezer and compare it to the one left out in the room. Based on the kinetic-molecular theory, what differences would you expect to see (a) immediately after taking the balloon from the freezer and (b) after the cold balloon warms to room temperature?

8-7b Effusion, Diffusion, and Graham's Law

The dependence of molecular speeds of gases on their molecular mass can be used to explain some interesting phenomena. One is **effusion**, the escape of gas molecules from a container through a tiny hole into a vacuum. *Lighter molecules effuse more readily than heavier ones.* For example, a balloon filled with He (lower molar mass, M) deflates faster than one filled with N_2, which consists of heavier molecules (greater M). According to

Graham's law, *the rate of effusion of a gas is inversely proportional to the square root of its molar mass.*

$$\text{Rate of effusion} \propto \frac{1}{\sqrt{M}}$$

The rates of effusion of two gases in a mixture at the same temperature and pressure can be expressed as a ratio

$$\frac{\text{Rate}_1}{\text{Rate}_2} = \frac{\sqrt{M_2}}{\sqrt{M_1}}$$

For a gaseous mixture of helium and nitrogen, the ratio is

$$\frac{\text{Rate(He)}}{\text{Rate(N}_2\text{)}} = \frac{\sqrt{M(\text{N}_2)}}{\sqrt{M(\text{He})}} = \frac{\sqrt{28 \text{ g/mol}}}{\sqrt{4.0 \text{ g/mol}}} = \sqrt{\frac{28}{4.0}} = \sqrt{7.0} = 2.6$$

Consequently, helium effuses 2.6 times more rapidly than nitrogen at the same temperature and pressure.

A related phenomenon is **diffusion**, the spread of gas molecules of one type through those of another type, for example, a pleasant odor spreading throughout air in a room. Diffusion is more complicated than effusion because the diffusing gas molecules of one type are colliding with the second type of gas molecules. Nevertheless, diffusion follows the same relationship between rate and molar mass as effusion. Diffusion of uranium hexafluoride gas was employed in the Manhattan Project during World War II to purify uranium for the atomic bomb. Gaseous $^{235}UF_6$ diffuses slightly faster than $^{238}UF_6$ gas molecules due to the differences in the masses of the two uranium isotopes.

EXERCISE 8.17

Graham's Law

Calculate the separation ratio of uranium-235 and uranium-238 hexafluorides based on their relative rates of diffusion. Are the two types of uranium hexafluorides separated readily by this method? Explain your answer.

8-8 The Behavior of Real (Non-ideal) Gases

The ideal gas law provides accurate predictions for the pressure, volume, temperature, and amount of a gas, provided the temperature is well above its boiling point and the pressure is near 1 atm. At STP (0 °C and 1 atm), most substances that are gases at room temperature deviate only slightly from ideal behavior. At much higher pressures or much lower temperatures, however, the ideal gas law does not work nearly as well. We can illustrate this departure from ideality by plotting PV/nRT for a gas as a function of pressure (Figure 8.12). For a gas that follows ideal gas behavior, $PV = nRT$, the ratio PV/nRT must equal 1. But Figure 8.12 shows that for $N_2(g)$ and $CH_4(g)$, the ratio PV/nRT deviates from 1, first dipping to values lower than 1 and then rising to values higher than 1 as pressure increases. Thus, the product of the measured pressure times the measured volume of the real gas is smaller than expected at medium pressures, and it is higher than expected at high pressures.

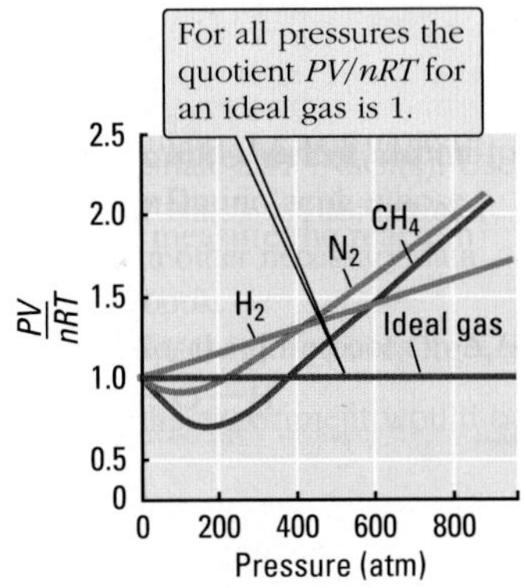

For all pressures the quotient PV/nRT for an ideal gas is 1.
2.5
2.0
1.5
1.0
0.5
0
PV/nRT
CH4
N2
H2
Ideal gas
0
200
400
600
800
Pressure (atm)

Figure 8.12 The non-ideal behavior of real gases compared with an ideal gas.

To see what causes these deviations, we must revisit two of the fundamental concepts of kinetic-molecular theory (← Sec. 8-2). The theory assumes that the molecules of a gas occupy no appreciable volume themselves and that gas molecules have no appreciable attraction for one another.

At STP, the volume occupied by a single molecule is very small relative to its share of the total gas volume. Recall that there are 6.02×10^{23} molecules in a mole and that

1 mol of a gas occupies 22.4 L (22.4×10^{-3} m^3) at STP (← **Sec. 8-3d**). The volume, V, that each molecule has to move around in is given by

$$V = \frac{22.4 \times 10^{-3}\ \text{m}^3}{6.02 \times 10^{23}\ \text{molecules}} = 3.72 \times 10^{-26}\ \text{m}^3/\text{molecule}$$

Assume that the volume available for a molecule to move around in is a sphere. The volume is given by $V = \frac{4}{3}\pi r^3$. Solving for r gives

$$r = \sqrt[3]{\frac{3V}{4\pi}} = \sqrt[3]{\frac{3(3.72 \times 10^{-26}\ \text{m}^3)}{4(3.14)}}$$

$$r = 2.07 \times 10^{-9}\ \text{m} = 2070\ \text{pm}$$

If this volume is assumed to be a sphere, then the radius, r, of the sphere is about 2000 pm. The radius of the smallest gas molecule, the helium atom, is 31 pm (← **Sec. 5-9a**) so a helium atom has a space to move around in that is similar to the room a pea has inside a basketball. Now suppose the pressure is increased significantly, to 1000 atm. The volume available to each molecule is now a sphere only about 200 pm in radius, which means the situation is now like that of a pea inside a sphere a bit larger than a table tennis ball.

A pea in a basketball corresponds roughly to the relative volume that gas molecules have to move about in without striking another gas molecule at 0 °C and 1 atm.
A pea in a table tennis ball corresponds roughly to the relative volume that gas molecules have to move about in without striking another gas molecule at 0 °C and 1000 atm.

The volume occupied by the gas molecules themselves is no longer negligible compared to the volume of the sphere. This violates the first postulate of the kinetic-molecular theory. In the kinetic-molecular theory and the ideal gas law, V is the volume available for the molecules to move around in, not the volume of the molecules themselves. However, the measured volume of the gas must include both. Therefore, at very high pressures, the measured volume is larger than predicted by the ideal gas law, and the value of PV/nRT is greater than 1.

At medium pressures, Figure 8.12 shows that the product PV for a real gas is smaller than the ideal gas value of nRT. The kinetic-molecular theory assumes that forces between molecules are negligible, but as pressure increases from STP, molecules become close enough to each other for noncovalent intermolecular forces to be important. Figure 8.13 shows how such forces cause molecules to hit container walls with less impact, reducing the measured pressure so that PV/nRT is less than 1. Thus,

- **the measured pressure for a real gas is less than the ideal pressure due to intermolecular attractive forces.**
- **the measured volume of a real gas is larger than the ideal volume due to the volume of the molecules themselves.**

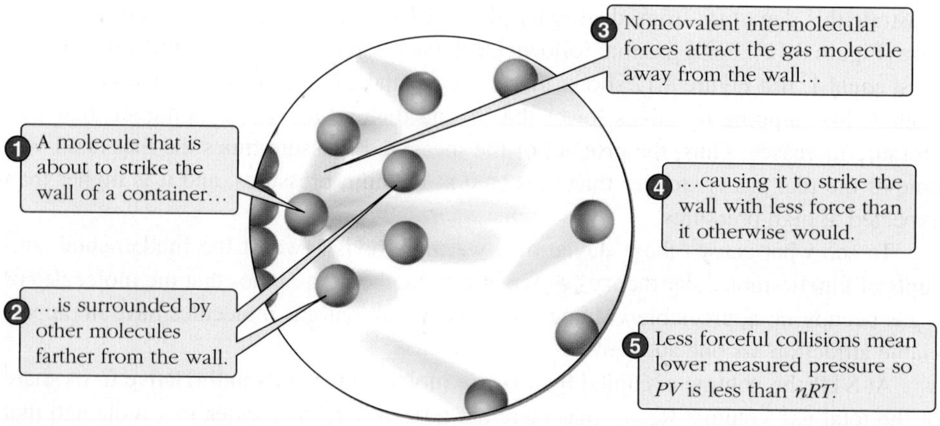
1 A molecule that is about to strike the wall of a container...
2 ...is surrounded by other molecules farther from the wall.
3 Noncovalent intermolecular forces attract the gas molecule away from the wall...
4 ...causing it to strike the wall with less force than it otherwise would.
5 Less forceful collisions mean lower measured pressure so PV is less than nRT.

Figure 8.13 Non-ideal gas behavior. Intermolecular forces make measured pressure lower.

The **van der Waals equation** predicts the behavior of a real gas quantitatively by taking account of intermolecular attraction and non-negligible molecular volume. It assumes that $P_{ideal} V_{ideal} = nRT$, but expresses the left-hand side in terms of *measured P* and *V*, which are corrected to give the ideal values.

$$\left(P_{measured} + \frac{n^2 a}{V^2_{measured}}\right)(V_{measured} - nb) = nRT$$

Correction for molecular attraction (adjusts measured *P* up)

Correction for volume of molecules (adjusts measured *V* down)

Table 8.5 Van der Waals Constants for Some Common Gases

Gas	$a\left(\frac{L^2\ atm}{mol^2}\right)$	$b\left(\frac{L}{mol}\right)$
He	0.034	0.0237
Ne	0.211	0.0171
Ar	1.35	0.0322
H_2	0.244	0.0266
N_2	1.39	0.0391
O_2	1.36	0.0318
Cl_2	6.49	0.0562
CO_2	3.59	0.0427
CH_4	2.25	0.0428
NH_3	4.17	0.0371
H_2O	5.46	0.0305

The measured volume is decreased by the factor nb, which accounts for the volume occupied by n moles of the gas molecules themselves. The van der Waals constant b has units of L/mol, and it is larger for larger molecules (see Table 8.5). The measured pressure is increased by the factor $(n^2/V^2)a$, which accounts for the reduction in pressure due to attractive forces between molecules. The van der Waals constant a has units of L^2 atm/mol^2; it is larger the more strongly the gas molecules attract one another. The a pressure correction term is proportional to the square of the number of molecules per unit volume, $(n/V)^2$, because the attractive forces are between *pairs* of molecules. The constants a and b are different for each gas (Table 8.5) and must be determined experimentally.

For an ideal gas, where the kinetic-molecular theory of gases holds, the attractive forces between molecules are negligible, so $a \cong 0$, and $P_{measured} = P_{ideal}$. The volume of the molecules themselves is negligible compared with the container volume, so $b \cong 0$ and $V_{measured} = V_{ideal}$. Therefore, for ordinary conditions of P and T, the van der Waals equation simplifies to the ideal gas law.

PROBLEM-SOLVING EXAMPLE 8.13

Using the van der Waals Equation

A 2.50-mol sample of CO_2 gas is in a sealed 2.00-L container at 0 °C. Calculate the pressure using the ideal gas law and using the van der Waals equation. Then, calculate the percentage difference between the two as a fraction of the ideal gas pressure.

Result Ideal P = 28.0 atm; van der Waals P = 24.0 atm; 14.3% difference

Analyze The amount, volume, and temperature are given. Constants a and b for the van der Waals equation must be looked up. Use the definition of percent: per 100.

Plan Calculate P using the ideal gas law and the van der Waals equation. Subtract one pressure from the other, divide the difference by the ideal gas pressure, and multiply by 100%.

Execute

$$P = \frac{nRT}{V} = \frac{(2.50\ \text{mol})(0.0821\ \text{L atm mol}^{-1}\ \text{K}^{-1})(273\ \text{K})}{2.00\ \text{L}} = 28.0\ \text{atm}$$

$$P = \left(\frac{nRT}{V - nb}\right) - \left(\frac{n^2 a}{V^2}\right)$$

$$P = \left(\frac{(2.50\ \text{mol})(0.0821\ \text{L atm mol}^{-1}\ \text{K}^{-1})\ (273\ \text{K})}{(2.00\ \text{L}) - (2.50\ \text{mol})(0.0427\ \text{L/mol})}\right) - \left(\frac{(2.50\ \text{mol})^2(3.59\ \text{L}^2\ \text{atm/mol}^2)}{(2.00\ \text{L})^2}\right) = 24.0\ \text{atm}$$

$$\text{Percentage difference in pressure} = \frac{28.0\ \text{atm} - 24.0\ \text{atm}}{28.0\ \text{atm}} \times 100\% = 14.3\%$$

The first term in the calculation of P, the one containing the constant b, equals 29.60 atm, which differs from the ideal gas pressure by 1.6 atm. The second term, containing the constant a, equals 5.61 atm. Thus, the term with a that depends on intermolecular forces accounts for most of the difference in pressure calculated by the van der Waals equation compared with the ideal gas law equation in this case.

PROBLEM-SOLVING PRACTICE 8.13

Calculate the percentage difference between the pressures predicted by the ideal gas law and the van der Waals equation if the volume in Problem-Solving Example 8.13 is increased by a factor of 10.0 to 20.0 L.

EXERCISE 8.18

Errors Caused by Deviations from Ideal Gas Behavior

In Problem-Solving Example 8.6 (← **Sec. 8-4**), the molar mass calculated using the ideal gas law, 139 g/mol, was slightly larger than that calculated using atomic weights, 137.9 g/mol, for the formula P_2F_4. Study the calculation in Problem-Solving Example 8.6 and explain why the amount is less than it should be.

CONCEPTUAL EXERCISE 8.19

van der Waals Constants

Look up the van der Waals constants, b, for H_2, N_2, O_2, and Cl_2. Based on the periodic table, predict atomic radii for H, N, O, and Cl. Use these values to explain the sizes of the b constants.

8-9 The Atmosphere

NASA

Earth.

Earth is enveloped by a few vertical miles of gaseous atoms and molecules that compose its atmosphere. Because of the atmosphere's mass, approximately 5.3×10^{15} metric tons (1 metric ton = 1000 kg = 2200 lb), Earth's surface experiences the atmosphere's pressure (← **Sec. 8-1**). The two major substances in our atmosphere are nitrogen, N_2, a rather unreactive gas, and oxygen, O_2, a highly reactive one. The composition of dry air at sea level is listed in Table 8.6. For every 100 L of air, 21 L are oxygen and 78 L are nitrogen;

Table 8.6 The Composition of Dry Air at Sea Level*

Gas	Percentage by Volume	Gas	Percentage by Volume
Nitrogen	78.080	Krypton	0.0001
Oxygen	20.947	Carbon monoxide‡	0.00001
Argon	0.934	Xenon	0.000008
Carbon dioxide†	0.039	Ozone‡	0.000002
Neon	0.00182	Ammonia	0.000001
Hydrogen	0.0010	Nitrogen dioxide‡	0.0000001
Helium	0.00052	Sulfur dioxide‡	0.00000002
Methane†	0.0002		

*The data in this table are for dry air because the percentage of water vapor in air varies from trace quantities in desert areas to as much as 4% over the oceans. On average, water makes up 1.5% of the atmosphere—the third most abundant substance and a very important component.

†The greenhouse gases carbon dioxide and methane are discussed in Section 8-11 as they relate to fuels and the burning of fuels for energy production.

‡Trace gases of environmental importance.

ESTIMATION | Thickness of Earth's Atmosphere

In the photograph of Earth nearby, Earth's diameter is about 5 cm. About 99.9% of the atmosphere is contained within 50 km of the Earth's surface. If you were to draw a circle around the photo properly relating the diameter of the Earth with the thickness of the atmosphere, how thick should the circle be drawn?

Earth's diameter is about 7900 miles, which is 12,700 km. Set up an equation that equates the ratio of the diameter of the photo to the diameter of Earth with the ratio of the sought line width (x) to the thickness of the atmosphere.

$$\frac{5 \text{ cm}}{12{,}700 \text{ km}} = \frac{x}{50 \text{ km}}$$

$$x = 0.02 \text{ cm} = 0.2 \text{ mm}$$

If the circle representing Earth is 5 cm in diameter, then the line width that properly shows the relative thickness of the atmosphere would be only 0.2 mm, less than the width of the period at the end of this sentence. The proportions are about the same as the skin on an apple in relation to the apple. Earth's atmosphere is quite thin.

except for water, which varies widely, each of the other gases is less than 1% by volume. Note that the atmosphere is our major source of nitrogen and our only source for most of the noble gases—neon, argon, krypton, and xenon.

8-9a The Troposphere and the Stratosphere

The mass of Earth's atmosphere, although large, is only about one millionth the total mass of Earth. The atmosphere's mass, 5.3×10^{15} metric tons, is equivalent to a cube of water 100 km (62 mi) on a side.

Earth's atmosphere can be roughly divided into layers, as shown in Figure 8.14. From the surface to an altitude of about 10 km, in the region named the **troposphere**, the temperature of the atmosphere decreases with increasing altitude. In this region the most violent mixing of air and the biggest variations in moisture content and temperature occur. Winds, clouds, storms, and precipitation are the result, the phenomena we know as weather. The troposphere is where we live. A commercial jet airplane flying at an altitude of about 10 km (approximately 33,000 ft) is still in the troposphere, although near its upper limits. The composition of the troposphere is roughly that of dry air near sea level (see Table 8.6), but the concentration of water vapor varies considerably, with an average of about 1.5%.

Just above the troposphere, from about 12 to 50 km above Earth's surface, is the **stratosphere**. The pressures in the stratosphere are extremely low, and little mixing occurs between the stratosphere and the troposphere. The lower limit of the stratosphere varies from night to day over the globe, and in the polar regions it may be as low as 8 to 9 km above Earth's surface. About 75% of the mass of the atmosphere is in the troposphere, and 99.9% of the atmosphere's mass is below 50 km, in the troposphere and stratosphere.

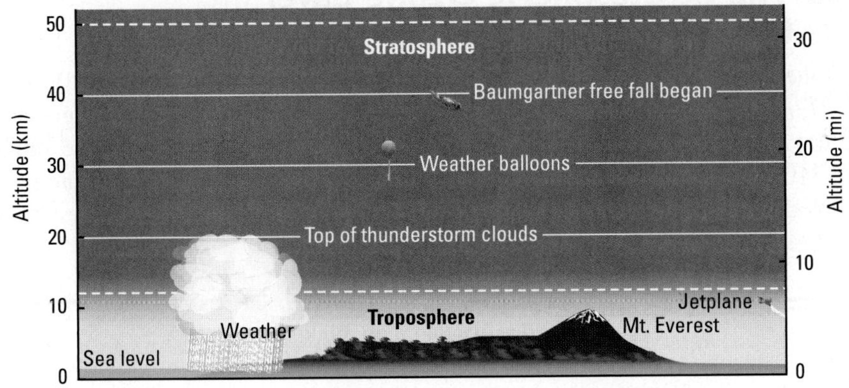

Altitude (km)
50
40
30
20
10
0
Altitude (mi)
30
20
10
0
Stratosphere
Baumgartner free fall began
Weather balloons
Top of thunderstorm clouds
Jetplane
Troposphere
Mt. Everest
Weather
Sea level

Figure 8.14 Regions of Earth's atmosphere.

Red Bull Stratos/EPA/Newscom

At an altitude of 39 km, near the top of the stratosphere, Felix Baumgartner steps from a capsule towed aloft by a weather balloon and begins his record-setting free fall (Oct. 14, 2012).

8-10 Ozone and Stratospheric Ozone Depletion

Ozone, O_3, is an important substance in the stratosphere, and its presence there is essential for life on Earth. A hybrid of two resonance structures, ozone has a bent (angular) geometry (← **Sec. 6-9**) with a bond angle of 116.8°.

Ozone resonance structures

The $\Delta_f H°$ of 142.3 kJ/mol shows that ozone is a higher-energy molecule than O_2 (which is assigned an enthalpy of formation of zero).

$$\tfrac{3}{2}\,O_2(g) \longrightarrow O_3(g) \qquad \Delta_f H° = 142.3 \text{ kJ/mol}$$

Ozone molecules are highly reactive and do not survive long after they form.

The *region of maximum ozone concentration in the upper atmosphere occurs at 25–30 km* and is called the **stratospheric ozone layer**. Under natural conditions, the stratospheric ozone concentration is kept at a relatively constant level by two competing, interrelated reactions—one that produces stratospheric ozone and an opposing one that destroys it at the same rate. The two linked processes constitute a natural cycle called the *Chapman cycle,* named for Sydney Chapman, the physicist who first proposed it in 1929.

Oxygen atoms are written $\cdot O \cdot$ because the electron configuration of an oxygen atom is [He] $2s^2 2p_x^2 2p_y^1 2p_z^1$ so $\cdot O \cdot$ is a free radical (← Sec. 6-10b) with two unpaired electrons, each written as a separate dot.

The bond energy of O_2 is 498 kJ/mol, so a calculation using $E = h\nu = hc/\lambda$ shows that the energy in a photon of $\lambda < 240$ nm is needed to break the O_2 bond.

Natural formation of O_3

$$O_2(g) \xrightarrow[\lambda < 240 \text{ nm}]{\text{UV radiation}} \cdot O \cdot (g) + \cdot O \cdot (g)$$

$$\cdot O \cdot (g) + O_2(g) \longrightarrow O_3(g)$$

Natural destruction of O_3

$$O_3 + \cdot O \cdot (g) \longrightarrow 2\,O_2(g)$$

Because ozone molecules and oxygen atoms are at very low concentrations in the stratosphere, the destruction reaction is relatively slow. However, it consumes ozone fast enough to maintain a balance with the ozone-forming reaction so that under normal conditions, stratospheric ozone concentration remains relatively constant. Even though approximately 4×10^8 tons of stratospheric ozone are formed and destroyed daily by these two processes, the actual amount of it in the stratosphere is surprisingly small. If all the stratospheric ozone molecules were brought together at sea level (STP = 1 atm, 273 K) and spread evenly, they would form a layer approximately only 3 mm thick on Earth's surface, about the thickness of three credit cards.

8-10a The Stratospheric Ozone Layer Protects Earth's Surface

Stratospheric ozone is important because it absorbs ultraviolet radiation. The UV *photons cause O_3 molecules to break apart,* a **photodissociation** reaction. This prevents damaging ultraviolet radiation from reaching Earth's surface. Wavelengths in the range of 200 to 310 nm are absorbed during the decomposition reaction. The oxygen atoms from the photodissociation of O_3 react with O_2 to regenerate O_3, so this reaction results in no net O_3 loss.

$$O_3(g) \xrightarrow[\lambda\ 200 \text{ to } 310 \text{ nm}]{\text{UV radiation}} \cdot O \cdot (g) + O_2(g)$$

Absorption of ultraviolet radiation in the stratosphere is *essential* for living things on Earth. Stratospheric ozone prevents 95% to 99% of the sun's ultraviolet radiation from reaching Earth's surface. Photons in this 200- to 310-nm range have enough energy to cause skin cancer in humans and damage to living plants. For every 1% decrease in the stratospheric ozone, an additional 2% of this damaging radiation reaches Earth's surface. Stratospheric ozone depletion therefore has the potential to drastically damage our environment.

F. Sherwood Rowland
1927–2012

HAL GARB/AFP/Getty Images

Born in Ohio, Sherwood Rowland entered Ohio Wesleyan University at age 16. He received his Ph.D. from the University of Chicago in 1952. He held faculty appointments at several institutions and joined the newly formed University of California at Irvine in 1964 as Chair of the Chemistry Department. In 1974 he and Mario Molina recognized that CFCs could deplete the ozone layer in the atmosphere. Their work was received skeptically at first, but it was confirmed by others in later experiments, notably in 1985 by satellite data showing an "ozone hole" over Antarctica. Rowland, Molina, and Paul Crutzen shared the 1995 Nobel Prize in Chemistry ". . . for their work in atmospheric chemistry, particularly concerning the formation and decomposition of ozone."

8-10b Chlorofluorocarbons

A **chlorofluorocarbon (CFC)** is a *small molecule with halogen atoms bonded to a central carbon.* CFCs play a major role in stratospheric ozone depletion. Specific examples of CFCs include $CFCl_3$ (CFC-11) and CF_2Cl_2 (CFC-12). These compounds are nonflammable, relatively inert, volatile yet readily liquefied, and nontoxic—an extremely useful set of properties. CFCs have been used as coolants for refrigeration, in foam plastics manufacture, in aerosol spray cans as propellants, and as industrial solvents. Unfortunately, the properties that make them useful are also the reason for their destructive effect on stratospheric ozone. Once gaseous CFCs are released into the atmosphere, they persist for a very long time in the troposphere because of their chemical nonreactivity. On average, a CF_2Cl_2 molecule survives roughly 100 years in the troposphere. Eventually, atmospheric mixing causes CFCs released into the troposphere to rise to the stratosphere, where they are decomposed by high-energy solar radiation. The decomposition products of CFCs participate in reactions in the stratosphere that result in a lowering of the concentration of ozone there as first proposed by S. Rowland and M. Molina in 1974.

Destruction of the stratospheric ozone layer by CFCs begins when a photon with sufficient energy breaks a carbon-chlorine bond in a CFC molecule. This produces a chlorine atom, as shown here using CFC-12 as an example.

$$CF_2Cl_2(g) \xrightarrow{h\nu} CF_2Cl\cdot(g) + Cl\cdot(g)$$

The chlorine atom, a free radical, then participates in what is called a *chain reaction mechanism.* It first combines with an ozone molecule, producing a chlorine monoxide, $ClO\cdot$, radical and an oxygen molecule.

Step 1: $Cl\cdot(g) + O_3(g) \longrightarrow ClO\cdot(g) + O_2(g)$

Thus, an ozone molecule has been destroyed. If this were the only reaction that particular CFC molecule caused, there would be little danger to the ozone layer. However, once Step 1 has occurred twice, the two $ClO\cdot$ radicals produced can react further.

The chlorine atom, Cl·, is a free radical due to its unpaired electron (← Sec. 6-10b). Because Cl· is regenerated in the chain reaction, Cl· accelerates the rate of ozone depletion.

Step 2: $ClO\cdot + ClO\cdot \longrightarrow ClOOCl$

Step 3: $ClOOCl \xrightarrow{h\nu} ClOO\cdot + Cl\cdot$

Step 4: $ClOO\cdot \xrightarrow{h\nu} Cl\cdot + O_2$

The net reaction obtained by adding two Step 1 reactions to Steps 2, 3, and 4 and canceling species that appear as both reactants and products is the conversion of two ozone molecules to three oxygen molecules.

$$\text{Net reaction:} \quad 2\,O_3(g) \longrightarrow 3\,O_2(g)$$

That is, this chain reaction increases the rate at which stratospheric ozone is destroyed, but it does not affect the rate at which ozone is formed. It is called a chain reaction because the reaction steps can repeat over and over. The two chlorine atoms that react when Step 1 occurs twice are regenerated in Steps 3 and 4, so there is no net change in

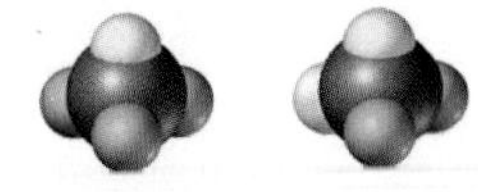

$CFCl_3$ (CFC-11) CF_2Cl_2 (CFC-12)

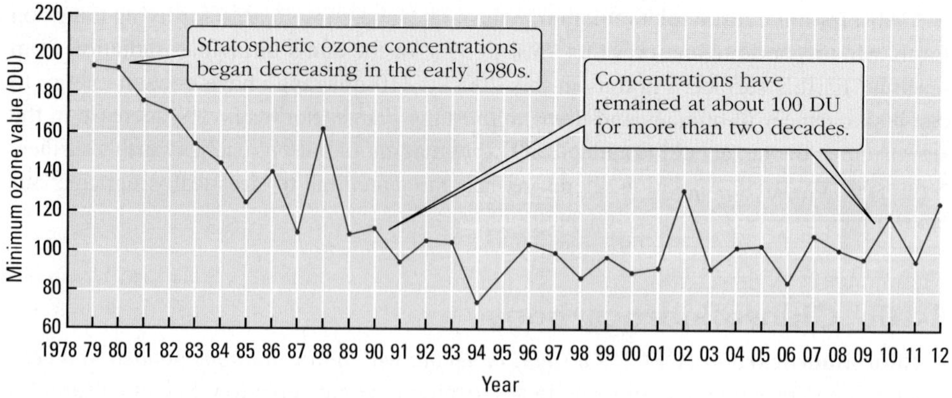
Stratospheric ozone concentrations began decreasing in the early 1980s.
Concentrations have remained at about 100 DU for more than two decades.
Minimum ozone value (DU)
220
200
180
160
140
120
100
80
60
1978 79 80 81 82 83 84 85 86 87 88 89 90 91 92 93 94 95 96 97 98 99 00 01 02 03 04 05 06 07 08 09 10 11 12
Year

Figure 8.15 Annual minimum stratospheric ozone concentrations, Antarctica, 1979–2012 in Dobson Units (DU). (Data from NASA: http://ozonewatch.gsfc.nasa.gov/meteorology/annual_data.html.)

concentration of chlorine atoms. It has been estimated that a single chlorine atom can destroy as many as 100,000 molecules of O_3 before it is inactivated or returned to the troposphere (probably as HCl). This sequence of ozone-destroying reactions upsets the balance in the stratosphere because ozone is being destroyed faster than it is being produced. As a consequence, the concentration of stratospheric ozone decreases.

8-10c Antarctic Stratospheric Ozone Hole

Stratospheric ozone concentration varies not only with altitude, but also with latitude. It is approximately 250 Dobson Units (DU) at the equator and increases to 350 DU or more at higher latitudes in the Northern and Southern hemispheres: 1 DU $= 2.69 \times 10^{16}$ molecules in a 1 cm^2 column extending from the surface of Earth to the top of the atmosphere. One DU is roughly equivalent to one ozone molecule for every billion atoms and molecules in air.

The most prominent manifestation of stratospheric ozone depletion is the **ozone "hole"**, *the seasonal reduction of ozone over Antarctica that occurs during late September or early October,* the beginning of the Antarctic Spring. (Seasons are reversed in the Southern Hemisphere.) The springtime stratospheric ozone concentration over Antarctica normally had been 350 DU. In 1985, British scientists reported a dramatic decrease in stratospheric ozone concentration over Antarctica to 124 DU (Figure 8.15). The ozone "hole" is a region throughout which stratospheric ozone is less than 220 DU; it does not mean that ozone is completely absent. The 2011 stratospheric ozone hole is shown in blue to purple in Figure 8.16, a representation of data taken by a total ozone mapping

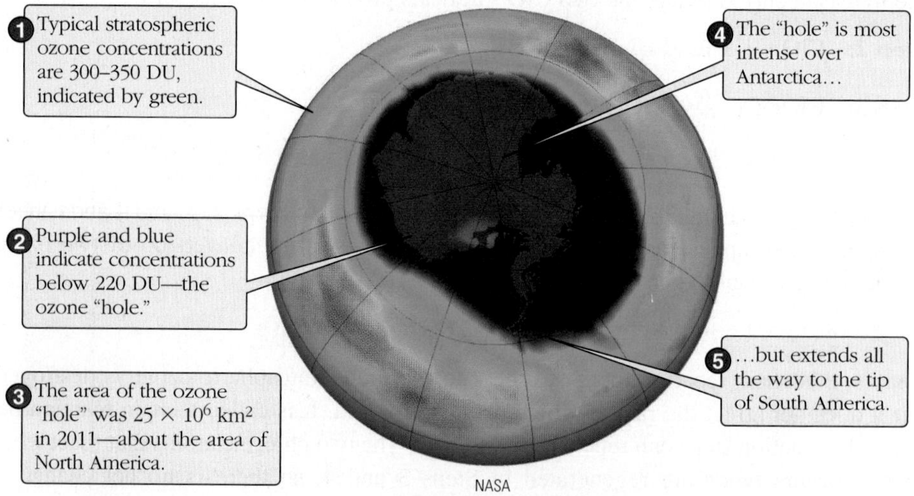
1 Typical stratospheric ozone concentrations are 300–350 DU, indicated by green.
2 Purple and blue indicate concentrations below 220 DU—the ozone "hole."
3 The area of the ozone "hole" was 25 × 10⁶ km² in 2011—about the area of North America.
4 The "hole" is most intense over Antarctica…
5 …but extends all the way to the tip of South America.
NASA

Figure 8.16 Antarctic stratospheric ozone hole, September 12, 2011. (*Source:* earthobservatory.nasa.gov/Features/WorldOfChange/ozone.php)

satellite (TOMS). Since 1992, the maximum area of the annual Antarctic ozone hole has equaled or exceeded the total area of North America, 24.7×10^6 km^2 (9.54×10^6 mi^2) in all but five years.

In 1987, using specially equipped aircraft flying over Antarctica, researchers obtained experimental evidence that depletion of Antarctic stratospheric ozone was correlated with the presence of Cl·, ClO·, and ClOOCl (Figure 8.17). This supported the chain reaction involving Cl· and ClO· radicals proposed by Rowland and Molina and implicated CFCs in stratospheric ozone depletion. However, the huge depletion of ozone in the Antarctic could not be explained solely by the reaction steps given earlier. In the dark Antarctic winter, a vortex of intensely cold air containing ice crystals, unique to the region, builds up. On the surfaces of these crystals additional reactions produce hydrogen chloride and chlorine nitrate, $ClONO_2$, which can react with each other to form chlorine molecules.

$$HCl(g) + ClONO_2(g) \longrightarrow Cl_2(g) + HNO_3(g)$$

When sunlight returns in the spring, solar photons dissociate the Cl_2 molecules into chlorine atoms, which can then participate in ozone destruction reactions.

$$Cl_2(g) \xrightarrow{h\nu} 2\,Cl\cdot(g)$$

The first direct confirmation of the relationship between stratospheric ozone depletion and increased ultraviolet intensity on Earth was reported in the September 10, 1999, issue of *Science* by scientists from the New Zealand National Institute of Water and Atmospheric Research. They found that ". . . over the past 10 years, peak levels of skin-frying and DNA-damaging ultraviolet (UV) rays have been increasing in New Zealand, just as the concentrations of protective stratospheric ozone have decreased." The Cancer Society of New Zealand has a SunSmart initiative that assists schools, businesses with outside workplaces, and recreation areas to develop policies and resources that help to reduce exposure to harmful sunlight during peak UV radiation hours (10 AM–4 PM).

8-10d An Arctic Ozone Hole?

Over the Arctic region, more variable annual meteorological conditions occur and temperatures in the stratosphere are usually not low enough to cause loss of ozone. Springtime ozone concentrations typically are approximately 400 DU. However, during March of 1990, 1996, 1997, and 2011, there were temperatures lower than −78 °C, the temperature at which the intense cold forms tiny ice crystals in polar stratospheric clouds. The crystals provide surfaces on which reactions produce diatomic chlorine molecules, just

Susan Solomon
1956–

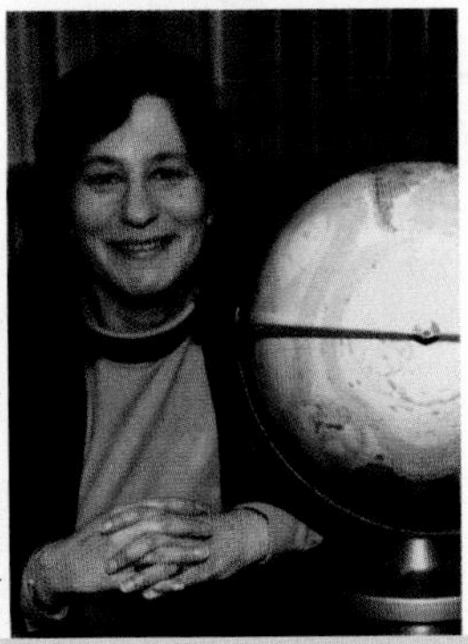

Carlye Calvin. Courtesy of Susan Solomon

A senior scientist at NOAA (National Oceanic and Atmospheric Administration), Susan Solomon was the first to propose a good explanation for the Antarctic ozone hole. Her chemist's intuition told her that the ice crystals that form during the Antarctic winter could provide a surface on which chemical reactions of CFC decomposition products could take place. In 1986, NSF chose Solomon (then 30 years old) to lead a team to Antarctica. Experiments performed during that visit showed that her cloud theory was correct and provided the first solid proof of a connection between CFCs and stratospheric ozone depletion. Solomon, a member of the National Academy of Sciences, was awarded a National Medal of Science in March 2000, the 2002 Weizmann Women & Science Award, and the 2004 Blue Planet Prize for "...achievements ... that have contributed to the resolution of global environmental problems." Solomon has said that her winters as a young girl in Chicago prepared her for her visits to Antarctica.

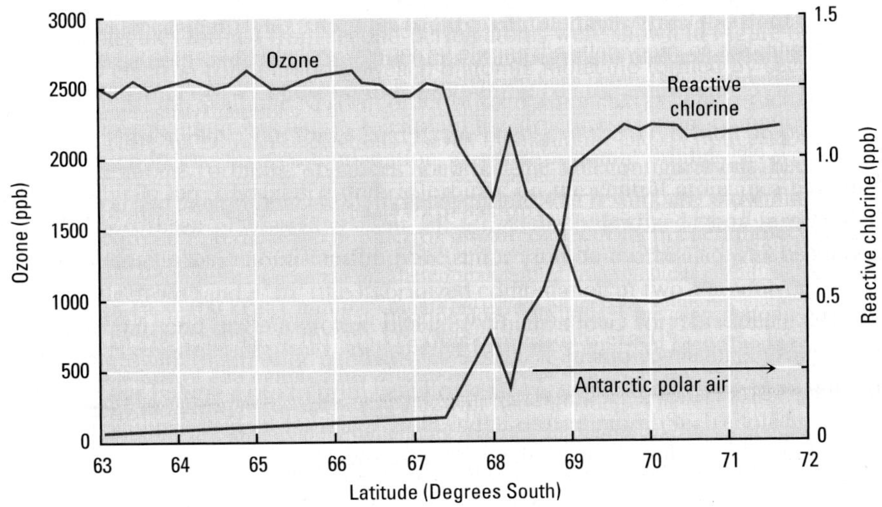

Ozone
Reactive chlorine
Antarctic polar air
Ozone (ppb)
Reactive chlorine (ppb)
Latitude (Degrees South)
3000
2500
2000
1500
1000
500
0
1.5
1.0
0.5
0
63
64
65
66
67
68
69
70
71
72

Figure 8.17 Antarctic stratospheric ozone concentration and reactive chlorine concentration. As the reactive chlorine concentration increases, the stratospheric ozone concentration decreases.

as they do over Antarctica. When sunlight reappears in the spring, the Cl_2 molecules photodissociate into chlorine atoms that cause the catalytic chain reaction through which ozone molecules are destroyed. The Arctic ozone depletion in March 2011 is notable because it was the most intense ever recorded. The ozone concentration dropped from 625 DU in January to 250 DU during March, a 60% decrease. Note, however, that at 250 DU, although less than the normal 400 DU springtime value, the Arctic stratospheric ozone concentration is still much greater than that over Antarctica during springtime there.

8-10e Dealing with the Problem

Because it is a global problem, stratospheric ozone depletion has led to global cooperation that has decreased the production and use of ozone-depleting CFCs. The Montreal Protocol on Substances that Deplete the Ozone Layer, first signed in 1987, was implemented, modified, and now has been ratified by 197 countries. The global production of CFCs fell from 1300 thousand tons in 1987 to fewer than 100 thousand tons in the past decade. By 1996, developed nations halted CFC production and use of CFCs in those countries was phased out completely in 2010. In spite of such laudable measures, however, the decrease in concentration of atmospheric CFCs will be slow due to their chemical inertness. CFCs are not broken down chemically in the troposphere, so they linger there a long time before reaching the stratosphere. Some have lifetimes in the atmosphere estimated to be about 100 years before decomposition occurs. Because of this, Antarctic stratospheric ozone levels are not expected to return to pre-1987 levels until 2060–2075.

Chemists have synthesized new compounds to replace CFCs, ones that do not contain ozone-depleting chlorine atoms, but creating such compounds has been a challenge. To be useful, these substitutes must have properties similar to those of CFCs—low toxicity, low boiling point, low chemical reactivity. The most recent substitutes are compounds known as **hydrofluorocarbons (HFCs)** in which hydrogen atoms replace chlorine atoms. Because of the hydrogen atoms, HFCs have much shorter atmospheric lifetimes than CFCs.

HFC-125 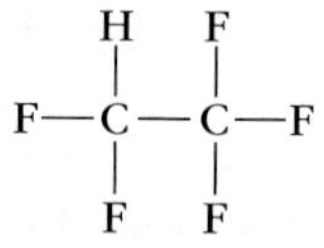 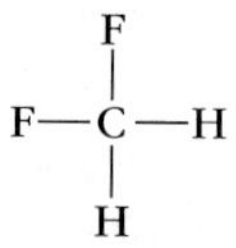 HFC-32

Another class of ozone-depleting compounds, halons, are similar to CFCs in that they contain chlorine or fluorine (not both), but they also contain bromine. The halon 1,2-dibromotetrafluoroethane, $C_2Br_2F_4$, (Halon-2402) has this Lewis structure:

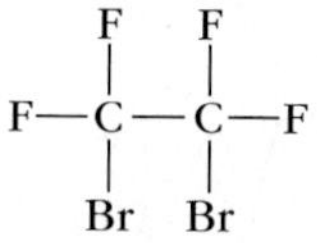 Halon-2402

Halons are used as fire suppressants where water cannot be used to put out a fire, such as on submarines and in aircraft. Halons are the last of the widely used ozone-depleting compounds to be phased out because it has been difficult to find adequate replacements for them.

In 1998, a Presidential Green Chemistry Challenge Award was presented to Pyrocool Technologies for its development of a fire extinguishing foam (FEF) that replaces halons in fighting large-scale fires. Following the collapse of the World Trade Center towers on September 11, 2001, a 0.4% aqueous solution of Pyrocool FEF was sprayed to control the spread of sublevel fires beneath the collapsed towers.

courtesy of Pyrocool Technologies

Pyrocool's fire extinguishing foam being used on subterranean fires in the rubble of the World Trade Center, September 30, 2001.

8-11 Greenhouse Gases and Global Warming

8-11a The Greenhouse Effect

An important function of our atmosphere is to moderate Earth's surface temperature. Most of the energy that heats Earth comes from the sun as electromagnetic radiation. Some of the electromagnetic radiation from the sun is reflected by the atmosphere back into space, and some is absorbed by the atmosphere. The remainder reaches the surface, warming its land and oceans. The warmed surfaces then reradiate this energy into the troposphere as infrared radiation (← **Secs. 5-1, 7-2d**). Carbon dioxide, water vapor, methane, and tropospheric ozone all absorb radiation in various portions of the infrared region. By absorbing this reradiated energy they warm the atmosphere, creating what is called the **greenhouse effect** (Figure 8.18). Thus, all four are "greenhouse gases." Such gases constitute an absorbing blanket that reduces the quantity of energy radiated back into space. Thanks to the greenhouse effect, Earth's average temperature is a comfortable 15 °C (59 °F). By comparison, the moon, with no moderating atmosphere but at about the same distance from the sun, has a surface mean temperature that fluctuates from approximately 107 °C when exposed to sunlight and −153 °C when in the dark.

The greenhouse effect derives its name by analogy with a botanical greenhouse. However, the warming in a botanical greenhouse is much more dependent on the fact that glass reduces convection than that it blocks infrared radiation from leaving the greenhouse.

Earth has such a vast reservoir of water in the oceans that human activity has a negligible influence on the concentration of water vapor in the atmosphere. In addition, methane is produced by natural processes in such large quantities that human contributions are negligible. Tropospheric ozone is present in such small concentrations that its contribution to the greenhouse effect is small. Among the four greenhouse gases, most attention is focused on CO_2.

Carbon dioxide is a greenhouse gas because it absorbs infrared radiation, causing C=O bonds in the molecule to stretch and bend, much like springs attached to balls. Stretching and compressing the C=O bonds requires more energy (shorter wavelength) than does bending them. Thus, C=O stretching and compressing occurs when infrared

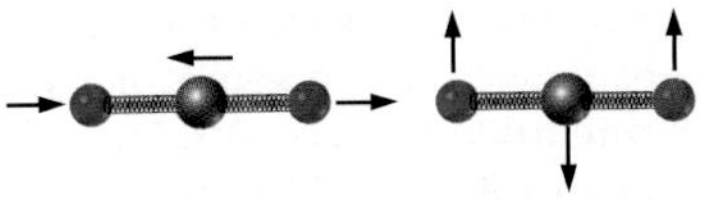
C=O
Stretching
O=C=O
Bending

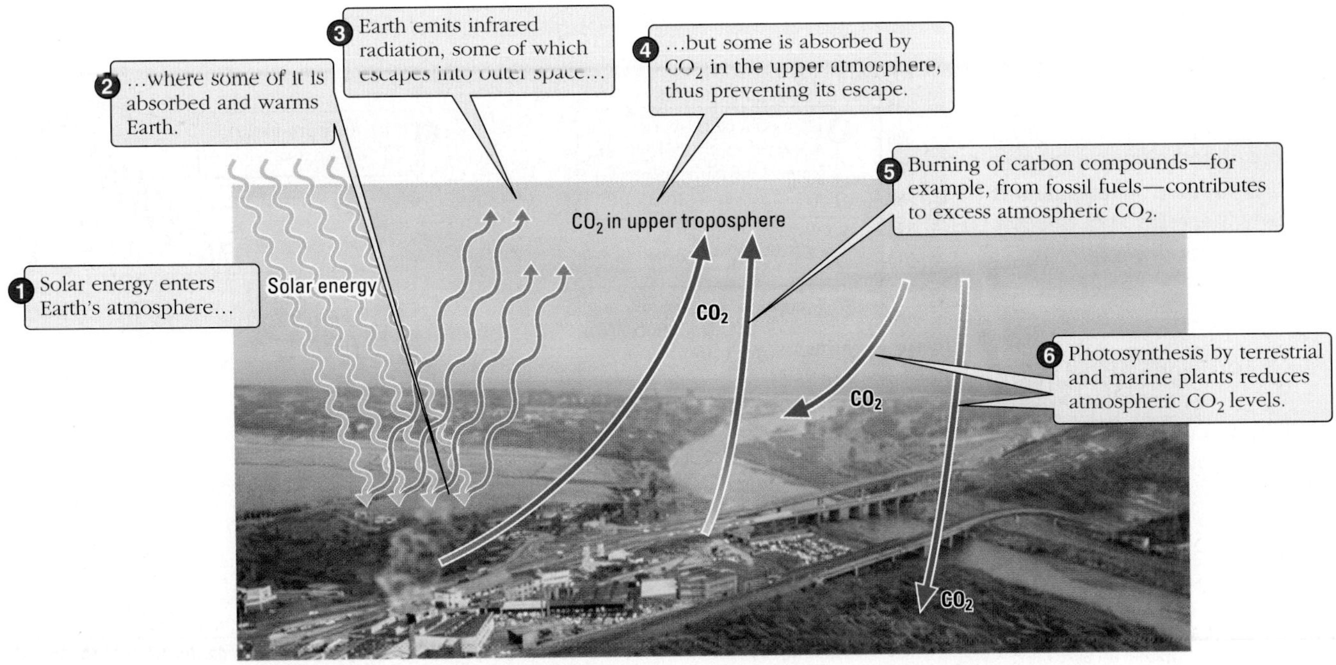
1 Solar energy enters Earth's atmosphere...
2 ...where some of it is absorbed and warms Earth.
3 Earth emits infrared radiation, some of which escapes into outer space...
4 ...but some is absorbed by CO_2 in the upper atmosphere, thus preventing its escape.
5 Burning of carbon compounds—for example, from fossil fuels—contributes to excess atmospheric CO_2.
6 Photosynthesis by terrestrial and marine plants reduces atmospheric CO_2 levels.
Solar energy
CO_2 in upper troposphere
CO_2
CO_2
CO_2

Figure 8.18 The greenhouse effect. Greenhouse gases form an effective barrier that prevents some heat from escaping Earth's surface.

Without the greenhouse effect, Earth's average temperature would be much lower.

1 μm = 1 × 10^{-6} m = 1 × 10^3 nm

Recall from Section 5-2 that energy and wavelength are inversely related.

Worldwide, nearly one third of all atmospheric CO_2 is released as a by-product from fossil fuel–burning electric power plants.

radiation with a wavelength of 4.257 μm is absorbed, whereas the C=O bending vibrations occur at lower energy (longer wavelength) when the molecule absorbs 15.000 μm infrared radiation.

Each year, combustion of fossil fuels (coal, petroleum, natural gas) worldwide puts billions of metric tons of carbon into the atmosphere as CO_2. About 45% is removed from the atmosphere by natural processes—some by plants during photosynthesis and the rest by dissolving in rainwater and the oceans to form hydrogen carbonates and carbonates.

$$CO_2(g) + H_2O(\ell) \longrightarrow H^+(aq) + HCO_3^-(aq)$$
$$HCO_3^-(aq) \longrightarrow H^+(aq) + CO_3^{2-}(aq)$$

The other 55% of the carbon dioxide from fossil fuel combustion remains in the atmosphere, increasing the global CO_2 concentration.

EXERCISE 8.20

Sources of CO_2

List as many natural sources of CO_2 as you can. List as many sources of CO_2 from human activities as you can.

Parts per million (ppm) is a convenient way to express low concentrations. One ppm means one part of something in one million parts. A CO_2 concentration of 360 ppm means that for every million molecules of air, 360 are CO_2 molecules.

Without human influences, the flow of carbon dioxide among the air, plants, animals, and the oceans would be roughly balanced. In 1750, during the preindustrial era, CO_2 concentration in the atmosphere was 277 parts per million (ppm). During the next 130 years, as the Industrial Revolution progressed, the concentration increased to 291 ppm, a 5% increase. Since 1900, however, the rate of increase of CO_2 has reflected the rapid increase in the use of fossil fuel combustion for industrial and domestic purposes, especially for motor vehicle transportation. Between 1959 and 2013, the atmospheric concentration of CO_2 increased from 316 to 400 ppm, an increase of 27% (Figure 8.19). Population pressure is also contributing heavily to increased CO_2 concentrations. In the Amazon region of Brazil, for example, forests are being cut and burned to create

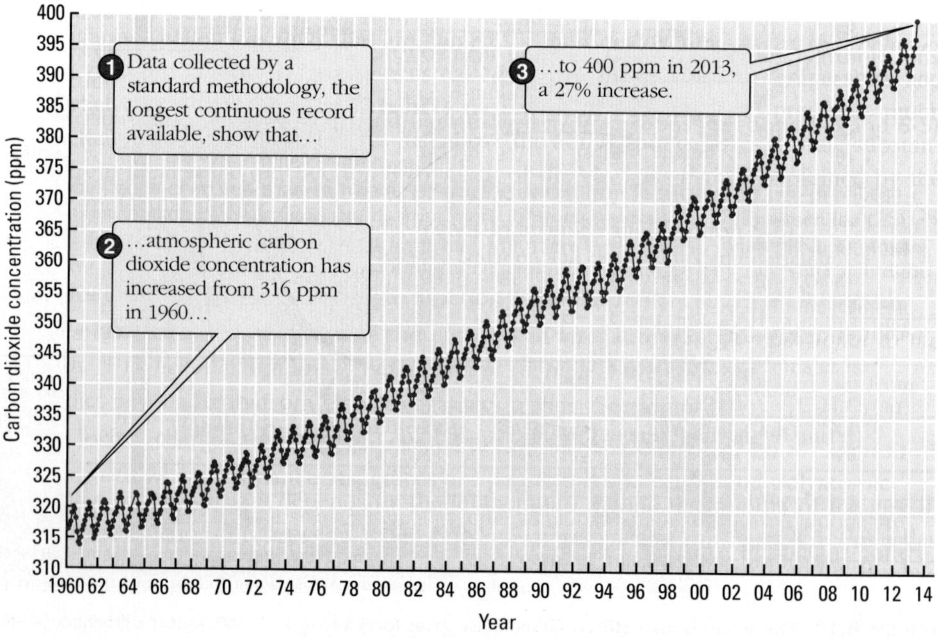

Carbon dioxide concentration (ppm)
400
395
390
385
380
375
370
365
360
355
350
345
340
335
330
325
320
315
310
1 Data collected by a standard methodology, the longest continuous record available, show that…
2 …atmospheric carbon dioxide concentration has increased from 316 ppm in 1960…
3 …to 400 ppm in 2013, a 27% increase.
1960 62 64 66 68 70 72 74 76 78 80 82 84 86 88 90 92 94 96 98 00 02 04 06 08 10 12 14
Year

Figure 8.19 Atmospheric carbon dioxide concentration measured at Mauna Loa, Hawaii. (*Source:* NOAA, http://www.esrl.noaa.gov/gmd/ccgg/trends/co2_data_mlo.html.)

cropland. This activity places a double burden on the natural CO_2 cycle, since there are fewer trees to use the CO_2 in photosynthesis and, at the same time, CO_2 is added to the atmosphere during burning.

Expectations are that the atmospheric CO_2 concentration will continue to increase at the current rate of about 1.5 ppm per year, due principally to industrialization, agricultural production, and an expanding global population. As a result, according to sophisticated computer models of climate change, the CO_2 concentration is projected to reach 550 ppm, double its preindustrial value, between 2030 and 2050. Perhaps the increase may not be so great if fuel costs increase substantially as fossil fuel supplies get tighter.

Marcelo Sayao/epa/Corbis
Clearing a Brazilian rain forest.

The Intergovernmental Panel on Climate Change (IPCC) is a prestigious multinational group of climate scientists organized by the United Nations. In its *IPCC Fourth Assessment Report* of 2007, the IPCC noted that, "Atmospheric concentrations of CO_2 (379 ppm) . . . in 2005 exceed by far the natural range over the last 650,000 years." The IPCC noted earlier that nearly 75% of CO_2 emissions from human activities in the past 20 years are due to the burning of fossil fuels.

To see how easily everyday activities affect the quantity of CO_2 being released into the atmosphere, consider a round-trip flight from New York to Los Angeles. Each passenger pays for about 200 gal jet fuel, which weighs 1400 lb. When burned, each pound of jet fuel produces about 3.14 lb carbon dioxide. So 4400 lb, or 2 metric tons, of carbon dioxide are produced per passenger during that trip.

EXERCISE 8.21

CO_2 Changes

Using data from Figure 8.19, calculate the percent increase in atmospheric CO_2 from
(a) 1960 to 1975. (b) 1975 to 1990. (c) 1990 to 2010.
Which period showed the greatest percent increase per year in CO_2?

CONCEPTUAL EXERCISE 8.22

Annual CO_2 Changes

During a year, atmospheric CO_2 concentration fluctuates, building up to a high value, then dropping to a low value before building up again (Figure 8.19). Explain what causes this fluctuation and when during the year the high and the low occur in the Northern Hemisphere.

CONCEPTUAL EXERCISE 8.23

Carbon Dioxide and Air Travel

Use an Internet search engine to find the annual airline passenger miles for a recent year and calculate how many tons of CO_2 were produced by those flights. Compare this quantity to the quantity of CO_2 produced for the same number of miles traveled in automobiles. (*Hint:* Assume a miles-per-gallon value and an average number of passengers per vehicle. Then, use the same numbers for gasoline as were used for jet fuel in the last paragraph in this section.)

8-11b Global Warming

During the 1860s, physicist John Tyndall first demonstrated experimentally that carbon dioxide absorbs infrared radiation. More than a century ago (1896), Svante Arrhenius, the noted Swedish chemist, estimated the effect that increasing atmospheric carbon dioxide could have on global temperature. He calculated that doubling the carbon dioxide concentration would increase the temperature by 4 °C. Disturbed by the prospects of

This section answers the question, "How is carbon dioxide involved in global warming?" that was posed in Chapter 1 (← Sec. 1-1).

Arrhenius had not foreseen the advent of motor vehicles and their impact on increasing atmospheric carbon dioxide.

Arrhenius's prediction, others described the possible scenario in picturesque terms: "We are evaporating our coal mines into the air."

CONCEPTUAL EXERCISE 8.24

Evaporating Coal Mines

Explain the quote about evaporating coal mines in terms of global warming.

The *enhanced greenhouse effect* is the process whereby gases in the atmosphere trap more than 80% of the thermal energy radiated from Earth's surface. **Global warming** is the term commonly used to describe *the increase in the average global temperature of the Earth's surface.* The vast majority of climate scientists attribute this increase to an increased greenhouse effect due to increasing concentrations of atmospheric CO_2. Figure 8.20 shows data collected from isotopic analysis of CO_2 in air bubbles trapped in ice cores taken from deep within the Antarctic ice, the oldest ice on the planet. The data span 160,000 years and show a remarkable correlation between fluctuations in CO_2 with that of global temperature; increases in CO_2 concentration matched increases of global temperature and decreases matched decreases. What is important to note from the graph is that CO_2 concentration is now higher than at any time over the past 160,000 years and it is climbing at a rapid rate. The average global temperature has risen 0.6 to 0.8 °C over the past 160 years.

Roger Ressmeyer/Corbis

Ice core samples from the Greenland Ice Sheet. The samples are analyzed for changes in their carbon dioxide concentration over many years.

Many computer models predict that increasing atmospheric CO_2 to 600 ppm will increase average global temperature, and most climate experts who do global warming research agree. What is uncertain is the extent of the temperature increase, with estimates varying from 1.5 to 4.5 °C (2.7 to 8.1 °F). The IPCC estimates a 1.0 to 3.5 °C (1.8 to 5.8 °F) increase by the year 2100. Warming by as little as 1.5 °C would produce the warmest climate seen on Earth in the past 6000 years; an increase of 4.5 °C would produce world temperatures higher than any since the Mesozoic era, the time of the dinosaurs.

Many scientists are concerned that rising temperatures are causing ice caps in Greenland and Antarctica to melt, raising the sea level. In conjunction with this, atmospheric currents will change and produce significant changes in weather and agricultural productivity. Since the 1940s, average summertime temperatures in Antarctica have increased 2.5 °C to just above 0 °C. Temperatures have risen sufficiently in the arctic regions that sea ice has receded and the long-sought Northwest Passage for shipping from the Atlantic to the Pacific via the arctic seas is now a reality.

Computer models used to predict future global temperature changes have become more sophisticated and accurate over the past decade. When factors in addition to

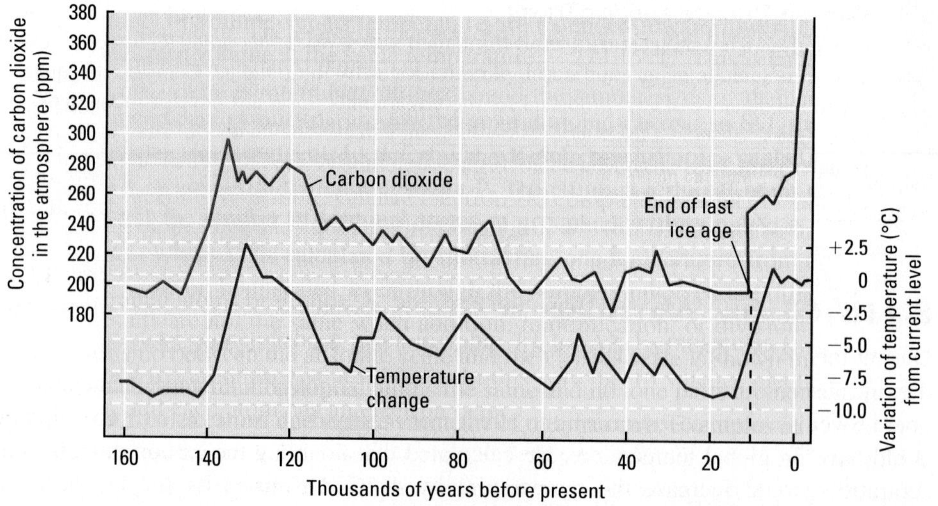

Concentration of carbon dioxide in the atmosphere (ppm)
380
360
340
320
300
280
260
240
220
200
180
Carbon dioxide
End of last ice age
Temperature change
+2.5
0
−2.5
−5.0
−7.5
−10.0
Variation of temperature (°C) from current level
160
140
120
100
80
60
40
20
0
Thousands of years before present

Figure 8.20 Global temperatures (blue) and carbon dioxide concentrations (red) over the past 160,000 years (ice core data).

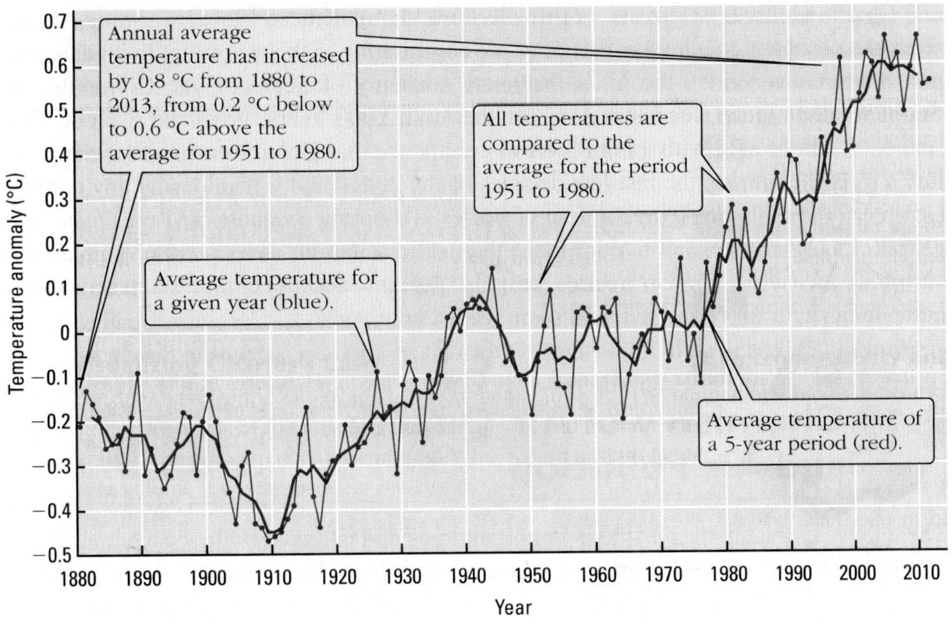
Annual average temperature has increased by 0.8 °C from 1880 to 2013, from 0.2 °C below to 0.6 °C above the average for 1951 to 1980.
All temperatures are compared to the average for the period 1951 to 1980.
Average temperature for a given year (blue).
Average temperature of a 5-year period (red).
Temperature anomaly (°C)
0.7
0.6
0.5
0.4
0.3
0.2
0.1
0
−0.1
−0.2
−0.3
−0.4
−0.5
1880
1890
1900
1910
1920
1930
1940
1950
1960
1970
1980
1990
2000
2010
Year

Figure 8.21 Global annual mean surface air temperature 1880-2013. (*Source NASA*: http://data.giss.nasa.gov/gistemp/graphs_v3/.)

greenhouse gases, such as the presence of aerosols and changes in the intensity of the sun's radiation are taken into account, the computer models predict temperature changes close to what has been observed since 1880 (Figure 8.21). Such work adds credibility to the predictions of temperature increases of 1.5 to 4.5 °C due to global warming.

Global warming has become an acknowledged worldwide problem. Concerns about global warming prompted delegates from 159 countries to convene in 1997 in Kyoto, Japan, where the Kyoto Protocol was negotiated. Goals were set to reduce greenhouse gas emissions below 1990 levels. Enough countries ratified the Kyoto Protocol to make it official in February 2005. The only industrialized country that did not ratify it was the United States. A United Nations Climate Change Conference was held from November 28 to December 11, 2011 in Durban, South Africa. The conference attempted to establish a new treaty to limit carbon emissions that would succeed the initial Kyoto Protocol, whose initial commitment period was to end in December 2012. Agreement on a new treaty could not be reached. Instead, the conference established that the terms of a future treaty are to be defined by 2015 and become effective legally in 2020. The participating nations at the conference agreed to continue the Kyoto Protocol in the interim. The 2011 conference is noteworthy because it marked the first time that developing countries such as China and India were part of the agreement.

The most obvious measure to reduce global warming is to control CO_2 emissions that result from human activities throughout the world. Given our addictive dependence on fossil fuels, however, getting and keeping these emissions under control will be very difficult. Alternatives to fossil fuels, such as solar power and nuclear energy, can help. One direct way to decrease the quantity of atmospheric CO_2 is to plant more trees and to replace those in deforested areas.

The projected extent of global warming remains a thorny issue. The uncertainty in the temperature changes attributable to global warming is related to three fundamental, but difficult, questions:

1. To what levels will CO_2 and other greenhouse gases rise during the next decades?
2. How responsive is Earth's climate to the warming created by the greenhouse effect?
3. To what extent is the current global warming being caused by human activities, or could it be part of a natural cycle of global temperature changes?

Credible climate scientists worldwide think that enhanced global warming has already occurred and that human activities are contributing to this warming. In 2006 an expert panel convened by the U. S. National Academy of Sciences, the Committee of Surface Temperature Reconstruction for the Last 2,000 Years, produced a report that stated, "It can be said with a high level of confidence that global mean surface temperature was higher during the last few decades of the 20th century than during any comparable period during the preceding four centuries." Adopting adequate measures to control global climate change will require more than valid scientific data on which to base the corrective measures. These measures must also take into account complex and sometimes competing political, economic, and social considerations.

8-12 Chemistry of Air Quality and Air Pollution

The quality of the air we breathe is important to us and is linked to the concentrations of air pollutants. An **air pollutant** is *a substance that degrades air quality*. If air-pollutant concentrations are high enough, they can adversely affect our health. Air pollutants come from both natural as well as anthropogenic (human-made) sources. Natural processes create air pollutants on a massive scale such as the ash, sulfur oxides, mercury vapor, hydrogen sulfide, and hydrogen chloride emitted from volcanoes. Decaying vegetation, odorous compounds from coniferous plants such as pine trees, methane gas from the digestive system of cows and other ruminants, and dinitrogen monoxide from decaying plant and animal proteins all add large quantities of pollutants to the troposphere. On the other hand, human activities such as driving automobiles, operating electricity-generating power plants, refining petroleum, and smelting ores to extract their minerals also produce significant quantities of air pollutants.

Note from Table 8.6 (← **Sec. 8-9**) that the tropospheric concentrations of sulfur dioxide, ozone, and nitrogen oxides are very low. It is common not to express such low concentrations as percentages by volume (parts per hundred, *pph*), but rather in terms of *parts per million* (*ppm*) or *parts per billion* (*ppb*). Because volume is proportional to the number of molecules of a gas, ppm and ppb units also give a ratio of molecules of one kind to those of another kind. For example, CO_2 at 0.039 percent by volume (Table 8.6) can be expressed as ppm by multiplying by 10,000: $0.039 \times 10^4 = 390$ ppm. This means that for every 1 million molecules in air, 390 of them are CO_2 molecules. That may not sound like much until you consider that just 1 cm^3 of air contains about 2.5×10^{19} molecules. Thus, if this 1 cm^3 of air contains 390 ppm CO_2, then it contains $(390/10^6) \times 2.5 \times 10^{19}$ molecules $= 9.8 \times 10^{15}$ CO_2 molecules. That's 9.8 million billion CO_2 molecules!

Percent means per 100 or divided by 100; thus 100% = 1. Parts per million means per million, that is, divided by 10^6, so 10^6 ppm = 1. Parts per billion means per billion or divided by 10^9, so 10^9 ppb = 1.

PROBLEM-SOLVING EXAMPLE 8.14

Air Quality Units

(a) A tropospheric air sample has an ozone concentration of 0.000002 volume percent. Convert this ozone concentration to ppb.

(b) An air sample contains sulfur dioxide at a concentration of 0.50 ppm, the maximum 3-h exposure limit permitted under the U. S. National Ambient Air Quality Standards. Convert this concentration to percent by volume.

Result (a) 20 ppb; (b) 5.0×10^{-5} %

Analyze Unit conversions are required. Percent means per 100 or divided by 100; ppm means per million or divided by 10^6; ppb means per billion or divided by 10^9.

Plan Use the definitions of percent, parts per million, and parts per billion to derive conversion factors. For example, 10^6 ppm $= 10^6/10^6 = 1$, so multiplying or dividing by 10^6 ppm is the same as multiplying or dividing by 1.

Execute

(a) 2×10^{-6} percent $= \dfrac{2 \times 10^{-6}}{100} \times 10^9$ ppb $= 2 \times 10^1$ ppb $= 20$ ppb

(b) 0.50 ppm $= \dfrac{0.50}{10^6} \times 100\% = 5.0 \times 10^{-5}\% = 0.000050\%$

☑ **Reasonable Result Check** A ppb value should be a larger number than a percent value, as is the case in part a. The reverse applies in going from 0.50 ppm to 0.000050 percent in part b. The results are reasonable.

PROBLEM-SOLVING PRACTICE 8.14

Make these conversions for atmospheric concentrations: (a) CO: 35 ppm to percent; (b) NO_2: 0.053 ppm to ppb.

We briefly will consider several tropospheric pollutants: sulfur dioxide; nitrogen oxides; and ozone. Each of these can be a hazard to human health when the pollutant's concentration exceeds a certain standard. In the United States, such limits are set by the National Ambient Air Quality Standards established by the U. S. Environmental Protection Agency (EPA). An air pollutant at a level below the standard for it indicates air that presumably is safe to breathe. The U. S. EPA has developed an air-quality index for each pollutant it regulates. An index value of 100 always corresponds to the U. S. National Ambient Air Quality Standard for the pollutant. The color-coded index is shown in a figure here and can be viewed at http://www.airnow.gov/. Of course, the levels of all other pollutants in the air need to be checked against each of their standards before considering that the air is truly healthful to breathe. An individual's exposure to an air pollutant depends on three factors: (1) *The concentration of the pollutant in air.* More toxic pollutants require setting lower concentration limits on them. (2) *Length of exposure.* Higher concentrations can be tolerated for a short period of time without harmful effects, but lower limits are required for longer exposure times. For example, the national standards for sulfur dioxide are 0.50 ppm for a 3-h average and 0.14 ppm for a 24-h average. (3) *Breathing rate and level of exertion.* Exercise increases breathing rate and thus, exposure to an air pollutant. Consequently, reducing or eliminating outdoor exercise may be required when air pollution levels are high.

8-12a Sulfur Dioxide

Sulfur dioxide, SO_2, is an air pollutant produced when sulfur or sulfur-containing compounds are burned in air, where SO_2 is a major contributor to acid rain and industrial smog. Many electric power plants in the United States burn coal to generate electricity, coal that contains from 1% to 4% weight percent sulfur in the form of the mineral pyrite, FeS_2, as well as sulfur in organic compounds making up the coal itself.

In the atmosphere, SO_2 is oxidized by ozone or oxygen to form sulfur trioxide, SO_3. Because of its strong affinity for water, SO_3 dissolves in aerosol droplets to form sulfuric acid, which contributes to acid precipitation:

$$SO_3(g) + H_2O(\ell) \longrightarrow H_2SO_4(aq)$$

Acid precipitation will be discussed in Section 15-3.

Air quality index

Index values	Health concern	Color code	Meaning
0-50	Good	Green	Air is satisfactory and poses little risk
51-100	Moderate	Yellow	Acceptable, may affect some people
101-150	Unhealthy for sensitive groups	Orange	Members of sensitive groups may experience health effects
151-200	Unhealthy	Red	Everyone may experience health effects
201-300	Very unhealthy	Purple	Warnings of emergency conditions
301-500	Hazardous	Maroon	Everyone may experience serious effects

Staff/MCT/Newscom

In high enough concentration, sulfur dioxide can be physiologically harmful to plants and animals. People with chronic respiratory difficulties such as bronchitis or asthma tend to be more sensitive to increased SO_2 concentration.

EXERCISE 8.25

Calculating SO_2 Emissions from Burning Coal

Large quantities of coal are burned in the United States to generate electricity. A 1000-MW coal-fired generating plant will burn 3.06×10^6 kg coal per hour. For coal that contains 4% sulfur by mass, calculate the mass of SO_2 released (a) per hour and (b) per year.

8-12b Nitrogen Oxides and Tropospheric Ozone

The common nitrogen oxides found in air, NO and NO_2, are collectively referred to as **NO_x**. These oxides are formed whenever nitrogen and oxygen, always present in air, are raised to high temperatures, such as in an automobile's internal combustion engine, where temperatures can reach 2500 °C. The NO produced reacts rapidly with atmospheric oxygen to produce NO_2. Such production can also occur around lightning bolts. Although NO and NO_2 generally wash out of the atmosphere during rain or other precipitation, the air near the surface in some urban areas can contain harmful levels of these gaseous oxides. In humans, bronchial constriction occurs when one breathes NO_2 at a concentration as low as 3 ppm for 1 hour; higher concentrations (150–250 ppm) at shorter exposure times cause lung damage that can be fatal.

Ronaldo SCHEMITD/AFP/ Getty Images

Early morning photochemical air pollution over Mexico City.

Recall from Section 8-10 that the ozone layer in the *stratosphere* forms a protective shield that prevents most (95–99%) of the sun's damaging UV radiation from reaching the Earth's surface. On the other hand, ozone in high enough concentration in the *troposphere* damages plants and animals. Ozone is produced in the troposphere through a complex sequence of reactions that involve O_2, NO_2, oxygen-containing organic compounds, hydrocarbons, and intense sunlight. Tropospheric ozone is a difficult pollutant to control, especially in urban areas, because it forms, in part, from the reactions among hydrocarbons and NO_2, which every gasoline-burning automobile emits to some degree. Thus, in cities with high tropospheric ozone concentrations, the underlying cause is almost always NO_x emissions from buses, automobiles, and trucks. Even at low levels, increased tropospheric ozone levels affect human health. In 2008, the U. S. EPA decreased the ozone standard to 0.075 ppm for an 8-h average exposure.

8-12c Air Quality Now?

The average air quality across the United States has improved significantly since the inception of the Clean Air Act in 1970. Data in Table 8.7 indicate the marked decline (negative percent change values) of SO_2, NO_2, and ozone concentrations. For example, the annual concentration of NO_2 decreased by 52% from 1980 to 2010. The data in the table come from the U. S. EPA using measurements from monitors across the country.

Table 8.7 Percent Changes in Air Pollution (Negative Numbers Are Decreases)

	1980 vs 2012	1990 vs 2012	2000 vs 2012
SO_2 (1-h)	−78%	−72%	−54%
NO_2 (annual)	−56%	−50%	−38%
O_3 (8-h)	−25%	−14%	−9%

epa.gov/airtrends/aqtrends.html

SUMMARY PROBLEM

In a typical automobile engine, a gasoline vapor-air mixture is compressed and ignited in the cylinders of the engine. This results in a combustion reaction that produces mainly carbon dioxide and water vapor. For simplicity, assume that the fuel is C_8H_{18} and has a density of 0.760 g/mL.

(a) Calculate the partial pressures of N_2 and O_2 in the air before it goes into the cylinder; assume the atmospheric pressure is 734 mmHg.

(b) Consider the case where the air, without any fuel added, is compressed in the cylinder to seven times atmospheric pressure, the compression ratio of many modern automobile engines. Calculate the partial pressures of N_2 and O_2 at this pressure.

(c) Now consider the case where 0.050 mL gasoline is added to the air in the cylinder just before compression and completely vaporized. Assume that the volume of the cylinder is 485 mL and the temperature is 150 °C. Calculate the partial pressure of the gasoline vapor.

(d) Calculate the amount (mol) of oxygen required to burn the gasoline in part (c) completely to CO_2 and H_2O.

(e) The combustion reaction in the cylinder creates temperatures in excess of 1200 K. Due to the high temperature, some of the nitrogen and oxygen in the air reacts to form nitrogen monoxide. If 10.% of the nitrogen is converted to NO, calculate the mass (g) of NO produced by this combustion.

(f) Hot-rod cars use another oxide of nitrogen, dinitrogen monoxide, to create an extra burst in power. When such a power boost is needed, dinitrogen monoxide gas is injected into the cylinders where it reacts with oxygen to form NO. Calculate the mass of dinitrogen monoxide that would have to be injected to form the same quantity of NO as produced in part (e). Assume that sufficient oxygen is present to do so.

HAVING STUDIED THIS CHAPTER . . .

. . . you should be able to:

- Convert from one gas pressure unit to another (Section 8-1). End-of-chapter Questions 10, 12
- State the fundamental postulates of the kinetic-molecular theory and apply them to explain gas behavior (Sections 8-2, 8-7). Questions 15, 61, 63
- Solve problems using the appropriate gas laws (Section 8-3). Questions 17, 19, 21, 25, 31
- Apply the ideal gas law to finding gas densities and molar masses (Section 8-4). Questions 33, 35, 37
- Calculate the quantities of gaseous reactants and products involved in chemical reactions (Section 8-5). Questions 39, 41, 45, 47
- Perform calculations using partial pressures and mole fractions of gases in mixtures (Section 8-6). Questions 50, 52, 54, 56
- Predict whether a gas will behave as a real (non-ideal) or an ideal gas (Section 8-8). Questions 67, 70
- Describe the major components and regions of the atmosphere (Section 8-9). Questions 73, 75
- Explain the main features of stratospheric ozone depletion and the role of CFCs in it (Section 8-10). Questions 78, 80
- Relate atmospheric CO_2 concentration to the greenhouse effect and to enhanced global warming (Section 8-11). Questions 83, 85
- Identify three tropospheric pollutants and the reactions that produce them (Section 8-12). Questions 87, 93

KEY TERMS

absolute temperature scale (Section 8-3b)

absolute zero (8-3b)

air pollutant (8-12)

atmosphere (atm) (8-1)

Avogadro's law (8-3c)

bar (8-1)

barometer (8-1)

Boyle's law (8-3a)

Charles's law (8-3b)

chlorofluorocarbon (CFC) (8-10b)

combined gas law (8-3d)

combining volumes, law of (8-5)

Dalton's law of partial pressures (8-6)

diffusion (8-7b)

effusion (8-7b)

global warming (8-11b)

Graham's law (8-7b)

greenhouse effect (8-11a)

hydrofluorocarbons (HFCs) (8-10e)

ideal gas (8-3)

ideal gas constant (*R*) (8-3d)

ideal gas law (8-3d)

Kelvin temperature scale (8-3)

kilopascal (kPa) (8-1)

millimeter of mercury (mmHg) (8-1)

mole fraction (*X*) (8-6)

newton (N) (8-1)

NO_x (8-12b)

ozone hole (8-10c)

partial pressure (8-6)

pascal (Pa) (8-1)

photodissociation (8-10a)

pound per square inch (psi) (8-1)

pressure (8-1)

standard molar volume (8-3d)

standard temperature and pressure (STP) (8-3d)

stratosphere (8-9a)

stratospheric ozone layer (8-10)

thermodynamic temperature scale (8-3b)

torr (Torr) (8-1)

troposphere (8-9a)

van der Waals equation (8-8)

QUESTIONS FOR REVIEW AND THOUGHT

Red-numbered questions have short answers at the back of this book in Appendix M and fully worked solutions in the *Student Solutions Manual.*

Review Questions

These questions test vocabulary and simple concepts.

1. Name the three gas laws and explain how they interrelate *P*, *V*, *T*, and *n*. Explain the relationships in words and with equations.
2. What are the conditions represented by STP?
3. What is the volume occupied by 1 mol of an ideal gas at STP?
4. What is the definition of pressure?
5. State Avogadro's law. Explain why two volumes of hydrogen react with one volume of oxygen to form two volumes of steam.
6. State Dalton's law of partial pressures. If the air we breathe is 78% N_2 and 21% O_2 on a mole basis, calculate the mole fraction of O_2. Calculate the partial pressure of O_2 if the total pressure is 720 mmHg.
7. Explain Boyle's law on the basis of the kinetic-molecular theory.
8. Explain why a gas at low temperature and high pressure does not obey the ideal gas equation as well as the same gas at high temperature and low pressure.
9. Gaseous water and carbon dioxide each absorb infrared radiation. Does either of them absorb ultraviolet radiation? Explain your answer.

Topical Questions

These questions are keyed to the major topics in the chapter. Usually a question that is answered at the back of this book is paired with a similar one that is not.

Gas Pressure (Section 8-1)

10. Gas pressures can be expressed in units of mmHg, atm, torr, and kPa. Convert these pressure values.
    (a) 720. mmHg to atm
    (b) 1.25 atm to mmHg
    (c) 542. mmHg to torr
    (d) 740. mmHg to kPa
    (e) 700. kPa to atm
11. Convert these pressure values.
    (a) 120. mmHg to atm
    (b) 2.00 atm to mmHg
    (c) 100. kPa to mmHg
    (d) 200. kPa to atm
    (e) 36.0 kPa to atm
    (f) 600. kPa to mmHg
12. Mercury has a density of 13.55 g/cm^3. A barometer is constructed using an oil with a density of 0.75 g/cm^3. If the atmospheric pressure is 1.0 atm, calculate the height in meters of the oil column in the barometer.
13. Why can't a hand-driven pump on a water well pull underground water from depths more than 33 ft? Would it help to have a motor-driven vacuum pump?
14. A scuba diver taking photos of a coral reef 60 ft below the ocean surface breathes out a stream of bubbles. What is the total gas pressure of these bubbles at the moment they are released? What is the gas pressure in the bubbles when they reach the surface of the ocean?

Kinetic-Molecular Theory (Section 8-2)

15. List the five basic postulates of the kinetic-molecular theory. Which assumption is incorrect at very high pressures? Which one is incorrect at low temperatures? Which assumption is probably most nearly correct?
16. Use the postulates of the kinetic-molecular theory to explain each phenomenon.
    (a) $Br_2(g)$ is reddish brown and transparent; $Br_2(\ell)$ is very dark brown and very little light passes through it.
    (b) When equal volumes of $Br_2(g)$ and $N_2(g)$ at the same T and P are brought into contact, they mix rapidly and the color is only half as dark as the initial Br_2 color.

The Behavior of Ideal Gases: Gas Laws (Section 8-3)

17. If 50.75 g of a gas occupies 10.0 L at STP, calculate the volume 129.3 g of the gas occupies at STP.
18. A sample of a gas has a pressure of 100. mmHg in a sealed 125-mL flask. This gas sample is transferred to another flask with a volume of 200. mL. Calculate the new pressure. Assume that the temperature remains constant.
19. Some butane, the fuel used in backyard grills, is placed in a sealed 3.50-L container at 25 °C; its pressure is 735 mmHg. You transfer the gas to a sealed 15.0-L container, also at 25 °C. Calculate the pressure of the gas in the larger container.
20. A sample of gas at 30. °C has a pressure of 2.0 atm in a sealed 1.0-L container. Calculate the pressure it will exert in a 4.0-L container. The temperature does not change.
21. Suppose you have a sample of CO_2 in a gas-tight syringe with a movable piston. The gas volume is 25.0 mL at a room temperature of 20. °C. Calculate the final volume of the gas if you hold the syringe in your hand to raise the gas temperature to 37 °C.
22. A sample of gas has a volume of 2.50 L at a pressure of 670. mmHg and a temperature of 80. °C. If the pressure remains constant but the temperature is decreased, the gas occupies 1.25 L. Determine this new temperature, in degrees Celsius.
23. A bicycle tire is inflated to a pressure of 3.74 atm at 15 °C. The tire is heated to 35 °C. Calculate the pressure in the tire. Assume the tire volume doesn't change.
24. An automobile tire is inflated to a pressure of 3.05 atm on a rather warm day when the temperature is 40. °C. The car is then driven to the mountains and parked overnight. The morning temperature is −5.0 °C. Calculate the gas pressure in the tire. Assume the volume of the tire doesn't change.
25. A sample of gas occupies 754 mL at 22 °C and a pressure of 165 mmHg. Calculate its volume if the temperature is raised to 42 °C and the pressure is raised to 265 mmHg. (The amount of gas does not change.)
26. A balloon is filled with helium to a volume of 1.05×10^3 L on the ground, where the pressure is 745 mmHg and the temperature is 20. °C.
    (a) Calculate the amount (mol) of helium in the balloon.
    (b) Calculate the volume of helium when the balloon ascends to a height of 2 miles, where the pressure is only 600. mmHg and the temperature is −33 °C.
27. Calculate the pressure exerted by 1.55 g Xe gas at 20. °C in a sealed 560-mL flask.
28. A 1.00-g sample of water is allowed to vaporize completely inside a sealed 10.0-L container. Calculate the pressure of the water vapor at a temperature of 150. °C.
29. Which of these gas samples contains the largest number of molecules and which contains the smallest?
    (a) 1.0 L H_2 at STP
    (b) 1.0 L N_2 at STP
    (c) 1.0 L H_2 at 27 °C and 760. mmHg
    (d) 1.0 L CO_2 at 0 °C and 800. mmHg
30. Ozone molecules attack rubber and cause cracks to appear. If enough cracks occur in a rubber tire, for example, it will be weakened, and the tread will wear away much faster. As little as 0.020 ppm O_3 will cause cracks to appear in rubber in about 1 hour. Assume that a 1.0-cm^3 sample of air containing 0.020 ppm O_3 is brought in contact with a sample of rubber that is 1.0 cm^2 in area. Calculate the number of O_3 molecules that are available to collide with the rubber surface. The temperature of the air sample is 25 °C and the pressure is 0.95 atm.
31. To find the volume of a flask, the flask is evacuated so it contains no gas. Next, 4.4 g CO_2 is introduced into the flask. On warming to 27 °C, the gas exerts a pressure of 730. mmHg. Calculate the volume of the flask in milliliters.
32. Determine the mass of helium required to fill a 5.0-L balloon to a pressure of 1.1 atm at 25 °C.

Gas Density, Molar Mass, Ideal Gas Law (Section 8-4)

33. Calculate the molar mass of a gas that has a density of 5.75 g/L at STP.
34. A 0.423-g sample of an unknown gas exerts a pressure of 0.965 atm in a 1.00-L container at 445.7 K. Calculate the molar mass of the gas.
35. Forty miles above Earth's surface the temperature is −23 °C, and the pressure is only 0.20 mmHg. Determine the density of air (molar mass = 29.0 g/mol) at this altitude.
36. A newly discovered gas has a density of 2.39 g/L at 23.0 °C and 715 mmHg. Determine the molar mass of the gas.
37. Consider two 5.0-L containers, each filled with gas at 25 °C. One container is filled with helium and the other with N_2. The density of gas in the two containers is the same. What is the relationship between the pressures in the two containers?
38. A hydrocarbon with the general formula C_xH_y is 92.26% carbon. Experiment shows that 0.293 g hydrocarbon fills a 185-mL flask at 23 °C with a pressure of 374 mmHg. Calculate the molecular formula for this compound.

Quantities of Gases in Chemical Reactions (Section 8-5)

39. When a commercial drain cleaner containing sodium hydroxide and small pieces of aluminum is poured, along with water, into a clogged drain, this reaction occurs:

$$2\ Al(s) + 2\ NaOH(aq) + 6\ H_2O(\ell) \longrightarrow 2\ NaAl(OH)_4(aq) + 3\ H_2(g)$$

If 6.5 g Al and excess NaOH are reacted, calculate the volume of H_2 gas produced at 742 mmHg and 22.0 °C.

40. Water can be made by combining gaseous O_2 and H_2. You begin with 1.5 L $H_2(g)$ at 360. mmHg and 23 °C. Calculate the volume in liters of $O_2(g)$ needed for complete reaction if the O_2 gas is also measured at 360. mmHg and 23 °C.
41. Gaseous silane, SiH_4, ignites spontaneously in air according to the equation

$$SiH_4(g) + 2\,O_2(g) \longrightarrow SiO_2(s) + 2\,H_2O(g)$$

If 5.2 L SiH_4 reacts with O_2, determine the volume in liters of O_2 required for complete reaction. Determine the volume of H_2O vapor produced. Assume all gases are measured at the same temperature and pressure.

42. A 0.05-g sample of the boron hydride, B_4H_{10}, is burned in pure oxygen to give B_2O_3 and H_2O.

$$2\,B_4H_{10}(s) + 11\,O_2(g) \longrightarrow 4\,B_2O_3(s) + 10\,H_2O(g)$$

Calculate the pressure of the gaseous water in a 4.25-L flask at 30. °C.

43. If 1.0×10^3 g uranium metal is converted to gaseous UF_6, calculate the pressure of UF_6 at 62 °C in a chamber that has a volume of 3.0×10^2 L.
44. Ten liters of F_2 gas at 1.00 atm and 100.0 °C reacts with 99.9 g $CaBr_2$ to form CaF_2 and bromine gas. Calculate the volume of Br_2 gas formed at this temperature and pressure.
45. Metal carbonates decompose to the metal oxide and CO_2 on heating according to this general equation.

$$M_x(CO_3)_y(s) \longrightarrow M_xO_y(s) + y\,CO_2(g)$$

You heat 0.158 g of a white, solid carbonate of a Group 2A metal and find that the evolved CO_2 has a pressure of 69.8 mmHg in a 285-mL flask at 25 °C. Determine the molar mass of the metal carbonate.

46. Nickel carbonyl, $Ni(CO)_4$, can be made by the room-temperature reaction of finely divided nickel metal with gaseous CO. This is the basis for purifying nickel on an industrial scale. If you have CO in a sealed 1.50-L flask at a pressure of 418 mmHg at 25.0 °C, calculate the maximum mass in grams of $Ni(CO)_4$ that can be made.
47. Assume that a car burns octane, C_8H_{18} ($d = 0.703$ g/cm^3).
    (a) Write the balanced equation for burning octane in air, forming CO_2 and H_2O.
    (b) The car has a fuel efficiency of 32 miles per gallon of octane; determine the volume of CO_2 at 25 °C and 1.0 atm that is generated when the car goes on a 10.-mile trip.
48. Follow the directions in the previous question, but use methanol, CH_3OH ($d = 0.791$ g/cm^3) as the fuel. Assume the fuel efficiency is 20. miles per gallon.
49. The build-up of excess carbon dioxide in the air of a submerged submarine is prevented by reacting CO_2 with sodium peroxide, Na_2O_2.

$$2\,Na_2O_2(s) + 2\,CO_2(g) \rightarrow 2\,Na_2CO_3(s) + O_2(g)$$

Calculate the mass of Na_2O_2 needed in a 24.0-h period per submariner if each exhales 240 mL CO_2 per minute at 21 °C and 1.02 atm.

Gas Mixtures and Partial Pressures (Section 8-6)

50. Calculate the total pressure of a mixture of 1.50 g H_2 and 5.00 g N_2 in a sealed 5.0-L vessel at 25 °C.
51. At 298 K, a sealed 750-mL vessel contains equimolar amounts of O_2, H_2, and He at a total pressure of 3.85 atm. Determine the partial pressure of the H_2 gas.
52. A sample of the atmosphere at a total pressure of 740. mmHg is analyzed to give these partial pressures: $P(N_2) = 575$ mmHg; $P(Ar) = 6.9$ mmHg; $P(CO_2) = 0.2$ mmHg; $P(H_2O) = 4.0$ mmHg. No other gases except O_2 have appreciable partial pressures. Calculate
    (a) the partial pressure of O_2.
    (b) the mole fraction of each gas.
    (c) the composition of this sample in percentage by volume. Compare your results with those of Table 8.6.
53. Gaseous CO exerts a pressure of 45.6 mmHg in a 56.0-L tank at 22.0 °C. This gas is released into a room with a volume of 2.70×10^4 L; determine the partial pressure of CO (in mmHg) in the room at 22 °C.
54. Three flasks are connected as shown. The starting conditions, with the stopcocks closed, are shown. Assume T does not change.
    (a) Determine the final pressure inside the system when all the stopcocks are open.
    (b) Calculate the partial pressure of each of the three gases. Assume that the connecting tube has negligible volume.

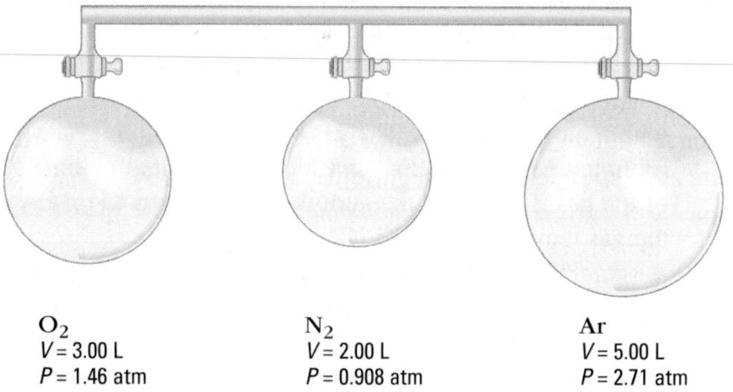
O2
V = 3.00 L
P = 1.46 atm
N2
V = 2.00 L
P = 0.908 atm
Ar
V = 5.00 L
P = 2.71 atm

55. The density of air at 20.0 km above Earth's surface is 92 g/m^3. The pressure is 42 mmHg and the temperature is −63 °C. Assuming the atmosphere contains only O_2 and N_2, calculate
    (a) the average molar mass of the air at 20.0 km.
    (b) the mole fraction of each gas.
56. Benzene has acute health effects. For example, it causes mucous membrane irritation at a concentration of 100 ppm and fatal narcosis at 20,000 ppm (by volume). Calculate the partial pressures in atmospheres at STP corresponding to these concentrations.
57. The mean fraction *by mass* of water vapor and cloud water in Earth's atmosphere is about 0.0025. Assume that the atmosphere contains two components: "air," with a molar mass of 29.2 g/mol, and water vapor. Determine the *mean* mole fraction of water vapor in Earth's atmosphere. Determine the *mean* partial pressure of water vapor. Why is this so much smaller than the typical partial pressure of water vapor at Earth's surface on a rainy summer day (25 mmHg)?

58. Acetylene can be made by reacting calcium carbide with water.

$$CaC_2(s) + 2\ H_2O(\ell) \longrightarrow C_2H_2(g) + Ca(OH)_2(aq)$$

Assume that you place 2.65 g CaC_2 in excess water and collect the acetylene over water. The volume of the acetylene and water vapor is 795 mL at 25.0 °C and a barometric pressure of 735.2 mmHg. Calculate the percent yield of acetylene. The vapor pressure of water at 25 °C is 23.8 mmHg.

59. Potassium chlorate, $KClO_3$, can be decomposed by heating.

$$2\ KClO_3(s) \longrightarrow 2\ KCl(s) + 3\ O_2(g)$$

If 465 mL gas was collected over water at a total pressure of 750. mmHg and a temperature of 25 °C, calculate the mass of O_2 collected.

Kinetic-Molecular Theory: Velocities of Gas Molecules (Section 8-7)

60. You are given two flasks of equal volume. Flask A contains H_2 at 0 °C and 1 atm pressure. Flask B contains CO_2 gas at 0 °C and 2 atm pressure. Compare these two samples with respect to each of these properties.
(a) Average kinetic energy per molecule
(b) Average molecular velocity
(c) Number of molecules

61. Place these gases in order of increasing average molecular speed at 25 °C: Kr, CH_4, N_2, and CH_2Cl_2.

62. Arrange these four gases in order of increasing average molecular speed at 25 °C: Cl_2, F_2, N_2, and O_2.

63. If equal amounts of the four inert gases Ar, Ne, Kr, and Xe are released at the same time at one end of a long, evacuated tube, which gas will reach the other end of the tube first? Explain your answer.

64. The reaction of SO_2 with Cl_2 to give dichlorine oxide is

$$SO_2(g) + 2\ Cl_2(g) \longrightarrow SOCl_2(g) + Cl_2O(g)$$

Place all molecules in the equation in order of increasing rate of effusion.

65. List all the gases in the atmosphere (Table 8.5) in order of *decreasing* average molecular speed. Concern has been expressed about one of these gases escaping into outer space because a significant fraction of its molecules have velocities large enough to break free from Earth's gravitational field. Which gas is it? (Assume the same T for all gases.)

The Behavior of Real (Non-ideal) Gases (Section 8-8)

66. From the density of liquid water and its molar mass, calculate the volume that 1 mol liquid water occupies. If water were an ideal gas at STP, what volume would a mole of water vapor occupy? Can we achieve the STP conditions for water vapor? Why or why not?

67. At low temperatures and very low pressures, gases behave ideally, but as the pressure is increased the product PV becomes less than the product nRT. Give a molecular-level explanation of this fact.

68. At high temperatures and low pressures, gases behave ideally, but as the pressure is increased the product PV becomes greater than the product nRT. Give a molecular-level explanation of this fact.

69. The densities of liquid noble gases and their normal boiling points are given in this table.

Gas	Normal Boiling Point (K)	Liquid Density (g/cm^3)
He	4.2	0.125
Ne	27.1	1.20
Ar	87.3	1.40
Kr	120.	2.42
Xe	165	2.95

Calculate the volume occupied by 1 mol of each of these liquids. Comment on any trend that you see. Determine the volume occupied by exactly 1 mol of each of these substances as an ideal gas at STP. Which gas would you expect to show the largest deviations from ideality at room temperature? Why?

70. Use the van der Waals constants in Table 8.5 to predict whether N_2 or CO_2 behaves more like an ideal gas at high pressures.

71. Calculate the pressure of 7.0 mol CO_2 in a sealed 2.00-L vessel at 50 °C using
(a) the ideal gas equation.
(b) the van der Waals equation.

72. Without looking at Table 8.5, predict which of these gases: Ne; N_2; H_2O; or CH_4
(a) has the largest van der Waals constant a.
(b) has the smallest van der Waals constant b.

The Atmosphere (Section 8-9)

73. Explain the major roles played by nitrogen in the atmosphere. Do the same for oxygen.

74. Beginning at Earth's surface and proceeding upward, name the first two layers or regions of the atmosphere. Describe, in general, the kinds of chemical reactions that occur in each layer.

75. (a) Calculate the volume of air in liters that you would inhale in 24 hours assuming that you inhaled 16 breaths per minute and each breath had a volume of approximately 0.50 L. (T = 18.0 °C; P = 0.970 atm.)
(b) Compare that total volume to the volume of air in a typical residence hall room, approximately 864 ft^3 (approx. 28 L/ft^3).
(c) Calculate the number of oxygen molecules you inhaled during that time.

76. At a spot 3,000 feet above sea level you take a sip of water through a straw before you begin a mountain hike. You take another sip when you reach the top at 10,400 ft. At which elevation is it easier to sip the water? Explain.

77. Felix Baumgartner, wearing a special pressurized suit, set a new skydiving record on October 14, 2012 by free falling from an altitude of 39 km, near the top of the stratosphere. Baumgartner was in a state of weightlessness for the first 25 s of his free fall. Explain why he was able to

gain maneuverability and ultimately deploy his parachute only after reaching the troposphere.

Stratospheric Ozone Depletion (Section 8-10)

78. Write the products for these reactions that take place in the stratosphere.
(a) $CF_3Cl \xrightarrow{h\nu}$ (b) $\cdot Cl + \cdot O \cdot \longrightarrow$
(c) $ClO\cdot + \cdot O \cdot \longrightarrow$
79. Can ozone form in the stratosphere at night? Explain why or why not.
80. The molecule CH_3F has much less ozone-depletion potential than the corresponding molecule CH_3Cl. Explain why.
81. Can CFCs catalyze the destruction of ozone in the stratosphere at night? Explain.
82. Are CFCs toxic? Compare the toxicity of CFCs with that of compounds used for refrigeration before CFCs were invented. Look up the toxicity of these compounds on the Internet.

Greenhouse Gases and Global Warning (Section 8-11)

83. What is the difference between the greenhouse effect and global warming? How are they related?
84. Name four greenhouse gases, and explain why they are called that.
85. Carbon dioxide is known to be a major contributor to the greenhouse effect. List some of its sources in our atmosphere and some of the processes that remove it. Currently, which predominates—the production of CO_2 or its removal?
86. Name a favorable effect of the global increase of CO_2 in the atmosphere.

Chemistry of Air Quality and Air Pollution (Section 8-12)

87. Define air pollution in terms of the kinds of pollutants, their sources, and the ways they are harmful.
88. Assume that limestone, $CaCO_3$, is used to remove 90.% of the sulfur from 4.0 metric tons of coal containing 2.0% S. The product is $CaSO_4$.

$$CaCO_3(s) + SO_3(g) \longrightarrow CaSO_4(s) + CO_2(g)$$

Calculate the mass of limestone required. Express your answer in metric tons.
89. Approximately 65 million metric tons of SO_2 enter the atmosphere every year from the burning of coal. If coal, on average, contains 2.0% S, calculate how many metric tons of coal were burned to produce this much SO_2. A 1000-MW power plant burns about 700. metric tons of coal per hour. Calculate the number of hours the quantity of coal will burn in one of these power plants.
90. Calculate the mass of gasoline that must be burned according to the reaction

$$C_8H_{18}(\ell) + 8.5\ O_2(g) \longrightarrow 8\ CO(g) + 9\ H_2O(g)$$

to raise the CO concentration to 1000. ppm in a garage that measures 7.00 m × 3.00 m × 3.00 m. (Assume STP conditions.)
91. What atmospheric reaction produces nitrogen monoxide, NO?
92. Give an example of a situation where atmospheric ozone is beneficial and an example of a situation where it is harmful. Explain how ozone is beneficial and how it is harmful.
93. The air pollutant sulfur dioxide, SO_2, is known to increase mortality in people exposed to it for 24 hours at a concentration of 0.175 ppm.
(a) Calculate the mole fraction of SO_2 when its mass fraction is 0.175 ppm.
(b) Calculate the partial pressure of SO_2 at the same mass fraction.
(c) Calculate the mass in micrograms of SO_2 in 1 m^3 of air at STP.

General Questions

These questions are not explicitly keyed to chapter topics; many require integration of several concepts.

94. HCl can be made by the direct reaction of H_2 and Cl_2 in the presence of light. Assume that 3.0 g H_2 and 140. g Cl_2 are mixed in a 10-L flask at 28 °C, and the flask is sealed.
Before the reaction:
(a) Calculate the partial pressures of the two reactants.
(b) Calculate the total pressure in the flask.
After the reaction:
(c) Calculate the total pressure in the flask.
(d) What reactant remains in the flask? Calculate the amount (mol) that remains.
(e) Calculate the partial pressure of each gas.
(f) Calculate the pressure inside the flask if the temperature is increased to 40. °C.
95. Worldwide, about 100. million metric tons of H_2S are produced annually from sources that include the oceans, bogs, swamps, and tidal flats. One of the major sources of SO_2 in the atmosphere is the oxidation of H_2S, produced by the decay of organic matter. The reaction in which H_2S molecules are oxidized to SO_2 involves O_3. Write an equation showing that one molecule of each reactant combines to form two product molecules, one of them being SO_2. Then, calculate the annual production in tons of H_2SO_4, assuming all of this SO_2 is converted to sulfuric acid.
96. Calculate the densities of Cl_2 and of SO_2 at 25 °C and 0.750 atm. Then, calculate the density of Cl_2 at 35 °C and 0.750 atm and the density of SO_2 at 25 °C and 2.60 atm.
97. The gas burner in a stove or furnace admits enough air so that methane gas can react completely with oxygen in the air according to the equation

$$CH_4(g) + 2\ O_2(g) \longrightarrow CO_2(g) + 2\ H_2O(g)$$

Air is one-fifth oxygen by volume. Both air and methane gas are supplied to the flame by passing them through separate small tubes. Compared with the tube for the methane gas, determine how much bigger the cross section of the tube for the air needs to be. Assume that both gases are at the same T and P.

98. You have 100 balloons of equal volume filled with a total of 26.8 g helium gas at 23.0 °C and 748 mmHg. The total volume of these balloons is 168 L. You are given 150 more balloons of the same size and 41.8 g He gas. The temperature and pressure remain the same. Determine by calculation whether you will be able to fill all the balloons with the He you have available.
99. The statement is made in Section 8-9 that the mass of Earth's atmosphere is 5.3×10^{15} metric tons. Perform calculations to show that this value is correct. 1 metric ton = 10^3 kg. The surface area of Earth is 5.1×10^8 km^2.
100. Argon is the most abundant atmospheric gas after nitrogen and oxygen. At a volume percent of 0.934, you might not consider argon as being "abundant". Calculate the mass (kg) of argon in the atmosphere. Assume the density of argon to be 1.6 g/L.

Applying Concepts

These questions test conceptual learning.

101. At 25 °C, the measured pressure of acetic acid vapor, $CH_3COOH(g)$, is significantly lower than that predicted by the ideal gas law. Explain this difference.
102. The air in a flask is evacuated by a high-quality vacuum system. The vacuum created corresponds to 1.0×10^{-8} Torr at 25 °C. Calculate the number of molecules of air per cm^3 remaining in the apparatus at this temperature and pressure.
103. If all the ozone, O_3, in the atmosphere could be isolated and brought to Earth's surface at standard temperature and pressure, the ozone would form a 3-mm thick layer around Earth. Assume that the volume of this ozone layer around Earth equals the thickness multiplied by the area, A: $A = 4\pi r^2$; r is the radius of Earth, 6.37×10^3 km.
    (a) Calculate the number of ozone molecules that would be in this layer.
    (b) To put this number of ozone molecules into perspective, compare it to the number of carbon dioxide molecules emitted annually by human activities such as the burning fossil fuels, cement production, and deforestation. It is estimated that 7.90×10^{15} g CO_2 is generated in this way.
104. The average kinetic energy of a gas molecule at 20.0 °C is 3.66×10^3 J/mol. Calculate the average velocity of carbon dioxide molecules in air at a temperature of 20.0 °C. Compare this average velocity with that of nitrogen molecules in the air at the same temperature. Calculate by what percent the average velocities differ. Explain the difference.
105. Consider a sample of N_2 gas under conditions in which it obeys the ideal gas law exactly. Which of these statements is/are true?
    (a) A sample of Ne(g) under the same conditions must obey the ideal gas law exactly.
    (b) The speed at which one particular N_2 molecule is moving changes from time to time.
    (c) Some N_2 molecules are moving more slowly than some of the molecules in a sample of $O_2(g)$ under the same conditions.
    (d) Some N_2 molecules are moving more slowly than some of the molecules in a sample of Ne(g) under the same conditions.
    (e) When two N_2 molecules collide, it is possible that both may be moving faster after the collision than they were before.
106. Which of these graphs best represents the distribution of molecular speeds for the gases acetylene, C_2H_2, and N_2? Both gases are in the same sealed flask with a total pressure of 750. mmHg. The partial pressure of N_2 is 500. mmHg.

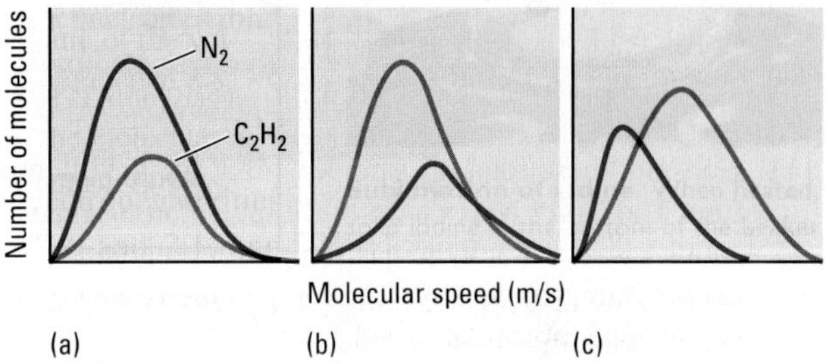

Number of molecules
N2
C2H2
Molecular speed (m/s)
(a)
(b)
(c)

107. Draw a graph representing the distribution of molecular speeds for the gases ethane, C_2H_6, and F_2 when both are in the same sealed flask with a total pressure of 720. mmHg and a partial pressure of 540. mmHg for F_2.
108. In this chapter Boyle's, Charles's, and Avogadro's laws were presented as word statements and mathematical relationships. Express each of these laws graphically.
109. Consider these four samples of helium (green spheres represent He atoms), all at the same temperature. The larger boxes have twice the volume of the smaller boxes. Rank the gas samples with respect to: (a) pressure, (b) density, (c) average kinetic energy, and (d) average molecular speed.

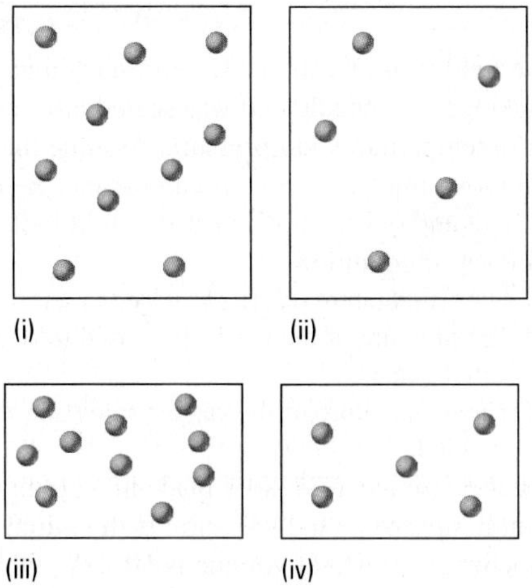

(i)
(ii)
(iii)
(iv)

110. Consider these four gas samples, all at the same temperature. The larger boxes have twice the volume of the smaller boxes. Rank the gas samples with respect to: (a) pressure, (b) density, (c) average kinetic energy, and

(d) average molecular speed. (Green spheres are He; violet spheres are Ne.)

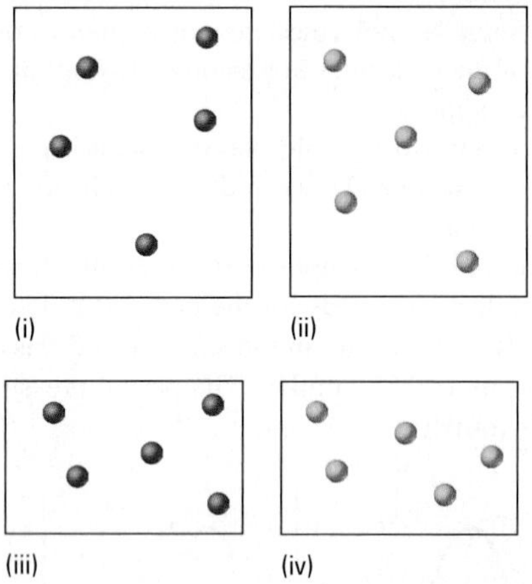
(i)
(ii)
(iii)
(iv)

111. Consider these four gas samples, all at the same temperature. The larger boxes have twice the volume of the smaller boxes. Rank the gas samples with respect to: (a) pressure, (b) density, (c) average kinetic energy, and (d) average molecular speed.

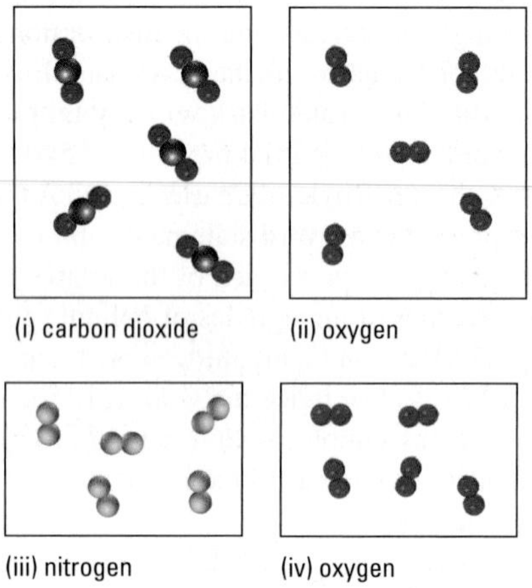
(i) carbon dioxide
(ii) oxygen
(iii) nitrogen
(iv) oxygen

112. The figure for Question 112 represents a gas collected in a syringe (the needle end was sealed after collecting) at room temperature and pressure. Assume that the plunger can move freely, but no gas can escape. Redraw the syringe and gas to show what it would look like under each set of conditions.
(a) The temperature of the gas is decreased by one half.
(b) The pressure of the gas is decreased to one half of its initial value.
(c) The temperature of the gas is tripled and the pressure is doubled.

113. A gas phase reaction takes place in a syringe at a constant temperature and pressure. If the initial volume is 40. cm^3 and the final volume is 60. cm^3, which of these general reactions took place? Explain your reasoning.
(a) $A(g) + B(g) \longrightarrow AB(g)$
(b) $2\,A(g) + B(g) \longrightarrow A_2B(g)$
(c) $2\,AB_2(g) \longrightarrow A_2(g) + 2\,B_2(g)$
(d) $2\,AB(g) \longrightarrow A_2(g) + B_2(g)$
(e) $2\,A_2(g) + 4\,B(g) \longrightarrow 4\,AB(g)$

Figure for Question 112.

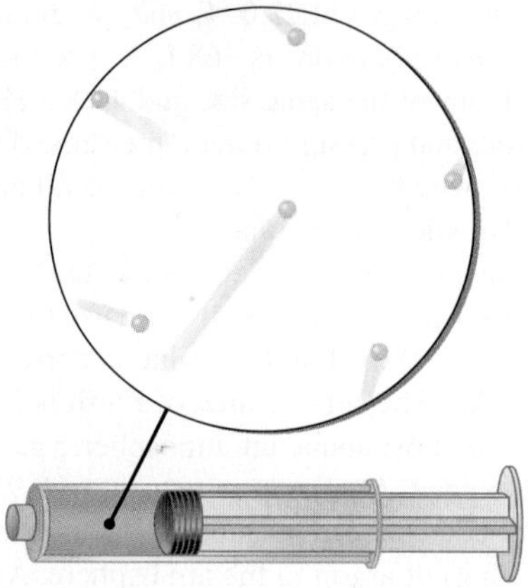

114. The gas molecules in the box undergo a reaction at constant temperature and pressure.

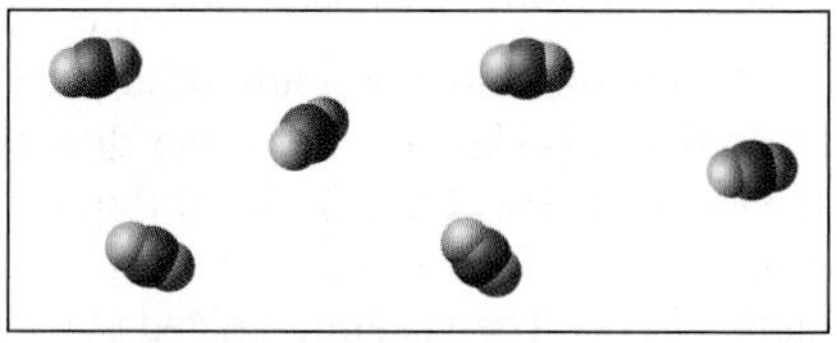

If the initial volume is 1.8 L and the final volume is 0.9 L, which of the boxes (a) through (e) could be the products of the reaction? Explain your reasoning.

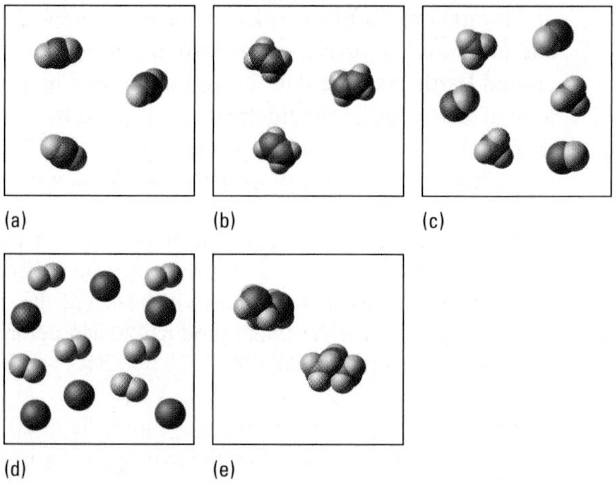
(a)
(b)
(c)
(d)
(e)

115. A substance is analyzed and found to contain 85.7% carbon and 14.3% hydrogen by mass. A gaseous sample of the substance is found to have a density of 1.87 g/L at STP.
(a) Calculate the molar mass of the compound.
(b) Determine the empirical and molecular formulas of the compound.
(c) Draw two possible Lewis structures for molecules of the compound.

116. A compound consists of 37.5% C, 3.15% H, and 59.3% F by mass. When 0.298 g of the compound is heated to 50. °C in an evacuated 125-mL flask, the pressure is observed to be 750. mmHg. The compound has three isomers.

(a) Calculate the molar mass of the compound.
(b) Determine the empirical and molecular formulas of the compound.
(c) Draw the Lewis structure for each isomer of the compound.

117. One very cold winter day you and a friend purchase a helium-filled balloon. As you leave the store and walk down the street, your friend notices the balloon is not as full as it was a moment ago in the store. He says the balloon is defective and he is taking it back. Do you agree with him? Explain why you do or do not agree.

More Challenging Questions

These questions require more thought and integrate several concepts.

118. A 2.69-g PCl_5 sample was completely vaporized in a 1.00-L flask at 250. °C. The resulting pressure in the flask was 1.00 atm. At this temperature, there is the possibility that some $PCl_5(g)$ decomposed to $PCl_3(g)$ and $Cl_2(g)$.
(a) Show calculations to determine whether any of the $PCl_5(g)$ decomposed.
(b) If some of the $PCl_5(g)$ decomposed, calculate the partial pressures of each of the three gaseous species under these experimental conditions.

119. The mean free path in a gas is the average distance traveled by a molecule (or atom of monoatomic gas) between its collision with another molecule (or atom). As expected, the mean free path depends on the number of molecules (or atoms) per cubic centimeter and upon the sizes of the molecules (or atoms) themselves. The mean free path can be calculated using the equation

$$\lambda = \frac{1}{\sqrt{2}\pi N\sigma^2}$$

in which σ represents the diameter of the atom or molecule, λ the distance traveled between collisions, and N, the number of molecules (or atoms) per cubic centimeter.
(a) Calculate the mean free path of gaseous argon atoms at 0.0 °C and 1.0 atm. The radius of an argon atom is 91 pm.
(b) Calculate how much greater the mean free path of argon atoms is compared to the diameter of the argon atom.
(c) Calculate the pressure (atm) required to change the mean free path of argon to 1.0 cm at 0.0 °C.

120. The relation between the average kinetic energy of a molecule, $\frac{1}{2}mv^2$, and the absolute temperature is

$$\frac{1}{2}mv^2 = \frac{3}{2}kT$$

m is the mass of the molecule; v is its average velocity; k is 1.38×10^{-23} J/K; T is the absolute temperature. $1\ \text{J} = 1\ \text{kg m}^2\ \text{s}^{-2}$. Calculate the average velocity of a nitrogen dioxide molecule in the atmosphere at 27.0 °C.

121. The reaction between the gases NH_3 and HBr produces NH_4Br, a white solid. The two gases are introduced simultaneously at opposite ends of an evacuated glass tube that is 1.0 m long. Calculate how far from the NH_3 end of the tube the white solid will form.

122. It is estimated that 50 g He are lost per second from the uppermost part of Earth's atmosphere. At this point, about 500 km above Earth's surface, there are so few atoms present that collisions between them are exceptionally rare and the helium atoms simply leave the top of the atmosphere and go into outer space.
(a) Use the equation given in question 120 to calculate the average velocity (km/s) of a helium atom at 1.0×10^3 K, the typical temperature of the atmosphere at 500 km above Earth's surface.
(b) To escape from the atmosphere, any object (including a molecule) must be traveling at 11 km/s or more. In light of the answer to part (a), explain how helium escapes from Earth's atmosphere.

123. A neon atom has a radius of 68 pm.
(a) Calculate the volume of 1 mol Ne atoms; $V = 4\pi r^3/3$.
(b) Calculate the fraction of the total volume of Ne(g) at 20. °C and 50. atm that is occupied by Ne atoms.

124. Acetic acid vapor contains both single acetic acid molecules and dimers of acetic acid in which two molecules of acetic acid are hydrogen bonded to each other. At 77 °C and 1.00 atm, the density of the vapor is 3.23 g/L. Calculate the percentage of acetic acid vapor that exists in the dimer form at these conditions. Does this percent increase or decrease with increasing temperature? Explain.

125. An ideal gas was contained in a glass vessel of unknown volume with a pressure of 0.960 atm. Some of the gas was withdrawn from the vessel and used to fill a 25.0-mL glass bulb to a pressure of 1.00 atm. The pressure of the gas remaining in the vessel of unknown volume was 0.882 atm. All the measurements were done at the same temperature. Determine the volume of the vessel.

126. You are holding two balloons, an orange balloon and a blue balloon, both at the same temperature and pressure. The orange balloon is filled with neon gas and the blue balloon is filled with argon gas. The orange balloon has twice the volume of the blue balloon. Determine the mass ratio of Ne to Ar in the two balloons.

127. A container of gas has a pressure of 550. Torr. A chemical change then occurs that consumes half of the molecules present at the start and produces two new molecules for each three consumed. Calculate the new pressure in the container if T and V are unchanged.

128. The effects of intermolecular interactions on gas properties depend on T and P. Do these effects become more or less significant when each change occurs? Why?
(a) A sealed container of gas is compressed to a smaller volume at constant temperature.
(b) A container of gas has more gas added into the same volume at constant temperature.
(c) The gas in a container of variable volume is heated at constant pressure.

129. Formaldehyde, CH_2O, is a volatile organic compound that is sometimes released from insulation used in home construction, and it can be trapped and build up inside

the home. When this happens, people exposed to the formaldehyde can suffer adverse health effects. The U. S. National Institute of Occupational Health and Safety (NIOSH) guideline for the maximum allowable concentration of formaldehyde in air in the workplace is 16 ppb (parts per billion) for an eight-hour average exposure.
(a) Determine the partial pressure of formaldehyde at the maximum allowable level of 16 ppb.
(b) Calculate how many molecules of formaldehyde are present in each cubic centimeter of air when formaldehyde is present at 16 ppb.
(c) Calculate how many total molecules of formaldehyde are present in a room: 15.0 ft long × 10.0 ft wide × 8.00 ft high (at 16 ppb).

Conceptual Challenge Problems

These rigorous, thought-provoking problems integrate conceptual learning with problem solving and are suitable for group work.

CP8.A (Section 8-3) Under what conditions would you expect to observe that the pressure of a confined gas at constant temperature and volume is *not* constant?

CP8.B (Section 8-3) Suppose that the gas constant, R, were defined as 1.000 L atm mol^{-1} deg^{-1} where the "deg" referred to a newly defined Basic temperature scale. Calculate the melting and boiling temperatures of water in degrees Basic (°B).

Liquids, Solids, and Materials

9

Panoramic/ZUMA Press/Newscom

The 102-foot, 89-ton Tûranor PlanetSolar is the first solar-powered ship to circle the world, doing so on an epic 19-month journey. The vessel's decks are covered by 825 solar panels that generate electricity to power the ship. Batteries provide back-up power at night and for up to three days when the sun is not shining. What are solar cells? How do they generate electricity when sunlight strikes their surface? What role does ultrapure silicon play in solar cells? What else is required in addition to the silicon? The nature and operation of solar cells is described in Section 9-10.

Gases and their behavior were described Chapter 8. In this chapter we consider liquids and solids. These are referred to as condensed states of matter because they can be formed by condensation of gases (such as the condensation of water vapor from air to form dew or frost). Condensed states of matter differ from gases at the nanoscale/molecular level in two important ways:

- At ordinary temperatures and pressures, the particles (atoms, molecules, or ions) that constitute a liquid or a solid are very much closer together than those of a gas. In fact, they are in contact with each other, whereas in a gas the particles are very far apart (← **Sec. 8-2**).

- There is no single equation analogous to the ideal gas law that describes the behavior of all liquids or solids; however, because particles are close together, noncovalent intermolecular forces have a major influence on physical properties (← **Sec. 7-6**).

To see that the molecules are much closer together in a liquid or a solid than in a gas, compare the volume of a mole of a gas with the volume of a mole of a liquid. For any gas at 1 atm and 25.0 °C, the volume of 1.00 mol gas can be calculated using the ideal gas law equation. For example, for krypton gas:

$$V_{Kr} = \frac{nRT}{P} = \frac{(1.00\ \text{mol})(0.0821\ \text{L atm mol}^{-1}\text{K}^{-1})(289\ \text{K})}{1.00\ \text{atm}} = 24.5\ \text{L}$$

$$= 2.45 \times 10^4\ \text{cm}^3$$

The molar volumes of a series of representative liquids at 1 atm and 25.0 °C are given in Table 9.1. These values can be calculated by dividing the molar mass by the density. For example, for liquid hexane at 1 atm and 25.0 °C,

$$V_m(\text{hexane}(\ell)) = \frac{\text{molar mass}}{\text{density}} = \frac{86.2\ \text{g/mol}}{0.655\ \text{g/cm}^3} = 132\ \text{cm}^3/\text{mol}$$

Thus, the volume of 1.00 mol hexane liquid is 132 cm^3, far smaller than the volume of 1.00 mol gas. As indicated by the data in Table 9.1, the molar volumes for liquids are all *much smaller* than for gases; this is also true for solids.

9-1 Liquids, Solids, and Intermolecular Forces

Despite the lack of an equation to describe them, the macroscale properties of condensed states of matter can be related to their nanoscale (molecular) structure. The atoms, molecules, or ions in liquids and solids are close enough to have strong interactions with each other, sufficient to condense a gas into a liquid and convert the liquid into a solid. These *noncovalent intermolecular forces* were described in Section 7-6 and are briefly reviewed here:

- *London forces (dispersion forces)* are present between molecules of *all* molecular substances and between atoms of noble gases. London forces result when a *momentary* uneven distribution of electrons relative to nuclei in one molecule induces *temporary* dipoles in adjacent molecules (← **Sec. 7-6a**). London forces are the *only* noncovalent intermolecular forces among nonpolar molecules. London forces range from approximately 0.05 to 40 kJ/mol. *The magnitude of London (dispersion) forces increases with increased number of electrons in a molecule.*
- *Dipole-dipole attraction occurs between polar molecules,* in addition to London forces. *Polar molecules have permanent dipoles,* which can align so that

Table 9.1 Molar Volumes of Liquids at 25 °C and 1 atm

Compound	Molar Mass (g/mol)	Density (g/cm^3)	Molar Volume (cm^3/mol)
Water, H_2O	18.0	1.00	18.0
Ethanol, C_2H_6O	46.1	0.785	58.7
Diethyl ether, $C_4H_{10}O$	74.1	0.708	105
1-butanol, $C_4H_{10}O$	74.1	0.806	91.9
Hexane, C_6H_{14}	86.2	0.655	132

opposite charges are close to each other, attracting polar molecules to each other (← **Figure 7.19, Sec. 7-6b**). Dipole-dipole forces range from 5 to 25 kJ/mol. *In general, the more polar a substance's molecules are, the stronger the intermolecular forces are, provided the London forces are similar* (that is, the molecules have similar numbers of electrons).

- *Hydrogen bonding, the attraction of a hydrogen atom in a molecule or part of a molecule X—H for another atom or group of atoms* (← **Sec. 7-6c**), is a stronger attraction than would be predicted by London forces and dipole forces alone. Strong hydrogen bonding occurs when a hydrogen atom bonded to F, O, or N (designated X) is attracted to a lone electron pair on a small, very electronegative atom, usually F, O, or N (designated :Z). The X—H · · · :Z angle is usually 180°. The H atom and the lone pair may be in two different molecules or in different parts of the same large molecule. Hydrogen bonds range from 10 to 40 kJ/mol. As we will see, hydrogen bonding plays a significant role in the properties of water.

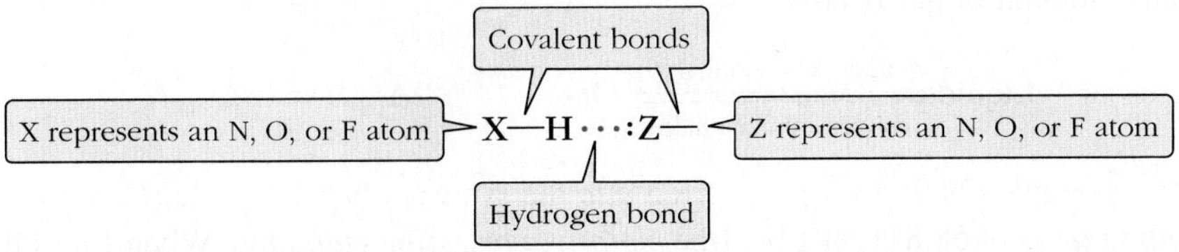
Covalent bonds
X represents an N, O, or F atom
X—H···:Z—
Z represents an N, O, or F atom
Hydrogen bond

The strengths of attractions due to London forces, dipole-dipole forces, and hydrogen bonding affect the physical properties of molecular liquids and solids. Stronger intermolecular attractions mean higher boiling point and melting point, for example. In both liquids and solids the particles are touching, which makes liquids and solids incompressible; there is little empty space between particles, so liquids and solids occupy a well-defined volume. In a liquid, particles are not held rigidly in place, so liquids flow and take the shape of their container. In a solid, atoms, molecules, or ions are constrained to specific positions in a crystal lattice, so a solid has a definite shape as well as a definite volume. Recall that in solid ionic compounds, very strong attractions among oppositely charged cations and anions result in high melting points (← **Sec. 2-6a**).

CONCEPTUAL **EXERCISE 9.1**

Noncovalent Intermolecular Forces

Describe the noncovalent intermolecular forces present in each substance; predict which has the highest boiling point. (a) CO; (b) CO_2; (c) CH_2O; (d) CH_3OH.

Answers to **EXERCISES** are provided at the back of this book in Appendix L.

EXERCISES that are labeled **CONCEPTUAL** are designed to test your understanding of one or more concepts; they usually involve qualitative rather than quantitative thinking.

9-2 Vaporization and Condensation

Like molecules of a gas, molecules of a liquid are in constant motion and have a range of kinetic energies at a given temperature, as shown in Figure 9.1. If a molecule at the surface of a liquid is moving upward *and* has sufficient energy to overcome the intermolecular attractive forces on it from other molecules in the liquid, the molecule can leave the surface and enter the gas phase. *Change of a liquid to a gas* is called **vaporization** or **evaporation** (Figure 9.1). Because work has to be done to separate molecules when they become farther apart in the gas phase, energy must be transferred into the system from the surroundings. Therefore vaporization is endothermic; the enthalpy change is called the vaporization enthalpy, $\Delta_{vap}H$ (← **Sec. 4-5d**). **Condensation** is the reverse *process in which a gas changes to a liquid;* molecules in the gas phase hit the liquid's surface, transfer energy to molecules in the liquid, and remain in the liquid phase. This is an exothermic process and the enthalpy change is the condensation enthalpy, $\Delta_{cond}H$ (← **Sec. 4-5d**). The energy transferred into the system upon vaporization of a given

Vaporization enthalpy is sometimes referred to as heat of vaporization.

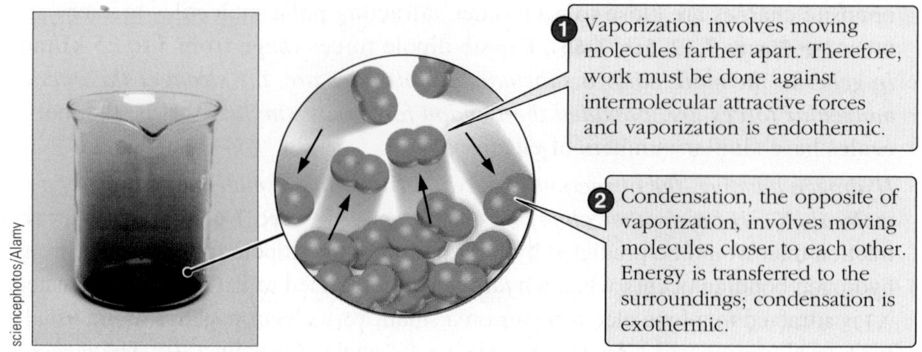
1 Vaporization involves moving molecules farther apart. Therefore, work must be done against intermolecular attractive forces and vaporization is endothermic.
2 Condensation, the opposite of vaporization, involves moving molecules closer to each other. Energy is transferred to the surroundings; condensation is exothermic.
sciencephotos/Alamy

Figure 9.1 Vaporization is endothermic; condensation is exothermic.

amount of liquid equals the energy transferred out of the system when condensation of the same amount of gas occurs.

$$\text{Liquid} \underset{\text{condensation enthalpy}}{\overset{\text{vaporization enthalpy}}{\rightleftharpoons}} \text{Gas} \qquad \Delta_{vap}H = -\Delta_{cond}H$$

For example, 29.96 kJ is transferred *into* the system when 1 mol $Br_2(\ell)$ vaporizes at its boiling point of 58.8 °C at 1 bar (the *molar vaporization enthalpy*). When 1 mol $Br_2(g)$ condenses at 58.8 °C and 1 bar, 29.96 kJ is transferred *out of* the system (the *molar condensation enthalpy*). Thus, the enthalpy changes for these two opposite processes *have equal magnitude, but opposite signs*:

$$Br_2(\ell) \longrightarrow Br_2(g) \qquad \Delta_r H° = \Delta_{vap}H = +29.96 \text{ kJ/mol}$$
$$Br_2(g) \longrightarrow Br_2(\ell) \qquad \Delta_r H° = \Delta_{cond}H = -29.96 \text{ kJ/mol}$$

CONCEPTUAL EXERCISE 9.2

Evaporative Cooling

In some countries where electric refrigeration is not readily available, drinking water is chilled by placing it in porous clay pots. Water slowly passes through the clay, and when it reaches the outer surface, it evaporates. Explain how this process cools the water inside the pot.

Table 9.2 illustrates the influence of noncovalent intermolecular forces on vaporization enthalpies. Larger intermolecular attractions make it more difficult for molecules at the surface of a liquid to escape into the gas above it, and correspondingly, make $\Delta_{vap}H$ larger. For example, consider three compounds: ethane, hydrogen chloride, and water. Their $\Delta_{vap}H$ values are C_2H_6, 15.7 kJ/mol; HCl, 17.5 kJ/mol; and H_2O, 40.7 kJ/mol. Ethane and hydrogen chloride have the same number of electrons so London forces should be similar, but hydrogen chloride is polar and its $\Delta_{vap}H$ is larger. Water has only 10 electrons, but it has a large dipole moment and, more importantly, strong hydrogen bonding, which makes its $\Delta_{vap}H$ the largest of the three.

PROBLEM-SOLVING EXAMPLE 9.1

Vaporization Enthalpy

To vaporize 50.0 g carbon tetrachloride, $CCl_4(\ell)$, requires 9.69 kJ at its normal boiling point of 76.7 °C and 1 bar pressure. Calculate the molar vaporization enthalpy of CCl_4 under these conditions.

Result $\Delta_{vap}H° = 29.8$ kJ/mol

Analyze The molar vaporization enthalpy is the enthalpy change per mole of $CCl_4(\ell)$ that vaporizes. The quantity vaporized is 50.0 g $CCl_4(\ell)$.

Plan Calculate the amount of $CCl_4(\ell)$ that vaporizes using the molar mass; then divide the enthalpy change by the amount.

Execute The molar mass of CCl_4 is 153.8 g/mol, so

$$n(\text{CCl}_4 \text{ vaporized}) = 50.0 \text{ g CCl}_4 \times \frac{1 \text{ mol CCl}_4}{153.8 \text{ g CCl}_4} = 0.325 \text{ mol CCl}_4$$

$$\Delta_{\text{vap}}H^\circ = \frac{9.69 \text{ kJ}}{0.325 \text{ mol}} = 29.8 \text{ kJ/mol}$$

☑ **Reasonable Result Check** We have approximately one-third mol $CCl_4(\ell)$, so the enthalpy required to vaporize this amount of CCl_4 should be approximately one third of the molar vaporization enthalpy, and it is.

PROBLEM-SOLVING PRACTICE 9.1

The molar vaporization enthalpy of benzene, C_6H_6, is 30.72 kJ/mol. Calculate the heat transfer of energy required to vaporize 25.0 g C_6H_6 at 1 bar and 80.1 °C, its boiling point.

PROBLEM-SOLVING PRACTICE answers are provided at the back of this book in Appendix K.

PROBLEM-SOLVING EXAMPLE 9.2

Using Vaporization Enthalpy

You empty the contents of a one-liter bottle of water into a pan at 100 °C and all the water evaporates. How much heat transfer was required to evaporate this water at 100 °C and 1 bar? The density of liquid water at 100 °C is 0.958 g/mL.

Table 9.2 Molar Vaporization Enthalpies for Some Common Substances*

Substance	Number of Electrons	$\Delta_{vap}H^\circ$ (kJ/mol)	Boiling Point (°C)
London forces only			
Ne	10	1.8	−246.0
Ar	18	6.5	−185.9
Xe	54	12.6	−107.1
F_2	18	6.54	−188.1
Cl_2	34	20.39	−34.6
Br_2	70	29.54	59.6
CH_4 (methane)	10	8.2	−161.5
CH_3—CH_3 (ethane)	18	14.7	−88.6
CH_3—CH_2—CH_3 (propane)	26	19.0	−42.1
CH_3—CH_2—CH_2—CH_3 (butane)	34	22.4	−0.5
London forces, dipole-dipole forces			
HCl	18	16.1	−84.8
HBr	36	17.6	−66.5
HI	54	19.8	−35.1
SO_2	32	24.9	−10.0
London forces, dipole-dipole forces, and hydrogen bonding			
HF	10	25.2	19.7
NH_3	10	23.3	−33.4
H_2O	10	40.7	100.0

*In this table molar vaporization enthalpies are reported at the normal boiling point, the temperature at which the vapor pressure of the liquid is 760 mmHg.

Result 2.16×10^3 kJ

Analyze The heat transfer of energy depends on the molar vaporization enthalpy of water and the amount of water vaporized. The volume of water that vaporizes is given. The density of water and the temperature of vaporization are given.

Plan Calculate the mass of water from the volume and density and use the molar mass of water to calculate the amount. Look up the molar vaporization enthalpy at the boiling point in Table 9.2 and use it to calculate the heat transfer.

Execute The molar mass of water is 18.02 g/mol.

$$n(\text{H}_2\text{O}) = 1.00 \times 10^3 \text{ mL water} \times \frac{0.958 \text{ g}}{1 \text{ mL}} \times \frac{1 \text{ mol}}{18.02 \text{ g}} = 53.16 \text{ mol H}_2\text{O}$$

The heat transfer of energy required for vaporization, q, is

$$q = 53.16 \text{ mol H}_2\text{O} \times \frac{40.7 \text{ kJ}}{1 \text{ mol}} = 2164 \text{ kJ} = 2.16 \times 10^3 \text{ kJ}$$

This quantity of energy transfer is equivalent to about one fourth the energy supplied by the daily food intake of an average person in the United States.

☑ **Reasonable Result Check** One liter of water is about 1 kg water. The molar mass is about 20 g/mol, so we have about $\frac{1000 \text{ g}}{20 \text{ g/mol}} = 50$ mol water. Each mole of water requires about 40 kJ to evaporate, so 50 mol $\times$ 40 kJ/mol is about 2000 kJ, which is close to our more carefully calculated result. The result is reasonable.

PROBLEM-SOLVING PRACTICE 9.2

What mass (g) of ethanol, $CH_3CH_2OH(\ell)$, can be vaporized at its boiling point of 78.4 °C by transfer of 500. kJ to the liquid? The $\Delta_{vap}H°$ of ethanol is 38.6 kJ/mol at this temperature.

9-3 Vapor Pressure

When a puddle dries after a rain or perspiration evaporates from the skin, a liquid, water, evaporates completely. A liquid, such as water or bromine, in an open container, evaporates until all the liquid is converted into gas (Figure 9.2).

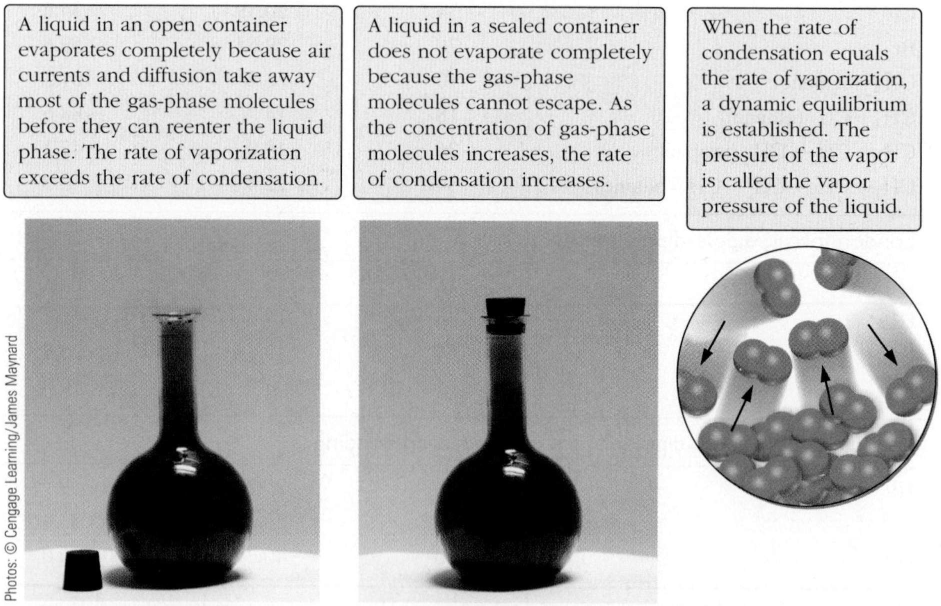
A liquid in an open container evaporates completely because air currents and diffusion take away most of the gas-phase molecules before they can reenter the liquid phase. The rate of vaporization exceeds the rate of condensation.
A liquid in a sealed container does not evaporate completely because the gas-phase molecules cannot escape. As the concentration of gas-phase molecules increases, the rate of condensation increases.
When the rate of condensation equals the rate of vaporization, a dynamic equilibrium is established. The pressure of the vapor is called the vapor pressure of the liquid.
Photos: © Cengage Learning/James Maynard

Figure 9.2 Vaporization and condensation in open and sealed containers.

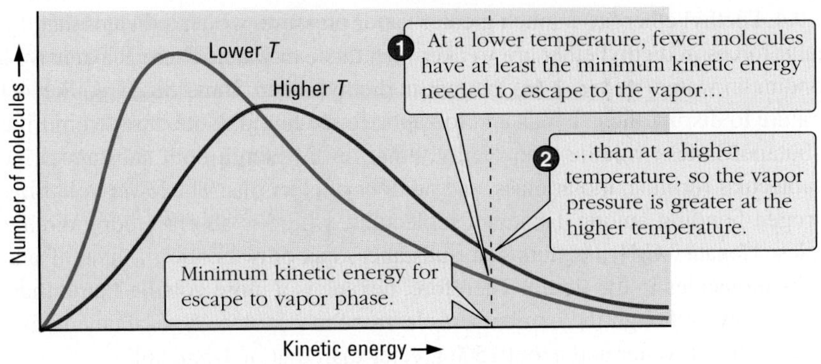

Lower T
Higher T
At a lower temperature, fewer molecules have at least the minimum kinetic energy needed to escape to the vapor...
...than at a higher temperature, so the vapor pressure is greater at the higher temperature.
Minimum kinetic energy for escape to vapor phase.
Number of molecules
Kinetic energy

Figure 9.3 Molecular kinetic energy and vapor pressure.

Figure 9.2 also shows that when a liquid is in a closed container, so that no molecules can escape, it does not evaporate completely. The rate of vaporization is initially much greater than the rate of condensation, but as the concentration of molecules in the gas phase increases, the rate of condensation also increases. Eventually, the rates of vaporization and condensation are equal and the two opposing processes balance each other. The system has reached a state of *dynamic equilibrium;* there is no *net* change in the amounts of liquid and vapor in the flask, but vaporization and condensation are still occurring. *The pressure of the vapor when liquid and vapor are in dynamic equilibrium* is known as the **equilibrium vapor pressure** *or* **vapor pressure** *of the liquid.*

The tendency of a liquid to vaporize, its *volatility*, increases with temperature. For example, hot water evaporates more quickly than cold water. This greater volatility can be explained by considering the distribution of energies of the molecules (← **Sec. 8-7**). As seen in Figure 9.3, molecules of a liquid have varying kinetic energies. Only a few molecules have kinetic energies large enough to escape from the liquid into the gas phase. Raising the temperature increases the fraction of molecules that have sufficient energy. Therefore, vaporization is faster at the higher temperature. Thus, *vapor pressure increases with increasing temperature*, as seen in Figure 9.4.

At a given temperature, the vapor pressures of the four liquids noted in Figure 9.4 differ due to differences in the strengths of their noncovalent intermolecular forces: a *liquid with stronger intermolecular attractions has a lower vapor pressure at a given temperature.* Consider the vapor pressures of diethyl ether and water at 30.0 °C as shown in

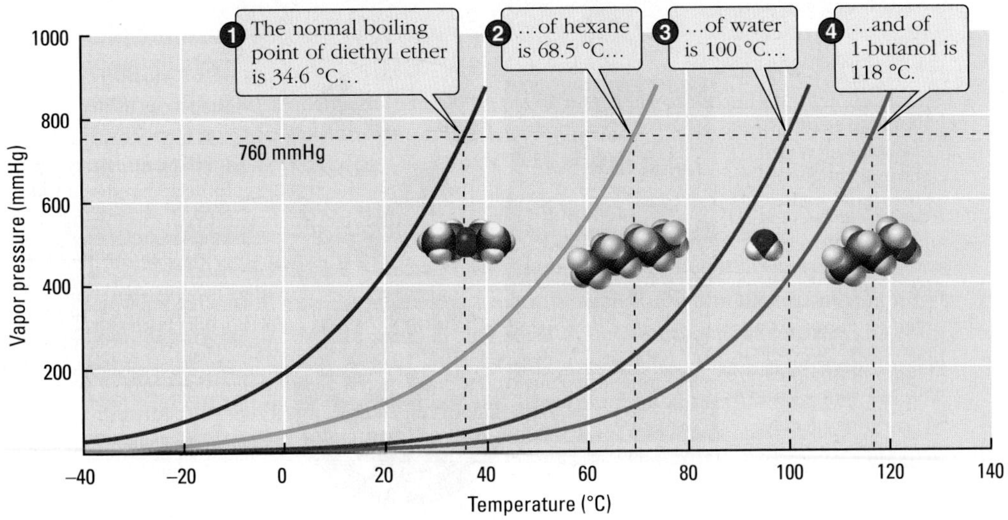

The normal boiling point of diethyl ether is 34.6 °C...
...of hexane is 68.5 °C...
...of water is 100 °C...
...and of 1-butanol is 118 °C.
760 mmHg
Vapor pressure (mmHg)
1000
800
600
400
200
–40
–20
0
20
40
60
80
100
120
140
Temperature (°C)

Figure 9.4 Vapor pressure curves for diethyl ether, hexane, water, and 1-butanol. Each curve represents temperatures and pressures where each pure liquid and its vapor are in equilibrium.

Figure 9.4. Diethyl ether has a much greater vapor pressure, which indicates that the intermolecular forces in diethyl ether are weaker than those in water. There is extensive hydrogen bonding in water (← **Sec. 7-6c**), but not in diethyl ether. Thus, water requires a higher temperature to disrupt those bonds and to vaporize the liquid. Note, also, from Figure 9.4 that 1-butanol is less volatile than diethyl ether, even though both substances have the same molecular formula, molar mass, and number of electrons. The lower volatility is due to hydrogen bonding among 1-butanol molecules, which is absent among diethyl ether molecules. Hexane, C_6H_{14}, a nonpolar substance, has only London forces of attraction among its molecules in the liquid. Therefore, hexane is a more volatile liquid than water or 1-butanol, even though the numbers of electrons in a molecule of hexane is five times greater than that of water and about 15% greater than that of 1-butanol.

9-3a Boiling Points

At a given temperature, liquids with weak intermolecular forces have higher vapor pressures than liquids with stronger intermolecular forces.

When a liquid is heated in an open container, it eventually boils (Figure 9.5). The **boiling point** *is the temperature at which the vapor pressure of the liquid equals the atmospheric pressure* on the surface of the liquid. Below the boiling point, only molecules at the surface of the liquid can go into the gas phase. At the boiling point, however, the vapor pressure equals the atmospheric pressure and bubbles of vapor form throughout the liquid. Due to their lower density, the bubbles rise through the liquid to the surface. *When the atmospheric pressure is 1 atm (760 mmHg), the boiling temperature is called the* **normal boiling point**. Normal boiling points are given in Figure 9.4 for the four liquids.

CONCEPTUAL EXERCISE 9.3

Estimating Boiling Points

Use Figure 9.4 to estimate the boiling points of: (a) hexane at 400 mmHg; (b) diethyl ether at 200 mmHg; (c) water at 400 mmHg.

CONCEPTUAL EXERCISE 9.4

Explaining Bubbles

One of your classmates believes that the bubbles in a boiling liquid are air bubbles. Explain to him what is wrong with that idea and what the bubbles actually are. Suggest an experiment to show that the bubbles are not air.

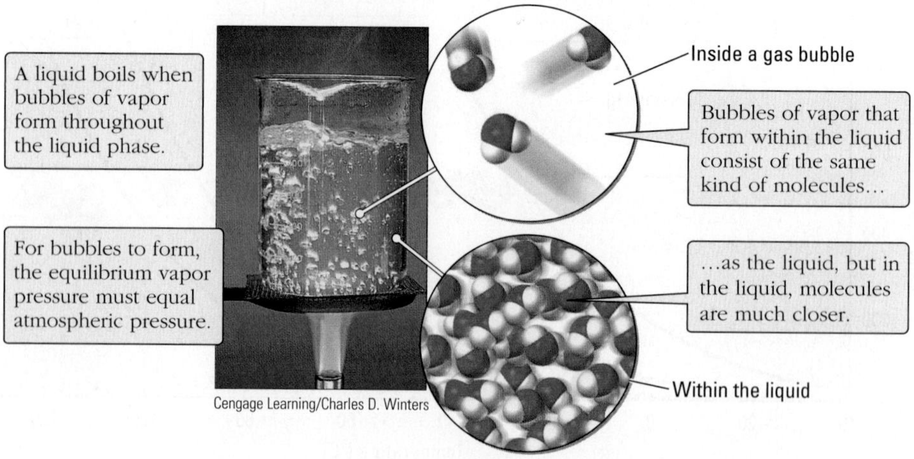
A liquid boils when bubbles of vapor form throughout the liquid phase.
For bubbles to form, the equilibrium vapor pressure must equal atmospheric pressure.
Inside a gas bubble
Bubbles of vapor that form within the liquid consist of the same kind of molecules...
...as the liquid, but in the liquid, molecules are much closer.
Within the liquid
Cengage Learning/Charles D. Winters

Figure 9.5 Water boiling. The gas has much lower density than the liquid so bubbles rise.

Note from Figure 9.4 that the boiling point decreases at lower pressure. For example, the boiling point of pure water is 100 °C at 1 atm. At higher altitudes, however, the atmospheric pressure is lower (← **Sec. 8-1**). In Denver, Colorado at 1609 m above sea level, the pressure is approximately 0.83 atm and water boils at about 94 °C (Table 9.3). Because of the decrease in boiling point with altitude, cooking and baking times are a bit longer in regions well above sea level, such as the Rocky Mountains.

Conversely, cooking times can be shortened by using a pressure cooker, a sealed steel pot in which food and water are heated to more than 100 °C at pressures greater than 1 atm. Porters who transport gear for high-mountain expeditions carry lightweight pressure cookers to use at base camps along the way. The principle of elevated boiling points is also used in autoclaves, devices that use high-pressure steam (2 atm); its higher temperature more rapidly sterilizes medical and laboratory instruments.

In *The Voyage of the Beagle*, Charles Darwin commented on this phenomenon when describing the difficulty of boiling potatoes at about 11,000 ft in the mountains of Mendoza.

All Natural
Fat Free
No Pudge!®
Fudge Brownie Mix
original
High Altitude
Increase oven temp to 365°F. Add 2 tablespoons of flour into dry brownie mix and mix per directions above. Bake for 35 to 40 minutes.

Courtesy Peter McGahey

9-3b The Clausius-Clapeyron Equation

We have seen that vaporization of a liquid is related to its vaporization enthalpy, vapor pressure, and temperature. The quantitative relationship among these three is given by the **Clausius-Clapeyron equation**:

$$\ln P = \frac{-\Delta_{vap}H}{RT} + C = \left(\frac{-\Delta_{vap}H}{R}\right)\left(\frac{1}{T}\right) + C$$

which relates the natural logarithm of the vapor pressure P of the liquid to the absolute temperature T, the universal gas constant R (8.314 J mol^{-1} K^{-1}), the molar vaporization enthalpy $\Delta_{vap}H$ (J/mol), and a constant C that is characteristic of the liquid. The form of the Clausius-Clapeyron equation shows that a graph of ln P versus $1/T$ is a straight line with its slope equal to $-\Delta_{vap}H/R$ and a y-axis intercept of C. That is, ln P equals C in the limit when T is very large and $1/T$ tends toward zero. Figure 9.6 shows a plot of ln P versus $1/T$ for diethyl ether, hexane, water, and 1-butanol. Figures 9.4 and 9.6 illustrate two different ways of expressing the relationship between T and P.

The Clausius-Clapeyron equation can be recast for two sets of pressure and temperature of a liquid to give

$$\ln\left(\frac{P_2}{P_1}\right) = \frac{-\Delta_{vap}H}{R}\left[\frac{1}{T_2} - \frac{1}{T_1}\right]$$

which can be used to calculate $\Delta_{vap}H$ from the two sets of data: T_1, T_2 and P_1, P_2. If $\Delta_{vap}H$ is known, then this form of the equation can be used to find the vapor pressure of a liquid at a new temperature given its vapor pressure at some other temperature.

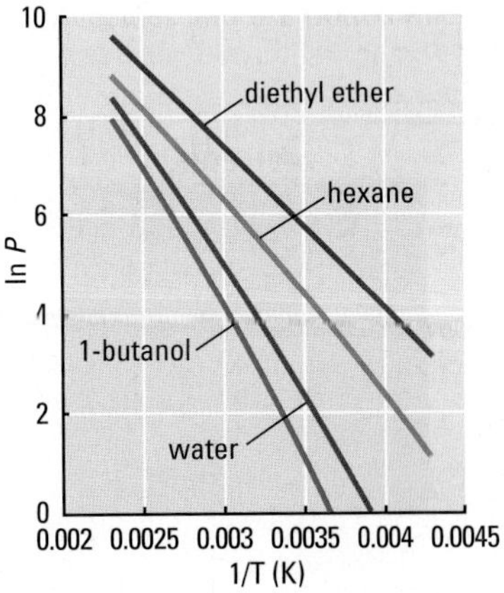
10
8
6
4
2
0
ln P
diethyl ether
hexane
1-butanol
water
0.002
0.0025
0.003
0.0035
0.004
0.0045
1/T (K)

Figure 9.6 Clausius-Clapeyron plots. Each slope = $-\Delta_{vap}H/R$.

Table 9.3 Boiling Points of Water at Various Altitudes

Locale	Elevation (m)	Approx. Atmos. Pressure (mmHg)	Approx. Boiling Point (°C)
Mt. Everest (Nepal; world's highest mountain)	8850	232	71
Denali/Mt. McKinley (Alaska; highest mountain in the United States)	6190	350	83
Mt. Blanc (Alps; highest mountain in Western Europe)	4810	413	85
Denver, CO	1609	631	94

Refrigerating fluids (refrigerants) take advantage of the fact that evaporation is an endothermic process (thermal energy is absorbed) and condensation is exothermic (thermal energy is transferred to the surroundings). As the liquid refrigerant passes through the coils of a refrigerator, the liquid refrigerant absorbs thermal energy from the refrigerator compartment, evaporating the liquid refrigerant and cooling the refrigerator's contents. The gaseous refrigerant is subsequently compressed back to a liquid, its released thermal energy transferred to the surroundings from the coils behind the refrigerator, and the cycle continues. To be effective coolants, refrigerants have relatively large vaporization enthalpies and normal boiling points just below room temperature (easily liquefiable), in addition to being nonflammable and nontoxic.

PROBLEM-SOLVING EXAMPLE 9.3

Clausius-Clapeyron Equation

Trichlorofluoromethane (CFC-11) is a refrigerant that was widely used in air conditioners before being phased out due to its role in stratospheric ozone depletion (← **Sec. 8-10b**). CFC-11 has a vapor pressure of 13.3 kPa at −23.0 °C and 43.7 kPa at 2.0 °C. Calculate $\Delta_{vap}H$ of CFC-11.

Result 27.2 kJ/mol

Analyze Pairs of vapor pressure and temperature data are given that can be used in the Clausius-Clapeyron equation to calculate the molar vaporization enthalpy.

Plan Convert temperatures to kelvins, use the vapor pressures given, use R = 8.314 J mol^{-1} K^{-1} so that the result is in J/mol, solve for $\Delta_{vap}H$, and calculate the result.

Execute

$$\ln\left[\frac{43.7 \text{ kPa}}{13.3 \text{ kPa}}\right] = \frac{-\Delta_{vap}H}{8.314 \text{ J mol}^{-1} \text{ K}^{-1}}\left[\frac{1}{275 \text{ K}} - \frac{1}{250 \text{ K}}\right]$$

$$\ln(3.286) = 1.19 = \frac{-\Delta_{vap}H}{8.314 \text{ J mol}^{-1} \text{ K}^{-1}}(-3.640 \times 10^{-4} \text{ K}^{-1})$$

$$\Delta_{vap}H = \frac{1.19 \times 8.314 \text{ J mol}^{-1} \text{ K}^{-1}}{(3.640 \times 10^{-4} \text{ K}^{-1})} = 27200 \text{ J/mol} = 27.2 \text{ kJ/mol}$$

☑ **Reasonable Result Check** CFC-11 contains 66 electrons and is a slightly polar molecule. Using Table 9.2 to find similar molecules, we expect CFC-11 to have $\Delta_{vap}H$ greater than HI (54 electrons, polar; 21.2 kJ/mol) and similar to Br_2 (70 electrons, nonpolar; 29.54 kJ/mol); the result is reasonable.

PROBLEM-SOLVING PRACTICE 9.3

Water is used as a coolant in pressurized-water nuclear power reactors. To prevent the water from boiling, it is kept at high temperature and pressure. Calculate the maximum temperature the water could reach without boiling in a pressurized-water nuclear reactor operating at 150 atm.

9-3c Critical Temperature and Critical Pressure

At a given temperature, applying pressure to a gas typically liquefies it. For example, consider a sample containing only water vapor at 100 °C and 20 mmHg. If the pressure is increased, some water vapor liquefies when the pressure reaches 760 mmHg; both liquid and gas phases are in equilibrium at this temperature and pressure. If the same sample is compressed at 110 °C, a greater pressure, 1075 mmHg, must be applied to liquefy some of the water vapor. Continuing to raise the temperature and the pressure on the sample ultimately reaches a point called the **critical point**, *the minimum temperature and pressure at which the liquid and vapor are indistinguishable; only one phase exists.*

The critical point for water is at 374 °C and 217.7 atm. The **critical temperature (T_c)** *is the temperature above which the liquid cannot exist, regardless of the pressure applied.* The **critical pressure (P_c)** *is the vapor pressure of the liquid at the critical temperature.* The uniform properties of liquid and vapor at the critical point are illustrated here for propane, C_3H_8.

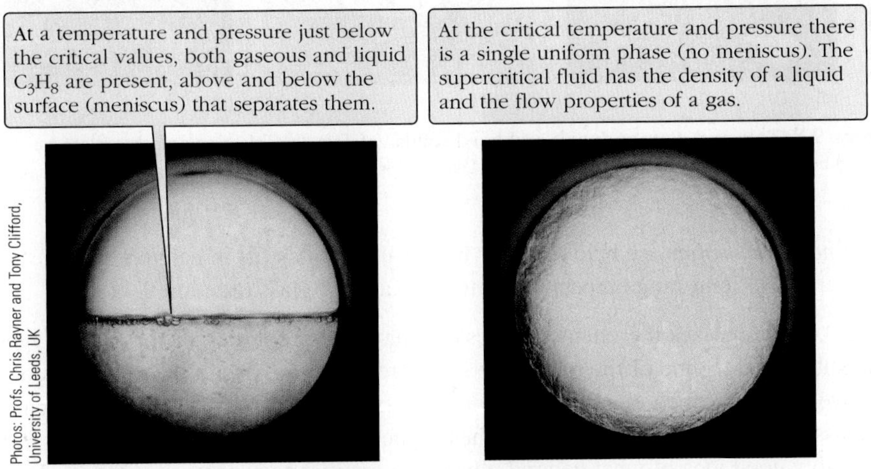
At a temperature and pressure just below the critical values, both gaseous and liquid C_3H_8 are present, above and below the surface (meniscus) that separates them.
At the critical temperature and pressure there is a single uniform phase (no meniscus). The supercritical fluid has the density of a liquid and the flow properties of a gas.
Photos: Profs. Chris Rayner and Tony Clifford, University of Leeds, UK

Propane, C_3H_8, just below the critical point and at the critical point.

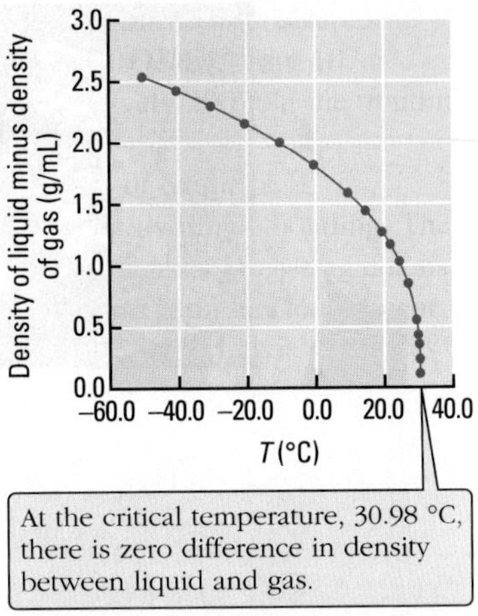
Density of liquid minus density of gas (g/mL)
3.0
2.5
2.0
1.5
1.0
0.5
0.0
−60.0 −40.0 −20.0 0.0 20.0 40.0
T (°C)
At the critical temperature, 30.98 °C, there is zero difference in density between liquid and gas.

Figure 9.7 Density of liquid CO_2 minus density of gaseous CO_2 at the critical pressure.

The difference between the density of a liquid and its vapor decreases dramatically as the temperature and pressure approach the critical point as shown in Figure 9.7 for carbon dioxide. The *density difference* falls from 2.5 g/mL at −50 °C to 0.1 g/mL at 30.97 °C, very close to the critical temperature, 30.98 °C. The density of CO_2 at its critical point, 30.98 °C and 7.375 MPa, is 0.4678 g/mL. At temperatures and pressures above the critical point, any substance exists as a **supercritical fluid** (SCF) that *has the density characteristics of a liquid, but the flow properties of a gas.* SCFs diffuse easily through many substances making SCFs excellent solvents. For example, supercritical CO_2, a nonpolar substance, is an excellent solvent for fats, oils, and other nonpolar substances. Supercritical carbon dioxide is used commercially as a solvent for dry cleaning clothing, to make decaffeinated coffee by extracting caffeine from coffee beans, and to remove fat from potato chips.

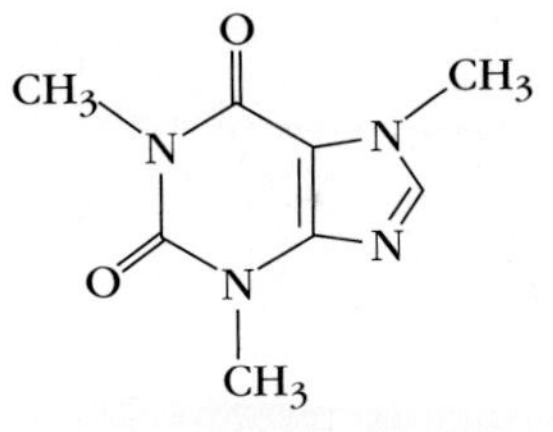
O
CH3
CH3
N
N
N
N
O
CH3

caffeine

Cengage Learning/Charles D. Winters

Coffee beans and the molecular structure of caffeine, which can be extracted from coffee using supercritical CO_2.

9-4 Solids and Changes of Phase

The relationship of nanoscale structure to macroscale properties is a central theme of this text. Nowhere is the influence of the nanoscale arrangement of atoms, molecules, or ions on properties more evident than in the study of solids. Chemists, in collaboration with physicists, engineers, and other scientists, explore such relationships as they work in the field of materials science. The nature of solid substances is determined by the type of forces holding their nanoscale particles together.

- *Ionic solids* are held together by electrostatic interactions between cations and anions. Example: table salt, NaCl (← **Sec. 2-6**)
- *Metallic solids* are held together by attractions among positively charged metal atom cores and valence electrons. Example: iron (Section 9-9)
- *Molecular solids* are held together by intermolecular interactions such as London forces, dipole-dipole forces, and hydrogen bonding. Example: ice (← **Secs. 2-7, 7-6**)
- *Network solids* consist of atoms bonded together into extremely large molecules by covalent bonds. Example: diamond (Section 9-7)

© Cengage Learning/James Maynard
(a)
© Cengage Learning/James Maynard
(b)
Cengage Learning/Charles D. Winters
(c)
© Cengage Learning/James Maynard
(d)

Figure 9.8 Talc and diamond: soft and hard solids. (a) Talc. (b) Talc is so soft that it can be crushed between one's fingers. (c) Diamond. (d) Diamond is hard enough to scratch glass.

- *Amorphous solids* are held together by covalent bonds (as in network solids), but there is no long-range repeating pattern. Example: glass (Section 9-11).

Table 9.4 summarizes the characteristics and physical properties of the major types of solid substances. Some of the nanoscale structures that give rise to the properties of solids have been described in earlier chapters; others will be described later in this chapter. By classifying a substance as one of these types of solid, you will be able to form a reasonably good idea of what general physical properties to expect, even for a substance that you have never encountered.

A solid is rigid, having its own shape rather than assuming the shape of its container as a liquid does. Solids have varying degrees of hardness that depend on the kinds of atoms, molecules, or ions in the solid and the types of forces that hold the particles together. For example, talc (soapstone, Figure 9.8a,b), which is used as a lubricant and in talcum powder, is one of the softest solids known. At the atomic level, talc consists of layered sheets containing silicon, magnesium, and oxygen atoms. Attractive forces between these sheets are very weak, so one sheet of talc can slide along another and be removed easily from the rest. In contrast, diamond (Figure 9.8c,d) is one of the hardest solids known. In diamond, each carbon atom is covalently bonded to four neighbors in a tetrahedral arrangement. Each of those neighbor atoms is in turn strongly bonded to four carbon atoms, and so on throughout the solid (a network solid, Section 9-7). Because of the number and strength of the bonds holding each carbon atom to its neighbors, diamond is so hard that it can scratch or cut almost any other solid. Diamonds are used in cutting tools and abrasives, which are more important commercially than gemstone diamonds.

Although all solids consist of atoms, molecules, or ions in relatively immobile positions, some solids exhibit greater regularity of structure than others.

- A **crystalline solid** has *an ordered, long-range arrangement of the individual particles, which is reflected in the planar faces and sharp angles of the crystals*. Examples: salt crystals, minerals, gemstones, and ice
- An **amorphous solid** *exhibits very little long-range order, yet it is hard and has a definite shape*. Examples: ordinary glass and organic polymers such as some forms of polyethylene and polystyrene

PROBLEM-SOLVING EXAMPLE 9.4

Types of Solids

What types of solids are these substances?

(a) Sucrose, $C_{12}H_{22}O_{11}$ (table sugar), has a melting point of about 185 °C. It has poor electrical conductance both as a solid and as a liquid.

(b) Solid Na_2SO_4 has a melting point of 884 °C and has low electrical conductivity that increases dramatically when the solid melts.

Table 9.4 Structures and Properties of Various Types of Solid Substances

	Ionic (← Sec. 2-6)	Metallic (Sections 9-6, 9-9)	Molecular (← Secs. 2-7, 7-6)	Network (Section 9-7)	Amorphous (glassy) (Section 9-11)
Examples	NaCl, K_2SO_4, $CaCl_2$, $CuSO_4 \cdot 5H_2O$	Iron, silver, copper, nickel, other metals and alloys	H_2, O_2, I_2, H_2O, CO_2, CH_4, CH_3OH, CH_3COOH	Graphite, diamond, quartz, feldspars, mica	Glass, polyethylene, nylon

Formed From	Reactive metals and reactive nonmetals	Elements that are metals	Nonmetals combined with nonmetals	Group IVA elements; many compounds containing silicon and oxygen	A broad range of elements and compounds
Structural Units	Positive and negative ions (some polyatomic); no discrete molecules	Metal atoms (positive metal ions surrounded by an electron sea)	Covalently bonded molecules	Atoms held in an infinite one-, two-, three-dimensional network	Covalently bonded networks of atoms or collections of large molecules with short-range order only
	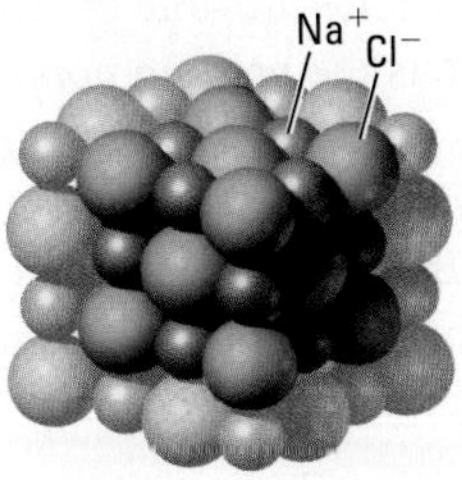 Na^+ Cl^-	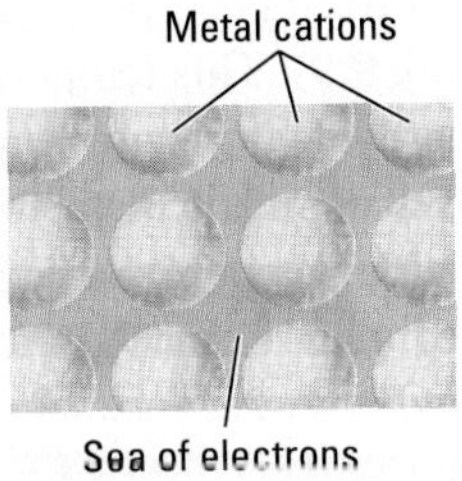 Metal cations Sea of electrons	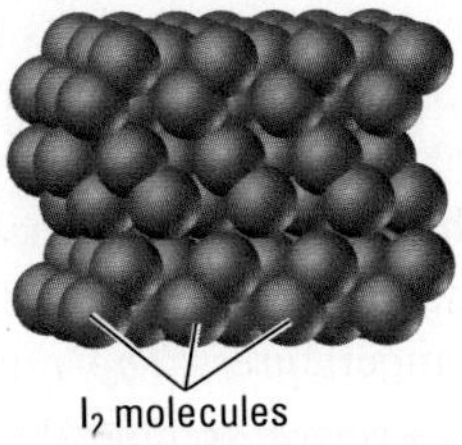 I_2 molecules	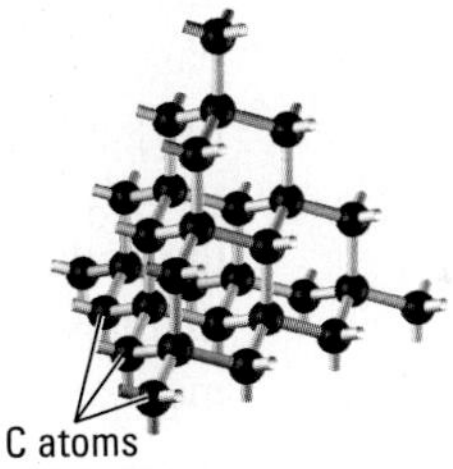 C atoms	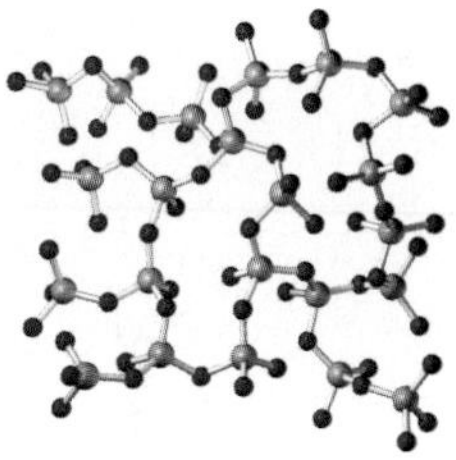
Forces Holding Units Together	Ionic bonding; attractions among charges on positive and negative ions	Metallic bonding; electrostatic attraction among metal ions and electrons	London forces, dipole-dipole forces, hydrogen bonds	Covalent bonds (directional electron-pair bonds)	Covalent bonds (directional electron-pair bonds)
Typical Properties	Hard; brittle; high melting point (700 to 3500 °C); poor electrical conductor as solid, but good conductor as liquid; often water soluble; poor conductor of heat	Malleable; ductile; good electrical conductor in solid and liquid; good heat conductor; wide range of hardness and melting point	Low to moderate melting point and boiling point (−250 to 600 °C); soft; poor electrical conductor as solid and liquid; poor conductor of heat	Wide range of hardness and melting point (three-dimensional bonding > two-dimensional bonding > one-dimensional bonding); poor electrical conductor, with some exceptions	Noncrystalline; wide temperature range for melting; poor electrical conductor, with some exceptions

Result (a) Molecular solid (b) Ionic solid

Analyze Physical properties are given for each substance and certain physical properties correlate with different types of substances.

Plan Compare the physical properties given for each substance with the characteristic physical properties of solids given in Table 9.4.

Execute

(a) Sucrose's properties correspond to those of a molecular solid since it has poor electrical conductance both as a solid and as a liquid.

(b) The high melting point, low electrical conductivity as a solid, and high electrical conductivity as a liquid are consistent with properties of an ionic solid.

☑ **Reasonable Result Check** Based on the chemical formulas it is reasonable to conclude that $C_{12}H_{22}O_{11}$ consists of molecules in which the atoms are covalently bonded and that Na_2SO_4 is an ionic compound composed of Na^+ and SO_4^{2-} ions.

PROBLEM-SOLVING PRACTICE 9.4

What types of solids are these substances?

(a) The hydrocarbon decane, $C_{10}H_{22}$, has a melting point of −31 °C and is a poor electrical conductor.

(b) Solid $MgCl_2$ has a melting point of 714 °C and conducts electricity only when melted.

9-4a Phase Changes of Solids: Melting and Freezing

When a solid is heated, its temperature increases until the solid either decomposes or begins to melt. A solid melts at a temperature where the kinetic energies of the molecules or ions are sufficiently high that the interparticle attractions in the solid are no longer strong enough to keep the particles in their fixed positions. The particles are able to move past each other, and the solid melts (Figure 9.9). This temperature is the *melting point* of the solid. Melting requires transfer of energy, the *fusion enthalpy,* from the surroundings into the system, so it is always endothermic. The *molar fusion enthalpy* (← **Sec. 4-5c**) is the enthalpy change required to melt 1 mol of a pure solid. Solids with high fusion enthalpies usually melt at high temperatures, and solids with low fusion enthalpies usually melt at low temperatures. *The reverse of melting*—called *solidification, freezing,* or

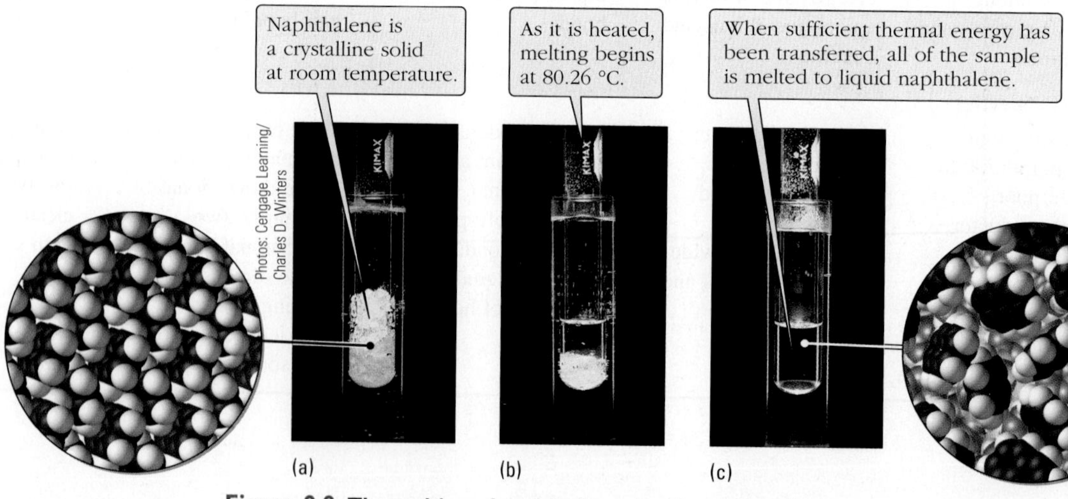
Naphthalene is a crystalline solid at room temperature.
As it is heated, melting begins at 80.26 °C.
When sufficient thermal energy has been transferred, all of the sample is melted to liquid naphthalene.
Photos: Cengage Learning/ Charles D. Winters
(a)
(b)
(c)

Figure 9.9 The melting of naphthalene, $C_{10}H_8$, at 80.26 °C.

Table 9.5 Melting Points and Fusion Enthalpies of Some Solids

Solid	Melting Point (°C)	Fusion Enthalpy (kJ/mol)	Intermolecular Noncovalent Forces
Molecular solids: Nonpolar molecules			
O_2	−219	0.445	These molecules have only London forces (which increase with the number of electrons).
F_2	−220	0.509	
Cl_2	−103	6.406	
Br_2	−7.2	10.794	
Molecular solids: Polar molecules			
HCl	−114	1.990	All of these molecules have London forces enhanced significantly by dipole-dipole forces. H_2O also has significant hydrogen bonding.
HBr	−87	2.406	
HI	−51	2.870	
H_2O	0	6.020	
H_2S	−86	2.395	
Ionic solids			
NaCl	800	28.16	All ionic solids have strong attractions between oppositely charged ions.
NaBr	747	26.23	
NaI	662	23.7	

crystallization—is always an exothermic process. The *molar crystallization enthalpy* has the same magnitude as the molar fusion enthalpy, but the opposite sign.

$$\text{Solid} \underset{\text{crystallization enthalpy}}{\overset{\text{fusion enthalpy}}{\rightleftharpoons}} \text{Liquid} \qquad \Delta_{\text{fus}}H = -\Delta_{\text{cryst}}H$$

Table 9.5 lists melting points and fusion enthalpies for examples of three classes of compounds: (a) nonpolar molecular solids; (b) polar molecular solids, some capable of hydrogen bonding; and (c) ionic solids. Solids composed of low molecular weight nonpolar molecules have the lowest melting temperatures, because their intermolecular attractions are weakest. These molecules are held to each other by London forces only (← **Sec. 9-1**), and they form solids with melting points so low that we seldom encounter them in the solid state at normal temperatures. Melting points and fusion enthalpies of nonpolar molecular solids increase with increasing number of electrons as the London forces become stronger. The ionic compounds in Table 9.5 have the highest melting points and fusion enthalpies because of the very strong ionic bonding that holds the oppositely charged ions together in the solid. The polar molecular solids have intermediate melting points.

PROBLEM-SOLVING EXAMPLE 9.5

Fusion Enthalpy

The molar fusion enthalpy of NaCl is 28.16 kJ/mol at its melting point. Calculate the thermal energy transfer into the system when 10.00 g NaCl melts.

Result 4.818 kJ

Analyze The energy transfer can be calculated from the amount of NaCl and the molar fusion enthalpy. The amount can be obtained from the mass.

Plan Calculate the amount from the mass using the molar mass of NaCl. Then calculate the energy transfer, q.

Execute The molar mass of NaCl is 58.443 g/mol, so

$$n(\text{NaCl}) = 10.00 \text{ g NaCl} \times \frac{1 \text{ mol NaCl}}{58.443 \text{ g NaCl}} = 0.1711 \text{ mol NaCl}$$

$$q = 0.1711 \text{ mol NaCl} \times \frac{28.16 \text{ kJ}}{1 \text{ mol NaCl}} = 4.818 \text{ kJ}$$

☑ **Reasonable Result Check** We have approximately one-sixth mol NaCl, so the enthalpy change required to melt this amount of NaCl should be approximately one sixth of the molar fusion enthalpy, and it is.

PROBLEM-SOLVING PRACTICE 9.5

Calculate the thermal energy transfer required to melt 0.500 mol NaI at its normal melting point.

EXERCISE 9.5

Thermal Energy Liberated upon Crystallization

Which transfers more thermal energy to the surroundings, the crystallization of 2 mol liquid bromine or the crystallization of 1 mol liquid water?

Figure 9.10 Dry Ice. When CO_2 sublimes, its cold vapors cause moisture to condense from the air, forming a white cloud of water droplets. The CO_2 vapors are denser than air, so they carry the cloud downward.

9-4b Phase Changes of Solids: Sublimation and Deposition

Atoms or molecules can escape directly from the solid to the gas phase; *change of a solid directly to a gas* is known as **sublimation**. The enthalpy change is the *sublimation enthalpy.* The reverse process, in which *a gas is converted directly to a solid,* is called **deposition**. The enthalpy change for this exothermic process (the deposition enthalpy) has the same magnitude as the sublimation enthalpy, but the opposite sign.

$$\text{Solid} \underset{\text{deposition enthalpy}}{\overset{\text{sublimation enthalpy}}{\rightleftharpoons}} \text{Gas} \qquad \Delta_{\text{sub}}H = -\Delta_{\text{dep}}H$$

One common substance that sublimes at normal atmospheric pressure is solid carbon dioxide (Dry Ice, Figure 9.10). Its vapor pressure reaches 1 atm at −78 °C, which is below its melting point. Thus solid carbon dioxide sublimes rather than melting. Because of the solid's high vapor pressure, liquid carbon dioxide can exist only at pressures much higher than 1 atm.

Have you noticed that snow outdoors and ice cubes in a frost-free refrigerator slowly disappear even if the temperature never gets above freezing? The sublimation enthalpy of ice is 51 kJ/mol, and its vapor pressure at 0 °C is 4.60 mmHg. Therefore, ice sublimes readily in air dry enough that the partial pressure of water vapor is below 4.60 mmHg (Figure 9.11). Given enough air passing over it, a sample of ice will sublime completely, leaving no trace behind. In a frost-free refrigerator, a current of dry air periodically blows across any ice formed in the freezer compartment, taking away water vapor (and hence the ice) without warming the freezer enough to thaw the food.

The piece of steak at the left is "freezer burned"; its surface has lost water by sublimation. The steak on the right is not burned.

In the reverse of sublimation, atoms or molecules in the gas phase can deposit (solidify directly) on the surface of a solid. Deposition is used to form thin coatings of metal atoms on surfaces. CD-ROM or DVD discs, for example, have shiny metallic surfaces of deposited aluminum or gold atoms. To make such discs, a metal filament is heated in a vacuum to a temperature at which metal atoms begin to sublime rapidly from the surface of the filament. The plastic disc is cooler than the filament, so the metal atoms in the gas phase quickly deposit on the cool surface. The purpose of the metal coating is to provide a reflective surface for the laser beam that reads the pits and lands (unpitted areas) containing the digital data or video information.

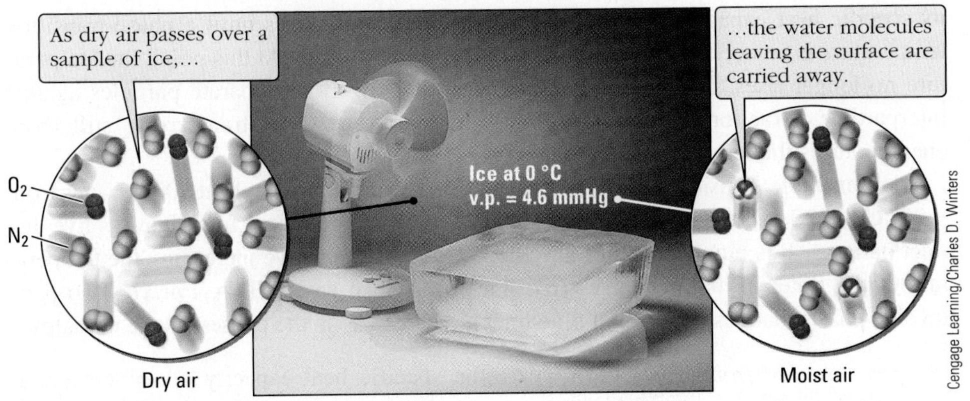
As dry air passes over a sample of ice,...
...the water molecules leaving the surface are carried away.
O_2
N_2
Ice at 0 °C
v.p. = 4.6 mmHg
Dry air
Moist air
Cengage Learning/Charles D. Winters

Figure 9.11 Ice subliming.

CONCEPTUAL EXERCISE 9.6

Frost-Free Refrigeration

Sometimes, because of high humidity, a frost-free refrigerator doesn't work as efficiently as it should. Explain why.

CONCEPTUAL EXERCISE 9.7

Purification by Sublimation

Sublimation is an excellent means of purification for compounds that will readily sublime. Explain how purification by sublimation works at the nanoscale.

9-4c Heating Curve

Figure 9.12 shows a **heating curve**, a *plot of temperature versus quantity of energy transferred to a sample*. Heating a solid or liquid increases its temperature as long as its phase does not change and the sample does not decompose. The size of this temperature increase is governed by the quantity of energy transferred to the substance, its mass, and

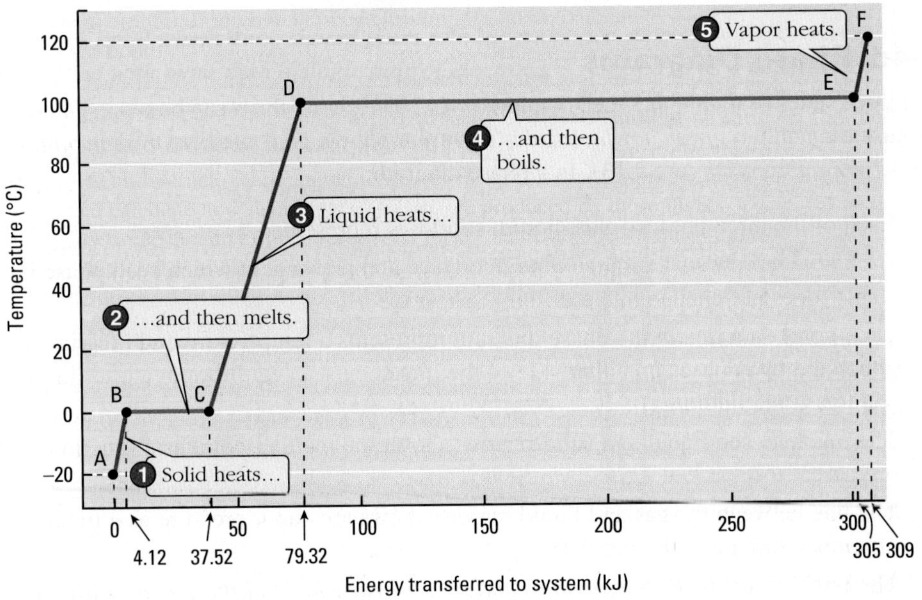
Temperature (°C)
120
100
80
60
40
20
0
−20
A
B
C
D
E
F
1 Solid heats...
2 ...and then melts.
3 Liquid heats...
4 ...and then boils.
5 Vapor heats.
0
4.12
50
37.52
79.32
100
150
200
250
300
305
309
Energy transferred to system (kJ)

Figure 9.12 Heating curve for 100. g water. Temperature is constant during each phase change.

its specific heat capacity (← **Sec. 4-4a**). The temperature rises until a phase-transition temperature is reached—for example, 100 °C for liquid water. At this point, the temperature no longer rises. Instead, the added energy does work to separate particles against interparticle attractions. When the phase change is complete, transferring still more energy causes the temperature to rise again.

For example, consider how much thermal energy is required to heat 100. g water from −20 °C to 120 °C, as illustrated in Figure 9.12. The heating curve for water has five distinct parts; each is numbered in Figure 9.12. Three parts involve increasing the temperature of the water in its three states (red lines), in which case $\Delta H° = cm\Delta T$. Two parts involve phase changes (horizontal blue lines), where $\Delta H°$ is the phase-change enthalpy.

1. *Heat the ice from −20 °C to 0 °C* (the specific heat capacity of solid water is 2.06 J g^{-1} °C^{-1}).

$$\Delta H° = (100.\ \text{g})(2.06\ \text{J g}^{-1}\ °\text{C}^{-1})[(0\ °\text{C}) - (-20.0\ °\text{C})] = 4120\ \text{J} = 4.12\ \text{kJ}$$

2. *Melt the ice at 0 °C* ($\Delta_{fus}H° = 6.020$ kJ/mol).

$$\Delta H° = 100.\ \text{g} \times \frac{1\ \text{mol}}{18.015\ \text{g}} \times \frac{6.020\ \text{kJ}}{1\ \text{mol}} = 33.4\ \text{kJ}$$

3. *Heat the water from 0 °C to 100 °C* (specific heat capacity of liquid water is 4.184 J g^{-1} °C^{-1}).

$$\Delta H° = (100.\ \text{g})(4.184\ \text{J g}^{-1}\ °\text{C}^{-1})[(100.\ °\text{C}) - (0\ °\text{C})] = 41{,}800\ \text{J} = 41.8\ \text{kJ}$$

4. *Boil the water at 100 °C* ($\Delta_{vap}H° = 40.7$ kJ/mol).

$$\Delta H° = 100.\ \text{g} \times \frac{1\ \text{mol}}{18.015\ \text{g}} \times \frac{40.7\ \text{kJ}}{1\ \text{mol}} = 226\ \text{kJ}$$

5. *Heat the water vapor from 100 °C to 120 °C* (specific heat capacity of water vapor is 1.84 J g^{-1} °C^{-1}).

$$\Delta H° = (100.\ \text{g})(1.84\ \text{J g}^{-1}\ °\text{C}^{-1})[(120.0\ °\text{C}) - (100.0\ °\text{C})] = 3680\ \text{J} = 3.68\ \text{kJ}$$

The total energy required to complete the transformation is 309 kJ, the sum of the five steps (Hess's law, ← **Sec. 4-9**). Notice that the largest portion of the energy, 226 kJ (73%), goes into vaporizing water at 100 °C to steam at 100 °C.

9-4d Phase Diagrams

The three states of matter and the six interconversions among them can be represented in a **phase diagram**, *a graph in which areas represent phases and curves represent equilibria between phases* (Figure 9.13). In a phase diagram:

- Each of the three phases—gas, liquid, solid—is represented by an area.
- Each area represents the range of temperature and pressure at which each phase is stable (does not change into a different phase).
- Each point on a line in the phase diagram represents a temperature and pressure at which the two phases on either side of the line coexist in equilibrium (solid/liquid, solid/gas, and liquid/gas).
- The line between liquid and solid regions shows the melting point as a function of pressure.
- The line between the gas and liquid regions shows the vapor pressure as a function of temperature (as in Figure 9.4).
- The **triple point** is *the temperature and pressure at which all three phases are in equilibrium.*

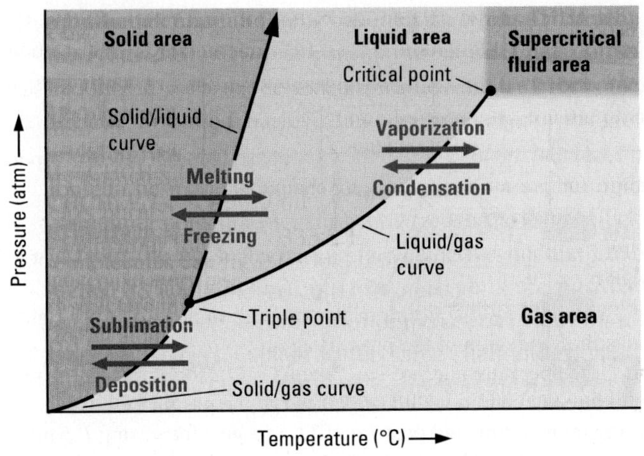
Solid area
Liquid area
Supercritical fluid area
Critical point
Solid/liquid curve
Vaporization
Melting
Condensation
Freezing
Liquid/gas curve
Sublimation
Triple point
Gas area
Deposition
Solid/gas curve
Pressure (atm)
Temperature (°C)

Figure 9.13 Generic phase diagram.

- The liquid/gas curve ends at the *critical point*, where the liquid and gas phases become indistinguishable.

Every pure substance that exists in all three phases has a characteristic phase diagram.

For CO_2 (Figure 9.14), as for most substances, the curve along which solid and liquid coexist in equilibrium has a positive slope. This reflects the fact that solid CO_2 is denser than liquid CO_2 (and most solids are denser than their liquid forms). The triple-point pressure is 5.2 atm; below this pressure the only equilibrium that can exist is between solid and gaseous CO_2. Thus, solid CO_2 sublimes when heated at atmospheric pressure. Liquid CO_2 can be produced only at pressures above 5.2 atm. Tanker trucks marked "liquid carbonic" carry liquid CO_2, pressurized above 5.2 atm. The smaller volume of liquid compared with gas makes it easier to transport CO_2 for making carbonated beverages.

The solid/liquid curve for water is an exception; it has a negative slope as discussed in Section 9-5.

PROBLEM-SOLVING EXAMPLE 9.6

Phase Diagrams

Use the phase diagram for CO_2 (Figure 9.14) to answer these questions.

(a) What is the temperature and pressure at the triple point?

(b) Starting at the triple point, what phase exists when the pressure is held constant and the temperature is increased to 65 °C?

(c) Starting at $T = -70$ °C and $P = 4$ atm, what phase change occurs when the pressure is held constant and the temperature is increased to -30 °C?

(d) Starting at $T = -30$ °C and $P = 5$ atm, what phase change occurs when the temperature is held constant and the pressure is increased to 10 atm?

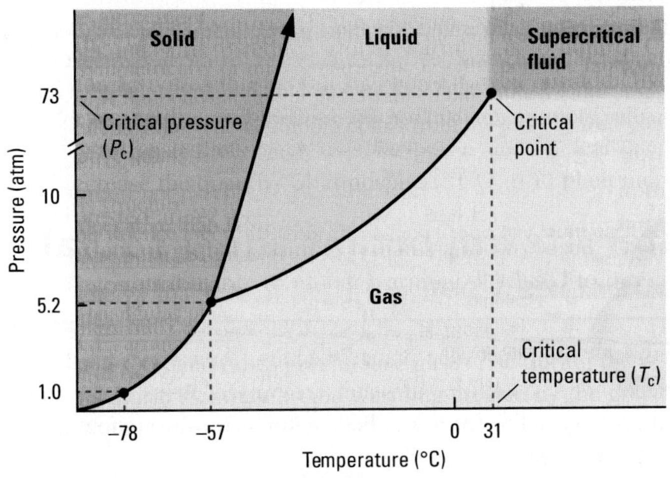
Solid
Liquid
Supercritical fluid
73
Critical pressure
(P_c)
Critical point
10
Pressure (atm)
5.2
Gas
Critical temperature (T_c)
1.0
−78
−57
0
31
Temperature (°C)

Figure 9.14 The phase diagram for carbon dioxide. The temperature and pressure scales are nonlinear. Note that CO_2 cannot be a liquid at 1 atm or any pressure less than 5.2 atm.

Result (a) $T = -57$ °C, $P = 5.2$ atm (b) Gas
(c) Solid to gas (sublimation) (d) Gas to liquid (condensation)

Analyze A phase diagram provides information about which phase is present at a given temperature and pressure.

Plan In each case, find the location on the CO_2 phase diagram that corresponds to the given temperature and pressure and trace any change in conditions.

Execute

(a) The phase diagram shows that the triple point occurs at a temperature of -57 °C and a pressure of 5.2 atm.
(b) Increasing T and holding P constant means moving somewhat to the right from the triple point. This is the gas region of the phase diagram.
(c) At the starting temperature and pressure, the CO_2 is a solid. Increasing temperature at constant pressure the conditions become those of the gas phase.
(d) At the starting temperature and pressure, CO_2 is a gas. Increasing P while holding T constant means moving upward, entering the liquid region.

PROBLEM-SOLVING PRACTICE 9.6

On the phase diagram for CO_2 (Figure 9.14), the point that corresponds to a temperature of -13 °C and a pressure of 7.5 atm is on the liquid-vapor curve where liquid phase and gas phase are in equilibrium. What phase exists when the pressure remains the same and the temperature is increased by several degrees?

EXERCISE 9.8

Using a Phase Diagram

Consider the phase diagram for CO_2 (Figure 9.14). A curve connects the triple point and the critical point; what is the name of this curve? What phase transition occurs when you traverse this curve at its midpoint going upward, that is, increasing the pressure at constant temperature? Downward? What phase transition occurs when you traverse this curve from left to right? Right to left?

The critical point for CO_2 is 72.79 atm and 30.98 °C. Above that T and P, CO_2 is a supercritical fluid. Under certain conditions, liquid CO_2 is present in carbon dioxide fire extinguishers. On a hot day, when $T > 31$ °C, the CO_2 becomes a supercritical fluid and cannot be heard sloshing. Under these conditions the only way to know whether CO_2 is in the extinguisher—without discharging it—is to weigh it and compare its mass with the mass of the empty container, which is usually indicated on a tag attached to the fire extinguisher.

EXERCISE 9.9

The Behavior of CO_2

If liquid CO_2 is slowly released to the atmosphere from a cylinder, what state will the CO_2 be in? If the liquid is suddenly released, as in the discharge of a CO_2 fire extinguisher, why is solid CO_2 seen? Can you explain this phenomenon on the basis of the phase diagram alone, or do you need to consider other factors?

9-5 Water: Its Important and Unusual Properties

"If there is magic on this planet, it is contained in water . . . Its substance reaches everywhere."
—Loren Eiseley

Earth is sometimes called the "blue planet" because three quarters of its surface is covered with oceans. Large quantities of water are present in its rocks and soils. Water is essential for almost every life form, has been a key actor throughout human history, and is a dominant factor in weather and climate.

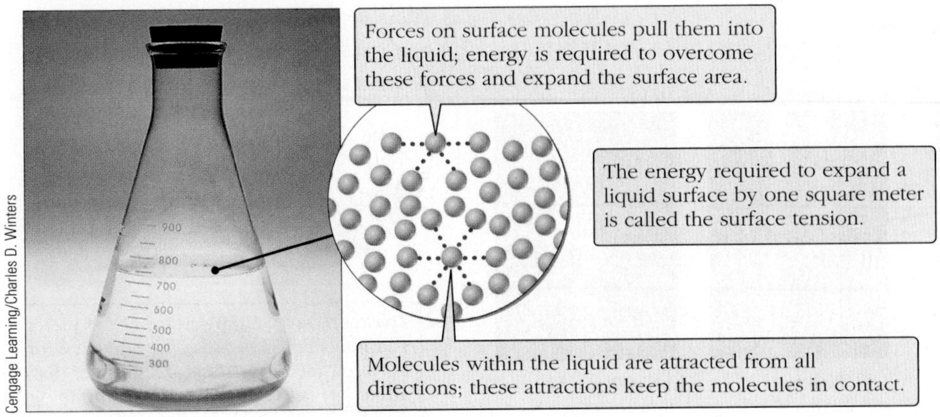
Forces on surface molecules pull them into the liquid; energy is required to overcome these forces and expand the surface area.
The energy required to expand a liquid surface by one square meter is called the surface tension.
Molecules within the liquid are attracted from all directions; these attractions keep the molecules in contact.
900
800
700
600
500
400
300
Cengage Learning/Charles D. Winters

Figure 9.15 Surface tension of a liquid.

In water and other liquids, particles are in contact and the volume is roughly constant. Thus all liquids have surfaces and distinctive surface properties. As illustrated in Figure 9.15 molecules in the interior of a liquid experience a uniform attraction in all directions; however, molecules at a horizontal surface are attracted from the sides and below, but not from above. Expanding the surface of a liquid requires energy to overcome these uneven attractions. *The energy required to expand a liquid surface by 1 m*^{2} *is its* **surface tension**. Because water has strong intermolecular attractions, it has high surface tension (7.92×10^{-2} J/m^2), sufficient to support a metal fishing fly or a water bug. Surface tension also causes rain droplets to be spherical and to bead up on waxy surfaces such as leaves. A sphere has less surface area than any other shape, so a spherical rain droplet has fewer surface molecules, minimizing the surface energy.

Fishing flies and water strider bugs "float" because they don't weigh enough to overcome water's surface tension.
Water beads up on leaves because a sphere has less surface area per unit volume than any other shape.
Douglas Allen/Getty Images
Hermann Eisenbeiss/Science Source
Ian Grainger/shutterstock.com

CONCEPTUAL EXERCISE 9.10

Surface Tension

Predict which liquid—glycerol, $HOCH_2CH(OH)CH_2OH$, or hexane, C_6H_{14}—has the greater surface tension. Explain your prediction.

When a glass tube with a small diameter and open at both ends, a *capillary tube,* is put into water, the water rises in the tube. *The rise of water in a capillary tube is called* **capillary action**. The glass walls of the tube are largely silicon dioxide, SiO_2, so water molecules form hydrogen bonds to oxygen atoms of the glass. This attractive force is stronger than the attractive forces among water molecules (also hydrogen bonding), so water creeps up the wall of the tube. Simultaneously, surface tension keeps the water's surface area small. The combination of the forces raises the water level in the tube and the liquid surface is concave upward. *The curved liquid surface is called a* ***meniscus*** (Figure 9.16a). Water rises in the tube until the force of the water-to-wall hydrogen bonds

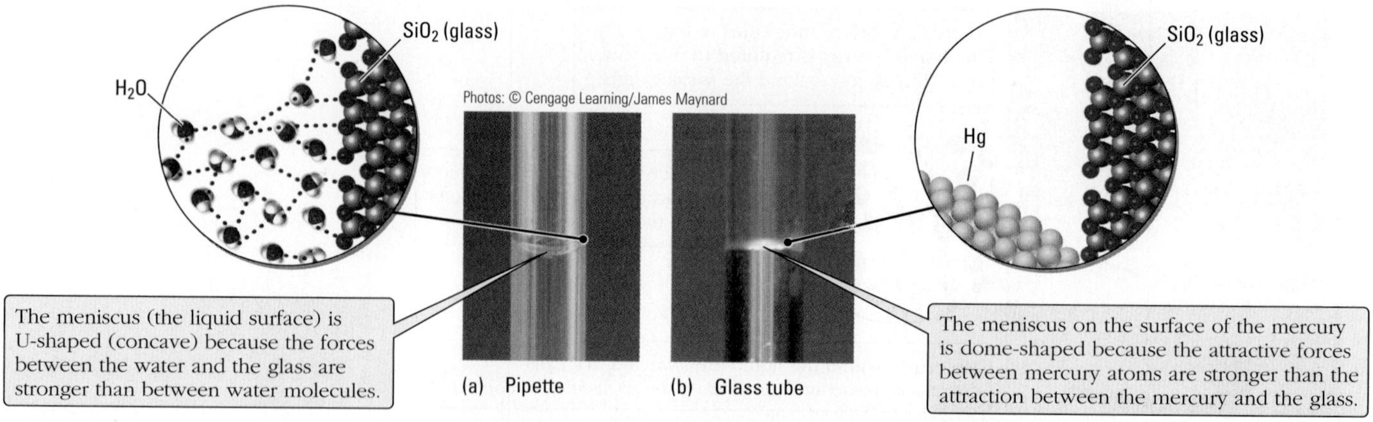
SiO_2 (glass)
H_2O
Photos: © Cengage Learning/James Maynard
SiO_2 (glass)
Hg
The meniscus (the liquid surface) is U-shaped (concave) because the forces between the water and the glass are stronger than between water molecules.
(a) Pipette
(b) Glass tube
The meniscus on the surface of the mercury is dome-shaped because the attractive forces between mercury atoms are stronger than the attraction between the mercury and the glass.

Figure 9.16 Meniscuses and noncovalent forces at surfaces.

is balanced by the pull of gravity. Capillary action is crucial to plant life because it helps water, with its dissolved nutrients, to move upward through plant tissues against the force of gravity. Certain athletic wear is made of materials that use capillary action to "wick away" moisture from the skin, keeping the athlete dry.

Mercury behaves oppositely from water. Mercury in a small-diameter glass tube has a dome-shaped (convex upward) meniscus because the attractive forces between mercury atoms are stronger than the attraction between the mercury and the glass wall (Figure 9.16b).

9-5a Unusual Properties of Water

We take water, that most common of liquids, for granted, but it is worthwhile to consider the many unique and wondrous properties water has:

- Ice is less dense than liquid water, so ice floats. This is in stark contrast with most other liquids, which exhibit the opposite behavior that the solid sinks in its liquid.
- It takes more energy to melt ice, heat water, and vaporize it per gram than for almost any other substance.
- Water has the highest thermal conductivity and surface tension of any molecular liquid.

These properties and their applications are given in greater detail in Table 9.6.

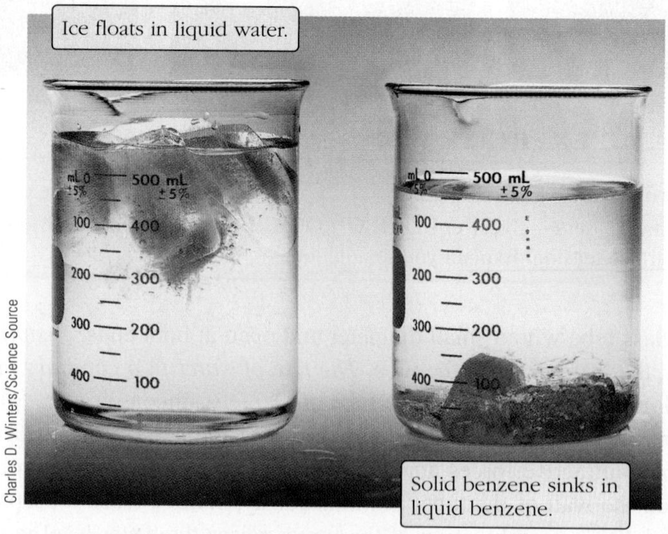
Ice floats in liquid water.
Charles D. Winters/Science Source
Solid benzene sinks in liquid benzene.

Most of water's unusual properties can be attributed to its molecules' unique capacity for hydrogen bonding. As seen in Figure 9.17, one water molecule can participate in four

Table 9.6 Unusual Properties of Water

Property	Comparison with Other Substances	Importance in Physical and Biological Environment
Specific heat capacity (4.18 J g^{-1} °C^{-1})	Highest of all liquids and solids except NH_3	Moderates temperature in the environment and in organisms; climate affected by movement of water (e.g., Gulf Stream)
Fusion enthalpy (333 J/g)	Highest of all molecular solids except NH_3	Freezing water releases large quantity of thermal energy; used to save crops from freezing by spraying them with liquid water
Vaporization enthalpy (2256 J/g)	Highest of all molecular substances	Condensation of water vapor in clouds releases large quantities of thermal energy, fueling storms
Surface tension (7.3×10^{-2} J/m^2)	Highest of all molecular liquids	Contributes to capillary action in plants; causes formation of spherical droplets; supports insects on water surfaces
Thermal conductivity (0.6 J s^{-1} m^{-1} °C^{-1})	Highest of all molecular liquids	Provides for transfer of thermal energy within organisms; rapidly cools organisms immersed in cold water, causing hypothermia

hydrogen bonds to other water molecules. When liquid water freezes, a three-dimensional network of water molecules forms to accommodate the maximum hydrogen bonding. In the crystal lattice of ice, the oxygen atoms lie at the corners of puckered, six-sided rings. Considerable open space is left within the rings, forming empty channels that run through the entire crystal lattice. This makes the solid's density less than the liquid's.

The six-sided symmetry of snowflakes corresponds to the symmetry of these hexagonal rings.

When ice melts to form liquid water, approximately 15% of the hydrogen bonds are broken and the rigid ice lattice collapses. This makes the density of liquid water greater than that of ice at the melting point. The density of ice at 0 °C is 0.917 g/cm^3, and that of liquid water at 0 °C is 0.9998 g/cm^3. The density difference is only about 10%, but it is enough so that ice floats on the surface of the liquid. This explains why about 90% of an iceberg is submerged and about 10% is above water.

This answers the question posed in Chapter 1 (← Sec. 1-1), "Why is only the tip of an iceberg (about 10%) above water?"

As the liquid water is warmed further, more hydrogen bonds break, and more empty space is filled by water molecules. The density continues to increase until a temperature

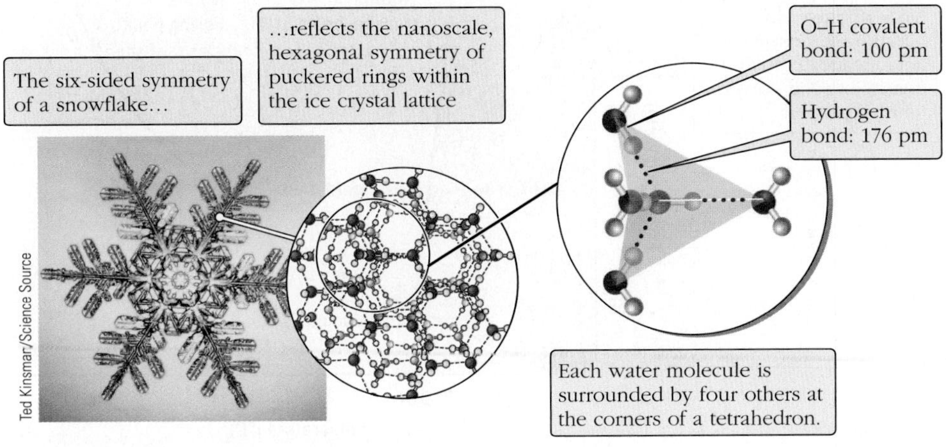
The six-sided symmetry of a snowflake...
...reflects the nanoscale, hexagonal symmetry of puckered rings within the ice crystal lattice
O–H covalent bond: 100 pm
Hydrogen bond: 176 pm
Each water molecule is surrounded by four others at the corners of a tetrahedron.
Ted Kinsmar/Science Source

Figure 9.17 The open-cage, hexagonal structure of ice.

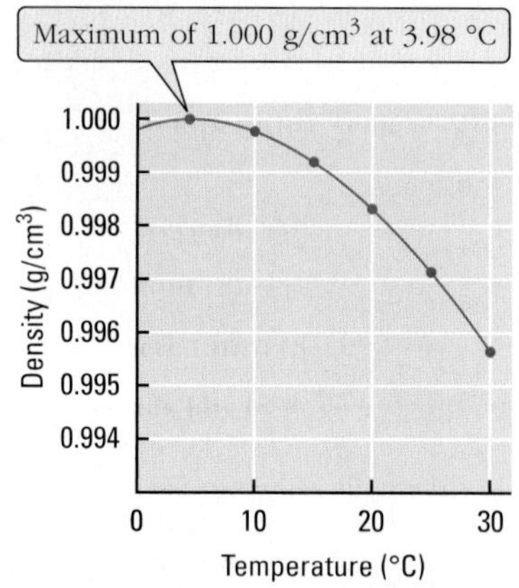
Maximum of 1.000 g/cm³ at 3.98 °C
1.000
0.999
0.998
0.997
0.996
0.995
0.994
Density (g/cm³)
0
10
20
30
Temperature (°C)

Figure 9.18 The density of liquid water between 0 °C and 30 °C. The density of ice at 0 °C and 1 bar is 0.917 g/cm^3, but the density of liquid water at those conditions is 0.9998 g/cm^3. Ice is about 10% less dense than water at 0 °C.

of 3.98 °C and a density of 1.000 g/cm^3 are reached. As the temperature rises beyond 3.98 °C, increased molecular motion causes the molecules to push each other aside more vigorously, the empty space between molecules increases, and the density decreases by about 0.0001 g/cm^3 for every 1 °C temperature rise above 3.98 °C (Figure 9.18).

Because of this unusual variation of density with temperature, when water in a lake is cooled to 3.98 °C, the higher density causes the cold water to sink to the bottom. Water cooled below 3.98 °C is less dense and stays on the surface, where it can be cooled even further. Consequently, the water at the bottom of the lake remains at 3.98 °C, while that on the surface freezes. Ice on the surface insulates the remaining liquid water from the cold air, and, unless the lake is quite shallow, not all of the water freezes. This allows fish and other organisms to survive without being frozen in the winter. When water at 3.98 °C sinks to the bottom of a lake in the fall, it carries with it dissolved oxygen. Nutrients from the bottom are brought to the surface by the water it displaces. This is called "turnover" of the lake. The same thing happens in the spring, when the ice melts and water on the surface warms to 3.98 °C. Spring and fall turnovers are essential to maintain nutrient and oxygen levels required by fish and other lake-dwelling organisms.

When water is heated, the increased molecular motion breaks additional hydrogen bonds. The strength of these intermolecular forces requires that considerable energy be transferred to raise the temperature of 1 g water by 1 °C. That is, water's specific heat capacity is quite large. Many hydrogen bonds are broken when liquid water vaporizes, because the molecules are completely separated. This gives rise to water's very large vaporization enthalpy, which helps humans to regulate body temperature by evaporation of sweat. As we have already described, hydrogen bonds are broken when ice melts, and this requires a large fusion enthalpy. Its larger-than-normal enthalpy changes upon vaporization and freezing, together with its large specific heat capacity, allow water to moderate climate and influence weather by a much larger factor than other liquids could. In the vicinity of a large body of water, summer temperatures do not get as high and winter temperatures do not get as low (at least until the water freezes over) as they do far away from water (← **Sec. 4-4a**). Seattle is farther north than Minneapolis, for example, but Seattle has a much more moderate climate because it borders the Pacific Ocean and Puget Sound.

9-5b Phase Diagram of Water

The phase diagram for water is shown in Figure 9.19. The temperature scale has been exaggerated to better illustrate some of the features of water's phase diagram. The liquid/gas curve, the line *AD*, is the same *vapor pressure curve* shown for water in Figure 9.4 (← **Sec. 9-3**). The triple point for water occurs at P = 4.58 mmHg and T = 0.01 °C.

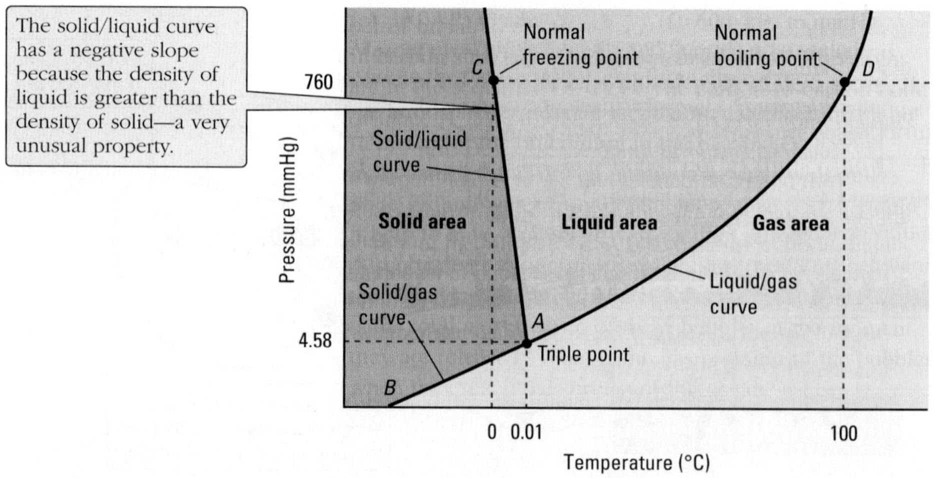
The solid/liquid curve has a negative slope because the density of liquid is greater than the density of solid—a very unusual property.
Normal freezing point
Normal boiling point
C
D
760
Solid/liquid curve
Solid area
Liquid area
Gas area
Pressure (mmHg)
Solid/gas curve
Liquid/gas curve
A
4.58
Triple point
B
0 0.01
100
Temperature (°C)

Figure 9.19 The phase diagram for water. Temperature and pressure scales are nonlinear.

You are most likely familiar with the common situation in which ice and liquid water are in equilibrium because of keeping food cool in an ice chest. If you fill an ice chest with ice, some of the ice melts to produce liquid. (Thermal energy is slowly transferred into the ice chest, no matter how well it is insulated.) The ice–liquid water mixture remains at approximately 0 °C until all of the ice has melted. This equilibrium between solid water and liquid water is represented in the phase diagram (Figure 9.19) by point *C*, at which $P = 760$ mmHg and $T = 0$ °C.

On a very cold day, ice can have a temperature well below 0 °C. Look along the temperature axis of the phase diagram for water and notice that below 1 atm and 0 °C the only equilibrium possible is between ice and water vapor. As a result, if the partial pressure of water in the air is low enough, ice will sublime on a cold day. Sublimation, which is endothermic, takes place more readily when solar radiation provides energy to the ice. The sublimation of ice allows snow or ice to gradually disappear even though the temperature does not climb above freezing (Figure 9.11, ← **Sec. 9-4b**).

The phase diagram for water is unusual in that the solid/liquid (melting point) curve *AC* slopes in the opposite direction from that for almost every other substance. The right-to-left, or negative, slope of the solid/liquid equilibrium line is a consequence of the lower density of ice compared to that of liquid water (which was discussed in the previous section). Thus, when ice and water are in equilibrium, one way to melt the ice is to apply greater pressure. This is evident from Figure 9.19. If you start at the normal freezing point (0 °C, 1 atm) and increase the pressure, you will move into the area of the diagram that corresponds to liquid water.

It was long thought that one's body weight applied to a very small area of an ice-skating blade caused sufficient pressure to form liquid water, thus making the ice under the blade slippery. However, more recent investigations show that the water molecules on the surface of ice are actually moving much more than in the main part of the solid—nearly as much as they do in liquid water. This is due to "surface melting" (Figure 9.20), which arises because the surface water molecules have interactions with fewer neighboring water molecules than do those molecules deeper in the solid. The ice lattice becomes less and less ordered closer to the surface because the surface molecules have the fewest hydrogen bonds holding them in place. Ice is made much more slippery when a thin coating of liquid water is present on its surface. This thin coat of liquid water acts as a lubricant to help the skates move easily over the ice. Although increasing the pressure on ice does cause it to melt, the effect is too small to contribute significantly in the context of ice skating.

Surface melting is just one of many examples where surfaces differ from bulk solids or liquids. *Surface science* explores behavior at surfaces and applies nanoscale ideas to explain anomalies like this one.

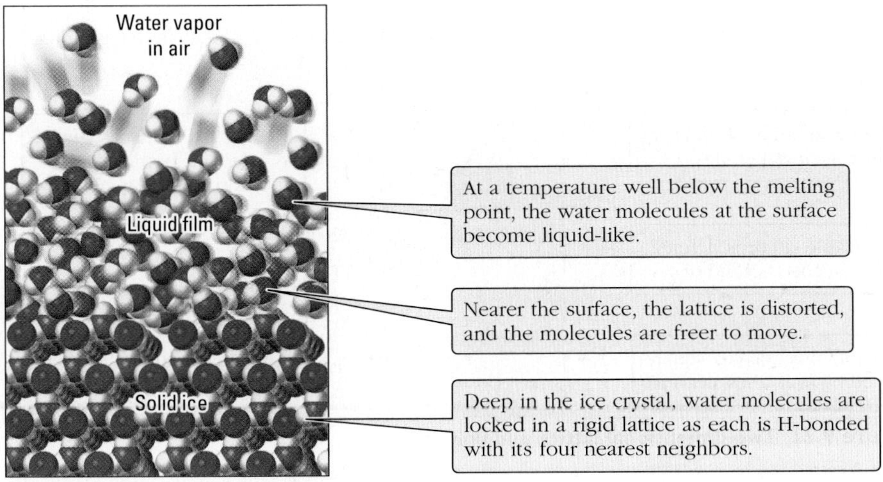
Water vapor in air
Liquid film
Solid ice
At a temperature well below the melting point, the water molecules at the surface become liquid-like.
Nearer the surface, the lattice is distorted, and the molecules are freer to move.
Deep in the ice crystal, water molecules are locked in a rigid lattice as each is H-bonded with its four nearest neighbors.

Figure 9.20 The surface of ice is slick due to a liquid-like film of water.

CONCEPTUAL EXERCISE 9.11

Lead into Gold?

Imagine a sample of gold and a sample of lead, each with a smoothly polished surface. The surfaces of these two samples are placed in contact with one another and held in place, under pressure, for about one year. After that time the two surfaces are analyzed. The gold surface is tested for the presence of lead, and the lead surface is tested for the presence of gold. Predict what the outcome of these two tests will be and explain what has happened, if anything.

9-6 Crystalline Solids

© Cengage Learning/James Maynard

Chromium alum crystal. The crystal's shape reflects its nanoscale structure.

Toward the end of the eighteenth century, scientists found that shapes of crystals can be used to identify minerals. The angles at which crystal faces meet are characteristic of a crystal's composition. The beautiful regularity of ice crystals, crystalline salts, and gemstones suggests that crystalline solids must have some *internal order.* The shape of each crystalline solid reflects the shape of its **crystal lattice**—*an orderly, repeating arrangement of points in three-dimensional space in which each point has surroundings identical to those of all other points.* The arrangement of atoms, molecules, or ions about each lattice point defines the position of each individual particle (← **Sec. 2-6b**). Hence, each atom, molecule, or ion is surrounded by neighbors in exactly the same arrangement. In the structures of metals and salts, atoms or ions usually occupy lattice points, but in some structures there are no particles located at the lattice points. A crystal is built up from a three-dimensional repetition of the same pattern, which gives the crystal long-range order throughout.

9-6a Unit Cells

A convenient way to describe and classify the repeating pattern of atoms, molecules, or ions in a crystal is to define a small segment of a crystal lattice as a **unit cell**—*a small part of the lattice that, when repeated along the directions defined by its edges, reproduces the entire crystal structure.* To help understand the idea of the unit cell, look at the simple two-dimensional array of circles shown in Figure 9.21. Each circle is centered on a lattice point. The same size circle is repeated over and over, but a circle is not the best unit cell because it gives no indication of its relationship to all the other circles. A better

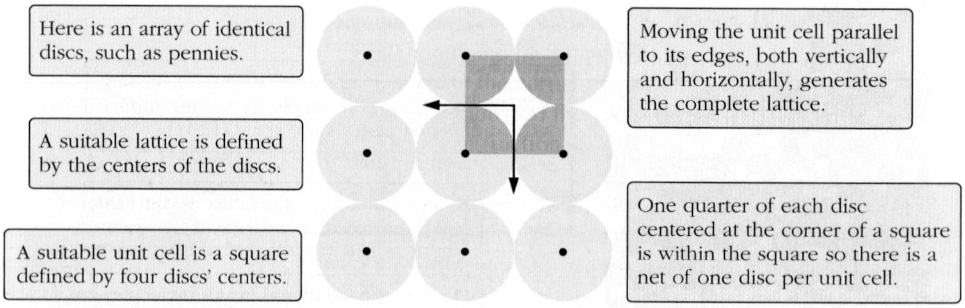
Here is an array of identical discs, such as pennies.
A suitable lattice is defined by the centers of the discs.
A suitable unit cell is a square defined by four discs' centers.
Moving the unit cell parallel to its edges, both vertically and horizontally, generates the complete lattice.
One quarter of each disc centered at the corner of a square is within the square so there is a net of one disc per unit cell.

Figure 9.21 Two-dimensional lattice and unit cell.

choice is to recognize that the lattice points at the centers of four adjacent circles lie at the corners of a square and to draw four lines connecting those lattice points. The square unit cell shown in Figure 9.21 results. Notice that each of four circles contributes one quarter of itself to the unit cell, so a net of one circle is located within the unit cell. When this two-dimensional unit cell is repeated by moving the square parallel to its edges (that is, when unit cells are placed to the left and right and above and below the first one and that process is repeated for each of the added cells), the two-dimensional lattice is generated. Notice that the corners of a unit cell (lattice points) are equivalent to each other and that collectively they define the crystal lattice.

The three-dimensional unit cells from which all known crystal lattices can be constructed fall into only seven categories. These seven types of unit cells differ because the edges (which connect lattice points) have different relative lengths and meet at different angles. We will discuss only one of these types—cubic unit cells—and only cubic unit cells composed of atoms or monoatomic ions. Such unit cells are quite common in nature and are easier to visualize than the other types of unit cells. The principles illustrated, however, apply to all unit cells and all crystal structures, including those composed of polyatomic ions and large molecules. In these more complicated cases the lattice points are often not occupied by atoms, molecules, or ions.

A **cubic unit cell** (Figure 9.22) *has edges of equal length that meet at 90° angles*; there are three types:

- *primitive cubic (pc)*
- *body-centered cubic (bcc)*
- *face-centered cubic (fcc)*

Many metals and ionic compounds crystallize in cubic unit cells. In metals, all three types of cubic unit cells have identical atoms centered at lattice points at each corner of the cube. When the cubes pack into three-dimensional space, an atom at a corner is

The 2011 Nobel Prize in Chemistry was awarded to Daniel Shechtman for his discovery and research of quasicrystals ("almost" crystals). In quasicrystals, the atoms in the solid are arranged in a pattern that doesn't quite repeat exactly, yet has long-range order. Quasicrystals have been found in a natural mineral and in surgical steel. Industrial applications being explored for these unusual solids include non-stick coatings and heat insulation.

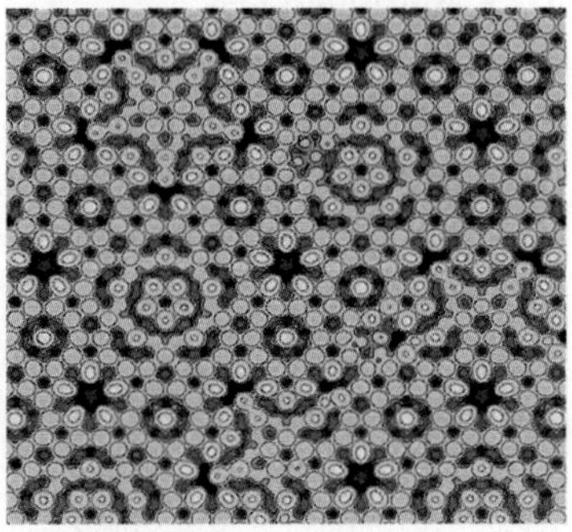

Dept of Energy/AFP/Getty Images

A quasicrystal pattern.

Examples of metals with cubic unit cells:

primitive cubic: Po

body-centered cubic: alkali metals, V, Cr, Mo, W, Fe

face-centered cubic: Ni, Cu, Ag, Au, Al, Pb

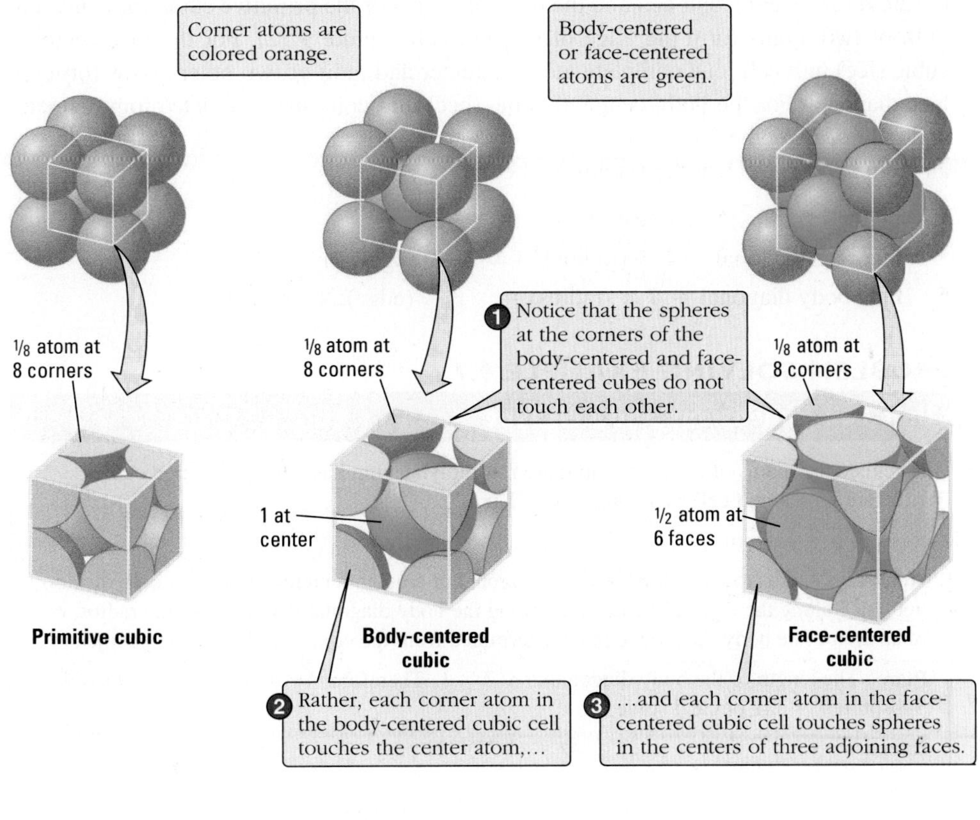
Corner atoms are colored orange.
Body-centered or face-centered atoms are green.
1/8 atom at 8 corners
1/8 atom at 8 corners
1 Notice that the spheres at the corners of the body-centered and face-centered cubes do not touch each other.
1/8 atom at 8 corners
1 at center
1/2 atom at 6 faces
Primitive cubic
Body-centered cubic
Face-centered cubic
2 Rather, each corner atom in the body-centered cubic cell touches the center atom,...
3 ...and each corner atom in the face-centered cubic cell touches spheres in the centers of three adjoining faces.

Figure 9.22 Three different types of cubic unit cells. Top row: space-filling atoms centered on lattice points. Bottom row: unit cells showing only the parts of atoms within the unit cell.

shared among eight cubes (Figure 9.22); thus only one eighth of each corner atom is actually within the unit cell.

In the primitive cubic (pc) unit cell, there are only lattice points at the corners of the cube. Since a cube has eight corners and since one eighth of the atom at each corner belongs to the unit cell, the net result is $8 \times \frac{1}{8} = 1$ *composite atom (or lattice point) per primitive cubic unit cell.*

In the body-centered cubic (bcc) unit cell, an additional lattice-point atom is at the center of the cube—entirely within the unit cell. This atom, combined with the net of one atom from the corners, gives a total of *two composite atoms (or lattice points) per body-centered cubic unit cell.*

pc: 1 atom/unit cell
bcc: 2 atoms/unit cell
fcc: 4 atoms/unit cell

In the face-centered cubic (fcc) unit cell, six atoms (lattice points) lie in the centers of the six faces of the cube. One half of each of these atoms (lattice points) belongs to the unit cell (Figure 9.22). In this case there is a net of $6 \times \frac{1}{2} = 3$ composite atoms (lattice points) per unit cell, plus the net of one atom (lattice point) contributed by the corners, for a total of *four atoms per face-centered cubic unit cell.* The number of atoms per unit cell helps to determine the density of a solid.

EXERCISE 9.12

Counting Atoms in Unit Cells of Metals

Crystalline polonium has a primitive cubic unit cell, lithium has a body-centered cubic unit cell, and calcium has a face-centered cubic unit cell. How many Po atoms belong to one unit cell? How many Li atoms belong to one unit cell? How many Ca atoms belong to one unit cell? Draw each unit cell. Indicate on your drawing what fraction of each atom lies within the unit cell.

9-6b Unit Cells and Density

For each of the three cubic unit cells, there is a close relationship between the radius of the atom forming the unit cell and the unit cell size. For the primitive cubic (pc) unit cell, radii of two atoms form the edge of the unit cell (Figure 9.22). For the face-centered cubic (fcc) unit cell, one complete atom diameter and radii of two other atoms form the face diagonal. For the body-centered cubic (bcc) unit cell, one complete atom diameter and radii of two other atoms form the body diagonal of the unit cell. If r is the radius of an atom and l is the length of a unit-cell edge,

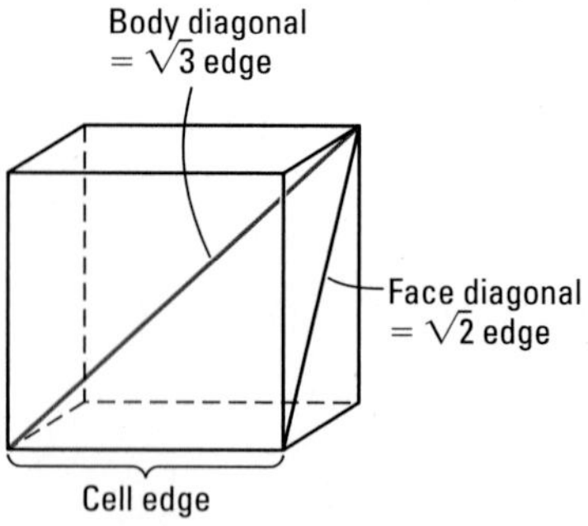

Body diagonal = √3 edge
Face diagonal = √2 edge
Cell edge

Dimensions of a cube unit cell. The edge, face diagonal, and body diagonal of the unit cell are directly related to the atomic radius for the pc, fcc, and bcc unit cells, respectively.

- **pc**: $2 \times \text{(radius)} = \text{(edge)}$; $2r = l$
- **fcc**: face diagonal $= 4 \times \text{(radius)} = \sqrt{2} \times \text{(edge)}$; $4r = \sqrt{2}\, l$
- **bcc**: body diagonal $= 4 \times \text{(radius)} = \sqrt{3} \times \text{(edge)}$; $4r = \sqrt{3}\, l$

PROBLEM-SOLVING EXAMPLE 9.7

Unit Cell Dimensions and Cell Type

Calculate the radius of a molybdenum atom if molybdenum has a body-centered cubic structure with a unit cell edge length of 321 pm.

Result $r = 139$ pm

Analyze Atoms touch along the body diagonal of a body-centered cubic unit cell; they do not touch along the edges. The distance along the body diagonal is four times the radius, r, of an atom. The body diagonal can be determined from the edge length, l, using geometry.

Plan The length of the body diagonal is $\sqrt{3} \times l$. Therefore, $4r = \sqrt{3} \times l = 1.73 \times l$. Substitute $l = 321$ pm and solve for r.

Execute

$$4r = (1.73)(321 \text{ pm})$$

$$r = \frac{(1.73)(321 \text{ pm})}{4} = 139 \text{ pm}$$

☑ **Reasonable Result Check** From Figure 9.22, note that the length of the unit cell edge in a bcc arrangement is longer than two atomic radii. In the case of molybdenum, $2 \times r = 278$ pm, which is less than the unit cell edge length, 321 pm, so the calculated radius is reasonable.

PROBLEM-SOLVING PRACTICE 9.7

Calculate the unit cell edge length of copper metal, which has an fcc arrangement. The radius of a copper atom is 126 pm.

The information given so far about unit cells allows us to check whether a proposed unit cell is reasonable. Because a unit cell can be replicated to give the entire crystal lattice, the unit cell must have the same density as the crystal. The density of a unit cell equals the mass of atoms within the cell divided by the volume of the cell. The volume can be calculated from the shape of the cell and the lengths of its edges.

PROBLEM-SOLVING EXAMPLE 9.8

Unit Cell Dimension and Density

Silver metal (molar mass = 107.86 g/mol) crystallizes in a face-centered cubic unit cell with a cell edge length of 409 pm. Use these data to calculate the density (g/cm^3) of metallic silver.

Result $d = 10.5\ \text{g/cm}^3$

Analyze The density of a unit cell is the same as the density of the substance. Density is mass per unit volume, so calculate the mass and volume of a unit cell.

Plan Calculate the unit cell volume, $V = l^3$, with $l = 409$ pm. Convert pm^3 to cm^3. Calculate the mass, m, of the unit cell from the fact that there are four atoms per fcc unit cell, the molar mass, and the Avogadro constant. Calculate $d = m/V$.

Execute

$$V(\text{unit cell}) = l^3 = (409\ \text{pm})^3 \times \left(\frac{1\ \text{m}}{10^{12}\ \text{pm}}\right)^3 \left(\frac{10^2\ \text{cm}}{1\ \text{m}}\right)^3 = 6.842 \times 10^{-23}\ \text{cm}^3$$

$$m(\text{unit cell}) = 4\ \text{Ag atoms} \times \frac{1\ \text{mol Ag}}{6.022 \times 10^{23}\ \text{Ag atoms}} \times \frac{107.86\ \text{g Ag}}{1\ \text{mol Ag}} = 7.164 \times 10^{-22}\ \text{g}$$

$$d = \frac{m}{V} = \frac{7.164 \times 10^{-22}\ \text{g}}{6.842 \times 10^{-23}\ \text{cm}^3} = 10.5\ \text{g/cm}^3$$

☑ **Reasonable Result Check** The calculated mass of silver in a unit cell is 10 times greater than its volume. Silver is near the bottom of the periodic table, where densities of metals are large; 10 times the density of water is reasonable.

PROBLEM-SOLVING PRACTICE 9.8

Vanadium metal crystallizes in a body-centered cubic unit cell. The radius of a vanadium atom is 134 pm. Calculate the density of vanadium metal, g/cm^3.

EXERCISE 9.13

Body-centered Cubic or Face-centered Cubic?

Strontium has as a cubic unit cell. The radius of a strontium metal atom is 215 pm and strontium metal has a density of 2.56 g/cm^3. Through calculations, determine whether strontium forms a body-centered or face-centered cubic unit cell.

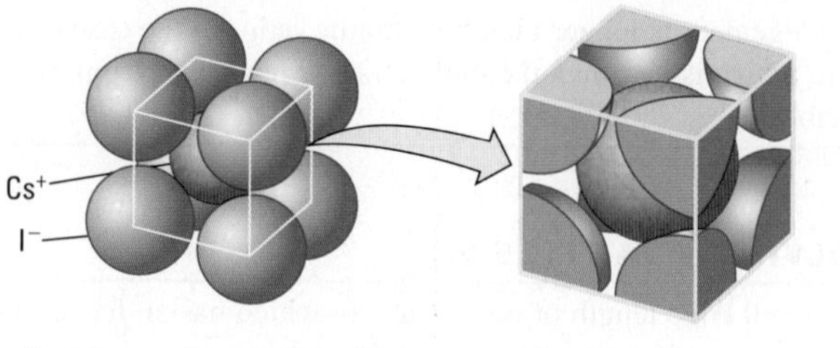
CsI unit cell
Cs+
I−

Figure 9.23 **Unit cell of the cesium iodide (CsI) crystal lattice.**

9-6c Ionic Crystal Structures

Remember that negative ions are usually larger than positive ions (← Sec. 5-10). Building an ionic crystal is a lot like placing golf balls (smaller positive ions) in the spaces among tennis balls (larger negative ions).

It is not appropriate to describe the CsI structure as bcc. In a bcc lattice, the atom or ion at the center of the unit cell must be the same as the atom or ion at the corner of the cell. In the CsI unit cell described here, there are iodide ions at the corners and a cesium ion at the center. Another, equally valid, unit cell for CsI has cesium ions at the corners and an iodide ion at the center.

Because the edge of a cube in a cubic lattice is surrounded by four cubes, one fourth of a spherical ion at the midpoint of an edge is within any one of the cubes.

The crystal structures of many ionic compounds can be described as primitive cubic or face-centered cubic lattices of spherical negative ions, with smaller positive ions occupying spaces (called holes) among the larger negative ions. The number and locations of the occupied holes are the keys to understanding the relation between the lattice structure and the formula of an ionic compound. The simplest example is an ionic compound in which the hole at the center of a primitive cubic unit cell (Figure 9.22) is occupied. The ionic compound cesium iodide, CsI, has such a structure. In it, each cube of I^- ions has a Cs^+ ion at its center (Figure 9.23). The spaces occupied by the Cs^+ ions are called cubic holes, and each Cs^+ has eight nearest neighbor I^- ions. The CsI unit cell contains one Cs^+ ion and one I^- ion for a formula of CsI.

The structure of sodium chloride, NaCl, is a very common ionic crystal lattice (Figures 9.24 and 9.25). If you look carefully at Figure 9.24, it is possible to determine the number of Na^+ and Cl^- ions in the NaCl unit cell. There is one eighth of a Cl^- ion at each corner of the unit cell (lattice point) and one half of a Cl^- ion at each lattice point in the middle of each face. The total number of Cl^- ions within the unit cell is

$\frac{1}{8}$ Cl^- per corner × 8 corners = 1 Cl^-

$\frac{1}{2}$ Cl^- per face × 6 faces = 3 Cl^-

Total of 4 Cl^- in a unit cell

There is one fourth of an Na^+ at the midpoint of each edge and a whole Na^+ in the center of the unit cell. For Na^+ ions, the total is

$\frac{1}{4}$ Na^+ per edge × 12 edges = 3 Na^+

1 Na^+ per center × 1 center = 1 Na^+

Total of 4 Na^+ in a unit cell

Thus, the unit cell contains four Na^+ and four Cl^- ions, a 1 : 1 ratio. This result agrees with the formula of NaCl for sodium chloride.

As shown in Figure 9.25, the NaCl crystal lattice consists of an fcc lattice of the larger Cl^- ions, in which Na^+ ions occupy so-called *octahedral* holes—octahedral

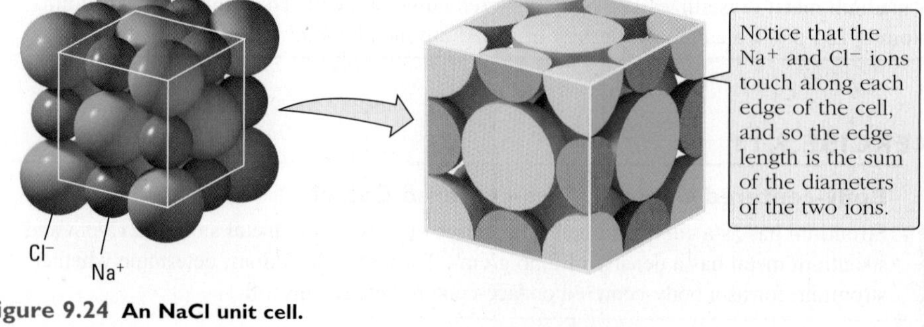
Notice that the Na+ and Cl− ions touch along each edge of the cell, and so the edge length is the sum of the diameters of the two ions.
Cl−
Na+

Figure 9.24 **An NaCl unit cell.**

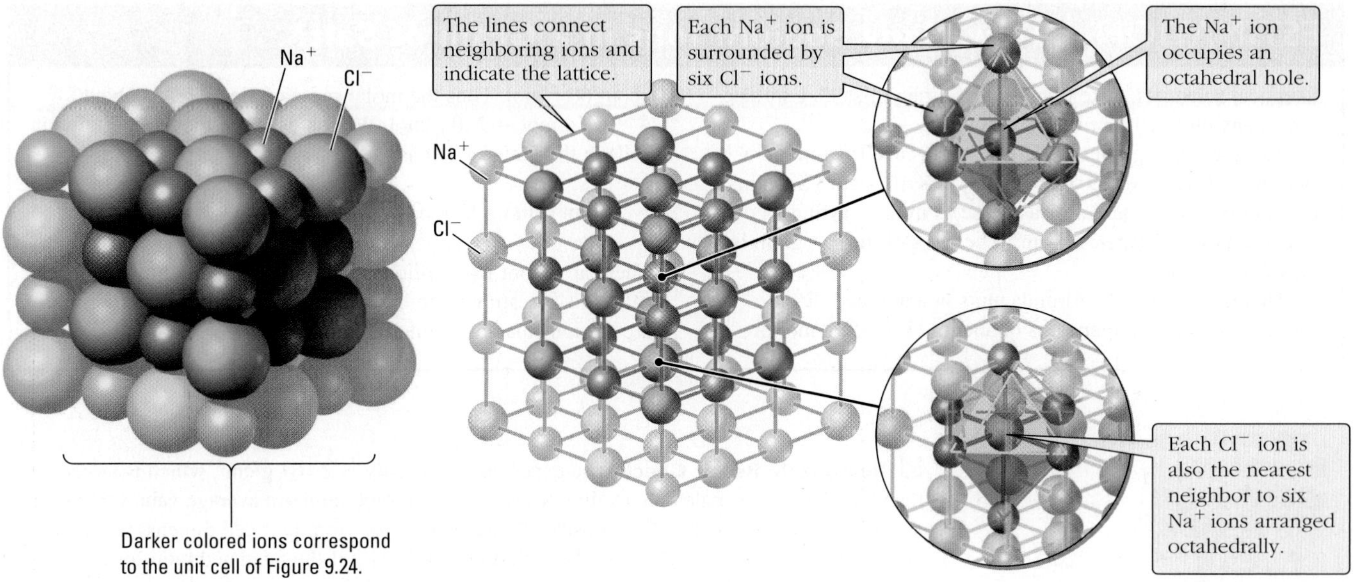

Na+
Cl−
The lines connect neighboring ions and indicate the lattice.
Each Na+ ion is surrounded by six Cl− ions.
The Na+ ion occupies an octahedral hole.
Na+
Cl−
Each Cl− ion is also the nearest neighbor to six Na+ ions arranged octahedrally.
Darker colored ions correspond to the unit cell of Figure 9.24.

Figure 9.25 The NaCl crystal lattice.

because each Na^+ ion is surrounded by six Cl^- ions at the corners of an octahedron. Likewise, each Cl^- ion is surrounded by six Na^+ ions. Figure 9.25 also shows a space-filling model of the NaCl lattice, in which each ion is drawn to scale based on its ionic radius.

EXERCISE 9.14

Formulas and Unit Cells

Cesium chloride has a cubic unit cell, like the cesium iodide unit cell in Figure 9.23. Show that the formula for the salt must be CsCl.

PROBLEM-SOLVING EXAMPLE 9.9

Calculating the Volume and Density of a Unit Cell

The fcc unit cell of NaCl is shown in Figure 9.24. The ionic radii for Na^+ and Cl^- are 116 pm and 167 pm, respectively. Calculate the density of NaCl (g/cm^3).

Result 2.14 g/cm^3

Analyze Density is mass/volume. Figure 9.24 shows four NaCl formula units (f.u.) in the cell. The mass of a unit cell is the mass of four NaCl units within the cell. The volume of a cubic unit cell can be determined from the cube of the edge length. Figure 9.24 shows that Na^+ and Cl^- ions touch along the edge.

Plan Calculate the unit cell mass from the four NaCl units, the molar mass of NaCl, and the Avogadro constant. Calculate the unit cell edge length as $2r(Na^+) + 2r(Cl^-)$. Calculate the volume of the unit cell in cm^3. Calculate density.

Execute

$$m(\text{unit cell}) = 4\ \text{NaCl units} \times \frac{1\ \text{mol}}{6.022 \times 10^{23}\ \text{units}} \times \frac{58.44\ \text{g}}{\text{mol}} = 3.882 \times 10^{-22}\ \text{g}$$

$$\text{Edge length} = l = 167\ \text{pm} + (2 \times 116\ \text{pm}) + 167\ \text{pm} = 566\ \text{pm}$$

$$V(\text{unit cell}) = (566\ \text{pm})^3 \times \left(\frac{1\ \text{m}}{10^{12}\ \text{pm}}\right)^3 \times \left(\frac{10^2\ \text{cm}}{1\ \text{m}}\right)^3 = 1.813 \times 10^{-22}\ \text{cm}^3$$

$$d = \frac{m}{V} = \frac{3.882 \times 10^{-22}\ \text{g}}{1.813 \times 10^{-22}\ \text{cm}^3} = 2.14\ \text{g/cm}^3$$

The general relationships among the unit cell type, ion or atom size, and solid density for most metallic or ionic solids are as follows:

Mass of 1 formula unit (e.g., NaCl) × number of formula units per unit cell (4) = mass of unit cell

Mass of unit cell ÷ unit cell volume (= edge^3 for NaCl) = density

ESTIMATION | How Many Sodium Ions in a Grain of Salt?

Skeptical scientists "take things with a grain of salt". Estimate how many unit cells are in a grain of salt.

Assume that a grain of salt, NaCl, is a cube 0.1 mm on a side. Its volume is thus $0.001\ mm^3 = 1 \times 10^{-6}\ cm^3$. Salt sinks in water but is nowhere near as dense as iron, so assume a density twice that of water, $2\ g/cm^3$. The mass of the salt grain is thus 2×10^{-6} g.

There are four NaCl formula units in a unit cell. Each formula unit has a molar mass of about (23 + 35) g/mol or about 60 g/mol. Thus the molar mass of a unit cell is about 4×60 g/mol = 240 g/mol. We can estimate the number of unit cells in the grain of salt as

$$N(\text{unit cells}) = 2 \times 10^{-6}\ g \times \frac{1\ mol}{240\ g} \times \frac{6 \times 10^{23}}{1\ mol} = 5 \times 10^{15}$$

There are about five million billion unit cells in a tiny crystal of salt. This is testimony to how small a unit cell is and how small are the ions that constitute the unit cell.

☑ **Reasonable Result Check** The experimental density is $2.164\ g/cm^3$, which is larger than but close to the calculated value. Remember that ionic radii are average values that fit a large range of compounds. The density of NaCl calculated from unit cell dimensions could easily have given a value closer to the experimental density if the tabulated radii for the Na^+ and Cl^- ions were slightly smaller.

PROBLEM-SOLVING PRACTICE 9.9

KCl has the same crystal structure as NaCl. Predict which has the larger unit cell, NaCl or KCl. Test your prediction by calculating the volume of the unit cell for KCl and the density of KCl, given the ionic radii K^+ = 152 pm and Cl^- = 167 pm.

Stockbyte/Getty Images

Spherical fruits in markets are stacked in closest-packed arrangements for efficiency.

In *abab* . . . packing, the third layer repeats the first layer; in *abcabc* . . . packing, the fourth layer repeats the first layer.

9-6d Closest Packing of Spheres

In crystalline solids, the atoms, molecules, or ions are often arranged as closely as possible, an arrangement known as **closest packing**. This maximizes their attractions and results in a stable structure. Closest packing is most easily illustrated for metals, where all atoms are identical and can be represented as spheres. The structure can be built up as a series of layers of atoms. In each layer of equal-sized spheres, as shown in Figure 9.26, each sphere is surrounded by six neighbors. A second layer can be put on top of the first layer so the spheres nestle into the depressions of the first layer. Then a third layer can be added. There are two ways to add the third layer, each of which yields a different structure, as shown in Figure 9.27. If the third layer is directly above the first layer, *an* abab. . . *arrangement results, called a* **hexagonal closest-packed** (hcp) structure. In this *abab*. . . structure, the centers of the spheres of the third layer are directly above the

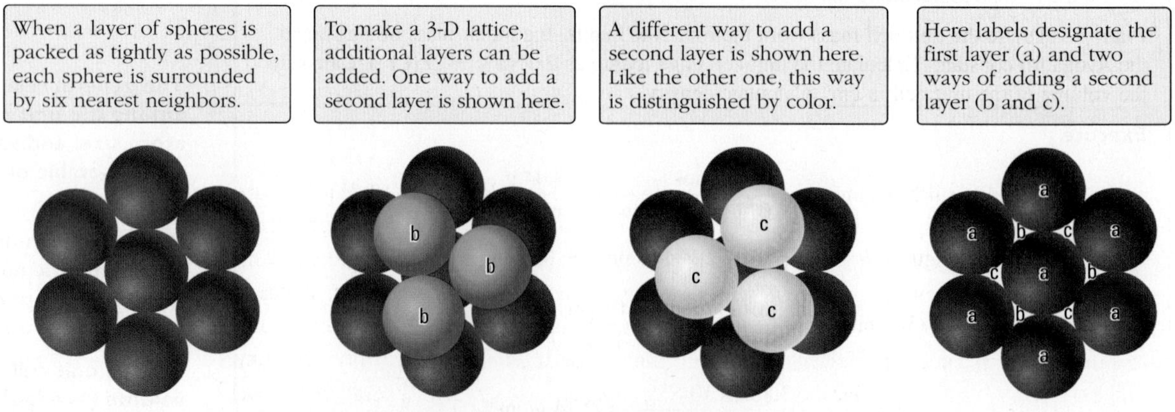
When a layer of spheres is packed as tightly as possible, each sphere is surrounded by six nearest neighbors.
To make a 3-D lattice, additional layers can be added. One way to add a second layer is shown here.
A different way to add a second layer is shown here. Like the other one, this way is distinguished by color.
Here labels designate the first layer (a) and two ways of adding a second layer (b and c).
b
c
a

Figure 9.26 Closest packing of spheres in layers.

Hexagonal closest packing. The second layer (b) is placed on the first (a). The third layer (a) is placed directly above the first. A fourth layer (b) is directly above the second, giving an abab repetition of layers.

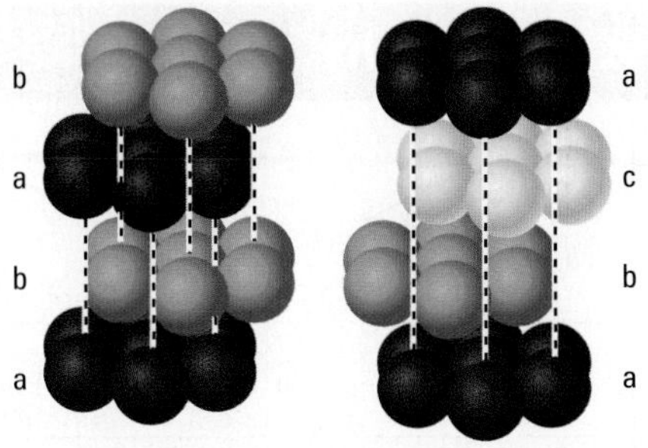
b
a
b
a
a
c
b
a

Cubic closest packing. The second layer (b) is placed on the first (a). The third layer (c) is placed above spaces in the first layer. A fourth layer (a) is directly above the first layer, giving an abcabc repetition.

Figure 9.27 Hexagonal closest packing and cubic closest packing of spheres.

centers of the spheres of the first layer (Figure 9.27). If the third layer is *not* directly above the first layer, *an* abcabc . . . *arrangement results, giving a* **cubic closest-packed** (ccp) structure (Figure 9.27). In this structure, the centers of the spheres of the third layer are directly above *spaces* (labeled "c" in Figure 9.26) between the spheres of the first layer. The centers of spheres in the fourth layer are directly above the centers of the spheres of the first layer. *The unit cell for a cubic closest-packed structure is face-centered cubic (fcc).* Therefore fcc and ccp designate the same structure.

Many repeating arrangements of layers other than abab and abcabc, such as abcbabcba, are possible; however, they rarely occur.

In each closest-packed structure (hcp and ccp), each sphere has 12 equidistant nearest-neighbor spheres (six in its layer and three each in the layers above and below); 74% of the total volume of the structure is occupied by spheres and 26% is empty. For the body-centered cubic structure only 68% of the volume is occupied, and for the primitive cubic structure only 52% is occupied.

A rigorous mathematical proof has shown the impossibility of packing spheres so that more than 74% of the total volume is occupied.

9-7 Network Solids

Some solids are composed of one or more kinds of nonmetal atoms connected by a network of covalent bonds. Such **network solids** really are *huge molecules in which all the atoms are connected to all the others via the network of covalent bonds*. Separate small molecules do not exist in a network solid.

The most important network solids are the *silicates*. Many bonding patterns exist among the silicates, but extended arrays of covalently bonded silicon and oxygen atoms are common. An example is the structure of quartz, SiO_2, which is shown nearby. Silicates are discussed in more detail in Chapter 19.

Silicon dioxide, quartz, has this structure. Each atom is connected to every other atom by a three-dimensional network of —O—Si—O— covalent bonds. Si atoms are gray; O atoms are red.

9-7a Diamond, Graphite, and Fullerenes

Two forms of carbon, diamond and graphite, are excellent examples of three-dimensional and two-dimensional covalent networks. Diamond, graphite, and a third form of carbon, fullerenes, are **allotropes**, *different forms of the same element in the same physical state at the same temperature and pressure.* Allotropes are possible because atoms of the same element can be connected in different ways when they form molecules or other structures.

Diamond You are probably familiar with diamond, a hard, colorless solid. Its structure is a *three-dimensional network* (Figure 9.28). The structure is built with sp^3-hybridized (tetrahedral) carbon atoms (← **Sec. 7-3c**), each bonded to four others by single covalent bonds. Thus each carbon atom is held firmly in place by strong covalent bonds. Because all of its valence electrons are localized in covalent bonds between carbon atoms, diamond does not conduct electricity. Diamond is one of the hardest materials and also one of the best conductors of heat known. It is also transparent to visible, infrared, and ultraviolet radiation.

TOOLS OF CHEMISTRY | X-ray Crystallography

X-ray crystallography is used to determine the arrangement of the atoms, molecules, or ions in crystalline samples. A powerful tool, it is routinely applied to solids as varied as crystalline minerals, proteins, and semiconductors. X-rays have wavelengths of about 1 nm, the same scale as atomic dimensions. When X-rays strike crystals, the X-rays are scattered in different directions and with different intensities due to interference (← **Sec. 5-2a**). When the interference is *constructive*, the waves combine to produce X-rays that can be observed indirectly because they expose a spot on photographic film or create an electrical signal in a detector. When the interference is *destructive*, the waves partly or entirely cancel each other, so little or no X-ray intensity is recorded.

Constructive interference. Two in-phase waves combine to produce a wave of greater amplitude.

Destructive interference. Two waves of equal amplitude that are exactly out of phase give zero amplitude.

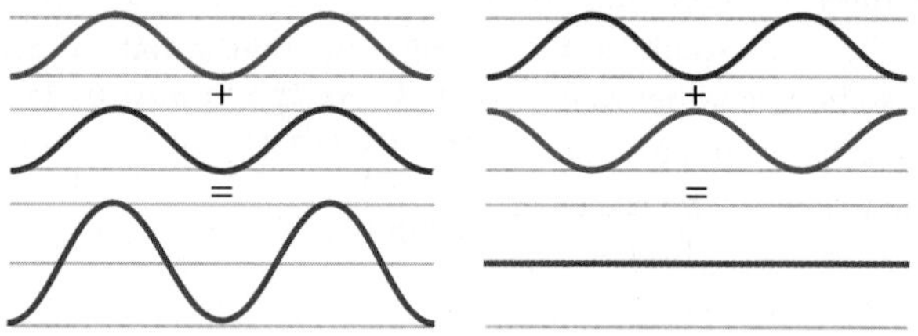

+
=
+
=

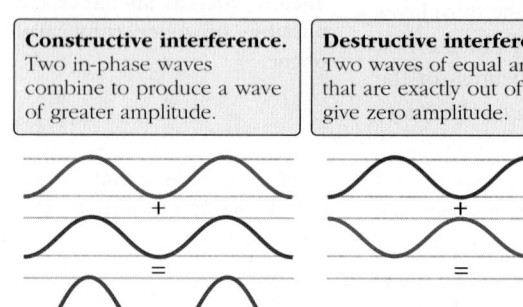

Interference effects create a diffraction pattern that depends on the X-ray wavelength and the spatial arrangement of atoms or ions in the crystal. The figure at the upper right shows how an equation called the Bragg equation relates the X-ray wavelength (λ), the spacing between atomic layers in the crystal (d), the angle of scattering (θ), and an integer (n, usually considered as 1):

$$n\lambda = 2d \sin \theta$$

Waves a and b reinforce each other and produce a spot or electrical signal when the Bragg equation is satisfied. Thus, the X-ray diffraction pattern can be related to the distance between layers of atoms or ions in the crystal.

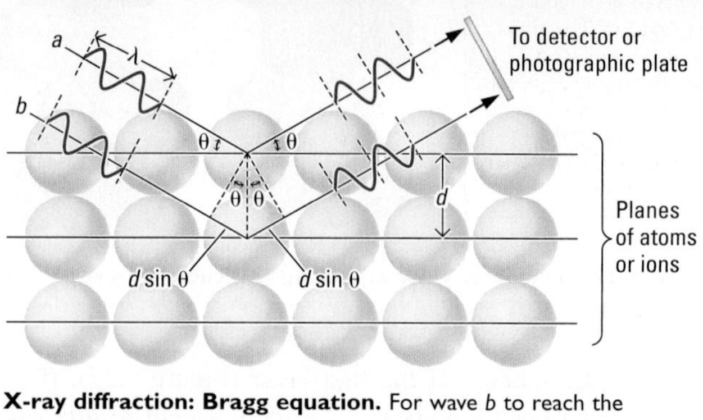

a
b
λ
θ
θ
θ θ
d
d sin θ
d sin θ
To detector or photographic plate
Planes of atoms or ions

X-ray diffraction: Bragg equation. For wave b to reach the detector in phase with wave a, b must travel farther by $2d \sin \theta$, which must be an integer multiple of the wavelength.

The intensity of the scattered X-ray beam depends on the electron density of the atoms in the crystal. A hydrogen atom is least effective in causing X-rays to scatter; heavy atoms such as lead or mercury are highly effective.

Photographic film was used to record the diffraction patterns in early instruments. Today, crystallographers use computer-controlled instruments, highly sensitive solid-state detectors, and other advanced instrumentation to measure the angles and intensities of X-ray beams diffracted by crystals.

In one of the most famous uses of X-ray crystallography, James Watson and Francis Crick in 1953 interpreted X-ray data obtained by Maurice Wilkins and Rosalind Franklin to discover the double-helix structure of DNA, the biological macromolecule that carries genetic information (← **Sec. 7-7b**). Watson, Crick, and Wilkins were awarded the Nobel Prize in Physiology or Medicine in 1962 for this discovery. (Franklin died in 1958 and therefore could not share the prize; Nobel Prizes are not awarded posthumously.)

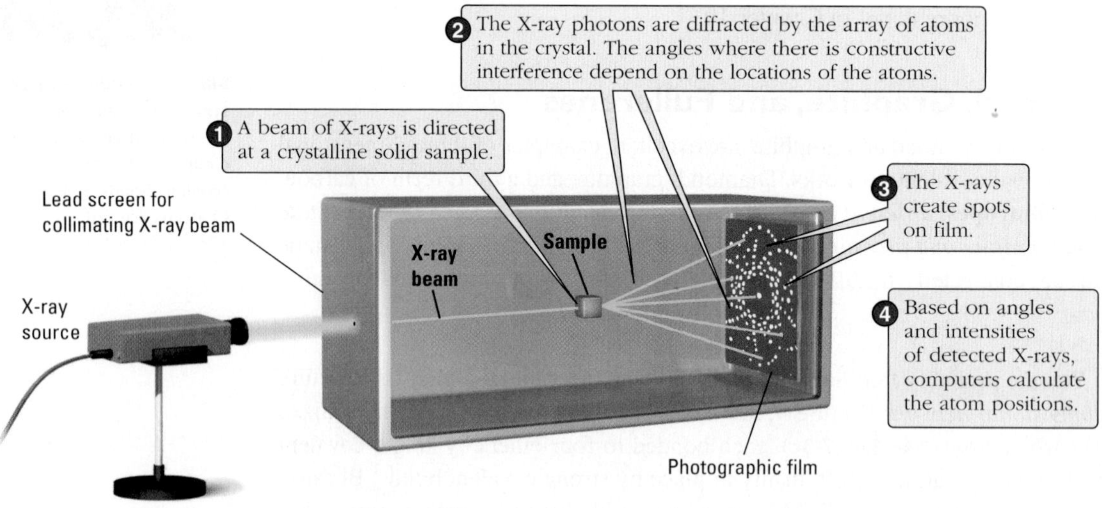

2 The X-ray photons are diffracted by the array of atoms in the crystal. The angles where there is constructive interference depend on the locations of the atoms.
1 A beam of X-rays is directed at a crystalline solid sample.
Lead screen for collimating X-ray beam
X-ray source
X-ray beam
Sample
3 The X-rays create spots on film.
4 Based on angles and intensities of detected X-rays, computers calculate the atom positions.
Photographic film

Schematic diagram of an early X-ray diffraction experiment in which photographic film was used to record the intensities of X-rays.

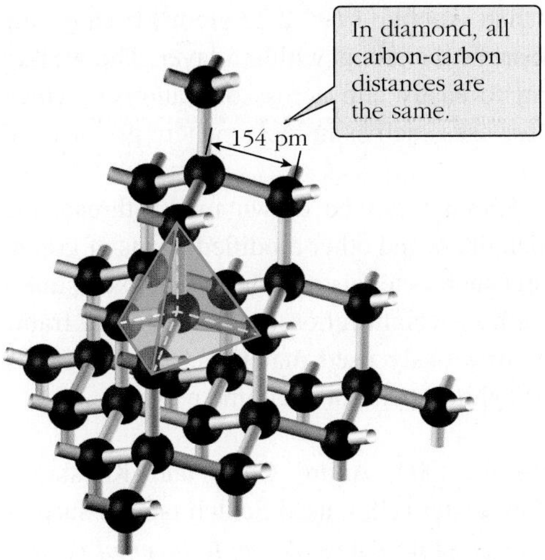
In diamond, all carbon-carbon distances are the same.
154 pm

Figure 9.28 The structure of diamond. Each carbon atom is covalently bonded to four other carbon atoms in an extended tetrahedral arrangement.

Graphite Graphite is a soft, black solid—very different from diamond. Graphite's name comes from the Greek *graphein,* meaning "to write," because one of its earliest uses was for writing on parchment. Artists today still draw with charcoal, an impure form of graphite, and we write with pencil leads that contain graphite. Graphite is an example of a *planar network solid* (Figure 9.29). Each carbon atom is covalently bonded by single bonds to three other carbon atoms. All carbon atoms are sp^2 hybridized and triangular planar (← **Sec. 7-3b**). Each sp^2 carbon atom has a third, unhybridized p orbital perpendicular to the triangular plane. These p orbitals overlap to form π bonds that are delocalized over the entire planar layer of carbon atoms. The layers consist of hexagonal rings of carbon atoms like those in benzene (← **Sec. 6-12c**). Each hexagon shares all six of its sides with other hexagons around it, forming a *two-dimensional network* resembling chicken wire. Because the π electrons are able to move freely around this network, graphite readily conducts electricity in the two directions parallel to the planar layers.

Within a plane, there are strong covalent bonds between carbon atoms, but attractions between the planes are due to noncovalent London forces and hence are weaker.

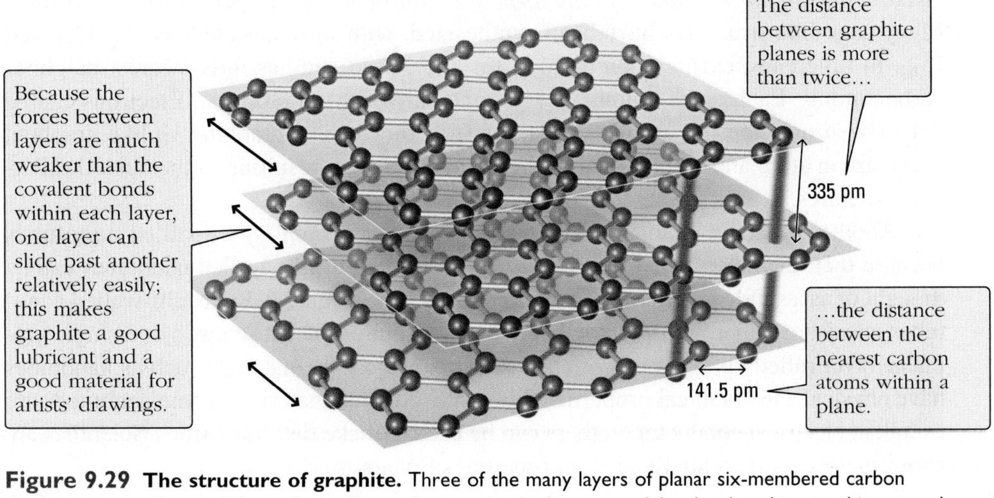
Because the forces between layers are much weaker than the covalent bonds within each layer, one layer can slide past another relatively easily; this makes graphite a good lubricant and a good material for artists' drawings.
The distance between graphite planes is more than twice...
335 pm
...the distance between the nearest carbon atoms within a plane.
141.5 pm

Figure 9.29 The structure of graphite. Three of the many layers of planar six-membered carbon rings are shown. Some of the carbon valence electrons in the layers are delocalized, making graphite a good conductor of electricity parallel to the layers.

cirkoglu/iStockphoto.com

Modern graphite-fiber tennis rackets.

Diamond (d = 3.51 g/cm^3) is much denser than graphite (d = 2.22 g/cm^3) because the graphite layers are more than twice as far apart as the atoms within a layer. The weaker attractive forces between planes allow them to easily slip across one another, which makes graphite an excellent solid lubricant for uses such as in locks, where greases and oils are undesirable.

Under carefully controlled conditions, graphite can be drawn into a thread and wrapped into a multistrand *graphite fiber*. Such fibers and other modified forms of graphite are used to make modern athletic equipment such as tennis rackets and lacrosse sticks. The fibers, which have extremely high strength-to-weight ratios, make the racket frame stiffer while lighter than fiberglass or other previously used materials. Lacrosse stick shafts and skis also incorporate high-grade graphite fibers for the same reasons.

Geim and Novoselov received the 2010 Nobel Prize in Physics for their research in studying the properties and potential applications of graphene.

Graphite can be considered to be stacks of graphene layers.

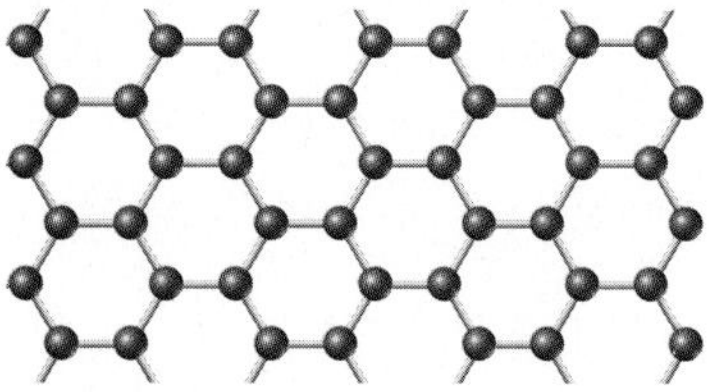

Graphene: Single Sheets of Graphite In 2004, Andre Geim and Konstantin Novoselov, physicists at the University of Manchester (UK), used Scotch tape to successively peel single layers of carbon atoms from graphite flakes. A *single layer of carbon atoms in a "chicken-wire" hexagonal arrangement is called* **graphene**, a two-dimensional material. You probably have created graphene when writing with a "lead" pencil. The "lead" is actually graphite, and when writing, you rub graphite onto the paper, likely occasionally peeling off a single layer of graphene among your markings. Graphene is also produced by the reduction of silicon carbide, SiC, at very high temperatures ($>$1300 K) and by several other methods.

Interest in graphene is widespread because of its unusual properties: It is very thin, only a single atom thick, yet it is extremely strong, about 200 times stronger than steel. Its surface can be coated to act as a very sensitive gas detector, such as one to detect explosives in luggage. As in a metal, electrons move through graphene extremely rapidly at room temperature because it is a "zero-gap" semiconductor (see Section 9-9). These properties make graphene highly valued for potential applications in electronic and digital devices, including the tantalizing possibility of graphene replacing silicon in computer chips (see Section 9-10). Electrons in graphene move about 1000 times faster than those in silicon, an advantage that could lead to smaller, more powerful computers.

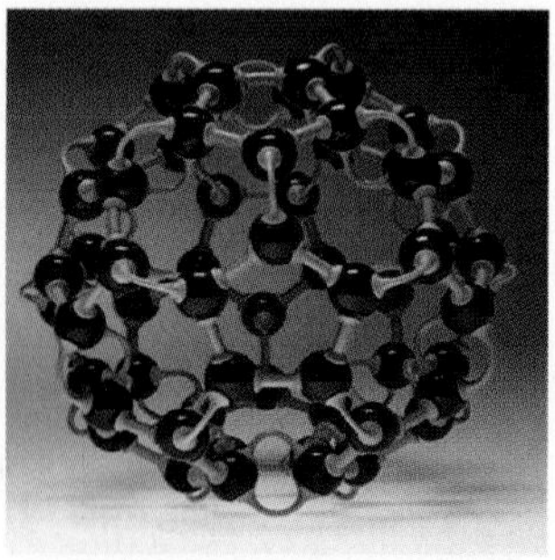

The C_{60} fullerene molecule, known as buckyball (above), has the same shape as a soccer ball (below).

Photos: Cengage Learning/Charles D. Winters

Fullerenes Diamond and graphite have been known for centuries, but it was only in 1985 that a third allotrope of carbon, buckyball, C_{60}, was discovered. Its structure, shown nearby, is composed of pentagons and hexagons of carbon atoms and is roughly spherical, similar to a soccer ball. Beginning in the 1980s, a variety of related carbon structures have been made, all classified as fullerenes. A **fullerene** *is a molecule that consists only of carbon and is in the form of a hollow sphere, distorted sphere, or tube.* Many cage-like structures have been synthesized, with formulas such as C_{70}, C_{84}, and C_{100}. In fullerenes, each carbon atom is sp^2 hybridized and has three nearest-neighbor carbon atoms. Each carbon atom also has one unhybridized p orbital. Electrons can be delocalized over these p atomic orbitals, just as they can in graphite. Unlike graphite, each carbon atom and its three bonds are not quite planar, resulting in the curved fullerene structures.

Another fullerene geometry is *elongated tubes of carbon atoms,* called **nanotubes** because their diameters are on the order of nanometers. Single-walled nanotubes can be thought of as a single graphene layer curved into a tube (Figure 9.30). Multi-walled nanotubes consist of either multiple graphene layers forming concentric tubes or a single graphene layer rolled up like a roll of chicken wire. Nanoscale materials such as nanotubes have physical and chemical properties different from bulk materials. Some nanotubes are excellent electrical conductors; others can be used to make tiny transistors. Scientists are exploring myriad potentially useful properties of nanotubes.

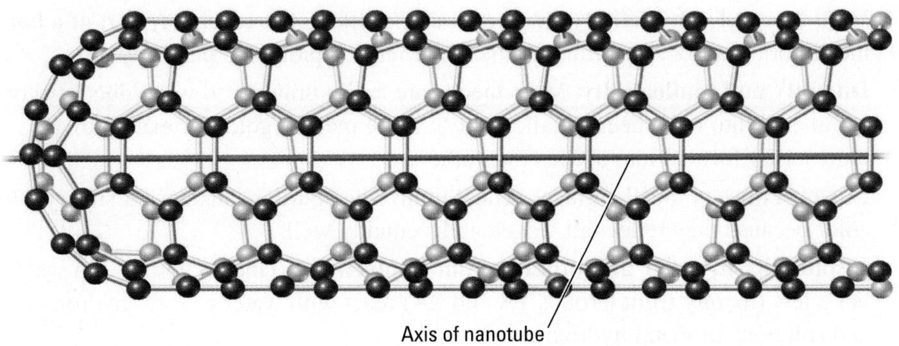
Axis of nanotube

Figure 9.30 A single-walled carbon nanotube is composed of hexagons of carbon.

9-8 Materials Science

Humans have long adapted or altered naturally occurring materials to use as clothing, structures, and devices. Periods of human history reflect the defining materials of the time—Stone Age, Bronze Age, and Iron Age. In early times, nanoscale structures were altered without any scientific understanding of the changes that took place. Only relatively recently has a fuller scientific understanding of materials been possible with the advent of a wide range of laboratory instruments that probe the structure of matter and materials at the nanoscale level. By *materials* we mean matter used to construct the devices and macroscale structures used by our highly developed technological society. **Materials science** *is the multidisciplinary scientific study of the relationships between the structure of materials and their physical and chemical properties. Structure* in this regard refers to the organization of the material from the nanoscale (electronic behavior; arrangements of atoms, molecules, and ions in solids) through to the macroscale (bulk physical properties). A major breakthrough in materials science occurred in the latter half of the 20th century when scientists and engineers developed the ability to design materials with properties customized to specific applications.

There are four major classes of materials:

- **Metals** Usually opaque and crystalline with shiny surface. Good conductors of thermal energy and electricity. Strong, malleable (able to be pounded into sheets) and ductile (able to be drawn into wires). A few are ferromagnetic (Fe, Ni, and Co), but most are not.
- **Ceramics** Clay-based; usually nonmetallic substances such as porcelain, brick, cement, and glass. Usually poor thermal conductors, but good thermal insulators. Glasses, a special class of ceramics, are strong, amorphous, and usually brittle.
- **Polymers** Natural (wool, silk, rubber) as well as synthetic (plastics). Typically are thermal and electrical insulators. Some are flexible; others are brittle. These are discussed in detail in Chapter 10.
- **Composites** Materials that may blend metallic, ceramic, and polymer components to achieve particular properties for specific applications.

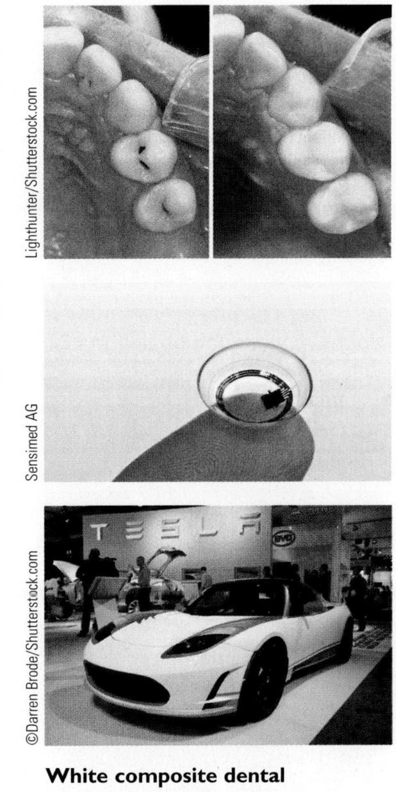
Lighthunter/Shutterstock.com
Sensimed AG
TESLA
©Darren Brode/Shutterstock.com

White composite dental restorations, soft contact lenses that report fluctuations in intraocular pressure, and the carbon-fiber body of the Tesla electric automobile all involve modern materials.

9-9 Metals, Semiconductors, and Insulators

All solid metals have these metallic properties:

- **High electrical conductivity** Metal wires carry electricity from power plants to homes and offices because electrons in metals are highly mobile.

Table 9.7 Fusion Enthalpies and Melting Points of Some Metals

Metal	$\Delta_{fusion}H°$ (kJ/mol)	Melting Point (°C)
Hg	2.3	−38.8
Ga	5.585	29.8*
Na	2.59	97.9
Li	3.0	180.5
Al	10.7	660.4
U	9.14	1132.1
Fe	13.8	1535.0
Ti	14.15	1670.
Cr	21.00	1907
W	52.31	3422

*Gallium metal will melt in the palm of your hand from the warmth of your body (37 °C). It happens that gallium is a liquid over the largest range of temperature of any metal. Its boiling point is approximately 2250 °C.

- **High thermal conductivity** We learn early in life not to touch any part of a hot metal pot because it will transfer thermal energy rapidly and painfully.
- **Ductility and malleability** Most metals are easily drawn into wire (ductility) or hammered into thin sheets (malleability); some metals (gold, for example) are more easily formed into shapes than others.
- **Luster** Polished metal surfaces reflect light. Most pure metals have a silvery white color because they reflect all wavelengths equally well.
- **Insolubility in water and other common solvents** No metal dissolves in water, but a few (mainly from Groups 1A and 2A) react with water to form hydrogen gas and solutions of metal hydroxides.

All metals are solids at room temperature except mercury (m.p. −38.8 °C). Many metals have high melting points and fusion enthalpies, which correlate with strong attractive forces among metal atoms. As seen from Figure 9.31 and Table 9.7, the transition metals, especially the third series, have very high melting points and exceptionally high fusion enthalpies. Tungsten (W) has the highest melting point (3422 °C) of any element except for carbon as graphite (3550 °C).

PROBLEM-SOLVING EXAMPLE 9.10

Using Fusion Enthalpy

Use data from Table 9.7 to calculate the thermal energy transfer needed to melt entirely a 25.0-g sample of tungsten at its melting point.

Result 7.11 kJ

Analyze The fusion enthalpies in Table 9.7 are per mole so the amount of tungsten is needed. The mass is given.

Plan Calculate the amount of tungsten from the mass using the molar mass. Then calculate the energy transfer using the fusion enthalpy.

Execute

$$\text{Energy transfer} = 25.0 \text{ g W} \times \frac{1 \text{ mol W}}{183.8 \text{ g W}} \times \frac{52.31 \text{ kJ}}{1 \text{ mol W}} = 7.11 \text{ kJ}$$

☑ **Reasonable Result Check** The sample is about 0.14 mol W, so the energy required should be about 0.14 of its molar fusion enthalpy, or about 7 kJ, which is close to the calculated result.

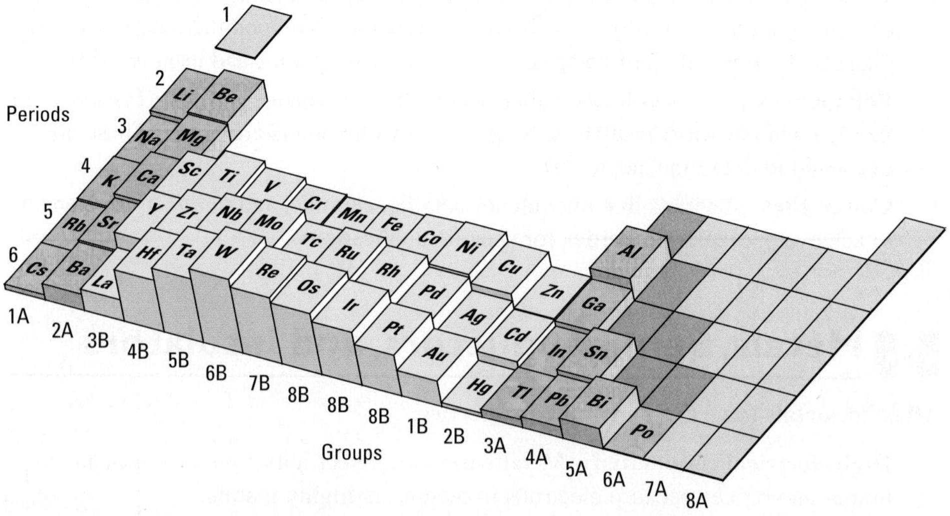

Periods
1
2
3
4
5
6
Li
Be
Na
Mg
K
Ca
Sc
Ti
V
Cr
Mn
Fe
Co
Ni
Cu
Zn
Ga
Al
Rb
Sr
Y
Zr
Nb
Mo
Tc
Ru
Rh
Pd
Ag
Cd
In
Sn
Cs
Ba
La
Hf
Ta
W
Re
Os
Ir
Pt
Au
Hg
Tl
Pb
Bi
Po
1A
2A
3B
4B
5B
6B
7B
8B
8B
8B
1B
2B
3A
4A
5A
6A
7A
8A
Groups

Figure 9.31 Relative fusion enthalpies for the metals in the periodic table. See Table 9.7 for some numerical values of fusion enthalpies.

PROBLEM-SOLVING PRACTICE 9.10

Lithium metal has a low fusion enthalpy (Table 9.7). Calculate what mass (g) of lithium is completely melted when 17.3 kJ is transferred to a Li sample at its melting point.

CONCEPTUAL EXERCISE 9.15

Cooling a Liquid Metal Until It Solidifies

The graph below is obtained when a liquid metal is cooled at a constant rate to the temperature at which it solidifies and the solid is cooled further. Account for the shape of this graph. Would all substances exhibit similar graphs?

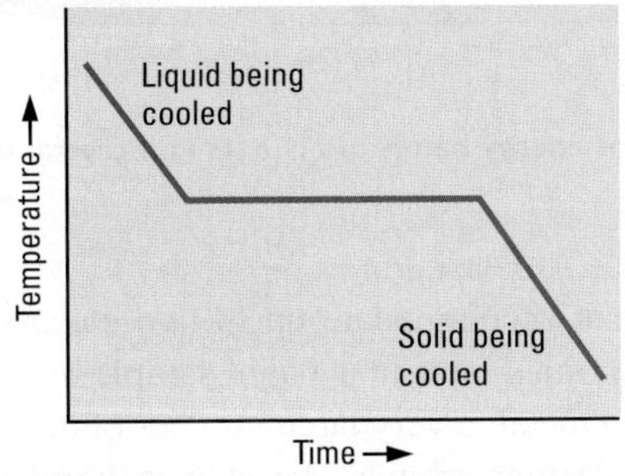
Liquid being cooled
Solid being cooled
Temperature
Time

CONCEPTUAL EXERCISE 9.16

Fusion Enthalpies and Electronic Configuration

Look in Appendix D and compare the electron configurations shown there with the fusion enthalpies for the metals shown in Table 9.7. Is there any correlation between these configurations and this property? Does strength of attraction among metal atoms correlate with number of valence electrons? Explain your answers.

9-9a Electrons in Metals, Semiconductors, and Insulators

Metals behave as though metal cations exist in a "sea" or "gas" of mobile electrons—the valence electrons of all the metal atoms. **Metallic bonding** *is the nondirectional attraction between positive metal ions and the surrounding sea of negative charge (valence electrons).* Each metal ion has a large number of near neighbors. The valence electrons are spread throughout the metal's crystal lattice, holding the positive metal ions together. When an electric field is applied to a metal, these valence electrons move toward the positive end of the field, and the metal conducts electricity.

The mobile valence electrons provide a uniform charge distribution in a metal lattice, so the positions of the positive ions can be changed without destroying the attractions among positive ions and electrons. Thus, most metals can be bent and drawn into wire. Conversely, when we try to deform an ionic solid, which consists of a lattice of positive and negative ions, the crystal cleaves or shatters (← **Sec. 2-6c**) because the balance of positive ions surrounded by negative ions, and vice versa, is disrupted.

Energy Bands To learn more about how bonding electrons behave in a metal, we can apply the ideas developed in earlier chapters regarding atomic and molecular orbitals. First, consider the arrangement of electrons in an individual atom far enough away from any neighbor so that no bonding occurs. In such an atom there are atomic orbitals, each of which can accommodate up to two electrons (← **Sec. 5-5**). In a lithium atom, the single valence electron is in a 2*s* atomic orbital (AO). If two lithium atoms are brought close together, the two valence AOs combine to form two molecular orbitals (MOs) (← **Sec. 6-12b**). The number of MOs equals the number of AOs from which the MOs

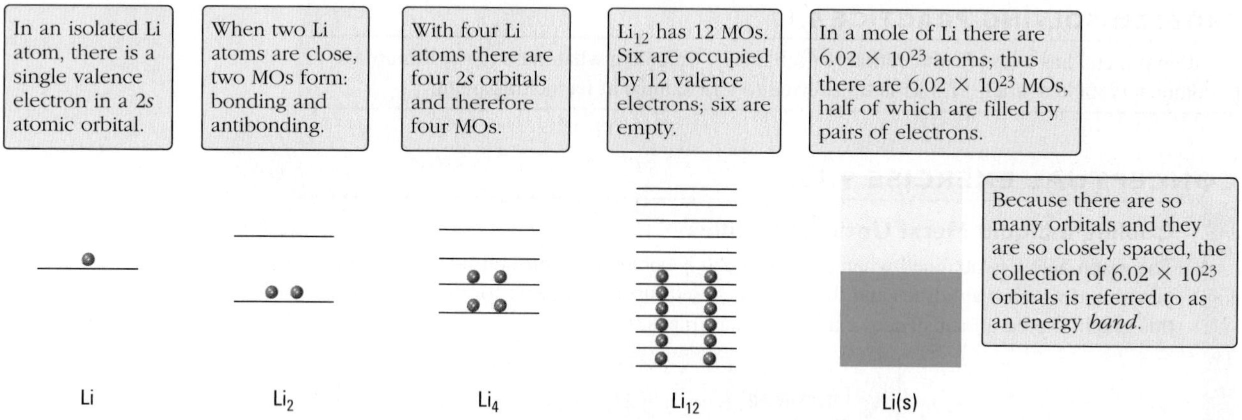

In an isolated Li atom, there is a single valence electron in a 2s atomic orbital.
When two Li atoms are close, two MOs form: bonding and antibonding.
With four Li atoms there are four 2s orbitals and therefore four MOs.
Li12 has 12 MOs. Six are occupied by 12 valence electrons; six are empty.
In a mole of Li there are 6.02 × 10^23 atoms; thus there are 6.02 × 10^23 MOs, half of which are filled by pairs of electrons.
Because there are so many orbitals and they are so closely spaced, the collection of 6.02 × 10^23 orbitals is referred to as an energy band.
Li
Li2
Li4
Li12
Li(s)

Figure 9.32 Formation of energy bands of orbitals in a crystal of lithium metal.

formed and each MO can accommodate up to two electrons. Figure 9.32 shows how Avogadro's number of orbitals form in a 1-mol sample of lithium. Similarly, in any laboratory scale sample of a metal, a very large number of atomic orbitals combine to form the same number of molecular orbitals. An **energy band** is *a large group of orbitals whose energies are closely spaced and whose average energy is the same as the energy of the corresponding orbital in an individual atom.* In some cases, energy bands for different types of electrons (*s*, *p*, *d*, and so on) overlap; in other cases, there is a gap between different energy bands.

Within each band, electrons fill the lowest energy orbitals first (Figure 9.32) and lower energy bands fill before higher energy bands. This is analogous to how electrons fill orbitals in atoms (← **Sec. 5-5**) or molecules (← **Sec. 6-12b**). Energy bands derived from completely filled atomic orbitals (core electrons) are completely filled. For example, in lithium the band derived from the 1*s* atomic orbitals has two electrons in every orbital. No electron can move from one orbital to another, because there is no empty spot for an electron to move to.

An *energy band containing valence electrons* is called the **valence band**. If the valence band is partially filled it requires very little added energy to excite a valence electron to a slightly higher energy orbital. Such a small increment of energy can be provided by applying an electric field, for example. The presence of low-energy, empty orbitals into which electrons can move allows the electrons to be mobile and to conduct an electric current.

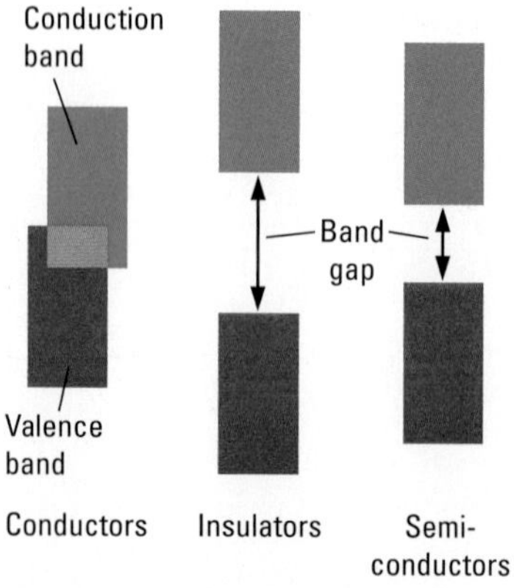

Conduction band
Band gap
Valence band
Conductors
Insulators
Semi-conductors

Figure 9.33 Differences in the energy bands of available orbitals in conductors, insulators, and semiconductors.

The *band with the next higher average energy above the valence band* is called the **conduction band**. In a metal, the valence band and the conduction band overlap, so electrons can move freely from the valence band to the conduction band; this explains the high electrical conductivity of metals. A *substance with overlapping valence and conduction bands* is an **electrical conductor** (or just **conductor**).

When the conduction band is close in energy to the valence band, the electrons can absorb a wide range of wavelengths in the visible region of the spectrum. As the excited electrons fall back to their lower energy states, they emit their extra energy as visible light, producing the luster characteristic of metals.

In a **semiconductor** a narrow *energy gap separates the valence band and the conduction band* (Figure 9.33). At very low temperatures, electrons remain in the filled lower energy valence band, and semiconductors are not good conductors. As temperature rises, more electrons have enough energy to jump across the band gap into the conduction band, so conduction increases. A sufficiently large electric field also can provide the energy needed for electrons to jump the band gap. This property of semiconductors—to switch from insulator to conductor with the application of an external electric field—is the basis for the operation of transistors, the cornerstone of modern electronics.

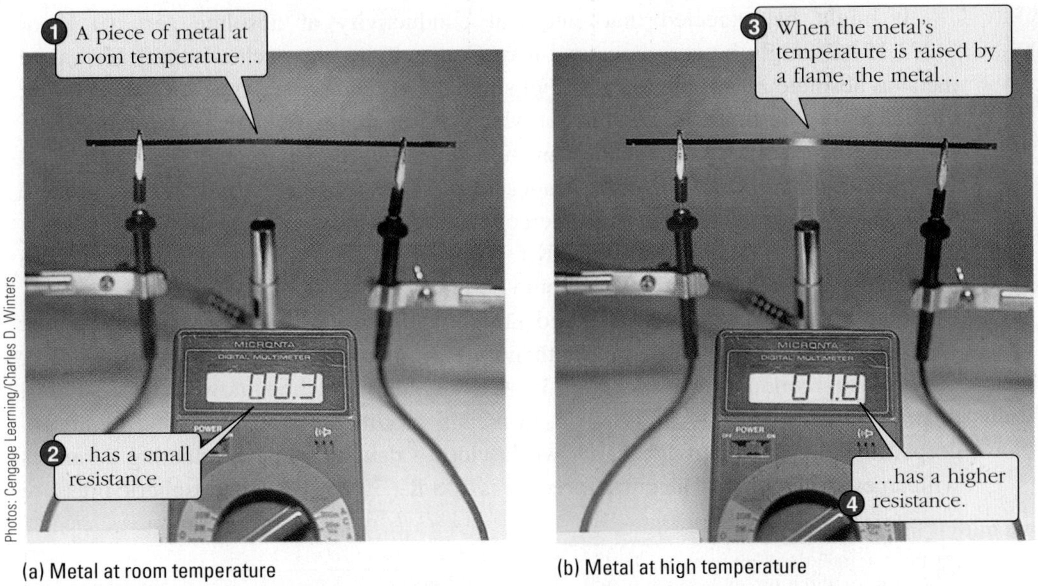
1 A piece of metal at room temperature...
2 ...has a small resistance.
3 When the metal's temperature is raised by a flame, the metal...
4 ...has a higher resistance.
(a) Metal at room temperature
(b) Metal at high temperature
Photos: Cengage Learning/Charles D. Winters

Figure 9.34 Electrical conductivity of a metal decreases (resistance increases) as temperature increases.

The energy-band theory also explains why some solids are **electrical insulators** (or just **insulators**), that is, *substances that do not conduct electricity*. In an insulator, there is a large band gap between the valence and conduction bands (Figure 9.33). Very few electrons have enough energy to move across the large gap from a filled lower energy band to an empty higher energy band, so no current flows through an insulator when an external electric field is applied.

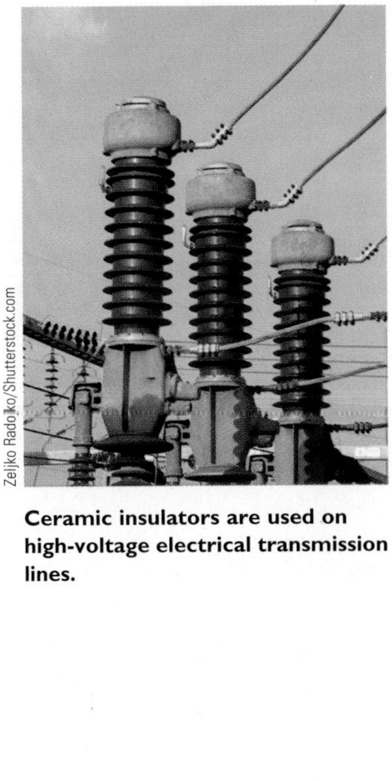
Zeljko Radojko/Shutterstock.com

Ceramic insulators are used on high-voltage electrical transmission lines.

9-9b Superconductors

The electrical conductivity of metals decreases with increasing temperature (Figures 9.34 and 9.35). This can be explained by considering the valence electrons as waves in the lattice of metal ions. Under the influence of an electrical voltage, the wave moves through the metal crystal where it encounters lattice positions at which metal ions are close enough to scatter the wave, analogous to the scattering of X-rays by atoms in a crystal (← **Sec. 9-7a**). The scattered electron wave moves off in another direction only to be scattered again and again. All of this scattering *lowers* the electrical conductivity of the metal. At higher temperatures, as the metal ions vibrate more, their lattice positions change causing even more scattering, further reducing the electrical conductivity.

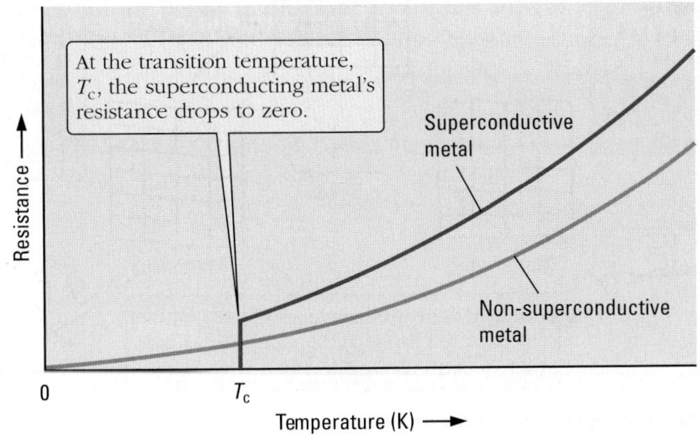
At the transition temperature, T_c, the superconducting metal's resistance drops to zero.
Superconductive metal
Non-superconductive metal
Resistance →
0
T_c
Temperature (K) →

Figure 9.35 Change in resistance with temperature of a superconducting metal and a non-superconducting metal.

magnet
superconductor
Yoav Levy/Phototake

Figure 9.36 Magnetic levitation. The magnet induces a current in the superconductor. This causes a magnetic field that repels the magnet, balancing the force of gravity, and holding the magnet above the superconductor.

It might be expected that electrical conductivity at absolute zero (0 K or −273.15 °C) might be very large and it is. Conductivity approaches infinity for a pure metal as absolute zero is approached. For some metals and special types of metal oxides, there is a low, but finite temperature, at which the conductivity increases to infinity (the resistance drops to zero). At that transition temperature, T_c, the material becomes a **superconductor** and *offers no resistance to the flow of an electrical current* (Figure 9.35). Once a current is started in a superconducting circuit, the current flows indefinitely. The highest T_c of a pure metal is 7.2 K (−266 °C) for fcc lead. Maintenance of such a low temperature requires expensive liquid helium.

The highest T_c achieved changed dramatically in 1986 when Alex Müller and Georg Bednorz discovered that a lanthanum-barium-copper oxide, a type of ceramic, became superconducting at 35 K (−238 °C), the highest T_c then known. The superconducting properties of $La_{(2-x)}Ba_xCuO_4$ (x means varying amounts of La and Ba are incorporated) led to intense research worldwide to develop superconducting ceramics with an even higher T_c. The current record is 138 K (−135 °C) at atmospheric pressure

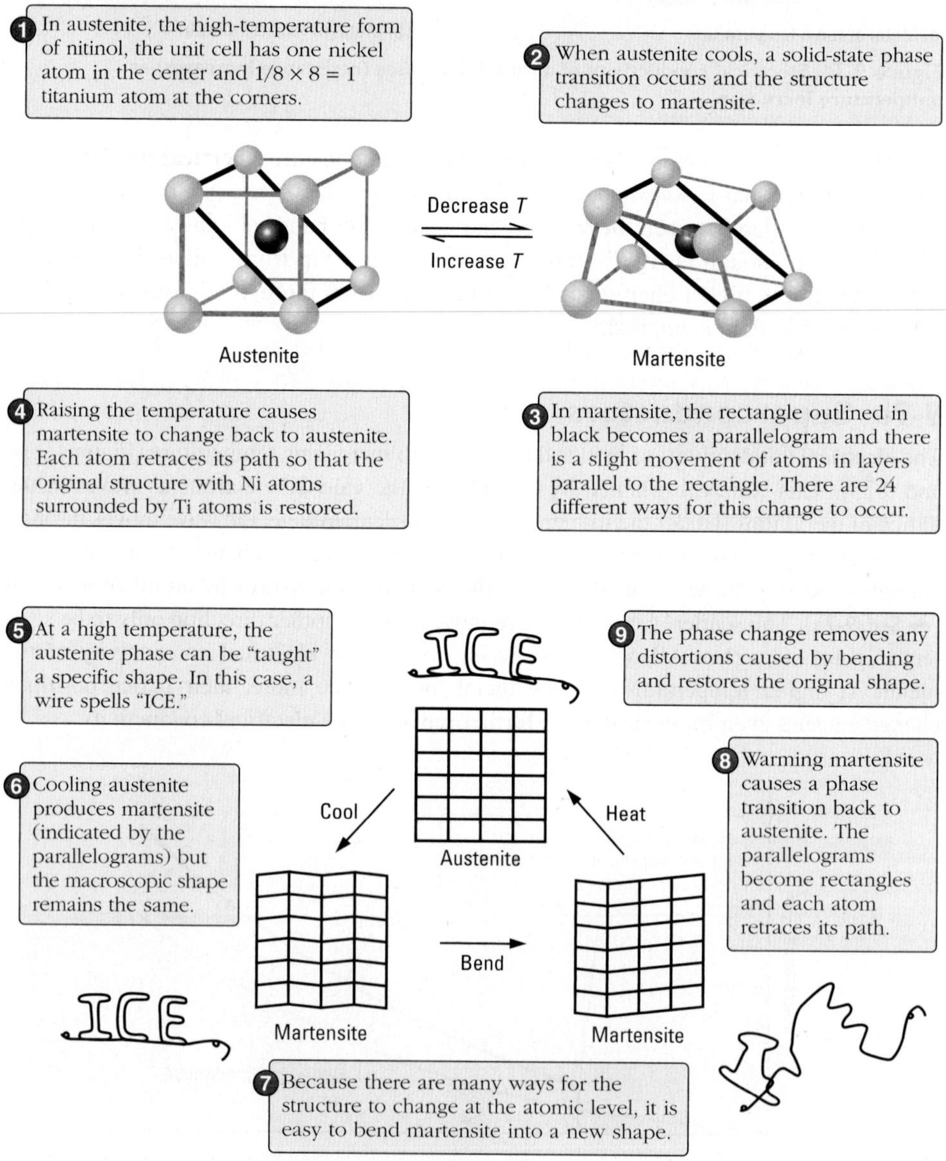
1 In austenite, the high-temperature form of nitinol, the unit cell has one nickel atom in the center and 1/8 × 8 = 1 titanium atom at the corners.
2 When austenite cools, a solid-state phase transition occurs and the structure changes to martensite.
Decrease T
Increase T
Austenite
Martensite
4 Raising the temperature causes martensite to change back to austenite. Each atom retraces its path so that the original structure with Ni atoms surrounded by Ti atoms is restored.
3 In martensite, the rectangle outlined in black becomes a parallelogram and there is a slight movement of atoms in layers parallel to the rectangle. There are 24 different ways for this change to occur.
5 At a high temperature, the austenite phase can be "taught" a specific shape. In this case, a wire spells "ICE."
ICE
9 The phase change removes any distortions caused by bending and restores the original shape.
6 Cooling austenite produces martensite (indicated by the parallelograms) but the macroscopic shape remains the same.
Cool
Heat
Austenite
8 Warming martensite causes a phase transition back to austenite. The parallelograms become rectangles and each atom retraces its path.
Bend
ICE
Martensite
Martensite
7 Because there are many ways for the structure to change at the atomic level, it is easy to bend martensite into a new shape.

Figure 9.37 Shape-memory property of nitinol.

for a complex ceramic with the composition $Hg_{0.8}Tl_{0.2}Ba_2Ca_2Cu_3O_{8.33}$. At 138 K, liquid nitrogen (BP = 77 K, −196 °C), not liquid helium, can be used to reach T_c, which is a major advantage because liquid nitrogen is available inexpensively from liquefaction of air.

Müller and Bednorz received the 1987 Nobel Prize in Physics for their discovery.

The excitement over high-temperature superconducting materials is understandable because such materials could be used to build more powerful electromagnets than those now used in MRI machines for medical diagnosis. More powerful electromagnets would permit higher energies to be maintained longer and create better MRI images. At present, the main barrier to wider application of superconductivity is the cost of cooling the electromagnets in such devices.

Because a superconductor repels magnetic materials (Figure 9.36), trains can be levitated above a track and travel at very high speeds without friction, except for air resistance. On December 31, 2002, the world's first maglev (magnetic levitation) train service ran at 430 km/h (267 mi/h) over a 30-km link from Pudong airport to Pudong (adjacent to Shanghai).

OTHK/Getty Images

Shanghai maglev train.

9-9c Alloys

An **alloy** is *a solid mixture of a metal with another substance, usually another metal.* Bronze, for example, is an alloy of copper and tin in which some of the copper atoms in the crystal lattice are replaced by tin atoms. The characteristic properties of an alloy reflect its composition. For example, brass, an alloy of copper and zinc, becomes harder as the percentage of zinc (which is harder than copper) increases.

Nitinol, an alloy that contains approximately a 1 : 1 mole ratio of nickel to titanium, has the unusual property of returning to its original shape at a higher temperature after being deformed at a lower temperature. Thus, the alloy is said to have a "shape-memory". A reversible structural change, called a phase transformation, occurs when martensite, the weaker, low-temperature form that was originally deformed, is warmed to the higher-temperature form (austenite). Figure 9.37 shows the structures of austenite and martensite and explains how shape-memory works. In addition to shape-memory, nitinol has superelasticity (pseudoelasticity), the property whereby austenite can be transformed into martensite by applying stress, such as pressure. When the pressure is removed, the nitinol returns to its original shape, much like a metal spring.

Nitinol is an acronym for _Ni_ckel _Ti_tanium _N_aval _O_rdnance _L_aboratory. The alloy was discovered by William Buehler and Frederick Wang at the U. S. Naval Ordnance Laboratory.

Because of its shape-memory ability, Nitinol is used extensively in a wide variety of applications in medical devices. Its superelasticity is useful in *self-expanding stents*, short metal tubes that are inserted into blood vessels to keep the blood vessels from collapsing. A collapsed nitinol stent in the martensite phase is placed into a collapsed vein. Upon warming to body temperature, a phase transformation expands the nitinol to its original tubular austenite form, allowing blood to flow smoothly through the vein. Nitinol is also widely used by orthodontists in wires for dental braces (Figure 9.38).

Approximately 1 billion U. S. dollar's worth of such stents are sold annually worldwide.

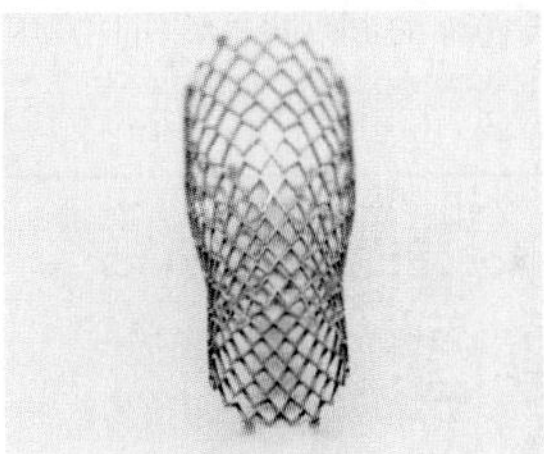
Alexey Kamenskiy/shutterstock.com
(a)

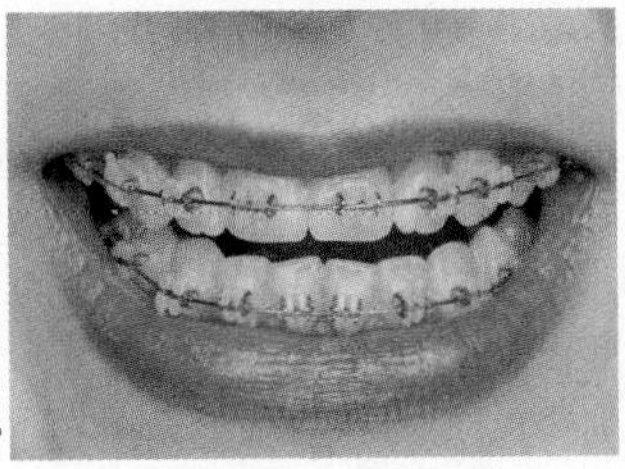
hightowernrw/Shutterstock.com
(b)

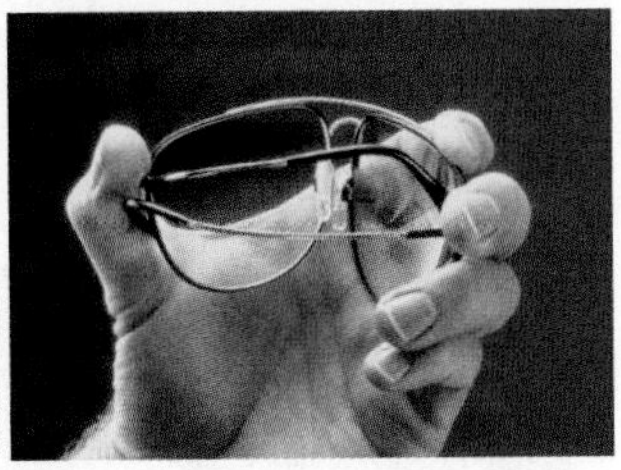
Science Source
(c)

Figure 9.38 Uses of nitinol shape-memory alloy. (a) Self-expanding stent inserted into a blood vessel; (b) dental braces; (c) self-repairing eyeglass frames.

9-10 Silicon and the Chip

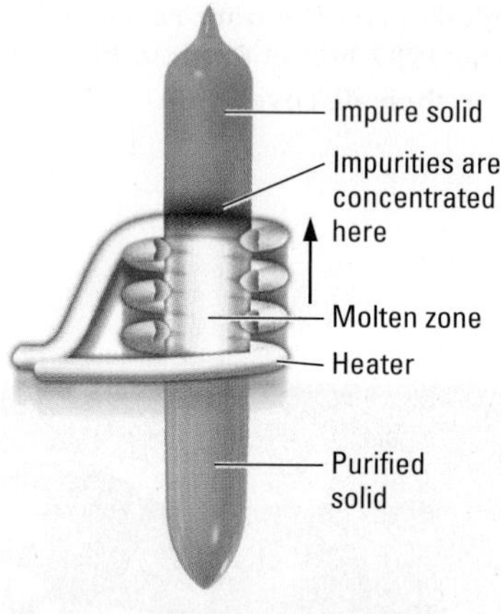

Impure solid
Impurities are concentrated here
Molten zone
Heater
Purified solid

Figure 9.39 Schematic of the zone refining process for silicon. As the heater moves upward, impurities are concentrated in the molten zone.

Silicon is known as an "intrinsic" semiconductor because the element itself is a semiconductor, although it must be highly purified before being used in semiconducting devices. Ultrapure silicon containing less than one part per billion of impurities is produced using **zone refining** (Figure 9.39). Because impurities are often more soluble in the liquid than solid phase, they move along the liquefied zone during zone refining. As the heated zone cools, the resolidified sample is purer than it was prior to melting. Multiple passes of the molten zone are required to achieve the very high purity necessary (at least 99.9999% pure).

Like all semiconductors, high-purity silicon fails to conduct an electric current until a certain electrical voltage is applied, but at higher voltages it conducts moderately well. Silicon's semiconducting properties can be improved dramatically by a process known as doping. **Doping** is *the addition of a tiny amount of some other element (a dopant) to an element.*

For example, consider what happens when a few atoms of a Group 5A element such as arsenic are substituted for a tiny fraction of the silicon atoms in purified solid silicon (Group 4A). Arsenic has five valence electrons, whereas silicon has four. Only four of the five valence electrons of As are used for bonding with four Si atoms, leaving one electron relatively free to move (Figure 9.40). This type of doped silicon is referred to as negative-type or *n-type silicon* because it has extra (negative) valence electrons. An ***n*-type semiconductor** is *one in which electrons are the charge carriers.*

A positive hole in an energy band of a semiconductor is a place where there is one less electron than normal and hence extra positive charge.

On the other hand, consider what happens when a small number of boron atoms (or atoms of some other Group 3A element) replace silicon atoms in purified solid silicon. Boron has only three valence electrons. This leaves a deficiency of one electron around the B atom, creating what is called a *hole* for every B atom added. Hence, silicon doped in this manner is referred to as positive-type or *p-type silicon* (Figure 9.40). A ***p*-type semiconductor** is *one in which holes are the charge carriers.* It is important to remember that such holes in semiconductors are the absence of electrons. Thus, the flow of holes in one direction is in reality the flow of electrons in the opposite direction.

When *p*-type and *n*-type semiconductors are brought together, a ***p-n* junction** results (Figure 9.41). Such a junction allows current to flow in one direction but not the other. When the two materials are joined, some of the excess electrons in the *n*-type material migrate across the junction and some of the holes in the *p*-type material migrate in the opposite direction. The result is the buildup of a negative charge on the *p*-type region and a positive charge on the *n*-type region. This charge buildup is called the *junction potential,* and it prevents the further migration of electrons or holes. However, if an external potential is applied to the *p-n* junction, one of the two effects shown in Figure 9.41 can result. In one case current flows, but if the connections are reversed there is no current flow.

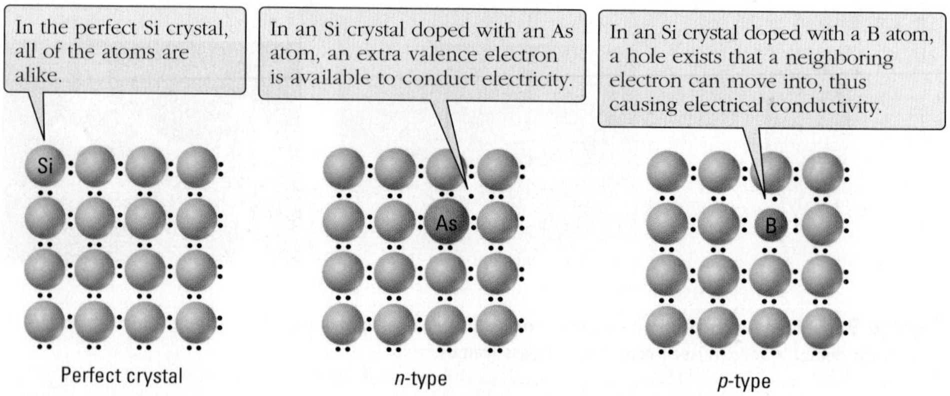

In the perfect Si crystal, all of the atoms are alike.
In an Si crystal doped with an As atom, an extra valence electron is available to conduct electricity.
In an Si crystal doped with a B atom, a hole exists that a neighboring electron can move into, thus causing electrical conductivity.
Si
As
B
Perfect crystal
n-type
p-type

Figure 9.40 Schematic drawing of semiconductor crystals derived from silicon.

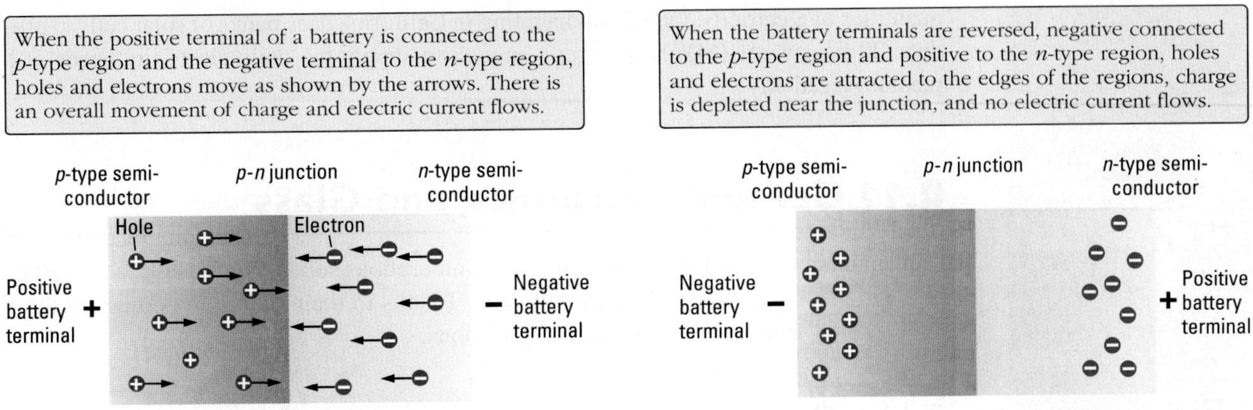
When the positive terminal of a battery is connected to the p-type region and the negative terminal to the n-type region, holes and electrons move as shown by the arrows. There is an overall movement of charge and electric current flows.
When the battery terminals are reversed, negative connected to the p-type region and positive to the n-type region, holes and electrons are attracted to the edges of the regions, charge is depleted near the junction, and no electric current flows.
p-type semi-conductor
p-n junction
n-type semi-conductor
Hole
Electron
Positive battery + terminal
– Negative battery terminal
Negative battery – terminal
+ Positive battery terminal

Figure 9.41 A *p-n* junction allows electric current to flow one way, but not the other.

These *p-n* junctions can be joined into larger composite structures to create transistors and integrated circuits. A single integrated circuit often contains thousands or even millions of transistors, as well as the circuits to carry the electrical signals. These devices, in the form of computer memories and central processing units (CPUs), permeate our society.

A *p-n* junction is also the basis of the **solar cell**, *a device for transforming solar energy into electric current* (Figure 9.42). When a photon strikes the *p-n* junction, the photon's energy excites an electron from the valence band to the conduction band, leaving a hole in the valence band. If the excited electron moves away from the *p-n* junction, it can migrate through the *n*-type region to an electrode and flow into an external electrical circuit. The circuit is completed when an electron moves to a second electrode that is connected to the *p*-type region and combines with a hole. Thus, light that strikes the *p-n* junction provides energy that causes an electrical current to flow. This process can continue as long as the silicon layers are exposed to sunlight and the circuit remains closed.

Solar cells are on the threshold of a major technological breakthrough, perhaps comparable to the computer chip. Although experimental solar-powered automobiles are now available and many novel applications of solar cells exist, the real breakthrough will come from the general use of banks of solar cells at utility power plants to produce huge

Courtesy of Intel Corporation

Intel Xeon E5-2600 microprocessor. The area shown is about 30 mm wide.

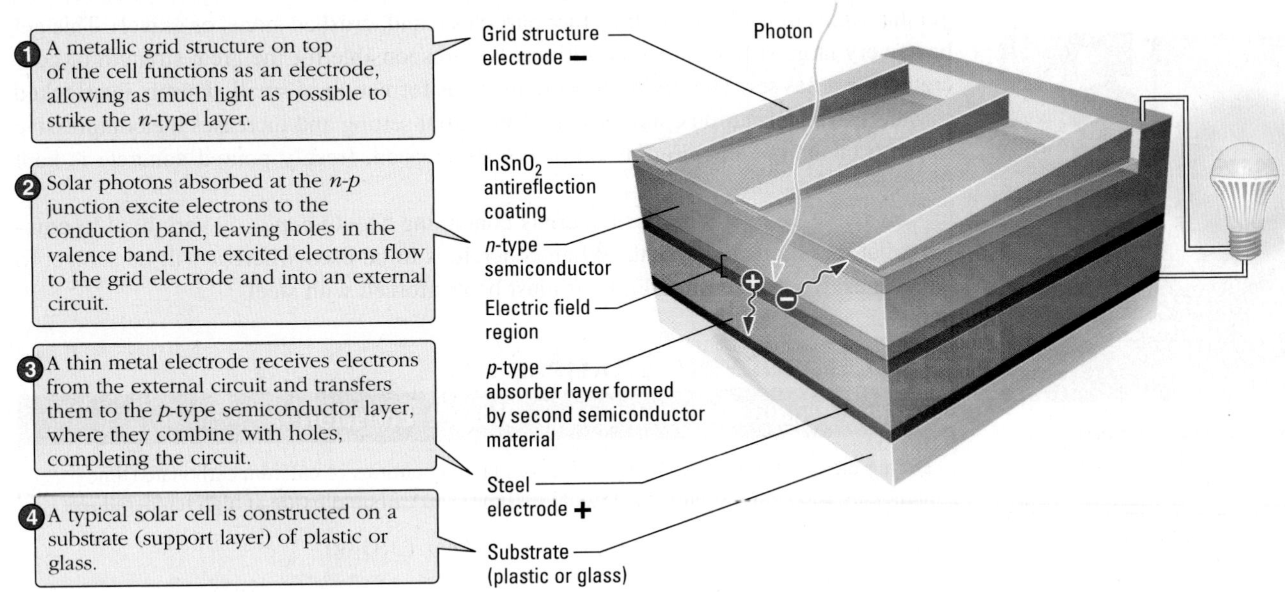
1 A metallic grid structure on top of the cell functions as an electrode, allowing as much light as possible to strike the n-type layer.
2 Solar photons absorbed at the n-p junction excite electrons to the conduction band, leaving holes in the valence band. The excited electrons flow to the grid electrode and into an external circuit.
3 A thin metal electrode receives electrons from the external circuit and transfers them to the p-type semiconductor layer, where they combine with holes, completing the circuit.
4 A typical solar cell is constructed on a substrate (support layer) of plastic or glass.
Grid structure electrode –
Photon
InSnO2 antireflection coating
n-type semiconductor
Electric field region
p-type absorber layer formed by second semiconductor material
Steel electrode +
Substrate (plastic or glass)

Figure 9.42 Typical photovoltaic cell using layers of doped silicon.

quantities of electricity. One plant operating in California uses banks of solar cells to produce 20 megawatts (MW) of power—enough to supply the daily electricity needs of a medium-sized city.

Cengage Learning/Charles D. Winters

Solar power can provide electricity in remote areas. This water pump in upstate New York is powered by solar-generated electricity.

An article on concrete is in *Journal of Chemical Education*, Vol. 83, 2006; pp. 1425–1427. An informative overview of cement chemistry is found in *Journal of Chemical Education*, Vol. 80, 2003; pp. 623–635.

An article describing uses for concrete is found in *Science News*, January 1, 2005; p. 7.

Compressive strength refers to a solid's resistance to shattering under high pressure.

Tensile strength refers to a substance's resistance to stretching.

9-11 Cement, Ceramics, and Glass

Cement, ceramics, and glass are examples of amorphous solids; that is, they lack crystalline structures with easily defined unit cells. Thanks to their useful properties, all three are extremely important in practical applications.

9-11a Cement

Cement consists of *microscopic particles of silicate clays that form a paste with water that reacts to form a solid mass*. Cement contains compounds with varying proportions of calcium, iron, aluminum, and silicon combined with oxygen. In the presence of water, cement forms hydrated particles with large surface areas that subsequently undergo recrystallization and reaction to bond to themselves as well as to the surfaces of bricks, stone, or other silicate materials.

Cement is made by heating a powdered mixture of calcium carbonate (limestone or chalk), silica (sand), aluminosilicate minerals (kaolin, clay, or shale), and iron oxide at a temperature of up to 1400 °C in a rotating kiln. As the materials pass through the kiln, they lose water and carbon dioxide and ultimately form a "clinker," in which they are partially fused (melted together). A small amount of calcium sulfate is added, and the cooled clinker is then ground to a very fine powder. A typical composition of cement is 60% to 67% CaO, 17% to 25% SiO_2, 3% to 8% Al_2O_3, up to 6% Fe_2O_3, and small amounts of magnesium oxide, magnesium sulfate, and oxides of potassium and sodium.

Cement is usually mixed with other substances. *Mortar* is a mixture of cement, sand, water, and lime. **Concrete** is *a mixture of cement, sand, and aggregate (crushed stone or pebbles)* in proportions that vary according to the application and the strength required.

In cement, the oxides are not separate ionic crystals; instead, an entire nanoscale structure forms that is a complex network of ions, each satisfying its charge requirements with ions of opposite charge. Many different reactions occur during the setting of cement. Various constituents react with water and with carbon dioxide in the air. The initial reaction of cement with water involves the hydrolysis of the calcium silicates, which forms a gel that sticks to itself and to the other particles (sand, crushed stone, or gravel). This gel has a very large surface area and ultimately is responsible for the great strength of concrete once it has set. The setting process involves formation of small, densely interlocked crystals. Their formation continues after the initial setting and increases the compressive strength of the cement (Figure 9.43). For this reason, freshly poured concrete is kept moist for several days.

Vadim Ratnikov/Shutterstock.com

Figure 9.43 Pouring concrete. After concrete has been poured, it sets. Water reacts with the cement to form strong bonds between the tiny particles originally in the dry cement.

Concrete, like many other materials containing Si—O bonds, is very hard to compress but lacks tensile strength. When concrete is to be used where it will be subject to tension, as in a bridge or building, it must be reinforced with steel.

PROBLEM-SOLVING EXAMPLE 9.11

Cement Production and Carbon Dioxide Emission

During the initial step in cement production, large quantities of calcium carbonate (limestone) are decomposed into calcium oxide (lime) and carbon dioxide at high temperatures:

$$CaCO_3(s) \longrightarrow CaO(s) + CO_2(g)$$

The United States produced 61 million metric tons of cement, which is approximately 63% CaO by mass, in 2010. Release of large quantities of carbon dioxide, a greenhouse gas (← **Sec. 8-11a**), during cement production is a concern. (a) Calculate the mass of CO_2 (kg) released per metric ton of cement; 1 metric ton = 1×10^3 kg. (b) Calculate the mass of CO_2 formed during the 2010 production of cement in the United States.

Result (a) 490 kg CO_2/1 ton cement (b) 3.0×10^{10} kg

Analyze Use the stoichiometric ratio of coefficients and the molar masses. Calculate the mass of CO_2 from the mass of CaO, which can be calculated from the mass of cement and the percent CaO in it. Then, calculate the mass of CO_2 produced per ton of cement.

Plan (a) To determine the mass of CO_2, first calculate the mass of CaO and the number of moles of it. Use the stoichiometric factor of 1 mol CO_2/1 mol CaO from the balanced chemical equation to calculate moles of CO_2 and then its mass. (b) Use the calculated conversion factor of kg CO_2/1 ton cement to calculate the kilograms of CO_2 formed.

Execute (a) The mass of CO_2 per metric ton of cement is

$$m(CO_2) = 1 \times 10^3 \text{ kg cement} \times \frac{63 \text{ kg CaO}}{100 \text{ kg cement}} \times \frac{1000 \text{ g}}{1 \text{ kg}} \times \frac{1 \text{ mol CaO}}{56.08 \text{ g CaO}} \times \frac{1 \text{ mol } CO_2}{1 \text{ mol CaO}} \times \frac{44.01 \text{ g } CO_2}{1 \text{ mol } CO_2} \times \frac{1 \text{ kg}}{1000 \text{ g}} = 4.9 \times 10^2 \text{ kg } CO_2$$

(b) Mass of CO_2 produced = 61×10^6 tons cement × (490 kg CO_2/1 ton cement)
= 3.0×10^{10} kg CO_2

☑ **Reasonable Result Check** CO_2 and CaO are produced in a 1 : 1 mole ratio. The molar mass of CO_2 is 44/56 = 0.78 times the molar mass of CaO. Thus, about 0.78 g CO_2 is produced per gram of CaO, or 0.78 kg CO_2 per kg CaO. The 61 million tons (61×10^9 kg) of cement is only 63% CaO, about 4×10^{10} kg CaO. The equivalent mass of CO_2 is $0.78 \times 4 \times 10^{10}$ kg = 3×10^{10} kg CO_2, so the answer is reasonable.

Concrete canoe. The University of Wisconsin-Madison concrete canoe team paddling its canoe.

AP Images/KEVIN P. CASEY

PROBLEM-SOLVING PRACTICE 9.11

Calculate the volume (L) of carbon dioxide formed during the 2010 production of cement in the United States. Assume 25 °C and 1 atm. The density of carbon dioxide gas is 1.799 g/L.

9-11b Ceramics

Ceramics are *materials fashioned from clay or other minerals at room temperature and then permanently hardened by heat in a baking ("firing") process* that binds the particles together.

Silicate ceramics include objects made from clays (aluminosilicates), such as pottery, bricks, and table china. China clay, or kaolin, is a mineral that is primarily kaolinite and is practically free of iron. (When present, iron imparts a red color to the clay.) China clay is white and particularly valuable in making fine pottery. Clays mixed with water form a moldable paste consisting of tiny silicate sheets that can easily slide past one another. When heated, the water is driven off, and new Si—O—Si bonds form so that the mass becomes permanently rigid.

Oxide ceramics are produced from powdered metal oxides such as alumina, Al_2O_3, and magnesia, MgO, by heating the solids under pressure, causing the particles to bind to one another and thereby form a rigid solid. Because it has a high electrical resistivity, alumina is used in spark plug insulators. High-density alumina has very high mechanical strength, so it is also used in armor plating and in high-speed cutting tools for machining metals. Magnesia is an insulator with a high melting point (2800 °C), so it is often used as insulation in electric heaters and electric stoves.

A third class of ceramics includes the *nonoxide ceramics* such as silicon nitride, Si_3N_4; silicon carbide, SiC; and boron nitride, BN. Heating the solids under pressure forms ceramics that are hard and strong, but brittle. Boron nitride has the same average number of electrons per atom as does elemental carbon and exists in the graphite structure or the diamond structure, making it comparable in hardness with diamond and more resistant to oxidation. For this reason, boron nitride cups and tubes are used to contain molten metals that are being evaporated. Silicon carbide (trade name Carborundum) is a widely used abrasive; its structure is the diamond structure with half of the C atoms replaced by Si atoms.

The one severely limiting problem in using ceramics is their brittleness. Ceramics deform very little before they fail catastrophically, with the failure resulting from a weak point in the bonding within the ceramic matrix. Such weak points are not consistent from sample to sample, so the failure is not readily predicted. Stress failure of ceramic composites occurs due to nanoscale irregularities resulting from impurities or disorder in the atomic arrangements, so much attention is now being given to using purer starting materials and more strictly controlling the processing steps. Adding fibers to ceramic composites makes them less susceptible to brittleness and sudden fracture.

9-11c Glasses

A **glass** is *an amorphous, clear solid formed from silicates or other oxides.* One of the more common glasses is soda-lime glass, which is clear and colorless if the purity of the ingredients is carefully controlled. Various substances can be added to color glass (Table 9.8).

Table 9.8 Substances Used to Color Glass

Substance	Color
Calcium fluoride	Milky white
Cobalt(II) oxide	Blue
Copper(I) oxide	Red, green, blue
Finely divided gold	Red
Iron(II) compounds	Green
Iron(III) compounds	Yellow
Manganese(IV) oxide	Violet
Tin(IV) oxide	Opaque
Uranium compounds	Yellow, green

The simplest glass is amorphous silica, SiO_2 (known as *vitreous silica*); it is prepared by melting and quickly cooling either quartz or cristobalite, both forms of SiO_2. Such glass is built up of corner-sharing SiO_4 tetrahedra linked into a three-dimensional network that lacks symmetry or long-range order. Another common glass is *borosilicate glass* or Pyrex, which contains SiO_2, B_2O_3, and Al_2O_3. It is made from melting boric acid, H_3BO_3; soda ash, Na_2CO_3; silica sand, SiO_2; alumina, Al_2O_3; and borax, $Na_2B_4O_7{\cdot}10H_2O$; and it is valuable because of its low expansion with temperature and high chemical durability.

It is important that glass be *annealed* properly during the manufacturing process. Annealing means cooling the glass slowly as it passes from a viscous liquid state to a solid at room temperature. If a glass is cooled too quickly, bonding forces become uneven because small regions of crystallinity develop. Poorly annealed glass may crack or shatter when subjected to mechanical shocks or sudden temperature changes. High-quality glass, such as that used in optics, must be annealed very carefully. For example, the 200-in mirror for the telescope at Mt. Palomar, California, was annealed from 500 °C to 300 °C over a period of a year.

Crystalline silica, SiO_2. There is a regular, repeating pattern in this crystalline solid. Red atoms are oxygen; gray atoms are silicon.

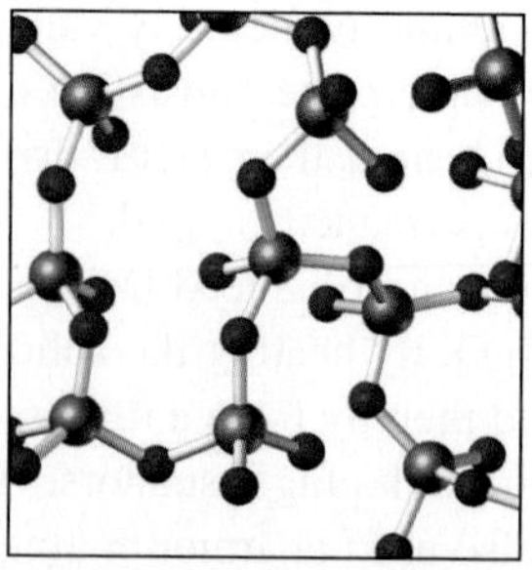

Vitreous silica, SiO_2. As in any amorphous solid, there is short-range order (Si atoms surrounded by four O atoms in this case), but no long-range repeating pattern.

Have you ever wondered about the glass screen on your smartphone and how it is made so it can take the physical abuse it does daily during ordinary use? The tough glass that makes the smartphone screen, now called "Gorilla Glass", was first developed in the late 1950s by researchers at Corning Glass Co. for possible use in auto and airplane windshields. The product was not commercialized then because safety glass for windshields ended up being made by laminating a layer of plastic film between two curved sheets of glass so that the glass does not shatter on impact.

Even though glass is very strong under compression, it is brittle under tension. The Corning scientists figured out a way to build some compression directly into the glass structure to reduce the possibility of breakage. To do so, sheets of glass are put into a bath of molten salts at 400 °C where sodium ions leave the glass and are replaced by potassium ions from the bath. Upon cooling, the potassium ions, which are larger than sodium ions (← **Sec. 5-10**), push against each other setting up a compressive stress on the surface of the glass. The thin, hard glass is an insulator. It is coated with a transparent electrical conductor, such as indium-tin oxide (ITO, a solid solution of about 90% In_2O_3 and 10% SnO_2). When a finger, which is also electrically conductive, touches the glass screen, there is a change in the quantity of charge where the finger touches. This changes the capacitance and can be detected and reported to the phone's microprocessor. Besides its durability, Gorilla Glass does not need a protective coating, so it can be made very thin, ideal for such an application.

cotesebastien/iStockphoto.com

A smartphone with a Gorilla Glass screen.

SUMMARY PROBLEM

Part I

Use the vapor pressure curves shown in Figure 9.4 and your knowledge of noncovalent intermolecular forces to answer these questions.

(a) Explain why hexane is less volatile than diethyl ether.

(b) At what temperature does 1-butanol have a vapor pressure of 500 mmHg?

(c) Explain why the boiling point of 1-butanol is greater than that of water.

(d) If you put a few drops of each substance in Figure 9.4 separately on your hand, which would evaporate immediately and which would remain liquid?

(e) The normal boiling point of 1-butanol is 117.3 °C and its vapor pressure is 389 mmHg at 100.0 °C. Calculate the $\Delta_{vap}H$ of 1-butanol.

Part II

Consider this phase diagram for an unknown substance, X.

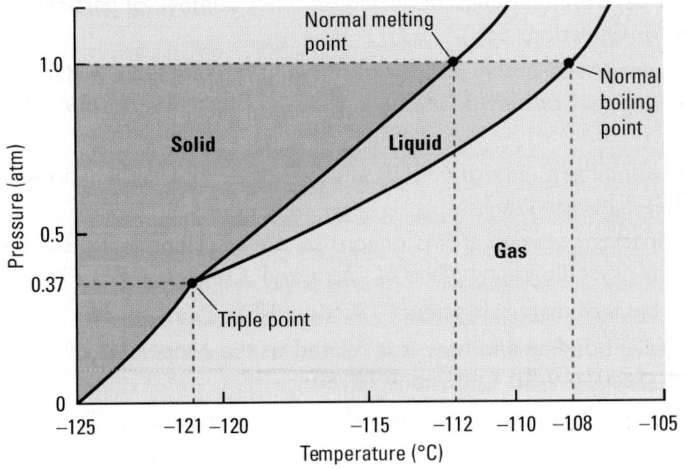
Normal melting point
Normal boiling point
Solid
Liquid
Gas
Triple point
Pressure (atm)
1.0
0.5
0.37
0
–125
–121 –120
–115
–112
–110
–108
–105
Temperature (°C)

(a) In what phase is X found at room temperature and 1.0 atm?

(b) In what phase does X exist at −114 °C and 0.75 atm?

(c) If the vapor pressure of a sample of X is 375 mmHg, what is the temperature of the liquid X?

(d) Identify a temperature and pressure at which X sublimes.

(e) At a pressure of 0.500 atm, determine the temperature at which X changes from a liquid to a gas.

(f) Which phase of X is more dense—solid or liquid? Explain your answer.

Part III

Consider the CsI unit cell shown in Figure 9.23.

(a) Determine how many Cs^+ ions are in a unit cell.

(b) Determine how many I^- ions are in a unit cell.

(c) Determine how many I^- ions share each face of a unit cell.

HAVING STUDIED THIS CHAPTER . . .

. . . you should be able to:

- Recall and apply knowledge of noncovalent intermolecular forces (Section 9-1). End-of-chapter Questions 11, 23, 33, 38
- Determine the energy transfer associated with phase changes (Sections 9-2, 9-4). Questions 15, 17, 35
- Relate vapor pressure to temperature and boiling point; describe how these properties are influenced by noncovalent intermolecular forces (Section 9-3). Questions 21, 24, 27
- Identify the significant features of phase diagrams and use them to interpret what occurs in a sample of matter when temperature and pressure change (Section 9-4). Question 43
- Understand critical temperature and critical pressure (Sections 9-3, 9-4). Question 45
- Differentiate among the major types of solids (Section 9-4). Questions 46, 48, 50
- Describe and explain the unusual properties of water (Section 9-5). Questions 52, 55
- Identify and describe unit cells; determine the net number of particles in a unit cell (Section 9-6). Questions 58, 61, 63
- Do calculations based on knowledge of simple unit cells and the dimensions of atoms and ions that occupy positions in those unit cells (Section 9-6). Questions 59, 65
- Explain the bonding in network solids and how it results in their properties (Section 9-7). Questions 67, 69
- Relate the spacing between atoms or ions to the wavelength and scattering angle used in X-ray crystallography (*Tools of Chemistry*) Questions 71, 73
- Explain the basis of materials science (Section 9-8). Question 75
- Explain metallic bonding and how it is related to the properties of metals and semiconductors (Section 9-9). Questions 78, 80
- Describe the phenomenon of superconductivity (Section 9-9). Question 82

- Describe *n*-type and *p*-type semiconductors and the use of *p-n* junctions (Section 9-10). Question 84
- Explain how the lack of regular structure in amorphous solids affects their properties (Section 9-11). Question 87

KEY TERMS

allotropes (Section 9-7a)
alloy (9-9c)
amorphous solid (9-4)
boiling point (9-3a)
capillary action (9-5)
cement (9-11a)
ceramics (9-11b)
Clausius-Clapeyron equation (9-3b)
closest packing (9-6d)
concrete (9-11a)
condensation (9-2)
conduction band (9-9a)
conductor (9-9a)
critical point (9-3c)
critical pressure (P_c) (9-3c)
critical temperature (T_c) (9-3c)
crystal lattice (9-6)
crystalline solids (9-4)
crystallization (9-4a)
cubic closest packed (9-6d)
cubic unit cell (9-6a)
deposition (9-4b)
dipole-dipole attraction (9-1)
doping (9-10)
electrical conductor (9-9a)
electrical insulator (9-9a)
energy band (9-9a)
equilibrium vapor pressure (9-3)
evaporation (9-2)
fullerene (9-7a)
glass (9-11c)
graphene (9-7a)
heating curve (9-4c)
hexagonal closest packed (9-6d)
hydrogen bonding (9-1)
insulator (9-9a)
ionic solid (9-4)
London (dispersion) forces (9-1)
materials science (9-8)
meniscus (9-5)
metallic bonding (9-9a)
metallic solid (9-4)
molecular solid (9-4)
***n*-type semiconductor** (9-10)
nanotubes (9-7a)
network solid (9-4, 9-7)
normal boiling point (9-3a)
***p-n* junction** (9-10)
***p*-type semiconductor** (9-10)
phase diagram (9-4d)
semiconductor (9-9a)
solar cell (9-10)
sublimation (9-4b)
superconductor (9-9b)
supercritical fluid (9-3c)
surface tension (9-5)
triple point (9-4d)
unit cell (9-6a)
valence band (9-9a)
vapor pressure (9-3)
vaporization (9-2)
X-ray crystallography (*Tools of Chemistry,* 9-7a)
zone refining (9-10)

QUESTIONS FOR REVIEW AND THOUGHT

Red-numbered questions have short answers at the back of this book in Appendix M and fully worked solutions in the *Student Solutions Manual.*

Review Questions

These questions test vocabulary and simple concepts.

1. Name three properties of solids that are different from those of liquids. Explain the differences for each.
2. What causes surface tension in liquids? Name a substance that has a very high surface tension. What kinds of intermolecular forces account for the high value?
3. Explain how the equilibrium vapor pressure of a liquid might be measured.
4. Define boiling point and normal boiling point.
5. Define the crystallization enthalpy of a substance. How is it related to the substance's fusion enthalpy?
6. Define sublimation.
7. Which processes are endothermic?
   (a) Condensation (b) Melting (c) Evaporation
   (d) Sublimation (e) Deposition (f) Freezing
8. Define the unit cell of a crystal.
9. Assuming the same substance could form crystals with its atoms or ions in either primitive cubic packing or

hexagonal closest packing, which form would have the higher density? Explain.

10. How does conductivity vary with temperature for (a) a metallic conductor, (b) a nonconductor, (c) a semiconductor, and (d) a superconductor? In your answer, begin at high temperatures and come down to low temperatures.

Topical Questions

These questions are keyed to the major topics in the chapter. Usually a question that is answered at the back of this book is paired with a similar one that is not.

Liquids, Solids, and Intermolecular Forces (Section 9-1)

11. Rank these substances in order of increasing noncovalent intermolecular attractions. For each substance, name the types of intermolecular attractions that occur.
(a) $CH_3CH_2CH_2CH_3$ (b) CH_3CH_3
(c) $CH_3CH_2CH_2OH$ (d) $CH_3CH_2OCH_3$
12. Rank these substances in order of increasing noncovalent intermolecular attractions. For each substance, name the types of intermolecular attractions that occur.
(a) Cl_2 (b) HF (c) F_2 (d) SO_2

Vaporization and Condensation (Section 9-2)

13. Explain on the molecular scale the processes of condensation and vaporization.
14. After exercising on a hot summer day and working up a sweat, you often become cool when you stop. What is the molecular-level explanation of this phenomenon?
15. The chlorofluorocarbon CCl_3F has a vaporization enthalpy of 24.8 kJ/mol. Calculate the heat energy transfer required to vaporize 1.00 kg of the compound.
16. The molar vaporization enthalpy of methanol is 38.0 kJ/mol at 25 °C. Calculate the heat energy transfer required to convert 250. mL of the alcohol from liquid to vapor. The density of CH_3OH is 0.787 g/mL at 25 °C.
17. Some camping stoves contain liquid butane, C_4H_{10}. They work only when the outside temperature is warm enough to allow the butane to have a reasonable vapor pressure (so they are not very good for camping in temperatures below about 0 °C). Assume the vaporization enthalpy of butane is 22.44 kJ/mol and the camp stove fuel tank contains 190. g liquid C_4H_{10}. Calculate the heat energy transfer required to vaporize all of the butane.
18. Mercury is highly toxic. Although it is a liquid at room temperature, it has a high vapor pressure and a low vaporization enthalpy (294 J/g). Calculate the heat energy transfer required to vaporize 0.500 mL mercury at 357 °C, its normal boiling point. The density of Hg(ℓ) is 13.6 g/mL. Compare your result with the energy transfer needed to vaporize 0.500 mL water. The molar vaporization enthalpy of H_2O is 40.7 kJ/mol.

Vapor Pressure (Section 9-3)

19. Use the concepts of noncovalent intermolecular forces to explain why vaporization is endothermic.
20. Give a molecular-level explanation of why the vapor pressure of a liquid increases with temperature.
21. Briefly explain the variations in the boiling points in this table. In your discussion be sure to mention the types of intermolecular forces involved.

Compound	Boiling Point (°C)
NH_3	−33.4
PH_3	−87.5
AsH_3	−62.4
SbH_3	−17

22. Explain the observation that 1-propanol, $CH_3CH_2CH_2OH$, has a boiling point of 97.2 °C, whereas a compound with the same empirical formula, ethyl methyl ether, $CH_3CH_2OCH_3$, boils at 7.4 °C.
23. Methanol, CH_3OH, has a normal boiling point of 64.7 °C and a vapor pressure of 100 mmHg at 21.2 °C. Formaldehyde, $H_2C{=}O$, has a normal boiling point of −19.5 °C and a vapor pressure of 100 mmHg at −57.3 °C. Explain why these two compounds have different boiling points and require different temperatures to achieve the same vapor pressure.
24. The highest mountain in the Western Hemisphere is Mt. Aconcagua, in the central Andes of Argentina (22,834 ft). Assume that atmospheric pressure decreases at a rate of 3.5 millibar every 100 feet. (a) Estimate the atmospheric pressure at the summit of Mt. Aconcagua. (b) Calculate the temperature at which water would boil at the summit.
25. The lowest sea-level barometric pressure ever recorded was 25.69 in mercury, recorded in a typhoon in the South Pacific. Suppose you were in this typhoon and, to calm yourself, boiled water to make yourself a cup of tea. At what temperature would the water boil? Use Figure 9.4 and remember that 1 atm is 760 mmHg and 1 in = 2.54 cm.
26. The vapor pressure curves for four substances are shown in the plot. Which one of these four substances has the greatest intermolecular attractive forces at 25 °C? Explain your answer.

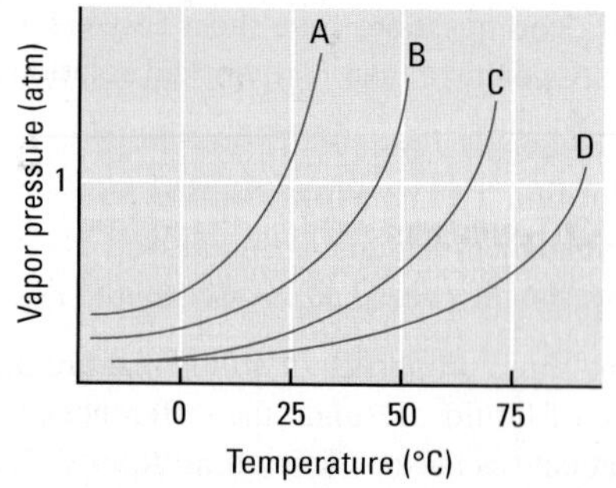
A
B
C
D
Vapor pressure (atm)
1
0
25
50
75
Temperature (°C)

27. A liquid has a $\Delta_{vap}H$ of 38.7 kJ/mol and a boiling point of 110 °C at 1 atm pressure. Calculate the vapor pressure of the liquid at 97 °C.

28. A liquid has a $\Delta_{vap}H$ of 44.0 kJ/mol and a vapor pressure of 370 mmHg at 90 °C. Calculate the vapor pressure of the liquid at 130 °C.
29. The vapor pressure of ethanol, C_2H_5OH, at 50.0 °C is 233 mmHg, and its normal boiling point at 1 atm is 78.3 °C. Calculate the $\Delta_{vap}H$ of ethanol.
30. Calculate the $\Delta_{vap}H$ for a substance whose vapor pressure doubled when its temperature was raised from 70.0 °C to 80.0 °C.

Solids and Changes of Phase (Section 9-4)

31. What does a low fusion enthalpy for a solid tell you about the solid (its bonding or type)?
32. What does a high melting point and a high fusion enthalpy tell you about a solid (its bonding or type)?
33. Which would you expect to have the higher fusion enthalpy, N_2 or I_2? Explain your choice.
34. The fusion enthalpy for H_2O is about 2.5 times larger than the fusion enthalpy for H_2S. What does this say about the relative strengths of the forces between the molecules in these two solids? Explain.
35. Calculate the total heat energy transfer required to change 0.50 mol ice at −5 °C to 0.50 mol steam at 100 °C. The solid ice and liquid water have heat capacities of 2.06 J g^{-1} $°C^{-1}$ and 4.184 J g^{-1} $°C^{-1}$, respectively. The fusion enthalpy for solid ice is 6.02 kJ/mol and the vaporization enthalpy of liquid water is 40.7 kJ/mol.
36. Calculate the heat energy transfer needed to melt a 36.00-g ice cube that is initially at −10 °C and bring it to room temperature (20 °C). (See Question 35 for data.)
37. The chlorofluorocarbon CCl_2F_2 was once used as a refrigerant. Calculate what mass of this substance must evaporate to freeze 2 mol water initially at 20 °C. The vaporization enthalpy for CCl_2F_2 is 289 J/g. The fusion enthalpy for solid ice is 6.02 kJ/mol and specific heat capacity for liquid water is 4.184 J g^{-1} $°C^{-1}$.
38. The ions of NaF and MgO all have the same number of electrons, and the internuclear distances are about the same (235 pm and 212 pm). Why, then, are the melting points of NaF and MgO so different (992 °C and 2825 °C, respectively)?
39. For the pair of compounds LiF and CsI, tell which compound is expected to have the higher melting point and briefly explain why.
40. Which of these substances has the highest melting point? The lowest melting point? Explain your choices briefly.
    (a) LiBr (b) CaO
    (c) CO (d) CH_3OH
41. Which of these substances has the highest melting point? The lowest melting point? Explain your choices briefly.
    (a) SiC (b) I_2
    (c) Rb (d) $CH_3CH_2CH_2CH_3$
42. During thunderstorms, very large hailstones can fall from the sky. To preserve some of these hailstones, you place them in the freezer compartment of your frost-free refrigerator. A friend, who is a chemistry student, tells you to put the hailstones in a tightly sealed plastic bag. Why?
43. In this phase diagram, make these identifications:

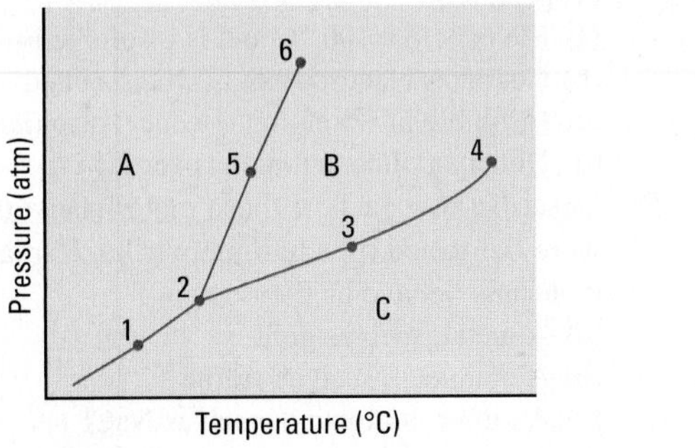

Pressure (atm)
Temperature (°C)
A
B
C
1
2
3
4
5
6

    (a) What phase is present in region A? Region B? Region C?
    (b) What phases are in equilibrium at point 1? Point 2? Point 3? Point 5?
44. Consult the phase diagram of CO_2 in Figure 9.14. What phase or phases are present under these conditions:
    (a) $T = -70$ °C and $P = 1.0$ atm.
    (b) $T = -40$ °C and $P = 15.5$ atm.
    (c) $T = -80$ °C and $P = 4.7$ atm.
45. At the critical point for carbon dioxide, the substance is very far from being an ideal gas. Prove this statement by calculating the density of an ideal gas in g/cm^3 at the conditions of the critical point and comparing it with the experimental value. Compute the experimental value from the fact that a mole of CO_2 at its critical point occupies 94 cm^3.
46. Classify each of these solids as ionic, metallic, molecular, network, or amorphous.
    (a) KF (b) I_2
    (c) SiO_2 (d) BN
47. Classify each of these solids as ionic, metallic, molecular, network, or amorphous.
    (a) Tetraphosphorus decaoxide (b) Brass
    (c) Ammonium phosphate (d) Graphite
48. On the basis of the description given, classify each of these solids as molecular, metallic, ionic, network, or amorphous, and explain your reasoning.
    (a) A brittle, yellow solid that melts at 113 °C; neither the solid nor the liquid conducts electricity
    (b) A soft, silvery solid that melts at 40 °C; both the solid and the liquid conduct electricity
    (c) A hard, colorless, crystalline solid that melts at 1713 °C; neither the solid nor the liquid conducts electricity
    (d) A soft, slippery solid that melts at 63 °C; neither the solid nor the liquid conducts electricity
49. On the basis of the description given, classify each of these solids as molecular, metallic, ionic, network, or amorphous, and explain your reasoning.
    (a) A soft, slippery solid that has no definite melting point but decomposes at temperatures above 250 °C; the solid does not conduct electricity
    (b) Violet crystals that melt at 114 °C and whose vapor irritates the nose; neither the solid nor the liquid conducts electricity
    (c) Hard, colorless crystals that melt at 2800 °C; the liquid conducts electricity, but the solid does not
    (d) A hard solid that melts at 3410 °C; both the solid and the liquid conduct electricity

50. What type of solid exhibits each of these sets of properties?
    (a) Melts below 100 °C and is insoluble in water
    (b) Conducts electricity only when melted
    (c) Insoluble in water and conducts electricity
    (d) Noncrystalline and melts over a wide temperature range
51. Describe how each of these materials would behave if it were deformed by a hammer strike. Explain why the materials behave as they do.
    (a) A metal, such as gold
    (b) A nonmetal, such as sulfur
    (c) An ionic compound, such as NaCl

Water: Its Important and Unusual Properties (Section 9-5)

52. For most substances, the density of the solid phase is larger than for the liquid phase, but for water the reverse is true. What is the molecular-scale reason for this property of water? Why is this property important?
53. Explain how the changes of the density of water with temperature causes "turnover" in a lake in the spring and fall. Explain why the turnover is important.
54. The surface tension of a liquid decreases with increasing temperature. Using the idea of intermolecular attractions, explain why this is so.
55. The boiling point of water is relatively high for a compound of such low molar mass. Explain why this is so.
56. Water can participate in hydrogen-bonding with other water molecules. In liquid water, how many hydrogen bonds does each water molecule engage in? What three-dimensional shape do these bonds assume?

Crystalline Solids (Section 9-6)

57. Each diagram given represents an array of like atoms that would extend indefinitely in two dimensions. Draw a two-dimensional unit cell for each array. How many atoms are in each unit cell?

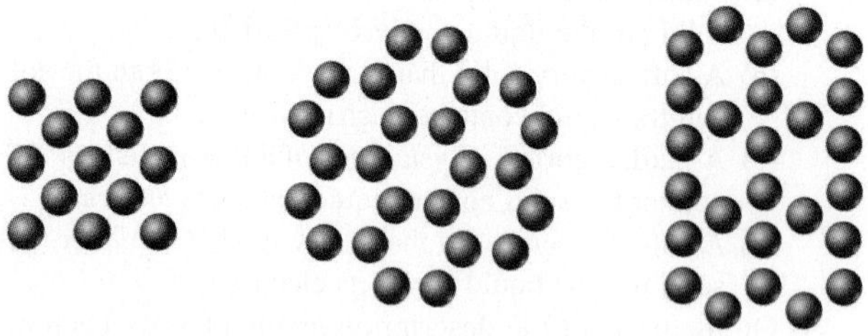

58. Name and draw the three cubic unit cells. Describe their similarities and differences.
59. Solid xenon forms crystals with a face-centered cubic unit cell that has an edge of 620 pm. Calculate the atomic radius of xenon.
60. Gold (atomic radius = 144 pm) crystallizes in an fcc unit cell. Calculate the length of a side of the cell.
61. Using the NaCl structure shown in Figure 9.24, how many unit cells share each of the Na^+ ions in the front face of the unit cell? How many unit cells share each of the Cl^- ions in this face?
62. The ionic radii of Cs^+ and Cl^- are 181 and 167 pm, respectively. What is the length of the body diagonal in the CsCl unit cell? What is the length of the side of this unit cell? (CsCl has the same unit cell as CsI, shown in Figure 9.23.)
63. Could $CaCl_2$ possibly have the NaCl structure? Explain your answer briefly.
64. You know that thallium chloride, TlCl, crystallizes in either a primitive cubic or a face-centered cubic lattice of Cl^- ions with Tl^+ ions in the holes. If the density of the solid is 7.00 g/cm^3 and the edge of the unit cell is 3.85×10^{-8} cm, determine the unit cell geometry.
65. Solid lithium has a body-centered cubic unit cell with the length of the edge of 351 pm at 20 °C. Calculate the density of lithium at this temperature.
66. Tungsten has a body-centered cubic unit cell and an atomic radius of 141 pm. Calculate the density of solid tungsten.

Network Solids (Section 9-7)

67. Explain why diamond is more dense than graphite.
68. Explain how two-dimensional and three-dimensional network solids differ on the nanoscale.
69. Determine, by looking up data in a reference such as the *Handbook of Chemistry and Physics,* whether the examples of network solids given in the text are soluble in water or other common solvents. Explain your answer in terms of the chemical bonding in network solids.
70. Explain why diamond is an electrical insulator and graphite is an electrical conductor.

Tools of Chemistry: X-Ray Crystallography (Section 9-7a)

71. For a clear diffraction pattern to be seen from a regularly spaced lattice, the radiation falling on the lattice must have a wavelength less than the lattice spacing. From the unit cell size of the NaCl crystal, estimate the maximum wavelength of the radiation that would be diffracted by this crystal. Calculate the frequency of the radiation and the energy associated with (a) one photon and (b) one mole of photons of the radiation. In what region of the spectrum is this radiation?
72. Taking the middle of the visible spectrum to be green light with a wavelength of 550 nm, calculate how many aluminum atoms (radius = 143 pm) touching their neighbors would make a straight line 550 nm long. Using this result, explain why an optical microscope using visible radiation will never be able to detect an individual aluminum atom (or any other atom, for that matter).
73. The second-order Bragg reflection ($n = 2$) from a copper crystal for X-rays with a wavelength of 166 pm is 27.35°. Calculate the spacing between the planes of copper atoms.
74. The first-order Bragg reflection ($n = 1$) from a NaCl crystal with a spacing of 282 pm is seen at 23.0°. Calculate the wavelength of the X-ray radiation used.

Materials Science (Section 9-8)

75. List the four major classes of materials and give one example of each.

76. Identify three items you use daily that have been developed recently using the principles of materials science.

Metals, Semiconductors, and Insulators (Section 9-9)

77. What is the principal difference between the orbitals that electrons occupy in individual, isolated atoms and the orbitals they occupy in solid metals?
78. In terms of band theory, what is the difference between a conductor and an insulator? Between a conductor and a semiconductor?
79. Name three properties of metals, and explain them by using a theory of metallic bonding.
80. Which substance has the greatest electrical conductivity? The smallest electrical conductivity? Explain your choice briefly.
    (a) Si (b) Ge (c) Ag (d) P_4
81. Which substance has the greatest electrical conductivity? The smallest electrical conductivity? Explain your choices briefly.
    (a) RbCl(ℓ) (b) NaBr(s) (c) Rb (d) Diamond
82. Define the term "superconductor." Give the chemical formulas of two kinds of superconductors and their associated transition temperatures.

Silicon and the Chip (Section 9-10)

83. Extremely high-purity silicon is required to manufacture semiconductors such as the memory chips found in calculators and computers. If a silicon wafer is 99.99999999% pure, approximately how many atoms of some other element are present per gram of high-purity silicon?
84. Explain why Group 3A and Group 5A elements are used to dope silicon to improve its semiconducting properties.
85. Explain the difference between *n*-type semiconductors and *p*-type semiconductors.

Cement, Ceramics, and Glass (Section 9-11)

86. Define the term "amorphous."
87. What makes a glass different from a crystalline solid such as SiO_2? Under what conditions could SiO_2 become glass-like?
88. A typical cement contains, by weight, 65% CaO, 20% SiO_2, 5% Al_2O_3, 6% Fe_2O_3, and 4% MgO. Determine the mass percent of each type of atom present. Then, determine an empirical formula of the material from the percent composition, setting the subscript of the least abundant element to 1.00.
89. Give two examples of (a) silicate ceramics, (b) oxide ceramics, and (c) nonoxide ceramics.

General Questions

These questions are not explicitly keyed to chapter topics; many require integration of several concepts.

90. Explain why, when you boil water in a pan, the water boils much faster when the pan has a lid on it than when it does not.
91. Will a closed container of water at 70 °C or an open container of water at the same temperature cool faster on a cold winter day? Explain why.
92. Given these properties: *Camphor:* colorless needles; density = 0.900 g/cm³ at 25 °C; sublimes at 204 °C; insoluble in water; very soluble in ethanol or ether. *Praseodymium chloride:* blue-green needle crystals; density = 4.02 g/cm³ at 25 °C; melting point 786 °C; boiling point 1700 °C; solubility 103.9 g/100 mL cold water, very soluble in hot water.
    (a) Is camphor an ionic or covalent compound? Explain your answer.
    (b) Is praseodymium chloride an ionic or covalent compound? Explain your answer.
93. Xenon has a triple point of 0.81 atm and −112 °C and a normal boiling point of −108 °C. If the pressure exerted on a xenon sample is 1.75 atm and the temperature is −105 °C, in what phase (physical state) does the xenon sample likely exist at these conditions? Explain your answer.
94. Ammonia has a triple point of 0.0604 atm and −77.8 °C; its normal boiling point is −33.4 °C. An ammonia sample is at 0.60 atm and −56.2 °C. In what phase is the ammonia sample likely to exist at this temperature and pressure? Explain your answer.
95. Use the vapor pressure curves for methyl ethyl ether, $CH_3OCH_2CH_3$; carbon disulfide, CS_2; and benzene, C_6H_6, to answer these questions.
    (a) What is the vapor pressure of methyl ethyl ether at 0 °C?
    (b) Which of these three liquids has the strongest intermolecular attractions?
    (c) At what temperature does benzene have a vapor pressure of 600 mmHg?
    (d) What are the normal boiling points of these three liquids?

Vapor pressure curves for Question 95.

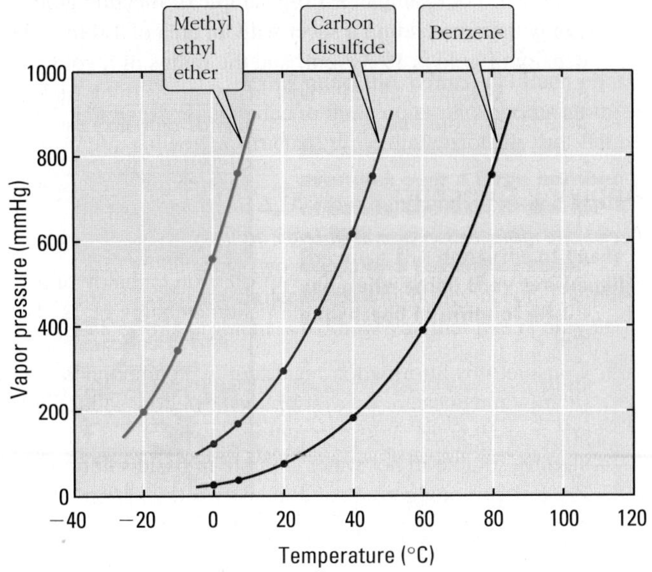

Methyl ethyl ether
Carbon disulfide
Benzene
1000
800
600
400
200
0
Vapor pressure (mmHg)
−40
−20
0
20
40
60
80
100
120
Temperature (°C)

96. Organic compounds with structures based on benzene, C_6H_6, can be formed by substituting an atom or a group of atoms for one of the hydrogens. Such substituted benzenes have their own properties, different from benzene and from each other. Explain the order of experimental boiling points for these four compounds.
   (a) C_6H_6 (80 °C) (b) C_6H_5Cl (131 °C)
   (c) C_6H_5Br (156 °C) (d) C_6H_5OH (182 °C)
97. The chlorofluorocarbon CCl_2F_2 was once used in air conditioners as the heat transfer fluid. Its normal boiling point is −30 °C, and its vaporization enthalpy is 165 J g^{-1}. The gas and the liquid have specific heat capacities of 0.61 J g^{-1} °C^{-1} and 0.97 J g^{-1} °C^{-1}, respectively. Calculate the heat energy transfer when 10.0 g CCl_2F_2 is cooled from 40 °C to −40 °C.
98. Liquid ammonia, $NH_3(\ell)$, was used as a refrigerant fluid before the discovery of the chlorofluorocarbons and is still widely used today. Its normal boiling point is −33.4 °C, and its vaporization enthalpy is 23.5 kJ/mol. The gas and liquid have specific heat capacities of 2.2 J g^{-1} K^{-1} and 4.7 J g^{-1} K^{-1}, respectively. Calculate the heat energy transfer required to raise the temperature of 10.0 kg liquid ammonia from −50.0 °C to 0.0 °C.
99. Sulfur dioxide, SO_2, is found in polluted air. What types of forces are responsible for binding SO_2 molecules to one another in the solid or liquid phase?
100. Using the information below, place the compounds listed in order of increasing intermolecular attractions. For each substance list all types of intermolecular forces that are important.

Compound	Normal Boiling Point (°C)
SO_2	−10
NH_3	−33.4
CH_4	−161.5
H_2O	100

101. Metallic gold is very malleable; that is, it can be hammered into very thin sheets, which are sometimes called gold leaf. For example, a 1.0-g sample of metallic gold can be hammered into a sheet with an area of 1.0 m^2. The density of gold is 19.3 g/cm^3 and the radius of a gold atom is 144 pm. Calculate how many atoms thick such a gold sheet would be.

Applying Concepts

These questions test conceptual learning.

102. Explain why, in general, the vaporization enthalpy of a liquid is much greater than the fusion enthalpy of its solid.
103. Consider this information regarding two compounds (common names are used). *Orpiment:* yellow solid; density = 3.49 g/cm^3 at 25 °C; melting point = 573 K; slightly soluble in hot water; soluble in basic solution. *Zeaxanthin:* orange-red solid; density = 0.93 g/cm^3 at 25 °C; melting point = 489 K; insoluble in water, soluble in benzene.
   (a) Is orpiment an ionic or molecular compound? Explain your answer.
   (b) Is zeaxanthin an ionic or molecular compound? Explain your answer.
104. Consider this information regarding two compounds. *Thallium azide:* yellow crystalline solid; melting point = 330 °C; slightly soluble in water, more soluble in hot water; insoluble in ethanol or diethyl ether. *Camphene:* colorless, cubic crystals; melting point = 51 °C; boiling point = 159 °C; insoluble in water; moderately soluble in ethanol; soluble in diethyl ether.
   (a) Is camphene an ionic or molecular compound? Explain your answer.
   (b) Is thallium azide an ionic or molecular compound? Explain your answer.
105. The phase diagram for red phosphorus is shown nearby.
   (a) Label the areas for the pure solid, liquid, and gas phases.
   (b) Label a point on the diagram at which red phosphorus liquid and vapor are in equilibrium.
   (c) Explain why solid red phosphorus cannot be melted in a container open to the atmosphere.
   (d) Identify all the phase changes that occur sequentially when conditions change from Point B to Point A.
106. Consider liquid water in equilibrium with its vapor at 100 °C. Estimate the number of water molecules per cm^3 in (a) the liquid (density = 0.958 g/cm^3 at 100 °C); (b) the vapor.
107. If you get boiling water at 100 °C on your skin, it burns. If you get 100 °C steam on your skin, it burns much more severely. Explain why this is so.
108. If water at room temperature is placed in a flask that is connected to a vacuum pump and the vacuum pump then lowers the pressure in the flask, we observe that the volume of the water decreases and the remaining water turns into ice. Explain what has happened.
109. The normal boiling point of SO_2 is 263.1 K and that of NH_3 is 239.7 K. At −40 °C, would you predict that ammonia has a vapor pressure greater than, less than, or equal to that of sulfur dioxide? Explain.

Phase diagram of red phosphorus for Question 105.

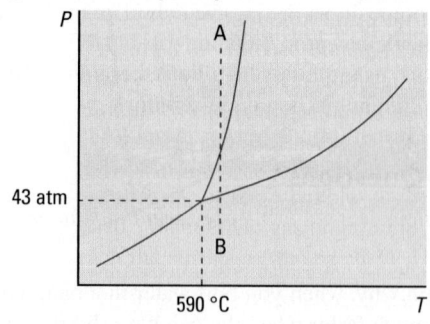

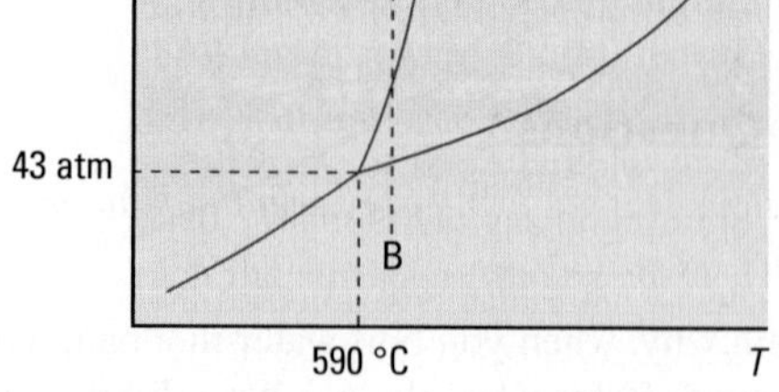
P
A
43 atm
B
590 °C
T

Nanoscale diagrams and phase diagram for Question 112.

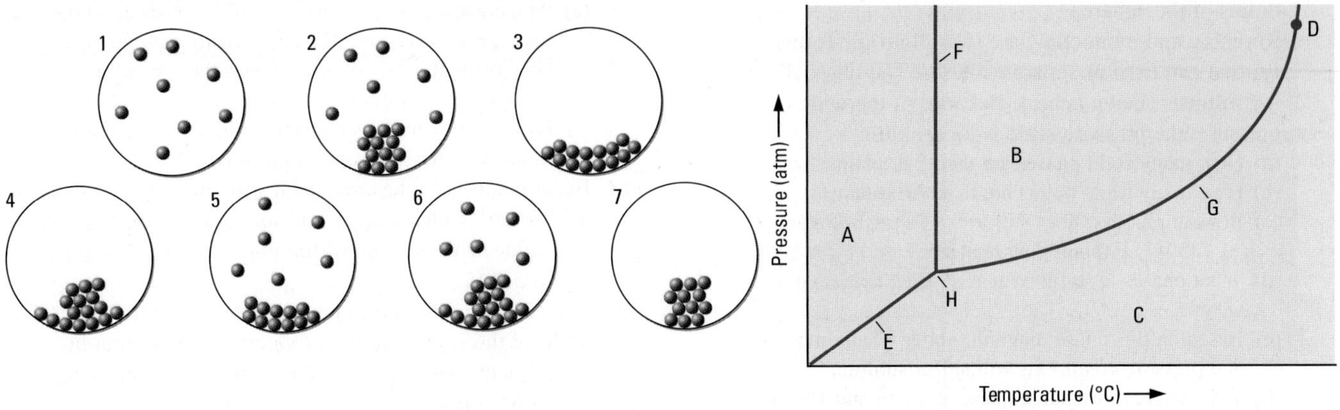
1
2
3
4
5
6
7
Pressure (atm)
Temperature (°C)
A
B
C
D
E
F
G
H

110. Butane is a gas at room temperature; however, if you look closely at a butane lighter you see it contains liquid butane. Explain how it is possible to have liquid butane present.

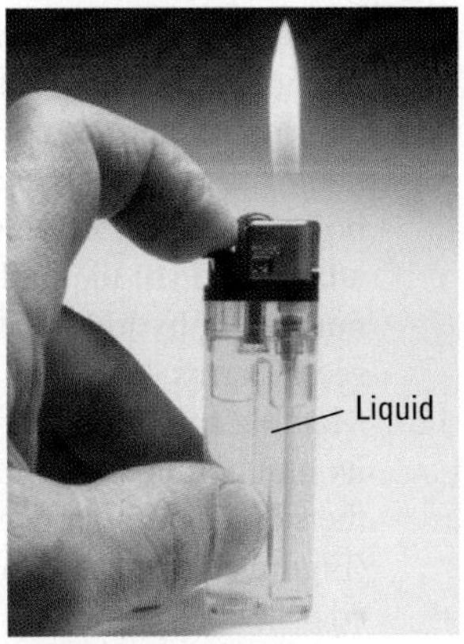
Liquid

Cengage Learning/Charles D. Winters

111. While camping with a friend in the Rocky Mountains, you decide to cook macaroni for dinner. Your friend says the macaroni will cook faster in the Rockies because the lower atmospheric pressure will cause the water to boil at a lower temperature. Do you agree with your friend? Explain your reasoning.
112. Examine the nanoscale diagrams and the phase diagram for Question 112. Match each particulate diagram (1 through 7) to its corresponding point (A through H) on the phase diagram.
113. Consider the phase diagram and heating-curve graphs for Question 113. Draw corresponding heating curves for T_1 to T_2 at pressures P_1 and P_2. Label each phase and phase change on your heating curves.
114. Consider three boxes of equal volume. One is filled with tennis balls, another with golf balls, and the third with marbles. If a closest-packing arrangement is used in each box, which one has the most occupied space? Which one has the least occupied space? (Disregard the difference in filling space at the walls, bottom, and top of the box.)

More Challenging Questions

These questions require more thought and integrate several concepts.

115. When spherical atoms are positioned in a unit cell, they don't occupy all of the space available in the unit cell. Some types of unit cells have more efficient packing than other types have. Consider two metals: one with a body-centered cubic unit cell; the other with a face-centered cubic unit cell. Calculate the volume of the atoms in each of these unit cells in comparison to the volume of the unit cell itself. From these data, calculate the fraction of space in each of the unit cells that is occupied by its

Phase diagram and heating-curve graphs for Question 113.

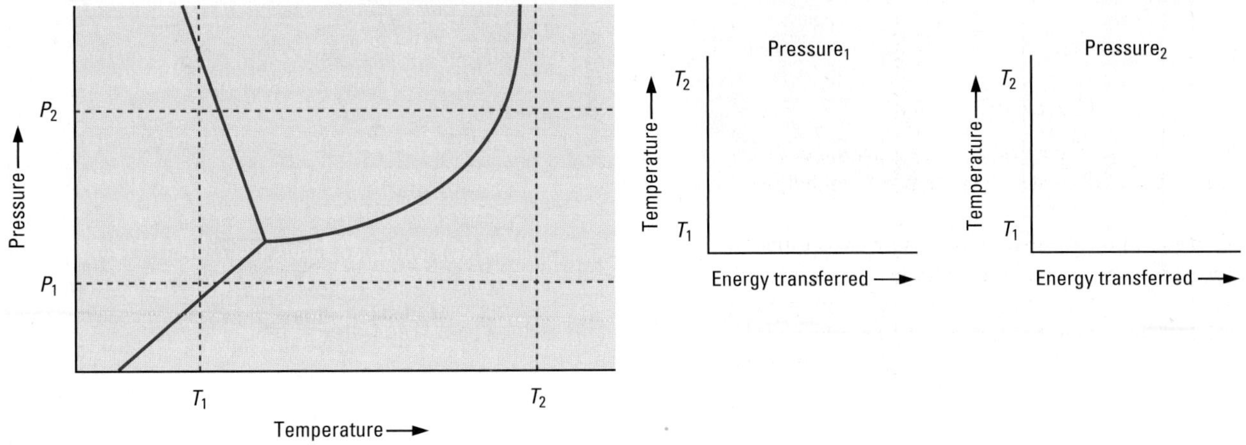
Pressure
Temperature
P2
P1
T1
T2
Pressure1
Pressure2
Temperature
Energy transferred

atoms. The volume of a sphere is $\frac{4}{3}\pi r^3$ where r is the radius of the sphere.

116. Rhombic and monoclinic are two allotropic forms of sulfur that can exist as separate phases. The phase diagram for sulfur is shown nearby. Because of the wide range of pressure, the pressure scale is logarithmic.
   (a) How many solid phases are there? Explain.
   (b) How many triple points are there? Explain.
   (c) In what phase(s) does sulfur exist(s) at 1 atm and 80 °C? At 125 °C? Explain your answers.
   (d) What phases is sulfur in at 151 °C? Explain your answer.
   (e) Based on this phase diagram, under what temperature and pressure conditions will sulfur sublime?
   (f) What is the most stable phase at 1 atm and 100 °C?
   (g) Identify *in sequence* the phases present at 10^{-4} atm as the temperature changes from 50 °C to 200 °C.
   (h) Determine the normal melting point of sulfur.

117. The phase diagram for water over a relative narrow pressure and temperature range is given in Figure 9.19. A phase diagram over a considerably wider range of temperature and pressure (kbar) is given nearby. This phase diagram illustrates the polymorphism of ice, the existence of a solid in more than one form. In this case, Roman numerals are used to designate each polymorphic form. For example, Ice I, ordinary ice, is the form that exists under ordinary pressures. The other forms exist only at higher pressures, in some cases extremely high pressure such as Ice VII and Ice VIII.
   (a) Using the phase diagram, give the approximate P and T conditions at the triple point for Ice III, Ice V, and liquid water.
   (b) Determine the approximate temperature and pressure for the triple point for Ices VI, VII, and VIII.
   (c) What is anomalously different about the fusion curves for Ice VI and Ice VII compared to that of Ice I?
   (d) What phases exist at 8 kbar and 20 °C?
   (e) At a constant temperature of −10 °C, start at 3 kbar and increase the pressure to 7 kbar. Identify all the phase changes that occur *sequentially* as these conditions change.
   (f) Explain why there is no triple point for the combination of Ice VII, Ice VIII, and liquid water.

118. Barium sulfide(s) has the NaCl structure and a density of 4.25 g/cm³. Calculate the interionic distance and compare this value with the sum of the ionic radii (Ba^{2+} = 149 pm; S^{2-} = 170 pm).

119. The Ne atom and the molecules HF, H_2O, NH_3, and CH_4 all have the same number of electrons. In a thought experiment, you can make HF from Ne by removing a single proton a short distance from the nucleus and having the electrons follow the new arrangement of nuclei so as to make a new chemical bond. You can do the same for each of the other molecules. (Of course, none of these thought experiments can actually be done because of the enormous energies required to remove protons from nuclei.) For all of these substances, make a plot of (a) the boiling point in kelvins versus the number of hydrogen atoms, and (b) the molar vaporization enthalpy versus the number of hydrogen atoms. Explain any trend that you see in terms of intermolecular forces.

120. Metallic lithium has a body-centered cubic structure, and its unit cell is 351 pm along an edge. Lithium iodide has the same crystal lattice structure as sodium chloride (Figure 9.24). The cubic unit cell is 600 pm along an edge.
   (a) Assume that the metal atoms in lithium touch along the body diagonal of the body-centered cubic unit cell, and estimate the radius of a lithium atom.
   (b) Assume that in lithium iodide the I^- ions touch along the face diagonal of the cubic unit cell and that the Li^+

Phase diagram for sulfur for Question 116.

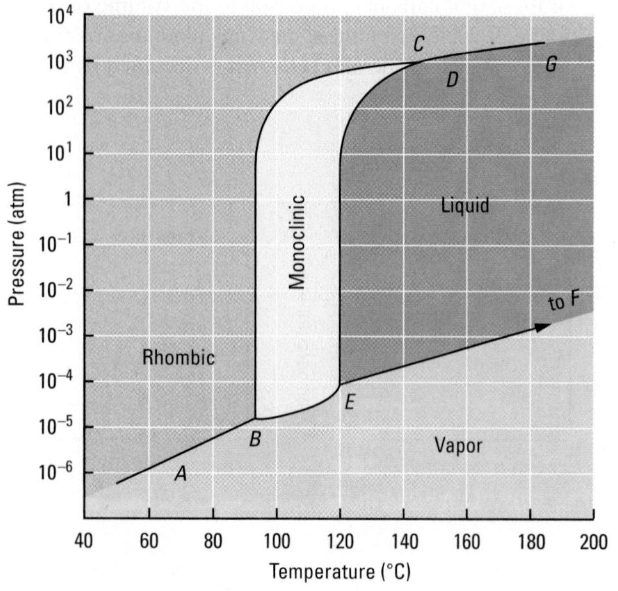
Pressure (atm)
10⁴
10³
10²
10¹
1
10⁻¹
10⁻²
10⁻³
10⁻⁴
10⁻⁵
10⁻⁶
C
G
D
Liquid
Monoclinic
Rhombic
to F
E
B
A
Vapor
40
60
80
100
120
140
160
180
200
Temperature (°C)

Phase diagram of water for Question 117.

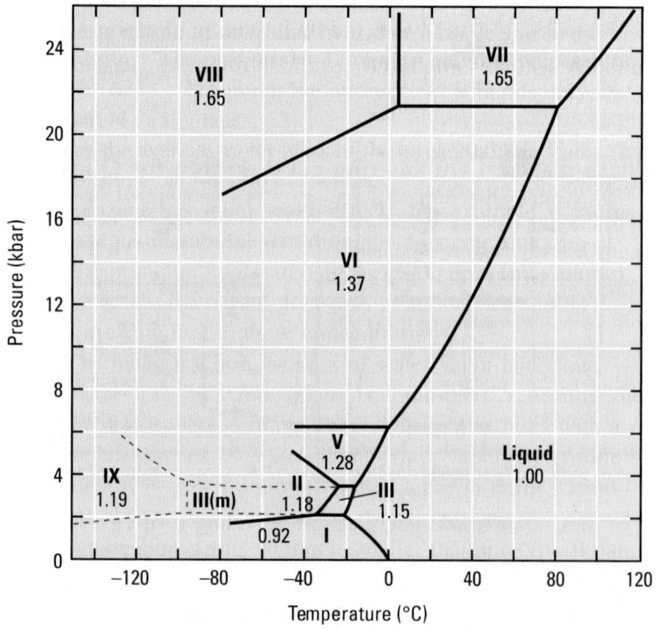
Pressure (kbar)
24
20
16
12
8
4
0
VIII
1.65
VII
1.65
VI
1.37
V
1.28
IX
1.19
III(m)
II
1.18
III
1.15
0.92
I
Liquid
1.00
−120
−80
−40
0
40
80
120
Temperature (°C)

and I^- ions touch along the edge of the cube; calculate the radius of an I^- ion and of an Li^+ ion.

(c) Compare your results in parts (a) and (b) for the radius of a lithium atom and a lithium ion. Are your results reasonable? If not, how could you account for the unexpected result? Could any of the assumptions that were made be in error? Explain.

121. Potassium chloride and rubidium chloride both have the sodium chloride structure (Figure 9.24). X-ray diffraction experiments indicate that their cubic unit cell dimensions are 629 pm and 658 pm, respectively.

(i) One mol KCl and 1 mol RbCl are ground together to a very fine powder in a mortar and pestle, and the X-ray diffraction pattern of the pulverized solid is measured. Two patterns are observed, each corresponding to a cubic unit cell—one with an edge length of 629 pm and one with an edge length of 658 pm. Call this Sample 1.

(ii) One mol KCl and 1 mol RbCl are heated until the entire mixture is molten and then cooled to room temperature. A single X-ray diffraction pattern indicates a cubic unit cell with an edge length of roughly 640 pm. Call this Sample 2.

(a) Suppose that Samples 1 and 2 were analyzed for their chloride content. What fraction of each sample is chloride? Could the samples be distinguished by means of chemical analysis?

(b) Interpret the two X-ray diffraction results in terms of the structures of the crystal lattices of Samples 1 and 2.

(c) What chemical formula should you write for Sample 1? For Sample 2?

(d) Suppose that you dissolved 1.00 g Sample 1 in 100 mL water in a beaker and did the same with 1.00 g Sample 2. Which sample would conduct electricity better, or would both be the same? What ions would be present in each solution at what concentrations?

122. In lithium chloride, the larger Cl^- ions form a face-centered cubic unit cell lattice with the Li^+ ions fitting between the Cl^- ions at the center of each edge of the unit cell. Sodium chloride has a similar structure. As shown in Figure 9.24, the Na^+ and Cl^- ions in sodium chloride are in a face-centered cubic unit cell in which the Na^+ and Cl^- ions touch along each cell edge. The Cl^- ions are also at the centers of each face. Whether cations can fit into a particular unit cell lattice of anions to create a stable structure depends on the relative sizes of the cations and anions.

(a) Use ionic radii to calculate the r_{cation}/r_{anion} ratio at which a cation just fits into a face-centered unit cell.

(b) How closely does sodium chloride meet this criterion?

(c) Comment on how well LiCl meets this criterion. Explain the difference between LiCl and NaCl in this regard.

123. Titanium metal crystallizes in a body-centered unit cell with a cell edge length of 330.6 pm. The density of titanium metal at 25.0 °C is 4.401 g/cm^3. Use these data to calculate the Avogadro constant.

Conceptual Challenge Problems

These rigorous, thought-provoking problems integrate conceptual learning with problem solving and are suitable for group work.

CP9.A (Section 9-2) The vaporization enthalpy of water depends on the temperature. At 100 °C this value is 40.7 kJ/mol, but at 25 °C it is 44.0 kJ/mol, a difference of 3.3 kJ/mol. List four enthalpy changes whose sum would equal this difference. Remember, the sum of the changes for a cyclic process must be zero because the system is returned to its initial state.

CP9.B (Section 9-3) For what reasons would you propose that two of the substances listed in Table 9.2 be considered better refrigerants for use in household refrigerators than the others listed there?

CP9.C (Section 9-4) A table of sublimation enthalpies is not given in Section 9-4, but the sublimation enthalpy of ice at 0 °C is given as 51 kJ/mol. Explain how this value was obtained. Tables 9.2 and 9.5 list the vaporization enthalpies and fusion enthalpies, respectively, for several substances. Determine from data in these tables the $\Delta_{sub}H$ for ice. Using the same method, estimate the sublimation enthalpies of HBr and HI at their melting points.

10 Fuels, Organic Chemicals, and Polymers

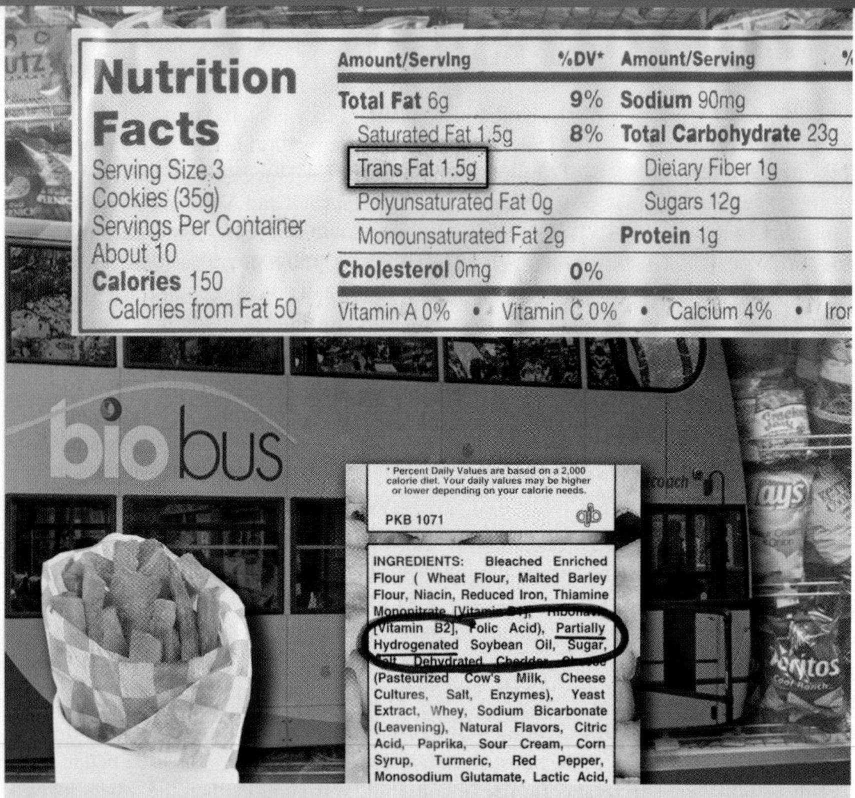

Nutrition Facts
Serving Size 3 Cookies (35g)
Servings Per Container About 10
Calories 150
Calories from Fat 50
Amount/Serving %DV* Amount/Serving
Total Fat 6g 9%
Saturated Fat 1.5g 8%
Trans Fat 1.5g
Polyunsaturated Fat 0g
Monounsaturated Fat 2g
Cholesterol 0mg 0%
Sodium 90mg
Total Carbohydrate 23g
Dietary Fiber 1g
Sugars 12g
Protein 1g
Vitamin A 0% • Vitamin C 0% • Calcium 4% •
bio bus
* Percent Daily Values are based on a 2,000 calorie diet. Your daily values may be higher or lower depending on your calorie needs.
PKB 1071
INGREDIENTS: Bleached Enriched Flour (Wheat Flour, Malted Barley Flour, Niacin, Reduced Iron, Thiamine
Vitamin B2], Folic Acid), Partially Hydrogenated Soybean Oil, Sugar,
(Pasteurized Cow's Milk, Cheese Cultures, Salt, Enzymes), Yeast Extract, Whey, Sodium Bicarbonate (Leavening), Natural Flavors, Citric Acid, Paprika, Sour Cream, Corn Syrup, Turmeric, Red Pepper, Monosodium Glutamate, Lactic Acid,

Photos: Peter McGahey; Richard Levine/Alamy; Cengage Learning/George Semple; Margoe Edwards/Shutterstock.com; Oramstock/Alamy.

Most dietary vegetable oils are high in *cis*-unsaturated fats (fatty acids). When such oils are partially hydrogenated for use in snack foods, some *cis* sites convert to the *trans* form. Health concerns have led to a proposed federal ban on *trans* fats in processed foods and commercial baked goods. Current federal law requires that nutrition labels indicate the mass of *trans* fats in fat-containing foods. Researchers are developing manufacturing plants to make oils high in unsaturated fats, but without *trans* fats, such as the oil used to cook the sweet potato fries shown above. There is also much work on converting used cooking fat to biodiesel fuel. What is an unsaturated fat and why does the difference between *cis* and *trans* forms of unsaturated fats create a health concern? (Section 10-5c) How are fatty acids converted to biodiesel? (Section 10-5e) This chapter discusses the molecular structures and reactivity of many types of organic compounds, including fats.

The combustion of *fossil fuels*—coal, natural gas, and petroleum—provides more than 80% of all energy used in the world. When these substances burn, the carbon they contain is released into the atmosphere as CO_2. Photosynthesis converts CO_2 back into other carbon-containing compounds. Many of the carbon compounds produced by photosynthesis are directly or indirectly very useful as energy sources for humans and animals. But such carbon compounds are not as convenient to use in power plants or automobiles as are fossil fuels, which are burned in prodigious quantities daily. As fossil fuels become scarcer, and if stringent conservation measures are not taken, conventional petroleum reserves could last for no more than into the last half of the 21st century. If this occurs, fuel costs will inevitably rise due to the increasing scarcity of petroleum.

Fossil fuels are also, by far, the largest source of starting material for the syntheses of most of the organic chemicals used to make consumer products such as plastics, pharmaceuticals, synthetic rubber, synthetic fibers, and hundreds of other products we rely on. For this reason, the organic chemical industry is often referred to as the petrochemical industry. In this chapter, we will discuss a few of the major classes of organic compounds and some of their reactions, especially those used to furnish energy and to make synthetic and natural polymers.

The great Russian chemist Dimitri Mendeleev recognized the importance of petroleum as a source from which to make valuable carbon compounds and not merely to be used as a fuel. On visiting the oil fields of Pennsylvania and Azerbaijan, he supposedly remarked that burning petroleum as a fuel "would be akin to firing up a kitchen stove with bank notes."

10-1 Petroleum

Petroleum is a complex mixture of alkanes, cycloalkanes, alkenes, and aromatic hydrocarbons formed underground over millions of years from the remains of plants and animals. More than 20,000 compounds, almost all of them hydrocarbons, are present in *crude oil,* the form of petroleum pumped from the ground. Crude oil's composition and color vary with the location in which it is found. Pennsylvania crude oils are primarily straight-chain hydrocarbons, whereas California crude oil contains a larger portion of aromatic hydrocarbons.

Petroleum contains about as many compounds as there are genes in the human body.

In 1859 the first commercial oil well was drilled at Titusville, Pennsylvania by the Seneca Oil Company in response to favorable reports about the likelihood of establishing a commercial oil drilling operation there. Dr. Benjamin Silliman, a Yale University chemistry professor, was commissioned to analyze petroleum samples from natural surface seepages in the area. In an 1855 report to the Pennsylvania Rock Oil Company he stated, ". . . In conclusion, gentlemen, it appears to me that . . . your Company have in your possession a raw material from which, by a simple and not expensive process, they may manufacture very valuable products." This classic example of understatement helped to launch the petroleum revolution.

The early uses for petroleum components were mainly for lubrication and as kerosene burned in lamps. The development of automobiles with internal combustion engines created the need for liquid fuels that would burn efficiently in these engines. To meet our need for gasoline and other petroleum products, it is necessary to refine crude oil—that is, to separate the various useful components from the complex mixture and to modify their properties.

10-1a Petroleum Refining

Substances that have sufficiently different boiling points can be separated by simple distillation. For example, Figure 10.1 shows how cyclohexane (b.p. 80.7 °C) can be separated from toluene (b.p. 110.6 °C). Petroleum contains thousands of different hydrocarbons and their separation as individual pure compounds is neither economically feasible nor necessary. Instead, petroleum refining uses *fractional distillation* to separate petroleum into groups of compounds that have similar boiling points (Figure 10.2). A **petroleum fraction** is *a mixture of many hundreds of hydrocarbons that have boiling points within a specified range*. Such separation is possible because the boiling point increases as the number of carbon atoms increases: Larger hydrocarbon molecules (larger number of electrons and greater polarizability) have greater noncovalent intermolecular forces and higher boiling points than smaller ones (← **Sec. 7-6**). These large molecules collect at the bottom of the fractional distillation tower. Small molecules with weak intermolecular forces are collected at the top.

The difference between simple distillation and fractional distillation is the degree of separation achieved.

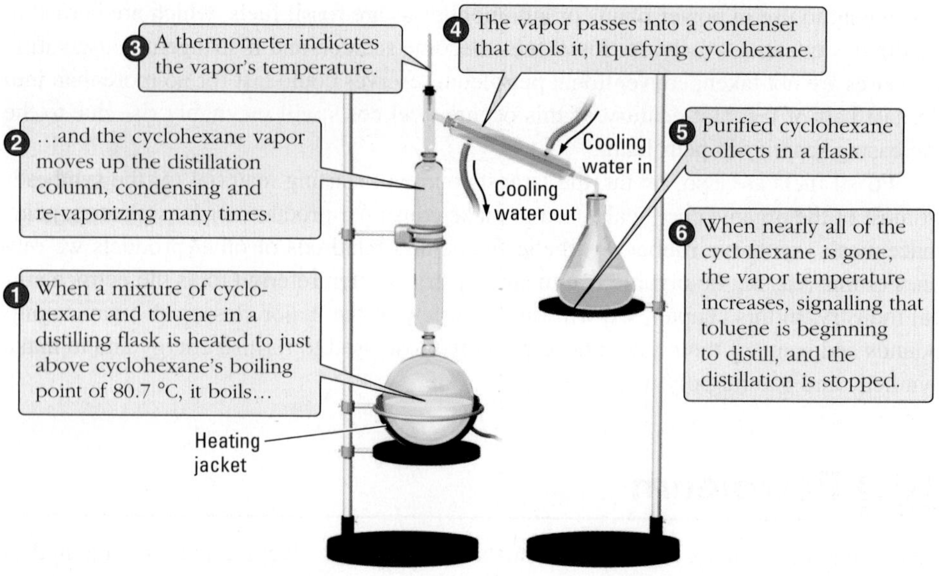
A thermometer indicates the vapor's temperature.
The vapor passes into a condenser that cools it, liquefying cyclohexane.
...and the cyclohexane vapor moves up the distillation column, condensing and re-vaporizing many times.
Cooling water in
Cooling water out
Purified cyclohexane collects in a flask.
When nearly all of the cyclohexane is gone, the vapor temperature increases, signalling that toluene is beginning to distill, and the distillation is stopped.
When a mixture of cyclohexane and toluene in a distilling flask is heated to just above cyclohexane's boiling point of 80.7 °C, it boils...
Heating jacket

Figure 10.1 Laboratory distillation. Cyclohexane and toluene can be separated because their boiling points differ.

Today, petroleum is seldom delivered in barrels, but rather is shipped in pipelines and ocean-going tankers. Nevertheless, the barrel (42 U. S. gallons) remains as the common unit of volume measure for petroleum.

The properties and consequently the uses of various fractions differ, as shown in Figure 10.2. Refined crude oil is used in these ways: fuel, 83% (gasoline, 47%; diesel and home heating oil, 24%; jet fuel, 9%; boiler fuel oil, 5%); lubricants, waxes, solvents, 9%; asphalt and road oil, 3%; and petrochemical feedstocks, 3%. Petrochemical feedstocks are the very important petroleum components needed as reactants to make fabrics, plastics, pharmaceuticals, and other synthetic organic chemicals on which we depend. A truly prodigious volume of refined crude oil is consumed daily worldwide, 91 million barrels per day in 2013, equivalent to about 1050 barrels or 44,000 gallons per second. The United States consumes about 21% of this.

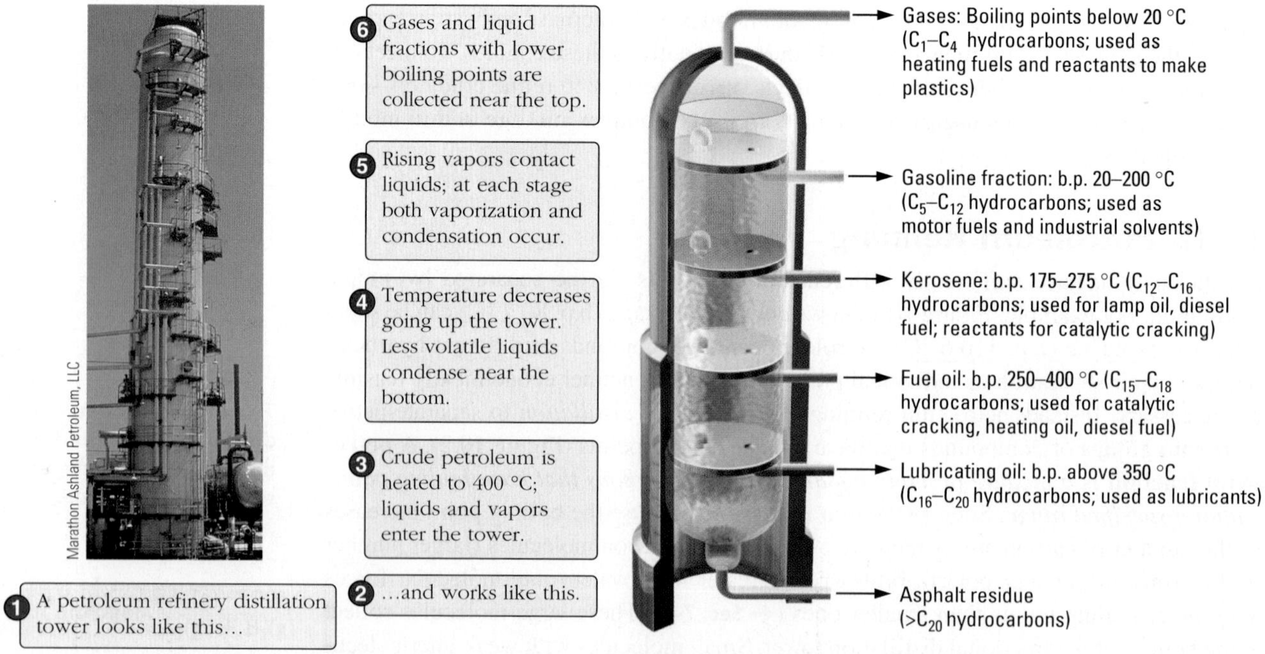
Marathon Ashland Petroleum, LLC
A petroleum refinery distillation tower looks like this...
...and works like this.
Crude petroleum is heated to 400 °C; liquids and vapors enter the tower.
Temperature decreases going up the tower. Less volatile liquids condense near the bottom.
Rising vapors contact liquids; at each stage both vaporization and condensation occur.
Gases and liquid fractions with lower boiling points are collected near the top.
Gases: Boiling points below 20 °C (C_1–C_4 hydrocarbons; used as heating fuels and reactants to make plastics)
Gasoline fraction: b.p. 20–200 °C (C_5–C_{12} hydrocarbons; used as motor fuels and industrial solvents)
Kerosene: b.p. 175–275 °C (C_{12}–C_{16} hydrocarbons; used for lamp oil, diesel fuel; reactants for catalytic cracking)
Fuel oil: b.p. 250–400 °C (C_{15}–C_{18} hydrocarbons; used for catalytic cracking, heating oil, diesel fuel)
Lubricating oil: b.p. above 350 °C (C_{16}–C_{20} hydrocarbons; used as lubricants)
Asphalt residue (>C_{20} hydrocarbons)

Figure 10.2 Distillation on an industrial scale separates petroleum into fractions.

10-1b Octane Number

The **octane number** of a gasoline is a measure of the gasoline's ability to burn smoothly and efficiently in an internal combustion engine; the higher the octane number, the smoother the combustion. A typical automobile engine uses the *gasoline fraction* of refined petroleum—a mixture of C_5 to C_{12} hydrocarbons. This hydrocarbon mixture contains relatively small molecules and has a fairly high autoignition temperature, the temperature at which the hydrocarbon vapor will ignite and burn without a spark or other source of ignition. Because of this fairly high autoignition temperature, a gasoline engine requires a source of ignition—a spark plug—to ignite its fuel. Once ignited, the fuel burns smoothly. If autoignition occurs, combustion is rough, which can reduce power or damage the engine.

Diesel engines have no spark plugs. Autoignition ignites the diesel fuel, which consists of larger hydrocarbon molecules that have lower autoignition temperatures. Because autoignition is associated with rougher combustion, diesel engines are noisier and must be stronger and heavier than gasoline engines.

The octane-number rating of a gasoline is determined by comparing its burning characteristics in a one-cylinder test engine with those obtained for mixtures of heptane and 2,2,4-trimethylpentane, a multibranched isomer of octane. Heptane does not burn smoothly and is assigned by general consensus an octane number of 0, whereas 2,2,4-trimethylpentane burns smoothly and is assigned an octane number of 100. A gasoline assigned an octane number of 87 has the same autoignition characteristics as a mixture of 13% heptane and 87% 2,2,4-trimethylpentane. This 87-octane rating corresponds to the octane number of regular unleaded gasoline currently available in the United States. Other, higher grades of gasoline available at gas stations have octane numbers of 89 (regular plus) and 92 (premium).

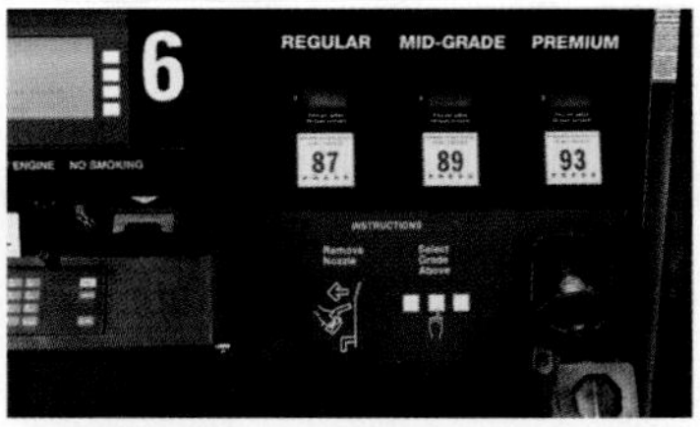
6
REGULAR
MID-GRADE
PREMIUM
87
89
93

© Cengage Learning/ Charles D. Winters

Typical octane ratings of commercially available gasoline.

Table 10.1 lists octane numbers of several substances. Straight-chain alkanes (← **Secs. 2-9a, 6-3**) are less thermally stable and burn less smoothly than branched-chain alkanes. For example, octane burns even less smoothly than heptane and therefore has a negative octane number; hexane rates at 25-octane. The gasoline fraction obtained directly from fractional distillation of petroleum ("straight-run" gasoline) has an octane number of 50 and is a poor motor fuel. It needs additional refining because it consists primarily of straight-chain hydrocarbons that autoignite too readily. The octane number of a gasoline can be increased by increasing the percentage of branched-chain and aromatic hydrocarbon components or by adding octane enhancers, some of which are listed in Table 10.1.

10-1c Catalytic Cracking

During petroleum refining, the percentage of each fraction collected can be adjusted to match the market demand. For example, there is greater demand for gasoline than

Table 10.1 Octane Numbers of Some Hydrocarbons and Gasoline Additives; Higher Octane Number Indicates Smoother Combustion

Name	Class of Compound	Octane Number
Octane	Alkane	−20
Heptane	Alkane	0
Hexane	Alkane	25
Pentane	Alkane	62
1-Pentene	Alkene	91
2,2,4-Trimethylpentane (isooctane)	Branched-chain alkane	100
Benzene	Aromatic hydrocarbon	106
Methanol	Alcohol	107
Ethanol	Alcohol	108
Tertiary-butyl alcohol	Alcohol	113
Toluene	Aromatic hydrocarbon	118

Since the method for determining octane numbers was established, fuels that are superior to 2,2,4-trimethylpentane have been developed and thus have octane numbers greater than 100.

for kerosene and diesel fuel. Demands also vary seasonally. In winter, the need for home heating oil is high; in summer, when more people take vacations, demand for gasoline is higher. In summer, refiners use chemical reactions in a process called "cracking" to convert some of the larger, kerosene-fraction molecules (C_{12} to C_{16}) into smaller molecules in the gasoline range (C_5 to C_{12}). **Catalytic cracking** *uses a catalyst, high temperature, and high pressure to break long-chain hydrocarbons into shorter-chain hydrocarbons* that include alkanes and alkenes, many in the gasoline range. A **catalyst** is *a substance that increases the rate of a chemical reaction without being consumed as a reactant is.* (Catalysts in chemical reactions are discussed in detail in Section 11-8.)

Catalytic cracking breaks large alkane molecules... ...into smaller molecules: an alkene and an alkane.

$$\underset{\text{An alkane}}{C_{16}H_{34}} \xrightarrow[\text{high temperature}]{\text{catalyst, pressure, and}} \underset{\text{An alkene}}{C_8H_{16}} + \underset{\text{An alkane}}{C_8H_{18}}$$

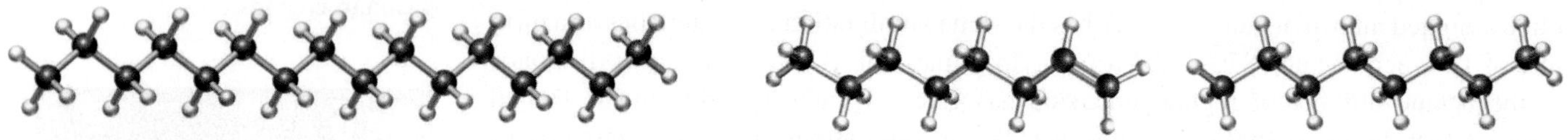

Because smaller hydrocarbons have higher octane numbers, and because alkenes have higher octane numbers than alkanes, catalytic cracking increases the octane number of gasoline. Catalytic cracking is beneficial in another way. It produces alkenes, which have C=C bonds and therefore are much more reactive than alkanes. Catalytic cracking of natural gas or petroleum produces acetylene, butylene, ethylene, and propylene (Table 10.2). These are then converted chemically into raw materials from which many kinds of commercial products are synthesized; examples are plastics and pharmaceuticals.

Table 10.2 Some Chemical Raw Materials and Commercial Products Made from Petroleum or Natural Gas

Unsaturated Hydrocarbon	Chemical Raw Materials	Commercial Products
 **Acetylene**	Neoprene	Synthetic rubber, paints, adhesives, fibers, solvents
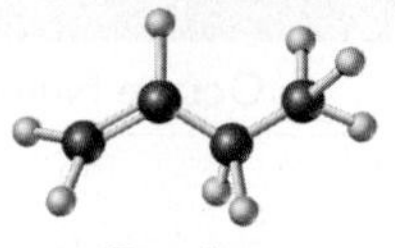 **Butylene**	Butadiene	Synthetic rubber, car bumpers, computer cases
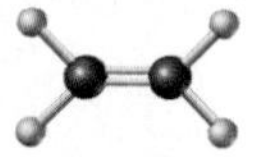 **Ethylene**	Ethyl alcohol, polyethylene, ethylene oxide	Plastics, pharmaceuticals, ethylene glycol, antifreeze, synthetic fibers
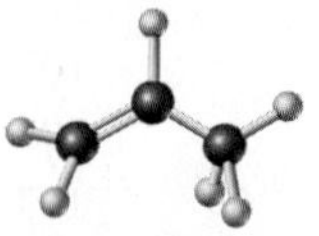 **Propylene**	Polypropylene	Plastics, pharmaceuticals, antifreeze, detergents, drycleaning fluid

10-1d Catalytic Reforming

The octane rating of gasoline can also be increased by **catalytic reforming**, which *converts straight-chain hydrocarbons to branched-chain hydrocarbons and aromatics.* In one type of catalytic reforming, certain catalysts, such as finely divided platinum on a support of Al_2O_3, enable the molecular structures of straight-chain hydrocarbons, which have low octane numbers, to change (re-form) into their branched-chain isomers, which have higher octane numbers.

$$CH_3CH_2CH_2CH_2CH_3 \xrightarrow{\text{catalyst}} CH_3\overset{\overset{\displaystyle CH_3}{|}}{C}HCH_2CH_3$$

pentane, 62 octane (C_5H_{12}) → 2-methylbutane, 94 octane (C_5H_{12})

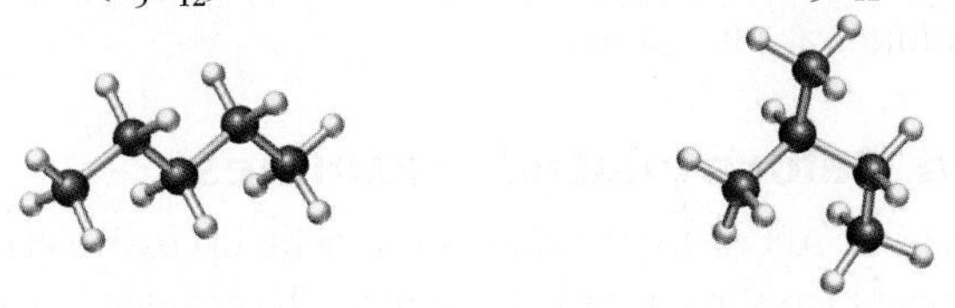

Catalytic reforming changes (re-forms) straight-chain hydrocarbon structures, converting them to branched-chain and aromatic structures that are more valuable.

By using different catalysts and petroleum mixtures, catalytic reforming also can produce aromatic hydrocarbons, which have very high octane numbers. For example, catalytic reforming converts a high percentage of gasoline, kerosene, and light oil fractions into a mixture of aromatic hydrocarbons including benzene, toluene, and xylenes, all of which have octane numbers above 100. An example is conversion of hexane into benzene. The hydrogen produced by catalytic reforming is also important for synthesizing ammonia, our main source of nitrogen fertilizer (see Section 12-8).

The additional cost of higher-octane gasoline is due to the extra processing required to make higher-octane compounds.

$$CH_3CH_2CH_2CH_2CH_2CH_3 \xrightarrow{\text{catalyst}} C_6H_6 + 4\ H_2$$

hexane, 25 octane (C_6H_{14}) → benzene, 106 octane (C_6H_6)

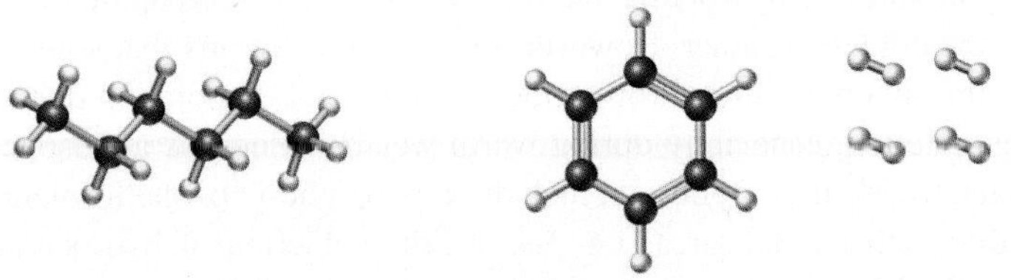

It is no exaggeration to state that our economy could not survive without petroleum and hydrocarbons. They provide fuels for transportation and raw materials to make the impressive array of consumer products shown in Table 10.2.

CONCEPTUAL EXERCISE 10.1

Rearranging Hydrocarbons

Heptane, C_7H_{16}, can be catalytically reformed to make toluene, $C_6H_5CH_3$, another seven-carbon molecule.

(a) How many hydrogen molecules are produced for every toluene molecule derived from heptane?
(b) Write a balanced chemical equation for this reaction.
(c) Why is it profitable to convert heptane into toluene?

Answers to **EXERCISES** are provided at the back of this book in Appendix L.

EXERCISES that are labeled **CONCEPTUAL** are designed to test your understanding of one or more concepts; they usually involve qualitative rather than quantitative thinking.

10-1e Octane Enhancers

The octane number of a given blend of gasoline can also be increased by adding octane enhancers to prevent autoignition. In the United States, prior to 1975, the most widely used anti-knock agent was tetraethyllead, $(C_2H_5)_4Pb$, but automobiles manufactured

Regrettably, many countries of the world still allow the use of leaded gasoline, putting millions of children at risk for lead poisoning from automobile exhaust fumes.

© Raymond Reuter/Sygma/Corbis

A catalytic converter. Platinum coats a honeycombed ceramic surface inside the catalytic converter. Reactions on the very large metal surface area remove pollutants from the exhaust gases.

since 1975 have been required to use lead-free gasoline. Because tetraethyllead can no longer be used in the United States and a few other countries, other octane enhancers are added to gasoline. These include toluene, 2-methyl-2-propanol (also called *tertiary*-butyl alcohol), methanol, and ethanol.

Unleaded gasoline has been required since 1975 because lead adversely affects catalytic converters. The exhaust emissions of internal combustion engines contain carbon monoxide, nitrogen oxides, and unburned hydrocarbons, all of which contribute to air pollution. To reduce urban air pollution, Congress passed the Clean Air Act of 1970, which required that 1975-model-year cars emit no more than 10% as much carbon monoxide and hydrocarbons as 1970 cars. The solution to lowering these emissions was a platinum-based catalytic converter, which accelerates conversion of carbon monoxide to carbon dioxide, more complete burning of hydrocarbons, and conversion of nitrogen oxides to N_2 and O_2. As little as two tanks of leaded gasoline can destroy the activity of a catalytic converter, so leaded gasoline was phased out.

10-1f Oxygenated and Reformulated Gasolines

The 1990 amendments to the Clean Air Act of 1970 require cities with excessive carbon monoxide pollution to use oxygenated gasoline during the winter. *Oxygenated gasoline* is gasoline to which organic compounds that contain oxygen, such as methanol, ethanol, and *tertiary*-butyl alcohol, have been added. Tests conducted by the U. S. Environmental Protection Agency (EPA) indicate that in cold weather, oxygenated gasoline burns more completely than nonoxygenated gasoline, thus potentially reducing carbon monoxide emissions in urban areas.

Oxygenated gasoline is produced by adding oxygen-containing organic compounds to refined gasoline. Reformulated gasoline, which contains such oxygenates, also requires changes in the refining process to alter its percentages of various hydrocarbons, particularly alkenes and aromatics.

Reformulated gasoline (RFG) is gasoline that is blended to burn more cleanly and has a lower volatility than ordinary gasoline. Reformulated gasoline is used in 18 states and the District of Columbia. From 1990 to 2006, federal regulations required that RFG contain a minimum of 2.0% oxygen by mass. In 2006, in order to give U. S. petroleum refiners more flexibility in producing RFG, the U. S. EPA did away with the oxygen requirement nationwide.

All gasolines are highly volatile and have vapors that can be ignited, enabling cars to start even in the coldest weather. However, this volatility means that some hydrocarbons get into the atmosphere as a result of accidental spills and evaporation during normal fillings at the gas station, especially during warm weather. Because atmospheric hydrocarbons contribute to urban air pollution, including tropospheric ozone formation, especially in heavy-traffic metropolitan areas (← **Sec. 8-12b**) reduction of hydrocarbon emissions contributes to the improvements in air quality reported in Table 8.7. During the summer tropospheric ozone season (from June 1 to September 15) federal regulations require that the vapor pressure of RFG meet a limit of 9.0 psi. More stringent requirements (7.8 psi) may apply in areas not meeting federal air quality guidelines.

Psi (pounds per square inch) is a pressure unit (← Sec. 8-1).

EXERCISE 10.2

Percent Oxygen in Ethanol

Calculate the percent oxygen by mass in ethanol, CH_3CH_2OH. Explain how ethanol was used to meet the former 2.0% oxygen requirement for RFG.

Engineering improvements, such as emission control systems, have succeeded in decreasing the emissions of ozone-forming nitrogen oxides and hydrocarbons from automobiles per mile traveled. Today's cars typically emit 70% less nitrogen oxides and 80% to 90% less hydrocarbons over their lifetimes compared with automobiles produced 50 years ago. Nevertheless, tropospheric ozone levels remain high for two reasons as noted in Figure 10.3. A significant portion of ozone-forming hydrocarbons come from a small percentage of vehicles (1%) with faulty emission-control systems that emit more than 10 g hydrocarbons per mile.

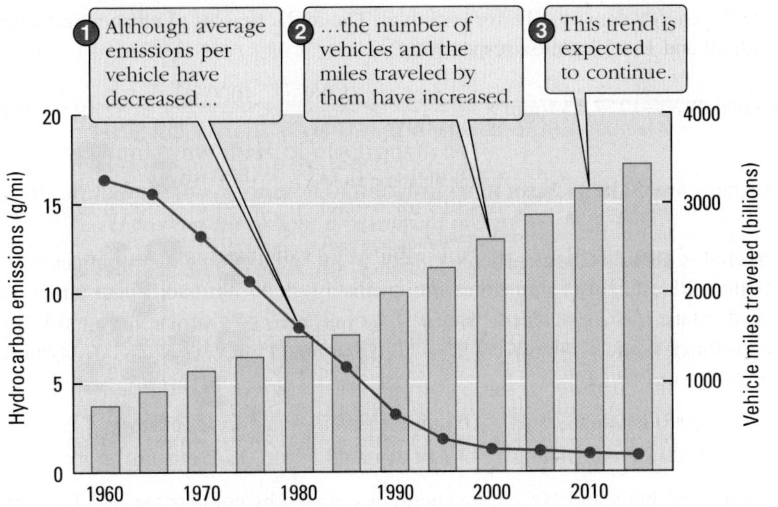
1 Although average emissions per vehicle have decreased...
2 ...the number of vehicles and the miles traveled by them have increased.
3 This trend is expected to continue.
Hydrocarbon emissions (g/mi)
Vehicle miles traveled (billions)
0
5
10
15
20
1000
2000
3000
4000
1960
1970
1980
1990
2000
2010

Figure 10.3 Average per-vehicle hydrocarbon emissions and mileage increases since 1960.

PROBLEM-SOLVING EXAMPLE 10.1

The Enthalpy of Combustion of Ethanol Compared to That of Octane

Calculate the standard combustion enthalpy of ethanol (kJ/mol) and compare the value with that of octane. Then, using the densities of the liquids, calculate the energy density (heat energy transfer per liter, ← **Sec. 4-11**) of each liquid fuel. The densities of octane and ethanol are 0.703 g/mL and 0.789 g/mL, respectively. The formation enthalpy of octane(g) is −208.447 kJ/mol.

Result

Ethanol: $\Delta_{comb}H° = -1277.4$ kJ/mol $= -2.19 \times 10^4$ kJ/L

Octane: $\Delta_{comb}H° = -5116.0$ kJ/mol $= -3.15 \times 10^4$ kJ/L

Analyze To calculate combustion enthalpy, use the balanced chemical equation, standard molar formation enthalpies for products and reactants from Appendix J, and Hess's law (← **Secs. 4-10, 4-9**). At the high temperatures of an internal combustion engine, all products and reactants are gases. Use calculated combustion enthalpies, molar masses, and densities to calculate energy densities in kJ/L.

Plan Write the balanced chemical equation for each combustion reaction. Apply the analysis given above.

Execute The balanced equation for the complete combustion of ethanol vapor is

$$C_2H_5OH(g) + 3\ O_2(g) \longrightarrow 2\ CO_2(g) + 3\ H_2O(g)$$

The enthalpy of combustion, $\Delta_{comb}H°$, for ethanol is

$$\Delta_{comb}H° = \{2[\Delta_f H°\ CO_2(g)] + 3[\Delta_f H°\ H_2O(g)]\} - \{1[\Delta_f H°\ C_2H_5OH(g)] + 3[\Delta_f H°\ O_2(g)]\}$$

$$= \{2(-393.509\text{ kJ/mol}) + 3(-241.818\text{ kJ/mol})\} - \{(-235.1\text{ kJ/mol}) + 0\}$$

$$= -1277.4\text{ kJ/mol}$$

The balanced combustion reaction for octane vapor is

$$C_8H_{18}(g) + \tfrac{25}{2}\ O_2(g) \longrightarrow 8\ CO_2(g) + 9\ H_2O(g)$$

$$\Delta_{comb}H° = \{8[\Delta_f H°\ CO_2(g)] + 9[\Delta_f H°\ H_2O(g)]\} - \{[\Delta_f H°\ C_8H_{18}(g)] + \tfrac{25}{2}\ [\Delta_f H°\ O_2(g)]\}$$

$$= \{8(-393.509\text{ kJ/mol}) + 9(-241.818\text{ kJ/mol})\} - \{(-208.447\text{ kJ/mol}) + 0)\}$$

$$= -5116.0\text{ kJ/mol}$$

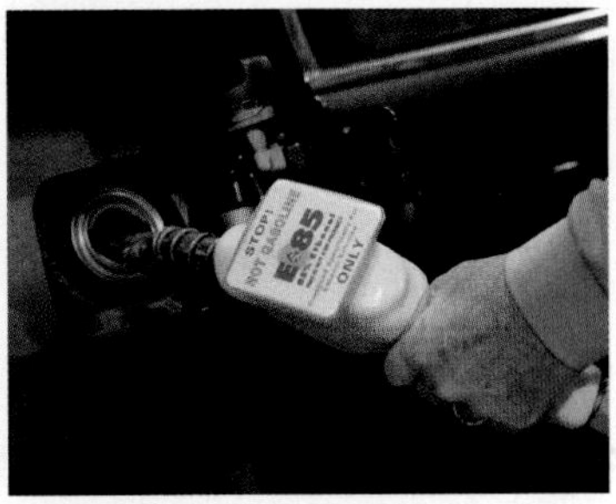
Carolina K. Smith MD/Shutterstock.com

Ethanol 85 (E85). E85 is a mixture of 85% ethanol and 15% gasoline by volume.

Calculate the energy density, ED, for each fuel. The molar masses of ethanol and octane are 46.069 g/mol and 114.23 g/mol, respectively.

$$\text{ED(ethanol)} = -1277.4 \text{ kJ/mol} \times \frac{1 \text{ mol}}{46.069 \text{ g}} \times \frac{0.789 \text{ g}}{\text{mL}} \times \frac{1000 \text{ mL}}{\text{L}} = -2.19 \times 10^4 \text{ kJ/L}$$

$$\text{ED(octane)} = -5116.0 \text{ kJ/mol} \times \frac{1 \text{ mol}}{114.23 \text{ g}} \times \frac{0.703 \text{ g}}{\text{mL}} \times \frac{1000 \text{ mL}}{\text{L}} = -3.15 \times 10^4 \text{ kJ/L}$$

☑ **Reasonable Result Check** There is about 17 mol ethanol in a liter of ethanol [(789 g ethanol/L ethanol)(1 mol ethanol/46 g ethanol) = 17 mol] and about 6 mol octane in a liter of octane [(703 g octane/L octane)(1 mol octane/114 g octane) = 6 mol]. But, nearly four times as much energy is liberated by burning a mole of octane. Applying this factor on a liter-to-liter basis:

$$\frac{5000 \text{ kJ/mol octane}}{1200 \text{ kJ/mol ethanol}} \times \frac{6 \text{ mol octane/L}}{17 \text{ mol ethanol/L}} = \frac{1.5 \text{ kJ/L octane}}{1.0 \text{ kJ/L ethanol}} = 1.5$$

The 1.5 indicates that about 50% more energy is released by combustion of 1 L octane than from 1 L ethanol. This is close to the 44% difference in the energy per liter calculated for the two fuels, $(3.15 \times 10^4/2.19 \times 10^4) \times 100\% = 1.44$.

PROBLEM-SOLVING PRACTICE answers are provided at the back of this book in Appendix K.

PROBLEM-SOLVING PRACTICE 10.1

Calculate the combustion enthalpy and the energy density of methanol using data from Appendix J. The density of methanol is 0.791 g/mL.

EXERCISE 10.3

Carbon Monoxide from Ethanol and from Toluene

(a) The combustion of C_2H_5OH can produce carbon monoxide and water. Use the balanced chemical equation for this reaction as the basis to calculate the mass in grams of CO produced per gram of ethanol burned.

(b) Repeat balancing the equation and the calculation this time using toluene, $C_6H_5CH_3$, instead of ethanol.

(c) From your answers, what conclusions can you draw about using ethanol to reduce carbon monoxide emissions?

10-2 U. S. Energy Sources and Consumption

As shown in Figure 10.4, the fossil fuels—coal, natural gas, and crude oil—are the predominant energy resource in the United States, accounting for approximately 78% of the total energy use. The major shift from 2007 to 2011 has been the increase in the percent of energy derived from natural gas compared with that from coal.

Figure 10.5 shows the sectors of our industrial society in which energy from all sources is used. In both 2007 and 2011, more than 90% of the energy was used to generate electrical power, for transportation, and in industry. From 2007 to 2011, patterns of energy use by sectors remained relatively unchanged.

10-2a Natural Gas

Natural gas is a mixture of hydrocarbons containing one to four carbon atoms and other gases trapped with petroleum in Earth's crust. In Europe and Japan, natural gas is essentially only methane. The natural gas found in North America is a mixture of C_1 to C_4 alkanes (70% to 90% methane, 0% to 20% ethane, propane, and butane, and 0% to 8% carbon dioxide) along with small and varying amounts of other gases, such as N_2, H_2S, and helium.

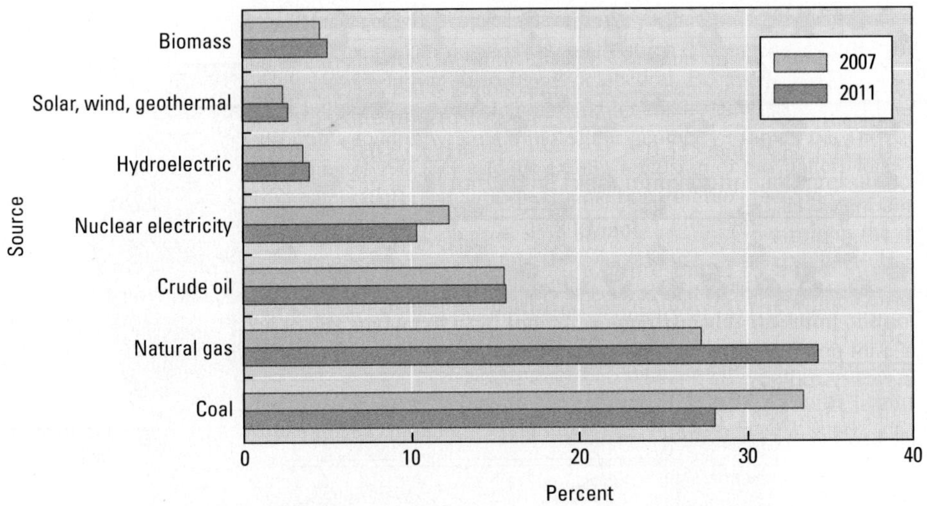
Biomass
Solar, wind, geothermal
Hydroelectric
Nuclear electricity
Crude oil
Natural gas
Coal
Source
2007
2011
0
10
20
30
40
Percent

Figure 10.4 U. S. energy resources 2007 and 2011. Total energy resource production: In 2007, 71.5 quadrillion kJ; in 2011, 83.3 quadrillion kJ (1 quadrillion kJ = 1.0×10^{15} kJ). (*Source:* Energy Information Administration, http://eia.doe.gov)

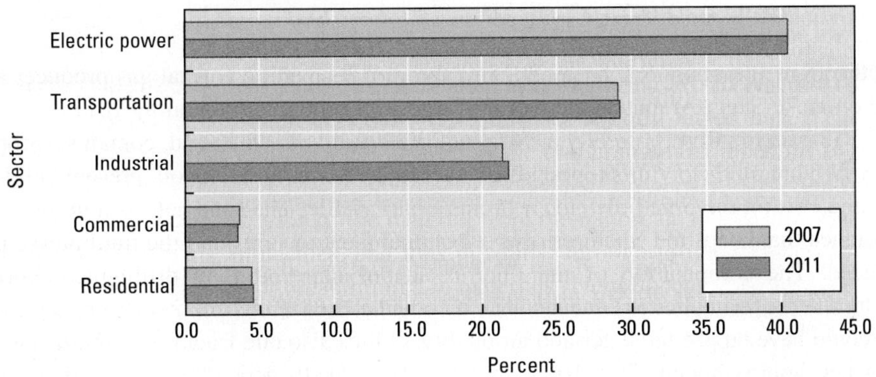
Electric power
Transportation
Industrial
Commercial
Residential
Sector
2007
2011
0.0
5.0
10.0
15.0
20.0
25.0
30.0
35.0
40.0
45.0
Percent

Figure 10.5 U. S. energy use by sector 2007 and 2011. (*Source:* Energy Information Administration, http://eia.doe.gov)

Until recently, natural gas was recovered mainly from vertically drilled oil wells and gas wells to which the liquid and gases migrated through the surrounding rock formation. In the late 1990s, a large-scale method known as *hydraulic fracturing* became economically feasible and more widely used to extract crude oil and natural gas from shale formations deep beneath the surface (1.5 to 6.0 km). **Hydraulic fracturing ("fracking")** *uses high pressure* (up to 100 MPa) *to force a fluid down a bored hole into the surrounding shale, fracturing it to release the natural gas and crude oil* (Figure 10.6). Because the deepest part of the drilling runs horizontally rather than vertically, more of the gas-bearing shale is exposed to the pressurized fluid, increasing the yield over that obtained from a conventional vertically drilled well. In the United States, three major shale deposits are currently being hydraulically fractured: the Barnett (Texas) and the Marcellus (Pennsylvania, Ohio, West Virginia) formations primarily yield natural gas; the Bakken (North Dakota) formation yields primarily crude oil. The U. S. Energy Information Administration (EIA) estimates that overall, 750 trillion cubic meters of natural gas and 24 billion barrels of crude oil are recoverable from these subterranean formations. The increased production of natural gas in the United States from hydraulic fracturing has been significant, from 1% in 2000 to more than 20% in 2010. The EIA predicts this percentage to increase to 46% by 2035. The International Energy Agency predicts that the United States, because of its enhanced production of natural gas and oil due to hydraulic

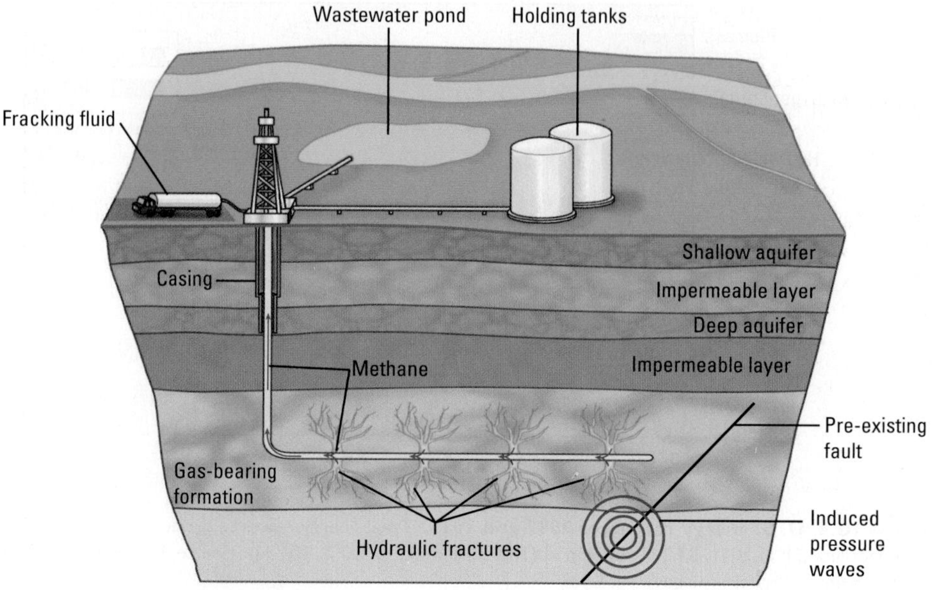
Wastewater pond
Holding tanks
Fracking fluid
Casing
Shallow aquifer
Impermeable layer
Deep aquifer
Impermeable layer
Methane
Pre-existing
fault
Gas-bearing
formation
Hydraulic fractures
Induced
pressure
waves

Figure 10.6 Hydraulic fracturing ("fracking").

fracturing, will surpass Russia by 2015 as the world's leading natural gas producer and will overtake Saudi Arabia as the world's largest oil producing nation by 2017.

The fracking fluid, typically a slurry of 90% water and 9.5% sand, contains a proprietary mixture of chemical compounds that act to decrease fluid friction, prevent build-up (scaling) within the bore hole, and serve as a bactericide. The sand acts as a *proppant*, a substance that keeps the fractures open after the injection stops and the fluid pressure is reduced. The composition of the fluid is altered depending on drilling conditions. Hydraulic fracturing has critics who are concerned about the environmental impact fracking could have on groundwater and air quality. Critics also cite fracking's significant use of water, approximately 5×10^6 L to 13×10^6 L/well, depending on the depth and geology of the well site; some of the water is reused. A 2011 study by a Massachusetts Institute of Technology (MIT) research team concluded, "The environmental impacts of shale development are challenging but manageable." ("The Future of Natural Gas: An Interdisciplinary MIT Study")

ESTIMATION | **Burning Oil**

A large oil refinery refines 400,000 barrels of crude oil daily. Assume that all the refined crude oil refined in a day is burned to produce electrical energy. Estimate how many toasters this amount of electrical energy would operate for a year assuming the conversion of thermal energy to electrical energy is 50% efficient (a typical value) and that each toaster uses 39 kWh (kilowatt hours) per year. 1.00 kWh = 3.60×10^3 kJ; the energy released by burning one barrel of crude oil is equivalent to 5.9×10^6 kJ.

The energy derived from burning the crude oil is approximately

$$E = (4 \times 10^5 \text{ barrels}) \times \left(\frac{6 \times 10^6 \text{ kJ}}{\text{barrel}}\right) = 2 \times 10^{12} \text{ kJ}$$

Assuming 50% efficiency in conversion from thermal to electrical energy and rounding values to one significant figure,

$$E_{\text{elec}} = (2 \times 10^{12} \text{ kJ}) \times \frac{50 \text{ kJ converted}}{100 \text{ kJ released}} \times \left(\frac{1 \text{ kWh}}{4 \times 10^3 \text{ kJ}}\right)$$

$$= 3 \times 10^8 \text{ kWh}$$

This quantity of electrical energy could operate almost 10 million toasters for a year.

$$\text{N(toasters)} = (3 \times 10^8 \text{ kWh}) \times \left(\frac{1 \text{ toaster}}{40 \text{ kWh}}\right) = 8 \times 10^6 \text{ toasters}$$

TOOLS OF CHEMISTRY | Gas Chromatography

Gas chromatography (GC) is *an important method to separate, identify, and determine the quantity of chemical compounds in complex mixtures.* Widely used in chemistry, biochemistry, environmental analysis, and forensic science, GC is applicable to compounds that have appreciable volatility and are thermally stable at temperatures up to several hundred degrees Celsius. We focus here on capillary column gas chromatography.

An analysis begins by injecting the sample into an inlet at the head of the GC column, where the sample is vaporized. A flow of inert carrier gas, usually helium, sweeps the vaporized sample along the column where the components are separated. The carrier gas stream containing the separated components passes through a detector, which allows identification of individual compounds in the gas stream. The resulting signal produces a *chromatogram*—a plot of concentration of each component versus time.

For capillary GC, the carrier gas usually flows at a rate in the range of 0.5 to 15 mL/min. A measured volume of sample (0.1 to 10 μL) is injected using a microsyringe. Capillary GC columns are made from specially purified silica tubes, typically 10 to 100 m long, that are drawn to have an inside diameter of 0.1 to 0.5 mm. The column is flexible, so it can be coiled and put inside an oven for temperature control. The inside of the column tube is coated with an immobilized liquid, the *stationary phase.* Many factors determine the coating used to separate a particular mixture, but ultimately the choice is dictated by the molecular structures of the stationary liquid phase material and the components being separated.

The detector must respond to very low concentrations of separated compounds in the flowing carrier gas. Two widely used detectors are: (1) flame-ionization detectors, which atomize and ionize the sample and measure the flow of current that results; and (2) thermal conductivity detectors, which measure the thermal conductivity of the gas stream. Mass spectrometers (← **Sec. 2-3**) are also used as detectors. The resulting chromatogram, a series of peaks, is a plot of the concentration of each component as a function of time. The position of each peak on the time (*x*) axis qualitatively identifies the components in the sample; the area under each peak provides a quantitative measure of the quantity of each component.

Compounds in a mixture are separated by GC due to their interactions with the stationary liquid phase on the column. Molecular structure dictates how quickly each component moves along the column to the detector. This depends on the magnitude of intermolecular attractions—London, dipole-dipole, and hydrogen bonding forces (← **Sec. 7-6**)—between molecules of each component of the mixture and of the stationary phase. Components having strong attractions for the stationary phase are retained longer on the column; those with weaker interactions reach the detector sooner. Thus, as the sample moves along the column, the components are progressively separated before they reach the detector.

If polar compounds are to be separated from a mixture, then a polar stationary phase works best, one that contains polar functional groups such as —CN, >C=O, and —OH. A nonpolar stationary phase contains hydrocarbon groups only. When the polarity of the stationary phase is matched well with that of the sample, the components exit from the column in order of their boiling points.

Capillary column GC can readily identify and measure components in gasoline. In the chromatogram below, 31 compounds ranging in size and complexity from butane and pentane to larger, ring compounds such as xylenes and toluene are separated from a gasoline sample.

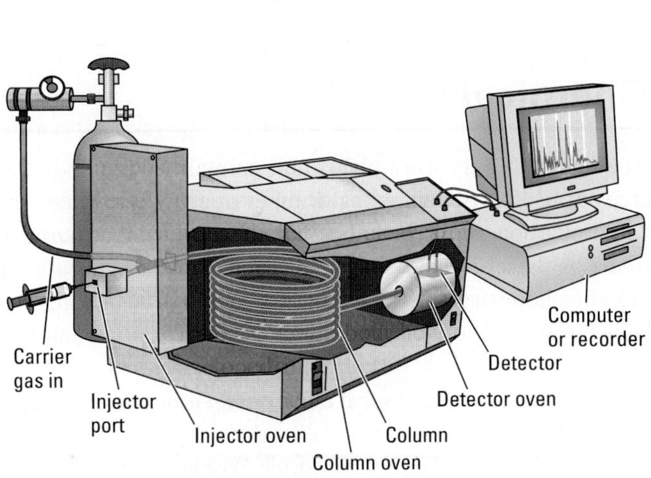

Carrier gas in
Injector port
Injector oven
Column
Column oven
Detector oven
Detector
Computer or recorder

Illustration of a gas chromatograph.

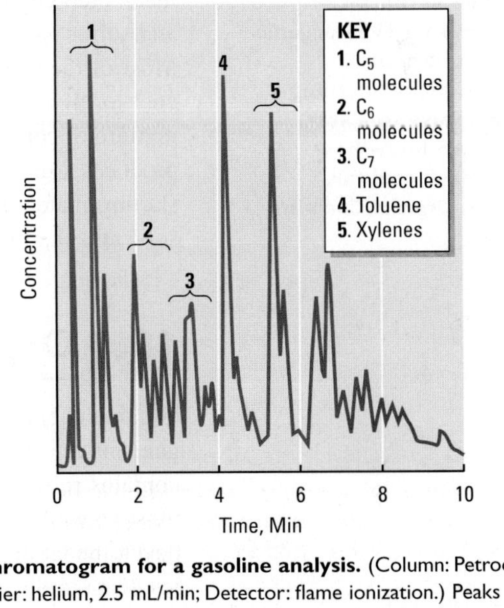

KEY
1. C_5 molecules
2. C_6 molecules
3. C_7 molecules
4. Toluene
5. Xylenes
1
2
3
4
5
Concentration
0
2
4
6
8
10
Time, Min

The chromatogram for a gasoline analysis. (Column: Petrocol A; Carrier: helium, 2.5 mL/min; Detector: flame ionization.) Peaks 1 are for components with five-carbon molecules, peaks 2 for six-carbon components, and peaks 3 for seven-carbon components. The last two peaks are for aromatic compounds: peak 4 indicates the presence of toluene, $C_6H_5CH_3$, and peak 5 the various xylenes, $C_6H_4(CH_3)_2$ (← **Sec. 6-11a**). These more complex aromatic hydrocarbons are held on the column longer than the nonaromatic hydrocarbons.

© Peter McGahey

This bus is powered by natural gas.

In the United States approximately 31% of natural gas is used industrially, 18% for home heating and cooling (approximately 50% of the homes), 12% in commercial applications, and 39% to produce electricity. Natural gas powers about 15 million vehicles worldwide, 112,000 in the United States. Test results indicate that vehicles powered by natural gas emit significantly less carbon monoxide, hydrocarbons, and nitrogen oxides per mile compared with gasoline-powered cars and trucks. Disadvantages of natural gas vehicles include the need for a pressurized gas tank and the lack of refueling stations that sell natural gas in liquefied form.

CONCEPTUAL EXERCISE 10.4

Natural Gas as a CO Source

The previous paragraph asserts that burning natural gas produces less carbon monoxide than burning gasoline. What factors need to be included in an analysis to determine whether this assertion is correct? Make appropriate assumptions and do calculations that either confirm or refute the assertion.

10-2b Coal

Known coal reserves are greater than petroleum reserves worldwide. The United States has 28% of the world's known coal reserves. Annually, about 90% of the U. S. coal production has been burned to produce electricity, although recently natural gas has been replacing coal at a significant rate due to the drop in natural gas prices. For example, on average, 39% of U. S. electricity generated in 2013 came from burning coal, down from 45% in 2009, and down significantly from the 50% average from 1990–2010. Burning coal is a major cause of air pollution because coal's 3–4% sulfur content leads to the production of atmospheric sulfur oxides; due to coal's high percentage of carbon, more CO_2 is produced per gram than for natural gas combustion.

The mass percent of carbon in coal varies from about 40% to more than 80%, depending on the type of coal.

Coal consists of a complex and irregular array of partially hydrogenated six-membered carbon rings and other structures, some of which contain oxygen, nitrogen, and sulfur atoms. Like petroleum, coal supplies raw materials to the chemical industry. Most of the useful compounds obtained from coal are aromatic hydrocarbons. Heating coal to high temperatures in the absence of air, a process called *pyrolysis,* produces a mixture of coke (mostly carbon), coal tar, and coal gas. Fractional distillation of coal tar produces aromatic hydrocarbons such as benzene, toluene, and xylene. They serve as the starting materials for a large variety of important commercial products such as paints, solvents, pharmaceuticals, plastics, dyes, and synthetic detergents.

The usefulness of coal tar to make other materials is summed up in this bit of doggerel from a 1905 organic chemistry textbook:

You can make anything
From a salve to a star
If you only know how
From black coal tar.
—Cornish, *Organic Chemistry,* 1905.

10-3 Organic Chemicals

Organic chemicals were once obtained only from plants and animals. A few living organisms are still direct sources of useful hydrocarbons. Rubber trees produce latex that contains rubber, a familiar hydrocarbon, and other plants produce an oil that burns almost as well in a diesel engine as diesel fuel. As important as these examples are, the development of synthetic organic chemistry has led to cheaper methods of making copies of naturally occurring substances and to many substances that have no counterpart in nature.

Prior to 1828, it was widely believed that chemical compounds found in living matter could not be made without living matter—a "vital force" was thought to be necessary for their synthesis. In 1828 a young German chemist, Friedrich Wöhler, contradicted the vital force theory when he prepared the organic compound urea, a major product in urine, by heating an aqueous solution of ammonium cyanate, a compound obtained from mineral sources. Soon thereafter, other chemists began to prepare more and more organic

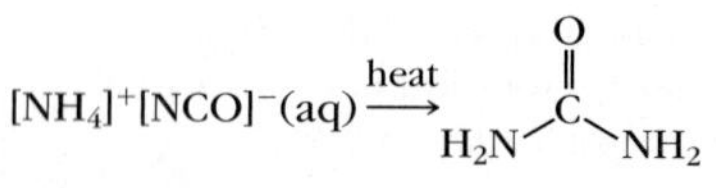
ammonium cyanate urea

chemicals in laboratory glassware—without involving an animal or plant for the synthesis.

Because about 85% of all known compounds are organic compounds, it is natural to ask: Why are there so many organic compounds, almost all of which contain carbon and hydrogen atoms (and commonly other kinds of atoms as well)? As discussed in earlier sections on bonding and isomerism (**← Secs. 6-3, 6-5**), two reasons are: (1) the ability of as many as thousands of carbon atoms to be linked to each other in a single molecule by stable C—C bonds, and (2) the occurrence of structural isomers. A third reason—the variety of functional groups that bond to carbon atoms—is further illustrated in this chapter. A **functional group** is *a distinctive grouping of atoms that, as part of an organic molecule, imparts specific properties to the molecule*. In addition to these three reasons, many molecules also display optical isomerism (**← Sec. 7-2f**), a very subtle type of molecular isomerism related to "handedness."

A table of functional groups and further information on how organic compounds are named is given in Appendix E.

All organic molecules can be viewed as derived from hydrocarbons, and many of them are prepared by chemists starting with hydrocarbons. Among the most useful of these compounds are alcohols, carboxylic acids, esters, amines, and the natural and synthetic polymers that can be made from them.

The alcohol, aldehyde, ketone, carboxylic acid, ester, amine, and amide functional groups are discussed in Sections 10-4 to 10-7.

10-4 Alcohols and Their Oxidation Products

Alcohols are a major class of organic compounds. Every **alcohol**, whether natural or synthetic, *contains the characteristic —OH functional group bonded directly to a carbon atom*. Some alcohols contain more than one —OH group. Examples of commercially important alcohols and their uses are listed in Table 10.3.

10-4a Methanol

Methanol, CH_3OH, the simplest of all alcohols, has just one carbon atom. Because methanol is so useful in making other products, more than 8 billion pounds of it are produced annually in the United States by the reaction of carbon monoxide with hydrogen in the presence of a catalyst at 250 °C.

$$CO(g) + 2\,H_2(g) \xrightarrow{\text{catalyst, 250 °C}} CH_3OH(g)$$

Methanol is sometimes called *wood alcohol* because the old method of producing it was by heating a hardwood such as beech, hickory, maple, or birch in the absence of air. Methanol is highly toxic. Drinking as little as 30 mL can cause death, and smaller quantities (10 to 15 mL) can cause blindness.

About 40% of methanol is used in the production of formaldehyde, HCHO, which is used to make plastics, embalming fluid, germicides, and fungicides, as well as other chemicals. Methanol is also used in jet fuels, antifreeze mixtures, and gasoline additives (**← Sec. 10-1e**).

Since methanol is relatively cheap, its potential as a fuel and as a starting material for the synthesis of other chemicals is receiving more attention. The technology for methanol-powered vehicles has existed for many years, particularly for racing cars that burn methanol because of its high octane rating (107). Methanol has both advantages and disadvantages as a gasoline replacement. As a motor fuel, methanol burns more cleanly than gasoline, but methanol has only about half the energy density of the same volume of gasoline. Therefore, twice as much methanol must be burned to give the same distance per tankful as gasoline. This disadvantage is partially compensated for by the fact that methanol costs about half as much to produce as gasoline.

Bernanrd Asset/Vandystadt/ Science Source

Some racing cars burn methanol.

Table 10.3 Some Important Alcohols

Condensed Formula	b.p. (°C)	Systematic Name	Common Name	Use
CH_3OH	65.0	Methanol	Methyl alcohol	Fuel, gasoline additive, making formaldehyde
CH_3CH_2OH	78.5	Ethanol	Ethyl alcohol	Beverages, gasoline additive, solvent
$CH_3CH_2CH_2OH$	97.4	1-propanol	Propyl alcohol	Industrial solvent
CH_3CHCH_3 OH	82.4	2-propanol	Isopropyl alcohol	Rubbing alcohol
CH_2CH_2 OH OH	198	1,2-ethanediol	Ethylene glycol	Antifreeze
$CH_2—CH—CH_2$ OH OH OH	290	1,2,3-propanetriol	Glycerol (glycerin)	Moisturizer in foods

PROBLEM-SOLVING EXAMPLE 10.2

Methanol and MTBE

Methanol reacts with 2-methylpropene to yield methyl *tert*-butyl ether (MTBE), which was formerly used as an octane enhancer and in reformulated gasolines.

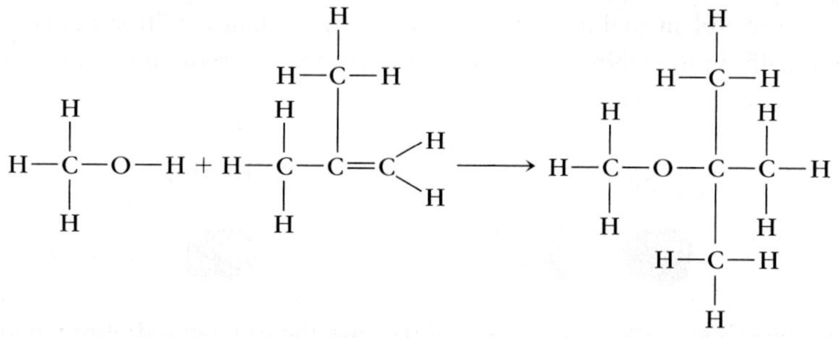

methanol 2-methylpropene methyl *tert*-butyl ether (MTBE)

Given the bond enthalpy data in the table, estimate the enthalpy change, $\Delta H°$, for the synthesis of MTBE. Is the reaction endothermic or exothermic?

Bond	Bond Enthalpy (kJ/mol)
C—H	416
C—O	336
O—H	467
C—C	356
C=C	598

Result −43.0 kJ/mol; exothermic.

Analyze During the reaction, bonds in the reactants are broken (endothermic process) and new bonds are formed in the products (exothermic). Bond enthalpies can be used to estimate the enthalpy change (← **Sec. 6-6b**).

Plan The chemical equation involving Lewis structures is balanced. Count the number of each kind of bond broken in the reactants and formed in the product. Multiply these numbers by the appropriate bond enthalpies, *D*, and sum them.

Execute

$$\Delta H° = \sum[(\text{number of bonds}) \times D(\text{bonds broken})] - \sum[(\text{number of bonds}) \times D(\text{bonds formed})]$$

$= \{[(3\ C{-}H)(416\ kJ/mol) + (1\ C{-}O)(336\ kJ/mol) + (1\ O{-}H)(467\ kJ/mol)] +$
$[(8\ C{-}H)(416\ kJ/mol) + (2\ C{-}C)(356\ kJ/mol) + (1\ C{=}C)(598\ kJ/mol)]\} -$
$\{[(12\ C{-}H)(416\ kJ/mol) + (2\ C{-}O)(336\ kJ/mol) + (3\ C{-}C)(356\ kJ/mol)]\}$
$= (6689\ kJ/mol) - (6732\ kJ/mol) = -43.0\ kJ/mol$

The negative sign indicates that the reaction is exothermic.

☑ **Reasonable Result Check** The net change in the reaction is the conversion of 1 C═C bond and 1 O—H bond into 1 C—H bond, 1 C—C bond, and 1 C—O bond. The other bonds are unchanged. The enthalpy change for this conversion is

$$\Delta H° = [(1\ C{=}C)(598\ kJ/mol) + (1\ O{-}H)(467\ kJ/mol)] - [(1\ C{-}H)(416\ kJ/mol) + (1\ C{-}C)(356\ kJ/mol) + (1\ C{-}O)(336\ kJ/mol)]$$

which equals −43.0 kJ/mol, the correct result.

PROBLEM-SOLVING PRACTICE 10.2

Using a table of average bond enthalpies, Table 6.2 (← **Sec. 6-6b**), estimate the enthalpy change for the industrial synthesis of methanol by the catalyzed reaction of carbon monoxide with hydrogen.

10-4b Ethanol

Ethanol, also called *ethyl alcohol* or *grain alcohol,* is the "alcohol" of alcoholic beverages. For millennia it has been prepared for this purpose by fermentation of carbohydrates (starch, sugars) from plant sources. For example, glucose is converted to ethanol and carbon dioxide by yeast in the absence of oxygen.

John Moore

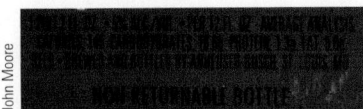

Alcohol concentration in a typical beer.

$$C_6H_{12}O_6 \xrightarrow{\text{Yeast}} 2\ C_2H_5OH + 2\ CO_2$$

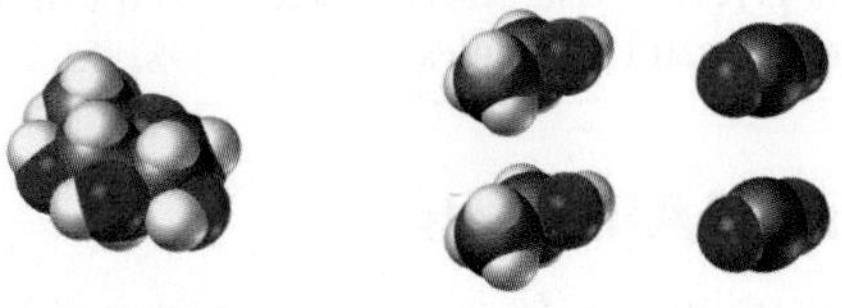

glucose ethanol

Fermentation eventually stops when the alcohol concentration reaches a level sufficient to inhibit the yeast cells. The "proof" of an alcoholic beverage is twice the volume percent of ethanol; 80-proof vodka, for example, contains 40% ethanol by volume. Ethanol is not as toxic as methanol, but one pint of pure ethanol, rapidly ingested, will kill most people. Ethanol is a depressant; the effects of various blood levels of alcohol are shown in Table 10.4. Rapid consumption of two 1-oz "shots" of 90-proof whiskey, two 12-oz beers, or two 4-oz glasses of wine can cause one's blood alcohol level to reach 0.05%.

In the United States, the federal tax on alcoholic beverages is $13.50 per gallon of 100-proof beverage, which is about 20 cents per ounce of ethanol in the beverage. Since the cost of producing ethanol is only about $1 per gallon, ethanol intended for industrial use must be *denatured* to avoid the beverage tax and to prevent people from drinking the alcohol. Ethanol is denatured by adding to it small amounts of a toxic substance, such as methanol or gasoline, that cannot be removed easily by chemical or physical means.

Ethanol, like methanol, is receiving increased attention as an alternative fuel. At present, most of it is used in a blend of 90% gasoline and 10% ethanol by mass (known as *gasohol* when introduced in the 1970s). Ethanol is also used as an oxygenated fuel to add to gasoline. The fermentation of corn, an abundant (but not limitless) source of carbohydrate, is used to produce ethanol for fuel use. The federal Renewable Fuel Standard (RFS), initiated in 2005 and expanded as part of the Energy Independence and Security Act of 2007, encourages development of biofuels, such as ethanol, to reduce dependence on foreign oil. The RFS mandates the production of 58 billion liters of corn-based

Table 10.4 Blood Alcohol Levels and Their Effects

% by Volume	Effect
0.05–0.15	Lack of coordination
0.08	Commonly defined point for "driving while intoxicated"
0.15–0.20	Obvious intoxication
0.30–0.40	Unconsciousness
0.50	Possible death

ethanol per year by 2022. Currently, 45% of the U. S. corn crop goes to this purpose. The corn from one acre of cropland can produce 400 gal of ethanol.

10-4c Hydrogen Bonding in Alcohols

Physical properties of liquid alcohols display the effects of hydrogen bonding among alcohol molecules. An alcohol structure can be considered as involving an alkyl group, such as CH_3— or CH_3CH_2—, substituted for one of the hydrogen atoms of water. The change from hydrogen to alkyl group reduces the number of hydrogen-bond donor sites and increases the size of the nonpolar, hydrocarbon part of the molecule. The boiling points listed in Table 10.3 for methanol, CH_3OH (65 °C), and ethanol, CH_3CH_2OH (78.5 °C), are lower than that of water, HOH (100 °C), because the alcohols have only one —OH hydrogen atom available for hydrogen bonding, but water has two. The higher boiling points of ethylene glycol (198 °C) and glycerol (290 °C) reflect the presence of two and three —OH groups per molecule. Ethanol is a liquid at room temperature, while propane, which has a similar number of electrons but no hydrogen bonding, is a gas. Alcohols with only a few carbon atoms are very water-soluble because of hydrogen bonding between water molecules and the —OH group.

CONCEPTUAL EXERCISE 10.5

Water Solubility of Alcohols

Predict the relative water solubility of these alcohols: CH_3OH; $C_5H_{11}OH$; and $C_{10}H_{21}OH$. Explain the reasons for your prediction.

CONCEPTUAL EXERCISE 10.6

One Alcohol in Another

Methanol dissolves in glycerol. Use Lewis structures to illustrate the hydrogen bonding between methanol and glycerol molecules.

10-4d Oxidation of Alcohols

Alcohols are classified according to the number of carbon atoms *directly* bonded to the —C—OH carbon. If there is no carbon atom or one carbon atom, the structure is a *primary* alcohol. With two carbon atoms, it is a *secondary* alcohol; with three, it is a *tertiary* alcohol. The use of R, R′, and R″ to represent alkyl groups indicates that the alkyl groups can be different.

In naming alcohols, the longest chain of carbon atoms is numbered so that the carbon with the —OH attached has the lowest possible number. A table of functional groups and more information on naming organic compounds is in Appendix E.

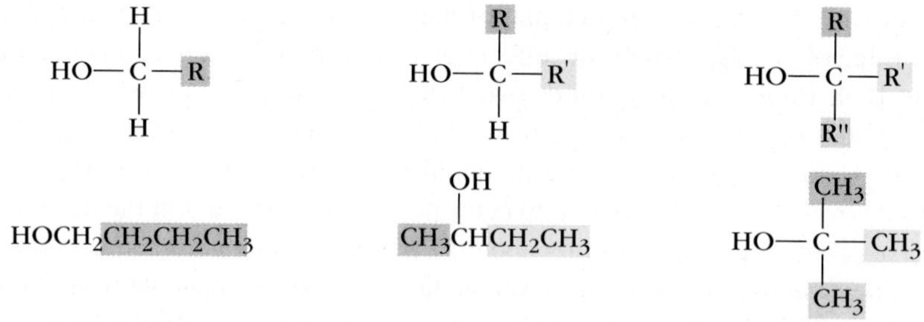

H
HO—C—R
H
R
HO—C—R'
H
R
HO—C—R'
R''
HOCH2CH2CH2CH3
OH
CH3CHCH2CH3
CH3
HO—C—CH3
CH3

1-butanol, a primary alcohol | **2-butanol, a secondary alcohol** | **2-methyl-2-propanol, a tertiary alcohol**

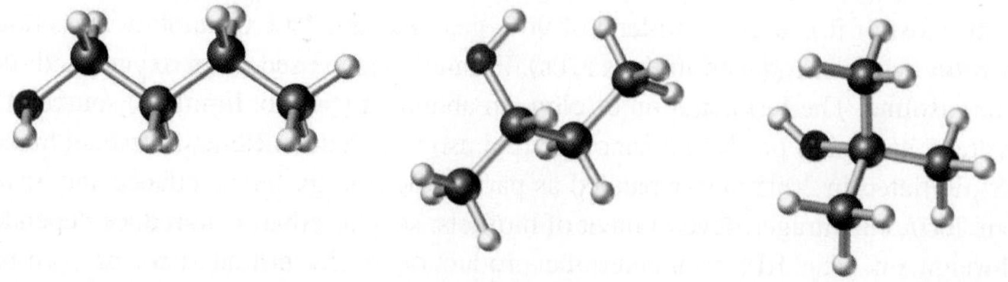

Controlled oxidation of primary alcohols initially produces compounds called aldehydes; aldehydes can be further oxidized to carboxylic acids.

$$\text{Primary alcohol} \xrightarrow{\text{oxidation}} \text{aldehyde} \xrightarrow{\text{oxidation}} \text{carboxylic acid}$$

Controlled oxidation of organic compounds usually involves either the loss of two hydrogen atoms or the gain of one oxygen atom. For example, stepwise oxidation of ethanol with an oxidizing agent such as aqueous potassium permanganate first produces acetaldehyde, where the number of H atoms is smaller by two than in ethanol. An **aldehyde** is *a substance that contains a —CHO functional group.*

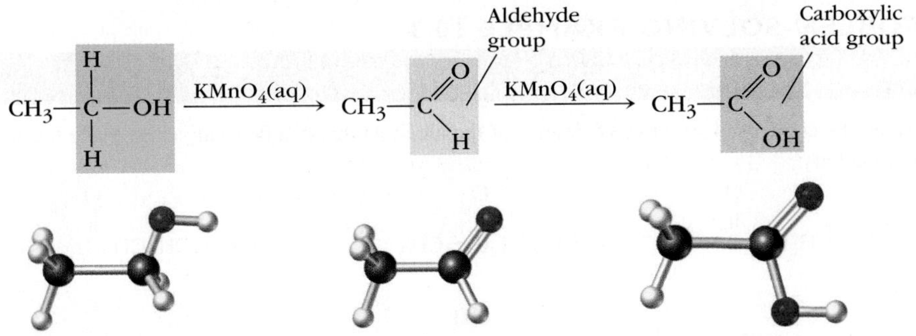
Aldehyde group
Carboxylic acid group
$KMnO_4(aq)$
$KMnO_4(aq)$

ethanol (a primary alcohol) acetaldehyde acetic acid

The acetaldehyde is then oxidized to acetic acid, a **carboxylic acid**, which *contains the —COOH functional group* (also represented as $—CO_2H$). When ethanol is ingested, enzymes in the liver produce the same products. Acetaldehyde, the intermediate product, contributes to the toxic effects of alcoholism. In the presence of oxygen in air, ethanol in wine is oxidized naturally to acetic acid, converting the wine from a beverage into vinegar—something more suitable for a salad dressing.

HEINZ
DISTILLED
WHITE
VINEGAR

Vinegar is 5% acetic acid in water.

CONCEPTUAL EXERCISE 10.7

Looking at the Oxidation of Primary Alcohols

Draw Lewis structures for ethanol and acetaldehyde. Describe the change in structure when ethanol is oxidized to acetaldehyde. Do the same for the oxidation of acetaldehyde to acetic acid. How do your descriptions differ?

Oxidation of a secondary alcohol involves loss of two hydrogen atoms and produces a **ketone**, which *contains the* $>C{=}O$ *functional group*. In the laboratory, an acidic solution of potassium dichromate, $K_2Cr_2O_7$, is a common oxidizing agent for this reaction. (Potassium permanganate would also work.)

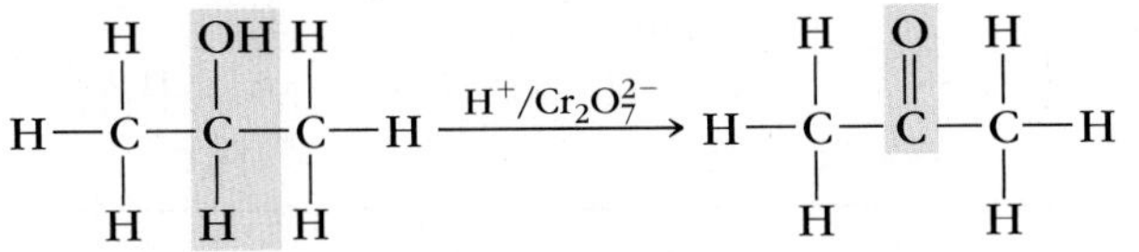
$H^+/Cr_2O_7^{2-}$

Secondary alcohol ⟶ ketone

Compound Class	Functional Group	Example
Aldehyde	O ‖ —CH	Formaldehyde O ‖ H—CH
Carboxylic acid	O ‖ —C—OH	Acetic acid O ‖ CH_3—C—OH
Ketone	O ‖ —C—	Acetone O ‖ CH_3—C—CH_3

Alcohols, aldehydes, carboxylic acids, and ketones are important biologically. Oxidation of a secondary alcohol to a ketone is important during strenuous exercise, when the lungs and circulatory system are unable to deliver sufficient oxygen to the muscles. Under these conditions, lactic acid builds up in muscle tissue, causing soreness and exhaustion. When you stop exercising, enzymes speed up oxidation of the secondary alcohol group in lactic acid to a ketone group, thereby converting the lactic acid to pyruvic acid. Pyruvic acid is an important biological compound that is further metabolized to provide energy.

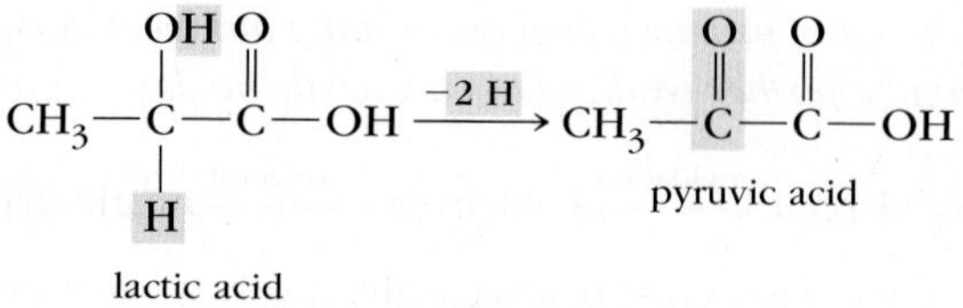
−2 H
lactic acid
pyruvic acid

Oxidation of organic compounds is usually the removal of two hydrogen atoms or the addition of one oxygen atom; reduction is usually the addition of two hydrogens or the removal of one oxygen.

Tertiary alcohols, which have no hydrogen atoms directly bonded to the carbon bearing the —OH group, are not oxidized easily.

PROBLEM-SOLVING EXAMPLE 10.3

Oxidation of Alcohols

Write the condensed structural formulas of the alcohols that can be oxidized to make these compounds:

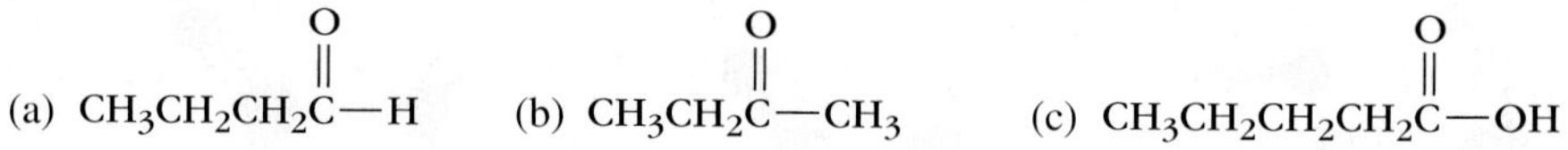

Result

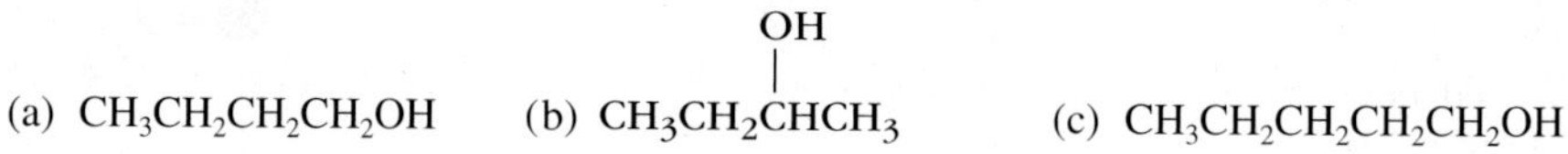
(a) $CH_3CH_2CH_2CH_2OH$
(b) $CH_3CH_2CHCH_3$
OH
(c) $CH_3CH_2CH_2CH_2CH_2OH$

Analyze Recognize in the structural formulas given whether the compound contains an aldehyde, a carboxylic acid, or a ketone functional group. Alcohols contain the —OH group. Primary alcohols are oxidized to aldehydes, which are then oxidized to carboxylic acids. Secondary alcohols are oxidized to ketones.

Plan An aldehyde or carboxylic acid group is formed when the —OH group is bonded to a carbon atom at the end of the molecule, that is, a primary alcohol. A ketone group is made when the —OH group is bonded to a carbon that is directly bonded to two other carbon atoms, that is, a secondary alcohol.

Execute

(a) The oxidation of the four-carbon primary alcohol, 1-butanol, will produce this aldehyde (see above).

(b) To produce this ketone, choose the secondary alcohol, 2-butanol, which has two carbons to the left and one carbon to the right of the C—OH group (see above).

(c) The oxidation of the five-carbon primary alcohol, 1-pentanol, will produce this carboxylic acid (see above).

PROBLEM-SOLVING PRACTICE 10.3

Draw the structural formula of the expected oxidation product of each compound.

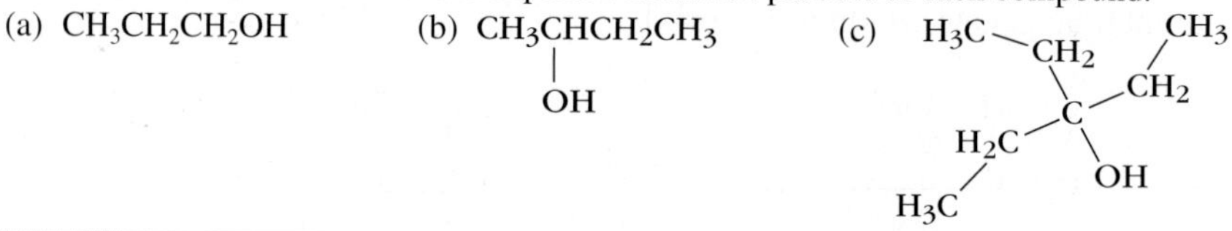
(a) $CH_3CH_2CH_2OH$
(b) $CH_3CHCH_2CH_3$
OH
(c)

CONCEPTUAL EXERCISE 10.8

Aldehydes and Combustion Products

Write the equation for the formation of an aldehyde by the oxidation of methanol. Critics of the use of methanol as a fuel or fuel oxygenate have cited the formation of this aldehyde as a major health threat. Use the Internet to find some of its toxic properties.

TOOLS OF CHEMISTRY | Nuclear Magnetic Resonance and Its Applications

Like the electron (← **Sec. 5-5d**), the nuclei of certain isotopes have quantized spin. For example, a ^{1}H nucleus (the hydrogen nucleus, a proton) can spin in either of two directions. In the absence of a magnetic field, the two spin states have the same energy. When a strong external magnetic field is applied, those ^{1}H nuclei with spins aligned with the external magnetic field are slightly lower in energy than those that are aligned against the magnetic field. The value of the energy difference, ΔE, is small enough that a photon of radio wave radiation (Figure 5.2, ← **Sec. 5-1**) has sufficient energy to change the direction of nuclear spin.

Consider the aligned hydrogen nuclei *absorbing* photons of a particular radio frequency ($h\nu$) that changes their spin to the less stable direction. When the nuclei return to the more stable spin direction, photons of the same radio frequency are *emitted* and can be measured with a radio receiver. This phenomenon, known as **nuclear magnetic resonance (NMR)**, provides an extremely valuable tool for studying molecular structure.

NMR is used extensively by chemists, especially organic chemists, because the radio frequency absorbed and then emitted depends on the chemical environment of the atoms with nuclear spin states in the sample. The study of hydrogen atoms by NMR (known as *proton* nuclear magnetic resonance, ^{1}H NMR) is an indispensable structural and analytical tool. Plots of the intensity of energy absorption versus the magnetic field strength are called *NMR spectra;* from them, chemists can deduce the kinds of atoms bonded to hydrogen as well as the number of hydrogen atoms present in a molecule.

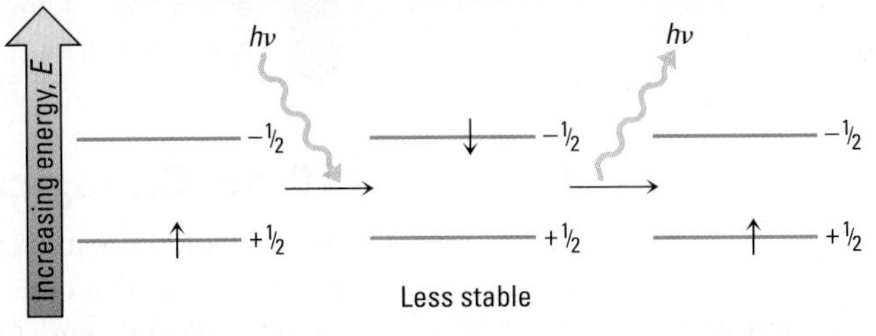

Increasing energy, E
hν
hν
−½
−½
−½
+½
+½
+½
Less stable

Absorption of radio frequency energy changes the direction and energy of spin of a proton. Emission of the energy returns the proton to its original spin.

There are more hydrogen atoms in the body than any other kind and H atoms give a strong NMR signal. Thus ^{1}H NMR can be used in medical imaging. MRI imaging is based on the time it takes for the protons in the unstable high-energy nuclear spin position to "relax" or return to the low-energy nuclear spin position. *Relaxation times* for protons in muscle, blood, and bone differ due to differences in the chemical environments. Based on these differences, a computer can produce a magnetic resonance image showing different tissues as different colors.

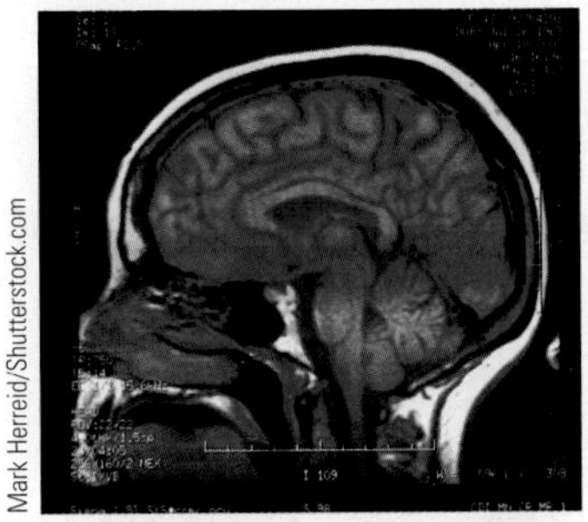

Mark Herreid/Shutterstock.com

A magnetic resonance imaging (MRI) scan of a normal human brain.

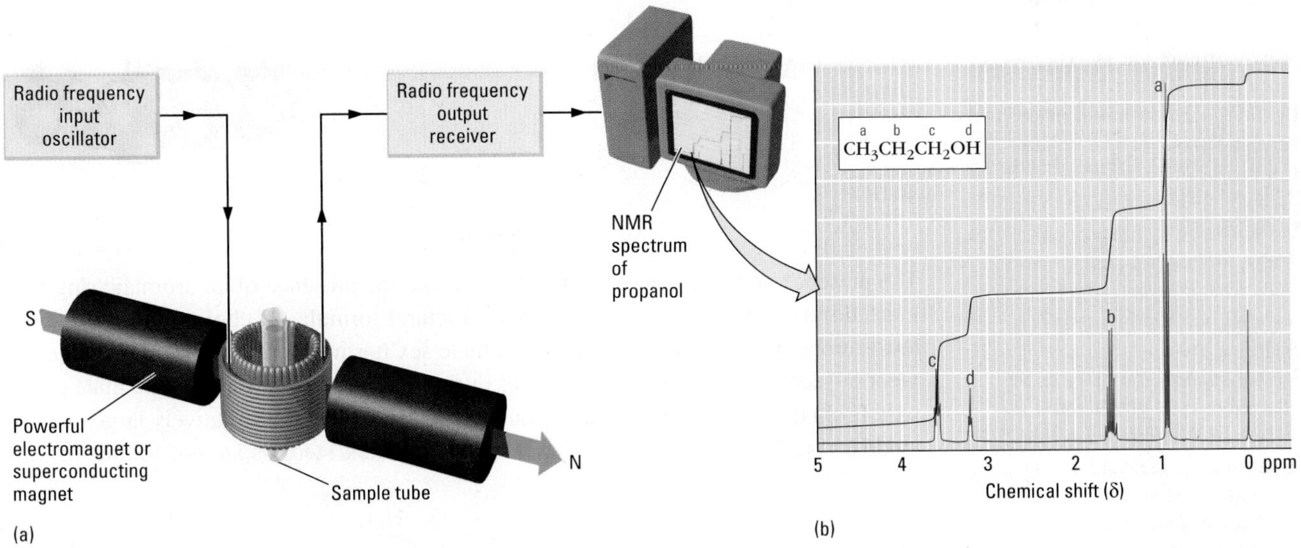

Radio frequency input oscillator
Radio frequency output receiver
NMR spectrum of propanol
S
N
Powerful electromagnet or superconducting magnet
Sample tube
(a)
a b c d
$CH_3CH_2CH_2OH$
a
b
c
d
5
4
3
2
1
0 ppm
Chemical shift (δ)
(b)

A modern NMR spectrometer. (a) Schematic diagram. (b) Spectrum of propanol. The spectrum shows four different types of protons (hydrogen nuclei)—a, b, c, d—present in different bonding environments in the molecule. The chemical shift (symbolized by δ and given in parts per million, ppm) is the position of the absorption relative to a defined zero point; chemical shifts depend on the strength of the magnetic field of the instrument and the other atoms near each proton.

CONCEPTUAL EXERCISE 10.9

Working Backward

Oxidation of an alcohol yields an aldehyde containing three carbon atoms. Further oxidation forms a three-carbon acid with the acid functional group on the number-one carbon of the chain. Write the structural formula and the name of the alcohol.

10-4e Biologically Important Alcohols

Cycloalkenes are nonaromatic ring compounds that contain at least one C=C double bond.

Many natural organic compounds are cyclic. They can be considered to be derivatives of hydrocarbons that consist of aromatic rings (← **Sec. 6-11**), cycloalkane rings (← **Sec. 6-3**), or cycloalkene rings. Often, two or more rings are fused, that is, the rings share carbon atoms. Steroids, all of which have the four-ring structure shown here, are an example of such fused-ring structures.

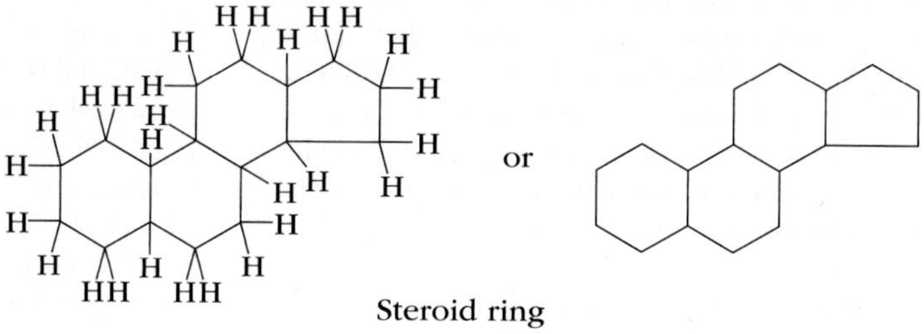
or

Steroid ring

Percy Lavon Julian
1899–1975

DePauw University Archives

After receiving a doctorate in chemistry in Vienna in 1931, Percy Julian spent 18 years as a research director in the chemical industry, and then directed his own research institute. He was granted more than 100 patents and was the first to synthesize hydrocortisone (a steroid) and to isolate from soybean oil the compounds used to make the first synthetic sex hormone (progesterone). These amazing accomplishments came from a person who had to leave his home in Montgomery, Alabama, after eighth grade for further studies—no more public education was available there for an African American man. Julian enrolled as a "subfreshman" at DePauw University in Indiana. On his first day, a white student welcomed him with a handshake. Julian later related his reaction: "In the shake of a hand my life was changed. I soon learned to smile and act like I believed they all liked me, whether they wanted to or not."

Many steroids—including cholesterol, estradiol and estrone (female sex hormones), and testosterone (male sex hormone)—contain alcohol functional groups. Cholesterol is the most abundant steroid in the human body, where it serves as a major component of cell membranes and as the starting point for the production of the steroid hormones. The elongated, rigid structure of cholesterol helps to make cell membranes sturdier. However, elevated levels of cholesterol in the blood are associated with *atherosclerosis,* a thickening of arterial walls that can lead to medical problems such as strokes and heart attacks.

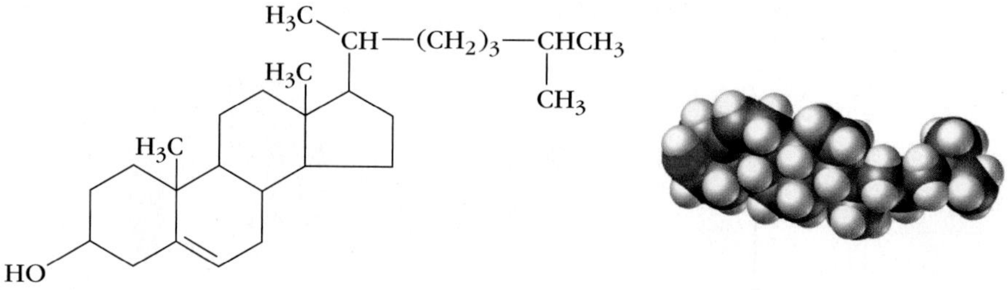
H3C
CH—(CH2)3—CHCH3
H3C
CH3
H3C
HO

cholesterol

The presence and position of the C=C bond and the presence of an aromatic ring distinguish different steroids, as seen from the structural formulas of cholesterol, estradiol, and testosterone. Estradiol and estrone are female sex hormones responsible for the development of secondary sexual characteristics during puberty; testosterone is the male counterpart. From their structural formulas you can see that steroids are relatively large molecules made up predominantly of carbon and hydrogen. Thus, steroids are not soluble in water.

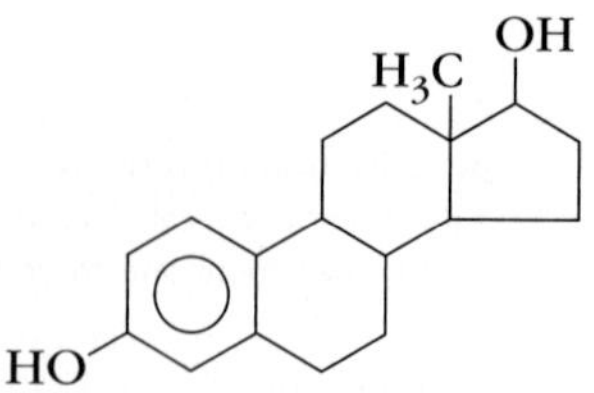
H3C
OH
HO

estradiol
(female hormone,
an estrogen)

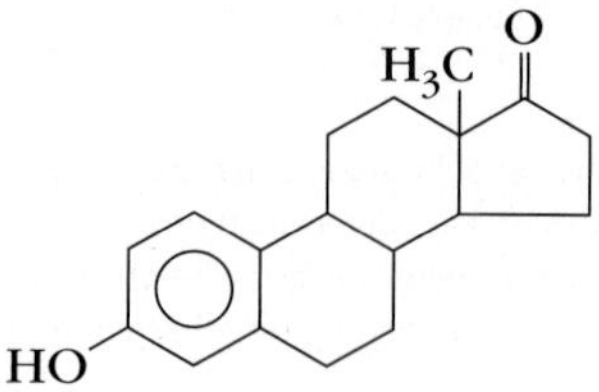
H3C
O
HO

estrone
(female hormone,
an estrogen)

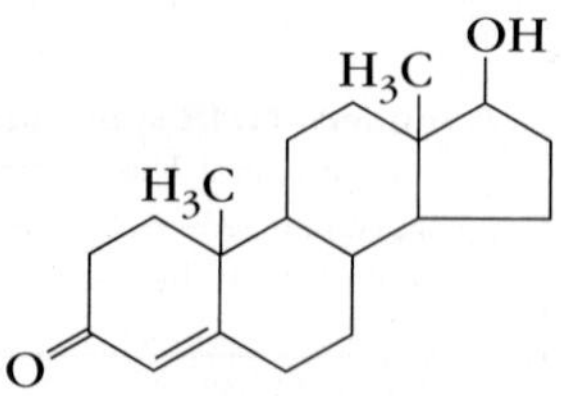
H3C
OH
H3C
O

testosterone
(male hormone)

CONCEPTUAL EXERCISE 10.10

A Closer Look

Look carefully at the structures for estradiol and estrone.

(a) What does the *-diol* suffix indicate in estradiol?

(b) What kind of alcohol group is on the five-membered ring in estradiol?

(c) What process could convert the alcohol group of the five-membered ring in estradiol to the functional group present in that same ring in estrone?

(d) Describe the differences in the molecular structures of the sex hormones testosterone (male) and estradiol (female).

EXERCISE 10.11

Performance-Enhancing Drugs

Sophisticated laboratory analyses have shown that cycling champion Lance Armstrong and other athletes used synthetic anabolic steroids to enhance their athletic performance. Androstenedione is such a steroid; its molecular structure is shown here. Compare this structure with the molecular structure of testosterone. In the body, what simple chemical process can convert testosterone to androstenedione?

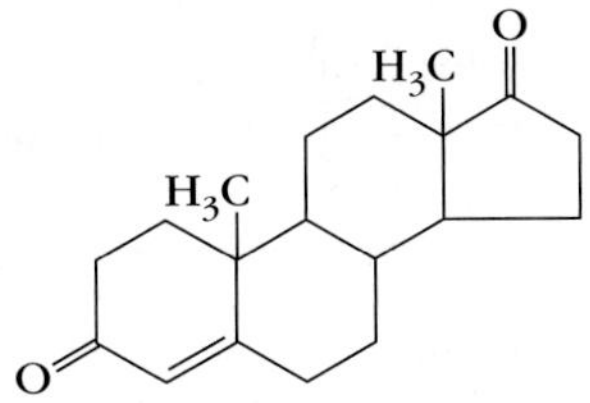
H3C
H3C
O
O

androstenedione

10-5 Carboxylic Acids and Esters

10-5a Carboxylic Acids

A **carboxylic acid** *contains the functional group* —C(=O)—OH *(—COOH or —CO_2H)* and can be prepared by the oxidation of an aldehyde or the complete oxidation of a primary alcohol (← **Sec. 10-4d**). All carboxylic acids react with bases to form salts; for example, sodium lactate is formed by the neutralization reaction (← **Sec. 3-4c**) of lactic acid and sodium hydroxide.

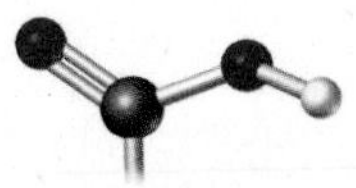
carboxylic
acid group

The carboxylic acid group is present in all organic acids. Acids, organic or inorganic, are an important class of compounds (← **Sec. 3-4**).

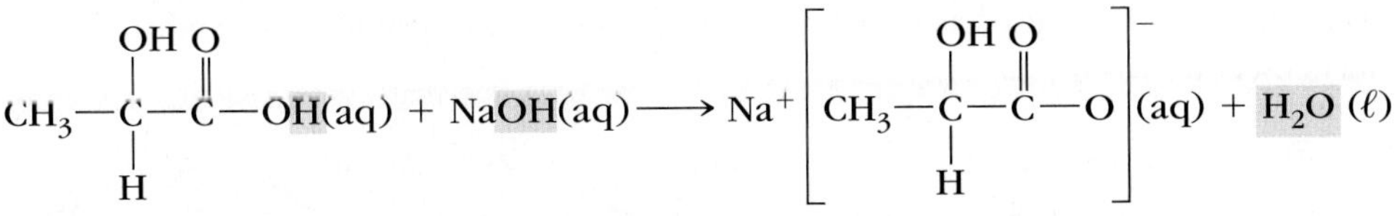
OH O
CH3—C—C—OH(aq) + NaOH(aq) ⟶ Na+ [CH3—C—C—O]− (aq) + H2O (ℓ)
H

lactic acid   sodium lactate

Carboxylic acid molecules are polar and form hydrogen bonds with each other. This hydrogen bonding results in relatively high boiling points for the acids—even higher than those of alcohols of comparable molecular size. For example, formic acid (46 g/mol) has a boiling point of 101 °C, while ethanol (46 g/mol) has a boiling point of only 78.5 °C. Both formic acid and ethanol form hydrogen bonds, but in formic acid more hydrogen bonds form.

Notice that the —OH part of the —COOH group has an acidic hydrogen because the —OH group is attached to a C=O group; the —OH group in an alcohol does not contain an acidic hydrogen.

EXERCISE 10.12

Hydrogen Bonding in Formic Acid

Based on the structural formula of formic acid in Table 10.5, draw as many ways as you can to form hydrogen bonds between formic acid molecules. Compare this with the number of ways hydrogen bonds can form between ethanol molecules.

A large number of carboxylic acids are found in nature and have been known for many years. As a result, some of the familiar carboxylic acids are almost always referred to by their common names (Table 10.5).

Carboxylic acids in foods. Citrus fruits contain citric acid as well as other naturally occurring acids. Carboxylic acids cause foods such as fruits, berries, and vinegar to have a tart, sour taste in addition to sweetness provided by sugars.

Table 10.5 Some Naturally Occurring Carboxylic Acids

Structure	Systematic Name	Common Name	b.p. (°C)	Natural Source
$HCOOH$ (H—C(=O)—OH)	Methanoic acid	Formic acid	101	Ants
CH_3COOH	Ethanoic acid	Acetic acid	118	Fermented fruit
CH_3CH_2COOH	Propanoic acid	Propionic acid	141	Dairy products
C_6H_5COOH	Benzoic acid	Benzoic acid	250	Berries
Structure		**Common Name**	**m.p. (°C)**	**Natural Source**
$HOOC—CH_2—C(OH)(COOH)—CH_2—COOH$		Citric acid	153	Citrus fruits
$HOOC—CH_2—CH(OH)—COOH$		Malic acid	131	Apples
$HOOC—CH(OH)—CH(OH)—COOH$		Tartaric acid	168–170	Grape juice, wine

In *Les Miserables*, Victor Hugo comments about naturally occurring acids when he writes, "Comrades, we will overthrow the government, as sure as there are fifteen acids intermediate between margaric acid and formic acid." Formic acid, the simplest organic acid, has just one carbon atom. Margaric acid, from which margarine is derived, has seventeen carbon atoms.

Three commercially important carboxylic acids—adipic acid, terephthalic acid, and phthalic acid—have two acid groups per molecule and are known as *dicarboxylic acids.* These three acids are used as starting materials to make vast quantities of synthetic polymers (Section 10-6).

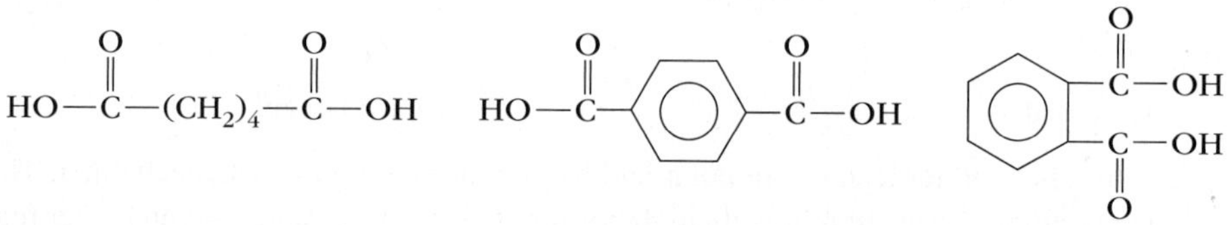

adipic acid terephthalic acid phthalic acid

10-5b Esters

Carboxylic acids react with alcohols in the presence of strong acids (such as sulfuric acid) to produce **esters**, which *contain the functional group*

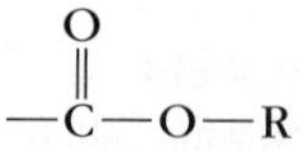

In an ester, the —OH of the carboxylic acid is replaced by the —OR group from the alcohol. The general equation for ester formation (esterification) is

Esterification is catalyzed by the presence of an acid (H^+ donor).

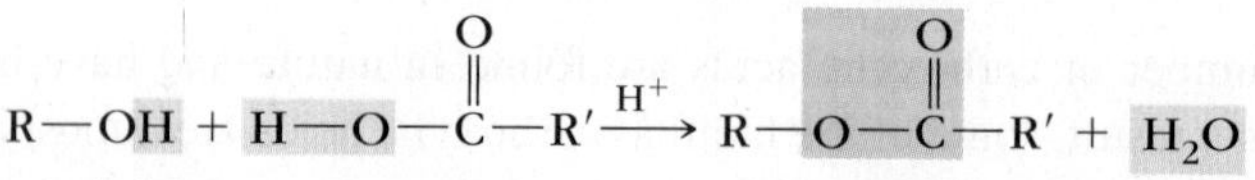

alcohol + organic acid ester + water

Esterification is an example of a condensation reaction. In a **condensation reaction**, *two molecules combine to form a larger molecule* (in esterification it is the ester) *while simultaneously producing a small molecule* (water in this case). For example, in 1897, Felix Hoffmann, a chemist at the Bayer Chemical Company in Germany, synthesized aspirin, the world's most common pain reliever, by esterifying an alcohol group on salicylic acid with acetic acid in strong acid solution to form acetylsalicylic acid, which is aspirin. The ester group makes aspirin less irritating to the stomach lining than salicylic acid. The carboxylic acid group originally in salicylic acid remains intact in aspirin.

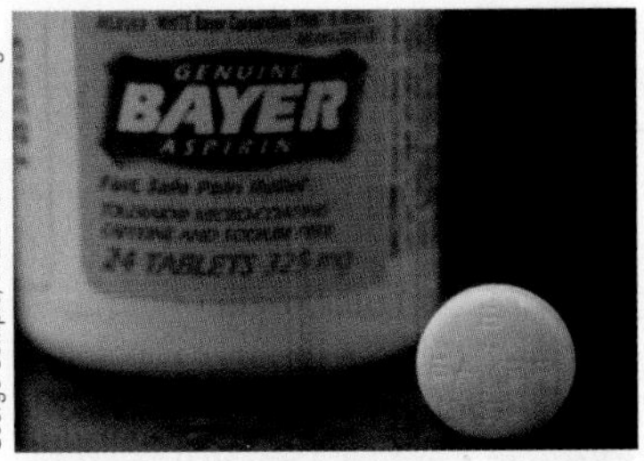
GENUINE
BAYER
ASPIRIN
24 TABLETS 325 mg
George Semple/Wadsworth Publishing

Aspirin (acetylsalicylic acid). The systematic name for aspirin is 2-(acetyloxy)-benzoic acid.

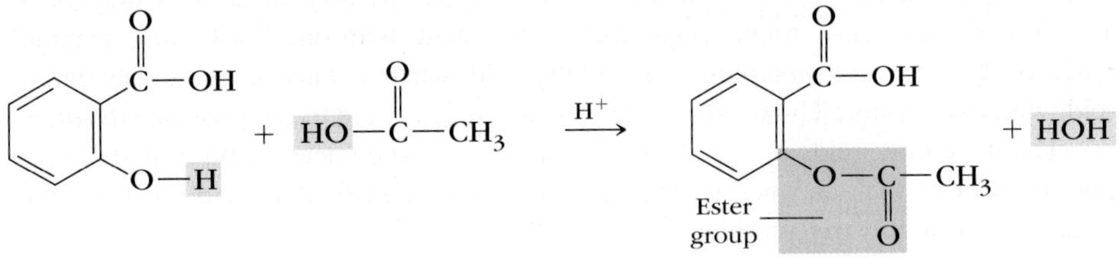
H+
Ester group
+ HOH

salicylic acid — acetic acid — aspirin (acetylsalicylic acid) — water

10-5c Triglycerides: Biologically Important Esters

One benefit of understanding organic molecules is in applying this understanding to the molecules involved in life—organic molecules that we eat, that provide the structure of our bodies, and that allow our bodies to function. Fats and oils are such molecules. *Fats* are solids at room temperature; *oils* are liquids at this temperature. Edible fats and oils are all **triglycerides** because they *share the common structural feature of a three-carbon backbone from glycerol to which three long-chain fatty acids are bonded*. Fatty acids usually have an even number of carbon atoms, ranging from 4 to 20. Triglycerides are formed by the esterification reaction of three moles of fatty acids with one mole of glycerol.

The discussion in this section answers the questions posed in Chapter 1 (← Sec. 1-1), "What is the difference between a saturated fat, an unsaturated fat, and a polyunsaturated fat? Why are trans fats harmful?"

Glycerol + 3 fatty acid ⟶ triglyceride + 3 H_2O

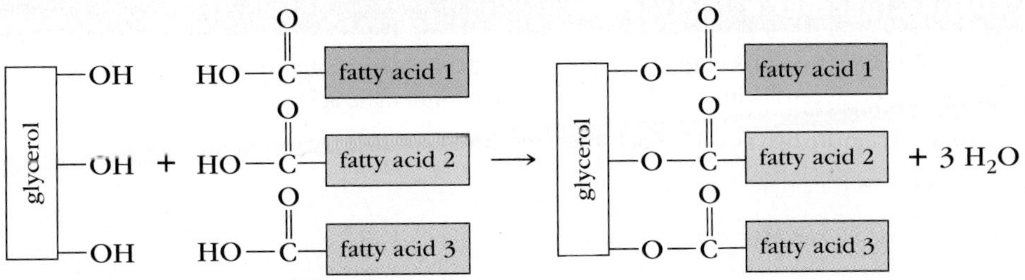
glycerol
fatty acid 1
fatty acid 2
fatty acid 3
glycerol
fatty acid 1
fatty acid 2
fatty acid 3
+ 3 H2O

For example, one glycerol molecule reacts with three molecules of stearic acid to produce tristearin, a very common animal fat, as shown here.

Glycerol + 3 stearic acid ⟶ tristearin + 3 H_2O

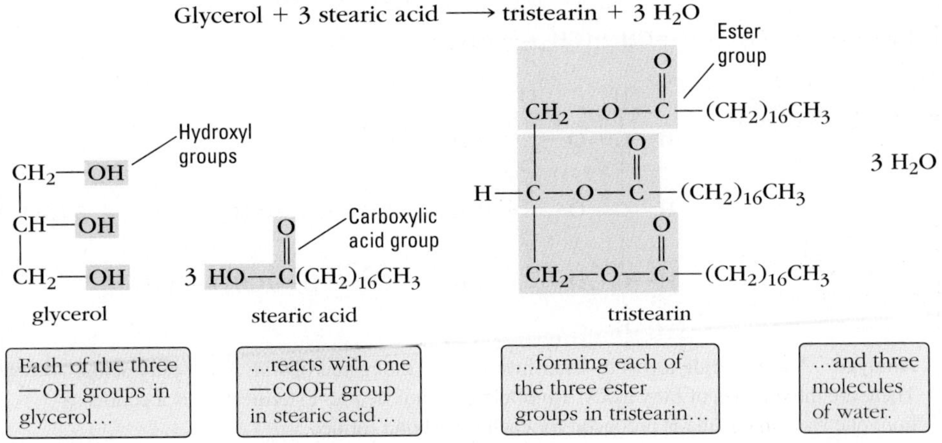
Ester group
Hydroxyl groups
Carboxylic acid group
3 H2O
glycerol
stearic acid
tristearin
Each of the three —OH groups in glycerol…
…reacts with one —COOH group in stearic acid…
…forming each of the three ester groups in tristearin…
…and three molecules of water.

FILIPPO BERIO
OLIVE OIL
Corn Oil
WESSON
CANOLA OIL
Peanut Oil

Charles D. Winters

Cooking oils are liquids and fats are solids at room temperature. Liquid oils are unsaturated; solid fats are saturated structures.

Stearic acid is representative of fatty acids, most of which are molecules with long chains of carbon atoms with their attached hydrogen atoms; in stearic acid there are 18 carbon atoms and 36 hydrogen atoms. At one end of the carbon chain is the —COOH group characteristic of all organic acids.

Animal and vegetable fats and oils vary considerably because their fatty acids differ in the length of the hydrocarbon chain and the number of double bonds in the chain. **Saturated fats**, such as tristearin, *contain only single C—C bonds in their hydrocarbon chains* (along with C—H bonds); they are usually solids. **Unsaturated fats** *contain one or more C═C double bonds in their hydrocarbon chains;* they are usually liquids, but may solidify just below room temperature. Oleic acid, with one C═C bond per molecule, is classified as a *monounsaturated* fatty acid, whereas fatty acids such as linoleic acid, with two or more C═C double bonds per molecule, are termed *polyunsaturated.*

The three fatty acids in a triglyceride can be the same (stearic acid in tristearin) or can be different such as when stearic acid, oleic acid, and linoleic acid condense with glycerol to form this triglyceride.

Saturated fats: No C═C

Monounsaturated fats: One C═C bond per molecule

Polyunsaturated fats: Several C═C bonds per molecule

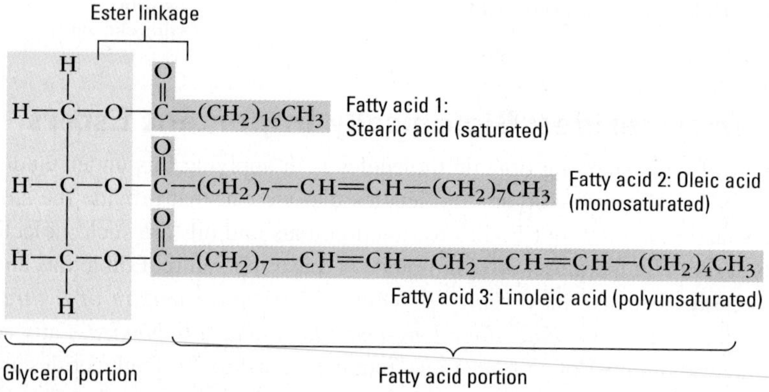
Ester linkage
Fatty acid 1: Stearic acid (saturated)
Fatty acid 2: Oleic acid (monosaturated)
Fatty acid 3: Linoleic acid (polyunsaturated)
Glycerol portion
Fatty acid portion

PROBLEM-SOLVING EXAMPLE 10.4

Putting Together a Triglyceride

Using structural formulas, write the equation for the formation of a triglyceride by the reaction of 1 mol glycerol with 1 mol stearic acid and 2 mol oleic acid.

Result There are two possible final structures; here is one.

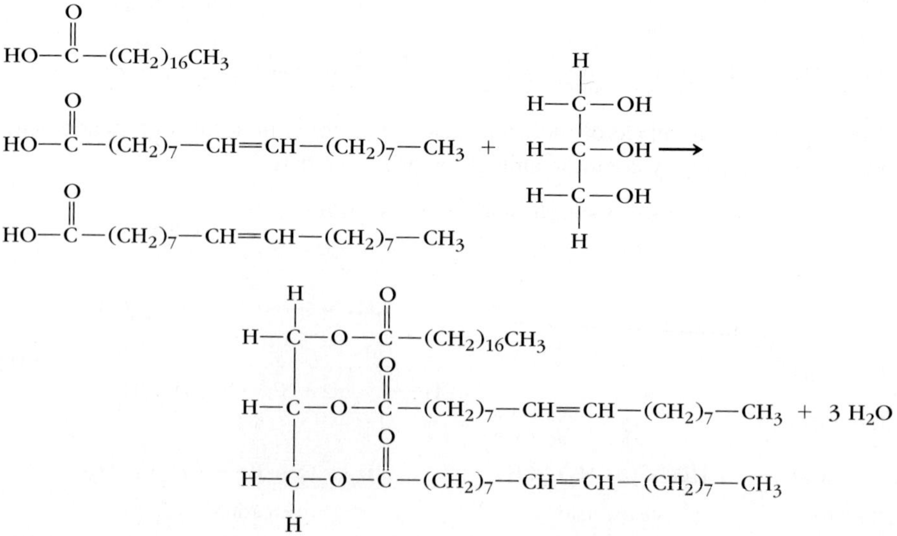

Analyze A triglyceride forms by esterification of 1 mol glycerol with 3 mol fatty acids. There are three moles of fatty acids: 2 mol oleic; 1 mol stearic. Esterification is a condensation: one water molecule is produced for each ester bond formed.

Plan Combine the structural formulas for the acids and glycerol to form three ester linkages and three water molecules, one for each ester bond.

Execute Stearic acid could condense with an —OH group on the end of the three-carbon glycerol chain or with an —OH group in the middle. The other two positions would be occupied by oleic acid. The Result depicts the first possibility.

PROBLEM-SOLVING PRACTICE 10.4

Describe the structure of the other triglyceride that could be obtained by reacting of 1 mol glycerol with 2 mol oleic acid and 1 mol stearic acid.

While we need fats in our diet, diets consisting of moderate amounts of fats and oils containing mono- and polyunsaturated fatty acids are considered better for good health than diets containing only saturated fats. In spite of this fact, there is a demand for solid or semisolid fats because of their texture and spreadability. Figure 10.7 indicates the relative proportions of saturated, monounsaturated, and polyunsaturated fats in some familiar cooking oils and solids.

The American Heart Association recommends that no more than 25% to 35% of dietary calories come from fat. Fats account for almost 40% of the calories in the average American's diet.

As seen from Figure 10.7, most dietary vegetable oils are high in unsaturated fatty acids, while most dietary animal fats contain mainly saturated fatty acids. In general, the greater the percentage of unsaturated fatty acids in a triglyceride, the lower its melting point. Thus, at room temperature, highly unsaturated fats are liquids (oils), whereas animal fats high in saturated fats are solids. The melting points of fats reflect the shape of the molecules. The C—C single bonds in the hydrocarbon portion of saturated fats are all the same, and they allow the molecule to adopt a rather linear shape, which fits nicely into a solid packing arrangement. The C═C double bonds in unsaturated fats create a different shape. Most natural unsaturated fats are in the *cis* configuration at the C═C double bond (← **Sec. 6-5a**), which puts a "kink" into the molecule, preventing *cis* unsaturated fat molecules from packing regularly into a solid. *Trans* fatty acid molecules are more linear, similar to those of saturated fats, raising health concerns about the use of *trans* fats in processed foods.

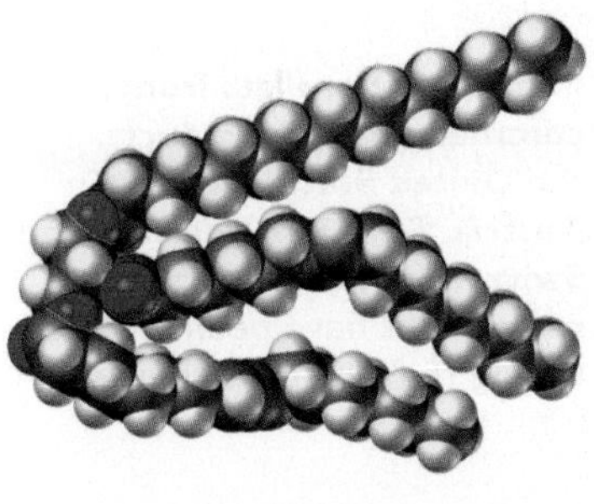

Triglyceride

A triglyceride. The fatty acids are (top) saturated, (middle) *cis*-monounsaturated, and (bottom) polyunsaturated. Note the "kink" in the middle structure due to the *cis* double bond.

Vegetable oils can be converted to semisolids by *hydrogenation*—reaction of the oil with H_2 in the presence of a metal catalyst such as palladium, which accelerates the reaction. The hydrogen reacts by addition to some of the C═C double bonds, converting

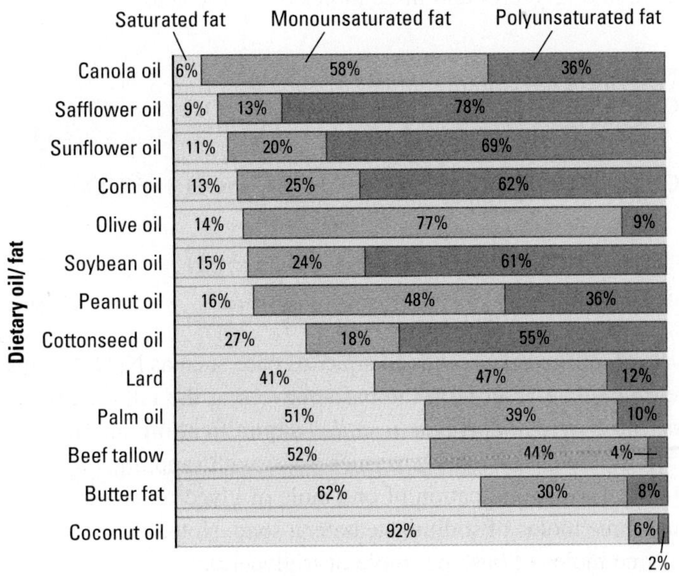

Saturated fat
Monounsaturated fat
Polyunsaturated fat
Dietary oil/fat
Canola oil 6% 58% 36%
Safflower oil 9% 13% 78%
Sunflower oil 11% 20% 69%
Corn oil 13% 25% 62%
Olive oil 14% 77% 9%
Soybean oil 15% 24% 61%
Peanut oil 16% 48% 36%
Cottonseed oil 27% 18% 55%
Lard 41% 47% 12%
Palm oil 51% 39% 10%
Beef tallow 52% 44% 4%
Butter fat 62% 30% 8%
Coconut oil 92% 6% 2%

Figure 10.7 Relative proportions of saturated, monounsaturated, and polyunsaturated fats in cooking oils and solid fats.

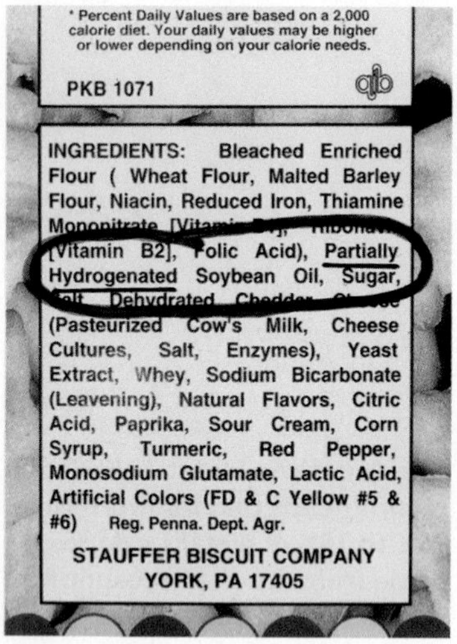
* Percent Daily Values are based on a 2,000 calorie diet. Your daily values may be higher or lower depending on your calorie needs.
PKB 1071
INGREDIENTS: Bleached Enriched Flour (Wheat Flour, Malted Barley Flour, Niacin, Reduced Iron, Thiamine
Vitamin B2], Folic Acid), Partially Hydrogenated Soybean Oil, Sugar,
(Pasteurized Cow's Milk, Cheese Cultures, Salt, Enzymes), Yeast Extract, Whey, Sodium Bicarbonate (Leavening), Natural Flavors, Citric Acid, Paprika, Sour Cream, Corn Syrup, Turmeric, Red Pepper, Monosodium Glutamate, Lactic Acid, Artificial Colors (FD & C Yellow #5 & #6) Reg. Penna. Dept. Agr.
STAUFFER BISCUIT COMPANY
YORK, PA 17405

George Semple

Snack food label. Most snack foods contain partially hydrogenated fats.

them to C—C single bonds, as shown below. Margarine, shortening (for example, Crisco®), and many snack foods contain partially hydrogenated fats, noted on the label as "partially hydrogenated soybean oil" (or cottonseed or other vegetable oils).

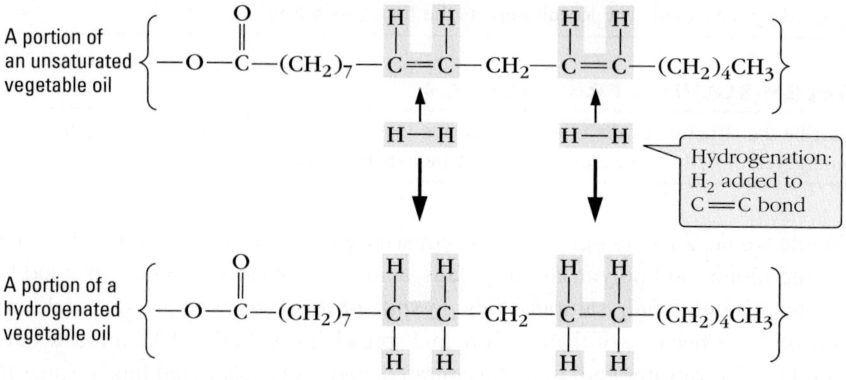
A portion of an unsaturated vegetable oil
H—H
H—H
Hydrogenation: H2 added to C=C bond
A portion of a hydrogenated vegetable oil

Diagram of a hydrogenation reaction.

In banning *trans* fats from commercial food products, the United States would join Austria, Denmark, Iceland, and Switzerland, where similar restrictions have been initiated during the past decade or so.

In November 2013, the U. S. Food and Drug Administration (FDA) proposed that partially hydrogenated oils, the source of dietary *trans* fats, be banned from snack foods, commercial baked goods, and other processed foods. Clear supporting evidence for the ban was cited, including a 2002 U. S. Institute of Medicine study that established no safe level for consumption of *trans* fats, whose dietary use is linked to heart disease. The proposed ban would, by FDA estimates, prevent 20,000 heart attacks and 7,000 deaths from heart disease annually. Some *trans* fats occur naturally, and so the FDA ban would apply only to *trans* fats that are added to foods.

10-5d Hydrolysis of Esters

Hydrolysis is the reverse of the condensation reaction by which an ester is formed.

Esters are not very reactive; their most important reaction is hydrolysis. In a **hydrolysis** reaction, *bonds are broken by their reaction with water, and the H— and —OH of water add to the atoms that were in the bond broken by hydrolysis*. Like other esters, triglycerides are hydrolyzed by strong acid or aqueous base. Triglycerides we eat are hydrolyzed during digestion in the same way by enzyme-catalyzed reactions. With strong acid hydrolysis, the three ester linkages in the triglyceride are broken; the products are glycerol and three moles of fatty acids, which is just the reverse of the ester formation condensation reaction. For example, acid hydrolysis of one mole of glycerol tristearate (tristearin) produces one mole of glycerol and three moles of stearic acid.

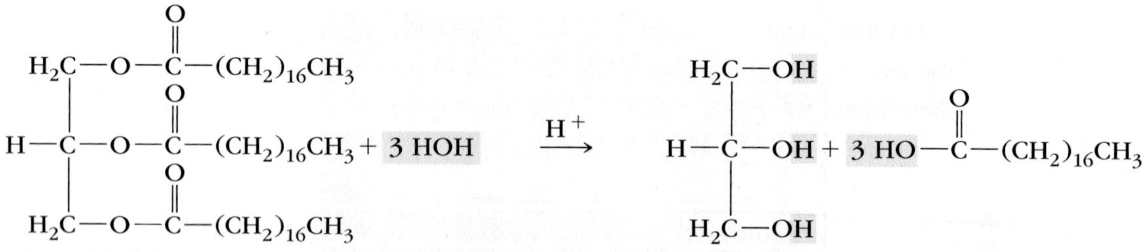

glycerol tristearate (tristearin) glycerol stearic acid

Hydrolysis of a triglyceride with an aqueous base such as NaOH forms glycerol; the fatty acids released then react with the base converting the fatty acids to their salts. *In aqueous base, the hydrolysis process is called* **saponification**, and the salts formed are called soaps. In biological systems, enzymes assist with saponification reactions that digest fats and oils. The saponification of one mole of glycerol tristearate forms one mole of glycerol and three moles of sodium stearate, a soap. Note that the saponification reaction requires three moles of base per mole of triglyceride.

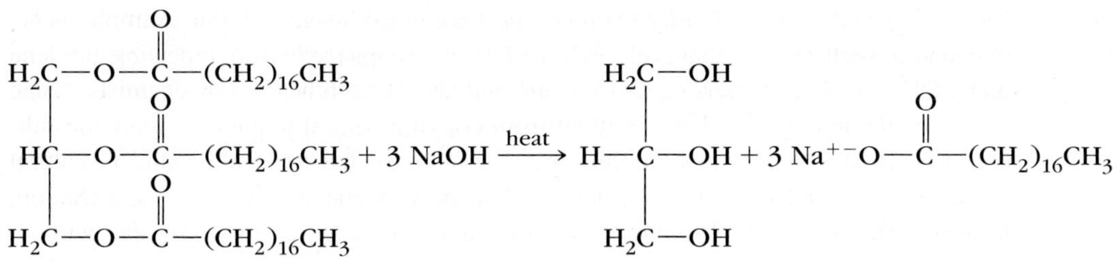

glycerol tristearate, a fat   NaOH   glycerol   sodium stearate, a soap

EXERCISE 10.13

Triglyceride Structures

(a) Draw the structural formula for the triglyceride formed when glycerol reacts with linoleic acid. Circle the ester linkages in this triester molecule.

(b) Using structural formulas, write the equation for the hydrolysis in aqueous NaOH of the triglyceride formed in part (a).

10-5e Biodiesel Fuel

Biodiesel is *a fuel made from biological sources, principally vegetable oils.* Waste vegetable oils, such as those from fast-food establishments and campus dining halls, can be used. Biodiesel fuel production involves the base-catalyzed *transesterification reaction* of triglycerides in vegetable oils with low molar mass alcohols, typically methanol or ethanol. The reaction produces glycerol (glycerine) and the biodiesel fuel, which is a mixture of esters of the fatty acids initially part of the triglyceride. Using ethanol as the alcohol, this is the reaction:

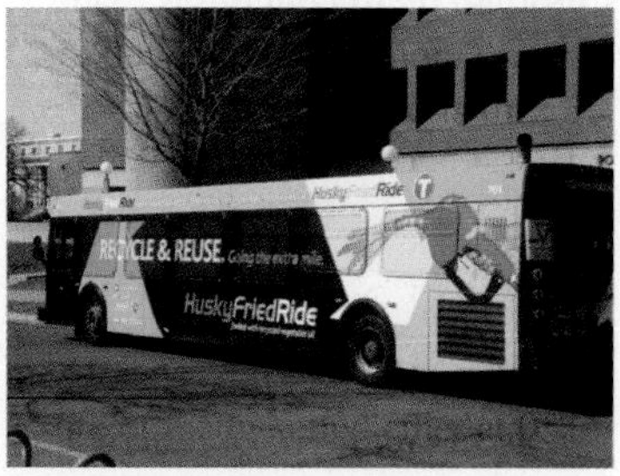
RECYCLE & REUSE
HuskyFriedRide

Courtesy of Don Neu

A bus powered by biodiesel fuel made from waste deep-fryer vegetable oil. The vegetable oil comes from the campus food service at St. Cloud State University (MN). The bus, termed the Husky Fried Ride, uses a mixture of 80% recycled vegetable oil biodiesel and 20% conventional diesel fuel.

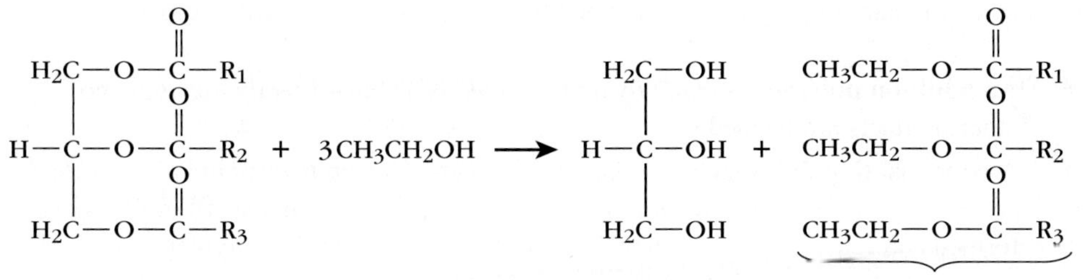

Triglyceride (in the vegetable oil) + 3 Ethanol ⟶ Glycerol + Ethyl esters of fatty acids (biodiesel fuel)

The waste oil is filtered to remove solid wastes and water is removed from the filtrate by heating. A measured sample of the cleaned oil is titrated (← **Sec. 3-12**) with a standardized solution of a strong base to evaluate the concentration of fatty acids in the oil, those not part of the triglycerides. Sufficient ethanol is used to provide the required stoichiometric ratio of three moles of ethanol per mole of triglyceride in the oil. A slight excess of NaOH is added as a catalyst. The NaOH/ethanol mixture is added to the oil and heated for several hours to complete the reaction. Once the reaction is complete, the more dense glycerol settles to the bottom and can be drawn off. The remaining biodiesel fuel is washed to remove the NaOH catalyst and any soaps that might have been formed during the reaction. After further drying, the biodiesel fuel is ready to use.

10-6 Synthetic Organic Polymers

Simple organic molecules are important, not only because of their own properties, but also because they are starting materials to make an important class of compounds called polymers. A **polymer** (*poly,* "many"; *mer,* "part") is *a large molecule composed of*

Nature makes many different polymers, including cellulose and starch in plants and proteins in both plants and animals.

smaller repeating units, usually arranged in a chain-like structure. For example, polyethylene consists of long chains of $—CH_2—CH_2—$ groups made from repeating ethylene units, $CH_2{=}CH_2$. Polymers occur in nature and also are synthesized by chemists. Some useful synthetic polymers have resulted from copying natural polymers. Synthetic rubber, used in almost every automobile tire, is a copy of the molecule found in natural rubber. Other useful synthetic polymers, such as polystyrene, nylon, Teflon, and Dacron, have no natural analogs. It is virtually impossible for us to get through a day without using a dozen or more synthetic organic polymers such as those used to make our clothes, package our foods, and build our computers, phones, and cars. You, a friend, or a family member may be alive because of a medical application of synthetic polymers. Synthetic polymers are so important that approximately 80% of the organic chemical industry is devoted to their production.

Charles D. Winters

Common household items made from plastic.

There are two broad categories of the synthetic polymers commonly known as "plastics." One type, called **thermoplastics**, *softens and flows when heated; when it is cooled, it hardens again.* Common plastic materials that undergo such reversible changes when heated and cooled are polyethylene (milk jugs), polystyrene (inexpensive sunglasses and toys), and polycarbonates (CD discs). The other general type is **thermosetting plastics**. *When first heated, a thermosetting plastic flows like a thermoplastic; however, when heated further, it forms a rigid structure that will not remelt.* Bowling balls, football helmets, and some kitchen countertops are examples of thermosetting plastics.

A macromolecule is a molecule with a very high molar mass.

Both synthetic and natural polymers are made by chemically joining many *small molecules*, each called a **monomer**, to form a *giant polymer molecule*, a **macromolecule**. Macromolecules have molar masses ranging from thousands to millions. In nature, polymerization reactions usually are controlled by enzymes, and in animal cells, biopolymer synthesis takes place rapidly at body temperature. Making synthetic polymers industrially often requires high temperatures and pressures and lengthy reaction times. Synthetic and natural polymers are formed by one of two main reaction types:

- An **addition polymer** is *made by adding monomer units directly together;* no other products are formed.
- A **condensation polymer** is *produced when monomer units combine to form the polymer and also release a small molecule,* usually water. One small molecule is formed and released for each new bond linkage formed in the polymer.

10-6a Addition Polymers

In addition polymerization, the *monomer units are added directly to each other,* hooking together like boxcars on a train to form the polymer. The monomers for making addition polymers usually contain one or more C=C double bonds.

Polyethylene The simplest monomer for addition polymerization is ethylene, $CH_2{=}CH_2$, which polymerizes to form *polyethylene*. When heated to 100 to 250 °C at a pressure of 1000 to 3000 atm in the presence of a catalyst, ethylene forms polyethylene chains with molar masses of up to many thousands. Polymerization involves three steps: initiation, propagation, and chain termination.

An organic peroxide, R—O—O—R, produces two free radicals, R—O·, each with an unpaired electron.

- **Initiation.** The first step, *initiation* of the polymerization, involves chemicals such as organic peroxides (R—O—O—R) that are unstable and easily break apart into free radicals, ·OR, each with an unpaired electron (← **Sec. 6-10b**). The free radicals react readily with molecules containing carbon-carbon double bonds to produce new free radicals.

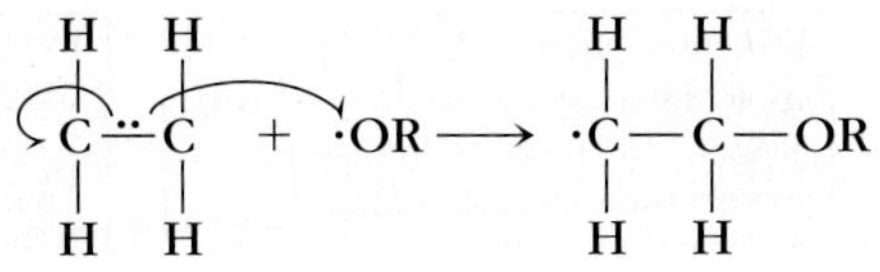

The two single-headed arrows in this equation each indicate movement of a single electron from the pi part of the double bond to form a new single bond or a new free radical.

- **Propagation.** The polyethylene chain begins to grow as the unpaired electron radical formed in the initiation step bonds to a double-bond electron in another ethylene molecule.

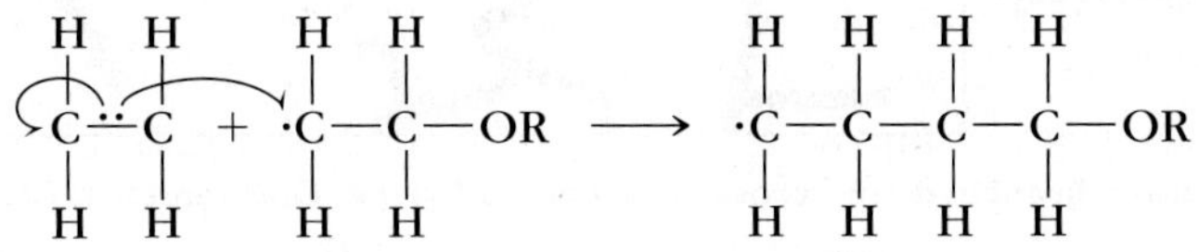

This forms another radical containing an unpaired electron that can bond with yet another ethylene molecule. The process continues, with the chain growing to form a huge polymer molecule.

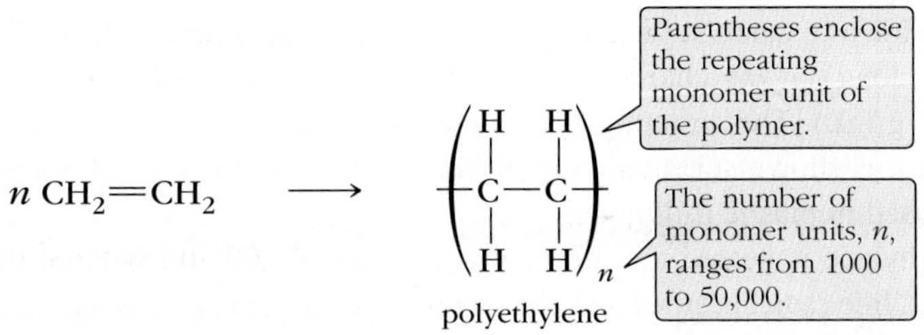
Parentheses enclose the repeating monomer unit of the polymer.
The number of monomer units, n, ranges from 1000 to 50,000.
polyethylene

The C=C double bonds in the ethylene monomer have been changed to C—C single bonds in the polyethylene chain.

- **Chain Termination.** Extension of the polymer chain stops when the supply of monomer is consumed or when the number of free radicals is depleted, such as when two free-radical chains combine, forming a new bond by pairing their unpaired electrons. The **degree of polymerization, *n*,** *is the number of repeating monomer units in the polymer chain.*

$$RO\!-\!(CH_2CH_2)_n\!-\!CH_2\cdot \;+\; \cdot CH_2\!-\!(CH_2CH_2)_{n^*}\!-\!OR' \longrightarrow RO\!-\!(CH_2CH_2)_n\!-\!CH_2CH_2\!-\!(CH_2CH_2)_{n^*}\!-\!OR'$$

n and n^* represent different numbers of CH_2CH_2 monomer units.

CONCEPTUAL EXERCISE 10.14

What Is at the Ends of the Polymer Chains?

What is attached at the ends of the polymer chains when all of the ethylene monomer molecules have been polymerized to form polyethylene?

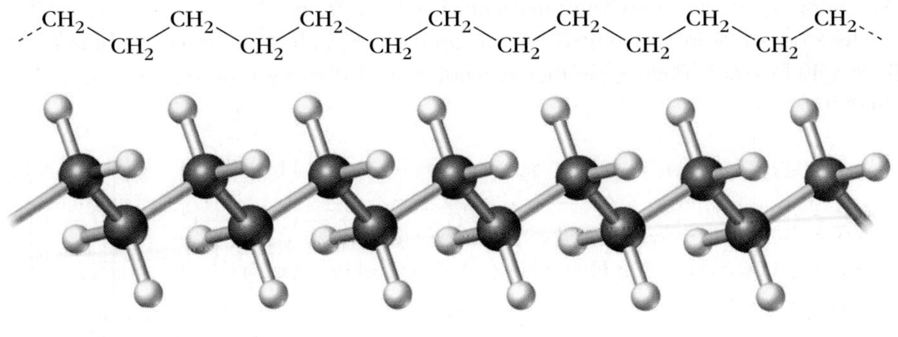

Figure 10.8 Two representations of a portion of the polymer chain in a polyethylene molecule. Notice that because carbon atoms are tetrahedral, polymer chains are zigzags.

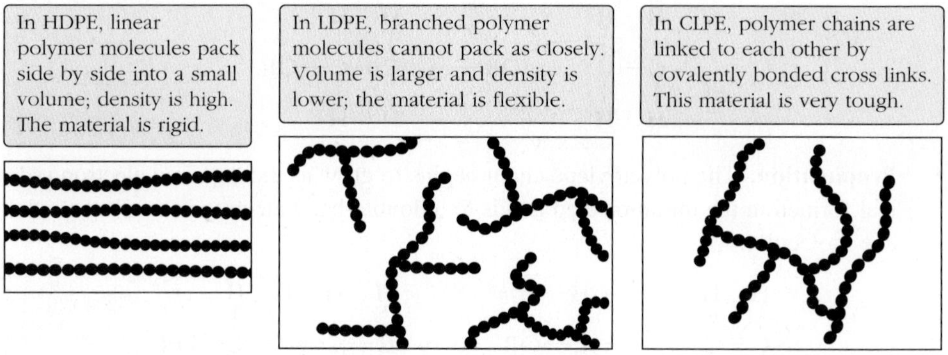
In HDPE, linear polymer molecules pack side by side into a small volume; density is high. The material is rigid.
In LDPE, branched polymer molecules cannot pack as closely. Volume is larger and density is lower; the material is flexible.
In CLPE, polymer chains are linked to each other by covalently bonded cross links. This material is very tough.

Figure 10.9 Linear, branched, and cross-linked polyethylene. Dots represent CH_2—CH_2 units.

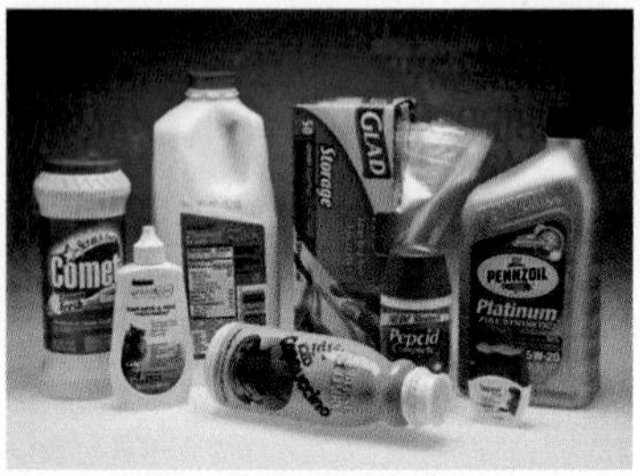
Charles D. Winters

The wide range of properties of different structural types of polyethylene leads to a wide variety of applications.

Bob Masini/Phototake

Manufacturing polyethylene.

Polyethylene is the world's most widely used synthetic polymer, partly because polyethylenes that have different molecular structures and hence different physical properties can be synthesized by changing pressures and catalytic conditions. For example, chromium oxide as a catalyst yields almost exclusively the linear polyethylene shown in Figure 10.8—long, zigzag chains with no branches that can pack closely giving a high density (0.97 g/mL). The structure of this material, referred to as high-density polyethylene (HDPE), is shown schematically in Figure 10.9. HDPE is hard, tough, and semirigid and is used in plastic milk jugs.

When ethylene is heated to 230 °C at a pressure of 200 atm without the chromium oxide catalyst, free radicals attack the chain at random positions, causing irregular branching (Figure 10.9). Branched polyethylene chains cannot pack as closely so the density is lower (0.92 g/mL). This material is called low-density polyethylene (LDPE). It is soft and flexible because of its weaker intermolecular forces and is used to make sandwich bags.

If the linear chains of polyethylene are treated in a way that causes short chains of —CH_2—CH_2— groups to connect adjacent chains, the result is cross-linked polyethylene (CLPE) (Figure 10.9). CLPE is a very tough material that is used for synthetic ice rinks and soft-drink bottle caps.

Other Addition Polymers Many different kinds of addition polymers are made from monomers in which one or more of the hydrogen atoms in ethylene have been replaced with a halogen atom or an organic group. Table 10.6 gives information on some of these monomers and their addition polymers. In each case, the replacement atom or group is represented by *X* in this general equation.

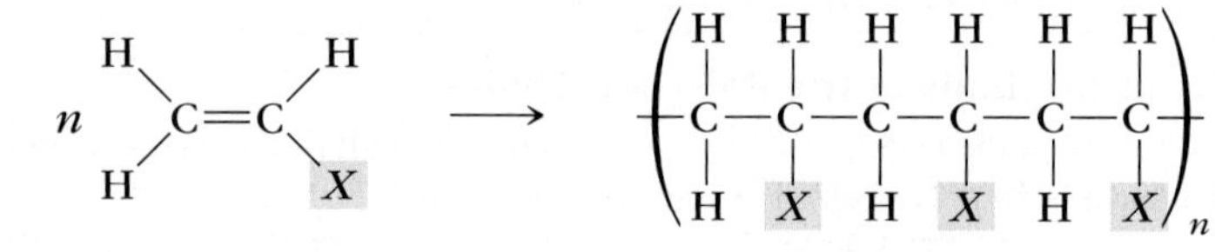

The growing chain

X can be an atom, such as Cl in vinyl chloride, or a group of atoms, such as —CH_3 in propylene or —CN in acrylonitrile. For example, vinyl chloride units combine to form poly(vinyl chloride). Each monomer is incorporated directly into the growing polymer chain.

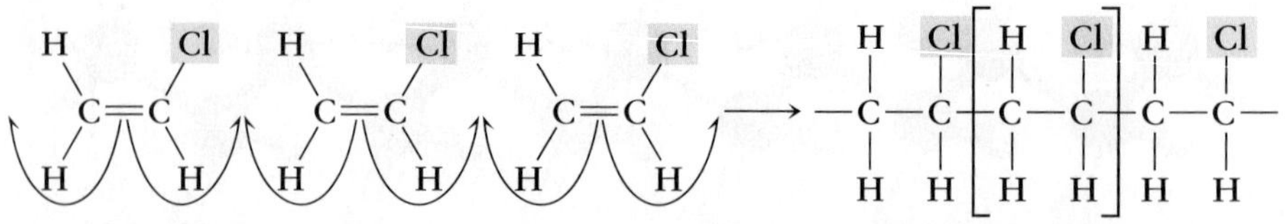

vinyl chloride

Repeating unit poly(vinyl chloride)

Table 10.6 Ethylene Derivatives That Undergo Addition Polymerization

Formula (X atom or group is highlighted)	Monomer Common Name	Polymer Name (trade names)	Uses
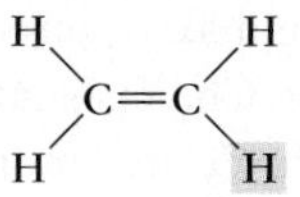	Ethylene	Polyethylene (Polythene)	Squeeze bottles, bags, films, toys and molded objects, electrical insulation
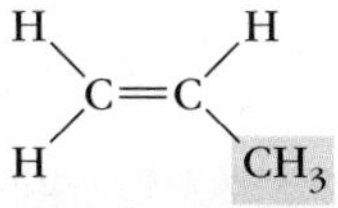	Propylene	Polypropylene (Herculon)	Bottles, films, indoor-outdoor carpets
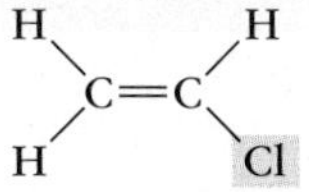	Vinyl chloride	Poly(vinyl chloride) (PVC)	Floor tile, raincoats, pipe
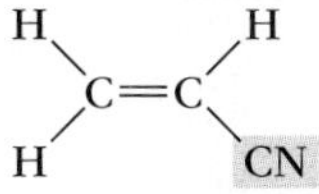	Acrylonitrile	Polyacrylonitrile (Orlon, Acrilan)	Rugs, fabrics
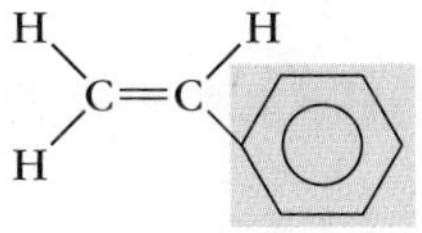	Styrene	Polystyrene (Styrene, Styrofoam, Styron)	Food and drink coolers, building material insulation
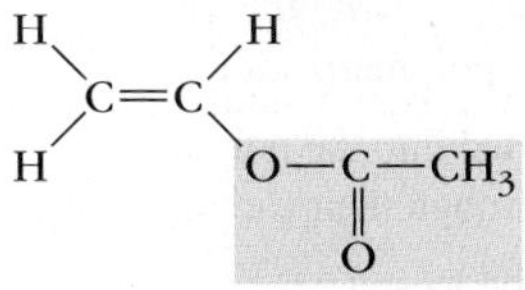	Vinyl acetate	Poly(vinyl acetate) (PVA)	Latex paint, adhesives, textile coatings
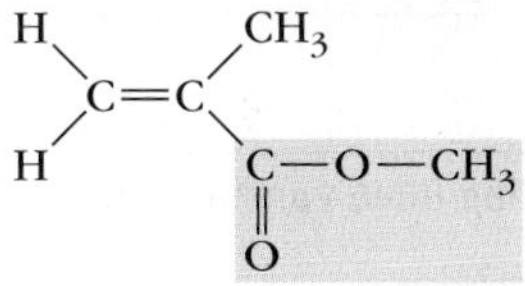	Methyl-methacrylate	Poly(methyl-methacrylate) (Plexiglas, Lucite)	High-quality transparent objects, latex paints, contact lenses
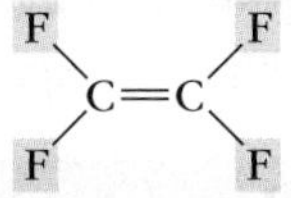	Tetrafluoro-ethylene	Polytetrafluoro-ethylene (Teflon)	Gaskets, insulation, bearings, cooking pan coatings

In polystyrene, n is typically about 5700. Polystyrene is a clear, hard, colorless, solid thermoplastic that can be molded easily at 250 °C. Nearly five billion pounds of polystyrene are used annually in the United States alone to make food containers, toys, electrical parts, and many other items. The variation in properties shown by polystyrene products is typical of synthetic polymers. For example, a clear polystyrene drinking glass that is brittle and breaks into sharp pieces somewhat like glass is quite different from an expanded polystyrene coffee cup that is soft and pliable (Figure 10.10).

X is an aromatic ring in styrene, the monomer for making polystyrene.

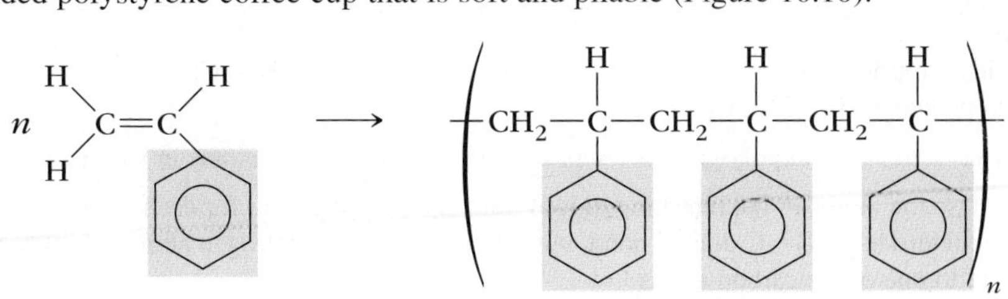

Styrene The growing chain

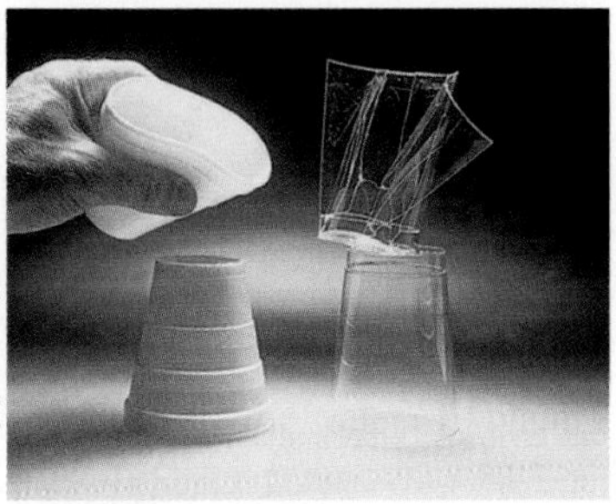

© Cengage Learning/Charles D. Winters

Figure 10.10 Polystyrene. Expanded polystyrene (Styrofoam®) coffee cup *(left)* is soft. Clear polystyrene cup *(right)* is brittle.

A major use of polystyrene is in the production of Styrofoam by "expansion molding." In this process, polystyrene beads are placed in a mold and heated with steam or hot air. The tiny beads contain 4–7% by weight of a low-boiling liquid such as pentane. The steam causes the low-boiling liquid to vaporize and expand the beads. As the foamed particles expand, they are molded into the shape of the mold cavity. Styrofoam is used for meat trays, coffee cups, and many kinds of packing materials.

The numerous variations in chain length, branching, and crosslinking make it possible to produce a variety of properties for each type of addition polymer. Chemists and chemical engineers can fine-tune the properties of the polymer to match the desired properties by appropriate selection of monomer and reaction conditions, thus accounting for the widespread and growing use of synthetic polymers.

PROBLEM-SOLVING EXAMPLE 10.5

Adding Up the Monomers

Kel-F is an addition polymer made from the monomer $FClC{=}CF_2$.

(a) Write the Lewis structure of this monomer.

(b) Write the Lewis structure of a portion of the Kel-F polymer containing three monomer units.

Result

(a) 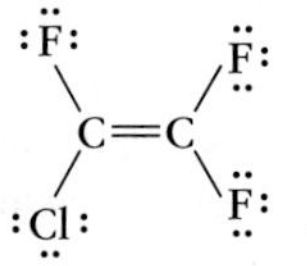

(b) 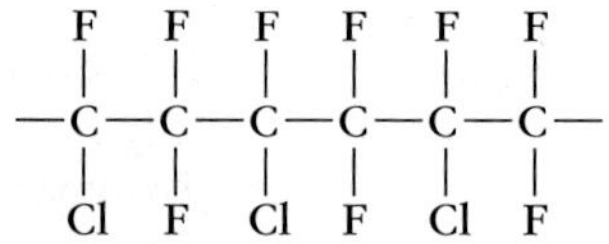

Analyze The condensed molecular formula is given; it can be used to write the Lewis structure. The monomer contains a C=C bond suggesting addition polymerization.

Plan In addition polymerization, all monomer atoms are incorporated into the polymer. Write the three-unit polymer chain as three monomer units joined through their carbon atoms.

Execute Carry out the plan to generate the structures shown in the Result.

PROBLEM-SOLVING PRACTICE 10.5

Draw the structural formula of a polymer chain formed by combining three vinyl acetate monomers. The structural formula of vinyl acetate is in Table 10.6.

PROBLEM-SOLVING EXAMPLE 10.6

Identify the Monomer

Use Table 10.6 to identify the monomer used to make each of the addition polymers, a portion of whose molecule is represented by the structures below.

PVC

(a) 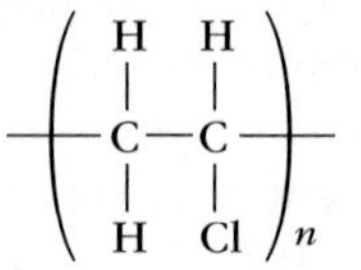

Acrilan

(b) 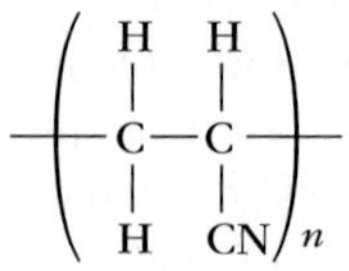

Polypropylene

(c) 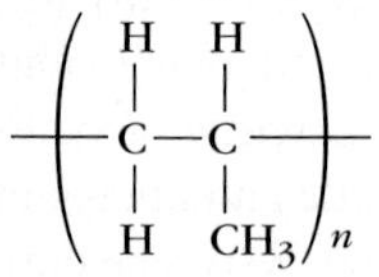

Result

(a) Vinyl chloride, $CH_2{=}CHCl$ (b) Acrylonitrile, $CH_2{=}CHCN$

(c) Propylene, $CH_2{=}CHCH_3$

Analyze An addition polymer has a repeating unit containing all atoms in its monomer.

Plan Use the atoms in the repeating unit to draw a Lewis structure for each monomer. Each structure has a C=C double bond, but differs in the non-hydrogen atom or group attached to one of the carbon atoms.

Execute
(a) $CH_2{=}CHCl$ (b) $CH_2{=}CHCN$
(c) $CH_2{=}CHCH_3$

PROBLEM-SOLVING PRACTICE 10.6

Draw the structure of the repeating unit for each of these addition polymers.
(a) Polypropylene (b) Poly(vinyl acetate) (c) Poly(methyl methacrylate)

Copolymers Many commercially important addition polymers are copolymers. A **copolymer** is *a polymer obtained by polymerizing a mixture of two or more different monomers.* A copolymer of styrene with butadiene is the most important synthetic rubber produced in the United States. About 1.1 million tons of styrene-butadiene rubber (SBR) are produced each year in the United States for making tires. A 3 : 1 mole ratio of butadiene to styrene is used to make SBR.

The suffix "diene" in butadiene indicates that there are two C=C bonds in the molecule.

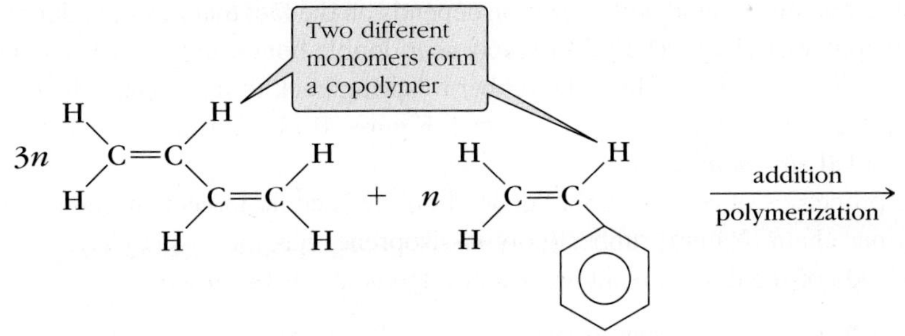
Two different monomers form a copolymer
addition
polymerization

1,3-butadiene styrene

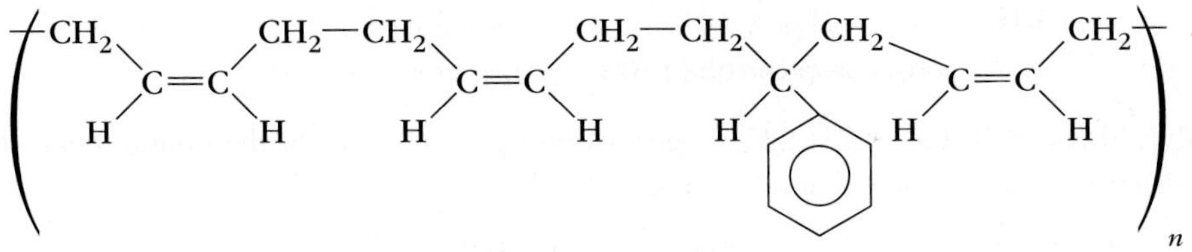

styrene-butadiene rubber (SBR)

Each 1,3-butadiene monomer molecule has two double bonds (four pi electrons). Only two of the double-bond electrons are involved in bonds between monomer units. Thus, one double bond remains in the polymer backbone for each 1,3-butadiene monomer.

Another important copolymer is made by polymerizing mixtures of acrylonitrile, butadiene, and styrene (ABS) to produce a sturdy material used in car bumpers and computer cases.

PROBLEM-SOLVING EXAMPLE 10.7

Macromolecular Masses

Polytetrafluoroethylene (Teflon) is an addition polymer made of chains of $CF_2{=}CF_2$ monomer units. This polymer has a molar mass of 1.0×10^6 g/mol. Calculate the degree of polymerization, n.

Result 10,000 monomer units

Analyze The degree of polymerization is the number of monomer units in the polymer. Teflon is an addition polymer, so all monomer atoms are present in the polymer. The molar mass of the polymer must be the number of monomer units times the molar mass of each monomer: $M_{polymer} = n \times M_{monomer}$. The molar mass of the polymer, $M_{polymer}$, is given.

Plan Each $CF_2{=}CF_2$ monomer unit has a molar mass of 100.0 g/mol. Rearrange the equation $M_{polymer} = n \times M_{monomer}$ and calculate n.

Execute $n = M_{polymer}/M_{monomer} = (1.0 \times 10^6 \text{ g/mol})/100.0 \text{ g/mol} = 1.0 \times 10^4$.

GRANDADAM/Science Source

Natural latex coming from a rubber tree. Natural rubber is an addition polymer whose monomer is 2-methyl-1,3-butadiene (isoprene).

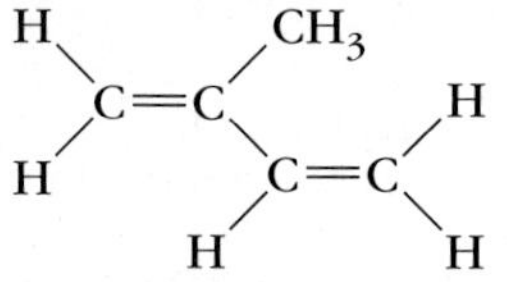

2-methyl-1,3-butadiene (isoprene)

PROBLEM-SOLVING PRACTICE 10.7

Saran wrap is a copolymer of vinyl chloride and 1,1-dichloroethene in a 1 : 1 mole ratio. Calculate the degree of polymerization of a polymer molecule that has a molar mass of 150,000 g/mol.

Natural and Synthetic Rubbers Natural rubber is a hydrocarbon whose monomer unit has the empirical formula C_5H_8. When rubber is decomposed in the absence of oxygen, the monomer 2-methyl-1,3-butadiene (isoprene) is obtained. Natural rubber occurs as *latex* (an emulsion of rubber particles in water) that oozes from rubber trees when the bark is cut. Precipitation of the rubber particles yields a gummy mass that is not only elastic and water-repellent but also very sticky, especially when warm. In 1839, after five years' work on natural rubber, Charles Goodyear (1800–1860) discovered that heating gum rubber with sulfur produces a material that is no longer sticky but is still elastic, water-repellent, and resilient. Heating with sulfur is called *vulcanization*.

Vulcanized rubber contains short chains of sulfur atoms that join (cross-link) the natural rubber polymer chains. Vulcanization depends on the fact that there are double bonds within the polymer chain: The sulfur reacts near double bonds and reduces the unsaturation. The sulfur cross links allow the rubber to spring back to its original shape and size after it is stretched and the stress is removed (Figure 10.11). Substances that behave this way are called *elastomers*.

The properties of natural rubber depend on the specific molecular geometry within the polymer chain. Natural rubber (poly-*cis*-isoprene) has the $—CH_2CH_2—$ groups on the same side of the double bond in a *cis* arrangement (← **Sec. 6-5a**),

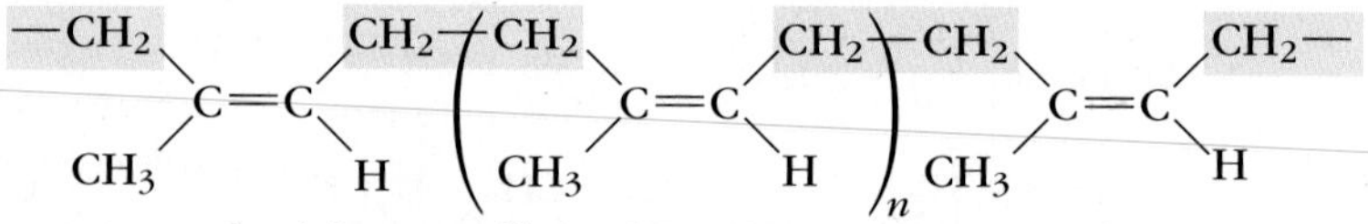

poly-*cis*-isoprene (the $—CH_2—CH_2—$ groups are *cis*)

Polyisoprene with the $—CH_2CH_2—$ groups on opposite sides of the double bond (the *trans* arrangement) also occurs in nature.

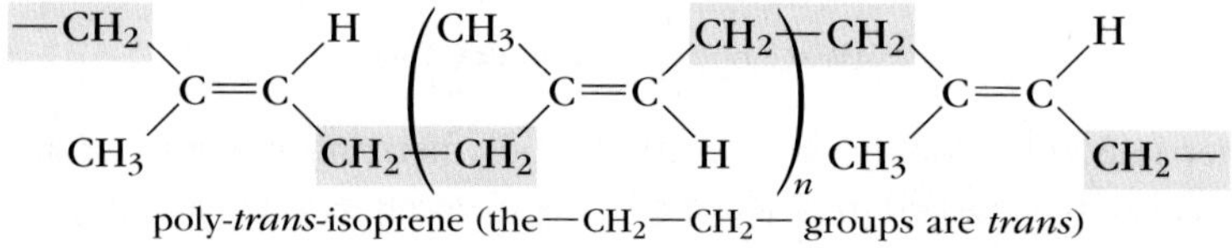

poly-*trans*-isoprene (the $—CH_2—CH_2—$ groups are *trans*)

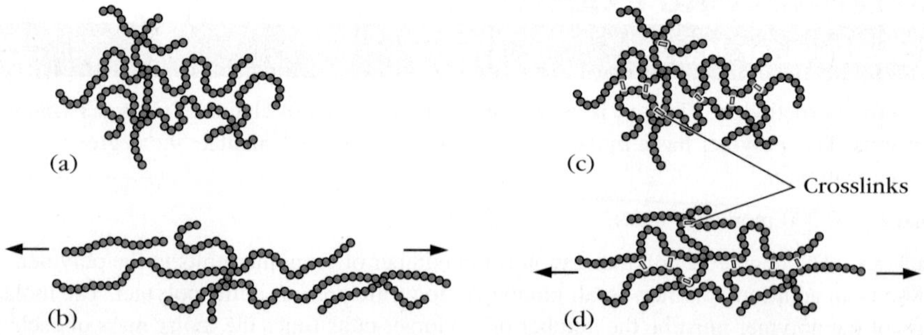
(a)
(b)
(c)
(d)
Crosslinks

Figure 10.11 Vulcanization. Each part illustrates several polymer chains: (a) and (b) are in non-vulcanized rubber; (c) and (d) are in vulcanized rubber. (b) and (d) illustrate the different effects of applying a horizontal stretching force to the rubber. Because the cross links (yellow) hold some of the polymer chains together, vulcanized rubber is stronger and harder to stretch (d) than non-vulcanized rubber (b). The diagram is schematic. In a real polymer there would be many more chains and each chain would be much longer. There would not be as many cross links. (*Source:* Adapted from *Journal of Chemical Education,* 85(10) p. 1324, 2008. Copyright © 2008, Division of Chemical Education, Inc. Reprinted by permission.)

In 1955, chemists at the Goodyear and Firestone companies almost simultaneously discovered how to prepare synthetic poly-*cis*-isoprene, a material structurally identical to natural rubber. Today, synthetic poly-*cis*-isoprene can be manufactured cheaply and is used when natural rubber is in short supply.

10-6b Condensation Polymers

Polyesters The reactions of alcohols with carboxylic acids to produce esters (← **Sec. 10-5b**) are examples of condensation reactions. Condensation polymers are formed by condensation polymerization reactions between monomers. Unlike addition polymerization reactions, condensation polymerization reactions do not depend on the presence of a C=C double bond in the reacting molecules. Rather, condensation polymerization reactions generally require *two different functional groups present in the same monomer or in two different monomers.* Each of the two monomers has a functional group at each end of the monomer. As the functional groups at each end of one monomer react with the groups of the other monomer, long-chain condensation polymers are produced.

For example, molecules with two carboxylic acid groups (—COOH), such as terephthalic acid, and other molecules with two alcohol groups, such as ethylene glycol, can react with each other at both ends to form ester linkages (shaded part of the structure). Water is the other product of this condensation polymerization reaction; it is produced during the formation of each ester linkage in the polymer.

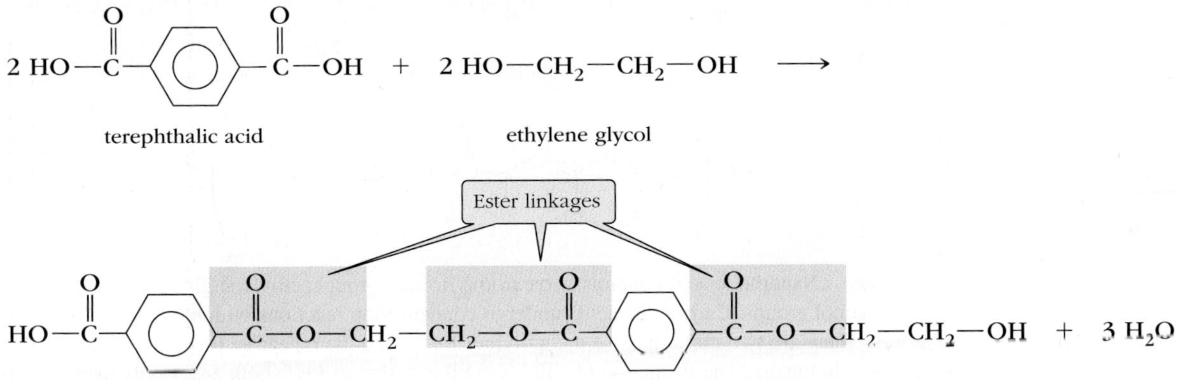
terephthalic acid
ethylene glycol
Ester linkages

The growing terephthalic acid-ethylene glycol polymer

Because this product has a —COOH group on one end and an —OH group on the other end, the —COOH group can react with an —OH group of another ethylene glycol molecule. Similarly, the remaining alcohol group on the ester can react with another terephthalic acid molecule. This process continues, forming long chains of poly(ethylene terephthalate), commonly known as PET. PET is a **polyester**, *a polymer made by linking monomers using ester linkages.*

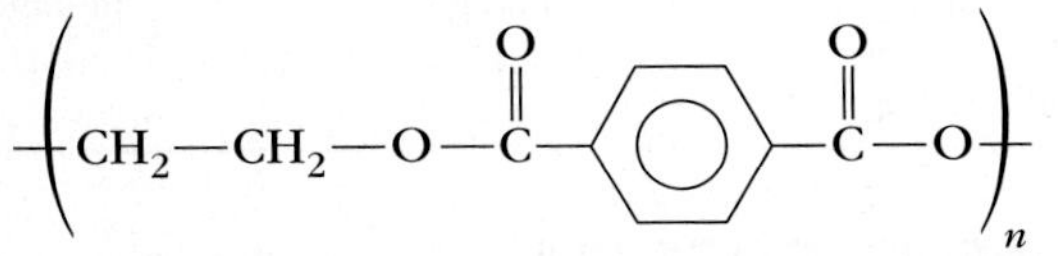

the repeating unit poly(ethylene terephthalate), PET

Each year more than 110 billion pounds of PET are produced globally for making beverage bottles, apparel, tire cord, food packaging, coatings for microwave and conventional ovens, and home furnishings. A variety of trade names is associated with these applications. PET textile fibers are marketed under such names as Dacron and Terylene. Films of the same polyester, when magnetically coated, are used to make TV tapes. This film, Mylar, has unusual strength and can be rolled into sheets one thirtieth the thickness

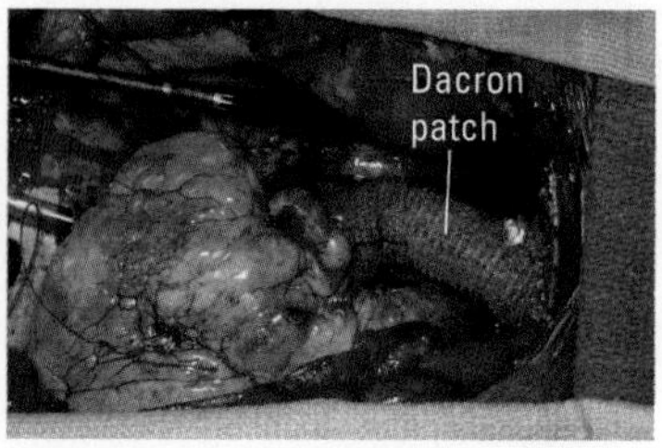
Dacron
patch

MICHAEL ENGLISH, M.D. Custom Medical Stock Photo/Newscom

Medical uses of Dacron. A Dacron patch is used to repair an aortic aneurysm site in a heart patient.

of a human hair. You have probably seen it in party balloons. The inert, nontoxic, noninflammatory, and non–blood-clotting characteristics of PET polymers make Dacron tubing an excellent substitute for human blood vessels in heart bypass operations. Dacron sheets are also used as a skin substitute for burn victims.

PROBLEM-SOLVING EXAMPLE 10.8

Condensation Polymerization

Poly(ethylene naphthalate) (PEN) is used for bar code labels. This condensation polymer is made by the reaction between naphthalic acid and ethylene glycol,

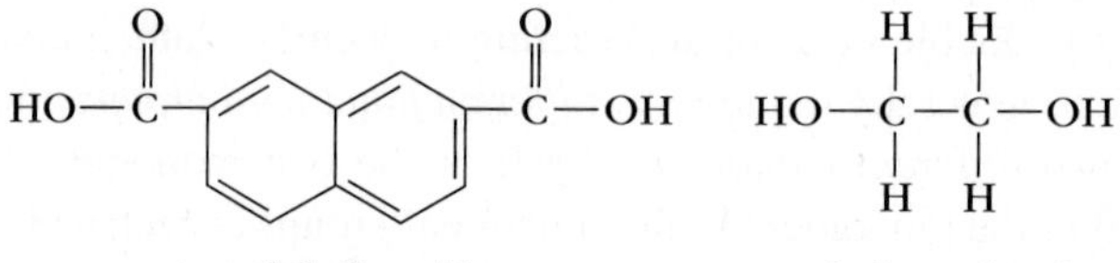

naphthalic acid   ethylene glycol

(a) Write the structural formula of the molecule formed after two ethylene glycol molecules have polymerized with two naphthalic acid molecules.
(b) Write the structural formula of the repeating unit of the polymer.

Result

(a)

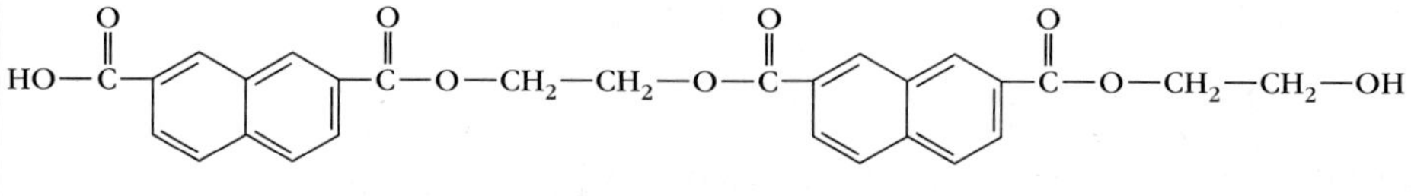

(b)

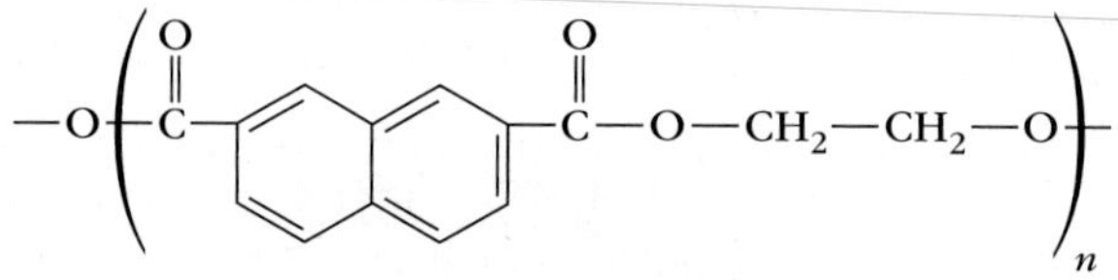

Analyze Naphthalic acid contains two carboxylic acid groups; ethylene glycol contains two alcohol groups. Carboxylic acids undergo condensation reactions with alcohols to form esters. There are two monomers so a copolymer forms and the repeating unit should consist of both monomers. The formation of PEN is similar to that of PET; both are copolymers and polyesters. A water molecule is produced for each ester linkage formed.

Plan Draw a polymer structure formed by condensation of —COOH and —OH groups. The repeating unit is the smallest unit that contains both monomers.

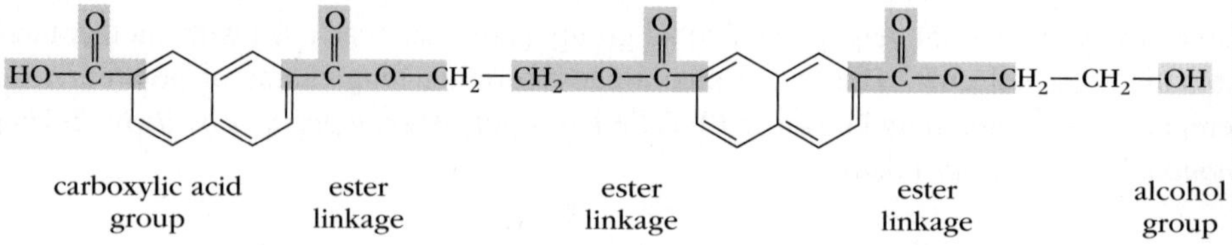
carboxylic acid
group
ester
linkage
ester
linkage
ester
linkage
alcohol
group

Execute See Result above.

PROBLEM-SOLVING PRACTICE 10.8

A condensation polymer can be made from the single monomer glycolic acid. A portion of the polymer is given below.

$$-O-CH_2-\overset{O}{\overset{\|}{C}}-O-CH_2-\overset{O}{\overset{\|}{C}}-O-CH_2-\overset{O}{\overset{\|}{C}}-O-CH_2-$$

Write the structural formula of glycolic acid.

Amines, Amides, and Polyamides An **amine** is *an organic compound* that can be thought of as *derived from the structure of ammonia,* NH_3. Amines are classified as primary, secondary, or tertiary amines according to how many of the H atoms in NH_3 are replaced by alkyl groups. There can be one, two, or three alkyl groups covalently bonded to the nitrogen atom, for example, as in methylamine, dimethylamine, and trimethylamine:

$$\underset{\text{methylamine}}{CH_3{-}\ddot{N}H_2} \qquad \underset{\text{dimethylamine}}{CH_3{-}\underset{\cdot\cdot}{N}(H){-}CH_3} \qquad \underset{\text{trimethylamine}}{CH_3{-}\underset{\cdot\cdot}{N}(CH_3){-}CH_3}$$

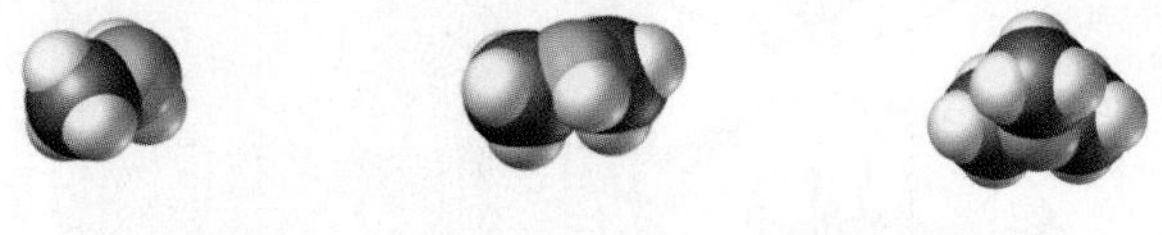

RNH_2 is primary; R_2NH is secondary; and R_3N is tertiary.

An important *condensation* reaction occurs when a primary or secondary amine reacts with a carboxylic acid at high temperature to produce a water molecule and form an *amide:*

$$\underset{\text{carboxylic acid}}{R{-}\overset{O}{\overset{\|}{C}}{-}OH} + \underset{\text{primary amine}}{H{-}\underset{H}{\underset{|}{N}}{-}R'} \xrightarrow{\text{heat}} \underset{\text{amide}}{R{-}\overset{O}{\overset{\|}{C}}{-}\underset{H}{\underset{|}{N}}{-}R'} + \underset{\text{water}}{H_2O}$$

The *functional group* $-\overset{O}{\overset{\|}{C}}-\overset{H}{\overset{|}{N}}-$ *is the* **amide group** or **amide linkage**.

Dr. Wallace Carothers of the DuPont Company discovered a useful and important type of condensation polymerization reaction that occurs when *diamines* (compounds containing two $-NH_2$ groups) react with *dicarboxylic acids* (compounds containing two $-COOH$ groups) to form polymers called polyamides or nylons. In February 1935 his research yielded a product known as nylon-66 prepared by copolymerizing adipic acid (a dicarboxylic acid) and hexamethylenediamine (a diamine). The overall equation is

$$n\ \underset{\text{adipic acid}}{HO{-}\overset{O}{\overset{\|}{C}}{-}(CH_2)_4{-}\overset{O}{\overset{\|}{C}}{-}OH} \; + \; n\ \underset{\text{hexamethylenediamine}}{H{-}\underset{H}{\underset{|}{N}}{-}(CH_2)_6{-}NH_2} \longrightarrow$$

$$HO{-}\overset{O}{\overset{\|}{C}}{-}(CH_2)_4{-}\overset{O}{\overset{\|}{C}}\left({-}\underset{H}{\underset{|}{N}}{-}(CH_2)_6{-}\underset{H}{\underset{|}{N}}{-}\overset{O}{\overset{\|}{C}}{-}(CH_2)_4{-}\overset{O}{\overset{\|}{C}}\right)_{n-1}{-}\underset{H}{\underset{|}{N}}{-}(CH_2)_6{-}\overset{H}{\overset{|}{N}}{-}H + (2n-1)\,H_2O$$

nylon-66

The *polymer contains many amide linkages;* hence it is known as a **polyamide**. Figure 10.12 shows how nylon-66 can be made in the laboratory from solutions of adipic acid and hexamethylene diamine.

The name of nylon-66 is based on the six carbon atoms in the diamine (hexamethylene diamine) and the six carbon atoms in the diacid (adipic acid), respectively.

Figure 10.13 illustrates another aspect of the structure of nylon—intermolecular hydrogen bonding—that explains why nylon makes such good fibers. To have good tensile strength, the chains of atoms in a polymer should be able to attract one another, but not so strongly that the polymer molecules cannot initially be extended to form the fibers. Linking the chains together with covalent bonds would be too strong. Hydrogen bonds, with a strength about one tenth that of a typical covalent bond, join the chains in the desired manner.

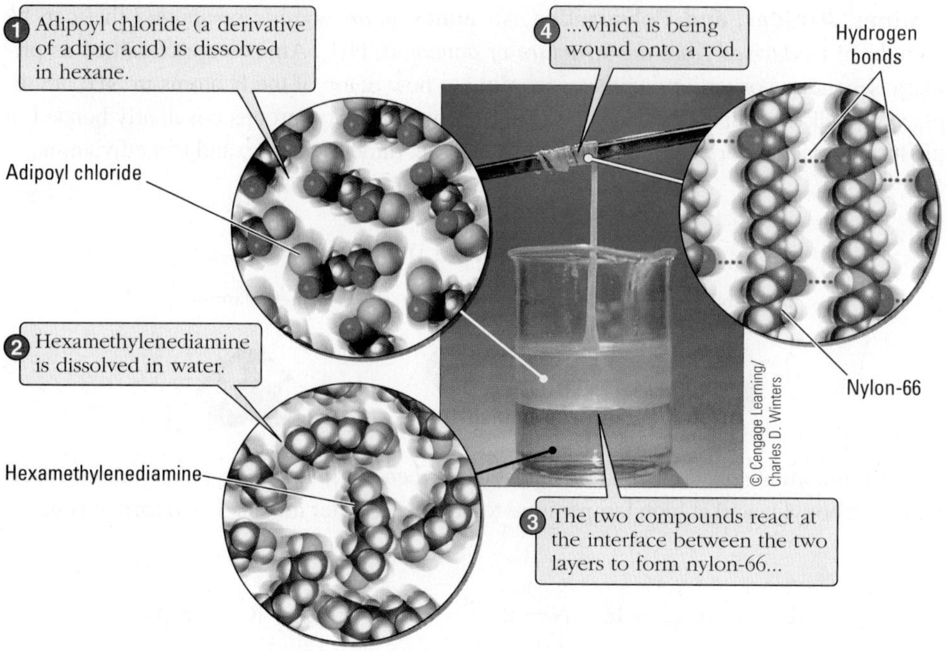
1 Adipoyl chloride (a derivative of adipic acid) is dissolved in hexane.
Adipoyl chloride
2 Hexamethylenediamine is dissolved in water.
Hexamethylenediamine
3 The two compounds react at the interface between the two layers to form nylon-66...
4 ...which is being wound onto a rod.
Hydrogen bonds
Nylon-66
© Cengage Learning/Charles D. Winters

Figure 10.12 Making nylon-66. Adipoyl chloride has the formula

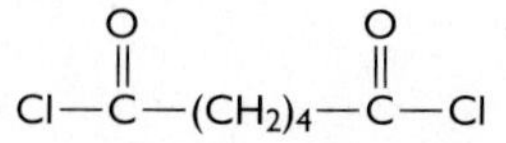

For clarity, solvent molecules (water and hexane) are not shown.

PROBLEM-SOLVING EXAMPLE 10.9

Kevlar, a Condensation Polyamide

Kevlar, used to make bulletproof vests and helmets, sturdy canoes, and baseball batting gloves, is made by polymerizing *p*-phenylenediamine and terephthalic acid.

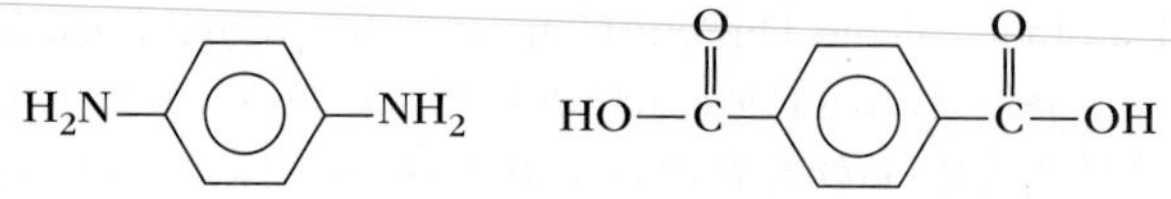

p-phenylenediamine terephthalic acid

(a) Using structural formulas, write the chemical equation for the formation of a segment of a Kevlar molecule containing three amide linkages.
(b) Identify the repeating unit for Kevlar.

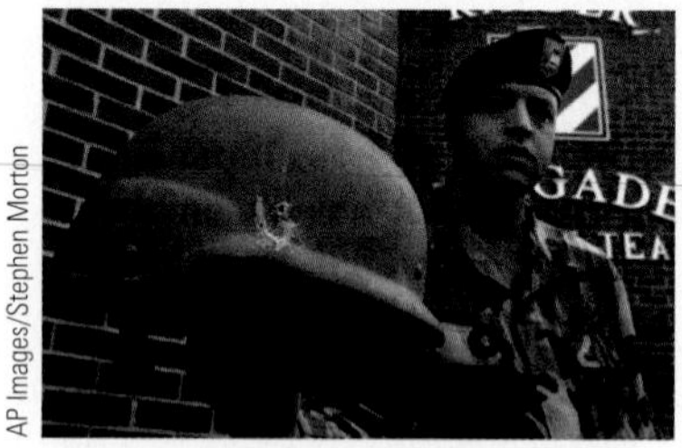
AP Images/Stephen Morton

A Kevlar helmet. A bullet creased, but did not penetrate, the helmet of this U. S. soldier.

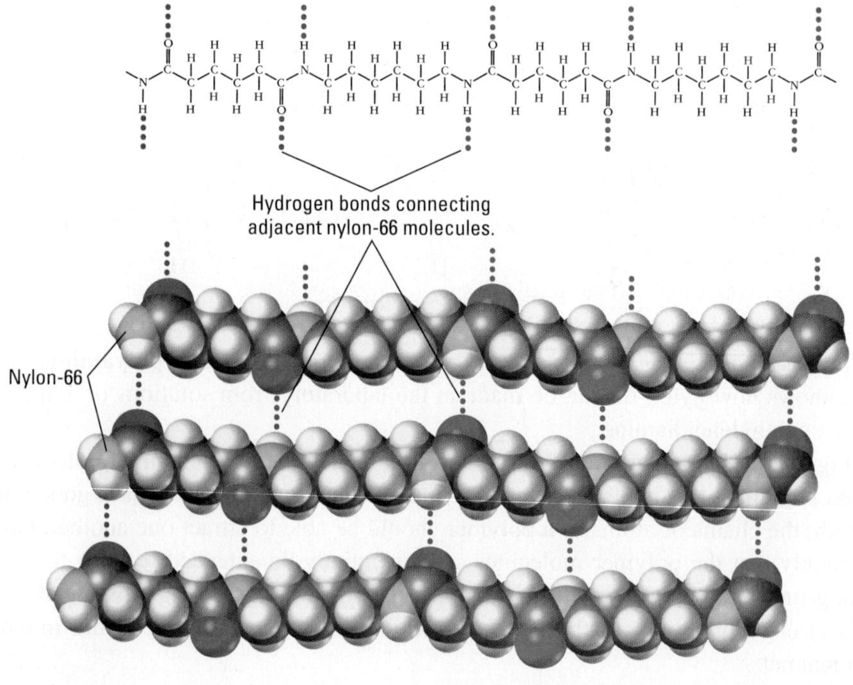
Hydrogen bonds connecting adjacent nylon-66 molecules.
Nylon-66

Figure 10.13 Hydrogen bonding in nylon-66.

Result

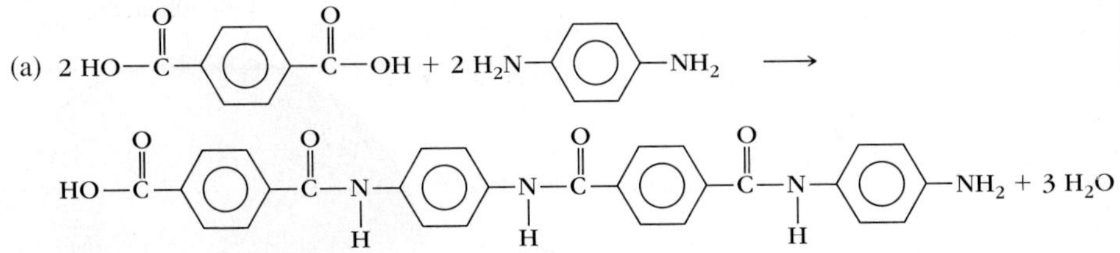

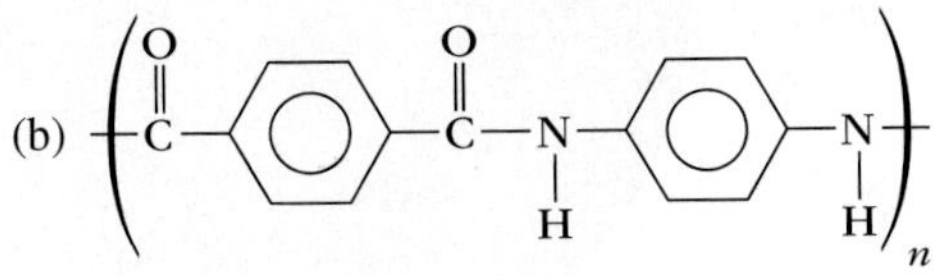

Analyze The reactants are a diamine and a diacid so they can react by condensation polymerization, forming a polyamide. The repeat unit must be derived from one diamine and one diacid. Forming an amide linkage releases one water molecule.

Plan Start with a —COOH group in the diacid. Make an amide linkage to an amine in the diamine by removing HOH. Then, react the amine on the other end of the diamine with a —COOH group of a second diacid. Continue this process.

Execute See Result.

PROBLEM-SOLVING PRACTICE 10.9

How many moles of water are formed from the condensation reaction between 6 mol *p*-phenylenediamine and 10 mol terephthalic acid?

CONCEPTUAL EXERCISE 10.15

Nylon from 3-Aminopropanoic Acid

Polyamides can also be formed from a single monomer that contains both an amine and a carboxylic acid group. For example, the compound 3-aminopropanoic acid can polymerize to form a nylon. Write the general formula for this polymer. Write a formula for the other product that is formed.

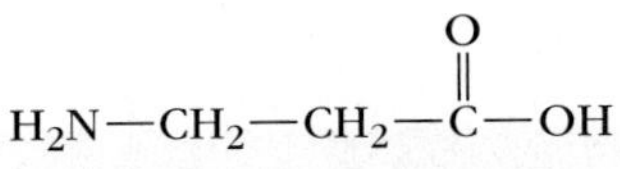

3-aminopropanoic acid

EXERCISE 10.16

Functional Groups

We have discussed several functional groups. Shown below is the structural formula of aspartame, an artificial sweetener. Identify each of the numbered functional groups.

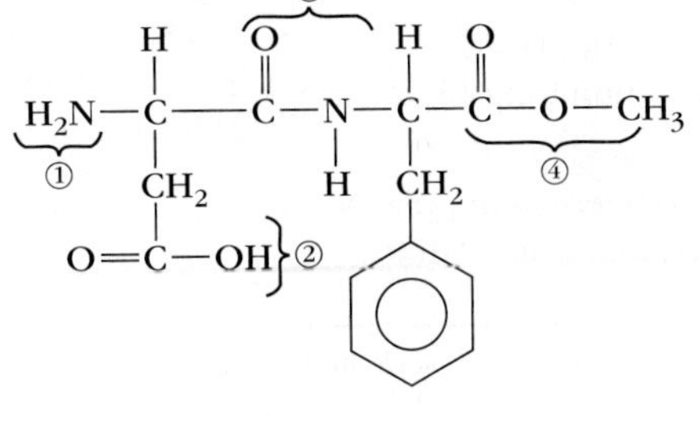

aspartame

Stephanie Louise Kwolek
1923–

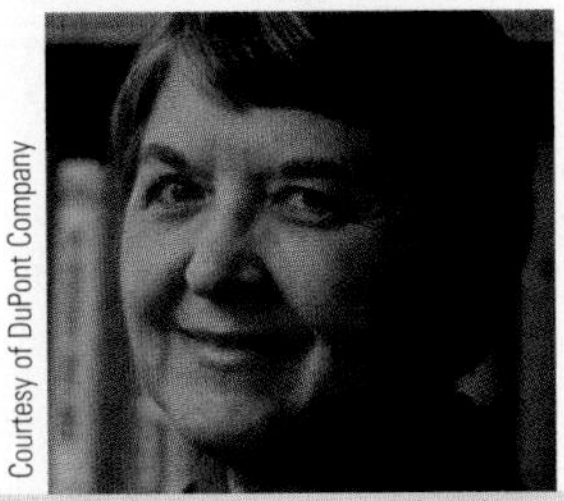

Courtesy of DuPont Company

In 1946, Stephanie Kwolek received a Bachelor of Science degree from Carnegie Tech (now Carnegie Mellon University). Although she wanted to study medicine, she couldn't afford it and decided to take a temporary job at DuPont. She liked her work so well that she stayed for 40 years, retiring in 1986. Kwolek is best known for the development of Kevlar fiber, which is five times stronger than steel on an equal weight basis. She has received many awards, including the Perkin Medal in 1997, considered one of the most prestigious awards a chemist can receive in the United States.

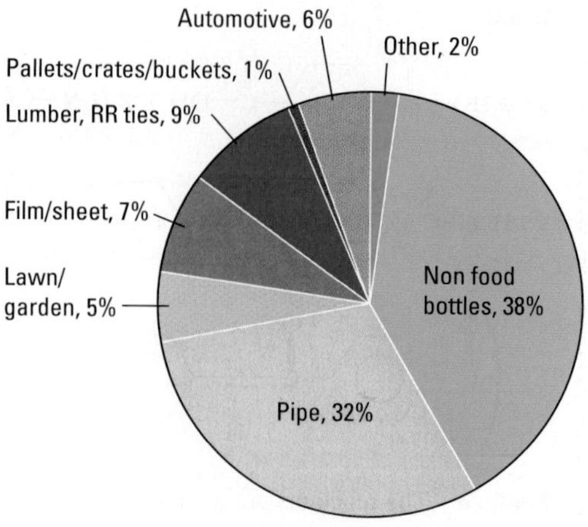
Automotive, 6%
Other, 2%
Pallets/crates/buckets, 1%
Lumber, RR ties, 9%
Film/sheet, 7%
Lawn/
garden, 5%
Non food
bottles, 38%
Pipe, 32%

Figure 10.14 Domestic recycled HDPE bottle end use, 2011. (*Source:* American Chemistry Council. 2011 United States National Post-Consumer Plastic Bottle Recycling report, Figure 7, p. 12.)

10-6c Recycling Plastics

I used to be a plastic bottle
Realimage/Alamy

A pen made from recycled PET.

More than 2.6 billion pounds of plastic bottles were recycled in the United States in 2011, up significantly from 600 million pounds in 1991. Almost all (98%) of these bottles were polyethylene terephthalate (PET) and high-density polyethylene (HDPE). Currently, these polymers are recycled in the United States at nearly a 30% rate. Recycled PET, the most recycled plastic, is used for carpet fibers, tennis balls, "fleece" for jackets, and insulation for ski jackets and sleeping bags. It takes only five recycled 2-L PET soft drink bottles to make the insulation for such a jacket; five such recycled bottles can also make a T-shirt. Recycled HDPE is used to make a variety of products (Figure 10.14). PET and HDPE are being replaced by polypropylene in manufacturing commercial food (yogurt, margarine) and takeout containers, as well as drink cups in fast food restaurants. Although these uses of polypropylene are growing, only about 2% is recycled because many recycling facilities can only recycle bottles, not cups.

Codes are stamped on plastic containers to help consumers identify and sort recyclable plastics (Table 10.7). Although there has been a dramatic increase in plastics recycling, further increases require a sufficient demand for products made partially or completely from recycled materials. Such use has been mandated in some areas,

Table 10.7 Plastic Container Codes

Code	Material and Uses	Code	Material and Uses
1 PETE	Polyethylene terephthalate (PET)*; beverage bottles, clothing	5 PP	Polypropylene; bottles, containers, auto battery cases
2 HDPE	High-density polyethylene; milk and water jugs, gasoline tanks, Tyvek wrap	6 PS	Polystyrene; thermal insulation, drinking glasses
3 PVC	Poly(vinyl chloride) (PVC)*; garden hoses, plumbing pipe	7 OTHER	All other resins and layered multimaterial; cell phone and computer casings
4 LDPE	Low-density polyethylene; plastic bags, toys, electrical insulation		

*Bottle codes are different from standard industrial identification to avoid confusion with registered trademarks.

Tyvek
Peter McGahey

A Tyvek-wrapped house under construction.

including California, which, since 1995, requires all HDPE packaging to contain 25% recycled HDPE. A significant challenge lies ahead in finding economically viable methods to recycle plastics (and metals) from the rapidly increasing and massive numbers of obsolete personal computers and mobile electronic devices. The U. S. EPA estimates that 225,000 *tons* of personal computers and 17,200 *tons* of mobile digital devices were discarded in 2010; about 40% of personal computers and 11% of mobile devices were recycled. To combat having such *electronic waste* ("*e waste*") end up in landfills or being incinerated, stricter laws are being put into place at the state and local levels regarding disposal and recycling of such waste.

10-7 Biopolymers: Polysaccharides and Proteins

Biopolymers—naturally occurring polymers—are an integral part of living things. Many advances in creating and understanding synthetic polymers have come from studying biopolymers. Cellulose and starch, made in plants by condensation polymerization reactions, resemble synthetic polymers in that the monomer molecules—glucose—are all alike. On the other hand, proteins, which are made by both plants and animals, differ from synthetic polymers because they include many different monomers. Also, the occurrences of the different monomers along the protein polymer chain are anything but regular. As a result, proteins are extremely complex condensation copolymers.

10-7a Monosaccharides to Polysaccharides

Nature makes an abundance of *compounds with the general formula* $C_x(H_2O)_y$, in which x and y are whole numbers, such as in glucose, $C_6H_{12}O_6$, where $x = 6$, $y = 6$. These compounds are variously known as **sugars**, **carbohydrates**, and mono-, di-, or polysaccharides (from the Latin *saccharum,* "sugar"—because they taste sweet). The simplest carbohydrates, such as glucose, are **monosaccharides**, also called simple sugars, because they *contain only a single saccharide molecule*. **Disaccharides** *consist of two monosaccharide units joined together,* such as sucrose (table sugar, glucose and fructose linked) or maltose (two glucose units linked). **Polysaccharides** are *polymers containing many monosaccharide units,* up to several thousand. This chart summarizes carbohydrate terminology.

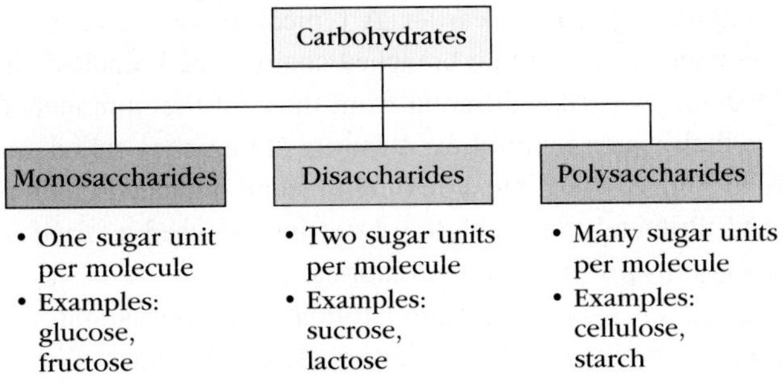
Carbohydrates
Monosaccharides
Disaccharides
Polysaccharides
• One sugar unit per molecule
• Examples: glucose, fructose
• Two sugar units per molecule
• Examples: sucrose, lactose
• Many sugar units per molecule
• Examples: cellulose, starch

The monomer units of disaccharides and polysaccharides are joined by a *C—O—C structure* called a **glycosidic linkage** between carbons 1 and 4 or 1 and 6 of adjacent monosaccharide units. The glycosidic linkages in disaccharides and polysaccharides are formed by a condensation reaction like the condensation reactions by which synthetic polymers are formed. A water molecule is released during the formation of each glycosidic bond from the reaction between the —OH groups of the monosaccharide monomers (Figure 10.15). Notice that maltose has two fewer hydrogen atoms and one less oxygen

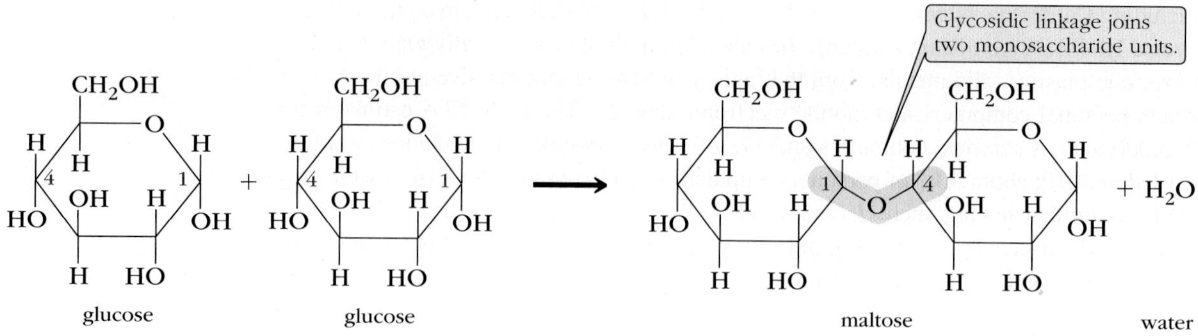
Glycosidic linkage joins two monosaccharide units.
glucose
glucose
maltose
water

Figure 10.15 Formation of maltose, a disaccharide, by condensation of glucose molecules.

atom than the combined formulas of two glucose molecules. This difference arises because a water molecule is a product of the condensation reaction.

Polysaccharides contain many monosaccharide monomers joined together via glycosidic linkages into a very large polymer. Starches and cellulose are the most abundant natural polysaccharides. D-glucose is the monomer in each of these polymers, which can contain as many as 5000 glucose units.

CONCEPTUAL EXERCISE 10.17

Sucrose Solubility

On a chemical basis, explain why table sugar (sucrose) is soluble in water.

10-7b Polysaccharides: Starches and Glycogen

Amylose turns blue-black when tested with iodine solution, whereas amylopectin turns red.

Plant starch is stored in protein-covered granules until glucose is needed for synthesis of new molecules or for energy production. If these granules are ruptured by high temperature, they yield two starches—*amylose* and *amylopectin.* Natural starches contain about 75% amylopectin and 25% amylose, both of which are polymers of glucose units joined by glycosidic linkages. Structurally, amylose is a condensation polymer with an average of about 200 glucose monomers per molecule arranged in a straight chain, like pearls on a necklace. Amylopectin is a branched-chain polymer analogous to the branched-chain synthetic polymers discussed earlier. A typical amylopectin molecule has about 1000 glucose monomers arranged into branched chains. The branched chains of glucose units give amylopectin properties different from those of the unbranched amylose. The main difference is their water solubilities. A family of enzymes called amylases helps to break down starches sequentially into a mixture of small branched-chain polysaccharides called dextrins and ultimately into glucose. Dextrins are used as food additives and in mucilage, paste, and finishes for paper and fabrics.

Animals store energy as fats rather than carbohydrates because fats have more energy per gram.

In animals, *glycogen* serves the same storage function as starch does in plants. Glycogen is stored in the liver and muscle tissues and provides glucose for "instant" energy until the process of fat metabolism can take over and serve as the energy source (← **Sec. 4-11**). The chains of D-glucose units in glycogen are more highly branched than those in amylopectin.

10-7c Cellulose, a Polysaccharide

Paper and cotton are nearly pure cellulose.

Cellulose is the most abundant organic compound on Earth, found as the woody part of trees and the supporting material in plants and leaves. Cotton is the purest natural form of cellulose. Like amylose, cellulose is composed of D-glucose monomer units. The

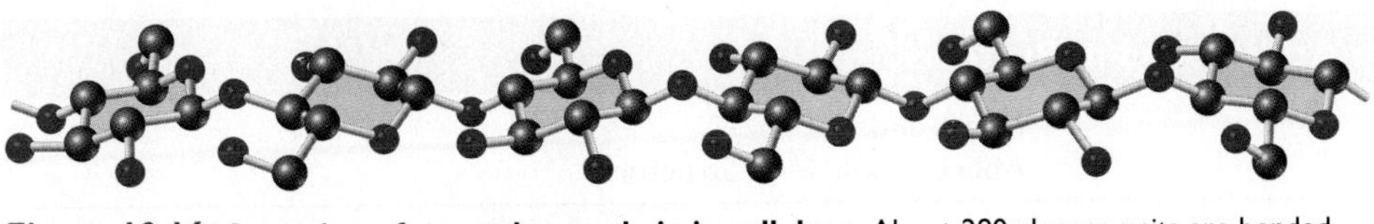

Figure 10.16 A portion of one polymer chain in cellulose. About 280 glucose units are bonded together to form a chain. Chains lie parallel to each other and are held together by hydrogen bonds. Glucose monomer units are highlighted in pink. H atoms are not shown.

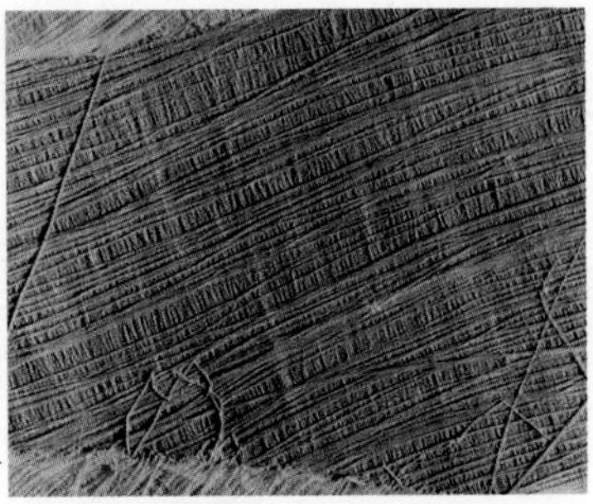

Biophoto Associates/Science Source

Parallel strands of cellulose in a plant fiber.

difference between the structures of cellulose and amylose lies in the orientation of the glycosidic linkages between the glucose monomer units. In cellulose, the —OH groups at carbons 1 and 4 are in the *trans* position, so the glycosidic linkages between glucose units alternate in direction; thus, every other glucose unit is turned over (Figure 10.16). In amylose, the —OH groups at carbons 1 and 4 are in the *cis* position, so all the glycosidic linkages are in the same direction. This subtle structural difference allows humans to digest starch, but not cellulose; we lack the enzyme necessary to break the *trans* glycosidic linkages in cellulose. However, termites, a few species of cockroaches, and ruminant mammals such as cows, sheep, goats, and camels do have the proper internal chemistry for this purpose. Because cellulose is so abundant, it would be advantageous if humans could use it, as well as starch, for food.

CONCEPTUAL EXERCISE 10.18

Digestion of Cellulose

Explain why humans cannot digest cellulose. Consult a reference on the Internet and explain why ruminant animals can digest cellulose.

CONCEPTUAL EXERCISE 10.19

What If Humans Could Digest Cellulose?

Think of some of the implications if humans could digest cellulose. What would be some desirable consequences? What would be some undesirable ones?

10-7d Amino Acids and Proteins

An **amino acid** *contains a carboxylic acid group and an amine group*. Proteins are copolymers of amino acids. The 20 naturally occurring amino acids from which proteins are made have the general formula

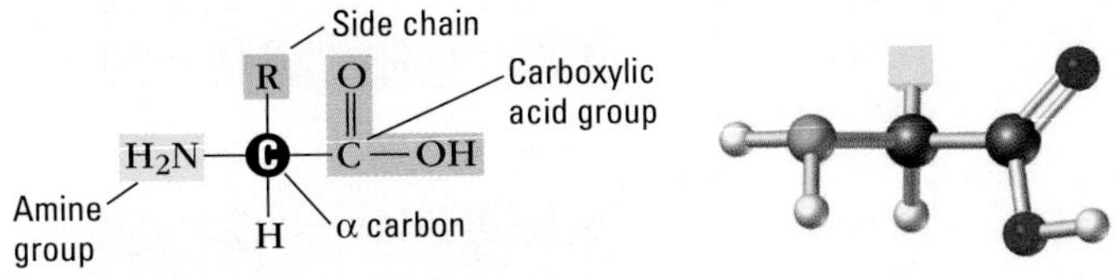
Side chain
R
O
Carboxylic
acid group
H2N
C
C—OH
Amine
group
H
α carbon

They are described as α-amino acids because the amine, $—NH_2$, group is attached to the **alpha (α) carbon**, *the first carbon next to the carboxylic acid,* —COOH, group. As shown in Table 10.8, the amino acids differ in that each has a different R group, called a *side chain*. Glycine, the simplest amino acid, has just hydrogen as its R group. Note from the table that some amino acid R groups contain only carbon and hydrogen, while others contain carboxylic acid, amine, or other functional groups. The amino acids are grouped in Table 10.8 according to whether the R group is nonpolar, polar, acidic, or basic. Each amino acid has a three-letter abbreviation for its name, also given in the table.

Table 10.8 Common L-Amino Acids Found in Proteins†

Name and Abbreviation	Structure	Name and Abbreviation	Structure
Nonpolar R groups			
Glycine Gly	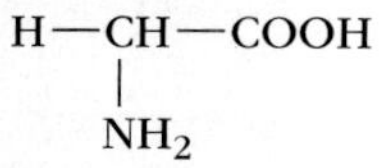	*Isoleucine Ile	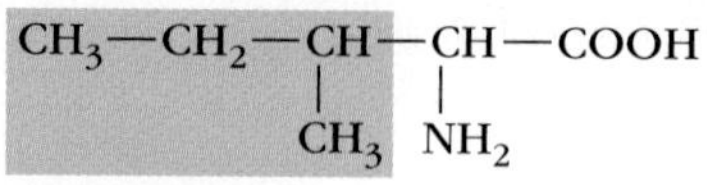
Alanine Ala	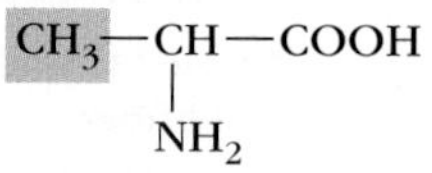	Proline Pro	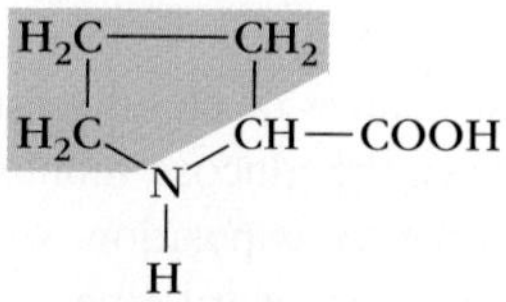
*Valine Val	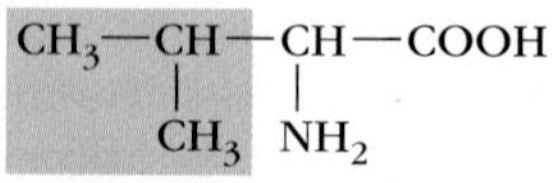	*Phenylalanine Phe	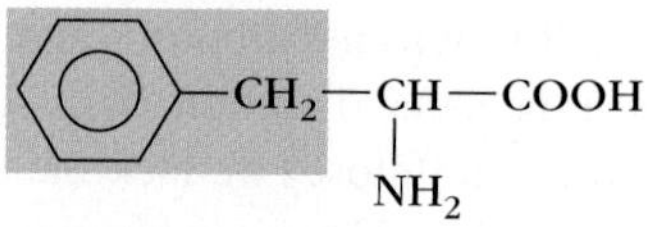
*Leucine Leu	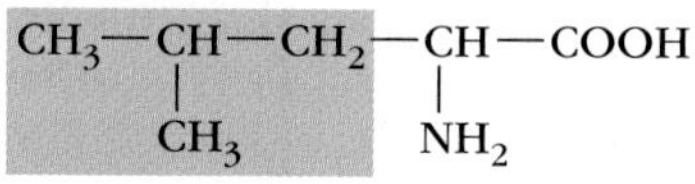	*Methionine Met	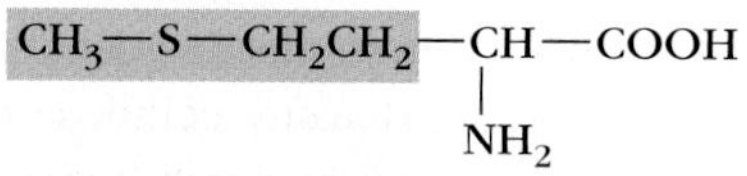
		*Tryptophan Trp	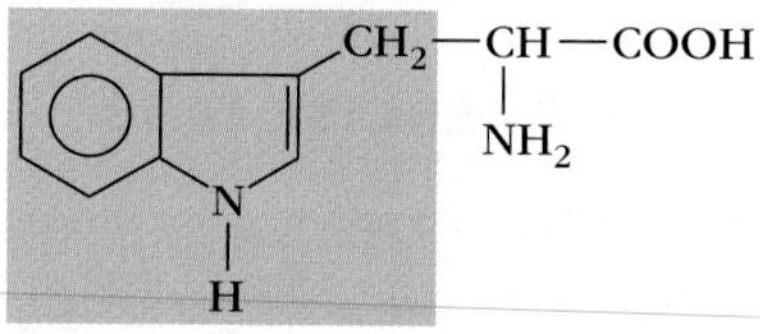
Polar but neutral R groups			
Serine Ser	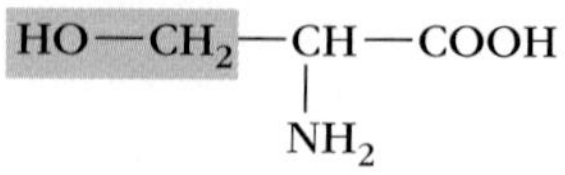	Asparagine Asn	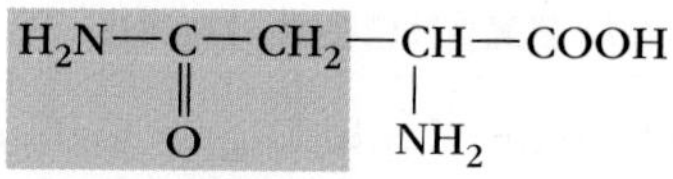
*Threonine Thr	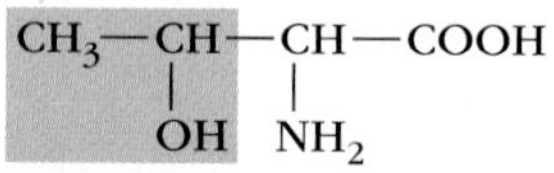	Glutamine Gln	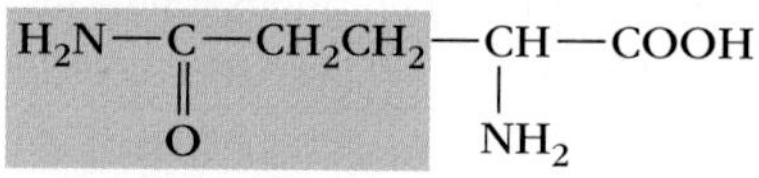
Cysteine Cys	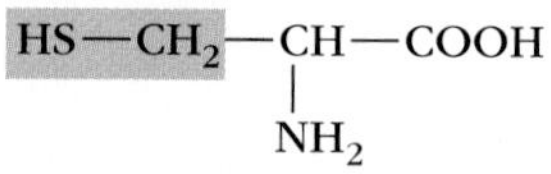	Tyrosine Tyr	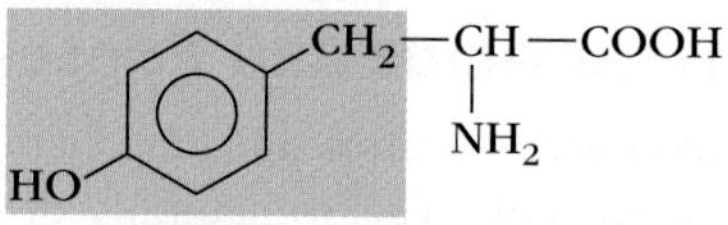
	Acidic R groups		**Basic R groups**
Glutamic acid Glu	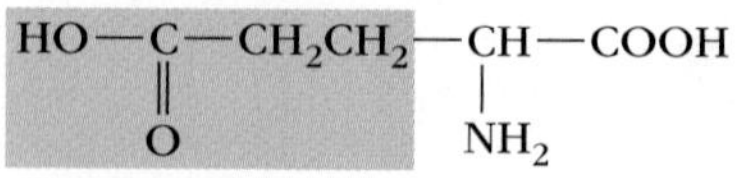	*Lysine Lys	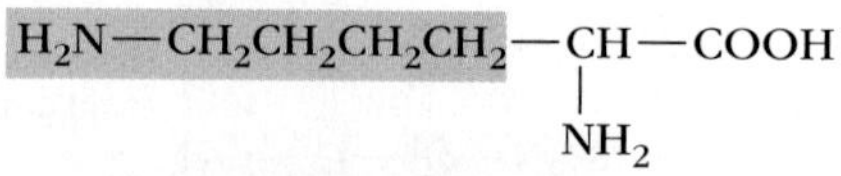
Aspartic acid Asp	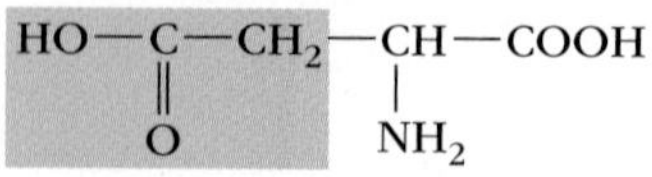	‡Arginine Arg	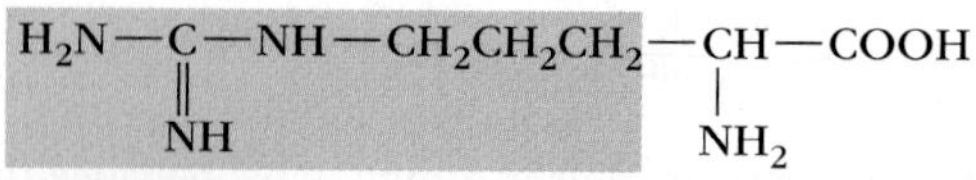
		*Histidine His	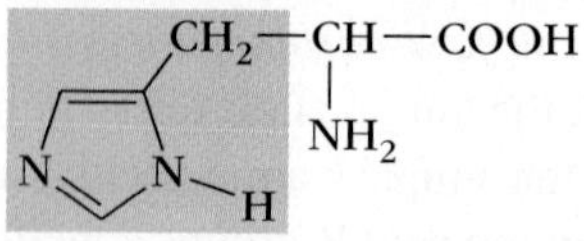

*Essential amino acids that must be part of the human diet; histidine is essential to juveniles, but not adults. The other amino acids can be synthesized by the body.
†The R group in each amino acid is highlighted.
‡Growing children also require arginine in their diet.

CONCEPTUAL EXERCISE 10.20

Hydrogen Bonding Between Amino Acids in Proteins

Choose two amino acids from Table 10.8 whose R groups can hydrogen-bond with one another if they were close together in a protein chain or in two adjacent protein chains. Then choose two whose R groups would not hydrogen-bond.

Like nylon, proteins are polyamides. The amide bond in a protein is formed by the *condensation* polymerization reaction between the amine group of one amino acid and the carboxylic acid group of another. *In proteins, the amide linkage is called a* **peptide linkage** or **peptide bond**. *A relatively small amino acid polymer (up to about 50 amino acids)* is known as a **polypeptide**. Proteins contain hundreds to thousands of amino acids bonded together.

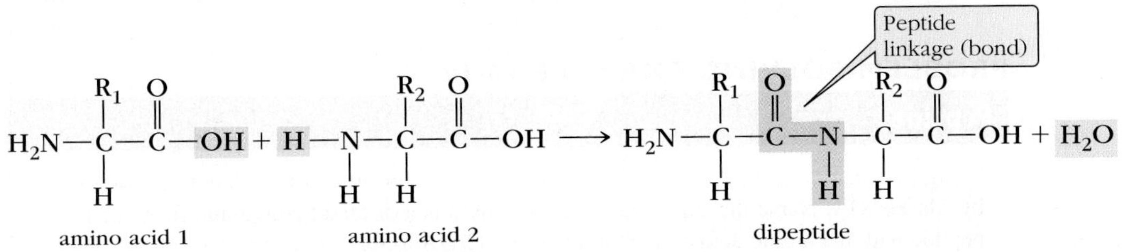
Peptide linkage (bond)
amino acid 1
amino acid 2
dipeptide

Any two amino acids can react to form two different dipeptides, depending on which amine and acid groups react from each amino acid. For example, glycine and alanine can react and join in either of these two ways:

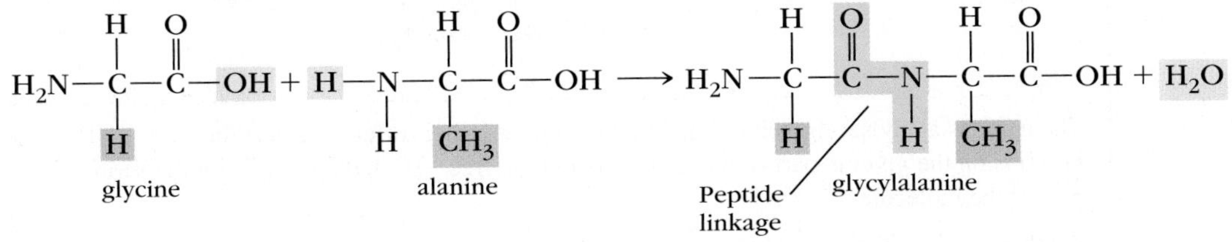
glycine
alanine
Peptide linkage
glycylalanine

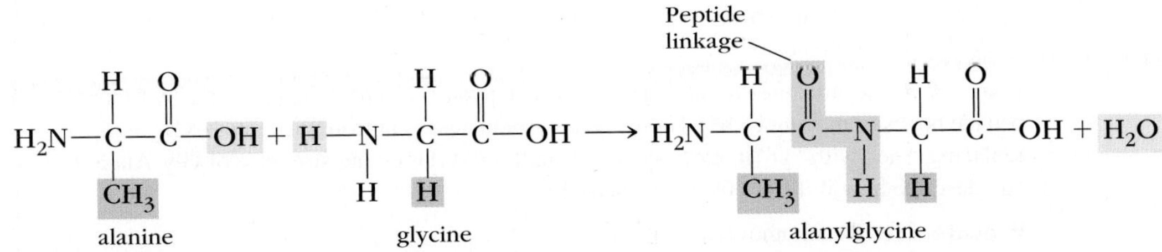
Peptide linkage
alanine
glycine
alanylglycine

Either dipeptide can react with another amino acid at each end. The extensive chains of amino acid units in proteins are built up by such condensation polymerization reactions. Polypeptide chains are named by starting at the $—NH_2$ end and naming each amino acid (with a "yl" ending) until the —COOH group is reached. Thus, as shown above, glycylalanine is different from alanylglycine.

All proteins, no matter how large, have a peptide backbone of amino acid units covalently bonded to each other through peptide linkages.

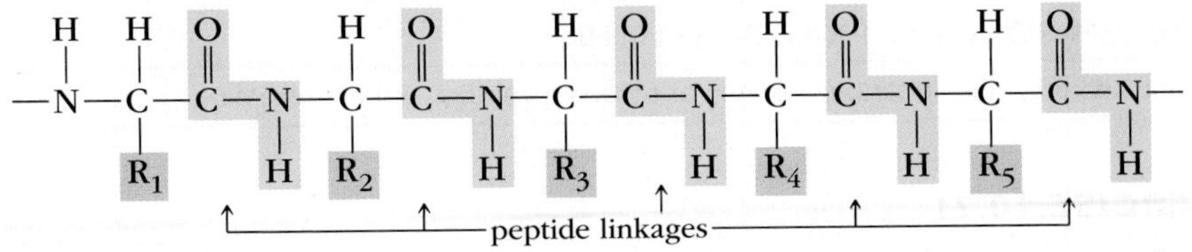
peptide linkages

As the number of amino acids in the chain increases, the number of variations quickly increases to a degree of complexity not usually found in synthetic polymers. As we have just seen, two different dipeptides can form by reacting two different amino acids. Six *tri*peptides

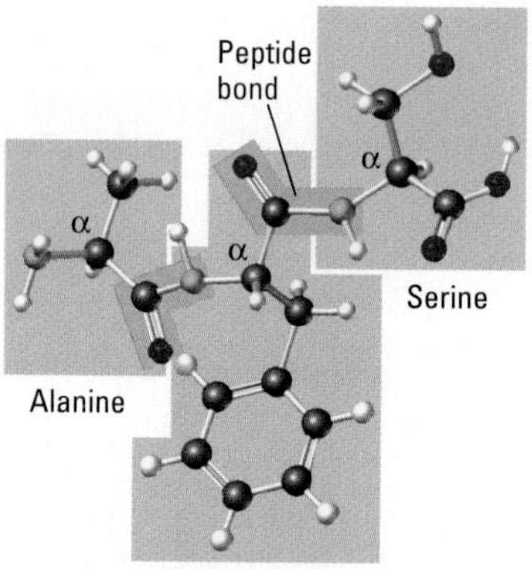
Peptide bond
α
α
α
Serine
Alanine
Phenylalanine
The Ala-Phe-Ser tripeptide

are possible if each of three different amino acids is used only once. For example, phenylalanine, Phe; alanine, Ala; and serine, Ser can be linked in these combinations:

Phe-Ala-Ser	Ser-Ala-Phe	Ala-Ser-Phe
Phe-Ser-Ala	Ser-Phe-Ala	Ala-Phe-Ser

If n different amino acids are present, the number of arrangements is $n!$ (n factorial). For four different amino acids, the number of different tetrapeptides is $4! = 4 \times 3 \times 2 \times 1 = 24$. If all 20 naturally occurring amino acids were each used once to form all possible polypeptides, there would be $20! = 2.43 \times 10^{18}$ (2.43 quintillion) unique 20-monomer polypeptide molecules. *Because a protein chain can also include more than one molecule of a given amino acid, the number of possible combinations is astronomical.* It is truly remarkable that of the many different proteins that could be made from a set of amino acids, a living cell makes only the relatively small number of proteins it needs.

PROBLEM-SOLVING EXAMPLE 10.10

Peptides

Using information from Table 10.8, draw the structural formula of the tripeptide represented by Ala-Ser-Gly. Name the tripeptide. Explain why it is a different compound from the tripeptide with the amino acids joined in the order Gly-Ala-Ser.

Result

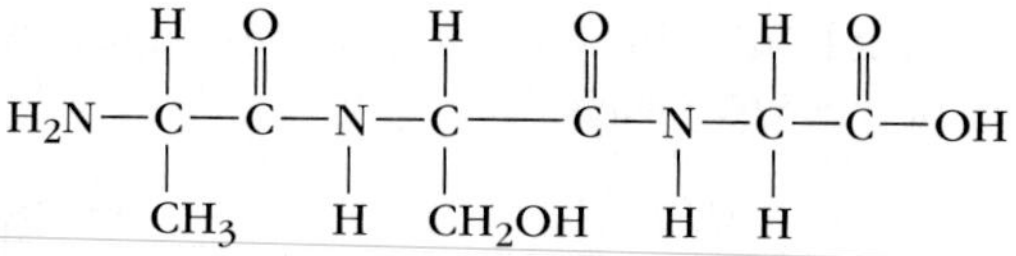
Ala-Ser-Gly

The name is alanylserylglycine. The structure Gly-Ala-Ser differs because the free —NH_2 group is on the glycine part of the molecule and the free —COOH group is on the serine part of the molecule.

Analyze The abbreviated name begins with the free H_2N— group and proceeds to the free —COOH group. Amino acids are connected by peptide bonds. The name of a polypeptide is derived from the names of the monomer amino acids.

Plan Write the structures for alanine, serine, and glycine. Connect the structures by forming peptide bonds (removing HOH). Name the polypeptide starting from the left (—NH_2) end, replacing "ine" with "yl" in each name except the last. Draw the structure of Gly-Ala-Ser and describe how it differs from Ala-Ser-Gly.

Execute See Result above.

Writing the structure of Gly-Ala-Ser shows how the two tripeptides are different.

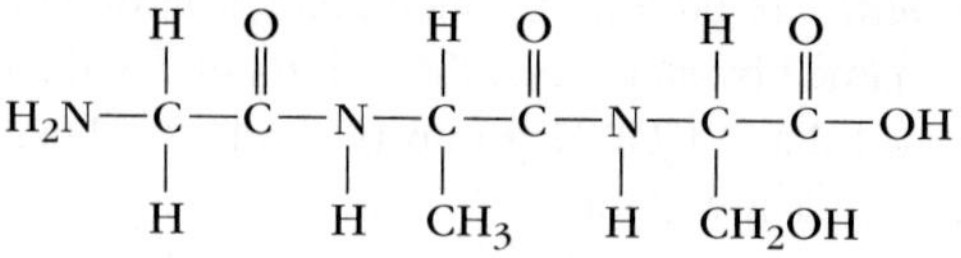
Gly-Ala-Ser

PROBLEM-SOLVING PRACTICE 10.10

Name and draw the structural formula of the tetrapeptide Lys-Phe-Ser-Ala.

EXERCISE 10.21

Peptide Sequences

Draw the structural formula of the tetrapeptide Ala-Ser-Phe-Cys. Identify each amino acid side chain as nonpolar, polar but neutral, acidic, or basic.

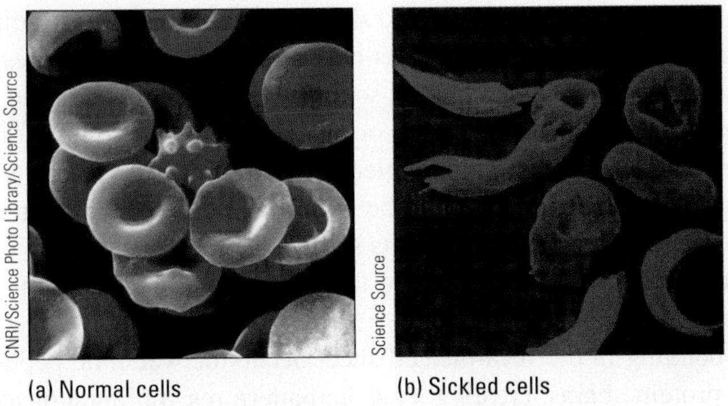
CNRI/Science Photo Library/Science Source
Science Source
(a) Normal cells
(b) Sickled cells

Figure 10.17 Red blood cells. Changing one amino acid in the primary protein structure of hemoglobin causes the disease sickle-cell anemia; symptoms are fatigue and pain.

The change in structure of hemoglobin that causes sickle-cell anemia answers the question, "How can a disease be caused or cured by a tiny change in a molecule?" that was posed in Chapter 1 (← Sec. 1-1).

10-7e Primary, Secondary, and Tertiary Protein Structure

The three-dimensional shape of a protein, and consequently the function it carries out in a living organism, is determined by the protein's **primary structure**—*the sequence of amino acids along the polymer chain*. The sequence of amino acids is reflected in the order of the R groups along the backbone. Even one amino acid out of place can create dramatic changes in the shape of a protein, which can lead to serious medical conditions. For example, the function of hemoglobin is to carry oxygen in red blood cells. Altering the primary structure of hemoglobin causes the disease sickle-cell anemia. Glutamic acid, the sixth amino acid in a 146–amino acid chain in hemoglobin, is replaced by valine. Replacement of the carboxylic acid side chain of glutamic acid with the nonpolar side chain of valine (Table 10.8) disturbs the attractions between hemoglobin molecules and causes them to gather into fibrous chains that distort the red blood cells into a sickle shape (Figure 10.17).

In addition to the sequence of amino-acid side chains, protein molecules have **secondary structure**, *regular, recurring structural patterns held in place by hydrogen bonding*. Figure 10.18 shows the two major types of secondary protein structure: the alpha-helix (α-helix) and the beta-pleated sheet (β-pleated sheet). The α-helix occurs

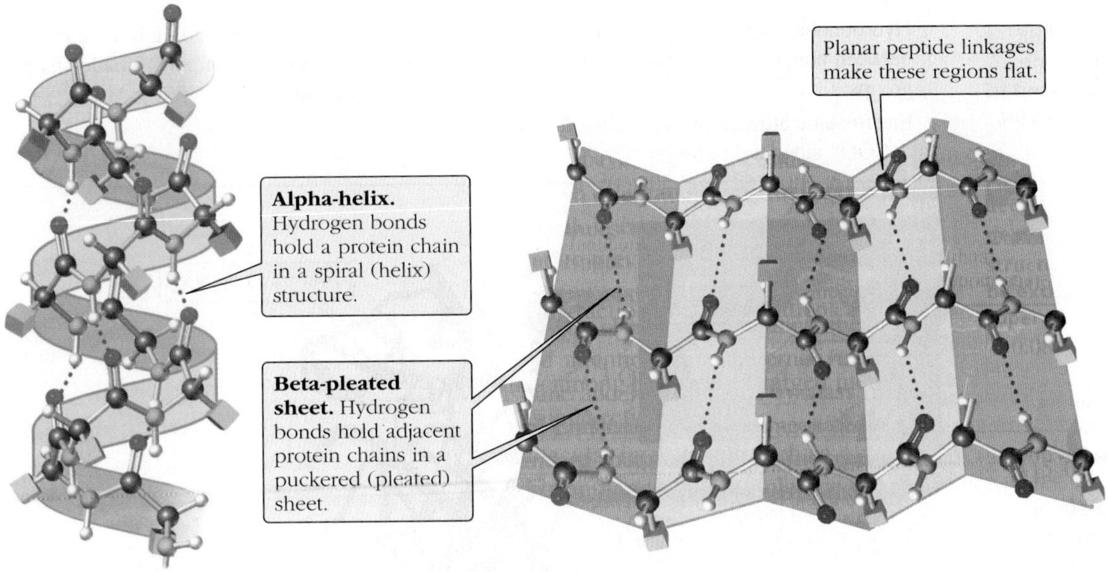
Planar peptide linkages make these regions flat.
Alpha-helix. Hydrogen bonds hold a protein chain in a spiral (helix) structure.
Beta-pleated sheet. Hydrogen bonds hold adjacent protein chains in a puckered (pleated) sheet.

Figure 10.18 Secondary protein structures: alpha-helix and beta-pleated sheet.

within a single protein chain. The β-pleated sheet involves two or more adjacent sections of protein chains; the two adjacent sections may be in separate molecules or within the same molecule where a protein chain folds so that different sections of the chain are parallel. In the alpha-helix, hydrogen bonding occurs between a hydrogen atom in an N—H bond in one peptide linkage with a lone electron pair on a C=O oxygen of a peptide bond four amino acid units farther down the backbone (Figure 10.18). In effect, the hydrogen in the hydrogen bond "looks over its shoulder," causing the protein to curl into a helix, much like the coiling of a spring. In the alpha-helix, the R groups on the amino acid units are on the outside of the helix.

Wool is primarily the protein keratin, which has an alpha-helix secondary structure. Fibroin, the principal protein in silk, is largely beta-pleated sheets.

Hydrogen bonding in the beta-pleated sheet occurs between the peptide backbones of neighboring protein chains, creating a zigzag pattern resembling a pleated sheet with the R groups above and below the sheet (Figure 10.18). The reason for the pleated, zigzag structure is that the six atoms involved in a peptide bond all are in the same plane. The peptide bond is a resonance hybrid of these two structures:

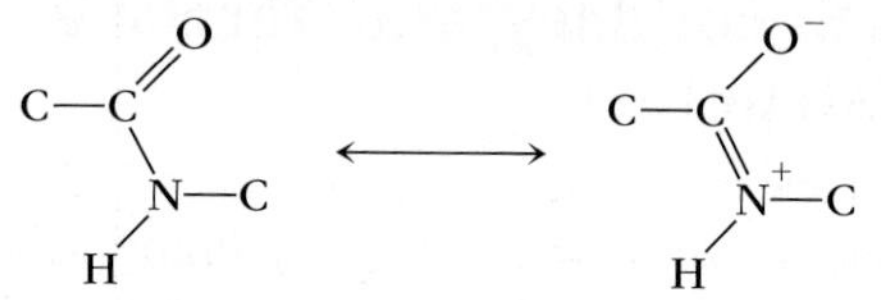

Because of the double bond in the right-hand structure, there is no free rotation around the C—N bond and both the C and N atoms have triangular-planar geometry. Thus, the only place where there can be a bend in a polypeptide chain is at the carbon atom to which the R group is attached and the sheets are pleated.

In many proteins, the backbone is folded into a globular structure as shown in Figure 10.19. Within the globular structure there may be regions of α-helix and β-pleated sheet. *The overall three-dimensional arrangement that accounts for all the twists and turns and folding of a protein* is called its **tertiary protein structure**. The twists and turns of the tertiary structure result in a protein molecule of maximum stability. Tertiary structure is determined by interactions among the side chains strategically placed along the backbone of the chain. These interactions include noncovalent forces of attraction and covalent bonding, as listed in this table and illustrated in Figure 10.20.

Noncovalent Attractions	Covalent Bonds
Hydrogen bonding between side-chain groups	Disulfide bonds, —S—S—
Hydrophobic (water-hating) interactions between nonpolar side-chain hydrocarbon groups	Coordinate covalent bonding between metal ions and electron pairs of side-chain N and O atoms
Electrostatic attractions between $—NH_3^+$ and $—COO^-$ side-chain groups	

Coordinate covalent bonding between metal ions and lone pairs of electrons is discussed in detail in Sections 14-9 and 20-6.

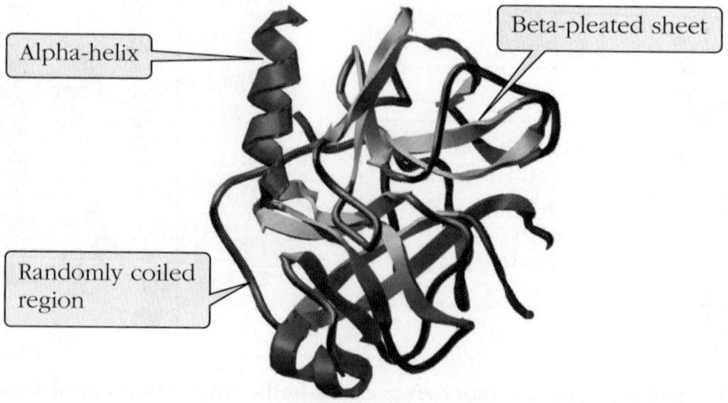
Alpha-helix
Beta-pleated sheet
Randomly coiled region

Figure 10.19 Protein folding in the enzyme chymotrypsin. Only the polypeptide backbone is shown. Alpha-helix regions are shown in blue, beta-pleated sheet regions are shown in green, and randomly coiled (neither alpha-helix nor beta-pleated sheet) regions are shown in copper.

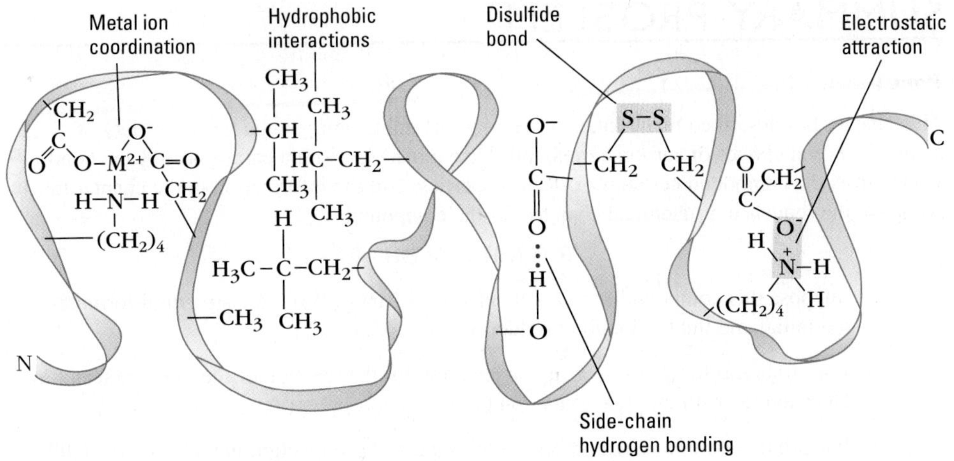
Metal ion coordination
Hydrophobic interactions
Disulfide bond
Electrostatic attraction
Side-chain hydrogen bonding
S–S
N
C

Figure 10.20 Noncovalent forces and bonds that stabilize protein tertiary structure. The tertiary structure of a protein is stabilized by several different forces.

The hydrogen bonding, hydrophobic interactions, and electrostatic attractions bring the groups closer together. Metal ions, such as Fe^{2+} in hemoglobin, are incorporated into a protein by the donation of lone pair electrons from oxygen or nitrogen atoms in side-chain groups to form coordinate covalent bonds to the metal ions. By loss of H, the —SH groups in nearby cysteine units are oxidized to form disulfide, —S—S—, covalent bonds that cross-link regions of the peptide backbone. The number and proximity of disulfide bonds help to limit the flexibility of a protein.

Proteins can be divided into two broad categories—fibrous proteins and globular proteins—that reflect differences in their tertiary structures. *Fibrous proteins,* such as hair, muscle fibers, and fingernails, have little tertiary structure and are rod-like, with the coils or sheets of protein aligned into parallel bundles, making for tough, water-insoluble materials. In contrast, *globular proteins* are highly folded, with hydrophilic (water-loving) side chains on the outside, making these proteins water-soluble. Hemoglobin and chymotrypsin (Figure 10.19) are globular proteins, as are most enzymes.

Because of the complexity and the variety of properties provided by the different R groups and associated molecules or ions, proteins are able to perform widely diverse functions in the body. Consider some of them.

Class of Protein	Example	Description of Function
Hormones	Insulin, growth hormone	Regulate vital processes according to needs
Muscle tissue	Myosin	Does mechanical work
Transport proteins	Albumin	Carry fatty acids and other hydrophobic molecules through the bloodstream
Clotting proteins	Fibrin	Form blood clots
Enzymes	Chymotrypsin	Catalyze biochemical reactions
Messengers	Endorphins	Transmit nerve impulses
Immune system proteins	Immunoglobulins	Fight off disease

SUMMARY PROBLEM

Part I

This chapter has described many kinds of carbon-containing compounds from a variety of sources. Several classes of carbon compounds have molecules with a carbon-to-oxygen bond, which gives the compound certain chemical properties. Use the information in the chapter to complete the sequence of chemical changes for the compound

$$CH_3(CH_2)_8CH_2OH$$

(a) Suppose this compound reacts with aqueous $KMnO_4$. Write the structural formulas of the initial and final oxidation products of this compound.

(b) Use structural formulas to write the equation for the reaction of the initial reactant in Question (a) with the final oxidation product in Question (a).

(c) Into what class of compounds does the product of the reaction in Question (b) fall?

(d) The final product in Question (b) reacts with aqueous sodium hydroxide. Write the equation for this reaction. What types of compounds are formed by this reaction? What would the products be if the reaction took place under acidic conditions instead of with aqueous base?

(e) The final oxidation product in Question (a) reacts with dimethylamine, $(CH_3)_2NH$. Write the structural formula of the product of this reaction. What class of compound is formed?

(f) Can the two reactants in Question (b) form a polymer? Explain.

Part II

The citric acid cycle is a series of steps involving the degradation and reformation of oxaloacetic acid. Two steps in the cycle involve a reaction studied in this chapter. One step converts isocitric acid into oxalosuccinic acid.

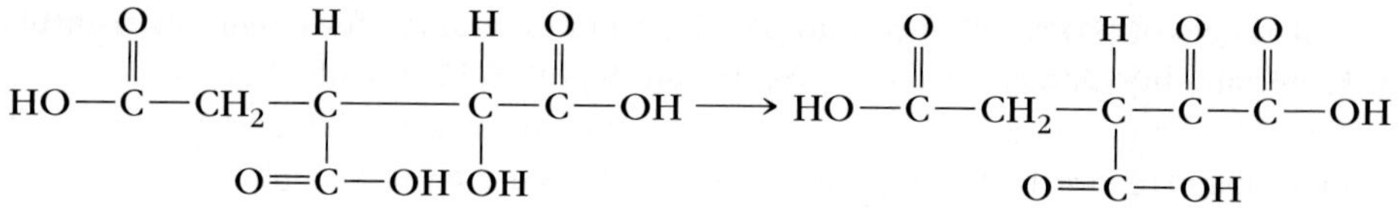

isocitric acid oxalosuccinic acid

In the last step of the cycle, L-malic acid is converted into oxaloacetic acid.

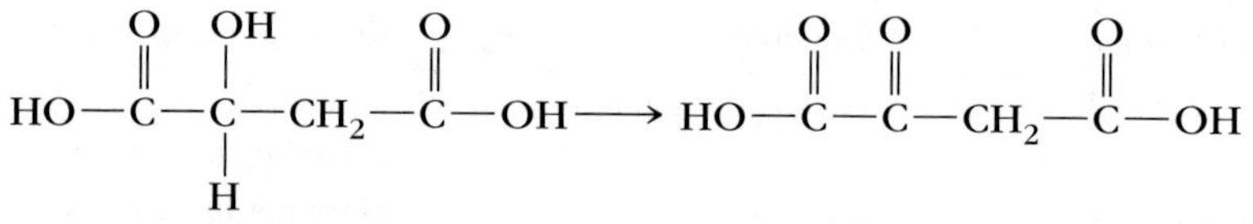

L-malic acid oxaloacetic acid

(a) Identify the general type of reaction in each step.

(b) What do the two steps have in common?

(c) Identify any chiral carbon atoms (← **Sec. 7-2f**) in the structures above in Part II.

Part III

Select any four amino acids from Table 10.8. Write the structural formulas of four different tetrapeptides that could be formed from these four amino acids.

HAVING STUDIED THIS CHAPTER . . .

. . . you should be able to:

- Describe petroleum refining, petroleum fractions, and the chemistry of increasing the octane number of gasoline (Section 10-1). End-of-chapter Questions 18, 21
- Describe the chemical nature of the major U. S. energy sources (Section 10-2). Questions 23, 25
- Explain how intermolecular forces apply in gas chromatography to separate and identify components of gaseous mixtures (*Tools of Chemistry,* Section 10-2). Question 28
- Recognize structures of primary, secondary, and tertiary alcohols (Section 10-4). Questions 30, 32
- Draw and name the structures of three functional groups produced by the oxidation of alcohols (Section 10-4). Questions 34, 36
- Give examples of the uses of some important alcohols (Section 10-4). Questions 38, 40
- Explain how nuclear magnetic resonance works and how it is used in chemical analysis and medical diagnosis (*Tools of Chemistry,* Section 10-4). Question 41
- List some properties of carboxylic acids, and write equations for the formation of esters from carboxylic acids and alcohols (Section 10-5). Questions 50, 52
- Use Lewis structures to illustrate triglyceride formation and hydrolysis (Section 10-5). Questions 43, 47
- Differentiate among saturated, monounsaturated, and polyunsaturated fats (Section 10-5). Question 45
- Identify or write the structures of the functional groups in alcohols, aldehydes, ketones, carboxylic acids, esters, and amines (Sections 10-4 to 10-6). Questions 34, 47, 66, 80
- Explain the formation of polymers by addition polymerization and give examples of synthetic polymers formed by such polymerization (Section 10-6). Questions 57, 59, 61
- Explain the formation of polymers by condensation polymerization and give examples of synthetic polymers formed by such polymerization (Section 10 6). Questions 64, 69
- Use chemical equations to show the formation of addition and condensation polymers, and identify the monomers that form them (Section 10-6). Questions 62, 74
- Identify the types of plastics being recycled most successfully (Section 10-6). Question 75
- Identify polysaccharides, their sources, the different ways they are linked, and the different uses resulting from these linkages (Section 10-7). Questions 83, 85, 87
- Illustrate the basics of protein structures, including primary, secondary, and tertiary structure, and explain how peptide linkages hold amino acids together in proteins (Section 10-7). Questions 78, 81

KEY TERMS

addition polymer (Section 10-6)

alcohol (10-4)

aldehyde (10-4d)

alpha (α) carbon (10-7d)

amide (10-6b)

amide group (10-6b)

amide linkage (10-6b)

amine (10-6b)

amino acid (10-7d)

biodiesel (10-5e)

carbohydrates (10-7a)

carboxylic acid (10-4d, 10-5a)

catalyst (10-1c)

catalytic cracking (10-1c)

catalytic reforming (10-1d)

condensation polymer (10-6)

condensation reaction (10-5b)

copolymer (10-6a)

degree of polymerization (*n*) (10-6a)

disaccharides (10-7a)

ester (10-5b)

functional group (10-3)

gas chromatography (GC) (*Tools of Chemistry,* Sec. 10-2a)

glycosidic linkage (10-7a)

hydraulic fracturing ("fracking") (10-2a)

hydrolysis (10-5d)

ketone (10-4d)

macromolecule (10-6)

monomer (10-6)

monosaccharides (10-7a)

nuclear magnetic resonance (NMR) (*Tools of Chemistry,* Sec. 10-4d)

octane number (10-1b)

peptide bond (10-7d)

peptide linkage (10-7d)

petroleum fraction (10-1a)

polyamide (10-6b)

polyester (10-6b)

polymer (10-6)

polypeptide (10-7d)

polysaccharides (10-7a)

primary structure (protein) (10-7e)

saponification (10-5d)

saturated fats (10-5c)

secondary structure (protein) (10-7e)

sugars (10-7a)

tertiary structure (protein) (10-7e)

thermoplastics (10-6)

thermosetting plastics (10-6)

triglycerides (10-5c)

unsaturated fats (10-5c)

QUESTIONS FOR REVIEW AND THOUGHT

Red-numbered questions have short answers at the back of this book in Appendix M and fully worked solutions in the *Student Solutions Manual.*

Review Questions

These questions test vocabulary and simple concepts.

1. Why is the organic chemical industry referred to as the petrochemical industry?
2. What products are produced by the petrochemical industry?
3. What is the difference between catalytic cracking and catalytic reforming?
4. Explain how the octane number of a gasoline is determined.
5. What is the difference between oxygenated gasoline and reformulated gasoline? Why are they being produced?
6. Table 10.1 lists several compounds with octane numbers above 100 and one compound with an octane number below zero. Explain why such values are possible.
7. Explain why methanol has a lower boiling point (65.0 °C) than water (100 °C).
8. Give two reasons why ethylene glycol has a higher boiling point than ethanol.
9. Outline the steps necessary to obtain 89-octane gasoline, starting with a barrel of crude oil.
10. What is the major difference between crude oil and coal as a source of hydrocarbons?
11. Write the structural formula of a representative compound for each of these classes of organic compounds: alcohols, aldehydes, ketones, carboxylic acids, esters, and amines.
12. Describe the structural feature a molecule must have to undergo addition polymerization.
13. What feature do all condensation polymerization reactions have in common?
14. Give examples of (a) a synthetic addition polymer, (b) a synthetic condensation polymer, and (c) a natural addition polymer.
15. Discuss which two plastics are currently being recycled the most successfully, and give examples of some products being made from these recycled plastics.
16. What is the difference between the formation of an addition polymer and a condensation polymer?

Topical Questions

These questions are keyed to the major topics in the chapter. Usually a question that is answered at the back of this book is paired with a similar one that is not.

Petroleum (Section 10-1)

17. What are petroleum fractions? What process is used to produce them?
18. (a) What is the boiling point range for the petroleum fraction containing the hydrocarbons that will provide fuel for your car?
    (b) Is the octane rating of "straight-run" gasoline obtained by fractional distillation of petroleum greater than 87? Explain your answer.
    (c) Would you use "straight-run" gasoline to fuel your car? Why or why not?
19. (a) Draw the Lewis structure for the hydrocarbon that is assigned an octane rating of 0.
    (b) Draw the Lewis structure for the hydrocarbon that is assigned an octane rating of 100.
    (c) What is the boiling point of each hydrocarbon?
20. Explain what is meant by this statement: "All gasolines are highly volatile."
21. What would be the advantage of removing the higher-octane components such as aromatics and alkenes from oxygenated gasolines?
22. Compare and contrast catalytic cracking and catalytic reforming. (a) How are they similar? (b) How do they differ?

U. S. Energy Sources and Consumption (Section 10-2)

23. Describe the major change that has occurred over the past few years in the percent of energy provided by fossil fuels in the United States.
24. What was the greatest energy source in the United States in 2007? In 2011?
25. Describe the major change that has occurred over the past few years in the source of natural gas in the United States.
26. Describe how hydraulic fracturing differs from conventionally drilled oil and gas wells.
27. Explain the trend in type of fuel used to generate electricity over the past five years.

Tools of Chemistry: Gas Chromatography (Section 10-2)

28. Do polar compounds appear earlier or later on a chromatogram when a nonpolar stationary phase is used? Explain your answer.
29. If you were to analyze an oxygenated gasoline using GC, would you use a less polar or more polar stationary phase than the one used for unoxygenated gasoline?

Alcohols and Their Oxidation Products (Section 10-4)

30. Give an example of (a) a primary alcohol, (b) a secondary alcohol, and (c) a tertiary alcohol. Draw Lewis structures for each example.
31. Classify each alcohol as primary, secondary, or tertiary.

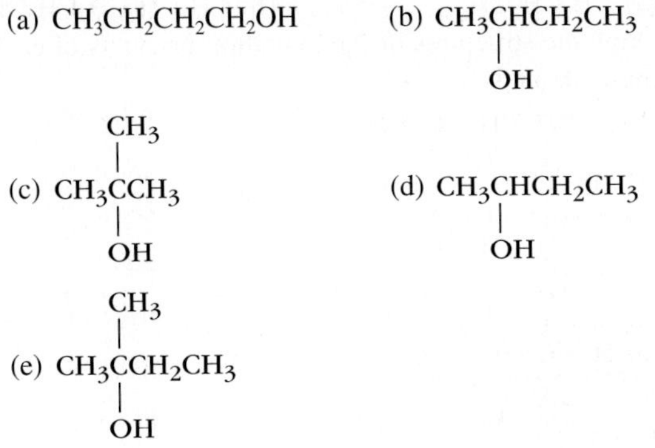

32. Classify each alcohol as primary, secondary, or tertiary.

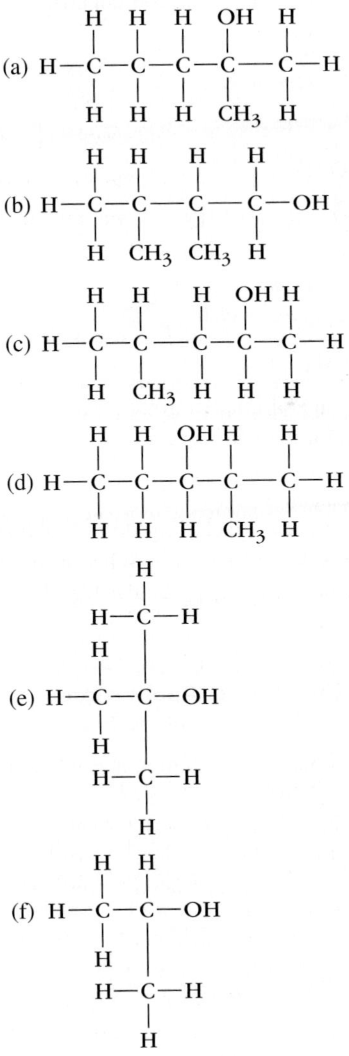

33. Explain what *oxidation* of organic compounds usually involves. What is meant by *reduction* of organic compounds?
34. Draw the structures of the first two oxidation products of each of these alcohols.
    (a) CH_3CH_2OH (b) $CH_3CH_2CH_2CH_2OH$
35. Draw the structures of the oxidation products of each of these alcohols.
    (a) 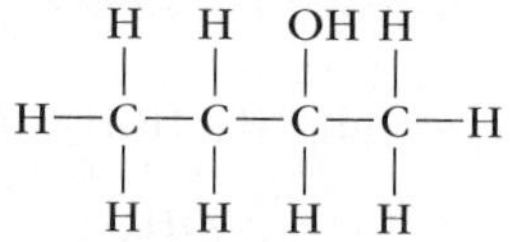
    (b) 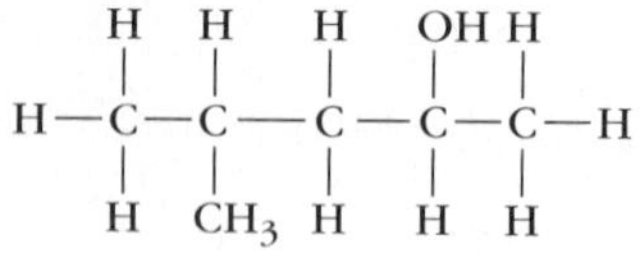
36. Write the condensed structural formula of the alcohols that can be oxidized to make these compounds.
    (a) 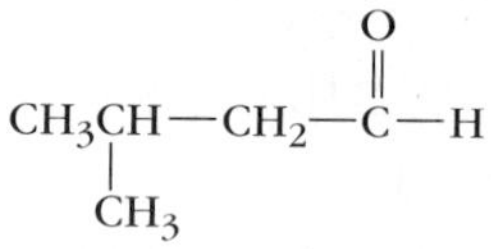
    (b) 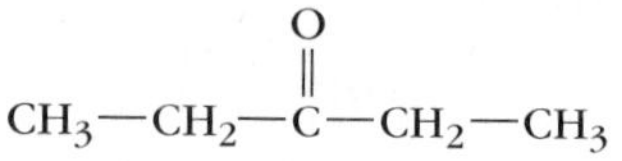
    (c) 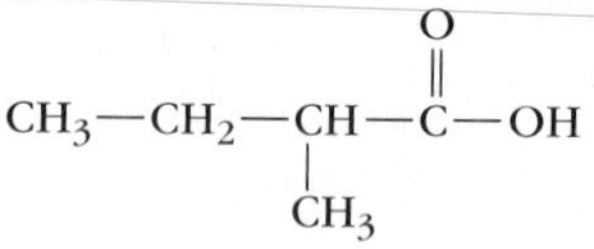
37. What is the percentage of ethanol in 90-proof vodka?
38. Explain how the common name *grain alcohol* for ethanol came about.
39. What is denatured alcohol? Why is it made?
40. Many biological molecules, including steroids and carbohydrates, contain many —OH groups. What need might biological systems have for this particular functional group?

Tools of Chemistry: NMR and Its Applications (Section 10-4d)

41. What kind of electromagnetic radiation is used in nuclear magnetic resonance (NMR)?
42. In magnetic resonance imaging (MRI), the intensity of the emitted signal is related to the ________ of hydrogen nuclei and the relaxation time is related to the type of tissue being examined.

Carboxylic Acids and Esters (Section 10-5)

43. Using structural formulas, write the equation for the formation of a triglyceride formed by the reaction of 1 mol glycerol with 2 mol stearic acid and 1 mol oleic acid.
44. Follow the directions given in Question 43, but form a triglyceride different from that obtained in Question 43.
45. For the three fatty acids given below,
    (a) classify each as saturated or mono-, di-, or polyunsaturated.
    (b) write a balanced equation for formation of a triglyceride that incorporates all three.
    (i) arachidonic acid
    $$CH_3{-}(CH_2)_4{-}(CH{=}CH{-}CH_2)_4{-}(CH_2)_2{-}COOH;$$
    (ii) nervonic acid
    $$CH_3{-}(CH_2)_7{-}CH{=}CH{-}(CH_2)_{13}{-}COOH;$$
    (iii) myristic acid
    $$CH_3{-}(CH_2)_{12}{-}COOH$$
46. For the three fatty acids given below,
    (a) classify each as saturated or mono-, di-, or polyunsaturated.
    (b) write a balanced equation for formation of a triglyceride that incorporates all three.
    (i) α-linolenic acid
    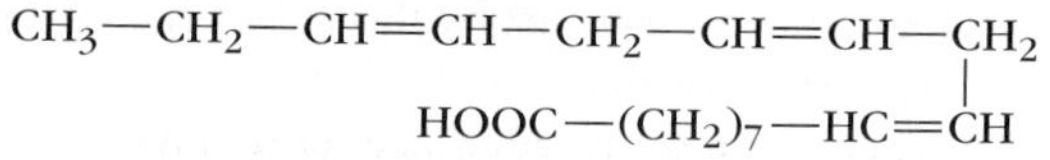
    (ii) lignoceric acid
    $$CH_3{-}(CH_2)_{22}{-}COOH$$
    (iii) palmitoleic acid
    $$CH_3{-}(CH_2)_5{-}CH{=}CH{-}(CH_2)_7{-}COOH$$
47. Melissyl cerotate has this structural formula:
    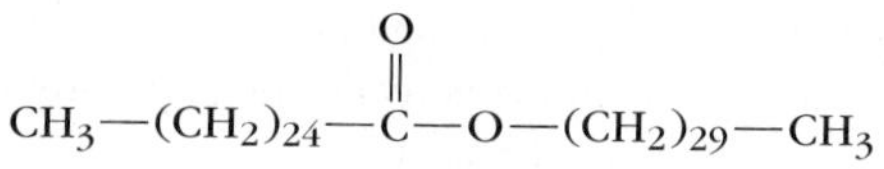
    (a) Identify the type of compound melissyl cerotate is.
    (b) Write the structural formulas of the compounds produced by the hydrolysis of melissyl cerotate.
48. Beeswax contains this compound:
    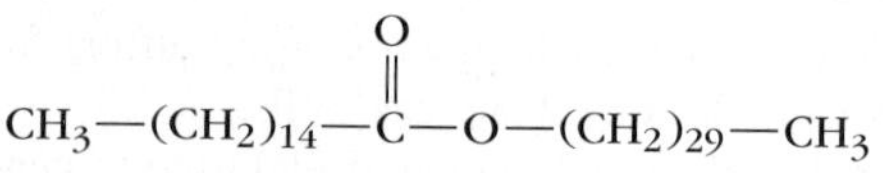
    (a) Identify what type of compound this is.
    (b) Write the structural formulas of the compounds produced by the hydrolysis of this compound.
49. Explain why the boiling points for carboxylic acids are higher than those for alcohols with comparable numbers of electrons.
50. Write the structural formula of the ester that can be formed from each reaction.
    (a) $CH_3COOH + CH_3CH_2OH$
    (b) $CH_3CH_2COOH + CH_3CH_2CH_2OH$
    (c) $CH_3CH_2COOH + CH_3OH$

51. Write the structural formula for the ester that can be produced by each reaction.
(a) Formic acid + methanol
(b) Butyric acid + ethanol
(c) Acetic acid + 1-butanol
(d) Propanoic acid + 2-propanol

52. Write the condensed formulas of the alcohol and acid that reacts to form each ester.
(a) $CH_3CH_2C(=O)—OCH_3$
(b) $HC(=O)—OCH_2CH_3$
(c) $CH_3C(=O)—OCH_2CH_3$

53. Explain why carboxylic acids are more soluble in water than are esters with the same number of electrons.

54. Explain why esters have lower boiling points than carboxylic acids with the same number of electrons.

Synthetic Organic Polymers (Section 10-6)

55. Give two examples of thermoplastics. What are the properties of thermoplastics when heated and cooled?

56. Give two examples of everyday items that are thermosetting plastics. What are the properties of thermosetting plastics when heated and cooled?

57. Draw the structure of the repeating unit in a polymer in which the monomer is
(a) 1-butene. (b) 1,1-dichloroethylene.
(c) vinyl acetate.

58. What is the principal structural difference between low-density and high-density polyethylene? Is polyethylene an addition or a condensation polymer?

59. Methyl methacrylate has the structural formula shown in Table 10.6 (**Sec. 10-6a**). When polymerized, it is very transparent, and it is sold in the United States under the trade names Lucite and Plexiglas.
(a) Use structural formulas to write the chemical equation for the formation of a segment of poly(methyl methacrylate) having four monomer units.
(b) Identify the repeating unit in the polymer.

60. Use structural formulas and data from Table 10.6 (**Sec. 10-6a**) to write the chemical equation for the formation of a polymer of five monomer units from each monomer.
(a) Acrylonitrile
(b) Propylene

61. Use structural formulas and data from Table 10.6 (**Sec. 10-6a**) to write the chemical equation for the formation of a polymer of five monomer units from each monomer.
(a) Vinylidene chloride, $H_2C{=}CCl_2$
(b) Tetrafluoroethylene

62. Write a structural formula for the monomer used to prepare each polymer.
(a) $—CH_2CH_2CH_2CH_2CH_2CH_2CH_2CH_2CH_2—$
(b) $—CH(CH_3)CH_2CH(CH_3)CH_2CH(CH_3)CH_2—$
(c) 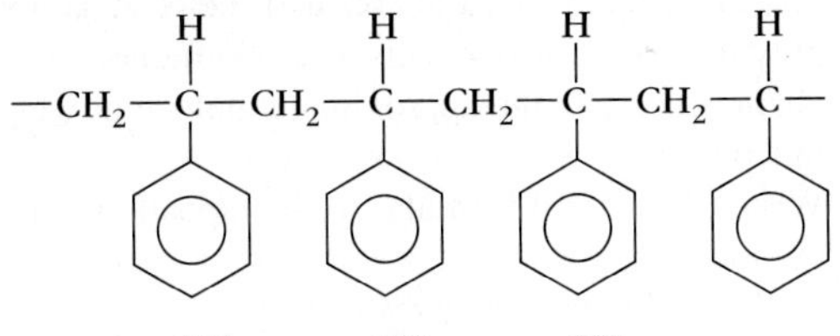
(d) $—CH_2—C(CH_3)(Cl)—CH_2—C(CH_3)(Cl)—CH_2—C(CH_3)(Cl)—$
(e) $—CH_2CH(C{=}O\,OC_2H_5)CH_2CH(C{=}O\,OC_2H_5)CH_2CH(C{=}O\,OC_2H_5)CH_2CH(C{=}O\,OC_2H_5)—$

63. Write the structural formula of four units of the polymer made from the reaction of $H_2N—(CH_2)_4—NH_2$ and HOOCCOOH. Indicate the repeating unit of the polymer.

64. 4-Hydroxybenzoic acid can form a condensation polymer. Use structural formulas to write the chemical equation for the formation of four units of the polymer. Indicate the repeating unit of the polymer.

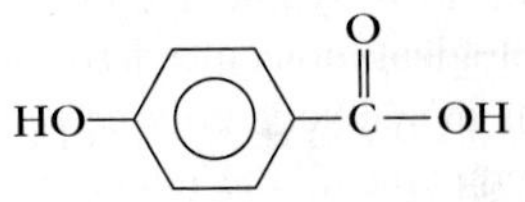

4-hydroxybenzoic acid

65. What are the two monomers used to make SBR?

66. Which functional groups must be present in a single monomer to form a polyester?

67. Name one important polyester polymer and its uses.

68. Polyamides are made by condensing which functional groups? Name the most common example of this class of synthetic polymers.

69. Draw structures of monomers that could form each of these condensation polymers.
(a) $—C(=O)—(CH_2)_8—C(=O)—NH(CH_2)_6NH—$
(b) 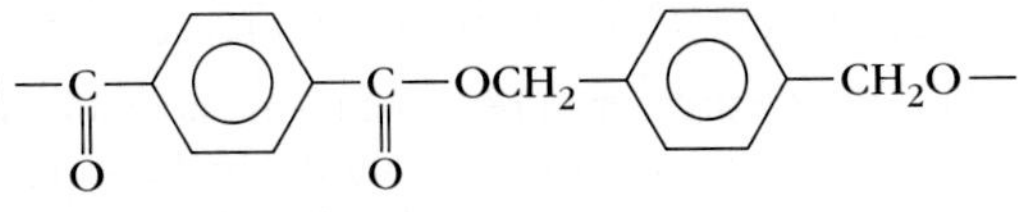

70. Orlon has this polymeric chain structure:

$$—CH_2—CH(CN)—CH_2—CH(CN)—CH_2—CH(CN)—$$

What is the monomer from which this structure can be made?

71. How many ethylene units are in a polyethylene molecule that has a molecular weight of approximately 42,000?
72. A sample of high molecular weight (HMW) polyethylene has a molecular weight of approximately 450,000. Determine its degree of polymerization.
73. Ultrahigh molecular weight polyethylene has high wear resistance and is used in conveyor belts. A sample of the polymer has approximately 150,000 monomer units per chain. Calculate the approximate molecular weight of this polymer.
74. Write a structural formula for the repeating unit of each polymer.
    (a) Natural rubber (poly-*cis*-isoprene)
    (b) Neoprene 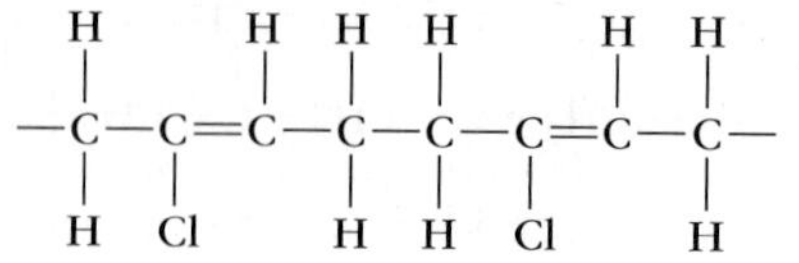
    (c) Polybutadiene
75. What are some major end uses for recycled PET and HDPE polymers?

Biopolymers: Polysaccharides and Proteins (Section 10-7)

76. State one major difference between proteins and synthetic polyamides.
77. How are amide linkages and peptide linkages similar? How are they different?
78. Which biological molecules have monomer units that are not all alike, as in synthetic copolymers?
79. Which biological molecules have monomer units that are all alike, as in synthetic polymers?
80. Identify and name all the functional groups in this tripeptide.

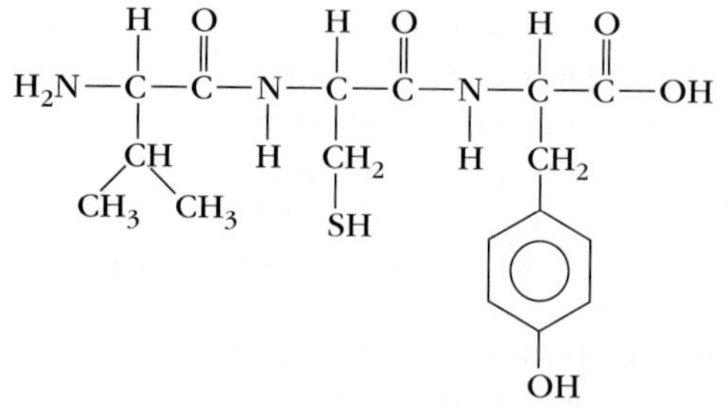

81. Draw the structural formula of alanylglycylphenylalanine.
82. Draw the structural formula of leucylmethionylalanylserine.
83. Explain the difference between (a) monosaccharides and disaccharides, and (b) disaccharides and polysaccharides.
84. What is the chief function of glycogen in animal tissue?
85. What polysaccharides yield only D-glucose upon complete hydrolysis?
86. (a) How do amylose and amylopectin differ?
    (b) How are they similar?
    (c) Are amylose and glycogen similar?
87. (a) Explain why humans can use glycogen but not cellulose for energy.
    (b) Why can cows digest cellulose?

General Questions

These questions are not explicitly keyed to chapter topics; many require integration of several concepts.

88. The total mass of carbon in living systems on Earth is estimated to be 7.5×10^{17} g; the total mass of carbon in all forms on Earth is estimated as 7.5×10^{22} g. Calculate the concentration of organic carbon as a fraction of total carbon. Express your answer as percent and in parts per million (ppm).
89. In his 1989 essay "The End of Nature," William McKibben states that ". . . a clean burning automobile engine will emit 5.5 lb of carbon in the form of carbon dioxide for every gallon of gasoline it consumes." Assume that the gasoline is C_8H_{18} and its density is 0.703 g/mL.
    (a) Write a balanced equation for the complete combustion of gasoline.
    (b) Is McKibben's assertion correct? Support your answer with calculations.
90. Compounds A and B both have the molecular formula C_2H_6O. The boiling points of compounds A and B are 78.5 °C and −23.7 °C, respectively. Use the table of functional groups in Appendix E and write the structural formulas and names of the two compounds.
91. Explain why ethanol, CH_3CH_2OH, is soluble in water in all proportions, but decanol, $CH_3(CH_2)_9OH$, is almost insoluble in water.
92. Nitrile rubber (Buna N) is a copolymer of two parts 1,3-butadiene to one part acrylonitrile. Draw the repeating unit of this polymer.
93. How are rubber molecules modified by vulcanization? Write a structure as part of your explanation.
94. Write the condensed structural formula for 3-ethyl-5-methyl-3-hexanol. Is this a primary, secondary, or tertiary alcohol?
95. Using structural formulas, write a reaction for the hydrolysis of a triglyceride that contains fatty acid chains, each consisting of 16 total carbon atoms.
96. Is the plastic wrap used in covering food a thermoplastic or thermosetting plastic? Explain.
97. Assume that a car burns pure octane,

$$C_8H_{18}\ (d = 0.703\ \text{g/cm}^3)$$

    (a) Write the balanced equation for burning octane in air, forming CO_2 and H_2O.
    (b) If the car has a fuel efficiency of 32 miles per gallon of octane, what volume of CO_2 at 25 °C and 1.0 atm is generated when the car goes on a 10.0-mile trip? (The volume of 1 mol $CO_2(g)$ at 25 °C and 1 atm is 24.5 L.)
98. Perform the same calculations as in Question 97, but use methanol, CH_3OH (d = 0.791 g/cm³), as the fuel. Assume the fuel efficiency is 20.0 miles per gallon.

99. Show structurally why glycogen forms granules when stored in the liver, but cellulose is found in cell walls as sheets.

100. Polytetrafluoroethylene (Teflon) is made by first treating HF with chloroform, then cracking the resultant difluorochloromethane.

$$CHCl_3 + 2\ HF \longrightarrow CHClF_2 + 2\ HCl$$

$$2\ CHClF_2 \xrightarrow[T]{\text{high}} F_2C{=}CF_2 + 2\ HCl$$

$$F_2C{=}CF_2 \xrightarrow{\text{peroxide catalyst}} \text{Teflon}$$

You want to make 1.0 kg Teflon. Calculate the mass of chloroform and HF required to make the starting material, $CHClF_2$. (Although it is not realistic, assume that each reaction step proceeds to a 100% yield.)

101. A 1.685-g sample of a hydrocarbon is burned completely to form 5.287 g carbon dioxide and 2.164 g water.
(a) Determine the empirical formula of the hydrocarbon.
(b) Identify whether it is an alkane or an alkene.
(c) Write a plausible Lewis structure for it.

102. Suppose 2.511 g of a hydrocarbon is burned completely to form 7.720 g carbon dioxide and 3.612 g water.
(a) Determine the empirical formula of the hydrocarbon.
(b) Identify whether it is an alkane or an alkene.
(c) Write a plausible Lewis structure for it.

Applying Concepts

These questions test conceptual learning.

103. Draw a structural formula for each of three possible repeating units of poly(vinyl chloride).

104. Two different compounds each have the formula $C_2H_4O_2$. Each contains a C=O group. Write the Lewis structure of each compound and identify the functional group or functional groups in each one.

105. Glycolic acid, $HOCH_2COOH$, and lactic acid form a copolymer used as absorbable stitches in surgery. Use structural formulas to write a chemical equation showing the formation of four units of the copolymer.

106. A particular condensation copolymer forms a silk-like synthetic fabric. A portion of the polymer's structure is shown below.

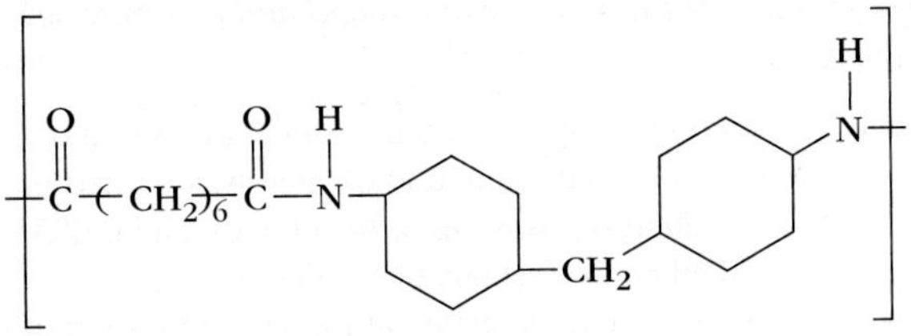

Write the structural formulas of the two monomers used to synthesize the polymer.

107. Hydrogen bonds can form between propanoic acid molecules and between 1-butanol molecules. Draw all the propanoic acid molecules that can hydrogen-bond to the one shown below. Draw all the 1-butanol molecules that can hydrogen-bond to the one shown below. Use dotted lines to represent the hydrogen bonds.

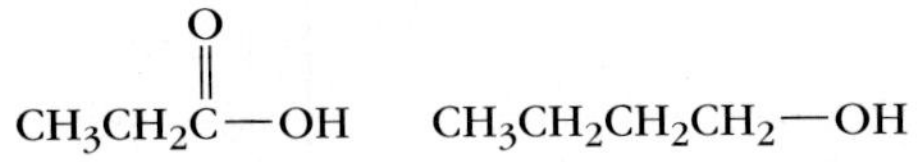

propanoic acid 1-butanol

Based on your drawings, which should have the higher boiling point, propanoic acid or 1-butanol? Explain your reasoning.

108. Both propanoic acid and ethyl methanoate form hydrogen bonds with water. Draw all the water molecules that can hydrogen-bond to these molecules. Use dotted lines to represent the hydrogen bonds.

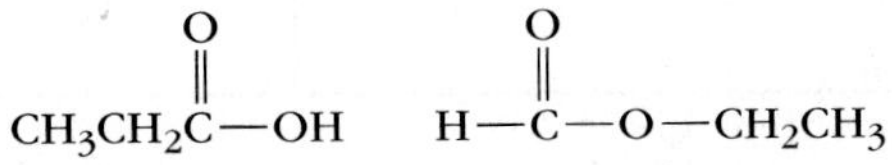

propanoic acid ethyl methanoate

Based on your drawings, which should be more soluble in water: propanoic acid or ethyl methanoate? Explain your reasoning.

109. What monomer formed this polymer?

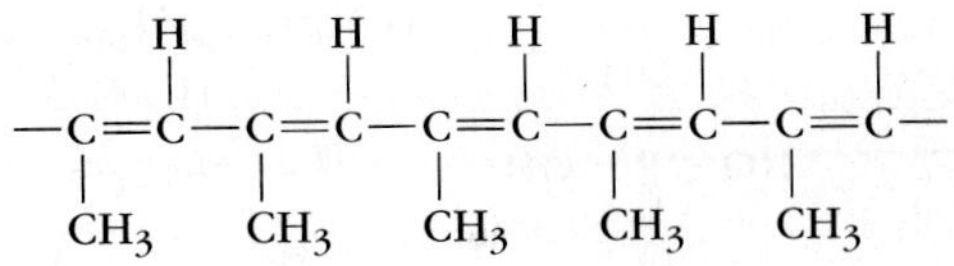

110. The illustrations below represent two different samples of polyethylene, each with the same number of monomer units (represented by dots •). Which one is high-density polyethylene and which is low-density polyethylene? Explain.

(a)

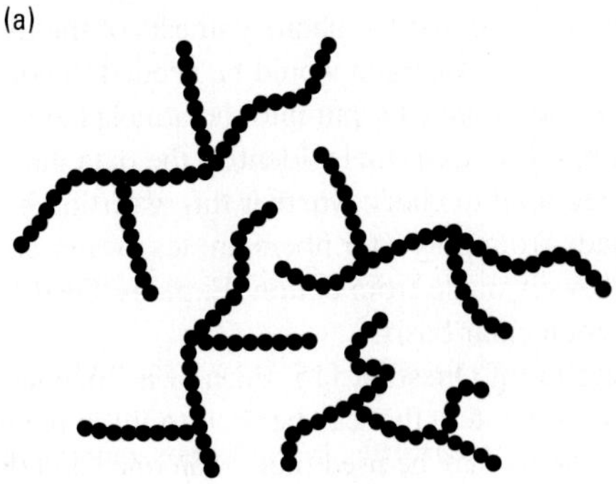

(b)

Grid for Question 115.

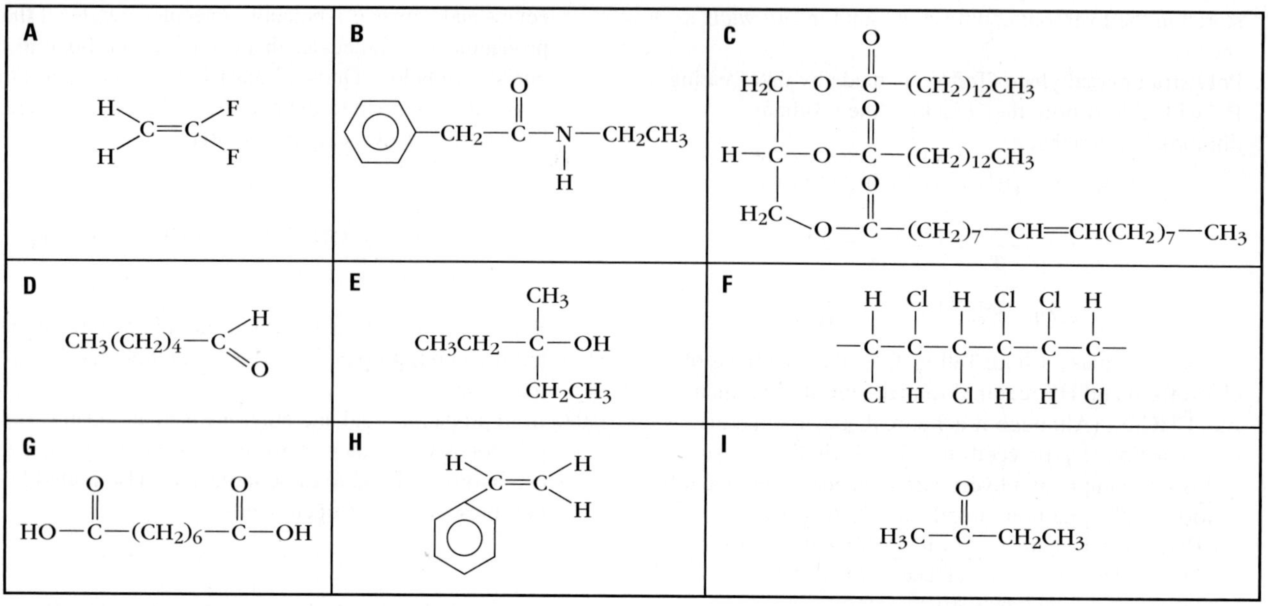

111. The backbone of a DNA molecule is a polymer of alternating sugar (deoxyribose) and phosphoric acid units held together by a phosphate ester bond. Draw a segment of the polymer consisting of at least two sugar and two phosphate units. Circle the phosphate ester bonds.

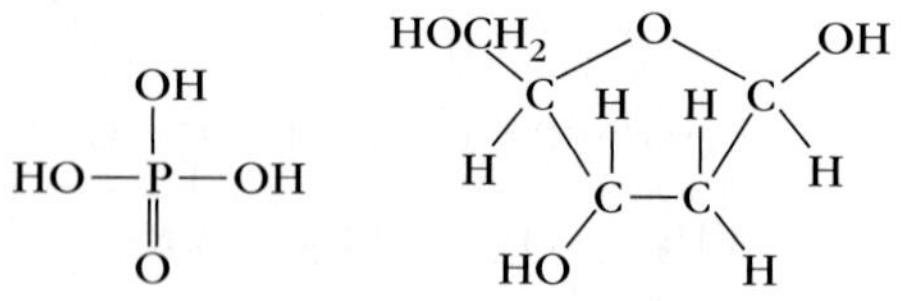

phosphoric acid   deoxyribose

112. Draw the structure of a molecule that could undergo a condensation reaction with itself to form a polyester. Draw a segment of the polymer consisting of at least five monomer units.
113. It has been asserted that the photosynthesis of the trees in a forest the size of Australia would be needed to compensate for the additional CO_2 put into the atmosphere each year from burning fossil fuels. Identify the data that would be required to check whether this assertion is valid.
114. Towels made from polyester fibers are less water absorbent than towels made from cotton. Explain this difference on a molecular basis.
115. Locate the grid for Question 115. Each of its nine lettered boxes contains an item that can be used to fill in the blanks that follow. Items may be used more than once and there may be more than one correct item for a blank. Place the letter(s) of the correct selection(s) on the appropriate line.
    (a) A triglyceride ________
    (b) Final oxidation product of a primary alcohol ________
    (c) Condensation polymerization monomer ________
    (d) Segment of an addition polymer molecule ________
    (e) Initial oxidation product of a primary alcohol ________
    (f) Contains an amide bond ________
    (g) Addition polymerization monomer ________
    (h) Contains an ester linkage ________
    (i) Final oxidation product of a secondary alcohol ________
    (j) A tertiary alcohol ________
116. Locate the grid for Question 116. Each of its nine lettered boxes contains an item that can be used to fill in the blanks that follow. Items may be used more than once and there may be more than one correct item for a blank. Place the letter(s) of the correct selection(s) on the appropriate line.
    (a) Can be oxidized to an aldehyde ________
    (b) Segment of a condensation polymer molecule ________
    (c) Can be oxidized to a carboxylic acid ________
    (d) A primary alcohol ________
    (e) A triglyceride ________
    (f) Addition polymerization monomer ________
    (g) Contains an amide linkage ________
    (h) Oxidized to form a ketone ________
    (i) Contains an ester linkage ________
    (j) Condensation polymerization monomer ________

More Challenging Questions

These questions require more thought and integrate several concepts.

117. A 0.0446-g sample of a pure liquid compound containing only carbon, hydrogen, and oxygen was burned in a combustion analysis forming 0.0979 g CO_2 and 0.0535 g H_2O.
    (a) Calculate the formula of the liquid.
    (b) Write a balanced equation for the combustion of the liquid.
    (c) Explain whether the chemical formula derived in part (a) is the empirical or molecular formula of the compound.
    (d) Write three possible Lewis structures for the compound and identify the functional groups in each.
    (e) When reacted with an acidic potassium dichromate solution, the pure compound forms acetone. Based on this, which of the structures written for part (d) is correct?

Grid for Question 116.

A $CH_3CH_2CH_2OH$	**B** $H_2N—(CH_2)_8—NH_2$	**C** 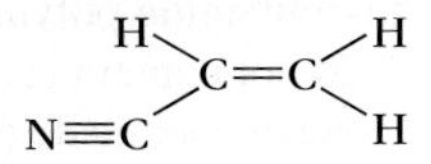
D 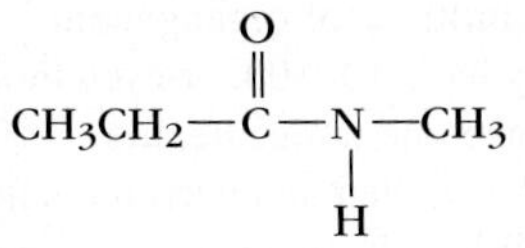	**E** 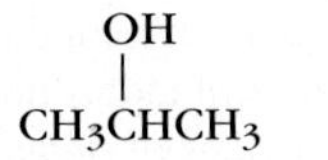	**F** 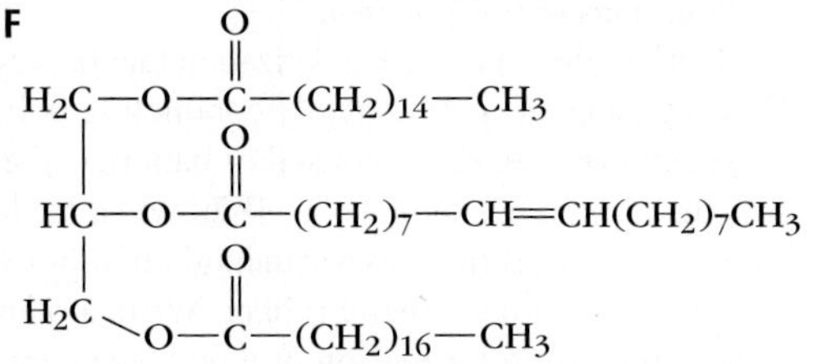
G 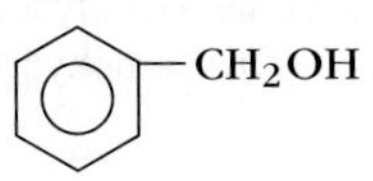	**H** 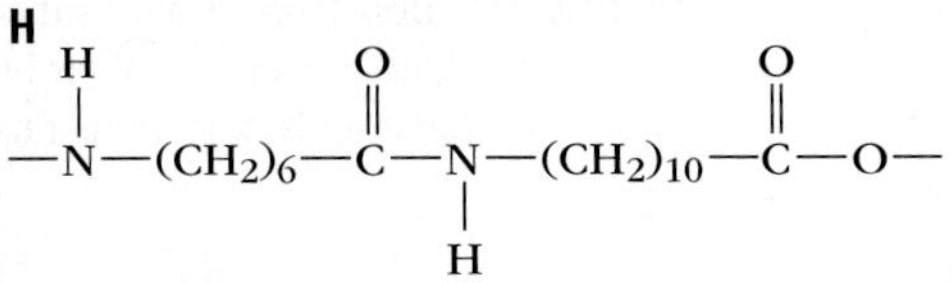	**I** 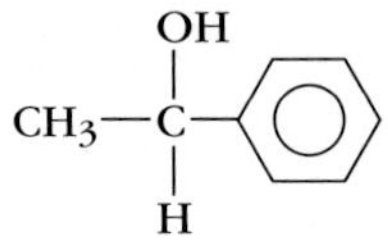

(f) When 10.00 g of the pure compound is burned, 333.8 kJ of thermal energy is released. Calculate the standard formation enthalpy of the compound.

118. Consider the compound C_3H_7NO in which there is no nitrogen-to-oxygen bond and there is an ethyl group.
(a) Write a Lewis structure for the compound.
(b) Draw two resonance structures for this molecule.
(c) Use VSEPR theory to predict the geometry around each carbon atom and the nitrogen atom in each resonance structure.
(d) Experimental evidence indicates that the nitrogen atom is triangular planar with bond angles close to 120°. Which prediction in part (c) is correct?
(e) Describe the hybridization of the nitrogen atom.

119. Ferritin is a protein that stores and releases iron. The iron is stored as Fe^{3+} and must be reduced to Fe^{2+} to be released. There are two types of channels through which Fe^{2+} could be released. One channel is lined with the amino acids glutamic and aspartic acid; the other channel is lined with the amino acid leucine (Table 10.8, **Sec. 10-7d**). Which is the more likely channel through which Fe^{2+} will leave ferritin? Explain your answer.

120. A hydrocarbon is 90.0% carbon and 10.0% hydrogen by mass; its molar mass is 40.1 g/mol. The hydrocarbon reacts with bromine by an addition reaction. It also undergoes hydrogenation with a nickel catalyst; 1.58 g of the hydrocarbon reacts with 1.77 L H_2 (measured at STP).
(a) From these data, derive the molecular formula of the hydrocarbon.
(b) Write the structural formulas of two possible isomers.
(c) For one of the isomers, use Lewis structures to write a balanced chemical equation for: (i) the bromination reaction; and (ii) the hydrogenation reaction.

121. 4-Octyne can be hydrogenated in two steps. The first step produces a *cis*-configuration product. The final step produces a saturated hydrocarbon.
(a) Use Lewis structures to write chemical equations for each step.
(b) Use bond enthalpies (Table 6.2, **Sec. 6-6b**) to calculate the standard enthalpy change for the overall reaction (both steps).

122. 3-Hexene reacts with HCl via an addition reaction.
(a) Use Lewis structures to write the chemical equation for this reaction.
(b) Use bond enthalpies (Table 6.2, **Sec. 6-6b**) to calculate the standard enthalpy change for this reaction.

123. In the laboratory, a student was given three bottles labeled X, Y, and Z. One of the bottles contained acetic acid, the second acetaldehyde, and the third ethanol. The student ran a series of experiments to determine the contents of each bottle. Substance X reacted with substance Y to form an ester under certain conditions and substance Y formed an acidic solution when dissolved in water. (a) Identify which compound is in which bottle, and write the Lewis structure for each compound. (b) Using structural formulas, write a chemical equation for the ester formation reaction and for the ionization of the acid. (c) The student confirmed the identifications by treating each compound with a strong oxidizing agent and found that compound X required roughly twice the amount of oxidizing agent as compound Z. Explain why this was the case.

124. Consider these data:

Hydrocarbon	Formula	Enthalpy of Combustion (kJ/mol)
Butane	C_4H_{10}	2853.9
Pentane	C_5H_{12}	3505.8
Hexane	C_6H_{14}	4159.5
Heptane	C_7H_{16}	4812.8
Octane	C_8H_{18}	5465.7

Graph these data with enthalpy of combustion on the *y*-axis and number of carbon atoms on the *x*-axis.
(a) Use the graph to estimate the enthalpy of combustion of propane, C_3H_8, nonane, C_9H_{20}, and hexadecane, $C_{16}H_{34}$.
(b) Use bond enthalpies to explain why the graph has a positive slope.

125. Explain why an amino acid has a higher boiling point than an amine with the same number of electrons.
126. Silicones are Si-containing polymers used in waterproof caulking. They do not contain carbon-to-carbon bonds.
   (a) Silly Putty is a polymer made from the monomer dimethylsilanol, $(CH_3)_2Si(OH)_2$. Write the structure of four units of this polymer.
   (b) Identify the type of polymerization that occurs.
127. Polyacetylene is an electrically conducting polymer of acetylene, $HC{\equiv}CH$. Now used in batteries, the polymer was discovered accidentally in 1970 when far too much catalyst was used in an experimental attempt to polymerize acetylene into synthetic rubber. Write a plausible Lewis structure for a portion of a polyacetylene molecule.

Conceptual Challenge Problems

These rigorous, thought-provoking problems integrate conceptual learning with problem solving and are suitable for group work.

CP10.A (Section 10-1) How are the boiling points of hydrocarbons during the distillation of petroleum related to their molecular size?

CP10.B (Section 10-4) Even though millions of organic compounds exist and each compound may have 10, 100, or even thousands of atoms bonded together to make one molecule, the reactions of organic compounds can be studied and even predicted for compounds yet to be discovered. What characteristic of organic compounds allows their reactions to be studied and predicted?

CP10.C (Section 10-5) What is the advantage of animals storing chemical potential energy in their bodies as triesters of glycerol and long-chain fatty acids, known as fats, instead of as carbohydrates?

Chemical Kinetics: Rates of Reactions

11

Eric Audras/ONOKY/Getty Images

Why does a movie move? A movie or video consists of a stream of individual images, each flashed for a very brief time on a screen. Because your eyes see each image even after it disappears, the individual images merge to make what appears to be a continuously moving view. In your eyes photons strike rhodopsin molecules, causing reactions that take less than 200 fs ($<200 \times 10^{-15}$ s). This sends signals that your brain interprets as an image. A much slower reaction turns off the signal. Rates of such reactions are directly responsible for the fact that we can enjoy videos. How do chemists define reaction rates? (Section 11-1) How are reaction rates measured? (Section 11-2) How do enzymes change reaction rates in the body? (Section 11-9) These and many other questions about rates of chemical reactions will be answered in this chapter.

Turn on the valve of a Bunsen burner in your laboratory, bring up a lighted match, and a rapid combustion reaction begins with a whoosh:

$$CH_4(g) + 2\ O_2(g) \longrightarrow CO_2(g) + 2\ H_2O(g) \qquad \Delta_r H° = -802.34\ \text{kJ/mol}$$

What would happen if you didn't put a lighted match in the methane-air stream? Nothing obvious. At room temperature the reaction of methane with oxygen is so slow that the two potential reactants can be mixed in a closed flask and stored unreacted for centuries. These facts about combustion of methane lie within the realm of **chemical kinetics**—*the study of the speeds of reactions and the nanoscale pathways or rearrangements by which atoms and molecules are transformed from reactants to products.*

Chemical kinetics is extremely important, because knowing about kinetics enables us to control many kinds of reactions in addition to combustion. In pharmaceutical chemistry an important problem is devising drugs that remain in their active form long enough to get to the site in the body where they are intended to act. Consequently, it is important to know whether a drug will react with other substances in the body and how long it will take to do so. In environmental chemistry, there was more than a decade of controversy over whether stratospheric ozone is being depleted by chlorofluorocarbons. Much of this debate hinged on verifying the sequence and rates of reactions by which stratospheric ozone is produced and consumed. Their careful studies of such reactions led to a Nobel Prize in Chemistry for Sherwood Rowland, Mario Molina, and Paul Crutzen (← **Sec. 8-10b**).

This chapter focuses on the factors that affect the speeds of reactions, the nanoscale basis for understanding those factors, and their importance in modern society, from industrial plants to cars to the cells of our bodies.

11-1 Reaction Rate

For a chemical reaction to occur, reactant molecules must come together so that their atoms can be exchanged or rearranged. Atoms and molecules are more mobile in the gas phase or in solution than in the solid phase, so reactions are often carried out in a mixture of gases or among solutes in a solution. For a **homogeneous reaction**, *one in which reactants and products are all in the same phase* (gas or solution, for example), four factors affect the speed of a reaction:

- The *properties* of reactants and products—in particular, their molecular structure and bonding
- The *concentrations* of the reactants and sometimes the products
- The *temperature* at which the reaction occurs
- The *presence* of a catalyst (Section 11-8) and, if one is present, its concentration

Many important reactions, including the ones in catalytic converters that remove air pollutants from automobile exhaust, are **heterogeneous reactions**. They *take place at a surface—at an interface between two different phases* (solid and gas, for example). The speed of a heterogeneous reaction depends on the four factors listed above as well as on the area and nature of the surface at which the reaction occurs. For example, very finely divided metal powder can burn very rapidly, whereas a pile of powder with much less surface exposed to oxygen in the air is difficult to ignite (Figure 11.1). The much more rapid reaction when greater surface is exposed has been responsible for explosions in grain elevators, sugar mills, and coal mines where finely divided, combustible solids in the air are exposed to a spark or flame.

The speed of any process is expressed as its **rate**, which is *the change in some measurable quantity per unit of time*. A car's rate of travel, for example, is found by measuring the change in its position, Δx, during a given time interval, Δt. Suppose you are driving on an interstate highway. If you pass mile marker 43 at 2:00 PM and mile marker 173 at 4:00 PM, $\Delta x = (173 - 43)$ mi $= 130$ mi and $\Delta t = 2.00$ h. You are traveling at an average rate of $\Delta x/\Delta t = 65$ mi/h. For a chemical process, the **reaction rate** is defined as *the change in concentration of a reactant or product per unit time.* (Time can be measured in seconds, hours, days, or whatever unit is most convenient for the speed of the reaction.)

Change in concentration is used (rather than change in total amount of reactant) because using change in concentration makes the rate independent of the volume of the reaction mixture.

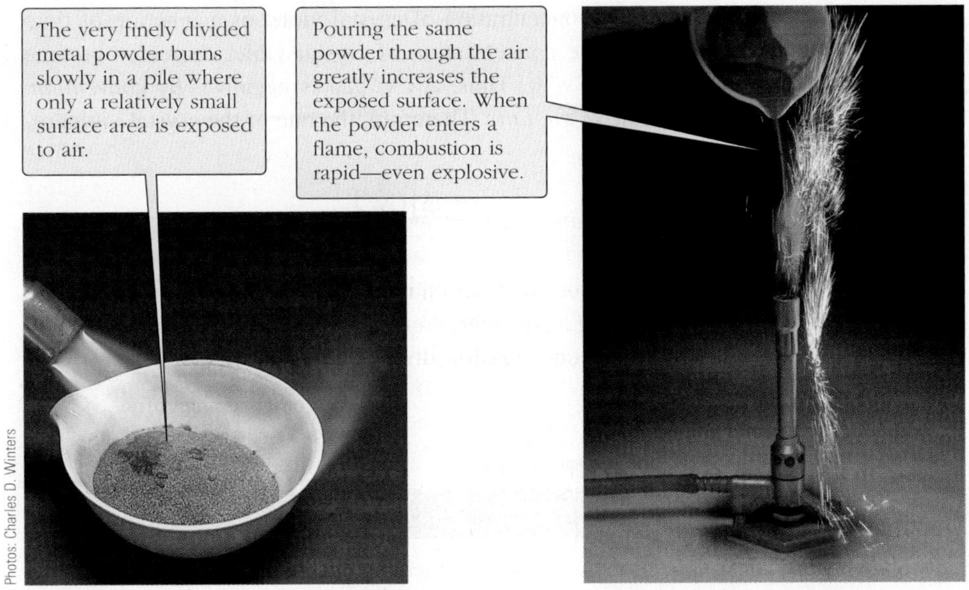
The very finely divided metal powder burns slowly in a pile where only a relatively small surface area is exposed to air.
Pouring the same powder through the air greatly increases the exposed surface. When the powder enters a flame, combustion is rapid—even explosive.
Photos: Charles D. Winters

Figure 11.1 Combustion of iron powder. The rate of combustion is greater when more surface area is exposed to air.

US Chemical Safety Board

Sugar mill explosion. Finely powdered sugar ignited and exploded, causing the destruction of this sugar mill.

As an example of measurements made in chemical kinetics, consider Figure 11.2, which shows the violet-colored dye, crystal violet, reacting with aqueous sodium hydroxide. The dye's violet color disappears over time, and the intensity of color can be used to determine the concentration of the dye. Crystal violet consists of polyatomic positive ions, which we abbreviate as Cv^+, and chloride ions, Cl^-. Its formula is CvCl. The beautiful violet $Cv^+(aq)$ ions combine with hydroxide ions, $OH^-(aq)$, to form a colorless, uncharged product abbreviated as CvOH(aq). The net ionic equation is

$$Cv^+(aq) + OH^-(aq) \longrightarrow CvOH(aq) \qquad [11.1]$$

The rate at which this reaction occurs can be calculated by dividing the change in concentration of crystal violet cation, $\Delta[Cv^+]$, by the elapsed time, Δt. (Concentration is indicated by square brackets.) For example, if the concentration of crystal violet is measured at some time t_1 to give $[Cv^+]_1$, and the measurement is repeated at a subsequent time t_2 to give $[Cv^+]_2$, then the rate of reaction is

Recall that the Greek letter Δ (delta) (← Sec. 4-3a) means to subtract the initial value of a quantity from the final value.

$$\text{Rate of change of concentration of } Cv^+ = \frac{\text{change of concentration of } Cv^+}{\text{elapsed time}}$$

$$= \frac{\Delta[Cv^+]}{\Delta t} = \frac{[Cv^+]_2 - [Cv^+]_1}{t_2 - t_1}$$

Photos: Charles D. Winters

Figure 11.2 Measurements in chemical kinetics. Aqueous crystal violet dye reacts with aqueous sodium hydroxide, forming a colorless product. The time for the color change is measured.

The experimentally measured concentration of crystal violet as a function of time is shown in Table 11.1. Because the concentration of crystal violet decreases as time increases, $[Cv^+]_2$ is smaller than $[Cv^+]_1$. Thus, $\Delta[Cv^+]/\Delta t$ is negative. *By convention, reaction rate is defined as a positive quantity.* Therefore, the rate of the crystal violet reaction is defined as

The negative sign converts a negative $\Delta[Cv^+]/\Delta t$ value to a positive reaction rate.

$$\text{Reaction rate} = -\frac{\Delta[Cv^+]}{\Delta t}$$

Table 11.1 also shows calculated values of reaction rates for each time interval. Because calculating the rate involves dividing a concentration difference by a time difference, the units of reaction rate are units of concentration divided by units of time, in this case mol/L divided by s, mol L^{-1} s^{-1}.

PROBLEM-SOLVING EXAMPLE 11.1

Calculating Average Rates

Using the data in the first two columns of Table 11.1, calculate $\Delta[Cv^+]$, Δt, and the average rate of reaction for each time interval given. Use the numbers given in the third column to check your results. (A good way to do so is to use a computer spreadsheet program.)

Result See the third column in Table 11.1.

Analyze The change in a quantity, represented by Δ, is the final value of the quantity minus the initial value. Reaction rate is $-\Delta[Cv^+]/\Delta t$.

Plan Use the definition of Δ to calculate $\Delta[Cv^+]$ and Δt. Then, calculate $-\Delta[Cv^+]/\Delta t$.

Execute The time interval from 80.0 s to 100.0 s provides an example of the calculation.

$$\Delta[Cv^+] = 0.232 \times 10^{-5}\ \text{mol/L} - 0.429 \times 10^{-5}\ \text{mol/L} = -0.197 \times 10^{-5}\ \text{mol/L}$$

$$\Delta t = 100.0\ \text{s} - 80.0\ \text{s} = 20.0\ \text{s}$$

$$\text{Rate} = -\frac{\Delta[Cv^+]}{\Delta t} = -\frac{(-0.197 \times 10^{-5}\ \text{mol/L})}{20.0\ \text{s}} = 0.985 \times 10^{-7}\ \text{mol L}^{-1}\ \text{s}^{-1}$$

Do the other calculations in a similar way.

☑ **Reasonable Result Check** The rates of reaction should be positive numbers and should have units of concentration divided by time (such as mol L^{-1} s^{-1}). Both of these conditions are met.

Table 11.1 Concentration-Time Data for Reaction of Crystal Violet with 0.10-M NaOH(aq) at 23 °C

Time, *t* (s)	Concentration of Crystal Violet Cation, $[Cv^+]$ (mol/L)	Average Rate (mol L^{-1} s^{-1})
0.0	5.000×10^{-5}	
		13.2×10^{-7}
10.0	3.680×10^{-5}	
		9.70×10^{-7}
20.0	2.710×10^{-5}	
		7.20×10^{-7}
30.0	1.990×10^{-5}	
		5.30×10^{-7}
40.0	1.460×10^{-5}	
		3.82×10^{-7}
50.0	1.078×10^{-5}	
		2.85×10^{-7}
60.0	0.793×10^{-5}	
		1.82×10^{-7}
80.0	0.429×10^{-5}	
		0.985×10^{-7}
100.0	0.232×10^{-5}	

PROBLEM-SOLVING PRACTICE 11.1

For the reaction of crystal violet with NaOH(aq), the measured rate of reaction is 1.27×10^{-6} mol L^{-1} s^{-1} when the concentration of crystal violet cation is 4.13×10^{-5} mol/L.

(a) Estimate how long it will take for the concentration of crystal violet to drop from 4.30×10^{-5} mol/L to 3.96×10^{-5} mol/L.

(b) Could you use the same method to make an accurate estimate of how long it would take for the concentration of crystal violet to drop from 4.30×10^{-5} mol/L to 0.43×10^{-5} mol/L? Explain why or why not.

PROBLEM-SOLVING PRACTICE answers are provided at the back of this book in Appendix K.

EXERCISE 11.1

Rates of Reaction

(a) From data in Table 11.1, calculate the rate of reaction for each time interval: (i) from 40.0 s to 60.0 s; (ii) from 20.0 s to 80.0 s; (iii) from 0.0 to 100.0 s.

(b) Use all of the data in the first two columns of Table 11.1 to draw a graph with time on the horizontal (x) axis and concentration on the vertical (y) axis. Draw a smooth curve through the data. On the graph, draw lines that correspond to $\Delta[Cv^+]/\Delta t$ for each interval.

(c) Why is the rate not the same for each time interval in part (a), even though the average time for each interval is 50.0 s? (That is, for interval i, the average time is (40.0 s + 60.0 s)/2 = 50.0 s.) Write an explanation of the reason for a friend who is taking this course, and ask your friend to evaluate what you have written.

Answers to **EXERCISES** are provided at the back of this book in Appendix L.

11-1a Reaction Rates and Stoichiometry

From the stoichiometry of the crystal violet reaction (Reaction 11.1), it is clear that for every mole of crystal violet that reacts, a mole of CvOH product is formed, because the coefficients of Cv^+(aq), OH^-(aq), and CvOH(aq) are the same. Thus, the rate of appearance of CvOH(aq) equals the rate of disappearance of Cv^+(aq). In many reactions, however, the coefficients are not all the same. For example, in the reaction

$$2\,NO_2(g) \longrightarrow 2\,NO(g) + O_2(g)$$

[11.2]

for every mole of O_2 formed, two moles of NO_2 react. Thus, the rate of disappearance of NO_2 is twice as great as the rate of appearance of O_2. We would like to define the rate of reaction in a way that does not depend on which substance's concentration change is measured. If we multiply $-\frac{\Delta[NO_2]}{\Delta t}$ by $\frac{1}{2}$ in the definition of the rate, then the rate is the same whether expressed in terms of $[O_2]$ or $[NO_2]$. Notice that $\frac{1}{2}$ is the reciprocal of the stoichiometric coefficient of NO_2 in Reaction 11.2. For the general reaction equation

$$a\,A + b\,B \longrightarrow c\,C + d\,D$$

where A, B, C, and D represent formulas of substances and a, b, c, and d are coefficients, the reaction rate can be defined uniformly in terms of each substance as

$$\text{Rate} = -\frac{1}{a}\frac{\Delta[A]}{\Delta t} = -\frac{1}{b}\frac{\Delta[B]}{\Delta t} = \frac{1}{c}\frac{\Delta[C]}{\Delta t} = \frac{1}{d}\frac{\Delta[D]}{\Delta t} \qquad [11.3]$$

Because reaction rates are related to Δconcentration/Δt as shown in Equation 11.3, it is important to know the exact chemical equation for which a rate is reported. If the coefficients in the equation are changed, for example, by doubling all of them, then the definition of reaction rate also changes.

That is, *the rate of change in concentration of any of the reactants or products is multiplied by the reciprocal of the stoichiometric coefficient to find the rate of reaction.* Because the concentrations of reactants decrease with time, their rates of change are given negative signs.

PROBLEM-SOLVING EXAMPLE 11.2

Rates and Stoichiometry

Reaction 11.2 shows the decomposition of $NO_2(g)$ to form NO(g) and $O_2(g)$.

(a) Define the rate of reaction in terms of the rate of change in concentration of each reactant and product.

(b) Suppose the rate of appearance of $O_2(g)$ is 0.023 mol L^{-1} s^{-1}. Calculate the rate of appearance of NO(g).

Result

(a) $\text{Rate} = \frac{1}{2}\frac{\Delta[\text{NO}]}{\Delta t} = \frac{\Delta[\text{O}_2]}{\Delta t} = -\frac{1}{2}\frac{\Delta[\text{NO}_2]}{\Delta t}$

(b) 0.046 mol L^{-1} s^{-1}

Analyze Rate of reaction is defined by Equation 11.3. This equation relates rates of change of concentrations of reactants and products.

Plan Write the rate of change of concentration of each substance in Reaction 11.2. Multiply each rate of change of concentration by the reciprocal of the stoichiometric coefficient. Place a negative sign in front of the rate for each reactant.

Execute

(a) See Result.

(b) Algebraically,

$$\frac{1}{2}\frac{\Delta[\text{NO}]}{\Delta t} = \frac{\Delta[\text{O}_2]}{\Delta t}$$

and

$$\frac{\Delta[\text{NO}]}{\Delta t} = 2 \times \frac{\Delta[\text{O}_2]}{\Delta t} = 2 \times 0.023 \text{ mol L}^{-1}\text{ s}^{-1} = 0.046 \text{ mol L}^{-1}\text{ s}^{-1}$$

PROBLEM-SOLVING PRACTICE 11.2

For the reaction

$$4\ \text{NO}_2(\text{g}) + \text{O}_2(\text{g}) \longrightarrow 2\ \text{N}_2\text{O}_5(\text{g})$$

(a) express the rate of formation of N_2O_5 in terms of the rate of disappearance of O_2.

(b) suppose the rate of disappearance of O_2 is 0.0037 mol L^{-1} s^{-1}. Calculate the rate of disappearance of NO_2.

If you are familiar with calculus, then you may recognize that in the limit of very small time intervals, Δ[A]/Δt, where A represents a substance, becomes the same as the derivative of concentration with respect to time. That is,

$$\lim_{\Delta t \to 0} \frac{\Delta[\text{A}]}{\Delta t} = \frac{d[\text{A}]}{dt}$$

This also means that the rate of reaction at any time can be found from the *slope* (at that time) of the tangent to a curve of concentration versus time, such as the curve in Figure 11.3. Appendix A-8 discusses how to determine the slope and intercept of a graph. The slope of a tangent to the graph at a given point is referred to as the derivative of the graph at that point and can be obtained using scientific graphing programs.

11-1b Average Rate and Instantaneous Rate

A reaction rate calculated from a change in concentration divided by a change in time is called the **average reaction rate** over the time interval from which it was calculated. For example, the average reaction rate at 23 °C for the crystal violet reaction over the interval from 0.0 to 10.0 s is 13.2×10^{-7} mol L^{-1} s^{-1}. The data in Table 11.1 indicate that as the concentration of Cv^+ decreases, the average rate also decreases. Most reactions are like this: The average rate becomes smaller as the concentration of one or more reactants decreases. Your results in Exercise 11.1 should have been different for each range of time over which you calculated. Because the average reaction rate changes over time, the rate you calculate depends on when, and for what range of time, you calculate. If you want to know the rate that corresponds to a particular concentration of Cv^+ (and therefore to a particular time after the reaction began), the average rate is not appropriate, because it depends on the size of the time interval.

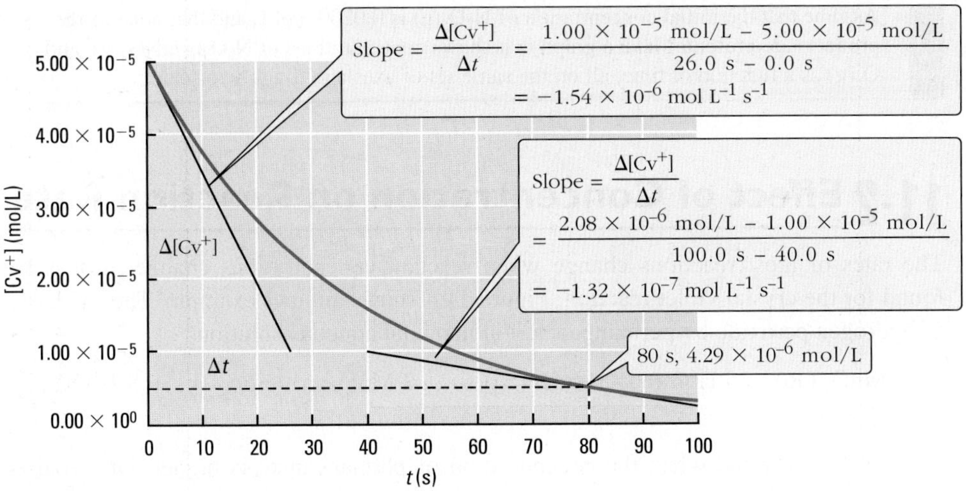

Slope = $\frac{\Delta[Cv^+]}{\Delta t} = \frac{1.00 \times 10^{-5}\ mol/L - 5.00 \times 10^{-5}\ mol/L}{26.0\ s - 0.0\ s}$
$= -1.54 \times 10^{-6}\ mol\ L^{-1}\ s^{-1}$
Slope = $\frac{\Delta[Cv^+]}{\Delta t}$
$= \frac{2.08 \times 10^{-6}\ mol/L - 1.00 \times 10^{-5}\ mol/L}{100.0\ s - 40.0\ s}$
$= -1.32 \times 10^{-7}\ mol\ L^{-1}\ s^{-1}$
80 s, 4.29×10^{-6} mol/L
$\Delta[Cv^+]$
Δt
$[Cv^+]$ (mol/L)
5.00×10^{-5}
4.00×10^{-5}
3.00×10^{-5}
2.00×10^{-5}
1.00×10^{-5}
0.00×10^{0}
0 10 20 30 40 50 60 70 80 90 100
t(s)

Figure 11.3 Instantaneous reaction rates. The experimentally measured concentration of Cv^+ is plotted as a function of time during the reaction in which Cv^+ reacts with OH^- in aqueous solution. The slopes at time 0 s and at 80 s are indicated on the graph. From these slopes the instantaneous rates 1.54×10^{-6} mol L^{-1} s^{-1} and 1.32×10^{-7} mol L^{-1} s^{-1} can be obtained.

The **instantaneous reaction rate** is *the rate at a particular time after a reaction has begun*. To obtain it, the rate must be calculated over a very small interval around the time or concentration for which the rate is desired. For example, to calculate the rate at which Cv^+ is reacting away when its concentration is 4.29×10^{-6} mol/L, you would need to calculate $\Delta[Cv^+]/\Delta t$ at exactly 80 s from the start of the reaction. A good way to do so is shown in Figure 11.3. ***The instantaneous rate is the slope of a line tangent to the concentration-time curve at the point corresponding to the specified concentration and time.*** For a particular concentration of the same reactant at the same temperature and the same concentrations of other species, the instantaneous rate has a specific value. As you saw in Exercise 11.1c, the value of the average rate depends on the size of the Δt used to calculate the average rate.

CONCEPTUAL EXERCISE 11.2

Instantaneous Rates

Instantaneous rates for the reaction of hydroxide ion with Cv^+ can be determined from the slope of the curve in Figure 11.3 at various concentrations. They are

(1) At 4.0×10^{-5} mol/L, rate = 12.3×10^{-7} mol L^{-1} s^{-1}
(2) At 3.0×10^{-5} mol/L, rate = 9.25×10^{-7} mol L^{-1} s^{-1}
(3) At 2.0×10^{-5} mol/L, rate = 6.16×10^{-7} mol L^{-1} s^{-1}
(4) At 1.5×10^{-5} mol/L, rate = 4.60×10^{-7} mol L^{-1} s^{-1}
(5) At 1.0×10^{-5} mol/L, rate = 3.09×10^{-7} mol L^{-1} s^{-1}

(a) What is the relationship between the rates in (1) and (3)? Between (2) and (4)? Between (3) and (5)?
(b) What is the relationship between the concentrations in each of these cases?
(c) Is the rate of the reaction proportional to the concentration of Cv^+? Explain your answer.

EXERCISES that are labeled **CONCEPTUAL** are designed to test your understanding of one or more concepts; they usually involve qualitative rather than quantitative thinking.

CONCEPTUAL EXERCISE 11.3

Graphing Concentrations versus Time

Consider the decomposition of $N_2O_5(g)$,

$$2\ N_2O_5(g) \longrightarrow 4\ NO_2(g) + O_2(g)$$

Assume that the initial concentration of $N_2O_5(g)$ is 0.0200 mol/L and that none of the products are present. Make a graph that shows concentrations of $N_2O_5(g)$, $NO_2(g)$, and $O_2(g)$ as a function of time, all on the same set of axes and roughly to scale.

11-2 Effect of Concentration on Reaction Rate

The rates of most reactions change when reactant concentrations change, just as we found for the crystal violet reaction. Figure 11.4 shows another example. The oxidation of hydrogen peroxide by permanganate ion in acidic aqueous solution

$$\underset{\text{purple}}{2\ MnO_4^-(aq)} + 5\ H_2O_2(aq) + 6\ H^+(aq) \longrightarrow 2\ Mn^{2+}(aq) + 5\ O_2(g) + 8\ H_2O(\ell)$$

is visibly more rapid when the concentration of permanganate is higher. One goal of chemical kinetics is to find out whether a reaction speeds up when the concentration of a reactant is increased and, if so, by how much.

11-2a The Rate Law

How the concentration of a reactant affects the rate can be determined by performing a series of experiments in which the concentration of that reactant is varied systematically (and temperature is held constant). Alternatively, a single experiment can be done in which concentration is determined continuously as a function of time. The latter approach gave the data for crystal violet shown in Table 11.1 and Figure 11.3, which you analyzed in Problem-Solving Example 11.1 and Exercise 11.2. You should have discovered that if the concentration of crystal violet cation is halved, the reaction rate is also halved. If the concentration of crystal violet cation is doubled, then the reaction rate is doubled. This leads to the expression

$$\text{Rate} \propto [Cv^+]$$

It says that the rate is directly proportional to (symbol $\propto$) the concentration of one of the reactants, crystal violet cation.

This proportionality can be changed to a mathematical equation by including a proportionality constant, *k*. *A mathematical equation that summarizes the relationship between reactant concentration and reaction rate* is called a **rate law** (or *rate equation*). For the crystal violet reaction the rate law is

$$\text{Rate} = k \times [Cv^+]$$

1 With dilute potassium permanganate, the reaction is slower...
2 ...than it is with concentrated potassium permanganate.
Photos: Charles D. Winters

Figure 11.4 Reaction of aqueous potassium permanganate with aqueous hydrogen peroxide. In both cases the temperature is the same.

The proportionality constant, k, is called the **rate constant**. The rate constant is independent of concentration, but it has different values at different temperatures, usually becoming larger as the temperature increases. The rate constant applies only to the specific reaction being studied and it applies at a specific temperature. Thus, the chemical equation and the temperature for the reaction should be given along with the rate constant. In this case we write

$$Cv^+(aq) + OH^-(aq) \longrightarrow CvOH(aq) \qquad k = 3.08 \times 10^{-2}\ s^{-1}\ (\text{at } 25\ °C)$$

11-2b Determining Rate Laws from Initial Rates

The relation between rate and concentration (the rate law) must be determined experimentally. One way to do so was illustrated in Exercise 11.3, but it is difficult to determine rates from tangents to a curve such as that in Figure 11.3. Another way is to measure initial rates. The **initial rate** of a reaction is *the instantaneous rate determined at the very beginning of the reaction.* A good approximation to the initial rate is to calculate $-\Delta[\text{reactant}]/\Delta t$ after no more than 2% of the limiting reactant has been consumed.

Many reactions can be started by mixing two different solutions or two different gas samples. Usually the concentrations of the reactants are known before they are mixed, so the initial rate corresponds to a known set of reactant concentrations. Several experiments can then be done in which initial concentrations are varied, and the change in the reaction rate can be correlated with changes in the concentration of each reactant. As an example, consider the reaction of a base with methyl acetate, CH_3COOCH_3, an ester (← **Sec. 10-5b**). This reaction produces acetate ion and methanol.

An advantage of measuring initial rates is that the concentrations of products are low early in the process. As a reaction proceeds, more and more products are formed. In some cases products can alter the rate; comparing initial rates with rates when products are present can reveal such a complication.

$$CH_3\overset{\overset{\displaystyle O}{\|}}{C}{-}O{-}CH_3 + OH^- \longrightarrow CH_3\overset{\overset{\displaystyle O}{\|}}{C}{-}O^- + CH_3OH \qquad [11.4]$$

methyl acetate — hydroxide ion — acetate ion — methanol

To control for the effect of temperature on rate, several experiments were done at the same temperature:

Experiment	Initial Concentration (mol/L) $[CH_3COOCH_3]$	$[OH^-]$	Initial Rate (mol L^{-1} s^{-1})
1	0.040	0.040	0.00022
	no change	× 2	× 2
2	0.040	0.080	0.00045
	× 2	no change	× 2
3	0.080	0.080	0.00090

Notice that in Experiments 1 and 2 the initial concentration of methyl acetate is the same. In Experiments 2 and 3 the initial concentration of hydroxide is the same.

To determine the rate law, compare two experiments in which only a single initial concentration changed. In Experiments 1 and 2, the $[CH_3COOCH_3]$ remained constant and the $[OH^-]$ doubled. The rate also doubled, which means that the rate is *directly proportional* to the $[OH^-]$. In Experiments 2 and 3, the $[OH^-]$ remained the same, the $[CH_3COOCH_3]$ doubled, and the rate doubled, indicating that the rate is also proportional to the $[CH_3COOCH_3]$. Therefore, the experimental data show that the rate is proportional to the *product* of the two concentrations, and the rate law is

Rate laws *must* be determined from experimental data such as these.

$$\text{Rate} = k\,[CH_3COOCH_3][OH^-]$$

This equation also tells us that doubling both initial concentrations at the same time would cause the rate to go up by a factor of 4, which it does from Experiment 1 to Experiment 3.

Another way to approach this problem involves proportions. As before, choose two experiments in which one concentration did not change. Then, calculate the ratio of the other concentrations and the ratio of rates. For the methyl acetate reaction, using Experiments 1 and 2 where the $[CH_3COOCH_3]$ was constant,

$$\frac{[OH^-]_2}{[OH^-]_1} = \frac{0.080 \text{ M}}{0.040 \text{ M}} = 2.0 \quad \text{and} \quad \frac{\text{rate}_2}{\text{rate}_1} = \frac{0.00045 \text{ mol L}^{-1}\text{ s}^{-1}}{0.00022 \text{ mol L}^{-1}\text{ s}^{-1}} = 2.0$$

it is clear that both the concentrations and the rates change in the same proportion. This same method could be applied to analyze results of Experiments 2 and 3, where the initial $[OH^-]$ was constant and the initial $[CH_3COOCH_3]$ changed.

Once the rate law is known, a value for k, the rate constant, can be found by substituting rate and initial concentration data for any one experiment into the rate law. For example, a value of k for the methyl acetate-hydroxide ion reaction could be obtained from data for the first experiment,

rate | $[CH_3COOCH_3]$ | $[OH^-]$

$$0.00022 \text{ mol L}^{-1}\text{ s}^{-1} = k\,(0.040 \text{ mol/L})(0.040 \text{ mol/L})$$

$$k = \frac{0.00022 \text{ mol L}^{-1}\text{ s}^{-1}}{(0.040 \text{ mol/L})(0.040 \text{ mol/L})} = 0.14 \text{ L mol}^{-1}\text{ s}^{-1}$$

Note that the units of the rate constant depend on the number of concentration terms multiplied. In this case the reaction has two concentration terms and the units of k are $\text{L mol}^{-1}\text{ s}^{-1}$.

A more precise value for k can be obtained by using all available experimental data—that is, by calculating a k for each experiment and then averaging the k values to obtain an overall result.

EXERCISE 11.4

Rates and Concentrations

Use the rate law for the reaction of methyl acetate with OH^- to predict the effect on the rate of reaction if the concentration of methyl acetate is doubled and the concentration of hydroxide ions is halved.

PROBLEM-SOLVING EXAMPLE 11.3

Rate Law from Initial Rates

Initial rates $\left(-\dfrac{\Delta[Cl_2]}{\Delta t}\right)$ for the reaction of nitrogen monoxide and chlorine

$$2\text{ NO(g)} + Cl_2\text{(g)} \longrightarrow 2\text{ NOCl(g)}$$

were measured at 27 °C starting with various concentrations of NO and Cl_2. These data were collected.

For the reaction as written in Problem-Solving Example 11.3,

$$\text{Rate} = -\frac{1}{2}\frac{\Delta[NO]}{\Delta t} = -\frac{\Delta[Cl_2]}{\Delta t} = +\frac{1}{2}\frac{\Delta[NOCl]}{\Delta t}$$

	Initial Concentrations (mol/L)		
Experiment	[NO]	$[Cl_2]$	Initial Rate ($\text{mol L}^{-1}\text{ s}^{-1}$)
1	0.020	0.010	8.27×10^{-5}
2	0.020	0.020	1.65×10^{-4}
3	0.020	0.040	3.31×10^{-4}
4	0.040	0.020	6.60×10^{-4}
5	0.010	0.020	4.10×10^{-5}

(a) Determine the rate law. (b) Determine the value of the rate constant k.

Result

(a) Rate = $k[Cl_2][NO]^2$ (b) $k = 2.1 \times 10^1$ mol^{-2} L^2 s^{-1}

Analyze Initial rates and concentrations are given; the rate law and rate constants are to be determined.

Plan To determine the effect of concentration of one reactant, analyze data from experiments where all other concentrations remain constant. Repeat this process for each reactant. Once the effect of each reactant is known, write the rate law. Use it to calculate *k* for each experiment; average the five *k* values.

Execute

(a) In Experiments 1, 2, and 3, the concentration of NO is constant, while the Cl_2 concentration increases from 0.010 to 0.020 to 0.040 mol/L. Each time $[Cl_2]$ is doubled, the initial rate also doubles. For example, when $[Cl_2]$ is doubled from 0.020 to 0.040 mol/L in Experiments 2 and 3, the initial rate doubles from 1.65×10^{-4} to 3.31×10^{-4} mol L^{-1} s^{-1}. The initial rate is directly proportional to $[Cl_2]$.

In Experiments 2, 4, and 5, $[Cl_2]$ is constant, while [NO] varies. In Experiments 2 and 4, [NO] is doubled, but the initial rate increases by a factor of 4, or 2^2.

$$\frac{\text{Experiment 4 rate}}{\text{Experiment 2 rate}} = \frac{6.60 \times 10^{-4}\ \text{mol L}^{-1}\ \text{s}^{-1}}{1.65 \times 10^{-4}\ \text{mol L}^{-1}\ \text{s}^{-1}} = \frac{4}{1} = \frac{2^2}{1}$$

This same result is found in Experiments 2 and 5. Thus, the initial rate is proportional to the *square* of [NO]. Therefore, the rate law is

$$\text{Rate} = k[Cl_2][NO]^2$$

(b) Once the rate law is known, the rate constant *k* can be calculated. For Experiment 1, for example,

$$8.27 \times 10^{-5}\ \text{mol L}^{-1}\ \text{s}^{-1} = k(0.010\ \text{mol/L})(0.020\ \text{mol/L})^2$$

$$k = \frac{8.27 \times 10^{-5}\ \text{mol L}^{-1}\ \text{s}^{-1}}{(0.010\ \text{mol/L})(0.020\ \text{mol/L})^2} = 2.1 \times 10^1\ \text{mol}^{-2}\ \text{L}^2\ \text{s}^{-1}$$

For Experiments 2, 3, 4, and 5, the rate constants are 2.1×10^1, 2.1×10^1, 2.1×10^1, and 2.0×10^1 mol^{-2} L^2 s^{-1}, respectively. The average of these values is the rate constant determined from this series of experiments, 2.1×10^1 mol^{-2} L^2 s^{-1}. It can be used to calculate the rate for any set of NO and Cl_2 concentrations at 27 °C.

☑ **Reasonable Result Check** The five calculated *k* values are nearly equal. If the rate law were incorrect, or if an error were made in one or more calculations, some *k* values would be quite different from the others.

When comparing one experiment with another, as is done in the Execute section of Problem-Solving Example 11.3, it is usually convenient to put the larger rate in the numerator. Otherwise fractions are obtained instead of whole numbers.

For this reaction the rate is proportional to one concentration and to the square of another. The rate constant equals the rate (units of mol L^{-1} s^{-1}) divided by three concentration terms multiplied together (units of mol^3 L^{-3}). Thus, the rate constant has units of

$$\frac{\text{mol L}^{-1}\ \text{s}^{-1}}{\text{mol}^3\ \text{L}^{-3}} = \text{mol}^{-2}\ \text{L}^2\ \text{s}^{-1}$$

PROBLEM-SOLVING PRACTICE 11.3

At 23 °C, these data were collected for the crystal violet reaction

$$Cv^+(aq) + OH^-(aq) \longrightarrow CvOH(aq)$$

Experiment	Initial Concentrations (mol/L) $[Cv^+]$	$[OH^-]$	Initial Rate (mol L^{-1} s^{-1})
1	4.3×10^{-5}	0.10	1.3×10^{-6}
2	2.2×10^{-5}	0.10	6.7×10^{-7}
3	1.1×10^{-5}	0.10	3.3×10^{-7}

(a) Is it possible to determine the complete rate law from the data given? Why or why not?

(b) Assume that the rate does not depend on the concentration of hydroxide ion. Determine the rate law.

(c) Calculate the rate constant, again assuming that the rate does not depend on the concentration of hydroxide ion.
(d) Calculate the initial rate of reaction when the concentration of crystal violet cation is 0.00045 M and the concentration of hydroxide ion is 0.10 M. Report your results in mol L^{-1} s^{-1}.
(e) Calculate the rate when the concentration of Cv^+ is half the initial value of 0.00045 M.

EXERCISE 11.5

Determining the Rate Law Using Logarithms

For the reaction in Problem-Solving Example 11.3, assume that the rate law is of the form Rate $= k[NO]^x[Cl_2]^y$. Show mathematically that by taking logarithms of both sides of the rate law and comparing Experiments 2 and 4, where the concentration of chlorine is the same, x, is given by

$$x = \log\left(\frac{\text{Rate}_4}{\text{Rate}_2}\right) \Big/ \log\left(\frac{[NO]_4}{[NO]_2}\right)$$

11-3 Rate Law and Order of Reaction

For many (but not all) homogeneous reactions, the rate law has the general form

The rate constant k always has the same value at a given temperature;

m and n are the orders of the reaction with respect to substances A and B.

$$\text{Rate} = k[A]^m[B]^n \ldots$$

where concentrations of substances, [A], [B], . . . are raised to powers, m, n, ... The substances A, B, . . . might be reactants, products, or catalysts. *The exponents* m, n, . . . *are usually positive whole numbers but might be negative numbers or fractions. These exponents define* the **order of the reaction** with respect to each reactant. If n is 1, for example, the reaction is *first*-order with respect to B; if m is 2, then the reaction is *second*-order with respect to A. *The sum of* m *and* n *(plus the exponents on any other concentration terms in the rate equation) gives* the **overall reaction order**. (The reaction in Problem-Solving Example 11.3 is first-order in Cl_2, second-order in NO, and third-order overall.) A very important point to remember is that ***the rate law and reaction orders must be determined experimentally.***

PROBLEM-SOLVING EXAMPLE 11.4

Reaction Order and Rate Law

For each reaction and experimentally determined rate law listed below, determine the order with respect to each reactant and the overall order.

(a) $2\ NO(g) + 2\ H_2(g) \longrightarrow N_2(g) + 2\ H_2O(g)$ Rate $= k[NO]^2[H_2]$

(b) $14\ H^+(aq) + 2\ HCrO_4^-(aq) + 6\ I^-(aq) \longrightarrow 2\ Cr^{3+}(aq) + 3\ I_2(aq) + 8\ H_2O(\ell)$
Rate $= k[HCrO_4^-][I^-]^2[H^+]^2$

(c) *cis*-2-butene(g) $\longrightarrow$ *trans*-2-butene(g) (catalytic concentration of I_2 present)
Rate $= k$[*cis*-2-butene]$[I_2]^{1/2}$

Result

(a) First-order in H_2, second-order in NO, third-order overall
(b) First-order in $HCrO_4^-$, second-order in I^-, second-order in H^+, fifth-order overall
(c) First-order in *cis*-2-butene, 0.5-order in I_2, 1.5-order overall

Analyze The order of a reaction with respect to a substance is the power to which that substance's concentration is raised in the rate law. The overall order is the sum of the orders for all substances in the rate law.

Plan Use the exponents in the rate law—not the stoichiometric coefficients—to determine the order for each substance; sum the individual orders to obtain the overall order.

Execute

(a) The rate law has a single term that is raised to the first power and another term that is squared, so the reaction is first-order in H_2, second-order in NO, and third-order overall.

(b) The rate law contains three terms. Since the $HCrO_4^-$ term is raised to the first power, the reaction is first-order in $HCrO_4^-$. The other two terms are squared, so the reaction is second-order in I^- and second-order in H^+. The exponents sum to five, so the reaction is fifth-order overall.

(c) In this case the rate of reaction depends on the concentration of the reactant and also on the square root ($\frac{1}{2}$ power) of the concentration of a catalyst, I_2. The reaction is therefore first-order in *cis*-2-butene, 0.5-order in I_2, and 1.5-order overall.

Notice that none of the reactions in Problem-Solving Example 11.4 has a rate law that can be derived correctly from the stoichiometric equation. For example, H_2 has a coefficient of 2 in Reaction (a), but the rate law involves $[H_2]$ to the first power.

PROBLEM-SOLVING PRACTICE 11.4

In Problem-Solving Example 11.3 the rate law for reaction of NO with Cl_2 was found to be

$$2\ NO(g) + Cl_2(g) \longrightarrow 2\ NOCl(g) \qquad \text{Rate} = k[NO]^2[Cl_2]$$

(a) Determine the order of the reaction with respect to NO and with respect to Cl_2.

(b) Suppose that you triple the concentration of NO and simultaneously decrease the concentration of Cl_2 by a factor of 8. Is the reaction faster or slower under the new conditions? How much faster or slower? (Assume that the temperature is the same in both sets of conditions.)

11-3a The Integrated Rate Law

Another approach to experimental determination of the rate law and rate constant for a reaction uses calculus to derive what is called the integrated rate law. As an example of the integrated rate law method, consider a hypothetical reaction in which a single substance A reacts to form products.

$$A \longrightarrow \text{products}$$

First-Order Reaction If the rate law is first-order, then

$$\text{Rate} = -\frac{\Delta[A]}{\Delta t} = k[A]$$

This expression can be transformed, using calculus, to the integrated first-order rate law,

$$\ln[A]_t = -kt + \ln[A]_0 \qquad [11.5]$$

where $[A]_t$ represents the concentration of A at time t, $[A]_0$ represents the initial concentration of A (when $t = 0$), and ln represents the natural logarithm function. (Logarithms are discussed in Appendix A-6.)

Equation 11.5 has the same form as the general equation for a straight line, $y = mx + b$, in which m is the slope and b is the y-intercept.

y-axis variable | slope | *x*-axis variable | *y*-intercept

$$\ln[A]_t = -k\,t + \ln[A]_0 \qquad [11.5]$$

$$y = m\,x + b$$

If the reaction is actually first-order, then a graph of ln[A] on the vertical (y) axis versus t on the horizontal (x) axis should be a straight line. A linear graph, such as the one in Figure 11.5a, is evidence that the reaction is first-order.

You do not have to know calculus to use the integrated rate law method. If you do know calculus, you can derive Equation 11.5 this way:

$$-\frac{d[A]}{dt} = k[A] \quad \text{and} \quad \frac{d[A]}{dt} = -k[A]$$

$$\int_{[A]_0}^{[A]_t} \frac{d[A]}{[A]} = -k\int_0^t dt$$

$$\ln[A]_t - \ln[A]_0 = -k(t_t - t_0)$$

If the reaction starts at time t_0, then $t_t - t_0 = t$, the elapsed time, and

$$\ln[A]_t - \ln[A]_0 = -kt \quad \text{or}$$

$$\ln[A]_t = -kt + \ln[A]_0$$

For a second-order reaction, calculus gives

$$-\frac{d[A]}{dt} = k[A]^2 \text{ and } \frac{d[A]}{[A]^2} = -k\,dt$$

$$\int_{[A]_0}^{[A]_t} \frac{d[A]}{[A]^2} = -k\int_0^t dt$$

$$-\frac{1}{[A]_t} - \left(-\frac{1}{[A]_0}\right) = -k(t_t - t_0)$$

If the reaction starts at time t_0, then $t_t - t_0 = t$, the elapsed time, and

$$\frac{1}{[A]_t} = kt + \frac{1}{[A]_0}$$

Second-Order Reaction For the same reaction

$$A \longrightarrow \text{products}$$

suppose that the rate depends on the square of the concentration of the reactant; that is, suppose the rate law is second-order.

$$\text{Rate} = k[A]^2$$

The integrated rate law derived using calculus is

$$\frac{1}{[A]_t} = kt + \frac{1}{[A]_0}$$

This equation is also of the form $y = mx + b$. If a reaction is second-order, a graph of $1/[A]_t$ versus t will be linear with slope $= k$ and y-intercept $= 1/[A]_0$. Such a straight-line graph indicates a second-order reaction (Figure 11.5b).

Zeroth-Order Reaction There are a few reactions for which the rate does not depend on the concentration of a reactant at all. These are called zeroth-order reactions, because the rate law can be written as a rate constant k times a concentration to the zeroth power. (Anything raised to the zeroth power equals 1.)

$$\text{Rate} = -\frac{\Delta[A]}{\Delta t} = k[A]^0 = k$$

For a zeroth-order reaction, no matter what the reactant concentration is, the rate is the same: it equals the rate constant, k. For a zeroth-order reaction you can derive the integrated form using algebra (no calculus required).

$$\Delta[A] = -k\Delta t$$
$$[A]_t - [A]_0 = -k(t_t - t_0) = -kt$$
$$[A]_t = -kt + [A]_0$$

Again, the equation is of the form $y = mx + b$. If a reaction is zeroth-order, graphing $[A_t]$ versus t gives a straight line with slope $= -k$ and y-intercept $= [A]_0$, as shown in Figure 11.5c.

To summarize these three situations, a rate law that involves powers of the reactant concentration can be written as

$$\text{Rate} = -\frac{\Delta[A]}{\Delta t} = k[A]^m$$

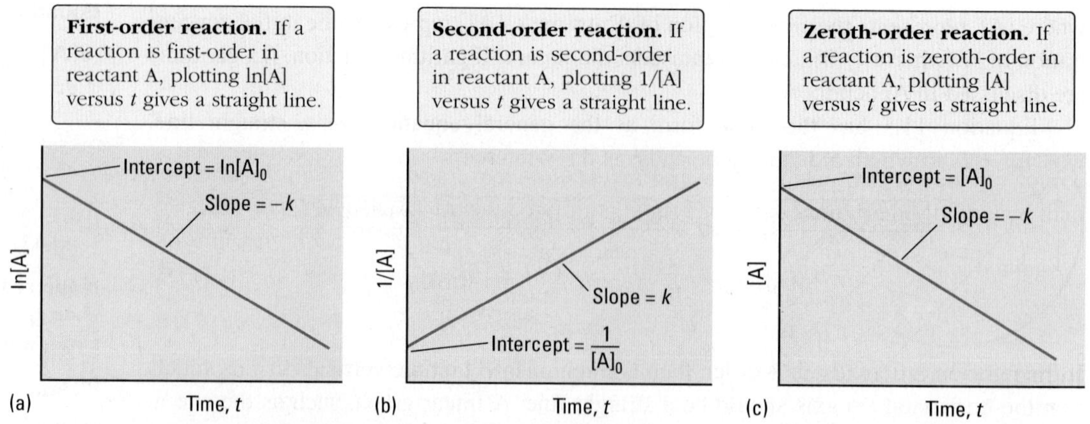
First-order reaction. If a reaction is first-order in reactant A, plotting ln[A] versus *t* gives a straight line.
Second-order reaction. If a reaction is second-order in reactant A, plotting 1/[A] versus *t* gives a straight line.
Zeroth-order reaction. If a reaction is zeroth-order in reactant A, plotting [A] versus *t* gives a straight line.
Intercept = ln[A]0
Slope = −k
ln[A]
(a)
Time, t
Slope = k
Intercept = 1/[A]0
1/[A]
(b)
Time, t
Intercept = [A]0
Slope = −k
[A]
(c)
Time, t

Figure 11.5 Integrated rate law graphs for reactions that are (a) first-order, (b) second-order, and (c) zeroth-order.

Table 11.2 Integrated Rate Laws

Order	Rate Equals	Integrated Rate Law*	Straight-line Plot	Slope of Plot	Units of $k^{\dagger}$
0	$k[A]^0 = k$	$[A]_t = -kt + [A]_0$	$[A]_t$ vs t	$-k$	conc. time^{-1}
1	$k[A]$	$\ln[A]_t = -kt + \ln[A]_0$	$\ln[A]_t$ vs t	$-k$	time^{-1}
2	$k[A]^2$	$\frac{1}{[A]_t} = kt + \frac{1}{[A]_0}$	$\frac{1}{[A]_t}$ vs t	k	conc.$^{-1}$ time^{-1}

*In the table, $[A]_0$ indicates the initial concentration of substance A, that is, the concentration of A at $t = 0$, the time when the reaction was started.

†If you know the units of k you can use the units to figure out whether the reaction is zeroth-order, first-order, or second-order because the units are different in each case.

where m is the order of the reaction. The integrated rate law depends on the value of m; the results for m equal to 0, 1, and 2 are given in Table 11.2.

To determine the order of reaction, then, we collect concentration-time data and make the three plots listed in Table 11.2 and shown in Figure 11.5. Only one (or perhaps none) of the plots will be a straight line. If one is straight, then it indicates the order and the rate constant can be calculated from its slope. The units of the rate constant are those of the slope of the line; they are given in the last column of Table 11.2 and depend on the order of the reaction.

PROBLEM-SOLVING EXAMPLE 11.5

Reaction Order and Rate Constant from Integrated Rate Law

These data were obtained for decomposition of cyclopentene at 825 K.

$$C_5H_8(g) \longrightarrow C_5H_6(g) + H_2(g)$$

Time (s)	$[C_5H_8]$ (mol/L)	Time (s)	$[C_5H_8]$ (mol/L)
0	0.0200	300.	0.0084
20.	0.0189	400.	0.0063
50.	0.0173	500.	0.0047
100.	0.0149	700.	0.0027
200.	0.0112	1000.	0.0011

Determine the order of the reaction and the rate constant.

Result The reaction is first-order in C_5H_8; $k = 2.88 \times 10^{-3}\ s^{-1}$.

Analyze Concentration-time data pairs are given. The integrated rate law method can be used to obtain order and rate constant from such data.

Plan Make zeroth-, first-, and second-order integrated rate law graphs.

Cyclopentene has this structure

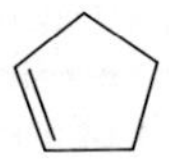

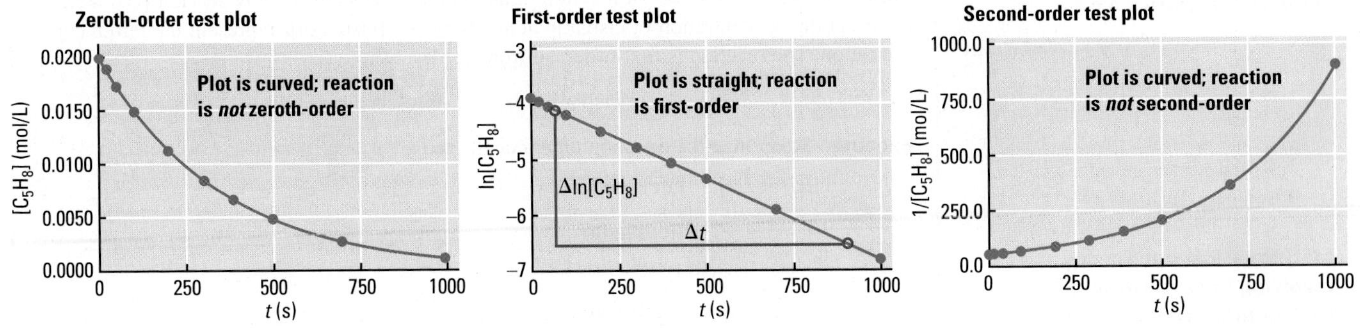

Zeroth-order test plot
Plot is curved; reaction is *not* zeroth-order
$[C_5H_8]$ (mol/L)
0.0200
0.0150
0.0100
0.0050
0.0000
0
250
500
750
1000
t (s)
First-order test plot
Plot is straight; reaction is first-order
$\ln[C_5H_8]$
−3
−4
−5
−6
−7
$\Delta\ln[C_5H_8]$
Δt
0
250
500
750
1000
t (s)
Second-order test plot
Plot is curved; reaction is *not* second-order
$1/[C_5H_8]$ (mol/L)
1000.0
750.0
500.0
250.0
0.0
0
250
500
750
1000
t (s)

Figure 11.6 Integrated rate law plots for cyclopentene decomposition reaction at 825 K.

Execute Figure 11.6. shows that zeroth-order and second-order plots are curved, whereas the first-order plot is a straight line. Thus, the reaction must be *first-order*.

From the first-order plot, calculate the slope using the points marked on the graph as red, open circles. The points are (910, −6.56) and (74, −4.15).

$$\text{Slope} = \frac{\{-6.56 - (-4.15)\}}{\{910 - 74\}\text{ s}} = -2.88 \times 10^{-3}\text{ s}^{-1}$$

When you calculate the slope, it is important to use two points on the straight line through the experimental data (open red circles in Figure 11.6), not two of the data points themselves. Making a graph is similar to averaging, because the straight line and its slope are based on all ten data pairs, not just two.

The slope, $-2.88 \times 10^{-3}\text{ s}^{-1}$, is the negative of the rate constant (see Table 11.2), which means that $k = 2.88 \times 10^{-3}\text{ s}^{-1}$.

☑ **Reasonable Result Check** The units are s^{-1}, which corresponds to the reciprocal time units indicated in Table 11.2 for a first-order rate constant.

PROBLEM-SOLVING PRACTICE 11.5

For the reaction of Cv^+ with OH^-, use the concentration-time data in Table 11.1 (← **Sec. 11-1**) to deduce the order of the reaction with respect to Cv^+. Determine the rate constant. (Assume that $[OH^-]$ is constant during the reaction.)

11-3b Calculating Concentration or Time from Rate Law

The rate law enables us to calculate the rate of reaction from the concentration of the reactants. The integrated rate law allows us to calculate the concentration of a reactant as a function of time.

Once the rate law has been determined experimentally, it provides a way to calculate the concentration of a reactant or product at any time after the reaction has begun. All that is needed is the integrated rate law (from Table 11.2), the value of the rate constant, and the initial concentration of reactant or product. These are related by the equations given in Table 11.2.

PROBLEM-SOLVING EXAMPLE 11.6

Calculating Concentrations

Cisplatin has the formula $Pt(NH_3)_2Cl_2$. Because one of the two chloride ions is replaced by a water molecule in the reaction, the cisplatin reactant has been written as cisplatinCl in the chemical equation to show where the chloride ion comes from.

The first-order rate constant is $1.87 \times 10^{-3}\text{ min}^{-1}$ at 37 °C (body temperature) for reaction of cisplatin, a cancer chemotherapy agent, with water. The reaction is

$$\text{cisplatinCl} + H_2O \longrightarrow \text{cisplatinOH}_2^+ + Cl^-$$

Suppose that the concentration of cisplatin in the bloodstream of a cancer patient is 4.73×10^{-4} mol/L. Calculate the concentration exactly 24 hours later.

Result 3.20×10^{-5} mol/L

Analyze The integrated rate law can be used to calculate concentration at any time after a reaction starts.

Plan Assume that the rate of reaction in the blood is the same as in water. Let [cisplatin] represent the concentration of cisplatin at any time and $[\text{cisplatin}]_0$ represent the initial concentration. The reaction is first-order, so from Table 11.2,

$$\ln[\text{cisplatin}] = -kt + \ln[\text{cisplatin}]_0$$

Execute Rearrange the equation algebraically to

$$\ln[\text{cisplatin}] - \ln[\text{cisplatin}]_0 = -kt$$

$$\ln\left\{\frac{[\text{cisplatin}]}{[\text{cisplatin}]_0}\right\} = -kt$$

$$\frac{[\text{cisplatin}]}{[\text{cisplatin}]_0} = \text{anti ln}(-kt) = e^{-kt}$$

Mathematical operations involving logarithms and exponentials (antilogarithms) are discussed in Appendix A-6.

$$[\text{cisplatin}] = [\text{cisplatin}]_0\, e^{-kt}$$
$$[\text{cisplatin}] = (4.73 \times 10^{-4}\ \text{mol/L})\, e^{-(1.87 \times 10^{-3}\ \text{min}^{-1} \times 24\ \text{h})}$$
$$[\text{cisplatin}] = (4.73 \times 10^{-4}\ \text{mol/L})\, e^{-(1.87 \times 10^{-3}\ \text{min}^{-1} \times 24\ \text{h} \times 60\ \text{min h}^{-1})}$$
$$[\text{cisplatin}] = (4.73 \times 10^{-4}\ \text{mol/L})(6.77 \times 10^{-2})$$
$$[\text{cisplatin}] = 3.20 \times 10^{-5}\ \text{mol/L}$$

☑ **Reasonable Result Check** If the drug is to be effective, it almost certainly needs to be in the body for some time—several minutes to hours or more. The concentration has dropped to a little less than 10% of its initial value in 24 hours, which is reasonable.

Because the solution to this example involves a ratio of concentrations in which the units cancel, the same approach can be taken in problems that involve the amount (moles), the mass, or the number of atoms or molecules at two different times.

PROBLEM-SOLVING PRACTICE 11.6

The first-order rate constant for decomposition of an insecticide in the environment is $3.43 \times 10^{-2}\ \text{d}^{-1}$. Calculate the time required for the concentration of insecticide to drop to $\frac{1}{10}$ of its initial value.

11-3c Half-Life

The **half-life ($t_{1/2}$)** of a reaction is *the time required for the concentration of a reactant A to fall to one half of its initial value.* That is, $[A]_{t_{1/2}} = \frac{1}{2}[A]_0$. For a *first-order reaction* the half-life has the same value, no matter what the initial concentration is. For other reaction orders this is *not* true.

The half-life is related to the first-order rate constant. To see how, use algebra to rearrange Equation 11.5 (which was associated with a reaction A ⟶ products):

$$\ln[A]_t = -kt + \ln[A]_0 \qquad [11.5]$$

$$\ln[A]_t - \ln[A]_0 = -kt$$

$$\ln[A]_{t_{1/2}} - \ln[A]_0 = -kt_{1/2} \qquad [11.6]$$

Remember that Equation 11.5 was derived for a reaction of the form A ⟶ products. Had the equation been 2 A ⟶ products, that is, had the coefficient of A been 2 in the chemical equation, the result would be slightly different. (See Question 121 at the end of this chapter.)

Because $[A]_{t_{1/2}} = \frac{1}{2}[A]_0$, Equation 11.5 can be rewritten as

$$\ln([A]_0/2) - \ln[A]_0 = -kt_{1/2}$$

and so

$$-kt_{1/2} = \ln[A]_0 - \ln(2) - \ln[A]_0 = -\ln 2$$

$$t_{1/2} = \frac{-\ln 2}{-k} = \frac{0.693}{k} \qquad [11.7]$$

This means that measuring the half-life of a first-order reaction determines the rate constant, and vice versa. Radioactive decay (Section 18-4) is a first-order process, and half-life is typically used to report the rate of decay of radioactive nuclei.

PROBLEM-SOLVING EXAMPLE 11.7

Half-Life and Rate Constant

In Problem-Solving Example 11.5, the rate constant for decomposition of cyclopentene at 825 K was found to be $2.88 \times 10^{-3}\ \text{s}^{-1}$. The reaction gave a linear first-order plot. Determine the half-life in seconds for this reaction.

Result 241 s

Analyze The half-life is to be calculated. For a first-order reaction A ⟶ products, the half-life is ln 2 divided by k (Equation 11.7).

Plan The reaction is first-order and is of the form A ⟶ products. Use Equation 11.7 to calculate $t_{1/2}$.

Execute

$$t_{1/2} = \frac{0.693}{k} = \frac{0.693}{2.88 \times 10^{-3}\ \text{s}^{-1}} = 2.41 \times 10^{2}\ \text{s} = 241\ \text{s}$$

☑ **Reasonable Result Check** The rate constant is about 3×10^{-3} s^{-1}. If the concentration of cyclohexene were 1.0 mol L^{-1}, then the reaction rate would be 0.003 s^{-1} × 1.0 mol L^{-1} = 0.003 mol L^{-1} s^{-1}. This means that 0.003 mol L^{-1} would react every second. For the concentration to drop to half of 1.0 mol L^{-1}, the change in concentration would be 0.500 mol L^{-1}. Therefore, it would take at least $\frac{0.500\ \text{mol L}^{-1}}{0.003\ \text{mol L}^{-1}\ \text{s}^{-1}} = 167$ s for the concentration to drop to half its initial value, and the half-life should be at least 167 s. Because the rate decreases as concentration decreases, it will take longer than 167 s. A half-life of 241 s is reasonable.

PROBLEM-SOLVING PRACTICE 11.7

From Figure 11.3 (← **Sec. 11-1b**) determine the time required for the concentration of Cv^+ to fall to one half the initial value. Verify that the same period is required for the concentration to fall from one half to one fourth the initial value. From this half-life, calculate the rate constant.

11-4 A Nanoscale View: Elementary Reactions

Macroscale experimental observations reveal that reactant concentrations, temperature, and catalysts can affect reaction rates. But how can we interpret such observations in terms of nanoscale models? We will use the *kinetic-molecular theory of matter,* which was first introduced in Section 1-9a and developed further in Section 8-2, together with the ideas about molecular structure developed in Chapters 6 and 7. These concepts provide a good basis for understanding how atoms and molecules move and chemical bonds are made or broken during the very short time it takes for reactant molecules to be converted into product molecules.

According to kinetic-molecular theory, molecules are in constant motion. In a gas or liquid they bump into one another; in a solid they vibrate about specific locations. Molecules also rotate, flex, or vibrate around or along the bonds that hold the atoms together.

ESTIMATION | Pesticide Decay

There are usually several different ways that a pesticide can decompose in an ecosystem. Thus, it is difficult to define a rate of decomposition and even more difficult to define a rate law. Often it is assumed that decomposition is first-order and an approximate half-life is reported.

Organochlorine pesticides such as DDT, lindane, and dieldrin may have half-lives as long as 10 years in the environment. The maximum contaminant level (MCL) for lindane is 0.2 ppb (parts per billion). Suppose that an ecosystem has been contaminated with lindane at a concentration of 200 ppb. How long would you have to wait before it would be safe to enter the ecosystem without protection from the pesticide?

The level of contamination is 200 ppb/0.2 ppb = 1000 times the MCL. Presumably it would be safe to wait until the level had dropped to 1/1000 of its initial value. You can estimate how long this would be by using powers of 2, because the number of half-lives required is n, where $(1/2)^n = 1/1000$. That is, as soon as 2^n exceeds 1000, you have waited long enough. Computer scientists, who deal with binary arithmetic, can easily tell you that $2^{10} = 1024$, so $n = 10$. You can verify this using your calculator's y^x-key, or you could simply raise 2 to a power until a value greater than 1000 was calculated. If you did not have a calculator, you could multiply 2 × 2 × 2 . . . in your head until you had enough factors to multiply out to a number bigger than 1000. Ten half-lives means 10 × 10 years, or 100 years, so after 100 years the ecosystem would be free of significant lindane contamination.

Collisions between pairs of molecules exchange energy and produce the transformations of structures that constitute chemical reactions.

There are only two important types of molecular transformations:

- A **unimolecular reaction** is *one in which the structure of a single particle (molecule or ion) rearranges to produce a different particle or particles.* A unimolecular reaction might involve breaking a bond (or bonds) and forming two new molecules, or it might involve rearrangement of one isomeric structure into another. A unimolecular reaction has a molecularity of one.
- A **bimolecular reaction** is *one in which two particles (atoms, molecules, or ions) collide and rearrange bonds to form products.* In a bimolecular reaction new bonds may be formed between the reactant particles and existing bonds may be broken. Sometimes the two particles combine to form a new, larger particle. Sometimes two or more new particles are formed from the two reactant particles. A bimolecular reaction has a molecularity of two.

All chemical reactions can be understood in terms of simple reactions such as those just described. Very complicated reactions can be built up from combinations of unimolecular and bimolecular reactions, just as complicated compounds can be built from chemical elements. For example, hundreds of such reactions are needed to understand how smog is produced in a city such as Los Angeles or to understand why chlorofluorocarbons can deplete stratospheric ozone (← **Sec. 8-10**). Like the chemical elements, *the simplest nanoscale reactions are building blocks, so each such reaction is referred to as an* **elementary reaction**. ***The equation for an elementary reaction shows exactly which molecules, atoms, or ions take part in the elementary reaction.*** The next two sections describe the two important types of elementary reactions in more detail.

EXERCISE 11.6

Unimolecular and Bimolecular Reactions

For each of these nanoscale molecular diagrams, write a balanced equation using chemical formulas. Which reactions are unimolecular? Which are bimolecular?

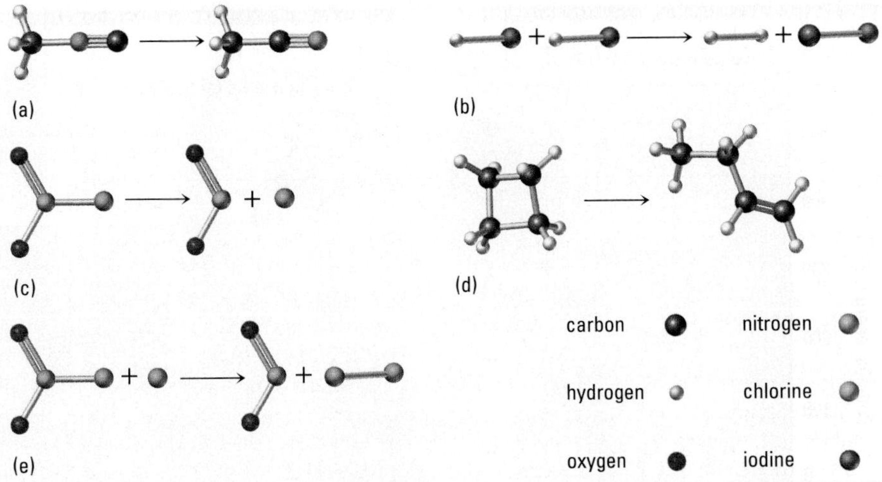
(a)
(b)
(c)
(d)
(e)
carbon
nitrogen
hydrogen
chlorine
oxygen
iodine

11-4a Unimolecular Reactions

An example of a unimolecular reaction is the conversion of *cis*-2-butene to *trans*-2-butene.

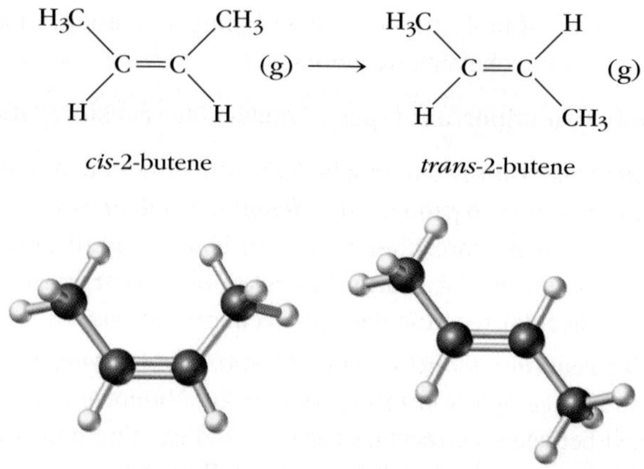
cis-2-butene
trans-2-butene

Cis-2-butene and *trans*-2-butene are *cis-trans* stereoisomers (← **Sec. 6-5a**), and all four carbon atoms are in the same plane. The difference between the two molecules is the orientation of the methyl groups, which are on the same side of the double bond in the plane of the flat molecule for the *cis* structure and on opposite sides for the *trans* structure. If we could grab one end of the molecule and twist it 180° around the axis of the double bond, we would get the other molecule. Thus, it is a reasonable hypothesis that the molecular pathway by which *cis*-2-butene changes to *trans*-2-butene involves twisting the molecule around the double bond (twisting one methyl group out of the plane of the molecule). The angle of twist around the double-bond axis measures the progress of the reaction on the nanoscale. The greater the angle, the less the molecule is like *cis*-2-butene and the more it is like *trans*-2-butene, until an angle of 180° is reached and it has become *trans*-2-butene. Figure 11.7 shows several structures with different angles of rotation, illustrating this process.

Twisting around the double bond to convert one isomer into the other requires that the reactant molecule have sufficient energy. Recall (← **Sec. 7-4**) that a double bond

Twisting in either direction requires the same increase in potential energy, so the energy for +30° equals that for −30°.

Figure 11.7 is similar to Figure 4.15 (← Sec. 4-7), which showed the energy change when bonds were broken and formed as H_2 and Cl_2 changed into HCl.

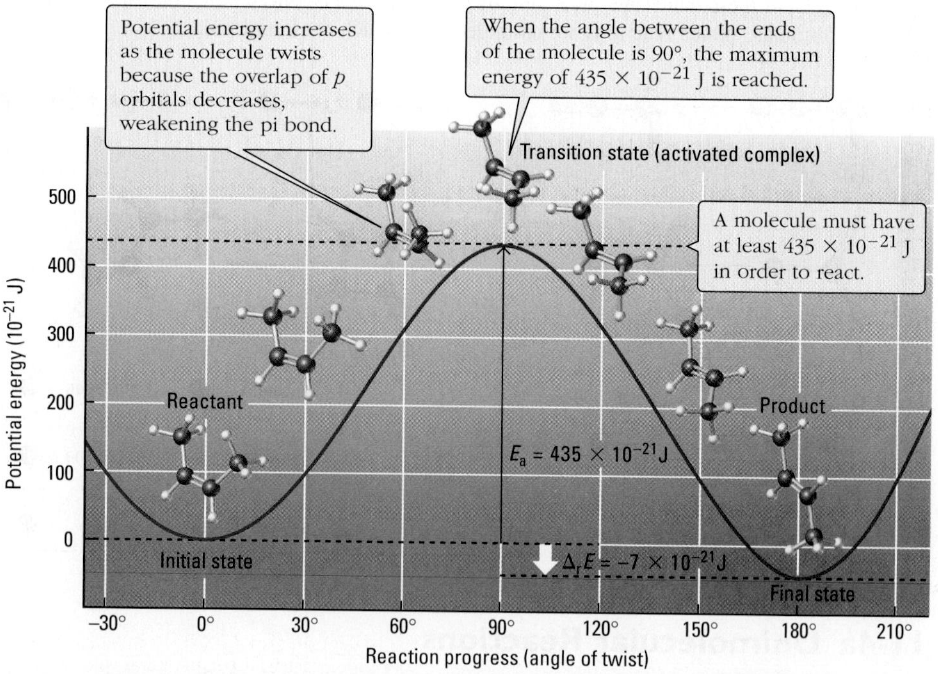
Potential energy increases as the molecule twists because the overlap of p orbitals decreases, weakening the pi bond.
When the angle between the ends of the molecule is 90°, the maximum energy of 435 × 10⁻²¹ J is reached.
Transition state (activated complex)
A molecule must have at least 435 × 10⁻²¹ J in order to react.
Reactant
Product
Initial state
Final state
Ea = 435 × 10⁻²¹ J
ΔrE = −7 × 10⁻²¹ J
Potential energy (10⁻²¹ J)
500
400
300
200
100
0
−30°
0°
30°
60°
90°
120°
150°
180°
210°
Reaction progress (angle of twist)

Figure 11.7 Energy diagram for the conversion of *cis*-2-butene to *trans*-2-butene. The progress of this first-order reaction (the change in the structure of this single molecule) is measured by the angle of twist.

consists of a sigma bond and a pi bond. The pi bond involves sideways overlap of one *p* orbital on each carbon atom. Such sideways overlap is greatest if the *p* orbitals are parallel to each other. Figure 11.8 shows how rotating *cis*-2-butene around the double bond by 90° decreases *p*-orbital overlap, effectively breaking the pi part of the double bond. This requires an increase in potential energy. Consequently, some kinetic energy must be converted to potential energy when one end of the *cis*-2-butene molecule twists relative to the other. At room temperature most of the molecules do not have enough energy to twist far enough to change *cis*-2-butene into *trans*-2-butene. Therefore, *cis*-2-butene can be kept in a sealed flask at room temperature for a long time without any appreciable quantity of *trans*-2-butene being formed. However, as the temperature is raised, more and more molecules have sufficient energy to react, and the reaction gets faster and faster.

Figure 11.7 shows potential energy versus the angle of twist in *cis*- and *trans*-2-butene. The potential energy is 435×10^{-21} J higher when one end of a *cis*-2-butene molecule is twisted by 90° from the initial flat molecule. This is similar to the increased potential energy that an object such as a car has at the top of a hill compared with its energy at the bottom. Just as a car cannot reach the top of a hill unless it has enough energy, a molecule cannot reach the top of the "hill" for a reaction unless it has enough energy. Notice that the top of the hill can be approached from either side, and from the top a twisted molecule can go downhill energetically to either the *cis* or the *trans* form. *The structure at the top of an energy diagram like this one* is called the **transition state** or **activated complex**. In this case it is a molecule that has been twisted so that the methyl groups are at a 90° angle.

Because molecules are very small, the energy required to twist one *cis*-2-butene molecule is very small. However, twisting a mole of molecules all at once would take a lot of energy. The energy required to reach the top of the "hill" is often reported per mole of molecules—that is, as $(435 \times 10^{-21}$ J/molecule$) \times$ (1 kJ/10^3 J) $\times$ $(6.022 \times 10^{23}$ molecules/mol$)$ = 262 kJ/mol.

CONCEPTUAL EXERCISE 11.7

Transition State

Methyl isonitrile reacts to form acetonitrile in a single-step elementary reaction.

$$CH_3NC(g) \longrightarrow CH_3CN(g)$$

During the reaction the nitrogen atom and one of the carbon atoms exchange places, but the rest of the molecule is unchanged. Suggest a structure for the transition state for this reaction. Draw this structure as a Lewis structure and as a ball-and-stick molecular model.

Almost every chemical reaction has an energy barrier that must be surmounted as reactant molecules change into product molecules. The heights of such barriers vary greatly—from almost zero to hundreds of kilojoules per mole. *At a given temperature,*

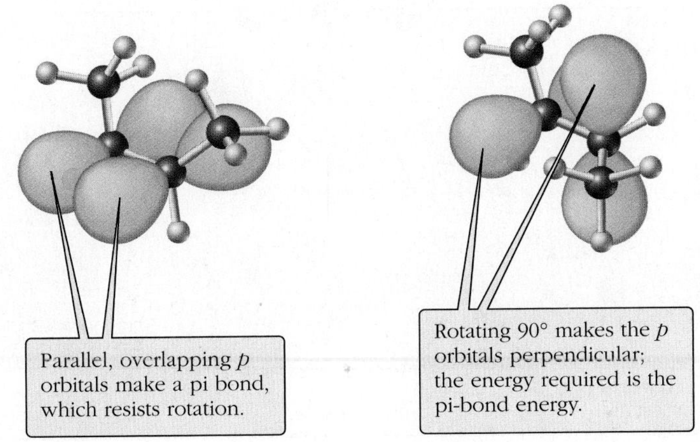
Parallel, overlapping *p* orbitals make a pi bond, which resists rotation.
Rotating 90° makes the *p* orbitals perpendicular; the energy required is the pi-bond energy.

Figure 11.8 Rotating around the double bond of *cis*-2-butene increases potential energy.

the higher the energy barrier, the slower the reaction. The *minimum energy required to surmount the barrier* is called the **activation energy, E_a,** for the reaction. For the *cis*-2-butene ⟶ *trans*-2-butene reaction the activation energy is 435×10^{-21} J/molecule, or 262 kJ/mol (see Figure 11.7).

Another interesting relationship shown in Figure 11.7 connects kinetics and thermodynamics. The energy of the product, one molecule of *trans*-2-butene, is 7×10^{-21} J *lower* than that of the reactant, one molecule of *cis*-2-butene. This means that the *cis* ⟶ *trans* reaction is *exothermic* by 7×10^{-21} J/molecule, which translates to 4 kJ/mol. Also, *cis*-2-butene is higher in energy by 7×10^{-21} J/molecule, so the reverse reaction requires that 4 kJ/mol be absorbed from the surroundings; it is *endothermic*. The height of the energy hill that must be climbed when the reverse reaction occurs is $(435 + 7) \times 10^{-21}$ J/molecule or 442×10^{-21} J/molecule (266 kJ/mol). Thus, the activation energy for the forward reaction is 4 kJ/mol less than that for the reverse reaction. For almost all reactions the activation energy for a forward reaction differs from the activation energy of the reverse reaction, and the difference is $\Delta_r E°$ for the reaction.

The generalization that higher activation energy results in slower reaction applies best if the reactions are similar. For example, it applies best to a group of reactions that all involve twisting around a double bond. It is less applicable when comparing one reaction that involves collision of two molecules with another reaction that involves twisting around a bond.

The actual relation is

$\Delta_r E° = E_a(\text{forward}) - E_a(\text{reverse})$.

Since $\Delta_r E°$ differs from $\Delta_r H°$ only when there is a change in volume, the difference in activation energies is often equated with the enthalpy change (← Sec. 4-5b).

11-4b Bimolecular Reactions

An example of a bimolecular process is the reaction of iodide ion, I^-, with methyl bromide, CH_3Br, in aqueous solution.

$$I^-(aq) + CH_3Br(aq) \longrightarrow ICH_3(aq) + Br^-(aq)$$

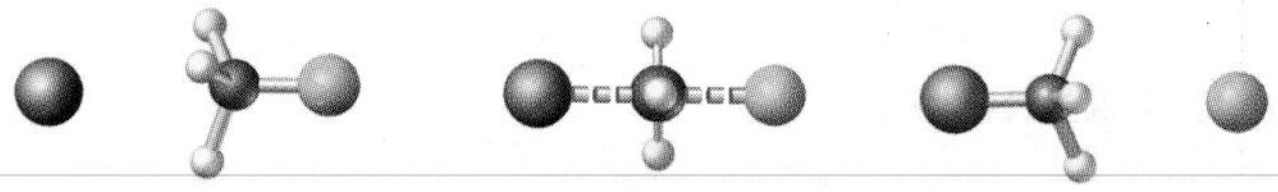

transition state

The equation for the elementary reaction shows that an iodide ion must collide with a methyl bromide molecule for the reaction to occur. The carbon-bromine bond does not break until after the iodine-carbon bond has begun to form. This makes sense, because just breaking a carbon-bromine bond would require a large increase in potential energy. Partially forming a carbon-iodine bond while the other bond is breaking reduces the

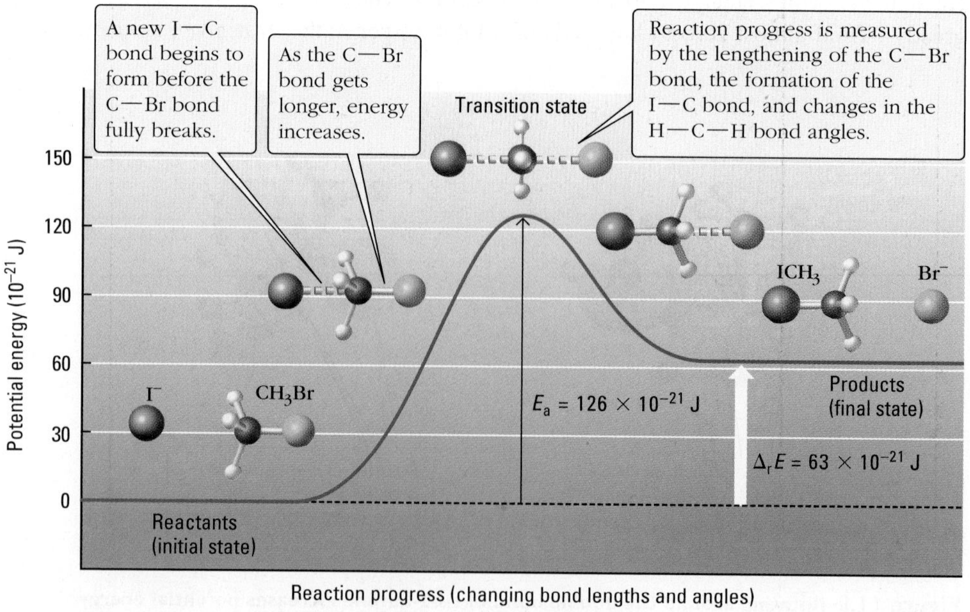
A new I—C bond begins to form before the C—Br bond fully breaks.
As the C—Br bond gets longer, energy increases.
Transition state
Reaction progress is measured by the lengthening of the C—Br bond, the formation of the I—C bond, and changes in the H—C—H bond angles.
150
120
90
60
30
0
Potential energy (10^-21 J)
I^-
CH3Br
ICH3
Br^-
Ea = 126 × 10^-21 J
Products (final state)
ΔrE = 63 × 10^-21 J
Reactants (initial state)
Reaction progress (changing bond lengths and angles)

Figure 11.9 Energy diagram for iodide-methyl bromide reaction.

In each of these three collisions the I^- strikes the CH_3Br on a side away from the C atom. Therefore these collisions are less likely to result in reaction than is...
...this one, where the I^- strikes the CH_3 side of the molecule.
CH_3Br
I^-

Figure 11.10 Steric factor. Orientation of molecules when they collide is important.

potential energy, making the activation energy hill lower than it would otherwise be. Figure 11.9 shows the energy diagram for this reaction.

The methyl bromide molecule has a tetrahedral shape (← **Sec. 7-2a**) that is distorted because the Br atom is much larger than the H atoms. Numerous experiments suggest that the reaction occurs most rapidly in solution when the I^- ion approaches the methyl bromide from the side of the tetrahedron opposite the bromine atom. That is, approach to only one of the four sides of CH_3Br can be effective, which limits reaction to only one fourth of all the collisions at most. This factor of one fourth is illustrated in Figure 11.10; it is called a **steric factor** because it *depends on the three-dimensional shapes of the reacting molecules.* For molecules much more complicated than methyl bromide, such geometry restraints mean that only a very small fraction of the total collisions can lead to reaction. No wonder some chemical reactions are slow.

The word *steric* comes from the same root as the prefix *stereo-*, which means "three-dimensional."

PROBLEM-SOLVING EXAMPLE 11.8

Reaction Energy Diagrams

A reaction by which ozone is destroyed in the stratosphere (← **Sec. 8-10**) is

$$O_3(g) + O(g) \longrightarrow 2\,O_2(g)$$

(O represents atomic oxygen, which is formed in the stratosphere when photons of ultraviolet light from the sun split oxygen molecules in two.) The activation energy for ozone destruction is 19 kJ/mol. Use standard formation enthalpies from Appendix J to calculate the enthalpy change for this reaction. Construct an energy diagram; draw vertical arrows to indicate the sizes of $\Delta_r H°$, E_a(forward), and E_a(reverse) for the reaction.

Result $\Delta_r H° = \Delta_r E° = -392$ kJ/mol of O_3 consumed; E_a(forward) = 19 kJ/mol; E_a(reverse) = 411 kJ/mol giving this energy diagram.

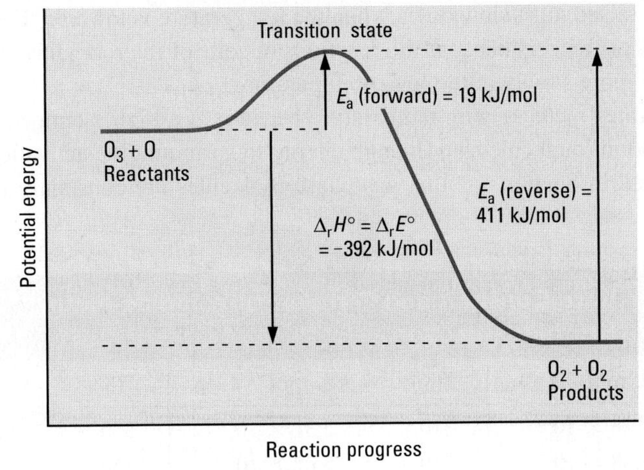
Transition state
E_a (forward) = 19 kJ/mol
$O_3 + O$
Reactants
E_a (reverse) =
411 kJ/mol
$\Delta_r H° = \Delta_r E°$
= −392 kJ/mol
Potential energy
$O_2 + O_2$
Products
Reaction progress

Because the amounts of gas-phase reactants and products are equal, there is no volume change at constant temperature, and $\Delta_r E° = \Delta_r H°$ for this reaction (← Sec. 4-5b).

Analyze An energy diagram indicates the energies of reactants, products, and transition states. The difference in energy of products compared to reactants is $\Delta_r H°$. The difference in energy of transition state compared to reactants is E_a. E_a(forward) − E_a(reverse) = $\Delta_r H°$.

Plan Apply the analysis to create the diagram.

Execute Standard formation enthalpies are 249.2 kJ/mol for ozone, 0 kJ/mol for diatomic oxygen, and 142.7 kJ/mol for atomic oxygen. Thus, $\Delta_r H° = 0 - (249.2 + 142.7)$ kJ/mol = -391.9 kJ/mol. The negative sign of $\Delta_r H°$ indicates that the reaction is exothermic, so the products must be lower in energy than the reactants by 391.9 kJ/mol. E_a(forward) = 19 kJ/mol so the transition state must be this much higher in energy than the reactants. Thus, the diagram in the Result can be drawn based on the E_a arrow and the $\Delta_r H°$ arrow.

$$E_a(\text{reverse}) = E_a(\text{forward}) - \Delta_r H° = [19 - (-391.9)]\ \text{kJ/mol} = 411\ \text{kJ/mol}$$

PROBLEM-SOLVING PRACTICE 11.8

For the reaction

$$Cl_2(g) + 2\,NO(g) \longrightarrow 2\,NOCl(g)$$

the activation energy is 18.9 kJ/mol. For the reaction

$$2\,NOCl(g) \longrightarrow 2\,NO(g) + Cl_2(g)$$

the activation energy is 98.1 kJ/mol. Draw a diagram similar to Figure 11.7 for the reaction of NO with Cl_2 to form NOCl. Is this reaction exothermic or endothermic? Explain.

CONCEPTUAL EXERCISE 11.8

Successful and Less Successful Collisions

The reaction

$$2\,NOCl(g) \longrightarrow 2\,NO(g) + Cl_2(g)$$

occurs in a single bimolecular step. Draw at least four possible ways that two NOCl molecules could collide, and rank them in order of greatest likelihood that a collision will be successful in producing products.

11-5 Temperature and Reaction Rate: The Arrhenius Equation

As a rough rule of thumb, the reaction rate increases by a factor of 2 to 4 for each 10-K rise in temperature.

The most common way to speed up a reaction is to increase the temperature. A mixture of methane and air can be ignited by a lighted match, which raises the temperature of the mixture of reactants. This increases the reaction rate, the heat transfer maintains the high temperature, and the reaction continues at a rapid rate. Reactions that speed up when the temperature is raised must slow down when the temperature is lowered. Foods are stored in refrigerators or freezers because the reactions in cells of microorganisms that produce spoilage occur more slowly at the lower temperature.

Reaction rates increase with temperature because at a higher temperature a greater fraction of reactant molecules has enough energy to surmount the activation energy barrier. You learned in Section 8-2 that gas-phase molecules are constantly in motion and

Table 11.3 Rate Constant and Temperature: Iodide-Methyl Bromide Reaction

T(K)	k(L mol^{-1} s^{-1})	T(K)	k(L mol^{-1} s^{-1})	T(K)	k(L mol^{-1} s^{-1})
273	4.18×10^{-5}	310.	2.31×10^{-3}	350.	6.80×10^{-2}
280.	9.68×10^{-5}	320.	5.82×10^{-3}	360.	1.41×10^{-1}
290.	2.00×10^{-4}	330.	1.39×10^{-2}	370.	2.81×10^{-1}
300.	8.60×10^{-4}	340.	3.14×10^{-2}		

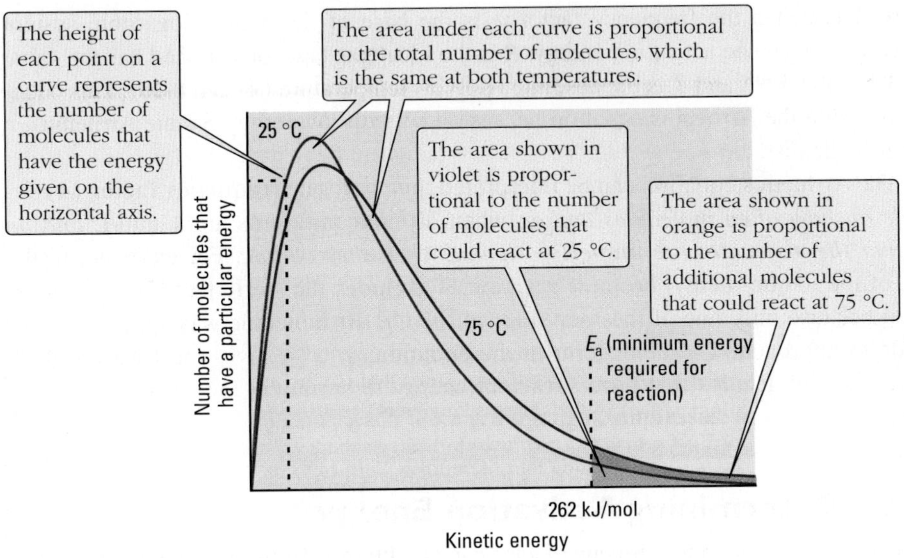

The height of each point on a curve represents the number of molecules that have the energy given on the horizontal axis.
The area under each curve is proportional to the total number of molecules, which is the same at both temperatures.
25 °C
The area shown in violet is proportional to the number of molecules that could react at 25 °C.
The area shown in orange is proportional to the number of additional molecules that could react at 75 °C.
75 °C
E_a (minimum energy required for reaction)
Number of molecules that have a particular energy
262 kJ/mol
Kinetic energy

Figure 11.11 Energy distribution curves. At 75 °C many more molecules than at 25 °C have energies of 262 kJ/mol or higher and can react. (This graph is not to scale; 262 kJ/mol should be much farther to the right.)

have a wide distribution of speeds and energies. Consider again the conversion of *cis*- to *trans*-2-butene and its activation barrier shown in Figure 11.7 (← **Sec. 11-4**). At room temperature relatively few *cis*-2-butene molecules have sufficient energy to surmount the energy barrier. However, as the temperature goes up, the number of molecules that have enough energy goes up rapidly, so the reaction rate increases rapidly. Figure 11.11 shows that at 75 °C, the number of molecules whose energy exceeds the activation energy is much higher than at 25 °C; the reaction rate is correspondingly higher at 75 °C.

A reaction is faster at a higher temperature because its rate constant is larger. That is, ***a rate constant is constant only for a given reaction at a given temperature.*** For example, for the reaction of iodide ion with methyl bromide, the data shown in Table 11.3 are found for the rate constant at different temperatures. When these data are graphed (Figure 11.12a), it is obvious that the rate constant increases very rapidly as temperature increases. A J-shaped graph like this is called an exponential curve. It can be represented by an equation involving a power of *e*, the base of the natural logarithm system:

$$k = Ae^{-E_a/RT} \qquad \textbf{[Arrhenius equation]}$$

Exponential curves also represent the growth of populations (such as human population) over time and are important in many other scientific fields. Logarithms and exponentials are discussed in Appendix A-6.

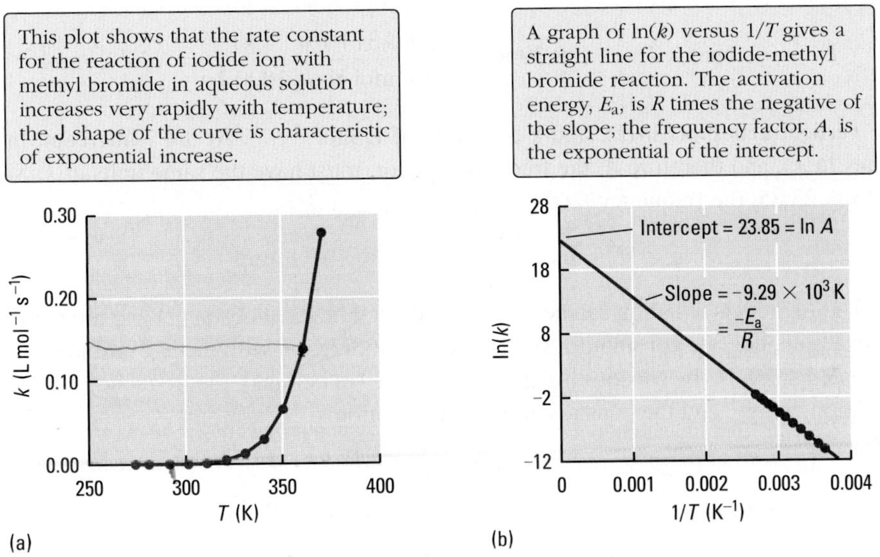

This plot shows that the rate constant for the reaction of iodide ion with methyl bromide in aqueous solution increases very rapidly with temperature; the J shape of the curve is characteristic of exponential increase.
0.30
0.20
0.10
0.00
k (L mol^{-1} s^{-1})
250
300
350
400
T (K)
(a)
A graph of ln(k) versus 1/T gives a straight line for the iodide-methyl bromide reaction. The activation energy, E_a, is R times the negative of the slope; the frequency factor, A, is the exponential of the intercept.
28
18
8
-2
-12
ln(k)
Intercept = 23.85 = ln A
Slope = −9.29 × 10^3 K = $\frac{-E_a}{R}$
0
0.001
0.002
0.003
0.004
1/T (K^{-1})
(b)

Figure 11.12 Effect of temperature on rate constant.

Because the line through the experimental data in part (b) has to be extrapolated a long way to reach the y-intercept, determining *A* accurately requires measuring *k* over a wide range of temperatures.

R is the constant found in the ideal gas law, $PV = nRT$ (← Sec. 8-3d). Here it is expressed in units of $J\ mol^{-1}\ K^{-1}$ instead of $L\ atm\ mol^{-1}\ K^{-1}$, so the numerical value is different.

where A is called the frequency factor, e is the base of the natural logarithm system (2.718 . . .), E_a is the activation energy, R is the ideal gas law constant and has the value 8.314 J mol^{-1} K^{-1}, and T is the absolute (Kelvin) temperature (← **Sec. 8-3b**). This equation is called the Arrhenius equation because it was discovered by Svante Arrhenius, a Swedish chemist.

The Arrhenius equation can be interpreted this way. The **frequency factor (A)** *depends on how often molecules collide* when all concentrations are 1 mol/L *and on whether the molecules are properly oriented when they collide.* For example, in the case of the iodide-methyl bromide reaction, A includes the steric factor of $\frac{1}{4}$ that resulted because only one of the four sides of the CH_3Br molecule was appropriate for iodide to approach. The other term in the equation, $e^{-E_a/RT}$, gives the fraction of all the reactant molecules that have sufficient energy to surmount the activation energy barrier.

11-5a Determining Activation Energy

The activation energy and frequency factor can be obtained from experimental measurements of rate constants as a function of temperature (such as those in Table 11.3). When a large number of experimental data pairs are given, a graph is usually a good way of obtaining information from the data. This is easier to do if the graph is linear. The activation energy (Arrhenius) equation can be modified by taking natural logarithms of both sides so that its graph is linear.

$$k = Ae^{-E_a/RT}$$

$$\ln(k) = \ln(A) + \ln(e^{-E_a/RT}) = \ln(A) + (-E_a/RT)$$

Rearranging this equation gives the equation of a straight line.

$$\ln(k) = -\frac{E_a}{R} \times \frac{1}{T} + \ln(A)$$

$$y = m \times x + b$$

That is, if we graph $\ln(k)$ on the vertical (y) axis and $1/T$ on the horizontal (x) axis, the result should be a straight line whose slope is $-E_a/R$ and whose y-intercept is $\ln(A)$. For the data in Table 11.3, such a graph is shown in Figure 11.12b. It is linear, its slope is -9.29×10^3 K, and the y-intercept is 23.85. Because the slope $= -E_a/R$, the activation energy can be calculated as

$$E_a = -(\text{slope}) \times R = -(-9.29 \times 10^3\ \text{K})\left(\frac{8.314\ \text{J}}{\text{mol K}}\right)\left(\frac{1\ \text{kJ}}{1000\ \text{J}}\right) = 77.2\ \text{kJ/mol}$$

The vertical axis plots $\ln(k)$, and k has units of L mol^{-1} s^{-1}. At the y-intercept, $\ln(A)$ equals $\ln(k)$, and therefore A, the frequency factor, must have the same units as k. Since $\ln(A) = 23.85$, the frequency factor, A, is

$$A = e^{23.85}\ \text{L mol}^{-1}\ \text{s}^{-1} = 2.28 \times 10^{10}\ \text{L mol}^{-1}\ \text{s}^{-1}$$

The Arrhenius equation can be used to calculate the rate constant at any temperature. For example, the rate constant for the reaction of iodide ion with methyl bromide can be calculated at 50. °C by substituting the temperature (in kelvins), the frequency factor, the activation energy, and the constant R into the equation.

$$k = Ae^{-E_a/RT} = (2.28 \times 10^{10}\ \text{L mol}^{-1}\ \text{s}^{-1})e^{(-77{,}200\ \text{J/mol})/(8.314\ \text{J K}^{-1}\ \text{mol}^{-1})(273.15\ +\ 50.)\text{K}}$$

$$= (2.28 \times 10^{10}\ \text{L mol}^{-1}\ \text{s}^{-1})e^{-28.7} = 7.56 \times 10^{-3}\ \text{L mol}^{-1}\ \text{s}^{-1}$$

As a means of calculating rate constants, the Arrhenius equation works best within the range of temperatures over which the activation energy and frequency factor were determined. (For the reaction of iodide with methyl bromide, that range was 273 to 370 K.)

PROBLEM-SOLVING EXAMPLE 11.9

Temperature Dependence of Rate Constant

The experimental rate constant for the reaction of iodide ion with methyl bromide is 7.70×10^{-3} L mol^{-1} s^{-1} at 50.00 °C and 4.18×10^{-5} L mol^{-1} s^{-1} at 0.00 °C. Calculate the frequency factor and activation energy.

Result $A = 1.83 \times 10^{10}$ L mol^{-1} s^{-1}; $E_a = 7.66 \times 10^4$ J mol^{-1}

Analyze Two rate constants and the temperatures at which they were measured are given. The frequency factor and activation are to be determined. The Arrhenius equation relates rate constant, temperature, frequency factor, and activation energy.

Plan This problem could be solved by making a graph of ln(k) versus $1/T$, but with only two data pairs, a graph is not the best way. Instead, write the Arrhenius equation twice, once for each data pair.

$$k_1 = Ae^{-E_a/RT_1} \qquad k_2 = Ae^{-E_a/RT_2}$$

Divide the first equation by the second to eliminate A.

$$\frac{k_1}{k_2} = \frac{Ae^{-E_a/RT_1}}{Ae^{-E_a/RT_2}} = e^{-E_a/RT_1} \times e^{+E_a/RT_2}$$

$$\frac{k_1}{k_2} = e^{\frac{E_a}{R}\left(\frac{1}{T_2} - \frac{1}{T_1}\right)}$$

Next, take the natural logarithm of both sides.

$$\ln\left(\frac{k_1}{k_2}\right) = \frac{E_a}{R}\left(\frac{1}{T_2} - \frac{1}{T_1}\right)$$

Execute Solve the equation for E_a.

$$E_a = \frac{R\ln\left(\frac{k_1}{k_2}\right)}{\frac{1}{T_2} - \frac{1}{T_1}} = \frac{(8.314 \text{ J mol}^{-1}\text{ K}^{-1})\ln\left[\frac{7.70 \times 10^{-3}}{4.18 \times 10^{-5}}\right]}{\frac{1}{273.15\text{ K}} - \frac{1}{323.15\text{ K}}}$$

$$= \frac{43.37 \text{ J mol}^{-1}\text{ K}^{-1}}{5.664 \times 10^{-4}\text{ K}^{-1}} = 7.66 \times 10^4 \text{ J mol}^{-1}$$

Finally, using the calculated value of E_a, solve one of the rate constant expressions for A.

$$k_1 = Ae^{-E_a/RT_1}$$

$$A = \frac{k_1}{e^{-E_a/RT_1}} = k_1 e^{E_a/RT_1}$$

$$= (7.70 \times 10^{-3}\text{ L mol}^{-1}\text{s}^{-1})e^{\left(\frac{7.66\times10^4\text{ J mol}^{-1}}{(8.314\text{ J mol}^{-1}\text{ K}^{-1})(323.15\text{ K})}\right)}$$

$$= (7.70 \times 10^{-3}\text{ L mol}^{-1}\text{ s}^{-1})(2.380 \times 10^{12}) = 1.83 \times 10^{10}\text{ L mol}^{-1}\text{ s}^{-1}$$

PROBLEM-SOLVING PRACTICE 11.9

Calculate the rate constant for the reaction of iodide ion with methyl bromide at a temperature of 75 °C.

Ahmed H. Zewail
1946–

AP Images/HO, HO

For his studies of the transition states of chemical reactions using femtosecond spectroscopy, Ahmed H. Zewail of the California Institute of Technology received the 1999 Nobel Prize in Chemistry. Zewail, who holds joint Egyptian and U. S. citizenship, pioneered the use of extremely short laser pulses—on the order of femtoseconds (10^{-15} s)—to study chemical kinetics. His technique has been called the world's fastest camera, and his research has enhanced understanding of many reactions, among them those involving rhodopsin and vision.

Be sure to use k_1 as the rate constant for T_1 and k_2 as the rate constant for T_2.

The Arrhenius equation enables us to calculate the rate constant as a function of temperature, and the rate law shows how the rate depends on concentration. To obtain a single equation that summarizes the effects of both temperature and concentration, substitute k from the Arrhenius equation into the rate law for the iodide-methyl bromide reaction, giving

Recall that the collision frequency contribution to the frequency factor is for 1-M concentrations.

$$\text{Rate} = k \times [\text{I}^-] \times [\text{CH}_3\text{Br}]$$

$$\text{Rate} = A \times e^{-E_a/RT} \times [\text{I}^-] \times [\text{CH}_3\text{Br}] \qquad [11.8]$$

Collision frequency × steric factor (A); Fraction of sufficiently energetic molecules ($e^{-E_a/RT}$); Concentrations of colliding molecules ($[\text{I}^-] \times [\text{CH}_3\text{Br}]$)

Equation 11.8 summarizes the effects of both temperature and concentration on rate of a reaction. The temperature effect depends primarily on the large increase in the number of sufficiently energetic collisions as the temperature increases, which shows up as larger values of k at higher temperatures. The effect of concentration is clearly indicated by the concentration terms in the rate law. If the rate law is known for a reaction, and if both the A and E_a values are known, then the rate can be calculated over a wide range of conditions.

EXERCISE 11.9

Activation Energy and Experimental Data

The frequency factor A is 6.31×10^8 L mol^{-1} s^{-1} and the activation energy is 10. kJ/mol for the gas-phase reaction

$$\text{NO(g)} + \text{O}_3\text{(g)} \longrightarrow \text{NO}_2\text{(g)} + \text{O}_2\text{(g)}$$

which is important in the chemistry of stratospheric ozone depletion.

(a) Calculate the rate constant for this reaction at 370. K.

(b) Assuming that this is an elementary reaction, calculate the rate of the reaction at 370. K if [NO] = 0.0010 M and $[\text{O}_3]$ = 0.00050 M.

11-6 Rate Laws for Elementary Reactions

An elementary reaction is a one-step process whose equation describes which nanoscale particles break apart, rearrange their positions, or collide to make a reaction occur. Therefore, from the equation for an elementary reaction, it is possible to figure out what the rate law and reaction order are, without doing an experiment. By contrast, when an equation represents a reaction that we do not understand at the nanoscale, rate laws and reaction orders must be determined experimentally (← **Secs. 11-2b, 11-3a**). The macroscale rate law can then be used to help develop a hypothesis about how a particular reaction takes place at the nanoscale.

11-6a Rate Law for a Unimolecular Reaction

In Section 11-4 we used the isomerization of *cis*-2-butene as an example of a reaction in which a single reactant molecule was converted to a product molecule or molecules—a unimolecular reaction (← **Sec. 11-4a**).

$$cis\text{-2-butene} \longrightarrow trans\text{-2-butene}$$

Suppose a flask contains 0.0050 mol/L *cis*-2-butene vapor at room temperature. The molecules have a wide range of energies, but only a few of them have enough energy at this temperature to get over the activation energy barrier. Thus, during a given period only a few molecules twist sufficiently to become *trans*-2-butene. Now, suppose that we double the concentration of *cis*-2-butene in the flask to 0.0100 mol/L, while keeping the temperature the same. The fraction of molecules with enough energy to cross over the barrier remains the same. However, because there are now twice as many molecules, twice as many must be crossing the barrier in any given time. Therefore, the rate of the *cis* ⟶ *trans* reaction is twice as great. That is, the reaction rate is proportional to the concentration of *cis*-2-butene, and the rate law must be

$$\text{Rate} = k[\textit{cis}\text{-2-butene}]$$

In the general case of any unimolecular elementary reaction,

$$\text{A} \longrightarrow \text{products} \qquad \text{the rate law is} \qquad \text{Rate} = k[\text{A}]$$

For any unimolecular reaction, the nanoscale mechanism predicts that a first-order rate law will be observed in a macroscale laboratory experiment.

Suppose that the fraction of molecules that have enough energy to react is 0.1%, or 0.001. If there are 10,000 molecules in a given volume, then 0.001 × 10,000 gives only 10 that have enough energy to react. If there are twice as many molecules in the same volume—that is, 20,000 molecules—then 0.001 × 20,000 gives 20 with enough energy to react, and the number reacting per unit volume (the rate) is twice as great.

11-6b Rate Law for a Bimolecular Reaction

A good example of a reaction in which two molecules collide [a bimolecular reaction, (← **Sec. 11-4b**)] is the gas-phase reaction of nitrogen monoxide and ozone that is involved in stratospheric ozone depletion and was mentioned in Exercise 11.9.

$$NO(g) + O_3(g) \longrightarrow NO_2(g) + O_2(g) \qquad [11.9]$$

The equation shows that the elementary reaction involves the collision of one NO molecule and one O_3 molecule. Because the molecules must collide to exchange atoms, the rate depends on the number of collisions per unit time.

Figure 11.13a represents one NO molecule (green sphere) and many O_3 molecules (purple spheres) in a tiny region within a flask where Reaction 11.9 is taking place. In a given time, the NO molecule collides with five O_3 molecules. If the concentration of NO molecules is doubled to two NO molecules in the same portion of the flask (Figure 11.13b), *each* NO molecule collides with five different O_3 molecules. Doubling the concentration of NO has doubled the number of collisions. This also doubles the rate, because the rate is proportional to the number of collisions. The number of collisions also

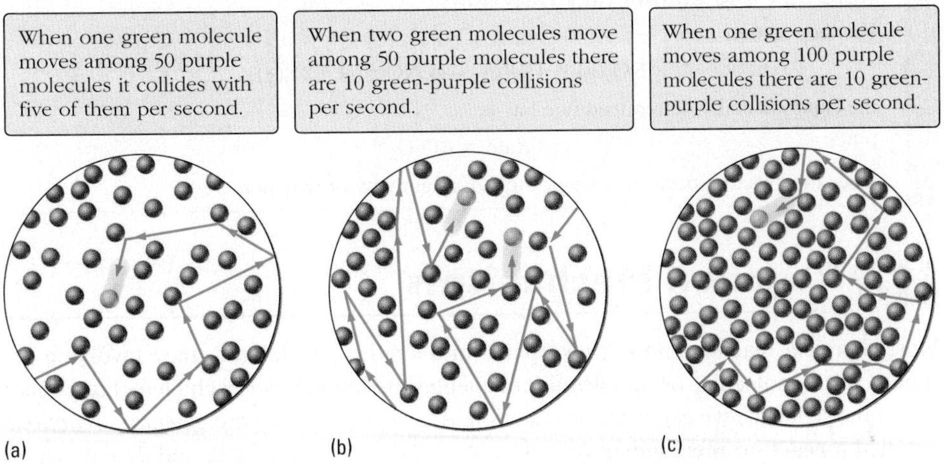
When one green molecule moves among 50 purple molecules it collides with five of them per second.
When two green molecules move among 50 purple molecules there are 10 green-purple collisions per second.
When one green molecule moves among 100 purple molecules there are 10 green-purple collisions per second.
(a)
(b)
(c)

Figure 11.13 Effect of concentration on frequency of bimolecular collisions. The number of collisions, and hence the reaction rate, is proportional to *both* the concentration of green molecules *and* the concentration of purple molecules.

In a real sample both the green and the purple molecules would be moving, but to make these diagrams clearer the motion of the purple molecules is not shown.

doubles when the O_3 concentration is doubled (Figure 11.13c). Thus, the rate law for this reaction must be

$$\text{Rate} = k[\text{NO}][\text{O}_3]$$

This description of the NO + O_3 reaction applies in general to bimolecular elementary reactions, even if the two molecules that must collide are of the same kind. That is, for the elementary reaction

$$\text{A} + \text{B} \longrightarrow \text{products} \qquad \text{the rate law is} \qquad \text{Rate} = k[\text{A}][\text{B}]$$

and for the elementary reaction

$$\text{A} + \text{A} \longrightarrow \text{products} \qquad \text{the rate law is} \qquad \text{Rate} = k[\text{A}]^2$$

For the NO + O_3 reaction the experimentally determined rate law is the same as the one we just derived by assuming the reaction occurs in one step. The chemical equation for the reaction is

$$\text{NO(g)} + \text{O}_3\text{(g)} \longrightarrow \text{NO}_2\text{(g)} + \text{O}_2\text{(g)}$$

and the experimental rate law is

$$\text{Rate} = k[\text{NO}][\text{O}_3]$$

This experimental observation suggests, but does not prove, that the reaction does take place in a single step. Other experimental evidence also suggests that this reaction is bimolecular.

In contrast, for the decomposition of hydrogen peroxide the chemical equation is

$$2\,\text{H}_2\text{O}_2\text{(aq)} \longrightarrow 2\,\text{H}_2\text{O}(\ell) + \text{O}_2\text{(g)}$$

and the experimental rate law is Rate = $k[H_2O_2]$. This rate law proves that this reaction *cannot* occur in a single step that involves collision of two H_2O_2 molecules. A single-step bimolecular reaction would have a second-order rate law, but the observed rate law is first-order. This means that more than a single elementary step is needed when hydrogen peroxide decomposes.

CONCEPTUAL **EXERCISE 11.10**

Rate Law and Elementary Reactions

For the reaction

$$\text{NO}_2\text{(g)} + \text{CO(g)} \longrightarrow \text{NO(g)} + \text{CO}_2\text{(g)}$$

the experimentally determined rate law is

$$\text{Rate} = k[\text{NO}_2]^2$$

Does this reaction occur in a single step? Explain why or why not.

11-7 Reaction Mechanisms

Most chemical reactions do not take place in a single step. Instead, they involve a sequence of unimolecular or bimolecular elementary reactions. For each elementary reaction in the sequence we can write an equation. *A set of equations for elementary reactions* is called a **reaction mechanism** (or just a *mechanism*). For example, iodide ion can be oxidized by hydrogen peroxide in acidic solution to form iodine and water according to this overall equation:

$$2\,\text{I}^-\text{(aq)} + \text{H}_2\text{O}_2\text{(aq)} + 2\,\text{H}^+\text{(aq)} \longrightarrow \text{I}_2\text{(aq)} + 2\,\text{H}_2\text{O}(\ell) \qquad [11.10]$$

When the acid concentration is between 10^{-3} M and 10^{-5} M, experiments show that the rate law is

$$\text{Rate} = k[I^-][H_2O_2]$$

The reaction is first-order in I^- and in H_2O_2; it is second-order overall.

Looking at Equation 11.10 for the oxidation of iodide ion by hydrogen peroxide, you might think that two iodide ions, one hydrogen peroxide molecule, and two hydrogen ions would all have to come together at once. However, the rate law corresponds with a bimolecular collision of I^- and H_2O_2. It is highly unlikely that at the same time five ions or molecules would all be at the same place, be properly oriented, and have enough energy to react. Instead, chemists who have studied this reaction propose that initially one H_2O_2 (HOOH) molecule and one I^- ion come together:

Step 1: $HOOH + I^- \xrightarrow{\text{slow}} OH^- + HOI$

This first step forms hypoiodous acid, HOI, and hydroxide ion, both known substances. The HOI then reacts with another I^- to form the product I_2:

Step 2: $HOI + I^- \xrightarrow{\text{fast}} OH^- + I_2$

In each of Steps 1 and 2, a hydroxide ion was produced. Because the solution is acidic, these OH^- ions react immediately with H^+ ions to form water. Two OH^- ions are formed so the reaction occurs twice:

Step 3: $OH^- + H^+ \xrightarrow{\text{fast}} H_2O$ $\qquad$ $OH^- + H^+ \xrightarrow{\text{fast}} H_2O$

Each of the three steps in this mechanism is an elementary reaction. Each has its own activation energy, E_a, and its own rate constant, k. When the three steps are summed (by putting all the reactants on the left, putting all the products on the right, and eliminating formulas that appear as both reactants and products), the overall stoichiometric equation (Equation 11.10) is obtained. ***Any valid mechanism must consist of a series of unimolecular or bimolecular elementary reaction steps that sum to the overall reaction.***

Step 1: $HOOH + I^- \xrightarrow{\text{slow}} HOI + OH^-$

Step 2: $HOI + I^- \xrightarrow{\text{fast}} I_2 + OH^-$

Step 3: $OH^- + H^+ \xrightarrow{\text{fast}} H_2O$

$OH^- + H^+ \longrightarrow H_2O$

Overall: $2\,I^- + HOOH + 2\,H^+ \longrightarrow I_2 + 2\,H_2O$ [11.10]

An analogy to the rate-limiting step is that no matter how quickly you shop in the supermarket, it seems that the time it takes to get out of the store depends on the rate at which you move through the checkout line.

Step 1 of the mechanism is slow, while Steps 2 and 3 are fast. Step 1 is called the **rate-limiting step** (or rate-determining step) because it is *the slowest step in the sequence and limits the rate* at which I_2 and H_2O can be produced. Steps 2 and 3 are rapid and therefore not rate-limiting. As soon as some HOI and OH^- are produced by Step 1, they are transformed into I_2 and H_2O by Steps 2 and 3. ***The rate of the overall reaction is limited by, and equal to, the rate of the slowest step in the mechanism.***

Step 1 is a bimolecular elementary reaction. Therefore, its rate must be first-order in HOOH and first-order in I^-. The rate law of the overall reaction is the rate law of the rate-limiting step so the mechanism predicts this rate law,

$$\text{Reaction rate} = k[\text{HOOH}][\text{I}^-]$$

which agrees with the experimentally observed rate law. ***A valid mechanism must correctly predict the experimentally observed rate law.***

If significant concentrations of an intermediate build up while a reaction is occurring, then reactants may disappear faster than products are formed (because buildup of the intermediate stores some of the used reactant before it is converted to final product). In such a case, the definition of reaction rate given in Equation 11.3 will not be correct. One way to detect formation of an intermediate is to notice that products are not formed as fast as reactants are used up.

The species HOI and OH^- are produced in Step 1 and used up in Step 2 or 3. In a mechanism, *atoms, molecules, or ions that are produced in one step and consumed in a later step (or later steps)* are called **reaction intermediates** (or just **intermediates**). Very small concentrations of HOI and OH^- are produced while the reaction is going on. Once the HOOH, the I^-, or both are used up, the intermediates HOI and OH^- are consumed by Steps 2 and 3 and disappear. HOI and OH^- are crucial to the reaction mechanism, but neither of them appears in the overall stoichiometric equation. If an experimenter is proficient enough to demonstrate that a particular intermediate was present, this provides additional evidence that a mechanism involving that intermediate is the correct one.

In summary, **a valid reaction mechanism should**

- **consist of a series of unimolecular or bimolecular elementary reactions;**
- **consist of reaction steps that sum to the overall reaction equation; and**
- **correctly predict the experimentally observed rate law.**

EXERCISE 11.11

Rate Law for an Elementary Reaction

Write the rate law for Step 2 of the mechanism for the reaction of hydrogen peroxide and iodide ion. Explain why the rate of Step 2 does not affect the overall rate.

11-7a Mechanisms with a Fast Initial Step

An experimental rate law should include only concentrations that can be measured experimentally. Usually this means concentrations of reactants (and perhaps products), but not the concentrations of intermediates. (The concentrations of intermediates usually are very small and therefore hard to measure.) The first step in the mechanism in the previous section was rate-limiting and involved two of the reactants, so it was easy to relate the overall rate to the concentrations of reactants. But what happens if the rate-limiting step is the second or a subsequent step?

Consider the reaction

$$2\,\text{NO(g)} + \text{Br}_2\text{(g)} \longrightarrow 2\,\text{NOBr(g)} \qquad \text{Rate (experimental)} = k[\text{NO}]^2[\text{Br}_2]$$

for which the currently accepted mechanism is

Step 1: (fast) $\text{NO(g)} + \text{Br}_2\text{(g)} \rightleftharpoons \text{NOBr}_2\text{(g)}$

Step 2: (slow) $NOBr_2(g) + NO(g) \longrightarrow 2\ NOBr(g)$

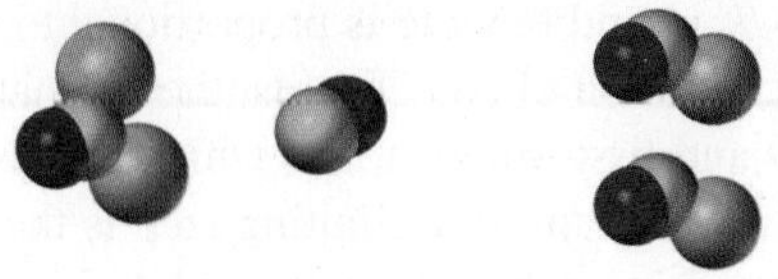

Because the second step is slow, $NOBr_2$ can often break apart and re-form NO + Br_2 before it reacts to form products. We say that Step 1 is reversible, which is indicated by the double arrow ($\rightleftharpoons$). Once $NOBr_2$ forms, it can react in either of two ways: back to the reactants, NO + Br_2, or forward (in Step 2) to form the product, NOBr.

The overall rate of the reaction is the rate of the slow (rate-limiting) step (Step 2):

$$\text{Rate} = k[NOBr_2][NO]$$

However, this is not a valid rate law to compare with experimental results, because an experimental rate law should include only concentrations that can be measured experimentally. (Usually, the concentration of an intermediate, such as $NOBr_2$, cannot be measured.) Therefore, we need a way to relate the concentration of $NOBr_2$ to the concentrations of the reactants, NO and Br_2. This can be done by recognizing that in the mechanism there are three reactions and three rate constants:

Step +1 (forward)	rate constant k_1
Step −1 (backward)	rate constant k_{-1}
Step 2	rate constant k_2

The concentration of $NOBr_2$ depends on all three reactions. It is increased by Step +1 and decreased by Step −1 and Step 2. Initially, the concentration of $NOBr_2$ builds up because Step +1 produces $NOBr_2$. However, when $[NOBr_2]$ gets big enough, the rates of Step −1 and Step 2 get bigger, and eventually the $[NOBr_2]$ reaches what is called a steady state—it neither increases nor decreases until the reaction is nearly over.

In the steady state, the rate of the reaction by which $NOBr_2$ is formed must equal the sum of the two rates by which it is used up; that is,

$$\begin{aligned}(\text{Rate of Step }+1) &= (\text{rate of Step }-1) + (\text{rate of Step 2})\\ k_1[NO][Br_2] &= k_{-1}[NOBr_2] + k_2[NOBr_2][NO] \qquad [11.11]\end{aligned}$$

Equation 11.11 can usually be simplified because Step 2 is much slower than either Step 1 or Step −1. This means that $k_2[NOBr_2][NO] \ll k_{-1}[NOBr_2]$ and $k_2[NOBr_2][NO]$ can be neglected in the calculation.

Suppose that

$k_{-1}[NOBr_2] = 0.20\ \text{mol L}^{-1}\ \text{s}^{-1}$

and

$k_2[NOBr_2][NO] = 0.00020\ \text{mol L}^{-1}\ \text{s}^{-1}$,

which is much smaller. Then

$k_{-1}[NOBr_2] + k_2[NOBr_2][NO] = (0.20 + 0.00020)\ \text{mol L}^{-1}\ \text{s}^{-1} = 0.20\ \text{mol L}^{-1}\ \text{s}^{-1}$.

That is, the rate for Step 2 is negligible and can be ignored in the calculation.

$$k_1[NO][Br_2] \cong k_{-1}[NOBr_2] \qquad [11.12]$$

Equation 11.12 can be rearranged algebraically to show that $[NOBr_2]$ is proportional to the concentration of each reactant.

$$[NOBr_2] = \frac{k_1}{k_{-1}}[NO][Br_2]$$

This allows the rate to be expressed in terms of the concentrations of reactants.

$$\begin{aligned}\text{Rate} &= k_2[NOBr_2][NO] = k_2\left(\frac{k_1}{k_{-1}}[NO][Br_2]\right)[NO]\\ &= \frac{k_1k_2}{k_{-1}}[NO]^2[Br_2] = k'[NO]^2[Br_2] \qquad [11.13]\end{aligned}$$

Equation 11.13 shows that the rate constant k' is actually a quotient of rate constants for three elementary reactions (k_1k_2/k_{-1}) and the rate is proportional to the concentration of Br_2 and to the square of the concentration of NO. That is, the mechanism predicts that the reaction is second-order in NO and first-order in Br_2, which agrees with the experimental rate law. For mechanisms in which the rate-limiting step is the second or a subsequent step, a mathematical relationship such as Equation 11.13 can usually be found that relates an overall rate constant, k', to the concentrations of the reactants and the rate constants for the steps up to and including the rate-limiting step.

11-7b Kinetics and Mechanism

Studying the kinetics of a chemical reaction involves collecting data on the concentrations of reactants as a function of time. From such data the rate law for the reaction and a rate constant can usually be obtained. The reaction can also be studied at several different temperatures to determine its activation energy. This allows us to predict how fast the macroscale reaction will be under a variety of experimental conditions, but it does not provide definitive information about the nanoscale mechanism by which the reaction takes place. A reaction mechanism is an educated guess—a hypothesis—about the way the reaction occurs. If the mechanism predicts correctly the overall stoichiometry of the reaction *and* the experimentally determined rate law, then it is a reasonable hypothesis. However, it is impossible to prove for certain that a mechanism is correct. Sometimes several mechanisms can agree with the same set of experiments. This is what makes kinetic studies one of the most interesting and rewarding areas of chemistry, but it also can provoke disputes among scientists who favor different possible mechanisms for the same reaction.

PROBLEM-SOLVING EXAMPLE 11.10

Rate Law and Reaction Mechanism

The gas-phase reaction between nitrogen monoxide and oxygen,

$$2\ NO(g) + O_2(g) \longrightarrow 2\ NO_2(g)$$

is found experimentally to obey the rate law

$$\text{Rate} = k[NO]^2[O_2]$$

Decide which of these mechanisms is (are) compatible with this information.

(a) $NO + NO \rightleftharpoons N_2O_2$	fast	(b) $NO + NO \longrightarrow NO_2 + N$	slow
$N_2O_2 + O_2 \longrightarrow 2\ NO_2$	slow	$N + O_2 \longrightarrow NO_2$	fast
(c) $NO + O \rightleftharpoons NO_2$	fast	(d) $NO + O_2 \rightleftharpoons NO_3$	fast
$NO_2 + NO \longrightarrow N_2O_3$	fast	$NO_3 + NO \longrightarrow 2\ NO_2$	slow
$N_2O_3 + O \longrightarrow 2\ NO_2$	slow		
(e) $NO + O_2 \longrightarrow NO_2 + O$	slow	(f) $2\ NO + O_2 \longrightarrow 2\ NO_2$	
$NO + O \longrightarrow NO_2$	fast		

Result Mechanisms (a) and (d) are compatible with the rate law and stoichiometry.

Analyze A valid reaction mechanism must (1) consist only of unimolecular steps, bimolecular steps, or both, (2) agree with the overall stoichiometry, and (3) predict the experimental rate law.

Plan Examine each mechanism to determine whether it meets these criteria. Eliminate mechanisms that do not. The remaining mechanism(s) may be correct.

Execute Mechanism (f) involves collision of three molecules: two NO and one O_2. It can be eliminated because it does not consist of unimolecular or bimolecular steps. All of the other mechanisms consist of bimolecular steps.

Mechanism (c) does not have O_2 as a reactant in the overall stoichiometry, so it can be eliminated. All other mechanisms predict the observed overall equation.

In mechanism (a) the first step is fast and reversible. Applying the idea that the rates are approximately equal for the forward and reverse reactions in that first step gives

$$k_1[NO]^2 = k_{-1}[N_2O_2] \text{ and } [N_2O_2] = \frac{k_1}{k_{-1}}[NO]^2$$

Since the overall rate equals the rate of the rate-limiting step,

$$\text{Rate} = k_2[N_2O_2][O_2] = k_2\frac{k_1}{k_{-1}}[NO]^2[O_2] = k'[NO]^2[O_2]$$

Consequently, mechanism (a) could be the actual mechanism.

The slow first step in mechanism (b) implies an overall rate $= k[NO]^2$, which eliminates it from consideration.

Continuing this kind of reasoning, mechanism (d) is seen to be a possibility, but mechanism (e) predicts rate $= k[NO][O_2]$ and therefore can be eliminated. Because there are still two possible mechanisms, (a) and (d), additional experiments need to be done to try to distinguish between them.

PROBLEM-SOLVING PRACTICE 11.10

The Raschig reaction produces the industrially important reducing agent hydrazine, N_2H_4, from ammonia, NH_3, and hypochlorite ion, OCl^-, in basic aqueous solution. A proposed mechanism is

Step 1: $NH_3(aq) + OCl^-(aq) \xrightarrow{\text{slow}} NH_2Cl(aq) + OH^-(aq)$

Step 2: $NH_2Cl(aq) + NH_3(aq) \xrightarrow{\text{fast}} N_2H_5^+(aq) + Cl^-(aq)$

Step 3: $N_2H_5^+(aq) + OH^-(aq) \xrightarrow{\text{fast}} N_2H_4(aq) + H_2O(\ell)$

(a) What is the overall stoichiometric equation?
(b) Which step is rate-limiting?
(c) What reaction intermediates are involved?
(d) What rate law is predicted by this mechanism?

EXERCISE 11.12

Rate Law and Mechanism

Consider the reaction mechanism

$$ICl(g) + H_2(g) \rightleftharpoons HI(g) + HCl(g) \quad \text{fast}$$
$$HI(g) + ICl(g) \longrightarrow HCl(g) + I_2(g) \quad \text{slow}$$

(a) What is the overall reaction equation?
(b) Derive the rate law predicted by this mechanism.
(c) Does the rate law depend on the concentration of one of the products of the reaction?
(d) Would the rate constant determined from the initial rate of this reaction equal the rate constant determined at a time when 80% of the reactants had been consumed? Explain why or why not.

11-8 Catalysts and Reaction Rate

Raising the temperature increases a reaction rate because it increases the fraction of molecules that are energetic enough to surmount the activation energy barrier. Increasing reactant concentrations can also increase the rate because it increases the number of molecules per unit volume. A third way to increase reaction rates is to add a catalyst.

Cengage Learning/Larry Cameron

(a) When 30% $H_2O_2(aq)$ is dropped onto liver, an enzyme catalyzes H_2O_2 decomposition to $O_2(g)$ and $H_2O(\ell)$.

Cengage Learning/Charles D. Winters

(b) Laboratory-scale explosion of a sample of ammonium nitrate, NH_4NO_3, catalyzed by Cl^- ions.

Bettmann/CORBIS

(c) Scene following the explosion of the ship *Grandcamp* in Texas City, Texas, April 16, 1947.

Figure 11.14 Catalysis in action.

For example, an aqueous solution of hydrogen peroxide can decompose to water and oxygen.

$$2\ H_2O_2(aq) \longrightarrow O_2(g) + 2\ H_2O(\ell)$$

At room temperature the rate of the decomposition reaction is exceedingly slow. If the hydrogen peroxide solution is stored in a cool, dark place in a clean plastic container, it is stable for months. However, in the presence of a manganese salt, an iodide-containing salt, or an enzyme (Section 11-9), the decomposition reaction can occur quite rapidly (Figure 11.14a).

Ammonium nitrate is used as fertilizer and is stable at room temperature. At higher temperatures and in the presence of chloride ion as a catalyst, however, ammonium nitrate can explode with tremendous force (Figure 11.14b). Approximately 600 people were killed in Texas City, Texas, in 1947 when workers tried to use salt water (which contains ~0.5-M Cl^-) to extinguish a fire in the ship *Grandcamp* and the ammonium nitrate cargo exploded (Figure 11.14c).

On April 17, 2013 a fertilizer plant in West, Texas caught fire. Ammonium nitrate in the plant exploded, leaving a 93-foot wide crater. The explosion killed 15 people and injured 200.

How does a catalyst or an enzyme help a reaction to go faster? It does so by participating in the reaction mechanism. That is, *the mechanism for a catalyzed reaction is different from the mechanism of the same reaction without the catalyst.* The rate-limiting step in the catalyzed mechanism has a lower activation energy and therefore is faster than the slow step for the uncatalyzed reaction. To see how this works, let us again consider conversion of *cis*- to *trans*-2-butene in the gas phase.

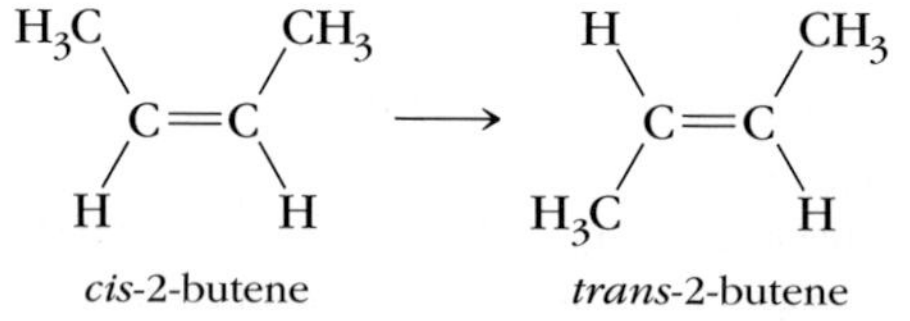

cis-2-butene → *trans*-2-butene

$$\text{Rate} = k\,[cis\text{-2-butene}]$$

If a trace of gaseous molecular iodine, I_2, is added to a sample of *cis*-2-butene, the iodine accelerates the change to *trans*-2-butene. The iodine is neither consumed nor produced in the overall reaction, so it does not appear in the overall balanced equation. However, because the reaction rate depends on the concentration of I_2, there is a term involving concentration of I_2 in the rate law for the catalyzed reaction.

$$\text{Rate} = k[cis\text{-2-butene}][I_2]^{1/2}$$

The exponent of $\frac{1}{2}$ for the concentration of I_2 in the rate law indicates dependence on the square root of the concentration. This usually means that only half a molecule—in this case a single iodine atom—is involved in the mechanism.

The rate of the conversion of *cis*- to *trans*-2-butene changes because the presence of I_2 somehow changes the reaction mechanism. The best hypothesis is that iodine molecules first dissociate to form iodine atoms.

Step 1: **I_2 dissociation**

$$\tfrac{1}{2}[\,I_2(g) \longrightarrow 2\,I(g)\,]$$

(This equation is multiplied by $\frac{1}{2}$ because only one of the two I atoms from the I_2 molecule is needed in subsequent steps of the mechanism.) An iodine atom then attaches to the *cis*-2-butene molecule, breaking half of the double bond between the two central carbon atoms and allowing the ends of the molecule to twist freely relative to each other.

Step 2: **Attachment of I atom to *cis*-2-butene**

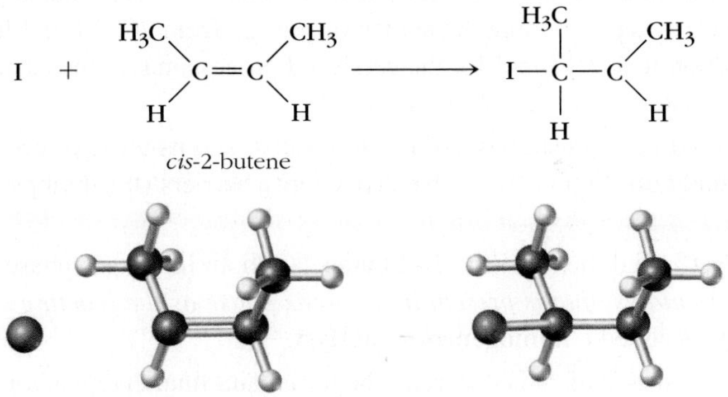
cis-2-butene

Step 3: **Rotation around the C—C bond**

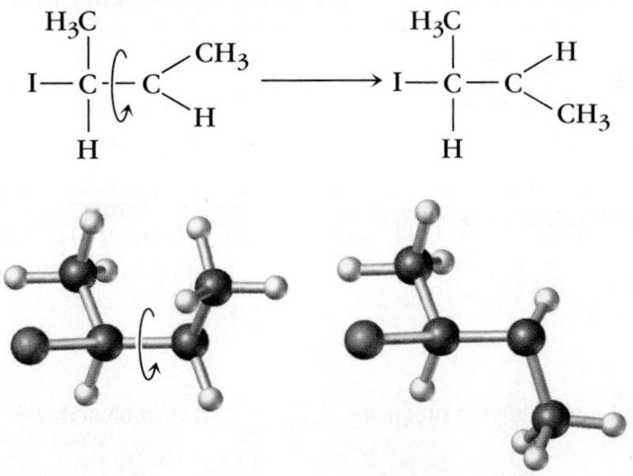

Step 4: **Loss of an I atom and re-formation of the carbon-carbon double bond**

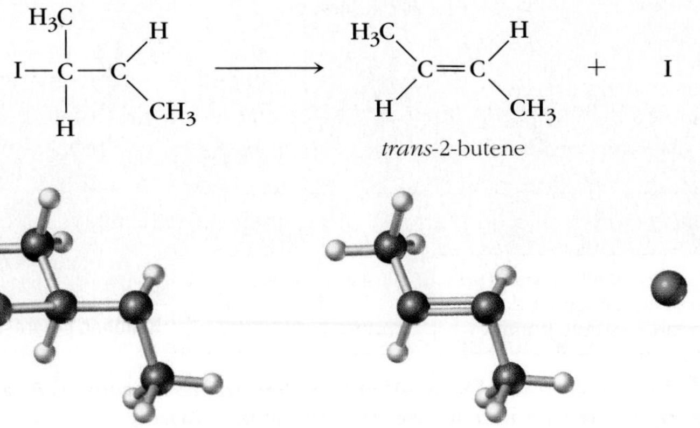
trans-2-butene

After the new double bond forms to give *trans*-2-butene and the iodine atom falls away, two iodine atoms come together to regenerate molecular iodine.

Step 5: **I_2 regeneration**

$$\tfrac{1}{2}[\,2\ I(g) \longrightarrow I_2(g)\,]$$

There are five important points concerning this mechanism:

- The I_2 dissociates to atoms and then re-forms. To an "outside" observer the concentration of I_2 is unchanged; I_2 is not involved in the balanced stoichiometric equation even though it has appeared in the mechanism. *A catalyst reacts in an early step of a mechanism and is regenerated later.*
- Figure 11.15 shows that the activation energy barrier is significantly lower for the catalyzed reaction (because the mechanism is different). Consequently the reaction rate is much faster. Dropping the activation energy from 262 kJ/mol for the uncatalyzed reaction to 115 kJ/mol for the catalyzed process makes the catalyzed reaction 10^{15} times faster at a temperature of 500. K.

Because $k = Ae^{-E_a/RT}$,

$$\frac{k_2}{k_1} = \frac{Ae^{-E_{a_2}/RT}}{Ae^{-E_{a_1}/RT}} = e^{-(E_{a_1} - E_{a_2})/RT}$$

$$= e^{\left(\frac{(262{,}000 - 115{,}000)\ \mathrm{J\ mol^{-1}}}{8.314\ \mathrm{J\ K^{-1}\ mol^{-1}} \times 500.\ \mathrm{K}}\right)}$$

$$= e^{3.536 \times 10^1} = 2.28 \times 10^{15}$$

- The catalyzed mechanism has five reaction steps, and its energy-versus-reaction-progress diagram (Figure 11.15) has five energy barriers (five humps appear in the curve).
- The catalyst I_2 and the reactant *cis*-2-butene are both in the gas phase during the reaction. *A catalyst that is present in the same phase as the reacting substance or substances* is called a **homogeneous catalyst**.
- Although the mechanism is different, the initial and final energies for the catalyzed reaction are the same as for the uncatalyzed reaction. This means that $\Delta_r E$ and $\Delta_r H$ are the same for the catalyzed as for the uncatalyzed reaction.

EXERCISE 11.13

Catalysis

The oxidation of thallium(I) ion by cerium(IV) ion in aqueous solution has the equation

$$2\ Ce^{4+}(aq) + Tl^{+}(aq) \longrightarrow 2\ Ce^{3+}(aq) + Tl^{3+}(aq)$$

The numbered steps in this figure correspond with the numbered steps in the preceding text.

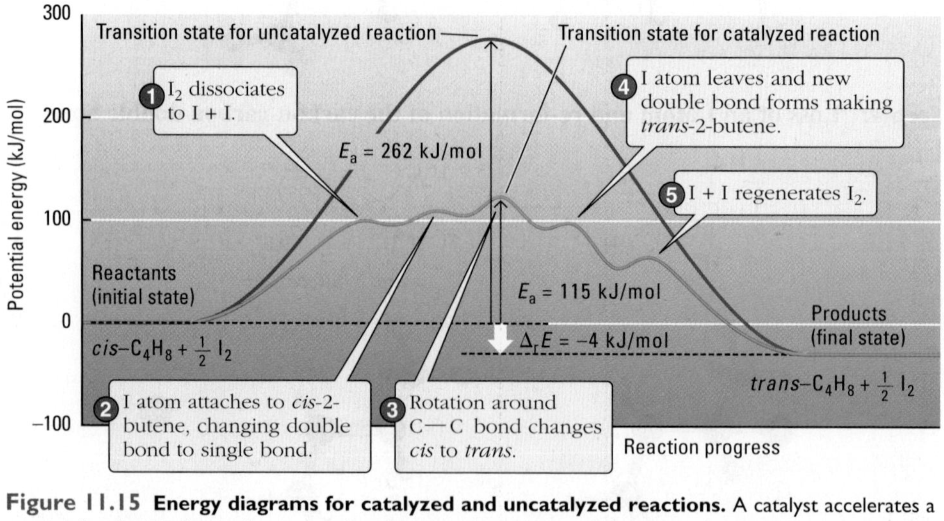

300
200
100
0
−100
Potential energy (kJ/mol)
Transition state for uncatalyzed reaction
Transition state for catalyzed reaction
1 I2 dissociates to I + I.
4 I atom leaves and new double bond forms making trans-2-butene.
Ea = 262 kJ/mol
5 I + I regenerates I2.
Reactants (initial state)
Ea = 115 kJ/mol
Products (final state)
cis–C4H8 + 1/2 I2
ΔrE = −4 kJ/mol
trans–C4H8 + 1/2 I2
2 I atom attaches to cis-2-butene, changing double bond to single bond.
3 Rotation around C—C bond changes cis to trans.
Reaction progress

Figure 11.15 Energy diagrams for catalyzed and uncatalyzed reactions. A catalyst accelerates a reaction by altering the mechanism so that the activation energy is reduced.

The accepted mechanism for this reaction is

Step 1: $Ce^{4+}(aq) + Mn^{2+}(aq) \longrightarrow Ce^{3+}(aq) + Mn^{3+}(aq)$

Step 2: $Ce^{4+}(aq) + Mn^{3+}(aq) \longrightarrow Ce^{3+}(aq) + Mn^{4+}(aq)$

Step 3: $Mn^{4+}(aq) + Tl^{+}(aq) \longrightarrow Mn^{2+}(aq) + Tl^{3+}(aq)$

(a) Verify that this mechanism predicts the overall reaction.
(b) Identify all intermediates in this mechanism.
(c) Identify the catalyst in this mechanism.
(d) Suppose that the first step in this mechanism is rate-limiting. Determine the rate law.
(e) Suppose that the second step in this mechanism is rate-limiting. Determine the rate law.

11-9 Enzymes: Biological Catalysts

Cengage Learning/Charles D. Winters

Oxidation of cellulose in wood.

Your body is a chemical factory of cells that can manufacture a broad range of compounds that are needed so that you can move, breathe, digest food, see, hear, smell, and even think. But, did you ever consider how the reactions that make those compounds are controlled? And how they can all occur reasonably quickly at the relatively low body temperature of 37 °C? Oxidation of glucose powers all the systems of your body, but you would not want it to take place in your body at the temperature it does when cellulose in wood burns in a fireplace. (Cellulose is a polymer of glucose (← **Sec. 10-7c**)). The chemical reactions of your body are catalyzed by enzymes. An **enzyme** is *a highly efficient catalyst for one or more chemical reactions in a living system.* The presence or absence of appropriate enzymes turns these reactions on or off by speeding them up or slowing them down. This allows your body to maintain nearly constant temperature and nearly constant concentrations of a variety of molecules and ions, an absolute necessity if you are to continue functioning.

Enzymes are usually proteins, but other biological macromolecules, such as RNA, can also increase the rates of reactions in living systems. Most enzymes are globular proteins, polymers of amino acids (← **Sec. 10-7d**) in which one or more long chains of amino acids fold into a nearly spherical shape. The shape of a globular protein is determined by noncovalent interactions (← **Sec. 10-7e**) among the amino acid components (hydrogen bonds, attractions of opposite ionic charges, and dipole-dipole forces), a few weak covalent bonds (such as disulfide bonds between cysteine side chains) (← **Sec. 10-7e**), and the fact that nonpolar (hydrophobic) amino acid side groups congregate in the middle of the molecule, avoiding the surrounding aqueous solution.

The 1989 Nobel Prize in Chemistry was given to Thomas Cech and Sidney Altman for the discovery that RNA molecules as well as proteins can be biological catalysts. In fact, the ribosome, the large protein/RNA complex that is responsible for all polymerization of amino acids into proteins, uses RNA to accomplish its reactions.

Enzymes are among the most effective catalysts known. They can increase reaction rates by factors of 10^9 to 10^{19}. For example, essentially every collision of the enzyme carbonic anhydrase with a carbonic acid molecule results in decomposition, and the enzyme can decompose about 36 million H_2CO_3 molecules every minute.

$$H_2CO_3(aq) \xrightarrow{\text{carbonic anhydrase}} CO_2(g) + H_2O(\ell)$$

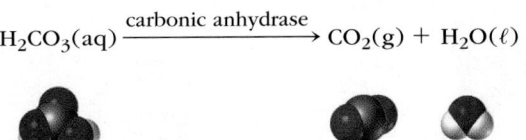

Most enzymes are highly specific catalysts. Some act on only one or two of the hundreds of different substances found in living cells. For example, carbonic anhydrase catalyzes only the decomposition of carbonic acid. Other enzymes can speed up several reactions, but usually these reactions are all of the same type.

Some enzymes can act as catalysts entirely on their own. Others require **cofactors**, *inorganic or organic molecules or ions that need to be present for an enzyme's catalytic activity to be fully available*. For example, many enzymes require nicotinamide adenine dinucleotide ion, NAD^+ (niacinamide ion). Molecules or ions that are cofactors are often derived from small quantities of minerals and vitamins in our diets (← **Sec. 1-14a**). If the cofactor needed for an enzyme to catalyze a reaction is not available because of dietary deficiency, that reaction does not occur when it is needed, and a bodily function is impaired.

11-9a Enzyme Activity and Specificity

A **substrate** is *a molecule whose reaction is catalyzed by an enzyme*. In some cases there may be more than one substrate, as when an enzyme catalyzes transfer of a group of atoms from one molecule to another. Enzyme catalysis is extremely effective and specific because the structure of the enzyme is finely tuned to minimize the activation energy barrier. Most enzymes have an **active site**, *a part of the enzyme molecule that interacts with the substrate.* The active site and substrate are held together by the same kinds of attractions that hold the enzyme in its globular structure. The nanoscale structure of an enzyme's active site is specifically suited to attract and bind a substrate molecule and to help the substrate react.

When a substrate binds to an enzyme, the structures of both molecules can change. Each structure adjusts to fit closely with the other, and the structures become complementary. *The change in shape of either the enzyme, the substrate, or both molecules when they bind* is called **induced fit**. Enzymes catalyze reactions of only certain molecules because the structures of most molecules are not close enough to the structure of the active site for an induced fit to occur. The induced fit of a substrate to an enzyme also can lower the activation energy for a reaction. For example, it may distort the substrate and stretch a bond that will be broken in the desired reaction. A schematic example of how this can work is shown in Figure 11.16. To see how it works in a specific case, consider the enzyme lysozyme, whose structure is shown in Figure 11.17 as a space-filling model with part of a substrate molecule in the active site. Lysozyme catalyzes hydrolysis reactions of polysaccharides (← **Sec. 10-7b**) found in bacterial cell walls. The reaction involved is

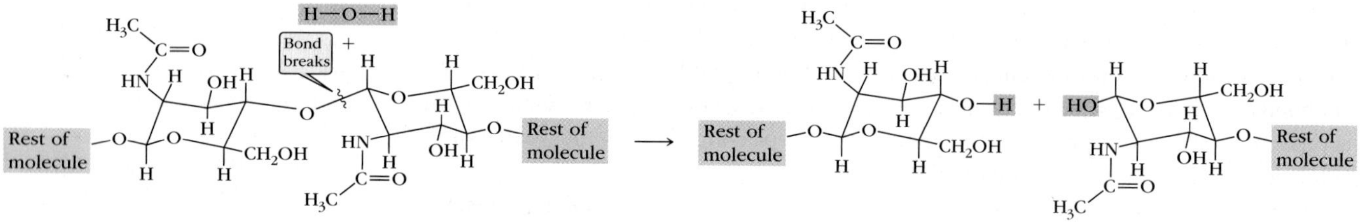

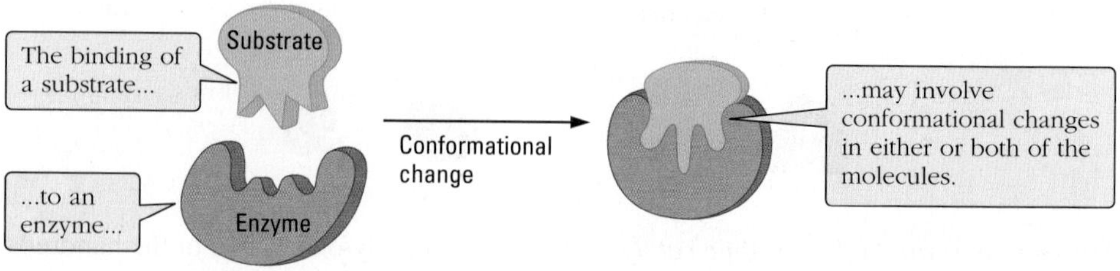
The binding of a substrate...
Substrate
...to an enzyme...
Enzyme
Conformational change
...may involve conformational changes in either or both of the molecules.

Figure 11.16 Induced fit of substrate to enzyme. Binding of a substrate to an enzyme may involve changing the shape of either or both molecules, thereby inducing them to fit together. In some cases a substrate molecule may be stretched or strained, helping bonds to break and reaction to occur.

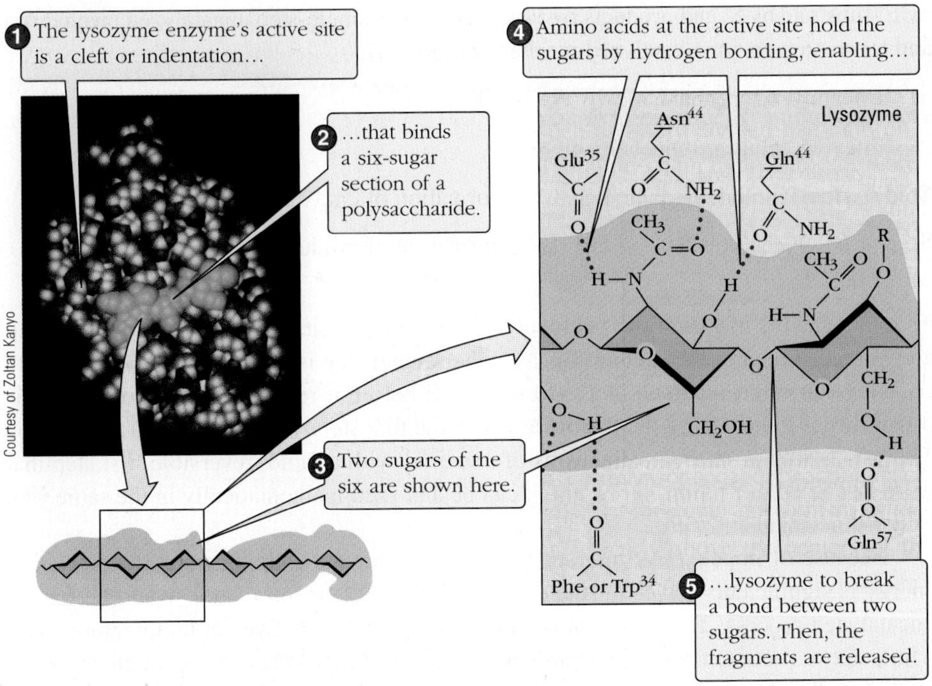
1 The lysozyme enzyme's active site is a cleft or indentation...
2 ...that binds a six-sugar section of a polysaccharide.
3 Two sugars of the six are shown here.
4 Amino acids at the active site hold the sugars by hydrogen bonding, enabling...
5 ...lysozyme to break a bond between two sugars. Then, the fragments are released.
Lysozyme
Asn44
Glu35
Gln44
Phe or Trp34
Gln57
CH2OH
Courtesy of Zoltan Kanyo

Figure 11.17 Lysozyme with substrate in the active site.

Hydrolysis is the opposite of condensation, the type of reaction by which most biopolymers are formed (← **Sec. 10-7d**). Breaking them into their building block molecules requires hydrolysis, so many important enzymes catalyze hydrolysis reactions.

In Section 10-5d, hydrolysis was described as a reaction in which a water molecule and some other molecule react, with both molecules splitting in two. In the lysozyme-catalyzed reaction, the H from the water ends up with one part of the substrate molecule, and the OH ends up with the other part.

The section of polysaccharide shown in Figure 11.17 fits nicely into the cleft along the surface of the lysozyme, but many other long-chain molecules, such as polypeptides, might fit there as well. Shape is important, but so are noncovalent attractions and their positioning so that the substrate can make the most effective use of them. The enlarged portion of Figure 11.17 shows many hydrogen bonds between enzyme and substrate. It should be clear that the specificity of the enzyme depends not only on the shape of the active site, but also on the positions of hydrogen-bonding groups and groups that participate in other noncovalent interactions so that they can adjust to complementary sites on the substrate.

To summarize, enzymes are extremely effective as catalysts for several reasons:

- Enzymes bring substrates into close proximity and hold them there while a reaction occurs.
- Enzymes hold substrates in the shape that is most effective for reaction.
- Enzymes can act as acids and bases during reaction, donating or accepting hydrogen ions from the substrate quickly and easily.
- The potential energy of a bond distorted by the induced fit of the substrate to the enzyme is already partway up the activation energy hill that must be surmounted for reaction to occur.
- Enzymes sometimes contain metal ions that are needed to help catalyze oxidation-reduction and hydrolysis reactions.

11-9b Enzyme Kinetics

An enzyme changes the mechanism of a reaction, as does any catalyst. The first step in the mechanism for any enzyme-catalyzed reaction is *binding the substrate to the enzyme*, which is referred to as formation of an **enzyme-substrate complex**. Representing enzyme by E, substrate by S, and products by P, we can write a single-step uncatalyzed mechanism and a two-step enzyme-catalyzed mechanism this way.

Uncatalyzed mechanism: $S \longrightarrow P$

Enzyme-catalyzed mechanism:

Step 1 (fast): $S + E \rightleftharpoons ES$ (formation of enzyme-substrate complex)

Step 2 (slow): $ES \longrightarrow P + E$ (formation of products and regeneration of enzyme)

That the enzyme is a catalyst is evident from the fact that it is a reactant in the first step and is regenerated in the second. Because the second step is slow, the enzyme-substrate complex can often separate and re-form S + E before it reacts to form products. This possibility is indicated by the double arrow in the first step. This mechanism for enzyme catalysis is similar to the mechanisms of reactions with a rapid, reversible first step that were discussed in Section 11-7a, and it can be analyzed mathematically in the same way as those mechanisms were.

Because of noncovalent interactions between enzyme and substrate, the activation energy is significantly lower for the enzyme-catalyzed reaction than it would be for the uncatalyzed process. This situation is shown in Figure 11.18. Even at temperatures only a little above room temperature, significant numbers of molecules have enough energy to surmount this lower barrier. Thus, enzyme-catalyzed reactions can occur reasonably quickly at body temperature.

11-9c Special Features of Enzyme Catalysis

Enzyme-catalyzed reactions obey the same principles of chemical kinetics that we discussed earlier in this chapter. Nevertheless, both the enzyme itself and the mechanism of enzyme catalysis have some special features that you should be aware of. First, because of the form of the mechanism, either the enzyme or the substrate may be the limiting reactant in the first step. If the substrate is limiting, increasing the concentration of substrate produces more enzyme-substrate complex and makes the reaction go faster. This is the expected behavior: Increasing the concentration of a reactant should increase the rate proportionately. If the enzyme becomes the limiting reactant, however, it can become completely converted to enzyme-substrate complex, leaving no enzyme available for additional substrate. If this happens, a further increase in the concentration of substrate will

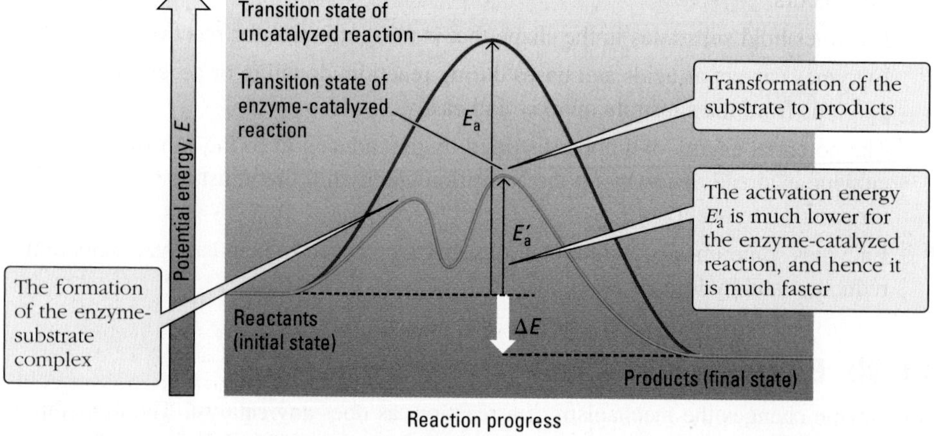
Transition state of uncatalyzed reaction
Transition state of enzyme-catalyzed reaction
Transformation of the substrate to products
E_a
The activation energy E_a' is much lower for the enzyme-catalyzed reaction, and hence it is much faster
E_a'
Potential energy, E
The formation of the enzyme-substrate complex
Reactants (initial state)
ΔE
Products (final state)
Reaction progress

Figure 11.18 Energy diagram for enzyme-catalyzed reaction. The purple curve is the energy profile for a typical reaction in a living system with no enzyme present. The green energy profile is drawn to the same scale for the same reaction with enzyme catalysis.

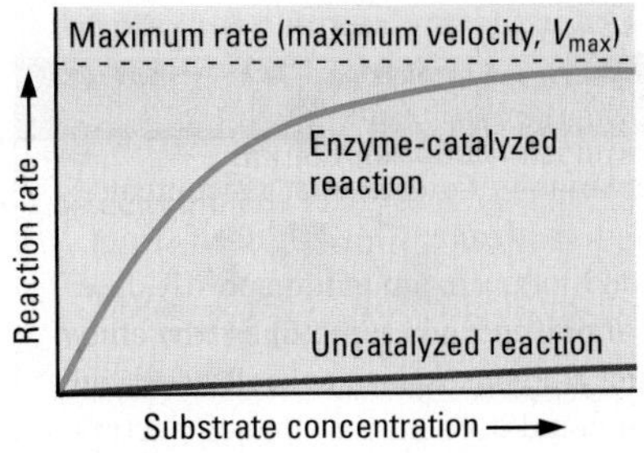
Maximum rate (maximum velocity, V_{max})
Enzyme-catalyzed reaction
Uncatalyzed reaction
Reaction rate
Substrate concentration

Figure 11.19 Maximum velocity for an enzyme-catalyzed reaction. Because there is only a limited quantity of enzyme available, increasing substrate concentration beyond the point at which the enzyme becomes the limiting reactant does not increase the rate further. There is a maximum rate (maximum velocity) for any enzyme-catalyzed reaction.

not increase the rate of reaction. As a result there is a *maximum rate* (those who study enzyme kinetics call this the maximum velocity) for an enzyme-catalyzed reaction. The behavior of rate with increasing substrate concentration is shown in Figure 11.19.

Enzyme-catalyzed reactions also behave unusually with respect to temperature. The rate does increase with increasing temperature, but if the temperature becomes high enough, there is a sudden decrease in rate, as shown in Figure 11.20. This happens because there is increased molecular and atomic motion as the temperature increases, and that motion can disrupt the secondary and tertiary structures of enzymes and other proteins. This *disruption of protein structure* is called **denaturation**. It occurs, for example, when an egg is boiled or fried. When a globular protein is denatured it loses its coiled globular structure, and its solubility and other properties change. Once an enzyme's structure has changed, the active site is no longer available, enzyme catalysis is seriously impaired, and the reaction rate falls to its uncatalyzed value. For many enzymes this happens only a little above 37 °C, which is body temperature for humans. Enzymes have evolved to produce maximum rates at body temperature, and slightly higher temperatures cause most of them to denature.

Charles D. Winters

Denaturation. Frying an egg causes protein in the white to denature and precipitate from the colorless, nearly clear original solution to form a white, flexible solid.

11-9d Inhibition of Enzymes

There is another way that the activity of an enzyme can be destroyed. Some molecules or ions, called **inhibitors**, can *fit an enzyme's active site, but remain there unreacted.* An inhibitor bound to an enzyme decreases its effective concentration, thereby decreasing the rate of the reaction that the enzyme catalyzes. If sufficient inhibitor becomes bound to an enzyme, the enzyme provides little catalytic effect because the concentration of available active sites becomes very small. An example of enzyme inhibition is the action of sulfa drugs on bacteria. Bacteria use *para*-aminobenzoic acid and an enzyme called dihydropteroate synthetase to synthesize folic acid, which is essential to their metabolism. Sulfa drugs bind to this enzyme, inhibit synthesis of folic acid, and destroy bacterial populations.

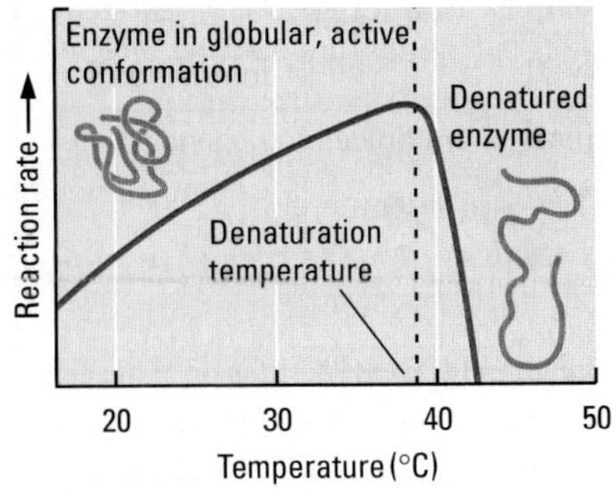
Enzyme in globular, active conformation
Denatured enzyme
Denaturation temperature
Reaction rate
20
30
40
50
Temperature (°C)

Figure 11.20 Enzyme activity destroyed by high temperature. At a temperature somewhat above normal body temperature, there is sufficient molecular motion to overcome the noncovalent interactions that maintain protein structure. This disrupts the structure of an enzyme, thereby destroying its catalytic activity. The process by which the enzyme structure becomes disrupted is called denaturation.

PROBLEM-SOLVING EXAMPLE 11.11

Enzyme Inhibition

The label of a container of methanol (methyl alcohol) invariably indicates that its contents are poisonous and should not be taken internally. Although methanol, CH_3OH, itself is not very toxic, it is labeled a poison because it is metabolized by the enzyme methanol oxidase to formaldehyde, $H_2C{=}O$, which is very toxic. Methanol poisoning is sometimes treated by giving the patient ethanol, CH_3CH_2OH, which inhibits the enzyme. Identify similarities and differences in the structures of methanol and ethanol that could account for ethanol's acting as an inhibitor.

Result Both molecules are alcohols and can hydrogen-bond. Methanol has three hydrogens on the carbon next to the —OH group. Ethanol has only two hydrogens on the carbon adjacent to the —OH group, and it has one more carbon and two more hydrogens. Because its shape is similar to that of methanol and because it can form hydrogen bonds of similar strength, ethanol should be able to bind to the active sites of methanol oxidase catalyst molecules. Probably because of the extra carbon atom or the difference in number of hydrogens adjacent to the —OH group, the catalyst is unable to speed up the oxidation of ethanol, and ethanol remains bound.

Analyze An inhibitor must have a structure similar to an enzyme's substrate, but must remain bonded to the enzyme without reacting. Condensed formulas for methanol and ethanol are given.

Plan Write structural formulas for methanol and ethanol; compare the structures.

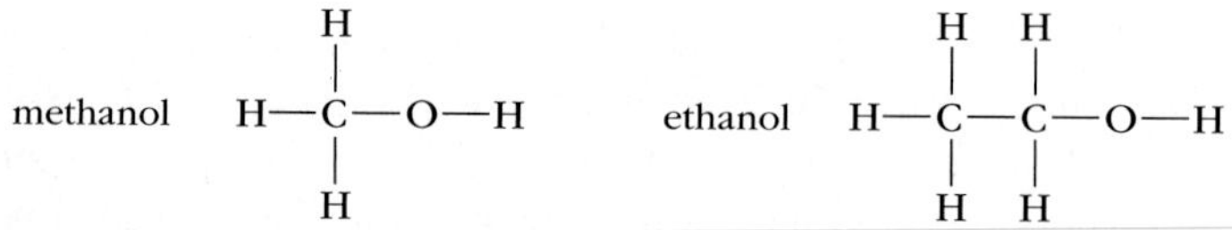
methanol
ethanol

Execute See Result.

PROBLEM-SOLVING PRACTICE 11.11

Bacteria need to use *p*-aminobenzoic acid to help synthesize folic acid in order to survive. Sulfa drugs interfere with this process. The structures of *p*-aminobenzoic acid and folic acid are

p-aminobenzoic acid

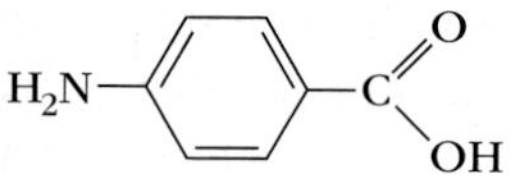

folic acid

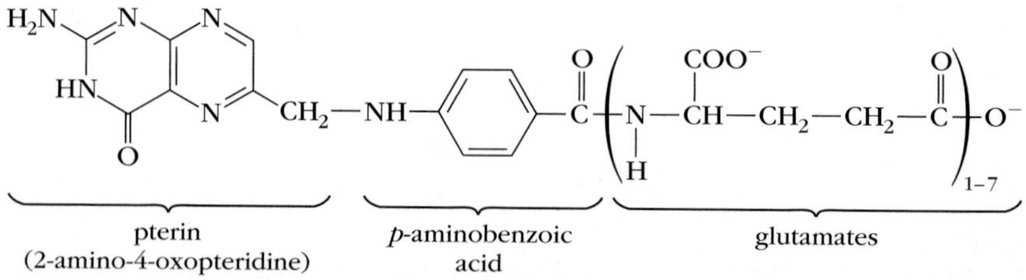
pterin
(2-amino-4-oxopteridine)
p-aminobenzoic
acid
glutamates
1–7

Which of these structures is most likely a sulfa drug? Explain your choice.

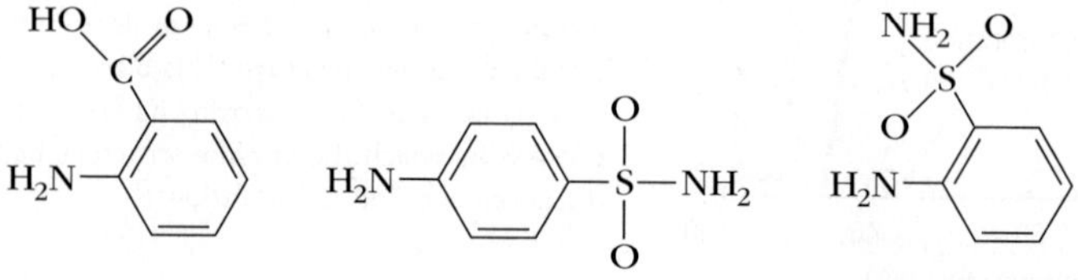

11-10 Catalysis in Industry

An expert in the field of industrial chemistry has said that every year more than one trillion dollars' worth of goods is manufactured with the aid of industrial catalysts. Without them, fertilizers, pharmaceuticals, fuels, synthetic fibers, solvents, and detergents would be in short supply. Indeed, 90% of all manufactured items use catalysts at some stage of production. The major areas of catalyst use are in petroleum refining, industrial production of chemicals, and environmental controls. In this section we provide a few examples of the many important industrial reactions that depend on catalysis.

Many industrial reactions use **heterogeneous catalysts**, *catalysts that are present in a different phase from that of the reactants being catalyzed.* Usually the catalyst is a solid and the reactants are in the gaseous or liquid phase. Heterogeneous catalysts are used in industry because they are more easily separated from the products and leftover reactants than are homogeneous catalysts. Catalysts for chemical processing are usually metal-based and often contain precious metals such as platinum and palladium. Worldwide, about \$15 billion worth of catalysts are used annually by the chemical industry.

11-10a Manufacture of Acetic Acid

The importance of acetic acid, CH_3COOH, in the organic chemicals industry is comparable to that of sulfuric acid in the inorganic chemicals industry; annual production of acetic acid worldwide exceeds 6 million metric tons. Acetic acid is used widely in industry to make plastics and synthetic fibers, as a fungicide, and as the starting material for preparing many dietary supplements. One way of synthesizing the acid is an excellent example of homogeneous catalysis: Aqueous rhodium(III) iodide is used to speed up the combination of carbon monoxide and methyl alcohol, both inexpensive chemicals, to form acetic acid.

$$\underset{\text{methyl alcohol}}{CH_3OH(aq)} + \underset{\text{carbon monoxide}}{CO(aq)} \xrightarrow{\text{RhI}_3\text{ catalyst}} \underset{\text{acetic acid}}{CH_3\overset{\overset{\displaystyle O}{\|}}{C}{-}OH(aq)}$$

The role of the rhodium(III) iodide catalyst in this reaction is to bring the reactants together and allow them to rearrange to form the products. Carbon monoxide and the methyl group from the alcohol become attached to the rhodium atom, which helps transfer the methyl group to the CO. The intermediate formed by this rearrangement reacts with solvent water to form acetic acid and the catalyst is regenerated.

11-10b Controlling Automobile Emissions

A major use of catalysts is in *emissions control* for both automobiles and power plants. This market uses very large quantities of platinum group metals: platinum, palladium, rhodium, and iridium. In 2012, approximately 9.5×10^4 kg platinum and 2.0×10^5 kg palladium were sold worldwide for automotive uses. Demand was also high for rhodium for this same purpose (about 2.4×10^4 kg). All three metals are also used in chemical processing as catalysts, and the petroleum industry uses platinum and rhodium to catalyze refining processes.

The purpose of the catalysts in a catalytic converter in the exhaust system of an automobile (Figure 11.21) is to ensure that the combustion of carbon monoxide and hydrocarbons is complete.

$$2\,CO(g) + O_2(g) \xrightarrow{\text{Pt-NiO catalyst}} 2\,CO_2(g)$$

$$\underset{\substack{\text{2,2,4-trimethylpentane,}\\ \text{a component of gasoline}}}{2\,C_8H_{18}(g)} + 25\,O_2(g) \xrightarrow{\text{Pt-NiO catalyst}} 16\,CO_2(g) + 18\,H_2O(g)$$

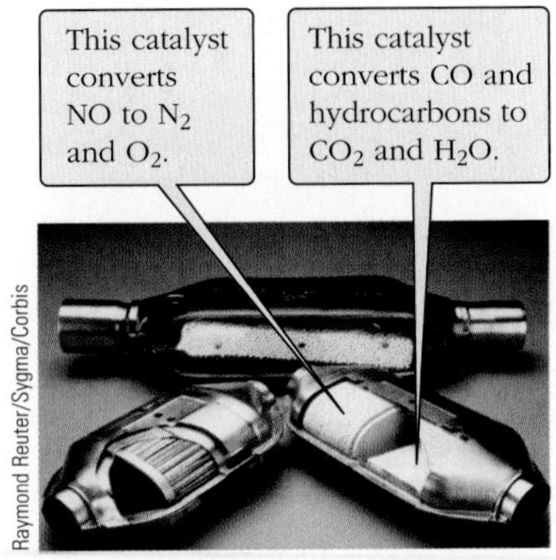
This catalyst converts NO to N_2 and O_2.
This catalyst converts CO and hydrocarbons to CO_2 and H_2O.
Raymond Reuter/Sygma/Corbis

Figure 11.21 Automobile catalytic converters. Catalytic converters are standard equipment on the exhaust systems of automobiles.

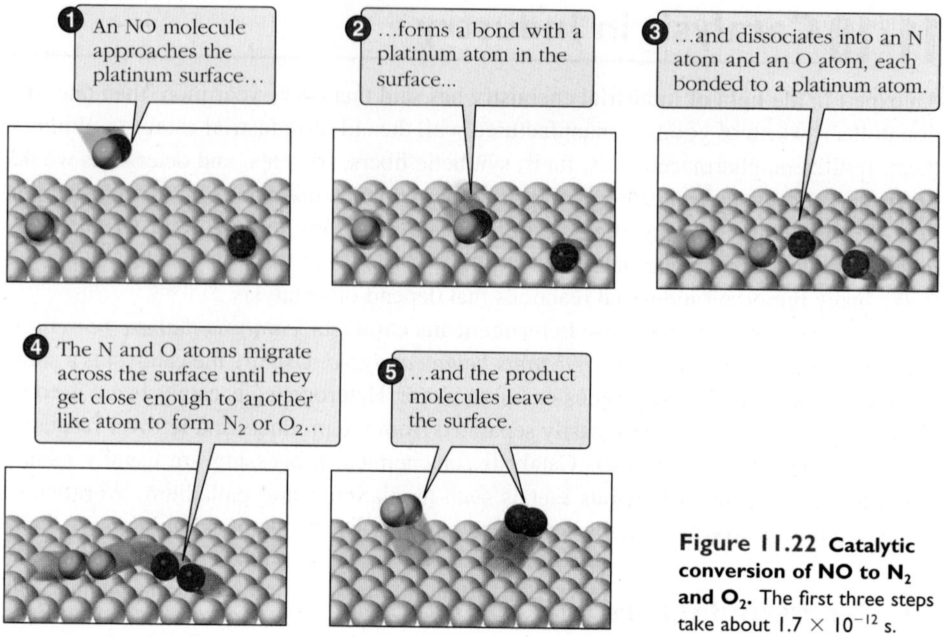
1 An NO molecule approaches the platinum surface...
2 ...forms a bond with a platinum atom in the surface...
3 ...and dissociates into an N atom and an O atom, each bonded to a platinum atom.
4 The N and O atoms migrate across the surface until they get close enough to another like atom to form N2 or O2...
5 ...and the product molecules leave the surface.

Figure 11.22 Catalytic conversion of NO to N_2 and O_2. The first three steps take about 1.7×10^{-12} s.

In addition, the catalytic converter converts nitrogen oxides to molecules that are less harmful to the environment. At the high temperature of combustion, some N_2 from air reacts with O_2 to give NO, a significant air pollutant. Nitrogen monoxide is unstable and should revert to N_2 and O_2 according to this equation:

$$2\ NO(g) \xrightarrow{\text{catalyst}} N_2(g) + O_2(g)$$

The rate of reversion of NO to N_2 and O_2 is slow, but catalysts have been developed that greatly speed this reaction. The role of the heterogeneous catalyst in the preceding reactions is probably to weaken the bonds of the reactants and to assist in product formation. For example, Figure 11.22 shows how NO molecules can dissociate into N and O atoms on the surface of a platinum metal catalyst.

11-10c Converting Methane to Liquid Fuel

Your home may be heated with natural gas, which consists largely of methane, CH_4. Although widely used, it is also widely wasted because much of it is found in geographical areas far removed from where fuels are consumed, and because transporting the flammable gas is expensive and dangerous. One solution to making methane useful is to convert it, where it is found, to a more readily transportable substance such as liquid methanol, CH_3OH. The methanol can then be used directly as a fuel, added to gasoline [as is currently done in some areas of the United States (← **Sec. 10-1e**)], or used to make other chemicals.

It has been known for some time that methane can be converted to carbon monoxide and hydrogen,

$$CH_4(g) + \tfrac{1}{2}\ O_2(g) \longrightarrow CO(g) + 2\ H_2(g)$$

and this mixture of gases can readily be turned into methanol in another step.

$$CO(g) + 2\ H_2(g) \longrightarrow CH_3OH(\ell)$$

Chemical engineers at the University of Minnesota discovered that methane can be converted to CO and H_2 under very mild conditions of temperature. They simply found the right catalyst, a heated, sponge-like ceramic disk coated with platinum or rhodium.

Rather than oxidizing the methane all the way to water and carbon dioxide, the process produces a hot mixture of CO and H_2, which can be converted in good yield to methanol. It is also possible to produce other partially oxidized hydrocarbons by a similar catalytic process.

CONCEPTUAL EXERCISE 11.14

Catalysis

Which of these statements is (are) true? If any are false, change the wording to make them true.

(a) The concentration of a homogeneous catalyst may appear in the rate law.
(b) A catalyst is always consumed in the overall reaction.
(c) A catalyst must always be in the same phase as the reactants.

SUMMARY PROBLEM

An excellent way to make highly pure nickel metal for use in specialized steel alloys is to decompose $Ni(CO)_4$ by heating it in a vacuum to slightly above room temperature.

$$Ni(CO)_4(g) \longrightarrow Ni(s) + 4\ CO(g)$$

The reaction is proposed to occur in four steps, the first of which is

$$Ni(CO)_4(g) \longrightarrow Ni(CO)_3(g) + CO(g)$$

Kinetic studies of this first-order decomposition reaction have been carried out between 47.3 °C and 66.0 °C to give the results in the table.*

Temperature (°C)	Rate Constant (s^{-1})
47.3	0.263
50.9	0.354
55.0	0.606
60.0	1.022
66.0	1.873

(a) Determine the activation energy for this reaction.

(b) $Ni(CO)_4$ is formed by the reaction of nickel metal with carbon monoxide. Suppose that 2.05 g CO is combined with 0.125 g nickel metal. Determine the maximum mass (g) of $Ni(CO)_4$ that can be formed.

Replacement of CO by another molecule in $Ni(CO)_4$ was studied in the nonaqueous solvents toluene and hexane to understand the general principles that govern the chemistry of such compounds.*

$$Ni(CO)_4(g) + P(CH_3)_3 \longrightarrow Ni(CO)_3P(CH_3)_3 + CO$$

A detailed study of the kinetics of the reaction led to the mechanism

Step 1: (slow) $\quad Ni(CO)_4 \longrightarrow Ni(CO)_3 + CO$

Step 2: (fast) $\quad Ni(CO)_3 + P(CH_3)_3 \longrightarrow Ni(CO)_3P(CH_3)_3$

(c) Which step in the mechanism is unimolecular? Which is bimolecular?

(d) Add the steps of the mechanism to show that the result is the balanced equation for the observed reaction.

(e) Is there an intermediate in this reaction? If so, what is it?

(f) It was found that doubling the concentration of $Ni(CO)_4$ increased the reaction rate by a factor of 2. Doubling the concentration of $P(CH_3)_3$ had no effect on the reaction rate. Based on this information, write the rate equation for the reaction.

(g) Does the experimental rate equation support the proposed mechanism? Why or why not?

*See Day, J. P., Basolo, F., and Pearson, R. G. *Journal of the American Chemical Society*, Vol. 90, 1968; p. 6933.

HAVING STUDIED THIS CHAPTER . . .

. . . you should be able to:

- Define reaction rate and calculate average rates (Section 11-1). End-of-chapter Questions 11, 13, 15
- Describe the effect that reactant concentrations have on reaction rate, and determine rate laws and rate constants from initial rates (Section 11-2). Questions 19, 21, 23, 25
- Determine reaction orders from a rate law, and use the integrated rate law method to obtain orders and rate constants (Section 11-3). Questions 28, 33, 35
- Calculate concentration at a given time, time to reach a certain concentration, and half-life for a reaction (Section 11-3). Questions 37, 39, 41
- Define and give examples of unimolecular and bimolecular elementary reactions (Section 11-4). Questions 42, 44
- Show by using an energy profile what happens as two reactant molecules interact to form product molecules (Sections 11-4 and 11-6). Question 61
- Define activation energy and frequency factor, and use them to calculate rate constants and rates under different conditions of temperature and concentration (Section 11-5). Questions 46, 48, 51, 53, 57
- Derive rate laws from the equations for unimolecular and bimolecular elementary reactions (Section 11-6). Questions 59, 67
- Define reaction mechanism; identify rate-limiting steps and intermediates (Section 11-7). Question 69
- Given several reaction mechanisms, decide which is (are) in agreement with experimentally determined stoichiometry and rate law (Section 11-7). Question 72
- Explain how a catalyst can speed up a reaction; draw energy profiles for catalyzed and uncatalyzed reaction mechanisms (Section 11-8). Questions 74, 76
- From overall stoichiometry and rate law, or from a reaction mechanism, determine whether a catalyst is involved (Section 11-8). Question 78
- Write a mechanism for a typical enzyme-catalyzed reaction and identify similarities and differences between enzyme-catalyzed reactions and other catalyzed reactions (Section 11-9). Question 82
- Distinguish homogeneous from heterogeneous catalysis and explain why industrial processes often involve heterogeneous catalysis. (Section 11-10). Questions 86, 87

KEY TERMS

activated complex (Section 11-4a)
activation energy (E_a) (11-4a)
active site (11-9a)
Arrhenius equation (11-5)
average reaction rate (11-1b)
bimolecular reaction (11-4)
chemical kinetics (Introduction)
cofactor (11-9)
denaturation (11-9c)
elementary reaction (11-4)
enzyme (11-9)
enzyme-substrate complex (11-9b)
frequency factor (A) (11-5)
half-life ($t_{1/2}$) (11-3c)
heterogeneous catalyst (11-10)
heterogeneous reaction (11-1)
homogeneous catalyst (11-8)
homogeneous reaction (11-1)
induced fit (11-9a)
inhibitor (11-9d)
initial rate (11-2b)
instantaneous reaction rate (11-1b)
intermediate (11-7)
order of reaction (11-3)
overall reaction order (11-3)
rate (11-1)
rate constant (k) (11-2a)

rate law (11-2a)
rate-limiting step (11-7)
reaction intermediate (11-7)
reaction mechanism (11-7)
reaction rate (11-1)
steric factor (11-4b)
substrate (11-9a)
transition state (11-4a)
unimolecular reaction (11-4)

QUESTIONS FOR REVIEW AND THOUGHT

Red-numbered questions have short answers at the back of this book in Appendix M and fully worked solutions in the *Student Solutions Manual.*

Review Questions

These questions test vocabulary and simple concepts.

1. Which of these is appropriate for determining the rate law for a chemical reaction?
   (a) Theoretical calculations based on balanced equations
   (b) Measuring the rate of the reaction as a function of the concentrations of the reacting species
   (c) Measuring the rate of the reaction as a function of temperature
2. Name at least three factors that affect the rate of a chemical reaction.
3. Using the rate law, rate $= k[A]^2[B]$, define the order of the reaction with respect to A and B and the overall reaction order.
4. Define the terms "unimolecular elementary reaction" and "bimolecular elementary reaction," and give an example of each.
5. Define the terms "activation energy" and "frequency factor." Write an equation that relates activation energy and frequency factor to reaction rate.
6. Write a one- or two-sentence definition in your own words for each term:

active site	substrate
cofactor	polypeptide
enzyme	polysaccharide
inhibition	protein

7. Write a one- or two-sentence definition in your own words for each term:

hydrolysis	globular protein
denaturation	induced fit
enzyme-substrate complex	maximum velocity

8. Explain the difference between a homogeneous and a heterogeneous catalyst. Give an example of each.

Topical Questions

These questions are keyed to the major topics in the chapter. Usually a question that is answered at the back of this book is paired with a similar one that is not.

Reaction Rate (Section 11-1)

9. Consider dissolving sugar as a simple process in which kinetics is important. Suppose that you dissolve an equal mass of each kind of sugar listed. Which dissolves the fastest? Which dissolves the slowest? Explain why in terms of rates of heterogeneous reactions. (If you are not sure which is fastest or slowest, try them all.)
   (a) Rock candy sugar (large sugar crystals)
   (b) Sugar cubes
   (c) Granular sugar
   (d) Powdered sugar
10. A cube of aluminum 1.0 cm on each edge is placed into 9-M NaOH(aq), and the rate at which H_2 gas is given off is measured.
   (a) Calculate by what factor this reaction rate will change if the aluminum cube is cut exactly in half and the two halves are placed in the solution. Assume that the reaction rate is proportional to the surface area, and that all of the surface of the aluminum is in contact with the NaOH(aq).
   (b) If you had to speed up this reaction as much as you could without raising the temperature, what would you do to the aluminum?
11. A compound called phenyl acetate reacts with water according to the equation

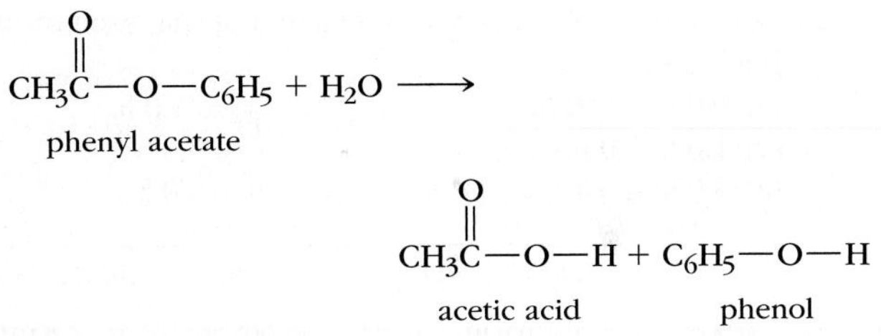
phenyl acetate
acetic acid
phenol

These data were collected at 5 °C.

Time (min)	[Phenyl acetate] (mol/L)
0.00	0.55
0.25	0.42
0.50	0.31
0.75	0.23
1.00	0.17
1.25	0.12
1.50	0.082

   (a) Make a graph of concentration as a function of time, describe the shape of the curve, and compare it with Figure 11.3 (← **Sec. 11-1b**).
   (b) Calculate the rate of change of the concentration of phenylacetate during the period from 0.20 min to 0.40 min, and then during the period from 1.2 min to 1.4 min. Compare the values and explain why one is smaller than the other.
   (c) Determine the rate of change of the phenol concentration during the period from 1.00 to 1.25 min.

12. Cyclobutane can decompose to form ethylene:

$$C_4H_8(g) \longrightarrow 2\ C_2H_4(g)$$

The cyclobutane concentration can be measured as a function of time by mass spectrometry (a graph is nearby).
   (a) Write an expression for the rate of reaction in terms of a changing concentration.
   (b) Calculate the average rate of reaction between 10. and 30. s.
   (c) Calculate the instantaneous rate of reaction after 20. s.
   (d) Calculate the initial rate of reaction.
   (e) Calculate the instantaneous rate of formation of ethylene 40. s after the start of the reaction.

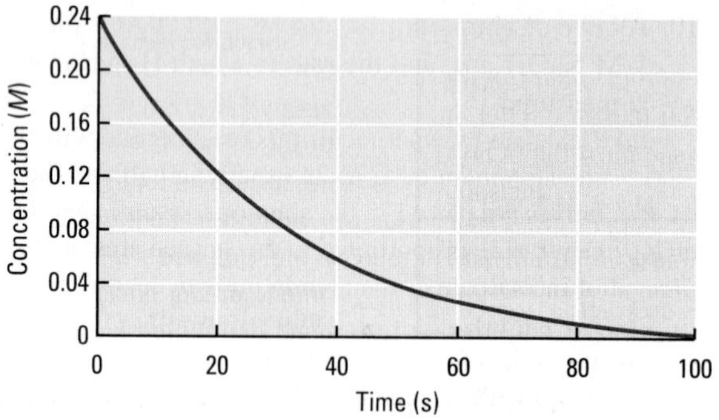

Concentration (M)
0.24
0.20
0.16
0.12
0.08
0.04
0
0
20
40
60
80
100
Time (s)

13. Using data given in the table for the reaction

$$N_2O_5 \longrightarrow 2\ NO_2 + \tfrac{1}{2}\ O_2$$

calculate the average rate of reaction during each of these intervals:
   (a) 0.00 to 0.50 h.   (b) 0.50 to 1.0 h.
   (c) 1.0 to 2.0 h.   (d) 2.0 to 3.0 h.
   (e) 3.0 to 4.0 h.   (f) 4.0 to 5.0 h.

Time (h)	$[N_2O_5]$ (mol/L)	Time (h)	$[N_2O_5]$ (mol/L)
0.00	0.849	3.00	0.352
0.50	0.733	4.00	0.262
1.00	0.633	5.00	0.196
2.00	0.472		

14. Using all your calculated rates from Question 13,
   (a) show that the reaction obeys the rate law

$$\text{Rate} = -\frac{\Delta[N_2O_5]}{\Delta t} = k[N_2O_5]$$

   (b) evaluate the rate constant k as an average of the values obtained for the six intervals.
   (c) calculate the reaction rate exactly 2.5 h from the start.

15. For the reaction

$$2\ NO_2(g) \longrightarrow 2\ NO(g) + O_2(g)$$

make qualitatively correct plots of the concentrations of $NO_2(g)$, $NO(g)$, and $O_2(g)$ versus time. Draw all three graphs on the same axes; assume that you start with $NO_2(g)$ at a concentration of 1.0 mol/L. Explain how you would determine, from these plots,
   (a) the initial rate of the reaction.
   (b) the final rate (that is, the rate as time approaches infinity).

16. For the reaction

$$O_3(g) + O(g) \longrightarrow 2\ O_2(g)$$

make qualitatively correct plots of the concentrations of $O_3(g)$, $O(g)$, and $O_2(g)$ versus time. Draw all three graphs on the same axes, assume that you start with $O_3(g)$ and $O(g)$, each at a concentration of 1.0 μmol/L. Explain how you would determine, from these plots,
   (a) the initial rate of the reaction.
   (b) the final rate (that is, the rate as time approaches infinity).

17. Express the rate of the reaction

$$2\ N_2H_4(\ell) + N_2O_4(\ell) \longrightarrow 3\ N_2(g) + 4\ H_2O(g)$$

in terms of
   (a) $\Delta[N_2O_4]$.   (b) $\Delta[N_2]$.

18. Ammonia is produced by the reaction between nitrogen and hydrogen gases.
   (a) Write a balanced equation using smallest whole-number coefficients for the reaction.
   (b) Write an expression for the rate of reaction in terms of $\Delta[NH_3]$.
   (c) The concentration of ammonia increases from 0.257 M to 0.815 M in 15.0 min. Calculate the average rate of reaction over this time interval.
   (d) Based on your result in part (c), calculate the rate of change of concentration of H_2 during the same time interval.

Effect of Concentration on Reaction Rate (Section 11-2)

19. If a reaction has the experimental rate law, Rate = $k[A]^2$, explain what happens to the rate when
   (a) the concentration of A is tripled.
   (b) the concentration of A is halved.

20. A reaction has the experimental rate law, Rate = $k[A]^2[B]$. If the concentration of A is doubled and the concentration of B is halved, what happens to the reaction rate?

21. The reaction of $CO(g) + NO_2(g)$ is second-order in NO_2 and zeroth-order in CO at temperatures less than 500 K.
   (a) Write the rate law for the reaction.
   (b) Determine how the reaction rate changes if the NO_2 concentration is halved.
   (c) Determine how the reaction rate changes if the concentration of CO is doubled.

22. Nitrosyl bromide, NOBr, is formed from NO and Br_2.

$$2\ NO(g) + Br_2(g) \longrightarrow 2\ NOBr(g)$$

Experiment shows that the reaction is first-order in Br_2 and second-order in NO.
   (a) Write the rate law for the reaction.
   (b) If the concentration of Br_2 is tripled, determine how the reaction rate changes.
   (c) Determine what happens to the reaction rate when the concentration of NO is doubled.

23. For the reaction of $Pt(NH_3)_2Cl_2$ with water,

$$Pt(NH_3)_2Cl_2 + H_2O \longrightarrow Pt(NH_3)_2(H_2O)Cl^+ + Cl^-$$

the rate law is Rate = $k[Pt(NH_3)_2Cl_2]$ with $k = 0.090\ h^{-1}$.

Red-numbered questions are answered in Appendix M

(a) Calculate the initial rate of reaction when the concentration of $Pt(NH_3)_2Cl_2$ is
(i) 0.010 M (ii) 0.020 M (iii) 0.040 M
(b) Determine how the rate of disappearance of $Pt(NH_3)_2Cl_2$ changes with its initial concentration.
(c) How is this related to the rate law?
(d) How does the initial concentration of $Pt(NH_3)_2Cl_2$ affect the rate of appearance of Cl^- in the solution?

24. The ester methyl acetate, CH_3COOCH_3, reacts with base to break one of the C—O bonds.

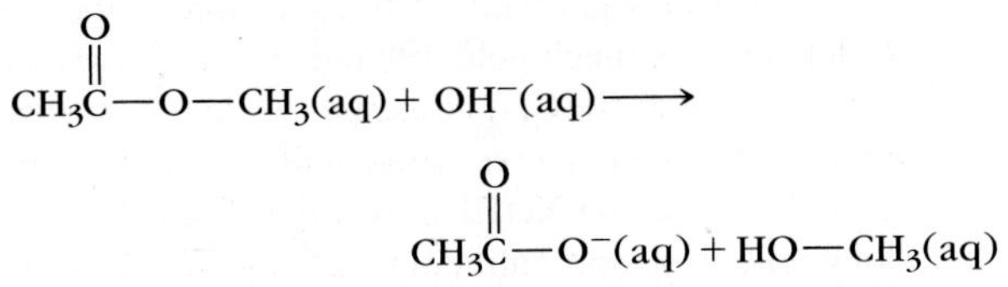

The rate law is rate $= k[CH_3COOCH_3][OH^-]$ where $k = 0.14\ L\ mol^{-1}\ s^{-1}$ at 25 °C.
(a) Calculate the initial rate at which the methyl acetate is converted to products when both reactants, CH_3COOCH_3 and OH^-, have a concentration of 0.025 M.
(b) Calculate the rate at which methyl alcohol, CH_3OH, initially appears in the solution.

25. Measurements of the initial rate of reaction between two compounds, triphenylmethyl hexachloroantimonate (substance **I**) and bis-(9-ethyl-3-carbazolyl)methane (substance **II**), in 1,2-dichloroethane at 40 °C yielded these data:

Initial Concentration × 10^5 (mol/L)		Initial Rate × 10^9 (mol $L^{-1}\ s^{-1}$)
[I]	[II]	
1.65	10.6	1.50
14.9	10.6	17.7
14.9	7.10	11.2
14.9	3.52	6.30
14.9	1.76	3.10
4.97	10.6	4.52
2.48	10.6	2.70

(a) Determine the order of the reaction with respect to substance **I** and substance **II**.
(b) Derive the rate law for this reaction.
(c) Calculate the rate constant k and express it in appropriate units.
(d) Calculate the initial rate of reaction when [**I**] = 8.3×10^{-5} mol/L and [**II**] = 6.78×10^{-5} mol/L.

26. For the reaction

$$2\ NO(g) + 2\ H_2(g) \longrightarrow N_2(g) + 2\ H_2O(g)$$

these data were obtained at 1100 K:

[NO] (mol/L)	$[H_2]$ (mol/L)	Initial Rate (mol $L^{-1}\ s^{-1}$)
5.00×10^{-3}	2.50×10^{-3}	3.0×10^{-3}
15.0×10^{-3}	2.50×10^{-3}	9.0×10^{-3}
15.0×10^{-3}	10.0×10^{-3}	3.6×10^{-2}

(a) Determine the order with respect to NO and with respect to H_2.
(b) Determine the overall order of this reaction.
(c) Write the rate law.
(d) Calculate the rate constant.
(e) Calculate the initial rate of this reaction at 1100 K when $[NO] = [H_2] = 8.0 \times 10^{-3}$ mol L^{-1}.

27. The transfer of an oxygen atom from NO_2 to CO has been studied at 540 K:

$$CO(g) + NO_2(g) \longrightarrow CO_2(g) + NO(g)$$

These data were collected:

Initial Rate (mol $L^{-1}\ h^{-1}$)	Initial Concentration (mol/L)	
	[CO]	$[NO_2]$
5.1×10^{-4}	0.35×10^{-4}	3.4×10^{-8}
5.1×10^{-4}	0.70×10^{-4}	1.7×10^{-8}
5.1×10^{-4}	0.18×10^{-4}	6.8×10^{-8}
1.0×10^{-3}	0.35×10^{-4}	6.8×10^{-8}
1.5×10^{-3}	0.35×10^{-4}	10.2×10^{-8}

(a) Write the rate law.
(b) Determine the reaction order with respect to each reactant.
(c) Calculate the rate constant and express it in appropriate units.

Rate Law and Order of Reaction (Section 11-3)

28. For each of these rate laws, state the reaction order with respect to the hypothetical substances A and B, and give the overall order.
(a) Rate $= k[A][B]^3$ (b) Rate $= k[A][B]$
(c) Rate $= k[A]$ (d) Rate $= k[A]^3[B]$

29. For each of the rate laws below, determine the order of the reaction with respect to the hypothetical substances X, Y, and Z. What is the overall order?
(a) Rate $= k[X][Y][Z]$ (b) Rate $= k[X]^2[Y]^{1/2}[Z]$
(c) Rate $= k[X]^{1.5}[Y]^{-1}$ (d) Rate $= k[X]/[Y]^2$

30. A reaction A + B ⟶ products is found to be second-order in B. Which rate equation cannot be correct?
(a) Rate $= k[A][B]$ (b) Rate $= k[A][B]^2$
(c) Rate $= k[B]^2$

31. The reaction

$$2\ NO(g) + 2\ H_2(g) \longrightarrow N_2(g) + 2\ H_2O(g)$$

is found to be first-order in $H_2(g)$. Which rate equation cannot be correct?
(a) Rate $= k[NO]^2[H_2]$ (b) Rate $= k[H_2]$
(c) Rate $= k[NO]^2[H_2]^2$

32. The decomposition of ammonia to nitrogen and hydrogen on tungsten at 1100 °C is zeroth-order with a rate constant of 2.5×10^{-4} mol L^{-1} min^{-1}.
(a) Write the rate expression.
(b) Calculate the rate when $[NH_3] = 0.075$ M.

33. For the reaction of phenyl acetate with water the concentration as a function of time was given in Question 11. Assume that the concentration of water does not change

during the reaction. Analyze the data from Question 11 to determine

(a) the rate law.
(b) the order of the reaction with respect to phenyl acetate.
(c) the rate constant.
(d) the rate of reaction when the concentration of phenyl acetate is 0.10 mol/L (assuming that the concentration of water is the same as in the experiments in the table in Question 11).

34. When phenacyl bromide and pyridine are both dissolved in methanol, they react to form phenacylpyridinium bromide.

$$C_6H_5{-}C({=}O){-}CH_2Br + C_5H_5N \longrightarrow C_6H_5{-}C({=}O){-}CH_2NC_5H_5^+ + Br^-$$

When equal concentrations of reactants were mixed in methanol at 35 °C, these data were obtained:

Time (min)	[Reactant] (mol/L)	Time (min)	[Reactant] (mol/L)
0	0.0385	500.	0.0208
100.	0.0330	600.	0.0191
200.	0.0288	700.	0.0176
300.	0.0255	800.	0.0163
400.	0.0220	1000.	0.0143

(a) Determine the rate law for this reaction.
(b) Determine the overall order of this reaction.
(c) Determine the rate constant for this reaction.
(d) Determine the rate constant for this reaction when the concentration of each reactant is 0.030 mol/L.

35. The compound *p*-methoxybenzonitrile *N*-oxide, which has the formula $CH_3OC_6H_4CNO$, reacts with itself to form a dimer—a molecule that consists of two *p*-methoxybenzonitrile *N*-oxide units connected to each other $(CH_3OC_6H_4CNO)_2$. The reaction can be represented as

$$A + A \longrightarrow B \quad \text{or} \quad 2\,A \longrightarrow B$$

where A represents *p*-methoxybenzonitrile *N*-oxide and B represents the dimer, $(CH_3OC_6H_4CNO)_2$. For the reaction in carbon tetrachloride at 40 °C with an initial concentration of 0.011 M, these data were obtained:

Time (min)	Percent Reaction	Time (min)	Percent Reaction
0	0	942	60.9
60.	9.1	1080.	64.7
120.	16.7	1212	66.6
215	26.5	1358	68.5
325	32.7	1518	70.3
565	47.3		

(a) Determine the rate law for the reaction.
(b) Determine the rate constant.
(c) Determine the order of the reaction with respect to A.

36. This question requires working with the equations of Table 11.2 (← **Sec. 11-3a**). Using an initial concentration $[A]_0$ of 1.0 mol/L and a rate constant k with a numerical value of 1.0 in appropriate units, make plots of [A] versus time over the time interval 0 to 5 s for each type of integrated rate law. Compare your results with Figure 11.5 (← **Sec. 11-3a**).

37. Radioactive gold-198 is used in the diagnosis of liver problems. ^{198}Au decays in a first-order process, emitting a β particle (electron). The half-life of this isotope is 2.7 days. You begin with a 5.6-mg sample of the isotope. Calculate how much gold-198 remains after 1.0 day.

38. The compound $Xe(CF_3)_2$ decomposes in a first-order reaction to elemental Xe with a half-life of 30. min. If you place 7.50 mg $Xe(CF_3)_2$ in a flask, calculate how long you must wait until only 0.25 mg $Xe(CF_3)_2$ remains.

39. The initial concentration of the reactant in a first-order reaction A ⟶ products is 0.64 mol/L and the half-life is 30. s.

(a) Calculate the concentration of the reactant 60. s after initiation of the reaction.
(b) Calculate how long it takes for the concentration of the reactant to drop to one-eighth its initial value.
(c) Calculate how long it takes for the concentration of the reactant to drop to 0.040 mol L^{-1}.

40. The compound SO_2Cl_2 decomposes in a first-order reaction

$$SO_2Cl_2(g) \longrightarrow SO_2(g) + Cl_2(g)$$

that has a half-life of 1.47×10^4 s at 600. K. If you begin with 1.6×10^{-3} mol of pure SO_2Cl_2 in a 2.0-L flask, calculate at what time the amount of SO_2Cl_2 will be 1.2×10^{-4} mol.

41. The first-order rate constant for the decomposition of a certain hormone in water at 25 °C is 3.42×10^{-4} day^{-1}.

(a) A 0.0200-M solution of the hormone is stored at 25 °C for two months. Calculate its concentration at the end of that period.
(b) Calculate how long it takes for the concentration of the solution to drop from 0.0200 M to 0.00350 M.
(c) Determine the half-life of the hormone.

A Nanoscale View: Elementary Reactions (Section 11-4)

42. Which of these reactions are unimolecular and elementary, which are bimolecular and elementary, and which are not elementary?

(a) $CH_4(g) + 2\,O_2(g) \longrightarrow CO_2(g) + 2\,H_2O(g)$
(b) $O_3(g) + O(g) \longrightarrow 2\,O_2(g)$
(c) $Mg(s) + 2\,H_2O(\ell) \longrightarrow H_2(g) + Mg(OH)_2(s)$
(d) $O_3(g) \longrightarrow O_2(g) + O(g)$

43. Which of these reactions are unimolecular and elementary, which are bimolecular and elementary, and which are not elementary?

(a) $HCl(g) + H_2O(g) \longrightarrow H_3O^+(g) + Cl^-(g)$
(b) $I^-(g) + CH_3Cl(g) \longrightarrow ICH_3(g) + Cl^-(g)$
(c) $C_2H_6(g) \longrightarrow C_2H_4(g) + H_2(g)$
(d) $N_2(g) + 3\,H_2(g) \longrightarrow 2\,NH_3(g)$
(e) $O_2(g) + O(g) \longrightarrow O_3(g)$

44. Assume that each gas-phase reaction occurs via a single bimolecular step. For which reaction would you expect the steric factor to be more important? Why?

$$Cl + O_3 \longrightarrow ClO + O_2 \quad \text{or} \quad NO + O_3 \longrightarrow NO_2 + O_2$$

45. Assume that each gas-phase reaction occurs via a single bimolecular step. For which reaction would you expect the steric factor to be more important? Why?

$$H_2C{=}CH_2 + H_2 \longrightarrow H_3C{-}CH_3 \quad \text{or}$$
$$(CH_3)_2C{=}CH_2 + HBr \longrightarrow (CH_3)_2CBr{-}CH_3$$

Temperature and Reaction Rate: The Arrhenius Equation (Section 11-5)

46. From Problem-Solving Example 11.8 (← Sec. 11-4b), where the energy profile of the ozone plus atomic oxygen reaction was derived, obtain the activation energy. Then, determine the ratio of the reaction rate for this reaction at 50. °C to the reaction rate at room temperature (25 °C). Assume that the initial concentrations are 1.00 M at both temperatures and $A = 4.82 \times 10^9\ M^{-1}\ s^{-1}$.
47. Suppose a reaction rate constant has been measured at two different temperatures, T_1 and T_2, and its values are k_1 and k_2, respectively.
    (a) Write the Arrhenius equation at each temperature.
    (b) By combining these two equations, derive an expression for the ratio of the two rate constants, k_1/k_2. Use this expression to answer the next four questions.
48. Suppose a chemical reaction has an activation energy of 76 kJ/mol, as in the example in Figure 11.12. Calculate by what factor the rate of the reaction at 50. °C is increased over its rate at 25 °C.
49. A chemical reaction has an activation energy of 30. kJ/mol. If you had to slow down this reaction a thousandfold by cooling it from room temperature (25 °C), calculate what the temperature would be.
50. Calculate the activation energy for a reaction if its rate constant is found to triple when the temperature is raised from 600. K to 610. K.
51. These data were obtained for the rate constant for reaction of an unknown compound with water:

T (°C)	k (s^{-1})
56.2	1.04×10^{-5}
78.2	1.45×10^{-4}

    (a) Calculate the activation energy and frequency factor for this reaction.
    (b) Estimate the rate constant of the reaction at a temperature of 100.0 °C.
52. *p*-Methylphenyl acetate reacts with imidazole to produce *p*-methylphenol and acetyl imidazole. The rate constants for this second-order reaction at two temperatures are given in the table.

T (°C)	k (L mol^{-1} s^{-1})
10.0	2.34×10^{-2}
60.0	1.52×10^{-1}

    (a) Calculate the activation energy and frequency factor for this reaction.
    (b) Estimate the rate constant for this reaction at a temperature of 100.0 °C.
53. For the reaction of iodine atoms with hydrogen molecules in the gas phase, these rate constants were obtained experimentally.

$$2\ I(g) + H_2(g) \longrightarrow 2\ HI(g)$$

T (K)	$10^{-5}\ k$ ($L^2\ mol^{-2}\ s^{-1}$)
417.9	1.12
480.7	2.60
520.1	3.96
633.2	9.38
666.8	11.50
710.3	16.10
737.9	18.54

    (a) Calculate the activation energy and frequency factor for this reaction.
    (b) Estimate the rate constant of the reaction at 400.0 K.
54. Make an Arrhenius plot and calculate the activation energy for the gas-phase reaction

$$2\ NOCl(g) \longrightarrow 2\ NO(g) + Cl_2(g)$$

T (K)	Rate Constant (L mol^{-1} s^{-1})
400.	6.95×10^{-4}
450.	1.98×10^{-2}
500.	2.92×10^{-1}
550.	2.60
600.	16.3

55. The activation energy E_a is 139.7 kJ mol^{-1} for the gas-phase reaction

$$HI + CH_3I \longrightarrow CH_4 + I_2$$

Calculate the fraction of the molecules whose collisions would be energetic enough to react at
    (a) 100. °C (b) 200. °C
    (c) 500. °C (d) 1000. °C
56. The activation energy E_a is 10. kJ/mol for the gas phase reaction

$$NO + O_3 \longrightarrow NO_2 + O_2$$

Calculate the fraction of the molecules whose collisions would be energetic enough to react at
    (a) 400. °C (b) 600. °C
    (c) 800. °C (d) 1000. °C
57. For the gas-phase reaction

$$CH_3CH_2I(g) \longrightarrow CH_2CH_2(g) + HI(g)$$

the activation energy E_a is 221 kJ/mol and the frequency factor A is $1.2 \times 10^{14}\ s^{-1}$. If the concentration of CH_3CH_2I is 0.012 mol/L, calculate the rate of the reaction at
    (a) 400. °C (b) 800. °C

Red-numbered questions are answered in Appendix M

58. For the gas-phase reaction

$$cis\text{-CHClCHCl(g)} \longrightarrow trans\text{-CHClCHCl(g)}$$

the activation energy E_a is 234 kJ/mol and the frequency factor A is $6.3 \times 10^{12}\ s^{-1}$. If the concentration of *cis*-CHClCHCl is 0.0043 mol/L, calculate the rate of the reaction at
(a) 400. °C (b) 800. °C

Rate Laws for Elementary Reactions (Section 11-6)

59. Write the rate law for each of these elementary reactions.
(a) $Cl(g) + ICl(g) \longrightarrow I(g) + Cl_2(g)$
(b) $CH_3N{=}NCH_3(g) \longrightarrow N_2(g) + C_2H_6(g)$
(c) $N_2O_4(g) \longrightarrow 2\ NO_2(g)$

60. Use the graph to answer these questions.

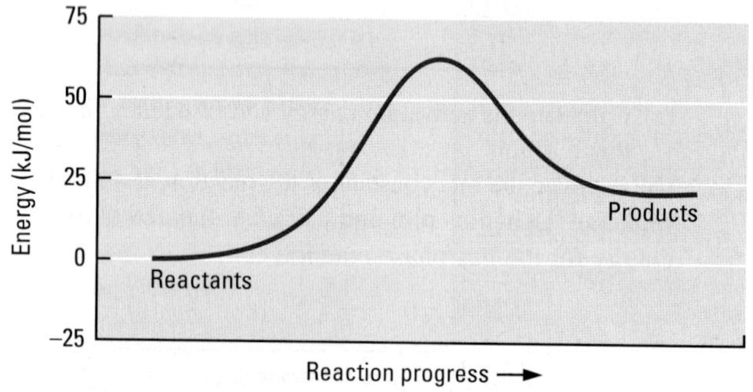
75
50
25
0
–25
Energy (kJ/mol)
Reactants
Products
Reaction progress

(a) Is the reaction exothermic or endothermic?
(b) Determine the approximate value of $\Delta_r E$ for the forward reaction.
(c) Calculate the approximate activation energy in each direction.
(d) A catalyst is found that lowers the activation energy of the reaction by 10 kJ/mol. How will this catalyst affect the rate of the reverse reaction?

61. Draw an energy versus reaction progress diagram (similar to the one in Question 60) for each of the reactions whose activation energy and enthalpy change are given below. Draw arrows to represent the activation energies of the forward and reverse reaction and $\Delta_r H°$.
(a) $\Delta_r H° = -145$ kJ mol^{-1}; $E_a = 75$ kJ mol^{-1}
(b) $\Delta_r H° = -70$ kJ mol^{-1}; $E_a = 65$ kJ mol^{-1}
(c) $\Delta_r H° = 70$ kJ mol^{-1}; $E_a = 85$ kJ mol^{-1}

62. Draw an energy versus reaction progress diagram (similar to the one in Question 60) for each of the reactions whose activation energy and enthalpy change are given below. Draw arrows to represent the activation energies of the forward and reverse reaction and $\Delta_r H°$.
(a) $\Delta_r H° = 105$ kJ mol^{-1}; $E_a = 175$ kJ mol^{-1}
(b) $\Delta_r H° = -43$ kJ mol^{-1}; $E_a = 95$ kJ mol^{-1}
(c) $\Delta_r H° = 15$ kJ mol^{-1}; $E_a = 55$ kJ mol^{-1}

63. Which of the reactions in Question 61 would (a) occur fastest? (b) occur slowest? (Assume equal temperatures, equal concentrations, equal frequency factors, and the same rate law for all reactions.)

64. Which of the reactions in Question 62 would (a) occur fastest? (b) occur slowest? (Assume equal temperatures, equal concentrations, equal frequency factors, and the same rate law for all reactions.)

65. For which of the reactions in Question 61 would the *reverse* reaction (a) be fastest? (b) be slowest? (Assume equal temperatures, equal concentrations, equal frequency factors, and the same rate law for all reactions.)

66. For which of the reactions in Question 62 would the *reverse* reaction (a) be fastest? (b) be slowest? (Assume equal temperatures, equal concentrations, equal frequency factors, and the same rate law for all reactions.)

67. Assuming that each reaction is elementary, predict the rate law.
(a) $NO(g) + NO_3(g) \longrightarrow 2\ NO_2(g)$
(b) $O(g) + O_3(g) \longrightarrow 2\ O_2(g)$
(c) $(CH_3)_3CBr(aq) \longrightarrow (CH_3)_3C^+(aq) + Br^-(aq)$
(d) $2\ HI(g) \longrightarrow H_2(g) + I_2(g)$

68. Assuming that each reaction is elementary, predict the rate law.
(a) $Br(g) + IBr(g) \longrightarrow I(g) + Br_2(g)$
(b) $Cl(g) + H_2(g) \longrightarrow HCl(g) + H(g)$
(c) $2\ NO_2(g) \longrightarrow N_2O_4(g)$
(d) Cyclopropane(g) $\longrightarrow$ propene(g)

Reaction Mechanisms (Section 11-7)

69. Experiments show that the reaction of nitrogen dioxide with fluorine

$$2\ NO_2(g) + F_2(g) \longrightarrow 2\ FNO_2(g)$$

has the rate law

$$\text{Rate} = k[NO_2][F_2]$$

and the reaction is thought to occur in two steps:

Step 1: $NO_2(g) + F_2(g) \longrightarrow FNO_2(g) + F(g)$

Step 2: $NO_2(g) + F(g) \longrightarrow FNO_2(g)$

(a) Show that the sum of this sequence of reactions gives the balanced equation for the overall reaction.
(b) Which step is rate-determining?

70. The reaction of $NO_2(g)$ and $CO(g)$ to form $CO_2(g)$ and $NO(g)$ is thought to occur in two steps:

Step 1: $NO_2(g) + NO_2(g) \longrightarrow NO(g) + NO_3(g)$ slow

Step 2: $NO_3(g) + CO(g) \longrightarrow NO_2(g) + CO_2(g)$ fast

(a) Show that the elementary steps add up to give the overall, stoichiometric equation.
(b) Determine the molecularity of each step.
(c) For this mechanism to be consistent with kinetic data, what must be the experimental rate equation?
(d) Identify any intermediates in this reaction.

71. For the reaction

$$2\ NO(g) + Cl_2(g) \longrightarrow 2\ NOCl(g)$$

the currently accepted mechanism is

$NO + Cl_2 \rightleftharpoons NOCl_2$ fast

$NOCl_2 + NO \longrightarrow 2\ NOCl$ slow

(a) Determine the rate law for this mechanism. (Be sure to express it in terms of concentrations of reactants or products of the overall reaction, not in terms of intermediates.)
(b) Suggest another mechanism that agrees with the same rate law.
(c) Suggest another mechanism that does not agree with the same rate law.

72. For the reaction

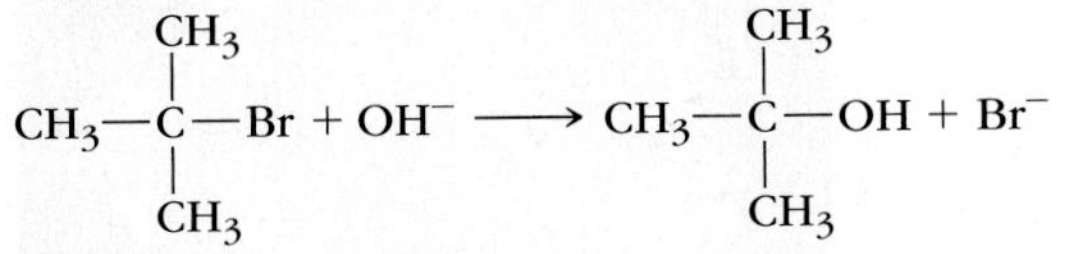

the rate law is

$$\text{Rate} = k[(CH_3)_3CBr]$$

Identify each mechanism that is compatible with the rate law.

(a) $(CH_3)_3CBr \longrightarrow (CH_3)_3C^+ + Br^-$ slow
$(CH_3)_3C^+ + OH^- \longrightarrow (CH_3)_3COH$ fast

(b) $(CH_3)_3CBr + OH^- \longrightarrow (CH_3)_3COH + Br^-$

(c) $(CH_3)_3CBr + OH^- \longrightarrow (CH_3)_2(CH_2)CBr^- + H_2O$ fast
$(CH_3)_2(CH_2)CBr^- \longrightarrow (CH_3)_2(CH_2)C + Br^-$ slow
$(CH_3)_2(CH_2)C + H_2O \longrightarrow (CH_3)_3COH$ fast

73. The mechanism for the reaction of CH_3OH and HBr is believed to involve two steps. The overall reaction is exothermic.

Step 1: $CH_3OH + H^+ \rightleftharpoons CH_3OH_2^+$ fast, endothermic

Step 2: $CH_3OH_2^+ + Br^- \longrightarrow CH_3Br + H_2O$ slow

(a) Write an equation for the overall reaction.
(b) Draw a reaction energy diagram for this reaction.
(c) Show that the rate law for this reaction is

$$\text{Rate} = k[CH_3OH][H^+][Br^-]$$

Catalysts and Reaction Rate (Section 11-8)

74. Which of these statements is (are) true? If a statement is false, reword it so that it becomes true.
(a) The concentration of a homogeneous catalyst may appear in the rate law.
(b) A catalyst is always consumed in the reaction.
(c) A catalyst must always be in the same phase as the reactants.
(d) A catalyst can change the course of a reaction and allow different products to be produced.

75. Hydrogenation reactions—processes in which H_2 is added to a molecule—are usually catalyzed. An excellent catalyst is a very finely divided metal suspended in the reaction solvent. Tell why finely divided rhodium, for example, is a much more efficient catalyst than a small block of the metal that has the same mass.

76. For this reaction mechanism,

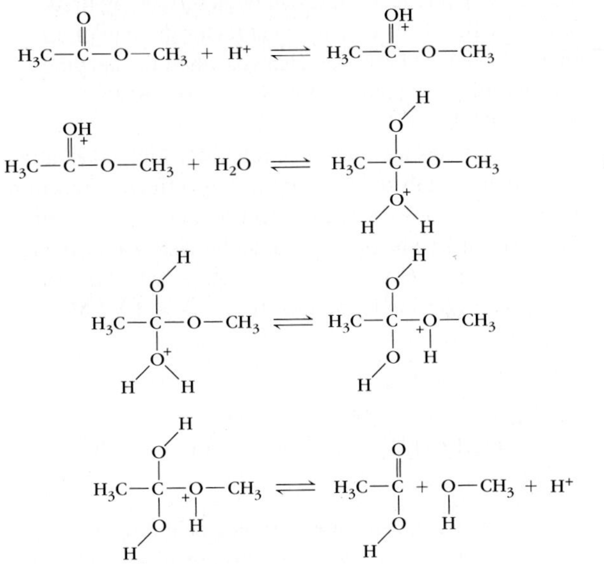
fast
slow
fast
fast

(a) write the chemical equation for the overall reaction.
(b) write the rate law for the reaction.
(c) is there a catalyst involved in this reaction? If so, what is it?
(d) identify all intermediates in the reaction.
(e) draw a reaction energy diagram for the reaction.

77. Which of these reactions appears to involve a catalyst? In those cases where a catalyst is present, tell whether it is homogeneous or heterogeneous.
(a) $CH_3COOCH_3(aq) + H_2O(\ell) \longrightarrow CH_3COOH(aq) + CH_3OH(aq)$
Rate $= k\,[CH_3COOCH_3][H^+]$
(b) $H_2(g) + I_2(g) \longrightarrow 2\,HI(g)$
Rate $= k[H_2][I_2]$
(c) $2\,H_2(g) + O_2(g) \longrightarrow 2\,H_2O(g)$
Rate $= k[H_2][O_2]$ (area of Pt surface)
(d) $H_2(g) + CO(g) \longrightarrow H_2CO(g)$
Rate $= k[H_2]^{1/2}[CO]$

78. In acid solution, methyl formate forms methyl alcohol and formic acid.

$$\underset{\text{methyl formate}}{HCO_2CH_3(aq)} + H_2O(\ell) \longrightarrow \underset{\text{formic acid}}{HCOOH(aq)} + \underset{\text{methyl alcohol}}{CH_3OH(aq)}$$

The rate law is Rate $= k[HCO_2CH_3][H^+]$. Why does H^+ appear in the rate law but not in the overall equation for the reaction?

Enzymes: Biological Catalysts (Section 11-9)

79. A biological catalyst lowers the activation energy of a reaction from 215 kJ/mol to 206 kJ/mol. Calculate by what factor the rate constant, *k*, would increase at 25 °C. Assume that the frequency factors (*A*) are the same for the uncatalyzed and catalyzed reactions.

80. When enzymes are present at very low concentration, their effect on reaction rate can be described by first-order kinetics. Calculate by what factor the rate of an enzyme-catalyzed reaction changes when the enzyme concentration is changed from 1.5×10^{-7} M to 4.5×10^{-6} M.

81. When substrates are present at relatively high concentration and are catalyzed by enzymes, the effect on reaction rate of changing substrate concentration can be described by zeroth-order kinetics. Calculate by what factor the rate of an enzyme-catalyzed reaction changes when the substrate concentration is changed from 1.5×10^{-2} M to 4.5×10^{-2} M.

82. The reaction

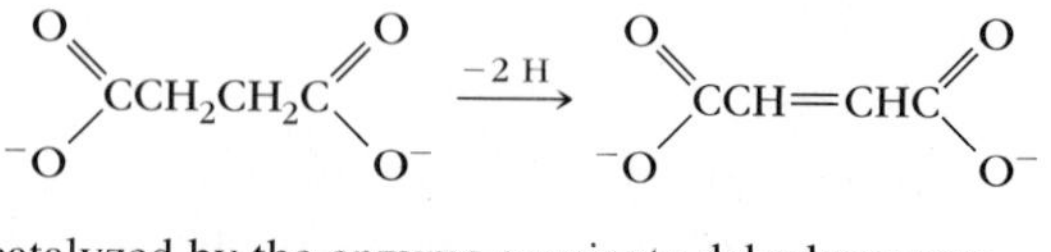

O
O
−2 H
O
O
CCH₂CH₂C
CCH=CHC
⁻O
O⁻
⁻O
O⁻

is catalyzed by the enzyme succinate dehydrogenase. When malonate ions or oxalate ions are added to the reaction mixture, the rate decreases significantly. Try to account for this observation in terms of the description of enzyme catalysis given in the text. The structures of malonate and oxalate ions are

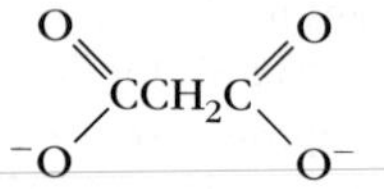

O
O
CCH₂C
⁻O
O⁻

malonate ion

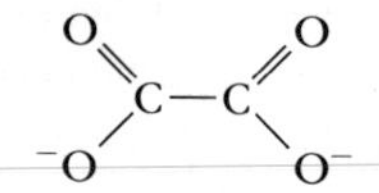

O
O
C—C
⁻O
O⁻

oxalate ion

83. Some enzymes can be inhibited by heavy metal ions such as Hg^{2+} and Pb^{2+}. These metal ions have a large affinity for sulfur-containing groups and can react with molecules such as CH_3CH_2SH to form compounds such as $Pb(CH_3CH_2S)_2$. Based on the structures of the amino acids given in Table 10.8 (← **Sec. 10-7d**), suggest at least one amino acid that is likely to be present in enzymes that are inhibited by heavy metals.

84. Many biochemical reactions are catalyzed by acids. A typical mechanism consistent with the experimental results (in which HA is the acid and X is the reactant) is

Step 1: reversible $HA(aq) \rightleftharpoons H^+(aq) + A^-(aq)$ fast

Step 2: reversible $X(aq) + H^+(aq) \rightleftharpoons XH^+(aq)$ fast

Step 3: $XH^+(aq) \longrightarrow$ products slow

(a) Derive the rate law from this mechanism.
(b) Determine the order of reaction with respect to HA.
(c) Determine how doubling the concentration of HA would affect the rate of the reaction.

Catalysis in Industry (Section 11-10)

85. In the first paragraph of Section 11-10, an expert in the field of industrial chemistry is quoted. Explain the expert's statement, in view of your understanding of the nature of catalysts. Why are catalysts so important?

86. Why are homogeneous catalysts harder to separate from products and leftover reactants than are heterogeneous reactants?

87. In an automobile catalytic converter the catalysis is accomplished on a surface consisting of platinum and other precious metals. The metals are deposited as a thin layer on a ceramic support that is a fine grid (see the photo).

Vicki Vellios Briner/The Patriot-News/Landov

(a) Why is the ceramic support arranged in the grid-like geometry?
(b) Why are the metals deposited on the ceramic surface instead of being used as strips or rods?

88. Find all examples of reactions described in this chapter that are catalyzed by metals. Are these metals main-group metals or transition metals? What type of chemical reactions are they: acid-base or oxidation-reduction? What conclusions can be drawn about metal-catalyzed chemical reactions from these examples?

General Questions

These questions are not explicitly keyed to chapter topics; many require integration of several concepts.

89. Draw a reaction energy diagram for an exothermic process. Mark the positions of reactants, products, and activated complex. Indicate the activation energies of the forward and reverse processes and explain how $\Delta_r E$ for the reaction can be calculated from the diagram.

90. Draw a reaction energy diagram for an endothermic process. Mark the positions of reactants, products, and activated complex. Indicate the activation energies of the forward and reverse processes and explain how $\Delta_r E$ for the reaction can be calculated from the diagram.

91. Under certain conditions, biphenyl, $C_{12}H_{10}$, can be produced by the decomposition of cyclohexane, C_6H_{12}:

$$2\ C_6H_{12} \longrightarrow C_{12}H_{10} + 7\ H_2$$

This table represents part of the data obtained in the kinetics experiment.

Time (s)	$[C_6H_{12}]$ (M)	$[C_{12}H_{10}]$ (M)	$[H_2]$ (M)
0.0	0.200	0.000	0.000
1.00	0.159	0.021	—
2.00	0.132	—	—
3.00	—	0.044	—

(a) Fill in the missing concentrations.
(b) Calculate the rate of reaction at 1.5 s.

92. The statements below relate to this reaction:

$$H_2(g) + I_2(g) \longrightarrow 2\ HI(g) \qquad \text{Rate} = k[H_2][I_2]$$

Determine which statements are true. If a statement is false, indicate why it is incorrect.
(a) The reaction must occur in a single step.
(b) This is a second-order reaction overall.
(c) Raising the temperature will cause the value of k to decrease.
(d) Raising the temperature lowers the activation energy for this reaction.
(e) If the concentrations of both reactants are doubled, the rate will double.
(f) Adding a catalyst in the reaction will cause the initial rate to increase.

93. Indicate whether each of these statements is true or false. Change the wording of each false statement to make it true.
(a) It is possible to change the rate constant for a reaction by changing the temperature.
(b) The reaction rate remains constant as a first-order reaction proceeds at a constant temperature.
(c) The rate constant for a reaction is independent of reactant concentrations.
(d) As a second-order reaction proceeds at a constant temperature, the rate constant changes.

94. For the reaction of NO and O_2 at 660 K,

$$2\ NO(g) + O_2(g) \longrightarrow 2\ NO_2(g)$$

Concentration (mol/L)		Rate of Disappearance
[NO]	$[O_2]$	of NO (mol L^{-1} s^{-1})
0.010	0.010	2.5×10^{-5}
0.020	0.010	1.0×10^{-4}
0.010	0.020	5.0×10^{-5}

(a) Determine the order with respect to each reactant.
(b) Write the rate equation for the reaction.
(c) Calculate the rate constant.
(d) Calculate the rate when [NO] = 0.025 mol/L and $[O_2]$ = 0.050 mol/L.
(e) If O_2 disappears at a rate of 1.0×10^{-4} mol L^{-1} s^{-1}, calculate the rate at which NO is consumed. Calculate the rate at which NO_2 is formed.

95. Nitryl fluoride is an explosive compound that can be made by oxidizing nitrogen dioxide with fluorine:

$$2\ NO_2(g) + F_2(g) \longrightarrow 2\ NO_2F(g)$$

Several kinetics experiments, all done at the same temperature and involving formation of nitryl fluoride, are summarized in this table:

	Initial Concentration (mol/L)			Initial Rate
Experiment	$[NO_2]$	$[F_2]$	$[NO_2F]$	(mol L^{-1} s^{-1})
1	0.0010	0.0050	0.0020	2.0×10^{-4}
2	0.0020	0.0050	0.0020	4.0×10^{-4}
3	0.0020	0.0020	0.0020	1.6×10^{-4}
4	0.0020	0.0020	0.0010	1.6×10^{-4}

(a) Write the rate law for the reaction.
(b) Determine what the order of the reaction is with respect to each reactant and each product.
(c) Calculate the rate constant k and express it in appropriate units.

96. The deep blue compound $CrO(O_2)_2$ can be made from the chromate ion by using hydrogen peroxide in an acidic solution.

$$HCrO_4^-(aq) + 2\ H_2O_2(aq) + H^+(aq) \longrightarrow CrO(O_2)_2(aq) + 3\ H_2O(\ell)$$

The kinetics of this reaction have been studied, and the rate equation is

$$\text{Rate of disappearance of } HCrO_4^- = k[HCrO_4^-][H_2O_2][H^+]$$

One of the mechanisms suggested for the reaction is

$$HCrO_4^- + H^+ \rightleftharpoons H_2CrO_4$$
$$H_2CrO_4 + H_2O_2 \longrightarrow H_2CrO(O_2)_2 + H_2O$$
$$H_2CrO(O_2)_2 + H_2O_2 \longrightarrow CrO(O_2)_2 + 2\ H_2O$$

(a) Give the order of the reaction with respect to each reactant.
(b) Show that the steps of the mechanism agree with the overall equation for the reaction.
(c) Which step in the mechanism is rate-limiting? Explain your answer.

97. Two mechanisms are proposed for the reaction

$$2\ NO(g) + O_2(g) \longrightarrow 2\ NO_2(g)$$

Mechanism 1: $NO + O_2 \rightleftharpoons NO_3$ fast
$NO_3 + NO \longrightarrow 2\ NO_2$ slow

Mechanism 2: $NO + NO \rightleftharpoons N_2O_2$ fast
$N_2O_2 + O_2 \longrightarrow 2\ NO_2$ slow

Show that each mechanism is consistent with the observed rate law: Rate = $k[NO]^2[O_2]$.

98. For a reaction involving the decomposition of a hypothetical substance Y, these data are obtained:

Rate (mol L^{-1} min^{-1})	0.288	0.245	0.202	0.158
[Y]	0.200	0.170	0.140	0.110

(a) Determine the order of the reaction.
(b) Write the rate law for the decomposition of Y.
(c) Calculate k for the experiment above.

99. The decomposition of N_2O_5 is first-order with a rate constant of 2.5×10^{-4} s^{-1}.
(a) Calculate the half-life for decomposition of N_2O_5.
(b) Calculate how long it takes for the concentration of N_2O_5 to drop to $\frac{1}{32}$ of its original value.
(c) Calculate how long it takes for the concentration of N_2O_5 to drop from 3.4×10^{-3} mol/L to 2.3×10^{-5} mol/L.

100. The rate constant for decomposition of azomethane at 425 °C is 0.68 s^{-1}.

$$CH_3N{=}NCH_3(g) \longrightarrow N_2(g) + C_2H_6(g)$$

(a) Based on the units of the rate constant, determine if this reaction is zeroth-, first-, or second-order.
(b) If 2.0 g azomethane is placed in a 2.0-L flask and heated to 425 °C, calculate the mass of azomethane that remains after 5.0 s.
(c) Calculate how long it takes for the mass of azomethane to drop from 2.0 g to 0.24 g.

(d) Calculate the mass of nitrogen that would be found in the flask after 0.50 s of reaction.

101. Cyclopropane isomerizes to propene when heated. Rate constants for the reaction

$$\text{cyclopropane} \longrightarrow \text{propene}$$

are $1.10 \times 10^{-4}\ s^{-1}$ at 470.0 °C and $1.02 \times 10^{-3}\ s^{-1}$ at 510.0 °C.

(a) Calculate the activation energy, E_a, for this reaction.

(b) Calculate how long it takes at 500. °C for the concentration of cyclopropane to drop from 0.10 M to 0.023 M.

Applying Concepts

These questions test conceptual learning.

102. This graph shows the change in concentration as a function of time for the reaction

$$2\ H_2O_2(g) \longrightarrow 2\ H_2O(g) + O_2(g)$$

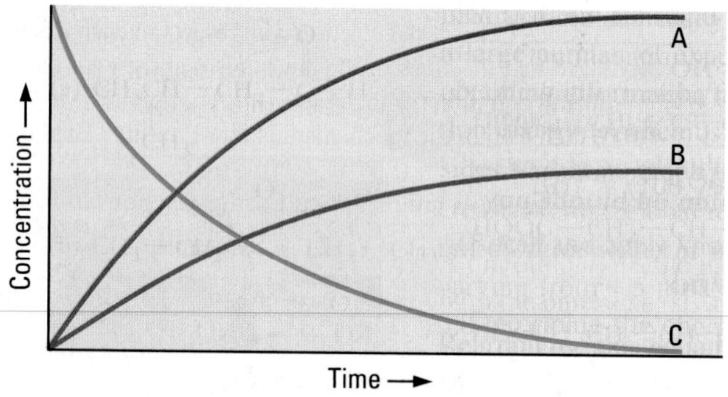
A
B
C
Concentration
Time

What substance does each of the curves A, B, and C represent?

103. Draw a graph similar to the one in Question 102 for the reaction

$$2\ N_2O_5(g) \longrightarrow 4\ NO_2(g) + O_2(g)$$

104. The left-most circle is a "snapshot" of the reactants at time = 0 for the reaction

$$H_2(g) + I_2(g) \longrightarrow 2\ HI(g)$$

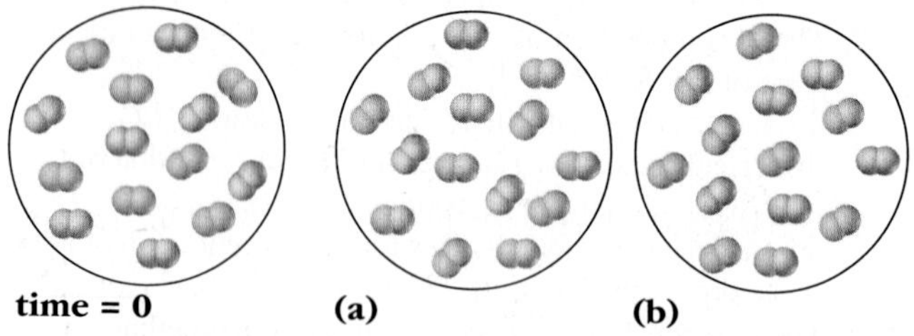
time = 0
(a)
(b)

Suppose the reaction is carried out at two different temperatures and that another snapshot is taken after a constant time has elapsed. Which of the other two snapshots, (a) or (b), corresponds to the lower temperature? Explain your answer.

105. Consider Question 104 again, only this time a catalyst is used instead of a lower temperature. Which snapshot, (a) or (b), corresponds to the presence of a catalyst? Explain your answer.

106. Initial rates for the reaction A + B + C ⟶ D + E were measured with various concentrations of A, B, and C as represented in these reaction snapshots. Based on these data, determine the rate law.

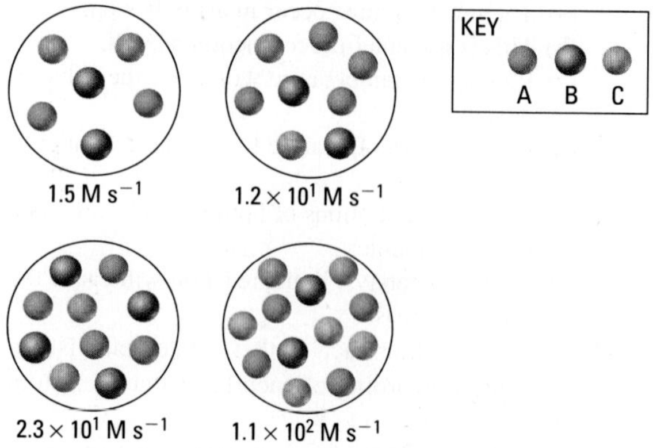
KEY
A B C
1.5 M s⁻¹
1.2 × 10¹ M s⁻¹
2.3 × 10¹ M s⁻¹
1.1 × 10² M s⁻¹

107. Examine the reaction energy diagram given here.

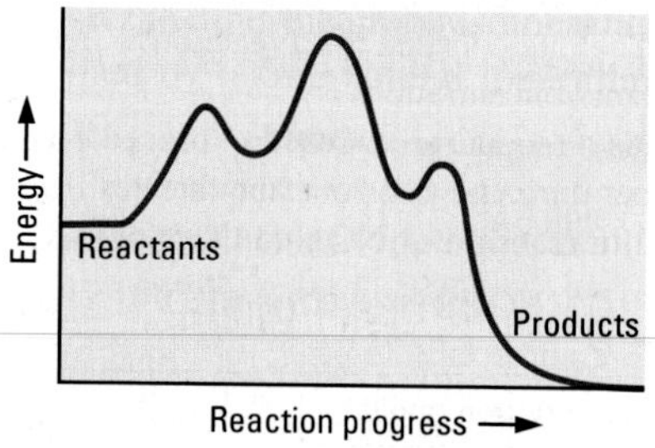
Energy
Reactants
Products
Reaction progress

(a) How many steps are in the mechanism for the reaction described by this diagram? Explain your answer.

(b) Is the overall reaction exothermic or endothermic? Explain your answer.

108. The rate of decay of a radioactive solid is independent of the temperature of that solid—at least for temperatures easily obtained in the laboratory. What does this observation imply about the activation energy for this process?

109. Platinum metal is used as a catalyst in the decomposition of NO(g) into $N_2(g)$ and $O_2(g)$. This graph shows the rate of the reaction as a function of NO concentration. Explain why the rate stops increasing.

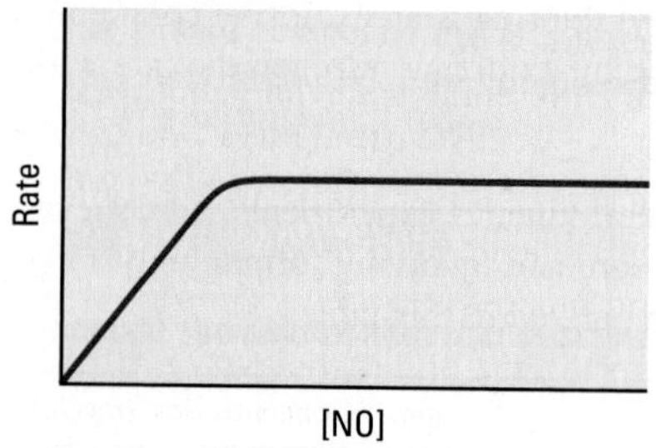
Rate
[NO]

110. Calculate the half-life of a first-order reaction if 30.5 s after the reaction starts the concentration of the reactant is 0.0451 M and 45.0 s after the reaction starts it is 0.0321 M. Calculate how many seconds after the start of the reaction it takes for the reactant concentration to decrease to 0.0100 M.

More Challenging Questions

These questions require more thought and integrate several concepts.

111. Which of these mechanisms (more than one may be chosen) is compatible with the rate law?

$$\text{Rate} = k[Cl_2]^{3/2}[CO]$$

(a) $\frac{1}{2}\,Cl_2 \rightleftharpoons Cl$	fast
$Cl + Cl_2 \rightleftharpoons Cl_3$	fast
$Cl_3 + CO \longrightarrow COCl_2 + Cl$	slow
$Cl \rightleftharpoons \frac{1}{2}\,Cl_2$	fast
(b) $Cl_2 + CO \longrightarrow CCl_2 + O$	slow
$O + Cl_2 \longrightarrow Cl_2O$	fast
$Cl_2O + CCl_2 \longrightarrow COCl_2 + Cl_2$	fast
(c) $\frac{1}{2}\,Cl_2 \rightleftharpoons Cl$	fast
$Cl + CO \rightleftharpoons COCl$	fast
$COCl + Cl_2 \longrightarrow COCl_2 + Cl$	slow
$Cl \rightleftharpoons \frac{1}{2}\,Cl_2$	fast
(d) $Cl_2 + CO \rightleftharpoons COCl + Cl$	fast
$COCl + Cl_2 \longrightarrow COCl_2 + Cl$	slow
$Cl + Cl \longrightarrow Cl_2$	fast

112. Let x represent the number of half-lives that have elapsed during the course of a first-order reaction. That is, the elapsed time t is x times the half-life $t_{1/2}$: $t = xt_{1/2}$.
(a) Show that at time t the concentration $[A]_t$ of reactant is related to the initial concentration $[A]_0$ by

$$[A]_t = [A]_0(\tfrac{1}{2})^x$$

(b) Use your result in part (a) to show that

$$\log\left(\frac{[A]_t}{[A]_0}\right) = \frac{t}{t_{1/2}}(-\log 2) = \frac{t}{t_{1/2}}(-0.301)$$

(c) Use your result in part (b) to show that for any first-order reaction a plot of $\log [A]_t$ versus t will be a straight line and that half-life can be obtained from the slope of the line.

113. Measurements of the initial rate of hydrolysis of benzenesulfonyl chloride in aqueous solution at 15 °C in the presence of fluoride ion yielded the results in the table for a fixed concentration of benzenesulfonyl chloride of 2×10^{-4} M. The reaction rate is known to be proportional to the concentration of benzenesulfonyl chloride.

$[F^-] \times 10^2$ (mol/L)	Initial Rate $\times 10^7$ (mol L^{-1} s^{-1})
0	2.4
0.5	5.4
1.0	7.9
2.0	13.9
3.0	20.2
4.0	25.2
5.0	32.0

Note that some reaction must be occurring in the absence of any fluoride ion, because at zero concentration of fluoride the rate is not zero. This residual rate should be subtracted from each observed rate to give the rate of the reaction being studied.
(a) Derive the complete rate law for the reaction.
(b) Calculate the rate constant k and express it in appropriate units.

114. Given these data and your result from Question 112, determine the half-life and the initial concentration of the reactant for the reaction

$$trans\text{-}CHClCHCl(g) \longrightarrow cis\text{-}CHClCHCl(g)$$

[*trans*-CHClCHCl] (mol/L)	Time (s)
9.23×10^{-4}	30
8.51×10^{-4}	60
7.86×10^{-4}	90
7.25×10^{-4}	120
6.19×10^{-4}	180
3.82×10^{-4}	360

115. When formic acid is heated, it decomposes to hydrogen and carbon dioxide in a first-order decay.

$$HCOOH(g) \longrightarrow CO_2(g) + H_2(g)$$

The rate of reaction is monitored by measuring the total pressure in the reaction container.

Time (s)	*P* (torr)
0	220
50	324
100	379
150	408
200	423
250	431
300	435

Calculate the rate constant and half-life, in seconds, for the reaction. At the start of the reaction (time = 0), only formic acid is present. (*Hint*: Find the partial pressure of formic acid [use Dalton's law of partial pressure (← **Sec. 8-6**) and the reaction stoichiometry to find P_{HCOOH} at each time]).

116. When heated, cyclobutane, C_4H_8, decomposes to ethylene, C_2H_4.

$$C_4H_8(g) \longrightarrow 2\,C_2H_4(g)$$

The activation energy, E_a, for this reaction is 262 kJ/mol.
(a) If the rate constant $k = 0.032\ s^{-1}$ at 800. K, calculate the value of k at 900. K.
(b) Calculate the cyclobutane concentration after 2 h at 850. K if the initial cyclobutane concentration was 0.0427 M.

117. For each reaction listed with its rate law, propose a reasonable mechanism.
(a) $CH_3COOCH_3(aq) + H_2O(\ell) \longrightarrow CH_3COOH(aq) + CH_3OH(aq)$
Rate $= k[CH_3COOCH_3][H^+]$
(b) $H_2(g) + I_2(g) \longrightarrow 2\,HI(g)$
Rate $= k[H_2][I_2]$

(c) $2\ H_2(g) + O_2(g) \longrightarrow 2\ H_2O(g)$
Rate $= k[H_2][O_2]$ (area of Pt surface)

(d) $H_2(g) + CO(g) \longrightarrow H_2CO(g)$
Rate $= k[H_2]^{1/2}[CO]$

118. Consider the reaction mechanism for iodine-catalyzed isomerization of *cis*-2-butene presented in Section 11-8. Use bond enthalpies in Table 6.2 (← **Sec. 6-6b**) to estimate the energy change for each step in the mechanism. (In Step 3 assume that rotation around the single bond requires about 5 kJ/mol.) Do your estimates agree with the energy values for the intermediates in Figure 11.15 (← **Sec. 11-8**)? Explain why or why not.

119. The iodine-catalyzed isomerization of *cis*-2-butene can be speeded up still further if a bright light with wavelength less than about 800 nm shines on the reaction mixture. Consider the mechanism presented in Section 11-8 and suggest a reason that the reaction is even faster in the presence of light than in the dark. Also explain why wavelengths longer than 800 nm are not effective in speeding up the reaction.

120. In a time-resolved picosecond spectroscopy experiment, Sheps, Crowther, Carrier, and Crim (*Journal of Physical Chemistry A*, Vol. 110, 2006; pp. 3087–3092) generated chlorine atoms in the presence of pentane. The pentane was dissolved in dichloromethane, CH_2Cl_2. The chlorine atoms are free radicals and are very reactive. After a nanosecond the chlorine atoms have reacted with pentane molecules, removing a hydrogen atom to form HCl and leaving behind a pentane radical with a single unpaired electron. The equation is

$$Cl \cdot (dcm) + C_5H_{12}(dcm) \longrightarrow HCl(dcm) + C_5H_{11} \cdot (dcm)$$

where (dcm) indicates that a substance is dissolved in dichloromethane. Measurements of the concentration of chlorine atoms were made as a function of time at three different concentrations of pentane in the dichloromethane. These results are shown in the table.

	Concentration of Chlorine Atoms (M) for Different C_5H_{12} Concentrations		
Time (ps)	0.15-M C_5H_{12}	0.30-M C_5H_{12}	0.60-M C_5H_{12}
100.0	4.42×10^{-5}	3.11×10^{-5}	1.77×10^{-5}
140.0	3.823×10^{-5}	2.39×10^{-5}	1.15×10^{-5}
180.0	3.38×10^{-5}	1.94×10^{-5}	8.48×10^{-6}
220.0	3.03×10^{-5}	1.49×10^{-5}	6.04×10^{-6}
260.0	2.68×10^{-5}	1.19×10^{-5}	4.12×10^{-6}
300.0	2.42×10^{-5}	9.45×10^{-6}	3.14×10^{-6}
340.0	2.08×10^{-5}	7.75×10^{-6}	2.38×10^{-6}
380.0	1.91×10^{-5}	6.35×10^{-6}	1.75×10^{-6}
420.0	1.71×10^{-5}	4.58×10^{-6}	1.61×10^{-6}
460.0	1.53×10^{-5}	3.77×10^{-6}	9.98×10^{-7}

(a) Determine the order of the reaction with respect to chlorine.
(b) Determine whether the reaction rate depends on the concentration of pentane in dichloromethane. If so, determine the order of the reaction with respect to pentane.
(c) Explain why the concentration of pentane in dichloromethane does not affect the data analysis that you performed in part (a).
(d) Write the rate law for the reaction and calculate the rate of reaction for a concentration of chlorine atoms equal to 1.0 μM and a pentane concentration of 0.23 M.
(e) Sheps, Crowther, Carrier, and Crim found that the rate of formation of HCl matched the rate of disappearance of Cl. From this they concluded that there were no intermediates and side reactions were not important. Explain the basis for this conclusion.

121. If you know some calculus, derive the integrated first-order rate law for the reaction

$$2\ A \longrightarrow \text{products}$$

by following these steps:
(i) Define the reaction rate in terms of [A].
(ii) Write the rate law in terms of [A], *k*, and *t*.
(iii) Separate variables in the rate law.
(iv) Integrate the rate law.
(v) Write the integrated equation in the form $y = mx + b$.
(vi) Derive the half-life as was done in Section 11-3c.

122. If you know some calculus, derive the integrated second-order rate law for the reaction

$$2\ A \longrightarrow \text{products}$$

Follow the same steps listed in Question 121.

Conceptual Challenge Problems

These rigorous, thought-provoking problems integrate conceptual learning with problem solving and are suitable for group work.

CP11.A (Section 11-5) A rule of thumb is that for a typical reaction, if concentrations are unchanged, a 10-K rise in temperature increases the reaction rate by two to four times. Use an average increase of three times to answer the questions below.
(a) What is the approximate activation energy of a "typical" chemical reaction at 298 K?
(b) If a catalyst increases a chemical reaction's rate by providing a mechanism that has a lower activation energy, then what change do you expect a 10-K increase in temperature to make in the rate of a reaction whose uncatalyzed activation energy of 75 kJ/mol has been lowered to one half this value (at 298 K) by addition of a catalyst?

CP11.B (Section 11-7) A sentence in an introductory chemistry textbook reads, "Dioxygen reacts with itself to form trioxygen, ozone, according to the equation, $3\ O_2 \longrightarrow 2\ O_3$." As a student of chemistry, what would you write to criticize this sentence?

CP11.C (Section 11-7) A classmate consults you about a problem concerning the reaction of nitrogen monoxide and dioxygen in the gas phase. She has been told that the reaction is second-order in nitrogen monoxide and first-order in dioxygen; hence, the rate law may be written as rate $= k[NO]^2[O_2]$. She has been asked to propose a mechanism for this reaction. She proposes that the mechanism is this single equation:

$$NO + NO + O_2 \longrightarrow NO_2 + NO_2$$

She asks your opinion about whether this is correct. What should you tell her to explain why the answer is correct or incorrect?

Structures for Conceptual Challenge Problem 11.D.

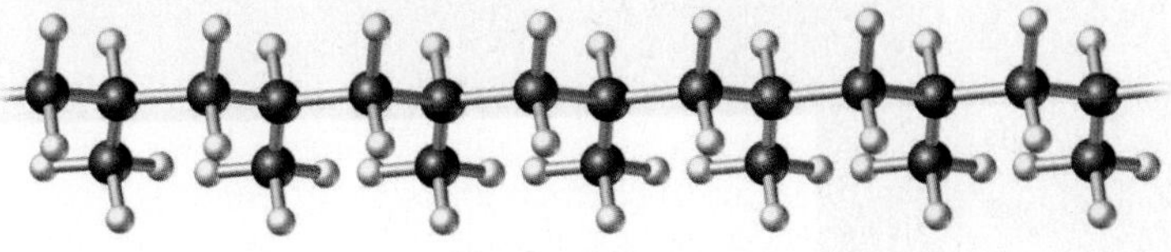

isotactic polypropylene

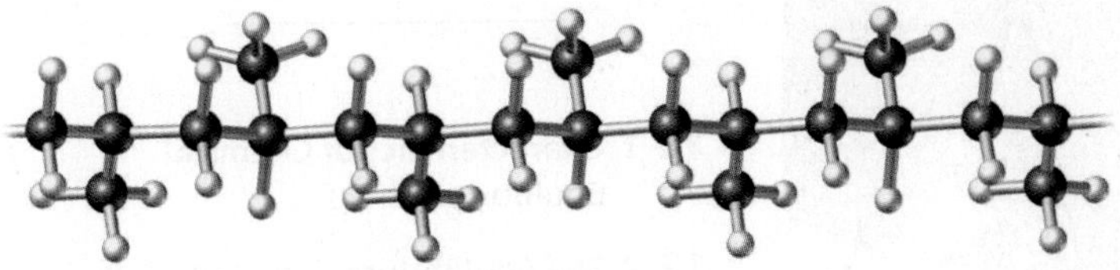

syndiotactic polypropylene

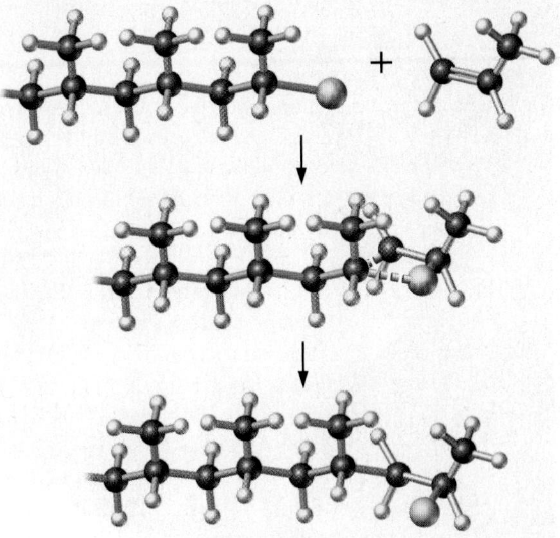
+

CP11.D (Section 11-8) Polypropylene is listed in Table 10.6 (← **Sec. 10-6a**) because it is an important type of plastic; over 3 million tons of it is produced every year in the United States. Properties of polypropylene can be changed by the way the polymer is made. For example, melting points between 130 °C and 160 °C can be obtained by using an appropriate catalyst to polymerize propylene (propene, C_3H_6). Two important types of polypropylene are isotactic, in which the methyl groups are all on the same side of the polymer chain, and syndiotactic, in which the methyl groups alternate between one side and the other of the chain.

Suppose that you are part of a team designing a new catalyst to polymerize propylene. Your catalyst will have a zirconium atom at the center of a structure consisting of carbon and hydrogen atoms. The zirconium atom will hold the growing polypropylene chain by bonding to one end of it. The metal atom will also attract a propylene molecule and bond to it before transferring the growing polypropylene chain to the other end of the new propylene molecule. The process is shown in the structures for Conceptual Challenge Problem 11.D.

What would be a reasonable shape for the rest of the catalyst molecule surrounding the metal atom so that isotactic polypropylene would be produced? It may help to build molecular models to see how each new propylene molecule needs to be added to the growing polymer chain to get all the methyl groups on the same side.

12 | Chemical Equilibrium

deoxygenated red blood cells
oxygenated red blood cells

Mario Colonel/Aurora/Getty Images; David M. Phillips/Science Source; CNRI/Science Photo Library/Science Source

At high altitudes, mountain climbers, such as these approaching a summit, carry tanks of oxygen to supplement the air they breathe. Atmospheric pressure is lower at high altitude (← **Sec. 8-1**) so the concentration of O_2 in the climbers' lungs is lower. This affects the chemical equilibrium by which hemoglobin molecules take up oxygen in the lungs and release it when they reach muscles or other tissues that require oxygen. This equilibrium can be written $Hb + 4\ O_2 \rightleftharpoons Hb(O_2)_4$, where Hb represents a hemoglobin molecule. As the O_2 concentration decreases, the equilibrium shifts to the left and less oxygen is carried in the bloodstream. Supplemental oxygen from tanks increases the O_2 concentration and shifts the equilibrium to the right allowing more oxygen to be carried. How can a chemical equilibrium be recognized? (Section 12-1) What factors shift a chemical equilibrium? (Section 12-7) These and other questions about chemical equilibrium will be answered in this chapter.

Chemical reactions that involve only pure solids or pure liquids are simpler than those that occur in the gas phase or in a solution. Either no reaction occurs between the solid and liquid reactants, or a reaction occurs in which at least one reactant is completely converted into products. [If one or more of the reactants are present in excess, those reactants will be left over, but the limiting reactant

(← **Sec. 3-7**) will be completely reacted away.] This happens because the concentration of a pure solid or a pure liquid does not change during a reaction, provided the temperature remains constant. If the initial concentrations of reactants are large enough to cause the reaction to occur, those same concentrations will be present throughout the reaction, and the reaction will not stop until the limiting reactant is used up.

When a reaction occurs in the gas phase or in a solution, concentrations of reactants decrease as the reaction takes place (← **Sec. 11-1b**). Eventually the concentrations decrease to the point at which conversion of reactants to products is no longer favored. Then, the concentrations of reactants and products stop changing, but none of the concentrations has become zero. At least a tiny bit (and often a lot) of each reactant and each product is present in the reaction mixture. Because the concentrations have stopped changing, it is often relatively easy to measure them, thus providing quantitative information about *how much* product can be obtained from the reaction. It is also possible to predict how changes in temperature, pressure, and concentrations will affect the quantity of product produced. This kind of information, combined with what you learned in Chapter 11 about factors that affect the rates of chemical reactions, enables us to predict how chemical reactions maintain living systems in balance (homeostasis) and which reactions will be useful for manufacturing a broad range of substances that enhance our quality of life.

As an example of the importance of such information, consider ammonia. The United States uses approximately 40 billion pounds of liquefied ammonia per year, mostly as fertilizer to provide nitrogen needed to support growth of a broad range of crops. Therefore, ammonia is a very important factor in providing people with food. Ammonia is synthesized directly from nitrogen and hydrogen by the *Haber-Bosch process* (Section 12-8).

The United States manufactures about 22 billion pounds of NH_3 per year and imports an additional 17 billion pounds.

$$N_2(g) + 3\,H_2(g) \rightleftharpoons 2\,NH_3(g) \qquad \Delta_r H^\circ = -92.2\ \text{kJ/mol}$$

The chemists and chemical engineers who operate ammonia manufacturing plants do their best to obtain the maximum quantity of ammonia with the minimum input of reactants and the minimum consumption of energy resources. The German chemist Fritz Haber won the 1918 Nobel Prize in Chemistry for research that showed how to determine the best conditions for carrying out this reaction. The German engineer Carl Bosch received the Nobel Prize in Chemistry in 1931 (together with Friedrich Bergius) for his pioneering chemical engineering work that enabled large-scale ammonia synthesis to be successful.

In this chapter you will learn the same principles that Haber used. With these principles you will be able to make both qualitative and quantitative predictions about how much product will be formed under a given set of reaction conditions.

12-1 Characteristics of Chemical Equilibrium

When the concentrations of reactants stop decreasing and the concentrations of products stop increasing, we say that a chemical reaction has reached equilibrium. In a **chemical equilibrium**, *there are finite concentrations of reactants and products, and these concentrations remain constant (unless reaction conditions change).* An equilibrium reaction always results in smaller amounts of products than the theoretical yield predicts (← **Sec. 3-8**), and sometimes the amount of products produced is tiny. The concentrations of reactants and products at equilibrium provide a quantitative way of determining how successful a reaction has been. *When products predominate over reactants*, the reaction is **product-favored**. *When the equilibrium mixture consists mostly of reactants* with very little product, the reaction is **reactant-favored**.

Answers to **EXERCISES** are provided at the back of this book in Appendix L.

EXERCISE 12.1

Concentrations of Pure Solids and Liquids

The introduction to this chapter states that at a given temperature the concentration of a pure solid or liquid does not depend on the quantity of substance present. Verify this assertion by calculating the concentration (in mol/L) of these solids and liquids at 20 °C. Obtain densities from Table 1.1 (← **Sec. 1-4d**).

(a) Aluminum (b) Benzene (c) Water (d) Gold

12-1a Equilibrium Is Dynamic

The *equi* in the word *equilibrium* means "equal." It refers to equal rates of forward and reverse reactions, not to equal quantities or concentrations of the substances involved. The *librium* part of the word comes from *libra*, meaning "balance." Chemical equilibrium is an equal balance between two reaction rates.

A set of double arrows, $\rightleftharpoons$, in an equation indicates a dynamic equilibrium in which forward and reverse reactions are occurring at equal rates; it also indicates that the reaction should be thought of in terms of the concepts of chemical equilibrium.

When a chemical reaction reaches equilibrium and concentrations of reactants and products remain constant, it appears that the reaction has stopped. This is true only of the net, macroscopic reaction, however. On the nanoscale level, both forward and reverse reactions continue, but *the rate of the forward reaction exactly equals the rate of the reverse reaction*. To emphasize that *chemical equilibrium involves a balance between opposite reactions*, it is often referred to as a **dynamic equilibrium**, and an equilibrium reaction is usually written with a double arrow ($\rightleftharpoons$) between reactants and products.

A good example of chemical equilibrium is provided by weak acids (← **Sec. 3-4a**), which ionize only partially in water.

Double arrows indicate dynamic equilibrium.

$$\underset{\text{acetic acid}}{CH_3COOH(aq)} \rightleftharpoons \underset{\text{acetate ion}}{CH_3COO^-(aq)} + \underset{\text{hydrogen ion}}{H^+(aq)}$$

After equilibrium has been reached at room temperature, more than 90% of the acetic acid remains in molecular form, CH_3COOH, and the equilibrium concentrations of acetate ions and hydrogen ions are each less than one tenth the concentration of acetic acid molecules. Nevertheless, both the forward and reverse reactions continue after equilibrium has been reached. Evidence to support this idea can be obtained by adding a tiny quantity of sodium acetate in which radioactive carbon-14 has been substituted into the CH_3COO^- ion to give $^{14}CH_3COO^-$. Almost immediately the radioactivity can be found in acetic acid molecules as well. This would not happen if the reaction had come to a halt, but it does happen because the reverse reaction,

$$^{14}CH_3COO^-(aq) + H^+(aq) \longrightarrow {}^{14}CH_3COOH(aq)$$

is still taking place. To a macroscopic observer nothing seems to be happening because the reverse reaction and the forward reaction are occurring at equal rates and there is no *net* change in the concentrations of reactants or products.

12-1b Equilibrium Is Independent of Direction of Approach

Another important characteristic of chemical equilibrium is that, *for a specific reaction at a specific temperature, the equilibrium state will be the same, no matter what the direction of approach to equilibrium*. As an example, consider again the synthesis of ammonia from N_2 and H_2.

$$N_2(g) + 3\,H_2(g) \rightleftharpoons 2\,NH_3(g)$$

Suppose that you introduce 1.0 mol $N_2(g)$ and 3.0 mol $H_2(g)$ into an empty (evacuated) 1.00-L container at 472 °C and seal the container. Some (but not all) of the N_2 reacts with the H_2 to form NH_3. After equilibrium is established, you would find that the concentration of H_2 has fallen from its initial value of 3.0 mol/L to an equilibrium value of 0.89 mol/L. You would also find equilibrium concentrations of 0.30 mol/L for N_2 and 1.4 mol/L for NH_3. This is shown schematically in Figure 12.1.

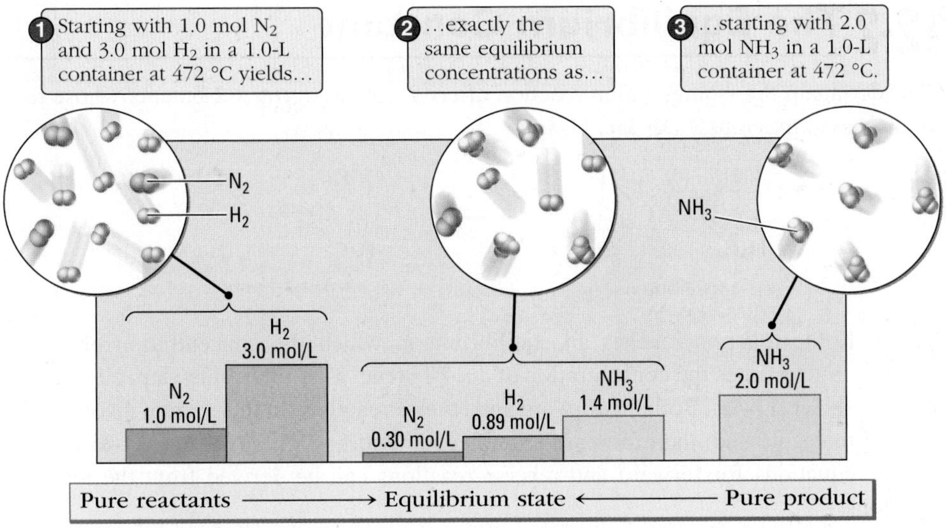
1 Starting with 1.0 mol N2 and 3.0 mol H2 in a 1.0-L container at 472 °C yields...
2 ...exactly the same equilibrium concentrations as...
3 ...starting with 2.0 mol NH3 in a 1.0-L container at 472 °C.
N2
H2
NH3
H2 3.0 mol/L
N2 1.0 mol/L
N2 0.30 mol/L
H2 0.89 mol/L
NH3 1.4 mol/L
NH3 2.0 mol/L
Pure reactants → Equilibrium state ← Pure product

Figure 12.1 The same equilibrium concentrations are achieved by starting with products as by starting with reactants. The reaction is $N_2 + 3\ H_2 \rightleftharpoons 2\ NH_3$.

Now, consider a second experiment at the same temperature in which you introduce 2.0 mol NH_3(g) into an empty 1.00-L container at 472 °C and seal the container. The 2.0 mol NH_3 consists of 2.0 mol N atoms and 6.0 mol H atoms—the same number of N atoms and H atoms contained in the 1.0 mol N_2 and 3.0 mol H_2 used in the first experiment. Because only NH_3 is present initially, the reverse reaction occurs, producing some N_2 and H_2. Measuring the concentrations at equilibrium reveals that the concentration of NH_3 dropped from the initial 2.0 mol/L to 1.4 mol/L, and the concentrations of N_2 and H_2 built up to 0.30 mol/L and 0.89 mol/L (see Figure 12.1). These equilibrium concentrations are the same as those achieved in the first experiment. Thus, *whether you start with reactants or products, the same equilibrium state is achieved*—as long as the number of atoms of each type, the volume of the container, and the temperature are the same.

12-1c Catalysts Do Not Affect Equilibrium Concentrations

Another important characteristic of chemical equilibrium is that *if a catalyst is present the same equilibrium state will be achieved, but more quickly*. A catalyst speeds up the forward reaction, but it also speeds up the reverse reaction equally. The overall effect is to produce exactly the same concentrations at equilibrium, whether or not a catalyst is in the reaction mixture. A catalyst can be used to speed up production of products in an industrial process, but it will not result in greater equilibrium concentrations of products, nor will it reduce the concentration of product present when the system reaches equilibrium.

CONCEPTUAL EXERCISE 12.2

Recognizing an Equilibrium State

A mixture of hydrogen gas and oxygen gas is maintained at 25 °C for 1 year. On the first day of each month the mixture is sampled and the concentrations of hydrogen and oxygen measured. In every case they are found to be 0.50 mol/L H_2 and 0.50 mol/L O_2; that is, the concentrations do not change over a long period. Is this mixture at equilibrium? If you think not, describe an experiment to prove it.

EXERCISES that are labeled **CONCEPTUAL** are designed to test your understanding of one or more concepts; they usually involve qualitative rather than quantitative thinking.

12-2 The Equilibrium Constant

Consider again the isomerization reaction of *cis*-2-butene to *trans*-2-butene, whose rate we discussed previously (← **Sec. 11-4a**).

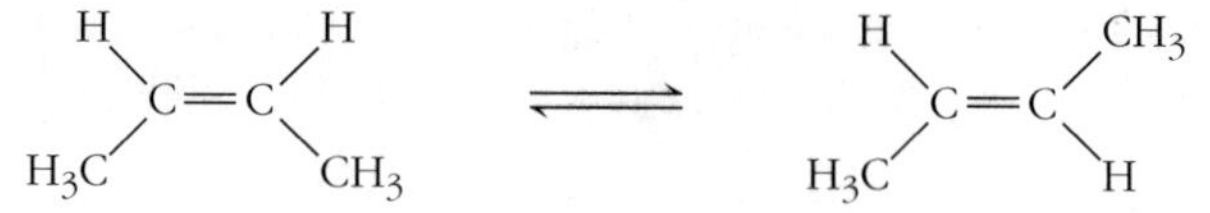

cis-2-butene *trans*-2-butene

At 500 K the reaction reaches an equilibrium in which the concentration of *trans*-2-butene is 1.65 times the concentration of *cis*-2-butene. This is a single-step, elementary process (← **Sec. 11-4a**). Both the forward and reverse reactions in this system involve only a single molecule and, therefore, are unimolecular and first-order (← **Sec. 11-6a**). Thus, the rate equations for forward and reverse reactions can be derived from the reaction equations.

$$\text{Rate}_{\text{forward}} = k_{\text{forward}}[cis\text{-2-butene}] \qquad \text{Rate}_{\text{reverse}} = k_{\text{reverse}}[trans\text{-2-butene}]$$

Suppose that we start with 0.100 mol *cis*-2-butene in a 5-L closed flask at 500 K. The *cis*-2-butene begins to react at a rate given by the forward rate equation. Initially no *trans*-2-butene is present, so the initial rate of the reverse reaction is zero. As the forward reaction proceeds, *cis*-2-butene is converted to *trans*-2-butene. The concentration of *cis*-2-butene decreases, so the forward rate decreases. As soon as some *trans*-2-butene has formed, the reverse reaction begins. As the concentration of *trans*-2-butene builds up, the reverse rate gets faster. The forward rate slows down (and the reverse rate speeds up) until the two rates are equal. At this time equilibrium has been achieved, and in the macroscopic system no further change in concentrations is observed (Figure 12.2).

On the nanoscale level, when equilibrium has been achieved, both reactions are still occurring, but the forward and reverse rates are equal. Therefore, we can equate the two rates to give

$$\text{Rate}_{\text{forward}} = \text{Rate}_{\text{reverse}}$$

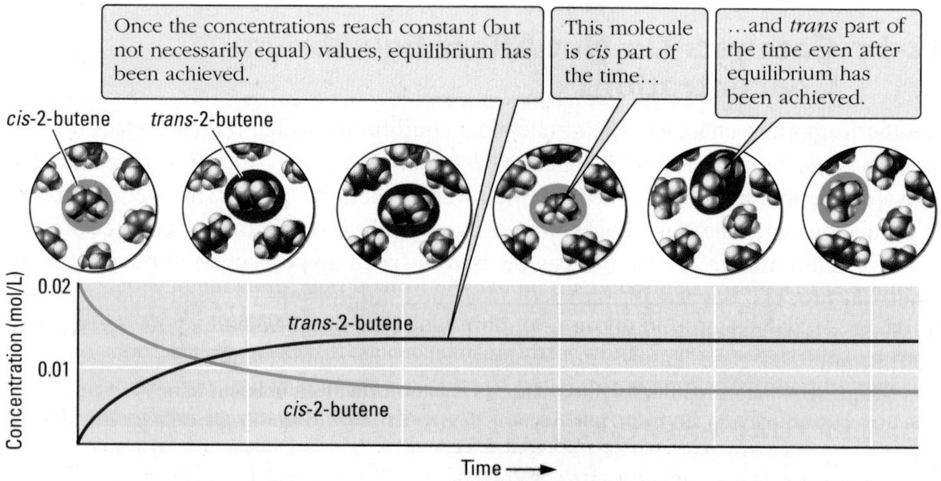
Once the concentrations reach constant (but not necessarily equal) values, equilibrium has been achieved.
This molecule is *cis* part of the time…
…and *trans* part of the time even after equilibrium has been achieved.
cis-2-butene
trans-2-butene
Concentration (mol/L)
0.02
0.01
trans-2-butene
cis-2-butene
Time →

Figure 12.2 Approach to equilibrium. The graph shows the concentrations of *cis*-2-butene and *trans*-2-butene as the *cis* compound reacts to form the *trans* compound at 500 K. The nanoscale diagrams above the graph show "snapshots" of the composition of a tiny portion of the reaction mixture. The same molecule has been highlighted in each diagram.

By substituting from the two previous rate equations we obtain

$$k_{\text{forward}}[cis\text{-2-butene}] = k_{\text{reverse}}[trans\text{-2-butene}]$$

(In this and subsequent equations in this chapter, square brackets are reserved to designate *equilibrium* concentrations; that is, [X] means the concentration of X *at equilibrium.* We designate concentrations for nonequilibrium situations by "conc. X".) Next, rearrange the equation so that both rate constants are on one side and both equilibrium concentrations are on the other side.

$$\frac{k_{\text{forward}}}{k_{\text{reverse}}} = \frac{[trans\text{-2-butene}]}{[cis\text{-2-butene}]}$$

This expression shows that the ratio of equilibrium concentrations is equal to a ratio of rate constants. Because a ratio of two constants is also a constant, the ratio of equilibrium concentrations must also be constant. We call this ratio K_c, where the capital letter K is used to distinguish it from lowercase ks of the rate constants, k_{forward} and k_{reverse}, and the subscript c indicates that it is a ratio of equilibrium *concentrations.* At 500 K the experimental value of K_c is 1.65 for the *cis* $\rightleftharpoons$ *trans* butene reaction.

$$K_c = \frac{k_{\text{forward}}}{k_{\text{reverse}}} = \frac{[trans\text{-2-butene}]}{[cis\text{-2-butene}]} = 1.65 \quad \text{(at 500 K)}$$

Because the values of the rate constants vary with temperature (← **Sec. 11-5**), *the value of* K_c *also varies with temperature.* For the butene isomerization reaction, K_c is 1.47 at 600 K and 1.36 at 700 K.

An **equilibrium constant (K_c)** *is a quotient of equilibrium concentrations of product and reactant substances that has a constant value for a given reaction at a given temperature.* Equilibrium constants can be used to answer three important questions about a reaction:

- When equilibrium has been achieved, do products predominate over reactants?
- Given initial concentrations of reactants and products, in which direction will the reaction go to achieve equilibrium?
- What concentrations of reactants and products are present at equilibrium?

If a reaction moves quickly to equilibrium, you can use equilibrium constants to determine the composition soon after reactants are mixed. Equilibrium constants are less valuable for slow reactions. Until equilibrium has been reached, only kinetics is capable of predicting the composition of a reaction mixture. One way to reach equilibrium faster is to use a catalyst; because the equilibrium concentrations are not affected by the presence of a catalyst, K_c for a given reaction has exactly the same value whether or not a catalyst is involved.

It is important to distinguish equilibrium concentrations, which do not change over time, from the changing concentrations of reactants and products before equilibrium because *only* equilibrium concentrations can be calculated from equilibrium constants.

When a reaction takes place by a mechanism that consists of a sequence of steps, the equilibrium constant expression can be obtained by multiplying together the rate constants for the forward reactions in all steps and then dividing by the rate constants for the reverse reactions in all steps. This process gives the same equilibrium constant expression that can be obtained from the coefficients of the balanced overall equilibrium equation.

When concentrations are large enough, the value of the equilibrium "constant" expressed in terms of concentrations does not remain constant, even at the same temperature. Attractions among molecules, and especially ions, cause them to behave differently as their concentrations become larger. To deal with this behavior, true equilibrium constants must be expressed in terms of corrected concentrations that are called *activities*.

CONCEPTUAL EXERCISE 12.3

Properties of Equilibrium

After a mixture of *cis*-2-butene and *trans*-2-butene has reached equilibrium at 600 K, where $K_c = 1.47$, half of the *cis*-2-butene is suddenly removed. Answer these questions:

(a) Is the new mixture at equilibrium? Explain why or why not.

(b) In the new mixture, which rate is faster, *cis* → *trans* or *trans* → *cis*? Or are both rates the same?

(c) In an equilibrium mixture, which concentration is larger, *cis*-2-butene or *trans*-2-butene?

(d) If the concentration of *cis*-2-butene at equilibrium is 0.10 mol/L, what will be the concentration of *trans*-2-butene?

12-2a Writing Equilibrium Constant Expressions

The **equilibrium constant expression** for any reaction is *a mathematical expression that relates the equilibrium concentrations of reactants and products.* It has concentrations of products in the numerator and concentrations of reactants in the denominator. Each concentration is raised to the power of its stoichiometric coefficient in the balanced equation. *The only concentrations that appear in an equilibrium constant expression are those of gases and of solutes in dilute solutions.* This is because these are the only concentrations that can change as a reaction occurs. *Concentrations of pure solids, pure liquids, and solvents in dilute solutions do not appear in equilibrium constant expressions.*

To illustrate how this works, consider the general equilibrium reaction

$$a\,\mathrm{A} + b\,\mathrm{B} \rightleftharpoons c\,\mathrm{C} + d\,\mathrm{D} \qquad [12.1]$$

By convention we write the equilibrium constant expression for this reaction as

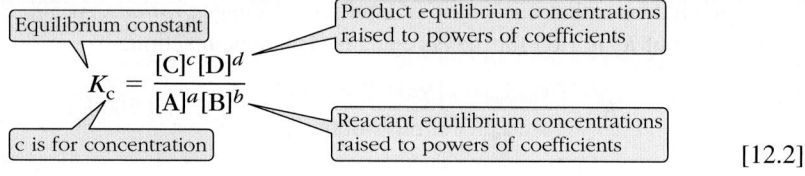

$$K_c = \frac{[\mathrm{C}]^c[\mathrm{D}]^d}{[\mathrm{A}]^a[\mathrm{B}]^b} \qquad [12.2]$$

Equilibrium constant

c is for concentration

Product equilibrium concentrations raised to powers of coefficients

Reactant equilibrium concentrations raised to powers of coefficients

The reaction of nitrogen with oxygen occurs in automobile engines and other high-temperature combustion processes where air is present.

Let's apply these ideas to the combination of nitrogen and oxygen gases to form nitrogen monoxide.

$$N_2(g) + O_2(g) \rightleftharpoons 2\ NO(g)$$

Because all substances in the reaction are gaseous, all concentrations appear in the equilibrium constant expression. The concentration of the product NO(g) is in the numerator and is squared because of its coefficient 2 in the balanced chemical equation. The concentrations of the reactants $N_2(g)$ and $O_2(g)$ are in the denominator. Each is raised to the first power because each has a coefficient of 1 in the balanced chemical equation.

$$\text{Equilibrium constant} = K_c = \frac{[NO]^2}{[N_2][O_2]}$$

12-2b Equilibria Involving Pure Liquids and Solids

This reaction occurs whenever a material that contains sulfur burns in air; it is responsible for a good deal of sulfur dioxide air pollution. It is also the first reaction in a sequence by which sulfur is converted to sulfuric acid, the number-one industrial chemical in the world.

As another example, consider the combustion of solid yellow sulfur, which consists of S_8 molecules. The combustion reaction produces sulfur dioxide gas.

$$\tfrac{1}{8}\ S_8(s) + O_2(g) \rightleftharpoons SO_2(g)$$

Because sulfur is a solid, the number of molecules per unit volume is fixed by the density of sulfur at any given temperature as shown in Exercise 12.1 (← **Sec. 12-1**). Therefore, the sulfur concentration is not changed either by reaction or by addition or removal of solid sulfur. It is an experimental fact that, as long as some solid sulfur is present, the equilibrium concentrations of O_2 and SO_2 are not affected by changes in the amount of sulfur. Therefore, the equilibrium constant expression for this reaction is properly written as

It is very important to remember that any substance designated as (s) or (ℓ) in the equilibrium equation does not appear in the equilibrium constant expression.

$$K_c = \frac{[SO_2(g)]}{[O_2(g)]} \qquad \text{At 25 °C, } K_c = 4.2 \times 10^{52}$$

Concentration of $S_8(s)$ is not included in K_c.

12-2c Equilibria in Dilute Solutions

Consider an aqueous solution of the weak base ammonia, which contains a small concentration of ammonium ions and hydroxide ions because ammonia reacts with water.

$$NH_3(aq) + H_2O(\ell) \rightleftharpoons NH_4^+(aq) + OH^-(aq)$$

If the concentration of ammonia molecules (and consequently of ammonium ions and hydroxide ions) is small, the number of water molecules per unit volume remains essentially the same as in pure water. Because the molar concentration of water is effectively constant for reactions involving dilute solutions, the concentration of water should not be included in the equilibrium constant expression. Thus, for this reaction we write

$$K_c = \frac{[NH_4^+][OH^-]}{[NH_3]}$$

and the concentration of water is not included in the denominator. At 25 °C, $K_c = 1.8 \times 10^{-5}$ for reaction of ammonia with water.

Notice that no units were included in the equilibrium constant for the ammonia ionization reaction. There are two concentrations in the numerator and only one in the denominator, which ought to give units of mol/L. However, if we always express concentrations in mol/L, the units of the equilibrium constant can be figured out from the equilibrium constant expression. Therefore, it is customary to omit the units from the equilibrium constant value, even if the equilibrium constant should have units. We follow that custom in this book.

PROBLEM-SOLVING EXAMPLE 12.1

Writing Equilibrium Constant Expressions

Write an equilibrium constant expression for each chemical equation.

(a) $4\ NO_2(g) + O_2(g) + 2\ H_2O(g) \rightleftharpoons 4\ HNO_3(g)$

(b) $BaSO_4(s) \rightleftharpoons Ba^{2+}(aq) + SO_4^{2-}(aq)$

(c) $6\ NH_3(aq) + Ni^{2+}(aq) \rightleftharpoons Ni(NH_3)_6^{2+}(aq)$

(d) $2\ BaO_2(s) \rightleftharpoons 2\ BaO(s) + O_2(g)$

(e) $NH_3(g) \rightleftharpoons NH_3(\ell)$

Result

(a) $K_c = \frac{[HNO_3]^4}{[NO_2]^4[O_2][H_2O]^2}$ (b) $K_c = [Ba^{2+}][SO_4^{2-}]$ (c) $K_c = \frac{[Ni(NH_3)_6^{2+}]}{[NH_3]^6[Ni^{2+}]}$

(d) $K_c = [O_2]$ (e) $K_c = \frac{1}{[NH_3]}$

Analyze To write an equilibrium constant expression, place equilibrium concentrations of products in the numerator and equilibrium concentrations of reactants in the denominator. Raise each concentration to the power of the stoichiometric coefficient of the species. Do not include pure solids, pure liquids, or solvents.

Plan In part (b), one species, $BaSO_4(s)$, is a pure solid and does not appear in the expression. In part (d), two species, $BaO_2(s)$ and $BaO(s)$, are solids and do not appear. In part (e), one species, $NH_3(\ell)$, is a pure liquid and does not appear.

Execute See Result.

PROBLEM-SOLVING PRACTICE 12.1

Write an equilibrium constant expression for each equation.

(a) $CaCO_3(s) \rightleftharpoons CaO(s) + CO_2(g)$

(b) $HCl(g) + LiH(s) \rightleftharpoons H_2(g) + LiCl(s)$

PROBLEM-SOLVING PRACTICE answers are provided at the back of this book in Appendix K.

(c) $CH_4(g) + H_2O(g) \rightleftharpoons CO(g) + 3\ H_2(g)$

(d) $CN^-(aq) + H_2O(\ell) \rightleftharpoons HCN(aq) + OH^-(aq)$

12-2d Equilibrium Constant Expressions for Related Reactions

Consider the equilibrium involving nitrogen, hydrogen, and ammonia.

$$N_2(g) + 3\ H_2(g) \rightleftharpoons 2\ NH_3(g) \qquad K_{c_1} = 3.5 \times 10^8 \quad (\text{at } 25\ °C)$$

Suppose we modify the equation so that 1 mol NH_3 is produced.

$$\tfrac{1}{2} N_2(g) + \tfrac{3}{2} H_2(g) \rightleftharpoons NH_3(g) \qquad K_{c_2} = ?$$

Coefficients half as big

Is the value of the equilibrium constant, K_{c_2}, for the second equation the same as the value of the equilibrium constant, K_{c_1}, for the first equation? To decide, write the equilibrium constant expression for each balanced equation.

Concentrations raised to powers half as big

$$K_{c_1} = \frac{[NH_3]^2}{[N_2][H_2]^3} = 3.5 \times 10^8 \quad \text{and} \quad K_{c_2} = \frac{[NH_3]}{[N_2]^{1/2}[H_2]^{3/2}} = ?$$

This makes it clear that K_{c_1} is the square of K_{c_2}; that is, $K_{c_1} = (K_{c_2})^2$. Therefore, the answer to our question about the value of K_{c_2} is

$$K_{c_2} = (K_{c_1})^{1/2} = (3.5 \times 10^8)^{1/2} = 1.9 \times 10^4$$

Whenever the stoichiometric coefficients of a balanced equation are multiplied by some factor, the equilibrium constant for the new equation (K$_{c_2}$ in this case) is the old equilibrium constant (K$_{c_1}$) raised to the power of the multiplication factor.

Suppose that we interchange products and reactants. What is the value of K_{c_3}, the equilibrium constant for the decomposition of ammonia to the elements, which is the reverse of the first equation? If we write the equilibrium constant expression for the reverse reaction,

$$2\ NH_3(g) \rightleftharpoons N_2(g) + 3\ H_2(g) \qquad K_{c_3} = ? = \frac{[N_2][H_2]^3}{[NH_3]^2}$$

Concentration of NH_3 is in denominator

we see that K_{c_3} is the reciprocal of K_{c_1}; that is, $K_{c_3} = 1/K_{c_1} = 1/(3.5 \times 10^8) = 2.9 \times 10^{-9}$.

The equilibrium constant for a reaction and that for its reverse are the reciprocals of one another. If a reaction has a very large equilibrium constant, the reverse reaction will have a very small one. That is, if a reaction is strongly product-favored, then its reverse is strongly reactant-favored. In the case of the production of ammonia from its elements at room temperature, the forward reaction has a large equilibrium constant, 3.5×10^8. As expected, the reverse reaction, decomposition of ammonia to its elements, has a small equilibrium constant, 2.9×10^{-9}.

EXERCISE 12.4

Manipulating Equilibrium Constants

The balanced equation for conversion of oxygen to ozone has a very small value of K_c.

$$3\ O_2(g) \rightleftharpoons 2\ O_3(g) \qquad K_c = 6.25 \times 10^{-58}$$

(a) Calculate the value of K_c if the equation is written as

$$\tfrac{3}{2}\ O_2(g) \rightleftharpoons O_3(g)$$

(b) Calculate the value of K_c for the conversion of ozone to oxygen.

$$2\ O_3(g) \rightleftharpoons 3\ O_2(g)$$

12-2e Equilibrium Constant for a Reaction That Combines Two or More Other Reactions

If two chemical equations can be combined to give a third, the equilibrium constant for the combined reaction can be obtained from the equilibrium constants for the two original reactions. For example, air pollution is produced when nitrogen monoxide forms from nitrogen and oxygen and then combines with additional oxygen to form nitrogen dioxide (← **Sec. 8-12b**).

(1) $N_2(g) + O_2(g) \rightleftharpoons 2\ NO(g)$ $$K_{c_1} = \frac{[NO]^2}{[N_2][O_2]}$$

(2) $2\ NO(g) + O_2(g) \rightleftharpoons 2\ NO_2(g)$ $$K_{c_2} = \frac{[NO_2]^2}{[NO]^2[O_2]}$$

The sum of these two equations is

(3) $N_2(g) + 2\ O_2(g) \rightleftharpoons 2\ NO_2(g)$

Sum of Equations 1 and 2

Product of equilibrium constants K_{c_1} and K_{c_2}

$$K_{c_3} = \frac{[NO_2]^2}{[N_2][O_2]^2} = \frac{[NO]^2}{[N_2][O_2]} \times \frac{[NO_2]^2}{[NO]^2[O_2]} = K_{c_1} \times K_{c_2}$$

If two chemical equations can be summed to give a third, the equilibrium constant for the overall equation equals the product of the two equilibrium constants for the equations that were summed. This is a powerful tool for obtaining equilibrium constants without having to measure them experimentally for each individual reaction.

PROBLEM-SOLVING EXAMPLE 12.2

Manipulating Equilibrium Constants

Given these equilibrium reactions and constants,

(1) $\frac{1}{4}\ S_8(s) + 2\ O_2(g) \rightleftharpoons 2\ SO_2(g)$ $\qquad K_{c_1} = 1.86 \times 10^{105}$

(2) $\frac{1}{8}\ S_8(s) + \frac{3}{2}\ O_2(g) \rightleftharpoons SO_3(g)$ $\qquad K_{c_2} = 1.77 \times 10^{53}$

calculate the equilibrium constant, K_{c_3}, for this reaction, which is important in the formation of acid-rain air pollution.

$$2\ SO_2(g) + O_2(g) \rightleftharpoons 2\ SO_3(g)$$

Result $K_{c_3} = 16.8$

Analyze Two equilibrium reactions and their associated equilibrium constants are given; the equilibrium constant for a third reaction is to be determined.

Plan Compare the given equations with the target equation. Manipulate each equation appropriately. Check that the equations sum to the target equation. Then, multiply the corresponding equilibrium constants.

Execute The target equation has SO_2 on the left, so Equation (1) needs to be reversed and we need to take the reciprocal of the equilibrium constant.

The target equation has SO_3 on the right. Equation (2) need not be reversed, but each coefficient must be multiplied by 2, which means squaring K_{c_2}.

Check that the equations sum to the correct equation and multiply the K values:

(1)′ $\quad 2\ SO_2(g) \rightleftharpoons \frac{1}{4}\ S_8(s) + 2\ O_2(g)$ $\qquad K'_{c_1} = \frac{1}{1.86 \times 10^{105}} = 5.38 \times 10^{-106}$

(2)′ $\quad \frac{1}{4}\ S_8(s) + 3\ O_2(g) \rightleftharpoons 2\ SO_3(g)$ $\qquad K'_{c_2} = (1.77 \times 10^{53})^2 = 3.13 \times 10^{106}$

$2\ SO_2(g) + O_2(g) \rightleftharpoons 2\ SO_3(g)$ $\qquad K_{c_3} = K'_{c_1} \times K'_{c_2} = 16.8$

Because K_{c_1} and K_{c_2} were known experimentally and could be used to calculate K_{c_3}, there is no need to measure K_{c_3} experimentally.

☑ **Reasonable Result Check** For Equation (1) the equilibrium constant was quite large, so for the reverse reaction it should be quite small, which it is. Also check the order of magnitude of the result by checking the powers of 10. In the square, the power of 10 should be doubled, which it is. The sum of the powers of 10 for the two equilibrium constants that were multiplied should be close to the power of 10 in the result, and it is.

PROBLEM-SOLVING PRACTICE 12.2

When carbon dioxide dissolves in water it reacts to produce carbonic acid, $H_2CO_3(aq)$, which can ionize in two steps.

$$H_2CO_3(aq) \rightleftharpoons HCO_3^-(aq) + H^+(aq) \qquad K_{c_1} = 4.2 \times 10^{-7}$$
$$HCO_3^-(aq) \rightleftharpoons CO_3^{2-}(aq) + H^+(aq) \qquad K_{c_2} = 4.8 \times 10^{-11}$$

Calculate the equilibrium constant for the reaction

$$H_2CO_3(aq) \rightleftharpoons CO_3^{2-}(aq) + 2\ H^+(aq)$$

12-2f Equilibrium Constants in Terms of Pressure

In a constant-volume system, when the concentration of a gas changes, the partial pressure of the gas (← **Sec. 8-6**) also changes. This follows from the ideal gas equation

$$PV = nRT$$

Solving for the partial pressure P_A of a gaseous substance, A,

[A] = n_A/V, the amount (mol) of A per unit volume

$$P_A = \frac{n_A}{V}RT = [A]RT \qquad [12.3]$$

Because K_c is related to K_P for the same gas-phase reaction, either can be used to calculate the composition of a gas-phase equilibrium mixture. Most examples in this chapter involve K_c, but the same rules apply to solving problems with K_P.

If all substances are in the gas phase, Equation 12.3 allows us to express the equilibrium constant for the general reaction in Equation 12.1 (← **Sec. 12-2a**) in a form similar to Equation 12.2, but in terms of partial pressures. The pressure **equilibrium constant (K_P)** is

$$K_P = \frac{P_C^c \times P_D^d}{P_A^a \times P_B^b} \qquad [12.4]$$

Product pressures raised to powers of coefficients

Reactant pressures raised to powers of coefficients

The subscript on K_P indicates that the equilibrium constant has been expressed in terms of partial pressures. For some gas-phase equilibria $K_c = K_P$; for many others it does not. Therefore, it is useful to be able to relate one type of equilibrium constant to the other. This can be done by combining Equations 12.2, 12.3, and 12.4 to give

$\Delta n = c + d - a - b$ is the sum of coefficients of *gaseous* products minus the sum of coefficients of *gaseous* reactants

$$K_P = \frac{P_C^c \times P_D^d}{P_A^a \times P_B^b} = \frac{\{[C]RT\}^c\{[D]RT\}^d}{\{[A]RT\}^a\{[B]RT\}^b} = \frac{[C]^c[D]^d}{[A]^a[B]^b}(RT)^{c+d-a-b} = K_c(RT)^{\Delta n}$$

As an example of this relation, consider the equilibrium

$$2\ NOCl(g) \rightleftharpoons 2\ NO(g) + Cl_2(g) \qquad K_c = 2.75 \times 10^{-9}\ \text{mol/L at 298 K}$$

Note that $K_c = K_p$ *only* when $\Delta n = 0$; that is, *only* when the number of gas molecules in the reactants is the same as in the products.

For this reaction $\Delta n = 2 + 1 - 2 = 3 - 2 = 1$, because there are three gas-phase product molecules and only two gas-phase reactants. Therefore,

$$K_P = K_c \times (RT)^1$$
$$= (2.75 \times 10^{-9}\ \text{mol/L}) \times (0.0821\ \text{L atm K}^{-1}\ \text{mol}^{-1}) \times (298\ \text{K}) = 6.74 \times 10^{-8}\ \text{atm}$$

EXERCISE 12.5

Relating K_c and K_P

For each of these reactions, calculate K_P from K_c.

(a) $N_2(g) + 3\ H_2(g) \rightleftharpoons 2\ NH_3(g)$	$K_c = 3.5 \times 10^{8}$	(at 25 °C)
(b) $2\ H_2(g) + O_2(g) \rightleftharpoons 2\ H_2O(g)$	$K_c = 3.5 \times 10^{81}$	(at 25 °C)
(c) $N_2(g) + O_2(g) \rightleftharpoons 2\ NO(g)$	$K_c = 1.6 \times 10^{-3}$	(at 2300 K)
(d) $2\ NO_2(g) \rightleftharpoons N_2O_4(g)$	$K_c = 1.7 \times 10^{2}$	(at 25 °C)

12-3 Determining Equilibrium Constants

To determine the value of an equilibrium constant it is necessary to know all of the equilibrium concentrations that appear in the equilibrium constant expression. This is most commonly done by allowing a system to reach equilibrium and then measuring the **equilibrium concentration** of one or more of the reactants or products. Algebra and stoichiometry are then used to obtain the numerical value of K_c.

12-3a Reaction Tables, Stoichiometry, and Equilibrium Concentrations

A systematic approach to calculations involving equilibrium constants involves making a table that shows initial conditions, changes that take place when a reaction occurs, and final (equilibrium) conditions. As an example, consider the colorless gas dinitrogen tetraoxide, $N_2O_4(g)$. When heated it dissociates to form red-brown $NO_2(g)$ according to the equation

$$N_2O_4(g) \rightleftharpoons 2\ NO_2(g)$$

Suppose that 2.00 mol $N_2O_4(g)$ is placed into an empty 5.00-L flask, the flask is sealed, and the N_2O_4 is heated to 400. K. Almost immediately a dark red-brown color appears, indicating that much of the colorless gas has been transformed into NO_2 (Figure 12.3). By measuring the intensity of color, it can be determined that the concentration of NO_2

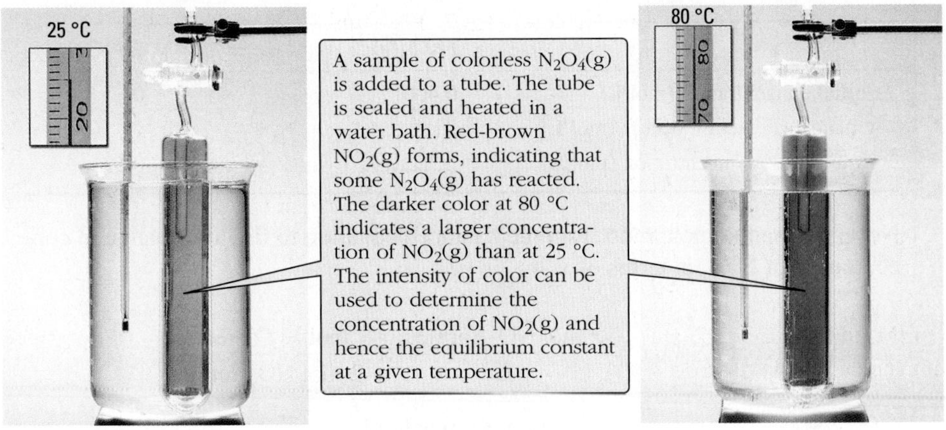
25 °C
80 °C
A sample of colorless N2O4(g) is added to a tube. The tube is sealed and heated in a water bath. Red-brown NO2(g) forms, indicating that some N2O4(g) has reacted. The darker color at 80 °C indicates a larger concentration of NO2(g) than at 25 °C. The intensity of color can be used to determine the concentration of NO2(g) and hence the equilibrium constant at a given temperature.

Photos: Charles D. Winters

Figure 12.3 Equilibrium mixtures of NO_2 and N_2O_4.

at equilibrium is 0.525 mol/L. To use this information to calculate the equilibrium constant, follow these steps:

1. ***Write the balanced equation for the equilibrium reaction. From it derive the equilibrium constant expression.*** The balanced equation and equilibrium constant expression are

$$N_2O_4(g) \rightleftharpoons 2\ NO_2(g) \qquad K_c = \frac{[NO_2]^2}{[N_2O_4]}$$

2. ***Set up a table containing initial concentration, change in concentration, and equilibrium concentration for each substance included in the equilibrium constant expression. Enter all known information into this reaction table.*** In this case the amount (moles) of $N_2O_4(g)$ and the volume of the flask were given, so the concentration of $N_2O_4(g)$ before any reaction occurs is (conc. N_2O_4) = 2.00 mol/5.00 L = 0.400 mol/L. Because the flask initially contained no NO_2, the initial concentration of NO_2 is zero. After the reaction took place and equilibrium was reached, the equilibrium concentration of NO_2 was measured as 0.525 mol/L. The reaction table looks like this:

Tables like this one are often called **ICE** tables from the initial letters of the labels on the rows: ***I****nitial*, ***C****hange*, and ***E****quilibrium*.

	$N_2O_4(g)$	$\rightleftharpoons$	$2\ NO_2(g)$
*I*nitial concentration (mol/L)	0.400		0
*C*hange as reaction occurs (mol/L)	______		______
*E*quilibrium concentration (mol/L)	______		0.525

3. ***Use* x *to represent the change in concentration of one substance. Use the stoichiometric coefficients in the balanced equilibrium equation to calculate the other changes in terms of* x.** When the reaction proceeds from left to right, the concentrations of reactants decrease. Therefore, the change in concentration of a reactant is negative. The concentrations of products increase, so the change in concentration of a product is positive. Usually it is best to begin with the reactant or product column that contains the most information. In this case that is the NO_2 column, where both initial and equilibrium concentrations are known. Therefore, let x represent the unknown change in concentration of NO_2.

	$N_2O_4(g)$	$\rightleftharpoons$	$2\ NO_2(g)$
*I*nitial concentration (mol/L)	0.400		0
*C*hange as reaction occurs (mol/L)	______		x
*E*quilibrium concentration (mol/L)	______		0.525

Next, we use the mole ratio from the balanced equation to find the change in concentration of N_2O_4 in terms of x.

For every 1 mol N_2O_4 that decomposes, 2 mol NO_2 forms. Therefore the mole ratio of N_2O_4:NO_2 is $\frac{1}{2}$:1.

$$\Delta(\text{conc. } N_2O_4) = \frac{x \text{ mol } NO_2 \text{ formed}}{\text{L}} \times \frac{1 \text{ mol } N_2O_4 \text{ reacted}}{2 \text{ mol } NO_2 \text{ formed}}$$

$$= \tfrac{1}{2}\, x \text{ mol } N_2O_4 \text{ reacted/L}$$

The sign of the change in concentration of N_2O_4 is *negative*, because the concentration of N_2O_4 *decreases*. The table becomes

	$N_2O_4(g)$	$\rightleftharpoons$	$2\ NO_2(g)$
*I*nitial concentration (mol/L)	0.400		0
*C*hange as reaction occurs (mol/L)	$-\frac{1}{2}x$		x
*E*quilibrium concentration (mol/L)	______		0.525

4. ***From initial concentrations and the changes in concentrations, calculate the equilibrium concentrations in terms of* x *and enter them in the table.*** The concentration of N_2O_4 at equilibrium, $[N_2O_4]$, is the initial 0.400 mol/L plus the change due to reaction ($-\frac{1}{2}x$ mol/L); that is, $[N_2O_4] = (0.400 - \frac{1}{2}x)$ mol/L. Similarly, the equilibrium concentration of NO_2 (which is already known to be 0.525 mol/L) is $0 + x$, and the table becomes

	$N_2O_4(g)$	$\rightleftharpoons$	$2\ NO_2(g)$
*I*nitial concentration (mol/L)	0.400		0
*C*hange as reaction occurs (mol/L)	$-\frac{1}{2}x$		x
*E*quilibrium concentration (mol/L)	$0.400 - \frac{1}{2}x$		$0.525 = 0 + x$

5. ***Use the simplest possible equation to solve for* x. *Then, use* x *to calculate the unknown you were asked to find.*** (Usually the unknown is K_c or a concentration.) In this case the simplest equation to solve for x is the last entry in the table, $0.525 = 0 + x$, and it is easy to see that $x = 0.525$. Calculate $[N_2O_4] = (0.400 - \frac{1}{2}x)$ mol/L $= (0.400 - \frac{1}{2} \times 0.525)$ mol/L $= 0.138$ mol/L. The problem stated that $[NO_2] = 0.525$ mol/L, so K_c is given by

$$K_c = \frac{[NO_2]^2}{[N_2O_4]} = \frac{(0.525\text{ mol/L})^2}{(0.138\text{ mol/L})} = 2.00 \quad (\text{at } 400.\text{ K})$$

6. ***Check your result to make certain it is reasonable.*** In this case the equilibrium concentration of product is larger than the concentration of reactant. Because the products are in the numerator of the equilibrium constant expression, we expect a value greater than 1, which is what we calculated.

PROBLEM-SOLVING EXAMPLE 12.3

Determining an Equilibrium Constant Value

Consider the gas-phase reaction

$$H_2(g) + I_2(g) \rightleftharpoons 2\ HI(g)$$

Suppose that a sealed flask containing H_2 and I_2 has been heated to 425 °C and the initial concentrations of H_2 and I_2 were each 0.0175 mol/L. With time, the concentrations of H_2 and I_2 decline and the concentration of HI increases. At equilibrium [HI] = 0.0276 mol/L. Use this experimental information to calculate the equilibrium constant.

Result $K_c = 56$

Analyze Initial concentrations of reactants and concentration of the product at equilibrium are given. The equilibrium constant is to be calculated. An ICE table helps keep track of changing concentrations as equilibrium is achieved.

Plan Follow the six-step procedure just described.

Note that products form at the expense of reactants.

The six steps and corresponding entries in the ICE table have been color coded to help you identify them.

Execute

1. ***Write the balanced equation and equilibrium constant expression.***

$$H_2(g) + I_2(g) \rightleftharpoons 2\,HI(g) \qquad K_c = \frac{[HI]^2}{[H_2][I_2]}$$

2. ***Construct a reaction (ICE) table (see Step 4); enter known information.***
3. ***Represent changes in concentration in terms of* x.**

The best choice is to enter x in the third column, because both initial and equilibrium concentrations of HI are known, and this provides a simple way to calculate x. Next, derive the rest of the concentration changes in terms of x. If the concentration of HI increases by a given quantity, the mole ratios say that the concentrations of H_2 and I_2 must decrease only half as much:

$$\frac{x \text{ mol HI produced}}{\text{L}} \times \frac{1 \text{ mol } H_2 \text{ consumed}}{2 \text{ mol HI produced}} = \frac{1}{2}x \text{ mol/L } H_2 \text{ consumed}$$

Because the coefficients of H_2 and I_2 are equal, each of their concentrations decreases by $\frac{1}{2}x$ mol/L. The entries in the table for change in concentration of H_2 and I_2 are negative, because their concentrations decrease.

4. ***Calculate equilibrium concentrations and enter them in the table.***

Color-coded entries in the table in this example correspond with color-coded steps in the Execute section preceding the table.

	$H_2(g)$	+ $I_2(g)$	$\rightleftharpoons$ 2 HI(g)
*I*nitial concentration (mol/L)	0.0175	0.0175	0
*C*hange as reaction occurs (mol/L)	$-\frac{1}{2}x$	$-\frac{1}{2}x$	x
*E*quilibrium concentration (mol/L)	$0.0175 - \frac{1}{2}x$	$0.0175 - \frac{1}{2}x$	$0.0276 = 0 + x$

5. ***Solve the simplest equation for* x.** The last row and column in the table contains $0.0276 = 0 + x$, which gives $x = 0.0276$. Substitute this value into the other two equations in the last row of the table to get the equilibrium concentrations, and substitute them into the equilibrium constant expression:

$$K_c = \frac{[HI]^2}{[H_2][I_2]} = \frac{(0.0276)^2}{\left[0.0175 - \left(\frac{1}{2} \times 0.0276\right)\right]\left[0.0175 - \left(\frac{1}{2} \times 0.0276\right)\right]}$$

$$= \frac{(0.0276)^2}{(0.0037)(0.0037)} = 56 \qquad \text{(at 425 °C)}$$

☑ **Reasonable Result Check** The equilibrium constant is larger than 1, so there should be more products than reactants when equilibrium is reached. The equilibrium concentration of the product (0.0276 mol/L) is larger than those of the reactants (0.0037 mol/L each).

PROBLEM-SOLVING PRACTICE 12.3

Measuring the conductivity of an aqueous solution in which 0.0200 mol CH_3COOH has been dissolved in 1.00 L of solution shows that 2.96% of the acetic acid molecules have ionized to CH_3COO^- ions and H^+ ions. Calculate the equilibrium constant for ionization of acetic acid and compare your result with the value given in Table 12.1.

Saying that 2.96% of the acetic acid molecules have ionized means that at equilibrium the concentration of acetate ions is 2.96/100 = 0.0296 times the initial concentration of acetic acid molecules.

12-4 The Meaning of the Equilibrium Constant

The numerical value of the equilibrium constant tells how far a reaction has proceeded by the time equilibrium has been achieved. Experimentally determined equilibrium constants for a few reactions are given in Table 12.1. These reactions occur to widely

Table 12.1 Selected Equilibrium Constants at 25 °C

Reaction	K_c	K_P
Gas-phase reactions		
$\frac{1}{8} S_8(s) + O_2(g) \rightleftharpoons SO_2(g)$	4.2×10^{52}	4.2×10^{52}
$2\ H_2(g) + O_2(g) \rightleftharpoons 2\ H_2O(g)$	3.2×10^{81}	1.3×10^{80}
$N_2(g) + 3\ H_2(g) \rightleftharpoons 2\ NH_3(g)$	3.5×10^{8}	5.8×10^{5}
$N_2(g) + O_2(g) \rightleftharpoons 2\ NO(g)$	4.5×10^{-31}	4.5×10^{-31}
$N_2(g) + O_2(g) \rightleftharpoons 2\ NO(g)$ at 2300 K	1.7×10^{-3}	1.7×10^{-3}
$H_2(g) + I_2(g) \rightleftharpoons 2\ HI(g)$	2.5×10^{1}	2.5×10^{1}
$2\ NO_2(g) \rightleftharpoons N_2O_4(g)$	1.7×10^{2}	7.0
$CH_4(g) + H_2O(g) \rightleftharpoons CO(g) + 3\ H_2(g)$	2.0×10^{-28}	1.25×10^{-25}
cis-2-butene(g) $\rightleftharpoons$ *trans*-2-butene(g)	3.2	3.2
Weak acids		
Formic: $HCOOH(aq) \rightleftharpoons H^+(aq) + HCOO^-(aq)$	3.0×10^{-4}	—
Acetic: $CH_3COOH(aq) \rightleftharpoons H^+(aq) + CH_3COO^-(aq)$	1.8×10^{-5}	—
Carbonic: $H_2CO_3(aq) \rightleftharpoons H^+(aq) + HCO_3^-(aq)$	4.3×10^{-7}	—
Weak base		
Ammonia: $NH_3(aq) + H_2O(\ell) \rightleftharpoons NH_4^+(aq) + OH^-(aq)$	1.8×10^{-5}	—
Very slightly soluble solids		
$CaCO_3(s) \rightleftharpoons Ca^{2+}(aq) + CO_3^{2-}(aq)$	2.8×10^{-9}	—
$AgCl(s) \rightleftharpoons Ag^+(aq) + Cl^-(aq)$	1.8×10^{-10}	—
$AgI(s) \rightleftharpoons Ag^+(aq) + I^-(aq)$	8.3×10^{-17}	—

differing extents, as shown by the wide range of K_c values. Equilibrium constants can also be used to calculate how much product will be present at equilibrium. There are three important cases to consider.

Case 1. $K_c \gg 1$: *Reaction is strongly product-favored; equilibrium concentrations of products are much greater than equilibrium concentrations of reactants.*

The symbol $\gg$ means "much greater than."

A large value of K_c *means that reactants have been converted almost entirely to products when equilibrium has been achieved.* That is, the products are strongly favored over the reactants. An example is the reaction of NO(g) with $O_3(g)$, which is one means by which ozone is destroyed in the stratosphere (← **Sec. 8-10**).

$$NO(g) + O_3(g) \rightleftharpoons NO_2(g) + O_2(g) \qquad K_c = \frac{[NO_2][O_2]}{[NO][O_3]} = 6 \times 10^{34} \text{ (at 25 °C)}$$

The very large value of K_c tells us that if 1 mol NO and 1 mol O_3 are mixed in a sealed flask at 25 °C and allowed to come to equilibrium, $[NO_2][O_2] \gg [NO][O_3]$. Virtually none of the reactants will remain, and essentially only NO_2 and O_2 will be found in the flask. For practical purposes, this reaction goes to completion, and it is not necessary to use the equilibrium constant to calculate the quantities of products that would be obtained. The simpler methods developed in Chapter 3 (← **Secs. 3-6 and 3-7**) would work just fine in this case.

The reaction goes essentially to completion when equilibrium has been reached. This could take a long time if the reaction is slow.

Case 2. $K_c \ll 1$: *Reaction is strongly reactant-favored; equilibrium concentrations of reactants are greater than equilibrium concentrations of products.*

The symbol $\ll$ means "much less than."

Conversely, *an extremely small* K_c *means that when equilibrium has been achieved, very little of the reactants has been transformed into products*. The reactants are favored over the products at equilibrium.

$$3\ O_2(g) \rightleftharpoons 2\ O_3(g) \qquad K_c = \frac{[O_3]^2}{[O_2]^3} = 6.25 \times 10^{-58} \text{ (at 25 °C)}$$

This means that $[O_3]^2 \ll [O_2]^3$ and if O_2 is placed in a flask at 25 °C and the flask is sealed, *very* little O_3 will be found when equilibrium is achieved. The concentration of O_2 would remain essentially unchanged. In the terminology of Chapter 3 (← **Secs. 3-3d, 3-5d**), we would write "N.R." and say that no reaction occurs.

The symbol ≅ means "approximately equal to."

Case 3. $K_c \cong 1$***: Equilibrium mixture contains significant concentrations of reactants and products; calculations are needed to determine equilibrium concentrations.***

If K_c *is neither extremely large nor extremely small, the equilibrium constant must be used to calculate how far a reaction proceeds toward products.* In contrast with the reactions in Case 1 and Case 2, dissociation of dinitrogen tetraoxide has neither a very large nor a very small equilibrium constant. At 391 K the value is 1.00, which means that significant concentrations of both N_2O_4 and NO_2 are present at equilibrium.

Recall that the equation for dissociation of dinitrogen tetraoxide is

$$N_2O_4(g) \rightleftharpoons 2\ NO_2(g)$$

$$K_c = 1.00 = \frac{[NO_2]^2}{[N_2O_4]}, \text{ so } [NO_2]^2 = [N_2O_4] \quad \text{(at 391 K)} \qquad [12.6]$$

What range of equilibrium constants represents this middle ground depends on how small a concentration is significant and on the form of the equilibrium constant. If the concentrations of N_2O_4 and NO_2 at equilibrium are both 1.0 mol/L, then the ratio of $[NO_2]^2/[N_2O_4]$ does equal the K_c value of 1.00 at 391 K. But, what if the concentrations were much smaller? Would they still be equal? You can verify this by using Equation 12.6. If the numeric value of the equilibrium concentration of NO_2 is 0.01, then the concentration of N_2O_4 must be $(0.01)^2$, which equals 0.0001. Thus, even though $K_c = 1.00$, the concentration of one substance can be much larger than the concentration of the other. This happens because there is a squared term in the numerator of the equilibrium constant expression and a term to a different power (the first power) in the denominator. Whenever the total of the exponents in the numerator differs from the total in the denominator, it becomes very difficult to say whether the equilibrium concentrations of the products exceed those of the reactants without doing a calculation.

By contrast, if the total of the exponents is the same, then if $K_c > 1$, products predominate over reactants, and if $K_c < 1$, reactants predominate over products. Examples in which this is true include

$$cis\text{-2-butene(g)} \rightleftharpoons trans\text{-2-butene(g)} \qquad K_c = \frac{[trans]}{[cis]} = 3.2 \quad \text{(at 25 °C)}$$

and

$$2\ HI(g) \rightleftharpoons H_2(g) + I_2(g) \qquad K_c = \frac{[H_2][I_2]}{[HI]^2} = 0.040 \quad \text{(at 25 °C)}$$

For the 2-butene *cis* $\rightleftharpoons$ *trans* equilibrium, Figure 12.4 shows that when K_c is small *cis*-2-butene predominates and when K_c is large *trans*-2-butene predominates.

You might wonder whether reactant-favored systems in which small quantities of products form are important. Many are. Examples include the acids and bases listed in Table 12.1. For acetic acid, the acidic ingredient in vinegar, the reaction is

$$CH_3COOH(aq) \rightleftharpoons H^+(aq) + CH_3COO^-(aq)$$

$$K_c = \frac{[H^+][CH_3COO^-]}{[CH_3COOH]} = 1.8 \times 10^{-5} \quad \text{(at 25 °C)}$$

The value of K_c for acetic acid is small, and at equilibrium the concentrations of the products (acetate ions and hydrogen ions) are small relative to the concentration of the reactant (acetic acid molecules). This confirms that acetic acid is a weak acid.

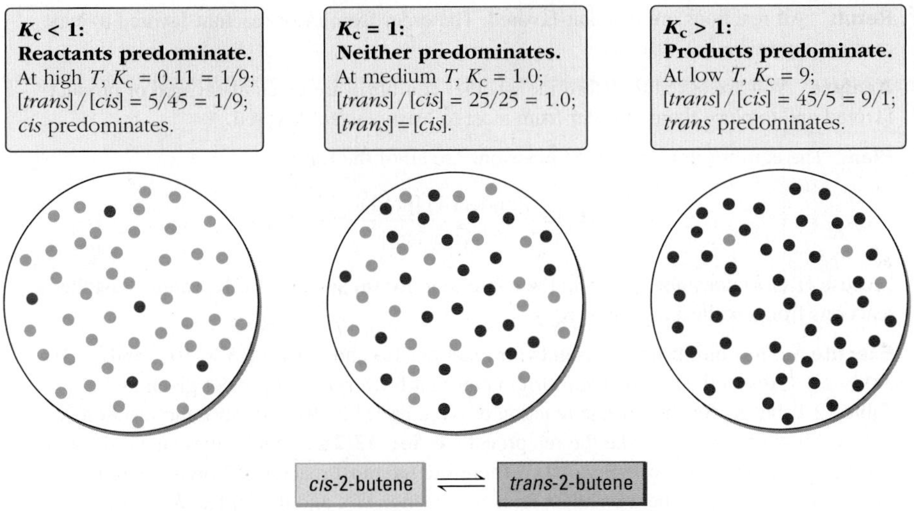
K_c < 1:
Reactants predominate.
At high T, $K_c = 0.11 = 1/9$;
[*trans*]/[*cis*] = 5/45 = 1/9;
cis predominates.
K_c = 1:
Neither predominates.
At medium T, $K_c = 1.0$;
[*trans*]/[*cis*] = 25/25 = 1.0;
[*trans*] = [*cis*].
K_c > 1:
Products predominate.
At low T, $K_c = 9$;
[*trans*]/[*cis*] = 45/5 = 9/1;
trans predominates.
cis-2-butene ⇌ *trans*-2-butene

Figure 12.4 Equilibrium constant size and concentrations of reactants and products. The smaller the equilibrium constant is, the smaller the fraction of product molecules becomes.

Nevertheless, vinegar tastes sour because a small percentage of the acetic acid molecules react with water to produce $H^+(aq)$.

If the form of the equilibrium constant is the same for two or more different reactions, then the degree to which each of those reactions is product-favored can be compared quantitatively by comparing the sizes of the K_c values. For example, the equilibrium reactions are similar and the equilibrium constant expressions all have the same form for formic acid, acetic acid, and carbonic acid, which can be represented as Hanion ($HCOOH$, CH_3COOH, or H_2CO_3):

$$\text{Hanion(aq)} \rightleftharpoons H^+(aq) + \text{anion}^-(aq) \qquad K_c = \frac{[H^+][\text{anion}^-]}{[\text{Hanion}]}$$

Anion$^-$ is $HCOO^-$, CH_3COO^-, or HCO_3^-. Data in Table 12.1 show that formic acid is stronger (has a larger K_c value) than acetic acid, and carbonic acid is the weakest of the three acids.

If a reaction has a large tendency to occur in one direction, then the reverse reaction has little tendency to occur. The equilibrium constant for the reverse of a strongly product-favored reaction is extremely small. Table 12.1 shows that combustion of hydrogen to form water vapor has an enormous equilibrium constant (3.2×10^{81}) and therefore is strongly product-favored. We say that it goes to completion.

The reverse reaction, decomposition of water to its elements,

$$2\ H_2O(g) \rightleftharpoons 2\ H_2(g) + O_2(g) \qquad K_c = \frac{[H_2]^2[O_2]}{[H_2O]^2} = 3.1 \times 10^{-82} \quad \text{(at 25 °C)}$$

is strongly reactant-favored, as indicated by the *very* small value of K_c.

PROBLEM-SOLVING EXAMPLE 12.4

Using Equilibrium Constants

Predict which reactions below are product-favored at 25 °C. List the reactions in order from most reactant-favored to least reactant-favored.

(a) $NH_3(aq) + H_2O(\ell) \rightleftharpoons NH_4^+(aq) + OH^-(aq)$

(b) $HCOOH(aq) \rightleftharpoons H^+(aq) + HCOO^-(aq)$

(c) $N_2O_4(g) \rightleftharpoons 2\ NO_2(g)$

Result All reactions are reactant-favored. The order from most reactant-favored to least reactant-favored is (a), (b), (c).

Analyze You are asked to determine whether equilibria are reactant-favored or product-favored and to place them in order from most to least reactant-favored.

Plan The equilibrium constant expressions are all of the form

$$K_c = \frac{[\text{product 1}]\,[\text{product 2}]}{[\text{reactant}]}$$

because $H_2O(\ell)$ does not appear in the expression for (a) so compare K_c values. List the reactions from smallest to largest K_c.

Execute The equilibrium constants for reactions (a) and (b) are 1.8×10^{-5} and 3.0×10^{-4}, respectively. The equilibrium constant for reaction (c) is not given in Table 12.1, but K_c for the reverse reaction is given as 1.7×10^2. Because the reaction is reversed it is necessary to take the reciprocal (← **Sec. 12-2d**), which gives an equilibrium constant for reaction (c) of 5.9×10^{-3}. Therefore the most reactant-favored reaction (smallest K_c) is (a), the next smallest K_c is for reaction (b), and the largest K_c is for reaction (c), which is the least reactant-favored.

PROBLEM-SOLVING PRACTICE 12.4

Suppose that solid AgCl and AgI are placed in 1.0 L water in separate beakers. Some, but not all, of each solid compound dissolves.

(a) In which beaker is the silver ion concentration, $[Ag^+]$, larger?

(b) Does the volume of water in which each compound dissolves affect the equilibrium concentration?

CONCEPTUAL EXERCISE 12.6

Manipulating Equilibrium Constants

The equilibrium constant is 1.8×10^{-5} for reaction of ammonia with water.

$$NH_3(aq) + H_2O(\ell) \rightleftharpoons NH_4^+(aq) + OH^-(aq)$$

(a) Is the equilibrium constant large or small for the reverse reaction, the reaction of ammonium ion with hydroxide ion to give ammonia and water?

(b) Calculate the value of K_c for the reaction of ammonium ions with hydroxide ions.

(c) What does the value of this equilibrium constant tell you about the extent to which a reaction can occur between ammonium ions and hydroxide ions?

(d) Predict what would happen if you added a 1.0-M solution of ammonium chloride to a 1.0-M solution of sodium hydroxide. What observations might allow you to test your prediction in the laboratory?

12-5 Using Equilibrium Constants

Because equilibrium constants have numeric values, they can be used to predict quantitatively in which direction a reaction will proceed and how far it will go. Such quantitative predictions involve the **reaction quotient (Q)**, which *has the same mathematical form as the equilibrium constant expression but is a ratio of actual concentrations in the mixture, instead of equilibrium concentrations.* For the general reaction

In the expression for Q, we have used (conc. N_2O_4) to represent the *actual* concentration of N_2O_4 at a given time. We use $[N_2O_4]$ to represent the *equilibrium* concentration of N_2O_4. When the reaction is at equilibrium (conc. N_2O_4) = $[N_2O_4]$.

$$a\,A + b\,B \rightleftharpoons c\,C + d\,D \qquad K_c = \frac{[C]^c[D]^d}{[A]^a[B]^b} \quad \text{and} \quad Q = \frac{(\text{conc. C})^c(\text{conc. D})^d}{(\text{conc. A})^a(\text{conc. B})^b}$$

K_c is a ratio of equilibrium concentrations, indicated by [A], [B], [C], and [D], whereas...

...Q is a ratio of actual concentrations, indicated by (conc. A), (conc. B), (conc. C), and (conc. D), with the reaction not necessarily at equilibrium.

12-5a Predicting the Direction of a Reaction

Consider a mixture of 50. mmol $NO_2(g)$ and 100. mmol $N_2O_4(g)$ at 25 °C in a sealed container with a volume of 10. L. Is the system at equilibrium? If not, in which direction will it react to achieve equilibrium? The reaction is

Recall that the metric prefix "m" means $\frac{1}{1000} = 10^{-3}$. Therefore, 1 mmol = 1×10^{-3} mol.

$$2\ NO_2(g) \rightleftharpoons N_2O_4(g) \qquad K_c = \frac{[N_2O_4]}{[NO_2]^2} = 1.7 \times 10^2 \text{ (at 25 °C)}$$

$$Q = \frac{(\text{conc. } N_2O_4)}{(\text{conc. } NO_2)^2} = \frac{(100. \times 10^{-3}\text{ mol/10. L})}{(50. \times 10^{-3}\text{ mol/10. L})^2} = \frac{1.0 \times 10^{-2}}{(5.0 \times 10^{-3})^2} = 4.0 \times 10^2$$

- *If Q is equal to* K_c, *then the reaction is at equilibrium.* The concentrations will not change.
- *If Q is less than* K_c, *then the concentrations of products are not as large as they would be at equilibrium.* The reaction will proceed from left to right to increase the product concentrations until they reach their equilibrium values.
- *If Q is greater than* K_c, *then the product concentrations are higher than they would be at equilibrium.* The reaction will proceed from right to left, increasing reactant concentrations, until equilibrium is achieved.

These relationships are shown schematically in Figure 12.5. In the case of the mixture of NO_2 and N_2O_4 just described, Q is greater than K_c, so, to establish equilibrium, some N_2O_4 will react to form NO_2. As the reverse reaction takes place, Q becomes smaller and eventually becomes equal to K_c.

PROBLEM-SOLVING EXAMPLE 12.5

Predicting Direction of Reaction

Consider this equilibrium, which is used industrially to generate hydrogen gas.

$$CH_4(g) + H_2O(g) \rightleftharpoons CO(g) + 3\ H_2(g) \qquad K_c = 3.01 \times 10^{-10} \text{ at 600 K}$$

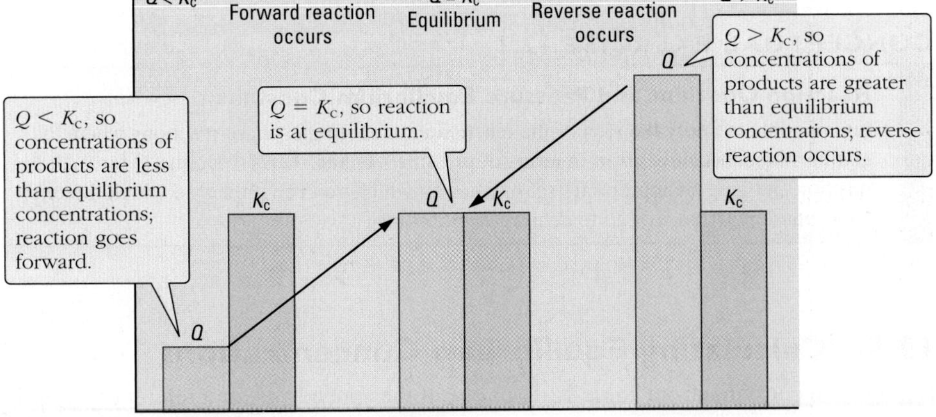

Q < Kc
Q = Kc
Q > Kc
Forward reaction occurs
Equilibrium
Reverse reaction occurs
Q < Kc, so concentrations of products are less than equilibrium concentrations; reaction goes forward.
Q = Kc, so reaction is at equilibrium.
Q > Kc, so concentrations of products are greater than equilibrium concentrations; reverse reaction occurs.
Q
Kc
Q
Kc
Q
Kc

Figure 12.5 Predicting direction of a reaction. The relative sizes of the reaction quotient, Q, and the equilibrium constant, K_c, determine in which direction a mixture of substances will react to achieve equilibrium. The reaction will proceed until $Q = K_c$.

If 1.0 mol CH_4, 1.0 mol H_2O, 2.0 mol H_2, and 0.50 mol CO gases are mixed in a sealed 10.0-L container at 25 °C, determine whether the concentration of H_2O is greater or less than 0.10 mol/L when equilibrium is reached.

Result Greater than 0.10 mol/L

Analyze Initial amounts of all substances involved in the equilibrium are given and the change in concentration of a reactant is to be determined. Q can be used to predict the direction a reaction will proceed.

Plan Calculate the initial concentration of each gas and thus evaluate Q. Then, compare Q with K_c.

Execute

$$(\text{conc. } CH_4) = \frac{1.0 \text{ mol}}{10.0 \text{ L}} = 0.10 \text{ mol/L} \qquad (\text{conc. } H_2) = \frac{2.0 \text{ mol}}{10.0 \text{ L}} = 0.20 \text{ mol/L}$$

$$(\text{conc. } H_2O) = \frac{1.0 \text{ mol}}{10.0 \text{ L}} = 0.10 \text{ mol/L} \qquad (\text{conc. CO}) = \frac{0.50 \text{ mol}}{10.0 \text{ L}} = 0.050 \text{ mol/L}$$

The units are defined by the equilibrium constant expression so leave them out.

$$Q = \frac{(\text{conc. CO})(\text{conc. } H_2)^3}{(\text{conc. } CH_4)(\text{conc. } H_2O)} = \frac{(0.050)(0.20)^3}{(0.10)(0.10)} = 0.040$$

which is much larger than 3.01×10^{-10}, the value of K_c. Because $Q > K_c$, the reverse reaction—reaction of CO with H_2 to form CH_4 and H_2O—occurs until the equilibrium concentrations are reached. The initial concentration of H_2O was 0.10 mol/L; when CO reacts with H_2, the H_2O concentration increases.

☑ **Reasonable Result Check** The concentrations are all approximately 10^{-1}, but the numerator has concentration to the fourth power and the denominator to the second power. This means that Q should be about 10^{-2}, and it is.

PROBLEM-SOLVING PRACTICE 12.5

For the equilibrium

$$2\ SO_2(g) + O_2(g) \rightleftharpoons 2\ SO_3(g) \qquad K_c = 245 \quad (\text{at } 1000 \text{ K})$$

the equilibrium concentrations are $[SO_2] = 0.102$, $[O_2] = 0.0132$, and $[SO_3] = 0.184$. The concentration of SO_2 is suddenly doubled. Show that the forward reaction takes place to reach a new equilibrium.

CONCEPTUAL EXERCISE 12.7

Reaction Quotient and Pressure Equilibrium Constant

Is it possible to apply the idea of the reaction quotient to gas-phase reactions where the equilibrium constant is given in terms of pressure (← **Sec. 12-2f**)? Define Q for such a reaction and give an appropriate set of rules by which you can predict in which direction a gas-phase reaction will go to achieve equilibrium.

12-5b Calculating Equilibrium Concentrations

Equilibrium constants from Table 12.1 (or one of the appendixes, or a reference compilation) can be used to calculate how much product is formed and how much of the reactants remains once a system has reached equilibrium. To verify our earlier statement that if an equilibrium constant is very large, essentially all of the reactants are converted to products (← **Sec. 12-4**), consider the reaction

$$\frac{1}{8}\,S_8(s) + O_2(g) \rightleftharpoons SO_2(g) \qquad K_c = 4.2 \times 10^{52} \quad \text{(at 25 °C)}$$

Suppose we place 4.0 mol O_2 and a large excess of sulfur in an empty, sealed 1.00-L flask and allow the system to reach equilibrium. We can calculate the quantity of O_2 left and the quantity of SO_2 formed at equilibrium by summarizing information in a table. Because S_8 is a solid and does not appear in the equilibrium constant expression, we do not need any entries under S_8 in the table.

	$\frac{1}{8}\,S_8(s)$	+	$O_2(g)$	$\rightleftharpoons$	$SO_2(g)$
*I*nitial concentration (mol/L)			4.0		0.0
*C*hange as reaction occurs (mol/L)			$-x$		$+x$
*E*quilibrium concentration (mol/L)			$4.0 - x$		$0.0 + x$

We know the concentrations of reactant and product before the reaction, but we do not know how many moles per liter of O_2 are consumed during the reaction, and so we designate this as x mol/L. (There is a minus sign in the table because O_2 is consumed.) Since the mole ratio is (1 mol SO_2)/(1 mol O_2), we know that x mol/L SO_2 is formed when x mol/L O_2 is consumed. To calculate the concentration of O_2, we take what was present initially, 4.0 mol/L, minus what was consumed in the reaction, x mol/L. The equilibrium concentration of SO_2 must be the initial concentration, 0 mol/L, plus what was formed by the reaction, x mol/L. Substituting into the equilibrium constant expression,

$$K_c = \frac{[SO_2]}{[O_2]} = \frac{x}{4.0 - x} = 4.2 \times 10^{52}$$

Solving algebraically for x (and following the usual rules for significant figures),

$$x = (4.2 \times 10^{52})(4.0 - x)$$

$$x = (16.8 \times 10^{52}) - (4.2 \times 10^{52})x$$

$$x + (4.2 \times 10^{52})x = 16.8 \times 10^{52}$$

Notice that $x + (4.2 \times 10^{52})x = (1 + 4.2 \times 10^{52})x$, which to a very good approximation is equal to $(4.2 \times 10^{52})x$. Thus,

Because 4.2×10^{52} is so much larger than 1, adding 1 to it makes no appreciable change in the very large number.

$$x = \frac{16.8 \times 10^{52}}{4.2 \times 10^{52}} = 4.0$$

The equilibrium concentration of SO_2 is $x = 4.0$ mol/L and that of O_2 is $(4.0 - x)$ mol/L, or 0 mol/L. That is, within the precision of our calculation, all the O_2 has been converted to SO_2. As stated earlier, a very large K_c value (4.2×10^{52} in this case) implies that essentially all of the reactants have been converted to products. The reaction is strongly product-favored and goes to completion, so the calculation could have been done using the methods in Sections 3-6 and 3-7.

The fact that samples of sulfur at 25 °C can be exposed to oxygen in the air for long periods without being converted to sulfur dioxide shows the importance of chemical kinetics. This reaction is very slow at room temperature, so only a faint odor of sulfur dioxide is noticeable in the vicinity of solid sulfur.

PROBLEM-SOLVING EXAMPLE 12.6

Calculating Equilibrium Concentrations

One way to generate hydrogen gas is to react carbon monoxide with steam.

$$CO(g) + H_2O(g) \rightleftharpoons CO_2(g) + H_2(g) \qquad K_c = 10.\ \text{at } 420\ °C$$

Suppose 2.5×10^{-3} mol/L CO and 2.5×10^{-3} mol/L H_2O are placed into a container. The container is sealed and heated to 420 °C. Determine the concentration of each of the four substances when equilibrium is reached.

Result [CO] = $[H_2O] = 6.0 \times 10^{-4}$ mol/L; $[CO_2] = [H_2] = 1.9 \times 10^{-3}$ mol/L

Analyze Concentrations of reactants are given and equilibrium concentrations are to be determined. A balanced equation and the corresponding equilibrium constant are given. No products are present initially so the reaction must go from left to right. An ICE table can be used to keep track of changing concentrations.

Plan Follow the procedure given earlier (← **Sec. 12-3a**).

Execute The chemical equation written in the statement of the problem shows that all of the mole ratios are 1 : 1. This tells us that equal amounts of CO and H_2O are consumed as the reaction proceeds to equilibrium. Because both substances are in the same flask, equal numbers of moles *per liter* (equal concentrations) of reactants must also be consumed. Designate each of these as $-x$. Mole ratios also tell us that if the concentration of CO decreases by x mol/L, then the concentration of H_2 (and the concentration of CO_2) must increase by the same quantity, x mol/L. Because their initial concentrations were the same and their mole ratio is 1 : 1, the equilibrium concentrations of the products H_2O and CO must be the same. Each is equal to the initial concentration plus the change (the change for reactants is $-x$, so this means subtracting x) as the reactants are consumed. The reaction table is

	CO	+	H_2O	$\rightleftharpoons$	CO_2	+ H_2
*I*nitial concentration (mol/L)	2.5×10^{-3}		2.5×10^{-3}		0.000	0.000
*C*hange as reaction occurs (mol/L)	$-x$		$-x$		$+x$	$+x$
*E*quilibrium concentration (mol/L)	$0.0025 - x$		$0.0025 - x$		x	x

Substituting these values into the expression for K_c, we have

$$K_c = 10.0 = \frac{[H_2][CO_2]}{[H_2O][CO]} = \frac{(x)(x)}{(0.0025 - x)(0.0025 - x)} = \frac{x^2}{(0.0025 - x)^2}$$

Solving this equation for x is not as difficult as it might seem at first glance. Because the right-hand side is a perfect square, we can take the square root of both sides, giving

$$\sqrt{K_c} = \sqrt{10.0} = 3.16 = \frac{x}{(0.0025 - x)}$$

and then solve for x.

$$x = (3.16)(0.0025 - x) = 0.0079 - 3.16x$$

$$4.16\,x = 0.0079; \qquad x = 1.9 \times 10^{-3}$$

The concentrations of the products are both equal to x mol/L—that is, to 1.9×10^{-3} mol/L—while the concentrations of the reactants that remain are both $(0.0025 - x)$ mol/L = $(0.0025 - 0.0019)$ mol/L = 6.0×10^{-4} mol/L. This result demonstrates quantitatively that when $K_c > 1$, a reaction is product-favored: at equilibrium the product concentrations are larger than the reactant concentrations.

☑ **Reasonable Result Check** When the calculated equilibrium concentrations are substituted into the equilibrium constant expression, the calculated K_c is the same as the value given in the statement of the problem.

$$K_c = \frac{[H_2][CO_2]}{[H_2O][CO]} = \frac{(1.9 \times 10^{-3})^2}{(6.0 \times 10^{-4})^2} = 10.$$

PROBLEM-SOLVING PRACTICE 12.6

The equilibrium constant for dissolving the insoluble substance gold(I) iodide, AuI(s), in aqueous solution is 1.6×10^{-23} at 25 °C. Write the equilibrium constant expression and calculate the concentration of $Au^+(aq)$ and $I^-(aq)$ ions in a solution in which 0.345 g AuI(s) is in equilibrium with the aqueous ions.

PROBLEM-SOLVING EXAMPLE 12.7

Q and Equilibrium Calculations

When colorless hydrogen iodide gas is heated to 745 K, a beautiful violet color appears. This shows that some iodine gas has been formed, which implies that some HI has been decomposed to its elements.

$$2\ HI(g) \rightleftharpoons H_2(g) + I_2(g) \qquad K_c = 0.0200 \quad \text{(at 745 K)}$$

Suppose that a mixture of 1.00 mol HI(g), 0.500 mol I_2(g), and 1.00 mol H_2(g) is sealed into a 10.0-L flask and heated to 745 K. Calculate the concentrations of all three substances when equilibrium has been achieved.

Result [HI] = 0.179 M, [I_2] = 0.0106 M, [H_2] = 0.0606 M

Analyze A balanced equation, initial amounts of all substances in the equilibrium, the volume of the container, and the value of K_c are given. Equilibrium concentrations are to be determined.

Plan Calculate initial concentrations of all substances from amounts and container volume. Calculate Q and compare with K_c to determine the direction of reaction. Use a reaction table and the K_c value to calculate equilibrium concentrations.

Execute (conc. HI) = (conc. H_2) = 1.00 mol/10.0 L = 0.100 mol/L.

(conc. I_2) = 0.500 mol/10.0 L = 0.0500 mol/L.

$$Q = \frac{(\text{conc. } H_2)(\text{conc. } I_2)}{(\text{conc. HI})^2} = \frac{(0.100\text{ M})(0.0500\text{ M})}{(0.100\text{ M})^2} = 0.500$$

Because $Q > K_c$, the concentrations of the products are larger than the equilibrium concentrations; some H_2 and I_2 will react to form HI. Therefore, it makes sense to let $-x$ mol/L be the change in concentration of H_2 and I_2 as the system reacts to reach equilibrium. (Note that the coefficients of H_2 and I_2 are the same so the change in concentration of I_2 is the same as the change in concentration of H_2.) Because the coefficient of HI is twice the coefficient of I_2, twice as many moles of HI must be formed as moles of I_2 that react; the change in concentration of HI is $2x$ mol/L. The reaction table is

	2 HI(g)	$\rightleftharpoons$	H_2(g)	+	I_2(g)
Initial concentration (mol/L)	0.100		0.100		0.0500
Change as reaction occurs (mol/L)	$2x$		$-x$		$-x$
Equilibrium concentration (mol/L)	$(0.100 + 2x)$		$(0.100 - x)$		$(0.0500 - x)$

Substitute the last line of the table into the equilibrium constant expression.

$$K_c = \frac{[H_2][I_2]}{[HI]^2} = \frac{(0.100 - x)(0.0500 - x)}{(0.100 + 2x)^2} = 0.0200$$

The ratio of terms involving x is not a perfect square so you cannot take a square root as was done in Problem-Solving Example 12.6. Multiply out the numerator and denominator to obtain

$$\frac{0.00500 - 0.150x + x^2}{0.0100 + 0.400x + 4x^2} = 0.0200$$

Multiply both sides by the denominator and then multiply out the terms.

$$0.00500 - 0.150x + x^2 = 0.0200 \times (0.0100 + 0.400x + 4x^2)$$
$$0.00500 - 0.150x + x^2 = 0.000200 + 0.00800x + 0.0800x^2$$

Collecting terms in x^2 and x, we have

$$0.9200x^2 - 0.1580x + 0.00480 = 0$$

This is a quadratic equation of the form $ax^2 + bx + c = 0$, where $a = 0.9200$, $b = -0.1580$, and $c = 0.004800$. The equation can be solved using the quadratic formula (Appendix A-7).

Richard Ramette

Iodine vapor has a beautiful violet color.

In solving this problem we could have chosen *x* to be the change in concentration of HI, in which case the changes in concentrations of H_2 and I_2 would each have been $-\frac{1}{2}x$.

$$x = \frac{-b \pm \sqrt{b^2 - 4ac}}{2a} = \frac{0.1580 \pm \sqrt{0.1580^2 - 4 \times 0.9200 \times 0.004800}}{2 \times 0.9200}$$

$$x = \frac{0.1580 - 0.08544}{1.840} = 0.03943$$

The other root, $x = 0.1323$, can be eliminated because, when subtracted from 0.0500, it gives a negative concentration of I_2 at equilibrium; this is clearly impossible. From the equilibrium concentration row of the table,

$$[HI] = (0.100 + 2x)\ \text{mol/L} = (0.100 + 2 \times 0.03943)\ \text{mol/L} = 0.179\ \text{mol/L}$$

$$[H_2] = (0.100 - x)\ \text{mol/L} = (0.100 - 0.03943)\ \text{mol/L} = 0.0606\ \text{mol/L}$$

$$[I_2] = (0.0500 - x)\ \text{mol/L} = (0.0500 - 0.03943)\ \text{mol/L} = 0.0106\ \text{mol/L}$$

☑ **Reasonable Result Check** Check the result by substituting the equilibrium concentrations into the equilibrium constant expression and verifying that the correct value of K_c (0.0200) results, which it does.

$$K_c = \frac{[H_2][I_2]}{[HI]^2} = \frac{(0.0606)(0.0106)}{(0.179)^2} = 0.0200$$

PROBLEM-SOLVING PRACTICE 12.7

Obtain the equilibrium constant for dissociation of dinitrogen tetraoxide to form nitrogen dioxide from Table 12.1. If 1.00 mol N_2O_4 and 0.500 mol NO_2 gases are initially placed in a container with a volume of 4.00 L and the container is sealed, calculate the concentrations of $N_2O_4(g)$ and $NO_2(g)$ present when equilibrium is achieved at 25 °C.

12-6 Shifting a Chemical Equilibrium: Le Chatelier's Principle

Le Chatelier is pronounced "luh SHOT lee ay."

Suppose you are an environmental engineer, biologist, or geologist, and you have just measured the concentration of hydrogen ion, H^+, in a lake. You know that the H^+ ions are involved in many different equilibrium reactions in the lake. How can you predict the influence of changing conditions? For example, what happens if there is a large increase in acid rainfall that has a hydrogen ion concentration different from that of the lake? Or, what happens if lime (calcium oxide), a strong base, is added to the lake? These questions and many others like them can be answered qualitatively by applying a useful guideline known as **Le Chatelier's principle:** ***If a system is at equilibrium and the conditions are changed so that it is no longer at equilibrium, the system will react to reach a new equilibrium in a way that partially counteracts the change.*** To adjust to a change, a system reacts in either the forward direction (producing more products) or the reverse direction (producing more reactants) until a new equilibrium state is achieved. Le Chatelier's principle applies to changes in the concentrations of reactants or products that appear in the equilibrium constant expression, the pressure or volume of a gas-phase equilibrium, and the temperature. *Changing conditions, thereby changing the equilibrium concentrations of reactants and products,* is called **shifting an equilibrium.** If the reaction occurs in the *forward direction,* we say that the equilibrium reaction has *shifted to the right.* If the system reacts in the *reverse direction,* the reaction has *shifted to the left.*

Henri Le Chatelier (1850–1936) was a French chemist who, as a result of his studies of the chemistry of cement, developed his ideas about how altering conditions affects an equilibrium system.

12-6a Changing Concentrations of Reactants or Products

If the concentration of a reactant or a product that appears in the equilibrium constant expression is changed, a system can no longer be at equilibrium because Q must have a different value from K_c. The equilibrium shifts to use up a substance that was added, or

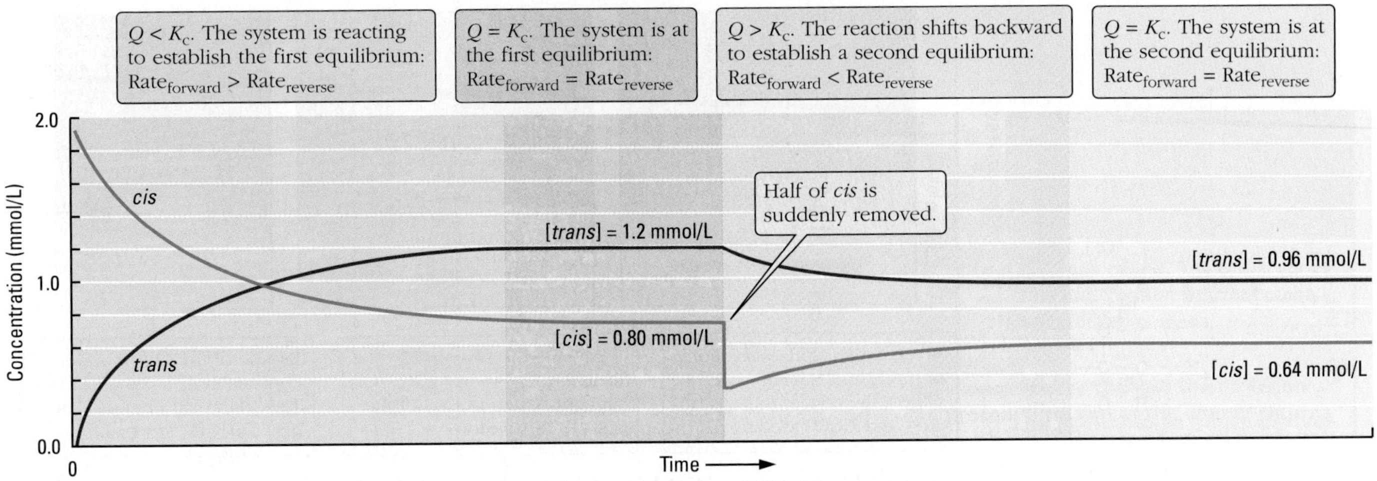
$Q < K_c$. The system is reacting to establish the first equilibrium: $Rate_{forward} > Rate_{reverse}$
$Q = K_c$. The system is at the first equilibrium: $Rate_{forward} = Rate_{reverse}$
$Q > K_c$. The reaction shifts backward to establish a second equilibrium: $Rate_{forward} < Rate_{reverse}$
$Q = K_c$. The system is at the second equilibrium: $Rate_{forward} = Rate_{reverse}$
2.0
1.0
0.0
Concentration (mmol/L)
cis
trans
[trans] = 1.2 mmol/L
[cis] = 0.80 mmol/L
Half of cis is suddenly removed.
[trans] = 0.96 mmol/L
[cis] = 0.64 mmol/L
0
Time

Figure 12.6 Approach to new equilibrium after a change in conditions. The reaction is *cis*-2-butene $\rightleftharpoons$ *trans*-2-butene. $K_c = [trans]/[cis]$ has the same value, 1.5, at all times at this temperature.

to replenish a substance that was removed. The shift will occur in whichever direction causes the value of Q to approach the value of K_c.

- If the concentration of a *reactant* is *increased,* the system will react in the *forward* direction.
- If the concentration of a *reactant* is *decreased,* the system will react in the *reverse* direction.
- If the concentration of a *product* is *increased,* the system will react in the *reverse* direction.
- If the concentration of a *product* is *decreased,* the system will react in the *forward* direction.

If temperature remains the same, the value of K_c also remains the same. Adding or removing a reactant or a product does not change K_c. If the concentration of the substance added or removed appears in the equilibrium constant expression, however, the value of Q changes. Because Q no longer equals K_c, the system is no longer at equilibrium and must react to achieve a new equilibrium.

To see why this happens, consider the simple equilibrium discussed in Section 12-2.

$$cis\text{-2-butene(g)} \rightleftharpoons trans\text{-2-butene(g)} \qquad K_c = \frac{[trans]}{[cis]} = 1.5 \quad (\text{at } 600\text{ K})$$

Suppose that 2 mmol *cis*-2-butene gas is placed into a 1.0-L container at 1000 K and the container is sealed. The forward reaction is faster than the reverse reaction until the concentration of *trans*-2-butene builds up to 1.5 times the concentration of *cis*-2-butene. Then, equilibrium is achieved. Now, suppose that half of the *cis*-2-butene is instantaneously removed from the container (see Exercise 12.3, ← **Sec. 12-2**). Because the concentration of *cis*-2-butene suddenly drops to half its former value, the forward reaction rate also drops to half its former value. The reverse rate is not affected, because the concentration of *trans*-2-butene has not changed. This means that the reverse reaction is twice as fast as the forward reaction, and *cis*-2-butene molecules are being formed twice as fast as they are reacting away. Therefore, the concentration of *cis*-2-butene increases and the concentration of *trans*-2-butene decreases until the forward and reverse rates are again equal. The graph in Figure 12.6 illustrates this situation.

The effect of changing concentration has many important consequences. For example, when the concentration of a substance needed by your body falls slightly, several enzyme-catalyzed chemical equilibria shift so as to increase the concentration of the essential substance. In industrial processes, reaction products are often continuously removed. This shifts one or more equilibria to produce more products and thereby maximize the yield of the reaction.

In nature, slight changes in conditions are responsible for effects such as the formation of limestone stalactites and stalagmites in caves and dissolving of coral reefs. Both of

Photos: Charles D. Winters
When clear, colorless solutions of $CaCl_2$ and $NaHCO_3$ are mixed, the concentrations of Ca^{2+} and HCO_3^- are large…
…causing the reverse of Reaction 12.7 to occur. $CaCO_3$ precipitates and CO_2 bubbles out of the solution.
Escape of the CO_2 reduces its concentration, shifting Reaction 12.7 to the left until the $CaCl_2$ and $NaHCO_3$ are gone.

Figure 12.7 Reaction of $CaCl_2(aq)$ with $NaHCO_3(aq)$.

these examples involve calcium carbonate, $CaCO_3$. Limestone, a form of calcium carbonate, is present in underground deposits, a leftover of the ancient oceans from which it precipitated long ago. Limestone dissolves when it reacts with an aqueous solution of CO_2.

$$CaCO_3(s) + CO_2(aq) + H_2O(\ell) \rightleftharpoons Ca^{2+}(aq) + 2\ HCO_3^-(aq) \qquad [12.7]$$

If groundwater that is saturated with CO_2 encounters a bed of limestone below Earth's surface, the forward reaction can occur until equilibrium is reached, and the water subsequently contains significant concentrations of aqueous Ca^{2+} and HCO_3^- ions in addition to dissolved CO_2.

Similarly, as the concentration of CO_2 in Earth's atmosphere increases (← **Sec. 8-11a**), more CO_2 dissolves in Earth's oceans, causing Reaction 12.7 to shift to the right. This causes coral reefs, which consist of limestone, to dissolve, robbing corals of their physical support and greatly changing coral ecosystems.

As with all equilibria, a reverse reaction is occurring in addition to the forward reaction. This reverse reaction can be demonstrated by mixing aqueous solutions of $CaCl_2$ and $NaHCO_3$ (salts containing the Ca^{2+} and HCO_3^- ions) in an open beaker (Figure 12.7). Bubbles of CO_2 gas and a precipitate of solid $CaCO_3$ form. Gaseous CO_2 escapes into the air, reducing the concentration of $CO_2(aq)$ and causing the equilibrium to shift to the left. Eventually, all of the dissolved Ca^{2+} and HCO_3^- ions are converted to gaseous CO_2, solid $CaCO_3$, and water.

Suppose that water containing dissolved CO_2, Ca^{2+}, and HCO_3^- contacts the air in a cave. Carbon dioxide bubbles out of the solution, the concentration of CO_2 decreases on the reactant side of Equation 12.7, and the equilibrium shifts toward the reactants. The reverse reaction forms $CO_2(aq)$, compensating partially for the reduced concentration of $CO_2(aq)$. Some of the calcium ions and hydrogen carbonate ions combine, and some $CaCO_3(s)$ precipitates (Figure 12.8) as a beautiful formation in the cave.

VLADJ55/Shutterstock.com

Figure 12.8 A limestone cave. Stalactites hanging from the ceiling and stalagmites growing from the floor both consist of limestone, $CaCO_3$.

CONCEPTUAL EXERCISE 12.8

Effect of Adding a Substance

Solid phosphorus pentachloride decomposes when heated to form gaseous phosphorus trichloride and gaseous chlorine. Write the equation for the equilibrium that is set up when solid phosphorus pentachloride is introduced into a container, the container is evacuated and sealed, and the solid is heated. Once the system has reached equilibrium at a given temperature, what is the effect on the equilibrium of each of these changes?

(a) Add chlorine to the container.
(b) Add phosphorus trichloride to the container.
(c) Add a small quantity of solid phosphorus pentachloride to the container.

EXERCISE 12.9

K Is a Constant

Use concentrations read from the graph in Figure 12.6 to show that K_c has the same value in the equilibrium regions both before and after half of the *cis*-2-butene is removed from the first equilibrium mixture.

12-6b Changing Volume or Pressure in Gaseous Equilibria

One way to change the pressure of a gaseous equilibrium mixture is to keep the volume constant and to add or remove one or more of the substances whose concentrations appear in the equilibrium constant expression. The effect of adding or removing a substance has just been discussed. We consider here other ways of changing pressure or volume.

The pressures of all substances in a gaseous equilibrium can be changed by changing the volume of the container. Consider the effect of tripling the pressure on the equilibrium

$$N_2O_4(g) \rightleftharpoons 2\ NO_2(g) \qquad K_c = \frac{[NO_2]^2}{[N_2O_4]}$$

by reducing the volume of the container to one third of its original value (at constant temperature). This situation is shown in Figure 12.9. Decreasing the volume increases the pressures of N_2O_4 and NO_2 to three times their equilibrium values. Decreasing the volume also increases the concentrations of N_2O_4 and NO_2 to three times their equilibrium values. Because $[NO_2]$ is squared in the equilibrium constant expression but $[N_2O_4]$ is not, tripling both concentrations increases the numerator of Q by $3^2 = 9$, but increases the denominator by only 3.

$$Q = \frac{(\text{conc. } NO_2)^2}{(\text{conc. } N_2O_4)} = \frac{(3 \times [NO_2])^2}{(3 \times [N_2O_4])} = \frac{9}{3} \times \frac{[NO_2]^2}{[N_2O_4]} = 3 \times K_c$$

Remember that the pressure of an ideal gas is proportional to the number of moles of gas, and therefore to the number of molecules of gas per unit volume (← Sec. 8-4).

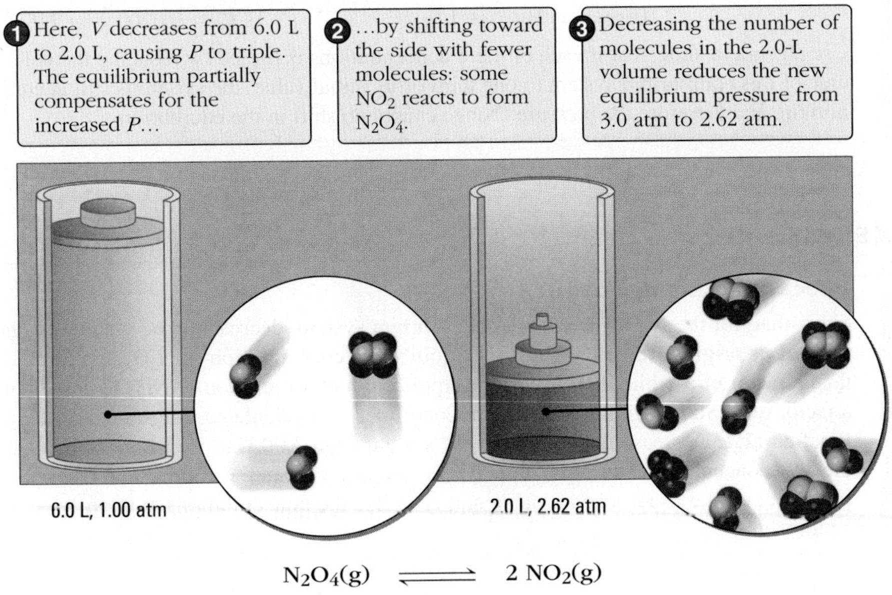
1 Here, V decreases from 6.0 L to 2.0 L, causing P to triple. The equilibrium partially compensates for the increased P...
2 ...by shifting toward the side with fewer molecules: some NO2 reacts to form N2O4.
3 Decreasing the number of molecules in the 2.0-L volume reduces the new equilibrium pressure from 3.0 atm to 2.62 atm.
6.0 L, 1.00 atm
2.0 L, 2.62 atm
N2O4(g) ⇌ 2 NO2(g)

Figure 12.9 Shifting an equilibrium by changing volume (and pressure).

Because Q is larger than K_c under the new conditions, the reaction should produce more reactant; that is, the equilibrium should shift to the left.

The same prediction is made using Le Chatelier's principle: The reaction should shift to partially compensate for the increase in pressure. That means decreasing the pressure, which can happen if the total number of gas-phase molecules decreases. In the case of the $N_2O_4 \rightleftharpoons 2\ NO_2$ equilibrium, the reverse reaction should occur, because one N_2O_4 molecule is produced for every two NO_2 molecules that react. A shift to the left reduces the number of gas-phase molecules and hence the pressure.

However, consider the situation with respect to another equilibrium we have already mentioned. At 425 °C all substances are in the gas phase.

$$2\ HI(g) \rightleftharpoons H_2(g) + I_2(g)$$

Suppose that the pressure of this system were tripled by reducing its volume to one third of the original volume. What would happen to the equilibrium? In this case, all the concentrations triple, but because equal numbers of moles of gaseous substances appear on both sides of the equation, Q still has the same numeric value as K_c. That is, the system is still at equilibrium, and no shift occurs. Thus, *changing pressure by changing the volume shifts an equilibrium only if the sum of the coefficients for gas-phase reactants is different from the sum of the coefficients for gas-phase products.*

An inert gas in this context is one that does not react with the other gases already present.

Finally, consider what happens if the pressure of the $N_2O_4 \rightleftharpoons 2\ NO_2$ equilibrium system is increased by adding an inert gas such as argon while retaining exactly the same volume. The total pressure of the system increases, but, because neither the amounts of N_2O_4 and NO_2 nor the volume change, the concentrations of N_2O_4 and NO_2 remain the same. Q still equals K_c, the system is still at equilibrium, and no shift occurs (or needs to). *Changing the pressure of an equilibrium system must change the concentration of at least one substance in the equilibrium constant expression if a shift in the equilibrium is to occur.*

CONCEPTUAL EXERCISE 12.10

Changing Volume Does Not Always Shift an Equilibrium

In Problem-Solving Example 12.7 we found that the equilibrium concentrations were [HI] = 0.179 M, $[I_2]$ = 0.0106 M, $[H_2]$ = 0.0606 M for this reaction:

$$2\ HI(g) \rightleftharpoons H_2(g) + I_2(g) \qquad K_c = 0.0200 \quad \text{(at 745 K)}$$

Use algebra to show that if each of these concentrations is tripled by reducing the volume of this equilibrium system to one third of its initial value, the system is still at equilibrium, and therefore the pressure change causes no shift in the equilibrium.

EXERCISE 12.11

Effect of Changing Volume

Show that, for the $N_2O_4 \rightleftharpoons 2\ NO_2$ equilibrium system, decreasing the volume to one third of its original value increases the equilibrium concentration of N_2O_4 by more than a factor of 3 while it increases the equilibrium concentration of NO_2 by less than a factor of 3. Start with the equilibrium conditions you calculated in Problem-Solving Practice 12.7. Decrease the volume of the system from 4.00 L to 1.33 L and calculate the new concentrations of N_2O_4 and NO_2 before any shift in the equilibrium takes place. Set up the usual reaction table and calculate the concentrations at the new equilibrium.

12-6c Changing Volume by Adding Solvent

Just as the concentrations (and pressures) of gas-phase reactants and products can be changed by changing the volume of a closed system, the concentrations of reactants and products in a solution can be changed by either adding solvent or removing solvent by evaporating it. For a reaction in which there is a change in the total number of solute particles, such a volume change will cause a shift in the equilibrium. For example, consider the reaction

$$NH_3(aq) + H_2O(\ell) \rightleftharpoons NH_4^+(aq) + OH^-(aq) \qquad K_c = \frac{[NH_4^+][OH^-]}{[NH_3]} = 1.8 \times 10^{-5}$$

Suppose that the volume of the solution is doubled by adding water. This reduces the concentrations of all solutes (NH_4^+, OH^-, and NH_3) to one half their previous values. For example, suppose that $[NH_4^+] = [OH^-] = 1.34 \times 10^{-3}$ M and $[NH_3] = 0.10$ M before the water was added. After the volume is doubled, (conc. NH_4^+) = (conc. OH^-) = 0.67×10^{-3} M and (conc. NH_3) = 0.050 M. This gives

$$Q = \frac{(0.67 \times 10^{-3})(0.67 \times 10^{-3})}{5.0 \times 10^{-2}} = 9.0 \times 10^{-6}$$

Thus, $Q < K_c$ immediately after the solution volume is doubled. Some reactant NH_3 must be converted to product NH_4^+ and OH^- to reach a new equilibrium, and the reaction shifts to the right. $Q < K_c$ because there are two concentrations in the numerator of K_c and only one in the denominator; that is, the halved concentrations are squared in the numerator, but only raised to the first power in the denominator. As in the case of the gas-phase equilibria discussed in the preceding section, if there had been equal numbers of solute particles on each side of the chemical equation, then after the solution volume doubled Q would still equal K_c, the system would still be at equilibrium, and no shift would occur. An example of a reaction in solution for which there would be no shift upon volume change is

$$Cu^{2+}(aq) + Zn(s) \rightleftharpoons Cu(s) + Zn^{2+}(aq)$$

For this reaction, there is one solute particle on each side of the equation so diluting the solution has no effect on Q; therefore the equilibrium does not shift.

12-6d Changing Temperature

When temperature changes, the values of most equilibrium constants also change, and systems will react to achieve new equilibria consistent with new values of K_c. You can make a qualitative prediction about the effect of temperature on an equilibrium if you know whether the reaction is exothermic or endothermic. As an example, consider the endothermic, gas-phase reaction of N_2 with O_2 to give nitrogen monoxide, NO.

For this and other endothermic reactions, the equilibrium constant increases as the temperature increases.

$$N_2(g) + O_2(g) \rightleftharpoons 2\,NO(g) \qquad \Delta_r H^\circ = 180.5 \text{ kJ/mol}$$

$$K_c = \frac{[NO]^2}{[N_2][O_2]}$$

$K_c = 4.5 \times 10^{-31}$ at 298 K

$K_c = 6.7 \times 10^{-10}$ at 900 K

$K_c = 1.7 \times 10^{-3}$ at 2300 K

In this case the equilibrium constant increases very significantly as the temperature increases. At 298 K the equilibrium constant is so small that essentially no reaction occurs. Suppose that a room-temperature equilibrium mixture were suddenly heated to 2300 K. What would happen? The equilibrium should shift to partially compensate for the temperature increase. Partial compensation occurs if the shift is in the endothermic direction, because that change would involve a transfer of energy into the reaction system, cooling the surroundings. For the $N_2 + O_2$ reaction the forward process is endothermic ($\Delta_r H° > 0$), so at the higher temperature N_2 and O_2 should react to produce more NO. At the new equilibrium, the concentration of NO should be higher and the concentrations of N_2 and O_2 should be lower. Since this makes the numerator in the K_c expression bigger and the denominator smaller, K_c should be larger at the higher temperature. This outcome corresponds with the experimental result.

The effect on K_c of raising the temperature of the $N_2 + O_2$ reaction leads us to a general conclusion:

- ***For an endothermic reaction, an increase in temperature always means an increase in*** $\mathbf{K_c}$***; an endothermic reaction becomes more product-favored at higher temperatures.***

When equilibrium is achieved at the higher temperature, the concentration of products is greater and that of the reactants is smaller. Likewise, and as illustrated by Problem-Solving Example 12.8, the opposite is true for an exothermic reaction:

- ***For an exothermic reaction, an increase in temperature always means a decrease in*** $\mathbf{K_c}$***; an exothermic reaction becomes less product-favored at higher temperatures.***

The effect of temperature on the reaction of N_2 with O_2 has important consequences. This reaction produces NO in Earth's atmosphere when lightning suddenly raises the temperature of the air along its path. Because the reverse reaction is slow at room temperature and because after the lightning bolt is over, the air rapidly cools to normal temperatures, much of the NO that is produced does not react to re-form N_2 and O_2 as it would at equilibrium. This situation provides one natural mechanism by which nitrogen in the air can be converted into a form that can be used by plants. (Converting nitrogen into a useful form is called nitrogen fixation; see Sections 19-3a and 19-6e). Humans have tried to use the same kind of process to produce NO and from it HNO_3 for use in manufacturing fertilizer. About a century ago scientists and many in the general public were worried that Earth's farmland could not grow enough food to support a growing population. Consequently, strenuous efforts were made to adjust the conditions of the $N_2 + O_2$ reaction so that significant yields of NO could be obtained. For several years at the end of the nineteenth century, a chemical plant at Niagara Falls, New York (where there was plentiful electric power), operated an electric arc process for fixing nitrogen. This electric arc plant was important because it was the first attempt to deal with the limitations on plant growth caused by lack of sufficient nitrogen in soils that had been heavily farmed.

Another consequence of the shift toward products in the $N_2 + O_2$ reaction at high temperatures is that automobile engines emit small concentrations of NO. The NO is rapidly oxidized to brown NO_2 in the air above cities, and the NO_2 in turn produces many further reactions that create air pollution problems. Air pollution involving NO was discussed in Section 8-12b. One of the functions of catalytic converters in automobiles is to speed up reduction of these nitrogen oxides back to elemental nitrogen (← Sec. 11-10b).

PROBLEM-SOLVING EXAMPLE 12.8

Le Chatelier's Principle

Consider an equilibrium mixture of nitrogen, hydrogen, and ammonia in which the reaction is

$$N_2(g) + 3\,H_2(g) \rightleftharpoons 2\,NH_3(g) \qquad \Delta_r H° = -92.2 \text{ kJ/mol} \quad \text{at } 25\ °C$$

For each change, (a) to (c), tell whether the value of K_c increases or decreases, and tell whether more NH_3 or less NH_3 is present at the new equilibrium established after the change.

(a) More H_2 is added (at a constant temperature of 25 °C and constant volume).
(b) The temperature is increased.
(c) The volume of the container is doubled (at constant temperature).

Result

(a) K_c stays the same; more NH_3 is present.
(b) K_c decreases; less NH_3 is present.
(c) K_c stays the same; less NH_3 is present.

Analyze An equilibrium equation and its associated $\Delta_r H°$ are given. The effect on the equilibrium of changes in conditions is to be predicted. Le Chatelier's principle is appropriate to make such predictions.

Plan Apply Le Chatelier's principle to each change.

Execute

(a) The temperature does not change so the value of K_c does not change. Adding H_2, a reactant, shifts the equilibrium toward the product, producing more NH_3. This can be seen in another way by considering the reaction quotient.

$$Q = \frac{(\text{conc. } NH_3)^2}{(\text{conc. } N_2)(\text{conc. } H_2)^3}$$

When more H_2 is added, the denominator becomes larger. This makes Q smaller than K_c, which predicts that the reaction should produce more product; that is, some of the added H_2 reacts with N_2 to make more NH_3. Notice that the concentration of N_2 decreases, because some reacts with the H_2, but the concentration of H_2 in the new equilibrium is greater than it was initially.

ESTIMATION | Generating Gaseous Fuel

The reaction of coke (mainly carbon) with steam, the water-gas reaction, was used for many years to generate gaseous fuel from coal. The thermochemical expression is

$$C(s) + H_2O(g) \rightleftharpoons CO(g) + H_2(g) \qquad \Delta_r H° = 131.293 \text{ kJ/mol}$$

The equilibrium constant K_P has the value 9.5×10^{-17} at 298 K, 1.9×10^{-7} at 500 K, 2.60×10^{-2} at 800 K, 18.8 at 1200 K, and 500. at 1600 K. Suppose that you have an equilibrium mixture in which the partial pressure of steam is 1.00 atm. Estimate the temperature at which the partial pressures of CO and H_2 also equal 1 atm, giving $K_P = 1$. (The reaction needs to be carried out at temperatures roughly this high to produce appreciable quantities of products.)

Graph A

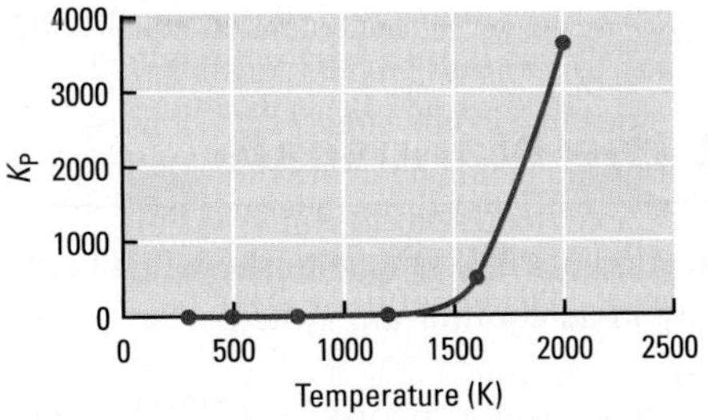
4000
3000
2000
1000
0
K_P
0
500
1000
1500
2000
2500
Temperature (K)

A good way to make this estimation is to use a spreadsheet program to plot the data and see which temperature corresponds to $K_P = 1$. This has been done in Graph A. It is clear that the range of values for the equilibrium constant is so wide that an accurate temperature cannot be read corresponding to $K_P = 1$. An estimate would be easier to make if the graph were a straight line. Because the values of K_P rise very rapidly, it is possible that taking the logarithm of each K_P value would generate a linear graph. This has been done in Graph B, where the natural logarithm of the equilibrium constant, $\ln(K_P)$, has been plotted on the vertical axis. Since $\ln(K_P) = 0$ when $K_P = 1$, we are looking for the temperature at which the graph crosses zero on the vertical axis. Although the graph is not linear, the appropriate temperature can be read from the graph and is slightly less than 1000 K.

Graph B

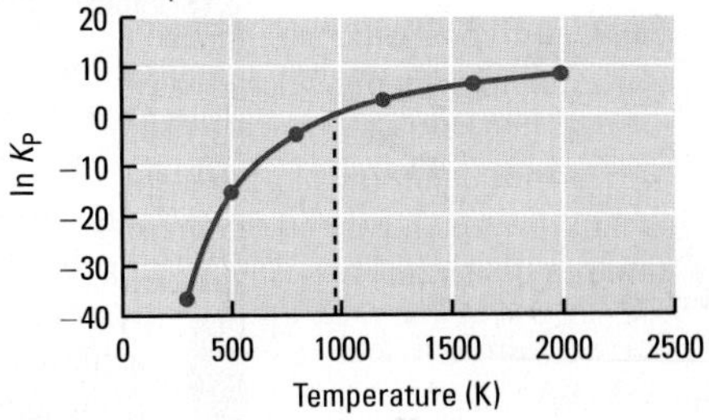
20
10
0
−10
−20
−30
−40
ln K_P
0
500
1000
1500
2000
2500
Temperature (K)

Experimenting with different functions of K_P on the vertical axis and different functions of T on the horizontal axis reveals that a linear graph is obtained when $\ln(K_P)$ is plotted against $1/T$, as in Graph C. (Note that the spreadsheet graph displays too many significant figures.) From this graph the equation of the straight line is found to be

$$\ln(K_P) = (-1.58 \times 10^4 \text{ K})\left(\frac{1}{T}\right) + 16.1$$

Substituting $K_P = 1$, that is, $\ln(K_P) = 0$, we find that

$$T = \frac{1.58 \times 10^4 \text{ K}}{16.1} = 980 \text{ K}$$

Graph C

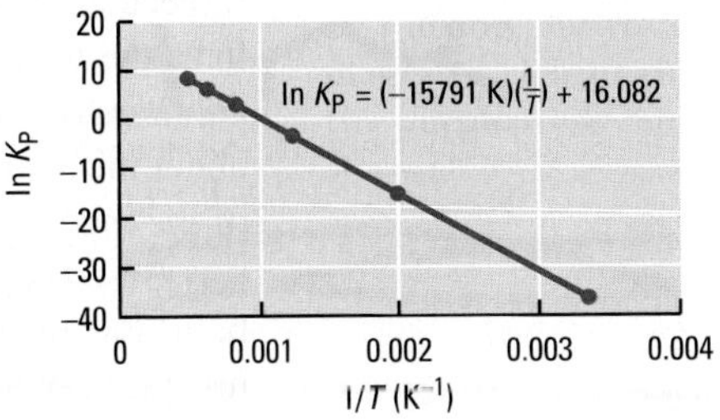
20
10
0
−10
−20
−30
−40
ln K_P
ln K_P = (−15791 K)($\frac{1}{T}$) + 16.082
0
0.001
0.002
0.003
0.004
1/T (K^{-1})

(b) The reaction is exothermic ($\Delta_r H° < 0$). Increasing the temperature shifts the equilibrium in the endothermic direction—that is, to the left (toward the reactants). This shift leads to a decrease in the NH_3 concentration, an increase in the concentrations of H_2 and N_2, and a decrease in the value of K_c.

(c) Because the temperature is constant, the value of K_c must be constant. Doubling the volume should cause the reaction to shift toward a greater number of gas-phase molecules—that is, toward the left. Doubling the volume would normally halve the pressure, but the shift of the equilibrium partially compensates for this effect, and the final equilibrium pressure is somewhat more than half the initial equilibrium pressure.

PROBLEM-SOLVING PRACTICE 12.8

Consider the equilibrium between N_2O_4 and NO_2 in a closed system.

$$N_2O_4(g) \rightleftharpoons 2\ NO_2(g)$$

Draw Lewis structures for the molecules involved in this equilibrium. Based on the bonding in the molecules, predict whether the reaction is exothermic or endothermic; hence, predict whether the concentration of N_2O_4 is larger in an equilibrium system at 25 °C or at 80 °C. Verify your prediction by looking at Figure 12.3.

CONCEPTUAL EXERCISE 12.12

Summarizing Le Chatelier's Principle

Construct a table to summarize your understanding of Le Chatelier's principle. Consider these changes in conditions:

(a) Addition of a reactant
(b) Removal of a reactant
(c) Addition of a product
(d) Removal of a product
(e) Increasing pressure by decreasing volume
(f) Decreasing pressure by increasing volume
(g) Increasing temperature
(h) Decreasing temperature

For each of these changes in conditions, indicate (1) how the reaction system changes to achieve a new equilibrium, (2) in which direction the equilibrium reaction shifts, and (3) whether the value of K_c changes. For some of these changes there are qualifications. For example, increasing pressure by decreasing volume does not always shift an equilibrium. List as many of these qualifications as you can.

12-7 Equilibrium at the Nanoscale

In Section 12-2, we used the isomerization of *cis*-2-butene to show that both forward and reverse reactions occur simultaneously in an equilibrium system.

$$cis\text{-2-butene(g)} \rightleftharpoons trans\text{-2-butene(g)} \qquad K_c = \frac{[trans]}{[cis]} = 2.0 \quad \text{(at 415 K)}$$

Because the equilibrium constant is 2.0, there are twice as many *trans* molecules as *cis* molecules in an equilibrium mixture at 415 K. In other words, two thirds of the molecules have the *trans* structure and one third have the *cis* structure. Because the molecules are continually reacting in both the forward and the reverse directions, another way to think about this situation is that each molecule is *trans* two thirds of the time and *cis* one third of the time.

Based on its molecular structure, you might think that a 2-butene molecule ought to be just as likely to be *cis* as *trans,* so why is the *trans* isomer favored at 415 K? In Figure 11.7 (← **Sec. 11-4a**) we noted that one molecule of the *trans* isomer is 7×10^{-21} J lower in energy than one molecule of the *cis* isomer. This difference is all important. For a mole of molecules, the difference in energy is $(-7 \times 10^{-21}\ \text{J})(6.022 \times 10^{23}\ \text{mol}^{-1}) = -4 \times 10^{3}\ \text{J/mol} = -4\ \text{kJ/mol}$, giving the thermochemical expression

$$cis\text{-2-butene(g)} \rightleftharpoons trans\text{-2-butene(g)} \qquad \Delta_r H^\circ = -4\ \text{kJ/mol}$$

Consider the rate constants for the forward and reverse reactions in the isomerization of 2-butene. Based on Figure 11.7, E_a(forward) = 262 kJ/mol and E_a(reverse) = 266 kJ/mol. This means that the rate constant for the reverse reaction is smaller than the rate constant for the forward reaction. At equilibrium the forward and reverse rates are equal, so

Larger k value and smaller concentration

Smaller k value and larger concentration

$$k_{\text{forward}} \times [cis] = k_{\text{reverse}} \times [trans]$$

If k_{reverse} is smaller than k_{forward}, then the concentration of *trans* must be larger than the concentration of *cis,* or the rates would not be equal. Because *cis*-2-butene is 4 kJ/mol higher in energy than *trans*-2-butene, it occurs only half as often as *trans*-2-butene at 415 K. We can generalize that ***in an equilibrium system, molecules that are higher in energy occur less often.*** We shall refer to this tendency as the energy effect on the position of equilibrium.

A second factor also affects how large an equilibrium constant is. It depends on how spread out, or dispersed, the total energy of products is compared with the total energy of reactants. This factor can be illustrated by the dissociation of dinitrogen tetraoxide.

$$N_2O_4(g) \rightleftharpoons 2\ NO_2(g) \qquad \Delta_r H^\circ = 57.2\ \text{kJ/mol}$$

In a constant-pressure system (such as a reaction at atmospheric pressure), the two moles of product molecules occupy twice the volume of the one mole of reactant molecules. This means that the energy of the NO_2 molecules is spread over twice the volume that would have been occupied by the N_2O_4 molecules. A thermodynamic quantity called **entropy** *provides a quantitative measure of how extensively energy is spread out when something happens.* Entropy will be defined more completely in Section 16-3. For now it suffices to say that ***in an equilibrium system, if there are more product molecules in the gas phase than reactant molecules in the gas phase, entropy favors the products.*** Although it depends on spreading out of energy, we shall refer to this tendency as the entropy effect on the position of equilibrium.

Despite the fact that the entropy factor favors NO_2 in the dissociation of N_2O_4, when data from Table 12.1 are used to calculate K_c at 25 °C (298 K) the result is $1/(1.7 \times 10^{2}) = 5.9 \times 10^{-3}$. This is because the dissociation of N_2O_4 is endothermic ($\Delta_r H^\circ = 57.2$ kJ/mol). The enthalpy of 2 mol NO_2 is 57.2 kJ higher than the enthalpy of 1 mol N_2O_4. From our first generalization that molecules with more energy occur less often in an equilibrium system, we expect that 2 NO_2 is less likely than N_2O_4 because of the energy effect. Therefore the equilibrium constant should be smaller than the entropy argument predicts. *Both the energy effect and the entropy effect must be taken into account to predict an equilibrium constant value.*

These ideas can also help us to understand the temperature dependence of the equilibrium constant. K_c values for the dissociation of N_2O_4 are given in the adjacent table. As the temperature rises, K_c becomes much larger. At high temperatures the molecules have lots of energy, and when the volume doubles, much more energy is dispersed. The energy difference between reactants and products ($\Delta_r H$) is smaller relative to the average energy per molecule.

$$N_2O_4(g) \rightleftharpoons 2\ NO_2(g)$$

T (K)	K_c
298	5.9×10^{-3}
350	1.3×10^{-1}
400	1.5×10^{0}
500	4.6×10^{1}
600	4.6×10^{2}

The higher the temperature is, the less important the energy effect ($\Delta_r H$) becomes and the more important the entropy effect (dispersal of energy) becomes in determining the position of equilibrium.

We have shown that if the number of gas-phase molecules increases when a reaction occurs, then the products are favored by entropy. This is not the only way for the products of a reaction to have greater entropy than the reactants, but it is one of the most important. In Chapter 16 we will show how entropy can be measured and tabulated. That discussion will enable us to use enthalpy changes and entropy changes for a reaction to calculate its equilibrium constant at a variety of temperatures.

PROBLEM-SOLVING EXAMPLE 12.9

Energy and Entropy Effects on Equilibria

For the equilibrium

$$CH_4(g) + H_2O(g) \rightleftharpoons CO(g) + 3\ H_2(g)$$

(a) Estimate whether entropy increases, decreases, or is the same when products form.
(b) Does the entropy effect favor reactants or products?
(c) Use data from Appendix J to calculate $\Delta_r H°$.
(d) Does the energy effect favor reactants or products?
(e) Is the reaction likely to be product-favored at high temperatures? Why or why not?

Result
(a) Entropy increases (b) Entropy favors products
(c) $\Delta_r H° = 206.10$ kJ/mol (d) Reactants
(e) Yes. Entropy favors products and the energy effect is less important at high temperatures.

Analyze A chemical equilibrium reaction is given and five questions are asked. Entropy increases if there are more gas-phase molecules in the products than the reactants. The side of an equilibrium with greater entropy is favored. $\Delta_r H°$ can be calculated from formation enthalpies, $\Delta_f H°$. The side of an equilibrium with lower energy (enthalpy) is favored. The entropy effect is more important than the energy effect at high temperatures.

Plan Apply the analysis to each question.

Execute
(a) Because there are 4 mol gas-phase products and only 2 mol gas-phase reactants, the products have higher entropy.
(b) The products have higher entropy and therefore are favored.
(c) $\Delta_r H° = \sum \{(\text{coefficient})\Delta_f H°(\text{product})\} - \sum\{(\text{coefficient})\Delta_f H°(\text{reactant})\}$

$$= \{(1)(-110.525 \text{ kJ/mol})\} - \{(1)(-74.81 \text{ kJ/mol}) + (1)(-241.818 \text{ kJ/mol})\}$$
$$= -110.525 \text{ kJ/mol} - (-316.628 \text{ kJ/mol})$$
$$= 206.10 \text{ kJ/mol}$$

(d) The reactants are lower in energy so they are favored by the energy effect.
(e) The reaction is product-favored at high temperatures because the energy effect favoring reactants becomes less important, and there is a relatively large entropy effect (four product molecules for every two reactant molecules in the gas phase).

☑ **Reasonable Result Check** In part (c) it is reasonable that the reaction is endothermic, because six bonds are broken and only four are formed when reactant molecules are changed into product molecules (← **Sec. 4-7a**).

PROBLEM-SOLVING PRACTICE 12.9

For the ammonia synthesis reaction

$$N_2(g) + 3\ H_2(g) \rightleftharpoons 2\ NH_3(g) \qquad \Delta_r H° = -92.2 \text{ kJ/mol}$$

(a) Does the entropy effect favor products? Explain your answer.
(b) Does the energy effect favor products? Explain your answer.
(c) Is the equilibrium concentration of $NH_3(g)$ greater at high or low temperature? Explain.

12-8 Controlling Chemical Reactions: The Haber-Bosch Process

The principles that allow us to control a reaction are based on our understanding of both equilibrium systems and the rates of chemical reactions. Some generalizations about equilibrium systems are

- ***A product-favored reaction has an equilibrium constant larger than 1.***
- ***If a reaction is exothermic, this energy factor favors the products.***
- ***If there is an increase in entropy when a reaction occurs, this entropy factor favors the products.***
- ***Product-favored reactions at low temperatures are usually exothermic.***
- ***Product-favored reactions at high temperatures are usually ones in which the entropy increases (energy becomes more dispersed).***

Using these general rules about equilibria, we can often predict whether a reaction is capable of yielding products. But it is also important that those products be produced rapidly. Recall these useful generalizations about reaction rates from Chapter 11.

- ***Reactions in the gas phase or in solution, where molecules of one reactant are completely mixed with molecules of another, occur more rapidly than do reactions between pure liquids or solids that do not dissolve in one another*** (← **Sec. 11-1**).
- ***Reactions occur more rapidly at high temperatures than at low temperatures*** (← **Sec. 11-5**).
- ***Reactions are faster when the reactant concentrations are high than when they are low*** (← **Sec. 11-2a**).
- ***Reactions between a solid and a gas, or between a solid and something dissolved in solution, are usually much faster when the solid particles are as small as possible*** (← **Sec. 11-1**).
- ***Reactions are faster in the presence of a catalyst. Often the right catalyst makes the difference between success and failure in industrial chemistry*** (← **Sec. 11-8**).

One of the best examples of the application of the principles of chemical reactivity is the chemical reaction used for the synthesis of ammonia from its elements. Even though Earth is bathed in an atmosphere that is about 80% N_2 gas, nitrogen cannot be used by most plants until it has been fixed—that is, converted into biologically useful forms. Although nitrogen fixation is done naturally by organisms such as cyanobacteria and some field crops such as alfalfa and soybeans, most plants cannot fix N_2. They must instead obtain nitrogen from cyanobacteria, some other organism, or fertilizer. Proper fertilization is especially important for recently developed varieties of wheat, corn, and rice that have resulted in much improved food production.

Direct combination of nitrogen and oxygen was used at the beginning of the twentieth century to provide fertilizer, but this process was not very efficient (← **Sec. 12-6d**). A much better way of manufacturing ammonia was devised by Fritz Haber and Carl Bosch, who chose the *direct synthesis of ammonia from its elements* as the basis for an industrial process.

$$N_2(g) + 3\,H_2(g) \rightleftharpoons 2\,NH_3(g)$$

Fritz Haber
1868–1934

The industrial chemical process by which ammonia is manufactured was developed by Fritz Haber, a chemist, and Carl Bosch, an engineer. Haber's studies in the early 1900s revealed that direct ammonia synthesis should be possible. In 1913 the engineering problems were solved by Bosch. Haber's contract with the manufacturer of ammonia called for him to receive 1 pfennig (one hundredth of a German mark—similar to a penny) per kilogram of ammonia, and he soon became not only famous, but also rich. In 1918 he was awarded a Nobel Prize in Chemistry for the ammonia synthesis, but the choice was criticized because of his role in developing the use of poison gases for Germany during World War I.

It is estimated that 40% to 60% of the nitrogen in the average human body has come from ammonia produced by the Haber-Bosch process and that increased agricultural productivity resulting from this process supports about 40% of the world's population.

CONCEPTUAL EXERCISE 12.13

Ammonia Synthesis

For the ammonia synthesis reaction, predict

(a) whether the reaction is exothermic or endothermic.
(b) whether the reaction product is favored by entropy.
(c) whether the reaction produces more products at low or high temperatures.
(d) what would happen if you tried to increase the rate of the reaction by increasing the temperature.

At first glance this reaction might seem to be a poor choice. In nature, hydrogen is available only in combined form—for example, in water or hydrocarbons—meaning that hydrogen must be extracted from these compounds at considerable expense in energy resources and money. As you discovered in Problem-Solving Practice 12.9, the ammonia synthesis reaction becomes less product-favored at higher temperatures. But, higher temperatures are needed for ammonia to be produced fast enough for the process to be efficient and economical. Nonetheless, the **Haber-Bosch process** (shown schematically in Figure 12.10) has been so well developed that ammonia is inexpensive (less than $600 per ton). For this reason it is widely used as a fertilizer and is often among the top five chemicals produced in the United States. In 2010, annual U. S. production of NH_3 by the Haber-Bosch process was approximately 22 billion pounds and about 17 billion pounds was imported that year.

Both the thermodynamics and the kinetics of the direct synthesis of ammonia have been carefully studied and fine-tuned by industry so that the maximum yield of product is obtained in a reasonable time and at a reasonable cost of both money and energy resources.

- The reaction is exothermic, and there is a decrease in entropy when it takes place. Therefore this reaction is predicted to be product-favored at low temperatures, but reactant-favored at high temperatures. (You should have made this prediction in Problem-Solving Practice 12.9, and it is in accord with Le Chatelier's principle for an exothermic process.)

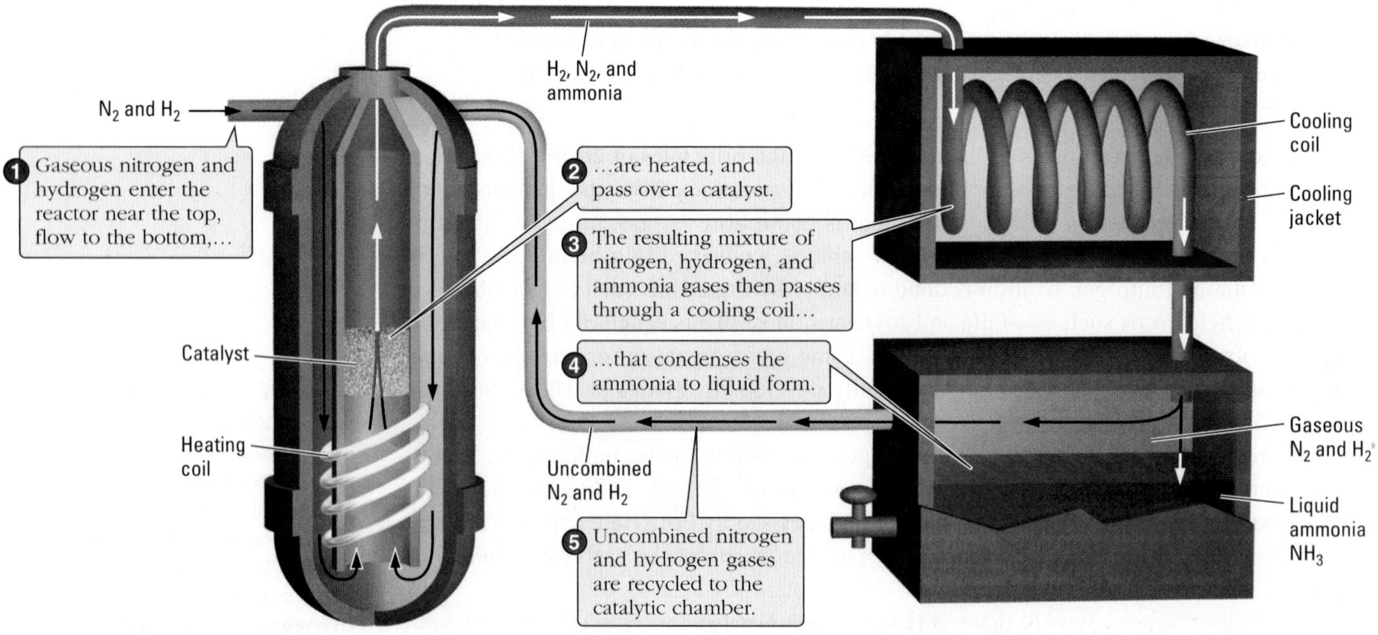
H2, N2, and ammonia
N2 and H2
1 Gaseous nitrogen and hydrogen enter the reactor near the top, flow to the bottom,...
2 ...are heated, and pass over a catalyst.
3 The resulting mixture of nitrogen, hydrogen, and ammonia gases then passes through a cooling coil...
4 ...that condenses the ammonia to liquid form.
5 Uncombined nitrogen and hydrogen gases are recycled to the catalytic chamber.
Catalyst
Heating coil
Uncombined N2 and H2
Cooling coil
Cooling jacket
Gaseous N2 and H2
Liquid ammonia NH3

Figure 12.10 Haber-Bosch process for synthesis of ammonia (schematic).

- To increase the equilibrium concentration of NH_3, the reaction is carried out at high pressure (200 atm). This does not change the value of K_c or K_P, but an increase in pressure can be compensated for by converting N_2 and H_2 to NH_3; 2 mol $NH_3(g)$ exerts less pressure than a total of 4 mol gaseous reactants $[N_2(g) + 3\ H_2(g)]$ in the same-sized container.
- Ammonia is continually liquefied and removed from the reaction vessel, which reduces the concentration of the product of the reaction and shifts the equilibrium toward the right.
- The reaction is quite slow at room temperature, so the temperature must be raised to increase the rate. Although the rate increases with increasing temperature, the equilibrium constant declines. Thus, the faster the reaction, the smaller the yield.
- The temperature cannot be raised too much in an attempt to increase the rate, but a rate increase can be achieved with a catalyst. An effective catalyst for the Haber-Bosch process is Fe_3O_4 mixed with KOH, SiO_2, and Al_2O_3. Since the catalyst is not effective below about 400 °C, the optimum temperature, considering all the factors controlling the reaction, is about 450 °C.

Making predictions about chemical reactivity is part of the challenge, the adventure, and the art of chemistry. Many chemists enjoy the challenge of making useful new materials, which usually means choosing to make them by reactions that we believe will be product-favored and reasonably rapid. Such predictions are based on the ideas outlined in this chapter and Chapter 11.

SUMMARY PROBLEM

On the first page of this chapter we described the equilibrium between oxygen from the air and hemoglobin in the bloodstream. This equilibrium enables oxygen to enter the bloodstream in the lungs, be transported to tissues where it is needed, and support metabolism in those tissues. In rapidly metabolizing tissues, oxygen combines with glucose to form carbon dioxide and water. Energy released by oxygen's reaction with glucose enables muscle tissues, for example, to make your arms or legs move.

Hemoglobin has a structure that consists of four protein chains (subunits), each folded into a tertiary structure (← **Sec. 10-7e**). The four subunits are held together by noncovalent attractions. Each subunit can bind to one oxygen molecule at a time. Thus, using Hb to represent a hemoglobin molecule, these four equilibrium reactions are involved in oxygen uptake:

$$Hb + O_2 \rightleftharpoons Hb(O_2) \qquad K_1 = 3.0$$

$$Hb(O_2) + O_2 \rightleftharpoons Hb(O_2)_2 \qquad K_2 = 6.8$$

$$Hb(O_2)_2 + O_2 \rightleftharpoons Hb(O_2)_3 \qquad K_3 = 1.5$$

$$Hb(O_2)_3 + O_2 \rightleftharpoons Hb(O_2)_4 \qquad K_4 = 720$$

The overall equilibrium reaction is

$$Hb + 4\ O_2 \rightleftharpoons Hb(O_2)_4 \qquad K_{overall}$$

Changing the concentration of $H^+(aq)$ in the bloodstream also affects the equilibrium because there are sites in a hemoglobin molecule where H^+ ions can bond. Thus, a more complete equation for the equilibrium is

$$HbH_2^{2+} + 4\ O_2 \rightleftharpoons Hb(O_2)_4 + 2\ H^+(aq) \qquad K_{complete}$$

1. Determine which of the first four reactions is most product-favored. Determine which reaction is least product-favored. Explain how you made these determinations.
2. Calculate the value of the overall equilibrium constant, $K_{overall}$, for the equilibrium reaction of hemoglobin with oxygen.
3. In the alveoli of the lungs, where oxygen enters the bloodstream, the partial pressure of oxygen is 100 mmHg. In capillaries of actively metabolizing muscle tissue, the partial

pressure of oxygen is 20 mmHg. Explain why hemoglobin absorbs oxygen in the lungs and releases it in muscles.

4. An important factor controlling the concentration of $H^+(aq)$ in the bloodstream is this equilibrium

$$H_2CO_3(aq) \rightleftharpoons H^+(aq) + HCO_3^-(aq) \qquad K_c = 4.3 \times 10^{-7}$$

Suppose that 0.24 mol H_2CO_3 is introduced into 250.0 mL water. Determine $[H^+]$ when equilibrium is reached.

5. The concentration of $H^+(aq)$ in the bloodstream is typically 4.0×10^{-8} M. Determine the ratio of $[H_2CO_3]$ to $[HCO_3^-]$ in the bloodstream.

6. Metabolic reactions produce $CO_2(aq)$, which reacts with water to form $H_2CO_3(aq)$. This fact contributes to the release of oxygen in actively metabolizing cells. Explain how.

7. Predict the sign of $\Delta_r H°$ for the overall reaction of hemoglobin with oxygen. Which direction does this equilibrium shift when temperature increases?

HAVING STUDIED THIS CHAPTER . . .

. . . you should be able to:

- Recognize a system at equilibrium and describe the properties of equilibrium systems (Section 12-1). End-of-chapter Question 8
- Describe the dynamic nature of equilibrium and the changes in concentrations of reactants and products that occur as a system approaches equilibrium (Sections 12-1, 12-2). Questions 10, 12
- Write equilibrium constant expressions, given balanced chemical equations (Section 12-2). Questions 14, 16, 18, 22
- Obtain equilibrium constant expressions for related reactions from the expression for one or more known reactions (Section 12-2). Questions 24, 26
- Calculate K_P from K_c or K_c from K_P for the same equilibrium (Section 12-2). Questions 28, 32
- Calculate a value of K_c for an equilibrium system, given information about initial concentrations and equilibrium concentrations (Section 12-3). Questions 34, 38, 42
- Make qualitative predictions about the extent of reaction based on equilibrium constant values—that is, predict whether a reaction is product-favored or reactant-favored based on the size of the equilibrium constant (Section 12-4). Questions 45, 47, 48
- Calculate concentrations of reactants and products in an equilibrium system if K_c and initial concentrations are known (Section 12-5). Questions 50, 54, 57
- Use the reaction quotient Q to predict in which direction a reaction will go to reach equilibrium (Section 12-5). Questions 59, 61
- Use Le Chatelier's principle to show how changes in concentrations, pressure or volume, and temperature shift chemical equilibria (Section 12-6). Questions 64, 66, 68, 76
- Use the change in enthalpy and the change in entropy qualitatively to predict whether products are favored over reactants (Section 12-7). Questions 78, 80
- List the factors affecting chemical reactivity, and apply them to predicting optimal conditions for producing products (Section 12-8). Question 84

KEY TERMS

chemical equilibrium (Section 12-1)
dynamic equilibrium (12-1a)
entropy (12-7)
equilibrium concentration (12-3)
equilibrium constant (K_c) (12-2)
equilibrium constant (K_P) (12-2f)
equilibrium constant expression (12-2a)
Haber-Bosch process (12-8)
Le Chatelier's principle (12-6)
product-favored (12-1)
reactant-favored (12-1)
reaction quotient (Q) (12-5)
shifting an equilibrium (12-6)

QUESTIONS FOR REVIEW AND THOUGHT

Red-numbered questions have short answers at the back of this book in Appendix M and fully worked solutions in the *Student Solutions Manual.*

Review Questions

These questions test vocabulary and simple concepts.

1. Define the terms *chemical equilibrium* and *dynamic equilibrium.*
2. If an equilibrium is product-favored, is its equilibrium constant large or small with respect to 1? Explain.
3. List three characteristics that you would need to verify in order to determine that a chemical system is at equilibrium.
4. Decomposition of ammonium dichromate is shown in the designated series of photos. In a closed container this process reaches an equilibrium state. Write a balanced chemical equation for the equilibrium reaction. How is the equilibrium affected if
   (a) more ammonium dichromate is added to the equilibrium system?
   (b) more water vapor is added?
   (c) more chromium(III) oxide is added?
5. For the equilibrium reaction in Question 4, write the expression for the equilibrium constant.
   (a) How would this equilibrium constant change if the total pressure on the system were doubled?
   (b) How would the equilibrium constant change if the temperature were increased?
6. Indicate whether each statement below is true or false. If a statement is false, rewrite it to produce a closely related statement that is true.
   (a) For a given reaction, the magnitude of the equilibrium constant is independent of temperature.
   (b) If there is an increase in entropy and a decrease in enthalpy when reactants in their standard states are converted to products in their standard states, the equilibrium constant for the reaction must be negative.
   (c) The equilibrium constant for the reverse of a reaction is the reciprocal of the equilibrium constant for the reaction itself.
   (d) For the reaction

$$H_2O_2(\ell) \rightleftharpoons H_2O(\ell) + \tfrac{1}{2}\,O_2(g)$$

   the equilibrium constant is one half the magnitude of the equilibrium constant for the reaction

$$2\,H_2O_2(\ell) \rightleftharpoons 2\,H_2O(\ell) + O_2(g)$$

Decomposition of ammonium dichromate, for Question 4.

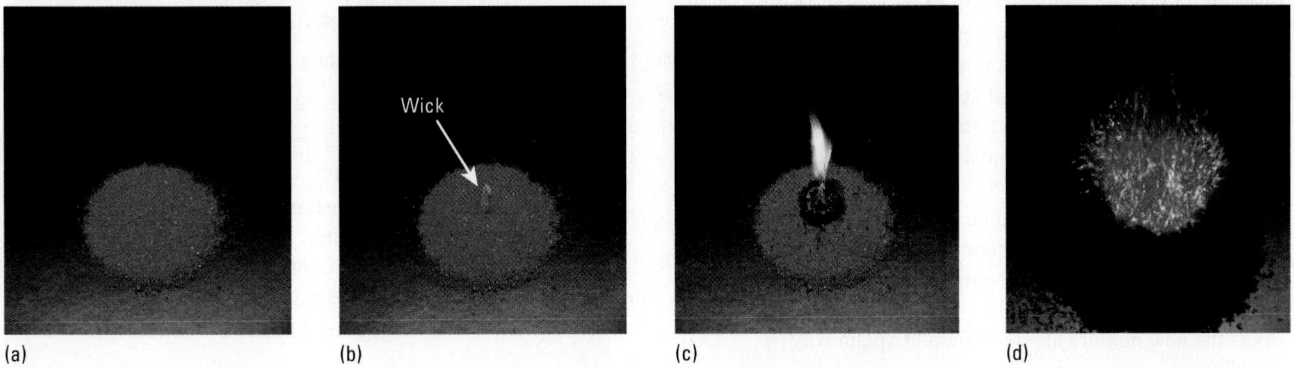
Wick
(a)
(b)
(c)
(d)

Photos: Cengage Learning/Charles D. Winters

Decomposition of $(NH_4)_2Cr_2O_7$. Orange, solid $(NH_4)_2Cr_2O_7$ (a) can be ignited by lighting a wick (b), which initiates decomposition (c) forming Cr_2O_3, the dark green solid in part (d), N_2 gas, and water vapor. Energy is transferred to the surroundings by the process.

7. If the reaction quotient is larger than the equilibrium constant, in what direction does the reaction proceed as it approaches equilibrium? What will happen if $Q < K$?

Topical Questions

These questions are keyed to the major topics in the chapter. Usually a question that is answered at the back of this book is paired with a similar one that is not.

Characteristics of Chemical Equilibrium (Section 12-1)

8. Think of an experiment you could do to demonstrate that the equilibrium

$$2\ NO_2(g) \rightleftharpoons N_2O_4(g)$$

is a dynamic process in which the forward and reverse reactions continue to occur after equilibrium has been achieved. Describe how such an experiment might be carried out.

9. Discuss this statement: "No true chemical equilibrium can exist unless reactant molecules are constantly changing into product molecules, and vice versa."
10. Suppose you place a large piece of ice into a well-insulated thermos with some water in it and it comes to equilibrium with part of the ice melted.
    (a) What is the temperature of the equilibrium system?
    (b) Is this a static or a dynamic equilibrium? Explain.
11. The atmosphere consists of about 80% N_2 and 20% O_2, yet there are many oxides of nitrogen that are stable and can be isolated in the laboratory.
    (a) Is the atmosphere at chemical equilibrium with respect to forming NO?
    (b) If not, why doesn't NO form? If so, how is it that NO can be made and kept in the laboratory for long periods?

The Equilibrium Constant (Section 12-2)

12. Consider the gas-phase reaction of $N_2 + O_2$ to give 2 NO and the reverse reaction of 2 NO to give $N_2 + O_2$, discussed in Section 12-2e. An equilibrium mixture of NO, N_2, and O_2 at 5000. K that contains equal concentrations of N_2 and O_2 has a concentration of NO about half as great. Make qualitatively correct plots of the concentrations of reactants and products versus time for these two processes, showing the initial state and the final dynamic equilibrium state. Assume a temperature of 5000. K. Don't do any calculations—just sketch how you think the plots should look.
13. After 0.1 mol pure *cis*-2-butene is allowed to come to equilibrium with *trans*-2-butene in a closed 5.0-L flask at 25 °C, another 0.1 mol *cis*-2-butene is suddenly added to the flask.
    (a) Is the new mixture at equilibrium? Explain why or why not.
    (b) In the new mixture, immediately after addition of the *cis*-2-butene, which rate is faster: *cis* → *trans* or *trans* → *cis*? Or are both rates the same?
    (c) After the second 0.1 mol *cis*-2-butene has been added and the system is at equilibrium, if the concentration of *trans*-2-butene is 0.01 mol/L, determine the concentration of *cis*-2-butene.
14. Write equilibrium constant expressions for these reactions. For gases, use either pressures or concentrations.
    (a) $3\ O_2(g) \rightleftharpoons 2\ O_3(g)$
    (b) $Fe(s) + 5\ CO(g) \rightleftharpoons Fe(CO)_5(g)$
    (c) $(NH_4)_2CO_3(s) \rightleftharpoons 2\ NH_3(g) + CO_2(g) + H_2O(g)$
    (d) $Ag_2SO_4(s) \rightleftharpoons 2\ Ag^+(aq) + SO_4^{2-}(aq)$
15. Write equilibrium constant expressions for these reactions:
    (a) $CH_4(g) + H_2O(\ell) \rightleftharpoons CO(g) + 3\ H_2(g)$
    (b) $4\ NH_3(g) + 5\ O_2(g) \rightleftharpoons 4\ NO(g) + 6\ H_2O(g)$
    (c) $BaCO_3(s) \rightleftharpoons BaO(s) + CO_2(g)$
    (d) $NH_3(g) + HCl(g) \rightleftharpoons NH_4Cl(s)$
16. Write the equilibrium constant expression for each reaction.
    (a) $2\ H_2O_2(g) \rightleftharpoons 2\ H_2O(g) + O_2(g)$
    (b) $PCl_3(g) + Cl_2(g) \rightleftharpoons PCl_5(g)$
    (c) $SiO_2(s) + 3\ C(s) \rightleftharpoons SiC(s) + 2\ CO(g)$
    (d) $H_2(g) + \frac{1}{8}\ S_8(s) \rightleftharpoons H_2S(g)$
17. Write the equilibrium constant expression for each reaction.
    (a) $N_2(g) + 2\ O_2(g) \rightleftharpoons N_2O_4(g)$
    (b) $SiH_4(g) + 2\ O_2(g) \rightleftharpoons SiO_2(s) + 2\ H_2O(g)$
    (c) $MgO(s) + SO_2(g) + \frac{1}{2}\ O_2(g) \rightleftharpoons MgSO_4(s)$
    (d) $2\ PbS(s) + 3\ O_2(g) \rightleftharpoons 2\ PbO(s) + 2\ SO_2(g)$
18. Write the expression for K_c for each reaction.
    (a) $PCl_5(s) \rightleftharpoons PCl_3(g) + Cl_2(g)$
    (b) $Co(H_2O)_6^{2+}(aq) + 4\ Cl^-(aq) \rightleftharpoons CoCl_4^{2-}(aq) + 6\ H_2O(\ell)$
    (c) $CH_3COOH(aq) \rightleftharpoons CH_3COO^-(aq) + H^+(aq)$
    (d) $2\ F_2(g) + H_2O(g) \rightleftharpoons OF_2(g) + 2\ HF(g)$
19. Write the equilibrium constant expression for each reaction.
    (a) The oxidation of ammonia with ClF_3 in a rocket motor

$$NH_3(g) + ClF_3(g) \rightleftharpoons 3\ HF(g) + \tfrac{1}{2}\ N_2(g) + \tfrac{1}{2}\ Cl_2(g)$$

    (b) The simultaneous oxidation and reduction of a chlorite ion

$$3\ ClO_2^-(aq) \rightleftharpoons 2\ ClO_3^-(aq) + Cl^-(aq)$$

    (c) $IO_3^-(aq) + 6\ OH^-(aq) + Cl_2(aq) \rightleftharpoons IO_6^{5-}(aq) + 2\ Cl^-(aq) + 3\ H_2O(\ell)$
20. Write the equilibrium constant expression for each of these heterogeneous systems.
    (a) $CaSO_4 \cdot 5H_2O(s) \rightleftharpoons CaSO_4 \cdot 3H_2O(s) + 2\ H_2O(g)$
    (b) $SiF_4(g) + 2\ H_2O(g) \rightleftharpoons SiO_2(s) + 4\ HF(g)$
    (c) $LaCl_3(s) + H_2O(g) \rightleftharpoons LaClO(s) + 2\ HCl(g)$

21. Write the equilibrium constant expression for each of these heterogeneous systems.
(a) $N_2O_4(g) + O_3(g) \rightleftharpoons N_2O_5(s) + O_2(g)$
(b) $C(s) + 2\ N_2O(g) \rightleftharpoons CO_2(g) + 2\ N_2(g)$
(c) $H_2O(\ell) \rightleftharpoons H_2O(g)$
22. Write a chemical equation for an equilibrium system that would lead to each expression (a–c) for K.

(a) $K = \dfrac{(P_{H_2S})^2(P_{O_2})^3}{(P_{SO_2})^2(P_{H_2O})^2}$ (b) $K = \dfrac{(P_{F_2})^{1/2}(P_{I_2})^{1/2}}{(P_{IF})}$

(c) $K = \dfrac{[Cl^-]^2}{(P_{Cl_2})[Br^-]^2}$

23. Consider this reaction at 122 °C:

$$2\ SO_3(g) \rightleftharpoons 2\ SO_2(g) + O_2(g)$$

(a) Write an equilibrium constant expression for the reaction and call the constant K_1.
(b) Write an equilibrium constant expression for the decomposition of 1 mol SO_3 to SO_2 and O_2 and call the constant K_2.
(c) Relate K_1 and K_2.
24. Consider these two equilibria involving $SO_2(g)$ and their corresponding equilibrium constants.

$$SO_2(g) + \tfrac{1}{2}\ O_2(g) \rightleftharpoons SO_3(g) \qquad K_{c_1}$$
$$2\ SO_3(g) \rightleftharpoons 2\ SO_2(g) + O_2(g) \qquad K_{c_2}$$

Which of these expressions correctly relates K_{c_1} to K_{c_2}?
(a) $K_{c_2} = K_{c_1}^2$ (b) $K_{c_2}^2 = K_{c_1}$
(c) $K_{c_2} = 1/K_{c_1}$ (d) $K_{c_2} = K_{c_1}$
(e) $K_{c_2} = 1/K_{c_1}^2$
25. The reaction of hydrazine, N_2H_4, with chlorine trifluoride, ClF_3, has been used in experimental rocket motors.

$$N_2H_4(g) + \tfrac{4}{3}\ ClF_3(g) \rightleftharpoons 4\ HF(g) + N_2(g) + \tfrac{2}{3}\ Cl_2(g)$$

How is the equilibrium constant, K_P, for this reaction related to K_P' for the reaction written this way?

$$3\ N_2H_4(g) + 4\ ClF_3(g) \rightleftharpoons 12\ HF(g) + 3\ N_2(g) + Cl_2(g)$$

(a) $K_P = K_P'$ (b) $K_P = 1/K_P'$
(c) $K_P^3 = K_P'$ (d) $K_P = (K_P')^3$
(e) $3K_P = K_P'$
26. At 627 °C, $K_c = 0.76$ for the reaction

$$2\ SO_2(g) + O_2(g) \rightleftharpoons 2\ SO_3(g)$$

Calculate K_c at 627 °C for
(a) synthesis of 1 mol sulfur trioxide gas.
(b) decomposition of 2 mol SO_3.
27. At 450 °C, the equilibrium constant K_c for the Haber-Bosch synthesis of ammonia is 0.16 for the reaction written as

$$3\ H_2(g) + N_2(g) \rightleftharpoons 2\ NH_3(g)$$

Calculate the value of K_c for the same reaction written as

$$\tfrac{3}{2}\ H_2(g) + \tfrac{1}{2}\ N_2(g) \rightleftharpoons NH_3(g)$$

28. For each reaction in Question 16, write the equilibrium constant expression for K_P.
29. For each reaction in Question 17, write the equilibrium constant expression for K_P.
30. Given these data at a certain temperature,

$$2\ N_2(g) + O_2(g) \rightleftharpoons 2\ N_2O(g) \qquad K = 1.2 \times 10^{-35}$$
$$N_2O_4(g) \rightleftharpoons 2\ NO_2(g) \qquad K = 4.6 \times 10^{-3}$$
$$\tfrac{1}{2}\ N_2(g) + O_2(g) \rightleftharpoons NO_2(g) \qquad K = 4.1 \times 10^{-9}$$

calculate K for the reaction between 1 mol dinitrogen oxide gas and oxygen gas to give dinitrogen tetraoxide gas.
31. Given these data at a certain temperature,

$$2\ H_2(g) + O_2(g) \rightleftharpoons 2\ H_2O(g) \qquad K_c = 3.2 \times 10^{81}$$
$$N_2(g) + 3\ H_2(g) \rightleftharpoons 2\ NH_3(g) \qquad K_c = 3.5 \times 10^{8}$$

calculate K_c for the reaction of ammonia with oxygen to give $N_2(g)$ and $H_2O(g)$.
32. The vapor pressure of water at 80. °C is 0.467 atm. Determine the value of K_c for the process

$$H_2O(\ell) \rightleftharpoons H_2O(g)$$

at this temperature.
33. The value of K_c for the reaction

$$N_2(g) + 3\ H_2(g) \rightleftharpoons 2\ NH_3(g)$$

is 2.00 at 400. °C. Determine the value of K_P for this reaction at this temperature using bars as pressure units.

Determining Equilibrium Constants (Section 12-3)

34. At elevated temperatures, BrF_5 establishes this equilibrium.

$$2\ BrF_5(g) \rightleftharpoons Br_2(g) + 5\ F_2(g)$$

The equilibrium concentrations of the gases at 1500 K are 0.0064 mol/L for BrF_5, 0.0018 mol/L for Br_2, and 0.0090 mol/L for F_2. Calculate K_c.
35. This reaction was examined at 250 °C.

$$PCl_5(g) \rightleftharpoons PCl_3(g) + Cl_2(g)$$

At equilibrium, $[PCl_5] = 4.2 \times 10^{-5}$ M, $[PCl_3] = 1.3 \times 10^{-2}$ M, and $[Cl_2] = 3.9 \times 10^{-3}$ M. Calculate the equilibrium constant K_c for the reaction.
36. At high temperature, hydrogen and carbon dioxide react to give water and carbon monoxide.

$$H_2(g) + CO_2(g) \rightleftharpoons H_2O(g) + CO(g)$$

Laboratory measurements at 986 °C show that there is 0.11 mol each of CO and water vapor and 0.087 mol each of H_2 and CO_2 at equilibrium in a sealed 1.0-L container. Calculate the equilibrium constant K_P for the reaction at 986 °C.

37. Carbon dioxide reacts with carbon to give carbon monoxide according to the equation

$$C(s) + CO_2(g) \rightleftharpoons 2\ CO(g)$$

At 700. °C, a 2.0-L sealed flask at equilibrium contains 0.10 mol CO, 0.20 mol CO_2, and 0.40 mol C. Calculate the equilibrium constant K_P for this reaction at the specified temperature.

38. A mixture of CO and Cl_2 is placed in a reaction flask that is then sealed. Initially [CO] = 0.0102 mol/L and $[Cl_2]$ = 0.00609 mol/L. When the reaction

$$CO(g) + Cl_2(g) \rightleftharpoons COCl_2(g)$$

reaches equilibrium at 600 K, $[Cl_2]$ = 0.00301 mol/L.
(a) Calculate the concentrations of CO and $COCl_2$ at equilibrium.
(b) Calculate K_c.

39. Carbon tetrachloride can be produced by this reaction:

$$CS_2(g) + 3\ Cl_2(g) \rightleftharpoons S_2Cl_2(g) + CCl_4(g)$$

Suppose 1.2 mol CS_2 and 3.6 mol Cl_2 are placed in a 1.00-L flask and the flask is sealed. After equilibrium has been achieved, the mixture contains 0.90 mol CCl_4. Calculate K_c.

40. Assume you place 0.010 mol $N_2O_4(g)$ in a sealed 2.0-L flask at 50. °C. After the system reaches equilibrium, $[N_2O_4]$ = 0.00090 M. Calculate the value of K_c for this reaction.

$$N_2O_4(g) \rightleftharpoons 2\ NO_2(g)$$

41. Nitrosyl chloride, NOCl, decomposes to NO and Cl_2 at high temperatures.

$$2\ NOCl(g) \rightleftharpoons 2\ NO(g) + Cl_2(g)$$

Suppose you place 2.00 mol NOCl in a 1.00-L flask, seal it, and raise the temperature to 462 °C. When equilibrium has been established, 0.66 mol NO is present. Calculate the equilibrium constant K_c for the decomposition reaction from these data.

42. Suppose 0.086 mol Br_2 is placed in a 1.26-L flask. The flask is sealed and heated to 1756 K, a temperature at which the Br_2 dissociates to atoms

$$Br_2(g) \rightleftharpoons 2\ Br(g)$$

If Br_2 is 3.7% dissociated at this temperature, calculate K_c.

43. H_2 gas and I_2 vapor are mixed in a flask. The flask is sealed and heated to 700 °C. The initial concentration of each gas is 0.0088 mol/L, and 78.6% of the I_2 has reacted when equilibrium is achieved according to the equation

$$H_2(g) + I_2(g) \rightleftharpoons 2\ HI(g)$$

Calculate K_c for this reaction.

44. Chemists carried out a study of the high temperature reaction of sulfur dioxide with oxygen in which a sealed reactor initially contained 0.0076-M SO_2, 0.0036-M O_2, and no SO_3. After equilibrium was achieved, the SO_2 concentration decreased to 0.0032 M. Calculate K_c at this temperature for

$$2\ SO_2(g) + O_2(g) \rightleftharpoons 2\ SO_3(g)$$

The Meaning of the Equilibrium Constant (Section 12-4)

45. Using the data of Table 12.1, predict which of these reactions is product-favored at 25 °C. Then, place all the reactions in order from most reactant-favored to most product-favored.
(a) $2\ NH_3(g) \rightleftharpoons N_2(g) + 3\ H_2(g)$
(b) $NH_4^+(aq) + OH^-(aq) \rightleftharpoons NH_3(aq) + H_2O(\ell)$
(c) $2\ NO(g) \rightleftharpoons N_2(g) + O_2(g)$

46. Using the data of Table 12.1, predict which of these reactions is product-favored at 25 °C. Then, place all the reactions in order from most reactant-favored to most product-favored.
(a) $2\ NO_2(g) \rightleftharpoons N_2O_4(g)$
(b) $H_2CO_3(aq) \rightleftharpoons HCO_3^-(aq) + H^+(aq)$
(c) $AgI(s) \rightleftharpoons Ag^+(aq) + I^-(aq)$

47. On the basis of the equilibrium constant values, choose the reactions in which the *reactants* are favored.
(a) $H_2O(\ell) \rightleftharpoons H^+(aq) + OH^-(aq)$ $K = 1.0 \times 10^{-14}$
(b) $[AlF_6]^{3-}(aq) \rightleftharpoons Al^{3+}(aq) + 6\ F^-(aq)$ $K = 2 \times 10^{-24}$
(c) $Ca_3(PO_4)_2(s) \rightleftharpoons 3\ Ca^{2+}(aq) + 2\ PO_4^{3-}(aq)$ $K = 1 \times 10^{-25}$
(d) $2\ Fe^{3+}(aq) + 3\ S^{2-}(aq) \rightleftharpoons Fe_2S_3(s)$ $K = 1 \times 10^{88}$

48. The equilibrium constants for dissolving silver sulfate and silver sulfide in water are 1.7×10^{-5} and 6×10^{-30}, respectively.
(a) Write the balanced dissociation reaction equation and the associated equilibrium constant expression for each process.
(b) Which compound is more soluble? Explain your answer.
(c) Which compound is less soluble? Explain your answer.

49. The equilibrium constants for dissolving calcium carbonate, silver nitrate, and silver chloride in water are 2.8×10^{-9}, 2.0×10^{2}, and 1.8×10^{-10}, respectively.
(a) Write the balanced dissociation reaction equation and the associated equilibrium constant expression for each process.
(b) Which compound is most soluble? Explain your answer.
(c) Which compound is least soluble? Explain your answer.

Using Equilibrium Constants (Section 12-5)

50. The hydrocarbon C_4H_{10} can exist in two gaseous forms: butane and 2-methylpropane. The value of K_c for conversion of butane to 2-methylpropane is 2.5 at 25 °C.

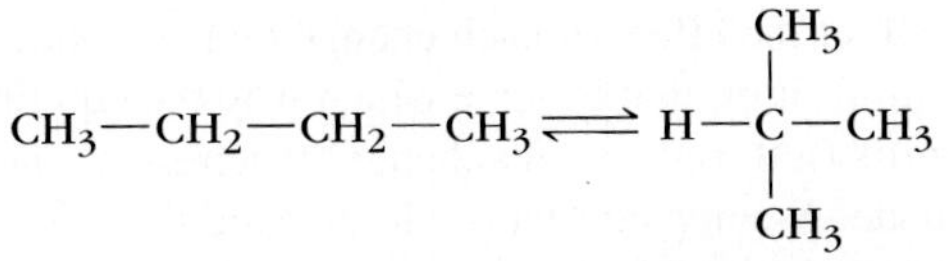
CH3
CH3—CH2—CH2—CH3 ⇌ H—C—CH3
CH3

butane 2-methylpropane

(a) Suppose that the initial concentrations of butane and 2-methylpropane are each 0.100 mol/L. Make up a table of initial concentrations, change in concentrations, and equilibrium concentrations for this reaction.
(b) Write the equilibrium constant expression in terms of x, the change in the concentration of butane, and then solve for x.
(c) If you place 0.017 mol butane in a 0.50-L sealed flask at 25 °C, calculate the equilibrium concentration of each isomer.

51. At room temperature, the equilibrium constant K_c for the reaction

$$2\ NO(g) \rightleftharpoons N_2(g) + O_2(g)$$

is 1.4×10^{30}.
(a) Is this reaction product-favored or reactant-favored? Explain your answer.
(b) In the atmosphere at room temperature the concentration of N_2 is 0.33 mol/L, and the concentration of O_2 is about 25% of that value. Calculate the equilibrium concentration of NO in the atmosphere produced by the reaction of N_2 and O_2.
(c) How does this affect your answer to Question 11?

52. The hydrocarbon cyclohexane, C_6H_{12}, can isomerize, changing into methylcyclopentane, a compound with the same molecular formula but a different molecular structure.

$$C_6H_{12}(g) \rightleftharpoons C_5H_9CH_3(g)$$

cyclohexane methylcyclopentane

The equilibrium constant K_c has been estimated to be 0.12 at 25 °C. If you place 3.79 g cyclohexane in an empty 2.80-L flask and seal the flask, calculate the mass of cyclohexane that is present when equilibrium is established.

53. Consider the equilibrium

$$N_2(g) + O_2(g) \rightleftharpoons 2\ NO(g)$$

At 2300 K the equilibrium constant $K_c = 1.7 \times 10^{-3}$. If 0.15 mol NO(g) is placed into an empty, sealed 10.0-L flask and heated to 2300 K, calculate the equilibrium concentrations of all three substances at this temperature.

54. The equilibrium constant, K_c, for the reaction

$$Br_2(g) + F_2(g) \rightleftharpoons 2\ BrF(g)$$

is 55.3. Calculate what the equilibrium concentrations of all these gases are if the initial concentrations of bromine and fluorine were both 0.220 mol/L. (Assume constant-volume conditions.)

55. The equilibrium constant K_c for the reaction

$$H_2(g) + I_2(g) \rightleftharpoons 2\ HI(g)$$

has the value 50.0 at 745 K.
(a) When 1.00 mol I_2 and 3.00 mol H_2 are allowed to come to equilibrium at 745 K in a sealed 10.00-L flask, calculate the amount (in moles) of HI produced.
(b) Calculate the amount of HI produced in a 5.00-L flask.
(c) Calculate the total amount of HI present at equilibrium if an additional 3.00 mol H_2 is added to the 10.00-L flask.

56. The equilibrium constant K_c for the *cis-trans* isomerization of gaseous 2-butene has the value 1.50 at 580. K.

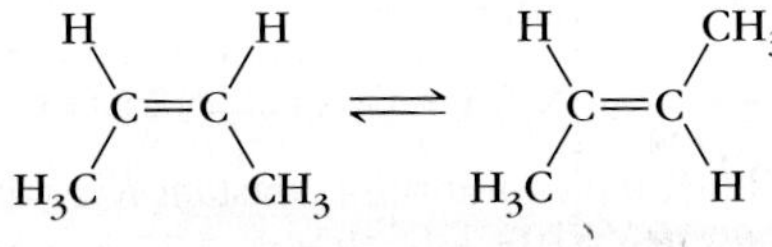
H
H
C=C
H3C
CH3
H
CH3
C=C
H3C
H

(a) Is the reaction product-favored at 580. K? Explain your answer.
(b) Calculate the amount (in moles) of *trans* isomer produced when 1 mol *cis*-2-butene is heated to 580. K in the presence of a catalyst in a sealed, 1.00-L flask and reaches equilibrium.
(c) What would the answer be if the flask had a volume of 10.0 L?

57. The equilibrium constant K_c for the reaction

$$CO(g) + H_2O(g) \rightleftharpoons CO_2(g) + H_2(g)$$

has the value 2.64×10^{-3} at 2300. K. If a mixture of 1.00 mol CO and 1.00 mol H_2O is allowed to come to equilibrium in a sealed, 1.00-L flask at 2300. K,
(a) calculate the final concentrations of all four species: CO, H_2O, CO_2, and H_2.
(b) calculate the equilibrium concentrations after an additional 1.00 mol each of CO and H_2O is added to the flask.

58. At 503 K the equilibrium constant K_c for the dissociation of N_2O_4,

$$N_2O_4(g) \rightleftharpoons 2\ NO_2(g)$$

has the value 40.0.
(a) Calculate the fraction of N_2O_4 left undissociated when 1.00 mol of this gas is heated to 503 K in a 10.0-L sealed container.
(b) If the volume is now reduced to 2.0 L, calculate the new fraction of N_2O_4 that is undissociated.
(c) Calculate all three equilibrium concentrations.

59. Consider the equilibrium at 25 °C

$$2\,SO_3(g) \rightleftharpoons 2\,SO_2(g) + O_2(g) \qquad K_c = 3.58 \times 10^{-3}$$

Suppose that 0.15 mol $SO_3(g)$, 0.015 mol $SO_2(g)$, and 0.0075 mol $O_2(g)$ are placed into a 10.0-L flask at 25 °C and the flask is sealed.
(a) Is the system at equilibrium?
(b) If the system is not at equilibrium, in which direction must the reaction proceed to reach equilibrium? Explain your answer.

60. At 2300 K the equilibrium constant for the formation of NO(g) is 1.7×10^{-3}.

$$N_2(g) + O_2(g) \rightleftharpoons 2\,NO(g)$$

(a) Analysis of the contents of a sealed flask at 2300 K shows that the concentrations of N_2 and O_2 are both 0.25 M and that of NO is 0.0042 M. Determine if the system is at equilibrium.
(b) If the system is not at equilibrium, in which direction does the reaction proceed?
(c) Calculate all three equilibrium concentrations.

61. Consider the equilibrium

$$N_2(g) + O_2(g) \rightleftharpoons 2\,NO(g)$$

At 2300 K the equilibrium constant $K_c = 1.7 \times 10^{-3}$. Suppose that 0.015 mol NO(g), 0.25 mol $N_2(g)$, and 0.25 mol $O_2(g)$ are placed into a 10.0-L flask, sealed, and heated to 2300 K.
(a) Determine whether the system is at equilibrium.
(b) If not, in which direction must the reaction proceed to reach equilibrium?
(c) Calculate the equilibrium concentrations of all three substances.

62. Consider the equilibrium

$$H_2(g) + I_2(g) \rightleftharpoons 2\,HI(g)$$

At 745 K the equilibrium constant $K_c = 50.0$. Suppose that 0.75 mol HI(g), 0.025 mol $H_2(g)$, and 0.025 mol $I_2(g)$ are placed into a sealed 20.0-L flask and heated to 745 K.
(a) Is the system at equilibrium?
(b) If not, in which direction must the reaction proceed to reach equilibrium?
(c) Calculate the equilibrium concentrations of all three substances.

63. The value of K_c is 3.7×10^{-23} at 25 °C for

$$C(graphite) + CO_2(g) \rightleftharpoons 2\,CO(g)$$

Describe what will happen if 3.5 mol CO and 3.5 mol CO_2 are mixed in a 1.5-L sealed graphite container with a suitable catalyst so that the reaction rate is rapid at this temperature.

Shifting a Chemical Equilibrium: Le Chatelier's Principle (Section 12-6)

64. K_p for this reaction is 0.16 at 25 °C:

$$2\,NOBr(g) \rightleftharpoons 2\,NO(g) + Br_2(g) \qquad \Delta_r H° = 16.3 \text{ kJ/mol}$$

Predict the effect of each change on the position of the equilibrium; that is, state which way the equilibrium shifts (left, right, or no change) when each change is made. Assume constant-volume conditions for parts (a), (b), and (c).
(a) Add more $Br_2(g)$.
(b) Remove some NOBr(g).
(c) Decrease the temperature.
(d) Increase the container volume.

65. The decomposition of NH_4HS is endothermic.

$$NH_4HS(s) \rightleftharpoons NH_3(g) + H_2S(g)$$

(a) Using Le Chatelier's principle, explain how increasing the temperature would affect the equilibrium.
(b) If more NH_4HS is added to a sealed flask in which this equilibrium exists, how is the equilibrium affected?
(c) What if some additional NH_3 is placed in a sealed flask containing an equilibrium mixture?
(d) What will happen to the partial pressure of NH_3 if some H_2S is removed from the flask?

66. Assume that the reaction

$$ClF_5(g) \rightleftharpoons ClF_3(g) + F_2(g)$$

is at equilibrium and the equilibrium conditions are changed as described. Indicate whether the forward or the reverse reaction rate is faster immediately after the change and explain your choice.
(a) Some $F_2(g)$ is added without changing the total volume.
(b) Some $ClF_3(g)$ is removed without changing the total volume.
(c) The total volume of the system is doubled.

67. Assume that the reaction

$$2\,HBr(g) \rightleftharpoons H_2(g) + Br_2(g)$$

is at equilibrium and the equilibrium conditions are changed as described. Indicate whether the forward or the reverse reaction rate is faster immediately after the change and explain your choice.
(a) Some HBr(g) is added without changing the total volume.
(b) Some $Br_2(g)$ is removed without changing the total volume.
(c) The total volume of the system is halved.

68. Hydrogen, bromine, and HBr in the gas phase are in equilibrium in a container of fixed volume.

$$H_2(g) + Br_2(g) \rightleftharpoons 2\,HBr(g) \qquad \Delta_r H° = -103.7 \text{ kJ/mol}$$

How will each of these changes affect the indicated quantities? Write "increase," "decrease," or "no change."

Change	$[Br_2]$	[HBr]	K_c	K_P
Some H_2 is added to the container.	___	___	___	___
The temperature of the gases in the container is increased.	___	___	___	___
The pressure of HBr is increased.	___	___	___	___

69. Nitrogen, oxygen, and nitrogen monoxide are in equilibrium in a container of fixed volume.

$$N_2(g) + O_2(g) \rightleftharpoons 2\ NO(g) \qquad \Delta_r H° = 180.5\ kJ/mol$$

How will each of these changes affect the indicated quantities? Write "increase," "decrease," or "no change."

Change	$[N_2]$	[NO]	K_c	K_P
Some NO is added to the container.	___	___	___	___
The temperature of the gases in the container is decreased.	___	___	___	___
The pressure of N_2 is decreased.	___	___	___	___

70. The equilibrium constant K_c for this reaction is 0.16 at 25 °C, and the standard reaction enthalpy is 16.1 kJ.

$$2\ NOBr(g) \rightleftharpoons 2\ NO(g) + Br_2(\ell)$$

Predict the effect of each of these changes on the position of the equilibrium; that is, state which way the equilibrium will shift (left, right, or no change) when each of these changes is made for a constant-volume system.
(a) Adding more Br_2
(b) Removing some NOBr
(c) Lowering the temperature

71. The formation of hydrogen sulfide from the elements is exothermic.

$$H_2(g) + \frac{1}{8}\ S_8(s) \rightleftharpoons H_2S(g) \qquad \Delta_r H° = -20.6\ kJ/mol$$

Predict the effect of each of these changes on the position of the equilibrium; that is, state which way the equilibrium will shift (left, right, or no change) when each change is made in a constant-volume system.
(a) Adding more sulfur
(b) Adding more H_2
(c) Raising the temperature

72. Heating a metal carbonate leads to decomposition.

$$BaCO_3(s) \rightleftharpoons BaO(s) + CO_2(g)$$

Predict the effect on the equilibrium of each change listed below. Answer by choosing (i) no change, (ii) shifts left, or (iii) shifts right. (Except for part (e), assume that the volume is constant.)
(a) Add $BaCO_3$
(b) Add CO_2
(c) Add BaO
(d) Raise the temperature
(e) Increase the volume of the reaction flask

73. Consider the equilibrium

$$PbCl_2(s) \rightleftharpoons Pb^{2+}(aq) + 2\ Cl^-(aq)$$

(a) Does the equilibrium concentration of aqueous lead(II) ion increase, decrease, or remain the same if some solid NaCl is added to a sealed flask? Explain your answer.
(b) Make a graph like the one in Figure 12.6 to illustrate what happens to each of the concentrations after the NaCl is added.

74. Consider the transformation of butane into 2-methylpropane (see Question 50). The system is originally at equilibrium at 25 °C in a 1.0-L sealed flask with [butane] = 0.010 M and [2-methylpropane] = 0.025 M. Suppose that 0.0050 mol 2-methylpropane is suddenly added to the flask, and the system shifts to a new equilibrium.
(a) Calculate the new equilibrium concentration of each gas.
(b) Make a graph like the one in Figure 12.6 to show how the concentrations of the isomers change when the 2-methylpropane is added.

75. Consider the system

$$4\ NH_3(g) + 3\ O_2(g) \rightleftharpoons 2\ N_2(g) + 6\ H_2O(\ell) \qquad \Delta_r H° = -1530.4\ kJ/mol$$

(a) How will the amount of ammonia at equilibrium be affected by
(i) removing $O_2(g)$ without changing the total gas volume?
(ii) adding $N_2(g)$ without changing the total gas volume?
(iii) adding water without changing the total gas volume?
(iv) expanding the container?
(v) increasing the temperature?
(b) Which of these changes (i to v) increases the value of K? Which decreases it?

76. Phosphorus pentachloride decomposes at high temperatures.

$$PCl_5(g) \rightleftharpoons PCl_3(g) + Cl_2(g)$$

An equilibrium mixture at some temperature consists of 3.120 g PCl_5, 3.845 g PCl_3, and 1.787 g Cl_2 in a sealed 1.00-L flask.
(a) If you add 1.418 g Cl_2 without changing the volume, how will the equilibrium be affected?
(b) Calculate the concentrations of all three substances when equilibrium is reestablished.

77. Predict whether the equilibrium for the photosynthesis reaction described by the equation

$$6\ CO_2(g) + 6\ H_2O(\ell) \rightleftharpoons C_6H_{12}O_6(s) + 6\ O_2(g) \qquad \Delta_r H° = 2801.69\ kJ/mol$$

would (i) shift to the right, (ii) shift to the left, or (iii) remain unchanged for each of these changes:
(a) decrease the concentration of CO_2 at constant volume.
(b) increase the partial pressure of O_2 at constant volume.
(c) remove one half of the $C_6H_{12}O_6$.
(d) decrease the total pressure by increasing the volume.
(e) increase the temperature.
(f) introduce a catalyst into a constant-volume system.

Equilibrium at the Nanoscale (Section 12-7)

78. For each of these reactions at 25 °C, indicate whether the entropy effect, the energy effect, both, or neither favors the reaction.
    (a) $N_2(g) + 3\ F_2(g) \rightleftharpoons 2\ NF_3(g)$ $\Delta_r H° = -249$ kJ/mol
    (b) $N_2F_4(g) \rightleftharpoons 2\ NF_2(g)$ $\Delta_r H° = 93.3$ kJ/mol
    (c) $N_2(g) + 3\ Cl_2(g) \rightleftharpoons 2\ NCl_3(g)$ $\Delta_r H° = 460$ kJ/mol
79. For each of these processes at 25 °C, indicate whether the entropy effect, the energy effect, both, or neither favors the process.
    (a) $C_3H_8(g) + 5\ O_2(g) \rightleftharpoons 3\ CO_2(g) + 4\ H_2O(g)$ $\Delta_r H° = -2045$ kJ/mol
    (b) $Br_2(g) \rightleftharpoons Br_2(\ell)$ $\Delta_r H° = -31$ kJ/mol
    (c) $2\ Ag(s) + 3\ N_2(g) \rightleftharpoons 2\ AgN_3(s)$ $\Delta_r H° = 618$ kJ/mol
80. For each of these chemical reactions, predict whether the equilibrium constant at 25 °C is greater than 1 or less than 1, or state that insufficient information is available. Also indicate whether each reaction is product-favored or reactant-favored.
    (a) $2\ NO(g) + O_2(g) \rightleftharpoons 2\ NO_2(g)$ $\Delta_r H° = -115$ kJ/mol
    (b) $2\ O_3(g) \rightleftharpoons 3\ O_2(g)$ $\Delta_r H° = -285$ kJ/mol
    (c) $N_2(g) + 3\ Cl_2(g) \rightleftharpoons 2\ NCl_3(g)$ $\Delta_r H° = 460$ kJ/mol
81. For each of these chemical reactions, predict whether the equilibrium constant at 25 °C is greater than 1 or less than 1, or state that insufficient information is available. Also indicate whether each reaction is product-favored or reactant-favored.
    (a) $2\ NaCl(s) \rightleftharpoons 2\ Na(s) + Cl_2(g)$ $\Delta_r H° = 823$ kJ/mol
    (b) $2\ CO(g) + O_2(g) \rightleftharpoons 2\ CO_2(g)$ $\Delta_r H° = -566$ kJ/mol
    (c) $3\ CO_2(g) + 4\ H_2O(g) \rightleftharpoons C_3H_8(g) + 5\ O_2(g)$ $\Delta_r H° = 2045$ kJ/mol

Controlling Chemical Reactions: The Haber-Bosch Process (Section 12-8)

82. Considering both the enthalpy effect and the entropy effect for the Haber-Bosch process, explain why choosing the temperature at which to run this reaction is very important.
83. Explain in your own words why it was so important to find a highly effective catalyst for the ammonia synthesis reaction before the Haber-Bosch process could become successful.
84. Ammonia is made in enormous quantities by the Haber-Bosch process. Sulfuric acid is made in even greater quantities by the *contact process*. A simplified version of this process can be represented by these three reactions.

$$S(s) + O_2(g) \rightleftharpoons SO_2(g)$$
$$2\ SO_2(g) + O_2(g) \rightleftharpoons 2\ SO_3(g)$$
$$SO_3(g) + H_2O(\ell) \rightleftharpoons H_2SO_4(\ell)$$

    (a) Use data from Appendix J to calculate $\Delta_r H°$ for each reaction.
    (b) Which reactions are exothermic? Which are endothermic?
    (c) In which of the reactions does entropy increase? In which does it decrease? In which does it stay about the same?
    (d) For which reaction(s) do low temperatures favor formation of products?
85. Lime, CaO(s), can be produced by heating limestone, $CaCO_3(s)$, to decompose it.
    (a) Write a balanced equation for the reaction.
    (b) Predict the sign of the enthalpy change for the reaction.
    (c) From the data in Appendix J, calculate $\Delta_r H°$ for this reaction at 25 °C to verify or contradict your prediction in part (b).
    (d) Predict the entropy effect for this reaction.
    (e) Is the reaction favored by entropy, energy, both, or neither?
    (f) Explain in terms of Le Chatelier's principle why limestone must be heated to make lime.

General Questions

These questions are not explicitly keyed to chapter topics; many require integration of several concepts.

86. Write equilibrium constant expressions, in terms of reactant and product concentrations, for each of these reactions.

$H_2O(\ell) \rightleftharpoons H^+(aq) + OH^-(aq)$ $K_c = 1.0 \times 10^{-14}$

$CH_3COOH(aq) \rightleftharpoons CH_3COO^-(aq) + H^+(aq)$ $K_c = 1.8 \times 10^{-5}$

$N_2(g) + 3\ H_2(g) \rightleftharpoons 2\ NH_3(g)$ $K_c = 3.5 \times 10^8$

Assume that all gases and solutes have initial concentrations of 1.0 mol/L. Then let the *first* reactant in each reaction change its concentration by $-x$.
    (a) Using the reaction table (ICE table) approach, write equilibrium constant expressions in terms of the unknown variable x for each reaction.
    (b) Which of these expressions yield quadratic equations?
    (c) How would you go about solving the others for x?
87. Write equilibrium constant expressions, in terms of reactant and product concentrations, for each of these reactions.

$2\ O_3(g) \rightleftharpoons 3\ O_2(g)$ $K_c = 7 \times 10^{56}$

$2\ NO_2(g) \rightleftharpoons N_2O_4(g)$ $K_c = 1.7 \times 10^2$

$HCOO^-(aq) + H^+(aq) \rightleftharpoons HCOOH(aq)$ $K_c = 5.6 \times 10^3$

$Ag^+(aq) + I^-(aq) \rightleftharpoons AgI(s)$ $K_c = 6.7 \times 10^{15}$

Assume that all gases and solutes have initial concentrations of 1.0 mol/L. Then, let the *first* reactant in each reaction change its concentration by $-x$.
    (a) Using the reaction table (ICE table) approach, write equilibrium constant expressions in terms of the unknown variable x for each reaction.
    (b) Which of these expressions yield quadratic equations?
    (c) How would you go about solving the others for x?

88. Consider the decomposition of ammonium hydrogen sulfide:

$$NH_4HS(s) \rightleftharpoons NH_3(g) + H_2S(g)$$

In a sealed flask at 25 °C are 10.0 g NH_4HS, ammonia with a partial pressure of 0.692 atm, and H_2S with a partial pressure of 0.0532 atm. When equilibrium is established, it is found that the partial pressure of ammonia has increased by 12.4%. Calculate K_P for the decomposition of NH_4HS at 25 °C.

89. The equilibrium constant K_c is 1.6×10^5 at 1297 K and 3.5×10^4 at 1495 K for the reaction

$$H_2(g) + Br_2(g) \rightleftharpoons 2\ HBr(g)$$

(a) Is $\Delta_r H°$ for this reaction positive or negative?
(b) Calculate K_c at 1297 K for the reaction

$$\tfrac{1}{2}\ H_2(g) + \tfrac{1}{2}\ Br_2(g) \rightleftharpoons HBr(g)$$

(c) Pure HBr is placed into an evacuated container of constant volume. The container is sealed and heated to 1297 K. Calculate the percentage of HBr that is decomposed to H_2 and Br_2 at equilibrium.

90. Many common nonmetallic elements exist as diatomic molecules at room temperature. When these elements are heated to 1500. K, the molecules break apart into atoms. A general equation for this type of reaction is

$$E_2(g) \rightleftharpoons 2\ E(g)$$

where E stands for an atom of each element. Equilibrium constants for dissociation of these molecules at 1500. K are

Species	K_c	Species	K_c
Br_2	8.9×10^{-2}	H_2	3.1×10^{-10}
Cl_2	3.4×10^{-3}	N_2	1×10^{-27}
F_2	7.4	O_2	1.6×10^{-11}

(a) Assume that 1.00 mol of each diatomic molecule is placed in a separate 1.0-L container, sealed, and heated to 1500. K. Calculate the equilibrium concentration of the atomic form of each element at 1500. K.
(b) From these results, predict which of the diatomic elements has the lowest bond dissociation energy, and compare your results with thermochemical calculations and with Lewis structures.

91. The total pressure for a mixture of N_2O_4 and NO_2 is 1.5 atm. If $K_P = 7.0$ (at 25 °C), calculate the partial pressure of each gas in the mixture.

$$2\ NO_2(g) \rightleftharpoons N_2O_4(g)$$

92. The chemistry of compounds composed of a transition metal and carbon monoxide has been an interesting area of research for more than 70 years. $Ni(CO)_4$ is formed by the reaction of nickel metal with carbon monoxide.
(a) Calculate the mass of $Ni(CO)_4$ that can be formed if you combine 2.05 g CO with 0.125 g nickel metal.
(b) An excellent way to make pure nickel metal is to decompose $Ni(CO)_4$ in a vacuum at a temperature slightly higher than room temperature. If the standard formation enthalpy of $Ni(CO)_4$ gas is −602.9 kJ/mol, calculate the enthalpy change for this decomposition reaction.

$$Ni(CO)_4(g) \longrightarrow Ni(s) + 4\ CO(g)$$

(c) Predict whether there is an increase or a decrease in entropy when this reaction occurs.
(d) In an experiment at 100. °C it is determined that with 0.010 mol $Ni(CO)_4(g)$ initially present in a sealed 1.0-L flask, only 0.000010 mol remains after decomposition.
(i) Calculate the equilibrium concentration of CO in the flask.
(ii) Calculate the value of the equilibrium constant K_c for this reaction at 100. °C.
(iii) Calculate the equilibrium constant K_P for this reaction at 100. °C.

93. A small sample of *cis*-dichloroethene in which one carbon atom is the radioactive isotope ^{14}C is added to an equilibrium mixture of the *cis* and *trans* isomers at a certain temperature. Eventually, 40% of the radioactive molecules are found to be in the *trans* configuration at any given time.
(a) Determine the value of K_c for the *cis* $\rightleftharpoons$ *trans* equilibrium.
(b) What would have happened if a small sample of radioactive *trans* isomer had been added instead of the *cis* isomer?

94. Solid ammonium iodide decomposes to ammonia and hydrogen iodide gases at sufficiently high temperatures.

$$NH_4I(s) \rightleftharpoons NH_3(g) + HI(g)$$

The equilibrium constant, K_P, for the decomposition at 673 K is 0.215. A 15.0-g sample of ammonium iodide is sealed in a 5.00-L flask and heated to 673 K.
(a) Calculate the total pressure in the flask at equilibrium.
(b) Calculate the amount (in moles) of ammonium iodide that decomposes.

95. These amounts of HI, H_2, and I_2 are introduced into a 10.00-L flask. The flask is sealed and heated to 745 K.

	n_{HI} (mol)	n_{H_2} (mol)	n_{I_2} (mol)
Case a	1.0	0.10	0.10
Case b	10.	1.0	1.0
Case c	10.	10.	1.0
Case d	5.62	0.381	1.75

The equilibrium constant for the reaction

$$2\ HI(g) \rightleftharpoons H_2(g) + I_2(g)$$

has the value 0.0200 at 745 K. In which cases does the concentration of HI increase as equilibrium is attained, and in which cases does the concentration of HI decrease?

96. These amounts of CO(g), H_2O(g), CO_2(g), and H_2(g) are introduced into a 10.00-L flask. The flask is sealed and heated to a very high temperature.

	n_{CO} (mol)	n_{H_2O} (mol)	n_{CO_2} (mol)	n_{H_2} (mol)
Case a	1.0	0.10	0.10	0.10
Case b	10.	1.0	1.0	1.0
Case c	10.	10.	1.0	1.0
Case d	5.62	0.381	1.75	1.75

The equilibrium constant for the reaction

$$CO(g) + H_2O(g) \rightleftharpoons CO_2(g) + H_2(g)$$

has the value $K_c = 4.00$ at this temperature. For which cases will the concentration of CO increase as equilibrium is attained, and in which cases will the concentration of CO decrease?

97. Carbonylbromide, $COBr_2$, can be formed by combining carbon monoxide and bromine gas.

$$CO(g) + Br_2(g) \rightleftharpoons COBr_2(g)$$

When equilibrium is established at 346 K, the partial pressures (in atm) of $COBr_2$, CO, and Br_2 are 0.12, 1.00, and 0.65, respectively.
(a) Calculate K_P at 346 K.
(b) Enough bromine condenses to decrease its partial pressure to 0.50 atm. Calculate the equilibrium partial pressures of all gases after equilibrium is re-established.

98. Mustard gas was used in chemical warfare in World War I. Mustard gas can be produced according to this reaction:

$$SCl_2(g) + 2\ C_2H_4(g) \rightleftharpoons S(CH_2CH_2Cl)_2(g)$$

An evacuated 5.00-L flask at 20.0 °C is filled with 0.258 mol SCl_2 and 0.592 mol C_2H_4 and sealed. After equilibrium is established, 0.0349 mol mustard gas is present.
(a) Calculate the partial pressure of each gas at equilibrium.
(b) Calculate K_c at 20.0 °C.

99. Limestone decomposes at high temperatures.

$$CaCO_3(s) \rightleftharpoons CaO(s) + CO_2(g)$$

At 1000. °C, $K_P = 3.87$. Pure $CaCO_3$ is placed into an empty 5.00-L flask. The flask is sealed and heated to 1000. °C. Calculate the mass of $CaCO_3$ that must decompose to achieve the equilibrium pressure of CO_2.

100. A sample of pure SO_3 weighing 0.8312 g was placed into a 1.00-L flask, sealed, and heated to 1100. K to decompose it partially.

$$2\ SO_3(g) \rightleftharpoons 2\ SO_2(g) + O_2(g)$$

If a total pressure of 1.295 atm was developed, calculate the value of K_c for this reaction at this temperature.

Applying Concepts

These questions test conceptual learning.

101. Two molecules of A react to form one molecule of B, as in the reaction

$$2\ A(g) \rightleftharpoons B(g)$$

Three experiments are done at different temperatures and equilibrium concentrations are measured. For each experiment, calculate the equilibrium constant, K_c.
(a) [A] = 0.74 mol/L, [B] = 0.74 mol/L
(b) [A] = 2.0 mol/L, [B] = 2.0 mol/L
(c) [A] = 0.01 mol/L, [B] = 0.01 mol/L
What can you conclude about this statement: "If the concentrations of reactants and products are equal, then the equilibrium constant is always 1.0."

102. Suppose that you have heated a mixture of *cis-* and *trans-*2-pentene to 600. K, and after 1 h you find that the composition is 40% *cis.* After 4 h the composition is found to be 42% *cis,* and after 8 h it is 42% *cis.* Next, you heat the mixture to 800. K and find that the composition changes to 45% *cis.* When the mixture is cooled to 600. K and allowed to stand for 8 h, the composition is found to be 42% *cis.* Is this system at equilibrium at 600. K? Or, would more experiments be needed before you could conclude that it was at equilibrium? If so, what experiments would you do?

103. In Table 12.1 (← **Sec. 12-3a**) the equilibrium constant for the reaction

$$\tfrac{1}{8}\ S_8(s) + O_2(g) \rightleftharpoons SO_2(g)$$

is given as 4.2×10^{52}. If this reaction is so product-favored, why can large piles of yellow sulfur exist in our environment (as they do in Louisiana and Texas)?

104. For the reaction

cis-2-butene ⇌ *trans*-2-butene

K_c is 1.65 at 500. K, 1.47 at 600. K, and 1.36 at 700. K. Predict whether the conversion from the *cis* to the *trans* isomer of 2-butene is exothermic or endothermic.

105. Figure 12.3 (← **Sec. 12-3a**) shows the equilibrium mixture of N_2O_4 and NO_2 at two different temperatures. Imagine that you can shrink yourself down to the size of the molecules in the two glass tubes and observe their behavior for a short period of time. Write a brief description of what you observe in each of the tubes.

106. Imagine yourself to be the size of ions and molecules inside a beaker containing this equilibrium mixture with a K_c greater than 1.

$$\underset{\text{pink}}{Co(H_2O)_6^{2+}(aq)} + 4\ Cl^-(aq) \rightleftharpoons \underset{\text{blue}}{CoCl_4^{2-}(aq)} + 6\ H_2O(\ell)$$

Write a brief description of what you observe around you before and after additional water is added to the mixture.

Diagrams for Questions 107 and 108.

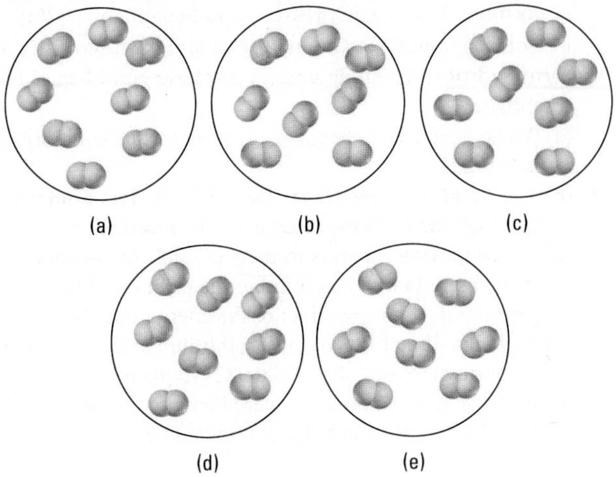
(a)
(b)
(c)
(d)
(e)

107. Which of the diagrams for Questions 107 and 108 represent equilibrium mixtures for the reaction

$$A_2(g) + B_2(g) \rightleftharpoons 2\,AB(g)$$

at temperatures where K_c is between 0.1 and 100?

108. Which of the diagrams for Questions 107 and 108 best represents an equilibrium mixture with an equilibrium constant of
(a) 0.44? (b) 4.0? (c) 36?

109. Draw a nanoscale (particulate) level diagram for an equilibrium mixture of

$$CO(g) + H_2O(g) \rightleftharpoons CO_2(g) + H_2(g) \qquad K_c = 4.00$$

110. The diagram represents an equilibrium mixture for the reaction

$$N_2(g) + O_2(g) \rightleftharpoons 2\,NO(g)$$

Estimate the equilibrium constant.

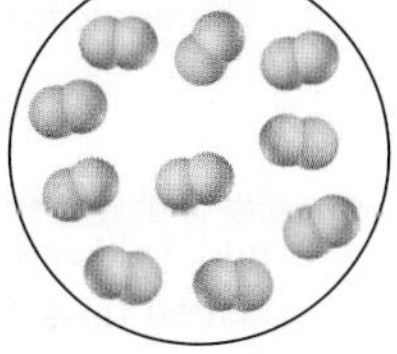

111. The equilibrium constant, K_c, is 1.05 at 350 K for the endothermic reaction

$$2\,CH_2Cl_2(g) \rightleftharpoons CH_4(g) + CCl_4(g)$$

Which of the diagrams for Question 111 represents an equilibrium mixture at 350 K?

112. For the reaction in Question 111, which diagram for Question 111 represents an equilibrium mixture at
(a) a temperature below 350 K?
(b) a temperature above 350 K?

113. A solid sample of benzoic acid, a carboxylic acid, is in equilibrium with an aqueous solution of benzoic acid. A tiny quantity of D_2O, water containing the isotope ^2H, deuterium, is added to the solution. The solution is allowed to stand at constant temperature for several hours, after which some of the solid benzoic acid is removed and analyzed. The benzoic acid is found to contain a tiny quantity of deuterium, D, and the formula of the deuterium-containing molecules is C_6H_5COOD. Explain how this can happen.

114. In a second experiment with the benzoic acid in Question 113, a tiny quantity of water that contains the isotope ^{18}O is added to the solution of benzoic acid in water. When some of the solid benzoic acid is analyzed, no ^{18}O is found in the benzoic acid. Compare this situation with the experiment involving deuterium, and explain how the results of the two experiments can differ as they do.

115. Samples of N_2O_4 can be prepared in which both nitrogen atoms are the heavier isotope ^{15}N. Designating this isotope as N*, we can write the formula of the molecules in such a sample as $O_2N^*{-}N^*O_2$ and the formula of typical N_2O_4 as $O_2N{-}NO_2$. When a tiny quantity of $O_2N^*{-}N^*O_2$ is introduced into an equilibrium mixture of N_2O_4 and NO_2, the ^{15}N immediately becomes distributed among both N_2O_4 and NO_2 molecules, and in the N_2O_4 it is invariably in the form $O_2N^*{-}NO_2$. Explain how this observation supports the idea that equilibrium is dynamic.

Diagrams for Question 111.

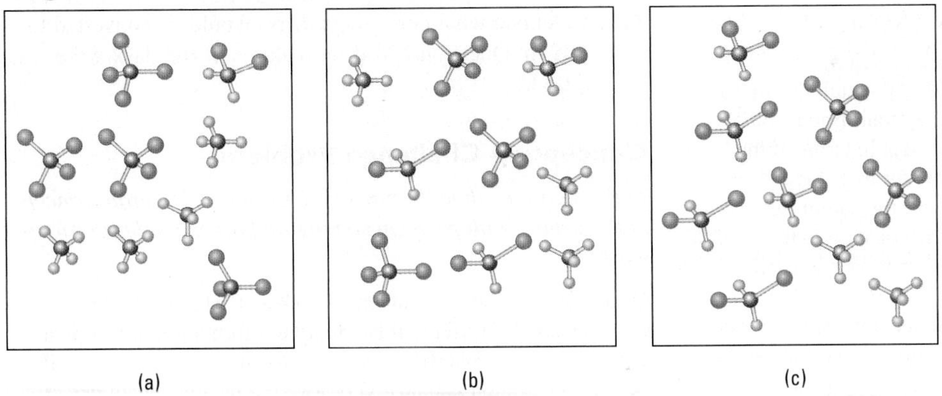
(a)
(b)
(c)

116. For the equilibrium

$$Co(H_2O)_6^{2+}(aq) + 4\ Cl^-(aq) \rightleftharpoons CoCl_4^{2-}(aq) + 6\ H_2O(\ell)$$

pink — blue

K_c is somewhat greater than 1. If water is added to a blue solution of $CoCl_4^{2-}(aq)$, the color changes from blue to pink.

(a) Does water appear in the equilibrium constant expression for this reaction?

(b) How can adding water shift the equilibrium to the left?

(c) Is this shift in the equilibrium in accord with Le Chatelier's principle? Why or why not?

More Challenging Questions

These questions require more thought and integrate several concepts.

117. A sealed 15.0-L flask at 300. K contains 64.4 g of a mixture of NO_2 and N_2O_4 in equilibrium. Calculate the total pressure in the flask. (For $2\ NO_2(g) \rightleftharpoons N_2O_4(g)$ $K_P =$ 6.67 at 300. K.)

118. The equilibrium constant K_c has a value of 3.30 at 760. K for the decomposition of phosphorus pentachloride,

$$PCl_5(g) \rightleftharpoons PCl_3(g) + Cl_2(g)$$

(a) Calculate the equilibrium concentrations of all three species arising from the decomposition of 0.75 mol PCl_5 in a sealed 5.00-L vessel.

(b) Calculate the equilibrium concentrations of all three species resulting from an initial mixture of 0.75 mol PCl_5 and 0.75 mol PCl_3 in a sealed 5.00-L vessel.

119. Use the fact that the equilibrium constant K_c equals the ratio of the forward rate constant divided by the reverse rate constant, together with the Arrhenius equation $k = Ae^{-E_a/RT}$, to show that a catalyst does not affect the value of an equilibrium constant even though the catalyst increases the rates of forward and reverse reactions. Assume that the frequency factors A for forward and reverse reactions do not change, and that the catalyst lowers the activation barrier for the catalyzed reaction.

120. Consider the reaction mechanism given for formation of NOBr in Section 11-7a.

Step 1: $NO(g) + Br_2(g) \rightleftharpoons NOBr_2(g)$ fast

Step 2: $NOBr_2(g) + NO(g) \longrightarrow 2\ NOBr(g)$ slow

(a) What is the overall stoichiometric equation?

(b) Assume that when the reaction reaches equilibrium, each step in the mechanism also reaches equilibrium. Write the equilibrium constant for each step in terms of the forward and reverse rate constants for that step.

(c) Use your result in part (b) to derive the equilibrium constant for the overall stoichiometric equation in terms of the rate constants for the forward and reverse reactions of each step in the mechanism.

(d) Does your result confirm the statement made in the margin in Section 12-2 that the equilibrium constant can be obtained by taking the product of the rate constants for all forward steps and dividing by the product of the rate constants for all reverse steps?

121. When a mixture of hydrogen and bromine is maintained at normal atmospheric pressure and heated above 200. °C in a closed container, the hydrogen and bromine react to form hydrogen bromide and a gas-phase equilibrium is established.

(a) Write a balanced chemical equation for the equilibrium reaction.

(b) Use bond enthalpies from Table 6.2 (**← Sec. 6-6b**) to estimate the enthalpy change for the reaction.

(c) Based on your answers to parts (a) and (b), which is more important in determining the position of this equilibrium, the entropy effect or the energy effect?

(d) In which direction will the equilibrium shift as the temperature increases above 200. °C? Explain.

(e) Suppose that the pressure were increased to triple its initial value. In which direction would the equilibrium shift?

(f) Why is the equilibrium not established at room temperature?

122. At 25 °C the vapor pressure of water is 0.03126 atm.

(a) Calculate K_P and K_c for

$$H_2O(\ell) \rightleftharpoons H_2O(g)$$

(b) Calculate the value of K_P for this same system at 100. °C.

(c) Suggest a general rule for calculating K_P for any liquid in equilibrium with its vapor at its normal boiling point.

123. A student studies the equilibrium

$$I_2(g) \rightleftharpoons 2\ I(g)$$

at a high temperature. She finds that the total pressure at equilibrium is 40.% greater than it was originally, when only I_2 was present at a pressure of 1.00 atm in the same sealed container. Calculate K_p.

124. The equilibrium constant K_c for the reaction

$$N_2(g) + 3\ H_2(g) \rightleftharpoons 2\ NH_3(g)$$

has the value 5.97×10^{-2} at 500. °C. If 1.00 mol N_2 gas and 1.00 mol H_2 gas are heated to 500. °C in a 10.00-L sealed flask together with a catalyst, calculate the percentage of N_2 converted to NH_3. (*Hint:* Assume that only a very small fraction of the reactants is converted to products. Obtain an approximate answer and use it to obtain a more accurate result.)

125. Calculate what percentage of N_2 would be converted to NH_3 in Question 124 if the volume of the flask were 5.00 L.

Conceptual Challenge Problems

These rigorous, thought-provoking problems integrate conceptual learning with problem solving and are suitable for group work.

Conceptual Challenge Problems CP12.A, CP12.B, CP12.C, CP12.D, and CP12.E are related to the information in this paragraph. Aqueous iron(III) ions, $Fe^{3+}(aq)$, are nearly colorless. If their concentration is 0.001 M or lower, a person cannot detect their color. Thiocyanate ions, $SCN^-(aq)$, are colorless also, but monothiocyanatoiron(III) ions, $Fe(SCN)^{2+}(aq)$, can be detected at very low concentrations because of their color. These ions are light amber in very dilute solutions, but as their concentration increases, the color intensifies and appears blood-red in more concentrated solutions. Suppose you prepared a stock solution by mixing equal volumes of 1.0×10^{-3}-M solutions of both iron(III) nitrate and potassium thiocyanate solutions. The equilibrium reaction is

$$\underset{\text{colorless}}{Fe^{3+}(aq)} + \underset{\text{colorless}}{SCN^-(aq)} \rightleftharpoons \underset{\text{amber}}{Fe(SCN)^{2+}(aq)}$$

CP12.A (Section 12-1) Describe how you would use 5-mL samples of the stock solution and additional solutions of 0.010-M $Fe^{3+}(aq)$ and 0.010-M $SCN^-(aq)$ to show experimentally that the reaction between $Fe^{3+}(aq)$ and $SCN^-(aq)$ does not go to completion but instead reaches an equilibrium state in which appreciable quantities of reactants and product are present. (Refer to the introductory paragraph for further information.)

CP12.B (Section 12-1) Suppose that you added one drop of 0.010-M $Fe^{3+}(aq)$ to a 5-mL sample of the stock solution, followed by 10 drops of 0.010-M $SCN^-(aq)$. You treated a second 5-mL sample of the stock solution by first adding 10 drops of 0.010-M $SCN^-(aq)$, followed by one drop of $Fe^{3+}(aq)$. How would the color intensity of these two solutions compare after the same quantities of the same solutions were added in reverse order? (Refer to the introductory paragraph for further information.)

CP12.C (Section 12-6) Predict what will happen if you add a small crystal of sodium acetate to a 5-mL sample of the stock solution (described in the introductory paragraph) so that some acetatoiron(III) ion, a coordination complex, is formed.

CP12.D (Section 12-6) Predict what will happen if you begin to add a 0.010-M solution of $Fe^{3+}(aq)$ drop by drop to a 5-mL sample of the stock solution (described in the introductory paragraph) until the total volume becomes 10. mL. (A 0.010-M solution of Fe^{3+} ions is pale yellow.) Predict what will happen if you do the same experiment but add 0.010-M $SCN^-(aq)$ to the stock solution. Predict what will happen if you mix 0.010-M $Fe^{3+}(aq)$ with 0.010-M $SCN^-(aq)$. Would the results of these three experiments be similar or different? Explain.

CP12.E (Section 12-6) Predict what will happen if you put a 5-mL sample of the stock solution (described in the introductory paragraph) in a hot water bath. Predict what will happen if the solution is placed in an ice bath.

CP12.F (Section 12-4) Consider the equilibrium reaction between dioxygen and trioxygen (ozone). Calculate the minimum volume of air (21% dioxygen by volume) at 1.00 atm and 25 °C that you would predict to have at least one molecule of trioxygen, if the only source of trioxygen were its formation from dioxygen, and if the atmospheric system were at equilibrium.

$$3\ O_2(g) \rightleftharpoons 2\ O_3(g) \qquad K_c = 6.3 \times 10^{-58} \quad \text{(at 25°C)}$$

(The volume of 1 mol air at 1 atm and 25 °C is 24.45 L.)

13 | The Chemistry of Solutes and Solutions

John W. Moore

Oceans are vast aqueous solutions containing dissolved solids, liquids, and gases. These crashing waves carry the dissolved materials into the air. What factors affect the solubility of solids, liquids, and gases in ocean water? (Sections 13-1 to 13-5) How is the composition of ocean water (or any other solution) described? (Section 13-6) What effects do solutes have on the properties of solutions? (Section 13-7) This chapter describes solutions and the nanoscale to macroscale links that govern solution behavior.

Every day we all encounter many solutions, such as soft drinks, juices, coffee, and gasoline. A *solution* is a homogeneous mixture of two or more substances (← **Sec. 1-7**). The component present in greatest quantity is usually called the *solvent;* the other components are *solutes* (← **Sec. 3-10**). In sweetened iced tea, for example, water is the solvent; sugar and soluble extracts of tea are the solutes.

Although solids dissolved in liquids or mixtures of liquids are the most common types of solutions, other kinds are possible. The possibilities, which encompass all three physical states of matter, are listed in Table 13.1. Chemistry

Table 13.1 Types of Solutions

Solution Type	Example
Gas in gas	Air—a mixture principally of N_2 and O_2, but containing other gases
Gas in liquid	Sparkling water—CO_2 in water
Gas in solid	Hydrogen in palladium metal
Liquid in liquid	Motor oil—a mixture of liquid hydrocarbons
Solid in liquid	Maple syrup—primarily sugar and water
Solid in solid	Alloys—bronze (Cu and Sn) or pewter (Sn, Sb, and Pb)

often focuses on liquid solutions, and in particular *aqueous solutions*, those in which water is the solvent, because water is the most important solvent on our planet.

In this chapter we explore in some detail the macroscale to nanoscale level connections between solutes and solvents to answer questions such as:

- Why does a particular solvent readily dissolve one kind of solute, but not another? Water, for example, dissolves salt, but does not dissolve gasoline.
- In what ways can the composition of a solution be expressed?
- Is energy transferred to or from the surroundings when a solute dissolves?
- How do changes in temperature or pressure affect the solubility of a solute in a given solvent? Why, for example, does a cold, carbonated beverage become "flat" when it is opened and warmed to room temperature?

We will apply what you have learned (← **Secs. 7-6a and 7-6b**) about intermolecular forces—London forces, dipole-dipole forces, and hydrogen bonding—to the interactions among solute and solvent molecules. We will also utilize the thermodynamic and equilibrium principles presented in Chapters 4 and 12, respectively. These principles help us to understand solubility and the effect that solutes have on the vapor pressures, melting points, and boiling points of solvents. We also discuss unwanted solutes in natural and polluted water.

13-1 Solubility and Intermolecular Forces

AP Images/U.S. Navy, Mass Communication Specialist 1st Class Michael B. Watkins

BP oil spill in Gulf of Mexico. The oil floats on top of the water, but does not dissolve in it.

An old adage correctly states that "oil and water don't mix." Chemists use a related saying: "Like dissolves like," where "like" refers to solutes and solvents whose molecules attract each other by similar types of intermolecular forces. By extension of this idea, substances with different types of intermolecular forces usually do not dissolve extensively in each other (Figure 13.1).

Consider, for example, dissolving the hydrocarbon octane, C_8H_{18}, in carbon tetrachloride, CCl_4 (Figure 13.2a). Both are nonpolar liquids so their molecules are attracted by London forces. These two *liquids dissolve in each other in any proportion* and are said

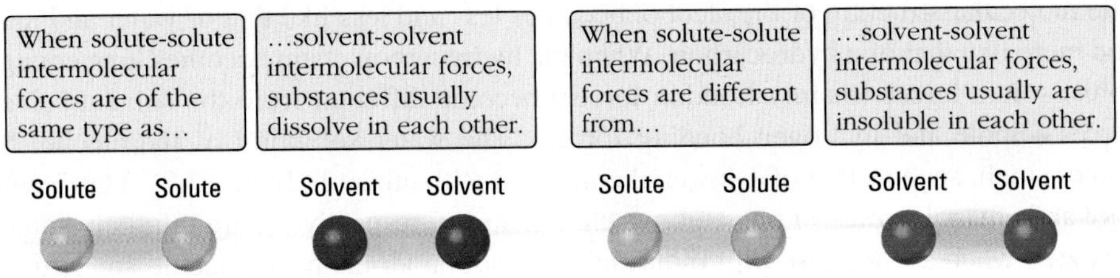
When solute-solute intermolecular forces are of the same type as...
...solvent-solvent intermolecular forces, substances usually dissolve in each other.
When solute-solute intermolecular forces are different from...
...solvent-solvent intermolecular forces, substances usually are insoluble in each other.
Solute
Solute
Solvent
Solvent
Solute
Solute
Solvent
Solvent

Figure 13.1 Like dissolves like.

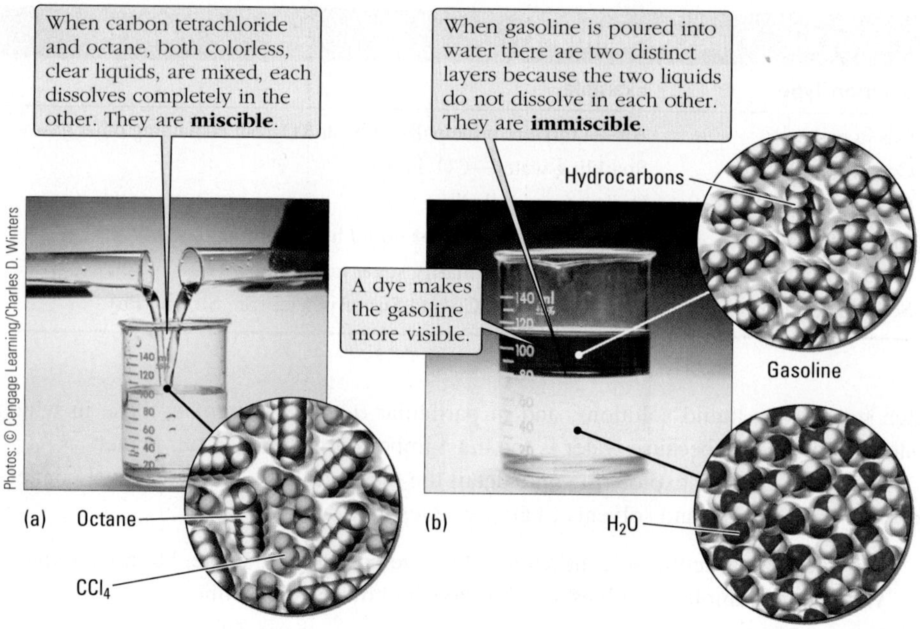
When carbon tetrachloride and octane, both colorless, clear liquids, are mixed, each dissolves completely in the other. They are **miscible**.
When gasoline is poured into water there are two distinct layers because the two liquids do not dissolve in each other. They are **immiscible**.
Hydrocarbons
A dye makes the gasoline more visible.
Gasoline
Photos: © Cengage Learning/Charles D. Winters
(a)
Octane
CCl_4
(b)
H_2O

Figure 13.2 **(a) Miscible and (b) immiscible liquids.**

to be **miscible**. In contrast, gasoline, a mixture of nonpolar hydrocarbons, is not miscible with water, a polar substance (Figure 13.2b). The nonpolar hydrocarbons have London forces only, but water molecules also have dipole and hydrogen-bonding forces. Thus, solute-solute intermolecular forces differ from solvent-solvent intermolecular forces and the liquids do not dissolve. *Liquids that do not dissolve appreciably in each other*, such as gasoline and water, are **immiscible**.

The discussion in this section answers the question posed in Chapter 1 (← Sec. 1-1), "Why do some substances dissolve in water but others do not?"

Because water is such an important solvent, substances with different types of intermolecular forces are often compared with water and referred to as hydrophilic (water loving) or hydrophobic (water fearing). In a **hydrophilic** substance or molecular structure *the primary intermolecular forces are dipole and hydrogen-bonding forces.* In a **hydrophobic** substance or molecular structure (← **Sec. 7-6d**) *the primary intermolecular forces are London forces.*

Table 13.2 Solubilities of Alcohols in Water, 20 °C

Name	Formula	Solubility (g/100 g)
Methanol	CH_3OH	Miscible
Ethanol	CH_3CH_2OH	Miscible
1-Propanol	$CH_3CH_2CH_2OH$	Miscible
1-Butanol	$CH_3CH_2CH_2CH_2OH$	7.9
1-Pentanol	$CH_3CH_2CH_2CH_2CH_2OH$	2.7
1-Hexanol	$CH_3CH_2CH_2CH_2CH_2CH_2OH$	0.6

The differing solubilities of various alcohols in water shown in Table 13.2 further illustrate the "like dissolves like" principle. Simple alcohols containing only a few carbon atoms, such as methanol and ethanol, dissolve completely in water. However, as the length of the carbon chain increases, solubility in water decreases.

Hydrogen-bonding forces are by far the most important intermolecular forces in pure water and pure ethanol (Figure 13.3). As the carbon chain in the alcohols in Table 13.2 becomes longer, London forces, which depend on the number of electrons, become more and more important. The molecular structure of the alcohol becomes less and less like that of water and more and more like that of a hydrocarbon. When the hydrocarbon chain becomes long enough, solute-solute forces (mainly London forces) become different from the solvent-solvent forces (dipole and hydrogen-bonding forces). The water solubility of the alcohol becomes small, such as that of 1-hexanol compared with ethanol (Table 13.2). Thus, methanol and ethanol, with just one and two carbon atoms, respectively, are infinitely soluble in water, whereas alcohols with more than six carbon atoms per molecule are virtually insoluble in water.

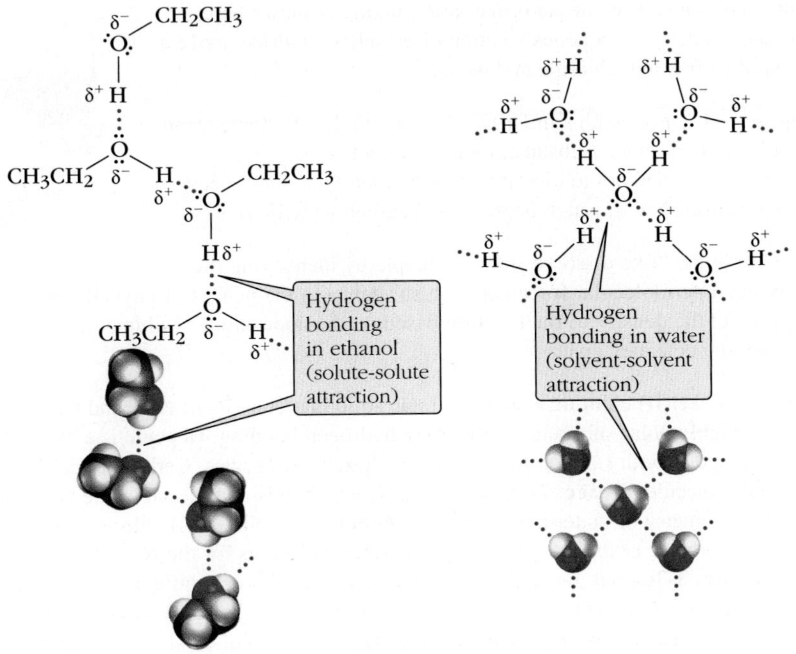
Hydrogen bonding in ethanol (solute-solute attraction)
Hydrogen bonding in water (solvent-solvent attraction)

Figure 13.3 Hydrogen bonding in pure ethanol and in water.

Here is a summary of the principle of "like dissolves like":

- ***Substances with similar intermolecular forces are likely to be soluble in each other.***
- ***Solutes do not readily dissolve in solvents whose intermolecular forces are quite different from their own.***

CONCEPTUAL EXERCISE 13.1

Predicting Water Solubility

How could the data in Table 13.2 be used to predict the solubility in water of 1-octanol or 1-decanol?

Answers to **EXERCISES** are provided at the back of this book in Appendix L.

EXERCISES that are labeled **CONCEPTUAL** are designed to test your understanding of one or more concepts; they usually involve qualitative rather than quantitative thinking.

CONCEPTUAL EXERCISE 13.2

Predicting Solubility

You have a sample of 1-octanol and a sample of methanol. Which is more water-soluble? Which is more soluble in gasoline? Explain your choices in terms of "like dissolves like."

PROBLEM-SOLVING EXAMPLE 13.1

Predicting Solubilities

A beaker initially contains three liquid layers as shown here. The colorless top layer is heptane, C_7H_{16} (density, d = 0.684 g/mL); the green middle layer is an aqueous solution of $NiCl_2$ (d = 1.10 g/mL); the colorless bottom layer is carbon tetrachloride, CCl_4 (d = 1.59 g/mL). The liquids in the beaker are stirred vigorously so that they are thoroughly mixed. Using the principle of "like dissolves like" and other information given, predict the appearance and composition of the contents of the beaker after the liquids are mixed and explain why the beaker's contents have that composition.

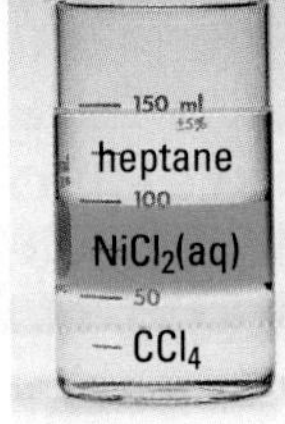
150 ml
heptane
100
NiCl2(aq)
50
CCl4

© Cengage Learning/Charles D. Winters

Result The appearance of the beaker after mixing is shown here. The beaker contains an aqueous solution of nickel(II) chloride above a solution of carbon tetrachloride and heptane.

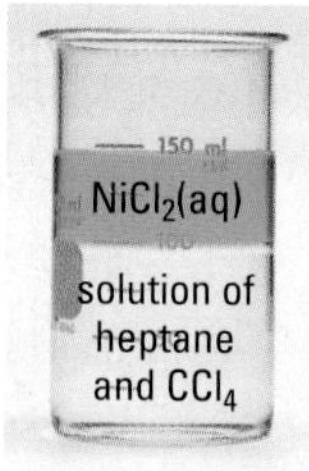

$NiCl_2(aq)$
solution of heptane and CCl_4

Analyze Substances with similar types of intermolecular forces tend to dissolve in each other. Substances with different types of intermolecular forces tend not to dissolve in each other. Four substances are in the beaker: $NiCl_2$, water, heptane, and carbon tetrachloride.

Plan Apply the "like dissolves like" principle by identifying the polarity and intermolecular forces of each substance in the beaker. If any substances dissolve, predict the density of the solution based on the approximate volume of each substance and the densities given.

Execute Nickel(II) chloride is a green, ionic solid that consists of Ni^{2+} and Cl^- ions. Water is a highly polar substance with strong hydrogen bonding. Heptane is a hydrocarbon and thus nonpolar with London forces between hexane molecules. Carbon tetrachloride is a tetrahedral molecule (← **Sec. 7-2a**) with polar C—Cl bonds, but the molecule is nonpolar due to the symmetry of its tetrahedral shape (← **Sec. 7-5**). Nickel(II) chloride should be soluble in water due to the attraction of polar water molecules for the Ni^{2+} cations and Cl^- anions (← **Sec. 3-3c**), but insoluble in heptane and CCl_4. Water should not dissolve in either heptane or CCl_4. Once the mixture is stirred, the nonpolar liquids dissolve in each other. Their volumes are nearly equal so the density of the solution should be approximately the average of their densities, that is, $(0.684 + 1.59)/2 = 1.14$ g/mL, which is greater than the density of the $NiCl_2$ solution. Thus, the nonpolar solution forms the bottom layer.

PROBLEM-SOLVING PRACTICE 13.1

Using the principle of "like dissolves like," predict whether
(a) ethylene glycol, $HOCH_2CH_2OH$, dissolves in gasoline.
(b) molecular iodine dissolves in carbon tetrachloride.
(c) motor oil dissolves in carbon tetrachloride.
Explain your predictions.

PROBLEM-SOLVING PRACTICE answers are provided at the back of this book in Appendix K.

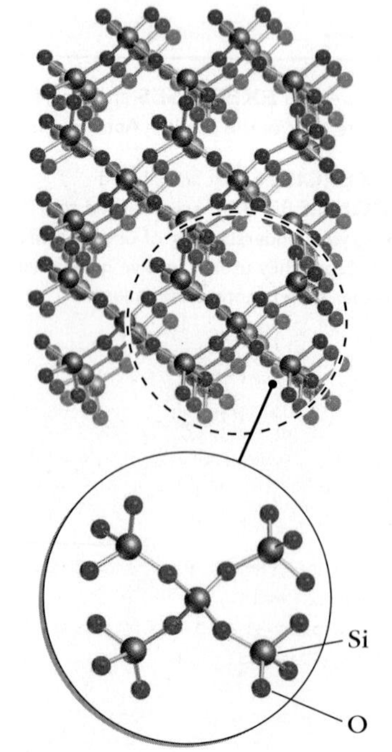

Si
O

A portion of a quartz, SiO_2, structure. The structure is based on SiO_4 tetrahedra linked through shared oxygen atoms.

Solids such as quartz, SiO_2, in which atoms are held together by an extensive network of covalent bonds (← **Sec. 9-7**) are usually not soluble in either polar or nonpolar solvents. In quartz, the silicon and oxygen atoms form SiO_4 tetrahedra linked through shared oxygen atoms. The links are strong covalent bonds that are not broken by weaker attractions to solvent molecules. Thus, quartz (and sand derived from it) is insoluble in water or any other solvent at room temperature. Sandy beaches do not dissolve in ocean water.

Whether a substance dissolves in water or in triglycerides—nonpolar fats or oily substances (← **Sec. 10-5c**)—plays an important role in our body chemistry. Vitamins, for example, are either water-soluble or fat-soluble. A major significance of this difference is the danger of overdosing on fat-soluble vitamins because they are stored in fatty tissues and may accumulate to harmful levels. By contrast, overdosing on water-soluble vitamins is difficult and uncommon because these vitamins are not stored in the body and any excess of them is excreted in urine.

PROBLEM-SOLVING EXAMPLE 13.2

Solubility and Noncovalent Forces

Use the structural formulas of niacin (nicotinic acid) and vitamin A to determine which vitamin is more soluble in water and which is more soluble in fat.

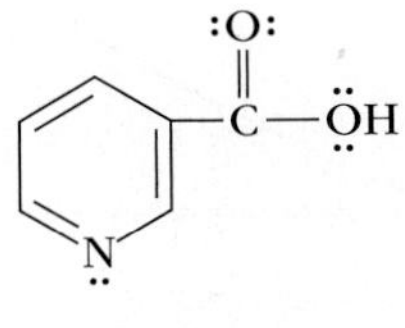

niacin
(nicotinic acid)

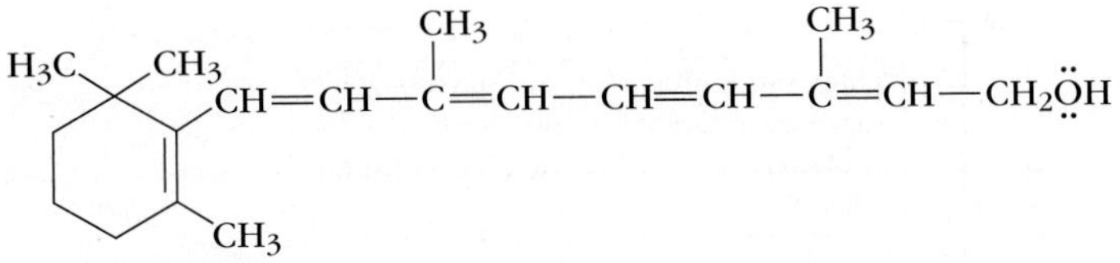

vitamin A

Result Niacin is water-soluble; vitamin A is fat-soluble.

Analyze Given structural formulas you are asked to determine whether substances are water-soluble or fat-soluble. Fats are nonpolar (hydrophobic).

Plan Examine the molecular structures; identify hydrophilic and hydrophobic regions.

Execute In niacin, the —COOH group and nitrogen atom portion are polar and can hydrogen bond; the hydrocarbon-like part of the molecule contains only five carbon atoms. Consequently, niacin is soluble in water and insoluble in fats. Vitamin A is mainly hydrocarbon-like, with a single alcohol group at one end. Therefore, vitamin A is insoluble in water and soluble in nonpolar fatty tissue.

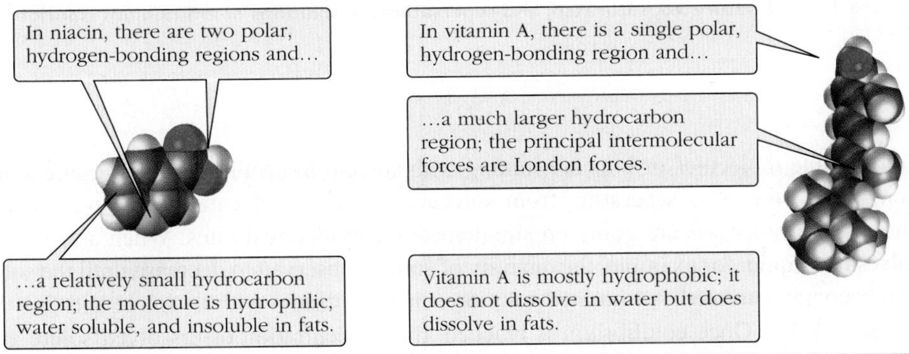

In niacin, there are two polar, hydrogen-bonding regions and...
...a relatively small hydrocarbon region; the molecule is hydrophilic, water soluble, and insoluble in fats.
In vitamin A, there is a single polar, hydrogen-bonding region and...
...a much larger hydrocarbon region; the principal intermolecular forces are London forces.
Vitamin A is mostly hydrophobic; it does not dissolve in water but does dissolve in fats.

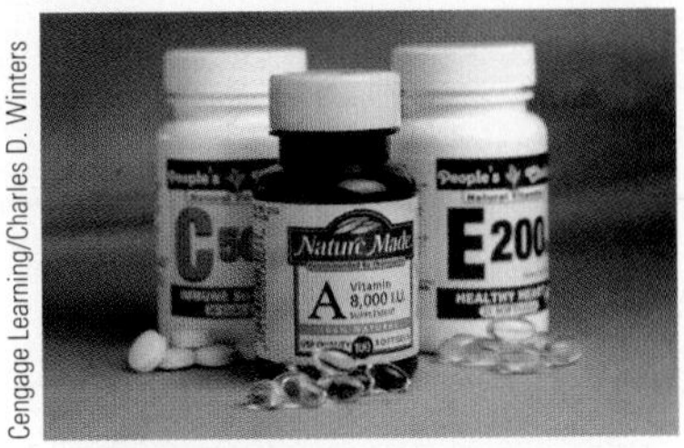

Cengage Learning/Charles D. Winters

Vitamins. Some are fat-soluble; some are water-soluble.

PROBLEM-SOLVING PRACTICE 13.2

Determine whether vitamin C, which has the structure shown here, is water-soluble.

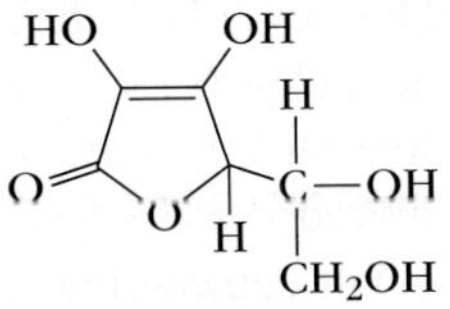

Vitamin C

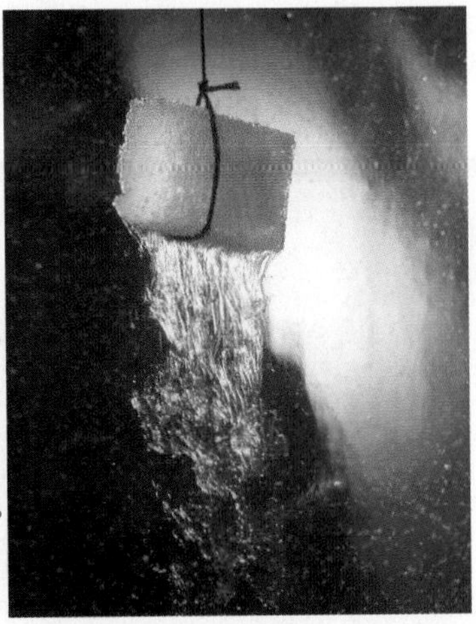

Richard Menga/Fundamental Photographs, NYC

A solute (sugar) dissolving in a solvent (water). As the solute dissolves, varying concentrations are observed as wavy lines.

13-2 Solubility and Equilibrium

In a few cases, such as ethanol and water, a solute and solvent can be combined in any proportions to form a solution. Usually, however, there is a limit to how much solute dissolves in a given mass of solvent and some solutes dissolve to a much greater extent than others. The **solubility** of a solute is *the maximum quantity of solute that dissolves in a given quantity of solvent when equilibrium is reached at a particular temperature.* The red curve in Figure 13.4 shows the solubility of ammonium chloride in water: each point on the curve represents the mass of NH_4Cl dissolved in 100. g H_2O at equilibrium at the temperature on the *x*-axis.

A **saturated solution** is one in which *the solute concentration equals its solubility* (all points *on* the curve in Figure 13.4). In a saturated solution there is a *dynamic equilibrium* (← **Sec. 12-1a**) between undissolved and dissolved solute particles.

$$\text{Solute} + \text{solvent} \rightleftharpoons \text{solute (in solution)} \qquad K_c = [\text{solute (in solution)}]$$

Remember that the concentration of a pure substance (solute or solvent) does not appear in the K_c expression.

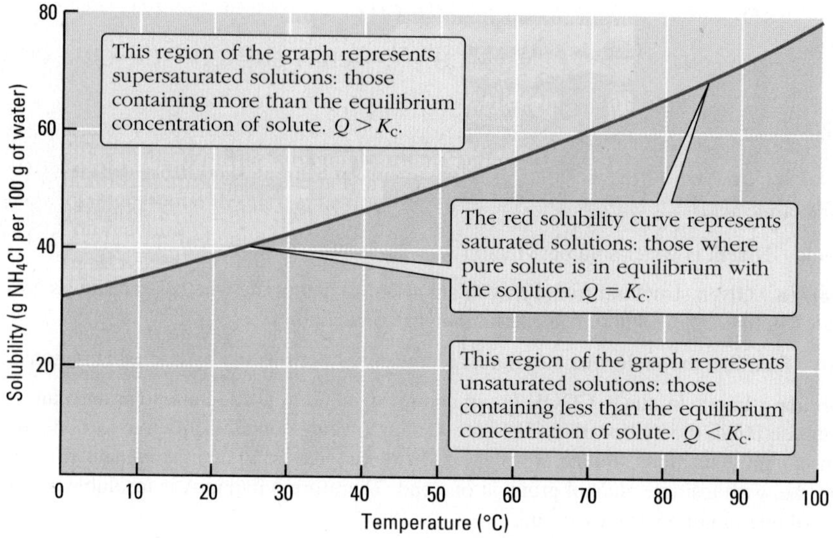
This region of the graph represents supersaturated solutions: those containing more than the equilibrium concentration of solute. $Q > K_c$.
The red solubility curve represents saturated solutions: those where pure solute is in equilibrium with the solution. $Q = K_c$.
This region of the graph represents unsaturated solutions: those containing less than the equilibrium concentration of solute. $Q < K_c$.
Solubility (g NH_4Cl per 100 g of water)
Temperature (°C)
0
20
40
60
80
10
30
50
70
90
100

Figure 13.4 Unsaturated, saturated, and supersaturated solutions of ammonium chloride.

Some solute molecules or ions are mixing with solvent molecules and going into solution, while others are separating from solvent molecules and entering the pure solute phase. Both processes are going on simultaneously at identical rates. When a solid dissolves in a liquid, for example, the quantity of solid is observed to decrease until the solution becomes saturated. A saturated solution is in equilibrium with its solute and $Q = K_c$ (← **Sec. 12-5a**). Once equilibrium is reached, the concentration of dissolved solute and the quantity of pure solute stop changing.

An **unsaturated solution** is *a solution in which the solute concentration is less than its solubility*; that is, *an unsaturated solution at a given temperature can accommodate additional solute* (the region *under* the curve in Figure 13.4). Thus, for an unsaturated solution, $Q < K_c$. If solute is in contact with an unsaturated solution at a given temperature, solute will dissolve until $Q = K_c$, the solution is saturated, and equilibrium has been achieved.

For some solutes, there is a third case. It is possible to prepare a **supersaturated solution**: *a solution that contains more than the equilibrium concentration of solute at a given temperature* (the region *above* the curve in Figure 13.4). For example, a supersaturated solution of ammonium chloride can be made by first making a saturated solution at 90 °C. Figure 13.4 shows that the solubility at 90 °C is 72 g NH_4Cl per 100 g water, so add 72 g solid NH_4Cl per 100 g water, heat the mixture to 90 °C, and stir until the solute all dissolves. Then, cool the saturated solution very slowly to 25 °C. If this is done carefully none of the NH_4Cl precipitates. The resulting solution is supersaturated because 72 g solute is dissolved in 100. g H_2O, whereas, at 25 °C, the solubility (Figure 13.4) is 40 g NH_4Cl per 100 g water. In equilibrium terms, $Q > K_c$ and excess solute should crystallize out of solution; however, precipitation is very slow. For the dissolved solute to form a crystal, several solute ions or molecules must move into an arrangement much like the appropriate crystal lattice positions, and it can take a long time for such an alignment to occur by chance. Some supersaturated solutions, such as honey or fudge, can remain so for days or months. However, precipitation of a solid from a supersaturated solution occurs rapidly if a tiny crystal of the solute is added to the solution (Figure 13.5). The lattice of the added crystal provides a template onto which more ions or molecules can be added.

Fudge involves a supersaturated sugar solution. In smooth fudge, the sugar remains uncrystallized; poor-quality fudge has a gritty texture because it contains crystallized excess sugar.

Sometimes other actions, such as stirring a supersaturated solution or scratching the inner walls of its container, cause solute to precipitate.

When a tiny seed crystal of sodium acetate...
...is added to a supersaturated sodium acetate solution...
...excess dissolved sodium acetate crystallizes.
Soon, numerous sodium acetate crystals can be seen. If the water were evaporated, solid sodium acetate would remain.

Photos: © Cengage Learning/Charles D. Winters

Figure 13.5 Supersaturated sodium acetate crystallizes when a seed crystal is added.

EXERCISE 13.3

Crystallizing Out of Solution

Using Figure 13.4, at 25 °C, calculate the mass (g) of excess NH_4Cl that would crystallize out of the supersaturated NH_4Cl solution just described.

EXERCISE 13.4

Solubility

Determine whether each of these masses of NH_4Cl dissolved in 100 g H_2O is an unsaturated, saturated, or supersaturated solution:

(a) 30 g NH_4Cl at 70 °C (b) 60 g NH_4Cl at 60 °C (c) 50 g NH_4Cl at 50 °C

13-3 Entropy, Enthalpy, and Dissolving Solutes

A solution forms when atoms, molecules, or ions of one kind mix with atoms, molecules, or ions of a different kind. In this section we apply to the solution process the ideas about enthalpy and entropy developed in Section 12-7. For any equilibrium, there is an *entropy effect* and an *enthalpy effect*: products (the solution) are favored if they have *greater entropy, lower enthalpy, or both.*

13-3a Entropy and Solution Formation

When solute and solvent particles mix to form solutions, as diagrammed in Figure 13.6, both have more space available to them. This spreads their energy over a larger volume and results in an increase in entropy. As described in Section 12-7, entropy is a quantitative measure of how extensively energy is spread out. *Processes in which the entropy of the system increases tend to be product-favored.* In many cases of solution formation, the only entropy effect is a large entropy increase as solvent and solute molecules mix. When the enthalpy change upon solution formation is rather small, this increase in entropy results in solubility or miscibility, such as when octane and carbon tetrachloride are mixed. In other cases, even though the enthalpy effect does not favor solution, the entropy increase can be large enough to cause the solution to form. This is true for some ionic solutes, such as ammonium nitrate, NH_4NO_3 (Section 13-2). Because molecules in gases are always much more dispersed than liquids, there is a significant decrease in entropy as gaseous solute molecules are brought closer together between solvent molecules. The entropy effect does not favor forming solid or liquid solutions with gaseous solutes.

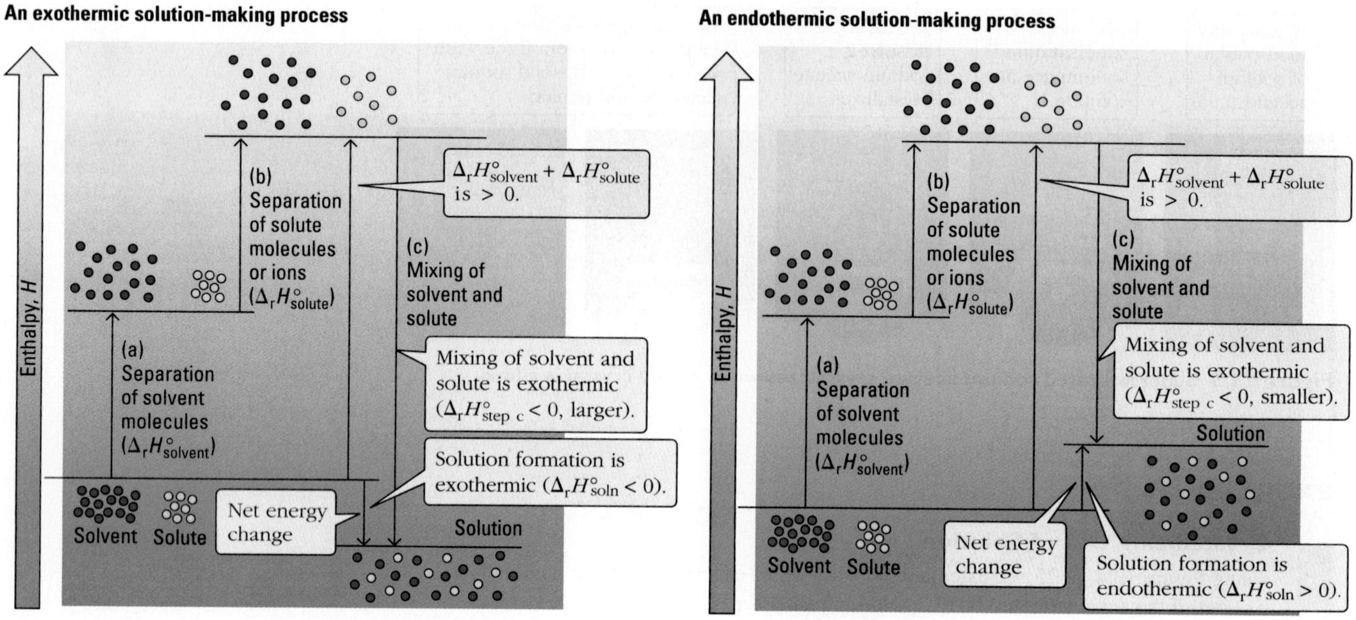

An exothermic solution-making process
Enthalpy, H
(b) Separation of solute molecules or ions ($\Delta_r H^\circ_{solute}$)
$\Delta_r H^\circ_{solvent} + \Delta_r H^\circ_{solute}$ is > 0.
(c) Mixing of solvent and solute
(a) Separation of solvent molecules ($\Delta_r H^\circ_{solvent}$)
Mixing of solvent and solute is exothermic ($\Delta_r H^\circ_{step\ c} < 0$, larger).
Solution formation is exothermic ($\Delta_r H^\circ_{soln} < 0$).
Net energy change
Solvent
Solute
Solution
An endothermic solution-making process
Enthalpy, H
(b) Separation of solute molecules or ions ($\Delta_r H^\circ_{solute}$)
$\Delta_r H^\circ_{solvent} + \Delta_r H^\circ_{solute}$ is > 0.
(c) Mixing of solvent and solute
Mixing of solvent and solute is exothermic ($\Delta_r H^\circ_{step\ c} < 0$, smaller).
Solution
(a) Separation of solvent molecules ($\Delta_r H^\circ_{solvent}$)
Solvent
Solute
Net energy change
Solution formation is endothermic ($\Delta_r H^\circ_{soln} > 0$).

Figure 13.6 The solution-formation process.

13-3b Enthalpy and Solution Formation

The enthalpy effect when a solution forms involves three steps (Figure 13.6).

- **Step (a)** The enthalpy of the collection of solvent molecules increases because intermolecular forces must be overcome when the molecules are separated. This is an *endothermic* process with a positive $\Delta_r H$ ($\Delta_r H > 0$).
- **Step (b)** The separation of solute ions or molecules is also an *endothermic* process ($\Delta_r H > 0$).
- **Step (c)** When solute and solvent particles mix, they attract each other and a decrease in enthalpy occurs; $\Delta_r H$ is negative ($\Delta_r H < 0$). Energy is released in this *exothermic* step.

The enthalpy of solution is also known as the heat of solution.

If the enthalpy of the solution (final state) is lower than that of the solute and solvent (initial state), forming the solution is favored by the enthalpy effect. As shown in Figure 13.6 (*left*), a net release of thermal energy to the surroundings occurs and the solution-making process is *exothermic*. When the enthalpy of the solution is greater than the enthalpies of the solute and solvent, the process is *endothermic* and the surroundings transfer thermal energy to the system (Figure 13.6, *right*). *The net energy change when a solution forms*, either exothermic or endothermic, is called the **enthalpy of solution**, $\Delta_r H_{soln}$.

$$\text{Enthalpy of solution} = \Delta_r H_{soln} = \Delta_r H_{step\ a} + \Delta_r H_{step\ b} + \Delta_r H_{step\ c}$$

When $\Delta_r H_{soln}$ is negative, the solution has lower energy than the pure solvent and solute and the enthalpy effect favors solution formation. However, if the solvent-solute forces (indicated by the magnitude of $\Delta_r H_{step\ c}$) are not strong enough to overcome the solute-solute and solvent-solvent attractions ($\Delta_r H_{step\ a} + \Delta_r H_{step\ b}$), dissolving is not favored by the enthalpy effect (← **Sec. 12-7**).

13-3c Dissolving Ionic Solids in Liquids

The enthalpy and entropy effects on solution are nicely illustrated by solutions of ionic solids. Sodium chloride is an ionic compound whose crystal lattice consists of Na^+ and Cl^- ions in a cubic array (← **Secs. 2-6b and 9-6c**). Strong electrostatic attractions between oppositely charged ions hold the ions tightly in the lattice. The enthalpy change when

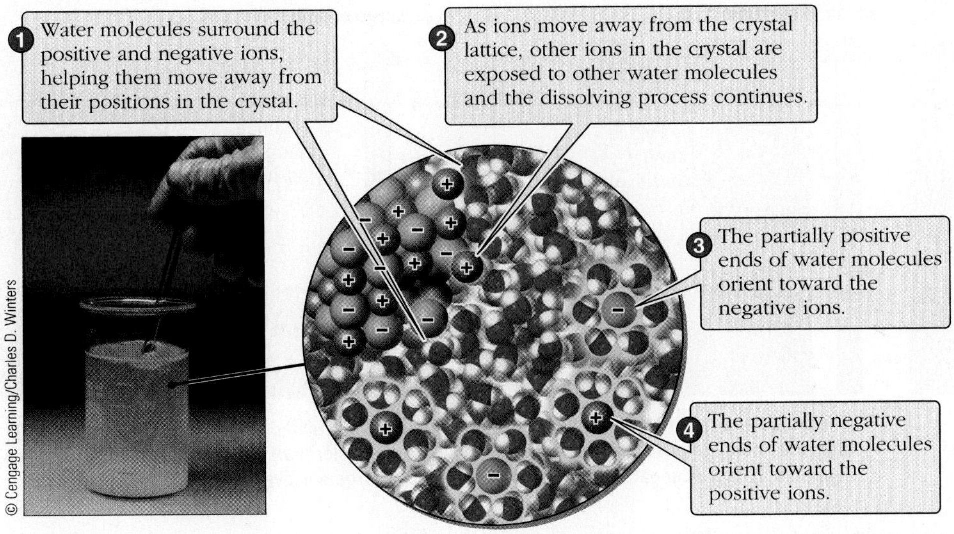
1 Water molecules surround the positive and negative ions, helping them move away from their positions in the crystal.
2 As ions move away from the crystal lattice, other ions in the crystal are exposed to other water molecules and the dissolving process continues.
3 The partially positive ends of water molecules orient toward the negative ions.
4 The partially negative ends of water molecules orient toward the positive ions.
© Cengage Learning/Charles D. Winters

Figure 13.7 A polar solvent, such as water, dissolves an ionic solid by solvating the ions.

1 mol Na^+ ions and 1 mol Cl^- ions in a crystal lattice are completely separated is the negative of the lattice energy (← **Sec. 5-13**). For NaCl, the lattice energy is quite large: −788 kJ/mol. Thus, 788 kJ must be supplied to 1 mol NaCl to separate the Na^+ and Cl^- ions. This accounts for sodium chloride's high melting point (800 °C). It is possible for solvent molecules to attract Na^+ ions in a crystal away from Cl^- ions, but they have to be the right kind of solvent molecules. Trying to dissolve NaCl (or any other ionic compound) with carbon tetrachloride or hexane (both nonpolar solvents) is a futile exercise because nonpolar molecules have very little attraction for ions. On the other hand, water and other highly polar solvents are attracted to ions and can dissolve NaCl.

Water is a good solvent for an ionic compound because water molecules are small and highly polar (← **Sec. 7-5**). As shown in Figures 13.7 and 13.8, the partially negative oxygen atoms of water molecules are attracted to cations and help pull them away from the crystal lattice, while the partially positive hydrogen atoms of other water molecules are attracted to the anions in the lattice and help pull them away from the lattice. *This process, in which water molecules surround cations and anions,* is called **hydration** (Figure 13.8). The *hydration enthalpy* is the enthalpy change when these new attractions form between ions and water-molecule dipoles. Energy is always required to separate ions from a crystal lattice (to overcome their attraction), and energy is always released when ions become hydrated because molecular dipoles attract the molecules to the ions.

A process similar to hydration occurs for any highly polar solvent. Positive ions are surrounded by the negative ends of solvent dipoles and negative ions are surrounded by the positive ends of solvent dipoles. *The process in which solvent molecules surround*

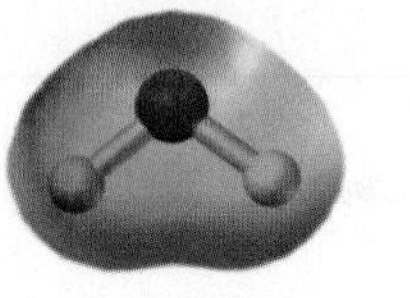

Electron density of a water molecule. The red area has high electron density and partial negative charge; the blue area has low electron density and partial positive charge.

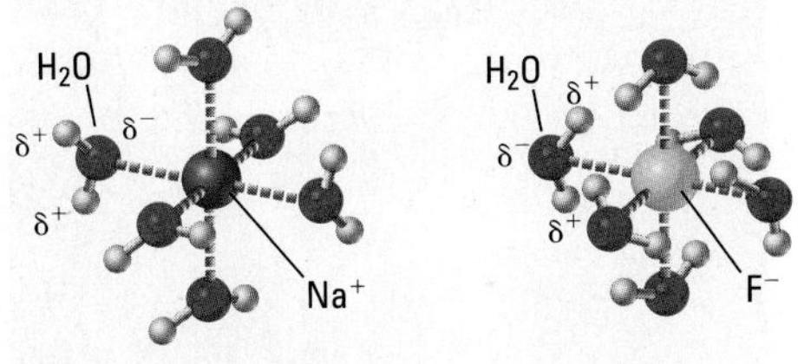
H2O
δ+
δ−
δ+
Na+
H2O
δ+
δ−
δ+
F−

Figure 13.8 Hydration of a sodium cation and a fluoride anion. The arrangement of water molecules around each ion is highly ordered. The dashed lines, - - - -, represent ion-dipole attraction between the central ion and the polar water molecules.

Hydration Enthalpies, $\Delta_r H$, for Selected Ions (kJ/mol)

Cations		Anions	
H^+	−1130	F^-	−483
Li^+	−558	Cl^-	−340
Na^+	−444	Br^-	−309
Mg^{2+}	−2003		
Ca^{2+}	−1557		
Al^{3+}	−4537		

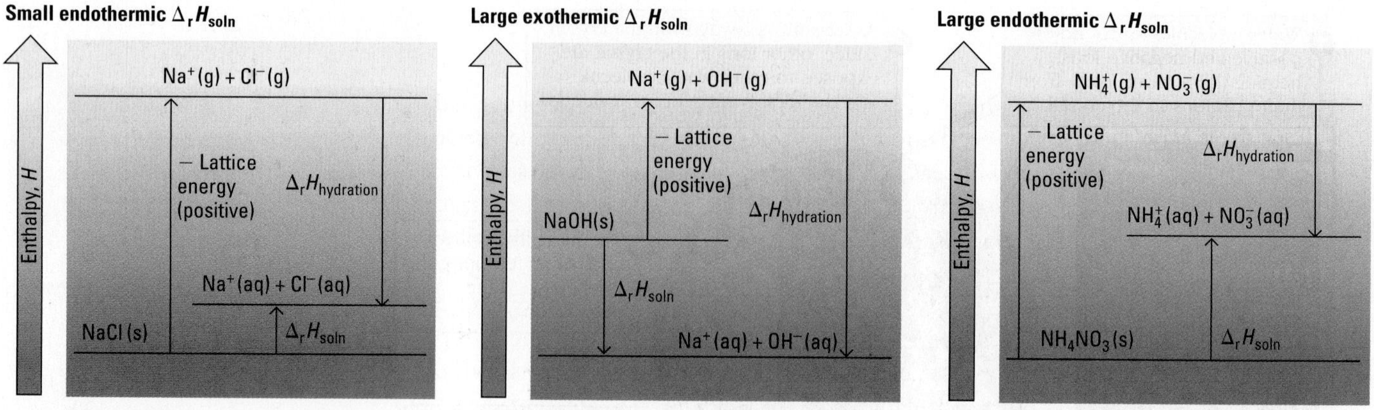

Small endothermic $\Delta_r H_{soln}$
Enthalpy, H
$Na^+(g) + Cl^-(g)$
− Lattice energy (positive)
$\Delta_r H_{hydration}$
$Na^+(aq) + Cl^-(aq)$
NaCl (s)
$\Delta_r H_{soln}$
Large exothermic $\Delta_r H_{soln}$
Enthalpy, H
$Na^+(g) + OH^-(g)$
− Lattice energy (positive)
NaOH(s)
$\Delta_r H_{hydration}$
$\Delta_r H_{soln}$
$Na^+(aq) + OH^-(aq)$
Large endothermic $\Delta_r H_{soln}$
Enthalpy, H
$NH_4^+(g) + NO_3^-(g)$
− Lattice energy (positive)
$\Delta_r H_{hydration}$
$NH_4^+(aq) + NO_3^-(aq)$
$NH_4NO_3(s)$
$\Delta_r H_{soln}$

(a) For NaCl, the lattice energy that must be overcome is larger than the energy released on hydration of the ions. $\Delta_r H_{soln}$ is positive (endothermic).

(b) The lattice energy for NaOH is much smaller than the energy released when the ions become hydrated. $\Delta_r H_{soln}$ is negative (highly exothermic).

(c) For NH_4NO_3, the hydration enthalpy is much smaller than the lattice energy. $\Delta_r H_{soln}$ has a large positive value (highly endothermic).

Figure 13.9 Enthalpies of solution of equal amounts of three different ionic compounds dissolving in equal volumes of water.

and stabilize ions in solution is called **solvation**. In general, because of the large enthalpy effects, ionic compounds are likely to dissolve in highly polar solvents but not likely to dissolve in nonpolar or slightly polar solvents.

Whether dissolving a particular ionic compound in water is exothermic or endothermic depends on the relative sizes of the lattice energy (← **Sec. 5-13**) of the ionic compound and the hydration enthalpies of its cations and anions. The relationship between the enthalpy of solution, the lattice energy of the ionic compound, and the hydration enthalpies of the ions is

$$\Delta_r H_{soln} = -\text{lattice energy} + \Delta_r H(\text{cation hydration}) + \Delta_r H(\text{anion hydration})$$

Figure 13.9 shows how lattice energy and the enthalpies of hydration combine to give the enthalpy of solution. The solubility rules for ionic compounds in water (← **Sec. 3-3c**) remind us that not all ionic compounds are highly water-soluble even though there are strong attractions between water molecules and ions. For some ionic compounds, the lattice energy is so large that water molecules cannot effectively pull ions away from the lattice. As a result, such compounds have large positive enthalpies of solution and usually are only slightly soluble.

Practical applications of endothermic and exothermic dissolution include cold packs containing NH_4NO_3 used to treat athletic injuries and hot packs containing $CaCl_2$ used to warm hands and foods.

Photos: © Cengage Learning/Charles D. Winters

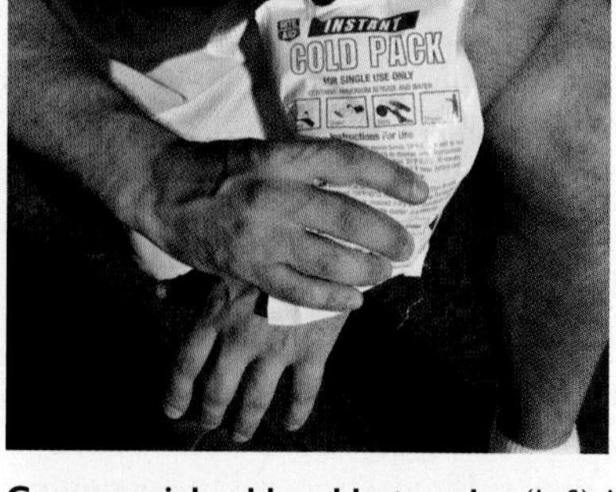

INSTANT
COLD PACK

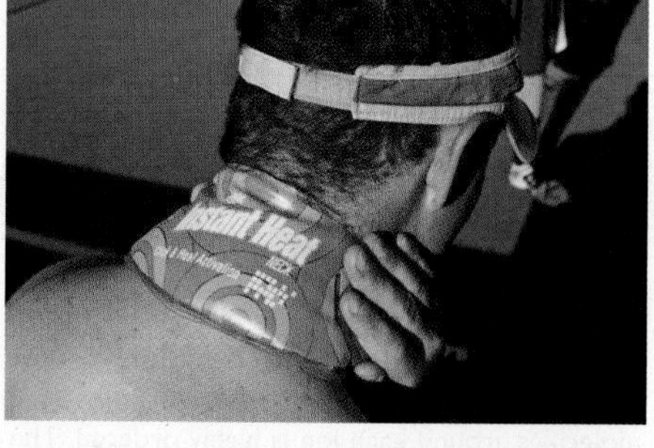

Instant Heat

Commercial cold and hot packs. (*Left*) Ammonium nitrate dissolving endothermically in water makes this cold pack cold. (*Right*) Calcium chloride dissolving exothermically in water makes this hot pack hot.

13-3d Entropy and the Dissolving of Ionic Compounds in Water

Both the spreading of energy introduced when ions separate from a crystal lattice and the spreading of energy introduced by the mixing of ions with solvent molecules favor the dissolving process. This entropy increase is counteracted by the ordering of solvent molecules around the ions, a negative entropy change (Figure 13.8). For 1+ and 1− charged ions, the overall entropy change is positive, and dissolving is favored. For some salts that contain 2+ or 3+ ions, the charges on the ions are so large and the ions so small that water molecules are aligned in a highly organized manner around the ions. When a large number of water molecules are locked into place by this strong hydration, the entropy of solution may be negative, which does not favor solubility. Calcium oxide is slightly soluble (only 0.131 g CaO dissolves in 100 mL water at 10 °C) and aluminum oxide, Al_2O_3, is insoluble due partly to this hydration entropy effect.

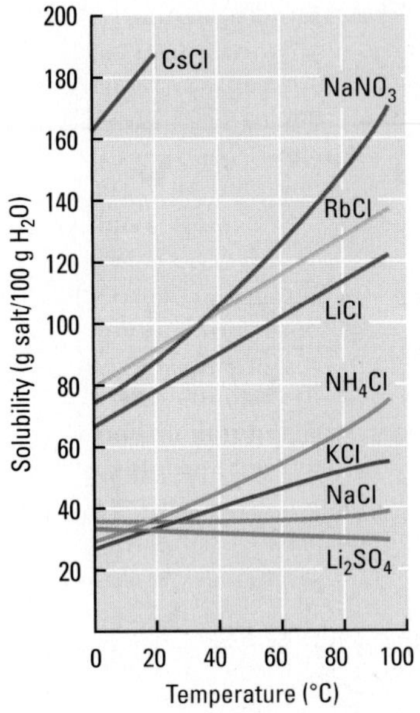
200
180
160
140
120
100
80
60
40
20
Solubility (g salt/100 g H2O)
CsCl
NaNO3
RbCl
LiCl
NH4Cl
KCl
NaCl
Li2SO4
0
20
40
60
80
100
Temperature (°C)

Figure 13.10 Solubility of ionic compounds versus temperature. The solubility of most ionic compounds in water increases with temperature.

Although generally useful, Le Chatelier's principle does not always correctly predict how the solubility of ionic solutes changes with temperature.

13-4 Temperature and Solubility

13-4a Solubility of Solids

Common experience tells us that more sugar (sucrose) can dissolve in hot coffee or hot tea than in cooler coffee or tea. This is an example of the fact that the aqueous solubility of most solid solutes, including ionic compounds, increases with increasing temperature (Figure 13.10). Although this generalization usually works, there are notable exceptions, such as Li_2SO_4 and $CaSO_4$.

13-4b Solubility of Gases

To understand how temperature affects gas solubility, we apply Le Chatelier's principle (← **Sec. 12-6**) to the equilibrium between a pure solute gas, the solvent, and a saturated solution of the gas.

$$\text{Gas} + \text{solvent} \rightleftharpoons \text{saturated solution} \qquad \text{Usually, } \Delta_r H_{soln} < 0 \text{ (exothermic)}$$

When a gas dissolves to form a saturated liquid solution, the process is almost always *exothermic*. Molecules that were relatively far apart in the gas phase are brought much closer in the solution to other molecules that attract them, lowering the potential energy of the system and releasing some energy to the surroundings.

If the temperature of a solution of a gas in a liquid increases, the equilibrium shifts in the direction that partially counteracts the temperature rise. That is, the equilibrium shifts to the left in the preceding equation. Thus, *a dissolved gas becomes less soluble with increasing temperature*. Conversely, cooling a solution of a gas that is at equilibrium with undissolved gas causes the equilibrium to shift to the right in the direction that liberates heat, so more gas dissolves. This is illustrated in Figure 13.11 with data for the solubility of oxygen gas in water.

Cooler water in contact with the atmosphere contains more dissolved oxygen at equilibrium than water at a higher temperature. For this reason fish seek out cooler (usually deeper) waters in the summer. Fish have an easier time obtaining oxygen when its concentration in the water is higher. The decrease in gas solubility as temperature increases makes *thermal pollution* a problem for aquatic life in rivers and streams. Natural heating of water by sunlight and by warmer air can usually be accommodated. But, excess heat from extended periods of very hot weather or from sources such as industrial facilities and electrical power plants can warm the water sufficiently that the concentration of dissolved oxygen is reduced to the point where some species of fish die.

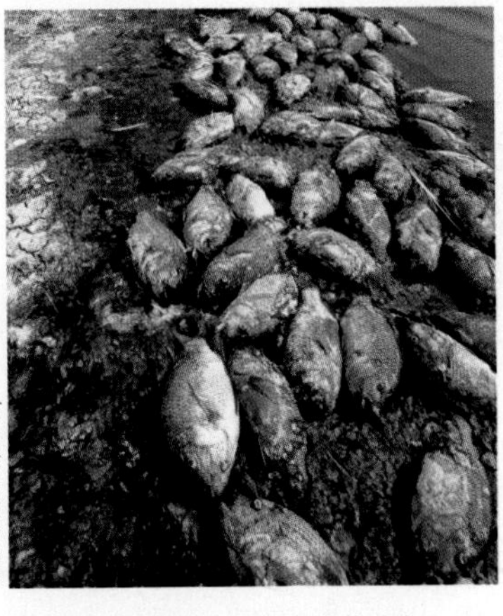

Mike Goldwater/Alamy

A fish kill caused by a lack of dissolved oxygen.

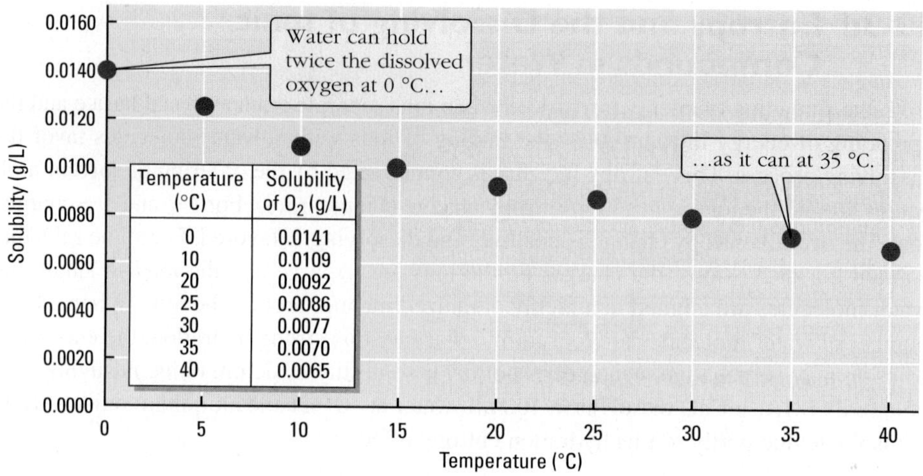
Water can hold twice the dissolved oxygen at 0 °C...
...as it can at 35 °C.
Solubility (g/L)
Temperature (°C)
0.0160
0.0140
0.0120
0.0100
0.0080
0.0060
0.0040
0.0020
0.0000
0
5
10
15
20
25
30
35
40

Temperature (°C)	Solubility of O_2 (g/L)
0	0.0141
10	0.0109
20	0.0092
25	0.0086
30	0.0077
35	0.0070
40	0.0065

Figure 13.11 Solubility of oxygen gas in water at various temperatures. The solubility of oxygen, like that of other gases, decreases with increasing temperature.

CONCEPTUAL EXERCISE 13.5

Carbonated Beverages

Explain why a carbonated beverage goes "flat" once it is opened and warms to room temperature.

CONCEPTUAL EXERCISE 13.6

Warming Water

Explain why water that has been used to cool a reactor in a nuclear power plant, and thus is at a relatively high temperature, must be cooled before it is put back into the lake or river from which it came.

CONCEPTUAL EXERCISE 13.7

Temperature and Solubility

If a substance has a positive enthalpy of solution, which would likely cause more of it to dissolve, hot solvent or cold solvent? Explain.

13-5 Pressure and Dissolving Gases in Liquids: Henry's Law

The partial pressure of a gas in a mixture of gases is the pressure that a pure sample of the gas would exert if it occupied the same volume as the mixture. Partial pressure is proportional to the mole fraction of the gas (← Sec. 8-6).

Pressure does not measurably affect the solubilities of solids or liquids in liquid solvents, but *the solubility of any gas in a liquid increases as the partial pressure of the gas increases.* A dynamic equilibrium is established when a gas is in contact with a liquid—the rate at which gas molecules enter the liquid phase equals the rate at which gas molecules escape from the liquid. If the partial pressure of a gas is increased, gas molecules strike the surface of the liquid more often, increasing the rate of dissolution of the gas. A new equilibrium is established when the rate of escape increases to match the rate of dissolution. The rate of escape is first order (← **Sec. 11-3**) in concentration of solute, so a higher rate of gas escape requires a higher concentration of solute gas molecules—that is, a higher gas solubility.

The fact that *solubility of a gas in a liquid is proportional to gas pressure* is known as **Henry's law**:

$$S_g = k_H P_g$$

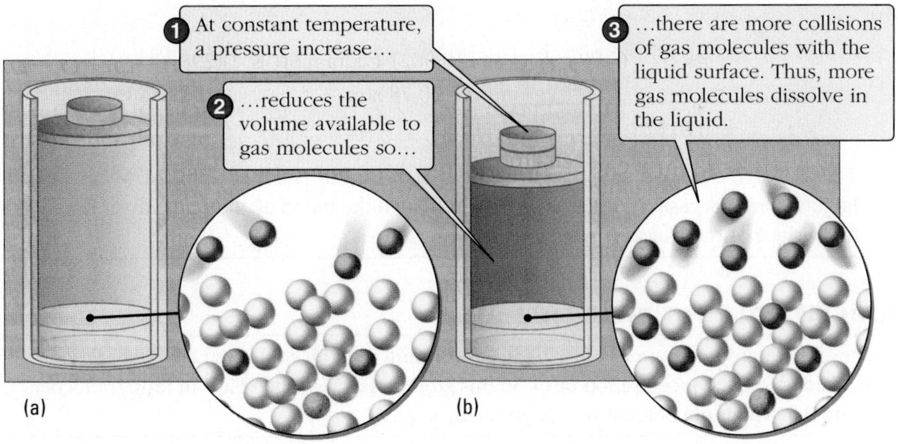
1 At constant temperature, a pressure increase...
2 ...reduces the volume available to gas molecules so...
3 ...there are more collisions of gas molecules with the liquid surface. Thus, more gas molecules dissolve in the liquid.
(a)
(b)

Figure 13.12 Henry's law. (a) A gas and a liquid solution of the gas under pressure in a closed container. (b) The pressure is increased at constant temperature, causing more gas to dissolve.

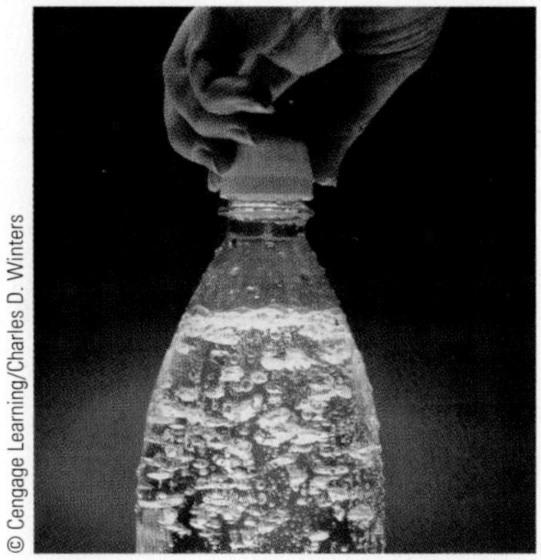
© Cengage Learning/Charles D. Winters

Henry's law. The greater the partial pressure of CO_2 over the soft drink in the bottle, the greater the concentration of dissolved CO_2. When the bottle is opened, the partial pressure of CO_2 drops and CO_2 bubbles out of the solution.

where S_g is the solubility of the gas in the liquid, and P_g is the pressure of the gas above the solution (or the partial pressure of the gas if the solution is in contact with a mixture of gases). The value of k_H, known as the Henry's law constant, depends on the identities of both the solute and the solvent and on the temperature (Table 13.3).

Figure 13.12 illustrates how gas solubility depends on pressure. The behavior of a carbonated drink when the cap is opened is an everyday illustration of the solubility of gases in liquids under pressure. The drink fizzes when opened because the partial pressure of CO_2 over the solution drops, the solubility of the gas decreases, and dissolved gas escapes from the solution.

PROBLEM-SOLVING EXAMPLE 13.3

Using Henry's Law

The "bends" is a condition that can occur in scuba divers due to the changes in the solubility of N_2 gas in the blood. The partial pressure of N_2 in the atmosphere is 0.78 atm. For a saturated solution, calculate

(a) the mass (g) of dissolved N_2 in 1 L blood at 25 °C.

(b) the volume (mL) of $N_2(g)$ that can be dissolved in 1 L blood at 25 °C.

Result (a) 1.40×10^{-2} g N_2 (b) 16 mL of N_2

Analyze

(a) Henry's law can be used to calculate the solubility of N_2 gas in water in mol/L. The molar mass of N_2 can be used to convert mol/L to g/L. Assume that the solubility of N_2 in blood is approximately that of N_2 in water and that the dissolved N_2 is in equilibrium with nitrogen in the air.

(b) The ideal gas law is appropriate to calculate the volume of N_2 gas (mL) from the amount (moles) dissolved in 1 L blood.

Plan

(a) Look up the Henry's law constant for N_2 in water at 25 °C. Convert atm to mmHg and apply Henry's law to calculate the solubility of N_2.

(b) Apply the ideal gas law to calculate the volume (mL) of N_2 gas dissolved in 1 L blood. Remember that the temperature must be in kelvins.

Execute The Henry's law constant is 8.42×10^{-7} mol L^{-1} mmHg^{-1}.

(a)

$$P_{N_2} = 0.78 \text{ atm} \times \frac{760 \text{ mmHg}}{1 \text{ atm}} = 593 \text{ mmHg}$$

$$S_{N_2} = k_H P_{N_2} = (8.42 \times 10^{-7} \text{ mol L}^{-1} \text{ mmHg}^{-1})(593 \text{ mmHg}) = 4.99 \times 10^{-4} \text{ mol/L}$$

$$m_{N_2} = (4.99 \times 10^{-4} \text{ mol/L})\left(\frac{28.02 \text{ g } N_2}{1 \text{ mol } N_2}\right) = 1.40 \times 10^{-2} \text{ g } N_2\text{/L}$$

Table 13.3 Henry's Law Constants for Water, 25 °C

Gas	k_H (mol L^{-1} mmHg^{-1})
N_2	8.42×10^{-7}
O_2	1.66×10^{-6}
CO_2	4.4×10^{-5}
He	4.6×10^{-7}
Air	8.6×10^{-7}

(b)

$$V = \frac{nRT}{P} = \frac{(4.99 \times 10^{-4}\ \text{mol})(0.0821\ \text{L atm mol}^{-1}\text{K}^{-1})(298\ \text{K})}{0.78\ \text{atm}}$$

$$= 0.0157\ \text{L} = 16\ \text{mL}$$

This is a significant volume of nitrogen gas dissolved in one liter of blood. When such a large volume of gaseous nitrogen is released from the blood as the scuba diver rises to the surface of the water, it can cause serious pain.

PROBLEM-SOLVING PRACTICE 13.3

Suppose that a trout stream at 25 °C is in equilibrium with air at normal atmospheric pressure. Calculate the concentration of O_2 in this stream. Express the result in milligrams per liter (mg/L). The mole percent of oxygen in air is 21%.

13-6 Expressing Solution Composition

The terms *unsaturated, saturated*, and *supersaturated* describe a solution with respect to the quantity of solute in a given quantity of solvent at a certain temperature. Sometimes a solution is described as *concentrated* or *dilute*. Although useful, these are all qualitative descriptors of solution composition. It is often important to specify the quantity of a solute in a given quantity of solution more precisely. There are several ways of doing this, including mass fraction, molarity, and molality. Often it is useful to be able to quantitatively describe quantities of solute across a wide range of concentrations, from rather large to very small. For example, many different units are used to express the quantities of unwanted—even potentially harmful—solute species in drinking water. Examples of unwanted species are lead, mercury, selenium, nitrate, and organic compounds. These units are also useful clinically in discussing solute concentrations in blood or urine.

13-6a Mass Fraction and Weight Percent

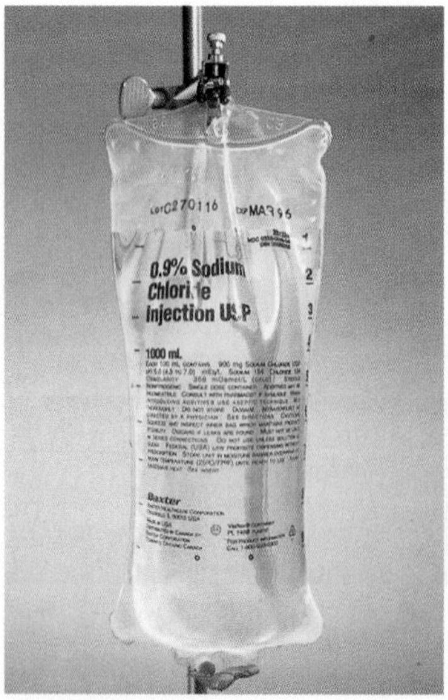
0.9% Sodium
Chloride
Injection USP
1000 mL

© Cengage Learning/Charles D. Winters

Sterile saline solution. Saline solutions like this one are routinely given to patients who have lost body fluids.

The **mass fraction** of a solute is *the fraction of the total mass of the solution that a solute contributes*—that is, the mass of a single solute divided by the total mass of all solutes *and* the solvent. Mass fraction is commonly expressed as a percentage and called mass percent or **weight percent**, which is *the mass fraction multiplied by 100%*. Weight percent is the same as the number of grams of solute per 100 g of solution. For example, the mass fraction of sucrose in a solution consisting of 25.0 g sucrose, 10.0 g fructose, and 300. g water is

$$\text{Mass fraction of sucrose} = \frac{\text{mass of sucrose}}{\text{mass of sucrose} + \text{mass of fructose} + \text{mass of water}}$$

$$= \frac{25.0\ \text{g}}{25.0\ \text{g} + 10.0\ \text{g} + 300.\ \text{g}} = 0.0746$$

The weight percent of sucrose in the solution is $(0.0746) \times 100\% = 7.46\%$.

PROBLEM-SOLVING EXAMPLE 13.4

Mass Fraction and Weight Percent

Sterile saline solutions containing NaCl in water are often used in medicine. Calculate the weight percent of NaCl in a solution made by dissolving 4.6 g NaCl in 500. g pure water.

Result 0.91%

Analyze Both masses are given in the same units (grams), so they can be used directly to calculate the weight percent of NaCl.

Plan Apply the definitions of mass fraction and weight percent to the solute and solution stated in the problem.

Execute

$$\text{Mass fraction of NaCl} = \frac{4.6 \text{ g NaCl}}{4.6 \text{ g NaCl} + 500. \text{ g } H_2O} = 0.0091$$

$$\text{Weight percent of NaCl} = \text{mass fraction of NaCl} \times 100\%$$
$$= 0.0091 \times 100\% = 0.91\%$$

☑ **Reasonable Result Check** The mass fraction is about 5 g NaCl in about 500 g solution, or about 0.01, which agrees with the more accurate result.

The symbol % means "per hundred," that is, divided by 100. Thus, 100% is $\frac{100}{100} = 1$. Weight percent could also be called parts per hundred.

PROBLEM-SOLVING PRACTICE 13.4

Calculate the weight percent of glucose in a solution containing 21.5 g glucose, $C_6H_{12}O_6$, in 750. g pure water.

EXERCISE 13.8

Mass Fraction and Weight Percent

Ringer's solution is used in physiology experiments. To make the solution, 1.00 L water is combined with 6.5 g NaCl, 0.20 g $NaHCO_3$, 0.10 g $CaCl_2$, and 0.10 g KCl. The density of water is 1.00 g/mL.

(a) Calculate the mass fraction and weight percent of $NaHCO_3$ in the solution.
(b) Determine which solute has the lowest mass fraction in the solution.

13-6b Parts per Million, Billion, and Trillion

Solutes in very dilute solutions have very low mass fractions. Consequently, the mass fraction in such solutions is often expressed in **parts per million (ppm)**. One part per million is equivalent to one gram of solute per one million grams of solution, or proportionally, one milligram of solute per one thousand grams of solution (1 mg/kg). Thus, a commercial bottled water with a calcium ion concentration of 66 ppm contains 66 mg Ca^{2+} per 1000 g solution, essentially one liter. For even smaller mass fractions, scientists often use **parts per billion (ppb)** (1 ppb = one gram of solute per one billion grams of solution or one microgram of solute per kilogram of solution, 1 μg/kg) and **parts per trillion (ppt)** (1 ppt = one gram of solute per one trillion grams of solution or one nanogram of solute in one kilogram of solution, 1 ng/kg). As the names imply, a mass fraction converts to parts per billion by multiplying by 10^9 ppb and to parts per trillion by multiplying by 10^{12} ppt.

1 ppm is equivalent to one penny in $10,000; 1 ppb is one penny in $10,000,000.

1 μg (microgram) = 10^{-6} g; 1 ng (nanogram) = 10^{-9} g.

The abbreviation ppb means per billion or $1/10^9 = 10^{-9}$. Thus, 10^9 ppb $= 10^9/10^9 = 1$. Multiplying by 10^9 ppb is the same as multiplying by 1.

PROBLEM-SOLVING EXAMPLE 13.5

ppm, ppb, and Mass Fraction

In Section 1-2 you learned that many water wells in the country of Bangladesh are contaminated with arsenic, a toxic material, with levels well above the World Health Organization's (WHO) guideline maximum value of 0.010 mg As/L. The WHO estimates that 28–35 million Bangladesh citizens have been exposed to high arsenic levels by drinking water from wells in which the arsenic concentration is at least 0.050 mg/L.

(a) Calculate the mass fraction of arsenic in a 0.050-mg/L solution. Assume that the solution is almost entirely water so that its density is 1.0 g/mL.
(b) Calculate the concentration of arsenic in ppb.

Result (a) 5.0×10^{-8} (b) 50. ppb

Analyze Density relates the mass to the volume of a sample. Mass fraction is the mass of solute divided by the total mass of solution. Mass fraction can be converted to ppb.

Plan

(a) Use the density to calculate the mass of 1 L solution and apply the definition of mass fraction.
(b) Multiply the mass fraction by 10^9 ppb.

Execute

(a) $m(\text{solution}) = V \times d = 1 \text{ L} \times 1.0 \text{ g/mL} \times 1000 \text{ mL/L} = 1.0 \times 10^3 \text{ g}$

$$\text{Mass fraction} = \left(\frac{0.050 \text{ mg arsenic}}{1 \times 10^3 \text{ g solution}}\right)\left(\frac{1.0 \times 10^{-3} \text{ g}}{1 \text{ mg}}\right) = 5.0 \times 10^{-8}$$

(b) $5.0 \times 10^{-8} \times 10^9 \text{ ppb} = 50. \text{ ppb}$

Notice that because very dilute aqueous solutions are essentially all water, 1 mL weighs 1 g. Thus, units of mg/kg are equivalent to units of mg/L.

☑ **Reasonable Result Check** Used correctly, the conversion factors change 0.050 mg/L directly to 5.0×10^{-5} g arsenic/L and 50 ppb arsenic.

PROBLEM-SOLVING PRACTICE 13.5

Drinking water may contain small quantities of selenium (Se). A sample of water contains 30. ppb Se; calculate the mass (μg) of Se in 100. mL of this water.

At the height of the Roman Empire, worldwide lead production was about 80,000 tons per year. Today it is about 3 million tons annually. Lead was first used for water pipes in ancient Rome. The Latin name for lead, *plumbum*, gave us the word "plumber."

CONCEPTUAL EXERCISE 13.9

Lead in Drinking Water

One drinking-water sample has a lead concentration of 20. ppb; another has a concentration of 0.0030 ppm.

(a) Determine which sample has the higher lead concentration.
(b) The current U. S. Environmental Protection Agency acceptable limit for lead in drinking water is 0.015 mg/L. Compare the lead concentration of each water sample with the acceptable limit.

EXERCISE 13.10

Bottled Water as a Magnesium Source

A 500-mL bottle of Evian bottled water contains 12 mg magnesium. The recommended daily allowance of magnesium for adult women is 320 mg/day. Calculate how many 1-L bottles of Evian a woman would have to drink to obtain her total daily allowance of magnesium solely in this way.

EXERCISE 13.11

Striking It Rich in the Oceans?

The mass fraction of gold in seawater is 1×10^{-3} ppm. Earth's oceans contain 3.5×10^{20} gal seawater. Estimate how many pounds of gold are in the oceans. 1 gal = 3.785 L; 1 lb = 454 g.

13-6c Molarity

As defined in Section 3-10a, the **molarity** of a solution is

$$\text{Molarity} = \frac{\text{amount of solute}}{\text{volume of solution}} = \frac{n}{V}$$

$$\left(\frac{\text{mol solute}}{\text{L solution}}\right) \times (\text{L solution}) = \text{mol solute}$$

Multiplying the volume of a solution (L) by its molarity yields the amount (mol) of solute in that volume of solution. For example, the amount of KNO_3 in 250. mL of 0.0200-M KNO_3 is

$$n(\text{KNO}_3) = 0.250\text{ L} \times \left(\frac{0.0200\text{ mol KNO}_3}{1\text{ L}}\right) = 5.00 \times 10^{-3}\text{ mol KNO}_3$$

from which the mass (g) can be determined using the molar mass of KNO_3.

$$m(\text{KNO}_3) = 5.00 \times 10^{-3}\text{ mol KNO}_3 \times \frac{101.1\text{ g KNO}_3}{1\text{ mol KNO}_3} = 0.506\text{ g KNO}_3$$

Thus, to make 250. mL of 0.0200-M KNO_3 solution, you would weigh 0.506 g KNO_3, put it into a 250-mL volumetric flask, and add to it sufficient water to bring the volume of the solution to 250. mL (← **Sec. 3-10a**). Note that as long as the *ratio* of amount of solute to volume of solution remains the same, the molarity of the solute does not change. For example,

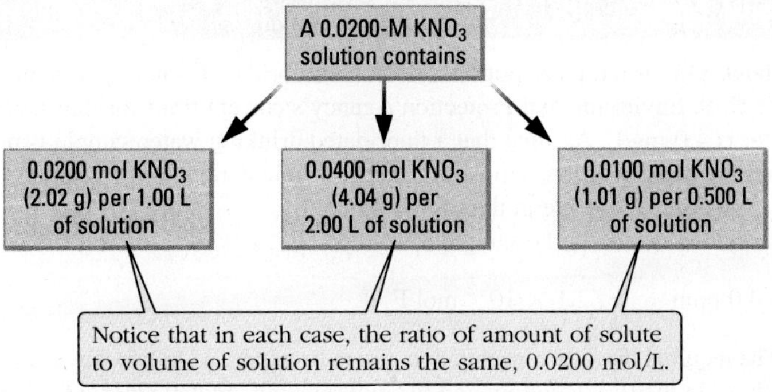
A 0.0200-M KNO3 solution contains
0.0200 mol KNO3 (2.02 g) per 1.00 L of solution
0.0400 mol KNO3 (4.04 g) per 2.00 L of solution
0.0100 mol KNO3 (1.01 g) per 0.500 L of solution
Notice that in each case, the ratio of amount of solute to volume of solution remains the same, 0.0200 mol/L.

PROBLEM-SOLVING EXAMPLE 13.6

Molarity

(a) Calculate the mass (g) of $NiCl_2$ needed to prepare 500. mL of 0.125-M $NiCl_2$.
(b) Calculate the volume (mL) of this solution required to prepare 250. mL of 0.0300-M $NiCl_2$.

Result (a) 8.10 g $NiCl_2$ (b) 60.0 mL

Analyze (a) Apply the definition of molarity to calculate amount of $NiCl_2$; molar mass relates amount to mass. (b) Apply the dilution relationship (← **Sec. 3-10c**); in a dilution the amount of solute ($NiCl_2$) remains the same.

Plan

(a) Multiply the volume (L) times the molarity to calculate amount; use the molar mass of $NiCl_2$, 129.6 g/mol, to calculate mass.
(b) To calculate the required volume of the more concentrated (undiluted) solution, use the relation

$$c_{conc} \times V_{conc} = c_{dil} \times V_{dil}$$

where c_{conc} and V_{conc} are the molarity and volume of the initial (undiluted) solution and c_{dil} and V_{dil} are the molarity and volume of the final (diluted) solution. Here,

$$c_{conc} = 0.125\text{ M} \qquad V_{conc} = \text{volume to be determined}$$
$$c_{dil} = 0.0300\text{ M} \qquad V_{dil} = 250\text{ mL} = 0.250\text{ L}$$

Execute

(a)

$$m(\text{NiCl}_2) = 0.500\text{ L} \times \frac{0.125\text{ mol NiCl}_2}{1\text{ L}} \times \frac{129.6\text{ g NiCl}_2}{1\text{ mol NiCl}_2} = 8.10\text{ g NiCl}_2$$

(b) Solve for V_{conc}.

$$V_{conc} = \frac{c_{dil} \times V_{dil}}{c_{conc}} = \frac{(0.0300\text{ M})(0.250\text{ L})}{0.125\text{ M}} = 0.0600\text{ L} = 60.0\text{ mL}$$

☑ **Reasonable Result Check** The molar mass of $NiCl_2$ is approximately 130 g/mol, so 1 L of a 0.125-M $NiCl_2$ solution would contain about 0.125 mol or 16 g $NiCl_2$. One half of a liter (500 mL) of that solution would contain one half as much $NiCl_2$, or about 8 g, which is close to the calculated value of 8.10 g. The diluted solution is 0.0300 M, which is only about one fourth the undiluted concentration. Therefore, in the dilution the volume must increase by a factor of about 4, from 60.0 mL to 250. mL.

PROBLEM-SOLVING PRACTICE 13.6

(a) Calculate the mass (g) of NaBr needed to prepare 250. mL of 0.0750-M NaBr.
(b) Calculate the volume (mL) of the solution in part (a) required to prepare 500. mL of 0.00150-M NaBr.

PROBLEM-SOLVING EXAMPLE 13.7

Molarity and ppm

Currently, about 60% of the U. S. population drinks fluoridated water from municipal water supplies. The U. S. Environmental Protection Agency's current limit for fluoride, F^-, in drinking water is 4.0 mg/L. Assume that a fluoridated drinking water sample is principally water and that the density of the sample is 1.0 g/mL. Calculate
(a) the mass fraction of fluoride in the sample in ppm.
(b) the molarity of fluoride in the sample.

Result (a) 4.0 ppm (b) 2.1×10^{-4} mol F^-/L

Analyze The required fluoride concentrations are to be expressed in two ways: (a) ppm, the mass (g) of fluoride in 10^6 g solution; and (b) molarity, the amount (mol) of F^- per unit volume (L) solution.

Plan Apply the definitions to calculate the values requested.
(a) Convert mg solute to g solute; use the density of 1.0 g/mL to calculate mass of solution. Divide mass of solute by mass of solution and multiply by 10^6 ppm.
(b) Use the molar mass to convert mass of fluoride to amount (mol); the volume of the solution is already expressed in liters.

Execute
(a)

$$\text{Mass fraction } (F^-) = \frac{4.0 \text{ mg } F^-}{1 \text{ L soln.}} \times \frac{1 \text{ g } F^-}{10^3 \text{ mg } F^-} \times \frac{1 \text{ L soln.}}{10^3 \text{ mL soln.}} \times \frac{1 \text{ mL soln.}}{1.0 \text{ g soln.}} \times 10^6 \text{ ppm}$$

$$= \frac{4.0 \text{ g } F^-}{10^6 \text{ g soln.}} \times 10^6 \text{ ppm} = 4.0 \text{ ppm}$$

(b)

$$c(F^-) = \frac{4.0 \text{ mg } F^-}{1 \text{ L soln.}} \times \frac{1 \text{ g } F^-}{10^3 \text{ mg } F^-} \times \frac{1 \text{ mol } F^-}{18.998 \text{ g } F^-} = \frac{2.1 \times 10^{-4} \text{ mol } F^-}{\text{L soln.}} = 2.1 \times 10^{-4} \text{ mol/L}$$

☑ **Reasonable Result Check** A liter of solution, which is about 1000 g of solution, contains about 0.004 g F^-. This is equivalent to 4 g F^- in 10^6 g solution or 4 ppm fluoride, close to the calculated result. A 4-g sample of F^- is approximately 0.2 mol F^-, so 4 mg F^- is approximately 0.0002 mol F^-, nearly equal to the calculated value. The results are reasonable.

PROBLEM-SOLVING PRACTICE 13.7

The concentration of magnesium ion in seawater is 0.0556 M, making magnesium the second most abundant cation in oceans. Express the Mg^{2+} concentration as (a) mass fraction and (b) ppm. Assume that the density of seawater is 1.03 g/mL.

PROBLEM-SOLVING EXAMPLE 13.8

Weight Percent and Molarity

Hydrochloric acid is sold as a concentrated aqueous solution with a density of 1.18 g/mL. The concentrated acid contains 38% HCl by mass. Calculate the molarity of hydrochloric acid in this solution.

Result 12.4 M

Analyze Molarity is to be determined, so calculate the amount (mol) of HCl and the volume (L) of the solution.

Plan A 38.0% hydrochloric acid solution means that 100. g solution contains 38.0 g HCl. Use the molar mass of HCl, 36.5 g/mol, to calculate the amount (mol) of HCl. Calculate the volume of the solution from its density.

Execute

$$\text{Amount(HCl)} = n(\text{HCl}) = 38.0 \text{ g HCl} \times \frac{1 \text{ mol HCl}}{36.5 \text{ g HCl}} = 1.04 \text{ mol HCl}$$

$$\text{Molarity} = c(\text{HCl}) = \frac{1.04 \text{ mol}}{100. \text{ g soln.}} \times \frac{1.18 \text{ g soln.}}{1 \text{ mL soln.}} \times \frac{1000 \text{ mL soln.}}{1 \text{ L soln.}} = 12.4 \frac{\text{mol}}{\text{L}} = 12.4 \text{ M}$$

☑ **Reasonable Result Check** The molar mass of HCl, 36.5 g/mol, is close to the number of grams of HCl in 100 g solution. The density of the solution is about 1 g/mL, so there is about 1 mol HCl in 100 mL of this solution. Consequently, there is about 10 mol HCl in 1 L solution—approximately a 10-M HCl solution. The calculated result is about 20% higher because we rounded the actual density (1.18 g/mL) to 1 g/mL.

PROBLEM-SOLVING PRACTICE 13.8

The density of a commercial 30.0% hydrogen peroxide, H_2O_2, solution is 1.11 g/mL at 25 °C. Calculate the molarity of hydrogen peroxide in this solution.

13-6d Molality

Solution composition can also be expressed in terms of the amount of solute in relation to the mass of the solvent. **Molality** (abbreviated *m*), is defined as the amount (mol) of solute per kilogram of *solvent* (not kilogram of solution).

$$\text{Molarity of solute A} = m(\text{A}) = \frac{\text{amount of solute A (mol)}}{\text{mass of solvent (kg)}}$$

For example, we can calculate the molality of a solution prepared by dissolving 0.413 g methanol, CH_3OH, in 1.50×10^3 g water by first determining the amount of methanol solute and the mass in kilograms of solvent, water.

$$n(\text{methanol}) = 0.413 \text{ g methanol} \times \frac{1 \text{ mol methanol}}{32.042 \text{ g methanol}} = 0.0129 \text{ mol methanol}$$

The mass of water is 1.50×10^3 g, which is 1.50 kg. Thus,

$$\text{Molality} = m(\text{methanol}) = \frac{0.0129 \text{ mol methanol}}{1.50 \text{ kg water}} = 8.60 \times 10^{-3} \text{ mol/kg}$$

Molality will be used in Section 13-7 when we deal with the effect of solute concentration on decreasing the freezing point and raising the boiling point of a solution.

PROBLEM-SOLVING EXAMPLE 13.9

Molarity and Molality

Automobile lead storage batteries contain an aqueous solution of sulfuric acid that has a density of 1.230 g/mL at 25 °C and contains 368 g H_2SO_4 per liter. Calculate

(a) the molarity of sulfuric acid in the solution.
(b) the molality of sulfuric acid in the solution.

Result (a) Molarity = 3.75 mol/L (b) Molality = 4.35 mol/kg

Analyze We are asked to express the solution composition in two ways: (a) molarity, and (b) molality. The density of solution and mass of solute per liter are given.

Plan Use the definitions of molarity and molality.

(a) Use molar mass to determine the amount of solute, H_2SO_4, from its mass. Divide by the volume of solution in liters.
(b) Determine the mass of solvent (water) by calculating the mass of 1 L solution from the solution density and subtracting the mass of H_2SO_4 in 1 L solution (368 g). Divide the amount of solute, which was calculated in part (a), by the mass of solvent. Express the result in mol/kg.

Molality and molarity are *not* the same, although the difference becomes negligibly small for dilute solutions, those less than 0.01 mol/L.

Execute

(a)

$$c(H_2SO_4) = \frac{368 \text{ g } H_2SO_4}{1 \text{ L soln.}} \times \frac{1 \text{ mol } H_2SO_4}{98.1 \text{ g } H_2SO_4} = 3.75 \text{ mol } H_2SO_4/\text{L soln.} = 3.75 \text{ M}$$

(b) First, calculate the mass of solution and mass of water.

$$\text{Mass of solution} = 1 \text{ L soln.} \times \frac{1.230 \text{ g soln.}}{1 \text{ mL soln.}} \times \frac{1000 \text{ mL}}{1 \text{ L}} = 1.230 \times 10^3 \text{ g soln.}$$

$$1230. \text{ g soln.} - 368 \text{ g } H_2SO_4 = 862 \text{ g water} = 0.862 \text{ kg water}$$

Next, calculate the molality of sulfuric acid in the solution.

$$\text{Molality}(H_2SO_4) = \frac{\text{amount solute (mol)}}{\text{mass solvent (kg)}} = \frac{3.75 \text{ mol } H_2SO_4}{0.862 \text{ kg water}} = 4.35 \text{ mol/kg}$$

☑ **Reasonable Result Check** The molar mass of sulfuric acid is about 100 g/mol so 368 g of the acid would be about 3.7 mol. There is about 4 mol of the acid in about a kilogram of water so the molality is about 4, close to the more accurate result of 4.35 mol/kg. Because the density of the solution is not 1.00 g/ml, the mass of solvent and the volume of the solution are not equal so the molarity and the molality differ.

PROBLEM-SOLVING PRACTICE 13.9

Calculate the molarity and the molality of NaCl in a 20.0% aqueous NaCl solution whose density is 1.148 g/mL at 25 °C.

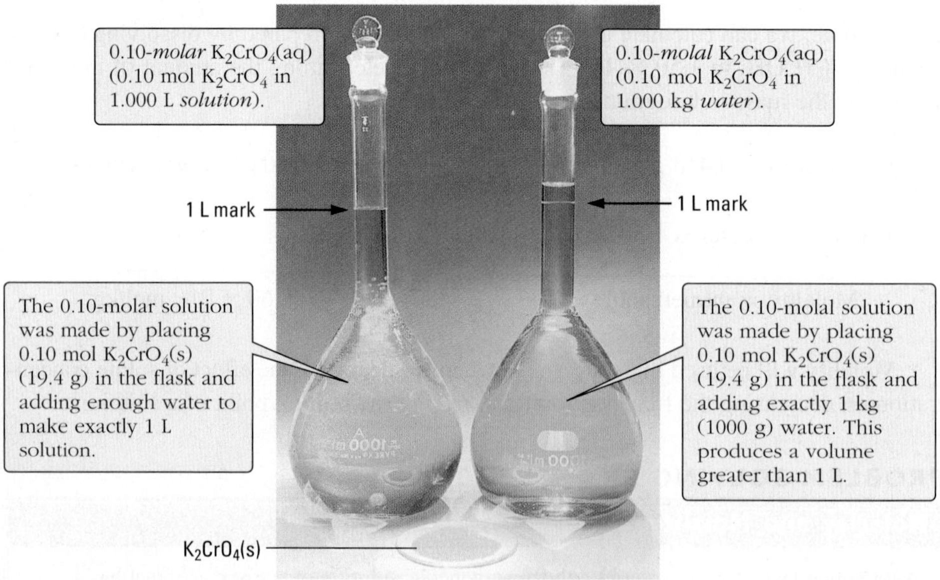

0.10-*molar* K_2CrO_4(aq) (0.10 mol K_2CrO_4 in 1.000 L *solution*).
0.10-*molal* K_2CrO_4(aq) (0.10 mol K_2CrO_4 in 1.000 kg *water*).
1 L mark
1 L mark
The 0.10-molar solution was made by placing 0.10 mol K_2CrO_4(s) (19.4 g) in the flask and adding enough water to make exactly 1 L solution.
The 0.10-molal solution was made by placing 0.10 mol K_2CrO_4(s) (19.4 g) in the flask and adding exactly 1 kg (1000 g) water. This produces a volume greater than 1 L.
K_2CrO_4(s)

Molarity and molality are not the same.

Table 13.4 Comparison of Expressions of Solution Composition

Name	Abbreviation	Name	Units
Mass fraction	None	Molarity	$\frac{\text{Amount solute (mol)}}{\text{volume } solution \text{ (L)}}$
Weight percent	Percent		
Parts per million	ppm		
Parts per billion	ppb	Molality	$\frac{\text{Amount solute (mol)}}{\text{mass } solvent \text{ (kg)}}$
Parts per trillion	ppt		

Table 13.4 summarizes the ways we have described solution composition.

EXERCISE 13.12

Solution Composition

A 12-oz (355-mL) Diet Coke contains 46.3 mg caffeine (M = 194.2 g/mol). The Diet Coke's density is 1.01 g/mL. For such a Diet Coke, calculate: (a) the mass fraction of caffeine in ppm; (b) the molarity of caffeine; and (c) the molality of caffeine.

CONCEPTUAL EXERCISE 13.13

Molality and Molarity

(a) What information is required to calculate the molality of a solution?
(b) What information is needed to calculate the molarity of a solution if the solution's composition is given in weight percent?

13-7 Colligative Properties of Solutions

Up to this point, we have discussed solutions in terms of the nature of the solute and the nature of the solvent. Some properties of solutions do not depend on the nature of the solute, but rather depend only on the *number* of dissolved solute particles—ions or molecules—per known quantity of solvent or solution.

In liquid solutions, solute molecules or ions disrupt solvent-solvent noncovalent attractions (← **Sec. 13-1**), causing changes in solvent properties that depend on these attractions. As a result, when a solute is added to a solvent, the freezing point of the resulting solution is lower and the boiling point is higher than those of the pure solvent. How much lower the freezing point is, or how much higher the boiling point is, depends on the nature of the solvent and the concentration of solute particles. **Colligative properties** of solutions are those that *depend only on the concentration of solute particles* (ions or molecules) *in the solution*, regardless of what kinds of particles are present. We consider four colligative properties that are important in the world around us: vapor pressure lowering, boiling-point elevation, freezing-point lowering, and osmotic pressure.

13-7a Vapor Pressure Lowering

In a closed container, a dynamic equilibrium exists between a pure liquid and its vapor—the rate at which molecules escape the liquid phase equals the rate at which vapor-phase molecules return to the liquid. This equilibrium gives rise to a *vapor pressure* (← **Sec. 9-3**) that depends on the temperature. **Vapor pressure lowering** is *the reduction in vapor pressure of a pure liquid when a solute is dissolved in the liquid.* Compare the vapor pressure of pure water with that of seawater (mainly an aqueous sodium chloride

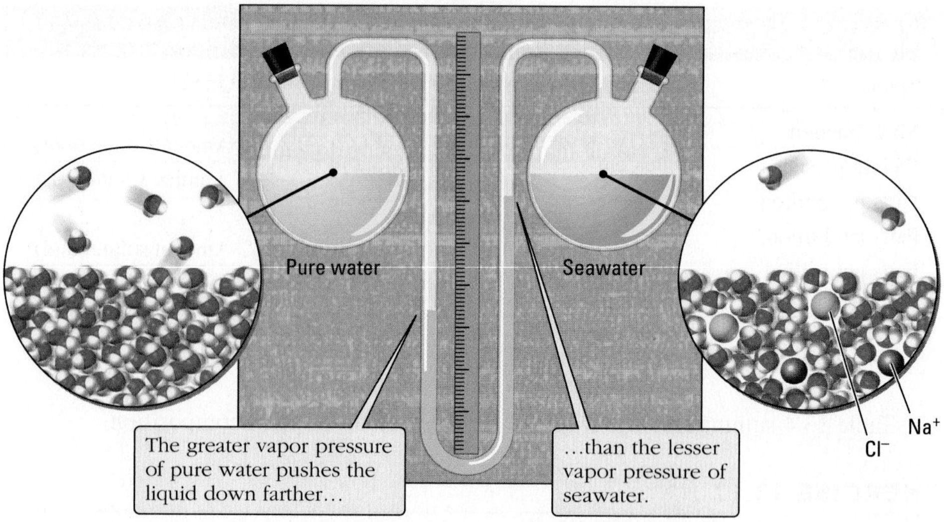
Pure water
Seawater
The greater vapor pressure of pure water pushes the liquid down farther...
...than the lesser vapor pressure of seawater.
Na+
Cl−

Figure 13.13 The vapor pressure of pure water and that of seawater. Seawater is an aqueous solution of NaCl and many other soluble salts. The vapor pressure over an aqueous solution is not as great as that over pure water at the same temperature.

Raoult's law is most accurate with dilute solutions. Deviations from Raoult's law occur when solute-solvent intermolecular forces are either much weaker or much stronger than the solvent-solvent and solute-solute intermolecular forces.

solution), as illustrated at the molecular scale in Figure 13.13. For an aqueous solution such as seawater, in which sodium ions and chloride ions (and many other kinds of ions and molecules) are dissolved, the vapor pressure of water in equilibrium with the solution is lower than for a sample of pure water.

According to **Raoult's law**, *in a solution of a nonvolatile solute the vapor pressure of pure solvent is lowered in proportion to the mole fraction of solvent.*

$$P_1 = X_1 P_1^0$$

Raoult's law can also be applied to solutions in which the solvent and the solute are both volatile so that an appreciable amount of each can be in the vapor above the solution. We will not consider such cases.

where P_1 is the vapor pressure of the solvent over the *solution,* P_1^0 is the vapor pressure of the *pure* solvent at the same temperature, and X_1 is the mole fraction (← **Sec. 8-6**) of *solvent* in the solution. For example, suppose you want to know the vapor pressure over an aqueous sucrose solution at 25 °C in which the mole fraction of water is 0.986. The vapor pressure of pure water at 25 °C is 23.76 mmHg. From these data, the vapor pressure of water over the solution can be calculated using Raoult's law.

$$P_{\text{water}} = (X_{\text{water}})(P^0_{\text{water}}) = (0.986)(23.76\text{ mmHg}) = 23.42\text{ mmHg}$$

The vapor pressure has been lowered by (23.76 − 23.42) mmHg = 0.34 mmHg. Therefore, the vapor pressure of the solution is only 98.6%, $\left(\frac{23.42}{23.76} \times 100\%\right)$ that of pure water at 25 °C.

PROBLEM-SOLVING EXAMPLE 13.10

Raoult's Law

ethylene glycol
$HOCH_2CH_2OH$

Ethylene glycol, $HOCH_2CH_2OH$, is used as an automobile coolant (antifreeze). Calculate the vapor pressure of water above a solution of 100.0 mL ethylene glycol and 100.0 mL water at 90 °C. Densities: ethylene glycol, 1.15 g/mL; water, 1.00 g/mL. The vapor pressure of pure water at 90 °C is 525.8 mmHg.

Result 395 mmHg

Analyze To determine the vapor pressure of water over the solution using Raoult's law, we must first calculate the mole fraction of water in the solution. Use the mole fraction of water and apply Raoult's law to calculate the vapor pressure of water above the solution.

Plan For the mole fraction, use densities to convert milliliters of ethylene glycol and of water to grams; then convert grams to moles using molar masses.

Execute

$$100.0 \text{ mL eth. gly.} \times \frac{1.15 \text{ g eth. gly.}}{1.00 \text{ mL eth. gly.}} \times \frac{1 \text{ mol eth. gly.}}{62.1 \text{ g eth. gly.}} = 1.85 \text{ mol eth. gly.}$$

$$100.0 \text{ mL water} \times \frac{1.00 \text{ g water}}{1.00 \text{ mL water}} \times \frac{1 \text{ mol water}}{18.0 \text{ g water}} = 5.56 \text{ mol water}$$

$$X_{\text{water}} = \frac{5.56}{5.56 + 1.85} = \frac{5.56}{7.41} = 0.750$$

Applying Raoult's law:

$$P_{\text{water}} = (X_{\text{water}})(P^0_{\text{water}}) = (0.750)(525.8 \text{ mmHg}) = 395 \text{ mmHg}$$

☑ **Reasonable Result Check** Because there are nearly three times the number of moles of water as there are moles of ethylene glycol, the mole fraction of water should be greater than the mole fraction of ethylene glycol, and it is (mole fraction of ethylene glycol = 1 − 0.750 = 0.250). Therefore, the vapor pressure of water over the solution should be about 75% that of pure water (395/525.8 × 100%), which it is. The result is reasonable.

Mole fraction of A

$$X_A = \frac{\text{moles of A}}{\text{total number of moles}} = \frac{\text{moles A}}{\text{moles A} + \text{moles B} + \ldots}$$

PROBLEM-SOLVING PRACTICE 13.10

The vapor pressure of an aqueous solution of urea, CH_4N_2O, is 291.2 mmHg at a measured temperature. The vapor pressure of pure water at that temperature is 355.1 mmHg. Calculate the mole fraction of each component.

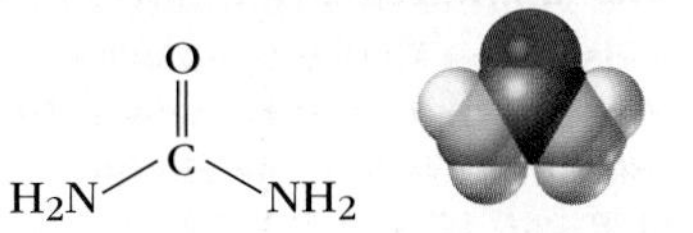
O
C
H2N
NH2
urea

EXERCISE 13.14

Vapor Pressure of a Mixture

Calculate the vapor pressure of water over a solution containing 50.0 g sucrose, $C_{12}H_{22}O_{11}$, and 100.0 g water at 45 °C. The vapor pressure of pure water at this temperature is 71.88 mmHg.

Entropy plays an important role in vapor pressure lowering. Compare the entropy change for the vaporization of pure water with that for the vaporization of a corresponding quantity of water from a sodium chloride solution (Figure 13.14). Because NaCl has

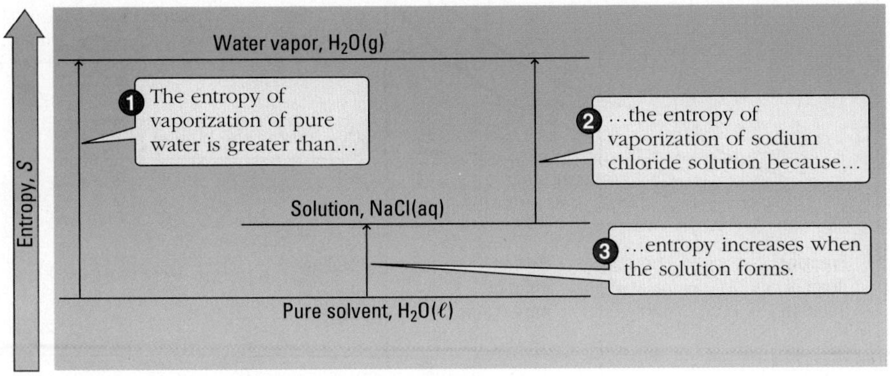
Entropy, S
Water vapor, H2O(g)
1 The entropy of vaporization of pure water is greater than...
2 ...the entropy of vaporization of sodium chloride solution because...
Solution, NaCl(aq)
3 ...entropy increases when the solution forms.
Pure solvent, H2O(ℓ)

Figure 13.14 Vapor pressure lowering and entropy.

a very low vapor pressure, very few sodium or chloride ions escape from the solution and the vapor in equilibrium with the salt water consists almost entirely of water molecules. Thus, the entropy of a given amount of the vapor is approximately the same for vaporization of pure water as for salt water. The entropy of the salt solution, however, is greater than that in pure water because Na^+ and Cl^- ions are more spread out in the solution.

Now consider what happens when water vaporizes from pure water and from the salt solution. As with any change from a liquid to a gas, there is a significant increase in entropy upon vaporization of either pure water or salt solution. But, as illustrated in Figure 13.14, the entropy increase is larger for vaporization of pure water. A larger entropy increase corresponds to a more product-favored process. Thus, at equilibrium, vaporization of pure water creates a higher concentration (and hence pressure) of water vapor than does vaporization of water from a salt solution.

13-7b Boiling-Point Elevation

The normal boiling point of any liquid is the temperature at which the vapor pressure reaches 760 mmHg (1 atm). The normal boiling point of pure water is 100 °C. As a result of vapor pressure lowering, the vapor pressure of an aqueous solution of a nonvolatile solute at 100 °C is less than 760 mmHg, that of pure water. For the solution to boil, it must be heated *above* 100 °C. The **boiling-point elevation**, ΔT_b, is *the difference between the normal boiling point of pure water and the higher boiling point of an aqueous solution of a nonvolatile nonelectrolyte solute* (Figure 13.15).

The volume of a solution, which is used to calculate molarity, expands or contracts when the solution is heated or cooled.

The increase in boiling point is proportional to the concentration of the solute. Molality, rather than molarity, is used in boiling-point-elevation determinations because the molality of a solution does not change as temperature changes, but the molarity of a solution does. Molality is based on the mass of solute *and* the mass of solvent, neither of which is affected by temperature changes. The boiling-point elevation can be calculated from this equation.

The units of the molal boiling-point-elevation constant are °C kg/mol, that is, °C kg mol⁻¹.

$$\Delta T_b = T_b(\text{solution}) - T_b(\text{solvent}) = K_b m_{\text{solute}}$$

where ΔT_b is the boiling-point elevation and K_b is the *molal boiling-point-elevation constant of the solvent.* K_b is different for different solvents. For example, K_b for water is

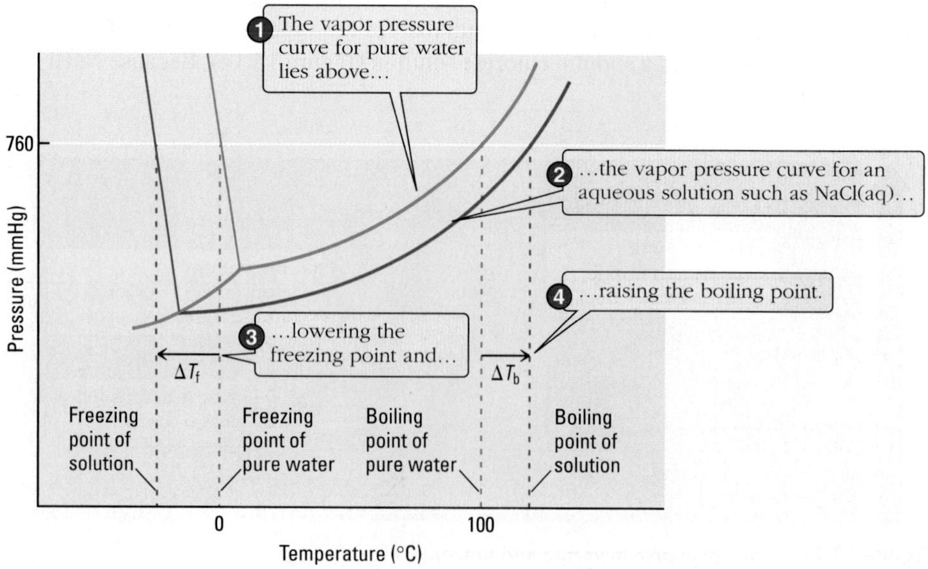
1 The vapor pressure curve for pure water lies above...
2 ...the vapor pressure curve for an aqueous solution such as NaCl(aq)...
3 ...lowering the freezing point and...
4 ...raising the boiling point.
760
Pressure (mmHg)
ΔT_f
ΔT_b
Freezing point of solution
Freezing point of pure water
Boiling point of pure water
Boiling point of solution
0
100
Temperature (°C)

Figure 13.15 Boiling-point elevation (ΔT_b) and freezing-point lowering (ΔT_f) for aqueous solutions. This is a phase diagram like the one for water in Figure 9.19 (← **Sec. 9-5b**).

0.512 °C kg/mol; K_b for benzene is 2.53 °C kg/mol. Some boiling-point-elevation constants are given in Table 13.5.

EXERCISE 13.15

Calculating the Boiling Point of a Solution

The boiling-point-elevation constant for benzene is 2.53 °C kg/mol. The boiling point of pure benzene is 80.10 °C. If a solute's concentration in benzene is 0.10 mol/kg, calculate the boiling point of the solution.

13-7c Freezing-Point Lowering

A pure liquid begins to freeze when the temperature is lowered to the substance's freezing point and the first few molecules cluster into a crystal lattice forming a tiny quantity of solid. As long as both solid and liquid phases are present and the temperature is at the freezing point, there is a dynamic equilibrium in which the rate of crystallization equals the rate of melting. When a *solution* freezes, a few molecules of *solvent* cluster together to form pure solid *solvent* (Figure 13.16), and a dynamic equilibrium is set up between solution and pure solid solvent.

The molecules or ions in the liquid in contact with the frozen solvent in a freezing solution are not all solvent molecules. This causes a slower rate at which particles move from solution to solid than the rate in the pure liquid solvent. To achieve dynamic equilibrium, a correspondingly slower rate of escape of molecules from the solid crystal lattice must occur. According to the kinetic-molecular theory, this slower rate occurs at a lower temperature, so the freezing point of the solution is lower than that of the pure liquid solvent (Figure 13.15).

The **freezing-point lowering**, ΔT_f, is proportional to the molality in the same way as the boiling-point elevation.

$$\Delta T_f = T_f(\text{solution}) - T_f(\text{solvent}) = K_f m_{\text{solute}}$$

As with K_b, the proportionality constant K_f depends only on the solvent. For water, K_f is −1.86 °C kg/mol; by comparison, K_f for benzene is −5.10 °C kg/mol and K_f for cyclohexane is −20.2 °C kg/mol. Some freezing-point-lowering constants are given in Table 13.5. Values are *negative* because the freezing point *decreases*.

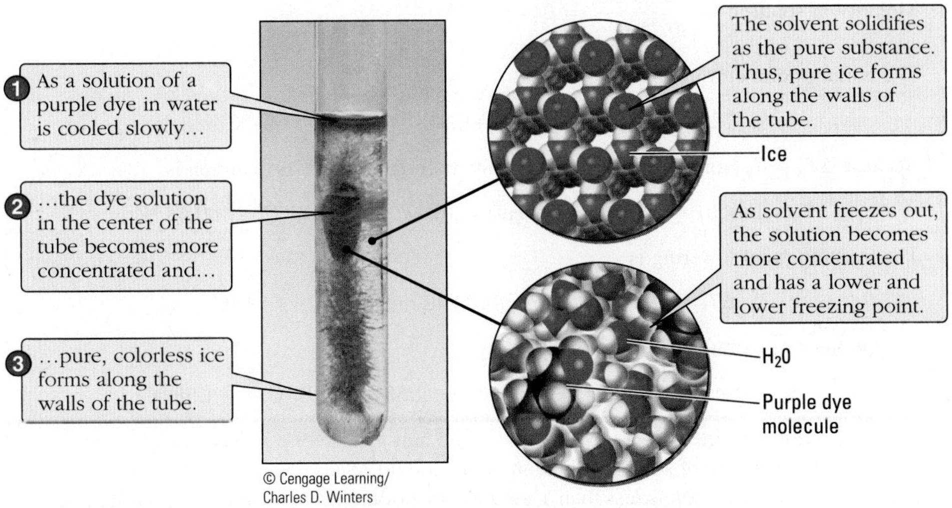
1 As a solution of a purple dye in water is cooled slowly…
2 …the dye solution in the center of the tube becomes more concentrated and…
3 …pure, colorless ice forms along the walls of the tube.
The solvent solidifies as the pure substance. Thus, pure ice forms along the walls of the tube.
Ice
As solvent freezes out, the solution becomes more concentrated and has a lower and lower freezing point.
H2O
Purple dye molecule
© Cengage Learning/ Charles D. Winters

Figure 13.16 Freezing a solution crystallizes pure solvent.

Table 13.5 Some Molal Boiling-Point, K_b, and Freezing-Point, K_f, Constants

Solvent	T_b (°C)	K_b (°C kg/mol)	T_f (°C)	K_f (°C kg/mol)
Benzene, C_6H_6	80.1	2.53	5.5	−5.10
Camphor, $C_{10}H_{16}O$	204	5.95	176	−37.7
Cyclohexane, C_6H_{12}	80.88	2.79	6.50	−20.2
Nitrobenzene, $C_6H_5NO_2$	210.8	5.24	5.7	−8.1
Water, H_2O	100.0	0.512	0.00	−1.86

Another practical application of freezing-point lowering is adding salt, NaCl, to ice when making homemade ice cream. Lowering the freezing temperature of the ice-salt water mixture freezes the ice cream more quickly.

Using ethylene glycol, $HOCH_2CH_2OH$, a relatively nonvolatile alcohol, in automobile cooling systems is a practical application of boiling-point elevation and freezing-point lowering. Ethylene glycol raises the boiling temperature of the coolant mixture of ethylene glycol and water to a level that prevents engine overheating in hot weather. Ethylene glycol also lowers the freezing point of the coolant, thereby keeping the solution from freezing in the winter.

PROBLEM-SOLVING EXAMPLE 13.11

Boiling-Point Elevation and Freezing-Point Lowering

Ethylene glycol is toxic and should not be allowed to get into drinking water supplies.

Calculate the boiling and freezing points of an aqueous solution containing 39.5 g ethylene glycol, $HOCH_2CH_2OH$, dissolved in 750. mL water. Assume the density of water to be 1.00 g/mL.

Result Boiling point = 100.43 °C; freezing point = −1.58 °C

Analyze Adding ethylene glycol (solute) to water (solvent) increases water's boiling point and lowers its freezing point. To use the equations for freezing-point and boiling-point changes, the molality of the solute must be determined.

Plan Use density to calculate the mass of water in kilograms, calculate the molality of the solution, and then apply the equations for freezing and boiling-point changes.

Execute

$$\text{Amount of ethylene glycol} = 39.5 \text{ g} \times \frac{1 \text{ mol}}{62.07 \text{ g}} = 0.636 \text{ mol}$$

$$\text{Mass of water} = 750. \text{ mL} \times \frac{1.00 \text{ g}}{\text{mL}} \times \frac{1 \text{ kg}}{10^3 \text{ g}} = 0.750 \text{ kg}$$

$$\text{Molality of solution} = \frac{0.636 \text{ mol}}{0.750 \text{ kg}} = 0.848 \text{ mol/kg}$$

The boiling-point elevation is

$$\Delta T_b = (0.512 \text{ °C kg/mol})(0.848 \text{ mol/kg}) = 0.43 \text{ °C}$$

Because $\Delta T_b = T_b(\text{solution}) - T_b(\text{solvent})$, the boiling point of the solution is

$$T_b(\text{solution}) = \Delta T_b + T_b(\text{solvent}) = (0.43 + 100.00) \text{ °C} = 100.43 \text{ °C}$$

The freezing-point lowering is

$$\Delta T_f = (-1.86 \text{ °C kg/mol})(0.848 \text{ mol/kg}) = -1.58 \text{ °C}$$

and the freezing point of the solution is

$$T_f(\text{solution}) = \Delta T_f + T_f(\text{solvent}) = (-1.58 + 0.00) \text{ °C} = -1.58 \text{ °C}$$

☑ **Reasonable Result Check** A little more than half a mole of ethylene glycol is in three-fourths kilogram of solvent, so the molality should be less than 1 mol/kg, which it is. Because the concentration is less than 1 mol/kg, the boiling point should be raised less than

0.512 °C and the freezing point should be lowered less than 1.86 °C, which they are. The results are reasonable.

PROBLEM-SOLVING PRACTICE 13.11

A water tank contains 6.50 kg water. Will the addition of 1.20 kg ethylene glycol be sufficient to prevent the solution from freezing if the temperature drops to −25 °C? Show calculations that support your result.

EXERCISE 13.16

Protection Against Freezing

Suppose that you are closing a cabin in the north woods for the winter and you do not want the water in the toilet tank to freeze. You know that the temperature might get as low as −30. °C, and you want to protect about 4.0 L water in the toilet tank from freezing. Calculate the volume of ethylene glycol (density = 1.113 g/mL; molar mass = 62.1 g/mol) you should add to the 4.0 L water.

PROBLEM-SOLVING EXAMPLE 13.12

Molar Mass from Freezing-Point Lowering

A researcher prepares a new compound and uses freezing-point-lowering measurements to determine the molar mass of the compound. The researcher dissolves 1.50 g of the new compound in 75.0 g pure cyclohexane. The freezing point of the solution is measured as 2.70 °C. Calculate the molar mass of the new compound.

Result 106 g/mol

Analyze The molar mass of the unknown compound can be calculated from its molality, which can be determined from the freezing-point lowering equation. This requires that we calculate the change in freezing point. Freezing points and freezing-point constants are in Table 13.5.

Plan Look up the freezing point and freezing-point constant of the solvent, cyclohexane, in Table 13.5. Calculate the freezing-point change and use it to calculate the molality of the solution. Convert the mass of cyclohexane from grams to kilograms. Rearrange the freezing-point equation and solve it for the molar mass, g/mol, of the unknown compound.

Execute The freezing point of cyclohexane is 6.50 °C and its freezing-point constant is −20.2 °C kg/mol. $\Delta T_f = (2.70 - 6.50)$ °C = − 3.80 °C.

$$\text{Molality} = \frac{\Delta T_f}{K_f} = \frac{-3.80\ ^\circ\text{C}}{-20.2\ ^\circ\text{C kg/mol}} = \frac{0.188 \text{ mol solute}}{1 \text{ kg solvent}}$$

The mass of solvent was 75.0 g, that is, 0.0750 kg, so

$$\text{Amount(solute)} = n(\text{solute}) = 0.0750 \text{ kg solvent} \times \frac{0.188 \text{ mol solute}}{1 \text{ kg solvent}} = 0.0141 \text{ mol}$$

$$\text{Molar mass} = \frac{\text{mass solute}}{\text{amount solute}} = \frac{1.50 \text{ g}}{0.0141 \text{ mol}} = 106 \text{ g/mol}$$

☑ **Reasonable Result Check** The 1.5-g compound in 0.0750 kg solvent is equivalent to 2 g per 0.1 kg or 20 g solute per 1 kg solvent. Because $m \cong 0.2$ mol/kg, so the 20 g solute in 1 kg solvent is about 0.2 mol (1/5 mol) solute. Thus, there is about (5×20) g = 100 g compound per mole. The result is reasonable.

PROBLEM-SOLVING PRACTICE 13.12

A student determines the freezing point to be 5.15 °C for a solution made from 0.180 g of a nonelectrolyte in 50.0 g benzene. Calculate the molar mass of the solute.

Jacobus Henricus van't Hoff
1852–1911

Edgar Fahs Smith Collection/University of Pennsylvania Libraries

Van't Hoff was one of the founders of physical chemistry, the branch of chemistry that applies the laws of physics to understand chemical phenomena. Van't Hoff conducted seminal experimental studies in chemical kinetics, chemical equilibrium, osmotic pressure, and chemical affinity. While still a graduate student, he proposed an explanation of optical isomerism based on the tetrahedral nature of the carbon atom (← **Sec. 7-2f**). Van't Hoff received the first Nobel Prize in Chemistry (1901) for fundamental discoveries including his work on the colligative properties of solutions.

13-7d Colligative Properties of Electrolytes

Experimentally, the vapor pressures of 1-molal aqueous solutions of sucrose, NaCl, and $CaCl_2$ are all less than that of water at the same temperature. This is expected because solutes lower the vapor pressure (vp) of the pure solvent. However,

$$\text{vp pure water} > \text{vp 1-molal sucrose} > \text{vp 1-molal NaCl} > \text{vp 1-molal } CaCl_2$$

The reason for this order is that colligative properties of dilute solutions are proportional to the concentration of solute *particles*. Due to their dissociation, electrolytes such as NaCl and $CaCl_2$ contribute more particles per mole than do nonelectrolytes such as sucrose, which do not dissociate. Whereas 1 mol sucrose contributes 1 mol particles (sucrose molecules) to a solution, 1 mol NaCl contributes 2 mol particles (1 mol Na^+ and 1 mol Cl^-), and 1 mol $CaCl_2$ produces 3 mol particles (1 mol Ca^{2+} and 2 mol Cl^-). Therefore, electrolytes have a greater effect on boiling and freezing points than nonelectrolytes have.

For solutions of electrolytes, the boiling-point-elevation equation can be written as

$$\Delta T_b = K_b\, m_{solute}\, i_{solute}$$

and the freezing-point-lowering equation becomes

$$\Delta T_f = K_f\, m_{solute}\, i_{solute}$$

The i_{solute} term is called the **van't Hoff factor**; it gives the *number of particles per formula unit of solute* and is named after Jacobus Henricus van't Hoff.

For nonelectrolytes, $i = 1$ because these molecular solutes, such as ethanol, sucrose, benzene, and carbon tetrachloride, *do* not *dissociate* in solution. *For soluble ionic solutes* (strong electrolytes), i *equals the number of ions per formula unit* of the ionic compound. In extremely dilute solutions $i_{solute} = 2$ for NaCl and $i_{solute} = 3$ for calcium chloride. These are theoretical i values; they assume that the ions act independently in solution, which is achieved *only* in extremely dilute solutions where cations and anions are widely separated by water molecules. For more concentrated solutions, the actual i_{solute} value must be determined experimentally because interactions between cations and anions cause i_{expt} to be less than i_{theor}. For example, in aqueous $MgSO_4$ solutions, $i_{theor} = 2$, whereas in 0.50 M $MgSO_4$, $i_{expt} = 1.07$ and in 0.005 M $MgSO_4$, $i_{expt} = 1.72$.

Another practical application of freezing-point lowering can be seen in areas where winter weather produces lots of frozen precipitation. To remove snow and particularly ice, roads and walkways are often salted. Although sodium chloride is usually used, calcium chloride is particularly good for this purpose because it has three ions per formula unit and dissolves exothermically. Not only is the freezing point of water lowered, but the heat of solution helps melt the ice.

This answers the question, "Why does salt help to clear ice and snow from roads?" that was posed in Chapter 1 (← Sec. 1-1).

CONCEPTUAL **EXERCISE 13.17**

Freezing-Point Lowering

The freezing point of a 2.0-mol/kg $CaCl_2$(aq) solution is measured as −4.78 °C. Calculate the i_{solute} factor and use it to approximate the degree of dissociation of $CaCl_2$ in this solution.

13-7e Osmotic Pressures of Solutions

A *membrane* is a thin layer of material that allows molecules or ions to pass through it. A **semipermeable membrane** *allows only certain kinds of molecules or ions to pass through while excluding others* (Figure 13.17). Examples of semipermeable membranes are animal bladders, cell membranes in plants and animals, and cellophane, a polymer derived from cellulose. When two solutions containing the same solvent are separated

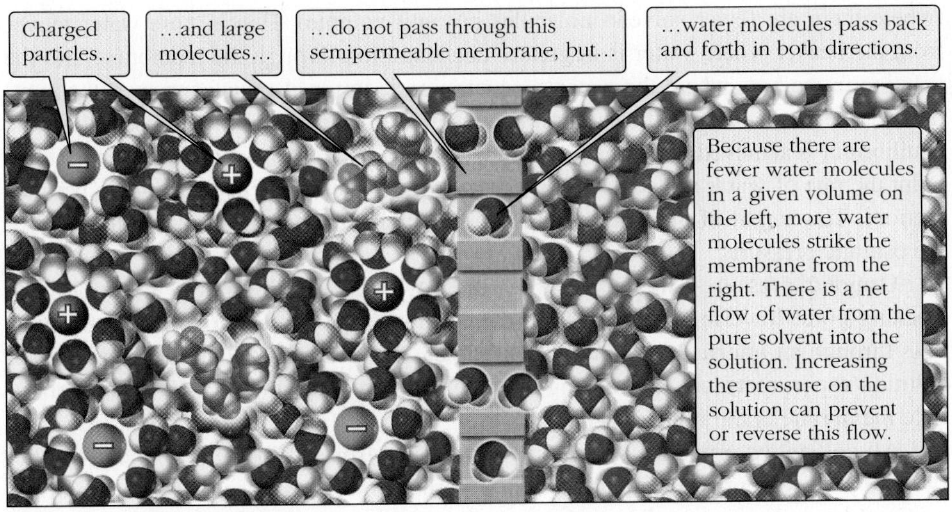
Charged particles...
...and large molecules...
...do not pass through this semipermeable membrane, but...
...water molecules pass back and forth in both directions.
Because there are fewer water molecules in a given volume on the left, more water molecules strike the membrane from the right. There is a net flow of water from the pure solvent into the solution. Increasing the pressure on the solution can prevent or reverse this flow.

Figure 13.17 Osmotic flow of a solvent through a semipermeable membrane to a solution. Although this semipermeable membrane is shown acting as a sieve, many membranes operate in different ways. The ultimate effect is the same.

by a membrane permeable only to solvent molecules, osmosis will occur. **Osmosis** is *the movement of a solvent through a semipermeable membrane from a region of higher solvent concentration (lower solute concentration) to a region of lower solvent concentration (higher solute concentration).* The **osmotic pressure (Π)** of a solution is *the pressure that must be applied to the solution to stop osmosis from a sample of pure solvent.*

Consider the osmosis example shown in Figure 13.18. A 5% aqueous sugar solution enclosed by a membrane permeable to water, but not sugar, molecules is submerged in pure water. Water flows into the bag by osmosis and raises the liquid level in the tube. When the bag is first submerged, more collisions of solvent molecules per unit area of the membrane occur on the pure solvent (water) side than there are on the solution side

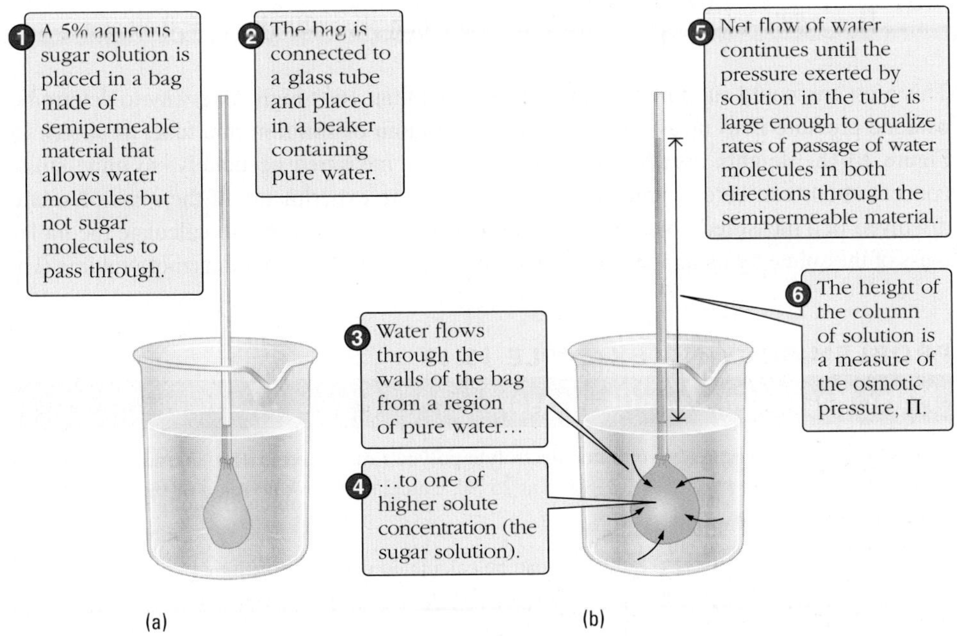
1 A 5% aqueous sugar solution is placed in a bag made of semipermeable material that allows water molecules but not sugar molecules to pass through.
2 The bag is connected to a glass tube and placed in a beaker containing pure water.
3 Water flows through the walls of the bag from a region of pure water...
4 ...to one of higher solute concentration (the sugar solution).
5 Net flow of water continues until the pressure exerted by solution in the tube is large enough to equalize rates of passage of water molecules in both directions through the semipermeable material.
6 The height of the column of solution is a measure of the osmotic pressure, Π.
(a)
(b)

Figure 13.18 Experimental demonstration of osmotic pressure.

(where there are fewer solvent molecules per unit volume). Hence, pure water moves from the beaker where water is in greater concentration through the membrane into the solution in the bag, where the water concentration is lower. As this continues, the number of water molecules in the solution increases and water rises in the tube. A dynamic equilibrium is achieved when the pressure in the bag equals the osmotic pressure. At this point the rate of water molecules passing through the membrane is the same in both directions. The height of the water column then remains unchanged, and is a measure of the osmotic pressure.

Osmotic pressure—like vapor pressure lowering, boiling-point elevation, and freezing-point lowering—results from the unequal rates at which solvent molecules pass through an interface or boundary. In the case of evaporation and boiling, it is the solution/vapor interface; for freezing, it is the solution/solid interface. The semipermeable membrane is the interface for osmosis.

All colligative properties can be understood in terms of differences in entropy between a pure solvent and a solution. This is perhaps most easily seen in the case of osmosis. When solvent and solute molecules mix, entropy usually increases. If pure solvent is added to a solution, a higher entropy state will be achieved as solvent and solute molecules diffuse among one another to form a more dilute solution. This increase in entropy favors mixing of solvent and solution. A semipermeable membrane prevents solute molecules from passing into pure solvent, so the only way mixing can occur (and entropy can increase) is for solvent to flow into the solution, and it does.

The osmotic pressure equation is similar to the ideal gas law equation, $PV = nRT$, which can be rearranged to $P = (n/V)RT = cRT$, where n/V is the molar concentration of the gas.

The more concentrated the solution, the more product-favored the mixing is and the greater is the pressure required to prevent it. Osmotic pressure (Π) is proportional to the *molarity* of the solution, *c*,

$$\Pi = cRTi$$

where *R* is the gas constant, *T* is the absolute temperature (in kelvins), and *i* is the van't Hoff factor, the number of particles per formula unit of solute.

Even though the solution concentration is small, osmotic pressure can be quite large. For example, the osmotic pressure of a 0.020-M solution of a molecular solute ($i = 1$) at 25 °C is

Note that molarity, not molality, is used to express solute concentration in this equation.

$$\Pi = cRTi = \left(\frac{0.020 \text{ mol}}{\text{L}}\right)(0.0821 \text{ L atm mol}^{-1}\text{K}^{-1})(298 \text{ K})(1) = 0.49 \text{ atm}$$

This pressure would support a water column more than 16 ft high. One way to determine osmotic pressure is to measure the height of a column of solution in a tube, as shown in Figure 13.18. Heights of a few centimeters can be measured accurately, so quite small concentrations can be determined by osmotic pressure experiments. If the mass of solute dissolved in a measured volume of solution is known, it is possible to calculate the molar mass of the solute by using the definition of molarity, $c = n/V$ = amount (mol)/volume (L).

Freezing-point-lowering and boiling-point-elevation measurements can also be used to find the molar mass in the same manner as shown in Problem-Solving Example 13.13 for osmotic pressure measurements. See Problem-Solving Example 13.12.

PROBLEM-SOLVING EXAMPLE 13.13

Molar Mass from Osmotic Pressure

A solution containing 2.50 g of a nonelectrolyte polymer dissolved in 150. mL solution has an osmotic pressure of 1.25×10^{-2} atm at 25 °C. Calculate the molar mass of the polymer.

Result 3.26×10^4 g/mol

Analyze The molarity of the solution can be calculated using the osmotic pressure equation. From it and the volume, the amount of polymer can be determined. The molar mass is mass divided by amount. The mass of polymer is given.

Plan Apply the osmotic pressure equation. Because the polymer is a nonelectrolyte, $i = 1$. Convert the volume of solution to liters so that units cancel with the ideal gas constant, $R = 0.0821 \text{ L atm K}^{-1} \text{ mol}^{-1}$.

Execute The molarity of the solution is

$$c = \frac{\Pi}{RTi} = \frac{1.25 \times 10^{-2}\ \text{atm}}{(0.0821\ \text{L atm mol}^{-1}\text{K}^{-1})(298\ \text{K})(1)} = 5.11 \times 10^{-4}\ \text{mol/L}$$

The volume of solution is 0.150 L, so the amount of polymer is

$$n(\text{polymer}) = (0.150\ \text{L})(5.11 \times 10^{-4}\ \text{mol/L}) = 7.67 \times 10^{-5}\ \text{mol}$$

The mass of this amount of polymer is 2.50 g so the average molar mass is

$$M(\text{polymer, average}) = \frac{2.50\ \text{g polymer}}{7.67 \times 10^{-5}\ \text{mol polymer}} = 3.26 \times 10^{4}\ \text{g/mol}$$

☑ **Reasonable Result Check** Because the molarity is low, the molar mass of the polymer must be relatively large to create an osmotic pressure of 1.25×10^{-2} atm. In Section 10-6, polymers were described as long chains of linked monomer units, so a molar mass of over 30,000 g/mol is not unreasonable for the polymer.

PROBLEM-SOLVING PRACTICE 13.13

At 25 °C, the osmotic pressure of a solution of 5.0 g of horse hemoglobin (a protein) in 1.0 L water is 1.8×10^{-3} atm. Calculate the molar mass of the hemoglobin.

Blood and other fluids inside living cells contain many different solutes, and the osmotic pressures of these solutions play an important role in the distribution and balance of solutes within the body. Dehydrated patients are often given water and nutrients intravenously. However, pure water cannot simply be dripped into a patient's veins (Figure 13.19). The water would flow into the red blood cells by osmosis, causing them to burst. *A solution that causes water to flow into cells* is called a **hypotonic** solution. To prevent cells from bursting, an isotonic intravenous solution must be used. An **isotonic** (or iso-osmotic) solution *has the same total concentration of solutes and therefore the same osmotic pressure* as the patient's blood. A solution of 0.9% sodium chloride is isotonic with fluids inside cells in the body. (See Problem-Solving Example 13.4, ← **Sec. 13-6a.**)

If an intravenous solution more concentrated than the solution inside a red blood cell were added to blood, the cell would lose water to the solution and shrivel up. *A solution that causes water to flow out of cells* is a **hypertonic** solution. Cell shriveling

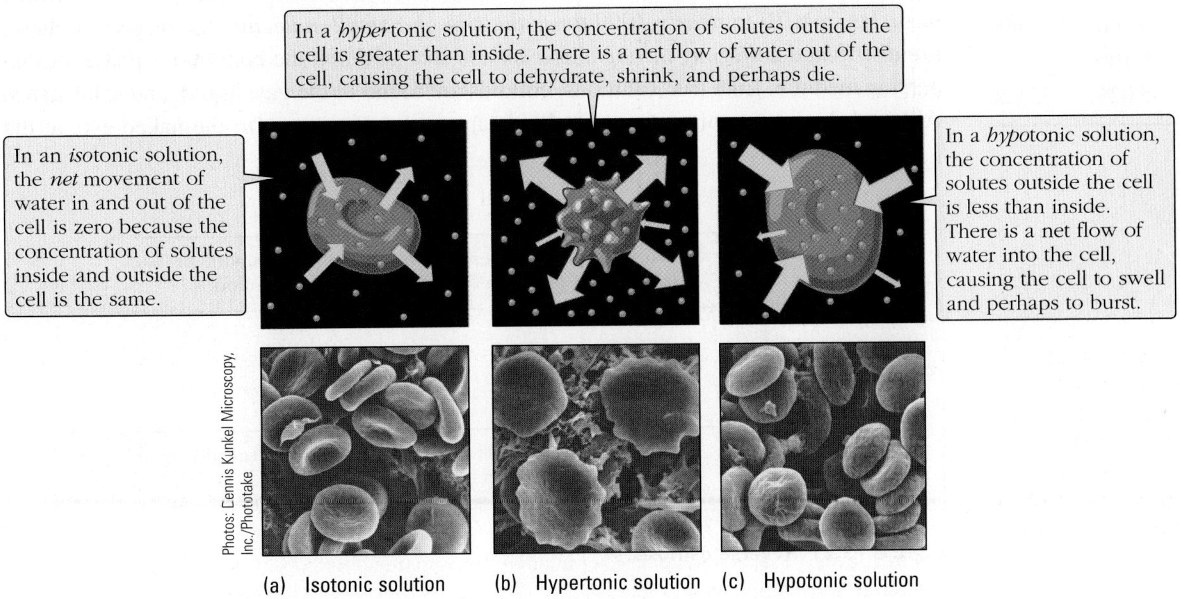
In a *hyper*tonic solution, the concentration of solutes outside the cell is greater than inside. There is a net flow of water out of the cell, causing the cell to dehydrate, shrink, and perhaps die.
In an *iso*tonic solution, the *net* movement of water in and out of the cell is zero because the concentration of solutes inside and outside the cell is the same.
In a *hypo*tonic solution, the concentration of solutes outside the cell is less than inside. There is a net flow of water into the cell, causing the cell to swell and perhaps to burst.
Photos: Lennis Kunkel Microscopy, Inc./Phototake
(a) Isotonic solution
(b) Hypertonic solution
(c) Hypotonic solution

Figure 13.19 Osmosis and the living cell. Isotonic, hypertonic, and hypotonic solutions.

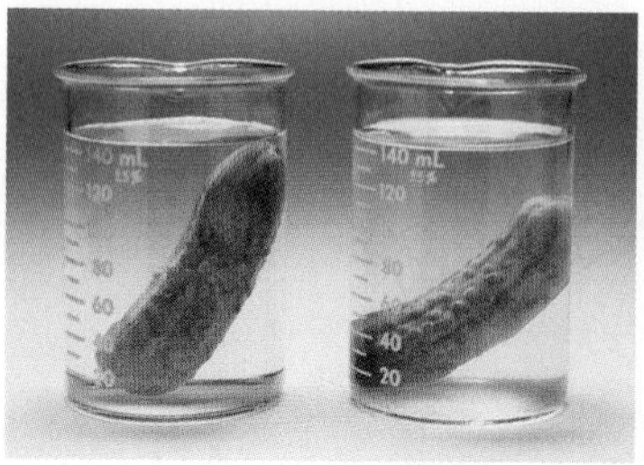
140 mL
140 mL

Figure 13.20 Osmosis in vegetables. A cucumber soaked in a concentrated salt solution (*right*) has lost much water and shrivels into a pickle. A cucumber soaked in pure water (*left*) is affected very little.

by osmosis happens when vegetables or meats are cured in *brine*, a concentrated solution of NaCl. If you put a fresh cucumber into brine, water will flow out of its cells and into the brine, leaving behind a shriveled vegetable (Figure 13.20). With proper spices added to the brine, a cucumber becomes a tasty pickle.

13-7f Reverse Osmosis

Reverse osmosis occurs when pressure greater than the osmotic pressure is applied and solvent is forced to flow through a semipermeable membrane from a concentrated solution to a dilute solution. In effect, the semipermeable membrane serves as a filter with very tiny pores through which only solvent can pass. Reverse osmosis can be used to remove small molecules or ions to obtain highly purified water. Seawater contains a high concentration of dissolved salts; its osmotic pressure is 24.8 atm. If a pressure greater than 24.8 atm is applied to a chamber containing seawater, water molecules can be forced to flow from seawater through a semipermeable membrane to a region containing purer water (Figure 13.21). Pressures up to 100 atm are used to provide reasonable rates of seawater purification. Seawater, which contains upward of 35,000 ppm (3.5%) of dissolved salts (Table 13.6), can be purified by reverse osmosis to between 400 and 500 ppm of solutes, which is well within the World Health Organization's limits for drinking water. Large reverse osmosis plants in places like the Persian Gulf countries and Florida can purify more than 100 million gallons of water per day. Nearly 50% of Saudi Arabia's fresh water is provided by one of the world's largest desalination plants at Jubail. Reverse osmosis purifies nearly 15 million gallons of brackish underground water daily for the city of Cape Coral, Florida. (Brackish water is a mixture of salt water and fresh water.) That facility is one of more than 100 in the state of Florida. Reverse osmosis is also used to further purify fresh water when ultrapure water is essential, as in the production of pharmaceuticals or in laboratory research.

Small reverse osmosis units are used to make ultrapure water for some "spotless" car washes.

Table 13.6 Ions Present in Seawater at 100 ppm or More

	Mass Fraction	
Ion	g/kg	ppm
Cl^-	19.35	19,350
Na^+	10.76	10,760
SO_4^{2-}	2.710	2710
Mg^{2+}	1.290	1290
Ca^{2+}	0.410	410
K^+	0.400	400
HCO_3^-, CO_3^{2-}	0.106	106
Total	35.026	35,026

13-8 Colloids

Around 1860 Thomas Graham found that starch, gelatin, glue, and egg albumin diffuse only very slowly in water and do not diffuse through a thin membrane, but sugar or salt do. He coined the word "colloid" to describe a class of substances distinctly different from sugar and salt and similar materials. A **colloid** is now understood to be *a mixture in which particles from 10 to nearly 1000 times the size of a small molecule*, the dispersed phase, *are distributed uniformly throughout a solvent-like medium*, the continuous phase, or dispersing medium. Like true solutions, colloids are found in the gas, liquid, and solid states. Although both true solutions and colloids appear homogeneous to the naked eye, at the

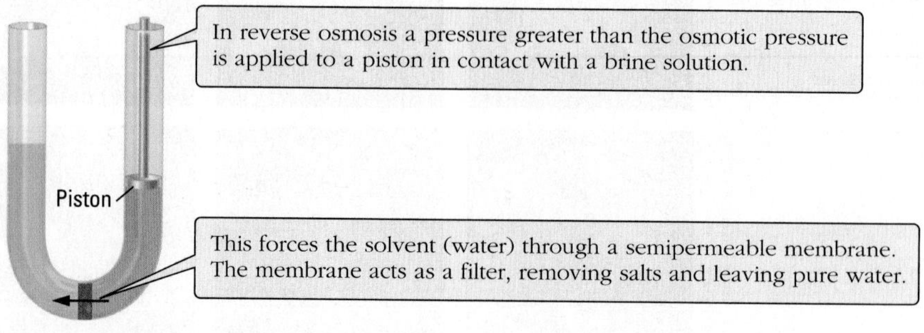
In reverse osmosis a pressure greater than the osmotic pressure is applied to a piston in contact with a brine solution.
Piston
This forces the solvent (water) through a semipermeable membrane. The membrane acts as a filter, removing salts and leaving pure water.

Figure 13.21 Reverse osmosis.

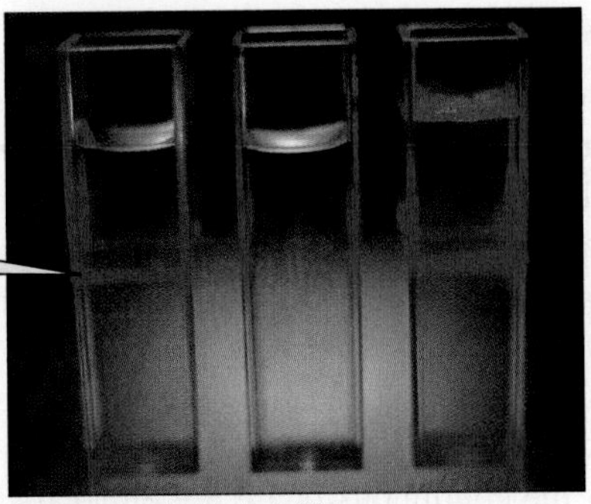
A narrow beam of red laser light passes through all three cuvets at this level.
The first and third cuvets contain colloidal dispersions and the laser beam is visible because of scattered light (Tyndall effect). The middle cuvet contains an NaCl solution. Na^+ and Cl^- are too small to scatter visible light, so the beam is not seen.
© Cengage Learning/Charles D. Winters

Figure 13.22 Tyndall effect: light scattering by a colloidal dispersion.

nanoscale level colloids are not homogeneous because the dispersed particles are larger than molecules and may be aggregations of many molecules. Colloidal particles are larger than those found in true solutions and smaller than the particles in suspensions.

	Smaller Particles ⟶		Larger Particles
	True Solution	Colloidal Dispersion	Suspension
Particles	Ions and molecules	Colloids	Large-sized particles
Particle size	0.2–2.0 nm	2–2000 nm	>2000 nm
Properties	Don't settle out on standing	Don't settle out on standing	Settle out on standing
	Not filterable	Not separable using ordinary filters	Filterable
Example	Seawater	Fog	River silt

Colloidal particles can be so large—up to 2000 nm in diameter—that they scatter visible light passing through the continuous medium. This *scattering of light by a colloid* is known as the **Tyndall effect**. Scattered light makes a light beam visible, as shown in Figure 13.22. Colloidal particles of dust and smoke in the air of a room can easily be observed in a beam of sunlight because they scatter the light; you have probably seen such a well-defined sunbeam many times. A common colloid, *fog,* consists of water droplets (the dispersed phase) in air (the continuous phase, and itself a solution). The fact that you cannot see through dense fog is another consequence of the Tyndall effect.

Galen Rowell/Corbis

The Tyndall effect. Shafts of light are visible through the mist in this forest because fog particles are large enough to scatter light.

Table 13.7 Types of Colloids

Continuous Phase	Dispersed Phase	Type	Examples
Gas	Liquid	Aerosol	Fog, clouds, aerosol sprays
Gas	Solid	Aerosol	Smoke, airborne viruses, automobile exhaust
Liquid	Gas	Foam	Shaving cream, whipped cream
Liquid	Liquid	Emulsion	Mayonnaise, milk, face cream
Liquid	Solid	Sol	Gold in water, milk of magnesia, mud
Solid	Gas	Foam	Foam rubber, sponge, pumice
Solid	Liquid	Gel	Jelly, cheese, butter
Solid	Solid	Solid sol	Milk glass, many alloys such as steel, some colored gemstones

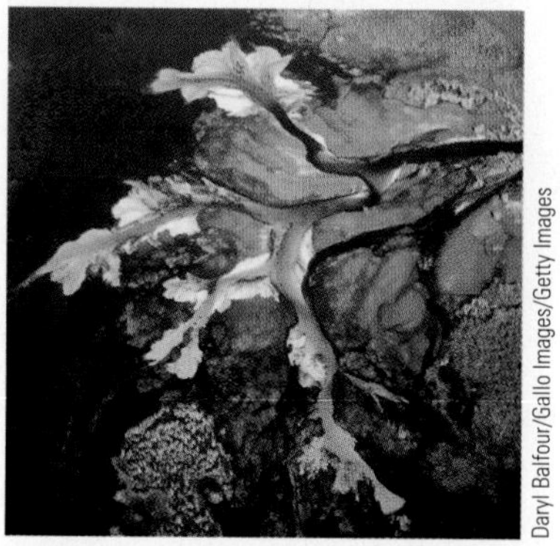
Daryl Balfour/Gallo Images/Getty Images

Figure 13.23 Aerial view of silt formation in a river delta from colloidal soil particles. Silt (tan color) forms at a river delta when colloidal particles in the river water meet the salt water in an ocean or a salt-water bay. The higher salt concentration causes the colloidal particles to coagulate.

13-8a Types of Colloids

Colloids are classified according to the state of the dispersed phase (solid, liquid, or gas) and the state of the continuous phase. Table 13.7 lists several types of colloids and some examples of each. Liquid-liquid colloids form only in the presence of an emulsifier—a third substance that coats and stabilizes the particles of the dispersed phase. *A colloidal dispersion involving an emulsifier* is called an **emulsion**. In mayonnaise, for example, egg yolk contains a protein that stabilizes the tiny drops of oil that are dispersed in the aqueous continuous phase. As you can see from examples listed in Table 13.7, colloids are very common in everyday life.

Colloids with water as the continuous phase are either hydrophilic or hydrophobic. In a hydrophilic colloid there is *a strong attraction between the dispersed phase and the continuous (aqueous) phase*. Hydrophilic colloids form when the molecules of the dispersed phase have multiple sites that interact with water through hydrogen bonding and dipole-dipole attractions. Proteins in aqueous media are hydrophilic colloids.

In a hydrophobic colloid there is a *lack of attraction between the dispersed phase and the continuous phase*. Although you might assume that such colloids would tend to separate quickly, hydrophobic colloids can be quite stable once they are formed. A colloidal dispersion (sol) of gold nanoparticles prepared in 1857 is still preserved in the British Museum.

A stable hydrophobic colloid coagulates when ions come into contact with the dispersed phase. Soil particles carried in rivers are hydrophobic. When river water containing large amounts of colloidal soil particles meets seawater with a high ion concentration, the colloidal particles coagulate to form silt. The deltas of the Mississippi and Nile Rivers have been and continue to be formed in this way (Figure 13.23).

13-9 Surfactants

A molecule that has a hydrophobic part and a hydrophilic part is called a **surfactant** (surface-active agent) because such molecules tend to act at the surface of a substance that is in contact with the solution that contains the surfactant. Soap, the classic surfactant, dates back to the Babylonians in 2800 BC. Chemically, soaps are salts of fatty acids and have always been made by the reaction of a fat with an alkali, a process known as *saponification* (← **Sec. 10-5d**).

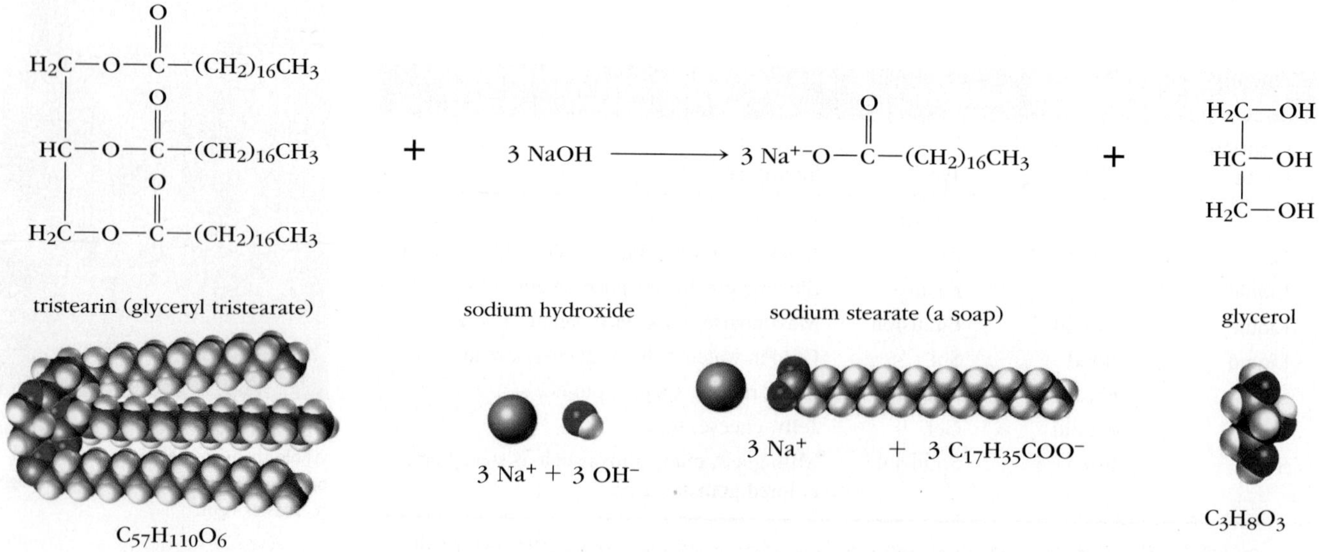
+
3 NaOH
3 Na⁺⁻O—C—(CH₂)₁₆CH₃
+
tristearin (glyceryl tristearate)
sodium hydroxide
sodium stearate (a soap)
glycerol
3 Na⁺ + 3 OH⁻
3 Na⁺ + 3 C₁₇H₃₅COO⁻
C₅₇H₁₁₀O₆
C₃H₈O₃

Sodium stearate is a typical soap. The long-chain hydrocarbon part of the molecule is hydrophobic, while the polar carboxylate group is hydrophilic.

$$\underbrace{CH_3CH_2CH_2CH_2CH_2CH_2CH_2CH_2CH_2CH_2CH_2CH_2CH_2CH_2CH_2CH_2CH_2}_{\text{Hydrophobic end}}\underbrace{\overset{\overset{\displaystyle O}{\|}}{C}{-}O^-Na^+}_{\text{Hydrophilic end}}$$

sodium stearate

Hand soaps are pure soaps to which dyes and perfumes are added.

Bile salts are important surfactants in the body that help to break up ingested fats. The hydrophobic portion of bile salts is incorporated into the surface of the ingested fat molecules, where it acts as an emulsifying agent, keeping fat molecules dispersed from each other. The hydrophilic portion of the bile salt keeps the emulsified fats in solution.

Water, oil, and a surfactant together form an emulsion. The surfactant acts as the emulsifying agent, such as in the case of bile salts emulsifying dietary fats.

A synthetic surfactant is called a **detergent**. Detergents are made from refined petroleum or coal products. Detergents have a long hydrocarbon chain that is hydrophobic and a polar end that is hydrophilic, somewhat like those of soaps. One common detergent is sodium lauryl sulfate, which is used in many shampoos.

$$CH_3CH_2CH_2CH_2CH_2CH_2CH_2CH_2CH_2CH_2CH_2CH_2OSO_3^-Na^+$$

sodium lauryl sulfate

CONCEPTUAL EXERCISE 13.18

Osmotic Pressure of a Soap Solution

Which solution has higher osmotic pressure, a 0.02-M sucrose solution or a 0.02-M soap solution, such as sodium stearate? Explain your result.

In water solutions, surfactants tend to aggregate to form hollow, colloid-sized, roughly spherical particles called *micelles* (Figure 13.24) that can transport various materials within them. The hydrophobic ends of the surfactant molecules point inward to the center of the micelle, while their hydrophilic heads point outward so that they interact with water molecules. Ordinary soap, a surfactant, forms micelles in water. Soaps cleanse because hydrophobic oil on clothing or skin joins to the hydrophobic centers of the soap micelles and is rinsed away (Figure 13.24). When ordinary soaps are used with hard water (Section 13-10b), the soaps react with Ca^{2+}, Mg^{2+}, and Fe^{2+} ions in the hard water to produce insoluble salts that form an unsightly scum around the bathtub or on clothes. Detergents do not form these insoluble salts, even with hard water.

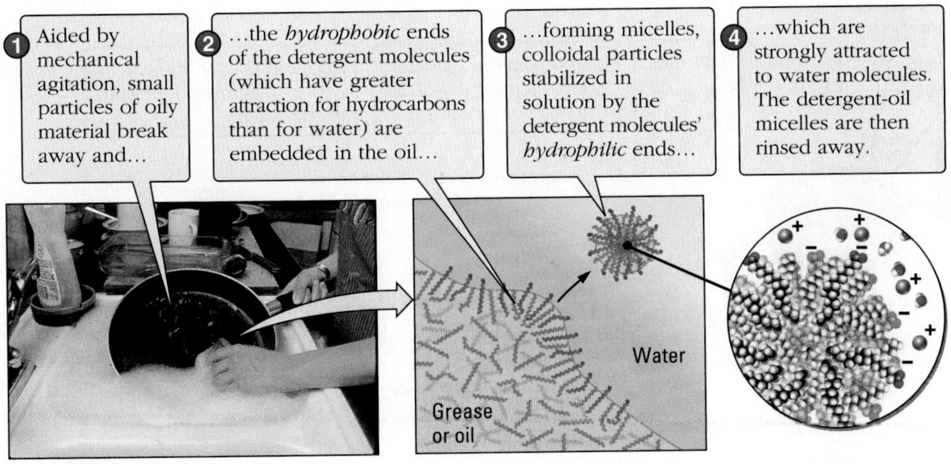

1 Aided by mechanical agitation, small particles of oily material break away and…
2 …the *hydrophobic* ends of the detergent molecules (which have greater attraction for hydrocarbons than for water) are embedded in the oil…
3 …forming micelles, colloidal particles stabilized in solution by the detergent molecules' *hydrophilic* ends…
4 …which are strongly attracted to water molecules. The detergent-oil micelles are then rinsed away.
Water
Grease or oil

© Cengage Learning/Charles D. Winters

Figure 13.24 The cleansing action of soaps and detergents depends on micelle formation.

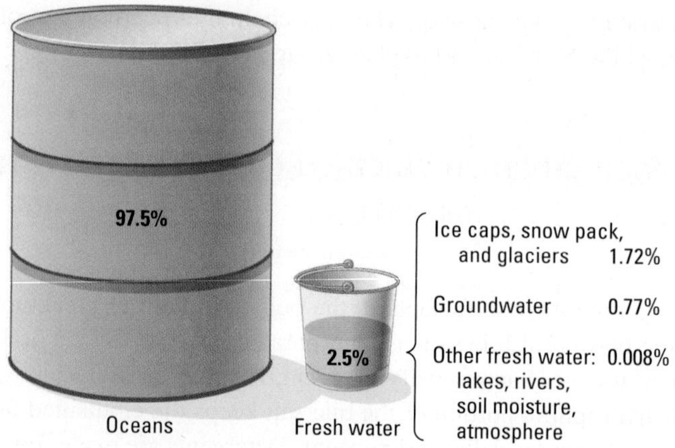
97.5%
2.5%
Oceans
Fresh water
Ice caps, snow pack, and glaciers 1.72%
Groundwater 0.77%
Other fresh water: 0.008%
lakes, rivers, soil moisture, atmosphere

Figure 13.25 Earth's water supply. Only 2.5% is fresh water, and less than 1% is available as groundwater or surface water.

13-10 Water: Natural, Clean, and Otherwise

"... Water, water everywhere, nor any drop to drink." ***The Rime of the Ancient Mariner*** **by Samuel Taylor Coleridge alludes to the fact that the dissolved salt in seawater renders it useless for drinking.**

More than 97% of water, the most abundant substance on Earth's surface, is found in the oceans, which cover about three quarters of Earth's surface (Figure 13.25). Glaciers, ice caps, and snow pack account for 1.72% of water. *Surface water*—lakes, rivers, streams, and reservoirs—makes up only a very small portion (0.008%). *Groundwater*, water that is held in large underground natural *aquifers*, makes up 0.77% of water.

Note from Figure 13.25 that fresh water makes up less than 3% of Earth's surface water, with most of it tied up in glaciers. Indeed, the amount of fresh water available to satisfy our demands (we can't live without it) is relatively limited. In the United States, a bit more than half (57%) of fresh water is used in industry and about a third (34%) in agriculture to irrigate crops; only 9% is available for domestic and municipal purposes. Even when sufficient water is available, the quality of the water needs to be assured before it is safe to drink (potable). Water must first be treated to make it potable.

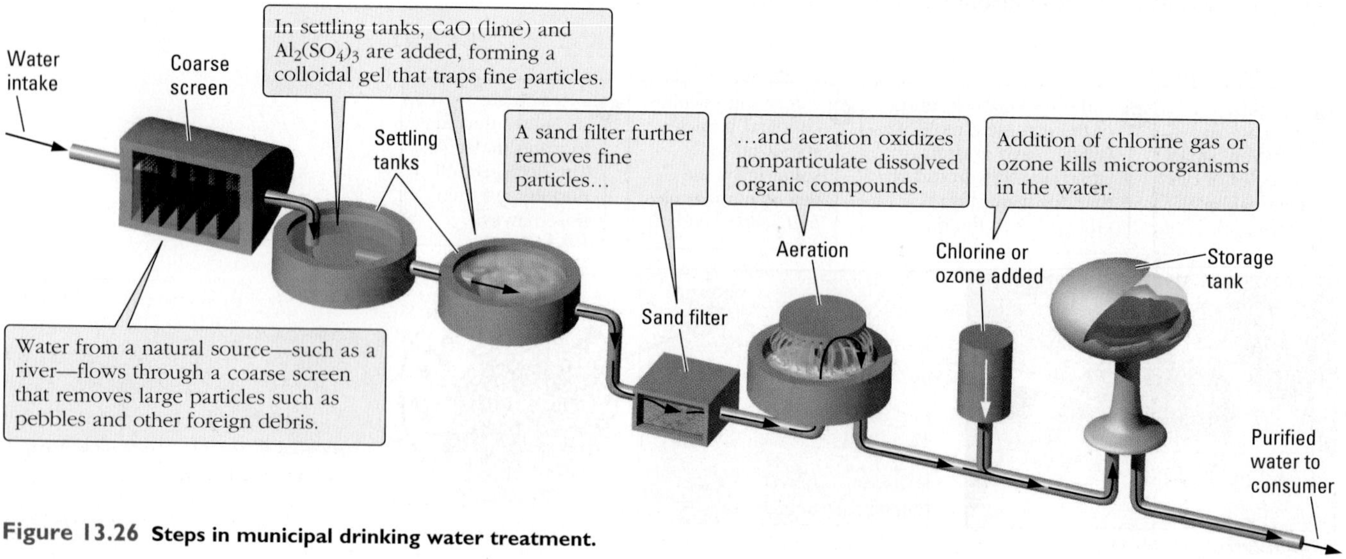
Water intake
Coarse screen
In settling tanks, CaO (lime) and $Al_2(SO_4)_3$ are added, forming a colloidal gel that traps fine particles.
Settling tanks
A sand filter further removes fine particles...
...and aeration oxidizes nonparticulate dissolved organic compounds.
Addition of chlorine gas or ozone kills microorganisms in the water.
Aeration
Chlorine or ozone added
Storage tank
Sand filter
Water from a natural source—such as a river—flows through a coarse screen that removes large particles such as pebbles and other foreign debris.
Purified water to consumer

Figure 13.26 Steps in municipal drinking water treatment.

13-10a Municipal Drinking Water Purification

Fresh water, even from natural sources, contains dissolved materials that must be removed entirely or partially to make water fit for domestic use, or for agricultural and industrial purposes. The Safe Water Drinking Act of 1974 established required standards of purity and safety for U. S. public water supplies. The U. S. EPA sets limits for contaminants that may be present in drinking water and requires continual monitoring of municipal water supplies.

Municipal water purification takes place in a series of steps (Figure 13.26). After a coarse filter (a screen) removes large objects such as tires, tree limbs, bottles, and so on, the water goes into a settling tank, where small clay and dirt particles settle out. To speed up the sedimentation, $Al_2(SO_4)_3$ (alum) and CaO (lime) are added. These compounds react to form a sticky gelatinous precipitate of aluminum hydroxide.

Calcium hydroxide (slaked lime) forms when lime, CaO, reacts with water.

$$Al_2(SO_4)_3(aq) + 3\ Ca(OH)_2(aq) \longrightarrow 2\ Al(OH)_3(s) + 3\ CaSO_4(aq)$$

The $Al(OH)_3$ gel collects suspended clay and dirt particles as it sinks slowly in the settling tank. Particles not settled out are removed by passing the water from the settling tank through a sand filter.

In the next step, the water is aerated—sprayed into the air to oxidize dissolved organic substances. To this point, nothing has been done to remove potentially harmful bacteria. These bacteria are killed in the final step by chemically treating the water. *Chlorination* is the most common bactericidal method used in the United States. Chlorination is done by adding chlorine gas, sodium hypochlorite, NaOCl, or calcium hypochlorite, $Ca(OCl)_2$. In all three cases, the antibacterial agent generated is HOCl, hypochlorous acid. In the case of Cl_2, the HOCl forms by the reaction of chlorine with water.

$$Cl_2(g) + H_2O(\ell) \longrightarrow HOCl(aq) + H^+(aq) + Cl^-(aq)$$

The extent of chlorination is adjusted so that between 0.075 and 0.600 ppm of HOCl remains in solution as the water leaves the treatment plant, which is sufficient to ensure that bacterial contamination does not occur before the water reaches the user.

Chlorination was first used for drinking water supplies in the early 1900s, with a resulting drop in the number of deaths in the United States caused by typhoid and other water-borne diseases from 35/100,000 population in 1900 to 3/100,000 population in 1930. Chlorination is the principal means of preventing water-borne diseases spread by bacteria, including cholera, typhoid, paratyphoid, and dysentery.

In spite of chlorination, most city water supplies are not bacteria-free, but only rarely do these surviving bacteria cause disease. In the United States the most common water-borne bacterial disease is giardiasis, a gastrointestinal disorder. Most often this disease is caused by bacteria in surface water that has leaked into drinking water supplies, but on occasion it can be traced to city water systems.

Chlorination is not without its own small risk because of by-products it forms. Even the best-designed purification systems allow some organic compounds to pass through; these then become chlorinated. In particular, humic acids, breakdown products of plant materials always present in surface water, react with residual HOCl. The reaction forms a class of compounds known as trihalomethanes (THMs), such as chloroform, $CHCl_3$. Most drinking water meets the current maximum contaminant level standard of 80 ppb THMs. Chloroform is the trihalomethane of chief concern because it is suspected of causing liver cancer. Information is still being evaluated about the seriousness of this potential threat, especially given the fact that the THM level in drinking water is normally much less than 80 ppb.

The 80 ppb standard for THMs was established in 1998 by the U. S. Environmental Protection Agency. The previous standard had been 100 ppb. The national average for THMs in drinking water is 51 ppb.

Gaseous ozone, O_3, is used in many European cities to disinfect municipal water supplies. Ozone is an even more effective bactericide than chlorine, so less of it is needed to purify the water. The disadvantages of ozone are that it must be generated on site, and that ozone does not remain in the water as long as chlorine, allowing the possibility of recontamination.

Disinfection of water by using ultraviolet radiation is becoming more popular. It is fast, is economical for small installations such as rural homes, and leaves no residual byproducts. Like ozonation, UV disinfection does not protect against bacterial contamination after water leaves the treatment site unless the appropriate amount of chlorine is added.

EXERCISE 13.19

Drinking Water

Selenium poisoning can occur in individuals who ingest more than 400 μg selenium per day. Calculate the mass of selenium ingested daily by an individual who drinks 3.0 qt water containing the maximum contaminant level of selenium, 0.050 ppm.

EXERCISE 13.20

Lead in Drinking Water

If the Pb^{2+} concentration in tap water is 0.0250 ppm, calculate how many liters of this water contain 100.0 μg Pb^{2+}.

Degrees of water hardness:
Soft water:
< 60 mg metal ion/L
Moderately hard:
60–120 mg/L
Hard: 120–180 mg/L
Very hard: > 180 mg/L

13-10b Hard Water: Natural Impurities

A relatively high concentration of Ca^{2+}, Mg^{2+}, Fe^{2+}, or Mn^{2+} ions imparts "hardness" to water. Water hardness is objectionable because it (1) causes precipitates (called scale) to form in boilers and hot-water pipes, (2) causes soaps to form insoluble curds, and (3) may impart a disagreeable taste to the water.

Water hardness is produced when surface water containing carbon dioxide trickles through limestone or dolomite, releasing calcium or magnesium ions as their soluble hydrogen carbonates.

$$\underset{\text{limestone}}{CaCO_3(s)} + CO_2(g) + H_2O(\ell) \longrightarrow Ca^{2+}(aq) + 2\ HCO_3^-(aq)$$

$$\underset{\text{dolomite}}{CaCO_3 \cdot MgCO_3(s)} + 2\ CO_2(g) + 2\ H_2O(\ell) \longrightarrow Ca^{2+}(aq) + Mg^{2+}(aq) + 4\ HCO_3^-(aq)$$

Such hard water can be softened by removing these ions by two principal methods: (1) the lime-soda process or (2) ion exchange.

In the lime-soda process, slaked lime, $Ca(OH)_2$, and soda, Na_2CO_3, are added to the water. Several reactions take place, which can be summarized as:

The second equation also applies if Mg^{2+} is substituted for Ca^{2+}.

In hard water **Added**

$$HCO_3^-(aq) + OH^-(aq) \longrightarrow CO_3^{2-}(aq) + H_2O(\ell)$$

$$Ca^{2+}(Aq) + Na_2CO_3(aq) \longrightarrow CaCO_3(s) + 2\ Na^+(aq)$$

$$Mg^{2+}(aq) + 2\ OH^-(aq) \longrightarrow Mg(OH)_2(s)$$

The lime-soda process works because calcium carbonate, $CaCO_3$, is much less soluble than calcium hydrogen carbonate, $Ca(HCO_3)_2$, and because magnesium hydroxide, $Mg(OH)_2$, is much less soluble than magnesium hydrogen carbonate, $Mg(HCO_3)_2$. The overall result of the lime-soda process is to precipitate almost all the calcium and magnesium ions and to leave sodium ions (which do not make water hard) as replacements.

Iron present as Fe^{2+} and manganese present as Mn^{2+} can be removed from water by aeration to produce higher oxidation states. If the water is neutral or slightly alkaline (either naturally or from the addition of lime), insoluble compounds $Fe(OH)_3$ and $MnO_2(H_2O)_x$ form and precipitate from solution.

Ion exchange is another way to remove ions that cause water hardness. In this kind of water softener, H^+ or Na^+ ions supplied by a polymer that contains negatively charged sites replace the ions that cause hard water. The hard-water ions become attached to the resin and H^+ or Na^+ ions become dissolved in the water. Home water treatment ion-exchange units usually replace hardness ions with Na^+ ions (Figure 13.27). The polymer, called an ion-exchange resin, contains numerous negatively charged $—SO_3^-$ functional groups that have Na^+ ions attached. When smaller, more highly charged ions like Mg^{2+} and Ca^{2+} in hard water flow over the resin, they displace Na^+ ions from the resin. The process happens many times, resulting in water that contains Na^+ ions in place of the Ca^{2+} and Mg^{2+} ions, which are bound to the resin. Two $—SO_3^-$ groups are required for every 2+ ion removed from solution; two Na^+ ions are released for every 2+ ion removed.

$$\underset{\text{ion-exchange resin}}{2\,(\text{Polymer—SO}_3^-)\text{Na}^+} + \underset{\text{from hard water}}{\text{Ca}^{2+}} \rightleftharpoons (\text{Polymer—SO}_3^-)_2\text{Ca}^{2+} + 2\,\text{Na}^+(aq)$$

Because Na^+ ions do not harden water, the resulting water is called "soft" water.

When all its sodium ions have been replaced, the ion-exchange resin becomes saturated with hard-water ions. The resin must be regenerated by treating it with a highly concentrated NaCl solution. This reverses the process given in the previous equation and the hard-water ions released from the resin are rinsed down the drain.

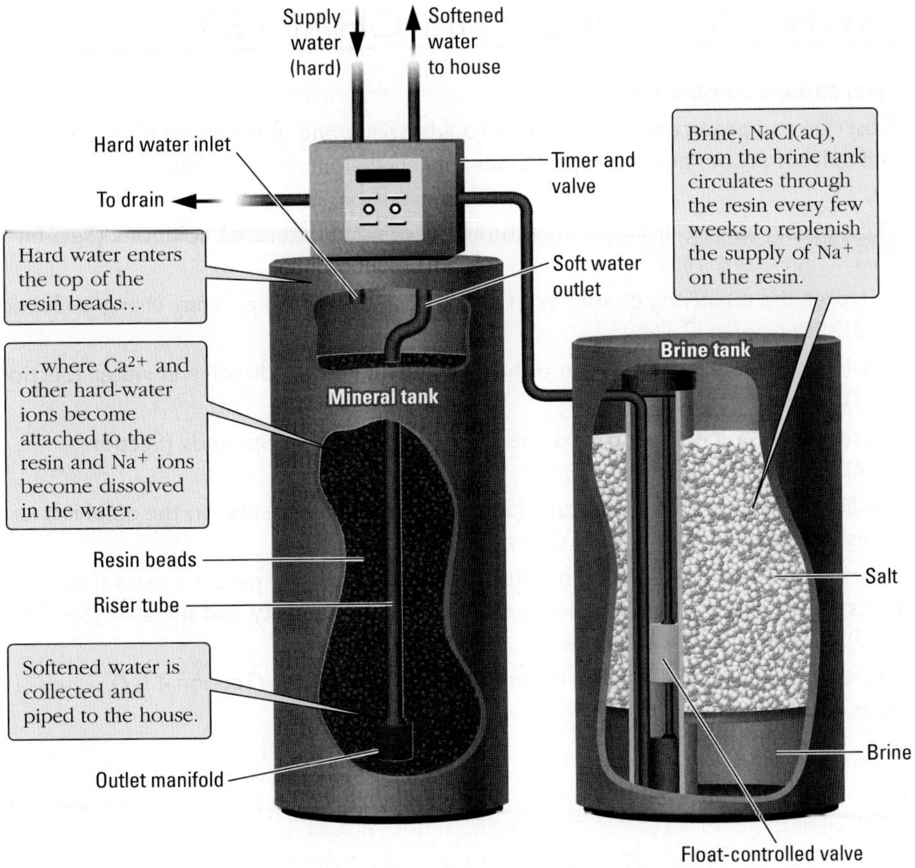
Supply water (hard)
Softened water to house
Hard water inlet
To drain
Timer and valve
Hard water enters the top of the resin beads...
Soft water outlet
Brine, NaCl(aq), from the brine tank circulates through the resin every few weeks to replenish the supply of Na+ on the resin.
Brine tank
Mineral tank
...where Ca2+ and other hard-water ions become attached to the resin and Na+ ions become dissolved in the water.
Resin beads
Riser tube
Salt
Softened water is collected and piped to the house.
Brine
Outlet manifold
Float-controlled valve

Figure 13.27 Schematic diagram of a home water softener based on ion exchange.

SUMMARY PROBLEM

You are asked to prepare three mixtures at 25 °C.

Mixture I: 25.0 g CCl_4 and 100. mL water

Mixture II: 15.0 g $CaCl_2$ in 125 mL water

Mixture III: 21 g ethylene glycol, $HOCH_2CH_2OH$, in 150. mL water

Answer these questions about these mixtures.

(a) Calculate the mass fraction of each mixture.

(b) Calculate the weight percent of each mixture.

(c) For which mixture or mixtures is a solution formed? (If a solution is formed, assume its density is 1.0 g/mL and answer the remaining questions.)

(d) Name the dissolved species in each solution and draw a diagram representing how the solvent (water) molecules interact with each solute.

(e) Express the composition of each solution in ppm.

(f) Calculate the molality of each solution.

(g) Calculate the vapor pressure of water in equilibrium with each solution.

(h) Calculate the boiling point of each solution.

(i) Calculate the freezing point of each solution.

(j) Calculate the osmotic pressure of each solution.

HAVING STUDIED THIS CHAPTER . . .

. . . you should be able to:

- Describe and predict the solubility of liquids, solids, and gases based on the solute and solvent properties and interactions (Section 13-1). End-of-chapter Questions 13, 15
- Differentiate among unsaturated, saturated, and supersaturated solutions (Section 13-2). Questions 18, 20
- Interpret the dissolving of solutes in terms of enthalpy and entropy changes (Section 13-3). Questions 22, 24
- Provide a nanoscale description of how ionic compounds dissolve in water (Section 13-3). Question 26
- Predict how temperature affects the solubility of ionic compounds (Section 13-4). Questions 27, 29
- Predict the effects of temperature (Section 13-4) and pressure on the solubility of gases in liquids (Section 13-5). Question 31
- Describe the compositions of solutions in terms of weight percent, mass fraction, parts per million, parts per billion, parts per trillion, molarity, and molality (Section 13-6). Questions 33, 37, 42, 44, 46, 52, 56, 58
- Interpret vapor pressure lowering in terms of Raoult's law (Section 13-7). Questions 61, 68

- Use molality to calculate the colligative properties: freezing-point lowering and boiling-point elevation (Section 13-7). Questions 66, 70, 75
- Compare and contrast the colligative properties of nonelectrolytes and electrolytes and apply their respective van't Hoff factor (Section 13-7). Questions 64, 76, 108
- Explain the phenomena of osmosis and reverse osmosis, and calculate osmotic pressure (Section 13-7). Questions 79, 80
- Describe the various kinds of colloids and their properties (Section 13-8). Questions 82, 84
- Explain how surfactants work (Section 13-9). Questions 86, 88
- Discuss Earth's fresh water supply and describe how drinking water is purified to make it potable (Section 13-10). Questions 90, 94, 98
- Describe what causes hard water and explain how water can be softened (Section 13-10). Question 96

KEY TERMS

boiling-point elevation (Section 13-7b)

colligative properties (13-7)

colloid (13-8)

detergent (13-9)

emulsion (13-8a)

enthalpy of solution (13-3b)

freezing-point lowering (13-7c)

Henry's law (13-5)

hydration (13-3c)

hydrophilic (13-1)

hydrophobic (13-1)

hypertonic (13-7e)

hypotonic (13-7e)

immiscible (13-1)

isotonic (13-7e)

mass fraction (13-6a)

miscible (13-1)

molality (13-6d)

molarity (13-6c)

osmosis (13-7e)

osmotic pressure (Π) (13-7e)

parts per billion (ppb) (13-6b)

parts per million (ppm) (13-6b)

parts per trillion (ppt) (13-6b)

Raoult's law (13-7a)

reverse osmosis (13-7f)

saturated solution (13-2)

semipermeable membrane (13-7e)

solubility (13-2)

solvation (13-3c)

supersaturated solution (13-2)

surfactant (13-9)

Tyndall effect (13-8)

unsaturated solution (13-2)

van't Hoff factor (i_{solute}) (13-7d)

vapor pressure lowering (13-7a)

weight percent (13-6a)

QUESTIONS FOR REVIEW AND THOUGHT

Red-numbered questions have short answers at the back of this book in Appendix M and fully worked solutions in the *Student Solutions Manual.*

Review Questions

These questions test vocabulary and simple concepts.

1. Explain why gasoline and motor oil are miscible, as in the fuel mixtures used in two-cycle lawn mower engines.
2. Describe the differences among solutions that are unsaturated, saturated, and supersaturated in terms of amount of solute, and in terms of Q and K_c.
3. State Henry's law. Name three factors that govern the solubility of a gas in a liquid.
4. In general, how does the water solubility of most ionic compounds change as the temperature is increased?
5. How does the solubility of gases in liquids change with increased temperature? Explain why.
6. Define molality. How does it differ from molarity?
7. Explain the difference between the mass fraction and the mole fraction of solute in a solution.
8. Explain why the vapor pressure of a solvent is lowered by the presence of a nonvolatile solute.
9. Explain the difference between (a) a hypotonic and an isotonic solution; (b) an isotonic and a hypertonic solution.
10. Explain how reverse osmosis works.
11. How do colloids differ from suspensions?
12. Explain why the Tyndall effect is not observed with solutions.

Topical Questions

These questions are keyed to the major topics in the chapter. Usually a question that is answered at the back of this book is paired with a similar one that is not.

Solubility and Intermolecular Forces (Section 13-1)

13. Why would the same solid readily dissolve in one liquid and be almost insoluble in another liquid? Give an example of such behavior.
14. Which of these general types of substances would you expect to dissolve readily in water? Explain why.
    (a) Alcohols (b) Hydrocarbons
    (c) Metals (d) Nonpolar molecules
    (e) Polar molecules (f) Salts
15. Beakers (a), (b), and (c) are representations of tiny sections (not to scale) of mixtures made from pure benzene and pure water. Select which beaker gives proper representation of the result when the two pure substances are mixed.

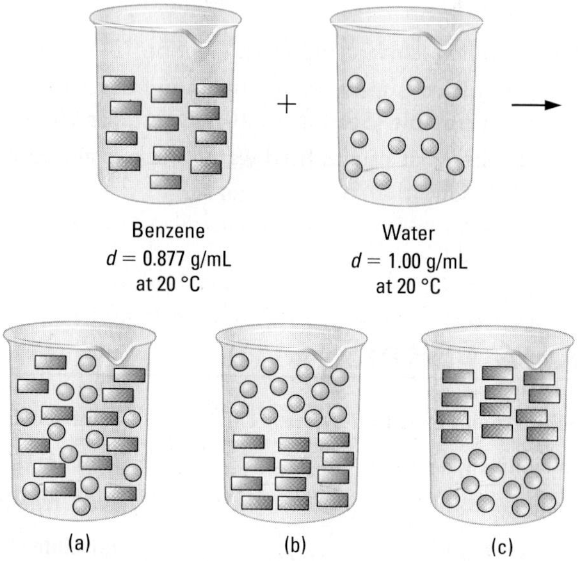

+
Benzene
d = 0.877 g/mL
at 20 °C
Water
d = 1.00 g/mL
at 20 °C
(a)
(b)
(c)

16. Beakers (a), (b), and (c) are representations of tiny sections (not to scale) of mixtures made from pure glycerol and pure cyclohexane. Select which beaker gives proper representation of the result when the two pure substances are mixed.

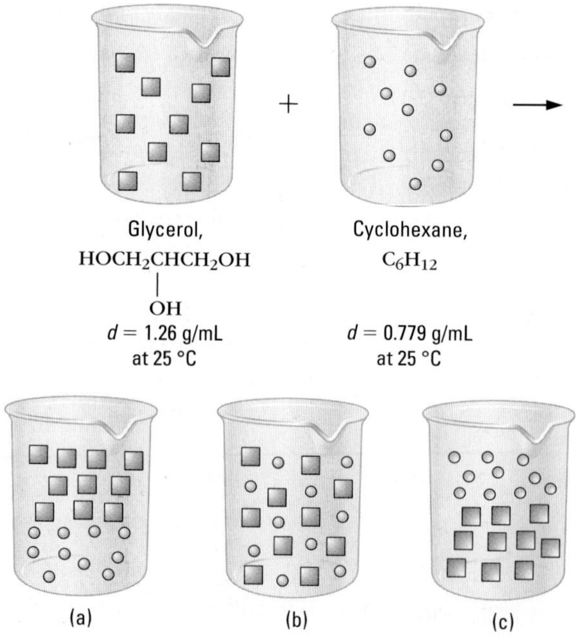

+
Glycerol,
$HOCH_2CHCH_2OH$ (with OH on the central C)
d = 1.26 g/mL
at 25 °C
Cyclohexane,
C_6H_{12}
d = 0.779 g/mL
at 25 °C
(a)
(b)
(c)

17. Simple acids such as formic acid, HCOOH, and acetic acid, CH_3COOH, are very soluble in water; however, fatty acids such as stearic acid, $CH_3(CH_2)_{16}COOH$, and palmitic acid, $CH_3(CH_2)_{14}COOH$, are water-insoluble. Based on what you know about the solubility of alcohols, explain the solubility of these organic acids.

Solubility and Equilibrium (Section 13-2)

18. Knowing that the solubility of oxalic acid at 25 °C is 1 g per 7 g water, describe how you would prepare 1 L of a saturated oxalic acid solution.
19. If a solution of a certain salt in water is saturated at some temperature and a few crystals of the salt are added to the solution, what happens? What happens if the same quantity of the same salt crystals is added to an unsaturated solution of the salt?
20. Determine whether each solution is unsaturated, saturated, or supersaturated at 25 °C by consulting Figure 13.4 (← **Sec. 13-2**).
    (a) 25.0 g $NH_4Cl(s)$ dissolved in 50.0 g H_2O
    (b) 25.0 g $NH_4Cl(s)$ dissolved in 75.0 g H_2O
21. For each solution in Question 20, calculate the mass (g) of NH_4Cl that would have to be added or that would have to precipitate to make a saturated solution.

Entropy, Enthalpy, and Dissolving Solutes (Section 13-3)

22. The lattice energy of $CaCl_2$ is −2258 kJ/mol, and the total enthalpy of hydration of its ions is −2175 kJ/mol. Determine whether the process of dissolving $CaCl_2$ in water is endothermic or exothermic.
23. The lattice energy of MgF_2 is −2961 kJ/mol and the total enthalpy of hydration of its ions is −2948 kJ/mol. Determine whether the process of dissolving MgF_2 in water is endothermic or exothermic.
24. Calculate the enthalpy of solution of $CaBr_2$ given that its lattice energy is −1984 kJ/mol and the total enthalpy of hydration of its ions is −1827 kJ/mol.
25. Given these data, calculate the enthalpy of hydration of I^-. Lattice energy of LiI = −759 kJ/mol; enthalpy of solution of LiI = −63.3 kJ/mol; hydration enthalpy of Li^+ = −558 kJ/mol.
26. Describe what happens when an ionic solid dissolves in water. Sketch an illustration that includes at least three positive ions, three negative ions, and a dozen or so water molecules in the vicinity of the ions.

Temperature and Solubility (Section 13-4)

27. Refer to Figure 13.10 (← **Sec. 13-4b**) to answer these questions.
    (a) Does a saturated solution occur when 65.0 g LiCl is present in 100 g H_2O at 40 °C? Explain your answer.
    (b) Consider a solution that contains 95.0 g LiCl in 100 g H_2O at 40 °C. Is the solution unsaturated, saturated, or supersaturated? Explain your answer.
    (c) Consider a solution that contains 50. g Li_2SO_4 in 200. g H_2O at 50 °C. Is this solution unsaturated, saturated, or supersaturated? Explain your answer.
28. Refer to Figure 13.10 to answer these questions.
    (a) A solution contains 25.0 g KCl in 100 g H_2O at 60 °C. Is the solution unsaturated, saturated, or supersaturated? Explain your answer.
    (b) An additional 25.0 g KCl is added to the solution in part (a); assume that the temperature does not change. Is the new solution unsaturated, saturated, or supersaturated? Explain your answer.
    (c) Consider a solution that contains 150. g CsCl in 100. g H_2O at 10 °C. Is this solution unsaturated, saturated, or supersaturated? Explain your answer.
29. A saturated solution of NH_4Cl was prepared by adding solid NH_4Cl to water until no more solid NH_4Cl would dissolve. The resulting mixture felt very cold and had a layer of undissolved NH_4Cl on the bottom. When the mixture reached room temperature, no solid NH_4Cl was present. Explain what happened. Was the solution still saturated? Explain your answer.

Pressure and Dissolving Gases in Liquids: Henry's Law (Section 13-5)

30. The partial pressure of O_2 in your lungs is about 100 mmHg. Calculate the concentration of O_2 (in grams per liter) that can dissolve in water at 37 °C when the O_2 partial pressure is 100. mmHg. The Henry's law constant for O_2 at 37 °C is 1.5×10^{-6} mol L^{-1} $mmHg^{-1}$.
31. The Henry's law constant for nitrogen in blood serum is approximately 8×10^{-7} mol L^{-1} $mmHg^{-1}$. Calculate the N_2 concentration in a diver's blood at a depth where the total pressure is 2.5 atm. The air the diver is breathing is 78% N_2 by volume.
32. The Henry's law constant for N_2 in water at 25 °C is 8.4×10^{-7} mol L^{-1} $mmHg^{-1}$. Calculate the solubility of N_2 in mol/L if its partial pressure is 1520 mmHg. Calculate the solubility when the N_2 partial pressure is 20. mmHg.

Expressing Solution Composition (Section 13-6)

33. Which is the highest solute concentration: 50 ppm, 500 ppb, or 0.05% by weight?
34. Estimate your concentration among students on campus in parts per million and parts per thousand.
35. Convert 73.2 ppm to weight percent.
36. Convert 2.5 ppm to weight percent.
37. Show mathematically how 1 ppb is equivalent to 1 μg/1 kg.
38. Show mathematically how 1 ppm is equivalent to 1 mg/1 kg.
39. Calculate the mass (g) of ethanol in 750. mL of a 12% ethanol solution. (Assume its density is the same as that of water.)
40. Calculate the mass (g) of sucrose in 1.0 kg of a 0.25% sucrose solution.
41. A sample of water contains 0.010 ppm lead ions, Pb^{2+}.
    (a) Calculate the mass of lead ions per liter in this solution. (Assume the density of the water solution is 1.0 g/mL.)
    (b) Calculate the mass fraction of lead in ppb.

42. A paint contains 200. ppm lead. Calculate what mass of lead (in grams) will be in 1.0 cm^2 of this paint (density = 10.0 lb/gal) when 1 gal is uniformly applied to a 500. ft^2 wall.
43. A liquid sample of lead-based paint contains 60.5 ppm lead. The density of the paint is 10.0 lb/gal. Calculate the mass of lead (in grams) that would be present in 50. gal of this paint.
44. Calculate the mass in grams of solute required to prepare each of these solutions.
    (a) 750. mL of 4.00-M NH_4Cl
    (b) 1.50 L of 0.750-M KCl
    (c) 150. mL of 0.350-M Na_2SO_4
45. Calculate the mass in grams of solute needed to prepare each of these solutions.
    (a) 250. mL of 0.50-M NaCl
    (b) 0.50 L of 0.15-M sucrose, $C_{12}H_{22}O_{11}$
    (c) 200. mL of 0.20-M $NaHCO_3$
46. Calculate the molarity of the solute in a solution containing
    (a) 14.2 g KCl in 250. mL solution.
    (b) 5.08 g K_2CrO_4 in 150. mL solution.
    (c) 0.799 g $KMnO_4$ in 400. mL solution.
    (d) 15.0 g $C_6H_{12}O_6$ in 500. mL solution.
47. Calculate the molarity of the solute in a solution containing
    (a) 6.18 g $MgNH_4PO_4$ in 250. mL solution.
    (b) 16.8 g $NaCH_3COO$ in 300. mL solution.
    (c) 2.50 g CaC_2O_4 in 750. mL solution.
    (d) 2.20 g $(NH_4)_2SO_4$ in 400. mL solution.
48. Concentrated sulfuric acid has a density of 1.84 g/cm^3 and is 18 M. Calculate the weight percent of H_2SO_4 in the solution.
49. Concentrated nitric acid is a 70.0% solution of nitric acid, HNO_3, in water. The density of the solution is 1.41 g/mL at 25 °C. Calculate the molarity of nitric acid in this solution.
50. A 0.6-mL teardrop contains 4 mg NaCl. Calculate the molarity of NaCl in the teardrop.
51. Consider a 13.0% solution of sulfuric acid, H_2SO_4, whose density is 1.090 g/mL.
    (a) Calculate the molarity of this solution.
    (b) To what volume should 100. mL of this solution be diluted to prepare a 1.10-M solution?
52. You want to prepare a 1.0 mol/kg solution of ethylene glycol, $C_2H_4(OH)_2$, in water. Calculate the mass of ethylene glycol you would need to mix with 950. g water.
53. You need a 0.050 mol/kg aqueous solution of methanol, CH_3OH. Calculate the mass of methanol you would need to dissolve in 500. g water to make this solution.
54. A 23.2% by weight aqueous solution of sucrose has a density of 1.127 g/mL. Calculate the molarity of sucrose in this solution.
55. Calculate the mass (g) of KI required to prepare 100. mL of 0.0200-M KI. How many milliliters of this solution are required to produce 250. mL of 0.00100-M KI?
56. A 12-oz (355-mL) Pepsi contains 38.9 mg caffeine (molar mass = 194.2 g/mol). Assume that the Pepsi, mainly water, has a density of 1.01 g/mL. For such a Pepsi, calculate: (a) its caffeine concentration in ppm; (b) its molarity of caffeine; and (c) the molality of caffeine.
57. A concentration of 0.05 to 0.07 mg F^-/kg body weight is the optimum range for maximum protection against tooth decay with a minimum of risk. Assume that in one day you eat these items, whose fluoride content is given in ppm: one 12-oz cup brewed coffee (0.91); one 3-oz hot dog (0.48); one 12-oz Diet Coke (0.60); one 12-oz glass tap water (0.71); one 4-oz chocolate ice cream cone (0.23). Use calculations to determine whether the cumulative fluoride concentration in what you ate that day falls within the optimum range. 1.0 kg = 2.2 lb.
58. Calculate the mass fraction and the weight percent of the solute in each of these solutions:
    (a) 20.7 g NaCl in 175 g H_2O.
    (b) 1.45 g ethanol in 10.0 g H_2O.
    (c) 20.0 g CS_2 in 45.0 g $CHCl_3$.
    (d) 4.00 mL benzene (d = 0.877 g/mL) in 120. g diethyl ether.
59. Calculate the mass fraction and the weight percent of the solute in each of these solutions:
    (a) 14.0 g K_2CrO_4 in 225 g H_2O.
    (b) 4.56 g ethanol in 50.0 g benzene.
    (c) 15.0 g methanol in 89.0 g ethanol.
    (d) 14.5 mL ethylene glycol (d = 1.11 g/mL) in 200. g H_2O.

Colligative Properties of Solutions (Section 13-7)

60. Which of these graphs correctly represents Raoult's law? Explain your choice.

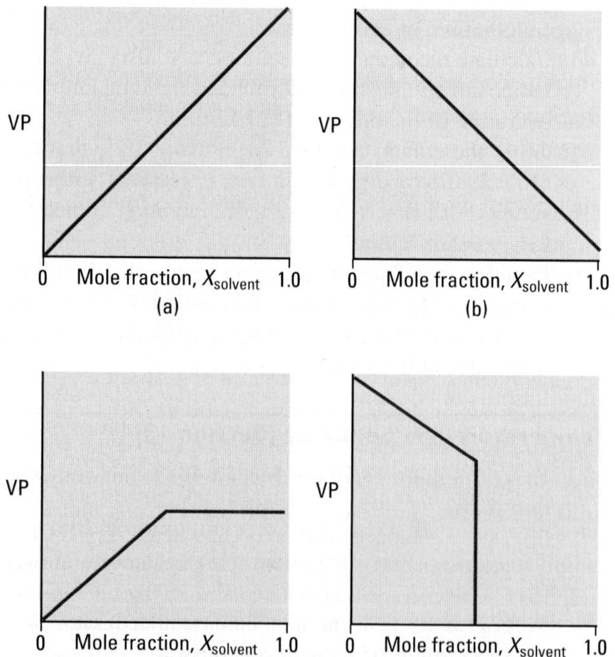
VP
0 Mole fraction, $X_{solvent}$ 1.0
(a)
VP
0 Mole fraction, $X_{solvent}$ 1.0
(b)
VP
0 Mole fraction, $X_{solvent}$ 1.0
(c)
VP
0 Mole fraction, $X_{solvent}$ 1.0
(d)

61. The two curves in this graph represent pure benzene and a solution in which benzene is the solvent. Which curve is for pure benzene and which for the solution? Explain your answer.

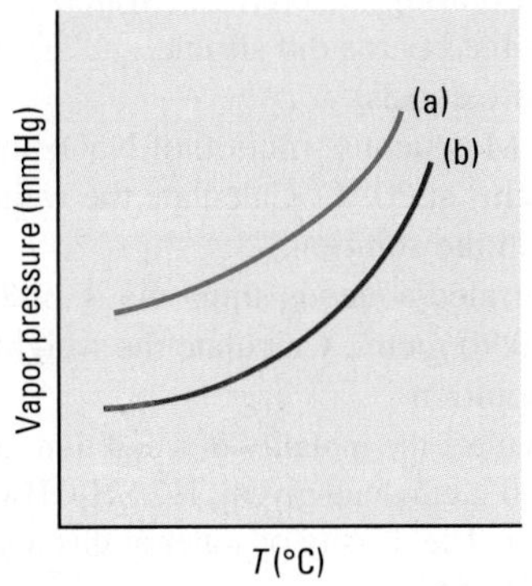
(a)
(b)
Vapor pressure (mmHg)
T (°C)

62. Calculate the boiling point of a solution composed of 15.0 g urea, $(NH_2)_2CO$, in 0.500 kg water.
63. Calculate the boiling point of a solution containing 0.200 mol of a nonvolatile nonelectrolyte solute in 100. g benzene. The normal boiling point of benzene is 80.10 °C, and K_b = 2.53 °C kg/mol.
64. List these aqueous solutions in order of decreasing freezing point.
    (a) 0.10 mol methanol/kg (b) 0.10 mol KCl/kg
    (c) 0.080 mol $BaCl_2$/kg (d) 0.040 mol Na_2SO_4/kg
    (Assume that all of the salts dissociate completely into their ions in solution.)
65. Place these aqueous solutions in order of increasing boiling point.
    (a) 0.10 mol KCl/kg (b) 0.10 mol glucose/kg
    (c) 0.080 mol $MgCl_2$/kg
    (Assume that all of the salts dissociate completely into their ions in solution.)
66. Calculate the freezing and boiling points (at 760 mmHg) of a solution of 4.00 g urea, $CO(NH_2)_2$, dissolved in 75.0 g water.
67. Calculate the boiling point and the freezing point of these solutions at 760 mmHg.
    (a) 20.0 g citric acid, $C_6H_8O_7$, in 100.0 g water
    (b) 3.00 g CH_3I in 20.0 g benzene (K_b benzene = 2.53 °C kg/mol; K_f benzene = −5.10 °C kg/mol)
68. At 60 °C the vapor pressure of pure water is 149.44 mmHg and that above an aqueous sucrose, $C_{12}H_{22}O_{12}$, solution is 119.55 mmHg. Calculate the mole fraction of water and the mass in grams of sucrose in the solution if the mass of water is 150. g.
69. Calculate the mass in grams of urea, $CO(NH_2)_2$, that must be added to 150. g water to give a solution whose vapor pressure is 2.5 mmHg less than that of pure water at 40 °C (vp H_2O at 40 °C = 55.34 mmHg).
70. The boiling point of benzene is increased by 0.65 °C when 5.0 g of an unknown organic compound (a nonelectrolyte) is dissolved in 100. g benzene. Calculate the approximate molar mass of the organic compound.
71. At 760 mmHg, a solution of 5.58 g glycerol in 40.0 g water has a boiling point of 100.777 °C. Calculate the molar mass of glycerol.
72. You add 0.255 g of an orange crystalline compound with an empirical formula of $C_{10}H_8Fe$ to 11.12 g benzene. The boiling point of the solution is 80.26 °C. Determine the molar mass and molecular formula of the compound.
73. The freezing point of *p*-dichlorobenzene is 53.1 °C, and its K_f is −7.10 °C kg/mol. A solution of 1.52 g of the drug sulfanilamide in 10.0 g *p*-dichlorobenzene freezes at 46.7 °C. Calculate the molar mass of sulfanilamide.
74. If you use only water and pure ethylene glycol, $HOCH_2CH_2OH$, in your car's cooling system, calculate what mass (in grams) of the glycol you must add to each quart of water to give freezing protection down to −31.0 °C. (1.00 L = 1.057 qt.)
75. Anthracene, a hydrocarbon obtained from coal, has an empirical formula of C_7H_5. To find its molecular formula you dissolve 0.500 g anthracene in 30.0 g benzene. The boiling point of the solution is 80.34 °C. Determine the molar mass and molecular formula of anthracene.
76. A 1.00 mol/kg aqueous sulfuric acid solution, H_2SO_4, freezes at −4.04 °C. Calculate *i*, the van't Hoff factor, for sulfuric acid in this solution.
77. Some ethylene glycol, $HOCH_2CH_2OH$, was added to your car's cooling system along with 5.0 kg water.
    (a) If the freezing point of the solution is −15.0 °C, determine what mass (g) of the glycol must have been added.
    (b) Calculate the boiling point of the coolant mixture.
78. Calculate the concentration of solute particles in human blood if the osmotic pressure is 7.63 atm at 37 °C, the temperature of the body.
79. The blood of cold-blooded animals and fish is isotonic with seawater. If seawater freezes at −2.3 °C, calculate the osmotic pressure of the blood of these animals at 20.0 °C. (Assume the density is that of pure water.)
80. The molar mass of a polymer was determined by measuring the osmotic pressure, 7.6 mmHg, of a benzene solution containing 5.0 g of the polymer dissolved in 1.0 L solution. Calculate the molar mass of the polymer. Assume a temperature of 298.15 K.
81. The osmotic pressure at 25 °C is 1.79 atm for a solution prepared by dissolving 2.50 g sucrose, empirical formula $C_{12}H_{22}O_{11}$, in enough water to give a solution volume of 100 mL. Use the osmotic pressure equation to show that the empirical formula for sucrose is the same as its molecular formula.

Colloids (Section 13-8)

82. Differentiate between the dispersed phase and the continuous phase of (a) soap suds; (b) milk; (c) airborne pollen grains; (d) margarine.
83. Differentiate between the dispersed phase and the continuous phase of (a) gelatin; (b) butter; (c) aerosol sprays; (d) mud.

84. Explain how globular proteins act as hydrophilic colloids in water.
85. Explain how silt forms at the delta of rivers when the river water meets seawater.

Surfactants (Section 13-9)

86. Sketch an illustration of a soap molecule. Based on its structure, why is it considered a surfactant?
87. Surfactant molecules have what common structural features?
88. Explain the role of micelles in the cleansing action of soaps.
89. Shampoos often contain sodium lauryl sulfate, a detergent. Explain why this detergent is an ingredient in shampoos.

Water: Natural, Clean, and Otherwise (Section 13-10)

90. Water is drawn from a well. Are its dissolved solutes the same type and concentration as those in surface water? Explain.
91. Where is most of the fresh water found on Earth?
92. Some dietitians recommend drinking six 8-oz glasses of water each day. If your drinking water contains the maximum contamination level for arsenic, 0.010 ppm, how much arsenic would you consume in a week following this recommendation?
93. The U. S. EPA acceptable limit for lead in drinking water is 0.015 ppm. If you drink six 8-oz. glasses of water each day and do not excrete any lead, calculate how long it would take for you to accumulate 0.50 mg Pb.
94. The maximum contamination level (MCL) for chlordane is 0.002 ppm. A sample of well water contained 5 ppb chlordane. Is the sample within the MCL for chlordane?
95. In a home, hard water containing 500. mg Ca^{2+}/gal passed through the Na^+-based ion-exchange water softener. If there was 200 gal. water and the ion-exchange resin operates at 100% efficiency, calculate the mass of Na^+ ions displaced from the resin.
96. How do the lime-soda and ion-exchange processes differ in treating hard water?
97. Explain how hard water produces "ring around the bathtub."
98. During municipal drinking water treatment, water is sprayed into the air. Why is this done?
99. Discuss the risks and benefits of using ozone to treat municipal drinking water.

General Questions

These questions are not explicitly keyed to chapter topics; many require integration of several concepts.

100. What is the difference between solubility and miscibility?
101. If 5 g solvent, 0.2 g solute A, and 0.3 g solute B are mixed to form a solution, calculate the weight percent of A.
102. A chemistry classmate tells you that a supersaturated solution is also saturated. Is the student correct? What would you tell the student about her/his statement?
103. In *The Rime of the Ancient Mariner* the poet Samuel Taylor Coleridge wrote, ". . . Water, water, everywhere/ And all the boards did shrink. . . ." Explain this effect in terms of osmosis.
104. A 10.0-M aqueous solution of NaOH has a density of 1.33 g/cm^3 at 20 °C. Calculate the weight percent of NaOH in the solution.
105. Concentrated aqueous ammonia is 14.8 M and has a density of 0.90 g/cm^3. Calculate the weight percent of NH_3 in the solution.
106. (a) Calculate the molality of a solution made by dissolving 115.0 g ethylene glycol, $HOCH_2CH_2OH$, in 500. mL water. The density of water at this temperature is 0.978 g/mL.
(b) Calculate the molarity of the solution.
107. Dimethylglyoxime (DMG) reacts with nickel(II) ion in aqueous solution to form a bright red compound. However, DMG is insoluble in water. To get it into aqueous solution where it can encounter Ni^{2+} ions, it must first be dissolved in a suitable solvent, such as ethanol. Suppose you dissolve 45.0 g DMG ($C_4H_8N_2O_2$) in 500. mL ethanol (C_2H_5OH; density = 0.7893 g/mL). Determine the molality and weight percent of DMG in this solution.
108. Arrange these aqueous solutions in order of decreasing freezing point. (Assume theoretical values for *i*.)
(a) 0.20 mol ethylene glycol/kg
(b) 0.12 mol Na_2SO_4/kg
(c) 0.10 mol NaBr/kg
(d) 0.12 mol KI/kg
109. Arrange these aqueous solutions in order of increasing boiling point. (Assume theoretical values for *i*.)
(a) 0.20 mol ethylene glycol/kg
(b) 0.12 mol K_2SO_4/kg
(c) 0.10 mol $BaCl_2$/kg
(d) 0.12 mol KBr/kg
110. The solubility of NaCl in water at 100 °C is 39.1 g/100. g water. Calculate the boiling point of a saturated solution NaCl.
111. The organic salt $[(C_4H_9)_4N][ClO_4]$ consists of the ions $(C_4H_9)_4N^+$ and ClO_4^-. The salt dissolves in chloroform. What mass (in grams) of the salt must have been dissolved if the boiling point of a solution of the salt in 25.0 g chloroform is 63.20 °C? The normal boiling point of chloroform is 61.70 °C and K_b = 3.63 °C kg mol^{-1}. Assume that the salt dissociates completely into its ions in solution.
112. A solution, prepared by dissolving 9.41 g $NaHSO_3$ in 1.00 kg water, freezes at −0.33 °C. From these data, decide which of these equations is the correct expression for the dissociation of the salt.
(a) $NaHSO_3(aq) \longrightarrow Na^+(aq) + HSO_3^-(aq)$
(b) $NaHSO_3(aq) \longrightarrow Na^+(aq) + H^+(aq) + SO_3^{2-}(aq)$

113. A 0.250-M sodium sulfate solution is added to a 0.200-M barium nitrate solution and 0.700 g barium sulfate precipitates.
(a) Write the balanced equation for this reaction.
(b) Calculate the minimum volume of barium nitrate solution that was used.
(c) Calculate the minimum volume of sodium sulfate needed to precipitate 0.700 g barium sulfate. Assume 100% yield.

114. The "proof" of an alcohol-containing beverage is twice the volume percentage of ethanol, C_2H_5OH (density = 0.789 g/mL), in the beverage. A bottle of 100-proof vodka is left overnight in the trunk of a car where the temperature is −15 °C.
(a) Use calculations to determine whether the vodka is frozen.
(b) Calculate the boiling point of the vodka.

115. A martini is a 5-oz (142-g) cocktail containing 30% by mass of alcohol. When the martini is consumed, about 15% of it passes directly into the bloodstream (blood volume = 7.0 L in an adult). Consider an adult who drinks two martinis with lunch. Estimate the blood alcohol concentration in this person after the two martinis have been consumed. An adult with a blood alcohol concentration of 3.0×10^{-4} g/mL or more is considered intoxicated. Is the person intoxicated?

116. A 0.100-mol/kg NaCl solution has a van't Hoff factor of 1.87. Calculate the freezing point, the boiling point, and the osmotic pressure at 25.00 °C of this solution, whose density is 1.01 g/mL.

Applying Concepts

These questions test conceptual learning.

117. Using these symbols,

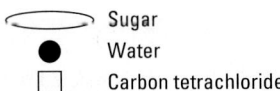

Sugar
Water
Carbon tetrachloride

draw nanoscale diagrams for the contents of a beaker containing
(a) water and sugar.
(b) carbon tetrachloride and sugar.

118. Using these symbols,

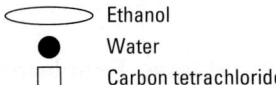

Ethanol
Water
Carbon tetrachloride

draw nanoscale diagrams for the contents of a beaker containing
(a) water and ethanol.
(b) water and carbon tetrachloride.

119. Refer to Figure 13.10 (← **Sec. 13-4b**) to determine whether these situations would result in an unsaturated, saturated, or supersaturated solution.
(a) 40. g NH_4Cl is added to 100. g H_2O at 80 °C.
(b) 100. g LiCl is dissolved in 100. g H_2O at 30 °C.
(c) 120. g $NaNO_3$ is added to 100. g H_2O at 40 °C.

120. Refer to Figure 13.10 (← **Sec. 13-4b**) to determine whether these situations would result in an unsaturated, saturated, or supersaturated solution.
(a) 120. g RbCl is added to 100. g H_2O at 50 °C.
(b) 30. g KCl is dissolved in 100. g H_2O at 70 °C.
(c) 20. g NaCl is dissolved in 50. g H_2O at 60 °C.

121. Complete this table.

Compound	Mass of Compound	Mass of Water	Mass Fraction of Solute	Weight Percent of Solute	ppm of Solute
Lye (NaOH)	______	125 g	0.375	______	______
Glycerol	33 g	200. g	______	______	______
Acetylene	0.0015 g	______	______	0.0009%	______

122. Complete this table.

Compound	Mass of Compound	Mass of Water	Mass Fraction of Solute	Weight Percent of Solute	ppm of Solute
Table salt	52 g	175 g	______	______	______
Glucose	15 g	______	______	______	7×10^4
Methane	______	100. g	______	0.0025%	______

123. What happens on the molecular level when a liquid freezes? What effect does a nonvolatile solute have on this process? Comment on the purity of water obtained by melting an iceberg.

124. If KI is added to a saturated solution of SrI_2, does the amount of solid SrI_2 present decrease, increase, or remain unchanged? What about the concentration of Sr^{2+} ions in solution?

125. In your own words, explain why
(a) seawater has a lower freezing point than fresh water.
(b) salt is added to the ice in an ice cream maker to freeze the ice cream faster.

126. Criticize these statements.
(a) A saturated solution is always a concentrated one.
(b) A 0.10-mol/kg sucrose solution and a 0.10-mol/kg KCl solution have the same osmotic pressure.

127. This illustration shows two beakers and their contents in a sealed container at some starting time, t_0. Both beakers contain the same solvent—pure solvent in one case, and solvent as part of a solution with a nonvolatile solute in the other case. Sketch the beakers and their contents in the sealed container to indicate any changes that have occurred after some time has passed.

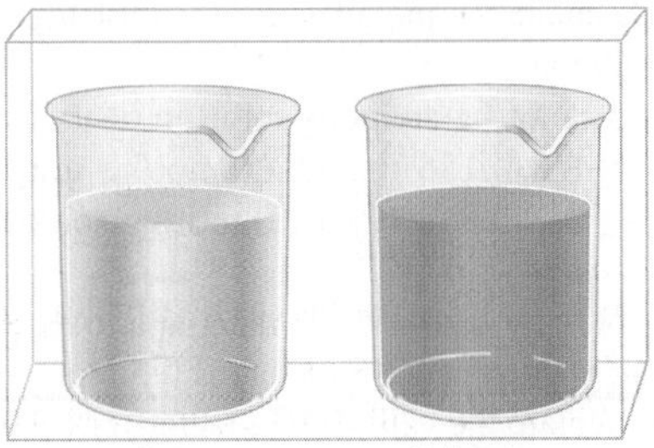
Pure solvent
Solvent + solute

More Challenging Questions

These questions require more thought and integrate several concepts.

128. In chemical research, newly synthesized compounds are often sent to commercial laboratories for analysis that determines the weight percent of C and H by burning the compound and collecting the evolved CO_2 and H_2O (← **Sec. 3-9**). The molar mass is determined by measuring the osmotic pressure of a solution of the compound. Calculate the empirical and molecular formulas of a compound, C_xH_yCr, given this information:
    (a) The compound contains 73.94% C and 8.27% H; the remainder is chromium.
    (b) At 25 °C, the osmotic pressure of 5.00 mg of the unknown dissolved in 100. mL of chloroform solution is 3.17 mmHg.
129. An osmotic pressure of 5.15 atm is developed by a solution containing 4.80 g dioxane (a nonelectrolyte) dissolved in 250. mL water at 15.0 °C. The empirical formula of dioxane is C_2H_4O. Use the osmotic pressure data to show that the empirical formula and the molecular formula of dioxane are not the same.
130. Osmosis is responsible for sap rising in trees. Calculate the approximate height to which it could rise if the sap were 0.13 M in sugar and the dissolved solids in the water outside the tree were at a concentration of 0.020 M. *Hint:* A column of liquid exerts a pressure directly proportional to its density.
131. The osmotic pressure of human blood at 37 °C is 7.63 atm. Calculate what the molarity of a glucose solution should be if it is to be safely administered intravenously.
132. A 0.109 mol/kg aqueous solution of formic acid, HCOOH, freezes at −0.210 °C. Calculate the percent dissociation of formic acid.
133. Consider these data for a solution of water and ethanol, C_2H_5OH.

Mass Percent Ethanol	Boiling Point (°C)
25.0	84.5
50.0	81.2
67.0	80.0
85.0	78.4

    (a) Plot these data and use the graph to determine the boiling point of a solution that is 59.0% ethanol by mass.
    (b) Calculate the molarity and the molality of the 67.0% solution.
134. A 0.050 mol/kg aqueous solution of iodic acid, HIO_3, freezes at −0.156 °C; a 0.20 mol/kg solution of the acid freezes at −0.542 °C; a 1.0 mol/kg solution freezes at −1.72 °C.
    (a) Plot these data and use the graph to determine the freezing point of a solution that is 0.5 mol iodic acid/kg.
    (b) Calculate the van't Hoff *i* factor for each solution.
    (c) Calculate the percent dissociation of iodic acid in each solution.
135. Consider these data for aqueous solutions of ammonium chloride, NH_4Cl.

Molality (mol/kg)	Freezing Point (°C)
0.0050	−0.0158
0.020	−0.0709
0.20	−0.678
1.0	−3.33

    (a) Plot these data and from the graph determine the freezing point of a 0.50 mol/kg ammonium chloride solution.
    (b) Calculate the van't Hoff *i* factor for each concentration. Explain any trend that you see.
    (c) Calculate the percent dissociation of ammonium chloride in each solution.
136. Maple syrup sap is 3% sugar (sucrose) and 97% water by mass. Maple syrup is produced by heating the sap to evaporate a certain amount of the water.
    (a) Describe what happens to the composition and boiling point of the solution as evaporation takes place.
    (b) A rule of thumb among maple syrup producers is that the finished syrup should boil about 4 °C higher than the original sap being boiled. Explain the chemistry behind this guideline.
    (c) If the finished product boils 4 °C higher than the original sap, calculate the concentration of sugar in the final product. Assume that sugar is the only solute and the operation is done at 1 atm pressure.
137. A 0.63% by weight aqueous tin(II) fluoride, SnF_2, solution is used as an oral rinse in dentistry to decrease tooth decay.
    (a) Calculate the SnF_2 concentration in ppm and ppb.
    (b) Calculate the molarity of SnF_2 in the solution.
    (c) Tin(II) fluoride is produced commercially by a series of steps that begin with reduction of cassiterite, SnO_2, the principal tin ore, with carbon.

$$SnO_2(s) + 2\,C(s) \longrightarrow Sn(s) + 2\,CO(g)$$

    Once purified, the tin is reacted with hydrogen fluoride vapor to produce SnF_2. Consider that one metric ton of cassiterite ore is reduced with sufficient carbon to tin metal in 80% yield. The tin is purified and reacted with sufficient hydrogen fluoride to produce SnF_2 in 94% yield. Calculate how many 250.-mL bottles of 0.63% SnF_2 solution could be prepared from the one metric ton of cassiterite by these steps.

Conceptual Challenge Problems

These rigorous, thought-provoking problems integrate conceptual learning with problem solving and are suitable for group work.

CP13.A (Section 13-6) Concentrations expressed in units of parts per million and parts per billion often have no meaning for people until they relate these small and large numbers to their own experiences.
(a) What time in seconds is 1 ppm of a year?
(b) What time in seconds is 1 ppb of a 70-year lifetime?

CP13.B (Section 13-4) Bodies of water with an abundance of nutrients that support a blooming growth of plants are said to be eutrophoric. In general, fish do not thrive for long in eutrophoric waters because little oxygen is available for them. Suppose someone asked you why this was true, given the fact that growing plants produce oxygen as a product of photosynthesis. How would you respond to this person's inquiry?

CP13.C (Section 13-7) Suppose that you want to produce the lowest temperature possible by using ice, sodium chloride, and water to chill homemade ice cream made in a 1.5-L metal cylinder surrounded by a coolant held in a wooden bucket. You have all the ice, salt, and water you want. How would you plan to do this?

14 | Acids and Bases

©Kostiantyn Ablazov/shutterstock.com

Citrus fruits such as oranges, grapefruit, lemons, and limes are often part of a balanced diet. The juices of these citrus fruits taste "tart" or "acidic" and are used to flavor drinks. What chemical species is common to citrus and many other fruit juices that makes them acidic? (Introduction) Not all citrus juices have the same acidity: lemon juice is much more acidic than orange juice, for example. How is this acidity difference measured and reported? (Sections 14-4 and 14-5) Other foods, such as egg whites, tea, and tofu (soy bean curd) are slightly basic (alkaline): the opposite of acidic. What chemical structures make a substance an acid; a base? (Sections 14-1, 14-2, 14-6) These and other questions about acids and bases will be discussed in this chapter.

It is difficult to overstate the importance of acids and bases. Aqueous solutions abound in our environment and in all living organisms and are almost always acidic or basic to some degree. Photosynthesis and respiration, the two most important biological processes on Earth, depend on acid-base reactions. Carbon dioxide, CO_2, is the most important acid-producing compound in nature. Rainwater is usually slightly acidic because of dissolved CO_2, and acid rain results from further acidification of rainwater by acids formed by the gaseous pollutants SO_2 and NO_2. The oceans are slightly basic, as are many ground and surface waters. Natural waters can also be acidic; the more acidic the water, the more easily metals such as lead can be dissolved from water pipes or soldered joints in older buildings.

Because lead is toxic, modern plumbing uses lead-free solder or plastic pipes and fittings.

Because of their importance in much of chemistry and biochemistry, the properties of acids and bases have been studied extensively. In 1776 Antoine Lavoisier proposed that oxygen made an acid acidic. He even derived the name *oxygen* from Greek words meaning "acid former." But in 1810, Humphry Davy showed experimentally that the gaseous compound HCl, which dissolves in water to give hydrochloric acid, contains only hydrogen and chlorine. It later became clear that hydrogen, not oxygen, is common to acids in aqueous solution. It was also shown that aqueous solutions of both acids and bases conduct an electrical current because acids and bases are electrolytes (← **Sec. 3-3b**)—they release ions into solution. In 1887, the Swedish chemist Svante Arrhenius proposed that acids *ionize in aqueous solution to produce hydrogen ions (protons) and anions;* bases *ionize to produce hydroxide ions and cations.* Today it is agreed that ***aqueous hydrogen ions, $H^+(aq)$, are responsible for the properties of acidic aqueous solutions, and aqueous hydroxide ions, $OH^-(aq)$, are responsible for the properties of basic aqueous solutions.***

Because of their small size and high charge density, hydrogen ions (protons) are always surrounded by clusters of water molecules in aqueous solution (← **Sec. 3-4a**).

14-1 Brønsted-Lowry Acids and Bases

A major problem with the Arrhenius acid-base definition is that certain substances, such as ammonia, NH_3, produce basic solutions and react with acids, yet contain no hydroxide ions. In 1923, J. N. Brønsted in Denmark and T. M. Lowry in England independently proposed a new way of defining acids and bases in aqueous solutions:

- **Brønsted-Lowry acids** are *hydrogen ion (proton) donors.*
- **Brønsted-Lowry bases** are *hydrogen ion (proton) acceptors.*

According to the Brønsted-Lowry definition, an acid can donate a hydrogen ion, H^+ (a proton), to another substance, while a base can accept an H^+ ion from another substance. In a Brønsted-Lowry **acid-base reaction,** *an acid donates an H^+ ion and a base accepts an H^+ ion.* To accept an H^+ and serve as a Brønsted-Lowry base, a molecule or ion must have an *unshared (lone) pair of electrons.* For example, in aqueous solutions, H^+ ions from acids such as nitric acid, HNO_3, react with water molecules, which accept protons to form hydronium ions, H_3O^+.

In this and subsequent chapters, a pink color is used to designate an acid; a blue color indicates a base.

Acid: H^+ donor		Base: H^+ acceptor				

$$HNO_3(aq) + H_2O(\ell) \longrightarrow H_3O^+(aq) + NO_3^-(aq)$$

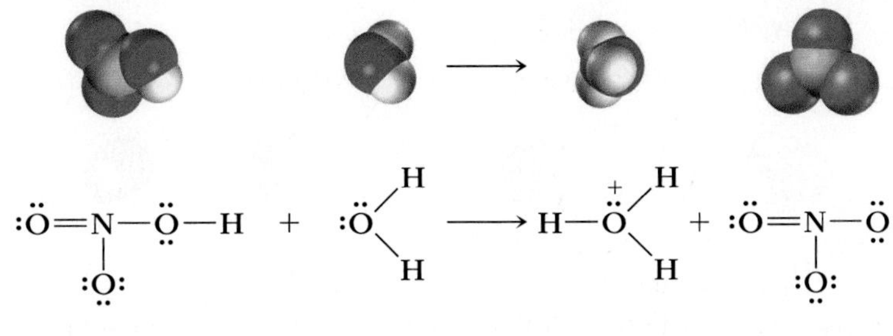

In reacting with an acid, water acts as a Brønsted-Lowry base by using an unshared electron pair to accept the H^+. Because nitric acid is a strong acid, it is 100% ionized in aqueous solution, and the equation above is strongly product-favored.

In Section 3-4a we described how water molecules surround H^+ ions in aqueous solution, forming clusters larger than H_3O^+. As a consequence, in previous chapters we have

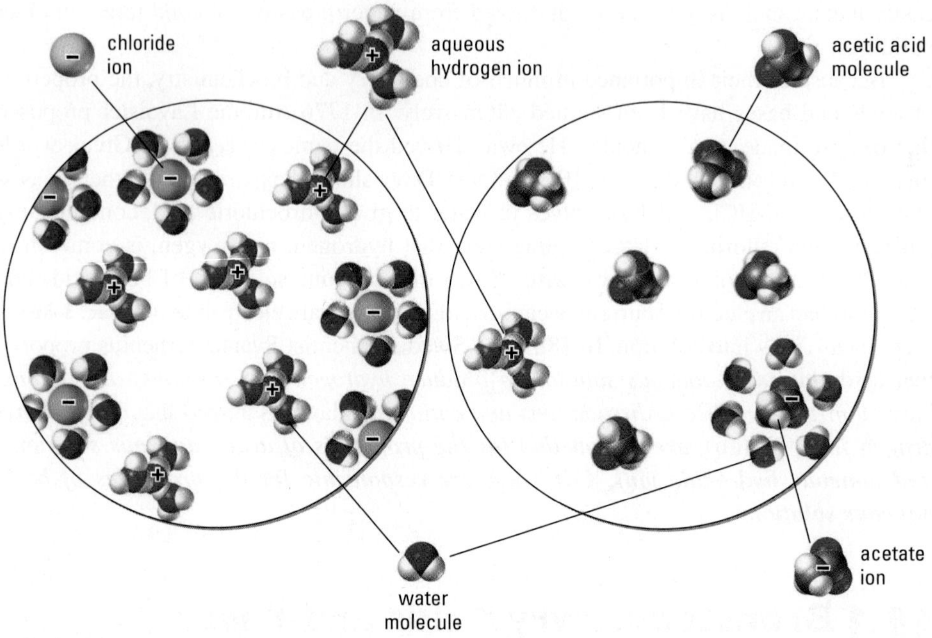

chloride ion
aqueous hydrogen ion
acetic acid molecule
acetate ion
water molecule

Ionization of acids in water. A strong acid (left) such as hydrochloric acid, HCl, is completely ionized in water; a weak acid (right) such as acetic acid, CH_3COOH, is only partially ionized in water. Water molecules not directly associated with ions have been omitted for clarity.

An ionic substance that dissolves in water is a strong electrolyte because it consists of ions (← Sec. 3-3b). A strong acid is a strong electrolyte because it reacts with water and is completely converted to ions.

abbreviated aqueous hydrogen ions as $H^+(aq)$. In this chapter, to emphasize the transfer of hydrogen ions, we shall represent clusters of water molecules around protons as $H_3O^+(aq)$.

In contrast, a weak acid does *not* ionize completely and therefore is a weak electrolyte; at equilibrium appreciable quantities of un-ionized acid molecules are present. For example, hydrofluoric acid, HF, is a weak acid, and its ionization in water is written as

$$\underset{\text{hydrofluoric acid, a weak acid}}{HF(aq)} + H_2O(\ell) \rightleftharpoons H_3O^+(aq) + F^-(aq)$$

The double arrow indicates an equilibrium between the reactants and the products (← **Sec. 12-1a**). Because ionization of HF is much less than 100%, this means that, at equilibrium, most HF molecules are un-ionized and there are relatively few hydronium and fluoride ions in solution. Thus, the ionization of *weak* acids is a reactant-favored process.

Ammonia, NH_3, is a compound consisting of electrically neutral molecules; the ammonium ion, NH_4^+, is a polyatomic ion.

Ammonia, NH_3, is a base because it accepts an H^+ from water (an acid) to form an ammonium ion, NH_4^+. Water, having donated an H^+, is converted into a hydroxide ion, OH^-.

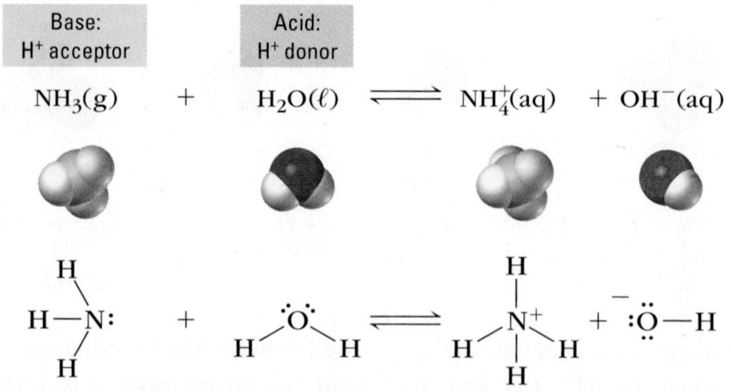

Base: H⁺ acceptor
Acid: H⁺ donor
NH3(g) + H2O(ℓ) ⇌ NH4+(aq) + OH−(aq)

Ammonia establishes an equilibrium with water, ammonium ions, and hydroxide ions and is therefore a weak base (← **Sec. 3-4b**). At equilibrium, the solution contains far more ammonia molecules than ammonium and hydroxide ions. Therefore, ammonium hydroxide is not an appropriate name for an aqueous solution of ammonia.

CONCEPTUAL EXERCISE 14.1

Brønsted-Lowry Acids and Bases

Identify each molecule or ion as a Brønsted-Lowry acid or base.

(a) HBr (b) Br^- (c) HNO_2 (d) CH_3NH_2

Answers to **EXERCISES** are provided at the back of this book in Appendix L.

EXERCISES that are labeled **CONCEPTUAL** are designed to test your understanding of one or more concepts; they usually involve qualitative rather than quantitative thinking.

CONCEPTUAL EXERCISE 14.2

Using Le Chatelier's Principle

Use Le Chatelier's principle (← **Sec. 12-6b**) to explain why the fraction of NH_3 molecules that reacts with water in a very dilute solution is larger than in a more concentrated solution.

14-1a Water's Role as Acid or Base

In aqueous solution, all Brønsted-Lowry acids and bases react with water molecules. As we have seen, a water molecule *accepts* an H^+ from an acid such as nitric acid, while a water molecule *donates* an H^+ to a base such as an ammonia molecule. According to the Brønsted-Lowry definitions, water serves as a base (an H^+ acceptor) when an acid is present and as an acid (an H^+ donor) when a base is present. Therefore, water displays both acid and base properties—*it can donate or accept H^+ ions,* depending on its chemical environment. The general reactions of water with acids, HA, and molecular bases, B, are

Water acting as a base, an H^+ acceptor:

$$\underset{\text{acid}}{HA} + \underset{\text{base}}{H_2O} \rightleftharpoons H_3O^+ + A^-$$

Water acting as an acid, an H^+ donor:

$$\underset{\text{base}}{B} + \underset{\text{acid}}{H_2O} \rightleftharpoons BH^+ + OH^-$$

If the base is an anion, A^-, it accepts H^+ from water to form AH.

A molecule or ion, like water, that can donate or accept H^+ is said to be **amphiprotic**.

EXERCISE 14.3

Acids and Bases

Complete these equations. (*Hint:* CH_3NH_2 and $(CH_3)_2NH$ are amines, so they are bases.)

(a) $HCN + H_2O \longrightarrow$ (b) $HBr + H_2O \longrightarrow$

(c) $CH_3NH_2 + H_2O \longrightarrow$ (d) $(CH_3)_2NH + H_2O \longrightarrow$

14-1b Conjugate Acid-Base Pairs

Whenever an acid donates H^+ to a base, a new acid and a new base are formed. *A pair of molecules or ions related to each other by the loss or gain of a single H^+* is called a **conjugate acid-base pair**. Every Brønsted-Lowry acid has its conjugate base, and every Brønsted-Lowry base has its conjugate acid.

We illustrate this using the reaction between acetic acid, CH_3COOH, and water (Figure 14.1). Acetic acid is an H^+ ion donor (an acid), and water is an H^+ ion acceptor (a base). The products of the reaction are a new acid, H_3O^+, and a new base, CH_3COO^-. In the reverse reaction, H_3O^+ acts as an acid (H^+ donor), and acetate ion as a base (H^+ acceptor). The structures of CH_3COOH and CH_3COO^- differ from one another by only a single H^+, just as the structures of H_2O and H_3O^+ do.

Removing an H^+ ion from the acid forms the conjugate base, which therefore has a charge that is one unit more negative than the acid. For example, hydrofluoric acid, HF, has the F^- (fluoride) ion as its conjugate base; HF and F^- are a conjugate acid-base *pair.*

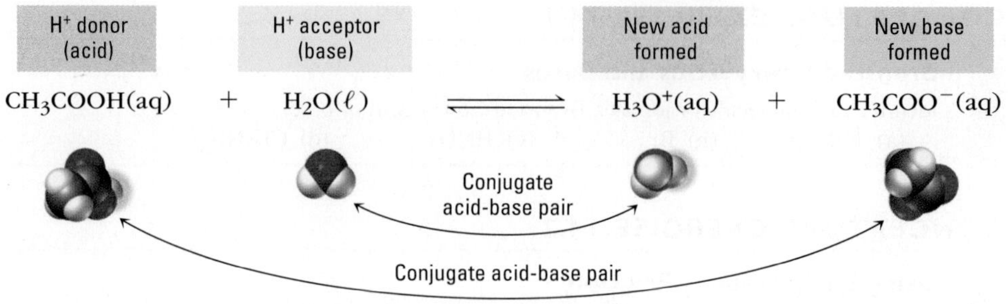

H⁺ donor (acid)
H⁺ acceptor (base)
New acid formed
New base formed
CH₃COOH(aq) + H₂O(ℓ) ⇌ H₃O⁺(aq) + CH₃COO⁻(aq)
Conjugate acid-base pair
Conjugate acid-base pair

Figure 14.1 Conjugate acid-base pairs. Transfer of an H^+ converts a conjugate acid into its conjugate base.

ACID Donates H^+ ⟶ ⟵ Accepts H^+ BASE

$$HF(aq) + H_2O(\ell) \rightleftharpoons H_3O^+(aq) + F^-(aq)$$

ACID base acid BASE

BASE Accepts H^+ ⟶ ⟵ Donates H^+ ACID

In the forward reaction, HF is a Brønsted-Lowry acid. It donates an H^+ to water, which, by accepting the H^+, acts as a Brønsted-Lowry base. In the reverse reaction, fluoride ion is the Brønsted-Lowry base, accepting an H^+ from H_3O^+, the acid. As noted in the equation, there are two conjugate acid-base pairs: (1) HF and F^- and (2) H_2O and H_3O^+. *One member of a conjugate acid-base pair is always a reactant and the other is always a product; they are never both products or both reactants.*

The conjugate acid formula can be derived from the formula of its conjugate base by adding an H^+ ion to the conjugate base, which makes the charge of the resulting conjugate acid one unit more positive than the conjugate base. Therefore, the conjugate acid of Cl^- ion is HCl; the conjugate acid of NH_3 is NH_4^+.

BASE Accepts H^+ ⟶ ⟵ Donates H^+ ACID

$$NH_3(aq) + H_2O(\ell) \rightleftharpoons NH_4^+(aq) + OH^-(aq)$$

BASE acid ACID base

ACID Donates H^+ ⟶ ⟵ Accepts H^+ BASE

PROBLEM-SOLVING EXAMPLE 14.1

Conjugate Acid-Base Pairs

Complete this table by identifying the correct conjugate acid or conjugate base.

Acid	Its Conjugate Base	Base	Its Conjugate Acid
HCOOH	________	CN^-	________
H_2S	________	________	HSO_4^-
PH_4^+	________	________	H_2SO_3
________	ClO^-	S^{2-}	________

Result

Acid	Its Conjugate Base	Base	Its Conjugate Acid
HCOOH	$HCOO^-$	CN^-	HCN
H_2S	HS^-	SO_4^{2-}	HSO_4^-
PH_4^+	PH_3	HSO_3^-	H_2SO_3
HClO	ClO^-	S^{2-}	HS^-

Analyze You are asked to find a conjugate base, given an acid, or a conjugate acid, given a base. In each of these cases, apply the relationship

ACID Donates $H^+ \longrightarrow$ BASE $\longleftarrow$ Accepts H^+

$$\text{Conjugate acid} \rightleftharpoons \text{Conjugate base}$$

Plan The conjugate acid can be identified by adding H^+ to the conjugate base; the conjugate base forms by loss of H^+ from the conjugate acid. For example, because CN^- has no H^+ to donate, CN^- must be a base; its conjugate acid, HCN, has an additional H^+. Likewise, HSO_3^- is the conjugate base of its conjugate acid, H_2SO_3.

Execute The other conjugate acid-base pairs can be worked out similarly. See table in Result.

PROBLEM-SOLVING PRACTICE 14.1

Complete the table.

Acid	Its Conjugate Base	Base	Its Conjugate Acid
$H_2PO_4^-$	______	______	HPO_4^{2-}
______	H^-	NH_2^-	______
HSO_3^-	______	ClO_4^-	______
HF	______	______	HBr

PROBLEM-SOLVING PRACTICE answers are provided at the back of this book in Appendix K.

EXERCISE 14.4

HSO_4^- as a Base

(a) Write the equation for HSO_4^- ion acting as a base in water.
(b) Identify the conjugate acid-base pairs in the equation.
(c) Is HSO_4^- amphiprotic? Explain your answer.

14-1c Relative Strengths of Acids and Bases in Conjugate Pairs

Strong acids are better H^+ donors than weak acids. For example, compare HCl, a strong acid, with HF, a weak acid.

$$HCl(aq) + H_2O(\ell) \longrightarrow H_3O^+(aq) + Cl^-$$
$$HF(aq) + H_2O(\ell) \rightleftharpoons H_3O^+(aq) + F^-(aq)$$

At concentrations <1.0 M the ionization of HCl is nearly 100%; essentially all of the HCl molecules react with water to form H_3O^+ and Cl^-. The Cl^- ion exhibits virtually no tendency to accept H^+ from H_3O^+; Cl^- is an extremely weak base. On the other hand, the reverse of the ionization of HF is significant; F^- ion readily accepts H^+ from H_3O^+ to form HF. Hydrofluoric acid is a weak acid that is mainly un-ionized (remains largely in its molecular form, HF). Thus, *stronger acids have weaker conjugate bases and weaker acids have stronger conjugate bases.* Correspondingly, strong bases are better H^+

acceptors than weak bases. Thus, *stronger bases have weaker conjugate acids and weaker bases have stronger conjugate acids.*

By measuring the extent to which various acids donate H^+ ions to water, chemists have developed an extensive tabulation of the relative strengths of acids and their conjugate bases. An abbreviated list is given in Figure 14.2. Study Figure 14.2 to verify these two important generalizations:

- ***As acid strength decreases, conjugate base strength increases; the weaker the acid, the stronger its conjugate base.***
- ***As base strength decreases, conjugate acid strength increases; the weaker the base, the stronger its conjugate acid.***

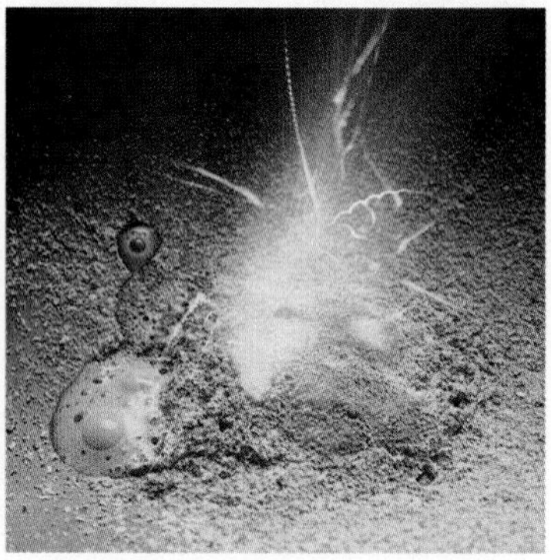
© Cengage Learning/Charles D. Winters

Hydride ion is strongly basic. The reaction of calcium hydride with water is highly exothermic because H^- is a strong base.

Notice from Figure 14.2 that a great many weak bases are anions, such as CN^- (cyanide), F^- (fluoride), and CH_3COO^- (acetate).

Knowing the relative acid and base strengths of the reactants, we can predict the direction of an acid-base reaction. ***The stronger acid and the stronger base will always react to form a weaker conjugate base and a weaker conjugate acid.*** Strong Brønsted-Lowry bases such as hydride ion, H^-; sulfide ion, S^{2-}; oxide ion, O^{2-}; amide ion, NH_2^-; and hydroxide ion, OH^-, readily accept H^+ ions, while weaker Brønsted-Lowry bases do so less readily. For example, the reaction of calcium hydride, CaH_2, with water is highly exothermic because of the extremely strong basic properties of the hydride ion, H^-, which avidly accepts H^+ from water to produce H_2, the extremely weak conjugate acid of hydride ion. In this reaction, hydride ion is a stronger base than OH^-, and water is a stronger acid than H_2, so the forward reaction is favored.

$$H^-(aq) + H_2O(\ell) \longrightarrow H_2(g) + OH^-(g)$$

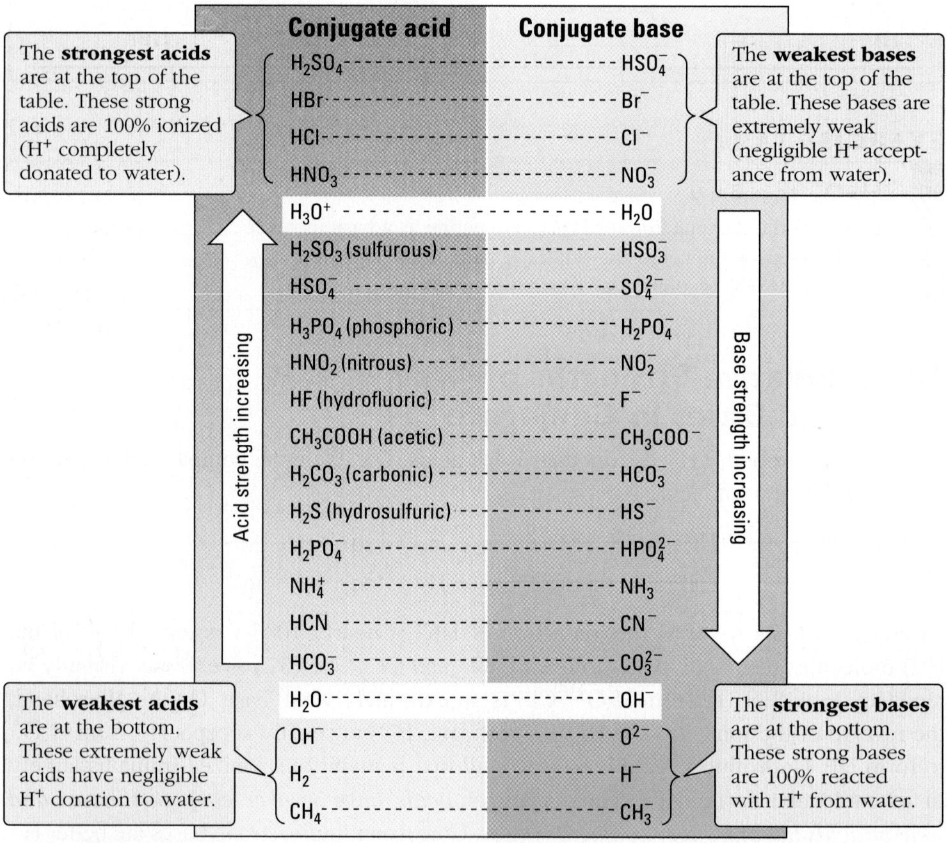

Conjugate acid
Conjugate base
The strongest acids are at the top of the table. These strong acids are 100% ionized (H+ completely donated to water).
The weakest bases are at the top of the table. These bases are extremely weak (negligible H+ acceptance from water).
H2SO4 — HSO4−
HBr — Br−
HCl — Cl−
HNO3 — NO3−
H3O+ — H2O
H2SO3 (sulfurous) — HSO3−
HSO4− — SO42−
H3PO4 (phosphoric) — H2PO4−
HNO2 (nitrous) — NO2−
HF (hydrofluoric) — F−
CH3COOH (acetic) — CH3COO−
H2CO3 (carbonic) — HCO3−
H2S (hydrosulfuric) — HS−
H2PO4− — HPO42−
NH4+ — NH3
HCN — CN−
HCO3− — CO32−
H2O — OH−
OH− — O2−
H2 — H−
CH4 — CH3−
Acid strength increasing
Base strength increasing
The weakest acids are at the bottom. These extremely weak acids have negligible H+ donation to water.
The strongest bases are at the bottom. These strong bases are 100% reacted with H+ from water.

Figure 14.2 Relative strengths of conjugate acids and bases in water.

We can apply information from Figure 14.2 to consider whether the forward or the reverse reaction is favored in an equilibrium. For example, consider the equilibrium

$$HSO_4^-(aq) + CO_3^{2-}(aq) \rightleftharpoons HCO_3^-(aq) + SO_4^{2-}(aq)$$

From Figure 14.2 we note that HSO_4^- is a stronger acid than HCO_3^- and CO_3^{2-} is a stronger base than SO_4^{2-}. Since acid-base reactions favor going from the stronger to the weaker member of each conjugate acid-base pair, the forward reaction is favored and H^+ is transferred from HSO_4^- to CO_3^{2-} to form hydrogen carbonate, HCO_3^-, and sulfate, SO_4^{2-}, ions.

CONCEPTUAL EXERCISE 14.5

Conjugate Acid-Base Strength

Use Figure 14.2 to predict whether the forward or reverse reaction is favored for the equilibrium

$$CH_3COOH(aq) + SO_4^{2-}(aq) \rightleftharpoons HSO_4^-(aq) + CH_3COO^-(aq)$$

14-2 Carboxylic Acids and Amines

Many weak acids such as acetic, lactic, and pyruvic acids are organic acids.

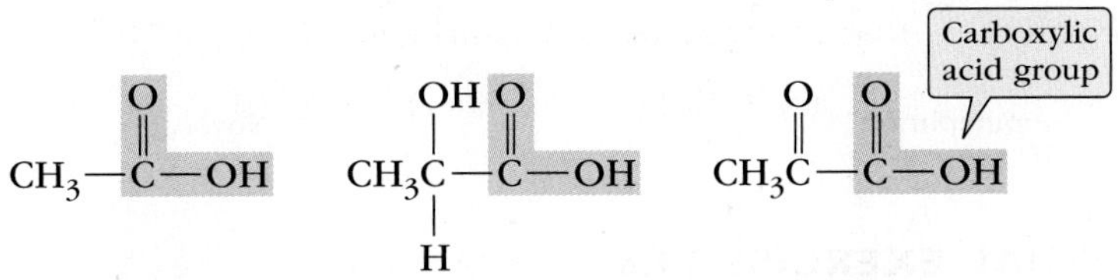
Carboxylic acid group

acetic acid   lactic acid   pyruvic acid

These acids all contain the carboxylic acid (abbreviated —COOH) functional group (← **Sec. 10-5a**). Although carboxylic acid molecules usually contain many other hydrogen atoms, *only the hydrogen atom bound to the oxygen atom of the carboxylic acid group is sufficiently positive to be donated as an H^+ ion in aqueous solution.* The two highly electronegative oxygen atoms of the carboxylic acid group pull electron density away from the hydrogen atom. As a result, the O—H bond of the acid is even more polar, its hydrogen atom has a larger partial positive charge, and is thus acidic. The C—H bonds in organic acids are relatively nonpolar and strong, and these hydrogen atoms are *not* acidic, as can be seen with butanoic acid, which has a single acidic H atom.

Note that although the carboxylic acid group is abbreviated —COOH, the oxygen atoms are not bonded to each other, but to carbon.

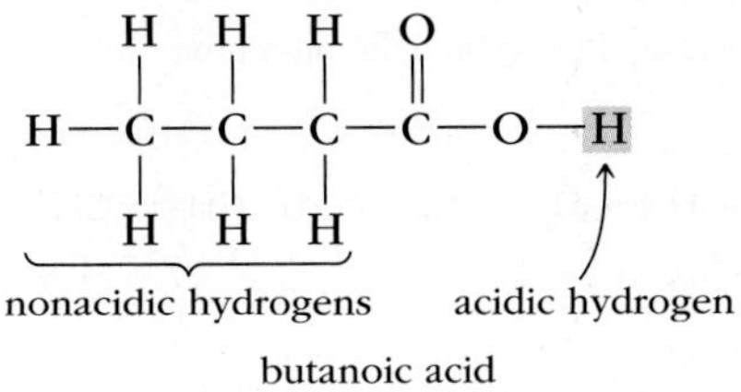
nonacidic hydrogens
acidic hydrogen

butanoic acid

Anions formed by loss of H^+ from a —COOH group, such as acetate ion, CH_3COO^-, from acetic acid, CH_3COOH, are stabilized by resonance (← **Sec. 6-9**).

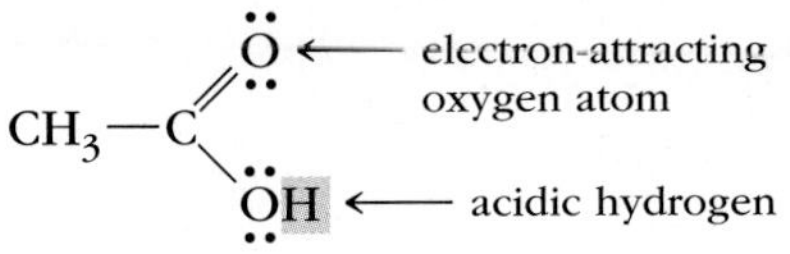
electron-attracting oxygen atom
acidic hydrogen

acetic acid

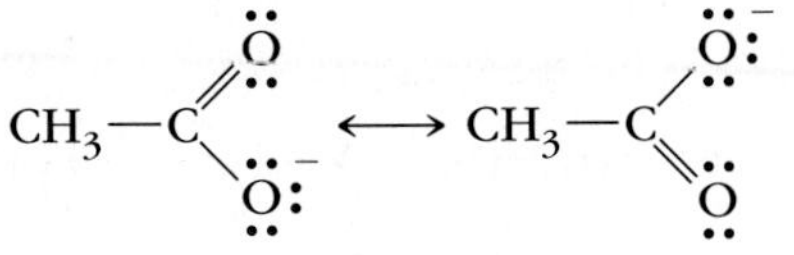

acetate ion

CH_3—$\ddot{N}H_2$

methylamine

An **amine**, such as methylamine, CH_3NH_2, is *a carbon compound that is related to ammonia,* NH_3 (← **Sec. 10-6b**). Like ammonia, amines have a nitrogen atom with three valence electron pairs in covalent bonds and one unshared electron pair. The lone pair of electrons can accept an H^+, and so, like ammonia, aqueous amines react as weak bases, accepting an H^+ from water.

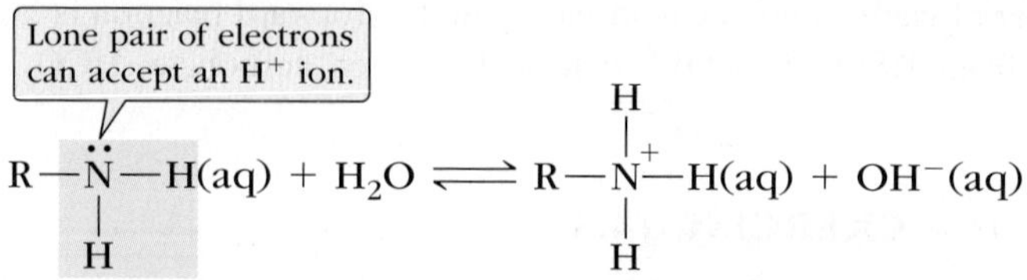
Lone pair of electrons can accept an H+ ion.

Many natural biochemicals and drugs are amines, such as epinephrine (adrenaline, a natural hormone) and Novocain (a local anesthetic).

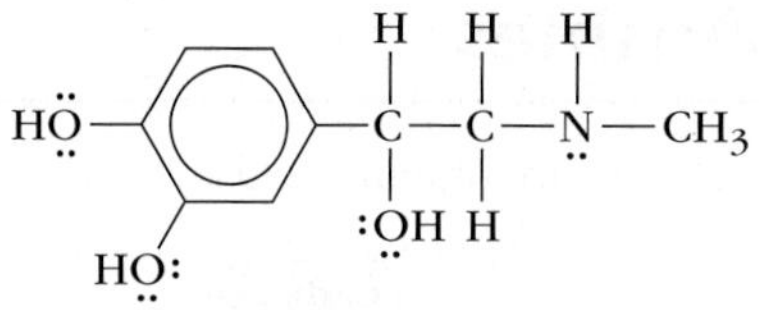

epinephrine

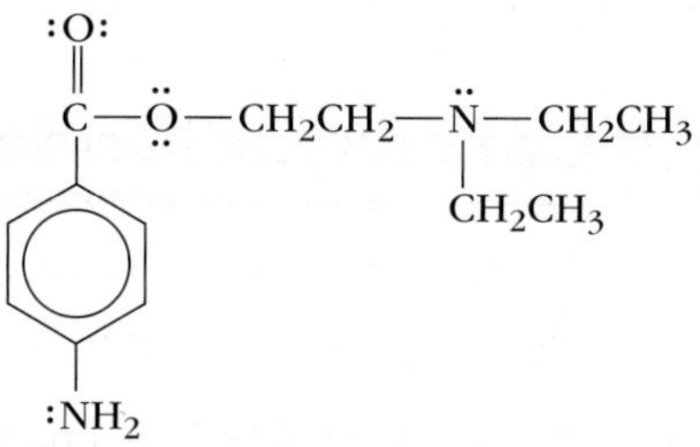

Novocain

McCormick
Coarse Ground
Black
Pepper

Black pepper contains piperidine.

CONCEPTUAL EXERCISE 14.6

Piperidine, an Amine

The structure of piperidine, a component of pepper, is shown here as a Lewis structure and a space-filling model. Write the equation for the reaction of piperidine (a) with water and (b) with hydrochloric acid.

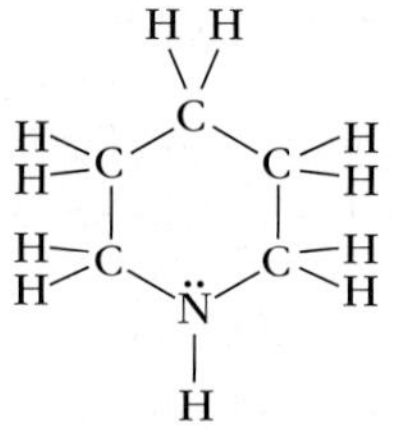

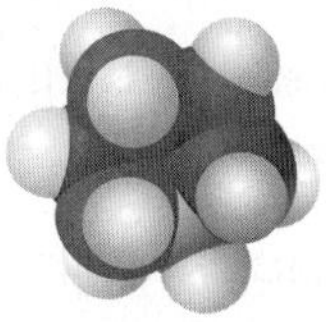

EXERCISE 14.7

Classifying Compounds

Identify each molecule as a carboxylic acid, an amine, both, or neither.

(a) 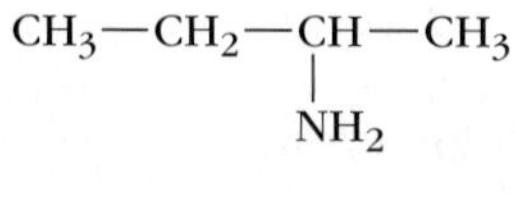

(b) 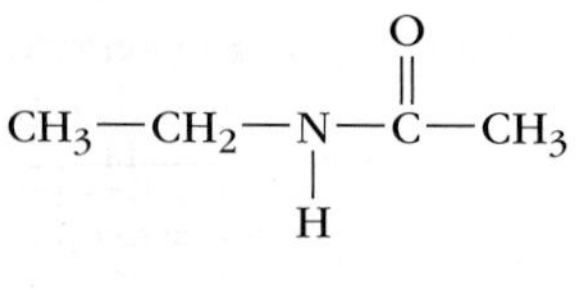

(c) 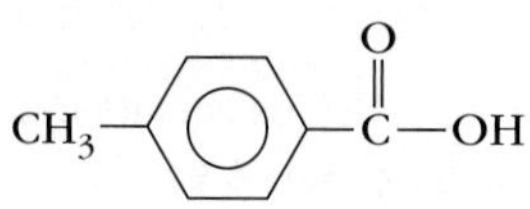

(d) 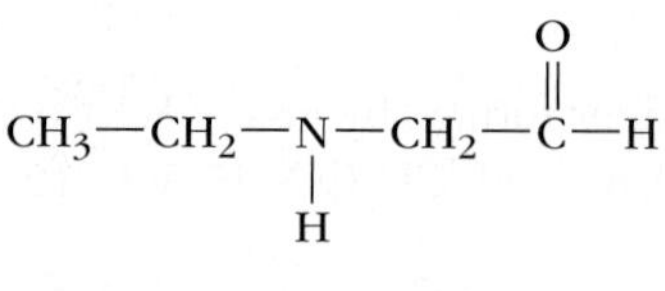

(e) 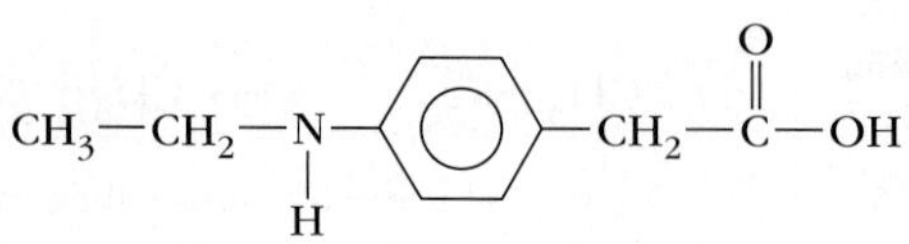

14-3 The Autoionization of Water

Even highly purified water conducts a very small electrical current, which indicates that pure water contains a very small concentration of ions. These ions are formed when *water molecules react to produce aqueous hydrogen ions and aqueous hydroxide ions* in a process called **autoionization**.

$$H_2O(\ell) + H_2O(\ell) \rightleftharpoons H_3O^+(aq) + OH^-(aq)$$

BASE acid ACID base

In this reaction, one water molecule serves as an H^+ acceptor (base) while the other is an H^+ donor (acid). The equilibrium between the water molecules and the hydronium and hydroxide ions is very reactant-favored. Therefore, the concentrations of these ions in pure water are *very* low. Nevertheless, autoionization of water is very important in understanding how acids and bases function in aqueous solutions. As in the case of any equilibrium reaction, an equilibrium constant expression can be written for the autoionization of water.

$$2\ H_2O(\ell) \rightleftharpoons H_3O^+(aq) + OH^-(aq) \qquad K_w = [H_3O^+][OH^-]$$

The equilibrium constant K_w is known as the **ionization constant for water**. From electrical conductivity measurements of pure water, we know that $[H_3O^+] = [OH^-] = 1.0 \times 10^{-7}$ M at 25 °C. Hence

$$K_w = [H_3O^+][OH^-] = (1.0 \times 10^{-7})(1.0 \times 10^{-7}) = 1.0 \times 10^{-14} \qquad \text{(at 25 °C)}$$

Like all equilibrium constant expressions, the one for K_w includes concentrations of solutes but not the concentration of the solvent, which in this case is water.

The equation $K_w = [H_3O^+][OH^-]$ ***applies to pure water and all aqueous solutions.*** Like other equilibrium constants, the value of K_w is temperature-dependent (Table 14.1).

At a given temperature, the product of the hydronium ion concentration times the hydroxide ion concentration is always the same. If the hydronium ion concentration increases (because an acid was added to the water, for example), then the hydroxide ion concentration must decrease, and vice versa. The equation also tells us that if we know one concentration, we can calculate the other.

In aqueous solutions, the H_3O^+ and OH^- concentrations are inversely related; as one increases, the other must decrease. Their product always equals 1.0×10^{-14} at 25 °C.

The relative concentrations of H_3O^+ and OH^- also indicate the acidic, neutral, or basic nature of the aqueous solution. **For all aqueous solutions** there are three possibilities.

Neutral solution: $[H_3O^+] = [OH^-]$ both equal to 1.0×10^{-7} M at 25 °C

Acidic solution: $[H_3O^+] > 1.0 \times 10^{-7}$ M; $[OH^-] < 1.0 \times 10^{-7}$ M at 25 °C

Basic (alkaline) solution: $[H_3O^+] < 1.0 \times 10^{-7}$ M; $[OH^-] > 1.0 \times 10^{-7}$ M at 25 °C

The term *alkaline* is also used to describe basic solutions.

When a solution has equal concentrations of $[H_3O^+]$ and $[OH^-]$, it is said to be **neutral**. If either an acid or a base is added to a neutral solution, the autoionization equilibrium between H_3O^+ and OH^- will be disturbed. Recall that according to Le Chatelier's principle (← **Sec. 12-6**), an equilibrium shifts in such a way as to offset the effect of any disturbance. When an acid is added, the concentration of H_3O^+ ions increases. To oppose this increase, some added H_3O^+ ions react with OH^- ions to form H_2O, thereby reducing the $[OH^-]$. When equilibrium is re-established, $[H_3O^+] > [OH^-]$ and the solution is *acidic*; however, the mathematical product $[H_3O^+][OH^-]$ is still equal to 1.0×10^{-14} at 25 °C. Similarly, if a base is added to water, some of the added OH^- ions react with H_3O^+ ions to form H_2O, thereby decreasing the $[H_3O^+]$. When equilibrium is re-established, $[H_3O^+] < [OH^-]$ and the solution is *basic*; the product $[H_3O^+][OH^-]$ still equals 1.0×10^{-14} at 25 °C.

Table 14.1 Temperature Dependence of K_w

T (°C)	K_w
10	0.29×10^{-14}
15	0.45×10^{-14}
20	0.68×10^{-14}
25	1.01×10^{-14}
30	1.47×10^{-14}
50	5.48×10^{-14}

PROBLEM-SOLVING EXAMPLE 14.2

Ionization Constant for Water

Determine which is more acidic at 25 °C: (a) a solution whose H_3O^+ concentration is 5.0×10^{-4} M; or (b) one that has an OH^- concentration of 3.0×10^{-8} M.

Result (a) A solution whose H_3O^+ concentration is 5.0×10^{-4} M is more acidic.

Analyze The greater the concentration of hydronium ions is, the greater the acidity of the solution is. When $[OH^-]$ is known, $[H_3O^+]$ can be calculated from K_w.

Plan For solution (b), calculate $[H_3O^+]$ from $[H_3O^+][OH^-] = 1.0 \times 10^{-14}$. Compare the calculated $[H_3O^+]$ with the $[H_3O^+]$ given for solution (a).

Execute For solution (b),

$$[H_3O^+][OH^-] = [H_3O^+](3.0 \times 10^{-8}) = 1.0 \times 10^{-14}$$

$$[H_3O^+] = \frac{1.0 \times 10^{-14}}{3.0 \times 10^{-8}} = 3.3 \times 10^{-7} \text{ M}$$

For solution (a), $[H_3O^+] = 5.0 \times 10^{-4}$ M; this is larger than 3.3×10^{-7} M, so solution (a) is more acidic.

☑ **Reasonable Result Check** In solution (a), $[H_3O^+]$ is much greater than 1×10^{-7} M. In solution (b), $[OH^-]$ is slightly less than 1×10^{-7} M, so solution (b)'s $[H_3O^+]$ is only slightly greater than 1×10^{-7} M. Thus, solution (a) is more acidic.

PROBLEM-SOLVING PRACTICE 14.2

Determine which is more basic: (a) a solution containing 2.0×10^{-5} mol HBr in 1.00 L solution, or (b) a solution that has an OH^- concentration of 5.0×10^{-9} M.

PROBLEM-SOLVING EXAMPLE 14.3

$[H_3O^+]$ and $[OH^-]$ Concentrations

Calculate:
(a) the hydroxide ion concentration at 25 °C in 0.10-M HCl.
(b) the hydronium ion concentration at 25 °C in 0.010-M KOH.
(c) the hydronium ion concentration at 25 °C in 0.010-M $Ba(OH)_2$.

Which solution is most acidic? Which solution is most basic?

Result (a) $[OH^-] = 1.0 \times 10^{-13}$ M (b) $[H_3O^+] = 1.0 \times 10^{-12}$ M
(c) $[H_3O^+] = 5.0 \times 10^{-13}$ M

The 0.10-M HCl solution is most acidic. The 0.010-M $Ba(OH)_2$ solution is most basic.

Analyze When $[OH^-]$ is known, $[H_3O^+]$ can be calculated from K_w and vice versa. HCl is a strong acid so it is 100% ionized. KOH and $Ba(OH)_2$ are soluble ionic compounds and strong bases so each is 100% dissociated. Because of the subscript 2 in the formula $Ba(OH)_2$ there is 2 mol OH^- in solution for every 1 mol $Ba(OH)_2$ that dissolves. At 25 °C, $[H_3O^+][OH^-] = 1.0 \times 10^{-14}$.

Plan From the concentration of each compound, use a stoichiometric ratio to calculate $[H_3O^+]$ or $[OH^-]$. Then, use the relation $[H_3O^+][OH^-] = 1.0 \times 10^{-14}$.

Execute

(a) $[H_3O^+] = 0.10 \text{ M} = 1.0 \times 10^{-1}$ M:

$$[H_3O^+][OH^-] = 1.0 \times 10^{-14} = (1.0 \times 10^{-1} \text{ M})\,[OH^-]$$

$$[OH^-] = \frac{1.0 \times 10^{-14}}{1.0 \times 10^{-1}} = 1.0 \times 10^{-13} \text{ M}$$

(b) $[OH^-] = 0.010 \text{ M} = 1.0 \times 10^{-2}$ M:

$$[H_3O^+] = \frac{1.0 \times 10^{-14}}{[OH^-]} = \frac{1.0 \times 10^{-14}}{1.0 \times 10^{-2}} = 1.0 \times 10^{-12} \text{ M}$$

(c) $[OH^-] = 2 \times 0.010 \text{ M} = 2.0 \times 10^{-2}$ M:

$$[H_3O^+] = \frac{1.0 \times 10^{-14}}{[OH^-]} = \frac{1.0 \times 10^{-14}}{2.0 \times 10^{-2}} = 5.0 \times 10^{-13} \text{ M}$$

Solution (a) has the largest $[H_3O^+]$ so it is most acidic. Solution (c) has the largest $[OH^-]$ so it is most basic.

☑ **Reasonable Result Check** In the acid solution, the hydroxide-ion concentration is very low; in the two basic solutions, the hydrogen-ion concentration is very low so the results are reasonable.

PROBLEM-SOLVING PRACTICE 14.3

Calculate the hydroxide ion concentration at 25 °C in 6.0-M HNO_3 and the hydronium ion concentration in 6.0-M NaOH.

14-4 The pH Scale

The $[H_3O^+]$ and $[OH^-]$ in an aqueous solution vary widely depending on the acid or base present and its concentration. In general, the $[H_3O^+]$ in aqueous solutions can range from about 10 mol/L down to about 10^{-15} mol/L. The $[OH^-]$ can also vary over the same range in aqueous solution.

Because these concentrations can be so small, they have very large negative exponents. It is more convenient to express these concentrations using logarithms. We define the **pH** of a solution as *the negative of the base 10 logarithm (log) of the hydronium ion concentration (expressed in mol/L).*

$$\mathbf{pH = -log[H_3O^+]}$$

The *negative* logarithm is used because it gives a positive pH value for small concentration values. Thus, the pH of pure water at 25 °C is given by

$$\begin{aligned} pH &= -\log[1.0 \times 10^{-7}] = -[\log(1.0) + \log(10^{-7})] \\ &= -[0 + (-7.00)] = 7.00 \end{aligned}$$

In terms of pH, for aqueous solutions at 25 °C we can write

Neutral solution:	pH = 7.00
Acidic solution:	pH < 7.00
Basic (alkaline) solution:	pH > 7.00

Figure 14.3 shows pH values and corresponding H_3O^+ and OH^- concentrations of some common solutions. Notice that, because $-\log(1 \times 10^{-x}) = x$,

Lemon juice: $[H_3O^+] = 1 \times 10^{-2}$ M; $pH = -\log(1 \times 10^{-2}) = 2$

Black coffee: $[H_3O^+] = 1 \times 10^{-5}$ M; $pH = -\log(1 \times 10^{-5}) = 5$

Keep in mind that, because pH involves a logarithm, a change of *one* in pH represents a *ten-fold* change in H_3O^+ concentration, a change of two in pH represent a 100-fold change, and so on. Thus, according to Figure 14.3, the $[H_3O^+]$ in lemon juice (pH = 2) is 1000 times *greater* than that in black coffee (pH = 5).

For solutions in which $[H_3O^+]$ or $[OH^-]$ has a value other than an exact power of 10 ($1, 1 \times 10^{-1}, 1 \times 10^{-2}, \ldots$) a calculator is convenient for finding the pH. For example, the pH is 2.35 for a solution that contains 0.0045 mol of the strong acid HNO_3 per liter of solution, that is, a solution whose hydrogen-ion concentration is 4.5×10^{-3} mol/L.

$$pH = -\log(4.5 \times 10^{-3}) = 2.35$$

If $[H_3O^+]$ is large enough, the pH can be negative; if it is small enough, the pH can be greater than 14. For example, when $[H_3O^+] = 2.0$, $pH = -\log(2.0) = -0.30$; when $[H_3O^+] = 2.0 \times 10^{-15}$, $pH = -\log(2.0 \times 10^{-15}) = 14.70$.

The p in pH is derived from French "puissance" meaning "power." Thus, pH is the "power of hydrogen."

The definition $pH = -\log[H_3O^+]$ is accurate only for very low concentrations of aqueous hydrogen ions (see McCarty, C. G.; Vitz, E. *Journal of Chemical Education*, Vol. 83, 2006; pp. 752–757). A more accurate definition is $pH = -\log a_{H^+}$, where a_{H^+} represents the *activity* of hydrogen ions. Activity is an effective concentration that has been corrected for noncovalent interactions among ions and molecules. A complete definition of pH is quite complicated (see goldbook.iupac.org/P04524.html and Galster, H. *pH Measurement: Fundamentals, Methods, Applications, Instrumentation;* VCH: New York, 1991).

See Appendix A-6 for more about using logarithms.

Notice that the concentration units, mol/L, are not included when the logarithm is taken. It is not possible to take the logarithm of a unit. The definition of pH applies only to concentrations expressed in mol/L.

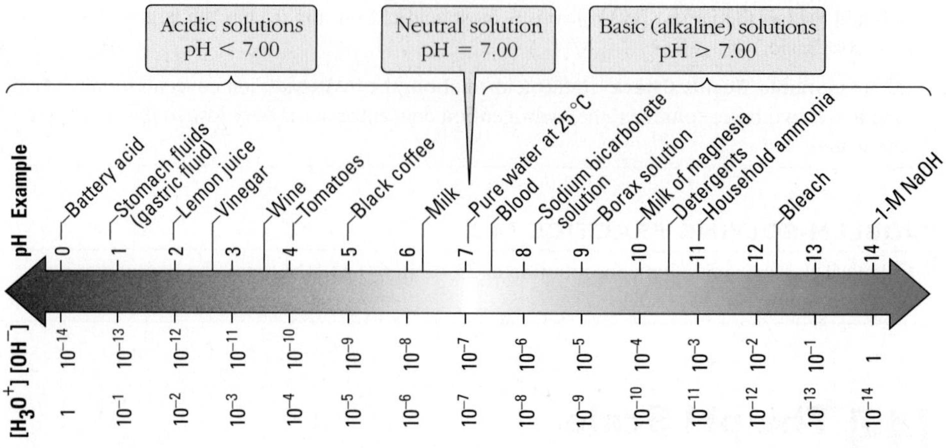

Acidic solutions
pH < 7.00
Neutral solution
pH = 7.00
Basic (alkaline) solutions
pH > 7.00
pH Example
Battery acid
Stomach fluids (gastric fluid)
Lemon juice
Vinegar
Wine
Tomatoes
Black coffee
Milk
Pure water at 25 °C
Blood
Sodium bicarbonate solution
Borax solution
Milk of magnesia
Detergents
Household ammonia
Bleach
1-M NaOH
[H3O+] [OH−]

Figure 14.3 The pH of some common aqueous solutions.

PROBLEM-SOLVING EXAMPLE 14.4

Calculating pH from $[H_3O^+]$

Calculate the pH of an aqueous HNO_3 solution that has a volume of 250. mL and contains 0.4649 g HNO_3.

Result pH = 1.530

Analyze The mass of HNO_3 and volume of solution are given. Nitric acid is a strong acid, so every mole that dissolves produces a mole of H_3O^+ and a mole of NO_3^- in solution. Molar mass relates mass to amount.

Plan Determine the amount of HNO_3, calculate the H_3O^+ concentration, and express this concentration as pH.

Execute

$$n(\text{HNO}_3) = 0.465 \text{ g HNO}_3 \times \frac{1 \text{ mol HNO}_3}{63.012 \text{ g HNO}_3} = 0.007380 \text{ mol HNO}_3$$

$$[\text{H}_3\text{O}^+] = \frac{0.007380 \text{ mol HNO}_3}{0.250 \text{ L}} = 0.0295 \text{ M}$$

$$\text{pH} = -\log(2.95 \times 10^{-2}) = 1.530$$

Note: This pH has been calculated to three significant figures (those to the right of the decimal point) because there are three significant digits in 250. mL, the least precise quantity. The digit or digits to the left of the decimal point in a pH represent a power of 10. Actual measurements of pH values are seldom this accurate.

☑ **Reasonable Result Check** If the $[H_3O^+]$ were 0.10 M, the pH would be 1.00; if the $[H_3O^+]$ were 0.010 M, the pH would be 2.00. The concentration, 0.0295 M, is between these two values so a pH between 1.00 and 2.00 is reasonable.

PROBLEM-SOLVING PRACTICE 14.4

Calculate the pH of a 0.040-M NaOH solution.

The hydronium ion concentration of a solution can be calculated from its pH value as shown in Problem-Solving Example 14.5.

PROBLEM-SOLVING EXAMPLE 14.5

Calculating $[H_3O^+]$ from pH

A hospital patient's blood sample has a pH of 7.40 at 25 °C.
(a) Calculate the sample's H_3O^+ concentration.
(b) Determine its OH^- concentration.
(c) Is the sample acidic, neutral, or basic?

Result (a) 4.0×10^{-8} M (b) 2.5×10^{-7} M (c) Basic

Analyze The pH is given. The definition of pH can be used to calculate the $[H_3O^+]$ and from it the $[OH^-]$. A pH < 7.0 is acidic; 7.0 is neutral; > 7.0 is basic.

Plan $pH = -\log[H_3O^+] = 7.40$, so $\log[H_3O^+] = -7.40$. Solve for the $[H_3O^+]$ and rearrange the $[H_3O^+][OH^-] = 1.0 \times 10^{-14}$ equation to solve for $[OH^-]$.

Execute (a) By the rules of logarithms (Appendix A-6), $10^{\log(x)} = x$, so we can write $10^{\log[H_3O^+]} = 10^{-pH} = [H_3O^+]$.

$$[H_3O^+] = 10^{-7.40} = 4.0 \times 10^{-8}\ M$$

(b)

$$[OH^{-1}] = \frac{1.0 \times 10^{-14}}{[H_3O^+]} = \frac{1.0 \times 10^{-14}}{4.0 \times 10^{-8}} = 2.5 \times 10^{-7}\ M$$

(c) Because the pH is greater than 7.0, the sample is basic.

☑ **Reasonable Result Check** Because the pH of 7.40 is slightly above 7.00, the hydroxide ion concentration should be a bit higher than 1.0×10^{-7} M, that of a neutral solution, which it is. Therefore, the sample is slightly basic.

PROBLEM-SOLVING PRACTICE 14.5

In a hospital laboratory the pH of a bile sample is measured as 7.90 at 25 °C.
(a) Calculate the H_3O^+ concentration.
(b) Is the sample acidic or basic?

CONCEPTUAL EXERCISE 14.8

pH of Solutions of Different Acids

Is the pH of a 0.1-M solution of HNO_3 the same as the pH of a 0.1-M solution of HCl? Explain.

EXERCISE 14.9

Acidity and pH

A sample of carbonated beverage has a pH of 2.10. A sample of lime juice has a hydronium ion concentration of 1.3×10^{-2} mol/L. Determine which sample is more acidic.

Using "p" before something to represent "take the negative log of" is a general convention. For example, $[OH^-]$ can be expressed as pOH.

$$\mathbf{pOH = -\log[OH^-]}$$

The $[OH^-]$ of pure water at 25 °C is 1.0×10^{-7} M, and therefore its pOH is

$$pOH = -\log(1 \times 10^{-7}) = -(-7.00) = 7.00$$

Because the values of $[H_3O^+]$ and $[OH^-]$ are related by the K_w expression, this equation holds for all aqueous solutions at 25 °C:

$$K_w = [H_3O^+][OH^-] = 1.0 \times 10^{-14}$$

Arnold Beckman
1900–2004

With his invention of the first electronic pH meter in 1934, Arnold Beckman revolutionized pH measurement. Beckman, a professor at the California Institute of Technology at the time, developed the instrument in response to a request from the California Fruit Growers' Association for a quicker, more accurate way to measure the acidity of lemon juice. He went on to found the highly successful Beckman Instrument Company, a firm that invented the first widely used infrared and ultraviolet spectrophotometers and other laboratory instruments. Arnold Beckman and his wife Mabel have donated millions of dollars to advance chemical research and education nationwide.

This equation can be rewritten by taking the negative logarithm of each side

$$-\log K_w = -\log[H_3O^+] + (-\log[OH^-]) = -\log(1.0 \times 10^{-14}) \quad \text{(at 25 °C)}$$

or

$$\mathbf{p}K_w = \mathbf{pH + pOH = 14.00}$$

The relation between pH and pOH can be used to find one value when the other is known. A 0.0010-M solution of the strong base NaOH, for example, has an OH^- concentration of 0.0010 M and a pOH given by

$$\text{pOH} = -\log(1.0 \times 10^{-3}) = 3.00$$

and therefore, at 25 °C,

$$\text{pH} = 14.00 - \text{pOH} = 14.00 - 3.00 = 11.00$$

EXERCISE 14.10

pOH and pH

Which solution is more basic, one that has a pH of 5.5 or one with a pOH of 8.5? Calculate the H_3O^+ concentration in each solution.

14-4a Measuring pH

The pH of a solution is readily measured using a pH meter (Figure 14.4). This device consists of a pair of electrodes (often in a single probe) that detect the H_3O^+ concentration of the test solution, convert it into an electrical signal, and display it directly as the pH value. The meter is initially calibrated using standard solutions of known pH. The pH of body fluids, soil, environmental and industrial samples, and other substances can be measured easily and accurately with a pH meter.

Acid-base indicators are a much older and less precise (but more convenient) method to determine pH. An indicator changes color within a narrow pH range, usually 1 to 2 pH units. Loss or gain of an H^+ ion changes the indicator's molecular structure so that it absorbs light in a different region of the visible spectrum. Consider the indicator bromthymol blue (Figure 14.4). At or below pH 6 it is yellow (this is called its acid form); at or above pH = 8 it is blue (its base form). At pH = 7, it is a mixture of 50% yellow and 50% blue, so it appears green. Thus, the pH of a sample that makes bromthymol blue turn green has a pH of about 7. If the indicator color is blue, the pH of the sample is at least 8, and it could be much higher. If the indicator is yellow, pH ≤ 6.

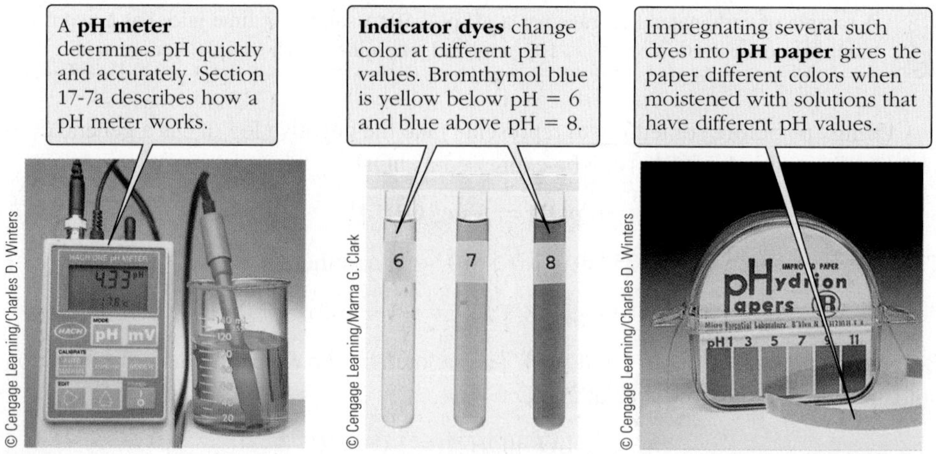
A **pH meter** determines pH quickly and accurately. Section 17-7a describes how a pH meter works.
Indicator dyes change color at different pH values. Bromthymol blue is yellow below pH = 6 and blue above pH = 8.
Impregnating several such dyes into **pH paper** gives the paper different colors when moistened with solutions that have different pH values.
6
7
8
© Cengage Learning/Charles D. Winters
© Cengage Learning/Marna G. Clark
© Cengage Learning/Charles D. Winters

Figure 14.4 Measuring pH. A pH meter, an indicator dye, and pH paper can determine pH.

Strips of paper impregnated with acid-base indicators are also used to test the pH of many substances. The color of the paper after it has been dampened by the solution to be tested is compared with a set of colors at known pHs.

14-5 Ionization Constants of Acids and Bases

You learned earlier that the greater the value of the equilibrium constant for a reaction, the more product-favored that reaction is (← **Sec. 12-1**). In an acid-base reaction, the stronger the reactant acid and the reactant base, the more product-favored the reaction. Consequently, the magnitude of equilibrium constants can tell us the relative strengths of weak acids and bases. For example, *the larger the equilibrium constant is for an acid's ionization, the stronger the acid is.*

14-5a Acid Ionization Constants

An ionization equation for the transfer to water of H^+ from any acid represented by the general formula HA is

$$\underset{\text{conjugate acid}}{HA(aq)} + H_2O(\ell) \rightleftharpoons H_3O^+(aq) + \underset{\text{conjugate base}}{A^-(aq)}$$

The **acid ionization constant expression** is *the equilibrium constant expression for the acid ionization equation*:

$$K_a = \frac{[H_3O^+][\text{conjugate base}]}{[\text{conjugate acid}]} = \frac{[H_3O^+][A^-]}{[HA]}$$

The equilibrium constant K_a is called the **acid ionization constant**. In the acid ionization constant expression:

Acid ionization constants are sometimes called acid dissociation constants.

- The *equilibrium* concentrations of conjugate base and hydronium ion appear in the numerator.
- The *equilibrium* concentration of *un-ionized* conjugate acid appears in the denominator.
- As with other equilibrium constant expressions, pure solids, pure liquids, and solvents, such as water, are not included.

As more acid ionizes, the [HA] denominator term in the acid ionization constant expression gets smaller and the numerator terms increase. Consequently, the ratio gets larger. *For strong acids* such as hydrochloric acid, the equilibrium is product-favored and *the acid ionization constant value is much larger than 1.* In contrast, weak acids such as acetic acid ionize to a much smaller extent, establishing equilibria in which significant concentrations of un-ionized weak acid molecules are still present in the solution. *All weak acids have* K_a *values less than 1* because the ionization of a weak acid is reactant-favored; *weak acids are weak electrolytes.*

The larger its acid ionization constant is, the stronger an acid is.

The common strong acids are
Nitric, HNO_3
Sulfuric, H_2SO_4
Perchloric, $HClO_4$
Hydrochloric, HCl
Hydrobromic, HBr
Hydroiodic, HI

PROBLEM-SOLVING EXAMPLE 14.6

Acid Ionization Constant Expressions

Write the ionization equation and ionization constant expression for each acid.
(a) HF (b) HBrO (c) $H_2PO_4^-$

Result

(a) $HF(aq) + H_2O(\ell) \rightleftharpoons H_3O^+(aq) + F^-(aq)$ $K_a = \frac{[H_3O^+][F^-]}{[HF]}$

(b) $HBrO(aq) + H_2O(\ell) \rightleftharpoons H_3O^+(aq) + BrO^-(aq)$ $K_a = \frac{[H_3O^+][BrO^-]}{[HBrO]}$

(c) $H_2PO_4^-(aq) + H_2O(\ell) \rightleftharpoons H_3O^+(aq) + HPO_4^{2-}(aq)$ $\qquad K_a = \frac{[H_3O^+][HPO_4^{2-}]}{[H_2PO_4^-]}$

Analyze An acid ionization equation represents the transfer of an H^+ ion from an acid to water creating a hydronium ion and the conjugate base of the acid. In the K_a expression, the product concentrations at equilibrium are divided by the reactant concentrations at equilibrium; $[H_2O]$ is not included.

Plan Write equations and equilibrium constant expressions based on the Analyze section.

Execute Use HF as an example. Parts (b) and (c) are worked out similarly:

$$HF(aq) + H_2O(\ell) \rightleftharpoons H_3O^+(aq) + F^-(aq) \qquad K_a = \frac{[H_3O^+][F^-]}{[HF]}$$

PROBLEM-SOLVING PRACTICE 14.6

Write the ionization equation and ionization constant expression for each acid:
(a) hydrazoic acid, HN_3 (b) formic acid, HCOOH (c) chlorous acid, $HClO_2$

That weak acids are only slightly ionized can be shown by measuring the pH of their aqueous solutions. The pH of a 0.10-M acetic acid solution is 2.88, which corresponds to a concentration of H_3O^+ of only 1.3×10^{-3} M. Compare this value with the 0.10-M concentration of H_3O^+ ions in a 0.10-M solution of HCl, a strong acid that has pH = 1.00. In a 0.10-M acetic acid solution, only 1.3% of the initial concentration of acetic acid is ionized:

Stronger acid than CH_3COOH

Stronger base than H_2O

$$CH_3COOH(aq) + H_2O(\ell) \rightleftharpoons H_3O^+(aq) + CH_3COO^-(aq) \qquad [14.1]$$

$$\%\ \text{ionization} = \frac{[H_3O^+]\ \text{at equilibrium}}{\text{initial acid conc.}} \times 100\% = \frac{1.3 \times 10^{-3}}{1.0 \times 10^{-1}} \times 100\% = 1.3\%$$

Therefore, almost 99% of the acetic acid remains in the un-ionized molecular form, CH_3COOH. This is why weak acids (and bases) are weak electrolytes.

In an acetic acid solution or an aqueous solution of any weak acid, two different bases compete for H^+ ions that can be donated from two different acids. In Equation 14.1, the two bases are water and acetate ion; the two acids are acetic acid and hydronium ion. Since acetic acid is a weak acid, its K_a is much less than 1 and the equilibrium favors the reactants. The acetate ion must be a stronger H^+ acceptor than the water molecule. Another way of looking at the same reaction is that the hydronium ion must be a stronger H^+ donor than the acetic acid molecule. Both of these statements are true. Recall from Section 14-1c that acid-base reactions favor going from the stronger to the weaker member of each conjugate acid-base pair. Thus, the acetic acid equilibrium is reactant-favored; a large fraction of the weak acid molecules are un-ionized at equilibrium.

14-5b Base Ionization Constants

A general equation analogous to that for the donation of H^+ to water by acids can be written for the *acceptance* of an H^+ *from* water by a molecular base, B, to form its conjugate acid, BH^+.

$$B(aq) + H_2O(\ell) \rightleftharpoons BH^+(aq) + OH^-(aq) \qquad [14.2]$$

conjugate base $\qquad$ conjugate acid

If the base B were NH_3, then BH^+ would be NH_4^+. The corresponding equilibrium constant expression is

$$K_b = \frac{[\text{conjugate acid}][OH^-]}{[\text{conjugate base}]} = \frac{[BH^+][OH^-]}{[B]}$$

The equilibrium constant K_b is called the **base ionization constant**, a term that can be misleading. Notice from the chemical reaction in Equation 14.2 that the base does not ionize. Rather, K_b and its equilibrium constant expression refer to *the reaction in which a base forms its conjugate acid (a cation) by removing an* H^+ *ion from water (leaving an anion,* OH^-*)*.

Many weak bases are anions (← Sec. 14-1c). When the base is an anion, A^-, the general equation for its reaction with water is

$$\underset{\text{conjugate base}}{A^-(aq)} + H_2O(\ell) \rightleftharpoons \underset{\text{conjugate acid}}{HA(aq)} + OH^-(aq)$$

If the base A^- were CH_3COO^-, then HA would be CH_3COOH. For acetate ion:

$$CH_3COO^-(aq) + H_2O(\ell) \rightleftharpoons CH_3COOH(aq) + OH^-(aq)$$

For a base that is an anion, the **base ionization constant expression** is

$$K_b = \frac{[\text{conjugate acid}][OH^-]}{[\text{conjugate base}]} = \frac{[HA][OH^-]}{[A^-]}$$

The magnitude of the K_b value indicates the extent to which the base removes H^+ ions from water to produce OH^- ions. The larger the base ionization constant, K_b, is

- the stronger the base is.
- the more product-favored is the H^+ transfer reaction from water to the base.
- the greater the OH^- concentration produced by the transfer reaction.

For a strong base, the base ionization constant is greater than 1. For a weak base, the base ionization constant is less than 1, sometimes considerably less than 1, because at equilibrium there is a significant concentration of unreacted weak conjugate base and a much smaller concentration of its conjugate acid and OH^- ions in solution.

PROBLEM-SOLVING EXAMPLE 14.7

Base Ionization

For each of these weak bases, write the equation for the reaction of the base with water and the companion K_b expression.

(a) C_5H_5N (b) NH_2OH (c) F^-

Result

(a) $C_5H_5N(aq) + H_2O(\ell) \rightleftharpoons C_5H_5NH^+(aq) + OH^-(aq)$ $K_b = \frac{[C_5H_5NH^+][OH^-]}{[C_5H_5N]}$

(b) $NH_2OH(aq) + H_2O(\ell) \rightleftharpoons NH_3OH^+(aq) + OH^-(aq)$ $K_b = \frac{[NH_3OH^+][OH^-]}{[NH_2OH]}$

(c) $F^-(aq) + H_2O(\ell) \rightleftharpoons HF(aq) + OH^-(aq)$ $K_b = \frac{[HF][OH^-]}{[F^-]}$

Analyze The general reaction is the same for each of the bases: The base removes an H^+ from water to form the corresponding conjugate acid.

Plan In parts (a) and (b), the base is a neutral molecule to which an H^+ is added, forming a positively charged conjugate acid. In part (c), the H^+ adds to a negatively charged ion, F^-, resulting in a conjugate acid with no net charge.

Execute See Result.

PROBLEM-SOLVING PRACTICE 14.7

Write the ionization equation and the K_b expression for each weak base.
(a) CH_3NH_2 (b) Phosphine, PH_3 (c) NO_2^-

14-5c Values of Acid and Base Ionization Constants

Table 14.2 summarizes the ionization constants for a number of acids and their conjugate bases. The ionization constants for strong acids (those above H_3O^+ in Table 14.2) and

Table 14.2 Ionization Constants for Some Acids and Their Conjugate Bases at 25 °C*

(Increasing Acid Strength: upward; Increasing Base Strength: downward)

Acid Name	Acid	$K_a = \frac{[H_3O^+][\text{conj base}]}{[\text{conj acid}]}$	Base Name	Base	$K_b = \frac{[\text{conj acid}][OH^-]}{[\text{conj base}]}$
Perchloric acid	$HClO_4$	Large	Perchlorate ion	ClO_4^-	Very small
Sulfuric acid	H_2SO_4	Large	Hydrogen sulfate ion	HSO_4^-	Very small
Hydrochloric acid	HCl	Large	Chloride ion	Cl^-	Very small
Nitric acid	HNO_3	≅20	Nitrate ion	NO_3^-	$\cong 5 \times 10^{-16}$
Hydronium ion	H_3O^+	1.0	Water	H_2O	1.0×10^{-14}
Sulfurous acid	H_2SO_3	1.7×10^{-2}	Hydrogen sulfite ion	HSO_3^-	5.9×10^{-13}
Hydrogen sulfate ion	HSO_4^-	1.1×10^{-2}	Sulfate ion	SO_4^{2-}	9.1×10^{-13}
Phosphoric acid	H_3PO_4	7.2×10^{-3}	Dihydrogen phosphate ion	$H_2PO_4^-$	1.4×10^{-12}
Hexaaquairon(III) ion	$Fe(H_2O)_6^{3+}$	6.76×10^{-3}	Pentaaquahydroxoiron(III) ion	$Fe(H_2O)_5OH^{2+}$	1.48×10^{-12}
Nitrous acid	HNO_2	7.1×10^{-4}	Nitrite ion	NO_2^-	1.4×10^{-11}
Hydrofluoric acid	HF	6.8×10^{-4}	Fluoride ion	F^-	1.5×10^{-11}
Formic acid	HCOOH	3.0×10^{-4}	Formate ion	$HCOO^-$	3.3×10^{-11}
Benzoic acid	C_6H_5COOH	1.2×10^{-4}	Benzoate ion	$C_6H_5COO^-$	8.3×10^{-11}
Acetic acid	CH_3COOH	1.8×10^{-5}	Acetate ion	CH_3COO^-	5.6×10^{-10}
Propanoic acid	CH_3CH_2COOH	1.4×10^{-5}	Propanoate ion	$CH_3CH_2COO^-$	7.1×10^{-10}
Hexaaquaaluminum ion	$Al(H_2O)_6^{3+}$	1.0×10^{-5}	Pentaaquahydroxoaluminum ion	$Al(H_2O)_5OH^{2+}$	1.0×10^{-9}
Carbonic acid	H_2CO_3	4.3×10^{-7}	Hydrogen carbonate ion	HCO_3^-	2.3×10^{-8}
Hydrosulfuric acid	H_2S	1×10^{-7}	Hydrogen sulfide ion	HS^-	1×10^{-7}
Hypochlorous acid	HClO	6.8×10^{-8}	Hypochlorite ion	ClO^-	1.5×10^{-7}
Dihydrogen phosphate ion	$H_2PO_4^-$	6.3×10^{-8}	Hydrogen phosphate ion	HPO_4^{2-}	1.6×10^{-7}
Hydrogen sulfite ion	HSO_3^-	6.3×10^{-8}	Sulfite ion	SO_3^{2-}	1.6×10^{-7}
Boric acid	$B(OH)_3(H_2O)$	5.8×10^{-10}	Tetrahydroxoborate ion	$B(OH)_4^-$	1.7×10^{-5}
Ammonium ion	NH_4^+	5.6×10^{-10}	Ammonia	NH_3	1.8×10^{-5}
Hydrocyanic acid	HCN	3.3×10^{-10}	Cyanide ion	CN^-	3.0×10^{-5}
Hydrogen carbonate ion	HCO_3^-	4.7×10^{-11}	Carbonate ion	CO_3^{2-}	2.1×10^{-4}
Hydrogen phosphate ion	HPO_4^{2-}	4.6×10^{-13}	Phosphate ion	PO_4^{3-}	2.2×10^{-2}
Water	H_2O	1.0×10^{-14}	Hydroxide ion	OH^-	1.0
Hydrogen sulfide ion	HS^-	1×10^{-19}	Sulfide ion	S^{2-}	1×10^5
Ethanol	C_2H_5OH	Very small	Ethoxide ion	$C_2H_5O^-$	Large
Ammonia	NH_3	Very small	Amide ion	NH_2^-	Large
Hydrogen	H_2	Very small	Hydride ion	H^-	Large
Methane	CH_4	Very small	Methide ion	CH_3^-	Large

*Taken from Högfeldt, E., Perrin, D. D. *Stability Constants of Metal-Ion Complexes,* 1st ed. Oxford; New Pergamon, 1979–1982. International Union of Pure and Applied Chemistry, Commission on Equilibrium.

strong bases (those below OH^- in Table 14.2) are too large to be measured easily. Fortunately, because their ionization reactions are virtually complete, these K_a and K_b values are hardly ever needed. For weak acids, K_a values show relative strengths quantitatively; for weak bases, K_b values do the same. Because the acids are listed in order of decreasing K_a values, Table 14.2 is analogous to Figure 14.2 (← **Sec. 14-1c**). The strongest acids are at the top and the strongest bases are at the bottom.

Consider acetic acid and boric acid. Boric acid is below acetic acid in Table 14.2, so boric acid must be a weaker acid than acetic acid; the K_a values tell us how much weaker. The K_a for boric acid is 5.8×10^{-10}; that for acetic acid is 1.8×10^{-5}, which shows that boric acid is somewhat more than 10^4 times weaker than acetic acid. In fact, boric acid is such a weak acid that a dilute solution of it can be used safely as an eyewash. Don't try that with acetic acid!

- **The smaller the K_a value is, the weaker the acid is.**
- **The smaller the K_b value is, the weaker the base is.**

CONCEPTUAL EXERCISE 14.11

Acid Strengths

The K_a of lactic acid is 1.5×10^{-4}; that of pyruvic acid is 3.2×10^{-3}.

(a) Which of these acids is the stronger acid?
(b) Which acid's ionization reaction is more reactant-favored?

14-5d K_a Values for Polyprotic Acids

So far we have concentrated on **monoprotic acids**, *acids that can donate a single H^+ per molecule*, such as hydrogen fluoride, HF, hydrogen chloride, HCl, and nitric acid, HNO_3.

Some acids, called **polyprotic acids**, *can donate more than one H^+ per molecule*. These include sulfuric acid, H_2SO_4; carbonic acid, H_2CO_3; and phosphoric acid, H_3PO_4. Oxalic acid, $H_2C_2O_4$ or HOOC—COOH, and other organic acids with two or more carboxylic acid, —COOH, groups are also polyprotic acids (Table 14.3).

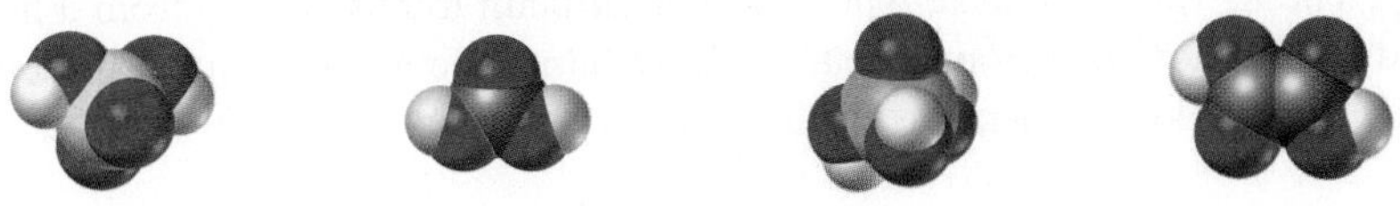

sulfuric acid carbonic acid phosphoric acid oxalic acid

In aqueous solution, a polyprotic acid donates its H^+ ions to water molecules in a stepwise manner. In the first step for sulfuric acid, hydrogen sulfate ion, HSO_4^-, is formed. Sulfuric acid is a strong acid, so this first ionization is essentially complete.

$$\text{First Ionization: } \underset{\text{ACID}}{H_2SO_4(aq)} + \underset{\text{base}}{H_2O(\ell)} \longrightarrow \underset{\text{acid}}{H_3O^+(aq)} + \underset{\text{BASE}}{HSO_4^-(aq)}$$

Hydrogen sulfate ion is the conjugate base of sulfuric acid.

Table 14.3 Polyprotic Acids

Acid Form	Conjugate Base Form
H_2S, hydrosulfuric acid	HS^-, hydrogen sulfide or bisulfide ion
H_3PO_4, phosphoric acid	$H_2PO_4^-$, dihydrogen phosphate ion
$H_2PO_4^-$, dihydrogen phosphate ion	HPO_4^{2-} monohydrogen phosphate ion
H_2CO_3, carbonic acid	HCO_3^- hydrogen carbonate or bicarbonate ion
$H_2C_2O_4$, oxalic acid	$HC_2O_4^-$, hydrogen oxalate ion
$C_3H_5O(COOH)_3$, citric acid	$C_3H_5O(COOH)_2COO^-$, hydrogen citrate ion

Many chemical reactions occur in steps that can be represented by individual equations. Sometimes only the overall equation is written.

In the next ionization step, hydrogen sulfate ion acting as an acid donates an H^+ ion to another water molecule. Hydrogen sulfate ion is a weak acid ($K_a < 1$) and, as with other weak acids, an equilibrium is established. Sulfate ion, SO_4^{2-} is the conjugate base of HSO_4^-, its conjugate acid.

When all acidic protons have been donated by a polyprotic acid, the result is a polyprotic base, such as SO_4^{2-}. A polyprotic base can accept more than one H^+ per molecule of base.

$$\text{Second Ionization: } \underset{\text{ACID}}{HSO_4^-(aq)} + \underset{\text{base}}{H_2O(\ell)} \rightleftharpoons \underset{\text{acid}}{H_3O^+(aq)} + \underset{\text{BASE}}{SO_4^{2-}(aq)}$$

CONCEPTUAL EXERCISE 14.12

Explaining Acid Strengths

Based on the charge on the hydrogen sulfate ion, explain why this ion is a weaker acid than H_2SO_4.

The weak acid H_3PO_4 can donate three H^+ ions per molecule through these three ionization reactions.

First ionization

$$H_3PO_4(aq) + H_2O(\ell) \rightleftharpoons H_3O^+(aq) + H_2PO_4^-(aq) \qquad K_a = 7.2 \times 10^{-3}$$

Second ionization

$$H_2PO_4^-(aq) + H_2O(\ell) \rightleftharpoons H_3O^+(aq) + HPO_4^{2-}(aq) \qquad K_a = 6.3 \times 10^{-8}$$

Third ionization

$$HPO_4^{2-}(aq) + H_2O(\ell) \rightleftharpoons H_3O^+(aq) + PO_4^{3-}(aq) \qquad K_a = 4.6 \times 10^{-13}$$

Both $H_2PO_4^-$ ion and HPO_4^{2-} ion are amphiprotic because they can gain or lose a proton.

The successive K_a values for the ionization of a polyprotic acid *decrease* by a factor of 10^4 to 10^5, indicating that each ionization step occurs to a *lesser* extent than the one before it. The $H_2PO_4^-$ ion ($K_a = 6.3 \times 10^{-8}$) is a much weaker acid than phosphoric acid ($K_a = 7.2 \times 10^{-3}$), and the HPO_4^{2-} ion ($K_a = 4.6 \times 10^{-13}$) is an even weaker acid than $H_2PO_4^-$. The K_a values indicate that it is more difficult to remove H^+ from a negatively charged $H_2PO_4^-$ ion than from a neutral H_3PO_4 molecule and even more difficult to remove H^+ from a doubly negative HPO_4^{2-} ion.

EXERCISE 14.13

Polyprotic Acids

Write equations for the stepwise ionization in aqueous solution of (a) oxalic acid and (b) citric acid. (Formulas for these acids are given in Table 14.3.)

14-6 Molecular Structure and Acid Strength

If all acids donate H^+ ions, why are some acids strong while others are weak? Why is there such a broad range of K_a values? To answer these questions we turn to the relationship of an acid's strength to its molecular structure. In doing so, we will consider a wide range of acids, from simple binary ones like HBr to structurally more complex ones containing oxygen, carbon, and other elements.

14-6a Factors Affecting Acid Strength

All acids have their acidic hydrogen bonded to some other atom, call it A, which can be bonded to other atoms as well. For the acid to transfer its hydrogen as an H^+ to water, the H—A bond must be broken and both electrons from the bond must remain with the A atom. That kind of bond breaking is more likely if the H—A bond is weak and polar. A strong H—A bond with very little polarity, such as the H—C bond in methane, CH_4, makes the hydrogens nonacidic, so methane is a very weak acid and is not a significant H^+ donor to water.

We consider first *binary acids,* such as HBr, acids that contain just hydrogen and one other element. In HBr, A is bromine. The H—Br bond is relatively weak and polar; HBr is a strong acid ($K_a \cong 10^8$). For acids with A atoms from the same group in the periodic table, for example HF, HCl, HBr, and HI, the H—A bond energies determine the relative acid strengths for binary acids. *As H—A bond energies decrease down a group, the H—A bond weakens, and binary acid strengths increase.*

$\overset{\delta^+ \quad \delta^-}{\longrightarrow}$
H—A

$\overset{\delta^+ \quad \delta^-}{\longrightarrow}$
H—Br

For a series of binary acids in which the A atoms are in the same period, H—A bond energies do not vary much, and therefore the H—A bond polarity is the principal factor affecting acid strengths. As the electronegativity of A increases across a period, the H—A bond polarity also gets larger due to greater electronegativity differences. Correspondingly, the hydrogen becomes more acidic, as seen for Period 3 nonmetals:

- The H—Si bond is relatively nonpolar and SiH_4 is nonacidic.
- The H—P bond is only slightly polar and the unshared electron pair on phosphorus makes PH_3 accept H^+ ions rather than donate them.
- The H—S bond is slightly polar and H_2S is a weak acid.
- The H—Cl bond is very polar and HCl is a strong acid.

14-6b Strengths of Oxoacids

Acids in which the acidic hydrogen is bonded directly to oxygen in an H—O— bond are called **oxoacids**. Three of the strong acids—nitric, HNO_3; perchloric, $HClO_4$; and sulfuric, H_2SO_4—are oxoacids.

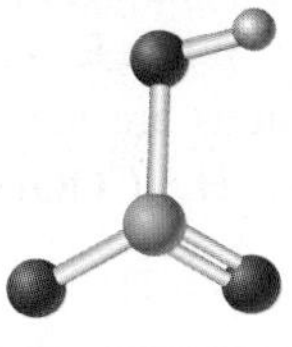
nitric acid

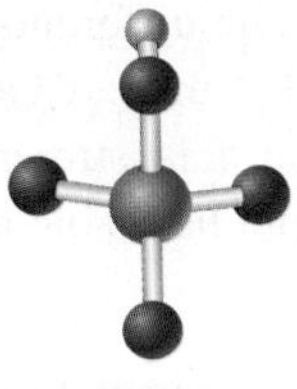
perchloric acid

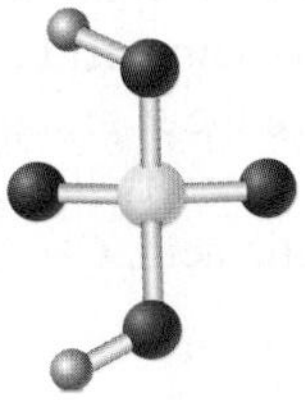
sulfuric acid

Like other oxoacids, they have at least one hydrogen atom bonded to an oxygen and have the general formula

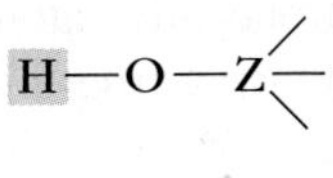

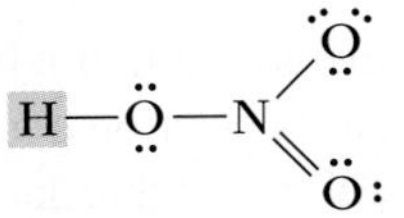
In nitric acid, HNO_3, Z is nitrogen.

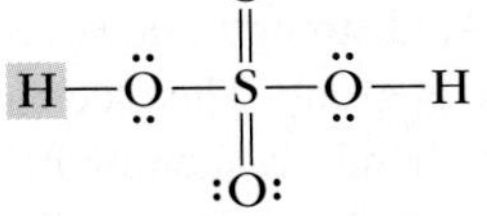
In sulfuric acid, H_2SO_4, Z is sulfur.

The nature of Z and other atoms that may be attached to it are important in determining the strength and polarity of the H—O bond and thus the strength of an oxoacid. In general, *acid strength decreases with the decreasing electronegativity of Z.* This is reflected in the differences among the K_a values of HOCl, HOBr, and HOI, as the electronegativity of the halogen decreases from chlorine (2.7) to bromine (2.6) to iodine (2.1).

Acid:	HOCl hypochlorous acid	HOBr hypobromous acid	HOI hypoiodous acid
	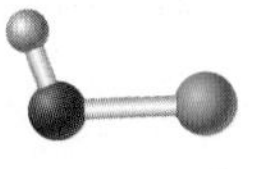	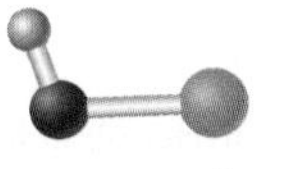	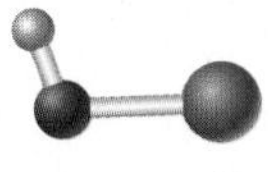
K_a:	6.8×10^{-8}	2.5×10^{-9}	2.3×10^{-11}

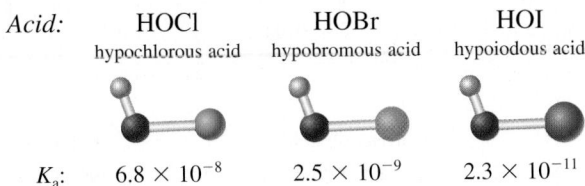

The number of oxygen atoms attached to Z also significantly affects the strength of the H—O bond and oxoacid strength: ***The K_a value increases by approximately 10^5 for each additional oxygen atom attached to Z.*** The terminal oxygen atoms (those not in an H—O bond) are sufficiently electronegative, along with Z, to withdraw electron density from the H—O bond. This weakens the bond and makes it more polar, promoting the transfer of an H^+ ion to water. The more terminal oxygen atoms present, the greater the electron density shift and the greater the acid strength. A particularly striking example of this trend is seen with the oxoacids of chlorine from the weakest, hypochlorous acid, HOCl, to the strongest, perchloric acid, $HClO_4$.

The chemical formula for perchloric acid can be considered as $HOClO_3$ as seen from its structural formula.

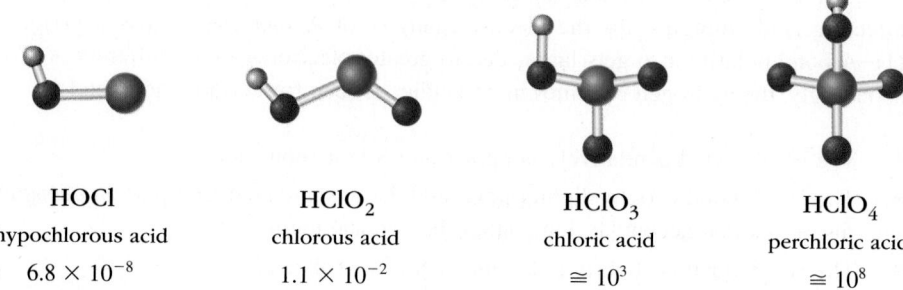

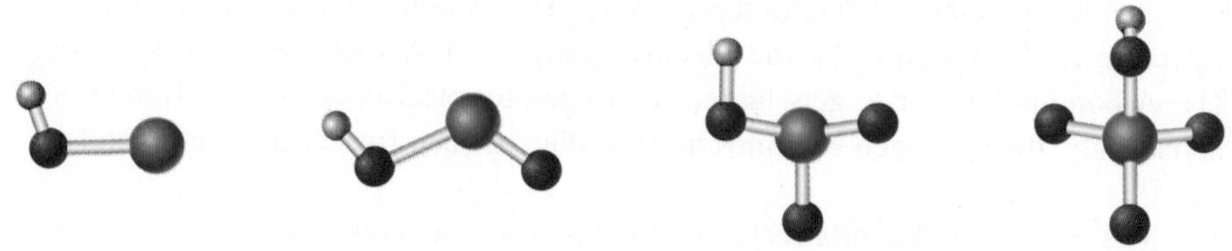

	HOCl	$HClO_2$	$HClO_3$	$HClO_4$
	hypochlorous acid	chlorous acid	chloric acid	perchloric acid
K_a:	6.8×10^{-8}	1.1×10^{-2}	$\cong 10^3$	$\cong 10^8$

To be a strong acid, an inorganic oxoacid must have *at least* two more oxygen atoms than acidic hydrogen atoms in the molecule. Thus, sulfuric acid is a strong acid. In contrast, phosphoric acid, H_3PO_4, has only four oxygen atoms for three hydrogen atoms and is a weak acid.

14-6c Strengths of Carboxylic Acids

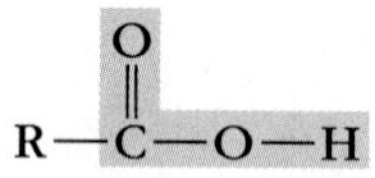

Carboxylic acid group

All carboxylic acids contain the carboxylic acid group, —COOH, although all such acids do not have the same acid strength. The differences in carboxylic acid strength are due to differences in the R group attached to the —COOH group. When R is simply a hydrocarbon group, there is little effect on acid strength; K_a values are similar, as seen by comparing acetic acid, CH_3COOH, and hexanoic acid, $CH_3(CH_2)_4COOH$.

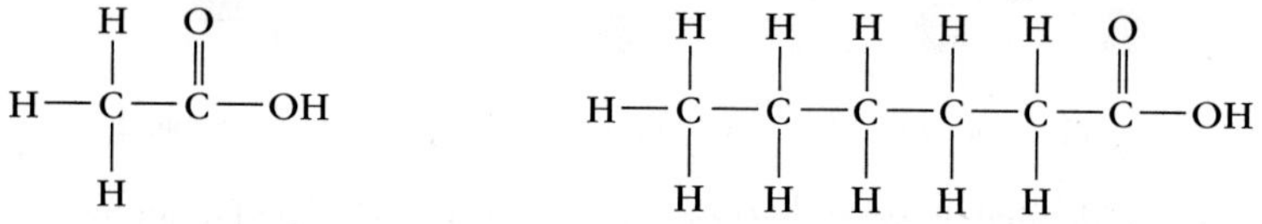

acetic acid
$K_a = 1.8 \times 10^{-5}$

hexanoic acid
$K_a = 1.4 \times 10^{-5}$

Acid strength, however, is affected by the addition of a highly electronegative atom to the R group. The electronegative atom causes electron density to shift away from the O—H bond, weakening it, making it more polar, and thus increasing the acid's strength. For example, replacing a hydrogen in the CH_3 group of acetic acid with chlorine, a more electronegative atom, forms chloroacetic acid, $ClCH_2COOH$. The K_a for this acid is 1.4×10^{-3}, nearly 100 times larger than that of acetic acid. Replacing all three hydrogens of the CH_3 group in acetic acid with chlorine atoms converts acetic acid to trichloroacetic acid, Cl_3CCOOH, which is about 10,000 times stronger than acetic acid and 100 times stronger than chloroacetic acid.

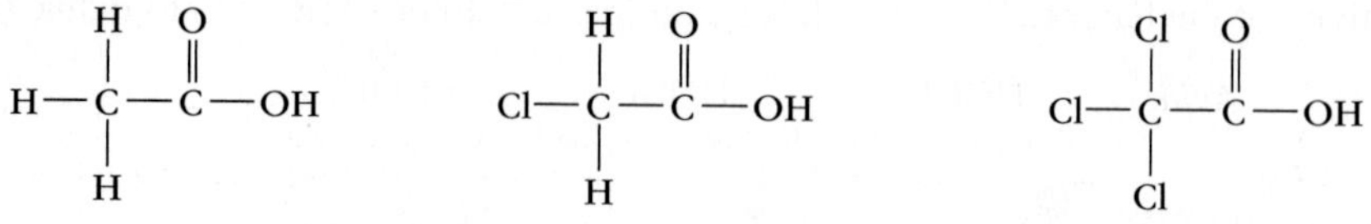

acetic acid
$K_a = 1.8 \times 10^{-5}$

chloroacetic acid
$K_a = 1.4 \times 10^{-3}$

trichloroacetic acid
$K_a = 2 \times 10^{-1}$

CONCEPTUAL EXERCISE 14.14

Molecular Structure and Acid Strength

Which has the larger K_a,

(a) fluorobenzoic acid, C_6H_4FCOOH, or benzoic acid, C_6H_5COOH?

(b) chloroacetic acid or bromoacetic acid, $BrCH_2COOH$?

Explain your prediction in each case.

CONCEPTUAL EXERCISE 14.15

Acid Strength and Molecular Structure

Consider acetic acid, CH_3COOH, and oxalic acid, HOOC—COOH. Which is the stronger acid? (Consider only the first ionization.) Explain your prediction.

14-6d Amino Acids and Zwitterions

Amino acids are the monomers from which proteins are assembled (← **Sec. 10-7d**). The general structural formula for amino acids is

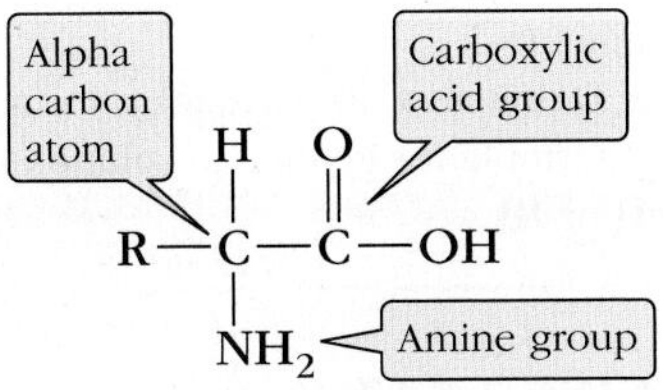
Alpha carbon atom
Carboxylic acid group
Amine group

The amino acids are called alpha (α) amino acids because the amine group is on the alpha carbon atom—the one next to the —COOH group.

Unlike acetic acid, CH_3COOH, and ethylamine, $CH_3CH_2NH_2$, which are liquids at 25 °C, amino acids are crystalline solids that usually melt above 200 °C. High melting points are characteristic of salts rather than simple organic compounds; the general structural formula for amino acids does not predict such properties.

An **amino acid** *has at least two functional groups: at least one acidic carboxylic acid group* and *at least one basic amine group*. From Table 14.2, K_a values for typical carboxylic acids, such as benzoic, acetic, and propanoic acids are about 10^{-4} to 10^{-5}. The K_a for methyl ammonium ion, $CH_3NH_3^+$, which is analogous to the conjugate acid of the amine group in an amino acid, is about 10^{-11}. That is, the carboxylic acid is a much stronger acid than the conjugate acid of the amine. Therefore, the carboxylic acid should react with the amine to produce its conjugate acid. An intramolecular Brønsted-Lowry acid-base reaction occurs in amino acids in which H^+ is transferred from the carboxylic acid group to the amine group. This resulting *structure contains both a negative charge* ($—COO^-$) *and a positive charge* ($—NH_3^+$) and is called a **zwitterion**. Because of the positive and negative charges, a zwitterion is highly polar and a salt-like substance. In the case of alanine the change is

The term "zwitterion" is derived from the German word *zwitter*, meaning "a hybrid."

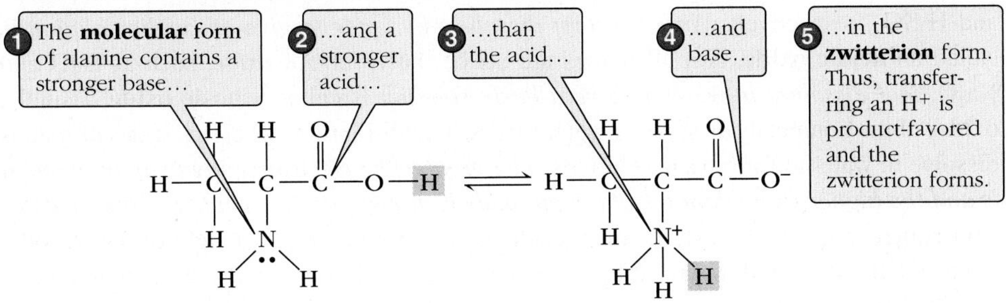
1 The **molecular** form of alanine contains a stronger base...
2 ...and a stronger acid...
3 ...than the acid...
4 ...and base...
5 ...in the **zwitterion** form. Thus, transferring an H+ is product-favored and the zwitterion forms.

The hydrogen ion is transferred from the carboxylic acid group to the lone pair on the nitrogen atom of the amine group.

The alanine zwitterion, for example, has no net charge, although regions of positive ($—NH_3^+$) and negative charge ($—COO^-$) exist in the molecule.

Alanine and other amino acids in solution can undergo additional acid-base reactions depending on the pH of the solution. Under acidic conditions, the H^+ concentration is large enough to add an H^+ ion to the negative carboxylate group, $—COO^-$, resulting in the formation of a —COOH group. Under these conditions, the amino acid has a net positive charge. In basic solution, the concentration of OH^- ions is sufficient to remove the H^+ attached to the nitrogen of the zwitterion, resulting in a net negative charge on alanine.

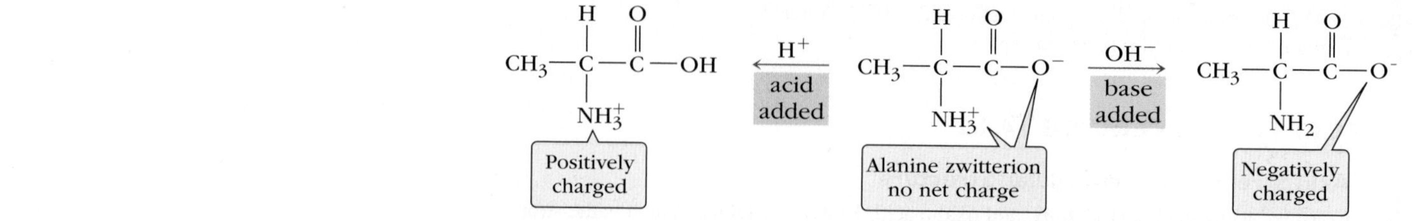

H⁺
acid added
OH⁻
base added
Positively charged
Alanine zwitterion no net charge
Negatively charged

The relative concentrations of the three alanine forms, or the analogous forms of any amino acid, depend on the pH of the body fluids or other aqueous solutions in which they are present.

CONCEPTUAL EXERCISE 14.16

Glycine and Its Zwitterion

Glycine is the simplest amino acid.

(a) Write the structural formula for its zwitterion form.

(b) Write the structural formula for glycine in solution at a pH of 2 and at a pH of 10.

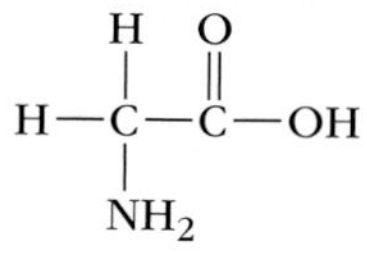

Glycine

CONCEPTUAL EXERCISE 14.17

H^+ Transfers

Consider this tripeptide, made from three amino acids. Draw a Lewis structure for the tripeptide at pH = 3, at pH = 7, and at pH = 11.

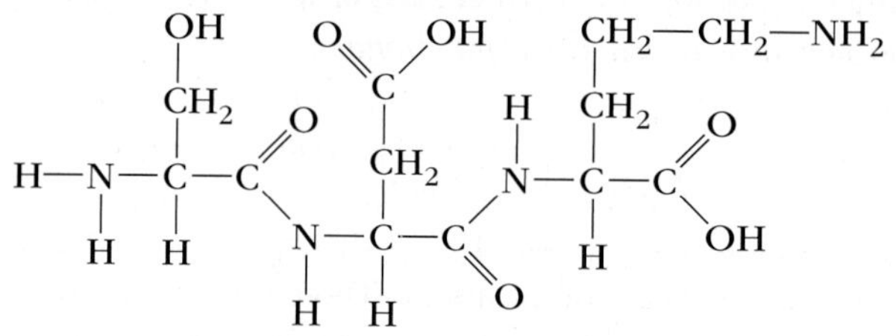

14-6e Acidic and Basic Oxides

In the introduction to this chapter we noted Antoine Lavoisier's hypothesis that all acids must contain oxygen. Such an idea is not surprising, because many oxides dissolve in water to form acids. Examples are CO_2 and SO_3, which react with water to form H_2CO_3 and H_2SO_4, respectively. *An oxide that dissolves in water to give an acidic solution* is called an **acidic oxide**. Not all oxides are acidic, however. *An oxide,* such as BaO and Na_2O, *that dissolves in water to give a basic solution* is called a **basic oxide**. Usually, oxides of the nonmetals dissolve in water to give acidic solutions whereas oxides of metals dissolve in water to form basic solutions. ***The greater the electronegativity of an element is and the higher its oxidation state in an oxide is, the greater the acidity of the oxide is.*** Consequently, elements with the most acidic oxides are in the upper right of the periodic table and elements with the most basic oxides are in the lower left of the periodic table.

CONCEPTUAL EXERCISE 14.18

Acidic and Basic Oxides

Rank these oxides from most acidic to most basic: SeO_2; CaO; SO_3; Al_2O_3; SO_2.

14-7 Problem Solving Using K_a and K_b

Calculations with K_a or K_b follow the same patterns as those of equilibrium calculations illustrated earlier (← **Sec. 12-3a**). Similar important relationships apply to these calculations.

- *Starting with only reactants, equilibrium can be achieved only if some amount of the reactants is converted to products; that is,* ***products are formed at the expense of reactants****.*
- *The chemical equilibrium equation for the ionization of the acid or the reaction of the base with water is the basis for the acid ionization or base ionization constant expression, respectively.*
- *The concentrations in the acid ionization or base ionization expression, expressed as molarity (mol/L), are those* ***at equilibrium****.*
- *The magnitude of the K_a or K_b value indicates how far the forward reaction occurs at equilibrium (K_a: H^+ donation* ***to*** *water by an acid; K_b: H^+ gain by a base* ***from*** *water).*

There are several experimental methods for determining acid or base ionization constants. The simplest is based on measuring the pH of an acid solution of known concentration. If both the acid concentration and the pH are known, the K_a for the acid can be calculated, as illustrated in Problem-Solving Example 14.8.

PROBLEM-SOLVING EXAMPLE 14.8

K_a from pH

Calculate the K_a of butanoic acid, $CH_3CH_2CH_2COOH$, a weak organic acid, if a 0.025-M butanoic acid solution has a pH of 3.21 at 25 °C.

Result $K_a = 1.6 \times 10^{-5}$

Analyze K_a of butanoic acid can be calculated if the equilibrium concentrations of all species are known. The K_a expression can be written based on the balanced chemical equation for the ionization of butanoic acid. The equilibrium concentration of H_3O^+ can be obtained from the pH. Butanoic acid ionizes to give equal concentrations of hydronium ions and butanoate ions. At equilibrium the concentration of un-ionized butanoic acid equals the original concentration minus the amount that dissociated.

Plan Use the definition of pH to calculate the equilibrium concentration H_3O^+. Set up a reaction (ICE) table to represent the initial and equilibrium concentrations of the species. Substitute the calculated equilibrium concentrations into the K_a expression to calculate the K_a value.

Execute The acid ionizes according to this balanced equation

$$CH_3CH_2CH_2COOH(aq) + H_2O(\ell) \rightleftharpoons H_3O^+(aq) + CH_3CH_2CH_2COO^-(aq)$$

The equilibrium concentration of H_3O^+ is $[H_3O^+] = 10^{-pH} = 10^{-3.21} = 0.00062$ M.

	$CH_3CH_2CH_2COOH$	H_3O^+	$CH_3CH_2COO^-$
Initial concentration (mol/L)	0.025	1.0×10^{-7} (from water)*	0
Change as reaction occurs (mol/L)	−0.00062	+0.00062	+0.00062
Equilibrium concentration (mol/L)	0.025 − 0.00062	0.00062	0.00062

*This concentration can be ignored because it is so small.

Substituting equilibrium concentrations into the K_a expression gives

$$K_a = \frac{[H_3O^+][CH_3CH_2CH_2COO^-]}{[CH_3CH_2CH_2COOH]}$$

$$= \frac{[0.00062][0.00062]}{0.0244} = 1.6 \times 10^{-5}$$

☑ **Reasonable Result Check** The fact that a 0.025-M butanoic acid solution has $[H_3O^+] = 0.00062$ M indicates that K_a is small and butanoic acid is a weak acid. Thus, the result makes sense; butanoic acid is only slightly ionized.

PROBLEM-SOLVING PRACTICE 14.8

Lactic acid is a monoprotic acid that occurs naturally in sour milk and also forms by metabolism in the human body. A 0.10-M aqueous solution of lactic acid, $CH_3CH(OH)COOH$, has a pH of 2.43. Calculate the value of K_a for lactic acid. Is lactic acid stronger or weaker than propanoic acid?

Acid-base ionization constants such as those in Table 14.2 can be used to calculate the pH of a solution of a weak acid or a weak base from its concentration.

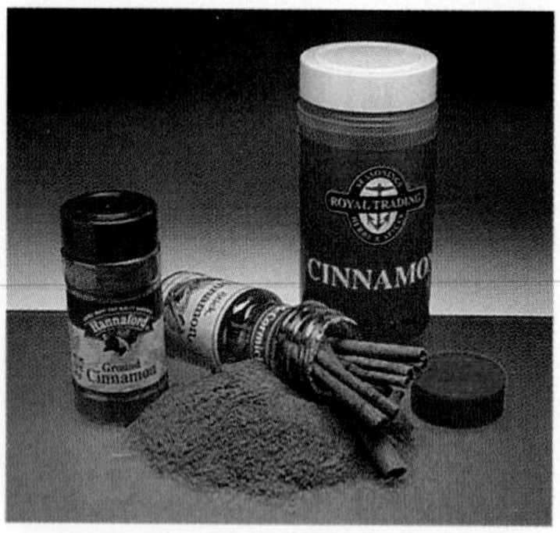
CINNAMO

Cinnamon.

PROBLEM-SOLVING EXAMPLE 14.9

pH from K_a

The flavor of cinnamon comes from cinnamaldehyde, which can be oxidized to *trans*-cinnamic acid. Calculate

(a) the pH of a 0.020-M solution of *trans*-cinnamic acid, $K_a = 3.6 \times 10^{-5}$ at 25 °C. (Abbreviate *trans*-cinnamic acid as HtCA.)

(b) the percent of the acid that has ionized in this solution.

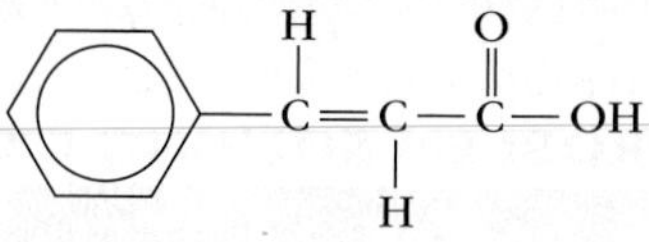

trans-cinnamic acid, $C_9H_8O_2$

Result (a) pH = 3.08 (b) 4.2% ionized

Analyze

(a) K_a and the initial concentration of the acid are given; pH is to be calculated. The equation for the ionization reaction can be used to obtain the K_a expression. A reaction (ICE) table can be used to relate equilibrium concentrations to initial concentrations. pH = $-\log[H_3O^+]$.

(b) Percent ionization is the fraction of acid that is ionized. The concentration of acid that is ionized equals the equilibrium concentration of H_3O^+.

Plan

(a) Write the ionization reaction and equilibrium constant expression. Organize the known information in an ICE table. Let x equal $[H_3O^+]$ because pH is to be calculated. At equilibrium, the tCA^- ion concentration is also equal to x because the reaction produces H_3O^+ and tCA^- in equal amounts.

(b) To calculate percent ionization, divide $[H_3O^+]$ at equilibrium by the initial acid concentration and multiply by 100%

Execute

(a) $HtCA(aq) + H_2O(\ell) \rightleftharpoons H_3O^+(aq) + tCA^-(aq)$ $\qquad K_a = \frac{[H_3O^+][tCA^-]}{[HtCA]}$

	HtCA	H_3O^+	tCA^-
*I*nitial concentration (mol/L)	0.020	1.0×10^{-7} (from water)*	0
*C*hange as reaction occurs (mol/L)	$-x$	$+x$	$+x$
*E*quilibrium concentration (mol/L)	$0.020 - x$	x	x

*This concentration can be ignored because it is so small.

Substituting equilibrium concentrations into the equilibrium constant expression gives

$$K_a = 3.6 \times 10^{-5} = \frac{[H_3O^+][tCA^-]}{[HtCA]} = \frac{(x)(x)}{0.020 - x}$$

Because K_a is very small, the reaction is reactant-favored and only a little product forms. Thus, the equilibrium concentrations of H_3O^+ and tCA^- are much smaller than the initial concentration of acid, and x must be quite small. Assume that when x is subtracted from 0.020, the result is almost exactly 0.020; that is, $0.020 - x \cong 0.020$. Then,

$$\frac{x^2}{0.020 - x} \cong \frac{x^2}{0.020} = 3.6 \times 10^{-5}$$

Solving for x gives

$$x = \sqrt{(3.6 \times 10^{-5})(0.020)} = \sqrt{7.2 \times 10^{-7}} = 8.5 \times 10^{-4} = [H_3O^+]$$

Now, we can check the assumption that x is small relative to 0.020:

$$0.020 - x = 0.020 - 8.5 \times 10^{-4} = 0.019$$

The value is not exactly equal to 0.020, but this calculation provides a more accurate value of $0.020 - x$ that can be substituted into the denominator of the equilibrium constant.

$$\frac{x^2}{0.020 - x} \cong \frac{x^2}{0.019} = 3.6 \times 10^{-5} \text{ and } x = \sqrt{0.019 \times 3.6 \times 10^{-5}} = 8.3 \times 10^{-4}$$

Finally, calculate the pH from this second, better value of x:

$$\text{pH} = -\log[H_3O^+] = -\log(8.3 \times 10^{-4}) = 3.08.$$

The solution is acidic (pH < 7.0).

(b)

$$\%\text{ ionization} = \frac{[H_3O^+]}{(HtCA)_{\text{initial}}} \times 100\% = \frac{8.3 \times 10^{-4}}{0.020} \times 100\% = 4.2\%$$

☑ **Reasonable Result Check** Its K_a value of 3.6×10^{-5} indicates that *trans*-cinnamic acid is a weak acid, similar to acetic acid ($K_a = 1.8 \times 10^{-5}$) in strength and should be only slightly ionized. That 0.020-M *trans*-cinnamic acid is only 4.2% ionized and has $[H_3O^+] = 8.3 \times 10^{-4}$ M and a pH = 3.08 is reasonable. Check the overall calculation by substituting calculated equilibrium values into K_a.

$$K_a = 3.6 \times 10^{-5} = \frac{[H_3O^+][tCA^-]}{[HtCA]} = \frac{(8.3 \times 10^{-4})(8.3 \times 10^{-4})}{0.020 - 8.3 \times 10^{-4}} = 3.6 \times 10^{-5}$$

The symbol "≅" means "is approximately equal to".

This method, calculating an approximate value of x and then using that value to calculate a more accurate value of x, is called the *method of successive approximations*. It is described in more detail in Appendix A-7.

PROBLEM-SOLVING PRACTICE 14.9

(a) Calculate the pH of a 0.015-M solution of hydrazoic acid, HN_3 ($K_a = 1.9 \times 10^{-5}$), at 25 °C.
(b) Calculate the percent ionization of the acid.

H—N̈=N=N̈:

hydrazoic acid

EXERCISE 14.19

The pH of Aqueous Nitrous Acid

Nitrous acid, HNO_2, has a $K_a = 7.1 \times 10^{-4}$ at 25 °C. Calculate the pH of a 0.495-M aqueous nitrous acid solution.

An analogous calculation can be done using K_b to find the pH of a solution of a weak base, such as piperidine, a component of pepper.

PROBLEM-SOLVING EXAMPLE 14.10

pH of a Weak Base from K_b

You wrote an equation for transfer of a proton to piperidine, $C_5H_{11}N$, in Exercise 14.6 (← **Sec. 14-2**). Piperidine, a nitrogen-containing base, has $K_b = 1.3 \times 10^{-3}$. Calculate the OH^- concentration and pH of a 0.025-M solution of piperidine.

Result $[OH^-] = 5.1 \times 10^{-3}$ M; pH = 11.71

Analyze K_b and the initial concentration of piperidine are given; the $[OH^-]$ and pH are to be determined. Piperidine is a base so H^+ ions transfer from water to piperidine, forming equal amounts of hydroxide ions and piperidinium ions, $C_5H_{11}NH^+$. The equation for the H^+ transfer reaction can be used to obtain the K_b expression. A reaction (ICE) table can be used to relate equilibrium concentrations to initial concentrations. $pOH = -\log[OH^-]$ and $pH = 14 - pOH$.

Piperidinium ion, $C_5H_{11}NH^+$, is the conjugate acid of piperidine; it is analogous to ammonium ion, NH_4^+.

Plan Write the balanced chemical equation and use it to write the K_b expression. Use the K_b expression, the K_b value, and an ICE table to calculate $[OH^-]$ and from it the pH. Let x be the change in concentration of OH^-. The equilibrium concentration of *unreacted* piperidine is its initial concentration, 0.025 mol/L, minus x.

Execute $$C_5H_{11}N(aq) + H_2O(\ell) \rightleftharpoons C_5H_{11}NH^+(aq) + OH^-(aq)$$

$$K_b = \frac{[C_5H_{11}NH^+][OH^-]}{[C_5H_{11}N]} = 1.3 \times 10^{-3}$$

	$C_5H_{11}N$	$C_5H_{11}NH^+$	OH^-
Initial concentration (mol/L)	0.025	0	1.0×10^{-7}*
Change as reaction occurs (mol/L)	$-x$	$+x$	$+x$
Equilibrium concentration (mol/L)	$(0.025 - x)$	x	x

*The low concentration can be ignored, as it was in the K_a calculations.

Substituting into the base ionization constant expression gives

$$K_b = \frac{[C_5H_{11}NH^+][OH^-]}{[C_5H_{11}N]} = \frac{x^2}{0.025 - x} = 1.3 \times 10^{-3} \quad [14.3]$$

Assume that piperidine is a weak enough base so that we can make the simplifying approximation that $0.025 - x \cong 0.025$. Then, calculate an approximate x.

$$x \cong \sqrt{(1.3 \times 10^{-3})(0.025)} = \sqrt{3.25 \times 10^{-5}} = 5.7 \times 10^{-3}$$

A general rule for K_a and K_b calculations is that, *if the calculated* x *value is greater than 5% of the initial concentration from which it is subtracted, the simplifying approximation is not accurate enough and should not be made.* In this case,

This guideline is called the *5% rule*.

$$\frac{x}{\text{initial concentration}} \times 100\% = \frac{5.7 \times 10^{-3}}{0.025} \times 100\% = 22.8\%$$

which is far greater than 5%, so the approximation is not justified. Under such circumstances, there are two ways to obtain a valid result: (1) carry out successive approximations, perhaps more than once; or (2) use the quadratic formula to obtain roots of Equation 14.3. We illustrate the latter method here.

Successive approximations gives these improving values of x:

$x_1 = \sqrt{0.025 \times 1.3 \times 10^{-3}} = 5.7 \times 10^{-3}$

$x_2 = \sqrt{0.0193 \times 1.3 \times 10^{-3}} = 5.0 \times 10^{-3}$

$x_3 = \sqrt{0.0200 \times 1.3 \times 10^{-3}} = 5.1 \times 10^{-3}$

Multiply both sides of the last equality in Equation 14.3 by $0.025 - x$ to give

$$x^2 = (0.025 - x)(1.3 \times 10^{-3}) = 3.3 \times 10^{-5} - (1.3 \times 10^{-3}x) \quad [14.4]$$

Rearrange Equation 14.4 into the quadratic form $ax^2 + bx + c = 0$ to give

$$x^2 + (1.3 \times 10^{-3}x) - (3.3 \times 10^{-5}) = 0 \quad [14.5]$$

If your calculator can obtain roots of a quadratic equation, then use your calculator to solve Equation 14.5 for x; if not, then use the quadratic formula.

Appendix A-7 reviews the use of the quadratic formula.

$$x = \frac{-(1.3 \times 10^{-3}) \pm \sqrt{(1.3 \times 10^{-3})^2 - (4 \times 1)(-3.3 \times 10^{-5})}}{2(1)}$$

$$= \frac{-(1.3 \times 10^{-3}) \pm \sqrt{1.34 \times 10^{-4}}}{2} = \frac{-(1.3 \times 10^{-3}) \pm (1.16 \times 10^{-2})}{2}$$

$$x = \frac{1.03 \times 10^{-2}}{2} = 5.1 \times 10^{-3} \quad \text{or} \quad x = \frac{-1.29 \times 10^{-2}}{2} = -6.4 \times 10^{-3}$$

Notice that the third, best value of x obtained from successive approximations was 5.1×10^{-3}, the same value obtained from the quadratic formula.

Therefore x, the OH^- concentration, equals 5.1×10^{-3} molar. (The negative root in the solution of the quadratic equation is disregarded because concentration cannot be negative; you can't have less than 0 mol/L of a substance.)

Finally, calculate the pOH and the pH.

$$\text{pOH} = -\log(5.1 \times 10^{-3}) = 2.29$$

$$\text{pH} = 14.00 - \text{pOH} = 14.00 - 2.29 = 11.71$$

Piperidine reacts sufficiently with water to generate a fairly basic solution.

☑ **Reasonable Result Check** Piperidine is not a strong base, so the pH should be less than that of a 0.025-M solution of a strong base like NaOH, which is 12.40.

$$[OH^-] = 0.025\text{ M}; \quad \text{pOH} = -\log(0.025) = 1.60; \quad \text{pH} = 12.40$$

A pH of 11.71 for 0.025-M piperidine is reasonable. Also, substitute the equilibrium concentrations into K_b and verify that 1.3×10^{-3} is calculated:

$$K_b = \frac{[C_5H_{11}NH^+][OH^-]}{[C_5H_{11}N]} = \frac{(5.1 \times 10^{-3})(5.1 \times 10^{-3})}{0.025 - 5.1 \times 10^{-3}} = 1.3 \times 10^{-3}$$

PROBLEM-SOLVING PRACTICE 14.10

Calculate the OH^- concentration and the pH of a 0.015-M solution of cyclohexylamine, $C_6H_{11}NH_2$. $K_b = 4.6 \times 10^{-4}$.

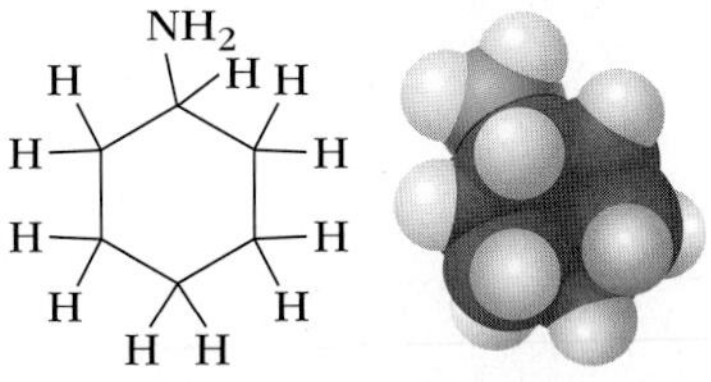
NH2
H
H H
H
H
H
H
H
H
H H

cyclohexylamine

14-7a Relationship between K_a and K_b Values

The right-hand side of Table 14.2 (← **Sec. 14-5c**) gives K_b values for the conjugate base of each acid. Try an experiment with these data: Multiply a few of the K_a values by K_b values for their conjugate bases. What do you find? Within a very small error you ought to find that $K_a \times K_b = 1.0 \times 10^{-14}$. This value is the same as K_w, the autoionization constant for water. To see why, multiply the equilibrium constant expressions for K_a and K_b.

$$K_a \times K_b = \left(\frac{[H_3O^+][A^-]}{[HA]}\right)\left(\frac{[HA][OH^-]}{[A^-]}\right)$$

Canceling like terms in the numerator and denominator of this expression gives

$$K_a \times K_b = \left(\frac{[H_3O^+]\cancel{[A^-]}}{\cancel{[HA]}}\right)\left(\frac{\cancel{[HA]}[OH^-]}{\cancel{[A^-]}}\right) = [H_3O^+][OH^-] = K_w$$

$K_a \times K_b = K_w$ is an inverse relationship; as K_a increases (stronger conjugate acid), K_b decreases (weaker conjugate base). This is noted in Table 14.2.

Taking the negative base-10 logarithm of both sides of this equation gives a second useful relation,

$$pK_a + pK_b = pK_w$$

At 25 °C, $K_w = 1.0 \times 10^{-14}$ and $pK_w = 14$, so $pK_a + pK_b = 14$. These two relations show that if you know K_a for an acid, you can find K_b for its conjugate base by using K_w. Furthermore, the larger the K_a, the smaller the K_b, and vice versa (because when multiplied they always have to give the same product, K_w). For example, K_a for HCN is 3.3×10^{-10}. The value of K_b for the conjugate base, CN^-, is

$$K_b \text{ (for CN}^-) = \frac{K_w}{K_a \text{ (for HCN)}} = \frac{1.0 \times 10^{-14}}{3.3 \times 10^{-10}} = 3.0 \times 10^{-5}$$

HCN has a relatively small K_a and lies fairly far down in Table 14.2, which means it is a relatively weak weak acid. However, CN^- is a fairly strong weak base; its K_b of 3.0×10^{-5} is nearly the same as the K_b for ammonia (1.8×10^{-5}), making CN^- a slightly stronger base than ammonia. *In general, if* $K_a > K_b$, *the acid is stronger than its conjugate base.* Alternatively, *if* $K_b > K_a$, *the conjugate base is stronger than its conjugate acid.* For example, hypochlorite ion, OCl^- ($K_b = 1.5 \times 10^{-7}$) is a stronger base than hypochlorous acid, HOCl, is an acid ($K_a = 6.8 \times 10^{-8}$).

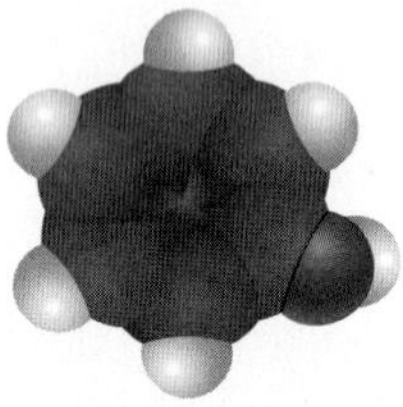

phenol, C_6H_5OH

EXERCISE 14.20

K_b from K_a

Phenol, or carbolic acid, C_6H_5OH, is a weak acid, $K_a = 1.3 \times 10^{-10}$. Calculate K_b for the phenolate ion, $C_6H_5O^-$. Which base in Table 14.2 (← **Sec. 14-5c**) is closest in strength to the phenolate ion? Explain how you made your choice.

14-8 Acid-Base Reactions of Salts

An exchange reaction between an acid and a base produces a salt plus water (← **Sec. 3-4c**). The salt's positive ion comes from the base and its negative ion comes from the acid. In the case of a metal hydroxide as a base, the salt-forming general reaction is

$$\underset{\text{acid}}{HX(aq)} + \underset{\text{base}}{MOH(aq)} \longrightarrow \underset{\text{salt}}{MX(aq)} + HOH(\ell)$$

Now that you know more about the Brønsted-Lowry acid-base definition and the strengths of acids and bases, it is useful to consider acid-base reactions and salt formation in more detail.

14-8a Salts of Strong Bases and Strong Acids

Strong acids react with strong bases to form *neutral* salts. Consider the reaction of the strong acid HCl with the strong base NaOH to form the salt NaCl. If the amounts of HCl and NaOH are in the correct stoichiometric ratio (1 mol HCl per 1 mol NaOH), this reaction occurs with the complete neutralization of the acidic properties of HCl and the basic properties of NaOH. The reaction can be described by an overall equation, a complete ionic equation, and a net ionic equation (← **Sec. 3-3e**). Each equation contains useful information.

$$HCl(aq) + NaOH(aq) \longrightarrow NaCl(aq) + H_2O(\ell) \quad \text{Overall equation}$$

$$H_3O^+(aq) + Cl^-(aq) + Na^+(aq) + OH^-(aq) \longrightarrow Na^+(aq) + Cl^-(aq) + 2\,H_2O(\ell) \quad \text{Complete ionic equation}$$

$$\underset{\text{ACID}}{H_3O^+(aq)} + \underset{\text{base}}{OH^-(aq)} \longrightarrow \underset{\text{BASE}}{H_2O(\ell)} + \underset{\text{acid}}{H_2O(\ell)} \quad \text{Net ionic equation}$$

- The overall equation shows the substances that were dissolved in the solutions that reacted or that could be recovered at the end of the reaction.
- The complete ionic equation indicates all of the ions present before and after the reaction.

- The net ionic equation emphasizes that a Brønsted-Lowry acid, H_3O^+, is reacting with a Brønsted-Lowry base, OH^-; the spectator ions, Na^+ and Cl^-, are omitted from the net ionic equation.

This reaction goes to completion because H_3O^+ is a strong acid, OH^- is a strong base, and water is a very weak acid and a very weak base.

The resulting solution contains only sodium ions and chloride ions, with a few more water molecules than before. The properties of the solution are the same as if it had been prepared by simply dissolving some NaCl(s) in water. The solution has a neutral pH because it contains no significant concentrations of acids or bases. The Cl^- ion is the conjugate base of a strong acid and hence is such a weak base that it does not react with water. The Na^+ ion also does not react as either an acid or a base with water. Examples of some other salts of this type are given in Table 14.4. These salts all form *neutral* solutions.

14-8b Salts of Strong Bases and Weak Acids

Strong bases react with weak acids to form *basic* salts. Suppose, for example, that 0.010 mol NaOH is added to 0.010 mol of the weak acid acetic acid in 1 L of solution. The three equations are

Overall equation $$CH_3COOH(aq) + NaOH(aq) \longrightarrow NaCH_3COO(aq) + H_2O(\ell)$$

Complete ionic equation $$CH_3COOH(aq) + Na^+(aq) + OH^-(aq) \longrightarrow CH_3COO^-(aq) + Na^+(aq) + H_2O(\ell)$$

Net ionic equation $$CH_3COOH(aq) + OH^-(aq) \longrightarrow CH_3COO^-(aq) + H_2O(\ell)$$

weak acid | strong base | base | acid

In this case, acetate ion, a weaker base than OH^-, has been formed by the reaction. Therefore, the solution is slightly basic (pH > 7), even though exactly the stoichiometric amount of acetic acid was added to the sodium hydroxide. The reaction that makes the solution basic is the reaction of water with acetate ion, a weak Brønsted-Lowry base.

$$CH_3COO^-(aq) + H_2O(\ell) \rightleftharpoons CH_3COOH(aq) + OH^-(aq)$$

This is a **hydrolysis** reaction, one in which a water molecule is split—in this case, into an H^+ ion and an OH^- ion. The H^+ ion is donated to the acetate ion to form acetic acid. The extent of hydrolysis is determined by the value of K_b for acetate ion.

All of the weak bases in Table 14.2 (← Sec. 14-5c), except for the very weak bases above water in the next to last column (NO_3^-, Cl^-, HSO_4^-, and ClO_4^-), undergo hydrolysis reactions in aqueous solution. *The larger the* K_b *value of a base, the more basic the solutions it produces.* The pH of a solution of a salt of a strong base and a weak acid can be calculated from K_b, as shown in Problem-Solving Example 14.11.

The term "hydrolysis" is derived from *hydro,* meaning "water," and *lysis,* meaning "to break apart." The hydrolysis reaction of a molecular compound results in the addition of H— and —OH to the molecules produced by breaking a covalent bond.

As K_b of the base increases, the pH of the solution increases.

Table 14.4 Some Salts Formed by Neutralization of Strong Acids with Strong Bases

	Strong Base		
Strong Acid	NaOH	KOH	$Ba(OH)_2$
	Salts	**Salts**	**Salts**
HCl	NaCl	KCl	$BaCl_2$
HNO_3	$NaNO_3$	KNO_3	$Ba(NO_3)_2$
H_2SO_4	Na_2SO_4	K_2SO_4	$BaSO_4$
$HClO_4$	$NaClO_4$	$KClO_4$	$Ba(ClO_4)_2$

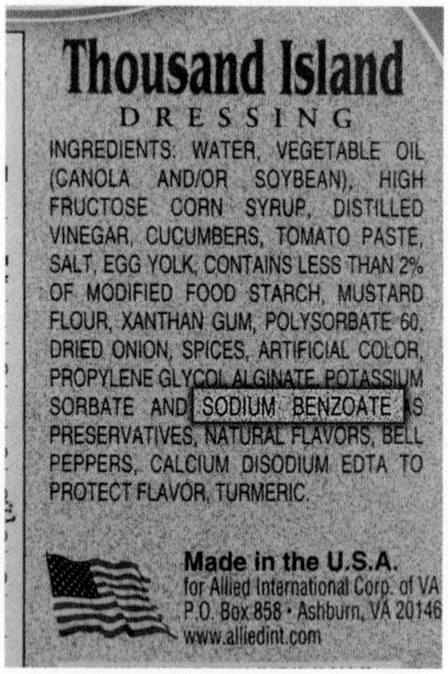
Thousand Island
DRESSING
INGREDIENTS: WATER, VEGETABLE OIL (CANOLA AND/OR SOYBEAN), HIGH FRUCTOSE CORN SYRUP, DISTILLED VINEGAR, CUCUMBERS, TOMATO PASTE, SALT, EGG YOLK, CONTAINS LESS THAN 2% OF MODIFIED FOOD STARCH, MUSTARD FLOUR, XANTHAN GUM, POLYSORBATE 60, DRIED ONION, SPICES, ARTIFICIAL COLOR, PROPYLENE GLYCOL ALGINATE, POTASSIUM SORBATE AND SODIUM BENZOATE AS PRESERVATIVES, NATURAL FLAVORS, BELL PEPPERS, CALCIUM DISODIUM EDTA TO PROTECT FLAVOR, TURMERIC.
Made in the U.S.A.
for Allied International Corp. of VA
P.O. Box 858 • Ashburn, VA 20146
www.alliedint.com

© Cengage Learning/Charles D. Winters

PROBLEM-SOLVING EXAMPLE 14.11

pH of a Salt Solution

Sodium benzoate, $NaC_7H_5O_2$, is used as a preservative in foods. Calculate the pH of a 0.025-M solution of $NaC_7H_5O_2$. (For benzoate ion, $K_b = 8.3 \times 10^{-11}$.)

Result 8.16

Analyze Sodium benzoate could be synthesized from a strong base, NaOH, and a weak acid, benzoic acid, $HC_7H_5O_2$. Thus, sodium benzoate is a basic salt. Soluble salts are completely dissociated so the solution contains Na^+ and $C_7H_5O_2^-$ ions. Na^+ does not react appreciably with water. $C_7H_5O_2^-$, the conjugate base of benzoic acid, is a weak base; its initial concentration and its K_b are given so $[OH^-]$ can be calculated. Use $[H_3O^+] \times [OH^-] = 1.0 \times 10^{-14}$ to calculate $[H_3O^+]$ and from it pH.

Plan Write the balanced chemical equation for the reaction of $C_7H_5O_2^-$ with water and use it to derive the K_b expression. Summarize the initial and equilibrium concentrations of benzoate ion, benzoic acid, and hydroxide ion in an ICE table. Let x be the equilibrium concentration of OH^- and of $C_7H_5O_2^-$ because they are formed in equal amounts. Calculate $[H_3O^+]$ from $[OH^-]$ and from it the pH.

Execute The balanced chemical equation and K_b expression are:

$$C_7H_5O_2^-(aq) + H_2O(\ell) \rightleftharpoons HC_7H_5O_2(aq) + OH^-(aq)$$

$$K_b = 8.3 \times 10^{-11} = \frac{[\text{conjugate acid}][OH^-]}{[\text{conjugate base}]} = \frac{[HC_7H_5O_2][OH^-]}{[C_7H_5O_2^-]}$$

	$C_7H_5O_2^-$	$HC_7H_5O_2$	OH^-
*I*nitial concentration (mol/L)	0.025	0	1.0×10^{-7} (from water)
*C*hange as reaction occurs (mol/L)	$-x$	$+x$	$+x$
*E*quilibrium concentration (mol/L)	$0.025 - x$	x	x

Benzoate ion has a very small K_b (8.3×10^{-11}) and thus is a very weak base, so it is reasonable to assume that x is negligibly small compared with 0.025, and $0.025 - x \cong 0.025$.

$$K_b = 8.3 \times 10^{-11} \cong \frac{x^2}{0.025}$$

Solving for x gives $x = 1.44 \times 10^{-6}$, which equals the hydroxide and benzoate ion concentrations. Because $0.025 - 1.44 \times 10^{-6} = 0.025$ (using the significant figures rules), our assumption that x is negligible compared with 0.025 is justified. Therefore, at equilibrium

$$[OH^-] = [HC_7H_5O_2] = 1.44 \times 10^{-6} \text{ mol/L}; [C_7H_5O_2^-] = 0.025 \text{ mol/L}$$

Finally, calculate the pH of the solution.

$$K_w = [H_3O^+][OH^-] = [H_3O^+](1.44 \times 10^{-6}) = 1.00 \times 10^{-14}$$

$$[H_3O^+] = \frac{1.00 \times 10^{-14}}{1.44 \times 10^{-6}} = 6.9 \times 10^{-9} \qquad pH = -\log(6.9 \times 10^{-9}) = 8.16$$

☑ **Reasonable Result Check** The reaction of benzoate ion with water produces hydroxide ions in addition to those from the dissociation of water. The excess hydroxide ions cause the solution to become basic, as indicated by the pH greater than 7. This is expected because the salt is formed from a strong base and a weak acid. To check the calculation, substitute equilibrium concentrations into K_b and verify that the result is 8.3×10^{-11}.

$$K_b = \frac{[HC_7H_5O_2][OH^-]}{[C_7H_5O_2^-]} = \frac{(1.44 \times 10^{-6})(1.44 \times 10^{-6})}{0.025} = 8.3 \times 10^{-11}$$

PROBLEM-SOLVING PRACTICE 14.11

Sodium carbonate is an environmentally benign paint stripper. It is water-soluble, and carbonate ion is a strong enough base to loosen paint so it can be scraped off. Calculate the pH of a 1.0-M solution of Na_2CO_3.

CONCEPTUAL EXERCISE 14.21

pH of Soap Solutions

Ordinary soaps are often sodium salts of fatty acids; fatty acids are weak organic acids (← **Sec. 10-5d**). Predict whether the pH of a soap solution is greater than or less than 7. Explain your prediction.

14-8c Salts of Weak Bases and Strong Acids

When a weak base reacts with a strong acid, the resulting salt solution is *acidic*. The conjugate acid of the weak base determines the pH of the solution. For example, suppose equal volumes of 0.10-M NH_3 and 0.10-M HCl are mixed. The reaction, shown in overall, complete ionic, and net ionic equations sequentially is

$$NH_3(aq) + HCl(aq) \longrightarrow NH_4Cl(aq)$$

$$NH_3(aq) + H_3O^+(aq) + Cl^-(aq) \longrightarrow NH_4^+(aq) + Cl^-(aq) + H_2O(\ell)$$

$$\underset{\text{weak base}}{NH_3(aq)} + \underset{\text{strong acid}}{H_3O^+(aq)} \longrightarrow \underset{\text{acid}}{NH_4^+(aq)} + \underset{\text{base}}{H_2O(\ell)}$$

As soon as it is formed, the weak acid NH_4^+ reacts with water and establishes an equilibrium. The resulting solution is slightly acidic, because the reaction produces hydronium ions.

$$NH_4^+(aq) + H_2O(\ell) \rightleftharpoons NH_3(aq) + H_3O^+(aq)$$

PROBLEM-SOLVING EXAMPLE 14.12

pH of Ammonium Nitrate Solution

Ammonium nitrate, NH_4NO_3, is a salt used in fertilizers and in making matches. The salt is made by the reaction of ammonia, NH_3, and nitric acid, HNO_3. Calculate the pH of a 0.15-M solution of ammonium nitrate at 25 °C; K_a $(NH_4^+) = 5.6 \times 10^{-10}$.

Result 5.04

Analyze Recognize that ammonium nitrate is the salt of a weak base (ammonia) and a strong acid (nitric acid), so the solution should be acidic, pH < 7.0. The initial concentration of NH_4NO_3 is given and the salt dissociates completely; the pH is to be calculated. The K_a value is very low indicating that ammonium ion is a very weak acid. Ammonium ions react with water to form ammonia (the conjugate base of ammonium ion) and hydronium ions, making the solution acidic.

Plan Write the balanced chemical equation for the equilibrium reaction of ammonium ion with water and from it derive the K_a expression. Use an ICE table to summarize the concentrations of ammonium ion, ammonia, and hydronium ions initially and at equilibrium. Let x equal the equilibrium concentration of H_3O^+ as well as that of NH_3 because they are formed in equal amounts. Calculate $[H_3O^+]$ and from it the pH.

Execute

$$NH_4^+(aq) + H_2O(\ell) \rightleftharpoons NH_3(aq) + H_3O^+(aq) \qquad K_a = \frac{[NH_3][H_3O^+]}{[NH_4^+]}$$

- **Strong base + strong acid yields a neutral salt (solution pH = 7.0)**
- **Strong base + weak acid yields a basic salt (solution pH > 7.0)**
- **Strong acid + weak base yields an acidic salt (solution pH < 7.0)**

	NH_4^+	NH_3	H_3O^+
*I*nitial concentration (mol/L)	0.15	0	1.0×10^{-7} (from water)
*C*hange as reaction occurs (mol/L)	$-x$	$+x$	$+x$
*E*quilibrium concentration (mol/L)	$0.15 - x$	x	x

$$K_a = \frac{[NH_3][H_3O^+]}{[NH_4^+]} = \frac{(x)(x)}{(0.15 - x)} = 5.6 \times 10^{-10}$$

Simplify the equation by approximating that $0.15 - x \cong 0.15$

$$K_a = \frac{(x)(x)}{(0.15)} \cong 5.6 \times 10^{-10}$$

and solve for x, the H_3O^+ and NH_3 concentrations. Calculate the pH

$$x^2 = (0.15)(5.6 \times 10^{-10}) \qquad x = 9.2 \times 10^{-6} = [H_3O^+]$$

$$pH = -\log[H_3O^+] = -\log(9.2 \times 10^{-6}) = 5.04$$

☑ **Reasonable Result Check** The pH of the solution is less than 7.0, which it should be. Substitute equilibrium concentrations into K_a and verify that the result is 5.6×10^{-10}.

$$K_a = \frac{[NH_3][H_3O^+]}{[NH_4^+]} = \frac{(9.2 \times 10^{-6})(9.2 \times 10^{-6})}{(0.15)} = 5.6 \times 10^{-10}$$

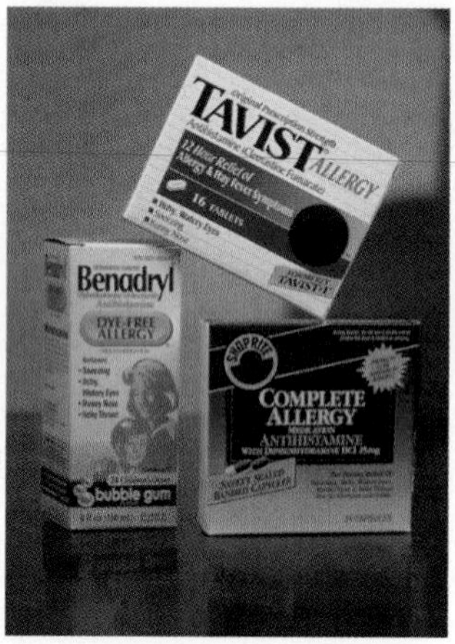
TAVIST ALLERGY
Benadryl
COMPLETE ALLERGY
ANTIHISTAMINE
bubble gum
© Cengage Learning/Charles D. Winters

Pharmaceutical amines.

PROBLEM-SOLVING PRACTICE 14.12

Calculate the pH of a 0.10-M aqueous solution of ammonium chloride, NH_4Cl at 25 °C.

Many drugs are high-molecular-weight amines that are weak bases (← **Sec. 14-2**). Such amines are not soluble in water, which limits the ways they can be administered and means that they are not soluble in body fluids such as blood plasma and cerebrospinal fluid. By reaction with hydrochloric acid, the amines are converted to water-soluble hydrochloride salts that can be administered by injection or dissolved in liquid oral medications. The resulting hydrochloride salts have the general formula BH^+Cl^-, where B represents the basic amine. This formula is like that of ammonium chloride, $NH_4^+Cl^-$. Two examples are

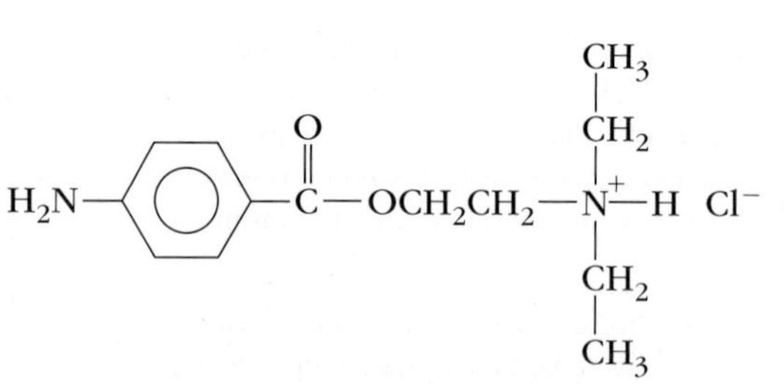

Novocain hydrochloride, a local anesthetic

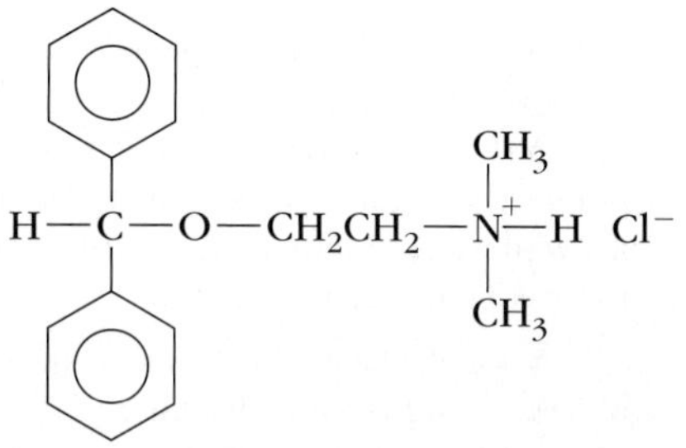

Diphenhydramine hydrochloride (Benadryl), an antihistamine

Because they are ionic, the amine hydrochloride salts are much more water-soluble than the amine form of a drug. For example, only 0.5 g Novocain dissolves in 100 g water, whereas 100 g Novocain hydrochloride dissolves in 100 g water.

EXERCISE 14.22

Forming a Drug Hydrochloride

To stymie illicit methamphetamine synthesis, phenylephrine is substituted for pseudoephedrine, an active ingredient in over-the-counter decongestants. Use structural formulas to write the equation for the formation of phenylephrine hydrochloride from phenylephrine.

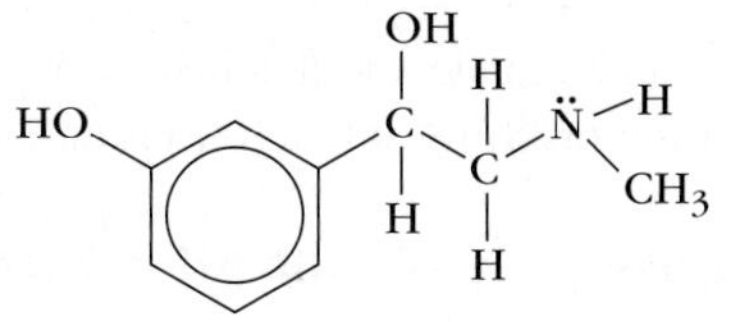
OH
H
HO
C
N
H
C
CH3
H
H

phenylephrine

CONCEPTUAL EXERCISE 14.23

Formal Charge and Drug Hydrochlorides

For phenylephrine hydrochloride, the amine hydrochloride drug in Exercise 14.22, calculate the formal charge (← **Sec. 6-8**) on each atom in the Lewis structure and thereby determine where the positive charge is located in the cation.

14-8d Salts of Weak Bases and Weak Acids

What is the pH of a solution of a salt such as NH_4F containing an acidic cation and a basic anion? The salt could be formed by the reaction of a weak acid and a weak base. There are two possible reactions that can determine the pH of the NH_4F solution: formation of H_3O^+ by H^+ transfer from the cation;

$$NH_4^+(aq) + H_2O(\ell) \rightleftharpoons H_3O^+(aq) + NH_3(aq) \qquad K_a(NH_4^+) = 5.6 \times 10^{-10}$$

and formation of OH^- by H^+ transfer from water to fluoride ion.

$$F^-(aq) + H_2O(\ell) \rightleftharpoons HF(aq) + OH^-(aq) \qquad K_b(F^-) = 1.5 \times 10^{-11}$$

Because $K_a(NH_4^+) > K_b(F^-)$, the reaction of ammonium ions with water to produce hydronium ions is the more product-favored reaction. Therefore, the resulting solution is slightly acidic.

In general, the K_a of the weak acid and the K_b of the weak base need to be considered to determine whether the aqueous solution of a salt of a weak acid and weak base is acidic or basic. Table 14.5 summarizes aqueous acid-base behavior of many ions.

Table 14.5 Acid-Base Properties of Typical Ions in Aqueous Solution

	Neutral		Basic			Acidic
Anions	Cl^-	NO_3^-	CH_3COO^-	CN^-	SO_4^{2-}	HSO_4^-
	Br^-	ClO_4^-	$HCOO^-$	PO_4^{3-}	HPO_4^{2-}	$H_2PO_4^-$
	I^-		CO_3^{2-}	HCO_3^-	SO_3^{2-}	HSO_3^-
			S^{2-}	HS^-	ClO^-	
			F^-	NO_2^-		
Cations	Li^+	Mg^{2+}		*None*		Al^{3+}
	Na^+	Ca^{2+}				NH_4^+
	K^+	Ba^{2+}				Transition metal ions

The acidity of aqueous Al^{3+} and transition metal ions is discussed in Section 14-9a.

CONCEPTUAL EXERCISE 14.24

Hydrolysis of a Salt of a Weak Acid and a Weak Base

Name a salt of a weak acid and a weak base where $K_a = K_b$. Predict the pH of a solution of this salt.

These generalizations can be made about acid-base neutralization reactions in aqueous solution and the pH of the resulting salt solutions at 25 °C.

- **Strong acid + strong base ⟶ salt with pH = 7 (neutral)**
- **Strong acid + weak base ⟶ salt with pH < 7 (acidic)**
- **Weak acid + strong base ⟶ salt with pH > 7 (basic)**
- **Weak acid + weak base ⟶ salt with pH determined by relative strengths of conjugate base and conjugate acid formed**

14-9 Lewis Acids and Bases

In 1923, when Brønsted and Lowry independently proposed their acid-base definition, Gilbert N. Lewis also was developing a new definition of acids and bases. By the early 1930s Lewis had proposed definitions of acids and bases that are more general than those of Brønsted and Lowry; the Lewis definitions are based on sharing of electron pairs rather than on H^+ ion transfers.

- **Lewis acid**: *a substance that can accept a pair of electrons to form a new bond.*
- **Lewis base**: *a substance that can donate a pair of electrons to form a new bond.*

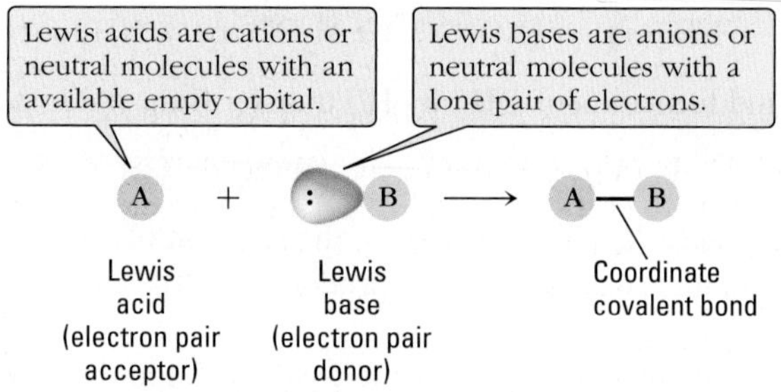
Lewis acids are cations or neutral molecules with an available empty orbital.
Lewis bases are anions or neutral molecules with a lone pair of electrons.
A
+
:
B
⟶
A
B
Lewis acid (electron pair acceptor)
Lewis base (electron pair donor)
Coordinate covalent bond

Those definitions mean that in the Lewis sense, an acid-base reaction occurs when a molecule (or ion) that has a lone pair of electrons that can be donated (a Lewis base) reacts with a molecule (or ion) that can accept an electron pair (a Lewis acid). *An electron-pair bond in which both electrons were originally associated with one of the bonded atoms* (the Lewis base) is called a **coordinate covalent bond** (← **Sec. 6-10a**).

A simple example of a Lewis acid-base reaction is the formation of a hydronium ion from H^+ and water. The H^+ ion has no electrons, while the water molecule has two lone pairs of electrons on the oxygen atom. One of the lone pairs can be shared between H^+ and oxygen, thus forming an O—H bond.

A Brønsted-Lowry base (H^+ ion acceptor) must also be a Lewis base because it donates an electron pair to bond with the H^+.

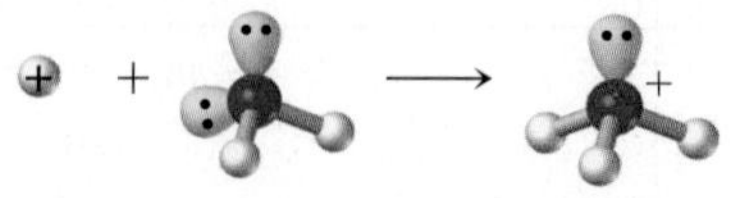

$$H^+ + H_2O \longrightarrow H_3O^+$$

hydronium ion

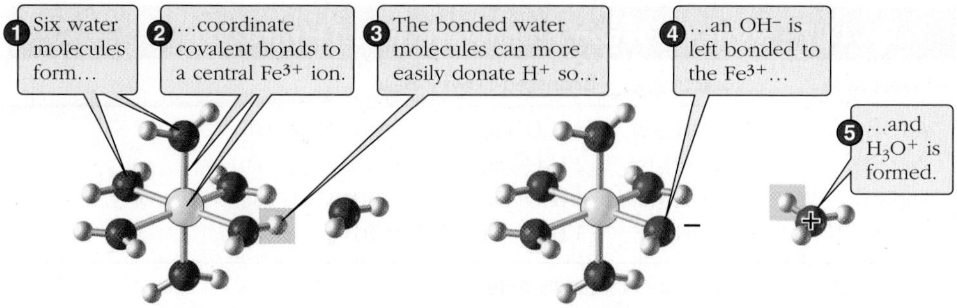
1 Six water molecules form…
2 …coordinate covalent bonds to a central Fe3+ ion.
3 The bonded water molecules can more easily donate H+ so…
4 …an OH− is left bonded to the Fe3+…
5 …and H3O+ is formed.
−
+

$$[Fe(H_2O)_6]^{3+}(aq) + H_2O(\ell) \rightleftharpoons [Fe(H_2O)_5(OH)]^{2+}(aq) + H_3O^+(aq)$$

Figure 14.5 Water molecules are more acidic when bonded to an Fe^{3+} ion in $[Fe(H_2O)_6]^{3+}$. Ions or molecules within brackets in the formulas are bonded directly to the metal ion, Fe^{3+}.

14-9a Positive Metal Ions as Lewis Acids

All metal cations are potential Lewis acids. Not only do the positively charged metal cations attract electrons, but all such cations have at least one empty orbital. This empty atomic orbital can accommodate an electron pair donated by a base, thereby forming a coordinate covalent bond. Consequently, metal ions readily form coordination complexes and also are hydrated in aqueous solution. When a metal ion such as Fe^{3+} becomes hydrated, one of the lone pairs on the oxygen atom in each of several water molecules forms a coordinate covalent bond to the metal ion; *the metal ion acts as a Lewis acid, and water acts as a Lewis base. A metal ion bonded to one or more Lewis bases by coordinate covalent bonds* forms a **complex ion**, for example, $[Fe(H_2O)_6]^{3+}$ (Figure 14.5). A hydrated metal ion (← **Sec. 13-3c**) often has six water molecules bonded to it, $[M(H_2O)_6]^{n+}$, where M represents a metal ion whose charge is $n+$.

Square brackets, [ ], surround all of the chemical species in the complex ion. Coordination complexes and complex ions are discussed in detail in Section 20-6a.

Hydrated metal ions, especially those of transition metal ions, are weak acids. Some K_a values are given in Table 14.6. As shown in Figure 14.5, there are M—O—H bonds in the hydrated metal ion. Transition metal ions and Al^{3+} ions have large enough charges and small enough sizes to attract the shared pair of the M—O bond to the metal ion. This attraction weakens the O—H bond making the hydrogen of the M—O—H bond more acidic than it would be in the O—H bond of a water molecule that is not bonded to a metal ion. Thus, the $[M(H_2O)_6]^{n+}$ complex ion can donate H^+ and the solution becomes acidic. The ionization reaction for a hydrated Fe^{3+} ion is given in Figure 14.5. The corresponding acid dissociation constant expression is

In this equilibrium constant expression, curly braces, { }, are used to indicate concentrations because square brackets indicate complex ions.

$$K_a = \frac{\{[Fe(H_2O)_5(OH)]^{2+}\} \times \{H_3O^+\}}{\{[Fe(H_2O)_6]^{3+}\}} = 6.3 \times 10^{-3}$$

Table 14.6 Ionization Constants for Hydrated Metal Complex Ions and Their Conjugate Bases at 25 °C

Acid	K_a	Base	K_b
$[Fe(H_2O)_6]^{3+}$	6.3×10^{-3}	$[Fe(H_2O)_5(OH)]^{2+}$	1.6×10^{-12}
$[Al(H_2O)_6]^{3+}$	7.9×10^{-6}	$[Al(H_2O)_5(OH)]^{2+}$	1.3×10^{-9}
$[Cu(H_2O)_6]^{2+}$	1.6×10^{-7}	$[Cu(H_2O)_5(OH)]^{1+}$	6.3×10^{-8}
$[Pb(H_2O)_6]^{2+}$	1.5×10^{-8}	$[Pb(H_2O)_5(OH)]^{1+}$	6.7×10^{-7}
$[Co(H_2O)_6]^{2+}$	1.3×10^{-9}	$[Co(H_2O)_5(OH)]^{1+}$	7.7×10^{-6}
$[Fe(H_2O)_6]^{2+}$	3.2×10^{-10}	$[Fe(H_2O)_5(OH)]^{1+}$	3.1×10^{-5}
$[Ni(H_2O)_6]^{2+}$	2.5×10^{-11}	$[Ni(H_2O)_5(OH)]^{1+}$	4.0×10^{-4}

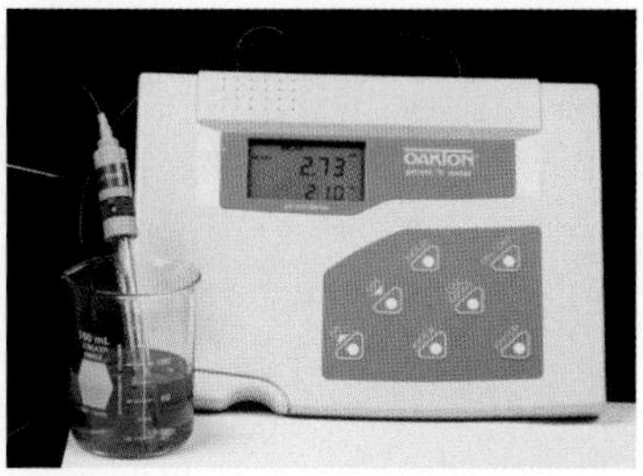
2.73
DAKTON

© Cengage Learning/James Maynard

Acidic pH of an aqueous copper(II) sulfate solution. The blue solution of this copper salt is acidic (pH = 2.73) due to hydrolysis of the $[Cu(H_2O)_6]^{2+}$ ion.

Table 14.7 Some Common Amphoteric Metal Hydroxides

Hydroxide	Reaction as a Brønsted-Lowry Base
$Al(OH)_3$	$Al(OH)_3(s) + 3\ H_3O^+(aq) \longrightarrow [Al(H_2O)_6]^{3+}(aq)$
$Zn(OH)_2$	$Zn(OH)_2(s) + 2\ H_3O^+(aq) \longrightarrow [Zn(H_2O)_4]^{2+}(aq)$
$Sn(OH)_4$	$Sn(OH)_4(s) + 4\ H_3O^+(aq) \longrightarrow [Sn(H_2O)_6]^{4+}(aq) + 2\ H_2O(\ell)$
$Cr(OH)_3$	$Cr(OH)_3(s) + 3\ H_3O^+(aq) \longrightarrow [Cr(H_2O)_6]^{3+}(aq)$
Hydroxide	**Reaction as a Lewis Acid**
$Al(OH)_3$	$Al(OH)_3(s) + OH^-(aq) \longrightarrow [Al(OH)_4]^-(aq)$
$Zn(OH)_2$	$Zn(OH)_2(s) + 2\ OH^-(aq) \longrightarrow [Zn(OH)_4]^{2-}(aq)$
$Sn(OH)_4$	$Sn(OH)_4(s) + 2\ OH^-(aq) \longrightarrow [Sn(OH)_6]^{2-}(aq)$
$Cr(OH)_3$	$Cr(OH)_3(s) + OH^-(aq) \longrightarrow [Cr(OH)_4]^-(aq)$

This K_a value shows that an aqueous $FeCl_3$ solution has about the same pH as a solution of phosphoric acid ($K_a = 7.2 \times 10^{-3}$) of equal concentration. The acidic properties of aqueous metal ions are important in the environment. For example, Al^{3+} ions in soils react with water to produce an acidic environment that can be detrimental to tree growth.

Metal ions (Lewis acids) form many complex ions with another Lewis base, ammonia, $:NH_3$. For example, silver ion readily forms a water-soluble, colorless complex ion in liquid ammonia or in aqueous ammonia.

$$Ag^+(aq) + 2\ :NH_3(aq) \longrightarrow [H_3N:Ag:NH_3]^+(aq)$$

This complex is so stable that AgCl, a compound that is almost completely insoluble in water, can be dissolved in aqueous ammonia.

$$AgCl(s) + 2\ :NH_3(aq) \longrightarrow [H_3N:Ag:NH_3]^+(aq) + Cl^-(aq)$$

The hydroxide ion, OH^-, is an excellent Lewis base and so it binds readily to metal cations to give metal hydroxides. An important feature of the chemistry of many metal hydroxides is that they are amphoteric. **Amphoteric** *describes a substance that can react as both a base and an acid.* Table 14.7 lists several amphoteric metal hydroxides. Aluminum hydroxide, for example, is amphoteric (Figure 14.6). It behaves as a Lewis acid when it dissolves in a basic solution by forming a complex ion containing one additional OH^- ion, a Lewis base.

$$\underset{\text{Lewis acid}}{Al(OH)_3(s)} + \underset{\text{Lewis base}}{OH^-(aq)} \rightleftharpoons \underset{\text{complex ion}}{[Al(OH)_4]^-(aq)}$$

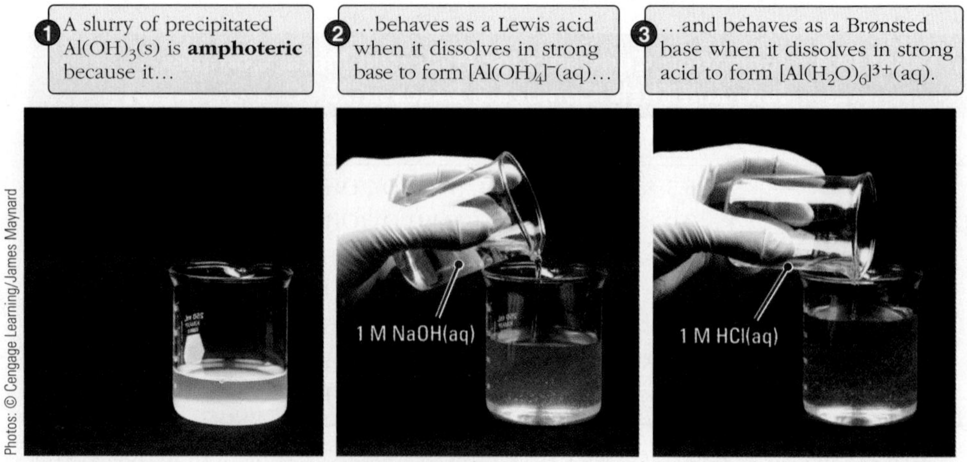
1 A slurry of precipitated Al(OH)3(s) is **amphoteric** because it...
2 ...behaves as a Lewis acid when it dissolves in strong base to form [Al(OH)4]−(aq)...
3 ...and behaves as a Brønsted base when it dissolves in strong acid to form [Al(H2O)6]3+(aq).
1 M NaOH(aq)
1 M HCl(aq)
Photos: © Cengage Learning/James Maynard

Figure 14.6 The amphoteric nature of $Al(OH)_3$.

Aluminum hydroxide behaves as a Brønsted-Lowry base when it reacts with a Brønsted-Lowry acid.

$$Al(OH)_3(s) + 3\ H_3O^+(aq) \rightleftharpoons [Al(H_2O)_6]^{3+}(aq)$$

14-9b Neutral Molecules as Lewis Acids

Lewis's ideas about acids and bases account nicely for the fact that oxides of nonmetals behave as acids (← **Sec. 14-6e**). Two important examples are carbon dioxide and sulfur dioxide, whose Lewis structures are

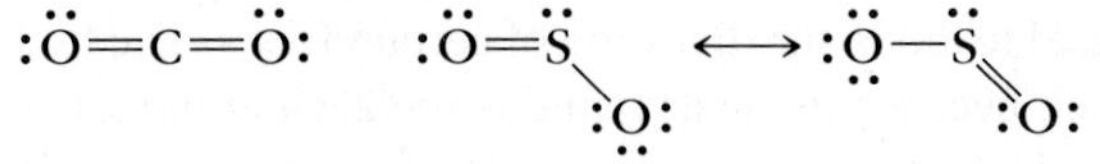

carbon dioxide sulfur dioxide

In each case, there is a double bond; an "extra" pair of electrons is being shared between an oxygen atom and the central atom. Because oxygen is highly electronegative, electrons in these bonds are attracted away from the central atom, which becomes slightly positively charged. This makes the central atom a likely site to attract a pair of electrons. A Lewis base such as OH^- can bond to the carbon atom in CO_2 to give hydrogen carbonate ion, HCO_3^-. This bonding displaces one double-bond pair of electrons onto an oxygen atom.

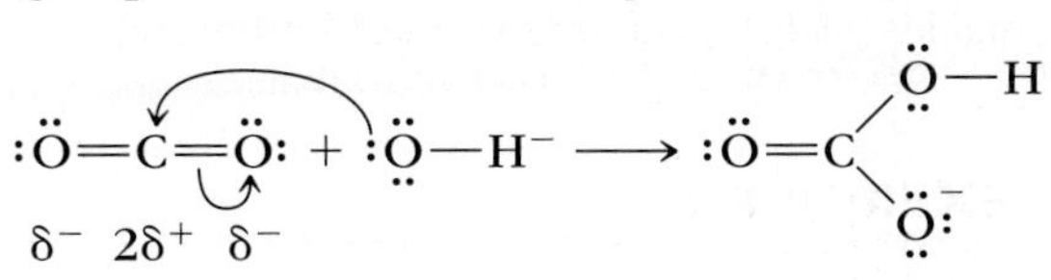
δ⁻ 2δ⁺ δ⁻

hydrogen carbonate ion

Carbon dioxide from the air can react to form sodium carbonate around the mouth of a bottle of sodium hydroxide. Sulfur dioxide can react similarly with hydroxide ion.

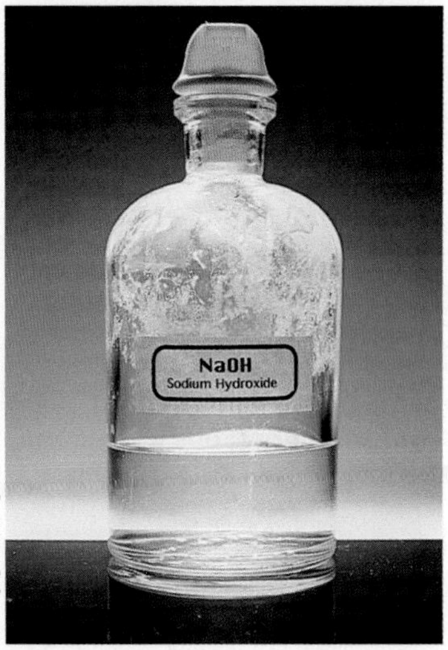
NaOH
Sodium Hydroxide

© Cengage Learning/Charles D. Winters

Carbon dioxide in air reacts with spilled base such as aqueous NaOH to form Na_2CO_3, the white solid coating the upper part of the bottle.

CONCEPTUAL EXERCISE 14.25

Lewis Acids and Bases

Predict whether each of these is a Lewis acid or a Lewis base. (*Hint:* Drawing a Lewis structure for a molecule or ion often helps to make such predictions.)

(a) PH_3 (b) BCl_3 (c) H_2S (d) NO_2 (e) Ni^{2+} (f) CO

EXERCISE 14.26

Salt of Weak Acid and a Weak Base

Use data given in the chapter to predict whether an aqueous solution of nickel(II) acetate has a pH < 7.0, $= 7.0$, or > 7.0. Explain your prediction.

EXERCISE 14.27

The pH of a Solution of $Ni(NO_3)_2$

When anhydrous nickel(II) nitrate dissolves in water, the Ni^{2+} ions become hydrated, forming $[Ni(H_2O)_6]^{2+}$ ions. Calculate the pH of a solution that is 0.15 M in nickel nitrate. K_a for $[Ni(H_2O)_6]^{2+}$ is 2.5×10^{-11}.

14-10 Additional Applied Acid-Base Chemistry

In addition to their uses in industry, various acids and bases find many applications around the home. Antacids are used to neutralize excess stomach acidity. Gardeners use acid salts such as sodium hydrogen sulfate, $NaHSO_4$, to help acidify soil and bases such

HEINZ
VINEGAR
CLUB
SODA

The pH of some household substances. The flasks contain club soda (left), vinegar (middle), and a household cleaner (right). The color of an acid-base indicator in each flask shows the acidity or alkalinity of the liquid—vinegar (red) is more acidic than club soda (orange); the cleaner (green) is much more basic than either of the other two liquids.

as lime, CaO, to make soil more basic. In the kitchen, baking soda and baking powders are used to make biscuit dough and cake batter rise. Mild acids and bases are used to clean everything from dishes and clothes to vehicles and the family dog.

14-10a Neutralizing Stomach Acidity

Human stomach fluids have a pH of approximately 1. This very acidic pH is caused by HCl secreted by thousands of cells in the stomach lining that specialize in transporting $H_3O^+(aq)$ and $Cl^-(aq)$ from the blood. The main purpose of this acid is to suppress the growth of bacteria and to aid in the digestion of certain foods. The hydrochloric acid does not harm the stomach because its inner lining is replaced at the rate of about half a million cells per minute. However, when too much food is eaten and the stomach is stretched, or when the stomach is irritated by very spicy food, some of its acidic contents can flow back into the esophagus (gastroesophageal reflux); the burning sensation that results is called *heartburn.*

An antacid is a base that is used to neutralize excess stomach acid. The recommended dose is the amount of the base required to neutralize *some,* but not all, of the stomach acid. Several antacids and their acid-base reactions are shown in Table 14.8. People who need to restrict the quantity of sodium (Na^+) in their diets should avoid sodium-containing antacids such as sodium hydrogen carbonate.

CONCEPTUAL EXERCISE 14.28

Strong Antacids?

Explain why strong bases such as NaOH or KOH are never used as antacids.

Chloride ions secreted by the stomach lining come mostly from the salty foods we eat and the salt we add to our foods.

PROBLEM-SOLVING EXAMPLE 14.13

Neutralizing Stomach Acid

Calculate the amount (mol) and mass (g) of HCl that could be neutralized by 0.750 g of the antacid $CaCO_3$.

Result 1.50×10^{-2} mol HCl; 0.547 g HCl

Analyze Use the balanced equation for the reaction given in Table 14.8. The equation shows that 1 mol $CaCO_3$ reacts with 2 mol HCl, $M(HCl) = 36.46$ g/mol.

Table 14.8 The Acid-Base Chemistry of Some Antacids

Compound	Reaction in Stomach	Examples of Commercial Products
Milk of magnesia: $Mg(OH)_2$ in water	$Mg(OH)_2(s) + 2\,H_3O^+(aq) \longrightarrow Mg^{2+}(aq) + 4\,H_2O(\ell)$	Phillips' Milk of Magnesia®
Calcium carbonate: $CaCO_3$	$CaCO_3(s) + 2\,H_3O^+(aq) \longrightarrow Ca^{2+}(aq) + 3\,H_2O(\ell) + CO_2(g)$	Tums®, Di-Gel®
Sodium hydrogen carbonate: $NaHCO_3$	$NaHCO_3(s) + H_3O^+(aq) \longrightarrow Na^+(aq) + 2\,H_2O(\ell) + CO_2(g)$	Baking soda, Alka-Seltzer®
Aluminum hydroxide: $Al(OH)_3$	$Al(OH)_3(s) + 3\,H_3O^+(aq) \longrightarrow Al^{3+}(aq) + 6\,H_2O(\ell)$	Amphojel®
Dihydroxyaluminum sodium carbonate: $NaAl(OH)_2CO_3$	$NaAl(OH)_2CO_3(s) + 4\,H_3O^+(aq) \longrightarrow Na^+(aq) + Al^{3+}(aq) + 7\,H_2O(\ell) + CO_2(g)$	Rolaids® (old formulation)

Plan Calculate the amount of calcium carbonate in 0.750 g $CaCO_3$. Use this amount and the stoichiometric mole ratio to first calculate the amount of HCl and from it the mass of HCl that is neutralized.

Execute

$$n(CaCO_3) = 0.750 \text{ g } CaCO_3 \times \frac{1 \text{ mol } CaCO_3}{100.1 \text{ g } CaCO_3} = 7.49 \times 10^{-3} \text{ mol } CaCO_3$$

$$n(HCl) = 7.49 \times 10^{-3} \text{ mol } CaCO_3 \times \frac{2 \text{ mol HCl}}{1 \text{ mol } CaCO_3} = 1.50 \times 10^{-2} \text{ mol HCl}$$

$$m(HCl) = 1.50 \times 10^{-2} \text{ mol HCl} \times \frac{36.46 \text{ g HCl}}{1 \text{ mol HCl}} = 0.547 \text{ g HCl}$$

☑ **Reasonable Result Check** Because the molar mass of $CaCO_3$ is about 100, the amount of $CaCO_3$ is about 0.0075 mol. Two moles of HCl are required to react with each mole of $CaCO_3$, so to neutralize 0.0075 mol $CaCO_3$ requires 0.0150 mol HCl, which is just over 0.5 g. The result is reasonable.

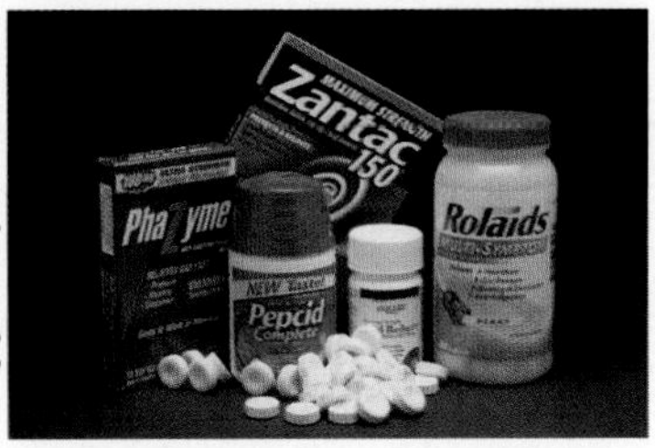
Zantac 150
Rolaids
Pepcid Complete
© Cengage Learning/Charles D. Winters

Commercial antacid products. These antacids neutralize excess stomach acid and relieve heartburn.

PROBLEM-SOLVING PRACTICE 14.13

Using the reactions in Table 14.8, determine which antacid, on a per gram basis, can neutralize the most stomach acid (assume 0.10-M HCl).

14-10b Acid-Base Chemistry in the Kitchen

Cooking and baking use chemical reactions, often involving acids and bases. Vinegar is an approximately 4%–5% aqueous acetic acid solution present in almost all salad dressings. Lemon juice, handy for flavoring tea and cooked fish and for making salad dressings, contains citric acid, as do all citrus fruits.

One of the most useful substances in the kitchen is carbon dioxide gas, produced by chemical reactions as it is needed. Pockets of the gas are generated in bread dough and cake batter. The expanding gas makes the resulting biscuits, breads, and cakes rise, producing lighter and more palatable baked goods.

ESTIMATION | Using an Antacid

Estimate how many Rolaids tablets (Table 14.8) it would take to neutralize the acidity in one glass (250 mL) of a regular cola drink. Assume the pH of the cola is 3.0. One Rolaids tablet contains 334 mg $NaAl(OH)_2CO_3$.

With a pH of 3.0, the cola has 1×10^{-3} mol acid per liter of cola, so 0.250 L cola has one fourth that much acid, or about 3×10^{-4} mol acid. To neutralize this amount of acid requires 3×10^{-4} mol base (1 mol base for every 1 mol acid). There are two bases in Rolaids—hydroxide ions and carbonate ions. Each mole of hydroxide neutralizes 1 mol acid, and each mole of carbonate neutralizes 2 mol acid.

$$H^+(aq) + OH^-(aq) \longrightarrow H_2O(\ell)$$

$$2\,H^+(aq) + CO_3^{2-}(aq) \longrightarrow H_2O(\ell) + CO_2(g)$$

Because each mole of $NaAl(OH)_2CO_3$ contains 2 mol OH^- ions and 1 mol CO_3^{2-} ions, 1 mol $NaAl(OH)_2CO_3$ neutralizes 4 mol acid. The molar mass of $NaAl(OH)_2CO_3$ is 144 g/mol, so one Rolaids tablet contains about 0.002 mol of the antacid.

$$\frac{0.334 \text{ g antacid}}{1 \text{ antacid tablet}} \times \frac{1 \text{ mol antacid}}{144 \text{ g antacid}} = 0.00232 \text{ mol antacid/tablet}$$

This tablet can neutralize four times that many moles of acid, or about 0.01 mol acid. To neutralize the 3×10^{-4} mol acid in the cola requires about 0.03 tablet.

$$3 \times 10^{-4} \text{ mol antacid} \times \frac{1 \text{ tablet}}{0.01 \text{ mol acid}} \approx 0.03 \text{ tablet}$$

It would take only a small portion of a tablet to do the job.

Various methods are used to generate CO_2. One is the addition of yeast, which causes bread dough to rise by catalyzing the fermentation of carbohydrates to produce ethyl alcohol and carbon dioxide.

$$C_6H_{12}O_6 \xrightarrow{\text{yeast}} 2\ CH_3CH_2OH + 2\ CO_2$$

Many commercial breads and homemade dinner rolls use this method to make their doughs rise.

Because CO_2 production by fermentation is slow, it is sometimes necessary to use another method, the reaction of a bicarbonate salt such as sodium hydrogen carbonate, $NaHCO_3$ (also known as *baking soda*), with acid. But which acid should be used? A weak acid is needed; if a strong acid were used, complete neutralization of the acid would be required to make the food safe to eat. Although vinegar could be used, it would impart an undesirable taste. Long ago it was discovered that lactic acid, $CH_3CH(OH)COOH$, present in milk and formed in larger quantities when milk sours to form buttermilk, is a good source of acid for reacting with hydrogen carbonate.

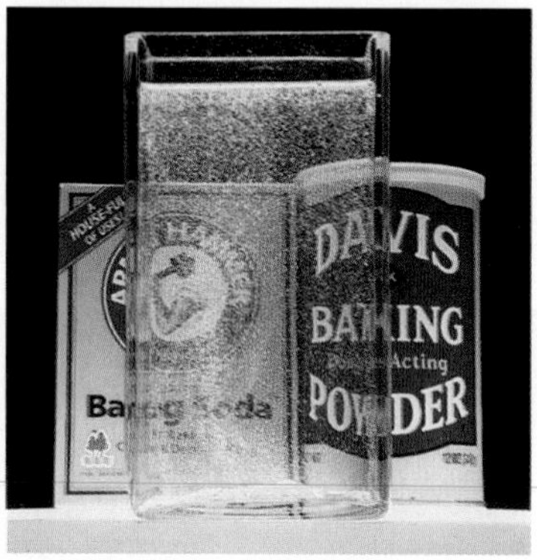

Action of baking powder. Baking powder contains the weak acid dihydrogen phosphate ion and the weak base hydrogen carbonate ion. When water is added, the acid and base react to produce carbon dioxide gas, which is seen bubbling out of the aqueous solution in the picture.

$$\underset{\text{lactic acid}}{CH_3\underset{\displaystyle |\atop \displaystyle OH}{C}HCOOH(aq)} + HCO_3^-(aq) \longrightarrow \underset{\text{lactate ion}}{CH_3\underset{\displaystyle |\atop \displaystyle OH}{C}HCOO^-(aq)} + H_2CO_3(aq)$$

$$\underset{\text{carbonic acid}}{H_2CO_3(aq)} \longrightarrow CO_2(g) + H_2O(\ell)$$

When buttermilk is not available, or when a different taste is desired, dihydrogen phosphate ion, $H_2PO_4^-$, is a convenient acid to react with the hydrogen carbonate ion. Baking powders are a mixture of sodium or potassium dihydrogen phosphate and sodium hydrogen carbonate. When dry, the two salts in baking powder do not react with one another. But, when mixed with water in the dough, they react to produce CO_2, a reaction that occurs even more quickly in a heated oven.

$$H_2PO_4^-(aq) \rightleftharpoons H^+(aq) + HPO_4^{2-}(aq)$$
$$H^+(aq) + HCO_3^-(aq) \longrightarrow H_2O(\ell) + CO_2(g)$$
$$\text{Net reaction:}\quad H_2PO_4^-(aq) + HCO_3^-(aq) \longrightarrow H_2O(\ell) + CO_2(g) + HPO_4^{2-}(aq)$$

EXERCISE 14.29

Acids and Muffins

Consider this list of ingredients for buttermilk blueberry muffins. There are two sources of acid and two sources of hydrogen carbonate, the reaction of which produces the CO_2 to make these muffins rise. What are the sources?

Buttermilk Blueberry Muffins

$2\frac{1}{2}$ cup flour	2 eggs, beaten
$1\frac{1}{2}$ tsp baking powder	1 cup buttermilk
$\frac{1}{2}$ tsp baking soda	3 oz butter
$\frac{1}{2}$ cup sugar	$1\frac{1}{2}$ cup blueberries
$\frac{1}{4}$ tsp salt	

14-10c Household Cleaners

Most cleaning compounds such as dishwashing detergents, scouring powders, laundry detergents, and oven cleaners are basic (sometimes highly basic). Synthetic detergents are derived from organic molecules designed to have even better cleaning action than soaps, but less reaction with the doubly positive ions (Mg^{2+} and Ca^{2+}) found in hard water.

The molecular structure of a synthetic detergent, like that of a soap, consists of a long oil-soluble (hydrophobic) group, and a water-soluble (hydrophilic) group (← **Sec. 13-9**). A typical synthetic detergent molecule looks like this:

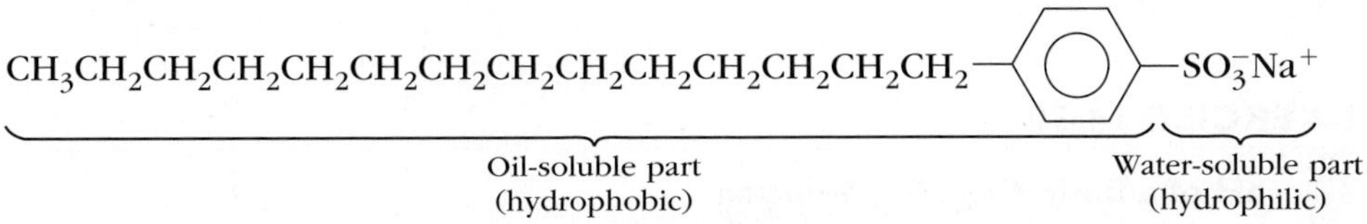
CH3CH2CH2CH2CH2CH2CH2CH2CH2CH2CH2CH2CH2CH2
SO3−Na+
Oil-soluble part
(hydrophobic)
Water-soluble part
(hydrophilic)

Typical hydrophilic groups in detergents include negatively charged sulfate, $—OSO_3^-$; sulfonate, $—SO_3^-$; and phosphate, $—OPO_3^{2-}$, groups. Compounds with these groups are called *anionic surfactants.* There are also *cationic surfactants,* almost all of which are quaternary ammonium halides in which a central nitrogen atom has four organic groups attached to it, one of which is a long hydrocarbon chain and another that frequently includes an —OH group. Some detergents are *nonionic;* they have an uncharged hydrophilic polar group attached to a large organic group of low polarity, for example,

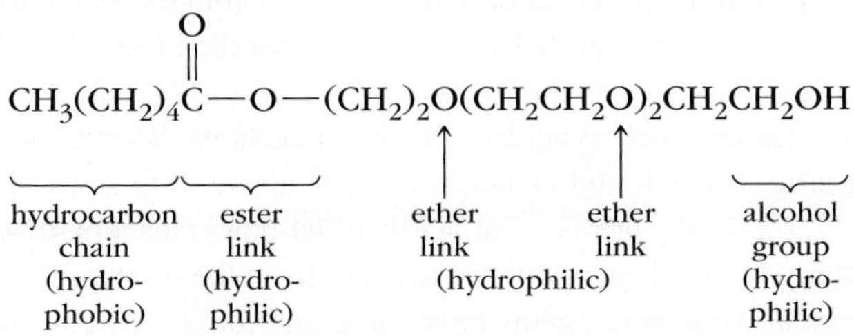
CH3(CH2)4C(=O)—O—(CH2)2O(CH2CH2O)2CH2CH2OH
hydrocarbon chain (hydrophobic)
ester link (hydrophilic)
ether link
ether link
(hydrophilic)
alcohol group (hydrophilic)

This molecule has the combination of properties needed for a detergent: a carbon chain that is oil-soluble and hydrophobic, and the ester and ether part of the molecule that is hydrophilic. Nonionic detergents have several advantages over ionic detergents. With no ionic groups, nonionic detergents cannot form salts with calcium, magnesium, and iron ions; consequently they are totally unaffected by hard water. For the same reason, nonionic detergents do not react with acids and may be used even in relatively strong acid solutions, which makes them useful in toilet bowl cleaners.

Many decades ago homemade lye soap was used to clean clothes as well as people's skins. This type of soap was made using either pure lye (NaOH, also called caustic soda) or sodium carbonate (Na_2CO_3, also called soda ash) and potassium carbonate, K_2CO_3 (also called potash) from wood ashes. Most of the soap made this way contained unreacted base, which was considered desirable because it helped raise the pH and break up the heavy soil particles common on fabrics in those days. In addition, the fabrics then were much more durable than fabrics today.

The closest thing to lye soap we see today is the typical dishwashing detergent. The first three ingredients in Table 14.9 contain anions that react with water to produce OH^-.

Table 14.9 Formulation of a Dishwashing Detergent

Ingredient	% by Weight
Sodium carbonate, Na_2CO_3	37.5
Sodium tripolyphosphate, $Na_5P_3O_{10}$	30
Sodium metasilicate, Na_2SiO_3	30
Low-foam surfactant	0.5
Sodium dichloroisocyanurate (Cl_2 source)	1.5
Other ingredients such as colorants	0.5

Together, these three salts produce solutions inside the dishwasher that have a pH near 12.5, high enough to quickly break away animal and vegetable oils from surfaces during the agitation cycle. The detergent helps to dissolve these oily particles and carry them away in the rinse water.

EXERCISE 14.30

pH of a Basic Cleaning Solution
Calculate the pH of 5.2-M aqueous sodium carbonate.

14-10d Corrosive Household Cleaners

Muriatic acid (hydrochloric acid) is used for cleaning bricks and concrete in new home construction or when remodeling is done. The strong acid should be handled with extreme caution.

Really tough cleaning jobs around the home require chemically aggressive cleaners. In those circumstances, either very acidic or highly basic cleaners are used to get rid of dirt, grease, or stains. Hydrochloric acid along with phosphoric acid and oxalic acid are used in toilet bowl cleaners to help get rid of stains. This combination of acids makes the pH of such cleaners very low, around 2. They should be handled cautiously because of their high acidity and reactivity. For example, labels on bottles of bleach warn against mixing bleach with other cleaners such as acidic toilet bowl cleaners. Bleach commonly contains sodium hypochlorite, NaOCl, and usually has a pH above 8; OCl^- and Cl^- are present in the bleach solution. In the presence of acidic toilet bowl cleaners, however, OCl^- and Cl^- are converted to toxic Cl_2 gas, which can erupt from the mixture. Bleach also should not be mixed with any cleaning agents containing ammonia. The chlorine in the bleach reacts with ammonia to produce fumes of chloramines such as NH_2Cl and $NHCl_2$, which can cause respiratory distress.

Drain and oven cleaners are at the other end of the pH spectrum, having very basic pHs of 12 or higher. Deposits of hair, grease, and fats build up inside pipes and eventually clog the drain. When this occurs, the drain has to be dismantled and cleaned (a messy job), or a drain cleaner is added to dissolve the clog.

Drain cleaners usually contain a strong base such as NaOH that reacts with fats and grease to form a soluble soap (← **Secs. 10-5d, 13-9**).

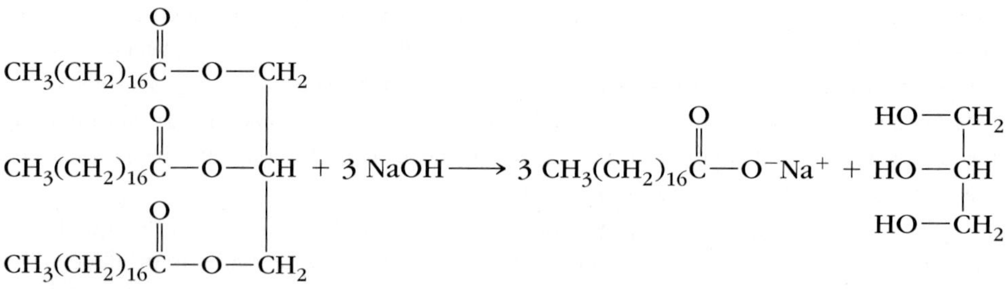

tristearin (glyceryl tristearate) sodium stearate (a soluble soap) glycerol

This reaction converts the grease or fat into water-soluble products (a sodium salt and glycerol) that are washed down the now-opened drain. The reaction is exothermic, and the released heat helps soften the grease or fats, which helps their removal. The strong base also decomposes and rinses away any hair trapped in the clog.

Grocery and home-improvement stores have an abundance of sodium hydroxide–based drain cleaners. Some are almost pure solid NaOH (pellets or flakes). Liquid drain cleaners are often 50% or more NaOH by weight in water. These solutions, being more dense than water, sink to the bottom of the drain trap to start working quickly. Although easy to use, the liquid cleaners become diluted when running water is put into the drain, reducing their efficiency. A particularly aggressive form of drain cleaner contains small bits of aluminum metal along with solid NaOH. In water, the aluminum reacts with the sodium hydroxide to produce hydrogen gas, whose bubbles help to unseat the clogged material. The net ionic equation for the reaction is

$$2\ Al(s) + 2\ OH^-(aq) + 6\ H_2O(\ell) \longrightarrow 3\ H_2(g) + 2\ [Al(OH)_4]^-(aq)$$

The hydrogen gas is flammable, so no flames or sparks should be present when this type of cleaner is used. If a pipe is weak and the clog very strong, sufficient gas pressure can build up to rupture the pipe (one way—but not a very desirable one—to unclog the drain).

Spray-on oven cleaners contain an NaOH solution mixed with a detergent and a propellant to apply the mixture to the soiled places. This mixture is sufficiently viscous to adhere to the oven surfaces long enough for the strongly basic solution to react with the baked-on food. If baked at a high temperature, the food has probably carbonized. If so, the cleaner will not be effective, and only scraping will remove the deposits.

You should always be cautious with household acids and bases. They are usually just as concentrated and harmful as industrial chemicals. All strongly acidic and basic solutions, both in the lab and in the home, can be hazardous. Acids, interestingly, are somewhat less dangerous than solutions of bases because the H_3O^+ ion tends to *denature* (← **Sec. 11-9c**) proteins in skin. The denatured proteins harden, forming a protective layer that prevents further attack by the acid, unless it is hot or highly concentrated. Basic solutions, by contrast, tend to dissolve proteins slowly and so produce little, if any, pain, causing considerable harm before any problem is noticed. If you should get acids or bases on your skin, wash with water for at least 15 min. If acid or base splashes into your eyes, have someone call a physician while you begin gently washing the affected area with lots of water. The international warning placard that is required for shipments of acids and bases in quantities of 1001 lb or more illustrates schematically the personal dangers of these substances.

CORROSIVE
8

A corrosive chemical warning placard. This type of placard is required by the U. S. Department of Transportation on loads of 1001 lb or more of acids or bases transported by highway or rail. Such a placard makes fairly clear the reactions with human skin and metals.

SUMMARY PROBLEM

Lactic acid, $CH_3CH(OH)COOH$, is a weak monoprotic acid with a melting point of 53 °C. It exists as two enantiomers (← **Sec. 7-2f**) that have slightly different K_a values. The D form has a K_a of 1.5×10^{-4} and the L form has a K_a of 1.6×10^{-4}. The D form is synthesized by some bacteria. The L form is produced in muscle cells during anaerobic metabolism in which glucose molecules are broken down into lactic acid and molecules of adenosine triphosphate (ATP) are formed. When lactic acid builds up too rapidly in muscle tissue, severe pain results.

(a) Which form of lactic acid (D or L) is the stronger acid? Explain your answer.

(b) Determine the pK_a that would be measured for a 50:50 mixture of the two forms of lactic acid in aqueous solution. $pK_a = -\log K_a$

(c) A solution of D-lactic acid is prepared. Use HL as a general formula for lactic acid, and write the equation for the ionization of lactic acid in water.

(d) If 0.100-M solutions of these two acids (D and L) were prepared, calculate what the pH of each solution would be.

(e) Before any lactic acid dissolves in the water, what reaction determines the pH?

(f) Calculate the pH of a solution made by dissolving 4.46 g D-lactic acid in 500. mL of water.

(g) Calculate the volume (mL) of 1.15-M NaOH(aq) required to completely neutralize 4.46 g of pure lactic acid.

(h) Calculate the pH of the solution when exactly enough NaOH was added to neutralize all of the lactic acid for (i) the D form; (ii) the L form; and (iii) a 50:50 mixture of the two forms.

HAVING STUDIED THIS CHAPTER . . .

. . . you should be able to:

- Use chemical equations to illustrate water's role in aqueous acid-base chemistry (Section 14-1). End-of-chapter Questions 9, 11
- Identify the conjugate acid-base pairs in a chemical reaction, and use the relationship between conjugate acid and base strengths (Section 14-1). Questions 13, 17
- Recognize how amines act as bases and how carboxylic acids ionize in aqueous solution (Section 14-2). Questions 21, 23
- Use the autoionization of water to calculate H^+ and OH^- concentrations in aqueous solutions of acids and bases (Section 14-3). Question 19
- Classify an aqueous solution as acidic, neutral, or basic based on its concentration of H_3O^+ or OH^- and its pH or pOH (Section 14-4). Questions 25, 33
- Calculate pH or pOH given $[H_3O^+]$ or $[OH^-]$, or calculate $[H_3O^+]$ or $[OH^-]$ given pH or pOH (Section 14-4). Questions 27, 29
- Estimate acid and base strengths from K_a and K_b values (Section 14-5). Questions 38, 46
- Write the ionization steps of polyprotic acids (Section 14-5). Question 51
- Describe the relationships between acid strength and molecular structure, and use them to predict the strength of an acid based on its molecular structure (Section 14-6). Question 48
- Explain the nature of zwitterions (Section 14-6). Questions 39, 41
- Calculate pH from K_a or K_b values and solution concentration (Section 14-7). Questions 54, 57, 58
- Predict whether a reaction occurs between an acid and a base, given K_a or K_b values (Sections 14-1 and 14-5). Questions 66, 68
- Describe the hydrolysis of aqueous salts and predict a particular salt solution's acidity or alkalinity (Section 14-8). Question 70
- Recognize Lewis acids and bases and describe how they react (Section 14-9). Questions 76, 85
- Describe the acidic behavior of hydrated metal ions (Section 14-9). Question 74
- Apply acid-base principles to the chemistry of antacids, kitchen chemistry, and household cleaners (Section 14-10). Questions 86, 91

KEY TERMS

acid-base reaction (Brønsted-Lowry) (Section 14-1)

acid ionization constant (K_a) (14-5a)

acid ionization constant expression (14-5a)

acidic oxide (14-6e)

acidic solution (14-3)

amine (14-2)

amino acid (14-6d)

amphiprotic (14-1a)

amphoteric (14-9a)

autoionization (14-3)

base ionization constant (K_b) (14-5b)

base ionization constant expression (14-5b)

basic oxide (14-6e)

basic (alkaline) solution (14-3)

Brønsted-Lowry acid (14-1)

Brønsted-Lowry base (14-1)

complex ion (14-9a)

conjugate acid-base pair (14-1b)

coordinate covalent bond (14-9)

hydrolysis (14-8b)

ionization constant for water (K_w) (14-3)

Lewis acid (14-9)

Lewis base (14-9)

monoprotic acids (14-5d)

neutral solution (14-3)

oxoacids (14-6b)

pH (14-4)

pK_w (14-4)

pOH (14-4)

polyprotic acids (14-5d)

zwitterion (14-6d)

QUESTIONS FOR REVIEW AND THOUGHT

Red-numbered questions have short answers at the back of this book in Appendix M and fully worked solutions in the *Student Solutions Manual.*

Review Questions

These questions test vocabulary and simple concepts.

1. Define a Brønsted-Lowry acid and a Brønsted-Lowry base.
2. Explain in your own words what 100% ionization means.
3. Write the chemical equation for the autoionization of water. Write the equilibrium constant expression for this reaction. What is the value of the equilibrium constant at 25 °C? What is this constant called?
4. When OH^- is the base in a conjugate acid-base pair, the acid is ________; when OH^- is the acid, the base is ________.
5. Designate the acid and the base on the left side of these equations, and designate the conjugate partner of each on the right side.
   (a) $HNO_3(aq) + H_2O(\ell) \longrightarrow H_3O^+(aq) + NO_3^-(aq)$
   (b) $NH_4^+(aq) + CN^-(aq) \longrightarrow NH_3(aq) + HCN(aq)$
6. Dissolving ammonium bromide in water gives an acidic solution. Write a balanced equation showing how that can occur.
7. Solution A has a pH of 8 and solution B a pH of 10. Which has the greater hydronium ion concentration? How many times greater is its concentration?
8. Contrast the main ideas of the Brønsted-Lowry and Lewis acid-base definitions. Name and write the formula for a substance that behaves as a Lewis acid but not as a Brønsted-Lowry acid.

Topical Questions

These questions are keyed to major topics in the chapter. Usually a question that is answered at the back of the book is paired with a similar one that is not.

Brønsted-Lowry Acids and Bases (Sections 14-1, 14-2)

9. Write a chemical equation to describe the proton transfer that occurs when each of these acids is added to water.
   (a) HCO_3^- (b) HCl
   (c) CH_3COOH (d) HCN
10. Write a chemical equation to describe the proton transfer that occurs when each of these acids is added to water.
   (a) HIO (b) $CH_3(CH_2)_4COOH$
   (c) HOOCCOOH (d) $CH_3NH_3^+$
11. Write a chemical equation to describe the proton transfer that occurs when each of these bases is added to water.
   (a) HSO_4^- (b) CH_3NH_2
   (c) I^- (d) $H_2PO_4^-$
12. Write a chemical equation to describe the proton transfer that occurs when each of these bases is added to water.
   (a) PO_4^{3-} (b) SO_3^{2-} (c) HPO_4^{2-}
13. Write the formula and name for the conjugate partner for each acid or base.
   (a) HI (b) NO_3^-
   (c) CO_3^{2-} (d) H_2CO_3
   (e) HSO_4^- (f) SO_3^{2-}
14. Write the formula and name for the conjugate partner for each acid or base.
   (a) CN^- (b) SO_4^{2-}
   (c) HS^- (d) S^{2-}
   (e) HSO_3^- (f) HCOOH (formic acid)
15. Which are conjugate acid-base pairs?
   (a) NH_2^- and NH_4^+ (b) NH_3 and NH_2^-
   (c) H_3O^+ and H_2O (d) OH^- and O^{2-}
   (e) H_3O^+ and OH^-
16. Which are conjugate acid-base pairs?
   (a) O^{2-} and H_3O^+ (b) H_3O^+ and O^{2-}
   (c) NH_2^- and NH_3 (d) NH_3 and NH_4^+
   (e) O^{2-} and H_2O
17. Identify the acid and the base that are reactants in each equation; identify the conjugate base and conjugate acid on the product side of the equation.
   (a) $HS^-(aq) + H_2O(\ell) \longrightarrow H_2S(aq) + OH^-(aq)$
   (b) $S^{2-}(aq) + NH_4^+(aq) \longrightarrow NH_3(g) + HS^-(aq)$
   (c) $HCO_3^-(aq) + HSO_4^-(aq) \longrightarrow H_2CO_3(aq) + SO_4^{2-}(aq)$
   (d) $NH_3(aq) + NH_2^-(aq) \longrightarrow NH_2^-(aq) + NH_3(aq)$
18. Identify the acid and the base that are reactants in each equation; identify the conjugate base and conjugate acid on the product side of the equation.
   (a) $CN^-(aq) + CH_3COOH(aq) \longrightarrow CH_3COO^-(aq) + HCN(aq)$
   (b) $O^{2-}(aq) + H_2O(\ell) \longrightarrow 2\ OH^-(aq)$
   (c) $HCO_2^-(aq) + H_2O(\ell) \longrightarrow HCOOH(aq) + OH^-(aq)$

pH and Autoionization of Water (Sections 14-3, 14-4)

19. Consider these four solutions:

Solution	$[H_3O^+]$ (M)	$[OH^-]$ (M)
D	2×10^{-3}	
E		2×10^{-7}
F	4×10^{-5}	
G		5×10^{-11}

   (a) Which solution has the highest H_3O^+ concentration?
   (b) Which solution has the highest OH^- concentration?
   (c) Which solution is closest to being a neutral solution?

20. Consider these four solutions:

Solution	$[H_3O^+]$ (M)	$[OH^-]$ (M)
W	5×10^{-6}	
X		2×10^{-4}
Y	4×10^{-2}	
Z		5×10^{-11}

(a) Which solution has the highest H_3O^+ concentration?
(b) One solution's H_3O^+ concentration is equal to the OH^- concentration of a different solution. Identify the two solutions and their concentrations.
(c) Which solution is closest to being a neutral solution?

21. Pyridine, C_5H_5N, is a weak base. Write a balanced chemical equation to represent why an aqueous solution of pyridine is basic.
22. Amantadine, $C_{10}H_{15}NH_2$, is a base used to treat Parkinson's disease. Write a balanced chemical equation to represent why an aqueous solution of amantadine is basic.
23. Pyruvic acid, $CH_3COCOOH$, is produced during aerobic respiration. Write a balanced chemical equation to represent why an aqueous solution of pyruvic acid is acidic.
24. Formic acid, HCOOH, is found in ants. Write a balanced chemical equation to represent why an aqueous solution of formic acid is acidic.
25. Milk of magnesia, $Mg(OH)_2$, has a pH of 10.5. Calculate the hydronium ion concentration of the solution. Is this solution acidic or basic?
26. A sample of coffee has a pH of 4.3. Calculate the hydronium ion concentration in this coffee. Is the coffee acidic or basic?
27. Calculate the pH of a solution that is 0.025-M in NaOH. Calculate the pOH of this solution.
28. Calculate the pH of a 0.0013-M solution of HNO_3. Calculate the pOH of this solution.
29. The hydronium ion concentration of a cyanoacetic acid solution is 0.032 M. Calculate its pOH.
30. A solution of benzyl amine, $C_7H_7NH_2$, has a hydroxide ion concentration of 2.4×10^{-3} M. Calculate the pH of the solution. Calculate its pOH.
31. A 1000.-mL solution of hydrochloric acid has a pH of 1.3. Calculate the mass (g) of HCl dissolved in the solution.
32. The pH of a $Ba(OH)_2$ solution is 10.66 at 25 °C. Calculate the hydroxide ion concentration of this solution. If the solution volume is 250. mL, calculate the mass (g) of $Ba(OH)_2$ that was used to make this solution.
33. Make these interconversions. In each case tell whether the solution is acidic or basic.

pH	$[H_3O^+]$ (M)	$[OH^-]$ (M)
(a) 1.00	________	________
(b) 10.5	________	________
(c) ________	1.8×10^{-4}	________
(d) ________	________	2.3×10^{-5}

34. Make these interconversions. In each case tell whether the solution is acidic or basic.

pH	$[H_3O^+]$ (M)	$[OH^-]$ (M)
(a) ________	6.1×10^{-7}	________
(b) ________	________	2.2×10^{-9}
(c) 4.67	________	________
(d) ________	2.5×10^{-2}	________
(e) 9.12	________	________

35. Figure 14.3 shows the pH of some common solutions. How many times more acidic or basic is each of these compared with a neutral solution?
(a) Milk (b) Seawater
(c) Blood (d) Battery acid
36. Figure 14.3 shows the pH of some common solutions. How many times more acidic or basic is each of these compared with a neutral solution?
(a) Black coffee (b) Household ammonia
(c) Baking soda (d) Vinegar
37. The measured pH of a sample of seawater is 8.30.
(a) Calculate the H_3O^+ concentration.
(b) Is the sample acidic or basic?

Acid-Base Strengths (Sections 14-5, 14-6)

38. Acid A has $K_a = 1 \times 10^{-5}$; Acid Z has $K_a = 5 \times 10^{-6}$; Base X has $K_b = 1 \times 10^{-4}$; Base Y has $K_b = 4 \times 10^{-5}$.
(a) Which acid is the stronger acid? Explain your answer.
(b) Which base is the stronger base? Explain your answer.
(c) Which base has the stronger conjugate acid? Explain your answer.
(d) Which acid has the weaker conjugate base? Explain your answer.
39. Valine is an amino acid with this Lewis structure:

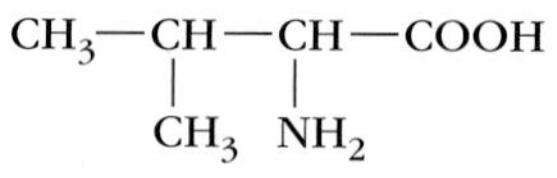

Write the Lewis structure for the zwitterion form of valine.

40. Leucine is an amino acid with this Lewis structure:

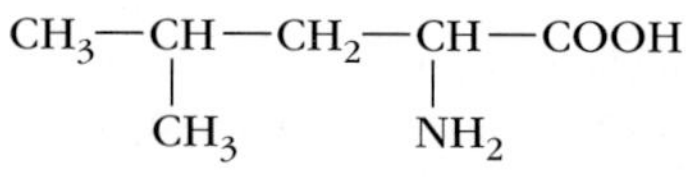

Write the Lewis structure for the zwitterion form of leucine.

41. Valine (see Question 39) can also exist as two other ionic forms: one has a net positive charge, the other has a net negative charge.
(a) Write the Lewis structure for each of these forms.
(b) Which form exists at high pH?
(c) Which exists at low pH?

42. Leucine (see Question 40) can also exist as two other ionic forms: one has a net positive charge, the other has a net negative charge.
(a) Write the Lewis structure for each of these forms.
(b) Which form exists at high pH?
(c) Which exists at low pH?
43. Write ionization equations and ionization constant expressions for these acids and bases.
(a) CH_3COOH (b) HCN
(c) SO_3^{2-} (d) PO_4^{3-}
(e) NH_4^+ (f) H_2SO_4
44. Write ionization equations and ionization constant expressions for these acids and bases.
(a) F^- (b) NH_3
(c) H_2CO_3 (d) H_3PO_4
(e) CH_3COO^- (f) S^{2-}
45. Which solution is more acidic?
(a) 0.10-M H_2CO_3 or 0.10-M NH_4Cl
(b) 0.10-M HF or 0.10-M $KHSO_4$
(c) 0.1-M $NaHCO_3$ or 0.1-M Na_2HPO_4
(d) 0.1-M H_2S or 0.1-M HCN
46. Which solution is more basic?
(a) 0.10-M NH_3 or 0.10-M NaF
(b) 0.10-M K_2S or 0.10-M K_3PO_4
(c) 0.10-M $NaNO_3$ or 0.10-M $NaCH_3COO$
(d) 0.10-M NH_3 or 0.10-M KCN
47. Without doing any calculations, assign each of these 0.10-M aqueous solutions to one of these pH ranges: pH 2; pH between 2 and 6; pH between 6 and 8; pH between 8 and 12; pH 12.
(a) HNO_2 (b) NH_4Cl (c) NaF
(d) $Mg(CH_3COO)_2$ (e) BaO (f) $KHSO_4$
(g) $NaHCO_3$ (h) $BaCl_2$
48. Based on formulas alone, which is the stronger acid?
(a) H_2CO_3 or H_2SO_4 (b) HNO_3 or HNO_2
(c) $HClO_4$ or H_2SO_4 (d) H_3PO_4 or $HClO_3$
(e) H_2SO_3 or H_2SO_4
49. Based on formulas alone, classify each of the following oxoacids as strong or weak.
(a) H_3PO_4 (b) H_2SO_4
(c) HClO (d) $HClO_4$
(e) HNO_3 (f) H_2CO_3
(g) HNO_2
50. Write balanced chemical equations that show phosphoric acid, H_3PO_4, ionizing stepwise as a polyprotic acid.
51. Write stepwise chemical equations for protonation or deprotonation of each of these polyprotic acids and bases in water.
(a) CO_3^{2-} (b) H_3AsO_4
(c) $NH_2CH_2COO^-$ (glycinate ion, a diprotic base)
52. Write stepwise chemical equations for protonation or deprotonation of each of these polyprotic acids and bases in water.
(a) H_2SO_3 (b) S^{2-}
(c) $NH_3CH_2COOH^+$ (glycinium ion, a diprotic acid)

Problem Solving Using K_a and K_b (Section 14-7)

53. Write the ionization equation for a weak acid and the equation for its conjugate base reaction with water. Show that adding these two equations gives the autoionization equation for water.
54. Calculate the pH of each solution in Question 47 to verify your prediction.
55. A 0.015-M solution of cyanic acid has a pH of 2.67. Calculate the ionization constant, K_a, of the acid.
56. Calculate the K_a of butyric acid if a 0.025-M butyric acid solution has a pH of 3.21.
57. The pH of a 0.10-M solution of propanoic acid, CH_3CH_2COOH, a weak organic acid, is measured at equilibrium and found to be 2.93 at 25 °C. Calculate the K_a of propanoic acid.
58. Calculate the equilibrium concentrations of H_3O^+, acetate ion, and acetic acid in a 0.20-M aqueous solution of acetic acid, CH_3COOH.
59. (a) Calculate the pH of a 0.050-M solution of benzoic acid, C_6H_5COOH; $K_a = 1.2 \times 10^{-4}$ at 25 °C.
(b) Calculate the percent of the acid that has ionized in this solution.
60. Calculate the pH of a 0.12-M aqueous solution of the base aniline, $C_6H_5NH_2$; $K_b = 3.9 \times 10^{-10}$.
61. Calculate the $[OH^-]$ and the pH of a 0.024-M methylamine solution; $K_b = 5.0 \times 10^{-4}$.
62. Amantadine, $C_{10}H_{15}NH_2$, is a weak base used in the treatment of Parkinson's disease. Its conjugate acid has $K_a = 7.9 \times 10^{-11}$. Calculate the pH of a 0.0010-M aqueous solution of amantadine at 25 °C.
63. Pyridine, C_5H_5N, is a weak base; its conjugate acid has $K_a = 6.3 \times 10^{-6}$. A 0.2-M solution of pyridine has pH = 8.5. Calculate the concentration of unreacted pyridine in this solution.
64. Lactic acid, $C_3H_6O_3$, occurs in sour milk as a result of the metabolism of certain bacteria. Calculate the pH of a solution of 56. mg lactic acid in 250. mL water. K_a for D-lactic acid is 1.5×10^{-4}.
65. Boric acid is a weak acid often used as an eyewash. K_a for boric acid is 5.8×10^{-10}. Calculate the pH of a 0.10-M solution of boric acid.

Acid-Base Reactions of Salts (Section 14-8)

66. Complete each of these reactions by filling in the blanks. Predict whether each reaction is product-favored or reactant-favored, and explain your reasoning.
(a) ______(aq) + Br^-(aq) ⇌ NH_3(aq) + HBr(aq)
(b) CH_3COOH(aq) + CN^-(aq) ⇌ ______(aq) + HCN(aq)
(c) ______(aq) + H_2O(ℓ) ⇌ NH_3(aq) + OH^-(aq)
67. Complete each of these reactions by filling in the blanks. Predict whether each reaction is product-favored or reactant-favored, and explain your reasoning.
(a) ______(aq) + HSO_4^-(aq) ⇌ HCN(aq) + SO_4^{2-}(aq)
(b) H_2S(aq) + H_2O(ℓ) ⇌ H_3O^+(aq) + ______(aq)
(c) H^-(aq) + H_2O(ℓ) ⇌ OH^-(aq) + ______(g)
68. Predict which of these acid-base reactions are product-favored and which are reactant-favored. In each case write a balanced equation for any reaction that might occur, even if the reaction is reactant-favored. Consult Table 14.2 if necessary.
(a) H_2O(ℓ) + HNO_3(aq) (b) H_3PO_4(aq) + H_2O(ℓ)
(c) CN^-(aq) + HCl(aq) (d) NH_4^+(aq) + F^-(aq)
69. Predict which of these acid-base reactions are product-favored and which are reactant-favored. In each case write a balanced equation for any reaction that might occur, even if the reaction is reactant-favored. Consult Table 14.2 if necessary.
(a) NH_4^+(aq) + HPO_4^{2-}(aq)
(b) CH_3COOH(aq) + OH^-(aq)
(c) HSO_4^-(aq) + $H_2PO_4^-$(aq)
(d) CH_3COOH(aq) + F^-(aq)
70. For each salt, predict whether an aqueous solution has a pH less than, equal to, or greater than 7. Explain your prediction.
(a) $AlCl_3$ (b) Na_2S (c) $NaNO_3$
71. For each salt, predict whether an aqueous solution has a pH less than, equal to, or greater than 7. Explain your prediction.
(a) Na_2HPO_4 (b) $(NH_4)_2S$ (c) KCH_3COO
72. Explain why $BaCO_3$ is soluble in aqueous HCl, but $BaSO_4$, which is used to make the intestines visible in X-ray photographs, remains sufficiently insoluble in the HCl in a human stomach so that poisonous barium ions do not get into the bloodstream.
73. Which of these substances has greater solubility at pH = 2 than at pH = 7?
(a) $Cu(OH)_2$ (b) $CuSO_4$ (c) $CuCO_3$
(d) CuS (e) $Cu_3(PO_4)_2$

Lewis Acids and Bases (Section 14-9)

74. Write a chemical equation that illustrates how the hydrated Ni^{2+} ion, $[Ni(H_2O)_6]^{2+}$, acts as an acid in aqueous solution.
75. The ionization constants of several kinds of hydrated metal ions are given in Table 14.6. Explain the significant difference between the K_a values of $[Fe(H_2O)_6]^{2+}$ and $[Fe(H_2O)_6]^{3+}$.
76. Which of these is a Lewis acid? A Lewis base?
(a) O^{2-} (b) CO_2 (c) H^-
77. Which of these is a Lewis acid? A Lewis base?
(a) Al^{3+} (b) H_2O (c) SCN^-
78. Which of these is a Lewis acid? A Lewis base?
(a) Cr^{3+} (b) SO_3 (c) CH_3NH_2
79. Which of these is a Lewis acid? A Lewis base?
(a) NH_3 (b) $BeCl_2$ (c) BCl_3
80. Identify the Lewis acid and the Lewis base in each reaction.
(a) H_2O(ℓ) + SO_2(aq) ⟶ H_2SO_3(aq)
(b) H_3BO_3(aq) + OH^-(aq) ⟶ $B(OH)_4^-$(aq)
(c) Cu^{2+}(aq) + 4 NH_3(aq) ⟶ $[Cu(NH_3)_4]^{2+}$(aq)
(d) 2 Cl^-(aq) + $SnCl_2$(aq) ⟶ $SnCl_4^{2-}$(aq)
81. Identify the Lewis acid and the Lewis base in each reaction.
(a) I_2(s) + I^-(aq) ⟶ I_3^-(aq)
(b) SO_2(g) + BF_3(g) ⟶ O_2SBF_3(s)
(c) Au^+(aq) + 2 CN^-(aq) ⟶ $[Au(CN)_2]^-$(aq)
(d) CO_2(g) + H_2O(ℓ) ⟶ H_2CO_3(aq)
82. Trimethylamine, $(CH_3)_3N$:, reacts readily with diborane, B_2H_6. The diborane dissociates to two BH_3 fragments, each of which can react with trimethylamine to form a complex, $(CH_3)_3N{:}BH_3$. Write an equation for this reaction and interpret it in terms of Lewis acid-base theory.
83. Draw a Lewis structure for ICl_3. Predict the shape of this molecule. Does it function as a Lewis acid or base when it reacts with chloride ion to form ICl_4^-? What is the structure of this ion?
84. Write the formula for the conjugate acid of $[Zn(OH)_4]^{2-}$.
85. Write the formula for the conjugate base of $[Zn(H_2O)_3(OH)]^{1+}$.

Additional Applied Acid-Base Chemistry (Section 14-10)

86. Double-acting baking powder contains two salts, sodium hydrogen carbonate and potassium dihydrogen phosphate, whose anions react in water to form CO_2 gas. Write a balanced chemical equation for the reaction. Which anion is the acid and which is the base?
87. Common soap is made by reacting sodium carbonate with stearic acid, $CH_3(CH_2)_{16}COOH$. Write a balanced equation for the reaction.
88. If 1 g of each antacid in Table 14.8 reacted with equal volumes of stomach acid, calculate which would neutralize the most stomach acid.
89. Why is it not a good idea to substitute dishwashing detergent for automobile-washing detergent?

90. Why do cleaning products containing sodium hydroxide feel slippery when you get them on your skin?
91. Some cooked fish have a "fishy" odor due to amines. Explain why putting lemon juice on the fish reduces this odor.

General Questions

These questions are not explicitly keyed to chapter topics; many require integration of several concepts.

92. Classify each of these as a strong acid, weak acid, strong base, weak base, amphiprotic substance, or neither acid nor base.
    (a) CH_3COOH (b) Na_2O (c) H_2SO_4
    (d) NH_3 (e) $Ba(OH)_2$ (f) $H_2PO_4^-$
93. Classify each of these as a strong acid, weak acid, strong base, weak base, amphiprotic substance, or neither acid nor base.
    (a) HCl (b) NH_4^+ (c) H_2O
    (d) CH_3COO^- (e) CH_4 (f) CO_3^{2-}
94. Several acids and their respective equilibrium constants are:

HF	$K_a = 6.8 \times 10^{-4}$
HS^-	$K_a = 1 \times 10^{-19}$
CH_3COOH	$K_a = 1.8 \times 10^{-5}$

    (a) Which is the strongest acid?
    (b) Which is the weakest acid?
    (c) Which acid has the weakest conjugate base?
    (d) Which acid has the strongest conjugate base?
95. State whether equal molar amounts of these would have a pH equal to 7, less than 7, or greater than 7.
    (a) A weak base and a strong acid react.
    (b) A strong base and a strong acid react.
    (c) A strong base and a weak acid react.
96. Ascorbic acid (vitamin C, $C_6H_8O_6$) is a diprotic acid ($K_{a_1} = 7.9 \times 10^{-5}$, $K_{a_2} = 1.6 \times 10^{-12}$). Calculate the pH of a solution that contains 5.0 mg acid per mL water. (Assume that only the first ionization is important in determining pH.)
97. Does the pH of the solution increase, decrease, or stay the same when you
    (a) Add solid ammonium chloride to 100. mL of 0.10-M NH_3?
    (b) Add solid sodium acetate to 50.0 mL of 0.015-M acetic acid?
    (c) Add solid NaCl to 25.0 mL of 0.10-M NaOH?
98. Does the pH of the solution increase, decrease, or stay the same when you
    (a) Add solid sodium oxalate, $Na_2C_2O_4$, to 50.0 mL of 0.015-M oxalic acid?
    (b) Add solid ammonium chloride to 100. mL of 0.016-M HCl?
    (c) Add 20.0 g NaCl to 1.0 L of 0.012-M sodium acetate, $NaCH_3COO$?
99. An aqueous ammonium iodide solution has a pH of 5.50. Calculate the ammonium iodide concentration of this solution.
100. Sodium hypochlorite, NaOCl, is used as a source of chlorine in some laundry bleaches, swimming pool disinfectants, and water treatment plants. Calculate the pH of a 0.010-M solution of NaOCl ($K_b = 1.5 \times 10^{-7}$).

Applying Concepts

These questions test conceptual learning.

101. Some commercial baking powders contain a dry mixture of aluminum sulfate and sodium hydrogen carbonate. When it is mixed with water and baking dough, a reaction occurs that produces a gas that makes the dough rise. Identify the gas and write balanced chemical equations for the reactions that create it. Explain the role of water in the process.
102. From the mid-1800s to the present, the average acidity of the oceans' surface water has increased 30% to its current pH of 8.1. Calculate what the pH was in the mid-1800s.
103. Carbon dioxide released from burning of fossil fuels contributes to the acidity of the oceans' surface water, whose current average pH is 8.1. It has been estimated that if CO_2 emissions are not curbed, the oceans' surface water pH could drop to 7.8 by the end of this century. Calculate the percent change in acidity if that were to happen.
104. Ammonium lactate, a salt, is used as a moisturizing agent in medicinal creams.
    (a) Identify the acid and the base that form ammonium lactate by a neutralization reaction.
    (b) Predict whether an ammonium lactate solution is acidic, neutral, or basic. Explain your answer.
    (c) A medicinal cream contains 12% ammonium lactate by weight. Calculate the amount (mol) of ammonium lactate in 1.5 g of this cream.
    (d) Calculate the total number of ammonium and lactate ions in the 1.5 g sample.
105. When a 0.1-M aqueous ammonia solution is tested with a conductivity apparatus (← **Sec. 3-3b**), the bulb glows dimly. When a 0.1-M hydrochloric acid solution is tested, the bulb glows brightly. As water is added to each of the solutions, predict whether the bulb glows brighter, stops glowing, or stays the same. Explain your reasoning.
106. Using Table 14.1, calculate what the pH of pure water is at 10 °C, 25 °C, and 50 °C. Classify the water at each temperature as either acidic, neutral, or basic.
107. When all the water is evaporated from a sodium hydroxide solution, solid sodium hydroxide is obtained. However, if you evaporate the water in an ammonium hydroxide solution, you will not produce solid ammonium hydroxide. Explain why. What will remain after the water is evaporated?

Diagrams for Question 109.

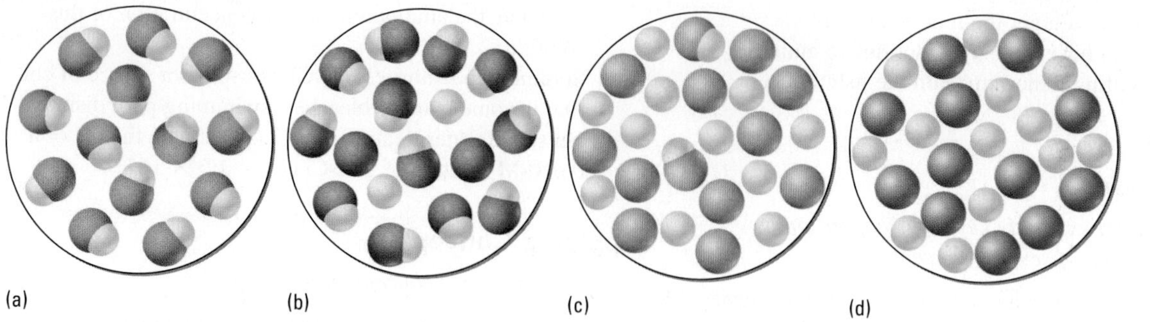
(a)
(b)
(c)
(d)

108. For each aqueous solution, predict what ions and molecules will be present. Without doing any calculations, list the ions and molecules in order of decreasing concentration.
   (a) HCl (b) $NaClO_4$ (c) HNO_2
   (d) NaClO (e) NH_4Cl (f) NaOH
109. The diagrams for Question 109 are nanoscale representations of different acids.
   (a) Which diagram best represents hydrochloric acid? (The yellow spheres are H^+ ions and the other spheres are Cl^- ions.)
   (b) Which diagram best represents acetic acid? (The yellow spheres are H^+ ions and the other spheres are CH_3COO^- ions.)
110. When asked to identify the conjugate acid-base pairs in the reaction

$$HCO_3^-(aq) + HSO_4^-(aq) \rightleftharpoons H_2CO_3(aq) + SO_4^{2-}(aq)$$

a student incorrectly wrote: "HCO_3^- is a base and HSO_4^- is its conjugate acid. H_2CO_3 is an acid and SO_4^{2-} is its conjugate base." Write a brief explanation to the student telling why the answer is incorrect.
111. Explain how the Arrhenius acid-base theory and the Brønsted-Lowry theory of acids and bases are explained by the Lewis acid-base theory.
112. Trichloroacetic acid, CCl_3COOH, has an acid dissociation constant of 2.0×10^{-1} at 25 °C.
   (a) Calculate the pH of a 0.0100-M trichloroacetic acid solution.
   (b) How many times greater is the H_3O^+ concentration in this solution than in a 0.0100-M acetic acid?
113. Explain why a 0.1-M NH_4NO_3 aqueous solution has a pH of 5.1.
114. Is the dissociation of water

$$2\ H_2O(\ell) \rightleftharpoons H_3O^+(aq) + OH^-(aq)$$

exothermic or endothermic? Refer to Table 14.1. Explain your answer.
115. Consider the nanoscale representations of aqueous solutions of three monoprotic acids—HX, HY, and HZ for Question 115. Water molecules are excluded for clarity.
   (a) Which acid has the largest K_a value? Explain your answer.
   (b) Which conjugate base—X^-, Y^-, or Z^-—has the largest K_b value? Explain your answer.
116. Consider the nanoscale representations of aqueous solutions of three monoprotic acids—HA, HB, and HC for Question 116. Water molecules are excluded for clarity.
   (a) Which acid has the smallest K_a value? Explain your answer.
   (b) Which conjugate base—A^-, B^-, or C^-—has the smallest K_b value? Explain your answer.

Aqueous solutions of three monoprotic acids—HX, HY, HZ; for Question 115.

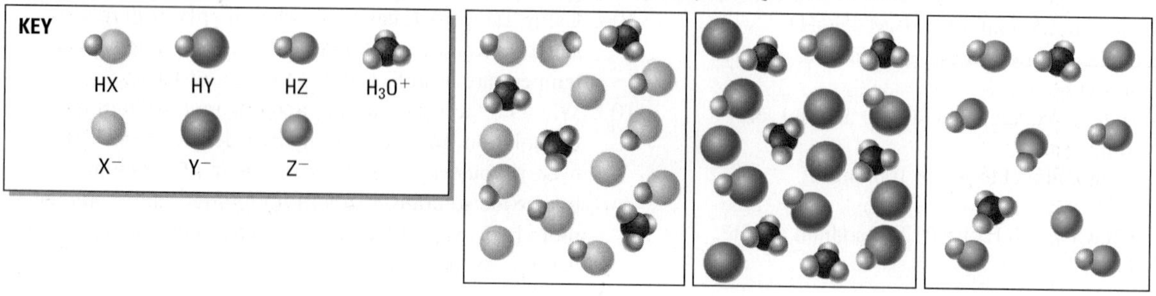
KEY
HX
HY
HZ
H3O+
X−
Y−
Z−

Aqueous solutions of three monoprotic acids—HA, HB, HC; for Question 116.

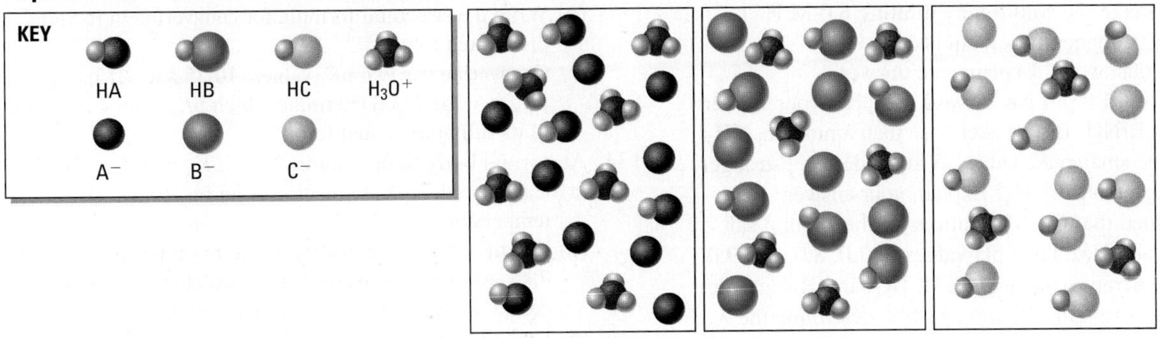
KEY
HA
HB
HC
H3O+
A−
B−
C−

117. Home gardeners spread aluminum sulfate powder around plants to increase the acidity of the soil. Describe the chemistry at the nanoscale level for such a soil treatment.
118. This grid has nine lettered boxes, each of which contains an item that is used to answer the questions that follow. Items may be used more than once and there may be more than one correct item in response to a question.

A K_2CO_3	**B** Cu^{2+}	**C** $[OH^-] = 3.0 \times 10^{-5}$ M
D H_2SO_3	**E** $H_2PO_4^-$	**F** 0.10-M LiOH
G $[H_3O^+] = 2.00 \times 10^{-8}$ M	**H** $NaNO_2$	**I** 0.010-M $HClO_4$

Place the letter(s) of the correct selection(s) on the appropriate line. Assume $T = 298$ K.
(a) Has pH $>$ 7.0 ________
(b) $[H_3O^+] > 1.0 \times 10^{-7}$ M ________
(c) A basic salt ________
(d) $K_a > K_b$ ________
(e) A Lewis acid ________
(f) Has pH = 2.00 ________
(g) A polyprotic acid ________

119. This grid has nine lettered boxes, each of which contains an item that is used to answer the questions that follow. Items may be used more than once and there may be more than one correct item in response to a question.

A NH_4Br	**B** Zn^{2+}	**C** 0.0010-M KOH
D 0.010-M HBr	**E** H_2S	**F** H_2O
G K_2HPO_4	**H** Na_3PO_4	**I** KF

Place the letter(s) of the correct selection(s) on the appropriate line. Assume $T = 298$ K.
(a) Has pH $<$ 7.0 ________
(b) Contains a conjugate acid or conjugate base of another item in the grid. ________
(c) Has pH = 11.00 ________
(d) A polyprotic acid ________
(e) $K_a > K_b$ ________
(f) A basic salt ________
(g) $[H_3O^+] < 1.0 \times 10^{-7}$ M ________
(h) A Lewis acid ________

More Challenging Questions

These questions require more thought and integrate several concepts.

120. The conductivity of BrF_3 is increased when AgF is added. Explain this increase in conductivity.
121. Calculate the pH of 0.00500 M $NH_4[B(OH)_4]$: (a) when the simplifying assumption regarding x is made; (b) by solving a quadratic equation; (c) using the method of successive approximations.
122. Calculate the pH of 0.010 M NH_4HCO_3.
123. Calculate the pH of 0.10 M $AlCl_3$.
124. If 1 g each of vinegar, lemon juice, and lactic acid react with equal masses of baking soda, which produces the most CO_2 gas?
125. A person claimed that his stomach ruptured when he took a teaspoonful of baking soda in a glass of water to relieve heartburn after a full meal (1 tsp = 5.0 g $NaHCO_3$). Assume that the pH of stomach acid is 1 and that the stomach has a volume of 1 L when expanded fully. Body temperature is 37 °C. Calculate the volume of carbon dioxide gas generated by the reaction of baking soda with stomach acids. Might his stomach have ruptured from this volume of CO_2?
126. A chilled carbonated beverage is opened and warmed to room temperature. Does the pH change and, if so, how does it change?

127. For an experiment a student needs a pH = 9.0 solution. He plans to make the solution by diluting 6.0-M HCl until the H_3O^+ concentration equals 1.0×10^{-9} M.
   (a) Will this plan work? Explain your answer.
   (b) Will diluting 1.00-M NaOH work? Explain your answer.
128. Explain why $BrNH_2$ is a weaker base than ammonia, NH_3. Which has the smaller K_b value? Will $ClNH_2$ be a stronger or weaker base than $BrNH_2$? Explain your answer.
129. It is determined that 0.1-M solutions of the sodium salts NaM, NaQ, and NaZ have pH values of 7.0, 8.0, and 9.0, respectively. Arrange the acids HM, HQ, and HZ in order of decreasing strength. Where possible, determine the K_a values of these acids.
130. Hydrogen peroxide, HOOH, is a powerful oxidizing agent that, in diluted form, is used as an antiseptic. For HOOH: $K_a = 2.1 \times 10^{-12}$; for OOH^-: $K_b = 4.8 \times 10^{-3}$.
   (a) Is hydrogen peroxide a strong or weak acid?
   (b) Is OOH^- a strong or weak base?
   (c) What is the relationship between HOOH and OOH^-?
   (d) Calculate the pH of 0.100-M aqueous hydrogen peroxide.
131. At 25 °C, a 0.10% aqueous solution of adipic acid, $C_5H_9O_2COOH$, has a pH of 3.2. A saturated solution of the acid, which contains 1.44 g acid per 100. mL of solution, has a pH = 2.7. Calculate the percent dissociation of adipic acid in each solution.
132. Niacin, a B vitamin, can act as an acid and as a base.
   (a) Write chemical equations showing the action of niacin as an acid in water. Symbolize niacin as HNc.
   (b) Write chemical equations showing the action of niacin as a base in water.
   (c) A 0.020-M aqueous solution of niacin has a pH = 3.26. Calculate the K_a of niacin.
   (d) Calculate the K_b of niacin.
   (e) Write the chemical equation that accompanies the K_b expression for niacin.
133. The loss of CO_2 from lysine forms cadaverine, an aptly named compound with an exceptionally revolting smell.

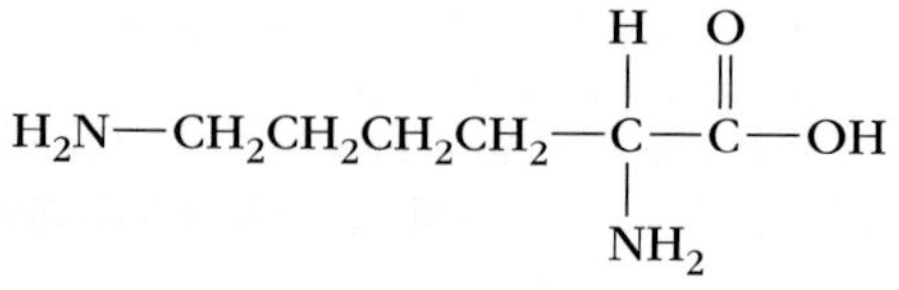

lysine

   (a) Write the structural formula for cadaverine.
   (b) Write the structural formula for cadaverine in its fully protonated form.
   (c) Cadaverine has two pK_a values: 10.25 and 9.13; $pK_a = -\log(K_a)$. Determine which pK_a is for cadaverine in its fully protonated form.
134. At normal body temperature, 37 °C, $K_w = 2.5 \times 10^{-14}$.
   (a) Calculate the pH of a neutral solution at this temperature.
   (b) The pH of blood at this temperature ranges from 7.35 to 7.45. Assuming a pH of 7.40, calculate the H_3O^+ concentration of the blood at this temperature.
   (c) By how much does the H_3O^+ concentration at this temperature differ from that at 25 °C?
135. The structural formula for lysine is given in Question 133. Explain why lysine has three pK_a values: 2.18, 8.95, and 10.53. $pK_a = -\log(K_a)$. Write a balanced net ionic equation for the reaction corresponding to each pK_a.

Conceptual Challenge Problems

These rigorous, thought-provoking problems integrate conceptual learning with problem solving and are suitable for group work.

CP14.A (Section 14-4) Determine the pH of water at 200 °C. Liquid water this hot must be under a pressure greater than 1.0 atm and might be found in a pressurized water reactor located in a nuclear power plant.

CP14.B (Section 14-5) Develop a set of rules by which you could predict the pH for solutions of strong or weak acids and strong or weak bases without using a calculator. Your predictions need to be accurate to ±1 pH unit. Assume that you know the concentration of the acid or base and that for the weak acids and bases you can look up the pK_a ($-\log K_a$) or K_a values. What rules would work to predict pH?

Additional Aqueous Equilibria

15

© hangingpixels/Shutterstock.com

Many natural phenomena occur in aqueous solutions. Stalactites, hanging from the ceiling of the cave in the photo, result from changes in the solubility of an ionic solid. What is the ionic solid that makes up a stalactite? What chemical reactions form stalactites? (Section 15-5a) What is acid rain and how does it affect the substance from which stalactites are made? (Sections 15-3 and 15-5a) In this chapter we consider aqueous acid-base and solubility equilibria and the factors that affect them.

In environments as different as the interior of red blood cells, coral reefs, ocean waters, and clouds high above Earth's surface, important interactions occur among solutes in aqueous solution. This chapter extends the discussion of aqueous solutions begun in Chapters 13 and 14 (conjugate acid-base behavior, acid-base neutralization, the link between solubility and precipitation) to quantitative aspects related to

- buffers, which are combinations of a weak acid and its conjugate base, or a weak base with its conjugate acid.
- acid-base titrations, which involve acid-base neutralization reactions.
- equilibria associated with solutions of slightly soluble salts.

15-1 Buffer Solutions

Adding a small amount of acid or base to pure water radically changes the pH. If 0.010 mol (1.0×10^{-2} mol) HCl is added to 1.0 L pure water, the pH changes from 7.00 to 2.00 because $[H_3O^+]$ changes from 1.0×10^{-7} M to 1.0×10^{-2} M. This pH change represents a *100,000-fold increase* in H_3O^+ concentration. Similarly, if 0.010 mol NaOH is added to 1.0 L pure water, the pH goes from 7.00 to 12.00, a *100,000-fold decrease* in $[H_3O^+]$ and a *100,000-fold increase* in $[OH^-]$. Most aquatic organisms could not survive such dramatic pH changes because they can live only within a narrow pH range. For example, if acid rain lowers the pH of a lake or stream sufficiently, fish such as trout may die.

Unlike pure water and aqueous solutions of NaOH or HCl, an aqueous solution called a **buffer solution** *maintains a nearly constant pH when limited amounts of base or acid are added to it*. All buffer solutions contain a **buffer**—*a chemical system that resists change in pH*. For example, a solution that contains 0.50 mol acetic acid, CH_3COOH, and 0.50 mol sodium acetate, $NaCH_3COO$, in 1.0 L solution is a buffer with a pH of 4.74. When 0.010 mol of a strong acid is added to it, the pH changes from 4.74 to 4.72, only 0.02 pH unit; adding 0.010 mol of a strong base to this buffer changes the pH from 4.74 to 4.76 (Figure 15.1). These slight pH changes are much less than those that occur when the same amounts of HCl or NaOH are added to water. How do buffers, such as the sodium acetate-acetic acid buffer, offset such additions of acid or base without significant change in pH?

We will discuss shortly how to calculate the pH of a buffer.

15-1a Buffer Action

A buffer maintains a relatively constant pH because the buffer contains both *a weak acid that can react with added base* and *a weak base that can react with added acid* (Figure 15.2). In addition, the acid and base components of a buffer solution must not react with each other. A conjugate acid-base pair, such as acetic acid and acetate ion (from sodium acetate), satisfies this second requirement: When the acid and base react with each other they just produce conjugate base and conjugate acid—no observable change occurs. For example, acetic acid reacts with acetate ion to form acetate ion and acetic acid.

$$\underset{\text{Conj acid}}{CH_3COOH(aq)} + \underset{\text{Conj base}}{CH_3COO^-(aq)} \rightleftharpoons \underset{\text{Conj base}}{CH_3COO^-(aq)} + \underset{\text{Conj acid}}{CH_3COOH(aq)}$$

A buffer usually contains approximately equal concentrations of a weak acid and its conjugate base, or a weak base and its conjugate acid.

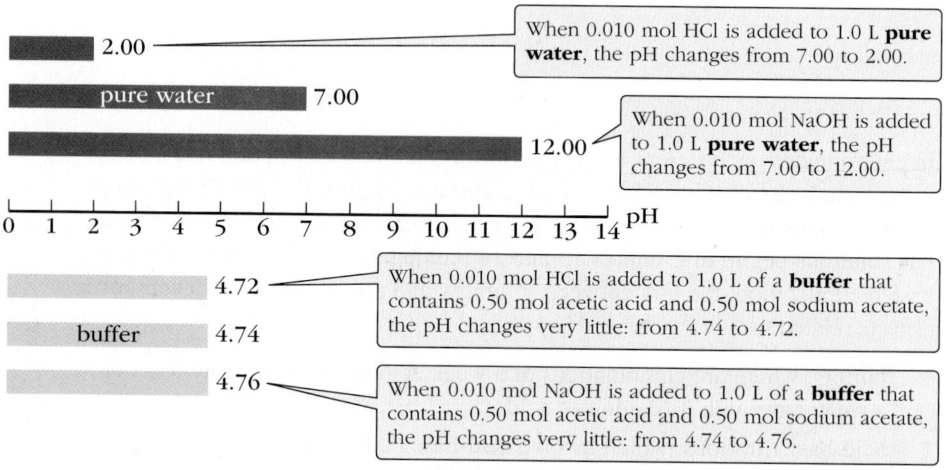
2.00
When 0.010 mol HCl is added to 1.0 L pure water, the pH changes from 7.00 to 2.00.
pure water
7.00
12.00
When 0.010 mol NaOH is added to 1.0 L pure water, the pH changes from 7.00 to 12.00.
0 1 2 3 4 5 6 7 8 9 10 11 12 13 14
pH
4.72
When 0.010 mol HCl is added to 1.0 L of a buffer that contains 0.50 mol acetic acid and 0.50 mol sodium acetate, the pH changes very little: from 4.74 to 4.72.
buffer
4.74
4.76
When 0.010 mol NaOH is added to 1.0 L of a buffer that contains 0.50 mol acetic acid and 0.50 mol sodium acetate, the pH changes very little: from 4.74 to 4.76.

Figure 15.1 pH changes. The pH of water changes dramatically when strong acid or base is added. The pH of a buffer changes hardly at all.

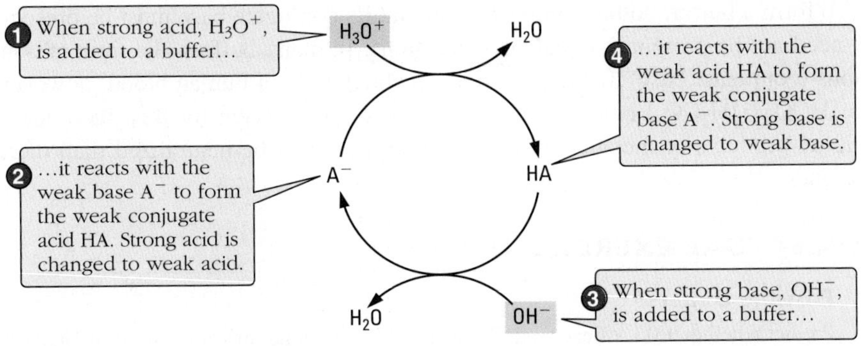
1 When strong acid, H_3O^+, is added to a buffer...
H_3O^+
H_2O
4 ...it reacts with the weak acid HA to form the weak conjugate base A^-. Strong base is changed to weak base.
2 ...it reacts with the weak base A^- to form the weak conjugate acid HA. Strong acid is changed to weak acid.
A^-
HA
H_2O
OH^-
3 When strong base, OH^-, is added to a buffer...

Figure 15.2 Buffer action. Adding either H_3O^+ or OH^- to a buffer changes pH only a little.

The blood of a mammal is an aqueous solution that maintains a constant pH, that is, a buffer. Consider human blood, whose normal pH is 7.40 ± 0.05. If the pH decreases below 7.35, a condition known as *acidosis* occurs; increasing the pH above 7.45 causes *alkalosis.* Both conditions can be life-threatening. Acidosis, for example, causes a decrease in oxygen transport by hemoglobin and depresses the central nervous system, leading in extreme cases to coma and death by creating weak and irregular cardiac contractions—symptoms of heart failure. To prevent such problems, your body must keep the pH of your blood nearly constant.

The term acidosis is used medically even though the pH is not less than 7.

Carbon dioxide provides the most important blood buffer (but not the only one). In solution, CO_2 reacts with water to form H_2CO_3, which ionizes to produce H_3O^+ and HCO_3^- ions. The equilibria are

$$CO_2(aq) + H_2O(\ell) \rightleftharpoons H_2CO_3(aq)$$
$$H_2CO_3(aq) + H_2O(\ell) \rightleftharpoons H_3O^+(aq) + HCO_3^-(aq)$$

Since H_2CO_3 is a weak acid and HCO_3^- is its conjugate weak base, together they constitute a buffer. The normal concentrations of H_2CO_3 and HCO_3^- in blood are 0.0025 M and 0.025 M, respectively—a 1 : 10 ratio. As long as the ratio of H_2CO_3 to HCO_3^- concentrations remains about 1 to 10, the pH of the blood remains near 7.4. (We will show how to calculate this pH in Problem-Solving Example 15.1.)

H_2CO_3: $K_a = 4.3 \times 10^{-7}$
HCO_3^-: $K_b = 2.3 \times 10^{-8}$

If a strong base such as NaOH is added to this buffer, carbonic acid—the conjugate acid in the buffer—will react with the added OH^-.

$$H_2CO_3(aq) + OH^-(aq) \longrightarrow HCO_3^-(aq) + H_2O(\ell)$$

$$K_c = \frac{1}{K_b(HCO_3^-)} = \frac{1}{2.3 \times 10^{-8}} = 4.3 \times 10^7$$

Here the equilibrium constant is $1/K_b$ of hydrogen carbonate ion, because the reaction is the reverse of the reaction of hydrogen carbonate ion with H_2O. Since OH^- is the strongest base that can exist in water solution, the reaction of hydroxide ions with carbonic acid is essentially complete, as indicated by the very large K_c value of 4.3×10^7.

If a strong acid such as HCl is added to this buffer, the HCO_3^- ion—the conjugate base in the buffer—reacts with the hydronium ions from HCl.

$$HCO_3^-(aq) + H_3O^+(aq) \longrightarrow H_2CO_3(aq) + H_2O(\ell)$$

$$K_c = \frac{1}{K_a(H_2CO_3)} = \frac{1}{4.3 \times 10^{-7}} = 2.3 \times 10^6$$

In this case, the equilibrium constant is $1/K_a$ of carbonic acid, because the reaction is the reverse of the ionization of carbonic acid. Since H_3O^+ is such a strong acid, the reaction between HCO_3^- and H_3O^+ is essentially complete and $K_c = 2.3 \times 10^6$.

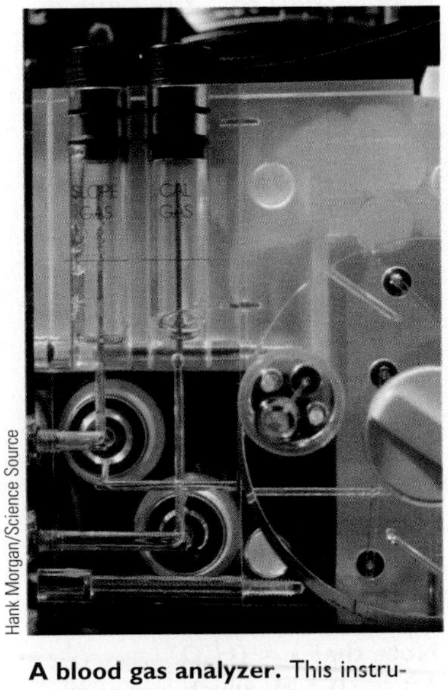
SLOPE GAS
CAL GAS

Hank Morgan/Science Source

A blood gas analyzer. This instrument measures CO_2 level, pH, and oxygen level in blood. These values are related and must be within a narrow range for good health.

To form a buffer, both a conjugate acid and its conjugate base must be present, but they need not be present in equal concentrations. In many buffers, the ratio of concentrations (conjugate base):(conjugate acid) is about 1:1. In human blood, however, the ratio $[HCO_3^-]:[H_2CO_3]$ is about 10:1. There is a good reason for this: there are more acidic by-products of metabolism in the blood that must be neutralized than there are basic ones.

CONCEPTUAL **EXERCISE 15.1**

Answers to **EXERCISES** are provided at the back of this book in Appendix L.

EXERCISES that are labeled **CONCEPTUAL** are designed to test your understanding of one or more concepts; they usually involve qualitative rather than quantitative thinking.

Possible Buffers?

Predict whether 1.0 L of each solution is a buffer: (a) equal amounts (mol) of HCl and NaCl; (b) equal amounts (mol) of KOH and KCl. Explain each prediction.

15-1b The pH of Buffer Solutions

If the K_a value and the concentrations of conjugate acid and conjugate base are known, there are two ways to calculate the pH of a buffer solution:

- Solve the K_a expression for $[H_3O]^+$ (Problem-Solving Example 15.1).
- Use the Henderson-Hasselbalch equation (Section 15-1c).

PROBLEM-SOLVING EXAMPLE 15.1

The pH of a Buffer from K_a

To mimic a blood buffer, a scientist prepared 1.00 L buffer containing 0.0025 mol carbonic acid and 0.025 mol hydrogen carbonate ion. Calculate the pH of the buffer. The K_a of carbonic acid is 4.3×10^{-7}.

Result 7.37

Analyze Carbonic acid (conjugate acid) and hydrogen carbonate ion (conjugate base) are a conjugate pair. The ionization-constant equation, K_a expression, and K_a value for carbonic acid can be used to calculate $[H_3O^+]$ and then pH.

Plan Set up an ICE table such as we have done for other equilibrium calculations. Let $x = [H_3O^+]$. Given the small value of K_a, assume that $x \ll 0.0025$; thus, $0.0025 - x \cong 0.0025$ and $0.025 + x \cong 0.025$. Substitute these values into the K_a expression to calculate $[H_3O^+]$ and then pH.

The symbol ≅ means "approximately equal to." The symbol ≪ means "much less than."

Execute

$$H_2CO_3(aq) + H_2O(\ell) \rightleftharpoons H_3O^+(aq) + HCO_3^-(aq)$$

$$K_a = 4.3 \times 10^{-7} = \frac{[H_3O^+][HCO_3^-]}{[H_2CO_3]}; \quad [H_3O^+] = K_a \times \frac{[H_2CO_3]}{[HCO_3^-]}$$

	H_2CO_3	+ H_2O ⇌	H_3O^+	+ HCO_3^-
*I*nitial concentration (mol/L)	0.0025		≅0	0.025
*C*hange as reaction occurs (mol/L)	$-x$		$+x$	$+x$
*E*quilibrium concentration (mol/L)	$0.0025 - x$		$+x$	$0.025 + x$

$$K_a = 4.3 \times 10^{-7} = \frac{x(0.025 + x)}{(0.0025 - x)} \cong \frac{x(0.025)}{(0.0025)}$$

$$x = (4.3 \times 10^{-7}) \times \frac{(0.0025)}{0.025} = 4.3 \times 10^{-8}$$

$$pH = -\log(4.3 \times 10^{-8}) = 7.37$$

Note that $x = [H_3O^+] = 4.3 \times 10^{-8}$ M, which is $\ll 0.0025$. Therefore, the assumption that $x \ll 0.0025$ was valid.

☑ **Reasonable Result Check** From the K_a expression, we see that when $[HCO_3^-] = [H_2CO_3]$, the $[H_3O^+] = K_a$. But, in this example, the concentration of HCO_3^-, the conjugate base, is ten times that of H_2CO_3, the conjugate acid. Therefore, the $[H_3O^+]$ should be ten times smaller than the K_a, which it is.

PROBLEM-SOLVING PRACTICE 15.1

Calculate the pH of blood containing 0.0020-M carbonic acid and 0.025-M hydrogen carbonate ion.

PROBLEM-SOLVING PRACTICE answers are provided at the back of this book in Appendix K.

15-1c The Henderson-Hasselbalch Equation

The *Henderson-Hasselbalch equation* can be used conveniently to calculate the pH of a buffer containing known concentrations of conjugate base and conjugate acid. The equation can also be applied to determine the ratio of conjugate base to conjugate acid concentrations needed to achieve a buffer of a given pH. This equation is derived by writing the acid ionization constant expression for a weak acid, HA, and solving for $[H_3O^+]$; A^- is the conjugate base of HA.

$$HA(aq) + H_2O(\ell) \rightleftharpoons H_3O^+(aq) + A^-(aq)$$

$$K_a = \frac{[H_3O^+][\text{conj base}]}{[\text{conj acid}]} = \frac{[H_3O^+][A^-]}{[HA]}$$

$$[H_3O^+] = K_a \times \frac{[\text{conj acid}]}{[\text{conj base}]} = K_a \times \frac{[HA]}{[A^-]}$$

This result is more useful when expressed in terms of pH. To do so, take the base 10 logarithm of each side of this equation to give

$$\log[H_3O^+] = \log K_a + \log\frac{[\text{conj acid}]}{[\text{conj base}]} = \log K_a + \log\frac{[HA]}{[A^-]}$$

Multiply both sides of the equation by -1 and use the fact that the negative of a logarithm is the logarithm of a reciprocal: $-\log(x) = \log(1/x)$. This gives

$$-\log[H_3O^+] = -\log K_a + \log\frac{[\text{conj base}]}{[\text{conj acid}]} = -\log K_a + \log\frac{[A^-]}{[HA]}$$

Use the definition of pH and define $-\log K_a$ as pK_a (analogous to the definition of pH); the resulting equation is the Henderson-Hasselbalch equation.

$$pH = pK_a + \log\frac{[\text{conj base}]}{[\text{conj acid}]} = pK_a + \log\frac{[A^-]}{[HA]}$$ **Henderson-Hasselbalch equation**

Because $-\log(x) = \log\left(\frac{1}{x}\right)$,

$$-\log\frac{[\text{conj acid}]}{[\text{conj base}]} = \log\frac{[\text{conj base}]}{[\text{conj acid}]}$$

and $-\log\frac{[HA]}{[A^-]} = \log\frac{[A^-]}{[HA]}$.

The Henderson-Hasselbalch equation can be used to determine the blood pH calculated in Problem-Solving Example 15.1; it gives the same results.

$$pH = pK_a + \log\frac{[\text{conj base}]}{[\text{conj acid}]} = -\log K_a + \log\frac{[HCO_3^-]}{[H_2CO_3]}$$

$$= -\log(4.3 \times 10^{-7}) + \log\left(\frac{0.025}{0.0025}\right) = -\log(4.3 \times 10^{-7}) + \log(10)$$

$$= 6.37 + \log(10) = 6.37 + 1.00 = 7.37$$

Note that because the concentration of conjugate base is ten times that of the conjugate acid, the pH should be greater than the pK_a by 1 pH unit, and it is.

Application of the Henderson-Hasselbalch equation has its limits. Two important rules for when it can be applied to buffer solutions are

1. The value of the $\frac{\text{[conj base]}}{\text{[conj acid]}}$ ratio *must* be between 0.1 and 10.
2. The [conj base] and [conj acid] must *each* exceed the K_a by a factor of 100 or more.

EXERCISE 15.2

Blood pH

Calculate the pH of blood containing 0.0025-M HPO_4^{2-}(aq) and 0.0015-M $H_2PO_4^-$(aq). K_a of $H_2PO_4^- = 6.3 \times 10^{-8}$. (Assume that this is the only blood buffer.)

PROBLEM-SOLVING EXAMPLE 15.2

A Buffer Solution and Its pH

Calculate the pH of a buffer containing 0.50 mol/L pyruvic acid, $CH_3COCOOH$, and 0.60 mol/L sodium pyruvate, $Na^+CH_3COCOO^-$. K_a of pyruvic acid $= 3.2 \times 10^{-3}$.

Result 2.57

Analyze Pyruvic acid and pyruvate ion are a conjugate acid-base pair. We can solve for the pH of the buffer in two ways. First, we can use the K_a expression to find the H_3O^+ concentration from which the pH can be calculated. Alternatively, we can use the Henderson-Hasselbalch equation to determine pH.

Plan To use the K_a expression, we first write the chemical equation for the equilibrium of pyruvic acid with pyruvate ions, and its corresponding K_a expression. Then, rearrange the equation to calculate $[H_3O^+]$ and from it, the pH. The Henderson-Hasselbalch equation calculates the pH directly, but the K_a must first be converted to pK_a; p$K_a = -\log K_a$.

Execute $CH_3COCOOH(aq) + H_2O(\ell) \rightleftharpoons H_3O^+(aq) + CH_3COCOO^-(aq)$

$$K_a = 3.2 \times 10^{-3} = \frac{[H_3O^+][CH_3COCOO^-]}{[CH_3COCOOH]} = \frac{[H_3O^+](0.60)}{(0.50)}$$

$$[H_3O^+] = \frac{3.2 \times 10^{-3}(0.50)}{0.60} = 2.67 \times 10^{-3}$$

$$pH = -\log(2.67 \times 10^{-3}) = 2.57$$

Applying the Henderson-Hasselbalch equation:

$$pH = pK_a + \log\frac{\text{[conj base]}}{\text{[conj acid]}} = -\log(3.2 \times 10^{-3}) + \log\frac{[CH_3COCOO^-]}{[CH_3COCOOH]}$$

$$pH = 2.49 + \log\left(\frac{0.60}{0.50}\right) = 2.49 + \log(1.2) = 2.49 + 0.079 = 2.57$$

☑ **Reasonable Result Check** The concentration of conjugate base is a bit greater than the concentration of conjugate acid. Therefore the pH should be slightly higher than the pK_a, which it is. The result is reasonable.

PROBLEM-SOLVING PRACTICE 15.2

Calculate the ratio of $[HPO_4^{2-}]$ to $[H_2PO_4^-]$ in blood at a normal pH of 7.40. Assume that this is the only buffer system present. For $H_2PO_4^-$, $K_a = 6.3 \times 10^{-8}$.

From the Henderson-Hasselbalch equation note that when the concentrations of conjugate base and conjugate acid are equal,

$$\frac{\text{[conj base]}}{\text{[conj acid]}} = \frac{[A^-]}{[HA]} = 1 \quad \log\frac{\text{[conj base]}}{\text{[conj acid]}} = \log\frac{[A^-]}{[HA]} = \log(1) = 0$$

Thus, pH = pK_a + log(1) = pK_a and ***a buffer's pH equals the pK_a of its weak acid when the concentrations of the acid and its conjugate base in the buffer are equal.***

In Problem-Solving Example 15.2, the [conj base]/[conj acid] ratio was > 1.0, so pH > pK_a.

15-1d Selecting an Appropriate Conjugate Acid-Base Buffer Pair

A buffer for maintaining a desired pH can be chosen readily by examining pK_a values, which are often tabulated along with K_a values. *Choose a conjugate acid-base pair whose conjugate acid has a pK_a near the desired pH.* Table 15.1 lists pK_a values of several common acids that could be used with their conjugate bases to prepare buffers over the pH range from 4 to 10.

To have comparable quantities of both acid and conjugate base in a buffer solution, the ratio of conjugate base to conjugate acid cannot get much smaller than 1 : 10 or much bigger than 10 : 1. ***The pH range of a buffer is limited to about one pH unit above or below the pK_a of the conjugate acid.*** In the carbonic acid/hydrogen carbonate case, for example, that would be a pH from 5.37 to 7.37 because pK_a = $-\log K_a = -\log 4.3 \times 10^{-7} = 6.37$. Other conjugate acid-base pairs, such as those in Table 15.1, can be used to prepare buffers with much different pH ranges, as determined by the pK_a value of the acid in the buffer (Table 15.1).

Additional K_a values are in Table 14.2 (← Sec. 14-5c) and Appendix G.

PROBLEM-SOLVING EXAMPLE 15.3

Selecting an Acid-Base Pair for a Buffer Solution of Known pH

You are doing an experiment that requires a buffer solution with a pH of 4.00. Available to you are solutions containing acetic acid, carbonic acid, lactic acid, sodium acetate, sodium hydrogen carbonate, and sodium lactate. Choose two of these solutions that are appropriate to make a pH 4.00 buffer solution. Determine the mole ratio of the compounds needed.

Result Lactic acid and sodium lactate, with a mole ratio of 1.0 mol lactic acid to 1.4 mol sodium lactate would work.

Table 15.1 Buffer Systems That Are Useful at Various pH Values

Desired pH	Weak Conjugate Acid	Weak Conjugate Base	K_a of Weak Conjugate Acid	pK_a
4	Lactic acid, $CH_3CHOHCOOH$	Lactate ion, $CH_3CHOHCOO^-$	1.4×10^{-4}	3.85
5	Acetic acid, CH_3COOH	Acetate ion, CH_3COO^-	1.8×10^{-5}	4.74
6	Carbonic acid, H_2CO_3	Hydrogen carbonate ion, HCO_3^-	4.3×10^{-7}	6.37
7	Dihydrogen phosphate ion, $H_2PO_4^-$	Monohydrogen phosphate ion, HPO_4^{2-}	6.3×10^{-8}	7.20
8	Hypochlorous acid, HClO	Hypochlorite ion, ClO^-	6.8×10^{-8}	7.17
9	Ammonium ion, NH_4^+	Ammonia, NH_3	5.6×10^{-10}	9.25
10	Hydrogen carbonate ion, HCO_3^-	Carbonate ion, CO_3^{2-}	4.7×10^{-11}	10.33

Taken from Högfeldt, E., Perrin, D. D., *Stability Constants of Metal-Ion Complexes,* 1st ed., Oxford; New York: Pergamon, 1979–1982. International Union of Pure and Applied Chemistry Commission on Equilibrium Data.

- **As K_a decreases, the pK_a increases.**
- **The weaker the acid, the larger its pK_a.**

Analyze To make a buffer, a conjugate acid-base pair is needed; the pK_a of the conjugate acid must be within ±1 of the desired pH. Available conjugate acid-base pairs are acetic acid/acetate ion, carbonic acid/carbonate ion, and lactic acid/lactate ion. Table 15.1 provides pK_a values; use it to select the conjugate acid-base pair with pK_a closest to pH = 4.00.

Plan Select an acid-base pair with pH = 4.00 ± 1. Use the Henderson-Hasselbalch equation to calculate the necessary conjugate acid and conjugate base concentrations.

Execute Lactic acid, $CH_3CHOHCOOH$, with $pK_a = 3.85$, is closest to the target pH of 4.00. Substituting this pK_a into the Henderson-Hasselbalch equation gives

$$pH = 4.00 = 3.85 + \log\frac{[CH_3CHOHCOO^-]}{[CH_3CHOHCOOH]}$$

$$\log\frac{[CH_3CHOHCOO^-]}{[CH_3CHOHCOOH]} = 4.00 - 3.85 = 0.15; \quad \frac{[CH_3CHOHCOO^-]}{[CH_3CHOHCOOH]} = 10^{0.15} = 1.41$$

The required concentration of lactate ion, $CH_3CHOHCOO^-$, is 1.41 times that of lactic acid, $CH_3CHOHCOOH$. Thus, the mole ratio of lactic acid to lactate ion is 1.0 mol lactic acid to 1.4 mol sodium lactate.

☑ **Reasonable Result Check** The target pH of 4.00 is greater than the pK_a, 3.85, so the desired buffer is more basic than an equimolar mixture of lactic acid and sodium lactate. Therefore, the concentration of the conjugate base, lactate ion, must be greater than that of the conjugate acid, lactic acid.

PROBLEM-SOLVING PRACTICE 15.3

Use the data in Table 15.1 to select a conjugate acid-base pair you could use to make buffer solutions having each of these hydrogen ion concentrations.

(a) 3.2×10^{-4} M (b) 5.0×10^{-5} M
(c) 7.0×10^{-8} M (d) 6.0×10^{-11} M

EXERCISE 15.3

Making a Buffer Solution

Calculate the mole ratio of sodium acetate and acetic acid needed to make a buffer of pH 4.68.

EXERCISE 15.4

Buffers and pH

Calculate the pH of these buffers.

(a) H_2CO_3 (0.10 M)/HCO_3^- (0.25 M) (b) $H_2PO_4^-$ (0.10 M)/HPO_4^{2-} (0.25 M)

15-1e The pH Change on Addition of Acid or Base to a Buffer

The pH of a buffer changes when acid or base is added to the buffer due to the changes in the amounts of conjugate acid and conjugate base in the buffer (Figure 15.2, Section 15-1a). **When acid (H_3O^+) is added to a buffer, the acid reacts with the conjugate base of the buffer to form its conjugate acid:**

$$\text{Conjugate base in buffer} + H_3O^+ \text{ added} \longrightarrow \text{conjugate acid} + \text{water}$$

When base (OH^-) is added to a buffer, the base reacts with the conjugate acid of the buffer, which is converted into its conjugate base:

$$\text{Conjugate acid in buffer} + OH^- \text{ added} \longrightarrow \text{conjugate base} + \text{water}$$

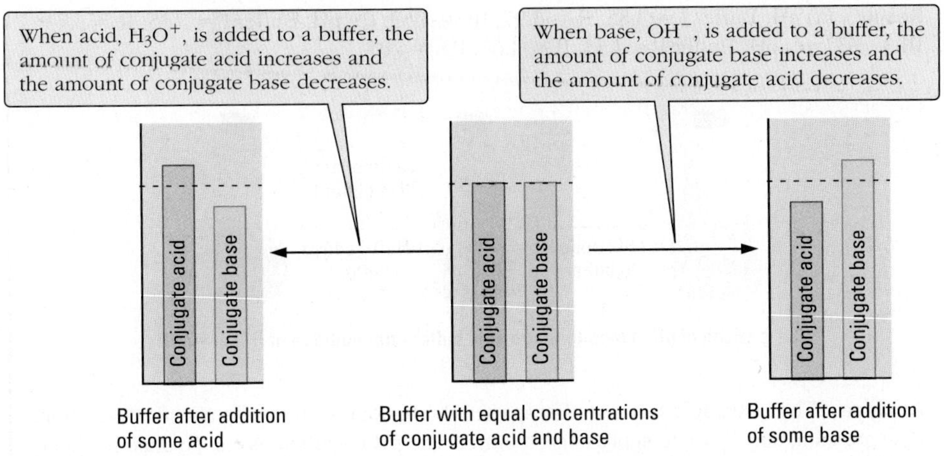
When acid, H_3O^+, is added to a buffer, the amount of conjugate acid increases and the amount of conjugate base decreases.
When base, OH^-, is added to a buffer, the amount of conjugate base increases and the amount of conjugate acid decreases.
Conjugate acid
Conjugate base
Conjugate acid
Conjugate base
Conjugate acid
Conjugate base
Buffer after addition of some acid
Buffer with equal concentrations of conjugate acid and base
Buffer after addition of some base

Figure 15.3 The effects of adding acid or base to a buffer. When H_3O^+ or OH^- ions are added to a buffer, the amounts of conjugate acid and conjugate base in the buffer change.

Figure 15.3 summarizes these relationships. For a buffer made from acetic acid and sodium acetate, the changes are

$$\underset{\text{Conjugate base of the buffer}}{CH_3COO^-(aq)} + \underset{\text{added}}{H_3O^+(aq)} \longrightarrow \underset{\text{Conjugate acid of the buffer}}{CH_3COOH(aq)} + H_2O(\ell)$$

$$\underset{\text{Conjugate acid of the buffer}}{CH_3COOH(aq)} + \underset{\text{added}}{OH^-(aq)} \longrightarrow \underset{\text{Conjugate base of the buffer}}{CH_3COO^-(aq)} + H_2O(\ell)$$

When acid or base is added to a buffer, the pH changes because of a change in the ratio [conjugate base]/[conjugate acid]. The extent of the pH change depends on the amount of acid or base added and on the amounts of conjugate acid or conjugate base remaining in the buffer to offset additional amounts of acid or base to be added. As shown in Problem-Solving Example 15.4, the change in pH of a buffer when acid or base is added to it can be calculated in two ways: (1) by using the K_a expression and the K_a value, or (2) by using the Henderson-Hasselbalch equation and pK_a.

PROBLEM-SOLVING EXAMPLE 15.4

pH Changes in a Buffer

Sodium lactate, $Na^+CH_3CHOHCOO^-$, and lactic acid, $CH_3CHOHCOOH$, are used to prepare three buffer solutions. Each of the three buffer solutions differs in its concentration of conjugate base (lactate ion) and conjugate acid (lactic acid). The concentrations are

Buffer	[Lactate$^-$], mol/L	[Lactic acid], mol/L
I	0.15	0.15
II	0.20	0.15
III	0.15	0.25

From Table 15.1, K_a lactic acid $= 1.4 \times 10^{-4}$; $pK_a = 3.85$. For 1.00 L of each buffer,

(a) calculate the initial pH of each buffer solution.

(b) calculate the pH of each buffer after 0.050 mol HCl has been added separately to each buffer (neglect volume changes).

(c) calculate the pH of each buffer after 0.10 mol NaOH has been added separately to each buffer (neglect volume changes).

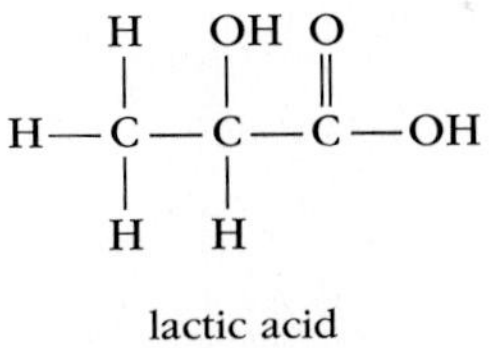
H OH O
H—C—C—C—OH
H H

lactic acid

Result (a) pH: Buffer I = 3.85; II = 3.97; III = 3.63; (b) pH: Buffer I = 3.55; II = 3.73; III = 3.37; (c) pH: Buffer I = 4.55; II = 4.63; III = 4.07

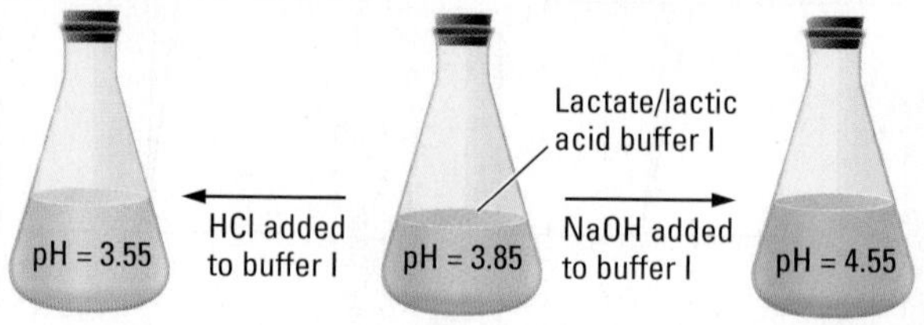
Lactate/lactic acid buffer I
HCl added to buffer I
NaOH added to buffer I
pH = 3.55
pH = 3.85
pH = 4.55

Comparison of pH of lactate/lactic acid buffer after addition of HCl or NaOH.

Analyze The lactic acid-lactate buffer is used throughout, so the same chemical equation, K_a expression, and K_a value apply to each buffer. The pH calculations can be done in either of two ways: (1) Use the K_a value and expression to calculate H_3O^+ concentration and, from it, the pH; or (2) apply the Henderson-Hasselbalch equation. In parts (b) and (c), the [conjugate base]/[conjugate acid] ratio changes as HCl and NaOH are added, respectively. Added HCl in part (b) reacts with lactate ion, the conjugate base, to form lactic acid, the conjugate acid. Correspondingly, in part (c), lactic acid reacts with any added NaOH to form lactate ion.

Plan In each case, the chemical equation for lactic acid dissociation and its related K_a expression and value are

$$CH_3CHOHCOOH(aq) + H_2O(\ell) \rightleftharpoons H_3O^+(aq) + CH_3CHOHCOO^-(aq)$$

$$K_a = 1.4 \times 10^{-4} = \frac{[H_3O^+][CH_3CHOHCOO^-]}{[CH_3CHOHCOOH]}$$

Solving for $[H_3O^+]$ gives:

$$[H_3O^+] = (1.4 \times 10^{-4})\frac{[CH_3CHOHCOOH]}{[CH_3CHOHCOO^-]}$$

The pH can also be calculated using the Henderson-Hasselbalch equation.

$$pH = pK_a + \log\frac{[CH_3CHOHCOO^-]}{[CH_3CHOHCOOH]}$$

Execute

(a) For Buffer I, $[CH_3CHOHCOO^-] = 0.15$ mol/L and $[CH_3CHOHCOOH] = 0.15$ mol/L; therefore, using the K_a expression, we have

$$[H_3O^+] = 1.4 \times 10^{-4}\ M \times \frac{0.15\ M}{0.15\ M} = 1.4 \times 10^{-4}\ M$$

$$pH = -\log(1.4 \times 10^{-4}) = 3.85$$

Using the Henderson-Hasselbalch equation gives

$$pH = pK_a + \log\frac{0.15\ M}{0.15\ M} = 3.85 + \log(1.00) = 3.85$$

The initial pH values of Buffers II and III can be calculated similarly.

(b) In each buffer, the 0.050 mol HCl added reacts with 0.050 mol lactate ion to form 0.050 mol lactic acid and water.

$$CH_3CHOHCOO^-(aq) + H_3O^+(aq) \rightleftharpoons CH_3CHOHCOOH(aq) + H_2O(\ell)$$

$CH_3CHOHCOO^-(aq)$	$H_3O^+(aq)$	$CH_3CHOHCOOH(aq)$
0.050 mol from buffer	0.050 mol added from HCl	0.050 mol formed by reaction

This changes the original [lactate]/[lactic acid] ratio in each buffer, so the pH changes because of the added HCl. For *Buffer I*:

	Amount Lactate⁻ (mol)	Amount Lactic Acid (mol)
Before reaction	0.15	0.15
After reaction	0.15 − 0.050 = 0.10	0.15 + 0.050 = 0.20

Addition of HCl changes the initial [lactate]/[lactic acid] ratio from $\frac{0.15}{0.15} = 1.0$ to $\frac{0.10}{0.20} = 0.50$, causing a change in pH.

$$[H_3O^+] = (1.4 \times 10^{-4})\frac{[CH_3CHOHCOOH]}{[CH_3CHOHCOO^-]} = (1.4 \times 10^{-4})\frac{0.20}{0.10} = 2.8 \times 10^{-4}$$

$$pH = -\log(2.8 \times 10^{-4}) = 3.55$$

The Henderson-Hasselbalch equation may be used to calculate the pH directly.

$$pH = pK_a + \log\frac{[CH_3CHOHCOO^-]}{[CH_3CHOHCOOH]} = 3.85 + \log\frac{(0.10)}{(0.20)} = 3.85 + (-0.30) = 3.55$$

The pH drops from 3.85 to 3.55. Note that 0.10 mol lactate remains to react with any additional HCl that might be added. The pH change in Buffer II and in Buffer III upon addition of HCl can be calculated similarly.

(c) To offset the addition of 0.10 mol NaOH requires 0.10 mol lactic acid from each buffer.

$$CH_3CHOHCOOH(aq) + OH^-(aq) \rightleftharpoons CH_3CHOHCOO^-(aq) + H_2O(aq)$$

0.10 mol from buffer	0.10 mol added from NaOH	0.10 mol formed in buffer by the reaction

This changes the original [lactate]/[lactic acid] ratio in each buffer, so the pH changes because of the added NaOH. For *Buffer I*:

	Amount Lactic Acid (mol)	Amount Lactate⁻ (mol)
Before reaction	0.15	0.15
After reaction	0.15 − 0.10 = 0.05	0.15 + 0.10 = 0.25

Addition of NaOH changes the initial [lactate]/[lactic acid] ratio in Buffer I from $\frac{0.15}{0.15} = 1.0$ to $\frac{0.25}{0.050} = 5.0$ and a change in pH occurs.

The new pH can be calculated from K_a as

$$[H_3O^+] = (1.4 \times 10^{-4})\frac{[CH_3CHOHCOOH]}{[CH_3CHOHCOO^-]} = (1.4 \times 10^{-4})\frac{0.05}{0.25} = 2.8 \times 10^{-5}$$

$$pH = -\log(2.8 \times 10^{-5}) = 4.55$$

Using the Henderson-Hasselbalch equation, the pH after NaOH addition is

$$pH = pK_a + \log\frac{[CH_3CHOHCOO^-]}{[CH_3CHOHCOOH]} = 3.85 + \log\left(\frac{0.25}{0.05}\right) = 3.85 + 0.70 = 4.55$$

The pH rises from 3.85 to 4.55. Note that 0.05 mol lactic acid remains to react with any additional NaOH that might be added. The pH change in Buffer II and in Buffer III upon addition of NaOH can be calculated similarly.

☑ **Reasonable Result Check** (a) The pH of the initial buffer, 3.85, is reasonable because the concentrations of lactate (conjugate base) and lactic acid (conjugate acid) are the same, therefore pH = pK_a. The initial pH of Buffer II should be a bit higher than 3.85 because the lactate/lactic acid ratio is > 1. The opposite is true of Buffer III: the initial lactate/lactic acid ratio is < 1 and the pH should be lower than 3.85. (b) When acid is added to the buffer, the pH should decrease and it does in each case. (c) The amount of NaOH added was greater than the amount of HCl, so the pH changes should be greater than for the addition of HCl, and they are. As expected upon addition of base to a buffer, the pH rose, but by less than 1 pH unit in each case. The results are reasonable.

PROBLEM-SOLVING PRACTICE 15.4

For the three lactate/lactic acid buffer solutions just given, calculate the pH when each quantity of acid or base is added separately to each buffer (neglect volume changes).
(a) 0.075 mol HCl (b) 0.065 mol NaOH

CONCEPTUAL EXERCISE 15.5

Blood Buffer Reaction

If an abnormally high CO_2 concentration is present in blood, which phosphorus-containing ion, $H_2PO_4^-$ or HPO_4^{2-}, can counteract the presence of excess CO_2? Explain your answer.

15-1f Buffer Capacity

When the conj base : conj acid ratio changes to 1 : 10, the pH *decreases* by 1 unit. When the ratio changes to 10 : 1, the pH *increases* by 1 unit.

The *amounts* of conjugate acid and conjugate base in the buffer solution determine the **buffer capacity**—*the amount of added acid or base that the buffer can accommodate without undergoing a pH change of more than 1 pH unit.*

- When all of the conjugate acid in a buffer has reacted with added base, adding just a little more base increases the pH significantly because there is no conjugate acid left in the buffer to react with the added base.
- If enough acid is added to a buffer to react with all of the buffer's conjugate base and excess unreacted acid remains, the pH decreases significantly.
- In each of these cases, the buffer capacity has been exceeded.

For example, 1 L of a buffer solution that is 0.25 M in CH_3COOH and 0.25 M in CH_3COO^- contains 0.25 mol CH_3COOH and 0.25 mol CH_3COO^-. If more than 0.25 mol H_3O^+ or 0.25 mol OH^- is added to this buffer, its buffer capacity is exceeded. Thus, the initial buffer *cannot* accommodate the addition of 0.30 mol of strong acid *or* 0.30 mol of strong base without undergoing a major change in pH. Such additions would use up all of the buffer's conjugate base or all of its conjugate acid, respectively, and exceed the buffer's capacity. The pH would drop or rise accordingly.

PROBLEM-SOLVING EXAMPLE 15.5

Buffer Capacity

A buffer is prepared using 0.25 mol $H_2PO_4^-$ and 0.15 mol HPO_4^{2-} in 500. mL of solution.
(a) Determine whether the buffer capacity will be exceeded if 6.2 g KOH is added. Calculate the pH of the solution after the addition.
(b) Determine whether the buffer capacity will be exceeded if 23.0 mL of 6.0-M HCl is added to the original buffer. Calculate the pH of the resulting solution.

Result (a) No, the buffer capacity will not be exceeded; pH = 7.47. (b) Yes, the buffer capacity is exceeded; pH = 5.61.

Analyze Buffer capacity is the quantity of added acid or base that a buffer can accommodate without undergoing a pH change of more than 1 pH unit. In this buffer, $H_2PO_4^-$ is the conjugate acid and HPO_4^{2-} is the conjugate base. Added OH^- ions, a base, react with $H_2PO_4^-$; added H_3O^+ ions, an acid, react with HPO_4^{2-}. There is a greater amount of conjugate acid than conjugate base in the initial buffer. Thus, this buffer will be able to react with a greater amount of added base than added acid without exceeding the buffer capacity. The volume of solution is 500. mL or 0.500 L and needs to be used to calculate concentrations. It is necessary to assume that when a solid is added to a solution there is no volume change and when two solutions are mixed the volumes are additive.

Plan Use the Henderson-Hasselbalch equation to calculate the initial pH of the buffer. From Table 15.1, the pK_a of $H_2PO_4^-$ is 7.20. For part (a), first calculate the amount of OH^- that will react with $H_2PO_4^-$ ions; 1 mol OH^- reacts with 1 mol $H_2PO_4^-$ and converts each mole of $H_2PO_4^-$ into HPO_4^{2-}, the conjugate base. Therefore, the initial [conjugate base]/[conjugate acid] ratio changes; use the resulting ratio and the Henderson-Hasselbalch equation to calculate the new pH. In part (b), calculate the amount of H_3O^+ ions that will react with HPO_4^{2-} ions; 1 mol H_3O^+ neutralizes 1 mol HPO_4^{2-} and converts each mole of HPO_4^{2-} into $H_2PO_4^-$, the conjugate acid. Therefore, the initial [conjugate base]/[conjugate acid] ratio changes; calculate the resulting pH using the Henderson-Hasselbalch equation.

Execute Calculate the initial pH of the buffer:

$$\text{pH} = 7.20 + \log\frac{[HPO_4^{2-}]}{[H_2PO_4^-]} = 7.20 + \log\frac{(0.15/0.500)}{(0.25/0.500)}$$

$$= 7.20 + \log(0.60) = 7.20 - 0.22 = 6.98$$

(a) The 6.2 g KOH contributes 0.11 mol OH^- to react with the conjugate acid.

$$6.2\text{ g KOH}\left(\frac{1\text{ mol KOH}}{56.1\text{ g KOH}}\right)\left(\frac{1\text{ mol OH}^-}{1\text{ mol KOH}}\right) = 0.11\text{ mol OH}^-$$

The 0.11 mol OH^- added to the buffer reacts with 0.11 mol $H_2PO_4^-$, to form 0.11 mol HPO_4^{2-}.

$$H_2PO_4^-(aq) + OH^- \longrightarrow HPO_4^{2-}(aq) + H_2O(\ell)$$

$H_2PO_4^-$	OH^-	HPO_4^{2-}
0.11 mol from buffer	0.11 mol from KOH	0.11 mol formed

The reaction changes the amounts of HPO_4^{2-} and $H_2PO_4^-$ remaining in the buffer:

	Amount HPO_4^{2-} (mol)	Amount $H_2PO_4^-$ (mol)
Before reaction	0.15	0.25
After reaction	0.15 + 0.11 = 0.26	0.25 − 0.11 = 0.14

The pH changes because the $[HPO_4^{2-}]/[H_2PO_4^-]$ ratio has changed:

$$\text{pH} = 7.20 + \log\frac{[HPO_4^{2-}]}{[H_2PO_4^-]} = 7.20 + \log\frac{(0.26/0.500)}{(0.14/0.500)}$$

$$= 7.20 + \log(1.9) = 7.20 + 0.27 = 7.47$$

Because the pH change is less than 1.0 (7.47 − 6.98 = 0.49) the buffer's capacity was *not exceeded* by adding 6.2 g KOH.

(b) Adding 0.0230 L of 6.0-M HCl provides:

$$(0.0230\text{ L})\left(\frac{6.0\text{ mol HCl}}{\text{L}}\right)\left(\frac{1\text{ mol H}_3\text{O}^+}{1\text{ mol HCl}}\right) = 1.4 \times 10^{-1}\text{ mol H}_3\text{O}^+$$

$$HPO_4^{2-}(aq) + H_3O^+(aq) \longrightarrow H_2PO_4^-(aq) + H_2O(\ell)$$

HPO_4^{2-}	H_3O^+	$H_2PO_4^-$
0.14 mol from buffer	0.14 mol from HCl	0.14 mol formed

	Amount HPO_4^{2-} (mol)	Amount $H_2PO_4^-$ (mol)
Before reaction	0.15	0.25
After reaction	0.15 − 0.14 = 0.01	0.25 + 0.14 = 0.39

The pH calculated using the K_a method is the same as determined here using the Henderson-Hasselbalch equation, even though the conjugate base : conjugate acid ratio is <0.1. This result is because

$$\left(\frac{0.01}{0.523} - x\right) \Big/ \left(\frac{0.39}{0.523} - x\right)$$

is essentially $\frac{(0.01/0.523)}{(0.39/0.523)}$

since $x \ll \left(\frac{0.01}{0.523}\right)$.

Thus, the composition of the buffer changes. The amount of HPO_4^{2-} *decreases* to 0.01 mol; the amount of $H_2PO_4^-$ *increases* from 0.25 to 0.39. The volume of the resulting solution is 0.523 L due to the addition of 0.0230 L HCl. The pH after addition of HCl is

$$\text{pH} = 7.20 + \log\frac{[HPO_4^{2-}]}{[H_2PO_4^-]} = 7.20 + \log\frac{(0.01 / 0.523)}{(0.39 / 0.523)}$$

$$= 7.20 + \log(0.026) = 7.20 - 1.59 = 5.61$$

The pH changed by 1.37 units, from 6.98 to 5.61, showing that the buffer capacity was *exceeded* by the addition of that much HCl. The final [conjugate base]/[conjugate acid] ratio was 0.026, smaller than 1 : 10.

☑ **Reasonable Result Check**

(a) Addition of base to a buffer should increase the pH to some extent depending on the amount of base added and the amount of conjugate acid initially in the buffer. The amount of KOH (0.11 mol) added was less than the amount of conjugate acid (0.25 mol) in the initial buffer. The [conjugate base]/[conjugate acid] ratio changed from 0.60 to 1.9, a change that did not exceed ten. Therefore, the buffer's capacity was not exceeded. The pH change should be less than 1 pH unit, which it is.

(b) The amount of acid added (0.14 mol) lowered the amount of conjugate base remaining in the buffer to 0.01 mol, while the amount of conjugate acid increased to 0.39 mol, thereby making the [conjugate base]/[conjugate acid] ratio greater than ten. Consequently, the pH rose by more than 1 pH unit and the buffer capacity was exceeded.

PROBLEM-SOLVING PRACTICE 15.5

Calculate the minimum mass (g) of KOH that would have to be added to the initial buffer in Problem-Solving Example 15.5 to exceed its buffer capacity.

15-2 Acid-Base Titrations

It is also possible to determine the concentration of the titrant, provided that the concentration of the solution to which it is added is known.

In Section 3-12 **titration** was described as a method by which the concentration of an acid or a base could be determined. An acid-base titration is carried out by slowly adding a measurable volume of an aqueous solution of a base (or acid) to a known volume of an aqueous acid (or base). If the concentration of one solution is known and both volumes are known, the concentration of the other solution can be determined. For example, a *standard solution* (← **Sec. 3-12**) of a base (known concentration) could be added from a buret to a known volume of acid whose concentration is to be determined, as in Figure 15.4. *The solution in the buret* is known as the **titrant**; its volume can be determined accurately by reading the buret before the titration begins and when the equivalence point is reached. The *equivalence point* (← **Sec. 3-12**) is reached when the stoichiometric amount of titrant has been added—the amount that exactly reacts with all the acid or base being titrated.

15-2a Detection of the Equivalence Point

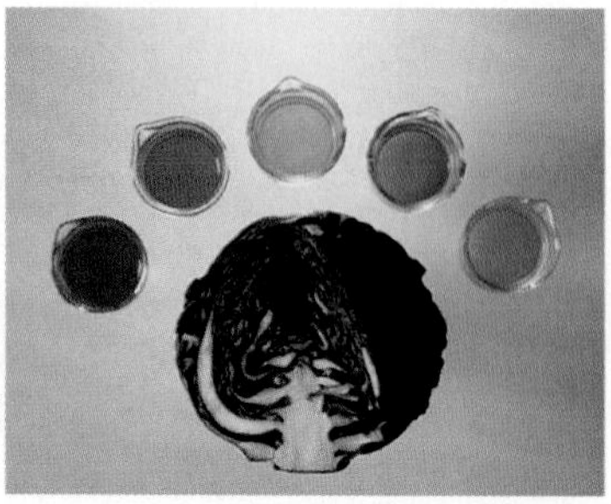

© Cengage Learning/Charles Steele

Red cabbage juice is a naturally occurring acid-base indicator. From left to right are solutions of pH 1, 4, 7, 10, and 13.

For an acid-base titration to be done successfully, a method is needed to detect the equivalence point. This can be done by using a pH meter (← **Sec. 14-4a**), a device that electronically measures the pH of the solution as the titration proceeds. Alternatively, the color change of an acid-base indicator (← **Sec. 14-4a**) can be used to detect when sufficient titrant has been added (Figure 15.4). The **end point** is *the point in a titration when an acid-base indicator changes color*. An indicator must be used that gives an end point very close to the equivalence point. The color changes of three acid-base indicators are shown in Figure 15.5.

The buret contains NaOH(aq) of known concentration; the flask contains an acid sample of unknown concentration and phenolphthalein indicator.
NaOH(aq) is added slowly to the flask; where the base has reacted with all of the acid the phenolphthalein indicator turns pink, its color in base.
Stirring the solution removes the pink color because there is enough acid to react with all of the added base. When enough base has been added to…
…react with all of the acid, stirring no longer causes the color to disappear. A very slightly pink solution indicates that the end point has been reached.
Photos: © Cengage Learning/James Maynard

Figure 15.4 Laboratory setup for titrating an acid sample (flask) with NaOH(aq) (buret).

Acid-base indicators, described in Section 14-4a, are typically weak organic acids (HIn) that differ in color from their conjugate bases (In^-).

$$\underset{\text{Color in acid}}{HIn(aq)} + H_2O(\ell) \rightleftharpoons H_3O^+(aq) + \underset{\text{Color in base}}{In^-(aq)}$$

Removal of an H^+ from the indicator molecule changes its structure so that the indicator molecule absorbs light in a different region of the visible spectrum, thus changing the color of the indicator. For methyl red the reaction is

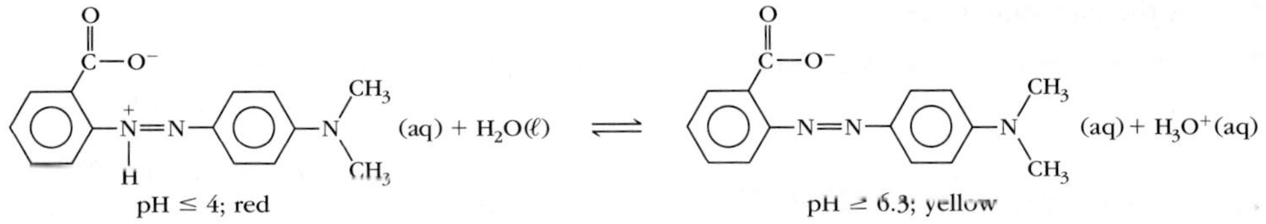

(aq) + H₂O(ℓ)
(aq) + H₃O⁺(aq)
pH ≤ 4; red
pH ≥ 6.3; yellow

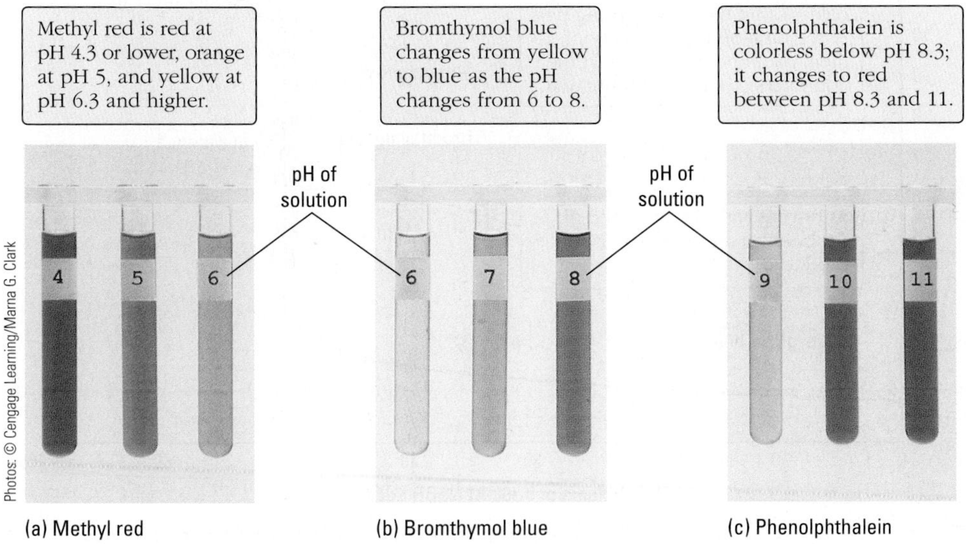

Methyl red is red at pH 4.3 or lower, orange at pH 5, and yellow at pH 6.3 and higher.
Bromthymol blue changes from yellow to blue as the pH changes from 6 to 8.
Phenolphthalein is colorless below pH 8.3; it changes to red between pH 8.3 and 11.
pH of solution
pH of solution
4
5
6
6
7
8
9
10
11
Photos: © Cengage Learning/Marna G. Clark
(a) Methyl red
(b) Bromthymol blue
(c) Phenolphthalein

Figure 15.5 Acid-base indicators. Acid-base indicators are compounds that change color in a particular pH range.

Methyl red changes color from red to yellow as the pH changes from 4 to 6.3 (Figure 15.5a); bromthymol blue changes from yellow to blue as the pH changes from 6 to 8 (Figure 15.5b).

If more than 90% of an indicator is in the acid form, the color observed is the acid color; if more than 90% of the indicator is in the base form, the color observed is the base color. This table summarizes the observed solution colors.

Ratio	Acid Form, HIn	Base Form, In^-	Color of Indicator Solution
$\frac{[HIn]}{[In^-]} \geq 10$	$\geq 90\%$	$\leq 10\%$	Color of HIn; conjugate acid color
$\frac{[HIn]}{[In^-]} \leq 0.1$	$\leq 10\%$	$\geq 90\%$	Color of In^-; conjugate base color
$\frac{[HIn]}{[In^-]} \cong 1$	$\cong 50\%$	$\cong 50\%$	Intermediate between conjugate acid and conjugate base colors

We will consider three types of acid-base titrations: (1) a strong acid titrated by a strong base (Section 15-2b); (2) a weak acid titrated by a strong base (Section 15-2c); and (3) a weak base titrated by a strong acid (Section 15-2d).

For each titration we will examine its **titration curve**, *a graph of pH as a function of the volume of titrant added*. In each case, we will be interested in the pH particularly at four stages of the titration:

- Prior to the addition of any titrant
- After addition of some titrant, but prior to the equivalence point
- At the equivalence point
- After more titrant has been added than at the equivalence point

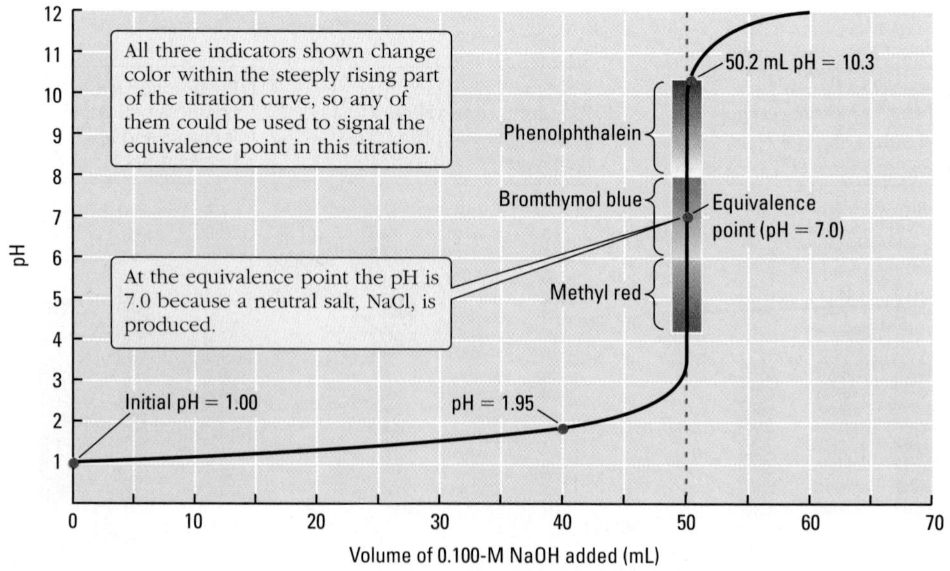
All three indicators shown change color within the steeply rising part of the titration curve, so any of them could be used to signal the equivalence point in this titration.
50.2 mL pH = 10.3
Phenolphthalein
Bromthymol blue
Equivalence point (pH = 7.0)
Methyl red
At the equivalence point the pH is 7.0 because a neutral salt, NaCl, is produced.
Initial pH = 1.00
pH = 1.95
pH
Volume of 0.100-M NaOH added (mL)

Figure 15.6 Curve for titration of a strong acid, 50.00 mL of 0.100-M HCl, with a strong base, 0.100-M NaOH.

15-2b Titration of a Strong Acid with a Strong Base

The titration curve for the titration of 50.0 mL of 0.100-M HCl, a strong acid, with 0.100-M NaOH, a strong base, as the titrant, is given in Figure 15.6. *The titration of a strong acid with a strong base produces a neutral salt and pH = 7.0 at the equivalence point.*

Prior to the equivalence point in a strong acid-strong base titration, the $[H_3O^+]$ decreases because some of the H_3O^+ has reacted with OH^- from the strong base. The volume of solution increases because of the added base solution. The $[H_3O^+]$ can be calculated for any volume of base added by the relation

$$[H_3O^+] = \frac{\text{initial amount acid (mol)} - \text{amount base added (mol)}}{\text{initial volume acid (L)} + \text{volume base added (L)}} \quad [15.1]$$

Equation 15.1 assumes that the sum of the initial volume and the added volume equals the exact volume of the mixed solutions.

Problem-Solving Example 15.6 illustrates pH calculations for the four points marked on the curve in Figure 15.6.

PROBLEM-SOLVING EXAMPLE 15.6

Titration of HCl with NaOH

A 0.100-M NaOH solution is used to titrate 50.0 mL of 0.100-M HCl. Calculate the pH of the solution at the four points in parts (a) through (d).

(a) Before any titrant is added
(b) After addition of 40.0 mL titrant
(c) After addition of 50.0 mL titrant
(d) After addition of 50.2 mL titrant
(e) Which of these indicators can be used to detect the equivalence point: methyl red, bromthymol blue, phenolphthalein?

Result

(a) 1.00 (b) 1.95 (c) 7.00 (d) 10.30
(e) All three: methyl red, bromthymol blue, and phenolphthalein

Analyze (a) Initially, only hydrochloric acid is present, so the pH can be calculated from the concentration of HCl, a strong acid and therefore completely ionized. (b) As NaOH is added the amount of H_3O^+ decreases due to the reaction of added OH^- ions with H_3O^+ ions in the acid:

$$\underset{\text{from HCl}}{H_3O^+(aq)} + \underset{\text{from NaOH}}{OH^-(aq)} \longrightarrow 2\,H_2O(\ell)$$

The $[H_3O^+]$ can be calculated prior to the equivalence point for any volume of base added by using Equation 15.1. (c) At the equivalence point, the amount (mol) of NaOH added exactly equals the initial amount (mol) of HCl; no un-neutralized HCl or unreacted NaOH is present; the solution contains only NaCl(aq). (d) Any NaOH added after the equivalence point will have no remaining HCl to neutralize it. (e) A suitable indicator must change color in the vertical part of the titration curve in Figure 15.6.

Plan (a) Calculate the pH based on the initial H_3O^+ concentration, which equals the initial concentration of HCl. (b) Use Equation 15.1 to calculate the pH after 40.0 mL of NaOH has been added. (c) The NaCl produced is a neutral salt (← **Sec. 14-8a**) so the pH is 7.00. (d) Calculate the hydroxide ion concentration of the excess (un-neutralized) NaOH, a strong base, and use it to determine the pH. (e) Select an indicator whose color changes along the vertical rise in the titration curve very close to the equivalence point (see Figure 15.6).

Execute

(a) Because HCl is a strong acid, the initial H_3O^+ concentration is 0.100 M and the pH is $-\log(0.100) = 1.00$.

We will round the pH values to two significant figures. See Appendix A-6 for treatment of significant figures when using logarithms.

(b) The initial 50.0-mL solution of HCl contains

$$n(H_3O^+) = (0.0500\ \text{L})(0.100\ \text{mol/L}) = 5.00 \times 10^{-3}\ \text{mol}\ H_3O^+$$

After 40.0 mL of 0.100-M NaOH has been added the amount of OH^- added is

$$n(OH^-) = (0.0400\ \text{L})(0.100\ \text{mol/L}) = 4.00 \times 10^{-3}\ \text{mol}\ OH^-$$

The $[H_3O^+]$ can be calculated from Equation 15.1:

$$[H_3O^+] = \frac{(5.00 \times 10^{-3}\ \text{mol}) - (4.00 \times 10^{-3}\ \text{mol})}{0.0500\ \text{L} + 0.0400\ \text{L}} = 1.11 \times 10^{-2}\ \text{M}$$

$$\text{pH} = -\log(1.11 \times 10^{-2}) = 1.95$$

(c) At the equivalence point, the 50.0 mL of 0.100-M NaOH added has exactly neutralized the 50.0 mL of 0.100-M HCl initially present in the flask. The pH is 7.00 because the NaCl produced is a neutral salt.

(d) Adding 50.2 mL of 0.100-M NaOH to the HCl solution puts 5.02×10^{-3} mol OH^- into the solution:

$$(0.0502\ \text{L})\left(\frac{0.100\ \text{mol NaOH}}{1\ \text{L}}\right)\left(\frac{1\ \text{mol OH}^-}{1\ \text{mol NaOH}}\right) = 5.02 \times 10^{-3}\ \text{mol OH}^-$$

In part (b), we calculated that 5.00×10^{-3} mol H_3O^+ was present initially; thus, 5.00×10^{-3} mol OH^- is required to neutralize all of the H_3O^+ in the initial sample. This leaves an additional 0.02×10^{-3} mol OH^-, now in 100.2 mL of solution, that is not neutralized. The pH of the solution is controlled by the concentration of *un-neutralized* OH^-.

$$[OH^-] = \frac{0.02 \times 10^{-3}\ \text{mol OH}^-}{0.1002\ \text{L}} = 2.0 \times 10^{-4}\ \text{M}$$

$$\text{pOH} = -\log(2.0 \times 10^{-4}) = 3.70; \qquad \text{pH} = 14.00 - \text{pOH} = 14.00 - 3.70 = 10.30$$

A volume of 0.2 mL is approximately 4 drops.

(e) Notice that the addition of just 0.2 mL of excess NaOH to the solution dramatically raises the pH (from 7.0 at the equivalence point to 10.3). There is a rapid, steep rise in pH very close to the equivalence point, so methyl red, bromthymol blue, or phenolphthalein could be used as the indicator (Figure 15.6).

PROBLEM-SOLVING PRACTICE 15.6

For the titration of 50.0 mL of 0.100-M HCl with 0.100-M NaOH, calculate the pH when these volumes of NaOH have been added:

(a) 10.0 mL (b) 25.00 mL (c) 45.0 mL (d) 50.5 mL

EXERCISE 15.6

Titration Curve

Draw the titration curve for the titration of 50.0 mL of 0.100-M NaOH using 0.100-M HCl as the titrant.

15-2c Titration of a Weak Acid with a Strong Base

As noted in Section 14-8b the reaction of a weak acid, such as acetic acid, with a strong base, like NaOH, produces a salt—sodium acetate in this case—that has a basic anion. Like other basic anions, the acetate ions react with water to produce hydroxide ions due to the hydrolysis reaction:

$$CH_3COO^-(aq) + H_2O(\ell) \rightleftharpoons CH_3COOH(aq) + OH^-(aq)$$

As a result, *when a weak acid is titrated with a strong base, the pH of the solution at the equivalence point is greater than 7.0 due to hydrolysis of the basic anion formed by the titration reaction.* A typical titration curve for a weak acid-strong base titration is shown in Figure 15.7 for the titration of 50.0 mL of 0.100-M acetic acid with 0.100-M NaOH. Problem-Solving Example 15.7 illustrates the calculations associated with the titration curve.

Notice in Figure 15.7 that the initial pH of 0.100-M acetic acid, 2.87, is higher than that of 0.100-M HCl, 1.00, in Figure 15.6. This is to be expected because acetic acid ($K_a = 1.8 \times 10^{-5}$) is a much weaker acid than HCl. Acetic acid is only slightly ionized

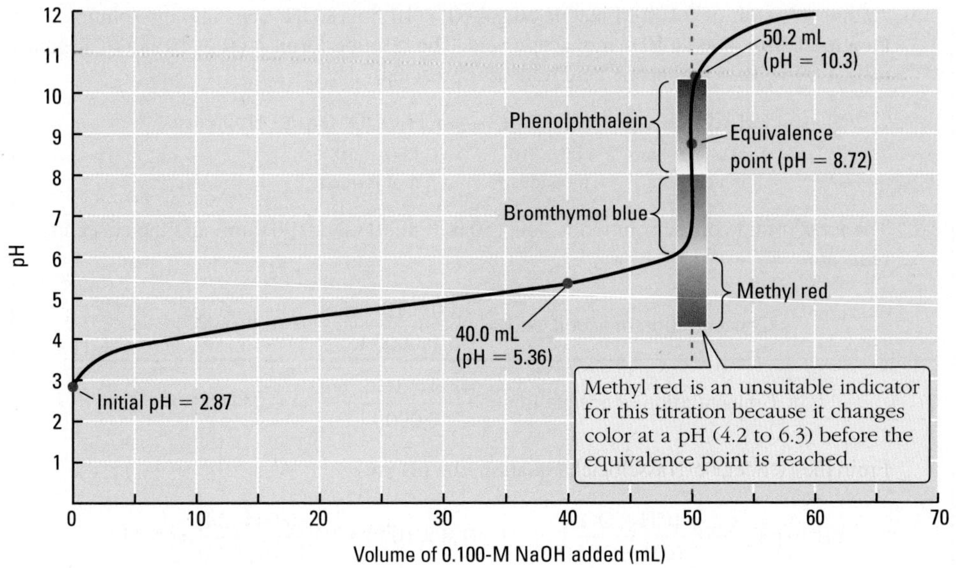

pH
Volume of 0.100-M NaOH added (mL)
50.2 mL (pH = 10.3)
Phenolphthalein
Equivalence point (pH = 8.72)
Bromthymol blue
Methyl red
40.0 mL (pH = 5.36)
Initial pH = 2.87
Methyl red is an unsuitable indicator for this titration because it changes color at a pH (4.2 to 6.3) before the equivalence point is reached.

Figure 15.7 Curve for titration of a weak acid, 50.0 mL of 0.100-M acetic acid, with a strong base, 0.100-M NaOH.

so its $[H_3O^+]$ is smaller and its pH is higher than for the same concentration of a strong acid like HCl. Also notice from Figure 15.7 that the rapidly rising portion of the titration curve near the equivalence point is shorter than it is for the titration of HCl with NaOH (Figure 15.6); the equivalence-point pH is 8.72 and the steeply rising part of the curve begins at about pH 6.5. This makes methyl red an unsuitable indicator because its color changes (pH 4.2 to 6.3) before the steep part of the curve is reached. Bromthymol blue or phenolphthalein is appropriate.

PROBLEM-SOLVING EXAMPLE 15.7

Titration of CH_3COOH with NaOH

A 0.100-M NaOH solution is used to titrate 50.0 mL of 0.100-M acetic acid ($K_a = 1.8 \times 10^{-5}$). Calculate the pH of the solution at these three points:

(a) After 40.0 mL of titrant has been added

(b) After 50.0 mL of NaOH has been added

(c) After 50.2 mL of NaOH has been added

Result (a) 5.34 (b) 8.72 (c) 10.30

Analyze We can calculate the amount of acetic acid present in the initial sample by multiplying its molarity (mol/L) times its volume (L). Because the concentrations of acid and base are both 0.10 M and 1 mol NaOH reacts with 1 mol acetic acid, 50.0 mL acetic acid requires 50.0 mL NaOH for complete reaction; the equivalence point is at 50.0 mL NaOH added. (a) Prior to the equivalence point, added NaOH reacts with acetic acid to produce sodium acetate and water. Sodium acetate and acetic acid form a buffer so the pH is controlled by the ratio [acetate ion]/[acetic acid]; the Henderson-Hasselbalch equation can be used to calculate the pH. (b) At the equivalence point all of the acetic acid has been converted to sodium acetate so we have a solution of sodium acetate, a basic salt (← **Sec. 14-8b**). The pH can be calculated from the concentration of sodium acetate and K_b(acetate ion). (c) Beyond the equivalence point there is excess NaOH so the $[OH^-]$ determines the pH.

Plan (a) Calculate the amount (mol) of acetate ion formed, the amount (mol) of acetic acid remaining, and the volume of solution. Divide each amount by the volume to calculate concentrations. Then, apply the Henderson-Hasselbalch equation. (b) Use the K_b expression and value for acetate ion to determine $[OH^-]$, the pOH, and the pH. Calculate the K_b for acetate ion from K_w and the K_a for acetic acid (← **Sec. 14-7a**).

Execute The initial 50.0-mL sample of 0.100-M acetic acid sample contains $(0.0500 \text{ L}) \times (0.100 \text{ mol/L}) = 5.00 \times 10^{-3}$ mol acetic acid.

(a) Adding 40.0 mL of 0.100-M NaOH puts 4.00×10^{-3} mol OH^- ions into the solution; they react with 4.00×10^{-3} mol acetic acid. The reaction forms 4.00×10^{-3} mol acetate ions; 1.00×10^{-3} mol acetic acid remains unreacted.

$$\underset{\substack{5.00 \times 10^{-3}\text{ mol} \\ \text{in acid soln}}}{CH_3COOH(aq)} + \underset{\substack{4.00 \times 10^{-3} \\ \text{mol added}}}{OH^-(aq)} \longrightarrow \underset{\substack{4.00 \times 10^{-3} \\ \text{mol formed}}}{CH_3COO^-(aq)} + H_2O(\ell)$$

The total volume of the solution is now (50.0 + 40.0) mL = 90.0 mL and the concentrations are

$$\text{Concentration of acetic acid} = \frac{1.00 \times 10^{-3}\text{ mol}}{0.0900\text{ L}} = 0.0111\text{ M}$$

$$\text{Concentration of acetate ion} = \frac{4.00 \times 10^{-3}\text{ mol}}{0.0900\text{ L}} = 0.0444\text{ M}$$

From the Henderson-Hasselbalch equation, the pH is

$$pH = pK_a + \log\frac{[CH_3COO^-]}{[CH_3COOH]} = -\log(1.8 \times 10^{-5}) + \log\frac{[0.0444\text{ M}]}{[0.0111\text{ M}]} = 5.35$$

pH can also be calculated using K_a:

$$[H_3O^+] = K_a\frac{[CH_3COOH]}{[CH_3COO^-]}$$

$$= (1.8 \times 10^{-5})\frac{(0.0111)}{(0.0444)}$$

$$= 4.5 \times 10^{-6}$$

$$pH = -\log(4.5 \times 10^{-6}) = 5.34$$

(b) At the equivalence point, the reaction has produced 5.00×10^{-3} mol acetate ion and the solution volume is 100. mL. The acetate-ion concentration is

$$c(CH_3COO^-) = \frac{5.00 \times 10^{-3}\text{ mol acetate}}{0.100\text{ L solution}} = 0.0500\text{ M}$$

Hydrolysis of acetate ion produces hydroxide ions (which make the solution basic) and the conjugate acid (acetic acid).

$$CH_3COO^-(aq) + H_2O(\ell) \rightleftharpoons CH_3COOH(aq) + OH^-(aq)$$

$$K_b = \frac{K_w}{K_a} = \frac{1.0 \times 10^{-14}}{1.8 \times 10^{-5}} = 5.6 \times 10^{-10}$$

Substituting into the K_b expression we let $x = [OH^-] = [CH_3COOH]$.

$$K_b = 5.6 \times 10^{-10} = \frac{[CH_3COOH][OH^-]}{[CH_3COO^-]} = \frac{x^2}{0.0500 - x}$$

Because K_b is small, we approximate $0.0500 - x$ to be 0.0500. Solving for x,

$$K_b = 5.6 \times 10^{-10} \approx \frac{x^2}{0.0500}; \quad x \approx 5.3 \times 10^{-6} = [OH^-]$$

$$pOH = -\log(5.3 \times 10^{-6}) = 5.28$$

$$pH = 14.00 - 5.28 = 8.72$$

This pH is considerably higher than equivalence-point pH of 7.00 for the NaOH-HCl titration (Problem-Solving Example 15.6), where neutral NaCl was the titration product.

(c) The pH beyond the equivalence point is controlled by the OH^- concentration from excess NaOH, a strong base. This is greater than the OH^- contributed by the hydrolysis of acetate ion. Therefore, the calculation for the pH beyond the equivalence point is like that for the titration with excess NaOH. An excess (un-neutralized) 0.2 mL NaOH has been added:

$$n(OH^-) = (0.2 \times 10^{-3}\text{ L})\left(\frac{0.100\text{ mol }OH^-}{\text{L}}\right) = 2 \times 10^{-5}\text{ mol }OH^-$$

$$[OH^-] = \frac{2 \times 10^{-5}\text{ mol }OH^-}{0.1002\text{ L}} = 2.0 \times 10^{-4}\text{ M}$$

$$pOH = -\log(2.0 \times 10^{-4}) = 3.70; \quad pH = 14.00 - pOH = 14.00 - 3.70 = 10.30$$

PROBLEM-SOLVING PRACTICE 15.7

Calculate the pH when these volumes of 0.100-M NaOH have been added when titrating 50.0 mL of 0.100-M acetic acid:
(a) 10.0 mL (b) 25.00 mL (c) 45.0 mL (d) 51.0 mL

As seen from Figures 15.6 and 15.7 and their associated Problem-Solving Examples, the titration curve for a strong base with a strong acid differs from that of a strong base with a weak acid of equal concentration in these ways:

- Before the titration, the initial pH of the solution is higher for the weak acid; the weaker the acid is, the higher the initial pH is.
- Very near the equivalence point the change in pH within the rapid rise of the curve is less for the weak acid; the weaker the acid is, the smaller the change in pH is.
- The pH at the equivalence point is higher for the weak acid; the weaker the acid is, the higher the pH of the equivalence point is.

A shorter rise in pH means that greater care is needed to select an appropriate indicator.

These features are illustrated in Figure 15.8.

One more feature of titration curves is discernible from Figure 15.8. When half as much titrant has been added as is needed to reach the equivalence point, half of the acid initially present has been converted into conjugate base.

	$HA(aq)$ +	$OH^-(aq) \longrightarrow$	$A^-(aq) + H_2O(\ell)$
Initially present	0.010 mol	≅ 0 mol	≅ 0 mol
After 5.0 mL NaOH added	0.005 mol	≅ 0 mol	0.005 mol

Thus, the amounts and concentrations of acid and conjugate base are equal when 5.0 mL of 1.0-M NaOH has been added in each titration of a weak acid in Figure 15.8. In our earlier discussion of the Henderson-Hasselbalch equation, we pointed out that when the concentrations of conjugate acid and conjugate base are equal, $pH = pK_a$ (← **Sec. 15-1c**). You can verify from Figure 15.8 that at the 5.0-mL point in each weak-acid titration $pH = -\log(K_a)$.

EXERCISE 15.7

Titration of Acetic Acid with NaOH

Use the K_a expression and value for acetic acid to calculate the pH after 30.0 mL of 0.100-M NaOH has been added to 50.0 mL of 0.100-M acetic acid.

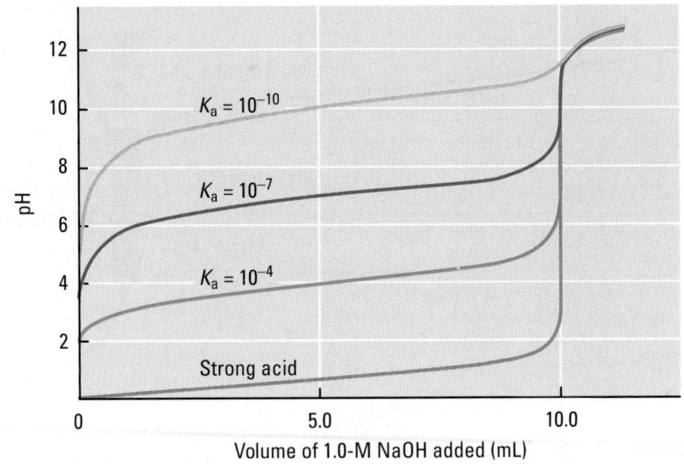

pH
12
10
8
6
4
2
0
5.0
10.0
Volume of 1.0-M NaOH added (mL)
$K_a = 10^{-10}$
$K_a = 10^{-7}$
$K_a = 10^{-4}$
Strong acid

Figure 15.8 The effect of acid strength on the shape of the titration curve. Each curve is for titration of 10.0 mL of a 1.0-M acid with 1.0-M NaOH.

CONCEPTUAL EXERCISE 15.8

Shape of the Titration Curve

Explain why the curve for the titration of acetic acid with NaOH in Figure 15.7 has a relatively flat region between ~10.0 and ~40.0 mL of NaOH added.

15-2d Titration of a Weak Base with a Strong Acid

The pH of 0.100-M NaOH is

$$pH = -\log\left(\frac{1.00 \times 10^{-14}}{1.00 \times 10^{-1}}\right) = 13.00$$

The titration curve for the titration of the weak base NH_3 with the strong acid HCl has the shape shown in Figure 15.9. The reaction produces ammonium chloride, NH_4Cl. Notice that *because NH_3 is a weak base, the starting pH is greater than 7.0, but less than it would be for 0.100-M NaOH (13.00).* Also notice that *the pH at the equivalence point, 5.28, is less than 7.0 because ammonium chloride is an acidic salt.*

$$NH_4^+(aq) + H_2O(\ell) \rightleftharpoons NH_3(aq) + H_3O^+(aq)$$

Beyond the equivalence point the pH continues to drop as excess acid is added. Although methyl red is a suitable indicator for this titration, phenolphthalein or bromthymol blue are *not* suitable because their color changes occur before the equivalence point.

15-2e Titration of a Polyprotic Acid with Base

Polyprotic acids—those with more than one ionizable hydrogen—react stepwise when titrated with bases, one step for each ionizable hydrogen. If the K_a values of the ionizable forms of the acid are sufficiently different, the titration curve has an equivalence point for each of the ionizable hydrogens removed from the acid molecule by titration. For example, maleic acid, HOOC—CH=CH—COOH, is a diprotic acid with two ionizable hydrogens, one from each —COOH group. Its titration with NaOH occurs in two steps:

$$HOOC{-}CH{=}CH{-}COOH(aq) + OH^-(aq) \rightleftharpoons HOOC{-}CH{=}CH{-}COO^-(aq) + H_2O(\ell)$$

$$HOOC{-}CH{=}CH{-}COO^-(aq) + OH^-(aq) \rightleftharpoons {}^-OOC{-}CH{=}CH{-}COO^-(aq) + H_2O(\ell)$$

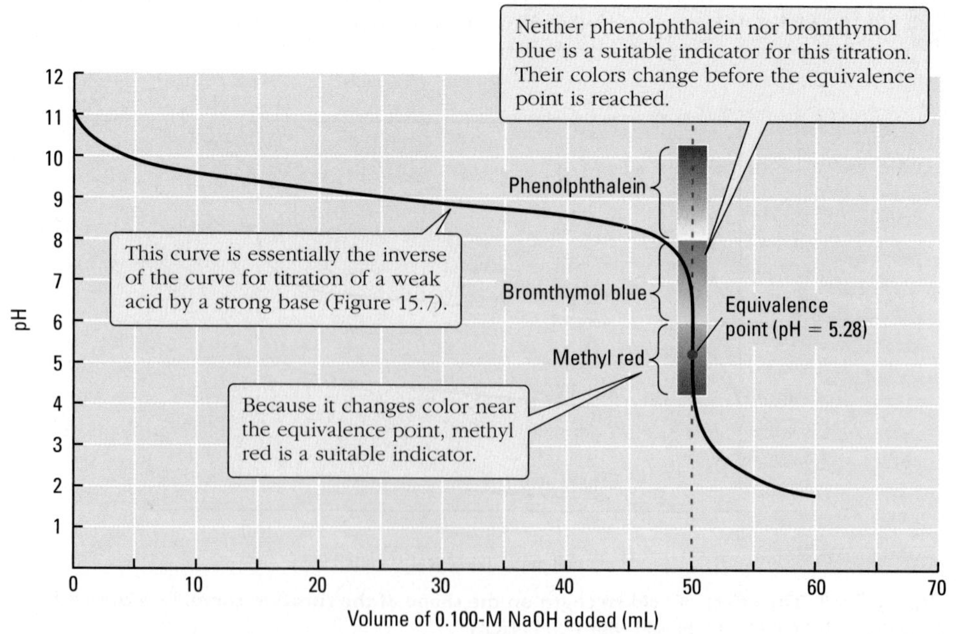
Neither phenolphthalein nor bromthymol blue is a suitable indicator for this titration. Their colors change before the equivalence point is reached.
Phenolphthalein
This curve is essentially the inverse of the curve for titration of a weak acid by a strong base (Figure 15.7).
Bromthymol blue
Equivalence point (pH = 5.28)
Methyl red
Because it changes color near the equivalence point, methyl red is a suitable indicator.
pH
12
11
10
9
8
7
6
5
4
3
2
1
0
10
20
30
40
50
60
70
Volume of 0.100-M NaOH added (mL)

Figure 15.9 Curve for titration of a weak base, 50.00 mL of 0.100-M NH_3, by a strong acid, 0.100-M HCl.

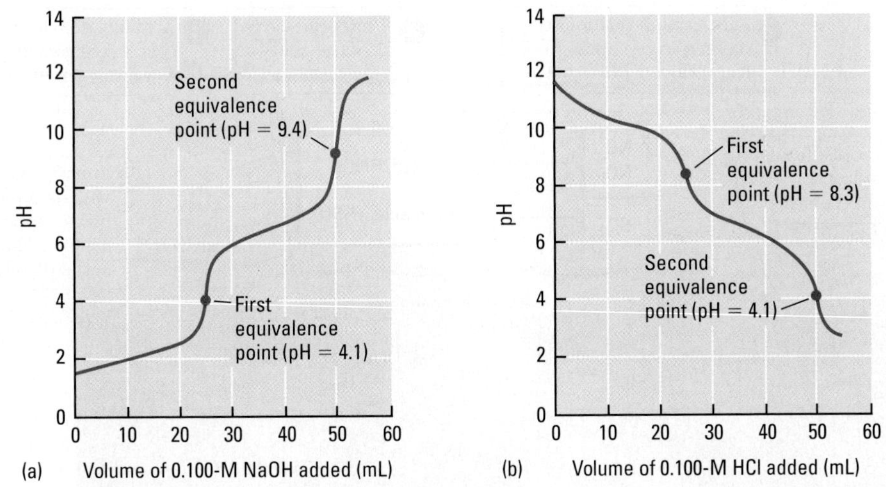

14
12
10
8
6
4
2
0
pH
Second equivalence point (pH = 9.4)
First equivalence point (pH = 4.1)
0 10 20 30 40 50 60
(a) Volume of 0.100-M NaOH added (mL)
First equivalence point (pH = 8.3)
Second equivalence point (pH = 4.1)
(b) Volume of 0.100-M HCl added (mL)

Figure 15.10 Curves for titration of a polyprotic acid and a polyprotic base. (a) Titration of 25.00 mL of 0.100-M maleic acid with 0.100-M NaOH. (b) Titration of 25.00 mL of 0.100-M sodium carbonate with 0.100-M HCl.

As shown in Figure 15.10, the two equivalence points occur at pH = 4.1 and pH = 9.4. Titration of a diprotic base, such as CO_3^{2-}, with acid also gives two equivalence points.

15-3 Acid Rain

The term *acid rain* was first used in 1872 by Robert Angus Smith, an English chemist and climatologist. In his book *Air and Rain,* Smith used the term to describe the acidic precipitation that fell on Manchester, England, in the mid-nineteenth century. **Acid rain** is now defined as *any precipitation with pH < 5.6, which is the pH of natural, unpolluted rain.* Although neutral water has a pH of 7, water in rain and in some lakes and rivers becomes acidified *naturally* from dissolved carbon dioxide, a normal component of the atmosphere. The carbon dioxide reacts reversibly with water to form a solution of carbonic acid, a weak acid, which ionizes into hydronium and hydrogen carbonate ions.

Although the term "acid rain" is commonly used, the more accurate term is "acid deposition," which takes into account acidic snow, sleet, rain, and fog.

$$2\,H_2O(\ell) + CO_2(g) \rightleftharpoons H_2CO_3(aq) + H_2O(\ell) \rightleftharpoons H_3O^+(aq) + HCO_3^-(aq)$$

The pH of water in equilibrium with CO_2 from the air is about 5.6, so *natural, unpolluted rainwater is slightly acidic.*

Several acidic, nonmetal oxides (← **Sec. 14-6e**) contribute to acid rain. Nitrogen dioxide, NO_2, from industrial and natural sources (← **Sec. 8-12b**), reacts with atmospheric water forming nitric acid, HNO_3, and nitrous acid, HNO_2.

$$2\,NO_2(g) + H_2O(\ell) \longrightarrow HNO_3(aq) + HNO_2(aq)$$

Atmospheric sulfur dioxide, SO_2, produced from burning sulfur-containing fossil fuels, reacts with water to produce sulfurous acid, H_2SO_3, and, if oxygen is present, sulfuric acid, H_2SO_4.

Approximately 125 million metric tons of SO_2 and 122 million metric tons of nitrogen oxides are emitted annually into the atmosphere from human activities, largely the burning of fossil fuels. Each year about 43 million metric tons of SO_2 and 59 million metric tons of nitrogen oxides are put into the atmosphere by natural sources.

$$SO_2(g) + H_2O(\ell) \longrightarrow H_2SO_3(aq)$$
$$2\,SO_2(g) + O_2(g) \longrightarrow 2\,SO_3(g)$$
$$SO_3(g) + H_2O(\ell) \longrightarrow H_2SO_4(aq)$$

The resulting acidic water droplets precipitate as rain or snow with a pH less than 5.6. Ice core samples taken in Greenland and dating back to 1900 contain sulfate, SO_4^{2-}, and

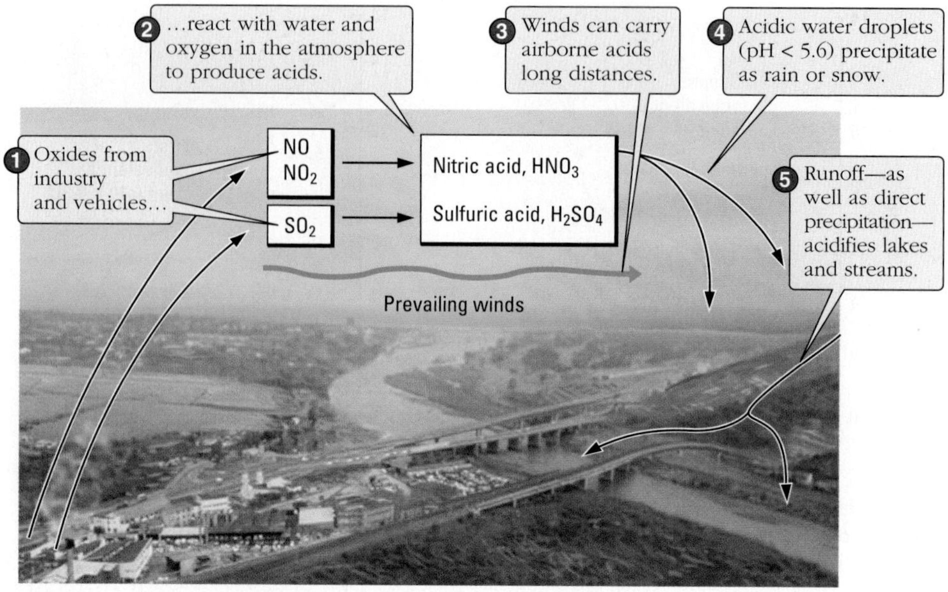

2 ...react with water and oxygen in the atmosphere to produce acids.
3 Winds can carry airborne acids long distances.
4 Acidic water droplets (pH < 5.6) precipitate as rain or snow.
1 Oxides from industry and vehicles...
NO
NO2
SO2
Nitric acid, HNO3
Sulfuric acid, H2SO4
5 Runoff—as well as direct precipitation—acidifies lakes and streams.
Prevailing winds

Figure 15.11 How acid deposition occurs.

nitrate, NO_3^-, ions indicating that acid rain has been commonplace, at least from 1900 onward.

Acid rain is a problem today due to the large amounts of these acidic oxides being put into the atmosphere by human activities every year (Figure 15.11). When such precipitation falls on areas without naturally occurring bases, such as limestone and other carbonate minerals that offset the acidity, serious environmental damage can occur. The average annual pH of precipitation falling on much of the northeastern United States and northeastern Europe is between 4.3 and 4.8 (Figure 15.12). In the past, rain in those areas has had pH values as low as 1.5. To further complicate matters, acid rain is an international problem because precipitation carried by winds does not observe international borders. Canadian residents are offended by the fact that much of the acid precipitation falling on Canadian cities and forests results from acidic oxides produced in the United States.

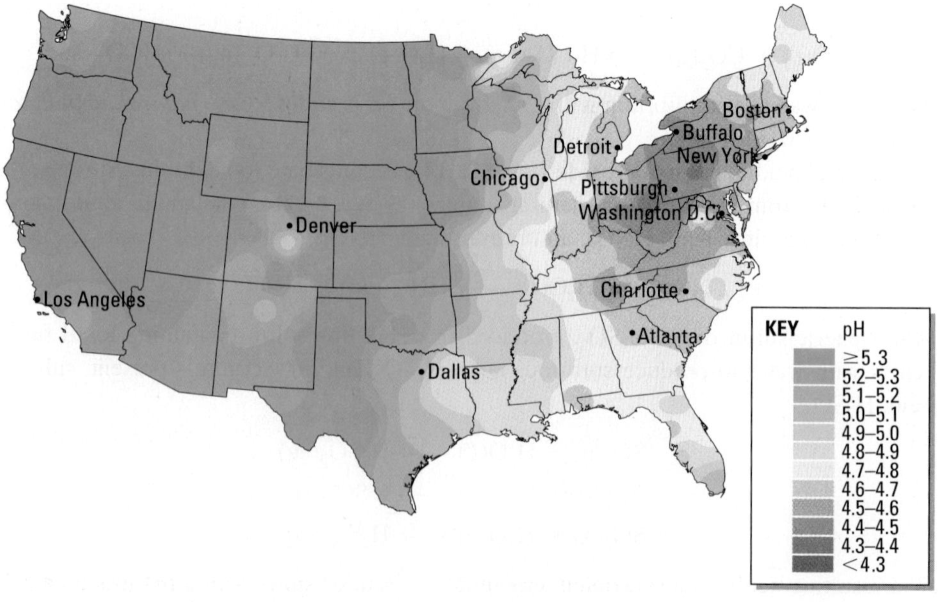

Boston
Buffalo
Detroit
New York
Chicago
Pittsburgh
Washington D.C.
Denver
Charlotte
Los Angeles
Atlanta
Dallas
KEY pH
≥5.3
5.2–5.3
5.1–5.2
5.0–5.1
4.9–5.0
4.8–4.9
4.7–4.8
4.6–4.7
4.5–4.6
4.4–4.5
4.3–4.4
<4.3

Figure 15.12 pH of precipitation within the contiguous United States.

(a) (b)

Photos: 2009 Dan Kunkle Lehigh Gap Nature Center

Figure 15.13 Acid rain damage and remediation. (a) An area damaged by acid rain and SO_2 emissions from a nearby zinc smelter, 1980. (b) The same area in 2007 with warm-season grasses that build and retain soil, growing in the damaged area without taking up toxic metal ions.

Acid rain can be caused by SO_2 produced by smelting sulfide-containing ores, such as ZnS, to form ZnO and SO_2. In addition to SO_2, the smelting releases toxic metal ions such as Zn^{2+}, Cd^{2+}, and Pb^{2+} into the atmosphere. Eventually, compounds of these ions precipitate into the surrounding soil. The photos in Figure 15.13 show acid rain damage to the area around the New Jersey Zinc Company smelting facility near Palmerton, PA between 1898 and 1980 and the significant remediation that has occurred recently on that same site.

15-4 Solubility Equilibria and the Solubility Product Constant, K_{sp}

Many ionic compounds, such as NH_4Cl and $NaNO_3$, are very soluble in water. Many other ionic compounds, however, are only modestly or slightly water-soluble; they produce *saturated* solutions of 0.001 M or less, far less in some cases (← **Sec. 3-3c**). Consider the case of a saturated aqueous solution of silver bromide, AgBr. When sufficient AgBr(s) is added to water, some of it dissolves to form a saturated solution, and some undissolved AgBr(s) is present. The dissolved AgBr dissociates, forming aqueous Ag^+ and Br^- ions in solution that are in equilibrium with the undissolved solid AgBr.

$$AgBr(s) \rightleftharpoons Ag^+(aq) + Br^-(aq)$$

This balanced chemical equation represents a solubility equilibrium (← **Sec. 12-2c**); we can derive an equilibrium constant from the chemical equation. *The equilibrium constant for a process in which a solid ionic compound dissolves in water* is called the **solubility product constant (K_{sp})**. The magnitude of K_{sp} indicates the extent to which the solid solute dissolves to give ions in solution. To evaluate the equilibrium constant, we first must write a **solubility product constant expression**. In general, the balanced chemical equation for dissolving a slightly soluble salt with the general formula A_xB_y is

The solubility product constant is commonly called just the solubility product.

$$A_xB_y(s) \rightleftharpoons x\,A^{n+}(aq) + y\,B^{m-}(aq)$$

This results in the general K_{sp} expression

$$K_{sp} = [A^{n+}]^x[B^{m-}]^y$$

A solubility product expression has the same general form that other equilibrium constant expressions have, except there is no denominator in the K_{sp} expression, because the reactant is always a pure solid.

For example, the equation for dissolving $Fe_2S_3(s)$ and its solubility product expression are

$$Fe_2S_3(s) \rightleftharpoons 2\,Fe^{3+}(aq) + 3\,S^{2-}(aq) \qquad K_{sp} = [Fe^{3+}]^2[S^{2-}]^3$$

Here $x = 2$, $y = 3$, $n = 3$, and $m = 2$. Notice that

- The chemical equation related to a solubility product constant expression has a solid solute compound as reactant and its aqueous ions as products.

- The concentration of the pure solid solute reactant is omitted from the K_{sp} expression because it remains unchanged as the reaction occurs.
- The K_{sp} value equals the product of the *equilibrium* molarities of the cation and the anion, each raised to the power given by the coefficient in the balanced chemical equation representing the solubility equilibrium.

PROBLEM-SOLVING EXAMPLE 15.8

Writing K_{sp} Expressions

Write the K_{sp} expression for each of these slightly soluble salts:
(a) $Fe(OH)_2$ (b) MgC_2O_4 (c) Ag_3PO_4

Result

(a) $K_{sp} = [Fe^{2+}][OH^-]^2$ (b) $K_{sp} = [Mg^{2+}][C_2O_4^{2-}]$ (c) $K_{sp} = [Ag^+]^3[PO_4^{3-}]$

Analyze To write the correct K_{sp} expression, we first must write the balanced equation for the dissociation of the solute and then substitute the equilibrium concentrations of the products appropriately into the K_{sp} expression where the cation and anion concentrations are multiplied.

Plan In each case, the working assumption is that, even though the solute is sparingly soluble, the amount of solute that does dissolve, although small, dissociates completely. Write the balanced chemical equation for the solute dissociation such that the equation has the correct coefficients for all terms and the charges of ions are shown correctly; the *sum* of the cation and the anion charges must be equal.

Execute

(a) The equilibrium reaction for the solubility of $Fe(OH)_2$ in water is

$$Fe(OH)_2(s) \rightleftharpoons Fe^{2+}(aq) + 2\ OH^-(aq)$$

In this example, the cation:anion relationship is 1:2; for every Fe^{2+} ion there are two OH^- ions produced, so the hydroxide ion concentration is raised to the power of 2 in the K_{sp} expression: $K_{sp} = [Fe^{2+}][OH^-]^2$.

(b) The equilibrium reaction for the solubility of MgC_2O_4 in water is

$$MgC_2O_4(s) \rightleftharpoons Mg^{2+}(aq) + C_2O_4^{2-}(aq)$$

Because the equilibrium chemical equation shows the cations and anions in a one-to-one relationship, the equilibrium constant expression is $K_{sp} = [Mg^{2+}][C_2O_4^{2-}]$.

(c) The equilibrium reaction for the solubility of Ag_3PO_4 in water is

$$Ag_3PO_4(s) \rightleftharpoons 3\ Ag^+(aq) + PO_4^{3-}(aq)$$

Because three Ag^+ ions are produced for every PO_4^{3-} ion, the K_{sp} expression is written $K_{sp} = [Ag^+]^3[PO_4^{3-}]$.

PROBLEM-SOLVING PRACTICE 15.8

Write the K_{sp} expression for each of these slightly soluble salts:
(a) CuBr (b) HgI_2 (c) $SrSO_4$

15-4a Solubility and K_{sp}

The solubility of a sparingly soluble solute and its solubility product constant, K_{sp}, are not the same thing, but they are related. The solubility is the amount of solute per unit volume of solution (mol/L) that dissolves to form a *saturated* solution. On the other hand, the solubility product constant is the equilibrium constant for the chemical equilibrium that exists between a solid ionic solute and its ions in a saturated solution. In general, solutes with very low solubility have very small K_{sp} values.

If the equilibrium concentrations of the ions are known, they can be used to calculate the K_{sp} value for the solute. For example, in a saturated AgCl solution at 10 °C, the

Table 15.2 K_{sp} Values for Some Slightly Soluble Salts

Compound	K_{sp} at 25 °C	Compound	K_{sp} at 25 °C
AgBr	5.0×10^{-13}	Hg_2Cl_2*	1.3×10^{-18}
AgCl	1.8×10^{-10}	Hg_2I_2*	4.5×10^{-29}
AgI	8.3×10^{-17}	Hg_2SO_4*	7.4×10^{-7}
Ag_2SO_4	1.4×10^{-5}	$Mn(OH)_2$	1.9×10^{-13}
AuI	1.6×10^{-23}	$PbBr_2$	4.0×10^{-5}
AuI_3	1×10^{-46}	$PbCl_2$	1.6×10^{-5}
$BaCrO_4$	1.2×10^{-10}	PbI_2	7.1×10^{-9}
$BaSO_4$	1.1×10^{-10}	$PbSO_4$	1.6×10^{-8}
$Fe(OH)_2$	8.0×10^{-16}	$SrSO_4$	3.2×10^{-7}
Hg_2Br_2*	5.6×10^{-23}		

*These compounds contain the diatomic Hg_2^{2+} ion.

molarities of Ag^+ and Cl^- each are experimentally determined to be 6.3×10^{-6} M. This means that K_{sp} at 10 °C is

$$K_{sp} = [Ag^+][Cl^-] = [6.3 \times 10^{-6}][6.3 \times 10^{-6}] = 4.0 \times 10^{-11}$$

The K_{sp} values for selected ionic compounds at 25 °C are listed in Table 15.2. A more extensive listing is in Appendix H.

PROBLEM-SOLVING EXAMPLE 15.9

Solubility and K_{sp}

The K_{sp} of $BaSO_4$ is 1.1×10^{-10} at 25 °C. Calculate the solubility of $BaSO_4$, in moles per liter.

Result 1.0×10^{-5} M

Analyze Based on its formula the dissociation of solid $BaSO_4$ forms Ba^{2+} ions and SO_4^{2-} ions in equal amounts. The balanced equation for dissociation can be used to write the correct K_{sp} expression. If S represents the solubility of $BaSO_4$, then at equilibrium the concentration of Ba^{2+} and SO_4^{2-} ions will each be S.

Plan Use the balanced chemical equation to write the K_{sp} expression. Substitute the solubility, S, into the K_{sp} expression and solve for S.

Execute $BaSO_4(s) \rightleftharpoons Ba^{2+}(aq) + SO_4^{2-}(aq) \qquad K_{sp} = [Ba^{2+}][SO_4^{2-}] = 1.1 \times 10^{-10}$

$$1.1 \times 10^{-10} = (S)(S) = S^2$$

$$S = \sqrt{1.1 \times 10^{-10}} = 1.0 \times 10^{-5}$$

Consequently, the aqueous solubility of $BaSO_4$ at 25 °C is 1.0×10^{-5} M.

PROBLEM-SOLVING PRACTICE 15.9

The K_{sp} of AgBr at 100 °C is 5×10^{-10}. Calculate the solubility of AgBr at that temperature in moles per liter.

A note of caution is in order. It might seem straightforward to calculate the solubility of an ionic compound from its K_{sp} (the calculation in Problem-Solving Example 15.9), or to do the reverse, that is, calculate the K_{sp} from the solubility (as we did for AgCl(s) just before Problem-Solving Example 15.9). However, this approach overlooks several complicating factors that can lead to incorrect results; we list these here using $PbCl_2$ as an example:

$PbCl_2(s) \rightleftharpoons PbCl^+(aq) + Cl^-(aq)$

$PbCl^+(aq) \rightleftharpoons Pb^{2+}(aq) + Cl^-(aq)$

$PbCl_2(s) \rightleftharpoons Pb^{2+}(aq) + 2\,Cl^-(aq)$

- Ionic solids such as $PbCl_2$ dissociate stepwise, so that $PbCl^+$ ions as well as Pb^{2+} and Cl^- ions are present in a $PbCl_2$ solution.
- Ion pairs, such as $PbCl^+Cl^-$ can exist, reducing the concentrations of unassociated Pb^{2+} and Cl^- ions.
- The solubilities of some solutes, such as metal hydroxides, depend on the acidity or alkalinity of the solution.
- Solutes containing anions such as CO_3^{2-} and PO_4^{3-} that react with water are more soluble than predicted by their K_{sp} values.

Solubilities calculated from K_{sp} values, and K_{sp} values calculated from solubilities, best agree with the experimentally measured solubilities of compounds with 1+ and 1− charged ions and ions that do not react with water.

PROBLEM-SOLVING EXAMPLE 15.10

Solubility and K_{sp}

A saturated aqueous solution of lead(II) iodate, $Pb(IO_3)_2$, contains 3.1×10^{-5}-M Pb^{2+} at 25 °C. Assuming that the ions do not react with water and that dissociation of the ions is complete, calculate the K_{sp} of lead(II) iodate at that temperature.

Result $K_{sp} = 1.2 \times 10^{-13}$

Analyze Based on its formula, when solid $Pb(IO_3)_2$ dissociates in solution, one mole of lead ions is produced for two moles of iodate ions. Thus, the iodate-ion concentration is twice the lead(II)-ion concentration, which is given. The dissociation equation and the K_{sp} expression can be written based on the formula.

Plan Write the chemical equation for the dissociation of the solute, derive the K_{sp} expression correctly from it, and then substitute concentrations of Pb^{2+} and IO_3^- ions into the expression to calculate K_{sp}.

Execute $Pb(IO_3)_2(s) \rightleftharpoons Pb^{2+}(aq) + 2\,IO_3^-(aq) \qquad K_{sp} = [Pb^{2+}][IO_3^-]^2$

The $[Pb^{2+}] = 3.1 \times 10^{-5}$ M so $[IO_3^-] = 2 \times (3.1 \times 10^{-5})$ M $= 6.2 \times 10^{-5}$ M.

$$K_{sp} = [Pb^{2+}][IO_3^-]^2 = (3.1 \times 10^{-5})(6.2 \times 10^{-5})^2 = 1.2 \times 10^{-13}$$

☑ **Reasonable Result Check** Given the fact that the iodate ion concentration is twice that of the lead(II) ion concentration, substituting these values correctly into the proper K_{sp} expression gives the calculated result, which is reasonable.

PROBLEM-SOLVING PRACTICE 15.10

A saturated solution of silver oxalate, $Ag_2C_2O_4$, contains 6.9×10^{-5}-M $C_2O_4^{2-}$ at 25 °C. Calculate the K_{sp} of silver oxalate at that temperature, assuming that the ions do not react with water and the dissociation is complete.

CONCEPTUAL EXERCISE 15.9

Solubility and Le Chatelier's Principle

At 25 °C, 0.014 g calcium carbonate dissolves in 100 mL water. Two equilibria are present in this solution.

(a) $CaCO_3(s) \rightleftharpoons Ca^{2+}(aq) + CO_3^{2-}(aq)$
(b) $CO_3^{2-}(aq) + H_2O(\ell) \rightleftharpoons HCO_3^- + OH^-(aq)$

Suppose Reaction (b) occurs to an appreciable extent. Use Le Chatelier's principle (← **Sec. 12-6**) to predict how Reaction (b) affects the solubility of $CaCO_3$.

15-5 Factors Affecting Solubility

The aqueous solubility of ionic compounds is affected by several factors, some of which have already been mentioned—temperature (← **Sec. 13-4b**), the formation of ion pairs, and competing equilibria. In this section we consider four other factors that affect the aqueous solubility of ionic compounds:

- The effect of acids and pH
- The presence of common ions
- The formation of complex ions
- Amphoterism

15-5a pH and Dissolving Slightly Soluble Salts Using Acids

As noted earlier in the discussion of solubility rules, many salts are only slightly soluble in water (← **Sec. 3-3c**). Some insoluble salts become more soluble if acid is added (pH is lowered); such salts contain a moderately basic ion. As an example, consider calcium carbonate, $CaCO_3$, which is found in minerals such as limestone and marble. $CaCO_3$ is not very soluble in pure water.

$$CaCO_3(s) \rightleftharpoons Ca^{2+}(aq) + CO_3^{2-}(aq) \qquad [15.2]$$

$$K_{sp} = 8.7 \times 10^{-9} \qquad S = \sqrt{8.7 \times 10^{-9}} = 9.3 \times 10^{-5}\ M$$

Because the solubility of calcium carbonate is so low, the equilibrium concentrations of Ca^{2+} and CO_3^{2-} must also be small. However, if acid is added to the solution, calcium carbonate dissolves and CO_2 is released from the solution. Adding acid adds hydronium ions, which react with carbonate and hydrogen carbonate ions.

$$CO_3^{2-}(aq) + H_3O^+(aq) \rightleftharpoons HCO_3^-(aq) + H_2O(\ell) \qquad [15.3]$$

$$HCO_3^-(aq) + H_3O^+(aq) \rightleftharpoons H_2CO_3(aq) + H_2O(\ell) \qquad [15.4]$$

$$H_2CO_3(aq) \rightleftharpoons CO_2(g) + H_2O(\ell) \qquad [15.5]$$

Reaction 15.3 is the reaction of a fairly strong weak base, CO_3^{2-}, with a strong acid, so nearly all of the carbonate is converted to hydrogen carbonate ion. Reaction 15.4 produces a product, carbonic acid, that is unstable and breaks down to CO_2 gas and water via Reaction 15.5, which has a large K_c value, $\cong 10^5$, and is a very product-favored reaction.

Reactions 15.2 through 15.5 are linked through carbonate, hydrogen carbonate, and carbonic acid:

- As CO_2 gas escapes from the solution, the H_2CO_3 concentration decreases (apply Le Chatelier's principle to Reaction 15.5).
- This shifts Reaction 15.4 to the right, decreasing the concentration of HCO_3^-.
- Decreasing the HCO_3^- concentration shifts Reaction 15.3 to the right, decreasing the concentration of CO_3^{2-}.
- To partially counteract this decrease in carbonate concentration, Reaction 15.2 shifts to the right, and more solid $CaCO_3$ dissolves.

Because the acidity of the solution determines the positions of equilibria 15.3 and 15.4, small changes in pH can cause limestone and marble to dissolve or precipitate (← **Sec. 12-6a**). Acid rain can dissolve a marble statue (Figure 15.14); it can also dissolve underground limestone deposits, creating massive cave formations. The impressive stalactite and stalagmite formations in caves that are shown on the first page of this chapter result from such changes. In addition, limestone has precipitated as layers of sedimentary rock on the ocean floor where a slight increase in the pH of seawater has occurred.

In general, insoluble salts containing anions that are Brønsted-Lowry bases dissolve in acidic solutions (those of low pH). This rule covers carbonates, sulfides (which produce $H_2S(g)$), phosphates, and other anions listed as bases in Table 14.5 (← **Sec. 14-8c**). The

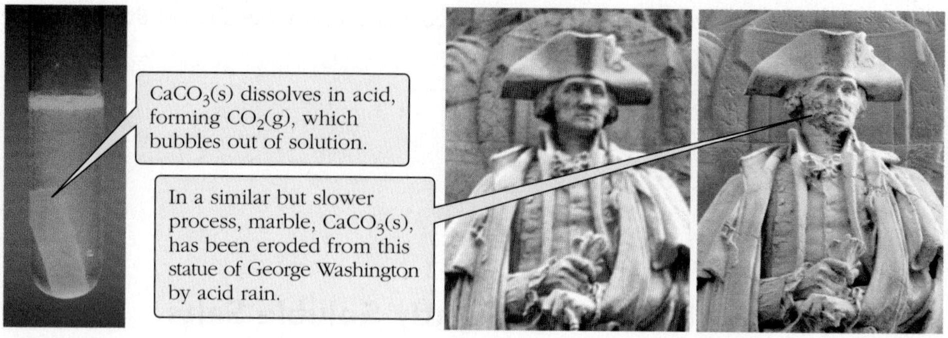

$CaCO_3$(s) dissolves in acid, forming CO_2(g), which bubbles out of solution.
In a similar but slower process, marble, $CaCO_3$(s), has been eroded from this statue of George Washington by acid rain.

Figure 15.14 Damage caused by acid rain. For more examples, see **http://pubs.usgs.gov/gip/acidrain/5.html and http://www.elmhurst.edu/~chm/vchembook/196buildings.html.**

principal exceptions to this rule are a few sulfides, such as HgS, CuS, and CdS, that have extremely low solubilities and therefore do not dissolve even when the pH is very low.

In contrast, an insoluble salt such as AgCl, which contains the conjugate base of a strong acid, is not soluble in strongly acidic solution because Cl^- is a very weak base and so does not react with H_3O^+.

15-5b Solubility and the Common Ion Effect

It is often desirable to remove a particular ion from solution by forming a precipitate of one of its insoluble compounds. For example, barium ions readily absorb X-rays and so are quite effective in making the intestinal tract visible when X-ray photographs are taken. But barium ions are poisonous and must not be allowed to dissolve in body fluids. Barium sulfate, an insoluble compound, can be ingested and used as an X-ray absorber, but both physician and patient want to be certain that the concentration of barium ions in solution is small enough that no harm is caused.

The solubility of $BaSO_4$ in water at 25 °C is 1.0×10^{-5} mol/L (calculated in Problem-Solving Example 15-9, Section 15-4a). In a saturated solution the concentration of Ba^{2+} ions and of sulfate ions is 1.0×10^{-5} M. The concentration of aqueous Ba^{2+} ions can be made much smaller by adding a soluble sulfate salt, such as Na_2SO_4. The solubility of $BaSO_4$ decreases because sulfate ions from Na_2SO_4 make the concentration of SO_4^{2-} ions much higher than it was from $BaSO_4$ alone. This shifts the solubility equilibrium to the left.

$$Na_2SO_4(s) \longrightarrow 2\,Na^+(aq) + SO_4^{2-}(aq)$$

sulfate from Na_2SO_4(s)

$$BaSO_4(s) \rightleftharpoons Ba^{2+}(aq) + SO_4^{2-}(aq)$$

sulfate from $BaSO_4$(s)

← Solubility of $BaSO_4$ decreases; additional solid $BaSO_4$ forms

In this case, the sulfate ion is called a "common ion" because it is common to both substances dissolved in the solution—barium sulfate and sodium sulfate. The common ion shifts the equilibrium to the left by what is called the **common ion effect**: *The presence of a second solute that provides a common ion lowers the solubility of an ionic compound.*

The common ion effect can be interpreted by using Le Chatelier's principle (← **Sec. 12-6**). Adding sodium sulfate to the saturated barium sulfate solution causes a stress (higher concentration of sulfate ions) on the equilibrium. To reduce the concentration of SO_4^{2-}, the equilibrium shifts to the left, reducing the concentration of Ba^{2+} ions as well, to form solid $BaSO_4$. The outcome is that less $BaSO_4$ is dissolved; its solubility is lower in the presence of sulfate, the common ion. Figure 15.15 shows another example of the common ion effect.

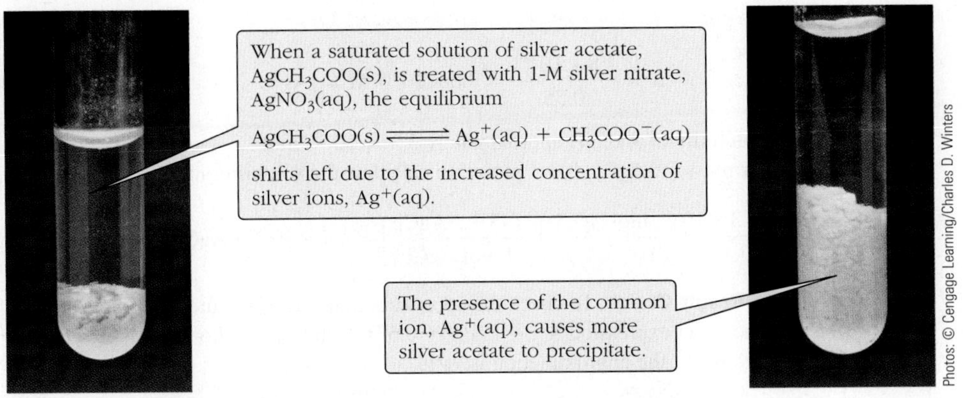
When a saturated solution of silver acetate, $AgCH_3COO(s)$, is treated with 1-M silver nitrate, $AgNO_3(aq)$, the equilibrium
$AgCH_3COO(s) \rightleftharpoons Ag^+(aq) + CH_3COO^-(aq)$
shifts left due to the increased concentration of silver ions, $Ag^+(aq)$.
The presence of the common ion, $Ag^+(aq)$, causes more silver acetate to precipitate.
Photos: © Cengage Learning/Charles D. Winters

Figure 15.15 The common ion effect. Silver ions are common to both solutes.

CONCEPTUAL EXERCISE 15.10

Common Ion Effect

Consider 0.0010-M solutions of these sparingly soluble solutes in equilibrium with their ions. Predict the effect on each equilibrium if a saturated solution of sodium iodide were added. Explain your prediction in each case.

$$AgI(s) \rightleftharpoons Ag^+(aq) + I^-(aq)$$
$$PbI_2(s) \rightleftharpoons Pb^{2+}(aq) + 2\,I^-(aq)$$

PROBLEM-SOLVING EXAMPLE 15.11

Common Ion Effect

The solubility of AgCl in pure water is 1.3×10^{-5} mol/L at 25 °C. If solid AgCl is added to a solution that is 0.55 M in NaCl, calculate what mass of AgCl dissolves per liter of solution. The K_{sp} of AgCl is 1.8×10^{-10} at 25 °C.

Result 4.7×10^{-8} g AgCl/L

Analyze The solution has two sources of chloride ions: AgCl and NaCl. The common ion effect predicts that the solubility of AgCl is less when NaCl is present than it would be in pure water. Because AgCl is the only source of Ag^+ ions in the solution, the $[Ag^+]$ equals S, the solubility of AgCl. S does not equal $[Cl^-]$ because most of the Cl^- comes from the NaCl, a highly soluble salt, not from AgCl.

Plan Write the chemical equation for the dissociation of AgCl(s) and the K_{sp} expression. Let S equal the silver ion concentration $[Ag^+]$ at equilibrium. Set up an ICE table. Substitute the equilibrium concentrations of Ag^+ and Cl^- from the ICE table into the K_{sp} equation to solve for $[Ag^+]$. Calculate the mass of AgCl that dissolved using the solubility, the volume of solution (1 L), and the molar mass of AgCl.

Execute The balanced chemical equation and ICE table are

	AgCl(s)	$\rightleftharpoons$	$Ag^+(aq)$	+	$Cl^-(aq)$
*I*nitial concentration (mol/L)			0		0.55
*C*hange as reaction occurs (mol/L)			$+S$		$+S$
*E*quilibrium concentration (mol/L)			S		$S + 0.55$

The total chloride ion concentration at equilibrium is what came from AgCl(s) *plus* what was already there (0.55 M) from the NaCl.

$$K_{sp} = 1.8 \times 10^{-10} = [Ag^+][Cl^-] = (S)(S + 0.55)$$

Because AgCl is not very soluble, assume that S is *very* small compared to 0.55; that is, assume that $(S + 0.55) \cong 0.55$. Therefore,

$$(S)(S + 0.55) \cong (S)(0.55) = K_{sp} = 1.8 \times 10^{-10}$$

The assumption that $S \ll 0.55$ is reasonable because we know that the solubility of AgCl in pure water equals only 1.3×10^{-5} mol/L. The NaCl present in solution further decreases the solubility of AgCl due to the presence of Cl^-, the common ion.

Solving for S, we get

$$S = \frac{1.8 \times 10^{-10}}{0.55} = 3.3 \times 10^{-10}\ \text{M} = [\text{Ag}^+]$$

Therefore, the solubility of AgCl is approximately 3.3×10^{-10} mol/L.

Use the molar mass for AgCl, 143.3 g/mol, to calculate the solubility in g/L.

$$\left(\frac{3.3 \times 10^{-10}\ \text{mol AgCl}}{1\ \text{L}}\right)\left(\frac{143.3\ \text{g AgCl}}{1\ \text{mol AgCl}}\right) = 4.7 \times 10^{-8}\ \text{g AgCl/L}$$

☑ **Reasonable Result Check** To check the approximation, substitute the approximate value of S into the exact expression $K_{sp} = (S)(S + 0.55)$. If this calculation results in the correct K_{sp}, 1.8×10^{-10}, the approximation is valid.

$$K_{sp} = (S)(S + 0.55) = (3.3 \times 10^{-10})(3.3 \times 10^{-10} + 0.55) = 1.8 \times 10^{-10}$$

We could also solve for S using the quadratic equation described in Appendix A-7 and used in Problem-Solving Example 14.10 (← **Sec. 14-7**). To two significant figures the quadratic equation gives the same value as the approximation.

PROBLEM-SOLVING PRACTICE 15.11

Calculate the solubility of $PbCl_2$ at 25 °C in a solution that is 0.50 M in NaCl.

PROBLEM-SOLVING EXAMPLE 15.12

pH and the Common Ion Effect

Manganese(II) hydroxide, $Mn(OH)_2$, is sparingly soluble in water: $K_{sp}(Mn(OH)_2) = 1.9 \times 10^{-13}$ at 25 °C. Calculate the solubility of manganese(II) hydroxide at that temperature in (a) pure water and (b) at a pH of 11.00.

Result (a) 3.6×10^{-5} M (b) 1.9×10^{-7} M

Analyze In part (a) no common ion is present. In part (b) where the initial pH is 11.00, there is a significant hydroxide ion concentration before any solid $Mn(OH)_2$ dissolves; hydroxide ion is the common ion. There are two sources of hydroxide ion: (1) the initial solution; and (2) the dissolution of solid $Mn(OH)_2$. Due to the common ion effect, the solubility of manganese(II) hydroxide in part (b) should be less than it is in part (a).

Plan Write the balanced chemical equation for the dissociation of solid $Mn(OH)_2$ and the K_{sp} expression. Let S equal the solubility of the solid solute, $Mn(OH)_2$. In this case, $S = [Mn^{2+}]$, because there is one mole per liter of Mn^{2+} for each mole per liter of $Mn(OH)_2$ that dissolves. The $[OH^-]$ is 2 S because there is 2 mol OH^- for 1 mol $Mn(OH)_2$. For part (a), set up an ICE table, substitute the equilibrium concentrations from it into the K_{sp} expression, and solve for S. For part (b), use the pH given to calculate the initial OH^- concentration. Then, set up an ICE table, substitute the equilibrium concentrations from it into the K_{sp} expression, and solve for S. Due to the common ion effect, $Mn(OH)_2$ will be even less soluble in a solution that already contains a significant concentration of hydroxide ions. Therefore, a simplifying assumption likely can be applied to solve for S.

Execute The chemical equilibrium and K_{sp} expression are

$$Mn(OH)_2(s) \rightleftharpoons Mn^{2+}(aq) + 2\ OH^-(aq) \qquad K_{sp} = [Mn^{2+}][OH^-]^2 = 1.9 \times 10^{-13}$$

(a) The ICE table is

	$Mn(OH)_2(s)$ $\rightleftharpoons$	$Mn^{2+}(aq)$ +	$2\ OH^-(aq)$
*I*nitial concentration (mol/L)		0	≅0*
*C*hange as reaction occurs (mol/L)		$+S$	$+2S$
*E*quilibrium concentration (mol/L)		$+S$	$+2S$

*We assume that the contribution of OH^- from the ionization of water is negligible.

Substitute into the K_{sp} expression and solve for S, the solubility of $Mn(OH)_2$.

$$K_{sp} = [Mn^{2+}][OH^-]^2 = (S)(2S)^2 = 4S^3 = 1.9 \times 10^{-13}$$

$$S = [Mn^{2+}] = \sqrt[3]{1.9 \times 10^{-13} / 4} = 3.6 \times 10^{-5}$$

The solubility of $Mn(OH)_2$ equals 3.6×10^{-5} M; the OH^- concentration is $2S$, which is 7.2×10^{-5} M.

(b) Before any $Mn(OH)_2$ is added to the solution, the pH is 11.00. Thus, initially,

$$[H_3O^+] = 1.0 \times 10^{-11} \text{ M and the } [OH^-] = \frac{1.0 \times 10^{-14}}{1.0 \times 10^{-11}} = 1.0 \times 10^{-3} \text{ M.}$$

When $Mn(OH)_2(s)$ is added, let S equal the concentration of $Mn(OH)_2$ that dissolves (the solubility) forming S mol/L of Mn^{2+} and $2S$ mol/L of OH^-. The OH^- concentration at equilibrium is $1.0 \times 10^{-3} + 2S$.

	$Mn(OH)_2(s) \rightleftharpoons$	$Mn^{2+}(aq)$ +	$2\ OH^-(aq)$
*I*nitial concentration (mol/L)		0	1.0×10^{-3}
*C*hange as addition of $Mn(OH)_2$ (mol/L)		$+S$	$+2S$
*E*quilibrium concentration (mol/L)		$+S$	$1.0 \times 10^{-3} + 2S$

Substitute equilibrium concentrations into the K_{sp} expression and solve for S.

$$(S)(1.0 \times 10^{-3} + 2S)^2 = 1.9 \times 10^{-13}$$

For the same reasons as in part (a) assume that $2S \ll 1.0 \times 10^{-3}$, which simplifies the equation to

$$(S)(1.0 \times 10^{-3})^2 \cong 1.9 \times 10^{-13}$$

$$S \cong \frac{1.9 \times 10^{-13}}{1.0 \times 10^{-6}} = 1.9 \times 10^{-7} \text{ M}$$

Thus, the presence of hydroxide as the common ion decreased the solubility of $Mn(OH)_2$ from that in pure water, 3.6×10^{-5} M, to 1.9×10^{-7} M in a starting solution of pH 11.00.

☑ **Reasonable Result Check** In part (a) the solubility is described by the expression $(S)(2S)^2 = 4S^3 = 1.9 \times 10^{-13}$. The result calculated using this expression is reasonable considering how very small the K_{sp} is. In part (b), check the approximation by substituting the approximate value of S into the actual expression $K_{sp} = (S)(1.0 \times 10^{-3} + 2S)^2$. Doing so we find that the product $(S)(1.0 \times 10^{-3} + 2S)^2$ equals the given K_{sp} value, so the approximation is legitimate.

PROBLEM-SOLVING PRACTICE 15.12

Calculate the solubility of $PbCl_2$ in (a) pure water and (b) 0.20-M NaCl. K_{sp} of $PbCl_2$ = 1.6×10^{-5}.

15-5c Complex Ion Formation

All metal cations are potential Lewis acids (← **Sec. 14-9**) because they can accept an electron pair donated by a Lewis base to form a complex ion. The reaction of Cu^{2+} ions with NH_3 is typical and is shown in Figure 15.16.

$$\underset{\text{Lewis acid}}{Cu^{2+}(aq)} + \underset{\text{Lewis base}}{4\ NH_3(aq)} \rightleftharpoons \underset{\text{complex ion}}{[Cu(NH_3)_4]^{2+}(aq)}$$

Many metal salts that are insoluble in water are brought into solution by complex ion formation with Lewis bases such as $S_2O_3^{2-}$, NH_3, OH^-, and CN^-. For example, the water solubility of AgBr is very low, 1.35×10^{-4} g/L, equivalent to 7.19×10^{-7} M.

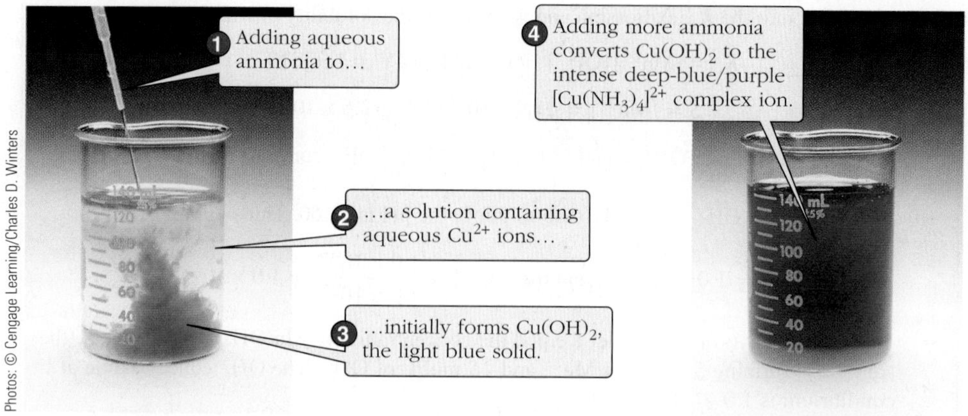
Photos: © Cengage Learning/Charles D. Winters
1 Adding aqueous ammonia to...
2 ...a solution containing aqueous Cu^{2+} ions...
3 ...initially forms $Cu(OH)_2$, the light blue solid.
4 Adding more ammonia converts $Cu(OH)_2$ to the intense deep-blue/purple $[Cu(NH_3)_4]^{2+}$ complex ion.

Figure 15.16 Formation of an ammonia complex ion from copper(II) ions.

However, AgBr dissolves readily in a sodium thiosulfate, $Na_2S_2O_3$, solution because the $[Ag(S_2O_3)_2]^{3-}$ complex ion forms (Figure 15.17).

$$AgBr(s) + 2\ S_2O_3^{2-}(aq) \rightleftharpoons [Ag(S_2O_3)_2]^{3-}(aq) + Br^-(aq)$$

The dissolving of AgBr in this way can be considered as the sum of two reactions—the solubility equilibrium of aqueous AgBr and the formation of the complex ion.

$$AgBr(s) \rightleftharpoons Ag^+(aq) + Br^-(aq)$$

$$Ag^+(aq) + 2\ S_2O_3^{2-}(aq) \rightleftharpoons [Ag(S_2O_3)_2]^{3-}(aq)$$

$$\textit{Net reaction:}\ AgBr(s) + 2\ S_2O_3^{2-}(aq) \rightleftharpoons [Ag(S_2O_3)_2]^{3-}(aq) + Br^-(aq)$$

The extent to which complex ion formation occurs can be evaluated from the magnitude of *the equilibrium constant for the formation of a complex ion*, the **formation constant, K_f**. For example, for $[Ag(S_2O_3)_2]^{3-}$, K_f is 2.0×10^{13}.

$$Ag^+(aq) + 2\ S_2O_3^{2-}(aq) \rightleftharpoons [Ag(S_2O_3)_2]^{3-}(aq)$$

$$K_f = \frac{\{[Ag(S_2O_3)_2]^{3-}\}}{[Ag^+][S_2O_3^{2-}]^2} = 2.0 \times 10^{13}$$

Formation constants for some metal complex ions are given in Table 15.3. The formation and structure of complex ions, which are very important in biochemistry and metallurgy, are considered in more detail in Chapter 20.

Silver bromide (white solid) is insoluble in water.
When aqueous sodium thiosulfate is added, the AgBr dissolves.
Ag^+ Br^-
Photos: © Cengage Learning/Charles D. Winters
(a)
(b)
$S_2O_3^{2-}$ (thiosulfate)
Coordinate covalent bonds

Figure 15.17 Sodium thiosulfate dissolves silver bromide. Water molecules and Na^+ ions have been omitted from the illustration for simplicity's sake.

PROBLEM-SOLVING EXAMPLE 15.13

Solubility and Complex Ion Formation

The K_{sp} of AgCl is 1.8×10^{-10}. The K_f of $[Ag(CN)_2]^-$ is 1.4×10^{20}. Use these data to show that AgCl dissolves in aqueous NaCN.

Result The net equilibrium constant is 2.5×10^{10}, which indicates that dissolving AgCl by complex ion formation is highly favored and AgCl will dissolve in NaCN.

Analyze If the equilibrium constant for the net chemical equation for dissolving AgCl by $[Ag(CN)_2]^-$ complex ion formation, K_{net}, is much greater than 1 the reaction is product-favored (← **Sec. 12-4**) and AgCl will dissolve in aqueous NaCN. If the K_{sp} and K_f equations can be combined to give the equation for dissolving AgCl, then K_{net} can be calculated from K_{sp} and K_f (← **Sec. 12-2e**).

Plan Write the chemical equations for the dissociation of AgCl and for the reaction of $Ag^+(aq)$ with $CN^-(aq)$. Combine them to get the equation for the net reaction to form $[Ag(CN)_2]^-(aq)$. Using the K_{sp} and K_f values associated with these reactions, calculate the K_{net} to determine whether $K_{net} \gg 1$.

Execute The net reaction for dissolving AgCl by $[Ag(CN)_2]^-$ complex ion formation is the sum of the K_{sp} and K_f equations.

$$AgCl(s) \rightleftharpoons Ag^+(aq) + Cl^-(aq) \qquad K_{sp} = 1.8 \times 10^{-10}$$

$$Ag^+(aq) + 2\,CN^-(aq) \rightleftharpoons [Ag(CN)_2]^-(aq) \qquad K_f = 1.4 \times 10^{20}$$

Net reaction: $$AgCl(s) + 2\,CN^-(aq) \rightleftharpoons [Ag(CN)_2]^-(aq) + Cl^-(aq) \qquad K_{net} = ?$$

Thus, $K_{net} = K_{sp} \times K_f = (1.8 \times 10^{-10})(1.4 \times 10^{20}) = 2.5 \times 10^{10}$. $K_{net} \gg 1$ so the net reaction is product-favored; AgCl is soluble in aqueous NaCN.

PROBLEM-SOLVING PRACTICE 15.13

The K_{sp} of AgCl is 1.8×10^{-10}. The K_f of $[Ag(S_2O_3)_2]^{3-}$ is 2.0×10^{13}. Use these data to show that dissolving AgCl by $[Ag(S_2O_3)_2]^{3-}$ complex ion formation is a product-favored process.

Table 15.3 Formation Constants for Some Complex Ions in Aqueous Solution*

Formation Equilibrium	K_f
$Ag^+ + 2\,CN^- \rightleftharpoons [Ag(CN)_2]^-$	1.4×10^{20}
$Ag^+ + 2\,S_2O_3^{2-} \rightleftharpoons [Ag(S_2O_3)_2]^{3-}$	2.0×10^{13}
$Ag^+ + 2\,NH_3 \rightleftharpoons [Ag(NH_3)_2]^+$	1.7×10^{7}
$Al^{3+} + 4\,OH^- \rightleftharpoons [Al(OH)_4]^-$	4.0×10^{36}
$Au^+ + 2\,CN^- \rightleftharpoons [Au(CN)_2]^-$	4.0×10^{36}
$Cd^{2+} + 4\,CN^- \rightleftharpoons [Cd(CN)_4]^{2-}$	1.6×10^{19}
$Cd^{2+} + 4\,NH_3 \rightleftharpoons [Cd(NH_3)_4]^{2+}$	9.4×10^{6}
$Co^{2+} + 6\,NH_3 \rightleftharpoons [Co(NH_3)_6]^{2+}$	1.3×10^{4}
$Cu^{2+} + 4\,NH_3 \rightleftharpoons [Cu(NH_3)_4]^{2+}$	3.3×10^{12}
$Fe^{2+} + 6\,CN^- \rightleftharpoons [Fe(CN)_6]^{4-}$	3.2×10^{32}
$Hg^{2+} + 4\,Cl^- \rightleftharpoons [HgCl_4]^{2-}$	1.3×10^{15}
$Ni^{2+} + 4\,CN^- \rightleftharpoons [Ni(CN)_4]^{2-}$	1.7×10^{30}
$Ni^{2+} + 6\,NH_3 \rightleftharpoons [Ni(NH_3)_6]^{2+}$	5.5×10^{8}
$Zn^{2+} + 4\,OH^- \rightleftharpoons [Zn(OH)_4]^{2-}$	4.5×10^{17}
$Zn^{2+} + 4\,NH_3 \rightleftharpoons [Zn(NH_3)_4]^{2+}$	2.9×10^{9}

*Taken from Högfeldt, E., Perrin, D. D., *Stability Constants of Metal-Ion Complexes,* 1st ed., Oxford; New York: Pergamon, 1979–1982. International Union of Pure and Applied Chemistry Commission on Equilibrium Data. Values for $[AgBr_2]^-$, $[Co(NH_3)_6]^{2+}$, and $[Zn(NH_3)_4]^{2+}$, from Patnaik, Pradyot, *Dean's Analytical Chemistry Handbook,* 2nd ed., New York, McGraw-Hill, 2004; Table 4.2.

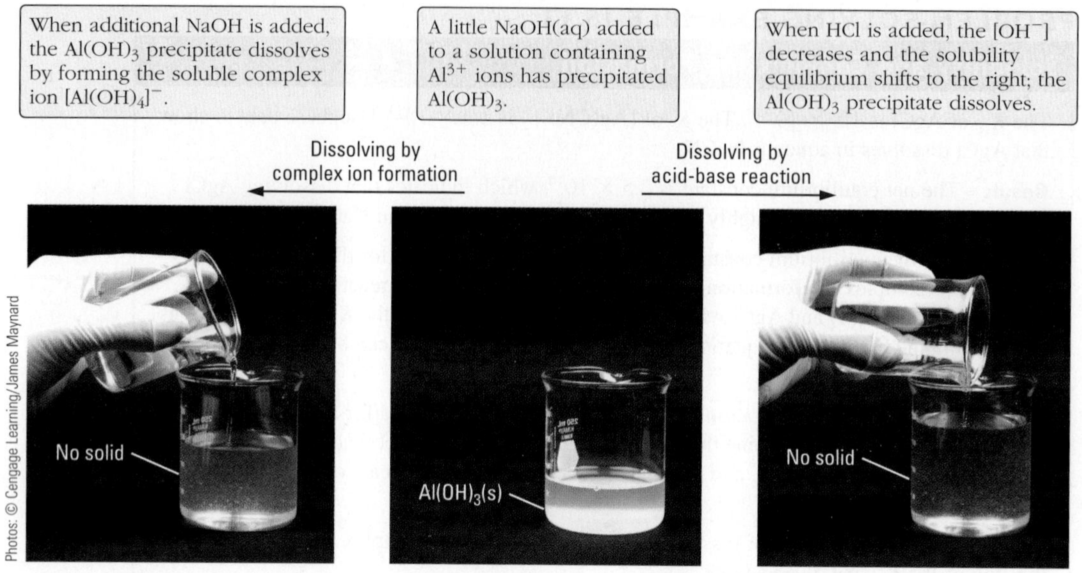
When additional NaOH is added, the $Al(OH)_3$ precipitate dissolves by forming the soluble complex ion $[Al(OH)_4]^-$.
A little NaOH(aq) added to a solution containing Al^{3+} ions has precipitated $Al(OH)_3$.
When HCl is added, the $[OH^-]$ decreases and the solubility equilibrium shifts to the right; the $Al(OH)_3$ precipitate dissolves.
Dissolving by complex ion formation
Dissolving by acid-base reaction
No solid
$Al(OH)_3(s)$
No solid
Photos: © Cengage Learning/James Maynard

$$[Al(OH)_4]^- \xleftarrow[\text{from NaOH(aq)}]{OH^-} Al(OH)_3(s) \xrightarrow[\text{from HCl(aq)}]{3\ H_3O^+} [Al(H_2O)_6]^{3+}$$

Figure 15.18 $Al(OH)_3$ is amphoteric: it dissolves in base and in acid.

15-5d Amphoterism

The majority of metal hydroxides are insoluble in water, but many dissolve in highly acidic or basic solutions. This is because these hydroxides are *amphoteric;* that is, they can react with both H_3O^+ ions and OH^- ions (← **Sec. 14-9a**). Aluminum hydroxide, $Al(OH)_3$, is an example of an amphoteric hydroxide (Figure 15.18). When it reacts with acid, aluminum hydroxide dissolves by acting as a base, donating OH^- ions to react with hydronium ions from the acid to form water molecules that bond to the aluminum ions.

$$Al(OH)_3(s) + 3\ H_3O^+(aq) \longrightarrow [Al(H_2O)_6]^{3+}(aq)$$

In highly basic solutions, $Al(OH)_3$ is dissolved through complex ion formation. In this case, Al^{3+} ions act as a Lewis acid by accepting electron pairs from OH^- ions, a Lewis base, to form $[Al(OH)_4]^-$.

$$Al(OH)_3(s) + OH^-(aq) \longrightarrow [Al(OH)_4]^-(aq)$$

15-6 Precipitation: Will It Occur?

Earlier, when writing net ionic equations, we used solubility rules to predict whether a precipitate forms when two solutions of salts are mixed (← **Sec. 3-3c**). Those rules apply to situations where the ions involved are at concentrations of 0.1 M or greater. Whether precipitation occurs when the ion concentrations are considerably less than 0.1 M depends on the concentrations of the ions in the resulting solution and the K_{sp} value for any precipitate that might form.

For example, AgBr might precipitate when a water-soluble silver salt, such as $AgNO_3$, is added to an aqueous solution of a bromide salt, such as KBr. The net ionic equation for the reaction is

$$Ag^+(aq) + Br^-(aq) \rightleftharpoons AgBr(s)$$

This is the reverse of the equation for K_{sp} of AgBr:

$$AgBr(s) \rightleftharpoons Ag^+(aq) + Br^-(aq)$$

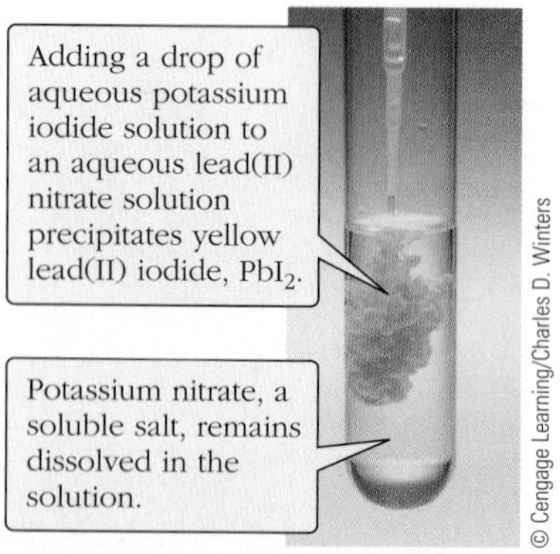
Adding a drop of aqueous potassium iodide solution to an aqueous lead(II) nitrate solution precipitates yellow lead(II) iodide, PbI_2.
Potassium nitrate, a soluble salt, remains dissolved in the solution.
© Cengage Learning/Charles D. Winters

Precipitation of lead(II) iodide.

To determine whether a precipitate forms, we use the *reaction quotient, Q* (← **Sec. 12-5a**). In this case the equilibrium constant expression is a product of ion concentrations, so Q is called the *ion product*. The Q expression has the same form as that for K_{sp}; however, the *initial* concentrations are used, *not those at equilibrium as in* K_{sp}. For AgBr the two expressions are

$$Q = (\text{conc. } Ag^+)(\text{conc. } Br^-) \qquad K_{sp} = [Ag^+][Br^-]$$

In this expression, (conc. Ag^+) represents the molarity of Ag^+ whether or not the system is at equilibrium. [Ag^+] represents molarity at *equilibrium*.

We compare the magnitude of the ion product, Q, with that of the solubility product constant, K_{sp}. Three cases are possible (Figure 15.19).

1. ***$Q < K_{sp}$*: *The solution is unsaturated and no precipitate forms.*** In this case, the solution contains ions at concentrations lower than required for equilibrium with the solid. An equilibrium is not established between a solid solute and its ions because no solid solute is present; more solute can be added to the solution before precipitation occurs. If a solid was present, some of the solid would dissolve (the equilibrium would shift to the right).
2. ***$Q > K_{sp}$*: *The solution contains a higher concentration of ions than it can hold at equilibrium; that is, the solution is supersaturated.*** To reach equilibrium, a precipitate forms, decreasing the concentration of ions until the ion product equals the K_{sp} (the equilibrium shifts to the left).
3. ***$Q = K_{sp}$*: *The solution is at equilibrium and at the point of precipitation; that is, the solution is saturated.***

Consider the case of two solutions, each made by combining $Pb(NO_3)_2(aq)$ and $Na_2SO_4(aq)$. In solution 1, the initial concentrations of Pb^{2+} and SO_4^{2-} are each 1.0×10^{-4} M. In solution 2, the initial concentrations are each 2.0×10^{-4} M, twice the concentrations of solution 1. In each of the two solutions, the reaction products are $NaNO_3$ and $PbSO_4$. The solubility rules indicate that $NaNO_3$ is soluble and remains in solution as Na^+ and NO_3^- ions, whereas $PbSO_4$ is insoluble (when concentrations are 0.1 M). Does a precipitate of $PbSO_4$ form in either or both solutions? K_{sp} of $PbSO_4 = 1.6 \times 10^{-8}$.

$$Q(\text{solution 1}) = (\text{conc. } Pb^{2+})(\text{conc. } SO_4^{2-}) = (1.0 \times 10^{-4}\text{ M})(1.0 \times 10^{-4}\text{ M})$$
$$= 1.0 \times 10^{-8}\text{; this is less than } K_{sp}\text{, which is } 1.6 \times 10^{-8}$$

$$Q(\text{solution 2}) = (\text{conc. } Pb^{2+})(\text{conc. } SO_4^{2-}) = (2.0 \times 10^{-4}\text{ M})(2.0 \times 10^{-4}\text{ M})$$
$$= 4.0 \times 10^{-8}\text{; this is greater than } K_{sp}\text{, which is } 1.6 \times 10^{-8}$$

Because Q of solution 1 is less than the K_{sp}, no precipitate forms. In contrast, Q of solution 2 exceeds the K_{sp}, and precipitation occurs.

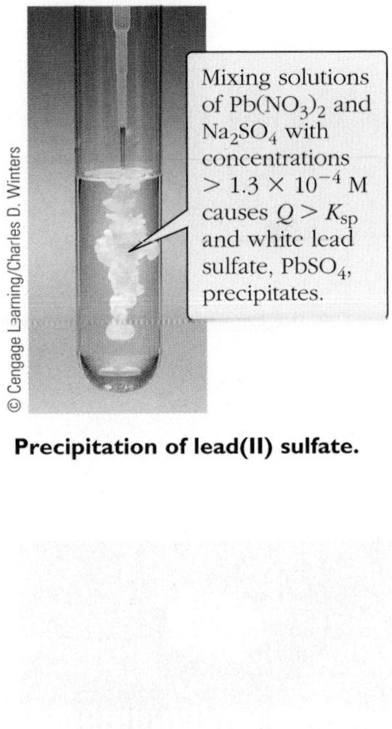
Mixing solutions of $Pb(NO_3)_2$ and Na_2SO_4 with concentrations $> 1.3 \times 10^{-4}$ M causes $Q > K_{sp}$ and white lead sulfate, $PbSO_4$, precipitates.
© Cengage Learning/Charles D. Winters

Precipitation of lead(II) sulfate.

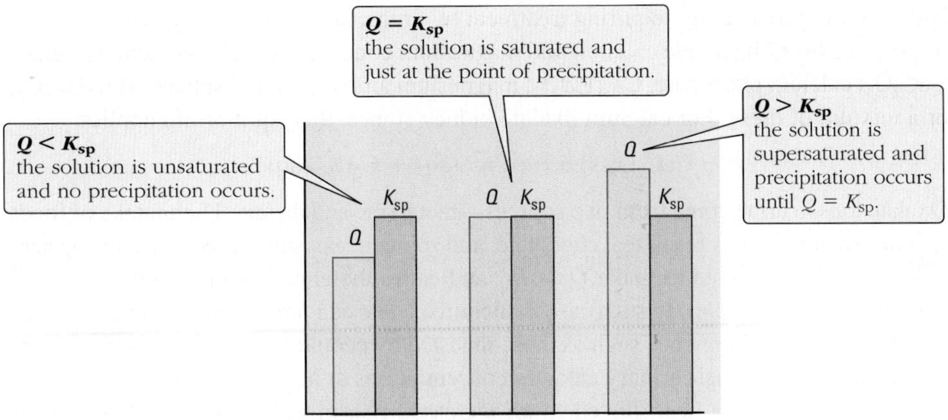
$Q = K_{sp}$ the solution is saturated and just at the point of precipitation.
$Q < K_{sp}$ the solution is unsaturated and no precipitation occurs.
$Q > K_{sp}$ the solution is supersaturated and precipitation occurs until $Q = K_{sp}$.
Q
K_{sp}
Q
K_{sp}
Q
K_{sp}

Figure 15.19 Predicting precipitation by comparing Q with K_{sp}.

PROBLEM-SOLVING EXAMPLE 15.14

Q, K_{sp}, and Precipitation

A chemistry student mixes 20.0 mL of 4.5×10^{-3}-M $AgNO_3$ with 10.0 mL of 7.5×10^{-2}-M $NaBrO_3$. Determine whether a precipitate forms. The K_{sp} of $AgBrO_3 = 6.7 \times 10^{-5}$.

Result Yes, $AgBrO_3$ precipitates.

Analyze Nitrate salts and sodium salts are soluble so if a precipitate forms it must be $AgBrO_3(s)$; the appropriate chemical equilibrium is

$$AgBrO_3(s) \rightleftharpoons Ag^+(aq) + BrO_3^-(aq)$$

For a precipitate to form, Q = (conc. Ag^+)(conc. BrO_3^-), must be greater than K_{sp}.

Plan Calculate the concentrations of Ag^+ and BrO_3^- ions, after mixing the solutions; then substitute them into the Q expression and compare Q with K_{sp}. Assume the volume of the mixed solution is the sum of the original solution volumes.

Execute The volume of the mixed solution is (0.020 + 0.010)L = 0.0300 L.

$$n(Ag^+) = (0.0200\ L)\left(\frac{4.5 \times 10^{-3}\ mol\ Ag^+}{1\ L}\right) = 9.0 \times 10^{-5}\ mol\ Ag^+$$

$$(conc.\ Ag^+) = \frac{9.0 \times 10^{-5}\ mol\ Ag^+}{0.0300\ L} = 3.0 \times 10^{-3}\ M\ Ag^+$$

$$n(BrO_3^-) = (0.0100\ L)\left(\frac{7.5 \times 10^{-2}\ mol\ BrO_3^-}{1\ L}\right) = 7.5 \times 10^{-4}\ mol\ BrO_3^-$$

$$(conc.\ BrO_3^-) = \left(\frac{7.5 \times 10^{-4}\ mol\ BrO_3^-}{0.0300\ L}\right) = 2.5 \times 10^{-2}\ M\ BrO_3^-$$

$$Q = (3.0 \times 10^{-3}\ M)(2.5 \times 10^{-2}\ M) = 7.5 \times 10^{-5}$$

which is greater than the $K_{sp} = 6.7 \times 10^{-5}$. Precipitation occurs until the ion product equals K_{sp}.

PROBLEM-SOLVING PRACTICE 15.14

(a) Determine whether AgCl precipitates from a solution containing 1.0×10^{-5}-M Ag^+ and 1.0×10^{-5}-M Cl^-. K_{sp} AgCl $= 1.8 \times 10^{-10}$.

(b) An AgCl precipitate forms from a solution that contains 1.0×10^{-5}-M Ag^+. Calculate the minimum Cl^- concentration required for precipitation to occur.

15-6a Kidney Stones—Common Ion Effect and Le Chatelier's Principle

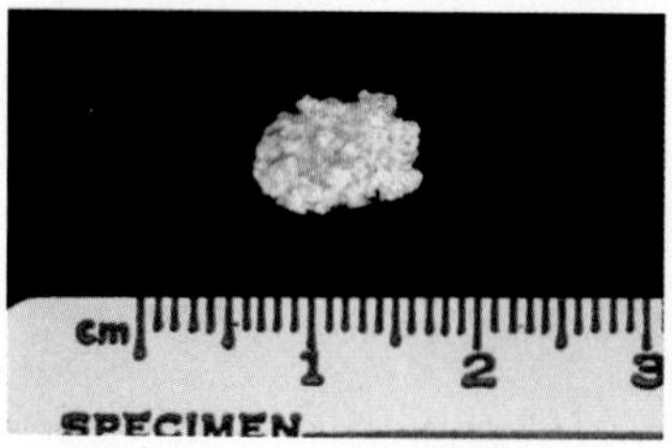
cm
1
2
3
SPECIMEN

Southern Illinois University/Science Source

Kidney stone. This kidney stone was surgically removed from a patient.

Many ions circulate in the human bloodstream; some combinations of ions can precipitate to form kidney stones. Such stones can become large enough to be extremely painful and even life-threatening, requiring treatment by drugs, lasers, or surgery. Kidney stones usually consist of insoluble calcium and magnesium compounds such as calcium oxalate, CaC_2O_4; calcium phosphate, $Ca_3(PO_4)_2$; magnesium ammonium phosphate, $MgNH_4PO_4$; or a mixture of these. For calcium oxalate kidney stones, this equilibrium applies:

$$CaC_2O_4(s) \rightleftharpoons Ca^{2+}(aq) + C_2O_4^{2-}(aq)$$

Leaving chocolate out of a diet seems far more punishment than forgoing spinach.

Oxalate ions in urine come from two sources—metabolic and dietary. High intake of foods rich in oxalate, such as black tea, chocolate, and spinach, can sufficiently raise the urinary concentration of oxalate to make $Q > K_{sp}$ and cause the equilibrium to shift to the left (Le Chatelier's Principle). In such cases, calcium oxalate can precipitate as a kidney stone. Taking calcium supplements, such as those ingested by people with osteoporosis or osteopenia, may increase their urinary calcium concentrations to high levels; continuing use of these supplements has been linked to an increased risk of kidney stones. Nutritionists advise taking such supplements with meals to reduce the risk. A high-sugar diet may also create kidney stones because excessive sugar promotes excretion of Ca^{2+} and Mg^{2+}, which increases the concentrations of these ions passing through the kidneys. This can cause kidney stone formation, such as through calcium phosphate precipitation:

$$3\ Ca^{2+}(aq) + 2\ PO_4^{3-}(aq) \longrightarrow Ca_3(PO_4)_2(s)$$

15-6b Selective Precipitation of Ions

If their solubilities are sufficiently different, ionic compounds can be precipitated selectively from solution. The more soluble compound remains in solution as the less soluble one starts to precipitate. For example, silver chloride, AgCl, and silver chromate, Ag_2CrO_4, are each only slightly soluble in water. The solubilities of these two solutes differ enough, however, that when they are both in the same solution, one can be precipitated, leaving the other behind in solution.

PROBLEM-SOLVING EXAMPLE 15.15

Selective Precipitation

Consider a solution containing 0.020-M Cl^- and 0.010-M CrO_4^{2-} ions to which Ag^+ ions are added slowly. Which precipitate forms first: AgCl or Ag_2CrO_4? $K_{sp}(AgCl) = 1.8 \times 10^{-10}$; $K_{sp}(Ag_2CrO_4) = 1.1 \times 10^{-12}$.

Result AgCl precipitates first.

Analyze Until a precipitate begins to form, the concentrations of chloride and chromate ions remain constant. The precipitate that forms first is the one where a smaller Ag^+ concentration causes the solubility product to be exceeded; that is, where the ion product, Q, just barely exceeds the K_{sp}.

Plan Write the dissociation reaction and equilibrium constant expression for each silver compound. Use the K_{sp} values and the anion concentrations given and rearrange the K_{sp} expression to solve for $[Ag^+]$ in each case. Compare the $[Ag^+]$ concentrations required for precipitation to occur; the silver salt that requires the smaller Ag^+ concentration precipitates first.

Execute To precipitate AgCl,

$$K_{sp}(AgCl) = [Ag^+][Cl^-] = 1.8 \times 10^{-10}$$

$$[Ag^+] = \frac{1.8 \times 10^{-10}}{[Cl^-]} = \frac{1.8 \times 10^{-10}}{2.0 \times 10^{-2}} = 9.0 \times 10^{-9}\ M$$

An Ag^+ concentration of slightly greater than 9.0×10^{-9} M gives $Q > K_{sp}$ and some AgCl precipitates from the solution.

To precipitate Ag_2CrO_4,

$$K_{sp} \text{ of } Ag_2CrO_4 = [Ag^+]^2[CrO_4^-] = 1.1 \times 10^{-12}$$

$$[Ag^+]^2 = \frac{1.1 \times 10^{-12}}{[CrO_4^{2-}]} = \frac{1.1 \times 10^{-12}}{1.0 \times 10^{-2}} = 1.1 \times 10^{-10}; \qquad [Ag^+] = 1.0 \times 10^{-5}$$

Silver chromate will precipitate when the Ag^+ concentration slightly exceeds 1.0×10^{-5} M. Because a *much* smaller concentration of Ag^+, 9.0×10^{-9} M, is required for AgCl precipitation, AgCl will precipitate before Ag_2CrO_4.

☑ **Reasonable Result Check** The Ag^+ concentration required to precipitate Ag_2CrO_4 is approximately 10,000 times greater than that for AgCl (1×10^{-5} M versus 9×10^{-9} M). Therefore, the result is reasonable: AgCl will precipitate first. In fact, the difference is so great that essentially all of the AgCl will precipitate before Ag_2CrO_4 precipitation begins.

PROBLEM-SOLVING PRACTICE 15.15

Hydrochloric acid is slowly added to a solution that is 0.10 M in Pb^{2+} and 0.010 M in Ag^+. Determine which precipitate is formed first, AgCl or $PbCl_2$.

SUMMARY PROBLEMS

In NaHCOO, $HCOO^-$ is the formate ion.

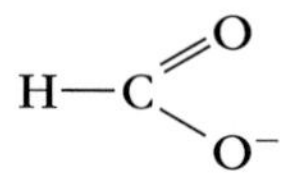

1. (a) Describe how to prepare a pH 3.70 buffer using formic acid, HCOOH, and sodium formate, NaHCOO.

   (b) Calculate the pH of this buffer after the addition of 0.0050 mol HCl.

   (c) Calculate the mass (g) of NaOH that could be added to the initial buffer before its buffer capacity is just exceeded.

2. Choose a weak-acid/weak-base conjugate pair from which you could prepare a buffer solution with pH = 7.5. Explain how you chose the conjugate pair. Calculate the ratio [weak base]/[weak acid] required to give pH = 7.5.

3. The K_a of nitrous acid, HNO_2, is 7.4×10^{-4}. In a titration, 50.0 mL of 1.00-M HNO_2 is titrated with 0.750-M NaOH.

   (a) Calculate the pH of the solution:

   (i) before the titration begins.

   (ii) when sufficient NaOH has been added to neutralize half of the nitrous acid originally present.

   (iii) at the equivalence point.

   (iv) when 0.05 mL NaOH less than that required to reach the equivalence point has been added.

   (v) when 0.05 mL NaOH more than that required to reach the equivalence point has been added.

   (b) Can bromthymol blue be used as the indicator for this titration? Explain your answer.

   (c) Is methyl red a satisfactory indicator? Explain why or why not.

   (d) Use data from part (a) to plot a graph of pH (*y*-axis) versus volume of titrant.

4. A 0.500-L solution contains 0.0010 mol Ag^+.

   (a) Calculate the minimum mass of NaCl that must be added to initiate precipitation of AgCl from the solution.

   (b) If excess Cl^- is added to the solution, the AgCl precipitate dissolves due to the formation of the complex ion $[AgCl_2]^-$; K_f of $[AgCl_2]^- = 1.8 \times 10^5$. Calculate the minimum amount (mol) of Cl^- that must be added to dissolve the precipitate.

HAVING STUDIED THIS CHAPTER . . .

. . . you should be able to:

- Explain how buffers maintain nearly constant pH, how to calculate their pH, how they are prepared, and the importance of buffer capacity (Section 15-1). End-of-chapter Questions 12, 21, 27
- Use the Henderson-Hasselbalch equation or the K_a expression to calculate the pH of a buffer and the pH change after acid or base has been added to a buffer (Section 15-1). Questions 25, 29
- Interpret acid-base titration curves and estimate or calculate the pH of the solution at various stages of the titration (Section 15-2). Questions 32, 35, 43
- Explain how acid rain is formed and its effects on the environment (Section 15-3). Questions 47, 49
- Relate a K_{sp} expression to its chemical equation (Section 15-4). Question 50
- Use the solubility of a slightly soluble solute to calculate its solubility product (Section 15-4). Questions 52, 54, 58

- Describe the factors affecting the aqueous solubility of ionic compounds and calculate their effect on solubility (Section 15-5). Questions 60, 62
- Apply Le Chatelier's principle to the common ion effect (Section 15-5). Question 64
- Use the solubility product to calculate the solubility of a sparingly soluble solute in pure water and in the presence of a common ion (Section 15-5). Questions 66, 70
- Describe the effect of complex ion formation on the solubility of a sparingly soluble ionic compound (Section 15-5). Questions 73, 75, 77
- Relate Q, the ion product, to K_{sp} to determine whether precipitation occurs and predict which of two sparingly soluble ionic solutes precipitates first (Section 15-6). Questions 80, 82

KEY TERMS

acid rain (Section 15-3)

buffer (15-1)

buffer capacity (15-1f)

buffer solution (15-1)

common ion effect (15-5b)

end point (15-2a)

formation constant (K_f) (15-5c)

Henderson-Hasselbalch equation (15-1c)

solubility product constant (K_{sp}) (15-4)

solubility product constant expression (15-4)

titrant (15-2)

titration (15-2)

titration curve (15-2a)

QUESTIONS FOR REVIEW AND THOUGHT

Red-numbered questions have short answers at the back of this book in Appendix M and fully worked solutions in the *Student Solutions Manual.*

Review Questions

These questions test vocabulary and simple concepts.

1. Define the term "buffer capacity".
2. What is the difference between the end point and the equivalence point in an acid-base titration?
3. What are the characteristics of a good acid-base indicator?
4. A strong acid is titrated with a strong base, such as KOH. Describe the changes in the composition of the solution as the titration proceeds: prior to the equivalence point, at the equivalence point, and beyond the equivalence point.
5. Repeat the description for Question 4, but use a weak acid rather than a strong one.
6. Use Le Chatelier's principle to explain why $PbCl_2$ is less soluble in 0.010-M $Pb(NO_3)_2$ than in pure water.
7. Describe what a complex ion is and give an example.
8. Define the term "amphoteric".
9. Distinguish between the ion product (Q) expression and the solubility product constant expression of a sparingly soluble solute.
10. Describe at least two ways that the solubility of a sparingly soluble metal hydroxide can be changed.

Topical Questions

These questions are keyed to the major topics in the chapter. Usually a question that is answered at the back of this book is paired with a similar one that is not.

Buffer Solutions (Section 15-1)

11. Briefly describe how a buffer solution can control the pH of a solution when strong acid is added and when strong base is added. Use NH_3/NH_4Cl as an example of a buffer and HCl and NaOH as the strong acid and strong base.
12. Identify each pair that could form a buffer.
    (a) HCl and CH_3COOH (b) NaH_2PO_4 and Na_2HPO_4
    (c) H_2CO_3 and $NaHCO_3$
13. Identify each pair that could form a buffer.
    (a) NaOH and NaCl (b) NaOH and NH_3
    (c) Na_3PO_4 and Na_2HPO_4
14. Many natural processes can be studied in the laboratory but only in an environment of controlled pH. Which of these combinations is the best to buffer the pH at approximately 7? Explain your choice.
    (a) H_3PO_4/NaH_2PO_4 (b) NaH_2PO_4/Na_2HPO_4
    (c) Na_2HPO_4/Na_3PO_4
15. Which of these combinations is the best to buffer the pH at approximately 9? Explain your choice.
    (a) $CH_3COOH/NaCH_3COO$ (b) HCl/NaCl
    (c) NH_3/NH_4Cl

16. Without doing calculations, determine the pH of a buffer made from equimolar amounts of these acid-base pairs.
(a) Nitrous acid and sodium nitrite
(b) Ammonia and ammonium chloride
(c) Formic acid and potassium formate
17. Without doing calculations, determine the pH of a buffer made from equimolar amounts of these acid-base pairs.
(a) Phosphoric acid and sodium dihydrogen phosphate
(b) Sodium monohydrogen phosphate and sodium dihydrogen phosphate
(c) Sodium phosphate and sodium monohydrogen phosphate
18. Select from Table 15.1 a conjugate acid-base pair that is suitable for preparing a buffer solution whose concentration of hydronium ions is
(a) 4.5×10^{-3} M. (b) 5.2×10^{-8} M.
(c) 8.3×10^{-6} M. (d) 9.7×10^{-11} M.
Explain your choices.
19. Select from Table 15.1 a conjugate acid-base pair that is suitable for preparing a buffer solution whose concentration (molarity) of hydronium ions is
(a) 3.55×10^{-4}. (b) 3.31×10^{-6}.
(c) 4.79×10^{-9}. (d) 7.08×10^{-11}.
Explain your choices.
20. Calculate the mass of sodium acetate, $NaCH_3COO$, you should add to 500. mL of a 0.150-M solution of acetic acid, CH_3COOH, to buffer a solution at a pH of 4.57.
21. Calculate the mass in grams of ammonium chloride, NH_4Cl, that would have to be added to 500. mL of 0.10-M NH_3 solution to have a pH of 9.00.
22. A buffer solution can be made from benzoic acid, C_6H_5COOH, and sodium benzoate, NaC_6H_5COO. Calculate the mass in grams of the acid you would have to mix with 14.4 g of the sodium salt to have a liter of solution with a pH of 3.88.
23. A buffer solution is prepared from 5.15 g NH_4NO_3 and 0.10 L of 0.15-M NH_3; calculate the pH of the solution.
24. You dissolve 0.425 g NaOH in 2.00 L of a solution that originally had $[H_2PO_4^-] = [HPO_4^{2-}] = 0.132$ M. Calculate the resulting pH.
25. A buffer solution is prepared by adding 0.125 mol ammonium chloride to 500. mL of 0.500-M aqueous ammonia. Calculate the pH of the buffer. If 0.0100 mol HCl gas is bubbled into 500. mL buffer and all of the gas dissolves, calculate the new pH of the solution.
26. If added to 1 L of 0.20-M acetic acid, CH_3COOH, which of these would form a buffer?
(a) 0.10 mol $NaCH_3COO$ (b) 0.10 mol NaOH
(c) 0.10 mol HCl (d) 0.30 mol NaOH
Explain your answers.
27. If added to 1 L of 0.20-M NaOH, which of these would form a buffer?
(a) 0.10 mol acetic acid (b) 0.30 mol acetic acid
(c) 0.20 mol HCl (d) 0.10 mol $NaCH_3COO$
Explain your answers.
28. Calculate the pH change when 10.0 mL of 0.100-M NaOH is added to 90.0 mL pure water, and compare the pH change with that when the same amount of NaOH solution is added to 90.0 mL of a buffer consisting of 1.00-M NH_3 and 1.00-M NH_4Cl. Assume that the volumes are additive. K_b of $NH_3 = 1.8 \times 10^{-5}$.
29. Calculate the pH change when 1.0 mL of 1.0-M NaOH is added to 0.100 L of a solution of
(a) 0.10-M acetic acid and 0.10-M sodium acetate.
(b) 0.010-M acetic acid and 0.010-M sodium acetate.
(c) 0.0010-M acetic acid and 0.0010-M sodium acetate.
30. Calculate the pH change when 1.0 mL of 1.0-M HCl is added to 0.100 L of a solution of
(a) 0.10-M acetic acid and 0.10-M sodium acetate.
(b) 0.010-M acetic acid and 0.010-M sodium acetate.
(c) 0.0010-M acetic acid and 0.0010-M sodium acetate.
31. A buffer consists of 0.20-M propanoic acid ($K_a = 1.4 \times 10^{-5}$) and 0.30-M sodium propanoate.
(a) Calculate the pH of this buffer.
(b) Calculate the pH after the addition of 1.0 mL of 0.10-M HCl to 0.010 L of the buffer.
(c) Calculate the pH after the addition of 3.0 mL of 1.0-M HCl to 0.010 L of the buffer.

Acid-Base Titrations (Section 15-2)

32. The titration curves for two acids with the same base are shown on this graph.

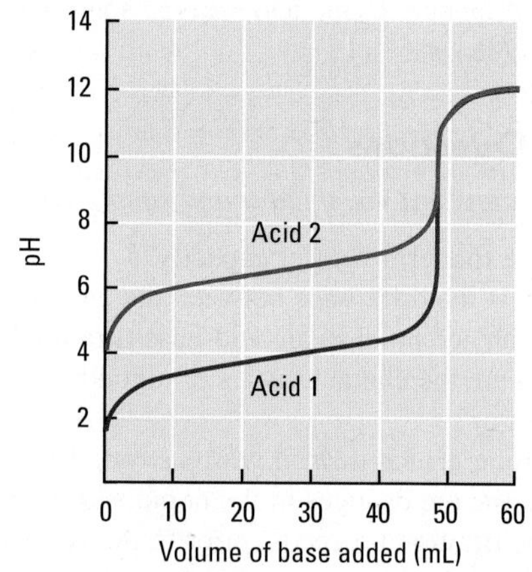
14
12
10
8
6
4
2
pH
Acid 2
Acid 1
0 10 20 30 40 50 60
Volume of base added (mL)

(a) Which is the curve for the weaker acid? Explain your choice.
(b) Give the approximate pH at the equivalence point for the titration of each acid.
(c) Explain why the pH at the equivalence point differs for each acid.
(d) Explain why the starting pH values of the two acids differ.
(e) Which indicator or indicators, phenolphthalein, bromthymol blue, or methyl red, could be used for the titration of Acid 1? For the titration of Acid 2? Explain your choices.

33. Explain why it is that the weaker the acid being titrated, the more alkaline the pH is at the equivalence point.
34. Sketch the titration curve for the titration of 20.0 mL of a 0.100-M solution of a strong acid by a 0.100-M weak base; that is, the base is the titrant. In particular, note the pH of the solution
   (a) prior to the titration.
   (b) when half the required volume of titrant has been added.
   (c) at the equivalence point.
   (d) 10.0 mL beyond the equivalence point.
35. Consider all acid-base indicators discussed in this chapter. Which of these indicators would be suitable for the titration of
   (a) NaOH with $HClO_4$. (b) acetic acid with KOH.
   (c) NH_3 solution with HBr. (d) KOH with HNO_3.
   Explain your choices.
36. Which of the acid-base indicators discussed in this chapter would be suitable for the titration of
   (a) HNO_3 with KOH. (b) KOH with acetic acid.
   (c) HCl with NH_3. (d) KOH with HNO_2.
   Explain your answers.
37. It required 22.6 mL of 0.0140-M $Ba(OH)_2$ solution to titrate a 25.0-mL sample of HCl to the equivalence point. Calculate the molarity of the HCl solution.
38. It took 12.4 mL of 0.205-M H_2SO_4 solution to titrate 20.0 mL of a sodium hydroxide solution to the equivalence point. Calculate the molarity of the original NaOH solution.
39. Vitamin C is a monoprotic acid. To analyze a vitamin C capsule weighing 0.505 g by titration took 24.4 mL of 0.110-M NaOH. Calculate the percentage of vitamin C, $C_6H_8O_6$, in the capsule. Assume that vitamin C is the only substance in the capsule that reacts with the titrant.
40. An acid-base titration was used to find the percentage of $NaHCO_3$ in 0.310 g of a powdered commercial product used to relieve upset stomachs. The titration required 14.3 mL of 0.101-M HCl to titrate the powder to the equivalence point. Assume that the $NaHCO_3$ in the powder is the only substance that reacted with the titrant. Calculate the percentage of $NaHCO_3$ in the powder.
41. Calculate the volume of 0.150-M HCl required to titrate to the equivalence point for each of these samples.
   (a) 25.0 mL of 0.175-M KOH
   (b) 15.0 mL of 6.00-M NH_3
   (c) 15.0 mL of propylamine, $CH_3CH_2CH_2NH_2$, which has a density of 0.712 g/mL
   (d) 40.0 mL of 0.0050-M $Ba(OH)_2$
42. Calculate the volume of 0.225-M NaOH required to titrate to the equivalence point for each of these samples.
   (a) 20.0 mL of 0.315-M HBr
   (b) 30.0 mL of 0.250-M $HClO_4$
   (c) 6.00 g of concentrated acetic acid, CH_3COOH, which is 99.7% pure
43. A 30.00-mL solution of 0.100-M benzoic acid, a monoprotic acid, is titrated with 0.100-M NaOH. The K_a of benzoic acid is 1.2×10^{-4}. Determine the pH after each of these volumes of titrant has been added:
   (a) 10.00 mL (b) 30.00 mL (c) 40.00 mL
44. The titration of 50.00 mL of 0.150-M HCl with 0.150-M NaOH (the titrant) is carried out in a chemistry laboratory. Calculate the pH of the solution after these volumes of the titrant have been added:
   (a) 0.00 mL (b) 25.00 mL (c) 49.9 mL
   (d) 50.00 mL (e) 50.1 mL (f) 75.00 mL
   Use the results of your calculations to plot a titration curve for this titration. On the curve indicate the position of the equivalence point.
45. The titration of 50.00 mL of 0.150-M NaOH with 0.150-M HCl (the titrant) is carried out in a chemistry laboratory. Calculate the pH of the solution after these volumes of the titrant have been added:
   (a) 0.00 mL (b) 25.00 mL (c) 49.9 mL
   (d) 50.00 mL (e) 50.1 mL (f) 75.00 mL
   Use the results of your calculations to plot a titration curve for this titration. On the curve indicate the position of the equivalence point.

Acid Rain (Section 15-3)

46. Explain why rain with a pH of 6.7 is not classified as acid rain.
47. Identify two oxides that are key producers of acid rain. Write chemical equations that illustrate how these oxides form acid rain.
48. Acid rain has been measured with a pH of 1.5. Calculate the hydrogen ion concentration of this rain.
49. Write a chemical equation that shows how limestone neutralizes acid rain.

Solubility Product (Section 15-4)

50. Write a balanced chemical equation for the equilibrium occurring when each of these solutes is added to water, then write the K_{sp} expression.
   (a) Ag_3AsO_4
   (b) Silver sulfate
   (c) Calcium phosphate
   (d) Manganese(III) hydroxide
   (e) Iron(II) carbonate
51. Write a balanced chemical equation for the equilibrium occurring when each of these solutes is added to water, then write the K_{sp} expression for each solute.
   (a) $BaCrO_4$ (b) $Mn(OH)_2$
   (c) Lead(II) carbonate (d) Nickel(II) hydroxide
   (e) Strontium phosphate (f) Mercury(I) sulfate
52. A saturated solution of silver arsenate, Ag_3AsO_4, contains 8.5×10^{-7} g Ag_3AsO_4 per mL. Calculate the K_{sp} of silver arsenate. Assume that there are no other reactions but the K_{sp} reaction.

53. The solubility of $PbBr_2$ is 2.2×10^{-2} g per 100.0 mL at 25 °C. Calculate the K_{sp} of $PbBr_2$, assuming that the solute dissociates completely into Pb^{2+} and Br^- ions and that these ions do not react with water.
54. At 20. °C, 2.03 g $CaSO_4$ dissolves per liter of water. From these data calculate the K_{sp} of calcium sulfate at 20. °C. Assume that there are no other reactions but the K_{sp} reaction.
55. The water solubility of strontium fluoride, SrF_2, is 0.011 g/100. mL. Calculate its solubility product constant. Assume that there are no reactions other than the K_{sp} reaction.
56. The solubility of silver chromate, Ag_2CrO_4, in water is 2.1×10^{-3} g/100. mL. Calculate the K_{sp} of silver chromate. Assume that there are no reactions other than the K_{sp} reaction.
57. Calculate the K_{sp} of HgI_2 given that its solubility in water is 2.0×10^{-10} M. Assume that there are no reactions other than the K_{sp} reaction.
58. The solubility of $PbCl_2$ in water is 1.62×10^{-2} M. Calculate the K_{sp} of $PbCl_2$. Assume that there are no reactions other than the K_{sp} reaction.
59. In a saturated CaF_2 solution at 25 °C, the calcium concentration is analyzed to be 9.1 mg/L. Use this value to calculate the K_{sp} of CaF_2 assuming that the solute dissociates completely into Ca^{2+} and F^- ions, and that neither ion reacts with water.

Factors Affecting Solubility (Section 15-5)

K_{sp} data are available in Appendix H.

60. Calculate the maximum concentration of Zn^{2+} in a solution of pH 10.00.
61. Determine the maximum concentration of Mn^{2+} in each solution:
    (a) pH = 7.81. (b) pH = 11.15.
62. Hydrochloric acid is added to dissolve 5.00 g $Mg(OH)_2$ in a liter of water. Calculate the pH required to just dissolve all of the $Mg(OH)_2$.

Common Ion Effect (Section 15-5b)

63. You have a saturated $Ca(OH)_2$ solution. What would you observe about the solution when these changes are made to separate samples of it:
    (a) The pH is increased.
    (b) Some 6 M NaOH is added to it.
    (c) Some 1 M HCl is added to it.
    (d) Some 1 M $CaCl_2$ is added to it.
64. Predict what effect each would have on this equilibrium:
$$PbCl_2(s) \rightleftharpoons Pb^{2+}(aq) + 2\ Cl^-(aq)$$
    (a) Addition of lead(II) nitrate solution
    (b) Addition of silver nitrate solution
    (c) Addition of NaCl solution
65. Predict what effect each would have on this equilibrium:
$$Cu^{2+}(aq) + 4\ NH_3(aq) \rightleftharpoons [Cu(NH_3)_4]^{2+}(aq)$$
    (a) Addition of copper(II) nitrate solution
    (b) Addition of HCl(aq)
    (c) Addition of 1 M NH_3
66. Calculate the molarity of Zn^{2+} ion in a saturated solution of $ZnCO_3$ that contains 0.25-M Na_2CO_3.
67. Calculate the solubility (mol/L) of $SrSO_4$ ($K_{sp} = 3.2 \times 10^{-7}$) in 0.010-M Na_2SO_4.
68. The solubility of $Mg(OH)_2$ in water is approximately 9.6 mg/L at a given temperature.
    (a) Calculate the K_{sp} of magnesium hydroxide.
    (b) Calculate the hydroxide concentration needed to precipitate Mg^{2+} ions such that no more than 5.0 μg Mg^{2+} per liter remains in the solution.
69. Iron(II) hydroxide, $Fe(OH)_2$, has a solubility in water of 6.0×10^{-1} mg/L at a given temperature.
    (a) From this solubility, calculate the K_{sp} of iron(II) hydroxide. Explain why the calculated K_{sp} differs from the experimental value of 8.0×10^{-16}.
    (b) Calculate the hydroxide concentration needed to precipitate Fe^{2+} ions such that no more than 1.0 μg Fe^{2+} per liter remains in the solution.
70. Calculate the solubility of $ZnCO_3$, $K_{sp} = 1.4 \times 10^{-11}$, in
    (a) water. (b) 0.050-M $Zn(NO_3)_2$.
    (c) 0.050-M K_2CO_3.
71. Calculate the Cl^- concentration (in mol/L) in a solution that is 0.05 M in $AgNO_3$ and contains some undissolved AgCl.

Complex Ion Formation (Section 15-5c)

72. When a few drops of 1×10^{-5}-M $AgNO_3$ are added to 0.01-M NaCl, a white precipitate immediately forms. When a few drops of 1×10^{-5}-M $AgNO_3$ are added to 5-M NaCl, no precipitate forms. Explain these observations.
73. Write the chemical equation for the formation of each complex ion and write its formation constant expression.
    (a) $[Ag(CN)_2]^-$ (b) $[Cd(NH_3)_4]^{2+}$
74. Write the chemical equation for the formation of each complex ion and write its formation constant expression.
    (a) $[CoCl_6]^{3-}$ (b) $[Zn(OH)_4]^{2-}$
75. Calculate the amount (mol) of $Na_2S_2O_3$ that must be added to dissolve 0.020 mol AgBr in 1.0 L water.
76. Gaseous ammonia is added to a 0.063-M solution of $AgNO_3$ until the aqueous ammonia concentration rises to 0.18 M. Calculate the concentrations of $[Ag(NH_3)_2]^+$ and Ag^+ in the solution.
77. Write chemical equations to illustrate the amphoteric behavior of
    (a) $Zn(OH)_2$. (b) $Sb(OH)_3$.

78. Write chemical equations to illustrate the amphoteric behavior of
(a) $Cr(OH)_3$. (b) $Sn(OH)_2$.

Precipitation: Will It Occur? (Section 15-6)

79. The Mohr titration has been used to determine chloride concentration in a sample. A standardized silver nitrate solution is used to precipitate chloride as silver chloride from the test solution. Potassium chromate is used as the indicator because it reacts with slight excess silver ion to form silver chromate, Ag_2CrO_4, a red precipitate. In practice, the silver chromate starts to precipitate just after the silver chloride precipitation is complete because at the equivalence point the only source of silver ions is from the AgCl precipitate dissociation. A 0.2651-g solid sample containing chloride was dissolved in sufficient water to form 20.00 mL of solution. Potassium chromate indicator was added to the solution and the titration required 31.32 mL of standardized 0.1041 M $AgNO_3$.
(a) Calculate the mass percent chloride in the sample.
(b) In a separate experiment it was determined that 20.00 mL of the potassium chromate indicator solution required 0.03 mL of the standardized $AgNO_3$ solution. Use this information to recalculate the mass percent chloride in the solid sample. *Ksp* values: $AgCl = 1.8 \times 10^{-10}$; $Ag_2CrO_4 = 1.1 \times 10^{-12}$.
80. A solution contains 0.020 M sodium phosphate and 0.050 M potassium sulfate. Thus, this solution contains two anion species—phosphate and sulfate. Barium chloride is added to the solution.
(a) Two precipitates are possible. Determine which forms first.
(b) Calculate the concentration of the first anion species when the second anion species starts to precipitate.
81. A solution contains 0.0040 M calcium chloride and 0.0020 M aluminum nitrate. Therefore, the solution contains two cation species—calcium and aluminum. Some sodium phosphate is added to the solution.
(a) Two precipitates are possible. Determine which forms first.
(b) Calculate the concentration of the first cation species when the second cation species starts to precipitate.
82. Solid sodium fluoride is slowly added to an aqueous solution that contains 0.0100-M Pb^{2+} and 0.0100-M Ca^{2+}.
(a) Determine which precipitates first, calcium fluoride or lead(II) fluoride.
(b) Calculate the percentage of Ca^{2+} or Pb^{2+} that has precipitated just prior to the precipitation of the compound that precipitates second.
83. Solid silver nitrate is slowly added to a solution containing 0.100-M NaBr, 0.0500-M Na_2CrO_4, and 0.00100-M Na_3PO_4. Determine in what order AgBr, Ag_2CrO_4, and Ag_3PO_4 precipitate.
84. These are the concentrations (mol/L) of selected cations in seawater:

Cation	Na^+	Mg^{2+}	Ca^{2+}	Al^{3+}	Fe^{2+}
Molarity mol/L	0.46	0.050	0.01	4×10^{-7}	2×10^{-7}

(a) Calculate the hydroxide concentration at which $Mg(OH)_2$ begins to precipitate.
(b) Determine whether any of the other cations will precipitate at this hydroxide concentration.
(c) When enough hydroxide is added to precipitate 50.% of the Mg^{2+}, calculate what percent of the other cations will be precipitated.

General Questions

These questions are not explicitly keyed to chapter topics; many require integration of several concepts.

85. A buffer solution was prepared by adding 4.95 g sodium acetate to 250. mL of 0.150-M acetic acid.
(a) What ions and molecules are present in the solution? List them in order of decreasing concentration.
(b) Calculate the pH of the buffer solution.
(c) Calculate the pH of 100. mL of the buffer solution if you add 80. mg NaOH. (Assume negligible change in volume.)
(d) Write a net ionic equation for the reaction that occurs to change the pH.
86. Calculate the mass in grams of NH_4Cl that must be added to 400. mL of a 0.93-M solution of NH_3 to prepare a pH = 9.00 buffer.
87. Calculate the relative concentrations of *o*-ethylbenzoic acid ($pK_a = 3.79$) and potassium *o*-ethylbenzoate that are needed to prepare a pH = 4.0 buffer.
88. Calculate the relative concentrations of the amine aniline ($pK_b = 9.41$) and anilinium chloride that are required to prepare a buffer of pH 5.00.
89. A solution contains 7.50 g KNO_2 per liter. Calculate the amount (moles) HNO_2 required to prepare 1.00 L of a buffer of pH 4.00. (Assume there is no volume change.)
90. Which of these buffers involving a weak acid HA has the greater resistance to change in pH? Explain your answer.
(i) $[HA] = 0.100\ M = [A^-]$
(ii) $[HA] = 0.300\ M = [A^-]$

91. (a) Calculate the pH of a 0.15-M acetic acid solution.
(b) You add 83 g sodium acetate to 1.50 L of the 0.15-M acetic acid solution. Calculate the new pH of the solution.

92. (a) Calculate the pH of a 0.050-M solution of HF.
(b) Calculate the pH of the solution after addition of 1.58 g NaF to 250. mL of the 0.050-M solution.

93. When 40.00 mL of a weak monoprotic acid solution is titrated with 0.100-M NaOH, the equivalence point is reached when 35.00 mL base has been added. After 20.00 mL NaOH solution has been added, the titration mixture has a pH of 5.75. Calculate the ionization constant of the acid.

94. Each of the solutions in the table has the same volume and the same concentration, 0.1 M.

Acid	pH	Acid	pH
HCl	1.0	Acetic	2.9
Formic	2.3	HCN	5.2

Which solution requires the greatest volume of 0.1-M NaOH to titrate to the equivalence point? Explain your answer.

95. What is the effect on the equilibrium if more solid AgCl is added to a solution saturated with Ag^+ and Cl^- ions?

Applying Concepts

These questions test conceptual learning.

96. You start with two 1.00-L samples: (1) pure water at pH = 7.00; and (2) lake water at pH = 6.96 due to the presence of 4.0×10^{-5} M H_2CO_3 and 1.6×10^{-4} M HCO_3^-. Then, 10.0 mL of 0.0010 M HCl was added to each sample.
(a) Calculate the resulting pH of the formerly pure water.
(b) Calculate the resulting pH of the lake water.
(c) Which sample did not act as a buffer? Explain your answer.

97. Lactic acid ($K_a = 1.4 \times 10^{-4}$) and pyruvic acid ($K_a = 3.2 \times 10^{-3}$) are very important in human metabolism. Calculate the [conjugate acid]/[conjugate base] ratio for each such that the two acid samples would have the same pH. Explain your results.

98. The average normal concentration of Ca^{2+} in urine is 5.33 g/L.
(a) Calculate the concentration of oxalate needed to precipitate calcium oxalate to initiate formation of a kidney stone. K_{sp} of calcium oxalate = 2.3×10^{-9}.
(b) Calculate the minimum phosphate concentration that would precipitate a calcium phosphate kidney stone. K_{sp} of calcium phosphate = 2.0×10^{-29}.

99. Explain why even though an aqueous acetic acid solution contains acetic acid and acetate ions, it cannot be a buffer.

100. Vinegar must contain at least 4% acetic acid (0.67 M). A 5.00-mL sample of commercial vinegar required 33.5 mL of 0.100-M NaOH to reach the equivalence point. Do calculations to determine whether the vinegar meets the legal requirement of at least 4% acetic acid.

101. An unknown acid is titrated with base, and the pH is 3.64 at the point where exactly half of the acid in the original sample has been neutralized. Calculate the value of the ionization constant of the acid.

102. When asked to prepare a carbonate buffer with pH = 10, a lab technician wrote this equation to determine the ratio of weak acid to conjugate base needed:

$$10 = 10.32 + \log \frac{[HCO_3^-]}{[H_2CO_3]}$$

What is wrong with this setup? If the technician prepared a solution containing equimolar concentrations of HCO_3^- and CO_3^{2-}, calculate the pH of the resulting buffer.

103. When you hold your breath, carbon dioxide gas is trapped in your body. Does this increase or decrease your blood pH? Does it lead to acidosis or alkalosis? Explain your answers.

104. Calcium fluoride, CaF_2, is used to fluoridate a municipal water supply. The water is extremely hard with a Ca^{2+} concentration of 0.070 M. Calculate the fluoride concentration in this solution. Calcium fluoride has $K_{sp} = 5.3 \times 10^{-9}$.

105. Apatite, $Ca_5(PO_4)_3OH$, is the mineral in teeth.

$$Ca_5(PO_4)_3OH(s) \rightleftharpoons 5\ Ca^{2+}(aq) + 3\ PO_4^{3-}(aq) + OH^-(aq)$$

(a) On a chemical basis explain why drinking milk strengthens young children's teeth.
(b) Sour milk contains lactic acid. Not removing sour milk from the teeth of young children can lead to tooth decay. Use chemical principles to explain why.

106. Calculate the maximum concentration of Mg^{2+} (molarity) that can exist in a solution of pH 12.00.

107. Choose the words that make this statement true: During a televised medical drama, a person went into cardiac arrest and stopped breathing. A doctor quickly injected sodium hydrogen carbonate solution into the heart. This would indicate that cardiac arrest leads to (acidosis or alkalosis) and that the sodium hydrogen carbonate helps to (increase or decrease) the pH. Explain your choices clearly.

108. The grid has six lettered boxes, each of which contains an item that may be used to answer the questions that follow. Items may be used more than once and there may be more than one correct item in response to a question.

A pH >7.0	**B** pH < 5.6	**C** $[AlF_6]^-$
D Large increase in pH	**E** NaOH and excess $H_2PO_4^-$	**F** $Q < K_{sp}$

Nanoscale representations for Question 110.

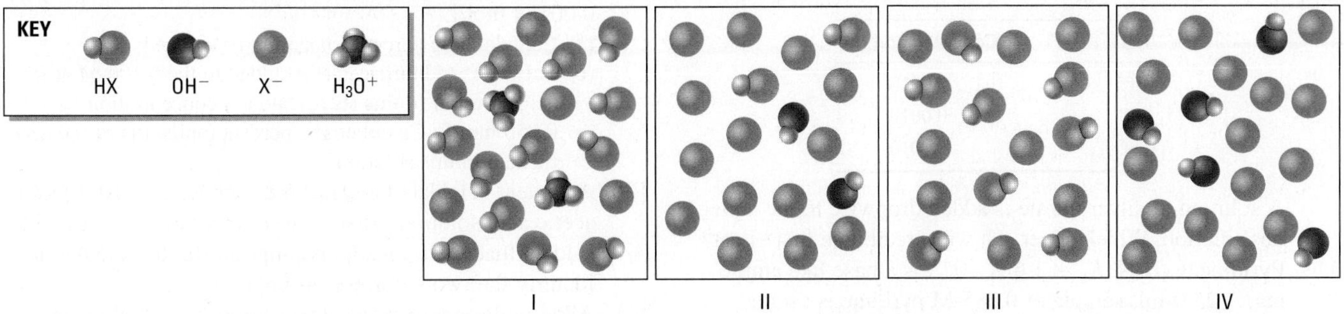
KEY
HX
OH⁻
X⁻
H₃O⁺
I
II
III
IV

Place the letter(s) of the correct selection(s) on the appropriate line.

(a) Add sufficient base to a buffer ________
(b) Acid precipitation ________
(c) Species formed by a Lewis acid-base reaction ________
(d) Condition required for no precipitation ________
(e) A buffer system ________
(f) Equivalence point of a weak acid/strong base titration ________
(g) Just prior to the equivalence point in the titration of lactic acid with KOH ________

109. The grid has six lettered boxes, each of which contains an item that may be used to answer the questions that follow. Items may be used more than once and there may be more than one correct item in response to a question.

A $[Cr(OH)_4]^-$	**B** $Q > K_{sp}$	**C** NaOH
D Decreasing pH	**E** $pH = pK_a$	**F** Sufficient KCl added to CuCl(aq)

Place the letter(s) of the correct selection(s) on the appropriate line.

(a) Could dissolve $Zn(OH)_2$ ________
(b) Could be used to prepare a buffer from HPO_4^{2-} ________
(c) Halfway to the equivalence point in the titration of a weak, monoprotic acid with strong base ________
(d) SO_2-related atmospheric phenomenon ________
(e) Species formed by a Lewis acid-base reaction ________
(f) General condition required for precipitation to occur ________
(g) [conj. base] = [conj. acid] in a buffer ________

110. Consider the nanoscale-level representations for Question 110 of the titration of the aqueous weak acid HX with aqueous NaOH, the titrant. Water molecules and Na^+ ions are omitted for clarity.
Which diagram corresponds to the situation:
(a) After a very small volume of titrant has been added to the initial HX solution?
(b) When enough titrant has been added to take the solution just past the equivalence point?
(c) Halfway to the equivalence point?
(d) At the equivalence point?

111. Consider the nanoscale-level representations for Question 111 of the titration of the aqueous strong acid HA with aqueous NaOH, the titrant. Water molecules and Na^+ ions are omitted for clarity.
Which diagram corresponds to the situation:
(a) After a very small volume of titrant has been added to the initial HA solution?
(b) Halfway to the equivalence point?
(c) When enough titrant has been added to take the solution just past the equivalence point?
(d) At the equivalence point?

Nanoscale representations for Question 111.

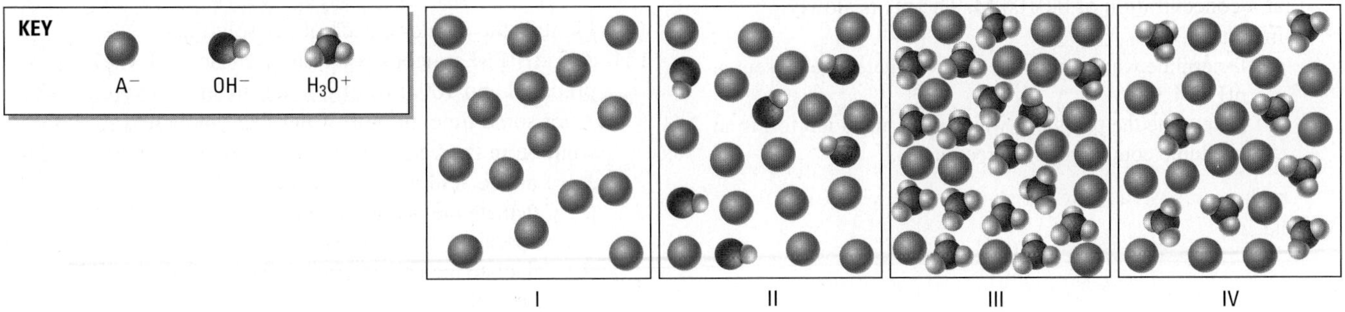
KEY
A⁻
OH⁻
H₃O⁺
I
II
III
IV

112. An aqueous solution contains these ions:

Ion	Concentration (M)
Br^-	0.0010
Cl^-	0.0010
CrO_4^{2-}	0.0010

A solution of silver nitrate is added dropwise to the aqueous solution. Which silver salt will precipitate first? Last?

113. Pyridine, C_5H_5N, $K_b = 1.6 \times 10^{-9}$, is a base like ammonia. A 25.0-mL sample of 0.085-M pyridine is titrated with 0.102-M HCl. The equivalence point occurs at 21.1 mL. Calculate the pH when 5.5 mL acid has been added.

114. You have 1.00 L of 0.10-M formic acid, HCOOH, whose $K_a = 3.0 \times 10^{-4}$. You want to bubble into the formic acid solution sufficient HCl gas to decrease the pH of the formic acid solution by 1.0 pH unit. Calculate the volume of HCl (liters) that must be used at STP to bring about the desired change in pH. Assume no volume change has occurred in the solution due to the addition of HCl gas.

115. One liter (1.0 L) of distilled water has an initial pH of 5.6 because it is in equilibrium with carbon dioxide in the atmosphere. One drop of concentrated hydrochloric acid, 12-M HCl, is added to the distilled water. Calculate the pH of the resulting solution. 20 drops = 1.0 mL.

116. You want to prepare a pH 4.50 buffer using sodium acetate and glacial acetic acid. You have on hand 300. mL of 0.100-M sodium acetate. Calculate the mass in grams of glacial acetic acid (d = 1.05 g/mL) you should use to prepare the buffer.

117. A 1.00-L solution contains 0.010-M F^- and 0.010-M SO_4^{2-}. Solid barium nitrate is slowly added to the solution.
   (a) Calculate the $[Ba^{2+}]$ when $BaSO_4$ begins to precipitate.
   (b) Calculate the $[Ba^{2+}]$ when BaF_2 starts to precipitate.
   Assume no volume change occurs. K_{sp} values: $BaSO_4$ = 1.1×10^{-10}; $BaF_2 = 1.0 \times 10^{-6}$.

More Challenging Questions

These questions require more thought and integrate several concepts.

118. An aqueous solution contains Ag^+, Fe^{3+}, and Zn^{2+}, each at a concentration of 0.0100 M. The pH is slowly increased.
   (a) Determine what precipitate or precipitates are present at pH = 4.35.
   (b) Determine the pH above which each metal ion forms an insoluble compound that precipitates.

119. A 0.100-M acetic acid solution has $[H_3O^+]$ = 0.00134 mol/L.
   (a) Calculate the percent ionization of the acid.
   (b) Sufficient sodium acetate is added to the 0.100-M acetic acid solution so that its acetate ion concentration is 0.050 mol/L. Calculate the percent ionization of the acetic acid in this solution.

120. A student calculated a pH of 5.89 for a 1.0×10^{-7} mol/L acetic acid solution. This answer is incorrect because the student made two invalid assumptions in the calculation. Identify the two incorrect assumptions.

121. Alkaliphiles are organisms that flourish in alkaline environments (pH 8 to 12) such as Octopus Spring in Yellowstone National Park and Mono Lake, CA. The cytoplasm (interior) of the cells of alkaliphiles contains polypeptides that are rich in amino acids such as lysine, arginine, and histidine that have positively charged side chains. Explain how such amino acids can buffer the cells against the alkaline external environment.

122. An experiment found that 0.0050 mol $Ca(OH)_2$ dissolved to form 0.100 L of a saturated aqueous solution.
   (a) Calculate the pH of the solution.
   (b) Calculate the K_{sp} of $Ca(OH)_2$. Explain why the calculated K_{sp} differs from the value of 5.5×10^{-6} given in Appendix H.

123. You are given four different aqueous solutions and told that they each contain NaOH, Na_2CO_3, $NaHCO_3$, or a mixture of these solutes. You do some experiments and gather these data about the samples.

   Sample A: Phenolphthalein is colorless in the solution.
   Sample B: The sample was titrated with HCl until the pink color of phenolphthalein disappeared, then methyl orange was added. The solution became pink. Methyl orange changes color from pH 3.01 (red) to pH 4.4 (orange).
   Sample C: Equal volumes of the sample were titrated with standardized acid. Using phenolphthalein as an indicator required 15.26 mL of standardized acid to change the phenolphthalein color. The other sample required 17.90 mL for a color change using methyl orange as the indicator.
   Sample D: Two equal volumes of the sample were titrated with standardized HCl. Using phenolphthalein as the indicator, it took 15.00 mL of acid to reach the equivalence point; using methyl orange as the indicator required 30.00 mL HCl to achieve neutralization.

   Identify the solute in each of the solutions.

124. A 0.0010-M aqueous solution of anisic acid, $C_8H_8O_3$, is prepared and 100. mL of it is left in an uncovered beaker. After some time, 50. mL water has evaporated from the solution in the beaker (assume no solute evaporated). The K_a of anisic acid is 3.38×10^{-5}.
   (a) Calculate the pH of the initial solution and the pH after evaporation has occurred.
   (b) Calculate the degree of dissociation of anisic acid in each of the solutions.

Titration curves for glutamic acid and lysine, for Question 125.

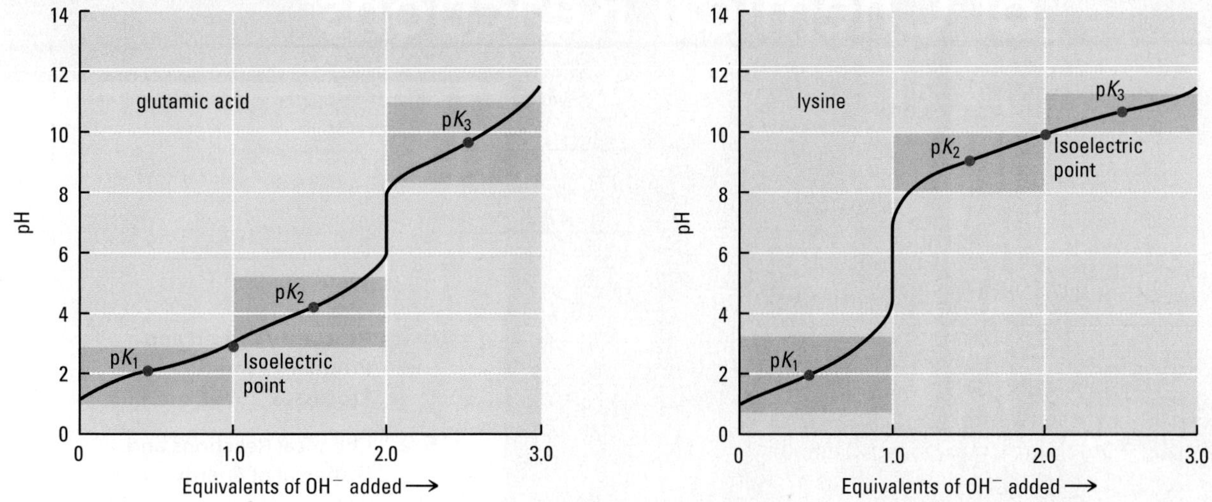
glutamic acid
pK1
pK2
pK3
Isoelectric point
pH
Equivalents of OH⁻ added
lysine
pK1
pK2
pK3
Isoelectric point
pH
Equivalents of OH⁻ added

125. Consider the titration curves for the amino acids glutamic acid and lysine for Question 125.
  (a) Using the structural formulas given in Table 10.8 (← **Sec. 10-7d**), write the structural formulas for the forms of each amino acid corresponding to points on the titration curve when 0, 1.0, 2.0, and 3.0 equivalents of OH^- have been added. An equivalent in this case is the amount of base added that neutralizes an amount of H_3O^+ ions equivalent to the amount of amino acid present.
  (b) The isoelectric point is the pH at which an amino acid has no net charge. Write the structural formulas of glutamic acid and lysine at their isoelectric points.
  (c) What relationship exists between the conjugate acid and conjugate base forms of these amino acids at each of the pK points noted on the titration curve?

126. An experiment requires the addition of 0.075 mol gaseous NH_3 to 1.0 L of 0.025-M $Mg(NO_3)_2$. Ammonium chloride, NH_4Cl, is added prior to the addition of the NH_3 to prevent precipitation of $Mg(OH)_2$. Calculate the minimum mass in grams of ammonium chloride that must be added. K_{sp} of $Mg(OH)_2 = 1.8 \times 10^{-11}$.

Conceptual Challenge Problems

These rigorous, thought-provoking problems integrate conceptual learning with problem solving and are suitable for group work.

CP15.A (Section 15-2) Suppose you were asked on a laboratory test to outline a procedure to prepare a buffered solution of pH 8.0 using hydrocyanic acid, HCN. You realize that a pH of 8.0 is basic, and you find that the K_a of hydrocyanic acid is 4.0×10^{-10}. What is your response?

CP15.B (Sections 15-4 and 15-5) Barium sulfate is swallowed to enhance X-ray studies of the gastrointestinal tract.
(a) Calculate the solubility of Ba^{2+} in mol/L in pure water. K_{sp} of $BaSO_4 = 1.1 \times 10^{-10}$.
(b) In the stomach, the HCl concentration is 0.10 M. Calculate the solubility of Ba^{2+} in mol/L in this solution. K_a of $HSO_4^- = 1.1 \times 10^{-2}$.
(c) The two calculations in parts (a) and (b) were done using data at 25 °C. Repeat part (b) at 37 °C, body temperature. At that temperature, K_{sp} of $BaSO_4 = 1.5 \times 10^{-10}$; K_a of $HSO_4^- = 7.1 \times 10^{-3}$.

16 Thermodynamics: Directionality of Chemical Reactions

© iStockphoto.com/rwittich

Rust is the enemy of iron and steel throughout the world. The remains of this ship will eventually change to an orange-red powder as a result of a chemical reaction: combination of iron, oxygen, and water to form hydrated iron(III) oxide. What characteristics of a chemical reaction determine whether it occurs of its own accord, as rusting does? (Section 16-5) If iron and steel change spontaneously to rust, how do we get them in the first place? (Section 16-8) Do the same factors that allow us to understand rusting apply to living systems? If so, how? (Section 16-9) This chapter will answer these and many other questions about how and why chemical reactions occur.

Many chemical reactions begin when the reactants come into contact and continue until at least one reactant, the limiting reactant (← **Sec. 3-7**), is completely converted to products. Some of these reactions, such as the rusting of iron at room temperature, happen slowly, but reactants are nevertheless converted completely to products. After many years and enough flaking of hydrated iron(III) oxide (rust) from its surface, a piece of iron exposed to air will rust completely.

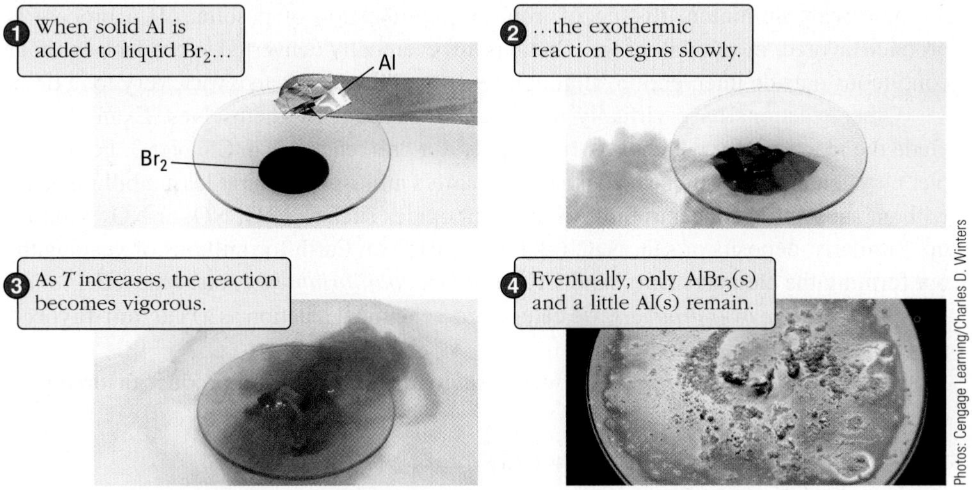
1 When solid Al is added to liquid Br_2...
Al
Br_2
2 ...the exothermic reaction begins slowly.
3 As T increases, the reaction becomes vigorous.
4 Eventually, only $AlBr_3(s)$ and a little Al(s) remain.
Photos: Cengage Learning/Charles D. Winters

Figure 16.1 The reaction of bromine with aluminum, $2\ Al(s) + 3\ Br_2(\ell) \longrightarrow 2\ AlBr_3(s)$.

An example of a reaction that is much faster than rusting is shown in Figure 16.1. Solid aluminum combines readily, and exothermically (← **Sec. 4-5a**), with liquid bromine to form solid aluminum bromide. Other reactions are quite slow. Gasoline reacts so slowly with air at room temperature that it can be stored safely for long periods. However, if its temperature is raised by a spark or flame, gasoline vapor burns rapidly and is essentially completely converted to CO_2 and H_2O.

Gasoline does "go bad" because air can slowly oxidize hydrocarbons to alkenes, which can polymerize and form insoluble gums. Therefore, gasoline is usually drained from lawn mowers or cars that are not used for many months or an antioxidant (such as N-nitrosodiethylamine) is added.

By contrast with the reactions just described, many chemical reactions do not occur by themselves. For example, table salt, NaCl, does not of its own accord decompose into sodium metal and chlorine gas. Neither does water change into hydrogen and oxygen all by itself. These reactions take place only if another process occurs simultaneously and transfers energy to them. (A significant portion of world energy resources is used to cause desirable reactions to occur—reactions that transform inexpensive, readily available substances into new substances with more useful properties, such as iron and steel, polymers, and medicines.) It is important to differentiate between a reaction that is so slow that it *appears* not to occur, such as air oxidation of gasoline, and one that *cannot* take place of its own accord, such as decomposition of sodium chloride. The principles of chemical kinetics in Chapter 11 can be applied to find ways to speed up a slow reaction, but they are of no use in dealing with a reaction that cannot occur by itself.

In Chapter 4 (← **Sec. 4-6**) you learned that thermal energy is transferred when most reactions occur. You also learned how to predict whether a reaction is exothermic or endothermic and how to calculate what quantity of energy transfer takes place as a reaction occurs. In this chapter you will learn how thermodynamics helps us to predict what happens when potential reactants are mixed. Are most or all of the reactants converted to products, as in the case of bromine and aluminum? Are some converted? Or virtually none?

16-1 Reactant-Favored and Product-Favored Processes

We prefer "product-favored" to "spontaneous" because some reactions do begin spontaneously, but produce only tiny quantities of products when equilibrium is reached. Also, the nonscientific usage of "spontaneous" implies a rapid change; if a product-favored reaction is very slow, it does not appear spontaneous at all.

In Chapter 12 (← **Sec. 12-1**) we introduced the idea that a chemical process can be described as reactant-favored or product-favored. *When products predominate over reactants at equilibrium,* we designate the reaction as a **product-favored** process. (Many scientists refer to such reactions as *spontaneous*.) Examples include the reaction of bromine with aluminum, rusting of iron, and combustion of gasoline. If a process is product-favored, most or all of the reactants are eventually converted to products without continuous outside intervention, although "eventually" may mean a very, very long time.

Other reactions have virtually no tendency to occur by themselves. Examples include the reactions for which we wrote "N.R." for "no reaction" in Chapter 3. For example, nitrogen and oxygen have coexisted in Earth's atmosphere for at least a billion years without substantial concentrations of nitrogen oxides such as N_2O, NO, or NO_2 building up. Similarly, deposits of salt, NaCl(s), have existed on Earth for millions of years without forming the elements Na(s) and Cl_2(g). *When equilibrium has been reached, if reactants predominate over products,* we categorize a chemical reaction as a **reactant-favored** process.

Although Earth's atmosphere is 78% N_2 and 21% O_2, the concentration of N_2O, the most abundant oxide of nitrogen in the atmosphere, is more than two million times smaller than the concentration of N_2.

A reactant-favored process is always *exactly the opposite* of a product-favored process. For example, the equation

$$2\ Na(s) + Cl_2(g) \longrightarrow 2\ NaCl(s)$$

describes a product-favored reaction, because sodium metal and chlorine gas react readily to produce sodium chloride. However, if we had written the same equation in the reverse direction

$$2\ NaCl(s) \longrightarrow 2\ Na(s) + Cl_2(g)$$

the system would be designated as reactant-favored. This equation represents decomposition of sodium chloride to form sodium and chlorine, a reaction that does not occur of its own accord. The designations "product-favored" and "reactant-favored" indicate the direction in which a chemical reaction will take place—either forward or backward based on a given equation.

Unless there is some continuous outside intervention, a reactant-favored process does not produce large quantities of products. What do we mean by continuous outside intervention? Usually it is some flow of energy. For example, if enough energy is provided to a sample of air to keep it at a very high temperature, small but significant quantities of NO can be formed from the N_2 and O_2. Such high temperatures are found in lightning bolts and in combustion reactions in electric power generating plants and automobile engines. A power plant or a large number of automobiles can produce enough NO and other nitrogen oxides, albeit at low concentrations, to cause significant air pollution problems (← **Sec. 8-12b**). Sodium chloride can be decomposed to its elements by continuously heating it to keep it molten and passing a direct electric current through it to separate the ions, carry out oxidation and reduction, and form the elements.

The air in the immediate vicinity of a lightning bolt can be heated enough to cause a small fraction of the nitrogen and oxygen to combine to form NO, but this reaction takes place only while the lightning is present. A similar reaction can occur in the engine of an automobile, but again only a small fraction of the air is converted to nitrogen oxides, and only while the temperature remains high (>1000 °C).

$$2\ NaCl(\ell) \xrightarrow[\text{current}]{\text{direct electric}} 2\ Na(\ell) + Cl_2(g)$$

The industrial process by which Na and Cl_2 are produced from NaCl is discussed in Sections 17-10 and 19-4a.

In each case, a reactant-favored process can be forced to produce products if sufficient energy is continuously supplied. This is in contrast to the situation for a product-favored process such as combustion of gasoline, which requires only a brief spark to initiate the reaction. Once started, gasoline combustion continues of its own accord without an additional continuous supply of energy from outside.

CONCEPTUAL EXERCISE 16.1

Reactant-Favored and Product-Favored Processes

Write a chemical equation for each process and classify each as reactant-favored or product-favored.

(a) A puddle of water evaporates on a summer day.
(b) Silicon dioxide (sand) decomposes to the elements silicon and oxygen.
(c) Paper, which is mainly cellulose $(C_6H_{10}O_5)_n$, burns at a temperature of 451 °F.
(d) A pinch of sugar dissolves in water at room temperature.

Answers to **EXERCISES** are provided at the back of this book in Appendix L.

EXERCISES that are labeled **CONCEPTUAL** are designed to test your understanding of one or more concepts; they usually involve qualitative rather than quantitative thinking.

16-2 Chemical Reactions and Dispersal of Energy

The fundamental rule that governs whether a process is product-favored is that ***energy spreads out (disperses) unless it is hindered from doing so.*** A simple example is the one-way transfer of energy from a hotter sample to a colder sample (← **Sec. 4-2b**). As a hot frying pan on a stove cools to room temperature, thermal energy that was concentrated in the pan spreads out over the atoms, molecules, and ions (particles) in the stove, the pan, and the surrounding air. When thermal equilibrium is reached, the room has become slightly warmer—energy has been dispersed and the process is product-favored.

Most exothermic reactions are product-favored at room temperature for a similar reason. When an exothermic reaction takes place, energy is transferred from the system to the surroundings. (See, for example, the reaction of bromine with aluminum in Figure 16.1.) Chemical potential energy that has been stored in bonds between relatively few particles (the reactants) spreads over many more particles as the surroundings (as well as the products) are heated. Therefore it is usually true that after an exothermic reaction, energy is more dispersed than it was before.

A chemical reaction system is usually defined as the collection of atoms that make up the reactants. These same atoms also make up the products, but there they are bonded in a different way. Everything else is designated as the surroundings (← Sec. 4-3).

16-2a Probability and Dispersal of Energy

Dispersal of energy occurs because the probability is much higher that energy will be spread over many particles than that it will be concentrated in a few. To better understand energy dispersal and probability, consider the hypothetical case of a very small sample of matter consisting of two atoms, A and B. Suppose that this sample contains two units of energy, each designated by *. The energy can be distributed over the two atoms in three ways: Atom A could have both units of energy, atom A and atom B could each have one unit, or atom B could have both units. Designate these three situations as

A**   A*B*   B**

Now suppose that atoms A and B come into contact with two other atoms, C and D, that have no energy. There are ten possibilities for distributing the two units of energy over four atoms.

A**   A*B*   A*C*   A*D*   B**   B*C*   B*D*   C**   C*D*   D**

Only three of these cases (A**, A*B*, and B**) have all the energy in atoms A and B, which was the initial situation. When all four atoms are in contact, there are seven chances out of ten that some energy will have transferred from A and B to C and D. Thus, there is a probability of 7/10 = 0.70 that the energy will become spread out over more than just the two atoms A and B.

The low probability that a lot of energy will be associated with only a few particles makes a substance with a lot of chemical potential energy valuable. Humans call substances such as coal, oil, and natural gas "energy resources" and sometimes fight wars over them because of their concentrated energy.

CONCEPTUAL EXERCISE 16.2

Probability of Energy Dispersal

Suppose that you have three units of energy to distribute over two atoms, A and B. Designate each possible arrangement. Now suppose that atoms A and B come into contact with three more atoms, C, D, and E. From the possible arrangements of energy over the five atoms, calculate the probability that all the energy will remain confined to atoms A and B.

Transfer of energy from a substance at a higher temperature to another substance at a lower temperature (← Sec. 4-2b) is an example of dispersal of energy over a larger number of particles.

The probability that energy will become dispersed becomes overwhelming when large numbers of atoms or molecules are involved. For example, suppose that atoms A and B had been brought into contact with a mole of other atoms. There would still be only three arrangements in which all the energy was associated with atoms A and B, but

1 There is a vacuum (no gas) in the upper flask.
2 The flasks are separated by a removable barrier.
3 The lower flask contains bromine vapor, Br2(g).
4 When the barrier is removed, bromine molecules rapidly rush into the upper flask.
5 Eventually bromine is evenly distributed throughout both flasks.
6 Gas molecules and energy are more dispersed after the gas expands to fill both containers.
Photos: © Cengage Learning/James Maynard

Figure 16.2 Gas expansion. Gas molecules and energy are dispersed when a gas expands.

there would be many, many more arrangements (more than 10^{47}) in which all the energy was associated with other atoms. In such a case it is essentially certain that energy will be transferred. ***If energy can be dispersed over a very much larger number of particles, it will be.***

16-2b Dispersal of Energy Accompanies Dispersal of Matter

Energy becomes more dispersed when a system consisting of atoms or molecules expands to occupy a larger volume. This kind of energy dispersal is illustrated by a characteristic property of gases: A gas expands until it fills a container. Recall that molecules in the gas phase are essentially independent of one another and that only weak forces attract the molecules to one another (← **Sec. 8-2**). Figure 16.2 illustrates expansion of a sample of bromine gas at room temperature.

Atoms or molecules of a material that is a solid at room temperature (like the glass flask) do not disperse, because there are strong attractive forces between them. Their tendency to disperse becomes more obvious if the temperature is raised so that they can vaporize.

That dispersal of atoms and molecules also involves dispersal of energy seems obvious, because energy accompanies the particles as they disperse. However, dispersal of energy refers to spreading of energy over a greater number of different *energy levels* (quantum levels, ← **Sec. 5-3**) of the atoms and molecules. When the volume of a gas-phase system increases, the energy levels associated with the motion of each atom or molecule get closer together; that is, there are more levels within the same range of energies. Therefore, at a given temperature, more different energy levels are accessible in the larger volume than in the smaller volume.

When matter spreads out, the number of ways of arranging the energy associated with that matter increases. The energy is more dispersed, and therefore the spreading out of matter is product-favored. This basic idea applies to mixing of different gases and dissolving of one substance in another, as well as to expansion of a gas. We have already mentioned it in connection with dissolving solutes (← **Sec. 13-3a**), and it is also useful in understanding colligative properties (← **Sec. 13-7**).

Expansion of a gas is product-favored, so the opposite process—compression of a gas—should be reactant-favored. If we wanted to reverse the expansion of a gas by concentrating all the particles into a smaller volume, a continuous outside influence such as a pump would be required—the pump could do work on the gas to force it into a less probable arrangement in which energy was less dispersed. The work done by the pump would be stored in the gas and could later be used for some other purpose.

To summarize, any physical or chemical process in which energy is concentrated in a few energy levels in the initial state and energy is dispersed over many energy levels in the final state is product-favored. Two important situations where this is true are

- an exothermic reaction, which disperses potential energy of chemical bonds to thermal energy of a much larger number of atoms or molecules, and
- a process where matter spreads out, which disperses energy as well as matter.

If *both* of these situations apply to a reaction, then it is *definitely product-favored*, because the final distribution of energy is more probable. On the other hand, a process that spreads out *neither energy nor matter* is *reactant-favored*—the initial substances will remain no matter how long we wait. If one of these situations applies but not the other, then quantitative information is needed to decide which effect is greater. The remainder of this chapter develops that quantitative information.

16-3 Measuring Dispersal of Energy: Entropy

The nanoscale dispersal of energy in a sample of matter is measured by a thermodynamic quantity called **entropy**, symbolized by S (← **Sec. 12-7**). Entropy changes can be measured with a calorimeter (← **Sec. 4-7a**), the same instrument used to measure the enthalpy change when a reaction occurs. For a process that takes place at constant temperature and pressure, the entropy change can be calculated by dividing the thermal energy transferred, q_{rev}, by the absolute temperature, T,

$$\Delta S = S_{final} - S_{initial} = \frac{q_{rev}}{T} \qquad [16.1]$$

The symbol Δ was defined in Section 4-3a. The quantity of energy transferred by heating, q, was defined in Section 4-3b.

The subscript "rev" has been added to q to indicate that the equation applies only to processes that can be reversed by a very small change in conditions. An example of such a process is melting of ice at 0 °C and normal atmospheric pressure. If the temperature is just a tiny bit below 0 °C, so that energy is transferred from the water to its surroundings, the water will freeze. If the temperature is a tiny bit above 0 °C, the ice will melt. *Any process for which a very small change in conditions can reverse its direction* is called a **reversible process**.

Another case where Equation 16.1 can be used is when energy is transferred to or from a thermal reservoir at constant temperature and pressure. A thermal reservoir is any large sample of matter, such as everything in your dorm room or the surroundings of a chemical reaction. For example, when a cup of hot coffee cools to room temperature, the quantity of energy transferred is small enough that the temperature of the thermal reservoir (your room) hardly changes. This small quantity of energy could be transferred into or out of your room reversibly. Therefore, to calculate the entropy change of your room you could measure room temperature (in kelvins), calculate how much energy transferred to the room as the coffee cooled, and divide energy transfer by temperature. You could not do the same calculation to determine the entropy change of the coffee, because its temperature changed during the process.

The absolute temperature scale was defined in Section 8-3b. Also called the Kelvin scale or thermodynamic scale, it should be used in all thermodynamic calculations involving temperature.

PROBLEM-SOLVING EXAMPLE 16.1

Calculating Entropy Change

The melting point of pure acetic acid is 16.6 °C at 1 bar. Its molar fusion enthalpy is 11.53 kJ/mol. Calculate $\Delta_r S°$ (in J K^{-1} mol^{-1}) for the process

$$CH_3COOH(s) \longrightarrow CH_3COOH(\ell)$$

Result $\Delta_r S° = 39.79 \text{ J K}^{-1} \text{ mol}^{-1}$

Analyze The process is melting acetic acid at its melting point of 16.6 °C, which is reversible. The pressure is 1 bar, the standard-state pressure (← **Sec. 4-6**), so $\Delta_r S° = \Delta_r S$. For a constant-pressure process, thermal energy transfer, q_{rev}, is the same as the enthalpy change, $\Delta_r H°$ (← **Sec. 4-5b**).

Entropy changes are usually reported in units of *joules* per kelvin per mole (J K^{-1} mol^{-1}), whereas enthalpy changes are usually given in *kilo*joules per mole (kJ/mol). This means that you need to be careful about the units to avoid being wrong by a factor of 1000.

Plan Use the equation $\Delta_r S = q_{rev}/T$. Note that a unit conversion is needed to obtain a result in J K^{-1} mol^{-1}.

Execute

$$\Delta_r S°(\text{melting acetic acid}) = \frac{q_{rev}}{T} = \frac{\Delta_r H°(\text{melting acetic acid})}{T} = \frac{\Delta_{fusion} H}{T}$$

$$= \frac{11.53 \text{ kJ/mol}}{(16.6 + 273.15) \text{ K}} = 3.979 \times 10^{-2} \text{ kJ K}^{-1} \text{ mol}^{-1} = 39.79 \text{ J K}^{-1} \text{ mol}^{-1}$$

☑ **Reasonable Result Check** The result is positive: Entropy increased when solid acetic acid was converted to liquid. Because larger entropy corresponds to greater spreading out of energy, and because molecules move greater distances and in more different ways in a liquid than in a solid (← **Sec. 1-9a**), this is reasonable.

PROBLEM-SOLVING PRACTICE
answers are provided at the back of this book in Appendix K.

PROBLEM-SOLVING PRACTICE 16.1

A chemical reaction transfers 30.8 kJ to a thermal reservoir that has a temperature of 45.3 °C before and after the energy transfer. Calculate the entropy change for the thermal reservoir.

CONCEPTUAL EXERCISE 16.3

The Importance of Absolute Temperature

Consider what would happen if the Celsius temperature scale were used when calculating entropy change by means of $\Delta_r S = q_{rev}/T$. Suppose, for example, that energy were transferred reversibly to $H_2O(s)$ at a temperature 10° below its melting point and we wanted to calculate the entropy change. Would the value calculated from the Celsius temperature agree with the fact that transfer of thermal energy to a sample always increases its entropy?

16-3a Absolute Entropy Values

In Chapter 4 we mentioned that there is no way to measure the total energy content of a sample of matter. Therefore, to summarize a large number of calorimetric measurements of enthalpy changes, we tabulated standard molar formation enthalpies in Table 4.2 (← **Sec. 4-10**). The standard formation enthalpy is the difference between the enthalpy of a substance in its standard state and the enthalpies of the elements that make up that substance, all in their standard states. The elements have enthalpy values, but we do not know what they are. Therefore, the standard enthalpies of formation of elements are by definition set to zero.

For entropy the situation is simpler, because it is possible to define conditions for which it is logical to assume the entropy of a substance has its lowest possible value—namely, zero. Measuring $\Delta_r S$ for a change from those zero-entropy conditions tells us the absolute entropy value for the substance under new conditions. This follows from the definition $\Delta_r S = S_{final} - S_{initial}$, because if $S_{initial} = 0$, then $\Delta_r S = S_{final}$, the absolute entropy of the substance. Because decreasing temperature corresponds to decreasing molecular motion, the minimum possible temperature can reasonably be expected to correspond to minimum motion and thus minimum dispersal of energy. Thus it is logical to assume that a perfect crystal of a substance at 0 K has an entropy value of zero. [In a perfect crystal every nanoscale particle is in exactly the right position in the crystal lattice (← **Sec. 9-6**), and there are no empty spaces or discontinuities.]

The idea that a perfect crystal of any substance at 0 K has minimum entropy is called the **third law of thermodynamics**. Even if a substance cannot be cooled to absolute zero, there are ways to estimate how much energy has been dispersed in the substance at temperatures just above 0 K. Thus, accurate entropy values can be obtained for many substances. To calculate the entropy change, start as close as possible to absolute zero, successively introduce small quantities of energy, and calculate $\Delta_r S$ from the equation $\Delta_r S = q_{rev}/T$ for each increment of energy at each temperature. Then, sum these entropy changes to give the total (or absolute) entropy of a substance at any desired temperature.

The results of such measurements for several substances at 298.15 K are given in Table 16.1. These are *standard molar entropy values*, and so they apply to 1 mol of each substance at the standard pressure of 1 bar and the specified temperature of 25 °C. The units are joules per kelvin per mole ($J\ K^{-1}\ mol^{-1}$). Because there is a real zero on the entropy scale, the values in Table 16.1 are not measured relative to elements in their most stable form under standard-state conditions. Therefore, absolute entropies can be determined for elements as well as compounds. ***The standard molar entropy of a substance at temperature* T *is a sum of the quantities of energy that must be dispersed in that substance at successive temperatures up to* T; *that is, it is* $\Delta_r S$ *from 0 K to* T.**

Although it is impossible to cool anything all the way to absolute zero, it is possible to get very close. Temperatures of a few nanokelvins can be achieved in a Bose-Einstein condensate—the coldest thing known to science. (See http://nobelprize.org/nobel_prizes/physics/laureates/2001/index.html)

The process of introducing small quantities of energy, calculating an entropy increase, and then summing these small entropy increases is actually done by measuring the heat capacity of a substance as a function of temperature and then using integral calculus to calculate the integral of the function q_{rev}/T between the limits of 0 K and the desired temperature.

16-3b Qualitative Guidelines for Entropy

Some useful guidelines can be drawn from the data given in Table 16.1.

- ***Entropies of gases are usually much larger than those of liquids, which in turn are usually larger than those of solids.*** Figure 16.3 shows why this is true. As an example, the entropies (in $J\ K^{-1}\ mol^{-1}$) of the halogens $I_2(s)$, $Br_2(\ell)$, and $Cl_2(g)$

Table 16.1 Some Standard Molar Entropy Values at 298.15 K*

Compound or Element	Entropy, $S°$ ($J\ K^{-1}\ mol^{-1}$)	Compound or Element	Entropy, $S°$ ($J\ K^{-1}\ mol^{-1}$)
C(graphite)	5.740	$O_2(g)$	205.138
C(g)	158.096	$NH_3(g)$	192.45
$CH_4(g)$	186.264	HCl(g)	186.908
$CH_3CH_3(g)$	229.60	$H_2O(g)$	188.825
$CH_3CH_2CH_3(g)$	269.9	$H_2O(\ell)$	69.91
$CH_3OH(\ell)$	126.8	Ca(s)	41.42
CO(g)	197.674	NaF(s)	51.5
$CO_2(g)$	213.74	MgO(s)	26.94
$F_2(g)$	202.78	NaCl(s)	72.13
$Cl_2(g)$	223.066	KOH(s)	78.9
$Br_2(\ell)$	152.231	$MgCO_3(s)$	65.7
$I_2(s)$	116.135	$NH_4NO_3(s)$	151.08
Ar(g)	154.7	NaCl(aq)	115.5
$H_2(g)$	130.684	$NH_4NO_3(aq)$	259.8
$N_2(g)$	191.61	KOH(aq)	91.6

*Data from Wagman, D. D., Evans, W. H., Parker, V. B., Schumm, R. H., Halow, I., Bailey, S. M., Churney, K. L., and Nuttall, R. "The NBS Tables of Chemical Thermodynamic Properties." *Journal of Physical and Chemical Reference Data*, Vol. 11, Suppl. 2, 1982.

Ludwig Boltzmann
1844–1906

S = k log W
LVDWIG
BOLTZMANN

Oesper Collection in the History of Chemistry. University of Cincinnati

An Austrian physicist, Ludwig Boltzmann, gave us the useful interpretation of entropy as probability. Engraved on his tombstone in Vienna is the equation

$$S = k \log W$$

It relates entropy, S, and thermodynamic probability, W—the number of different nanoscale arrangements of energy that correspond to a given macroscale system. The proportionality constant, k, is called Boltzmann's constant, and log stands for the natural logarithm (in modern symbolism, ln).

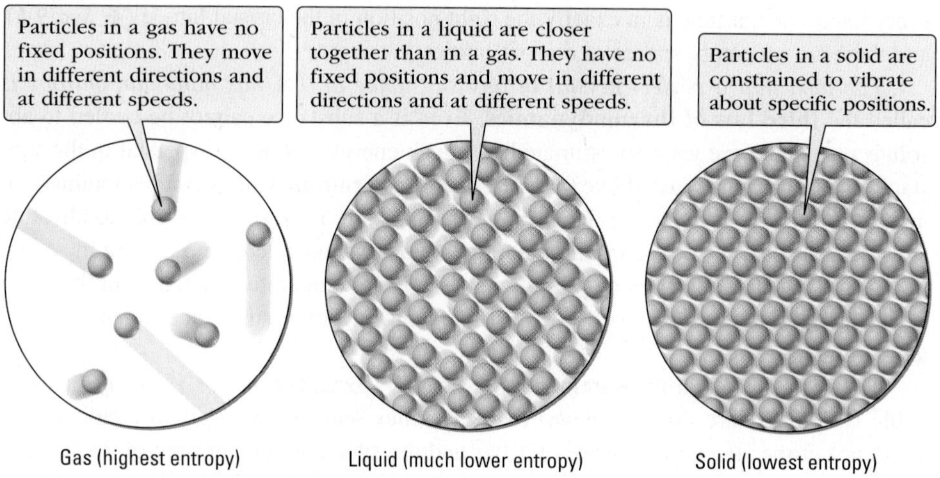
Particles in a gas have no fixed positions. They move in different directions and at different speeds.
Particles in a liquid are closer together than in a gas. They have no fixed positions and move in different directions and at different speeds.
Particles in a solid are constrained to vibrate about specific positions.
Gas (highest entropy)
Liquid (much lower entropy)
Solid (lowest entropy)

Figure 16.3 Entropies of gas, liquid, and solid phases. The entropy of each phase is related to the freedom of motion of the particles.

For a more detailed look at estimating entropy changes, see Craig, N. C. *Journal of Chemical Education*, Vol. 80, 2003; pp. 1432–1436.

are 116.1, 152.2, and 223.1, respectively. Similarly, the entropies of C(s, graphite) and C(g) are 5.7 and 158.1.

- ***Entropies of more complex molecules are larger than those of simpler molecules,*** especially in a series of closely related compounds. In a more complicated molecule there are more ways for the atoms to move about in three-dimensional space, and hence there is greater entropy. For example, the entropies (in J K^{-1} mol^{-1}) of the gases methane, CH_4; ethane, CH_3CH_3; and propane, $CH_3CH_2CH_3$, are 186.26, 229.6, and 269.9, respectively. To control for any effect of molar mass, compare the gases Ar, CO_2, and $CH_3CH_2CH_3$, which have entropies of 154.7, 213.74, and 269.9, respectively (Figure 16.4).
- ***Entropies of ionic solids that have similar formulas are larger when the attractions among the ions are weaker.*** The weaker such forces are, the easier it is for ions to vibrate about their lattice positions and the greater the entropy is. The entropy of NaF(s) is 51.5 J K^{-1} mol^{-1}, and that of MgO(s) is 26.94 J K^{-1} mol^{-1};

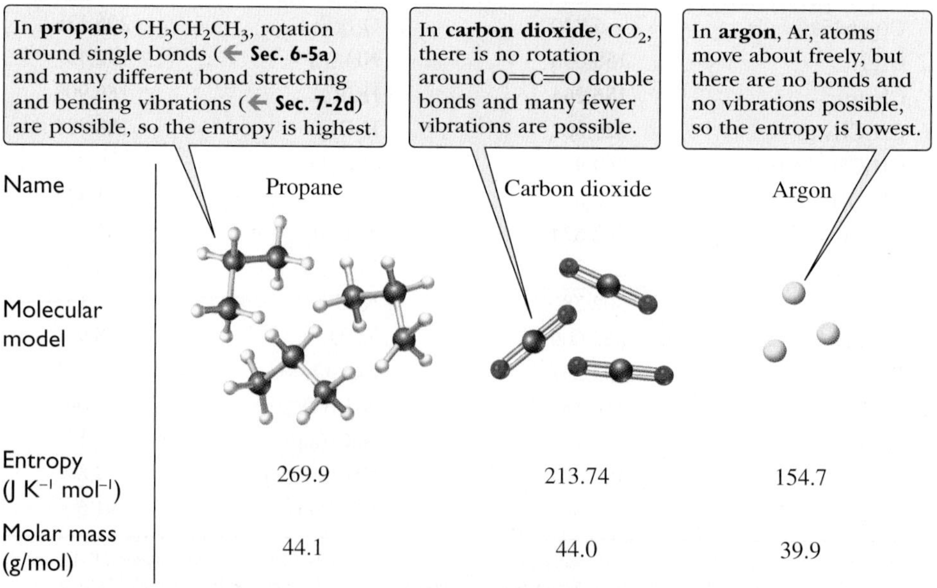
In propane, CH3CH2CH3, rotation around single bonds (← Sec. 6-5a) and many different bond stretching and bending vibrations (← Sec. 7-2d) are possible, so the entropy is highest.
In carbon dioxide, CO2, there is no rotation around O=C=O double bonds and many fewer vibrations are possible.
In argon, Ar, atoms move about freely, but there are no bonds and no vibrations possible, so the entropy is lowest.
Name
Propane
Carbon dioxide
Argon
Molecular model
Entropy (J K-1 mol-1)
269.9
213.74
154.7
Molar mass (g/mol)
44.1
44.0
39.9

Figure 16.4 Entropy increases with molecular complexity. Three particles for each of three gases are shown. The molar masses of all three substances are nearly the same.

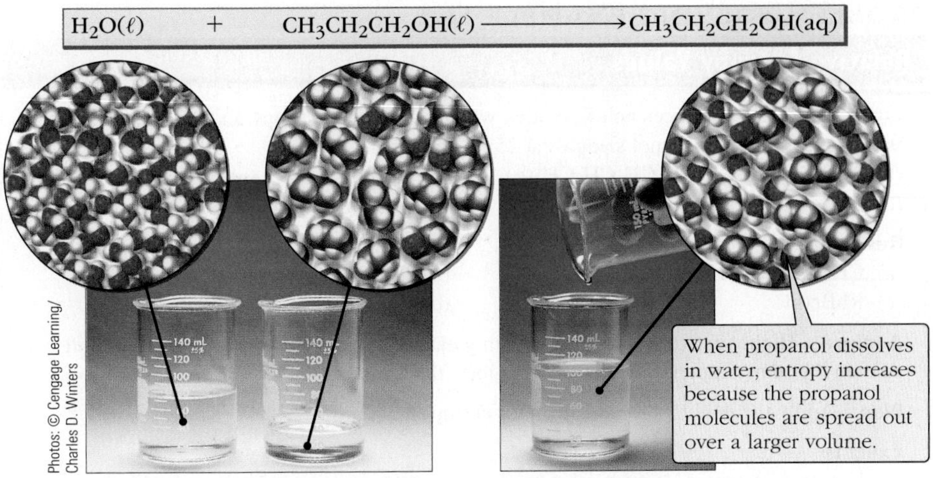
$H_2O(\ell)$ + $CH_3CH_2CH_2OH(\ell)$ → $CH_3CH_2CH_2OH(aq)$
When propanol dissolves in water, entropy increases because the propanol molecules are spread out over a larger volume.
Photos: © Cengage Learning/ Charles D. Winters

Figure 16.5 Entropy and dissolving. Entropy usually increases when a solid or liquid dissolves in a liquid solvent because solute particles become dispersed among solvent particles.

Na^+ and F^-, with unit positive and negative charges, attract each other less than Mg^{2+} and O^{2-}, each of which has two units of charge (← **Sec. 2-6a**); therefore NaF(s) has higher entropy. NaF(s) and NaCl(s) have entropies of 51.5 and 72.13 J K^{-1} mol^{-1}. Chloride ions, Cl^-, are larger than fluoride ions, F^-, and attractions are smaller when the ions are farther apart.

- ***Entropy usually increases when a pure liquid or solid dissolves in a solvent.*** Energy usually becomes more dispersed when different kinds of molecules mix together and occupy a larger volume (Figure 16.5). An example is $NH_4NO_3(s)$ and $NH_4NO_3(aq)$ with standard molar entropies of 151.08 J K^{-1} mol^{-1} and 259.8 J K^{-1} mol^{-1}, respectively. [Some ionic compounds dissolving in water are exceptions to this generalization because small ions with large charges are strongly hydrated (← **Sec. 13-3c**). The water molecules surrounding such ions are constrained in a rigid structure and are no longer free to move and rotate; thus entropy decreases.]
- ***Entropy decreases when a gas dissolves in a liquid.*** Although gas molecules are dispersed among solvent molecules in solution, the very large entropy of the gas phase is lost when the widely separated gas particles become crowded together with solvent particles in the liquid solution (Figure 16.6).

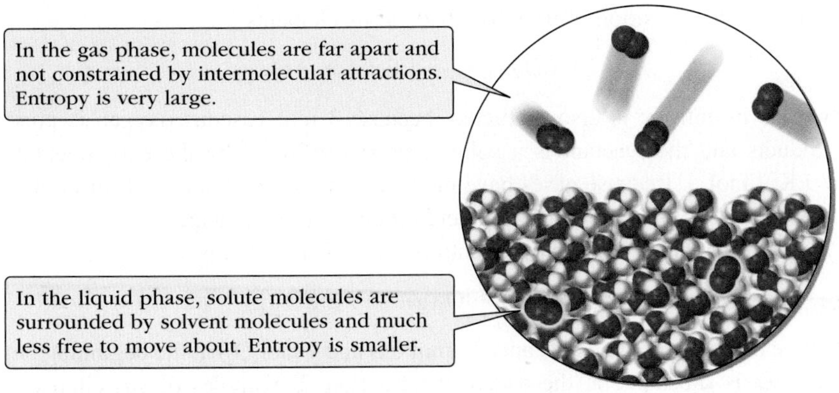
In the gas phase, molecules are far apart and not constrained by intermolecular attractions. Entropy is very large.
In the liquid phase, solute molecules are surrounded by solvent molecules and much less free to move about. Entropy is smaller.

Figure 16.6 When a gas dissolves in a liquid, entropy decreases.

PROBLEM-SOLVING EXAMPLE 16.2

Relative Entropy Values

For each pair of substances below, predict which has greater entropy and give a reason for your choice. (Assume 1-mol samples at 25 °C and 1 bar.)

(a) $H_2C{=}CH_2(g)$ or $CH_3CH_2CH_2CH_3(g)$ (b) $CO_2(aq)$ or $CO_2(g)$
(c) LiF(s) or RbBr(s) (d) $N_2(\ell)$ or $N_2(s)$

Result

(a) $CH_3CH_2CH_2CH_3(g)$ (b) $CO_2(g)$
(c) RbBr(s) (d) $N_2(\ell)$

Analyze Higher entropy corresponds with greater dispersal of energy. The samples all contain the same amount of substance and have the same T and P.

Plan Apply the qualitative guidelines for entropy.

Execute

(a) Larger, more complex molecules have greater entropy than similar smaller ones, so the entropy of $CH_3CH_2CH_2CH_3(g)$ is greater.
(b) Molecules of a gas are free to move, rotate, and vibrate, so when a gas dissolves in a liquid, the entropy decreases; therefore $CO_2(g)$ has greater entropy.
(c) These are ionic solids, both with 1+ and 1− ions. The attractive forces are greater the closer the ions are to each other, and the ions are smaller in LiF because ionic radii increase going down the periodic table (← **Sec. 5-10**). Therefore, RbBr has greater entropy.
(d) Entropy increases from solid to liquid to gas for the same substance, so $N_2(\ell)$ has greater entropy.

PROBLEM-SOLVING PRACTICE 16.2

In each case, predict which of the two substances has greater entropy, assuming 1-mol samples at 25 °C and 1 bar. Check your prediction by looking up each substance's absolute entropy in Table 16.1.

(a) C(g) or C(s, graphite) (b) Ca(s) or Ar(g) (c) KOH(s) or KOH(aq)

16-3c Predicting Entropy Changes

Predicting entropy changes for chemical processes is usually easier than predicting enthalpy changes. For gas phase reactions, the enthalpy-change guideline is that having more bonds or stronger bonds (or both) in the products gives a negative $\Delta_r H°$ (← Sec. 4-7a); however, a table of bond enthalpies is usually needed to tell which bonds are stronger.

The general guidelines about entropy can be used to predict whether an increase or decrease in entropy occurs when reactants are converted to products. For example, in both of the processes

$$H_2O(s) \longrightarrow H_2O(\ell) \quad \text{and} \quad H_2O(\ell) \longrightarrow H_2O(g)$$

an entropy increase is expected. Water molecules in the solid phase are more restricted and their energy is more localized than in the liquid, and water molecules in the liquid cannot move, rotate, or vibrate as freely as they can in the gas. This is confirmed by the data in Table 16.1, where $S°(H_2O(g)) > S°(H_2O(\ell))$.

For the decomposition of iron(III) oxide to its elements,

$$2\ Fe_2O_3(s) \longrightarrow 4\ Fe(s) + 3\ O_2(g)$$

an increase in entropy is also predicted, because 3 mol gaseous oxygen is present in the products and the reactant is a solid. This is confirmed by the experimental $\Delta_r S°$, 549.7 J K^{-1} mol^{-1}. Because gases have much higher entropy than solids or liquids, gaseous substances are most important in determining entropy changes.

An example where a decrease in entropy can be predicted is

$$2\ CO(g) + O_2(g) \longrightarrow 2\ CO_2(g)$$

Here there is 3 mol gaseous substance (2 mol CO and 1 mol O_2) at the beginning but only 2 mol gaseous substance at the end of the reaction. Two moles of gas almost always has less entropy than three moles of gas, so $\Delta_r S°$ is negative (experimentally, $\Delta_r S° = -173.0 \text{ J K}^{-1} \text{ mol}^{-1}$). Another example in which entropy decreases is the process

$$Ag^+(aq) + Cl^-(aq) \longrightarrow AgCl(s)$$

Here the reactant ions are free to move about among water molecules in aqueous solution, but those same ions are held in a crystal lattice in the solid, a situation with greater constraint and thus less spreading out of energy.

Predicting an entropy decrease when ions precipitate from aqueous solution does not always work, especially when the ions have more than single positive or negative charges, such as Mg^{2+} or SO_4^{2-}. The higher the charge on an ion the more tightly water-molecule dipoles are attracted to it. When water molecules are tightly attracted around an ion their motion is restricted and entropy is smaller.

CONCEPTUAL EXERCISE 16.4

Predicting Entropy Changes

For each process, predict whether entropy increases or decreases, and explain how you arrived at your prediction.

(a) $2\ CO_2(g) \longrightarrow 2\ CO(g) + O_2(g)$

(b) $NaCl(s) \longrightarrow NaCl(aq)$

(c) $MgCO_3(s) \longrightarrow MgO(s) + CO_2(g)$

16-4 Calculating Entropy Changes

The standard molar entropy values given in Table 16.1 can be used to calculate entropy changes for physical and chemical processes. Assume that each reactant and each product is at the standard pressure of 1 bar and at the temperature given (298.15 K). The number of particles of each substance is specified by its stoichiometric coefficient in the equation for the process. Multiply the entropy of each product substance by the number of particles of that product and add the entropies of all products. Calculate the total entropy of the reactants in the same way and subtract it from the total entropy of the products. This is summarized in the equation

$$\Delta_r S° = \Sigma[(\text{coefficient of product}) \times S°(\text{product})] - \Sigma[(\text{coefficient of reactant}) \times S°(\text{reactant})] \quad [16.2]$$

Notice that the equation for calculating $\Delta_r S°$ has the same form as that for calculating $\Delta_r H°$ for a reaction (← Sec. 4-10).

Note that this calculation gives the entropy change for the chemical reaction *system* only. It tells whether the energy of the atoms that make up the system is more dispersed or less dispersed after the reaction. It does *not* account for any entropy change in the surroundings.

PROBLEM-SOLVING EXAMPLE 16.3

Calculating an Entropy Change from Tabulated Values

The reaction

$$CO(g) + 2\ H_2(g) \longrightarrow CH_3OH(\ell)$$

is being evaluated as a possible way to manufacture liquid methanol, $CH_3OH(\ell)$, for use in motor fuel. Calculate $\Delta_r S°$ for the reaction.

Result $\Delta_r S° = -332.3 \text{ J K}^{-1} \text{ mol}^{-1}$

Analyze A balanced chemical equation is given and $\Delta_r S°$ is to be calculated.

Plan Use Equation 16.2 and data from Table 16.1.

Execute $\Delta_r S° = \Sigma[(\text{coefficient of product}) \times S°(\text{product})] - \Sigma[(\text{coefficient of reactant}) \times S°(\text{reactant})]$

$$= [1 \times S°\{CH_3OH(\ell)\}] - [1 \times S°\{CO(g)\} + 2 \times S°\{H_2(g)\}]$$

$$= [1 \times (126.8 \text{ J K}^{-1} \text{ mol}^{-1})] - [1 \times 197.7 + 2 \times 130.7] \text{ J K}^{-1} \text{ mol}^{-1}$$

$$= -332.3 \text{ J K}^{-1} \text{ mol}^{-1}$$

Note that because all substances, including elements, have nonzero absolute entropy values, elements as well as compounds must be included in this calculation.

☑ **Reasonable Result Check** There is a decrease in entropy, which is reasonable because 3 mol gaseous reactants is converted to 1 mol liquid-phase product. The product molecule is more complicated, but the fact that it is a liquid makes its entropy much smaller than that of 1 mol gaseous carbon monoxide and 2 mol gaseous hydrogen.

PROBLEM-SOLVING PRACTICE 16.3

Calculate the entropy change for each of these processes, thereby verifying the predictions made in Conceptual Exercise 16.4.

(a) $2\,CO_2(g) \longrightarrow 2\,CO(g) + O_2(g)$ (b) $NaCl(s) \longrightarrow NaCl(aq)$

(c) $MgCO_3(s) \longrightarrow MgO(s) + CO_2(g)$

16-5 Entropy and the Second Law of Thermodynamics

A great deal of experience with many chemical reactions and other processes in which energy is transformed and transferred is consistent with the conclusion that ***whenever a product-favored chemical or physical process occurs, energy becomes more dispersed***. This is summarized in the **second law of thermodynamics**, which states that *the total entropy of the universe (a system plus its surroundings) is continually increasing*. Evaluating whether the total entropy increases during a proposed chemical reaction allows us to predict whether reactants will form appreciable quantities of products.

Predicting whether a reaction is product-favored can be done in three steps:

Steps 1 and 2 can be carried out by assuming that reactants under standard conditions of 1 bar and a specified temperature are converted to products under the same standard conditions.

1. Calculate the entropy change that results from transfer of energy between system and surroundings ($\Delta_r S_{surroundings}$).
2. Calculate the entropy change that results from dispersal of energy within the system ($\Delta_r S_{system}$).
3. Add these two results to get the overall entropy change of the universe:

$$\Delta_r S_{universe} = \Delta_r S_{system} + \Delta_r S_{surroundings} \qquad [16.3]$$

Let us apply these steps to the reaction between carbon monoxide and hydrogen

$$CO(g) + 2\,H_2(g) \longrightarrow CH_3OH(\ell)$$

A larger tabulation of standard formation enthalpies and standard entropies is available in Appendix J.

We will consider the effect of changing temperature later in this chapter. The effects of changing other conditions can often be predicted qualitatively by using Le Chatelier's principle (← Sec. 12-6).

for which we calculated the entropy change in Problem-Solving Example 16.3. If the reaction is product-favored, it would be a good way to produce methanol for use as automotive fuel. The reactants can be obtained from plentiful resources: coal and water. We base our prediction upon having all substances at 1 bar and 298.15 K (25 °C). Then, standard formation enthalpies from Table 4.2 (← **Sec. 4-10**) and standard entropies from Table 16.1 apply. If the entropy of the universe is predicted to be higher after the product has been produced, then the reaction is product-favored under these conditions and might be useful. If not, perhaps some other conditions could be used, or perhaps we should consider some other reaction altogether.

Step 1: ***Calculate the reaction's dispersal of energy to, or concentration of energy from, the surroundings by calculating $\Delta_r H°$ and assuming that this quantity***

of thermal energy is transferred reversibly. The entropy change for the surroundings can be calculated as

$$\Delta_r S^\circ_{surroundings} = \frac{q_{rev}}{T} = \frac{\Delta_r H_{surroundings}}{T} = \frac{-\Delta_r H^\circ_{system}}{T} = \frac{-\Delta_r H^\circ}{T}$$

The minus sign in this equation comes from the fact that the direction of energy transfer for the surroundings is opposite from the direction of energy transfer for the system. For an exothermic reaction (negative $\Delta_r H^\circ$) there is an increase in entropy of the surroundings, a fact that we have already mentioned. For the proposed methanol-producing reaction, $\Delta_r H^\circ$ (calculated from data in Appendix J) is -128.1 kJ/mol, so the entropy change is

$$\Delta_r S^\circ_{surroundings} = \frac{-(-128.1 \text{ kJ/mol})}{298 \text{ K}} \times \frac{1000 \text{ J}}{\text{kJ}} = 430. \text{ J K}^{-1} \text{ mol}^{-1}$$

Step 2: ***Calculate the entropy change for dispersal of energy within the system*** (that is, for the atoms involved in the methanol-producing reaction). This entropy change can be evaluated from the absolute entropies of the products and reactants as described in the previous section and has already been calculated in Problem-Solving Example 16.3 to be

$$\Delta_r S^\circ_{system} = \Delta_r S^\circ = -332.3 \text{ J K}^{-1} \text{ mol}^{-1}$$

Step 3: ***Calculate the total entropy change for the system and the surroundings, $\Delta_r S^\circ_{universe}$.*** Because the universe includes both system and surroundings, $\Delta_r S^\circ_{universe}$ is the sum of the entropy change for the system and the entropy change for the surroundings. (We assume that nothing else but our reaction happens, so there are no other entropy changes.) This total entropy change is

$$\Delta_r S^\circ_{universe} = \Delta_r S^\circ_{surroundings} + \Delta_r S^\circ_{system} = \frac{-\Delta_r H^\circ_{system}}{T} + \Delta_r S^\circ_{system}$$

$$= (430. - 332.3) \text{ J K}^{-1} \text{ mol}^{-1}$$

$$= 98 \text{ J K}^{-1} \text{ mol}^{-1}$$

Calculation of the entropy change for the universe is summarized in the entropy diagram at the right.

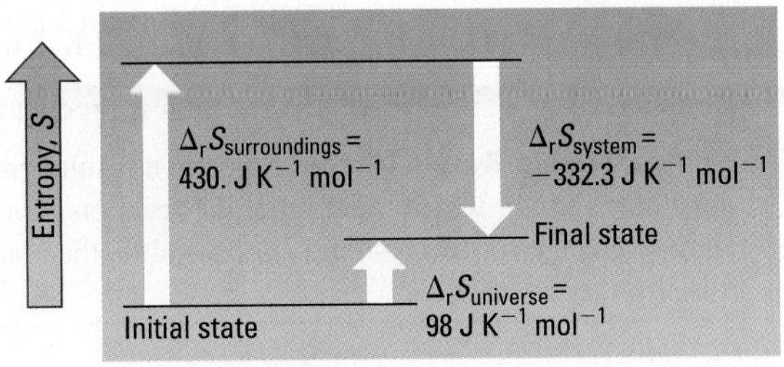

Entropy, S
ΔrSsurroundings = 430. J K−1 mol−1
ΔrSsystem = −332.3 J K−1 mol−1
Final state
ΔrSuniverse = 98 J K−1 mol−1
Initial state

Combination of carbon monoxide and hydrogen to form methanol increases the entropy of the universe. Because the process is product-favored, it could be useful for manufacturing methanol.

CONCEPTUAL EXERCISE 16.5

Effect of Temperature on Entropy Change

The reaction of carbon monoxide with hydrogen to form methanol is quite slow at room temperature. As a general rule, reactions go faster at higher temperatures. Suppose that you tried to speed up this reaction by increasing the temperature.

(a) Assuming that $\Delta_r H^\circ$ does not change very much as the temperature changes, what effect would increasing the temperature have on $\Delta_r S^\circ_{surroundings}$?

(b) Assuming that $\Delta_r S^\circ$ for a reaction system does not change much as the temperature changes, what effect would increasing the temperature have on $\Delta_r S^\circ_{universe}$?

PROBLEM-SOLVING EXAMPLE 16.4

Calculating $\Delta_r S^\circ_{universe}$

When gasoline burns, one reaction is combustion of octane with oxygen from air.

$$2\ C_8H_{18}(g) + 25\ O_2(g) \longrightarrow 16\ CO_2(g) + 18\ H_2O(\ell)$$

Calculate $\Delta_r S^\circ_{universe}$ for this reaction, thereby confirming that the reaction is product-favored. (Assume that reactants and products are at 298.15 K and 1 bar.)

Result $\Delta_r S^\circ_{universe} = 35{,}591\ J\ K^{-1}\ mol^{-1}$

Analyze A balanced chemical equation for the reaction is given. Data for standard formation enthalpies and standard entropies are needed to use Equation 16.3.

Plan Obtain needed data from Appendix J. Calculate $\Delta_r H^\circ$ and $\Delta_r S^\circ$ by subtracting the sum of reactant values from the sum of product values. Calculate $\Delta_r S_{surroundings}$ as $-\Delta_r H^\circ/T$ and use Equation 16.3 to calculate $\Delta_r S_{universe}$.

Execute

$$\begin{aligned}\Delta_r H^\circ &= \Sigma\{(\text{coefficient of product}) \times \Delta_f H^\circ(\text{product})\} - \\ &\quad \Sigma\{(\text{coefficient of reactant}) \times \Delta_f H^\circ(\text{reactant})\} \\ &= \{16 \times (-393.509\ \text{kJ/mol}) + 18 \times (-285.830\ \text{kJ/mol})\} - \\ &\quad \{2 \times (-208.447\ \text{kJ/mol}) + 25 \times (0\ \text{kJ/mol})\} \\ &= -11{,}024.2\ \text{kJ/mol}\end{aligned}$$

$$\begin{aligned}\Delta_r S^\circ &= \Sigma\{(\text{coefficient of product}) \times S^\circ(\text{product})\} - \\ &\quad \Sigma\{(\text{coefficient of reactant}) \times S^\circ(\text{reactant})\} \\ &= \{16 \times (213.74\ J\ K^{-1}\ mol^{-1}) + 18 \times (69.91\ J\ K^{-1}\ mol^{-1})\} - \\ &\quad \{2 \times (466.835\ J\ K^{-1}\ mol^{-1}) + 25 \times (205.138\ J\ K^{-1}\ mol^{-1})\} \\ &= -1383.9\ J\ K^{-1}\ mol^{-1}\end{aligned}$$

$$\begin{aligned}\Delta_r S^\circ_{universe} &= \Delta_r S^\circ_{system} + \Delta_r S^\circ_{surroundings} = \Delta_r S^\circ + \frac{-\Delta_r H^\circ}{T} \\ &= (-1383.9\ J\ K^{-1}\ mol^{-1}) + \frac{-(-11{,}024.2\ \text{kJ/mol})}{298.15\ K} \\ &= -1383.9\ J\ K^{-1}\ mol^{-1} + 36.975\ kJ\ K^{-1}\ mol^{-1} \\ &= -1383.9\ J\ K^{-1}\ mol^{-1} + 36{,}975\ J\ K^{-1}\ mol^{-1} = 35{,}591\ J\ K^{-1}\ mol^{-1}\end{aligned}$$

Notice that for $-\Delta_r H^\circ/T$ the units were kilojoules per kelvin per mole ($kJ\ K^{-1}\ mol^{-1}$), while for $\Delta_r S^\circ$ the units were joules per kelvin per mole ($J\ K^{-1}\ mol^{-1}$). Therefore it was necessary to convert kilojoules to joules by multiplying the second term by 1000 J/kJ.

Because $\Delta_r S^\circ_{universe}$ is positive, the reaction is product-favored.

☑ **Reasonable Result Check** A positive result is reasonable, because you know that gasoline burns in air, which means that the reaction is product-favored. Even though the entropy change for the system is unfavorable, the reaction is highly product-favored because it is strongly exothermic.

PROBLEM-SOLVING PRACTICE 16.4

Determine whether the synthesis of ammonia from nitrogen and hydrogen is product-favored at 298.15 K and 1 bar.

CONCEPTUAL EXERCISE 16.6

Variation of $\Delta_r H^\circ$ and $\Delta_r S^\circ$ with Temperature

Suppose that the combustion of gaseous octane is carried out at 150 °C.

(a) How would this affect the chemical equation for the reaction?

(b) What effect would the change in the chemical equation have on $\Delta_r H^\circ$ and $\Delta_r S^\circ$ for the reaction?

(c) When is it definitely *not* safe to assume that $\Delta_r H^\circ$ and $\Delta_r S^\circ$ for a reaction will be almost the same over a broad range of temperatures?

Predictions like the ones we have just made by calculating $\Delta_r S^\circ_{universe}$ can also be made qualitatively, without calculating, if we know whether a reaction is exothermic and if we can predict whether energy is dispersed within the system when the reaction takes place.

- ***A reaction is certain to be product-favored if it is exothermic* and *the entropy of the product atoms, molecules, and ions is greater than the entropy of the reactants ($\Delta_r S_{system} > 0$).***
- ***A reaction is certain to be reactant-favored if it is endothermic* and *there is a decrease in entropy of the system ($\Delta_r S_{system} < 0$).***

There are two other possible cases, as indicated in Table 16.2, but they are more difficult to predict without quantitative information.

An example of the first type is reactions of carbonates with acids (← **Sec. 3-4e**). These are product-favored: They are exothermic and produce gases (entropy increase). Reaction of limestone with hydrochloric acid is typical.

$$CaCO_3(s) + 2\,HCl(aq) \longrightarrow CaCl_2(aq) + H_2O(\ell) + CO_2(g) \qquad \text{(exothermic)}$$

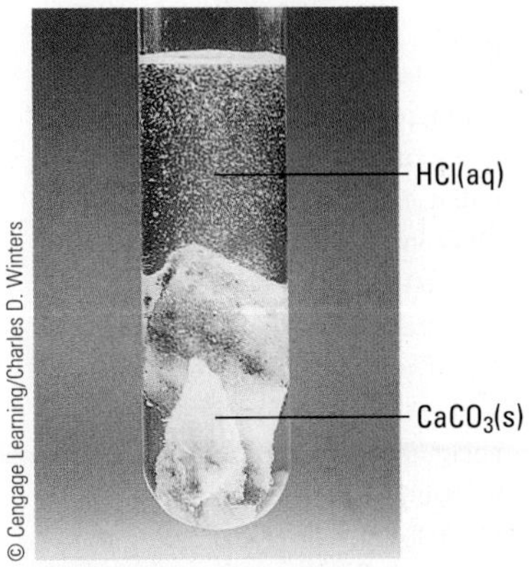
HCl(aq)
CaCO3(s)
© Cengage Learning/Charles D. Winters

A product-favored reaction. Carbonates react rapidly with acid to produce carbon dioxide and water.

Similarly, combustion reactions of hydrocarbons such as butane, $CH_3CH_2CH_2CH_3$, are product-favored because they are exothermic and produce a larger number of gaseous product molecules than there were gaseous reactant molecules.

$$2\,CH_3CH_2CH_2CH_3(g) + 13\,O_2(g) \longrightarrow 8\,CO_2(g) + 10\,H_2O(g) \qquad \text{(exothermic)}$$

But what about a reaction such as the production of ethylene, $CH_2{=}CH_2$, from ethane, CH_3CH_3? Although entropy is predicted to increase (one gas-phase molecule forms two), the reaction is very endothermic. (That the reaction is endothermic might be predicted on the basis of a decrease from seven bonds in the reactant molecule to six bonds—five single, one double—in the products.)

$$CH_3CH_3(g) \longrightarrow H_2(g) + CH_2{=}CH_2(g)$$

$$\Delta_r H^\circ = +137 \text{ kJ/mol};\ \Delta_r S^\circ = +121 \text{ J K}^{-1}\text{ mol}^{-1}$$

Enthalpy change predicts that this process is reactant-favored, while entropy change predicts the opposite. Which is more important? It depends on the temperature.

Calculating $\Delta_r S^\circ_{surroundings}$ $(= -\Delta_r H^\circ/T)$ requires dividing the enthalpy change by the temperature. Because $\Delta_r H^\circ$ stays pretty much the same at different temperatures, the higher the temperature, the smaller the absolute value of $\Delta_r S^\circ_{surroundings}$ $(= -\Delta_r H^\circ/T)$. At room temperature $\Delta_r S^\circ_{surroundings}$ is usually bigger in absolute value than $\Delta_r S^\circ_{system}$, so exothermic reactions are expected to be product-favored and endothermic reactions (like this one) are expected to be reactant-favored. The ethylene-producing reaction is reactant-favored at 25 °C, because $\Delta_r S^\circ_{universe} = -339$ J K^{-1} mol^{-1}. To make a successful industrial process, chemical engineers have designed plants that carry out this reaction at about 1000 °C. At this higher temperature, $\Delta_r S^\circ_{surroundings}$ is smaller in magnitude than $\Delta_r S^\circ_{system}$. Thus $\Delta_r S^\circ_{universe} = 13$ J K^{-1} mol^{-1}, and products are predicted to predominate over reactants.

At 25 °C:

$$\Delta_r S^\circ_{surroundings} = \frac{-137 \text{ kJ/mol}}{298 \text{ K}} = -460 \text{ J K}^{-1}\text{ mol}^{-1}$$

$$\Delta_r S^\circ_{system} = 121 \text{ J K}^{-1}\text{ mol}^{-1}$$

$$\Delta_r S^\circ_{universe} = (-460 + 121) \text{ J K}^{-1}\text{ mol}^{-1} = -339 \text{ J K}^{-1}\text{ mol}^{-1}$$

At 1000 °C:

$$\Delta_r S^\circ_{surroundings} = \frac{-137 \text{ kJ/mol}}{(1000 + 273) \text{ K}} = -108 \text{ J K}^{-1}\text{ mol}^{-1}$$

$$\Delta_r S^\circ_{system} = 121 \text{ J K}^{-1}\text{ mol}^{-1}$$

$$\Delta_r S^\circ_{universe} = 13 \text{ J K}^{-1}\text{ mol}^{-1}$$

Table 16.2 Predicting Whether a Reaction Is Product-Favored (at constant *T* and *P*)

Sign of $\Delta_r H_{system}$	Sign of $\Delta_r S_{system}$	Product-Favored?
Negative (exothermic)	Positive	Yes
Negative (exothermic)	Negative	Yes at low *T*; no at high *T*
Positive (endothermic)	Positive	No at low *T*; yes at high *T*
Positive (endothermic)	Negative	No

Josiah Willard Gibbs
1839–1903

Beinecke Rare Book and Manuscript Library, Yale University

Gibbs was the son of a Yale professor of sacred literature. Like many of his forebears Gibbs studied at Yale, where he received his Ph.D. in 1863. His interest in both mathematics and engineering was reflected in the title of his thesis, "On the Form of the Teeth of Wheels in Spur Gearing." He spent three years in Europe, studying in Paris, Berlin, and Heidelberg, before becoming the first professor of mathematical physics at Yale. His work was not well known until long after he published it, because his published articles were compactly written and very abstract. They were also profound and influential, providing the basis for chemical thermodynamics. (See **http://www.aip.org/history/gap/Gibbs/Gibbs.html** for a more detailed biography.)

EXERCISE 16.7

Predicting the Direction of a Reaction

Using data from Appendix J, complete the table and then classify each reaction into one of the four types in Table 16.2. Predict whether each reaction is product-favored or reactant-favored at room temperature.

Reaction	$\Delta_r H°$, 298 K (kJ/mol)	$\Delta_r S°$, 298 K (J K^{-1} mol^{-1})
(a) $C_2H_4(g) + 3\ O_2(g) \longrightarrow 2\ H_2O(\ell) + 2\ CO_2(g)$	______	______
(b) $2\ Fe_2O_3(s) + 3\ C(\text{graphite}) \longrightarrow 4\ Fe(s) + 3\ CO_2(g)$	______	______
(c) $C(\text{graphite}) + O_2(g) \longrightarrow CO_2(g)$	______	______
(d) $2\ Ag(s) + 3\ N_2(g) \longrightarrow 2\ AgN_3(s)$	______	______

EXERCISE 16.8

Product- or Reactant-Favored?

(a) Is the combination reaction of hydrogen gas and chlorine gas to give hydrogen chloride gas (at 1 bar) predicted to be product-favored or reactant-favored at 298 K?
(b) Calculate the value for $\Delta_r S°_{\text{universe}}$ for the reaction in part (a).

16-6 Gibbs Free Energy

Calculations of the sort done in the previous section would be simpler if we did not have to separately evaluate the entropy change of the surroundings from T and a table of $\Delta_f H°$ values and the entropy change of the system from a table of $S°$ values. To simplify such calculations, a new thermodynamic function was defined by J. Willard Gibbs. It is now called **Gibbs free energy**, symbol G, and is defined as *the negative of temperature times entropy change of the universe*: $\Delta_r G_{\text{system}} = -T\Delta_r S_{\text{universe}}$. Because of the minus sign, if the entropy of the universe increases, the Gibbs free energy of the system must decrease. That is, ***a decrease in Gibbs free energy of a system is characteristic of a process that is product-favored at constant temperature and pressure.***

In the previous section we showed that the total entropy change accompanying a chemical reaction carried out at constant temperature and pressure is

$$\Delta_r S_{\text{universe}} = \Delta_r S_{\text{surroundings}} + \Delta_r S_{\text{system}} = \frac{-\Delta_r H_{\text{system}}}{T} + \Delta_r S_{\text{system}}$$

Combining this algebraically with Gibbs's definition of free energy, we have

$$\Delta_r G_{\text{system}} = -T\Delta_r S_{\text{universe}} = -T\left[\frac{-\Delta_r H_{\text{system}}}{T} + \Delta_r S_{\text{system}}\right] = \Delta_r H_{\text{system}} - T\Delta_r S_{\text{system}}$$

or, under standard-state conditions.

$$\Delta_r G° = \Delta_r H° - T\Delta_r S° \qquad [16.4]$$

For $\Delta_r H°$, $\Delta_r S°$, and $\Delta_r G°$ you can assume that the values apply to the system—that is, $\Delta_r G° = \Delta_r G°_{\text{system}}$—unless a subscript is attached to indicate that a value is for the surroundings.

This equation summarizes the ideas about chemical equilibrium that were developed in Section 12-7. A *negative* value of $\Delta_r G°$ indicates that a reaction is *product-favored*, and the equation says that two conditions will make $\Delta_r G°$ more negative: (1) if the reaction is exothermic, $\Delta_r H°$ is negative, thereby favoring the products; and (2) if the products have greater entropy than the reactants, then $\Delta_r S°$ is positive and the $-T\Delta_r S°$ term is negative, which favors the products. Because $\Delta_r S°$ is multiplied by T, the entropy change of the system is more important at higher temperatures.

In Chapter 12 (← Sec. 12-7) we stated that if a reaction is exothermic or involves an increase in entropy of the system products are favored. We also said that the entropy effect becomes more important the higher the temperature is.

CONCEPTUAL **EXERCISE 16.9**

Predicting Whether a Process Is Product-Favored

Make a table similar to Table 16.2, but add a new column for the sign of $\Delta_r G°$. Based on the value of $\Delta_r G°$, predict whether each type of reaction is product-favored. If there is insufficient information, indicate whether products are favored more at high temperatures or at low temperatures. Check your results against Table 16.2.

The Gibbs free energy change provides a way of predicting whether a reaction is product-favored that depends only on the system—that is, the chemical substances undergoing reaction. Therefore, we can tabulate values of the standard Gibbs free energy of formation, $\Delta_f G°$, for a variety of substances, and from these values calculate

$$\Delta_r G° = \Sigma[(\text{coefficient of product}) \times \Delta_f G°\,(\text{product})] - \Sigma[(\text{coefficient of reactant}) \times \Delta_f G°\,(\text{reactant})] \quad [16.5]$$

for a great many reactions. The calculation is similar to using $\Delta_f H°$ values from Table 4.2 or Appendix J to calculate $\Delta_r H°$ for a reaction (← **Sec. 4-10**). As was the case for $\Delta_f H°$ values, there are no $\Delta_f G°$ values for elements in their standard states, because forming an element from itself constitutes no change at all. Appendix J contains a table that includes $\Delta_f G°$ values for many compounds.

It is important to realize that $\Delta_r G°$ varies significantly as the temperature changes (because of the $-T\Delta_r S°$ term). Therefore, values of $\Delta_r G°$ calculated from Equation 16.5 apply only to the temperature specified in the table of $\Delta_f G°$ values. Appendix J specifies a temperature of 25 °C.

PROBLEM-SOLVING EXAMPLE 16.5

Using Standard Gibbs Free Energies of Formation

Calculate the standard Gibbs free energy change for the combustion of octane using values of $\Delta_f G°$ from Appendix J. Assume that the initial and final states have the same temperature and pressure so that Equation 16.5 applies.

Result $\Delta_r G° = -10{,}611$ kJ/mol

Analyze Combustion means converting a substance to CO_2 and H_2O. This information can be used to write a balanced equation for the reaction. $\Delta_r G°$ can be calculated from $\Delta_f G°$ values for reactants and products. Elements in their standard states have $\Delta_f G° = 0$ (just as they have $\Delta_f H° = 0$).

Plan Write a balanced equation for the combustion reaction, look up $\Delta_f G°$ values in Appendix J, and use Equation 16.5.

Execute The balanced equation and $\Delta_f G°$ values are

$$2\ C_8H_{18}(g) + 25\ O_2(g) \longrightarrow 18\ H_2O(\ell) + 16\ CO_2(g)$$

$\Delta_f G°$ (kJ/mol): 16.72 0 −237.1 −394.4

$$\begin{aligned}\Delta_r G° &= \Sigma[(\text{coefficient of product}) \times \Delta_f G°(\text{product})] - \Sigma[(\text{coefficient of reactant}) \times \Delta_f G°(\text{reactant})]\\ &= [18 \times \Delta_f G°\{H_2O(\ell)\} + 16 \times \Delta_f G°\{CO_2(g)\}] - [2 \times \Delta_f G°\{C_8H_{18}(g)\} - 25 \times \Delta_f G°\{O_2(g)\}]\\ &= [18 \times (-237.1\ \text{kJ/mol}) + 16 \times (-394.4\ \text{kJ/mol})] - [2 \times (16.72\ \text{kJ/mol}) - 25 \times (0\ \text{kJ/mol})]\\ &= -10{,}611\ \text{kJ/mol}\end{aligned}$$

The Gibbs free energy change, $\Delta_r G°$, is a large negative number, indicating that the reaction is product-favored under standard-state conditions.

☑ **Reasonable Result Check** A large, negative Gibbs free energy change is reasonable for a combustion reaction, which, once initiated, occurs of its own accord.

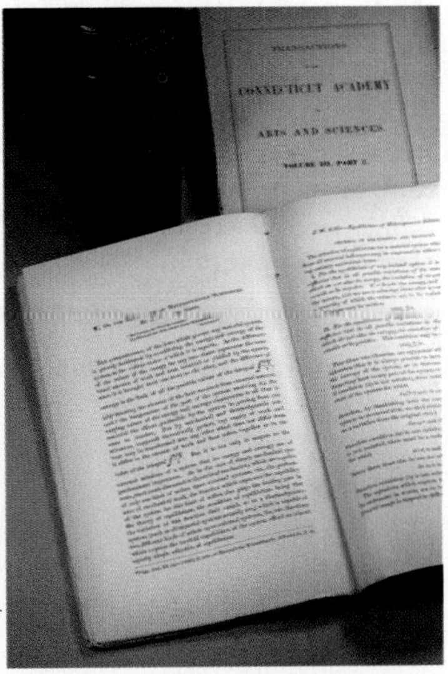

Othmer Library of Chemical History, Chemical Heritage Foundation. Photo by James R. Voelkel

Scientific communication. This is the first page of Josiah Willard Gibbs's paper describing results of his theoretical research on chemical equilibrium, entropy, and thermodynamics. The paper was 320 pages long and filled the entire third volume of the *Transactions of the Connecticut Academy of Arts and Sciences*. Because this journal was not read by most scientists, the publication went unnoticed for a long time, even though it was crucial for advancing our understanding of thermodynamics. This underscores how important it is to communicate scientific results widely.

PROBLEM-SOLVING PRACTICE 16.5

In the text we concluded that the reaction to produce methanol from CO and H_2 is product-favored.

$$CO(g) + 2\ H_2(g) \longrightarrow CH_3OH(\ell)$$

(a) Verify this result by calculating $\Delta_r G°$ from $\Delta_r H°$ and $\Delta_r S°$ for the system. Use values of $\Delta_f H°$ and $S°$ from Appendix J.

(b) Compare your result in part (a) with the calculated value of $\Delta_r G°$ obtained from $\Delta_f G°$ values from Appendix J.

(c) Is the sign of $\Delta_r G°$ positive or negative? Is the reaction product-favored? At all temperatures?

16-6a The Effect of Temperature on Reaction Direction

Many reactions are product-favored at some temperatures and reactant-favored at other temperatures. Thus, it might be possible to make such a reaction produce products by increasing or decreasing the temperature. There is a simple, approximate way to estimate the temperature at which a reactant-favored process becomes product-favored.

In Exercises 16.5 and 16.6 (← **Sec. 16-5**) we developed the idea that $\Delta_r H°$ and $\Delta_r S°$ have nearly constant values over a broad range of temperatures, provided that each of the substances involved in a chemical reaction remains in the same state of matter (solid, liquid, or gas). Because $\Delta_r H°$ and $\Delta_r S°$ are nearly constant, the T on the right-hand side of the equation $\Delta_r G° = \Delta_r H° - T\Delta_r S°$ implies that $\Delta_r G°$ must vary with temperature. It also implies that if we know $\Delta_r H°$ and $\Delta_r S°$ at one temperature, we can estimate $\Delta_r G°$ over a range of temperatures. For example, suppose we want to know whether the reaction

The effect of temperature on reaction direction can be seen in Equation 16.3 (← Sec. 16-5). Because

$$\Delta_r S_{surroundings} = -\Delta_r H_{system}/T$$

and $\Delta_r H$ is nearly constant over a broad range of temperatures, the larger T is the smaller (and therefore less important) the entropy change of the surroundings becomes relative to the entropy change of the system. At high temperatures $\Delta_r S_{system}$ governs the directionality of the reaction.

$$2\ HgO(s) \longrightarrow 2\ Hg(\ell) + O_2(g)$$

produces products at a temperature of 350. °C and a pressure of 1 bar. To find out, calculate $\Delta_r H°$ and $\Delta_r S°$ at 298 K and then estimate $\Delta_r G°$, assuming that $\Delta_r H°$ and $\Delta_r S°$ have the same values at 350. °C (623 K) that they do at 298 K. The boiling point of mercury is 357 °C, and mercury(II) oxide does not melt until well above 500. °C, so the substances are all in the same states at 350. °C that they were at 25 °C.

The notation "$\Delta_r S°$(298.15 K)" designates the entropy change at a temperature of 298.15 K.

$$\begin{aligned}\Delta_r S°\ (298.15\text{ K}) &= \{2\text{ mol} \times S°\ [Hg(\ell)] + (1\text{ mol}) \times S°\ [O_2(g)]\} - \\ &\quad \{2\text{ mol} \times S°\ [HgO(s)]\} \\ &= \{2 \times 76.02\text{ J K}^{-1}\text{ mol}^{-1} + 205.138\text{ J K}^{-1}\text{ mol}^{-1}\} - \\ &\quad \{2 \times 70.29\text{ J K}^{-1}\text{ mol}^{-1}\} \\ &= 216.60\text{ J K}^{-1}\text{ mol}^{-1}\end{aligned}$$

$$\begin{aligned}\Delta_r H°\ (298.15\text{ K}) &= \{2\text{ mol} \times \Delta_f H°[Hg(\ell)] + 1\text{ mol} \times \Delta_f H°[O_2(g)]\} - \\ &\quad \{2\text{ mol} \times \Delta_f H°[HgO(s)]\} \\ &= \{0\text{ kJ/mol} + 0\text{ kJ/mol}\} - \{2 \times (-90.83)\text{ kJ/mol}\} \\ &= 181.66\text{ kJ/mol}\end{aligned}$$

$$\begin{aligned}\Delta_r G°\ (623\text{ K}) &= \Delta_r H°\ (298.15\text{ K}) - T \times \Delta_r S°\ (298.15\text{ K}) \\ &= 181.66\text{ kJ/mol} - 623\text{ K} \times 216.60\text{ J K}^{-1}\text{ mol}^{-1} \\ &= 181.66\text{ kJ/mol} - 134{,}942\text{ J/mol} \\ &= 181.66\text{ kJ/mol} - 134.9\text{ kJ/mol} = 46.8\text{ kJ/mol}\end{aligned}$$

Note that 134,942 J/mol has been converted to 134.9 kJ/mol in the last step.

Because $\Delta_r G°$ is positive, the reaction is not product-favored at 623 K. But at 623 K, $\Delta_r G°$ has a smaller positive value than at 298 K ($\Delta_r G° = -2 \times \Delta_f G°$ (HgO[s]) = 117.1 kJ/mol at 298 K). That is, the reaction is closer to being product-favored at 623 K than it is at room temperature. Because the calculated values of $\Delta_r H°$ and $\Delta_r S°$ are both positive, the reaction is expected to be product-favored at high temperatures, but not at low temperatures. Heating to an even higher temperature than 623 K does decompose mercury(II) oxide (see Figure 16.7).

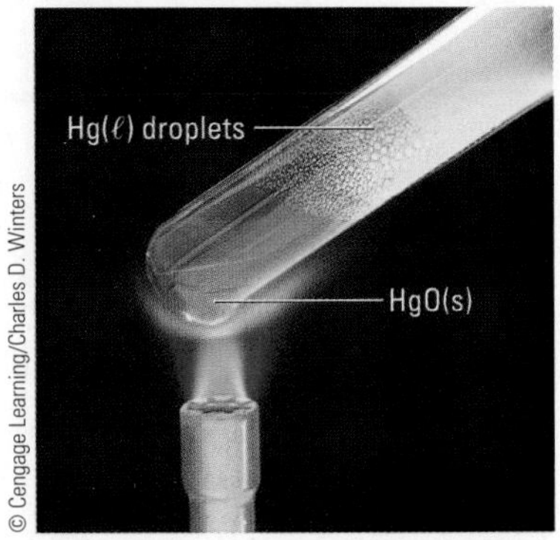
Hg(ℓ) droplets
HgO(s)
© Cengage Learning/Charles D. Winters

Figure 16.7 Decomposition of HgO(s). When heated, red mercury(II) oxide decomposes to liquid mercury metal and oxygen gas. Shiny droplets of mercury can be seen where they have condensed in the cooler part of the test tube.

CONCEPTUAL **EXERCISE 16.10**

High-Temperature Decomposition

Suppose that a sample of HgO(s) is heated above the boiling point of mercury (357 °C).

(a) Could you use the same method to estimate the Gibbs free energy change that was used in the preceding paragraph? Why or why not? (*Hint:* Write the chemical equation for the process that would occur if the temperature were 400 °C.)

(b) At 400 °C, would you expect the reaction to be more or less product-favored than it was at 350 °C? Give two reasons for your choice.

For reactions such as decomposition of mercury(II) oxide that are reactant-favored at low temperatures and product-favored at high temperatures, it is possible to calculate the minimum temperature to which the system must be heated to make the reaction product-favored. Below that temperature $\Delta_r G°$ is positive, and above that temperature $\Delta_r G°$ is negative. Therefore, $\Delta_r G°$ must equal zero at the desired temperature. Because $\Delta_r G° = \Delta_r H° - T\Delta_r S°$ (Equation 16.4), we can set $\Delta_r H° - T\Delta_r S° = 0$. Solving for T gives

$$T(\text{at which } \Delta_r G° \text{ changes sign}) = \frac{\Delta_r H°}{\Delta_r S°} \qquad [16.6]$$

As an example, the contributions of $\Delta_r H°$ and $-T\Delta_r S°$ to $\Delta_r G°$ for the decomposition of silver(I) oxide are shown graphically in Figure 16.8. Equation 16.6 also applies to reactions that are product-favored at low temperatures and reactant-favored at high temperatures. For such reactions Equation 16.6 gives the temperature *below* which the reaction is product-favored. Heating the system above this temperature will probably result in insufficient quantities of products being produced.

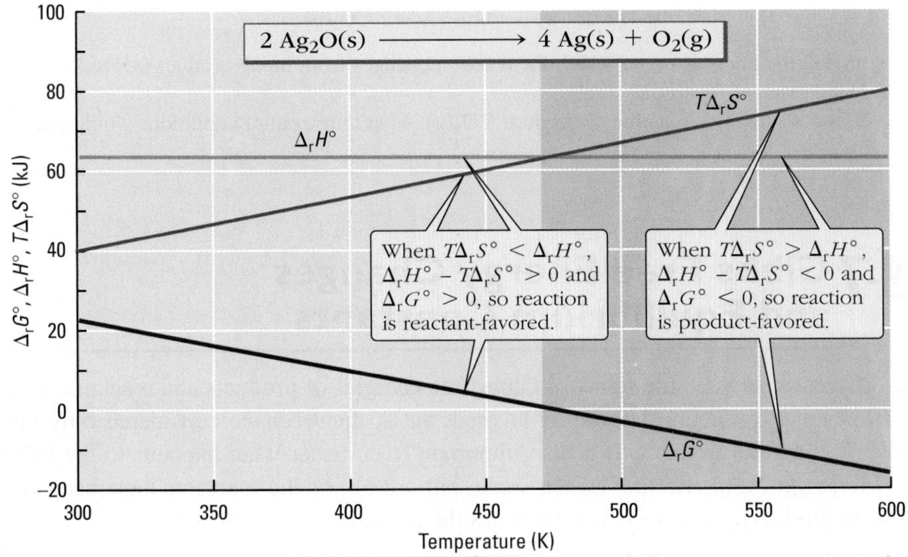
2 Ag2O(s) → 4 Ag(s) + O2(g)
TΔrS°
ΔrH°
ΔrG°
When TΔrS° < ΔrH°, ΔrH° – TΔrS° > 0 and ΔrG° > 0, so reaction is reactant-favored.
When TΔrS° > ΔrH°, ΔrH° – TΔrS° < 0 and ΔrG° < 0, so reaction is product-favored.
ΔrG°, ΔrH°, TΔrS° (kJ)
100
80
60
40
20
0
–20
300
350
400
450
500
550
600
Temperature (K)

Figure 16.8 Effect of temperature on reaction spontaneity. The two terms that contribute to the Gibbs free energy change, $\Delta_r H°$ and $T\Delta_r S°$, are plotted as a function of temperature for the reaction of silver(I) oxide, $Ag_2O(s)$, to form silver metal and oxygen gas.

PROBLEM-SOLVING EXAMPLE 16.6

Effect of Temperature on Gibbs Free Energy Change

The reaction of nitrogen gas with hydrogen gas to form ammonia gas is product-favored at room temperature. Calculate the maximum temperature at which this reaction is product-favored under standard-state conditions.

Result 464.0 K (191.6 °C)

Analyze We need to find the temperature at which the sign of $\Delta_r G°$ changes; that is, the temperature at which $\Delta_r G° = 0$. $\Delta_r G°$ can be estimated by calculating $\Delta_r H°$ and $\Delta_r S°$ at 25 °C and assuming they remain nearly constant as T changes.

Plan Write a balanced chemical equation for the reaction. Use values from Appendix J to calculate $\Delta_r H°$ and $\Delta_r S°$ at 25 °C. Then, use Equation 16.6 to calculate the desired temperature.

Execute

$$N_2(g) + 3\ H_2(g) \longrightarrow 2\ NH_3(g)$$

$$\begin{aligned}\Delta_r H° &= [2 \times (-46.11\ \text{kJ/mol})] - [1 \times (0\ \text{kJ/mol}) + 3 \times (0\ \text{kJ/mol})] \\ &= -92.22\ \text{kJ/mol}\end{aligned}$$

$$\begin{aligned}\Delta_r S° &= [2 \times (192.45\ \text{J K}^{-1}\ \text{mol}^{-1})] - \\ &\quad [1 \times (191.61\ \text{J K}^{-1}\ \text{mol}^{-1}) + 3 \times (130.684\ \text{J K}^{-1}\ \text{mol}^{-1})] \\ &= -198.76\ \text{J K}^{-1}\ \text{mol}^{-1}\end{aligned}$$

$$T = \frac{-92.22\ \text{kJ/mol}}{-198.76\ \text{J K}^{-1}\ \text{mol}^{-1}} \times \frac{1000\ \text{J}}{1\ \text{kJ}} = 464.0\ \text{K}$$

Note that $\Delta_r H°$ has units of kJ/mol, whereas $\Delta_r S°$ has units of J K^{-1} mol^{-1}; this necessitates a unit conversion so that the energy units can cancel.

☑ **Reasonable Result Check** Two moles of gaseous products are produced from four moles of gaseous reactants so the entropy of the system should decrease. Six bonds are formed from four bonds; although the N≡N bond is quite strong, the N—H bonds are also strong and the enthalpy change should be negative. Reactants and products are already gases so increasing T cannot cause any phase changes; the approximation of nearly constant $\Delta_r S°$ and $\Delta_r H°$ is reasonable.

PROBLEM-SOLVING PRACTICE 16.6

For the reaction

$$2\ CO(g) + O_2(g) \longrightarrow 2\ CO_2(g)$$

(a) predict the temperature at which the reaction changes from being product-favored to being reactant-favored.
(b) if you wanted this reaction to produce $CO_2(g)$, what temperature conditions would you choose?

16-7 Gibbs Free Energy Changes and Equilibrium Constants

The difference between the standard Gibbs free energies of products and reactants determines whether a reaction is product-favored, but so far we have considered only pure reactants and pure products. It is also important to consider what happens to the Gibbs free energy *during a reaction* (when some, but not all, of the reactants have been converted to products) because that determines the position of equilibrium.

16-7a Variation of Gibbs Free Energy During a Reaction

One way to remove carbon dioxide from air is to pass the air over solid sodium hydroxide.

$$NaOH(s) + CO_2(g) \longrightarrow NaHCO_3(s) \qquad \Delta_r G°\ (\text{at } 25\ °C) = -77.2\ \text{kJ/mol} \qquad [16.7]$$

Suppose that this reaction is carried out so that the pressure remains at 1 bar and the temperature remains at 25 °C throughout the change from NaOH and CO_2 to $NaHCO_3$. Then, all three substances will remain in their standard states throughout the reaction. The concentration of CO_2, which is proportional to its pressure, remains constant, and the concentrations of the two solids also remain constant, as you showed in Exercise 12.1 (← **Sec. 12-1**).

When the reaction is halfway over, half of the reactants have been converted to products, so 0.5 mol NaOH(s) and 0.5 mol CO_2(g) have been converted to 0.5 mol $NaHCO_3$(s). The equation for what has happened so far is

$$0.5\ NaOH(s) + 0.5\ CO_2(g) \longrightarrow 0.5\ NaHCO_3(s)$$
$$\Delta_r G° \text{ (at 25 °C)} = (0.5)(-77.2 \text{ kJ/mol}) = -38.6 \text{ kJ/mol}$$

Recall that a thermochemical expression refers to the exact number of moles of reactants and products indicated by the coefficients.

At this point we say that the extent of reaction is 0.5. The **extent of reaction**, which we will represent by x, *is the fraction of the reactants that has been converted to products.* For the general equation

$$a\,A + b\,B \longrightarrow c\,C + d\,D$$

if z mol reactant A has been consumed, then the extent of reaction is z/a. If y mol B has been consumed, the extent of reaction is y/b. If w mol D has formed, the extent of reaction is w/d. For a reaction such as the combination of NaOH with CO_2, in which all substances remain in their standard states throughout the chemical change, an extent of reaction of x corresponds to a Gibbs free energy change of x times the Gibbs free energy change for the complete reaction. This is shown graphically in Figure 16.9.

If we start with a mol A, then the maximum quantity of A that can be consumed is a mol A. Therefore $z \leq a$ and $x = z/a$ is always a fraction. Its maximum value is 1.

CONCEPTUAL EXERCISE 16.11

Gibbs Free Energy and Extent of Reaction

For Reaction 16.7, $NaOH(s) + CO_2(g) \longrightarrow NaHCO_3(s)$

(a) Calculate the Gibbs free energy change when the extent of reaction is 0.10, 0.40, and 0.80.

(b) Verify the statement in the text that an extent of reaction of x corresponds to a Gibbs free energy change of x times $\Delta_r G°$ for the complete reaction.

(c) Show that the slope of the line in Figure 16.9 is $\Delta_r G°$.

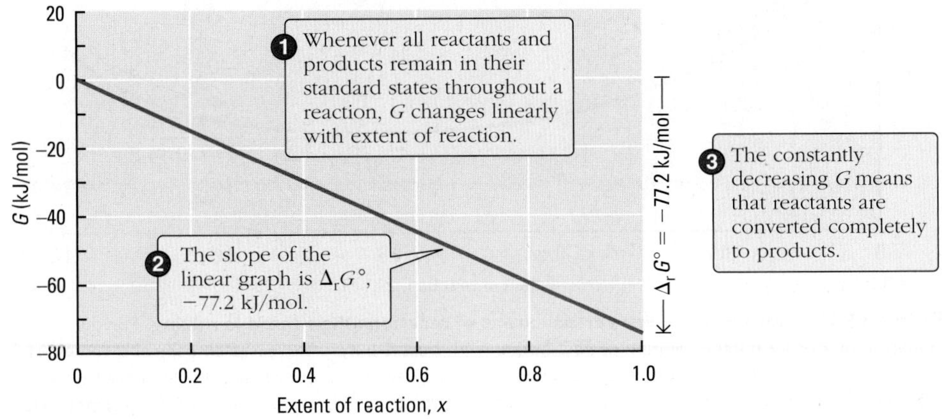
20
0
−20
−40
−60
−80
G (kJ/mol)
0
0.2
0.4
0.6
0.8
1.0
Extent of reaction, x
1 Whenever all reactants and products remain in their standard states throughout a reaction, G changes linearly with extent of reaction.
2 The slope of the linear graph is ΔrG°, −77.2 kJ/mol.
ΔrG° = −77.2 kJ/mol
3 The constantly decreasing G means that reactants are converted completely to products.

Figure 16.9 Gibbs free energy as a function of extent of reaction for the reaction of NaOH(s) with CO_2(g). The Gibbs free energy of the reactants has been arbitrarily set to zero so that it is easier to see how big the differences in Gibbs free energy are.

16-7b Reactions That Reach Equilibrium

When a reaction occurs in which two or more substances form a solution, the situation is different. Such reactions reach an equilibrium state in which both reactants and products are present. Consider an equilibrium reaction we discussed in Chapter 12 (← **Sec. 12-2**), the isomerization of *cis*-2-butene at constant pressure.

$$cis\text{-2-butene(g)} \rightleftharpoons trans\text{-2-butene(g)} \qquad \Delta_r G° \text{ (at 500 K)} = -2.08 \text{ kJ/mol}$$

If we start with 1.0 mol *cis*-2-butene at 500 K and 1 bar, then as soon as some *trans*-2-butene forms it will mix with (dissolve in) the *cis*-2-butene. When one gas dissolves in the other, each becomes less concentrated. Because the total pressure remains at 1 bar in the constant-pressure process, the partial pressure of each gas must be less than 1 bar, which means that neither gas is in its standard state, and the method you used in Exercise 16.11 to calculate $\Delta_r G$ as a function of extent of reaction no longer applies. However, we can tell something about how G varies as the reaction proceeds. Mixing two ideal gases involves no change in enthalpy, but it does involve an increase in entropy because each gas expands to fill the entire container, thereby causing increased dispersal of energy. (This is in addition to any difference in entropy between products and reactants.) Because $\Delta_r S_{dilution}$ is positive and $\Delta_r H_{dilution}$ is zero, then $\Delta_r G_{dilution}$ must be negative. Therefore, in Figure 16.10, there is *not* a straight line from reactants to products. Because there is a negative component ($\Delta_r G_{dilution}$) in addition to x times $\Delta_r G°$, there is a curved line below where the straight line would have been.

Recall that one of the assumptions on which the ideal gas law is based is that there are no forces of attraction among the molecules. Therefore there is no enthalpy change when the gases mix.

The graph in Figure 16.10 is steeper at both ends than a straight line from $G_{reactants}$ to $G_{products}$ and passes through a minimum at $x = 0.62$. This value of x means that the fraction of *trans* molecules is 0.62 and the fraction of *cis* molecules is $1 - 0.62 = 0.38$. Since both *cis*- and *trans*-2-butene occupy the same container, their concentrations and partial pressures are proportional to the number of molecules. Therefore, the equilibrium constant is

$$K_p = K_c = \frac{P_{trans}}{P_{cis}} = \frac{[trans]}{[cis]} = \frac{0.62}{0.38} = 1.6$$

If there is the same number of gas-phase molecules in the reactants as in the products, $K_c = K_P$.

This is the same value that was reported in Chapter 12 (← **Sec. 12-2a**). That is, the equilibrium concentrations correspond to the minimum in the graph of G versus extent of

When $x = 0.62$, the system is at equilibrium and $Q = K$. When x differs from 0.62, $Q \neq K$ and the system must react to reach equilibrium.

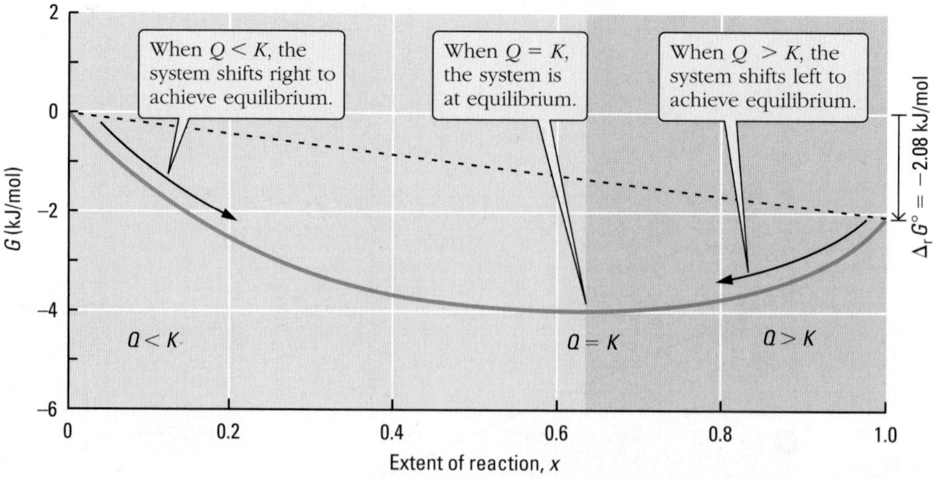
When Q < K, the system shifts right to achieve equilibrium.
When Q = K, the system is at equilibrium.
When Q > K, the system shifts left to achieve equilibrium.
Q < K
Q = K
Q > K
G (kJ/mol)
2
0
−2
−4
−6
0
0.2
0.4
0.6
0.8
1.0
Extent of reaction, x
ΔrG° = −2.08 kJ/mol

Figure 16.10 Gibbs free energy versus extent of reaction when there is mixing. For the isomerization of *cis*-2-butene, dilution of *cis*-2-butene with *trans*-2-butene increases entropy and decreases Gibbs free energy. This causes the graph of Gibbs free energy versus extent of reaction to curve below a straight line from reactants to products. As in Figure 16.9, the total Gibbs free energy of the reactants has been set arbitrarily to zero.

reaction. This correspondence makes sense, because any set of concentrations that differs from $x = 0.62$ has a higher value of G and therefore could change to the equilibrium concentrations with a decrease in Gibbs free energy (which corresponds to an increase in entropy of the universe).

At the far left of Figure 16.10 the graph drops faster than a straight line from $G_{reactants}$ to $G_{products}$. The slope at any point on the curve differs from the slope of the straight line ($\Delta_r G°$) by a factor that depends on the reaction quotient, Q (← **Sec. 12-5**).

$$\text{Slope} = \Delta_r G° + RT \ln Q \qquad [16.8]$$

At equilibrium the slope is zero, and $Q = K°$, where the superscript indicates the standard equilibrium constant. The **standard equilibrium constant ($K°$)** is *similar to the equilibrium constant as defined in Chapter 12* (← **Sec. 12-2**) *except that each concentration is divided by the standard-state concentration of 1 mol/L and each pressure is divided by the standard-state pressure of 1 bar.* That is, even if the equilibrium constant has units (of concentration or pressure), $K°$ is unitless. Substituting into Equation 16.8 gives

$$\text{Slope} = 0 = \Delta_r G° + RT \ln K°$$

which rearranges to

$$\Delta_r G° = -RT \ln K° \qquad [16.9]$$

If the reaction occurs in solution, $K°$ has the same form (and the same value) as the concentration equilibrium constant. For gases, because the standard state involves pressure, $K°$ relates pressures, not concentrations, and has the same value as K_P.

Equation 16.9 indicates that if $K°$ is larger than 1, then $\ln K°$ is positive and $\Delta_r G°$ is negative because of the minus sign. Both of these conditions, a negative $\Delta_r G°$ and $K° > 1$, indicate that the reaction is *product-favored* under standard-state conditions. Conversely, if $K° < 1$, then $\ln K°$ is negative and $\Delta_r G°$ must be positive, indicating a *reactant-favored* system.

$\Delta_r G°$ is the difference in Gibbs free energy between products in their standard states and reactants in their standard states. For example, for ionization of formic acid in water,

$$\underset{1\ \text{mol/L}}{HCOOH(aq)} \rightleftharpoons \underset{1\ \text{mol/L}}{HCOO^-(aq)} + \underset{1\ \text{mol/L}}{H^+(aq)} \qquad \Delta_r G° = 21.3\ \text{kJ/mol}$$

the Gibbs free energy change of 21.3 kJ/mol is for converting 1 mol HCOOH(aq) at a concentration of 1 mol/L into 1 mol $HCOO^-$(aq) and 1 mol H^+(aq), each at a concentration of 1 mol/L. Because $\Delta_r G°$ is positive, we predict that the process is reactant-favored, which means that the concentrations of products will not reach 1 mol/L. This agrees with the fact that formic acid is a weak acid, is only slightly ionized, and therefore has an equilibrium constant much smaller than 1.

Relation between $\Delta_r G°$ and $K°$ at 25 °C

$\Delta_r G°$ (kJ/mol)	$K°$
200	9×10^{-36}
100	3×10^{-18}
10	2×10^{-2}
1	7×10^{-1}
0	1
−1	1.5
−10	6×10^{1}
−100	3×10^{17}
−200	1×10^{35}

$K°$	$\Delta_r G°$ ($-RT \ln K°$)	Product-Favored?
<1	Positive	No
>1	Negative	Yes
=1	0	Neither

PROBLEM-SOLVING EXAMPLE 16.7

Gibbs Free Energy and Equilibrium Constant

In the preceding paragraph you learned that $\Delta_r G° = 21.3$ kJ/mol at 25 °C for the reaction

$$HCOOH(aq) \rightleftharpoons HCOO^-(aq) + H^+(aq)$$

Use this information to calculate the equilibrium constant, K_a, for ionization of formic acid in aqueous solution at 25 °C.

Result $K_a = K_c = K° = 1.8 \times 10^{-4}$

Analyze $\Delta_r G°$ is related to $K°$ by Equation 16.9. Because the reaction occurs in aqueous solution the standard state is 1 mol/L, the standard equilibrium constant involves concentrations divided by 1 mol/L, and $K° = K_c = K_a$.

For formic acid in water, the equilibrium constant is K_a, a ratio of concentrations (← Sec. 14-5a).

Plan Solve Equation 16.9 for $K°$. Substitute appropriate values and calculate $K°$.

Execute $$\Delta_r G° = -RT \ln K°$$

Divide both sides of the equation by $-RT$:

$$-\Delta_r G°/RT = \ln K°$$

Make use of the properties of logarithms (see Appendix A-6). Remove the natural logarithm function by using each side of the equation as an exponent of e.

$$e^{-\Delta_r G°/RT} = e^{\ln K°} = K°$$

Now we can substitute the known values into the equation.

$$K° = e^{-\Delta_r G°/RT} = e^{-(21.3 \text{ kJ/mol})(1000 \text{ J/kJ})/(8.314 \text{ J K}^{-1} \text{ mol}^{-1})(298 \text{ K})} = 1.8 \times 10^{-4}$$

Thus, the positive value of $\Delta_r G°$ results in a value of $K°$ less than 1 and, indeed, indicates a reactant-favored system (see Figure 16.11).

☑ **Reasonable Result Check** Formic acid is a weak acid and is not expected to have a very large K_a. The value calculated appears to be reasonable.

Make certain to check units in calculations like this one. Standard Gibbs free energy changes involve kilojoules, and the gas constant *R* involves joules, so a unit conversion is needed.

PROBLEM-SOLVING PRACTICE 16.7

For each of these reactions, evaluate $K°$ at 298 K from the standard Gibbs free energy change. If necessary, obtain data from Appendix J to calculate $\Delta_r G°$. Check your results against the K_c and K_P values in Table 12.1 (← **Sec. 12-4**). For which of these reactions is $K_c = K°$?

(a) $CaCO_3(s) \rightleftharpoons Ca^{2+}(aq) + CO_3^{2-}(aq)$

(b) $H_2CO_3(aq) \rightleftharpoons HCO_3^-(aq) + H^+(aq)$

(c) $2\ NO_2(g) \rightleftharpoons N_2O_4(g)$

Remember also that $\Delta_r G°$ can be calculated from the equation

$$\Delta_r G° = \Delta_r H° - T\Delta_r S° \quad [16.4]$$

If we know or can estimate changes in enthalpy and entropy for a reaction, then we can calculate or estimate the standard Gibbs free energy change and hence the equilibrium constant. And because $\Delta_r H°$ and $\Delta_r S°$ have nearly constant values over a wide range of

Assuming constant values of $\Delta_r H°$ and $\Delta_r S°$ over a broad range of temperatures does not work well for reactions involving ions in aqueous solution, such as (a) and (b) in Problem-Solving Practice 16.7, because the extent of hydration of the ions varies with temperature.

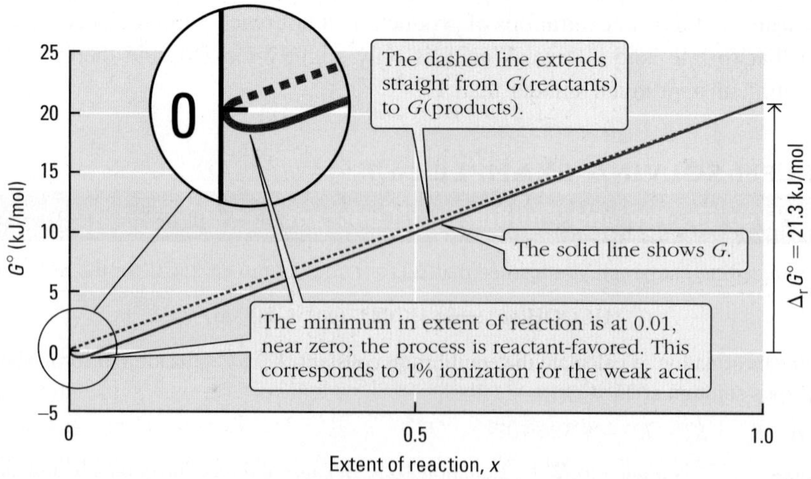

25
20
15
10
5
0
–5
G° (kJ/mol)
0
The dashed line extends straight from G(reactants) to G(products).
The solid line shows G.
The minimum in extent of reaction is at 0.01, near zero; the process is reactant-favored. This corresponds to 1% ionization for the weak acid.
$\Delta_r G° = 21.3$ kJ/mol
0
0.5
1.0
Extent of reaction, x

Figure 16.11 Graph of G as a function of extent of reaction for ionization of formic acid, a reactant-favored process (Problem-Solving Example 16.7).

temperatures, we can estimate equilibrium constants at a variety of temperatures, not just at 25 °C. Problem-Solving Example 16.8 shows how to do this.

PROBLEM-SOLVING EXAMPLE 16.8

Estimating $K°$ at Different Temperatures

Calculate values of $\Delta_r H°$ and $\Delta_r S°$ for the reaction

$$N_2(g) + O_2(g) \rightleftharpoons 2\ NO(g)$$

From these data, estimate the value of $\Delta_r G°$ and hence the value of $K°$ at (a) 298 K, (b) 1000. K, and (c) 2300. K.

Result

(a) $\Delta_r G° = 173.1$ kJ/mol; $K° = 4.5 \times 10^{-31}$
(b) $\Delta_r G° = 155.7$ kJ/mol; $K° = 7.3 \times 10^{-9}$
(c) $\Delta_r G° = 123.5$ kJ/mol; $K° = 1.57 \times 10^{-3}$

Analyze Data from which to calculate $\Delta_r H°$ and $\Delta_r S°$ values are in Appendix J. Equation 16.4 can be used to estimate $\Delta_r G°$ at different temperatures. All species are gaseous, so no phase changes can occur as T increases.

Plan Calculate $\Delta_r H°$ and $\Delta_r S°$ values at 298 K. At each temperature, use $\Delta_r G° = \Delta_r H° - T\Delta_r S°$. Then, calculate $K°$ as was done in Problem-Solving Example 16.7.

Execute

$$\begin{aligned}\Delta_r S°\ (298\ \text{K}) &= [(2 \times 210.761) - (191.61 + 205.138)]\ \text{J K}^{-1}\ \text{mol}^{-1}\\ &= 24.774\ \text{J K}^{-1}\ \text{mol}^{-1}\end{aligned}$$

$$\begin{aligned}\Delta_r H° &= [(2 \times 90.25) - (0 + 0)]\ \text{kJ/mol}\\ &= 180.5\ \text{kJ/mol}\end{aligned}$$

Parts (a) and (b) are done similarly to part (c), which is shown here:

$$\begin{aligned}\text{(c)}\ \Delta_r G° = \Delta_r H° - T\Delta_r S° &= 180{,}500\ \text{J/mol} - (2300.\ \text{K})(24.774\ \text{J K}^{-1}\ \text{mol}^{-1})\\ &= 1.235 \times 10^5\ \text{J} = 123.5\ \text{kJ/mol}\end{aligned}$$

$$K° = e^{-\Delta_r G°/RT} = e^{-(1.235 \times 10^5\ \text{J mol}^{-1})/(8.314\ \text{J mol}^{-1}\ \text{K}^{-1})(2300.\ \text{K})} = 1.57 \times 10^{-3}$$

☑ **Reasonable Result Check** The reaction is endothermic, so Le Chatelier's principle predicts that the equilibrium will shift toward products as the temperature increases. This is reflected by the increasing values of $K°$ as the temperature rises.

This reaction is of great importance because it can take place to a significant extent in high-temperature combustion processes. If the temperature is high enough, nitrogen and oxygen react to form the air pollutant NO.

PROBLEM-SOLVING PRACTICE 16.8

For the ammonia synthesis reaction,

$$N_2(g) + 3\ H_2(g) \rightleftharpoons 2\ NH_3(g)$$

estimate the equilibrium constant at (a) 298 K, (b) 450. K, and (c) 800. K.

16-7c Gibbs Free Energy Changes under Nonstandard-State Conditions

In previous sections we calculated $\Delta_r G°$ for reactions in which reactants in their standard states were converted to products in their standard states. However, substances usually are not at a pressure of 1 bar or at a concentration of 1 mol/L. How do we calculate $\Delta_r G$ if the reactants and products are not at standard concentration or standard pressure? A simple adjustment can be made to $\Delta_r G°$ to account for the difference between actual pressures or concentrations and standard-state pressures or concentrations. The equation from which $\Delta_r G$ (for *nonstandard-state* conditions) can be calculated from $\Delta_r G°$ (for *standard-state* conditions) is

$$\Delta_r G = \Delta_r G° + RT \ln Q \qquad [16.10]$$

For the standard state (concentration of 1 mol/L or pressure of 1 bar), Q = 1. Substituting this into Equation 16.10, we get ln Q = ln(1) = 0 and $\Delta_r G = \Delta_r G°$, the correct value for the standard state.

According to Equation 16.10, the bigger Q becomes, the more positive the correction factor $RT \ln Q$ becomes, and the more positive $\Delta_r G$ becomes. This makes sense, because the larger Q is, the more the concentrations (or pressures) of products exceed those of reactants. According to Le Chatelier's principle, increasing the concentrations of products (or decreasing the concentrations of reactants) causes the reaction to shift in the reverse direction. A shift toward more reactants is expected, because the more positive $\Delta_r G$ is, the more reactant-favored a process is.

To determine the change in Gibbs free energy when reactants at nonstandard-state concentrations or pressures are converted to products at nonstandard-state concentrations or pressures, we first write an appropriate chemical equation, then calculate $\Delta_r G°$ and Q, and finally use Equation 16.10 to correct the value of $\Delta_r G°$ for the nonstandard-state conditions, giving $\Delta_r G$.

PROBLEM-SOLVING EXAMPLE 16.9

Gibbs Free Energy Change for Nonstandard-State Conditions

For the ammonia synthesis reaction at 25 °C, calculate the change in Gibbs free energy if 1 mol $N_2(g)$ at 0.230 bar and 3 mol $H_2(g)$ at 0.420 bar are converted to 2 mol $NH_3(g)$ at 1.45 bar.

Result $\Delta_r G = -21.0$ kJ/mol

Analyze The Gibbs free energy change can be calculated from $\Delta_r G°$ and Q, both of which can be determined based on a balanced chemical equation.

Plan Write a balanced equation, then calculate $\Delta_r G°$ and Q. Finally, use Equation 16.10 to calculate $\Delta_r G$.

Execute

$$N_2(g) + 3\,H_2(g) \rightleftharpoons 2\,NH_3(g)$$

$$\Delta_r G° = [2 \times (-16.45 \text{ kJ/mol})] - [0 + 0] = -32.90 \text{ kJ/mol}$$

$$Q = \frac{P^2_{NH_3}}{P_{N_2}P^3_{H_2}} = \frac{(1.45)^2}{(0.230)(0.420)^3} = 123.4$$

$$\Delta_r G = \Delta_r G° + RT \ln Q = -32.90 \text{ kJ/mol} + (8.314 \text{ J mol}^{-1} \text{ K}^{-1})(298.15 \text{ K}) \times \ln(123.4)$$

$$= -32.90 \text{ kJ/mol} + 11936 \text{ J/mol}$$

$$= -32.90 \text{ kJ/mol} + 11.936 \text{ kJ/mol} = -21.0 \text{ kJ/mol}$$

☑ **Reasonable Result Check** The nonstandard-state conditions involve a concentration of ammonia (the product) well above standard pressure and concentrations of nitrogen and hydrogen (the reactants) well below standard pressure. According to Le Chatelier's principle, if an equilibrium is disturbed by increasing concentrations of products or decreasing concentrations of reactants (both of which apply here), then the equilibrium will shift toward the left—that is, in a reactant-favored direction. The value of $\Delta_r G$ (−21.0 kJ/mol) is negative, but it is less negative than the value of $\Delta_r G°$ (−32.90 kJ/mol). A less negative $\Delta_r G$ corresponds to a less product-favored (that is, more reactant-favored) process, which corresponds with the qualitative prediction from Le Chatelier's principle.

PROBLEM-SOLVING PRACTICE 16.9

Calculate $\Delta_r G$ at 25 °C for a reaction in which $Ca^{2+}(aq)$ combines with $CO_3^{2-}(aq)$ to form a precipitate of $CaCO_3(s)$; the concentrations of $Ca^{2+}(aq)$ and $CO_3^{2-}(aq)$ are 0.023 M and 0.13 M, respectively.

16-8 Gibbs Free Energy, Maximum Work, and Energy Resources

An important interpretation of the Gibbs free energy is that ***the magnitude of $\Delta_r G$ represents the maximum useful work that can be done by a product-favored system on its surroundings (at constant temperature and pressure). $\Delta_r G$ also represents the minimum work that must be done to cause a reactant-favored process to occur.*** Consider the product-favored reaction of hydrogen with oxygen to form liquid water under standard conditions.

The "useful" work represented by $\Delta_r G$ does not include work done by expansion of a system, pushing back the atmosphere.

$$2\ H_2(g) + O_2(g) \rightleftharpoons 2\ H_2O(\ell) \qquad \Delta_r G^\circ = -474.258\ \text{kJ/mol}$$

This thermochemical expression tells us that for every 2 mol $H_2O(\ell)$ produced, as much as 474.258 kJ of work could be done. The negative sign of $\Delta_r G^\circ$ tells us that the work is done on the surroundings. (Because the system has less Gibbs free energy after the reaction than before it, the surroundings has more energy.) Even if the reactants and the products are not at standard pressure or concentration, $\Delta_r G$ still equals $-w_{max}$, the maximum work the system can do on its surroundings.

$$\Delta_r G = -w_{max}\ \text{(work done on the surroundings)} \qquad [16.11]$$

It is important to remember that w_{max} is the maximum work *on the surroundings*. Because w_{max} is for the surroundings, the sign of w_{max} is opposite to that of $\Delta_r G$.

Now consider the decomposition of water to form hydrogen and oxygen, which is the reverse of the previous reaction.

$$2\ H_2O(\ell) \rightleftharpoons 2\ H_2(g) + O_2(g) \qquad \Delta_r G^\circ = 474.258\ \text{kJ/mol}$$

The positive value of $\Delta_r G^\circ$ indicates that this process is reactant-favored. Because the Gibbs free energy of the products is greater than the Gibbs free energy of the reactant, at least 474.258 kJ must be supplied for every 2 mol $H_2O(\ell)$ that decomposes. This 474.258 kJ is the minimum work that must be done to change 2 mol liquid water into hydrogen gas and oxygen gas. One way to supply this work is to use a direct electric current to carry out electrolysis of the water. In general, a continuous supply of energy is required to cause a reactant-favored process, such as decomposition of liquid water, to happen.

Because transformations of energy from one form to another are not 100% efficient, we seldom observe anything close to the maximum quantity of useful work given by the value of $\Delta_r G^\circ$.

PROBLEM-SOLVING EXAMPLE 16.10

Gibbs Free Energy Change and Maximum Work

Predict whether each reaction is product-favored or reactant-favored at 25 °C and 1 bar. For each product-favored reaction, calculate the maximum useful work the reaction could do. For each reactant-favored process, calculate the minimum work needed to cause it to occur.

(a) $2\ Al_2O_3(s) \longrightarrow 4\ Al(s) + 3\ O_2(g)$ (b) $Cl_2(g) + Mg(s) \longrightarrow MgCl_2(s)$

Result

(a) Reactant-favored; at least 3164.6 kJ/mol must be supplied
(b) Product-favored; can do up to 591.79 kJ/mol of useful work

Analyze The magnitude of the Gibbs free energy change gives the maximum work that a product-favored reaction can do or the minimum work required to make a reactant-favored process occur.

Plan Use data from Appendix J to calculate $\Delta_r G^\circ$ for each reaction. Based on the sign, determine whether the reaction can do work or work is required.

Execute

(a) $\Delta_r G^\circ = 0 + 0 - 2(-1582.3)$ kJ/mol = 3164.6 kJ/mol; at least 3164.6 kJ/mol is required.
(b) $\Delta_r G^\circ = -591.79$ kJ/mol $- 0 - 0 = -591.79$ kJ/mol; up to 591.79 kJ/mol useful work can be done.

☑ **Reasonable Result Check** Reaction (a) is decomposition of an oxide to a metal and oxygen. Because metals are good reducing agents and oxygen is a strong oxidizing agent, the reverse of this reaction is likely to be product-favored, which would make Reaction (a) reactant-favored. This result agrees with the calculation. Reaction (b) is combination of an alkaline-earth element with a halogen, which should form a stable ionic compound. Therefore Reaction (b) should be product-favored, which agrees with the calculation. In both cases the value of $\Delta_r G°$ is large, which also is expected based on the arguments just given.

PROBLEM-SOLVING PRACTICE 16.10

Predict whether each reaction is reactant-favored or product-favored at 298 K and 1 bar, and calculate the minimum work that would have to be done to force it to occur, or the maximum work that could be done by the reaction.

(a) $2\ CO_2(g) \longrightarrow 2\ CO(g) + O_2(g)$ (b) $4\ Fe(s) + 3\ O_2(g) \longrightarrow 2\ Fe_2O_3(s)$

16-8a Coupling Reactant-Favored Processes with Product-Favored Processes

A dead battery in an electric car will not charge itself. Charging a battery is a reactant-favored process. But a battery can be charged if it is connected to a charger that is, in turn, powered by electricity generated in a power plant that burns coal. Coal, which is mainly carbon, burns in air according to the equation

$$C(s) + O_2(g) \longrightarrow CO_2(g) \qquad \Delta_r G° = -394.4\ \text{kJ/mol}$$

If enough coal is burned, the large negative Gibbs free energy change for its combustion more than offsets the positive Gibbs free energy change of the battery-charging process. An overall decrease in Gibbs free energy occurs, even though there is an increase in the battery-charging system we are interested in. Once a battery has been charged, the chemicals in the battery can undergo a product-favored reaction, supplying electricity to make the car go. Some of the Gibbs free energy lost when the coal was burned was stored in the car's battery for use later (Figure 16.12).

Charging a battery is an example of coupling a product-favored reaction with a reactant-favored process to cause the latter to take place. Both processes occur at the same time and in a way that allows the Gibbs free energy released by the product-favored reaction to be used by the reactant-favored reaction. Other examples include obtaining aluminum or iron from their ores; synthesizing large, complicated molecules from simple reactants to make medicines, plastics, and other useful materials; and keeping a house cool on a day when the outside temperature is above 100 °F. All involve decreasing entropy in the region of our interest, but all can be made to occur provided that there is a larger increase in entropy at a power plant or somewhere else.

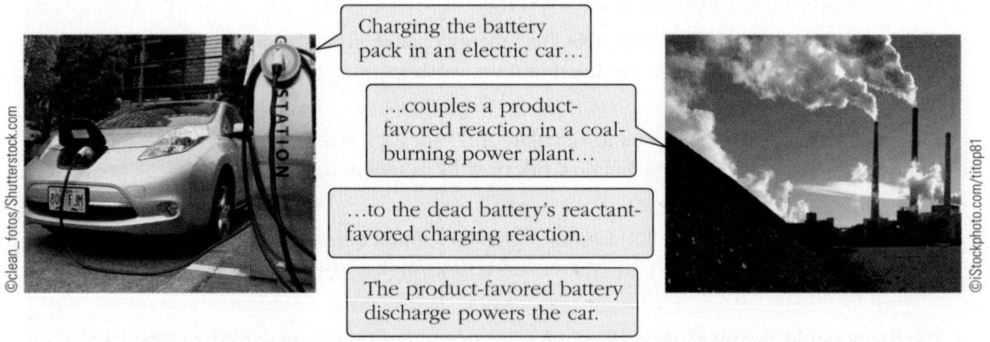
Charging the battery pack in an electric car...
...couples a product-favored reaction in a coal-burning power plant...
...to the dead battery's reactant-favored charging reaction.
The product-favored battery discharge powers the car.
STATION
©clean_fotos/Shutterstock.com
©iStockphoto.com/titop81

Figure 16.12 Coupled chemical reactions. Charging the battery in an electric car.

The Gibbs free energy change indicates a chemical reaction's capacity to drive a different reactant-favored system to produce products. The word "free" in the name indicates not "zero cost," but rather "available." ***Gibbs free energy is available to do useful tasks that would not happen on their own.*** Another way of saying this is that Gibbs free energy is a measure of the *quality* of the energy contained in a chemical reaction system. If it contains a lot of Gibbs free energy, a chemical system can do a lot of work for us; the energy is of high quality—potentially useful to humankind. As the system's reactants transform into products, that available free energy can do useful work, but *only* if the reaction is coupled to some other, reactant-favored process we want to carry out. If systems are *not* coupled, then the Gibbs free energy released by a reaction is wasted as thermal energy. All living things depend on coupling of chemical reactions to provide for their energy needs.

CONCEPTUAL EXERCISE 16.12

Coupling Reactions

One way to produce iron metal is to reduce iron(III) oxide with aluminum. This is called the thermite reaction and is shown here. You can think of the reaction as occurring in two steps. The first is the loss of oxygen from iron(III) oxide,

(i) $Fe_2O_3(s) \longrightarrow 2\ Fe(s) + \frac{3}{2}\ O_2(g)$

The second is the combination of aluminum with the oxygen,

(ii) $2\ Al(s) + \frac{3}{2}\ O_2(g) \longrightarrow Al_2O_3(s)$

© Cengage Learning/Charles Steele

Thermite reaction.

(a) Calculate the enthalpy, entropy, and Gibbs free energy changes for each step. Decide whether each step is product- or reactant-favored. Comment on the signs of $\Delta_r H°$, $\Delta_r S°$, and $\Delta_r G°$ for each step.

(b) What is the overall net reaction that occurs when aluminum is combined with iron(III) oxide? What are the enthalpy, entropy, and Gibbs free energy changes for the overall reaction? Is it product- or reactant-favored? Comment on the signs of $\Delta_r H°$, $\Delta_r S°$, and $\Delta_r G°$ for the overall reaction.

(c) Discuss briefly how coupling Reaction (i) with Reaction (ii) affects our ability to obtain iron metal from iron(III) oxide by reacting it with aluminum.

(d) Suggest a reaction other than oxidation of aluminum that might be used to reduce iron(III) oxide to iron. Test your selection by calculating the Gibbs free energy change for the coupled system.

16-9 Gibbs Free Energy and Biological Systems

Have you ever thought about how unlikely it is that a human being can exist? Your body contains about 100 trillion (10^{14}) cells, all working together to make you what you are. Each of those cells contains trillions of molecules, and many of the molecules (such as insulin, Figure 16.13) contain hundreds to tens of thousands of atoms arranged in exactly the appropriate way. Those molecules and cells are arranged in structures such as organs, bones, and skin that provide for all the functions of your body and that determine its overall shape and size. When necessary, you can synthesize molecules on very short notice. For example, it does not take long to generate the surge of adrenalin your body makes when you are scared. Your body is a very highly organized system, which means that its entropy must be very low. This in turn means that, thermodynamically speaking, you are very, very improbable. How can it be, then, that you exist? How can all the molecules from which you are made be synthesized and organized into the organs and other tissues of your body?

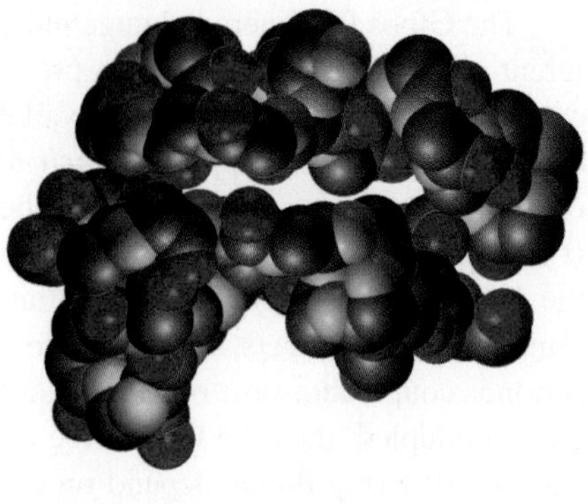

Figure 16.13 Insulin, a protein. Like all protein molecules, the insulin molecule is highly ordered. It contains 51 amino acids (← **Sec. 10-7d**) and 777 atoms connected in exactly the correct order and folded into exactly the molecular shape needed for its function in the metabolism of glucose. (Hydrogen atoms are not shown.)

16-9a Human Metabolism and Gibbs Free Energy

The answer lies in the coupling of reactions described in the preceding section. Because you are a very-low-entropy system, you must be very high in Gibbs free energy. Your body obtains that Gibbs free energy by oxidizing the food you eat with oxygen you inhale. In the processes of metabolism, foods are oxidized, and the Gibbs free energy released by their oxidation causes other reactions to occur that store Gibbs free energy in specific molecules within your body. Later those molecules can release the Gibbs free energy to cause muscles to contract, nerve signals to be sent, important molecules to be synthesized, and other processes to occur. **Metabolism** refers to *all of the chemical changes that occur as food nutrients are processed by an organism to release Gibbs free energy and form the complex chemical constituents of living cells.* **Nutrients** are *the chemical raw materials needed for survival of an organism.*

As an example of metabolism, consider the single nutrient glucose, also known as dextrose or blood sugar (← **Sec. 10-7b**). Glucose can be oxidized to carbon dioxide and water according to the equation

The prime symbol (′) on $\Delta_r G^{\circ\prime}$ indicates that the value of the Gibbs free energy change is for pH = 7, the same concentration of H^+(aq) ions as in a typical cell. When aqueous solutions are involved, $\Delta_r G^\circ$ values in tables such as Appendix J refer to H^+ concentrations of 1 mol/L, but such a high concentration of acid would destroy most cells. Consequently, biochemists have calculated a set of $\Delta_r G^{\circ\prime}$ values that apply to solutions at pH = 7. These values are usually reported for a temperature of 37 °C, human body temperature.

$$C_6H_{12}O_6(aq) + 6\ O_2(g) \longrightarrow 6\ CO_2(g) + 6\ H_2O(\ell) \qquad \Delta_r G^{\circ\prime} = -2870 \text{ kJ/mol}$$

Thus, a large quantity of Gibbs free energy can be released when glucose is oxidized. This reaction, which is strongly product-favored, is an **exergonic** reaction—*a reaction that releases Gibbs free energy.* The same quantity of Gibbs free energy is available whether glucose is burned in air or reacts in your body. However, burning glucose would release all of the Gibbs free energy as thermal energy. This process would not be appropriate in your body because it would raise the temperature rapidly, which in turn would kill cells. Instead, your body makes use of a large number of reactions that allow the Gibbs free energy to be released in small steps and stored in small quantities for use later.

Gibbs free energy is temporarily stored in your body in small packets through formation of adenosine triphosphate (ATP) from adenosine diphosphate (ADP). The structures of ADP and ATP are closely related and are shown in Figure 16.14. The energy-storage reaction, a condensation reaction, is

$$ADP^{3-}(aq) + H_2PO_4^-(aq) \longrightarrow ATP^{4-}(aq) + H_2O(\ell) \qquad \Delta_r G^{\circ\prime} = 30.5 \text{ kJ/mol}$$

The words "exergonic" and "endergonic" have nearly the same prefixes as "exothermic" and "endothermic." In both cases *ex* means "out" and *end* means "into." *Thermic* indicates that thermal energy is released or taken up. *Ergonic* indicates that Gibbs free energy is released or taken up.

This is an example of an **endergonic** reaction—*one that consumes Gibbs free energy and is therefore reactant-favored.* In a typical bacterial cell, this reaction takes place 38 times for each molecule of glucose that is oxidized. In human cells, it takes place 32 times for each molecule of glucose that is oxidized. That is, in a human cell the Gibbs free energy released by the exergonic glucose oxidation is used to force the endergonic, reactant-favored process of forming ATP from ADP to occur 32 times. The overall process is

$$C_6H_{12}O_6(aq) + 6\ O_2(g) + 32\ ADP^{3-}(aq) + 32\ H_2PO_4^-(aq) \longrightarrow$$
$$6\ CO_2(g) + 32\ ATP^{4-}(aq) + 38\ H_2O(\ell) \qquad \Delta_r G^{\circ\prime} = -1894 \text{ kJ/mol}$$

Because it is exergonic, this overall reaction must be product-favored, and therefore appreciable quantities of products can be obtained.

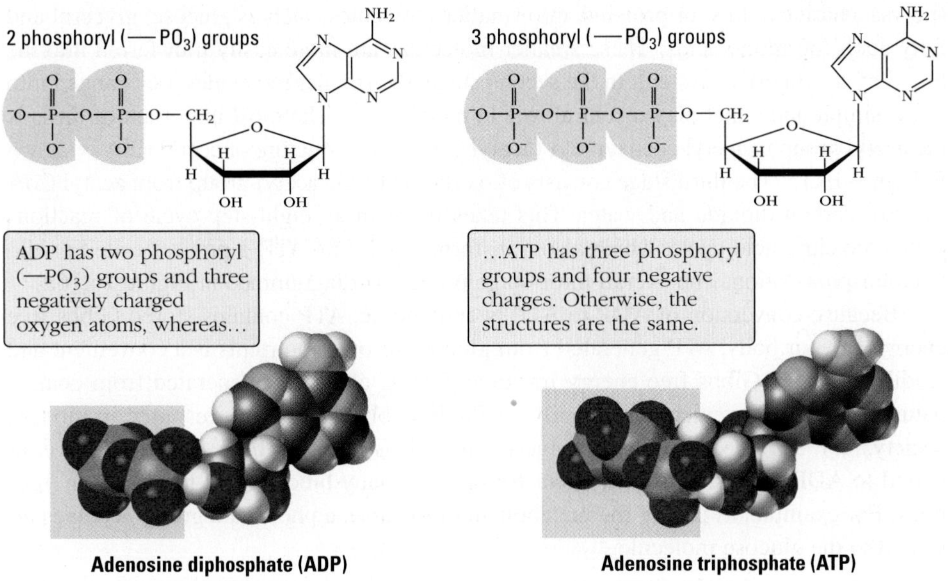
2 phosphoryl (—PO3) groups
3 phosphoryl (—PO3) groups
ADP has two phosphoryl (—PO3) groups and three negatively charged oxygen atoms, whereas…
…ATP has three phosphoryl groups and four negative charges. Otherwise, the structures are the same.
Adenosine diphosphate (ADP)
Adenosine triphosphate (ATP)

Figure 16.14 Gibbs free energy is stored when ADP is transformed to ATP.

EXERCISE 16.13

Coupled Metabolic Reactions

Add the Gibbs free energy change for oxidation of glucose to the appropriate Gibbs free energy change for 32 conversions of ADP to ATP, thereby verifying that the Gibbs free energy change of −1894 kJ/mol for the overall reaction is correct. What happens to the 1894 kJ/mol of Gibbs free energy released by the overall reaction?

Because ATP is high in Gibbs free energy, it is said to be a high-energy molecule (or ion). Sometimes the bonds in ATP are called high-energy bonds, but this description is a misnomer. Actually, the bonds have low bond energies and can undergo hydrolysis easily to form ADP and release Gibbs free energy.

The metabolic process by which the Gibbs free energy contained in nutrients is stored in ATP is far more complicated than the overall equation given previously makes it seem. Metabolism can be divided into three stages that were first clearly identified by Hans Krebs. The first stage, digestion, breaks apart large molecules, such as carbohydrates

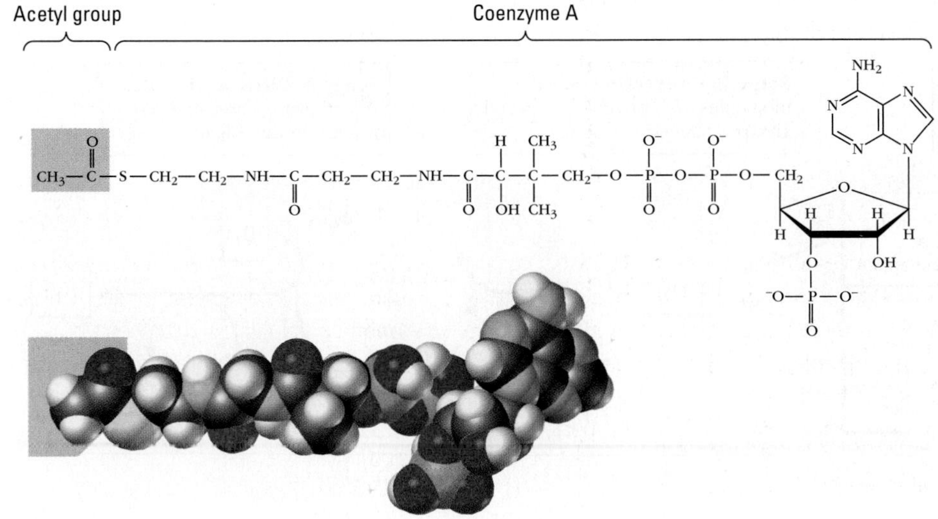
Acetyl group
Coenzyme A

Figure 16.15 Structure of acetyl coenzyme A. Structural formula and space-filling model of acetyl coenzyme A. The acetyl group at the far left end of the molecule can be formed from glucose that came originally from starch. (This figure is discussed on the next page.)

(polysaccharides), fats, or proteins, into smaller molecules, such as glucose, glycerol and fatty acids, or amino acids. These smaller molecules are more easily transferred into the blood by the digestive system. In the second stage, the smaller molecules are changed into a few simple units that play a central role in metabolism. The most important of these is the acetyl group in acetyl coenzyme A (acetyl CoA). The structure of acetyl CoA is shown in Figure 16.15. The third stage consists of oxidation of the acetyl group from acetyl CoA to form carbon dioxide and water. This takes place in an eight-step cycle of reactions called the citric acid cycle, which also transforms ADP into ATP, a process called oxidative phosphorylation. The overall three-stage process is diagrammed in Figure 16.16.

The citric acid cycle is also known as the Krebs cycle or the tricarboxylic acid (TCA) cycle.

Conversion of ATP to ADP is the source of the energy that causes muscles to contract. This answers the question, "Where does the energy come from to make my muscles work?" that was posed in Chapter 1 (← Sec. 1-1).

Because conversion of ADP to ATP is endergonic, ATP contains stored Gibbs free energy. In your body, ATP generated from glucose or other nutrients is a convenient and readily available Gibbs free energy resource, just as electricity generated from coal or natural gas is a convenient and readily available Gibbs free energy resource in modern society. ATP can release Gibbs free energy in packets of 30.5 kJ/mol for each ATP converted to ADP. This size is convenient for driving many biochemical processes in your body. For example, as part of the metabolism of glucose, a phosphate group becomes attached to the glucose molecule.

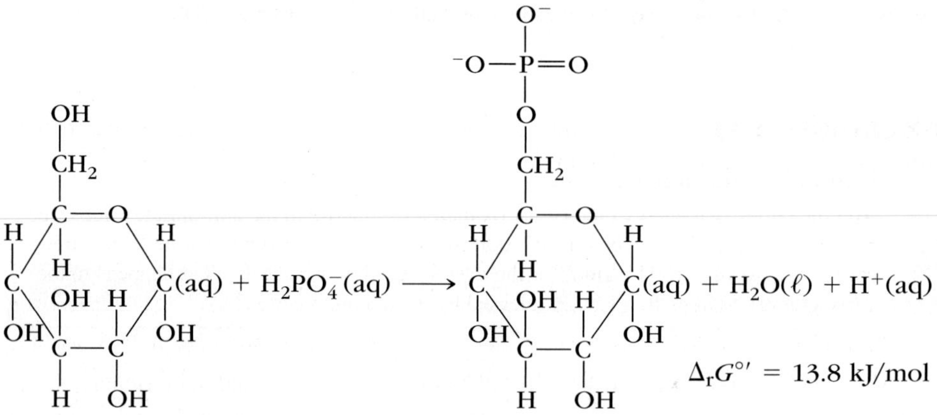

(aq) + $H_2PO_4^-$(aq) →
(aq) + $H_2O(\ell)$ + H^+(aq)
$\Delta_r G^{\circ\prime} = 13.8$ kJ/mol

glucose   glucose 6-phosphate^{2-}

The "6" in glucose 6-phosphate indicates that the phosphate group has been added to the oxygen atom attached to carbon number 6 of glucose, the only carbon atom not in the ring structure.

This reaction is endergonic by 13.8 kJ/mol and, therefore, is not product-favored. It does not occur unless forced to do so.

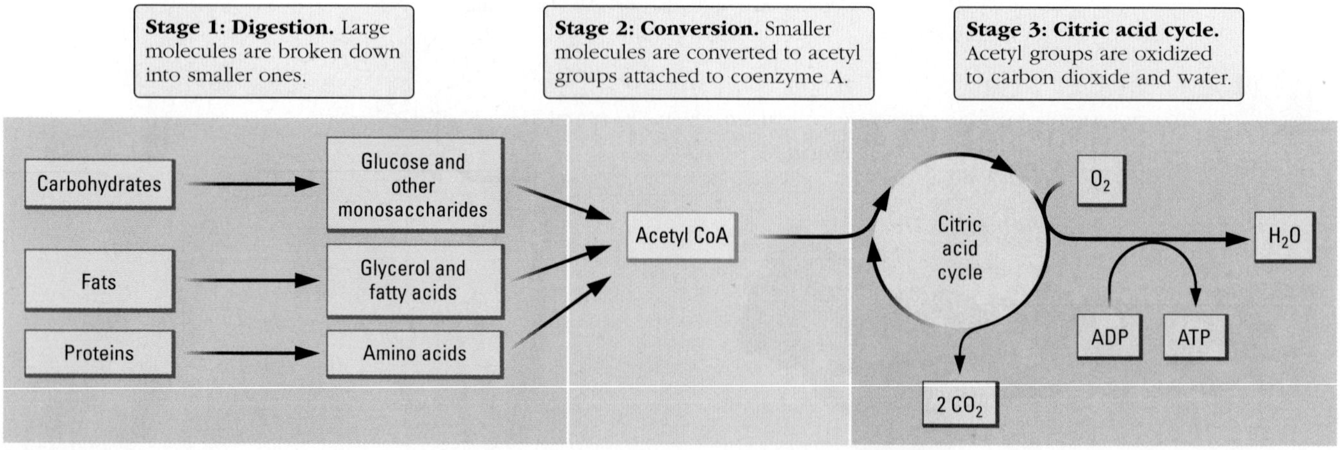

Stage 1: Digestion. Large molecules are broken down into smaller ones.
Stage 2: Conversion. Smaller molecules are converted to acetyl groups attached to coenzyme A.
Stage 3: Citric acid cycle. Acetyl groups are oxidized to carbon dioxide and water.
Carbohydrates
Fats
Proteins
Glucose and other monosaccharides
Glycerol and fatty acids
Amino acids
Acetyl CoA
Citric acid cycle
O_2
H_2O
ADP
ATP
2 CO_2

Figure 16.16 Extraction of Gibbs free energy from nutrients is a three-stage process.

The endergonic reaction can be caused to occur by coupling it to the transformation of ATP to ADP.

$$ATP^{4-}(aq) + H_2O(\ell) \longrightarrow ADP^{3-}(aq) + H_2PO_4^-(aq) \qquad \Delta_rG°' = -30.5 \text{ kJ/mol}$$

$H_2PO_4^-$ produced by this reaction can react with glucose to form glucose 6-phosphate, coupling the two reactions directly. Also, water produced in the first reaction is used up in the second one. The overall process is

$$\text{Glucose(aq)} + ATP^{4-}(aq) \longrightarrow$$
$$\text{glucose 6-phosphate}^{2-}(aq) + ADP^{3-}(aq) + H^+(aq)$$
$$\Delta_rG°' = (-30.5 + 13.8)\text{ kJ/mol} = -16.7 \text{ kJ/mol}$$

The negative value of $\Delta_rG°'$ indicates that the overall process is exergonic and product-favored. Thus the ATP $\longrightarrow$ ADP transformation can cause glucose to undergo a condensation reaction with dihydrogen phosphate. The 16.7 kJ/mol of Gibbs free energy released appears as thermal energy transferred from the system to its surroundings.

In biochemistry the convention is to write these equations in a shorthand notation that indicates that they are coupled. The process just described is represented as:

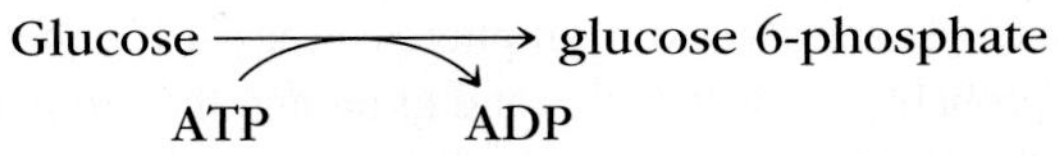
Glucose
glucose 6-phosphate
ATP
ADP

The curved line indicates that the transformation of ATP to ADP occurs simultaneously with the glucose reaction and that the two are coupled.

PROBLEM-SOLVING EXAMPLE 16.11

Biochemical Standard State

Many biochemical processes involve reactions that take place at a temperature of 37 °C and a pH of 7 in body fluids. Under these conditions the Gibbs free energy change is specified as $\Delta_rG°'$, where the prime specifies that all substances are at their standard-state concentrations except for H^+(aq), which is at a biological concentration of about 10^{-7} mol/L (pH = 7). Calculate $\Delta_rG°$ (standard-state conditions of 1 mol/L H^+(aq)) for this reaction.

$$\text{Glucose(aq)} + ATP^{4-}(aq) \longrightarrow \text{glucose 6-phosphate}^{2-}(aq) + ADP^{3-}(aq) + H^+(aq)$$
$$\Delta_rG°' = -16.7 \text{ kJ/mol}$$

Result $\Delta_rG° = 24.8$ kJ/mol

Analyze The $\Delta_rG°'$ value differs from $\Delta_rG°$ because one of the concentrations (that of H^+(aq)) has a nonstandard value of 10^{-7} mol/L. That is, $\Delta_rG°'$ is Δ_rG for conditions such that every concentration is 1 mol/L except for the concentration of H^+(aq).

Plan Set $\Delta_rG = \Delta_rG°'$, calculate Q, and use Equation 16.10 (← **Sec. 16-7c**) to calculate $\Delta_rG°$.

Execute

$$\Delta_rG°' = -16.7 \text{ kJ/mol} = \Delta_rG = \Delta_rG° + RT \ln Q$$
$$\Delta_rG° = -16.7 \text{ kJ/mol} - RT \ln Q$$

$$Q = \frac{(\text{conc. glucose 6-phosphate}^{2-})(\text{conc. } ADP^{3-})(\text{conc. } H^+)}{(\text{conc. glucose})(\text{conc. } ATP^{4-})}$$

$$= \frac{(1)(1)(1 \times 10^{-7})}{(1)(1)} = 1 \times 10^{-7}$$

$$\Delta_rG° = -16.7 \text{ kJ/mol} - RT \ln(1 \times 10^{-7})$$
$$= -16.7 \text{ kJ/mol} - (8.314 \text{ J mol}^{-1} \text{ K}^{-1})\{(273 + 37) \text{ K}\}(-16.12)$$
$$= -16.7 \text{ kJ/mol} + 41542 \text{ J/mol} = -16.7 \text{ kJ/mol} + 41.5 \text{ kJ/mol} = 24.8 \text{ kJ/mol}$$

☑ **Reasonable Result Check** Using Le Chatelier's principle, we predict less shift toward products for a system in which the concentration of $H^+(aq)$, a product, is 1 mol/L than there would be for a system in which the concentration of $H^+(aq)$ is 1×10^{-7} mol/L. Less shift toward products means a more positive $\Delta_r G$, and the value of $\Delta_r G^\circ$ is indeed more positive than $\Delta_r G^{\circ\prime}$.

PROBLEM-SOLVING PRACTICE 16.11

Is $\Delta_r G^\circ$ larger than, smaller than, or the same size as $\Delta_r G^{\circ\prime}$ for this reaction?

$$C_6H_{12}O_6(aq) + 6\ O_2(g) \longrightarrow 6\ CO_2(g) + 6\ H_2O(\ell) \qquad \Delta_r G^{\circ\prime} = -2870\ \text{kJ/mol}$$

Explain why you chose the response you did.

16-9b Photosynthesis and Gibbs Free Energy

You may be wondering where the nutrients you take into your body get the Gibbs free energy they so obviously have. The answer is from solar energy via photosynthesis. **Photosynthesis** is *a series of reactions in a green plant that combines carbon dioxide with water to form carbohydrates and oxygen.* The carbohydrates and other constituents you consume in vegetables are derived from photosynthesis. If you eat meat, the animal from which it came probably ate vegetables and grain and therefore derived its nutrients from plant photosynthesis. The overall reaction in photosynthesis is just the opposite of oxidation of glucose.

$$6\ CO_2(g) + 6\ H_2O(\ell) \longrightarrow C_6H_{12}O_6(aq) + 6\ O_2(g) \qquad \Delta_r G^{\circ\prime} = 2870\ \text{kJ/mol}$$

Photosynthesis is endergonic and can occur only because of an influx of energy in the form of sunlight. That is, the energy in the photons of sunlight causes this reactant-favored process to form appreciable quantities of products, and the sunlight's energy is stored as Gibbs free energy in the glucose and oxygen that are formed. This process is diagrammed in Figure 16.17.

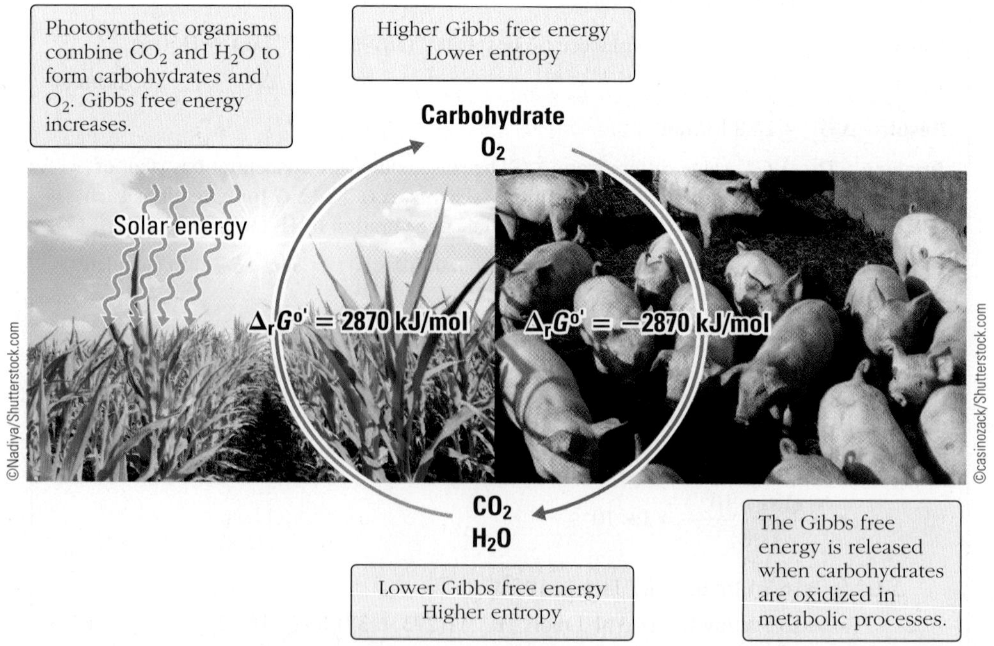

Photosynthetic organisms combine CO2 and H2O to form carbohydrates and O2. Gibbs free energy increases.
Higher Gibbs free energy
Lower entropy
Carbohydrate
O2
Solar energy
ΔrG°′ = 2870 kJ/mol
ΔrG°′ = −2870 kJ/mol
©Nadiya/Shutterstock.com
©casinozack/Shutterstock.com
CO2
H2O
Lower Gibbs free energy
Higher entropy
The Gibbs free energy is released when carbohydrates are oxidized in metabolic processes.

Figure 16.17 Solar energy storage by photosynthesis. The Gibbs free energy in the foods animals eat is derived from solar energy via photosynthesis.

Organisms that can carry out photosynthesis are called *phototrophs* (literally, "light-feeders") because they can use sunlight to supply needed energy. Phototrophs include all green plants, all algae, and some groups of bacteria. The phototrophs capture light by means of photosynthetic pigment systems and store the light energy in molecules such as glucose that have relatively high Gibbs free energies. Nearly all other organisms belong to the class of *chemotrophs* (literally, "chemical-feeders"), which must depend on the chemicals created by the phototrophs for their energy. All animals, fungi, and most bacteria are chemotrophs. A world composed only of chemotrophs would not last long because without the phototrophs, food supplies would disappear almost immediately. Without sunlight and its ability to drive a reactant-favored system to form products (carbohydrate and oxygen), organisms such as humans and indeed almost the entire biosphere of planet Earth could not exist.

Some organisms, the chemautotrophs, which are found near deep-ocean volcanic vents, do not depend on phototrophs for their energy supply.

Both phototrophs and chemotrophs make use of the Gibbs free energy stored during photosynthesis by using oxidation of glucose to drive a large number of conversions of ADP to ATP and then using the ATP to couple to desired endergonic reactions and force them to occur. Thus, ATP is the minute-to-minute energy currency of living cells. The Gibbs free energy released in these reactions contributes to synthesis of molecules needed by the cell, causes some desirable process such as muscle contraction, or is dissipated as thermal energy. If more Gibbs free energy is taken in than the organism needs, then the excess Gibbs free energy can be stored long term through the synthesis of fats, which have approximately twice as much Gibbs free energy as an equal mass of carbohydrate.

It is significant that when ATP reacts and causes other reactions to occur, the product, ADP, is very similar to the reactant ATP. ADP can easily be recycled to form ATP. A reasonable estimate of the quantity of ATP converted to ADP during one day in the life of an average human is 117 mol. The molar mass of the sodium salt of ATP, 551 g/mol, allows us to calculate the mass of ATP converted to ADP every day:

$$m(\text{ATP}) = 117 \text{ mol} \times \frac{551 \text{ g}}{1 \text{ mol}} = 64{,}500 \text{ g ATP}$$

This is 64.5 kg, which is close to the 70-kg body weight of an average person. Obviously ATP is not a long-term storage molecule for Gibbs free energy. Instead it is recycled from ADP as needed and used almost immediately for some necessary process. The typical 70-kg human body contains only 50 g ATP and ADP. If we actually had to take in 64.5 kg ATP per day to provide Gibbs free energy, it would be a very expensive habit. The price of ATP from a laboratory supplier is currently about \$50 per gram, which puts the cost of supplying each of us with our daily energy currency at more than three million dollars!

PROBLEM-SOLVING EXAMPLE 16.12

Coupling of Biological Reactions

ATP undergoes hydrolysis with release of Gibbs free energy according to the equation

(i) Adenosine triphosphate + $H_2O(\ell)$ $\longrightarrow$
adenosine diphosphate + dihydrogen phosphate $\Delta_r G°'$ (i) = −30.5 kJ/mol

Other organophosphates undergo similar hydrolysis reactions. For creatine phosphate and glycerol 3-phosphate the hydrolysis reactions are

(ii) Creatine phosphate + $H_2O(\ell)$ $\longrightarrow$ creatine + dihydrogen phosphate
$\Delta_r G°'$ (ii) = −43.1 kJ/mol

(iii) Glycerol 3-phosphate + $H_2O(\ell)$ $\longrightarrow$ glycerol + dihydrogen phosphate
$\Delta_r G°'$ (iii) = −9.7 kJ/mol

Predict whether each reaction, (a) and (b) that follows, is product-favored and, if it is, calculate the maximum work that could be done if the reaction took place as written.

(a) Adenosine triphosphate + creatine ⟶ creatine phosphate + adenosine diphosphate
(b) Glycerol + adenosine triphosphate ⟶ glycerol 3-phosphate + adenosine diphosphate

Result

(a) Not product-favored
(b) Product-favored; $\Delta_r G^{\circ\prime} = -20.8$ kJ/mol, so up to 20.8 kJ of work could be done on the surroundings per mole of reaction.

Analyze The same procedure as for Hess's law calculations (← **Sec. 4-9**) can be applied to calculate Gibbs free energy changes. If $\Delta_r G^{\circ\prime}$ for a reaction is negative, the process is product-favored. The maximum work that can be done is $-\Delta_r G^{\circ\prime}$.

Plan Write overall reactions that couple two of the reactions for which $\Delta_r G^{\circ\prime}$ values are known. Then, sum the $\Delta_r G^{\circ\prime}$ values to obtain $\Delta_r G^{\circ\prime}$ for the desired reaction.

Execute Use part (a) as an example. (The calculation for part (b) follows a similar procedure.) Because ATP is a reactant in the desired equation, use Reaction (i) as written.

(i) Adenosine triphosphate + $H_2O(\ell)$ ⟶
adenosine diphosphate + dihydrogen phosphate $\quad \Delta_r G^{\circ\prime}(\text{i}) = -30.5$ kJ/mol

Because creatine phosphate is a product in the desired reaction, reverse Reaction (ii) and change the sign of $\Delta_r G^{\circ\prime}$.

Reverse of (ii) creatine + dihydrogen phosphate ⟶
creatine phosphate + $H_2O(\ell)$ $\quad \Delta_r G^{\circ\prime} = -\Delta_r G^{\circ\prime}(\text{ii}) = +43.1$ kJ/mol

The overall reaction is

Adenosine triphosphate + creatine ⟶ adenosine diphosphate + creatine phosphate

$$\Delta_r G^{\circ\prime} = \Delta_r G^{\circ\prime}(\text{i}) - \Delta_r G^{\circ\prime}(\text{ii}) = -30.5 \text{ kJ/mol} + 43.1 \text{ kJ/mol} = 12.6 \text{ kJ/mol}$$

Therefore, the process (a) is reactant-favored, not product-favored, and at least 12.6 kJ/mol would have to be supplied to force it to occur.

PROBLEM-SOLVING PRACTICE 16.12

ATP, creatine phosphate, and glycerol 3-phosphate could be thought of as phosphate donors (just as Brønsted-Lowry acids can be thought of as proton donors).
(a) Which of the three substances is the strongest phosphate donor?
(b) Which is the weakest?
(c) Explain your choices.

EXERCISE 16.14

Recycling of ATP

From the figures given previously for the daily quantity of ATP converted to ADP by an average human and the quantity of ATP and ADP actually present in the body, calculate the number of times each ADP molecule must be recycled to ATP each day.

16-10 Conservation of Gibbs Free Energy

When a ton of coal is burned its energy has not been used up. The law of conservation of energy (← **Sec. 4-2**) summarizes many experiments whose results verify that energy cannot be destroyed. When coal is burned in a power plant, its chemical energy is changed to an equal quantity of energy in other forms. These are electrical energy, which can be very useful, and thermal energy in the immediate surroundings of the plant, especially in the gases going up the smokestack, which is much less useful. However, an *energy resource* (Figure 16.18) has been used up: the coal's ability to store energy and release it

Figure 16.18 Energy resources are like material resources. Like other natural resources, energy resources are high quality and concentrated. An analogy is a material resource such as a diamond, which is valuable because it consists of a beautiful single crystal with carbon atoms arranged in a specific way. If you ground a diamond into dust and spread the dust over the area of a city block, the diamond would be nearly worthless, because it would require tremendous expense to collect the carbon and convert it back to diamond. Similarly, an energy resource is valuable not for the energy it contains, but because that energy is concentrated and available to do useful work.

to do work. When coal burns in air, some of the Gibbs free energy that was in the coal and the oxygen that combined with it has been used up. This fact is indicated by the negative value of $\Delta_r G°$ for the combustion of coal. The same is true of any other product-favored reaction.

What we commonly refer to as **energy conservation** *is actually conservation of useful energy: Gibbs free energy.* Energy conservation does not mean conserving energy—nature takes care of conserving energy automatically. But, nature does not automatically conserve Gibbs free energy. Substances with high Gibbs free energies are energy resources, and it is their *useful* energy that we must take pains to conserve. Once a product-favored reaction with a negative $\Delta_r G°$ has taken place, it cannot be reversed, thereby restoring the Gibbs free energy of its reactants, without coupling the reverse reaction with some other product-favored reaction. That is, once we have used an energy resource, it cannot be restored, except by using some other energy resource. Analysis of chemical systems in terms of Gibbs free energy can lead to important insights into how energy resources can be conserved effectively.

The molecules of a bathtub full of water at room temperature are moving in a variety of ways and therefore have a lot of energy, but that energy cannot easily be used to boil water or to generate electricity. The energy in the water is not useful in the same sense that energy stored in a fuel and oxygen is useful.

By comparing Gibbs free energy changes calculated using the equations in this chapter with the actual loss of Gibbs free energy in industrial processes, environmentalists and industrialists can suggest ways to minimize loss of Gibbs free energy. For example, there is a very large quantity of Gibbs free energy stored in aluminum metal and oxygen gas compared with aluminum oxide, Al_2O_3, from an ore. This can be seen from the thermochemical expression

$$2\ Al_2O_3(s) \longrightarrow 4\ Al(s) + 3\ O_2(g) \qquad \Delta_r G° = 3164.6\ kJ/mol$$

which shows that the Gibbs free energy of 4 mol Al(s) and 3 mol O_2(g) is 3164.6 kJ higher than the Gibbs free energy of 2 mol Al_2O_3(s). If 4 mol Al(s) is oxidized to aluminum oxide, 3164.6 kJ of Gibbs free energy is lost—energy that was expended to manufacture the aluminum from its ore is wasted if the aluminum is oxidized. It is not surprising, then, that major programs for recycling aluminum operate throughout the United States. A similar statement can be made about almost every metal: Once reduced from their ores, metals are storehouses of Gibbs free energy that should be maintained in their reduced forms to avoid repeating the expenditure of Gibbs free energy needed to separate them from chemical combination with oxygen.

16-10a Energy Conservation and Coupled Reactions

The previous section mentioned that in a typical human cell, oxidation of 1 mol glucose to carbon dioxide and water can cause 32 conversions of adenosine diphosphate to adenosine triphosphate (← **Sec. 16-9a**). In Exercise 16.13, you calculated the overall

ESTIMATION | Gibbs Free Energy and Automobile Travel

To illustrate the sizable energy resources consumed by automobile travel, estimate the quantity of Gibbs free energy consumed when a car makes a 1000-mile round trip.

Assume that the typical car averages 25 miles per gallon and that combustion of gasoline can be approximated by combustion of octane. (Gasoline is a mixture of hydrocarbons and most gasoline contains 10% or more ethanol, but we are making an estimate.) Because the trip is a round trip, the car ends up exactly where it started, which means that no useful thermodynamic work has been done. Therefore all of the Gibbs free energy released by combustion of the fuel is lost. The combustion reaction is

$$C_8H_{18}(\ell) + \tfrac{25}{2}\,O_2(g) \longrightarrow 8\,CO_2(g) + 9\,H_2O(\ell) \qquad \Delta_r G° = -5295.74 \text{ kJ/mol}$$

Fuel economy of 25 miles per gallon means that four gallons of fuel are used in 100 miles and 40 gallons in 1000 miles. One gallon is four quarts and a quart is about a liter, so the volume of octane is about 40 gal × 4 L/gal = 160 L. The density of gasoline must be less than that of water, because gasoline floats on water. Assume that it is about 80% as big, that is, 0.8 g/mL or 800 g/L. The 160 L fuel weighs about 160 L × 800 g/L = 128,000 g.

The molar mass of octane, C_8H_{18}, is about 8 × 12 + 18 = 114 g/mol. To make the arithmetic easier, round this value to 100 g/mol. Then 128,000 g octane corresponds to 1280 mol octane. The Gibbs free energy released by combustion is about 5000 kJ for every mole of octane burned. Since 1280 × 5000 = 6,400,000 kJ, more than 6 million kilojoules of useful energy is consumed for every 1000 miles a car is driven. Most of us drive ten times that far every year, and there are a lot of cars in the United States. The energy resources consumed by automobile travel are huge.

change in Gibbs free energy when 1 mol glucose is metabolized and 32 mol ADP is transformed into 32 mol ATP.

$$C_6H_{12}O_6(aq) + 6\,O_2(g) \longrightarrow 6\,CO_2(g) + 6\,H_2O(\ell) \qquad \Delta_r G°' = -2870 \text{ kJ}$$

$$32\,ADP^{3-}(aq) + 32\,H_2PO_4^-(aq) \longrightarrow 32\,ATP^{4-}(aq) + 32\,H_2O(\ell) \qquad \Delta_r G°' = 976.0 \text{ kJ}$$

$$C_6H_{12}O_6(aq) + 6\,O_2(g) + 32\,ADP^{3-}(aq) + 32\,H_2PO_4^-(aq) \longrightarrow 6\,CO_2(g) + 32\,ATP^{4-}(aq) + 38\,H_2O(\ell) \qquad \Delta_r G°' = -1894 \text{ kJ}$$

Some of the Gibbs free energy released when the glucose is oxidized does useful work by causing synthesis of ATP, the energy-storage medium of living cells. However, nearly two thirds of the Gibbs free energy does no useful work and ends up as thermal energy. That is, about two thirds of the original Gibbs free energy is lost, and one third is stored in ATP. The energy-storage process has an efficiency of about 33%, because only about 33% of the Gibbs free energy change actually does useful work.

The more efficient a biochemical process (or an industrial process) is, the less Gibbs free energy is lost and the more energy is used productively. Efficiency is defined as the useful work done per 100 units of energy input. Current energy-conversion systems have a broad range of efficiencies. For example, the generators in a large electrical generating plant convert about 99% of the mechanical energy input to electric energy output; based on the energy of the coal combustion reaction, however, the overall efficiency of such a plant is only about 40%. An incandescent electric light bulb converts only about 5% of the electrical energy it receives into light energy, but a fluorescent light converts four times as much—about 20%. The higher the energy efficiency of the devices we use, the less Gibbs free energy is lost.

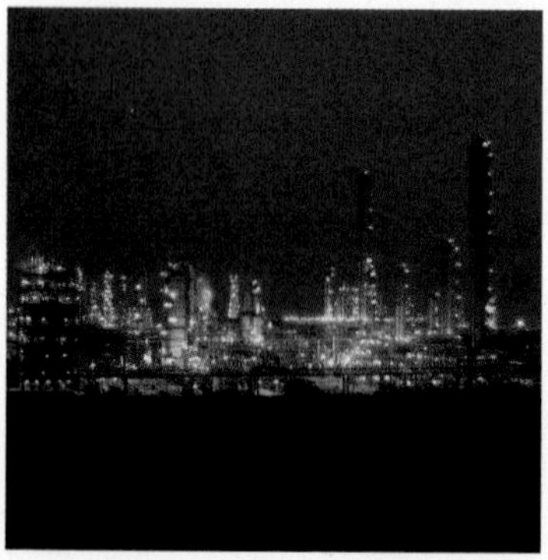

BASF SE

Figure 16.19 A chemical manufacturing plant that produces ethylene. From the ethylene the plant also manufactures other products, such as polyethylene, that are used in many consumer items (← **Sec. 10-6a**).

Like ATP in your body, many compounds can store Gibbs free energy. An example is ethylene. About 50 million pounds of this gas is produced in the United States every year from the dehydrogenation of ethane in chemical plants like the one shown in Figure 16.19.

$$C_2H_6(g) \longrightarrow H_2(g) + C_2H_4(g) \qquad \Delta_r G° = 100.97 \text{ kJ/mol}$$

When a mole of hydrogen and a mole of ethylene are produced from a mole of ethane, at least 100.97 kJ of Gibbs free energy must be supplied from an external source. This Gibbs free energy is then stored in the hydrogen and ethylene. Ethylene production is the largest single consumer of Gibbs free energy in the chemical industry, so there has been great interest in improving the process to save energy and money. Since 1960 the Gibbs free energy requirement per pound of ethylene produced has declined by 60%. Even so, the energy used to make 1 mol ethylene from 1 mol ethane (about 400 kJ) is four times the minimum required (100.97 kJ) because manufacture of ethylene is only 25% efficient. This is largely due to inefficiencies in energy transfer from external sources to the reaction system.

It is important to recognize that completely eliminating consumption of Gibbs free energy is impossible. Whenever anything happens, whether a chemical reaction or a physical process, the final state must have less Gibbs free energy than was available initially. This is the same as saying that the entropy of the universe must have increased during the change. This statement is true of any system in which the initial substances are changed into something new—any product-favored system. Thus, losses of Gibbs free energy are inevitable. The aim of energy conservation is to minimize—not eliminate—them. This can be done by maximizing the efficiency of coupling exergonic reactions to endergonic processes we want to cause to occur. The ideas of thermodynamics help us figure out how to accomplish that goal and are the most powerful tool we have for conserving energy while maintaining a high standard of living.

16-11 Thermodynamic and Kinetic Stability

Chemists often say that substances are "stable," but what exactly does that statement mean? Usually it means that the substance in question does not decompose or react with other substances that normally come in contact with it. Most chemists, for example, would say that the aluminum can that holds the soda you drink is stable. It will be around for quite a long time. The fact that aluminum cans do not decompose rapidly is one of the reasons you are encouraged to recycle them instead of throwing them away. Some aluminum cans have emerged almost unchanged from landfills after 40 or 50 years.

©Dmitry Bruskov/Shutterstock.com

Iron displays kinetic stability. When exposed to the oxygen in air, iron rusts, iron is not thermodynamically stable under these conditions. The many iron objects we use—tools, cars, bridges, pipes—all will eventually rust completely, but the reaction rate is slow enough that we can use them for many years.

Strictly speaking, there are two kinds of stability. We discussed one of them earlier in this chapter. A substance is *thermodynamically stable* if it does not undergo product-favored reactions. Such reactions disperse energy and increase the entropy of their surroundings. Although we just said it was stable, the aluminum in a soda can is actually *thermodynamically unstable* compared with its oxide, because its reaction with oxygen in air has a negative Gibbs free energy change.

$$4\ Al(s) + 3\ O_2(g) \longrightarrow 2\ Al_2O_3(s) \qquad \Delta_r G^\circ = -3164.6\ kJ/mol$$

The aluminum exhibits a different kind of stability—it is *kinetically stable.* Although it has the potential to undergo a product-favored oxidation reaction, this reaction proceeds so slowly that the can remains essentially unchanged for a long time. This happens because a thin coating of aluminum oxide forms on the surface of the aluminum and prevents oxygen from reaching the rest of the aluminum atoms below the can's surface. If we grind the aluminum into a fine powder and throw it into a flame, the powder burns and the evolved heat leads to an entropy increase in the little piece of the universe around the burning metal. The reason an aluminum can is stable (does not oxidize away) is that the oxidation is slow (kinetics), not that formation of the oxide would not occur of its own accord (thermodynamics).

Another substance that is *thermodynamically unstable* but *kinetically stable* is diamond. If you look up the data in Appendix J, you will find that the conversion of diamond to graphite has a negative Gibbs free energy change; therefore that conversion is product-favored. But, diamonds don't change into graphite. Engagement rings contain diamonds precisely because the diamond (like the love it represents) is expected to last for a long time. It does so because there is a very high activation energy barrier (← **Sec. 11-4a**) for the change from the diamond structure to the graphite structure. When a chemist says something is stable, it usually means that it is kinetically stable—only an activation energy barrier prevents it from reacting fast enough for us to see a change.

PROBLEM-SOLVING EXAMPLE 16.13

Thermodynamic and Kinetic Stability

Whenever air is heated to a very high temperature, the reaction between nitrogen and oxygen to form nitrogen monoxide occurs. It is an important source of nitrogen-containing air pollutants that can be formed in the cylinders of an automobile engine.

(a) Write a balanced equation with minimum whole-number coefficients for the equilibrium reaction of N_2 with O_2 to form NO.

(b) Is this reaction product-favored at room temperature? That is, is NO thermodynamically stable compared to N_2 and O_2?

(c) Estimate the temperature at which the standard equilibrium constant for this reaction equals 1.

(d) If NO is formed at high temperature in an automobile engine, why does it not all change back to N_2 and O_2 when the mixture of gases enters the exhaust system and its temperature falls?

(e) How might the concentration of NO in automobile exhaust be decreased?

Result

(a) $N_2(g) + O_2(g) \rightleftharpoons 2\ NO(g)$ (b) No (c) 7286 K

(d) The reverse reaction is too slow. (e) Use a suitable catalyst.

Analyze From a balanced equation $\Delta_r G°$ can be calculated. If $K°$ is known $\Delta_r G°$ can be calculated. $\Delta_r G°$ as a function of T can be estimated from $\Delta_r G° = \Delta_r H° - T\Delta_r S°$, and $\Delta_r H°$ and $\Delta_r S°$ do not vary much with T. Therefore, T can be estimated from $\Delta_r G°$. Chemical kinetics can be used to explain kinetic stability; chemical thermodynamics can be used to explain thermodynamic stability.

Plan Apply the analysis to each part of the problem.

Execute

(a) See Result.

(b) Calculate $\Delta_r G°$ at 25 °C using data from Appendix J.
$\Delta_r G° = 2\{\Delta_f G°\ [NO(g)]\} = 2(86.55\ kJ/mol) = 173.10\ kJ/mol.$
Because $\Delta_r G° > 0$, the reaction is not product-favored.

(c) If $K° = 1$, then $\Delta_r G° = -RT \ln K° = -RT \ln(1) = 0$.
Because $\Delta_r G° = \Delta_r H° - T\Delta_r S° = 0$, $\Delta_r H° = T\Delta_r S°$ and $T = \Delta_r H°/\Delta_r S°$.
Using data from Appendix J gives

$$\Delta_r H° = 2\{\Delta_f H°\ [NO(g)]\} = 2(90.25\ kJ/mol) = 180.50\ kJ/mol,\ \text{and}$$

$$\begin{aligned}\Delta_r S° &= 2\{S°[NO(g)]\} - \{S°[N_2(g)] + S°[O_2(g)]\}\\ &= 2(210.76\ J\ K^{-1}\ mol^{-1}) - 191.61\ J\ K^{-1}\ mol^{-1} - 205.138\ J\ K^{-1}\ mol^{-1}\\ &= 24.772\ J\ K^{-1}\ mol^{-1}\end{aligned}$$

$$\begin{aligned}\text{Therefore } T &= (180.50\ kJ/mol)/(24.772\ J\ K^{-1}\ mol^{-1})\\ &= (180{,}500\ J/mol)/(24.772\ J\ K^{-1}\ mol^{-1}) = 7286\ K\end{aligned}$$

(d) When the mixture of gases, which contains some NO as well as N_2 and O_2, leaves the cylinders of the automobile engine and enters the exhaust system, the gas mixture cools very rapidly to a temperature below 500 K. The reverse reaction should occur, according to thermodynamics, but it does not. The activation energy for the decomposition of NO is quite high, because NO contains a double bond, and it is very difficult to separate the two atoms (which must be done to form N_2 and O_2). Therefore, the reaction rate is greatly affected by temperature. At low temperatures the reverse reaction is very slow, so significant concentrations of NO exist in automobile exhaust, even though N_2 and O_2 are thermodynamically stable compared to NO.

(e) With a suitable catalyst, decomposition of NO to its elements can take place at appreciable rates even at relatively low temperatures. Catalytic converters are installed in the exhaust systems of cars partly to reduce the concentration of NO. Because N_2 and O_2 are

Automobile emissions. NO emitted from automobile exhausts is a major cause of air pollution (← **Sec. 8-12b**).

more stable thermodynamically, it is reasonable to use a catalyst to speed up their formation.

☑ **Reasonable Result Check** It is reasonable that $\Delta_r G°$ for the reaction of N_2 with O_2 is positive, because N_2 and O_2 are the principal components of the atmosphere where their partial pressures are close to standard pressure and they do not react with each other. It is reasonable that $\Delta_r S°$ for the reaction is small and positive. The total number of gas-phase molecules does not change, but the product molecules have two different atoms, and both reactant molecules have two atoms that are the same, making the product molecules slightly more probable. It is reasonable that the reaction is endothermic, because the reactant molecules have a triple bond and a double bond, and the product molecules have two double bonds. The bonds broken are therefore expected to be stronger than the bonds formed.

PROBLEM-SOLVING PRACTICE 16.13

All of these substances are stable with respect to decomposition to their elements at 25 °C. Which are kinetically stable and which are thermodynamically stable?

(a) $MgO(s)$ (b) $N_2H_4(\ell)$ (c) $C_2H_6(g)$ (d) $N_2O(g)$

Finally, think about whether you yourself are stable (thermodynamically or kinetically). From a thermodynamic standpoint, most of the substances you are made of are unstable with respect to oxidation to carbon dioxide, water, and other substances. That is, based on Gibbs free energy changes, most of the substances that you are made of should undergo product-favored reactions that would completely destroy them. Your proteins, fats, carbohydrates, and even DNA should spontaneously change into much smaller, simpler molecules. Fortunately for you, the reactions by which this change could happen are very slow at room temperature and body temperature. Only when enzymes catalyze those reactions do they occur with reasonable speed. It is the combination of thermodynamic instability and kinetic stability that allows those enzymes to control the reactions in your body or in any living organism. Were it not for the kinetic stability of a wide variety of substances, everything would be quickly converted to a small number of very thermodynamically stable substances. Life and the environment as we know them would then be impossible.

Another example of kinetic versus thermodynamic stability is gasoline and other fuels. We want a fuel to be thermodynamically unstable compared with CO_2 and H_2O, but kinetically stable so that it will not burn unless we want it to burn.

The roles of chemical thermodynamics and chemical kinetics in determining chemical reactivity can be summarized this way:

- ***Thermodynamics tells whether a reaction can produce predominantly products under standard conditions and, if it does, how much useful work can be accomplished by coupling the reaction to another process.***
- ***Thermodynamics tells how to calculate the standard equilibrium constant and allows quantitative prediction of how much product is formed for reactions that involve dilution of substances in the gas phase or in solution.***
- ***Thermodynamics can also be used to predict what happens under nonstandard conditions.***
- ***Kinetics tells how fast a given reaction goes and indicates how we can control the rate of reaction.***

Together, thermodynamics and kinetics provide the intellectual foundation on which modern chemical industries are based and the principles that are the basis for our fundamental understanding of physiology and medicine.

SUMMARY PROBLEM

In a blast furnace for making iron from iron ore, large quantities of coke (which is mainly carbon) are dumped into the top of the furnace along with iron ore (which can be assumed to be Fe_2O_3) and limestone (which is used to help remove impurities from the iron). The overall process is

$$2\,Fe_2O_3(s) + 3\,C(s) \longrightarrow 4\,Fe(s) + 3\,CO_2(g)$$

This reaction can be thought of as a combination of several individual steps.

$$2\,Fe_2O_3(s) \longrightarrow 4\,FeO(s) + O_2(g)$$
$$2\,FeO(s) \longrightarrow 2\,Fe(s) + O_2(g)$$
$$2\,C(s) + O_2(g) \longrightarrow 2\,CO(g)$$
$$2\,CO(g) + O_2(g) \longrightarrow 2\,CO_2(g)$$

(a) Calculate the enthalpy change for each step, assuming a temperature of 25 °C. Which steps are exothermic and which are endothermic?

(b) Based on the equations, predict which of the individual steps involves an increase and which a decrease in the entropy of the system.

(c) Based on your results in parts (a) and (b), what can you say about whether each step is reactant-favored or product-favored at room temperature? At a much higher temperature (>1000 K)?

(d) Calculate the entropy change and the Gibbs free energy change for each reaction step, assuming a temperature of 25 °C.

(e) Keeping in mind the equation $\Delta_r G° = \Delta_r H° - T\Delta_r S°$ and the fact that the enthalpy change and entropy change for a reaction do not vary much with temperature, what is the slope of a graph of $\Delta_r G°$ versus T for each of the reactions? For which of the reactions does $\Delta_r G°$ become more negative as the temperature increases? For which does it become more positive? Does this agree with what you predicted in part (c)?

(f) For which of these reactions might the assumption of nearly constant $\Delta_r H°$ and $\Delta_r S°$ not be valid as the temperature increases from 25 °C? For each reaction you choose, explain why the assumption might not be correct.

(g) Use your results from previous parts of this problem to estimate the Gibbs free energy change for each reaction at a temperature of 1500 K.

(h) Which of the two iron oxides is more easily reduced at 1500 K? Which of the reactions involving carbon compounds is more product-favored at 1500 K? What chemical reactions are taking place in the hottest part of the blast furnace?

(i) In portions of the furnace where the temperature is about 800 K, would the same reactions be occurring as in the higher-temperature part of the furnace? Why or why not?

(j) Show that the individual steps can be combined to give the overall reaction. From the enthalpy, entropy, and Gibbs free energy changes already calculated, calculate these changes for the overall reaction.

(k) In a typical blast furnace every kilogram of iron produced requires 2.5 kg iron ore, 1 kg coke, and nearly 6 kg air (to provide oxygen for oxidation of the coke to heat the furnace and to combine with carbon in the coke, forming CO(g)). How much Gibbs free energy would be destroyed if the coke were simply burned to form carbon dioxide? Given the quantity of iron produced in a typical furnace, determine how much Gibbs free energy is stored by coupling the oxidation of coke to the reduction of iron oxides. Calculate what percentage of the Gibbs free energy available from combustion of coke is wasted per kilogram of iron produced.

HAVING STUDIED THIS CHAPTER . . .

. . . you should be able to:

- Understand and apply appropriately the terms "product-favored" and "reactant-favored" (Section 16-1). End-of-chapter Question 13
- Explain why there is a higher probability that energy will be dispersed than that it will be concentrated in a small number of nanoscale particles (Section 16-2). Questions 15, 17
- Use qualitative rules to predict the sign of the entropy change for a process (Section 16-3). Questions 19, 23, 27
- Calculate the entropy change for a process occurring at constant temperature (Section 16-3). Questions 31, 33
- Calculate the entropy change for a chemical reaction, given standard entropy values for elements and compounds (Section 16-4). Questions 35, 37
- Use entropy and enthalpy changes to predict whether a reaction is product-favored (Section 16-5). Questions 41, 43, 49, 51
- Describe the connection between enthalpy and entropy changes for a reaction and the Gibbs free energy change; use this relation to estimate quantitatively how temperature affects whether a reaction is product-favored (Section 16-6). Questions 53, 55, 59, 61
- Calculate the Gibbs free energy change for a reaction from values given in a table of standard molar Gibbs free energies of formation (Section 16-6). Questions 57, 63, 65
- Relate the standard Gibbs free energy change and the standard equilibrium constant for the same reaction and be able to calculate one from the other (Section 16-7). Questions 67, 69, 73
- Describe how a reactant-favored system can be coupled to a product-favored system so that a desired reaction can be carried out (Section 16-8). Questions 75, 77, 80
- Explain how biological systems make use of coupled reactions to maintain the high degree of order found in all living organisms; give examples of coupled reactions that are important in biochemistry (Section 16-9). Questions 82, 84
- Explain the relationship between Gibbs free energy and energy conservation (Sections 16-8 and 16-10). Question 86
- Distinguish between thermodynamic stability and kinetic stability and describe the effect of each on whether a reaction is useful in producing products (Section 16-11). Questions 88, 90

KEY TERMS

endergonic (Section 16-9a)

energy conservation (16-10)

entropy (16-3)

exergonic (16-9a)

extent of reaction (16-7a)

Gibbs free energy (16-6)

metabolism (16-9a)

nutrients (16-9a)

photosynthesis (16-9b)

product-favored (16-1)

reactant-favored (16-1)

reversible process (16-3)

second law of thermodynamics (16-5)

standard equilibrium constant ($K°$) (16-7b)

third law of thermodynamics (16-3a)

QUESTIONS FOR REVIEW AND THOUGHT

Red-numbered questions have short answers at the back of this book in Appendix M and fully worked solutions in the *Student Solutions Manual.*

Review Questions

These questions test vocabulary and simple concepts.

1. Define the terms "product-favored system" and "reactant-favored system." Give one example of each.
2. What are the two ways that a final chemical state of a system can be more probable than its initial state?
3. Define the term "entropy," and give an example of a sample of matter that has zero entropy. What are the units of entropy? How do they differ from the units of enthalpy?
4. State five useful qualitative rules for predicting entropy changes when chemical or physical changes occur.
5. State the second law of thermodynamics.
6. In terms of values of $\Delta_r H^\circ$ and $\Delta_r S^\circ$, under what conditions can you be sure that a reaction is product-favored? When can you be sure that it is not product-favored?
7. Define the Gibbs free energy change of a chemical reaction in terms of its enthalpy and entropy changes. Why is the Gibbs free energy change especially useful in predicting whether a reaction is product-favored?
8. Why are materials whose reactions release large quantities of Gibbs free energy useful to society? Give two examples of such materials.
9. Define the terms "endergonic" and "exergonic."
10. Define these important biochemistry terms: metabolism, nutrients, ATP, ADP, coupled reactions, photosynthesis.
11. Describe two ways to cause reactant-favored reactions to form products.
12. Describe the process by which sunlight is employed to convert high-entropy, low-Gibbs-free-energy substances into low-entropy, high-Gibbs-free-energy substances.

Topical Questions

These questions are keyed to major topics in the chapter. Usually a question that is answered at the back of this book is paired with a similar one that is not.

Reactant-Favored and Product-Favored Processes (Section 16-1)

13. For each process, write a chemical equation and classify the process as reactant-favored or product-favored.
    (a) Water decomposes to its elements, hydrogen and oxygen.
    (b) Gasoline spilled on the ground evaporates (use octane, C_8H_{18}, to represent gasoline).
    (c) Sugar dissolves in water at room temperature.
14. For each process, write a chemical equation and classify the process as reactant-favored or product-favored.
    (a) Carbon dioxide gas decomposes to its elements, carbon and oxygen.
    (b) The steel (mostly iron) body of an automobile rusts.
    (c) Gasoline reacts with oxygen to form carbon dioxide and water (use octane, C_8H_{18}, to represent gasoline).

Chemical Reactions and Dispersal of Energy (Section 16-2)

15. Suppose you flip a coin.
    (a) What is the probability that the coin will come up heads?
    (b) What is the probability that it will come up tails?
    (c) If you flip the coin 100 times, what is the most likely number of heads and tails you will see?
16. Suppose you make a tetrahedron and put numbers 1, 2, 3, and 4 on each of the four sides. You toss the tetrahedron in the air and observe it after it comes to rest.
    (a) What is the probability that the tetrahedron will come to rest with the numbers 2, 3, and 4 visible?
    (b) What is the probability that the tetrahedron will come to rest with the numbers 1, 2, and 3 visible?
    (c) If you toss the tetrahedron 100 times, what is the most likely number of times you will see a 1 after it comes to rest?
17. Consider two equal-sized flasks connected as shown in the figure.

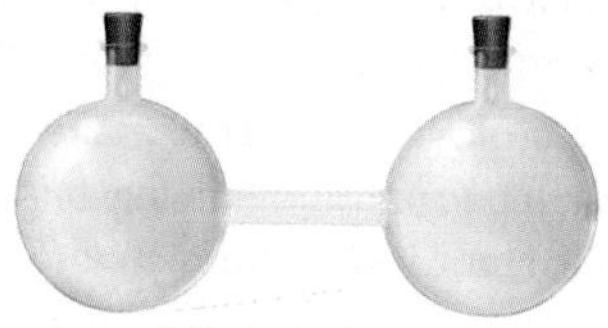

    (a) Suppose you put one molecule inside. What is the probability that the molecule will be in flask A? What is the probability that it will be in flask B?
    (b) If you put 100 molecules into the two-flask system, what is the most likely arrangement of molecules? Which arrangement has the highest entropy?
18. Suppose you have four identical molecules labeled 1, 2, 3, and 4. Draw 16 simple two-flask diagrams as in the figure for Question 17, and draw all possible arrangements of the four molecules in the two flasks. How many of these arrangements have two molecules in each flask? How many have no molecules in one flask? From these results, what is the most probable arrangement of molecules? Which arrangement has the highest entropy?

Measuring Dispersal of Energy: Entropy (Section 16-3)

19. For each process, tell whether the entropy change of the system is positive or negative.
    (a) Water vapor (the system) deposits as ice crystals on a cold windowpane.
    (b) A can of carbonated beverage loses its fizz. (Consider the beverage but not the can as the system. What happens to the entropy of the dissolved gas?)
    (c) A glassblower heats glass (the system) to its softening temperature.
20. For each process, tell whether the entropy change of the system is positive or negative.
    (a) Water boils.
    (b) A teaspoon of sugar dissolves in a cup of coffee. (The system consists of both sugar and coffee.)
    (c) Calcium carbonate precipitates out of water in a cave to form stalactites and stalagmites. (Consider only the calcium carbonate to be the system.)
21. For each situation described in Question 13, predict whether the entropy of the system increases or decreases.
22. For each situation described in Question 14, predict whether the entropy of the system increases or decreases.
23. For each pair of items, predict which has the higher entropy, and explain why.
    (a) Item 1, a sample of solid CO_2 at −78 °C, or item 2, CO_2 vapor at 0 °C
    (b) Item 1, solid sugar, or item 2, the same sugar dissolved in a cup of tea
    (c) Item 1, a 100-mL sample of pure water and a 100-mL sample of pure alcohol, or item 2, the same samples of water and alcohol after they had been poured together and stirred
24. For each pair of items, predict which has the higher entropy, and explain why.
    (a) Item 1, a sample of pure silicon (to be used in a computer chip), or item 2, a piece of silicon having the same mass but containing a trace of some other element, such as B or P
    (b) Item 1, an ice cube at 0 °C, or item 2, the same mass of liquid water at 0 °C
    (c) Item 1, a sample of pure I_2 solid at room temperature, or item 2, the same mass of iodine vapor at room temperature
25. Comparing the formulas or states for each pair of substances, predict which has the higher entropy per mole at the same temperature, and explain why.
    (a) NaCl(s) or CaO(s)
    (b) Cl_2(g) or P_4(g)
    (c) NH_4NO_3(s) or NH_4NO_3(aq)
26. From each pair of substances, select the one having the larger standard molar entropy at 25 °C. Give reasons for your choice.
    (a) Ga(s) or Ga(ℓ)
    (b) AsH_3(g) or Kr(g)
    (c) NaF(s) or MgO(s)
27. Without doing a calculation, predict whether the entropy change is positive or negative when each reaction occurs in the direction it is written.
    (a) $C_2H_4(g) + H_2(g) \longrightarrow C_2H_6(g)$
    (b) $CH_3OH(\ell) + \frac{3}{2}\,O_2(g) \longrightarrow CO_2(g) + 2\,H_2O(g)$
    (c) $N_2(g) + 3\,H_2(g) \longrightarrow 2\,NH_3(g)$
    (d) $CaCO_3(s) \longrightarrow CaO(s) + CO_2(g)$
28. Without doing a calculation, predict whether the entropy change is positive or negative when each reaction occurs in the direction it is written.
    (a) $CH_3OH(\ell) \longrightarrow CO(g) + 2\,H_2(g)$
    (b) $Br_2(\ell) + H_2(g) \longrightarrow 2\,HBr(g)$
    (c) $C_3H_8(g) \longrightarrow C_2H_4(g) + CH_4(g)$
    (d) $Ag^+(aq) + I^-(aq) \longrightarrow AgI(s)$
29. Without consulting a table of standard molar entropies, predict whether $\Delta_r S^\circ_{system}$ is positive or negative for each of these reactions.
    (a) $2\,CO(g) + O_2(g) \longrightarrow 2\,CO_2(g)$
    (b) $2\,H_2(g) + O_2(g) \longrightarrow 2\,H_2O(\ell)$
    (c) $2\,O_3(g) \longrightarrow 3\,O_2(g)$
30. Without consulting a table of standard molar entropies, predict whether $\Delta_r S^\circ_{system}$ is positive or negative for each of these reactions.
    (a) $2\,NH_3(g) \longrightarrow N_2(g) + 3\,H_2(g)$
    (b) $2\,Na(s) + Cl_2(g) \longrightarrow 2\,NaCl(s)$
    (c) $H_2(g) + I_2(s) \longrightarrow 2\,HI(g)$
31. Calculate the entropy change, $\Delta_r S^\circ$, for the vaporization of ethanol, C_2H_5OH, at the boiling point of 78.3 °C. The heat of vaporization of the alcohol is 39.3 kJ/mol.

$$C_2H_5OH(\ell) \longrightarrow C_2H_5OH(g) \qquad \Delta_r S^\circ = ?$$

32. Diethyl ether, $(C_2H_5)_2O$, was once used as an anesthetic. Calculate the entropy change, $\Delta_r S^\circ$, for the vaporization of ether if its heat of vaporization is 26.0 kJ/mol at the boiling point of 35.0 °C.
33. Calculate $\Delta_r S^\circ$ for each substance when the quantity of thermal energy indicated is transferred reversibly to the system at the temperature specified. Assume that you have enough of each substance so that its temperature remains constant as the thermal energy is transferred.
    (a) H_2(g), 0.775 kJ/mol, 295 K
    (b) KCl(s), 500. kJ/mol, 500. K
    (c) N_2(g), 2.45 kJ/mol, 1000. K
34. Calculate $\Delta_r S^\circ$ for each of these substances when the quantity of thermal energy indicated is transferred reversibly to the system at the temperature specified. Assume that you have enough of each substance so that its temperature remains constant as the thermal energy is transferred.
    (a) NaCl(s), 5.00 kJ/mol, 500. K
    (b) N_2O(g), 0.30 kJ/mol, 300. K

Calculating Entropy Changes (Section 16-4)

35. Check your predictions in Question 27 by calculating the entropy change for each reaction. Standard entropies not in Table 16.1 can be found in Appendix J.
36. Check your predictions in Question 28 by calculating the entropy change for each reaction. Standard entropies not in Table 16.1 can be found in Appendix J.
37. Check your predictions in Question 29 by calculating the entropy change for each reaction. Standard entropies not in Table 16.1 can be found in Appendix J.
38. Check your predictions in Question 30 by calculating the entropy change for each reaction. Standard entropies not in Table 16.1 can be found in Appendix J.

Entropy and the Second Law of Thermodynamics (Section 16-5)

39. Calculate $\Delta_r S^\circ_{system}$ at 25 °C for the reaction

$$C_2H_4(g) + H_2O(g) \longrightarrow C_2H_5OH(\ell)$$

Can you tell from the result of this calculation whether this reaction is product-favored? If you cannot tell, what additional information do you need? Obtain the needed information and determine whether the reaction is product-favored.

40. Calculate $\Delta_r S^\circ_{system}$ at 25 °C for the reaction

$$C_6H_6(\ell) + 4\ H_2(g) \longrightarrow C_6H_{14}(\ell)$$

Can you tell from the result of this calculation whether this reaction is product-favored? If you cannot tell, what additional information do you need? Obtain the needed information and determine whether the reaction is product-favored.

41. Is this reaction predicted to favor the products at low temperatures, at high temperatures, or both? Explain your answer briefly.

$$Mg(s) + \tfrac{1}{2}\ O_2(g) \longrightarrow MgO(s) \qquad \Delta_r H^\circ = -601.70 \text{ kJ/mol}$$

42. Is this reaction predicted to favor the products at low temperatures, at high temperatures, or both? Explain your answer briefly.

$$MgCO_3(s) \longrightarrow MgO(s) + CO_2(g) \quad \Delta_r H^\circ = 100.59 \text{ kJ/mol}$$

43. Explain briefly why the exothermic combustion of propane is product-favored.

$$C_3H_8(g) + 5\ O_2(g) \longrightarrow 3\ CO_2(g) + 4\ H_2O(g)$$

44. Explain briefly why the exothermic reaction of a metal carbonate with an acid is product-favored.

$$CuCO_3(s) + H_2SO_4(aq) \longrightarrow CuSO_4(aq) + CO_2(g) + H_2O(\ell)$$

45. Sodium reacts violently with water according to the equation

$$Na(s) + H_2O(\ell) \longrightarrow NaOH(aq) + \tfrac{1}{2}\ H_2(g)$$

(a) Predict the signs of $\Delta_r H^\circ$ and $\Delta_r S^\circ$ for the reaction.
(b) Verify your predictions with calculations.

46. Once ignited, magnesium reacts vigorously with oxygen in air according to the equation

$$2\ Mg(s) + O_2(g) \longrightarrow 2\ MgO(s)$$

(a) Predict the signs of $\Delta_r H^\circ$ and $\Delta_r S^\circ$ for the reaction.
(b) Verify your predictions with calculations.

47. Hydrogen burns in air with considerable heat transfer to the surroundings. Consider the decomposition of water to gaseous hydrogen and oxygen. Without doing any calculations, and basing your prediction on the enthalpy change and the entropy change, is this reaction product-favored at 25 °C? Explain your answer briefly.
48. Hydrogen gas combines with chlorine gas in an exothermic reaction to form HCl(g). Consider the decomposition of gaseous hydrogen chloride to hydrogen and chlorine. Without doing any calculations, and basing your prediction on the enthalpy change and the entropy change, is this reaction product-favored at 25 °C? Explain your answer briefly.
49. For each reaction, calculate $\Delta_r H^\circ$ and $\Delta_r S^\circ$ and predict whether the reaction is always product-favored, product-favored only at low temperatures, product-favored only at high temperatures, or never product-favored.
(a) $Fe_2O_3(s) + 2\ Al(s) \longrightarrow 2\ Fe(s) + Al_2O_3(s)$
(b) $N_2(g) + 2\ O_2(g) \longrightarrow 2\ NO_2(g)$
50. For each reaction, calculate $\Delta_r H^\circ$ and $\Delta_r S^\circ$ and predict whether the reaction is always product-favored, product-favored only at low temperatures, product-favored only at high temperatures, or never product-favored.
(a) $C_6H_{12}O_6(s) + 6\ O_2(g) \longrightarrow 6\ CO_2(g) + 6\ H_2O(\ell)$
(b) $MgO(s) + C(s, graphite) \longrightarrow Mg(s) + CO(g)$
51. Determine whether the combustion of ethane, C_2H_6, is product-favored at 25 °C.

$$C_2H_6(g) + \tfrac{7}{2}\ O_2(g) \longrightarrow 2\ CO_2(g) + 3\ H_2O(\ell)$$

(a) Calculate $\Delta_r S_{universe}$. Required values of $\Delta_f H^\circ$ and S° are in Appendix J.
(b) Verify your result by calculating the value of $\Delta_r G^\circ$ for the reaction.
(c) Do your calculated answers in parts (a) and (b) agree with your preconceived idea of this reaction?

52. The reaction of magnesium with water can be used as a means for heating food.

$$Mg(s) + 2\ H_2O(\ell) \longrightarrow Mg(OH)_2(s) + H_2(g)$$

Determine whether this reaction is product-favored at 25 °C.
(a) Calculate $\Delta_r S_{universe}$. See Appendix J for the needed data.
(b) Verify your result by calculating $\Delta_r G^\circ$ for the reaction.

Gibbs Free Energy (Section 16-6)

53. Use a mathematical equation to show how the statement leads to the conclusion cited: If a reaction is exothermic (negative $\Delta_r H$) and if the entropy of the system increases (positive $\Delta_r S$), then $\Delta_r G$ must be negative, and the reaction is product-favored.

54. Use a mathematical equation to show how the statement leads to the conclusion cited: If $\Delta_r H$ and $\Delta_r S$ have the same sign, then the magnitude of T determines whether $\Delta_r G$ is negative and whether the reaction is product-favored.
55. Predict whether the reaction given is product-favored or reactant-favored by calculating $\Delta_r G°$ from the entropy and enthalpy changes for the reaction at 25 °C.

$$H_2(g) + CO_2(g) \longrightarrow H_2O(g) + CO(g)$$
$$\Delta_r H° = 41.17 \text{ kJ/mol} \qquad \Delta_r S° = 42.08 \text{ J K}^{-1} \text{ mol}^{-1}$$

56. Predict whether this reaction is product-favored at 25 °C by calculating the change in standard Gibbs free energy from the entropy and enthalpy changes.

$$H_2(g) + I_2(g) \rightleftharpoons 2\ HI(g)$$
$$\Delta_r H° = 52.96 \text{ kJ/mol} \qquad \Delta_r S° = 21.81 \text{ J K}^{-1} \text{ mol}^{-1}$$

57. If this reaction were product-favored, it would be a good way to make pure silicon, crucial in the semiconductor industry, from sand (SiO_2).

$$SiO_2(s) + C(s) \longrightarrow Si(s) + CO_2(g)$$

Calculate $\Delta_r G°$ from data in Appendix J and decide whether the reaction is a good choice to produce silicon at 25 °C.

58. From data in Appendix J, calculate $\Delta_r G°$ for the reactions of sand with hydrogen fluoride and hydrogen chloride. Explain why hydrogen fluoride attacks glass, whereas hydrogen chloride does not.

$$SiO_2(s) + 4\ HF(g) \longrightarrow SiF_4(g) + 2\ H_2O(g)$$
$$SiO_2(s) + 4\ HCl(g) \longrightarrow SiCl_4(g) + 2\ H_2O(g)$$

59. If a system falls within the second or third category in Table 16.2 (← **Sec. 16-5**), then there must be a temperature at which it shifts from being reactant-favored to being product-favored. For each reaction, obtain data from Appendix J and calculate what that temperature is.
    (a) $CO(g) + 2\ H_2(g) \rightleftharpoons CH_3OH(\ell)$
    (b) $2\ Fe_2O_3(s) + 3\ C(s, \text{graphite}) \rightleftharpoons 4\ Fe(s) + 3\ CO_2(g)$
60. If a system falls within the second or third category in Table 16.2 (← **Sec. 16-5**) then there must be a temperature at which it shifts from being reactant-favored to being product-favored. For each reaction, obtain data from Appendix J and calculate what that temperature is.
    (a) $2\ H_2O(g) \rightleftharpoons 2\ H_2(g) + O_2(g)$
    (b) $2\ Ag_2O(s) \longrightarrow 4\ Ag(s) + O_2(g)$
61. Estimate $\Delta_r G°$ at 2000. K for each reaction in Question 59.
62. Estimate $\Delta_r G°$ at 2000. K for each reaction in Question 60.
63. Many metal carbonates can be decomposed to the metal oxide and carbon dioxide by heating.

$$CaCO_3(s) \longrightarrow CaO(s) + CO_2(g)$$

    (a) Calculate the enthalpy, entropy, and Gibbs free energy changes for this reaction at 25.00 °C.
    (b) Is it product-favored or reactant-favored?
    (c) Based on the signs of $\Delta_r H°$ and $\Delta_r S°$, predict whether the reaction is product-favored at all temperatures.
    (d) Predict the lowest temperature at which appreciable quantities of products can be obtained.
64. Some metal oxides, such as lead(II) oxide, can be decomposed to the metal and oxygen simply by heating.

$$PbO(s) \longrightarrow Pb(s) + \tfrac{1}{2}\ O_2(g)$$

    (a) Is the decomposition of lead(II) oxide product-favored at 25 °C? Explain.
    (b) If not, can it become so if the temperature is raised?
    (c) As the temperature increases, at what temperature does the reaction first become product-favored?
65. Use the experimentally determined thermochemical expression

$$CaC_2(s) + 2\ H_2O(\ell) \longrightarrow C_2H_2(g) + Ca(OH)_2(aq) \qquad \Delta_r G° = -119.282 \text{ kJ/mol}$$

and data from Appendix J to calculate $\Delta_f G°$ for $Ca(OH)_2(aq)$ at 25 °C. Compare your result with the value in Appendix J.

66. Use the thermochemical expression

$$PCl_3(g) + Cl_2(g) \longrightarrow PCl_5(g) \qquad \Delta_r G° = -37.2 \text{ kJ/mol}$$

and data from Appendix J to calculate $\Delta_f G°$ for $PCl_5(g)$.

Gibbs Free Energy Changes and Equilibrium Constants (Section 16-7)

67. Use data from Appendix J to obtain the equilibrium constant K_P for each reaction at 298.15 K.
    (a) $2\ HCl(g) \rightleftharpoons H_2(g) + Cl_2(g)$
    (b) $N_2(g) + O_2(g) \rightleftharpoons 2\ NO(g)$
68. Use data from Appendix J to obtain the equilibrium constant K_P for each of these reactions at 298 K.
    (a) $CH_4(g) + 2\ O_2(g) \rightleftharpoons CO_2(g) + 2\ H_2O(g)$
    (b) $2\ NO_2(g) \rightleftharpoons N_2O_4(g)$
69. Ethylene reacts with hydrogen to produce ethane.

$$H_2C{=}CH_2(g) + H_2(g) \rightleftharpoons H_3C{-}CH_3(g)$$

    (a) Using the data in Appendix J, calculate $\Delta_r G°$ for the reaction at 25 °C. Is the reaction product-favored under standard conditions?
    (b) Calculate K_P from $\Delta_r G°$. Comment on the connection between the sign of $\Delta_r G°$ and the magnitude of K_P.
70. Use the data in Appendix J to calculate $\Delta_r G°$ and K_P at 25 °C for the reaction

$$2\ HBr(g) + Cl_2(g) \rightleftharpoons 2\ HCl(g) + Br_2(\ell)$$

Comment on the connection between the sign of $\Delta_r G°$ and the magnitude of K_P.

71. For each reaction, estimate $K°$ at the temperature indicated.
    (a) $2\ H_2(g) + O_2(g) \rightleftharpoons 2\ H_2O(g)$ at 800. K
    (b) $2\ SO_2(g) + O_2(g) \rightleftharpoons 2\ SO_3(g)$ at 500. K
    (c) $2\ HF(g) \rightleftharpoons H_2(g) + F_2(g)$ at 2000. K
72. For each reaction, estimate $K°$ at the temperature indicated.
    (a) $H_2(g) + I_2(g) \rightleftharpoons 2\ HI(g)$ at 500. K
    (b) $N_2(g) + 3\ H_2(g) \rightleftharpoons 2\ NH_3(g)$ at 400. K
    (c) $CO(g) + 3\ H_2(g) \rightleftharpoons CH_4(g) + H_2O(g)$ at 800. K

73. For each reaction, an equilibrium constant at 298 K is given. Calculate $\Delta_r G°$ for each reaction.
(a) $Br_2(\ell) + H_2(g) \rightleftharpoons 2\ HBr(g)$ $\quad K_P = 4.4 \times 10^{18}$
(b) $H_2O(\ell) \rightleftharpoons H_2O(g)$ $\quad K_P = 3.17 \times 10^{-2}$
(c) $N_2(g) + 3\ H_2(g) \rightleftharpoons 2\ NH_3(g)$ $\quad K_c = 3.5 \times 10^{8}$

74. For each reaction, an equilibrium constant at 298 K is given. Calculate $\Delta_r G°$ for each reaction.
(a) $\frac{1}{8}\ S_8(s) + O_2(g) \rightleftharpoons SO_2(g)$ $\quad K_p = 4.2 \times 10^{52}$
(b) $2\ H_2(g) + O_2(g) \rightleftharpoons 2\ H_2O(g)$ $\quad K_c = 3.3 \times 10^{81}$
(c) $CH_4(g) + H_2O(g) \rightleftharpoons CO(g) + 3\ H_2(g)$
$K_c = 2.0 \times 10^{-28}$

Gibbs Free Energy, Maximum Work, and Energy Resources (Section 16-8)

75. Which of these reactions are capable of being harnessed to do useful work at 298 K and 1 bar? Which require that work be done to make them occur?
(a) $2\ C_6H_6(\ell) + 15\ O_2(g) \longrightarrow 12\ CO_2(g) + 6\ H_2O(g)$
(b) $2\ NF_3(g) \longrightarrow N_2(g) + 3\ F_2(g)$
(c) $TiO_2(s) \longrightarrow Ti(s) + O_2(g)$

76. Which of these reactions are capable of being harnessed to do useful work at 298 K and 1 bar? Which require that work be done to make them occur?
(a) $Al_2O_3(s) \longrightarrow 2\ Al(s) + \frac{3}{2}\ O_2(g)$
(b) $2\ CO(g) + O_2(g) \longrightarrow 2\ CO_2(g)$
(c) $C_2H_6(g) \longrightarrow C_2H_4(g) + H_2(g)$

77. For each of the reactions in Question 75 that requires work to be done, calculate the minimum mass of graphite that would have to be oxidized to $CO_2(g)$ to provide the necessary work.

78. For each of the reactions in Question 76 that requires work to be done, calculate the minimum mass of hydrogen gas that would have to be burned to form water vapor to provide the necessary work.

79. To obtain a metal from its ore, the decomposition of the metal oxide to form the metal and oxygen is often coupled with oxidation of coke (carbon) to carbon monoxide. For each metal oxide listed, write a balanced equation for the decomposition of the oxide and for the overall reaction when the decomposition is coupled to oxidation of coke to carbon monoxide. Calculate the overall value of $\Delta_r G°$ for each coupled reaction at 25 °C. Which of the metals could be obtained from these ores at 25 °C by this method?
(a) $CuO(s)$ (b) $Ag_2O(s)$
(c) $HgO(s)$ (d) $MgO(s)$
(e) $PbO(s)$

80. From which of the metal oxides in Question 79 could the metal be obtained by coupling reduction of the oxide with oxidation of coke to carbon monoxide at 800 °C?

81. From which of the metal oxides in Question 79 could the metal be obtained by coupling reduction of the oxide with oxidation of coke to carbon monoxide at 1500 °C?

Gibbs Free Energy and Biological Systems (Section 16-9)

82. The molecular structure shown is of one form of glucose, $C_6H_{12}O_6$.

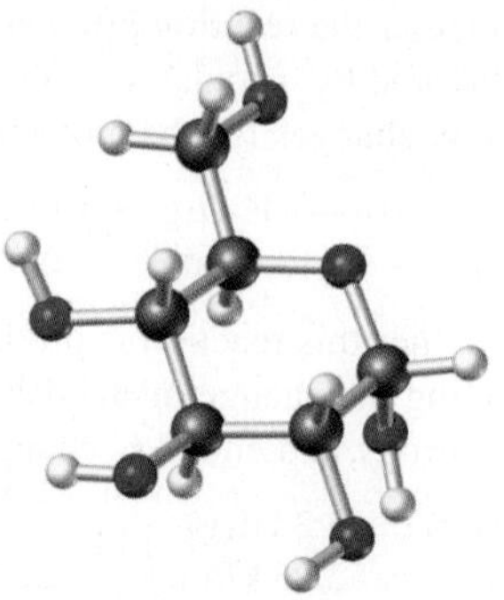

glucose

Glucose can be oxidized to carbon dioxide and water according to the equation

$$C_6H_{12}O_6(s) + 6\ O_2(g) \longrightarrow 6\ CO_2(g) + 6\ H_2O(g)$$

(a) Using the method described in Section 6-6a for estimating enthalpy changes from bond energies, estimate $\Delta_r H°$ for the oxidation of this form of glucose. Make a list of all bonds broken and all bonds formed in this process.
(b) Compare your result with the experimental value of −2816 kJ/mol for combustion of glucose. Why might there be a difference between this value and the one you calculated in part (a)?

83. Another step in the metabolism of glucose, which occurs after the formation of glucose 6-phosphate, is the conversion of fructose 6-phosphate to fructose 1,6-bisphosphate ("bis" means *two*):

$$\text{Fructose 6-phosphate(aq)} + H_2PO_4^-(aq) \longrightarrow$$
$$\text{fructose 1,6-bisphosphate(aq)} + H_2O(\ell) + H^+(aq)$$

(a) This reaction has a Gibbs free energy change of +16.7 kJ/mol of fructose 6-phosphate. Is it endergonic or exergonic?
(b) Write the equation for the formation of 1 mol ADP from ATP, for which $\Delta_r G° = -30.5$ kJ/mol.
(c) Couple these two reactions to get an exergonic process; write its overall chemical equation, and calculate the Gibbs free energy change.

84. In muscle cells under the condition of vigorous exercise, glucose is converted to lactic acid ("lactate"), $CH_3CHOHCOOH$, by the chemical reaction

$$C_6H_{12}O_6 \longrightarrow 2\ CH_3CHOHCOOH \quad \Delta_r G°' = -197 \text{ kJ/mol}$$

(a) If all of the Gibbs free energy from this reaction were used to convert ADP to ATP, calculate how many moles of ATP could be produced per mole of glucose.
(b) The actual reaction involves the production of 3 mol ATP per mole of glucose. Calculate the $\Delta_r G°$ for this overall reaction.
(c) Is the overall reaction in part (b) reactant-favored or product-favored?

85. The biological oxidation of ethanol, C_2H_5OH, is also a source of Gibbs free energy.
(a) Does the oxidation of 1 g ethanol give more or less energy than the oxidation of 1 g glucose? (*Hint:* Write the balanced equation for the production of carbon dioxide and water from ethanol and oxygen, and use Appendix J.)
(b) Comment on potential problems of replacing glucose with ethanol in your diet.

Conservation of Gibbs Free Energy (Section 16-10)

86. What are the resources human society uses to supply Gibbs free energy? (*Hint:* Consider information you learned in Section 4-11 as well as in this chapter.)
87. For one day, keep a log of all the activities you undertake that consume Gibbs free energy. Distinguish between Gibbs free energy provided by nutrient metabolism and that provided by other energy resources.

Thermodynamic and Kinetic Stability (Section 16-11)

88. Billions of pounds of acetic acid are made each year, much of it by the reaction of methanol with carbon monoxide. (Assume $T = 298$ K.)

$$CH_3OH(\ell) + CO(g) \longrightarrow CH_3COOH(\ell)$$

(a) By calculating the standard Gibbs free energy change, $\Delta_r G°$, for this reaction, show that it is product-favored.
(b) Determine the standard Gibbs free energy change, $\Delta_r G°$, for the reaction of acetic acid with oxygen to form gaseous carbon dioxide and liquid water.
(c) Based on this result, is acetic acid thermodynamically stable compared with $CO_2(g)$ and $H_2O(\ell)$?
(d) Is acetic acid kinetically stable compared with $CO_2(g)$ and $H_2O(\ell)$?
89. Determine the standard Gibbs free energy change, $\Delta_r G°$, for the reactions of liquid methanol, of CO(g), and of ethyne, $C_2H_2(g)$, with oxygen gas to form gaseous carbon dioxide and (if hydrogen is present) liquid water at 298 K. Use your calculations to decide which of these substances are kinetically stable and which are thermodynamically stable: $CH_3OH(\ell)$, CO(g), $C_2H_2(g)$, $CO_2(g)$, $H_2O(\ell)$.
90. There are millions of organic compounds known, and new ones are being discovered or made at a rate of more than 100,000 compounds per year. Organic compounds burn readily in air at high temperatures to form carbon dioxide and water. Several classes of organic compounds are listed, with a simple example of each. Write a balanced chemical equation for the combustion in O_2 of each of these compounds, and then use the data in Appendix J to show that each reaction is product-favored at room temperature.

Class of Organics	Simple Example
Aliphatic hydrocarbons	Methane, CH_4
Aromatic hydrocarbons	Benzene, C_6H_6
Alcohols	Methanol, CH_3OH

From these results, it is reasonable to hypothesize that *all* organic compounds are thermodynamically unstable in an oxygen atmosphere (that is, their room-temperature reaction with $O_2(g)$ to form $CO_2(g)$ and $H_2O(\ell)$ is product-favored). If this hypothesis is true, how can organic compounds exist on Earth?
91. Actually, the carbon in $CO_2(g)$ is thermodynamically unstable with respect to the carbon in calcium carbonate (limestone). Verify this by determining the standard Gibbs free energy change for the reaction of lime, CaO(s), with $CO_2(g)$ to make $CaCO_3(s)$.

General Questions

These questions are not explicitly keyed to chapter topics; many require integration of several concepts.

92. The standard molar entropy of methanol vapor, $CH_3OH(g)$, is 239.8 J K^{-1} mol^{-1}.
(a) Calculate the entropy change for the vaporization of 1 mol methanol (use data from Table 16.1 or Appendix J).
(b) Calculate the enthalpy of vaporization of methanol, assuming that $\Delta_r S°$ doesn't depend on temperature and taking the boiling point of methanol to be 64.6 °C.
93. The standard molar entropy of iodine vapor, $I_2(g)$, is 260.7 J K^{-1} mol^{-1} and the standard molar enthalpy of formation is 62.4 kJ/mol.
(a) Calculate the entropy change for vaporization of 1 mol of solid iodine (use data from Table 16.1 or Appendix J).
(b) Calculate the enthalpy change for sublimation of iodine.
(c) Assuming that $\Delta_r S°$ does not change with temperature, estimate the temperature at which iodine would sublime (change directly from solid to gas).
94. This problem will help you understand the dependence of the U. S. economy on energy. Refer to the Energy Consumption Data for Question 94 and use the Internet to convert from quads (quadrillion Btu) to joules. Calculate the energy resources (J) used by the petroleum and coal products industries
(a) in one year.
(b) in one day.
(c) in one second.

Energy Consumption Data for Question 94.

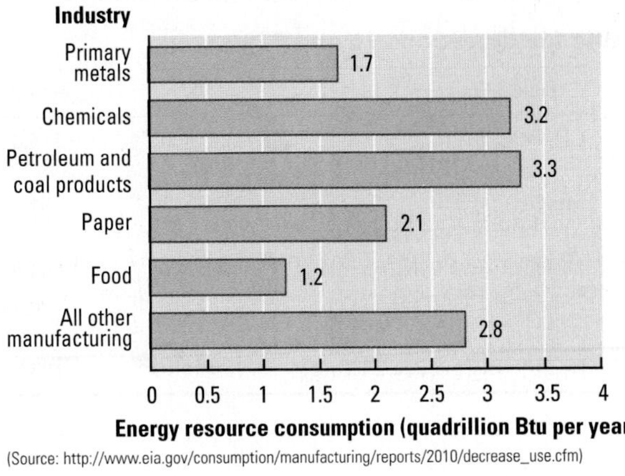
Industry
Primary metals
Chemicals
Petroleum and coal products
Paper
Food
All other manufacturing
1.7
3.2
3.3
2.1
1.2
2.8
0
0.5
1
1.5
2
2.5
3
3.5
4
Energy resource consumption (quadrillion Btu per year)

(Source: http://www.eia.gov/consumption/manufacturing/reports/2010/decrease_use.cfm)

(d) Remembering that 1 watt is the expenditure of 1 joule every second, calculate the average power needs of the petroleum and coal products industries in watts.
(e) Assuming a U. S. population of 317 million people, calculate the power needed by the agriculture, mining, and construction industries *per person in the United States.*

95. Suppose you signed a contract to provide to the petroleum and coal products industries the energy they use each year (see Question 94) by eating glucose and giving them the resulting energy from its oxidation in your body.
(a) Calculate how much glucose you would have to eat each day to meet your contract. Assume that it is someone else's job to figure out how to get the energy stored in your ATP to the industries!
(b) An Olympic sprinter uses energy at the rate of 700 to 900 watts in a sprint. Compare this figure with the one you calculated in part (a), and draw conclusions about the feasibility of fulfilling your contract.

96. The Data Table for Questions 96 and 97 provides data at 25 °C for five reactions. For which (if any) of the reactions 1 through 5 is
(a) K_P greater than K_c?
(b) the reaction product-favored?
(c) there only a single concentration in the K_c expression?
(d) there an increase in the concentrations of products when the temperature increases?
(e) there a change in the sign of $\Delta_r G°$ if water is liquid instead of gas?

97. The Data Table for Questions 96 and 97 provides data at 25 °C for five reactions. For which (if any) of the reactions 1 through 5 is
(a) K_P less than K_c?
(b) there a decrease in the concentrations of products when the pressure increases?
(c) the value of $\Delta_r S°$ positive?
(d) the sign of $\Delta_r G°$ dependent on temperature?

98. Mercury is a poison, and its vapor is readily absorbed through the lungs. Therefore it is important that the partial pressure of mercury be kept as low as possible in any area where people could be exposed to it (such as a dentist's office). The relevant equilibrium reaction is

$$Hg(\ell) \rightleftharpoons Hg(g)$$

For Hg(g), $\Delta_f H° = 61.4$ kJ/mol, $S° = 175.0 \text{ J K}^{-1} \text{ mol}^{-1}$, and $\Delta_f G° = 31.8$ kJ/mol. Use data from Appendix J and these values to evaluate the vapor pressure of mercury at different temperatures. (Remember that concentrations of pure liquids and solids do not appear in the equilibrium constant expression, and for gases $K°$ involves pressures in bars.)
(a) Calculate $\Delta_r G°$ for vaporization of mercury at 25 °C.
(b) Write the equilibrium constant expression for vaporization of mercury.
(c) Calculate $K°$ for this reaction at 25 °C.
(d) Determine the vapor pressure of mercury at 25 °C.
(e) Estimate the temperature at which the vapor pressure of mercury reaches 10 mmHg.

Data Table for Questions 96 and 97.

Reaction	Chemical Equation	K_c	$\Delta_r H°$ (kJ)
1	$CH_3OH(g) + H_2(g) \rightleftharpoons CH_4(g) + H_2O(g)$	3.6×10^{20}	−115.4
2	$Mg(OH)_2(s) \rightleftharpoons MgO(s) + H_2O(g)$	1.24×10^{-5}	81.1
3	$2\ CH_4(g) \rightleftharpoons C_2H_6(g) + H_2(g)$	9.5×10^{-13}	64.9
4	$2\ H_2(g) + CO(g) \rightleftharpoons CH_3OH(g)$	3.76	90.7
5	$H_2(g) + Br_2(g) \rightleftharpoons 2\ HBr(g)$	1.9×10^{24}	−103.7

Applying Concepts

These questions test conceptual learning.

99. A friend says that the boiling point of water is twice that of cyclopentane, which boils at 50 °C. Write a brief statement about the validity of this observation.

100. Using the second law of thermodynamics, explain why it is very difficult to unscramble an egg. Who was Humpty-Dumpty? Why did his moment of glory illustrate the second law of thermodynamics?

101. Appendix J lists standard molar entropies $S°$, not standard entropies of formation $\Delta_f S°$. Why is this possible for entropy but not for internal energy, enthalpy, or Gibbs free energy?

102. When calculating $\Delta_r S°$ from $S°$ values, it is necessary to look up all substances, including elements in their standard state, such as $O_2(g)$, $H_2(g)$, and $N_2(g)$. When calculating $\Delta_r H°$ from $\Delta_f H°$ values, however, elements in their standard state can be ignored. Why is the situation different for $S°$ values?

103. Use Appendix J to look up the standard formation enthalpies of the oxides of at least two alkaline-earth metals, one metal from Group IIIA, and at least three transition metals. Based on what you find, are oxidations of metals usually exothermic or endothermic? Is oxidation of a metal at 25 °C usually reactant-favored or product-favored? Propose an explanation for your conclusions.

104. Explain how the entropy of the universe increases when an aluminum metal can is made from aluminum ore. The first step is to extract the ore, which is primarily a form of Al_2O_3, from the ground. After it is purified by freeing it from oxides of silicon and iron, aluminum oxide is changed to the metal by an input of electrical energy.

$$2\ Al_2O_3(s) \xrightarrow{\text{electrical energy}} 4\ Al(s) + 3\ O_2(g)$$

105. Suppose that at a certain temperature T, a chemical reaction is found to have a standard equilibrium constant $K°$ of 1.0. Indicate whether each statement is true or false and explain why.
(a) The enthalpy change for the reaction, $\Delta_r H°$, is zero.
(b) The entropy change for the reaction, $\Delta_r S°$, is zero.
(c) The Gibbs free energy change for the reaction, $\Delta_r G°$, is zero.
(d) $\Delta_r H°$ and $\Delta_r S°$ have the same sign.
(e) $\Delta_r H°/T = \Delta_r S°$ at the temperature T.

106. When you eat a candy bar, how does your body store the Gibbs free energy that is released during oxidation of the sugars (glucose and other carbohydrates) in the candy bar? What was the original source of the Gibbs free energy needed to synthesize the sugars before they went into the candy bar?

107. Explain how biological systems make use of coupled reactions to maintain the high degree of order found in all living organisms.

108. How can kinetically stable substances exist at all if they are not thermodynamically stable?

109. Criticize this statement: Provided it occurs at an appreciable rate, any chemical reaction for which $\Delta_r G < 0$ will proceed until all reactants have been converted to products.

110. Reword the statement in Question 109 so that it is always true.

111. Assuming that the particles shown in these three figures represent molecules that have the same molar mass, in which case is the entropy
(a) highest? (b) lowest?

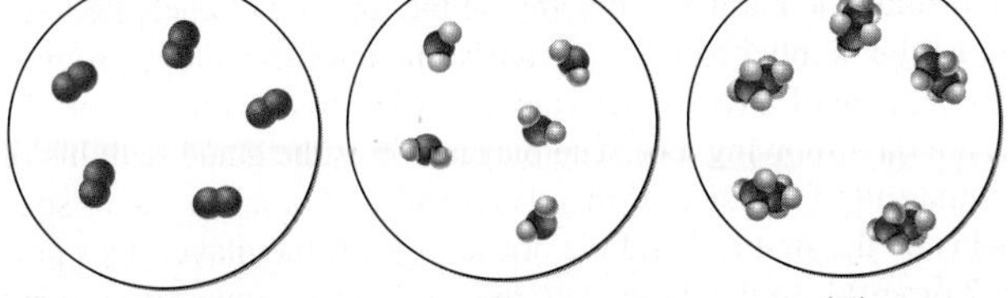

(i) (ii) (iii)

112. Assuming that the spheres shown in these three figures represent atoms of the same element, which case represents a situation where the entropy is
(a) highest? (b) lowest?

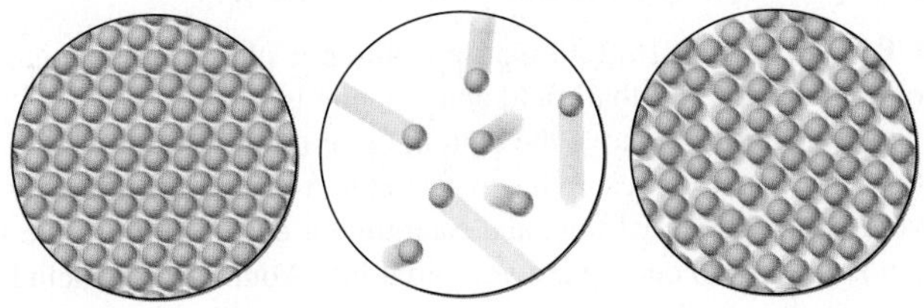

(i) (ii) (iii)

113. Calculate the entropy change for formation of exactly 1 mol of each of these gaseous hydrocarbons under standard conditions from carbon (graphite) and hydrogen. What trend do you see in these values? Does $\Delta_r S°$ increase or decrease on adding H atoms?
(a) acetylene, $C_2H_2(g)$ (b) ethylene, $C_2H_4(g)$
(c) ethane, $C_2H_6(g)$

114. Calcium hydroxide, $Ca(OH)_2(s)$, can be dehydrated to form lime, CaO, by heating. Without doing any calculations, and basing your prediction on the enthalpy change and the entropy change, is this reaction product-favored at 25 °C? Explain your answer briefly.

More Challenging Questions

These questions require more thought and integrate several concepts.

115. This is a group project: Estimate or look up, to the nearest order of magnitude,
(a) the number of kilograms of CH_3OH made each year.
(b) the number of kilograms of CO in the entire atmosphere.
(c) the number of kilograms of CH_3COOH made each year.
(d) the number of kilograms of H_2O on Earth.
(e) the number of kilograms of CO_2 in the atmosphere.
What do these facts tell you about the difference between kinetic stability and thermodynamic stability?

116. From data in Appendix J, estimate
(a) the boiling point of bromine.
(b) the boiling point of tin(IV) chloride.

117. From data in Appendix J, estimate
(a) the boiling point of titanium(IV) chloride.
(b) the boiling point of carbon disulfide, CS_2, which is a liquid at 25 °C and 1 bar.

118. Nitric oxide and chlorine combine at 25 °C to produce nitrosyl chloride, NOCl.

$$2\ NO(g) + Cl_2(g) \longrightarrow 2\ NOCl(g)$$

(a) Calculate the equilibrium constant K_P for the reaction.
(b) Is the reaction product-favored or reactant-favored?
(c) Calculate the equilibrium constant K_c for the reaction.

119. Hydrogen for use in the Haber-Bosch process for ammonia synthesis is generated from natural gas by the reaction

$$CH_4(g) + H_2O(g) \rightleftharpoons CO(g) + 3\ H_2(g)$$

(a) Calculate $\Delta_r G°$ for this reaction at 25 °C.
(b) Calculate K_P for the reaction at 25 °C.
(c) Is the reaction product-favored at 25 °C? If not, at what temperature does it become so?
(d) Estimate K_c for the reaction at 1000. K.

120. It would be very useful if we could use the inexpensive carbon in coal to make more complex organic molecules such as gaseous or liquid fuels. The formation of methane from coal and water is reactant-favored and thus cannot occur unless there is some energy transfer from outside. This problem examines the feasibility of other reactions using coal and water.
   (a) Write three balanced equations for the reactions of coal (carbon) and steam to make ethane gas, $C_2H_6(g)$, propane gas, $C_3H_8(g)$, and liquid methanol, $CH_3OH(\ell)$, with carbon dioxide as a by-product.
   (b) Using the data in Appendix J, calculate $\Delta_r H°$, $\Delta_r S°$, and $\Delta_r G°$ for each reaction, and then comment on whether any of them would be a feasible way to make the stated products.
121. You are exploring the marketing possibilities of a scheme by which every family in the United States produces enough water for its own needs by the combustion of hydrogen and oxygen. Would the release of Gibbs free energy from the combination of hydrogen and oxygen be sufficient to supply the family's energy needs? Do not try to collect the actual data you would use, but define the problem well enough so that someone else could collect the necessary data and do the calculations that would be needed.
122. Quite often a graph of ln $K°$ versus $1/T$ is a straight line. Use Equation 16.9 (← **Sec. 16-7b**) to show how $\Delta_r H°$ and $\Delta_r S°$ can be determined from such a graph. Does the fact that such a graph is straight tell you anything about the dependence of $\Delta_r H°$ and $\Delta_r S°$ on temperature? Explain your answer.
123. Assuming that $\Delta_r H°$ and $\Delta_r S°$ do not vary with temperature, use Equation 16.9 (← **Sec. 16-7b**) to derive a formula relating $K_1°$ at temperature T_1 to $K_2°$ at temperature T_2.
124. Without consulting tables of $\Delta_f H°$, $S°$, or $\Delta_f G°$ values, predict which of these reactions is
   (i) always product-favored.
   (ii) product-favored at low temperatures, but not product-favored at high temperatures.
   (iii) not product-favored at low temperatures, but product-favored at high temperatures.
   (iv) never product-favored.
   (a) $2\ NO_2(g) \longrightarrow N_2O_4(g)$
   (b) $C_5H_{12}(g) + 8\ O_2(g) \longrightarrow 5\ CO_2(g) + 6\ H_2O(g)$
   (c) $P_4(g) + 10\ F_2(g) \longrightarrow 4\ PF_5(g)$
   (*Hint:* Use the qualitative rules regarding bond enthalpies in Section 4-7 to predict the sign of $\Delta_r H°$.)
125. Using the reactions

$$2\ H_2(g) + O_2(g) \longrightarrow 2\ H_2O(\ell)$$
$$2\ H_2(g) + O_2(g) \longrightarrow 2\ H_2O(g)$$

as an example, explain why it may be incorrect to assume for reactions involving solids or liquids that $\Delta_r S°$ and $\Delta_r H°$ do not change appreciably with increasing temperature.

126. For the reaction

$$CH_3OH(\ell) + \tfrac{3}{2}\ O_2(g) \longrightarrow 2\ H_2O(\ell) + CO_2(g)$$

the value of $\Delta_r G°$ is −702.35 kJ/mol at 25 °C. Other data at 25 °C are

	$\Delta_f H°$ (kJ/mol)	$S°$ (J mol^{-1} K^{-1})
$CH_3OH(\ell)$	−238.66	126.8
$H_2O(\ell)$	−285.83	69.91
$CO_2(g)$	−393.509	213.74

Calculate the standard molar entropy, $S°$, for $O_2(g)$.

127. The standard equilibrium constant is 2.1×10^9 for this reaction at 25 °C

$$Zn^{2+}(aq) + 4\ NH_3(aq) \rightleftharpoons Zn(NH_3)_4^{2+}(aq)$$

   (a) Calculate $\Delta_r G°$ at this temperature.
   (b) If standard-state concentrations of the reactants and products are combined, in which direction will the reaction proceed?
   (c) Calculate $\Delta_r G$ when $[Zn(NH_3)_4^{2+}] = 0.010$ M, $[Zn^{2+}] = 0.0010$ M, and $[NH_3] = 3.5 \times 10^{-4}$ M.

Conceptual Challenge Problems

These rigorous, thought-provoking problems integrate conceptual learning with problem solving and are suitable for group work.

CP16.A (Section 16-2) Suppose that you are invited to play a game as either the "player" or the "house." A pair of dice is used to determine the winner. Each die is a cube having a different number, one through six, showing on each face. The player rolls two dice and sums the numbers showing on the top side of each die to determine the number rolled. Obviously, the number rolled has a minimum value of 2 (both dice showing a 1) and a maximum of 12 (both dice showing a 6). The player begins the game with his or her initial roll of the dice. If the player rolls a 7 or an 11, he or she wins on the first roll and the house loses. If the player does not roll a 7 or an 11 on the initial roll, then whatever number was rolled is called the point, and the player must roll again. For the player to win, he or she must roll the point again before either a 7 or an 11 is rolled. Should the player roll a 7 or an 11 before rolling the point a second time, the house wins. Which would you choose to be, player or house? Explain clearly in terms of the probabilities of rolling the dice why you chose the role you did.

CP16.B (Section 16-2) Suppose a button is placed in the middle of an American football field and a penny is flipped to decide which direction to move the button, up or down the field. Each time the penny comes up heads, the button is moved 10 cm toward your opponent's goal line; and each time it comes up tails, the button is moved 10 cm toward your goal line. Your friend concludes that after many flips of the penny the button is likely to remain within 10 cm of the middle of the field, because numerous flips of the penny will produce heads just as often as tails. You doubt this because you know that perfume molecules and particles diffuse away from their original source, even though, like the button, they are just as likely to be hit from one direction as from any other by the moving molecules around them. How would you explain the error of your friend's conclusion about the movement of the button?

CP16.C (Section 16-3) When thermal energy is transferred to a substance at its standard melting point or boiling point, the substance melts or vaporizes, but its temperature does not change while it is doing so. It is clear then that temperature cannot be a measure of "how much energy is in a sample of matter" or the "intensity of energy in a sample of matter." In Figure 16.3 (← **Sec. 16-3b**) we noted that atoms and molecules are not stationary, but rather are in constant motion. When heated, their motion increases. If this is true, what can you infer that temperature measures about a sample of matter?

CP16.D (Section 16-10) Suppose that you are a member of an environmental group and have been assigned to evaluate various ways of delivering milk to consumers with respect to Gibbs free energy conservation. Think of all the ways that milk could be delivered, the kinds of containers that could be used, and the ways they could be transported. Consider whether the containers could be reused (refilled) or recycled. Define the problem in terms of the kinds of information you would need to collect, how you would analyze the information, and which criteria you would use to decide which systems are more efficient in use of Gibbs free energy. Do not try to collect the actual data you would use, but define the problem well enough so that someone could collect the necessary data based on your statement of the problem.

CP16.E (Section 16-11) Consider planet Earth as a thermodynamic system. Is Earth thermodynamically or kinetically stable? Discuss your choice, providing as many arguments as you can to support it.

17 | Electrochemistry and Its Applications

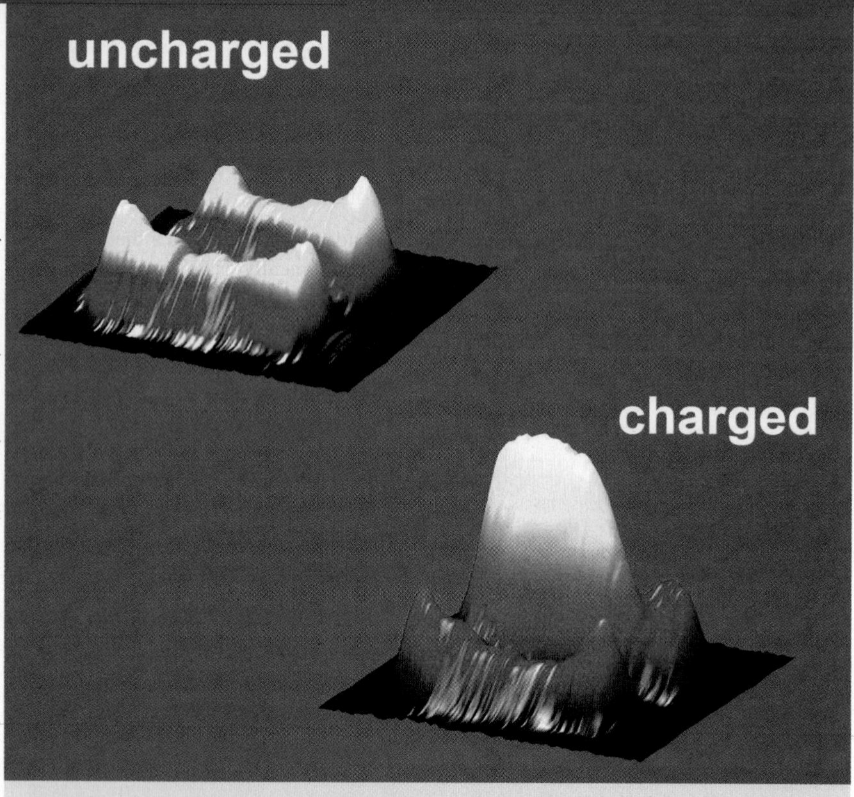
uncharged
charged

Chem. Eng. News, 2012, 90 (8) 19 Feb 2012, pp 40–41. Reproduced with permission of Alexej Jerschow

If you have a laptop computer, tablet computer, or smart phone, it probably contains a lithium-ion battery. The lifetime of such batteries can be shortened if deposits of lithium form on an electrode during charging or discharging. The photo above was produced by a team of researchers who used lithium-7 magnetic resonance imaging (MRI) to visually monitor growth of lithium deposits as a battery was charged. These studies will likely contribute to development of longer-lived, safer lithium-ion batteries. Batteries generate electric current from product-favored oxidation-reduction reactions. How do such reactions make a battery? (Section 17-3) How do chemists predict the voltage of a battery and how long it will take to discharge? (Sections 17-4 and 17-7) Why are lithium-ion batteries so common and how do they differ from other batteries? (Section 17-8) This chapter answers these and many other questions about the ways chemistry and electricity are related.

It is estimated that there are approximately 6.8 billion cell phones in use today among the world's 7.2 billion people (NYTimes.com, June 11, 2013). Each cell phone contains a *battery*, an apparatus that provides electric current generated by a product-favored oxidation-reduction (redox) reaction (← **Sec. 16-1**). A cell phone is also likely to contain parts in which one metal is electroplated onto another ("chrome" or gold-plated parts). In electroplating, an electric current causes a redox reaction to occur. Batteries and electroplating are examples of

electrochemistry, *the study of the relationship between chemical changes and electricity*. In a battery, the voltage depends on the relative strengths of the oxidizing and reducing agents used as reactants. The quantity of current depends on the number of electrons transferred in the redox reaction and on the amounts of substances that react. In electroplating, the quantity of electric charge transferred determines how much chemical reaction takes place. The applied voltage determines which redox reactions can occur.

17-1 Redox Reactions

Many chemical reactions are redox reactions in which the reactants can be atoms, ions, or molecules. How can we determine whether a reaction involves oxidation-reduction? In Chapter 3, we used these guidelines:

You may want to review the definitions of oxidation and reduction in Section 3-5 and the rules for assigning oxidation numbers in Section 3-5c.

- ***the presence of strong oxidizing or reducing agents as reactants*** (Table 3.3, ← **Sec. 3-5b**).
- ***a change in oxidation number of one or more elements*** (← **Sec. 3-5c**). To look for a change in oxidation number you have to determine the oxidation number of each element as it appears in a reactant or a product.
- ***the presence of an uncombined element as a reactant or product*** (← **Sec. 3-5c**). Producing a free element or incorporating one into a compound almost always results in a change in oxidation number.

An uncombined element is always assigned an oxidation number of 0.

EXERCISE 17.1

The Nitrogen Cycle: A Review of Oxidation Numbers

The nitrogen cycle is a series of chemical pathways involving the conversion of atmospheric N_2 into other nitrogen-containing species. Identify the oxidation number of nitrogen in each of these species in the nitrogen cycle: (a) N_2; (b) NO_3^-; (c) NH_3; (d) NH_4^+.

Answers to **EXERCISES** are provided at the back of this book in Appendix L.

We can apply the guidelines given above to the reaction between magnesium metal and hydrochloric acid (Figure 17.1a); the oxidation numbers of the species are shown above their symbols.

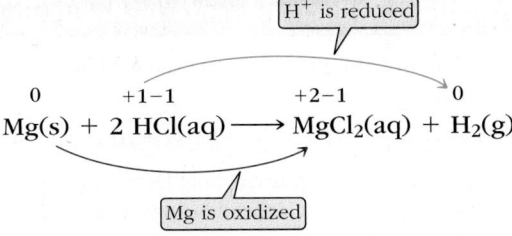

H^+ is reduced

$$\overset{0}{Mg(s)} + 2\ \overset{+1\,-1}{HCl(aq)} \longrightarrow \overset{+2\,-1}{MgCl_2(aq)} + \overset{0}{H_2(g)}$$

Mg is oxidized

In this book, oxidation numbers are written as +1, +2, and so forth, whereas charges on ions are written as 1+, 2+, and so forth.

This is a redox reaction as indicated by:

- the presence of a strong reducing agent, the alkaline-earth metal, Mg.
- the presence of Mg and H_2, elements in their "free" state (uncombined with another element).
- changes in the oxidation numbers of the reactants: Mg(s) is oxidized (increases in oxidation number from 0 to +2); H^+ ions from HCl are reduced (decrease in oxidation number from +1 to 0).

All redox reactions involve the *simultaneous* transfer of electrons from a reducing agent to an oxidizing agent. For a given reaction, oxidizing agents and reducing agents are always *reactants*, never products.

© Cengage Learning/James Maynard
Cengage Learning/Charles D. Winters
(a)
(b)

Figure 17.1 Two dramatic redox reactions. (a) Magnesium metal reacting with hydrochloric acid inflates a balloon and (b) iron(III) oxide reacting with aluminum metal—the thermite reaction.

- **Oxidizing agents are reduced** (gain electrons). In this case, HCl is the oxidizing agent because its H^+ ions remove electrons from magnesium metal, causing the metal to be oxidized; simultaneously, H^+ ions gain these electrons and are reduced to H atoms in uncharged H_2 molecules.
- **Reducing agents are oxidized** (lose electrons). Magnesium metal is the reducing agent here because it is oxidized to Mg^{2+} by losing electrons to H^+ ions.

Redox reactions, such as this one, involve a flow of electrons. If the electron flow from a product-favored redox reaction occurs along a pathway that completes an electric circuit, useful electrical work can be done by the system.

PROBLEM-SOLVING EXAMPLE 17.1

Identifying Oxidizing and Reducing Agents in Redox Reactions

In the thermite reaction (Figure 17.1b), iron(III) oxide and aluminum metal react to give iron metal and aluminum oxide:

$$Fe_2O_3(s) + 2\,Al(s) \longrightarrow 2\,Fe(s) + Al_2O_3(s)$$

Give one piece of evidence that this is a redox reaction. Determine what is reduced, what is oxidized, what is the oxidizing agent, what is the reducing agent.

Result Oxidation number changes: Fe, +3 to 0; Al, 0 to +3. The Fe^{3+} in iron(III) oxide is reduced to metallic iron. The aluminum metal is oxidized to Al^{3+} ions. The oxidizing agent is Fe^{3+} in iron(III) oxide and the reducing agent is aluminum metal.

Analyze A redox reaction can be identified by looking for strong oxidizing or reducing agents in the reactants, by looking for changes in oxidation numbers, or by looking for the presence of an uncombined element as a reactant or product. Reduction involves a decrease in oxidation number; oxidation involves an increase. The oxidizing agent is the reactant that is reduced and the reducing agent is the reactant that is oxidized.

Plan Notice that Al(s) is a reactant and Fe(s) is a product. Assign oxidation numbers to see which atoms increase and which decrease.

Execute Because there are uncombined elements this is a redox reaction.

Oxidation numbers for reactants: Oxygen in compounds normally is −2 (← **Sec. 3-5c**). Because the sum of the oxidation numbers of all atoms in a formula must equal the charge on the formula, each iron in Fe_2O_3 must be +3. The oxidation number for metallic Al is 0, as it is for all uncombined elements.

Oxidation numbers for products: Oxygen is −2, so Al in Al_2O_3 must be +3. Iron is uncombined, so its oxidation number is 0. Thus, the oxidation numbers are

$$\overset{+3\ -2}{Fe_2O_3(s)} + 2\,\overset{0}{Al(s)} \longrightarrow 2\,\overset{0}{Fe(s)} + \overset{+3\ -2}{Al_2O_3(s)}$$

The oxidation number of iron decreased, while the oxidation number of aluminum increased. Thus, Fe^{3+} in Fe_2O_3 was reduced (+3 to 0) and aluminum was oxidized (0 to +3). Consequently, Fe^{3+} in Fe_2O_3 is the oxidizing agent and Al metal is the reducing agent. Oxygen is neither reduced nor oxidized in this reaction; its oxidation number remains −2.

PROBLEM-SOLVING PRACTICE 17.1

Give the oxidation number for each atom and identify the oxidizing and reducing agents in these balanced chemical equations.

(a) $2\,Fe(s) + 3\,Cl_2(g) \longrightarrow 2\,FeCl_3(s)$

(b) $2\,H_2(g) + O_2(g) \longrightarrow 2\,H_2O(\ell)$

(c) $Cu(s) + 2\,NO_3^-(aq) + 4\,H^+(aq) \longrightarrow Cu^{2+}(aq) + 2\,NO_2(g) + 2\,H_2O(\ell)$

(d) $C(s) + O_2(g) \longrightarrow CO_2(g)$

(e) $6\,Fe^{2+}(aq) + Cr_2O_7^{2-}(aq) + 14\,H^+(aq) \longrightarrow 6\,Fe^{3+}(aq) + 2\,Cr^{3+}(aq) + 7\,H_2O(\ell)$

PROBLEM-SOLVING PRACTICE answers are provided at the back of this book in Appendix K.

17-2 Half-reactions and Redox Reactions

Consider the redox reaction between zinc metal and copper(II) ions shown in Figure 17.2. The net ionic equation is

reduction

$$Zn(s) + Cu^{2+}(aq) \longrightarrow Zn^{2+}(aq) + Cu(s)$$

oxidation

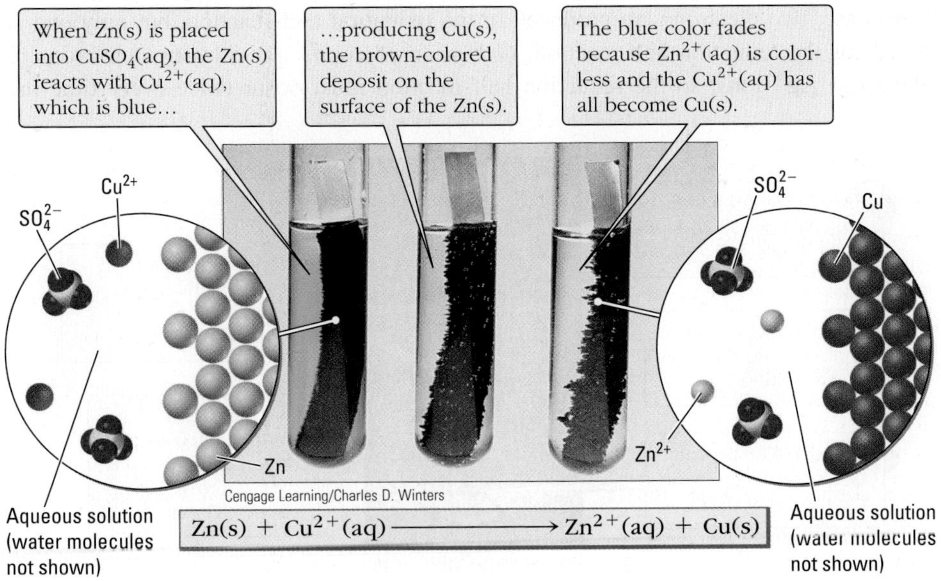
When Zn(s) is placed into CuSO4(aq), the Zn(s) reacts with Cu2+(aq), which is blue...
...producing Cu(s), the brown-colored deposit on the surface of the Zn(s).
The blue color fades because Zn2+(aq) is colorless and the Cu2+(aq) has all become Cu(s).
Cu2+
SO42−
Zn
Zn2+
SO42−
Cu
Cengage Learning/Charles D. Winters
Aqueous solution (water molecules not shown)
Zn(s) + Cu2+(aq) → Zn2+(aq) + Cu(s)
Aqueous solution (water molecules not shown)

Figure 17.2 An oxidation-reduction reaction. Cu^{2+}(aq) reacts with Zn(s) to produce Cu(s) and Zn^{2+}(aq). Water molecules are not shown so that the ions in solution are clearly discernible.

Zinc metal is oxidized to Zn^{2+} ions, and Cu^{2+} ions are reduced to copper metal. As Cu^{2+} ions are converted to Cu(s) in this half-reaction, the blue color of the solution becomes less intense and metallic copper forms on the zinc surface.

To see more clearly how electrons are transferred, this overall reaction can be thought of as the result of two simultaneous half-reactions. A **half-reaction** *shows the reactants, products, and electrons transferred for either an oxidation or a reduction.* In this case, there is one half-reaction for the oxidation of Zn and one half-reaction for the reduction of Cu^{2+}. The oxidation half-reaction

$$Zn(s) \longrightarrow Zn^{2+}(aq) + 2\,e^-$$

shows that each zinc atom loses two electrons when it is oxidized to a Zn^{2+} ion. These two electrons are accepted by a Cu^{2+} ion in the reduction half-reaction,

$$Cu^{2+}(aq) + 2\,e^- \longrightarrow Cu(s)$$

The overall reaction is the sum of the oxidation and reduction half-reactions.

Note that the sum of the charges on the left side of the reaction equals the sum of the charges on the right side, even for half-reactions. Otherwise, conservation of matter would be violated—electrons would be created or destroyed.

Oxidation half-reaction: $Zn(s) \longrightarrow Zn^{2+}(aq) + 2\,e^-$

Reduction half-reaction: $Cu^{2+}(aq) + 2\,e^- \longrightarrow Cu(s)$

Overall reaction: $Zn(s) + Cu^{2+}(aq) \longrightarrow Zn^{2+}(aq) + Cu(s)$

Notice that no electrons appear in the equation for the overall reaction because the number of electrons produced by the oxidation half-reaction must equal the number of electrons gained by the reduction half-reaction. This must be true in any redox reaction. Otherwise, electrons would be created or destroyed, violating the laws of conservation of matter (← **Sec. 1-10**) and conservation of electrical charge (← **Sec. 3-3c**).

Consider another example (Figure 17.3). A copper metal screen is placed in a solution of silver nitrate. As the reaction proceeds, the colorless solution gradually turns blue, and fine, silvery, hair-like crystals form on the copper screen. Knowing that Cu^{2+} ions in aqueous solution appear blue, we can conclude that the copper metal is being oxidized to Cu^{2+}. Reduction must also be taking place, so it is reasonable to conclude that the hair-like crystals of silver result from the reduction of Ag^+ ions to metallic silver. The two half-reactions are

Oxidation half-reaction: $Cu(s) \longrightarrow Cu^{2+}(aq) + 2e^-$

Reduction half-reaction: $Ag^+(aq) + e^- \longrightarrow Ag(s)$

In this case, two electrons are produced in the oxidation half-reaction, but only one is needed for the reduction half-reaction. *One* atom of copper provides two electrons to reduce *two* Ag^+ ions, so the reduction half-reaction must occur twice every time the

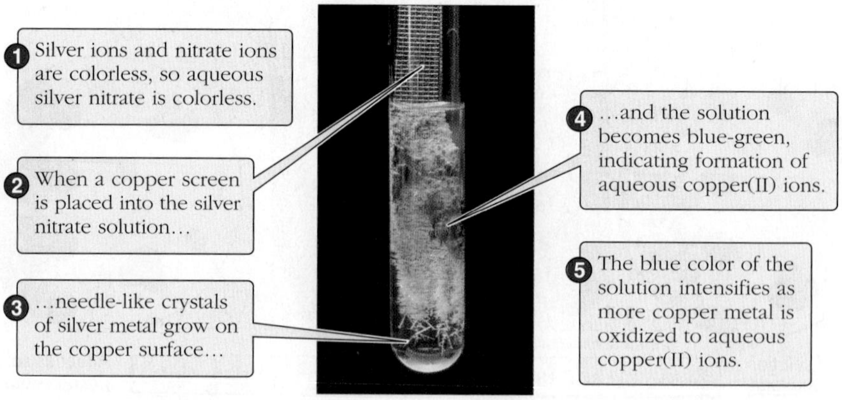
1 Silver ions and nitrate ions are colorless, so aqueous silver nitrate is colorless.
2 When a copper screen is placed into the silver nitrate solution...
3 ...needle-like crystals of silver metal grow on the copper surface...
4 ...and the solution becomes blue-green, indicating formation of aqueous copper(II) ions.
5 The blue color of the solution intensifies as more copper metal is oxidized to aqueous copper(II) ions.
Cengage Learning/Charles D. Winters

Figure 17.3 Copper metal reacting with a solution of silver nitrate.

oxidation half-reaction occurs once. To indicate this relationship, we multiply the reduction half-reaction by 2.

Reduction half-reaction × *2:* $2\ Ag^{+}(aq) + 2\ e^{-} \longrightarrow 2\ Ag(s)$

Adding this reduction half-reaction to the oxidation half-reaction gives the overall equation

Reduction half-reaction × *2:*	$2\ Ag^{+}(aq) + 2\ e^{-} \longrightarrow 2\ Ag(s)$
Oxidation half-reaction:	$Cu(s) \longrightarrow Cu^{2+}(aq) + 2\ e^{-}$
Overall reaction:	$Cu(s) + 2\ Ag^{+}(aq) \longrightarrow Cu^{2+}(aq) + 2\ Ag(s)$

The method shown here is a general one. *An overall equation can always be generated by writing oxidation and reduction half-reactions, using coefficients to adjust the half-reaction equations so the number of electrons released by the oxidation equals the number gained by the reduction, and then adding the two half-reactions to give the equation for the overall reaction.* A more detailed description of how to balance equations for redox reactions in both acidic and basic aqueous solutions is given in Appendix F.

PROBLEM-SOLVING EXAMPLE 17.2

Determining Half-reactions from Net Redox Reactions

Aluminum metal undergoes a redox reaction with $Zn^{2+}(aq)$ to produce $Al^{3+}(aq)$ and zinc metal.

(unbalanced equation) $Al(s) + Zn^{2+}(aq) \longrightarrow Al^{3+}(aq) + Zn(s)$

Write the oxidation half-reaction, write the reduction half-reaction, and combine them to give the balanced equation for the overall reaction.

Result

Oxidation half-reaction:	$Al(s) \longrightarrow Al^{3+}(aq) + 3\ e^{-}$
Reduction half-reaction:	$Zn^{2+}(aq) + 2\ e^{-} \longrightarrow Zn(s)$
Overall reaction:	$2\ Al(s) + 3\ Zn^{2+}(aq) \longrightarrow 2\ Al^{3+}(aq) + 3\ Zn(s)$

Analyze The oxidation half-reaction must involve an increase in oxidation number; the reduction half-reaction must involve a decrease in oxidation number.

Plan Assign oxidation numbers. Find the atom or ion whose oxidation number increases and write a half-reaction involving that atom or ion and electrons. Find the atom or ion whose oxidation number decreases and write a half-reaction. Then sum the half-reactions, equalizing the number of electrons lost and gained.

Execute The oxidation number of aluminum increases from 0 to +3, so aluminum is oxidized. The half-reaction is

$$Al(s) \longrightarrow Al^{3+}(aq) + 3\ e^{-}$$

The three electrons on the right side balance the 3+ charge of the aluminum ion.

The oxidation number of zinc ion decreases from +2 to 0, so zinc ion is reduced.

$$Zn^{2+}(aq) + 2\ e^{-} \longrightarrow Zn(s)$$

The two electrons on the left side balance the 2+ charge of the zinc ion.

These two half-reactions include different numbers of electrons. Multiply the first by 2 and the second by 3, so that six electrons appear in each half-reaction.

$2\ [Al(s) \longrightarrow Al^{3+}(aq) + 3\ e^{-}]$	gives	$2\ Al(s) \longrightarrow 2\ Al^{3+}(aq) + 6\ e^{-}$
$3\ [Zn^{2+}(aq) + 2\ e^{-} \longrightarrow Zn(s)]$	gives	$3\ Zn^{2+}(aq) + 6\ e^{-} \longrightarrow 3\ Zn(s)$
Overall reaction:		$2\ Al(s) + 3\ Zn^{2+}(aq) \longrightarrow 2\ Al^{3+}(aq) + 3\ Zn(s)$

☑ **Reasonable Result Check** The Al atoms and Zn atoms balance. There is a total charge of 3 × 2+ = 6+ on the left and 2 × 3+ = 6+ on the right. The result is reasonable.

PROBLEM-SOLVING PRACTICE 17.2

Write oxidation and reduction half-reactions for these unbalanced redox equations. For each, show that the sum of the half-reactions is the overall reaction.
(a) $Cd(s) + Cu^{2+}(aq) \longrightarrow Cd^{2+}(aq) + Cu(s)$
(b) $Mg(s) + Cr^{3+}(aq) \longrightarrow Mg^{2+}(aq) + Cr(s)$
(c) $Fe^{3+}(aq) + Cu^{+}(aq) \longrightarrow Fe^{2+}(aq) + Cu^{2+}(aq)$

EXERCISES that are labeled **CONCEPTUAL** are designed to test your understanding of one or more concepts; they usually involve qualitative rather than quantitative thinking.

CONCEPTUAL EXERCISE 17.2

Electrons Lost Equal Electrons Gained

Explain why the number of electrons lost must always equal the number gained in a redox reaction.

17-3 Voltaic Cells

It is easy to see by the color changes in the two redox reactions shown in Figures 17.2 and 17.3 that these reactions favor the formation of products—as soon as the reactants are mixed, changes begin to occur. All product-favored reactions release Gibbs free energy (← **Sec. 16-6**)—energy that can do useful work. A voltaic cell is a way to capture that useful work as electrical work: By forcing electrons through an electrical circuit a voltaic cell can, for example, light a flashlight bulb. A voltaic cell is one type of **electrochemical cell**, *an apparatus in which a chemical reaction and electricity interact.*

In a redox reaction, electrons are *transferred* from one kind of atom, molecule, or ion to another. *In a* **voltaic cell**, *an oxidizing agent and a reducing agent are arranged in such a way that they can react only if electrons flow through an outside conductor.* Voltaic cells are commonly called *batteries*, although, strictly speaking, a **battery** is *two or more voltaic cells connected in series electrically*. Every voltaic cell consists of two **half-cells**, so named because *half of the overall reaction occurs in each half-cell*. The **anode** is *the half-cell where oxidation occurs*; the **cathode** is *the half-cell where reduction occurs*. It is easy to remember which is which: ***a****node* and ***o****xidation* begin with vowels; ***c****athode* and ***r****eduction* begin with consonants. Figure 17.4 diagrams how a voltaic cell (flashlight battery) does work (lights a bulb) by causing a flow of electrons around a circuit.

The voltaic cell is named for the Italian scientist Alessandro Volta. In about 1800 Volta made a repeating stack of a zinc disk, a silver disk, a porous paper disk soaked in salt water (an electrolyte), a zinc disk, and so on. Volta showed that this "pile" produced electric current and he later showed that any two different metals and an electrolyte would generate electricity. For example, you can generate a voltage by sticking a strip of zinc and a strip of copper into a grapefruit!

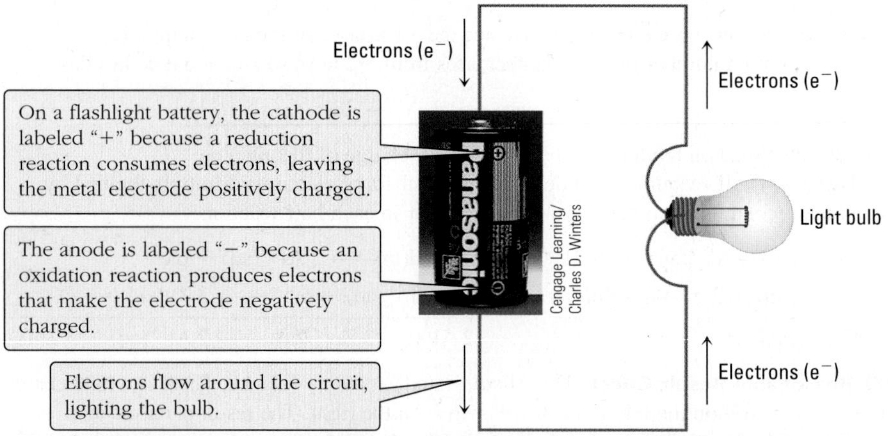
Electrons (e⁻)
Electrons (e⁻)
On a flashlight battery, the cathode is labeled "+" because a reduction reaction consumes electrons, leaving the metal electrode positively charged.
Panasonic
Cengage Learning/ Charles D. Winters
Light bulb
The anode is labeled "−" because an oxidation reaction produces electrons that make the electrode negatively charged.
Electrons (e⁻)
Electrons flow around the circuit, lighting the bulb.

Figure 17.4 A battery. Voltaic cells generate a flow of electrons around an electrical circuit.

Figure 17.5 diagrams how a voltaic cell can be made from the reaction of Zn with Cu^{2+} shown previously in Figure 17.2. The anode half-cell contains a strip of Zn metal in a 1-M solution of $ZnSO_4$. The cathode half-cell contains a strip of Cu metal in a 1-M solution of $CuSO_4$. The metallic zinc and copper strips are called electrodes. An **electrode** *conducts electrical current (electrons) into or out of something*. An electrode is most often a metal plate or wire, but it can also be a piece of graphite or some other electrical conductor; some electrodes, such as graphite, do not react with the solution they contact.

In the voltaic cell diagrammed in Figure 17.5, electrons are released at the anode by the oxidation half-reaction

Anode reaction: $Zn(s) \longrightarrow Zn^{2+}(aq) + 2\ e^-$

Electrons flow from the anode through the filament in the bulb, causing it to glow, and eventually travel to the cathode, where they react with copper(II) ions in the reduction half-reaction

Cathode reaction: $Cu^{2+}(aq) + 2\ e^- \longrightarrow Cu(s)$

If nothing else but electron flow took place, the concentration of Zn^{2+} ions in the anode compartment would increase as zinc metal became oxidized, building up positive charge in the anode solution. The concentration of Cu^{2+} ions in the cathode compartment

0.95

Cengage Learning/Charles D. Winters

A grapefruit battery. Alessandro Volta demonstrated that two different metals immersed in an electrolyte generate a voltage. Zinc and copper electrodes in the acidic solution within a grapefruit produce 0.95 V. To learn how the grapefruit battery works, see Goodisman, J. *Journal of Chemical Education,* Vol. 78, 2001; p. 516.

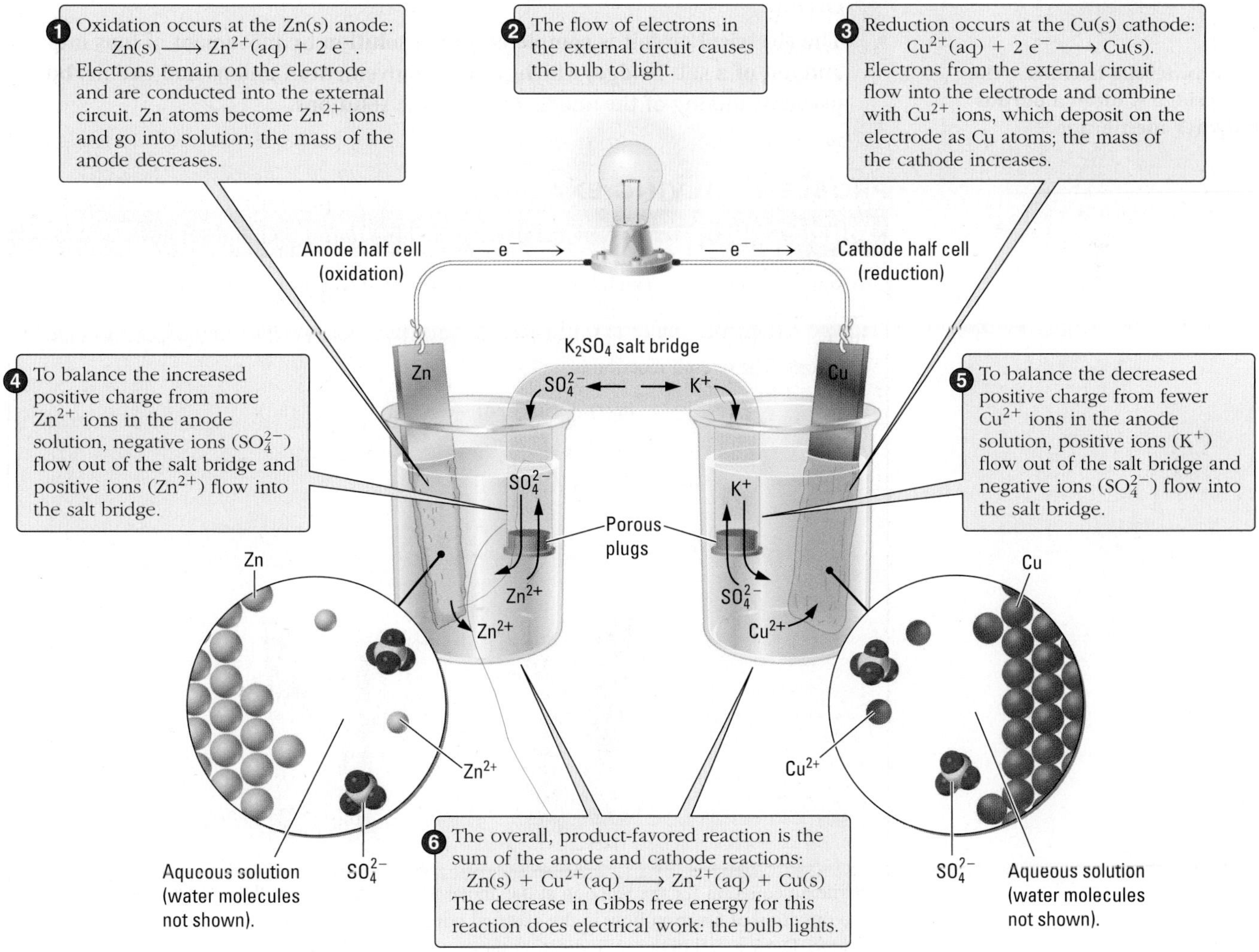
1 Oxidation occurs at the Zn(s) anode: $Zn(s) \longrightarrow Zn^{2+}(aq) + 2\ e^-$. Electrons remain on the electrode and are conducted into the external circuit. Zn atoms become Zn^{2+} ions and go into solution; the mass of the anode decreases.
2 The flow of electrons in the external circuit causes the bulb to light.
3 Reduction occurs at the Cu(s) cathode: $Cu^{2+}(aq) + 2\ e^- \longrightarrow Cu(s)$. Electrons from the external circuit flow into the electrode and combine with Cu^{2+} ions, which deposit on the electrode as Cu atoms; the mass of the cathode increases.
Anode half cell (oxidation)
$— e^- \longrightarrow$
$— e^- \longrightarrow$
Cathode half cell (reduction)
K_2SO_4 salt bridge
SO_4^{2-}
K^+
Zn
Cu
4 To balance the increased positive charge from more Zn^{2+} ions in the anode solution, negative ions (SO_4^{2-}) flow out of the salt bridge and positive ions (Zn^{2+}) flow into the salt bridge.
5 To balance the decreased positive charge from fewer Cu^{2+} ions in the anode solution, positive ions (K^+) flow out of the salt bridge and negative ions (SO_4^{2-}) flow into the salt bridge.
SO_4^{2-}
K^+
Porous plugs
Zn^{2+}
Zn^{2+}
SO_4^{2-}
Cu^{2+}
Zn
Cu
Zn^{2+}
Cu^{2+}
Aqueous solution (water molecules not shown).
SO_4^{2-}
6 The overall, product-favored reaction is the sum of the anode and cathode reactions: $Zn(s) + Cu^{2+}(aq) \longrightarrow Zn^{2+}(aq) + Cu(s)$ The decrease in Gibbs free energy for this reaction does electrical work: the bulb lights.
SO_4^{2-}
Aqueous solution (water molecules not shown).

Figure 17.5 A voltaic cell.

would decrease as Cu^{2+} ions became reduced to metallic copper. There would be more SO_4^{2-} ions than Cu^{2+} ions, so the cathode solution would become negatively charged. Because of this charge imbalance, the flow of electrons in the wires would stop.

The charge imbalance can be avoided by using a salt bridge to connect the two compartments. In Figure 17.5 the salt bridge is an upside-down U-shaped tube containing a solution of K_2SO_4 confined by a porous plug at each end. A **salt bridge** *is a solution of a salt arranged so that the bulk of that solution cannot flow into the half-cell solutions, but ions can pass freely.* If you think of the cell as an electrical circuit, the salt bridge completes the circuit. As electrons (negative charges) flow through the wire from the zinc electrode to the copper electrode, negative ions (SO_4^{2-} in this case) move into the salt bridge from the cathode solution and out of the salt bridge into the anode solution. That is, both electrons and negative ions move around a complete circuit in a clockwise direction. In the salt bridge, positive ions move in the opposite direction, helping to conduct current. Zn^{2+} ions enter the salt bridge from the anode solution and K^+ ions move into the cathode solution. This flow of ions offsets the charge imbalance and completes an electrical circuit, allowing current to flow.

All voltaic cells and batteries have these characteristics:

- **The oxidation-reduction reaction that occurs must be product-favored.**
- **Oxidation occurs at the anode.**
- **Reduction occurs at the cathode.**
- **Electrons move from the anode to the cathode through an external electrical circuit.**
- **The electrical circuit is completed in the solution by movement of ions into and out of a salt bridge, which permits movement of anions and cations but prevents mixing of the anode and cathode solutions.**

In commercial batteries, the salt bridge is often a porous polymer membrane.

PROBLEM-SOLVING EXAMPLE 17.3

Voltaic Cells

A simple voltaic cell is assembled with Fe(s) and $Fe(NO_3)_2$(aq) in one compartment and Cu(s) and $Cu(NO_3)_2$(aq) in the other compartment. An external wire and light bulb connects the two electrodes, and a salt bridge containing aqueous $NaNO_3$ solution connects the two solutions. The overall reaction is

$$Fe(s) + Cu^{2+}(aq) \longrightarrow Cu(s) + Fe^{2+}(aq)$$

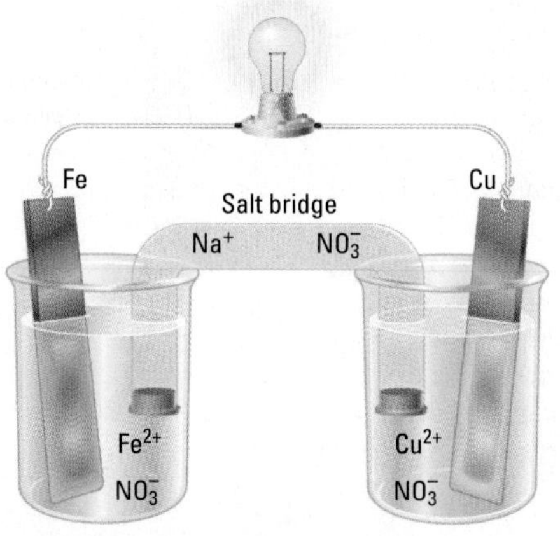
Fe
Cu
Salt bridge
Na+
NO3-
Fe2+
NO3-
Cu2+
NO3-

Write an equation for the reaction at the anode and the reaction at the cathode. Determine the direction of electron flow in the external wire and the direction of ion flow in the salt bridge. Draw a cell diagram, indicating the anode, the cathode, and the directions of electron flow in the external circuit and ion flow in the salt bridge.

Result

Anode reaction: $Fe(s) \longrightarrow Fe^{2+}(aq) + 2\,e^-$

Cathode reaction: $Cu^{2+}(aq) + 2\,e^- \longrightarrow Cu(s)$

The electrons flow through the wire from the anode to the cathode. Anions move from right to left, into and out of the salt bridge. Cations move from left to right, into and out of the salt bridge. The completed cell diagram is shown below.

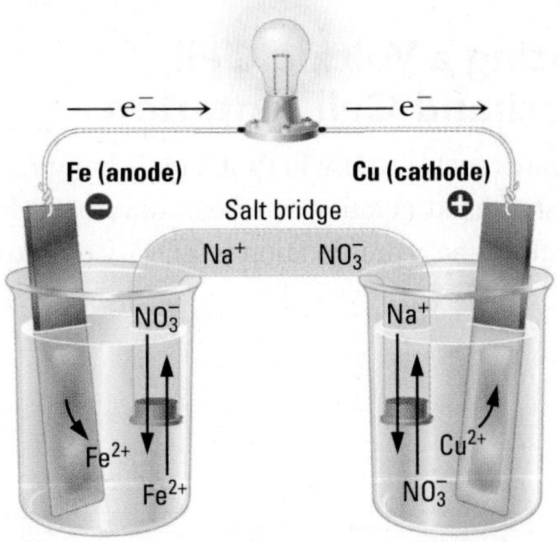
e⁻
e⁻
Fe (anode)
Cu (cathode)
Salt bridge
Na⁺
NO₃⁻
NO₃⁻
Na⁺
Fe²⁺
Fe²⁺
Cu²⁺
NO₃⁻

Analyze Given the overall cell reaction and a diagram of the setup of a voltaic cell, the half-reactions and directions of electron and ion flow are to be determined. Use the characteristics of voltaic cells in Figure 17.5 to draw the cell diagram.

Plan Use oxidation numbers to determine what is oxidized, what is the product of the oxidation, what is reduced, and what is the product of the reduction in the overall equation. Assign the oxidation reaction to be the anode half-cell and the reduction to be the cathode. Electrons released by oxidation at the anode travel to the cathode. Show negative ions moving around the electrical circuit in the same direction (clockwise or counterclockwise) as electrons and positive ions moving in the opposite direction.

Anode $\xrightarrow{e^-\text{ flow}}$	Cathode
oxidation	reduction
(loss of e^- s)	(gain of e^- s)
$Fe \rightarrow Fe^{2+} + 2e^-$	$Cu^{2+} + 2e^- \rightarrow Cu$

Execute The equation with oxidation numbers is

$$\overset{0}{Fe}(s) + \overset{+2}{Cu}{}^{2+}(aq) \longrightarrow \overset{0}{Cu}(s) + \overset{+2}{Fe}{}^{2+}(aq)$$

Anode reaction (oxidation: oxidation number ↑): $Fe(s) \longrightarrow Fe^{2+}(aq) + 2\,e^-$

Cathode reaction (reduction: oxidation number ↓): $Cu^{2+}(aq) + 2\,e^- \longrightarrow Cu(s)$

In the external circuit electrons flow from the anode (oxidation is loss of electrons and the electrons lost by Fe atoms enter the external circuit) to the cathode (reduction is gain of electrons). Because positive ions (Fe^{2+}) are being formed in the anode half-cell, some of the excess positive ions (Fe^{2+}) enter the salt bridge and some negative ions (NO_3^-) leave the salt bridge; this keeps the total charge from positive ions equal to the total charge from negative ions in the anode compartment. Because positive ions (Cu^{2+}) are being converted to uncharged Cu atoms in the cathode half-cell, negative ions begin to outnumber positive ions. This build up of negative charge is counteracted by movement of excess negative ions (NO_3^-) into the salt bridge and movement of positive ions (Na^+) out of the salt bridge; this keeps the total charge from positive ions equal to the total charge from negative ions in the cathode cell.

To maintain electrical neutrality, ions move into or out of the salt bridge solution. Moving two NO_3^- ions out of the salt bridge solution is electrically equivalent to moving one Fe^{2+} ion into the salt bridge solution.

The cell diagram is partly given in the statement of the problem. Based on the discussion above, the left compartment contains the anode and the right contains the cathode. Electrons flow from anode to cathode in the external circuit. Negative ions move from the cathode half-cell into the salt bridge and out of the salt bridge into the anode half-cell; positive ions move in the opposite direction. (Note that the negative ions move clockwise through the circuit, as do the electrons; positive ions move in the opposite direction, counterclockwise.)

PROBLEM-SOLVING PRACTICE 17.3

A voltaic cell is assembled to use this overall reaction.

$$Ni(s) + 2\,Ag^+(aq) \longrightarrow Ni^{2+}(aq) + 2\,Ag(s)$$

(a) Write half-reactions for this cell, and indicate which is the oxidation reaction and which is the reduction reaction.
(b) Name the electrodes at which these reactions take place.
(c) Determine the direction of flow of electrons in an external wire and electrical device connected between the electrodes.
(d) If a salt bridge connecting the two electrode compartments contains KNO_3, in what direction do the K^+ ions and the NO_3^- ions move?

17-3a Representing a Voltaic Cell with Shorthand Cell Notation

Drawing voltaic cell diagrams such as those in Problem-Solving Example 17.3 is quite time-consuming. Therefore a shorthand notation has been developed for representing a voltaic cell. For the cell shown below, the representation is $Zn(s)|Zn^{2+}(aq)||Cu^{2+}(aq)|Cu(s)$.

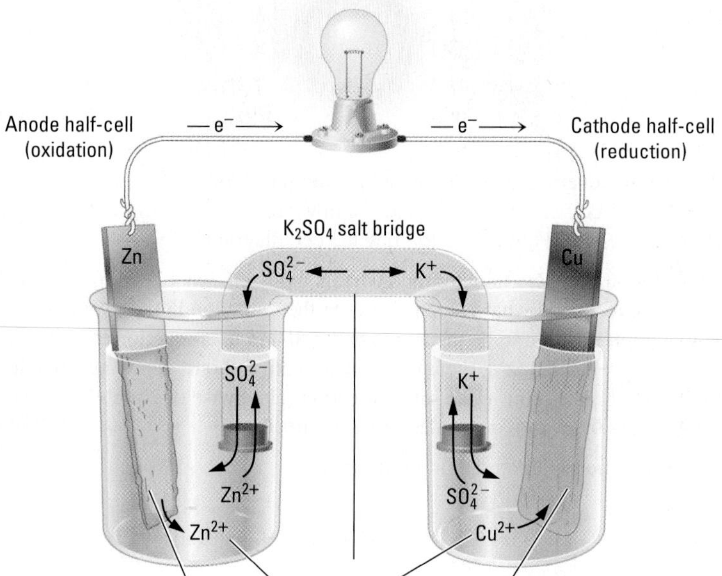
Anode half-cell (oxidation)
e⁻
e⁻
Cathode half-cell (reduction)
K₂SO₄ salt bridge
Zn
Cu
SO₄²⁻
K⁺
SO₄²⁻
K⁺
Zn²⁺
SO₄²⁻
Zn²⁺
Cu²⁺

Shorthand: $Zn(s)\ |\ Zn^{2+}(aq)\ ||\ Cu^{2+}(aq)\ |\ Cu(s)$

Cell reaction: $Zn(s) + Cu^{2+}(aq) \longrightarrow Zn^{2+}(aq) + Cu(s)$

In the shorthand notation, the anode half-cell is represented on the left, and the cathode half-cell on the right. The electrodes are written on the extreme left (anode, Zn) and extreme right (cathode, Cu). A single vertical line denotes a phase boundary (between the Zn(s) electrode and the aqueous solution, for example). A double vertical line denotes the salt bridge or porous barrier that separates half-cells. Within each half-cell the reactants are written first, followed by the products. The electron flow in the external circuit is from left to right (anode to cathode).

CONCEPTUAL EXERCISE 17.3

Battery Design

Devise an internal on-off switch for a battery that would not be a part of the flow of electrons.

CONCEPTUAL EXERCISE 17.4

Electrons and Ions in a Voltaic Cell

For the voltaic cell in Problem-Solving Example 17.3, assume that 3 mol e^- flows through the external circuit. If half of the charge build up in the anode compartment is balanced by movement of ions out of the salt bridge and half is balanced by movement of ions into the salt bridge, which ions move and how many moles of each?

17-4 Voltaic Cells and Cell Potential

The flow of electrons from the anode to the cathode of a voltaic cell is caused by the *difference in electrical potential energy* between the two electrodes. Just as water flows downhill in response to a difference in gravitational potential energy, so an electric current flows from an electrode of higher electrical potential energy to an electrode of lower potential energy. The moving water can do work; so can the electric current. For example, the current could run a motor.

The quantity of electrical work done is proportional to the quantity of electric charge that goes from higher to lower potential energy as well as to the size of the potential energy difference.

This is similar to comparing the quantity of work a few drops of water can do when falling 100 m with that possible when a few tons of water fall the same distance.

$$\text{Electrical work} = \text{electrical charge} \times \text{potential energy difference}$$

Electrical charge is measured in coulombs. The charge on a single electron is very small (-1.6022×10^{-19} C), so it takes 6.24×10^{18} electrons to produce 1 coulomb of charge. A **coulomb, C,** *is the quantity of charge that passes a fixed point in an electrical circuit when a current of 1 ampere flows for 1 second.* The **ampere, A,** is *the SI unit of electric current.*

$$\frac{1.6022 \times 10^{-19}\,\text{C}}{1\,e^-} \times 6.24 \times 10^{18}\,e^- = 1\,\text{C}$$

An average lightning strike generates 10,000 to 30,000 amperes.

$$1 \text{ coulomb} = 1 \text{ ampere} \times 1 \text{ second} \quad \text{or} \quad 1 \text{ ampere} = 1 \text{ coulomb/1 second}$$

EXERCISE 17.5

Electrical Charges

Which has the larger charge, 1.0 C or Avogadro's number (1.0 mol) of electrons?

Electrical potential energy difference is measured in volts. The **volt, V,** *is defined such that one joule of work is performed when one coulomb of charge moves through a potential difference of one volt*:

When a single electron moves through a potential of 1 V, the work done is one electron-volt, abbreviated eV.

$$1 \text{ volt} = \frac{1 \text{ joule}}{1 \text{ coulomb}} \quad \text{or} \quad 1 \text{ joule} = 1 \text{ volt} \times 1 \text{ coulomb}$$

The electrical potential energy difference of a voltaic cell, commonly *called its* **cell potential** or **cell voltage**, shows how much work a cell can do for each coulomb of charge that the chemical reaction transfers. The cell potential can be measured with a voltmeter.

The cell potential of a voltaic cell depends on the chemical characteristics of the substances that make up the cell, and on their partial pressures if they are gases or their concentrations if they are solutes in solution. It also varies with temperature. The quantity of charge (coulombs) depends on how much of each substance reacts. Consider the 1.5-V batteries shown in Figure 17.6. They have the same voltage because they have electrodes with the same potential energy difference. However, a larger battery is capable of far more work than a smaller one, because the larger one contains larger quantities of oxidizing and reducing agents. In this section and the next, we consider how cell potential depends on the materials from which a cell is made. In Section 17-6, we will consider the question of how much electrical work a cell can do.

DURACELL
Panasonic
Energizer

Cengage Learning/Charles D. Winters

Figure 17.6 Dry cell batteries. The larger batteries can do more work because they contain greater amounts of the oxidizing and reducing agents.

17-4a Cell Potential and Half-cells

The potential for a voltaic cell is measured as the potential difference between the cathode and the anode. That is, the positive terminal of a voltmeter is connected to the cathode, the negative terminal is connected to the anode, and the meter displays the difference in potential between cathode and anode. Thus,

$$E_{\text{cell}} = E_{\text{cathode}} - E_{\text{anode}}$$

It is important that no chemical reaction occurs during measurement of a cell potential; the voltmeter must have a very high electrical resistance so that no current flows.

Every redox reaction can be thought of as the sum of two half-reactions—a reduction half-reaction and an oxidation half-reaction. Every voltaic cell consists of two half-cells—a cathode half-cell where reduction occurs and an anode half-cell where oxidation occurs. Because a cell potential is calculated as the difference between two half-cells, it is convenient to assign a potential to every possible half-cell; from such half-cell potentials, the cell potential for any voltaic cell could be calculated.

The convention of assigning potentials to half-cells is similar to the convention of tabulating standard enthalpies of formation; in both cases a relatively small table of data can provide information about a large number of different reactions.

CONCEPTUAL EXERCISE 17.6

Cell Potentials and Half-cells

Is it reasonable to conclude that a potential could be assigned to each half-cell in a voltaic cell, based on these data for three voltaic cells? Explain.

$Zn(s)\,|\,Zn^{2+}(aq)\,\|\,Cu^{2+}(aq)\,|\,Cu(s)$	cell potential = 1.10 V
$Zn(s)\,|\,Zn^{2+}(aq)\,\|\,Ag^{+}(aq)\,|\,Ag(s)$	cell potential = 1.56 V
$Cu(s)\,|\,Cu^{2+}(aq)\,\|\,Ag^{+}(aq)\,|\,Ag(s)$	cell potential = 0.46 V

Because the cell potential for a given pair of half-cells varies with partial pressures and concentrations, **standard-state conditions** are defined as *conditions where all reactants and products are present as pure solids, pure liquids, gases at 1 bar pressure, or solutes at 1-M concentration. A cell potential measured under standard-state conditions* is a **standard potential ($E°$)**. Unless specified otherwise, $E°$ values can be assumed to be for 25 °C (298 K). By definition, *standard cell potentials for product-favored reactions are positive. If the cell potential is negative, the reaction is reactant-favored; that is, the reverse reaction is product-favored.* For standard-state conditions, the equation above becomes

$$E°_{\text{cell}} = E°_{\text{cathode}} - E°_{\text{anode}}$$

For example, consider $Zn(s)\,|\,Zn^{2+}(aq)\,\|\,Cu^{2+}(aq)\,|\,Cu(s)$, the first voltaic cell we encountered. The cell is shown in Figure 17.5 (← **Sec. 17-3**); it is based on the reaction shown in Figure 17.2 (← **Sec. 17-2**). The half-reactions and the overall reaction are

Anode: $Zn(s) \longrightarrow Zn^{2+}(aq, 1\ M) + 2\ e^-$

Cathode: $Cu^{2+}(aq, 1\ M) + 2\ e^- \longrightarrow Cu(s)$

Overall: $Zn(s) + Cu^{2+}(aq, 1\ M) \longrightarrow Zn^{2+}(aq, 1\ M) + Cu(s)$

The experimentally determined half-cell potentials are:

$$E°_{Zn^{2+}(aq)|Zn(s)} = -0.76\ V$$

$$E°_{Cu^{2+}(aq)|Cu(s)} = +0.34\ V$$

$$E°_{\text{cell}} = E°_{\text{cathode}} - E°_{\text{anode}} = E°_{Cu^{2+}(aq)|Cu(s)} - E°_{Zn^{2+}(aq)|Zn(s)}$$

$$= (+0.34\ V) - (-0.76\ V) = +1.10\ V$$

Figure 17.2 shows that the overall reaction is product-favored. Earlier we stated that the cell potential is positive for a product-favored reaction. This is confirmed by the positive $E°_{\text{cell}}$ value just calculated.

A very important point about ***half-cell potentials*** is that they ***apply to half-cells (that is, electrodes), not to half-reactions***. That is, the value $E°_{Zn^{2+}(aq)|Zn(s)} = -0.76$ V, which is measured with no reaction occurring, applies to the $Zn^{2+}(aq)\,|\,Zn(s)$ half-cell: a Zn(s) electrode in a beaker containing 1-M $Zn^{2+}(aq)$ (together with negative ions to balance the positive charge) connected by a salt bridge and an external circuit to another half-cell. It does not matter whether the half-reaction in the half-cell when current is drawn is an oxidation or a reduction.

17-4b Standard Half-cell Potentials, the Standard Hydrogen Electrode, and Cell Potentials

It is important to note that it is not possible to measure the potential for just a single half-cell. Only a *difference* in potential energy between two half-cells can be measured. This dilemma is resolved by choosing a half-cell that can serve as a reference standard. All other half-cells are then compared to that standard half-cell. The half-cell chosen as the standard is the **standard hydrogen electrode** *in which hydrogen gas at a pressure of 1 bar is bubbled over an inert platinum metal electrode immersed in 1-M aqueous acid* (Figure 17.7).

$$\mathrm{Pt}|\mathrm{H_2(g,\ 1\ bar)}|\mathrm{H^+(aq,\ 1\ M)}$$

This half-cell is *arbitrarily assigned* a potential of exactly 0 V. By convention, when half-cell potentials are measured, the standard hydrogen electrode is always connected to the negative terminal on the voltmeter. Under these conditions, *because the potential of the standard hydrogen electrode is 0 V, the overall cell potential equals the potential of the other electrode.*

To see how this works, consider the cell

$$\mathrm{Pt}|\mathrm{H_2(g,\ 1\ bar)}|\mathrm{H^+(aq,\ 1\ M)}||\mathrm{Cu^{2+}(aq)}|\mathrm{Cu(s)}$$

This cell is shown in Figure 17.8. The difference in potential between the voltmeter's positive lead and its negative lead is +0.34 V. This means that the standard half-cell potential is +0.34 V for the $Cu^{2+}(aq)|Cu(s)$ half-cell.

The positive meter reading indicates that, *if the cell produced current,* electrons (negative particles) would enter the negative lead of the meter. Therefore, oxidation would be taking place at the standard hydrogen electrode and it would be the anode. Thus, the copper electrode must be the cathode, where reduction takes place. The cell reaction is the sum of these half-cell reactions.

Remember that during measurement of a cell potential, no current flows and no chemical reaction takes place.

Anode: $$\mathrm{H_2(g,\ 1\ bar)} \longrightarrow \mathrm{2\ H^+(aq,\ 1\ M) + 2\ e^-}$$

Cathode: $$\mathrm{Cu^{2+}(aq,\ 1\ M) + 2\ e^-} \longrightarrow \mathrm{Cu(s)}$$

Overall: $$\mathrm{H_2(g,\ 1\ bar) + Cu^{2+}(aq,\ 1\ M)} \longrightarrow \mathrm{2\ H^+(aq,\ 1\ M) + Cu(s)}$$

$$E^\circ_{\text{cell}} = E^\circ_{\text{cathode}} - E^\circ_{\text{anode}} = E^\circ_{\text{Cu(s)|Cu}^{2+}\text{(aq, 1M)}} - 0\ \text{V} = +0.34\ \text{V}$$

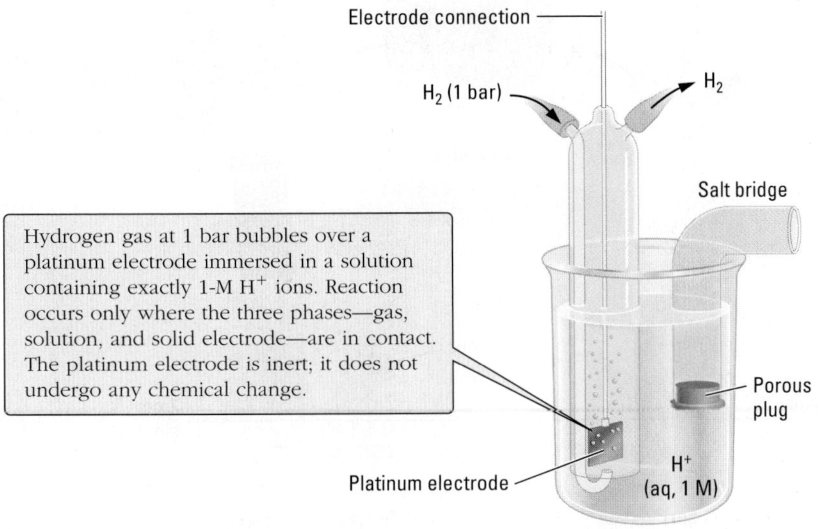
Electrode connection
H₂ (1 bar)
H₂
Salt bridge
Hydrogen gas at 1 bar bubbles over a platinum electrode immersed in a solution containing exactly 1-M H⁺ ions. Reaction occurs only where the three phases—gas, solution, and solid electrode—are in contact. The platinum electrode is inert; it does not undergo any chemical change.
Porous plug
H⁺ (aq, 1 M)
Platinum electrode

Figure 17.7 The standard hydrogen electrode. E° = 0 V for this electrode, by definition.

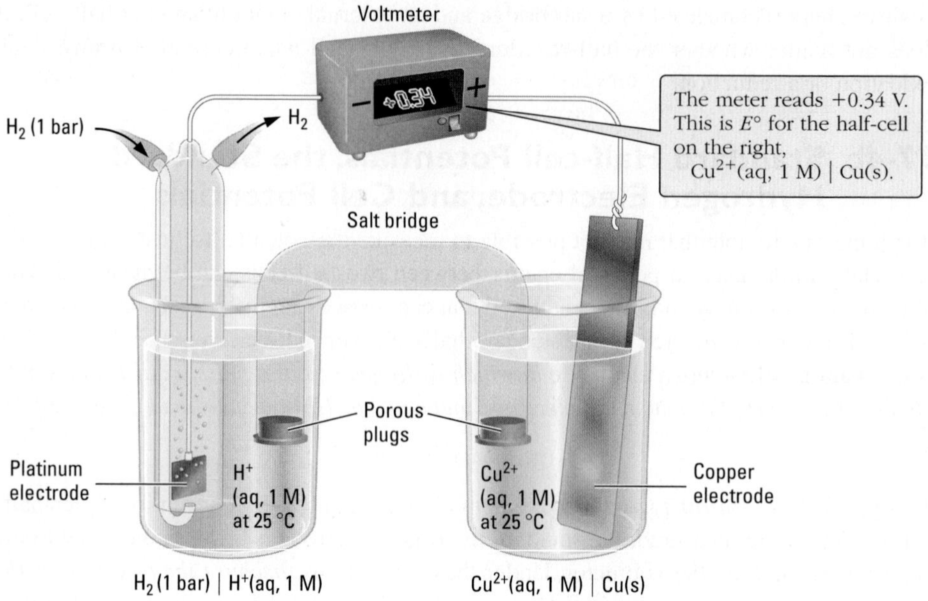
Voltmeter
+0.34
H$_2$ (1 bar)
H$_2$
The meter reads +0.34 V. This is $E°$ for the half-cell on the right, Cu^{2+}(aq, 1 M) | Cu(s).
Salt bridge
Porous plugs
Platinum electrode
H^+ (aq, 1 M) at 25 °C
Cu^{2+} (aq, 1 M) at 25 °C
Copper electrode
H$_2$ (1 bar) | H^+(aq, 1 M)
Cu^{2+}(aq, 1 M) | Cu(s)

Figure 17.8 A voltaic cell involving a standard hydrogen electrode and a Cu^{2+}(aq)|Cu(s) half-cell.

As another example, consider this cell, which is shown in Figure 17.9.

$$\text{Pt}|\text{H}_2(\text{g, 1 bar})|\text{H}^+(\text{aq, 1 M})||\text{Zn}^{2+}(\text{aq})|\text{Zn(s)}$$

In this case, the voltmeter indicates that $E°_{cell} = -0.76$ V. This means that the standard half-cell potential for the Zn^{2+}(aq)|Zn(s) half-cell is -0.76 V.

The convention for the shorthand cell notation is that the anode is on the left and the cathode on the right (← **Sec. 17-3a**). Thus, *for the cell as written above and diagrammed in Figure 17.9,* we assume that, if current were drawn from the cell, oxidation would occur at the standard hydrogen electrode and reduction at the Zn electrode. Therefore the half-reactions and overall reaction *when current is drawn from the cell as written* are

Anode: $\text{Zn}^{2+}(\text{aq, 1 M}) + 2\,e^- \longrightarrow \text{Zn(s)}$

Cathode: $\text{H}_2(\text{g, 1 bar}) \longrightarrow 2\,\text{H}^+(\text{aq, 1 M}) + 2\,e^-$

Overall: $\text{Zn}^{2+}(\text{aq, 1 M}) + \text{H}_2(\text{g, 1 bar}) \longrightarrow \text{Zn(s)} + 2\,\text{H}^+\,(\text{aq, 1 M})$

$$E°_{cell} = E°_{cathode} - E°_{anode} = E°_{\text{Zn(s)}|\text{Zn}^{2+}(\text{aq, 1 M})} - 0\text{ V} = -0.76\text{ V}$$

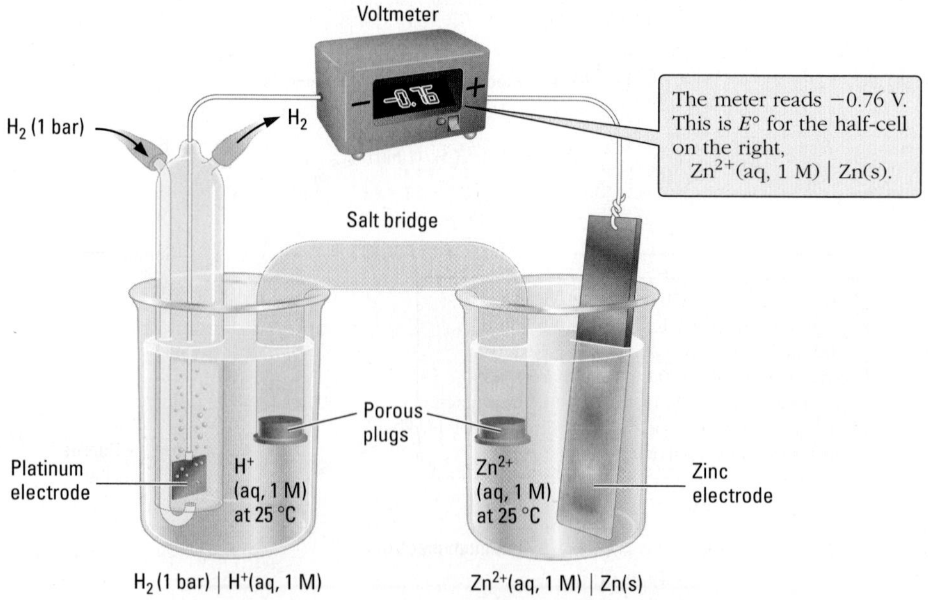
Voltmeter
-0.76
H$_2$ (1 bar)
H$_2$
The meter reads −0.76 V. This is $E°$ for the half-cell on the right, Zn^{2+}(aq, 1 M) | Zn(s).
Salt bridge
Porous plugs
Platinum electrode
H^+ (aq, 1 M) at 25 °C
Zn^{2+} (aq, 1 M) at 25 °C
Zinc electrode
H$_2$ (1 bar) | H^+(aq, 1 M)
Zn^{2+}(aq, 1 M) | Zn(s)

Figure 17.9 A voltaic cell involving a standard hydrogen electrode and a Zn^{2+}(aq)|Zn(s) half-cell.

Thus, we assign $E° = -0.76$ V to the $Zn(s)|Zn^{2+}$(aq, 1 M) half-cell.

The negative voltmeter reading, that is, the negative cell potential, indicates that this overall cell reaction is reactant-favored and the reverse reaction would occur if the cell were set up and current flowed through the external circuit. Thus, the overall reaction that would actually occur is

$$Zn(s) + 2\,H^+(aq) \longrightarrow Zn^{2+}(aq) + H_2(g)$$

CONCEPTUAL EXERCISE 17.7

Determining What Happens in a Voltaic Cell

Devise an experiment that would show that Zn is being oxidized when current is drawn from the voltaic cell shown in Figure 17.9.

Standard potentials of many different half-cells can be measured by comparing them with the standard hydrogen electrode. For example, in a cell consisting of the $Ag^+|Ag$ half-cell connected to a standard hydrogen electrode, a reading of +0.7991 V is obtained when the standard hydrogen electrode is connected to the negative voltmeter lead. This means that the standard hydrogen electrode must be the anode and the reactions are

$$H_2(g, 1\ bar) \longrightarrow 2\,H^+(aq, 1\ M) + 2\,e^- \qquad E°_{anode} = 0\ V$$

$$2\,Ag^+(aq, 1\ M) + 2\,e^- \longrightarrow 2\,Ag(s) \qquad E°_{cathode} = ?$$

$$H_2(g, 1\ bar) + 2\,Ag^+(aq, 1\ M) \longrightarrow 2\,H^+(aq, 1\ M) + 2\,Ag(s) \qquad E°_{cell} = +0.7991\ V$$

The potential for the Ag^+(aq, 1 M)|Ag(s) half-cell is therefore +0.7991 V.

PROBLEM-SOLVING EXAMPLE 17.4

Determining a Half-cell Potential

Under standard-state conditions, the voltaic cell shown generates a potential difference of $E°_{cell} = 0.36$ V at 25 °C. The overall cell reaction is

$$Zn(s) + Cd^{2+}(aq, 1\ M) \longrightarrow Zn^{2+}(aq, 1\ M) + Cd(s)$$

The standard half-cell potential for $Zn(s)|Zn^{2+}$(aq, 1 M) is −0.76 V.

(a) Determine which electrode is the anode and which is the cathode.

(b) Show the direction of electron flow through the circuit outside the cell when current flows, and complete the cell diagram.

(c) Calculate the standard half-cell potential for $Cd^{2+}(aq)|Cd(s)$.

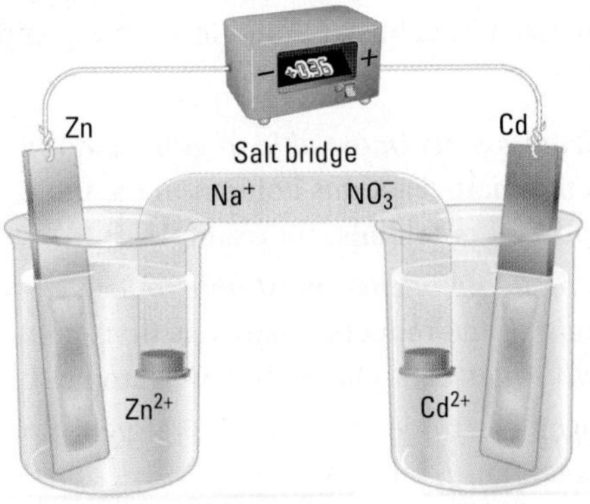

−
+
Zn
Cd
Salt bridge
Na+
NO3−
Zn2+
Cd2+

Result

(a) Zinc is the anode, and cadmium is the cathode.

(b) The completed cell diagram is shown on the next page.

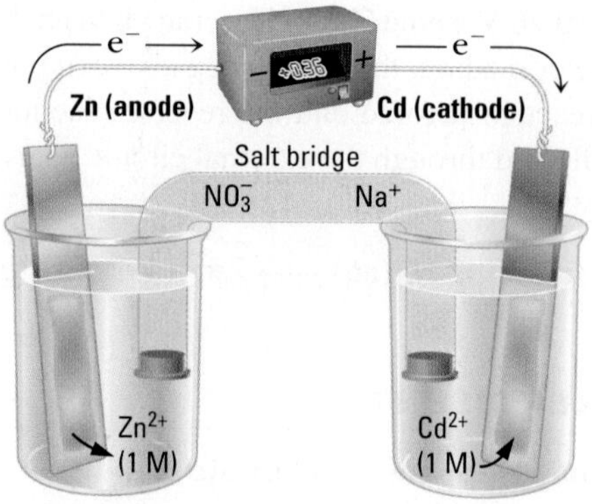
e⁻
+0.36
e⁻
Zn (anode)
Cd (cathode)
Salt bridge
NO3⁻
Na⁺
Zn2+
(1 M)
Cd2+
(1 M)

(c) The standard half-cell potential for $Cd^{2+}(aq)|Cd(s)$ is -0.40 V.

Analyze Oxidation occurs at the anode, reduction at the cathode. Electrons released by oxidation flow from the anode through the circuit to the cathode. Standard cell potential is calculated as $E^\circ_{cell} = E^\circ_{cathode} - E^\circ_{anode}$.

Plan Use oxidation numbers to determine what is oxidized and what is reduced. Draw the cell diagram, label the anode and cathode, and indicate electron flow from anode to cathode. Use the standard cell potential, the standard half-cell potential for Zn, and the equation $E^\circ_{cell} = E^\circ_{cathode} - E^\circ_{anode}$ to calculate the desired standard half-cell potential.

Execute See the figure in the Result section.

$$Zn(s) \longrightarrow Zn^{2+}(aq, 1\text{ M}) + 2\,e^- \qquad E^\circ_{anode} = -0.76\text{ V (anode)}$$

$$Cd^{2+}(aq, 1\text{M}) + 2\,e^- \longrightarrow Cd(s) \qquad E^\circ_{cathode} = ?\text{ V (cathode)}$$

$$Zn(s) + Cd^{2+}(aq, 1\text{M}) \longrightarrow Zn^{2+}(aq, 1\text{ M}) + Cd(s) \qquad E^\circ_{cell} = +0.36\text{ V}$$

$$E^\circ_{cathode} = E^\circ_{cell} + E^\circ_{anode} = 0.36\text{ V} + (-0.76\text{ V}) = -0.40\text{ V}$$

PROBLEM-SOLVING PRACTICE 17.4

Given this reaction, its standard potential, and the standard half-cell potential of 0.34 V for the $Cu^{2+}|Cu$ half-cell, calculate E° for the $Fe(s)|Fe^{2+}(aq)$ half-cell.

$$Fe(s) + Cu^{2+}(aq, 1\text{ M}) \longrightarrow Fe^{2+}(aq, 1\text{ M}) + Cu(s) \qquad E^\circ_{cell} = +0.78\text{ V}$$

17-5 Using Standard Half-cell Potentials

A half-cell potential measured under standard conditions by comparison with a standard hydrogen electrode is called a **standard half-cell potential** or a **standard electrode potential**. The results of a great many measurements of cell potentials such as the ones just described are summarized in Table 17.1 and in a more complete table in Appendix I. Note from Table 17.1 that:

- The strongest *oxidizing agents* (*most able to gain electrons and be reduced*) are at *top left* of the reduction half-reactions column in the table. They have the most *positive* standard half-cell potentials, for example, $F_2(g)$.
- The most powerful *reducing agents* (*most able to lose electrons and be oxidized*) are at the *bottom right* of the reduction half-reactions column in the table. They have the most *negative* half-cell potentials, for example, Li(s).
- The standard hydrogen electrode serves as the reference point for all half-cells; its standard half-cell potential, E°, is 0 V.

All the half-cells above the hydrogen half-cell have a positive standard half-cell potential because *when paired with the standard hydrogen electrode, the corresponding half-reactions occur as reductions*. For example, Cu^{2+} ions are reduced to copper metal by $H_2(g)$ when current flows in the cell shown in Figure 17.8. In this case, the standard half-cell potential is +0.34 V.

$$Cu^{2+}(aq, 1\ M, 25\ °C) + 2\ e^- \longrightarrow Cu(s) \qquad E° = +0.34\ V$$

Half-cells below the hydrogen half-cell have a *negative* standard half-cell potential because when paired with the standard hydrogen electrode, their corresponding half-reactions occur as oxidations. For example, Zn is oxidized by $H^+(aq)$ to $Zn^{2+}(aq)$ when current flows in the cell diagrammed in Figure 17.9. In this case, the standard half-cell potential is −0.76 V.

$$Zn^{2+}(aq, 1\ M, 25\ °C) + 2\ e^- \longrightarrow Zn(s) \qquad E° = -0.76\ V$$

Table 17.1 Standard Half-cell Potentials in Aqueous Solution at 25 °C*

Reduction Half-reaction			Half-cell	E° (V)
Oxidizing Agent		Reducing Agent		
$F_2(g) + 2\ e^-$	$\longrightarrow$	$2\ F^-(aq)$	$F_2(g)\vert F^-(aq)\vert Pt$	+2.87
$H_2O_2(aq) + 2\ H^+(aq) + 2\ e^-$	$\longrightarrow$	$2\ H_2O(\ell)$	$H_2O_2(aq), H^+(aq), H_2O(\ell)\vert Pt$	+1.763
$PbO_2(s) + SO_4^{2-}(aq) + 4\ H^+(aq) + 2\ e^-$	$\longrightarrow$	$PbSO_4(s) + 2\ H_2O(\ell)$	$PbO_2(s)\vert SO_4^{2-}(aq), H^+(aq)\vert PbSO_4(s)\vert Pb$	+1.690
$Au^{3+}(aq) + 3\ e^-$	$\longrightarrow$	$Au(s)$	$Au^{3+}(aq)\vert Au(s)$	+1.52
$MnO_4^-(aq) + 8\ H^+(aq) + 5\ e^-$	$\longrightarrow$	$Mn^{2+}(aq) + 4\ H_2O(\ell)$	$MnO_4^-(aq), H^+(aq), Mn^{2+}(aq)\vert Pt$	+1.51
$Cr_2O_7^{2-}(aq) + 14\ H^+(aq) + 6\ e^-$	$\longrightarrow$	$2\ Cr^{3+}(aq) + 7\ H_2O(\ell)$	$Cr_2O_7^{2-}(aq), H^+(aq), Cr^{3+}(aq)\vert Pt$	+1.36
$Cl_2(g) + 2\ e^-$	$\longrightarrow$	$2\ Cl^-(aq)$	$Cl_2(g)\vert Cl^-(aq)\vert Pt$	+1.358
$O_2(g) + 4\ H^+(aq) + 4\ e^-$	$\longrightarrow$	$2\ H_2O(\ell)$	$O_2(g)\vert H^+(aq)\vert Pt$	+1.229
$Br_2(\ell) + 2\ e^-$	$\longrightarrow$	$2\ Br^-(aq)$	$Br_2(\ell)\vert Br^-(aq)\vert Pt$	+1.066
$NO_3^-(aq) + 4\ H^+(aq) + 3\ e^-$	$\longrightarrow$	$NO(g) + 2\ H_2O(\ell)$	$NO_3^-(aq), H^+(aq)\vert NO(g)\vert Pt$	+0.96
$OCl^-(aq) + H_2O(\ell) + 2\ e^-$	$\longrightarrow$	$Cl^-(aq) + 2\ OH^-(aq)$	$OCl^-(aq), Cl^-(aq), OH^-(aq)\vert Pt$	+0.89
$Hg^{2+}(aq) + 2\ e^-$	$\longrightarrow$	$Hg(\ell)$	$Hg^{2+}(aq)\vert Hg(\ell)$	+0.8535
$Ag^+(aq) + e^-$	$\longrightarrow$	$Ag(s)$	$Ag^+(aq)\vert Ag(s)$	+0.7991
$Hg_2^{2+}(aq) + 2\ e^-$	$\longrightarrow$	$2\ Hg(\ell)$	$Hg_2^{2+}(aq)\vert Hg(\ell)$	+0.7960
$Fe^{3+}(aq) + e^-$	$\longrightarrow$	$Fe^{2+}(aq)$	$Fe^{3+}(aq), Fe^{2+}(aq)\vert Pt$	+0.771
$I_2(s) + 2\ e^-$	$\longrightarrow$	$2\ I^-(aq)$	$I_2(s)\vert I^-(aq)\vert Pt$	+0.535
$O_2(g) + 2\ H_2O(\ell) + 4\ e^-$	$\longrightarrow$	$4\ OH^-(aq)$	$O_2(g)\vert OH^-(aq)\vert Pt$	+0.401
$Cu^{2+}(aq) + 2\ e^-$	$\longrightarrow$	$Cu(s)$	$Cu^{2+}(aq)\vert Cu(s)$	+0.340
$Sn^{4+}(aq) + 2\ e^-$	$\longrightarrow$	$Sn^{2+}(aq)$	$Sn^{4+}(aq), Sn^{2+}(aq)\vert Pt$	+0.15
$2\ H^+(aq) + 2\ e^-$	$\longrightarrow$	$H_2(g)$	$H^+(aq)\vert H_2(g)\vert Pt$	0
$Sn^{2+}(aq) + 2\ e^-$	$\longrightarrow$	$Sn(s)$	$Sn^{2+}(aq)\vert Sn(s)$	−0.1375
$Ni^{2+}(aq) + 2\ e^-$	$\longrightarrow$	$Ni(s)$	$Ni^{2+}(aq)\vert Ni(s)$	−0.25
$PbSO_4(s) + 2\ e^-$	$\longrightarrow$	$Pb(s) + SO_4^{2-}(aq)$	$PbSO_4(s)\vert SO_4^{2-}(aq)\vert Pb(s)$	−0.3505
$Cd^{2+}(aq) + 2\ e^-$	$\longrightarrow$	$Cd(s)$	$Cd^{2+}(aq)\vert Cd(s)$	−0.403
$Fe^{2+}(aq) + 2\ e^-$	$\longrightarrow$	$Fe(s)$	$Fe^{2+}(aq)\vert Fe(s)$	−0.44
$Zn^{2+}(aq) + 2\ e^-$	$\longrightarrow$	$Zn(s)$	$Zn^{2+}(aq)\vert Zn(s)$	−0.763
$2\ H_2O(\ell) + 2\ e^-$	$\longrightarrow$	$H_2(g) + 2\ OH^-(aq)$	$H_2O(\ell), OH^-(aq)\vert H_2(g)\vert Pt$	−0.8277
$Al^{3+}(aq) + 3\ e^-$	$\longrightarrow$	$Al(s)$	$Al^{3+}(aq)\vert Al(s)$	−1.676
$Mg^{2+}(aq) + 2\ e^-$	$\longrightarrow$	$Mg(s)$	$Mg^{2+}(aq)\vert Mg(s)$	−2.356
$Na^+(aq) + e^-$	$\longrightarrow$	$Na(s)$	$Na^+(aq)\vert Na(s)$	−2.714
$K^+(aq) + e^-$	$\longrightarrow$	$K(s)$	$K^+(aq)\vert K(s)$	−2.925
$Li^+(aq) + e^-$	$\longrightarrow$	$Li(s)$	$Li^+(aq)\vert Li(s)$	−3.045

Data from Bard, A. J., Parsons, R., and Jordan, J. *Standard Potentials in Aqueous Solution.* New York: Marcel Dekker: 1985. International Union of Pure and Applied Chemistry Commission on Electrochemistry and Electroanalytical Chemistry.

*In volts (V) versus the standard hydrogen electrode. For cases where an electrode is needed to conduct electrons into or out of the half-cell, an electrode that does not react with the solutions is required; Pt is specified in most cases.

The negative standard half-cell potential indicates that, when paired with the standard hydrogen electrode, the *reverse* of the reaction in Table 17.1 is favored; zinc metal donates electrons and becomes oxidized.

$$Zn(s) \longrightarrow Zn^{2+}(aq,\ 1\ M,\ 25\ °C) + 2\ e^-$$

Here are some important points to notice about Table 17.1:

1. ***Each half-reaction is written as a reduction.*** Thus, a species on the left-hand side of each half-reaction is in a higher oxidation state than the species on the right-hand side. *Oxidizing agents are on the left side; reducing agents are on the right side.*
2. ***Each half-reaction can occur in either direction.*** A given substance can react at the anode or the cathode, depending on the conditions. For example, we have already seen cases in which H_2 is oxidized to H^+ and others in which H^+ is reduced to H_2 by different reactants.
3. ***The more positive the value of the standard half-cell potential, E°, the more easily the substance on the left-hand side of a half-reaction can be reduced.*** When a substance is easy to reduce, it is a strong oxidizing agent. Thus, $F_2(g)$ is the best oxidizing agent in the table, and Li^+ is the poorest oxidizing agent in the table. Other strong oxidizing agents are at the top left of the table:

$$H_2O_2(aq),\ PbO_2(s),\ Au^{3+}(aq),\ Cl_2(g),\ O_2(g)$$

Recall that an oxidizing agent is reduced when it oxidizes something else.

4. ***The less positive the value of the standard half-cell potential, E°, the less likely the reaction will occur as a reduction, and the more likely an oxidation (the reverse reaction) will occur.*** The farther down we go in the table, the better the reducing (electron-donating) ability of the atom, ion, or molecule on the right. Thus, Li(s) is the strongest reducing agent in the table, and F^- is the weakest reducing agent in the table. Other strong reducing agents are alkali and alkaline-earth metals at the lower right of the reduction half equation column in the table.

When they are involved in redox reactions, $F_2(g)$ is always reduced and Li(s) is always oxidized.

5. ***Under standard-state conditions, any species on the left of a half-reaction will oxidize any species that is below it on the right side of the table.*** For example, we can apply this rule to predict that $Br_2(\ell)$ will oxidize Mg(s). The overall reaction is found by adding the half-reactions, and the cell potential can be calculated as $E°_{cell} = E°_{cathode} - E°_{anode}$.

$$Br_2(\ell) + 2\ e^- \longrightarrow 2\ Br^-(aq) \qquad E°_{cathode} = +1.07\ V$$
$$Mg(s) \longrightarrow Mg^{2+}(aq) + 2\ e^- \qquad E°_{anode} = -2.36\ V$$
$$Br_2(\ell) + Mg(s) \longrightarrow Mg^{2+}(aq) + 2\ Br^-(aq) \qquad E°_{cell} = +3.43\ V$$

$$E°_{cell} = E°_{cathode} - E°_{anode} = 1.07\ V - (-2.36\ V) = +3.43\ V$$

The very large positive cell potential denotes a very product-favored reaction.

6. ***Standard half-cell potentials depend on the nature and concentration of reactants and products, but not on the quantity of each that reacts.*** Changing the stoichiometric coefficients for a half-reaction does *not* change the value of $E°$. For example, the $Fe^{3+}|Fe^{2+}$ half-cell has $E° = +0.771$ V whether the reaction is written as

$$Fe^{3+}(aq,\ 1\ M) + e^- \longrightarrow Fe^{2+}(aq,\ 1\ M) \qquad E° = +0.771\ V$$

or

$$2\ Fe^{3+}(aq,\ 1\ M) + 2\ e^- \longrightarrow 2\ Fe^{2+}(aq,\ 1\ M) \qquad E° = +0.771\ V$$

This fact about half-cell potentials may seem unusual at first, but consider that a half-cell potential is energy per unit charge (1 volt = 1 joule/1 coulomb). When a half-reaction is multiplied by some number, both the energy and the charge are multiplied by that number. Thus, the ratio of the energy to the charge (the potential) does not change.

In this respect, $E°$ values differ from $\Delta_r H°$ and $\Delta_r G°$ values, which do depend on the coefficients in thermochemical equations.

7. ***Standard half-cell potentials do not depend on the direction in which a half-cell reaction is written.*** The mixture of atoms, ions, or molecules specified in the short-

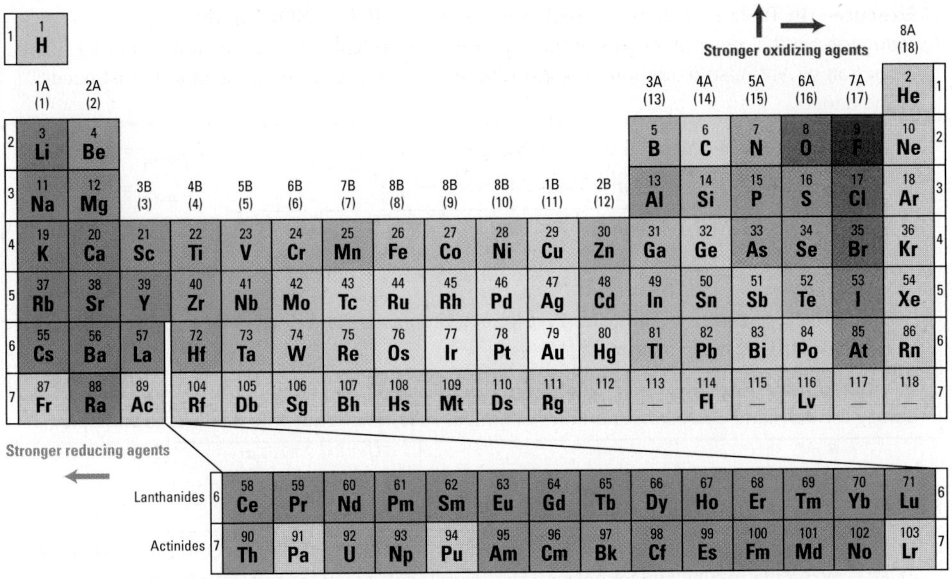
Stronger oxidizing agents
Stronger reducing agents
1A (1)
2A (2)
3B (3)
4B (4)
5B (5)
6B (6)
7B (7)
8B (8)
8B (9)
8B (10)
1B (11)
2B (12)
3A (13)
4A (14)
5A (15)
6A (16)
7A (17)
8A (18)
1 H
2 He
3 Li 4 Be 5 B 6 C 7 N 8 O 9 F 10 Ne
11 Na 12 Mg 13 Al 14 Si 15 P 16 S 17 Cl 18 Ar
19 K 20 Ca 21 Sc 22 Ti 23 V 24 Cr 25 Mn 26 Fe 27 Co 28 Ni 29 Cu 30 Zn 31 Ga 32 Ge 33 As 34 Se 35 Br 36 Kr
37 Rb 38 Sr 39 Y 40 Zr 41 Nb 42 Mo 43 Tc 44 Ru 45 Rh 46 Pd 47 Ag 48 Cd 49 In 50 Sn 51 Sb 52 Te 53 I 54 Xe
55 Cs 56 Ba 57 La 72 Hf 73 Ta 74 W 75 Re 76 Os 77 Ir 78 Pt 79 Au 80 Hg 81 Tl 82 Pb 83 Bi 84 Po 85 At 86 Rn
87 Fr 88 Ra 89 Ac 104 Rf 105 Db 106 Sg 107 Bh 108 Hs 109 Mt 110 Ds 111 Rg 112 — 113 — 114 Fl 115 — 116 Lv 117 — 118 —
Lanthanides 58 Ce 59 Pr 60 Nd 61 Pm 62 Sm 63 Eu 64 Gd 65 Tb 66 Dy 67 Ho 68 Er 69 Tm 70 Yb 71 Lu
Actinides 90 Th 91 Pa 92 U 93 Np 94 Pu 95 Am 96 Cm 97 Bk 98 Cf 99 Es 100 Fm 101 Md 102 No 103 Lr

Figure 17.10 Variation of standard half-cell potential among selected elements. Dark blue indicates the most negative potentials; bright red indicates the most positive potentials; gray indicates no data. Note that nonmetals have more positive standard half-cell potentials than metals.

hand description of a half-cell generates a particular electrical potential. Whether the corresponding half-cell reaction occurs as an oxidation or a reduction is determined by which other half-cell it is connected to in a cell. When a half-cell is coupled with another half-cell that is above it in Table 17.1, the half-cell reaction is an oxidation (the reverse of the half-reaction in Table 17.1); when a half-cell is coupled with another half-cell below it in Table 17.1, the half-cell reaction is a reduction. The half-cell potential does not change sign.

8. ***In the periodic table, elements that are the strongest oxidizing agents are in the upper right; elements that are the strongest reducing agents are on the left.*** Figure 17.10 shows trends in standard half-cell potentials for selected elements.

The preceding guidelines and the table of standard half-cell potentials enable us to *predict* whether redox reactions occur and to calculate E°_{cell}.

PROBLEM-SOLVING EXAMPLE 17.5

Does a Redox Reaction Occur?

As shown, copper metal, a transition element, reacts readily with nitric acid. Gold is also a transition metal. Use standard half-cell potentials to determine whether metallic gold reacts with 1-M nitric acid, HNO_3 at 25 °C.

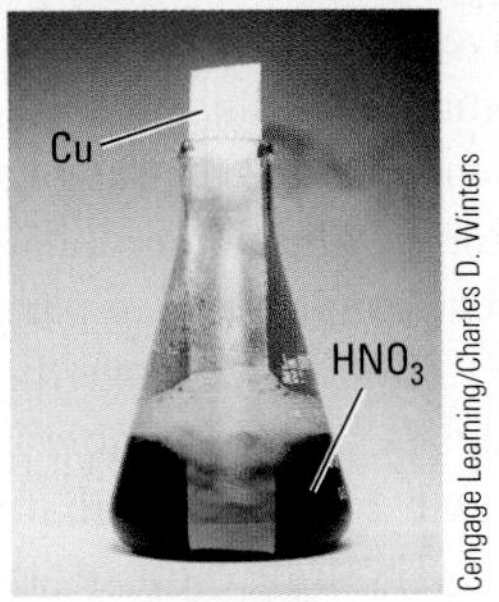
Cu
HNO3
Cengage Learning/Charles D. Winters

Cu + HNO_3

Result The cell potential, E°_{cell}, is −0.56 V. A negative cell potential indicates a reactant-favored reaction. Therefore, gold does not react with 1-M nitric acid.

Analyze The overall reaction must have Au(s) and HNO_3, that is, H^+(aq) and NO_3^-(aq), as reactants. The cathode half-reaction must be a reduction. A balanced equation can be written by combining half-cell reactions so that electrons are conserved. A positive standard cell potential indicates a product-favored reaction under standard conditions at a given temperature; a negative potential indicates a reactant-favored process.

Plan Use balanced half-cell reactions, their associated half-cell potentials, and $E^\circ_{cell} = E^\circ_{cathode} - E^\circ_{anode}$ to write a chemical equation and calculate the cell potential. Refer to Table 17.1 for half-reactions and potentials.

Execute In Table 17.1 there is a half-cell reaction involving $NO_3^-(aq)$ that occurs as a reduction, so that reaction occurs at the cathode; there is no half-cell reaction involving reduction of Au(s), so oxidation of Au(s) (the reverse of a reaction in the table) must occur at the anode.

$$NO_3^-(aq) + 4\,H^+(aq) + 3\,e^- \longrightarrow NO(g) + 2\,H_2O(\ell) \qquad E^\circ_{cathode} = +0.96\text{ V}$$
$$Au(s) \longrightarrow Au^{3+}(aq) + 3\,e^- \qquad E^\circ_{anode} = +1.52\text{ V}$$
$$NO_3^-(aq) + 4\,H^+(aq) + Au(s) \longrightarrow NO(g) + 2\,H_2O(\ell) + Au^{3+}(aq) \qquad E^\circ_{cell} = ?$$

$$E^\circ_{cell} = E^\circ_{cathode} - E^\circ_{anode} = (+0.96\text{ V}) - (+1.52\text{ V}) = -0.56\text{ V}$$

The negative cell potential indicates that the reaction as written does not occur; gold does not react with 1-M nitric acid at 25 °C.

☑ **Reasonable Result Check** Through the centuries, gold metal has been prized for its ability to withstand corrosion and other chemical reactions. Thus, the result is reasonable.

PROBLEM-SOLVING PRACTICE 17.5

Look at Table 17.1 and determine which two half-reactions would produce the largest value of E°_{cell}. Write the two half-reactions and the overall cell reaction, and calculate the E°_{cell} for the reaction.

CONCEPTUAL EXERCISE 17.8

Using $E°$ Values

Transporting chemicals is of great practical and economic importance. Suppose that you have a large volume of 1-M aqueous mercury(II) chloride solution, $HgCl_2$, that needs to be transported. A driver brings a tanker truck made of aluminum to the loading dock. Is it okay to load the truck with your solution? Explain your answer fully.

Stainless steel is an alloy of iron.

Standard half-cell potentials can be used to explain an annoying experience many of us have had—a pain in a tooth when a filling is touched with a stainless steel fork or a piece of aluminum foil. A common material for dental fillings is a dental amalgam—tin and silver dissolved in mercury to form solid solutions having compositions approximating Ag_2Hg_3, Ag_3Sn, and Sn_xHg (where x ranges from 7 to 9). All of these compounds can undergo electrochemical reactions; for example,

$$3\,Hg_2^{2+}(aq) + 4\,Ag(s) + 6\,e^- \longrightarrow 2\,Ag_2Hg_3(s) \qquad E^\circ = +0.85\text{ V}$$
$$Sn^{2+}(aq) + 3\,Ag(s) + 2\,e^- \longrightarrow Ag_3Sn(s) \qquad E^\circ = -0.05\text{ V}$$

The $E°$ values in Table 17.1 indicate that both iron and aluminum have much more negative half-cell potentials and, therefore, are much better reducing agents than any of the amalgam fillings. Consequently, if a piece of iron or aluminum comes in contact with a dental filling, the saliva and gum tissue act as a salt bridge, resulting in a voltaic cell. The iron or aluminum donates electrons, producing a tiny electrical current that activates a nerve and results in pain.

Before we leave this discussion of Table 17.1, consider again the activity series of metals shown in Table 3.4 (← **Sec. 3-5d**), which contains many of the same elements as Table 17.1. Notice that the most active metal, lithium, at the top of Table 3.4 is at the very bottom right of Table 17.1. That is because Table 17.1 is arranged by *half-cell potential,* and lithium ion has the lowest tendency to be reduced. Table 3.4, on the other hand, lists the metals in order of activity, that is, their tendency to be oxidized. Because oxidation is the opposite of reduction, lithium metal is in opposite positions in the two tables.

CONCEPTUAL EXERCISE 17.9

Predicting Redox Reactions Using E° Values

Consider the reduction half-reactions in this table:

Half-reaction	Half-cell	E° (V)
$Cl_2(g) + 2\,e^- \longrightarrow 2\,Cl^-(aq)$	$Cl_2(g) \mid Cl^-(aq)$	+1.36
$I_2(s) + 2\,e^- \longrightarrow 2\,I^-(aq)$	$I_2(s) \mid I^-(aq)$	+0.535
$Pb^{2+}(aq) + 2\,e^- \longrightarrow Pb(s)$	$Pb^{2+}(aq) \mid Pb(s)$	−0.125
$V^{2+}(aq) + 2\,e^- \longrightarrow V(s)$	$V^{2+}(aq) \mid V(s)$	−1.13

(a) Which is the weakest oxidizing agent?
(b) Which is the strongest oxidizing agent?
(c) Which is the strongest reducing agent?
(d) Which is the weakest reducing agent?
(e) Does Pb(s) reduce V^{2+}(aq) to V(s)?
(f) Does I_2(s) oxidize Cl^-(aq) to Cl_2(g)?
(g) Name the molecules or ions in the table that can be reduced by Pb(s).

CONCEPTUAL EXERCISE 17.10

Predicting E° Values

Lead is the first element above hydrogen in Table 3.4 (← **Sec. 3-5d**) and antimony is the first element below hydrogen, but Pb and Sb are not listed in Table 17.1. Indicate where they would appear in Table 17.1 and, based on values from the table, estimate half-cell potentials for reduction of their positive ions to metal atoms.

17-6 E°_{cell}, Gibbs Free Energy, and K°

The sign of E°_{cell} indicates whether a redox reaction is product-favored (positive E°) or reactant-favored (negative E°). In Chapter 16 you learned another way to decide whether a reaction is product-favored: The change in standard Gibbs free energy, $\Delta_r G^\circ$, must be negative (← **Sec. 16-6**). Because both E°_{cell} and $\Delta_r G^\circ$ tell something about whether a reaction will occur, it is not surprising that they are related.

The "free" in Gibbs free energy indicates that it is energy available to do work (← **Sec. 16-8**). The electrical work that a cell can do depends on the amounts of reactants that react in the cell reaction and on the cell voltage, E°. The electrical work can be calculated by multiplying the quantity of electrical charge transferred times the electrical potential difference, E°. A greater amount of reactants reacting means more electrons (charge) transferred and hence more work. The quantity of charge is given by the number of electrons transferred in the overall reaction, n, multiplied by the amount of reaction that occurs, times the electrical charge per mole of electrons.

$$\text{Quantity of charge} = \text{number of electrons transferred} \times \text{amount of reaction} \times \text{charge per mole of electrons}$$

The charge on 1 mol e^- can be calculated from the charge on one electron and the Avogadro constant.

$$\text{Charge on 1 mol } e^- = \left(\frac{1.60218 \times 10^{-19}\ \text{C}}{e^-}\right)\left(\frac{6.02214 \times 10^{23}\ e^-}{1\ \text{mol } e^-}\right)$$

$$= 9.6485 \times 10^4\ \text{C/mol } e^-$$

Michael Faraday
1791–1867

As an apprentice to a London bookbinder, Michael Faraday became fascinated by science when he was a boy. At age 21 he was appointed as a laboratory assistant at the Royal Institution and became its director within 12 years. A skilled experimenter in chemistry and physics, he made many discoveries, the most important of which was electromagnetic induction, the basis of modern electromagnetic technology. Faraday built the first electric motor, generator, and transformer. A popular speaker and educator, he also performed chemical and electrochemical experiments and first isolated benzene, C_6H_6.

The quantity 9.6485×10^4 C/mol of electrons (sometimes rounded to 96,500 C/mol) *is known as the* **Faraday constant, *F***, in honor of Michael Faraday, who was the first to explore the quantitative aspects of electrochemistry.

The electrical work that can be done by a cell is equal to the Faraday constant (F) multiplied by the number of electrons transferred in the overall reaction (n) and by the cell potential (E°_{cell}).

$$\text{Electrical work done by a cell} = nFE^\circ_{cell}$$

It is possible to operate a cell in such a way that the maximum electrical work is obtained from the cell reaction. Because $\Delta_r G = -w_{max}$ (← **Sec. 16-8**), that is, because the Gibbs free energy change is minus the maximum work that can be done by a chemical reaction, we can write (for standard-state conditions)

$$\Delta_r G^\circ = -w_{max} = -nFE^\circ_{cell}$$

The negative sign in the right-most term of the equation reflects the fact that $\Delta_r G^\circ$ *is always negative for a product-favored process, but* E°_{cell} *is always positive for a product-favored process.* Thus, these values must have opposite signs.

We can use this equation to calculate $\Delta_r G^\circ$ for the $Zn|Zn^{2+}||Cu^{2+}|Cu$ cell, that is, the maximum work that the cell can do. For the reaction

$$Cu^{2+}(aq) + Zn(s) \longrightarrow Cu(s) + Zn^{2+}(aq) \qquad E^\circ_{cell} = +1.10 \text{ V}$$

two electrons are transferred for each copper ion reduced; $n = 2$. The Gibbs free energy change when this quantity of reactants is converted is

$$\Delta_r G^\circ = -nFE^\circ = -(2)\left(\frac{9.65 \times 10^4 \text{ C}}{\text{mol e}^-}\right)(1.10 \text{ V})\left(\frac{1 \text{ J}}{1 \text{ V} \times 1 \text{ C}}\right)\left(\frac{1 \text{ kJ}}{10^3 \text{ J}}\right) = -212 \text{ kJ/mol}$$

The large positive E°_{cell} and the large negative Gibbs free energy change indicate a significantly product-favored reaction.

PROBLEM-SOLVING EXAMPLE 17.6

Determining E°_{cell} and $\Delta_r G^\circ$

Zinc–air batteries are used in hearing aids and other devices that require low mass and small size. Such batteries use a paste of metallic zinc and aqueous KOH contained in a metal cup, which serves as the anode. Air vents and a membrane permeable to oxygen from air allow oxygen to flow over a metal cathode. The half-reactions are

Anode: $Zn(s) + 2\ OH^-(aq) \longrightarrow ZnO(s) + H_2O(\ell) + 2\ e^- \qquad E^\circ_{anode} = -1.25 \text{ V}$

Cathode: $O_2(g) + 2\ H_2O(\ell) + 4\ e^- \longrightarrow 4\ OH^-(aq) \qquad E^\circ_{cathode} = +0.401 \text{ V}$

Write a balanced overall equation for the cell reaction. Calculate E°_{cell} and $\Delta_r G^\circ$ for the reaction in such a battery under standard-state conditions at 25 °C.

Result $2\ Zn(s) + O_2(g) \longrightarrow 2\ ZnO(s)$; $E^\circ_{cell} = +1.65$ V;
$\Delta_r G^\circ = -6.37 \times 10^5 \text{ J} = -6.37 \times 10^2 \text{ kJ}$.

Analyze The balanced overall equation can be obtained by combining half-equations so that electrons are conserved. The cell potential is $E^\circ_{cell} = E^\circ_{cathode} - E^\circ_{anode}$.

$$\Delta_r G^\circ = -nFE^\circ_{cell}$$

Plan The half-equations involve different numbers of electrons, so multiply the anode half-equation by 2 to equalize electrons lost with the 4 e⁻ gained in the cathode equation. Then, sum the half-equations to get the overall equation. Calculate the cell potential and $\Delta_r G^\circ$ using the appropriate equations.

Execute

Note that the standard half-cell potentials are *not* multiplied by the balancing coefficients.

$$2 \times [Zn(s) + 2\ OH^-(aq) \longrightarrow ZnO(s) + H_2O(\ell) + 2\ e^-] \qquad E^\circ_{anode} = -1.25 \text{ V}$$

$$O_2(g) + 2\ H_2O(\ell) + 4\ e^- \longrightarrow 4\ OH^-(aq) \qquad E^\circ_{cathode} = +0.401 \text{ V}$$

$$2\ Zn(s) + O_2(g) \longrightarrow 2\ ZnO(s) \qquad E^\circ_{cell} = ?$$

$$E^\circ_{cell} = E^\circ_{cathode} - E^\circ_{anode} = (+0.401 \text{ V}) - (-1.25 \text{ V}) = +1.65 \text{ V}$$

In the overall equation, four electrons are transferred, thus $n = 4$.

$$\Delta_r G^\circ = -(4)\left(\frac{9.65 \times 10^4\ \text{C}}{1\ \text{mol e}^-}\right)(1.65\ \text{V})\left(\frac{1\ \text{J}}{1\ \text{V} \times 1\ \text{C}}\right)\left(\frac{1\ \text{kJ}}{1000\ \text{J}}\right) = -6.37 \times 10^2\ \text{kJ/mol}$$

☑ **Reasonable Result Check** Because the overall reaction powers a battery, producing electric energy, the reaction must be product-favored. The cell potential should be positive and the Gibbs free energy change should be negative; they are.

PROBLEM-SOLVING PRACTICE 17.6

Using standard half-cell potentials, determine whether this reaction is product-favored as written at 25 °C.

$$Hg^{2+}(aq) + 2\ I^-(aq) \longrightarrow Hg(\ell) + I_2(s)$$

EXERCISE 17.11

Apple iPod Cell Potential

An Apple iPod model is advertised as being able to operate for 36 hours before needing to be recharged. The iPod battery is rated at 650 mA. To operate the iPod for 36 h without recharging the battery, assuming 100% efficiency, would require 3.0×10^5 J of energy. Calculate the voltage of the battery.

17-6a $\Delta_r G^\circ$, E°_{cell}, and K°

We have seen that the magnitude of the standard Gibbs free energy change is directly proportional to the E°_{cell} for a voltaic cell at standard conditions

$$\Delta_r G^\circ = -nFE^\circ_{cell}$$

The magnitude of the standard Gibbs free energy change is also directly proportional to the logarithm of the equilibrium constant of a reaction as shown in Equation 16.8 (← **Sec. 16-7b**).

$$\Delta_r G^\circ = -RT \ln K^\circ$$

Putting these two equations together yields

$$-nFE^\circ_{cell} = -RT \ln K^\circ$$

which, when solved for E°_{cell} or solved for K° yields

$$E^\circ_{cell} = \frac{RT}{nF} \ln K^\circ \quad \text{or} \quad \ln K^\circ = \frac{nFE^\circ_{cell}}{RT}$$

Thus, by measuring E°_{cell}, the values of $\Delta_r G^\circ$ and K° can be calculated. The relationships linking $\Delta_r G^\circ$, E°_{cell}, and K° are summarized in Figure 17.11.

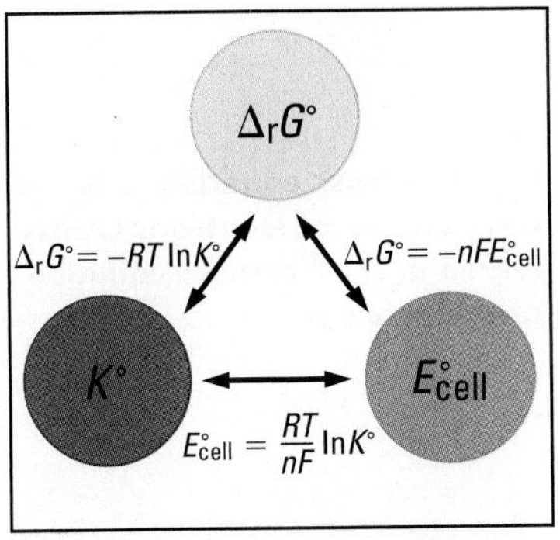

ΔrG°
ΔrG° = −RT ln K°
ΔrG° = −nFE°cell
K°
E°cell
E°cell = RT/nF ln K°

Figure 17.11 The relationships linking $\Delta_r G^\circ$, E°_{cell}, and K°. Given any one of the values, the other two can be calculated.

PROBLEM-SOLVING EXAMPLE 17.7

Equilibrium Constant for a Redox Reaction

Calculate the equilibrium constant K_c for this reaction at 25 °C.

$$Fe(s) + Cd^{2+}(aq) \rightleftharpoons Fe^{2+}(aq) + Cd(s)$$

Result $K_c = K^\circ = 20$

Analyze The reaction occurs in aqueous solution, so $K_c = K^\circ$ (← **Sec. 16-7b**). K° can be calculated from E° for a cell based on the overall reaction given.

Plan Determine the half-reactions that combine to give the overall reaction. Look up half-cell potentials in Table 17.1, calculate the cell potential, and calculate K° from

$$\ln K^\circ = \frac{nFE^\circ_{cell}}{RT}$$

Execute The two half-reactions and the overall reaction are

$$\begin{array}{rlr} Fe(s) \longrightarrow Fe^{2+}(aq) + 2\,e^- & & E^\circ_{anode} = -0.44\text{ V} \\ Cd^{2+}(aq) + 2\,e^- \longrightarrow Cd(s) & & E^\circ_{cathode} = -0.40\text{ V} \\ \hline Fe(s) + Cd^{2+}(aq) \longrightarrow Fe^{2+}(aq) + Cd(s) & & E^\circ_{cell} = +0.04\text{ V} \end{array}$$

Two electrons are transferred.

$$\ln K^\circ = \frac{nFE^\circ_{cell}}{RT} = \frac{(2)\left(\frac{96{,}485\text{ C}}{1\text{ mol e}^-}\right)(0.04\text{ V})}{(8.314\text{ J K}^{-1}\text{ mol}^{-1})(298\text{ K})} = 3.12 \quad\text{and}\quad K^\circ = e^{3.12} = 20$$

Note that the units cancel in this calculation because $1\text{ C} \times 1\text{ V} = 1\text{ J}$. The K° value is larger than 1, which shows that the reaction is product-favored as written. Because the reaction occurs in aqueous solution, $K_c = K^\circ = 20$.

☑ **Reasonable Result Check** The value for E°_{cell} is positive, which indicates a product-favored reaction, as does $K^\circ > 1$.

PROBLEM-SOLVING PRACTICE 17.7

Using the standard half-cell potentials listed in Table 17.1, calculate the equilibrium constant for this reaction at 25 °C.

$$Ni(s) + Cd^{2+}(aq) \longrightarrow Ni^{2+}(aq) + Cd(s)$$

17-7 Effect of Concentration on Cell Potential: The Nernst Equation

As product-favored reactions proceed in voltaic cells reactants are converted to products, the concentrations of the reactants decrease continuously, and the cell potential drops. Eventually, the cell potential drops to zero when the reactants and products are at equilibrium. Thus, the voltage of a voltaic cell is related to the concentration of reactants and products of its chemical reaction. To determine how the cell potential depends on concentrations, we start with the relationship between Gibbs free energy change and concentration (← **Sec. 16-7c**).

$$\Delta_r G = \Delta_r G^\circ + RT\ln Q$$

where Q is the reaction quotient (← **Sec. 12-5**). Q has the same form as the equilibrium constant, but it refers to any reaction mixture, whether or not it is at equilibrium.

Because $\Delta_r G = -nFE_{cell}$ and $\Delta_r G^\circ = -nFE^\circ_{cell}$,

$$-nFE_{cell} = -nFE^\circ_{cell} + RT\ln Q$$

where E°_{cell} is the potential at standard-state concentrations and pressures and E_{cell} is the potential at nonstandard conditions. Solving for E_{cell} gives the relationship we seek: It is called the **Nernst equation**.

In the Nernst equation, *n* is the number of electrons transferred in the *balanced equation* of the process.

$$E_{cell} = E^\circ_{cell} - \frac{RT}{nF}\ln Q$$

If the concentrations in Q all equal 1 M, which is the standard state, then $Q = 1$ and $\ln(1) = 0$, so the Nernst equation reduces to $E_{cell} = E^\circ_{cell}$.

The Nernst equation is used to calculate the potential produced by a voltaic cell under *nonstandard* conditions. It can also be used to calculate the concentration of a reactant or product in an electrochemical reaction from the measured value of the potential produced.

PROBLEM-SOLVING EXAMPLE 17.8

Using the Nernst Equation

Consider this electrochemical reaction:

$$Zn(s) + Ni^{2+}(aq) \longrightarrow Zn^{2+}(aq) + Ni(s)$$

The standard cell potential $E^\circ_{cell} = 0.51$ V. Calculate the cell potential at 25 °C if the Ni^{2+} concentration is 5.0 M and the Zn^{2+} concentration is 0.050 M.

Result $E_{cell} = 0.57$ V

Analyze Q can be obtained from the balanced equation and the concentrations given. The equation indicates that one Ni^{2+} forms one Ni, so $n = 2$. E°_{cell} is given.

Plan Calculate the cell potential using the Nernst equation.

Execute At 298 K,

$$E_{cell} = E^\circ_{cell} - \frac{RT}{nF}\ln\frac{(\text{conc. } Zn^{2+})}{(\text{conc. } Ni^{2+})} = 0.51\text{ V} - \frac{(8.314\text{ J mol}^{-1}\text{ K}^{-1})(298\text{ K})}{2\,(96{,}485\text{ C mol}^{-1})}\ln\frac{(0.050\text{ M})}{(5.0\text{ M})}$$

$$E_{cell} = 0.51\text{ V} - (1.284 \times 10^{-2}\text{ V})(-4.60) = 0.57\text{ V}$$

☑ **Reasonable Result Check** The concentration of Ni^{2+} is 5.0 M, which is larger than the standard-state value of 1.0 M, and the concentration of Zn^{2+} is 0.050 M, much smaller than the standard-state value. Each of these departures from standard-state conditions tends to make the potential slightly larger than the standard cell potential ($E^\circ_{cell} = 0.51$ V), and it is.

Note that the *larger* the concentrations of *reactants* are, the *larger* the cell voltage is; the *smaller* the concentrations of *products* are, the *larger* the cell voltage is.

PROBLEM-SOLVING PRACTICE 17.8

Calculate the cell potential for the $Zn(s) + Ni^{2+}(aq)$ reaction (at the same temperature) if (conc. Zn^{2+}) = 3.0 M and (conc. Ni^{2+}) = 0.010 M.

17-7a Concentration Cells: Measuring pH

When a more concentrated solution contacts a more dilute solution of the same solute, the concentration of the former solution decreases and the concentration of the latter increases until both concentrations are equal. This product-favored process can generate electric current in a **concentration cell**, *a voltaic cell that has the same chemical species in the anode and cathode compartments, but with different concentrations*. The voltage produced by the difference in concentrations can be calculated using the Nernst equation. As current flows, the concentration increases in the more dilute compartment and decreases in the more concentrated compartment so the potential difference decreases.

The H^+ concentration in a solution can be measured using a concentration cell based on the $H_2(g)|H^+(aq)$ half-reaction. The cathode compartment contains a standard hydrogen electrode with a 1-M H^+ concentration (← **Fig. 17.7, Sec. 17-4b**); the anode compartment contains the same type of electrode in contact with a solution of unknown H^+ concentration. Due to the difference in H^+ concentrations, a voltage is created. From the measured voltage, the pH can be calculated. Because of the complexity of the standard hydrogen electrode, this type of experiment is very cumbersome. The pH meter, which uses the same principles, is a much more convenient way to measure pH routinely.

The **pH meter** *is a special concentration cell that either has two electrodes* (Figure 17.12) *or a single probe that combines the two* (Figure 17.13). The glass *indicator* electrode contains a 1-M HCl solution plus an Ag|AgCl half-cell. The very thin glass membrane tip of the electrode is sensitive to H^+ concentration differences. Thus, the glass electrode measures the difference between its internal H^+ concentration (half-cell 1) relative to the H^+ concentration in a test solution (half-cell 2). The second electrode, a silver-silver chloride electrode, serves as the *reference* electrode. It consists of a silver

A standard calomel electrode is sometimes used as the reference electrode. This electrode consists of a platinum metal wire dipping into a paste of Hg_2Cl_2 (calomel), liquid mercury, and a saturated KCl solution.

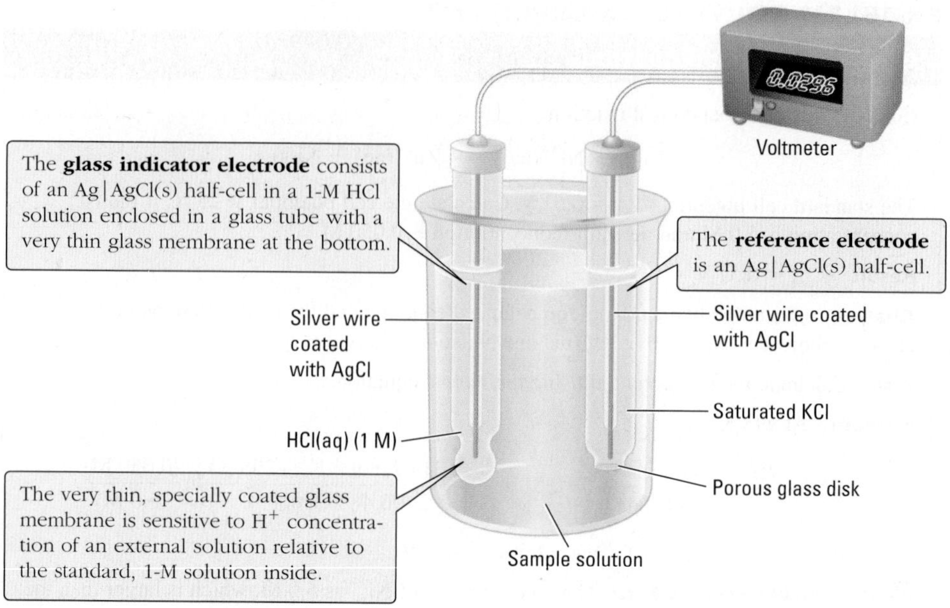
Voltmeter
The **glass indicator electrode** consists of an Ag | AgCl(s) half-cell in a 1-M HCl solution enclosed in a glass tube with a very thin glass membrane at the bottom.
The **reference electrode** is an Ag | AgCl(s) half-cell.
Silver wire coated with AgCl
Silver wire coated with AgCl
Saturated KCl
HCl(aq) (1 M)
Porous glass disk
The very thin, specially coated glass membrane is sensitive to H+ concentration of an external solution relative to the standard, 1-M solution inside.
Sample solution

Figure 17.12 The electrodes of a two-electrode commercial pH meter.

chloride-coated silver wire dipping into a saturated KCl solution. The overall cell can be represented as

$$\text{Ag(s)}\,|\,\text{AgCl(s)}\,|\,\text{Cl}^-\text{(aq), H}^+\text{ (1.0 M)}\,\overset{\text{glass membrane}}{|}\,\text{H}^+\text{ (unknown conc.)}\,\|\,\text{Cl}^-\text{(aq)}\,|\,\text{AgCl(s)}\,|\,\text{Ag(s)}$$

The voltage difference between the glass indicator electrode and the reference electrode is converted electronically to give the pH of the test solution.

The two electrodes are connected by a salt bridge and the half-reactions are

Indicator (glass) electrode: $\text{Ag(s)} + \text{Cl}^-\text{(aq)} \rightarrow \text{AgCl(s)} + e^-$

$\text{H}^+\text{ (1.0 M)} \rightarrow \text{H}^+\text{ (unknown)}$

Reference electrode: $\text{AgCl(s)} + e^- \rightarrow \text{Ag(s)} + \text{Cl}^-\text{(aq)}$

The half-cell potentials of the two AgCl half-cells cancel and therefore make no contribution to the overall cell potential. That potential is governed only by the difference in H^+ concentration.

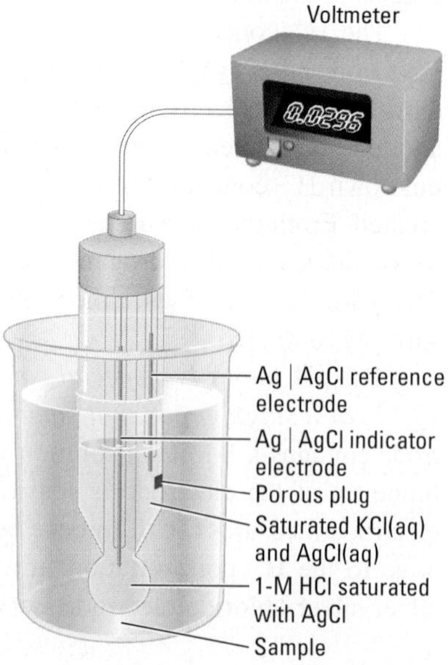
Voltmeter
Ag | AgCl reference electrode
Ag | AgCl indicator electrode
Porous plug
Saturated KCl(aq) and AgCl(aq)
1-M HCl saturated with AgCl
Sample

Figure 17.13 A pH meter with a combination pH probe, which contains both an indicator and a reference electrode.

PROBLEM-SOLVING EXAMPLE 17.9

Measuring pH

When the pH of an unknown solution is measured at 25.0 °C, a pH meter registers a voltage of 0.177 V. Calculate (a) the pH and (b) the H^+ concentration.

Result pH = 2.99; H^+ concentration = 1.02×10^{-3} M

Analyze E_{cell} is given and E°_{cell} is zero because for standard-state conditions there is no reaction. The H^+ concentration in the indicator half-cell is 1 M. Assume that the unknown concentration is less than 1 M. It can be calculated using the standard-state concentration (1 M), the Nernst equation, and pH = −log(conc. H^+).

Plan Write half-equations and an overall equation. From the overall equation determine n and write the expression for Q. Substitute into the Nernst equation and solve for the unknown concentration. Take the negative log to calculate pH.

Execute (a) The two half-reactions are

Reduction: $2\,H^+\,(1\,M) + 2\,e^- \longrightarrow H_2(g,\,1\,bar)$

Oxidation: $H_2(g,\,1\,bar) \longrightarrow 2\,H^+\,(unknown\,M) + 2\,e^-$

Overall: $2\,H^+(1\,M) \longrightarrow 2\,H^+\,(unknown\,M)$ $\quad E_{cell} = 0.177$ V

The Nernst equation at 25 °C is

$$E_{cell} = E^\circ_{cell} - \frac{(8.314\ \text{J mol}^{-1}\ \text{K}^{-1})(298\ \text{K})}{2\ (96{,}485\ \text{C mol}^{-1})}\ln\frac{(\text{conc. H}^+)^2_{unknown}}{(\text{conc. H}^+)^2_{standard}}$$

Under standard-state conditions, (conc. H^+) = 1 M and $E^\circ_{cell} = 0$, so

$$E_{cell} = -1.284 \times 10^{-2}\ \text{V} \times \ln\,(\text{conc. H}^+)^2_{unknown}$$
$$E_{cell} = -1.284 \times 10^{-2}\ \text{V} \times 2.303 \times \log\,(\text{conc. H}^+)^2_{unknown}$$

Because $\log x^2 = 2 \log x$,

$$E_{cell} = -1.284 \times 10^{-2}\ \text{V} \times 2.303 \times 2 \times \log\,(\text{conc. H}^+)_{unknown} = 0.177\ \text{V}$$
$$= 0.591\ \text{V} \times -\log\,(\text{conc. H}^+)_{unknown} = 0.177\ \text{V}$$

By definition, pH = −log (conc. H^+), so

$$0.0591\ \text{V} \times \text{pH} = 0.177\ \text{V} \quad \text{and} \quad \text{pH} = \frac{0.177\ \text{V}}{0.0591\ \text{V}} = 2.99$$

(b) $(\text{conc. H}^+)_{unknown} = 10^{-pH} = 10^{-2.99} = 1.02 \times 10^{-3}$ M.

☑ **Reasonable Result Check** The general relationship between the variables is pH = E_{cell}/0.0591 V, so using approximate values gives 0.18/0.06 ≅ 3, which is close to our more exact result.

Recall that ln(x) = 2.303 log(x)

PROBLEM-SOLVING PRACTICE 17.9

In a laboratory experiment, a student uses a pH meter to measure the pH of an unknown solution, which is found to be 7.30. Calculate the corresponding E_{cell}.

17-8 Common Batteries

A **primary battery** is *a battery in which the electrochemical reactions cannot easily be reversed.* Such batteries are "single use" because when the reactants are consumed, the battery is "dead" and no longer useful. A good example of a primary battery is the *alkaline battery*, which is typically used in flashlights. It produces 1.54 V by the oxidation of zinc metal as the anode reaction under alkaline (basic) conditions. Electrons released at the anode travel through the external circuit and reduce manganese(IV) oxide at the cathode.

Anode, oxidation: $Zn(s) + 2\,OH^-(aq) \longrightarrow ZnO(s) + H_2O(\ell) + 2\,e^-$

Cathode, reduction: $MnO_2(s) + H_2O(\ell) + e^- \longrightarrow MnO(OH)(s) + OH^-(aq)$

©Huguette Roe/Shutterstock.com

"Dead" batteries ready for recycling.

In the "button-sized" alkaline batteries used in calculators and watches, silver(I) oxide replaces the manganese(IV) oxide used in larger alkaline batteries.

In contrast, a **secondary battery** *uses a reversible electrochemical reaction, which makes it rechargeable.* As a secondary battery discharges, the oxidation products remain at the anode and the reduction products remain at the cathode. When the electron flow is reversed in the recharging process, the reactants are regenerated. Lead-acid storage batteries, nickel-cadmium and nickel-metal hydride batteries, and lithium-ion batteries are examples of secondary batteries.

The process of charging a secondary battery involves electrolysis, which is discussed in Section 17-11.

17-8a Lead-Acid Storage Batteries

The lead-acid battery was first described to the French Academy of Science in 1860 by Gaston Planté.

The *familiar automobile battery*, the **lead-acid storage battery**, consists of six cells (Figure 17.14). Each cell contains porous metallic lead plates as the anode and lead(IV) oxide plates as the cathode. The electrodes are immersed in a 4.5-M aqueous sulfuric acid solution. When the battery produces an electric current, lead metal is oxidized to lead(II) sulfate at the anode and lead(IV) oxide is reduced to lead(II) sulfate at the cathode.

$$Pb(s) + PbO_2(s) + 2\,H^+(aq) + 2\,HSO_4^-(aq) \longrightarrow 2\,PbSO_4(s) + 2\,H_2O(\ell) \qquad E^\circ_{cell} = 2.041\text{ V} \qquad [17.1]$$

Each cell generates 2.0 V. The six cells are connected in series, which makes their voltages additive. This produces a total potential of 12 V.

The lead-acid battery can be recharged because the lead(II) sulfate formed at the electrodes is insoluble. It remains on the electrode surface, available for the reverse, recharging reaction at both electrodes. The recharging reaction is

The number of electrons (quantity of charge) that a battery can move from the anode to the cathode is proportional to the amount of reactants that react.

$$2\,PbSO_4(s) + 2\,H_2O(\ell) \longrightarrow Pb(s) + PbO_2(s) + 2\,H^+(aq) + 2\,HSO_4^-(aq) \qquad E^\circ_{cell} = -2.041\text{ V} \qquad [17.2]$$

As seen from Figure 17.15, when a car is started the battery discharges, providing power. The battery is recharged by electric current generated by an alternator during normal driving. Figure 17.15 shows the half-cell reactions that occur during charging and discharging modes.

As the battery recharges, a small amount of aqueous hydrogen ions is reduced to hydrogen gas at the cathode and, simultaneously, a small amount of water is oxidized to oxygen gas at the anode.

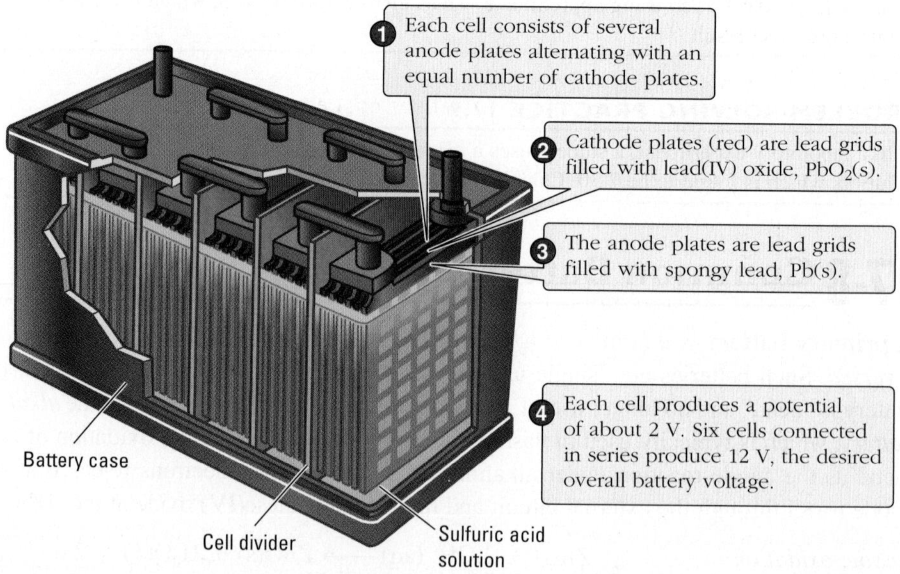
1 Each cell consists of several anode plates alternating with an equal number of cathode plates.
2 Cathode plates (red) are lead grids filled with lead(IV) oxide, PbO2(s).
3 The anode plates are lead grids filled with spongy lead, Pb(s).
4 Each cell produces a potential of about 2 V. Six cells connected in series produce 12 V, the desired overall battery voltage.
Battery case
Cell divider
Sulfuric acid solution

Figure 17.14 Lead-acid storage battery. Six 2-V cells connected in series produce 12 V.

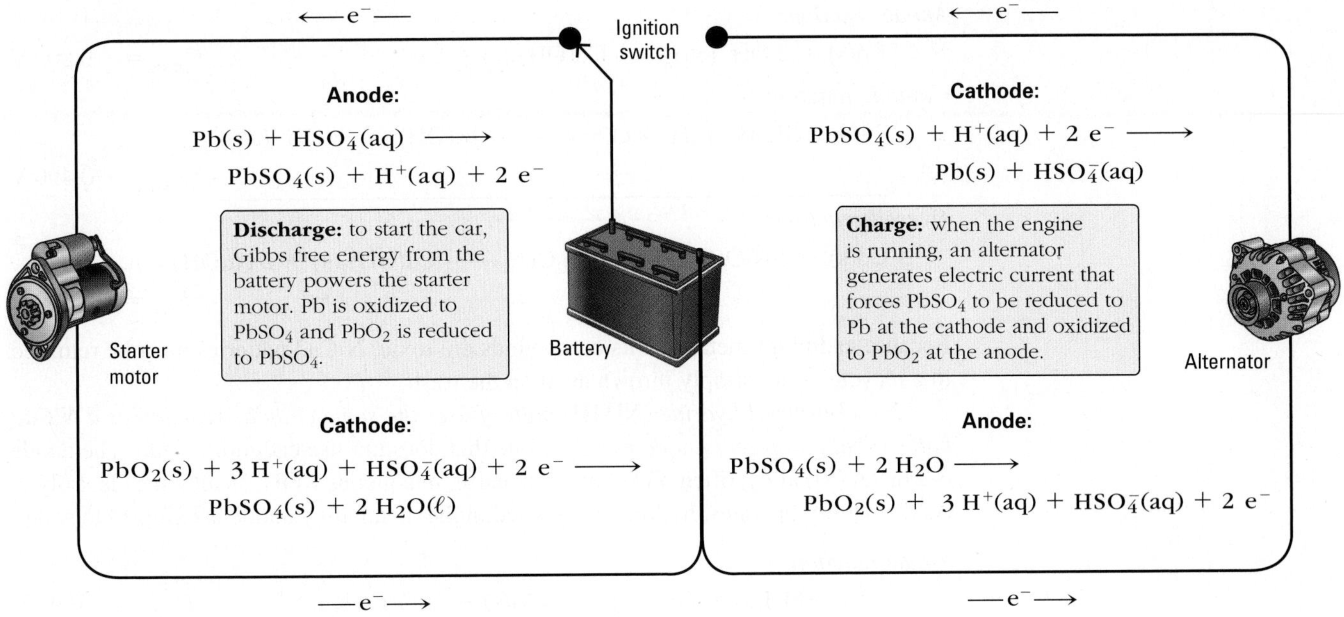

←e⁻—
Ignition switch
←e⁻—
Anode:
Pb(s) + HSO₄⁻(aq) ⟶ PbSO₄(s) + H⁺(aq) + 2 e⁻
Discharge: to start the car, Gibbs free energy from the battery powers the starter motor. Pb is oxidized to PbSO₄ and PbO₂ is reduced to PbSO₄.
Starter motor
Battery
Cathode:
PbO₂(s) + 3 H⁺(aq) + HSO₄⁻(aq) + 2 e⁻ ⟶ PbSO₄(s) + 2 H₂O(ℓ)
Cathode:
PbSO₄(s) + H⁺(aq) + 2 e⁻ ⟶ Pb(s) + HSO₄⁻(aq)
Charge: when the engine is running, an alternator generates electric current that forces PbSO₄ to be reduced to Pb at the cathode and oxidized to PbO₂ at the anode.
Alternator
Anode:
PbSO₄(s) + 2 H₂O ⟶ PbO₂(s) + 3 H⁺(aq) + HSO₄⁻(aq) + 2 e⁻
—e⁻⟶
—e⁻⟶

Figure 17.15 **Discharge and recharge reactions for a lead-acid storage battery.**

Cathode reaction: $4\ H^+(aq) + 4\ e^- \longrightarrow 2\ H_2(g)$

Anode reaction: $2\ H_2O(\ell) \longrightarrow O_2(g) + 4\ H^+(aq) + 4\ e^-$

This hydrogen-oxygen mixture inside the battery, if accidentally ignited, can explode. Consequently, no sparks or open flames should be near a lead-acid storage battery, even the sealed types.

When a car with a dead battery is jump-started, the last jumper cable connection should be to the car's frame—well away from the battery—to avoid a spark igniting any H_2 in the battery.

As the battery discharges, sulfuric acid is consumed in both the anode and cathode reactions as seen from Equation 17.1 and Figure 17.15. This decrease in sulfuric acid concentration diminishes the cell potential, which can be calculated from the Nernst equation (← **Sec. 17-7**). In this case

$$E_{\text{cell}} = E^\circ_{\text{cell}} - \left(\frac{RT}{nF}\right)\ln Q = E^\circ_{\text{cell}} - \left(\frac{RT}{2F}\right)\ln\left(\frac{1}{[H^+]^?[HSO_4^-]^?}\right)$$

$$= 2.041\text{ V} - (0.0128\text{ V})\ln\left(\frac{1}{[H^+]^2[HSO_4^-]^2}\right)$$

It is important to note here the *logarithmic* relation between Q and the cell potential. Due to this type of relationship, the cell potential does not decrease appreciably below 2.0 V until the battery is approximately 97% discharged. On the other hand, once that point is reached, the cell voltage plummets to zero resulting in a "dead" battery.

EXERCISE 17.12

Cell Potential Change

Calculate the decrease in cell potential when the sulfuric acid concentration in a lead-acid storage battery decreases from 0.500 M to 0.100 M.

17-8b Nickel-Based Batteries

Nickel-cadmium (**NiCad**) batteries are *small, lightweight secondary batteries* used in cordless appliances and other applications. The batteries are rechargeable due to the fact that insoluble nickel and cadmium hydroxides produced at the electrodes stay on the electrodes as the reactions occur in basic solution.

Anode reaction:

$$\mathrm{Cd(s) + 2\ OH^-(aq) \longrightarrow Cd(OH)_2(s) + 2\ e^-} \qquad E^\circ_{\text{anode}} = -0.809\ \mathrm{V}$$

Cathode reaction:

$$\mathrm{2\ [NiO(OH)(s) + H_2O(\ell) + e^- \longrightarrow Ni(OH)_2(s) + OH^-(aq)]} \qquad E^\circ_{\text{cathode}} = +0.490\ \mathrm{V}$$

Overall reaction:

$$\mathrm{Cd(s) + 2\ NiO(OH)(s) + 2\ H_2O(\ell) \longrightarrow Cd(OH)_2(s) + 2\ Ni(OH)_2(s)} \qquad E^\circ_{\text{cell}} = +1.299\ \mathrm{V}$$

Because cadmium metal and its compounds are toxic, NiCad batteries must be returned to a recycler—not simply thrown away in the trash.

A *nickel-metal hydride* (**NiMH**) *battery uses the same cathode reaction as a NiCad battery, but a different anode reaction*, one that does not use cadmium metal. The anode is a metal (M) alloy, often nickel or a rare earth, in aqueous KOH, an alkaline electrolyte. As the battery operates, hydrogen absorbed in the metal alloy anode is oxidized to water.

Anode reaction:

$$\mathrm{MH(s) + OH^-(aq) \longrightarrow M(s) + H_2O(\ell) + e^-} \qquad E^\circ_{\text{anode}} = -0.91\ \mathrm{V}$$

Overall reaction:

$$\mathrm{MH(s) + NiO(OH)(s) \longrightarrow M(s) + Ni(OH)_2(s)} \qquad E^\circ_{\text{cell}} = +1.4\ \mathrm{V}$$

The Toyota Prius, a hybrid vehicle, combines a gasoline engine with an electric motor powered by an assembly of nickel-metal hydride batteries. The batteries are recharged when the gasoline engine is running and when the brakes are used (regenerative braking), converting the car's kinetic energy into electrical energy.

Batteries
Spencer Grant/PhotoEdit

Toyota Prius NiMH battery assembly.

EXERCISE 17.13

Recharging a NiCad Battery

Write equations for the electrode reactions that occur when a NiCad battery is recharged. Identify the anode and cathode half-reactions.

17-8c Lithium-Ion Batteries

GM Corp

Chevrolet Volt.

The Chevrolet Volt automobile is a plug-in hybrid electric vehicle (PHEV) that recharges its batteries by plugging into the existing electric grid (electrical sockets). PHEVs require significantly more electrical energy per unit volume and mass of their batteries than can be supplied by NiMH batteries. Thus, PHEV manufacturers use lithium-ion batteries, which have a large voltage (3.4 V per cell), have a very high energy output for their mass, and are rechargeable. These advantages lead to the use of lithium-ion batteries in laptop computers, cell phones, and digital cameras, in addition to PHEVs.

In a **lithium-ion battery** (Figure 17.16) *both anode and cathode are layered structures that allow Li^+ ions to migrate in and out.* In one type of lithium-ion battery the anode is a specially treated form of graphite and the cathode is a cobalt oxide in which oxide ions form a cubic closest-packed structure (← **Sec. 9-6d**) with cobalt ions between every other layer of oxide ions. Anode and cathode are separated by an electrolyte (salt bridge) consisting of an organic solvent and a salt, such as $LiPF_6$. ($LiPF_6$ consists of Li^+ ions and PF_6^- ions.)

In a fully charged battery, lithium ions occupy tiny spaces between the graphene layers of the graphite and there are no lithium ions in the cathode (oxide electrode). To balance the positive charges of the Li^+, there are negative charges in the anode associated with groups of carbon atoms, represented by C_6 in the equations that follow.

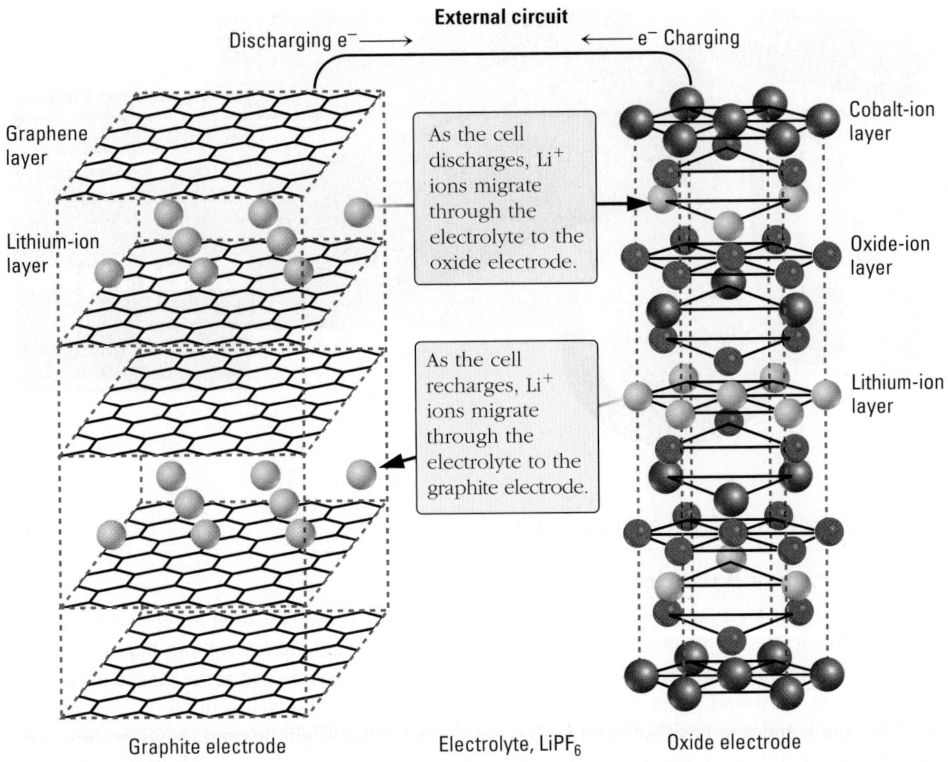
External circuit
Discharging e⁻ ⟶
⟵ e⁻ Charging
Graphene layer
Lithium-ion layer
As the cell discharges, Li+ ions migrate through the electrolyte to the oxide electrode.
As the cell recharges, Li+ ions migrate through the electrolyte to the graphite electrode.
Cobalt-ion layer
Oxide-ion layer
Lithium-ion layer
Graphite electrode
Electrolyte, LiPF6
Oxide electrode

Figure 17.16 Lithium-ion battery discharge and recharge operation. Source: Panasonic (http://industrial.panasonic.com/www-data/pdf/ACA4000/ACA4000PE3.pdf).

This section answers the question, "How does my cell phone battery work?" that was posed in Chapter 1 (← Sec. 1-1).

When a lithium-ion battery discharges (provides current), the anode reaction is

Anode reaction: $LiC_6(s) \longrightarrow Li^+(\text{electrolyte}) + C_6(s) + e^-$

Electrons are conducted through the graphite to a wire and into the external circuit. Lithium ions migrate through the electrolyte to the cathode.

The cathode reaction involves reduction of Co^{4+}, a strong oxidizing agent, to Co^{3+} by electrons from the external circuit. If nothing else happened in the cathode, the battery could not supply current because the total negative charge of oxide ions would be greater than the total positive charge of cobalt ions. To make up for the loss of positive charge in the cathode, Li^+ ions migrate from the anode and occupy spaces between layers of oxide ions that are not already occupied by cobalt ions. Each Li^+ ion balances one unit of positive charge. When the battery is fully discharged there are no lithium ions in the anode and the formula of the cathode material is $LiCoO_2$, which contains Co^{3+} ions.

Cathode reaction: $Li^+(\text{electrolyte}) + e^- + CoO_2(s) \longrightarrow LiCoO_2(s)$

Overall reaction: $LiC_6(s) + CoO_2(s) \longrightarrow C_6(s) + LiCoO_2(s) \qquad E^\circ_{\text{cell}} = +3.5\text{ V}$

Recharging a lithium-ion battery involves reversing the reactions just described. A flow of electrons from the external circuit into the graphite electrode creates negatively charged groups of carbon atoms. These attract Li^+ ions from the electrolyte into the spaces between the graphene layers. The external circuit removes electrons from the cobalt oxide electrode. Combined with the flow of Li^+ ions into the electrolyte, this changes the $LiCoO_2$ back to CoO_2.

Examples of several of the types of batteries we have discussed are shown in Figure 17.17.

Esther S. Takeuchi
1953–

FABIANO/SIPA/Newscom

Esther S. Takeuchi, an extraordinarily prolific inventor, holds more than 150 patents. She studies the mechanisms of battery systems and develops new battery materials for implantable medical devices. One device she developed, a lithium/silver vanadium oxide battery, powers the majority of implanted cardiac defibrillators. A double major with bachelor's degrees in history and chemistry from the University of Pennsylvania and a Ph.D. in chemistry from The Ohio State University, she is currently a SUNY Distinguished Professor in the Stony Brook University departments of chemistry and materials science and engineering, as well as Chief Scientist in Brookhaven National Laboratory's Global and Regional Solutions Directorate. Takeuchi, a member of the National Academy of Engineering, received the National Medal of Technology and Innovation in 2009 and has been inducted into the National Inventor's Hall of Fame.

SUPER
ALKALINE
GP
PX625A
CHINA
ALKALINE CELL
DieHard

Figure 17.17 Primary and secondary batteries. Alkaline primary batteries come as cylinders, rectangular solids, or "buttons". Lead-acid car batteries and lithium-ion laptop batteries are secondary batteries.

David J. Green—electrical/Alamy; © Mino Surkala/Shutterstock.com; Charles D. Winters/Science Source; Kim Kyung-Hoon/Reuters/Landov

CONCEPTUAL EXERCISE 17.14

Emergency Batteries

You are stranded on an island and need to communicate your location to receive help. You have a battery-powered radio transmitter, but the lead batteries are discharged. There is a swimming pool nearby and you find a tank of chlorine gas and some plastic tubing that can withstand being exposed to chlorine. There is an iron chain-link fence around the pool, and you have some tools. Devise a battery that might be used to power the radio using these items.

17-9 Fuel Cells

A **fuel cell** is *a voltaic cell that converts the chemical energy of its fuels directly into electricity*. Although it functions like a battery, a fuel cell differs in that its reactants are continually supplied from an *external* source. In a hydrogen-oxygen fuel cell with a proton-exchange membrane, hydrogen gas is oxidized in the anode chamber and oxygen gas is reduced in the cathode chamber. The two chambers are separated by a semipermeable proton-exchange membrane (PEM) (Figure 17.18). The membrane is a thin plastic sheet that allows the passage of protons (H^+ ions), *but not electrons*. The electrons are forced to pass through an external circuit where they can do useful work. The two half-reactions and the overall reaction are given in Figure 17.18. The $H_2|O_2$ fuel cell produces

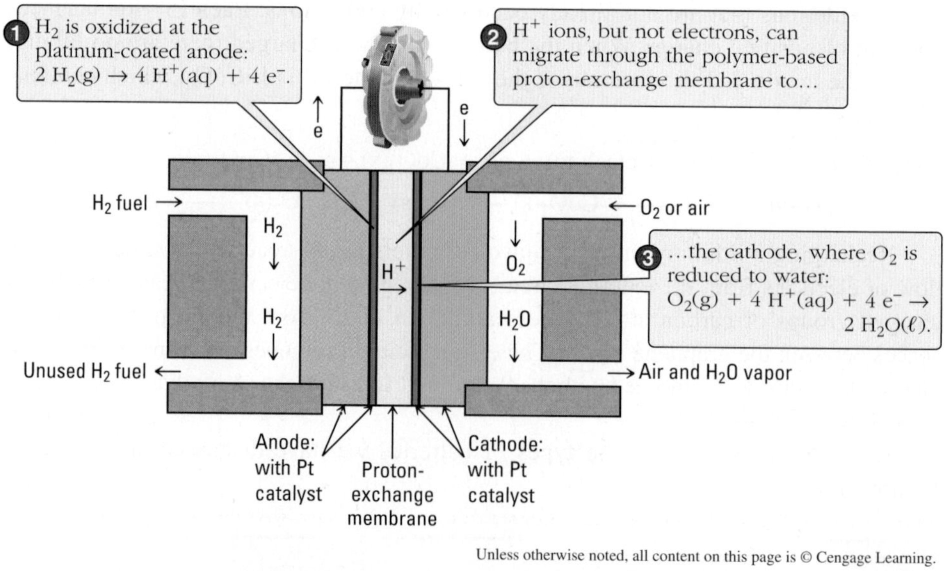
1 H₂ is oxidized at the platinum-coated anode: 2 H₂(g) → 4 H⁺(aq) + 4 e⁻.
2 H⁺ ions, but not electrons, can migrate through the polymer-based proton-exchange membrane to…
3 …the cathode, where O₂ is reduced to water: O₂(g) + 4 H⁺(aq) + 4 e⁻ → 2 H₂O(ℓ).
e
e
H₂ fuel
H₂
H₂
Unused H₂ fuel
H⁺
O₂ or air
O₂
H₂O
Air and H₂O vapor
Anode: with Pt catalyst
Proton-exchange membrane
Cathode: with Pt catalyst

Figure 17.18 A proton-exchange membrane $H_2|O_2$ fuel cell. $2\ H_2(g) + O_2(g) \rightarrow 2\ H_2O(\ell)$.

approximately 1 V. Several hundred fuel cells can be stacked in series to obtain higher voltages. Such a stacked device (the size of a suitcase) can power an automobile.

Pure hydrogen is required for PEM fuel cells, which creates concerns. Pure hydrogen is difficult to store and distribute and it is flammable. Hydrogen does not occur in nature as H_2, so it has to be manufactured. Currently most H_2 is produced as a by-product of petroleum refining (← **Sec. 10-1d**) or by treating methane with steam, so a lot of fossil-fuel energy is required to manufacture H_2. A device called a *reformer* can convert a pure hydrocarbon (such as natural gas), or an alcohol such as methanol, into hydrogen, which can then be fed into a fuel cell. The disadvantage to this approach is that the efficiency of a fuel cell with a reformer is lower than that of a regular fuel cell and products other than water are produced.

17-10 Electrolysis—Causing Reactant-Favored Redox Reactions to Occur

Reactant-favored redox systems (positive $\Delta_r G°$) can be made to produce products by **electrolysis**, *a process in which electrons are forced into the electrochemical system from an external source of direct electric current*, such as a battery. Electrolysis reactions are reactant-favored electrochemical reactions that are forced to occur by the flow of electricity. Recharging a battery is a good example of electrolysis. Electrolytic processes are extremely important industrially for the production and purification of many metals, including copper and aluminum (Sections 20-3 and 19-4), and in electroplating processes that produce a thin coating of metal on many different kinds of items.

Lysis means "splitting," so electrolysis means "splitting with electricity."

Electrolysis is used in industrial production of sodium metal and chlorine gas by decomposition of sodium chloride. At room temperature, the reaction is

$$2\ NaCl(s) \longrightarrow 2\ Na(s) + Cl_2(g) \qquad \Delta_r G° = 384.138\ kJ/mol;\ E° = -1.99\ V$$

The reaction is clearly reactant-favored and work must be done to cause it to occur. In the industrial process, the work is supplied by an external source of direct current. A pair of electrodes dips into molten NaCl (Figure 17.19), which consists of Na^+ and Cl^- ions that are free to move. At the surface of the cathode, Na^+ ions from the liquid are reduced to $Na(\ell)$ by electrons supplied by the external circuit. As the concentration of $Na^+(\ell)$ near the cathode decreases, other Na^+ ions move through the liquid to take their place. At the surface of the anode, Cl^- ions are oxidized to $Cl_2(g)$ and the electrons released are removed from the anode by the external power supply.

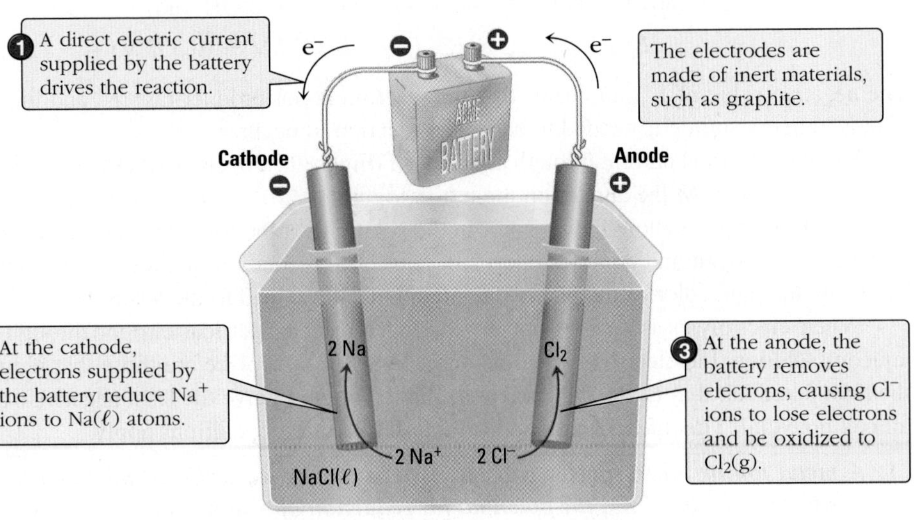
1 A direct electric current supplied by the battery drives the reaction.
e⁻
e⁻
The electrodes are made of inert materials, such as graphite.
ACME BATTERY
Cathode
Anode
2 At the cathode, electrons supplied by the battery reduce Na⁺ ions to Na(ℓ) atoms.
2 Na
Cl₂
3 At the anode, the battery removes electrons, causing Cl⁻ ions to lose electrons and be oxidized to Cl₂(g).
2 Na⁺
2 Cl⁻
NaCl(ℓ)

Figure 17.19 Electrolysis of molten sodium chloride with inert electrodes.

Cathode, reduction:	$2\ Na^+(\text{in melt}) + 2\ e^- \longrightarrow 2\ Na(\ell)$
Anode, oxidation:	$2\ Cl^-(\text{in melt}) \longrightarrow Cl_2(g) + 2\ e^-$
Overall reaction:	$2\ Na^+(\text{in melt}) + 2\ Cl^-(\text{in melt}) \longrightarrow 2\ Na(\ell) + Cl_2(g)$

To cause the reaction to occur, the external source of direct current must be supplied at a potential higher than the negative potential of the reaction.

The electrolysis of molten salts is an energy-intensive process because energy is needed to melt the salt as well as to cause the anode and cathode reactions to take place. It is used because there are few other ways to isolate very reactive elements like sodium.

What happens if a direct current is passed through an *aqueous solution* of a salt, such as potassium iodide, KI, rather than through the molten salt? To predict the outcome of the electrolysis we must first decide what is in the solution that can be oxidized and what can be reduced. For KI(aq), the solution contains K^+ ions, I^- ions, and H_2O molecules. Potassium is already in its highest common oxidation state in K^+, so it cannot be oxidized. However, both the I^- ion and the H_2O could be oxidized. The possible anode half-reaction oxidations and their half-cell potentials are

$$2\ I^-(aq) \longrightarrow I_2(s) + 2\ e^- \qquad E° = 0.535\ V$$

$$2\ H_2O(\ell) \longrightarrow O_2(g) + 4\ H^+(aq) + 4\ e^- \qquad E° = 1.229\ V$$

Whenever two or more electrochemical reactions are possible at the same electrode, use Table 17.1 (← **Sec. 17-5**) *or Appendix I to decide which reaction is more likely to occur under standard-state conditions.* Considering the two possible anode reactions, item 4 in the discussion of Table 17.1 indicates that the less positive the half-cell potential, the more likely a half-reaction is to occur as an oxidation. That is, the farther toward the bottom of Table 17.1 a half-reaction is, the more likely it is to occur as an oxidation. Oxidation of I^- to I_2 has the less positive $E°$, so this is the more likely anode reaction.

Since I^- is the lowest common oxidation state of iodine, there are only two species that can be reduced at the cathode: K^+ ions and water molecules. The possible cathode half-reaction reductions and their half-cell potentials are

$$K^+(aq) + e^- \longrightarrow K(s) \qquad E° = -2.925\ V$$

$$2\ H_2O(\ell) + 2\ e^- \longrightarrow H_2(g) + 2\ OH^-(aq) \qquad E° = -0.8277\ V$$

Point 3 in the discussion of Table 17.1 states that the more positive the half-cell potential, the more easily a substance on the left-hand side of a half-reaction can be reduced. Since $E°$ for H_2O is more positive (less negative) than $E°$ for K^+, H_2O is more likely to be reduced. Therefore, in aqueous KI solution, the overall reaction and cell potential are

$$2\ I^-(aq) + 2\ H_2O(\ell) \longrightarrow I_2(s) + H_2(g) + 2\ OH^-(aq)$$

$$E°_{cell} = E°_{cathode} - E°_{anode} = -0.8277\ V - (0.535\ V) = -1.363\ V$$

The negative value of $E°_{cell}$ indicates that the reaction is not product-favored and that an external energy source is needed to cause the reaction to occur.

Passing electrical current through aqueous KI (Figure 17.20) shows that this predicted equation is correct. At the anode (on the left in Figure 17.20), the I^- ions are oxidized to I_2, which produces a yellow-brown color in the solution. At the cathode, water is reduced to gaseous hydrogen and aqueous hydroxide ions. The formation of excess OH^- ions is shown by the pink color of the phenolphthalein indicator added to the solution.

When electrolysis is carried out by passing a direct electrical current through an aqueous solution, the electrode reactions most likely to take place are those that require the least potential, that is, the half-reactions that combine to give the least negative overall cell potential. This means that in aqueous solution these conditions apply.

1. ***A metal ion or other species can be reduced if it has a half-cell potential more positive than −0.8277 V, the potential for reduction of water.*** Table 17.1 shows that many metal ions are in this category. If a species has a half-cell potential more

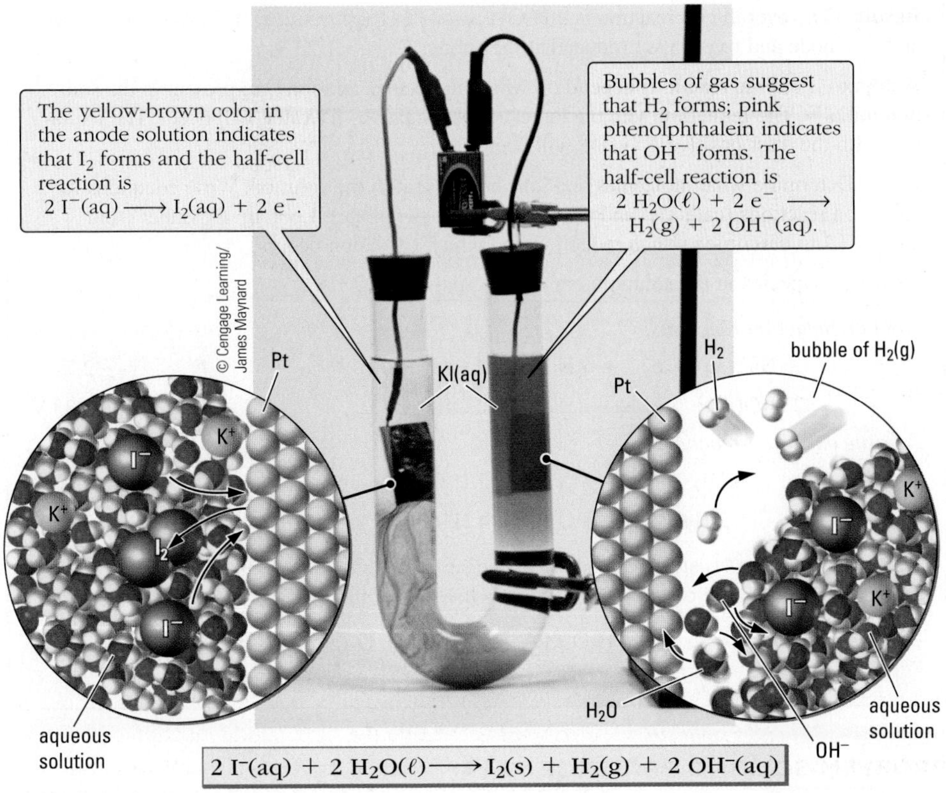

The yellow-brown color in the anode solution indicates that I_2 forms and the half-cell reaction is $2\ I^-(aq) \longrightarrow I_2(aq) + 2\ e^-$.
Bubbles of gas suggest that H_2 forms; pink phenolphthalein indicates that OH^- forms. The half-cell reaction is $2\ H_2O(\ell) + 2\ e^- \longrightarrow H_2(g) + 2\ OH^-(aq)$.
© Cengage Learning/ James Maynard
Pt
KI(aq)
Pt
H_2
bubble of $H_2(g)$
K^+
I^-
I_2
I^-
K^+
K^+
I^-
I^-
K^+
H_2O
aqueous solution
aqueous solution
OH^-
$2\ I^-(aq) + 2\ H_2O(\ell) \longrightarrow I_2(s) + H_2(g) + 2\ OH^-(aq)$

Figure 17.20 The electrolysis of aqueous potassium iodide. Aqueous KI is in both sides of the cell, and both electrodes are platinum.

negative than −0.8277 V, then water will preferentially be reduced to $H_2(g)$ and OH^- ions. Metal ions in this latter category include Na^+, K^+, Mg^{2+}, and Al^{3+}. Consequently, producing these metals from their ions requires electrolysis of a molten salt with no water present.

2. ***A species can be oxidized in aqueous solution if it has a half-cell potential less positive than 1.229 V, the potential for reduction of $O_2(g)$ to water.*** If the half-cell potential is less positive than for reduction of $O_2(g)$, then oxidation of the species on the right-hand side of a half-reaction is more likely than oxidation of water. Most of the half-equations in Table 17.1 are in this category. If a species has a half-cell potential more positive than 1.229 V (that is, if its half-reaction is above the water-oxygen half-reaction in Table 17.1), water will be oxidized preferentially. For example, $F^-(aq)$ cannot be oxidized electrolytically to $F_2(g)$ because water will be oxidized to $O_2(g)$ instead.

The potential that must be applied to an electrolysis cell is always somewhat greater than the potential calculated from standard half-cell potentials. An *overvoltage* is required, which is an additional potential needed to overcome limitations in the electron transfer rate at the interface between electrode and solution. Another factor affecting the necessary potential is that concentrations and pressures are usually not at standard-state conditions. Redox reactions that involve the formation of O_2 or H_2 are especially prone to have large overvoltages. Because overvoltages cannot be predicted accurately, the only way to determine with certainty which half-reaction will occur in an electrolysis cell when two possible reactions have similar standard half-cell potentials is to perform the experiment.

PROBLEM-SOLVING EXAMPLE 17.10

Electrolysis of Aqueous NaOH

Predict the results of passing a direct electrical current through a 1-M aqueous solution of NaOH. Calculate the standard cell potential.

Result The overall cell reaction is $2\ H_2O(\ell) \longrightarrow 2\ H_2(g) + O_2(g)$. Hydrogen is produced at the cathode and oxygen is produced at the anode. $E° = -1.23$ V.

Analyze Possible reactions depend on which molecules and ions are present in the solution initially. The reduction with the largest (least negative) $E°$ value will occur. The oxidation with the least positive $E°$ value will occur.

Plan Determine what molecules and ions are present in the solution. Write equations for reduction reactions (cathode) and oxidation reactions (anode). Look up $E°$ values from Table 17.1 to determine which reduction and which oxidation occurs.

Execute Species in the solution: Na^+, OH^-, and H_2O.

Possible reduction reactions:

$$Na^+(aq) + e^- \longrightarrow Na(s) \qquad E°_{cathode} = -2.71\ V$$

$$2\ H_2O(\ell) + 2\ e^- \longrightarrow H_2(g) + 2\ OH^-(aq) \qquad E°_{cathode} = -0.83\ V$$

Possible oxidation reactions:

$$4\ OH^-(aq) \longrightarrow O_2(g) + 2\ H_2O(\ell) + 4\ e^- \qquad E°_{anode} = +0.40\ V$$

$$2\ H_2O(\ell) \longrightarrow O_2(g) + 4\ H^+(aq) + 4\ e^- \qquad E°_{anode} = +1.229\ V$$

Water is reduced to H_2 at the cathode (least negative $E°$); OH^- is oxidized at the anode (least positive $E°$). The overall cell reaction is the sum of the two half-equations:

Overall reaction: $$2\ H_2O(\ell) \longrightarrow 2\ H_2(g) + O_2(g)$$

$$E°_{cell} = E°_{cathode} - E°_{anode} = (-0.83\ V) - (+0.40\ V) = -1.23\ V$$

PROBLEM-SOLVING PRACTICE 17.10

Predict the results of passing a direct electrical current through (a) molten NaBr, (b) aqueous NaBr, and (c) aqueous $SnCl_2$.

The electrolysis of aqueous NaCl (brine) is an extremely important industrial reaction in the chlor-alkali process. This process is the major commercial source of chlorine gas and sodium hydroxide; it is described in detail in Section 19-4b.

Table 17.2 compares and contrasts the properties of electrochemical cells.

Table 17.2 Summary of Properties of Electrochemical Cells

Common to All Electrochemical Cells (Voltaic and Electrolytic)	
Two electrodes are in contact with a conducting solution or porous membrane and an external electrical circuit.	
The overall reaction can be divided into an oxidation half-reaction and a reduction half-reaction.	
Oxidation occurs at the anode; reduction occurs at the cathode.	
Specific to a Voltaic Cell	**Specific to an Electrolytic Cell**
A *product-favored redox reaction* causes electrons to flow through the external circuit.	A source of direct electric current in the external circuit forces a *reactant-favored redox reaction* to occur.
Oxidation at the anode releases electrons, which flow into the external circuit; reduction at the cathode uses electrons, which are removed from the external circuit.	Electrons are forced into the cathode from the external circuit, causing reduction; electrons are removed from the anode by the external circuit, causing oxidation.
Half-reactions usually occur in *separate half-cells* connected by a salt bridge or porous membrane.	Half-reactions often occur in the *same container* with no separation of oxidation and reduction.
The *electrodes often participate in the redox reaction;* the anode usually loses mass and the cathode gains mass.	The *electrodes sometimes do not participate in the redox reaction;* often their main function is to conduct electrons into and out of the cell.

CONCEPTUAL **EXERCISE 17.15**

Making F_2 Electrolytically

In 1886, Henri Moissan was the first to prepare F_2 by the electrolysis of F^- ions. He electrolyzed KF dissolved in pure HF. No water was present, so only F^- ions were available at the anode. What was produced at the cathode? Write the half-equations for the oxidation and the reduction reactions, and then write the overall cell reaction.

17-11 Counting Electrons

When a direct electric current is passed through an aqueous solution of the soluble salt $AgNO_3$, metallic silver is produced at the cathode. One mole of electrons is required for every mole of Ag^+ reduced.

$$Ag^+(aq) + e^- \longrightarrow Ag(s)$$

For an aqueous copper(II), 2 mol electrons is required to produce 1 mol metallic copper from 1 mol copper(II) ions.

$$Cu^{2+}(aq) + 2\ e^- \longrightarrow Cu(s)$$

Each of these balanced half-reactions is like any other balanced chemical equation. That is, each illustrates the fact that both matter and charge are conserved in chemical reactions. Thus, if you could measure the amount (mol) of electrons flowing through an electrolysis cell, you could determine the amount (mol) of silver or copper produced. Conversely, if you knew the amount of silver or copper produced, you could calculate the amount of electrons that had passed through the circuit.

The amount of electrons transferred during a redox reaction is usually determined experimentally by measuring the current flowing in the external electrical circuit during a given time. The product of the current and the time interval equals the electric charge that flowed through the circuit.

$$\text{Charge} = \text{current} \times \text{time}$$

If the current is measured in amperes, A, and the time in seconds, s, the charge has units of coulombs, C:

$$1 \text{ ampere} \times 1 \text{ second} = 1 \text{ coulomb/second} \times 1 \text{ second} = 1 \text{ coulomb}$$

The Faraday constant (96,485 C/mol of electrons) can then be used to find the amount of electrons from a known charge in coulombs. This information is of practical significance in chemical analysis and synthesis.

Figure 17.21 shows the relationship between quantity of charge used and the quantities of substances that are oxidized or reduced during electrolysis.

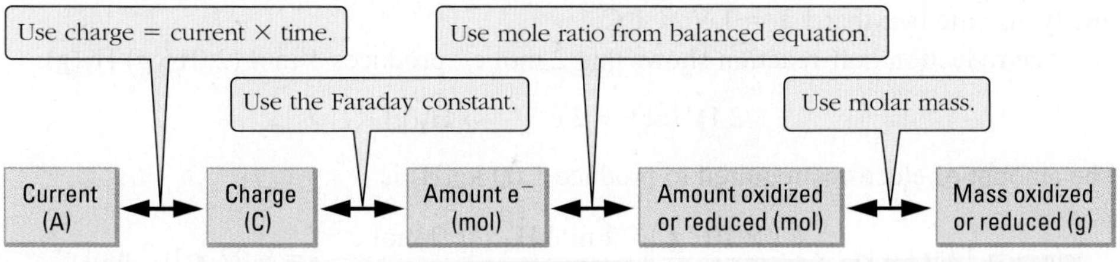
Use charge = current × time.
Use mole ratio from balanced equation.
Use the Faraday constant.
Use molar mass.
Current (A)
Charge (C)
Amount e⁻ (mol)
Amount oxidized or reduced (mol)
Mass oxidized or reduced (g)

Figure 17.21 Relationships in electrolysis calculations.

Large electric currents, like those needed to run a hair dryer or refrigerator, are measured in amperes (1 A = 1 C/s). Smaller currents, in the milliampere (mA) range, are more commonly used in laboratory electrolysis experiments.

PROBLEM-SOLVING EXAMPLE 17.11

Using the Faraday Constant

An aqueous solution containing Cu^{2+} ions is electrolyzed. Calculate the mass of copper deposited at the cathode when an electric current of 15 mA is applied for 1.0 h. Assume the process is 100% efficient and that some $Cu^{2+}(aq)$ remains in the solution when the current stops.

Result 0.018 g Cu

Analyze The mass of Cu can be calculated from the amount of Cu, which can be obtained from the amount of electrons passing through the circuit and the number of electrons transferred for each Cu atom.

Plan Write a balanced half-reaction for reduction of Cu^{2+} ions and determine n. Calculate the charge transferred from the product of current and time. Use the Faraday constant to calculate the amount of electrons, a stoichiometric ratio to calculate the amount of Cu, and the molar mass to calculate the mass.

Execute

$$Cu^{2+}(aq) + 2\,e^- \longrightarrow Cu(s)$$

$$\text{Charge} = 15 \times 10^{-3}\text{ A} \times 60\text{ min/h} \times 60\text{ s/min} = 15 \times 10^{-3}\text{ C/s} \times 3600\text{ s} = 54\text{ C}$$

$$m(\text{Cu}) = (54\text{ C})\left(\frac{1\text{ mol e}^-}{9.65 \times 10^4\text{C}}\right)\left(\frac{1\text{ mol Cu}}{2\text{ mol e}^-}\right)\left(\frac{63.5\text{ g Cu}}{1\text{ mol Cu}}\right) = 0.018\text{ g Cu}$$

PROBLEM-SOLVING PRACTICE 17.11

In the commercial production of sodium metal by electrolysis, the cell operates at 7.0 V and a current of 25×10^3 A. Calculate the mass of metallic sodium that can be produced in 1 h.

CONCEPTUAL EXERCISE 17.16

How Many Electrons?

Which requires the most electrons?

(a) Making 1 mol Al from Al^{3+}
(b) Making 2 mol Na from Na^+
(c) Making 2 mol Cu from Cu^{2+}

17-11a Electrolytic Production of Hydrogen

Hydrogen can be produced by the electrolysis of water to which a drop or two of sulfuric acid has been added to make the solution electrically conductive. The overall electrochemical reaction is

$$2\,H_2O(\ell) \longrightarrow 2\,H_2(g) + O_2(g) \qquad E^\circ_{cell} = -1.23\text{ V}$$

Oxygen is produced at the anode and hydrogen at the cathode. The minimum potential required for this reaction is 1.23 V (1.23 J/C), but in practice overvoltage requires a higher potential of about 2 V.

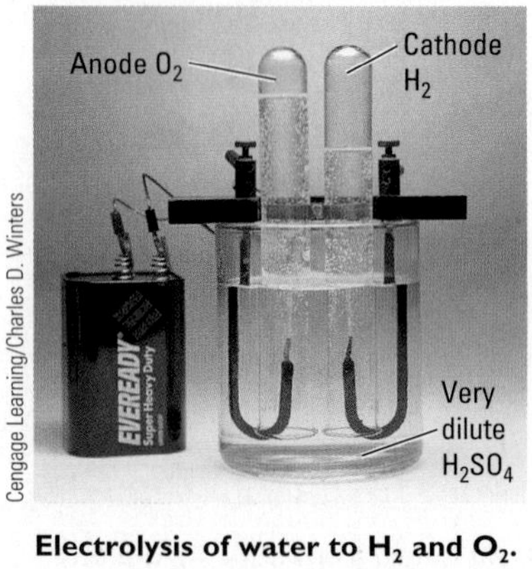
Anode O_2
Cathode H_2
Very dilute H_2SO_4
EVEREADY
Cengage Learning/Charles D. Winters

Electrolysis of water to H_2 and O_2.

Let's consider how much electrical energy is required to produce 1.00 kg $H_2(g)$ (about 11,200 L at STP) and at what cost. First, use the reduction half-reaction to find the amount of electrons; then, use the Faraday constant to calculate the charge in coulombs; finally, use the fact that 1 J = 1 V × 1 C.

The reduction half-reaction shows that 2 mol e^- produces 1 mol (2.016 g) $H_2(g)$.

$$2\,H^+(aq) + 2\,e^- \longrightarrow H_2(g)$$

The amount of electrons required to produce 1.00 kg H_2 is

$$n(e^-) = 1.00\text{ kg H}_2 \times \left(\frac{1 \times 10^3\text{ g}}{\text{kg}}\right)\left(\frac{1\text{ mol H}_2}{2.016\text{ g H}_2}\right)\left(\frac{2\text{ mol e}^-}{1\text{ mol H}_2}\right) = 9.92 \times 10^2\text{ mol e}^-$$

Next, calculate the charge transferred using the Faraday constant.

$$\text{Charge} = (9.92 \times 10^2 \text{ mol e}^-) \times \left(\frac{9.65 \times 10^4 \text{ C}}{1 \text{ mol e}^-}\right) = 9.57 \times 10^7 \text{ C}$$

Calculate the energy (in joules) from the charge and the cell potential.

$$\text{Energy} = \text{charge} \times \text{potential} = (9.57 \times 10^7 \text{ C})(1.23 \text{ V}) = 1.18 \times 10^8 \text{ J}$$

Convert joules to kilowatt-hours (kWh), the unit we see on an electric bill.

The kilowatt-hour (kWh) is a unit of energy: 1 kWh = 3.60×10^6 J.

$$\text{Energy} = 1.18 \times 10^8 \text{ J} \times \frac{1 \text{ W}}{1 \text{ J/s}} \times \frac{1 \text{ kW}}{1000 \text{ W}} \times \frac{1 \text{ h}}{3.60 \times 10^3 \text{ s}} = 32.7 \text{ kWh}$$

At a rate of 10 cents per kilowatt-hour, the production of 1.00 kg hydrogen would cost \$3.27.

Hydrogen holds great promise as a fuel in our economy because it is a gas and can be transported through pipelines, it burns without producing pollutants, and it could be used in fuel cells to generate electricity on demand. Both water and sulfuric acid are in plentiful supply. The major problem with producing hydrogen in quantities large enough to meet the nation's energy demands is finding a cheap enough source of electricity to decompose water to its elements.

EXERCISE 17.17

Calculations Based on Electrolysis

In the production of aluminum metal, Al^{3+} is reduced to Al metal using currents of about 50,000 A and a low voltage of about 4.0 V. Calculate how much energy (in kilowatt-hours) is required to produce 2000. metric tons of aluminum metal.

CONCEPTUAL EXERCISE 17.18

How Much Energy?

Think of a battery you just purchased at the store as an energy source that can deliver some quantity of energy. Name the two pieces of information you need to calculate the energy (J) this battery can deliver. Which one is obviously available as you read the label on the battery? Devise a means of determining the other information needed.

17-11b Electroplating

If a metal or other electrical conductor serves as the cathode in an electrolysis cell, the conductor can be plated with another metal to decorate it or protect it against corrosion. To plate an object with copper, for example, we make the object's surface conducting and use the object as the cathode in an electrolysis cell containing a solution of Cu^{2+} ions from a soluble copper salt. The object becomes coated with metallic copper, and the coating thickens as the electrolysis continues. If the plated object is a metal, it will conduct electricity by itself. If the object is a nonmetal, its surface can be lightly dusted with graphite powder to make it conducting.

Precious metals such as gold are often electroplated onto cheaper metals such as copper to make jewelry. By controlling the current and duration of the known plating reaction, it is possible to calculate the mass of gold metal that is plated onto the cathode surface. For example, the gold medals awarded during the 2012 Olympics are actually a thin layer of gold plated on silver and copper. Each medal weighs 400 g, only 1.5% by mass of which is gold, just 6.0 g. Assuming that gold is selling for \$1300 per ounce, that's more than \$250 worth of gold. The 6.0 g of gold could be plated onto the medal by

Martin Bache/Alamy

Gold medal from 2012 Olympics.

immersing the medal into an aqueous $AuCl_3$ solution and making the medal the cathode. The circuit is completed by immersing an inert electrode in the solution and connecting the electrode to the positive terminal of a battery operating at a current of 2.0 A. How long (minutes) will it take to plate the 6.0 g of gold, assuming 100% efficiency?

The reduction reaction at the cathode is

$$Au^{3+}(aq) + 3\ e^- \longrightarrow Au(s)$$

The time required can be calculated as

$$\text{Time} = 6.0\ \text{g Au}\left(\frac{1\ \text{mol Au}}{197\ \text{g Au}}\right)\left(\frac{3\ \text{mol e}^-}{1\ \text{mol Au}}\right)\left(\frac{96{,}500\ \text{C}}{1\ \text{mol e}^-}\right)\left(\frac{1\ \text{s}}{2.0\ \text{C}}\right)\left(\frac{1\ \text{min}}{60\ \text{s}}\right) = 74\ \text{min}$$

EXERCISE 17.19

Electroplating Copper

Calculate the mass of copper in a 2012 Olympic gold medal if it takes 12.1 min at the current of 100. A to plate this mass of copper at the cathode using this reaction: $Cu^{2+}(aq) + 2\ e^- \longrightarrow Cu(s)$.

17-12 Corrosion: Undesirable Product-Favored Redox Reactions

Corrosion is *the oxidation of a metal that is exposed to the environment.* Visible corrosion on the steel supports of a bridge, for example, indicates possible structural failure. Corrosion is so commonplace that about 25% of the annual U. S. production of steel is used to replace metal lost to corrosion.

Corrosion reactions are invariably product-favored, which means that $E°$ for the reaction is positive and $\Delta_r G°$ is negative. Corrosion of iron, for example, takes place quite

ESTIMATION | Making an Aluminum Baseball Bat

Although major league baseball players are not permitted to use them, aluminum bats are used widely in all other baseball programs. Such aluminum bats are actually alloys that contain mostly aluminum with very small quantities of other metals added to increase strength and other desirable properties. Approximately how much energy (in kWh) is required to manufacture a typical 800-g Al bat?

The aluminum metal is produced by using an electric current to reduce Al^{3+} ions to Al(s). Let's assume that the bat contains only aluminum and the reduction process is 80% efficient. The reduction of Al^{3+} is run commercially at 50,000 A and 4 V (4 J/C); 1 kWh = 3.60×10^6 J.

The charge required to generate 800 g Al from sufficient Al^{3+} ions is

$$\text{Charge} = (800\ \text{g Al})\left(\frac{1\ \text{mol Al}}{26.98\ \text{g Al}}\right)\left(\frac{3\ \text{mol e}^-}{1\ \text{mol Al}}\right)\left(\frac{96{,}500\ \text{C}}{1\ \text{mol e}^-}\right)$$
$$= 8.6 \times 10^6\ \text{C}$$

If the process were 100% efficient, the quantity of energy used would be

$$E = (8.6 \times 10^6\ \text{C})(4\ \text{J/C})\left(\frac{1\ \text{kWh}}{3.60 \times 10^6\ \text{J}}\right) = 9.6\ \text{kWh}$$

However, because the process is only 80% efficient, the actual quantity of energy used is

$$E = \left(\frac{9.6\ \text{kWh}}{0.80}\right) = 12\ \text{kWh}$$

This quantity of energy would light 1000 13-W compact fluorescent light bulbs for almost an hour.

It should be noted here (without doing the actual calculations) that making the baseball bat from recycled aluminum rather than from aluminum ore would require about 95% less energy.

readily and is difficult to prevent. It produces the red-brown substance we call rust, which is hydrated iron(III) oxide ($Fe_2O_3 \cdot xH_2O$, where x varies from 2 to 4). The rust that forms when iron corrodes does not adhere to the surface of the metal, so it can easily flake off and expose fresh metal surface to corrosion (Figure 17.22). The corrosion of aluminum, a metal that is even more reactive than iron, is also very product-favored. However, the aluminum oxide that forms as a result of corrosion adheres tightly as a thin coating on the surface of the metal, creating a protective coating that prevents further corrosion.

Cengage Learning/Charles D. Winters

Figure 17.22 Rusting. Formation of rust destroys the structural integrity of objects made of iron and steel. Given time, this chain will completely change to rust.

For corrosion of a metal (M) to occur, the metal must have an anodic area where the oxidation can occur. The general reaction is

Anode reaction: $$M(s) \longrightarrow M^{n+} + n\,e^-$$

There must also be a cathodic area where electrons are consumed. Frequently, the cathode reactions are reductions of oxygen or water.

Cathode reactions: $$O_2(g) + 2\,H_2O(\ell) + 4\,e^- \longrightarrow 4\,OH^-(aq)$$
$$2\,H_2O(\ell) + 2\,e^- \longrightarrow 2\,OH^-(aq) + H_2(g)$$

Anodic areas may occur at cracks in the oxide coating that protects the surfaces of many metals (such as aluminum) or around impurities. Cathodic areas may occur at the metal oxide coating, at less reactive metallic impurity sites, or around other metal compounds trapped at the surface, such as sulfides or carbides.

The other requirements for corrosion are electrical conduction between the anode and the cathode and an electrolyte in contact with both anode and cathode. Both requirements are easily fulfilled—the metal itself is the conductor, and ions dissolved in moisture from the environment provide the electrolyte.

In the corrosion of iron, the anodic reaction is the oxidation of metallic iron (Figure 17.23). If both water and O_2 gas are present, the cathode reaction is the reduction of oxygen, giving the overall reaction

Anode reaction: $$2\,[Fe(s) \longrightarrow Fe^{2+}(aq) + 2\,e^-]$$
Cathode reaction: $$O_2(g) + 2\,H_2O(\ell) + 4\,e^- \longrightarrow 4\,OH^-(aq)$$
Overall reaction: $$2\,Fe(s) + O_2(g) + 2\,H_2O(\ell) \longrightarrow \underset{\text{iron(II) hydroxide}}{2\,Fe(OH)_2(s)}$$

In the presence of an ample supply of oxygen and water, as in open air or flowing water, the iron(II) hydroxide is oxidized to the hydrated red-brown iron(III) oxide seen in Figure 17.22.

$$4\,Fe(OH)_2(s) + O_2(g) \longrightarrow \underset{\text{red-brown}}{2\,Fe_2O_3 \cdot 2H_2O(s)}$$

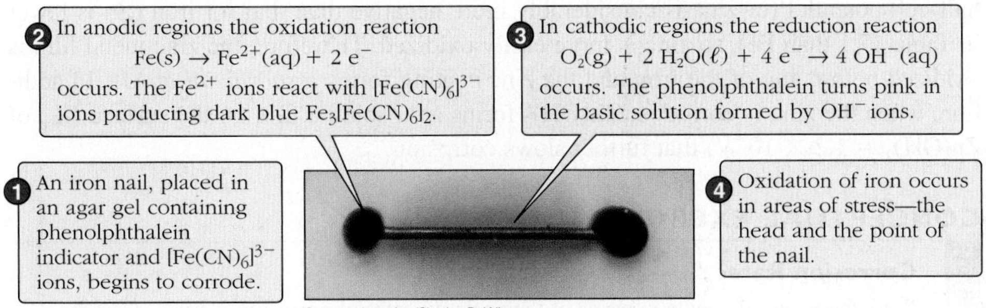
2 In anodic regions the oxidation reaction
Fe(s) → Fe2+(aq) + 2 e−
occurs. The Fe2+ ions react with [Fe(CN)6]3− ions producing dark blue Fe3[Fe(CN)6]2.
3 In cathodic regions the reduction reaction
O2(g) + 2 H2O(ℓ) + 4 e− → 4 OH−(aq)
occurs. The phenolphthalein turns pink in the basic solution formed by OH− ions.
1 An iron nail, placed in an agar gel containing phenolphthalein indicator and [Fe(CN)6]3− ions, begins to corrode.
4 Oxidation of iron occurs in areas of stress—the head and the point of the nail.
Cengage Learning/Charles D. Winters

Figure 17.23 A corroding iron nail has anodic and cathodic regions.

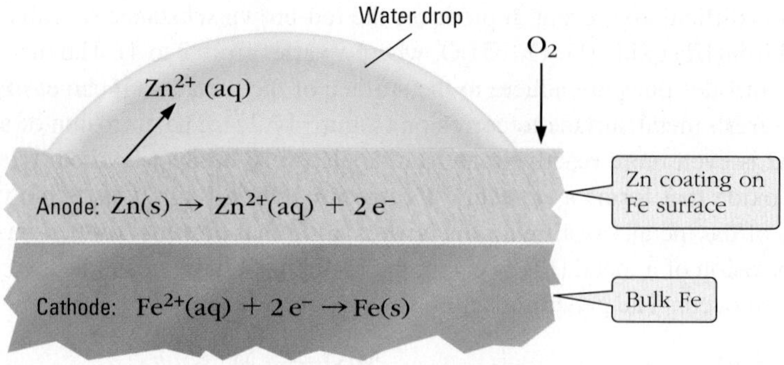
Water drop
O_2
Zn^{2+} (aq)
Anode: $Zn(s) \rightarrow Zn^{2+}(aq) + 2\,e^-$
Zn coating on Fe surface
Cathode: $Fe^{2+}(aq) + 2\,e^- \rightarrow Fe(s)$
Bulk Fe

Figure 17.24 Galvanizing provides cathodic protection of an iron-containing object. The iron is coated with a film of zinc, a metal more easily oxidized than iron. The zinc acts as the anode and forces iron to become the cathode, thereby preventing the corrosion of the iron.

This hydrated iron oxide is the familiar rust you see on iron and steel objects and the substance that colors the water red in some mountain streams and home water pipes.

Other substances in air and water can hasten corrosion. Dissolved metal salts, such as the chlorides of sodium and calcium from sea air or from salt spread on roadways in the winter, function as electrolyte salt bridges between anodic and cathodic regions, thus speeding up corrosion reactions.

CONCEPTUAL EXERCISE 17.20

Do All Metals Corrode?

Do all metals corrode as readily as iron? Name three metals that you would expect to corrode about as readily as iron (ignoring the inhibiting effect of oxide coatings on the surface), and name three metals that do not corrode readily. Name a use for each of the three noncorroding metals. Explain why metals fall into these two groups.

17-12a Corrosion Protection

How can metal corrosion be prevented? The general approaches are to: (1) inhibit the anodic process; (2) inhibit the cathodic process; or (3) do both. The most common method is **anodic inhibition**, which *directly limits or prevents the oxidation half-reaction* by painting the metal surface, coating it with grease or oil, or allowing a thin film of metal oxide to form. This prevents oxygen and water from reaching the metal surface.

Cathodic protection is accomplished by *forcing the metal to become the cathode instead of the anode*. Usually, this goal is achieved by attaching another, more readily oxidized metal to the metal being protected. The most common example involves galvanized iron, iron that has been coated with a thin film of zinc (Figure 17.24). The standard half-cell potential for zinc is considerably more negative than that for iron (Zn is lower in Table 17.1 than Fe), so zinc is more easily oxidized. Therefore, the zinc metal film is oxidized before any of the iron and the zinc coating forms a *sacrificial anode.* In addition, when the zinc is corroded, $Zn(OH)_2$ forms an insoluble film on the surface (K_{sp} of $Zn(OH)_2 = 1.2 \times 10^{-17}$) that further slows corrosion.

Cengage Learning/Charles D. Winters

Galvanized objects. A thin coating of zinc helps prevent the oxidation of iron.

CONCEPTUAL EXERCISE 17.21

Corrosion Rates

Rank these environments in terms of their relative rates of corrosion of iron. Place the fastest first. Explain your answer.

(a) Moist clay
(b) Sand by the seashore
(c) The surface of the moon
(d) Desert sand in Arizona

SUMMARY PROBLEM

At the turn of the 20th century, electric automobiles were more popular than gasoline-fueled cars. Thomas Edison, the prolific American inventor, was interested in electric automobiles and developed an iron-nickel (Fe-Ni) battery to power them. He claimed that, due to their higher energy density, his batteries were superior to the lead-acid batteries then being used in automobiles. Today, the Edison battery is available in various voltages up to 48 V. It is used in London subway locomotives, in New York City subway cars, for standalone locations that are off the electric power grid, and for backup emergency power. The battery has an NiO(OH) cathode, an iron anode, and uses a mixture of KOH and LiOH as the electrolyte. The half-reactions are

Cathode: $NiO(OH)(s) + H_2O(\ell) + e^- \longrightarrow Ni(OH)_2(s) + OH^-(aq)$

Anode: $Fe(s) + 2\ OH^-(aq) \longrightarrow Fe(OH)_2(s) + 2\ e^-$

The E°_{cell} is 1.37 V at 25 °C, which drops to 1.2 V when operating.

(a) Assign an oxidation number to nickel in the reactant and product of the cathode reaction.

(b) Identify what is oxidized and what is reduced.

(c) Write a balanced chemical equation for the overall cell reaction.

(d) Calculate the cell voltage at 25 °C when the hydroxide ion concentration is 3.5 M.

(e) An AA iron-nickel battery delivers a constant current of 750 mA at 1.2 V.

(1) Calculate how long it will take the battery to convert 17.1 g NiO(OH) to $Ni(OH)_2$, assuming 75% efficiency.

(2) Calculate the mass (g) of metallic iron converted to $Fe(OH)_2$ during this time.

(3) The battery provides energy at a cost of 6.0×10^{-3} kilowatt-hour (kWh) per dollar; calculate the cost of operating the battery in part (1).

(f) An AA iron-nickel battery is recharged at 65% efficiency using a constant current of 1.50 A at 1.65 V. Calculate:

(1) The number of kilowatt-hours (kWh) of electricity used to recharge the battery.

(2) If the electricity used to recharge the battery costs $0.11 per kWh, calculate the cost of recharging the battery.

Edison also developed a nickel-zinc battery for use in electric automobiles. The battery is now used in some cellphones and digital cameras. This battery operates in much the same way as the iron-nickel battery; both use NiO(OH) as the cathode. The E°_{cell} for the nickel-zinc cell is 1.65 V at 25 °C.

(a) Calculate the $\Delta_r G^\circ$ (kJ/mol) for each of the battery reactions.

(b) Which battery—the iron-nickel or nickel-zinc—has the more product-favored reaction at standard conditions? Explain your answer.

(c) Does the nickel-zinc battery have a very small or very large equilibrium constant under standard conditions? Explain your answer.

HAVING STUDIED THIS CHAPTER . . .

. . . you should be able to:

- Identify the oxidizing and reducing agents in a redox reaction (Section 17-1). End-of-chapter Question 6
- Write equations for oxidation and reduction half-reactions, and use them to balance the overall equation (Section 17-2). Questions 10, 14
- Identify and describe the functions of the parts of a voltaic cell; describe the direction of electron movement outside the cell and the direction of the ion movement inside the cell (Section 17-3). Question 21

- Use shorthand cell notation to represent a voltaic cell (Section 17-3). Question 23
- Define standard half-cell potentials; use them to predict whether a reaction is product-favored as written (Sections 17-4 and 17-5). Questions 27, 31, 32
- Calculate $\Delta_r G°$ and $K°$ from the value of $E°$; calculate $E°$ from $\Delta_r G°$ or $K°$ for a redox reaction (Section 17-6). Questions 36, 38, 42
- Explain how product-favored electrochemical reactions can be used to do useful work (Section 17-6).
- Explain how the Nernst equation relates concentrations of redox reactants to E_{cell} and use the Nernst equation to calculate the potentials of cells that are not at standard conditions (Section 17-7). Questions 46, 49, 51
- Describe the chemistry of the lead-acid storage battery, the NiCad and NiMH batteries, and the Li-ion battery (Section 17-8). Questions 52, 54, 71
- Describe how a fuel cell works, and indicate its similarities to and differences from a battery (Section 17-9). Questions 56, 58
- Use standard half-cell potentials to predict the products of electrolysis of an aqueous salt solution (Section 17-10). Questions 61, 63
- Calculate the quantity of product formed at an electrode during an electrolysis reaction, given the current passing through the cell and the time during which the current flows (Section 17-11). Questions 65, 73
- Explain how electroplating works (Section 17-11). Question 75
- Describe how corrosion occurs and how it can be prevented (Section 17-12). Questions 78, 80

KEY TERMS

ampere (A) (Section 17-4)
anode (17-3)
anodic inhibition (17-12a)
battery (17-3)
cathode (17-3)
cathodic protection (17-12a)
cell potential (17-4)
cell voltage (17-4)
concentration cell (17-7a)
corrosion (17-12)
coulomb (C) (17-4)
electrochemical cell (17-3)
electrochemistry (Introduction)
electrode (17-3)
electrolysis (17-10)
Faraday constant (F) (17-6)
fuel cell (17-9)
half-cell (17-3)
half-reaction (17-2)
lead-acid storage battery (17-8a)
lithium-ion battery (17-8c)
Nernst equation (17-7)
NiCad and NiMH batteries (17-8b)
pH meter (17-7a)
primary battery (17-8)
salt bridge (17-3)
secondary battery (17-8)
standard-state conditions (17-4a)
standard electrode potential ($E°$) (17-5)
standard half-cell potential ($E°$) (17-5)
standard hydrogen electrode (17-4b)
standard potential ($E°$) (17-4a)
volt (V) (17-4)
voltaic cell (17-3)

QUESTIONS FOR REVIEW AND THOUGHT

Red-numbered questions have short answers at the back of this book in Appendix M and fully worked solutions in the *Student Solutions Manual.*

Review Questions

These questions test vocabulary and simple concepts.

1. Make a drawing showing the principal parts of
   (a) a voltaic cell: show the anode, the cathode, the direction of electron movement outside the cell, and the direction of ion movement inside the cell.
   (b) a standard hydrogen electrode: describe the components of the electrode and explain how it works.
2. Explain how product-favored electrochemical reactions can be used to do useful work.
3. Explain how reactant-favored electrochemical reactions can be induced to make products.
4. Explain how electroplating works.
5. Identify each statement as true or false. Rewrite each false statement to make it true.
   (a) Oxidation always occurs at the anode of an electrochemical cell.
   (b) The anode of a discharging voltaic cell is the site of reduction and is negative.
   (c) Standard-state conditions for electrochemical cells are a concentration of 1.0 M for dissolved species and a pressure of 1 bar for gases.
   (d) The potential of a voltaic cell does not change with temperature.
   (e) All product-favored oxidation-reduction reactions have a standard cell potential E°_{cell}, with a negative sign.

Topical Questions

These questions are keyed to the major topics in the chapter. Usually a question that is answered at the back of this book is paired with a similar one that is not.

Redox Reactions (Section 17-1)

6. In this reaction, assign an oxidation number to each atom in reactants and products. Identify which substance is oxidized and which is reduced. Identify the oxidizing agent and the reducing agent.

$$8\ H^+(aq) + MnO_4^-(aq) + 5\ Fe^{2+}(aq) \longrightarrow 5\ Fe^{3+}(aq) + Mn^{2+}(aq) + 4\ H_2O(\ell)$$

7. In each of these reactions assign an oxidation number to each atom in reactants and products. Identify which substance is oxidized and which is reduced. Identify the oxidizing agent and the reducing agent.
   (a) $Fe(s) + Br_2(\ell) \longrightarrow FeBr_2(s)$
   (b) $2\ Al(s) + 3\ Cl_2(g) \longrightarrow 2\ AlCl_3(s)$
   (c) $8\ HI(aq) + H_2SO_4(aq) \longrightarrow H_2S(aq) + 4\ I_2(s) + 4\ H_2O(\ell)$
   (d) $H_2O_2(aq) + 2\ Fe^{2+}(aq) + 2\ H^+(aq) \longrightarrow 2\ Fe^{3+}(aq) + 2\ H_2O(\ell)$
   (e) $FeS(s) + 3\ NO_3^-(aq) + 4\ H^+(aq) \longrightarrow 3\ NO(g) + SO_4^{2-}(aq) + Fe^{3+}(aq) + 2\ H_2O(\ell)$
8. Choose four elements: a metal that is a representative element, a transition metal, a nonmetal, and a metalloid. Using the index to this text, find a chemical reaction in which each element occurs as a reactant. Assign oxidation numbers to all elements on the reactant and product sides, and identify the oxidizing agent and the reducing agent.
9. Answer Question 8 again, but this time find a chemical reaction in which each element is produced.

Half-reactions and Redox Reactions (Section 17-2)

10. Write half-reactions for these changes:
    (a) Oxidation of cadmium to Cd^{2+} ions
    (b) Reduction of Fe^{3+} ions to Fe metal
    (c) Reduction of Sn^{4+} ions to Sn^{2+} ions
    (d) Reduction of chlorine to Cl^- ions
    (e) Oxidation of sulfur dioxide to sulfate ions in acidic solution
11. Write half-reactions for these changes:
    (a) Reduction of MnO_4^- ion to Mn^{2+} ion in acid solution
    (b) Reduction of $Cr_2O_7^{2-}$ ion to Cr^{3+} ion in acid solution
    (c) Oxidation of chlorine gas to ClO^- ions
    (d) Reduction of hydrogen peroxide to water in acidic solution
    (e) Oxidation of nitrous acid to nitrate ions in acidic solution
12. For the reaction in Question 6, write balanced half-reactions.
13. For each reaction in Question 7, write balanced half-reactions.
14. For the reaction in Question 6, combine the balanced half-reactions you wrote in Question 12 to give a balanced overall equation.
15. For each reaction in Question 7, combine the balanced half-reactions you wrote in Question 13 to give a balanced overall equation.
16. Balance these redox reactions, and identify the oxidizing agent and the reducing agent.
    (a) $CO(g) + O_3(g) \longrightarrow CO_2(g)$
    (b) $H_2(g) + Cl_2(g) \longrightarrow HCl(g)$
    (c) $H_2O_2(aq) + Ti^{2+}(aq) \longrightarrow H_2O(\ell) + TiO_2(s)$ in acidic solution
    (d) $Cl^-(aq) + MnO_4^-(aq) \longrightarrow Cl_2(g) + MnO_2(s)$ in acidic solution
    (e) $FeS_2(s) + O_2(g) \longrightarrow Fe_2O_3(s) + SO_2(g)$
    (f) $O_3(g) + NO(g) \longrightarrow O_2(g) + NO_2(g)$
    (g) $Zn(s) + HgO(s) \longrightarrow Zn(OH)_2(s) + Hg(\ell)$ in basic solution

17. Balance these redox reactions, and identify the oxidizing agent and the reducing agent.
(a) $FeO(s) + O_3(g) \longrightarrow Fe_2O_3(s)$
(b) $P_4(s) + Br_2(\ell) \longrightarrow PBr_5(\ell)$
(c) $H_2O_2(aq) + Co^{2+}(aq) \longrightarrow H_2O(\ell) + Co^{3+}(aq)$ in acidic solution
(d) $Cl^-(aq) + Cr_2O_7^{2-}(aq) \longrightarrow Cl_2(g) + Cr^{3+}(aq)$ in acidic solution
(e) $MnO_4^-(aq) + Zn(s) \longrightarrow MnO_2(s) + Zn(OH)_2(s)$ in basic solution
(f) $H_2CO(g) + O_2(g) \longrightarrow CO_2(g) + H_2O(\ell)$
(g) $C_3H_8(g) + O_2(g) \longrightarrow CO_2(g) + H_2O(\ell)$

Voltaic Cells (Section 17-3)

18. For the reaction $Cu^{2+}(aq) + Zn(s) \longrightarrow Cu(s) + Zn^{2+}(aq)$, why can't you generate electric current by placing a piece of copper metal and a piece of zinc metal in a solution containing $CuCl_2(aq)$ and $ZnCl_2(aq)$?
19. Explain the function of a salt bridge in a voltaic cell.
20. Tell whether this statement is true or false. If false, rewrite it to make it a correct statement: The value of an electrode potential changes when the half-reaction is multiplied by a factor. That is, $E°$ for $Li^+ + e^- \longrightarrow Li$ is different from that for $2\ Li^+ + 2\ e^- \longrightarrow 2\ Li$.
21. A voltaic cell is assembled with Pb(s) and $Pb(NO_3)_2(aq)$ in one compartment and Zn(s) and $ZnCl_2(aq)$ in the other. An external wire connects the two electrodes, and a salt bridge containing $KNO_3(aq)$ connects the two solutions.
(a) In the product-favored reaction, zinc metal is oxidized to Zn^{2+}. Write a balanced net ionic equation for this reaction.
(b) Which half-reaction occurs at each electrode? Which is the anode and which is the cathode?
(c) Draw a diagram of the cell, indicating the direction of electron movement outside the cell and of ion movement within the cell.
22. A voltaic cell is assembled with Sn(s) and $Sn(NO_3)_2(aq)$ in one compartment and Ag(s) and $AgNO_3(aq)$ in the other. An external wire connects the two electrodes, and a salt bridge containing $KNO_3(aq)$ connects the two solutions.
(a) In the product-favored reaction, Ag^+ is reduced to silver metal. Write a balanced net ionic equation for this reaction.
(b) Which half-reaction occurs at each electrode? Which is the anode and which is the cathode?
(c) Draw a diagram of the cell, indicating the direction of electron movement outside the cell and of ion movement within the cell.
23. Draw a diagram of each cell. Label the anode, the cathode, the species in each half-cell solution, the direction of electron movement in an external circuit, and the direction of movement of ions within the cell.
(a) $Cu(s)|Cu^{2+}(aq)||Fe^{2+}(aq)|Fe(s)$
(b) $Pt(s)|H_2O_2(aq), H^+(aq)||Fe^{2+}(aq), Fe^{3+}(aq)|Pt(s)$

Voltaic Cells and Cell Potential (Section 17-4)

24. You light a 25-W light bulb with the current from a 12-V lead-acid storage battery. Calculate how much energy the light bulb utilized after 1.0 h of operation. Calculate how many coulombs passed through the bulb. Assume 100% efficiency. (1 W = 1 J/s.)
25. Copper can reduce silver ion to metallic silver, a reaction that could, in principle, be used in a battery.

$$Cu(s) + 2\ Ag^+(aq) \longrightarrow Cu^{2+}(aq) + 2\ Ag(s)$$

(a) Write equations for the half-reactions involved.
(b) Which half-reaction is an oxidation and which is a reduction? Which half-reaction occurs in the anode compartment and which takes place in the cathode compartment?
26. Chlorine gas can oxidize zinc metal in a reaction that has been suggested as the basis of a battery. Write the half-reactions involved. Label which is the oxidation half-reaction and which is the reduction half-reaction.

Using Standard Half-cell Potentials (Section 17-5)

27. In Table 17.1 (← Sec. 17-5), identify (a) the strongest oxidizing agent; (b) the strongest reducing agent; (c) the weakest oxidizing agent; and (d) the weakest reducing agent.
28. One of the most energetic redox reactions is that between F_2 gas and lithium metal.
(a) Write the half-reactions involved. Label the oxidation half-reaction and the reduction half-reaction.
(b) Using data from Table 17.1, calculate $E°_{cell}$ for this reaction.
29. Using the standard half-cell potentials in Table 17.1, place these elements in order of increasing ability to function as reducing agents:
(a) Cl_2 (b) Fe
(c) Ag (d) Na
(e) H_2
30. Using the standard half-cell potentials in Table 17.1, place these elements in order of increasing ability to function as oxidizing agents:
(a) O_2 (b) H_2O_2
(c) $PbSO_4$ (d) H_2O
31. Calculate the value of $E°_{cell}$ for each of these reactions. Decide whether each is product-favored.
(a) $I_2(s) + Mg(s) \longrightarrow Mg^{2+}(aq) + 2\ I^-(aq)$
(b) $Ag(s) + Fe^{3+}(aq) \longrightarrow Ag^+(aq) + Fe^{2+}(aq)$
(c) $Sn^{2+}(aq) + 2\ Ag^+(aq) \longrightarrow Sn^{4+}(aq) + 2\ Ag(s)$
(d) $2\ Zn(s) + O_2(g) + 2\ H_2O(\ell) \longrightarrow 2\ Zn^{2+}(aq) + 4\ OH^-(aq)$
32. Consider these half-reactions:

Half-reaction	$E°$ (V)
$Au^{3+}(aq) + 3\ e^- \longrightarrow Au(s)$	1.52
$Pt^{2+}(aq) + 2\ e^- \longrightarrow Pt(s)$	1.118
$Co^{2+}(aq) + 2\ e^- \longrightarrow Co(s)$	−0.277
$Mn^{2+}(aq) + 2\ e^- \longrightarrow Mn(s)$	−1.18

(a) Which is the weakest oxidizing agent?
(b) Which is the strongest oxidizing agent?
(c) Which is the strongest reducing agent?
(d) Which is the weakest reducing agent?
(e) Will Co(s) reduce Pt^{2+}(aq) to Pt(s)?
(f) Will Pt(s) reduce Co^{2+}(aq) to Co(s)?
(g) Which ions can be reduced by Co(s)?

33. Consider these half-reactions:

Half-reaction	$E°$ (V)
$Ce^{4+}(aq) + e^- \longrightarrow Ce^{3+}(aq)$	1.72
$Ag^+(aq) + e^- \longrightarrow Ag(s)$	0.80
$Hg_2^{2+}(aq) + 2\,e^- \longrightarrow 2\,Hg(\ell)$	0.80
$Sn^{2+}(aq) + 2\,e^- \longrightarrow Sn(s)$	−0.14
$Ni^{2+}(aq) + 2\,e^- \longrightarrow Ni(s)$	−0.25
$Al^{3+}(aq) + 3\,e^- \longrightarrow Al(s)$	−1.68

(a) Which is the weakest oxidizing agent?
(b) Which is the strongest oxidizing agent?
(c) Which is the strongest reducing agent?
(d) Which is the weakest reducing agent?
(e) Will Sn(s) reduce Ag^+(aq) to Ag(s)?
(f) Will Hg(ℓ) reduce Sn^{2+}(aq) to Sn(s)?
(g) Name the ions that can be reduced by Sn(s).
(h) Which metals can be oxidized by Ag^+(aq)?

34. In principle, a battery could be made from aluminum metal and chlorine gas.
(a) Write a balanced equation for the reaction that would occur in a battery using Al^{3+}(aq)|Al(s) and Cl_2(g)|Cl^-(aq) half-cells.
(b) Identify the half-reaction at the anode and at the cathode. Do electrons flow from the Al electrode when the cell does work? Explain.
(c) Calculate the standard potential, $E°_{cell}$, for the battery.

$E°_{cell}$, Gibbs Free Energy, and $K°$ (Section 17-6)

35. Choose the correct answers: In a product-favored chemical reaction, the standard cell potential, $E°_{cell}$, is (greater/less) than zero, and the Gibbs free energy change, $\Delta_r G°$, is (greater/less) than zero.

36. Hydrazine, N_2H_4, can be used as the reducing agent in a fuel cell.

$$N_2H_4(aq) + O_2(aq) \longrightarrow N_2(g) + 2\,H_2O(\ell)$$

(a) If $\Delta_r G°$ for the reaction is −598 kJ, calculate the value of $E°$ expected for the reaction.
(b) Suppose the equation is written with all coefficients doubled. Determine $\Delta_r G°$ and $E°$ for this new reaction.

37. The standard cell potential for the oxidation of Mg by Br_2 is 3.42 V.

$$Br_2(\ell) + Mg(s) \longrightarrow Mg^{2+}(aq) + 2\,Br^-(aq)$$

(a) Calculate $\Delta_r G°$ for this reaction.
(b) Suppose the equation is written with all coefficients doubled. Determine $\Delta_r G°$ and $E°$ for this new equation.

38. The standard cell potential, $E°$, for the reaction of Zn(s) and Cl_2(g) is 2.12 V. Write the chemical equation for the reaction of 1 mol zinc. Calculate the standard Gibbs free energy change, $\Delta_r G°$, for this reaction.

39. For each of the reactions in Question 31, compute the Gibbs free energy change, $\Delta_r G°$.

40. Calculate the equilibrium constant K_c and $\Delta_r G°$ for the reaction between Cd(s) and Cu^{2+}(aq).

41. Calculate the equilibrium constant K_c and $\Delta_r G°$ for the reaction between I_2(s) and Br^-(aq).

42. Calculate the equilibrium constant K_c and $\Delta_r G°$ for the reaction between Ag(s) and Zn^{2+}(aq).

43. Calculate the equilibrium constant K_c and $\Delta_r G°$ for the reaction between Cl_2(g) and Br^-(aq).

44. Consider a voltaic cell with the reaction given below. As the cell reaction proceeds, what happens to the values of E_{cell}, $\Delta_r G$, and K_c? Explain your answers.

$$Cu^{2+}(aq, 1\ M) + Zn(s) \longrightarrow Cu(s) + Zn^{2+}(aq, 1\ M) \qquad E°_{cell} = 1.10\ V$$

45. Calculate the equilibrium constant K_c for this reaction.

$$Ni(s) + Co^{2+}(aq) \rightleftharpoons Ni^{2+}(aq) + Co(s) \qquad E°_{cell} = -0.027\ V$$

Effect of Concentration on Cell Potential: The Nernst Equation (Section 17-7)

46. Consider the voltaic cell

$$Zn(s) + Cd^{2+}(aq) \longrightarrow Zn^{2+}(aq) + Cd(s)$$

operating at 298 K.
(a) Calculate the $E°_{cell}$ for this cell.
(b) If E_{cell} = 0.390 and (conc. Cd^{2+}) = 2.00 M, calculate the (conc. Zn^{2+}).
(c) If (conc. Cd^{2+}) = 0.068 M and (conc. Zn^{2+}) = 1.00 M, calculate the E_{cell}.

47. Consider the voltaic cell

$$2\,Ag^+(aq) + Cd(s) \longrightarrow 2\,Ag(s) + Cd^{2+}(aq)$$

operating at 298 K.
(a) Calculate the $E°_{cell}$ for this cell.
(b) If (conc. Cd^{2+}) = 2.0 M and (conc. Ag^+) = 0.25 M, calculate E_{cell}.
(c) If E_{cell} = 1.25 V and (conc. Cd^{2+}) = 0.100 M, calculate (conc. Ag^+).

48. Consider a voltaic cell with the reaction

$$H_2(g) + Sn^{4+}(aq) \longrightarrow 2\,H^+(aq) + Sn^{2+}(aq)$$

operating at 298 K.
(a) Calculate the $E°_{cell}$ for this cell.
(b) Calculate the E_{cell} for P_{H_2} = 1.0 bar, (conc. Sn^{2+}) = 6.0×10^{-4} M, (conc. Sn^{4+}) = 5.0×10^{-4} M, and pH = 3.60.

49. Calculate the cell potential of a concentration cell that contains two hydrogen electrodes if the cathode contacts a solution with pH = 7.8 and the anode contacts a solution with (conc. H^+) = 0.05 M.

50. Calculate the potential of a cell with one electrode made from zinc metal immersed in a solution where (conc. Zn^{2+}) = 0.010 M and the other electrode is a standard hydrogen electrode.
51. For a voltaic cell with the reaction

$$Pb(s) + Sn^{2+}(aq) \longrightarrow Pb^{2+}(aq) + Sn(s)$$

calculate the ratio of concentrations of lead and tin ions when $E_{cell} = 0$.

Common Batteries (Section 17-8)

52. Describe the advantages and disadvantages of lead-acid storage batteries.
53. Describe the advantages and disadvantages of Li-ion batteries.
54. NiCad batteries are rechargeable and are commonly used in cordless appliances. Although such batteries actually function under basic conditions, imagine a voltaic cell using the setup in the diagram shown.

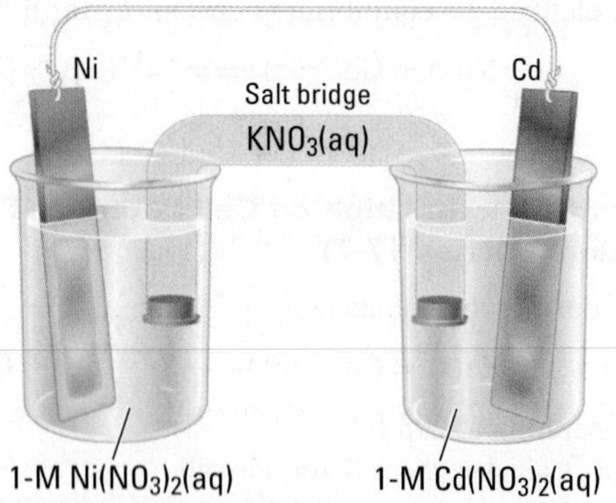

Ni
Cd
Salt bridge
KNO3(aq)
1-M Ni(NO3)2(aq)
1-M Cd(NO3)2(aq)

(a) Write a balanced net ionic equation depicting the reaction occurring in the cell.
(b) What is oxidized? What is reduced? What is the reducing agent and what is the oxidizing agent?
(c) Which is the anode and which is the cathode?
(d) Calculate $E°$ for the cell.
(e) Indicate the direction of electron flow in the external wire.
(f) If the salt bridge contains KNO_3, toward which compartment will the NO_3^- ions migrate?

55. Consider the NiCad cell in Question 54.
(a) If the concentration of Cd^{2+} is reduced to 0.010 M, and (conc. Ni^{2+}) = 1.0 M, is E_{cell} smaller or larger than when the concentration of $Cd^{2+}(aq)$ was 1.0 M? Explain your answer in terms of Le Chatelier's principle.
(b) Begin with 1.0 L of each of the solutions, both initially 1.0 M in dissolved species. Each electrode weighs 50.0 g at the start. If 0.050 A is drawn from the battery, calculate how long the battery can operate.

Fuel Cells (Section 17-9)

56. How does a fuel cell differ from a battery?
57. Describe the principal parts of an $H_2|O_2$ fuel cell. Write a balanced equation for the reaction at the cathode; at the anode. Give the formula of the product of the fuel cell reaction.
58. Hydrazine, N_2H_4, has been proposed as the fuel in a fuel cell in which oxygen is the oxidizing agent. The reactions are

$$N_2H_4(aq) + 4\ OH^-(aq) \longrightarrow N_2(g) + 4\ H_2O(\ell) + 4\ e^-$$
$$O_2(g) + 2\ H_2O(\ell) + 4\ e^- \longrightarrow 4\ OH^-(aq)$$

(a) Which reaction occurs at the anode and which at the cathode?
(b) What is the overall cell reaction?
(c) If the cell is to produce 0.50 A of current for 50.0 h, calculate what mass in grams of hydrazine must be present.
(d) Calculate what mass (g) of O_2 must be available to react with the mass of N_2H_4 determined in part (c).

Electrolysis—Causing Reactant-Favored Redox Reactions to Occur (Section 17-10)

59. Consider the electrolysis of water in the presence of very dilute H_2SO_4. What species is produced at the anode? At the cathode? What are the relative amounts of the species produced at the two electrodes?
60. Write chemical equations for the electrolysis of molten salts of three different alkali halides to produce the corresponding halogens and alkali metals.
61. From Table 17.1 write down all of the aqueous metal ions that can be reduced by electrolysis to the corresponding metal.
62. From Table 17.1 write down all of the aqueous species that can be oxidized by electrolysis, and determine the products.
63. Identify the products of the electrolysis of a 1-M aqueous solution of NaBr. What species are present in the solution? What is formed at the cathode? What is formed at the anode?
64. For each of these solutions, tell what reactions take place at the anode and at the cathode during electrolysis.
(a) $NiBr_2(aq)$ (b) $NaI(aq)$
(c) $CdCl_2(aq)$ (d) $CuI_2(aq)$
(e) $MgF_2(aq)$ (f) $HNO_3(aq)$

Counting Electrons (Section 17-11)

65. A current of 0.015 A is passed through a solution of $AgNO_3$ for 155 min. Calculate the mass of silver deposited at the cathode.
66. A current of 1.0 mA is passed through a solution containing $Ag^+(aq)$. Calculate the mass of silver in the solution if all the silver was deposited as Ag metal in 14.5 min.
67. A current of 2.50 A is passed through a solution of $Cu(NO_3)_2$ for 2.00 h. Calculate the mass of copper deposited at the cathode.
68. A current of 0.0125 A is passed through a solution of $CuCl_2$ for 2.00 h. Calculate the mass of copper deposited at the cathode and the volume of Cl_2 gas (in mL at STP) produced at the anode.

69. The major reduction half-reaction occurring in the cell in which molten Al_2O_3 and molten aluminum salts are electrolyzed is $Al^{3+}(\ell) + 3\,e^- \longrightarrow Al(s)$. The cell operates at 5.0 V and 1.0×10^5 A. Calculate the mass (g) of aluminum metal produced in 8.0 h.
70. The vanadium(II) ion can be produced by electrolysis of a vanadium(III) salt in solution. Calculate how long you must carry out an electrolysis if you wish to convert completely 0.125 L of 0.0150-M $V^{3+}(aq)$ to $V^{2+}(aq)$ using a current of 0.268 A.
71. The reactions occurring in a lead-acid storage battery are given in Section 17-8a. A typical battery might be rated at 50. ampere-hours (A-h). This means that it has the capacity to deliver 50. A for 1.0 h or 1.0 A for 50. h. If it does deliver 1.0 A for 50. h, calculate the mass of lead consumed.
72. An effective battery can be built using the reaction between Al metal and O_2 from the air. If the Al anode of this battery consists of a 3-oz piece of aluminum (84 g), determine how long (h) the battery can produce 1.0 A before going dead.
73. Assume that the anode reaction for the lithium battery is

$$LiC_6(s) \longrightarrow Li^+(electrolyte) + C_6(s) + e^-$$

and the anode reaction for the lead-acid storage battery is

$$Pb(s) + HSO_4^-(aq) \longrightarrow PbSO_4(s) + 2\,e^- + H^+(aq)$$

Compare the masses of metals consumed when each of these batteries supplies a current of 1.0 A for 10. min.

74. A hydrogen-oxygen fuel cell operates on the simple reaction

$$2\,H_2(g) + O_2(g) \longrightarrow 2\,H_2O(\ell)$$

If the cell is designed to produce 1.5 A of current, determine how long it can operate if there is an excess of oxygen and only sufficient hydrogen to fill a 1.0-L tank at 200. bar pressure at 25 °C.

75. Calculate how long it would take to electroplate a metal surface with 0.500 g nickel metal from a solution of Ni^{2+} with a current of 4.00 A.
76. Calculate how much current is required to electroplate a metal surface with 0.400 g chromium metal from a solution of Cr^{3+} in 1.00 h.

Corrosion: Undesirable Product-Favored Redox Reactions (Section 17-12)

77. Explain how rust is formed from iron materials by corrosion.
78. Why does iron corrode faster in salt water than in fresh water?
79. Name one common metal that does not corrode readily under normal conditions.
80. Why does coating a steel object with chromium stop corrosion of the iron?
81. Explain how galvanizing iron stops corrosion of the underlying iron.

General Questions

These questions are not explicitly keyed to chapter topics; many require integration of several concepts.

82. Does 1.0-M nitric acid, HNO_3, oxidize metallic gold to form a 1-M Au^{3+} solution? Explain why or why not.
83. A 12-V automobile battery consists of six cells of the type described in Section 17-8a. The cells are connected in series so that the same current flows through all of them. Calculate the theoretical minimum electrical potential difference needed to recharge an automobile battery. (Assume standard-state concentrations.) How does this compare with the maximum voltage that could be delivered by the battery? Assuming that the lead plates in an automobile battery each weigh 2.50 kg and that sufficient PbO_2 is available, calculate the maximum possible work that could be obtained from the battery.
84. Three electrolytic cells are connected in series, so that the same current flows through all of them for 20. min. In cell A, 0.0234 g Ag plates out from a solution of $AgNO_3(aq)$; cell B contains $Cu(NO_3)_2(aq)$; cell C contains $Al(NO_3)_3(aq)$. Calculate what mass of Cu will plate out in cell B. Calculate what mass of Al will plate out in cell C.
85. Fluorinated organic compounds are important commercially; they are used as herbicides, flame retardants, and fire-extinguishing agents, among other things. A reaction such as

$$CH_3SO_2F + 3\,HF \longrightarrow CF_3SO_2F + 3\,H_2$$

is actually carried out electrochemically in liquid HF as the solvent.

(a) Draw the structural formula for CH_3SO_2F. (S is the "central" atom with the O atoms, F atom, and CH_3 group bonded to it.) What is the geometry around the S atom? What are the O—S—O and O—S—F bond angles?

(b) If you electrolyze 150. g CH_3SO_2F, determine how many grams of HF are required and how many grams of each product can be isolated.

(c) Is H_2 produced at the anode or the cathode of the electrolysis cell?

(d) A typical electrolysis cell operates at 8.0 V and 250 A. Calculate how many kilowatt-hours of energy one such cell consumes in 24 h.

86. Fluorine, F_2, is made by the electrolysis of anhydrous HF.

$$2\,HF(\ell) \longrightarrow H_2(g) + F_2(g)$$

Typical electrolysis cells operate at 4000 to 6000 A and 8 to 12 V. A large-scale plant can produce about 9.0 metric tons of F_2 gas per day.

(a) Calculate the mass (g) of HF consumed.

(b) Using the conversion factor of 3.60×10^6 J/kWh, calculate how much energy in kilowatt-hours is transferred to a cell operating at 6.0×10^3 A at 12 V for 24 h.

87. What reaction takes place when a 1.0-M solution of $Cr_2O_7^{2-}$ is added to a 1.0-M solution of HBr?
88. An electric current of 2.00 A was passed through a platinum salt solution for 3.00 hours, and 10.9 g of metallic platinum was formed at the cathode. Determine the charge on the platinum ions in the solution.
89. You wish to electroplate a copper surface having an area of 1200 mm² with a 1.0-μm-thick coating of silver from a solution of $Ag(CN)_2^-$ ions. If you use a current of 150.0 mA, calculate how much electrolysis time you should use. The density of metallic silver is 10.5 g/cm³.
90. In a mercury battery, the anode reaction is

$$Zn(s) + 2\ OH^-(aq) \longrightarrow ZnO(aq) + H_2O(\ell) + 2\ e^-$$

and the cathode reaction is

$$HgO(s) + H_2O(\ell) + 2\ e^- \longrightarrow Hg(\ell) + 2\ OH^-(aq)$$

The cell potential is 1.35 V. Calculate how many hours such a battery can provide power at a rate of 4.0×10^{-4} watt (1 watt = 1 J s^{-1}) if 1.25 g HgO is available.

Applying Concepts

These questions test conceptual learning.

91. Four metals A, B, C, and D exhibit these properties:
   (a) Only A and C react with 1.0-M HCl to give H_2 gas.
   (b) When C is added to solutions of ions of the other metals, metallic A, B, and D are formed.
   (c) Metal D reduces B^{n+} ions to give metallic B and D^{n+} ions.

   On the basis of this information, arrange the four metals in order of increasing ability to act as reducing agents.
92. The table below lists the cell potentials for the ten possible voltaic cells assembled from the elements A, B, C, D, and E and their respective ions in solutions. Using the data in the table, establish a standard half-cell potential table similar to Table 17.1. Assign a half-cell potential of 0.00 V to the element that falls in the middle of the series.

	A(s) in A^{n+}(aq)	B(s) in B^{n+}(aq)
E(s) in E^{n+}(aq)	+0.21 V	+0.68 V
D(s) in D^{n+}(aq)	+0.35 V	+1.24 V
C(s) in C^{n+}(aq)	+0.58 V	+0.31 V
B(s) in B^{n+}(aq)	+0.89 V	—
	C(s) in C^{n+}(aq)	**D(s) in D^{n+}(aq)**
E(s) in E^{n+}(aq)	+0.37 V	+0.56 V
D(s) in D^{n+}(aq)	+0.93 V	—
C(s) in C^{n+}(aq)	—	—
B(s) in B^{n+}(aq)	—	—

93. When the voltaic cell shown here runs for several hours, the green solution gets lighter and the yellow solution gets darker.

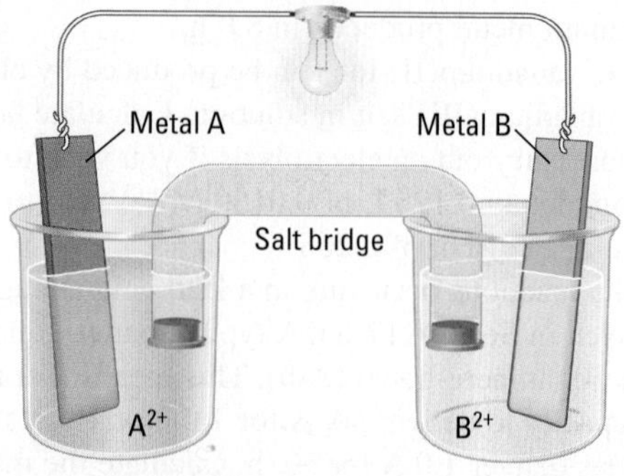
Metal A
Metal B
Salt bridge
A2+
B2+

   (a) Determine what is oxidized and what is reduced.
   (b) Identify the oxidizing agent and the reducing agent.
   (c) Identify the anode and the cathode.
   (d) Write equations for the half-reactions.
   (e) Which metal electrode gains mass?
   (f) What is the direction of the electron transfer through the external wire?
   (g) If the salt bridge contains KNO_3(aq), into which solution will the K^+ ions migrate?
94. An electrolytic cell is set up with Cd(s) in $Cd(NO_3)_2$(aq) and Zn(s) in $Zn(NO_3)_2$(aq). Initially both electrodes weigh 5.00 g. After running the cell for several hours the electrode in the left compartment weighs 4.75 g.
   (a) Which electrode is in the left compartment?
   (b) Does the mass of the electrode in the right compartment increase, decrease, or stay the same? If the mass changes, what is the new mass?
   (c) Does the volume of the electrode in the right compartment increase, decrease, or stay the same? If the volume changes, what is the new volume? (The density of Cd is 8.65 g/cm³.)
95. Using data from Appendix I, show why
   (a) Co^{3+} is not stable in aqueous solution.
   (b) Fe^{2+} is not stable in air.
96. When H_2O_2 is mixed with Fe^{2+}, which redox reaction occurs—the oxidation of Fe^{2+} to Fe^{3+} or the reduction of Fe^{2+} to Fe? Determine what the E°_{cell} values are for the voltaic cells corresponding to the two reactions.
97. The permanganate ion MnO_4^- can be reduced to the manganese(II) ion Mn^{2+} in aqueous acidic solution, and the half-cell potential for this half-cell reaction is 1.51 V. If this half-cell is combined with a $Zn^{2+}|Zn$ half-cell to form a voltaic cell at standard conditions,
   (a) Write the chemical equation for the half-reaction occurring at the anode.
   (b) Write the chemical equation for the half-reaction occurring at the cathode.
   (c) Write the overall balanced equation for the reaction.
   (d) Calculate the cell potential.

98. Consider two different electrolytic cells; one cell contains aqueous Zn^{2+} and the other contains Cr^{3+}. The initial metal ion concentration is the same in each cell and the metal ions are reduced to the metal during the electrolysis. Each cell operates at the same current. *Without doing calculations,* predict which cell has the greater mass of metal deposited after 5 min. Explain your prediction.

More Challenging Questions

These questions require more thought and integrate several concepts.

99. The K_{sp} of $Cu(IO_3)_2$ is 1.4×10^{-7}. Calculate the $E°$ for $Cu(IO_3)_2(s) + 2\,e^- \longrightarrow Cu(s) + 2\,IO_3^-(aq)$.
100. Use calculations to determine whether (a) oxygen in air can oxidize silver metal in acidic solution; and (b) oxygen in air can oxidize silver metal in basic solution. In each case, write the balanced cell reaction equation and calculate the standard electrical potential.
101. To measure the Ag^+ concentration, 25.00 mL of a silver-containing solution is titrated with 0.015 M KI at 25 °C by using a silver electrode immersed in the test solution and the electrical potential measured against a standard hydrogen electrode. It required 16.7 mL of the KI solution to reach the equivalence point, where the potential was 0.325 V.
 (a) Calculate the molarity of Ag^+ in the solution.
 (b) Calculate the K_{sp} of AgI.
102. The K^+ concentration inside a nerve cell differs from that outside the cell. Calculate the difference in electrical potential required for the K^+ concentration inside a nerve cell to be 45 times greater than that outside the cell. Assume that the potential difference is due only to the K^+ concentration difference.
103. A pH meter is standardized using a pH 9.40 buffer; the cell potential is +0.060 V. When the buffer is replaced with a solution of unknown $H^+(aq)$ concentration, the cell potential is +0.22 V. Calculate the pH of the test solution.
104. Consider these *unbalanced* equations for two reactions:

$$NO_3^-(aq) + H^+(aq) + Hg(\ell) \longrightarrow Hg_2^{2+}(aq) + NO(g) + H_2O(\ell)$$

$$Hg^{2+}(aq) + Br^-(aq) \longrightarrow Hg_2^{2+}(aq) + Br_2(\ell)$$

 For each reaction:
 (a) balance the equation.
 (b) draw a cell diagram.
 (c) calculate the standard cell potential.
 (d) calculate the $\Delta_r G°$.
 (e) determine whether each reaction is product-favored. Explain your reason.
105. The standard potential of a $Zn|Zn^{2+}\|Cu^{2+}|Cu$ voltaic cell is 1.103 V. As the cell operates, the cell potential changes due to concentration changes of Zn^{2+} and Cu^{2+}. Calculate the ratio (conc. Zn^{2+})/(conc. Cu^{2+}) when the cell potential is 0.050 V.
106. In an electrolytic cell, a 10.0-A direct current passes through an aqueous copper(II) nitrate solution and 9.50 g metallic copper plates out.
 (a) Calculate how long it took for this mass of copper to be deposited at the cathode. Assume 100% efficiency.
 (b) A gas is produced at the anode and collected. Identify the gas and calculate its volume. The gas was collected at 25 °C and 0.945 atm.
107. Consider a voltaic cell to study the reaction of chromium and zinc. The cell consists of the usual array of a salt bridge, appropriate wiring, and two half-cells: (1) a chromium metal electrode dipping into an aqueous solution of Cr^{2+} ions; and (2) a zinc metal electrode dipping into an aqueous solution of Zn^{2+} ions.
 (a) Calculate the cell potential at standard conditions.
 (b) Calculate $\Delta_r G°_{cell}$.
 (c) The concentration of metal ions is changed to Cr^{2+} = 0.050 M and Zn^{2+} = 0.010 M. Calculate $\Delta_r G_{cell}$ and E_{cell}.
108. If Cl_2 and Br_2 are added to an aqueous solution that contains Cl^- and Br^-, what product-favored reaction will occur?
109. This reaction occurs in a cell with $H_2(g)$ pressure of 1.0 atm and (conc. Cl^-) = 1.0 M at 25 °C; the measured E_{cell} = 0.34 V. Calculate the pH of the solution.

$$H_2(g) + 2\,AgCl(s) \longrightarrow 2\,H^+(aq) + 2\,Cl^-(aq) + 2\,Ag(s)$$

110. E_{cell} = 0.013 V for a voltaic cell with this reaction at 25 °C.

$$Sn(s) + Pb^{2+}(aq) \longrightarrow Sn^{2+}(aq) + Pb(s)$$

 (a) Calculate the equilibrium constant K_c for the reaction.
 (b) If a solution with (conc. Pb^{2+}) = 1.1 M had excess tin metal added to it, calculate the equilibrium concentrations of Sn^{2+} and Pb^{2+}.
111. A student wanted to measure the copper(II) concentration in an aqueous solution. For the cathode half-cell she used a silver electrode with a 1.00-M solution of $AgNO_3$. For the anode half-cell she used a copper electrode dipped into the aqueous sample. The cell gave E_{cell} = 0.62 V at 25 °C. Calculate the copper(II) ion concentration of the solution.

Conceptual Challenge Problems

These rigorous, thought-provoking problems integrate conceptual learning with problem solving and are suitable for group work.

CP17.A (Section 17-6) Most automobiles run on internal combustion engines, in which the energy used to run the vehicle is obtained from the combustion of gasoline. The main component of gasoline is octane, C_8H_{18}. An automobile manufacturer has announced a chemical method for generating hydrogen gas from gasoline and proposes to develop a car in which an $H_2|O_2$ fuel cell powers an electric propulsion motor, thus eliminating the internal combustion engine with its problems (for example, the generation of unwanted by-products that pollute the air). The hydrogen for the fuel cell would be directly generated from gasoline on board the vehicle. There are two steps in this hydrogen generation process:

(i) Partial oxidation of octane by oxygen to carbon monoxide and hydrogen

(ii) Combination of carbon monoxide with additional gaseous water to form carbon dioxide and more hydrogen (the water-gas shift reaction)

(a) Write the chemical equation for the complete combustion of 1 mol octane.

(b) Write balanced chemical equations for the two-step hydrogen generation process. Calculate the amount (mol) of H_2 produced per mole of octane. (Remember that water is a reactant in the two-step process.)

(c) By combining these equations, show that the *overall* reaction is the same as in the combustion of octane.

(d) Assuming that the entire Gibbs free energy change of the $H_2|O_2$ fuel cell reaction is available for use by the electric propulsion motor, calculate the energy produced by a fuel cell when it consumes all of the hydrogen produced from 1 mol of octane. Compare this energy with the Gibbs free energy change for the combustion of 1 mol of octane. (*Note:* The Gibbs free energy of formation, $\Delta_f G°$, for $C_8H_{18}(\ell)$ is 6.707 kJ/mol.)

CP17.B (Section 17-4) People obtain energy by oxidizing food. Glucose, a typical foodstuff, is a carbohydrate. When metabolized, glucose is oxidized to water and carbon dioxide.

$$C_6H_{12}O_6(aq) + 6\ O_2(g) \longrightarrow 6\ CO_2(g) + 6\ H_2O(\ell)$$

The combustion enthalpy of glucose is 2.80×10^3 kJ/mol, which means that as glucose is oxidized, its electrons lose 2.80×10^3 kJ/mol as they give up potential energy in a complicated series of chemical steps.

(a) Assume that a person requires 2400 Calories per day and that this energy is obtained solely from the oxidation of glucose. Calculate the amount (mol) of O_2 a person must breathe each day to react with this much glucose.

(b) Each mole of O_2 requires 4 mol electrons, regardless of whether the O atoms become part of CO_2 or H_2O. Determine the average electric current (C/s) in a human body using the quantity of energy described in part (a) per day.

(c) Use the answer from part (b) and calculate the electrical potential this current flows through in a day to produce the 2400 Calories.

CP17.C (Section 17-10) A piece of chromium metal is attached to a battery and the metal is dipped into 50. mL of 0.30-M KOH solution in a 250-mL beaker. A stainless steel electrode is connected to the other electrode of the battery and dipped into the same solution. A steady current of 0.50 A is maintained for exactly 2 hours. Several samples of a gas formed at the stainless steel electrode during the electrolysis are captured, and all are found to ignite in air. After the electrolysis, the chromium electrode is weighed and found to have decreased in mass by 0.321 g. The mass of the stainless steel electrode does not change.

After electrolysis, the KOH solution is neutralized with nitric acid to a pH of slightly less than 7, then is heated and reacted with 0.151-M lead(II) nitrate solution. As the lead(II) nitrate solution is added, a yellow precipitate quickly forms from the hot solution. The formation of precipitate stops after 40.4 mL of the lead(II) nitrate solution has been added. The yellow solid is then filtered, dried, and weighed. Its mass is 1.97 g.

(a) Calculate electrical charge that passed through the cell.

(b) Determine the amount (mol) of Cr that reacted.

(c) What is the oxidation state of the Cr after reacting?

(d) Assuming that the yellow compound that precipitates from the solution during the titration contains both Pb and Cr, what do you conclude to be the ratio of the numbers of atoms of Pb and Cr?

(e) If the yellow compound contains an element other than Pb and Cr, what is it and how much of it is in the compound? Determine the formula of the yellow compound.

Nuclear Chemistry

18

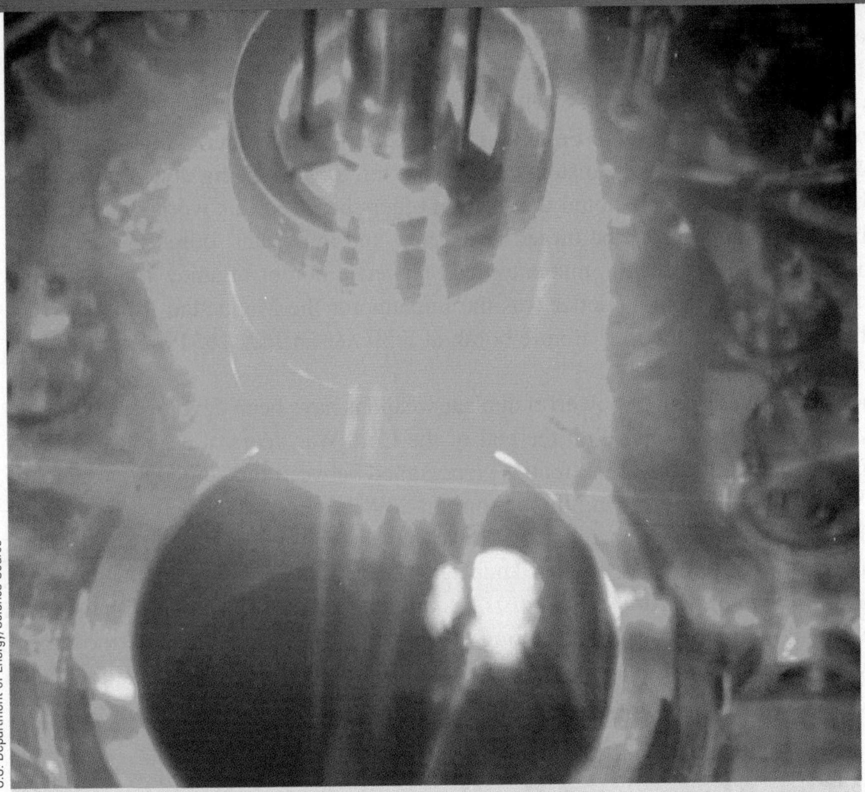

U.S. Department of Energy/Science Source

A fuel rod is being removed from a nuclear reactor at Oak Ridge National Laboratory. Commercial nuclear reactors produce about 19% of the electricity in the United States. What are the components of a nuclear power reactor and what nuclear process provides the energy to produce electricity? (Section 18-6) Can a nuclear power reactor explode like an atomic bomb? (Section 18-6) How does nuclear radiation affect human health? How are such effects measured? (Section 18-8) These and many other aspects of unstable (radioactive) atomic nuclei are discussed in this chapter.

Nuclear chemistry, a subject that bridges chemistry and physics, has a significant impact on our society. No matter what your reason for taking a college course in chemistry—to prepare for a career in science or engineering or simply to gain knowledge as a concerned citizen—you should know about nuclear chemistry. Radioactive isotopes are widely used in medicine. PET (positron emission tomography) scans depend on radioactivity. Your room may be protected by a smoke detector that uses a radioactive element as part of its sensor, and research in all fields of science employs radioactive isotopes and their compounds. The national security of the United States since World War II has depended on nuclear weapons, and more than 30 nations use nuclear reactors to produce electricity. This chapter considers several aspects of nuclear chemistry: changes in atomic nuclei and their effects; fusion of nuclei and the energy that can be derived from such changes; units used to measure radioactivity; and beneficial applications of radioactive isotopes.

On August 2, 1939, with the world hovering on the brink of World War II, Albert Einstein sent a letter to President Franklin D. Roosevelt. In this letter, which profoundly changed the course of history, Einstein called attention to work being done on the physics of the atomic nucleus. He noted that he and others believed this work suggested the possibility that "uranium may be turned into a new and important source of energy . . . and [that it was] conceivable . . . that extremely powerful bombs of a new type may thus be constructed. . . ." Einstein's letter was the impetus for the Manhattan Project, which led to the detonation of the first atomic bomb at 5:30 AM on July 16, 1945, in the New Mexico desert.

Einstein's original letter can be seen at http://www.fdrlibrary.marist.edu/psf/box5/a64a01.html.

The Alsos Digital Library for Nuclear Issues has a website at http://alsos.wlu.edu with a section on the Manhattan Project.

The Bulletin of the Atomic Scientists can be found online at http://www.thebulletin.org.

Since World War II, more powerful nuclear weapons have been developed and stockpiled by a number of nations. With the end of the Cold War, fears of a nuclear disaster receded, and nuclear disarmament treaties were signed between the United States and the former Soviet Union for removing the plutonium-239 and other nuclear fuel from existing nuclear warheads. Unfortunately, those fears have been replaced by the concern that other nations have developed or acquired nuclear weapons. For many years, the respected journal *The Bulletin of the Atomic Scientists* has used the symbol of a clock with its hands near the fateful midnight hour (representing nuclear annihilation) to illustrate the danger faced by the world from atomic weapons. Even with the end of the Cold War, the hands have moved back only a little from midnight.

18-1 The Nature of Radioactivity

Many minerals, called phosphors, glow for some time after being exposed to sunlight or ultraviolet light. In 1896 French physicist Antoine Henri Becquerel was studying this phenomenon, called *phosphorescence*, when he made an important and totally unexpected observation that led him to the accidental discovery of radioactivity. While waiting for a sunny day, Becquerel stored a photographic plate wrapped in black paper along with a uranium salt (a material known to phosphoresce) in a dark drawer. To his amazement, the image of the uranium salt appeared on the plate that had been in the drawer, unexposed to sunlight. Becquerel realized that radiation from the uranium salt had penetrated the black paper and exposed the photographic plate even though the uranium salt had not been stimulated by sunlight.

Becquerel performed many more related experiments and found that pure uranium metal produced the same emissions as uranium salts did, but even more strongly. This result would be expected if the radiation were the property of the metal and not dependent on its form of chemical combination. But no pure metal was known to phosphoresce, which mystified Becquerel. What was the source of this radiation? It turned out that the radiation had nothing to do with phosphorescence. Failing to find the reasons for the emissions, Becquerel gave the project to his graduate student Marie Curie. She and her husband Pierre, a physicist, studied the phenomenon intensively and termed it *radioactivity*.

For more on experiments done by Becquerel and the Curies, see Walton, H. F. *Journal of Chemical Education*, Vol. 69, 1992; p. 10.

One of Marie Curie's first findings was to confirm Becquerel's observation that uranium metal itself was radioactive and that the degree to which a uranium-containing sample was radioactive depended on the percentage of uranium present. When she tested pitchblende, a common ore containing uranium and other metals (such as lead, bismuth,

Table 18.1 Characteristics of α, β, and γ Emissions

Name	Symbol	Charge	Mass (g/particle)	Penetrating Power*
Alpha	$^{4}_{2}He^{2+}$, $^{4}_{2}\alpha$, $^{4}_{2}He$	2+	6.64×10^{-24}	0.03 mm
Beta	$^{0}_{-1}e$, $^{0}_{-1}\beta$	1−	9.11×10^{-28}	2 mm
Gamma	$^{0}_{0}\gamma$, γ	0	0	10 cm

*Distance at which half the radiation has been stopped by water.

In the atomic representation $^{4}_{2}He$, the subscript denotes the atomic number (number of protons) of the atom, and the superscript denotes the mass number (number of protons plus number of neutrons) of the atom (← Sec. 2-2c).

and copper), she was surprised to find that it was even more radioactive than pure uranium. Only one explanation was possible: pitchblende contained an element (or elements) more radioactive than uranium. Eventually, the Curies discovered the element they named *polonium* after Marie's homeland of Poland. They also discovered radium, another highly radioactive element.

In England at about the same time, Sir J. J. Thomson and his graduate student Ernest Rutherford were studying the radiation from uranium and thorium (← **Sec. 2-1a**). Rutherford found that "There are present at least two distinct types of radiation—one that is readily absorbed, which will be termed for convenience alpha (α) radiation, and the other of a more penetrative character, which will be termed beta (β) radiation." **Alpha (α) radiation**, he discovered, *is composed of positively charged particles that, when passed through an electric field, were attracted to the negative side of the field* (← **Sec. 2-1a**). Indeed, his later studies showed that an **alpha (α) particle** is *a helium nucleus, $^{4}_{2}He^{2+}$, that has been ejected at high speed from a radioactive atom's nucleus* (Table 18.1). Alpha particles have limited penetrating power and can be absorbed by skin, clothing, or several sheets of ordinary paper.

In the same experiment with electric fields, Rutherford found that **beta (β) radiation** must be *composed of negatively charged particles, because the beam of beta radiation was attracted to the electrically positive plate of an electric field.* Later work by Becquerel showed that these particles have an electric charge and mass equal to those of an electron. Thus, a **beta (β) particle** is *an electron that has been ejected at high speed from a radioactive atom's nucleus.* Beta particles are more penetrating than alpha particles (Table 18.1), and a $\frac{1}{8}$ inch thick piece of aluminum is necessary to stop them. Beta particles can penetrate 1 to 2 cm into living bone or tissue.

Encouraged by the Curies, Becquerel returned to the study of radiation. He found that the radiation from uranium was affected by magnetic fields and consisted of two kinds of particles, which we now know to be alpha and beta particles.

A third type of radiation was later discovered by P. Villard, a Frenchman, who named it *gamma (γ) radiation*, using the third letter in the Greek alphabet in keeping with Rutherford's scheme. Unlike alpha and beta radiation, which consist of charged particles, **gamma (γ) radiation** *is a form of electromagnetic radiation and is not deflected by an electric field.* Gamma radiation is the most penetrating, and it can pass completely through the human body. Thick layers of lead or concrete are required to stop a beam of gamma rays.

18-2 Nuclear Reactions

18-2a Equations for Nuclear Reactions

Ernest Rutherford found that radium not only emits alpha particles but also produces the radioactive gas radon. Such observations led Rutherford and Frederick Soddy, in 1902, to propose the revolutionary theory that *radioactivity is the result of a natural change of a radioactive isotope of one element into an isotope of a different element*, that is, a **nuclear reaction** or *transmutation*. In a nuclear reaction, an unstable nucleus (the *parent* *nucleus*) spontaneously emits radiation and is converted (decays) into a more stable nucleus of a different element (the *daughter product*). Thus, a nuclear reaction results in a change in atomic number (number of protons) and, in some cases, a change in mass number as well. For example, the reaction of radium studied by Rutherford can be written as

$$^{226}_{88}Ra \longrightarrow {}^{4}_{2}He + {}^{222}_{86}Rn$$

When a radioactive atom decays, the emission of a charged particle leaves behind a charged atom. Thus, when radium-226 decays, it gives a helium-4 cation ($^{4}_{2}He^{2+}$) and a radon-222 anion ($^{222}_{86}Rn^{2-}$). By convention, the ion charges are not shown in balanced nuclear equations.

In a chemical change, the atoms in molecules and ions are rearranged, but atoms are neither created nor destroyed: The number of atoms remains the same. In a nuclear reaction the total number of *nucleons* remains the same. A **nucleon** is *a nuclear particle—a proton or neutron.* During a nuclear reaction, one nucleon can change into a different nucleon along with the release of energy. A proton can change to a neutron or a neutron can change to a proton, but the total number of nucleons remains the same.

- **The sum of the mass numbers of reacting nuclei must equal the sum of the mass numbers of the nuclei produced.**
- **The sum of the atomic numbers of the products must equal the sum of the atomic number of the reactants.**

Recall (← Sec. 2-2c) that the atomic number is the number of protons in an atom's nucleus (that is, the total positive charge on the nucleus), and the mass number is the sum of protons and neutrons in a nucleus.

These principles can be verified for the preceding nuclear equation.

	$^{226}_{88}Ra$ radium-226	$\longrightarrow$	$^{4}_{2}He$ alpha particle	+	$^{222}_{86}Rn$ radon-222
Mass number:	226	$\longrightarrow$	4	+	222
Atomic number:	88	$\longrightarrow$	2	+	86

18-2b Alpha and Beta Particle Emission

One way a radioactive isotope can decay is to eject an alpha particle from the parent nucleus. This is illustrated by the radium-226 reaction above and by the conversion of uranium-234 to thorium-230 by alpha emission.

	$^{234}_{92}U$ uranium-234 (parent nucleus)	$\longrightarrow$	$^{4}_{2}He$ alpha particle	+	$^{230}_{90}Th$ thorium-230 (daughter product)
Mass number:	234	$\longrightarrow$	4	+	230
Atomic number:	92	$\longrightarrow$	2	+	90

In alpha emission, *the atomic number of the parent nucleus decreases by two units and its mass number decreases by four units for each alpha particle emitted.*

Emission of a beta particle is another way for a radioactive isotope to decay. For example, loss of a beta particle by uranium-239 (parent nucleus) to form neptunium-239 (daughter product) is represented by

Note that in beta decay the mass number is unchanged.

	$^{239}_{92}U$ uranium-239	$\longrightarrow$	$^{0}_{-1}e$ beta particle	+	$^{239}_{93}Np$ neptunium-239
Mass number:	239	$\longrightarrow$	0	+	239
Atomic number:	92	$\longrightarrow$	−1	+	93

How does a nucleus, composed only of protons and neutrons, increase its number of protons by ejecting an electron during beta emission? It is generally accepted that a series of reactions is involved, but the net process is

If a neutron changes to a proton, conservation of charge requires that a negative particle (the beta particle) be created.

$$^{1}_{0}n \longrightarrow {}^{0}_{-1}e + {}^{1}_{1}p$$

neutron → electron + proton

where we use the symbol p for a proton and n for a neutron. In this overall process, a neutron is converted to a proton and a beta particle is released. Therefore, *the ejection of a beta particle always means that a different element is formed because a neutron has been converted into a proton. The new element (daughter product) has an atomic number one unit greater than that of the decaying (parent) nucleus.* The mass number does not change, however, because no proton or neutron has been emitted.

In many cases, the emission of an alpha or beta particle results in the formation of a product nucleus that is also unstable and therefore radioactive. The new radioactive product may undergo a number of successive transformations until a stable, nonradioactive nucleus is finally produced. Such *a series of successive nuclear reactions* is called a **radioactive series**. One such series begins with uranium-238 and ends with lead-206, as illustrated in Figure 18.1. The first step in the series is

A nucleus formed as a result of alpha or beta emission is often in an excited state and therefore emits a gamma ray.

$$^{238}_{92}\text{U} \longrightarrow {}^{4}_{2}\text{He} + {}^{234}_{90}\text{Th}$$

The final step, the conversion of polonium-210 to lead-206, is

$$^{210}_{84}\text{Po} \longrightarrow {}^{4}_{2}\text{He} + {}^{206}_{82}\text{Pb}$$

PROBLEM-SOLVING EXAMPLE 18.1

Radioactive Series

An intermediate species in the uranium-238 decay series shown in Figure 18.1 is polonium-218. It emits an alpha particle, followed by emission of a beta particle. Write the nuclear equations for these two reactions.

Result $^{218}_{84}\text{Po} \longrightarrow {}^{4}_{2}\text{He} + {}^{214}_{82}\text{Pb}$ $\quad ^{214}_{82}\text{Pb} \longrightarrow {}^{0}_{-1}\text{e} + {}^{214}_{83}\text{Bi}$

Analyze During alpha particle emission, the atomic number decreases by two and the mass number decreases by four. Beta emission causes the atomic number of the daughter product to be one greater and leaves the mass number unchanged.

Plan Consider the sequence: alpha emission by polonium-218 produces lead-214. This product undergoes beta emission to form bismuth-214.

Execute

$$\underset{\text{polonium-218}}{^{218}_{84}\text{Po}} \longrightarrow {}^{4}_{2}\text{He} + \underset{\text{lead-214}}{^{214}_{82}\text{Pb}}$$

$$\underset{\text{lead-214}}{^{214}_{82}\text{Pb}} \longrightarrow {}^{0}_{-1}\text{e} + \underset{\text{bismuth-214}}{^{214}_{83}\text{Bi}}$$

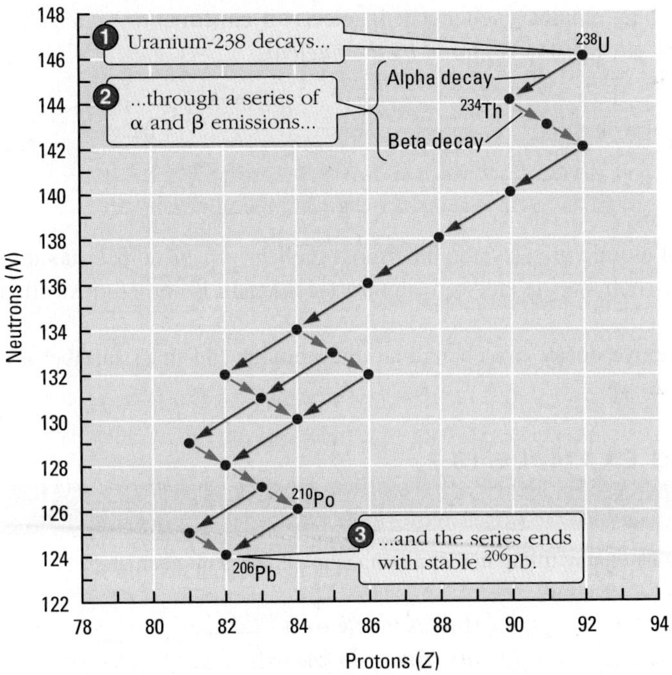

1 Uranium-238 decays...
2 ...through a series of α and β emissions...
3 ...and the series ends with stable 206Pb.
Alpha decay
Beta decay
238U
234Th
210Po
206Pb
Neutrons (N)
Protons (Z)
148
146
144
142
140
138
136
134
132
130
128
126
124
122
78
80
82
84
86
88
90
92
94

Figure 18.1 The $^{238}_{92}$U decay series.

☑ **Reasonable Result Check** Alpha emission decreases the atomic number by two and decreases the mass number by four. Beta emission increases the atomic number by one and leaves the mass number unchanged. Therefore, one alpha and one beta emission would leave the atomic number one less and decrease the mass number by four, which is what our systematic analysis found.

PROBLEM-SOLVING PRACTICE answers are provided at the back of this book in Appendix K.

PROBLEM-SOLVING PRACTICE 18.1

(a) Write an equation showing the emission of an alpha particle by an isotope of neptunium, $^{237}_{93}Np$, to produce an isotope of protactinium.

(b) Write an equation showing the emission of a beta particle by sulfur-35, $^{35}_{16}S$, to produce an isotope of chlorine.

Answers to **EXERCISES** are provided at the back of this book in Appendix L.

EXERCISE 18.1

Radioactive Decay Series

The actinium decay series begins with uranium-235, $^{235}_{92}U$, and ends with lead-207, $^{207}_{82}Pb$. The first five steps involve the successive emission of α, β, α, β, and α particles. Identify the radioactive isotope produced in each of the steps, beginning with uranium-235.

18-2c Other Types of Radioactive Decay

The positron was discovered by Carl Anderson in 1932. It is sometimes called an "anti-electron," one of a group of particles that have become known as "antimatter." Contact between an electron and a positron leads to mutual annihilation of both particles with production of two high-energy photons (gamma rays). This process is the basis of positron emission tomography (PET) scanning to detect tumors (Sec. 18-9c).

In addition to emission of alpha, beta, or gamma radiation, other nuclear decay processes are known. Some nuclei decay by emission of a **positron**, $^{0}_{+1}e$ or β^+, which is effectively *a positively charged electron.* For example, positron emission by polonium-207 leads to the formation of bismuth-207.

	$^{207}_{84}Po$	⟶	$^{0}_{+1}e$	+	$^{207}_{83}Bi$
	polonium-207		positron		bismuth-207
Mass number:	207	⟶	0	+	207
Atomic number:	84	⟶	+1	+	83

Notice that this process is the opposite of beta decay, because positron decay leads to a *decrease* in atomic number. Like beta decay, positron decay does not change the mass number because no proton or neutron is ejected.

Another nuclear process is *electron capture,* in which the atomic number is reduced by one but the mass number remains unchanged. In **electron capture**, *an inner-shell electron (for example, a 1s electron) is captured by the nucleus.*

	$^{7}_{4}Be$	+	$^{0}_{-1}e$	⟶	$^{7}_{3}Li$
	beryllium-7		electron		lithium-7
Mass number:	7	+	0	⟶	7
Atomic number:	4	+	−1	⟶	3

In the old nomenclature of atomic physics, the innermost shell ($n = 1$ principal quantum number) was called the K-shell, so the electron capture mechanism is sometimes called *K-capture*.

The four ways radioactive decay can change atomic number and mass number are summarized in Figure 18.2.

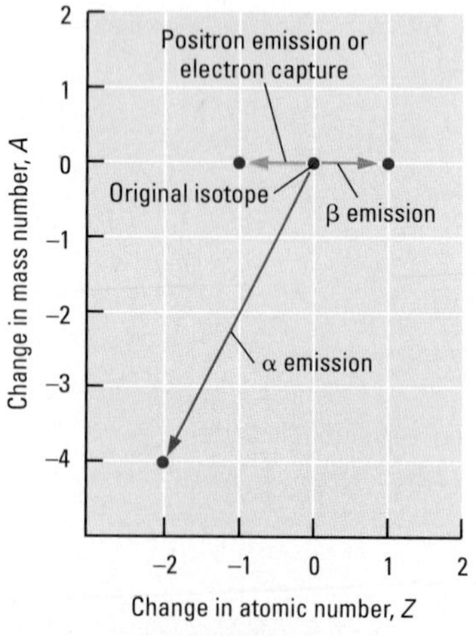
Positron emission or electron capture
Original isotope
β emission
α emission
Change in mass number, A
Change in atomic number, Z
2
1
0
−1
−2
−3
−4
−2
−1
0
1
2

Figure 18.2 Effects on Z and A of four radioactive decay processes.

PROBLEM-SOLVING EXAMPLE 18.2

Nuclear Equations

Complete these nuclear equations by filling in the missing symbol, mass number, and atomic number of the product species.

(a) $^{18}_{9}F \longrightarrow {}^{18}_{8}O + $ ______

(b) $^{26}_{13}Al + {}^{0}_{-1}e \longrightarrow$ ______

(c) $^{208}_{79}Au \longrightarrow {}^{208}_{80}Hg + $ ______

(d) $^{218}_{84}Po \longrightarrow {}^{4}_{2}He + $ ______

Result

(a) ${}^{0}_{+1}e$ (b) ${}^{26}_{12}Mg$ (c) ${}^{0}_{-1}e$ (d) ${}^{214}_{82}Pb$

Analyze In a nuclear equation the sum of mass numbers and the sum of atomic numbers must be the same on each side of the arrow.

Plan In each case, deduce the missing species by comparing the atomic numbers and mass numbers of the reactants and products.

Execute

(a) The missing particle has a mass number of 0 and a charge of +1, so it must be a positron, ${}^{0}_{+1}e$. When the positron is included in the equation, the atomic mass is 18 on each side, and the atomic numbers sum to 9 on each side.

(b) The missing nucleus must have a mass number of $26 + 0 = 26$ and an atomic number of $13 - 1 = 12$, so it is ${}^{26}_{12}Mg$.

(c) The missing particle has a mass number of 0 and a charge of −1, so it must be a beta particle, ${}^{0}_{-1}e$.

(d) The missing nucleus has a mass number of $218 - 4 = 214$ and an atomic number of $84 - 2 = 82$, so it is ${}^{214}_{82}Pb$.

PROBLEM-SOLVING PRACTICE 18.2

Complete these nuclear equations by filling in the missing symbol, mass number, and atomic number of the product species.

(a) ${}^{11}_{6}C \longrightarrow {}^{11}_{5}B + ?$ (b) ${}^{35}_{16}S \longrightarrow {}^{35}_{17}Cl + ?$

(c) ${}^{30}_{15}P \longrightarrow {}^{0}_{+1}e + ?$ (d) ${}^{22}_{11}Na \longrightarrow {}^{0}_{-1}e + ?$

EXERCISE 18.2

Nuclear Reactions

Aluminum-26 can undergo either positron emission or electron capture. Write the balanced nuclear equation for each case.

18-3 Stability of Atomic Nuclei

The naturally occurring isotopes of elements from hydrogen to bismuth are shown in Figure 18.3, where the radioactive isotopes are represented by orange circles and the stable (nonradioactive) isotopes are represented by purple and green circles. You may be surprised that so few stable isotopes exist. Why are there not hundreds more? To investigate this question, we systematically examine the elements, starting with hydrogen.

In its simplest and most abundant form, hydrogen has only one nuclear particle, a single proton. In addition, the element has two other naturally occurring isotopes: nonradioactive deuterium, with one proton and one neutron, ${}^{2}_{1}H = D$; and radioactive tritium, with one proton and two neutrons, ${}^{3}_{1}H = T$. Helium, the next element, has two protons and two neutrons in its most stable isotope. At the end of the actinide series is element 103, lawrencium, one isotope of which has 154 neutrons and a mass number of 257. From hydrogen to lawrencium, except for ${}^{1}_{1}H$ and ${}^{3}_{2}He$, *the mass numbers of stable isotopes are always at least twice as large as the atomic number*. In other words, except for ${}^{1}_{1}H$ and ${}^{3}_{2}He$, every stable isotope of every element has a nucleus containing *at least* one neutron for every proton. Apparently the tremendous *repulsive* forces between the positively charged protons in the nucleus are moderated by the presence of neutrons, which have no electrical charge. Figure 18.3 illustrates a number of principles:

- **For light elements up to Ca ($Z = 20$), the stable isotopes usually have equal numbers of protons and neutrons ($N = Z$), or perhaps one more neutron than protons.** Examples include ${}^{7}_{3}Li$, ${}^{12}_{6}C$, ${}^{16}_{8}O$, and ${}^{32}_{16}S$.

- **Beyond calcium the neutron-to-proton ratio becomes increasingly greater than 1.** The region containing stable isotopes deviates more and more from the line $N = Z$ so apparently more neutrons are needed for nuclear stability in the heavier elements. For example, whereas one stable isotope of Fe has 26 protons and 30 neutrons ($N/Z = 1.15$), one stable isotope of platinum has 78 protons and 117 neutrons ($N/Z = 1.50$).
- **For elements beyond bismuth-209 (83 protons and 126 neutrons), all nuclei are unstable and radioactive.** Furthermore, the rate of disintegration becomes greater as the nuclear mass increases. For example, half of a $^{238}_{92}U$ sample decays in 4.5 billion years, whereas half of a $^{256}_{103}Lr$ sample decays in only 27 seconds.

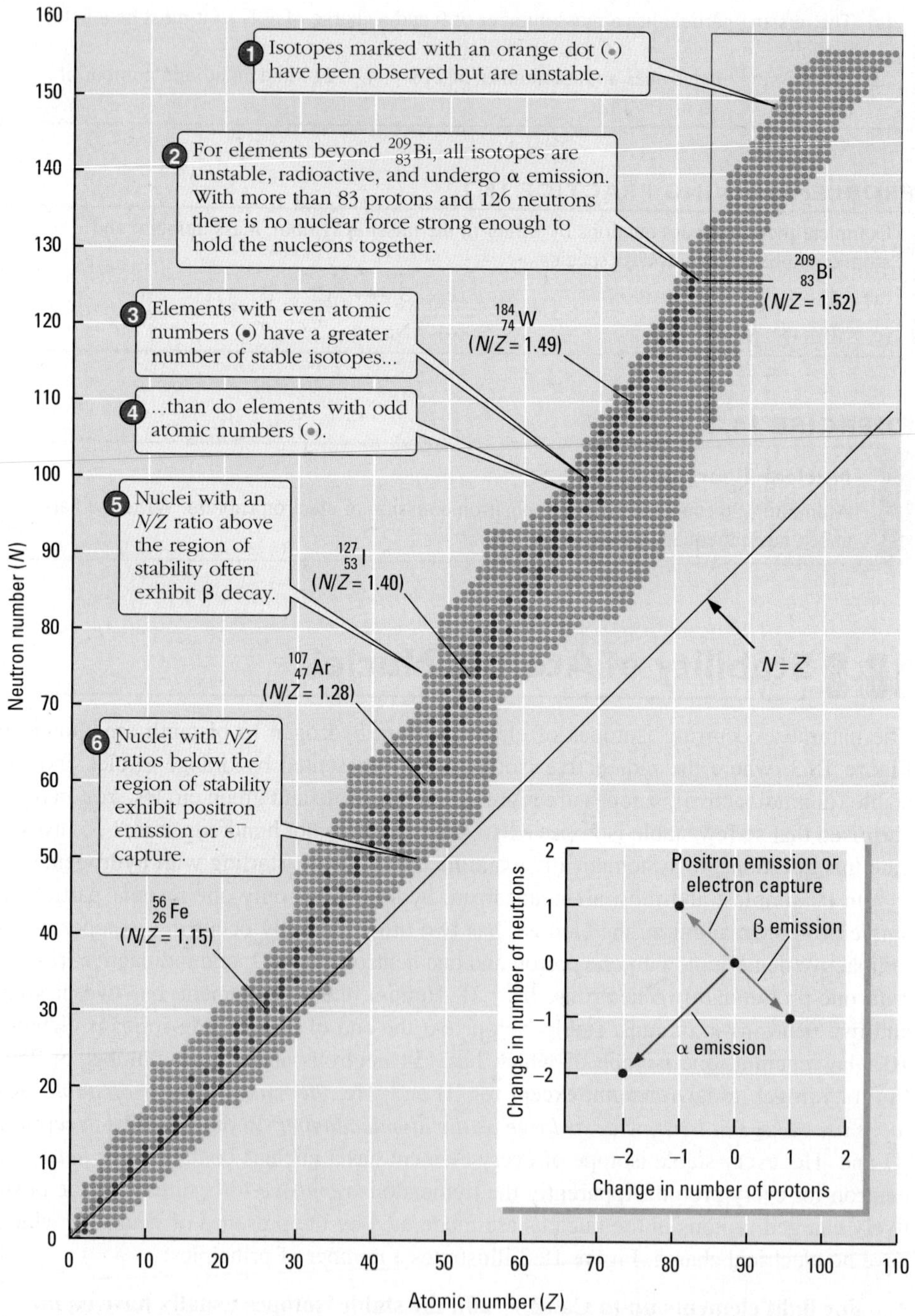

1 Isotopes marked with an orange dot (•) have been observed but are unstable.
2 For elements beyond $^{209}_{83}Bi$, all isotopes are unstable, radioactive, and undergo α emission. With more than 83 protons and 126 neutrons there is no nuclear force strong enough to hold the nucleons together.
3 Elements with even atomic numbers (•) have a greater number of stable isotopes...
4 ...than do elements with odd atomic numbers (•).
5 Nuclei with an N/Z ratio above the region of stability often exhibit β decay.
6 Nuclei with N/Z ratios below the region of stability exhibit positron emission or e⁻ capture.
$^{209}_{83}Bi$ (N/Z = 1.52)
$^{184}_{74}W$ (N/Z = 1.49)
$^{127}_{53}I$ (N/Z = 1.40)
$^{107}_{47}Ar$ (N/Z = 1.28)
$^{56}_{26}Fe$ (N/Z = 1.15)
N = Z
Neutron number (N)
Atomic number (Z)
Positron emission or electron capture
β emission
α emission
Change in number of neutrons
Change in number of protons

Figure 18.3 Nuclear stability. There is a narrow region of stable isotopes (purple and green).

- **Elements with an even atomic number have a greater number of stable isotopes than do those with an odd atomic number.**
- **For elements with an odd atomic number, the most stable isotope has an even number of neutrons.**
- **Of the nearly 270 stable isotopes shown in Figure 18.3, approximately 160 have an even number of neutrons *and* an even number of protons. Only about 110 have an odd number of either protons or neutrons. Only four isotopes ($^{2}_{1}H$, $^{6}_{3}Li$, $^{10}_{5}B$, and $^{14}_{7}N$) have odd numbers of *both* protons and neutrons.**

18-3a The Peninsula of Stability and Type of Radioactive Decay

The narrow region of stable isotopes in Figure 18.3 (the purple and green circles) is sometimes called the *peninsula of stability* in a "sea of instability." Any nucleus (the orange circles) not on this peninsula will decay in such a way that the nucleus can come ashore on the peninsula. The inset in the chart can help us predict what type of decay will be observed.

The nuclei of all elements beyond Bi ($Z = 83$) are unstable—that is, radioactive—and most decay by *alpha particle emission*. As shown in the inset, α emission results in a move down and to the left, bringing the newly formed isotope closer to the peninsula of stability. For example, americium, the radioactive element used in smoke alarms, decays in this manner.

$$^{243}_{95}Am \longrightarrow {}^{4}_{2}He + {}^{239}_{93}Np$$

Beta emission occurs in isotopes that have too many neutrons to be stable—that is, isotopes above the peninsula of stability in Figure 18.3. When beta decay converts a neutron to a proton and an electron is ejected, the atomic number increases by one, and the mass number remains constant.

$$^{60}_{27}Co \longrightarrow {}^{0}_{-1}e + {}^{60}_{28}Ni$$

Lighter nuclei—below the peninsula of stability—have too few neutrons. They attain stability by *positron emission* or by *electron capture*, because these processes have the effect of converting a proton to a neutron in one step.

$$^{13}_{7}N \longrightarrow {}^{0}_{+1}e + {}^{13}_{6}C$$

$$^{41}_{20}Ca + {}^{0}_{-1}e \longrightarrow {}^{41}_{19}K$$

Positron emission or electron capture is observed for elements with atomic numbers ranging from 4 to greater than 100; as Z increases, electron capture becomes more likely than positron emission.

PROBLEM-SOLVING EXAMPLE 18.3

Nuclear Stability

Write a nuclear equation for decay of each of these unstable isotopes.

(a) Silicon-32, $^{32}_{14}Si$ (b) Titanium-43, $^{43}_{22}Ti$

(c) Plutonium-239, $^{239}_{94}Pu$ (d) Manganese-56, $^{56}_{25}Mn$

Result

(a) $^{32}_{14}Si \longrightarrow {}^{0}_{-1}e + {}^{32}_{15}P$ (b) $^{43}_{22}Ti \longrightarrow {}^{0}_{+1}e + {}^{43}_{21}Sc$ or $^{43}_{22}Ti + {}^{0}_{-1}e \longrightarrow {}^{43}_{21}Sc$

(c) $^{239}_{94}Pu \longrightarrow {}^{4}_{2}He + {}^{235}_{92}U$ (d) $^{56}_{25}Mn \longrightarrow {}^{0}_{-1}e + {}^{56}_{26}Fe$

Analyze The ratio of protons to neutrons indicates the likelihood of a particular mode of nuclear decay (see Figure 18.3). Isotopes beyond $^{209}_{83}Bi$ decay by α emission.

Plan If there are excess neutrons, beta emission is probable. If there are excess protons, either electron capture or positron emission is probable.

Execute See Result for nuclear equations.

(a) Silicon-32 has excess neutrons, so beta decay is expected.
(b) Titanium-43 has excess protons, so either positron emission or electron capture is probable.
(c) Plutonium-239 has an atomic number greater than 83, so alpha decay is probable.
(d) Manganese-56 has excess neutrons, so beta decay is expected.

PROBLEM-SOLVING PRACTICE 18.3

Write a nuclear equation for decay of each unstable isotope.
(a) $^{42}_{19}K$ (b) $^{234}_{92}U$ (c) $^{20}_{9}F$

18-3b Binding Energy

As demonstrated by Ernest Rutherford's alpha particle scattering experiment (← **Sec. 2-2**), the nucleus of the atom is extremely small. Yet the nucleus can contain up to 83 protons, each of which is positively charged and repels all of the others, before becoming unstable. This suggests that there must be a very strong, short-range nuclear binding force that overcomes the electrostatic repulsive force of the protons. A measure of the force holding the nucleus together is the nuclear **binding energy**, *the energy change, ΔE, that occurs if the protons and neutrons in a nucleus are completely separated.* For example, a deuterium nucleus consists of one proton and one neutron. If a mole of deuterium nuclei is separated into a mole of protons and a mole of neutrons, the energy change (the binding energy) is more than 200 million kJ.

The nuclear binding energy is similar to the bond energy for a chemical bond (← Sec. 6-6b). The binding energy is the energy that must be supplied to separate all of the particles (protons and neutrons) that make up the atomic nucleus; the bond energy is the energy that must be supplied to separate two atoms linked by a chemical bond. In both cases the energy change is positive, because work must be done to separate the particles, but nuclear binding energy is much greater than chemical bond energy.

$$^{2}_{1}H \longrightarrow {}^{1}_{1}H + {}^{1}_{0}n$$

$$\text{Binding energy} = E_b = \Delta E = 2.15 \times 10^8 \text{ kJ/mol} \quad [18.1]$$

The reverse of this highly endothermic process, formation of a mole of deuterium nuclei from a mole of protons and a mole of neutrons, is highly exothermic. The energy released is more than 200 million kJ, the equivalent of exploding 73 tons of TNT. A deuterium nucleus is far more stable than an isolated proton and an isolated neutron.

Because of the enormous energy changes in nuclear reactions, it is possible to observe the corresponding changes in mass predicted by Albert Einstein's 1905 theory of special relativity. Einstein stated that mass and energy are simply different manifestations of the same quantity and the energy of a body is equivalent to its mass times the square of the speed of light, $E = mc^2$. Whenever there is a change in energy, there is a corresponding change in mass that is given by the equation

$$\Delta E = (\Delta m)c^2 \quad [18.2]$$

Careful measurements of the isotopic mass of deuterium and of the masses of a proton and a neutron show that the mass of a nucleus is less than the sum of the masses of the constituent protons and neutrons. For a deuterium nucleus,

$$\underset{2.013553 \text{ g/mol}}{^{2}_{1}H} \longrightarrow \underset{1.007276 \text{ g/mol}}{^{1}_{1}H} + \underset{1.008665 \text{ g/mol}}{^{1}_{0}n}$$

Change in mass $= \Delta m =$ sum of masses of products − mass of reactant

$$\Delta m = 1.007276 \text{ g/mol} + 1.008665 \text{ g/mol} - 2.013553 \text{ g/mol}$$

$$\Delta m = 2.015941 \text{ g/mol} - 2.013553 \text{ g/mol}$$

$$\Delta m = 0.002388 \text{ g/mol} = 2.388 \times 10^{-6} \text{ kg/mol}$$

This change in mass is equivalent to a change in energy of

1 J = 1 kg m²/s²

$$\Delta E = (2.388 \times 10^{-6}\ \text{kg/mol})(2.998 \times 10^{8}\ \text{m/s})^2$$
$$= 2.146 \times 10^{11}\ \text{J/mol} = 2.146 \times 10^{8}\ \text{kJ/mol}$$

This is the ΔE value (the binding energy) given in Equation 18.1 for the change in energy when a mole of deuterium nuclei is separated into a mole of protons and a mole of neutrons. Because the energy of a nucleus is less than the energy of the separated protons and neutrons, the mass of a nucleus is always less than the sum of the masses of its constituent protons and neutrons.

Consider another example, the formation of a helium-4 nucleus from two protons and two neutrons.

$$2\ {}^{1}_{1}\text{H} + 2\ {}^{1}_{0}\text{n} \longrightarrow {}^{4}_{2}\text{He} \qquad E_b = 2.73 \times 10^{9}\ \text{kJ/mol}\ {}^{4}_{2}\text{He nuclei}$$

This binding energy, E_b, is even larger than that for deuterium because when more particles combine there is a greater change in mass/energy. To compare nuclear stabilities more appropriately, nuclear scientists *divide the binding energy of a nucleus by the number of nucleons (the number of protons plus the number of neutrons)*; this is called the **binding energy per nucleon**. Each ${}^{4}_{2}$He nucleus contains four nucleons. Therefore,

$$E_b \text{ per nucleon} = \frac{2.73 \times 10^{9}\ \text{kJ}}{\text{mol}\ {}^{4}_{2}\text{He nuclei}} \times \frac{1\ \text{mol}\ {}^{4}_{2}\text{He nuclei}}{4\ \text{mol nucleons}}$$
$$= 6.83 \times 10^{8}\ \text{kJ/mol nucleons}$$

The greater the binding energy per nucleon is, the greater the stability of the nucleus is. Binding energies per nucleon are known for many nuclei; they are plotted as a function of mass number in Figure 18.4. It is very interesting and important that the point of maximum stability occurs at iron-56, ${}^{56}_{26}$Fe. This means that *all nuclei are thermodynamically unstable with respect to iron-56.* That is, very heavy nuclei can split, or *fission*, to form smaller, more stable nuclei with atomic numbers nearer to that of iron, while simultaneously releasing enormous quantities of energy (Section 18-6). In contrast, two very light

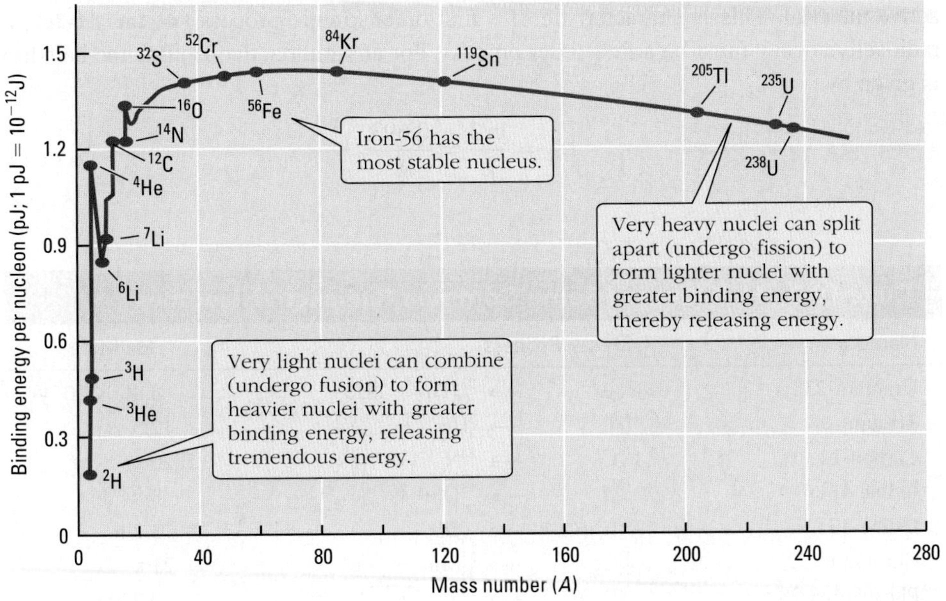
Binding energy per nucleon (pJ; 1 pJ = 10⁻¹² J)
Mass number (A)
0
0.3
0.6
0.9
1.2
1.5
0
40
80
120
160
200
240
280
²H
³He
³H
⁶Li
⁷Li
⁴He
¹²C
¹⁴N
¹⁶O
³²S
⁵²Cr
⁵⁶Fe
⁸⁴Kr
¹¹⁹Sn
²⁰⁵Tl
²³⁵U
²³⁸U
Iron-56 has the most stable nucleus.
Very heavy nuclei can split apart (undergo fission) to form lighter nuclei with greater binding energy, thereby releasing energy.
Very light nuclei can combine (undergo fusion) to form heavier nuclei with greater binding energy, releasing tremendous energy.

Figure 18.4 Binding energy per nucleon. Nuclei at the top of the curve are most stable.

nuclei can come together and undergo *nuclear fusion* exothermically to form heavier nuclei (Section 18-7). Because of its high nuclear stability, *iron is the most abundant of the heavier elements in the universe.*

EXERCISE 18.3

Binding Energy

Calculate the binding energy, in kJ/mol, for lithium-6, $^{6}_{3}Li$. The necessary masses are $^{1}_{1}H$ = 1.007276 g/mol, $^{1}_{0}n$ = 1.008665 g/mol, and $^{6}_{3}Li$ = 6.015123 g/mol. Is the binding energy greater than or less than that for helium-4? Compare the binding energy per nucleon of lithium-6 and helium-4. Which nucleus is more stable?

CONCEPTUAL EXERCISE 18.4

Binding Energy: Fission and Fusion

By interpreting the shape of the curve in Figure 18.4, predict which is more exothermic per gram—fission or fusion. Explain your prediction.

EXERCISES that are labeled **CONCEPTUAL** are designed to test your understanding of one or more concepts; they usually involve qualitative rather than quantitative thinking.

18-4 Rates of Disintegration Reactions

Cobalt-60 is radioactive and is used as a source of β particles and γ rays to treat malignancies in the human body. One-half of a sample of cobalt-60 changes via beta decay into nickel-60 in 5.26 years (Table 18.2). On the other hand, iodine-131, which is used to treat thyroid gland malfunction, decays much more rapidly; half of the radioactive iodine-131 decays in 8.04 days. These two radioactive isotopes are clearly different in their rates of decay.

18-4a Half-Life

The relative instability of a radioactive isotope is expressed as its **half-life ($t_{1/2}$)**, the *time required for one half of a given quantity of the isotope to undergo radioactive decay.* For all radioactive isotopes, the half-life is found to be independent of the quantity of radioactive material. This is characteristic of a first-order kinetic process (← **Sec. 11-3c**) so radioactive decay must be a first-order process. For any radioactive isotope the half-life is given by

$$t_{1/2} = \frac{\ln 2}{k} = \frac{0.693}{k}$$

The relationship $t_{1/2} = \frac{0.693}{k}$ was introduced in the context of kinetics of first-order reactions (← Sec. 11-3c).

Table 18.2 Half-Lives of Some Common Radioactive Isotopes

Name	Decay Process	Half-Life
Uranium-238	$^{238}_{92}U \longrightarrow ^{234}_{90}Th + ^{4}_{2}He$	4.46×10^{9} yr
Tritium	$^{3}_{1}H \longrightarrow ^{3}_{2}He + ^{0}_{-1}e$	12.3 yr
Carbon-14	$^{14}_{6}C \longrightarrow ^{14}_{7}N + ^{0}_{-1}e$	5730 yr
Iodine-131	$^{131}_{53}I \longrightarrow ^{131}_{54}Xe + ^{0}_{-1}e$	8.04 d
Iodine-123	$^{123}_{53}I + ^{0}_{-1}e \longrightarrow ^{123}_{52}Te$	13.2 h
Chromium-57	$^{57}_{24}Cr \longrightarrow ^{57}_{25}Mn + ^{0}_{-1}e$	21 s
Phosphorus-28	$^{28}_{15}P \longrightarrow ^{28}_{14}Si + ^{0}_{+1}e$	0.270 s
Strontium-90	$^{90}_{38}Sr \longrightarrow ^{90}_{39}Y + ^{0}_{-1}e$	28.8 yr
Cobalt-60	$^{60}_{27}Co \longrightarrow ^{60}_{28}Ni + ^{0}_{-1}e$	5.26 yr

where k is the first-order rate constant for the decay. As illustrated by Table 18.2, isotopes have widely varying half-lives. Some take years, even millennia, for half of the sample to decay (^{238}U, ^{14}C), whereas others decay to half the original number of atoms in seconds or less (^{57}Cr, ^{28}P). Half-lives are reported using whatever time unit is most appropriate—anything from years to seconds.

As an example, consider plutonium-239, an isotope formed in nuclear reactors, which decays by alpha emission.

$$^{239}_{94}Pu \longrightarrow\, ^{4}_{2}He + \,^{235}_{92}U$$

The half-life of plutonium-239 is 24,100 years. Thus, half of the quantity of $^{239}_{94}Pu$ present at any given time has disintegrated 24,100 years later. For example, if we begin with 1.00 g $^{239}_{94}Pu$, 0.500 g of the isotope remains after 24,100 years. After 48,200 years (two half-lives), only half of the 0.500 g, or 0.250 g, remains. After 72,300 years (three half-lives), only half of the 0.250 g, or 0.125 g, is still present. The amounts of $^{239}_{94}Pu$ present at various times are illustrated in Figure 18.5. All radioactive isotopes follow this type of decay curve.

PROBLEM-SOLVING EXAMPLE 18.4

Half-Life

Iodine-131, used to treat hyperthyroidism, has a half-life of 8.04 days.

$$^{131}_{53}I \longrightarrow\, ^{131}_{54}Xe + \,^{0}_{-1}e \qquad t_{1/2} = 8.04 \text{ days}$$

For a sample containing 10.0 μg of iodine-131, calculate the mass (μg) of the isotope that remains after 32.2 days.

Result 0.625 μg

Analyze First, we find the number of half-lives in the given 32.2-day time period. Use this number of half-lives to calculate the mass of iodine-131 remaining.

Plan Because the time period is 32.2 days and the half-life is 8.04 days, there are exactly four half-lives during this period.

Execute This means that the initial quantity of 10.0 μg is reduced by half four times.

$$10.0\ \mu g \times 1/2 \times 1/2 \times 1/2 \times 1/2 = 10.0\ \mu g \times 1/16 = 0.625\ \mu g$$

After 32.2 days, only one sixteenth (0.625 μg/10.0 μg) of the original ^{131}I remains.

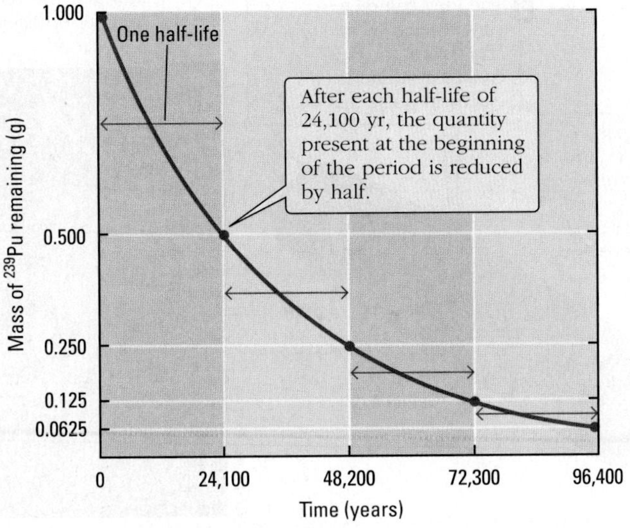
1.000
One half-life
After each half-life of 24,100 yr, the quantity present at the beginning of the period is reduced by half.
0.500
0.250
0.125
0.0625
Mass of ²³⁹Pu remaining (g)
0
24,100
48,200
72,300
96,400
Time (years)

Figure 18.5 The radioactive decay of 1.00 g plutonium-239.

☑ **Reasonable Result Check** After the passage of four half-lives, the remaining mass of ^{131}I should be a small fraction of the initial mass, and it is.

PROBLEM-SOLVING PRACTICE 18.4

Strontium-90 is a radioisotope ($t_{1/2}$ = 29 years) produced in atomic bomb explosions. Its long half-life and tendency to concentrate in bone marrow by replacing calcium make it particularly dangerous to people and animals.

(a) Strontium-90 decays through beta emission. Write a balanced nuclear equation showing the other product of decay.

(b) A ^{90}Sr sample emits 2000 β particles per minute. Calculate the number of half-lives and years necessary to reduce the emission to 125 β particles per minute.

EXERCISE 18.5

Half-Lives

The radioactivity of formerly highly radioactive isotopes is essentially negligible after ten half-lives. Calculate the percentage of the original radioisotope that remains after this amount of time (ten half-lives).

18-4b Rate of Radioactive Decay

The half-life of a radioactive element can be determined by measuring its *rate of decay*, that is, the number of atoms that disintegrate in a given time. Each nuclear disintegration emits radiation: an alpha particle; a beta particle; a gamma ray; or some combination of emissions that can be detected with a device such as a Geiger counter (Figure 18.6). A Geiger counter detects radioactive emissions as they ionize a gas to form free electrons and cations that can be attracted to a pair of electrodes. Each disintegration generates a pulse of current; the pulses are counted, and the number of pulses per unit time is the output of the Geiger counter. The rate of pulses, that is, *the number of disintegrations per unit time* is called the **activity (*A*)** of the sample.

The **curie (Ci)** is commonly used as a unit of activity. One curie represents a decay rate of 3.7×10^{10} disintegrations per second (s^{-1}), which is the decay rate of 1 g radium. One

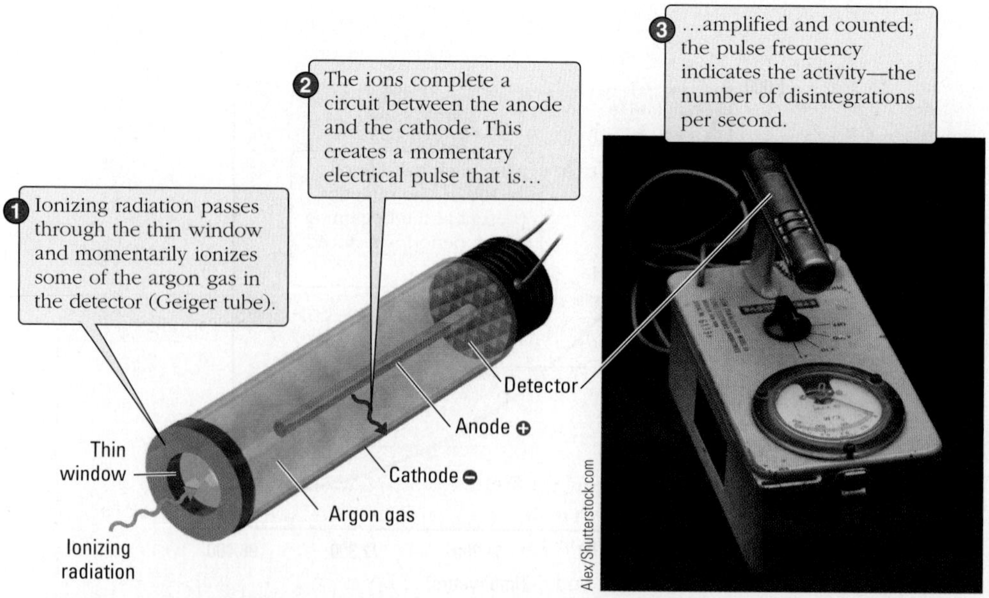
1 Ionizing radiation passes through the thin window and momentarily ionizes some of the argon gas in the detector (Geiger tube).
2 The ions complete a circuit between the anode and the cathode. This creates a momentary electrical pulse that is...
3 ...amplified and counted; the pulse frequency indicates the activity—the number of disintegrations per second.
Thin window
Ionizing radiation
Argon gas
Cathode ⊖
Anode ⊕
Detector
Alex/Shutterstock.com

Figure 18.6 A Geiger counter measures ionizing radiation produced by radioactive decays.

millicurie, mCi, $= 10^{-3}$ Ci $= 3.7 \times 10^7$ s^{-1}. Another unit of radioactivity is the **becquerel (Bq)**; 1 becquerel is equal to one nuclear disintegration per second (1 Bq $= 1$ s^{-1}).

As noted in Section 18-4a, radioactive decay is a first-order process. Its rate is directly proportional to the number of radioactive atoms present (N). This proportionality can be expressed as a rate law (Equation 18.3) in which A is the activity of the sample and k is the first-order rate constant or *decay constant* characteristic of that radioisotope.

The *curie* was named for Pierre Curie by his wife, Marie; the *becquerel* honors Henri Becquerel. The unit for curie and becquerel is s^{-1} because each is a number (of disintegrations) per second.

$$\text{Rate of radioactive decay} = \text{activity} = A = kN \qquad [18.3]$$

Suppose the activity of a sample is measured at some time t_0 and then measured again after a few minutes, hours, or days. If the initial activity is A_0 at t_0, then a second measurement at a later time t will detect a smaller activity A. Using Equation 18.3, the ratio of the activity A at some time t to the activity at the beginning of the experiment A_0 must equal the ratio of the number of radioactive atoms N that are present at time t to the number present at the beginning of the experiment, N_0.

$$\frac{A}{A_0} = \frac{kN}{kN_0} \quad \text{or} \quad \frac{A}{A_0} = \frac{N}{N_0}$$

Thus, either A/A_0 or N/N_0 expresses the fraction of radioactive atoms remaining in a sample after some time has elapsed.

The fraction of radioactive atoms still present in a sample after some time has elapsed, N/N_0, can be calculated using the integrated rate equation for a first-order reaction (← **Sec. 11-3a**).

$$\ln N = -kt + \ln N_0 \quad \text{which rearranges to} \quad \ln\frac{N}{N_0} = -kt \qquad [18.4]$$

Equation 18.4 can be derived from Equation 18.3 using calculus (← Sec. 11-3a).

Equation 18.4 can also be stated in terms of the fraction of radioactive atoms still present in the sample at time t.

$$\ln\frac{A}{A_0} = -kt \qquad [18.5]$$

Notice the negative sign in Equations 18.4 and 18.5. The ratio N/N_0 is less than 1 because N is always less than N_0. This means that the logarithm of N/N_0 is negative, and the other side of the equation has a compensating negative sign because k and t are always positive. The same applies to the ratio A/A_0.

As radioactive atoms decay, N becomes a smaller and smaller fraction of N_0.

As we have seen, the half-life of an isotope is inversely proportional to the first-order rate constant k:

$$t_{1/2} = \frac{0.693}{k}$$

Thus, the half-life can be found by calculating k from Equation 18.5 using N and N_0 from laboratory measurements over the time period t, as illustrated in Problem-Solving Example 18.5.

PROBLEM-SOLVING EXAMPLE 18.5

Half-Life

A sample of ^{24}Na initially undergoes 3.50×10^4 disintegrations per second (s^{-1}). After 24.0 h, its disintegration rate has fallen to 1.16×10^4 s^{-1}. Calculate the half-life of ^{24}Na.

Result 15.0 h

Analyze In exactly one half-life period, the disintegration rate (activity) decreases to one half of what it was originally. We use Equation 18.5 relating activity (disintegration rate) at time zero and time t with the decay constant k.

Plan Using the data provided (A, A_0, and the time, t) and Equation 18.4, calculate the value of k. Use this value to calculate the half-life.

Execute

$$\ln\left(\frac{1.16 \times 10^4\ s^{-1}}{3.50 \times 10^4\ s^{-1}}\right) = \ln(0.331) = -k(24.0\ h)$$

$$k = -\frac{\ln(0.331)}{24.0\ h} = -\left(\frac{-1.104}{24.0\ h}\right) = 0.0460\ h^{-1}$$

From k we determine $t_{1/2}$.

$$t_{1/2} = \frac{0.693}{k} = \frac{0.693}{0.0460\ h^{-1}} = 15.0\ h$$

☑ **Reasonable Result Check** The activity (disintegration rate) fell to between one half and one quarter of its initial value in 24 h, so the half-life must be less than 24 h, and this agrees with our more accurate calculation.

PROBLEM-SOLVING PRACTICE 18.5

The decay of iridium-192, a radioisotope used in cancer radiation therapy, has a rate constant of $9.3 \times 10^{-3}\ d^{-1}$.

(a) Calculate the half-life of ^{192}Ir.

(b) Calculate the fraction of an ^{192}Ir sample that remains after 100. days.

PROBLEM-SOLVING EXAMPLE 18.6

Time and Radioactivity

A 1.00-mg sample of ^{131}I ($t_{1/2}$ = 8.04 days) has an initial disintegration rate of $4.7 \times 10^{12}\ s^{-1}$. Calculate the time required for the disintegration rate of the sample to fall to $2.9 \times 10^{11}\ s^{-1}$.

Result 776 h

Analyze We use the half-life of ^{131}I to calculate the decay constant, k. First, convert the known half-life from days to hours. Then, use this half-life value to calculate the decay constant, k. We use this k value in the equation relating disintegration rate to time.

Plan The initial disintegration rate is $A_0 = 4.7 \times 10^{12}\ s^{-1}$, and the disintegration rate after the elapsed time is $A = 2.9 \times 10^{11}\ s^{-1}$. We can use Equation 18.5 to calculate the time t, required to decrease the disintegration rate to $2.9 \times 10^{11}\ s^{-1}$.

Execute

$$t_{1/2} = 8.04\ \text{days} \times \frac{24\ h}{1\ \text{day}} = 193\ h$$

$$k = \frac{0.693}{t_{1/2}} = \frac{0.693}{193\ h} = 3.59 \times 10^{-3}\ h^{-1}$$

Note that both disintegration rates are given in disintegrations per second, so we can use them as provided. If we converted them both to disintegrations per hour, we would get the same numerical result for the ratio.

$$\ln\left(\frac{2.9 \times 10^{11}\ s^{-1}}{4.7 \times 10^{12}\ s^{-1}}\right) = -kt = -(3.59 \times 10^{-3}\ h^{-1})t$$

$$t = \frac{-2.785}{-3.59 \times 10^{-3}\ h^{-1}} = 776\ h$$

☑ **Reasonable Result Check** The disintegration rate has fallen by approximately a factor of sixteen $((4.7 \times 10^{12})/(2.9 \times 10^{11}) = 16.2)$, so the elapsed time must be approximately four half-lives of ^{131}I, and it is.

PROBLEM-SOLVING PRACTICE 18.6

In 1921 the women of America honored Marie Curie by giving her a gift of 1.00 g pure radium, which is now in Paris at the Curie Institute of France (see http://www.northnet.org/stlawrenceaauw/curie.htm). The principal isotope, ^{226}Ra, has a half-life of 1.60×10^3 years. Calculate the mass (g) of radium-226 that remains today.

18-4c Carbon-14 Dating

In 1947 Willard Libby developed a technique for measuring the age of archaeological objects using radioactive carbon-14. Carbon is an important building block of all living systems, and all organisms contain the three isotopes of carbon: ^{12}C, ^{13}C, and ^{14}C. The first two isotopes are stable (nonradioactive) and have been present for billions of years. Carbon-14, however, is radioactive and decays to nitrogen-14 by beta emission.

$$^{14}_{6}C \longrightarrow ^{0}_{-1}e + ^{14}_{7}N$$

The half-life of ^{14}C is known by experiment to be 5.73×10^3 years. The number of carbon-14 atoms (N) in a carbon-containing sample can be measured from the activity of the sample (A). If the number of carbon-14 atoms originally in the sample (N_0) can be determined, or if the initial activity (A_0) can be determined, the age of the sample can be found from Equation 18.4 or 18.5.

Bettmann/CORBIS

Willard Libby and his apparatus for carbon-14 dating. Willard Libby won the 1960 Nobel Prize in Chemistry for his discovery of radiocarbon dating.

This method of age determination clearly depends on knowing how much ^{14}C was originally in the sample. The answer to this question comes from work by physicist Serge Korff, who discovered in 1939 that ^{14}C is continually generated in the upper atmosphere. High-energy cosmic rays collide with gas molecules in the upper atmosphere and cause them to eject neutrons. These free neutrons collide with nitrogen atoms to produce carbon-14.

$$^{14}_{7}N + ^{1}_{0}n \longrightarrow ^{14}_{6}C + ^{1}_{1}H$$

Throughout the *entire* atmosphere, only about 7.5 kg ^{14}C is produced per year. However, this relatively small quantity of radioactive carbon is incorporated into CO_2 and other carbon compounds and then distributed worldwide as part of the cycling of carbon throughout Earth's environment. The continual formation of ^{14}C, transfer of the isotope within the oceans, atmosphere, and biosphere, and decay of living matter keep the supply of ^{14}C constant.

Plants absorb carbon dioxide from the atmosphere and convert it into food via photosynthesis (← **Sec. 16-9b**). In this way, the ^{14}C becomes incorporated into living tissue, where radioactive ^{14}C atoms react chemically the same way as nonradioactive ^{12}C atoms. The beta decay activity of carbon-14 in *living* plants and animals and in the air is constant at 15.3 disintegrations per minute per gram ($min^{-1}\ g^{-1}$) of carbon-14. When a plant or animal dies, however, carbon-14 disintegration continues *without the ^{14}C being replaced*. Consequently, the ^{14}C activity of the dead plant or animal material decreases with the passage of time. The smaller the activity of carbon-14 in the plant or animal, the longer the time between the death of the organism and the present. Assuming that ^{14}C activity in living organisms was about the same hundreds or thousands of years ago as it is now, measurement of the ^{14}C beta activity of an artifact can be used to date an article containing carbon. The slight fluctuations of the ^{14}C activity in living plants for the past several thousand years have been measured by studying growth rings of long-lived trees, and the carbon-14 dates of objects can be corrected accordingly.

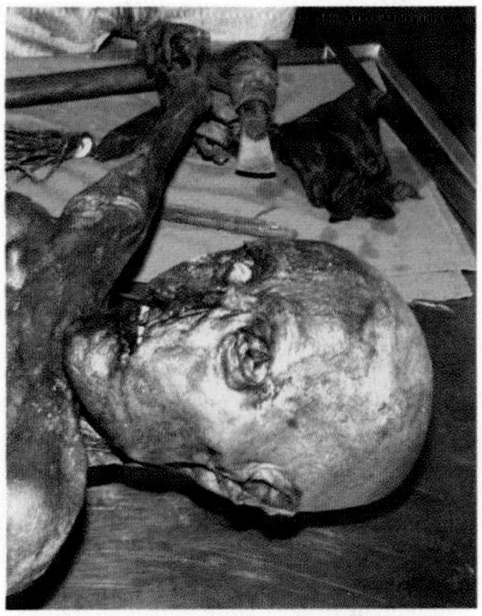

Reuters/CORBIS

The Ice Man. This human mummy was found in 1991 in glacial ice high in the Alps. Carbon-14 dating determined that he lived about 5300 years ago. The mummy is exhibited at the South Tyrol Museum of Archaeology in Bolzano, Italy.

The time scale accessible to carbon-14 dating is determined by the value of the half-life of ^{14}C: 5.73×10^3 yr. Therefore, this method for dating objects can be extended back approximately 50,000 years. This span of time is almost nine half-lives, during which the number of disintegrations per minute per gram of carbon would fall by a factor of about

$(\frac{1}{2})^9 = 1.95 \times 10^{-3}$ from about 15.3 min^{-1} g^{-1} to about 0.030 min^{-1} g^{-1}, which is a disintegration rate so low that it is difficult to measure accurately.

PROBLEM-SOLVING EXAMPLE 18.7

Carbon-14 Dating

Charcoal fragments found in a prehistoric cave in Lascaux, France, had a measured disintegration rate of 1.90 min^{-1} g^{-1} carbon. Calculate the approximate age of the charcoal.

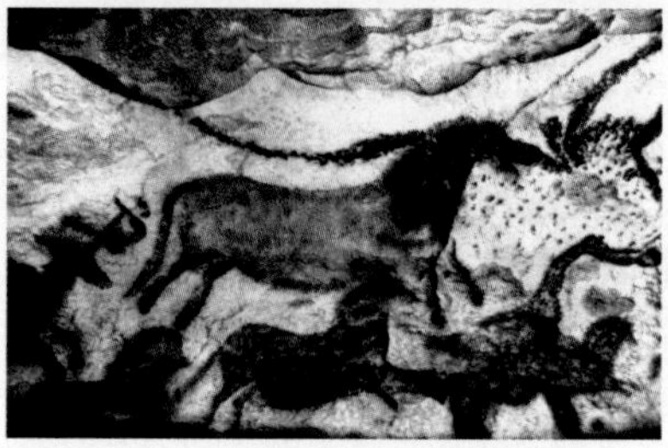
Images-of-france/Alamy

Cave paintings from Lascaux, France.

Result 1.73×10^4 yr

Analyze The current activity, 1.90 min^{-1} g^{-1} carbon, is given; to calculate t from Equation 18.5, the rate constant and the initial activity are needed. The rate constant can be obtained from the half-life and the initial activity is given in the text.

Plan First calculate k, the rate constant, using the half-life of carbon-14, 5.73×10^3 yr. Then, use Equation 18.4 to calculate the time; A is the activity of the charcoal (1.90 min^{-1} g^{-1}) and A_0 is the activity of the carbon-14 in living material (15.3 min^{-1} g^{-1}).

Execute Calculate k.

$$k = \frac{0.693}{t_{1/2}} = \frac{0.693}{5.73 \times 10^3 \text{yr}} = 1.21 \times 10^{-4}\ \text{yr}^{-1}$$

Calculate the time, t.

$$\ln\left(\frac{A}{A_0}\right) = \ln\left(\frac{1.90\ \text{min}^{-1}\ \text{g}^{-1}}{15.3\ \text{min}^{-1}\ \text{g}^{-1}}\right) = -kt$$

$$\ln(0.124) = -(1.21 \times 10^{-4}\ \text{yr}^{-1})t$$

$$t = \frac{-2.09}{-1.21 \times 10^{-4}\ \text{yr}^{-1}} = 1.73 \times 10^4\ \text{yr}$$

Thus, the charcoal is approximately 17,300 years old.

☑ **Reasonable Result Check** The disintegration rate has fallen by a factor of about eight from the rate for living material, so about three half-lives have elapsed. This agrees with our calculated result.

PROBLEM-SOLVING PRACTICE 18.7

Tritium, ^3H ($t_{1/2}$ = 12.3 yr), is produced in the atmosphere and incorporated in living plants in much the same way as ^{14}C. Estimate the age of a sealed sample of Scotch whiskey that has a tritium content 0.60 times that of the water in the area where the whiskey was produced.

EXERCISE 18.6

Radiochemical Dating

The radioactive decay of uranium-238 to lead-206 provides a method of radiochemically dating ancient rocks by using the ratio of ^{206}Pb atoms to ^{238}U atoms in a sample. Using this method, a moon rock was found to have a $^{206}Pb/^{238}U$ ratio of 100/109, that is, 100 ^{206}Pb atoms for every 109 ^{238}U atoms. No other lead isotopes were present in the rock, indicating that all of the ^{206}Pb was produced by ^{238}U decay. Estimate the age of the moon rock. The half-life of ^{238}U is 4.46×10^9 years.

CONCEPTUAL EXERCISE 18.7

Detecting Fraudulently Fortified Wine

Ethanol, C_2H_5OH, is produced by the fermentation of grains or by the reaction of water with ethylene, which is made from petroleum. The alcohol content of wines can be increased fraudulently beyond the usual 12% available from fermentation by adding ethanol produced from ethylene. How can carbon dating techniques be used to differentiate the ethanol sources in these wines?

18-5 Artificial Transmutations

In the course of his experiments, Rutherford found in 1919 that alpha particles ionize atomic hydrogen, knocking off an electron from each atom. Using atomic nitrogen instead, he found that bombardment with alpha particles *produced protons*. He correctly concluded that the alpha particles had knocked a proton out of the nitrogen nucleus and that a nucleus of another element had been produced. In other words, nitrogen had undergone a *transmutation* to oxygen.

$$^{4}_{2}\text{He} + {}^{14}_{7}\text{N} \longrightarrow {}^{17}_{8}\text{O} + {}^{1}_{1}\text{H}$$

Rutherford had proposed that protons and neutrons are the fundamental building blocks of nuclei. Although his search for the neutron was not successful, it was later found by James Chadwick in 1932 as a product of the alpha particle bombardment of beryllium.

$$^{9}_{4}\text{Be} + {}^{4}_{2}\text{He} \longrightarrow {}^{12}_{6}\text{C} + {}^{1}_{0}\text{n}$$

Changing one element into another by alpha particle bombardment has its limitations. Before a positively charged particle (such as an alpha particle) can be captured by a positively charged nucleus, the two particles must have sufficient kinetic energy to overcome the very large electrostatic repulsion between the two positively charged particles. But the neutron is electrically neutral, so Enrico Fermi (in 1934) reasoned that a nucleus would not oppose its entry. By this approach, nearly every element has since been transmuted, and a number of *transuranium elements* (elements beyond uranium) have been prepared. For example, plutonium-239 forms americium-241 by neutron bombardment.

$$^{239}_{94}\text{Pu} + {}^{1}_{0}\text{n} \longrightarrow {}^{240}_{94}\text{Pu}$$

$$^{240}_{94}\text{Pu} + {}^{1}_{0}\text{n} \longrightarrow {}^{241}_{94}\text{Pu}$$

$$^{241}_{94}\text{Pu} \longrightarrow {}^{241}_{95}\text{Am} + {}^{0}_{-1}\text{e}$$

Of the elements currently known, all those with atomic numbers of 92 or less exist in nature except Tc and Pm. The transuranium elements, those with atomic numbers greater than 92, are all synthetic. Up to element 101, mendelevium, all the elements can be made by bombarding the nucleus of a lighter element with small particles such as $^{4}_{2}\text{He}$ or $^{1}_{0}\text{n}$. Beyond element 101, special techniques using heavier particles are required and are still being developed. For example, the International Union of Pure and Applied Chemistry (IUPAC) has acknowledged the *validated* syntheses of the highest atomic numbered (super-heavy) elements—111 roentgenium (Rg), 112 copernicium (Cn), and most recently, 114 flerovium (Fl) in 2009 and 116 livermorium (Lv) in 2010. Flerovium-289 was produced by bombarding a plutonium-244 target with calcium-48 nuclei.

$$^{48}_{20}\text{Ca} + {}^{244}_{94}\text{Pu} \longrightarrow {}^{289}_{114}\text{Fl} + 3\,{}^{1}_{0}\text{n}$$

Glenn Seaborg
1912–1999

University of California, Lawrence Berkeley National Laboratory

A pioneer in developing radioisotopes for medical use (Section 18-9b), Glenn Seaborg was the first to produce iodine-131, used subsequently to treat his mother's abnormal thyroid condition. As a result of Seaborg's further research, it became possible to predict accurately the properties of many of the as-yet-undiscovered transuranium elements. In a remarkable 21-year span (1940–1961), Seaborg and his colleagues synthesized ten new transuranium elements (plutonium to lawrencium). He received the Nobel Prize in 1951 for his creation of new elements. In the 1990s Seaborg was honored by having element 106 named for him.

The bombardment of curium-248 with calcium-48 nuclei produced livermorium-293.

$$^{48}_{20}\text{Ca} + ^{248}_{96}\text{Cm} \longrightarrow ^{293}_{116}\text{Lv} + 3\,^{1}_{0}\text{n}$$

Preliminary evidence (2004) indicates that element 115 was formed by bombardment of an americium-243 target with calcium-48 nuclei. After nearly a month of bombardment, four atoms of element 115 were detected. Through alpha decay, these atoms formed element 113. In September 2013, an independent synthesis of element 115 was reported, but the element's discovery has not yet been verified by IUPAC. Elements 117 (2010) and 118 (2006) have been synthesized using calcium-48 nuclei to bombard berkelium and californium targets, respectively. Their synthesis adds credibility to the likely existence of the so-called "island of stability", a region of the periodic table in which nuclei of super-heavy elements ($Z > 112$) become more stable with an increase in their number of neutrons. Neither element 117 nor element 118 has yet been officially recognized by IUPAC. Until an element's discovery has been verified by IUPAC, placeholder names (and symbols) based on the atomic number are used. These names are ununtrium (Uut), ununpentium (Uup), ununseptium (Uus), and ununoctium (Uuo) for elements 113, 115, 117, and 118, respectively.

EXERCISE 18.8

Nuclear Transmutations

Balance these equations for nuclear reactions, indicating the symbol, the mass number, and the atomic number of the remaining product.

(a) $^{16}_{8}\text{O} + ^{1}_{1}\text{H} \longrightarrow ^{4}_{2}\text{He} + ?$

(b) $^{64}_{28}\text{Ni} + ^{209}_{83}\text{Bi} \longrightarrow ^{1}_{0}\text{n} + ?$

(c) $^{27}_{13}\text{Al} + ? \longrightarrow ^{1}_{0}\text{n} + ^{30}_{15}\text{P}$

CONCEPTUAL EXERCISE 18.9

Element Synthesis

In 1998, researchers in Dubna, Russia synthesized element 112, copernicium, by bombardment of uranium-238 nuclei with calcium-48 nuclei. The copernicium-283 isotope was produced along with neutrons. Write a balanced nuclear equation to represent this synthesis.

18-6 Nuclear Fission

Element 109 (Mt) is named in honor of Lise Meitner.

In 1938 the nuclear chemists Otto Hahn and Fritz Strassmann were confounded when they isolated barium from a sample of uranium that had been exposed to a beam of neutrons. Further work by Lise Meitner, Otto Frisch, Niels Bohr, and Leó Szilárd confirmed that the uranium-235 nucleus had formed barium by the capture of a neutron followed by **nuclear fission**, *a nuclear reaction in which a nucleus splits into two lighter nuclei* (Figure 18.7).

1 A neutron strikes a $^{235}_{92}$U nucleus...
2 ...causing it to fission into...
3 ...a $^{92}_{36}$Kr nucleus...
4 ...a $^{141}_{56}$Ba nucleus...
5 ...and three neutrons.

Figure 18.7 A neutron causes a $^{235}_{92}$U nucleus to undergo fission (Equation 18.6).

$$^{235}_{92}\text{U} + ^{1}_{0}\text{n} \longrightarrow ^{236}_{92}\text{U} \longrightarrow ^{141}_{56}\text{Ba} + ^{92}_{36}\text{Kr} + 3\,^{1}_{0}\text{n} \qquad \Delta E = -2 \times 10^{10}\ \text{kJ/mol} \qquad [18.6]$$

This nuclear equation shows that a single neutron as reactant produces three neutrons among the products. The fact that the fission reaction produces more neutrons than are required to begin the process is important. Each of the product neutrons is capable of inducing another fission reaction, so three neutrons would induce three fissions, which would release nine neutrons to induce nine more fissions, from which 27 neutrons are obtained, and so on. Because the neutron-induced fission of uranium-235 is extremely rapid, this sequence of reactions can lead to an explosive chain reaction, as illustrated in Figure 18.8.

Nuclear fission of uranium-235 produces a variety of products. Thirty-four elements have been detected among the fission products, including those shown in Figure 18.8. If the quantity of uranium-235 is small, then most of the neutrons escape without hitting a nucleus. Because so few neutrons are captured by ^{235}U nuclei the chain reaction cannot be sustained. *The minimum mass of fissionable material required for a self-sustaining chain reaction* is termed the **critical mass**. In an atomic bomb, two small masses of uranium-235, neither capable of sustaining a chain reaction, are brought together rapidly to form one mass greater than the critical mass. A nuclear fission explosion results.

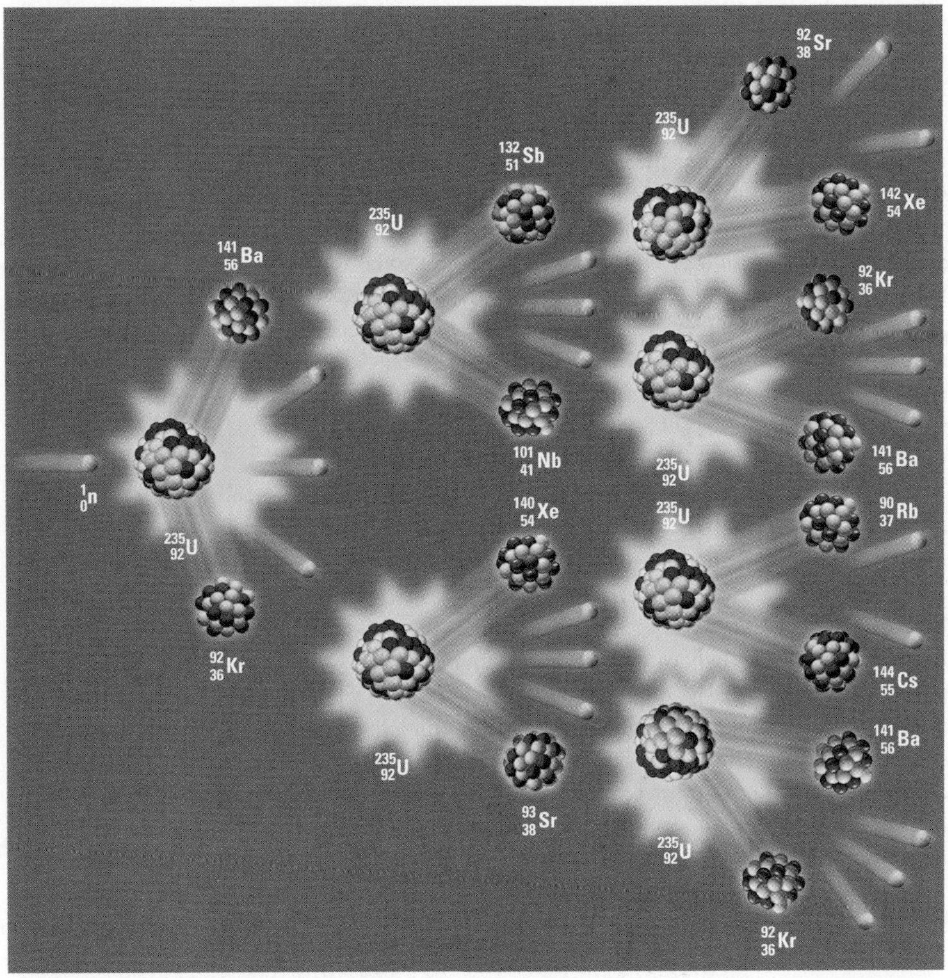

1 0 n
235 92 U
141 56 Ba
92 36 Kr
132 51 Sb
101 41 Nb
140 54 Xe
93 38 Sr
92 38 Sr
142 54 Xe
92 36 Kr
141 56 Ba
90 37 Rb
144 55 Cs
141 56 Ba
92 36 Kr

Figure 18.8 A self-propagating nuclear chain reaction initiated by capture of a neutron. Each fission event produces two lighter nuclei plus two or three neutrons.

18-6a Nuclear Reactors

A **nuclear reactor** is *a device by which a nuclear reaction can be carried out in a safe and controlled manner.* In a nuclear fission reactor, the rate of fission is controlled by inserting rods of cadmium or other neutron absorbers into the *reactor core,* the part of the reactor that contains fissionable material. The rods absorb neutrons that would otherwise cause fission reactions. The rate of the overall reaction can be increased by withdrawing the control rods, or decreased by inserting them. The materials that control the numbers of neutrons (by absorbing them) or control their energy (by absorbing some of their energy) are known as *moderators.*

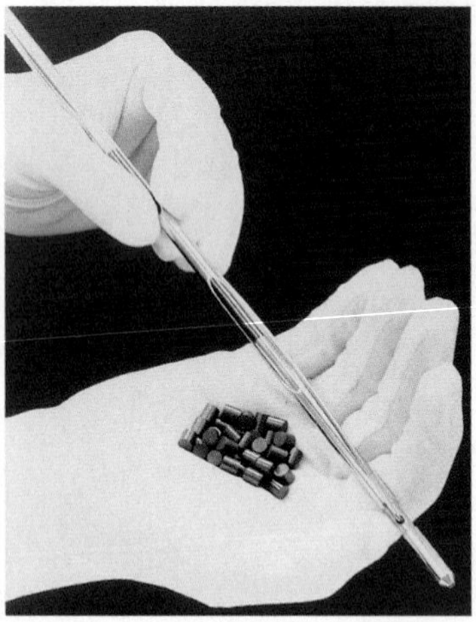

Science Source

Uranium oxide pellets used in nuclear fuel rods. Note that protective gloves are worn when handling the pellets.

Nuclear fission reactors are the thermal energy source in *nuclear power plants* (Figure 18.9). The nuclear reactor fuel is a fissionable isotope, for example, uranium-235 in the form of uranium dioxide, UO_2, pellets. The pellets, which are about the size of the eraser on a pencil, are placed end-to-end in metal alloy tubes, which are then assembled into stainless steel–clad bundles. As pointed out earlier, once a fission reaction is started, it can be sustained as a chain reaction. However, a source of neutrons is needed to initiate the chain reaction. One means of generating these initial neutrons is a neutron source, such as beryllium-9, and a heavy, alpha-emitting element, such as plutonium or americium. The heavy element emits alpha particles; when they strike a beryllium-9 nucleus, the two nuclei combine to form a carbon-12 nucleus and a neutron is emitted.

$$^{238}_{94}\text{Pu} \longrightarrow {}^{4}_{2}\text{He} + {}^{234}_{92}\text{U}$$

$$^{4}_{2}\text{He} + {}^{9}_{4}\text{Be} \longrightarrow {}^{12}_{6}\text{C} + {}^{1}_{0}\text{n}$$

These neutrons then initiate the nuclear fission of uranium-235 in the reactor core.

The tremendous energy transferred by the nuclear fission reaction in the reactor core heats the primary coolant, a substance with a very high heat capacity, usually water. The primary coolant is at a pressure of more than 150 atm, so it does not boil, even though the temperature is higher than its normal boiling point (← **Sec. 9-3a**). The hot primary coolant is pumped in a closed loop from the nuclear reaction vessel to a heat exchanger, where heat transfer to water that runs the steam generators lowers the temperature of the primary coolant, which then returns to the reactor to be reheated. This closed loop separates the nuclear reactor from the rest of the power plant.

Conventional (non-nuclear) power plants burn fossil fuel (coal or natural gas) to transfer thermal energy that produces steam to drive a turbine.

In the heat exchanger, water in a second loop (the secondary coolant) is vaporized to steam to turn the steam generator turbines. The steam strikes the large turbine blades, causing the turbine to spin. The turbine shaft is connected to the generator, which is surrounded by a magnetic field. The rapid spinning of the turbine shaft in a magnetic field produces electricity.

The efficiency of a steam generator depends on the difference between the highest and lowest temperature of the coolant. Thus, after striking the turbine blades, the steam must be cooled and condensed so that the heating/cooling cycle can be repeated to create additional electricity. To do so, cooling water is pumped from a neighboring river or lake to the secondary coolant loop. Enormous amounts of outside cooling water are needed to condense the vast quantity of steam produced by such power plants. For example, the nuclear power reactor at the Entergy Arkansas Unit 1 uses 750,000 gal of cooling water per minute. To avoid reducing the concentration of dissolved oxygen in its source and thereby harming fish (← **Sec. 13-4b**), the cooling water itself must then be cooled. In many nuclear power plants, such cooling is done by passing the water through large hourglass-shaped cooling towers, in which some water evaporates and cools the rest. Cooling towers are sometimes mistaken for the nuclear reactors themselves.

Cooling towers are also used by fossil fuel–burning power plants.

Not all nuclei can be made to fission on colliding with a neutron, but ^{235}U and ^{239}Pu are both fissionable isotopes. Natural uranium contains an average of only 0.72% of the fissionable ^{235}U isotope. More than 99% of the natural element is nonfissionable

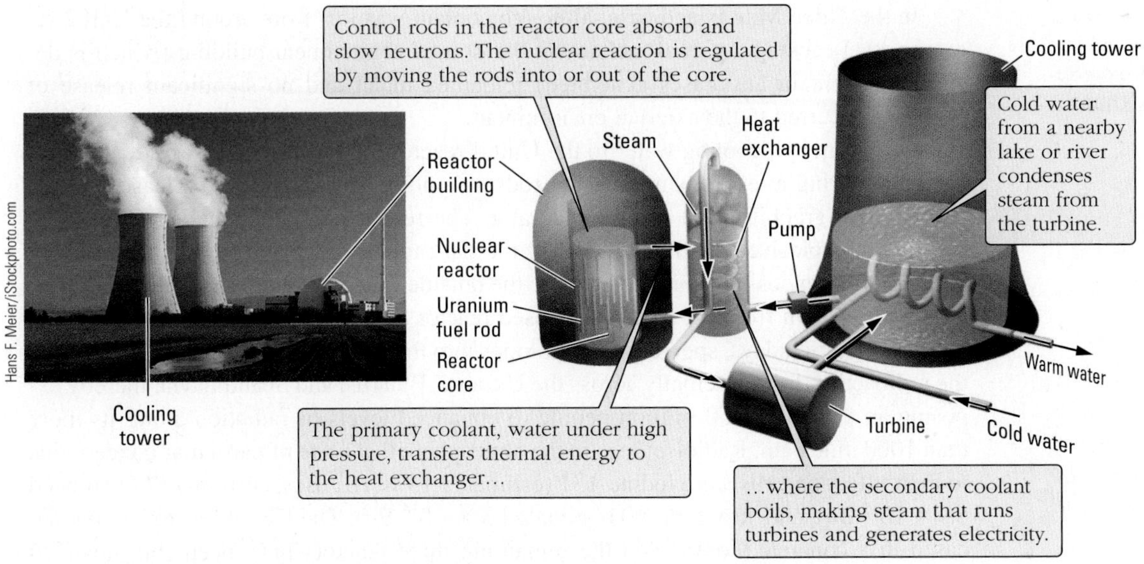
Control rods in the reactor core absorb and slow neutrons. The nuclear reaction is regulated by moving the rods into or out of the core.
Cooling tower
Cold water from a nearby lake or river condenses steam from the turbine.
Heat exchanger
Steam
Reactor building
Nuclear reactor
Uranium fuel rod
Reactor core
Pump
Warm water
Cold water
Turbine
Cooling tower
The primary coolant, water under high pressure, transfers thermal energy to the heat exchanger…
…where the secondary coolant boils, making steam that runs turbines and generates electricity.
Hans F. Meier/iStockphoto.com

Figure 18.9 Photograph and schematic diagram of a nuclear power plant.

uranium-238. Because the percentage of ^{235}U in natural uranium is too small to sustain a chain reaction, nuclear power fuel must be enriched to about 3% uranium-235. To accomplish this goal, some of the ^{238}U isotope in a sample is effectively discarded, thereby raising the concentration of ^{235}U. As the nuclear fission reaction proceeds, the percentage of ^{235}U decreases and the percentage of fission products increases. Approximately one third of the fuel rods in a nuclear reactor are replaced annually because fission by-products absorb neutrons, reducing the efficiency of the fission reactions.

Uranium-238 can fission, but only when bombarded by fast neutrons, unlike those in nuclear reactors. Thus, we consider uranium-238 to be nonfissionable in the context of nuclear reactors.

Because the total mass of uranium-235 in the fuel rods of a nuclear power plant is lower than the critical mass needed for an atomic bomb, the reactor core *cannot* undergo an uncontrolled explosive chain reaction.

Nuclear fission fuels have extremely large energy density (← **Sec. 4-11**). For example, fission of 1.0 kg (2.2 lb) uranium-235 releases 9.0×10^{13} J, the equivalent of exploding 33,000 tons (33 kilotons) TNT. Each UO_2 fuel pellet used in a nuclear reactor has the energy equivalent to burning 136 gallons oil, 2.5 tons wood, or 1 ton coal.

Discussion in this section answers the question, "Can a nuclear power plant explode like an atomic bomb?" that was posed in Chapter 1 (← Sec. 1-1). Uranium is considered to be "weapons-grade" if its ^{235}U content is greater than 90%. Even in reactors using weapons-grade uranium, the ^{235}U is still too dispersed to produce uncontrolled fission.

EXERCISE 18.10

Energy of Nuclear Fission

Burning 1.0 kg high-grade coal produces 2.8×10^4 kJ/mol, whereas fission of 1.0 mol uranium-235 generates 2.1×10^{10} kJ/mol. Calculate how many metric tons of coal (1 metric ton = 10^3 kg) are needed to produce the same quantity of energy as that released by the fission of 1.0 kg uranium-235. (Assume that the processes have equal efficiency.)

18-6b Significant Nuclear Power Plant Accidents

Due to faulty technology, design flaws, operator errors, and natural forces there have been three major commercial nuclear power plant accidents, each different in its own way.

- 1979: Three Mile Island (TMI) near Harrisburg, PA
- 1986: near Chernobyl in the Ukraine
- 2011: Fukushima Daiichi near Fukushima, Japan

The domed roof of a containment building is designed so that it can withstand exceedingly high pressure without rupture.

In the Three Mile Island case, although coolant was lost from around the Unit 2 reactor vessel causing a partial meltdown, the domed containment building (which is designed to contain any gases generated) remained intact and no significant release of radiation occurred to the external environment.

At Chernobyl, cooling water to the Unit 4 reactor (one of four) was accidentally interrupted during a test causing the fuel rods to rupture and steam to be released, which reacted with graphite used as the moderator. The reaction produced hydrogen gas that caught fire explosively, sufficient to blow the flat (non-domed) roof off the containment building, thus exposing the melted core to the outside. Large quantities of radioactive material were blown from the molten, exposed reactor core into the atmosphere. The reactor burned for ten days, spewing radioisotopes into the air where prevailing winds carried the radioactive dust principally across the Ukraine, Belarus, and Scandinavia, thereby exposing an estimated 250 million people to enhanced levels of radiation along its more than 1000 mile path. Radioisotopes in the atmosphere that were of particular concern due to their effect on cells were iodine-131 (estimated 13×10^{17} Bq), cesium-137 (estimated 85×10^{15} Bq), and strontium-90 (estimated 3×10^{14} Bq). The Unit 4 reactor is now encased in a concrete "tomb" and the remaining three reactors have been shut down. It should be pointed out that no commercial nuclear reactors in the United States have the design flaws of the Chernobyl reactors.

The Fukushima Daiichi Nuclear Power Plant (FDNPP) accident, one whose magnitude is on the level of Chernobyl, might be considered as the result of a "perfect storm" of factors. Although designed to withstand a magnitude 8.2 earthquake, the FDNPP was overwhelmed by a magnitude 9.0 earthquake 80 miles from the reactors. Together with its associated tsunami, the earthquake caused severe damage. Generators that supplied electric power to the cooling system pumps were destroyed causing loss of coolant water around the reactor cores. Hydrogen formed by the reaction of steam with the alloy covering the nuclear fuel rods ignited and exploded. The explosion destroyed the outer containment buildings of two reactors and severely damaged a third, causing a massive release of radioactive debris. During the next 25 days, an estimated 150×10^{15} Bq of iodine-131 and $\sim 13 \times 10^{15}$ Bq of cesium-137 were released into the atmosphere. Due to the prevailing westerly winds, nearly 80% was deposited over the northern Pacific Ocean; the remainder was deposited on land, mostly northwest of the site. During this same period, water contaminated with an estimated 4×10^{15} Bq of cesium-137 also leaked from the failed reactors into the ocean. The most contaminated land area lies within 20 km of the FDNPP. Annual radiation doses there exceed 50 mSv, well above the 5 mSv limit for human occupancy and 20 mSv for evacuation. (The millisievert, mSv, is a unit that measures radiation damage to human health.)

Digital Globe/ABACAUSA.COM/Newscom

Fukushima Daiichi Nuclear Power Plant during the accident, March 2011.

In the wake of the FDNPP accident, Japan has abandoned plans to build 14 new reactors, which would have produced 50% of the country's electricity. Germany and Switzerland followed suit. The 17 power reactors in Germany are to be shut down by 2022; plans are to close the five Swiss reactors by 2034.

18-6c Nuclear Power Pros and Cons

There is, of course, substantial controversy surrounding the use of nuclear power plants, and not just in the United States. Nuclear energy proponents argue that the health of our economy and our standard of living depend on inexpensive, reliable, and safe sources of electrical energy. Just within the past few years the demand for electric power has at times exceeded the supply in the United States, and many believe nuclear power plants should be built to meet that demand. Nuclear power plants can be the source of "clean" energy, in that they do not pollute the atmosphere with ash, smoke, or oxides of sulfur, nitrogen, or carbon as coal-fired plants do. In addition, nuclear plants help to ensure that our

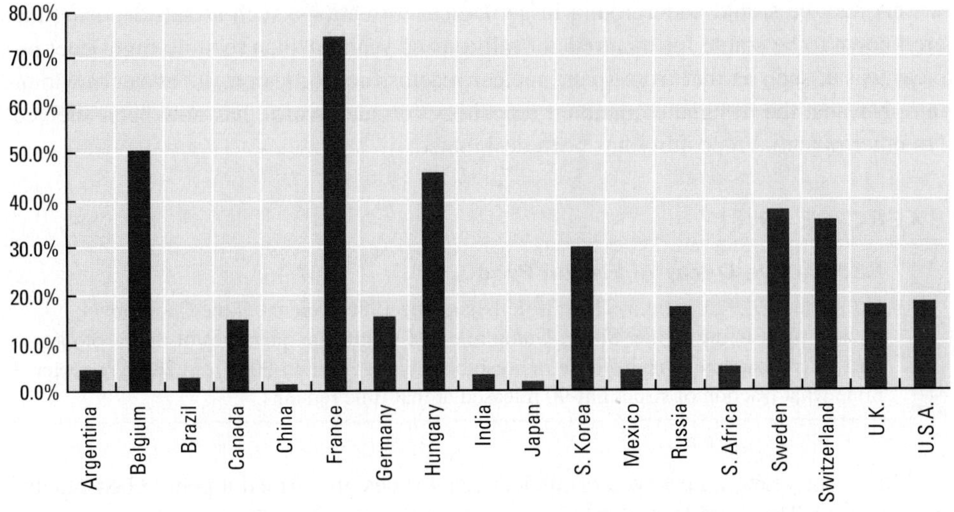
80.0%
70.0%
60.0%
50.0%
40.0%
30.0%
20.0%
10.0%
0.0%
Argentina
Belgium
Brazil
Canada
China
France
Germany
Hungary
India
Japan
S. Korea
Mexico
Russia
S. Africa
Sweden
Switzerland
U.K.
U.S.A.

Figure 18.10 The fraction of electricity generated by nuclear power in selected countries. About 19% of the electricity in the United States is produced by nuclear power. Data from http://www.world-nuclear.org/info/Facts-and-Figures/Nuclear-generation-by-country/#.Ua-dk_nVB8E.

supplies of fossil fuels will not be depleted as quickly in the near future, and they reduce our dependency on buying such fuels from other countries. Currently, more than 100 nuclear power plants operate in the United States, supplying 19% of our nation's electric energy. Worldwide, about 435 nuclear power plants in 31 countries produce 14% of the world's electricity and 12 countries use nuclear power to generate at least 25% of their electricity. As shown in Figure 18.10 there is a wide range in percentage of electricity produced by nuclear plants. In France, for example, 58 nuclear plants generate three out of every four kilowatts of electricity (75%). In contrast, China's 17 reactors generate only 2% of its electricity and Iran's single nuclear plant generates only 0.6% of its electricity.

Five reactors at four different nuclear plants in CA, FL, VT, and WI were scheduled to be shut down (decommissioned) in 2013.

No new nuclear power plants were built in the United States between March 1979, the time of the Three Mile Island accident, and March 2013. In February 2012, the U. S. Nuclear Regulatory Commission (NRC) approved a construction license for two new nuclear power reactors, Vogtle 3 and 4, to be built near Waynesboro, Georgia. This was quickly followed by NRC approval for two additional plants, Summer 2 and 3, to be constructed near Jenkinsville, South Carolina. On March 9, 2013, construction began at the Summer 2 site, the first nuclear power reactor construction in over three decades in the United States. Three days later, construction began at the Vogtle 3 site. To reduce construction time, plans call for large sections of each plant to be prefabricated, shipped to the site, and welded together there. Current schedules call for all four plants to be generating electricity by 2018.

A major problem associated with nuclear power plants is the highly radioactive fission products in fuel rods that have become depleted of ^{235}U (spent fuel). In the United States, tens of thousands of tons of spent fuel waste are being stored, and the quantity is growing steadily. Although some fission products are put to various uses (Section 18-9), many are unsuitable as a nuclear fuel or for other purposes. Because these products are often highly radioactive and some have long half-lives (plutonium-239, $t_{1/2} = 24{,}100$ yr), proper, long-term disposal is essential. One approach is to encase high-level radioactive wastes in a glassy material, a process called vitrification. In 1996, the U. S. Department of Energy Savannah River site near Augusta, Georgia began encapsulating radioactive waste in glass. A mixture of glass particles and radioactive waste is heated to 1200 °C. The molten mixture is poured into stainless steel canisters, cooled, and stored, producing a volume of about 2 m^3 per nuclear reactor per year. Eventually, such high-level nuclear

US Department of Energy

These stainless steel canisters (2 feet in diameter and 10 feet tall) hold high-level radioactive waste that has been vitrified into a glassy solid.

wastes may be stored underground in geological formations, such as salt deposits, that are known to be stable for hundreds of millions of years. A plan to store high-level nuclear waste, such as that from spent nuclear reactor fuel rods, in a site at Yucca Mountain, Nevada, the designated national repository for such waste, has now been shelved. No other potential repository has been designated.

EXERCISE 18.11

Radioactive Decay of Fission Products

Unlike the 1979 incident at Three Mile Island, the 1986 accident at the Chernobyl nuclear plant in the former Soviet Union released significant quantities of radioisotopes into the atmosphere. One of those radioisotopes was strontium-90 ($t_{1/2}$ = 28.8 yr). Determine what fraction of strontium-90 released at that time remains.

In recent years, a new type of nuclear reactor design, termed a pebble bed reactor, has emerged. These are high-temperature, gas-cooled reactors. The fuel elements, "pebbles" the size of a tennis ball, contain approximately 4% fissionable uranium encased in silicon carbide and specially treated (pyrolytic) graphite, which act as the neutron moderators. The core of such a reactor contains hundreds of thousands of these pebbles. They are cooled by a flow of inert gas (helium, nitrogen, carbon dioxide), which is heated and then used to drive turbines to generate power. The gas temperature is in the range of 700 °C to 900 °C, which provides up to 50% thermal efficiency for power production. Pebbles are withdrawn from the pebble bed, tested to see whether they are spent, and replaced with new ones as needed. In this design, the fuel, its containment, and the moderator are all together in the pebble. This design has been advanced as being inherently safe. Even if the gas flow stops, the reactor core cools rather than increasing its temperature. One criticism of the design is that it produces more radioactive waste than traditional designs. A current pebble bed reactor research project is China's HTR-10 reactor at the Institute of Nuclear and New Energy Technology (INET) at Tsinghua University near Beijing. This research reactor has been running since 2000. China's energy plans for the next decade include a major move into the use of nuclear power.

18-7 Nuclear Fusion

Tremendous amounts of energy are generated by **nuclear fusion**, *a nuclear reaction in which very light nuclei combine to form heavier nuclei.* One of the best examples is the fusion of hydrogen nuclei (protons) to give helium nuclei.

$$4\,{}^{1}_{1}\text{H} \longrightarrow {}^{4}_{2}\text{He} + 2\,{}^{0}_{+1}\text{e} \qquad \Delta E = -2.5 \times 10^{9}\ \text{kJ/mol}$$

As shown in Figure 18.4 (← **Sec. 18-3b**), the helium nucleus produced by this reaction is more stable (has higher binding energy per nucleon) than the reactant hydrogen nuclei. This fusion reaction is the source of the energy from our sun and other stars, and it is the beginning of the synthesis of all of the elements in the universe (Section 19-1). Temperatures of 10^6 to 10^7 K, found in the core and radioactive zone of the sun, are required to bring the positively charged ${}^{1}_{1}\text{H}$ nuclei together with enough kinetic energy to overcome nuclear repulsions and to react.

Deuterium can also be fused to give helium-3,

$$ {}^{2}_{1}\text{H} + {}^{2}_{1}\text{H} \longrightarrow {}^{3}_{2}\text{He} + {}^{1}_{0}\text{n} \qquad \Delta E = -3.2 \times 10^{8}\ \text{kJ/mol}$$

which can undergo further fusion with a proton to give helium-4.

$$ {}^{1}_{1}\text{H} + {}^{3}_{2}\text{He} \longrightarrow {}^{4}_{2}\text{He} + {}^{0}_{+1}\text{e} \qquad \Delta E = -1.9 \times 10^{9}\ \text{kJ/mol}$$

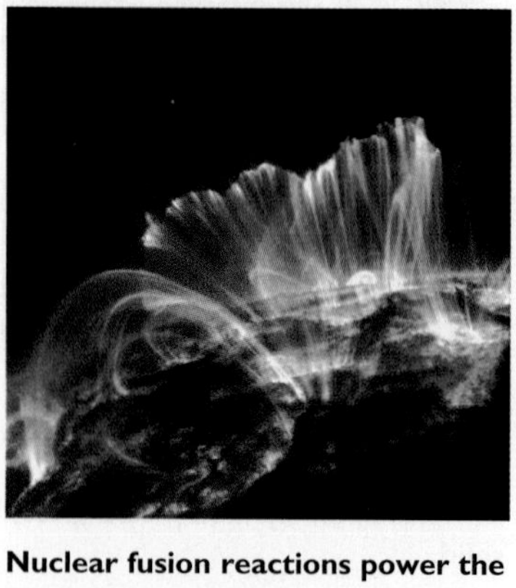
NASA/GSFC

Nuclear fusion reactions power the sun, whose surface is shown here.

Each of these reactions releases an enormous quantity of energy, so it has been the dream of many nuclear physicists to harness them to provide energy for the people of the world.

The activation energy (← Sec. 11-4a) for nuclear fusion is very large because there is very strong repulsion between two positive nuclei when they are brought very close together (Coulomb's law, ← Sec. 2-6a). Thus, very high temperatures are needed for nuclear fusion to occur.

Nuclear fusion is appealing as a commercial energy source because hydrogen isotopes are available (from water), and fusion products are usually nonradioactive or have short half-lives, eliminating the problem of disposal of high-level radioactive waste. However, controlling a nuclear fusion reaction for peaceful commercial uses has proven to be extraordinarily difficult. Three critical requirements must be met to achieve controlled fusion. First, the temperature must be high enough for fusion to occur sufficiently rapidly. The fusion of deuterium and tritium, for example, requires a temperature of at least 100 million K. At the very high temperatures that allow fusion reactions to occur rapidly, atoms do not exist as such. Instead, there is a **plasma**, which *consists of unbound nuclei and electrons*. Second, the plasma must be confined long enough to generate a net output of energy. Third, the energy must be recovered in some usable form.

Magnetic "bottles" (enclosures bounded by magnetic fields) can confine a plasma so that controlled fusion has been achieved. But the energy generated by the fusion has been less than that required to produce the magnetic bottle and control the fusion reaction. Using more energy to produce less energy is not a commercially appealing investment. Thus, commercial fusion reactors are not likely in the near future without a dramatic breakthrough in fusion technology.

Confining a plasma is one of the biggest problems in developing controlled nuclear fusion.

Hydrogen bombs are based on fusion. To achieve the high temperatures required for the nuclear fusion reaction of the hydrogen bomb, a nuclear fission bomb (atomic bomb) is first set off. In one type of hydrogen bomb, lithium-6 deuteride (LiD, a solid salt) is placed around a ^{235}U or ^{239}Pu fission bomb, and the nuclear fission reaction is set off to initiate the process. Lithium-6 nuclei absorb neutrons produced by the fission and split into tritium and helium.

$$^{1}_{0}n + ^{6}_{3}Li \longrightarrow ^{3}_{1}H + ^{4}_{2}He$$

The temperature reached by the fission of uranium or plutonium is high enough to bring about the fusion of tritium and deuterium, accompanied by the release of 1.7×10^9 kJ per mole of ^3H. A 20-megaton hydrogen bomb usually contains about 300 lb LiD, as well as a considerable mass of plutonium and uranium.

EXERCISE 18.12

Nuclear Fusion

Complete the equations for these nuclear fusion reactions.

(a) $^{7}_{3}Li + ^{1}_{1}H \longrightarrow ^{1}_{0}n + \_\_\_\_$ (b) $^{2}_{1}H + \_\_\_\_ \longrightarrow ^{4}_{2}He + ^{1}_{1}H$

18-8 Nuclear Radiation: Effects and Units

The use of nuclear energy and radiation carries both risks and benefits. It can be used to harm (nuclear armaments) or to cure (radioisotopes in medicine). Alpha, beta, and gamma radiation disrupt normal cell processes in living organisms by interacting with key biomolecules (← **Sec. 10-7**), breaking their covalent bonds, and producing energetic free radicals (← **Sec. 6-10b**) and ions that can lead to further disruptive reactions. The potential for serious radiation damage to humans has been well documented, beginning with the biological effects of the atomic bombs exploded at Hiroshima and Nagasaki, Japan, in 1945 at the close of World War II. However, controlled exposure to nuclear radiation can be beneficial in destroying malignant tissue, as in radiation therapy for

All technologies carry risks as well as benefits. In the 1800s railroads were new and the poet William Wordsworth wrote of their risks and benefits in terms of "Weighing the mischief with the promised gain. . . ."

treating some cancers. Assessing risks and benefits requires measurements that depend on carefully defined units.

18-8a Radiation Units

The SI unit of radioactivity is the becquerel; 1 Bq = 1 s^{-1}. Another common unit of radioactivity is the curie; 1 Ci = 3.70×10^{10} s^{-1}. However, to measure the effects of radiation on living organisms, units are needed for radiation dosages that take into account the energy absorbed by tissue when radiation passes into it.

To quantify radiation and its effects, particularly on humans, several units have been developed. The SI unit of absorbed radiation dose is the **gray (Gy)**, which is *equal to the absorption of 1 J per kilogram of material*; 1 Gy = 1 J/kg. The rad (*r*adiation *a*bsorbed *d*ose) also measures the quantity of radiation energy *absorbed;* 1 **rad** *represents a dose of* 1.00×10^{-2} *J absorbed per kilogram of material*. Thus, 1 Gy = 100 rad. Another unit is the **roentgen (R)**, which *corresponds to the deposition of* 93.3×10^{-7} *J per gram of tissue*.

The roentgen is named to honor Wilhelm Röntgen, the German physicist who discovered X-rays.

The biological effects of radiation per rad or gray differ with the kind of radiation. This difference is made quantitative using the **rem** (*r*oentgen *e*quivalent in *m*an), which is *radiation dose in rads multiplied by a quality factor*:

$$\text{Effective dose in rems} = \text{quality factor} \times \text{dose in rads}$$

The quality factor depends on the type of radiation and other factors. It is arbitrarily set as 1 for beta and gamma radiation. It is between 10 and 20 for alpha particles, depending on total dose, dose rate, and type of tissue. Because one rem is a large quantity of radiation, the millirem (mrem) is commonly used (1 mrem = 10^{-3} rem). The SI unit of effective dose is the **sievert (Sv)**, which is *defined similarly to the rem, except that the absorbed dose is in grays*, not rads. Consequently, 1 Sv = 100 rem.

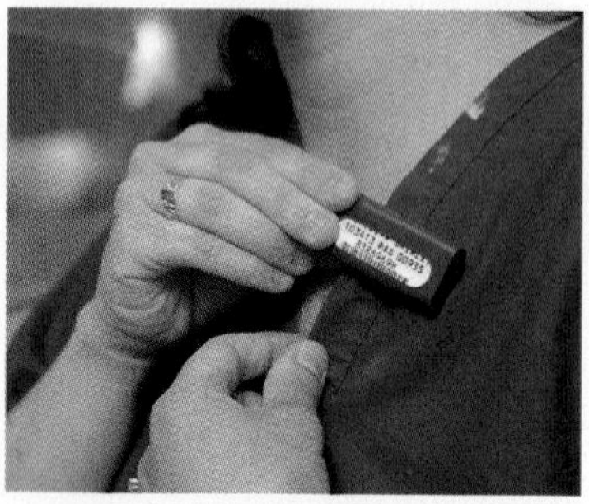

Cliff Moore/Science Source

Individuals who work where there is potential danger from exposure to excessive nuclear radiation wear film badges to monitor their radiation dose.

18-8b Background Radiation

Humans are constantly exposed to natural and artificial **background radiation**, estimated to be collectively about 360 mrem per year (Figure 18.11), well below 500 mrem, the U. S. federal government's background radiation standard for the general public. Note that *most* background radiation, about 300 mrem per year (82%), comes from *natural* background radiation sources: cosmic radiation and radioactive elements and minerals found naturally in soil, water, air, and materials around and within us. The remaining 18% comes from artificial sources.

The greatest quantity of natural background radiation, 55%, comes from radon, a by-product of radium decay; radon is discussed in Section 18-8c. Cosmic radiation, emitted by the sun and other stars, continually impinges on Earth and accounts for about 8% of natural background radiation. The remainder comes from radioactive isotopes such as ^{40}K. Potassium is present to the extent of about 0.3 g per kilogram of soil and is essential to all living organisms. We all acquire some radioactive potassium from the foods we eat. For example, a hamburger contains 112 μg ^{40}K, giving 29 disintegrations per second (29 Bq); a hot dog contains 9 μg ^{40}K and gives off 3 Bq; a serving of french fries has 76 μg ^{40}K and gives off 20 Bq. Other radioactive elements found in some abundance on Earth include thorium-232 and uranium-238. Approximately 8% of the natural background radiation arises from Th-232 and U-238 in rocks and soil. Thorium, for example, is found to the extent of 12 g per 1000 kg soil.

Burning fossil fuels (coal and oil) releases large quantities of naturally occurring radioactive isotopes originally in the fossil fuel and adds significantly to background radiation. Far more thorium and uranium are released annually into the atmosphere from fossil fuel-burning plants than from nuclear power plants, which emit essentially no thorium or uranium.

On average, roughly 15% of our annual exposure comes from medical procedures such as diagnostic X-rays and the use of radioactive compounds to trace the body's functions. Consumer products account for 3% of our total annual exposure. Contrary to popular belief, less than 1% comes from sources such as the radioactive fission products that

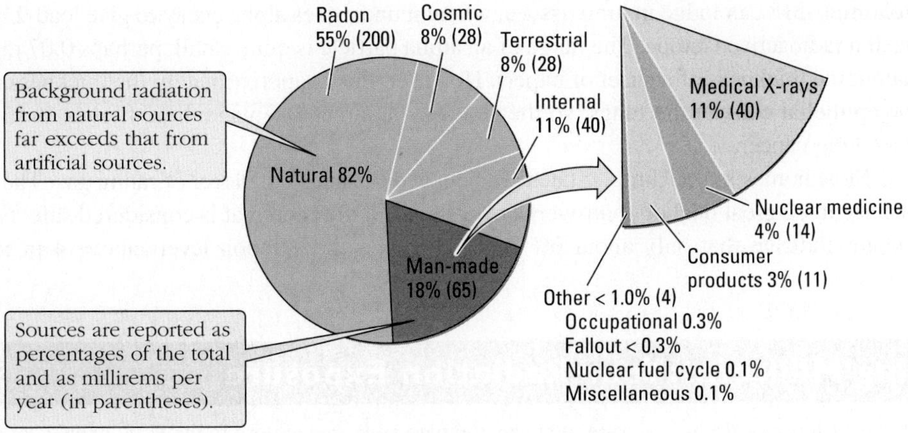
Radon
55% (200)
Cosmic
8% (28)
Terrestrial
8% (28)
Internal
11% (40)
Medical X-rays
11% (40)
Background radiation from natural sources far exceeds that from artificial sources.
Natural 82%
Nuclear medicine
4% (14)
Consumer products 3% (11)
Man-made
18% (65)
Other < 1.0% (4)
Occupational 0.3%
Fallout < 0.3%
Nuclear fuel cycle 0.1%
Miscellaneous 0.1%
Sources are reported as percentages of the total and as millirems per year (in parentheses).

Figure 18.11 Sources of average background radiation exposure in the United States.

remain from testing nuclear explosives in the atmosphere, nuclear power plants and their wastes, nuclear weapons manufacture, and nuclear fuel processing.

A report published by the U. S. National Research Council concludes that there is no safe level of radiation for humans, although the risks of low-dose radiation are small. The researchers studied doses of radiation of 0.1 Sv (10,000 mrem) or more, which is much greater than the 360 mrem per year natural background level discussed above. Rules for nuclear workers and others who are systematically exposed to elevated radiation levels may be affected by this type of research finding.

There are no observable physiologic effects from a single dose of radiation less than 25 rem (25×10^3 mrem).

EXERCISE 18.13

Radioactivity of a Common Food

Bananas are radioactive because they contain potassium-40, a beta and gamma emitter with a half-life of 1.26×10^9 yr. A typical banana contains 500. mg K^+; the natural abundance of ^{40}K is 0.0117%. Calculate the disintegration rate (Bq) for a typical banana. Compare your result with the disintegration rate (3.0×10^4 Bq) of ^{243}Am in a typical household smoke detector. (The molar mass of ^{40}K is 39.96 g/mol.)

18-8c Radon

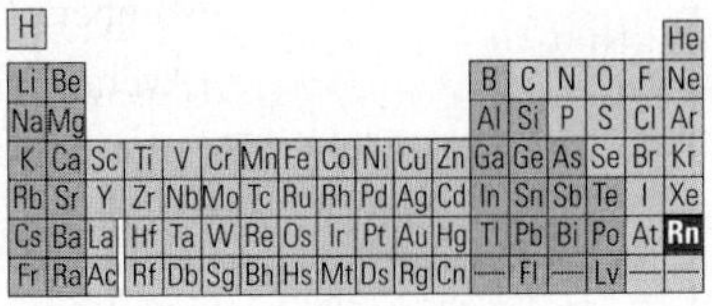
H
He
Li Be
B C N O F Ne
Na Mg
Al Si P S Cl Ar
K Ca Sc Ti V Cr Mn Fe Co Ni Cu Zn Ga Ge As Se Br Kr
Rb Sr Y Zr Nb Mo Tc Ru Rh Pd Ag Cd In Sn Sb Te I Xe
Cs Ba La Hf Ta W Re Os Ir Pt Au Hg Tl Pb Bi Po At Rn
Fr Ra Ac Rf Db Sg Bh Hs Mt Ds Rg Cn — Fl — Lv — —

Radon, one of the noble gases in the same periodic table group as helium, neon, argon, krypton, and xenon, accounts for 55% of natural background radiation. Radon-222 is produced in the decay series of naturally occurring uranium-238 (← **Sec. 18-2b**). Other isotopes of radon are products of other decay series.

Radon occurs naturally in our environment. Because it comes from natural uranium deposits, the quantity of radon depends on the nature of the rocks and soil in a given locality. Furthermore, because the gas is chemically inert, it is not trapped by chemical processes in the soil or water. Thus, radon gas is free to seep up from the ground and into underground mines or into homes through pores in concrete block walls, cracks in basement floors or walls, and around pipes. Radon-222 decays to give polonium-218, a radioactive, heavy-metal element that is not a gas but, unlike radon-222, is not chemically inert.

Although chemically inert, radon is problematic because as a gas it can be inhaled. If radon is inhaled, decay to polonium occurs deep in the lungs.

$$^{222}_{86}\text{Rn} \longrightarrow {}^{4}_{2}\text{He} + {}^{218}_{84}\text{Po} \qquad t_{1/2} = 3.82 \text{ days}$$

$$^{218}_{84}\text{Po} \longrightarrow {}^{4}_{2}\text{He} + {}^{214}_{82}\text{Pb} \qquad t_{1/2} = 3.10 \text{ minutes}$$

Polonium-218 can lodge in lung tissues, where it undergoes alpha decay to give lead-214, itself a radioactive isotope. The range of an alpha particle is quite small, perhaps 0.07 mm (about the thickness of a sheet of paper). However, this is approximately the thickness of the epithelial cells of the lungs, so the alpha radiation can damage these tissues and induce lung cancer.

An online radiation calculator is available at http://www.epa.gov/radiation/understand/calculate.html.

Most homes in the United States are believed to have some level of radon gas. There is currently a great deal of controversy over the level of radon that is considered safe. Estimates indicate that only about 6% of U. S. homes have radon levels above 4 pCi/L

ESTIMATION | Counting Millirems: Your Radiation Exposure

In 1987 the Committee on Biological Effects of Ionizing Radiation of the U. S. National Academy of Sciences issued a report that contained a survey for an individual to evaluate his or her exposure to ionizing radiation. The table below is adapted from this report and updated. By adding up your exposure, you can compare your annual dose to the U. S. annual average of 360 mrem.

Source: Based on the BEIR Report III. National Academy of Sciences, Committee on Biological Effects of Ionizing Radiation. *The Effects on Populations of Exposure to Low Levels of Ionizing Radiation.* Washington, DC: National Academy of Sciences, 1987.

Common Sources of Radiation		Your Annual Dose (mrem)
Where you live	**Location:** Cosmic radiation at sea level	27
	For your elevation (in feet), add this number of mrem	______
	Elevation / *mrem*: 1000 – 2; 2000 – 5; 3000 – 9; 4000 – 15; 5000 – 21; 6000 – 29; 7000 – 40; 8000 – 53; 9000 – 70	
	Ground: U. S. average	26
	Radon: U. S. average	200
	House construction: For stone, concrete, or masonry building, add 7; for wood, add 30	______
What you eat, drink, and breathe	**Radioisotopes** in the body from	
	Food, air, water: U. S. average	40
	Weapons test fallout	4
How you live	**X-ray and radiopharmaceutical diagnosis**	
	Number of chest X-rays ____ × 10	______
	Number of lower gastrointestinal tract X-rays ____ × 500	______
	Number of radiopharmaceutical examinations ____ × 300	______
	(Average dose to total U. S. population = 53 mrem)	
	Jet plane travel: For each 2500 miles add 1 mrem	______
	TV viewing: Number of hours per day ____ × 0.15	______
How close you live to a nuclear plant	**At site boundary:** Average number of hours per day ____ × 0.2	______
	One mile away: Average number of hours per day ____ × 0.02	______
	Five miles away: Average number of hours per day ____ × 0.002	______
	Over five miles away: None	______
	Note: Maximum allowable dose determined by "as low as reasonably achievable" (ALARA) criteria established by the U. S. Nuclear Regulatory Commission. Experience shows that your actual dose is substantially less than these limits.	
	Your total annual dose in mrem:	______

Elevation table:

Elevation	*mrem*	*Elevation*	*mrem*	*Elevation*	*mrem*
1000	2	4000	15	7000	40
2000	5	5000	21	8000	53
3000	9	6000	29	9000	70

Compare your annual dose with the U. S. annual average of 360 mrem.

(picocurie per liter) of air, the action level standard set by the U. S. Environmental Protection Agency (1 pCi = 1×10^{-12} Ci). Some believe that 1.5 pCi/L is more likely the average level and that only about 2% of the homes will contain more than 8 pCi/L. To test for the presence of radon, you can purchase testing kits of various kinds. If your home shows higher levels of radon gas than 4 pCi/L, you should probably have it tested further and perhaps take corrective actions such as sealing cracks around the foundation and in the basement. But, keep in mind the relative risks involved. A 1.5 pCi/L level of radon leads to a lung cancer risk about the same as the risk of your dying in an accident in your home.

PRO-LAB
RATED #1
PROFESSIONAL
RADON GAS
TEST KIT
• SAFE, EASY TO USE
• TWO DETECTORS FOR ACCURATE TESTING*
• MEETS ALL EPA AND STATE REQUIREMENTS
$15.00
Cengage Learning/Charles D. Winters

A commercially available kit to test for radon gas in a residence.

EXERCISE 18.14

Radon Levels

A home is contaminated with radon-222, $t_{1/2}$ = 3.82 days. The activity is measured as 16 pCi. If the home's basement is sealed to prevent additional radon from entering, calculate the time required for the activity to drop to

(a) 4 pCi, the EPA action level.
(b) 1.5 pCi, approximately the U. S. average.

18-9 Applications of Radioactivity

18-9a Food Irradiation, Cold Pasteurization

In some parts of the world, spoilage of stored food may claim up to 50% of the food crop. In developed nations, this figure is lowered considerably by refrigeration, canning, and preservatives. Still, there are problems with food spoilage, and food preservation costs are a substantial fraction of the final cost of food. Food and grains can be preserved by gamma irradiation, also known as cold pasteurization. Contrary to some popular opinion, such irradiation does *not* make foods radioactive, any more than a dental X-ray makes you radioactive.

The radura, the international symbol for irradiated food.

Food irradiation with gamma rays from ^{60}Co or ^{137}Cs sources is allowed in 40 countries and is endorsed by the World Health Organization and the American Medical Association. Astronauts' food has been preserved by gamma irradiation. The United States and several other countries require that foods preserved by irradiation be labeled with the international symbol for irradiated food, the radura.

Bacteria, molds, and yeasts are killed or their growth is retarded by irradiation with gamma rays. As a result, the shelf life of irradiated foods during refrigeration is prolonged in much the same way that heat pasteurization protects milk. In recent years, outbreaks of food-borne illnesses caused by new types of harmful bacteria or inappropriate food handling have heightened interest in the benefits of irradiation as a safety measure, especially for use with meat.

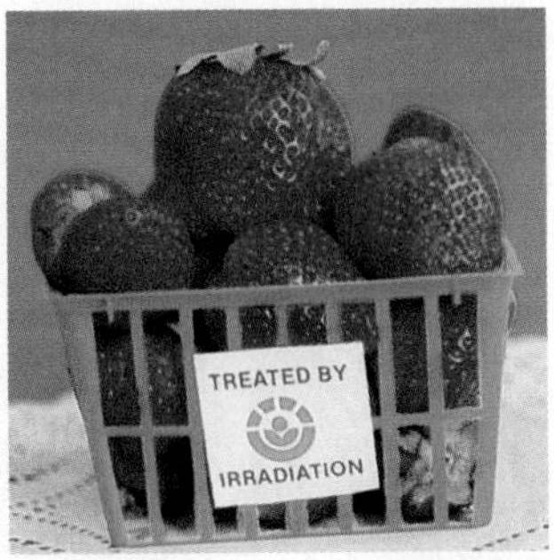
TREATED BY
IRRADIATION
Courtesy of Food Technology Service, Inc.

Strawberries preserved by gamma irradiation.

The U. S. Food and Drug Administration (FDA) has approved irradiation of meat, poultry, and a variety of fresh fruits and vegetables and spices. Except for spices, however, irradiated foods are not yet widely available.

The FDA permits irradiation up to 300 kilorads for the pasteurization of meat and poultry. Radiation levels in the 1- to 5-Mrad range (1 Mrad = 10^6 rads) sterilize, killing every living organism. Foods irradiated at these levels will keep indefinitely when sealed in plastic or aluminum foil packages. However, FDA approval is unlikely for irradiation sterilization of foods in the near future because of potential problems caused by as-yet-undiscovered, but possible, "unique radiolytic products." For example, irradiation sterilization might produce a substance that is capable of causing genetic damage. To prove or disprove the presence of these substances, animal feeding studies using foods sterilized by irradiation are now being conducted in the United States.

In addition to preserving foods, gamma irradiation is used to sterilize bandages, contact lens solutions, and many cosmetics.

Table 18.3 Radioisotopes Used as Tracers

Name	Isotope	Half-Life	Use
Carbon-14	^{14}C	5730 years	CO_2 for photosynthesis research
Tritium	^{3}H	12.33 years	Tag hydrocarbons
Sulfur-35	^{35}S	87.2 days	Tag pesticides, measure air flow
Phosphorus-32	^{32}P	14.3 days	Measure phosphorus uptake by plants

18-9b Radioactive Tracers

The chemical behavior of a radioisotope is essentially identical to that of the nonradioactive isotopes of the same element. Compounds containing radioactive atoms undergo chemical reactions in exactly the same way as compounds containing the same nonradioactive atoms. Therefore, chemists can use radioactive isotopes as **tracers**, *radioisotopes used to track the pathway of a radioactive element in a chemical or biological process.* To use a tracer, a chemist prepares a reactant compound in which one of the elements consists of both radioactive and stable (nonradioactive) isotopes, and uses the compound in a reaction (or feeds it to an organism). After the reaction, the chemist measures the radioactivity of the products (or determines which parts of the organism contain the radioisotope) using a Geiger counter or other radiation detector. Several radioisotopes commonly used as tracers are listed in Table 18.3.

Melvin Calvin used ^{14}C to monitor the uptake and release of $^{14}CO_2$ to determine the basic biochemical pathways of photosynthesis. This groundbreaking work earned him the 1961 Nobel Prize in Chemistry.

For example, plants take up phosphorus-containing compounds from the soil through their roots. The use of the radioactive phosphorus isotope ^{32}P, a beta emitter, provides a way to detect the uptake of phosphorus by a plant, as well as to measure the speed of uptake under various conditions. Plant biologists can grow hybrid strains of plants that quickly absorb phosphorus, an essential nutrient. They can test the new plants by measuring their uptake of the radioactive ^{32}P tracer. This type of research can lead to faster-maturing crops, better yields per acre, and more food or fiber at less expense.

18-9c Medical Imaging

Using radioactive isotopes to diagnose and treat disease is called **nuclear medicine**. In the diagnosis of internal disorders such as tumors, physicians need information on the locations of abnormal tissue. This identification is done by imaging, a technique in which the radioisotope, either alone or combined with some other substance, accumulates at the site of the disorder. There, the radioisotope decays, emitting its characteristic radiation, which is then detected. Modern medical diagnostic instruments not only determine where the radioisotope is located in the patient's body, but also construct an image that shows where the radioisotope is concentrated in the body.

Eight of the most common diagnostic radioisotopes are described in Table 18.4. Most are created in particle accelerators, devices in which heavy, charged nuclear particles bombard other target atoms. Each of these radioisotopes produces gamma radiation, which, in low doses, is less harmful to tissue than internal ionizing radiations such as beta or alpha particles because gamma rays pass through the tissue without being absorbed. When combined with special carrier compounds, these radioisotopes can be made to accumulate in specific areas of the body. For example, the pyrophosphate ion, $P_4O_7^{4-}$, can bond to the technetium-99m radioisotope. Together they accumulate in the skeletal structure where abnormal bone metabolism is occurring (Figure 18.12). The technetium-99m radioisotope is metastable, as denoted by the letter m; this term means that the nucleus is in an excited state and can lose energy by changing to a lower energy version of the same isotope and emitting a photon (γ-ray).

The metastable ^{99m}Tc is in an excited nuclear energy state. This is analogous to a H atom in which an electron is in an excited atomic energy state (← Sec. 5-3a).

$$^{99m}Tc \longrightarrow {}^{99}Tc + \gamma$$

Images produced by detecting the gamma rays often pinpoint bone tumors.

Table 18.4 Diagnostic Radioisotopes

Radio Isotope	Name	Mode of Decay	Half-Life (hours)	Diagnostic Site
^{99m}Tc*	Technetium-99m	Gamma	6.0	Thyroid, as $^{99m}TcO_4^-$
^{18}F	Fluorine-18	Positron (97%); electron capture (3%)	1.83	Tumors
^{133}Xe	Xenon-133	Beta	126	Pulmonary ventilation
^{111}In	Indium-111	Electron capture	67.3	Central nervous system
^{201}Tl	Thallium-201	Electron capture	73.0	Cardiac function
^{13}N	Nitrogen-13	Alpha and positron	0.17	Cardiovascular
^{51}Cr	Chromium-51	Gamma	665	Kidney function
^{68}Ga	Gallium-68	Positron (90%); electron capture (10%)	1.13	Tumors, abscesses

*Technetium-99m is the most commonly used diagnostic radioisotope; the m stands for "metastable."

EXERCISE 18.15

Comparing Half-Lives

Chromium-51 is a radioisotope ($t_{1/2}$ = 27.7 days) used to evaluate the lifetime of red blood cells; the radioisotope iron-59 ($t_{1/2}$ = 44.5 days) is used to assess bone marrow function. A hospital laboratory has 80. mg iron-59 and 100. mg chromium-51. After 90 days, which radioisotope is present in greater mass?

Positron emission tomography (PET) is a form of nuclear imaging that uses positron emitters, such as carbon-11, fluorine-18, nitrogen-13, or oxygen-15. These radioisotopes are neutron-deficient, have short half-lives, and therefore must be prepared in a cyclotron immediately before use. When they decay, a proton is converted into a neutron, a positron, and a neutrino; the neutrino is usually not shown in the equation.

The neutrino, first observed experimentally in 1956, is a subatomic particle with zero electrical charge and a mass smaller than that of an electron.

$$^{1}_{1}p \longrightarrow {}^{1}_{0}n + {}^{0}_{+1}e$$

Because matter is essentially transparent to neutrinos, they escape undetected. However, the positron, ${}^{0}_{+1}e$, travels on average less than a few millimeters before it encounters an electron, ${}^{0}_{-1}e$, and undergoes antimatter-matter annihilation.

$${}^{0}_{+1}e + {}^{0}_{-1}e \longrightarrow 2\,\gamma$$

The annihilation event produces two gamma rays that move in opposite directions and are detected by detectors located 180° apart in the PET scanner. Several million annihilation gamma rays can be detected within a circular field around the subject over approximately 10 minutes. Computerized signal-averaging techniques applied to these data generate an image of the region of tissue containing the radioisotope (Figure 18.13).

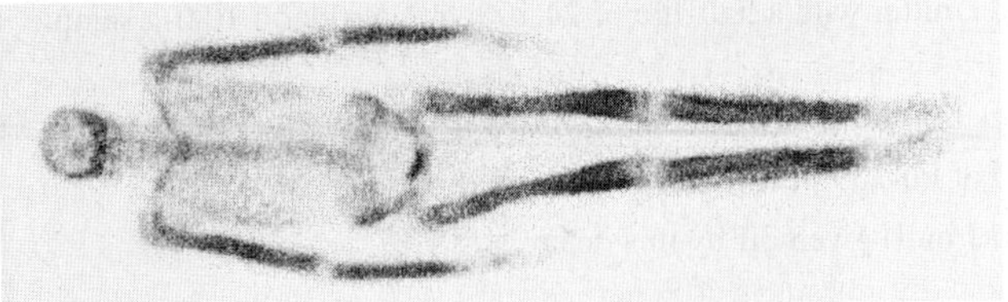

Susan Law Cain/iStockphoto.com

Figure 18.12 A whole-body scan. Phosphate with technetium-99m was injected into the blood and then absorbed by the bones and kidneys. This picture was taken three hours after injection.

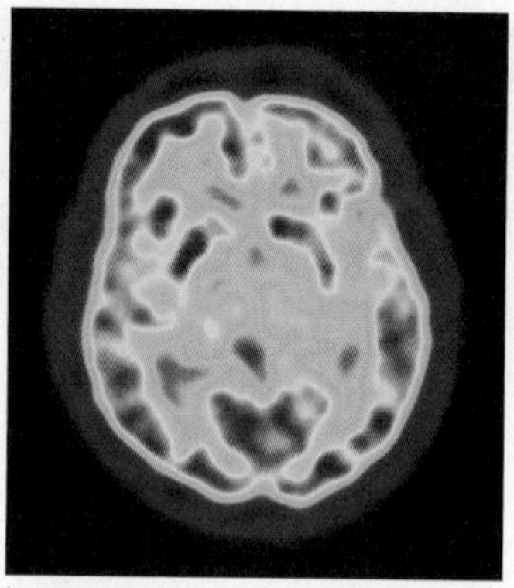

1. This PET image of an axial section through a normal human brain was obtained by injecting the patient with a derivative of glucose that contained fluorine-18.

2. Glucose is the energy source for the brain (←Sec. 16-9a) so glucose must be supplied to regions that are most actively metabolizing.

3. The glucose derivative containing fluorine-18 accumulates in active areas of the brain. These areas are colored red and orange in this image.

4. Fluorine-18 emits positrons, which react with electrons generating two γ-rays that are recorded by a circular detector and converted to a color image by computer.

Image courtesy of Dr. Jens Langner [1]. It can be found on Wikipedia [2]. [1] http://jens-langner.de/
[2] http://commons.wikipedia.org/wiki/File:PET-image.jpg

Figure 18.13 PET (positron emission tomography) scan of a normal human brain.

18-9d Medical Treatment Involving Radiation

Paradoxically, high-energy radiation, which can kill healthy cells, is used therapeutically to kill malignant, cancerous cells—those exhibiting rapid, uncontrolled growth. Because they divide more rapidly than normal cells, malignant cells are more susceptible to radiation damage. Thus, malignant cells are more likely to be killed than normal cells. For external radiation therapy, a narrow beam of high-energy gamma radiation from a cobalt-60 or cesium-137 source is focused on the cancerous cells. Internal radiation therapy uses gamma-emitting salts of radioisotopes such as ^{192}Ir ($t_{1/2}$ = 73.8 days). The radioactive salts are encapsulated in platinum or gold "seeds" or needles and surgically implanted into the body. Because the thyroid gland uses iodine, thyroid cancer can be treated internally by oral administration of a sodium iodide solution containing a relatively high concentration of radioactive iodine-131 as iodide ion.

SUMMARY PROBLEM

1. Plutonium-239 can be produced in so-called "breeder reactors" through a sequence of nuclear reactions that involve absorption of a neutron by uranium-238 followed by two consecutive beta emissions. Write a balanced nuclear equation for each step of this three-step process.

2. Plutonium-239 is converted to fissionable uranium-235 by alpha emission. Write the balanced nuclear equation for this process.

3. When uranium-235 undergoes fission in a nuclear reactor, neutrons and a mixture of by-products form; one pair of products is barium-144 and krypton-89. Write the balanced nuclear equation for this fission reaction.

4. The fission of plutonium-239 can be represented by this equation:

$$^{239}_{94}\text{Pu} + {}^{1}_{0}\text{n} \longrightarrow {}^{146}_{58}\text{Ce} + {}^{90}_{38}\text{Sr} + 2\,{}^{0}_{-1}\text{e} + 4\,{}^{1}_{0}\text{n}$$

   (a) Calculate the mass (g) of plutonium-239 that would need to react to produce 100. kJ of energy.

   (b) Calculate how many plutonium-239 atoms would have to fission to produce this quantity of energy.

5. Plutonium-239 is an alpha emitter with a half-life of 24,100 yr. Consider a 10.0-g sample of pure Pu-239.

   (a) Write the balanced equation for the nuclear decay of Pu-239.

   (b) Calculate the mass (g) of Pu-239 decay in 1000. yr.

   (c) Calculate the energy (kJ/mol) given off by this decay.

(d) Calculate the time (s) required for the plutonium-239 activity to decrease to 10.0% of its initial activity.

(e) Calculate the radiation dosage (rem) received by a 70.0-kg person exposed to the initial sample of Pu-239 for 30.0 days. Assume that the quality factor in this case is 20.

(f) Deep underground burial is one method proposed for long-term storage of high-level nuclear waste such as plutonium-239 from nuclear weapons and spent fuel rods from commercial nuclear power reactors. Comment on factors that need to be considered for such burial and storage.

HAVING STUDIED THIS CHAPTER . . .

. . . you should be able to:

- Characterize the three major types of radiation observed in radioactive decay: alpha (α), beta (β), and gamma (γ) (Section 18-1). End-of-chapter Question 10
- Write a balanced equation for a nuclear reaction or transmutation (Section 18-2). Questions 14, 16, 20
- Decide whether a particular radioactive isotope will decay by α, β, positron emission, or electron capture (Sections 18-2, 18-3). Question 21
- Calculate the binding energy for a particular isotope and understand what this energy means in terms of nuclear stability (Section 18-3). Question 23
- Calculate the half-life of a radioactive isotope ($t_{1/2}$) or decay constant (k) from the activity of a sample at two different times, or use the half-life or decay constant to find the time required for an isotope to decay to a particular activity (Section 18-4). Questions 28, 32, 36, 38
- Describe transmutation of elements and write balanced nuclear equations for transmutation reactions (Section 18-5). Questions 40, 42
- Describe nuclear chain reactions, nuclear fission, and nuclear fusion (Sections 18-6, 18-7). Questions 49, 55, 57
- Describe the basic functioning of a nuclear power reactor (Section 18-6). Question 53
- Describe some sources of background radiation and the units used to measure radiation (Section 18-8). Questions 59, 63
- Give examples of some uses of radioisotopes (Section 18-9). Questions 65, 67

KEY TERMS

activity (*A*) (Section 18-4b)
alpha (α) particle (18-1)
alpha (α) radiation (18-1)
background radiation (18-8b)
becquerel (Bq) (18-4b)
beta (β) particle (18-1)
beta (β) radiation (18-1)
binding energy (18-3b)
binding energy per nucleon (18-3b)
critical mass (18-6)
curie (Ci) (18-4b)
electron capture (18-2c)
gamma (γ) radiation (18-1)
gray (Gy) (18-8a)
half-life, $t_{1/2}$ (18-4a)
nuclear fission (18-6)
nuclear fusion (18-7)
nuclear medicine (18-9c)
nuclear reaction (18-2a)
nuclear reactor (18-6a)
nucleon (18-2a)
plasma (18-7)
positron (18-2c)
rad (18-8a)
radioactive series (18-2b)
rem (18-8a)
roentgen (R) (18-8a)
sievert (Sv) (18-8a)
tracers (18-9b)

QUESTIONS FOR REVIEW AND THOUGHT

Red-numbered questions have short answers at the back of this book in Appendix M and fully worked solutions in the *Student Solutions Manual.*

Review Questions

These questions test vocabulary and simple concepts.

1. Compare nuclear and chemical reactions in terms of changes in reactants, type of products formed, and conservation of matter and energy.
2. What is meant by the "peninsula of stability"?
3. Describe the basis for and define the binding energy of a nucleus.
4. If the mass number of an isotope is much greater than twice the atomic number, what type of radioactive decay might you expect?
5. If the number of neutrons in an isotope is much less than the number of protons, what type of radioactive decay might you expect?
6. Define critical mass and chain reaction.
7. What is the difference between nuclear fission and nuclear fusion? Illustrate your answer with an example of each.
8. Use the Internet to locate the nuclear reactor power plant nearest to your campus residence. Do you consider the reactor to pose a threat to your health and safety? If so, why? If not, why not?
9. Name at least two uses of radioactive isotopes (outside of their use in power reactors and weapons).

Topical Questions

These questions are keyed to the major topics in the chapter. Usually a question that is answered at the back of this book is paired with a similar one that is not.

The Nature of Radioactivity (Section 18-1)

10. Complete the table.

	Symbol	Mass	Charge
α particle	______	______	______
β particle	______	______	______
γ radiation	______	______	______

11. Complete the table.

	Ionizing Power	Penetrating Power
α particle	______	______
β particle	______	______
γ radiation	______	______

Nuclear Reactions (Section 18-2)

12. By what processes do these transformations occur?
    (a) Thorium-230 to radium-226
    (b) Cesium-137 to barium-137
    (c) Potassium-38 to argon-38
    (d) Zirconium-97 to niobium-97
13. By what processes do these transformations occur?
    (a) Uranium-238 to thorium-234
    (b) Iodine-131 to xenon-131
    (c) Nitrogen-13 to carbon-13
    (d) Bismuth-214 to polonium-214
14. Fill in the mass number, atomic number, and symbol for the missing particle in each nuclear equation.
    (a) $^{242}_{94}\text{Pu} \longrightarrow {}^{4}_{2}\text{He} + \square$
    (b) $\square \longrightarrow {}^{32}_{16}\text{S} + {}^{0}_{-1}\text{e}$
    (c) $^{252}_{98}\text{Cf} + \square \longrightarrow 3\,{}^{1}_{0}\text{n} + {}^{259}_{103}\text{Lr}$
    (d) $^{55}_{26}\text{Fe} + \square \longrightarrow {}^{55}_{25}\text{Mn}$
    (e) $^{15}_{8}\text{O} \longrightarrow \square + {}^{0}_{+1}\text{e}$
15. Fill in the mass number, atomic number, and symbol for the missing particle in each nuclear equation.
    (a) $\square \longrightarrow {}^{22}_{10}\text{Ne} + {}^{0}_{+1}\text{e}$
    (b) $^{122}_{53}\text{I} \longrightarrow {}^{122}_{54}\text{Xe} + \square$
    (c) $^{210}_{84}\text{Po} \longrightarrow \square + {}^{4}_{2}\text{He}$
    (d) $^{195}_{79}\text{Au} + \square \longrightarrow {}^{195}_{78}\text{Pt}$
    (e) $^{241}_{94}\text{Pu} + {}^{16}_{8}\text{O} \longrightarrow 5\,{}^{1}_{0}\text{n} + \square$
16. Write a balanced nuclear equation for each word statement.
    (a) Magnesium-28 undergoes β emission.
    (b) When uranium-238 is reacted with carbon-12, four neutrons are emitted and a new element forms.
    (c) Hydrogen-2 and helium-3 react to form helium-4 and another particle.
    (d) Argon-38 forms by positron emission.
    (e) Platinum-175 forms osmium-171 by spontaneous radioactive decay.
17. Write a balanced nuclear equation for each word statement.
    (a) Einsteinium-253 combines with an alpha particle to form a neutron and a new element.
    (b) Nitrogen-13 undergoes positron emission.
    (c) Iridium-178 captures an electron to form a stable nucleus.
    (d) A proton and boron-11 fuse, forming three identical particles.
    (e) Nobelium-252 and six neutrons form when carbon-12 collides with a transuranium isotope.
18. One radioactive series that begins with uranium-235 and ends with lead-207 undergoes this sequence of emission reactions: α, β, α, β, α, α, α, α, β, β, α. Identify the radioisotope produced in each of the *first five steps.*
19. One radioactive series that begins with uranium-235 and ends with lead-207 undergoes this sequence of emission reactions: α, β, α, β, α, α, α, α, β, β, α. Identify the radioisotope produced in each of the *last six steps.*

20. Radon-222 is unstable and its presence in homes may constitute a health hazard. It decays by this sequence of emissions: α, α, β, β, α, β, β, α. Write the sequence of nuclear reactions leading to the final product nucleus, which is stable.

Stability of Atomic Nuclei (Section 18-3)

21. Write a nuclear equation for the type of decay each of these unstable isotopes is most likely to undergo.
(a) Neon-19 (b) Thorium-230
(c) Bromine-82 (d) Polonium-212
22. Write a nuclear equation for the type of decay each of these unstable isotopes is most likely to undergo.
(a) Silver-114 (b) Sodium-21
(c) Radium-226 (d) Iron-59
23. Boron has two stable isotopes, ^{10}B (abundance = 19.78%) and ^{11}B (abundance = 80.22%). Calculate the binding energies per nucleon of these two nuclei and compare their stabilities.

$$5\ {}^1_1H + 5\ {}^1_0n \longrightarrow {}^{10}_{5}B$$

$$5\ {}^1_1H + 6\ {}^1_0n \longrightarrow {}^{11}_{5}B$$

The required masses (in g/mol) are ${}^1_1H$ = 1.00783; ${}^1_0n$ = 1.00867; ${}^{10}_{5}B$ = 10.01294; and ${}^{11}_{5}B$ = 11.00931.
24. Calculate the binding energy in kJ per mole of P for the formation of ${}^{30}_{15}P$ and ${}^{31}_{15}P$

$$15\ {}^1_1H + 15\ {}^1_0n \longrightarrow {}^{30}_{15}P$$

$$15\ {}^1_1H + 16\ {}^1_0n \longrightarrow {}^{31}_{15}P$$

Which is the more stable isotope? The required masses (in g/mol) are ${}^1_1H$ = 1.00783; ${}^1_0n$ = 1.00867; ${}^{30}_{15}P$ = 29.97832; and ${}^{31}_{15}P$ = 30.97376.
25. The most abundant isotope of uranium is ^{238}U, which has an isotopic mass of 238.0508 g/mol. Calculate its nuclear binding energy in kJ/mol and its binding energy per nucleon.
26. Calculate the nuclear binding energy in kJ/mol and binding energy per nucleon of chlorine-35, which has an isotopic mass of 34.9689 g/mol.
27. Calculate the nuclear binding energy and binding energy per nucleon in kJ/mol of iodine-127, which has an isotopic mass of 126.9045 g/mol.

Rates of Disintegration Reactions (Section 18-4)

28. Sodium-24 is a diagnostic radioisotope used to measure blood circulation time. Calculate the mass (mg) of a 20-mg sample of sodium-24 that remains after 1 day and 6 hours if sodium-24 has $t_{1/2}$ = 15 hours.
29. Iron-59 in the form of iron(II) citrate is used in iron metabolism studies. Its half-life is 44.5 days. If you start with 0.56 mg iron-59, calculate the mass (mg) that remains after 1 year.
30. Iodine-131 is used in the form of sodium iodide to treat cancer of the thyroid.
(a) The isotope decays by ejecting a β particle. Write a balanced equation to show this process.
(b) The isotope has a half-life of 8.04 days. If you begin with 25.0 mg of radioactive $Na^{131}I$, calculate the mass (mg) that remains after 32.2 days.
31. Phosphorus-32 is used in the form of $Na_2H^{32}PO_4$ in the treatment of chronic myeloid leukemia.
(a) The isotope decays by emitting a β particle. Write a balanced equation to show this process.
(b) The half-life of ^{32}P is 14.3 days. If you begin with 9.6 mg radioactive $Na_2H^{32}PO_4$, calculate what mass (mg) remains after 28.6 days.
32. Calculate the half-life of a radioisotope if it decays to 12.5% of its radioactivity in 12 years.
33. After 2 hours, tantalum-172 has $\frac{1}{16}$ of its initial radioactivity. Calculate its half-life (s).
34. Radioisotopes of iodine are widely used in medicine. For example, iodine-131 ($t_{1/2}$ = 8.04 days) is used to treat thyroid cancer. If you ingest a sample of $Na^{131}I$, calculate the time required for the isotope to decrease to 5.0% of its original activity.
35. The noble gas radon has been the focus of much attention because it may be found in homes. Radon-222 emits α particles and has a half-life of 3.82 days.
(a) Write a balanced equation to show this process.
(b) Calculate the time required for a sample of radon to decrease to 10.0% of its original activity.
36. A sample of wood from a Thracian chariot found in an excavation in Bulgaria has a ^{14}C activity of 11.2 disintegrations per minute per gram. Estimate the age of the chariot and the year it was made. ($t_{1/2}$ for ^{14}C is 5.73×10^3 years, and the activity of ^{14}C in living material is 15.3 disintegrations per minute per gram.)
37. A piece of charred bone found in the ruins of a Native American village has a $^{14}C/^{12}C$ ratio of 0.72 times that found in living organisms. Calculate the age of the bone fragment. (See Question 36 for required data on carbon-14.)
38. Calculate the time (s) required for a sample of plutonium-239 with a half-life of 2.4×10^4 years to decay to 1% of its original activity.
39. A 1.00-g sample of wood from an archaeological site gave 4100 disintegrations of ^{14}C in a 10-hour measurement. In the same time, a 1.00-g modern sample gave 9200 disintegrations. Calculate the age of the wood.

Artificial Transmutations (Section 18-5)

40. Fluorine-18, a diagnostic radioisotope used to evaluate cardiovascular conditions, is made by reaction of oxygen-18 with a proton. A neutron is also produced in addition to fluorine-18. Write a balanced nuclear equation to represent this process.

41. Nitrogen-13, a diagnostic radioisotope used to detect tumors, is made by reaction of oxygen-16 with a proton. An alpha particle is also produced in addition to nitrogen-13. Write a balanced nuclear equation to represent this process.
42. In 2006, Alexander Litvinenko, a vocal critic of the Putin government in Russia, was poisoned in London with a lethal dose of polonium-210 injected by Russian agents. Polonium-210 is synthesized by the uptake of a neutron by bismuth-209. Write a balanced nuclear equation to represent this process.
43. In 1998, researchers in Dubna, Russia, synthesized element 112, copernicium, by reaction of uranium-238 nuclei with calcium-48 nuclei. The copernicium-283 isotope was produced along with neutrons. Write a balanced nuclear equation to represent this synthesis.
44. The synthesis of ${}^{292}_{116}Lv$ has been attempted by trying to merge calcium-40 and californium-249 nuclei, but has not yet succeeded. Write a balanced nuclear equation to represent this attempted synthesis.
45. There are two isotopes of americium, both with half-lives sufficiently long to allow the handling of large quantities. Americium-241, an alpha emitter, has a half-life of 433 years; it is used in gauging the thickness of materials and in smoke detectors. This isotope is formed from ${}^{239}Pu$ by absorption of two neutrons followed by emission of a β particle. Write a balanced equation for this process.
46. Americium-240 is made by reacting plutonium-239 atoms with α particles. In addition to ${}^{240}Am$, the products are a proton and two neutrons. Write a balanced equation for this process.
47. To synthesize the heavier transuranium elements, one must react a lighter nucleus with a relatively large particle. If you know that the products of such a reaction are californium-246 and four neutrons, what particle must have reacted with uranium-238 atoms?
48. The officially named element with the highest atomic number is livermorium, ${}^{293}_{116}Lv$, named to honor the long history of the synthesis of post-uranium elements at the Lawrence Livermore National Laboratory in Berkeley, CA. In attempts to make elements with a higher atomic number than 116, reactions have been attempted between californium-249 and calcium-48. Determine the atomic number of the element that would be formed.

Nuclear Fission and Fusion (Sections 18-6, 18-7)

49. Elements are formed by nuclear fusion reactions. One step in this process is so-called "helium burning", the fusion of two helium-4 nuclei to produce beryllium-8. Write a balanced nuclear equation to represent this process.
50. Elements are formed by nuclear fusion reactions including so-called "carbon burning", the fusion of two carbon-12 nuclei to produce sodium-23 and a proton. Write a balanced nuclear equation to represent this process.
51. Elements are formed by nuclear fusion reactions such as so-called "carbon and oxygen burning", the fusion of a carbon-12 nucleus with an oxygen-16 nucleus to produce silicon-28. Write a balanced nuclear equation to represent this process.
52. Under proper conditions, fusion of a helium-3 nucleus with a proton produces helium-4 plus a neutron. Write a balanced nuclear equation to represent this process.
53. Name the fundamental parts of a nuclear fission reactor and describe their functions.
54. Explain why it is easier for a nucleus to capture a neutron than for a nucleus to capture a proton.
55. Determine the missing product in each of these fission equations.
    (a) ${}^{235}_{92}U + {}^{1}_{0}n \longrightarrow \square + {}^{93}_{38}Sr + 3\,{}^{1}_{0}n$
    (b) ${}^{235}_{92}U + {}^{1}_{0}n \longrightarrow \square + {}^{132}_{51}Sb + 3\,{}^{1}_{0}n$
    (c) ${}^{235}_{92}U + {}^{1}_{0}n \longrightarrow \square + {}^{141}_{56}Ba + 3\,{}^{1}_{0}n$
56. Explain why no commercial fusion reactors are in operation today.
57. The average energy output of a barrel of oil is 5.9×10^6 kJ/barrel. Fission of 1 mol ${}^{235}U$ releases 2.1×10^{10} kJ/mol. Calculate the number of barrels of oil needed to produce the same energy as 1.0 lb ${}^{235}U$.
58. A concern in the nuclear power industry is that, if nuclear power becomes more widely used, there may be serious shortages in worldwide supplies of fissionable uranium. One solution is to build breeder reactors that manufacture more fuel than they consume. One such cycle works as follows:
    (i) A ${}^{238}U$ nucleus collides with a neutron to produce ${}^{239}U$.
    (ii) ${}^{239}U$ decays by β emission ($t_{1/2}$ = 24 minutes) to give an isotope of neptunium.
    (iii) The neptunium isotope decays by β emission to give a plutonium isotope.
    (iv) The plutonium isotope is fissionable. On its collision with a neutron, fission occurs and gives energy, at least two neutrons, and other nuclei as products.

    Write an equation for each of these steps, and explain how this process can be used to breed more fuel than the reactor originally contained and still produce energy.

Nuclear Radiation: Effects and Units (Section 18-8)

59. In 2012, a radiation reading of 73 sieverts (Sv) was recorded in the containment structure at the Fukushima Daiichi reactor 2. Assuming a quality factor of 20. convert this reading to:
    (a) rads.
    (b) grays (Gy).
    (c) mrem.
60. The average annual dose of ionizing radiation in the United States is 360 mrem. Convert this value to
    (a) sieverts (Sv).
    (b) rads (assume a quality factor of 10.).
    (c) grays (Gy).

61. Two common units of radiation used in newspaper and news magazine articles are the rad and rem. What does each measure? Which would you use in an article describing the damage an atomic bomb would inflict on a human population? What relationship does the gray have with these units?
62. Which electrical power plant—fossil fuel or nuclear—exposes a community to more nuclear radiation? Explain why.
63. Explain how our own bodies are sources of nuclear radiation.
64. Describe the main source of radiation exposure during jet plane travel.

Applications of Radioactivity (Section 18-9)

65. Why are foods irradiated with gamma rays instead of alpha or beta particles?
66. X-rays and PET scans are two medical imaging techniques. How are they similar and how are they different?
67. To measure the volume of the blood system of an animal, this experiment was done. A 1.0-mL sample of an aqueous solution containing tritium with an activity of $2.0 \times 10^6\ s^{-1}$ was injected into the bloodstream. After time was allowed for complete circulatory mixing, a 1.0-mL blood sample was withdrawn and found to have an activity of $1.5 \times 10^4\ s^{-1}$. Calculate the volume of the circulatory system. (The half-life of tritium is 12.3 years, so this experiment assumes that only a negligible quantity of tritium has decayed during the experiment.)
68. Cobalt-60 is a therapeutic radioisotope used in treating certain cancers. A sample of cobalt-60 initially disintegrates at a rate of $4.3 \times 10^6\ s^{-1}$ and after 21.2 years the rate has dropped to $2.6 \times 10^5\ s^{-1}$. Calculate the half-life of cobalt-60.

General Questions

These questions are not explicitly keyed to chapter topics; many require integration of several concepts.

69. Complete these nuclear equations.
    (a) $^{214}Bi \longrightarrow \square + {}^{214}Po$
    (b) $4\ {}^1_1H \longrightarrow \square$ + 2 positrons
    (c) ^{249}Es + neutron $\longrightarrow$ 2 neutrons + $\square + {}^{161}Gd$
    (d) $^{220}Rn \longrightarrow \square$ + alpha particle
    (e) ^{68}Ge + electron $\longrightarrow \square$
70. Complete these nuclear equations.
    (a) $\square$ + neutron $\longrightarrow$ 2 neutrons + $^{137}Tc + {}^{97}Zr$
    (b) $^{45}Ti \longrightarrow \square$ + positron
    (c) $\square \longrightarrow$ beta particle + ^{59}Co
    (d) ^{24}Mg + neutron $\longrightarrow \square$ + proton
    (e) $^{131}Cs + \square \longrightarrow {}^{131}Xe$
71. Radioactive nitrogen-13 has a half-life of 10. minutes. Calculate the mass of this isotope that remains after an hour in a sample that originally contained 96 mg.
72. The half-life of molybdenum-99 is 67.0 hours. Calculate how much of a 1.000-mg sample of ^{99}Mo is left after 335 hours. Determine how many half-lives it underwent.
73. The oldest known fossil cells were found in South Africa. The fossil has been dated by the reaction

$$^{87}_{37}Rb \longrightarrow {}^{87}_{38}Sr + {}^{0}_{-1}e \qquad t_{1/2} = 4.9 \times 10^{10}\ \text{years}$$

    If the ratio of the present quantity of ^{87}Rb to the original quantity is 0.951, calculate the age of the fossil cells.
74. Balance these equations used for the synthesis of transuranium elements.
    (a) $^{238}_{92}U + {}^{14}_{7}N \longrightarrow \square + 5\ {}^1_0n$
    (b) $^{238}_{92}U + \square \longrightarrow {}^{249}_{100}Fm + 5\ {}^1_0n$
    (c) $^{253}_{99}Es + \square \longrightarrow {}^{256}_{101}Md + {}^1_0n$
    (d) $^{246}_{96}Cm + \square \longrightarrow {}^{254}_{102}No + 4\ {}^1_0n$
    (e) $^{252}_{98}Cf + \square \longrightarrow {}^{257}_{103}Lr + 5\ {}^1_0n$
75. On December 2, 1942, the first man-made self-sustaining nuclear fission chain reactor was operated by Enrico Fermi and others under the University of Chicago stadium. In June 1972, natural fission reactors, which operated billions of years ago, were discovered in Oklo, Gabon. At present, natural uranium contains 0.72% $^{235}_{92}U$. How many years ago did natural uranium contain 3.0% $^{235}_{92}U$, sufficient to sustain a natural reactor? ($t_{1/2}$ for $^{235}_{92}U = 7.04 \times 10^8$ years.)
76. Element 117 was synthesized by collision of calcium-48 nuclei with berkelium-249 nuclei. Two isotopes of element 117 were formed—one when three neutrons were emitted; the other when four neutrons were emitted for each 117 nucleus formed.
    (a) Write a balanced nuclear equation to represent the formation of each isotope.
    (b) The lighter isotope of element 117 undergoes three successive alpha particle emissions. Write a series of balanced nuclear equations to represent this sequence.
    (c) The heavier isotope of element 117 undergoes six successive alpha particle emissions. Write a series of balanced nuclear equations to represent this sequence.
    (d) Identify the group of the periodic table to which element 117 belongs.
    (e) What name is commonly used for that group?
77. Element 118 was first synthesized by collision of calcium-48 nuclei with californium-249 nuclei. Three neutrons were emitted for each 118 nucleus formed.
    (a) Write a balanced nuclear equation to represent this process.
    (b) The element 118 nucleus underwent three successive alpha particle emissions. Write a series of balanced nuclear equations to represent this sequence.
    (c) Identify the group of the periodic table to which element 118 belongs.
    (d) What name is commonly used for that group?

78. Chemical elements are formed in the Sun by step-wise nuclear fusion reactions. In one step of the process, ^3He (molar mass 3.0160297) is formed by the fusion of a deuteron, ^2_1H, and a proton, ^1_1H.
(a) Calculate ΔE (in kilojoules per gram) of deuterium fused.
(b) Water contains approximately 0.016% deuterium. The total mass of water in the world's oceans is 1.4×10^{24} g. Estimate how much energy would be available by the fusion of hydrogen atoms with all the deuterium atoms in seawater.
(c) The world's annual energy requirement is approximately 4.7×10^{17} kJ. Calculate the percent of deuterium in seawater that would need to be fused to meet this energy requirement.

79. In June 2009, researchers at the Joint Institute of Nuclear Research in Dubna, Russia, began continuous bombardment of a 22-mg target of berkelium-249 atoms with calcium-48 nuclei in an attempt to synthesize element 117. The bombardment continued until January 2010 when evidence indicated that a few atoms of the new element had been synthesized. The half-life of berkelium-249 is 330 days. Assuming that the reaction period was 240 days, calculate the mass of berkelium-249 remaining.

80. A 60-kg woman has an average of 145 g of potassium in her body. Of this potassium, 0.0117% is radioactive potassium-40, which has a half-life of 1.2×10^9 yr. Calculate the radioactivity (Bq) emitted from the potassium-40.

81. The combustion of methane gas and the conversion of sodium-23 to sodium-24 by adding a neutron are both exothermic, but they are very different processes. The first is a chemical reaction; the latter is a nuclear transformation.
(a) Use standard formation enthalpy data to calculate the combustion enthalpy (kJ/mol) of methane and convert this to the energy per molecule.
(b) Use the equation $E = (\Delta m)c^2$ to calculate the energy change for the conversion of sodium-23 to sodium-24 in terms of kJ/mol and convert this to energy per nucleus. (Molar masses: ^{23}Na, 22.9897697; ^{24}Na, 23.9909633.)
(c) Compare the energy of the two reactions in terms of energy per gram.

82. The mass of the Sun is about 2×10^{30} kg. The Sun is mostly hydrogen and it emits energy at a rate of about 4×10^{26} J/s, principally by the fusion of protons into helium-4 nuclei. Calculate how long it would take for the Sun to lose 50% of its mass of ^1H at this rate.

Applying Concepts

These questions test conceptual learning.

83. In Chapter 12 it was stated that chemical equilibrium is dynamic. Consider this equilibrium

$$CH_3COOH(aq) + H_2O(\ell) \rightleftharpoons CH_3COO^-(aq) + H_3O^+(aq)$$

Choose a radioisotope that could be used as a tracer to demonstrate that this equilibrium is dynamic. Explain how an experiment using the tracer could show the equilibrium is dynamic.

84. Radioactive isotopes are often used as "tracers" to follow an atom through a chemical reaction. Acetic acid reacts with methanol, CH_3OH, by eliminating a molecule of H_2O to form methyl acetate, CH_3COOCH_3. Explain how you would use the radioactive isotope ^{18}O to show whether the oxygen atom in the water product comes from the —OH of the acid or the —OH of the alcohol.

$$\underset{\text{acetic acid}}{CH_3COOH} + \underset{\text{methanol}}{CH_3OH} \longrightarrow \underset{\text{methyl acetate}}{CH_3COOCH_3} + H_2O$$

85. If a radioisotope is used for diagnosis (e.g., detecting cancer), it should decay by gamma radiation. However, if its use is therapeutic (e.g., treating cancer), it should decay by alpha or beta radiation. Explain why in terms of ionizing and penetrating power.

86. During the Three Mile Island incident, people in central Pennsylvania were concerned that strontium-90 (a beta emitter) released from the reactor could become a health threat (it did not). Where would this isotope collect in the body? If so, what types of problems could it cause?

87. Classify the isotopes ^{17}Ne, ^{20}Ne, and ^{23}Ne as stable or unstable. What type of decay would you expect the unstable isotope(s) to have?

88. This demonstration was carried out to illustrate the concept of a nuclear chain reaction. Explain the connections between the demonstration and the reaction.

Eighty mousetraps are arranged side by side in eight rows of ten traps each. Each trap is set with two rubber stoppers for bait. A small plastic mouse is tossed into the middle of the traps, setting off one trap, which in turn sets off two traps and so on until all the traps are sprung.

89. Most students have no trouble understanding that 1.5 g of a 24-g sample of a radioisotope would remain after 8 h if it had $t_{1/2} = 2$ h. What they don't always understand is where the other 22.5 g went. How would you explain this disappearance to another student?

90. Nuclear chemistry is a topic that raises many debatable issues. Briefly discuss your views on the following.
(a) Twice a year the general public is allowed to visit the Trinity Site in Alamogordo, New Mexico, where the first atomic bomb was tested. If you had the opportunity to do so, would you visit the site? Explain your answer.
(b) Now that the Cold War has ended, should the United States stockpile nuclear weapons? Explain your answer.
(c) The practice of sterilizing food by irradiation is controversial in the United States. What are some of the possible advantages and disadvantages of this practice? Explain your answer.

More Challenging Questions

These questions require more thought and integrate several concepts.

91. The activation energy and the temperature required for fusion of two deuterons (deuterium nuclei) can be estimated based on Coulomb's law (← **Sec. 2-6a**). Coulomb's law can be used to derive an equation from which the energy required to bring the two nuclei close enough to form an alpha particle can be calculated. The equation is $E = (2.307 \times 10^{-28}\text{ J m}) \times Q_1 \times Q_2/r$, where Q_1 and Q_2 are the charges on the nuclei in units of proton charge (Q for a proton is $+1$) and r is the distance between the nuclei. A reasonable estimate for r is the radius of an alpha particle, 1×10^{-14} m.
    (a) Estimate the activation energy for fusion of two deuterons.
    (b) Assuming that one deuteron is not moving, use the equation for kinetic energy (← **Sec. 4-1**) to calculate the minimum speed for the second deuteron that would produce a collision with greater than the fusion activation energy.
    (c) The high temperature required for nuclear fusion can be estimated by assuming that the minimum speed for the second deuteron calculated in part (b) is the average speed of all deuterons in a plasma and using an equation derived from the kinetic-molecular theory of gases (← **Sec. 8-2**). The equation is $v_{ave} = (3RT/M)^{1/2}$, where R is the gas constant, T is the absolute temperature, and M is the molar mass. Use this equation to calculate the minimum temperature required for deuterium fusion.
92. All radioactive decays are first order. Why is this so?
93. A sample of the alpha emitter $^{222}_{86}Rn$ had an initial activity, A_0, of 7.00×10^4 Bq. After 10.0 days its activity, A, had fallen to 1.15×10^4 Bq. Calculate the decay constant and half-life of radon-222.
94. When a wine was analyzed for its tritium, ^3_1H, content, it was found to contain 1.45% of the tritium originally present when the wine was produced. Determine the age of the wine. ($t_{1/2}$ of ^3_1H = 12.3 years.)
95. A chemist is setting up an experiment using $^{47}_{20}Ca$, which has a half-life of 4.5 days. She needs 10.0 μg of the calcium. Calculate the minimum mass (μg) of $^{47}_{20}CaCO_3$ she must order if the delivery time is 50 hours.
96. To determine the age of the charcoal found at an archaeological site, this sequence of experiments was done: The charcoal sample was burned in excess oxygen and the CO_2 obtained was captured by bubbling it through lime water, $Ca(OH)_2$, to form a precipitate of $CaCO_3$. This precipitate was filtered, dried, and weighed. A sample of 1.14 g $CaCO_3$ produced 2.17×10^{-2} Bq from carbon-14. Modern carbon-14 produces 15.3 disintegrations per minute per gram. Calculate the age of the charcoal. (The half-life of carbon-14 is 5730 years.)

Conceptual Challenge Problems

These rigorous, thought-provoking problems integrate conceptual learning with problem solving and are suitable for group work.

CP18.A (Section 18-4) The half-life for the alpha decay of uranium-238 to thorium-234 is 4.5×10^9 years, which happens to be the estimated age of Earth.
(a) Calculate how many atoms were decaying per second in a 1.0-g sample of uranium-238 that existed 1.0×10^6 years ago.
(b) Explain how you would find the number of atoms now decaying per second in this sample.

CP18.B (Section 18-4) If Earth is 4.5×10^9 years old and the quantity of radioactivity in a sample becomes smaller with time, how is it possible for there to be any radioactive elements on Earth that have half-lives less than a few million years?

CP18.C (Section 18-4) Using experiments based on a sample of living wood, a nuclear chemist estimates that the uncertainty of her measurements of the carbon-14 radioactivity in the sample is 1.0%. The half-life of carbon-14 is 5730 years.
(a) Determine how long a sample of wood must be separated from a living tree before the chemist's radioactivity measurements on the sample provide credible evidence for the time when the tree died.
(b) Suppose that the chemist's uncertainty in the radioactivity of carbon-14 continues to be 1.0% of the radioactivity of living wood. Determine how long a sample of wood must be dead before the chemist's measurements support the claim that the time since the wood was separated from the tree is not changing.

CP18.D (Section 18-8) You have read that alpha radiation is the least penetrating type of radiation, followed by beta radiation. Gamma radiation penetrates matter well, and thick samples of matter are required to contain gamma radiation. Knowing these facts, what can you correctly deduce about the harmful effects of these three types of radiation on living tissue?

CP-18.E (Section 18-8) Death will probably occur within weeks for a 150-lb person who receives 500,000 mrem of radiation over a short time, an exposure that is 1000 times the U. S. government's standard for 1 year (500 mrem/yr). A student realizes that 500,000 mrem is 500 rem, and that 500 rem has the effect of depositing 317 J in the body of the 150-lb person. The student is puzzled. How can the deposition of only 317 J from nuclear radiation—much less energy than that deposited by cooling a cup of coffee 1 °C within a person's body—have such a disastrous effect on the person? Describe how you would explain this to the puzzled student.

19 | The Chemistry of the Main-Group Elements

NASA/Alamy

A Hubble Space Telescope photograph of Cassiopeia A, the remnants of the youngest supernova in the Milky Way galaxy. Chemical elements are formed when a star explodes as a supernova. Bright green indicates areas richest in oxygen; red and purple indicate sulfur; and blue areas are mainly hydrogen and nitrogen. What processes form chemical elements in exploding supernovas? (Section 19-1) How do heavier elements form from lighter elements? (Section 19-1) Which elements are most abundant in Earth's crust? (Section 19-2) How are elements obtained from their natural sources? (Sections 19-3 to 19-5) These questions are answered in this chapter and much other information is provided about ten selected main-group elements.

Elements have been discussed throughout this book. Their names, symbols, physical properties, and chemical reactivity have been noted, demonstrating their enormous diversity. Some elements, such as sodium and fluorine, react violently with many other substances; other elements, the noble gases, are so quiescent that they enter into very few or no chemical combinations. In spite of this diversity, elements in each group of the periodic table have predictable chemical similarities based on their number of valence electrons.

A fundamental question not yet discussed concerns the origin of the elements: How did they form? How, for example, did magnesium atoms acquire a different number of protons from atoms of calcium or helium? This chapter will answer such questions. It will also describe how 10 selected main-group elements—those in Groups 1A through 8A—are extracted from natural sources, the chemical principles associated with the extraction processes, and the properties of those three metallic and seven nonmetallic elements. The chapter concludes with an overview of the main-group elements, those in Groups 1A through 8A, describing their properties and uses, and relating these to the periodic table. The transition elements, all metals, are discussed in Chapter 20.

"I now mean by Elements. . . certain Primitive and Simple, or perfectly unmingled bodies; which not being made of other bodies, or of one another, are the ingredients of which all those call'd perfectly mixt Bodies are immediately compounded, and into which they are ultimately resolved. . ."

Robert Boyle, *The Sceptical Chymist*, 1661

19-1 Formation of the Elements

Cosmologists—scientists who study the formation of the universe—use spectral evidence as well as knowledge of nuclear reactions to develop theories about the origin of the universe. Cosmologists theorize that about 13.8 billion years ago all the matter in the universe was contained in a pinpoint-sized region that expanded at inconceivably high temperatures, estimated to be about 10^{32} K, in what is called the "Big Bang." This produced a universe expanding so quickly that in one second it cooled to 10^9 K, a temperature at which the fundamental subatomic particles formed—neutrons, protons, and electrons. Within two hours, the temperature had dropped to 10^7 K, temperatures suitable for the formation of light nuclei—^{2}H, ^{3}He, and ^{4}He. Over a much longer time gravitational attraction caused some matter to contract and form stars where temperatures became high enough for nuclear reactions to occur.

Expansion is a cooling process; contraction is a heating process.

19-1a Nuclear Burning

The elements from hydrogen to iron are formed within stars by **nuclear burning**, *a sequence of nuclear fusion reactions* (← **Sec. 18-7**) not to be confused with chemical combustion. The *fusion of protons (hydrogen-1 nuclei) to form helium-4 nuclei* is called **hydrogen burning**.

This section answers part of the question, "How many chemical elements are known and how are they formed?" that was posed in Chapter 1 (← Sec. 1-1).

$$4\,{}^1_1\text{H} \longrightarrow {}^4_2\text{He} + 2\,{}^{\;\,0}_{+1}\text{e} + 2\,\gamma$$

After billions of years of hydrogen burning, the star contracts and the core becomes sufficiently dense and hot for **helium burning**, *the fusion of helium-4 nuclei*, to occur.

Terms such as "nuclear burning" arise from the fact that nuclear fusion reactions are highly exothermic and that we think of the sun and other stars as burning in the sky.

$${}^4_2\text{He} + {}^4_2\text{He} \longrightarrow {}^8_4\text{Be}$$

Beryllium-8 is unstable, but by fusing with another helium-4 nucleus, beryllium-8 is converted to stable carbon-12.

$${}^8_4\text{Be} + {}^4_2\text{He} \longrightarrow {}^{12}_{\;6}\text{C}$$

The low natural abundance of beryllium is evidence of the instability of beryllium-8 (← **Sec. 18-3a**).

When helium burning stops, the star contracts further with the result that the temperature becomes high enough for heavier nuclei to form by fusion. Starting with carbon-12, three successive fusions with helium-4 nuclei form oxygen-16, then neon-20, and then magnesium-24. The process continues up to the formation of calcium-40. Carbon and oxygen burning also occurs:

Nuclear fusion reactions have very high activation energy barriers (← Sec. 11-4a) because two positively charged nuclei repel each other strongly. Thus, high temperatures and high kinetic energies of colliding particles are necessary to overcome the high activation energy barrier. The greater the nuclear charge, the higher the energy barrier and the higher the temperature required to overcome the barrier.

$${}^{12}_{\;6}\text{C} + {}^{12}_{\;6}\text{C} \longrightarrow {}^{23}_{11}\text{Na} + {}^1_1\text{H}$$

$${}^{12}_{\;6}\text{C} + {}^{16}_{\;8}\text{O} \longrightarrow {}^{28}_{14}\text{Si}$$

Starting with silicon-28, fusion reactions build up heavier nuclei all the way to iron-56 and nickel-58, which are very stable nuclei with the highest binding energies per nucleon

(← **Sec. 18-3b**). Elements heavier than iron and nickel are less stable and cannot be formed by such nuclear fusion reactions.

Answers to **EXERCISES** are provided at the back of this book in Appendix L.

EXERCISE 19.1

Fusing Nuclei

Write balanced nuclear equations representing the formation of oxygen-16, neon-20, and magnesium-24 starting from carbon-12 and helium-4 nuclei.

19-1b Formation of Heavier Elements

Figure 19.1 indicates the amount of time a star spends in the various stages of elemental synthesis in relation to the central temperature of the star. Following helium burning, elements heavier than iron form by neutron capture in massive stars in which the core collapses rapidly. Stable nuclei such as those of iron decompose into neutrons and protons, and the protons are converted into additional neutrons by capturing electrons. The result is the formation of a neutron star whose outer layers explode away as a supernova.

Heavier elements form during supernova generation by one of two processes. In the *s* (slow) *process,* the slow capture of neutrons takes place over thousands of years. Because this capture shifts the neutron/proton ratio, eventually the nuclei become beta emitters. As noted in Section 18-2b, beta emission causes an increase in the atomic number, so a new element is formed from the parent nucleus. Such is the case for ${}^{98}_{42}Mo$, for example, which is converted to technetium-99 by this two-step process.

$$ {}^{98}_{42}Mo + {}^{1}_{0}n \longrightarrow {}^{99}_{42}Mo $$

$$ {}^{99}_{42}Mo \longrightarrow {}^{99}_{43}Tc + {}^{0}_{-1}e $$

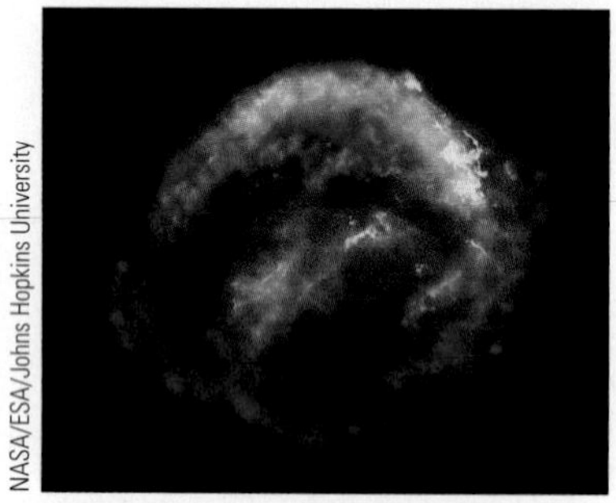

NASA/ESA/Johns Hopkins University

Kepler's supernova remnant. A bubble-shaped cloud of gas and dust envelops Kepler's supernova remnant. This fast-moving material from an exploded star (supernova) is partly composed of newly synthesized chemical elements.

Isotopes with mass numbers up to 209 can form by the *s process.*

Some elements, including the radioactive actinides, are produced by the very rapid *r process,* which occurs during the explosive stage of a star. Because the *r* process is very fast, new elements can be produced from nuclei with very short half-lives—too short to be produced by the *s* process. In the *r* process, a nucleus captures many neutrons in an extremely short time (0.01 to 10 s) in a series of reactions that produce a nucleus much heavier than the original one. Suppose, for example, that ${}^{130}_{48}Cd$ is produced in the *r* process. This isotope is highly unstable because it has far too many neutrons; ${}^{116}_{48}Cd$ is the most stable known isotope of cadmium. The ${}^{130}_{48}Cd$ nucleus can undergo a rapid series of beta decays, increasing in atomic number until it reaches ${}^{130}_{52}Te$, the most abundant isotope of tellurium.

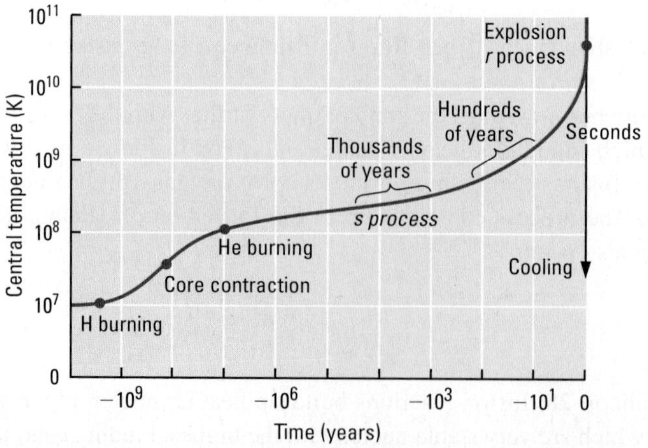

Central temperature (K)
10^11
10^10
10^9
10^8
10^7
0
Explosion
r process
Hundreds of years
Seconds
Thousands of years
s process
He burning
Cooling
Core contraction
H burning
−10^9
−10^6
−10^3
−10^1
0
Time (years)

Figure 19.1 The time scale for various stages of elemental syntheses in stars.

EXERCISE 19.2

Cadmium-130 Decay

Cadmium-130 decays to tellurium-130 through four consecutive beta decays. Write balanced equations for these reactions, starting with cadmium-130 and ending with tellurium-130.

19-2 Terrestrial Elements

In this section and Sections 19-3 to 19-5, we describe how nitrogen, oxygen, sulfur, sodium, chlorine, magnesium, aluminum, phosphorus, bromine, and iodine are obtained from their natural sources; we also consider the properties and uses of these elements.

Nitrogen and oxygen are obtained from the atmosphere. Ocean water is treated to extract commercial quantities of magnesium, bromine, and sodium chloride. Earth's crust is an indispensable source of most of the other elements. Figure 19.2 illustrates the relation between Earth's crust, mantle, and core and indicates the most important elements in each. Compared to the diameter of Earth, the crust, which extends from the surface to a depth of about 35 km, is equivalent to the thickness of the skin of an apple compared to the apple.

The average composition of Earth's crust is also given in Figure 19.2. All the elements shown in the pie chart are found in compounds. Because of their reactivity these elements do not exist in their "free" (elemental) form in Earth's crust. Note the preponderance of oxygen and silicon; they are the major components of silicate minerals, clays, and sand. Aluminum is the most abundant metal ion, followed by ions of iron and several alkali and alkaline-earth metals—calcium, sodium, potassium, and magnesium.

Most elements in the crust of Earth are in chemically combined forms as complex ionic solids known as minerals. A **mineral** is commonly defined as *a naturally occurring inorganic compound with a characteristic composition and crystal structure*. The major chemical form in which each element occurs on Earth's surface as a source of the element is shown in Figure 19.3. In particular, note the predominance of oxygen-containing

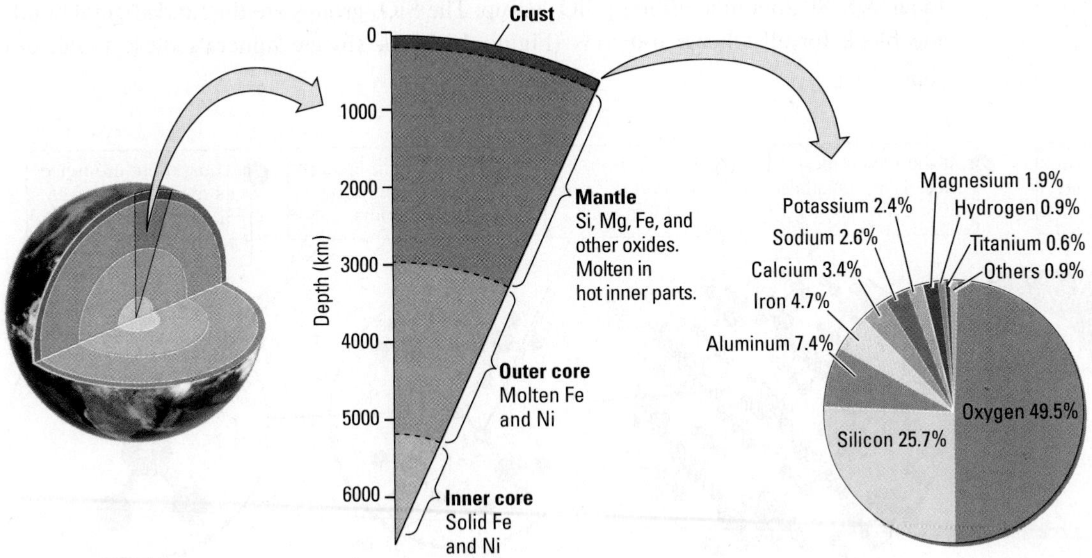
Crust
0
1000
2000
3000
4000
5000
6000
Depth (km)
Mantle
Si, Mg, Fe, and other oxides. Molten in hot inner parts.
Outer core
Molten Fe and Ni
Inner core
Solid Fe and Ni
Magnesium 1.9%
Potassium 2.4%
Hydrogen 0.9%
Sodium 2.6%
Titanium 0.6%
Calcium 3.4%
Others 0.9%
Iron 4.7%
Aluminum 7.4%
Oxygen 49.5%
Silicon 25.7%

Figure 19.2 Cross section and elemental composition of Earth. Most of the iron on Earth is in the core and the mantle. The crust, which is very thin compared with the mantle and the core, is 75% silicon and oxygen. Percentages of other major elements in the crust are in the pie chart.

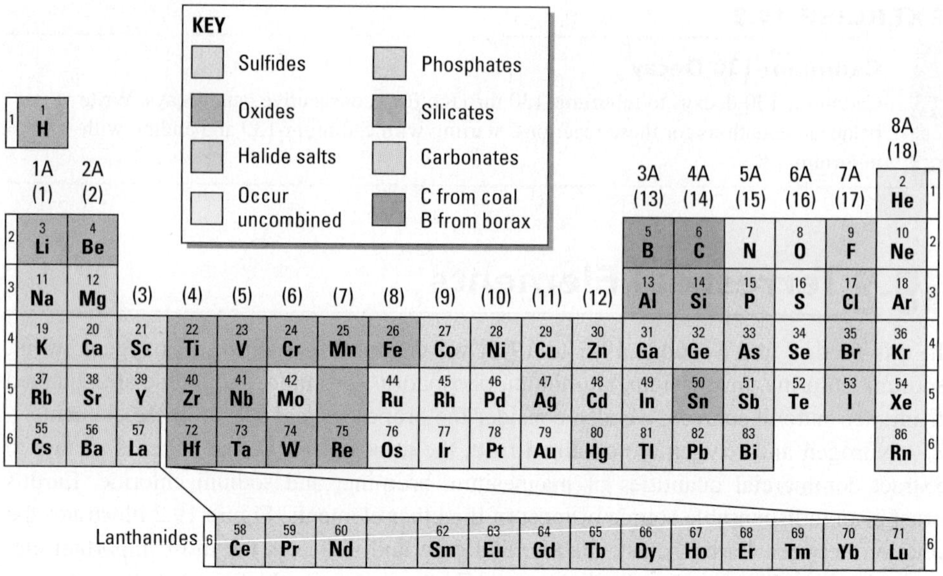
KEY
Sulfides
Phosphates
Oxides
Silicates
Halide salts
Carbonates
Occur uncombined
C from coal
B from borax
1A (1) 2A (2) (3) (4) (5) (6) (7) (8) (9) (10) (11) (12) 3A (13) 4A (14) 5A (15) 6A (16) 7A (17) 8A (18)
1 H 2 He
3 Li 4 Be 5 B 6 C 7 N 8 O 9 F 10 Ne
11 Na 12 Mg 13 Al 14 Si 15 P 16 S 17 Cl 18 Ar
19 K 20 Ca 21 Sc 22 Ti 23 V 24 Cr 25 Mn 26 Fe 27 Co 28 Ni 29 Cu 30 Zn 31 Ga 32 Ge 33 As 34 Se 35 Br 36 Kr
37 Rb 38 Sr 39 Y 40 Zr 41 Nb 42 Mo 44 Ru 45 Rh 46 Pd 47 Ag 48 Cd 49 In 50 Sn 51 Sb 52 Te 53 I 54 Xe
55 Cs 56 Ba 57 La 72 Hf 73 Ta 74 W 75 Re 76 Os 77 Ir 78 Pt 79 Au 80 Hg 81 Tl 82 Pb 83 Bi 86 Rn
Lanthanides
58 Ce 59 Pr 60 Nd 62 Sm 63 Eu 64 Gd 65 Tb 66 Dy 67 Ho 68 Er 69 Tm 70 Yb 71 Lu

Figure 19.3 Main types of minerals in Earth's crust used as sources of elements.

compounds, either binary ones such as MgO and TiO_2, or more complex ones such as carbonates, phosphates, and silicates. The preponderance of such compounds is testimony to the abundance of oxygen and silicon in Earth's surface. The lanthanides and naturally occurring actinides also form oxide minerals. Many transition metals and heavier elements of Groups 3A (13) to 6A (16) are found as sulfides, such as ZnS and Sb_2S_3.

19-2a Silica and Silicates

Silicon and oxygen, the two most abundant elements in Earth's crust, are combined in the crust as silica and silicate minerals. *Silica* is pure SiO_2. Its most common form is α-quartz, which is a major component of many rocks such as granite and sandstone. Alpha-quartz also occurs as a pure rock crystal (Figure 19.4) and in several less pure forms. As shown in Figure 19.4, such minerals all contain SiO_4 groups in which four oxygen atoms are arranged tetrahedrally around a central Si atom and each oxygen is bonded to another Si atom in a different SiO_4 group. The SiO_4 groups are the fundamental building block for all silicate minerals (Figure 19.5). In silicate minerals these tetrahedra

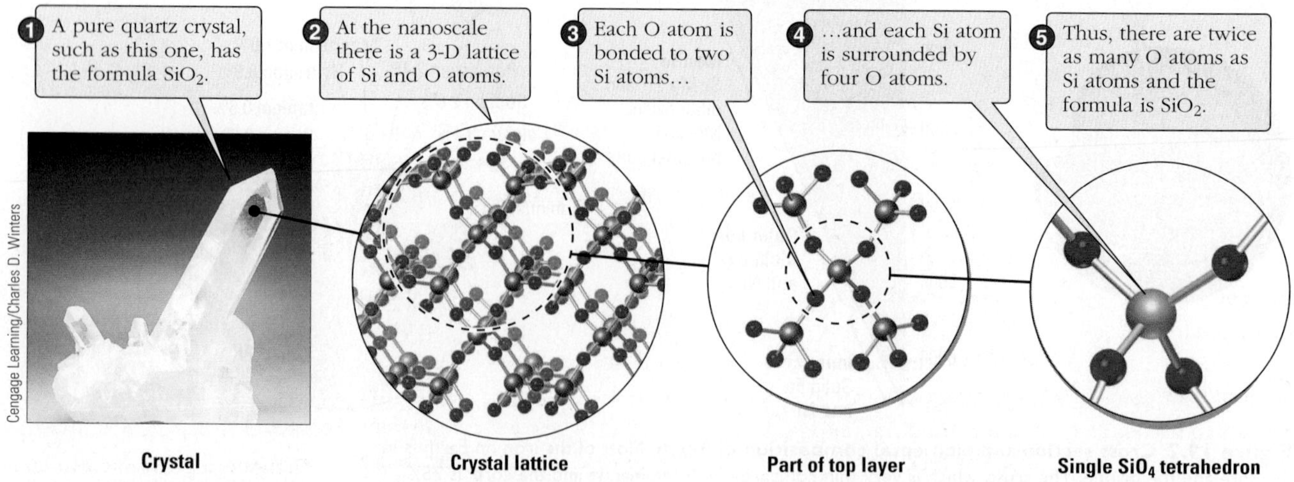
1 A pure quartz crystal, such as this one, has the formula SiO₂.
2 At the nanoscale there is a 3-D lattice of Si and O atoms.
3 Each O atom is bonded to two Si atoms...
4 ...and each Si atom is surrounded by four O atoms.
5 Thus, there are twice as many O atoms as Si atoms and the formula is SiO₂.
Cengage Learning/Charles D. Winters
Crystal
Crystal lattice
Part of top layer
Single SiO₄ tetrahedron

Figure 19.4 A pure quartz crystal (pure SiO_2). Quartz crystals are used as oscillators in watches, radios, and computers.

Class	Independent tetrahedra	Single chains; pyroxenes	Double chains; amphiboles	Sheet silicates; mica
Unit composition	SiO_4^{4-}	$(SiO_3^{2-})_n$	$(Si_4O_{11}^{6-})_n$	$(Si_2O_5^{2-})_n$
Arrangement of SiO_4 tetrahedra	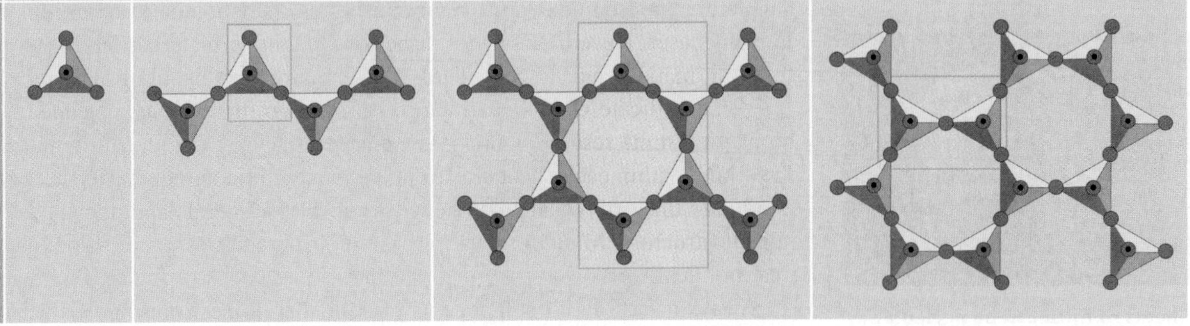			

Figure 19.5 Silicate structures. These structures are all based on the tetrahedral SiO_4 unit. The repeating unit of each structure is shown with a tan background.

typically share one or more oxygen atoms to form chains, sheets, rings, and three-dimensional networks such as the structure of quartz.

An individual SiO_4 group in which none of the oxygens is shared with another silicon has a charge of 4−. An example is found in the mineral forsterite, Mg_2SiO_4, which contains two Mg^{2+} ions for each SiO_4^{4-} ion to balance the charge. The simple silicate mineral willemite, Zn_2SiO_4, and the gemstone garnet contain discrete SiO_4^{4-} units that do not share oxygens.

CONCEPTUAL EXERCISE 19.3

Charge on a Silicate Ion

Draw a Lewis structure for SiO_4^{4-}, predict the structure based on VSEPR theory, calculate the formal charge on each atom, and verify that the overall charge is the sum of the formal charges.

EXERCISES that are labeled **CONCEPTUAL** are designed to test your understanding of one or more concepts; they usually involve qualitative rather than quantitative thinking.

Condensed silicates contain SiO_4 tetrahedra in which oxygen atoms are shared. *Pyroxenes* contain extended chains of linked SiO_4 tetrahedra, each sharing two oxygen atoms (Figure 19.5). The repeating unit in the pyroxene polymer is SiO_3, so pyroxene appears to contain the metasilicate ion, SiO_3^{2-}; a typical formula is Na_2SiO_3. Pyroxenes are abundant in the ocean's floor and Earth's mantle. If two pyroxene chains are laid side by side, they can link by sharing oxygen atoms in adjoining chains to form a type of silicate called an *amphibole,* with the typical repeating unit $Si_4O_{11}^{6-}$ (Figure 19.5).

An excellent example of an amphibole is crocidolite, one form of *asbestos.* What is called "asbestos" is not a single substance; rather, this term applies broadly to a family of naturally occurring hydrated silicates that crystallize as fibers. Asbestos minerals are generally subdivided into two forms: serpentine and amphibole fibers. Approximately 4 million tons of the serpentine form of asbestos, chrysotile, are mined each year, chiefly in Canada and the former Soviet Union. Chrysotile is the only form widely used commercially in the United States. Another form, the amphibole crocidolite, is mined in small quantities, mainly in South Africa. The two asbestos minerals differ greatly in composition, color, shape, solubility, and persistence in human tissue. This last property is important in determining their toxicity. Crocidolite is blue, is relatively insoluble, and persists in tissue. Its long, thin, straight fibers can penetrate narrow lung passages. In contrast, chrysotile is white, and it tends to be soluble and disappear in tissue. Chrysotile fibers are curly, so they ball up like yarn and are more easily rejected by the body.

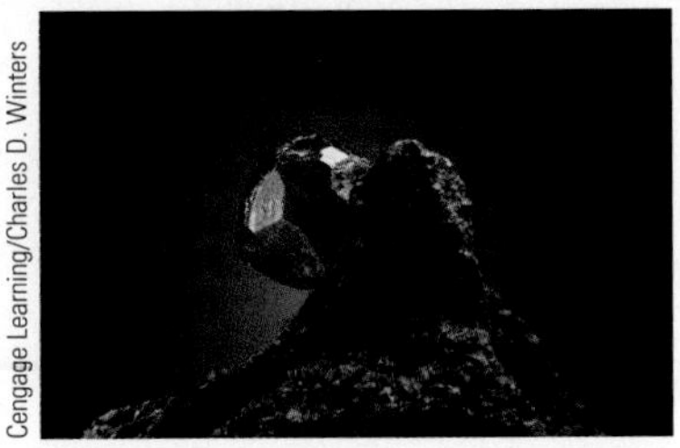

Cengage Learning/Charles D. Winters

Garnet is a gemstone containing tetrahedral SiO_4^{4-} ions.

Long-term occupational exposure to certain asbestos minerals can lead to lung cancer. Although some disagreement persists in the medical and scientific communities, evidence strongly suggests that amphiboles such as crocidolite are much more potent cancer-causing agents than the serpentines such as chrysotile. Most asbestos in public buildings is the chrysotile type, so initiatives to remove asbestos insulation may be misguided overreaction in many cases. Nevertheless, most asbestos-containing materials have been removed from the U. S. market, and strict standards now exist for their handling and use.

Cengage Learning/Charles D. Winters

Sheets of mica can be peeled away from each other using a razor blade because the nanoscale sheets (Figure 19.5) are weakly bonded to each other.

When silicate chains continue to link in two dimensions, extended sheets of SiO_4 tetrahedral units result (Figure 19.5), characterized by the repeating unit $Si_2O_5^{2-}$. All of the atoms within each sheet are strongly covalent bonded, but each sheet is only weakly bonded to those above and below it. Various clay minerals and mica have this sheet-like silicate structure. Mica, for example, is used to prepare "metallic"-looking paint on new automobiles.

Clays are essential components of soils that come from the weathering of igneous rocks. Since early in human history, clays have been used for pottery, bricks, tiles, and writing materials. Clays are actually *aluminosilicates.* In an aluminosilicate, some of the tetrahedral groups are AlO_4 instead of SiO_4, and some O atoms are shared between an Al atom and a Si atom. An example is feldspar, $KAlSi_3O_8$, a component of many rocks. When Al^{3+} ions are replaced by other 3+ metal ions, the clay may become colored. For example, a red clay contains Fe^{3+} ions in place of some Al^{3+} ions.

Artists who work with clay first wet it and then mold the clay into a shape. Water molecules strongly interact with the oxygen atoms as well as the metal ions near the surface of clay particles, and the silicate layers slide over one another, making the clay pliable. After the clay has been formed into the desired shape, the object is heated to remove the water. Bonds form between exposed oxygen atoms and aluminum or silicon atoms on the surfaces of adjacent particles, which causes the clay to harden. Too much water in the wet clay can make it unstable. This instability occurs not only in clay for pottery, but also on a much larger scale in nature. During very heavy rains, entire hillsides of clay can shift and slide downhill, causing massive destruction of property (Figure 19.6).

AP Images/Nick UT

Figure 19.6 A mudslide caused by shifting clay. During heavy rains, clays become saturated with water, allowing the aluminosilicate layers to slide over each other, which causes the ground to shift.

CONCEPTUAL EXERCISE 19.4

Linking Tetrahedra

Explain how the silicate unit in pyroxenes has the general formula SiO_3^{2-}, not SiO_4^{4-}.

19-2b Methods for Obtaining Pure Elements

Relatively few elements are available in nature directly in their uncombined form: nitrogen and oxygen as diatomic molecules in the atmosphere, the noble gases, mercury, gold, silver, copper, and sulfur. For these elements, only purification is needed, although in several cases the elemental form is not the least expensive source of the element. All other elements needed in elemental form for practical applications must be extracted from their compounds. The types of extraction methods used are listed in Table 19.1. Metal cations in minerals must be reduced to their elemental form. Therefore, chemical and electrochemical oxidation-reduction reactions are needed in the production of metals.

In the subsequent sections of this chapter we describe how some elements are extracted from their naturally occurring forms by

- physical methods (Section 19-3: nitrogen, oxygen, and sulfur),
- electrochemical redox reactions (Section 19-4: sodium, chlorine, magnesium, and aluminum), and
- chemical redox reactions (Section 19-5: phosphorus, bromine, and iodine).

An **ore** is *a mineral that contains a sufficiently high concentration of an element to make its extraction profitable*. Because not all elements are used to the same extent, and

Table 19.1 Methods for Extraction of Elements from Their Ores

Extraction Method	Examples of Elements Extracted by This Method
Carbon reduction of oxide	Si, Fe, Sn
Oxidation of anion with Cl_2	Br, I
Reaction of sulfide with O_2	Cu, Hg
Conversion of sulfide to oxide, followed by reduction with C	Zn, Pb
Reduction of metal halide with sodium or other highly reactive metal	K, Ti, Cr, Cs, U
Reduction of halide or oxide with H_2	B, Ni, Mo, W
Electrolysis of solution or molten salt	H, Li, F, Na, Ca, Al, Cl

the quantity used can vary with market demands, a metal is extracted from its ore in response to such demands. At present, known reserves of some common elements such as aluminum and iron are sufficient to last hundreds of years at the current rate of use, whereas the known reserves of other widely used elements, such as copper, tin, and lead, are rather slim (Table 19.2). The United States does not have major reserves of several critical metals, for example, the chromium and manganese needed for making steel and other alloys. Thus, we must import and stockpile such metals.

19-3 Extraction by Physical Methods: Nitrogen, Oxygen, and Sulfur

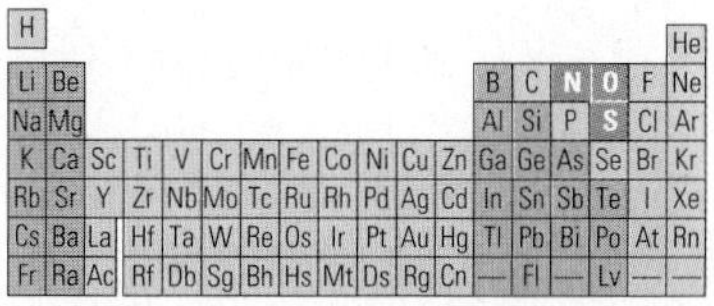

19-3a Elements from the Atmosphere

The composition of the atmosphere given in Table 19.3 shows that nitrogen is by far the most abundant component, its concentration being nearly four times that of oxygen, the next most abundant component. The gases of the atmosphere can be separated from one another by liquefying and fractionally distilling air in a process similar to the separation of petroleum fractions (← **Sec. 10-1a**), except at much lower temperatures and higher pressures.

Table 19.2 Known Reserves of Selected Elements

Element	Reserves (10^9 kg)	Lifetime (yr)	Locations of Major Reserves
Al	20,000	220	Australia, Brazil, Guinea
Fe	66,000	120	Australia, Canada, CIS*
Mn	800	100	CIS,* Gabon, S. Africa
Cr	400	100	CIS,* S. Africa, Zimbabwe
Cu	300	36	Chile, CIS,* USA, Zaïre
Zn	150	21	Australia, Canada, USA
Pb	71	20	Australia, Canada, CIS,* USA
Ni	47	55	Canada, CIS,* Cuba, New Caledonia
Sn	5	28	Brazil, China, Indonesia, Malaysia
U	2.8	58	Australia, CIS,* S. Africa, USA

*No individual breakdown is available for nations constituting the Commonwealth of Independent States (formerly the USSR).

Table 19.3 Composition of Clean, Dry Air at Sea Level

Component	Percent by Volume	Component	Percent by Volume
N_2	78.080	He, Ne, Kr, Xe	0.002
O_2	20.947	CH_4	0.00015*
Ar	0.934	H_2	0.00005
CO_2	0.039*	All others combined	$<$0.00004

*Variable: changes with season of the year

Cengage Learning/Charles D. Winters

The Joule-Thomson effect. When the tab is opened on a can of carbonated beverage, the gases in the liquid are expelled rapidly enough to cool the water vapor to a liquid in the vicinity of the mouth of the can. The water vapor cools to form a tiny, visible "cloud."

At low temperatures and high pressures, gases no longer behave ideally and the attractive forces between molecules cause the gases in air to condense to liquids. Because of their different boiling points, the liquid components can then be separated from one another by distillation. Before pure oxygen and nitrogen can be obtained from air, water vapor and carbon dioxide are removed (Figure 19.7). The dry air is compressed by more than 100 times normal atmospheric pressure, cooled to room temperature, and allowed to expand into a chamber. This expansion produces a cooling effect (the *Joule-Thomson effect*) because energy is required to overcome intermolecular forces as the molecules move farther apart. The expanding gas absorbs kinetic energy from the motion of its own molecules, which cools the gas. If this expansion is repeated and controlled properly, the expanding air cools to the point of liquefaction.

The temperature of the liquid air is usually well below the normal boiling points of nitrogen (−195.8 °C), oxygen (−183 °C), and argon (−186 °C). The very cold liquid air is again allowed to vaporize partially. Because N_2 is more volatile and has a lower boiling point than O_2 or Ar, the N_2 evaporates first and the remaining liquid becomes more concentrated in O_2 and Ar. This process, known as the *Linde process,* produces high-purity nitrogen (>99.5%) and oxygen (99.5%). Further processing produces pure Ar.

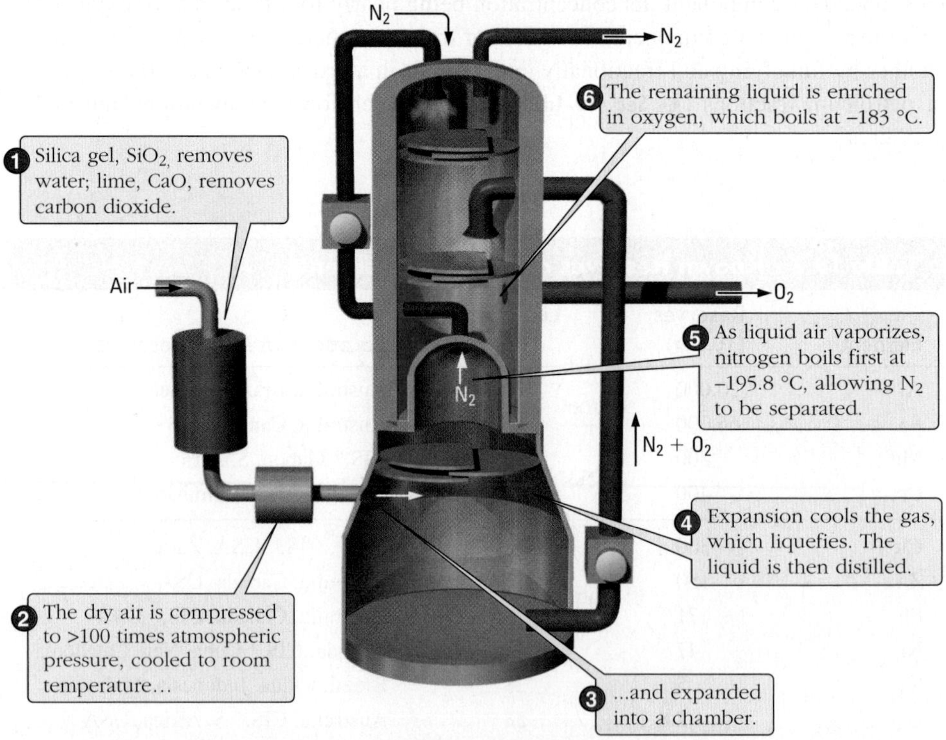
N₂
N₂
6 The remaining liquid is enriched in oxygen, which boils at −183 °C.
1 Silica gel, SiO₂, removes water; lime, CaO, removes carbon dioxide.
Air
O₂
5 As liquid air vaporizes, nitrogen boils first at −195.8 °C, allowing N₂ to be separated.
N₂
N₂ + O₂
4 Expansion cools the gas, which liquefies. The liquid is then distilled.
2 The dry air is compressed to >100 times atmospheric pressure, cooled to room temperature...
3 ...and expanded into a chamber.

Figure 19.7 Fractional distillation of air. Air can be liquefied by using low temperatures and high pressure. The components of the liquefied air are then separated by distillation.

EXERCISE 19.5

Liquefied Gases

Liquid air has been proposed as a means of storing energy and powering automobiles (http://www.telegraph.co.uk/motoring/green-motoring/10087205/Liquid-Air-the-future-of-motoring.html). When the liquid boils, the expanding volume of the gas can power a car. A cryogenic flask contains 5.0 L of liquid air, which has a density of 0.87 g/mL. Assuming the air contains 78% N_2, 21% O_2, and 1% Ar, calculate the volume this air occupies at STP after it boils.

19-3b Sulfur

Sulfur, the element known biblically as brimstone, is a bright yellow solid. Very pure sulfur has been obtained from large deposits in salt domes along the coast of the Gulf of Mexico in the United States and Mexico and in underground deposits in Poland. Such sulfur deposits are believed to have been formed by bacterial reduction of sulfur in the naturally occurring mineral gypsum, $CaSO_4 \cdot 2H_2O$. Millions of tons of sulfur have been recovered from such deposits by the *Frasch process*, developed in the 1890s by Herman Frasch, a petroleum engineer. Most sulfur is now produced by extracting it from petroleum and natural gas. Removing the sulfur avoids the formation of sulfur dioxide, an atmospheric pollutant produced when petroleum burns. High-sulfur natural gas from Alberta, Canada, an especially large source of recovered sulfur, has now displaced the Frasch process as the chief source of sulfur. Outside the United States much sulfur is obtained from pyrite minerals.

Paul Thompson Images/Alamy

Sulfur. Large piles of sulfur are in a harbor awaiting shipment.

19-4 Extraction by Electrolysis: Sodium, Chlorine, Magnesium, and Aluminum

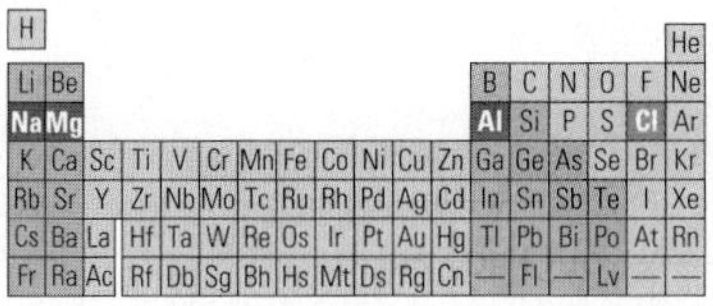
H He
Li Be B C N O F Ne
Na Mg Al Si P S Cl Ar
K Ca Sc Ti V Cr Mn Fe Co Ni Cu Zn Ga Ge As Se Br Kr
Rb Sr Y Zr Nb Mo Tc Ru Rh Pd Ag Cd In Sn Sb Te I Xe
Cs Ba La Hf Ta W Re Os Ir Pt Au Hg Tl Pb Bi Po At Rn
Fr Ra Ac Rf Db Sg Bh Hs Mt Ds Rg Cn — Fl — Lv — —

Chapter 17 described how electrolysis is used to force reactant-favored chemical reactions to occur (← **Sec. 17-10**). Electrolysis is applied commercially on a vast scale to extract the elements sodium, magnesium, aluminum, and chlorine from their natural sources. These reactive elements exist naturally only in ionic form. Consequently, the metals must be obtained by reduction of their cations from compounds, and chlorine must be oxidized from Cl^- to Cl_2.

19-4a Sodium

Sodium metal was discovered by Humphry Davy in 1807; he electrolyzed molten NaOH. The half-reactions are

Anode, oxidation: $4\,OH^-(\text{in melt}) \longrightarrow O_2(g) + 2\,H_2O(g) + 4\,e^-$

Cathode, reduction: $4\,Na^+(\text{in melt}) + 4\,e^- \longrightarrow 4\,Na(\text{in melt})$

Overall reaction:

$$4\,Na^+(\text{in melt}) + 4\,OH^-(\text{in melt}) \longrightarrow 4\,Na(\text{in melt}) + O_2(g) + 2\,H_2O(g)$$

By the early 1900s, commercial uses for sodium metal had increased so that a large-scale production method was needed. In 1921, the *Downs process* was developed to meet this demand. In a Downs cell, molten NaCl is electrolyzed at 7 to 8 V and 25,000 to 40,000 A (Figure 19.8). The cell is filled with a 1 : 3 mixture of NaCl and $CaCl_2$. Pure NaCl is not used because of its high melting point (800 °C). Mixing the two salts lowers the melting point of the mixture to approximately 600 °C.

In the Downs cell, sodium metal is produced at a cathode made of copper or iron that surrounds a cylindrical graphite anode. Directly over the cathode is an inverted trough through which the molten sodium flows (sodium melts at 97.8 °C); liquid sodium is less

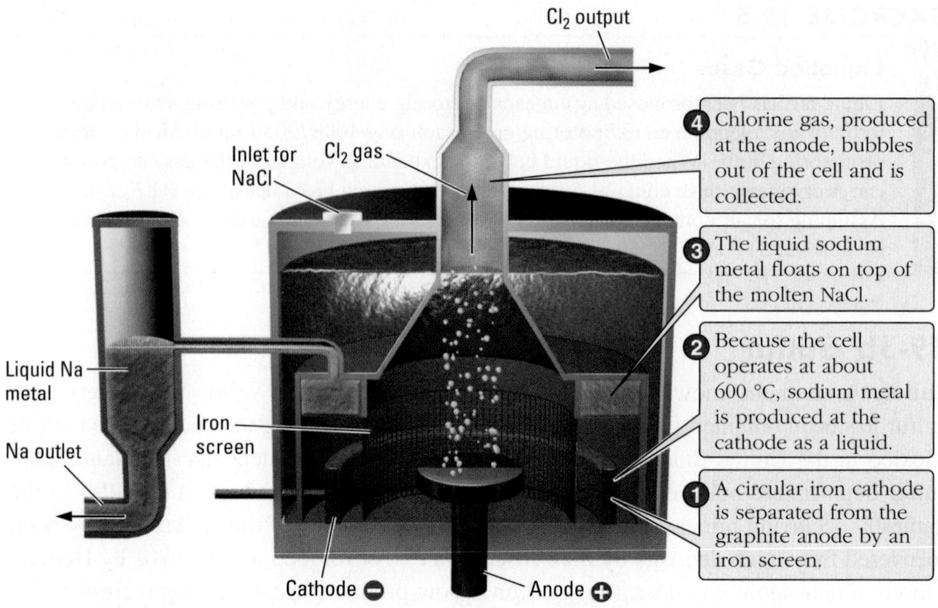
Cl_2 output
Chlorine gas, produced at the anode, bubbles out of the cell and is collected.
Inlet for NaCl
Cl_2 gas
The liquid sodium metal floats on top of the molten NaCl.
Because the cell operates at about 600 °C, sodium metal is produced at the cathode as a liquid.
Liquid Na metal
Iron screen
Na outlet
A circular iron cathode is separated from the graphite anode by an iron screen.
Cathode
Anode

Figure 19.8 The Downs cell for the electrolysis of molten NaCl.

dense than the molten mixture and therefore floats on top of it. Gaseous chlorine, the other product of the electrolysis, passes through an inverted cone of nickel metal extending through the molten salt mixture and is collected, cooled, and liquefied.

Anode, oxidation:	$Cl^-(\text{in melt}) \longrightarrow \frac{1}{2} Cl_2(g) + e^-$
Cathode, reduction:	$Na^+(\text{in melt}) + e^- \longrightarrow Na(\ell)$
Overall reaction:	$Na^+(\text{in melt}) + e^- \longrightarrow Na(\ell) + \frac{1}{2} Cl_2(g)$

PROBLEM-SOLVING EXAMPLE 19.1

Titanium Production

Assume that the annual production of sodium metal in the United States is 76,000 tons. If half of this amount were used to produce titanium from $TiCl_4$, calculate the mass (ton) of titanium that could be produced.

$$TiCl_4(\ell) + 4\,Na(\ell) \longrightarrow Ti(\ell) + 4\,NaCl(\ell)$$

Result 2.0×10^4 tons

Analyze Half of the sodium produced is 38,000 tons. Use the sodium-to-titanium mole ratio to calculate the amount of titanium, from which we can then find the mass of titanium.

Plan From the balanced equation we see that there is a 4 mol sodium : 1 mol titanium ratio. Use this ratio and the appropriate conversion factors to calculate the mass (tons) of titanium produced.

Execute Calculate the amount of sodium and from it the amount of titanium.

$$n(\text{Na}) = 3.8 \times 10^4 \text{ ton Na}\left(\frac{2000 \text{ lb Na}}{1 \text{ ton Na}}\right)\left(\frac{454 \text{ g Na}}{1 \text{ lb Na}}\right)\left(\frac{1 \text{ mol Na}}{23.0 \text{ g Na}}\right) = 1.5 \times 10^9 \text{ mol Na}$$

$$n(\text{Ti}) = 1.5 \times 10^9 \text{ mol Na}\left(\frac{1 \text{ mol Ti}}{4 \text{ mol Na}}\right) = 3.8 \times 10^8 \text{ mol Ti}$$

and

$$m(\text{Ti}) = 3.8 \times 10^8 \text{ mol Ti}\left(\frac{47.9 \text{ g Ti}}{1 \text{ mol Ti}}\right)\left(\frac{1 \text{ lb Ti}}{454 \text{ g Ti}}\right)\left(\frac{1 \text{ ton Ti}}{2000 \text{ lb Ti}}\right) = 2.0 \times 10^4 \text{ ton Ti}$$

☑ **Reasonable Result Check** It takes 4 mol Na to produce 1 mol Ti and the molar mass of Ti (48 g/mol) is somewhat more than twice the molar mass of Na (23 g/mol) so the mass of Ti should be about half (23/48) the mass of Na. Therefore, 38,000 tons Na should produce about 19,000 tons Ti; the result is reasonable.

PROBLEM-SOLVING PRACTICE 19.1

Calculate the mass (ton) of sodium chloride produced under the same conditions as in Problem-Solving Example 19.1.

PROBLEM-SOLVING PRACTICE answers are provided at the back of this book in Appendix K.

Manufacturing facilities in the United States can produce about 76,000 tons of sodium metal per year. Many are located near Niagara Falls, New York, because of the relatively low-cost electricity available from hydroelectric plants.

PARK KIBAE/SIPA/Newscom

Hydroelectric power at Niagara Falls. Water diverted around Niagara Falls flows through a hydroelectric plant downstream from the falls. The falling water spins turbines that generate electricity.

EXERCISE 19.6

The Downs Cell

Calculate how many tons of sodium can be produced in one day by a Downs cell operating at 2.0×10^4 A. Determine how many tons of Cl_2 are produced in this same time.

19-4b Chlorine and Sodium Hydroxide

Most chlorine is produced by the electrolysis of aqueous sodium chloride in the **chlor-alkali process**; the alkali produced by this process is sodium hydroxide. This process produces approximately 25 billion pounds of sodium hydroxide and a similar quantity of chlorine annually in the United States. These large quantities testify to the usefulness of these two products. The oxidizing and bleaching ability of chlorine is utilized in many industrial and everyday applications, and this element is a raw material in the manufacture of chlorine-containing chemicals. Sodium hydroxide is the base of choice in many industrial chemistry applications because it is inexpensive. It is also used widely to produce soaps, detergents, and other compounds.

The chlor-alkali process electrolyzes saturated aqueous NaCl (saturated brine), as illustrated in Figure 19.9. Chloride ions are oxidized at the anode, and water is reduced at the cathode. The anode and cathode compartments are separated by a special polymeric membrane that allows only cations to pass through it. The brine is added to the

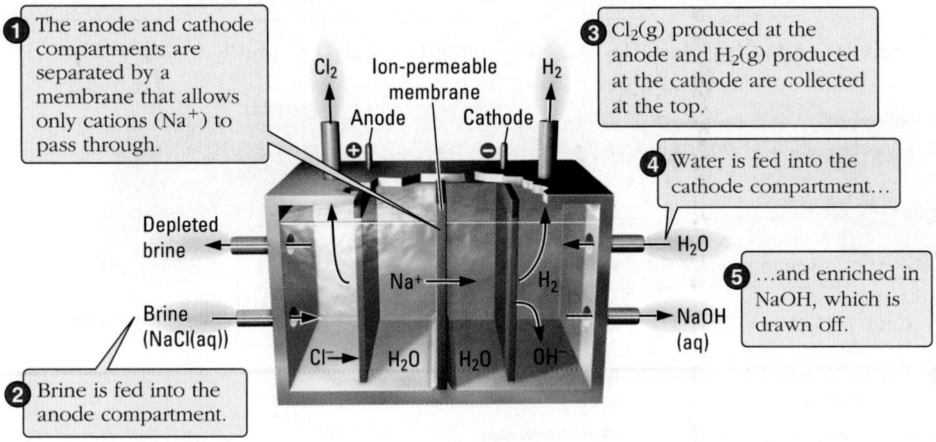

1 The anode and cathode compartments are separated by a membrane that allows only cations (Na^+) to pass through.
Cl_2
Ion-permeable membrane
H_2
Anode
Cathode
3 $Cl_2(g)$ produced at the anode and $H_2(g)$ produced at the cathode are collected at the top.
4 Water is fed into the cathode compartment…
Depleted brine
H_2O
Na^+
H_2
5 …and enriched in NaOH, which is drawn off.
Brine (NaCl(aq))
NaOH (aq)
Cl^-
H_2O
H_2O
OH^-
2 Brine is fed into the anode compartment.

Figure 19.9 A membrane cell used in the chlor-alkali process.

anode compartment, and sodium ions pass through the membrane into the cathode compartment. The half-reactions are

Anode, oxidation:	$2\ Cl^-(aq) \longrightarrow Cl_2(g) + 2\ e^-$
Cathode, reduction:	$2\ H_2O(\ell) + 2\ e^- \longrightarrow 2\ OH^-(aq) + H_2(g)$
Overall reaction:	$2\ Cl^-(aq) + 2\ H_2O(\ell) \longrightarrow Cl_2(g) + 2\ OH^-(aq) + H_2(g)$

The reduction potential of Na^+ is more negative than that of water, so water, not sodium ions, is reduced.

The anode is specially treated titanium, and the cathode is stainless steel or nickel. The membrane is not permeable to water and acts as a salt bridge. In the anode compartment, chloride ions are oxidized to Cl_2, reducing the concentration of negative ions. To maintain charge balance, sodium ions must migrate from the anode compartment to the cathode compartment, where OH^- ions are being produced by the cathode reaction. The resulting NaOH solution in the cathode compartment is 21% to 30% NaOH by weight.

The membrane cell was developed to replace the mercury cell that had been used previously in the chlor-alkali process. A major problem with mercury cells is the environmental damage caused by loss of mercury during normal operation of the cells. In the past, when mercury cells were cleaned, mercury was routinely allowed to run into neighboring bodies of water.

EXERCISE 19.7

NaOH Production

A chlor-alkali membrane cell operates at 2.00×10^4 A for 100. hours. Calculate the mass (ton) of NaOH produced. (Assume 100% efficiency.)

19-4c Magnesium from Seawater

$Mg(OH)_2$ ($K_{sp} = 1.8 \times 10^{-11}$) is much less soluble than $Ca(OH)_2$ ($K_{sp} = 5.5 \times 10^{-6}$) so a solution of $Ca(OH)_2$ has a high enough OH^- concentration to precipitate $Mg(OH)_2$.

With a concentration of 1.35 mg Mg^{2+} per liter, the oceans provide a nearly limitless supply of magnesium, containing approximately 6 thousand tons of Mg per cubic mile. As with other reactive metals, converting the metal ion to the metal atom is not product-favored, so electrolysis is required to extract magnesium metal.

The *Dow process*, which is used to reduce Mg^{2+} ions in seawater into magnesium metal, is summarized in Figure 19.10 together with chemical equations for the reactions. It involves using $Ca(OH)_2$ to precipitate Mg^{2+} from seawater as its insoluble hydroxide ($K_{sp} = 1.8 \times 10^{-11}$). The magnesium hydroxide is filtered and neutralized by

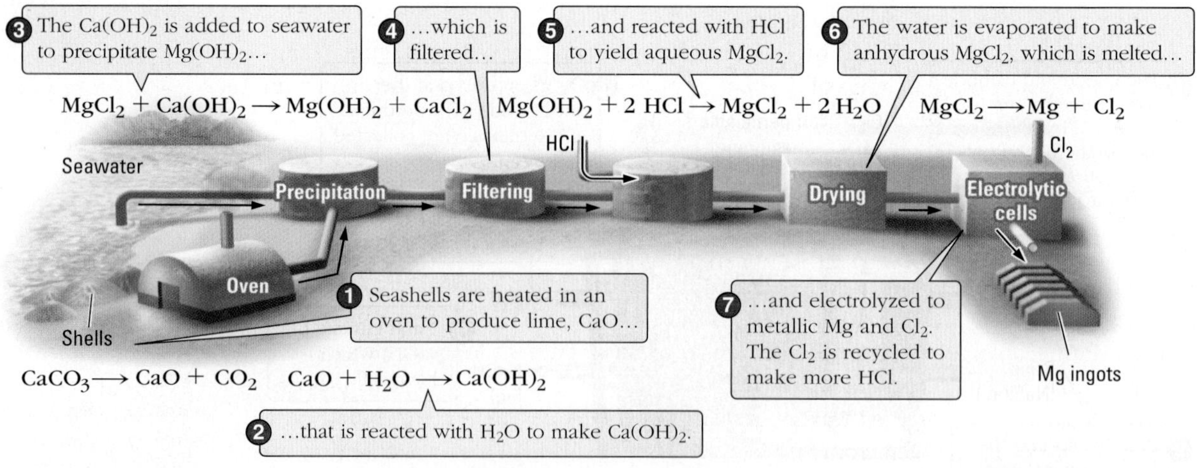
3 The Ca(OH)2 is added to seawater to precipitate Mg(OH)2...
4 ...which is filtered...
5 ...and reacted with HCl to yield aqueous MgCl2.
6 The water is evaporated to make anhydrous MgCl2, which is melted...
MgCl2 + Ca(OH)2 → Mg(OH)2 + CaCl2
Mg(OH)2 + 2 HCl → MgCl2 + 2 H2O
MgCl2 → Mg + Cl2
HCl
Cl2
Seawater
Precipitation
Filtering
Drying
Electrolytic cells
Oven
Shells
1 Seashells are heated in an oven to produce lime, CaO...
7 ...and electrolyzed to metallic Mg and Cl2. The Cl2 is recycled to make more HCl.
Mg ingots
CaCO3 → CaO + CO2
CaO + H2O → Ca(OH)2
2 ...that is reacted with H2O to make Ca(OH)2.

Figure 19.10 The steps for extracting magnesium metal from seawater.

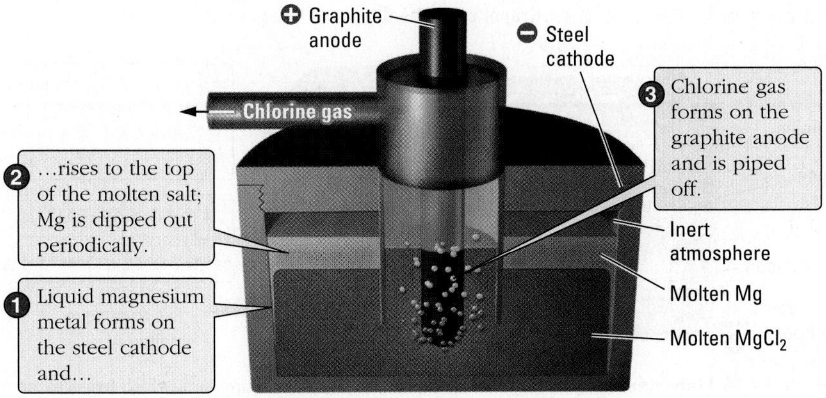
Graphite anode
Steel cathode
Chlorine gas
Chlorine gas forms on the graphite anode and is piped off.
...rises to the top of the molten salt; Mg is dipped out periodically.
Inert atmosphere
Molten Mg
Liquid magnesium metal forms on the steel cathode and...
Molten $MgCl_2$

Figure 19.11 Electrolysis of molten magnesium chloride.

hydrochloric acid, another inexpensive chemical, to produce $MgCl_2$. Magnesium chloride is melted and electrolyzed in a steel pot, which serves as the cathode (Figure 19.11). The electrode reactions are

Anode, oxidation:	$2\ Cl^-(\text{in melt}) \longrightarrow Cl_2(g) + 2\ e^-$
Cathode, reduction:	$Mg^{2+}(\text{in melt}) + 2\ e^- \longrightarrow Mg(\text{in melt})$
Overall reaction:	$Mg^{2+}(\text{in melt}) + 2\ Cl^-(\text{in melt}) \longrightarrow Mg(\text{in melt}) + Cl_2(g)$

Molten magnesium is less dense than molten $MgCl_2$ and floats at the surface, where the metal can be removed. Chlorine produced at the anode is converted to HCl by mixing Cl_2 with methane from natural gas and burning the mixture.

$$4\ Cl_2(g) + 2\ CH_4(g) + O_2(g) \longrightarrow 2\ CO(g) + 8\ HCl(g)$$

The HCl is recycled to neutralize $Mg(OH)_2$, forming more $MgCl_2$.

19-4d Aluminum Production

Aluminum is the most abundant metal in Earth's surface (7.4%), where it is present as Al^{3+} ions, from which the metal must be obtained by reduction. Aluminum was first isolated in metallic form in 1825 by an expensive and potentially dangerous method—using metallic sodium or potassium to reduce Al^{3+} ions in aluminum chloride, $AlCl_3$. Thus, metallic aluminum was very expensive and considered to be a precious metal, like gold or platinum. An early use of aluminum was in jewelry, including the Danish King Frederik VII's dress helmet. In the 1855 Exposition in Paris, some of the first pieces of aluminum metal produced were displayed along with the French crown jewels. In 1884, a 2.8-kg aluminum cap, produced by sodium reduction, topped the Washington Monument as ornamentation and the tip of a lightning rod system. At that time, the aluminum cap cost about the same as the same mass of silver.

Napoleon II saw the advantages of using aluminum for military purposes because of its low density, and he commissioned studies on improving its production. Near the town of Les Baux, France, was a ready source of the aluminum-containing ore bauxite ($Al_2O_3 \cdot 3H_2O$ combined with oxides of Si, Fe, and other elements); but, how could aluminum be extracted readily from the ore? In 1886, Paul Héroult, a Frenchman, conceived an electrochemical process to do so that is still used today. In a curious coincidence, Charles Martin Hall, an American, independently came up with the identical process two months earlier. Hence, the commercial method is known as the *Hall-Héroult process*. Just five years after the process was first used to produce aluminum commercially, the price of the metal plummeted from $12 per kilogram, a substantial sum at that time, to 70 cents per kilogram. What was once a precious metal soon became commonplace.

Charles Martin Hall
1863–1914

Bettmann/Corbis

While a student at Oberlin College (OH), Charles Martin Hall became intrigued with trying to separate aluminum from its ores cheaply. When just 22 years old, using batteries and a blacksmith's forge, Hall succeeded in reducing Al_2O_3 dissolved in cryolite to metallic aluminum. To take advantage of his discovery, he formed what has become the Aluminum Corporation of America (ALCOA), an enterprise that made Hall a multimillionaire. Although Paul Héroult filed for a U. S. patent on the process about a month before Hall, Hall received precedence because he demonstrated the electrolytic extraction of aluminum metal on February 9, 1886, about two months before Héroult's French patent filing. The two men had an amicable relationship. Born in the same year, they also shared the same year of death; Héroult died just eight days after Hall.

Paul Louis-Toussaint Héroult
1863–1914

William Haynes Portrait Collection, Chemical Heritage Foundation

While at a boarding school near Paris, Paul Héroult became interested in extracting aluminum from its ores after reading a book on electrolytic reduction of Al^{3+} ions. The 22-year-old Héroult convinced his mother to give him 50,000 francs, a large sum of money at the time, to buy a 400-A, 30-V battery to use in electrolytic reduction of aluminum ores. Like Charles Martin Hall, Héroult used molten cryolite to dissolve the ores and produced metallic aluminum. He filed his French patent on April 23, 1886 and his U. S. patent May 22, 1886, just months before Hall filed his patent on July 9, 1886.

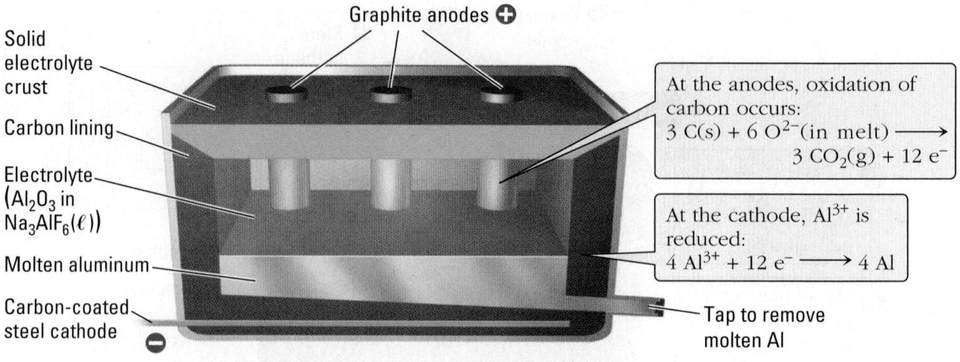
Graphite anodes ⊕
Solid electrolyte crust
Carbon lining
Electrolyte (Al_2O_3 in $Na_3AlF_6(\ell)$)
Molten aluminum
Carbon-coated steel cathode ⊖
At the anodes, oxidation of carbon occurs:
$3\ C(s) + 6\ O^{2-}(\text{in melt}) \longrightarrow 3\ CO_2(g) + 12\ e^-$
At the cathode, Al^{3+} is reduced:
$4\ Al^{3+} + 12\ e^- \longrightarrow 4\ Al$
Tap to remove molten Al

Figure 19.12 A Hall-Héroult process electrolytic cell. Molten aluminum, which has greater density than molten Al_2O_3 in cryolite, is drawn off from the bottom of the cell into molds.

In the Hall-Héroult process, metallic aluminum is obtained by electrolysis of Al_2O_3 dissolved in molten cryolite, Na_3AlF_6. The cryolite allows the electrolysis to be carried out at a lower temperature (1000 °C) than would be required for molten Al_2O_3 (m.p. 2030 °C). The aluminum oxide-cryolite mixture is electrolyzed in a cell using carbon anodes and a carbon cell lining that serves as the cathode (Figure 19.12). The half-reactions for extracting aluminum are

Anode, oxidation: $3\ C(s) + 6\ O^{2-}(\text{in melt}) \longrightarrow 3\ CO_2(g) + 12\ e^-$

Cathode, reduction: $4\ Al^{3+}(\text{in melt}) + 12\ e^- \longrightarrow 4\ Al(\text{in melt})$

Overall reaction: $4\ Al^{3+}(\text{melt}) + 3\ C(s) + 6\ O^{2-}(\text{melt}) \longrightarrow 4\ Al(\text{melt}) + 3\ CO_2(g)$

As the cell operates, molten aluminum deposits on the cathode and sinks to the bottom of the cell, from which it is removed periodically. Such cells operate at a very low voltage of 4.0 to 5.5 V, but at a very high current of 50,000 to 150,000 A.

Aluminum production uses extremely large quantities of electricity, so aluminum production plants are located near hydroelectric power sources, such as those in the Pacific Northwest, because electricity from hydroelectric plants is usually less expensive than that from fossil fuel power plants. Production of each kilogram of aluminum requires about 13 to 16 kWh (4.7×10^4 to 5.8×10^4 kJ) of electric energy; additional energy is required to melt the cryolite. Because of the high energy cost to extract aluminum metal from its ore, there is widespread interest in recycling aluminum beverage containers and other aluminum objects. It takes far less energy to melt and process recycled aluminum than to generate new metal from bauxite. You could run your television set for three hours on the energy saved by recycling just one aluminum can!

BotfR/Alamy

A power-generating dam in the Pacific Northwest.

PROBLEM-SOLVING EXAMPLE 19.2

Aluminum Production

If electricity costs \$0.100 per kilowatt-hour (kWh), calculate the cost of the electricity to produce 1.00 ton aluminum in a Hall-Héroult cell operating at 5.00 V. 1 kWh = 3.60×10^6 J; 1 V = 1 J/C.

Result \$1350

Analyze The reduction reaction is $Al^{3+} + 3\ e^- \longrightarrow Al$ and so 1 mol aluminum requires 3 mol electrons. Calculate the amount (mol) of aluminum metal produced. Use it and Faraday's constant to calculate the charge (C) required. Use the charge, the voltage, and the conversion factor between volts and kWh to calculate the amount of electricity used.

Plan Apply the above analysis to calculate the cost of the electricity used taking into account that 1 kWh costs \$0.10.

Execute First, calculate the amount of aluminum produced.

$$n(\text{Al}) = 1.00 \text{ ton Al}\left(\frac{2000 \text{ lb Al}}{1 \text{ ton Al}}\right)\left(\frac{454 \text{ g Al}}{1 \text{ lb Al}}\right)\left(\frac{1 \text{ mol Al}}{26.98 \text{ g Al}}\right) = 3.37 \times 10^4 \text{ mol Al}$$

1 mol aluminum requires 3 mol electrons. Therefore,

$$\text{Total charge} = 3.37 \times 10^4 \text{ mol Al} \times \frac{3 \text{ mol e}^-}{1 \text{ mol Al}} \times \frac{9.65 \times 10^4 \text{ C}}{1 \text{ mol e}^-} = 9.74 \times 10^9 \text{ C}$$

The electrical energy in kilowatt-hours is

$$E_{\text{electrical}} = 9.74 \times 10^9 \text{ C} \times \frac{5.00 \text{ J}}{1 \text{ C}} \times \frac{1 \text{ kWh}}{3.60 \times 10^6 \text{ J}} = 1.35 \times 10^4 \text{ kWh}$$

$$\text{Cost} = 1.35 \times 10^4 \text{ kWh} \times \frac{\$0.100}{1 \text{ kWh}} = \$1.35 \times 10^3 \text{ or } \$1350$$

☑ **Reasonable Result Check** One ton of aluminum represents more than 30,000 mol of the metal, a substantial amount. In addition, each mole of Al requires three moles of electrons, a large amount of charge. The energy required is significant, more than 10,000 kWh. Even though the cost of the electricity is relatively low, just \$0.10/kWh, the high cost, more than \$1,000, is due to the large amount of aluminum produced. The answer is reasonable.

PROBLEM-SOLVING PRACTICE 19.2

Calculate the time required to produce 1.00 ton Al in a Hall-Héroult cell operating at 1.00×10^5 A.

19-5 Extraction by Chemical Oxidation-Reduction: Phosphorus, Bromine, and Iodine

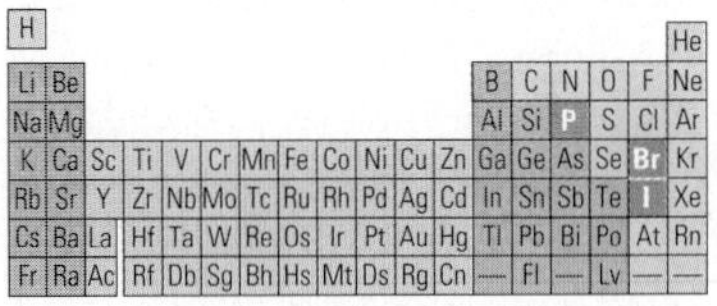

19-5a Phosphorus

Elemental phosphorus is extracted from phosphate-bearing rock by heating the rock with sand (SiO_2) and coke in an electric furnace (Figure 19.13). At 1400 to 1500 °C, a redox

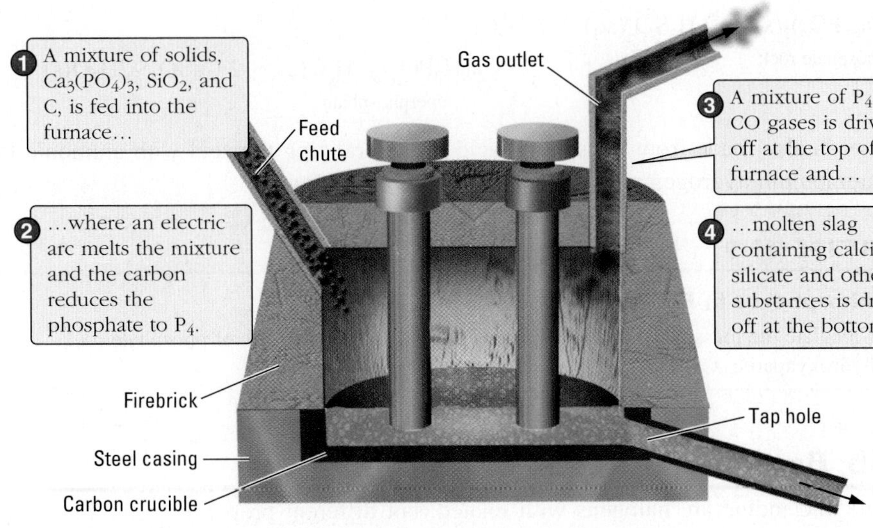

1 A mixture of solids, $Ca_3(PO_4)_3$, SiO_2, and C, is fed into the furnace…
Gas outlet
Feed chute
3 A mixture of P_4 and CO gases is driven off at the top of the furnace and…
2 …where an electric arc melts the mixture and the carbon reduces the phosphate to P_4.
4 …molten slag containing calcium silicate and other substances is drawn off at the bottom.
Firebrick
Tap hole
Steel casing
Carbon crucible

Figure 19.13 The production of phosphorus in an electric furnace.

An electric arc furnace operates on the same principle as electric arc welding, but on a much larger scale.

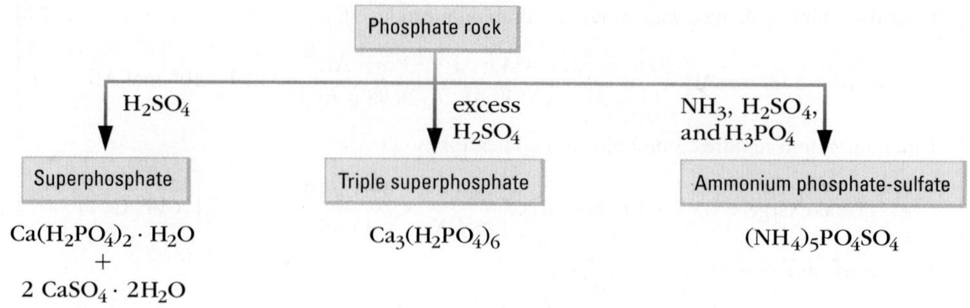
Phosphate rock
H_2SO_4
excess H_2SO_4
NH_3, H_2SO_4, and H_3PO_4
Superphosphate
Triple superphosphate
Ammonium phosphate-sulfate
$Ca(H_2PO_4)_2 \cdot H_2O$ + $2\ CaSO_4 \cdot 2H_2O$
$Ca_3(H_2PO_4)_6$
$(NH_4)_5PO_4SO_4$

Figure 19.14 Schematic diagram showing fertilizers produced from phosphate rock.

reaction produces gaseous phosphorus, which evaporates from the mixture, leaving behind liquid calcium silicate.

$$2\ Ca_3(PO_4)_2(\ell) + 10\ C(s) + 6\ SiO_2(\ell) \longrightarrow P_4(g) + 10\ CO(g) + 6\ CaSiO_3(\ell)$$

The mixture of phosphorus vapor and carbon monoxide gas is passed through water, where the phosphorus condenses and the CO bubbles out.

CONCEPTUAL EXERCISE 19.8

Phosphorus Extraction

The extraction of phosphorus from phosphate rock involves oxidation and reduction. Identify which element is oxidized and which is reduced.

About 90% of the elemental phosphorus produced is oxidized subsequently in air to P_4O_{10}, which reacts with water to produce phosphoric acid, H_3PO_4.

$$P_4(s) + 5\ O_2(g) \longrightarrow P_4O_{10}(s)$$
$$P_4O_{10}(s) + 6\ H_2O(\ell) \longrightarrow 4\ H_3PO_4(aq)$$

Some phosphoric acid is used in soft drinks, baking powder, and detergents.

The principal use of phosphate rock is to make fertilizers directly rather than to produce elemental phosphorus (Figure 19.14). Phosphate rock is reacted with sulfuric acid and converted into a soluble fertilizer. The mixture of hydrated calcium dihydrogen phosphate and hydrated calcium sulfate is called "superphosphate."

$$\underset{\text{phosphate rock}}{Ca_3(PO_4)_2(s)} + 2\ H_2SO_4(aq) + 5\ H_2O(\ell) \longrightarrow \underset{\text{superphosphate}}{Ca(H_2PO_4)_2 \cdot H_2O(s) + 2\ CaSO_4 \cdot 2H_2O(s)}$$

Fertilizer is also made from phosphoric acid by neutralizing the acid with ammonia to form ammonium hydrogen phosphate, $(NH_4)_2HPO_4$.

EXERCISE 19.9

Phosphorus in Phosphate Rock

Calculate the mass percent of phosphorus present in another form of phosphate rock, hydroxyapatite, $Ca_5(PO_4)_3OH$.

19-5b Bromine and Iodine

Bromine and iodine are halogens with related, but different properties. Like the other halogens, they are too reactive to be found uncombined in nature. Consequently, Br_2 and I_2 are produced by the oxidation of their anions.

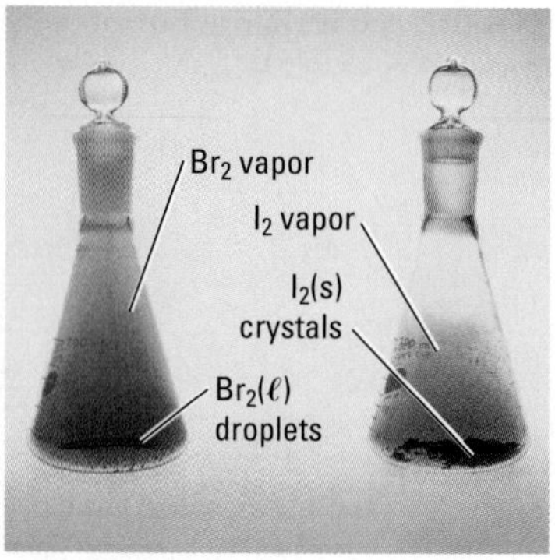
Br_2 vapor
I_2 vapor
I_2(s) crystals
$Br_2(\ell)$ droplets

Bromine and iodine.

Bromine and iodine are produced from seawater or brines (underground natural salt water deposits) by treating the solution with chlorine gas, which oxidizes Br^- to Br_2 and I^- to I_2. This is a case of a more reactive halogen, chlorine, displacing a less reactive one, bromide or iodide, from solution (Figure 19.15).

$$Cl_2(g) + 2\ Br^-(aq) \longrightarrow Br_2(\ell) + 2\ Cl^-(aq) \qquad E° = +0.292\ V$$
$$Cl_2(g) + 2\ I^-(aq) \longrightarrow I_2(aq) + 2\ Cl^-(aq) \qquad E° = +0.823\ V$$

The large positive voltages indicate both as very product-favored reactions.

Another source of iodine is iodate ions, IO_3^-, in Chilean ore deposits. Iodate is converted to I_2 in a two-step process using hydrogen sulfite ions.

$$2\ IO_3^-(aq) + 6\ HSO_3^-(aq) \longrightarrow 2\ I^-(aq) + 6\ HSO_4^-(aq)$$
$$5\ I^-(aq) + IO_3^-(aq) + 3\ H^+(aq) + 3\ HSO_4^-(aq) \longrightarrow 3\ I_2(aq) + 3\ SO_4^{2-}(aq) + 3\ H_2O(\ell)$$

PROBLEM-SOLVING EXAMPLE 19.3

Oxidation-Reduction Reactions

Identify the oxidizing and reducing agents in the first step of the extraction of iodine from IO_3^--bearing ores.

Result Oxidizing agent: IO_3^-; reducing agent: HSO_3^-

Analyze Recall from Chapter 17 that reduction involves a decrease in oxidation number due to a gain of electrons. Oxidation is an increase in oxidation number, indicating a loss of electrons. A reducing agent donates the electrons and is oxidized; its oxidation number increases. Conversely, the oxidizing agent gains electrons and is reduced; its oxidation number decreases.

Plan Assign oxidation numbers and assess their changes. Use these changes to determine the oxidizing and reducing agents.

Execute The +4 oxidation number (state) of sulfur in HSO_3^- is increased to +6 in HSO_4^-. This is oxidation, an increase in oxidation number, indicating a loss of electrons. Thus, the reducing agent, HSO_3^-, is oxidized (+4 sulfur to +6 sulfur); simultaneously, iodine in the oxidizing agent, IO_3^-, is reduced (from +5 to −1).

PROBLEM-SOLVING PRACTICE 19.3

Identify the oxidizing and reducing agents in the second step of the extraction of I_2 from Chilean ores.

CONCEPTUAL EXERCISE 19.10

Bromine Conversion

Use the terms *oxidation, reduction, oxidizing agent,* and *reducing agent* to explain the extraction of bromine from brines.

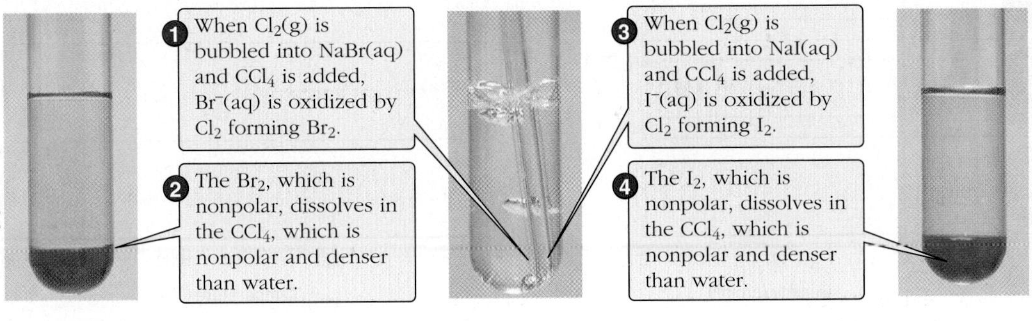

1 When $Cl_2(g)$ is bubbled into NaBr(aq) and CCl_4 is added, $Br^-(aq)$ is oxidized by Cl_2 forming Br_2.
2 The Br_2, which is nonpolar, dissolves in the CCl_4, which is nonpolar and denser than water.
3 When $Cl_2(g)$ is bubbled into NaI(aq) and CCl_4 is added, $I^-(aq)$ is oxidized by Cl_2 forming I_2.
4 The I_2, which is nonpolar, dissolves in the CCl_4, which is nonpolar and denser than water.

Photos: Cengage Learning/Charles D. Winters

Figure 19.15 Displacement of Br_2 and I_2 by Cl_2.

Herbert H. Dow
1866–1930

Courtesy of Dow Historical Collection, Chemical Heritage Foundation Collections

Herbert H. Dow was the first to produce bromine by the electrolysis of brine (1891). In the 1920s, the demand for bromine rose sharply in order to make ethylene dibromide, which was starting to be used in the higher octane gasoline required by high-performance automobile engines. Dow realized that ethylene dibromide demand would be so great that brine sources could not supply enough bromine. He told the head of General Motors that to meet the demand, "... we'll have to go to sea and extract bromine from ocean water."* Herbert Dow died four years before achieving this goal, which was accomplished by his son Willard.

*Brandt, E. N. *Chemical Heritage*, Vol. 18, Number 3, Fall 2000; p. 39.

80 ml
±5%
60
40
20

Cengage Learning/Charles D. Winters

Chunks of metallic potassium in oil. A layer of oil prevents the potassium from reacting with oxygen in the air.

EXERCISE 19.11

Iodine Reaction

Calculate $E°$ for the reaction of I_2(s) with Br^-(aq). What does the value of $E°$ indicate about using I_2(s) to oxidize Br^-(aq) to $Br_2(\ell)$?

19-6 A Periodic Perspective: The Main-Group Elements

This section describes some of the properties and uses of selected main-group elements, Groups 1A–8A (1, 2, 13–18). For each group, there is a general overview and at least one element is highlighted for more detailed description.

19-6a Group 1A(1): The Alkali Metals

The alkali metals make up the leftmost group of the periodic table. Their densities, melting points, boiling points, atomic radii, and ionic radii are given in Figure 19.16. The densities are low because the alkali metals have relatively large atomic radii compared to their molar masses. Weak metallic bonding (← Sec. 9-9a) is responsible for their softness and relatively low melting points.

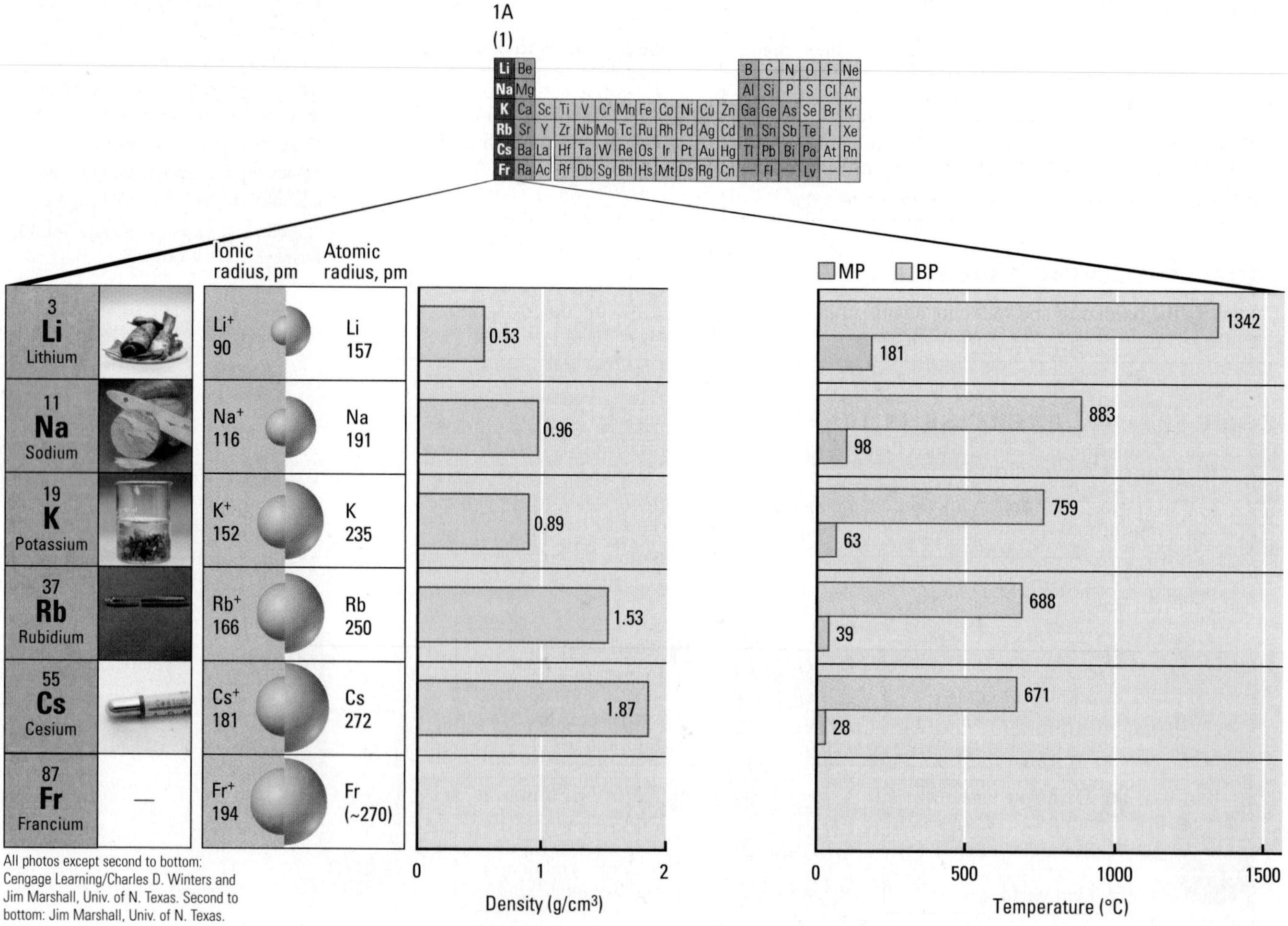
1A
(1)
Li Be B C N O F Ne
Na Mg Al Si P S Cl Ar
K Ca Sc Ti V Cr Mn Fe Co Ni Cu Zn Ga Ge As Se Br Kr
Rb Sr Y Zr Nb Mo Tc Ru Rh Pd Ag Cd In Sn Sb Te I Xe
Cs Ba La Hf Ta W Re Os Ir Pt Au Hg Tl Pb Bi Po At Rn
Fr Ra Ac Rf Db Sg Bh Hs Mt Ds Rg Cn — Fl — Lv — —
Ionic radius, pm
Atomic radius, pm
MP
BP
3 Li Lithium
Li+ 90
Li 157
0.53
1342
181
11 Na Sodium
Na+ 116
Na 191
0.96
883
98
19 K Potassium
K+ 152
K 235
0.89
759
63
37 Rb Rubidium
Rb+ 166
Rb 250
1.53
688
39
55 Cs Cesium
Cs+ 181
Cs 272
1.87
671
28
87 Fr Francium
—
Fr+ 194
Fr (~270)
0
1
2
Density (g/cm3)
0
500
1000
1500
Temperature (°C)

All photos except second to bottom: Cengage Learning/Charles D. Winters and Jim Marshall, Univ. of N. Texas. Second to bottom: Jim Marshall, Univ. of N. Texas.

Figure 19.16 Group 1A(1) elements, the alkali metals: Li, Na, K, Rb, Cs, Fr.

Table 19.4 Reactions of Alkali Metals

Group 1A Metal (M)	Combining Substance	Reaction
Li	Oxygen	$4\,Li(s) + O_2(g) \longrightarrow 2\,Li_2O(s)$
Na	Oxygen	$2\,Na(s) + O_2(g) \longrightarrow Na_2O_2(s)$
K, Rb, Cs	Oxygen	$M(s) + O_2(g) \longrightarrow MO_2(s)$
All	Halogens	$2\,M(s) + X_2 \longrightarrow 2\,MX(s)$; $X = F, Cl, Br, I$
All	Sulfur	$2\,M(s) + S(s) \longrightarrow M_2S(s)$
Li	Nitrogen	$6\,Li(s) + N_2(g) \longrightarrow 2\,Li_3N(s)$
All	Water	$2\,M(s) + 2\,H_2O(\ell) \longrightarrow 2\,M^+(aq) + 2\,OH^-(aq) + H_2(g)$

The chemical behavior of the alkali metals is dominated by loss of the ns^1 outer electron leading solely to the formation of M^+ cations. Therefore, most alkali-metal compounds are ionic, except for organometallic compounds containing an alkali metal-to-carbon bond. The Group 1A elements react with air, water, and most nonmetals. The reactions of the heavier alkali metals are particularly vigorous, even explosive. Reactions of alkali metals with oxygen, sulfur, the halogens, and water are summarized in Table 19.4. The uses of some alkali metal compounds are given in Table 19.5.

The product of the reaction of an alkali metal with oxygen is dependent on the alkali metal, as seen from Table 19.4. Lithium is the only Group 1A metal that reacts directly with oxygen to form in good yield the normal oxide, M_2O, containing M^+ metal ions and O^{2-} oxide ions. In contrast, sodium reacts directly with oxygen to form predominantly sodium peroxide, Na_2O_2, which contains Na^+ and O_2^{2-}, peroxide, ions. The remaining alkali metals produce the metal superoxide, MO_2, in which M^+ and superoxide O_2^- ions are present. The peroxide and superoxide ions contain two covalently bonded oxygen atoms, with superoxide ions having one fewer electron than peroxide ions. Potassium superoxide is a convenient source of oxygen used in emergency breathing apparatus, such as for firefighters and miners in rescue circumstances where the concentration of oxygen is low. Water vapor in the breath reacts with superoxide ions to produce oxygen and potassium hydroxide; the latter removes exhaled carbon dioxide.

The molecular orbital theory (← Sec. 6-12) can be used to describe bonding in O_2^{2-} and O_2^- ions.

Self-contained self-rescue devices are described in Chapter 8 (← Sec. 8-5).

$$4\,KO_2(s) + 2\,H_2O(\ell) \longrightarrow 4\,KOH(s) + 3\,O_2(g)$$
$$2\,KOH(s) + CO_2(g) \longrightarrow K_2CO_3(s) + H_2O(\ell)$$

Table 19.5 Uses of Alkali Metals and Some of Their Compounds

Element or Compound	Uses
Lithium	Lithium-ion batteries for computers, cell phones
Lithium carbonate, Li_2CO_3	Treatment of bipolar disorder
Sodium	Nuclear reactor coolant, manufacture of Ti
Sodium chloride, NaCl	Production of sodium metal, chlorine, NaOH
Sodium hydroxide, NaOH	Soaps and detergents, pulp and paper industry, bleach preparation, widely used industrial base
Sodium carbonate, Na_2CO_3	Glass manufacturing, water softening, detergents, reduction of SO_2 stack gas emission
Sodium hydrogen carbonate, $NaHCO_3$	Baking powder, baking soda, fire extinguishers, pharmaceuticals
Potassium nitrate, KNO_3	Gunpowder, fireworks; strong oxidizing agent
Potassium superoxide, KO_2	Oxygen source in emergency breathing apparatus
Rubidium and cesium	Photoelectric cells

Cengage Learning/Charles D. Winters

Figure 19.17 Reaction of potassium with water. When water is dripped onto potassium metal, a violent reaction occurs.

EXERCISE 19.12

Lewis Structures

Write the Lewis structures for oxide, peroxide, and superoxide ions. Write the molecular orbital diagrams for oxygen, peroxide ion, and superoxide ion.

Halogens react directly with alkali metals to produce stable binary halide salts whose lattice energies are substantial (← **Sec. 5-13**). Examples include NaCl, KBr, and CsI. Likewise, the Group 1A metals all react directly with sulfur to form ionic sulfides with the general formula M_2S. Lithium is the only Group 1A metal that reacts directly with nitrogen gas, forming an ionic nitride, Li_3N.

Water reacts vigorously with the alkali metals, especially the heavier ones (Figure 19.17), to produce hydrogen gas plus a solution of the metal hydroxide. The reaction of sodium with water is a good example:

$$Na(s) + 2\ H_2O(\ell) \longrightarrow 2\ Na^+(aq) + 2\ OH^-(aq) + H_2(g)$$

This reaction is highly exothermic: $\Delta_r H° = -368.6$ kJ/mol.

Because sodium metal is a strong reducing agent, it is used to obtain other metals from their metal halides. In particular, titanium, an element essential in aircraft production, can be prepared from its chloride by reduction with sodium.

$$TiCl_4(s) + 4\ Na(s) \longrightarrow Ti(s) + 4\ NaCl(s)$$

A major use for sodium metal was in the production of tetraethyllead, $Pb(C_2H_5)_4$, once used as an octane enhancer in leaded gasoline. Although leaded gasoline is still used in some countries, tetraethyllead is banned as a gasoline additive in the United States. Consequently, sodium production has declined.

Liquid sodium has high thermal conductivity and an anomalously high heat capacity. Metallic sodium has a low melting point and can be liquefied easily. These properties make liquid sodium an excellent heat-exchange liquid in some types of nuclear reactors (← **Sec. 18-6a**).

19-6b Group 2A(2): The Alkaline-Earth Metals

Like their Group 1A neighbors, the alkaline-earth metals (Group 2A) are silvery white, ductile, and malleable metals that are a bit harder than the alkali metals. The densities, melting points, boiling points, atomic radii, and ionic radii of the alkaline-earth metals are given in Figure 19.18. The Group 2A elements, like the adjacent Group 1A elements, show regular changes in properties down the group. Chemical reactivity increases down the group.

Using pulsed-laser vaporization of beryllium metal, researchers recently created and characterized Be_2 in its ground state for the first time.

The alkaline-earth metals are characterized chemically by the loss of the ns^2 outer electrons to yield 2+ ions. Beryllium is the exception, forming no predominantly ionic compounds due to the high charge density of Be^{2+} ions, which are found only as hydrated

Table 19.6 Some Reactions of Alkaline-Earth Elements

Group 2A Metal (M)	Combining Substance	Reaction
Be, Mg, Ca	Oxygen	$2\ M(s) + O_2(g) \longrightarrow 2\ MO(s)$
Sr, Ba	Oxygen	$M(s) + O_2(g) \longrightarrow MO_2(s)$
All	Halogens	$M(s) + X_2 \longrightarrow MX_2(s)$; X = F, Cl, Br, I
All (high temp.)	Nitrogen	$3\ M(s) + N_2(g) \longrightarrow M_3N_2(s)$
Ca, Sr, Ba	Water	$M(s) + 2\ H_2O(\ell) \longrightarrow M(OH)_2(aq) + H_2(g)$
Mg, Ca, Sr, Ba	Hydrogen	$M(s) + H_2(g) \longrightarrow MH_2(s)$
Mg, Ca, Sr, Ba	Carbon	$M(s) + 2\ C(s) \longrightarrow MC_2(s)$

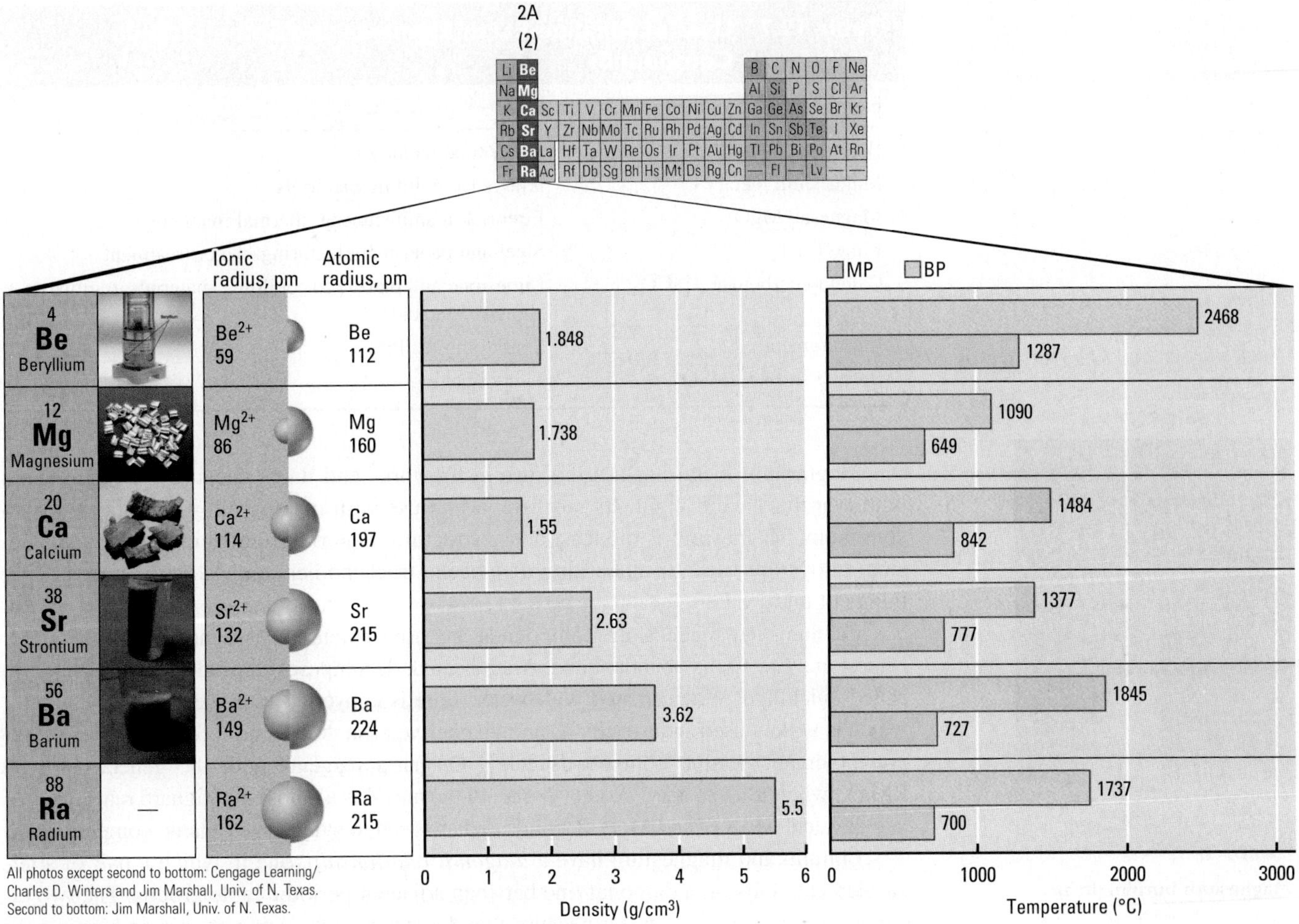
2A
(2)
Ionic radius, pm
Atomic radius, pm
MP
BP
4 Be Beryllium
Be2+ 59
Be 112
1.848
2468
1287
12 Mg Magnesium
Mg2+ 86
Mg 160
1.738
1090
649
20 Ca Calcium
Ca2+ 114
Ca 197
1.55
1484
842
38 Sr Strontium
Sr2+ 132
Sr 215
2.63
1377
777
56 Ba Barium
Ba2+ 149
Ba 224
3.62
1845
727
88 Ra Radium
—
Ra2+ 162
Ra 215
5.5
1737
700
0 1 2 3 4 5 6
Density (g/cm3)
0 1000 2000 3000
Temperature (°C)

All photos except second to bottom: Cengage Learning/ Charles D. Winters and Jim Marshall, Univ. of N. Texas. Second to bottom: Jim Marshall, Univ. of N. Texas.

Figure 19.18 Group 2A(2) elements, the alkaline earths.

ions such as $[Be(H_2O)_4]^{2+}$. Anhydrous beryllium compounds are covalent and many are polymeric solids, such as $BeCl_2$, which has a Cl-bridging chain structure. Beryllium and its compounds are poisonous.

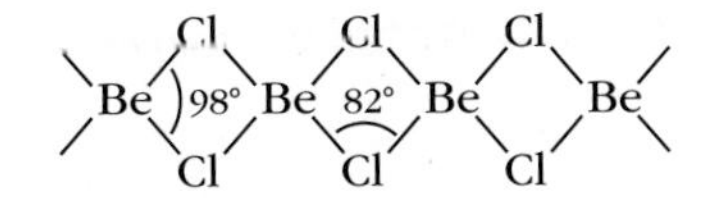
Cl
Be
98°
82°
Cl

The polymeric structure of solid beryllium chloride.

Some reactions of the alkaline-earth metals are summarized in Table 19.6. Like the alkali metals, the lighter alkaline-earth metals, Mg and Ca, react with oxygen to form oxides, while the heavier metals, Sr and Ba, form peroxides. Beryllium oxide forms directly only at temperatures higher than 600 °C. The oxides, except BeO, react with water to form the corresponding hydroxides. The vigorousness of the reaction of the metals with water increases down the group: Beryllium does not react; magnesium does so only with steam above 100 °C; calcium and strontium react slowly with liquid water at room temperature, but barium reacts rapidly. All of the Group 2A elements react directly with nitrogen to form nitrides, which react with water to form aqueous hydroxides and release ammonia. Direct halogenation of the metals forms ionic halide salts of the general formula MX_2. The alkaline-earth metals, except beryllium, react directly with hydrogen to form hydrides, MH_2, and with carbon to form carbides, MC_2.

The toxicity of Be compounds is considered to be due to beryllium displacing Mg^{2+} ions from Mg-based enzymes.

The carbide ion, C_2^{2-}, has the Lewis structure $:C{\equiv}C:^{2-}$.

EXERCISE 19.13

Group 2A Compounds

Write the formulas for these compounds.

(a) Calcium oxide (b) Barium peroxide (c) Strontium nitride
(d) Calcium carbide

Table 19.7 Uses of Alkaline-Earth Elements and Some of Their Compounds

Element or Compound	Uses
Beryllium metal	X-ray tube windows
Magnesium metal	Alloys for building materials
Magnesia, MgO	Firebrick manufacturing, thermal insulator
Lime, CaO	Steel and paper manufacturing, water treatment
Calcium carbonate, $CaCO_3$	Limestone and marble for building materials, toothpaste abrasive, antacid
Barium metal	Spark plugs (alloy)
Barium sulfate, $BaSO_4$	X-ray imaging

Cengage Learning/Charles D. Winters

Magnesium burning in air.

Lithium is the only Group 1A element that forms a nitride by direct reaction with N_2.

Magnesium metal has a minor use in fireworks and flares because the metal burns with a brilliant white light. Its most important use is in making alloys, principally with aluminum. Magnesium is the least dense structural material; lightweight, strong magnesium alloys are used to make aircraft wheels, truck bodies, and ladders, among other things (Table 19.7).

Calcium and magnesium compounds are used extensively, as noted in Table 19.7. Limestone, $CaCO_3$, is abundant and, when heated, decomposes to produce lime, CaO, also called quicklime. When treated with water, lime is converted into slaked lime, $Ca(OH)_2$, which is widely used industrially as an inexpensive base. Magnesium forms a series of important organometallic compounds called Grignard reagents with the general formula RMgX, where R is an alkyl group (← **Sec. 10-4c**) and X is a halogen. Grignard reagents contain covalent Mg—C and Mg—X bonds and are used to synthesize organic compounds.

Lithium and magnesium have a *diagonal relationship*, one in which a pair of similar elements falls on a diagonal line between adjacent periodic table groups. The two elements are chemically similar because they have nearly the same atomic and ionic radii (Li 157 pm, Mg 160 pm; Li^+ 90 pm, Mg^{2+} 86 pm). For example, each of these metals reacts with nitrogen directly to form a nitride.

EXERCISE 19.14

Lighting Things Up

When magnesium metal burns in air, magnesium nitride and magnesium oxide are produced.

(a) Write the formula for magnesium nitride.
(b) Write a balanced chemical equation for the formation of magnesium nitride from the elements.

19-6c Group 3A(13): Boron, Aluminum, Gallium, Indium, Thallium

All elements of Group 3A have an ns^2np^1 outer electron configuration leading to +3 as the stable oxidation state except thallium, for which the +1 oxidation state is the more stable. *Where multiple oxidation states are possible for elements in a group, the lower oxidation state is usually the more favored in the heavier elements of the group.*

Some of the physical and atomic properties of the elements in Group 3A are given in Figure 19.19. The elements of this group exhibit a wider range of properties than those in Group 1A or 2A. Boron, the first member of the group, is a metalloid, an anomaly in a group where all other elements are silvery white metals. Aluminum is more representative of the group and for that reason is discussed in more detail here. The extraction of aluminum metal from bauxite ore was described in Section 19-4d. Aluminum is the most

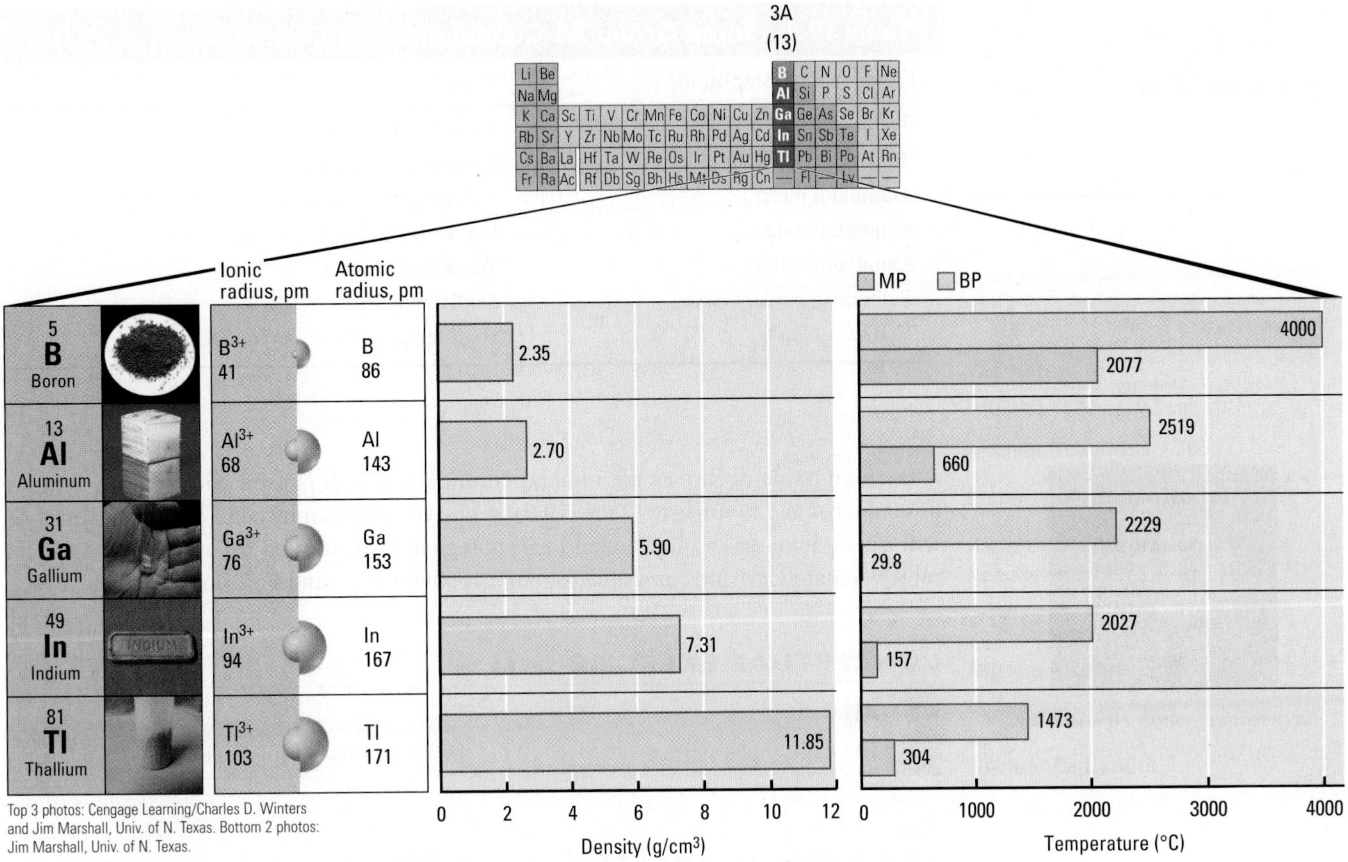
3A
(13)
Li Be B C N O F Ne
Na Mg Al Si P S Cl Ar
K Ca Sc Ti V Cr Mn Fe Co Ni Cu Zn Ga Ge As Se Br Kr
Rb Sr Y Zr Nb Mo Tc Ru Rh Pd Ag Cd In Sn Sb Te I Xe
Cs Ba La Hf Ta W Re Os Ir Pt Au Hg Tl Pb Bi Po At Rn
Fr Ra Ac Rf Db Sg Bh Hs Mt Ds Rg Cn — Fl — Lv — —
Ionic radius, pm
Atomic radius, pm
MP
BP
5 B Boron
B^{3+} 41
B 86
2.35
2077
4000
13 Al Aluminum
Al^{3+} 68
Al 143
2.70
660
2519
31 Ga Gallium
Ga^{3+} 76
Ga 153
5.90
29.8
2229
49 In Indium
INDIUM
In^{3+} 94
In 167
7.31
157
2027
81 Tl Thallium
Tl^{3+} 103
Tl 171
11.85
304
1473
0 2 4 6 8 10 12
Density (g/cm³)
0 1000 2000 3000 4000
Temperature (°C)

Top 3 photos: Cengage Learning/Charles D. Winters and Jim Marshall, Univ. of N. Texas. Bottom 2 photos: Jim Marshall, Univ. of N. Texas.

Figure 19.19 Group 3A(13) elements.

abundant metal ion, Al^{3+}, in Earth's crust, exceeded in elemental abundance only by oxygen and silicon. As with Li and Mg, a diagonal relationship exists between Be and Al and between B and Si due to similarities in electronegativity and effective nuclear charge (← **Sec. 5-9c**) within each pair of elements. Table 19.8 summarizes some reactions of the Group 3A elements.

Boron and hydrogen form an extensive series of covalently bonded hydrides called *boranes* with the general formulas B_nH_{n+4} or B_nH_{n+6} such as B_2H_6 and B_5H_{11}, respectively. These compounds contain boron atoms bridged by hydrogen atoms in what is described as a three-center-two-electron bond.

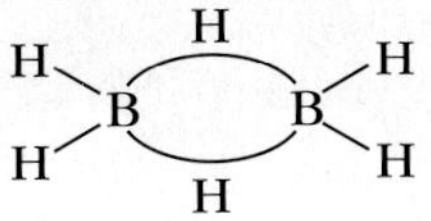

The B—H—B bridge bonding in diborane, B_2H_6.

Aluminum is an economically important, useful metal because of its low density ($2.70\ g/cm^3$) and high strength when alloyed. It can be fashioned into wire, food wrapping sheets, stepladders, aircraft and automotive parts, and many other useful items. Aluminum metal resists corrosion because a transparent, chemically inactive film of aluminum oxide clings avidly to the metal's surface and protects the metal beneath it from further oxidation.

$$4\ Al(s) + 3\ O_2(g) \longrightarrow 2\ Al_2O_3(s)$$

Table 19.8 Some Reactions of Group 3A Elements

Group 3A Element (M)	Combining Substance	Reaction
Al, Ga, In	Oxygen	$4\ M(s) + 3\ O_2(g) \longrightarrow 2\ M_2O_3(s)$
B, Al, Ga, In	Halogens	$2\ M(s) + 3\ X_2 \longrightarrow 2\ MX_3$; X = F, Cl, Br, I
Ga, In (high temperature)	Water	$2\ M(s) + 6\ H_2O(g) \longrightarrow 2\ M^{3+}(aq) + 6\ OH^-(aq) + 3\ H_2(g)$

Table 19.9 Some Group 3A Compounds and Their Uses

Element or Compound	Uses
Boron oxide	Borosilicate glass
Boric acid, H_3BO_3	Eyewash, astringent
Aluminum metal	Foil wrap, alloys, structural material
Aluminum oxide	Pigments, fireworks, refractory bricks, toothpaste
Aluminum sulfate	Water purification
Gallium arsenide, GaAs	Semiconductor
$Tl_2Ba_2Ca_2Cu_3O_{10}$	High-temperature superconductor

Gemstones: sapphire and ruby.

Aluminum oxide occurs as the mineral corundum, which is used widely as an abrasive in sandpaper and toothpaste (Table 19.9). A number of precious gems are primarily Al_2O_3 with small amounts of other metal ions strategically substituted for aluminum ions. Red rubies contain Cr^{3+} ions and blue sapphires contain Fe^{2+} and Fe^{3+} ions.

CONCEPTUAL EXERCISE 19.15

Al_4^{4-}

The Al_4^{4-} ion has been synthesized. Write the Lewis structure of this ion.

19-6d Group 4A(14): Carbon, Silicon, Germanium, Tin, Lead

The relative uniformity of the Group 1A alkali metals is absent from the Group 4A elements, which display the full range of element types from nonmetals to metals. Carbon, the first member, is a nonmetal; silicon and germanium are both metalloids; and tin and lead are metals. Some physical and atomic properties of the Group 4A elements are shown in Figure 19.20.

The Group 4A elements all have an ns^2np^2 outer electron configuration. Promotion of the ns^2 electrons into empty np orbitals allows for hybridization and, through electron sharing, the formation of four bonds as in compounds such as CH_4, $SiBr_4$, $SnCl_4$, and $Pb(C_2H_5)_4$. The bonding in Group 4A compounds shifts from predominantly covalent in earlier members of the group to more ionic with tin and lead. The lower oxidation state (+2) is more important for tin and lead than the +4 state. Sn(II) and Pb(II) compounds are white, crystalline solids, whereas the Sn(IV) and Pb(IV) analogs are volatile liquids consisting of covalently bonded molecules. The np^2 electrons are used to form the lower oxidation state compounds. In such cases, the ns^2 electrons are not involved in bonding, a phenomenon sometimes referred to as the *"inert pair effect."* Sn(II) compounds are reducing agents, being converted to Sn(IV) by oxidizing agents. In contrast, Sn(IV) compounds are oxidizing agents that are reduced to Sn(II) by reducing agents.

SnCl₄(ℓ)
PbCl₄(ℓ)
SnCl₂(s)
PbCl₂(s)

Chlorides of tin and lead. Compounds of Sn(IV) and Pb(IV) have properties of covalent substances; compounds of Sn(II) and Pb(II) have ionic properties.

Carbon compounds, and to a lesser extent silicon compounds, exhibit **catenation**, in which *bonds between atoms of the same element form chains or rings*. Hydrocarbons containing carbon-to-carbon bonds exemplify this phenomenon (← **Sec. 2-9**). The chemistry of carbon and its compounds has been described in Chapters 2 and 10. Silicon chemistry was discussed in Section 19-2a.

Tin and lead have been known since ancient times; lead lined the aqueducts to ancient Rome. In contrast, germanium was not discovered until 1886 after Mendeleev predicted its expected properties in 1871, calling the proposed element *ekasilicon.* Like diamond and silicon, germanium has a covalent network structure and was used in the

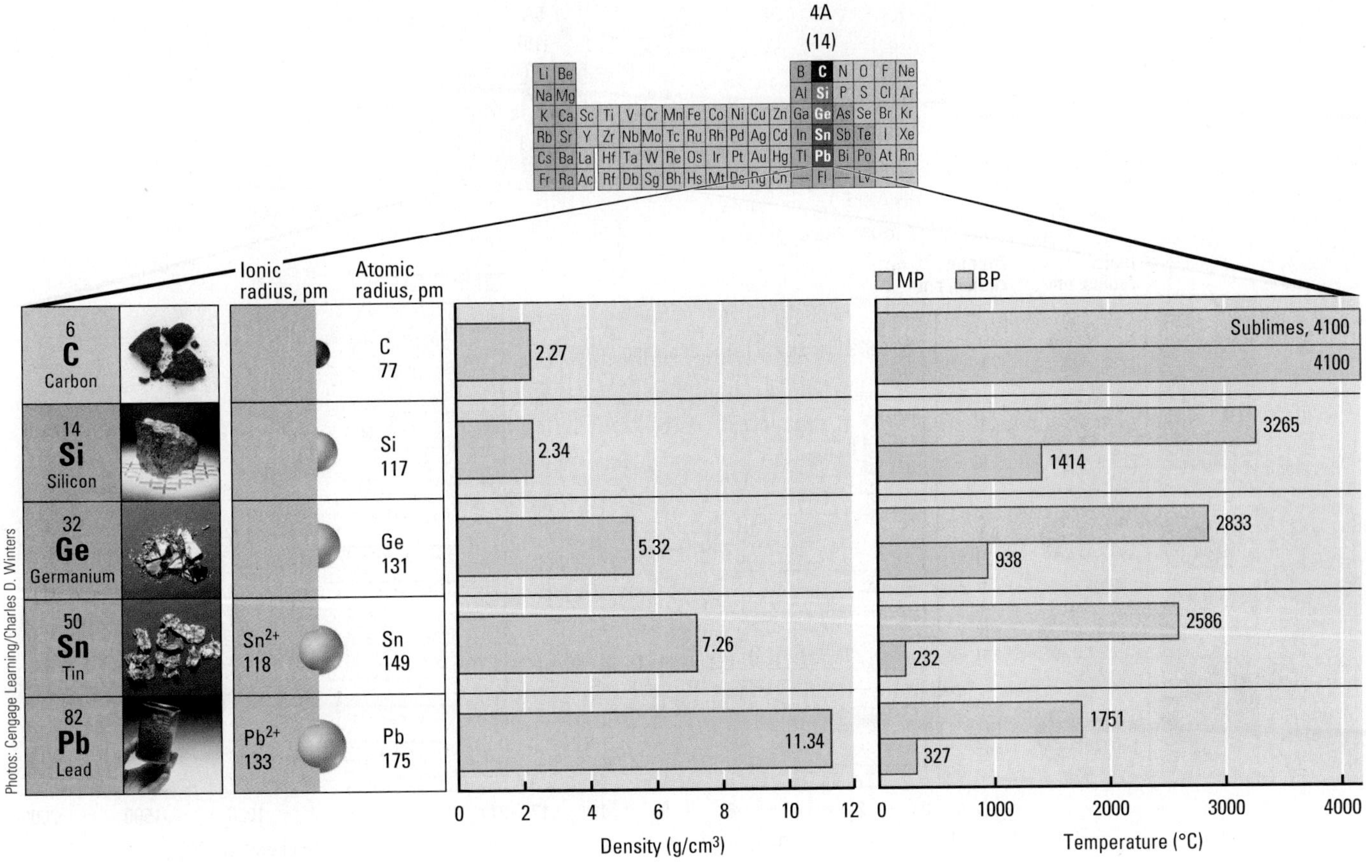
4A
(14)
Ionic radius, pm
Atomic radius, pm
MP
BP
6 C Carbon
14 Si Silicon
32 Ge Germanium
50 Sn Tin
82 Pb Lead
C 77
Si 117
Ge 131
Sn²⁺ 118
Sn 149
Pb²⁺ 133
Pb 175
2.27
2.34
5.32
7.26
11.34
Sublimes, 4100
4100
3265
1414
2833
938
2586
232
1751
327
0 2 4 6 8 10 12
Density (g/cm³)
0 1000 2000 3000 4000
Temperature (°C)
Photos: Cengage Learning/Charles D. Winters

Figure 19.20 **Group 4A(14) elements.**

first transistors due to its semiconductor properties. Because ultrapure silicon is cheaper and more rugged, it has replaced germanium for this application.

Tin is used to make pewter, an alloy of 85% tin and the remainder a combination of copper, zinc and antimony, or lead. Bronze, an alloy of tin (20%) and copper (80%), revolutionized tool making and weaponry because bronze can be fabricated into a sharp, hard edge. Its use for this purpose ushered in the Bronze Age. Elemental copper occurs naturally. Tin was available because cassiterite, SnO_2, its ore, can be reduced easily using a charcoal fire.

Aksenova Natalya/Shutterstock.com
F. Jimenez Meca/Shutterstock.com

Articles made of bronze (left) and pewter (right).

$$SnO_2(s) + 2\,C(s) \longrightarrow Sn(\ell) + 2\,CO(g)$$

The molten tin was recovered readily as it flowed from the fire.

Lead was used to produce tetraethyllead $(C_2H_5)_4Pb$, as a gasoline additive to enhance octane rating (← **Sec. 10-1b**), but this use has been phased out in the United States due to the release of toxic lead compounds into the atmosphere and the fact that lead destroys the catalytic effect of automobile catalytic converters. Lead reacts with oxygen and carbon dioxide to form an oxide or carbonate coating on the catalyst that prevents reactants from contacting the metal. Lead reacts with and dissolves slowly in water. Because lead compounds are toxic, they must not be allowed in water used for human consumption.

19-6e Group 5A(15): Nitrogen, Phosphorus, Arsenic, Antimony, Bismuth

Like the elements in Group 4A, the lightest to heaviest members of Group 5A range from typical nonmetals (N and P) to metalloids (As and Sb) and then to a metal (Bi). Some physical properties of the Group 5A elements are given in Figure 19.21.

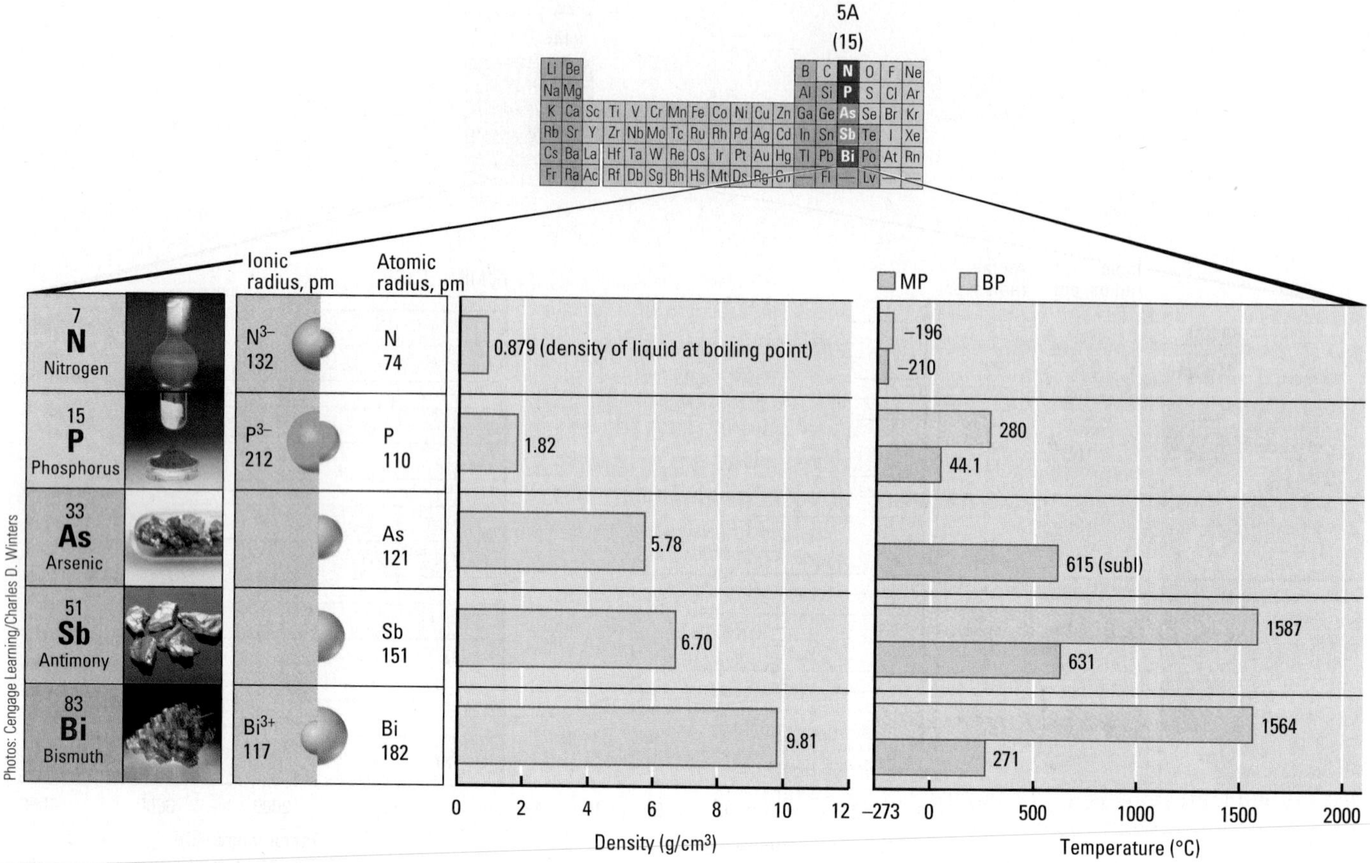
5A
(15)
Ionic radius, pm
Atomic radius, pm
7 N Nitrogen
15 P Phosphorus
33 As Arsenic
51 Sb Antimony
83 Bi Bismuth
N3– 132
P3– 212
Bi3+ 117
N 74
P 110
As 121
Sb 151
Bi 182
0.879 (density of liquid at boiling point)
1.82
5.78
6.70
9.81
Density (g/cm3)
MP
BP
–196
–210
280
44.1
615 (subl)
1587
631
1564
271
Temperature (°C)
Photos: Cengage Learning/Charles D. Winters

Figure 19.21 Group 5A(15) elements.

All of these elements have an ns^2np^3 outer electron configuration. Except for nitrogen, all can form pentavalent compounds such as PCl_5 and $BiCl_5$. Nitrogen forms only trivalent compounds such as NH_3 and NCl_3, because the nitrogen atom is too small to accommodate five bonding pairs of electrons around it.

Cryogen, from the Greek word *kryos* meaning "icy cold."

Uses of Nitrogen and Its Compounds Liquid nitrogen, b.p. = −196 °C, is a **cryogen**, *a substance that can maintain very low temperatures.* It is used in cryosurgery, for example, to cool an area of skin prior to removal of a wart or other unwanted or pathogenic tissue. Because of its low boiling point and inertness, liquid nitrogen has found wide use in frozen-food preparation and preservation during transit. Because nitrogen is chemically unreactive at room temperature, it is used as an inert atmosphere for applications such as welding.

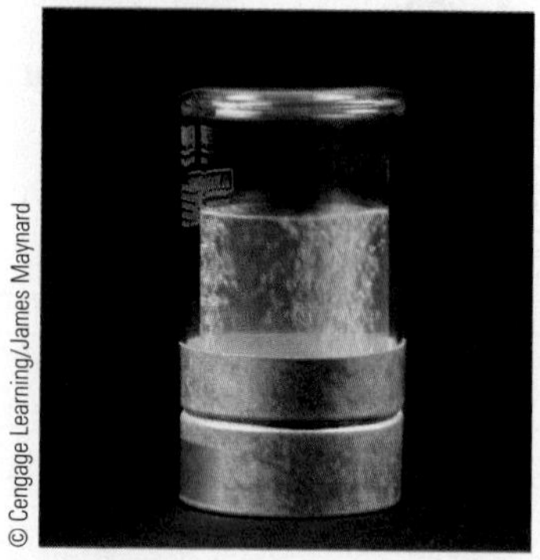
© Cengage Learning/James Maynard

Liquid nitrogen. Even when insulated by a transparent vacuum flask, liquid nitrogen boils vigorously at room temperature.

Nitrogen, phosphorus, and potassium are primary nutrients for plants. Although bathed in an atmosphere containing abundant nitrogen, most plants are unable to use the air directly as a supply of this vital element due to the energy required to break the N≡N triple bond, one of the strongest bonds known. **Nitrogen fixation** is *the process of changing atmospheric nitrogen into water-soluble compounds that can be absorbed through plant roots and assimilated by a plant.* Nitrogen fixation is part of the **nitrogen cycle** (Figure 19.22), a *natural cycle of chemical pathways involving nitrogen and its compounds.* In nitrogen fixation, nitrogen-fixing bacteria convert N_2 into NH_3. Ammonia is converted by other bacteria into nitrate, NO_3^-, which is used by plants. When an organism dies, bacteria reverse the process by converting nitrate to N_2 and organic nitrogen compounds to ammonia.

Most plants thrive on soils rich in nitrates, but many plants that grow in swamps, where there is a lack of oxidized materials, can use reduced forms of nitrogen such as the

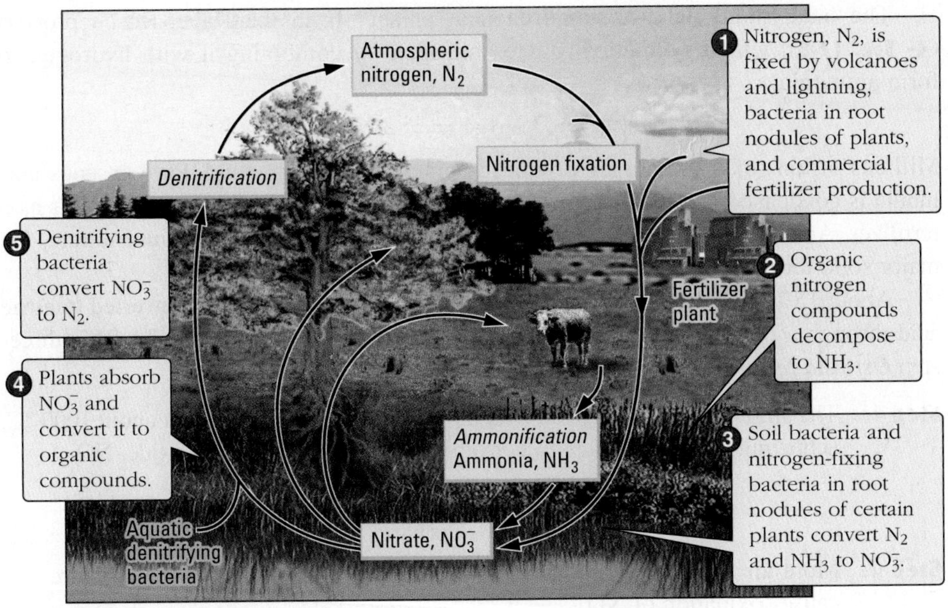

Atmospheric nitrogen, N_2
1 Nitrogen, N_2, is fixed by volcanoes and lightning, bacteria in root nodules of plants, and commercial fertilizer production.
Denitrification
Nitrogen fixation
5 Denitrifying bacteria convert NO_3^- to N_2.
2 Organic nitrogen compounds decompose to NH_3.
Fertilizer plant
4 Plants absorb NO_3^- and convert it to organic compounds.
Ammonification Ammonia, NH_3
3 Soil bacteria and nitrogen-fixing bacteria in root nodules of certain plants convert N_2 and NH_3 to NO_3^-.
Aquatic denitrifying bacteria
Nitrate, NO_3^-

Figure 19.22 Natural nitrogen chemical pathways: the nitrogen cycle.

ammonium ion. The nitrate ion is the most highly oxidized form of combined nitrogen, and the ammonium ion is the most reduced form of nitrogen.

CONCEPTUAL EXERCISE 19.16

Chemically Combined Nitrogen

Verify the statement that the nitrate ion is the most highly oxidized form of combined nitrogen and that the ammonium ion is the most reduced form of combined nitrogen.

Nitrogen is fixed by natural processes on a massive scale in two ways. In the first method, nitrogen is oxidized under highly energetic conditions in the discharge of lightning or, to a lesser extent, in a fire. The initial atmospheric reaction is that of nitrogen with oxygen to form nitrogen monoxide, NO, a colorless, reactive gas.

$$N_2(g) + O_2(g) \rightleftharpoons 2\,NO(g) \qquad K_c = 1.7 \times 10^{-3} \text{ (at 2300 K)}$$

Once formed, nitrogen monoxide is easily oxidized in air to nitrogen dioxide, NO_2, which dissolves in water to form nitrous acid, HNO_2 and nitric acid, HNO_3.

$$H_2O(\ell) + 2\,NO_2(g) \longrightarrow HNO_2(aq) + HNO_3(aq)$$

nitrous acid nitric acid

These acids are readily soluble in rain, clouds, or ground moisture, and thus increase nitrogen concentration in soil. They also contribute to the formation of acid rain (← **Sec. 15-3**).

In the second natural method of nitrogen fixation, bacteria that live on the roots of plants called *legumes,* such as clover, beans, and peas, convert atmospheric nitrogen into ammonia. This complex series of reactions depends on enzyme catalysis. Under ideal conditions, legume fixation can add more than 100 lb of nitrogen per acre of soil in one growing season.

U. S. production (2012):
NH_3, 9.1 × 10^9 kg;
NH_4NO_3, 2.0 × 10^9 kg

The main industrial use of nitrogen at present is in the Haber-Bosch process (← **Sec. 12-8**), which synthetically fixes nitrogen by combining it with hydrogen to form ammonia.

$$N_2(g) + 3\,H_2(g) \rightleftharpoons 2\,NH_3(g)$$

Millions of tons of ammonia are produced annually by this method. Pure gaseous ammonia is condensed and the liquid anhydrous ammonia is applied directly to fields as a fertilizer. Ammonia is also reacted with nitric acid to produce ammonium nitrate, the major solid fertilizer in the world.

About 15% of the ammonia made by the Haber-Bosch process is converted to nitric acid through a process developed by a German chemist, Wilhelm Ostwald. This three-step *Ostwald process* is carried out at pressures of 1 to 10 atm.

Step 1: The ammonia is burned in air over a platinum-rhodium catalyst at about 1000 °C, achieving a greater than 95% conversion of ammonia to nitric oxide, NO.

$$4\,NH_3(g) + 5\,O_2(g) \xrightarrow[\text{catalyst}]{1000\ °C} 4\,NO(g) + 6\,H_2O(g)$$

Step 2: More air is added to the gaseous mixture, which lowers the temperature and causes oxidation of NO.

$$2\,NO(g) + O_2(g) \rightleftharpoons 2\,NO_2(g)$$

Step 3: The nitrogen dioxide produced in the second step is passed through water to produce nitric acid.

$$3\,NO_2(g) + H_2O(\ell) \longrightarrow 2\,HNO_3(aq) + NO(g)$$

The resulting aqueous solution is about 60% nitric acid by mass. The anhydrous acid is produced by adding sulfuric acid and boiling the mixture to distill nearly pure nitric acid from it.

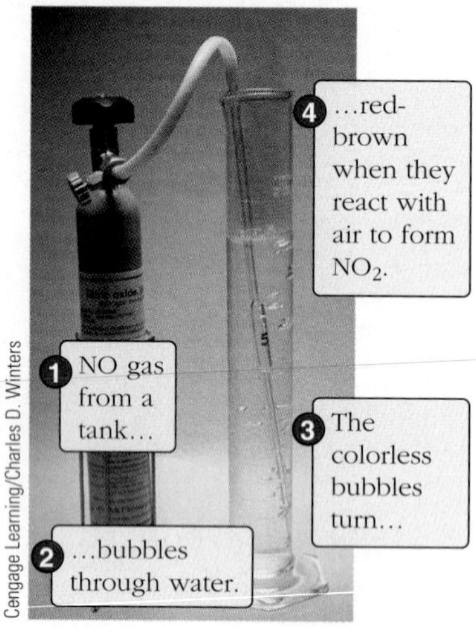
4 ...red-brown when they react with air to form NO_2.
1 NO gas from a tank...
3 The colorless bubbles turn...
2 ...bubbles through water.
Cengage Learning/Charles D. Winters

Colorless NO reacts with air forming red-brown NO_2.

PROBLEM-SOLVING EXAMPLE 19.4

Another Nitrogen-Hydrogen Compound

Compounds of nitrogen and hydrogen other than ammonia exist. For example, *trans*-tetrazene, N_4H_4, has four nitrogen atoms in a chain with the terminal nitrogens each bonded to two hydrogen atoms. Write the Lewis structure of *trans*-tetrazene.

Result

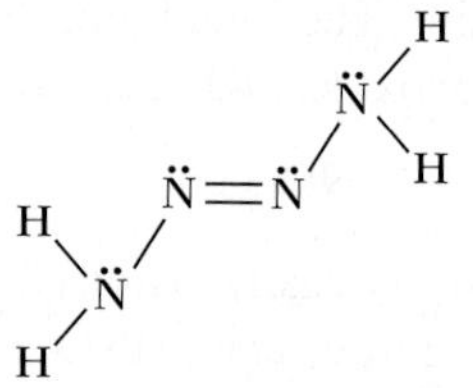

Analyze The terminal nitrogen atoms are each bonded to two hydrogen atoms. The remaining two nitrogen atoms are between the two terminal nitrogen atoms. The *trans* prefix indicates that the —NH_2 groups must be across the plane of a double bond.

Plan The interior nitrogen atoms have a N=N double bond between them. Distribute the valence electrons so that each nitrogen atom has an octet of electrons; hydrogen atoms have only a single bond and no lone pairs.

Execute The correct Lewis structure is shown above.

PROBLEM-SOLVING PRACTICE 19.4

Dinitrogen pentaoxide is the acid anhydride of nitric acid.

(a) Write the Lewis structure of dinitrogen pentaoxide.

(b) Write the balanced chemical equation for the formation of nitric acid from the reaction of dinitrogen pentaoxide with water.

PROBLEM-SOLVING EXAMPLE 19.5

Producing Nitric Acid

Consider the second step in the Ostwald process.

$$2\ NO(g) + O_2(g) \rightleftharpoons 2\ NO_2(g) \qquad \Delta_r H^\circ = -113.0\ kJ/mol$$

What happens to the yield of NO_2 at equilibrium if
(a) the pressure is increased by decreasing the volume?
(b) the temperature is increased?
(c) a catalyst is added?

Result
(a) The yield increases.
(b) The yield decreases.
(c) No effect on yield.

Analyze Each case represents a stress applied to an equilibrium that might affect the yield of NO_2. Le Chatelier's principle predicts the effect on the equilibrium.

Plan In each case, apply Le Chatelier's principle (← **Sec. 12-6**).

Execute
(a) The yield would increase because the pressure change favors the formation of fewer moles of gas: 3 mol gaseous reactants, 2 mol gaseous products.
(b) An increase in temperature favors the endothermic reverse reaction, decreasing the yield.
(c) A catalyst will speed up both the forward and reverse reactions equally, so it has no effect on yield.

PROBLEM-SOLVING PRACTICE 19.5

Use Le Chatelier's principle to explain why in the manufacturing of nitric acid
(a) lowering the temperature after Step 1 favors NO_2 formation.
(b) coupling Step 2 with Step 3 of the Ostwald process favors the formation of nitric acid.

A Lifesaving Use of N_2: Automobile Air Bags Air bags save the lives of thousands of motorists annually in the United States. This is done through the application of a simple decomposition reaction of sodium azide, an ionic compound containing sodium ions and azide ions, N_3^-, which decomposes rapidly to liberate N_2 gas. An uninflated air bag has a small cylinder containing a carefully formulated mixture of the solids sodium azide, NaN_3, potassium nitrate, KNO_3, and silicon dioxide, SiO_2. When a car decelerates rapidly, as in a collision, a sensor sends an electrical signal to the mixture, rapidly igniting and decomposing the sodium azide and releasing nitrogen gas, which inflates the air bag.

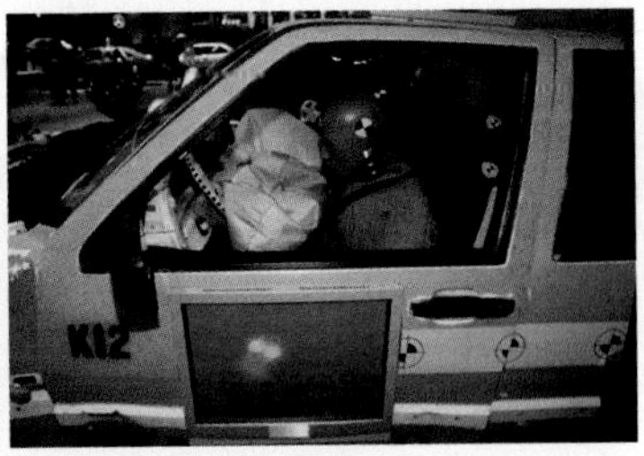
Richard Olivier/Corbis

Expanding air bags.

$$2\ NaN_3(s) \longrightarrow 2\ Na(s) + 3\ N_2(g)$$

Residual sodium metal must be removed because it could react vigorously with water. Removal in this case is achieved by reacting the residual sodium with potassium nitrate in a reaction that produces additional nitrogen gas to inflate the air bag.

In addition to being used in automobile air bags, azides are used as explosives.

$$10\ Na(s) + 2\ KNO_3(s) \longrightarrow K_2O(s) + 5\ Na_2O(s) + N_2(g)$$

The energy released by these reactions melts the solid products and the silicon dioxide (sand), fusing them into an unreactive glass.

$$K_2O(s) + Na_2O(S) + SiO_2(s) \xrightarrow{heat} glass$$

EXERCISE 19.17

Expanding Air Bags

Calculate the volume of N_2 released in an air bag at STP when 150. g sodium azide decomposes.

The Discovery of Phosphorus, 1775, William Pether Engraving after Joseph Wright painting. Fisher Collections. Chemical Heritage Foundation. Photo by Will Brown.

The discovery of phosphorus in 1669 by Hennig Brand by extraction from urine. In a very limited oxygen supply, white phosphorus glows with a greenish light, the source of the term "phosphorescence."

Phosphorus and Its Compounds Phosphorus has two main allotropes, *white* phosphorus and *red* phosphorus, which have very different properties (Figure 19.23). White phosphorus is highly reactive, igniting spontaneously in air at room temperature. For this reason, white phosphorus is stored under water. The waxy, nonpolar, solid white phosphorus is soft and easily cut, reflecting the fact that it consists of P_4 tetrahedra held together by weak noncovalent intermolecular forces (← **Sec. 7-6**). Because it is nonpolar, white phosphorus is not soluble in water, but dissolves readily in nonpolar liquids such as carbon disulfide, CS_2, or hexane, C_6H_{14}. Red phosphorus, unlike white phosphorus, does not ignite in air at room temperature and is much less toxic. A third, less common allotrope called black phosphorus is produced by heating white phosphorus at high pressure.

Although white phosphorus is toxic, phosphorus is an essential dietary mineral because of the many ways it is used by the body. Phosphorus is part of phosphate groups that link alternately with deoxyribose units to form the backbone of the DNA double helix (← **Sec. 7-7**). Phosphate anhydride linkages, which have this structure,

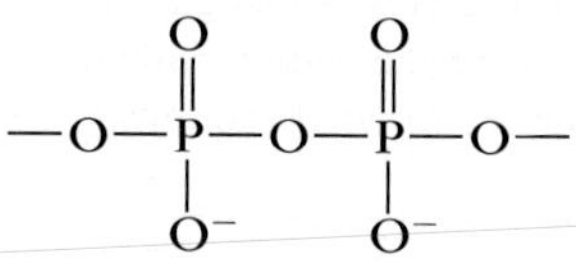

are responsible for how cellular energy is stored in ATP (← **Sec. 16-9a**).

Tooth enamel and bone contain the mineral hydroxyapatite, $Ca_5(PO_4)_3OH$. Water fluoridation reduces tooth decay because fluoride ions from the fluoridated water substitute for OH^- ions in tooth enamel to form fluoroapatite, $Ca_5(PO_4)_3F$, which is more resistant to decay than hydroxyapatite.

PROBLEM-SOLVING EXAMPLE 19.6

Phosphorus Pentachloride

Although the empirical formula for phosphorus pentachloride is PCl_5, in the solid state the compound actually consists of PCl_4^+ and PCl_6^- ions. Write the Lewis structure for each of these ions and predict its shape.

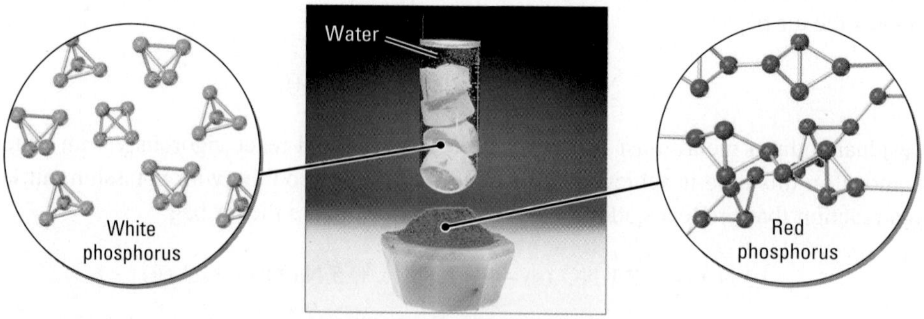
Water
White phosphorus
Red phosphorus
Cengage Learning/Charles D. Winters

Figure 19.23 Phosphorus allotropes. White phosphorus reacts with air at room temperature, and therefore must be stored under water. On the other hand, red phosphorus does not react with air at room temperature.

Result

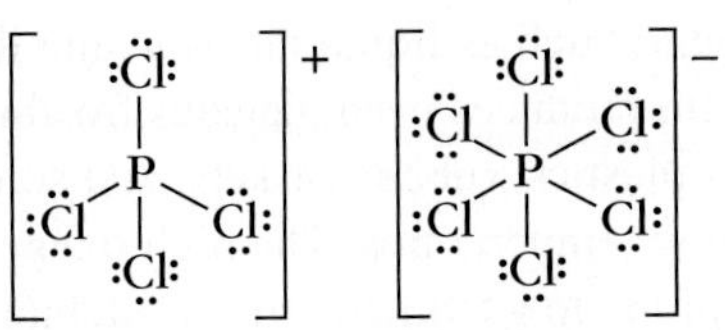

Tetrahedral   Octahedral

Analyze Count the number of valence electrons and allow for the charge of the ion in each case. Phosphorus as a central atom can exceed an octet of electrons.

Plan The PCl_4^+ ion has 32 valence electrons; PCl_6^- has 48 valence electrons. The PCl_4^+ ion is an example of an AX_4E_0 type and is tetrahedral; PCl_6^- is an AX_6E_0 type and is octahedral.

Execute The Lewis structures are shown in the Result.

PROBLEM-SOLVING PRACTICE 19.6

Pure phosphoric acid, H_3PO_4, occurs only as a solid. When it melts, phosphoric acid units gradually lose water. Write the Lewis structure for the other product of this dehydration.

Arsenic, Antimony, and Bismuth Arsenic and antimony each have a metallic allotropic form and one or more amorphous allotropes. Amorphous arsenic is yellow and, like white phosphorus, is soluble in carbon disulfide, where it exists as tetrahedral As_4 molecules. Antimony and arsenic are used to harden lead alloys such as those used in automobile batteries and in bullets. Arsenic compounds are poisonous, a fact often put to use in mystery novels. A 0.1 g or greater dose of As_2O_3 is fatal to humans. All of the trihalides of arsenic and antimony are known, such as AsF_3, a colorless liquid, and SbI_3, a red crystalline solid. By contrast, only four pentahalides are known: AsF_5, $AsCl_5$, SbF_5, and $SbCl_5$. That the other pentahalides have not been found is likely due to the strong oxidizing nature of arsenic's +5 oxidation state.

Bismuth is a silvery white metal that often appears yellowish because of an oxide coating on its surface. Its chief use is in low-melting alloys for automatic sprinkler systems and electrical fuses. One physical property of bismuth is worth noting: It is one of the few substances that expands on freezing. Thus, bismuth was used in low-melting alloys to make printing type that expands on solidification when cast in a mold. The expanded metal gives a very sharp edge to the type, which produces a clear ink image on the printed page. Printing type made of these alloys was first used in the Middle Ages shortly after the initial Gutenberg printing press was developed (1440). Bismuth is commonly found in the lower +3 oxidation state, whereas the lighter elements (P, As, and Sb) are often found in the higher +5 oxidation state. This is another example of the inert pair effect, as noted previously for the heavier Group 3A and 4A elements.

19-6f Group 6A(16): Oxygen, Sulfur, Selenium, Tellurium, Polonium

Group 6A is the oxygen family of elements; S, Se, and Te are also known as the *chalcogens*. Some physical properties of the Group 6A elements are in Figure 19.24. Like the two preceding groups, Group 6A is diverse, consisting of true nonmetals (O, S, and Se), a metalloid (Te), and a metal (Po, a rare, radioactive element).

The term *chalcogen* is derived from the Greek word *khalkos* for copper, possibly because copper ores contain compounds of these elements.

Properties and Uses of Oxygen Most of the oxygen produced by fractional distillation of liquid air is used as an oxidizing agent and in steel making (Section 20-2a), although some is used in rocket propulsion and in controlled oxidation reactions of other types. Liquid oxygen (LOX) can be shipped and stored at its boiling temperature of

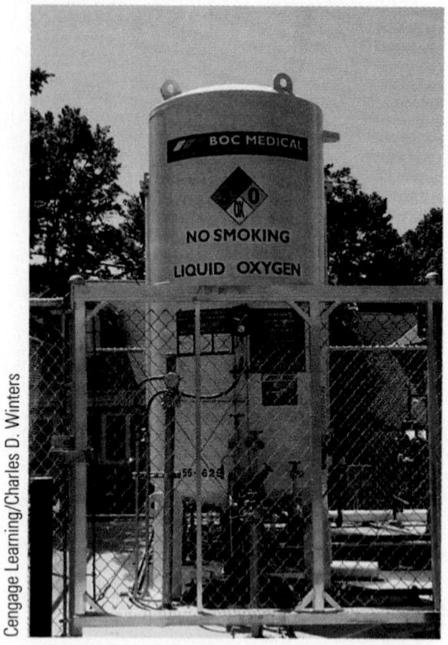
BOC MEDICAL
NO SMOKING
LIQUID OXYGEN

A cryogenic container of liquid oxygen.

−183 °C under atmospheric pressure. Cryogens, such as liquid nitrogen and liquid oxygen, present special hazards. Contact with them produces instantaneous frostbite of flesh and brittleness in structural materials such as plastics, rubber gaskets, and some metals, causing materials to fracture easily at these low temperatures. The high oxygen concentration present in liquid oxygen can, in spite of the low temperature, accelerate oxidation reactions to the point of explosion. For this reason, contact between liquid oxygen and substances that will ignite and burn in air must be prevented.

Special cryogenic containers holding liquid oxygen incorporate huge vacuum-walled bottles much like those used to carry hot soup or hot coffee. These special containers can be seen outside hospitals and industrial complexes, on highways and railroads, and even aboard ocean-going vessels. In hospitals, as well as in homes, supplemental oxygen is used to help patients who have difficulty breathing.

Most atmospheric oxygen comes from photosynthesis, in which green plants use energy from sunlight to convert water and carbon dioxide into glucose and oxygen. The concentration of atmospheric oxygen has upper and lower limits that are essential for health and safety. Should the concentration exceed 25%, the rates of oxidation reactions would increase significantly, potentially endangering us by the increased rates of oxygen-requiring metabolic processes. With too little atmospheric oxygen, less than 17%, we would suffocate.

Properties and Uses of Sulfur Sulfur exists in two common allotropic forms—rhombic (mp 115 °C) and monoclinic (mp 119 °C), both consisting of S_8 rings in the solid. When sulfur is heated above 150 °C, the S_8 rings break open, forming chains that

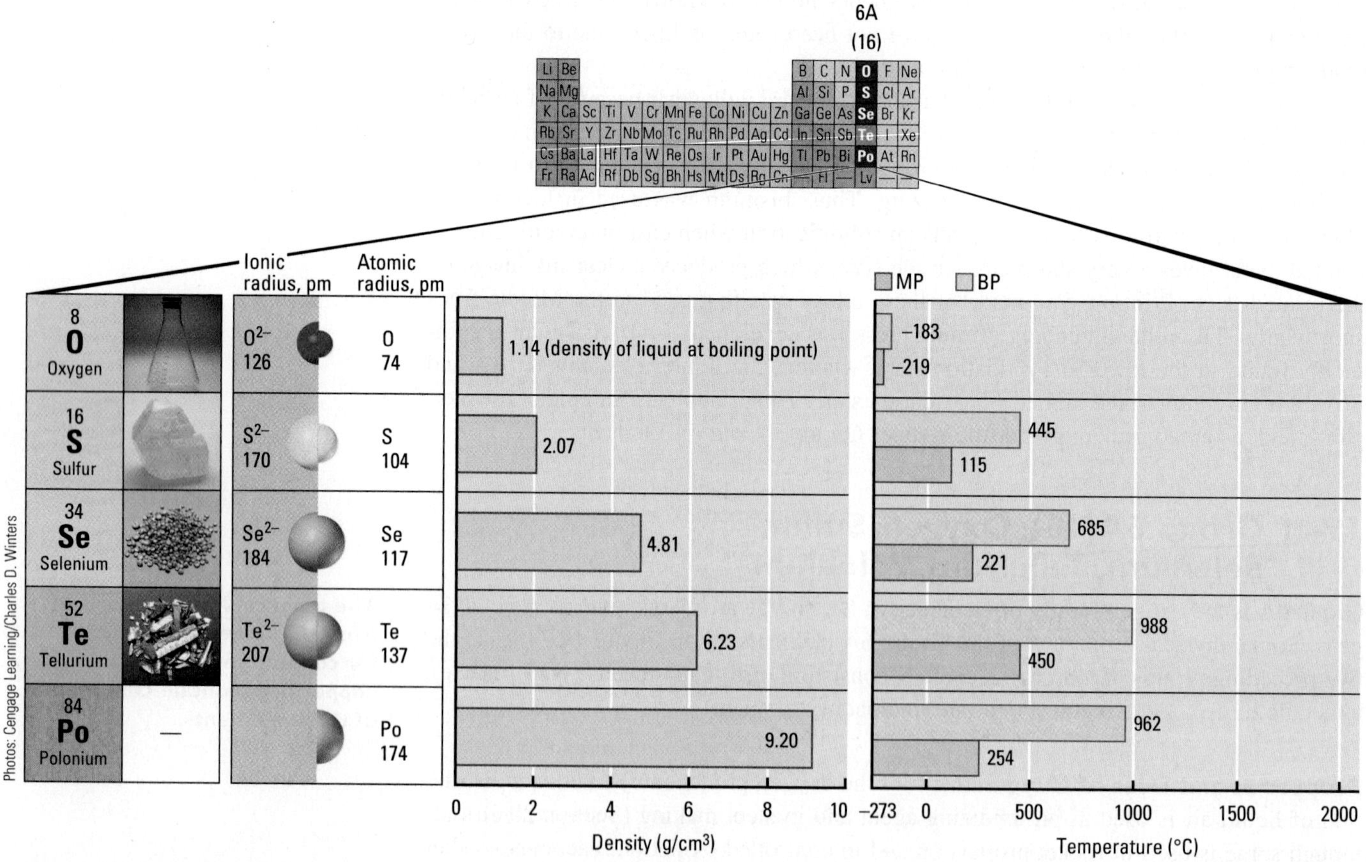
6A
(16)
Ionic radius, pm
Atomic radius, pm
8 O Oxygen
O2– 126
O 74
1.14 (density of liquid at boiling point)
16 S Sulfur
S2– 170
S 104
2.07
34 Se Selenium
Se2– 184
Se 117
4.81
52 Te Tellurium
Te2– 207
Te 137
6.23
84 Po Polonium
—
Po 174
9.20
MP
BP
−183
−219
445
115
685
221
988
450
962
254
0 2 4 6 8 10
Density (g/cm3)
−273 0 500 1000 1500 2000
Temperature (°C)
Photos: Cengage Learning/Charles D. Winters

Figure 19.24 Group 6A(16) elements.

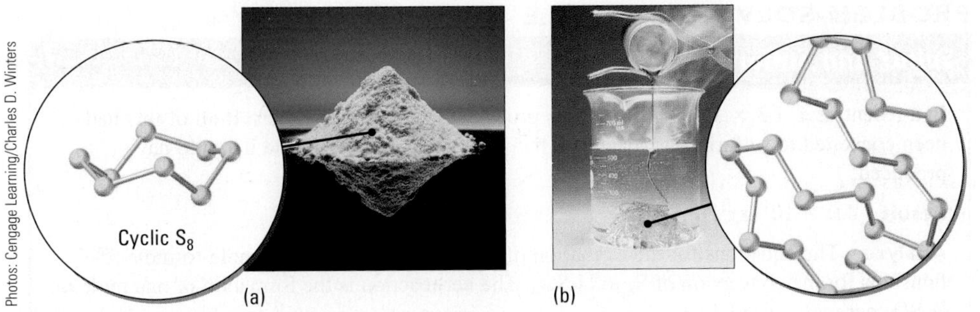
Photos: Cengage Learning/Charles D. Winters
Cyclic S_8
(a)
(b)

Figure 19.25 Sulfur allotropes. (a) At room temperature, sulfur is a bright yellow solid. At the nanoscale it consists of rings of eight sulfur atoms. (b) When melted, the rings break open to form long chains.

become entangled, thereby increasing the viscosity of the molten sulfur. Upon continued heating, the color of sulfur changes from yellow to dark red because of unpaired electrons at the ends of the chains. If heated to 210 °C and poured into cold water, the sulfur forms an uncrystallized polymer called "plastic sulfur," which reverts back to the common crystalline forms at 25 °C (Figure 19.25).

Sulfur is a critical element in the body, necessary for the formation of methionine, an essential amino acid (← **Sec. 10-7d**). Sulfur atoms form —S—S— disulfide linkages between chains of amino acids; the disulfide linkages help to create the essential molecular shapes of proteins and enzymes (← **Sec. 10-7e**). Sulfur is also used to cross-link polymer chains in the vulcanization of rubber (← **Sec. 10-6a**). The sulfur helps to align the polymer chains, which makes the rubber more elastic and prevents it from becoming sticky in warm weather.

Sulfuric Acid Production Most sulfur is used to produce sulfuric acid, the workhorse industrial chemical used in steel production, in automobile batteries, in the petroleum industry, and in the manufacture of fertilizers, plastics, drugs, dyes, and many other products. Because sulfuric acid costs less to make than any other acid, it is the first to be considered when an acid is needed in an industrial process.

Sulfur is converted to sulfuric acid in four steps, collectively called the *contact process*. In the first step, sulfur is burned in air to give mostly sulfur dioxide.

$$S_8(s) + 8\,O_2(g) \longrightarrow 8\,SO_2(g)$$

The SO_2 is then converted to SO_3 over a heated catalyst, such as platinum metal or vanadium(V) oxide.

$$2\,SO_2(g) + O_2(g) \xrightarrow{\text{catalyst}} 2\,SO_3(g)$$

The next step converts the sulfur trioxide to sulfuric acid by the addition of water. The best way to do this is to pass the SO_3 into H_2SO_4 to form pyrosulfuric acid, $H_2S_2O_7$, and then to dilute the $H_2S_2O_7$ with water. The net reaction is 1 mol H_2SO_4 for every 1 mol SO_3.

$$SO_3(g) + H_2SO_4(\ell) \longrightarrow H_2S_2O_7(\ell)$$
$$H_2S_2O_7(\ell) + H_2O(\ell) \longrightarrow 2\,H_2SO_4(aq)$$
$$\textit{Net reaction: } SO_3(g) + H_2O(\ell) \longrightarrow H_2SO_4(aq)$$

Sulfur dioxide for the contact process can also be obtained as a by-product from copper or lead smelting. Unless this sulfur dioxide is recovered, it pollutes the atmosphere (← **Sec. 8-12a**).

PROBLEM-SOLVING EXAMPLE 19.7

Sulfur and Sulfuric Acid

In a recent year, 1.3×10^{10} kg sulfur was produced in the United States. If all of this had been converted to sulfuric acid, calculate the mass (kg) of sulfuric acid it would have produced.

Result 4.0×10^{10} kg H_2SO_4

Analyze The equations for the formation of sulfuric acid provide the mole-to-mole relationships for the conversion of S_8 to H_2SO_4. The net reaction is the formation of one mole of H_2SO_4 per mole of SO_3.

Plan Calculate the moles of SO_3 and from it, determine the mass of sulfuric acid produced.

Execute Calculate the amount of SO_3 from mass of S_8.

$$n(SO_3) = 1.3 \times 10^{10} \text{ kg } S_8 \left(\frac{10^3 \text{ g } S_8}{1 \text{ kg } S_8}\right)\left(\frac{1 \text{ mol } S_8}{256.5 \text{ g } S_8}\right)\left(\frac{8 \text{ mol } SO_2}{1 \text{ mol } S_8}\right)\left(\frac{1 \text{ mol } SO_3}{1 \text{ mol } SO_2}\right)$$

$$= 4.05 \times 10^{11} \text{ mol } SO_3$$

From this amount, calculate the mass of sulfuric acid.

$$m(H_2SO_4) = 4.05 \times 10^{11} \text{ mol } SO_3 \left(\frac{1 \text{ mol } H_2SO_4}{1 \text{ mol } SO_3}\right)\left(\frac{98.08 \text{ g } H_2SO_4}{1 \text{ mol } H_2SO_4}\right)\left(\frac{1 \text{ kg } H_2SO_4}{10^3 \text{ g } H_2SO_4}\right)$$

$$= 4.0 \times 10^{10} \text{ kg } H_2SO_4$$

☑ **Reasonable Result Check** Sulfuric acid is approximately 1/3 sulfur by mass (32 g S/98 g H_2SO_4). Thus, 1 kg sulfur forms about 3 kg sulfuric acid, and 1.3×10^{10} kg S forms about 3.9×10^{10} kg sulfuric acid, which is close to the calculated value of 4.0×10^{10} kg H_2SO_4.

PROBLEM-SOLVING PRACTICE 19.7

Calculate the mass (kg) of SO_3 produced from 1.3×10^{10} kg S_8.

19-6g Group 7A(17): The Halogens

The halogens exhibit a trend in physical states not found in any other group: The lightest halogen, fluorine, is a gas and the heaviest, iodine, is a solid at room temperature. Some physical properties of the Group 7A elements are given in Figure 19.26. At room temperature, fluorine and chlorine, the first two elements, are diatomic pale yellow and yellow-green gases, respectively; bromine, the next element, is a reddish-brown liquid; and iodine is a violet-black solid. These physical states reflect the increasing strengths of noncovalent intermolecular forces with increasing numbers of electrons in the diatomic molecules (← **Sec. 7-6a**). Astatine is an intensely radioactive element with isotopes of very short half-lives, accounting for the scarcity of astatine in nature.

All halogens have an ns^2np^5 outer electron configuration that leads to a gain of one electron to form a 1− ion or a sharing of one electron in a pair to complete an octet. Fluorine is extremely reactive, evidence of its exceptionally high electronegativity, the highest of any element, and of the very weak F—F bond in F_2 (bond enthalpy 158 kJ/mol). The halogens are oxidizing agents in many reactions including the oxidation of a heavier halide ion by a lighter halogen. For example, chlorine oxidizes bromide ions to diatomic bromine.

$$Cl_2(g) + 2\ Br^-(aq) \longrightarrow 2\ Cl^-(aq) + Br_2(aq)$$

This reaction is used to extract bromine from seawater, which contains 65 ppm bromide ions (Section 19-5b). Some uses of the halogens and their compounds are given in Table 19.10.

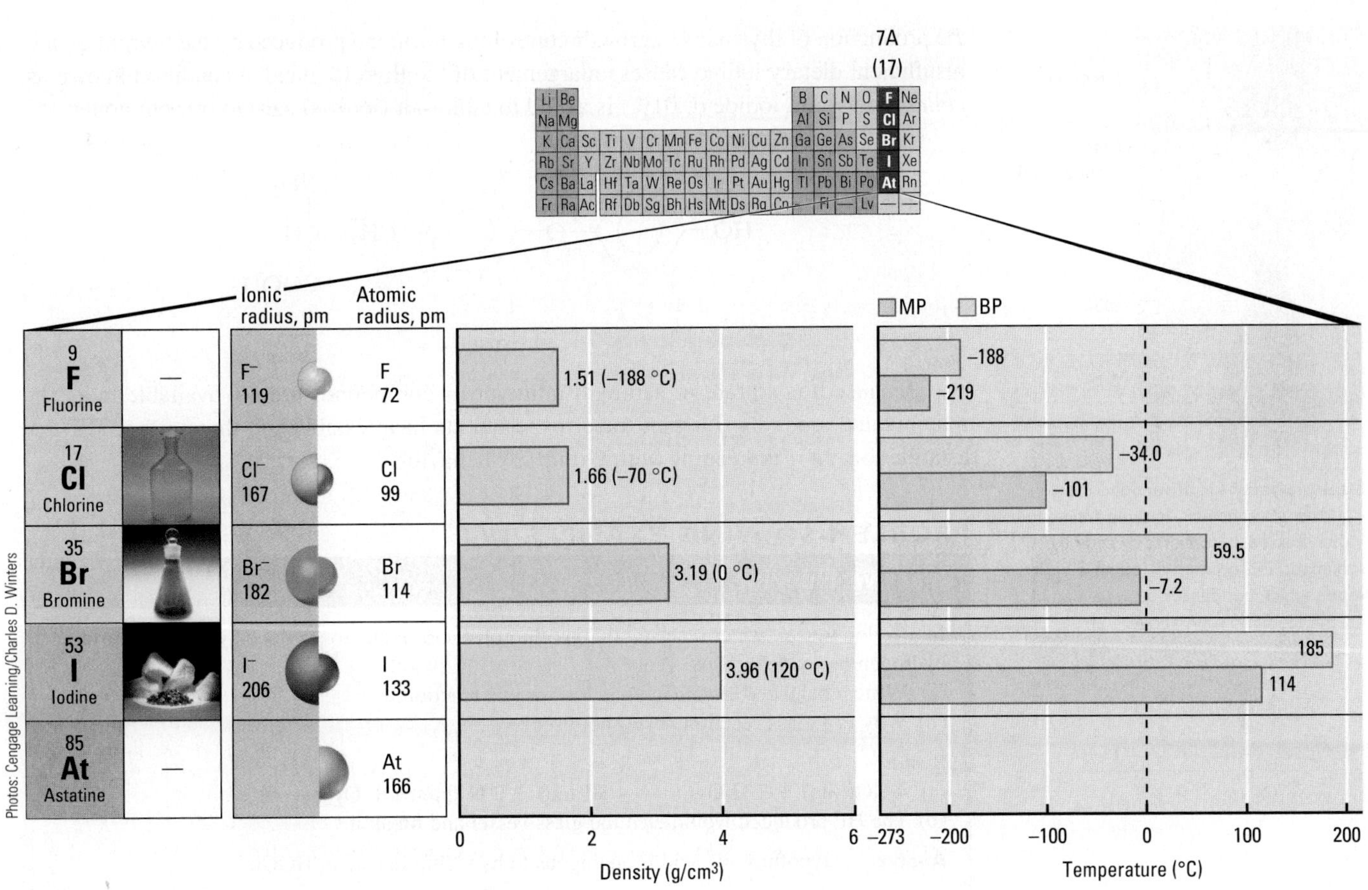
7A
(17)
Ionic radius, pm
Atomic radius, pm
9
F
Fluorine
F⁻
119
F
72
1.51 (–188 °C)
17
Cl
Chlorine
Cl⁻
167
Cl
99
1.66 (–70 °C)
35
Br
Bromine
Br⁻
182
Br
114
3.19 (0 °C)
53
I
Iodine
I⁻
206
I
133
3.96 (120 °C)
85
At
Astatine
At
166
Density (g/cm³)
MP
BP
–188
–219
–34.0
–101
59.5
–7.2
185
114
Temperature (°C)
Photos: Cengage Learning/Charles D. Winters

Figure 19.26 Group 7A(17) elements, the halogens.

Chlorine, a toxic gas with an irritating odor, is the most important halogen used in industry. Chlorine is used to purify water (← **Sec. 13-10a**), to bleach paper and textiles, to manufacture herbicides, insecticides, and other chlorinated organic compounds, to produce polyvinyl chloride, and to extract titanium metal from its ores.

Bromine is the only nonmetal that is a liquid at room temperature. It is used to prepare methyl bromide, CH_3Br, an efficient pesticide.

Iodine is the only common halogen that is a solid at room temperature. It is a violet-black metallic-looking solid that sublimes to a violet-colored vapor. Iodine was discovered by burning dried seaweed, which contains a relatively high concentration of iodide ions. Iodine is an essential dietary mineral for humans because the iodide ions are necessary for

Table 19.10 Uses of Some Halogen Compounds

Halogen Compound	Uses
Hydrogen fluoride	Frosting light bulbs and television tubes, production of uranium hexafluoride
Uranium hexafluoride, UF_6	Separation of fissionable U-235 from U-238 during processing of uranium ores
Hydrochloric acid, HCl	Magnesium manufacturing, manufacture of vinyl chloride and chlorinated solvents; human stomach acid (0.1 M)
Ammonium perchlorate, NH_4ClO_4	Solid propellant for Space Shuttle
Methyl bromide, CH_3Br	Pesticide
Potassium iodide, KI	Salt additive to prevent goiter, a thyroid condition

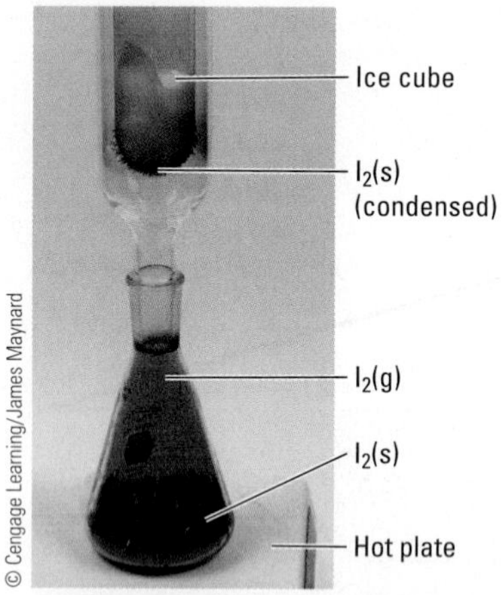
Ice cube
$I_2(s)$ (condensed)
$I_2(g)$
$I_2(s)$
Hot plate
© Cengage Learning/James Maynard

Sublimation of iodine. Solid iodine is a dark gray, metallic-looking solid. When heated, it converts directly by sublimation into gaseous iodine, a violet vapor, which condenses back to a solid on the ice-filled tube above.

the production of thyroxine, a growth-controlling hormone produced by the thyroid gland. Insufficient dietary iodine causes enlargement of the thyroid gland, a condition known as *goiter.* Potassium iodide (0.01%) is added to table salt (iodized salt) to prevent goiter.

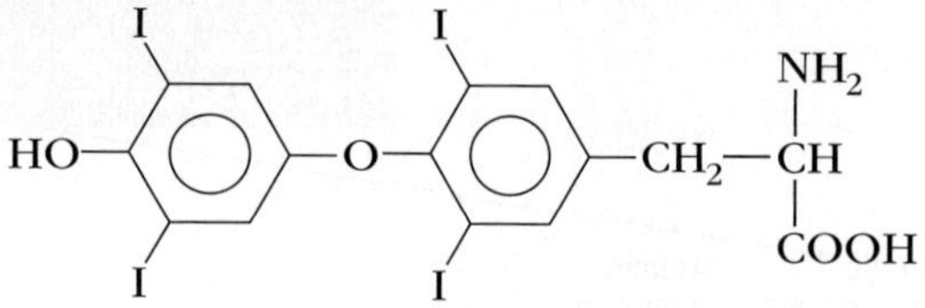
I
I
NH$_2$
HO
O
CH$_2$—CH
COOH
I
I

thyroxine

Because it is so rare in nature, astatine or its compounds are not available in easily manipulated amounts. Elegant tracer experiments have established the existence of the astatide ion, At^-, in keeping with Group 7A behavior.

PROBLEM-SOLVING EXAMPLE 19.8

Reactive Fluorine

In a Teflon vessel, water reacts with hypofluorous acid, HOF, to produce hydrogen fluoride, hydrogen peroxide, and oxygen.
(a) Write a balanced chemical equation for this reaction.
(b) Why is a glass reaction vessel not used?

Result
(a) $4\ HOF(aq) + 2\ H_2O(\ell) \longrightarrow 4\ HF(aq) + 2\ H_2O_2(aq) + O_2(g)$
(b) The HF produced would etch the glass vessel and break it.

Analyze Hypofluorous acid is analogous to hypochlorous acid, HOCl.

Plan Balance the equation by writing the correct formulas and coefficients for the reactants and products.

Execute
(a) $4\ HOF(aq) + 2\ H_2O(\ell) \longrightarrow 4\ HF(aq) + 2\ H_2O_2(aq) + O_2(g)$
(b) Teflon is used because it is nonreactive. Glass would react with the HF produced by the reaction.

PROBLEM-SOLVING PRACTICE 19.8

Classify each of these equations as being either a redox reaction or a non-redox reaction. If a redox reaction occurs, also identify the oxidizing agent and the reducing agent.
(a) $2\ NaF(s) + H_2SO_4(aq) \longrightarrow 2\ HF(g) + Na_2SO_4(aq)$
(b) $S_8(s) + 24\ F_2(g) \longrightarrow 8\ SF_6(g)$

19-6h Group 8A(18): The Noble Gases

Some physical properties of the Group 8A elements are given in Figure 19.27. Although they were once called the "rare gases," they are not rare. Helium is relatively low in terrestrial abundance but is the second most abundant element in the universe. Argon makes up nearly 1% of Earth's atmosphere.

Evidence of helium was first obtained in 1868 when a new yellow line was observed spectroscopically from Earth during an eclipse of the sun. Proof of the existence of terrestrial helium first came from observation of the same line in the spectrum of gases released during the eruption of Mount Vesuvius in 1882. Subsequently, the remaining noble gases except radon were isolated from liquid air and characterized by Rayleigh, Ramsay, and Travers between 1895 and 1898. Radon was first isolated and identified in 1902 by Rutherford and Soddy. These findings led to the addition of a new Group, 8A, to the periodic table.

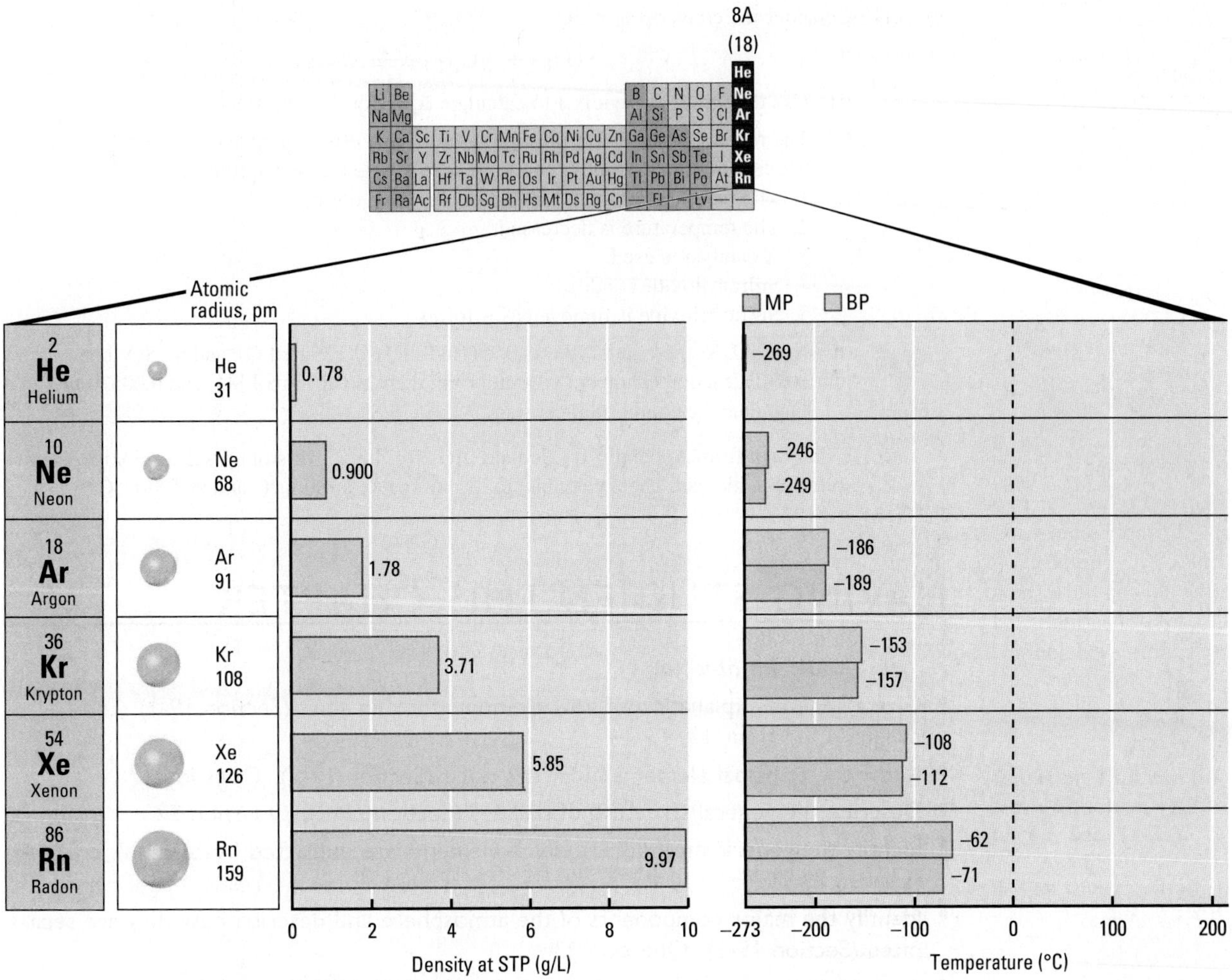
8A
(18)
Atomic radius, pm
MP
BP
2 He Helium
10 Ne Neon
18 Ar Argon
36 Kr Krypton
54 Xe Xenon
86 Rn Radon
He 31
Ne 68
Ar 91
Kr 108
Xe 126
Rn 159
0.178
0.900
1.78
3.71
5.85
9.97
Density at STP (g/L)
−269
−246
−249
−186
−189
−153
−157
−108
−112
−62
−71
Temperature (°C)

Figure 19.27 Group 8A(18) elements, the noble gases.

The reluctance of Group 8A elements to form compounds is to be expected given their complete octets of ns^2np^6 outer electrons (only ns^2 for helium). In fact, it was a long-standing accepted chemical truism that these elements, then called the "inert elements," formed no compounds. That view was overthrown dramatically in 1962 with the synthesis of a series of xenon compounds. Since then, compounds of xenon containing oxygen and fluorine have been synthesized, as well as fluorides of krypton and radon. Spectroscopic evidence exists for the argon compound HArF.

Helium is used as an inert atmosphere in welding, as a nitrogen substitute in gas mixtures for underwater diving, as a liquid cryogen, and to fill meteorological balloons. Argon is used as a protective atmosphere for arc-welding aluminum, in airport runway lights, and as the gas in incandescent light bulbs, where it decreases the evaporation of the tungsten metal filament.

SUMMARY PROBLEM

(a) In the United States annual consumption of crude oil is about 24 barrels per person (1 barrel = 42 gallons; density of crude oil = 0.83 g/mL). Assume that the crude oil is 3% sulfur by mass and that all of the sulfur was removed from the crude oil before it was used. Calculate the volume (L) of SO_2 at 25 °C and 1 atm that was prevented from entering the atmosphere from your share of crude oil.

(b) Consider the conversion of $SO_2(g)$ to $SO_3(g)$.

$$SO_2(g) + \frac{1}{2} O_2(g) \rightleftharpoons SO_3(g)$$

(i) Use data from Appendix J to calculate $\Delta_r H°$ for this reaction.

(ii) The reaction reaches equilibrium. Identify the effect on the concentration of sulfur trioxide produced if these conditions are applied to the equilibrium.
1. The pressure is increased by reducing the volume.
2. The temperature is decreased.
3. A catalyst is used.
4. Sulfur dioxide is added.
5. Sulfur trioxide is removed as it forms.

(iii) At 1000. K, there initially was 0.250 mol SO_2, 0.125 mol O_2, and no SO_3 in a 10.0-L reaction chamber. At equilibrium there is 0.136 mol SO_3 and 0.00570 mol O_2. Calculate the equilibrium constant, K_c.

(c) Using data from Appendix J, calculate $\Delta_r G°$ for the conversion of sulfur dioxide gas to sulfur trioxide gas. Then, calculate the value for K_P at 800. °C and at 1000. °C.

HAVING STUDIED THIS CHAPTER . . .

. . . you should be able to:

- Give a general explanation of how elements form in stars (Section 19-1). End-of-chapter Questions 13, 15
- Know the principal elements in Earth's crust (Section 19-2). Question 17
- Describe the general structure of silicates (Section 19-2). Question 22
- Identify the general methods by which elements are extracted from Earth's crust (Section 19-2).
- Identify the major components of the atmosphere and describe how they are separated (Section 19-3). Question 23
- Apply chemical principles to the processes for extracting and purifying elements, and the reactions they undergo (Sections 19-3 to 19-5). Questions 25, 27
- Describe how electrolysis is used to obtain sodium, chlorine, magnesium, and aluminum (Section 19-4). Questions 29, 35
- Explain how chemical redox reactions are used to extract bromine, iodine, and phosphorus from compounds (Section 19-5). Question 38
- Explain how sulfuric acid and nitric acid are produced (Section 19-6). Question 40
- Interpret the periodic trends among the main-group elements; describe reactions and properties of some main-group elements and their compounds (Section 19-6).

KEY TERMS

catenation (Section 19-6d)

chlor-alkali process (19-4b)

cryogen (19-6e)

helium burning (19-1a)

hydrogen burning (19-1a)

mineral (19-2)

nitrogen cycle (19-6e)

nitrogen fixation (19-6e)

nuclear burning (19-1a)

ore (19-2b)

QUESTIONS FOR REVIEW AND THOUGHT

Red-numbered questions have short answers at the back of this book in Appendix M and fully worked solutions in the *Student Solutions Manual.*

Review Questions

These questions test vocabulary and simple concepts.

1. What is meant by hydrogen burning and helium burning in relation to the formation of elements?
2. Identify the most abundant nonmetallic element in Earth's crust. Identify the most abundant metallic element in Earth's crust.
3. Describe the difference between an ore and a mineral.
4. Differentiate among pyroxenes, amphiboles, and silica.
5. Explain how the silicate unit in mica and other sheet silicates has the general formula $Si_2O_5^{2-}$.
6. Identify two major differences between white phosphorus and red phosphorus.
7. Identify
   (a) two elements obtained from the atmosphere.
   (b) two elements obtained from the sea.
   (c) two elements obtained from Earth's crust.
8. Why are nitrogen and oxygen important industrial chemicals?
9. Describe how nature fixes nitrogen. Why is nitrogen fixation necessary?
10. Identify two uses of phosphate rock.
11. Describe the structural changes that occur in sulfur as it is heated from room temperature to 210 °C.
12. Why is phosphate rock not applied directly as a phosphorus fertilizer?

Topical Questions

These questions are keyed to the major topics in the chapter. Usually a question that is answered at the back of this book is paired with a similar one that is not.

Formation of the Elements (Section 19-1)

13. Write balanced equations to represent nuclear fusion reactions for
    (a) the conversion of oxygen-16 into neon-20.
    (b) carbon "burning" to form neon-20.
14. Write balanced equations to represent nuclear fusion reactions for
    (a) the conversion of calcium-44 to titanium-48.
    (b) the conversion of oxygen-16 to sulfur-31.
15. Describe how heavier elements are formed from lighter elements.
16. Explain the differences between the *s process* and the *r process* of element formation.

Terrestrial Elements (Section 19-2)

17. Identify six elements that are found in nature in uncombined form.
18. Identify the most abundant metalloid element in Earth's crust.
19. Describe the difference between amphibole and serpentine forms of asbestos in terms of their potential as health hazards.
20. Identify four different anions that are common in minerals in Earth's crust. Give an example of each type of mineral.
21. Give a simple explanation for the abundance of clay minerals in Earth's crust.
22. Explain how the silicate unit in amphiboles has the general formula $Si_4O_{11}^{6-}$.

Extraction by Physical Methods: N_2, O_2, S (Section 19-3)

23. What physical property allows the separation of nitrogen, oxygen, and argon from each other in fractional distillation of air?
24. Describe two different methods for obtaining pure sulfur.
25. Identify the elements obtained by
    (a) the Linde process. (b) the Frasch process.
26. A 500-g sample of gaseous argon is collected at −185 °C and 5.0 atm. Calculate its volume at this temperature and pressure.

Extraction by Electrolysis: Na, Cl_2, Mg, Al (Section 19-4)

27. Identify the substance or substances produced by each of these commercial processes and write a balanced chemical equation for the main reaction of the process.
    (a) Downs (b) Chlor-alkali
    (c) Dow (d) Hall-Héroult
28. Write balanced equations for the recovery of magnesium from seawater. Begin with the precipitation of magnesium hydroxide by addition of calcium hydroxide to seawater.
29. (a) Write the balanced chemical equation for the electrolysis of aqueous NaCl.
    (b) In 2002, 8.98×10^9 kg NaOH and 1.14×10^{10} kg chlorine were produced in the United States. Does the ratio of these masses agree with the ratio of masses from the balanced chemical equation? If not, what does that suggest about the ways that NaOH and Cl_2 are produced?
30. Briefly explain why different products are obtained from the electrolysis of molten NaCl and the electrolysis of aqueous NaCl.

31. To produce magnesium metal, 1000. kg molten $MgCl_2$ are electrolyzed.
    (a) At which electrode is magnesium produced?
    (b) What is produced at the other electrode?
    (c) Calculate the amount (mol) of electrons used in the process.
    (d) An industrial process uses 8.4 kWh per pound of Mg. Calculate the energy (J) required per mole of magnesium.
32. A Downs cell operates at 7.0 V and 4.0×10^4 A.
    (a) Calculate how much Na(s) and $Cl_2(g)$ can be produced in 24 hours by such a cell.
    (b) Assuming 100% efficiency, calculate the energy consumption (kWh) of this cell.
33. Calculate how much energy (kWh) is required to prepare a ton of sodium in a typical Downs cell operating at 25,000 A and 7.0 V.
34. Calculate the mass in grams of aluminum that can be produced when 6.0×10^4 A is passed through a series of 100 Hall-Héroult electrolytic cells operating at 85% efficiency for 24 hours.
35. Calculate what mass in grams of aluminum can be produced from the electrolysis of molten $AlCl_3$ in an electrolytic cell operating at 100. A for 2.00 hr.

Extraction by Chemical Oxidation-Reduction: P, Br_2, I_2 (Section 19-5)

36. Phosphine, PH_3, is a hydride of phosphorus as ammonia, NH_3, is of nitrogen. Explain why phosphine boils at a much lower temperature than NH_3 in spite of having a molar mass twice that of ammonia.
37. Given:

$$F_2(g) + 2\,e^- \longrightarrow 2\,F^-(aq) \qquad E° = +2.87\text{ V}$$
$$Cl_2(g) + 2\,e^- \longrightarrow 2\,Cl^-(aq) \qquad E° = +1.36\text{ V}$$

determine whether this reaction is product-favored:

$$Cl_2(g) + 2\,F^-(aq) \longrightarrow 2\,Cl^-(aq) + F_2(g)$$

38. Given:

$$Br_2(\ell) + 2\,e^- \longrightarrow 2\,Br^-(aq) \qquad E° = +1.07\text{ V}$$
$$I_2(s) + 2\,e^- \longrightarrow 2\,I^-(aq) \qquad E° = +0.54\text{ V}$$

determine whether this reaction is product-favored:

$$Br_2(g) + 2\,I^-(aq) \longrightarrow 2\,Br^-(aq) + I_2(s)$$

39. Given:

$$F_2(g) + 2\,e^- \longrightarrow 2\,F^-(aq) \qquad E° = +2.87\text{ V}$$
$$Br_2(\ell) + 2\,e^- \longrightarrow 2\,Br^-(aq) \qquad E° = +1.07\text{ V}$$

determine whether this reaction is product-favored:

$$F_2(g) + 2\,Br^-(aq) \longrightarrow 2\,F^-(aq) + Br_2(\ell)$$

A Periodic Perspective: The Main-Group Elements (Section 19-6)

40. Identify the substance or substances produced by each of these commercial processes and write balanced chemical equations for all reactions in each process.
    (a) Ostwald (b) Contact
    (c) Haber-Bosch
41. Phosphorus forms two oxides, P_4O_6 and P_4O_{10}.
    (a) Determine the oxidation number of phosphorus in each of these compounds.
    (b) Write Lewis structures for each compound.
42. Bipolar disorder can be treated with daily 1 to 2 g doses of lithium carbonate. Calculate what mass of lithium each dosage contains.
43. A human body contains approximately 5 L of blood. A bipolar disorder patient receives 2.0 g lithium carbonate as a daily dose. Calculate the concentration of Li^+ (mmol/L) that this dosage provides.
44. Metallic tin was obtained in ancient times by the reaction of the principal tin ore, cassiterite, SnO_2, with carbon from a charcoal fire:

$$SnO_2(s) + 2\,C(s) \longrightarrow 2\,CO(g) + Sn(s)$$

Identify the oxidizing and reducing agents in this reaction.

45. Galena, PbS, is the principal lead ore. To produce lead metal, galena is first reacted with oxygen; the oxide is then reacted with carbon.
    (a) Write balanced chemical equations for these two reactions.
    (b) Is either of these reactions a redox reaction? If so, explain your answer by identifying the oxidizing and reducing agent in each case.
46. Which of these is soluble in carbon disulfide, CS_2?
    (a) water (b) benzene, C_6H_6
    (c) white phosphorus (d) iodine
    (e) ethanol, CH_3CH_2OH
    Explain your answers.

General Questions

These questions are not explicitly keyed to chapter topics; many require integration of several concepts.

47. Complete this table.

Formula	Name	Oxidation State of Nitrogen
______	Nitrogen	______
NH_3	______	______
______	Nitrous acid	______
______	Nitrogen dioxide	______
NH_4^+	______	______
______	Ammonium nitrate	______

48. Complete this table.

Formula	Name	Oxidation State of Phosphorus
________	Phosphorus	________
$(NH_4)_2HPO_4$	________	________
________	Phosphoric acid	________
________	Tetraphosphorus decaoxide	________
$Ca_3(PO_4)_2$	________	________
________	Calcium dihydrogen phosphate	________

49. Molten NaCl is electrolyzed in a Downs cell operating at 1.00×10^4 A for 24 hr.
(a) Calculate the mass in grams of sodium produced.
(b) Calculate the volume of Cl_2 in liters that is collected from the outlet tube at 20. °C and 15 atm.
50. Bauxite, the principal source of aluminum oxide, contains 55% Al_2O_3. Calculate how much bauxite is required to produce the 5.0×10^6 tons of Al produced annually by electrolysis.
51. Write a plausible Lewis structure for azide ion, N_3^-.
52. Write a plausible Lewis structure for P_4O_{10}.
53. Write the chemical equation for the
(a) combustion of white phosphorus.
(b) reaction of the combustion product with water.
54. Estimate the temperature at which the conversion of white phosphorus to red phosphorus occurs.

$\Delta_r H° = -70.4$ kJ/mol; $\Delta_r S° = -73.16$ J mol^{-1} K^{-1} at 25 °C.

55. There are two common oxides of sulfur. Name these oxides, and write chemical equations for the reaction of each with water. Identify the products.
56. What raw materials are used to produce sulfuric acid? Write chemical equations to represent the steps in the contact process to produce sulfuric acid.
57. Write Lewis structures for all the resonance forms of sulfuric acid.
58. Write Lewis structures for all the resonance forms of nitric acid.
59. Iodine trichloride, ICl_3, is an interhalogen compound.
(a) Write the Lewis structure of ICl_3.
(b) Does the central atom have more than an octet of valence electrons?
(c) Using VSEPR theory, predict the molecular shape of ICl_3.
60. At some temperature, a gaseous mixture in a 1.00-L vessel originally contained 1.00 mol SO_2 and 5.00 mol O_2. When equilibrium was reached, 77.8% of the SO_2 had been converted to SO_3. Calculate the equilibrium constant (K_c) for this reaction at this temperature.
61. White phosphorus is soluble in carbon disulfide. At 10. °C, 800. g P_4 dissolves in 100. g CS_2. The density of carbon disulfide at 10. °C is 1.26 g/mL.
(a) Calculate the molarity of phosphorus in the solution.
(b) Calculate the molality of phosphorus in the solution.
62. Drug stores sell 3% aqueous hydrogen peroxide that is used as an antiseptic. Hydrogen peroxide, H_2O_2, decomposes to water and oxygen. Calculate the volume of oxygen produced if 250. mL of 3% hydrogen peroxide decomposes fully at 750. mmHg and 22 °C.
63. Calculate the volume of concentrated (98%) sulfuric acid that is needed to produce two tons of phosphoric acid from the reaction of sulfuric acid with sufficient phosphate-bearing rock. Density of conc. sulfuric acid = 1.84 g/mL.

$$Ca_3(PO_4)_2(s) + 3\,H_2SO_4(\ell) \longrightarrow 2\,H_3PO_4(\ell) + 3\,CaSO_4(s)$$

Applying Concepts

These questions test conceptual learning.

64. The K_{sp} of $Ca(OH)_2$ is 5.5×10^{-6}; that of $Mg(OH)_2$ is 1.8×10^{-11}. Calculate the equilibrium constant for the reaction

$$Ca(OH)_2(s) + Mg^{2+}(aq) \longrightarrow Ca^{2+}(aq) + Mg(OH)_2(s)$$

Use it to explain why this reaction can be used commercially to extract magnesium from seawater.
65. Commercial concentrated nitric acid contains 69.5 mass percent HNO_3 and has a density of 1.42 g/mL.
(a) Calculate the molarity of this solution.
(b) Calculate what volume of the concentrated acid must be used to prepare 10.0 L of 6.00-M HNO_3.
66. The compound nitrosyl azide, N_4O, is a covalent compound with an NNNNO atomic arrangement. Write a plausible Lewis structure for this compound.
67. Hydrazoic acid, HN_3, is very explosive in its pure state but can be studied in aqueous solution. The acid is prepared by the reaction of hydrazine with nitrous acid.

$$N_2H_4(\ell) + HNO_2(aq) \longrightarrow HN_3(aq) + 2\,H_2O(\ell)$$

(a) Determine the oxidation states of nitrogen in the compounds in this reaction.
(b) What is the oxidizing agent in this reaction?
(c) The K_a of hydrazoic acid is 1.0×10^{-5} at 25 °C. Calculate the pH of a 0.010-M solution of HN_3.
68. (a) Write two plausible resonance structures for hydrazoic acid, HN_3.
(b) Use bond enthalpy data (Table 6.2, ← **Sec. 6-6b**) to calculate the $\Delta_f H°$ for each resonance form; $\Delta_f H°$ = 218.0 kJ/mol for H(g) and 472.7 kJ/mol for N(g). (See Appendix J.)

69. Given the reaction

$$Cl_2(g) + H_2O(\ell) \rightleftharpoons H^+(aq) + Cl^-(aq) + HOCl(aq)$$

(a) identify the oxidizing agent and the reducing agent.
(b) write the equilibrium constant expression for the reaction.
(c) calculate the concentration of HOCl in equilibrium with $Cl_2(g)$ at 1.0 atm. $K_c = 2.7 \times 10^{-5}$.

70. Dinitrogen trioxide, N_2O_3, is a blue liquid formed by the reaction of NO_2 and NO.
(a) Write a balanced chemical equation for the formation of N_2O_3.
(b) Write the Lewis structure of N_2O_3 and any plausible resonance forms.
(c) Predict the O—N—O and the N—N—O bond angles.

71. (a) Write the resonance forms of SO_3.
(b) Predict the molecular shape of SO_3 and the O—S—O bond angle.

72. Iodine can be produced by the oxidation of iodide ion with permanganate ion.

$$MnO_4^-(aq) + 2\ I^-(aq) + 8\ H^+(aq) \longrightarrow I_2(s) + Mn^{2+}(aq) + 4\ H_2O(\ell)$$

Excess HI is added to 0.200 g MnO_4^-. Assuming 100% yield, calculate the mass (g) of iodine produced.

73. In a Downs cell, molten NaCl is electrolyzed to sodium metal and chlorine gas.

$$2\ NaCl(\ell) \longrightarrow 2\ Na(\ell) + Cl_2(g)$$

$\Delta_r H°$ and $\Delta_r S°$ for the reaction are +820 kJ/mol and +180 J K^{-1} mol^{-1}, respectively.
(a) Calculate $\Delta_r G°$ at 600. °C, the electrolysis temperature.
(b) Calculate the voltage required for the electrolysis.

74. A 425-gal tank of water contains 175 g NaI. Calculate the volume (L) of chlorine gas at 758 mmHg and 25 °C required to convert all the iodide to iodine.

75. Use the phase diagram for sulfur for Question 75. The solid forms of sulfur are rhombic and monoclinic.
(a) A triple point is defined as a temperature and pressure where three phases are in equilibrium. How many triple points does sulfur have? Indicate the approximate temperature and pressure at each.
(b) Which physical states are present at equilibrium under these conditions?
(i) 10^{-2} atm and 80 °C
(ii) 10^{-1} atm and 140 °C
(iii) 10^{-3} atm and 110 °C
(iv) 10^{-4} atm and 160 °C

76. Consult the phase diagram for sulfur for Question 75. Describe the conditions of temperature and pressure under which monoclinic sulfur sublimes.

Phase diagram for sulfur, for Question 75.

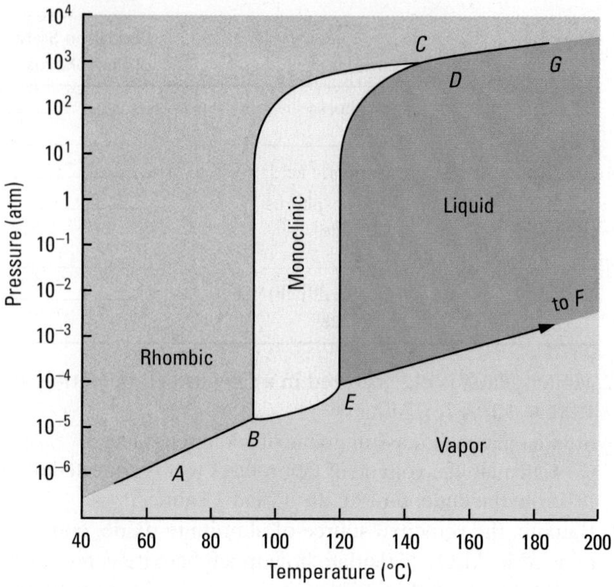

Pressure (atm)
10^4
10^3
10^2
10^1
1
10^-1
10^-2
10^-3
10^-4
10^-5
10^-6
C
D
G
Liquid
Monoclinic
to F
Rhombic
E
B
Vapor
A
40
60
80
100
120
140
160
180
200
Temperature (°C)

77. In the laboratory, small quantities of bromine can be produced by the reaction of hydrobromic acid with MnO_2. The unbalanced equation for the reaction is

$$MnO_2(s) + Br^-(aq) \longrightarrow Mn^{2+}(aq) + Br_2(\ell)$$

(a) Balance the equation.
(b) If the reaction has 100% yield, calculate the amount of bromide ions that reacts to produce 6.50 g bromine.
(c) Calculate the mass (g) of MnO_2 required in part (b).

78. Lapis lazuli is an aluminum silicate whose brilliant blue color is due to the presence of S_3^- ions. Write a plausible Lewis structure for this ion.

Tom McHugh/Science Source

Lapis lazuli jewelry.

79. Bromine is prepared by bubbling 0.240 mol gaseous chlorine into 0.500 L of a solution that is 0.500 M in bromide ion.
(a) Determine the limiting reactant.
(b) Calculate the theoretical amount (mol) of bromine that could be produced.
(c) If 0.124 mol bromine is produced, calculate the percent yield.

80. Magnesium can be extracted from dolomite, a mineral that contains 13.2% Mg, 21.7% Ca, 13.0% C, and 52.1% O. Determine the simplest formula for this ionic compound.
81. A natural brine found in Arkansas has a bromide ion concentration of 5.00×10^{-3} M. If 210. g Cl_2 were added to 1.00×10^3 L of the brine,
    (a) determine the limiting reactant.
    (b) calculate the theoretical yield of Br_2 (d = 3.12 g/mL).
82. Chlorine gas was first prepared by Carl Wilhelm Scheele in 1774 by the reaction of sodium chloride, manganese(IV) oxide, and sulfuric acid. In addition to chlorine, the reaction produces water, sodium sulfate, and manganese(II) sulfate. Write the balanced equation for this reaction.
83. Mercury(II) azide, $Hg(N_3)_2$, is an unstable compound used as a detonator in blasting caps. Calculate the volume (L) of nitrogen produced at 1 atm and 25 °C when 2.50 g mercury azide decomposes to liquid mercury and nitrogen.

More Challenging Questions

These questions require more thought and integrate several concepts.

84. Use the Clausius-Clapeyron equation and the phase diagram in Question 75 to calculate the sublimation enthalpy (J/mol) of monoclinic sulfur.
85. At 20. °C the vapor pressure of white phosphorus is 0.0254 mmHg; at 40. °C it is 0.133 mmHg. Use the Clausius-Clapeyron equation to estimate the heat of sublimation (J/mol) of white phosphorus.
86. The density of sulfur vapor at 700. °C and 1.00 atm is 0.8012 g/L. Determine the molecular formula of sulfur in the vapor.
87. Assume that the radius of Earth is 6400 km, the crust is 50. km thick, the density of the crust is 3.5 g/cm^3, and 25.7% of the crust is silicon by mass. Calculate the total mass of silicon in the crust of Earth.
88. A 5.00-g sample of white phosphorus is burned in excess oxygen and the product is dissolved in sufficient water to form 250. mL of solution.
    (a) Write the balanced chemical equation for the burning of phosphorus in excess oxygen.
    (b) Calculate the pH of the resulting solution.
    (c) An excess of aqueous calcium nitrate is added to the solution causing a white precipitate to form. Write a balanced chemical equation for this reaction and calculate the mass of precipitate formed.
    (d) An excess of zinc is added to the remaining solution. The reaction generates a colorless gas. Identify the gas and calculate its volume at STP.
89. Bromine, Br_2, reacts vigorously with hydrogen, H_2, to form hydrogen bromide. At STP, 100. mL hydrogen gas reacts with a stoichiometric amount of bromine. The resulting hydrogen bromide is dissolved in sufficient water to form 250. mL solution. Calculate the pH of the solution.
90. A solid-fuel rocket booster is used to lift a rocket from its launching pad. The solid fuel contains a mixture of ammonium perchlorate, NH_4ClO_4, and powdered aluminum metal that reacts in the presence of an Fe_2O_3 catalyst.

$$3\ NH_4ClO_4(s) + 3\ Al(s) \longrightarrow Al_2O_3(s) + AlCl_3(s) + 6\ H_2O(g) + 3\ NO(g)$$

    (a) Which element is oxidized? Which is reduced?
    (b) Use the data in Appendix J and calculate the enthalpy change for the reaction. For $NH_4ClO_4(s)$, $\Delta_f H° = -295$ kJ/mol.
91. At 1 atm and approximately 1800 °C, 50% of P_4 is dissociated into 2 P_2. If equilibrium is established under these conditions, calculate the equilibrium constant.
92. The reaction for the production of white phosphorus is exothermic; $\Delta_r H° = -3060.$ kJ/mol.

$$2\ Ca_3(PO_4)_2(s) + 6\ SiO_2(s) + 10\ C(s) \longrightarrow 6\ CaSiO_3(s) + 10\ CO(g) + P_4(g)$$

Calculate the $\Delta_f H°$ of $CaSiO_3(s)$. For $Ca_3(PO_4)_2(s)$, $\Delta_f H° = -4138$ kJ/mol. The sublimation enthalpy of phosphorus is 13.06 kJ/mol.

93. A typical electric furnace using 500. V and a certain current produces four tons (4000. kg) phosphorus per hour using the reaction

$$2\ Ca_3(PO_4)_2(s) + 6\ SiO_2(s) + 10\ C(s) \longrightarrow 6\ CaSiO_3(s) + 10\ CO(g) + P_4(g)$$

Calculate the current needed to produce this quantity of phosphorus.

94. The Gibbs free energy of formation, $\Delta_f G°$, of HI is +1.70 kJ/mol at 25 °C. Calculate the equilibrium constant for the reaction $HI(g) \rightleftharpoons \frac{1}{2} H_2(g) + \frac{1}{2} I_2(g)$.
95. Predict whether the formation of ions of the alkaline-earth elements requires less energy or more energy than formation of ions of the alkali metals. Explain your answer.
96. Use a Born-Haber cycle (← **Sec. 5-13**) to calculate the lattice energy of MgF_2 using these thermodynamic data.

$$\Delta_{sub}H\ Mg(s) = +146\ kJ/mol;$$
$$B.E.\ F_2(g) = +158\ kJ/mol;$$
$$I.E._1\ Mg(g) = +738\ kJ/mol;$$
$$I.E._2\ Mg^+(g) = +1450\ kJ/mol;$$
$$E.A.\ F(g) = -328\ kJ/mol;$$
$$\Delta_f H°\ MgF_2(s) = -1124\ kJ/mol.$$

Compare this lattice energy with that of SrF_2, −2496 kJ/mol. Explain the difference in the values in structural terms.

97. Aluminum metal reacts rapidly and completely with hydrochloric acid, but only incompletely with nitric acid. Explain.
98. Elemental analysis of a borane indicates this composition: 84.2% B and 15.7% H. The compound has a molar mass of 76.7 g/mol. Determine the molecular formula of the borane.
99. The reaction of iodine with excess liquid chlorine produces the interhalogen compound I_2Cl_6.
    (a) Write the Lewis structure of this molecule.
    (b) Use VSEPR theory to predict the structure of the molecule.
100. Hydroxyapatite is the important compound in tooth enamel. It dissociates according to the equation

$$Ca_5(PO_4)_3OH(s) \rightleftharpoons 5\ Ca^{2+}(aq) + 3\ PO_4^{3+}(aq) + OH^-(aq)$$

Children drink milk to obtain calcium, but the fermentation of the milk produces lactic acid, which remains on the teeth.
    (a) Use the dissociation equation to explain how drinking milk helps babies to produce "strong" teeth.
    (b) Explain why the lactic acid inhibits formation of strong teeth.

Chemistry of Selected Transition Elements and Coordination Compounds

20

David Toase/Photodisc Green/Getty Images

For centuries, stained glass artists have added transition metal oxides to colorless molten glass to make colored glass. Some oxides and their colors are: Cu_2O, red; Cr_2O_3, green; Co_2O_3, blue; and MnO_2, violet. We see vivid, beautiful colors because only part of the white light striking the stained glass is transmitted. Why are only some visible wavelengths transmitted and does the same transition metal ion always transmit the same color? (Section 20-7) How are transition metals used in our society and how are they obtained and purified? (Sections 20-2 to 20-4) What are coordination compounds and why do most of them involve transition metal ions? (Section 20-6). This chapter focuses on transition metals and their compounds.

It is hard to overstate how important metals have been to the development of civilizations. Transitions from the Stone Age to the Bronze Age to the Iron Age and to the Computer Age have been marked by humans' ability to extract metals from ores and to process the metals into tools and objects useful in industry, warfare, and dwellings. In this chapter we consider the transition metals, the *d*-block elements of the periodic table. Some of these elements and their compounds are of major economic importance. The precious metals gold, silver, and platinum are transition elements used in coinage and jewelry. Others, such as iron and its alloy, steel, are valuable for their structural uses.

We first consider the transition metals in overview and then look more closely at a few of them, including iron, the most economically important. The chapter closes by describing coordination compounds. In such compounds, ions or molecules are bonded to transition metal ions or atoms. These compounds run the gamut from being responsible for the vivid colors of famous oil paintings to their role in significant biomolecules such as hemoglobin, vitamin B_{12}, and critical enzymes.

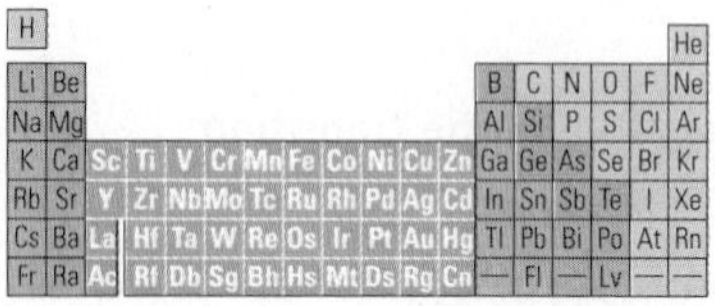

The transition elements in Period 7 beginning with rutherfordium, element 104, are all radioactive. They have been made synthetically in *very* small amounts, just several atoms in some cases, and, therefore, little is currently known about their properties.

20-1 Properties of the Transition (*d*-Block) Elements

The four series of *d*-block elements called the *transition elements* are in the center of the periodic table in Periods 4 through 7. These elements lie between the very reactive *s*-block metals and the less reactive *p*-block metals. Compared to the representative elements (*s* block and *p* block), there is a slow, steady transition in properties from one transition metal to the next, hence the name transition elements. These generalities apply to the transition elements:

- All are metals that conduct electricity well, but to varying degrees.
- Most are ductile (able to be drawn into a wire) and malleable (able to be hammered into thin sheets).
- Except for gold and copper, they are silvery-white or bluish.
- They usually have higher melting and boiling points than the main-group elements; tungsten has the highest melting point of any metal, 3410 °C. Mercury is an exception, being the only liquid metal at room temperature.
- They usually have high densities; osmium (22.61 g/mL) and iridium (22.65 g/mL) are the most dense, even more dense than gold (19.3 g/mL).
- Many form brightly colored compounds, as solids and in aqueous solution.
- Some are paramagnetic; a few are ferromagnetic (← **Sec. 5-8a**).
- All form complex ions (Section 20-6).
- Most have several oxidation states; the scandium (+3) and zinc (+2) groups are exceptions.

Oxidation states (oxidation numbers) are described in Section 3-5c.

$Co^{2+}(aq)$
$Ni^{2+}(aq)$
$Cu^{2+}(aq)$

Cengage Learning/Charles D. Winters

The transition elements at the beginning of each row in the periodic table are quite different in their chemical behavior from those at the end of each series. Such differences in chemical behavior can be attributed to the number and distribution of *d*-orbital electrons (Table 20.1). At the left side of the series, members of the scandium group of elements (Sc, Y, La) are reactive metals like the alkaline-earth metals, their predecessors in each period. On the other hand, the zinc family members (Zn, Cd, Hg) are not like other transition elements in that the zinc family members have filled $(n - 1)\,d$ and ns sublevels. In fact, the elements of the zinc family are sometimes not classified as transition elements.

20-1a Electron Configurations

All transition elements have the electron configuration

$$[\text{noble gas}]\,(n - 1)d^x ns^y$$

Table 20.1 Outermost Electron Configurations of *d*-Block Elements

Deviations (marked in color) occur when a nonstandard configuration is more stable.

Configuration	$(n-1)d\,ns^2$	$(n-1)d^2\,ns^2$	$(n-1)d^3\,ns^2$	$(n-1)d^4\,ns^2$	$(n-1)d^5\,ns^2$	$(n-1)d^6\,ns^2$	$(n-1)d^7\,ns^2$	$(n-1)d^8\,ns^2$	$(n-1)d^9\,ns^2$	$(n-1)d^{10}\,ns^2$
First series:	21 **Sc** $3d^1\,4s^2$	22 **Ti** $3d^2\,4s^2$	23 **V** $3d^3\,4s^2$	24 **Cr** $3d^5\,4s^1$	25 **Mn** $3d^5\,4s^2$	26 **Fe** $3d^6\,4s^2$	27 **Co** $3d^7\,4s^2$	28 **Ni** $3d^8\,4s^2$	29 **Cu** $3d^{10}\,4s^1$	30 **Zn** $3d^{10}\,4s^2$
Second series:	39 **Y** $4d^1\,5s^2$	40 **Zr** $4d^2\,5s^2$	41 **Nb** $4d^4\,5s^1$	42 **Mo** $4d^5\,5s^1$	43 **Tc** $4d^5\,5s^2$	44 **Ru** $4d^7\,5s^1$	45 **Rh** $4d^8\,5s^1$	46 **Pd** $4d^{10}$	47 **Ag** $4d^{10}\,5s^1$	48 **Cd** $4d^{10}\,5s^2$
Third series:	57 **La** $5d^1\,6s^2$	72 **Hf** $4f^{14}5d^2\,6s^2$	73 **Ta** $4f^{14}5d^3\,6s^2$	74 **W** $4f^{14}5d^4\,6s^2$	75 **Re** $4f^{14}5d^5\,6s^2$	76 **Os** $4f^{14}5d^6\,6s^2$	77 **Ir** $4f^{14}5d^7\,6s^2$	78 **Pt** $4f^{14}5d^9\,6s^1$	79 **Au** $4f^{14}5d^{10}\,6s^1$	80 **Hg** $4f^{14}5d^{10}\,6s^2$

4f-elements intervene

where n is the period number (4 through 7), x is the number of d electrons (1 through 10), and y is the number of s electrons (1 or 2, except in palladium), as summarized in Table 20.1. The number of d electrons increases from left to right across a transition metal series. Elements in the zinc group (Zn, Cd, Hg) at the end of the series have filled $(n-1)d^{10}$ sublevels. In the preceding group, the elements Cu, Ag, and Au also have filled $(n-1)d^{10}$ sublevels, along with half-filled ns^1 sublevels, as described below.

The progressive filling of the d atomic orbitals from Sc to Zn in the first series is not uniform, as seen in Table 20.2. The first three elements of the series—Sc, Ti, and V—have [Ar]$3d^14s^2$, [Ar]$3d^24s^2$, and [Ar]$3d^34s^2$ electron configurations, respectively. The filling sequence changes at chromium, which has a ground-state electron configuration of [Ar]$3d^54s^1$ with *two* half-filled sublevels, which is a lower energy state than the expected [Ar]$3d^44s^2$ configuration. The sequence reverts back to the expected one from manganese through nickel, with the pairing of $3d$ electrons. It changes again with

Table 20.2 Atomic Orbital Occupancy of the First Transition Series Elements

Element	Partial Atomic Orbital Diagram: 3*d*					4*s*	4*p*			Unpaired Electrons
Sc	↑					↑↓				1
Ti	↑	↑				↑↓				2
V	↑	↑	↑			↑↓				3
Cr*	↑	↑	↑	↑	↑	↑				6
Mn	↑	↑	↑	↑	↑	↑↓				5
Fe	↑↓	↑	↑	↑	↑	↑↓				4
Co	↑↓	↑↓	↑	↑	↑	↑↓				3
Ni	↑↓	↑↓	↑↓	↑	↑	↑↓				2
Cu*	↑↓	↑↓	↑↓	↑↓	↑↓	↑				1
Zn	↑↓	↑↓	↑↓	↑↓	↑↓	↑↓				0

*Exceptions to normal orbital occupancy sequence.

copper, which has a filled $3d^{10}$ sublevel and a half-filled $4s^1$ sublevel (Table 20.2). This configuration is more stable than the expected $[Ar]3d^94s^2$ configuration. The first series ends with zinc, which has filled 3*d* *and* 4*s* sublevels, $[Ar]3d^{10}4s^2$.

When transition metal atoms lose electrons to form cations, the* ns *electrons are lost before the* (n − *1*)d *electrons. Thus, both of the 4*s* electrons are lost when Fe^{2+} forms. When Fe^{3+} forms, one of the 3*d* electrons is lost as well as both of the 4*s* electrons. Fe^{2+} and Fe^{3+} ions differ in their number of 3*d* electrons; each ion has lost both of the original two 4*s* electrons.

Fe atom		Fe^{2+} ion + 2 e^-
$[Ar]3d^64s^2$	$\longrightarrow$	$[Ar]3d^64s^0$
3*d*: ↑↓ ↑ ↑ ↑ ↑ 4*s*: ↑↓		3*d*: ↑↓ ↑ ↑ ↑ ↑ 4*s*: (empty)

Fe atom		Fe^{3+} ion + 3 e^-
$[Ar]3d^64s^2$	$\longrightarrow$	$[Ar]3d^54s^0$
3*d*: ↑↓ ↑ ↑ ↑ ↑ 4*s*: ↑↓		3*d*: ↑ ↑ ↑ ↑ ↑ 4*s*: (empty)

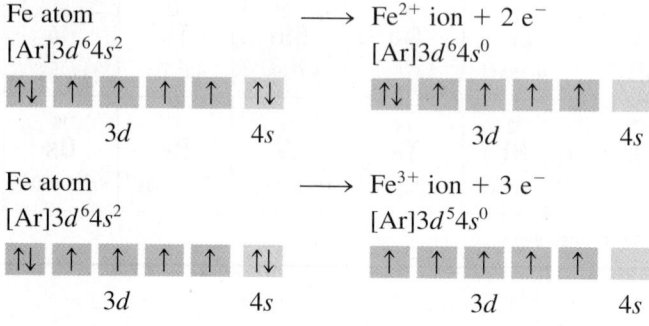

Magnetic susceptibility measurements (Figure 20.1) confirm the electron configurations of the first row and other transition metal ions. The *magnetic moment* is a value calculated from the measured paramagnetism of a sample and indicates the number of unpaired electrons in the sample (← **Sec. 5-8a**). As seen from Figure 20.2, the greater the number of unpaired electrons, the greater the magnetic moment of the substance. Figure 20.2 includes experimental evidence confirming that the 4*s* electrons are removed first, leaving Fe^{2+} and Fe^{3+} with four and five unpaired electrons, respectively. If the 3*d* electrons were removed first, these ions would have four (Fe^{2+}) and three (Fe^{3+}) unpaired electrons.

PROBLEM-SOLVING EXAMPLE 20.1

Transition-Metal Ion Electron Configurations

Use partial atomic orbital box diagrams to explain the number of unpaired electrons shown in Figure 20.2 for (a) Ti^{3+}, (b) Cr^{2+}, and (c) Cu^{2+}.

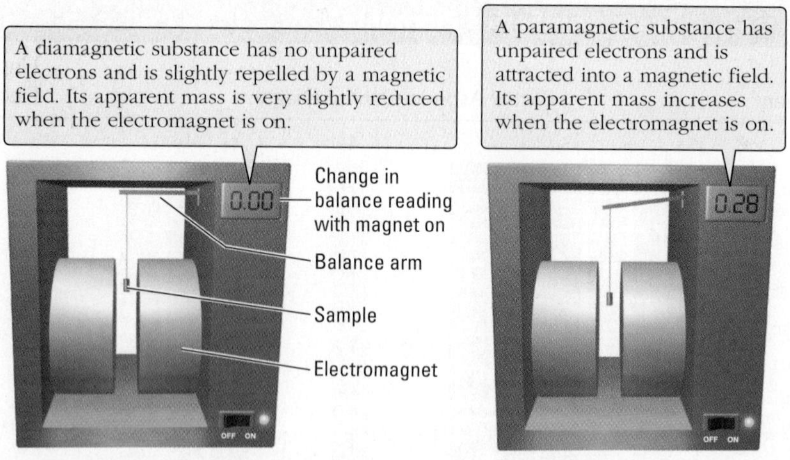
A diamagnetic substance has no unpaired electrons and is slightly repelled by a magnetic field. Its apparent mass is very slightly reduced when the electromagnet is on.
A paramagnetic substance has unpaired electrons and is attracted into a magnetic field. Its apparent mass increases when the electromagnet is on.
0.00
Change in balance reading with magnet on
Balance arm
Sample
Electromagnet
0.28
OFF ON
OFF ON

Figure 20.1 Measurement of magnetic properties.

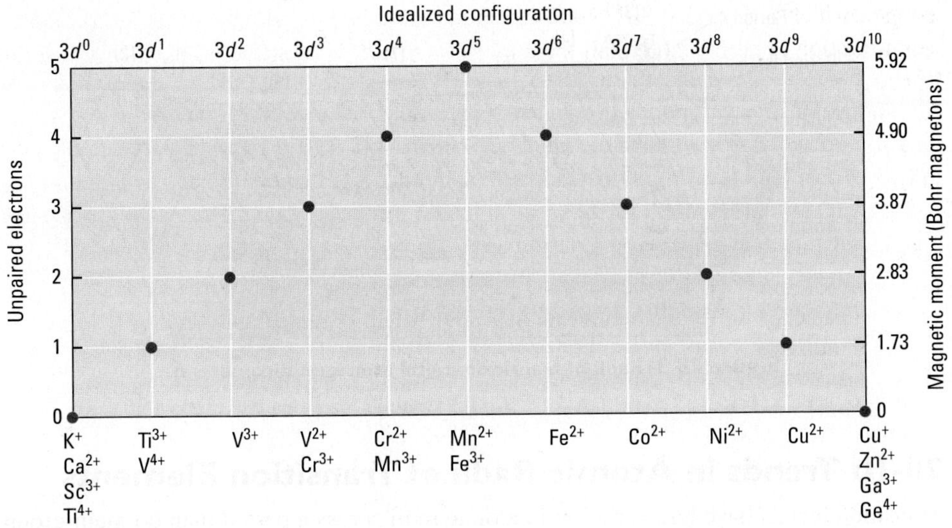
Idealized configuration
3d0 3d1 3d2 3d3 3d4 3d5 3d6 3d7 3d8 3d9 3d10
Unpaired electrons
0 1 2 3 4 5
Magnetic moment (Bohr magnetons)
0 1.73 2.83 3.87 4.90 5.92
K+ Ca2+ Sc3+ Ti4+
Ti3+ V4+
V3+
V2+ Cr3+
Cr2+ Mn3+
Mn2+ Fe3+
Fe2+
Co2+
Ni2+
Cu2+
Cu+ Zn2+ Ga3+ Ge4+

Figure 20.2 The number of unpaired electrons in the first-row transition-metal ions and their magnetic moments. (The Bohr magneton is a unit that measures paramagnetism.)

Result

(a) $[Ar]3d^14s^0$

3d
4s

(b) $[Ar]3d^44s^0$

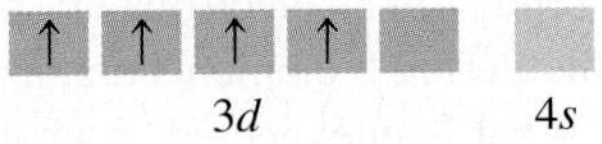
3d
4s

(c) $[Ar]3d^94s^0$

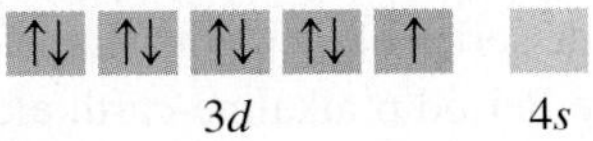
3d
4s

Analyze When an ion is formed from a Period 4 transition metal, 4*s* electrons are removed before 3*d* electrons.

Plan Use a partial atomic orbital box diagram to keep track of removal of 4*s* and then 3*d* electrons from an atom and to determine the number of unpaired electrons in the ion.

Execute

(a) Ti atom $\longrightarrow$ Ti^{3+} ion + $3\ e^-$

$[Ar]3d^24s^2$ $\qquad$ $[Ar]3d^14s^0$

3d
4s

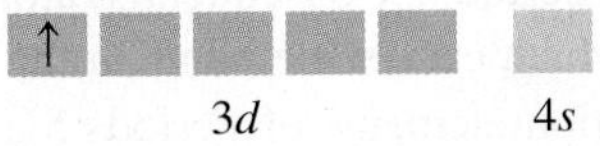
3d
4s

(b) Cr atom $\longrightarrow$ Cr^{2+} ion + $2\ e^-$

$[Ar]3d^54s^1$ $\qquad$ $[Ar]3d^44s^0$

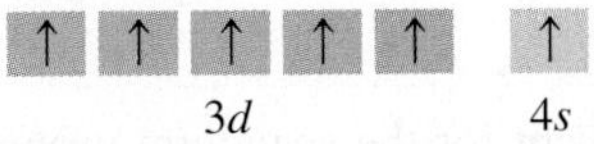
3d
4s

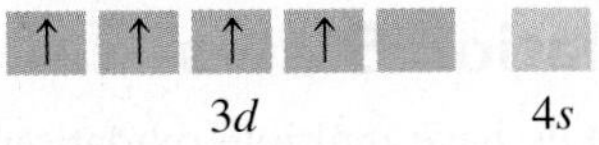
3d
4s

(c) Cu atom $\longrightarrow$ Cu^{2+} ion + $2\ e^-$

$[Ar]3d^{10}4s^1$ $\qquad$ $[Ar]3d^94s^0$

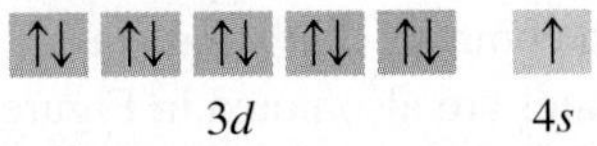
3d
4s

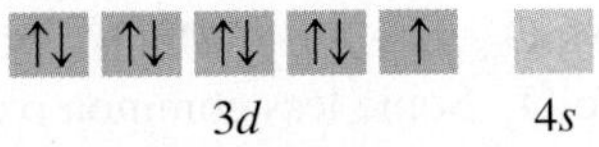
3d
4s

PROBLEM-SOLVING PRACTICE 20.1

Use partial atomic orbital box diagrams to explain the number of unpaired electrons shown in Figure 20.2 for (a) Mn^{2+} and (b) Cr^{3+}.

PROBLEM-SOLVING PRACTICE answers are provided at the back of this book in Appendix K.

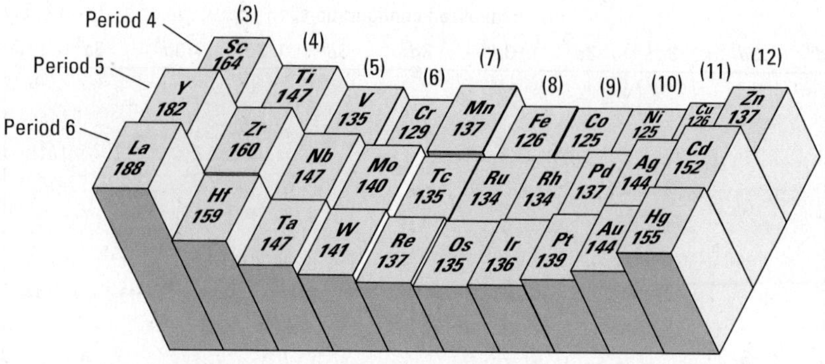

Period 4
Period 5
Period 6
(3)
(4)
(5)
(6)
(7)
(8)
(9)
(10)
(11)
(12)
Sc 164
Ti 147
V 135
Cr 129
Mn 137
Fe 126
Co 125
Ni 125
Cu 126
Zn 137
Y 182
Zr 160
Nb 147
Mo 140
Tc 135
Ru 134
Rh 134
Pd 137
Ag 144
Cd 152
La 188
Hf 159
Ta 147
W 141
Re 137
Os 135
Ir 136
Pt 139
Au 144
Hg 155

Figure 20.3 Radii of transition metal atoms (in picometers).

20-1b Trends in Atomic Radii of Transition Elements

Transition metals have less variation in atomic radii across a period than do main-group elements. Across a row of transition elements, the atomic radii decrease steadily and then increase slightly; Mn is an exception (Figure 20.3). The decrease occurs because as each *d* electron is added, there is also an increase of one unit of nuclear charge (a proton). Repulsions among the electrons are not sufficient to counteract the added attraction of the nucleus to the electrons, so the atomic radius decreases. Toward the end of each transition-metal series, the radii increase slightly due to several factors, including electron-electron repulsion as electrons are paired in *d* atomic orbitals.

The radii of the Period 5 elements are, as expected, greater than those of the Period 4 transition metals. What is unexpected is what occurs with the atomic radii in going from the Period 5 to Period 6. Instead of Period 6 radii being larger, they are nearly same as those of Period 5. This is a consequence of the lanthanide series of elements (La, element 57, to Lu, element 71) intervening between barium, the Period 6 alkaline-earth element, and hafnium, the next transition element of Period 6. In the lanthanide series atoms, the effective nuclear charge (← **Sec. 5-9c**) builds up. This causes a decrease in their atomic radii because all the additional electrons go into the 4*f* atomic orbitals, which do not effectively screen valence electrons from the increasing nuclear charge. The increased nuclear charge pulls the valence electrons closer to the nucleus, decreasing the atomic radii. This *decrease in atomic radii across the lanthanides* is known as the **lanthanide contraction**. It just offsets the expected size increase going from Period 5 transition elements to those of Period 6. Consequently, atoms of Periods 5 and 6 transition elements are of similar size, which causes elements in the same group to have similar chemical properties. The transition elements of Periods 5 and 6 occur together in ores, and because of their chemical similarities, they are very difficult to separate from each other.

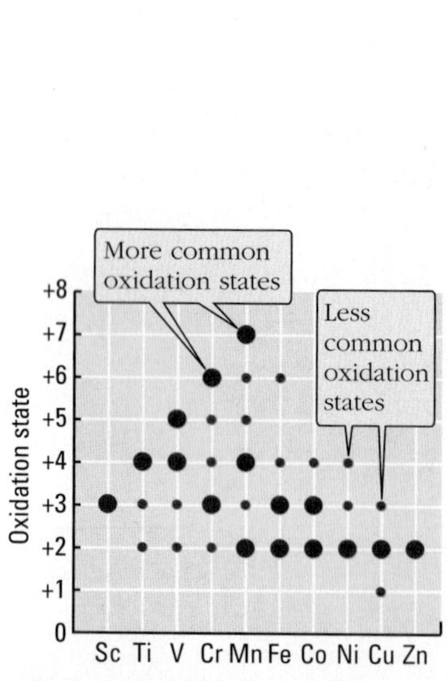

More common oxidation states
Less common oxidation states
Oxidation state
+8
+7
+6
+5
+4
+3
+2
+1
0
Sc Ti V Cr Mn Fe Co Ni Cu Zn

Figure 20.4 Oxidation states of first transition series elements. Note that the greatest number of oxidation states occurs with elements in the middle of the first transition series.

20-1c Oxidation States

All transition metals have multiple oxidation states except for the scandium group (+3) and zinc group (+2) elements. The oxidation states of the first transition series elements are listed in Figure 20.4. Manganese, for example, has three common oxidation states: +2 in Mn^{2+}, +4 in MnO_2, and +7 in MnO_4^-. Iron has two common oxidation states: +2 in FeO and +3 in Fe_2O_3. Some less common oxidation states are also noted in Figure 20.4.

Transition metals that form 2+ ions do so, in general, by losing two *ns* electrons before losing any $(n-1)d$ electrons. Higher oxidation states involve losing $(n-1)d$ electrons as well. The maximum oxidation state for the first five elements in the first series—Sc through Mn—equals the total of the $(n-1)d$ plus *ns* electrons. Thus, the maximum oxidation state is +6 for chromium, electron configuration $3d^5\,4s^1$; it is found in CrO_4^{2-} and $Cr_2O_7^{2-}$. The maximum oxidation state of manganese is +7, which is found in MnO_4^-. These high oxidation states make $Cr_2O_7^{2-}$ (in acidic solution) and MnO_4^- strong oxidizing agents. In general, compounds in which the transition metal has a low oxidation state tend to be ionic, whereas compounds with transition metals in high oxidation states are relatively covalent. Thus, $MnCl_2$ (mp 650 °C) is an ionic solid containing Mn^{2+} and Cl^- ions. On the other hand, MnO_4^- is a polyatomic ion containing covalent Mn—O bonds.

Cr^0 [Ar]$3d^54s^1$
Mn^0 [Ar]$3d^54s^2$

Ions of transition elements with partially filled *d* atomic orbitals can accept or donate electrons, a property that makes them effective components of catalysts. In iron compounds, for example, iron can be present as Fe^{2+} (reduced form) or Fe^{3+} (oxidized form). Each iron ion acts as an electron shuttle, losing or gaining electrons between the oxidized and reduced forms when catalyzing electron transfer reactions, such as the production of ammonia by the Haber-Bosch process (← **Sec. 12-8**).

EXERCISE 20.1

Oxidation States of Transition Metals

Determine the oxidation state of the transition metal(s) in each compound.

(a) $Na_2V_4O_{11}$ (b) $KAgF_4$ (c) $CaTiO_3$ (d) $MnAl_2O_4$

Answers to **EXERCISES** are provided at the back of this book in Appendix L.

EXERCISE 20.2

Oxidation State

Determine the oxidation state of copper in $K_2Ca[Cu(NO_2)_6]$.

20-2 Iron and Steel: Pyrometallurgy

Iron is the most abundant transition metal and the second most abundant metallic element in Earth's crust (4.7%) (← **Sec. 19-2**). Pure iron is a silvery-white, rather soft metal. Its great commercial importance comes with the addition of very small amounts of carbon or other alloying elements to it to form steel.

Iron occurs naturally in compounds containing Fe^{2+} and Fe^{3+} ions. In an aqueous solution exposed to air, Fe^{2+}(aq) is oxidized to Fe^{3+}(aq).

Fe^{2+}(aq) is the hydrated ion, $[Fe(H_2O)_6]^{2+}$.

$$4\ Fe^{2+}(aq) + O_2(g) + 4\ H^+(g) \longrightarrow 4\ Fe^{3+}(aq) + 2\ H_2O(\ell)$$

Aqueous Fe^{3+} reacts with water to form a hydrated oxide known as *rust*.

$$2\ Fe^{3+}(aq) + 4\ H_2O(\ell) \longrightarrow Fe_2O_3 \cdot H_2O(s) + 6\ H^+(aq)$$

Iron reacts with weakly oxidizing acids such as HCl and acetic acid to form the pale green $[Fe(H_2O)_6]^{2+}$ ion.

$$Fe(s) + 2\ H^+(aq) \longrightarrow Fe^{2+}(aq) + H_2(g) \qquad E° = +0.44\ V$$

When reacted with strongly oxidizing acids such as dilute nitric acid, the metal is oxidized directly to Fe^{3+}.

$$Fe(s) + 4\ H^+(aq) + NO_3^-(aq) \longrightarrow Fe^{3+}(aq) + NO(g) + 2\ H_2O \qquad E° = +1.00\ V$$

The principal iron ores are hematite, Fe_2O_3, and magnetite, Fe_3O_4, which are found in large deposits in Minnesota, Russia, Brazil, India, and Australia. Iron production involves steps to concentrate and purify the ores. Then, iron ions in the oxide ores are reduced to the metal by using carbon in the form of coke as the reducing agent at high

Iron also occurs in nature as the pyrite, FeS_2. This mineral is not used as an ore because in steel making it is difficult to remove all the sulfur, which makes the steel brittle.

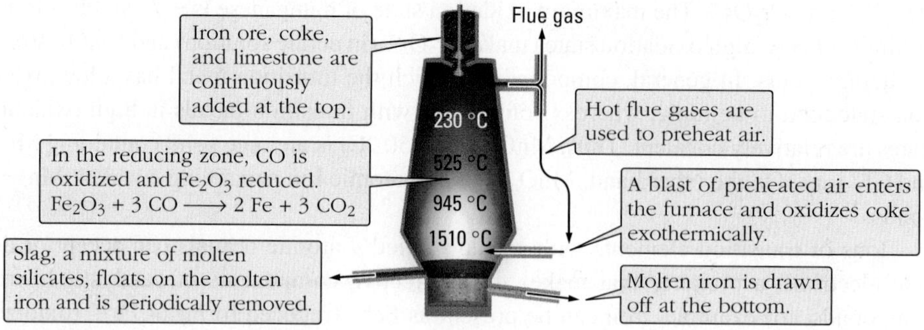
Flue gas
Iron ore, coke, and limestone are continuously added at the top.
230 °C
525 °C
945 °C
1510 °C
Hot flue gases are used to preheat air.
In the reducing zone, CO is oxidized and Fe_2O_3 reduced.
$Fe_2O_3 + 3\ CO \longrightarrow 2\ Fe + 3\ CO_2$
A blast of preheated air enters the furnace and oxidizes coke exothermically.
Slag, a mixture of molten silicates, floats on the molten iron and is periodically removed.
Molten iron is drawn off at the bottom.

Figure 20.5 Diagram of a blast furnace. Iron ore is reduced to iron in a blast furnace.

temperatures in a blast furnace (Figure 20.5). *The extraction of a metal from its ore using chemical reactions carried out at high temperatures* is called **pyrometallurgy**.

To extract metallic iron, a mixture of iron ore, coke, and limestone, $CaCO_3$, is fed into the top of the blast furnace, and a blast of heated air or oxygen is forced up from the bottom of the furnace. The coke reacts exothermically with the heated air, producing a high temperature that speeds up the iron-forming reactions, which makes the process economical. The iron ore is reduced to metallic iron by these reactions

$$2\ C(s) + O_2(g) \longrightarrow 2\ CO(g) \qquad \text{exothermic}$$

$$Fe_2O_3(s) + 3\ CO(g) \longrightarrow 2\ Fe(s) + 3\ CO_2(g) \qquad \text{exothermic}$$

Limestone is added to remove silica-containing impurities in the ore.

$$CaCO_3(s) \longrightarrow CaO(s) + CO_2(g) \qquad \text{endothermic}$$

$$CaO(s) + SiO_2(s) \longrightarrow CaSiO_3(\ell) \qquad \text{endothermic}$$

Calcium silicate and other metal silicates form *slag*, which is a liquid at the temperature of the blast furnace and is immiscible with molten iron. The molten iron, which contains a substantial concentration of dissolved carbon and smaller concentrations of other impurities, is more dense than the slag and collects at the bottom of the furnace (Figure 20.5).

The iron produced by a blast furnace is *pig iron*, a material that is brittle due to impurities of as much as 4.5% carbon, 1.7% manganese, 0.3% phosphorus, 0.04% sulfur, and 1% silicon. The main material causing brittleness is cementite, Fe_3C, an iron carbide formed at the temperatures of the blast furnace.

$$3\ Fe(s) + C(s) \longrightarrow Fe_3C(s)$$

Molten pig iron that is poured into molds of a desired shape is called *cast iron.* It can be used directly to make molded automobile engine blocks, brake drums, transmission housings, and the like. Cast iron is too brittle for most structural uses. To convert cast iron or pig iron into **steel**, a much stronger material, the phosphorus, sulfur, and silicon impurities must be removed and the carbon content reduced to about 1.3%.

20-2a Steel

Many iron alloys are known collectively as *steels,* each with its own particular structural properties. One of the most common is carbon steel, an iron alloy containing 0.5 to 1.3% carbon. To convert pig iron to steel, the excess carbon is oxidized away using oxygen. Thus, whereas extracting iron from an ore is a *reduction* process, steel making is an *oxidation* process. One of several techniques used to make steel is the *basic oxygen process* (Figure 20.6), in which pure oxygen is blown through a water-cooled ceramic tube that

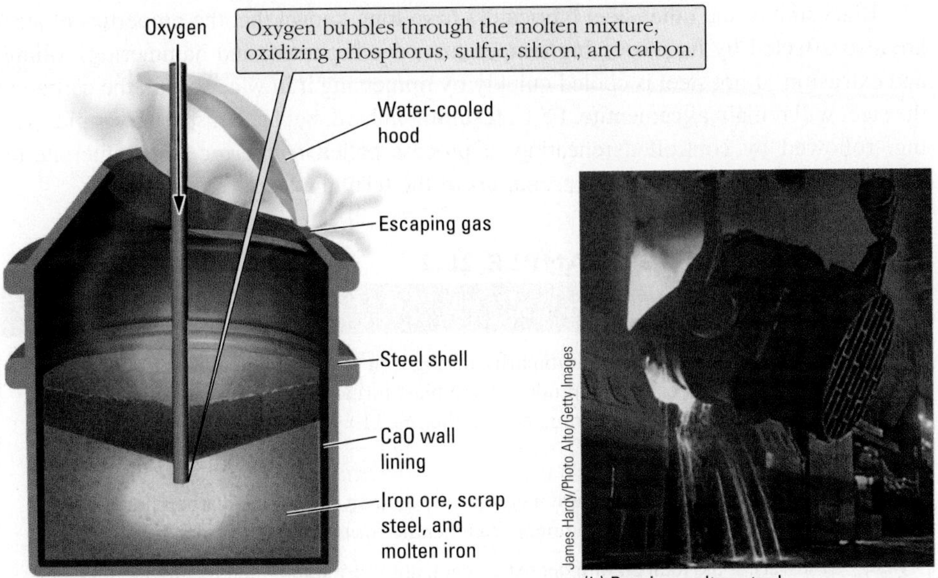
Oxygen
Oxygen bubbles through the molten mixture, oxidizing phosphorus, sulfur, silicon, and carbon.
Water-cooled hood
Escaping gas
Steel shell
CaO wall lining
Iron ore, scrap steel, and molten iron
James Hardy/Photo Alto/Getty Images

(a) Schematic diagram (b) Pouring molten steel

Figure 20.6 A basic oxygen furnace. Much of the steel manufactured today is produced by the basic oxygen process.

is pushed below the surface of molten, impure iron. At the high temperatures of the melt (1900 °C), the dissolved carbon reacts rapidly with the oxygen to form gaseous carbon monoxide and carbon dioxide, which are vented. The scale of the basic oxygen process operation is impressive. About 200 tons of molten pig iron, 100 tons of scrap iron, and 20 tons of limestone are loaded into the furnace at a time. The steel is produced within an hour using such a process.

The composition of steel is varied by adding silicon, chromium, manganese, molybdenum, nickel, or other metals to give the steel specific physical, chemical, and mechanical properties. Table 20.3 lists the composition and uses of some common steel alloys. Magnetic alloys can be permanent magnets, such as those in audio speakers and older computer hard drives, or temporary magnets such as those in electric motors, generators, and transformers. Alnico is the general name for a series of popular permanent magnets containing Al, Ni, Co, Fe, and sometimes Cu and Ti. Alnico V, for example, contains 51% Fe, as well as four other elements—14% Ni, 24% Co, 8% Al, and 3% Cu.

Cengage Learning/
Charles D. Winters

An Alnico magnet picking up iron or steel objects.

Table 20.3 Some Steels and Their Uses

Name	Composition	Properties	Uses
Carbon steel	98.7% Fe, 1.3% C	Hard	Sheet steel, tools
Manganese steel	10–18% Mn, 90–82% Fe, 0.5% C	Hard, resistant to wear	Railroad rails, safes, armor plate
Stainless steel	14–18% Cr, 7–9% Ni, 79–73% Fe, 0.2% C	Resistant to corrosion	Cutlery, instruments
Nickel steel	2–4% Ni, 98–96% Fe, 0.5% C	Hard, elastic, resistant to corrosion	Drive shafts, gears, cables
Duriron	12–15% Si, 88–85% Fe, 0.85% C	Resistant to corrosion by acids	Pipes
High-speed steel	14–20% W, 86–80% Fe, 0.5% C	Retains temper when hot	High-speed cutting tools

The artist Michelangelo wrote about using fire to transform substances:

"It is with fire that blacksmiths iron subdue
Unto fair form, the image of their thought . . ."
—Sonnet 59

Blacksmiths and other steel fabricators have long known that the properties of steel are also affected by its processing temperature, cooling rates, and hammering, rolling, and extrusion. If hot steel is cooled quickly by immersing it in water or oil, the carbon in the steel will remain as cementite, Fe_3C, resulting in hard, but brittle, steel. By rapid cooling, followed by controlled reheating, a process called *tempering,* the cementite-to-graphite ratio is adjusted and the properties of the resultant steel varied further.

PROBLEM-SOLVING EXAMPLE 20.2

Cementite

Cementite, Fe_3C, is produced as an impurity during iron production in a blast furnace. To decrease the amount of cementite produced, the blast furnace temperature is decreased. Is the formation of cementite exothermic or endothermic? Explain your result.

Result Endothermic

Analyze Lowering the temperature increases the amount of iron and carbon not converted to cementite, thus decreasing the amount of cementite formed.

Plan If lowering the temperature decreases cementite formation, then raising the temperature increases the amount of cementite formed, which indicates that cementite formation is an endothermic reaction.

Execute

$$3\ Fe(s) + C(s) \rightleftharpoons Fe_3C \qquad \text{endothermic} \qquad (\Delta_r H > 0)$$

PROBLEM-SOLVING PRACTICE 20.2

In a blast furnace at approximately 1000 °C, much of the carbon is converted into cementite, Fe_3C, during the production of iron, which is used subsequently to produce steel. If the temperature of the steel is decreased rapidly, cementite is trapped in the iron to produce a steel with a high cementite concentration causing the steel to be brittle. Explain why this is the case. If the steel is cooled slowly, what would the effect be on the properties of the steel?

PROBLEM-SOLVING EXAMPLE 20.3

Iron Production

(a) Calculate the mass (g) of carbon monoxide needed to form 1.00×10^3 kg iron from hematite, Fe_2O_3, and from magnetite, Fe_3O_4.
(b) Calculate the mass (g) of carbon in the form of coke needed to prepare the total amount of CO required for the two reductions in part (a).

Result

(a) 7.52×10^5 g CO for hematite and 6.69×10^5 g CO for magnetite
(b) 6.09×10^5 g C

Analyze

(a) For both reactions, write the balanced chemical equation and use its stoichiometric factors to relate the mass of iron to the mass of carbon monoxide needed.
(b) Write the balanced chemical equation and use its stoichiometric factors to relate the mass of carbon needed.

Plan

(a) Use a mole ratio from the equation for the reduction of hematite to iron

$$Fe_2O_3(s) + 3\ CO(g) \longrightarrow 2\ Fe(s) + 3\ CO_2(g)$$

and a mole ratio from the equation for the reduction of magnetite to iron.

$$Fe_3O_4(s) + 4\ CO(g) \longrightarrow 3\ Fe(s) + 4\ CO_2(g)$$

(b) The carbon monoxide is produced from carbon by the reaction

$$2\ C(s) + O_2(g) \longrightarrow 2\ CO(g)$$

Execute

(a) Applying the information from the balanced chemical equation, the mass of CO required to form 1.00×10^3 kg Fe from hematite is

$$m(\text{CO}) = 1.00 \times 10^3 \text{ kg Fe}\left(\frac{10^3 \text{ g Fe}}{1 \text{ kg Fe}}\right)\left(\frac{1 \text{ mol Fe}}{55.84 \text{ g Fe}}\right)\left(\frac{3 \text{ mol CO}}{2 \text{ mol Fe}}\right)\left(\frac{28.00 \text{ g CO}}{1 \text{ mol CO}}\right)$$
$$= 7.52 \times 10^5 \text{ g CO}$$

The CO required to reduce magnetite can be calculated the same way based on the balanced equation for reduction of magnetite.

$$m(\text{CO}) = 1.00 \times 10^3 \text{ kg Fe}\left(\frac{10^3 \text{ g Fe}}{1 \text{ kg Fe}}\right)\left(\frac{1 \text{ mol Fe}}{55.84 \text{ g Fe}}\right)\left(\frac{4 \text{ mol CO}}{3 \text{ mol Fe}}\right)\left(\frac{28.00 \text{ g CO}}{1 \text{ mol CO}}\right)$$
$$= 6.69 \times 10^5 \text{ g CO}$$

(b) From part (a), the total mass of CO required is $(7.52 \times 10^5 \text{ g}) + (6.69 \times 10^5 \text{ g}) = 1.42 \times 10^6$ g. The required amount of carbon is

$$m(\text{C}) = 1.42 \times 10^6 \text{ g CO}\left(\frac{1 \text{ mol CO}}{28.00 \text{ g CO}}\right)\left(\frac{2 \text{ mol C}}{2 \text{ mol CO}}\right)\left(\frac{12.01 \text{ g C}}{1 \text{ mol C}}\right) = 6.09 \times 10^5 \text{ g C}$$

☑ **Reasonable Result Check**

(a) Hematite: 10^3 kg iron is equivalent to about

$$10^6 \text{ g Fe} \times \frac{1 \text{ mol Fe}}{56 \text{ g Fe}} = 1.8 \times 10^4 \text{ mol Fe}$$

which requires 1.5 times that amount of CO, about 2.7×10^4 mol CO. This is equivalent to approximately 8×10^5 g CO.

Magnetite: Using the same approach, we estimate that magnetite requires 7×10^5 mol CO. The estimated values for CO needed for each ore are close to the calculated values, so the results are reasonable.

(b) The sum of the estimated masses of CO is about 1.5×10^6 g CO. This should be more than twice the mass of C, so 6×10^5 g C is reasonable.

PROBLEM-SOLVING PRACTICE 20.3

Calculate the total mass of iron ore needed to produce 1.00×10^3 kg iron by the process described in Problem-Solving Example 20.3.

ESTIMATION | Steeling Automobiles

Steel is the most recycled consumer product, given that over 96% of steel from automobiles is recycled. Iron and steel make up approximately 60% of the weight of an automobile. The Steel Recycling Institute estimates that 18 million tons of steel scrap from automobiles was recycled in 2012, the equivalent of 18 million automobiles. Estimate how much iron ore (tons) containing approximately 2% Fe_3O_4 did not have to be mined that year due to the use of recycled steel. For this estimation problem, assume that the steel is 100% iron.

Approximately 18×10^6 tons of iron (steel) is recycled. Each 10^3 tons iron ore contains 20 tons Fe_3O_4 in which there are 168 tons Fe per 232 tons Fe_3O_4.

$$m(\text{ore}) = 18 \times 10^6 \text{ tons iron} \times \frac{232 \text{ tons Fe}_3\text{O}_4}{168 \text{ tons iron}} \times \frac{10^3 \text{ tons iron ore}}{20 \text{ tons Fe}_3\text{O}_4} = 1 \times 10^9 \text{ tons ore}$$

About one billion tons of ore did not have to be mined, transported, and processed because recycled steel was used instead. This saved a substantial quantity of material and energy.

EXERCISE 20.3

High-Speed Steel

Ultrahigh-speed steel is used in some saw blades. Such a saw blade, weighing 500. g, contains 0.6% C, 4.0% Cr, 18% W, 1.0% Mo, 1.5% V, and 6.0% Co along with iron.

(a) Calculate the mass of W and of Co in the saw blade.

(b) Determine which alloying metal is present in the greatest mole fraction.

20-3 Copper: A Coinage Metal

Copper metal is sometimes found in nature, and evidence suggests that such naturally occurring metallic copper was known and used during the Stone Age. As early as 10,000 years ago, the metal was hammered into useful items such as coins, jewelry, tools, and weapons. Nearly five millennia later, the Bronze Age was ushered in when humans learned how to alloy copper with tin to make bronze.

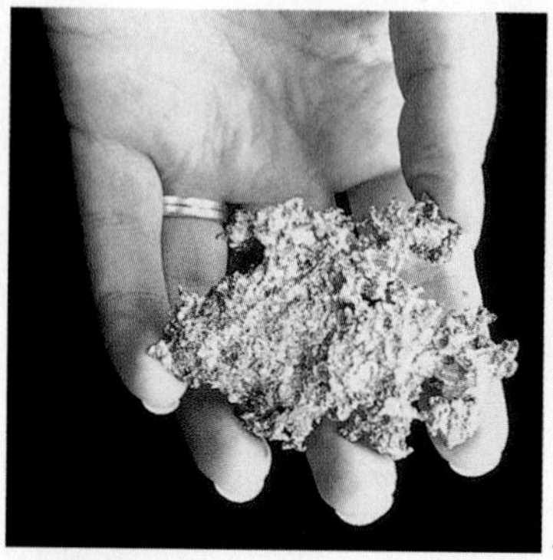

Cengage Learning/Charles D. Winters

A sample of nearly pure native copper.

20-3a The Metallurgy of Copper

Native copper, metallic copper that is found in nature uncombined chemically, is not available in sufficient supply to meet current demands for the metal, and so chemical methods have been developed to extract copper metal from its ores. The principal ores are chalcocite, Cu_2S, and chalcopyrite, $CuFeS_2$, which occur in conjunction with two iron sulfides, FeS_2 and FeS.

Modern methods to extract copper from its ores begin with crushing the ore and separating it from rocks. The ore is then heated in air to temperatures high enough to drive off the sulfur as sulfur dioxide, a process called *roasting*.

Tim Roberts Photography/Shutterstock.com

An open-pit mine near Green Valley, Arizona, from which copper ore is obtained.

$$3\ FeS_2(s) + 8\ O_2(g) \xrightarrow{\text{heating}} Fe_3O_4(s) + 6\ SO_2(g)$$

$$2\ CuFeS_2(s) + O_2(s) \xrightarrow{\text{heating}} Cu_2S(s) + 2\ FeS(s) + SO_2(g)$$

Copper is separated from the iron by melting the Cu_2S and Fe_3O_4 mixture and then combining it with oxygen and SiO_2 to form a liquid iron silicate slag and molten Cu_2S. The slag is less dense than the molten copper(I) sulfide and floats on it; from there the slag can be drawn off periodically.

The conversion of copper(I) sulfide to metallic copper takes place in a process similar to the basic oxygen process for steel making. After the iron silicate slag is removed, air is blown through the molten Cu_2S, converting it to Cu_2O, which reacts with the remaining Cu_2S to form copper metal and sulfur dioxide.

$$2\ Cu_2S(\ell) + 3\ O_2(g) \longrightarrow 2\ Cu_2O(\ell) + 2\ SO_2(g)$$
$$2\ Cu_2O(\ell) + Cu_2S(\ell) \longrightarrow 6\ Cu(\ell) + SO_2(g)$$

The resulting copper, called *blister copper,* is about 96% to 99.5% copper; it can be further purified by electrolysis.

The electrorefining of copper is carried out in large electrolysis cells (← **Sec. 17-10**). Anodes of impure blister copper alternate with cathodes that are very thin sheets of pure copper (Figure 20.7). The electrolyte is $CuSO_4(aq)$ and $H_2SO_4(aq)$. The electrode reactions are

Anode:	$Cu(s, \text{impure blister copper}) \longrightarrow Cu^{2+}(aq) + 2\ e^-$
Cathode:	$Cu^{2+}(aq) + 2\ e^- \longrightarrow Cu(s, \text{pure})$
Overall Reaction:	$Cu(s, \text{impure blister copper}) \longrightarrow Cu(s, \text{pure})$

By controlling the voltage, only copper and those impurities (zinc, lead, iron) in the blister copper anode that have a low enough half-cell potential (are more easily oxidized

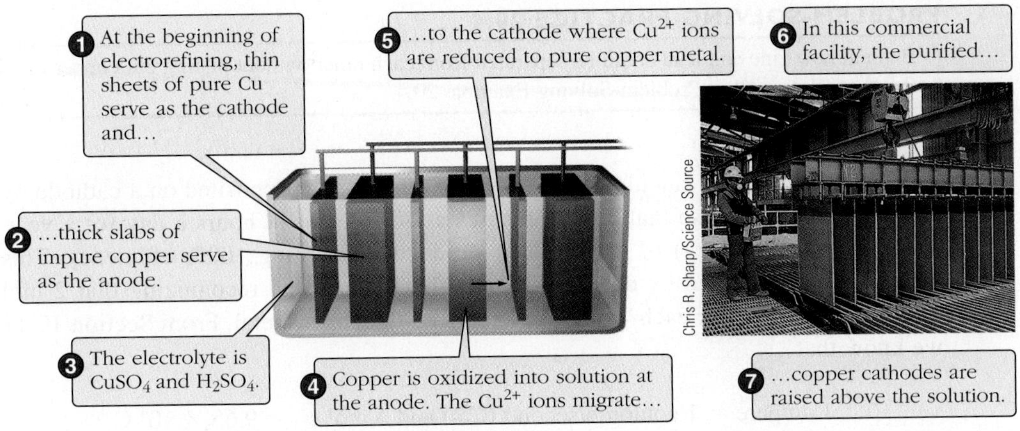
1 At the beginning of electrorefining, thin sheets of pure Cu serve as the cathode and...
2 ...thick slabs of impure copper serve as the anode.
3 The electrolyte is $CuSO_4$ and H_2SO_4.
4 Copper is oxidized into solution at the anode. The Cu^{2+} ions migrate...
5 ...to the cathode where Cu^{2+} ions are reduced to pure copper metal.
6 In this commercial facility, the purified...
7 ...copper cathodes are raised above the solution.
Chris R. Sharp/Science Source

Figure 20.7 **Blister copper is refined (purified) in electrolytic cells.**

than Cu) are oxidized and dissolved in the electrolyte. Any less easily oxidized metal impurities, such as metallic gold and silver, drop from the anode as solid metals as the anode is consumed during electrolysis, forming an *anode sludge.* The voltage is regulated so that only copper, the least electropositive of the metals in the electrolyte, is plated onto the pure copper cathode. Electrorefined copper is greater than 99.9% pure, a purity required for copper used in electrical applications. Usually, enough gold, silver, and platinum are recovered from the anode sludge to pay for the cost of copper electrorefining.

PROBLEM-SOLVING EXAMPLE 20.4

Copper Electrorefining

Copper is electrorefined by removing impurities such as lead, silver, gold, and zinc. Based on these standard half-cell potentials, explain how copper can be separated by electrolysis from these impurities.

	$E°$(V)
$Ag^+(aq) + e^- \longrightarrow Ag(s)$	+0.799
$Au^+(aq) + e^- \longrightarrow Au(s)$	+1.83
$Cu^{2+}(aq) + 2\,e^- \longrightarrow Cu(s)$	+0.340
$Pb^{2+}(aq) + 2\,e^- \longrightarrow Pb(s)$	−0.125
$Zn^{2+}(aq) + 2\,e^- \longrightarrow Zn(s)$	−0.763

Result At the potential that just oxidizes blister copper from the anode (−0.340 V relative to a standard hydrogen electrode), zinc and lead are also oxidized. Silver and gold are not oxidized and form the anode sludge. This potential is not sufficient to reduce Zn^{2+} and Pb^{2+}, so only copper forms at the cathode.

Analyze Consider what reactions could occur at the impure blister copper anode and the pure copper cathode. Adjust the cell potential so that only copper and metals that are more easily oxidized than copper will dissolve.

Plan Ag^+ and Au^+ have more positive standard half-cell potentials than Cu^{2+}, so Ag^+ and Au^+ are easier to reduce than Cu^{2+}; therefore, Cu is easier to oxidize than Ag or Au. Pb^{2+} and Zn^{2+} are more difficult to reduce than Cu^{2+}; therefore, Pb and Zn are easier to oxidize than Cu, and so a potential that oxidizes Cu at the anode will also oxidize Zn and Pb.

Execute Because the other metals that dissolve are easier to oxidize than copper, their ions are harder to reduce and, therefore, copper(II) ions are reduced at the cathode. At the potential that just oxidizes blister copper from the anode (−0.340 V relative to a standard hydrogen electrode), zinc and lead are also oxidized. This potential that just oxidizes Cu will not oxidize Ag or Au, which will form the anode sludge; Cu^{2+}, Zn^{2+}, and Pb^{2+} will be available in solution for reduction at the cathode. Because Pb^{2+} and Zn^{2+} are more difficult to reduce than Cu^{2+}, only Cu^{2+} reacts at the cathode, forming pure Cu.

PROBLEM-SOLVING PRACTICE 20.4

Explain how zinc and lead could be separated from each other without plating out copper in the electrolytic cell in Problem-Solving Example 20.4.

As an example of calculating the mass of pure copper deposited on a cathode by electrorefining, suppose that a cell operates at 200. A for 24 hours a day for a year. Calculate the mass (kg) of pure copper produced, assuming 100% efficiency. This mass can be calculated by determining the charge used and recognizing that 2 mol electrons is needed for each 1 mol Cu^{2+} reduced to copper metal. From Section 17.11 we know that

$$1 \text{ ampere} = 1 \text{ coulomb/second (C/s) and } 1 \text{ mol } e^- = 9.65 \times 10^4 \text{ C}$$

$$\text{Charge} = 365 \text{ days} \times 200. \text{ A} \times \left(\frac{1 \text{ C/s}}{1 \text{ A}}\right)\left(\frac{24 \text{ h}}{1 \text{ day}}\right)\left(\frac{3600 \text{ s}}{1 \text{ h}}\right) = 6.31 \times 10^9 \text{ C}$$

$$n(\text{Cu}) = 6.31 \times 10^9 \text{ C}\left(\frac{1 \text{ mol } e^-}{9.65 \times 10^4 \text{ C}}\right)\left(\frac{1 \text{ mol Cu}}{2 \text{ mol } e^-}\right) = 3.27 \times 10^4 \text{ mol Cu}$$

$$m(\text{Cu}) = 3.27 \times 10^4 \text{ mol Cu}\left(\frac{63.55 \text{ g Cu}}{1 \text{ mol Cu}}\right)\left(\frac{1 \text{ kg Cu}}{10^3 \text{ g Cu}}\right) = 2.08 \times 10^3 \text{ kg Cu}$$

EXERCISE 20.4

Refining Copper

Calculate the mass of pure copper deposited during electrorefining in a cell operating at 250. A for 12.0 h. Assume the process is 100% efficient.

20-3b Bronze and Brass

The Bronze Age began about 3000 BC with the discovery that bronze formed when tin combined with copper. This discovery was likely accidental, brought on by the fact that copper ores and tin ores are often found together. In a stroke of good fortune, some tin ore was present during the reduction of copper ore by the charcoal of a wood fire, and tin ions were also reduced to tin metal. The heated mixture of the resulting copper and tin metals formed bronze. The advantage of bronze is that it is sufficiently hard to keep a cutting edge, something copper cannot do. Bronze usually contains 7% to 10% tin, but bronzes with as much as 20% tin are known. Because of its properties, all early civilizations that produced bronze used it to create weapons as well as works of art and everyday items such as coins.

On prolonged exposure to moist air, copper or bronze, by a redox reaction, forms a green outer coating (a patina) of copper(II) hydroxycarbonate, $Cu_2(OH)_2CO_3$, often seen on statuary, such as the Statue of Liberty, and on copper-clad roofs.

$$\underset{\text{copper or copper in bronze}}{2\,\text{Cu(s)}} + \text{O}_2\text{(g)} + \text{CO}_2\text{(g)} + \text{H}_2\text{O(g)} \longrightarrow \underset{\text{patina: copper(II) hydroxycarbonate}}{\text{Cu}_2\text{(OH)}_2\text{CO}_3\text{(s)}}$$

For you design/Shutterstock.com

Ancient bronze coins.

Brass is an alloy made of varying proportions of copper and zinc (20% to 45% Zn), which becomes harder as the percentage of zinc increases. Because brass is easy to forge, cast, and stamp, it is widely used for pipes, valves, and fittings.

20-3c Uses and Reactions of Copper

Copper is used in all U. S. coins. By 1982, the price of copper had risen to where it was costing the U. S. Treasury Department more than 1 cent to make a penny. Since then, to conserve copper and reduce costs, the penny has been made of 97.5% zinc and 2.5% copper, with the zinc core sandwiched between two thin layers of copper. Silver-colored coins actually contain no silver. The U. S. nickel is made of a copper (75%) and nickel (25%) alloy of uniform composition throughout the coin. The other silver-colored coins are a "sandwich" made of a pure copper core covered with a thin layer of a copper and nickel alloy (91.67% copper). The Sacagawea dollar coin, which looks like gold, contains no gold.

Metallic copper is not attacked by most acids, although it does react with nitric acid. The nitrate ion, NO_3^-, can oxidize copper. In dilute nitric acid, NO_3^- is reduced to NO; concentrated nitric acid yields NO_2.

$$3\ Cu(s) + 2\ NO_3^-(aq) + 8\ H^+(aq) \longrightarrow 3\ Cu^{2+}(aq) + 2\ NO(g) + 4\ H_2O(\ell)$$

$$Cu(s) + 2\ NO_3^-(aq) + 4\ H^+(aq) \longrightarrow Cu^{2+}(aq) + 2\ NO_2(g) + 2\ H_2O(\ell)$$

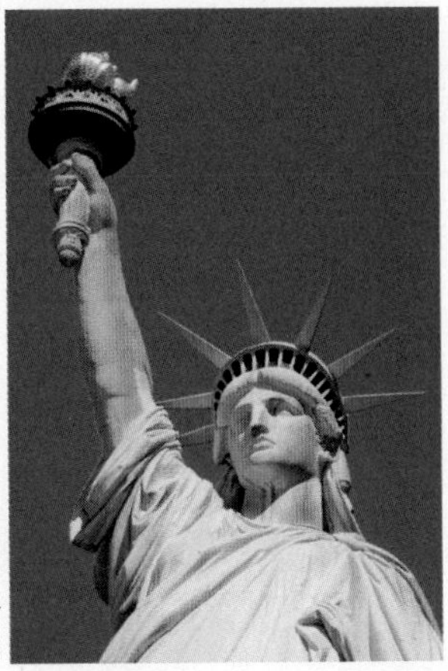

ErickN/Shutterstock.com

Patina on the Statue of Liberty. The green color is a copper(II) hydroxycarbonate coating (patina).

Copper(II) sulfate pentahydrate, $CuSO_4 \cdot 5H_2O$, is the most widely used copper compound. Commonly called blue vitriol, this compound is used to kill algae and fungi. In the solid it contains the hydrated copper(II) complex ion, $[Cu(H_2O)_4]^{2+}$. The fifth water molecule is bound to the sulfate ion through hydrogen bonding. The water of hydration can be removed by gentle heating or by putting the hydrate into a desiccator. Many aqueous solutions containing copper(II) ions are blue due to the presence of the $[Cu(H_2O)_6]^{2+}$ ion. The color of aqueous copper(II) ions changes if a Lewis base other than water is bonded to the Cu^{2+} ion (this is described in more detail in Section 20-7c).

Cengage Learning/ Charles D. Winters

Pre-1982 (left) and post-1982 (right) pennies. Since 1982 the penny has been copper-clad zinc (not solid copper): note the silvery-appearing zinc core when the copper cladding has been filed away.

Copper also exists in a +1 oxidation state, Cu^+; however, aqueous copper chemistry involves principally Cu^{2+} because Cu^+ in aqueous solutions disproportionates (undergoes both oxidation and reduction) into Cu and Cu^{2+}.

$$Cu^+(aq) + e^- \longrightarrow Cu(s) \qquad E^\circ = +0.52\ V$$

$$Cu^+(aq) \longrightarrow Cu^{2+}(aq) + e^- \qquad E^\circ = +0.15\ V$$

$$2\ Cu^+(aq) \longrightarrow Cu(s) + Cu^{2+}(aq) \qquad E^\circ = +0.37\ V$$

In a disproportionation reaction, the same substance is oxidized and reduced.

In basic solution, however, Cu^{2+} can be reduced to copper(I) oxide, Cu_2O. This reaction was once used to test for glucose in urine. When heated, a basic solution containing blue aqueous Cu^{2+} and a reducing sugar such as glucose react to form a brick-red

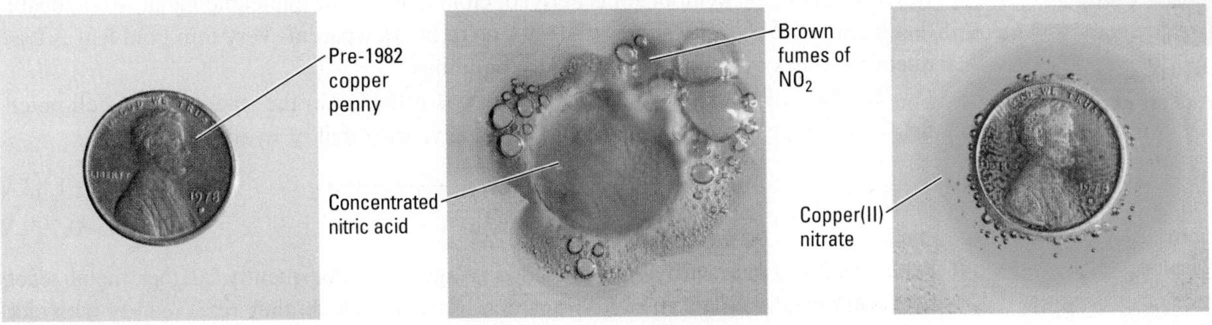
Pre-1982
copper
penny
Brown
fumes of
NO_2
Concentrated
nitric acid
Copper(II)
nitrate

Photos: John Moore

The reaction of metallic copper with concentrated nitric acid produces brown fumes of NO_2 and blue copper(II) nitrate solution.

Photos: Cengage Learning/Charles D. Winters

Hydrated copper(II) sulfate (blue)

Partially anhydrous copper(II) sulfate (white)

Hydrated copper(II) sulfate is converted to anhydrous copper(II) sulfate by heating, which drives off the waters of hydration and causes the color to change.

precipitate of Cu_2O. The reducing sugar is represented here with its aldehyde functional group, RCHO.

$$\underset{\text{blue}}{2\,Cu^{2+}(aq)} + \underset{\text{reducing sugar}}{RCHO(aq)} + 5\,OH^-(aq) \longrightarrow \underset{\text{brick red}}{Cu_2O(s)} + RCOO^-(aq) + 3\,H_2O(\ell)$$

Cengage Learning/Charles D. Winters

Samples of red copper(I) oxide, Cu_2O, and black copper(II) oxide, CuO.

At elevated temperatures copper reacts with oxygen. Below 1000 °C it forms black copper(II) oxide, CuO. Above 1000 °C copper reacts with oxygen to form red copper(I) oxide, Cu_2O, which is found in the mineral cuprite.

20-4 Silver and Gold: The Other Coinage Metals

Gabriel Eckert/iStockphoto.com

Gold can be hammered into ultrathin sheets for decorative purposes.

Silver and gold occur in elemental (uncombined) form in nature, although such sources have dwindled. In the past, gold prospectors, such as the forty-niners in the 1849 California gold rush, panned for gold. They simply swirled gold-bearing sand and gravel from streams in a pan. Because of its high density (19.3 g/cm^3), gold settles out from the sand (about 2.5 g/cm^3) and other rocky impurities. Today, the gold content in such deposits is much too low for panning to be commercially effective.

Early civilizations highly prized silver and gold for their luster, corrosion resistance, and workability in making jewelry, art objects, and coins. Silver is the best metallic conductor of heat and electricity. Long synonymous with wealth and power, gold was thought to be a part of the sun, thus its Latin name *aurum,* meaning "bright dawn," from which its chemical symbol Au is derived. Gold is the most malleable metal, so malleable that it can be rolled into sheets thin enough to be transparent. Very thin gold leaf is used to cover domes of churches and capitol buildings.

The lack of reactivity of gold and silver is reflected in the standard half-cell potentials of their ions, which are much more positive than that of hydrogen (0.00 V).

$$Au^+(aq) + e^- \longrightarrow Au(s) \qquad E^\circ = +1.83\text{ V}$$
$$Ag^+(aq) + e^- \longrightarrow Ag(s) \qquad E^\circ = +0.799\text{ V}$$

Both metal ions are fairly strong oxidizing agents. Consequently, neither metal reacts with weak oxidizing acids such as hydrochloric acid, nor do they react readily with oxygen at normal temperatures. When heated in air, gold does not react with oxygen, even at high temperatures. Silver, however, slowly forms silver(I) oxide, Ag_2O, which is unstable and decomposes back to the elements when heated strongly. Silver reacts with oxidizing acids such as nitric acid, HNO_3, forming $Ag^+(aq)$. However, it takes aqua regia, a 3 : 1 mixture of concentrated HCl and HNO_3, to dissolve gold.

The ability of aqua regia (Latin, "royal water") to dissolve gold has been known since the 1300s.

$$3\ Ag(s) + 4\ H^+(aq) + NO_3^-(aq) \longrightarrow 3\ Ag^+(aq) + NO(g) + 2\ H_2O(\ell)$$

$$Au(s) + 6\ H^+(aq) + 3\ NO_3^-(aq) + 4\ Cl^-(aq) \longrightarrow AuCl_4^-(aq) + 3\ NO_2(g) + 3\ H_2O(\ell)$$

The oxidation of metallic gold to Au^{3+} is highly unfavorable, but the reaction in aqua regia occurs because as Au^{3+} ions form, they combine with Cl^- ions to form the complex ion, $AuCl_4^-$, in a highly product-favored reaction.

$$Au^{3+}(aq) + 4\ Cl^-(aq) \longrightarrow AuCl_4^-(aq) \qquad K_c = 1 \times 10^{38}$$

Complex ions such as $AuCl_4^-$ were briefly described in Section 15-5c and will be discussed in more detail in Section 20-6.

EXERCISE 20.5

Putting Silver and Gold into Solution

(a) Identify the oxidizing and reducing agents in the reaction of nitric acid with silver.

(b) Do the same for the reaction of gold with aqua regia.

Gold and silver are both relatively soft metals whose hardness is increased by alloying with other metals. Sterling silver, for example, is 92.5% silver and 7.5% copper. The proportion of gold in its alloys is expressed in carats. Pure gold (24 carats) is too soft to be used in jewelry. It is alloyed with copper or other metals to make the 18-carat and 14-carat gold for jewelry, which are 75% (18/24 × 100%) and 58% (14/24 × 100%) gold, respectively.

As noted earlier (← **Sec. 20-3a**), silver and gold are by-products of the electrorefining of copper. They are also obtained from ores. Silver is obtained from its principal ore argentite, Ag_2S, by cyanide extraction. The ore is ground and put into a 0.5% solution of NaCN.

$$Ag_2S(s) + 4\ CN^-(aq) \longrightarrow 2\ Ag(CN)_2^-(aq) + S^{2-}(aq)$$

The formation of the $Ag(CN)_2^-$ complex ion brings the insoluble silver sulfide into solution.

$$\begin{array}{r} Ag_2S(s) \rightleftharpoons 2\ Ag^+(aq) + S^{2-}(aq) \\ 2\ Ag^+(aq) + 4\ CN^-(aq) \rightleftharpoons 2\ Ag(CN)_2^-(aq) \\ \hline Ag_2S(s) + 4\ CN^-(aq) \longrightarrow 2\ Ag(CN)_2^-(aq) + S^{2-}(aq) \end{array}$$

Powdered zinc, a good reducing agent, is added to the solution to convert Ag^+ to metallic silver.

$$Zn(s) + 2\ Ag(CN)_2^-(aq) \longrightarrow 2\ Ag(s) + Zn(CN)_4^{2-}(aq)$$

The silver can be further purified by electrolytic refining.

Like silver, gold is obtained from ores by cyanide extraction followed by reduction with zinc.

In all of history, the amount of gold that has been mined is relatively small, not quite enough to fill four Olympic-sized swimming pools.

$$4\ Au(s) + 8\ CN^-(aq) + O_2(g) + 2\ H_2O(\ell) \longrightarrow 4\ Au(CN)_2^-(aq) + 4\ OH^-(aq)$$

$$Zn(s) + 2\ Au(CN)_2^-(aq) \longrightarrow 2\ Au(s) + Zn(CN)_4^{2-}(aq)$$

The cyanide extraction of silver and gold depends on the formation of the $Ag(CN)_2^-$ and $Au(CN)_2^-$ complex ions. Because of the toxicity of CN^-, waste cyanide solutions must be disposed of properly. Improper disposal, dumping the solutions near the processing site, has caused serious environmental damage.

20-5 Chromium

Chromium is characteristic of the middle transition elements, which exhibit multiple oxidation states (← **Sec. 20-1c**). In compounds, chromium has oxidation states from +2 to +6 because of its five 3*d* electrons and one 4*s* electron, although +2 and +3 are the most common oxidation states. The +6 oxidation state occurs in CrO_4^{2-} and $Cr_2O_7^{2-}$ ions (Figure 20.8). The +4 and +5 oxidation states are uncommon.

Chromium is a hard, brittle metal that is extremely corrosion resistant due to a chromium oxide surface layer that passivates and protects the metal beneath from further oxidation. In this regard, chromium resembles aluminum (← **Sec. 19-6c**). Chromium is used in stainless steel and is plated on truck bumpers to give them a bright surface. Its oxides are used in magnetic recording tapes, CrO_2, in abrasives, Cr_2O_3, and as a glass pigment, Cr_2O_3.

The chief chromium ore is chromite, $FeCr_2O_4$, which is treated with lime, CaO, oxygen, and sodium carbonate to form sodium chromate. Water is added to remove the soluble sodium chromate produced.

$$4\,FeCr_2O_4(s) + 8\,CaO(s) + 8\,Na_2CO_3(s) + 7\,O_2(g) \longrightarrow 8\,Na_2CrO_4(s) + 2\,Fe_2O_3(s) + 8\,CaCO_3(s)$$

Sodium chromate is used widely in chrome plating and as an intermediate in the formation of many chromium compounds. One of these compounds is chromium(III) oxide, from which chromium metal is obtained by reduction with aluminum metal.

$$2\,CrO_4^{2-}(aq) + 3\,SO_2(g) + H_2O(\ell) \longrightarrow Cr_2O_3(s) + 3\,SO_4^{2-}(aq) + 2\,H^+(aq)$$

$$Cr_2O_3(s) + 2\,Al(s) \longrightarrow 2\,Cr(s) + Al_2O_3(s)$$

EXERCISE 20.6

Production of Chromium Metal

Use data from Appendix J to calculate the enthalpy change and the Gibbs free energy change for the reduction of chromium(III) oxide by aluminum.

In aqueous solution, yellow chromate ions, CrO_4^{2-}, and orange dichromate ions, $Cr_2O_7^{2-}$, exist in a highly pH-dependent equilibrium.

$$H^+(aq) + \underset{\text{yellow}}{CrO_4^{2-}(aq)} \rightleftharpoons HCrO_4^-(aq)$$

$$2\,HCrO_4^-(aq) \rightleftharpoons \underset{\text{orange}}{Cr_2O_7^{2-}(aq)} + H_2O(\ell)$$

$HCrO_4^-$ $Cr_2O_7^{2-}$ H_2O

Notice from the second equilibrium that dichromate is formed by a condensation reaction in which two $HCrO_4^-$ units are joined and a water molecule forms.

The net equilibrium is

$$2\,H^+(aq) + 2\,CrO_4^{2-}(aq) \rightleftharpoons Cr_2O_7^{2-}(aq) + H_2O(\ell) \qquad K_c = 4 \times 10^{14}$$

From the very large value of K_c, you should recognize that in acid, chromate is converted to dichromate; chromate is stable in basic or neutral solution.

Compounds with chromium oxidation state +3, Cr(III):
CrCl3 (green) and...
...Cr(NO3)3 (violet).
Compounds with chromium oxidation state +6, Cr(VI):
K2CrO4 (chromate, yellow) and...
...K2Cr2O7 (dichromate, orange).
Cr(NO3)3(aq)
CrCl3(aq)
K2CrO4(aq)
K2Cr2O7
Cengage Learning/ Charles D. Winters

Figure 20.8 Chromium compounds in solution. The color changes with the oxidation state of chromium and the chemical environment of the metal ion.

CONCEPTUAL EXERCISE 20.7

Applying Le Chatelier's Principle

Use Le Chatelier's principle to explain how the addition of acid or base shifts the equilibrium to favor chromate or dichromate.

EXERCISES that are labeled **CONCEPTUAL** are designed to test your understanding of one or more concepts; they usually involve qualitative rather than quantitative thinking.

In highly acidic solution the dichromate ion is a powerful oxidizing agent.

$$Cr_2O_7^{2-}(aq) + 14\ H^+(aq) + 6\ e^- \longrightarrow 2\ Cr^{3+}(aq) + 7\ H_2O(\ell) \qquad E° = 1.36\ V$$

Dichromate ion is sufficiently strong to oxidize alcohols to aldehydes or ketones (← **Sec. 10-4d**) and aldehydes to carboxylic acids; for example, dichromate oxidizes acetaldehyde to acetic acid.

$$Cr_2O_7^{2-}(aq) + 3\ CH_3CHO(aq) + 8\ H^+(aq) \longrightarrow 2\ Cr^{3+}(aq) + 3\ CH_3COOH(aq) + 4\ H_2O(\ell)$$

The oxidizing strength of dichromate decreases as the pH increases, as shown in Problem-Solving Example 20.5.

PROBLEM-SOLVING EXAMPLE 20.5

Dichromate and pH

Use the Nernst equation (← **Sec. 17-7**) to calculate the E_{cell} for the reduction of dichromate ion by iodide ion at a pH of 4.0 and 25 °C with all concentrations other than H^+ equal to 1.0 M. $E°_{cell}$ is 0.825 V.

Result $E_{cell} = 0.273$ V

Analyze The balanced equation for the reaction and the Nernst equation can be used to calculate the cell potential at the nonstandard conditions.

Plan The balanced equation for the reaction is

$$Cr_2O_7^{2-}(aq) + 6\ I^-(aq) + 14\ H^+(aq) \longrightarrow 2\ Cr^{3+}(aq) + 3\ I_2(aq) + 7\ H_2O(\ell)$$

The Nernst equation is

$$E_{cell} = E°_{cell} - \frac{RT}{nF}\ln Q = 0.825\ V - \frac{0.0257\ V}{6}\ln\frac{[Cr^{3+}]^2[I_2]^3}{[Cr_2O_7^{2-}][I^-]^6[H^+]^{14}}$$

In the equation, 6 mol e^- are transferred so $n = 6$. All concentrations are 1.0 M except $[H^+]$, which is $10^{-pH} = 10^{-4}$ M.

Appendix F describes how to balance redox equations such as this one.

Six moles of electrons are transferred per mole of $Cr_2O_7^{2-}$:

$$2\ Cr^{6+} + 6\ e^- \longrightarrow 2\ Cr^{3+}$$
$$6\ I^- \longrightarrow 3\ I_2 + 6\ e^-$$

Execute In this case, the Nernst equation becomes

$$E_{cell} = 0.825 \text{ V} - \frac{0.0257 \text{ V}}{6} \ln \frac{1}{(1.0 \times 10^{-4})^{14}}$$

$$= 0.825 \text{ V} - \frac{0.0257 \text{ V}}{6} \ln(1.0 \times 10^{56})$$

$$= 0.825 \text{ V} - 0.552 \text{ V} = 0.273 \text{ V}$$

☑ **Reasonable Result Check** Changing the $[H^+]$ from 1.0 M (the standard-state concentration) to 1.0×10^{-4} M (pH = 4.0) decreases the cell potential by 0.552 V. This large change results because of the large coefficient (14) of H^+ in the balanced equation. Dichromate ion is a much weaker oxidizing agent at pH = 4.0 than at standard-state conditions.

PROBLEM-SOLVING PRACTICE 20.5

At what pH does $E_{cell} = 0.00$ V for the reduction of dichromate by iodide ion in acid solution, assuming standard-state concentrations of all species except H^+ ion?

20-6 Coordinate Covalent Bonds: Complex Ions and Coordination Compounds

In Section 6-10a, the formation of F_3BNH_3 was described as occurring by the sharing of a lone pair from NH_3 with BF_3. This type of covalent bond, in which one atom contributes *both* electrons for the shared pair, is called a *coordinate covalent bond.* Atoms with lone pairs of electrons, such as nitrogen, phosphorus, and sulfur, can use those lone pairs to form coordinate covalent bonds. For example, the formation of the ammonium ion from ammonia results from formation of a coordinate covalent bond between H^+ and the lone pair of electrons of nitrogen in NH_3.

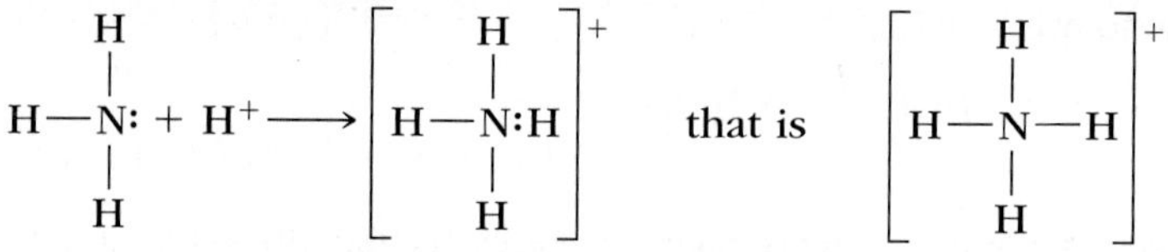

Once the coordinate covalent bond is formed, it is impossible to distinguish which of the N—H bonds it is; all four N—H bonds are equivalent.

20-6a Metals and Coordination Compounds

Much of the chemistry of *d*-block transition metals is related to their ability to form coordinate covalent bonds with molecules or ions that have lone pair electrons. Transition-metal atoms or ions with vacant *d* atomic orbitals can act as Lewis acids and accept the lone pairs into those orbitals (← **Sec. 14-9a**).

One of the reasons that water is a good solvent for ionic compounds is that ions in aqueous solution are surrounded by water molecules (← **Sec. 3-3a**); we now consider this in more detail. For example, the Ni^{2+} ion in aqueous solution is surrounded by six water molecules. This type of ion, in which several molecules or ions are connected to a central metal ion or atom by coordinate covalent bonds, is known as a *complex ion. Molecules or ions bonded to a central metal ion or atom by coordinate covalent bonds* are called **ligands**, from the Latin verb *ligare*, "to bind."

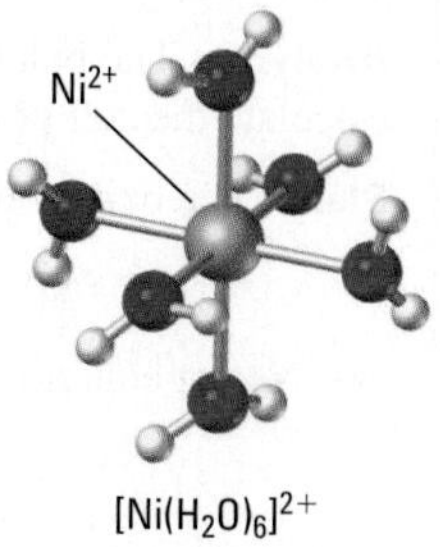
Ni2+
$[Ni(H_2O)_6]^{2+}$

Ligands act as Lewis bases, electron pair donors; transition-metal ions are Lewis acids, electron pair acceptors (← Sec. 14-9a).

Each ligand (a water molecule in this example) has one or more atoms with lone pairs that can form coordinate covalent bonds to the metal ion. To write the formula of a complex ion, the ligand formulas are placed in parentheses following the metal ion. The entire complex ion formula is enclosed by brackets, and the ionic charge, if any, is a superscript outside the brackets. For the nickel(II) complex ion with six water ligands this gives $[Ni(H_2O)_6]^{2+}$.

The charge of a complex ion is determined by the charges of the metal ion and the charges of its ligands. In $[Ni(H_2O)_6]^{2+}$ the water ligands have no net charge, so the charge of the complex ion is that of the Ni^{2+} ion. In the complex ion formed by Ni^{2+} with four chloride ions, $[NiCl_4]^{2-}$, the net 2− charge results from the 4− charge of four chloride ions and the 2+ charge of the nickel ion.

Compounds that contain metal ions surrounded by ligands are called **coordination compounds**. Usually, complex ions are combined with oppositely charged ions *(counter ions)* to form neutral compounds. Coordination compounds are usually brightly colored as solids or in solution (Figure 20.8, two left flasks). The complex ion part of a coordination compound's formula is enclosed in brackets; counter ions are written outside the brackets, as in the formula $[Ni(H_2O)_6]Cl_2$ of the compound consisting of chloride counter ions with the $[Ni(H_2O)_6]^{2+}$ complex ion. The two Cl^- ions compensate for the 2+ charge of the complex ion. $[Ni(H_2O)_6]Cl_2$ is an ionic compound analogous to $CaCl_2$, which also contains a 2+ cation and two Cl^- ions. In some cases, no compensating ions are needed outside the brackets for a coordination compound. For example, the anticancer drug $[Pt(NH_3)_2Cl_2]$ (cisplatin) is a coordination compound containing NH_3 and Cl^- ligands coordinated to a central Pt^{2+} ion. The two Cl^- ions compensate for the charge of the Pt^{2+} ion, resulting in a neutral coordination compound rather than a complex ion.

When the word *coordinated* is used in chemistry, such as in "the chloride ions in $[NiCl_4]^{2-}$ are coordinated to the nickel ion," it means that coordinate covalent bonds have been formed.

PROBLEM-SOLVING EXAMPLE 20.6

Coordination Compounds

For the coordination compound $K_3[Fe(CN)_6]$, identify
(a) the central metal ion.
(b) the ligands.
(c) the formula and charge of the complex ion and the charge of the central metal ion.

Result
(a) Iron (b) Six cyanide ions, CN^- (c) $[Fe(CN)_6]^{3-}$, Fe^{3+}

Analyze Apply the guidelines and concepts described previously.

Plan In a formula, the complex ion is enclosed in square brackets; within the brackets, ligands coordinated to the central metal ion are enclosed by parentheses.

Execute
(a) The iron ion is the central metal ion, as shown by its placement inside the brackets.
(b) Cyanide ions, CN^-, are coordinated to the central iron ion.
(c) The charge on three potassium ions is $(3 \times 1+) = 3+$. Therefore, the compensating charge of the complex ion must be 3−, arising from the 6− charge of six cyanide ions $(6 \times 1- = 6-)$, combined with the 3+ charge of the central iron(III) ion: $(6-) + (3+) = 3-$.

PROBLEM-SOLVING PRACTICE 20.6

For the coordination compound $[Cu(NH_3)_4]SO_4$, identify
(a) the counter ion.
(b) the central metal ion.
(c) the ligands.
(d) the formula and charge of the complex ion.

CONCEPTUAL EXERCISE 20.8

Coordination Complex Ion

In a complex ion, a central Cr^{3+} ion is bonded to two ammonia molecules, two water molecules, and two hydroxide ions. Determine the formula and the net charge of this complex ion.

20-6b Naming Complex Ions and Coordination Compounds

In early times, coordination compounds, like other compounds, were known by common names such as "roseo cobaltic chloride" and "Zeise's salt". Today, a systematic nomenclature is applied to complex ions and coordination compounds. This nomenclature system indicates the central metal ion and its oxidation state, as well as the number and kinds of ligands. Here we provide some basic aspects of the system.

In naming any coordination compound or complex ion, the *ligands are written first, in alphabetical order. The name and oxidation state (a Roman numeral in parentheses) of the metal ion are given last.* Greek prefixes *di, tri, tetra,* and so on are used to *denote the number of times each ligand appears in the formula.* Such prefixes are ignored when determining the alphabetical order of the ligands. The names of ligands are derived from the names of the parent ions or molecules; for example, the names of anions end in "o" instead of "ide". Table 20.4 lists the names and formulas of some common ligands.

Consider the coordination compound $[Co(NH_3)_3(OH)_3]$, which is named triamminetrihydroxocobalt(III). From Table 20.4, we see that the name and formula indicate that three ammonia molecules (triammine) and three hydroxide ions (trihydroxo) are bonded to a central cobalt ion in the +3 oxidation state (roman numeral III). The three hydroxide ions carry a total 3− charge, ammonia molecules have no net charge, and the cobalt ion has a 3+ charge, so the compound has no net charge.

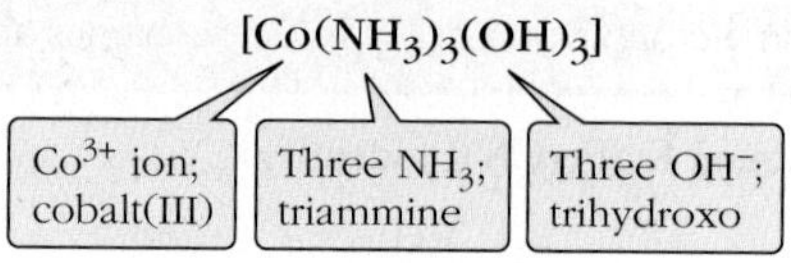
[Co(NH3)3(OH)3]
Co3+ ion; cobalt(III)
Three NH3; triammine
Three OH−; trihydroxo

triamminetrihydroxocobalt(III)

Counter ions offset the charge of the complex ion.

Next, consider $[Fe(H_2O)_2(NH_3)_4]Cl_3$, a coordination compound that consists of a complex *cation,* $[Fe(H_2O)_2(NH_3)_4]^{3+}$, and three chloride ions as counter ions. As for any ionic compound, the *complex cation is always named first, followed by the name of the anionic counter ions.* The compound's name is tetraamminediaquairon(III) chloride. From Table 20.4, we see that the ligands are ammine (NH_3, four of them) and aqua (H_2O,

Table 20.4 Names and Formulas of Some Common Ligands

Neutral Ligand	Ligand Name	Anionic Ligand	Ligand Name
NH_3	Ammine	Br^-	Bromo
CO	Carbonyl	CO_3^{2-}	Carbonato
$H_2NCH_2CH_2NH_2$	Ethylenediamine, en	Cl^-	Chloro
H_2O	Aqua	CN^-	Cyano
		F^-	Fluoro
		OH^-	Hydroxo
		$C_2O_4^{2-}$	Oxalato, ox
		NCS^-	Isothiocyanato
		SCN^-	Thiocyanato

two of them). *For complex cations, the metal ion and its oxidation state follow the names of the ligands.*

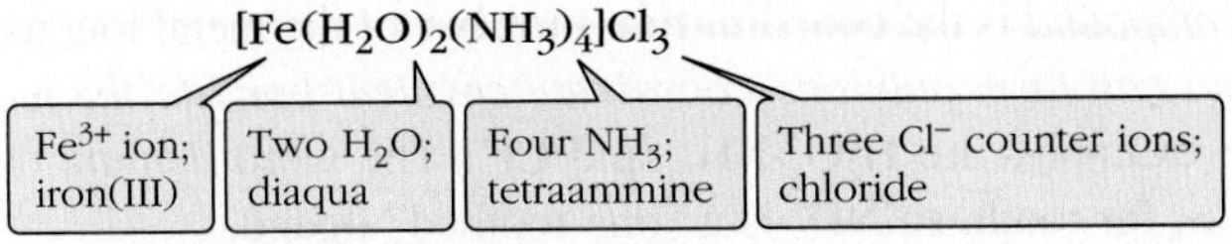
$[Fe(H_2O)_2(NH_3)_4]Cl_3$
Fe^{3+} ion; iron(III)
Two H_2O; diaqua
Four NH_3; tetraammine
Three Cl^- counter ions; chloride

tetraamminediaquairon(III) chloride

The compound $K_2[PtCl_4]$ contains a complex *anion,* $[PtCl_4]^{2-}$, and two K^+ ions as counter ions and is named potassium tetrachloroplatinate(II). As with any ionic compound, the *cation is named first, followed by the anion name.* For complex anions, the central metal ion's name ends in *-ate* followed by its oxidation state in parentheses.

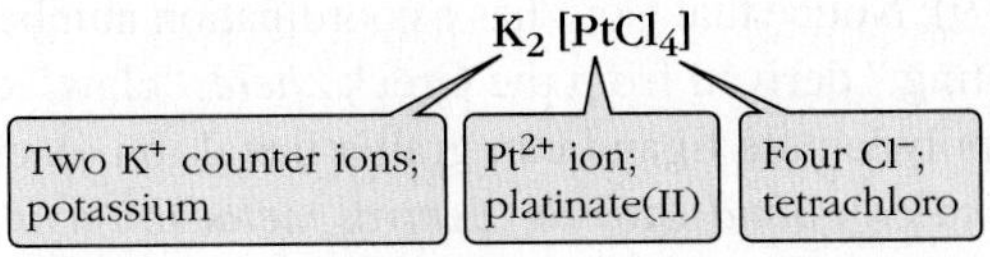
$K_2\ [PtCl_4]$
Two K^+ counter ions; potassium
Pt^{2+} ion; platinate(II)
Four Cl^-; tetrachloro

potassium tetrachloroplatinate(II)

PROBLEM-SOLVING EXAMPLE 20.7

Formulas and Names of Coordination Compounds

(a) Write the formula of diamminetriaquahydroxochromium(III) nitrate.
(b) Name $K[Cr(NH_3)_2(C_2O_4)_2]$.

Result

(a) $[Cr(NH_3)_2(H_2O)_3(OH)](NO_3)_2$
(b) Potassium diamminedioxalatochromate(III)

Analyze (a) Use the names and formulas of the ligands in Table 20.4; chromium(III) in the name indicates a Cr^{3+} ion in a complex cation. (b) Use the names and formulas of the ligands in Table 20.4; chromate(III) in the name indicates a Cr^{3+} ion as part of a complex anion.

Plan Compound (a) contains a complex cation, and compound (b) contains a complex anion.

Execute

(a)

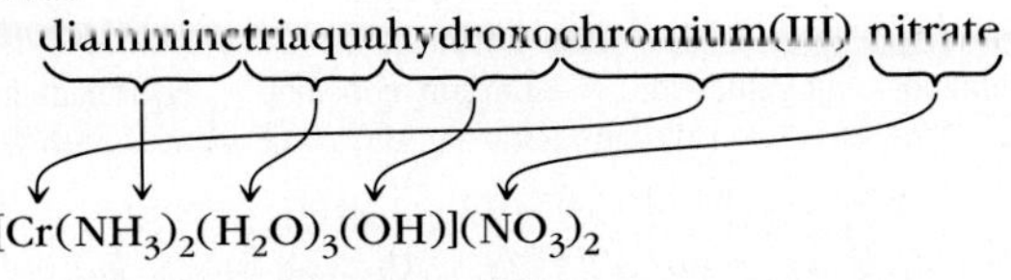
diamminetriaquahydroxochromium(III) nitrate
$[Cr(NH_3)_2(H_2O)_3(OH)](NO_3)_2$

(b)

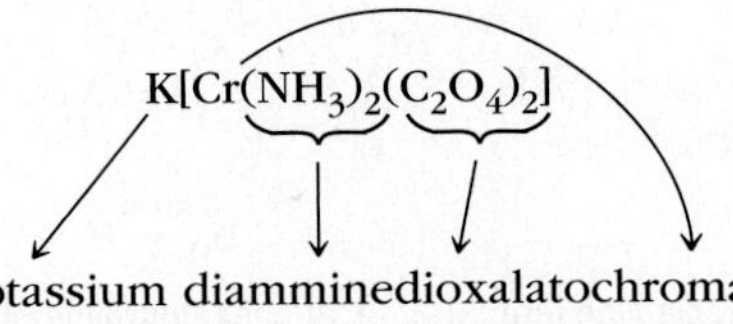
$K[Cr(NH_3)_2(C_2O_4)_2]$

potassium diamminedioxalatochromate(III)

PROBLEM-SOLVING PRACTICE 20.7

(a) Name this coordination compound: $[Ag(NH_3)_2]NO_3$.
(b) Write the formula of pentaaquaisothiocyanatoiron(III) chloride.

CONCEPTUAL EXERCISE 20.9

Coordination Compounds

$CaCl_2$ and $[Ni(H_2O)_6]Cl_2$ have the same formula type, MCl_2. Give the formula of an ionic compound that is not a coordination compound and has a formula analogous to $K_2[NiCl_4]$.

20-6c Types of Ligands and Coordination Number

Coordination Number	Examples
2	$[Ag(NH_3)_2]^+$, $[AuCl_2]^-$
4	$[NiCl_4]^{2-}$, $[Pt(NH_3)_4]^{2+}$
6	$[Fe(H_2O)_6]^{2+}$, $[Co(NH_3)_6]^{3+}$

The number of coordinate covalent bonds between the ligands and the central metal ion in a coordination compound is the **coordination number** of the metal ion, usually 2, 4, or 6.

A ligand that can form only one coordinate covalent bond to the metal is a **monodentate ligand**; examples are H_2O, NH_3, and Cl^-. The word *dentate* derives from the Latin word *dentis,* for tooth, so NH_3 is a "one-toothed" ligand.

A ligand that can form two or more coordinate covalent bonds to the same metal ion is called a **polydentate ligand**. Such ligands have two or more atoms with lone pairs separated by several intervening atoms. A **bidentate ligand** *forms two coordinate covalent bonds to the central metal ion.* A good example is 1,2-diaminoethane, $H_2NCH_2CH_2NH_2$, commonly called ethylenediamine and abbreviated *en.* When a lone pair of electrons from each nitrogen atom in en coordinates to the same metal ion, a stable five-membered ring forms (Figure 20.9). Notice that Co^{3+} has a coordination number of 6 in $[Co(en)_3]^{3+}$.

The word "chelating," derived from the Greek *chele*, "claw," describes the pincer-like way in which a polydentate ligand can grab a metal ion. A **chelating ligand** is *a polydentate ligand and can bond with the central metal ion using two or more electron pairs.* Examples of monodentate and polydentate (chelating) ligands are shown in Figure 20.10.

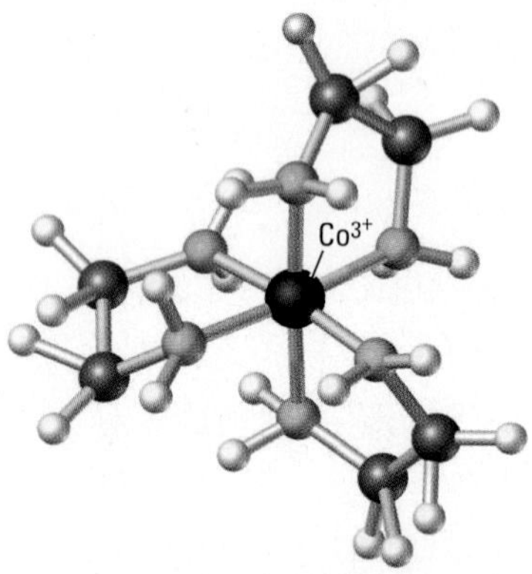

Co3+

Figure 20.9 The $[Co(en)_3]^{3+}$ complex ion. Cobalt ion (Co^{3+}) forms a six-coordinate complex ion with three bidentate ethylenediamine ligands.

PROBLEM-SOLVING EXAMPLE 20.8

Chelating Agents

Two ethylenediamine (en) ligands and two chloride ions form a complex ion with Co^{3+}.
(a) Write the formula of this complex ion.
(b) Determine the coordination number of the Co^{3+} ion.
(c) Write the formula of the coordination compound formed by Cl^- counter ions and the Co^{3+} complex ion.

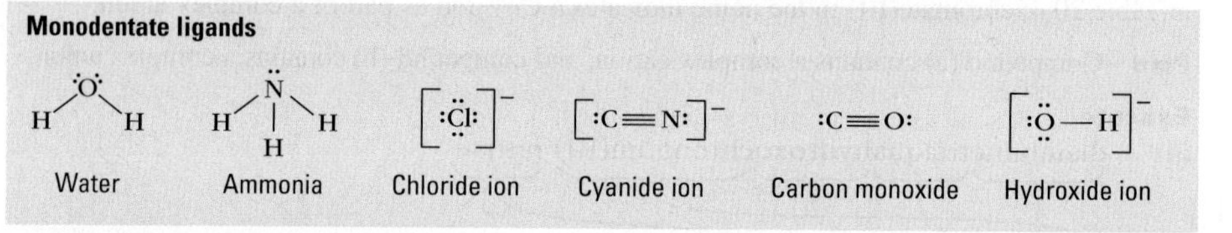

Monodentate ligands
Water
Ammonia
Chloride ion
Cyanide ion
Carbon monoxide
Hydroxide ion

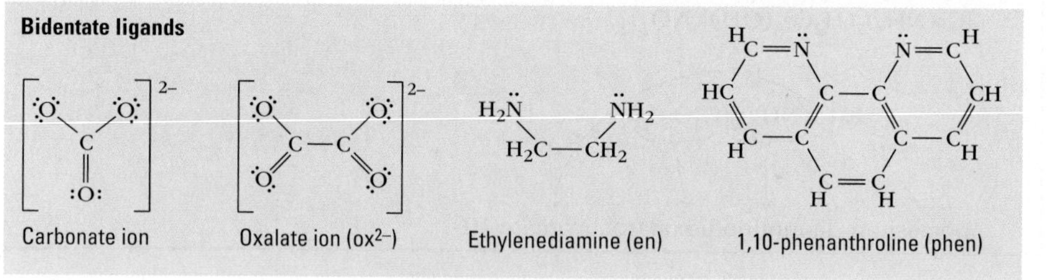

Bidentate ligands
Carbonate ion
Oxalate ion (ox2–)
Ethylenediamine (en)
1,10-phenanthroline (phen)

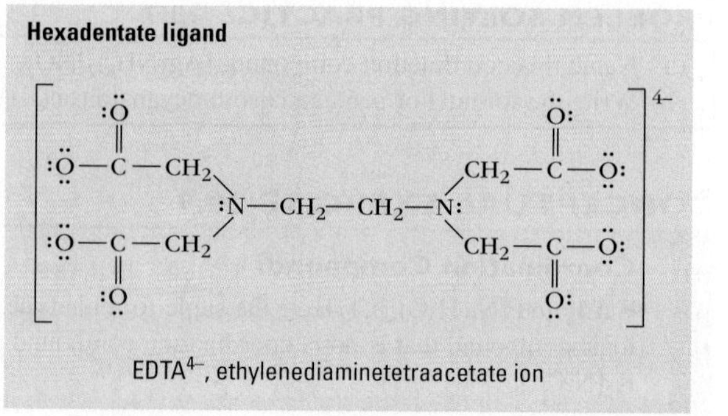

Hexadentate ligand
EDTA4–, ethylenediaminetetraacetate ion

Figure 20.10 Monodentate, bidentate, and hexadentate ligands. Chelating ligands have two lone pairs (bidentate) or up to six lone pairs (hexadentate) that can be shared with a central metal ion.

Result (a) $[Co(en)_2Cl_2]^+$ (b) 6 (c) $[Co(en)_2Cl_2]Cl$

Analyze Recall that ethylenediamine (en) is a bidentate ligand that forms two coordinate covalent bonds per en molecule. Sufficient counter ions of the proper charge are required to balance the net charge on the complex ion.

Plan

(a) Ethylenediamine is a neutral molecule, each chloride ion is 1−, and cobalt has a 3+ charge, so the complex ion will have a 1+ charge.
(b) The coordination number is the number of coordinate covalent bonds to the central Co^{3+} ion.
(c) The 1+ charge of the complex ion requires one chloride ion as a counter ion.

Execute

(a) Two en molecules and two chloride ions are bonded to the central cobalt ion, so the formula of the complex ion is $[Co(en)_2Cl_2]^+$. Cobalt has a 3+ charge, ethylenediamine is a neutral ligand, and each chloride ion is 1−. Therefore, the charge on the complex ion is $(3+) + 2(0) + 2(1-) = 1+$.
(b) The coordination number is 6 because there are six coordinate covalent bonds to the central Co^{3+} ion—two from each bidentate ethylenediamine and one from each monodentate chloride ion.
(c) $[Co(en)_2Cl_2]Cl$

Note that Cl_2 in the complex ion's formula represents two chloride ions, *not* a diatomic chlorine molecule.

PROBLEM-SOLVING PRACTICE 20.8

The dimethylglyoximate anion (abbreviated DMG^-),

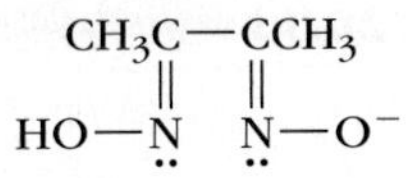
CH3C—CCH3
HO—N N—O−

is a bidentate ligand used to test for the presence of nickel. It reacts with Ni^{2+} to form a beautiful red solid in which the Ni^{2+} has a coordination number of 4. DMG^- coordinates to Ni^{2+} by the lone pairs on the nitrogen atoms (Figure 20.11).

(a) How many DMG^- ions are needed to satisfy a coordination number of 4 on the central Ni^{2+} ion?
(b) What is the net charge after coordination occurs?
(c) How many atoms are in the ring formed by one DMG^- and one Ni^{2+}?

CONCEPTUAL EXERCISE 20.10

Chelating and Complex Ions

Oxalate ion forms a complex ion with Mn^{2+} by coordinating at the oxygen lone pairs (see Figure 20.10).

(a) How many oxalate ions are needed to satisfy a coordination number of 6 on the central Mn^{2+} ion?
(b) What is the charge on this complex ion?
(c) How many atoms are in the ring formed between one oxalate ion and the central metal ion?

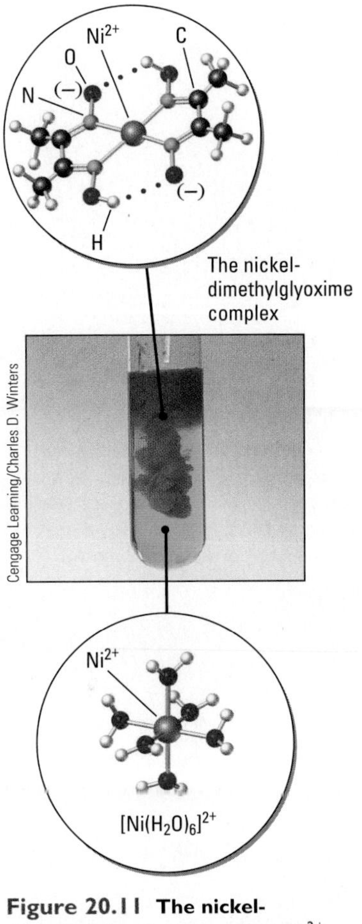
Ni2+
O
C
N
(−)
(−)
H
The nickel-dimethylglyoxime complex
Cengage Learning/Charles D. Winters
Ni2+
[Ni(H2O)6]2+

Figure 20.11 The nickel-dimethylglyoxime complex. Ni^{2+} ions react with the dimethylglyoximate anion (DMG^-) to form a beautiful red solid.

For metals that display a coordination number of 6, an especially effective ligand is the *hexadentate* ethylenediaminetetraacetate ion (abbreviated $EDTA^{4-}$; see Figure 20.10). A **hexadentate ligand** *has six lone pair donor atoms that can coordinate to a single metal ion.* In the case of EDTA there are four O atoms and two N atoms that encapsulate and firmly bind a metal ion, so $EDTA^{4-}$ is an excellent chelating ligand. $EDTA^{4-}$ is often added to commercial salad dressing to remove traces of metal ions from solution, because these metal ions could otherwise accelerate the oxidation of oils in the product and make them rancid. Another use of $EDTA^{4-}$ is in bathroom cleansers, where it removes hard water deposits of insoluble $CaCO_3$ and $MgCO_3$ by chelating Ca^{2+} or Mg^{2+} ions, making them soluble and allowing them to be rinsed away. EDTA is also used in the treatment of lead and mercury poisoning because it chelates these metals and aids in their removal from the body (Figure 20.12).

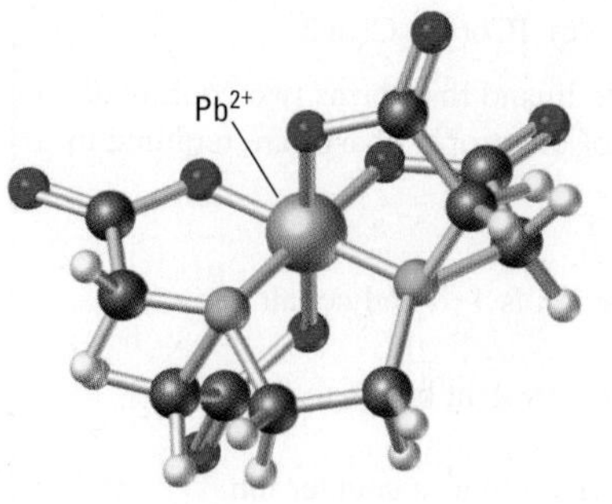
Pb^{2+}

(a)

(b)

Figure 20.12 EDTA is an extremely effective chelating ligand. (a) The structure of the EDTA complex of Pb^{2+}. (b) Some household products containing EDTA. EDTA forms complexes with and makes soluble ions such as Pb^{2+} and Ca^{2+}, enabling them to be removed from our bodies or from household surfaces.

Coordination compounds of *d*-block transition metals are often colored, and the colors of the complexes of a given transition-metal ion depend on both the metal ion and the ligand (Figure 20.13). Many transition-metal coordination compounds are used as pigments in paints and dyes. For example, Prussian blue, $Fe_4[Fe(CN)_6]_3$, a deep-blue pigment known for hundreds of years, is the "bluing agent" in engineering blueprints. The origin of colors in coordination compounds will be discussed in Section 20.7c.

CONCEPTUAL EXERCISE 20.11

Complex Ions

Prussian blue contains two kinds of iron ions. What is the charge of the iron in

(a) the complex ion $[Fe(CN)_6]^{4-}$?

(b) the iron ion not in the complex ion?

20-6d Geometry of Coordination Compounds and Complex Ions

The geometry of a complex ion or coordination compound is dictated by the arrangement of the electron donor atoms of the ligands attached to the central metal ion. Although other geometries are possible, we will discuss only the four most common ones, those associated with coordination numbers of 2, 4, and 6. To simplify matters, we will consider only monodentate ligands, L, bonded to a central metal ion, M^{n+}.

Coordination Number = 2, ML_2^{n+} All such complex ions have a *linear* geometry with the two ligands on opposite sides of the central metal ion to give an L—M—L bond angle of 180°, such as that in $[Ag(NH_3)_2]^+$. Other examples are $[CuCl_2]^-$ and $[Au(CN)_2]^-$, the complex ion used to extract gold from ores (← **Sec. 20-4**).

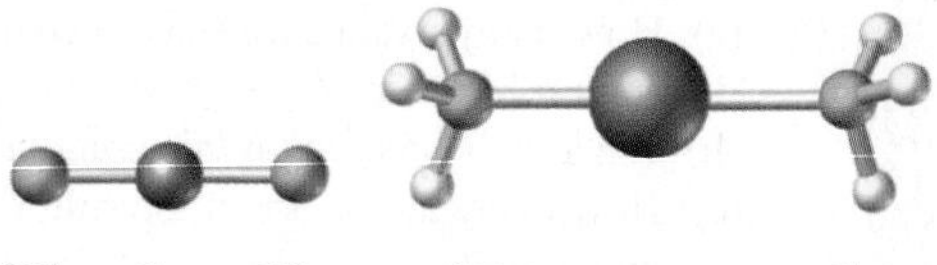

$[Cl—Cu—Cl]^-$ $[H_3N—Ag—NH_3]^+$

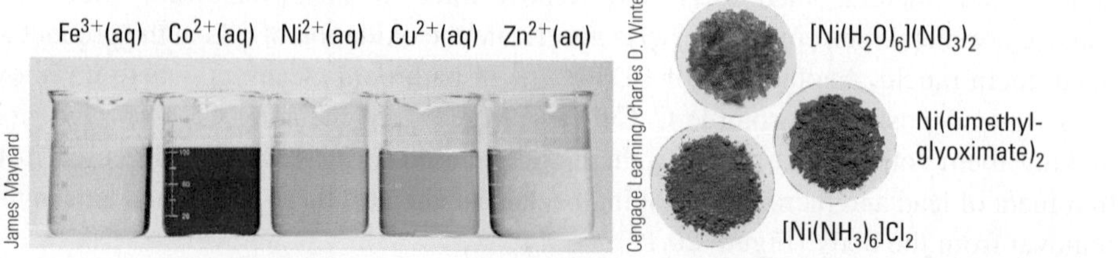
Fe^{3+}(aq)
Co^{2+}(aq)
Ni^{2+}(aq)
Cu^{2+}(aq)
Zn^{2+}(aq)
[Ni(H$_2$O)$_6$](NO$_3$)$_2$
Ni(dimethyl-glyoximate)$_2$
[Ni(NH$_3$)$_6$]Cl$_2$

Figure 20.13 Colors of transition-metal ions in aqueous solutions and as solids.

Coordination Number = 4, ML_4^{n+} Four-coordinate complex ions have either tetrahedral or square planar geometries. In the *tetrahedral* case, the four monodentate ligands are at the corners of a tetrahedron, such as in $[Zn(NH_3)_4]^{2+}$; L—M—L bond angle is 109.5°. In *square planar* geometry, the ligands lie in a plane at the corners of a square as in $[Ni(CN)_4]^{2-}$ and $[Pt(NH_3)_4]^{2+}$ ions; L—M—L bond angles are 90° and 180°.

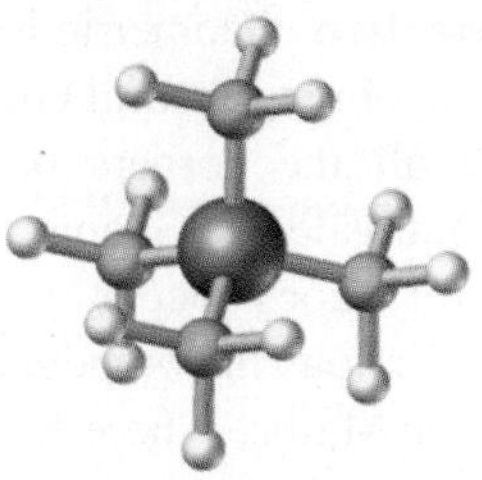

Tetrahedral
$[Zn(NH_3)_4]^{2+}$

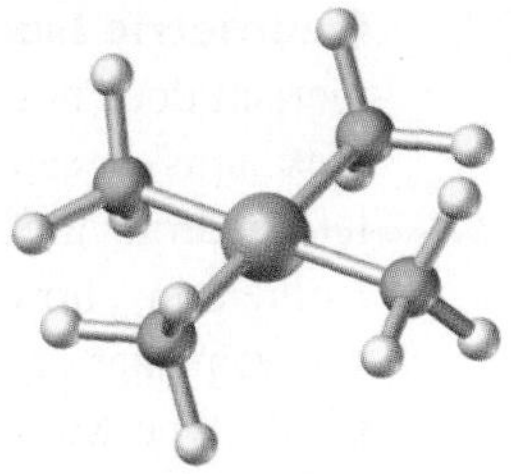

Square planar
$[Pt(NH_3)_4]^{2+}$

Coordination Number = 6, ML_6^{n+} Octahedral geometry is characteristic of this coordination number. The six ligands are at the corners of an octahedron with the central metal ion at its center. Octahedral geometry can be regarded as derived from a square planar geometry by adding two ligands, one above and one below the square plane. Two common octahedral complex ions are $[Co(NH_3)_6]^{3+}$ and $[Fe(CN)_6]^{3-}$, in which the six ligands are equidistant from the central metal ion and all six ligand sites are equivalent; L—M—L bond angles are 90° and 180°.

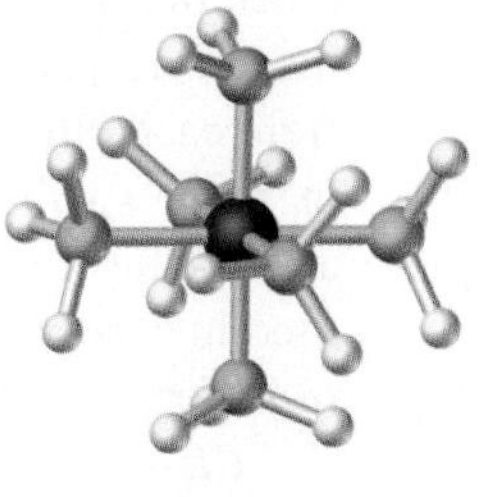

Octahedral
$[Co(NH_3)_6]^{3+}$

20-6e Isomerism in Coordination Compounds and Complex Ions

Various types of isomerism have been discussed previously with regard to organic compounds. *Constitutional isomerism* occurs with molecules that have the same molecular formula but differ in the way their atoms are connected together, such as occurs with butane and 2-methylpropane (← **Sec. 2-9a**). *Stereoisomerism* is a second general category of isomerism in which the isomers have the same bond connections, but the atoms are arranged differently in space. One type of stereoisomerism is *geometric isomerism,* such as that found in *cis*- and *trans*-1,2-dichloroethene (← **Sec. 6-5a**). The other type of stereoisomerism is *optical isomerism,* which occurs when mirror images are nonsuperimposable (← **Sec. 7-2f**). Constitutional, geometric, and optical isomers also occur with coordination complex ions and coordination compounds.

Linkage Isomerism, a Type of Constitutional Isomerism Linkage isomerism occurs when a ligand can bond to the central metal using either of two different electron-donating atoms. Thiocyanato $(SCN)^-$ and isothiocyanato $(NCS)^-$ ligands are an example: Sulfur bonds with a metal ion in the first case and nitrogen in the second, as illustrated for the two Co^{3+} complex ions here.

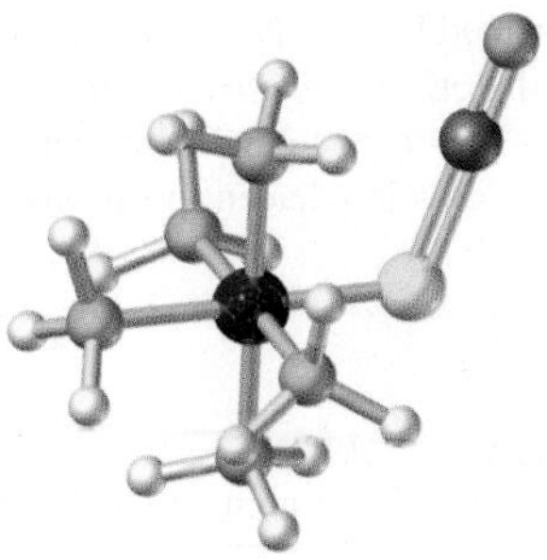

pentaammine-
thiocyanatocobalt(III) ion

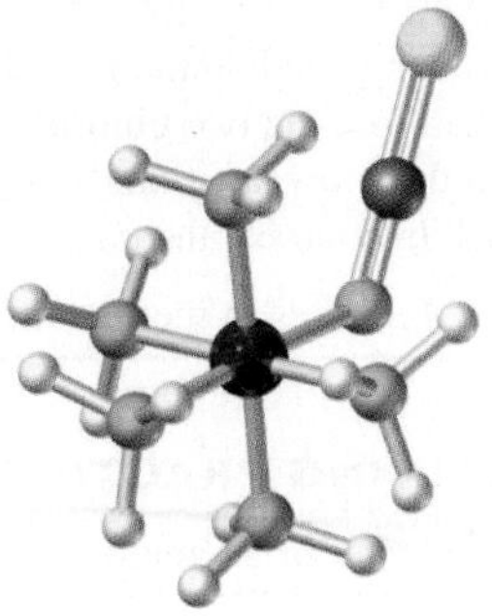

pentaammine-
isothiocyanatocobalt(III) ion

Alfred Werner
1866–1919

Science Photo Library/Science Source

In 1893, while still a young professor, Alfred Werner published a revolutionary paper about transition-metal compounds. He asserted that transition-metal ions could exhibit a secondary valence as well as a primary one, such as in $CoCl_3 \cdot 6NH_3$ (now written as $[Co(NH_3)_6]Cl_3$). The primary valence was represented by the ionic bonds between Co^{3+} and the chloride ions; the secondary valence was represented by the coordinate covalent bonds between the metal ion and six NH_3 molecules, what we now call the coordination sphere around the central metal ion. Werner also made the inspired proposal that the ammonia molecules were octahedrally coordinated around the Co^{3+} ion, thereby laying the foundation for understanding the geometry of complex ions. For his groundbreaking work, Werner received the 1913 Nobel Prize in Chemistry.

Geometric Isomerism Geometric isomerism does not exist in tetrahedral complex ions because all the corners of a tetrahedron are equivalent. Geometric isomerism, however, does occur with square planar complex ions in compounds of the type Ma_2b_2 or Ma_2bc, where M is the central metal ion and a, b, and c are *different* ligands. The square planar coordination compound $[Pt(NH_3)_2Cl_2]$, an Ma_2b_2 type, occurs in two geometric forms. The *cis*-$[Pt(NH_3)_2Cl_2]$ isomer has the chloride ligands as close as possible at 90° to one another. In *trans*-$[Pt(NH_3)_2Cl_2]$, the chloride ions are as far apart as possible, directly across the square plane of the molecule at 180° from each other. These two isomers differ in water solubility, color, melting point, and chemical reactivity. The *cis* isomer is used in cancer chemotherapy, whereas the *trans* form is not effective against cancer.

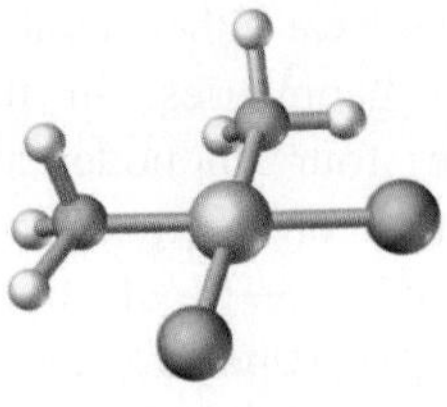
cis-$[Pt(NH_3)_2Cl_2]$

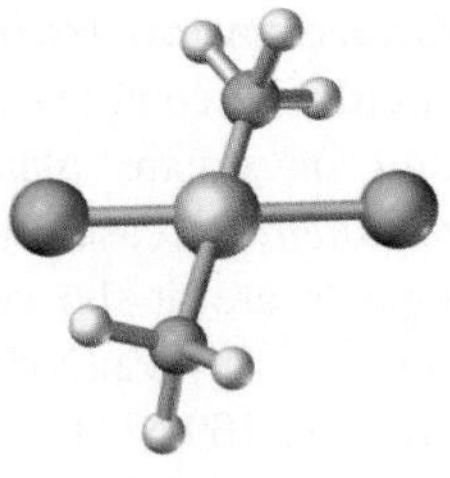
trans-$[Pt(NH_3)_2Cl_2]$

Cis-trans isomerism is also possible in octahedral complex ions and compounds, as illustrated with $[Co(NH_3)_4Cl_2]^+$. In this complex ion the *cis* isomer has the chloride ions adjacent to each other; the *trans* isomer has them opposite each other. The differences in properties are striking, particularly the color. The *cis* isomer is violet, whereas the *trans* form is green.

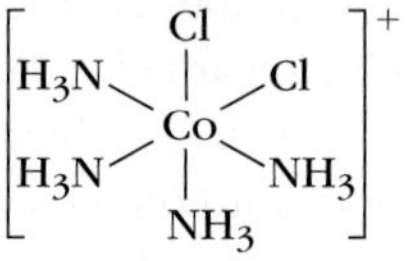
Cl
H3N
Cl
Co
H3N
NH3
NH3
+

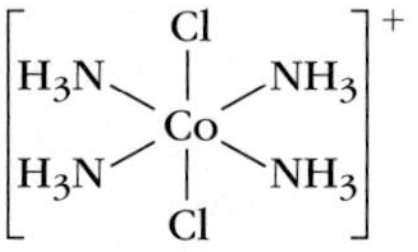
Cl
H3N
NH3
Co
H3N
NH3
Cl
+

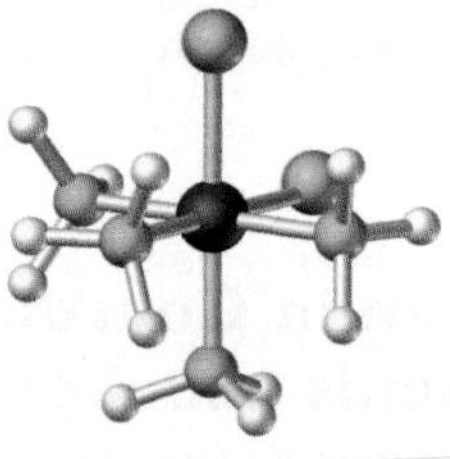
cis-$[Co(NH_3)_4Cl_2]^+$
(violet)

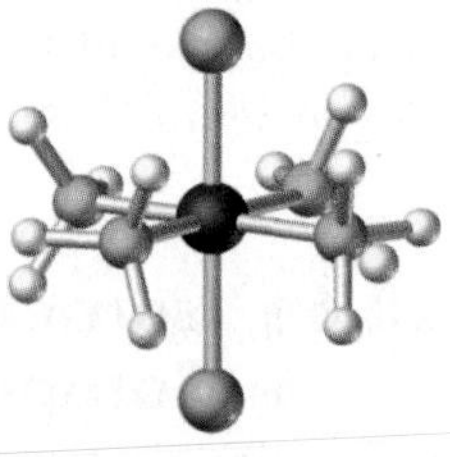
trans-$[Co(NH_3)_4Cl_2]^+$
(green)

PROBLEM-SOLVING EXAMPLE 20.9

Geometric Isomerism

How many geometric isomers are there for $[Co(en)_2Cl_2]^+$?

Result Only two geometric isomers are possible, *cis* and *trans*.

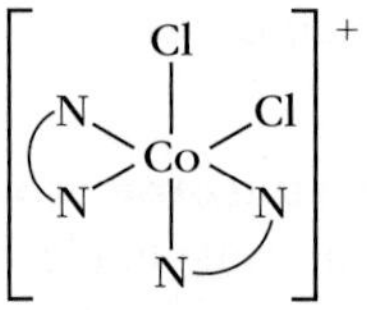
Cl
N
Cl
Co
N
N
N
+
cis

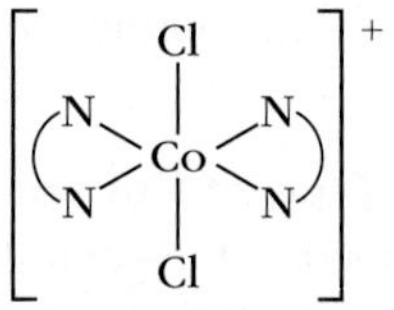
Cl
N
N
Co
N
N
Cl
+
trans

Analyze The two chloride ions can occupy either the *trans* (opposite) positions or *cis* (adjacent) positions. Each en ligand must have both N atoms bonded at a 90° angle because the molecule is not long enough to reach to 180°.

Plan The two ethylenediamine ligands (en), represented here as N⌒N, occupy four sites around the cobalt ion; the two chloride ions occupy the other two sites. In one case, the two Cl^- ions are in the *trans* positions, that is, one at the "top" of the octahedron and the other at the "bottom." In contrast, the *cis* isomer has the Cl^- ions in adjacent positions.

Execute See structures in the Result.

PROBLEM-SOLVING PRACTICE 20.9

Determine the number of geometric isomers for the square planar compound $[Pt(NH_3)_2ClBr]$.

EXERCISE 20.12

Geometric Isomerism

Determine the number of isomers possible for $[Co(NH_3)_3Cl_3]$. Write the structural formula of each isomer.

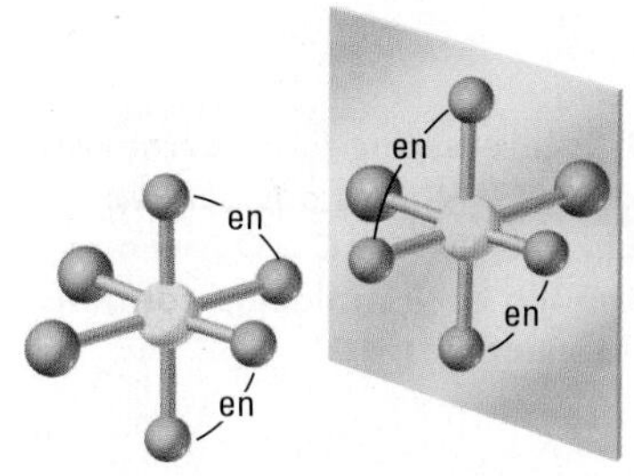
en
en
en
en
$[Cr(en)_2Cl_2]^+$

Figure 20.14 Optical isomerism in $[Cr(en)_2Cl_2]^+$. The ion on the *left* cannot be superimposed on its mirror image *(right)*.

Optical Isomerism Optical isomers are mirror images that are not superimposable. Such nonsuperimposable mirror images are known as *enantiomers* (← **Sec. 7-2f**). An example of optical isomerism in a complex ion is $[Cr(en)_2Cl_2]^+$. There are two enantiomers, as shown in Figure 20.14. No matter how they are twisted and turned, the two mirror images cannot be superimposed.

Optical isomerism is not possible for square planar complexes based on the geometry around the metal ion; the mirror images are superimposable. Although optical isomers of tetrahedral complex ions with four different ligands are theoretically possible, no such stable complexes are known.

20-6f Coordination Compounds and Life

Bioinorganic chemistry, the study that applies chemical principles to inorganic ions and compounds in biological systems, is a rapidly growing field centered mainly around coordination compounds. This is because the very existence of living systems depends on many coordination compounds in which metal ions are chelated to the nitrogen and oxygen atoms in proteins and especially in enzymes. Copper-containing proteins, for example, transport oxygen in crabs, lobsters, and snails and give the blood of these organisms its blue color.

In humans, molecular oxygen, O_2, is transported through the circulatory system by hemoglobin, a very large protein (molar mass of about 64,500 g/mol) in red blood cells. Hemoglobin is blue but becomes red when oxygenated. This is why arterial blood is bright red (high O_2 concentration) and blood in veins is bluish (low O_2 concentration).

A hemoglobin molecule carries four O_2 molecules, each of which forms a coordinate covalent bond to one of four Fe^{2+} ions (Figure 20.15). Each Fe^{2+} ion is at the center of one heme, a nonprotein part of the hemoglobin molecule that consists of four linked nitrogen-containing rings. Bound in this way, molecular oxygen is carried to the cells, where it is released as needed by breaking the Fe—O_2 coordinate covalent bond.

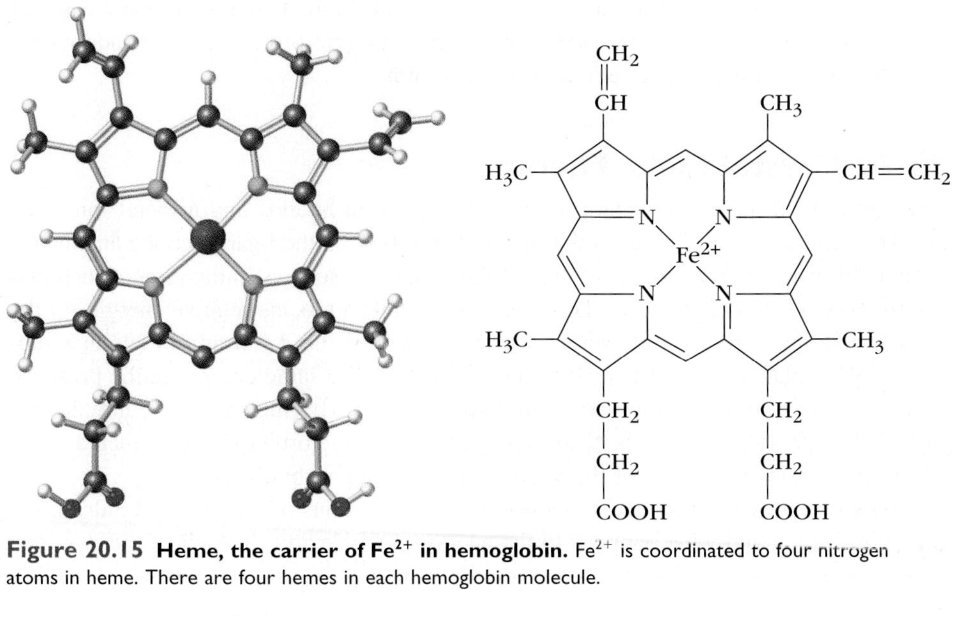
CH_2
CH
CH_3
H_3C
CH=CH_2
N
N
Fe^{2+}
N
N
H_3C
CH_3
CH_2
CH_2
CH_2
CH_2
COOH
COOH

Figure 20.15 Heme, the carrier of Fe^{2+} in hemoglobin. Fe^{2+} is coordinated to four nitrogen atoms in heme. There are four hemes in each hemoglobin molecule.

It is interesting (and fortunate) that N≡N does not behave chemically like C≡O, even though each molecule contains 14 electrons.

Other substances that can donate an electron pair can also bond to the Fe^{2+} in heme. Carbon monoxide is such a ligand and forms an exceptionally strong Fe^{2+}—CO bond, about 200 times stronger than the O_2—Fe^{2+} bond. Therefore, when a person breathes in CO, it displaces O_2 from hemoglobin and prevents red blood cells from carrying oxygen. The initial effect is drowsiness. But if CO inhalation continues, cells deprived of oxygen can no longer function and the person suffocates.

Structures similar to the oxygen-carrying unit in hemoglobin are also found in other biologically important compounds, including such diverse ones as myoglobin and vitamin B-12. Myoglobin, like hemoglobin, contains Fe^{2+} and carries and stores molecular oxygen, principally in muscles. At the center of a vitamin B-12 molecule is a Co^{3+} ion bonded to the same type of group that Fe^{2+} bonds to in hemoglobin. Vitamin B-12 is the only known dietary use of cobalt, but it makes cobalt an essential mineral (← **Sec. 1-14a**).

The dietary necessity of zinc for humans has become established only since the 1980s. Zinc, in the form of Zn^{2+} ions, is essential to the functioning of several hundred enzymes, including those that catalyze the breaking of P—O—P bonds in adenosine triphosphate (ATP), an important energy-releasing compound in cells (← **Sec. 16-9a**).

Copper ranks third among biologically important transition-metal ions in humans, after iron and zinc. Although we usually excrete any dietary excess of copper, a genetic defect causes Wilson's disease, a condition in which Cu^{2+} accumulates in the liver and brain. Fortunately, Wilson's disease can be treated by administering chelating agents that coordinate excess Cu^{2+} ions, allowing them to be excreted harmlessly.

20-7 Crystal-Field Theory: Color and Magnetism in Coordination Compounds

Bright colors are characteristic of many coordination compounds (Figure 20.13). Any theory about bonding in such compounds needs to address the origin of their colors. One such approach is the crystal-field theory developed by Hans Bethe and J. H. van Vleck. Crystal-field theory explains color as originating from electron transitions between two sets of *d* atomic orbitals in a complex ion, similar to the bright line atomic spectra that originate from electron transitions among atomic orbitals in elements. In many cases of complex ions, the energy difference between the *d* atomic orbitals corresponds with a wavelength within the visible region of the spectrum.

20-7a Crystal-Field Theory

The crystal-field theory considers the bonding between ligands and a metal ion to be primarily electrostatic. It assumes that the electron pairs on the ligands create an electrostatic field around the *d* atomic orbitals of the metal ion, such as in the case of the octahedral $[Fe(CN)_6]^{4-}$ complex ion. This electrostatic field alters the relative energies of the various *d* atomic orbitals. Before this interaction occurs, the *d* atomic orbitals in a sublevel of the isolated metal ion, such as the 3*d*, all have the same energy. In the presence of ligands, the *d* atomic orbitals split into two groups of differing energy (Figure 20.16). The higher-energy pair consists of the $d_{x^2-y^2}$ and d_{z^2} atomic orbitals and is labeled *e*; the lower-energy trio is made up of the d_{xy}, d_{yz}, and d_{xz} atomic orbitals, labeled t_2.

N
C
NC
CN
Fe2+
NC
CN
C
N
4−

The $[Fe(CN)_6]^{4-}$ complex ion. The six CN^- ions are arranged octahedrally around the central Fe^{2+} ion.

Why does this splitting occur? Consider the orientations of *d* atomic orbitals on the central metal ion and what happens to the energy of an electron in each of the *d* atomic orbitals when ligands approach a metal ion. From Figure 20.17 note that two of the five

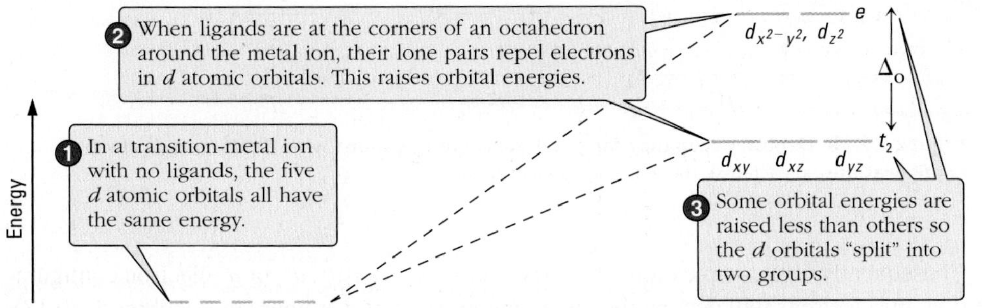
2 When ligands are at the corners of an octahedron around the metal ion, their lone pairs repel electrons in d atomic orbitals. This raises orbital energies.
e
$d_{x^2-y^2}$, d_{z^2}
Δ_o
1 In a transition-metal ion with no ligands, the five d atomic orbitals all have the same energy.
Energy
t_2
d_{xy} d_{xz} d_{yz}
3 Some orbital energies are raised less than others so the d orbitals "split" into two groups.

Figure 20.16 The splitting of *d* atomic orbitals in an octahedral field of ligands.

d atomic orbitals—the $d_{x^2-y^2}$ and the d_{z^2}—have their lobes of maximum electron density directly *along* the *x*-, *y*-, and *z*-axes. In contrast, the other three *d* atomic orbitals—d_{xy}, d_{yz}, and d_{xz}—have their lobes of maximum electron density aligned *between* the *x*-, *y*-, and *z*-axes rather than directly along them. Thus, when CN^- ligands are oriented along the *x*-, *y*-, and *z*-axes to form an octahedral complex, the electrostatic repulsion between ligands and electrons in the $d_{x^2-y^2}$ and d_{z^2} atomic orbitals is greater than between ligands and electrons in orbitals not directly along the axes—the d_{xy}, d_{yz}, and d_{xz} atomic orbitals. An electron in a $d_{x^2-y^2}$ or d_{z^2} orbital, therefore, has higher energy than an electron in a d_{xy}, d_{yz}, or d_{xz}. This gives the *d*-orbital energy levels shown in Figure 20.16. *The energy difference between the* t_2 *and* e *sets of* d *atomic orbitals* is known as the **crystal-field splitting energy** (**Δ**). For an octahedral field of ligands, as in $Fe(CN)_6^{4-}$, a subscript o is added, Δ_o.

20-7b Electron Configurations and Magnetic Properties of Coordination Complex Ions

The electron configurations and magnetic properties of transition-metal ions can be interpreted by using crystal-field theory. Applying Hund's rule (← **Sec. 5-7a**), electrons occupy the vacant atomic orbitals of lowest energy first. If the vacant atomic orbitals all have the same energy, electrons occupy those orbitals unpaired with their spins parallel.

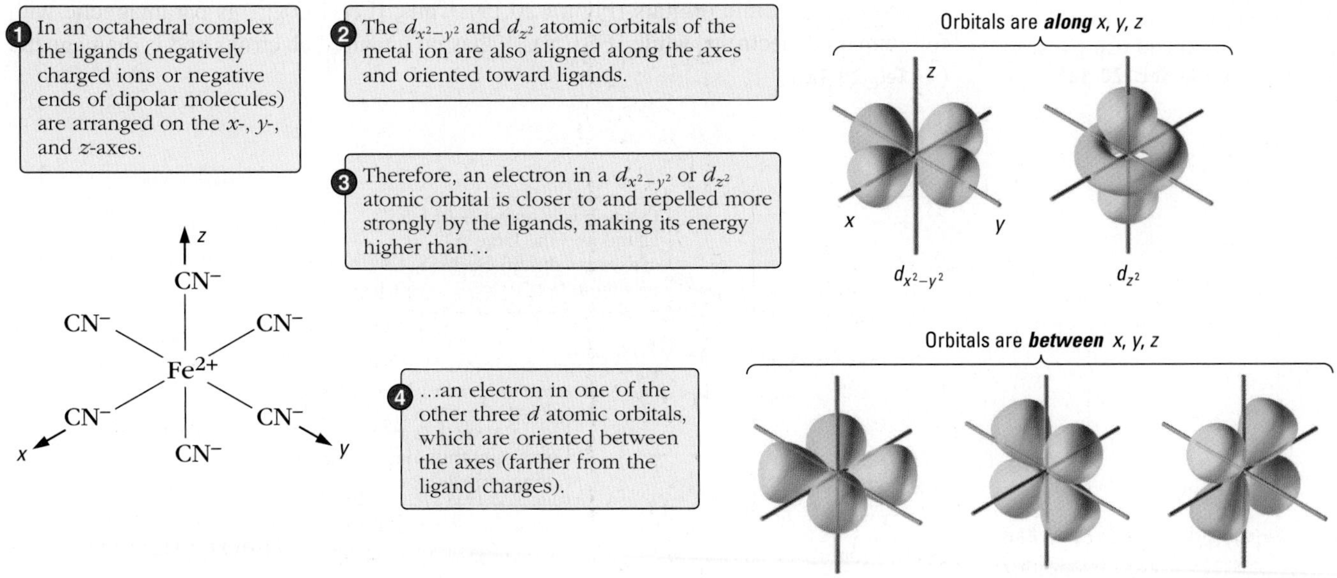
1 In an octahedral complex the ligands (negatively charged ions or negative ends of dipolar molecules) are arranged on the x-, y-, and z-axes.
2 The $d_{x^2-y^2}$ and d_{z^2} atomic orbitals of the metal ion are also aligned along the axes and oriented toward ligands.
3 Therefore, an electron in a $d_{x^2-y^2}$ or d_{z^2} atomic orbital is closer to and repelled more strongly by the ligands, making its energy higher than…
4 …an electron in one of the other three d atomic orbitals, which are oriented between the axes (farther from the ligand charges).
z
CN^-
Fe^{2+}
x
y
Orbitals are along x, y, z
$d_{x^2-y^2}$
d_{z^2}
Orbitals are between x, y, z
d_{xy}
d_{yz}
d_{xz}

Figure 20.17 The origin of crystal-field splitting energy.

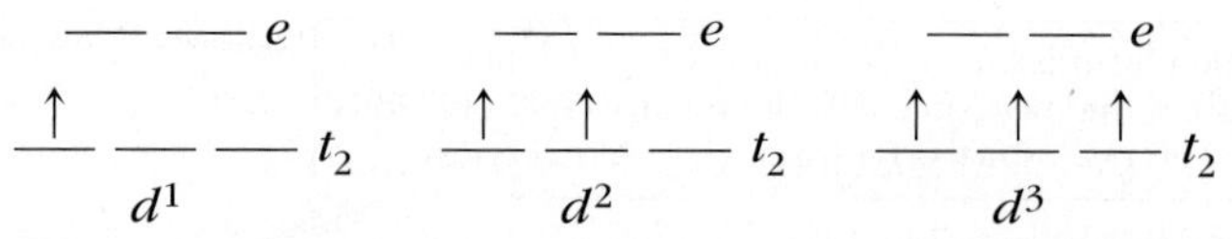
e
e
e
t2
t2
t2
d1
d2
d3

Figure 20.18 Orbital occupancy for octahedral complex ions with d^1, d^2, and d^3 electron configurations. In each case the electrons are unpaired.

Consequently, in a complex ion where the metal ion has a d^1, d^2, or d^3 electron configuration, the electrons will occupy the lower-energy t_2 set of d atomic orbitals (Figure 20.18).

We might expect that the trend of unpaired electrons would continue with the filling of the set of e higher-energy d atomic orbitals for metal ions with d^4 and d^5 electron configurations. This is true in some, but not all, cases. *Whether these electrons are unpaired or paired depends on the magnitude of the crystal-field splitting energy, Δ_o, the energy gap between the metal ion's two sets of* d *atomic orbitals.* That crystal-field splitting energy depends on the metal ion and the ligands. The relative effect of the ligands is given by the **spectrochemical series**, *a listing of ligands in the order of their crystal-field splitting energy.* The spectrochemical series is determined experimentally. An abbreviated spectrochemical series for octahedral complexes is

$$Cl^- < F^- < H_2O < NCS^- < NH_3 < en < NO_2^- < CN^-$$

weak field ⟶ increasing Δ_o ⟶ strong field

Ligands such as CN^-, NO_2^-, and en that cause a large Δ_o are called *strong-field* ligands; those at the other end of the series, such as Cl^- and F^-, with smaller Δ_o are termed *weak-field* ligands.

Consider two complex ions, $[Fe(CN)_6]^{4-}$ and $[Fe(H_2O)_6]^{2+}$, each containing the Fe^{2+} ion with six 3d electrons. As seen from the spectrochemical series, cyanide ion, CN^-, is a strong-field ligand with a Δ_o large enough to cause the six d electrons to pair in the three lower-energy t_2 atomic orbitals (Figure 20.19). In $[Fe(H_2O)_6]^{2+}$, water is a weak-field ligand with a Δ_o sufficiently smaller that four of the six d electrons remain unpaired. Using Hund's rule for the aqua complex ion, the first five electrons occupy each of the five d atomic orbitals individually, with pairing occurring when the sixth electron is added into one of the t_2 atomic orbitals. The result is four electrons occupying the t_2 atomic orbitals and the remaining two electrons unpaired in the higher-energy set of d_{z^2} and $d_{x^2-y^2}$ atomic orbitals (Figure 20.19). Thus, $[Fe(H_2O)_6]^{2+}$ is paramagnetic with four unpaired electrons, while $[Fe(CN)_6]^{4-}$ has no unpaired electrons and is diamagnetic (← **Sec. 20-1a**).

How paramagnetic a metal ion is can be measured using a special balance (← Sec. 20-1a).

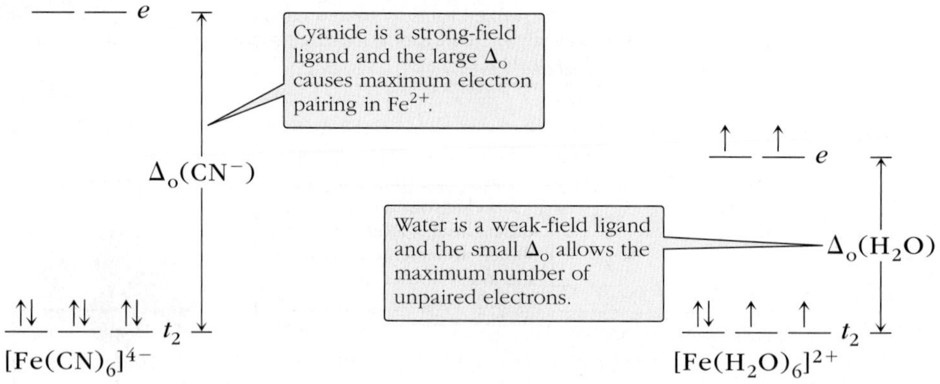
e
Cyanide is a strong-field ligand and the large Δo causes maximum electron pairing in Fe2+.
Δo(CN−)
t2
[Fe(CN)6]4−
e
Water is a weak-field ligand and the small Δo allows the maximum number of unpaired electrons.
Δo(H2O)
t2
[Fe(H2O)6]2+

Figure 20.19 Ligands affect the number of unpaired electrons in these two complex ions.

High-spin and Low-spin Complexes There are many other cases like this, where the same metal ion has greater or smaller numbers of unpaired electrons depending on the ligand. They occur when the metal ion contains between four and seven valence *d* electrons. *The complex ion with the greater number of unpaired electrons* is known as the **high-spin complex**; the **low-spin complex** *contains the lesser number of unpaired electrons*. High-spin complexes are expected with weak-field ligands where the crystal-field splitting energy Δ_o is small. The opposite applies to low-spin complexes in which strong-field ligands cause maximum pairing of electrons in the set of three t_2 atomic orbitals due to large Δ_o.

High-spin: maximum number of unpaired electrons. $[Fe(H_2O)_6]^{2+}$ is high-spin.

Low-spin: minimum number of unpaired electrons. $[Fe(CN)_6]^{4-}$ is low-spin.

PROBLEM-SOLVING EXAMPLE 20.10

High-spin and Low-spin Complex Ions

Experimental data indicate that $[Co(CN)_6]^{3-}$ is diamagnetic and $[CoF_6]^{3-}$ is paramagnetic.
(a) Use the crystal-field model to illustrate the $3d$ electron configuration of the cobalt ion in each complex.
(b) How many unpaired electrons are in the $3d$ atomic orbitals of the paramagnetic complex?
(c) Which is the low-spin complex?

Result (a) See the diagram (b) Four (c) $[Co(CN)_6]^{3-}$

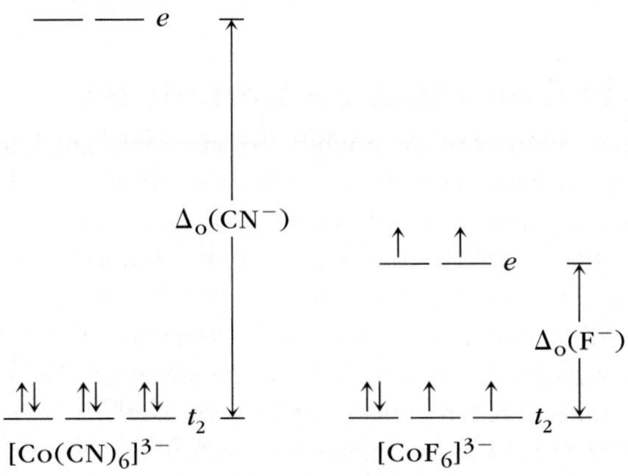
e
$\Delta_o(CN^-)$
t_2
$[Co(CN)_6]^{3-}$
e
$\Delta_o(F^-)$
t_2
$[CoF_6]^{3-}$

Analyze In each case, we first must determine the number of $3d$ electrons of Co^{3+}. Recall that in forming ions, transition metals lose *ns* electrons before losing $(n-1)d$ electrons (← **Sec. 20-1a**). We then use the spectrochemical series to determine the relative crystal field strengths of cyanide and fluoride ions as ligands. Strong-field ligands cause maximum pairing of electrons (low spin).

Plan Cobalt metal atoms have the $[Ar]4s^23d^7$ electron configuration and lose three electrons to form Co^{3+} ions. In this case, two $4s$ electrons are lost plus one $3d$ electron to give Co^{3+} ion an $[Ar]3d^6$ electron configuration.

Execute
(a) From the spectrochemical series, we note that cyanide ion is a strong-field ligand that will cause maximum pairing of electrons (low spin). In contrast, fluoride ion, a weak-field ligand, is not sufficiently strong to cause maximum *d* electron pairing; instead, the maximum number of unpaired electrons occurs (high spin) as shown in the Result above.
(b) In $[CoF_6]^{3-}$, there are four unpaired $3d$ electrons. (See Result above.)
(c) Because $[Co(CN)_6]^{3-}$ is diamagnetic, it contains no unpaired electrons and must be the low-spin complex of these two.

PROBLEM-SOLVING PRACTICE 20.10

A Cr^{2+} ion contains four $3d$ electrons. How many unpaired electrons are in a high-spin octahedral complex of this metal ion? A low-spin complex of this ion? Is either complex paramagnetic? Explain your result.

CONCEPTUAL EXERCISE 20.13

High- and Low-spin Complexes

Explain why Ni^{2+} ions cannot form high- and low-spin complexes.

Tetrahedral and Square Planar Complexes Up to this point we have focused on the application of crystal-field theory to octahedral complexes. It also can be applied to tetrahedral and square planar complexes, which have different crystal-field splitting patterns (Figure 20.20). The *d* atomic orbital splitting pattern for tetrahedral complexes is the opposite of the octahedral pattern. In tetrahedral complexes the $d_{x^2-y^2}$ and d_{z^2} atomic orbitals are lower in energy by Δ_t than the set of d_{xz}, d_{xy}, and d_{yz} atomic orbitals. In addition, for a given ligand the value of Δ_t is less than the value of Δ_o.

Remember that the d_{z^2} atomic orbital has a donut of electron density in the *x-y* plane (Figure 20.17, Section 20-7b).

In square planar complexes, we assume that ligands are in a plane containing the *x*- and *y*-axes. Because the $d_{x^2-y^2}$ atomic orbital points along these axes, it is the highest in energy, followed by the d_{xy} atomic orbital and then the d_{z^2} atomic orbital. The d_{xz} and d_{yz} atomic orbitals are of equal and lowest energy (Figure 20.20). The crystal-field splitting energy for square planar complexes, Δ_{sp}, is the energy difference between the $d_{x^2-y^2}$ and d_{xy} atomic orbitals. A square planar complex can be considered as forming by removing the two ligands along the *z*-axis from an octahedral complex.

20-7c Color in Coordination Complexes

Most transition-metal complexes are brightly colored and explaining their color is a major success of crystal-field theory. The colors arise because of *d*-to-*d* transitions. A ***d*-to-*d* transition** is *a transition of an electron from a lower-energy* d *atomic orbital to a higher-energy* d *atomic orbital* of the metal ion. In a coordination complex, the energy difference between these sets of *d* atomic orbitals typically corresponds to photon wavelengths within the visible region of the spectrum (400 nm to 700 nm), thus giving rise to a color when some wavelengths of visible light are absorbed. The color we see is the color of light that is transmitted; *it is the complementary color of the light absorbed.* On the color wheel in Figure 20.21 complementary colors are directly across from each other. Therefore, a complex appears yellow, for example, because it absorbs blue-violet light, which is across the wheel from yellow.

Coordination compounds with metal ions whose *d* atomic orbitals are filled, such as Zn^{2+}, or have no *d* electrons, such as Sc^{3+}, are not colored.

We can apply crystal-field theory to understand the origin of the vivid purple (red-violet) color of the $[Ti(H_2O)_6]^{3+}$ complex ion. This case is particularly simple because

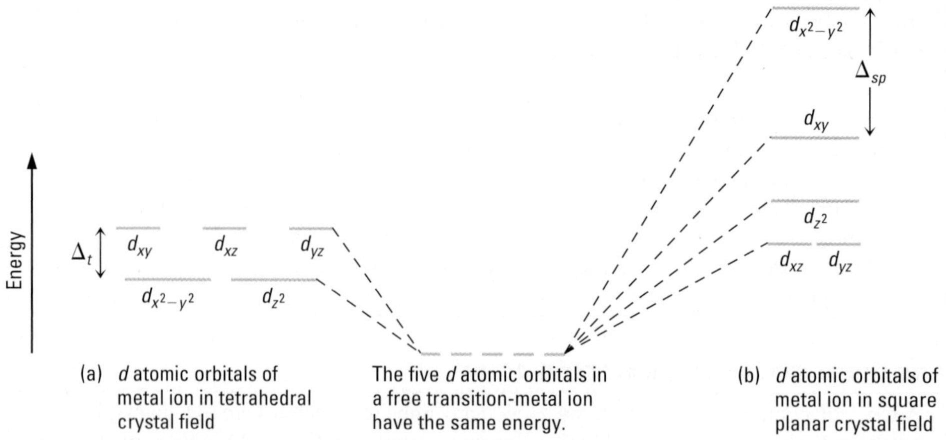
Energy
Δ_t
d_{xy}
d_{xz}
d_{yz}
$d_{x^2-y^2}$
d_{z^2}
$d_{x^2-y^2}$
Δ_{sp}
d_{xy}
d_{z^2}
d_{xz}
d_{yz}
(a) *d* atomic orbitals of metal ion in tetrahedral crystal field
The five *d* atomic orbitals in a free transition-metal ion have the same energy.
(b) *d* atomic orbitals of metal ion in square planar crystal field

Figure 20.20 Splitting of metal ion *d* atomic orbitals in (a) tetrahedral and (b) square planar complexes.

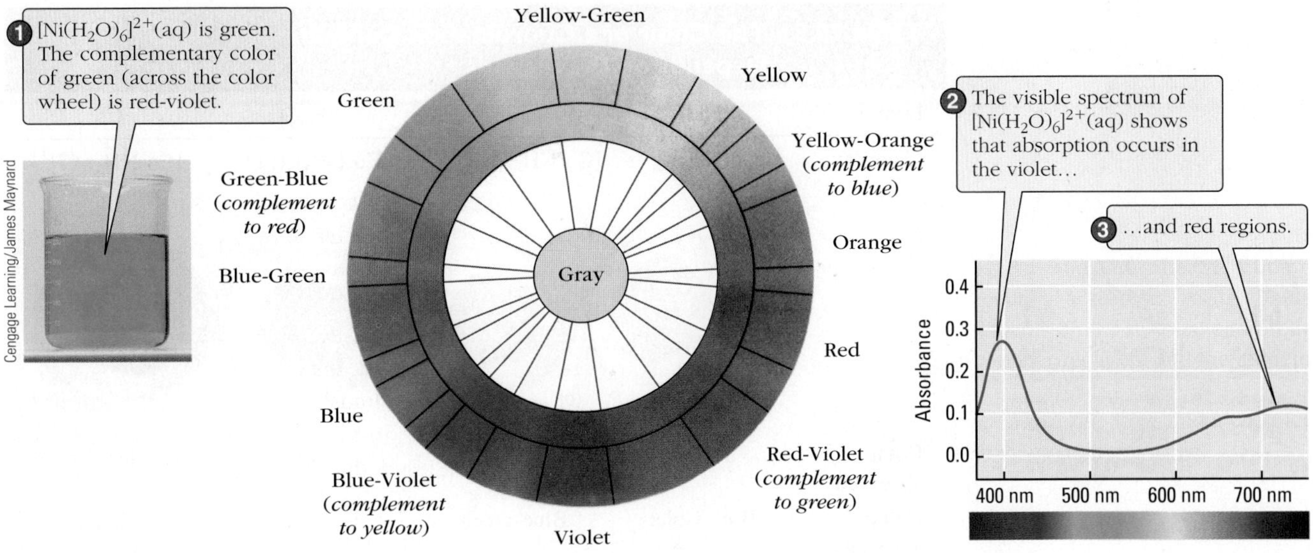
1 $[Ni(H_2O)_6]^{2+}(aq)$ is green. The complementary color of green (across the color wheel) is red-violet.
Cengage Learning/James Maynard
Yellow-Green
Yellow
Green
Yellow-Orange (complement to blue)
Green-Blue (complement to red)
Orange
Blue-Green
Gray
Red
Blue
Red-Violet (complement to green)
Blue-Violet (complement to yellow)
Violet
2 The visible spectrum of $[Ni(H_2O)_6]^{2+}(aq)$ shows that absorption occurs in the violet…
3 …and red regions.
Absorbance
0.4
0.3
0.2
0.1
0.0
400 nm
500 nm
600 nm
700 nm

Figure 20.21 The color wheel. The observed color of a coordination complex is complementary to the color of light absorbed.

Ti^{3+} has only a single 3*d* electron. The complex ion absorbs light at 510 nm, the green region of the visible spectrum, raising the *d* electron from a lower-energy t_2 atomic orbital to a higher-energy *e* atomic orbital (Figure 20.22).

$$\begin{array}{ccc} \text{— —}\ e & E = \Delta_o \longrightarrow & \uparrow\text{— —}\ e \\ \uparrow\text{— — —}\ t_2 & = \dfrac{hc}{\lambda} & \text{— — —}\ t_2 \end{array}$$

d-to-*d* transition of 3*d* electron of $[Ti(H_2O)_6]^{3+}$

In this case, the energy difference is a direct measure of Δ_o. We can calculate Δ_o, the crystal-field splitting energy, for the octahedral $[Ti(H_2O)_6]^{3+}$ complex ion by relating it to the wavelength and energy of the absorbed photon. For one photon of 510-nm light,

$$E = \frac{hc}{\lambda} = \frac{(6.626 \times 10^{-34}\text{ J s})(2.998 \times 10^{8}\text{ m/s})}{510 \times 10^{-9}\text{ m}} = 3.90 \times 10^{-19}\text{J}$$

This translates into 235 kJ/mol, which is the splitting energy, Δ_o, the energy separating the t_2 and *e* sets of *d* atomic orbitals.

$$E = 3.90 \times 10^{-19}\text{ J} \times \frac{1\text{ kJ}}{10^3\text{ J}} \times \frac{6.022 \times 10^{23}}{1\text{ mol}} = 2.35 \times 10^2\text{ kJ/mol}$$

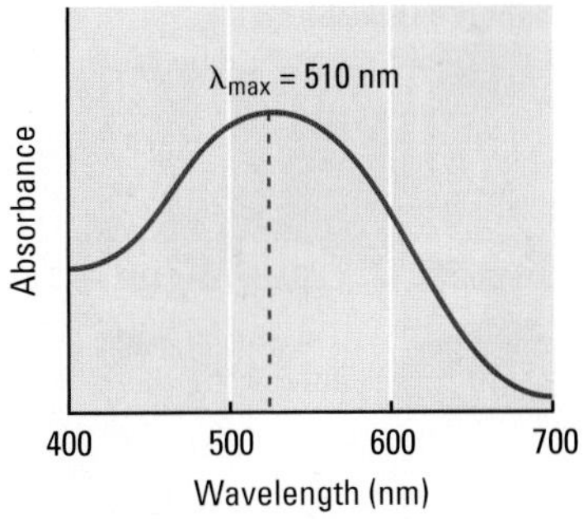
λ_{max} = 510 nm
Absorbance
400
500
600
700
Wavelength (nm)

Figure 20.22 The absorption spectrum of $[Ti(H_2O)_6]^{3+}$. The absorption of 510-nm wavelength light causes a *d*-to-*d* transition of the 3*d* electron of $[Ti(H_2O)_6]^{3+}$.

20-7d Color and the Spectrochemical Series

The color of a coordination complex depends on the splitting energy, which depends on the metal ion and the field strength of its ligands, as given by the spectrochemical series (← **Sec. 20-7b**). *Stronger-field ligands cause a larger splitting energy.* Consequently, complexes with such ligands absorb light at shorter wavelengths compared to those with weaker-field ligands, as seen vividly in Table 20.5 for a series of Co^{3+} complexes. Note from the table that as weaker-field ligands are substituted for NH_3 in the complex, the wavelength of the absorbed radiation becomes longer, indicating a lower splitting energy.

Table 20.5 The Color of a Coordination Complex Depends on the Ligands

Ligand	NH_3	NCS^-	H_2O	Cl^-
Complex	$[Co(NH_3)_6]^{3+}$	$[Co(NH_3)_5NCS]^{2+}$	$[Co(NH_3)_5H_2O]^{3+}$	$[Co(NH_3)_5Cl]^{2+}$
Color observed	Yellow	Orange	Red	Reddish purple
Color absorbed	Blue-violet	Blue-green	Green-blue	Green
Approximate wavelength absorbed (nm)	430	470	500	522

(a)
(b)
(c)
(d)

Cengage Learning/Charles D. Winters

CONCEPTUAL EXERCISE 20.14

Color and Electron Configuration

An aqueous solution of $[Cr(H_2O)_6]^{2+}$ is blue. Predict whether the 3*d* electrons of the central chromium ion are in a low-spin or high-spin configuration.

SUMMARY PROBLEM

Part I

Solid iron(II) sulfide, FeS, is roasted to form either $Fe_2O_3(s)$ or $Fe_3O_4(s)$.

(a) Write balanced equations for the roasting of FeS(s) to $Fe_2O_3(s)$ and to $Fe_3O_4(s)$.

(b) Use thermochemical data from Appendix J to calculate the enthalpy change for these reactions at 25 °C, given $\Delta_f H° = -101.671$ kJ/mol for FeS(s).

(c) Use thermochemical data from Appendix J to calculate the Gibbs free energy change for these reactions at 25 °C, given $S° = 82.81$ J mol^{-1} K^{-1} for FeS(s). Which reaction is more product-favored at this temperature?

(d) Calculate the Gibbs free energy change for the conversion of FeS to Fe_2O_3 and to Fe_3O_4 at 600 °C. Which reaction is more product-favored at this temperature?

Part II

Iron(III) forms a deep red coordination compound, $[Fe(acac)_3]$, with acetylacetonate ions, $acac^-$.

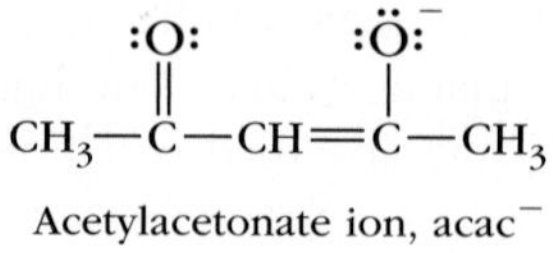

Acetylacetonate ion, $acac^-$

The acetylacetonate anion is a bidentate ligand that furnishes lone pairs from the C=O and C—O oxygens.

(a) Calculate the formal charge (← **Sec. 6-8**) on each atom in $acac^-$ to verify that one oxygen atom has a negative charge.

(b) Draw Lewis structures for all plausible resonance forms of $acac^-$. Explain why both oxygen atoms bond equally strongly to the iron(III) ion.

(c) Write a structural formula for $[Fe(acac)_3]$.

(d) Give the coordination number of Fe^{3+} in this compound.

(e) Account for the fact that there is no net charge on $[Fe(acac)_3]$.

(f) Explain why no counter ions are needed for this compound.

HAVING STUDIED THIS CHAPTER . . .

. . . you should be able to:

- Recognize the general properties of transition metals (Section 20-1).
- Write electron configurations and orbital box diagrams for transition metals and their ions (Section 20-1). End-of-chapter Questions 13, 84, 86
- Explain why most transition metals have two or more common oxidation states (Section 20-1). Questions 15, 23
- Explain trends in transition-metal atomic radii (Section 20-1). Question 27
- Describe how iron ore is processed into iron and then into steel (Section 20-2). Questions 29, 31
- Discuss how copper is extracted from its ores and purified (Section 20-3). Questions 33, 91
- Discuss the chemistry of gold and silver (Section 20-4). Questions 37, 39
- Discuss the chemistry of chromium (Section 20-5). Questions 41, 67
- Explain the coordinate covalent bonding of ligands in coordination compounds and complexes (Section 20-6). Questions 43, 45, 52, 56, 58
- Interpret the names and formulas of coordination complex ions and compounds (Section 20-6). Questions 59, 61
- Discuss isomerism in coordination compounds and complex ions (Section 20-6). Questions 63, 67, 94
- Give examples of coordination compounds and their uses (Section 20-6).
- Describe crystal-field theory and its applications to interpreting colors and magnetic properties of coordination compounds (Section 20-7). Questions 71, 76, 79, 81

KEY TERMS

bidentate ligand (Section 20-6c)
chelating ligand (20-6c)
coordination compound (20-6a)
coordination number (20-6c)
crystal-field splitting energy, Δ (20-7a)
***d*-to-*d* transition** (20-7c)
hexadentate ligand (20-6c)
high-spin complex (20-7b)
lanthanide contraction (20-1b)
ligands (20-6a)
low-spin complex (20-7b)
monodentate ligand (20-6c)
polydentate ligand (20-6c)
pyrometallurgy (20-2)
spectrochemical series (20-7b)
steel (20-2)

QUESTIONS FOR REVIEW AND THOUGHT

Red-numbered questions have short answers at the back of this book in Appendix M and fully worked solutions in the *Student Solutions Manual.*

Review Questions

These questions test vocabulary and simple concepts.

1. What is the primary reducing agent in the production of iron from its ores? Write a balanced chemical equation for this reduction process.
2. Why is limestone necessary in the blast furnace reduction of iron ore?
3. What is the difference between pig iron and cast iron?
4. Explain the purpose of each of these materials in the blast-furnace conversion of iron ore to iron.
   (a) Air (b) Limestone (c) Coke
5. Identify what is produced by each of these processes or operations.
   (a) Blast furnace
   (b) Basic oxygen process
   (c) Roasting
6. Identify a typical use for Cr, Cu, Fe, Au, and Ag.
7. Name three transition metals that are found "free" in nature.
8. What is the lanthanide contraction? Why does it occur?
9. In general, how do the atomic radii change across the first transition series (Period 4)?
10. What is the distinguishing chemical feature of a ligand?
11. Distinguish between
    (a) a monodentate and a bidentate ligand.
    (b) a *cis* and a *trans* isomer.
    (c) a coordination compound and a coordination complex ion.
    (d) a geometric and an optical isomer.
12. Define these words or phrases and give an example for each.
    (a) Coordination compound (b) Complex ion
    (c) Ligand (d) Chelate
    (e) Bidentate ligand

Topical Questions

These questions are keyed to the major topics in the chapter. Usually a question that is answered at the back of this book is paired with a similar one that is not.

Transition (*d*-block) Elements (Section 20-1)

13. Write electron configurations for the common oxidation states of
    (a) silver. (b) gold.
14. Write electron configurations for the 2+ ions of
    (a) iron. (b) copper. (c) chromium.
15. Of the two more common oxidation states of chromium in Figure 20.4, which are paramagnetic?
16. Which Period 4 transition-metal ions are isoelectronic with
    (a) Zn^{2+} (b) Mn^{2+}
    (c) Cr^{3+} (d) Fe^{3+}
17. Arrange these in order of decreasing strength as oxidizing agents: $Cr_2O_7^{2-}$ (in acid), Cr^{2+}, Cr^{3+}. Explain the trend.
18. Arrange these in order of decreasing strength as oxidizing agents: Mn^{2+}, MnO_4^-, MnO_2. Explain the trend.
19. Write a balanced equation to represent
    (a) the reduction of Fe_2O_3 with carbon monoxide in a blast furnace.
    (b) the production of hydrogen gas when hydrochloric acid reacts with an iron nail.
20. Write a balanced equation to represent
    (a) the roasting of Cu_2S to copper metal.
    (b) the reduction of Fe_2O_3 with aluminum.
21. Balance this redox reaction (in acidic solution).
    $$Fe(s) + NO_3^-(aq) \longrightarrow Fe^{3+}(aq) + NO_2(g)$$
22. Balance this redox reaction (in acidic solution).
    $$Cu(s) + NO_3^-(aq) \longrightarrow Cu^{2+}(aq) + NO_2(g)$$
23. Determine the oxidation state (oxidation number) of the transition metal in each compound.
    (a) V_2O_5 (b) $K_2Cr_2O_7$
    (c) MnO_2 (d) OsO_4
24. Determine the oxidation state of iron in $KFe[Fe(CN)_6]$. Explain your answer.
25. Explain why the atomic radii of transition metals vary less across a period than do main-group elements.
26. Explain why the radii of Period 5 and Period 6 transition elements are nearly the same.
27. Predict which element in the pair has the larger atomic radius.
    (a) Cu or Ag (b) Ti or Cr (c) W or Hg
28. Predict which element in the pair has the smaller atomic radius.
    (a) Pt or Ni (b) Ti or Ni (c) Zr or Rh

Iron and Steel (Section 20-2)

29. Describe how extracting iron from its ore is done by reduction, but producing steel is an oxidation process.
30. Describe the major differences between cast iron and steel.
31. Describe the production of steel by the basic oxygen process.

32. Identify the elements used, in addition to iron and carbon, to make
    (a) duriron.
    (b) high-speed steel.
    (c) permanent magnets.

Copper (Section 20-3)

33. Describe how the composition of brass differs from that of bronze.
34. Describe native copper. Why is it important?
35. Describe how blister copper is electrorefined.
36. Explain what a reducing sugar is in terms of aqueous copper chemistry.

Silver, Gold, and Chromium (Sections 20-4, 20-5)

37. Compare the action of concentrated nitric acid on silver with that on gold.
38. Describe how the percent of gold in its alloys is expressed in carats.
39. Describe why other metals are added to silver and gold.
40. (a) Identify the oxidizing agent in this equation:

$$Zn(s) + 2\ Au(CN)_2^-(aq) \longrightarrow 2\ Ag(s) + Zn(CN)_4^{2-}$$

(b) Identify the reducing agent in this equation:

$$4\ Au(s) + 8\ CN^-(aq) + O_2(g) + 2\ H_2O(\ell) \longrightarrow 4\ OH^-(aq) + 4\ Au(CN)_2^-(aq)$$

41. Give the oxidation state (number) of chromium in each of these compounds.
    (a) $FeCr_2O_4$ (b) Cr_2O_5
    (c) $K_2[CrF_6]$ (d) $[Cr(en)_3]Cl_2$
42. Use data from Appendix J to determine whether this reaction is product-favored.

$$2\ Al(s) + Cr_2O_3(s) \longrightarrow 2\ Cr(s) + Al_2O_3(s)$$

Coordination Compounds (Section 20-6)

43. In a complex ion, a central Cr^{3+} ion is bonded to two ammonia molecules, three water molecules, and a hydroxide ion.
    (a) Give the formula and charge of the complex ion.
    (b) Identify a single counter ion that could be used with the complex ion to form an uncharged compound.
44. In a complex ion, a central ruthenium ion, Ru(III), is bonded to six ammonia molecules.
    (a) Give the formula and net charge for this complex ion.
    (b) How many chloride ions are needed to balance the net charge on this complex ion?
    (c) Write the formula for the compound formed by this complex ion and the chloride ions that are not part of it.
45. Consider the complex ion $[Co(C_2O_4)_2Cl_2]^{3-}$.
    (a) Identify the ligands and their charges (if any).
    (b) Determine the charge on the central metal ion.
    (c) Determine the formula and charge of the complex ion in which the $C_2O_4^{2-}$ ions were replaced by NH_3 molecules.
46. Consider the complex ion $[Cr(NH_3)_2(H_2O)_2Br_2]^+$.
    (a) Identify the ligands and their charges (if any).
    (b) Determine the charge on the central metal ion.
    (c) Determine the formula of the sulfate salt of this cation.
47. Determine the charge of the central metal ion in each case.
    (a) $[Zn(H_2O)_3(OH)]^+$
    (b) $[Pt(NH_3)_3Cl_3]^+$
    (c) $[Cr(CN)_6]^{3-}$
48. For coordination compounds $Na_3[IrCl_6]$ and $[Mo(CO)_4]Br_2$, identify in each case
    (a) the ligands.
    (b) the central metal ion and its charge.
    (c) the formula and charge of the complex ion.
    (d) any ions not part of the complex ion.
49. Give the coordination number of the central metal ion in
    (a) $[Pt(en)_2]^{2+}$. (b) $[Cu(C_2O_4)_2]^{2-}$.
50. Give the coordination number of the central metal ion in
    (a) $[Ni(en)_2(NH_3)_2]^{2+}$. (b) $[Fe(en)(C_2O_4)Cl_2]^-$.
51. Write a structural formula for the coordination compound $[Cr(en)(NH_3)_2I_2]$, and give the coordination number for the central Cr^{2+} ion.
52. Give the formula of each of these coordination compounds formed with Pt^{2+}.
    (a) Two ammonia molecules and two bromide ions
    (b) One ethylenediamine molecule and two nitrite ions, NO_2^-
    (c) One chloride ion, one bromide ion, and two ammonia molecules
53. Give the charge on the central metal ion in each of these.
    (a) $[VCl_6]^{4-}$ (b) $[Sc(H_2O)_3Cl_3]$
    (c) $[Mn(NO)(CN)_5]^{3-}$ (d) $[Ni(en)_2(NH_3)_2]^{2+}$
54. Identify the coordination number of the metal ion in these coordination complexes.
    (a) $[FeCl_4]^-$ (b) $[PtBr_4]^{2-}$
    (c) $[Mn(en)_3]^{2+}$ (d) $[Cr(NH_3)_5H_2O]^{3+}$
55. Using structural formulas, show how the carbonate ion can be either a monodentate or bidentate ligand to a transition-metal cation.
56. Classify each ligand as monodentate, bidentate, and so on.
    (a) $(CH_3)_3P$
    (b) 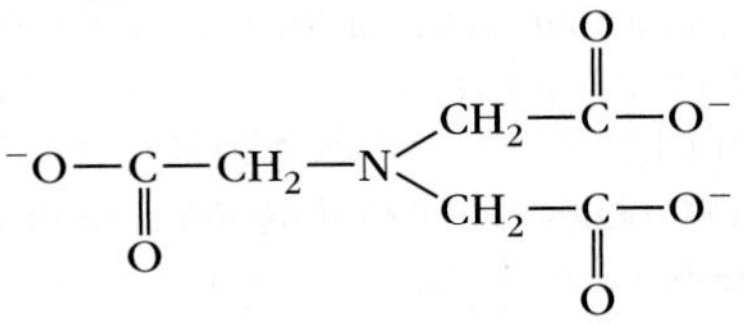
    (c) $H_2N{-}(CH_2)_2{-}NH{-}(CH_2)_2{-}NH_2$
    (d) H_2O
57. Which of these would be expected to be effective chelating agents? Explain your answer.
    (a) CH_3CH_2OH (b) $H_2N{-}(CH_2)_3{-}NH_2$
    (c) 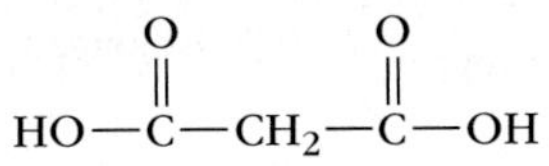 (d) PH_3

58. Give the formula of a simple ionic (noncoordination) compound analogous to $[Rh(en)_3]Cl_3$.

Naming Complex Ions and Coordination Compounds (Section 20-6)

59. Write the formula for
    (a) potassium diaquadioxalatocobaltate(III).
    (b) diamminetriaquahydroxochromium(II) nitrate.
    (c) ammonium tetrachlorocuprate(II).
60. Write the formula for
    (a) tetrachloroethylenediaminecobaltate(III).
    (b) triaquatrifluorocobalt(III).
61. Write the name corresponding to each formula.
    (a) $[MnCl_4]^{2-}$
    (b) $K_3[Fe(C_2O_4)_3]$
    (c) $[Pt(NH_3)_2(CN)_2]$
62. Write the name corresponding to each formula.
    (a) $[Fe(H_2O)_5(OH)]^{2+}$ (b) $[Mn(en)_2Cl_2]$

Geometry of Coordination Complexes (Section 20-6)

63. Sketch the geometry of
    (a) *cis*-$[Pt(H_2O)_2Cl_2]$. (b) *trans*-$[Cr(H_2O)_4Cl_2]^+$.
64. Sketch the geometry of
    (a) *cis*-$[Cu(H_2O)_2Br_4]^{2-}$. (b) *trans*-$[Ni(NH_3)_2(en)_2]^{2+}$.
65. The acetylacetonate ion (acac)$^-$

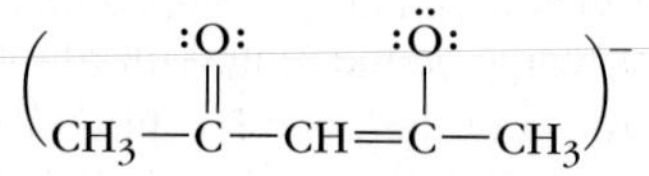

forms a complex with Co^{3+}. Sketch the geometry of $[Co(acac)_3]$.

66. The ligand 1,2-diaminocyclohexane

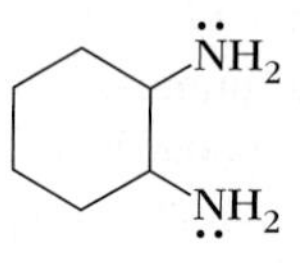

is abbreviated "dach." Sketch the geometry of *cis*-$[Pd(H_2O)_2(dach)]^{2+}$.

67. Which of these octahedral coordination complexes can exhibit geometric isomerism?
    (a) $[Cr(H_2O)_3Cl_3]$ (b) $[Cr(H_2O)_4Cl_2]^+$
68. Which of these octahedral coordination complexes can exhibit geometric isomerism?
    (a) $[Pt(H_2O)_2Cl_2Br_2]$ (b) $[Pt(H_2O)_2Cl_3Br]$
69. Draw the possible geometric isomers, if any.
    (a) $[Co(H_2O)_4Cl_2]^+$ (b) $[Pt(NH_3)Cl_3]^-$
    (c) $[Co(H_2O)_3Cl_3]$ (d) $[Co(en)_2(NH_3)Br]^{2+}$
70. Draw the possible geometric isomers, if any, of
    (a) $[Ni(NH_3)_4Cl_2]$.
    (b) $[Pt(NH_3)_2(SCN)Br]$. (The S in SCN is bonded to Pt^{2+}.)
    (c) $[Co(en)Cl_4]^-$.

Crystal-Field Theory and Magnetic Properties of Complex Ions (Section 20-7)

71. Draw the crystal-field splitting diagrams and put in the *d* electrons for these octahedral complexes. In those cases where they are possible, draw diagrams for both low-spin and high-spin cases.
    (a) $[Cr(H_2O)_6]^{2+}$ (b) $[Mn(H_2O)_6]^{2+}$
    (c) $[FeF_6]^{3-}$ (d) $[Cr(en)_3]^{3+}$
72. Using the spectrochemical series, predict the actual number of unpaired electrons in each complex in Question 71 for which the possibility of low-spin and high-spin forms exist.
73. Fe^{3+} forms octahedral complexes with NCS^- and with NO_2^- ligands. One complex displays a greater paramagnetism than the other.
    (a) Write the formula for each of these complex ions.
    (b) Use the spectrochemical series to predict whether the complex ions are high-spin or low-spin.
    (c) Identify which complex ion is more paramagnetic.
    (d) Draw the crystal-field splitting diagram, including *d* electrons, for each complex ion.
74. Explain why Cr^{2+} forms high-spin and low-spin octahedral complexes, but Cr^{3+} does not.
75. How many unpaired electrons are in the high-spin and low-spin octahedral complexes of Cr^{2+}?
76. Use crystal-field theory to explain why some Co^{3+} octahedral complexes are diamagnetic and others are paramagnetic.
77. Use crystal-field theory to explain why some octahedral Co^{2+} complexes are more paramagnetic than others.
78. Use crystal-field theory to explain why Cu^{2+} does not form high-spin and low-spin octahedral complexes.

Crystal-Field Theory and Color in Complex Ions (Section 20-7)

79. An aqueous solution of $[Rh(C_2O_4)_3]^{3-}$ is yellow. Predict the approximate wavelength and predominant color of light absorbed by the complex.
80. An aqueous solution of $[Ni(NH_3)_6]^{2+}$ is purple. Predict the approximate wavelength and predominant color of light absorbed by the complex.
81. As discussed in Section 20-7c, an aqueous solution of $[Ti(H_2O)_6]^{3+}$ is purple (red-violet). Predict how the value of Δ_o would change if all H_2O ligands were replaced by CN^- ligands; by Cl^- ligands. How would the color change in each case?
82. An aqueous solution of $[Rh(C_2O_4)_3]^{3-}$ is yellow; that of $[Rh(en)_3]^{3+}$ is a different color. Oxalate is to the right of en in the spectrochemical series. Predict what the change in color likely will be from $[Rh(C_2O_4)_3]^{3-}$ to $[Rh(en)_3]^{3+}$. In which direction does the absorbed wavelength change?

83. A solution of a complex ion absorbs visible light at a wavelength of 540 nm.
   (a) What is the color of the solution?
   (b) Calculate the energy of an absorbed photon in joules and in kJ/mol.

General Questions

These questions are not explicitly keyed to chapter topics; many require integration of several concepts.

84. Give the electron configuration of
   (a) Cr^{2+}. (b) Zn^{2+}. (c) Co^{2+}. (d) Mn^{4+}.
85. Give the electron configuration of
   (a) Ti^{3+}. (b) V^{2+}. (c) Ni^{3+}. (d) Cu^{+}.
86. Write a partial atomic orbital box diagram and determine the number of unpaired electrons for each species in Question 84.
87. Write a partial atomic orbital box diagram and determine the number of unpaired electrons for each species in Question 85.
88. Assuming 100% recovery of the metal, which would yield the greater mass of copper?
   (a) One kilogram of an ore containing 3.60 mass % azurite, $Cu(OH)_2 \cdot 2\ CuCO_3$
   (b) One kilogram of an ore containing 4.95 mass % chalcopyrite, $CuFeS_2$
89. Calculate the mass of copper that is electroplated from a $CuSO_4$ solution using an electric current of 2.50 A for 5.00 h. Assume 100% efficiency.
90. Copper metal is obtained directly by roasting covellite, CuS.
   (a) Write a balanced equation for this process.
   (b) Assume that the roasting is 90.0% efficient. Calculate how many tons of SO_2 are released into the air by roasting 500. tons of covellite.
91. Determine what mass of SO_2 is produced when 1.0 ton of chalcocite, Cu_2S, is roasted to Cu_2O.
92. Determine the coordination number of the central metal ion in
   (a) $[Ni(en)Cl_2]$. (b) $[Mo(CO)_4Br_2]$.
   (c) $[Cd(CN)_4]^{2-}$. (d) $[Co(CN)_5(OH)]^{3-}$.
93. Determine the coordination number of the central metal ion in
   (a) $[Pt(NH_3)_2Br_2]$. (b) $[Fe(CN)_6]^{3-}$.
   (c) $[Ti(H_2O)Cl_5]^{2-}$. (d) $[Mn(C_2O_4)_3]^{4-}$.
94. Draw structures for as many octahedral complexes as you can for the formula $Co(NH_3)_4Cl_2Br$.
95. Draw structures for all possible octahedral complexes of Co^{3+} using only ethylenediamine and/or Cl^- as ligands.
96. In your own words explain why
   (a) $H_2N—(CH_2)_3—NH_2$ is a bidentate ligand.
   (b) AgCl dissolves in NH_3.
   (c) there are no geometric isomers of tetrahedral complexes.
97. Determine whether each statement is true or false. If it is false, correct the statement.
   (a) The coordination number of the Fe^{3+} ion in $[Fe(H_2O)_4(C_2O_4)]^+$ is five.
   (b) Cu^+ has two unpaired electrons.
   (c) The net charge of a coordination complex of Cr^{3+} with two NH_3, one en, and two H_2O is 2+.
98. Determine whether each statement is true or false. If it is false, correct the statement.
   (a) In $[Pt(NH_3)_4Cl_4]$, platinum has a 4+ charge and a coordination number of six.
   (b) In general, Cu^{2+} is more stable than Cu^+ in aqueous solutions.
99. The metal ion in $[Pt(NH_3)_2(C_2O_4)]$ is surrounded by a square planar array of coordinating atoms.
   (a) Give the oxidation number of the central metal ion.
   (b) Draw the structural formula of this coordination compound.

Applying Concepts

These questions test conceptual learning.

100. Chromium(III) forms three different compounds with water and chloride ions, all of which have the same composition: 19.51% Cr, 39.92% Cl, and 40.57% water. One of the compounds is violet and dissolves in water to give a complex ion with a 3+ charge plus three chloride ions. All three chloride ions precipitate immediately as AgCl when $AgNO_3$ is added to the solution. Draw the structural formula of this complex ion.
101. Iron nails are put into Fe^{2+} aqueous solutions to reduce any Fe^{3+} that forms back to Fe^{2+}. Write a balanced chemical equation for this preventative reaction.
102. Use VSEPR theory to predict the shape and bond angles around chromium in
   (a) chromate ions. (b) dichromate ions.
103. The structure of cyclam is

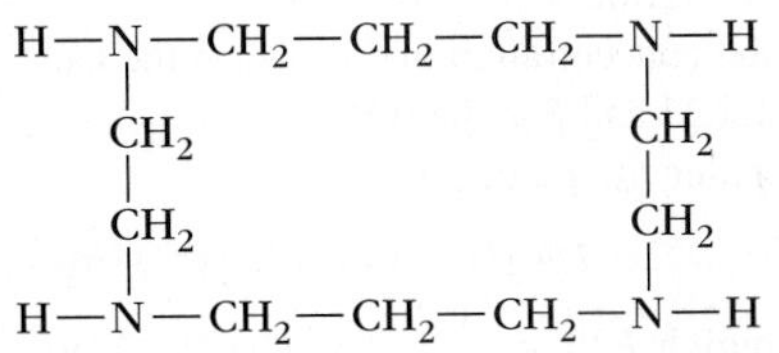
H—N—CH2—CH2—CH2—N—H
CH2
CH2
CH2
CH2
H—N—CH2—CH2—CH2—N—H

   Cyclam can act as a ligand. How many coordinate covalent bonds can one cyclam molecule form with a central metal ion?
104. The compound 1,10-phenanthroline is a chelating agent used in analytical chemistry. Its isomer 4,7-phenanthroline is not. Use these structural formulas to explain this difference.

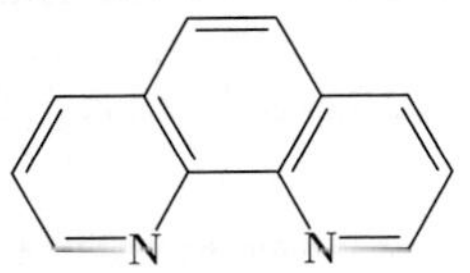
N
N

1,10-phenanthroline

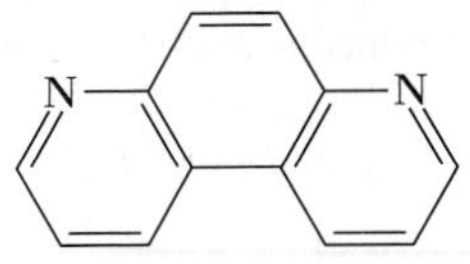
N
N

4,7-phenanthroline

105. Two different isomers are known with the formula $[Pt(py)_2Cl_2]$, where py represents pyridine, an uncharged monodentate ligand in which an N atom bonds to the metal ion. There is, however, only one structure known for $[Pt(phen)Cl_2]$, where phen represents 1,10-phenanthroline, an uncharged bidentate ligand (Question 104). Draw the structural formulas of all three molecules and explain why there are isomers in one case, but not the other.

106. An electrochemical cell is made by immersing a strip of chromium into a 1.0-M solution of Cr^{3+} and a strip of gold into a 1.0-M solution of Au^{3+}. The half-cells are connected by a salt bridge. A wire and light bulb complete the circuit.
   (a) Write the balanced chemical equation for the reaction that is product-favored.
   (b) Calculate the cell potential.
   (c) Draw a sketch of the cell and indicate the anode, cathode, and direction of electron flow.

107. Repeat the directions for Question 106 using a cell constructed of a strip of nickel immersed in a 1.0-M Ni^{2+} solution and a strip of silver dipping into a 1.0-M Ag^+ solution.

108. Calculate $\Delta_r G°$ for the reduction of Fe_2O_3 with CO gas at 25 °C and at 1000 °C. What application does this have to the conversion of iron ore to iron in a blast furnace?

109. To determine the percent iron in an ore, a 1.500-g sample of the ore containing Fe^{2+} is titrated to the equivalence point with 18.6 mL of 0.05012-M $KMnO_4$. The products of the titration are Fe^{3+} and Mn^{2+}. Calculate the weight percent of iron in the ore.

110. Consider the reaction

$$2\ Cu^+(aq) \longrightarrow Cu^{2+}(aq) + Cu(s)$$

for which $E°_{cell} = +0.37$ V. Use the Nernst equation to calculate
   (a) E when the Cu^{2+} concentration is equal to the Cu^+ concentration $= 1 \times 10^{-4}$ M.
   (b) the concentration of Cu^+ when the Cu^{2+} concentration = 1.0 M and $E = 0.00$ V.

111. Consider the reaction

$$2\ Ag^+(aq) \longrightarrow Ag(s) + Ag^{2+}(aq)$$

for which $E°_{cell} = -1.18$ V. Use the Nernst equation to calculate
   (a) E when the Ag^+ concentration $= 1 \times 10^{-4}$ M, which is five times the concentration of Ag^{2+}.
   (b) the concentration of Ag^{2+} when the Ag^+ concentration = 1.0 M and $E = 0.00$ V.

112. Use the Nernst equation to calculate E_{cell} for $Cr_2O_7^{2-}$ oxidation of Cl^- in 6.0-M H^+ when the concentration of $Cr_2O_7^{2-}$ = concentration of Cr^{3+} = 0.10 M, and all other concentrations = 1.0 M.

113. If 1.00 mol of each compound is dissolved in a separate sample of water sufficient to dissolve the compound, how many moles of ions are present in each solution?
   (a) $[Pt(en)Cl_2]$
   (b) $Na[Cr(en)_2(SO_4)_2]$
   (c) $K_3[Au(CN)_4]$
   (d) $[Ni(H_2O)_2(NH_3)_4]Cl_2$

114. For each of the compounds in Question 113, state which it would most likely resemble in colligative properties and conductivity: $(NH_2)_2CO$ (urea), KCl, K_2SO_4, or K_3PO_4.

115. In aqueous solution, $[Cr(NH_3)_6]Cl_3$ is yellow, but aqueous $[Cr(NH_3)_5Cl]Cl_2$ is purple. Explain the difference in colors.

116. Early coordination chemists relied on close experimental observation to determine the formulas of coordination compounds. They found, for example, that aqueous $BaCl_2$ did not cause precipitation when added to a solution of a Co^{3+}-containing coordination compound, but precipitation occurred when aqueous silver nitrate was added to a solution of the coordination compound. The coordination compound was known to contain one Co^{3+} ion, one sulfate ion, one chloride ion, and four ammonia molecules. Write the structural formula of the coordination compound that is consistent with the experimental results.

More Challenging Questions

These questions require more thought and integrate several concepts.

117. The bidentate oxalate ion, $C_2O_4^{2-}$, forms octahedral complexes with Fe^{3+} and Ru^{3+} ions.
   (a) Write the structural formula for each complex.
   (b) The ruthenium complex is low-spin; the iron complex is high-spin. Write the *d*-orbital splitting diagram for each metal ion.
   (c) Which complex has the higher Δ_o? Explain your answer.

118. Analysis of a coordination compound gives these results: 22.0% Co, 31.4% N, 6.78% H, and 39.8% Cl. One mole of the compound dissociates in water to form 4 mol ions.
   (a) Determine the formula of the compound.
   (b) Write the equation for its dissociation in water.

119. A chemist synthesizes two coordination compounds. One compound decomposes at 210 °C, the other at 240 °C. When analyzed, the compounds give the same mass percent data: 52.6% Pt, 7.6% N, 1.63% H, and 38.2% Cl. Both compounds contain a 4+ central metal ion.
   (a) Determine the simplest formula of the compounds.
   (b) Draw structural formulas for the complexes present.

120. A coordination compound has the simplest formula $PtN_2H_6Cl_2$ with a molar mass of about 600 g/mol. It contains a complex cation and a complex anion. Draw its structural formula.

121. The glycinate ion (gly) is $H_2NCH_2CO_2^-$. It can act as a ligand coordinating through the nitrogen and one of the oxygens. Using $\widehat{NO}$ to represent glycinate ion, draw structural formulas for four stereoisomers of $[Co(gly)_3]$.

122. Five-coordinate coordination complexes are known, including $[CuCl_5]^{3-}$ and $[Ni(CN)_5]^{3-}$. Write the structural formulas and identify a plausible geometry for these complexes.
123. Predict the number of unpaired electrons in a square planar transition metal ion with seven *d* electrons.
124. A coordination compound has the empirical formula $Fe(H_2O)_4(CN)_2$. Its paramagnetism is the equivalent of 2.67 unpaired electrons per Fe ion. Explain how this is possible.
125. Two different compounds are known with the formula $Pd(py)_2Cl_2$, but there is only one compound with the formula $Zn(py)_2Cl_2$. The symbol py is for pyridine, a monodentate ligand. Explain the differences in the Pd and Zn compounds.
126. An octahedral coordination complex ion is formed by the combination of an Fe^{3+} ion and det ligands (det is $H_2NCH_2CH_2NHCH_2CH_2NH_2$). Write a structural formula for the complex ion.

APPENDIX A Problem Solving and Mathematical Operations

In this book we provide many illustrations of problem solving and many problems for practice. Some are numerical problems that must be solved by mathematical calculations. Others are conceptual problems that must be solved by applying an understanding of the principles of chemistry. Often, it is necessary to use chemical concepts to relate what we know about matter at the nanoscale to the properties of matter at the macroscale. The problems throughout this book are representative of the kinds of problems that chemists and other scientists must regularly solve to pursue their goals, although our problems are often not as difficult as those encountered in the real world.

Problem solving is not a simple skill that can be mastered with a few hours of study or practice. Because there are many different kinds of problems and many different kinds of people who are problem solvers, no hard and fast rules are available that are guaranteed to lead you to solutions. The general guidelines presented in this appendix are, however, helpful in getting you started on any kind of problem and in checking whether your answers are correct. The problem-solving skills you develop in a chemistry course can later be applied to difficult and important problems that may arise in your profession, your personal life, or the society in which you live.

In getting a clear picture of a problem and asking appropriate questions regarding the problem, you need to keep in mind all the principles of chemistry and other subjects that you think may apply. In many real-life problems, not enough information is available for you to arrive at an unambiguous solution; in such cases, try to look up or estimate what is needed and then forge ahead, noting assumptions you have made. Often the hardest part is deciding which principle or idea is most likely to help solve the problem and what information is needed. To some degree this can be a matter of luck or chance. Nevertheless, in the words of Louis Pasteur, "In the field of observation chance only favors those minds which have been prepared." The more practice you have had, and the more principles and facts you can keep in mind, the more likely you are to be able to solve the problems that you face.

A-1 General Problem-Solving Strategies

"The mere formulation of a problem is often far more essential than its solution, which may be merely a matter of mathematical or experimental skill. To raise new questions, new possibilities, to regard old problems from a new angle, requires creative imagination and marks real advances in science."

—Albert Einstein

This book advocates a four-step strategy for approaching and solving problems.

Step 1: **Analyze the problem.** Carefully review the information contained in the problem. What is the problem asking you to find? What key principles are involved? What known information is necessary for solving the problem and what information is there just to place the question in context? Organize the information to see what is necessary and to see the relationships among the known data. Try writing the information in an organized way. If the information is numerical, be sure to include proper units. Can you picture the situation under consideration? Try sketching it and including any relevant dimensions in the sketch.

Step 2: **Plan a solution.** Have you solved a problem of this type before? If you recognize the new problem as similar to ones you know how to solve, you can use the same method that worked before. Try reasoning backward from the units of what is being sought. What data are needed to find an answer in those units?

Can the problem be broken down into smaller pieces, each of which can be solved separately to produce information that can be assembled to solve the entire problem? When a problem can be divided into simpler problems, it often helps to write down a plan that lists the simpler problems and the order in which those problems must be put together to arrive at an overall solution. Many major problems in chemical research have to be solved in this way. In problems in this book we have mostly provided the needed numerical data, but in the laboratory, the first aspect of solving a problem is often devising experiments to gather the data or searching databases to find needed information.

If you are still unsure about what to do, do something anyway. It may not be the right thing to do, but as you work on it, the way to solve the problem may become apparent, or you may see what is wrong with your initial approach, thereby making clearer what a good plan would be.

Step 3: **Execute the plan.** Carefully write down each step of a mathematical problem, being sure to keep track of the units. Do the units cancel to give you the answer in the desired units? Don't skip steps. Don't do any except the simplest steps in your head. Once you've written down the steps for a mathematical problem, check what you've written—is it all correct? Students often say they got a problem wrong because they "made a stupid mistake." Teachers—and textbook authors—make mistakes, too. These errors usually arise because they don't take the time to write down the steps of a problem clearly and correctly. In solving a mathematical problem, remember to apply the principles of dimensional analysis and significant figures. Dimensional analysis was introduced in Section 1-5; it is described in more detail in Appendix A-2. Section 1-5b introduced significant figures and this topic is developed fully in Appendix A-3.

Step 4: ☑ **Check the answer to see whether it is reasonable.** As a final check of your solution to any problem, ask yourself whether the answer is reasonable: Are the units of a numerical answer correct? Is a numerical answer of about the right size? Don't just copy a result from your calculator without thinking about whether it makes sense.

Suppose you have been asked to convert 100. yards to a distance in meters. Using dimensional analysis and some well-known factors for converting from the English system to the metric system, you could write

$$\text{Distance} = 100.\text{ yd} \times \frac{3\text{ ft}}{1\text{ yd}} \times \frac{12\text{ in}}{1\text{ ft}} \times \frac{2.54\text{ cm}}{1\text{ in}} \times \frac{1\text{ m}}{100\text{ cm}} = 91.4\text{ m}$$

To check that a distance of 91.4 m is about right, recall that a yard is a little shorter than a meter. Therefore, 100 yd should be a little less than 100 m. If you had mistakenly divided instead of multiplied by 3 ft/yd in the first step, your final answer would have been a little more than 10 m. This is equivalent to only about 30 ft, and you probably know a 100-yd football field is longer than that.

Here is an example that shows how this problem-solving strategy can be applied to a simple problem. The problem has been taken verbatim from Chapter 1 (← **Sec. 1-5c**).

PROBLEM-SOLVING EXAMPLE A.1

Density

In an old movie, thieves are shown running off with pieces of gold bullion that are about a foot long and have a square cross section of about six inches. The volume of each piece of gold is 7000 mL. Is what the movie shows physically possible? [Hint: calculate the mass of gold and express the result in pounds (lb). 1 lb = 454 g.]

Result Probably not; 1.4×10^5 g; 300 lb

Strategy and Explanation A good approach to problem solving is to (1) analyze the problem, (2) plan a solution, (3) execute the plan, and (4) check your result to see whether it is reasonable. (These four steps are described in more detail in Appendix A-1.)

Analyze the problem.

Step 1: *Analyze the problem.* You are asked to calculate the mass of the gold, and you know the volume is 7000 mL (one significant figure).

Plan a solution.

Step 2: *Plan a solution.* Density relates mass and volume and is the appropriate proportionality factor, so look up the density in a table. Mass is proportional to volume, so the volume either has to be multiplied by the density or divided by the density. Use the units to decide which. Use the information that 1 lb = 454 g to obtain a conversion factor for the units.

Execute the plan.

Step 3: *Execute the plan.* According to Table 1.1, (← **Sec. 1-4d**), the density of gold is 19.32 g/mL. Setting up the calculation so that the unit (milliliter) cancels gives

$$7000\text{ mL} \times \frac{19.32\text{ g}}{1\text{ mL}} = 1.35 \times 10^5\text{ g}$$

This can be converted to pounds

$$1.35 \times 10^5 \not{g} \times \frac{1 \text{ lb}}{454 \not{g}} = 297 \text{ lb, which rounds to 300 lb}$$

Notice that the result is expressed to one significant figure, because the volume was given to only one significant figure and only multiplications and divisions were done. The intermediate result, 1.35×10^5 g, was expressed to more significant figures because rounding should only be done at the end of a calculation.

☑ **Reasonable Result Check** The units are mass units, so they are reasonable. Gold is nearly 20 times denser than water. A liter (1000 mL) of water is about a quart, and a quart of water (two pints) weighs about two pounds. Seven liters (7000 mL) of water should weigh 14 lb, and 20 times 14 gives 280 lb, which rounds to 300 lb, so 300 lb is reasonable. The movie is not—few people could run while carrying a 300-lb object!

Check that the result is reasonable.

A-2 Numbers, Units, and Quantities

Many scientific problems require you to use mathematics to calculate a result or draw a conclusion. Therefore, knowledge of mathematics and its application to problem solving is important. However, one aspect of scientific calculations is often absent from pure mathematical work: Science deals with *measurements* in which an unknown quantity is compared with a standard or unit of measure. For example, using a balance to determine the mass of an object involves comparing the object's mass with standard masses, usually in multiples or fractions of one gram; the result is reported as some number of grams, say 4.356 g. *Both the number and the unit are important.* If the result had been 123.5 g, this would clearly be different, but a result of 4.356 oz (ounces) would also be different, because the unit "ounce" is different from the unit "gram." A *result that describes the quantitative measurement of a property,* such as 4.356 g, is called a **quantity** (or **physical quantity**), and it consists of a number and a unit. Chemical problem solving requires calculating with quantities. Notice that whether a quantity is large or small depends on the units as well as the number; the two quantities 123.5 g and 4.356 oz, for example, represent the *same* mass.

A quantity is always treated as though the number and the units are multiplied; that is, 4.356 g can be handled mathematically as $4.356 \times$ g. Using this simple rule, you will see that calculations involving quantities follow the normal rules of algebra and arithmetic: $5 \text{ g} + 7 \text{ g} = 5 \times \text{g} + 7 \times \text{g} = (5 + 7) \times \text{g} = 12 \text{ g}$; or $6 \text{ g} \div 2 \text{ g} = (6 \not{g})/(2 \not{g}) = 3$. (Notice that in the second calculation the unit g appears in both the numerator and the denominator and cancels, leaving a pure number, 3.) Treating units as algebraic entities has the advantage that *if a calculation is set up correctly, the units will cancel or multiply together so that the final result has appropriate units.* For example, if you measured the size of a sheet of paper and found it to be 8.5 in by 11 in, the area *A* of the sheet could be calculated as area = length × width = 11 in × 8.5 in = 94 in^2, or 94 square inches. If a calculation is set up incorrectly, the units of the result will be inappropriate. *Using units to check whether a calculation has been properly set up* is called **dimensional analysis** (← **Sec. 1-5**).

This idea of using algebra on units as well as numbers is useful in all kinds of situations. For example, suppose you are having a party for some friends who like pizza. A large pizza consists of 12 slices and costs \$15.75. You expect to need 36 slices of pizza and want to know how much you will have to spend. A strategy for solving the problem is first to figure out how many pizzas you need and then to figure out the cost in dollars. This solution could be diagrammed as

$$\text{Slices} \xrightarrow[\text{step 1}]{\text{slices per pizza}} \text{pizzas} \xrightarrow[\text{step 2}]{\text{dollars per pizza}} \text{dollars}$$

Step 1: Find the number of pizzas required by dividing the number of slices per pizza into the number of slices, thus converting "units" of slices to "units" of pizzas:

$$\text{Number of pizzas} = 36 \text{ slices}\left(\frac{1 \text{ pizza}}{12 \text{ slices}}\right) = 3 \text{ pizzas}$$

If you had multiplied the number of slices times the number of slices per pizza, the result would have been labeled pizzas × slices^2, which does not make sense. In other words, the labels indicate whether multiplication or division is appropriate.

Strictly speaking, slices and pizzas are not units in the same sense that a gram is a unit. Nevertheless, labeling things this way will often help you keep in mind what a number refers to—pizzas, slices, or dollars in this case.

Step 2: Find the total cost by multiplying the cost per pizza by the number of pizzas needed, thus converting "units" of pizzas to "units" of dollars:

$$\text{Total cost} = 3 \text{ pizzas}\left(\frac{\$15.75}{1 \text{ pizza}}\right) = \$47.25$$

Notice that in each step you have multiplied by a factor that allowed the initial units to cancel algebraically, giving the answer in the desired units. A factor such as (1 pizza/12 slices) or ($15.75/pizza) is referred to as a proportionality factor (← **Sec. 1-5c**). This name **proportionality factor** indicates that it is *a factor that comes from a proportion.* For instance, in the pizza problem you could set up the proportion

$$\frac{x \text{ pizzas}}{36 \text{ slices}} = \frac{1 \text{ pizza}}{12 \text{ slices}} \quad \text{or} \quad x \text{ pizzas} = 36 \text{ slices}\left(\frac{1 \text{ pizza}}{12 \text{ slices}}\right) = 3 \text{ pizzas}$$

A proportionality factor such as (1 pizza/12 slices) is also called a *conversion factor,* which indicates that it converts one kind of unit or label to another; in this case the label "slices" is converted to the label "pizzas."

Many everyday scientific problems involve proportionality. For example, the bigger the volume of a solid or liquid substance is, the bigger its mass is. When the volume is zero, the mass is also zero. These facts indicate that mass, m, is directly proportional to volume, V, or, symbolically,

$$m \propto V$$

where the symbol $\propto$ means "is proportional to." Whenever a proportion is expressed this way, it can also be expressed as an equality by using a proportionality constant; for example,

$$m = d \times V$$

In this case, the proportionality constant, d, is called the density of the substance. This equation embodies the definition of density as mass per unit volume, because it can be rearranged algebraically to

$$d = \frac{m}{V}$$

As with any algebraic equation involving three variables, it is possible to calculate any one of the three quantities m, V, or d, provided the other two are known. If density is wanted, simply use the definition: Density is mass per unit volume. If mass or volume is to be calculated, the density can be used as a proportionality factor.

Suppose that you are going to buy a ton of gravel and want to know how big a bin you will need to store it. You know the mass of gravel and want to find the volume of the bin; this implies that density will be useful. If the gravel is primarily limestone, you can assume that its density is about the same as for limestone and look it up. Limestone has the chemical formula $CaCO_3$ and its density is 2.7 kg/L. However, these mass units are different from the units for mass of gravel—namely, tons. Therefore you need to recall or look up the mass of 1 ton (exactly 2000 pounds [lb]) and the fact that there are 2.20 pounds per kilogram. This provides enough information to calculate the volume needed. Here is a "roadmap" plan for the calculation:

$$\text{Mass of gravel in tons} \xrightarrow[\text{step 1}]{\text{change units}} \text{mass of gravel in kilograms} \xrightarrow[\text{step 2}]{\text{density}} \text{volume of bin}$$

Step 1: Figure out how many kilograms of gravel are in a ton.

$$m_{\text{gravel}} = 1 \text{ ton} = 2000 \text{ lb} = 2000 \text{ lb}\left(\frac{1 \text{ kg}}{2.20 \text{ lb}}\right) = 909 \text{ kg}$$

The fact that there are 2.20 pounds per kilogram implies two proportionality factors: (2.20 lb/1 kg) and (1 kg/2.20 lb). The latter was used because it results in appropriate cancellation of units.

Step 2: Use the density to calculate the volume of 909 kg of gravel.

$$V_{\text{gravel}} = \frac{m_{\text{gravel}}}{d_{\text{gravel}}} = \frac{909 \text{ kg}}{2.7 \text{ kg/L}} = 909 \text{ kg}\left(\frac{1 \text{ L}}{2.7 \text{ kg}}\right) = 340 \text{ L}$$

In this step we used the definition of density, solved algebraically for volume, substituted the two known quantities into the equation, and calculated the result. However, it is quicker simply to remember that mass and volume are related by a proportionality factor called density and to use the units of the quantities to decide whether to multiply or divide by that factor. In this case we divided mass by density because the units kilograms canceled, leaving a result in liters, which is a unit of volume.

Also, it is quicker and more accurate to solve a problem like this one by using a single setup. Then all the calculations can be done at once, and no intermediate results need to be written down. The "roadmap" plan presented previously can serve as a guide to the single-setup solution, which looks like this:

$$V_{\text{gravel}} = 1\text{ ton}\left(\frac{2000\text{ lb}}{1\text{ ton}}\right)\left(\frac{1\text{ kg}}{2.20\text{ lb}}\right)\left(\frac{1\text{ L}}{2.7\text{ kg}}\right) = 340\text{ L}$$

To calculate the result, then, you would enter 2000 on your calculator, divide by 2.20, and divide by 2.7. Such a setup makes it easy to see what to multiply and divide by, and the calculation goes more quickly when it can be entered into a calculator all at once.

The liter is not the most convenient volume unit for this problem, however, because it does not relate well to what we want to find out—how big a bin to make. A liter is about the same volume as a quart, but whether you are familiar with liters, quarts, or both, 300 of them is not easy to visualize. Let's convert liters to something we can understand better. A liter is a volume equal to a cube one tenth of a meter (1 dm) on a side; that is, $1\text{ L} = 1\text{ dm}^3$. Consequently,

Notice that because the unit being converted (dm^3) was raised to the third power, the conversion factor $(1\text{ m}/10\text{ dm})^3$ also had to be raised to the third power.

$$340\text{ L} = 340\text{ L}\left(\frac{1\text{ dm}^3}{1\text{ L}}\right)\left(\frac{1\text{ m}}{10\text{ dm}}\right)^3 = 340\text{ dm}^3\left(\frac{1\text{ m}^3}{1000\text{ dm}^3}\right) = 0.34\text{ m}^3$$

Thus, the bin would need to have a volume of about one third of a cubic meter; that is, it could be a meter wide, a meter long, and about a third of a meter high and it would hold the ton of gravel.

One more thing should be noted about this example. We don't need to know the volume of the bin very precisely, because being off a bit will make very little difference; it might mean getting a little too much wood to build the bin or not making the bin quite big enough and having a little gravel spill out, but this isn't a big deal. In other cases, such as calculating the quantity of fuel needed to get a space shuttle into orbit, being off by a few percent could be a life-or-death matter. Because it is important to know how precise data are and to be able to evaluate how important precision is, scientific results usually indicate their precision. The simplest way to do so is by means of significant figures.

A-3 Precision, Accuracy, and Significant Figures

As mentioned in the preceding section, measurement consists of comparing the item to be measured with a unit; that is, of counting how many units correspond with the item. Usually a measuring device is calibrated with some kind of scale that indicates the units. Then, the object being measured is compared with the scale. In reading a scale, you should find the smallest scale division, determine how many of those divisions correspond with the object, and then estimate to the best of your ability between those smallest scale units. For example, consider rulers A and B shown in Figure A.1. Ruler A is marked every centimeter. If you were measuring the length of a metal rod with ruler A, the smallest scale unit is a whole centimeter. You could estimate between the centimeter graduations, but your estimate would not be exact. Therefore you might report your measurement of the length of the rod in Figure A.1 as 7.7 cm. This number has two significant figures. The **significant figures** in a reported result include *all digits that are known exactly plus one more digit that is estimated.* In this case, where the length of the rod is between 7 and 8 cm, the first figure (7) is known exactly, but the second figure (.7) has been estimated; thus, there are two significant figures (sig figs).

The **precision** of a measurement *indicates how well several determinations of the same measurement agree.* Some devices can make more precise measurements than others. Compare ruler A with ruler B, which is marked every millimeter. Clearly you could measure at least to the nearest millimeter with ruler B, and probably you could estimate to the nearest 0.2 mm or so. This would give a result such as 76.7 mm or 7.67 cm, each of which has three sig figs. Thus, ruler B allows

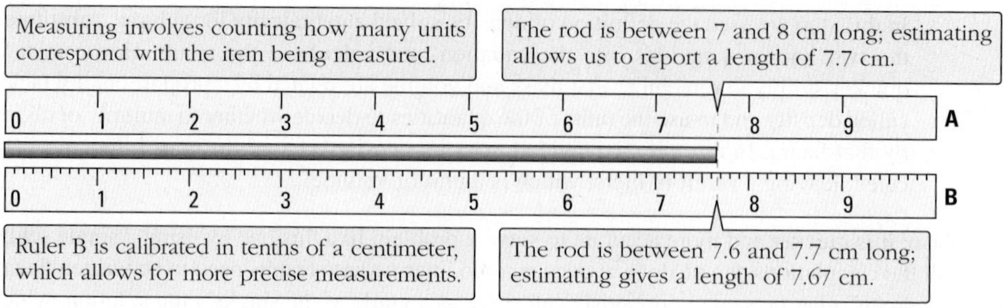
Measuring involves counting how many units correspond with the item being measured.
The rod is between 7 and 8 cm long; estimating allows us to report a length of 7.7 cm.
0 1 2 3 4 5 6 7 8 9
A
0 1 2 3 4 5 6 7 8 9
B
Ruler B is calibrated in tenths of a centimeter, which allows for more precise measurements.
The rod is between 7.6 and 7.7 cm long; estimating gives a length of 7.67 cm.

Figure A.1 Two 10-cm rulers measuring the same metal rod.

more precise measurement than ruler A. If several different people used ruler B to measure the length of an iron bar, their measurements would agree more closely than if they used ruler A.

Precision is also illustrated by the results of throwing darts at a bull's-eye. In Figure A.2a, the darts are scattered all over the board; the dart thrower was apparently not very skillful (or threw the darts from a long distance away from the board), and the precision of their placement on the board is low. This is analogous to the results that would be obtained by several people using ruler A. In part (b), the darts are all clustered together, indicating much better reproducibility on the part of the thrower—that is, greater precision. This is analogous to measurements made using ruler B.

Notice also that in Figure A.2b every dart has come very close to the bull's-eye; we describe this result by saying that the thrower has been quite **accurate**—*the average* of all throws *is very close to the accepted value*, the bull's-eye. Figure A.2c illustrates that it is possible to be precise without being accurate—the dart thrower has consistently missed the bull's-eye, but all darts are clustered very precisely around the wrong point on the board. This third case is like an experiment with some flaw (either in its design or in a measuring device) that causes all results to differ from the correct value by the same quantity. An example is measuring length with ruler C, on which the scale is incorrectly labeled. The precision (reproducibility) of measurements with ruler C might be just as good as with ruler B, but the accuracy would be poor because all items longer than 1 cm would be off by a centimeter.

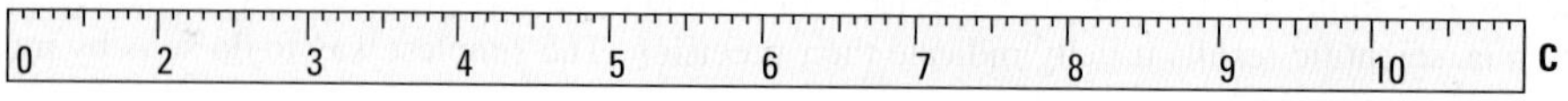
0 2 3 4 5 6 7 8 9 10
C

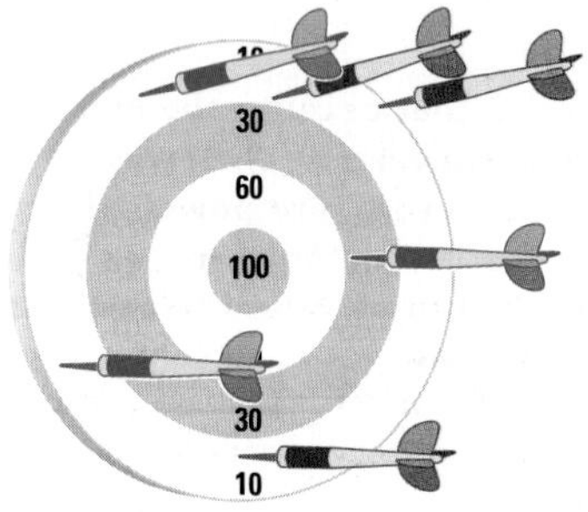
30
60
100
30
10

(a) Poor precision

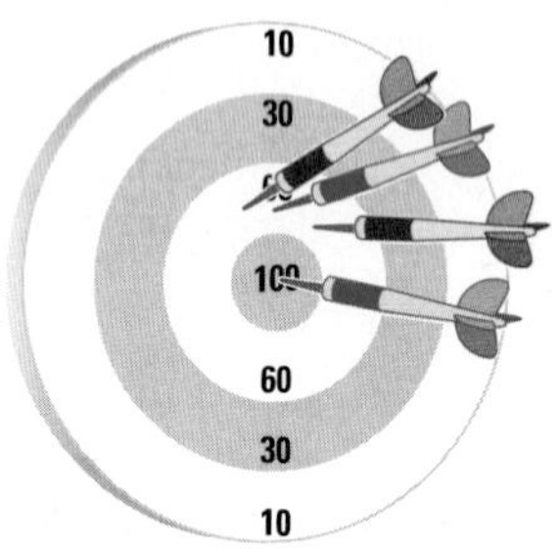
10
30
60
30
10

(b) Good precision; good accuracy

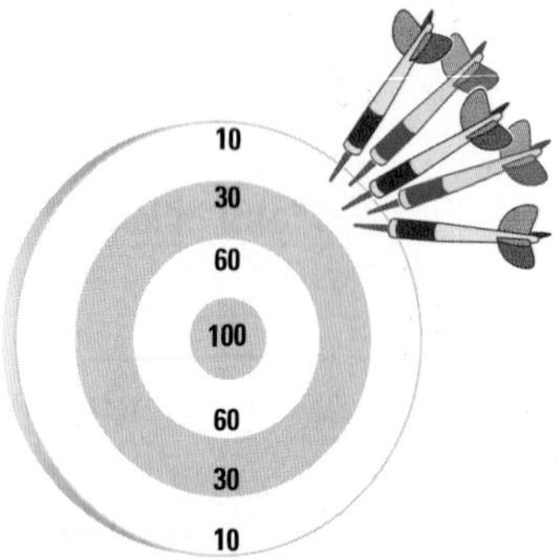
10
30
60
100
60
30
10

(c) Good precision; poor accuracy

Figure A.2 Precision and accuracy.

In the laboratory we attempt to set up experiments so that the greatest possible accuracy can be obtained. As a further check on accuracy, results may be compared among different laboratories so that any flaw in experimental design or measurement can be detected. For each individual experiment, several measurements are usually made and their precision determined. In most cases, better precision is taken as an indication of better experimental work, and it is necessary to know the precision to compare results among different experimenters. If two different experimenters both had results like those in Figure A.2a, for example, their average values could differ quite a lot before they would say that their results did not agree within experimental error.

In most experiments several different kinds of measurements must be made, and some can be done more precisely than others. It is common sense that *a calculated result can be no more precise than the least precise piece of information that went into the calculation.* This is where the rules for significant figures come in. In the last example in the preceding section, the quantity of gravel was described as "a ton." Usually gravel is measured by weighing an empty truck, putting some gravel in the truck, weighing the truck again, and subtracting the weight of the truck from the weight of the truck plus gravel. The quantity of gravel is not adjusted if there is a bit too much or a bit too little, because that would be a lot of trouble and gravel is inexpensive. If you asked for a ton, you might end up with as much as 2200 pounds or as little as 1800 pounds. In terms of significant figures this would be expressed as 2.0×10^3 lb.

The quantity 2.0×10^3 lb has two significant figures; it designates a quantity in which the 2 is taken to be exactly right but the 0 is not known precisely. (In this case, the number could be as large as 2.2 or as small as 1.8, so the 0 clearly is not exactly right.) Usually, in a number that represents a scientific measurement, the last digit on the right is taken to be inexact, but all digits farther to the left are assumed to be exact. When you do calculations using such numbers, you must follow some simple rules so that calculated results will reflect the precision of all the measurements that went into the calculations. Here are the rules:

Rule 1: To *determine the number of significant figures* in a measurement, read the number from left to right and count all digits, starting with the first digit that is *not* zero.

Example	Number of Significant Figures
1.23 g	3
7.64×10^{-2} m	3; for a number written in scientific notation, all digits preceding the 10^x term are significant.
0.00123 g	3; the zeros to the left of the 1 simply locate the decimal point. The number of significant figures is more obvious if you write numbers in scientific notation; thus, $0.00123 = 1.23 \times 10^{-3}$.
2.0 g and 0.020 g	2; both have two significant digits. When a number is greater than 1, *all zeros to the right of the decimal point are significant.* For a number less than 1, only zeros to the right of the first significant digit are significant.
100 g	1; in numbers that do not contain a decimal point, "trailing" zeros may or may not be significant. To eliminate possible confusion, the practice followed in this book is to include a decimal point if the zeros are significant. Thus, 100. has three significant digits, while 100 has only one. Alternatively, we write in scientific notation 1.00×10^2 (three significant digits) or 1×10^2 (one significant digit).
100 cm/m	Infinite number of significant figures, because this is a defined quantity. There are *exactly* 100 centimeters in one meter.
$\pi = 3.1415926...$	The value of π is known to a greater number of significant figures than any data you will ever use in a calculation.

The number π is now known to 1,011,196,691 digits. It is doubtful that you will need that much precision in this course—or ever.

Rule 2: When *adding or subtracting,* the number of decimal places in the answer should be equal to the number of decimal places in the number with the *fewest* places. Suppose you add three numbers:

0.12	2 significant figures	2 decimal places
1.6	2 significant figures	1 decimal place
+ 10.976	5 significant figures	3 decimal places
12.696		

This sum should be reported as 12.7, a number with one decimal place, because 1.6 has only one decimal place. The number 12.7 has three sig figs.

Rule 3: In *multiplication or division,* the number of significant figures in the answer should be the same as that in the quantity with the fewest significant figures.

$$\frac{0.01208}{0.0236} = 0.512 \text{ or, in scientific notation, } 5.12 \times 10^{-1}$$

Because 0.0236 has only three significant figures, while 0.01208 has four, the answer is limited to three significant figures.

Rule 4: When *rounding a number* to reduce the number of digits, follow these rules:

- The last digit retained is left unchanged if the following digit is less than 5 (4.327 rounded to two significant digits is 4.3).
- The last digit retained is increased by 1 if the following digit is greater than 5 or is a 5 followed by other non-zero digits (4.573 rounded to two significant digits is 4.6).
- If the last digit retained is followed by only a single digit 5 or by a 5 followed by zeros, then the last digit retained is increased by 1 if it is *odd* and remains the same if it is *even* (4.75 rounded to two significant digits is 4.8; 4.850 rounded to two significant digits is 4.8).

The examples in the table further illustrate these rules.

Full Number	Number Rounded to Three Significant Figures
12.696	12.7
16.249	16.2
18.350	18.4
18.45	18.4
18.351	18.4

One last word regarding significant figures and calculations. In working problems on a hand-held electronic calculator, you should do the calculation using all the digits allowed by the calculator and round only at the end of the problem. Rounding in the middle can introduce small errors (called *rounding errors* or *round-off errors*). If your answers do not quite agree with those in the Appendices of this book, rounding errors may be the source of the disagreement.

The ppm unit stands for "parts per million." If a substance is present with a concentration of 1 ppm by mass, there is 1 gram of the substance in 1 million grams of the sample.

Now let us consider a problem that is of practical importance and that makes use of all the rules. Suppose you discover that young children are eating chips of paint that flake off a wall in an old house. The paint contains 200. ppm lead by mass (200. milligrams of Pb per kilogram of paint). Suppose that a child eats five such chips. How much lead has the child gotten from the paint?

As stated, this problem does not include enough information for a solution to be obtained; however, some reasonable assumptions can be made, and they can lead to experiments that could be used to obtain the necessary information. The statement does not say how big the paint chips are. Let's assume that they are 1.0 cm by 1.0 cm so that the area is 1.0 cm^2. Then, eating five chips means eating 5.0 cm^2 of paint. (This assumption could be improved by measuring similar chips from the same place.) Because the concentration of lead is reported in units of mass of lead per mass of paint, we need to know the mass of 5.0 cm^2 of paint. This could be determined by measuring the areas of several paint chips and determining the mass of each. Suppose that the results of such measurements were those given in the table.

Mass of Chip (mg)	Area of Chip (cm^2)	Mass per Unit Area (mg/cm^2)
29.6	2.34	12.65
21.9	1.73	12.66
23.6	1.86	12.69

$$\text{Average mass per unit area} = \frac{(12.65 + 12.66 + 12.69)\ \text{mg/cm}^2}{3}$$

$$= 12.67\ \text{mg/cm}^2 = 12.7\ \text{mg/cm}^2$$

The average has been rounded to three significant figures because each experimentally measured mass and area has three significant figures. (Notice that more than three significant figures were kept in the intermediate calculations so as not to lose precision.) Now we can use this information to calculate how much lead the child has consumed.

$$m_{\text{paint}} = 5.0\ \text{cm}^2\ \text{paint}\left(\frac{12.7\ \text{mg paint}}{1\ \text{cm}^2\ \text{paint}}\right)\left(\frac{1\ \text{g}}{1000\ \text{mg}}\right)\left(\frac{1\ \text{kg}}{1000\ \text{g}}\right)$$

$$= 6.35 \times 10^{-5}\ \text{kg paint}$$

$$m_{\text{Pb}} = 6.35 \times 10^{-5}\ \text{kg paint}\left(\frac{200.\ \text{mg Pb}}{1\ \text{kg paint}}\right) = 1.27 \times 10^{-2}\ \text{mg Pb}$$

$$= 1.3 \times 10^{-2}\ \text{mg Pb} = 0.013\ \text{mg Pb}$$

The final result was rounded to two significant figures because there were only two significant figures in the initial 5.0-cm^2 area of the paint chip. This is quite adequate precision, however, for you to determine whether this quantity of lead is likely to harm the child.

The methods of problem solving presented here have been developed over time and represent a good way of keeping track of the precision of results, the units in which those results were obtained, and the correctness of calculations. These methods are not the only way that such goals can be achieved, but they do work well. We recommend that you include units in all calculations and check that they cancel appropriately. It is also important not to overstate the precision of results by keeping too many significant figures. By solving many problems, you should be able to develop your problem-solving skills so that they become second nature and you can use them without thinking about the mechanics. You can then devote all your thought to the logic of a problem solution.

A-4 Electronic Calculators

The advent of inexpensive electronic calculators has made calculations in introductory chemistry much more straightforward. You are well advised to purchase a calculator that has the capability of performing calculations in scientific notation, has both base 10 and natural logarithms, and is capable of raising any number to any power and of finding any root of any number. In the discussion below, we will point out in general how these functions of your calculator can be used. You should practice using your calculator to carry out arithmetic operations and make certain that you are able to use all of its functions correctly.

The directions for calculator use in this section are given for calculators using "algebraic" logic. Such calculators are the most common type used by students in introductory courses. For calculators using RPN logic (such as those made by Hewlett-Packard), the procedure will differ slightly.

Although electronic calculators have simplified calculations greatly, they have also forced us to focus on significant figures. A calculator easily handles eight or more significant figures, but real laboratory data are rarely known to this precision. Therefore, if you have not already done so, review Appendix A-3 on significant figures, precision, and rounding numbers.

The mathematical skills required to read and study this textbook successfully involve algebra, some geometry, scientific notation, logarithms, and solving quadratic equations. The next three sections review the last three of these topics.

A-5 Exponential or Scientific Notation

In exponential or scientific notation, a number is expressed as a product of two numbers: $N \times 10^n$. The first number, N, is called the digit term and is a number between 1 and 10. The second number, 10^n, the exponential term, is some integer power of 10. For example, 1234 would be written in scientific notation as 1.234×10^3 or 1.234 multiplied by 10 three times.

$$1234 = 1.234 \times 10^1 \times 10^1 \times 10^1 = 1.234 \times 10^3$$

Conversely, a positive number less than 1, such as 0.01234, would be written as 1.234×10^{-2}. This notation tells us that 1.234 should be divided twice by 10 to obtain 0.01234.

$$0.01234 = \frac{1.234}{10^1 \times 10^1} = 1.234 \times 10^{-1} \times 10^{-1} = 1.234 \times 10^{-2}$$

Some other examples of scientific notation are

$10{,}000 = 1 \times 10^4$	$12{,}345 = 1.2345 \times 10^4$
$1000 = 1 \times 10^3$	$1234 = 1.234 \times 10^3$
$100 = 1 \times 10^2$	$123 = 1.23 \times 10^2$
$10 = 1 \times 10^1$	$12 = 1.2 \times 10^1$
$1 = 1 \times 10^0$	(any number to the zeroth power = 1)
$1/10 = 1 \times 10^{-1}$	$0.12 = 1.2 \times 10^{-1}$
$1/100 = 1 \times 10^{-2}$	$0.012 = 1.2 \times 10^{-2}$
$1/1000 = 1 \times 10^{-3}$	$0.0012 = 1.2 \times 10^{-3}$
$1/10{,}000 = 1 \times 10^{-4}$	$0.00012 = 1.2 \times 10^{-4}$

When converting a number to scientific notation, notice that the exponent n is positive if the number is greater than 1 and negative if the number is less than 1. The value of n is the number of places by which the decimal was shifted to obtain the number in scientific notation.

$$1\,2\,3\,4\,5. = 1.2345 \times 10^4$$

Decimal shifted 4 places to the left. Therefore, n *is positive and equal to 4.*

$$0.0\,0\,1\,2 = 1.2 \times 10^{-3}$$

Decimal shifted 3 places to the right. Therefore, n *is negative and equal to 3.*

If you wish to convert a number in scientific notation to the usual form, the procedure just described is simply reversed.

$$6.273 \times 10^2 = 627.3$$

Decimal point shifted 2 places to the right, because* n *is positive and equal to 2.

$$006.273 \times 10^{-3} = 0.006273$$

Decimal point shifted 3 places to the left, because* n *is negative and equal to 3.

To enter a number in scientific notation into a calculator, first enter the number itself. Then, press the EE (Enter Exponent) key followed by n, the power of 10. For example, to enter 6.022×10^{23} you press these keys in succession:

6 . 0 2 2 EE 2 3 .

(Do not enter the number using the multiplication key, the number 10, and the EE key. This will result in a number that is 10 times bigger than you want. For example, pressing these keys

6 . 0 2 2 × 1 0 EE 2 3 (incorrect)

enters the number $6.022 \times 10 \times 10^{23} = 6.022 \times 10^{24}$, because EE 2 3 means 10^{23}.)

There are two final points concerning scientific notation. First, if you are used to working on a computer you may be in the habit of writing a number such as 1.23×10^3 as 1.23E3, or 6.45×10^{-5} as 6.45E−5. Second, some electronic calculators allow you to convert numbers readily to scientific notation. If you have such a calculator, you can change a number shown in the usual form to scientific notation by pressing an appropriate key or keys.

A-5a Raising Numbers in Scientific Notation to Powers

Electronic calculators usually have two methods of raising a number to a power. To square a number, enter the number and then press the "x^2" key. To raise a number to any power, use the "y^x" key. For example, to raise 1.42×10^2 to the 4th power, that is, to find $(1.42 \times 10^2)^4$,

On some calculators, Steps (a) and (c) may be interchanged.

(a) Enter 1.42×10^2.
(b) Press "y^x".
(c) Enter the desired power, 4 (the number 4 should appear on the display).
(d) Press "=". The result, $4.0658689\ldots \times 10^8$, will appear on the display. (The number of digits depends on the calculator.)

As a final step, express the number with the correct number of significant figures (4.07×10^8 in this case).

A-5b Taking Roots of Numbers in Scientific Notation

To take a square root on an electronic calculator, enter the number and then press the "$\sqrt{x}$" key. To find a higher root of a number, such as the fourth root of 5.6×10^{-10},

(a) Enter 5.6×10^{-10}.
(b) Press the "$\sqrt[x]{y}$" key. (On most calculators, the sequence you actually use is to press "2ndF" and then "y^x". Alternatively, you may have to press "INV" and then "y^x".)
(c) Enter the desired root, 4 (the number 4 should appear in the display).
(d) Press "=". The result, $4.864598558\ldots \times 10^{-3}$, will appear on the display. Round to 4.9×10^{-3}.

A general procedure for finding any root is to use the "y^x" key. For a square root, x is 0.5 (or $\frac{1}{2}$), whereas it is 0.33 (or $\frac{1}{3}$) for a cube root, 0.25 (or $\frac{1}{4}$) for a fourth root, and so on.

A-6 Logarithms

There are two types of logarithms used in this text: common logarithms (abbreviated log), whose base is 10, and natural logarithms (abbreviated ln), whose base is e ($e = 2.7182818284$). A logarithm is the power to which the base must be raised to produce the number:

$$\log x = n \qquad \text{where } x = 10^n$$

$$\ln x = m \qquad \text{where } x = e^m$$

Most equations in chemistry and physics were developed in natural or base e logarithms, and this practice is followed in this text. The relation between log and ln is

$$\ln x = 2.303 \log x$$

Logarithms to the base 10 are needed when dealing with pH.

Whether based on 10 or e, logarithms are used in the same manner. Here we describe the use of common logarithms. The common logarithm of a number is the power to which you must raise 10 to obtain the number. For example, the log of 100 is 2, because you must raise 10 to the power 2 to obtain 100. Other examples are

$$\log 1000 = \log (10^3) = 3$$
$$\log 10 = \log (10^1) = 1$$
$$\log 1 = \log (10^0) = 0$$
$$\log 1/10 = \log (10^{-1}) = -1$$
$$\log 1/10{,}000 = \log (10^{-4}) = -4$$

To obtain the common logarithm of a number other than a simple power of 10, use an electronic calculator. For example,

$$\log 2.10 = 0.3222, \text{ which means that } 10^{0.3222} = 2.10$$
$$\log 516 = 2.7126, \text{ which means that } 10^{2.7126} = 516$$
$$\log 0.3125 = -0.5051, \text{ which means that } 10^{-0.5051} = 0.3125$$

To check this result on your calculator, enter the number and then press the "log" key.

To obtain the natural logarithm of each number, enter the number and press "ln".

$$\ln 2.10 = 0.7419, \text{ which means that } e^{0.7419} = 2.10$$
$$\ln 516 = 6.2461, \text{ which means that } e^{6.2461} = 516$$

A-6a Significant Figures and Logarithms

The mantissa (digits to the right of the decimal point in the logarithm) should have as many significant figures as the number whose log was found. (So that you could more clearly see the result obtained with a calculator or a table, this rule was not strictly followed in the previous section.) Thus, expressed to the correct number of significant figures, $\log 2.10 = 0.322$ (three significant figures in 2.10 and three figures to the right of the decimal point in 0.322). Also, $\log 5.16 \times 10^3 = 3.713$ (note that the left-most 3 in the logarithm is not significant but rather tells the power of 10 in 5.16×10^3). Finally, for numbers smaller than 1, logarithms are always negative; for example, $\log 1.8 \times 10^{-5} = -4.74$.

***Nomenclature of Logarithms:* The number to the left of the decimal in a logarithm is called the *characteristic*, and the number to the right of the decimal is called the *mantissa*.**

A-6b Obtaining Antilogarithms

If you are given the logarithm of a number and need to find the number from it, you need to obtain the "antilogarithm" or "antilog" of the number. There are two common procedures used by electronic calculators to do this:

Procedure A	Procedure B
(a) Enter the log or ln (a number).	(a) Enter the log or ln (a number).
(b) Press 2ndF.	(b) Press INV.
(c) Press 10^x or e^x.	(c) Press log or ln x.

Test one or the other of these procedures with these examples.

EXERCISE A.1

Find the number whose log is 5.234.

Recall that $\log x = n$, where $x = 10^n$. In this case $n = 5.234$. Enter that number in your calculator and find the value of 10^n, the antilog. In this case,

$$10^{5.234} = 10^{0.234} \times 10^5 = 1.71 \times 10^5$$

Notice that the characteristic (5) sets the decimal point; it is the power of 10 in the exponential form. The mantissa (0.234) gives the value of the number x. A mantissa of 0.234 corresponds to $10^{0.234} = 1.71$.

EXERCISE A.2

Find the number whose ln is −3.456.

$$e^{-3.456} = 3.16 \times 10^{-2}$$

A-6c Mathematical Operations Using Logarithms

Logarithms are exponents so operations involving them follow the same rules as the use of exponents. Thus, when two numbers are multiplied their logarithms are added.

$$\log xy = \log x + \log y$$

When two numbers are divided, their logarithms are subtracted.

$$\log \frac{x}{y} = \log x - \log y$$

For example, taking the log of both sides of this expression

$$K_a = \frac{[H_3O^+][CH_3COCOO^-]}{[CH_3COCOOH]}$$

gives this one.

$$\log K_a = \log [H_3O^+] + \log [CH_3COCOO^-] - \log [CH_3COCOOH]$$

Logarithms can also be applied to powers and roots as shown in these two equations:

$$\log x^y = y(\log x)$$

$$\log \sqrt[y]{x} = \log x^{1/y} = \frac{1}{y} \log x$$

Thus, for an equation such as this,

$$\text{Rate} = k[A]^x$$

the logarithmic form is this,

$$\log \text{Rate} = \log k + x \log [A]$$

A-7 Quadratic Equations

Algebraic equations of the form $ax^2 + bx + c = 0$ are called *quadratic equations*. The coefficients a, b, and c may be either positive or negative. The two roots of the equation may be found using the *quadratic formula*.

$$x = \frac{-b \pm \sqrt{b^2 - 4ac}}{2a}$$

As an example, solve the equation $5x^2 - 3x - 2 = 0$. Here $a = 5$, $b = -3$, and $c = -2$. Therefore,

$$x = \frac{3 \pm \sqrt{(-3)^2 - 4(5)(-2)}}{2(5)} = \frac{3 \pm \sqrt{9 - (-40)}}{10} = \frac{3 \pm \sqrt{49}}{10} = \frac{3 \pm 7}{10}$$

$$x = 1 \quad \text{and} \quad x = -0.4$$

How do you know which of the two roots is the correct answer? You have to decide in each case which root has physical significance. However, for most problems in this book, negative values are not appropriate.

Many calculators have a built-in quadratic-equation solver. If your calculator offers this capability, it will be convenient to use it, and you should study the manual until you know how to use your calculator to obtain the roots of a quadratic equation.

When you have solved a quadratic expression, you should always check your values by substituting them into the original equation. In the example above, we find that $5(1)^2 - 3(1) - 2 = 0$ and that $5(-0.4)^2 - 3(-0.4) - 2 = 0$.

You will encounter quadratic equations in the chapters on chemical equilibria, particularly in Chapters 12, 14, and 15. Here you may be faced with solving an equation such as

$$1.8 \times 10^{-4} = \frac{x^2}{0.0010 - x}$$

This equation can certainly be solved by using the quadratic formula or your calculator (to give $x = 3.4 \times 10^{-4}$). However, you may find the *method of successive approximations* to be especially convenient. Here you begin by making a reasonable approximation of x. This approximate value is substituted into the original equation, and this expression is solved to give what is hoped to be a more correct value of x. This process is repeated until the answer converges on a particular value of x—that is, until the value of x derived from two successive approximations is the same.

Step 1: Assume that x is so small that $(0.0010 - x) \cong 0.0010$. This means that

$$x^2 = (1.8 \times 10^{-4})(0.0010)$$

$$x = 4.2 \times 10^{-4} \text{ (to 2 significant figures)}$$

Step 2: Substitute the value of x from Step 1 into the denominator (but not the numerator) of the original equation and again solve for x.

$$x^2 = (1.8 \times 10^{-4})(0.0010 - 0.00042)$$

$$x = 3.2 \times 10^{-4}$$

Step 3: Repeat Step 2 using the value of x found in that step.

$$x = \sqrt{(1.8 \times 10^{-4})(0.0010 - 0.00032)} = 3.5 \times 10^{-4}$$

Step 4: Continue by repeating the calculation, using the value of x found in the previous step. Step 4 gives $x = 3.4 \times 10^{-4}$.

Step 5: $x = \sqrt{(1.8 \times 10^{-4})(0.0010 - 0.00034)} = 3.4 \times 10^{-4}$

Here we find that iterations after the fourth step give the same value for x, indicating that we have arrived at a valid answer (and the same one obtained from the quadratic formula).

Some final thoughts on using the method of successive approximations. First, in some cases this method does not work. Successive steps may give answers that are random or that diverge from the correct value. For quadratic equations of the form $K = x^2/(C - x)$, the method of approximations will work only as long as $K < 4C$ (assuming one begins with $x = 0$ as the first guess; that is, $K \cong x^2/C$). This will always be true for weak acids and bases.

Second, values of K in the equation $K = x^2/(C - x)$ are usually known only to two significant figures. Therefore, we are justified in carrying out successive steps until two answers are the same to two significant figures.

Finally, if your calculator does not automatically obtain roots of quadratic equations, we highly recommend this method. If your calculator has a memory function, successive approximations can be carried out easily and very rapidly. Even without a memory function, the method of successive approximations is much quicker than solving a quadratic equation by hand.

A-8 Graphing

When analyzing experimental data, chemists and other scientists often graph the data to see whether the data agree with a mathematical equation. If the equation does fit the data (often indicated by a linear graph), then the graph can be used to obtain numerical values (parameters) that can be used in the equation to predict information not specifically included in the data set. For

Data Table: Masses and Volumes of Aluminum Samples

Mass (g)	Volume (mL)	Mass (g)	Volume (mL)	Mass (g)	Volume (mL)
2.03	0.760	18.02	6.68	32.84	12.17
5.27	1.95	21.83	8.10	36.27	13.43
9.57	3.54	25.17	9.32	36.77	13.62
11.46	4.25	30.08	11.14	39.36	14.58
14.96	5.55				

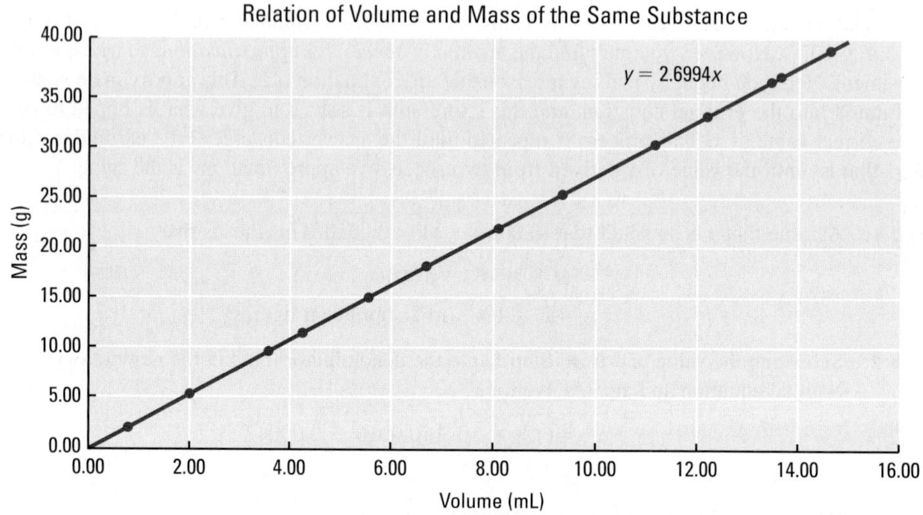
Relation of Volume and Mass of the Same Substance
y = 2.6994x
Mass (g)
40.00
35.00
30.00
25.00
20.00
15.00
10.00
5.00
0.00
0.00
2.00
4.00
6.00
8.00
10.00
12.00
14.00
16.00
Volume (mL)

Graph 1. Example graph of volume and mass of different samples of the same substance.

example, suppose that you measured the masses and the volumes of several samples of aluminum. Your data set might look like the Data Table. When these data are plotted with volume along the horizontal (x) axis and mass on the vertical (y) axis, Graph 1 is obtained.

Notice the features of this graph. It has a title that describes its contents, it has a label for each axis that specifies the quantity plotted and the units used, the vertical grid lines are equally spaced, the horizontal grid lines are equally spaced, and the numbers associated with those grid lines have the same difference between successive grid lines. (The differences are 2.00 mL on the horizontal axis and 5.00 g on the vertical axis.) The experimental points are indicated by circular markers, and

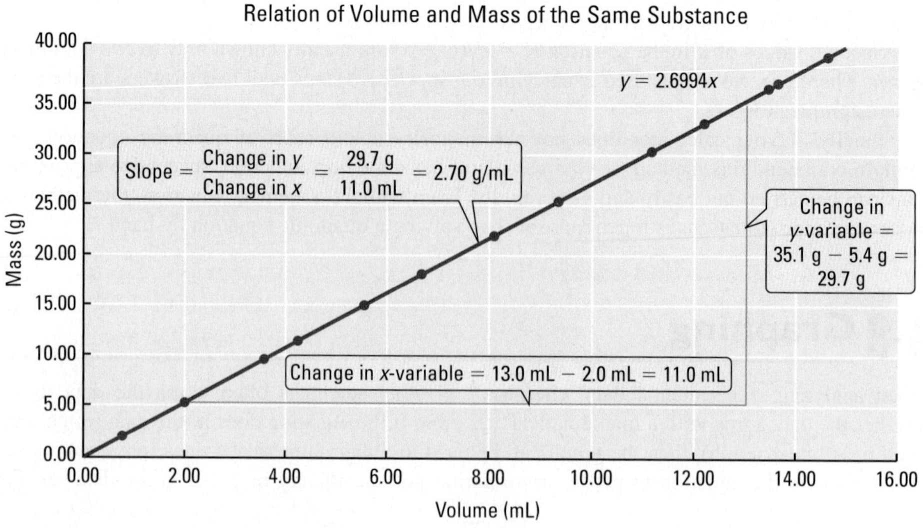
Relation of Volume and Mass of the Same Substance
y = 2.6994x
Slope = Change in y / Change in x = 29.7 g / 11.0 mL = 2.70 g/mL
Change in y-variable = 35.1 g − 5.4 g = 29.7 g
Change in x-variable = 13.0 mL − 2.0 mL = 11.0 mL
Mass (g)
40.00
35.00
30.00
25.00
20.00
15.00
10.00
5.00
0.00
0.00
2.00
4.00
6.00
8.00
10.00
12.00
14.00
16.00
Volume (mL)

Graph 2. Determining the slope of a graph.

a straight line has been drawn through the points. Because the line passes through all of the points and through the origin (point 0, 0), the data can be represented by the equation $y = 2.70x$, where 2.70 is the slope of the line. (On the graph, the slope is indicated by 2.6994; in the equation, it has been rounded to 2.70 because some of the data have only three significant figures.)

The slope of a graph such as this one can be obtained by choosing two points on the straight line (not two data points). The points should be far apart to obtain the most precise result. The slope is then given by the change in the y-axis variable divided by the change in the x-axis variable from the first point to the second. An example is shown on Graph 2. From the calculation of the slope, it is more obvious that the slope has only three significant figures. For this graph, the slope represents mass divided by volume—that is, density. A good way to measure density is to measure the mass and volume of several samples of the same substance and then plot the data. The resulting graph should be linear and should pass very nearly through the origin. The density can be obtained from its slope. When a graph passes through the origin, we say that the y-axis variable is directly proportional to the x-axis variable.

When a graph does not pass through the origin, there is an intercept as well as a slope. The intercept is the value of the y-axis variable when the x-axis variable equals zero. A good example is the relation between Celsius and Fahrenheit temperatures. Suppose you have measured a series of temperatures using both a Celsius thermometer and a Fahrenheit thermometer. Your data set might look something like this table.

Temperature (°C)	Temperature (°F)	Temperature (°C)	Temperature (°F)
−5.23	22.586	63.59	146.462
13.54	56.372	74.89	166.802
32.96	91.328	88.02	190.436
48.34	119.012	105.34	221.612

When graphed, these data look like Graph 3. In this case, although the data are on a straight line, that line does not pass through the origin. Instead, it intersects the vertical line corresponding to $x = 0$ (0 °C) at a value of 32 °F. This makes sense, because the normal freezing point of water is 0 °C, which is the same temperature as 32 °F. This value, 32 °F, is the intercept.

If we determine the slope by starting at the intercept value of 32 °F and going to 212° F, the normal boiling point of water, the change in temperature on the Fahrenheit scale is 180 °F. The corresponding change on the Celsius scale is 100 °C, from which we can determine that the slope is 180 °F/100 °C. Thus the equation relating Celsius and Fahrenheit temperatures is

$$y \quad = \quad m \quad \times \quad x \quad + \quad b$$

$$\text{temperature °F} = \frac{180\ \text{°F}}{100\ \text{°C}} \times \text{temperature °C} + 32\ \text{°F}$$

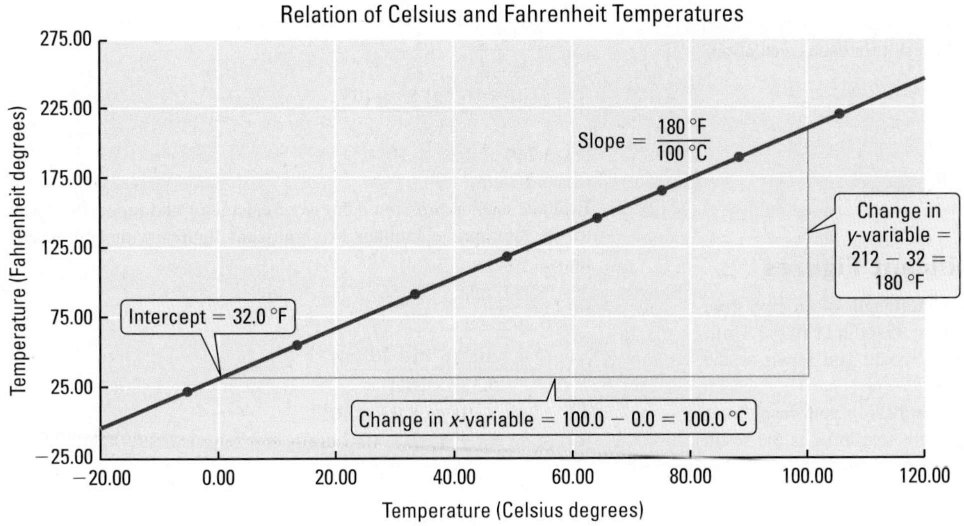
Relation of Celsius and Fahrenheit Temperatures
Temperature (Fahrenheit degrees)
275.00
225.00
175.00
125.00
75.00
25.00
−25.00
−20.00
0.00
20.00
40.00
60.00
80.00
100.00
120.00
Temperature (Celsius degrees)
Slope = 180 °F / 100 °C
Change in y-variable = 212 − 32 = 180 °F
Intercept = 32.0 °F
Change in x-variable = 100.0 − 0.0 = 100.0 °C

Graph 3. Example graph with non-zero intercept.

QUESTIONS FOR APPENDIX A

Red-numbered questions have short answers at the back of this book in Appendix M and fully worked solutions in the *Student Solutions Manual.*

General Problem-Solving Strategies

1. List four steps that can be used for guidance in solving problems. Choose a problem that you are interested in solving, and apply the four steps to that problem.
2. You are asked to study a lake in which fish are dying and determine the cause of their deaths. Suggest three things you might do to define this problem.
3. You have calculated the area in square yards of a carpet whose dimensions were originally given to you as 96 in by 72 in. Suggest at least one way to check that the results of your calculation are reasonable. (12 in = 1 ft; 3 ft = 1 yd)

Numbers, Units, and Quantities

4. When a calculation is done and the units for the answer do not make sense, what can you conclude about the solution to the problem? What would you do if you were faced with a situation like this?
5. The term "quantity" (or "physical quantity") has a specific meaning in science. What is that meaning, and why is it important?
6. To measure the length of a pencil, you use a tape measure calibrated in inches with marks every sixteenth of an inch. Which of these results would be a suitable record of your observation? Why would the other results be unsuitable? (1/16 in = 0.0625 in)
   (a) 8.38 ft (b) 8.38 m (c) 8.38 in
   (d) 8.38 (e) 0.698 ft
7. To measure the inseam length of a pair of slacks, you use a tape measure calibrated in centimeters with marks every tenth of a centimeter. Which of these results would be a suitable record of your observation? Why would the other results be unsuitable?
   (a) 75.0 cm (b) 75.0 m (c) 75.0
   (d) 75.0 in (e) 750 mm
8. What is wrong with each calculation? How would you carry out each calculation correctly? What is the correct result for each calculation?
   (a) $4.32 \text{ g} + 5.63 \text{ g} = 9.95 \text{ g}^2$
   (b) $5.23 \text{ g} \times \dfrac{4.87 \text{ g}}{1.00 \text{ mL}} = 25.5 \text{ g}$
   (c) $3.57 \text{ cm}^3 \times \dfrac{1 \text{ m}}{100 \text{ cm}} = 3.57 \times 10^{-2} \text{ m}^3$
9. What is wrong with each calculation? How would you carry out each calculation correctly? What is the correct result for each calculation?
   (a) $7.86 \text{ g} - 5.63 \text{ g} = 2.23 \text{ g}^2$
   (b) $7.37 \text{ mL} \times \dfrac{1.00 \text{ mL}}{2.23 \text{ g}} = 3.30 \text{ mL}$
   (c) $9.26 \text{ m}^3 \times \dfrac{100 \text{ cm}}{1 \text{ m}} = 9.26 \times 10^2 \text{ m}^3$

Precision, Accuracy, and Significant Figures

10. These measurements were reported for the length of an eight-foot pole: 95.31 in; 96.44 in; 96.02 in; 95.78 in; 95.94 in (1 ft = 12 in).
    (a) Based on these measurements, what would you report as the length of the pole?
    (b) How many significant figures should appear in your result?
    (c) Assuming that the pole is exactly eight feet long, is the result accurate?
11. These measurements were reported for the mass of a sample in the laboratory: 32.54 g; 32.67 g; 31.98 g; 31.76 g; 32.05 g.
    (a) Based on these measurements, what would you report as the mass of the sample?
    (b) How many significant figures should appear in your result?
    (c) The sample is exactly balanced by a weight known to have a mass of 35.0 g. Is the result accurate?
12. How many significant figures are in each quantity?
    (a) 3.274 g (b) 0.0034 L
    (c) 43,000 m (d) 6200. ft
13. How many significant figures are in each quantity?
    (a) 0.2730 g (b) 8.3 g/mL
    (c) 300 m (d) 2030.0 dm^3
14. Round each number to four significant figures.
    (a) 43.3250 (b) 43.3165 (c) 43.3237
    (d) 43.32499 (e) 43.3150 (f) 43.32501
15. Round each number to three significant figures.
    (a) 88.3520 (b) 88.365 (c) 88.45
    (d) 88.5500 (e) 88.2490 (f) 88.4501
16. Evaluate each expression and report the result to the appropriate number of significant figures.
    (a) $\dfrac{4.47}{0.3260}$ (b) $\dfrac{4.03 + 3.325}{29.75}$
    (c) $\dfrac{8.234}{5.673 - 4.987}$
17. Evaluate each expression and report the result to the appropriate number of significant figures.
    (a) $\dfrac{9.316}{6.211}$ (b) $\dfrac{4.441 - 3.52}{1.782}$
    (c) $\dfrac{16.82}{4.266} + 6.21$

Exponential or Scientific Notation

18. Without using a calculator, express each number in scientific notation and with the appropriate number of significant figures.
    (a) 76,003 (b) 0.00037 (c) 34,000
19. Without using a calculator, express each number in scientific notation and with the appropriate number of significant figures.
    (a) 49,002 (b) 0.0234 (c) 23,400
20. Evaluate each expression using your calculator and report the result to the appropriate number of significant figures with appropriate units.
    (a) $\dfrac{0.7346 \text{ g}}{304.2 \text{ g}}$
    (b) $\dfrac{(3.45 \times 10^{-3})(1.83 \times 10^{12})}{23.4}$
    (c) $3.240 - 4.33 \times 10^{-3}$
    (d) $(4.87 \text{ cm})^3$
21. Evaluate each expression using your calculator and report the result to the appropriate number of significant figures with appropriate units.
    (a) $\dfrac{893.0 \text{ mL}}{0.2032 \text{ mL}}$
    (b) $\dfrac{(5.4 \times 10^{3})(8.36 \times 10^{-12})}{5.317 \times 10^{-3}}$
    (c) $3.240 \times 10^5 - 8.33 \times 10^3$
    (d) $(4.87 \text{ cm} + 7.33 \times 10^{-1} \text{ cm})^3$

Logarithms

22. Use your calculator to find the logarithm of each number and report the logarithm to the appropriate number of significant figures.
    (a) log(0.7327) (b) ln(34.5)
    (c) $\log(6.022 \times 10^{23})$ (d) $\ln(6.022 \times 10^{23})$
    (e) $\log\left(\dfrac{8.34 \times 10^{-5}}{2.38 \times 10^{3}}\right)$
23. Use your calculator to find the logarithm of each number and report the logarithm to the appropriate number of significant figures.
    (a) log(54.3) (b) ln(0.0345)
    (c) $\log(4.344 \times 10^{-3})$ (d) $\ln(8.64 \times 10^{4})$
    (e) $\ln\left(\dfrac{4.33 \times 10^{24}}{8.32 \times 10^{-2}}\right)$
24. Use your calculator to evaluate each expression and report the result to the appropriate number of significant figures.
    (a) antilog(0.7327) (b) antiln(34.5)
    (c) $10^{2.043}$ (d) $e^{3.20 \times 10^{-4}}$
    (e) exp(4.333/3.275)
25. Use your calculator to evaluate each expression and report the result to the appropriate number of significant figures.
    (a) antilog(87.2) (b) antiln(0.0034)
    (c) $e^{2.043}$ (d) $10^{(3.20 \times 10^{-4})}$
    (e) $\exp(4.3 \times 10^{3}/8.314)$

Quadratic Equations

26. Find the roots of each quadratic equation.
    (a) $3.27x^2 + 4.32x - 2.83 = 0$
    (b) $x^2 + 4.32 = 4.57x$
27. Find the roots of each quadratic equation.
    (a) $8.33x^2 - 2.32x - 7.53 = 0$
    (b) $4.3x^2 - 8.37 = -2.22x$

Graphing

28. Graph these data involving mass and volume, label the graph appropriately, and determine whether m is directly proportional to V.

V (mL)	m (g)
0.347	0.756
1.210	2.638
2.443	5.326
7.234	15.76
11.43	24.90

29. Graph these data for heat transfers during a reaction, label the graph appropriately, and determine whether the heat evolved is directly proportional to the amount of reactant consumed.

Amount of Reactant (mol)	Heat Evolved (J)
94.2	43.2
70.7	32.5
65.7	30.1
34.2	15.7
54.3	24.9

APPENDIX B Units, Equivalences, and Conversion Factors

B-1 Units of the International System (SI)

The metric system was begun by the French National Assembly in 1790 and has undergone many modifications since its inception. The International System of Units, or *Système International* (SI), which represents an extension of the metric system, was adopted by the 11th General Conference on Weights and Measures in 1960. It is constructed from seven base units, each of which represents a particular physical quantity (Table B.1). More information about the SI is available at **http://physics.nist.gov/cuu/Units/index.html**.

The first five units listed in Table B.1 are particularly useful in chemistry. They are defined as follows:

1. The *meter* is the length of the path traveled by light in a vacuum during a time interval of 1/299,792,458 of a second.
2. The *kilogram* represents the mass of a platinum-iridium block kept at the International Bureau of Weights and Measures in Sevres, France.
3. The *second* is the duration of 9,192,631,770 periods of a certain line in the microwave spectrum of cesium-133.
4. The *kelvin* is 1/273.16 of the temperature interval between absolute zero and the triple point of water (the temperature at which liquid water, ice, and water vapor coexist).
5. The *mole* is the amount of substance that contains as many elementary entities (atoms, molecules, ions, other particles, or groups of such particles) as there are atoms in exactly 0.012 kg of carbon-12.

Decimal fractions and multiples of metric and SI units are designated by using the prefixes listed in Table B.2. The prefix *kilo-*, for example, means that a unit is multiplied by 10^3.

$$1 \text{ kilogram} = 1 \times 10^3 \text{ gram} = 1000 \text{ gram}$$

The prefix *centi-* means that the unit is multiplied by the factor 10^{-2}.

$$1 \text{ centigram} = 1 \times 10^{-2} \text{ gram} = 0.01 \text{ gram}$$

Prefixes are used to give units a size appropriate to what is being measured. It is more convenient and understandable to report the diameter of an atom (say, 1.00×10^{-10} m) as 100 pm than as 100×10^{-12} m. Following Table B.2 is a list of units for measuring distances, from very small to very large.

Table B.1 SI Fundamental Units

Physical Quantity	Name of Unit	Symbol
Length	Meter	m
Mass	Kilogram	kg
Time	Second	s
Temperature	Kelvin	K
Amount of substance	Mole	mol
Electric current	Ampere	A
Luminous intensity	Candela	cd

Table B.2 Prefixes for Metric and SI Units*

Factor	Prefix	Symbol	Factor	Prefix	Symbol
10^{24}	yotta-	Y	10^{-1}	*deci-*	d
10^{21}	zetta-	Z	10^{-2}	*centi-*	c
10^{18}	exa-	E	10^{-3}	*milli-*	m
10^{15}	peta-	P	10^{-6}	*micro-*	μ
10^{12}	tera-	T	10^{-9}	*nano-*	n
10^{9}	giga-	G	10^{-12}	*pico-*	p
10^{6}	mega-	M	10^{-15}	femto-	f
10^{3}	*kilo-*	k	10^{-18}	atto-	a
10^{2}	hecto-	h	10^{-21}	zepto-	z
10^{1}	deca-	da	10^{-24}	yocto-	y

*The prefixes most commonly used in chemistry are shown in italics.

yoctometer (ym)	0.000000000000000000000001 meter
zeptometer (zm)	0.000000000000000000001 meter
attometer (am)	0.000000000000000001 meter
femtometer (fm)	0.000000000000001 meter
picometer (pm)	0.000000000001 meter
nanometer (nm)	0.000000001 meter
micrometer (μm)	0.000001 meter
millimeter (mm)	0.001 meter
centimeter (cm)	0.01 meter
decimeter (dm)	0.1 meter
meter (m)	1 meter
decameter (dam)	10 meters
hectometer (hm)	100 meters
kilometer (km)	1000 meters
megameter (Mm)	1,000,000 meters
gigameter (Gm)	1,000,000,000 meters
terameter (Tm)	1,000,000,000,000 meters
petameter (Pm)	1,000,000,000,000,000 meters
exameter (Em)	1,000,000,000,000,000,000 meters
zettameter (Zm)	1,000,000,000,000,000,000,000 meters
yottameter (Ym)	1,000,000,000,000,000,000,000,000 meters

Table B.3 Derived SI Units

Physical Quantity	Name of Unit	Symbol	Definition	Expressed in Fundamental Units
Area	Square meter	m^2	—	
Volume	Cubic meter	m^3	—	
Density	Kilogram per cubic meter	kg/m^3	—	
Force	Newton	N	$\frac{\text{(kilogram)(meter)}}{\text{(second)}^2}$	$kg\ m\ s^{-2}$
Pressure	Pascal	Pa	$\frac{\text{(newton)}}{\text{(meter)}^2}$	$N/m^2 = kg\ m^{-1}\ s^{-2}$
Energy	Joule	J	(newton)(meter)	$kg\ m^2\ s^{-2}$
Electric charge	Coulomb	C	(ampere)(second)	A s
Electric potential difference	Volt	V	$\frac{\text{(joule)}}{\text{(coulomb)}}$	$J\ A^{-1}\ s^{-1} = kg\ m^2\ s^{-3}\ A^{-1}$

In the International System of Units, all physical quantities are represented by appropriate combinations of the base units listed in Table B.1. The result is a derived unit for each kind of measured quantity. Derived units that are often used in chemistry are listed in Table B.3. It is easy to see that the derived unit for area is length × length = meter × meter = square meter, m^2, or that the derived unit for volume is length × length × length = meter × meter × meter = cubic meter, m^3. More complex derived units are arrived at by a similar kind of combination of units. Units such as the joule, which measures energy, have been given simple names that represent the combination of fundamental units by which they are defined.

B-2 Conversion of Units for Physical Quantities

The result of a measurement is a physical quantity, which consists of a number and a unit. Algebraically, a physical quantity can be treated as if the number is multiplied by the unit. To convert a physical quantity from one unit of measure to another requires a conversion factor (proportionality factor) based on equivalences between units of measure such as those given in Table B.4. (See Appendix A-2 for more about physical quantities and proportionality factors.) Each equivalence provides two conversion factors that are the reciprocals of each other. For example, the equivalence between a quart and a liter, 1 quart = 0.9463 liter, gives

$$\frac{1\text{ quart}}{0.9463\text{ liter}} \qquad \text{There is 1 quart per 0.9463 liter.}$$

$$\frac{0.9463\text{ liter}}{1\text{ quart}} \qquad \text{There is 0.9463 liter per 1 quart.}$$

The method of canceling units described in Appendix A-2 provides the basis for choosing which conversion factor is needed: It is always the one that allows the unit being converted to be canceled and leaves the new unit uncanceled.

To convert 2 quarts to liters:

$$2\text{ quarts} \times \frac{0.9463\text{ liter}}{1\text{ quart}} = 1.893\text{ liters}$$

To convert 2 liters to quarts:

$$2\text{ liters} \times \frac{1\text{ quart}}{0.9463\text{ liter}} = 2.113\text{ quarts}$$

Because of the definitions of Celsius degrees and Fahrenheit degrees, conversions between these temperature scales are a bit more complicated. Both units are based on the properties of water. The Celsius unit is defined by assigning 0 °C as the freezing point of pure water and 100 °C as its boiling point, when the pressure is exactly 1 atm. The size of the Fahrenheit degree is equally arbitrary. Fahrenheit defined 0 °F as the freezing point of a solution in which he had dissolved the maximum quantity of ammonium chloride (because this was the lowest temperature he could reproduce reliably), and he intended 100 °F to be the normal human body temperature (but this value turned out to be 98.6 °F). Today, the reference points are set at exactly 32 °F and 212 °F (the freezing and boiling points of pure water, at 1 atm). The number of units between these two Fahrenheit temperatures is 180 °F. Thus, the Celsius degree is almost twice as large as the Fahrenheit degree; it takes only 5 Celsius degrees to cover the same temperature range as 9 Fahrenheit degrees.

$$\frac{100\ ^\circ\text{C}}{180\ ^\circ\text{F}} = \frac{5\ ^\circ\text{C}}{9\ ^\circ\text{F}}$$

This relationship is the basis for converting a temperature on one scale to a temperature on the other. If t_C is the numerical value of the temperature in °C and t_F is the numerical value of the temperature in °F, then

$$t_C = \left(\tfrac{5}{9}\right)(t_F - 32)$$

$$t_F = \left(\tfrac{9}{5}\right) t_C + 32$$

Table B.4 Common Units of Measure

Mass and Weight

1 pound = 453.59 grams = 0.45359 kilogram

1 kilogram = 1000 grams = 2.205 pounds

1 gram = 10 decigrams = 100 centigrams = 1000 milligrams

1 gram = 6.022×10^{23} unified atomic mass units

1 unified atomic mass unit = 1.6605×10^{-24} grams

1 short ton = 2000 pounds = 907.2 kilograms

1 long ton = 2240 pounds

1 metric tonne = 1000 kilograms = 2205 pounds

Length

1 inch = 2.54 centimeters (exactly)

1 mile = 5280 feet = 1.609 kilometers

1 yard = 36 inches = 0.9144 meter

1 meter = 100 centimeters = 39.37 inches = 3.281 feet = 1.094 yards

1 kilometer = 1000 meters = 1094 yards = 0.6215 miles

1 Angstrom = 1.0×10^{-10} meter = 0.10 nanometer = 100 picometers
= 1.0×10^{-8} centimeter = 3.937×10^{-9} inch

Volume

1 quart = 0.946353 liter

1 liter = 1.05669 quarts

1 liter = 1 cubic decimeter = 10^3 cubic centimeters = 10^{-3} cubic meter

1 milliliter = 1 cubic centimeter = 0.001 liter = 1.05669×10^{-3} quart

1 cubic foot = 28.31685 liters = 29.9221 quarts = 7.48052 gallons

Force and Pressure

1 atmosphere = 760.0 millimeters of mercury = 1.01325×10^5 pascals
= 14.70 pounds per square inch

1 bar = 10^5 pascals = 0.98692 atmosphere

1 Torr = 1 millimeter of mercury

1 pascal = 1 kg m^{-1} s^{-2} = 1 N/m^2 = 9.8692×10^{-6} atmospheres = 7.5006×10^{-3} mmHg

Energy

1 joule = 1×10^7 ergs

1 thermochemical calorie = 4.184 joules = 4.184×10^7 ergs
= 4.129×10^{-2} liter-atmospheres
= 2.611×10^{19} electron volts

1 erg = 1×10^{-7} joules = 2.3901×10^{-8} calorie

1 electron volt = 1.6022×10^{-19} joule = 1.6022×10^{-12} erg = 96.48 kJ/mol*

1 liter-atmosphere = 24.217 calories = 101.325 joules = 1.01325×10^9 ergs

1 British thermal unit = 1055.06 joules = 1.05506×10^{10} ergs = 252.165 calories

Temperature

0 K = −273.15 °C

If t_K is the numerical value of the temperature in kelvins, t_C is the numerical value of the temperature in °C, and t_F is the numerical value of the temperature in °F, then

$$t_K = t_C + 273.15$$
$$t_C = \left(\tfrac{5}{9}\right)(t_F - 32)$$
$$t_F = \left(\tfrac{9}{5}\right)t_C + 32$$

*The other units in this line are per particle and must be multiplied by 6.022×10^{23} to be strictly comparable.

For example, to show that your normal body temperature of 98.6 °F corresponds to 37.0 °C, use the first equation.

$$t_C = \left(\tfrac{5}{9}\right)(t_F - 32) = \left(\tfrac{5}{9}\right)(98.6 - 32) = \left(\tfrac{5}{9}\right)(66.6) = 37.0$$

Thus, body temperature is 37.0 °C.

Laboratory work is almost always done using Celsius units, and we rarely need to make conversions to and from Fahrenheit degrees. It is best to try to calibrate your senses to Celsius units; to help you do so, it is useful to know that water freezes at 0 °C, a comfortable room temperature is about 22 °C, your body temperature is 37 °C, and the hottest water you could leave your hand in for some time is about 60 °C.

QUESTIONS FOR APPENDIX B

Red-numbered questions have short answers at the back of this book in Appendix M and fully worked solutions in the *Student Solutions Manual.*

Units of the International System

1. Which SI unit accompanied by which prefix would be most convenient for describing each quantity?
   (a) Mass of this book
   (b) Volume of a glass of water
   (c) Thickness of this page
2. Which SI unit accompanied by which prefix would be most convenient for describing each quantity?
   (a) Distance from New York to San Francisco
   (b) Mass of a glass of water
   (c) Area of this page
3. Explain the difference between an SI base (fundamental) unit and a derived SI unit.
4. What is the official SI definition of the mole? Describe in your own words each part of the definition and explain why each part of the definition is important.

Conversion of Units for Physical Quantities

5. Express each quantity in SI base (fundamental) units. Use exponential (scientific) notation whenever it is needed.
   (a) 475 pm (b) 56 Gg (c) 4.28 μA
6. Express each quantity in SI base (fundamental) units. Use exponential (scientific) notation whenever it is needed.
   (a) 32.5 ng (b) 56 Mm (c) 439 pm
7. Express each quantity in SI base (fundamental) units. Use exponential (scientific) notation whenever it is needed.
   (a) 8.7 nm^2 (b) 27.3 aJ (c) 27.3 μN
8. Express each quantity in SI base (fundamental) units. Use exponential (scientific) notation whenever it is needed.
   (a) 56.3 cm^3 (b) 5.62 MJ (c) 33.4 kV
9. Express each quantity in the units indicated. Use scientific notation.
   (a) 1.00 kg in pounds
   (b) 2.45 ton in kilograms
   (c) 1 L in cubic inches (in^3)
   (d) 1 atm in pascals and in bars
10. Express each quantity in the units indicated. Use scientific notation.
    (a) 24.3 u in grams
    (b) 87.3 mL in cubic feet (ft^3)
    (c) 24.7 dg in ounces
    (d) 1.02 bar in millimeters of mercury (mmHg) and in torr
11. Express each temperature in Fahrenheit degrees.
    (a) 37 °C (b) −23.6 °C (c) −40.0 °C
12. Express each temperature in Celsius degrees.
    (a) 180. °F (b) −40.0 °F (c) 28.3 °F

APPENDIX C Physical Constants* and Sources of Data

Quantity	Symbol	SI Units	Traditional Units
Acceleration of gravity	g_n	9.806 65 m/s^2	980.665 cm/s^2
Atomic mass unit ($\frac{1}{12}$ the mass of ^{12}C atom)	u or amu	1.660 538 921(73) × 10^{-27} kg	1.660 538 921(73) × 10^{-24} g
Avogadro constant	N_A, L	6.022 141 29(27) × 10^{23} particles/mol	6.022 141 29(27) × 10^{23} particles/mol
Bohr radius	a_0	5.291 772 1092(17) × 10^{-11} m	0.529 177 210 92(17) Å
Boltzmann constant	k	1.380 6488(13) × 10^{-23} J/K	1.380 6488(13) × 10^{-16} erg/K
Charge-to-mass ratio of electron	e/m_c	−1.758 820 088(39) × 10^{11} C/kg	−1.758 820 088(39) × 10^{8} C/g
Elementary charge (electron or proton charge)	e	1.602 176 565(35) × 10^{-19} C	1.602 176 565(35) × 10^{-19} C
Electron rest mass	m_c	9.109 382 91(40) × 10^{-31} kg	9.109 382 91(40) × 10^{-28} g
Faraday constant	F	96 485.3365(21) C/mol e$^-$ 96 485.3365(21) J V^{-1} mol^{-1}	96 485.3365(21) C/mol e$^-$ 23.060 549 kcal V^{-1} mol^{-1}
Gas constant	R	8.314 4621(75) dm^3 Pa mol^{-1} L^{-1} 8.314 4621(75) J mol^{-1} K^{-1}	0.082 057 L atm mol^{-1} K^{-1} 1.9872 cal mol^{-1} K^{-1}
Molar volume (1 atm, 273.15 K)	V_m	22.414 × 10^{-3} m^3/mol 22.414 dm^3/mol	22.414 L/mol
Neutron rest mass	m_n	1.674 927 351(74) × 10^{-27} kg 1.008 664 916 u	1.674 927 351(74) × 10^{-24} g 1.008 664 916 amu
Planck's constant	h	6.626 069 57(29) × 10^{-34} J s	6.626 069 57(29) × 10^{-27} erg s
Proton rest mass	m_p	1.672 621 777(74) × 10^{-27} kg 1.007 276 467 u	1.672 621 777(74) × 10^{-24} g 1.007 276 467 amu
Rydberg constant	R_∞	1.097 373 156 8539(55) × 10^{7} m^{-1} 2.179 872 171(96) × 10^{-18} J	1.097 303 156 8539(55) × 10^{5} cm^{-1} 2.179 872 171(96) × 10^{-11} erg
Speed of light (in a vacuum)	c, c_0	2.997 924 58 × 10^{8} m/s	2.997 927 58 × 10^{10} cm/s 186 282 mile/s

*Data from the National Institute for Standards and Technology (**http://physics.nist.gov/cuu/Constants/index.html**) reference on constants, units, and uncertainty. The uncertainty in each number is given in parentheses. For example, the notation 1.234 56(12) means that the value of the quantity is 1.234 56 ± 0.000 12; that is, the value can range from 1.234 44 up to 1.234 68. Spaces are used between numbers to make reading easier.

Online Sources

- SIRCh: Selected Internet Resources for Chemistry
  http://en.wikibooks.org/wiki/Chemical_Information_Sources/SIRCh
- SIRCh: Physical Property Information
  http://en.wikibooks.org/wiki/Chemical_Information_Sources/Physical_Property_Searches
- Thermodex. University of Texas at Austin.
  http://www.lib.utexas.edu/thermodex/
- How Many? A Dictionary of Units of Measurement.
  http://www.unc.edu/~rowlett/units/index.html
- NIST Chemistry Web Book
  http://webbook.nist.gov/chemistry
- Hawley's Condensed Chemical Dictionary (Internet Version)
  http://onlinelibrary.wiley.com/book/10.1002/9780470114735

Print Sources

- Lide, D. R., Ed. *CRC Handbook of Chemistry and Physics,* 94th ed., Boca Raton, FL: CRC Press, 2013–2014.
- O'Neil, M. J., Heckelman, P. E., Koch, C. B., and Roman, K. J. Eds. *The Merck Index: An Encyclopedia of Chemicals, Drugs, and Biologicals,* 15th ed., Rahway, NJ: Merck & Co., 2006, 2013.
- Speight, J., Ed. *Lange's Handbook of Chemistry,* 16th ed., New York: McGraw-Hill, 2004.
- Perry, R. H., and Green, D. W., Eds. *Perry's Chemical Engineers' Handbook,* 8th ed., New York: McGraw-Hill, 2007.
- Zwillinger, D., Ed. *CRC Standard Mathematical Tables and Formulae,* 32nd ed., Boca Raton, FL: CRC Press, 2011.
- Lewis, R. J., Sr. *Hawley's Condensed Chemical Dictionary,* 15th ed., New York: Wiley, 2007.

APPENDIX D Ground-State Electron Configurations of Atoms*

Z	Element	Configuration	Z	Element	Configuration	Z	Element	Configuration
1	H	$1s^1$	40	Zr	[Kr] $4d^2 5s^2$	79	Au	[Xe] $4f^{14} 5d^{10} 6s^1$
2	He	$1s^2$	41	Nb	[Kr] $4d^4 5s^1$	80	Hg	[Xe] $4f^{14} 5d^{10} 6s^2$
3	Li	[He] $2s^1$	42	Mo	[Kr] $4d^5 5s^1$	81	Tl	[Xe] $4f^{14} 5d^{10} 6s^2 6p^1$
4	Be	[He] $2s^2$	43	Tc	[Kr] $4d^5 5s^2$	82	Pb	[Xe] $4f^{14} 5d^{10} 6s^2 6p^2$
5	B	[He] $2s^2 2p^1$	44	Ru	[Kr] $4d^7 5s^1$	83	Bi	[Xe] $4f^{14} 5d^{10} 6s^2 6p^3$
6	C	[He] $2s^2 2p^2$	45	Rh	[Kr] $4d^8 5s^1$	84	Po	[Xe] $4f^{14} 5d^{10} 6s^2 6p^4$
7	N	[He] $2s^2 2p^3$	46	Pd	[Kr] $4d^{10}$	85	At	[Xe] $4f^{14} 5d^{10} 6s^2 6p^5$
8	O	[He] $2s^2 2p^4$	47	Ag	[Kr] $4d^{10} 5s^1$	86	Rn	[Xe] $4f^{14} 5d^{10} 6s^2 6p^6$
9	F	[He] $2s^2 2p^5$	48	Cd	[Kr] $4d^{10} 5s^2$	87	Fr	[Rn] $7s^1$
10	Ne	[He] $2s^2 2p^6$	49	In	[Kr] $4d^{10} 5s^2 5p^1$	88	Ra	[Rn] $7s^2$
11	Na	[Ne] $3s^1$	50	Sn	[Kr] $4d^{10} 5s^2 5p^2$	89	Ac	[Rn] $6d^1 7s^2$
12	Mg	[Ne] $3s^2$	51	Sb	[Kr] $4d^{10} 5s^2 5p^3$	90	Th	[Rn] $6d^2 7s^2$
13	Al	[Ne] $3s^2 3p^1$	52	Te	[Kr] $4d^{10} 5s^2 5p^4$	91	Pa	[Rn] $5f^2 6d^1 7s^2$
14	Si	[Ne] $3s^2 3p^2$	53	I	[Kr] $4d^{10} 5s^2 5p^5$	92	U	[Rn] $5f^3 6d^1 7s^2$
15	P	[Ne] $3s^2 3p^3$	54	Xe	[Kr] $4d^{10} 5s^2 5p^6$	93	Np	[Rn] $5f^4 6d^1 7s^2$
16	S	[Ne] $3s^2 3p^4$	55	Cs	[Xe] $6s^1$	94	Pu	[Rn] $5f^6 7s^2$
17	Cl	[Ne] $3s^2 3p^5$	56	Ba	[Xe] $6s^2$	95	Am	[Rn] $5f^7 7s^2$
18	Ar	[Ne] $3s^2 3p^6$	57	La	[Xe] $5d^1 6s^2$	96	Cm	[Rn] $5f^7 6d^1 7s^2$
19	K	[Ar] $4s^1$	58	Ce	[Xe] $4f^1 5d^1 6s^2$	97	Bk	[Rn] $5f^9 7s^2$
20	Ca	[Ar] $4s^2$	59	Pr	[Xe] $4f^3 6s^2$	98	Cf	[Rn] $5f^{10} 7s^2$
21	Sc	[Ar] $3d^1 4s^2$	60	Nd	[Xe] $4f^4 6s^2$	99	Es	[Rn] $5f^{11} 7s^2$
22	Ti	[Ar] $3d^2 4s^2$	61	Pm	[Xe] $4f^5 6s^2$	100	Fm	[Rn] $5f^{12} 7s^2$
23	V	[Ar] $3d^3 4s^2$	62	Sm	[Xe] $4f^6 6s^2$	101	Md	[Rn] $5f^{13} 7s^2$
24	Cr	[Ar] $3d^5 4s^1$	63	Eu	[Xe] $4f^7 6s^2$	102	No	[Rn] $5f^{14} 7s^2$
25	Mn	[Ar] $3d^5 4s^2$	64	Gd	[Xe] $4f^7 5d^1 6s^2$	103	Lr	[Rn] $5f^{14} 7s^2 7p^1$
26	Fe	[Ar] $3d^6 4s^2$	65	Tb	[Xe] $4f^9 6s^2$	104	Rf	[Rn] $5f^{14} 6d^2 7s^2$
27	Co	[Ar] $3d^7 4s^2$	66	Dy	[Xe] $4f^{10} 6s^2$	105	Db	[Rn] $5f^{14} 6d^3 7s^2$
28	Ni	[Ar] $3d^8 4s^2$	67	Ho	[Xe] $4f^{11} 6s^2$	106	Sg	[Rn] $5f^{14} 6d^4 7s^2$
29	Cu	[Ar] $3d^{10} 4s^1$	68	Er	[Xe] $4f^{12} 6s^2$	107	Bh	[Rn] $5f^{14} 6d^5 7s^2$
30	Zn	[Ar] $3d^{10} 4s^2$	69	Tm	[Xe] $4f^{13} 6s^2$	108	Hs	[Rn] $5f^{14} 6d^6 7s^2$
31	Ga	[Ar] $3d^{10} 4s^2 4p^1$	70	Yb	[Xe] $4f^{14} 6s^2$	109	Mt	[Rn] $5f^{14} 6d^7 7s^2$
32	Ge	[Ar] $3d^{10} 4s^2 4p^2$	71	Lu	[Xe] $4f^{14} 5d^1 6s^2$	110	Ds	[Rn] $5f^{14} 6d^9 7s^1$
33	As	[Ar] $3d^{10} 4s^2 4p^3$	72	Hf	[Xe] $4f^{14} 5d^2 6s^2$	111	Rg	[Rn] $5f^{14} 6d^{10} 7s^1$
34	Se	[Ar] $3d^{10} 4s^2 4p^4$	73	Ta	[Xe] $4f^{14} 5d^3 6s^2$	112	Cn	[Rn] $5f^{14} 6d^{10} 7s^2$
35	Br	[Ar] $3d^{10} 4s^2 4p^5$	74	W	[Xe] $4f^{14} 5d^4 6s^2$	113	—	[Rn] $5f^{14} 6d^{10} 7s^2 7p^1$
36	Kr	[Ar] $3d^{10} 4s^2 4p^6$	75	Re	[Xe] $4f^{14} 5d^5 6s^2$	114	Fl	[Rn] $5f^{14} 6d^{10} 7s^2 7p^2$
37	Rb	[Kr] $5s^1$	76	Os	[Xe] $4f^{14} 5d^6 6s^2$	115	—	[Rn] $5f^{14} 6d^{10} 7s^2 7p^3$
38	Sr	[Kr] $5s^2$	77	Ir	[Xe] $4f^{14} 5d^7 6s^2$	116	Lv	[Rn] $5f^{14} 6d^{10} 7s^2 7p^4$
39	Y	[Kr] $4d^1 5s^2$	78	Pt	[Xe] $4f^{14} 5d^9 6s^1$	117	—	[Rn] $5f^{14} 6d^{10} 7s^2 7p^5$
						118	—	[Rn] $5f^{14} 6d^{10} 7s^2 7p^6$

*All electron configurations in this table are from experimental determinations except for elements number 103 through 118, which are predicted theoretically. For elements with atomic number greater than 103 only a few atoms can be produced at a time and it is very difficult to determine electron configurations.

APPENDIX E Naming Simple Organic Compounds

The systematic nomenclature for organic compounds was proposed by the International Union of Pure and Applied Chemistry (IUPAC). The IUPAC set of rules provides different names for the more than 50 million known organic compounds and allows names to be assigned to new compounds as they are synthesized. Many organic compounds also have *common* names. Usually the common name came first and is widely known. Many consumer products are labeled with the common name, and when only a few isomers are possible, the common name adequately identifies the product for the consumer. However, as illustrated in Section 2-9a, a system of common names quickly fails when several structural isomers are possible.

E-1 Hydrocarbons

The name of each member of the hydrocarbon classes has two parts. The first part, called the prefix (*meth-*, *eth-*, *prop-*, *but-*, and so on), reflects the number of carbon atoms. When more than four carbons are present, the Greek or Latin number prefixes are used: *pent-*, *hex-*, *hept-*, *oct-*, *non-*, and *dec-*. The second part of the name, called the suffix, tells the class of hydrocarbon. Alkanes have carbon-carbon single bonds, alkenes have carbon-carbon double bonds, and alkynes have carbon-carbon triple bonds.

E-1a Unbranched Alkanes and Alkyl Groups

The names of the first 20 unbranched (straight-chain) alkanes are given in Table E.1.

Alkyl groups are named by dropping *-ane* from the parent alkane and adding *-yl*. Examples are ethyl, CH_3CH_2—, and octyl, $CH_3CH_2CH_2CH_2CH_2CH_2CH_2CH_2$—.

E-1b Branched-Chain Alkanes

Rules for naming branched-chain alkanes are:

1. *Find the longest continuous chain of carbon atoms; it determines the parent name for the compound.* For example, this compound has two methyl groups attached to a *heptane* parent; the longest continuous chain contains seven carbon atoms.

$$\begin{array}{l} CH_3CH_2CH_2\underset{\displaystyle CH_3}{\underset{|}{CH}}CH_2\underset{\displaystyle CH_3}{\underset{|}{CH}}CH_3 \end{array}$$

Table E.1 Names of Unbranched Alkanes

CH_4	Methane	C_8H_{18}	Octane	$C_{15}H_{32}$	Pentadecane
C_2H_6	Ethane	C_9H_{20}	Nonane	$C_{16}H_{34}$	Hexadecane
C_3H_8	Propane	$C_{10}H_{22}$	Decane	$C_{17}H_{36}$	Heptadecane
C_4H_{10}	Butane	$C_{11}H_{24}$	Undecane	$C_{18}H_{38}$	Octadecane
C_5H_{12}	Pentane	$C_{12}H_{26}$	Dodecane	$C_{19}H_{40}$	Nonadecane
C_6H_{14}	Hexane	$C_{13}H_{28}$	Tridecane	$C_{20}H_{42}$	Eicosane
C_7H_{16}	Heptane	$C_{14}H_{30}$	Tetradecane	$C_{40}H_{82}$	Tetracontane

The longest continuous chain may not be obvious from the way the formula is written, especially for the straight-line format that is commonly used (← **Sec. 6-3**). For example, the longest continuous chain of carbon atoms in this chain is *eight,* not *four* or *six.*

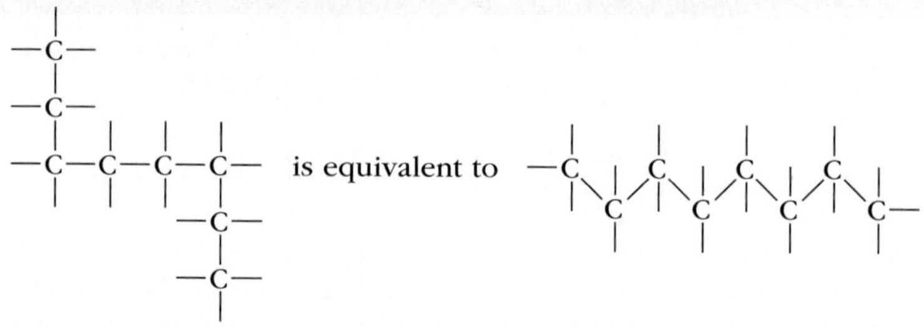
is equivalent to

2. *Number the longest chain beginning with the end of the chain nearest the branching. Use these numbers to designate the location of the attached group. When two or more groups are attached to the parent, give each group a number corresponding to its location on the parent chain.* For example, the name of

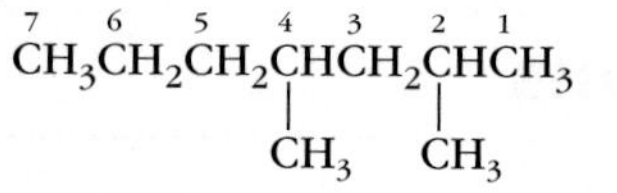

is 2,4-dimethylheptane. The name of

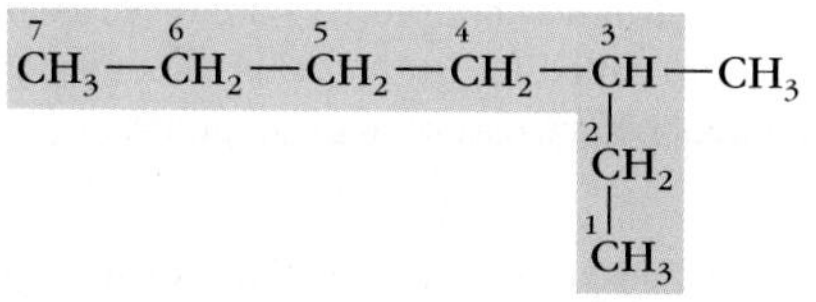

is 3-methylheptane, not 5-methylheptane or 2-ethylhexane.

3. *When two or more substituents are identical, indicate this by the use of the prefixes di-, tri-, tetra-, and so on. Positional numbers of the substituents should have the smallest possible sum.*

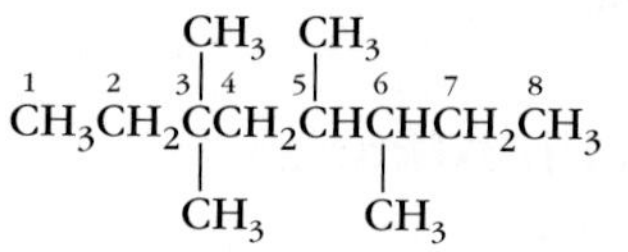

The correct name of this compound is 3,3,5,6-tetramethyloctane.

4. *If there are two or more different groups, the groups are listed alphabetically.*

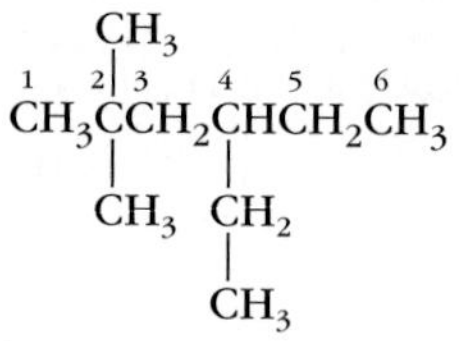

The correct name of this compound is 4-ethyl-2,2-dimethylhexane. Note that the prefix *di-* is ignored in determining alphabetical order.

E-1c Alkenes

Alkenes are named by using the prefix to indicate the number of carbon atoms and the suffix *-ene* to indicate one or more double bonds. The systematic names for the first two members of the alkene series are *ethene* and *propene.*

$$CH_2{=}CH_2 \qquad CH_3CH{=}CH_2$$

When groups, such as methyl or ethyl, are attached to carbon atoms in an alkene, the longest hydrocarbon chain is numbered from the end that will give the double bond the lowest number and then numbers are assigned to the attached groups. For example, the name of

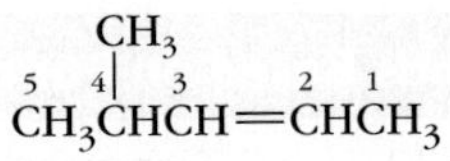

is 4-methyl-2-pentene. See Section 6-5a for a discussion of *cis-trans* isomers of alkenes.

E-1d Alkynes

The naming of alkynes is similar to that of alkenes, with the lowest number possible being used to locate the triple bond. For example, the name of

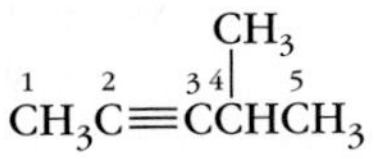

is 4-methyl-2-pentyne.

E-1e Benzene Derivatives

Monosubstituted benzene derivatives are named by using a prefix for the substituent. Some examples are

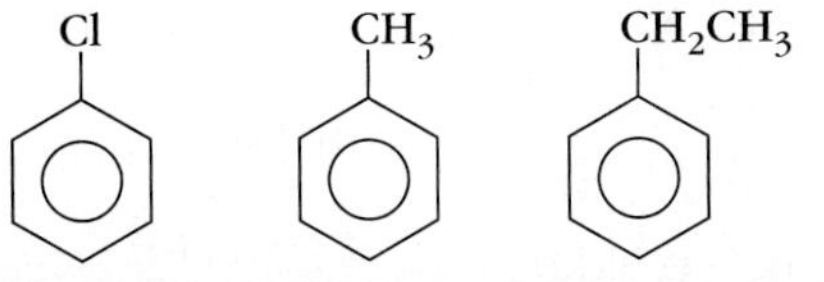

chlorobenzene methylbenzene (toluene) ethylbenzene

Three isomers are possible when two groups are substituted for hydrogen atoms on the benzene ring. The relative positions of the substituents are indicated either by the prefixes *ortho-*, *meta-*, and *para-* (abbreviated *o-*, *m-*, and *p-*, respectively) or by numbers. For example,

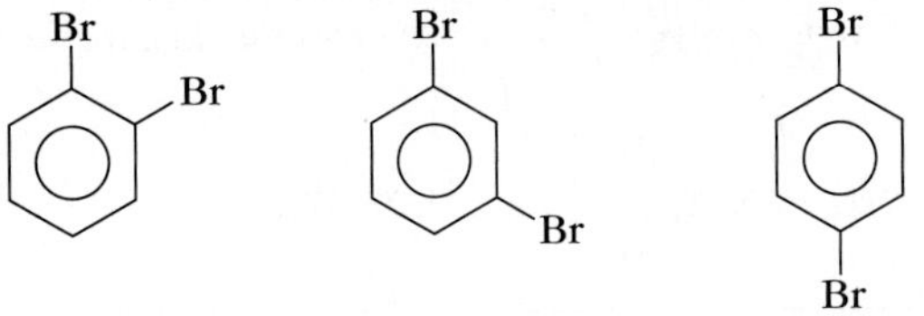

1,2-dibromobenzene (*o*-dibromobenzene) 1,3-dibromobenzene (*m*-dibromobenzene) 1,4-dibromobenzene (*p*-dibromobenzene)

The dimethylbenzenes are called *xylenes.*

If more than two groups are attached to the benzene ring, numbers must be used to identify the positions. The benzene ring is numbered to give the lowest possible numbers to the substituents.

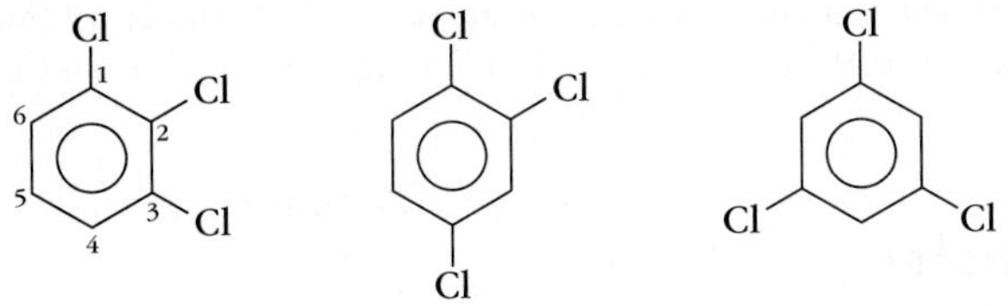

1,2,3-trichlorobenzene 1,2,4-trichlorobenzene 1,3,5-trichlorobenzene

E-2 Functional Groups

An atom or group of atoms that defines the structure of a specific class of organic compounds and determines their properties is called a *functional group* (← **Sec. 10-3**). The millions of organic compounds include classes of compounds that are obtained by replacing hydrogen atoms of hydrocarbons with functional groups, (← **Secs. 10-3, 10-4, and 10-5**). The important functional groups are shown in Table E.2.

Table E.2 Classes of Organic Compounds Based on Functional Groups*

General Formulas of Class Members	Class Name	Typical Compound	Compound Name	Common Use of Sample Compound
R—X	Halide	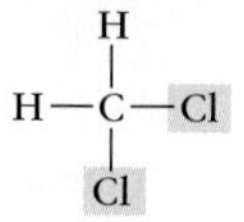	Dichloromethane (methylene chloride)	Solvent
R—OH	Alcohol	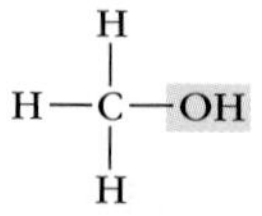	Methanol (wood alcohol)	Solvent
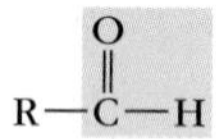	Aldehyde	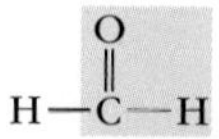	Methanal (formaldehyde)	Preservative
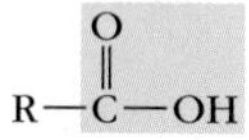	Carboxylic acid	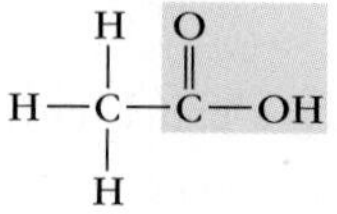	Ethanoic acid (acetic acid)	Vinegar
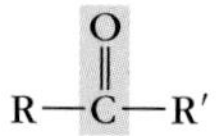	Ketone	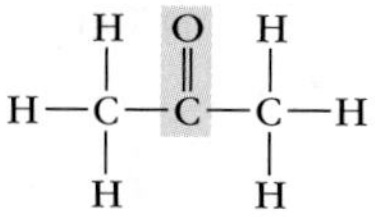	Propanone (acetone)	Solvent
R—O—R′	Ether	C_2H_5—O—C_2H_5	Diethyl ether (ethyl ether)	Anesthetic
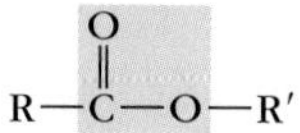	Ester	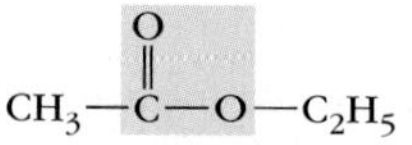	Ethyl ethanoate (ethyl acetate)	Solvent in fingernail polish
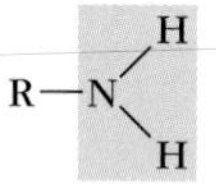	Amine	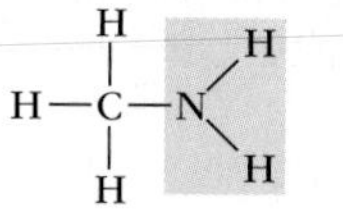	Methylamine	Tanning hides (foul odor)
	Amide	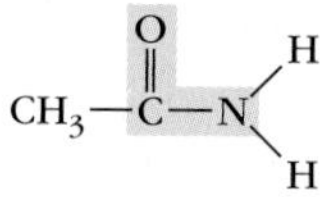	Acetamide	Plasticizer

*R stands for an H or a hydrocarbon group such as —CH_3 or —C_2H_5. R′ could be a different group from R.

The "R" attached to the functional group represents the hydrocarbon framework with one hydrogen removed for each functional group added. The IUPAC system provides a systematic method for naming all members of a given class. For example, alcohols end in *-ol* (methan*ol*); aldehydes end in *-al* (methan*al*); carboxylic acids end in *-oic* (ethan*oic* acid); and ketones end in *-one* (propan*one*).

E-2a Alcohols

Isomers are also possible for molecules containing functional groups. For example, three different alcohols are obtained when a hydrogen atom in pentane is replaced by —OH, depending on which hydrogen atom is replaced. The rules for naming the "R" or hydrocarbon framework are the same as those for hydrocarbon compounds.

$CH_3CH_2CH_2CH_2CH_2OH$ 1-pentanol

$CH_3CH_2CH_2CHCH_3$
|
OH
2-pentanol

$CH_3CH_2CHCH_2CH_3$
|
OH
3-pentanol

Compounds with one or more functional groups (Table E.2) and alkyl substituents are named so as to give the functional groups the lowest numbers. For example, the correct name of

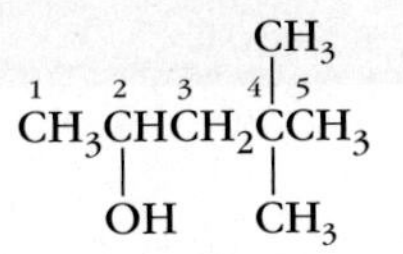

is 4,4-dimethyl-2-pentanol.

E-2b Aldehydes and Ketones

The systematic names of the first three aldehydes are methanal, ethanal, and propanal.

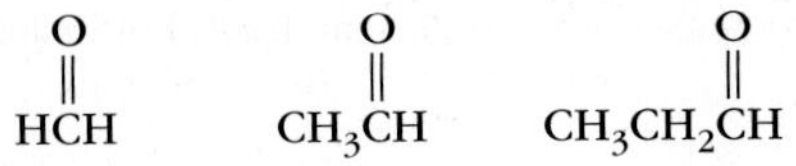

methanal (formaldehyde) ethanal (acetaldehyde) propanal (propionaldehyde)

For ketones, a number is used to designate the position of the carbonyl group, and the chain is numbered in a way that gives the carbonyl carbon the smallest number.

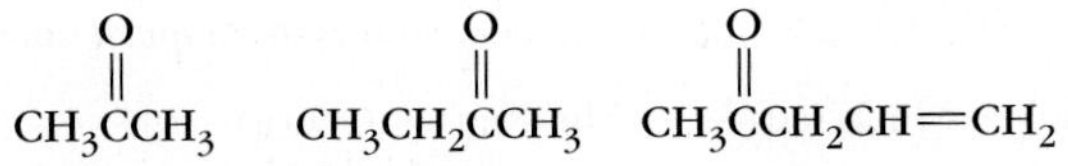

2-propanone (acetone) 2-butanone (methyl ethyl ketone) 4-penten-2-one

E-2c Carboxylic Acids

The systematic names of carboxylic acids are obtained by dropping the final *e* of the name of the corresponding alkane and adding *-oic acid.* For example, the name of

$$CH_3CH_2CH_2CH_2CH_2COOH$$

is hexanoic acid. Other examples are

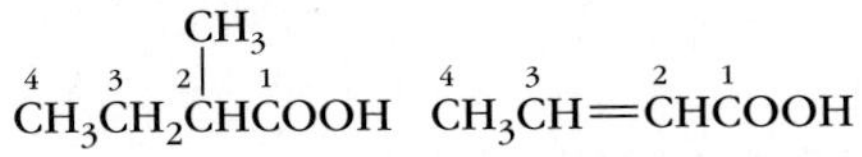

2-methylbutanoic acid 2-butenoic acid

E-2d Esters

The systematic names of esters are derived from the names of the alcohol and the acid used to prepare the ester. The general formula for esters is

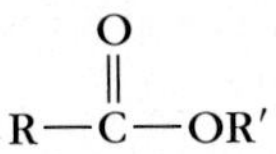

As shown in Section 10-5, the $R{-}\overset{\overset{O}{\|}}{C}$ comes from the acid and the OR′ comes from the alcohol. The alcohol part is named first, followed by the name of the acid changed to end in *-ate.* For example,

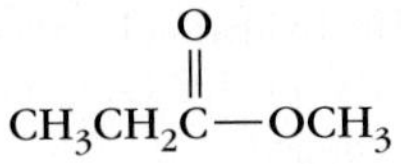

is named methyl propanoate and

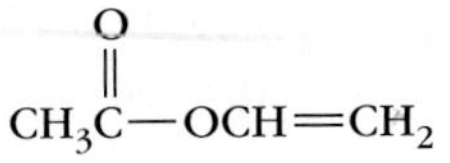

is named ethenyl ethanoate.

APPENDIX F Balancing Oxidation-Reduction Reactions

In Section 17-2 we showed that two half-reactions could be summed to give an overall reaction. This is straightforward if you know what the half-reactions are, but often you cannot find half-reaction equations in a table of data (such as Appendix G). Redox equations and half-reactions often involve water, hydrogen ions, or hydroxide ions as reactants or products. It is difficult to tell by observing the unbalanced equation how many H_2O, H^+, and OH^- are involved and whether they will be reactants or products—or even present at all. Fortunately, there are systematic ways to figure this out.

F-1 Acidic Solution

Consider the reaction of permanganate ion with oxalic acid in an acidic aqueous solution. The products are manganese(II) ions and carbon dioxide, so the *unbalanced* equation is

Oxalic acid, HOOC—COOH, is the simplest organic acid containing two carboxylic acid groups.

$$\underset{\text{oxalic acid}}{H_2C_2O_4(aq)} + \underset{\text{permanganate ion}}{MnO_4^-(aq)} \longrightarrow Mn^{2+}(aq) + CO_2(g) \qquad \text{(unbalanced equation)}$$

If you try to balance this equation by trial and error, you will almost certainly have a hard time balancing hydrogen and oxygen. You have probably already noticed that no hydrogen-containing species appears on the product side of the unbalanced equation. Because the reaction takes place in an aqueous acidic solution, water and hydrogen ions can be involved, but how many of each? Generating the balanced equation for a reaction like this one is best done by following a series of steps. In each step you use what you know about the oxidation and reduction half-reactions, as well as conservation of matter and conservation of electrical charge. Problem-Solving Example F.1 illustrates the steps that produce a balanced equation for a redox reaction occurring in acidic solution.

PROBLEM-SOLVING EXAMPLE F.1

Balancing Redox Equations for Reactions in Acidic Solutions

Balance the equation for the reaction of oxalic acid with acidic potassium permanganate in aqueous solution. The products of this reaction are CO_2 and Mn^{2+} ions.

Result $5\,H_2C_2O_4(aq) + 6\,H^+(aq) + 2\,MnO_4^-(aq) \longrightarrow 10\,CO_2(g) + 2\,Mn^{2+}(aq) + 8\,H_2O(\ell)$

Analyze If the reaction is a redox reaction, it is best to follow a systematic approach. In a redox reaction, oxidation numbers change.

Plan Follow the series of steps that are listed in the Execute section.

Execute

Step 1: ***Determine whether the reaction is an oxidation-reduction process. If it is, then determine what is reduced and what is oxidized.*** This is a redox reaction because the oxidation number of Mn changes from +7 in MnO_4^- to +2 in Mn^{2+}, so the Mn in MnO_4^- is reduced. The oxidation number of each C changes from +3 in $H_2C_2O_4$ to +4 in CO_2, so the C in $H_2C_2O_4$ is oxidized. The oxidation numbers of H (+1) and O (−2) are unchanged.

Step 2: ***Break the overall unbalanced equation into half-reactions.***

$$H_2C_2O_4(aq) \longrightarrow CO_2(g) \qquad \text{(oxidation half-reaction)}$$
$$MnO_4^-(aq) \longrightarrow Mn^{2+}(aq) \qquad \text{(reduction half-reaction)}$$

(a)
(b)

Reaction of oxalic acid with permanganate in acidic solution. The purple color of the permanganate (a) diminishes as the reaction proceeds to completion (b).

In acidic solution, balance O by adding H_2O, and balance H by adding H^+.

Step 3: ***Balance the atoms in each half-reaction.*** First, balance all atoms except for O and H, then balance O by adding H_2O and balance H by adding H^+. (Hydroxide ion, OH^-, cannot be used here because the reaction occurs in an acidic solution and the OH^- concentration is very low.)

Oxalic acid half-reaction: First, balance the carbon atoms in the half-reaction.

$$H_2C_2O_4(aq) \longrightarrow 2\ CO_2(g)$$

This step balances the O atoms as well (no H_2O needed here), so only H atoms remain to be balanced. Because the product side is deficient by two H, we put 2 H^+ there.

$$H_2C_2O_4(aq) \longrightarrow 2\ CO_2(g) + 2\ H^+(aq)$$

Permanganate half-reaction: The Mn atoms are already balanced, but the oxygen atoms are not balanced until H_2O is added. Adding 4 H_2O on the product side takes care of the needed oxygen atoms.

$$MnO_4^-(aq) \longrightarrow Mn^{2+}(aq) + 4\ H_2O(\ell)$$

Now there are 8 H atoms on the right and none on the left. To balance hydrogen atoms, 8 H^+ are placed on the left side of the half-reaction.

$$8\ H^+(aq) + MnO_4^-(aq) \longrightarrow Mn^{2+}(aq) + 4\ H_2O(\ell)$$

Step 4: ***Balance the half-reactions for charge using electrons (e^-).***

Oxalic acid half-reaction: There is a net charge of 0 on the left side and 2+ on the right. The reactants have lost two electrons when they form products. To show this fact, 2 e^- should appear on the right side.

$$H_2C_2O_4(aq) \longrightarrow 2\ CO_2(g) + 2\ H^+(aq) + 2\ e^-$$

This confirms that $H_2C_2O_4$ is the reducing agent (it loses electrons and is oxidized). The loss of two electrons is also in keeping with the increase in the oxidation number of each of two C atoms by 1, from +3 to +4. The 2 e^- also balance the charge on the product side of the equation.

Permanganate half-reaction: The MnO_4^- half-reaction has a charge of 7+ on the left and 2+ on the right. Therefore, to achieve a net 2+ charge on each side, 5 e^- must appear on the left. The gain of electrons shows that MnO_4^- is the oxidizing agent; it is reduced.

$$5\ e^- + 8\ H^+(aq) + MnO_4^-(aq) \longrightarrow Mn^{2+}(aq) + 4\ H_2O(\ell)$$

Step 5: ***Multiply the half-reactions by appropriate factors so that the oxidation half-reaction produces as many electrons as the reduction half-reaction accepts.*** In this case, one half-reaction involves two electrons, and the other half-reaction involves five electrons. It takes ten electrons to balance the overall reaction. The oxalic acid half-reaction must be multiplied by 5, and the permanganate half-reaction by 2.

$$5\ [H_2C_2O_4(aq) \longrightarrow 2\ CO_2(g) + 2\ H^+(aq) + 2\ e^-]$$
$$2\ [5\ e^- + 8\ H^+(aq) + MnO_4^-(aq) \longrightarrow Mn^{2+}(aq) + 4\ H_2O(\ell)]$$

Step 6: ***Add the half-reactions to give the overall reaction and cancel equal amounts of reactants and products that appear on both sides of the arrow.***

$$5\ H_2C_2O_4(aq) \longrightarrow 10\ CO_2(g) + 10\ H^+(aq) + \cancel{10\ e^-}$$
$$\cancel{10\ e^-} + 16\ H^+(aq) + 2\ MnO_4^-(aq) \longrightarrow 2\ Mn^{2+}(aq) + 8\ H_2O(\ell)$$

$$5\ H_2C_2O_4(aq) + 16\ H^+(aq) + 2\ MnO_4^-(aq) \longrightarrow 10\ CO_2(g) + 10\ H^+(aq) + 2\ Mn^{2+}(aq) + 8\ H_2O(\ell)$$

Because 16 H^+ appear on the left and 10 H^+ appear on the right, 10 H^+ are canceled, leaving 6 H^+ on the left.

$$5\ H_2C_2O_4(aq) + 6\ H^+(aq) + 2\ MnO_4^-(aq) \longrightarrow 10\ CO_2(g) + 2\ Mn^{2+}(aq) + 8\ H_2O(\ell)$$

Step 7: ***Check the balanced equation to make sure both atoms and charge are balanced.***
Atom balance: Each side of the equation has 2 Mn, 28 O, 10 C, and 16 H atoms.
Charge balance: Each side has a net charge of 4+.
On the left side, $(6 \times 1+) + (2 \times 1-) = 4+$.
On the right side, $2\,(2+) = 4+$.

PROBLEM-SOLVING PRACTICE F.1

Balance this equation for the reaction of Zn with $Cr_2O_7^{2-}$ in acidic aqueous solution.

$$Zn(s) + Cr_2O_7^{2-}(aq) \longrightarrow Cr^{3+}(aq) + Zn^{2+}(aq)$$

F-2 Basic Solution

For redox reactions that occur in basic solutions, the final electrochemical reaction must be completed with water and OH^- ions rather than water and H^+ ions that we used for acidic solutions. However, during the balancing process, the half-reactions can be treated as if they occurred in acidic solution. Then, at the end of the series of steps, the H^+ ions can be neutralized by adding an equal number of OH^- ions to both sides of the electrochemical equation and, if necessary, canceling water molecules that appear as both reactants and products. Problem-Solving Example F.2 illustrates how to balance a redox reaction in basic solution.

PROBLEM-SOLVING EXAMPLE F.2

Balancing Redox Equations for Reactions in Basic Solutions

In a nickel-cadmium (nicad) battery, cadmium metal and nickel(III) oxide, Ni_2O_3, react in alkaline solution to form $Cd(OH)_2$ and $Ni(OH)_2$. Write the balanced equation for this reaction.

Result $Cd(s) + Ni_2O_3(s) + 3\,H_2O(\ell) \longrightarrow Cd(OH)_2(s) + 2\,Ni(OH)_2(s)$

Analyze If the reaction is a redox reaction, it is best to follow a systematic approach. In a redox reaction, oxidation numbers change.

Plan Follow the systematic steps in the Execute section.

Execute

Step 1: ***Determine whether the reaction is an oxidation-reduction process. Then determine what is reduced and what is oxidized.*** This is a redox reaction because the oxidation number of Cd changes from 0 in Cd metal to +2 in $Cd(OH)_2$, so Cd metal is oxidized. The oxidation number of each Ni changes from +3 in Ni_2O_3 to +2 in $Ni(OH)_2$, so the Ni is reduced.

Step 2: ***Break the overall unbalanced equation into half-reactions.***

$$Cd(s) \longrightarrow Cd(OH)_2(s) \qquad \text{(oxidation half-reaction)}$$
$$Ni_2O_3(s) \longrightarrow Ni(OH)_2(s) \qquad \text{(reduction half-reaction)}$$

Step 3: ***Balance the atoms in each half-reaction.*** First, balance all atoms except the O and H atoms. Then, balance O by adding H_2O and balance H by adding H^+; that is, balance each half-reaction as if it were in an *acidic* solution.

Cd half-reaction: In the Cd half-reaction, the Cd atoms are balanced. Adding two water molecules on the left balances the two O atoms on the right, but this leaves four H atoms on the left and only two on the right. Adding two H^+ ions on the right balances H atoms in this half-reaction.

$$2\,H_2O(\ell) + Cd(s) \longrightarrow Cd(OH)_2(s) + 2\,H^+(aq)$$

Ni_2O_3 half-reaction: For the Ni_2O_3 half-reaction, a coefficient of 2 is needed for $Ni(OH)_2$ because there are two Ni atoms on the left. To balance the four O atoms

and four H atoms now on the right with the three O atoms on the left requires one water molecule and two H^+ ions on the left.

$$2\ H^+(aq) + H_2O(\ell) + Ni_2O_3(s) \longrightarrow 2\ Ni(OH)_2(s)$$

Step 4: ***Balance the half-reactions for charge using electrons.***

Cd half-reaction: The Cd half-reaction has two positive charges on the right, which requires 2 e^- as a product.

$$2\ H_2O(\ell) + Cd(s) \longrightarrow Cd(OH)_2(s) + 2\ H^+(aq) + 2e^- \quad \text{(balanced)}$$

Ni_2O_3 half-reaction: The Ni_2O_3 half-reaction requires 2 e^- as a reactant.

$$2\ H^+(aq) + H_2O(\ell) + Ni_2O_3(s) + 2\ e^- \longrightarrow 2\ Ni(OH)_2(s) \quad \text{(balanced)}$$

Step 5: ***Multiply the half-reactions by appropriate factors so that the reducing agent produces as many electrons as the oxidizing agent accepts.*** The Cd half-reaction produces two electrons, and the Ni_2O_3 half-reaction accepts two, so the number of electrons lost in the oxidation equals the number gained in the reduction.

Step 6: ***Because H^+ does not exist at any appreciable concentration in a basic solution, remove H^+ by adding an appropriate amount of OH^- to both sides of each half-reaction. H^+ and OH^- react to form H_2O.***

Cd half-reaction: Add two OH^- ions to each side to get

$$2\ OH^-(aq) + 2\ H_2O(\ell) + Cd(s) \longrightarrow Cd(OH)_2(s) + 2\ H_2O(\ell) + 2\ e^-$$

(On the product side, two OH^- ions plus two H^+ ions form two H_2O molecules.)

Ni_2O_3 half-reaction: Add two OH^- ions to each side to get

$$3\ H_2O(\ell) + Ni_2O_3(s) + 2\ e^- \longrightarrow 2\ Ni(OH)_2(s) + 2\ OH^-(aq)$$

Step 7: ***Add the half-reactions to give the overall reaction, and cancel reactants and products that appear in equal amounts on both sides of the reaction arrow.***

$$\cancel{2\ OH^-}(aq) + \cancel{2\ H_2O}(\ell) + Cd(s) \longrightarrow Cd(OH)_2(s) + \cancel{2\ H_2O}(\ell) + \cancel{2\ e^-}$$
$$3\ H_2O(\ell) + Ni_2O_3(s) + \cancel{2\ e^-} \longrightarrow Ni(OH)_2(s) + \cancel{2\ OH^-}(aq)$$
$$\overline{Cd(s) + Ni_2O_3(s) + 3\ H_2O(\ell) \longrightarrow Cd(OH)_2(s) + 2\ Ni(OH)_2(s)}$$

Step 8: ***Check the final result to make sure both atoms and charge are balanced.*** The equation is balanced. In the final equation, there are no net charges on either side of the reaction arrow, and the numbers of atoms of each kind on each side of the reaction arrow are equal.

PROBLEM-SOLVING PRACTICE F.2

In a basic solution, aluminum metal forms $[Al(OH)_4]^-$ ion as the metal reduces NO_3^- ion to NH_3. Write the balanced equation for this reaction.

QUESTIONS FOR APPENDIX F

Red-numbered questions have short answers at the back of this book in Appendix M and fully worked solutions in the *Student Solutions Manual.*

1. Write half-reactions for these changes:
   (a) Oxidation of nickel to Ni^{2+} ion
   (b) Reduction of H^+ ion to hydrogen gas
   (c) Reduction of Fe^{3+} ion to Fe^{2+} ion
   (d) Reduction of bromine to Br^- ion
   (e) Oxidation of nitrogen dioxide to nitrate ion in acidic solution
2. Write half-reactions for these changes:
   (a) Reduction of CrO_4^{2-} ion to Cr^{3+} ion in basic solution
   (b) Reduction of $Cr_2O_7^{2-}$ ion to Cr^{3+} ion in acid solution
   (c) Oxidation of chlorine gas to ClO^- ions in basic solution
   (d) Oxidation of water to hydrogen peroxide in acidic solution
   (e) Oxidation of nitric oxide to nitrite ion in acidic solution
3. Balance this redox reaction in a basic solution:
   $$Cd(s) + NO_3^-(aq) \longrightarrow [Cd(OH)_4]^{2-}(aq) + NH_3(aq)$$
4. Balance this redox reaction in a basic solution:
   $$NO_2^-(aq) + Sc(s) \longrightarrow NH_3(aq) + [Sc(OH)_6]^-(aq)$$

5. Balance these redox reactions.
   (a) $NO(g) + O_3(g) \longrightarrow NO_2(g)$
   (b) $H_2(g) + P_2(g) \longrightarrow PH_3(g)$
   (c) $H_2O_2(aq) + Fe^{2+}(aq) \longrightarrow H_2O(\ell) + Fe^{3+}(aq)$ in acidic solution
   (d) $Br^-(aq) + MnO_4^-(aq) \longrightarrow Br_2(g) + MnO_2(s)$ in basic solution
   (e) $CH_3OH(aq) + Cr_2O_7^{2-}(aq) \longrightarrow CH_2O(aq) + Cr^{3+}(aq)$ in acidic solution
   (f) $As_2O_3(s) + NO_3^-(aq) \longrightarrow H_3AsO_4(aq) + NO(g)$ in acidic solution

6. Balance these redox reactions.
   (a) $FeO(s) + O_2(g) \longrightarrow Fe_3O_4(s)$
   (b) $I^-(aq) + ClO^-(aq) \longrightarrow I_3^-(aq) + Cl^-(aq)$ in acidic solution
   (c) $Pb(s) + PbO_2(s) + HSO_4^-(aq) \longrightarrow PbSO_4(s)$ in acidic solution
   (d) $Al(s) + MnO_4^-(aq) \longrightarrow MnO_2(s) + Al(OH)_4^-(aq)$ in basic solution
   (e) $Cr(s) + CrO_4^{2-}(aq) \longrightarrow Cr(OH)_3(s)$ in basic solution
   (f) $Zn(Hg)$ (amalgam) $+ HgO(s) \longrightarrow ZnO(s) + Hg(\ell)$ in basic solution

APPENDIX G Ionization Constants for Weak Acids and Weak Bases at 25 °C*

Acid	Formula and Ionization Equation	K_a
Acetic	$CH_3COOH + H_2O \rightleftharpoons H_3O^+ + CH_3COO^-$	1.8×10^{-5}
Arsenic	$H_3AsO_4 + H_2O \rightleftharpoons H_3O^+ + H_2AsO_4^-$	$K_1 = 6.17 \times 10^{-3}$
	$H_2AsO_4^- + H_2O \rightleftharpoons H_3O^+ + HAsO_4^{2-}$	$K_2 = 1.17 \times 10^{-7}$
	$HAsO_4^{2-} + H_2O \rightleftharpoons H_3O^+ + AsO_4^{3-}$	$K_3 = 3.09 \times 10^{-12}$
Benzoic	$C_6H_5COOH + H_2O \rightleftharpoons H_3O^+ + C_6H_5COO^-$	1.2×10^{-4}
Boric	$B(OH)_3(H_2O) + H_2O \rightleftharpoons H_3O^+ + B(OH)_4^-$	5.8×10^{-10}
Carbonic	$H_2CO_3 + H_2O \rightleftharpoons H_3O^+ + HCO_3^-$	$K_1 = 4.3 \times 10^{-7}$
	$HCO_3^- + H_2O \rightleftharpoons H_3O^+ + CO_3^{2-}$	$K_2 = 4.7 \times 10^{-11}$
Chlorous	$HClO_2 + H_2O \rightleftharpoons H_3O^+ + ClO_2^-$	1.1×10^{-2}
Citric	$H_3C_6H_5O_7 + H_2O \rightleftharpoons H_3O^+ + H_2C_6H_5O_7^-$	$K_1 = 1.4 \times 10^{-3}$
	$H_2C_6H_5O_7^- + H_2O \rightleftharpoons H_3O^+ + HC_6H_5O_7^{2-}$	$K_2 = 4.5 \times 10^{-5}$
	$HC_6H_5O_7^{2-} + H_2O \rightleftharpoons H_3O^+ + C_6H_5O_7^{3-}$	$K_3 = 1.5 \times 10^{-6}$
Formic	$HCOOH + H_2O \rightleftharpoons H_3O^+ + HCOO^-$	3.0×10^{-4}
Hydrazoic	$HN_3 + H_2O \rightleftharpoons H_3O^+ + N_3^-$	1.0×10^{-5}
Hydrocyanic	$HCN + H_2O \rightleftharpoons H_3O^+ + CN^-$	3.3×10^{-10}
Hydrofluoric	$HF + H_2O \rightleftharpoons H_3O^+ + F^-$	6.8×10^{-4}
Hydrogen peroxide	$H_2O_2 + H_2O \rightleftharpoons H_3O^+ + HO_2^-$	2.1×10^{-12}
Hydrosulfuric†	$H_2S + H_2O \rightleftharpoons H_3O^+ + HS^-$	$K_1 = 1 \times 10^{-7}$
	$HS^- + H_2O \rightleftharpoons H_3O^+ + S^{2-}$	$K_2 = 1 \times 10^{-19}$
Hypochlorous	$HOCl + H_2O \rightleftharpoons H_3O^+ + OCl^-$	6.8×10^{-8}
Nitrous	$HNO_2 + H_2O \rightleftharpoons H_3O^+ + NO_2^-$	7.4×10^{-4}
Oxalic	$H_2C_2O_4 + H_2O \rightleftharpoons H_3O^+ + HC_2O_4^-$	$K_1 = 5.5 \times 10^{-2}$
	$HC_2O_4^- + H_2O \rightleftharpoons H_3O^+ + C_2O_4^{2-}$	$K_2 = 1.4 \times 10^{-4}$
Phenol	$HC_6H_5O + H_2O \rightleftharpoons H_3O^+ + C_6H_5O^-$	1.7×10^{-10}
Phosphoric	$H_3PO_4 + H_2O \rightleftharpoons H_3O^+ + H_2PO_4^-$	$K_1 = 7.2 \times 10^{-3}$
	$H_2PO_4^- + H_2O \rightleftharpoons H_3O^+ + HPO_4^{2-}$	$K_2 = 6.3 \times 10^{-8}$
	$HPO_4^{2-} + H_2O \rightleftharpoons H_3O^+ + PO_4^{3-}$	$K_3 = 4.6 \times 10^{-13}$
Phosphorous	$H_3PO_3 + H_2O \rightleftharpoons H_3O^+ + H_2PO_3^-$	$K_1 = 2.4 \times 10^{-2}$
	$H_2PO_3^- + H_2O \rightleftharpoons H_3O^+ + HPO_3^{2-}$	$K_2 = 2.9 \times 10^{-7}$
Propanoic	$CH_3CH_2COOH + H_2O \rightleftharpoons H_3O^+ + CH_3CH_2COO^-$	1.33×10^{-5}
Selenic	$H_2SeO_4 + H_2O \rightleftharpoons H_3O^+ + HSeO_4^-$	$K_1 =$ very large
	$HSeO_4^- + H_2O \rightleftharpoons H_3O^+ + SeO_4^{2-}$	$K_2 = 2.2 \times 10^{-2}$
Selenous	$H_2SeO_3 + H_2O \rightleftharpoons H_3O^+ + HSeO_3^-$	$K_1 = 2.5 \times 10^{-3}$
	$HSeO_3^- + H_2O \rightleftharpoons H_3O^+ + SeO_3^{2-}$	$K_2 = 1.6 \times 10^{-9}$
Sulfuric	$H_2SO_4 + H_2O \rightleftharpoons H_3O^+ + HSO_4^-$	$K_1 =$ very large
	$HSO_4^- + H_2O \rightleftharpoons H_3O^+ + SO_4^{2-}$	$K_2 = 1.1 \times 10^{-2}$
Sulfurous	$H_2SO_3 + H_2O \rightleftharpoons H_3O^+ + HSO_3^-$	$K_1 = 1.7 \times 10^{-2}$
	$HSO_3^- + H_2O \rightleftharpoons H_3O^+ + SO_3^{2-}$	$K_2 = 6.3 \times 10^{-8}$
Tellurous	$H_2TeO_3 + H_2O \rightleftharpoons H_3O^+ + HTeO_3^-$	$K_1 = 1.9 \times 10^{-3}$
	$HTeO_3^- + H_2O \rightleftharpoons H_3O^+ + TeO_3^{2-}$	$K_2 = 4.6 \times 10^{-10}$

Base	Formula and Ionization Equation	K_b
Ammonia[‡]	$NH_3 + H_2O \rightleftharpoons NH_4^+ + OH^-$	1.77×10^{-5}
Aniline[§]	$C_6H_5NH_2 + H_2O \rightleftharpoons C_6H_5NH_3^+ + OH^-$	3.9×10^{-10}
Dimethylamine[§]	$(CH_3)_2NH + H_2O \rightleftharpoons (CH_3)_2NH_2^+ + OH^-$	5.8×10^{-4}
Ethylenediamine	$(CH_2)_2(NH_2)_2 + H_2O \rightleftharpoons (CH_2)_2(NH_2)_2H^+ + OH^-$	$K_1 = 7.8 \times 10^{-5}$
	$(CH_2)_2(NH_2)_2H^+ + H_2O \rightleftharpoons (CH_2)_2(NH_2)_2H_2^{2+} + OH^-$	$K_2 = 2.1 \times 10^{-8}$
Hydrazine	$N_2H_4 + H_2O \rightleftharpoons N_2H_5^+ + OH^-$	$K_1 = 1.2 \times 10^{-6}$
	$N_2H_5^+ + H_2O \rightleftharpoons N_2H_6^{2+} + OH^-$	$K_2 = 1.3 \times 10^{-15}$
Hydroxylamine	$NH_2OH + H_2O \rightleftharpoons NH_3OH^+ + OH^-$	9.3×10^{-9}
Methylamine	$CH_3NH_2 + H_2O \rightleftharpoons CH_3NH_3^+ + OH^-$	5.0×10^{-4}
Pyridine	$C_5H_5N + H_2O \rightleftharpoons C_5H_5NH^+ + OH^-$	1.6×10^{-9}
Trimethylamine[§]	$(CH_3)_3N + H_2O \rightleftharpoons (CH_3)_3NH^+ + OH^-$	6.2×10^{-5}

*Taken from Högfeldt, E., and Perrin, D. D. *Stability Constants of Metal-Ion Complexes,* 1st ed. Oxford; New York: Pergamon, 1979–1982. International Union of Pure and Applied Chemistry Commission on Equilibrium Data.

[†]See Myers, R. "The New Low Value for the Second Dissociation Constant for H_2S." *Journal of Chemical Education,* Vol. 63, 1986; pp. 687–690.

[‡]This value from Read, A. J. *Journal of Solution Chemistry,* Vol. 11, No. 9, 1982, pp. 649–664.

[§]These values are from Meites, L., Ed. *Handbook of Analytical Chemistry,* 1st ed. New York: McGraw-Hill, 1963.

APPENDIX H Solubility Product Constants for Some Inorganic Compounds at 25 °C*,†

Substance	K_{sp}
Aluminum Compounds	
$AlAsO_4$	1.6×10^{-16}
$Al(OH)_3$ amorphous	1.3×10^{-33}
$AlPO_4$	6.3×10^{-19}
Barium Compounds	
$Ba_3(AsO_4)_2$	8.0×10^{-15}
$BaCO_3$	5.1×10^{-9}
BaC_2O_4	1.6×10^{-7}
$BaCrO_4$	1.2×10^{-10}
BaF_2	1.0×10^{-6}
$Ba(OH)_2$	5×10^{-3}
$Ba_3(PO_4)_2$	3.4×10^{-23}
$BaSeO_4$	3.5×10^{-8}
$BaSO_4$	1.1×10^{-10}
$BaSO_3$	8×10^{-7}
BaS_2O_3	1.6×10^{-5}
Bismuth Compounds	
$BiAsO_4$	4.4×10^{-10}
$BiOCl$[‡]	1.8×10^{-31}
$BiO(OH)$	4×10^{-10}
$Bi(OH)_3$	4×10^{-31}
BiI_3	8.1×10^{-19}
$BiPO_4$	1.3×10^{-23}
Cadmium Compounds	
$Cd_3(AsO_4)_2$	2.2×10^{-33}
$CdCO_3$	5.2×10^{-12}
$Cd(CN)_2$	1.0×10^{-8}
$Cd_2[Fe(CN)_6]$	3.2×10^{-17}
$Cd(OH)_2$ fresh	2.5×10^{-14}
Calcium Compounds	
$Ca_3(AsO_4)_2$	6.8×10^{-19}
$CaCO_3$	2.8×10^{-9}
$CaCrO_4$	7.1×10^{-4}
$CaC_2O_4 \cdot H_2O$[§]	4×10^{-9}
CaF_2	5.3×10^{-9}
$Ca(OH)_2$	5.5×10^{-6}
$CaHPO_4$	1×10^{-7}
$Ca_3(PO_4)_2$	2.0×10^{-29}
$CaSeO_4$	8.1×10^{-4}
$CaSO_4$	9.1×10^{-6}
$CaSO_3$	6.8×10^{-8}
Chromium Compounds	
$CrAsO_4$	7.7×10^{-21}
$Cr(OH)_2$	2×10^{-16}
$Cr(OH)_3$	6.3×10^{-31}
$CrPO_4 \cdot 4H_2O$ green[§]	2.4×10^{-23}
$CrPO_4 \cdot 4H_2O$ violet[§]	1.0×10^{-17}
Cobalt Compounds	
$Co_3(AsO_4)_2$	7.6×10^{-29}
$CoCO_3$	1.4×10^{-13}
$Co(OH)_2$ fresh	1.6×10^{-15}
$Co(OH)_3$	1.6×10^{-44}
$CoHPO_4$	2×10^{-7}
$Co_3(PO_4)_2$	2×10^{-35}
Copper Compounds	
$CuBr$	5.3×10^{-9}
$CuCl$	1.2×10^{-6}
$CuCN$	3.2×10^{-20}
CuI	1.1×10^{-12}
$CuOH$	1×10^{-14}
$CuSCN$	4.8×10^{-15}
$Cu_3(AsO_4)_2$	7.6×10^{-36}
$CuCO_3$	1.4×10^{-10}
$Cu_2[Fe(CN)_6]$	1.3×10^{-16}
$Cu(OH)_2$	2.2×10^{-20}
$Cu_3(PO_4)_2$	1.3×10^{-37}
Gold Compounds	
$AuCl$	2.0×10^{-13}
AuI	1.6×10^{-23}
$AuCl_3$	3.2×10^{-25}
$Au(OH)_3$	5.5×10^{-46}
AuI_3	1×10^{-46}
Iron Compounds	
$FeCO_3$	3.2×10^{-11}
$Fe(OH)_2$	8.0×10^{-16}
$FeC_2O_4 \cdot 2H_2O$[§]	3.2×10^{-7}
$FeAsO_4$	5.7×10^{-21}
$Fe_4[Fe(CN)_6]_3$	3.3×10^{-41}
$Fe(OH)_3$	4×10^{-38}
$FePO_4$	1.3×10^{-22}

(continued)

Substance	K_{sp}	Substance	K_{sp}
Lead Compounds		***Nickel Compounds***	
$Pb_3(AsO_4)_2$	4.0×10^{-36}	$Ni_3(AsO_4)_2$	3.1×10^{-26}
$PbBr_2$	4.0×10^{-5}	$NiCO_3$	6.6×10^{-9}
$PbCO_3$	7.4×10^{-14}	$2\ Ni(CN)_2 \rightleftharpoons Ni^{2+} + [Ni(CN)_4]^{2-}$	1.7×10^{-9}
$PbCl_2$	1.6×10^{-5}		
$PbCrO_4$	2.8×10^{-13}	$Ni_2[Fe(CN)_6]$	1.3×10^{-15}
PbF_2	2.7×10^{-8}	$Ni(OH)_2$ fresh	2.0×10^{-15}
$Pb(OH)_2$	1.2×10^{-15}	NiC_2O_4	4×10^{-10}
PbI_2	7.1×10^{-9}	$Ni_3(PO_4)_2$	5×10^{-31}
PbC_2O_4	4.8×10^{-10}	***Silver Compounds***	
$PbHPO_4$	1.3×10^{-10}	Ag_3AsO_4	1.0×10^{-22}
$Pb_3(PO_4)_2$	8.0×10^{-43}	$AgBr$	5.0×10^{-13}
$PbSeO_4$	1.4×10^{-7}	Ag_2CO_3	8.1×10^{-12}
$PbSO_4$	1.6×10^{-8}	$AgCl$	1.8×10^{-10}
$Pb(SCN)_2$	2.0×10^{-5}	Ag_2CrO_4	1.1×10^{-12}
Magnesium Compounds		$AgCN$	1.2×10^{-16}
$Mg_3(AsO_4)_2$	2.1×10^{-20}	$Ag_2Cr_2O_7$	2.0×10^{-7}
$MgCO_3$	3.5×10^{-8}	$Ag_4[Fe(CN)_6]$	1.6×10^{-41}
$MgCO_3 \cdot 3H_2O$§	2.1×10^{-5}	$AgOH$	2.0×10^{-8}
$MgC_2O_4 \cdot 2H_2O$§	1×10^{-8}	AgI	8.3×10^{-17}
MgF_2	6.5×10^{-9}	Ag_3PO_4	1.4×10^{-16}
$Mg(OH)_2$	1.8×10^{-11}	Ag_2SO_4	1.4×10^{-5}
$Mg_3(PO_4)_2$	10^{-23} to 10^{-27}	Ag_2SO_3	1.5×10^{-14}
$MgSeO_3$	1.3×10^{-5}	$AgSCN$	1.0×10^{-12}
$MgSO_3$	3.2×10^{-3}	***Strontium Compounds***	
$MgNH_4PO_4$	2.5×10^{-13}	$Sr_3(AsO_4)_2$	8.1×10^{-19}
Manganese Compounds		$SrCO_3$	1.1×10^{-10}
$Mn_3(AsO_4)_2$	1.9×10^{-29}	$SrCrO_4$	2.2×10^{-5}
$MnCO_3$	1.8×10^{-11}	$SrC_2O_4 \cdot H_2O$§	1.6×10^{-7}
$Mn_2[Fe(CN)_6]$	8.0×10^{-13}	$Sr_3(PO_4)_2$	4.0×10^{-28}
$Mn(OH)_2$	1.9×10^{-13}	$SrSO_4$	3.2×10^{-7}
$MnC_2O_4 \cdot 2H_2O$§	1.1×10^{-15}	$SrSO_3$	4×10^{-8}
Mercury Compounds		***Tin Compounds***	
Hg_2Br_2	5.6×10^{-23}	$Sn(OH)_2$	1.4×10^{-28}
Hg_2CO_3	8.9×10^{-17}	$Sn(OH)_4$	1×10^{-56}
$Hg_2(CN)_2$	5×10^{-40}	***Zinc Compounds***	
Hg_2Cl_2	1.3×10^{-18}	$Zn_3(AsO_4)_2$	1.3×10^{-28}
Hg_2CrO_4	2.0×10^{-9}	$ZnCO_3$	1.4×10^{-11}
Hg_2I_2	4.5×10^{-29}	$Zn_2[Fe(CN)_6]$	4.0×10^{-16}
Hg_2SO_4	7.4×10^{-7}	$Zn(OH)_2$	1.2×10^{-17}
Hg_2SO_3	1.0×10^{-27}	ZnC_2O_4	2.7×10^{-8}
$Hg_2(OH)_2$	2.0×10^{-24}	$Zn_3(PO_4)_2$	9.0×10^{-33}
$Hg(OH)_2$	3.0×10^{-26}		

*Taken from Patnaik, P. *Dean's Analytical Chemistry Handbook,* 2nd ed. New York: McGraw-Hill, 2004, Table 4.2.

†No metal sulfides are listed in this table because sulfide ion is such a strong base that the usual solubility product equilibrium equation does not apply. See Myers, R. J. *Journal of Chemical Education,* Vol. 63, 1986; pp. 687–690.

‡Taken from Meites, L., Ed. *Handbook of Analytical Chemistry,* 1st ed. New York: McGraw-Hill, 1963.

§Because $[H_2O]$ does not appear in equilibrium constants for equilibria in aqueous solution in general, it does not appear in the K_{sp} expressions for hydrated solids.

APPENDIX I Standard Electrode Potentials in Aqueous Solution at 25 °C*

Acidic Solution	Standard Electrode Potential, $E°$ (volts)
$F_2(g) + 2e^- \longrightarrow 2\ F^-(aq)$	2.87
$Co^{3+}(aq) + e^- \longrightarrow Co^{2+}(aq)$	1.92
$Au^+(aq) + e^- \longrightarrow Au(s)$	1.83
$H_2O_2(aq) + 2\ H^+(aq) + 2\ e^- \longrightarrow 2\ H_2O(\ell)$	1.763
$Ce^{4+}(aq) + e^- \longrightarrow Ce^{3+}(aq)$	1.72
$Pb^{4+}\ (aq) + 2\ e^- \longrightarrow Pb^{2+}(aq)$	1.69
$PbO_2(s) + SO_4^{2-}\ (aq) + 4\ H^+(aq) + 2\ e^- \longrightarrow PbSO_4(s) + 2\ H_2O(\ell)$	1.690
$NiO_2(s) + 4\ H^+(aq) + 2\ e^- \longrightarrow Ni^{2+}(aq) + 2\ H_2O(\ell)$	1.68
$2\ HClO(aq) + 2\ H^+(aq) + 2\ e^- \longrightarrow Cl_2(g) + 2\ H_2O(\ell)$	1.63
$Au^{3+}(aq) + 3\ e^- \longrightarrow Au(s)$	1.52
$MnO_4^-\ (aq) + 8\ H^+(aq) + 5\ e^- \longrightarrow Mn^{2+}(aq) + 4\ H_2O(\ell)$	1.51
$BrO_3^-(aq) + 6\ H^+(aq) + 5\ e^- \longrightarrow \frac{1}{2}\ Br_2(aq) + 3\ H_2O(\ell)$	1.478
$2\ ClO_3^-(aq) + 12\ H^+(aq) + 10\ e^- \longrightarrow Cl_2(g) + 6\ H_2O(\ell)$	1.47
$Cr_2O_7^{2-}(aq) + 14\ H^+(aq) + 6\ e^- \longrightarrow 2\ Cr^{3+}(aq) + 7\ H_2O(\ell)$	1.36
$Cl_2(g) + 2\ e^- \longrightarrow 2\ Cl^-(aq)$	1.358
$N_2H_5^+(aq) + 3\ H^+(aq) + 2\ e^- \longrightarrow 2\ NH_4^+(aq)$	1.275
$MnO_2(s) + 4\ H^+(aq) + 2\ e^- \longrightarrow Mn^{2+}(aq) + 2\ H_2O(\ell)$	1.23
$O_2(g) + 4\ H^+(aq) + 4\ e^- \longrightarrow 2\ H_2O(\ell)$	1.229
$ClO_4^-(aq) + 2\ H^+(aq) + 2\ e^- \longrightarrow ClO_3^-(aq) + H_2O(\ell)$	1.201
$IO_3^-(aq) + 6\ H^+(aq) + 5\ e^- \longrightarrow \frac{1}{2}\ I_2(aq) + 3\ H_2O(\ell)$	1.195
$Pt^{2+}(aq) + 2\ e^- \longrightarrow Pt(s)$	1.188
$Br_2(\ell) + 2\ e^- \longrightarrow 2\ Br^-(aq)$	1.066
$[AuCl_4]^-(aq) + 3\ e^- \longrightarrow Au(s) + 4\ Cl^-(aq)$	1.00
$NO_3^-(aq) + 4\ H^+(aq) + 3\ e^- \longrightarrow NO(g) + 2\ H_2O(\ell)$	0.96
$NO_3^-(aq) + 3\ H^+(aq) + 2\ e^- \longrightarrow HNO_2(aq) + H_2O(\ell)$	0.94
$Pd^{2+}(aq) + 2\ e^- \longrightarrow Pd(s)$	0.915
$2\ Hg^{2+}(aq) + 2\ e^- \longrightarrow Hg_2^{2+}(aq)$	0.9110
$Hg^{2+}(aq) + 2\ e^- \longrightarrow Hg(\ell)$	0.8535
$SbCl_6^-(aq) + 2\ e^- \longrightarrow SbCl_4^-(aq) + 2\ Cl^-(aq)$	0.84†
$Ag^+(aq) + e^- \longrightarrow Ag(s)$	0.7991
$Hg_2^{2+}(aq) + 2\ e^- \longrightarrow 2\ Hg(\ell)$	0.7960
$Fe^{3+}(aq) + e^- \longrightarrow Fe^{2+}(aq)$	0.771
$[PtCl_4]^{2-}(aq) + 2\ e^- \longrightarrow Pt(s) + 4\ Cl^-(aq)$	0.758
$[PtCl_6]^{2-}(aq) + 2\ e^- \longrightarrow [PtCl_4]^{2-}(aq) + 2\ Cl^-(aq)$	0.726
$O_2(g) + 2\ H^+(aq) + 2\ e^- \longrightarrow H_2O_2(aq)$	0.695
$TeO_2(s) + 4\ H^+(aq) + 4\ e^- \longrightarrow Te(s) + 2\ H_2O(\ell)$	0.604
$H_3AsO_4(aq) + 2\ H^+(aq) + 2\ e^- \longrightarrow HAsO_2(aq) + 2\ H_2O(\ell)$	0.560
$I_2(s) + 2\ e^- \longrightarrow 2\ I^-(aq)$	0.535
$Cu^+(aq) + e^- \longrightarrow Cu(s)$	0.521
$[RhCl_6]^{3-}(aq) + 3\ e^- \longrightarrow Rh(s) + 6\ Cl^-(aq)$	0.5
$Cu^{2+}(aq) + 2e^- \longrightarrow Cu(s)$	0.340
$Hg_2Cl_2(s) + 2e^- \longrightarrow 2\ Hg(\ell) + 2\ Cl^-(aq)$	0.27

(continued)

Acidic Solution	Standard Electrode Potential, $E°$ (volts)
$AgCl(s) + e^- \longrightarrow Ag(s) + Cl^-(aq)$	0.222
$Cu^{2+}(aq) + e^- \longrightarrow Cu^+(aq)$	0.159
$SO_4^{2-}(aq) + 4\,H^+(aq) + 2\,e^- \longrightarrow H_2SO_3(aq) + H_2O(\ell)$	0.158
$Sn^{4+}(aq) + 2e^- \longrightarrow Sn^{2+}(aq)$	0.15
$S(s) + 2\,H^+(aq) + 2\,e^- \longrightarrow H_2S(aq)$	0.144
$AgBr(s) + e^- \longrightarrow Ag(s) + Br^-(aq)$	0.0713
$2\,H^+(aq) + 2\,e^- \longrightarrow H_2(g)$ (reference electrode)	0
$N_2O(g) + 6\,H^+(aq) + H_2O(\ell) + 4\,e^- \longrightarrow 2\,NH_3OH^+(aq)$	−0.05
$HgS(s, black) + 2\,H^+(aq) + 2\,e- \longrightarrow Hg(\ell) + H_2S(g)$	−0.085
$Se(s) + 2\,H^+(aq) + 2\,e^- \longrightarrow H_2Se(aq)$	−0.115
$Pb^{2+}(aq) + 2\,e^- \longrightarrow Pb(s)$	−0.125
$Sn^{2+}(aq) + 2\,e^- \longrightarrow Sn(s)$	−0.1375
$AgI(s) + e^- \longrightarrow Ag(s) + I^-(aq)$	−0.1522
$[SnF_6]^{2-}(aq) + 4\,e^- \longrightarrow Sn(s) + 6\,F^-(aq)$	−0.200
$Ni^{2+}(aq) + 2\,e^- \longrightarrow Ni(s)$	−0.25
$Co^{2+}(aq) + 2\,e^- \longrightarrow Co(s)$	−0.277
$Tl^+(aq) + e^- \longrightarrow Tl(s)$	−0.3363
$PbSO_4(s) + 2\,e^- \longrightarrow Pb(s) + SO_4^{2-}(aq)$	−0.3505
$Cd^{2+}(aq) + 2\,e^- \longrightarrow Cd(s)$	−0.403
$Cr^{3+}(aq) + e^- \longrightarrow Cr^{2+}(aq)$	−0.424
$Fe^{2+}(aq) + 2\,e^- \longrightarrow Fe(s)$	−0.44
$2\,CO_2(g) + 2\,H^+(aq) + 2\,e^- \longrightarrow (COOH)_2(aq)$	−0.481
$TiO_2(s) + 4\,H^+(aq) + 2\,e^- \longrightarrow Ti^{2+}(aq) + 2\,H_2O(\ell)$	−0.502
$Ga^{3+}(aq) + 3\,e^- \longrightarrow Ga(s)$	−0.53
$Cr^{3+}(aq) + 3\,e^- \longrightarrow Cr(s)$	−0.74
$Zn^{2+}(aq) + 2\,e^- \longrightarrow Zn(s)$	−0.763
$Cr^{2+}(aq) + 2\,e^- \longrightarrow Cr(s)$	−0.90
$V^{2+}(aq) + 2\,e^- \longrightarrow V(s)$	−1.13
$Mn^{2+}(aq) + 2\,e^- \longrightarrow Mn(s)$	−1.18
$Zr^{4+}(aq) + 4\,e^- \longrightarrow Zr(s)$	−1.55
$Al^{3+}(aq) + 3\,e^- \longrightarrow Al(s)$	−1.676
$H_2(g) + 2\,e^- \longrightarrow 2\,H^-(aq)$	−2.25
$Mg^{2+}(aq) + 2\,e^- \longrightarrow Mg(s)$	−2.356
$Na^+(aq) + e^- \longrightarrow Na(s)$	−2.714
$Ca^{2+}(aq) + 2\,e^- \longrightarrow Ca(s)$	−2.84
$Sr^{2+}(aq) + 2\,e^- \longrightarrow Sr(s)$	−2.89
$Ba^{2+}(aq) + 2\,e^- \longrightarrow Ba(s)$	−2.92
$Rb^+(aq) + e^- \longrightarrow Rb(s)$	−2.925
$K^+(aq) + e^- \longrightarrow K(s)$	−2.925
$Li^+(aq) + e^- \longrightarrow Li(s)$	−3.045

Basic Solution	Standard Electrode Potential, $E°$ (volts)
$ClO^-(aq) + H_2O(\ell) + 2\,e^- \longrightarrow Cl^-(aq) + 2\,OH^-(aq)$	0.89
$OOH^-(aq) + H_2O(\ell) + 2\,e^- \longrightarrow 3\,OH^-(aq)$	0.867
$2\,NH_2OH(aq) + 2\,e^- \longrightarrow N_2H_4(aq) + 2\,OH^-(aq)$	0.73
$ClO_3^-(aq) + 3\,H_2O(\ell) + 6\,e^- \longrightarrow Cl^-(aq) + 6\,OH^-(aq)$	0.622
$MnO_4^-(aq) + 2\,H_2O(\ell) + 3\,e^- \longrightarrow MnO_2(s) + 4\,OH^-(aq)$	0.60
$MnO_4^-(aq) + e^- \longrightarrow MnO_4^{2-}(aq)$	0.56
$NiO_2(s) + 2\,H_2O(\ell) + 2\,e^- \longrightarrow Ni(OH)_2(s) + 2\,OH^-(aq)$	0.49
$Ag_2CrO_4(s) + 2\,e^- \longrightarrow 2\,Ag(s) + CrO_4^{2-}(aq)$	0.4491
$O_2(g) + 2\,H_2O(\ell) + 4\,e^- \longrightarrow 4\,OH^-(aq)$	0.401
$ClO_4^-(aq) + H_2O(\ell) + 2\,e^- \longrightarrow ClO_3^-(aq) + 2\,OH^-(aq)$	0.374
$Ag_2O(s) + H_2O(\ell) + 2\,e^- \longrightarrow 2\,Ag(s) + 2\,OH^-(aq)$	0.342
$2\,NO_2^-(aq) + 3\,H_2O(\ell) + 4\,e^- \longrightarrow N_2O(g) + 6\,OH^-(aq)$	0.15
$N_2H_4(aq) + 2\,H_2O(\ell) + 2\,e^- \longrightarrow 2\,NH_3(aq) + 2\,OH^-(aq)$	0.10
$HgO(s) + H_2O(\ell) + 2\,e^- \longrightarrow Hg(\ell) + 2\,OH^-(aq)$	0.0977
$O_2(g) + H_2O(\ell) + 2\,e^- \longrightarrow OOH^-(aq) + OH^-(aq)$	0.0649
$[Co(NH_3)_6]^{3+}(aq) + e^- \longrightarrow [Co(NH_3)_6]^{2+}(aq)$	0.058
$NO_3^-(aq) + H_2O(\ell) + 2\,e^- \longrightarrow NO_2^-(aq) + 2\,OH^-(aq)$	0.01
$MnO_2(s) + 2\,H_2O(\ell) + 2\,e^- \longrightarrow Mn(OH)_2(s) + 2\,OH^-(aq)$	−0.05
$CrO_4^{2-}(aq) + 4\,H_2O(\ell) + 3\,e^- \longrightarrow Cr(OH)_3(s) + 5\,OH^-(aq)$	−0.11
$Cu_2O(s) + H_2O(\ell) + 2\,e^- \longrightarrow 2\,Cu(s) + 2\,OH^-(aq)$	−0.365
$FeO_2^-(aq) + H_2O(\ell) + e^- \longrightarrow HFeO_2^-(aq) + OH^-(aq)$	−0.69
$HFeO_2^-(aq) + H_2O(\ell) + 2\,e^- \longrightarrow Fe(s) + 3\,OH^-(aq)$	−0.8
$2\,H_2O(\ell) + 2\,e^- \longrightarrow H_2(g) + 2\,OH^-(aq)$	−0.8277
$2\,NO_3^-(aq) + 2\,H_2O(\ell) + 2\,e^- \longrightarrow N_2O_4(g) + 4\,OH^-(aq)$	−0.86
$SO_4^{2-}(aq) + H_2O(\ell) + 2\,e^- \longrightarrow SO_3^{2-}(aq) + 2\,OH^-(aq)$	−0.936
$N_2(g) + 4\,H_2O(\ell) + 4\,e^- \longrightarrow N_2H_4(aq) + 4\,OH^-(aq)$	−1.16
$Zn(OH)_2(s) + 2\,e^- \longrightarrow Zn(s) + 2\,OH^-(aq)$	−1.246
$[Zn(OH)_4]^{2-}(aq) + 2\,e^- \longrightarrow Zn(s) + 4\,OH^-(aq)$	−1.285
$Cr(OH)_3(s) + 3\,e^- \longrightarrow Cr(s) + 3\,OH^-(aq)$	−1.33
$[Zn(CN)_4]^{2-}(aq) + 2\,e^- \longrightarrow Zn(s) + 4\,CN^-(aq)$	−1.34
$SiO_3^{2-}(aq) + 3\,H_2O(\ell) + 4\,e^- \longrightarrow Si(s) + 6\,OH^-(aq)$	−1.69

*Taken from Bard, A. J., Parsons, R., and Jordan, J. *Standard Potentials in Aqueous Solution.* New York: Marcel Dekker, 1985. International Union of Pure and Applied Chemistry, Commission on Electrochemistry and Electroanalytical Chemistry.

†Taken from Brown, R. A., and Swift, E. H. *Journal of the American Chemical Society,* Vol. 71, 1949; pp. 2719–2723.

APPENDIX J Selected Thermodynamic Values at 298.15 K*

Species	$\Delta_f H°$ (kJ/mol)	$S°$ (J K^{-1} mol^{-1})	$\Delta_f G°$ (kJ/mol)
Aluminum			
Al(s)	0	28.33	0
Al^{3+}(aq)	−531	−321.7	−485
$AlCl_3$(s)	−704.2	110.67	−628.8
Al_2O_3(s, corundum)	−1675.7	50.92	−1582.3
Argon			
Ar(g)	0	154.843	0
Ar(aq)	−12.1	59.4	16.4
Barium			
$BaCl_2$(s)	−858.6	123.68	−810.4
BaO(s)	−553.5	70.42	−525.1
$BaSO_4$(s)	−1473.2	132.2	−1362.2
$BaCO_3$(s)	−1216.3	112.1	−1137.6
Beryllium			
Be(s)	0	9.5	0
$Be(OH)_2$(s)	−902.5	51.9	−815
Bromine			
Br(g)	111.884	175.022	82.396
Br_2(ℓ)	0	152.231	0
Br_2(g)	30.907	245.463	3.110
Br_2(aq)	−2.59	130.5	3.93
Br^-(aq)	−121.55	82.4	−103.96
BrCl(g)	14.64	240.10	−0.98
BrF_3(g)	−255.6	292.53	−229.43
HBr(g)	−36.40	198.695	−53.45
Calcium			
Ca(s)	0	41.42	0
Ca(g)	178.2	154.884	144.3
Ca^{2+}(g)	1925.9	—	—
Ca^{2+}(aq)	−542.83	−53.1	−553.58
CaC_2(s)	−59.8	69.96	−64.9
$CaCO_3$(s, calcite)	−1206.92	92.9	−1128.79
$CaCl_2$(s)	−795.8	104.6	−748.1
CaF_2(s)	−1219.6	68.87	−1167.3
CaH_2(s)	−186.2	42	−147.2
CaO(s)	−635.09	39.75	−604.03
CaS(s)	−482.4	56.5	−477.4
$Ca(OH)_2$(s)	−986.09	83.39	−898.49
$Ca(OH)_2$(aq)	−1002.82	−74.5	−868.07
$CaSO_4$(s)	−1434.11	106.7	−1321.79

Species	$\Delta_f H°$ (kJ/mol)	$S°$ (J K^{-1} mol^{-1})	$\Delta_f G°$ (kJ/mol)
Carbon			
C(s, graphite)	0	5.74	0
C(s, diamond)	1.895	2.377	2.9
C(g)	716.682	158.096	671.257
CCl_4(ℓ)	−135.44	216.4	−65.21
CCl_4(g)	−102.9	309.85	−60.59
$CHCl_3$(ℓ)	−134.47	201.7	−73.66
$CHCl_3$(g)	−103.14	295.71	−70.34
CH_4(g, methane)	−74.81	186.264	−50.72
C_2H_2(g, ethyne)	226.73	200.94	209.2
C_2H_4(g, ethene)	52.26	219.56	68.15
C_2H_6(g, ethane)	−84.68	229.6	−32.82
C_3H_8(g, propane)	−103.8	269.9	−23.49
C_4H_{10}(g, butane)	−126.148	310.227	−16.985
C_6H_6(ℓ, benzene)	49.03	172.8	124.5
C_6H_{14}(ℓ, hexane)	−198.782	296.018	−4.035
C_8H_{18}(g, octane)	−208.447	466.835	16.718
C_8H_{18}(ℓ, octane)	−249.952	361.205	6.707
CH_3OH(ℓ, methanol)	−238.66	126.8	−166.27
CH_3OH(g, methanol)	−200.66	239.81	−161.96
CH_3OH(aq, methanol)	−245.931	133.1	−175.31
C_2H_5OH(ℓ, ethanol)	−277.69	160.7	−174.78
C_2H_5OH(g, ethanol)	−235.1	282.7	−168.49
C_2H_5OH(aq, ethanol)	−288.3	148.5	−181.64
$C_6H_{12}O_6$(s, glucose)	−1274.4	235.9	−917.2
CH_3COO^-(aq)	−486.01	86.6	−369.31
CH_3COOH(aq)	−485.76	178.7	−396.46
CH_3COOH(ℓ)	−484.5	159.8	−389.9
CO(g)	−110.525	197.674	−137.168
CO_2(g)	−393.509	213.74	−394.359
H_2CO_3(aq)	−699.65	187.4	−623.08
HCO_3^-(aq)	−691.99	91.2	−586.77
CO_3^{2-}(aq)	−677.14	−56.9	−527.81
$HCOO^-$(aq)	−425.55	92.0	−351.0
HCOOH(aq)	−425.43	163	−372.3
HCOOH(ℓ)	−424.72	128.95	−361.35
CS_2(g)	117.36	237.84	67.12
CS_2(ℓ)	89.70	151.34	65.27
$COCl_2$(g)	−218.8	283.53	−204.6
Cesium			
Cs(s)	0	85.23	0
Cs^+(g)	457.964	—	—
CsCl(s)	−443.04	101.17	−414.53
Chlorine			
Cl(g)	121.679	165.198	105.68
Cl^-(g)	−233.13	—	—
Cl^-(aq)	−167.159	56.5	−131.228
Cl_2(g)	0	223.066	0
Cl_2(aq)	−23.4	121	6.94
HCl(g)	−92.307	186.908	−95.299
HCl(aq)	−167.159	56.5	−131.228

(continued)

Species	$\Delta_f H°$ (kJ/mol)	$S°$ (J K^{-1} mol^{-1})	$\Delta_f G°$ (kJ/mol)
Chlorine, continued			
ClO_2(g)	102.5	256.84	120.5
Cl_2O(g)	80.3	266.21	97.9
ClO^-(aq)	−107.1	42.0	−36.8
HClO(aq)	−120.9	142	−79.9
ClF_3(g)	−163.2	281.61	−123.0
Chromium			
Cr(s)	0	23.77	0
Cr_2O_3(s)	−1139.7	81.2	−1058.1
$CrCl_3$(s)	−556.5	123	−486.1
Copper			
Cu(s)	0	33.15	0
CuO(s)	−157.3	42.63	−129.7
$CuCl_2$(s)	−220.1	108.07	−175.7
$CuSO_4$(s)	−771.36	109	−661.8
Fluorine			
F_2(g)	0	202.78	0
F(g)	78.99	158.754	61.91
F^-(g)	−255.39	—	—
F^-(aq)	−332.63	−13.8	−278.79
HF(g)	−271.1	173.779	−273.2
HF(aq, un-ionized)	−320.08	88.7	−296.82
HF(aq, ionized)	−332.63	−13.8	−278.79
Hydrogen†			
H_2(g)	0	130.684	0
H_2(aq)	−4.2	57.7	17.6
HD(g)	0.318	143.801	−1.464
D_2(g)	0	144.960	0
H(g)	217.965	114.713	203.247
H^+(g)	1536.202	—	—
H^+(aq)	0	0	0
OH^-(aq)	−229.994	−10.75	−157.244
H_2O(ℓ)	−285.83	69.91	−237.129
H_2O(g)	−241.818	188.825	−228.572
H_2O_2(ℓ)	−187.78	109.6	−120.35
H_2O_2(aq)	−191.17	143.9	−134.03
HO_2^-(aq)	−160.33	23.8	−67.3
HDO(ℓ)	−289.888	79.29	−241.857
D_2O(ℓ)	−294.600	75.94	−243.439
Iodine			
I_2(s)	0	116.135	0
I_2(g)	62.438	260.69	19.327
I_2(aq)	22.6	137.2	16.40
I(g)	106.838	180.791	70.25
I^-(g)	−197	—	—
I^-(aq)	−55.19	111.3	−51.57
I_3^-(aq)	−51.5	239.3	−51.4
HI(g)	26.48	206.594	1.70
HI(aq, ionized)	−55.19	111.3	−51.57

Species	$\Delta_f H°$ (kJ/mol)	$S°$ (J K^{-1} mol^{-1})	$\Delta_f G°$ (kJ/mol)
Iodine, continued			
IF(g)	−95.65	236.17	−118.51
ICl(g)	17.78	247.551	−5.46
ICl(ℓ)	−23.89	135.1	−13.58
IBr(g)	40.84	258.773	3.69
ICl_3(s)	−89.5	167.4	−22.29
Iron			
Fe(s)	0	27.28	0
FeO(s, wustite)	−266.27	57.49	−245.12
Fe_2O_3(s, hematite)	−824.2	87.4	−742.2
Fe_3O_4(s, magnetite)	−1118.4	146.4	−1015.4
$FeCl_2$(s)	−341.79	117.95	−302.3
$FeCl_3$(s)	−399.49	142.3	−344
FeS_2(s, pyrite)	−178.2	52.93	−166.9
$Fe(CO)_5$(ℓ)	−774	338.1	−705.3
Lead			
Pb(s)	0	64.81	0
$PbCl_2$(s)	−359.41	136	−314.1
PbO(s, yellow)	−217.32	68.7	−187.89
PbS(s)	−100.4	91.2	−98.7
$PbSO_4$(s)	−919.94	148.57	−813.14
$PbCl_4$(ℓ)	−329.3	—	—
PbO_2(s)	−277.4	68.6	−217.33
Lithium			
Li(s)	0	29.12	0
Li^+(g)	685.783	—	—
LiOH(s)	−484.93	42.8	−438.95
LiOH(aq)	−508.48	2.8	−450.58
LiCl(s)	−408.61	59.33	−384.37
Magnesium			
Mg(s)	0	32.68	0
Mg^{2+}(aq)	−466.85	−138.1	−454.8
$MgCl_2$(g)	−400.4	—	—
$MgCl_2$(s)	−641.32	89.62	−591.79
$MgCl_2$(aq)	−801.15	−25.1	−717.1
MgO(s)	−601.70	26.94	−569.43
$Mg(OH)_2$(s)	−924.54	63.18	−833.51
MgS(s)	−346	50.33	−341.8
$MgSO_4$(s)	−1284.9	91.6	−1170.6
$MgCO_3$(s)	−1095.8	65.7	−1012.1
Mercury			
Hg(ℓ)	0	76.02	0
$HgCl_2$(s)	−224.3	146	−178.6
HgO(s, red)	−90.83	70.29	−58.539
HgS(s, red)	−58.2	82.4	−50.6
Nickel			
Ni(s)	0	29.87	0
NiO(s)	−239.7	37.99	−211.7
$NiCl_2$(s)	−305.332	97.65	−259.032

(continued)

Species	$\Delta_f H°$ (kJ/mol)	$S°$ (J K^{-1} mol^{-1})	$\Delta_f G°$ (kJ/mol)
Nitrogen			
N_2(g)	0	191.61	0
N_2(aq)	−10.8	—	—
N(g)	472.704	153.298	455.563
NH_3(g)	−46.11	192.45	−16.45
NH_3(aq)	−80.29	111.3	−26.50
NH_4^+(aq)	−132.51	113.4	−79.31
N_2H_4(ℓ)	50.63	121.21	149.34
NH_4Cl(s)	−314.43	94.6	−202.87
NH_4Cl(aq)	−299.66	169.9	−210.52
NH_4NO_3(s)	−365.56	151.08	−183.87
NH_4NO_3(aq)	−339.87	259.8	−190.56
NO(g)	90.25	210.761	86.55
NO_2(g)	33.18	240.06	51.31
N_2O(g)	82.05	219.85	104.20
N_2O_4(g)	9.16	304.29	97.89
N_2O_4(ℓ)	−19.50	209.2	97.54
NOCl(g)	51.71	261.69	66.08
HNO_3(ℓ)	−174.10	155.60	−80.71
HNO_3(g)	−135.06	266.38	−74.72
HNO_3(aq)	−207.36	146.4	−111.25
NO_3^-(aq)	−205.0	146.4	−108.74
NF_3(g)	−124.7	260.73	−83.2
Oxygen[†]			
O_2(g)	0	205.138	0
O_2(aq)	−11.7	110.9	16.4
O(g)	249.170	161.055	231.731
O_3(g)	142.7	238.93	163.2
OH^-(aq)	−229.994	−10.75	−157.244
Phosphorus			
P_4(s, white)	0	164.36	0
P_4(s, red)	−70.4	91.2	−48.4
P(g)	314.64	163.193	278.25
PH_3(g)	5.4	210.23	13.4
PCl_3(g)	−287.0	311.78	−267.8
PCl_3(ℓ)	−319.7	217.1	−272.3
PCl_5(s)	−443.5	—	—
P_4O_{10}(s)	−2984.0	228.86	−2697.7
H_3PO_4(s)	−1279	110.5	−1119.1
Potassium			
K(s)	0	64.18	0
KF(s)	−567.27	66.57	−537.75
KCl(s)	−436.747	82.59	−409.14
KCl(aq)	−419.53	159.0	−414.49
KBr(s)	−393.798	95.90	−380.66
KI(s)	−327.900	106.32	−324.892
$KClO_3$(s)	−397.73	143.1	−296.25
KOH(s)	−424.764	78.9	−379.08
KOH(aq)	−482.37	91.6	−440.5

Species	$\Delta_f H°$ (kJ/mol)	$S°$ (J K^{-1} mol^{-1})	$\Delta_f G°$ (kJ/mol)
Silicon			
Si(s)	0	18.83	0
$SiBr_4$(ℓ)	−457.3	277.8	−443.8
SiC(s)	−65.3	16.61	−62.8
$SiCl_4$(g)	−657.01	330.73	−616.98
SiH_4(g)	34.3	204.62	56.9
SiF_4(g)	−1614.94	282.49	−1572.65
SiO_2(s, quartz)	−910.94	41.84	−856.64
Silver			
Ag(s)	0	42.55	0
Ag^+(aq)	105.579	72.68	77.107
Ag_2O(s)	−31.05	121.3	−11.2
AgCl(s)	−127.068	96.2	−109.789
AgI(s)	−61.84	115.5	−66.19
AgN_3(s)	308.8	104.2	376.2
$AgNO_3$(s)	−124.39	140.92	−33.41
$AgNO_3$(aq)	−101.8	219.2	−34.16
Sodium			
Na(s)	0	51.21	0
Na(g)	107.32	153.712	76.761
Na^+(g)	609.358	—	—
Na^+(aq)	−240.12	59.0	−261.905
NaF(s)	−573.647	51.46	−543.494
NaF(aq)	−572.75	45.2	−540.68
NaCl(s)	−411.153	72.13	−384.138
NaCl(g)	−176.65	229.81	−196.66
NaCl(aq)	−407.27	115.5	−393.133
NaBr(s)	−361.062	86.82	−348.983
NaBr(aq)	−361.665	141.4	−365.849
NaI(s)	−287.78	98.53	−286.06
NaI(aq)	−295.31	170.3	−313.47
NaOH(s)	−425.609	64.455	−379.494
NaOH(aq)	−470.114	48.1	−419.15
$NaClO_3$(s)	−365.774	123.4	−262.259
$NaHCO_3$(s)	−950.81	101.7	−851.0
Na_2CO_3(s)	−1130.68	134.98	−1044.44
Na_2SO_4(s)	−1387.08	149.58	−1270.16
Sulfur			
S(s, monoclinic)	0.33	—	—
S(s, rhombic)	0	31.80	0
S(g)	278.805	167.821	238.250
S^{2-}(aq)	33.1	−14.6	85.8
S_2Cl_2(g)	−18.4	331.5	−31.8
SF_6(g)	−1209	291.82	−1105.3
SF_4(g)	−774.9	292.03	−731.3
H_2S(g)	−20.63	205.79	−33.56
H_2S(aq)	−39.7	121	−27.83
HS^-(aq)	17.6	62.8	12.08
SO_2(g)	−296.830	248.22	−300.194
SO_3(g)	−395.72	256.76	−371.06

(continued)

Species	$\Delta_f H°$ (kJ/mol)	$S°$ (J K^{-1} mol^{-1})	$\Delta_f G°$ (kJ/mol)
Sulfur, continued			
$SOCl_2(g)$	−212.5	309.77	−198.3
$SO_4^{2-}(aq)$	−909.27	20.1	−744.53
$H_2SO_4(\ell)$	−813.989	156.904	−690.003
$H_2SO_4(aq)$	−909.27	20.1	−744.53
$HSO_4^-(aq)$	−887.34	131.8	−755.91
Tin			
Sn(s, white)	0	51.55	0
Sn(s, gray)	−2.09	44.14	0.13
$SnCl_2(s)$	−325.1	—	—
$SnCl_4(\ell)$	−511.3	258.6	−440.1
$SnCl_4(g)$	−471.5	365.8	−432.2
$SnO_2(s)$	−580.7	52.3	−519.6
Titanium			
Ti(s)	0	30.63	0
$TiCl_4(\ell)$	−804.2	252.34	−737.2
$TiCl_4(g)$	−763.2	354.9	−726.7
$TiO_2(s)$	−939.7	49.92	−884.5
Uranium			
U(s)	0	50.21	0
$UO_2(s)$	−1084.9	77.03	−1031.7
$UO_3(s)$	−1223.8	96.11	−1145.9
$UF_4(s)$	−1914.2	151.67	−1823.3
$UF_6(g)$	−2147.4	377.9	−2063.7
$UF_6(s)$	−2197.0	227.6	−2068.5
Zinc			
Zn(s)	0	41.63	0
$ZnCl_2(s)$	−415.05	111.46	−369.398
ZnO(s)	−348.28	43.64	−318.3
ZnS(s, sphalerite)	−205.98	57.7	−201.29

*Taken from Wagman, D. D., Evans, W. H., Parker, V. B., Schumm, R. H., Halow, I., Bailey, S. M., Churney, K. L., and Nuttall, R. "The NBS Tables of Chemical Thermodynamic Properties." *Journal of Physical and Chemical Reference Data,* Vol. 11, Suppl. 2, 1982.

†Many hydrogen-containing and oxygen-containing compounds are listed only under other elements; for example, HNO_3 appears under nitrogen.

APPENDIX K Answers to Problem-Solving Practice Problems

Chapter 1

In Answer 1.1 we have included headings for these problem-solving steps: *Analyze the problem, Plan a solution, Execute the plan,* and *Check your result.* To save space, we provide results only for subsequent problems.

1.1 (1) *Analyze the problem:* You are asked to find the volume of the sample, and you know the mass.

(2) *Plan a solution:* Density relates mass and volume and is the appropriate conversion factor, so look up the density in a table. Volume is proportional to mass, so the mass has to be either multiplied by the density or multiplied by the reciprocal of the density. Use the units to decide which.

(3) *Execute the plan:* According to Table 1.1, the density of benzene is 0.880 g/mL. Setting up the calculation so that the unit (grams) cancels gives

$$V_{\text{benzene}} = 4.33 \text{ g} \times \frac{1 \text{ mL}}{0.880 \text{ g}} = 4.92 \text{ mL}$$

Notice that the result is expressed to three significant figures, because both the mass and the density had three significant figures.

(4) *Check your result:* Because the density is a little less than 1.00 g/mL, the volume in milliliters should be a little larger than the mass in grams. The calculated answer, 4.92 mL, is a little larger than the mass, 4.33 g.

1.2 Substance A must be a mixture because some of it dissolves and some, substance B, does not.

Substance C is the soluble portion of substance A. Because all of substance C dissolves in water there is no way to determine how many components it has. Additionally, it is not possible to determine whether the one or more components themselves are elements or compounds. Therefore it is not possible to say whether C is an element, a compound, or a mixture.

The only thing we know about substance B is that it is insoluble in water. We do not know whether it is one insoluble substance, or more than one insoluble substance. Additionally, we do not know whether the substance or substances of B are elements or compounds. Therefore it is not possible to say whether B is an element, a compound, or a mixture.

1.3

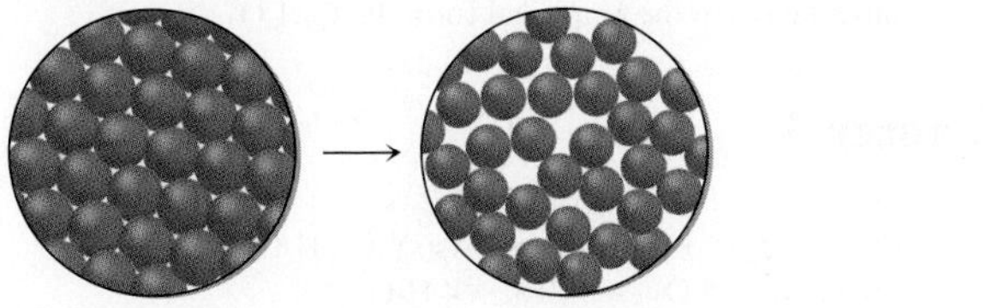

Cu(s) ⟶ Cu(ℓ)

Solid copper at the macroscale is a reddish orange-colored metal; it conducts electricity and retains its shape unless bent or hammered. Liquid copper is similar to the solid in color and electrical conductivity but is fluid and adopts the shape of its container.

Chapter 2

2.1 $\text{Diameter} = \dfrac{1.0 \text{ mm}}{3.97 \times 10^6 \text{ atoms}} \times \dfrac{10^9 \text{ pm}}{1 \text{ mm}} = 252 \text{ pm};$

Radius = (0.500)(252 pm) = 126 pm

2.2 $r = 5.15 \times 10^{-15} \text{ m} \left(\dfrac{10^2 \text{ cm}}{1 \text{ m}}\right) = 5.15 \times 10^{-13} \text{ cm}$

$$V = \frac{4}{3}\pi r^3 = 4.189(5.15 \times 10^{-13} \text{ cm})^3 = 5.72 \times 10^{-37} \text{ cm}^3$$

$$d = \frac{m}{V} = \frac{3.27 \times 10^{-22} \text{ g}}{5.72 \times 10^{-37} \text{ cm}^3} = 5.72 \times 10^{14} \text{ g/cm}^3$$

2.3 (a) A phosphorus atom ($Z = 15$) with 16 neutrons has $A = 31$.

(b) A neon-22 atom has $A = 22$ and $Z = 10$, so there are 10 protons and 10 electrons; the number of neutrons is $A - Z = 22 - 10 = 12$ neutrons.

(c) The periodic table shows that the element with 82 protons is lead. The mass number of this isotope of lead is 82 + 125 = 207, so the correct symbol is $^{207}_{82}Pb$.

2.4 All magnesium isotopes have 12 protons, $Z = 12$. The isotope with 12 neutrons has $A = 24$ and the notation is $^{24}_{12}Mg$; the isotope with 13 neutrons has $A = 25$ and notation $^{25}_{12}Mg$; the isotope with 14 neutrons has $A = 26$ and $^{26}_{12}Mg$.

2.5 Calculate the weighted average of the isotopic masses.

$$m_{\text{ave}} = \left(\frac{7.500 \text{ atoms } ^6\text{Li}}{100 \text{ Li atoms}}\right)\left(\frac{6.015121 \text{ u}}{1 \text{ atom } ^6\text{Li}}\right) + \left(\frac{92.50 \text{ atoms } ^7\text{Li}}{100 \text{ Li atoms}}\right)\left(\frac{7.016003 \text{ u}}{1 \text{ atom } ^7\text{Li}}\right) = 6.941 \text{ u per Li atom}$$

2.6 (a) A Ca^{4+} charge is unlikely because calcium is in Group 2A, the elements of which lose two electrons to form 2+ ions.

(b) Cr^{2+} is possible because chromium is a transition metal ion that forms 2+ and 3+ ions.

(c) Strontium is a Group 2A metal and forms 2+ ions; thus, a Sr^- ion is highly unlikely.

(d) Selenium is a Group 6A nonmetal that gains two electrons to form 2− ions; loss of two electrons from a selenium atom to form an Se^{2+} ion is unlikely.

2.7 (a) $CaSO_3$ contains one Ca^{2+} ion and one sulfite ion, SO_3^{2-}.

(b) $Mg_3(PO_4)_2$ contains three Mg^{2+} ions and two phosphate ions, PO_4^{3-}.

2.8 (a) $MgBr_2$ (b) Li_2CO_3 (c) CuI, CuI_2 (d) NH_4Cl

2.9 (a) KNO_2 is potassium nitrite.

(b) $NaHSO_3$ is sodium hydrogen sulfite.

(c) $Mn(OH)_2$ is manganese(II) hydroxide.

(d) $Mn_2(SO_4)_3$ is manganese(III) sulfate.

(e) Ba_3N_2 is barium nitride.

(f) LiH is lithium hydride.

2.10 (a) KH_2PO_4 (b) CuOH (c) NaClO

(d) NH_4ClO_4 (e) $CrCl_3$ (f) $FeSO_3$

2.11 (a) $C_{10}H_{11}N_5O_{13}P_3$ (b) $C_{18}H_{27}NO_3$

(c) $C_4H_{10}O$

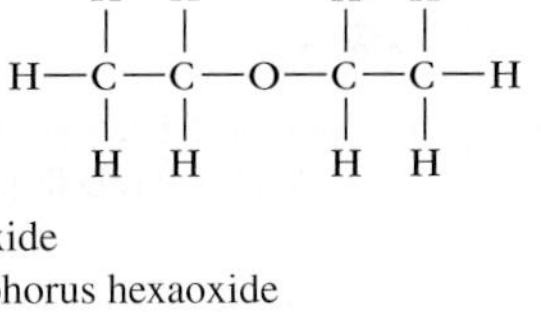

2.12 (a) Sulfur dioxide

(b) Tetraphosphorus hexaoxide

(c) Carbon tetrachloride

2.13 (a) $n_{Ti} = 4.00 \text{ g Ti} \times \dfrac{1 \text{ mol Ti}}{47.87 \text{ g Ti}} = 8.36 \times 10^{-2} \text{ mol Ti}$

(b) $m_{Ag} = 3.00 \times 10^{-2} \text{ mol Ag} \times \dfrac{107.9 \text{ g Ag}}{1 \text{ mol Ag}} = 3.24 \text{ g Ag}$

2.14 (a) The molar mass of $Ca(NO_3)_2$ is 164.09 g/mol.

$$n_{Ca(NO_3)_2} = 12.0 \text{ g} \times \frac{1 \text{ mol}}{164.09 \text{ g}} = 7.31 \times 10^{-2} \text{ mol}$$

(b) The molar mass of $KMnO_4$ is 158.03 g/mol.

$$n_{KMnO_4} = 12.0 \text{ g} \times \frac{1 \text{ mol}}{158.03 \text{ g}} = 7.59 \times 10^{-2} \text{ mol}$$

(c) The molar mass of nickel(II) chloride hexahydrate is 237.68 g/mol.

$$n_{NiCl_2 \cdot 6H_2O} = 12.0 \text{ g} \times \frac{1 \text{ mol}}{237.68 \text{ g}} = 5.05 \times 10^{-2} \text{ mol}$$

2.15 (a) The molar mass of sucrose, $C_{12}H_{22}O_{11}$, is 342.3 g/mol.

$$m_{sucrose} = 5.00 \times 10^{-3} \text{ mol} \times \frac{342.3 \text{ g}}{1 \text{ mol}} = 1.71 \text{ g}$$

(b) $m_{ACTH} = 3.00 \times 10^{-6} \text{ mol} \times \dfrac{4600 \text{ g}}{1 \text{ mol}} \times \dfrac{10^3 \text{ mg}}{1 \text{ g}} = 14 \text{ mg}$

2.16 (a) The molar mass of aspirin is 180.16 g/mol and there is 4 mol O per 1 mol aspirin.

$$N_{O\ atoms} = 12.0 \text{ g asp.} \times \frac{1 \text{ mol asp.}}{180.16 \text{ g asp.}} \times \frac{4 \text{ mol O}}{1 \text{ mol asp.}} \times \frac{6.02 \times 10^{23} \text{ O atoms}}{1 \text{ mol O}} = 1.60 \times 10^{23} \text{ O atoms}$$

The molar mass of calcium phosphate is 310.18 g/mol and there is 8 mol O per 1 mol calcium phosphate.

$$N_{O\ atoms} = 12.0 \text{ g } Ca_3(PO_4)_2 \times \frac{1 \text{ mol } Ca_3(PO_4)_2}{310.18 \text{ g } Ca_3(PO_4)_2} \times \frac{8 \text{ mol O}}{1 \text{ mol } Ca_3(PO_4)_2} \times \frac{6.02 \times 10^{23} \text{ O atoms}}{1 \text{ mol O}}$$

$$= 1.86 \times 10^{23} \text{ O atoms}$$

(b) The molar mass of manganese(III) sulfate is 398.06 g/mol and there is 12 mol O per 1 mol manganese(III) sulfate.

$$m_{Mn_2(SO_4)_3} = 6.022 \times 10^{23} \text{ O atoms} \times \frac{1 \text{ mol O}}{6.022 \times 10^{23} \text{ O atoms}} \times \frac{1 \text{ mol } Mn_2(SO_4)_3}{12 \text{ mol O}} \times \frac{398.06 \text{ g } Mn_2(SO_4)_3}{1 \text{ mol } Mn_2(SO_4)_3}$$

$$= 33.17 \text{ g } Mn_2(SO_4)_3$$

The molar mass of caffeine is 194.19 g/mol and there is 2 mol O per 1 mol caffeine.

$$m_{caffeine} = 6.022 \times 10^{23} \text{ O atoms} \times \frac{1 \text{ mol O}}{6.022 \times 10^{23} \text{ O atoms}} \times \frac{1 \text{ mol caffeine}}{2 \text{ mol O}} \times \frac{194.19 \text{ g caffeine}}{1 \text{ mol caffeine}}$$

$$= 97.10 \text{ g caffeine}$$

2.17 The mass of Si in 1 mol SiO_2 is 28.0855 g. The mass of O in 1 mol SiO_2 is 2 mol × 15.9994 g/mol = 31.9988 g. The mass of SiO_2 in 1 mol SiO_2 is 60.08 g.

$$\% \text{ Si in } SiO_2 = \frac{28.0855 \text{ g}}{60.08 \text{ g}} \times 100\% = 46.7\%$$

$$\% \text{ O in } SiO_2 = \frac{31.9988 \text{ g}}{60.08 \text{ g}} \times 100\% = 53.3\%$$

2.18 The molar mass of $Na_2SO_4 \cdot 10H_2O$ is [2(22.989) + (32.07) + 4(16.00) + 20(1.008) + 10(16.00)] g/mol = 322.20 g/mol.

$$\% \text{ Na} = \frac{45.98 \text{ g Na}}{322.20 \text{ g hydrate}} \times 100\% = 14.27\%$$

$$\% \text{ S} = \frac{32.07 \text{ g S}}{322.20 \text{ g hydrate}} \times 100\% = 9.95\%$$

$$\% \text{ H} = \frac{20.16 \text{ g H}}{322.20 \text{ g hydrate}} \times 100\% = 6.26\%$$

$$\% \text{ O} = \frac{64.00 \text{ g O} + 160.0 \text{ g O}}{322.20 \text{ g hydrate}} \times 100\% = 69.5\%$$

2.19 The amount of each element in 100 g compound is

$$n_N = 87.42 \text{ g N} \times \frac{1 \text{ mol N}}{14.0067 \text{ g N}} = 6.241 \text{ mol N}$$

$$n_H = 12.58 \text{ g H} \times \frac{1 \text{ mol H}}{1.0079 \text{ g H}} = 12.48 \text{ mol H}$$

The mole ratio is

$$\frac{12.48 \text{ mol H}}{6.241 \text{ mol N}} = \frac{2.000 \text{ mol H}}{1.000 \text{ mol N}}$$

The simplest (empirical) formula is therefore NH_2, which has molar mass = [1(14.0067 + 2(1.0079)] g/mol = 16.02 g/mol. The experimental molar mass is 32.05 g/mol, twice the molar mass calculated from the empirical formula. Therefore the molecular formula is N_2H_4, twice the empirical formula.

2.20 The amount of each element in 100.0 g of vitamin C is

$$n_C = 40.9 \text{ g C} \times \frac{1 \text{ mol C}}{12.011 \text{ g C}} = 3.405 \text{ mol C}$$

$$n_H = 4.58 \text{ g H} \times \frac{1 \text{ mol H}}{1.0079 \text{ g H}} = 4.544 \text{ mol H}$$

$$n_O = 54.5 \text{ g O} \times \frac{1 \text{ mol O}}{15.9994 \text{ g O}} = 3.406 \text{ mol O}$$

The mole ratios are the same for H : C and H : O.

$$\frac{4.544 \text{ mol H}}{3.405 \text{ mol C}} = \frac{1.335 \text{ mol H}}{1.000 \text{ mol C}} = \frac{3}{3} \times \frac{1.335 \text{ mol H}}{1.000 \text{ mol C}} = \frac{4.005 \text{ mol H}}{3.000 \text{ mol C}}$$

This gives $C_3H_4O_3$ for the empirical formula. The empirical formula weight is (3)(12.011) + (4)(1.0079) + (3)(15.9994) = 88.06 g/mol. The actual molar mass, however, is 176.13 g/mol (= 2 × 88.06 g/mol) so the molecular formula, $C_6H_8O_6$, must be twice the empirical formula, $C_3H_4O_3$.

Chapter 3

3.1 (a) $2\,Cr(s) + 3\,Cl_2(g) \rightarrow 2\,CrCl_3(s)$
(b) $As_2O_3(s) + 3\,H_2(g) \rightarrow 2\,As(s) + 3\,H_2O(\ell)$
(c) $C_3H_8 + 5\,O_2 \rightarrow 3\,CO_2 + 4\,H_2O$
(d) $C_2H_5OH + 3\,O_2 \rightarrow 2\,CO_2 + 3\,H_2O$

3.2 (a) $Na_2Cr_2O_7(aq) + 2\,AgNO_3(aq) \rightarrow Ag_2Cr_2O_7(s) + 2\,NaNO_3(aq)$
(b) $2\,NaHCO_3(aq) + CaCl_2(aq) \rightarrow CaCO_3(s) + H_2CO_3(aq) + 2\,NaCl(aq)$

3.3 (a) $(NH_4)_2S(aq) \rightarrow 2\,NH_4^+(aq) + S^{2-}(aq)$
(b) $KMnO_4(aq) \rightarrow K^+(aq) + MnO_4^-(aq)$
(c) $K_2C_2O_4(aq) \rightarrow 2\,K^+(aq) + C_2O_4^{2-}(aq)$
(d) $Li_2CO_3(aq) \rightarrow 2\,Li^+(aq) + CO_3^{2-}(aq)$
(e) $Co(SCN)_2(aq) \rightarrow Co^{2+}(aq) + 2\,SCN^-(aq)$

3.4 (a) NaF is soluble.
(b) $Ca(CH_3COO)_2$ is soluble.
(c) $SrCl_2$ is soluble.
(d) MgO is not soluble.

(e) $PbCl_2$ is not soluble.
(f) HgS is not soluble.

3.5 (a) This reaction forms insoluble nickel hydroxide and aqueous sodium chloride.

$$NiCl_2(aq) + 2\ NaOH(aq) \longrightarrow Ni(OH)_2(s) + 2\ NaCl(aq)$$

(b) This reaction forms aqueous potassium bromide and a precipitate of calcium carbonate.

$$K_2CO_3(aq) + CaBr_2(aq) \longrightarrow CaCO_3(s) + 2\ KBr(aq)$$

3.6 (a) $BaCl_2(aq) + Na_2SO_4(aq) \longrightarrow BaSO_4(s) + 2\ NaCl(aq)$

$$Ba^{2+}(aq) + SO_4^{2-}(aq) \longrightarrow BaSO_4(s)$$

(b) $(NH_4)_2S(aq) + FeCl_2(aq) \longrightarrow FeS(s) + 2\ NH_4Cl(aq)$

$$Fe^{2+}(aq) + S^{2-}(aq) \longrightarrow FeS(s)$$

(c) $(NH_4)_2SO_4(aq) + KCl(aq) \longrightarrow$ N.R.

3.7 Any of the strong acids in Table 3.2 is a strong electrolyte. Any weak acid or base in Table 3.2 is a weak electrolyte. Any organic compound that yields no ions on dissolution is a nonelectrolyte.

3.8 $H_3PO_4(aq) + 3\ NaOH(aq) \rightarrow Na_3PO_4(aq) + 3\ H_2O(\ell)$

3.9 (a) Sulfuric acid and magnesium hydroxide
(b) Carbonic acid and strontium hydroxide

3.10 $2\ HI(aq) + Ca(OH)_2(aq) \longrightarrow CaI_2(aq) + 2\ H_2O(\ell)$

$$2\ H^+(aq) + 2\ I^-(aq) + Ca^{2+}(aq) + 2\ OH^-(aq) \longrightarrow Ca^{2+}(aq) + 2\ I^-(aq) + 2\ H_2O(\ell)$$

$$H^+(aq) + OH^-(aq) \longrightarrow H_2O(\ell)$$

3.11 The oxidation numbers of Fe and Sb are 0 (Rule 1). The oxidation numbers in Sb_2S_3 are +3 for Sb^{3+} and −2 for S^{2-} (Rules 2 and 4). The oxidation numbers in FeS are +2 for Fe^{2+} and −2 for S^{2-} (Rules 2 and 4). Sb has been reduced; Fe has been oxidized. Fe is the reducing agent; Sb_2S_3 is the oxidizing agent.

3.12 Reactions (a) and (b) occur. Aluminum is above copper and chromium in Table 3.4; therefore, aluminum is oxidized and acts as the reducing agent in reactions (a) and (b). In reaction (a), Cu^{2+} is reduced and Cu^{2+} is the oxidizing agent. In reaction (b), Cr^{3+} is the oxidizing agent and is reduced to Cr metal. Reactions (c) and (d) do not occur because Pt is below H_2 and cannot reduce H^+, and Au is below Ag and cannot reduce Ag^+.

3.13 0.433 mol hematite needs 0.433 × 3 = 1.30 mol CO. The molar mass of CO is 28.01 g/mol, so 1.30 mol × 28.01 g/mol = 36.4 g CO.

3.14 Based on the balanced equation given, the amount of Mg that reacts equals the amount of $MgCl_2$ formed.

$$n(Mg) = 6.46\ g\ MgCl_2 \times \frac{1\ mol\ MgCl_2}{95.21\ g\ MgCl_2} \times \frac{1\ mol\ Mg}{1\ mol\ MgCl_2}$$

$$= 6.785 \times 10^{-2}\ mol\ Mg$$

$$m(Mg) = 6.785 \times 10^{-2}\ mol\ Mg \times \frac{24.30\ g\ Mg}{1\ mol\ Mg} = 1.65\ g\ Mg$$

$$\%\ Mg\ in\ impure\ sample = \frac{1.65\ g\ Mg}{1.72\ g\ sample} \times 100\% = 95.9\%$$

3.15 The reaction occurs until the limiting reactant is all reacted. The balanced equation is

$$N_2O_4(g) + 2\ N_2H_4(g) \rightarrow 3\ N_2(g) + 4\ H_2O(g)$$

Thus, 1 mol N_2O_4 requires 2 mol N_2H_4; that is, 3 mol N_2H_4 requires 1.5 mol N_2O_4. Because only 1.25 mol N_2O_4 is available, N_2O_4 is the limiting reactant and must be used to calculate the amount of water formed.

$$n(H_2O) = 1.25\ mol\ N_2O_4 \times \frac{4\ mol\ H_2O}{1\ mol\ N_2O_4} \times = 5.00\ mol\ H_2O$$

3.16 Determine the limiting reactant as the reactant that would produce the smaller mass of nitrogen:

$$m(N_2,\ from\ N_2O_4) = 200.\ g\ N_2O_4 \times \frac{1\ mol\ N_2O_4}{92.02\ g\ N_2O_4} \times \frac{3\ mol\ N_2}{1\ mol\ N_2O_4} \times \frac{28.01\ g\ N_2}{1\ mol\ N_2} = 183\ g\ N_2$$

$$m(N_2,\ from\ N_2H_4) = 100.\ g\ N_2H_4 \times \frac{1\ mol\ N_2H_4}{32.05\ g\ N_2H_4} \times \frac{3\ mol\ N_2}{2\ mol\ N_2H_4} \times \frac{28.01\ g\ N_2}{1\ mol\ N_2} = 131\ g\ N_2$$

Therefore, hydrazine, N_2H_4, is the limiting reactant and 131 g N_2 is produced.

3.17 (a) Determine the limiting reactant as the reactant that would produce the smaller mass of SiC:

$$m(SiC,\ from\ SiO_2) = 100.\ g\ SiO_2 \times \frac{1\ mol\ SiO_2}{60.08\ g\ SiO_2} \times \frac{1\ mol\ SiC}{1\ mol\ SiO_2} \times \frac{40.10\ g\ SiC}{1\ mol\ SiC} = 66.7\ g\ SiC$$

$$m(SiC,\ from\ C) = 100.\ g\ C \times \frac{1\ mol\ C}{12.01\ g\ C} \times \frac{1\ mol\ SiC}{3\ mol\ C} \times \frac{40.10\ g\ SiC}{1\ mol\ SiC} = 111\ g\ C$$

SiO_2 is the limiting reactant and 66.7 g SiC is produced.
(b) The mass of C that reacted is less than the initial mass of C; it can be calculated from the mass of the limiting reactant:

$$m(C\ reacted) = 100.\ g\ SiO_2 \times \frac{1\ mol\ SiO_2}{60.08\ g\ SiO_2} \times \frac{3\ mol\ C}{1\ mol\ SiO_2} \times \frac{12.01\ g\ C}{1\ mol\ C} = 59.9\ g\ C$$

The mass of C remaining is 100. g − 59.9 g = 40. g.

3.18 To make 1.00 kg CH_3OH with 85.0% yield will require using enough reactant to produce (1000 g)/0.850, or 1176 g CH_3OH.

$$n(CH_3OH) = 1176\ g\ CH_3OH \times \frac{1\ mol}{32.04\ g} = 36.70\ mol\ CH_3OH$$

$$n(H_2) = 36.70\ mol\ CH_3OH \times \frac{2\ mol\ H_2}{1\ mol\ CH_3OH} = 73.40\ mol\ H_2$$

$$m(H_2) = 73.40\ mol\ H_2 \times \frac{2.016\ g\ H_2}{1\ mol\ H_2} = 148\ g\ H_2$$

3.19 Calculate the mass of Cu_2S you should have produced and compare it with the mass actually produced.

$$n(Cu) = 2.50\ g\ Cu \times \frac{1\ mol\ Cu}{63.546\ g\ Cu} = 3.93 \times 10^{-2}\ mol\ Cu$$

$$n(Cu_2S) = 3.93 \times 10^{-2}\ mol\ Cu \times \frac{8\ mol\ Cu_2S}{16\ mol\ Cu} = 1.97 \times 10^{-2}\ mol\ Cu_2S$$

$$m(Cu_2S) = 1.97 \times 10^{-2}\ mol\ Cu_2S \times \frac{159.16\ g\ Cu_2S}{1\ mol\ Cu_2S} = 3.14\ g\ Cu_2S$$

$$Percent\ yield = \frac{2.53\ g\ Cu_2S}{3.14\ g\ Cu_2S} \times 100\% = 80.6\%$$

Your synthesis met the standard.

3.20 $n(Na_2SO_4) = 36.0\ g\ Na_2SO_4 \times \frac{1\ mol\ Na_2SO_4}{142.0\ g\ Na_2SO_4}$

$$= 0.254\ mol\ Na_2SO_4$$

$$Molarity = \frac{0.254\ mol\ Na_2SO_4}{0.750\ L\ solution} = 0.339\ mol/L = 0.339\ M$$

3.21 (a) 1.00 L of 0.125-M Na_2CO_3 contains 0.125 mol Na_2CO_3.

$$m(Na_2CO_3) = 0.125 \text{ mol} \times \frac{105.99 \text{ g}}{1 \text{ mol}} = 13.2 \text{ g } Na_2CO_3$$

Prepare the solution by adding 13.2 g Na_2CO_3 to a 1.00-L volumetric flask, dissolving the Na_2CO_3 and mixing thoroughly, and adding pure water until the solution volume is 1.00 L.

(b) 500. mL of 0.215-M $KMnO_4$ contains 1.70 g $KMnO_4$.

$$m(KMnO_4) = 500. \text{ mL} \times \frac{1 \text{ L}}{1000 \text{ mL}} \times \frac{0.0215 \text{ mol } KMnO_4}{1 \text{ L solution}} \times \frac{158.0 \text{ g } KMnO_4}{1 \text{ mol } KMnO_4} = 1.70 \text{ g } KMnO_4$$

Put 1.70 g $KMnO_4$ into a 500.-mL volumetric flask, add water, mix thoroughly, and add water to the 500.-mL mark on the flask.

3.22 $$V(\text{conc}) = \frac{0.150 \text{ M} \times 0.0500 \text{ L}}{0.500 \text{ M}} = 0.0150 \text{ L} = 15.0 \text{ mL}$$

3.23 $$V = 1.2 \times 10^{10} \text{ kg NaOH} \times \frac{1000 \text{ g}}{1 \text{ kg}} \times \frac{1 \text{ mol NaOH}}{40.0 \text{ g NaOH}} \times \frac{2 \text{ mol NaCl}}{2 \text{ mol NaOH}} \times \frac{58.4 \text{ g NaCl}}{1 \text{ mol NaCl}} \times \frac{1.0 \text{ L brine}}{360 \text{ g NaCl}} = 4.9 \times 10^{10} \text{ L}$$

3.24 Calculate the amount of AgBr; from it, calculate the amount and volume of $Na_2S_2O_3$. $M(\text{AgBr}) = 187.77$ g/mol.

$$V(Na_2S_2O_3) = 50.0 \text{ mg AgBr} \times \frac{1 \text{ g}}{1000 \text{ mg}} \times \frac{1 \text{ mol AgBr}}{187.77 \text{ g AgBr}} \times \frac{2 \text{ mol } Na_2S_2O_3}{1 \text{ mol AgBr}} \times \frac{1 \text{ L}}{0.0150 \text{ mol } Na_2S_2O_3} \times \frac{1000 \text{ mL}}{1 \text{ L}} = 35.5 \text{ mL}$$

3.25 $H_2SO_4(aq) + 2\ NaOH(aq) \longrightarrow Na_2SO_4(aq) + 2\ H_2O(\ell)$

$$n(\text{NaOH}) = (0.0413 \text{ L})(0.100 \text{ mol/L}) = 0.00413 \text{ mol NaOH}$$

$$n(H_2SO_4) = 0.00413 \text{ mol NaOH} \times \frac{1 \text{ mol } H_2SO_4}{2 \text{ mol NaOH}} = 0.002065 \text{ mol } H_2SO_4$$

$$\text{Molarity} = \frac{0.002065 \text{ mol } H_2SO_4}{20.0 \text{ mL solution}} \times \frac{1000 \text{ mL}}{1 \text{ L}} = 0.103 \text{ mol/L}$$

Chapter 4

4.1 (a) $160 \text{ Cal} \times \frac{1000 \text{ cal}}{\text{Cal}} \times \frac{4.184 \text{ J}}{\text{cal}} = 6.7 \times 10^5 \text{ J}$

(b) $16 \text{ kJ} \times \frac{1 \text{ kcal}}{4.184 \text{ kJ}} = 3.8 \text{ kcal}$

4.2 $\Delta E = -2400 \text{ J} = q + w = -1.89 \text{ kJ} + w$
$w = -2400 \text{ J} + 1.89 \text{ kJ} = -2.4 \text{ kJ} + 1.89 \text{ kJ} = -0.5 \text{ kJ}$

4.3 $q = c \times m \times \Delta T = c \times m \times (T_{\text{final}} - T_{\text{initial}})$

$$T_{\text{final}} = T_{\text{initial}} + \frac{q}{c \times m} = 5\ °\text{C} + \frac{24{,}100 \text{ J}}{(0.902 \text{ J g}^{-1}\ °\text{C}^{-1})(250. \text{ g})} = 5\ °\text{C} + 106.8\ °\text{C} = 112\ °\text{C}$$

4.4 Because no work is done, $\Delta E = q$. Assume that the tea has a mass of 250. g and that its specific heat capacity is the same as that of water, namely, 4.184 J g^{-1} °C^{-1}.

$\Delta E = q = (\text{mass}) \times (\text{specific heat capacity}) \times (\text{change in temperature})$

$\Delta E = q = (250. \text{ g}) \times (4.184 \text{ J g}^{-1}\ °\text{C}^{-1}) \times [(65.0 - 37.0)\ °\text{C}] = 2.93 \times 10^4 \text{ J} = 2.93 \times 10^1 \text{ kJ}$

4.5 $q_{\text{water}} = -q_{\text{iron}}$
$(4.184 \text{ J g}^{-1}\ °\text{C}^{-1})(1000. \text{ g})(32.8\ °\text{C} - 20.0\ °\text{C}) = -(0.451 \text{ J g}^{-1}\ °\text{C}^{-1})(400. \text{ g})(32.8\ °\text{C} - T_i)$

$T_i = (297 + 32.8)\ °\text{C} = 330.\ °\text{C}$

4.6 $1.00 \text{ g K(s)} \times \frac{1 \text{ mL}}{0.86 \text{ g}} = 1.16 \text{ mL};\ 1.00 \text{ g K}(\ell) \times \frac{1 \text{ mL}}{0.82 \text{ g}} = 1.22 \text{ mL}$

The change in volume is $(1.22 - 1.16) \text{ mL} = 0.06 \text{ mL}$.

$$w = 0.06 \text{ mL} \times \frac{0.10 \text{ J}}{1 \text{ mL}} = 6 \times 10^{-3} \text{ J}$$

$$\Delta H = \frac{14.6 \text{ cal}}{1 \text{ g}} \times 1.00 \text{ g} \times \frac{4.184 \text{ J}}{1 \text{ cal}} = 61.1 \text{ J}$$

$$\Delta E = \Delta H + w = 61.1 \text{ J} + (6 \times 10^{-3} \text{ J}) = 61.1 \text{ J}$$

4.7 (a) $10.0 \text{ g } I_2 \times \frac{1 \text{ mol } I_2}{253.8 \text{ g } I_2} \times \frac{62.4 \text{ kJ}}{1 \text{ mol } I_2} = 2.46 \text{ kJ}$

(b) $3.42 \text{ g } I_2 \times \frac{1 \text{ mol } I_2}{253.8 \text{ g } I_2} \times \frac{-62.4 \text{ kJ}}{1 \text{ mol } I_2} = -0.841 \text{ kJ} = -841 \text{ J}$

(c) This process is the reverse of the one in part (a), so $\Delta H°$ is negative. Thus, the process is exothermic. The quantity of energy transferred is 841 J.

4.8 $AgNO_3(aq) + NaCl(aq) \longrightarrow AgCl(s) + NaNO_3(aq)$ $\quad \Delta_r H° = -69$ kJ/mol

First, determine the limiting reactant:

$$n(\text{NaCl}) = (1.00 \text{ mol NaCl/L})(0.0150 \text{ L}) = 0.0150 \text{ mol NaCl}$$

$$n(AgNO_3) = (0.873 \text{ mol } AgNO_3\text{/L})(0.0250 \text{ L}) = 0.0218 \text{ mol } AgNO_3$$

From the balanced equation, the reactants are in a 1:1 mole ratio. Thus, NaCl is the limiting reactant and will produce 0.0150 mol AgCl.

$$(0.0150 \text{ mol AgCl})(-69 \text{ kJ/mol}) = -1.04 \text{ kJ}$$

4.9 $\Delta T = (25.43 - 20.64)\ °\text{C} = 4.79\ °\text{C}$

$$\Delta E_{\text{calorimeter}} = \frac{877 \text{ J}}{°\text{C}} \times 4.79\ °\text{C} = 4.200 \times 10^3 \text{ J} = 4.200 \text{ kJ}$$

$$\Delta E_{\text{water}} = 832 \text{ g} \times \frac{4.184 \text{ J}}{\text{g}\ °\text{C}} \times 4.79\ °\text{C} = 16.67 \text{ kJ}$$

$$\Delta E_{\text{reaction}} = -(\Delta E_{\text{calorimeter}} + \Delta E_{\text{water}}) = -(4.200 + 16.67) \text{ kJ} = -20.87 \text{ kJ}$$

$$20.87 \text{ kJ} \times \frac{1 \text{ kcal}}{4.184 \text{ kJ}} \times \frac{1 \text{ Cal}}{1 \text{ kcal}} = 4.99 \text{ Cal}$$

Since metabolizing the Fritos chip corresponds to oxidizing it, the result of 4.99 Cal verifies the statement that one chip provides 5 Cal.

4.10 The total volume of the initial solutions is 200. mL, which corresponds to 200. g of solution. The quantities of reactants are 0.10 mol $H^+(aq)$ and 0.050 mol $OH^-(aq)$, so 0.050 mol H_2O is formed.

$$0.050 \text{ mol } H_2O \times \frac{-58.7 \text{ kJ}}{1 \text{ mol } H_2O} = -2.94 \text{ kJ}$$

Because $\Delta_r H°$ is negative, energy is transferred to the water, and its temperature will rise.

$$\Delta T = \frac{q}{c \times m} = \frac{2.94 \times 10^3 \text{ J}}{(4.184 \text{ J g}^{-1}\ °\text{C}^{-1})(200. \text{ g})} = 3.5\ °\text{C}$$

The final temperature will be $(20.4 + 3.5)\ °\text{C} = 23.9\ °\text{C}$.

4.11 The balanced chemical equation is

$$2\ Fe_3O_4(s) \longrightarrow 6\ FeO(s) + O_2(g)$$

The equations given in the problem can be arranged so that when added they give this balanced equation:

$2 \times [Fe_3O_4(s) \longrightarrow 3\ Fe(s) + 2\ O_2(g)] \qquad \Delta_r H° = 2(1118.4\ kJ/mol)$

$6 \times [Fe(s) + \frac{1}{2}\ O_2(g) \longrightarrow FeO(S) \qquad \Delta_r H° = 6(-272.0\ kJ/mol)$

So

$$\Delta_r H° = 2236.8\ kJ/mol - 1632.0\ kJ/mol = 604.8\ kJ/mol$$

4.12 (a) $\frac{1}{2}\ N_2(g) + \frac{3}{2}\ H_2(g) \longrightarrow NH_3(g) \qquad \Delta_f H° = -46.11\ kJ/mol$

(b) $C(graphite) + \frac{1}{2}\ O_2(g) \longrightarrow CO(g) \qquad \Delta_f H° = -110.525\ kJ/mol$

4.13 For the reaction given,

$$\Delta_r H° = 6 \times \Delta_f H°\{CO_2(g)\} + 5 \times \Delta_f H°\{H_2O(g)\} - 2 \times \Delta_f H°\{C_3H_5(NO_3)_3(\ell)\}$$
$$= \{6(-393.509) + 5(-241.818) - 2(-364)\}\ kJ/mol$$
$$= -2.84 \times 10^3\ kJ/mol$$

For 10.0 g nitroglycerin (nitro),

$$q = 10.0\ g \times \frac{1\ mol\ nitro}{227.09\ g} \times \frac{-2.84 \times 10^3\ kJ}{2\ mol\ nitro} = -62.5\ kJ$$

(The 2 mol nitro in the last factor comes from the coefficient of 2 associated with nitroglycerin in the chemical equation.)

4.14 $SO_2(g) + \frac{1}{2}\ O_2(g) \longrightarrow SO_3(g) \qquad \Delta_r H° = ?$

$$\Delta_r H° = \Delta_f H°\{SO_3(g)\} - [\Delta_f H°\{SO_2(g)\} + \tfrac{1}{2}\ \Delta_f H°\{O_2(g)\}]$$
$$= -395.72\ kJ/mol - (-296.830 + 0)\ kJ/mol$$
$$= -98.89\ kJ$$

4.15 Represent fuel value as FV; represent energy density as ED.

(a) $C_8H_{18}(\ell) + \frac{25}{2}\ O_2(g) \longrightarrow 8\ CO_2(g) + 9\ H_2O(g)$

$$\Delta_r H° = [8(-393.509) + 9(-241.818) - (-249.952)]\ kJ/mol$$
$$= -5074.48\ kJ/mol$$

$$FV = \frac{5074.48\ kJ}{1\ mol\ C_8H_{18}} \times \frac{1\ mol}{114.23\ g} = 44.423\ kJ/g\ C_8H_{18}$$

$$ED = \frac{44.423\ kJ}{1\ g\ C_8H_{18}} \times \frac{0.703\ g}{1\ mL} \times \frac{1000\ mL}{1\ L} = 3.12 \times 10^4\ kJ/L\ C_8H_{18}$$

(b) $N_2H_4(\ell) + O_2(g) \longrightarrow N_2(g) + 2\ H_2O(g)$

$$\Delta_r H° = [2(-241.818) - 50.63]\ kJ/mol = -534.26\ kJ/mol$$

$$FV = \frac{534.26\ kJ}{1\ mol\ N_2H_4} \times \frac{1\ mol}{32.045\ g} = 16.672\ kJ/g\ N_2H_4$$

$$ED = \frac{16.672\ kJ}{1\ g\ N_2H_4} \times \frac{1.004\ g}{1\ mL} \times \frac{1000\ mL}{1\ L} = 1.67 \times 10^4\ kJ/L\ N_2H_4$$

$16.672\ kJ/g\ N_2H_4 \times 1004\ g/L = 1.67 \times 10^4\ kJ/L\ N_2H_4$

(c) $CH_3OH(\ell) + \frac{3}{2}\ O_2(g) \longrightarrow CO_2(g) + H_2O(g)$

$$\Delta_r H° = 1 \times (-393.509\ kJ/mol) + 2 \times (-241.818\ kJ/mol)] - [1 \times (-238.66\ kJ/mol)]$$
$$= -638.49\ kJ/mol$$

Fuel value: (638.49 kJ/mol)(1 mol/32.0406 g) = 19.928 kJ/g

Energy density: (19.928 kJ/g)(792 g/L) = 1.58×10^4 kJ/L

Octane has the highest fuel value and the highest energy density.

Chapter 5

5.1 (a) $\lambda = \frac{c}{\nu} = \frac{2.998 \times 10^8\ m/s}{9.99 \times 10^{14}\ s^{-1}} = 3.00 \times 10^{-7}\ m$

(b) $3.00 \times 10^{-7}\ m \times \frac{1\ nm}{10^{-9}\ m} = 300.\ nm$

5.2 (a) One photon of ultraviolet radiation has more energy because ν is larger in the UV spectral region than in the microwave region.
(b) One photon of blue light has more energy because the blue portion of the visible spectrum has a higher frequency than the green portion of the visible spectrum.

5.3 Any from $n_{hi} > 8$ to $n_{lo} = 2$

5.4 (a) $$\nu = \frac{-2.179 \times 10^{-18}\ J}{h}\left(\frac{1}{n_f^2} - \frac{1}{n_i^2}\right)$$
$$= \left(\frac{-2.179 \times 10^{-18}\ J}{6.626 \times 10^{-34}\ Js}\right)\left(\frac{1}{4^2} - \frac{1}{6^2}\right)$$
$$= (-3.289 \times 10^{15}\ s^{-1})\left(\frac{1}{16} - \frac{1}{36}\right)$$
$$= (-3.289 \times 10^{15}\ s^{-1})(3.47 \times 10^{-2})$$
$$= -1.14 \times 10^{14}\ s^{-1}$$

The negative sign indicates that energy is emitted.

$$\lambda = 2.63 \times 10^{-6}\ m$$

(b) Longer than that of the $n = 7$ to $n = 4$ transition.

5.5 $$\lambda = \frac{h}{mv} = \frac{6.626 \times 10^{-34}\ J\ s}{1.67 \times 10^{-27}\ kg \times 2.998 \times 10^8\ m/s \times 0.10}$$
$$= 1.32 \times 10^{-14}\ m$$

5.6 (a) 6*d* (b) 5 (c) 2, 1, 0, −1, −2

5.7 (a) [Ne] $3s^23p^2$ (b) [Ne] 3*s* [↑↓] 3*p* [↑][↑][ ]

5.8 The electron configurations for :S̈e· and :T̈e· are [Ar] $4s^23d^{10}4p^4$ and [Kr] $5s^24d^{10}5p^4$, respectively. Elements in the same main group have similar electron configurations.

5.9 (a) S^{2-} (b) Ca^{2+}

5.10 The ground-state Cu atom has a configuration [Ar] $4s^13d^{10}$. When it loses one electron, it becomes the Cu^+ ion with configuration [Ar] $3d^{10}$. There is an added stability for the completely filled set of 3*d* orbitals.

5.11 B < Mg < Na < K

5.12 (a) Cs^+
(b) La^{3+}. All three ions have the same number of electrons. Cs^+ has the least number of protons and La^{3+} has the greatest number of protons so the electron cloud is (a) largest for Cs^+ and (b) smallest for La^{3+}.

5.13 F > N > P > Na. In the periodic table, Na is to the left of P so Na is larger; P is below N so P is larger; N is to the left of F so N is larger; F is the smallest.

Chapter 6

6.1 (a) 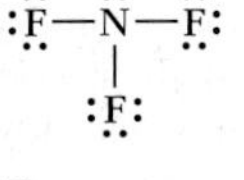 (b) 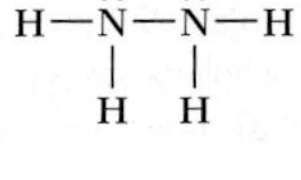

(c) 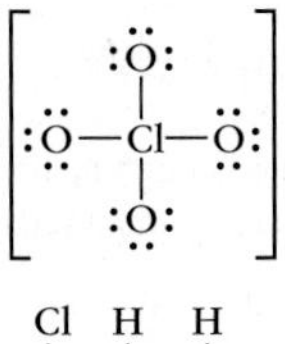

6.2 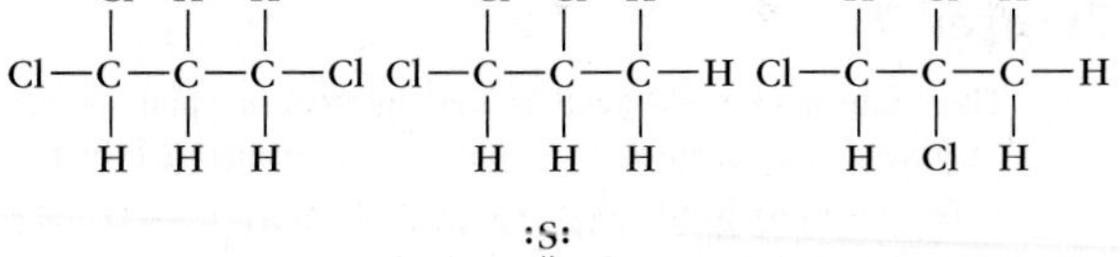

6.3 (a) $[:N{\equiv}O:]^+$ (b) 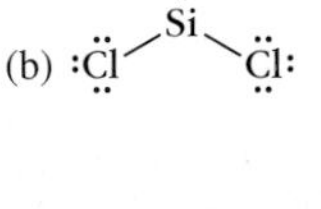

6.4 Only (c) can have geometric isomers. Molecules in (a) and (b) each have the same two groups on one of the double-bonded carbons.

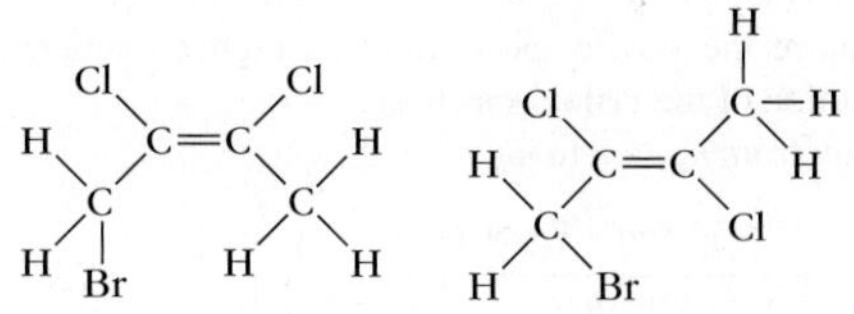

cis-1-bromo-2,3-dichloro-2-butene   *trans*-1-bromo-2,3-dichloro-2-butene

6.5 (a) Si is a larger atom than S.
(b) Br is a larger atom than Cl.
(c) The greater electron density in the triple bond brings the N≡O atoms closer together than the smaller electron density in the N═O double bond does.

6.6 $\Delta_r H$ = [(4 C—H)(416 kJ/mol) + (2 O═O)(498 kJ/mol)] − [(4 O—H)(467 kJ/mol) + (2 C═O)(803 kJ/mol)] = (2660 kJ/mol) − (3474 kJ/mol) = −814 kJ/mol

6.7 (a) B—Cl is more polar; $\overset{\delta^+}{B}$—$\overset{\delta^-}{Cl}$; N—H is more polar; $\overset{\delta^-}{N}$—$\overset{\delta^+}{H}$.

6.8 The other Lewis structure is :O≡N—N̈:

	O	N	N
Valence electrons	6	5	5
Lone pair electrons	2	0	6
$\frac{1}{2}$ shared electrons	3	4	1
Formal charge	+1	+1	−2

This resonance structure is not preferred to either of the previous structures. The third structure has the higher formal charge and most negative charge on N, not O, the more electronegative atom.

6.9 The N-to-O bond length in NO_2^- is 124 pm. From Table 6.1, N—O is 136 pm; N═O is 115 pm. Thus, the nature of the bond in NO_2^- is between that of a N—O single bond and a N═O double bond.

6.10 (a) :F̤—Be—F̤:   (b) :Ö—C̈l—Ö:

(c) 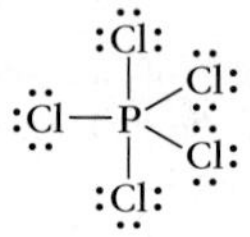   (d) $[H—B—H]^+$

(e) 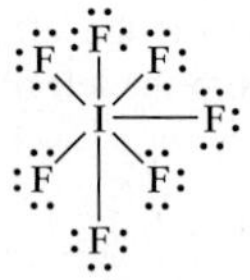

BeF_2—not an octet around the central Be atom; ClO_2—an odd number of electrons around Cl; PCl_5—more than four electron pairs around the central phosphorus atom; BH_2^+—only two electron pairs around the central B atom; IF_7—iodine has seven shared electron pairs.

6.11 Ten valence electrons. The bond order is (8 − 2)/2 = 3. No unpaired electrons. See Table 6.4.

6.12 O_2^-: Bond order = (8 − 5)/2 = 1.5. There is one unpaired electron. O_2^{2-}: Bond order = (8 − 6)/2 = 1; there are no unpaired electrons.

Chapter 7

7.1 There are two Be—F bonds in BeF_2 and no lone pairs on beryllium. Therefore, the electron-pair and the molecular geometry are the same, linear, with 180° Be—F bond angles.

7.2

Central Atom (underscored)	Bond Regions	Lone Pairs	Electron-Region Geometry	Molecular Geometry	Bond Angle
$\underline{Br}O_3^-$	3	1	Tetrahedral	Triangular pyramid	<109.5°
$\underline{Se}F_2$	2	2	Tetrahedral	Angular	<109.5°
$\underline{N}O_2^-$	2	1	Triangular planar	Angular	<120°

7.3 (a) ClF_2^-: triangular bipyramidal electron-pair geometry and linear molecular geometry
(b) XeO_3: tetrahedral electron-pair geometry and triangular pyramidal molecular geometry

7.4 The chiral centers are identified by an asterisk in the structural formula. Each of those carbon atoms has four different groups or atoms attached to it.

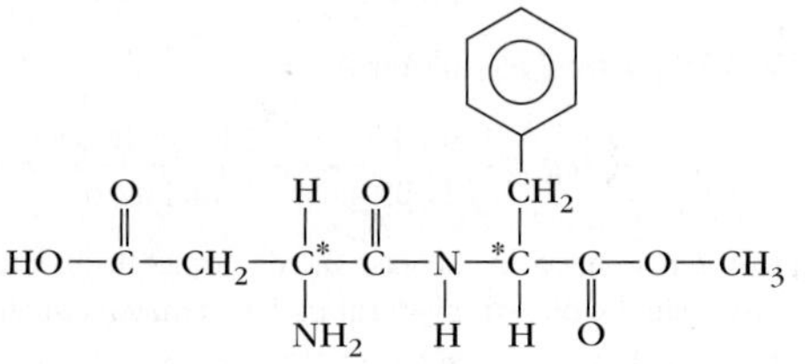

7.5 (a) $BeCl_2$: *sp* hybridization (two bonding electron pairs around the central Be atom), linear geometry

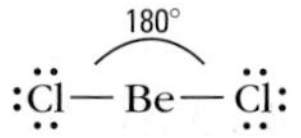

(b) The central N atom has four bonding pairs and no lone pairs in sp^3 hybridized orbitals on N giving tetrahedral electron-pair and molecular geometries.

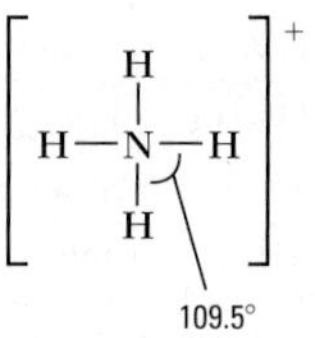

(c) Each carbon is sp^3 hybridized with no lone pairs, so the electron-pair and molecular geometries are both tetrahedral. The sp^3 hybridized oxygen atom has two bonding pairs, each in single bonds to a carbon atom plus two lone pairs giving it a tetrahedral electron-pair geometry and an angular molecular geometry.

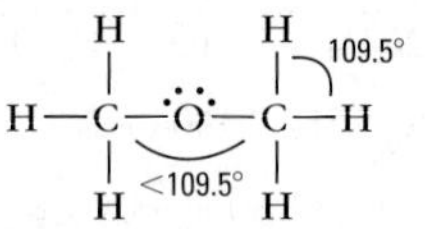

(d) The central boron atom has four single bonds, one to each fluorine atom, and no lone pairs resulting in tetrahedral electron-pair and molecular geometries. The boron is sp^3 hybridized to accommodate the four bonding pairs.

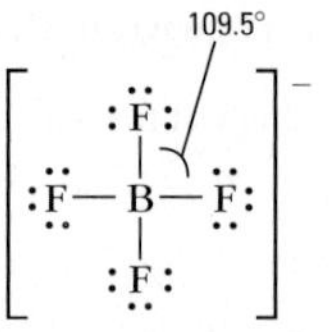

7.6 (a) In HCN, the *sp* hybridized carbon atom is sigma bonded to H and to N, as well as having two pi bonds to N. The sigma and two pi bonds form the C≡N triple bond. The nitrogen is *sp* hybridized with a sigma and two pi bonds to carbon; a lone pair is in the non-bonding *sp* hybrid orbital on N.

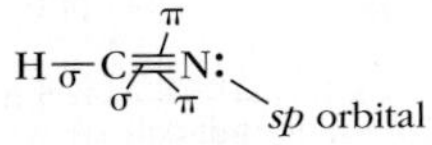

π
H σ C N:
σ π
sp orbital

(b) The double-bonded carbon and nitrogen are both sp^2 hybridized. The sp^2 hybrid orbitals on C form sigma bonds to H and to N; the unhybridized *p* orbital on C forms a pi bond with the unhybridized *p* orbital on N. The sp^2 hybrid orbitals on N form sigma bonds to carbon and to H; the N lone pair is in the nonbonding sp^2 hybrid orbital. (See diagram.)

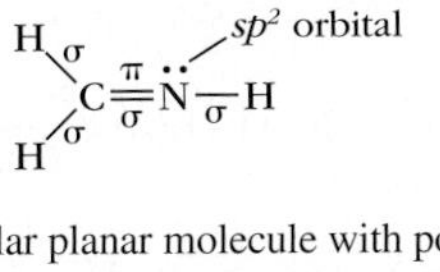

sp² orbital
H σ
π
C N σ H
σ
σ
H

7.7 (a) $BFCl_2$ is a triangular planar molecule with polar B—F and B—Cl bonds. The molecule is polar because the B—F bond is more polar than the B—Cl bonds, resulting in a net dipole. The B—Cl regions are partially positive; the B—F region is partially negative.
(b) NH_2Cl is a triangular pyramidal molecule with polar N—H bonds. (N—Cl is a slightly polar bond; N and Cl have nearly the same electronegativity.) It is a polar molecule because the N—H dipoles do not cancel and produce a net dipole. The partial positive charge is on each H; the partial negative charge is on the N.
(c) SCl_2 is an angular polar molecule. The S—Cl bond dipoles do not cancel each other because they are not symmetrically arranged due to the two lone pairs on S. The partial positive charge resides on the S; the partial negative charges are on each Cl.

7.8 (a) London forces between Kr atoms must be overcome for krypton to melt.
(b) The C—H covalent bonds in propane must be broken to release C and H atoms; the H atoms covalently bond to form H_2.

7.9 (a) London forces occur between N_2 molecules.
(b) CO_2 is nonpolar, and London forces occur between it and polar water molecules. CO_2 can also hydrogen bond to water (H from water and lone pair on the O of CO_2).
(c) London forces occur between the two molecules, but the principal intermolecular forces are the hydrogen bonds between the H on NH_3 with the lone pairs on the OH oxygen, and the hydrogen bonds between the H on the oxygen in CH_3OH and lone pair on nitrogen in NH_3.

Chapter 8

8.1 (a) (100. psi)(101.325 kPa/14.7 psi) = 689 kPa
(b) (135 mmHg)(1.01325 × 10^5 Pa/760 mmHg) = 1.80 × 10^4 Pa
(c) (690 Torr)(14.7 psi/760 mmHg) = 13.3 psi;
(690 Torr)(1 atm/760 Torr)(1.01325 bar/1 atm) = 0.920 bar

8.2 $$V = \frac{nRT}{P} = \frac{(2.64\text{ mol})(0.0821\text{ L atm mol}^{-1}\text{ K}^{-1})(304\text{ K})}{0.640\text{ atm}} = 103\text{ L}$$

8.3 $$V_2 = \frac{P_1V_1T_2}{P_2T_1} = \frac{(710\text{ mmHg}) \times (21\text{ mL}) \times (299.6\text{ K})}{(740\text{ mmHg}) \times (295.4\text{ K})} = 20.\text{ mL}$$

$$V_2 = \frac{(21\text{ mL})(299.6\text{ K})}{(295.4\text{ K})} = 21\text{ mL}$$

8.4 $$V_2 = \frac{P_1V_1}{P_2} = \frac{(1.00\text{ atm})(400.\text{ mL})}{0.750\text{ atm}} = 533\text{ mL}$$

8.5 $$d_{SO_2} = \frac{PM}{RT} = \frac{(2.60\text{ atm})(64.06\text{ g/mol})}{(0.0821\text{ L atm mol}^{-1}\text{ K}^{-1})(25.0 + 273.15)\text{ K}}$$
$$= 6.80\text{ g/L}$$

8.6 $$M = \frac{nRT}{PV} = \frac{(20.0\text{ mol})(0.0821\text{ L atm mol}^{-1}\text{ K}^{-1})(283\text{ K})}{(1.41\text{ atm})(2.50\text{ L})}$$
$$= 132\text{ g/mol}$$

The gas is xenon.

8.7 V_{NO} = (1.0 L O_2)(2 L NO/1 L O_2) = 2.0 L NO

8.8 It was assumed that all of the N_2 was from Equation 8.3. Thus,

$$n_{Na} = 2.46\text{ mol N}_2 \times \frac{2\text{ mol Na}}{3\text{ mol N}_2} = 1.64\text{ mol Na}$$

$$m_{KNO_3} = 1.64\text{ mol Na} \times \frac{2\text{ mol KNO}_3}{10\text{ mol Na}} \times \frac{101.1\text{ g KNO}_3}{1\text{ mol KNO}_3}$$
$$= 33.2\text{ g KNO}_3$$

$$n_{\text{additional N}_2} = 1.64\text{ mol Na} \times \frac{1\text{ mol N}_2}{10\text{ mol Na}} = 0.164\text{ mol N}_2$$

8.9 $$\text{Amount of N}_2 = 7.0\text{ g N}_2 \times \frac{1\text{ mol N}_2}{28.10\text{ g N}_2} = 0.25\text{ mol N}_2$$

$$\text{Amount of H}_2 = 6.0\text{ g H}_2 \times \frac{1\text{ mol H}_2}{2.02\text{ g H}_2} = 3.0\text{ mol H}_2$$

Total amount of gas = (3.0 + 0.25) mol = 3.25 mol

$$P_{total} = \frac{(3.25\text{ mol})(0.0821\text{ L atm mol}^{-1}\text{ K}^{-1})(773\text{ K})}{5.0\text{ L}} = 41.3\text{ atm}$$

$$X_{N_2} = \frac{0.25\text{ mol}}{3.25\text{ mol}} = 0.077 \qquad X_{H_2} = \frac{3.0\text{ mol}}{3.25\text{ mol}} = 0.92$$

$$P_{N_2} = \frac{(0.25\text{ mol})(0.0821\text{ L atm mol}^{-1}\text{ K}^{-1})(773\text{ K})}{5.0\text{ L}} = 3.2\text{ atm}$$

$$P_{H_2} = \frac{(3.0\text{ mol})(0.0821\text{ L atm mol}^{-1}\text{ K}^{-1})(773\text{ K})}{5.0\text{ L}} = 38\text{ atm}$$

8.10 For NO:

$$n = \frac{(1.0\text{ atm})(4.0\text{ L})}{(0.0821\text{ L atm mol}^{-1}\text{ K}^{-1})(298\text{ K})} = 0.163\text{ mol NO}$$

For O_2:

$$n = \frac{(0.40\text{ atm})(2.0\text{ L})}{(0.0821\text{ L atm mol}^{-1}\text{ K}^{-1})(298\text{ K})} = 0.0327\text{ mol O}_2$$

All the O_2 is used.

$$0.0327\text{ mol O}_2 \times \frac{2\text{ mol NO}}{1\text{ mol O}_2} = 0.0654\text{ mol NO used}$$

0.163 mol − 0.0654 mol = 0.0976 mol NO remains

$$0.0327\text{ mol O}_2 \times \frac{2\text{ mol NO}_2}{1\text{ mol O}_2} = 0.0654\text{ mol NO}_2\text{ formed}$$

$$n_{total} = 0.0976 + 0.0654 = 0.163\text{ mol of gas}$$

$$P_{total} = \frac{nRT}{V} = \frac{(0.163\text{ mol})(0.0821\text{ L atm mol}^{-1}\text{ K}^{-1})(298\text{ K})}{6.0\text{ L}}$$
$$= 0.665\text{ atm}$$

8.11 $P_{H_2} = P_{total} - P_{water}$ = 740 mmHg − 23.8 mmHg = 716 mmHg

$$n = \frac{PV}{RT} = \frac{(719/760\text{ atm})(0.260\text{ L})}{(0.0821\text{ L atm mol}^{-1}\text{ K}^{-1})(296\text{ K})} = 0.0101\text{ mol H}_2$$

0.0101 mol × 2.0158 g/mol = 0.0204 g = 20.4 mg H_2

8.12 (a) The temperature remains constant, so the average energy of the gas molecules remains constant. If the volume is decreased, then the gas molecules must hit the walls more frequently, and the pressure is increased.

(b) The temperature remains constant, so the average energy of the gas molecules remains constant. The addition of more molecules within a fixed volume must mean that the molecules hit walls more frequently, so the pressure is increased.

8.13 $$P = \frac{nRT}{V} = \frac{(2.50\text{ mol})(0.0821\text{ L atm mol}^{-1}\text{ K}^{-1})(273\text{ K})}{20.0\text{ L}}$$
$$= 2.80\text{ atm}$$

$$P = \left(\frac{nRT}{V - nb}\right) - \left(\frac{n^2a}{V^2}\right)$$

$$P = \left(\frac{(2.50\text{ mol})(0.0821\text{ L atm mol}^{-1}\text{ K}^{-1})(273\text{ K})}{(20.0\text{ L}) - (2.50\text{ mol})(0.0427\text{ L/mol})}\right) - \left(\frac{(2.50\text{ mol})^2(3.59\text{ L}^2\text{ atm/mol}^2)}{(20.0\text{ L})^2}\right) = 2.76\text{ atm}$$

$$\text{Percentage difference in pressure} = \frac{2.80\text{ atm} - 2.76\text{ atm}}{2.80\text{ atm}} \times 100\% = 1.43\%$$

8.14 (a) 35 ppm = $(35/10^6) \times 100\% = 3.5 \times 10^{-3}\%$
(b) $(0.053/10^6) \times 10^9 = 53$ ppb

Chapter 9

9.1 $$q = 25.0\text{ g benzene} \times \frac{1\text{ mol benzene}}{78.0\text{ g benzene}} \times \frac{30.72\text{ kJ}}{1\text{ mol benzene}} = 9.85\text{ kJ}$$

9.2 The molar mass of ethanol is 46.05 g/mol.

$$m_{\text{ethanol}} = 500.\text{ kJ} \times \frac{1\text{ mol}}{38.6\text{ kJ}} \times \frac{40.05\text{ g}}{1\text{ mol}} = 597\text{ g}$$

9.3 $\Delta_{\text{vap}}H° = 40.7$ kJ/mol $H_2O(\ell)$

$$\ln\left(\frac{150\text{ atm}}{1\text{ atm}}\right) = \frac{-40.7 \times 10^3\text{ J/mol}}{8.314\text{ J mol}^{-1}\text{ K}^{-1}}\left[\frac{1}{T_2} - \frac{1}{T_1}\right]$$

$$5.01 = (-4.90 \times 10^3\text{ K})\left[\frac{1}{T_2} - \frac{1}{373\text{ K}}\right] = -4.90 \times 10^3\text{ K}\left[\frac{1}{T_2} - 0.00268\text{ K}^{-1}\right]$$

$$5.01 = \left[\frac{-4.90 \times 10^3\text{ K}}{T_2} + 13.13\right];\ T_2 = \left[\frac{-4.90 \times 10^3\text{ K}}{-8.12}\right] = 603\text{ K}$$

Therefore, the maximum temperature of water without boiling is 602 K.

9.4 (a) Solid decane is a molecular solid.
(b) Solid $MgCl_2$ is composed of Mg^{2+} and Cl^- ions and is an ionic solid.

9.5 0.500 mol × 23.7 kJ/mol = 11.8 kJ

9.6 The gas phase

9.7 Copper is fcc. Face diagonal $= 4r \times 126\text{ pm} = 504\text{ pm} = \sqrt{2} \times l$

$l = (504\text{ pm}/1.414) = 356\text{ pm}$

9.8 Vanadium is bcc and has two atoms per unit cell. The density equals the mass of the unit cell (grams) divided by its volume (cm^3).

$$m_{\text{uc}} = \frac{50.94\text{ g V}}{1\text{ mol V}} \times \frac{1\text{ mol V}}{6.022 \times 10^{23}\text{ V atoms}} \times \frac{2\text{ V atoms}}{\text{unit cell}}$$
$$= 1.69 \times 10^{-22}\text{ g/unit cell}$$

Calculate the length of the cell edge, l from the dimensions of the cube diagonal. Use the value of l to calculate the unit cell volume, l^3:

Cube diagonal $= 4 \times r = 4 \times 134\text{ pm} = 536\text{ pm} = \sqrt{3} \times l$

$l = (536\text{ pm}/1.732) = 310\text{ pm}$;

$V_{\text{uc}} = l^3 = (310\text{ pm})^3(10^{-10}\text{ cm/1 pm})^3 = 2.98 \times 10^{-23}\text{ cm}^3\text{/unit cell}$

$$d_{\text{uc}} = \frac{1.69 \times 10^{-22}\text{ g/unit cell}}{2.98 \times 10^{-23}\text{ cm}^3\text{/unit cell}} = 5.67\text{ g/cm}^3$$

9.9 The edge of the KCl unit cell is $2 \times 152\text{ pm} + 2 \times 167\text{ pm} = 638$ pm.
The unit cell of KCl is larger than that of NaCl.

$$V_{\text{unit cell}} = (638\text{ pm})^3 = 2.597 \times 10^8\text{ pm}^3 \times \left(\frac{10^{-10}\text{ cm}}{\text{pm}}\right)^3 = 2.597 \times 10^{-22}\text{ cm}^3$$

$$m_{\text{unit cell}} = (4\text{ formula units KCl}) \times \frac{1\text{ mol}}{6.022 \times 10^{23}\text{ formula units}} \times \frac{74.55\text{ g KCl}}{1\text{ mol KCl}} = 4.952 \times 10^{-22}\text{ g}$$

$$d = \frac{m}{v} = \frac{4.952 \times 10^{-22}\text{ g}}{2.597 \times 10^{-22}\text{ cm}^3} = 1.91\text{ g/cm}^3$$

9.10 $$m_{\text{Li}} = 17.3\text{ kJ} \times \frac{1\text{ mol Li}}{3.0\text{ kJ}} \times \frac{6.94\text{ g Li}}{1\text{ mol Li}} = 40.\text{ g Li}$$

9.11 $$V_{CO_2} = 3.0 \times 10^{13}\text{ g CO}_2 \times \frac{1\text{ L CO}_2}{1.799\text{ g CO}_2} = 1.7 \times 10^{13}\text{ L CO}_2$$

Chapter 10

10.1 The balanced combustion reaction for methanol vapor is

$$CH_3OH(g) + \tfrac{3}{2}\,O_2(g) \longrightarrow CO_2(g) + 2\,H_2O(g)$$

Using Hess's law, we see that the heat of combustion of methanol vapor is

$$\Delta_{\text{comb}}H = [\Delta_fH°\,CO_2(g)] + 2[\Delta_fH°\,H_2O(g)] - 1[\Delta_fH°\,CH_3OH(g)]$$
$$= -393.509\text{ kJ/mol} + 2(-241.818\text{ kJ/mol}) - (-200.66\text{ kJ/mol})$$
$$= -676.49\text{ kJ/mol}$$

The energy density of methanol is

$$\text{ED} = \frac{-676.49\text{ kJ}}{1\text{ mol}} \times \frac{1\text{ mol}}{32.04\text{ g}} \times \frac{0.791\text{ g}}{1\text{ mL}} \times \frac{1000\text{ mL}}{1\text{ L}}$$
$$= -1.67 \times 10^4\text{ kJ/L}$$

10.2 $CO(g) + 2\,H_2(g) \longrightarrow CH_3OH(g)$

$$\Delta_rH° = \{(1\text{ C}{\equiv}\text{O})(1073\text{ kJ/mol}) + (2\text{ H—H})(436\text{ kJ/mol})\} - \{(3\text{ C—H})(416\text{ kJ/mol}) + (1\text{ C—O})(336\text{ kJ/mol}) + (1\text{ O—H})(467\text{ kJ/mol})\}$$
$$= (1945\text{ kJ/mol}) - (2051\text{ kJ/mol}) = -106\text{ kJ/mol}$$

10.3 (a) The first oxidation product of $CH_3CH_2CH_2OH$ is the aldehyde.

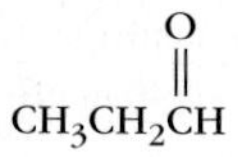

The second oxidation product of $CH_3CH_2CH_2OH$ is the acid.

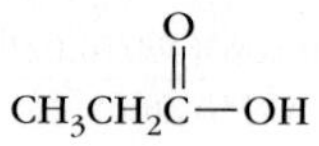

(b) The oxidation product of this secondary alcohol is the ketone.

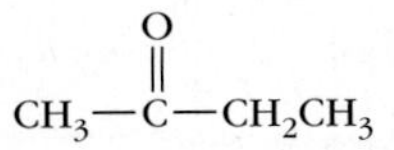

(c) This is a tertiary alcohol, which is not oxidized easily.

10.4 In this case, stearic acid would be on carbon 2 of glycerol where it would be flanked by oleic acids at carbons 1 and 3 of glycerol. See Problem-Solving Example 10.4 (← **Sec. 10-5c**) for the structural formulas of stearic and oleic acids.

10.5

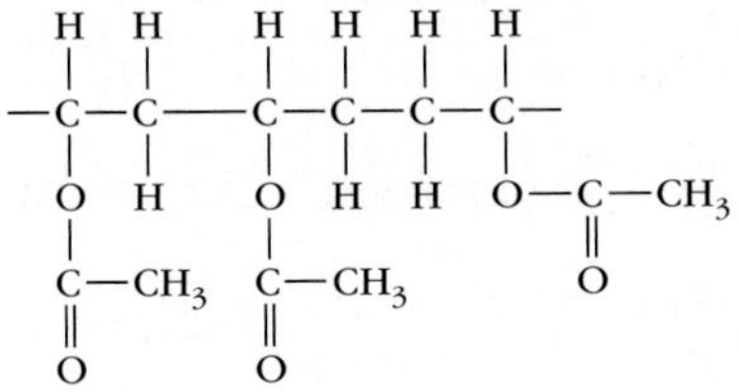

10.6 (a)

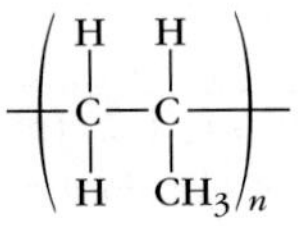

(b)

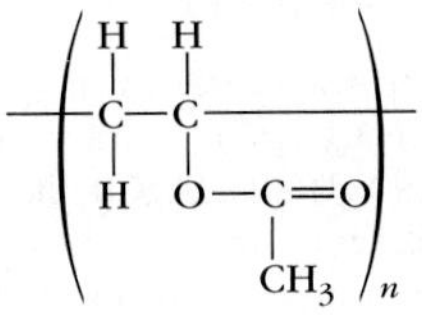

(c)

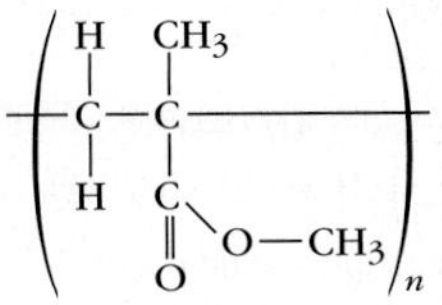

10.7 $M_{\text{co-monomer}} = M\ C_2H_3Cl + M\ C_2H_2Cl_2 =$

$62.5\ \text{g/mol} + 97.0\ \text{g/mol} = 159.5\ \text{g/mol}$

$n = (M_{\text{polymer}}/M_{\text{co-monomer}}) = (150{,}000/160) = 938$

10.8

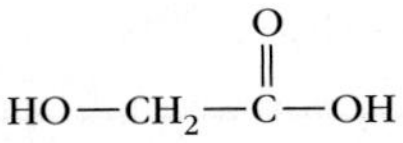

10.9 Twelve moles of water

10.10 The name of the tetrapeptide is lysylphenylalanylserylalanine.

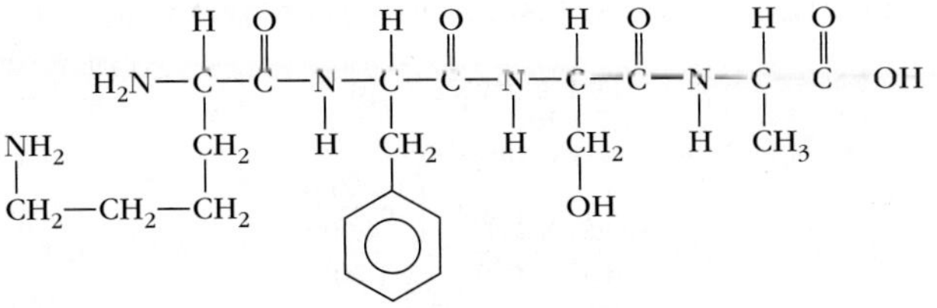

Chapter 11

11.1 (a) $\text{Rate} = \dfrac{-\Delta[Cv^+]}{\Delta t} = 1.27 \times 10^{-6}\ \text{mol L}^{-1}\ \text{s}^{-1}$

$$\Delta t = \frac{-\Delta[Cv^+]}{1.27 \times 10^{-6}\ \text{mol L}^{-1}\ \text{s}^{-1}}$$

$$= \frac{(4.30 \times 10^{-5} - 3.96 \times 10^{-5})\ \text{mol/L}}{1.27 \times 10^{-6}\ \text{mol L}^{-1}\ \text{s}^{-1}}$$

$$= 2.7\ \text{s}$$

(b) No. The rate of reaction depends on the concentration of Cv^+ and, therefore, becomes slower as the reaction progresses. Therefore, the method used in part (a) works only over a small range of concentrations.

11.2 (a) The balanced chemical equation shows that for every mole of O_2 consumed two moles of N_2O_5 are produced. Therefore, the rate of formation of N_2O_5 is twice the rate of disappearance of O_2.

(b) Four moles of NO_2 are consumed for every mole of O_2 consumed. Therefore, if O_2 is consumed at the rate of $0.0037\ \text{mol L}^{-1}\ \text{s}^{-1}$ the rate of disappearance of NO_2 is four times this rate.

$$4 \times (0.0037\ \text{mol L}^{-1}) = 0.015\ \text{mol L}^{-1}\ \text{s}^{-1}$$

11.3 (a) The effect of $[OH^-]$ on the rate of reaction cannot be determined, because the $[OH^-]$ is the same in all three experiments.

(b) $\text{Rate} = k[Cv^+]$

(c) $k_1 = \dfrac{\text{rate}}{[Cv^+]} = \dfrac{1.3 \times 10^{-6}\ \text{mol L}^{-1}\ \text{s}^{-1}}{4.3 \times 10^{-5}\ \text{mol/L}}$

$= 3.0 \times 10^{-2}\ \text{s}^{-1}$

$k_2 = 3.0 \times 10^{-2}\ \text{s}^{-1}$

$k_3 = 3.0 \times 10^{-2}\ \text{s}^{-1}$

$$k = \frac{k_1 + k_2 + k_3}{3} = 3.0 \times 10^{-2}\ \text{s}^{-1}$$

(d) $\text{Rate} = k[Cv^+]$

$= (3.0 \times 10^{-2}\ \text{s}^{-1})(0.00045\ \text{mol/L})$

$= 1.4 \times 10^{-5}\ \text{mol L}^{-1}\ \text{s}^{-1}$

(e) $\text{Rate} = (3.0 \times 10^{-2}\ \text{s}^{-1})(0.05 \times 0.00045\ \text{mol/L})$

$= 6.8 \times 10^{-6}\ \text{mol L}^{-1}\ \text{s}^{-1}$

11.4 (a) The order of the reaction with respect to each chemical is the exponent associated with the concentration of that chemical. The reaction is second-order with respect to NO and it is first-order with respect to Cl_2.

(b) Tripling the concentration of NO will make the reaction go $3^2 = 9$ times faster, but decreasing the concentration of Cl_2 by a factor of 8 will make the reaction go at 1/8 the initial rate. If these two changes are made simultaneously the relevant factor will be $9/8 = 1.13$, so the reaction will occur 13% faster than it did under the initial conditions.

11.5 Make three plots of the data.

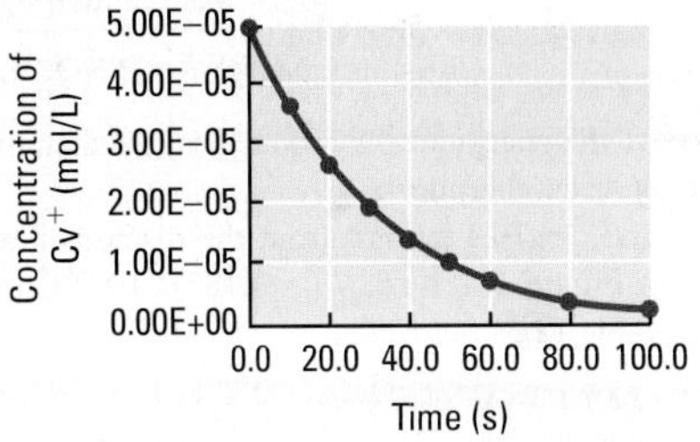
5.00E–05
4.00E–05
3.00E–05
2.00E–05
1.00E–05
0.00E+00
Concentration of Cv+ (mol/L)
0.0 20.0 40.0 60.0 80.0 100.0
Time (s)

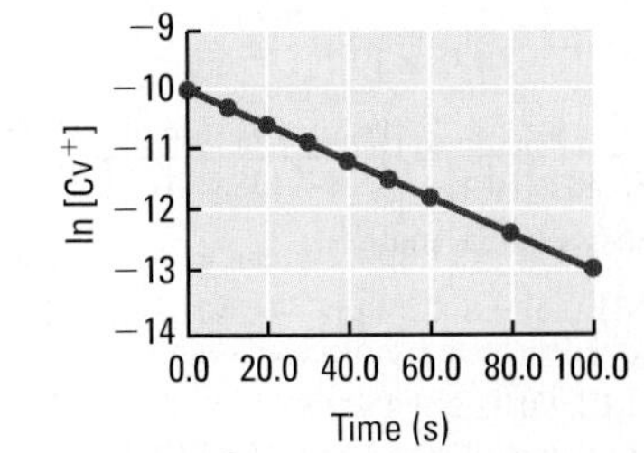
−9
−10
−11
−12
−13
−14
ln [Cv+]
0.0 20.0 40.0 60.0 80.0 100.0
Time (s)

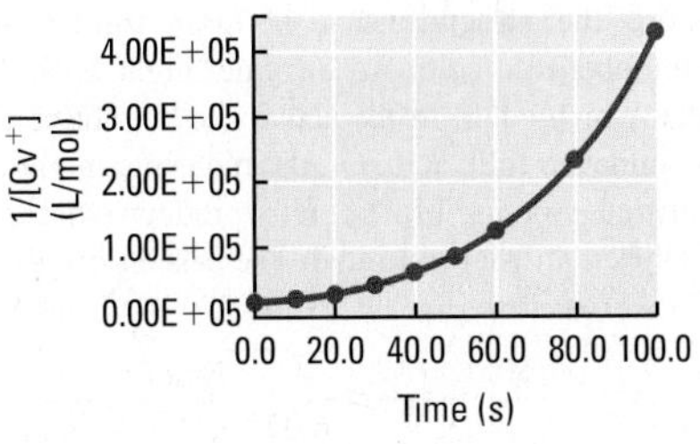
4.00E+05
3.00E+05
2.00E+05
1.00E+05
0.00E+05
1/[Cv+] (L/mol)
0.0 20.0 40.0 60.0 80.0 100.0
Time (s)

The first-order plot is a straight line and the other plots are curved, so the reaction is first-order. The slope of the first-order plot is $-0.0307\ \text{s}^{-1}$, so $k = -\text{slope} = 0.031\ \text{s}^{-1}$.

11.6 Use the integrated first-order rate law from Table 11.2.

$$\ln[A]_t = -kt + \ln[A]_0$$

$$\ln\frac{[A]_t}{[A]_0} = -kt$$

$$t = -\frac{1}{k}\ln\frac{[A]_t}{[A]_0} = -\left(\frac{1}{3.43\times10^{-2}\ \text{d}^{-1}}\right)\ln\left(\frac{0.1}{1.0}\right)$$

$$= -(29.15\ \text{d})(-2.303) = 67.1\ \text{d}$$

11.7 In Figure 11.3 the $[Cv^+]$ falls from 5.00×10^{-5} M to 2.5×10^{-5} M in 23 s. The $[Cv^+]$ falls from 2.5×10^{-5} M to 1.25×10^{-5} M between 23 s and 46 s. The two times are equal, so $t_{1/2} = 23$ s.

$$k = \frac{0.693}{t_{1/2}} = \frac{0.693}{23\ \text{s}} = 3.0\times10^{-2}\ \text{s}^{-1}$$

11.8

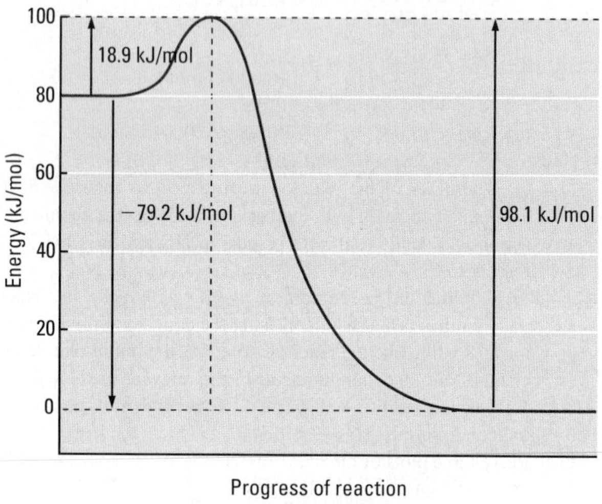

18.9 kJ/mol
−79.2 kJ/mol
98.1 kJ/mol
Energy (kJ/mol)
100
80
60
40
20
0
Progress of reaction

$$Cl_2(g) + 2\ NO(g) \longrightarrow 2\ NOCl(g)$$

Because the energy of the products is less than that of reactants, the reaction is exothermic.

11.9 Obtain $E_a = 76.3$ kJ/mol from the discussion and analysis of the data in Figure 11.11 and $k = 4.18 \times 10^{-5}$ L mol^{-1} s^{-1} at 273 K from Table 11.3.

$$\frac{k_1}{k_2} = e^{\left[\frac{E_a}{R}\left(\frac{1}{T_2}-\frac{1}{T_1}\right)\right]} = e^{\left[\frac{76{,}300\ \text{J mol}^{-1}}{8.314\ \text{J K}^{-1}\ \text{mol}^{-1}}\left(\frac{1}{348\ \text{K}}-\frac{1}{275\ \text{K}}\right)\right]}$$

$$= e^{-7.245} = 7.138\times10^{-4}$$

$$k_2 = \frac{k_1}{7.138\times10^{-4}} = \frac{4.18\times10^{-5}\ \text{L mol}^{-1}\ \text{s}^{-1}}{7.138\times10^{-4}}$$

$$= 5.86\times10^{-2}\ \text{L mol}^{-1}\ \text{s}^{-1}$$

11.10 (a) $2\ NH_3(aq) + OCl^-(aq) \longrightarrow N_2H_4(aq) + Cl^-(aq) + H_2O(\ell)$

(b) Step 1

(c) NH_2Cl, OH^-, $N_2H_5^+$

(d) Rate = rate of step 1 = $k[NH_3][OCl^-]$

11.11 Choose the structure that is most similar to the structure of *p*-aminobenzoic acid. An enzyme might be inhibited by this molecule, which could fit the active site but not be converted to a product similar to folic acid. Or, the molecule might react in an enzyme-catalyzed process, producing a product whose biological function was different from folic acid. The best choice is

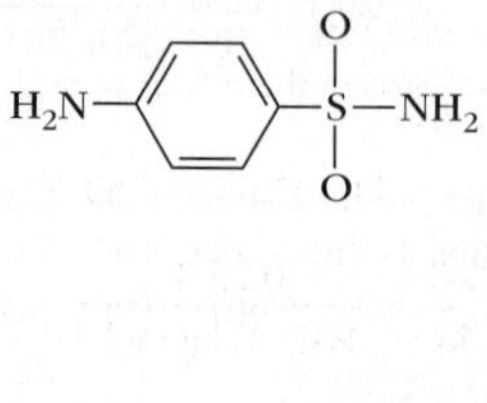

Chapter 12

12.1 (a) $K_c = [CO_2(g)]$ (b) $K_c = \dfrac{[H_2]}{[HCl]}$

(c) $K_c = \dfrac{[CO][H_2]^3}{[CH_4][H_2O]}$ (d) $K_c = \dfrac{[HCN][OH^-]}{[CN^-]}$

12.2 $K_c = K_{c_1} \times K_{c_2} = (4.2 \times 10^{-7})(4.8 \times 10^{-11})$
$= 2.0 \times 10^{-17}$

12.3 $[CH_3COO^-] = \dfrac{2.96}{100} \times 0.0200\ \text{mol/L} = 5.92\times10^{-4}$ M

$[H^+] = 5.92 \times 10^{-4}$ M

$[CH_3COOH] = \dfrac{100-2.96}{100}\times 0.200\ \text{mol/L} = 1.94\times10^{-2}$ M

$$K_c = \frac{[H^+][CH_3COO^-]}{[CH_3COOH]} = \frac{(5.92\times10^{-4})(5.92\times10^{-4})}{1.94\times10^{-2}}$$

$$= 1.81\times10^{-5}$$

The result agrees with the value in Table 12.1.

12.4 (a) $K_c(AgCl) = 1.8 \times 10^{-10}$; $K_c(AgI) = 1.5 \times 10^{-16}$. Because $K_c(AgI) < K_c(AgCl)$, the concentration of silver ions is larger in the beaker of AgCl.

(b) Unless all of the solid AgCl or AgI dissolves (which would mean that there was no equilibrium reaction), the concentrations at equilibrium are independent of the volume.

12.5 $Q = \dfrac{(\text{conc. }SO_3)^2}{(\text{conc. }SO_2)^2(\text{conc. }O_2)} = \dfrac{(0.184)^2}{(0.102\times2)^2(0.0132)} = 61.6$

Because $Q < K_c$, the forward reaction should occur.

12.6

	$AuI(s) \rightleftharpoons$	$Au^+(aq)$	$+ I^-(aq)$
Initial concentration (mol/L)		0	0
Change as reaction occurs (mol/L)		$+x$	$+x$
Equilibrium concentration (mol/L)		x	x

$K_c = 1.6 \times 10^{-23} = [Au^+][I^-] = x^2$

$x = \sqrt{1.6\times10^{-23}} = 4.0 \times 10^{-12} = [Au^+] = [I^-]$

12.7 The reaction is the reverse of the one in Table 12.1, so

$$K_c = \frac{1}{1.7\times10^2} = 5.9\times10^{-3}$$

$$Q = \frac{(\text{conc. }NO_2)^2}{(\text{conc. }N_2O_4)} = \frac{\left(\frac{0.500\ \text{mol}}{4.00\ \text{L}}\right)^2}{\left(\frac{1.00\ \text{mol}}{4.00\ \text{L}}\right)} = 6.25\times10^{-2}$$

Because $Q > K_c$, the reaction should go in the reverse direction. Therefore, let x be the change in concentration of N_2O_4, giving the ICE table:

	N_2O_4	$\rightleftharpoons$	$2\ NO_2$
Initial concentration (mol/L)	$\frac{1.00}{4.00} = 0.250$		$\frac{0.500}{4.00} = 0.125$
Change as reaction occurs (mol/L)	x		$-2x$
Equilibrium concentration (mol/L)	$0.250 + x$		$0.125 - 2x$

$$K_c = 5.9\times10^{-3} = \frac{(0.125-2x)^2}{0.250+x} = \frac{(1.56\times10^{-2}) - 0.500x + 4x^2}{0.250+x}$$

$(1.48 \times 10^{-3}) + (5.9 \times 10^{-3})x = (1.56 \times 10^{-2}) - 0.500x + 4x^2$

$4x^2 - 0.5059x + (1.412 \times 10^{-2}) = 0$

$$x = \frac{-(-0.5059) \pm \sqrt{(-0.5059)^2 - 4 \times 4 \times 1.412 \times 10^{-2}}}{2 \times 4}$$

$$= \frac{0.5059 \pm \sqrt{3.001 \times 10^{-2}}}{8}$$

$$= \frac{0.5059 \pm 0.1732}{8}$$

$x = 8.49 \times 10^{-2}$ or $x = 4.16 \times 10^{-2}$

If $x = 8.49 \times 10^{-2}$, then $[N_2O_4] = 0.250 + x = 0.335$ and $[NO_2] = 0.125 - 2x = 0.125 - (2 \times 8.49 \times 10^{-2}) = -0.0448$. A negative concentration is impossible, so x must be 4.16×10^{-2}. Then

$$[N_2O_4] = 0.250 + 0.0416 = 0.292$$

$$[NO_2] = 0.125 - (2 \times 0.0416) = 0.0418$$

As predicted by Q, the reverse reaction has occurred, and the concentration of N_2O_4 has increased.

12.8 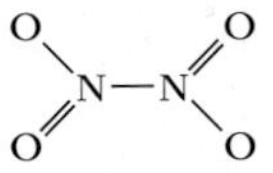 and 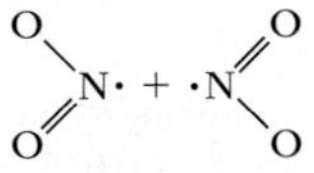

Because a bond is broken and because bond breaking is always endothermic (← **Sec. 4-7a**), the reaction must be endothermic. Increasing temperature shifts the equilibrium in the endothermic direction. Figure 12.3 shows that at a higher temperature there is a greater concentration of brown NO_2.

12.9 (a) There are more moles of gas-phase reactants than products, so entropy favors the reactants.
(b) The reaction is exothermic, so the energy effect favors the products.
(c) As T increases, the reaction shifts in the endothermic direction, which is toward reactants. The entropy effect also becomes more important at high T, and it favors reactants. There is a greater concentration of NH_3 at low temperature.

Chapter 13

13.1 (a) Ethylene glycol molecules are polar and attracted to each other by dipole-dipole attractions and hydrogen bonding. They will not dissolve in gasoline, a nonpolar substance.
(b) Molecular iodine and carbon tetrachloride are nonpolar; therefore iodine should dissolve readily in carbon tetrachloride.
(c) Motor oil contains a mixture of nonpolar hydrocarbons that will dissolve in carbon tetrachloride, a nonpolar solvent.

13.2 The —OH groups attached to the ring and to the side chain of vitamin C hydrogen bond to water molecules. The oxygen atoms in and on the ring also form hydrogen bonds to water. Thus, vitamin C is water soluble.

13.3 Use Henry's law, the Henry's law constant, and the fact that air is only 21 mol percent oxygen, a 0.21 mole fraction of oxygen.

$$P_{O_2} = (1.0 \text{ atm})\left(\frac{760 \text{ mmHg}}{1 \text{ atm}}\right)(0.21) = 160. \text{ mmHg}$$

$$S_g = k_H P_g = \left(1.66 \times 10^{-6} \frac{\text{mol/L}}{\text{mmHg}}\right)(160. \text{ mmHg})$$

$$= 2.66 \times 10^{-4} \text{ mol/L}$$

$$(2.66 \times 10^{-4} \text{ mol/L})\left(\frac{32.00 \text{ g } O_2}{1 \text{ mol } O_2}\right) = 0.0085 \text{ g/L or } 8.5 \text{ mg/L}$$

13.4 Total mass is 750 + 21.5 = 771.5 g.

$$\text{Weight percent glucose} = \left(\frac{21.5 \text{ g}}{771.5 \text{ g}}\right) \times 100\% = 2.79\%$$

13.5 $$\left(\frac{30 \text{ g Se}}{10^9 \text{ g } H_2O}\right)\left(\frac{1 \text{ g } H_2O}{1 \text{ mL } H_2O}\right)\left(\frac{10^6\ \mu\text{g Se}}{1 \text{ g Se}}\right) = 3.0 \times 10^{-2}\ \mu\text{g Se/mL } H_2O$$

Mass Se in 100. mL water =

$$\frac{3.0 \times 10^{-2}\ \mu\text{g Se}}{1 \text{ mL } H_2O} \times 100. \text{ mL } H_2O = 3.0\ \mu\text{g Se}$$

13.6 (a) $$m_{NaBr} = 0.250 \text{ L} \times \frac{0.0750 \text{ mol NaBr}}{1 \text{ L}} \times \frac{102.9 \text{ g NaBr}}{1 \text{ mol NaBr}}$$

$$= 1.93 \text{ g NaBr}$$

(b) $$V_c = \frac{c_d \times V_d}{c_c} = \frac{(0.00150 \text{ M})(0.500 \text{ L})}{0.0750 \text{ M}}$$

$$= 0.0100 \text{ L} = 10.0 \text{ mL}$$

13.7 (a) $$\text{Mass fraction Mg} = \frac{0.0556 \text{ mol } Mg^{2+}}{1 \text{ L soln}} \times \frac{24.3 \text{ g } Mg^{2+}}{1 \text{ mol } Mg^{2+}} \times \frac{1 \text{ L soln}}{1.03 \times 10^3 \text{ g soln}} = 0.00131 \frac{\text{g } Mg^{2+}}{\text{g soln}}$$

(b) 0.00131×10^6 ppm = 1310 ppm

13.8 $$\text{Molarity } H_2O_2 = \frac{\text{moles } H_2O_2}{\text{L solution}}$$

$$= \frac{30.0 \text{ g } H_2O_2}{100. \text{ g solution}} \times \frac{1.11 \text{ g solution}}{1 \text{ mL solution}} \times \frac{10^3 \text{ mL solution}}{1 \text{ L solution}} \times \frac{1 \text{ mol } H_2O_2}{34.0 \text{ g } H_2O_2}$$

$$= \frac{3.33 \times 10^4 \text{ mol } H_2O_2}{3.40 \times 10^3 \text{ L}} = 9.79 \text{ M}$$

13.9 A 20.0% solution contains 20.0 g solute in 100.0 g solution, so there is 80.0 g (0.0800 kg) solvent.

Thus,

$$\text{Molarity} = \frac{20.0 \text{ g NaCl}}{100.0 \text{ g soln}} \times \frac{1.148 \text{ g soln}}{1.000 \text{ mL soln}} \times \frac{1 \text{ mol NaCl}}{58.44 \text{ g soln}} \times \frac{10^3 \text{ mL}}{1 \text{ L}} = 3.93 \text{ mol NaCl/L}$$

$$\text{Molality, } m = \left(\frac{20.0 \text{ g NaCl}}{0.0800 \text{ kg } H_2O}\right)\left(\frac{1 \text{ mol NaCl}}{58.5 \text{ g NaCl}}\right)$$

$$= 4.3 \text{ mol NaCl/kg } H_2O$$

13.10 $P_{water} = (X_{water})(P^\circ_{water})$

$$291.2 \text{ mmHg} = (X_{water})(355.1 \text{ mmHg})$$

$$X_{water} = \left(\frac{291.2 \text{ mmHg}}{355.1 \text{ mmHg}}\right) = 0.8201$$

$$X_{urea} = 1.000 - X_{water} = 1.000 - 0.8201 = 0.180$$

13.11 $$\text{Molality} = \frac{1.20 \text{ kg ethylene glycol}}{6.50 \text{ kg } H_2O} \times \frac{1000 \text{ g}}{1 \text{ kg}} \times \frac{1 \text{ mol ethylene glycol}}{62.068 \text{ g ethylene glycol}} = 2.97 \text{ mol/kg } H_2O$$

Next, calculate the freezing point depression of a 2.97-mol/kg solution.

$$\Delta T_f = (1.86 \text{ °C kg mol}^{-1})(2.97 \text{ mol/kg}) = 5.52 \text{ °C}$$

This solution will freeze at −5.52 °C, so this quantity of ethylene glycol will not protect the 6.5 kg water in the tank if the temperature drops to −25 °C.

13.12 F.P. benzene = 5.50 °C; ΔT_f = 5.50 °C − 5.15 °C = 0.35 °C

$$\text{Molality} = \frac{\Delta T_f}{K_f} = \frac{0.35 \text{ °C}}{5.10 \text{ °C kg mol}^{-1}} = 0.0686 \text{ mol/kg}$$

$$\frac{0.0686 \text{ mol solute}}{1 \text{ kg benzene}} \times 0.0500 \text{ kg benzene} = 0.00343 \text{ mol solute}$$

$$\frac{0.180 \text{ g solute}}{0.00343 \text{ mol solute}} = 53 \text{ g/mol}$$

13.13 Hemoglobin is a molecular substance, so the factor i is 1.

$$c = \frac{\Pi}{RTi} = \frac{1.8 \times 10^{-3} \text{ atm}}{(0.0821 \text{ L atm mol}^{-1} \text{ K}^{-1})(298 \text{ K})(1)}$$

$$= 7.36 \times 10^{-5} \text{ mol/L}$$

$$M_{\text{hemoglobin}} = \frac{5.0 \text{ g}}{7.36 \times 10^{-5} \text{ mol}} = 6.8 \times 10^{4} \text{ g/mol}$$

Chapter 14

14.1

Acid	Its Conjugate Base	Base	Its Conjugate Acid
$H_2PO_4^-$	HPO_4^{2-}	PO_4^{3-}	HPO_4^{2-}
H_2	H^-	NH_2^-	NH_3
HSO_3^-	SO_3^{2-}	ClO_4^-	$HClO_4$
HF	F^-	Br^-	HBr

14.2 (a) $[H_3O^+] = 2.0 \times 10^{-5}$ M
(b) $[H_3O^+] = (1.0 \times 10^{-14})/(5.0 \times 10^{-9}) = 2.0 \times 10^{-6}$ M
Solution (a) has the larger hydrogen-ion concentration; it is more acidic.

14.3 Assume that nitric acid and sodium hydroxide, both strong electrolytes, are 100% ionized and that $[H_3O^+][OH^-] = 1.0 \times 10^{-14}$.

For 6.0-M HNO_3: $[H_3O^+][OH^-] = 1.0 \times 10^{-14} = (6.0)[OH^-]$
$[OH^-] = (1.0 \times 10^{-14})/(6.0) = 1.7 \times 10^{-15}$ M

For 6.0 M NaOH: $[H_3O^+][OH^-] = 1.0 \times 10^{-14} = [H_3O^+](6.0)$
$[H_3O^+] = (1.0 \times 10^{-14})/(6.0) = 1.7 \times 10^{-15}$ M

14.4 In a 0.040-M solution of NaOH, the $[OH^-]$ is 0.040 M because the NaOH is 100% dissociated; $[H_3O^+] = (1.0 \times 10^{-14})/(0.040) = 2.5 \times 10^{-13}$; pH $= -\log(2.5 \times 10^{-13}) = 12.60$.

14.5 (a) $[H_3O^+] = 10^{-7.90} = 1.3 \times 10^{-8}$ M
(b) A pH of 7.90 is basic.

14.6 (a) $HN_3(aq) + H_2O(\ell) \rightleftharpoons H_3O^+(aq) + N_3^-(aq)$

$$K_a = \frac{[H_3O^+][N_3^-]}{[HN_3]}$$

(b) $HCOOH(aq) + H_2O(\ell) \rightleftharpoons H_3O^+(aq) + HCOO^-(aq)$

$$K_a = \frac{[H_3O^+][HCOO^-]}{[HCOOH]}$$

(c) $HClO_2(aq) + H_2O(\ell) \rightleftharpoons H_3O^+(aq) + ClO_2^-(aq)$

$$K_a = \frac{[H_3O^+][ClO_2^-]}{[HClO_2]}$$

14.7 (a) $CH_3NH_2(aq) + H_2O(\ell) \rightleftharpoons CH_3NH_3^+ + OH^-(aq)$

$$K_b = \frac{[CH_3NH_3^+][OH^-]}{[CH_3NH_2]}$$

(b) $PH_3(aq) + H_2O(\ell) \rightleftharpoons PH_4^+(aq) + OH^-(aq)$

$$K_b = \frac{[PH_4^+][OH^-]}{[PH_3]}$$

(c) $NO_2^-(aq) + H_2O(\ell) \rightleftharpoons HNO_2(aq) + OH^-(aq)$

$$K_b = \frac{[HNO_2][OH^-]}{[NO_2^-]}$$

14.8 Setting up a small ICE table for lactic acid, HLa:

	$HLa + H_2O \rightleftharpoons$	H_3O^+ +	La^-
Initial concentration (mol/L)	0.10	10^{-7}	0
Change due to reaction (mol/L)	$-x$	$+x$	$+x$
Equilibrium concentration (mol/L)	$0.10 - x$	x	x

But $x = 10^{-2.43} = 3.7 \times 10^{-3}$ because $x = [H_3O^+]$. Substituting in the K_a expression,

$$K_a = \frac{[H_3O^+][La^-]}{[HLa]} = \frac{(3.7 \times 10^{-3})^2}{0.10 - (3.7 \times 10^{-3})}$$

$$= \frac{1.4 \times 10^{-5}}{0.1} = 1.4 \times 10^{-4}$$

Lactic acid is a stronger acid than propanoic acid, $K_a = 1.4 \times 10^{-5}$.

14.9 (a) Using the same methods as shown in the example. First approximation:

$$\frac{x^2}{0.015} = 1.9 \times 10^{-5}$$

Solving for x, which is $[H_3O^+]$, we get

$$x = \sqrt{(1.9 \times 10^{-5})(0.015)} = 5.2 \times 10^{-4} \text{ M} = [H_3O^+].$$

Second approximation:

$$0.015 - x = 0.0145;\ x = 5.2 \times 10^{-4}$$

The pH of this solution is $-\log(5.2 \times 10^{-4}) = 3.27$.

(b) $$\%\text{ ionization} = \frac{[H_3O^+]}{[HN_3]_{\text{initial}}} \times 100\% = \frac{5.2 \times 10^{-4}}{0.015} \times 100\% = 3.5\%$$

14.10 In such cases use the K_b expression and value to calculate $[OH^-]$ and then pOH from $[OH^-]$. Calculate pH from 14 − pOH.

$C_6H_{11}NH_2(aq) + H_2O(\ell) \rightleftharpoons C_6H_{11}NH_3^+(aq) + OH^-(aq)$

$$K_b = \frac{[C_6H_{11}NH_3^+][OH^-]}{[C_6H_{11}NH_2]} = 4.6 \times 10^{-4}$$

$$K_b = \frac{[C_6H_{11}NH_3^+][OH^-]}{[C_6H_{11}NH_2]} = \frac{x^2}{0.015 - x} = 4.6 \times 10^{-4}$$

$x^2 = (0.015 - x)(4.6 \times 10^{-4}) = (6.9 \times 10^{-6}) - (4.6 \times 10^{-4}\, x)$

Solve the quadratic equation for x.

$$x = \frac{4.8 \times 10^{-3}}{2} = 2.4 \times 10^{-3} = [OH^-]$$

pOH $= -\log(2.4 \times 10^{-3}) = 2.62$; pH $= 14.00 - 2.62 = 11.38$

14.11 Using the same methods as those used in the example, letting $x = [OH^-]$ and $[HCO_3^-]$, and using the value of 2.1×10^{-4} for K_b for CO_3^{2-}, we get

$$\frac{x^2}{1.0} = 2.1 \times 10^{-4} \qquad x = \sqrt{2.1 \times 10^{-4}} = 1.45 \times 10^{-2}$$

pOH = 1.84 and pH = 12.16

14.12 NH_4Cl dissolves by dissociating into NH_4^+ and Cl^- ions. The ammonium ions react with water to produce an acidic solution.

$$NH_4Cl(s) \rightleftharpoons NH_4^+(aq) + Cl^-(aq)$$

$$NH_4^+(aq) + H_2O(\ell) \rightleftharpoons H_3O^+(aq) + NH_3(aq)$$

$$\text{The } K_a \text{ of } NH_4^+ = \frac{K_w}{K_b} = \frac{1.0 \times 10^{-14}}{1.8 \times 10^{-5}};$$

$$K_a = 5.6 \times 10^{-10} = \frac{[H_3O^+][NH_3]}{[NH_4^+]}$$

	$NH_4^+ + H_2O \rightleftharpoons H_3O^+ + NH_3$		
Initial	0.10	0	0
Change	$-x$	$+x$	$+x$
Equilibrium	$0.10 - x$	x	x

$$5.6 \times 10^{-10} = \frac{(x)(x)}{(0.10 - x)}$$

Assume $0.10 - x = 0.10$ because K_a is so small. Thus,

$(5.6 \times 10^{-10})(0.10) = x^2$; $x = [H_3O^+] = 7.48 \times 10^{-6}$ M.

$pH = -\log[H_3O^+] = -\log(7.48 \times 10^{-6}) = 5.13$

14.13 The molar masses and moles of acid per gram for the five antacids are:

	Formula Weight	Mol Acid/Gram
$Mg(OH)_2$	58.32	1 mol acid/29.16 g antacid
$CaCO_3$	100.10	1 mol acid/50.05 g antacid
$NaHCO_3$	84.00	1 mol acid/84.00 g antacid
$Al(OH)_3$	78.0034	1 mol acid/26.00 g antacid
$NaAl(OH)_2CO_3$	143.99	1 mol acid/36.00 g antacid

Of these antacids, $Al(OH)_3$ neutralizes the most stomach acid per gram of antacid.

Chapter 15

15.1 $K_a = 4.3 \times 10^{-7}$

$$= \frac{[H_3O^+][HCO_3^-]}{[H_2CO_3]} = \frac{[H_3O^+](0.025)}{0.0020} = [H_3O^+] \times 12.5$$

$$[H_3O^+] = \frac{4.3 \times 10^{-7}}{12.5} = 3.4 \times 10^{-8}$$

$pH = -\log(3.4 \times 10^{-8}) = 7.46$

15.2 $7.40 = 7.20 + \log(\text{ratio}) = 7.20 + \log\frac{[HPO_4^{2-}]}{[H_2PO_4^-]}$

$$\log\frac{[HPO_4^{2-}]}{[H_2PO_4^-]} = 7.40 - 7.20 = 0.20$$

$$\frac{[HPO_4^{2-}]}{[H_2PO_4^-]} = 10^{0.20} = 1.6$$

Therefore, $[HPO_4^{2-}] = 1.6 \times [H_2PO_4^-]$.

15.3 (a) Lactic acid-lactate (b) Acetic acid-acetate

(c) Hypochlorous acid-hypochlorite; $H_2PO_4^-/HPO_4^{2-}$

(d) Hydrogen carbonate-carbonate

15.4 Using the Henderson-Hasselbalch equation and the same methods as shown in the example,

(a) Buffer I:

$$pH = 3.85 + \log\left(\frac{(0.15 - 0.075)}{(0.15 + 0.075)}\right) = 3.85 - 0.48 = 3.37$$

Buffer II:

$$pH = 3.85 + \log\left(\frac{(0.20 - 0.075)}{(0.15 + 0.075)}\right) = 3.85 - 0.26 = 3.59$$

Buffer III:

$$pH = 3.85 + \log\left(\frac{(0.15 - 0.075)}{(0.25 + 0.075)}\right) = 3.85 - 0.64 = 3.21$$

(b) Buffer I:

$$pH = 3.85 + \log\left(\frac{(0.15 + 0.065)}{(0.15 - 0.065)}\right) = 3.85 + 0.64 = 4.25$$

Buffer II:

$$pH = 3.85 + \log\left(\frac{(0.20 + 0.065)}{(0.15 - 0.065)}\right) = 3.85 + 0.49 = 4.34$$

Buffer III:

$$pH = 3.85 + \log\left(\frac{(0.15 + 0.065)}{(0.25 - 0.065)}\right) = 3.85 + 0.065 = 3.92$$

15.5 From Problem-Solving Example 15.5 the initial pH of the buffer is 6.98. Adding KOH increases the pH, so solve for the ratio of (conjugate base)/(conjugate acid) that corresponds with pH = 7.98.

$$\log\frac{[HPO_4^{2-}]}{[H_2PO_4^-]} = pH - pK_a = 7.98 - 7.20 = 0.78$$

$$\frac{[HPO_4^{2-}]}{[H_2PO_4^-]} = 6.02$$

Let x = amount KOH added; then,

$$\frac{\frac{0.15}{0.500} + \frac{x}{0.500}}{\frac{0.25}{0.500} - \frac{x}{0.500}} = 6.02$$

and $0.15 + x = 6.02(0.25 - x)$

$x = 0.19$ mol

$$0.19 \text{ mol OH}^- = 0.19 \text{ mol KOH} \times \frac{56 \text{ g KOH}}{1 \text{ mol KOH}} = 11 \text{ g.}$$

Thus, slightly more than 11 g KOH will exceed the buffer capacity.

15.6 (a) $[H_3O^+] = \dfrac{(5.00 \times 10^{-3}) - (1.00 \times 10^{-3})}{0.0500 + 0.0100}$

$$= \frac{4.00 \times 10^{-3}}{0.0600} = 6.67 \times 10^{-2} \text{ M}$$

$pH = 1.176 = 1.18$

(b) $[H_3O^+] = \dfrac{(5.00 \times 10^{-3}) - (0.00250)}{0.0500 + 0.0250}$

$$= \frac{2.50 \times 10^{-3}}{0.0750} = 3.33 \times 10^{-2} \text{ M}$$

$pH = -\log(3.33 \times 10^{-2}) = 1.48$

(c) $[H_3O^+] = \dfrac{(5.00 \times 10^{-3}) - (0.00450)}{0.0500 + 0.0450}$

$$= \frac{5.00 \times 10^{-4}}{0.0950} = 5.26 \times 10^{-3}$$

$pH = -\log(5.26 \times 10^{-3}) = 2.28$

(d) $[OH^-] = \dfrac{0.05 \times 10^{-3} \text{ mol OH}^-}{0.0500 \text{ L} + 0.0505 \text{ L}} = 5.0 \times 10^{-4}$ mol/L

$pOH = -\log(5.0 \times 10^{-4}) = 3.30$

$pH = 14.00 - 3.30 = 10.70$

15.7 (a) Adding 10.0 mL of 0.100-M NaOH is adding

(0.100 mol/L) (0.0100 L) = 0.00100 mol OH^-,

which neutralizes 0.00100 mol acetic acid, converting it to 0.00100 mol acetate ion.

$$pH = pK_a + \log\frac{[\text{acetate}]}{[\text{acetic acid}]}$$

$$= 4.74 + \log\frac{(0.00100/0.0600)}{(0.00400/0.0600)}$$

$$= 4.74 + \log(0.25) = 4.74 + (-0.602) = 4.14$$

(b) $pH = 4.74 + \log\dfrac{(0.00250/0.0750)}{(0.00250/0.0750)} = 4.74 + \log(1) = 4.74$

(c) $pH = 4.74 + \log\dfrac{(0.00450/0.0950)}{(0.00050/0.0950)}$

$$= 4.74 + \log(9) = 4.74 + 0.95 = 5.70$$

(d) $[OH^-] = \dfrac{0.10 \times 10^{-3}\text{ mol OH}^-}{0.0500\text{ L} + 0.0510\text{ L}} = 9.9 \times 10^{-4}\text{ mol/L}$

$\text{pOH} = -\log(9.99 \times 10^{-4}) = 3.00$

$\text{pH} = 14.00 - 3.00 = 11.00$

15.8 (a) $K_{sp} = [Cu^+][Br^-]$ (b) $K_{sp} = [Hg^{2+}][I^-]^2$
(c) $K_{sp} = [Sr^{2+}][SO_4^{2-}]$

15.9 $AgBr(s) \rightleftharpoons Ag^+(aq) + Br^-(aq)$

$K_{sp} = [Ag^+][Br^-] = 5.0 \times 10^{-13} = x^2$; x = solubility

$x = \sqrt{5.0 \times 10^{-13}} = 7.1 \times 10^{-7}$ M

15.10 $Ag_2C_2O_4(s) \rightleftharpoons 2\,Ag^+(aq) + C_2O_4^{2-}(aq)$

$K_{sp} = [Ag^+]^2[C_2O_4^{2-}]$

$K_{sp} = [Ag^+]^2[C_2O_4^{2-}] = (1.4 \times 10^{-4})^2\,(6.9 \times 10^{-5})$
$= 1.4 \times 10^{-12}$

15.11

	$PbCl_2 \rightleftharpoons$ Pb^{2+}	+ 2 Cl^-
Initially (mol/L)	0	0.50
Change due to dissolving (mol/L)	$+S$	$0.50 + 2S$
Equilibrium (mol/L)	S	0.50 (ignore $2S$ because S will be small)

$K_{sp} = (S)(0.50)^2 = 1.6 \times 10^{-5}$

$S = [Pb^{2+}] = \dfrac{1.6 \times 10^{-5}}{(0.50)^2} = 6.4 \times 10^{-5}$ mol/L

15.12 (a) $PbCl_2(s) \rightleftharpoons Pb^{2+}(aq) + 2\,Cl^-(aq)$
$K_{sp} = [Pb^{2+}][Cl^-]^2 = 1.6 \times 10^{-5}$. Let S equal the solubility of lead chloride, which equals $[Pb^{2+}]$.

$K_{sp} = 1.6 \times 10^{-5} = (S)(2S)^2 = 4S^3$

$S = \sqrt[3]{\dfrac{1.6 \times 10^{-5}}{4}} = \sqrt[3]{4.0 \times 10^{-6}} = 1.6 \times 10^{-2}$ mol/L

$= (4.0 \times 10^{-6})^{1/3} = 1.6 \times 10^{-2}$ mol/L

(b) Let solubility of $PbCl_2 = [Pb^{2+}] = S$; $[Cl^-] = 0.20$ M.

$[Pb^{2+}] = S = \dfrac{1.6 \times 10^{-5}}{(0.20)^2} = 4.0 \times 10^{-4}$ mol/L

This is less than that in pure water due to the common ion effect of the presence of chloride ion.

15.13 $AgCl(s) \rightleftharpoons Ag^+(aq) + Cl^-(aq)$ $K_{sp} = 1.8 \times 10^{-10}$
$Ag^+(aq) + 2\,S_2O_3^{2-}(aq) \rightleftharpoons Ag(S_2O_3)_2^{3-}(aq)$ $K_f = 2.0 \times 10^{13}$

Net reaction:

$AgCl(s) + 2\,S_2O_3^{2-}(aq) \rightleftharpoons Ag(S_2O_3)_2^{3-}(aq) + Cl^-(aq)$

Therefore, the equilibrium constant for the net reaction is $K_{sp} \times K_f = (1.8 \times 10^{-10})(2.0 \times 10^{13}) = 3.6 \times 10^3$. Because K_{net} is much greater than 1, the net reaction is product-favored, and AgCl is much more soluble in an $Na_2S_2O_3$ solution than it is in water.

15.14 (a) $Q = (1.0 \times 10^{-5})(1.0 \times 10^{-5}) = 1.0 \times 10^{-10} < K_{sp}$; no precipitation.
(b) For precipitation to occur, $Q \geq K_{sp}$; Q = (conc. Ag^+) $\times$ (conc. Cl^-); $K_{sp} = [Ag^+][Cl^-] = 1.8 \times 10^{-10}$;

$[Cl^-] = \dfrac{1.8 \times 10^{-10}}{1.0 \times 10^{-5}} = 1.8 \times 10^{-5}$ M

the minimum Cl^- concentration for AgCl precipitation.

15.15 AgCl precipitates first. $[Cl^-]$ needed to precipitate AgCl:

$[Cl^-] = \dfrac{1.8 \times 10^{-10}}{1.0 \times 10^{-2}} = 1.8 \times 10^{-8}$ M

$[Cl^-]$ needed to precipitate $PbCl_2$:

$[Cl^-] = \sqrt{\dfrac{1.6 \times 10^{-5}}{1.0 \times 10^{-1}}} = 1.3 \times 10^{-2}$ M

Chapter 16

16.1 $\Delta S = q_{rev}/T = (30.8 \times 10^3\text{ J})/(273.15 + 45.3)\text{ K}$
$= (30.8 \times 10^3\text{ J})/(318.45\text{ K})$
$\Delta S = 96.7\text{ J K}^{-1}$

16.2 (a) C(g) has higher $S°$, 158.096 J K^{-1} mol^{-1} versus 5.740 J K^{-1} mol^{-1} for C (graphite).
(b) Ar(g) has higher $S°$, 154.7 J K^{-1} mol^{-1} versus 41.42 J K^{-1} mol^{-1} for Ca(s).
(c) KOH(aq) has higher $S°$, 91.6 J K^{-1} mol^{-1} versus 78.9 J K^{-1} mol^{-1} for KOH(s).

16.3 (a) $\Delta_r S° = 2 \times S°(CO(g)) + 1 \times S°(O_2(g)) - 2 \times S°(CO_2(g))$
$= \{2 \times (197.674) + (205.138) - 2 \times (213.74)\}\text{ J K}^{-1}\text{ mol}^{-1}$
$= 173.01\text{ J K}^{-1}\text{ mol}^{-1}$
(b) $\Delta_r S° = 1 \times S°(NaCl(aq)) - 1 \times S°(NaCl(s))$
$= (115.5 - 72.13)\text{ J K}^{-1}\text{ mol}^{-1} = 43.4\text{ J K}^{-1}\text{ mol}^{-1}$
(c) $\Delta_r S° = 1 \times S°(MgO(s)) + 1 \times S°(CO_2(s)) - 1 \times S°(MgCO_3(s))$
$= (26.94 + 213.74 - 65.7)\text{ J K}^{-1}\text{ mol}^{-1}$
$= 175.0\text{ J K}^{-1}\text{ mol}^{-1}$

16.4 $N_2(g) + 3\,H_2(g) \longrightarrow 2\,NH_3(g)$

$\Delta_r H° = 2 \times \Delta_f H°(NH_3(g))$
$= 2\,(-46.11\text{ kJ/mol}) = -92.22\text{ kJ/mol}$

$\Delta_r S° = 2 \times S°(NH_3(g)) - 1 \times S°(N_2(g)) - 3 \times S°(H_2(g))$
$= \{2(192.45) - (191.61) - 3(130.684)\}\text{ J K}^{-1}\text{ mol}^{-1}$
$= -198.76\text{ J K}^{-1}\text{ mol}^{-1}$

$\Delta_r S°_{universe} = \dfrac{-\Delta_r H°}{T} + \Delta_r S° = \dfrac{92.2\text{ kJ/mol}}{298.15\text{ K}} + (-198.76\text{ J K}^{-1}\text{ mol}^{-1})$

$= \dfrac{92{,}200\text{ J/mol}}{298.15\text{ K}} - 198.76\text{ J K}^{-1}\text{ mol}^{-1} = 110.5\text{ J K}^{-1}\text{ mol}^{-1}$

The process is product-favored.

16.5 (a) $\Delta_r H° = \{(-238.66) - (-110.525)\}$ kJ/mol
$= -128.14$ kJ/mol
$\Delta_r S° = \{(126.8) - 197.674 - 2 \times 130.684\}\text{ J K}^{-1}\text{ mol}^{-1}$
$= -332.2\text{ J K}^{-1}\text{ mol}^{-1}$
$\Delta_r G° = \Delta_r H° - T\Delta_r S°$
$= -128.14 \times 10^3\text{ J/mol} - 298.15\text{ K} \times (-332.2\text{ J K}^{-1}\text{ mol}^{-1})$
$= -29.09 \times 10^3\text{ J/mol} = -29.09$ kJ/mol.
(b) $\Delta_r G° = [-166.27 - (-137.168)]$ kJ/mol $= -29.10$ kJ/mol. The two results agree.
(c) $\Delta_r G°$ is negative. The reaction is product-favored at 298.15 K. Because $\Delta_r S°$ is negative, at very high temperatures the reaction becomes reactant-favored.

16.6 (a) $T = \Delta_r H°/\Delta_r S° = (-565{,}968\text{ J/mol})/(-173.01\text{ J K}^{-1}\text{ mol}^{-1})$
$= 3271$ K
(b) The reaction is exothermic and therefore is product-favored at temperatures lower than 3271 K.

16.7 (a) $\Delta_r G° = \{-553.58 - 527.81 - (-1128.79)\}$ kJ/mol
$= 47.40$ kJ/mol
$K° = e^{-\Delta_r G°/RT} = e^{-(47.40\text{ kJ/mol})/(8.314\text{ J K}^{-1}\text{ mol}^{-1})(298\text{ K})}$
$= e^{-(47{,}400\text{ J/mol})/(8.314\text{ J K}^{-1}\text{ mol}^{-1})(298\text{ K})}$
$= e^{-19.13} = 4.9 \times 10^{-9}$ (close to K_c)
(b) $K° = e^{-14.66} = 4.3 \times 10^{-7}$ (agrees with K_c)
(c) $K° = e^{-(-1.909)} = 6.75$ (agrees with K_P)
For reactions (a) and (b), $K_c = K°$.

16.8 (a) At 298 K,
$\Delta_r G° = 2 \times (-16.45\text{ kJ/mol}) = -32.9$ kJ/mol
$K° = e^{-(-32{,}900\text{ J/mol})/(8.314\text{ J K}^{-1}\text{ mol}^{-1})(298\text{ K})} = e^{13.28}$
$= 5.8 \times 10^5$
(b) At 450. K,
$\Delta_r G° = \Delta_r H° - T\Delta_r S°$
$= -92.22\text{ kJ/mol} - (450.\text{ K})(-0.19876\text{ kJ K}^{-1}\text{ mol}^{-1})$
$= -2.78$ kJ/mol
$K° = e^{-(-2780\text{ J/mol})(8.314\text{ J K}^{-1}\text{ mol}^{-1})(450.\text{ K})} = 2.10$

(c) At 800. K,

$$\Delta_r G^\circ = -92.22\ \text{kJ/mol} - (800.\ \text{K})(-0.19876\ \text{kJ K}^{-1}\ \text{mol}^{-1}) = 66.79\ \text{kJ/mol}$$

$$K^\circ = e^{-(66{,}790\ \text{J/mol})/(8.314\ \text{J K}^{-1}\ \text{mol}^{-1})(800.\ \text{K})} = 4.3 \times 10^{-5}$$

16.9 $\Delta_r G = \Delta_r G^\circ + RT \ln Q$

$\Delta_r G$ was calculated in Problem-Solving Practice 16.7 to be 47.40 kJ/mol for the reverse of this reaction. Therefore,

$\Delta_r G^\circ = -47.40$ kJ/mol

$$\Delta_r G = -47.40\ \text{kJ/mol} + (8.314\ \text{J K}^{-1}\ \text{mol}^{-1})(298\ \text{K}) \ln\left(\frac{1}{(0.023)(0.13)}\right)$$

$$= -47.40\ \text{kJ/mol} + 1.44 \times 10^4\ \text{J/mol}$$

$$= -47.40\ \text{kJ/mol} + 14.4\ \text{kJ/mol} = -33.0\ \text{kJ/mol}$$

16.10 (a) $\Delta_r G^\circ = 2(-137.168\ \text{kJ/mol}) - 2(-394.359\ \text{kJ/mol}) = 514.382$ kJ/mol. The reaction is reactant-favored, and at least 514.382 kJ/mol work must be done to make it occur.

(b) $\Delta_r G^\circ = 2(-742.2\ \text{kJ/mol}) = -1484.4$ kJ/mol. The reaction is product-favored and could do up to 1484.4 kJ/mol of useful work.

16.11 $\Delta_r G^{\circ\prime} = \Delta_r G^\circ$ for this reaction because none of the reactants or products requires a standard state different from 1 bar or 1 mol/L.

16.12 (a) The strongest phosphate donor has the most negative $\Delta_r G^{\circ\prime}$ for its reaction with water to produce dihydrogen phosphate. The $\Delta_r G^{\circ\prime}$ values are given in Problem-Solving Example 16.12. Creatine phosphate, at -43.1 kJ/mol, has the most negative value and is the strongest phosphate donor.

(b) Glycerol 3-phosphate has the least negative $\Delta_r G^{\circ\prime}$ with value -9.7 kJ/mol; it therefore is the weakest phosphate donor.

(c) See parts (a) and (b) for explanation.

16.13 (a) $\Delta_f G^\circ(\text{MgO(s)}) = -569.43$ kJ/mol, so formation of MgO(s) is product-favored and MgO(s) is thermodynamically stable.

(b) $\Delta_f G^\circ(\text{N}_2\text{H}_4(\ell)) = 149.34$ kJ/mol; kinetically stable.

(c) $\Delta_f G^\circ(\text{C}_2\text{H}_6(\text{g})) = -32.82$ kJ/mol; thermodynamically stable.

(d) $\Delta_f G^\circ(\text{N}_2\text{O(g)}) = 104.20$ kJ/mol; kinetically stable.

Chapter 17

17.1 Reducing agents are indicated by "red" and oxidizing agents are indicated by "ox." Oxidation numbers are shown above the symbols for the elements.

(a) $2\,\overset{0}{\underset{\text{red}}{\text{Fe}}}(s) + 3\,\overset{0}{\underset{\text{ox}}{\text{Cl}_2}}(g) \longrightarrow 2\,\overset{+3\;-1}{\text{FeCl}_3}(s)$

(b) $2\,\overset{0}{\underset{\text{red}}{\text{H}_2}}(g) + \overset{0}{\underset{\text{ox}}{\text{O}_2}}(g) \longrightarrow 2\,\overset{+1\;-2}{\text{H}_2\text{O}}(\ell)$

(c) $\overset{0}{\underset{\text{red}}{\text{Cu}}}(s) + 2\,\overset{+5\;-2}{\underset{\text{ox}}{\text{NO}_3^-}}(aq) + 4\,\overset{+1}{\text{H}^+}(aq) \longrightarrow \overset{+2}{\text{Cu}^{2+}}(aq) + 2\,\overset{+4\;-2}{\text{NO}_2}(g) + 2\,\overset{+1\;-2}{\text{H}_2\text{O}}(\ell)$

(d) $\overset{0}{\underset{\text{red}}{\text{C}}}(s) + \overset{0}{\underset{\text{ox}}{\text{O}_2}}(g) \longrightarrow \overset{+4\;-2}{\text{CO}_2}(g)$

(e) $6\,\overset{+2}{\underset{\text{red}}{\text{Fe}^{2+}}}(aq) + \overset{+6\;-2}{\underset{\text{ox}}{\text{Cr}_2\text{O}_7^{2-}}}(aq) + 14\,\overset{+1}{\text{H}^+}(aq) \longrightarrow 6\,\overset{+3}{\text{Fe}^{3+}}(aq) + 2\,\overset{+3}{\text{Cr}^{3+}}(aq) + 7\,\overset{+1\;-2}{\text{H}_2\text{O}}(\ell)$

17.2 (a) Oxidation half-reaction: $\text{Cd(s)} \longrightarrow \text{Cd}^{2+}(\text{aq}) + 2\,\text{e}^-$
Reduction half-reaction: $\text{Cu}^{2+}(\text{aq}) + 2\,\text{e}^- \longrightarrow \text{Cu(s)}$
The sum of the two half-reactions is
$\text{Cd(s)} + \text{Cu}^{2+}(\text{aq}) \longrightarrow \text{Cd}^{2+}(\text{aq}) + \text{Cu(s)}$

(b) Oxidation half-reaction: $\text{Mg(s)} \longrightarrow \text{Mg}^{2+}(\text{aq}) + 2\,\text{e}^-$
Reduction half-reaction: $\text{Cr}^{3+}(\text{aq}) + 3\,\text{e}^- \longrightarrow \text{Cr(s)}$
The sum of the two half-reactions is
$3\,\text{Mg(s)} + 2\,\text{Cr}^{3+}(\text{aq}) \rightarrow 3\,\text{Mg}^{2+}(\text{aq}) + 2\,\text{Cr(s)}$

(c) Oxidation half-reaction: $\text{Cu}^+(\text{aq}) \longrightarrow \text{Cu}^{2+}(\text{aq}) + \text{e}^-$
Reduction half-reaction: $\text{Fe}^{3+}(\text{aq}) + \text{e}^- \longrightarrow \text{Fe}^{2+}(\text{aq})$
The sum of the two half-reactions is
$\text{Fe}^{3+}(\text{aq}) + \text{Cu}^+(\text{aq}) \longrightarrow \text{Fe}^{2+}(\text{aq}) + \text{Cu}^{2+}(\text{aq})$

17.3 (a) $\text{Ni(s)} \longrightarrow \text{Ni}^{2+}(\text{aq}) + 2\,\text{e}^-$ (This is the oxidation half-reaction.)

$2\,\text{Ag}^+(\text{aq}) + 2\,\text{e}^- \longrightarrow 2\,\text{Ag(s)}$ (This is the reduction half-reaction.)

(b) The oxidation of Ni takes place at the anode and the reduction of Ag^+ takes place at the cathode.

(c) Electrons would flow through an external circuit from the anode (where Ni is oxidized) to the cathode (where Ag^+ ions are reduced).

(d) Nitrate ions would flow through the salt bridge to the anode compartment. Potassium ions would flow into the cathode compartment.

17.4 Oxidation half-reaction:

$\text{Fe(s)} \longrightarrow \text{Fe}^{2+}(\text{aq, 1M}) + 2\,\text{e}^-$ (anode)

Reduction half-reaction:

$\text{Cu}^{2+}(\text{aq, 1 M}) + 2\,\text{e}^- \longrightarrow \text{Cu(s)}$ (cathode)

$$E^\circ_{\text{cell}} = +0.78\ \text{V} = E^\circ_{\text{cathode}} - E^\circ_{\text{anode}}$$

Because $E^\circ_{\text{cathode}} = +0.34$ V, E°_{anode} must be -0.44 V.

17.5 $\text{F}_2(\text{g}) + 2\,\text{e}^- \longrightarrow 2\,\text{F}^-(\text{aq})$ $\quad E^\circ_{\text{cathode}} = +2.87$ V
$2\,\text{Li(s)} \longrightarrow 2\,\text{Li}^+(\text{aq}) + 2\,\text{e}-$ $\quad E^\circ_{\text{anode}} = -3.045$ V

$2\,\text{Li(s)} + \text{F}_2(\text{g}) \longrightarrow 2\,\text{Li}^+(\text{aq}) + 2\,\text{F}^-(\text{aq})$

$$E^\circ_{\text{cell}} = E^\circ_{\text{cathode}} - E^\circ_{\text{anode}} = +2.87 - (-3.045) = +5.92\ \text{V}$$

17.6 The two half-reactions are

$\text{Hg}^{2+}(\text{aq}) + 2\,\text{e}^- \longrightarrow \text{Hg}(\ell)$ $\quad E^\circ_{\text{cathode}} = +0.854$ V
$2\,\text{I}^-(\text{aq}) \longrightarrow \text{I}_2(\text{s}) + 2\,\text{e}^-$ $\quad E^\circ_{\text{anode}} = +0.535$ V

$$E^\circ_{\text{cell}} = E^\circ_{\text{cathode}} - E^\circ_{\text{anode}} = +0.854 - 0.535 = +0.319\ \text{V}$$

The reaction is product-favored as written.

17.7 $\text{Ni(s)} \longrightarrow \text{Ni}^{2+}(\text{aq}) + 2\,\text{e}^-$ $\quad E^\circ_{\text{anode}} = -0.25$ V
$\text{Cd}^{2+}(\text{aq}) + 2\,\text{e}^- \longrightarrow \text{Cd(s)}$ $\quad E^\circ_{\text{cathode}} = -0.40$ V

$\text{Ni(s)} + \text{Cd}^{2+}(\text{aq}) \longrightarrow \text{Ni}^{2+}(\text{aq}) + \text{Cd(s)}$

$$E^\circ_{\text{cell}} = E^\circ_{\text{cathode}} - E^\circ_{\text{anode}} = (-0.40\ \text{V}) - (0.25\ \text{V}) = -0.15\ \text{V}$$

$$\ln K^\circ = \frac{nFE^\circ_{\text{cell}}}{RT} = \frac{(2)(96485\ \text{C/mol e}^-)(-0.15\ \text{V})}{(8.314\ \text{J mol}^{-1}\ \text{K}^{-1})(298\ \text{K})} = -11.7$$

$$K^\circ = e^{-11.7} = 8.3 \times 10^{-6}$$

17.8 $$E_{\text{cell}} = E^\circ_{\text{cell}} - \frac{RT}{nF} \ln \frac{(\text{conc. Zn}^{2+})}{(\text{conc. Ni}^{2+})}$$

$$= 0.51\ \text{V} - \frac{(8.314\ \text{J mol}^{-1}\ \text{K}^{-1})(298\ \text{K})}{2(96{,}485\ \text{C mol}^{-1})} \ln \frac{(3.0\ \text{M})}{(0.010\ \text{M})}$$

$$E_{\text{cell}} = 0.51\ \text{V} - (1.284 \times 10^{-2}\ \text{V})(5.70) = 0.44\ \text{V}$$

17.9 At pH = 7.30, $[\text{H}^+] = 10^{-\text{pH}} = 10^{-7.30} = 5.0 \times 10^{-8}$ mol/L

$$E_{\text{cell}} = E^\circ_{\text{cell}} - \frac{(8.314\ \text{J K}^{-1}\ \text{mol}^{-1})(298\ \text{K})}{(2)(96485\ \text{C/mol})} \ln \frac{(\text{conc. H}^+)^2_{\text{unk}}}{(\text{conc. H}^+)^2_{\text{std}}}$$

$$E_{\text{cell}} = (-1.284 \times 10^{-2}\ \text{V})(-33.6) = 0.43\ \text{V}$$

17.10 (a) The net cell reaction would be

$$2\,\text{Na}^+(\text{aq}) + 2\,\text{Br}^-(\text{aq}) \longrightarrow 2\,\text{Na(s)} + \text{Br}_2(\ell)$$

Sodium ions would be reduced at the cathode and bromide ions would be oxidized at the anode.

(b) H_2 would be produced at the cathode for the same reasons given in Problem-Solving Example 17.10. That reaction is

$$2\,\text{H}_2\text{O}(\ell) + 2\,\text{e}^- \longrightarrow \text{H}_2(\text{g}) + 2\,\text{OH}^-(\text{aq})$$

At the anode, two reactions are possible: the oxidation of water and the oxidation of Br^- ions.

$2\,\text{H}_2\text{O}(\ell) \longrightarrow \text{O}_2(\text{g}) + 4\,\text{H}^+(\text{aq}) + 4\,\text{e}^-$ $\quad E^\circ = 1.229$ V
$2\,\text{Br}^-(\text{aq}) \longrightarrow \text{Br}_2(\ell) + 2\,\text{e}^-$ $\quad E^\circ = 1.07$ V

Bromide ions will be oxidized to Br_2 because that potential is smaller. The net cell reaction is

$$2\ H_2O(\ell) + 2\ Br^-(aq) \longrightarrow Br_2(\ell) + H_2(g) + 2\ OH^-(aq)$$

(c) Sn metal will be formed at the cathode because its reduction potential (−0.14 V) is less negative than the potential for the reduction of water. O_2 will form at the anode because the $E°$ value for the oxidation of water is smaller than the $E°$ value for the oxidation of Cl^-. The net cell reaction is

$$2\ Sn^{2+}(aq) + 2\ H_2O(\ell) \longrightarrow 2\ Sn(s) + O_2(g) + 4\ H^+(aq)$$

17.11 First, calculate the quantity of charge:

$$\text{Charge} = (25 \times 10^3\ A)(1\ h)\left(\frac{60\ s}{1\ min}\right)\left(\frac{60\ min}{1\ h}\right)$$
$$= 9.0 \times 10^7\ A \cdot s = 9.0 \times 10^7\ C$$

Then, calculate the mass of Na:

$$\text{Mass of Na} = (9.0 \times 10^7\ C) \times \left(\frac{1\ mol\ e^-}{96{,}500\ C}\right)\left(\frac{1\ mol\ Na}{1\ mol\ e^-}\right)\left(\frac{22.99\ g\ Na}{1\ mol\ Na}\right)$$
$$= 2.1 \times 10^4\ g\ Na$$

Chapter 18

18.1 (a) $^{237}_{93}Np \rightarrow {}^4_2He + {}^{233}_{91}Pa$ (b) $^{35}_{16}S \rightarrow {}^{0}_{-1}e + {}^{35}_{17}Cl$

18.2 (a) $^{11}_{6}C \rightarrow {}^{11}_{5}B + {}^{0}_{1}e$ (b) $^{35}_{16}S \rightarrow {}^{35}_{17}Cl + {}^{0}_{-1}e$

(c) $^{30}_{15}P \rightarrow {}^{0}_{+1}e + {}^{30}_{14}Si$ (d) $^{22}_{11}Na \rightarrow {}^{0}_{-1}e + {}^{22}_{12}Mg$

18.3 (a) $^{42}_{19}K \rightarrow {}^{0}_{-1}e + {}^{42}_{20}Ca$ (b) $^{234}_{92}U \rightarrow {}^{4}_{2}He + {}^{230}_{90}Th$

(c) $^{20}_{9}F \rightarrow {}^{0}_{-1}e + {}^{20}_{10}Ne$

18.4 (a) $^{90}_{38}Sr \rightarrow {}^{0}_{-1}e + {}^{90}_{39}Y$

(b) It takes 4 half-lives (4 × 29 y = 116 y) for the activity to decrease to 125 beta particles emitted per minute:

Number of Half-lives	Change of Activity	Total Elapsed Time (y)
1	2000 to 1000	29
2	1000 to 500	58
3	500 to 250	87
4	250 to 125	116

18.5 (a) $t_{1/2} = \dfrac{0.693}{9.3 \times 10^{-3}\ d^{-1}} = 75\ d$

(b) $\ln(\text{fraction remaining}) = -k \times t$
$= -(9.3 \times 10^{-3}\ d^{-1}) \times (100\ d)$
$= -0.930$

Fraction of iridium-192 remaining $= e^{-0.930} = 0.39$. Therefore, 39% of the original iridium-192 remains.

18.6 $k = \dfrac{0.693}{1.60 \times 10^3\ y} = 4.33 \times 10^{-4}\ y^{-1}$

As of 2013: $\ln(\text{fraction remaining}) = -kt$
$= -(4.33 \times 10^{-4}\ y^{-1}) \times 92\ y$
$= -3.98 \times 10^{-2}$

Fraction of radium-226 remaining $= e^{-0.0398} = 0.961$.

Therefore, 96.1% of the original radium-226 remains; mass of the original radium remaining = 0.961 × 1.00 g = 0.961 g.

18.7 $\ln(0.60) = -0.51 = -k \times t$

$$k = \frac{0.693}{t_{1/2}} = \frac{0.693}{12.3\ y} = 0.0563\ y^{-1}$$

$$t = \frac{-0.51}{-0.0563\ y^{-1}} = 9.1\ y$$

Chapter 19

19.1 $m_{NaCl} = 1.5 \times 10^9\ mol\ Na \times \dfrac{4\ mol\ NaCl}{4\ mol\ Na} \times \dfrac{58.5\ g\ NaCl}{1\ mol\ NaCl} \times \dfrac{1\ lb\ NaCl}{454\ g\ NaCl} \times \dfrac{1\ ton\ NaCl}{2000\ lb\ NaCl} = 9.7 \times 10^4\ ton\ NaCl$

19.2 $n_{Al} = 1.00\ ton\ Al \times \dfrac{2000\ lb}{1\ ton} \times \dfrac{454\ g}{1\ lb} \times \dfrac{1\ mol}{26.98\ g}$
$= 3.37 \times 10^4\ mol\ Al$

$Al^{3+} + 3\ e^- \rightarrow Al$; therefore, 3 mol electrons are needed to produce 1 mol Al.

$\text{Amount of electrons} = 3.37 \times 10^4\ mol\ Al \times \dfrac{3\ mol\ e^-}{1\ mol\ Al}$
$= 1.01 \times 10^5\ mol\ e^-$

$\text{Charge} = 1.01 \times 10^5\ mol\ e^- \times \dfrac{9.65 \times 10^4\ C}{1\ mol\ e^-}$
$= 9.75 \times 10^9\ C$

$\text{Time} = 9.75 \times 10^9\ C \times \dfrac{1\ s}{1.00 \times 10^5\ C} = 9.75 \times 10^4\ s$
$= 1.625 \times 10^3\ min = 27.1\ h$

19.3 I^- is oxidized to I_2; IO_3^- is reduced to I_2. IO_3^- is the oxidizing agent, and I^- is the reducing agent.

19.4 (a)

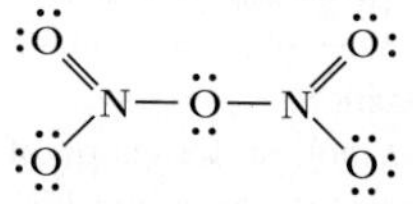

O
O
N
O
N
O
O

(b) $N_2O_5 + H_2O \rightarrow 2\ HNO_3$

19.5 (a) The formation of NO_2 from NO is exothermic. Thus, lowering the temperature favors the forward reaction (NO_2 formation).
(b) By reacting with water, NO_2 is converted to HNO_3 thereby removing NO_2 from the reaction mixture.

19.6

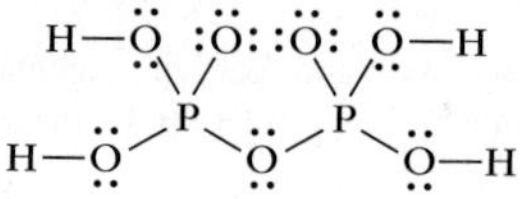

H—O
O
O
O—H
P
P
H—O
O
O—H

19.7 $S_8(s) + 12\ O_2(g) \longrightarrow 8\ SO_3(g)$

$m_{SO_3} = 1.3 \times 10^{10}\ kg\ S_8 \times \dfrac{1000\ g\ S_8}{1\ kg\ S_8} \times \dfrac{1\ mol\ S_8}{256\ g\ S_8} \times \dfrac{8\ mol\ SO_3}{1\ mol\ S_8} \times \dfrac{80\ g\ SO_3}{1\ mol\ SO_3} \times \dfrac{1\ kg}{1000\ g} = 3.3 \times 10^{10}\ kg\ SO_3$

19.8 (a) Non-redox reaction; no change in oxidation numbers
(b) Redox reaction; F_2 is the oxidizing agent; S_8 is the reducing agent.

Chapter 20

20.1 (a) Mn [Ar]$3d^5 4s^2$

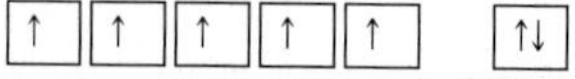

3*d* 4*s*

Mn^{2+} [Ar]$3d^5 4s^0$

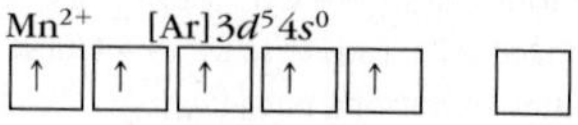

3*d* 4*s*

(b) Cr [Ar]$3d^5 4s^1$

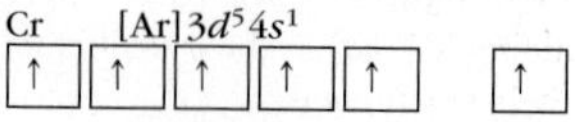

3*d* 4*s*

Cr^{3+} [Ar]$3d^3 4s^0$

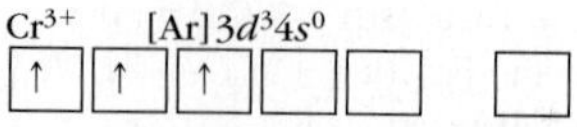

3*d* 4*s*

20.2 Cooling the iron slowly would shift the equilibrium to favor the reverse reaction, the conversion of cementite to iron and carbon (graphite).

20.3 Assume that 50% of the iron comes from each ore.
It requires 1 mol Fe_2O_3 to produce 2 mol Fe; that is, 159.7 g Fe_2O_3 yields 111.7 g Fe. Thus,

$$m_{Fe_2O_3} = 5.00 \times 10^2 \text{ kg Fe} \times \frac{159.7 \text{ kg } Fe_2O_3}{111.7 \text{ kg Fe}} = 7.15 \times 10^2 \text{ kg } Fe_2O_3$$

It requires 1 mol Fe_3O_4 to produce 3 mol Fe; that is, 231.5 g Fe_3O_4 yields 167.5 g Fe. Thus,

$$m_{Fe_3O_4} = 5.00 \times 10^2 \text{ kg Fe} \times \frac{231.5 \text{ kg } Fe_3O_4}{167.5 \text{ kg Fe}} = 6.91 \times 10^2 \text{ kg } Fe_3O_4$$

Total = 7.15×10^2 kg + 6.91×10^2 kg = 1.41×10^3 kg

20.4 The standard half-cell potentials are $Pb^{2+}(aq)|Pb(s)$, $E° = -0.125$ V, $Zn^{2+}(aq)|Zn(s)$, $E° = -0.763$ V, and $Cu^{2+}(aq)|Cu(s)$, $E° = +0.340$ V. Thus, Zn is most easily oxidized from the impure anode. If the anode is set to a potential between +0.763 V and +0.125 V relative to a standard hydrogen electrode, only Zn will be oxidized and only $Zn^{2+}(aq)$ will be in solution. The $Zn^{2+}(aq)$ can then be plated out onto the cathode, separating the Zn. Next, Pb can be oxidized and plated out, leaving Cu in the impure anode.

20.5 $$E_{cell} = 0.00 \text{ V} = 0.825 \text{ V} - 0.00985 \text{ V} \log\left(\frac{1}{[H^+]^{14}}\right)$$

$$-0.825 \text{ V} = -0.00985 \text{ V} \log\frac{1}{[H^+]^{14}}$$

$$\log\frac{1}{[H^+]^{14}} = \frac{-0.825 \text{ V}}{-0.00985 \text{ V}} = 83.7$$

$$-14 \log[H^+] = 83.7$$

$$-\log[H^+] = 83.7/14 = 5.98 = \text{pH}$$

20.6 (a) SO_4^{2-} (b) Cu^{2+} (c) NH_3 (d) $[Cu(NH_3)_4]^{2+}$

20.7 (a) Diamminesilver(I) nitrate (b) $[Fe(H_2O)_5(NCS)]Cl_2$

20.8 (a) Two (b) Zero (c) Five

20.9 Two

20.10 High-spin: four unpaired electrons; low-spin: two unpaired electrons; both complexes are paramagnetic due to their unpaired electrons.

Appendix F

F.1 ***Step 1.*** This is an oxidation-reduction reaction. It is obvious that Zn is oxidized by its change in oxidation state.
Step 2. The half-reactions are

$$Zn(s) \longrightarrow Zn^{2+}(aq) \quad \text{(This is the oxidation reaction.)}$$

$$Cr_2O_7^{2-}(aq) \longrightarrow 2\,Cr^{3+}(aq) \quad \text{(This is the reduction reaction.)}$$

Step 3. Balance the atoms in the half-reactions. The atoms are balanced in the Zn half-reaction. We need to add water and H^+ in the $Cr_2O_7^{2-}$ half-reaction. Fourteen H^+ ions are required on the right to combine with the seven O atoms.

$$Cr_2O_7^{2-}(aq) + 14\,H^+(aq) \longrightarrow 2\,Cr^{3+}(aq) + 7\,H_2O(\ell)$$

Step 4. Balance the half-reactions for charge. Write the Zn half-reaction as

$$Zn(s) \longrightarrow Zn^{2+}(aq) + 2\,e^-$$

and write the $Cr_2O_7^{2-}$ half-reaction as

$$Cr_2O_7^{2-}(aq) + 14\,H^+(aq) + 6\,e^- \longrightarrow 2\,Cr^{3+}(aq) + 7\,H_2O(\ell)$$

Step 5. Multiply the half-reactions by factors to make the number of electrons gained equal to the number lost.

$$3\,[Zn(s) \longrightarrow Zn^{2+}(aq) + 2\,e^-]$$

$$1\,[Cr_2O_7^{2-}(aq) + 14\,H^+(aq) + 6\,e^- \longrightarrow 2\,Cr^{3+}(aq) + 7\,H_2O(\ell)]$$

Step 6. Add the two half-reactions, canceling the electrons.

$$3\,Zn(s) \longrightarrow 3\,Zn^{2+}(aq) + 6\,e^-$$

$$Cr_2O_7^{2-}(aq) + 14\,H^+(aq) + 6\,e^- \longrightarrow 2\,Cr^{3+}(aq) + 7\,H_2O(\ell)$$

$$Cr_2O_7^{2-}(aq) + 3\,Zn(s) + 14\,H^+(aq) \longrightarrow 2\,Cr^{3+}(aq) + 3\,Zn^{2+}(aq) + 7\,H_2O(\ell)$$

Step 7. Everything checks.

F.2 ***Step 1.*** This is an oxidation-reduction reaction. The wording of the question says Al reduces NO_3^- ion. Al is oxidized.
Step 2. The half-reactions are:

$$Al(s) \longrightarrow Al(OH)_4^-(aq) \quad \text{(This is the oxidation reaction.)}$$

$$NO_3^-(aq) \longrightarrow NH_3(aq) \quad \text{(This is the reduction reaction.)}$$

Step 3. Balance the atoms in the half-reactions. For the Al half-reaction, add four H^+ ions on the right and four water molecules on the left.

$$Al(s) + 4\,H_2O(\ell) \longrightarrow Al(OH)_4^-(aq) + 4\,H^+(aq)$$

For the NO_3^- half-reaction,

$$NO_3^-(aq) + 9\,H^+(aq) \longrightarrow NH_3(aq) + 3\,H_2O(\ell)$$

Step 4. Balance the half-reactions for charge. Put 3 e^- on the right in the Al half-reaction

$$Al(s) + 4\,H_2O(\ell) \longrightarrow Al(OH)_4^-(aq) + 4\,H^+(aq) + 3\,e^-$$

and put 8 e^- on the left side of the NO_3^- half-reaction.

$$NO_3^-(aq) + 9\,H^+(aq) + 8\,e^- \longrightarrow NH_3(aq) + 3\,H_2O(\ell)$$

Step 5. Multiply the half-reactions by factors to make the electrons gained equal to those lost.

$$8\,[Al(s) + 4\,H_2O(\ell) \longrightarrow Al(OH)_4^-(aq) + 4\,H^+(aq) + 3\,e^-]$$

$$3\,[NO_3^-(aq) + 9\,H^+(aq) + 8\,e^- \longrightarrow NH_3(aq) + 3\,H_2O(\ell)]$$

Step 6. Remove $H^+(aq)$ ions by adding an appropriate amount of OH^-. For the Al half-reaction, add 32 OH^- ions to get

$$8\,Al(s) + 32\,OH^-(aq) + 32\,H_2O(\ell) \longrightarrow 8\,Al(OH)_4^-(aq) + 32\,H_2O(\ell) + 24\,e^-$$

For the NO_3^- half-reactions, add 27 OH^- ions to get

$$3\,NO_3^-(aq) + 27\,H_2O(\ell) + 24\,e^- \longrightarrow 3\,NH_3(aq) + 9\,H_2O(\ell) + 27\,OH^-(aq)$$

Step 7. Add both half-reactions and cancel the electrons.

$$8\,Al(s) + 32\,OH^-(aq) + 32\,H_2O(\ell) \longrightarrow 8\,Al(OH)_4^-(aq) + 32\,H_2O(\ell) + 24\,e^-$$

$$3\,NO_3^-(aq) + 27\,H_2O(\ell) + 24\,e^- \longrightarrow 3\,NH_3(aq) + 9\,H_2O(\ell) + 27\,OH^-(aq)$$

$$3\,NO_3^-(aq) + 8\,Al(s) + 59\,H_2O(\ell) + 32\,OH^-(aq) \longrightarrow 8\,Al(OH)_4^-(aq) + 3\,NH_3(aq) + 27\,OH^-(aq) + 41\,H_2O(\ell)$$

Step 8. Make a final check. Since there are OH^- ions and water molecules on both sides of the equation, cancel them. This gives the final balanced equation.

$$3\,NO_3^-(aq) + 8\,Al(s) + 18\,H_2O(\ell) + 5\,OH^-(aq) \longrightarrow 8\,Al(OH)_4^-(aq) + 3\,NH_3(aq)$$

(This is a fairly complicated equation to balance. If you balanced this one with a minimum of effort, your understanding of balancing redox equations is rather good. If you had to struggle with one or more of the steps, go back and repeat them.)

APPENDIX L Answers to Exercises

Chapter 1

1.1 (a) These temperatures can be compared to the boiling point of water, 212 °F or 100 °C. So 110 °C is a higher temperature than 180 °F.
(b) These temperatures can be compared to normal body temperature, 98.6 °F or 37.0 °C. So 36 °C is a lower temperature than 100 °F.
(c) This temperature can be compared to normal body temperature, 37.0 °C. Because body temperature is above the melting point, gallium held in one's hand will melt.

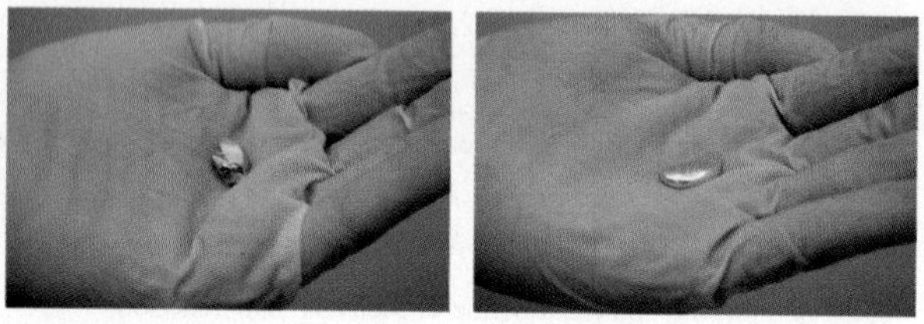

Photos: Courtesy of James H. Maynard/University of Wisconsin

A sample of gallium melts in a gloved hand.

1.2 The liquid densities in Section 1-4d indicate that kerosene is the top layer, linseed oil is the middle layer, and water is the bottom layer.
(a) Because the least dense liquid is the top layer and the densest liquid is the bottom layer, the densities increase in the order kerosene, linseed oil, water.
(b) If linseed oil is added to the tube, the top and bottom layers remain the same, but the middle layer becomes larger.
(c) If kerosene is now added to the tube, the top layer will grow, but the middle and bottom layers remain the same. The order of levels does *not change*. Density does not depend on the quantity of material present. So, no matter how much of each liquid is present, the densities increase in the order kerosene, linseed oil, water.

1.3 (a) Properties: blue (qualitative), melts at 99 °C (quantitative)
Change: melting
(b) Properties: white, cubic (both qualitative)
Change: none
(c) Properties: mass of 0.123 g, melts at 327 °C (both quantitative)
Change: melting
(d) Properties: colorless, vaporizes easily (both qualitative), boils at 78 °C, density of 0.789 g/mL (both quantitative)
Changes: vaporizing, boiling

1.4 $$1\text{ mL} = 1\text{ mL} \times \frac{1\text{ L}}{1000\text{ mL}} \times \frac{1\text{ dm}^3}{1\text{ L}} \times \left(\frac{1\text{ m}}{10\text{ dm}}\right)^3 \times \left(\frac{100\text{ cm}}{1\text{ m}}\right)^3$$
$$= 1\text{ cm}^3$$

1.5 (a) (i) 3; (ii) 3; (iii) 2; (iv) 1; (v) infinite number (defined quantity); (vi) 4; (vii) far greater than you will ever use in a calculation
(b) (i) 12.7; (ii) 12.2; (iii) 18.4; (iv) 18.4; (v) 24.8; (vi) 18.4
(c) (i) 15.89; (ii) 4.5; (iii) 2.085; (iv) 632 (round at the end of the calculation)

1.6 Physical change: boiling water
Chemical changes: combustion of propane, cooking the egg

1.7 (a) Homogeneous mixture (solution)
(b) Heterogeneous mixture (contains carbon dioxide gas bubbles in a solution of sugar and other substances in water)
(c) Heterogeneous mixture of dirt and oil
(d) Element; diamond is pure carbon
(e) Modern quarters (since 1965) are composed of a pure copper core (that can be seen when they are viewed side-on) and an outer layer of 75% Cu, 25% Ni alloy, so they are heterogeneous matter. Pre-1965 quarters are fairly pure silver (90%).
(f) Compound; contains carbon, hydrogen, and oxygen

1.8 (a) Energy from the sun warms the ice and the water molecules vibrate more; eventually they break away from their fixed positions in the solid and liquid water forms. As the temperature of the liquid increases, some of the molecules have enough energy to become widely separated from the other molecules, forming water vapor (gas).
(b) Some of the water molecules in the clothes have enough speed and energy to escape from the liquid state and become water vapor; these molecules are carried away from the clothes by breezes or air currents. Eventually nearly every water molecule in the clothes vaporizes, and the clothes become dry.
(c) Water molecules from the air come into contact with the cold glass, and their speeds are decreased, allowing them to become liquid. As more and more molecules enter the liquid state, droplets form on the glass.
(d) Some water molecules escape from the liquid state, forming water vapor. As more and more molecules escape, the ratio of sugar molecules to water molecules becomes larger and larger, and eventually some sugar molecules start to stick together. As more and more sugar molecules stick to each other, a visible crystal forms. Eventually all of the water molecules escape, leaving sugar crystals behind.

1.9 (a) Tellurium, Te, Earth (Latin *tellus* means Earth); uranium, U, for Uranus; neptunium, Np, for Neptune; and plutonium, Pu, for Pluto. (Mercury, like the planet Mercury, is named for a Roman god.)
(b) Californium, Cf
(c) Curium, Cm, for Marie Curie; and meitnerium, Mt, for Lise Meitner
(d) Scandium, Sc, for Scandinavia; gallium, Ga, for France (Latin *Gallia* means France); germanium, Ge, for Germany; ruthenium, Ru, for Russia; europium, Eu, for Europe; polonium, Po, for Poland; francium, Fr, for France; americium, Am, for America; californium, Cf, for California
(e) H, He, C, N, O, F, Ne, P, S, Cl, Ar, Se, Br, Kr, I, Xe, At, Rn

1.10 The question should ask: Which of the following men does not have an *element* named after him? The correct answer to the question, after it is properly posed, is Isaac Newton.

1.11
1. C, Si, Ge, Sn, Pb, Fl
2. 18
3. (a) group 4B(4), period 4; (b) group 4A(14), period 3; (c) group 4B(4), period 7; (d) group 6A(16), period 4; (e) group 8B(8), period 4
4. (a) 7A(17) (b) 2A(2) (c) 8A(18)
5. Elements consisting of diatomic molecules are H_2, N_2, O_2, and the halogens.

1.12
1. (a) 13 metals: potassium (K), calcium (Ca), scandium (Sc), titanium (Ti), vanadium (V), chromium (Cr), manganese (Mn), iron (Fe), cobalt (Co), nickel (Ni), copper (Cu), zinc (Zn), and gallium (Ga)
(b) Three nonmetals: selenium (Se), bromine (Br), and krypton (Kr)
(c) Two metalloids: germanium (Ge) and arsenic (As)

2. (a) Groups 1A (except hydrogen), 2A, 1B, 2B, 3B, 4B, 5B, 6B, 7B, 8B
   (b) Groups 7A and 8
   (c) None
3. Period 6

1.13 (a) Carbon, nitrogen, oxygen, phosphorus, hydrogen, selenium, sulfur, chlorine, fluorine, iodine (b) Calcium and magnesium (c) Chloride, fluoride, and iodide (d) Iron, copper, zinc, vanadium (also chromium, manganese, cobalt, nickel, and molybdenum).

Chapter 2

2.1 The movement of the comb though your hair removes some electrons, leaving slight charges on your hair and the comb. The charges must sum to zero; therefore, one must be slightly positive and one must be slightly negative, so they attract each other.

2.2 (a) A nucleus is about one ten-thousandth as large as an atom, so $100\ m \times (1 \times 10^{-4}) = 1 \times 10^{-2}\ m = 1$ cm. (b) Many everyday objects are about 1 cm in size—for example, a grape.

2.3 The statement is wrong because two atoms that are isotopes always have the same number of protons. It is the number of neutrons that varies from one isotope of an element to another.

2.4 Because the most abundant isotope is magnesium-24 (78.99%), the atomic weight of magnesium is closer to 24 than to 25 or 26, the mass numbers of the other magnesium isotopes, which make up approximately 21% of the remaining mass. The simple arithmetic average is $(24 + 25 + 26)/3 = 25$, which is larger than the atomic weight. In the arithmetic average, the relative abundance of each magnesium isotope is 33%, far less than the actual percent abundance of magnesium-24, and much more than the natural percent abundances of magnesium-25 and magnesium-26.

2.5 There is no reasonable pair of values of the mass numbers for Ga that has an average value of 69.72.

2.6 Ionic: (a), (b), and (e) contain a cation and an anion; (c), (d), and (f) do not contain a cation and anion.

2.7 (a) Non-ionic—soft solid composed of two nonmetals; (b) ionic—contains a cation, NH_4^+, and an anion, $C_2O_4^{2-}$; (c) ionic—very high melting point and brittle; (d) ionic—contains a metal cation and an oxide anion.

2.8 Ionic: (b), (d), and (e); molecular: (a), (c), and (f)

2.9 Propylene glycol structural formula:

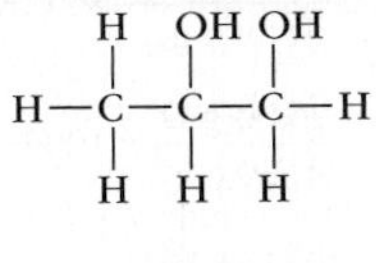

Condensed formula:

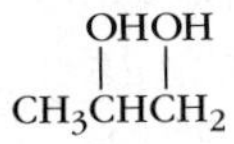

Molecular formula: $C_3H_8O_2$

2.10 (a) CS_2 (b) PCl_3 (c) SBr_2 (d) SeO_2 (e) OF_2 (f) XeO_3

2.11 (a)

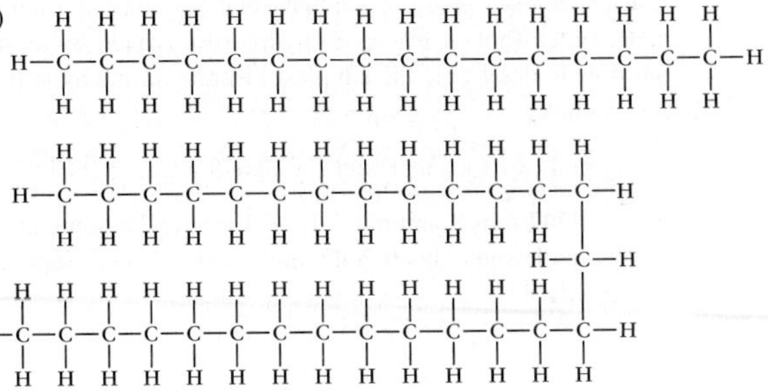

(b) $C_{14}H_{30}$ has 30 hydrogen atoms.
(c) If n is the number of carbon atoms, then the number of hydrogens should be $2n + 2$. For $C_{16}H_{34}$, $16 \times 2 + 2 = 34$. For $C_{28}H_{58}$, $2 \times 28 + 2 = 58$. For $C_{14}H_{30}$, $2 \times 14 + 2 = 30$.

2.12 (a) As the number of carbon atoms increases, the boiling point increases. (b) No. (c) For each CH_2 unit added to a hydrocarbon chain, the addition changes the size of the molecule to a greater relative extent for a short chain than it does for a long chain; the intermolecular forces (and hence the boiling point) depend on the size of and number of electrons in the molecule.

2.13 The structural and condensed formulas for three constitutional isomers of five-carbon alkanes (pentanes) are

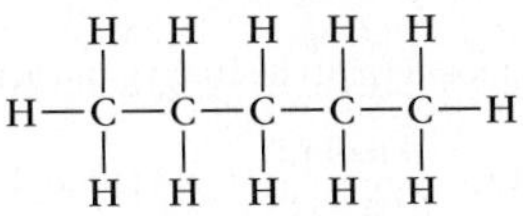 $CH_3CH_2CH_2CH_2CH_3$

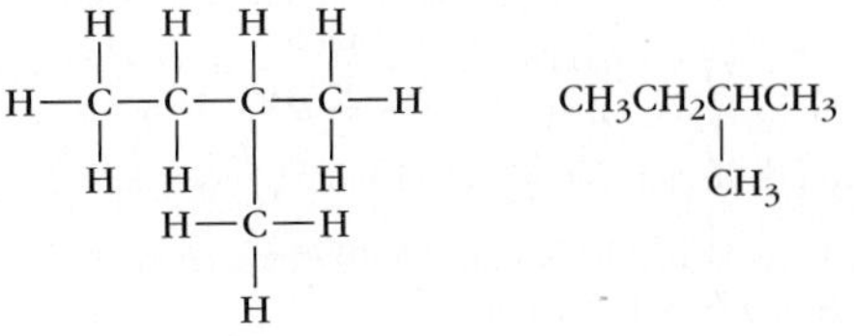

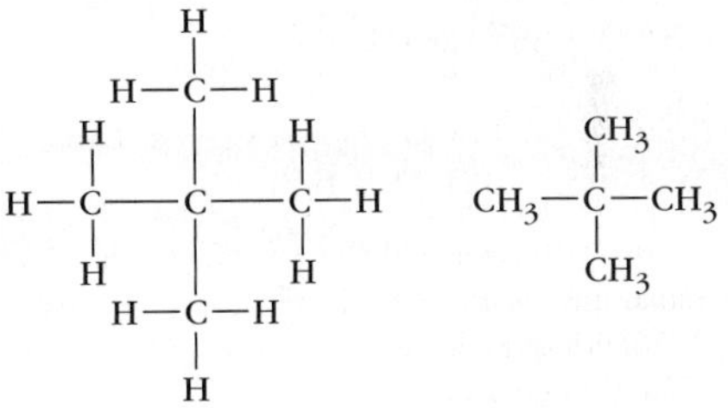

2.14

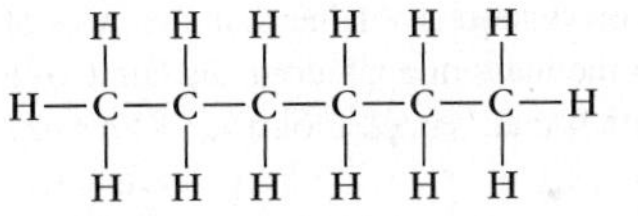 $CH_3CH_2CH_2CH_2CH_2CH_3$

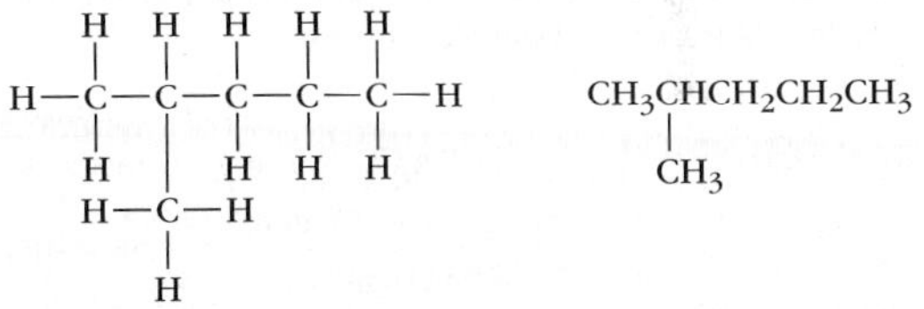

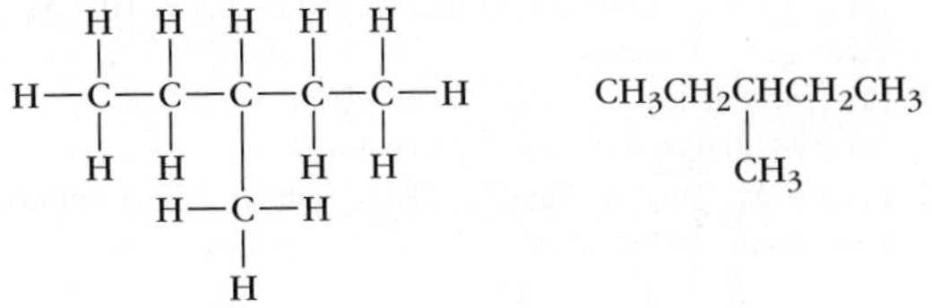

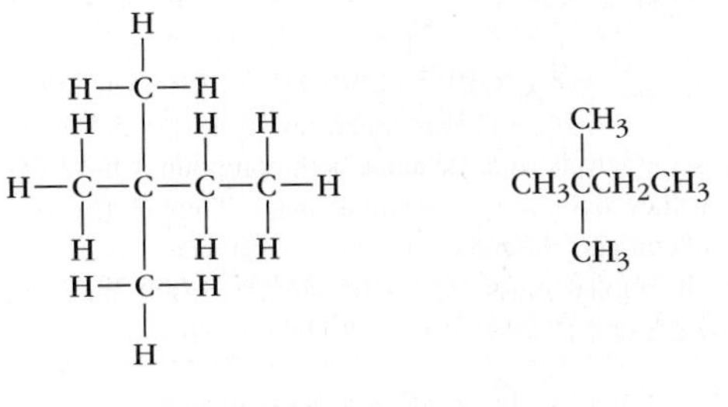

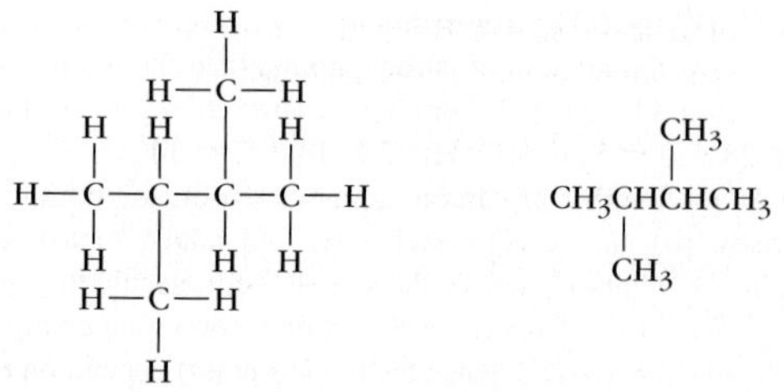

2.15 The fact that iodine consists of diatomic molecules does not affect the number of iodine atoms in a sample. The problem required equal numbers of *atoms* of I and Al, so the number of I_2 molecules was not needed.

2.16 Start by calculating the amount (mol) in 10.00 g of each element.

$$n_{Li} = 10.00 \text{ g Li} \times \frac{1 \text{ mol Li}}{6.941 \text{ g Li}} = 1.441 \text{ mol Li}$$

$$n_{Ir} = 10.00 \text{ g Ir} \times \frac{1 \text{ mol Ir}}{192.22 \text{ g Ir}} = 0.05202 \text{ mol Ir}$$

Multiply the amount of each element by Avogadro's number.

$$N_{Li} = 1.441 \text{ mol Li} \times 6.022 \times 10^{23} \text{ atoms/mol} = 8.678 \times 10^{23} \text{ atoms Li}$$

$$N_{Ir} = 0.05202 \text{ mol Ir} \times 6.022 \times 10^{23} \text{ atoms/mol} = 3.133 \times 10^{22} \text{ atoms Ir}$$

Find the difference.

$$(8.678 \times 10^{23}) - (0.3133 \times 10^{23}) = 8.365 \times 10^{23} \text{ more atoms of Li than Ir}$$

2.17 The molar masses are
$M_{Al} = 26.9815$ g/mol
$M_P = 30.9738$ g/mol
$M_{Cl} = 35.453$ g/mol
Because the mass of an Al atom is smaller than the mass of a P atom, which is smaller than the mass of a Cl atom, a sample of given mass contains more Al atoms than samples of P and Cl with the same mass.

For example, suppose each sample has a mass of 27.0 g. Using chlorine as an example,

$$N_{\text{Cl atoms}} = 27.0 \text{ g } Cl_2 \times \frac{1 \text{ mol } Cl_2}{71.0 \text{ g } Cl_2} \times \frac{2 \text{ mol Cl atoms}}{1 \text{ mol } Cl_2 \text{ molecules}} \times \frac{6.022 \times 10^{23} \text{ Cl atoms}}{1 \text{ mol Cl atoms}} = 4.58 \times 10^{23} \text{ Cl atoms}$$

Similar calculations for Al and P_4 yield 6.03×10^{23} Al atoms and 5.25×10^{23} P atoms.

2.18 (a) 174.18 g/mol (b) 386.66 g/mol
(c) 398.06 g/mol (d) 194.19 g/mol

2.19 (a) Epsom salt is $MgSO_4 \cdot 7H_2O$, which has a molar mass of 246 g/mol.

(b) $n_{\text{Epsom salt}} = 20. \text{ g} \times \frac{1 \text{ mol}}{246 \text{ g}} = 8.1 \times 10^{-2}$ mol Epsom salt

(c) $N_{\text{O atoms}} = (8.1 \times 10^{-2} \text{ mol})(6.022 \times 10^{23} \text{ formula units/mol}) \times (11 \text{ O atoms/formula unit}) = 5.4 \times 10^{23}$ O atoms

2.20 The statement is true. Because both compounds have the same formula, they have the same molar mass. Thus, 100 g of each compound contains the same amount of substance.

2.21 (a) SF_6 molar mass is 146.06 g/mol; 1.000 mol SF_6 contains 32.07 g S and 18.9984 × 6 = 113.99 g F. The mass percents are

$$\frac{32.07 \text{ g S}}{146.06 \text{ g } SF_6} \times 100\% = 21.96\% \text{ S}$$

$$100.0\% - 21.96\% = 78.04\% \text{ F}$$

(b) $C_{12}H_{22}O_{11}$ has a molar mass of 342.3 g/mol; 1.000 mol $C_{12}H_{22}O_{11}$ contains

$$(12.011 \text{ g/mol}) \times 12 \text{ mol} = 144.13 \text{ g C}$$
$$(1.0079 \text{ g/mol}) \times 22 \text{ mol} = 22.174 \text{ g H}$$
$$(15.9994 \text{ g/mol}) \times 11 \text{ mol} = 175.99 \text{ g O}$$

The mass percents of the three elements are

$$\frac{144.13 \text{ g C}}{342.3 \text{ g}} \times 100\% = 42.11\% \text{ C}$$

$$\frac{22.174 \text{ g H}}{342.3 \text{ g}} \times 100\% = 6.478\% \text{ H}$$

$$\frac{175.99 \text{ g O}}{342.3 \text{ g}} \times 100\% = 51.41\% \text{ O}$$

(c) $Al_2(SO_4)_3$ molar mass is 342.15 g/mol; 1.000 mol $Al_2(SO_4)_3$ contains

$$(26.9815 \text{ g/mol}) \times 2 \text{ mol} = 53.963 \text{ g Al}$$
$$(32.065 \text{ g/mol}) \times 3 \text{ mol} = 96.195 \text{ g S}$$
$$(15.9994 \text{ g/mol}) \times 12 \text{ mol} = 191.99 \text{ g O}$$

The mass percents of the three elements are

$$\frac{53.963 \text{ g Al}}{342.15 \text{ g}} \times 100\% = 15.77\% \text{ Al}$$

$$\frac{96.195 \text{ g S}}{342.15 \text{ g}} \times 100\% = 28.12\% \text{ S}$$

$$\frac{191.99 \text{ g O}}{342.15 \text{ g}} \times 100\% = 56.11\% \text{ O}$$

(d) $U(OTeF_5)_6$ molar mass is 1669.6 g/mol; 1.000 mol $U(OTeF_5)_6$ contains 238.0289 g U and

$$(15.9994 \text{ g/mol}) \times 6 \text{ mol} = 96.00 \text{ g O}$$
$$(127.60 \text{ g/mol}) \times 6 \text{ mol} = 765.6 \text{ g Te}$$
$$(18.9984 \text{ g/mol}) \times 30 \text{ mol} = 570.0 \text{ g F}$$

The mass percents of the four elements are

$$\frac{238.0289 \text{ g U}}{1669.6 \text{ g}} \times 100\% = 14.26\% \text{ U}$$

$$\frac{96.00 \text{ g O}}{1669.6 \text{ g}} \times 100\% = 5.750\% \text{ O}$$

$$\frac{765.6 \text{ g Te}}{1669.6 \text{ g}} \times 100\% = 45.86\% \text{ Te}$$

$$\frac{570.0 \text{ g F}}{1669.6 \text{ g}} \times 100\% = 34.14\% \text{ F}$$

2.22 $n_{\text{water}} = 9.903 \text{ g} \times \frac{1 \text{ mol}}{18.015 \text{ g}} = 0.5497$ mol

Use the masses given to calculate the amounts of calcium, sulfur, and oxygen (not in waters of hydration). Divide by least common multiple to determine the simplest formula for the hydrate. Calcium, for example:

$$n_{Ca} = 11.013 \text{ g Ca (1 mol Ca/40.078 g Ca)} = 0.2748 \text{ mol Ca.}$$

$n_S = 0.2747$ mol S and $n_O = 1.0994$ mol O. Dividing all, including 0.5497 mol water, by 0.2747 mol S yields the simplest formula, $CaSO_4 \cdot 2H_2O$.

Chapter 3

3.1 (a) One mole of washing soda reacts with one mole of sulfuric acid to produce one mole each of carbon dioxide gas, liquid water, and sodium sulfate.
(b) Each side of the balanced equation contains 2 mol Na^+, 1 mol SO_4^{2-}, 1 mol carbon, 2 mol hydrogen, and 3 mol oxygen.

3.2 (a) The total mass of reactants {4 Fe(s) + 3 O_2(g)} must equal the total mass of products {2 Fe_2O_3(s)}, which is 2.50 g.
(b) The stoichiometric coefficients are 4, 3, and 2.
(c) $$N_{Fe} = 1.000 \times 10^4 \text{ O atoms} \times \frac{1\ O_2 \text{ molecule}}{2 \text{ O atoms}} \times \frac{4 \text{ Fe atoms}}{3\ O_2 \text{ molecules}} = 6.667 \times 10^3 \text{ Fe atoms}$$

3.3 One example is the reaction between barium hydroxide and iron(II) sulfate:
$$Ba^{2+}(aq) + 2\,OH^-(aq) + Fe^{2+}(aq) + SO_4^{2-}(aq) \rightarrow BaSO_4(s) + Fe(OH)_2(s)$$

3.4 $H_3PO_4(aq) \rightleftharpoons H_2PO_4^-(aq) + H^+(aq)$
$H_2PO_4^-(aq) \rightleftharpoons HPO_4^{2-}(aq) + H^+(aq)$
$HPO_4^{2-}(aq) \rightleftharpoons PO_4^{3-}(aq) + H^+(aq)$

3.5 (a) Hydrogen ions and perchlorate ions:
$$HClO_4(aq) \rightarrow H^+(aq) + ClO_4^-(aq)$$
(b) $Ca(OH)_2(aq) \rightarrow Ca^{2+}(aq) + 2\,OH^-(aq)$

3.6 (a) $H^+(aq) + Cl^-(aq) + K^+(aq) + OH^-(aq) \rightarrow H_2O(\ell) + K^+(aq) + Cl^-(aq)$
$H^+(aq) + OH^-(aq) \rightarrow H_2O(\ell)$
(b) $2\,H^+(aq) + SO_4^{2-}(aq) + Ba^{2+}(aq) + 2\,OH^-(aq) \rightarrow 2\,H_2O(\ell) + BaSO_4(s)$
Because $BaSO_4$ is insoluble, the complete ionic equation and the net ionic equation are the same.
(c) $CH_3COOH(aq) + Na^+(aq) + OH^-(aq) \rightarrow Na^+(aq) + CH_3COO^-(aq) + H_2O(\ell)$
$CH_3COOH(aq) + OH^-(aq) \rightarrow H_2O(\ell) + CH_3COO^-(aq)$

3.7 $Al(OH)_3(s) + 3\,H^+(aq) + 3\,Cl^-(aq) \rightarrow 3\,H_2O(\ell) + Al^{3+}(aq) + 3\,Cl^-(aq)$
$Al(OH)_3(s) + 3\,H^+(aq) \rightarrow 3\,H_2O(\ell) + Al^{3+}(aq)$

3.8 $S^{2-}(aq) + 2\,H^+(aq) \rightarrow H_2S(g)$

3.9 (a) The products are aqueous sodium sulfate, water, and carbon dioxide gas.
$$Na_2CO_3(aq) + H_2SO_4(aq) \rightarrow Na_2SO_4(aq) + H_2O(\ell) + CO_2(g)$$
$$2\,H^+(aq) + CO_3^{2-}(aq) \rightarrow H_2O(\ell) + CO_2(g)$$
(b) The products are aqueous iron(II) chloride and hydrogen sulfide gas.
$$FeS(s) + 2\,HCl(aq) \rightarrow FeCl_2(aq) + H_2S(g)$$
$$2\,H^+(aq) + FeS(s) \rightarrow H_2S(g) + Fe^{2+}(aq)$$
(c) The products are aqueous potassium chloride, water, and sulfur dioxide gas.
$$K_2SO_3(aq) + 2\,HCl(aq) \rightarrow 2\,KCl(aq) + H_2O(\ell) + SO_2(g)$$
$$2\,H^+(aq) + SO_3^{2-}(aq) \rightarrow H_2O(\ell) + SO_2(g)$$

3.10 (a) Aqueous sodium chloride and insoluble barium oxalate are produced.
$$BaCl_2(aq) + Na_2C_2O_4(aq) \rightarrow BaC_2O_4(s) + 2\,NaCl(aq)$$
$$Ba^{2+}(aq) + C_2O_4^{2-}(aq) \rightarrow BaC_2O_4(s)$$
(b) Nitric acid reacts with strontium hydroxide, a base, to produce water and strontium nitrate, a soluble salt.
$$2\,HNO_3(aq) + Sr(OH)_2(s) \rightarrow Sr(NO_3)_2(aq) + H_2O(\ell)$$
$$Sr(OH)_2(s) + 2\,H^+(aq) \rightarrow Sr^{2+}(aq) + 2\,H_2O(\ell)$$
(c) Products are a precipitate of nickel(II) phosphate and aqueous ammonium chloride.
$$2\,(NH_4)_3PO_4(aq) + 3\,NiCl_2(aq) \rightarrow Ni_3(PO_4)_2(s) + 6\,NH_4Cl(aq)$$
$$2\,PO_4^{3-}(aq) + 3\,Ni^{2+}(aq) \rightarrow Ni_3(PO_4)_2(s)$$
(d) Products are aqueous potassium chlorate and water.
$$KOH(aq) + HClO_4(aq) \rightarrow H_2O(\ell) + KClO_4(aq)$$
$$OH^-(aq) + H^+(aq) \rightarrow H_2O(\ell)$$

3.11 (a) $Br_2(aq) + 2\,I^-(aq) \rightarrow 2\,Br^-(aq) + I_2(aq)$. Bromine is the oxidizing agent and is reduced to bromide; iodide ion is the reducing agent and is oxidized to iodine.
(b) $Br_2(\ell) + 2\,K(s) \rightarrow 2\,KBr(s)$. Bromine is the oxidizing agent and is reduced to bromide ions; potassium metal is the reducing agent and is oxidized to potassium ions.

3.12 (a) This is not a redox reaction. Nitric acid is a strong oxidizing agent, but here it serves as an acid.
(b) In this redox reaction, chromium metal (Cr) is oxidized (loses electrons) to form Cr^{3+} ions in Cr_2O_3; oxygen (O_2) is reduced (gains electrons) to form oxide ions, O^{2-}. Oxygen is the oxidizing agent, and chromium is the reducing agent.
(c) This is an acid-base reaction, but not a redox reaction; there are no strong oxidizing or reducing agents present.
(d) Copper is oxidized and chlorine is reduced in this redox reaction, in which copper is the reducing agent and chlorine is the oxidizing agent.

3.13 (a) $\overset{+2}{Mg_3}\overset{-3}{N_2}$ (b) $\overset{+3}{Cl}\overset{-1}{F_3}$ (c) $\overset{-3}{N}\overset{+1}{H_4}\overset{+5}{N}\overset{-2}{O_3}$

3.14 (a) $CH_3CH_2OH(\ell) + 3\,O_2(g) \rightarrow 3\,H_2O(\ell) + 2\,CO_2(g)$; redox
(b) $2\,Fe(s) + 6\,HNO_3(aq) \rightarrow 2\,Fe(NO_3)_3(aq) + 3\,H_2(g)$; redox
(c) $AgNO_3(aq) + KBr(aq) \rightarrow AgBr(s) + KNO_3(aq)$; precipitation, not redox

3.15 Four additional stoichiometric ratios (still others are possible):
$$\frac{2 \text{ mol } N_2H_4}{3 \text{ mol } N_2} \quad \frac{4 \text{ mol } H_2O}{1 \text{ mol } N_2O_4} \quad \frac{2 \text{ mol } N_2H_4}{4 \text{ mol } H_2O} \quad \frac{3 \text{ mol } N_2}{4 \text{ mol } H_2O}$$
The mole ratio 2 mol N_2H_4/3 mol N_2 means that for every 3 mol N_2 formed 2 mol N_2H_4 must react.
The mole ratio 4 mol H_2O/1 mol N_2O_4 means that for every 1 mol N_2O_4 that reacts 4 mol H_2O is formed.
The mole ratio 2 mol N_2O_4/4 mol H_2O means that for every 4 mol H_2O produced 2 mol N_2H_4 must react.
The mole ratio 3 mol N_2/4 mol H_2O means that for every 4 mol H_2O produced 3 mol N_2 must be produced.

3.16 $$n_{H_2O} = 0.300 \text{ mol } CH_4 \times \frac{2 \text{ mol } H_2O}{1 \text{ mol } CH_4} = 0.600 \text{ mol } H_2O$$
$$m_{H_2O} = 0.600 \text{ mol } H_2O \times \frac{18.02 \text{ g } H_2O}{1 \text{ mol } H_2O} = 10.8 \text{ g } H_2O$$

3.17 (a) $C_{12}H_{22}O_{11} + 12\,O_2 \rightarrow 12\,CO_2 + 11\,H_2O$
(b) $$m_{O_2} = 35.0 \text{ g sugar} \times \frac{1 \text{ mol sugar}}{342.3 \text{ g sugar}} \times \frac{12 \text{ mol } O_2}{1 \text{ mol sugar}} \times \frac{32.0 \text{ g } O_2}{1 \text{ mol } O_2} = 39.3 \text{ g } O_2$$
$$m_{H_2O} = 35.0 \text{ g sugar} \times \frac{1 \text{ mol sugar}}{342.3 \text{ g sugar}} \times \frac{11 \text{ mol } H_2O}{1 \text{ mol sugar}} \times \frac{18.02 \text{ g } H_2O}{1 \text{ mol } H_2O} = 20.3 \text{ g } H_2O$$

$$m_{CO_2} = 35.0 \text{ g sugar} \times \frac{1 \text{ mol sugar}}{342.3 \text{ g sugar}} \times \frac{12 \text{ mol } CO_2}{1 \text{ mol sugar}} \times \frac{44.01 \text{ g } CO_2}{1 \text{ mol } CO_2} = 54.0 \text{ g } CO_2$$

3.18 (a) Because the masses of products are to be determined, use the mass method.

$$m_{NH_3}(\text{if urea limiting}) = 300. \text{ g urea} \times \frac{1 \text{ mol urea}}{60.05 \text{ g urea}} \times \frac{2 \text{ mol } NH_3}{1 \text{ mol urea}} \times \frac{17.03 \text{ g } NH_3}{1 \text{ mol } NH_3} = 170.2 \text{ g } NH_3$$

$$m_{NH_3}(\text{if water limiting}) = 100. \text{ g } H_2O \times \frac{1 \text{ mol } H_2O}{18.02 \text{ g } H_2O} \times \frac{2 \text{ mol } NH_3}{1 \text{ mol } H_2O} \times \frac{17.03 \text{ g } NH_3}{1 \text{ mol } NH_3} = 189.0 \text{ g } NH_3$$

Because it produces less NH_3, urea is limiting. Use urea to calculate the mass of carbon dioxide.

$$m_{CO_2}(\text{urea limiting}) = 300. \text{ g urea} \times \frac{1 \text{ mol urea}}{60.05 \text{ g urea}} \times \frac{1 \text{ mol } CO_2}{1 \text{ mol urea}} \times \frac{44.00 \text{ g } CO_2}{1 \text{ mol } CO_2} = 219.8 \text{ g } CO_2$$

(b) $$m_{H_2O}(\text{reacted}) = 300. \text{ g urea} \times \frac{1 \text{ mol urea}}{60.05 \text{ g urea}} \times \frac{1 \text{ mol } H_2O}{1 \text{ mol urea}} \times \frac{18.02 \text{ g } H_2O}{1 \text{ mol } H_2O} = 90.0 \text{ g } H_2O$$

m_{H_2O} remaining $= 100.0 \text{ g} - 90.0 \text{ g} = 10.0 \text{ g}$

3.19 (1) Impure reactants; (2) inaccurate weighing of reactants and products

3.20 First, write the balanced chemical equation:

$$C(s) + 2\,H_2O(g) \rightarrow 2\,H_2(g) + CO_2(g)$$

Use the expression for atom economy:

$$\text{Atom economy} = \frac{\sum(\text{atomic weights atoms in useful product})}{\sum(\text{atomic weights atoms in reactants})}$$

$$= \frac{2 \times (2 \times 1.008 \text{ u})}{12.011 \text{ u} + 2 \times (2 \times 1.008 \text{ u}) + (2 \times 15.999 \text{ u})} = 0.08393$$

Percent atom economy = Atom economy × 100% = 8.393%

3.21 (a) Convert milligrams to grams. Calculate the mass of each element in the compound. For carbon it is

$$m_{C \text{ in sample}} = 0.491 \text{ g } CO_2 \times \frac{1 \text{ mol } CO_2}{44.01 \text{ g } CO_2} \times \frac{1 \text{ mol C}}{1 \text{ mol } CO_2} \times \frac{12.01 \text{ g C}}{1 \text{ mol C}} = 0.134 \text{ g C}$$

Using the mass of water, do a similar calculation to determine the mass of hydrogen in the sample, 0.0112 g H. Calculate the mass of oxygen in the sample by difference:

$$0.175 \text{ g phenol} - 0.134 \text{ g C} - 0.0112 \text{ g H} = 0.0298 \text{ g O}.$$

Calculate the amount of each element. For carbon:

$$n_{C \text{ in sample}} = 0.134 \text{ g C} \times \frac{1 \text{ mol}}{12.01 \text{ g C}} = 0.0112 \text{ mol C}$$

Similar calculations yield 0.0111 mol H and 0.00186 mol O. Divide each by the least common multiple, 0.00186, to determine the relative amount of each element in the compound.
(0.0112 mol C/0.00186 mol O) = 6.02 mol C/1 mol O;
(0.0111 mol H/0.00186 mol O) = 5.97 mol H/1 mol O;
(0.00186 mol O/0.00186 mol O) = 1.00 mol O.
The empirical formula is C_6H_6O.

(b) To determine the molecular formula the molar mass is needed. The molar mass of phenol is 94 g/mol, which corresponds to the empirical formula molar mass.

3.22 Assume that all of the carbon and oxygen in the 1.90 g carbon dioxide came only from the decomposition of the calcium carbonate. Calculate the amount (mol) of carbon and of oxygen in the CO_2. Likewise, calculate the amount (mol) of calcium and oxygen in the 2.40 g CaO. Add the amounts of oxygen in CaO and CO_2. Divide the amounts of calcium, carbon, and oxygen by the least common multiple to calculate the simplest (empirical formula) of calcium carbonate, $CaCO_3$.

3.23 Molar mass of cholesterol = 386.7 g/mol

$$n_{cholesterol} = 240 \text{ mg} \times \frac{1 \text{ g}}{10^3 \text{ mg}} \times \frac{1 \text{ mol}}{386.7 \text{ g}} = 6.21 \times 10^{-4} \text{ mol cholesterol}$$

$$c_{cholesterol} = \frac{6.21 \times 10^{-4} \text{ mol}}{0.100 \text{ L}} = 6.2 \times 10^{-3} \text{ M}$$

3.24 (a) $$n_{Al(NO_3)_3} = 6.37 \text{ g } Al(NO_3)_3 \times \frac{1 \text{ mol } Al(NO_3)_3}{213.0 \text{ g } Al(NO_3)_3} = 0.0299 \text{ mol } Al(NO_3)_3;$$

$$c_{Al(NO_3)_3} = \frac{0.0299 \text{ mol}}{0.250 \text{ L}} = 0.120 \text{ M } Al(NO_3)_3$$

(b) Molarity: Al^{3+} = 0.120 M; NO_3^- = 3(0.120 M) = 0.360 M

3.25 If the description of solution preparation is always worded in terms of adding enough solvent to make a specific volume of solution, then any possible expansion or contraction has no effect on the molarity of the solution. The denominator of the definition of molarity is liters of *solution.*

3.26 The amount of HCl in the concentrated solution is (6.0 mol/L) × (0.100 L) = 0.60 mol HCl. The amount of HCl in the dilute solution is (1.20 mol/L)(0.500 L) = 0.60 mol HCl.

3.27 The molarity could be increased by evaporating some of the solvent.

3.28 $$n_{NaCl} = 0.0193 \text{ L} \times \frac{0.200 \text{ mol } AgNO_3}{1 \text{ L}} \times \frac{1 \text{ mol } Ag^+}{1 \text{ mol } AgNO_3} \times \frac{1 \text{ mol } Cl^-}{1 \text{ mol } Ag^+} \times \frac{1 \text{ mol NaCl}}{1 \text{ mol } Cl^-} = 3.86 \times 10^{-3} \text{ mol NaCl}$$

$$c_{NaCl} = \frac{3.86 \times 10^{-3} \text{ mol NaCl}}{0.0250 \text{ L}} = 0.154 \text{ M NaCl}$$

3.29 The oxidation number of manganese in $KMnO_4$ is +7; in $MnSO_4$ it is +2. Therefore, Mn in $KMnO_4$ is the oxidizing agent and is reduced from +7 to +2. The oxidation number of oxygen in H_2O_2 is −1; in O_2 it is 0. Thus, the oxygen in hydrogen peroxide is the reducing agent and is oxidized from −1 to 0.

Chapter 4

4.1 You transfer some mechanical energy to the ball to accelerate it upward. The ball's potential energy increases the higher it gets, but its kinetic energy decreases by an equal quantity, and eventually it stops rising and begins to fall. As it falls, some of the ball's potential energy changes to kinetic energy, and the ball goes faster and faster until it hits the floor. When the ball hits the floor, some of its kinetic energy is transferred to the atoms, molecules, or ions that make up the floor, causing them to move faster. Much of the ball's kinetic energy is used to compress the ball, squeezing together and deforming the molecules within the ball. When those molecules move apart and their original shape is restored, the ball bounces up, but not to the full height from which it was dropped. Eventually all of the ball's energy is transferred, and the ball stops moving. The nanoscale particles in the floor and the ball (and some in the air that the ball fell through) are moving faster on average, and the temperature of the floor, the ball, and the air is slightly higher. The energy has spread out over a much larger number of particles.

4.2

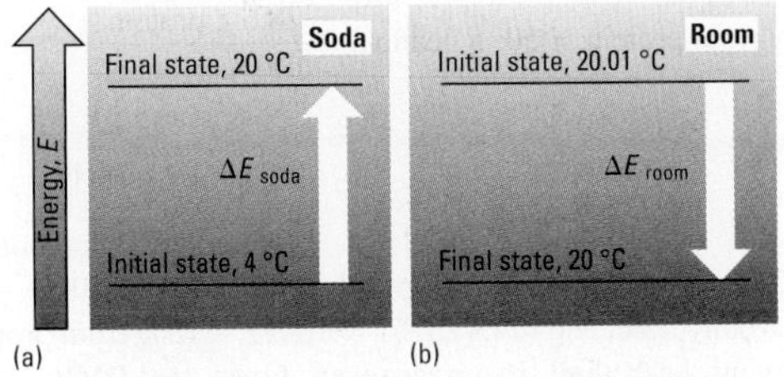
Soda
Final state, 20 °C
ΔE soda
Initial state, 4 °C
Energy, E
(a)
Room
Initial state, 20.01 °C
ΔE room
Final state, 20 °C
(b)

4.3 (a) Heat transfer = (25.0 °C − 1.0 °C) (1.5 kJ/1.0 °C) = 36 kJ
(b) The system is the can and the liquid it contains.
(c) The surroundings includes the air and other materials in contact with the can, or close to the can.
(d) ΔE is negative because the system transferred energy to the surroundings as it cooled; $\Delta E = -36$ kJ.
(e)

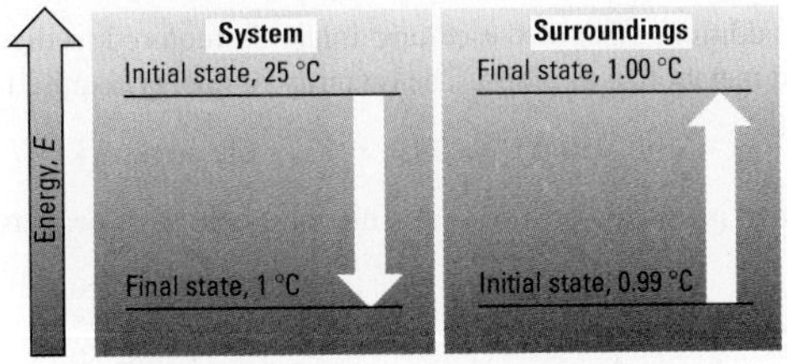
System
Initial state, 25 °C
Final state, 1 °C
Energy, E
Surroundings
Final state, 1.00 °C
Initial state, 0.99 °C

The surroundings are warmed very slightly, say from 0.99 °C to 1.00 °C, by the heat transfer from the can of soda.

4.4 The same quantity of energy is transferred out of each beaker and the mass of each sample is the same. Therefore the sample with the smaller specific heat capacity will cool more. Look up the specific heat capacities in Table 4.1. Because glass has a larger specific heat capacity than carbon, the carbon will cool more and therefore will have the lower temperature.

4.5 The calculation for Al is given as an example.

$$\text{Molar heat capacity} = \frac{0.902\ \text{J}}{\text{g}\ °\text{C}} \times \frac{26.98\ \text{g}}{1\ \text{mol}} = 24.3\ \text{J mol}^{-1}\ °\text{C}^{-1}$$

Metal	Molar Heat Capacity (J mol^{-1} °C^{-1})	Metal	Molar Heat Capacity (J mol^{-1} °C^{-1})
Al	24.3	Cu	24.5
Fe	25.2	Au	25.2

The molar heat capacities of most metals are about 25 J mol^{-1} °C^{-1}. This rule does not work for ethanol or other compounds listed in Table 4.2.

4.6 $T_1 = 0$ °C.
333 J = 1.00 g × (0.451 J g^{-1} °C) × (T_2 − 0); T_2 = 738 °C.

4.7 (a) Because the heat of vaporization is almost seven times larger than the heat of fusion, the temperature stays constant at 100 °C almost seven times longer than it stays constant at 0 °C. It stays constant at 0 °C for slightly less time than it takes to heat the water from 0 °C to 100 °C (see graph). Because the heating is at a constant rate, time is proportional to quantity of energy transferred.

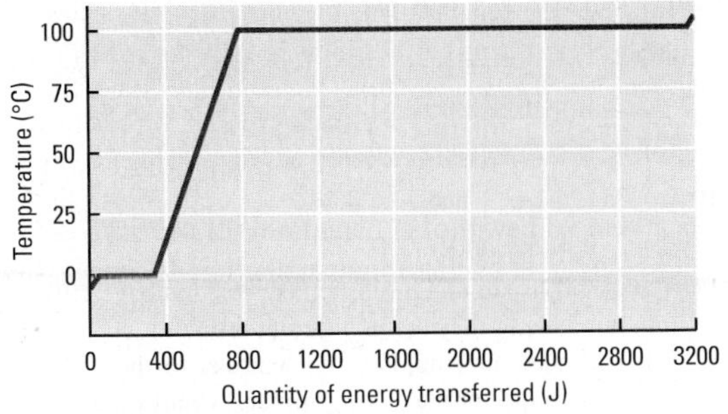
Temperature (°C)
100
75
50
25
0
0
400
800
1200
1600
2000
2400
2800
3200
Quantity of energy transferred (J)

(b) The mass of water is half as great as in part (a), so each process takes half as long. A graph to the same scale as in part (a) begins at 105 °C and reaches −5 °C with half the quantity of energy transferred.

4.8 Heat of fusion: 237 g × 333 J/g = 78.9 kJ
Heating liquid: 237 g × 4.184 J g^{-1} °C^{-1} × 100.0 °C = 99.2 kJ
Heat of vaporization: 237 g × 2260 J/g = 536 kJ
Total heating = (78.9 + 99.2 + 536) kJ = 714 kJ

4.9 The negative sign indicates an exothermic process in which energy is transferred from the system to the surroundings.

4.10 Because of heats of fusion and heats of vaporization, the enthalpy change is different when a reactant or product is in a different state.

4.11 (a) 0.50 mol

$$n_{\text{reaction}} = 1\ \text{mol NH}_3 \times \frac{1\ \text{mol reaction}}{2\ \text{mol NH}_3} = 0.5\ \text{mol reaction}$$

(b) 0.125 mol

$$n_{\text{reaction}} = 0.375\ \text{mol H}_2 \times \frac{1\ \text{mol reaction}}{3\ \text{mol H}_2} = 0.125\ \text{mol reaction}$$

(c) 0.158 mol

$$n_{\text{reaction}} = 4.42\ \text{g N}_2 \times \frac{1\ \text{mol N}_2}{28.013\ \text{g N}_2} \times \frac{1\ \text{mol reaction}}{1\ \text{mol N}_2}$$
$$= 0.158\ \text{mol reaction}$$

4.12 $2\ H_2(g) + O_2(g) \rightarrow 2\ H_2O(g)$ $\quad \Delta_r H° = -483.6$ kJ/mol
Using the coefficients of the balanced chemical equation, 0.5000 mol H_2 reacts with 0.2500 mol O_2 to produce 0.5000 mol H_2O. The thermal energy released by these amounts of reactants, one fourth of $\Delta_r H° = -120.9$ kJ.

4.13 Yes, it would violate the first law of thermodynamics. According to the supposition, we could create energy by starting with 2 mol HCl, breaking all the molecules apart, recombining the atoms to form 1 mol H_2 and 1 mol Cl_2, and then reacting the H_2 and Cl_2 to give 2 HCl.

$2\ HCl \rightarrow \text{atoms} \rightarrow H_2 + Cl_2$ $\quad \Delta H° = +185$ kJ

$H_2 + Cl_2 \rightarrow 2\ HCl$ $\quad \Delta H° = -190.$ kJ

The net effect of these two processes is that there is still 2 mol HCl, but 5 kJ of energy has been created. This is impossible according to the first law of thermodynamics.

4.14 (a) In the reaction $2\ HF \rightarrow H_2 + F_2$, there are two bonds in the two reactant molecules and two bonds in the two product molecules. Because the reaction is endothermic, the bonds in the reactant molecules must be stronger than in the products.
(b) For the reaction $2\ H_2O \rightarrow 2\ H_2 + O_2$, there are four bonds in the two reactant molecules but only three bonds in the three product molecules. The reaction is endothermic because more bonds are broken than are formed.

4.15 $C_6H_{12}O_6(s) + 6\ O_2(g) \rightarrow 6\ CO_2(g) + 6\ H_2O(\ell)$
Because the volume of any ideal gas is proportional to the amount (moles) of gas, and because there are 6 mol of gaseous reactant and 6 mol of gaseous product, there will be very little change in volume. Almost no work will be done, and $\Delta H \cong \Delta E$.

4.16 In Problem-Solving Example 4.10, the reaction produced 0.25 mol NaCl and the heat transfer associated with the reaction caused 500 mL of solution to warm by 7 °C.
(a) Here the reaction produces 0.20 mol NaCl by neutralizing 0.20 mol NaOH and the heat transfer warms 400. mL water. So there is less heat transfer, but it will be heating a smaller volume. The quantity of reaction is 0.20 mol/0.25 mol = 0.80 as much, so the heat transfer is 0.80 as much. The quantity of water to be heated is 400. mL/500. mL = 0.800 as much. Therefore the combined effect on temperature is the same as in Problem-Solving Example 4.10, and the temperature change associated with this process will also be 7 °C.

(b) Here the limiting reactant is 0.10 mol NaOH. (Only half of the H_2SO_4 reacts.) The heat transfer from the reaction warms 200. mL water. So there is less heat transfer, but it will heat a smaller volume. The quantity of reaction is 0.10 mol/0.25 mol = 0.40 as much, so the heat transfer is 0.40 as much. The quantity of water is 200./400. = 0.40 as much. Therefore once again the combined effect on temperature is the same as in Problem-Solving Example 4.10, and the temperature change associated with this process will also be 7 °C.

4.17 $NH_4NO_3(s) \rightarrow NH_4^+(aq) + NO_3^-(aq)$. The total mass of solution = 150. g water + 7.07 g NH_4NO_3 = 157.1 g

$$q_P = -c \times m \times \Delta T$$
$$= -(4.184\ \text{J g}^{-1}\ °\text{C})(157.1\ \text{g})(19.2\ °\text{C} - 22.3\ °\text{C})$$
$$= 2.04 \times 10^3\ \text{J}$$

This is the quantity of thermal energy transferred when 7.07 g solid NH_4NO_3 dissolves, not that for 1 mol NH_4NO_3. Calculating that quantity:

$$n_{NH_4NO_3} = 7.07\ \text{g} \times \frac{1\ \text{mol}}{80.05\ \text{g}} = 8.83 \times 10^{-2}\ \text{mol}$$

$$\Delta H = \frac{2.04 \times 10^3\ \text{J}}{8.83 \times 10^{-2}\ \text{mol}} \times \frac{1\ \text{kJ}}{1000\ \text{J}} = 23.1\ \text{kJ/mol}$$

4.18 $N_2(g) \rightarrow N_2(g)$

(a) The product is the same as the reactant, so there is no change—nothing happens.

(b) Because product and reactant are the same, $\Delta_f H° = 0$.

4.19 The fuel values and also the food values are: CH_4 = 50.013 kJ/g; $C_6H_6O_6$ = 15.551 kJ/g; $C_{59}H_{110}O_6$ = 42.36 kJ/g; and C_2H_5OH = 26.81 kJ/g. Methane is not a food (at least for mammals) and so will not be considered. Ethanol, C_2H_5OH, is also not considered to be a food, but it is metabolized and will be considered. The percent oxygen by mass in glucose is (96.0 g O/180. g glucose) × 100% = 53.3% O. Likewise, tristearin is 10.8% oxygen and ethanol is 34.8% oxygen. Because of the presence of oxygen in them, each of the compounds can be considered as being partially oxidized prior to combustion. However, due to its lower percent oxygen, tristearin is the least oxidized in that way and correspondingly releases the largest amount of energy of the three upon combustion.

4.20 (a) The average BMR is given in the text as 1750 Cal/day for a 70-kg male at rest. So the BMR is

$$\frac{1750\ \text{Cal}}{\text{d}} \times \frac{1000\ \text{cal}}{\text{Cal}} \times \frac{4.184\ \text{J}}{\text{cal}} = 7.32 \times 10^6\ \text{J/day}$$

$$(7.322 \times 10^6\ \text{J/day}) \times (1\ \text{day/24 h}) \times (1\ \text{h/60 min}) \times (1\ \text{min/60 s}) = 84.75\ \text{W}$$

(b) The average BMR for a 70-kg male playing basketball is 7 times the above or 593.2 W. Typical incandescent light bulbs are in the range of 50 to 250 W.

4.21 (a) Data from Table 4.3 show that 100. g of unsalted peanuts contains 50.0 g fat, 21.4 g carbohydrate, and 28.6 g protein. The fat provides 38 kJ/g and the carbohydrate and protein provide 17 kJ/g each. Therefore, the energy provided is

$$\text{Energy from fat} = 50.\ \text{g peanuts} \times \frac{50.0\ \text{g fat}}{100.\ \text{g peanuts}} \times \frac{38\ \text{kJ}}{1\ \text{g fat}} = 950\ \text{kJ}$$

$$\text{Energy from carbohydrate} = 50.\ \text{g peanuts} \times \frac{21.4\ \text{g carbo}}{100.\ \text{g peanuts}} \times \frac{17\ \text{kJ}}{1\ \text{g carbo}} = 182\ \text{kJ}$$

$$\text{Energy from protein} = 50.\ \text{g peanuts} \times \frac{28.6\ \text{g protein}}{100.\ \text{g peanuts}} \times \frac{17\ \text{kJ}}{1\ \text{g protein}} = 243\ \text{kJ}$$

The total caloric intake is 1375 kJ, and the fractions are (950/1375) × 100% = 69% from fat, (182/1375) × 100% = 13% from carbohydrate, and (243/1375) × 100% = 18% from protein.

(b) Playing basketball requires seven times the BMR—that is, 7 × 1525 Cal/day.

$$\text{Energy required} = 7 \times \frac{1525\ \text{Cal}}{\text{day}} \times \frac{1\ \text{day}}{24\ \text{h}} \times \frac{1\ \text{h}}{60\ \text{min}} \times 30\ \text{min} \times \frac{4.184\ \text{kJ}}{1\ \text{Cal}} = 930.\ \text{kJ}$$

In addition, 10% of the caloric intake is required to digest, absorb, and metabolize the food. The quantity of energy required is thus

$$930.\ \text{kJ} + (0.10 \times 1375\ \text{kJ}) = 1068\ \text{kJ}$$

which is less than the caloric value of the 50. g of peanuts.

Chapter 5

5.1 Wavelength and frequency are inversely related. Therefore, low-frequency radiation has long-wavelength radiation.

5.2 (a) Ultraviolet; (b) 1.20×10^{15} Hz has shorter wavelength than 7.898×10^{14} Hz radiation.

5.3 For 4.275 μm radiation:

$$E = \frac{hc}{\lambda} = \frac{(6.626 \times 10^{-34}\ \text{J s})(2.998 \times 10^8\ \text{m/s})}{4.257\ \mu\text{m} \times \left(\frac{1\ \mu\text{m}}{10^{-6}\ \text{m}}\right)} = 4.666 \times 10^{-20}\ \text{J}$$

A similar calculation for 15.00 μm indicates 1.324×10^{-20} J. Thus, 4.257 μm radiation has the greater amount of energy.

5.4 (a) 656.4 nm red; 486.2 nm green; 434.1 nm blue; 410.2 nm violet

(b) $$E_{\text{red line}} = \frac{(6.626 \times 10^{-34}\ \text{J s})(2.998 \times 10^8\ \text{m/s})}{656.4 \times 10^{-9}\ \text{m}} = 3.026 \times 10^{-19}\ \text{J}$$

The energies for green (4.086×10^{-19} J), blue (4.575×10^{-19} J), and violet (4.843×10^{-19} J) can be calculated in the same way.

5.5 In a sample of excited hydrogen gas there are many atoms, and each can exist in one of the excited states possible for hydrogen. The observed spectral lines are a result of all the possible transitions of all these hydrogen atoms.

5.6 $$(2.179 \times 10^{-18}\ \text{J/photon}) \times \left(\frac{1\ \text{kJ}}{10^3\ \text{J}}\right)\left(\frac{6.022 \times 10^{23}\ \text{photons}}{1\ \text{mol}}\right) = 1312\ \text{kJ/mol photons}$$

5.7 (a) 5*d* (b) 4*f* (c) 6*p*

5.8 (a) The $n = 3$ level can have only three types of sublevels—*s*, *p*, and *d*.

(b) The $n = 2$ level can have only *s* and *p* sublevels, not *d* sublevels ($\ell < 2$).

5.9 (a) 3, 0, 0, $+\frac{1}{2}$; 3, 0, 0, $-\frac{1}{2}$

(b) 2, 1, 1, $+\frac{1}{2}$; 2, 1, 1, $-\frac{1}{2}$; 2, 1, 0, $+\frac{1}{2}$; 2, 1, 0, $-\frac{1}{2}$; 2, 1, −1, $+\frac{1}{2}$; 2, 1, −1, $-\frac{1}{2}$

(c) 4, 2, −2, $+\frac{1}{2}$; 4, 2, −2, $-\frac{1}{2}$

5.10 (a) 3, 1, −1, $+\frac{1}{2}$ and 3, 1, 0, $+\frac{1}{2}$

(b) 3, 1, 1, $+\frac{1}{2}$ and 3, 0, 0, $+\frac{1}{2}$

5.11 The electrons are in a *p* subshell ($\ell = 1$) but have different m_ℓ quantum numbers so they are in different orbitals.

5.12 (a) The maximum number of electrons in the $n = 3$ level is 18 (2 electrons per orbital). The orbitals would be designated 3*s*, $3p_x$, $3p_y$, $3p_z$, $3d_{z^2}$, $3d_{xy}$, $3d_{yz}$, $3d_{xz}$, and $3d_{x^2-y^2}$.

(b) The maximum number of electrons in the $n = 5$ level is 50. The orbitals would be designated $5s$, $5p_x$, $5p_y$, $5p_z$, $5d_{yz}$, $5d_{xz}$, $5d_{xy}$, $5d_{z^2}$, $5d_{x^2-y^2}$, and the seven $5f$ orbitals and the nine $5g$, which are not designated by name in the text.

5.13 $n = 6$; number of atomic orbitals $= 2\ell + 1 = 2(5) + 1 = 11$

5.14 The $[Ar]3d^4 4s^2$ configuration for chromium has four unpaired electrons, and the $[Ar]3d^5 4s^1$ configuration has six unpaired electrons.

5.15 For the chlorine atom, $n = 3$, and there are seven electrons in this highest energy level. The configuration is

3s 3p

[↑↓] [↑↓] [↑↓] [↑]. For the selenium atom, the highest energy level is $n = 4$, and there are six electrons in the $n = 4$ level.

4s 4p

The configuration is [↑↓] [↑↓] [↑] [↑].

5.16 (a) VF_5 (V^{5+} has no unpaired electrons); (b) VF_2 (V^{2+} has three unpaired electrons); (c) VF_3 (V^{3+} has two unpaired electrons)

5.17 KCl < LiF < SrSe < CaS < CaO. Melting point increases with increasing ionic charge and decreasing ionic radius.

Chapter 6

6.1 (a) 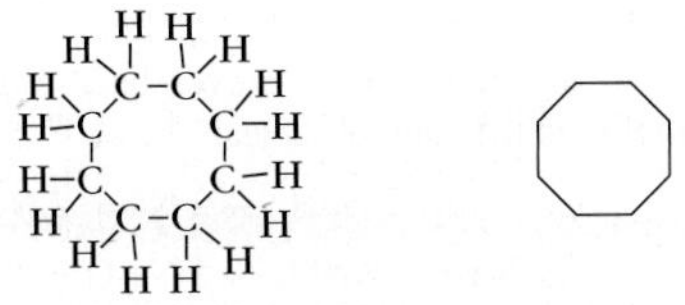 C_8H_{16}

(b) 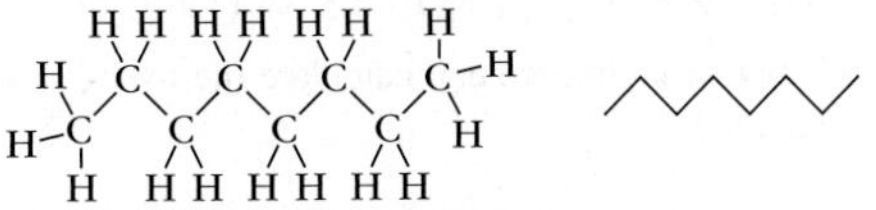 C_8H_{18}

6.2 N_2 has only 10 valence electrons. The Lewis structure shown has 14 valence electrons.

6.3 None of the structures is correct. (a) is incorrect because sulfur does not have an octet of electrons (it has only six); (b) is incorrect because, although it shows the correct number of valence electrons (26), there is a double bond between F and N rather than a single bond with a lone pair on N; (c) is incorrect because the left carbon has five bonds; (d) is incorrect because OCCl should have 17 valence electrons, not 18 as shown.

6.4 (a) C_5H_{10} (b) Four

6.5 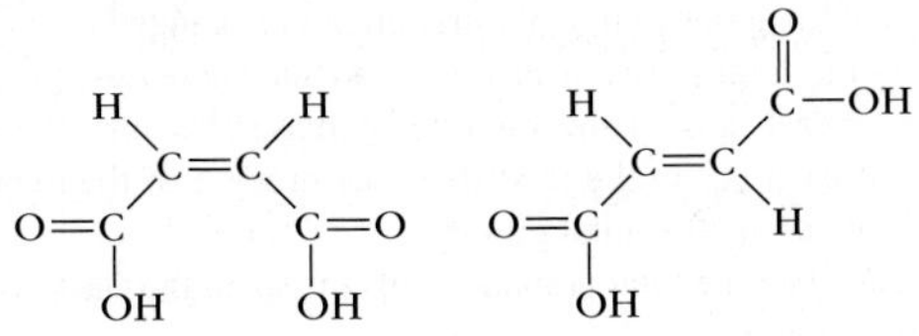

maleic acid (the *cis* isomer) fumaric acid (the *trans* isomer)

6.6 C—N > C═N > C≡N. The order of decreasing bond energy is the reverse order: C≡N > C═N > C—N. See Tables 6.1 and 6.2.

6.7 (a) The electronegativity difference between sodium and chlorine is 1.8, sufficient to cause electron transfer from sodium to chlorine to form Na^+ and Cl^- ions. Molten NaCl conducts an electric current, indicating the presence of ions.

(b) There is an electronegativity difference of 1.4 in BrF, which is sufficient to form a polar covalent bond, but not great enough to cause electron transfer leading to ion formation.

6.8 The Lewis structure of hydrazine is

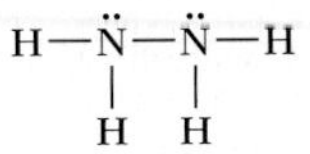

	H	H	N	N	H	H
Valence electrons	1	1	5	5	1	1
Lone pair electrons	0	0	2	2	0	0
$\frac{1}{2}$ shared electrons	1	1	3	3	1	1
Formal charge	0	0	0	0	0	0

6.9 Atoms cannot be rearranged to derive a resonance structure. There is no N-to-O bond in cyanate ion; therefore, such an arrangement cannot be a resonance structure of cyanate ion.

6.10 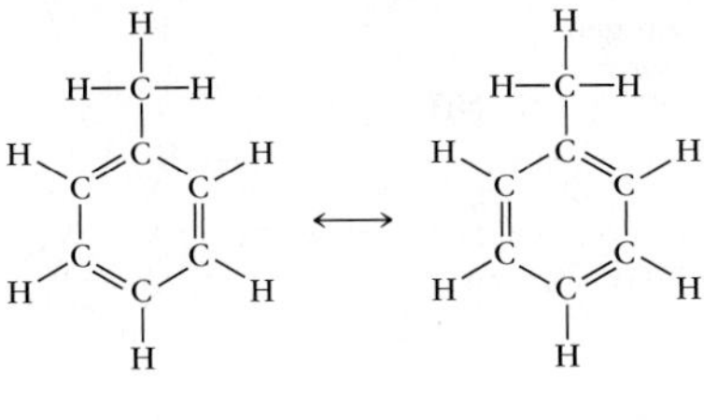

6.11 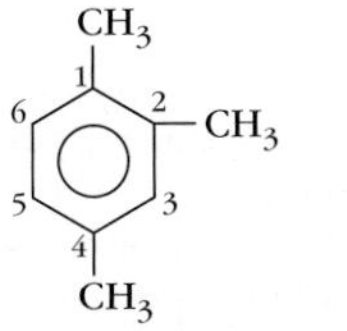

1,2,4-trimethylbenzene

Both 1,5,6-trimethylbenzene and 1,2,3-trimethylbenzene have all three methyl groups on adjacent C atoms.

6.12 1. (a) 20 carbon atoms and 30 hydrogen atoms
(b) The carbon atom at the top of the six-membered ring
(c) Five C═C double bonds

2. (a) $C_{29}H_{50}O_2$
(b) No C═C double bonds
(c) The H—O bond

Chapter 7

7.1 When the central atom has no lone pairs

7.2 The triangular bipyramidal shape has five regions of electron density; three situated in equatorial positions 120° apart and the remaining two in axial positions. The square pyramidal shape is different because it involves six regions of electron density arranged in an octahedron; one is a lone pair, four are used to bond atoms in a square plane, and the sixth region bonds an atom directly above the central atom and equidistant from the other four.

7.3 (a) $H_2C{=}CH_2$ Triangular planar around each carbon atom
(b) $CH_3{-}O{-}CH_3$ Tetrahedral around each carbon atom; angular around the oxygen
CH_3CH_2OH Tetrahedral around the terminal carbon atom and the interior carbon atom; angular around the oxygen

7.4 Pi bonding is not possible for a carbon atom with sp^3 hybridization because the atom has no unhybridized $2p$ orbitals. All of its $2p$ orbitals have been hybridized.

7.5 (a) Bromine is more electronegative than iodine, and the H—Br bond is more polar than the H—I bond.
(b) Chlorine is more electronegative than the other two halogens; therefore, the C—Cl bond is more polar than the C—Br and C—I bonds.

7.6 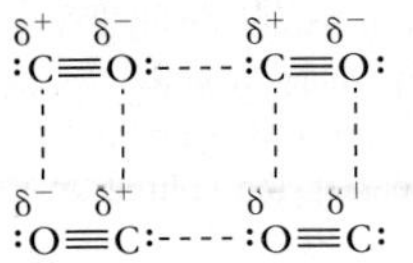 ---- dipole-dipole forces

7.7 The F—H · · · F—H hydrogen bond is the strongest because the electronegativity difference between H and F produces a more polar F—H bond than does the lesser electronegativity difference between O and H or N and H in the O—H or N—H bonds.

7.8 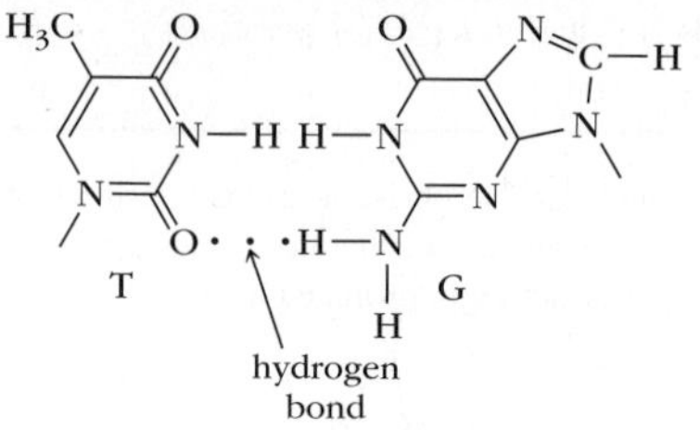
H₃C
T
G
hydrogen bond

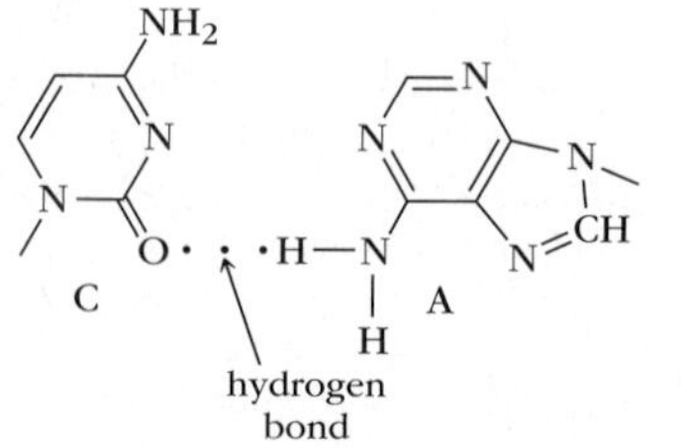
NH₂
C
A
hydrogen bond

7.9 Replication would be very difficult because strong, covalent bonds would have to be broken. DNA replicates easily because it is weaker hydrogen bonds, not covalent bonds, that occur between the base pairs and hold the two DNA strands together.

Chapter 8

8.1 (a) 25.17 inHg (25.4 mmHg/1 inHg)(1 Torr/1 mmHg) = 639.3 Torr
26.64 inHg (25.4 mmHg/1 inHg)(1 Torr/1 mmHg) = 676.7 Torr
(b) 639.3 Torr (101.325×10^5 Pa/760 Torr)(1 mbar/10^2 Pa) = 852.3 mbar
676.7 Torr (101.325×10^5 Pa/760 Torr)(1 mbar/10^2 Pa) = 902.2 mbar
(c) 639.3 Torr (101.325 kPa/760 Torr) = 85.23 kPa
676.7 Torr (101.325 kPa/760 Torr) = 90.22 kPa
(d) 639.3 Torr (1 atm/760 Torr) = 0.841 atm
676.7 Torr (1 atm/760 Torr) = 0.890 atm

8.2 (0.973 atm)(760 mmHg/1 atm)(1 m/1000 mm) = 0.7395 mHg
Because the density of Hg is 13.55 g/mL and the density of H_2O is 0.998 g/mL, a column of H_2O must be 13.55/0.998 = 13.56 times higher than a column of Hg that exerts the same pressure.
(13.56 mH_2O/1 mHg)(0.7395 mHg) = 10.04 mH_2O
(10.04 mH_2O)(3.281 ft/1 m) = 32.94 ftH_2O

8.3 First, gas molecules are far apart. This allows most light to pass through. Second, molecules are much smaller than the wavelengths of visible light. This means that the waves are not reflected or diffracted by the molecules.

8.4 As more gas molecules are added to a container of fixed volume, there will be more collisions of all of the gas molecules with the container walls. This causes the observed pressure to rise.

8.5 The gas in the shock absorbers will be more highly compressed. The gas molecules will be closer together. The gas molecules will collide with the walls of the shock absorber more often, and the pressure exerted will be larger.

8.6 Increasing the temperature of a gas causes the gas molecules to move faster, on average. This means that each collision with the container walls involves greater force, because, on average, a molecule is moving faster and hits the wall harder. If the container remained the same (constant volume), there would also be more collisions with the container wall because faster-moving molecules would hit the walls more often. Increasing the volume of the container, on the other hand, requires that the faster-moving molecules must travel a greater distance before they strike the container walls. Increasing the volume enough would just balance the greater numbers of harder collisions caused by increased temperature. To maintain a constant volume requires that the pressure increases to match the greater pressure due to more and harder collisions of gas molecules with the walls.

8.7 1. Increase the pressure.
2. Decrease the temperature.
3. Increase P and decrease T at the same time.

8.8 Density of He = 1.66×10^{-4} g/mL.
Density of Li = 0.53 g/mL.
Because the density of He is so much less than that of Li, the atoms in a sample of He must be much farther apart than the atoms in a sample of Li. This idea is in keeping with the general principle of the kinetic-molecular theory that the particles making up a gas are far from one another.

8.9 A 50-50 mixture of N_2 and O_2 would have less N_2 in it than does air. Because O_2 molecules have greater mass than N_2 molecules, this 50-50 mixture has greater density than air.

8.10 Upon braking, the air molecules continue to move forward and get compressed slightly against the windshield causing the density of the "forward" air to increase a bit. Helium in the balloon is less dense than air and the balloon moves backward toward the less-dense air region in the back of the car.

8.11
$$2\ NaN_3 \rightarrow 2\ Na + 2\ N_2$$
$$10\ Na + 2\ KNO_3 \rightarrow 5\ Na_2O + K_2O + N_2$$
$$\overline{2\ NaN_3 + 8\ Na + 2\ KNO_3 \rightarrow 5\ Na_2O + K_2O + 4\ N_2}$$

Using this net equation, we can calculate the overall amount of nitrogen generated.

$$n_{N_2} = 107\ g\ NaN_3 \times \frac{1\ mol\ NaN_3}{65.01\ g\ NaN_3} \times \frac{4\ mol\ N_2}{2\ mol\ NaN_3} = 3.29\ mol\ N_2$$

The volume of nitrogen at the stated conditions is

$$V = \frac{nRT}{P} = \frac{(3.29\ mol)(0.0821\ L\ atm\ mol^{-1}\ K^{-1})(26.0 + 273.1)K}{1.10\ atm}$$
$$= 73.4\ L$$

8.12 (a) If lowering the temperature causes the volume to decrease, by $PV = nRT$, the pressure can be assumed to be constant. The value of n is unchanged. Since both P and n remain unchanged, the partial pressures of the gases in the mixture remain unchanged.
(b) When the total pressure of a gas mixture increases, the partial pressure of each gas in the mixture increases because the partial pressure of each gas in the mixture is the product of the mole fraction for that gas and the total pressure.

8.13 We can calculate the total amount (mol) of gas in the flask from the given information.

$$n = \frac{PV}{RT} = \frac{\left(\frac{626}{760}\right)\ atm\ (0.355\ L)}{(0.0821\ L\ atm\ mol^{-1}\ K^{-1})(308\ K)} = 0.01156\ mol\ gas$$

The amount (mol) of Ne is

$$0.146\ g\ Ne \times \frac{1\ mol\ Ne}{20.18\ g/mol} = 0.007235\ mol\ Ne$$

We find the amount (mol) of Ar by subtraction.

$$0.01156\ mol\ gas - 0.007235\ mol\ Ne = 0.004325\ mol\ Ar$$
$$0.004325\ mol\ Ar \times 39.948\ g/mol = 0.173\ g\ Ar$$

8.14 All have the same kinetic energy at the same temperature.

8.15 For a sample of helium, the plot would look like the curve marked He in Figure 8.11. When an equal number of argon molecules, which are heavier, are added to the helium, the distribution of molecular speeds would look like the sum of the curves marked He and O_2 in Figure 8.11, except that the curve for Ar would have its peak a little to the left of the O_2 curve.

8.16 (a) The balloon placed in the freezer will be smaller than the one kept at room temperature because its sample of helium is colder.
(b) Upon warming, the helium balloon that had been in the freezer will be either the same size as the balloon kept at room temperature or perhaps slightly larger because there is a greater chance that He atoms leaked out of the room-temperature balloon during the time the other balloon was kept in the freezer. This would be caused by the faster-moving He atoms in the room-temperature balloon having more chances to escape from tiny openings in the balloon's walls.

8.17 $$\frac{\text{Rate } ^{235}\text{U}}{\text{Rate } ^{238}\text{U}} = \frac{\sqrt{M_{^{238}\text{UF}_6}}}{\sqrt{M_{^{235}\text{UF}_6}}} = \sqrt{\frac{352}{349}} = 1.004$$

The separation ratio of 1.004 indicates a low separation of 0.4% for each "pass" through a diffusion barrier making the separation of the isotopes difficult.

8.18 The value of n depends directly on the measured pressure, P. Intermolecular attractions in a real gas would cause the measured P to be slightly smaller than for an ideal gas. The lower value of P would cause the calculated amount (mol) to be somewhat smaller. Using this slightly smaller value of n in the denominator would cause the calculated molar mass to be a little larger than it should be.

8.19 See Table 8.5 for the van der Waals constants. The periodic table trends for atomic radii are: H < O < N < Cl. The trend in the van der Waals constants is an increase in the value of the constant as the molecular diameter increases. This means that there is a greater volume correction in the van der Waals equation for heavier gases than for lighter ones.

8.20 Natural sources: animal respiration, forest fires, decay of cellulose materials, partial digestion of carbohydrates, volcanoes; Human sources: burning fossil fuels, burning agricultural wastes and refined cellulose products such as paper, decay of carbon compounds in landfills

8.21 (a) 4.8% (b) 7.0% (c) 10.0%. The greatest percent increase per year (0.50% per year) occurred between 1990 and 2010.

8.22 The fluctuations occur because of the seasons. Photosynthesis, which uses CO_2, is greatest during the spring and summer, accounting for lower CO_2 levels.

8.23 For this calculation, we can use the figure of 450×10^9 passenger miles. Using the ratio of 2×10^3 kg CO_2/3000 passenger miles, we can calculate the CO_2 released.

$$\text{Quantity of CO}_2 = 450 \times 10^9 \text{ passenger mi} \times \frac{2 \times 10^3 \text{ kg CO}_2}{3 \times 10^3 \text{ passenger mi}} = 3 \times 10^{11} \text{ kg CO}_2$$

If a typical automobile gets 20 mi/gal of fuel and about 1.5 passengers are transported for every mile the automobile travels, then an automobile gets 30 passenger miles per gallon. The number of gallons used is

$$\text{Volume of gasoline} = 450 \times 10^9 \text{ passenger mi} \times \frac{1 \text{ gal gasoline}}{30 \text{ passenger mi}} = 2 \times 10^{10} \text{ gal gasoline}$$

Assume the gasoline produces about the same mass of CO_2 per gallon as does jet fuel, or 2×10^3 kg CO_2/200 gal.

$$\text{Quantity of CO}_2 \text{ from gasoline} = 2 \times 10^{10} \text{ gal gasoline} \times \frac{2 \times 10^3 \text{ kg CO}_2}{200 \text{ gal gasoline}} = 2 \times 10^{11} \text{ kg CO}_2$$

So the numbers are about the same for these two modes of transportation.

8.24 The quote meant that burning coal converted its carbon into carbon dioxide in the air, thus increasing atmospheric CO_2 concentration.

8.25 Mass of S burned per hour = $(3.06 \times 10^6 \text{ kg/h})(0.04) = 1 \times 10^5$ kg/h

Mass of SO_2 per hour =

$$(1 \times 10^5 \text{ kg/h})\left(\frac{64.06 \text{ kg SO}_2}{32.07 \text{ kg S}}\right) = 2 \times 10^5 \text{ kg/h}$$

Mass of SO_2 per year = $(2 \times 10^5 \text{ kg/h})(8760 \text{ h/yr}) = 2 \times 10^9$ kg/yr

Chapter 9

9.1 (a) Dipole-dipole and London forces
(b) London forces only
(c) Dipole-dipole and London forces
(d) Hydrogen bonding, dipole-dipole forces, and London forces. Methanol has the highest boiling point.

9.2 The evaporating water carries with it thermal energy from the water inside the pot. In addition, a large quantity of thermal energy is required to cause the water to evaporate. Much of this thermal energy comes from the water inside the pot.

9.3 (a) 50 °C (b) 0 °C (c) 82 °C

9.4 Bubbles form within a boiling liquid when the vapor pressure of the liquid equals the pressure of the surroundings of the liquid sample. The bubbles are actually filled with vapor of the boiling liquid. One way to prove this would be to trap some of these bubbles and allow them to condense. They would condense to form the liquid that had boiled.

9.5 Two moles of liquid bromine crystallizing liberates 21.59 kJ of energy. One mole of liquid water crystallizing liberates 6.02 kJ of energy.

9.6 High humidity conditions make the evaporation of water or the sublimation of ice less favorable. Under these conditions, the sublimation of ice required to make the frost-free refrigerator work is less favorable, so the defrost cycle is less effective.

9.7 The impurity molecules are less likely to be converted from the solid phase to the vapor phase. This causes them to be left behind as the molecules that sublime go into the gas phase and then condense at some other place. The molecules that condense are almost all of the same kind, so the sublimed sample is much purer than the original.

9.8 The curve is the vapor pressure curve. Upward: condensation. Downward: vaporization. Left to right: vaporization. Right to left: condensation.

9.9 (a) If liquid CO_2 is slowly released from a cylinder of CO_2, gaseous CO_2 is formed. The temperature remains constant (at room temperature) because there is time for energy to be transferred from the surroundings to separate the CO_2 molecules from their intermolecular attractions. This can be seen from the phase diagram as the phase changes from liquid to vapor as the pressure decreases.
(b) If the pressure is suddenly released, the attractive forces between a large number of CO_2 molecules must be overcome, which requires energy. This energy comes from the surroundings as well as from the CO_2 molecules themselves, causing the temperature of both the surroundings and the CO_2 molecules to decrease. On the phase diagram for CO_2, a decrease in both temperature and pressure moves into a region where only solid CO_2 exists.

9.10 Glycerol due to hydrogen bonding among —OH groups of adjacent molecules.

9.11 It is predicted that a small concentration of gold will be found in the lead and that a small concentration of lead will be found in the gold. This will occur because of the movement of the metal atoms with time, as predicted by the kinetic molecular theory.

9.12 One Po atom belongs to its unit cell. Two Li atoms belong to its unit cell. Four Ca atoms belong to its unit cell.

Primitive cubic

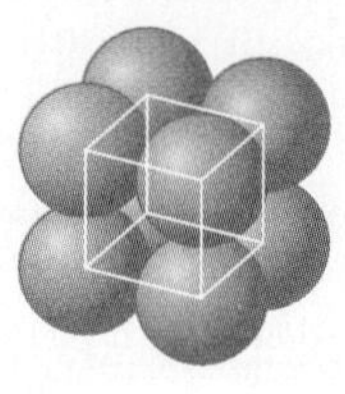

1 Po atom
Each of the 8 atoms contributes 1/8 to unit cell

Body-centered cubic

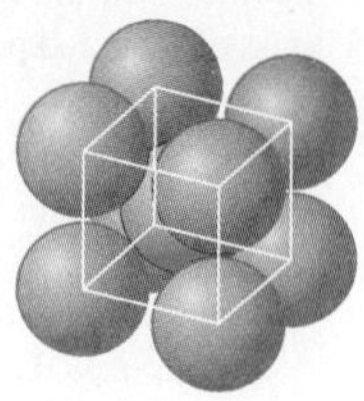

2 Li atoms
8/8 from corner atoms
+ 1 atom at center

Face-centered cubic

4 Ca atoms
8/8 from corner atoms
+ 6/2 from each atom on the 6 faces contributing 1/2 atom

9.13 Assume the unit cell is bcc. Calculate the density. If the calculated density agrees with the experimental density of Sr, the structure is bcc; if not, the structure is fcc.
For bcc, there are two atoms per unit cell and the spherical atoms touch along the body diagonal. Thus

$$\text{Body diagonal} = 4r = \sqrt{3}\, l$$

where r is the radius of a Sr atom and l is the length of the cube edge.

$$\text{Body diagonal} = 4r = 4(215\text{ pm}) = 860\text{ pm}$$

$$l = 4r/\sqrt{3} = (860\text{ pm})/(1.713) = 497\text{ pm}$$

$$V_{\text{unit cell}} = l^3 = (497\text{ pm})^3 \times \left(\frac{1 \times 10^{-12}\text{ m}}{1\text{ pm}}\right)^3 \times \left(\frac{1\text{ cm}}{1 \times 10^{-2}\text{ m}}\right)^3$$

$$= 1.23 \times 10^{-22}\text{ cm}^3$$

$$m_{\text{unit cell}} = 2\text{ Sr atoms} \times \frac{1\text{ mol}}{6.022 \times 10^{23}\text{ Sr atoms}} \times \frac{87.62\text{ g}}{1\text{ mol}}$$

$$= 2.91 \times 10^{-22}\text{ g}$$

$$d_{\text{unit cell}} = \frac{m_{\text{unit cell}}}{V_{\text{unit cell}}} = 2\,\frac{2.91 \times 10^{-22}\text{ g}}{1.23 \times 10^{-22}\text{ cm}^3} = 2.37\text{ g/cm}^3$$

This is not the density of strontium (2.58 g/cm^3). Therefore, the unit cell for strontium is fcc, not bcc.

9.14 Each Cs^+ ion at the center of the cube has eight Cl^- ions as its neighbors. One eighth of each Cl^- ion belongs to that Cs^+ ion. So the formula for this salt must be a 1 : 1 ratio of Cs^+ ions to Cl^- ions, or CsCl.

9.15 Cooling a liquid above its freezing point causes the temperature to decrease. When the liquid begins to solidify, energy is released as atoms, molecules, or ions move closer together to form in the solid crystal lattice. This causes the temperature to remain constant until all the molecules in the liquid have positioned themselves in the lattice. Further cooling then causes the temperature to decrease. The shape of this curve is common to all substances that can exist as liquids.

9.16 Increasing strength of metallic bonding is related to increasing numbers of valence electrons. In the transition metals, the presence of *d*-orbital electrons causes stronger metallic bonding. Beyond a half-filled set of *d*-orbitals, however, extra electrons have the effect of decreasing the strength of metallic bonding.

Chapter 10

10.1 (a) 4 H_2
(b) $C_7H_{16} \rightarrow C_7H_8 + 4\ H_2$
(c) Toluene has a much higher octane number and can be used as an octane enhancer.

10.2 Mass percent oxygen in ethanol = (16.0 g/46.0 g) × 100% = 35%. Ethanol was blended with gasoline, a hydrocarbon, in sufficient proportion to account for at least 2% oxygen by mass in the blended fuel.

10.3 (a) $C_2H_5OH + 2\ O_2 \rightarrow 2\ CO + 3\ H_2O$

$$\frac{\text{g CO}}{\text{g ethanol}} = \frac{2(28.0\text{ g})}{46.0\text{ g}} = 1.22\text{ g CO/g ethanol}$$

(b) $C_7H_8 + \frac{11}{2}\ O_2 \rightarrow 7\ CO + 4\ H_2O$

$$\frac{\text{g CO}}{\text{g toluene}} = \frac{7(28.0\text{ g})}{92.0\text{ g}} = 2.13\text{ g CO/g toluene}$$

(c) Ethanol produces less CO per gram than does toluene.

10.4 Assume that the same quantity of energy is required to drive one mile with each fuel. Assume that natural gas is only methane and that gasoline is octane, C_8H_{18}. Assume that the same fraction of carbon in each fuel is converted to CO. The combustion equations are

$$CH_4(g) + \tfrac{3}{2}\ O_2(g) \rightarrow CO(g) + 2\ H_2O(g) \qquad \Delta_r H° = -515.751\text{ kJ/mol}$$

$$C_8H_{18}(\ell) + \tfrac{17}{2}\ O_2(g) \rightarrow 8\ CO(g) + 9\ H_2O(g) \qquad \Delta_r H° = -2811.61\text{ kJ/mol}$$

For methane there is 1 mol CO produced for every 515.751 kJ provided to move the vehicle; thus,

$$\frac{m_{CO}}{\text{kJ}} = \frac{1\text{ mol CO}}{515.751\text{ kJ}} \times \frac{28.0\text{ g CO}}{1\text{ mol}} = 5.43 \times 10^{-2}\text{ g CO/kJ}$$

For octane there is 8 mol CO for every 2811.61 kJ; thus,

$$\frac{m_{CO}}{\text{kJ}} = \frac{8\text{ mol CO}}{2811.61\text{ kJ}} \times \frac{28.0\text{ g CO}}{1\text{ mol}} = 7.97 \times 10^{-2}\text{ g CO/kJ}$$

More than 40% more CO is produced per mile for octane compared with methane (7.97/5.43 = 1.47 = 147% or 47% more). This is mainly due to the larger ratio of carbon atoms to hydrogen atoms in octane (8/18 = 0.44) than in methane (1/4 = 0.25).

10.5 The order is: $CH_3OH > C_5H_{11}OH \gg C_{10}H_{21}OH$. There is intermolecular hydrogen bonding in each. However, as the carbon chain lengthens, the number of electrons increases and London forces increase, eventually sufficient to overcome the effectiveness of hydrogen bonding.

10.6

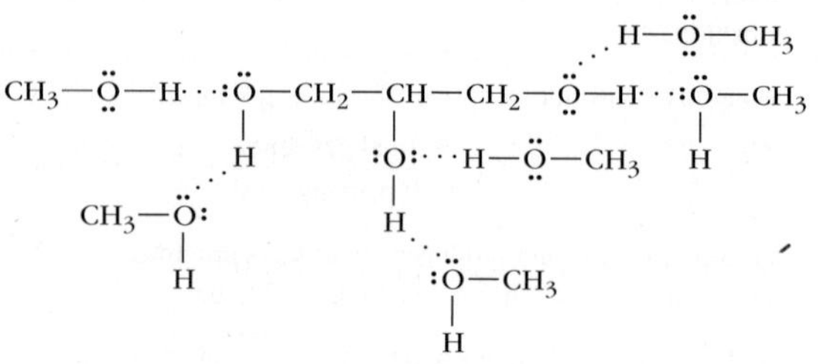

10.7 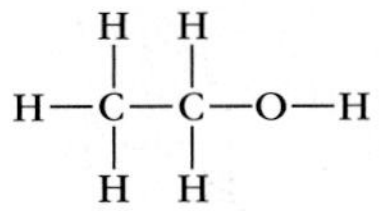

ethanol

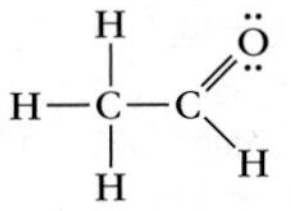

acetaldehyde

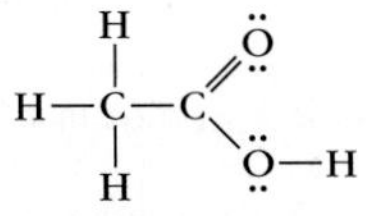

acetic acid

The acetaldehyde molecule has two fewer hydrogen atoms compared with the ethanol molecule. Loss of hydrogen is oxidation. The acetaldehyde molecule is more oxidized than the ethanol molecule. Comparing the formulas for acetaldehyde and acetic acid, the hydrogen atoms are the same, but the acetic acid molecule has one additional oxygen atom. Gain of oxygen is oxidation. So the acetic acid molecule is more oxidized than the acetaldehyde molecule.

10.8 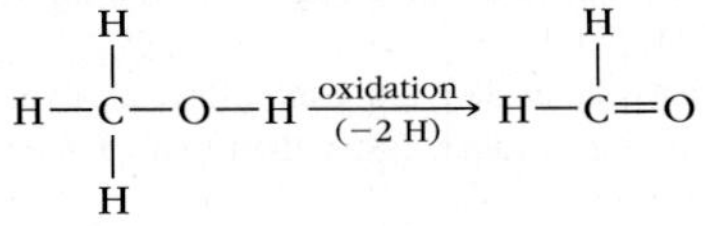

10.9 $CH_3CH_2CH_2OH$, 1-propanol

10.10 (a) Estradiol contains two alcohol groups (—OH)

(b) Secondary alcohol

(c) Oxidation (removal of two hydrogens)

(d) Estradiol: Two alcohol groups, aromatic ring, one $—CH_3$ group
Testosterone: One alcohol group; one ketone group; C=C double bond in ring; two $—CH_3$ groups

10.11 Conversion of a secondary alcohol on the five-membered ring to a ketone by oxidation (removal of 2 H atoms)

10.12 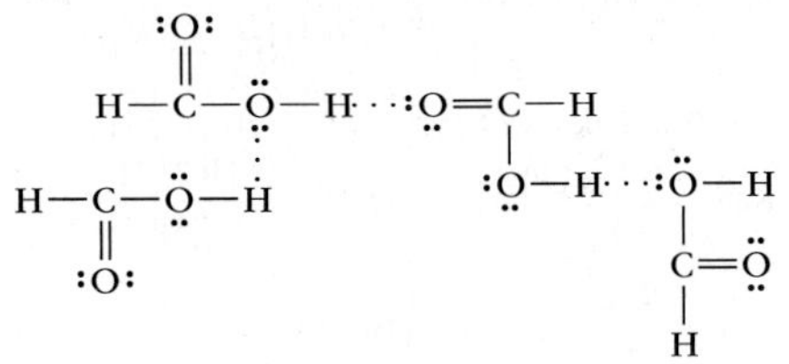

10.13 (a)

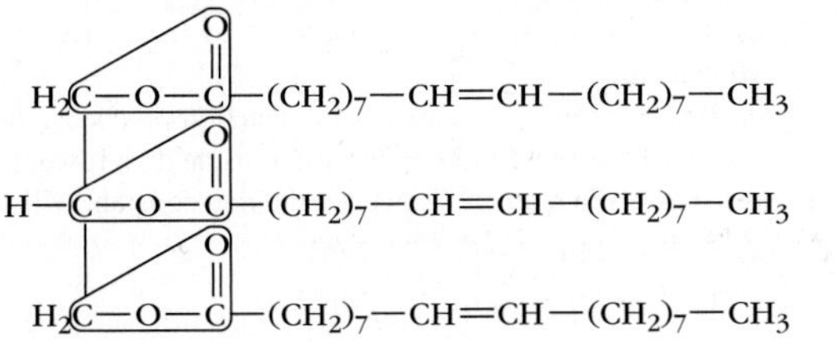

(b)

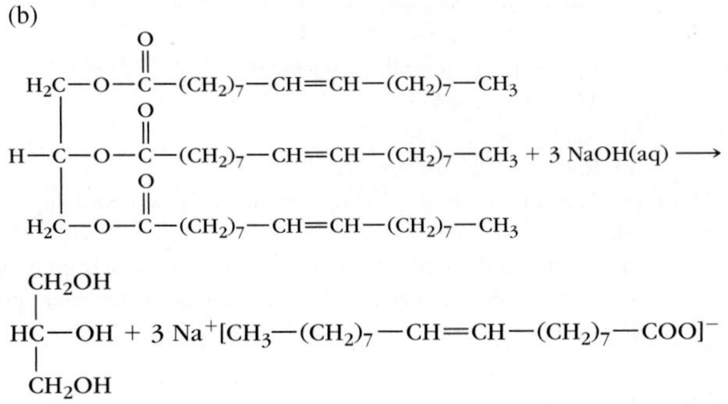

10.14 The ends of the chains are possibly occupied by the OR groups from the initiator molecules.

10.15 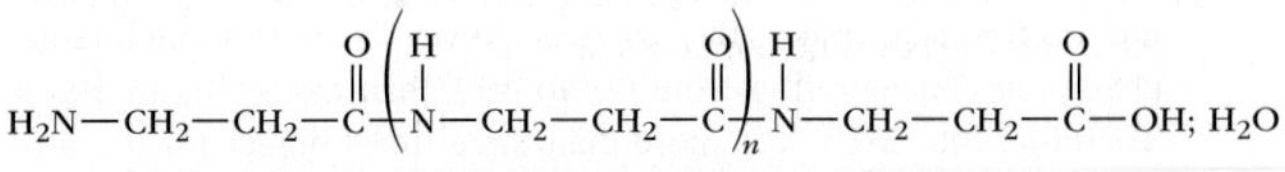

10.16 (1) Amine (2) Carboxylic acid (3) Amide (4) Ester

10.17 The —OH groups in this molecule allow it to be extensively hydrogen-bonded with solvent water molecules.

10.18 Cellulose contains glucose molecules linked together by *trans* 1,4 linkages. Humans lack the enzymes needed to catalyze breaking of *trans* 1,4 linkages. Ruminant animals have large colonies of bacteria and protozoa that live in the forestomach and digest cellulose.

10.19 If humans could digest cellulose, then common plants that are easy to grow could become food. There might be less reliance upon cultivation of plants for food. In addition, the entire plant could be used for food rather than just certain parts eaten and the other parts wasted. On the other hand, in times of famine, there might not be enough cellulose to go around. Destroying trees and other plants for food might cause enlargements of desert regions and the disappearance of entire species of plants.

10.20 Serine and glutamine can hydrogen-bond to one another if they were close in two adjacent protein chains because they have polar groups containing H atoms in their R groups. Glycine and valine would not because they have no additional polar groups in their R groups. (Other correct choices are possible.)

10.21 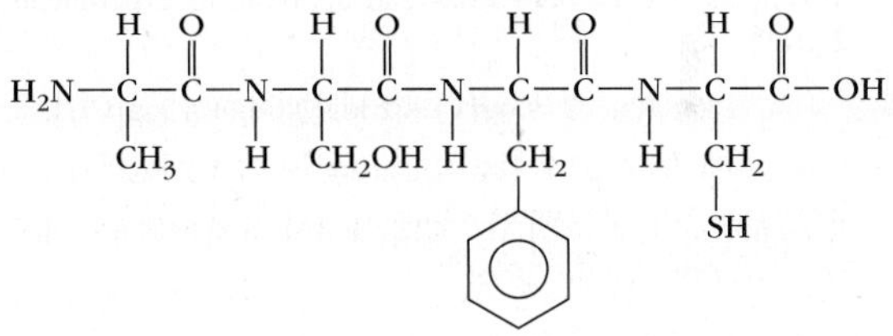

nonpolar — polar neutral — nonpolar — polar neutral

Chapter 11

11.1 (a)

i. $$\text{Rate} = -\frac{\Delta[\text{Cv}^+]}{\Delta t}$$

$$= -\frac{(0.793 \times 10^{-5} - 1.46 \times 10^{-5})\ \text{mol/L}}{(60.0 - 40.0)\ \text{s}}$$

$$= \frac{6.67 \times 10^{-6}\ \text{mol/L}}{20.0\ \text{s}} = 3.3 \times 10^{-7}\ \text{mol L}^{-1}\ \text{s}^{-1}$$

ii. $$\text{Rate} = -\frac{(0.429 \times 10^{-5} - 2.71 \times 10^{-5})\ \text{mol/L}}{(80.0 - 20.0)\ \text{s}}$$

$$= 3.8 \times 10^{-7}\ \text{mol L}^{-1}\ \text{s}^{-1}$$

iii. $$\text{Rate} = -\frac{(0.232 \times 10^{-5} - 5.00 \times 10^{-5})\ \text{mol/L}}{(100.0 - 0.0)\ \text{s}}$$

$$= 4.8 \times 10^{-7}\ \text{mol L}^{-1}\ \text{s}^{-1}$$

(b) 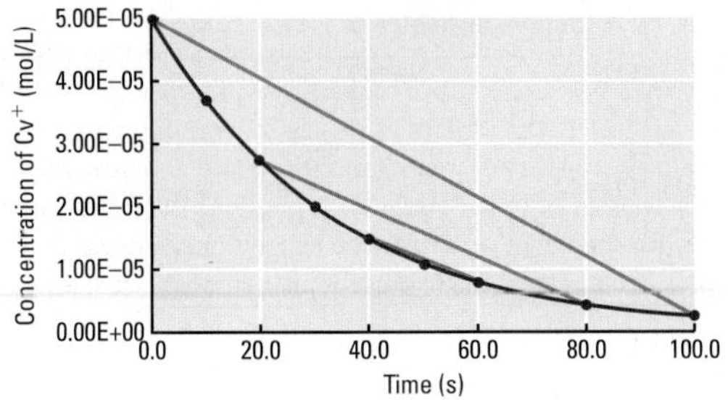
5.00E–05
4.00E–05
3.00E–05
2.00E–05
1.00E–05
0.00E+00
Concentration of Cv+ (mol/L)
0.0
20.0
40.0
60.0
80.0
100.0
Time (s)

(c) The rate is faster when the concentration of Cv^+ is larger. As the reaction takes place, the rate gets slower. There is a much larger change in concentration from 0 s to 50 s than from 50 s to 100 s. Therefore, the $\Delta[Cv^+]$ is more than three times bigger for the time range from 20 s to 80 s than it is for the time range from 40 s to 60 s, even though Δt is exactly three times larger.

11.2 (a) Rate (1) is twice rate (3); rate (2) is twice rate (4); rate (3) is twice rate (5).
(b) In each case, the $[Cv^+]$ is twice as great when the rate is twice as great.
(c) Yes, the rate doubles when $[Cv^+]$ doubles.

11.3
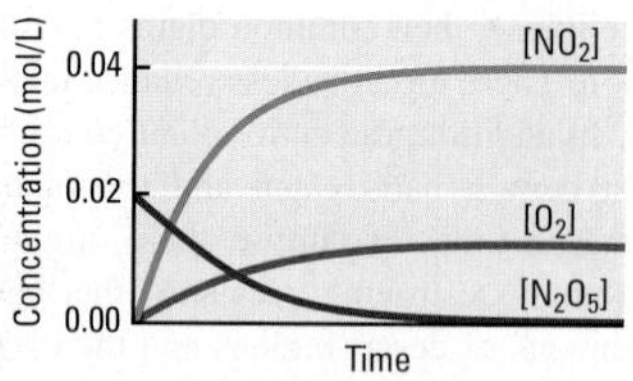
Concentration (mol/L)
0.04
0.02
0.00
Time
[NO2]
[O2]
[N2O5]

11.4 $\text{Rate}_1 = k[CH_3COOCH_3][OH^-]$
$\text{Rate}_2 = k(2[CH_3COOCH_3])(\tfrac{1}{2}[OH^-]) = k[CH_3COOCH_3][OH^-]$
The rate is unchanged.

11.5
$$\text{Rate} = k[NO]^x[Cl_2]^y$$

Taking the log of both sides and applying to experiments 4 and 2 gives

$$\log(\text{Rate}_4) = \log(k) + x\log[NO]_4 + y\log[Cl_2]_4$$
$$\log(\text{Rate}_2) = \log(k) + x\log[NO]_2 + y\log[Cl_2]_2$$

Recognizing that $[Cl_2]_4 = [Cl_2]_2$ and then subtracting the second equation from the first gives

$$\log(\text{Rate}_4) - \log(\text{Rate}_2) = x\log[NO]_4 - x\log[NO]_2$$

Using the properties of logarithms gives

$$\log\{\text{Rate}_4/\text{Rate}_2\} = x\{\log[NO]_4/[NO]_2\}$$

So

$$x = \log\{\text{Rate}_4/\text{Rate}_2\}/\{\log[NO]_4/[NO]_2\} \quad \text{QED}$$

Now we can insert the actual numbers from experiments 4 and 2:

$$x = [\log(6.60\times10^{-4}/1.65\times10^{-4})]/\log(0.04/0.02)$$
$$x = \log 4/\log 2 = 0.602/0.301 = 2$$

11.6

(a)	$CH_3NC \rightarrow CH_3CN$	unimolecular
(b)	$2\,HI \rightarrow H_2 + I_2$	bimolecular
(c)	$NO_2Cl \rightarrow NO_2 + Cl$	unimolecular
(d)	$C_4H_8 \rightarrow C_4H_8$	unimolecular
(e)	$NO_2Cl + Cl \rightarrow NO_2 + Cl_2$	bimolecular

11.7
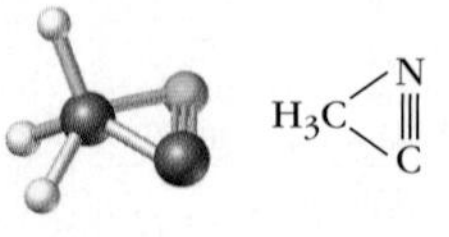
H3C
N
C

The Lewis structure has five pairs of electrons around one C atom, which would not be stable. However, this is a transition state, which is, by definition, unstable.

11.8 (1)
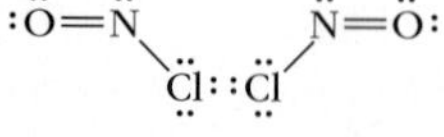

(2)
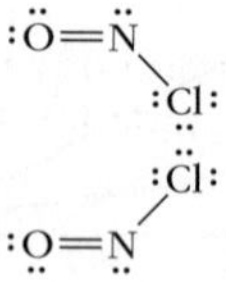

(3)
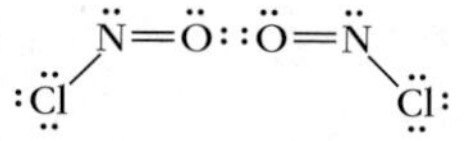

(4)
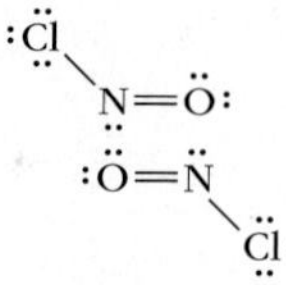

(1) and (2) are much more likely to result in a reaction than are (3) and (4).

11.9 (a) $k = Ae^{-E_a/RT} = (6.31\times10^8\ \text{L mol}^{-1}\ \text{s}^{-1})\ e^{\frac{-10{,}000\ \text{J/mol}}{(8.314\ \text{J K}^{-1}\ \text{mol}^{-1})(370\ \text{K})}}$

$k = 2.4\times10^7\ \text{L mol}^{-1}\ \text{s}^{-1}$

(b) $\text{Rate} = k[NO][O_3]$
$= (2.4\times10^7\ \text{L mol}^{-1}\ \text{s}^{-1})(1.0\times10^{-3}\ \text{mol/L}) \times (5.0\times10^{-4}\ \text{mol/L})$
$= 12\ \text{mol L}^{-1}\ \text{s}^{-1}$

11.10 The reaction does not occur in a single step. If it did, the rate law would be $\text{Rate} = k[NO_2][CO]$.

11.11 $\text{Rate} = k[HOI][I^-]$. The rate of Step 2 doesn't affect the overall rate because the rate of Step 2 is much faster than that of Step 1, the slow, rate-determining step.

11.12 (a) $2\,ICl(g) + H_2(g) \rightarrow 2\,HCl(g) + I_2(g)$
(b) $\text{Rate} = k_2[HI][ICl]$
However, HI is an intermediate. Assume that the concentration of HI reaches a steady state.

$$\text{Rate of Step } {+1} = \text{rate of Step } {-1} + \text{rate of Step 2}$$

Also assume that Step 2 is much slower than Step +1 and Step −1. Then

$$k_1[ICl][H_2] = k_{-1}[HI][HCl]$$
$$[HI] = \frac{k_1[ICl][H_2]}{k_{-1}[HCl]}$$
$$\text{Rate} = k_2[ICl][HI] = k_2[ICl] \times \frac{k_1[ICl][H_2]}{k_{-1}[HCl]}$$
$$= \frac{k_1k_2}{k_{-1}}[ICl]^2[H_2][HCl]^{-1}$$

(c) The rate is inversely proportional to the concentration of the product, HCl.
(d) Because [HCl] increases as the reaction proceeds, the rate of reaction will decrease more quickly over time than it would if $[HCl]^{-1}$ were not in the rate law. However, the rate constant will not change.

11.13 (a)
$$Ce^{4+} + Mn^{2+} \rightarrow Ce^{3+} + Mn^{3+}$$
$$Ce^{4+} + Mn^{3+} \rightarrow Ce^{3+} + Mn^{4+}$$
$$\underline{Mn^{4+} + Tl^{+} \rightarrow Mn^{2+} + Tl^{3+}}$$
$$2\,Ce^{4+} + Tl^{+} \rightarrow 2\,Ce^{3+} + Tl^{3+}$$
(b) Intermediates are Mn^{3+} and Mn^{4+}.
(c) The catalyst is Mn^{2+}.
(d) $\text{Rate} = k[Ce^{4+}][Mn^{2+}]$.
(e) $\text{Rate} = k[Ce^{4+}]^2[Mn^{2+}][Ce^{3+}]^{-1}$.

11.14 (a) The concentration of a homogeneous catalyst *must* appear in the rate law.
(b) A catalyst does not appear in the equation for an overall reaction.
(c) A *homogeneous* catalyst must always be in the same phase as the reactants.

Chapter 12

12.1 (a) $(\text{conc. Al}) = \frac{2.70\ \text{g}}{\text{mL}} \times \frac{1000\ \text{mL}}{1\ \text{L}} \times \frac{1\ \text{mol}}{26.98\ \text{g}}$
$= 100.\ \text{mol/L}$

(b) $(\text{conc. benzene}) = \frac{0.880\text{ g}}{\text{mL}} \times \frac{1000\text{ mL}}{1\text{ L}} \times \frac{1\text{ mol}}{78.11\text{ g}}$

$= 11.3\text{ mol/L}$

(c) $(\text{conc. water}) = \frac{0.998\text{ g}}{\text{mL}} \times \frac{1000\text{ mL}}{1\text{ L}} \times \frac{1\text{ mol}}{18.02\text{ g}}$

$= 55.4\text{ mol/L}$

(d) $(\text{conc. Au}) = \frac{19.32\text{ g}}{\text{mL}} \times \frac{1000\text{ mL}}{1\text{ L}} \times \frac{1\text{ mol}}{196.97\text{ g}}$

$= 98.1\text{ mol/L}$

12.2 The mixture is not at equilibrium, but the reaction is so slow that there is no change in concentrations. You could show that the system was not at equilibrium by providing a catalyst or by raising the temperature to speed up the reaction.

12.3 (a) The new mixture is not at equilibrium because the quotient (conc. *trans*)/(conc. *cis*) no longer equals the equilibrium constant. Because (conc. *cis*) was halved, the quotient is twice K_c.
(b) The rate *trans* $\rightarrow$ *cis* remains the same as before, because (conc. *trans*) did not change. The rate *cis* $\rightarrow$ *trans* is only half as great, because (conc. *cis*) is half as great as at equilibrium.
(c) At 600 K, K_c is 1.47. Thus, [*trans*] = 1.47[*cis*].
(d) 0.15 mol/L

12.4 (a) If the coefficients of an equation are halved, the numerical value for the new equilibrium constant is the square root of the previous equilibrium constant. So the new equilibrium constant is $(6.25 \times 10^{-58})^{1/2} = 2.50 \times 10^{-29}$.
(b) If a chemical equation is reversed, the value for the new equilibrium constant is the reciprocal of the previous equilibrium constant. So the new equilibrium constant is $1/(6.25 \times 10^{-58}) = 1.60 \times 10^{57}$.

12.5 (a) $K_P = K_c \times (RT)^{\Delta n} = (3.5 \times 10^8\text{ L}^2\text{ mol}^{-2}) \times \{(0.082057\text{ L atm K}^{-1}\text{ mol}^{-1})(298\text{ K})\}^{-2}$

$= 5.8 \times 10^5\text{ atm}^{-2}$

(b) $K_P = (3.2 \times 10^{81}\text{ L mol}^{-1} \times \{(0.082057\text{ L atm mol}^{-1}\text{ K}^{-1})(298\text{ K})\}^{-1}$

$= 1.3 \times 10^{80}\text{ atm}^{-1}$

(c) $K_P = K_c = 1.6 \times 10^{-3}$

(d) $K_P = (1.7 \times 10^2\text{ L mol}^{-1}) \times \{(0.082057\text{ L atm mol}^{-1}\text{ K}^{-1})(298\text{ K})\}^{-1}$

$= 7.0\text{ atm}^{-1}$

12.6 (a) Because K_c for the forward reaction is small, K_c' for the reverse reaction is large.

(b) $K_c' = \frac{1}{K_c} = \frac{1}{1.8 \times 10^{-5}} = 5.6 \times 10^4$

(c) Ammonium ions and hydroxide ions should react, using up nearly all of whichever is the limiting reactant.

(d) $NH_4^+(aq) + OH^-(aq) \rightleftharpoons NH_3(aq) + H_2O(\ell)$
$NH_3(aq) \rightleftharpoons NH_3(g)$
You might detect the odor of $NH_3(g)$ above the solution. A piece of moist red litmus paper above the solution would turn blue.

12.7 Q should have the same mathematical form as K_P, so for the general Equation 12.1

$$Q_P = \frac{P_C^c \times P_D^d}{P_A^a \times P_B^b}$$

The rules for Q_P and K_P are analogous to those for Q and K_c:
If $Q_P > K_P$, then the reverse reaction occurs.
If $Q_P = K_P$, then the system is at equilibrium.
If $Q_P < K_P$, the forward reaction occurs.

12.8 $PCl_5(s) \rightleftharpoons PCl_3(g) + Cl_2(g)$
(a) Adding $Cl_2(g)$ shifts the equilibrium to the left.
(b) Adding $PCl_3(g)$ to the container shifts the equilibrium to the left.
(c) Because $PCl_5(s)$ does not appear in the K_c expression, adding some will not affect the equilibrium.

12.9 For the first equilibrium, [*trans*] = 1.2 mmol/L and [*cis*] = 0.80 mmol/L; this gives K_c = 1.2/0.80 = 1.5. For the second equilibrium, [*trans*] = 0.96 mmol/L and [*cis*] = 0.64 mmol/L; this gives K_c = 0.96/0.64 = 1.5, the same value. K does not change unless the temperature changes.

12.10 $Q = \frac{(\text{conc. H}_2)(\text{conc. I}_2)}{(\text{conc. HI})^2} = \frac{(0.0606 \times 3)(0.0106 \times 3)}{(0.179 \times 3)^2}$

$= \frac{(0.0606)(0.0106)(9)}{(0.179)^2(9)} = 0.020 = K_c$

Because $Q = K_c$, the system is at equilibrium under the new conditions. No shift is needed and none occurs.

12.11 From Problem-Solving Practice 12.7, $[NO_2]$ = 0.0418 mol/L and $[N_2O_4]$ = 0.292 mol/L.
Decreasing the volume from 4.00 to 1.33 L increases the concentrations as

$$(\text{conc. NO}_2) = 0.0418 \times \frac{4.00}{1.33} = 0.1257\text{ mol/L}$$

$$(\text{conc. N}_2\text{O}_4) = 0.292 \times \frac{4.00}{1.33} = 0.8782\text{ mol/L}$$

$$Q = \frac{(\text{conc. NO}_2)^2}{(\text{conc. N}_2\text{O}_4)} = \frac{(0.1257)^2}{0.8782} = 1.80 \times 10^{-2}$$

which is greater than K_c. Therefore, the equilibrium should shift to the left. Let x be the change in concentration of NO_2, (which should be negative).

	N_2O_4	$\rightleftharpoons$	$2\ NO_2$
Initial concentration (mol/L)	0.8782		0.1257
Change as reaction occurs (mol/L)	$-\frac{1}{2}x$		x
New equilibrium concentration (mol/L)	$0.8782 - \frac{1}{2}x$		$0.1257 + x$

$$K_c = 5.9 \times 10^{-3} = \frac{(0.1257 + x)^2}{0.8782 - 0.500x}$$

$$= \frac{(1.580 \times 10^{-2}) + (2.514 \times 10^{-1})x + x^2}{0.8782 - 0.500x}$$

$$5.18 \times 10^{-3} - (2.95 \times 10^{-3})x = (1.580 \times 10^{-2}) + (2.514 \times 10^{-1})x + x^2$$

$$x^2 + 0.2544x + (1.062 \times 10^{-2}) = 0$$

$$x = \frac{-0.2544 \pm \sqrt{6.472 \times 10^{-2} - (4 \times 1 \times 1.062 \times 10^{-2})}}{2}$$

$$x = \frac{-0.2544 \pm 0.1491}{2}$$

$x = -0.0526$ or $x = -0.2018$

The second root is mathematically reasonable, but results in a negative value for the $[NO_2]$. The new equilibrium concentrations are

$$[NO_2] = 0.1257 - 0.0526 = 0.0731\text{ mol/L}$$

$$[N_2O_4] = 0.8782 - \tfrac{1}{2}(-0.0526) = 0.9045\text{ mol/L}$$

Compared to the initial equilibrium, the concentrations have changed by

$$NO_2\text{: } \frac{0.0731}{0.0418} = 1.75 \qquad N_2O_4\text{: } \frac{0.9045}{0.292} = 3.10$$

The concentration of N_2O_4 did increase by more than a factor of 3. The concentration of NO_2 increased by 1.75, which is less than a factor of 3.

12.12

	How Reaction System Changes	Equilibrium Shifts	Change in K_c?
(a) Add reactant	Some reactants consumed	To right	No
(b) Remove reactant	More reactants formed	To left	No
(c) Add product	More reactants formed	To left	No
(d) Remove product	More products formed	To right	No
(e) Increase *P* by decreasing *V*	Total pressure decreases	Toward fewer gas molecules	No
(f) Decrease *P* by increasing *V*	Total pressure increases	Toward more gas molecules	No
(g) Increase *T*	Heat transfer into system	In endothermic direction	Yes
(h) Decrease *T*	Heat transfer out of system	In exothermic direction	Yes

If a substance is added or removed, the equilibrium is affected only if the substance's concentration appears in the equilibrium constant expression, or if its addition or removal changes concentrations that appear in the equilibrium constant expression. Changing pressure by changing volume affects an equilibrium only for gas-phase reactions in which there is a difference in the amount of gaseous reactants and products.

12.13 (a) The reaction is exothermic. $\Delta_r H° = -92.22$ kJ/mol.
(b) The reaction is not favored by entropy.
(c) The reaction produces more products at low temperatures.
(d) If you increase *T* the reaction will go faster, but a smaller amount of products will be produced.

Chapter 13

13.1 The data in Table 13.2 indicate that the solubility of alcohols decreases as the hydrocarbon chain lengthens. Thus, 1-octanol is less soluble in water than 1-heptanol and 1-decanol should be even less soluble than 1-octanol.

13.2 Methanol is more water soluble than is octanol, but octanol is more soluble in gasoline. The octanol molecule is more hydrocarbon-like and this explains its solubility in gasoline. The methanol molecule is more water-like and this explains its greater solubility in water.

13.3 72 g − 38 g = 34 g NH_4Cl would crystallize from solution at 20 °C.

13.4 (a) Unsaturated (b) Supersaturated (c) Saturated

13.5 The solubility of CO_2 decreases with increasing temperature, and the beverage loses its carbonation, causing it to go "flat."

13.6 Putting back water that is too warm would decrease the solubility of oxygen in the lake or river, thereby decreasing the oxygen concentration. This could cause a fish kill if the oxygen concentration dropped sufficiently.

13.7 Hot solvent would cause more of the solute to dissolve because Le Chatelier's principle states that at higher temperature an equilibrium will shift in the endothermic direction.

13.8 (a) Mass fraction of $NaHCO_3$ =

$$\frac{0.20}{1000 + 6.5 + 0.20 + 0.1 + 0.10} = 2.0 \times 10^{-4}$$

Weight percent = $2.0 \times 10^{-4} \times 100\% = 2.0 \times 10^{-2}\%$

(b) KCl and $CaCl_2$ each have the lowest mass fraction, 0.015.

13.9 (a) The 20. ppb sample has the higher lead concentration. (The other sample is 3.0 ppb lead.)
(b) 0.015 mg/L is equivalent to 0.015 ppm, which is 15 ppb. The 20-ppb sample exceeds the EPA limit; the 3.0-ppb sample does not.

13.10 $\frac{320 \text{ mg}}{1 \text{ d}} \times \frac{0.500 \text{ L}}{12 \text{ mg}} = \frac{13 \text{ L}}{\text{d}} = 13$ bottles per day

13.11 $3.5 \times 10^{20} \text{ gal} \times \frac{3.785 \text{ L}}{\text{gal}} \times \frac{1 \times 10^{-3} \text{ mg Au}}{1 \text{ L}}$

$= 1.3 \times 10^{18}$ mg Au $= 1.3 \times 10^{15}$ g Au

$$1.3 \times 10^{15} \text{ g Au} \times \frac{1 \text{ lb Au}}{454 \text{ g Au}} = 3 \times 10^{12} \text{ lb Au}$$

13.12 (a) $355 \text{ mL} \times \frac{1.01 \text{ g}}{\text{mL}} = 359$ g solution;

$$\frac{46.3 \times 10^{-3} \text{ g caffeine}}{359 \text{ g solution}} = \frac{1.29 \times 10^{-4} \text{ g caffeine}}{1 \text{ g solution}} = \frac{129 \text{ g caffeine}}{10^6 \text{ g solution}} = 129 \text{ ppm}$$

(b) $c = \frac{\text{moles solute}}{\text{L solution}}$

$$= \frac{(46.3 \times 10^{-3} \text{ g caffeine}) \times \left(\frac{1 \text{ mol caffeine}}{194.2 \text{ g caffeine}}\right)}{0.355 \text{ L}}$$

$= 6.71 \times 10^{-4}$ M

(c) $m = \frac{\text{moles solute}}{\text{kg solution}}$

$$= \frac{(46.3 \times 10^{-3} \text{ g caffeine}) \times \left(\frac{1 \text{ mol caffeine}}{194.2 \text{ g caffeine}}\right)}{0.359 \text{ kg solution}}$$

$= 6.64 \times 10^{-4}$ m

13.13 (a) Moles of solute and kilograms of solvent
(b) Molar mass of the solute and the density of the solution

13.14 $50.0 \text{ g sucrose} \times \left(\frac{1 \text{ mol sucrose}}{342 \text{ g sucrose}}\right) = 0.146$ mol sucrose

$$100.0 \text{ g water} \times \left(\frac{1 \text{ mol water}}{18.0 \text{ g water}}\right) = 5.56 \text{ mol water}$$

$$X_{H_2O} = \frac{5.56}{5.56 + 0.146} = \frac{5.56}{5.706} = 0.974$$

Applying Raoult's law:

$$P_{\text{water}} = (X_{\text{water}})(P^0_{\text{water}}) = (0.974)(71.88 \text{ mmHg}) = 70.0 \text{ mmHg}$$

13.15 $\Delta T_b = (2.53 \text{ °C kg mol}^{-1})(0.10 \text{ mol/kg}) = 0.25$ °C. The boiling point of the solution is 80.10 °C + 0.25 °C = 80.35 °C.

13.16 First, calculate the required molality of the solution that would have a freezing point of −30. °C.

$$\Delta T_f = -30.\ \text{°C} = (-1.86 \text{ °C} \cdot \text{kg mol}^{-1}) \times m$$

$$m = \frac{-30.\ \text{°C}}{-1.86 \text{ °C kg mol}^{-1}} = 16.1 \text{ mol/kg}$$

To protect 4.0 kg of water from this freezing temperature, you would need 4.0 × 16.1 mol of ethylene glycol, or 64 mol.

$$64 \text{ mol}\left(\frac{62.1 \text{ g}}{\text{mol}}\right)\left(\frac{1 \text{ mL}}{1.113 \text{ g}}\right)\left(\frac{1 \text{ L}}{1000 \text{ mL}}\right) = 3.6 \text{ L}$$

13.17 $\Delta T_f = K_f \times m \times i$;

$$-4.78 \text{ °C} = (-1.86 \text{ °C kg mol}^{-1})(2.0 \text{ mol/kg})(i)$$

$$i = \frac{-4.78 \text{ °C}}{(2.0 \text{ mol/kg})(-1.86 \text{ °C kg mol}^{-1})} = 1.28$$

Degree of dissociation: If completely dissociated, 1 mol $CaCl_2$ should yield 3 mol ions

$$CaCl_2 \longrightarrow Ca^{2+} + 2\ Cl^-$$

and i should be 3. The degree of dissociation in this solution is $1.28/3 = 0.427 \cong 43\%$.

13.18 The 0.02-M solution of ordinary soap contains more particles because a soap is a salt of a fatty acid, whereas sucrose is a nonelectrolyte. Therefore, its osmotic pressure is larger.

13.19 $$\frac{3\text{ qt}}{\text{d}} \times \frac{1\text{ L}}{1.06\text{ qt}} \times \frac{0.050\text{ mg Se}}{\text{L}} \times \frac{10^3\ \mu\text{g}}{1\text{ mg}} = \frac{142\ \mu\text{g Se}}{\text{d}}$$

13.20 0.0250 ppm Pb means that for every liter (1 kg) of water, there is 0.0250 mg, or 25.0 μg of Pb. Using this factor

$$\text{Volume of water} = 100.0\ \mu\text{g Pb}\left(\frac{1\text{ L}}{25.0\ \mu\text{g Pb}}\right) = 4.00\text{ L}$$

Chapter 14

14.1 (a) Acid (b) Base
(c) Acid (d) Base

14.2 The equilibrium reaction is

$$NH_3(aq) + H_2O(\ell) \rightleftharpoons NH_4^+(aq) + OH^-(aq)$$

Diluting the solution decreases the concentration of species in solution and by Le Chatelier's principle, the equilibrium shifts to counteract this change. There are two aqueous species on the right and one on the left so the equilibrium shifts right.

14.3 (a) $H_3O^+(aq) + CN^-(aq)$
(b) $H_3O^+\ (aq) + Br^-\ (aq)$
(c) $CH_3NH_3^+(aq) + OH^-(aq)$
(d) $(CH_3)_2NH_2^+(aq) + OH^-(aq)$

14.4 (a) $HSO_4^-(aq) + H_2O(\ell) \rightarrow H_2SO_4(aq) + OH^-(aq)$
(b) H_2SO_4 is the conjugate acid of HSO_4^-; water is the conjugate acid of OH^-.
(c) Yes; it can be an H^+ donor or acceptor.

14.5 The reverse reaction is favored because HSO_4^- is a stronger acid than CH_3COOH.

14.6 (a)

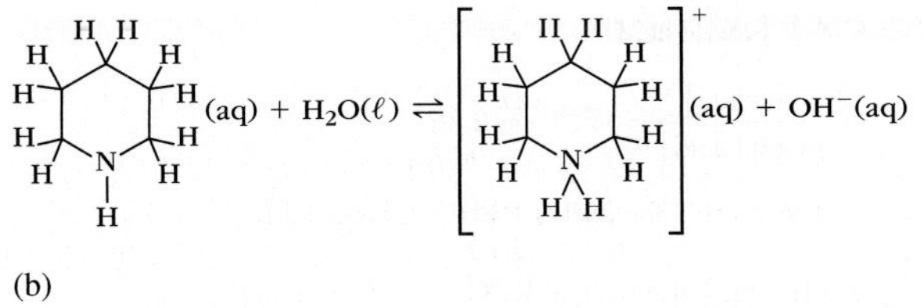

(b)

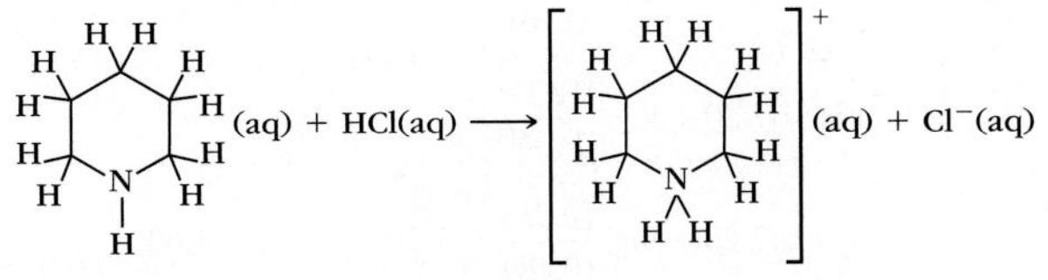

14.7 (a) Amine (b) Neither (c) Acid
(d) Amine (e) Amine and acid

14.8 The pH values of 0.1-M solutions of these two strong acids would be essentially the same because they both are 100% ionized, resulting in $[H_3O^+]$ values that are the same.

14.9 Lime juice is more acidic: pH $= -\log\ [H_3O^+] = -\log\ (1.3 \times 10^{-2}) = 1.89$; pH carbonated beverage is 2.10 (less acidic).

14.10 Because pH + pOH = 14.0, both solutions have a pOH of 8.5. The $[H_3O^+] = 10^{-pH} = 10^{-5.5} = 3 \times 10^{-6}$ M.

14.11 Pyruvic acid is the stronger acid as indicated by its larger K_a value. Lactic acid's ionization reaction is more reactant-favored (less acid ionizes).

14.12 Being negatively charged, the HSO_4^- ion has a lower tendency to lose a positively charged proton because of the electrostatic attractions of opposite charges.

14.13 (a) Step 1: $HOOC{-}COOH(aq) + H_2O(\ell) \rightleftharpoons H_3O^+(aq) + HOOC{-}COO^-(aq)$
Step 2: $HOOC{-}COO^-(aq) + H_2O(\ell) \rightleftharpoons H_3O^+(aq) + {}^-OOC{-}COO^-(aq)$
(b) Step 1: $C_3H_5O(COOH)_3(aq) + H_2O \rightleftharpoons H_3O^+(aq) + C_3H_5O(COOH)_2COO^-(aq)$
Step 2: $C_3H_5O(COOH)_2COO^-(aq) + H_2O \rightleftharpoons H_3O^+(aq) + C_3H_5O(COOH)(COO)_2^{2-}(aq)$
Step 3: $C_3H_5O(COOH)(COO)_2^{2-}(aq) + H_2O \rightleftharpoons H_3O^+(aq) + C_3H_5O(COO)_3^{3-}(aq)$

14.14 (a) Fluorobenzoic acid (b) Chloroacetic acid
In both cases, the more electronegative halogen atom increases electron withdrawal from the acidic hydrogen, thereby increasing its partial positive charge.

14.15 Oxalic acid is the stronger acid because it has a greater number of oxygens.

14.16 (a)

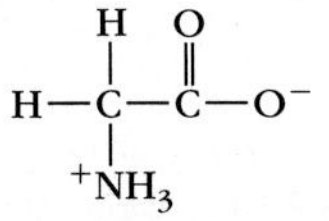

(b)

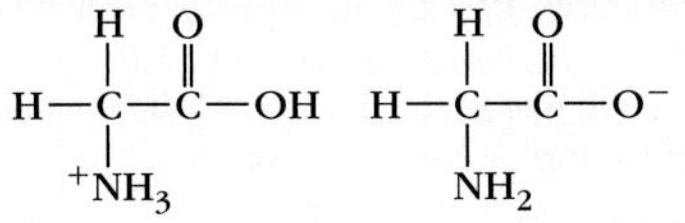

at pH 2 at pH 10

14.17 At pH 3

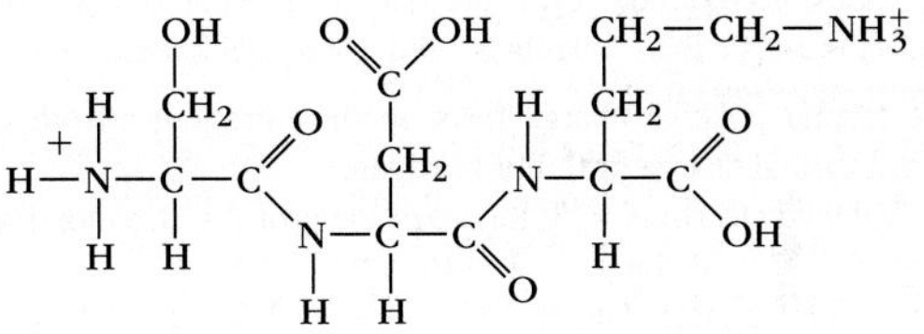

At pH 7

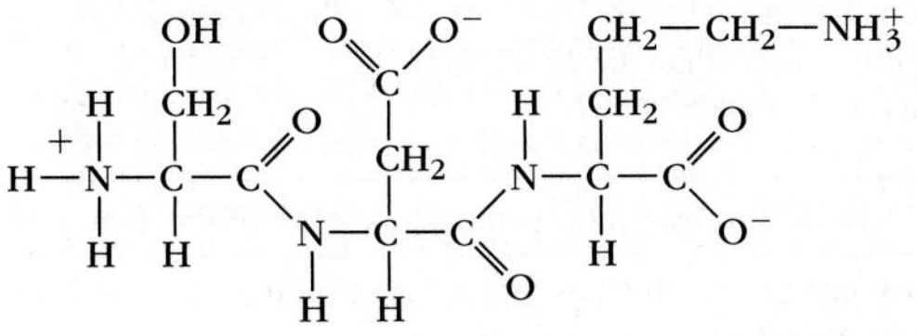

At pH 11

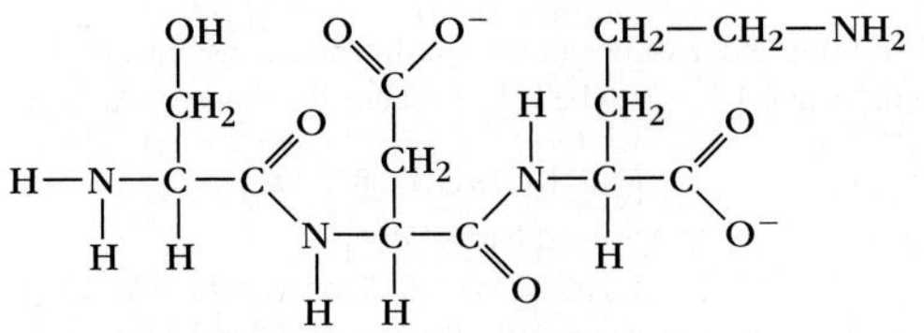

14.18 Nonmetal oxides are acidic; metal oxides are basic: SO_3, SO_2, SeO_2, Al_2O_3, CaO.

14.19 Set up an ICE table for the dissociation of nitrous acid.

	$HNO_2(aq)$	$\rightleftharpoons$ $H^+(aq)$	+ NO_2^-
Initial conc. (M)	0.495	0	0
Conc. change (M)	$-x$	$+x$	$+x$
Equil. conc. (M)	$0.495 - x$	$+x$	$+x$

Substitute the values into the equilibrium constant expression.

$$K_a = 7.1 \times 10^{-4} = \frac{[H^+][NO_2^-]}{[HNO_2]} = \frac{(x)(x)}{(0.495 - x)} = \frac{x^2}{(0.495 - x)}$$

Assume that x is small compared to 0.495 and the equation simplifies to

$$K_a = 7.1 \times 10^{-4} = \frac{[H^+][NO_2^-]}{[HNO_2]} = \frac{(x)(x)}{(0.495 - x)} \cong \frac{x^2}{(0.495)}$$

$x^2 \cong (7.1 \times 10^{-4})(0.495) = 3.51 \times 10^{-4}$

$x = 1.87 \times 10^{-2}$ M = $[H^+]$

A second and third approximation gives

$x = 1.84 \times 10^{-2} = [H^+]$ and

pH = $-\log(1.84 \times 10^{-2}) = 1.73$.

14.20 $K_b = \frac{1.0 \times 10^{-14}}{K_a} = \frac{1.0 \times 10^{-14}}{1.3 \times 10^{-10}} = 7.7 \times 10^{-5}$; carbonate; by comparing K_b values

14.21 The pH of soaps is >7 due to the reaction of the conjugate base in the soap with water to form a basic solution.

14.22 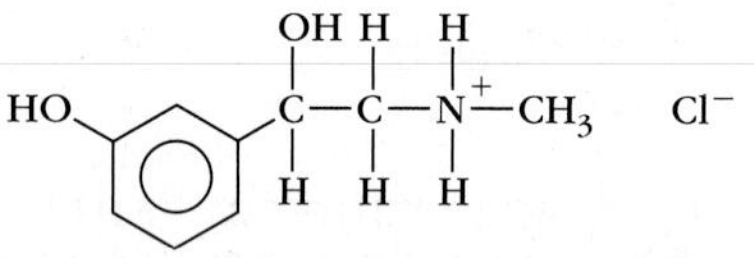

14.23 The formal positive charge rests on the nitrogen; all other atoms have a formal charge of 0. For nitrogen:
Valence electrons = 5; lone-pair electrons = 0; $\frac{1}{2}$ bonding electrons = 4; formal charge = 5 − (0 + 4) = +1.

14.24 Ammonium acetate. The pH of a solution of this salt will be 7.

14.25 (a) Lewis base (b) Lewis acid
(c) Lewis acid and base (d) Lewis acid
(e) Lewis acid (f) Lewis base

14.26 The solution is slightly basic. The K_b of acetate ion, 5.6×10^{-10}, is slightly greater than the K_a of $Ni(H_2O)_6^{2+}$, 2.5×10^{-11}.

14.27 The reaction table is

	$H_2O(\ell)$ + $Ni(H_2O)_6^{2+}(aq)$	$\rightleftharpoons$ $Ni(H_2O)_5(OH)^+(aq)$	+ $H_3O^+(aq)$
Initial (mol/L)	0.15	0	10^{-7}
Change (mol/L)	$-x$	$+x$	$+x$
Equilibrium (mol/L)	$0.15 - x$	x	x

Substituting these values in the equilibrium constant expression, and simplifying $0.15 - x$ to be 0.15 because the value of K_a is so small

$$K_a = \frac{[Ni(H_2O)_5(OH)^+][H_3O^+]}{[Ni(H_2O)_6^{2+}]}$$

$$= \frac{(x)(x)}{(0.15 - x)} \cong \frac{x^2}{0.15} = 2.5 \times 10^{-11}$$

Solving for x, which is the $[H_3O^+]$

$$x = \sqrt{(0.15)(2.5 \times 10^{-11})} = 1.94 \times 10^{-6}$$

So, the pH of this solution = $-\log(1.94 \times 10^{-6}) = 5.71$.

14.28 Strong bases would cause damage to tissue.

14.29 Baking powder and baking soda are sources of hydrogen carbonate; buttermilk and baking powder are sources of acid.

14.30 Set up an ICE table for the hydrolysis reaction.

	H_2O + CO_3^{2-}	$\rightleftharpoons$ HCO_3^-	+ OH^-
Initial conc. (mol/L)	5.2	0	10^{-7}
Conc. change due to reaction (mol/L)	$-x$	$+x$	$+x$
Equilibrium conc. (mol/L)	$5.2 - x$	x	x

Using the K_b expression and substituting the values from the table

$$K_b = \frac{K_w}{K_a(HCO_3^-)} = \frac{[HCO_3^-][OH^-]}{[CO_3^{2-}]} = \frac{x^2}{5.2 - x} \cong \frac{x^2}{5.2} = 2.1 \times 10^{-4}$$

$x = [OH^-] = \sqrt{(5.2)(2.1 \times 10^{-4})} = 3.3 \times 10^{-2}$

pOH = $-\log(3.3 \times 10^{-2}) = 1.48$

pH = 14.00 − 1.48 = 12.52

Chapter 15

15.1 HCl and NaCl: no; has no significant H^+ acceptor (Cl^- is a very poor base).
KOH and KCl: no; has no H^+ donor.

15.2 $\text{pH} = 7.20 + \log\frac{(0.0025)}{(0.0015)} = 7.20 + \log(1.67)$

$= 7.20 + 0.22 = 7.42$

15.3 $\text{pH} = pK_a + \log\frac{[\text{acetate}]}{[\text{acetic acid}]}$

$4.68 = 4.74 + \log\frac{[\text{acetate}]}{[\text{acetic acid}]}$

$\log\frac{[\text{acetate}]}{[\text{acetic acid}]} = 4.68 - 4.74 = -0.06$

$\frac{[\text{acetate}]}{[\text{acetic acid}]} = 10^{-0.06} = 0.86$

Therefore, [acetate] = 0.86 × [acetic acid].

15.4 (a) $\text{pH} = 6.37 + \log\frac{[0.25]}{[0.10]} = 6.37 + 0.398 = 6.77$

(b) $\text{pH} = 7.20 + \log\frac{[HPO_4^{2-}]}{[H_2PO_4^-]}$

$= 7.20 + \log\frac{(0.25)}{(0.10)} = 7.20 + 0.398 = 7.60$

15.5 Because CO_2 reacts to form an acid, H_2CO_3, the phosphate ion that is the stronger base, HPO_4^{2-}, will be used to counteract excess CO_2.

15.6

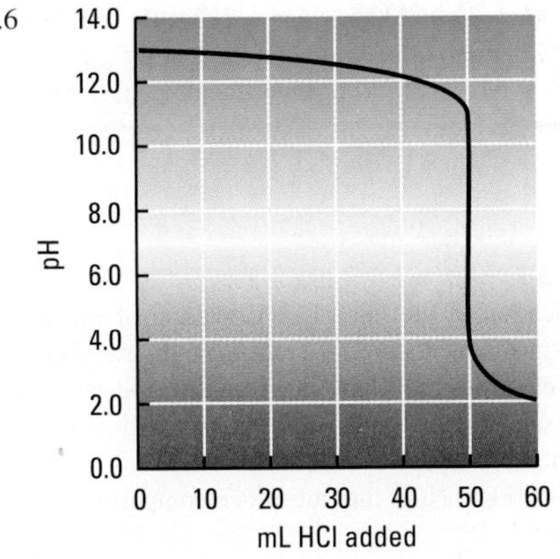
14.0
12.0
10.0
8.0
6.0
4.0
2.0
0.0
pH
0 10 20 30 40 50 60
mL HCl added

15.7 The addition of 30.0 mL of 0.100-M NaOH neutralizes 30.0 mL of 0.100-M acetic acid, forming 0.00300 mol of acetate ions, which is in 80.0 mL of solution. There is (0.0200 L)(0.100 mol/L) = 0.00200 mol acetic acid that is unreacted.

$$K_a = 1.8 \times 10^{-5} = \frac{[H^+][C_2H_3O_2^-]}{[HC_2H_3O_2]} = \frac{[H^+] \times \left(\frac{0.00300 \text{ mol}}{0.0800 \text{ L}}\right)}{\left(\frac{0.00200 \text{ mol}}{0.0800 \text{ L}}\right)}$$

$$1.8 \times 10^{-5} = \frac{[H^+] \times (0.0375)}{(0.025)} = [H^+] \times 1.5$$

$$[H^+] = \frac{1.8 \times 10^{-5}}{1.5} = 1.2 \times 10^{-5}$$

$$pH = -\log(1.2 \times 10^{-5}) = 4.92$$

15.8 As NaOH is added, it reacts with acetic acid to form sodium acetate. After 20.0 mL NaOH has been added, just less than half of the acetic acid has been converted to sodium acetate; when 30.0 mL NaOH has been added, just over half of the acetic acid has been neutralized. Thus, after 20.0 mL and 30.0 mL base have been added, the solution contains approximately equal amounts of acetic acid and acetate ion, its conjugate base, which acts as a buffer.

15.9 Because Reaction (b) occurs to an appreciable extent, CO_3^{2-} is removed from the solution. CO_3^{2-} is a product of reaction (a) and removing it causes reaction (a) to shift right. This causes additional $CaCO_3(s)$ to dissolve.

15.10 The excess iodide would stress the equilibria involving $I^-(aq)$, shifting them to the left. Some AgI and some PbI_2 would precipitate from solution.

Chapter 16

16.1 (a) $H_2O(\ell) \rightarrow H_2O(g)$ Product-favored
(b) $SiO_2(s) \rightarrow Si(s) + O_2(g)$ Reactant-favored
(c) $(C_6H_{10}O_5)_n(s) + 6n\ O_2(g) \rightarrow 6n\ CO_2(g) + 5n\ H_2O(g)$ Product-favored
(d) $C_{12}H_{22}O_{11}(s) \rightarrow C_{12}H_{22}O_{11}(aq)$ Product-favored

16.2 A*** A**B* A*B** B***

If C, D, and E are added, there are many more arrangements in addition to these:

A*B*C*	A*B*D*	A*B*E*	A*C*D*	A*C*E*
A*D*E*	B*C*D*	B*C*E*	B*D*E*	C*D*E*
A**C*	A**D*	A**E*	B**C*	B**D*
B**E*	C**A*	C**B*	C**D*	C**E*
D**A*	D**B*	D**C*	D**E*	E**A*
E**B*	E**C*	E**D*	C***	D***
E***				

There are 35 possible arrangements, but only 4 of them have the energy confined to atoms A and B. The probability that all energy remains with A and B is thus 4/35 = 0.114, or a little more than 11%.

16.3 Using Celsius temperature and $\Delta_r S = q_{rev}/T$, if the temperature were −10 °C, the value of $\Delta_r S$ would be negative, in disagreement with the fact that transfer of energy to a sample should increase molecular motion and, hence, entropy.

16.4 (a) The reactant is a gas. The products are also gases, but the number of molecules has increased, so entropy is greater for products. (Entropy increases.)
(b) The reactant is a solid. The product is a solution. Mixing sodium and chloride ions among water molecules results in greater entropy for the product. (Entropy increases.)
(c) The reactant is a solid. The products are a solid and a gas. The much larger entropy of the gas results in greater entropy for the products. (Entropy increases.)

16.5 (a) Because $\Delta_r S_{surroundings} = -\Delta_r H/T$ at a given temperature, the larger the value of T, the smaller the value of $\Delta_r S_{surroundings}$.
(b) If $\Delta_r S_{system}$ does not change much with temperature, then $S_{universe}$ must also get smaller. In this case, because $\Delta_r S_{system}$ is negative, $\Delta_r S_{universe}$ would become negative at a high enough temperature.

16.6 (a) The reaction would have gaseous water as a product.
(b) Both $\Delta_r H°$ and $\Delta_r S°$ would change. $\Delta_r H° = -10{,}232.197$ kJ and $\Delta_r S° = 756.57$ J K^{-1} mol^{-1}.
(c) If any of the reactants or products change to a different phase (s, ℓ, or g) over the range of temperature, $\Delta_r H°$ and $\Delta_r S°$ will change significantly at the temperature of the phase transition.

16.7

Reaction	$\Delta_r H°$ 298 K (kJ)	$\Delta_r S°$, 298 K (J K^{-1} mol^{-1})
(a)	−1410.94	−267.67
(b)	467.87	558.3
(c)	−393.509	2.862
(d)	617.6	−451.5

Reaction (a) is product-favored at low T (room temperature) and reactant-favored at high T.
Reaction (b) is reactant-favored at low T, product-favored at high T.
Reaction (c) is product-favored at all values of T.
Reaction (d) is reactant-favored at all values of T.

16.8 (a) $\Delta_r S°_{system} = 2 \times S°(HCl(g)) - 1 \times S°(H_2(g)) - 1 \times S°(Cl_2(g))$
$= (2 \times 186.908 - 130.684 - 223.066)$ J K^{-1} mol^{-1}
$= 20.066$ J K^{-1} mol^{-1}

$\Delta_r S°_{surroundings} = -\Delta_r H°/T = -\Delta_r H°/T$
$= -[2 \times (-92.307 \text{ kJ/mol})]/298.15$ K
$= 619.20$ J K^{-1} mol^{-1}

(b) $\Delta_r S°_{universe} = (619.20 + 20.066)$ J K^{-1} mol^{-1}
$= 639.26$ J K^{-1} mol^{-1} Product-favored

16.9

Sign of $\Delta_r H°$	Sign of $\Delta_r S°$	Sign of $\Delta_r G°$	Product-favored?
Negative (exothermic)	Positive	Negative	Yes
Negative (exothermic)	Negative	Depends on T	Yes at low T; no at high T
Positive (endothermic)	Positive	Depends on T	No at low T; yes at high T
Positive (endothermic)	Negative	Positive	No

16.10 (a) At 400 °C the equation is

$$2\ HgO(s) \rightarrow 2\ Hg(g) + O_2(g)$$

Because Hg(g) is a product, instead of Hg(ℓ), both $\Delta_r H°$ and $\Delta_r S°$ will have significantly different values above 356 °C from their values below 356 °C. Therefore, the method of estimating $\Delta_r G°$ would not work above 356 °C.

(b) At 400 °C the entropy change should be more positive, which would make the reaction more product-favored. Because $\Delta_r H°$ and $\Delta_r S°$ are both positive, the reaction is product-favored at high temperatures.

16.11 (a) If the extent of reaction is 0.10, then 0.10 mol of NaOH has reacted with 0.10 mol of CO_2 to produce 0.10 mol of $NaHCO_3$.

$$\begin{aligned}\Delta_r G°(0.10 \text{ extent}) &= 0.10 \times \Delta_f G°(NaHCO_3(s)) - \\ &\quad 0.10 \times \Delta_f G° (NaOH(s)) - \\ &\quad 0.10 \times \Delta_f G° (CO_2(g)) \\ &= -0.10 \times 851.0 \text{ kJ/mol} + \\ &\quad 0.10 \times 379.484 \text{ kJ/mol} + \\ &\quad 0.10 \times 394.359 \text{ kJ/mol} \\ &= -7.72 \text{ kJ/mol}\end{aligned}$$

Similarly,

$$\Delta_r G°(0.40 \text{ extent}) = 0.40[-72.2 \text{ kJ}] = -30.9 \text{ kJ}$$
$$\Delta_r G°(0.80 \text{ extent}) = 0.80[-72.2 \text{ kJ}] = -61.8 \text{ kJ}$$

(b) In each case, $\Delta_r G°(x \text{ extent}) = x\Delta_r G°(\text{full extent})$, which verifies the statement.

(c) Because $\Delta_r G°(x \text{ extent}) = x\Delta_r G°(\text{full extent})$,

$$y = xm + b$$

where $b = 0$ and $m = \Delta_r G°(\text{full extent}) = \text{slope}$.

16.12 (a)
$$\begin{aligned}\Delta_r S°(i) &= \{2 \times (27.28) + \tfrac{3}{2} \times (205.138) - (87.40)\} \text{ J K}^{-1}\text{mol}^{-1} \\ &= 274.87 \text{ J K}^{-1}\text{ mol}^{-1}\end{aligned}$$

$$\Delta_r H°(i) = -\Delta_f H°(Fe_2O_3(s)) = 824.2 \text{ kJ/mol}$$

$$\Delta_r G°(i) = -\Delta_f G°(Fe_2O_3(s)) = 742.2 \text{ kJ/mol}$$

$$\begin{aligned}\Delta_r S°(ii) &= \{50.92 - 2 \times (28.33) - \tfrac{3}{2} \times (205.138)\} \text{ J K}^{-1}\text{mol}^{-1} \\ &= -313.45 \text{ J K}^{-1}\text{ mol}^{-1}\end{aligned}$$

$$\Delta_r H°(ii) = \Delta_f H°(Al_2O_3(s)) = -1675.7 \text{ kJ/mol}$$

$$\Delta_r G°(ii) = \Delta_f G°(Al_2O_3(s)) = -1582.3 \text{ kJ/mol}$$

Step i is reactant-favored. Step ii is product-favored. For step (i) $\Delta_r S°$ is positive but $\Delta_r H°$ and $\Delta_r G°$ are both negative; for step (ii) all signs are opposite from step (i).

(b) Net reaction

$$Fe_2O_3(s) + 2\,Al(s) \rightarrow 2\,Fe(s) + Al_2O_3(s)$$

$$\begin{aligned}\Delta_r S° &= 274.87 \text{ J K}^{-1}\text{mol}^{-1} + (-313.45 \text{ J K}^{-1}\text{mol}^{-1}) \\ &= -38.6 \text{ J K}^{-1}\text{mol}^{-1}\end{aligned}$$

$$\begin{aligned}\Delta_r H° &= 824.2 \text{ J K}^{-1}\text{mol}^{-1} + (-1675.7 \text{ J K}^{-1}\text{mol}^{-1}) \\ &= -851.5 \text{ J K}^{-1}\text{mol}^{-1}\end{aligned}$$

$$\begin{aligned}\Delta_r G° &= 742.2 \text{ J K}^{-1}\text{mol}^{-1} + (-1582.3 \text{ J K}^{-1}\text{mol}^{-1}) \\ &= -840.1 \text{ J K}^{-1}\text{mol}^{-1}\end{aligned}$$

The net reaction has negative $\Delta_r G°$ and is therefore product-favored. For the *net* reaction, $\Delta_r S°$, $\Delta_r H°$, and $\Delta_r G°$ are all negative.

(c) If the two reactions are coupled, it is possible to obtain iron from iron(III) oxide even though that reaction is not product-favored by itself. The large negative $\Delta_r G°$ for formation of $Al_2O_3(s)$ makes the overall $\Delta_r G°$ negative for the coupled reactions.

(d) $Mg(s) + \tfrac{1}{2}\,O_2(g) \rightarrow MgO(s)$

$$\Delta_r G° = \Delta_f G°(MgO(s)) = -569.43 \text{ kJ}$$

Coupling the reactions, we have

$$Fe_2O_3(s) \rightarrow 2\,Fe(s) + \tfrac{3}{2}O_2(g) \quad \Delta_r G_1° = 742.2 \text{ kJ}$$

$$3 \times [Mg(s) + \tfrac{1}{2}\,O_2(g) \rightarrow MgO(s)]$$
$$\Delta_r G_2° = 3(-569.43) \text{ kJ} = -1708.29 \text{ kJ}$$

$$Fe_2O_3(s) + 3\,Mg(s) \rightarrow 2\,Fe(s) + 3\,MgO(s)$$
$$\Delta_r G_3° = -966.1 \text{ kJ}$$

16.13 $\Delta_r G° = -2870 \text{ kJ} + 32 \times (30.5 \text{ kJ}) = -1894 \text{ kJ}$. The 1894 kJ of Gibbs free energy is transformed into thermal energy.

16.14 64,500 g ATP/50 g ATP = 1290 times each ADP must be recycled to ATP on average each day.

Chapter 17

17.1 (a) 0 (b) +5 (c) −3 (d) −3

17.2 This is an application of the law of conservation of matter. If the number of electrons gained were different from the number of electrons lost, some electrons must have been created or destroyed.

17.3 Removal of the salt bridge would effectively switch off the flow of electricity from the battery.

17.4 Fe^{2+} into the salt bridge; NO_3^- into the anode compartment; 1.5 mol NO_3^-; 0.75 mol Fe^{2+}.

17.5 Avogadro's number of electrons is 96,485 coulombs of charge, so it is 96,485 times as large as one coulomb of charge.

17.6 Notice that the cell potential for the zinc-silver cell is the sum of the cell potentials for the zinc-copper and copper-silver cells. This suggests that the copper half cell might have the same potential in each case and that a cell potential could be calculated as a difference of half-cell potentials.

17.7 The zinc anode could be weighed before the battery was put into use. After a period of time, the zinc anode could be dried and reweighed. A loss in mass would be interpreted as being caused by the loss of Zn atoms from the surface through oxidation.

17.8 No, because Hg^{2+} ions can oxidize Al metal to Al^{3+} ions. The net cell reaction is

$$2\,Al(s) + 3\,Hg^{2+}(aq) \rightarrow 2\,Al^{3+}(aq) + 3\,Hg(\ell)$$
$$E_{cell} = +2.51 \text{ V}$$

17.9 For this table,
(a) V^{2+} ion is the weakest oxidizing agent.
(b) Cl_2 is the strongest oxidizing agent.
(c) V is the strongest reducing agent.
(d) Cl^- is the weakest reducing agent.
(e) No, E_{cell} for that reaction would be <0.
(f) No, E_{cell} for that reaction would be <0.
(g) Pb can reduce I_2 and Cl_2.

17.10 In Table 17.1, Sb would be above H_2, and Pb would be below H_2. For Sb, the electrode potential would be between 0.00 and 0.340 V; for Pb, the value would be between 0.00 and −0.1375 V.

17.11 Electrical work = (quantity of charge) × (electrical potential difference)

$$\text{Charge} = 650 \text{ mA} \times \frac{1 \text{ A}}{1000 \text{ mA}} \times \frac{1 \text{ C/s}}{1 \text{ A}} \times 36 \text{ h} \times \frac{3600 \text{ s}}{1 \text{ h}} = 8.42 \times 10^4 \text{ C}$$

$$\text{Pot. Diff.} = \frac{3.0 \times 10^5 \text{ J}}{8.42 \times 10^4 \text{ C}} \times \frac{1 \text{ V} \times 1 \text{ C}}{1 \text{ J}} = 3.6 \text{ V}$$

17.12 $E_{cell} = 2.041 \text{ V} - (0.0128 \text{ V})\ln\left(\dfrac{1}{(0.500)^2(0.500)^2}\right) = 2.006 \text{ V}$

The cell potential for 0.100 M sulfuric acid can be calculated similarly; $E_{cell} = 1.923$ V. The cell potential decreases by 0.083 V.

17.13 During charging, the reactions at each electrode are reversed. At the electrode that is the anode during discharge, the charging reaction is

$$Cd(OH)_2(s) + 2\,e^- \rightarrow Cd(s) + 2\,OH^-(aq)$$

This is reduction, so this electrode is now a cathode.
At the electrode that is the cathode during discharge, the charging reaction is

$$Ni(OH)_2 + OH^-(aq) \rightarrow NiO(OH)(s) + H_2O(\ell) + e^-$$

This is oxidation, so this electrode is now an anode.

17.14 Remove the lead cathodes and as much sulfuric acid as you can from the discharged battery. Use some wire from the chain-link fence and construct a battery with Cl_2 gas flowing across a piece of iron wire. The two half-reactions would be

$$Cl_2(g) + 2\,e^- \rightarrow 2\,Cl^-(aq) \qquad E° = 1.36\text{ V}$$
$$Pb(s) + SO_4^{2-}(aq) \rightarrow PbSO_4(s) + 2\,e^- \qquad E° = -0.356\text{ V}$$
$$E_{cell} = 1.36 - (-0.356) = 1.71\text{ V}$$

17.15 Potassium metal was produced at the cathode.
Oxidation reaction: $2\,F^-(\text{molten}) \rightarrow F_2(g) + 2\,e^-$
Reduction reaction: $2[K^+(\text{molten}) + e^- \rightarrow K(\ell)]$
Net cell reaction:
$2\,K^+(\text{molten}) + 2\,F^-(\text{molten}) \rightarrow 2\,K(\ell) + F_2(g)$

17.16 Reaction (c) making 2 mol of Cu from Cu^{2+} would require 4 Faradays of electricity. Two F are required for part (b), and 3 F are required for part (a).

17.17 First, calculate how many coulombs of electricity are required to make this much aluminum.

$$(2000.\text{ ton Al})\left(\frac{1000.\text{ kg Al}}{1\text{ ton Al}}\right)\left(\frac{10^3\text{ g}}{1\text{ kg}}\right)\left(\frac{1\text{ mol Al}}{26.982\text{ g Al}}\right)\times$$
$$\left(\frac{3\text{ mol e}^-}{1\text{ mol Al}}\right)\left(\frac{96{,}485\text{ C}}{1\text{ mol e}^-}\right) = 2.146\times10^{13}\text{ C}$$

Next, using the product of charge and voltage, calculate how many joules are required; then convert to kilowatt-hours.

$$\text{Energy} = (2.146\times10^{13}\text{ C})(4.0\text{ V})\left(\frac{1\text{ J}}{(1\text{ C})(1\text{ V})}\right)\times$$
$$\left(\frac{1\text{ kWh}}{3.60\times10^6\text{ J}}\right) = 2.4\times10^7\text{ kWh}$$

17.18 To calculate how much energy is stored in a battery, you need the voltage and the number of coulombs of charge the battery can provide. The voltage is usually given on the battery label. To determine the number of coulombs available, you would have to disassemble the battery and determine the masses of the chemicals at the cathode and anode.

17.19 $$m_{Cu} = 12.1\text{ min}\times\frac{60\text{ s}}{1\text{ min}}\times\frac{100.\text{ C}}{1\text{ s}}\times\frac{1\text{ mol e}^-}{96{,}500\text{ C}}\times\frac{1\text{ mol Cu}}{2\text{ mol e}^-}\times$$
$$\frac{63.55\text{ g Cu}}{1\text{ mol Cu}} = 23.9\text{ g Cu}$$

17.20 No, not all metals corrode as easily. Three metals that would corrode about as readily as Fe and Al are Zn, Mg, and Cd. Three metals that do not corrode as readily as Fe and Al are Cu, Ag, and Au. These three metals are used in making coins and jewelry. Metals fall into these two broad groups because of their relative ease of oxidation compared with the oxidation of H_2. In Table 17.1, you can see this breakdown easily.

17.21 (b) > (a) > (d) > (c)
Sand by the seashore, (b), would contain both moisture and salts, which would aid corrosion. Moist clay, (a), would contain water but less dissolved salts. If an iron object were embedded within the clay, the clay's impervious nature might prevent oxygen from getting to the iron, which would also lower the rate of corrosion. Desert sand in Arizona, (d), would be quite dry, and this low-moisture environment would not lead to a rapid rate of corrosion. On the moon, (c), there would be a lack of moisture and oxygen. This would lead to a very low rate of corrosion.

Chapter 18

18.1 ${}^{235}_{92}\text{U} \rightarrow {}^{4}_{2}\text{He} + {}^{231}_{90}\text{Th}$ ${}^{231}_{90}\text{Th} \rightarrow {}^{0}_{-1}\text{e} + {}^{231}_{91}\text{Pa}$
${}^{231}_{91}\text{Pa} \rightarrow {}^{4}_{2}\text{He} + {}^{227}_{89}\text{Ac}$ ${}^{227}_{89}\text{Ac} \rightarrow {}^{0}_{-1}\text{e} + {}^{227}_{90}\text{Th}$
${}^{227}_{90}\text{Th} \rightarrow {}^{4}_{2}\text{He} + {}^{223}_{88}\text{Ra}$

18.2 ${}^{26}_{13}\text{Al} \rightarrow {}^{0}_{+1}\text{e} + {}^{26}_{12}\text{Mg}$ ${}^{26}_{13}\text{Al} + {}^{0}_{-1}\text{e} \rightarrow {}^{26}_{12}\text{Mg}$

18.3 Mass difference = $\Delta m = -0.032700$ g/mol
$\Delta E = (-3.2700\times10^{-5}\text{ kg/mol})(2.998\times10^{8}\text{ m/s})^2$
$= -2.9390\times10^{12}\text{ J/mol} = -2.9390\times10^{9}\text{ kJ/mol}$
E_b per nucleon $= 5.150\times10^{8}$ kJ/nucleon
E_b for ^{6}Li is smaller than E_b for ^{4}He; therefore, helium-4 is more stable than lithium-6.

18.4 From the graph it can be seen that the binding energy per nucleon increases more sharply for the fusion of lighter elements than it does for heavy elements undergoing fission. Therefore, fusion is more exothermic per gram than fission.

18.5 $(\frac{1}{2})^{10} = 9.8\times10^{-4}$; this is equivalent to 0.098% of the radioisotope remaining.

18.6 All the lead came from the decay of ^{238}U; therefore, at the time the rock was dated, $N = 100$ and $N_0 = 109 + 100 = 209$. The decay constant, k, can be determined:

$$k = \frac{0.693}{4.46\times10^9\text{ y}} = 1.55\times10^{-10}\text{ y}^{-1}$$

The age of the rock (t) can be calculated using Equation 18.4:

$$\ln\frac{100}{209} = -(1.55\times10^{-10}\text{ y}^{-1})\times t$$
$$t = 4.76\times10^{9}\text{ y}$$

18.7 Ethylene is derived from petroleum, which was formed millennia ago. The half-life of ^{14}C is 5730 y, and thus much of ethylene's ^{14}C would have decayed and would be much less than that of the ^{14}C alcohol produced by fermentation.

18.8 (a) ${}^{16}_{8}\text{O} + {}^{1}_{1}\text{H} \rightarrow {}^{4}_{2}\text{He} + {}^{13}_{7}\text{N}$
(b) ${}^{64}_{28}\text{Ni} + {}^{209}_{83}\text{Bi} \rightarrow {}^{1}_{0}\text{n} + {}^{272}_{111}\text{Rg}$
(c) ${}^{27}_{13}\text{Al} + {}^{4}_{2}\text{He} \rightarrow {}^{1}_{0}\text{n} + {}^{30}_{15}\text{P}$

18.9 ${}^{238}_{92}\text{U} + {}^{48}_{20}\text{Ca} \rightarrow {}^{283}_{112}\text{Cn} + 3\,{}^{1}_{0}\text{n}$

18.10 Burning a metric ton of coal produces 2.8×10^{7} kJ of energy.

$$\left(\frac{2.8\times10^4\text{ kJ}}{1.0\text{ kg}}\right)\left(\frac{10^3\text{ kg}}{\text{metric ton}}\right) = 2.8\times10^7\text{ kJ of energy}$$

The fission of 1.0 kg of ^{235}U produces

$$\frac{2.1\times10^{10}\text{ kJ}}{0.235\text{ kg }{}^{235}\text{U}} = 8.93\times10^{10}\text{ kJ}$$

It would require burning 3.2×10^{3} metric tons of coal to equal the amount of energy from 1.0 kg of ^{235}U:

$$8.93\times10^{10}\text{ kJ from }{}^{235}\text{U}\times\frac{1\text{ metric ton coal}}{2.8\times10^7\text{ kJ}} = 3.2\times10^3\text{metric tons}$$

18.11 $$k = \frac{0.693}{28.8\text{ yr}} = 2.406\times10^{-2}\text{ yr}$$

$$\ln(\text{fraction}) = \ln\left(\frac{N}{N_0}\right) = -kt = -(2.406\times10^{-2}\text{ yr})(27\text{ yr})$$
$$= -0.6496$$
$$\text{fraction} = e^{-0.6496} = 0.522 = 52\%$$

18.12 (a) ${}^{7}_{3}\text{Li} + {}^{1}_{1}\text{H} \rightarrow {}^{1}_{0}\text{n} + {}^{7}_{4}\text{Be}$
(b) ${}^{2}_{1}\text{H} + {}^{3}_{2}\text{He} \rightarrow {}^{4}_{2}\text{He} + {}^{1}_{1}\text{H}$

18.13 $k = (0.693)/(1.26\times10^{9}\text{ yr}) = 5.50\times10^{-10}\text{ yr}^{-1}$
(500. mg K-40) (0.000117) = 5.85×10^{-2} mg = 5.85×10^{-5} g K-40
(5.85×10^{-5} g K-40)(6.022×10^{23} K-40 atoms/39.96 g K-40) = 8.82×10^{17} K-40 atoms
Disintegration rate = $A = kN$
= ($5.50\times10^{-10}\text{ yr}^{-1}$)($8.82\times10^{17}$ K-40 atoms)
= 4.85×10^{8} K-40 atoms/yr
(4.85×10^{8} K-40 atoms/yr)(1 yr/3.15×10^{7} s) = 15 disintegrations/s = 15 Bq. The smoke detector is much more radioactive.

18.14 $k = 0.693/3.82\text{ d} = 0.181\text{ d}^{-1}$
(a) $\ln(4/16) = \ln(0.25) = -1.39 = -(0.181\text{ d}^{-1})(t);\ t = 7.7\text{ d}$
(b) $\ln(1.5/16) = \ln(0.0938) = -2.37 = -(0.181\text{ d}^{-1})(t);\ t = 13\text{ d}$

18.15 Iron-59 $k = \dfrac{0.693}{44.5\text{ d}} = 1.557 \times 10^{-2}\text{ d}^{-1}$

Chromium-51 $k = \dfrac{0.693}{27.7\text{ d}} = 0.02502\text{ d}^{-1}$

Fractions remaining:

^{59}Fe: $\ln(\text{fraction}) = -(1.557 \times 10^{-2}\text{ d}^{-1}) \times 90\text{ d}$

$\text{fraction} = e^{-1.401} = 0.246;\ 80\text{ mg} \times 0.246 = 19.7\text{ mg left}$

^{51}Cr: $\ln(\text{fraction}) = -(0.02502\text{ d}^{-1}) \times 90\text{ d};\ 10.5\text{ mg left}$

Alternatively, consider the fact that 90 days is approximately two half-lives of ^{59}Fe. Therefore, approximately $\frac{3}{4}$ of it (about 60 mg) has decayed after 90 days, and about 20 mg remains. In that same time, ^{51}Cr has undergone more than three half-lives, so that less than $\frac{1}{8}$ remains (less than 12.5 mg).

Chapter 19

19.1 $^{12}_{6}C + ^{4}_{2}He \rightarrow ^{16}_{8}O$
$^{16}_{8}O + ^{4}_{2}He \rightarrow ^{20}_{10}Ne$
$^{20}_{10}Ne + ^{4}_{2}He \rightarrow ^{24}_{12}Mg$

19.2 $^{130}_{48}Cd \rightarrow ^{0}_{-1}e + ^{130}_{49}In$
$^{130}_{49}In \rightarrow ^{0}_{-1}e + ^{130}_{50}Sn$
$^{130}_{50}Sn \rightarrow ^{0}_{-1}e + ^{130}_{51}Sb$
$^{130}_{51}Sb \rightarrow ^{0}_{-1}e + ^{130}_{52}Te$

19.3 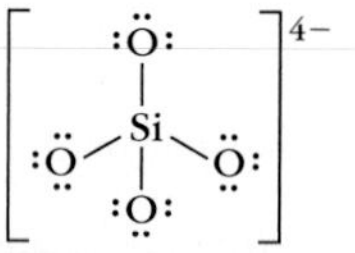

:Ö:
Si
:Ö
Ö:
:Ö:
4−

The ion is tetrahedral.
Formal charges: Si = 0; each oxygen is −1 for an overall formal charge of −4, the net charge on the ion is 4−.

19.4 Two of the four oxygens in an SiO_4 unit are shared with other SiO_4 tetrahedra. Therefore, for each SiO_4 unit,

1 Si + 2 oxygen not shared + (2 oxygen shared) × $\frac{1}{2}$ gives $SiO_{2+1} = SiO_3^{2-}$

19.5 The "molar mass" of air = (28.0)(0.78) + (32.0)(0.21) + (40.0)(0.010) = 29 g/mol

$$n_{air} = 5.0\text{ L} \times \frac{0.87\text{ g}}{1\text{ mL}} \times \frac{1000\text{ mL}}{1\text{ L}} \times \frac{1\text{ mol}}{29\text{ g}} = 1.5 \times 10^2\text{ mol}$$

$$V_{air} = \frac{nRT}{P} = \frac{(1.5 \times 10^2\text{ mol})(0.0821\text{ L atm mol}^{-1}\text{ K}^{-1})(273\text{ K})}{1.0\text{ atm}}$$

$$= 3.4 \times 10^3\text{ L}$$

19.6 $2\text{ NaCl}(\ell) \rightarrow 2\text{ Na}(\ell) + Cl_2(g)$

Charge = current × time

$$= 2.0 \times 10^4\text{ A} \times 24\text{ h} \times \frac{3600\text{ s}}{\text{h}} = 1.7 \times 10^9\text{ C}$$

$$1.7 \times 10^9\text{ C} \times \frac{1\text{ mol e}^-}{9.65 \times 10^4\text{ C}} = 1.8 \times 10^4\text{ mol e}^-$$

$$1.8 \times 10^4\text{ mol e}^- \times \frac{1\text{ mol Na}}{1\text{ mol e}^-} \times \frac{23.0\text{ g Na}}{1\text{ mol Na}} \times \frac{1\text{ lb}}{454\text{ g}} \times \frac{1\text{ ton}}{2000\text{ lb}} = 0.45\text{ tons Na}$$

$$1.8 \times 10^4\text{ mol e}^- \times \frac{1\text{ mol Cl}_2}{2\text{ mol e}^-} \times \frac{70.9\text{ g Cl}_2}{1\text{ mol Cl}_2} \times \frac{1\text{ lb}}{454\text{ g}} \times \frac{1\text{ ton}}{2000\text{ lb}} = 0.70\text{ tons Cl}_2$$

19.7 $\text{Charge} = 2.00 \times 10^4\text{ A} \times 100.\text{ h} \times \dfrac{3600\text{ s}}{\text{h}} = 7.20 \times 10^9\text{ C}$

$$m_{NaOH} = 7.20 \times 10^9\text{ C} \times \frac{1\text{ mol e}^-}{9.65 \times 10^4\text{ C}} \times \frac{1\text{ mol NaOH}}{1\text{ mol e}^-} \times \frac{40.00\text{ g NaOH}}{1\text{ mol NaOH}} \times \frac{1\text{ lb}}{454\text{ g}} \times \frac{1\text{ ton}}{2000\text{ lb}}$$

$$= 3.29\text{ tons NaOH}$$

19.8 Phosphorus in $Ca_3(PO_4)_2$ has an oxidation state of +5; it is reduced to 0 in P_4. Carbon is oxidized to CO (oxidation state changes from 0 to −2).

19.9 $\dfrac{93\text{ g}}{502\text{ g}} \times 100\% = 19\%\text{ P}$

19.10 Br^- (reducing agent) is oxidized to Br_2; Cl_2 (oxidizing agent) is reduced to Cl^-.

19.11 $I_2(s) + 2\text{ e}^- \rightarrow 2\text{ I}^-(aq)\quad E° = 0.535\text{ V}$
$Br_2(\ell) + 2\text{ e}^- \rightarrow 2\text{ Br}^-(aq)\quad E° = 1.066\text{ V}$
$I_2(s) + 2\text{ Br}^-(aq) \rightarrow 2\text{ I}^-(aq) + Br_2(\ell)$
$E°_{cell} = E°_{cathode} - E°_{anode} = (0.535 - 1.066)\text{ V} = -0.531\text{ V}$

$I_2(s)$ is below $Br^-(aq)$ so iodine will not oxidize bromide ion to bromine, as seen by the negative cell voltage. Rather, bromine will oxidize iodide to iodine.

19.12 Oxide, O^{2-}, [:Ö:]$^{2-}$; peroxide, O_2^{2-}, [:Ö:Ö:]$^{2-}$; superoxide, O_2^-, [:Ö:Ö·]$^-$

MO diagrams:

Peroxide ion, O_2^{2-}: $\sigma_{2s}(\uparrow\downarrow)\ \sigma^*_{2s}(\uparrow\downarrow)\ \pi_{2p}(\uparrow\downarrow)\ \pi_{2p}(\uparrow\downarrow)\ \sigma_{2p}(\uparrow\downarrow)$
$\pi^*_{2p}(\uparrow\downarrow)\ \pi^*_{2p}(\uparrow\downarrow)$

Superoxide ion, O_2^-: $\sigma_{2s}(\uparrow\downarrow)\ \sigma^*_{2s}(\uparrow\downarrow)\ \pi_{2p}(\uparrow\downarrow)\ \pi_{2p}(\uparrow\downarrow)\ \sigma_{2p}(\uparrow\downarrow)$
$\pi^*_{2p}(\uparrow\downarrow)\ \pi^*_{2p}(\uparrow)$

19.13 (a) CaO (b) BaO_2 (c) Sr_3N_2 (d) CaC_2

19.14 (a) Mg_3N_2 (b) $3\text{ Mg}(s) + N_2(g) \rightarrow Mg_3N_2(s)$

19.15 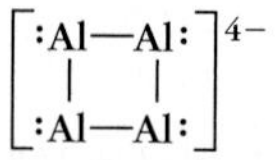

:Al—Al:
:Al—Al:
4−

19.16 NO_3^-; oxidation number of oxygen is −2; oxidation number of N is +5. Nitrogen is in group 5A so its maximum oxidation number should be +5.
NH_4^+; oxidation number of hydrogen is +1; oxidation number of nitrogen is −3. N can gain three electrons for a noble-gas configuration so the minimum oxidation number should be −3.

19.17 $n(N_2) = 150.\text{ g NaN}_3 \times \dfrac{1\text{ mol NaN}_3}{65.0\text{ g NaN}_3} \times \dfrac{3\text{ mol N}_2}{2\text{ mol NaN}_3} = 3.46\text{ mol N}_2$

$$V = \frac{nRT}{P} = \frac{(3.46)(0.0821)(273)}{(1)}\text{ L} = 77.6\text{ L}$$

Chapter 20

20.1 (a) V^{+5} (b) Ag^{+3} (c) Ti^{+4} (d) Mn^{+2}

20.2 Cu^{+2}

20.3 (a) $500.\text{ g} \times \dfrac{18\text{ g W}}{100\text{ g}} = 90.0\text{ g W}$

$$500.\text{ g} \times \frac{6.0\text{ g Co}}{100\text{ g}} = 30.0\text{ g Co}$$

(b) Iron is present in the greatest mole fraction.

$$500.\ \text{g} \times \frac{68.9\ \text{g Fe}}{100\ \text{g}} \times \frac{1\ \text{mol Fe}}{55.845\ \text{g Fe}} = 6.17\ \text{mol}$$

The amount of each other metal can be calculated similarly, giving 0.25 mol C, 0.38 mol Cr, 0.49 mol W, 0.052 mol Mo, 0.15 mol V, and 0.51 mol Co. The total amount of metal is 8.00 mol.

$$\frac{6.17}{8.00} = 0.771\ \text{mole fraction Fe}$$

20.4 $Cu^{2+}(aq) + 2\ e^- \rightarrow Cu(s)$

$$m_{Cu} = 12.0\ \text{h}\left(\frac{3600\ \text{s}}{\text{h}}\right)\left(\frac{250\ \text{C}}{\text{s}}\right)\left(\frac{1\ \text{mol e}^-}{9.65 \times 10^4\ \text{C}}\right) \times$$

$$\left(\frac{1\ \text{mol Cu}}{2\ \text{mol e}^-}\right)\left(\frac{63.55\ \text{g Cu}}{1\ \text{mol Cu}}\right) = 3.56 \times 10^3\ \text{g Cu}$$

20.5 (a) NO_3^- is the oxidizing agent; silver metal is the reducing agent.
(b) NO_3^- is the oxidizing agent; gold is the reducing agent.

20.6 $2\ Al(s) + Cr_2O_3(s) \rightarrow Al_2O_3(s) + 2\ Cr(s)$

$$\begin{aligned}\Delta_f H^\circ &= \Delta_f H^\circ(Al_2O_3(s)) - \Delta_f H^\circ(Cr_2O_3(s))\\ &= (-1675.7\ \text{kJ}) - (-1139.7\ \text{kJ})\\ &= -536.0\ \text{kJ/mol}\ Cr_2O_3\end{aligned}$$

$$\begin{aligned}\Delta G^\circ &= \Delta_f G^\circ(Al_2O_3(s)) - \Delta_f G^\circ(Cr_2O_3(s))\\ &= (-1582.3\ \text{kJ}) - (-1058.1\ \text{kJ})\\ &= -524.2\ \text{kJ/mol}\ Cr_2O_3\end{aligned}$$

20.7 The addition of acid (H^+, a reactant) shifts the equilibrium to the right, converting CrO_4^{2-} to $Cr_2O_7^{2-}$. Added base reacts with H^+ to form water, causing the equilibrium to shift to the left, converting $Cr_2O_7^{2-}$ to CrO_4^{2-}.

20.8 $[Cr(H_2O)_2(NH_3)_2(OH)_2]^+$

20.9 Should be $M_2^+X^{2-}$; such as K_2SO_4

20.10 (a) Three (b) 4− (c) Five

20.11 (a) Fe^{2+} (b) Fe^{3+}

20.12 Two isomers

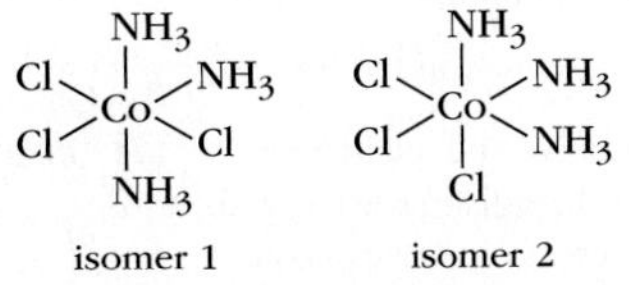
NH3
Cl
Co
NH3
Cl
Cl
NH3
isomer 1
NH3
Cl
Co
NH3
Cl
NH3
Cl
isomer 2

20.13 Ni^{2+} ions have a $3d^8$ configuration, which has three pairs of electrons in the t_2 orbitals and an unpaired electron in each of the two e orbitals. No other electron configuration is possible, so high- and low-spin Ni^{2+} complex ions are not formed.

20.14 The complementary color of blue is yellow-orange (Figure 20.21) so the complex absorbs longer wavelengths of visible light. Also, H_2O is a weaker-field ligand in the spectrochemical series. Both indicate a smaller energy difference, Δ_o, between t_2 and e orbitals. This smaller Δ_o allows all electrons to remain unpaired so the complex is high spin.

APPENDIX M Answers to Selected Questions for Review and Thought

Chapter 1

9. (a) Qualitative (b) Quantitative and qualitative
   (c) Quantitative and qualitative (d) Qualitative
11. Bromine is a reddish brown liquid.
    Sulfur is a chalky yellowish solid.
    No property is in common. Physical phase, shape, color, and appearance are all different.
13. The solid will melt because your body temperature of 37 °C is above the melting point of 29.76 °C.
15. 0.00283 kg; Ca and F
17. No, 23.4 mi/hr $<$ 25 mi/hr
19. (a) Four (b) Three (c) Four
    (d) Four (e) Three (f) Four
21. (a) 1.9 g/mL (b) 291.2 cm^3
    (c) 0.0217 (d) 5.21×10^{-5}
23. Copper
25. Aluminum
27. 3.9×10^3 g
29. (a) Physical (b) Chemical
    (c) Chemical (d) Physical
31. (a) Chemical (b) Chemical (c) Physical
33. (a) Forcing a chemical reaction to occur
    (b) Causing work to be done
    (c) Causing work to be done
    (d) Forcing a chemical reaction to occur
35. Heterogeneous; use a magnet
37. (a) Evaporate the water.
    (b) Use a magnet.
    (c) Dissolve the sucrose, separate the solution from the solid magnesium, then evaporate the water.
39. (a) A compound that decomposed
    (b) A compound that decomposed
41. (a) Heterogeneous mixture
    (b) Pure compound
    (c) Element
    (d) Homogeneous mixture
43. (a) No; the substance reacted with H_2 and formed two other substances one of which is water, so it must have contained O and the red-orange substance
    (b) Maybe; the black powder might have contained three elements, two of which remain in the red-orange substance, so you cannot tell
45. Nanoscale
47. Carbon dioxide molecules crowded within the unopened can escape through the opening.
49. A large quantity of heat energy must be transferred to the large stable molecule to increase the atomic motion enough to break bonds that can allow formation of caramelized products.
51. All atoms in the reactants must be accounted for in the product, so there would be no change in the collective mass.
53. If two compounds contain the same elements, and samples of those two compounds both contain the same number of atoms of one element, then the ratio of the atoms of the other element will be a small whole number.
55.

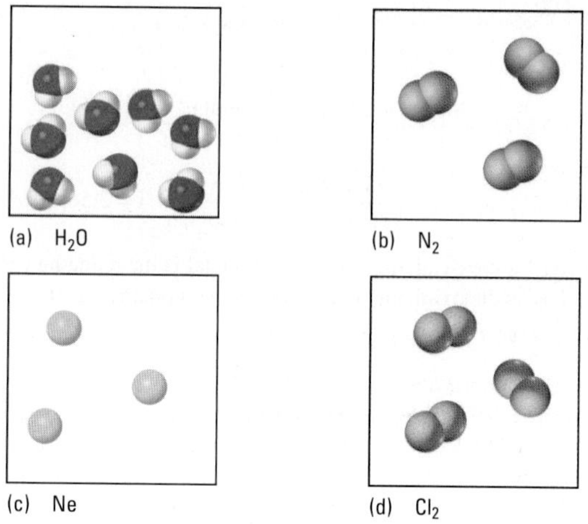
(a) H_2O
(b) N_2
(c) Ne
(d) Cl_2

57. $S(s) + O_2(g) \longrightarrow SO_2(g)$

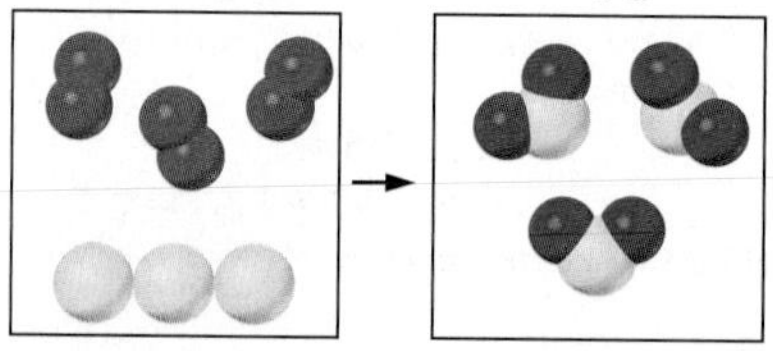

A yellow solid combines with a colorless, odorless gas to form a gas with a choking odor.

59. $I_2(s) \longrightarrow I_2(g)$

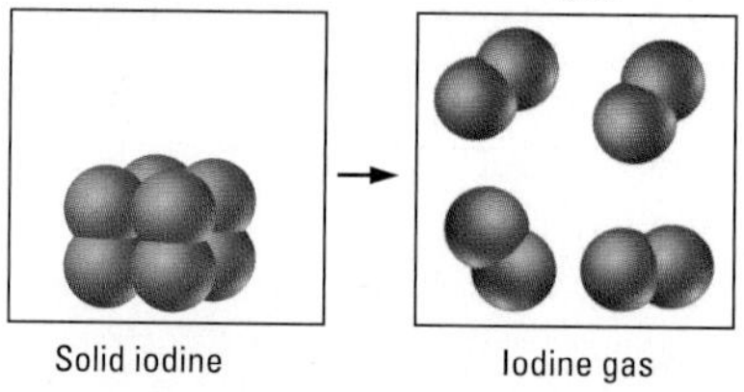
Solid iodine
Iodine gas

A dark colored solid changes to a violet colored gas.

61. (a) Iron, Fe; gold, Au (b) Hydrogen, H; oxygen, O
    (c) Boron, B; silicon, Si (d) Beryllium (Be); magnesium (Mg)
    These are examples; many other answers are also correct.
65. Six Group 4A elements; nonmetal: carbon (C), metalloids: silicon (Si) and germanium (Ge), and metals: tin (Sn), lead (Pb), and flerovium (Fl).
67. (a) I (b) In (c) Ir (d) Fe
69. (a)* Mg (b)* Na (c) C
    (d) S (e) I (f) Mg
    (g) Kr (h) O (i)* Ge
    *Other answers exist for (a), (b), and (i).
75. (a) Mass is a quantitative physical property.
    Colors are qualitative physical properties.
    The reaction with a dye is a qualitative chemical property.
    (b) Mass is a quantitative physical property. Identities of reactants are qualitative physical properties. The chemical reaction occurring is a qualitative chemical property.
77. Garden needs 3.0 ft^3, more than 1.45 ft^3 bag. The density of mulch in the compressed bale is larger than the density when scattered on the lawn so the volume is larger when scattered.

79. 0.197 nm; 197 pm
81. Atoms of calcium are smaller than the atoms of potassium.
83. 508 m
85. Teflon is at the bottom in liquid perfluorohexane. Polyvinyl chloride floats on the perfluorohexane. The next layer up is liquid water with pieces of HDPE plastic floating on it. The liquid hexane is the top layer.
88. (a) K (b) Ar (c) Cu (d) Ge (e) H (f) Ca (g) Br (h) P
89. (a) Lavender area (b) Gray and blue areas (c) Orange and lavender areas
91. Se and S are from the same periodic group.
93. To prove that something is an element requires many tests that all have negative results.
95. (a) Nickel, lead, and magnesium (b) Titanium
96. 6.02×10^{-29} m^3
100. (a) Bromobenzene sample (b) Gold sample (c) Lead sample
102. (a) 2.7×10^2 mL ice (b) Deformed, overflowing, or broken
104. (a) Water (b) Ethanol will float on the water (c) Ethanol and water dissolve in each other and float on bromobenzene layer
106. Drawing (b)
111. 0.7527 g Ag, 0.2473 g Cl, 0.8854 g I
113. (a) 3×10^{24} molecules (b) fraction $= 2 \times 10^{-20}$ (c) 300 molecules
115. 7.056 g/cm^3 $\neq$ 7.917 g/cm^3, different densities, so they are not the same metal
119. (a) Copper (b) 120 mL
122. (a) 32.1 g sulfur (b) 29.8 g zinc sulfide
124. No, the samples contain variable percentages of iron.
126. 3.1 L

Chapter 2

7.

Name	Electric Charge (C)	Mass (g)	Deflected by Electric Field?
Proton	—	—	Yes
—	3.2044×10^{-19}	6.6447×10^{-24}	Yes
Electron	—	9.1094×10^{-28}	Yes

9. 40,000 cm
11. (a) $^{67}_{34}$Se (b) $^{72}_{36}$Kr (c) $^{72}_{36}$Kr (d) $^{67}_{34}$Se
14. Ions
16.

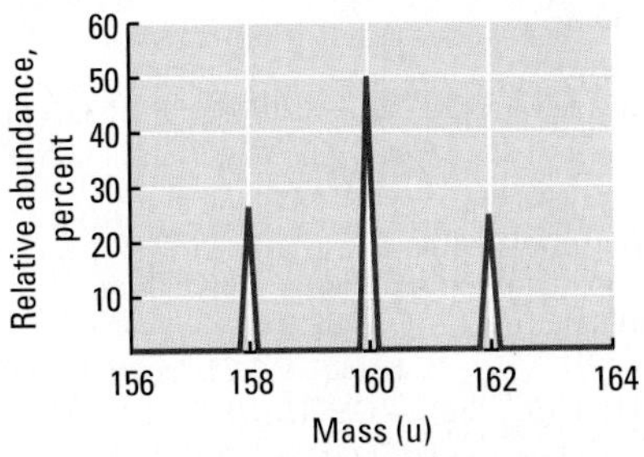

60
50
40
30
20
10
Relative abundance, percent
156
158
160
162
164
Mass (u)

18. The number of neutrons is different by three.
20. 27 electrons, 27 protons, and 33 neutrons
22. 78.92 u
24. (a) $^{23}_{11}$Na (b) $^{39}_{18}$Ar (c) $^{69}_{31}$Ga
26. $(0.07500 \times 6.015121$ u$) + (0.9250 \times 7.016003$ u$) = 6.941$ u
28. 60.12% ^{69}Ga, 39.88% ^{71}Ga
30. (a) Li^+ (b) Sr^{2+} (c) Al^{3+} (d) Zn^{2+}
32. (a) 2+ (b) 3− (c) 2+ or 3+ (d) 2−
34. CoO, Co_2O_3
36. Choices (c) and (d) are correct. (a) $AlCl_3$, (b) NaF
38. Choices (b), (c), and (e) are ionic, because the compounds contain metals and nonmetals.
40. $BaSO_4$, barium ion, 2+, sulfate, 2−; $Mg(NO_3)_2$, magnesium ion, 2+, nitrate, 1−; $NaCH_3COO$, sodium ion, 1+, acetate, 1−
42. (a) $Ni(NO_3)_2$ (b) $NaHCO_3$ (c) LiClO (d) $Mg(ClO_3)_2$ (e) $CaSO_3$
44. (a) $(NH_4)_2CO_3$ (b) CaI_2 (c) $CuBr_2$ (d) $AlPO_4$
46. (a) Potassium sulfide (b) Nickel(II) sulfate (c) Ammonium phosphate (d) Aluminum hydroxide (e) Cobalt(III) sulfate
48. MgO; MgO has higher ionic charges and smaller ion sizes than NaCl.
50. (a) Ionic (b) Molecular (c) Molecular (d) Ionic
52. (a) Structural:

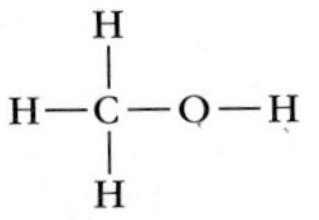

Molecular: CH_4O

(b) Structural:

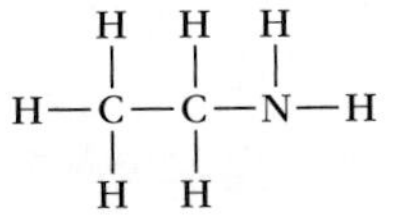

Molecular: C_2H_7N

(c) Structural:

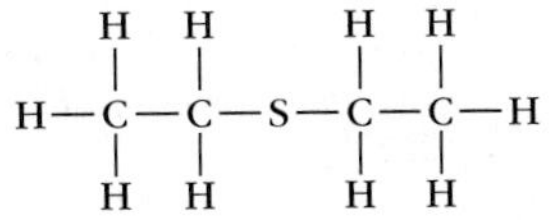

Molecular: $C_4H_{10}S$

(d) Structural:

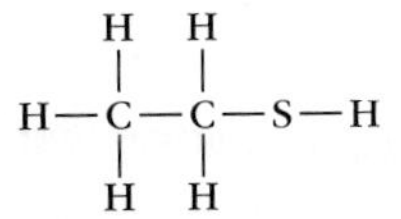

Molecular: C_2H_6S

54. (a) C_7H_{16} (b) C_3H_3N
56. (a) 1 Ca, 2 C, 4 O (b) 8 C, 8 H (c) 2 N, 8 H, 1 S, 4 O (d) 1 Pt, 2 N, 6 H, 2 Cl (e) 4 K, 1 Fe, 6 C, 6 N
58. (a) Sulfur dioxide (b) Carbon tetrachloride (c) Tetraphosphorus decasulfide (d) Sulfur tetrafluoride
60. (a) NI_3 (b) CS_2 (c) N_2O_4 (d) SeF_6
62. Three
64. (a) The same number of atoms of each kind (b) Different bonding arrangements
66. 42 hydrogen atoms
68. 2×10^8 years
70. (a) 27 g B (b) 0.48 g O_2 (c) 6.98×10^{-2} g Fe (d) 2.61×10^3 g He
72. (a) 1.9998 mol Cu (b) 0.499 mol Ca (c) 0.6208 mol Al (d) 3.1×10^{-4} mol K (e) 2.1×10^{-5} mol Am
74. 4.131×10^{23} Cr atoms
76. (a) 12.63 g (b) 7.689×10^{20} molecules (c) 1.538×10^{21} N atoms (d) 14.4 g N

78. (a) 41.7 pennies
(b) 9.28×10^{-4} mol Cu
(c) 5.59×10^{20} Cu atoms

80.

	CH_3OH	Carbon	Hydrogen	Oxygen
Amount of substance (mol)	1 mol	1 mol	4 mol	1 mol
No. of molecules or atoms	6.022×10^{23} molecules	6.022×10^{23} atoms	2.409×10^{24} atoms	6.022×10^{23} atoms
Molar mass (grams per mol methanol)	32.0417	12.0107	4.0316	15.9994

82. (a) 0.0312 mol (b) 0.0101 mol (c) 0.0125 mol
(d) 0.00406 mol (e) 0.00599 mol

84. (a) 151.1622 g/mol (b) 0.0352 mol (c) 25.1 g

87. (a) 239.3 g/mol PbS, 86.60% Pb, 13.40% S
(b) 30.0688 g/mol C_2H_6, 79.8881% C, 20.1119% H
(c) 60.0518 g/mol CH_3COOH, 40.0011% C, 6.7135% H, 53.2854% O
(d) 80.0432 g/mol NH_4NO_3, 34.9979% C, 5.0368% H, 59.9654% O

89. 245.745 g/mol, 25.858% Cu, 22.7992% N, 5.74197% H, 13.048% S, 32.5528% O

91. (a) $C_{10}H_{12}NO$ (b) $C_{20}H_{24}N_2O_2$

94. One

96. 6

98. $C_4H_8N_2O_2$

102. 0.995 g Pt

105. 89 tons/yr

107. 0.038 mol

109. (a) Iodine monobromide (b) Bromine trifluoride
(c) Diiodine hexachloride (d) Chlorine pentafluoride
(e) Iodine heptafluoride

111. (a) 28.8515% N (b) 6.57×10^{20} molecules
(c) 5.26×10^{21} C atoms (d) Nine times greater

113. (a) Ionic: (ii), (iv), and (vi)
Molecular: (i), (v), (vii)
No compound: (iii)
(b) (i) BrCl, bromine monochloride;
(ii) Li_2Te, lithium telluride;
(iii) No compound;
(iv) MgF_2, magnesium fluoride,
(v) NF_3, nitrogen trifluoride;
(vi) In_2S_3, indium sulfide;
(vii) $SeBr_2$, selenium dibromide

115. (a) 1.66×10^{-3} mol (b) 0.346 g

117.

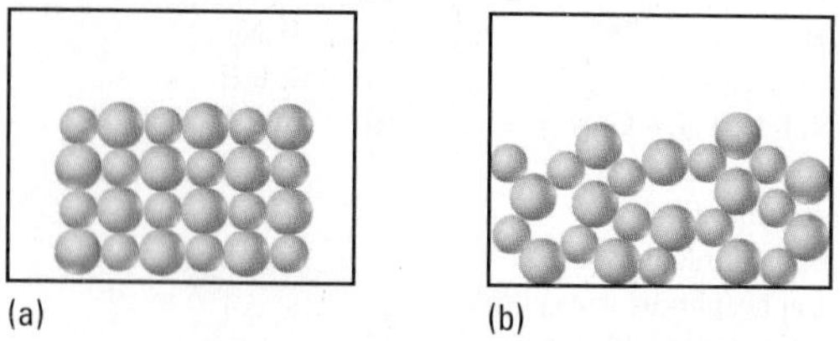

(a) (b)

119. (a) Not possible; mass number and atomic number wrong
(b) Possible
(c) Not possible; mass number wrong
(d) Not possible; mass number wrong
(e) Possible
(f) Not possible; mass number wrong

121. ^{39}K

123. (a) 1 mol of Cl_2 (b) 1 mol of O_2
(c) one nitrogen molecule (d) 6.03210^{23} molecules of F_2
(e) 20.3 grams of neon (f) 159.8 grams of Br_2
(g) 9.6 grams of Li (h) 58.9 g CO
(i) 6.02210^{23} calcium atoms (j) Same

126. (a) Three (b) (i) and (iv), (ii) and (iii), (v) and (vi)

127. Tl_2CO_3, Tl_2SO_4

130. (a) ^{79}Br–^{79}Br, ^{79}Br–^{81}Br, ^{81}Br–^{81}Br
(b) 78.918 g/mol, 80.916 g/mol
(c) 79.92 g/mol
(d) 50.1% ^{79}Br, 49.9% ^{81}Br

132. 75.0% sulfate

134. 6.73% ^{41}K and 0.01% ^{40}K

136. 71.6% Fe, 18.5% Cr, and 8.96% Ni

140. I: K_2O, II: KO_2, III: KO_3, IV: K_2O_2

141. (a) I: XeF_4, II: XeF_2, III: XeF_6
(b) I: xenon tetrafluoride
II: xenon difluoride
III: xenon hexafluoride

145. 45.0% $CaCl_2$

147. (a) 2− (b) Al_2X_3 (c) Se

Chapter 3

10.

	NH_3	O_2	NO	H_2O
No. of molecules	4	5	4	6
No. of atoms	16	10	8	18
Amount of molecules (mol)	4	5	4	6
Mass (g)	68.1216	159.9940	120.0244	108.0912
Total mass of reactants (g)	228.1156		—	
Total mass of products (g)	—		228.1156	

12. $A_2 + 2\ B \rightarrow 2\ AB$

15. (a) Box (a) (b) Box (c)

18. $X_2 + Y_2 \rightarrow XY_2 + X$

20. (a) $UO_2(s) + 4\ HF(\ell) \rightarrow UF_4(s) + 2\ H_2O(\ell)$
(b) $B_2O_3(s) + 6\ HF(\ell) \rightarrow 2\ BF_3(s) + 3\ H_2O(\ell)$
(c) $BF_3(g) + 3\ H_2O(\ell) \rightarrow 3\ HF(\ell) + H_3BO_3(s)$

22. (a) $H_2NCl(aq) + 2\ NH_3(g) \rightarrow NH_4Cl(aq) + N_2H_4(aq)$
(b) $(CH_3)_2N_2H_2(\ell) + 2\ N_2O_4(g) \rightarrow 3\ N_2(g) + 4\ H_2O(g) + 2\ CO_2(g)$
(c) $CaC_2(s) + 2\ H_2O(\ell) \rightarrow Ca(OH)_2(s) + C_2H_2(g)$

24. (a) $C_6H_{12}O_6 + 6\ O_2 \rightarrow 6\ CO_2 + 6\ H_2O$
(b) $C_5H_{12} + 8\ O_2 \rightarrow 5\ CO_2 + 6\ H_2O$
(c) $2\ C_7H_{14}O_2 + 19\ O_2 \rightarrow 14\ CO_2 + 14\ H_2O$
(d) $C_2H_4O_2 + 2\ O_2 \rightarrow 2\ CO_2 + 2\ H_2O$

25. (a) K^+ and OH^- (b) K^+ and SO_4^{2-}
(c) Na^+ and NO_3^- (d) NH_4^+ and Cl^-

27. (a) and (d)

29. (a) Soluble, K^+ and HPO_4^{2-}
(b) Soluble, Na^+ and ClO^-
(c) Soluble, Mg^{2+} and Cl^-
(d) Soluble, Ca^{2+} and OH^-
(e) Soluble, Al^{3+} and Br^-

31. Box (b)

33. Complete ionic: $2\ K^+(aq) + CO_3^{2-}(aq) + Cu^{2+}(aq) + 2\ NO_3^-(aq) \rightarrow CuCO_3(s) + 2\ K^+(aq) + 2\ NO_3^-(aq)$;
net ionic: $CO_3^{2-}(aq) + Cu^{2+}(aq) \rightarrow CuCO_3(s)$;
precipitate is copper(II) carbonate.

35. (a) $Ca(OH)_2(s) + CoCl_2(aq) \rightarrow CaCl_2(aq) + Co(OH)_2(s)$
complete ionic: $Ca(OH)_2(s) + Co^{2+}(aq) + 2\ Cl^-(aq) \rightarrow Ca^{2+}(aq) + 2\ Cl^-(aq) + Co(OH)_2(s)$
net ionic: $Ca(OH)_2(s) + Co^{2+}(aq) \rightarrow Ca^{2+}(aq) + Co(OH)_2(s)$
(b) $BaCl_2(aq) + Na_2CO_3(aq) \rightarrow BaCO_3(s) + 2\ NaCl(aq)$
complete ionic: $Ba^{2+}(aq) + 2\ Cl^-(aq) + 2\ Na^+(aq) + CO_3^{2-}(aq) \rightarrow BaCO_3(s) + 2\ Na^+(aq) + 2\ Cl^-(aq)$
net ionic: $Ba^{2+}(aq) + CO_3^{2-}(aq) \rightarrow BaCO_3(s)$
(c) $2\ Na_3PO_4(aq) + 3\ Ni(NO_3)_2(aq) \rightarrow Ni_3(PO_4)_2(s) + 6\ NaNO_3(aq)$
complete ionic:
$6\ Na^+(aq) + 2\ PO_4^{3-}(aq) + 3\ Ni^{2+}(aq) + 6\ NO_3^-(aq) \rightarrow Ni_3(PO_4)_2(s) + 6\ Na^+(aq) + 6\ NO_3^-(aq)$
net ionic: $2\ PO_4^{3-}(aq) + 3\ Ni^{2+}(aq) \rightarrow Ni_3(PO_4)_2(s)$

38. $Pb(NO_3)_2(aq) + 2\ KCl(aq) \rightarrow PbCl_2(s) + 2\ KNO_3(aq)$;
reactants: lead(II) nitrate and potassium chloride;
products: lead(II) chloride and potassium nitrate
40. (a) Base, strong, K^+ and OH^-
(b) Base, strong, Mg^{2+} and OH^-
(c) Acid, weak, H^+ and ClO^-
(d) Acid, strong, H^+ and Br^-
(e) Base, strong, Li^+ and OH^-
(f) Acid, weak; small amounts of H^+, HSO_3^-, and SO_3^{2-}
42. (a) HNO_2; NaOH;
complete ionic: $HNO_2(aq) + Na^+(aq) + OH^-(aq) \rightarrow H_2O(\ell) + Na^+(aq) + NO_2^-(aq)$
net ionic: $HNO_2(aq) + OH^-(aq) \rightarrow H_2O(\ell) + NO_2^-(aq)$
(b) H_2SO_4; $Ca(OH)_2$;
complete ionic and net ionic: $H^+(aq) + HSO_4^-(aq) + Ca(OH)_2(s) \rightarrow 2\ H_2O(\ell) + CaSO_4(s)$
(c) HI; NaOH;
complete ionic: $H^+(aq) + I^-(aq) + Na^+(aq) + OH^-(aq) \rightarrow H_2O(\ell) + Na^+(aq) + I^-(aq)$
net ionic: $H^+(aq) + OH^-(aq) \rightarrow H_2O(\ell)$
(d) H_3PO_4; $Mg(OH)_2$;
complete ionic and net ionic: $2\ H_3PO_4(aq) + 3\ Mg(OH)_2(s) \rightarrow 6\ H_2O(\ell) + Mg_3(PO_4)_2(s)$
44. (a) Precipitation reaction; products are NaCl and MnS;
$MnCl_2(aq) + Na_2S(aq) \rightarrow 2\ NaCl(aq) + MnS(s)$
(b) Precipitation reaction; products are NaCl and $ZnCO_3$;
$Na_2CO_3(aq) + ZnCl_2(aq) \rightarrow 2\ NaCl(aq) + ZnCO_3(s)$
(c) Gas-forming reaction; products are $KClO_4$, H_2O, and CO_2;
$K_2CO_3(aq) + 2\ HClO_4(aq) \rightarrow 2\ KClO_4(aq) + H_2O(\ell) + CO_2(g)$
46. $MnCO_3(aq) + 2\ HCl(aq) \rightarrow H_2O(\ell) + CO_2(g) + MnCl_2(aq)$
47. (a) Strong electrolyte (b) Weak electrolyte
(c) Strong electrolyte (d) Strong electrolyte
49. (a) −1 (b) +1 (c) +3 (d) +5 (e) +7
52. (a) Precipitation (b) Oxidation-reduction
(c) Acid-base neutralization
55. Substances (b), (c), and (d)
57. (a) Most: Groups 1A and 2A;
Least: right side of the transition elements
(b) No
(c) Yes, $Pb(s) + 2\ AgNO_3(aq) \rightarrow Pb(NO_3)_2(aq) + 2\ Ag(s)$
(d) Al(s) > Pb(s) > Ag(s)
59. 1.1 mol O_2, 35 g O_2, 1.0×10^2 g NO_2
61.

$(NH_4)_2PtCl_6$	Pt	HCl
—	5.428 g	5.410 g
0.02782 mol	0.02782 mol	0.1484 mol

63. (a) $4\ Fe(s) + 3\ O_2(g) \rightarrow 2\ Fe_2O_3(s)$
(b) 7.98 g (c) 2.40 g
65. (a) $CCl_2F_2 + 2\ Na_2C_2O_4 \rightarrow C + 4\ CO_2 + 2\ NaCl + 2\ NaF$
(b) 170. g $Na_2C_2O_4$ (c) 112 g CO_2
67. (a) 699 g (b) 526 g
69. (a) Cl_2 is limiting. (b) 5.08 g Al_2Cl_6
(c) 1.67 g Al unreacted
71. (a) CH_4 (b) 188 g (c) 739 g
73. 699 g, 93.5%
74. 56.0%
76. 5.3 g SCl_2
78. $C_3H_6O_2$
80. (a) $C_9H_{11}NO_4$ (b) $C_9H_{11}NO_4$
82. (a) 0.254 M Na_2CO_3
(b) 0.508 M Na^+, 0.254 M CO_3^{2-}
84. 5.08×10^3 mL
86. Method (b), because it is the only one with the correct concentration
88. 39.4 g $NiSO_4 \cdot 6H_2O$
90. 121 mL HNO_3
93. (a) Step (ii) is not correct. (Steps (iii) and (iv) show correct calculations but use the wrong numbers.)
(b) 3.94×10^{-3} g citric acid
95. 16.1% $H_2C_2O_4$
97. 104.0 g/mol
98. (b) Cu
100. SiH_4
102. KOH; KOH
104. (a) K^+ is a spectator ion.
(b) Elements are changing oxidation states.
(c) MnO_4^- is the oxidizing agent.
H_2S is the reducing agent.
107. (a) Acid-base reaction; H_3PO_4 is the acid; NaOH is the base.
(b) Acid-base reaction; CO_2 and H_2O combine as the acid; NH_3 is the base.
(c) Redox reaction; oxidizing agent is Ti; reducing agent is Mg.
(d) Gas-forming reaction.
108. (a) $CaF_2(s) + H_2SO_4(aq) \rightarrow 2\ HF(g) + CaSO_4(s)$; calcium fluoride, sulfuric acid, hydrogen fluoride, calcium sulfate
(b) Precipitation
(c) Carbon tetrachloride, antimony(V) pentachloride, hydrogen monochloride
(d) CCl_3F
109. 6.28% impurity
111. 99.7% CH_3OH, 0.3% C_2H_5OH
116. For $LiCl(aq) + AgNO_3(aq)$
(a) Before: clear and colorless solutions; after: insoluble white AgCl precipitate in a clear colorless solution of $LiNO_3$
(b)
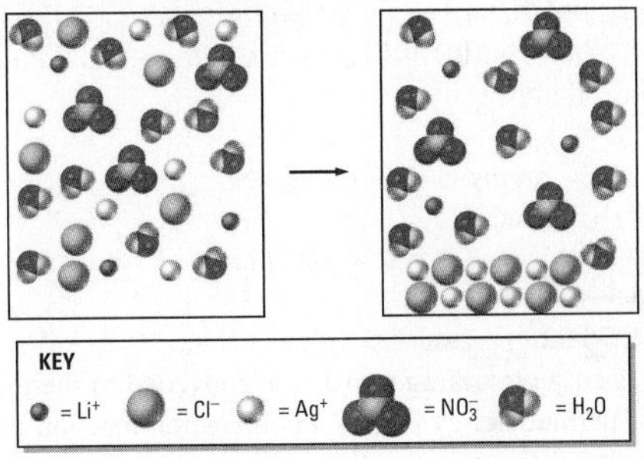
KEY
= Li+
= Cl−
= Ag+
= NO3−
= H2O
(c) $Li^+(aq) + Cl^-(aq) + Ag^+(aq) + NO_3^-(aq) \rightarrow AgCl(s) + Li^+(aq) + NO_3^-(aq)$
For $NaOH(aq) + HCl(aq)$
(a) Before: clear and colorless; after: clear and colorless
(b)
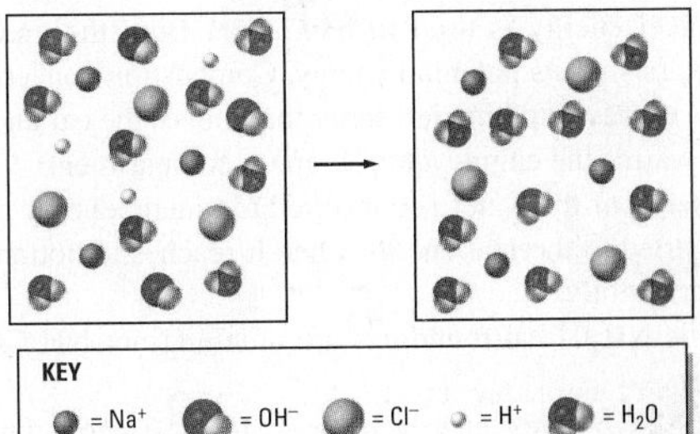
KEY
= Na+
= OH−
= Cl−
= H+
= H2O
(c) $Na^+(aq) + OH^-(aq) + H^+(aq) + Cl^-(aq) \rightarrow H_2O(\ell) + Na^+(aq) + Cl^-(aq)$
118. 2 butane molecules react with 13 diatomic oxygen molecules to produce 8 carbon dioxide molecules and 10 water molecules.
2 mol of gaseous butane molecules reacts with 13 mol of gaseous diatomic oxygen molecules to produce 8 mol of gaseous carbon dioxide molecules and 10 mol of liquid water molecules.
120. Ag^+, Cu^{2+}, and NO_3^-
123. Equation (b)
124. $A_2 + 4\ BC_2 \rightarrow 4\ C_2 + 2\ AB_2$

126. When the metal mass is less than 1.2 g, the metal is the limiting reactant.
When the metal mass is greater than 1.2 g, the bromine is the limiting reactant.
128. (a) Box 2 (b) Box 3 (c) Box 1
131. (a) Combine H_2SO_4 and $Ba(OH)_2$.
(b) Combine Na_2SO_4 and $Ba(NO_3)_2$.
(c) Combine $H_2SO_4(aq)$ and $BaCO_3(s)$.
133. Use HCl; precipitate forms, if Pb^{2+} is present; no precipitate, if Ba^{2+} is present
135. Solution (d)
137. (a) and (d) are correct.
140. (a) Groups A and B: $Ag^+(aq) + Cl^-(aq) \rightarrow AgCl(s)$;
Groups C and D: $Ag^+(aq) + Br^-(aq) \rightarrow AgBr(s)$
(b) Silver halide product is the same for A and B and different from C or D.
(c) Curve has upward slope while the Ag^+ is the limiting reactant. Bromide has greater mass than chloride, so the curve levels out at greater masses for the bromides.
142. $H_2(g) + 3\ Fe_2O_3(s) \rightarrow H_2O(\ell) + 2\ Fe_3O_4(s)$
144. 44.9 u
145. 0 g $AgNO_3$, 9.82 g Na_2CO_3, 6.79 g Ag_2CO_3, 4.19 g $NaNO_3$
148. 0.28 M NaCl
150. Combine the solutions pair-wise. HCl forms a precipitate with $AgNO_3$, bubbles of CO_2 with Na_2CO_3, and a warmer solution with NaOH.
$AgNO_3$ forms a precipitate with all three other solutions.
Na_2CO_3 forms CO_2 with HCl and a precipitate with $AgNO_3$.
NaOH forms a warmer solution with HCl and a precipitate with $AgNO_3$.
153. (a) $AsO_4^{3-}(aq) + FeCl_3(aq) \rightarrow 3\ Cl^-(aq) + FeAsO_4(s)$
(b) $AsO_4^{3-}(aq) + Fe^{3+}(aq) \rightarrow FeAsO_4(s)$
(c) 4.2 mL (d) 0.12 g
155. 84.5%

Chapter 4

9. (a) 399 Cal (b) 5.0×10^6 J/day
11. 1.04×10^4 J
13. 3.60×10^6 J, \$0.02
15. (a) Kinetic energy of striking a match is converted to thermal energy by friction. Thermal energy causes a combustion reaction in the fuse and later in the chemical propellants, converting chemical potential energy into heat energy, light energy, kinetic energy, and potential energy as the speed and altitude increase. When the rocket explodes, more chemical potential energy is converted into light, heat, and kinetic energy.
(b) Mechanical energy is used to pump fuel from the underground storage tank, raising its potential energy. Combustion converts chemical potential energy into kinetic energy that moves the car and thermal energy that warms the engine and passenger compartment.
17. Potential energy of the water is converted to kinetic energy as it falls, then is converted to thermal energy when it reaches to bottom, warming the water slightly.
19. (a) System is NH_4Cl; surroundings are anything not NH_4Cl, including the water.
(b) The goal is to study energy released to the surroundings when dissolving the ionic compound in water.
(c) Thermal energy is transferred from the surroundings to the system. There is no material transfer, though water interacts strongly with dissolved ions.
(d) Endothermic
21.

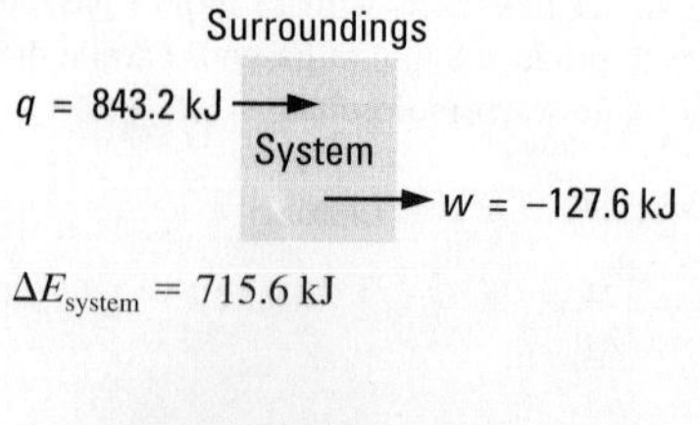
Surroundings
q = 843.2 kJ
System
w = −127.6 kJ

$\Delta E_{system} = 715.6$ kJ

23. Process (a)
25. Copper. The specific heat capacity for Al is greater than for Cu, so more energy is required to heat the Al by 1 °C and the Cu warms faster.
27. 136 J mol^{-1} K^{-1}
29. 0.25 J g^{-1} °C^{-1}
31. 330. °C
33. Gold
35. $\Delta T_{surroundings}$ = positive, ΔE_{system} = negative
37. 9.98 kJ
39. 273 J
41. (a) Negative (b) Positive
43. 49.3 kJ
45. Endothermic
47. $CaO(s) + H_2O(\ell) \rightarrow Ca^{2+}(aq) + 2\ OH^-(aq)$; exothermic
49. (a) -2.1×10^2 kJ (b) −33 kJ
51. (a) 4 mol rxn (b) 0.115 mol rxn (c) 0.0788 mol rxn
53. -3.3×10^4 kJ
55. -1.45×10^3 kJ/mol
57. 2.35×10^3 kJ
59. HF
61. For F_2:
(a) 594 kJ/mol (b) −1132 kJ/mol (c) −538 kJ/mol;
For Cl_2:
(a) 678 kJ/mol (b) −862 kJ/mol (c) −184 kJ/mol
(d) Reaction of F_2 with H_2 is more exothermic.
63. Endothermic; energy must be transferred into the system to work against the force holding the atoms together.
64. 23.9 °C
66. (a) 1.4×10^4 J transferred (other answers are possible)
(b) −42 kJ/mol
68. $\Delta_r E$ per mole of glucose is -2.80×10^3 kJ/mol
70. The experiment is described in Section 4-8a.
The combustion (bomb) calorimeter apparatus is shown in Figure 4.15.
72. $\Delta_r H° = -43.6$ kJ/mol
74. −320. kJ/mol
76. (a) $2\ Al(s) + \frac{3}{2}\ O_2(g) \rightarrow Al_2O_3(s)$ $\Delta_f H° = -1675.7$ kJ/mol
(b) $Ti(s) + 2\ Cl_2(g) \rightarrow TiCl_4(\ell)$ $\Delta_f H° = -804.2$ kJ/mol
(c) $N_2(g) + 2\ H_2(g) + \frac{3}{2}\ O_2(g) \rightarrow NH_4NO_3(s)$
$\Delta_f H° = -365.56$ kJ/mol
77. (a) −1675.7 kJ/mol (b) −205.4 kJ
79. −584 kJ/mol
81. (a) 178.32 kJ/mol (b) −595.2 kJ/mol (c) 44 kJ/mol
83. 103.6 kJ/mol
85. 41.2 kJ
87. 1.2×10^2 g CH_4
89. 44.422 kJ/g octane > 19.927 kJ/g methanol
91. 2.2 hours walking
93. 4.82×10^4 J
95. 75.4 g
97. −96.4 kJ/mol
101. (a) −36.03 kJ/mol (b) 1.18×10^4 kJ evolved
103. Endothermic; exothermic
105. (a) $X(\ell)$ (b) The fusion enthalpy (c) Positive
107. Gold
108. Substance A
110. Greater than; because a larger mass of the same substance contains a larger quantity of thermal energy at a given temperature.
112. Enthalpy change is an extensive property. The given equation produces 2 mol SO_3. Formation enthalpy is for the production of 1 mol SO_3, so the enthalpy values must be different by a factor of two.
114. $\Delta_r E = 310$ J; $w = 0$ J
116. $\Delta_f H°(OF_2) = 24.6$ kJ/mol
119. Mass of a tennis ball (from www.usta.com/2011_tennis_ball_specifications/) is between 56.0 g and 59.4 g; use the lower value.
$m_{tennis\ ball} = (56.0\ g)(1\ kg/1000\ g) = 0.0560$ kg

$v_{\text{tennis ball}} =$
(130 mi/h)(1 h/60 min)(1 min/60 s)(1 km/0.6214 mi)(1000 m/1 km) = 58.1 m/s

$E_k = \frac{1}{2}mv^2 = (0.5)(0.0560\text{ kg})(58.1\text{ m/s})^2 = 94.6\text{ kg m}^2\text{ s}^{-2} = 94.6\text{ J}$

$m_{\text{peanuts}} = (94.6\text{ J})(1\text{ g}/23.91\text{ kJ})(1\text{ kJ}/1000\text{ J}) = 4.0 \times 10^{-3}\text{ g}$

120. (a) 26.6 °C, temperature leveled off from Experiment 3 on
(b) Ascorbic acid is limiting in Experiments 1, 2, and 3; sodium hydroxide is limiting in Experiments 3, 4, and 5
(c) One; equal quantities of each reactant are present in Experiment 3 at the stoichiometric equivalence point
122. Results will vary depending on sources used:
(a) Approximately 4.6×10^9 kJ
(b) Approximately 1.0 metric kilotons TNT
(c) 15 kilotons for Hiroshima bomb, 20 kilotons for Nagasaki; largest nuclear weapon ever = 57 megatons; approximately 6.7% Hiroshima bomb or 5.0% Nagasaki bomb
(d) Same energy as approximately 3.1 seconds of a mature hurricane

Chapter 5

11. (a) Visible green light (b) 5.71×10^{14} Hz
13. (a) Radio waves have less energy than infrared.
(b) Microwaves are higher frequency than radio waves.
15. 4.4×10^{-19} J/photon
17. 1.11×10^{15} Hz, 7.36×10^{-19} J/photon
20. Photons of this light are too low in energy. Increasing the intensity only increases the number of photons, not their individual energy.
22. No; insufficient energy
27. The higher-energy state to the lower-energy state; difference
28. (a) Absorbed (b) Emitted (c) Absorbed (d) Emitted
31. (a), (b), (d)
32. 4.576×10^{-19} J absorbed, 434.0 nm
34. $\Delta E = -6.18 \times 10^{-19}$ J; λ = 320.5 nm; ultraviolet
37. Moon, bowling ball, baseball, neon atom, electron
39. It shows where the electron can be found in space and does not require the simultaneous knowledge of the electron's position and momentum.
41. (a) (iii) and (iv)
(b) (i) and (iii)
43. (a) First electron: $n = 1, \ell = 0, m_\ell = 0, m_s = +\frac{1}{2}$; second electron: $n = 1, \ell = 0, m_\ell = 0, m_s = -\frac{1}{2}$; third electron: $n = 2, \ell = 0, m_\ell = 0, m_s = +\frac{1}{2}$; fourth electron: $n = 2, \ell = 0, m_\ell = 0, m_s = -\frac{1}{2}$; fifth electron: $n = 2, \ell = 1, m_\ell = 1, m_s = +\frac{1}{2}$ (The fifth electron could have different m_ℓ and m_s values.)
(b) $n = 3, \ell = 0, m_\ell = 0, m_s = +\frac{1}{2}$ and $n = 3, \ell = 0, m_\ell = 0, m_s = -\frac{1}{2}$
(c) $n = 3, \ell = 2, m_\ell = 2, m_s = +\frac{1}{2}$ (Different m_ℓ and m_s values are also possible.)
45. (a) Cannot occur, m_ℓ too large (b) Can occur
(c) Cannot occur, m_s cannot be 1
(d) Cannot occur, ℓ must be less than n (e) Can occur
47. (a) $n = 4, \ell = 0, m_\ell = 0, m_s = +\frac{1}{2}$
(b) $n = 3, \ell = 1, m_\ell = 1, m_s = -\frac{1}{2}$
(c) $n = 3, \ell = 2, m_\ell = 0, m_s = +\frac{1}{2}$
49. 18 elements, all possible orbital electron combinations are already used
50.

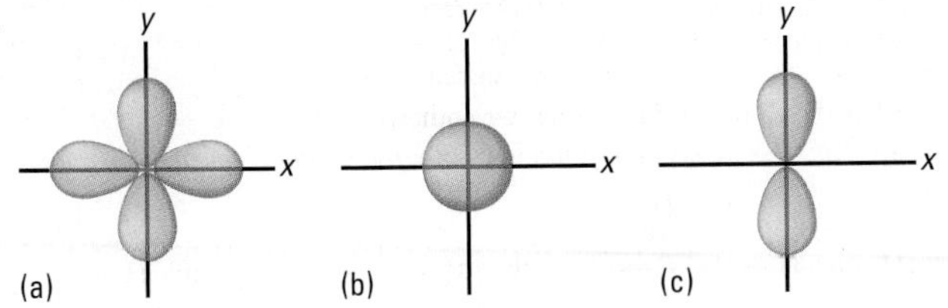
y
x
(a)
y
x
(b)
y
x
(c)

52. Four subshells
54. 4s orbital is filled before the 3d orbitals in Ti; atoms lose the outermost shell electrons first so 4s electrons are lost in Cr^{2+}
56. $_{32}Ge$: $1s^22s^22p^63s^23p^63d^{10}4s^24p^2$
58. Oxygen; Group 6A has 6 valence electrons with valence electron configuration of ns^2np^4.
60. (a) Cs^+, Se^{2-} (b) Cs^+, Se^{2-}
61. (a) 4s orbital must be full.
(b) Orbital labels must be 3, not 2; electrons in 3p subshell should be in separate orbitals with parallel spin.
(c) Electrons should have been removed from the 5p subshell, not the 4d subshell.
62. (a) V

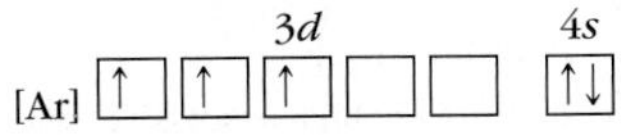
3d
4s
[Ar]

(b) V^{2+}

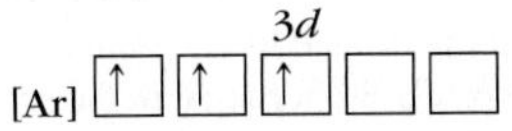
3d
[Ar]

(c) V^{4+}

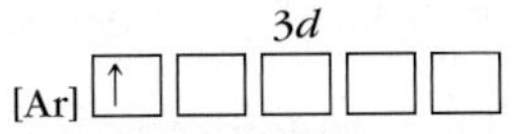
3d
[Ar]

64. Mn: $1s^22s^22p^63s^23p^63d^54s^2$; it has 5 unpaired electrons:

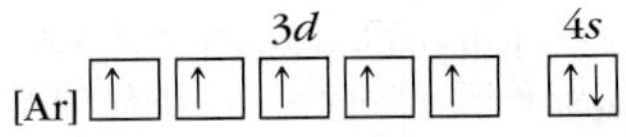
3d
4s
[Ar]

Mn^{2+}: $1s^22s^22p^63s^23p^63d^5$; it has 5 unpaired electrons:

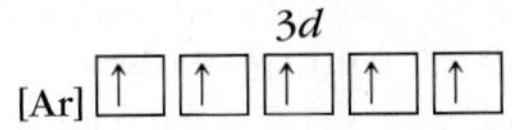
3d
[Ar]

Mn^{3+}: $1s^22s^22p^63s^23p^63d^4$; it has 4 unpaired electrons:

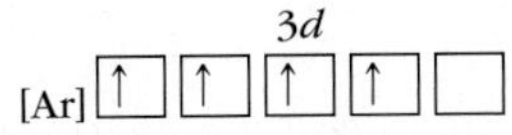
3d
[Ar]

67. (a) $[Xe]4f^76s^2$ or $[Xe]4f^65d^16s^2$
(b) $[Xe]4f^{14}6s^2$ or $[Xe]4f^{13}5d^16s^2$
68. (a) ·Sr· (b) :B̈r· (c) ·Ga· (d) ·S̈b·
70. (a) [Ar] (b) [Ar]
(c) [Ne]; Ca^{2+} and K^+ are isoelectronic.
72. $_{50}Sn$: $[Kr]4d^{10}5s^25p^2$; $_{50}Sn^{2+}$: $[Kr]4d^{10}5s^2$, $_{50}Sn^{4+}$: $[Kr]4d^{10}$
74. (a) Zn (b) Cr
76. In both paramagnetic and ferromagnetic substances, atoms have unpaired spins and thus are attracted to magnets. Ferromagnetic substances retain their aligned spins after an external magnetic field has been removed, so they can function as permanent magnets. Paramagnetic substances lose their aligned spins after a time and, therefore, cannot be used as permanent magnets.
77. (a) TiO_2; it is diagmagnetic (b) TiO; it is paramagnetic
79. (a) Cl has one more proton in its nucleus than S and adding one electron gives a noble-gas configuration making its electron affinity greater than S.
(b) Be has a full subshell, so it takes more energy to ionize it than B.
(c) F electron is in a lower-energy 2p, whereas Cl electron is in a higher-energy 3p orbital.
(d) N has a half-filled subshell, so N takes more energy to ionize than O.
(e) Br is smaller than I, so an electron added to Br is closer to the nucleus than in I.
81. B, C, N, O, F
83. P < Ge < Ca < Sr < Rb
85. (a) Rb smaller (b) O smaller (c) Br smaller
(d) Ba^{2+} smaller (e) Ca^{2+} smaller
87. (c)
89. Na; it must have one valence electron.
90. (a) Al (b) Al (c) Al < B < C

92. Adding a negative electron to a negatively charged ion requires additional energy to overcome the coulombic charge repulsion.
95. −862 kJ/mol
97. LiCl, because it is composed of the smallest ions and all ions have single charges.
99. (a) He (b) Sc (c) Na
102. (a) F < O < S; in the periodic table S is below O, so S is larger and F is to the right of O, so F is smaller.
(b) S; S is closest to the top right corner of the periodic table.
104. (a) Sulfur (b) Radium (c) Nitrogen
(d) Ruthenium (e) Copper
106. (a) Z (b) Z
108. In^{4+}, Fe^{6+}, and Sn^{5+}; very high successive ionization energies
110. (a) B, D, I (b) E (c) D (d) A, G (e) H, C
112. (a) Directly proportional to, not inversely proportional to
(b) Inversely proportional to *the square of* the principle quantum number, not inversely proportional to the principle quantum number
(c) Before, not as soon as
(d) Wavelength, not frequency
115. (a) $[Rn]5f^{14}6d^{10}7s^27p^68s^2$ (b) Magnesium
(c) XO, XCl_2
121. (a) Br^- product has a noble-gas electron configuration, so it is very stable.
(b) Mg atom has all full subshells.
(c) P atom has a half-full *p*-subshell.
(d) Cl electron comes from a 3*p* orbital and Br comes from a 4*p* orbital.
123. XCl
125. (a) Ground state (b) Could be ground state or excited state
(c) Excited state (d) Impossible
(e) Excited state (f) Excited state
127. $[Rn]5f^{14}5g^{18}6d^{10}6f^{14}7s^27p^67d^{10}8s^28p^2$, Group 4A
128. (a) Increase, decrease
(b) Helium
(c) 5 and 13
(d) He has only two electrons; there are no electrons left.
(e) The first electron is a valence electron, but the second electron is a core electron.
(f) $Mg^{2+}(g) \rightarrow Mg^{3+}(g) + e^-$
130. (a) 3.35 (b) 3.35 (c) 10.85
133. (a) 4.34×10^{-19} J (b) 6.54×10^{14} Hz
135. 2.18×10^{-18} J
137. To remove an electron from a N atom requires disrupting a half-filled subshell (p^3), which is relatively stable, so the ionization of O requires less energy than the ionization of N.
139. (a) Ir (b) $[Rn]7s^25f^{14}6d^7$
141. (a) Transition metals (b) $[Rn]7s^25f^{14}6d^{10}$

Chapter 6

13. At 25 pm, proton-proton repulsion and electron-electron repulsion destabilize the molecule. At 100 pm, the atoms are not close enough for optimal overlap of orbitals and sharing of electrons.
15. (a) :C̈l—F̈: (b) H—S̈e—H

(c) 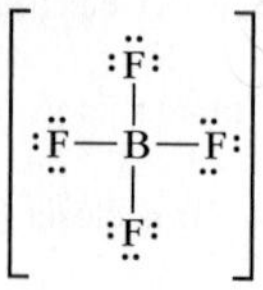 (d) 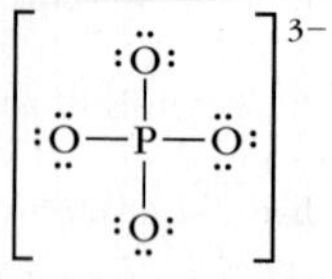

17. (a) 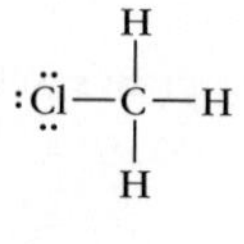 (b) 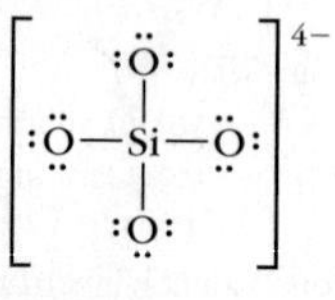

(c) 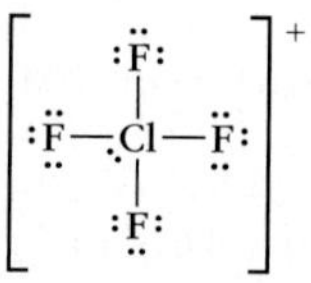 (d) 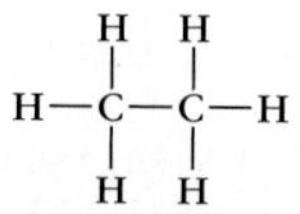

19. (a) 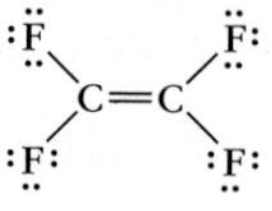 (b) 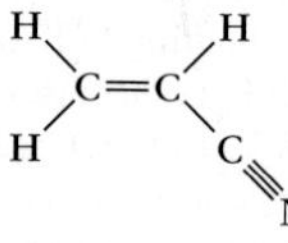

21. (a) Incorrect; the F atoms are missing electrons.
(b) Incorrect; the structure has 10 electrons but needs 12 electrons.
(c) Incorrect; one hydrogen atom has more than two electrons. One hydrogen atom is completely missing.
(d) Incorrect; the structure has three too many electrons; carbon has nine electrons. The single electron should be deleted. Oxygen has ten electrons; delete one pair of electrons.
(e) Incorrect; the structure has 16 electrons but needs 18 electrons. The N atom doesn't follow the octet rule. It needs another pair of electrons in the form of a lone pair.
23. Four branched-chain compounds

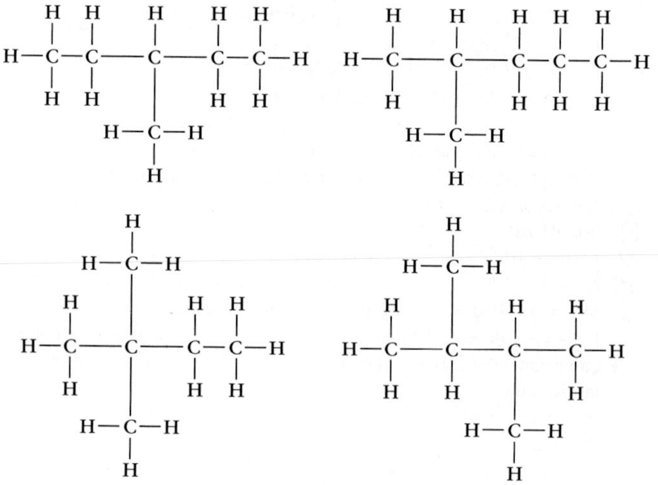

25. (a) Alkyne (b) Alkane (c) Alkene
27. (a) No
(b) Yes

cis

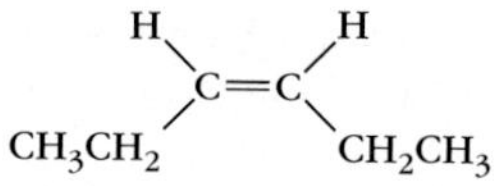

trans

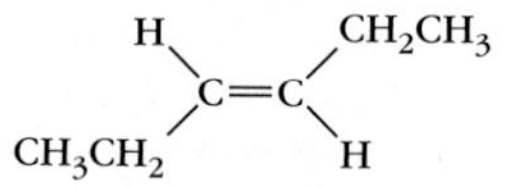

(c) Yes

cis

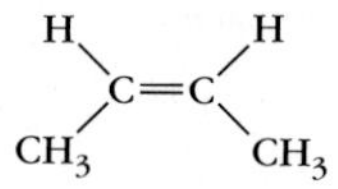

trans

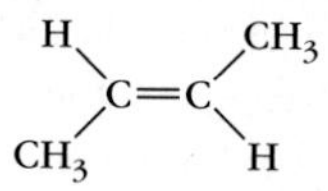

(d) No
29. (a) B—Cl (b) C—O (c) P—O (d) C=O
31. (a) Si—F
34. −92 kJ; the reaction is exothermic
37. (a) C—O (δ^+ δ^-) (b) B—O (δ^+ δ^-) (c) P—N (δ^+ δ^-) (d) B—Cl (δ^+ δ^-)
39. (a) All the bonds in urea are somewhat polar.
(b) The most polar bond is C=O; the O end is partially negative.
41. (a) 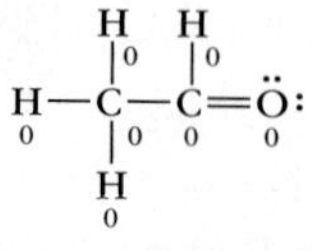 (b) 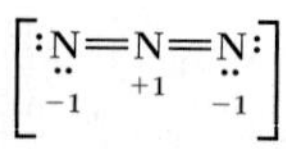 (c) 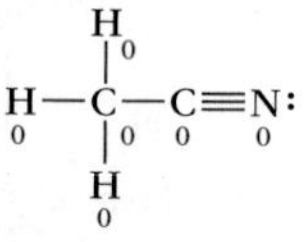

43. 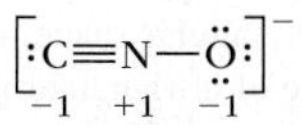

45. 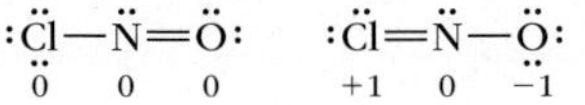

49. (a)

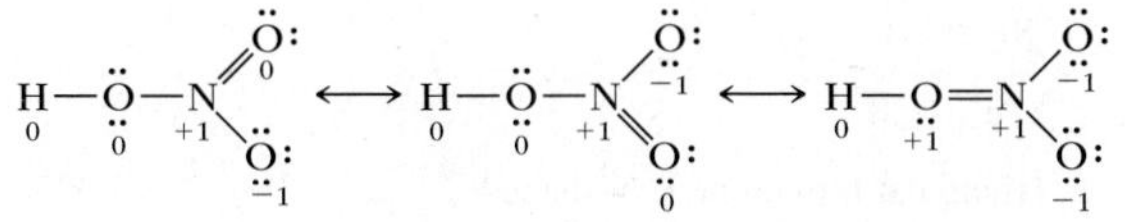

(b)

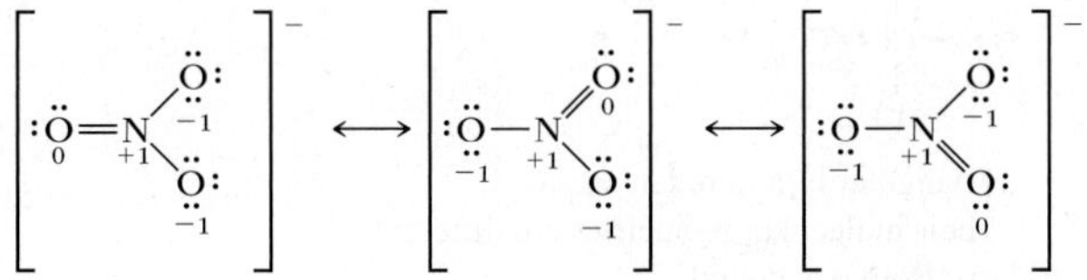

51. 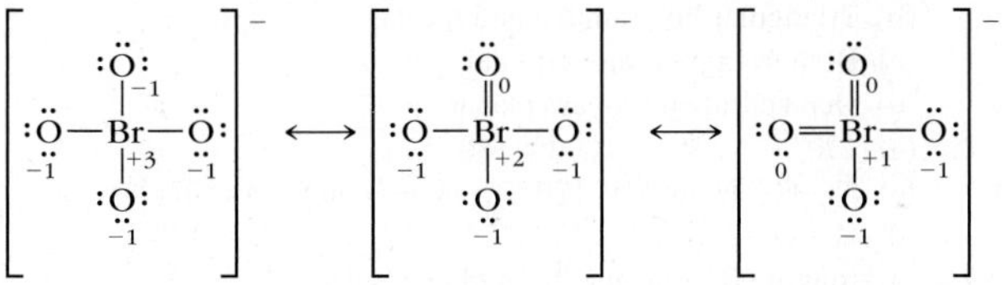

Most plausible

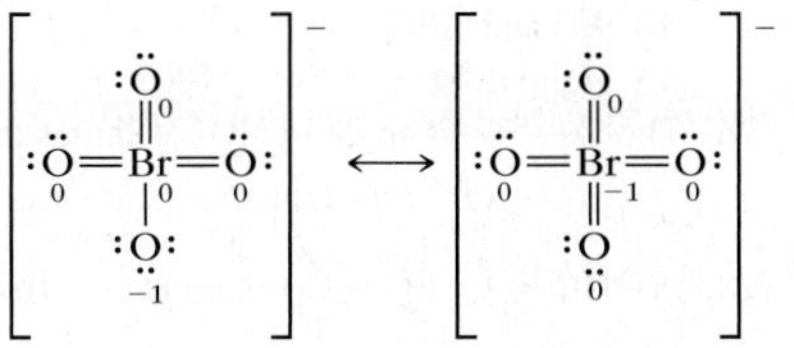

53. 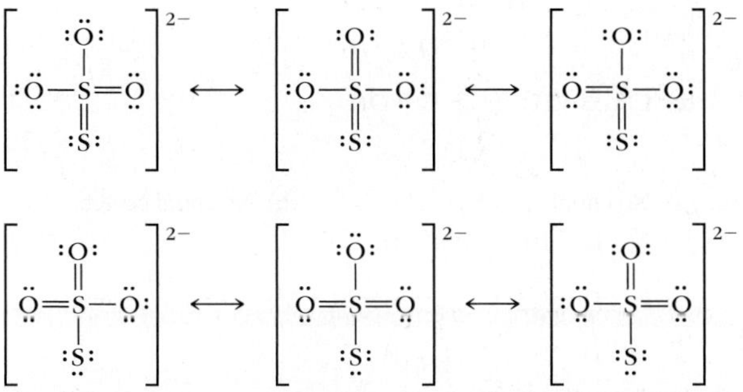

All of these structures have minimum formal charges and make equal contributions to the resonance hybrid. There are additional structures where there is only one S=O double bond and where there are no S=O double bonds; these have higher formal charges and, based on formal charges, make smaller contributions.

54. CO_3^{2-} has longer C—O bonds.

58. (a) (b) (c) 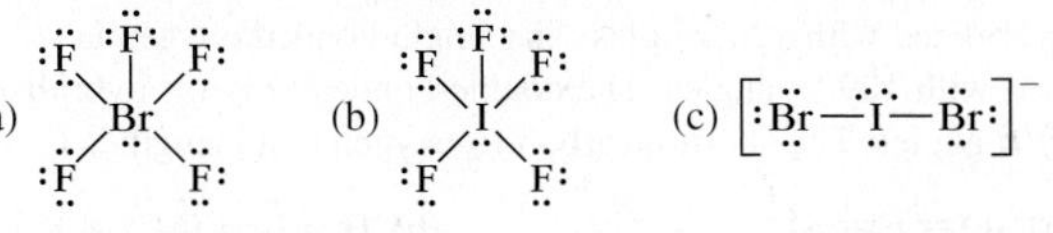

60. (c) B has only six electrons (d) Odd number of electrons

63. Against; with C=C and C—C, the bond lengths would be different.

65. 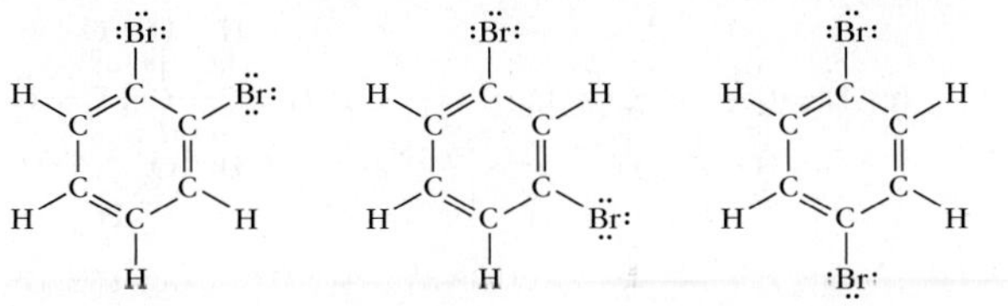

ortho-dibromobenzene *meta*-dibromobenzene *para*-dibromobenzene

67. (a)

σ_{2s}	σ^*_{2s}	π_{2p}	π_{2p}	σ_{2p}	π^*_{2p}	π^*_{2p}	σ^*_{2p}
(↑↓)	(↑↓)	(↑↓)	(↑↓)	(↑↓)	(↑↓)	(↑↓)	( )

1 bond; no unpaired electrons

(b)

σ_{2s}	σ^*_{2s}	π_{2p}	π_{2p}	σ_{2p}	π^*_{2p}	π^*_{2p}	σ^*_{2p}
(↑↓)	(↑↓)	(↑)	( )	( )	( )	( )	( )

0.5 bond; 1 unpaired electron

(c)

σ_{2s}	σ^*_{2s}	π_{2p}	π_{2p}	σ_{2p}	π^*_{2p}	π^*_{2p}	σ^*_{2p}
(↑)	( )	( )	( )	( )	( )	( )	( )

0.5 bond; 1 unpaired electron

(d)

σ_{2s}	σ^*_{2s}	π_{2p}	π_{2p}	σ_{2p}	π^*_{2p}	π^*_{2p}	σ^*_{2p}
(↑↓)	(↑↓)	(↑↓)	(↑↓)	(↑↓)	(↑)	( )	( )

2.5 bond; 1 unpaired electron

69. (a)

σ_{2s}	σ^*_{2s}	π_{2p}	π_{2p}	σ_{2p}	π^*_{2p}	π^*_{2p}	σ^*_{2p}
(↑↓)	(↑↓)	(↑↓)	(↑↓)	( )	( )	( )	( )

bond order = 2; no unpaired electrons

(b)

σ_{2s}	σ^*_{2s}	π_{2p}	π_{2p}	σ_{2p}	π^*_{2p}	π^*_{2p}	σ^*_{2p}
(↑↓)	(↑↓)	(↑↓)	(↑↓)	(↑↓)	( )	( )	( )

bond order = 3; no unpaired electrons

71. Si—F; Si is in the third period and less electronegative than C, S, or O.

73. Yes; "close" means similar electronegativities, therefore covalent bonds. "Far apart" means different electronegativities, therefore ionic bonds.

74. (a) The C=C is shorter. (b) The C=C is stronger.
(c) C≡N (δ^+ δ^-)

76. (a) (b) (c) (d) 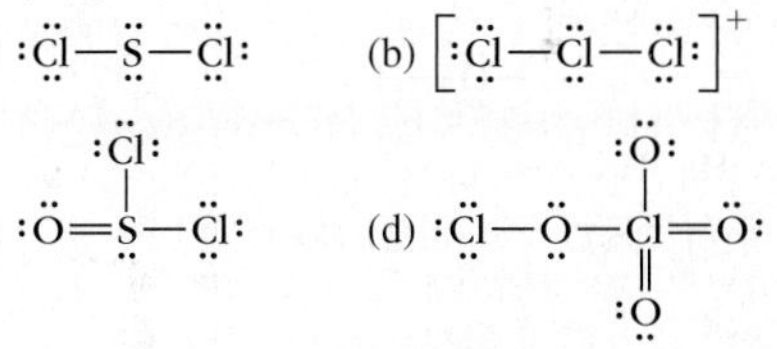

80. Forgot to subtract one electron for the positive charge.

83. Atoms are not bonded to the same atoms.

84. (a) (b) 

85. Cl: 2.7, S: 2.3, Br: 2.6, Se: 2.4, As: 2.1

87. (a) Box 7 (b) Box 1 (c) Boxes 4 and 9
(d) Box 10 (e) Box 12 (f) Box 8

89. NF_5; you cannot expand the octet of N.

90. :F—N=O: (a) N—F (b) N—O (c) F—N

92. (a) C—O (b) C≡N (c) C—O

94. 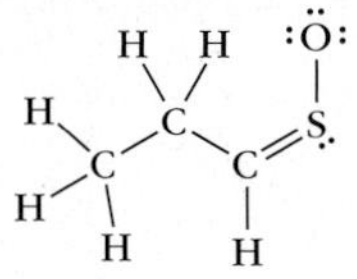

Other Lewis structures are possible, for example,

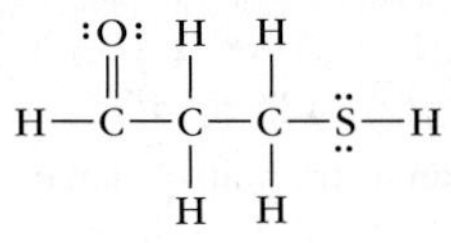

95. 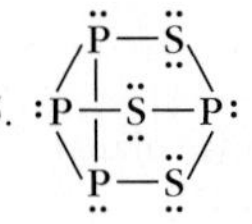

100. (a) 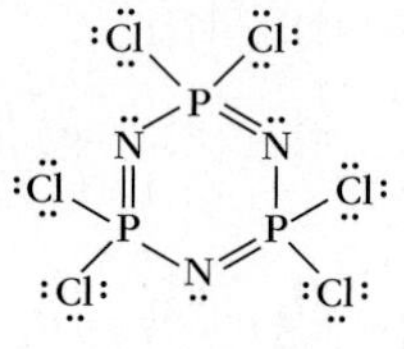 (b) 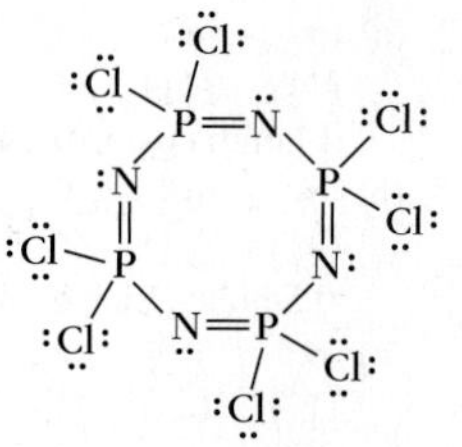

101. The NON structure has higher formal charges and positive formal charge on the higher–EN O atom.

104. (a) 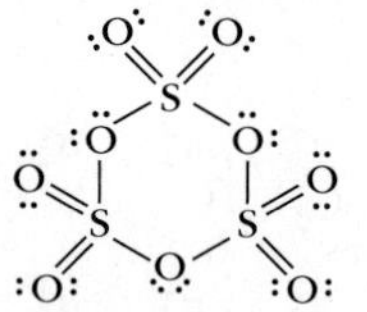 (b) 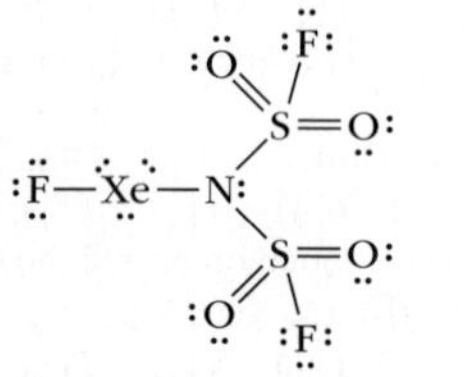

105. $H-C\equiv C-C\equiv N:$

109. 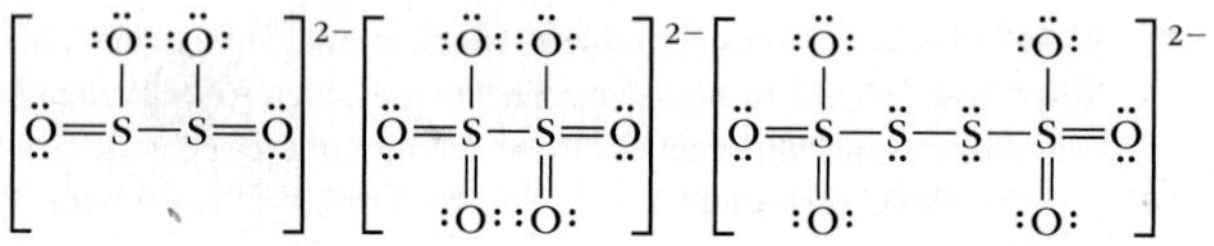

111. (a) 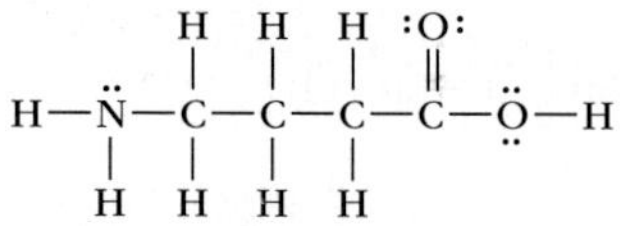

(b) C—C (c) C=O (d) C=O

Chapter 7

13. (a) H—Be—H Linear

(b) 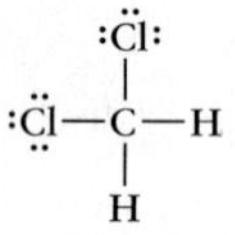 Tetrahedral

(c) 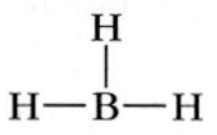 Triangular planar

(d) 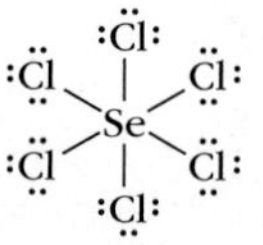 Octahedral

(e) 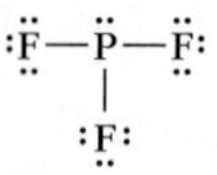 Triangular pyramidal

15. (a) 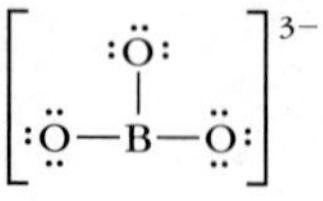 Triangular planar, triangular planar

(b) 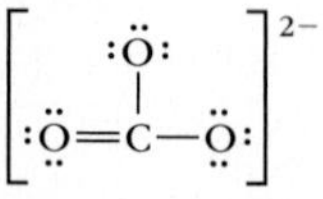 Triangular planar, triangular planar

(c) 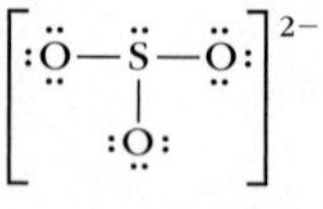 Tetrahedral, triangular pyramidal

(d) 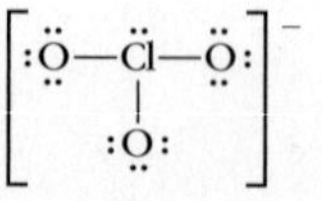 Tetrahedral, triangular pyramidal

Three atoms bonded to the central atom gives triangular shape; however, structures with 26 electrons have the shape of a triangular pyramidal and structures with 24 electrons have the shape of a triangular planar.

17. 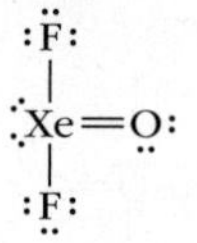

Triangular bipyramidal, T-shaped

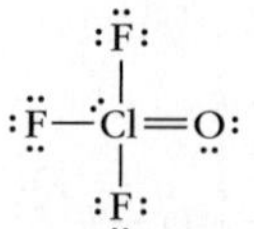

Triangular bipyramidal, seesaw

Their molecular geometries are different.

19. (a) Both octahedral
(b) Triangular bipyramidal and seesaw
(c) Both triangular bipyramidal
(d) Octahedral and square planar

22. (a) 120° (b) 120°
(c) H—C—H angle is 120°, C—C—N angle is 180°
(d) H—O—N angle is 109.5°, O—N—O angle is 120°

24. (a) four at 90°, one at 120°, and one at 180°
(b) 90° and 120°
(c) eight at 90° and two at 180°

26. The O—N—O angle in NO_2^+ is larger than in NO_2.

28. (a) 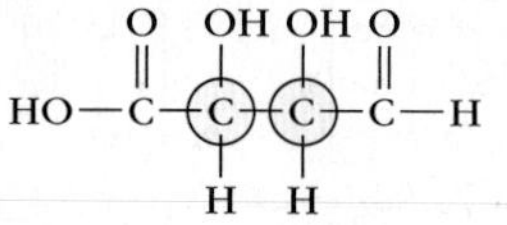 (b) No chiral centers

(c) 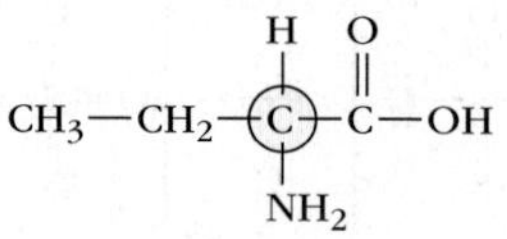

30. (a) No chiral centers (b) No chiral centers

(c) 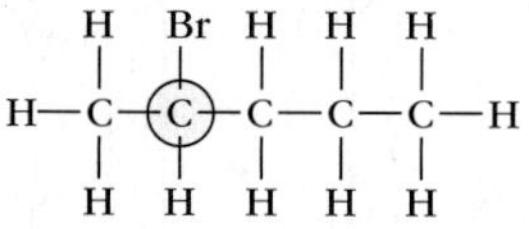 (d) No chiral centers

32. Tetrahedral, sp^3 hybridized carbon atom

34. The central S atom in SCl_2 has sp^3 hybridization; the central C atom in OCS has sp hybridization.

35. (a) Methyl carbon: sp^3 carboxyl carbon: sp^2
(b) Methyl carbon: 109.5°, carboxyl carbon: 120°

37. The N atom is sp^3 hybridized with 109.5° angles. The first two carbons are sp^3 hybridized with 109.5° angles. The third carbon is sp^2 hybridized with 120° angles. The single-bonded oxygen is sp^3 hybridized with 109.5° angles. The double-bonded O is sp^2 hybridized with 120° angles. The shortest carbon-to-oxygen bond length is C=O.

39. (a) 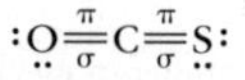 (b) 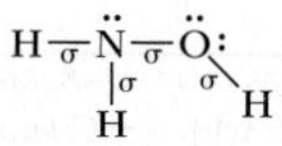

(c) 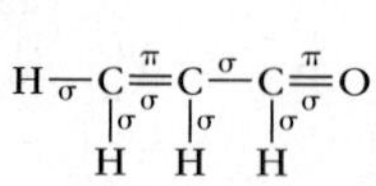 (d) 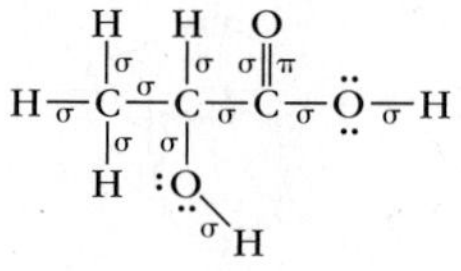

41. 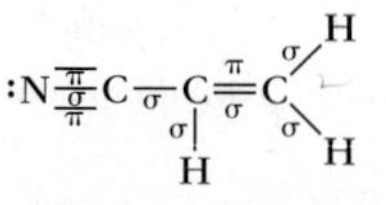

(a) Six (b) Three (c) *sp* (d) *sp* (e) Both sp^2

43. (a) Nonpolar
(b) Polar; the H-side of the molecule is more positive.
(c) Polar; the H-side of the molecule is more positive.
(d) Nonpolar
45. (a) The Br—F bond has a larger electronegativity difference.
(b) The H—O bond has a larger electronegativity difference.

46. (a) 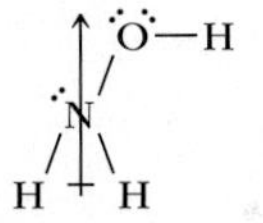 (b) 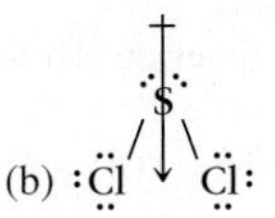

48.

Noncovalent Interaction	Strength	Example
Ion-ion	Greatest strength	Na^+ interaction with F^-
Ion-dipole	High strength	Na^+ ions in H_2O
Dipole-dipole	Medium strength	H_2O interaction with H_2O
Dipole-induced dipole	Low strength	H_2O interaction with CH_4
Induced dipole-induced dipole	Lowest strength	CH_4 interaction with CH_4

49. Water molecules can hydrogen bond to the lone pairs on dimethyl ether, but do not do so with the lone pairs on S in dimethyl sulfide.
51. Wax molecules interact using London forces. Water molecules interact using hydrogen bonding. The water molecules interact much more strongly with other water molecules; hence beads form as the water tries to avoid contact with the surface of the wax.
53. (c), (d), and (e)
55. (b) < (c) < (d) < (a); Hydrogen bonding interactions are stronger than dipole-dipole interactions and London forces are the weakest. London forces increase as the number of electrons increases.
57. Vitamin C is capable of forming hydrogen bonds with water.
60. (a) London forces (b) London forces
(c) Intramolecular (covalent) forces
(d) Dipole-dipole force (e) Hydrogen bonding
61.

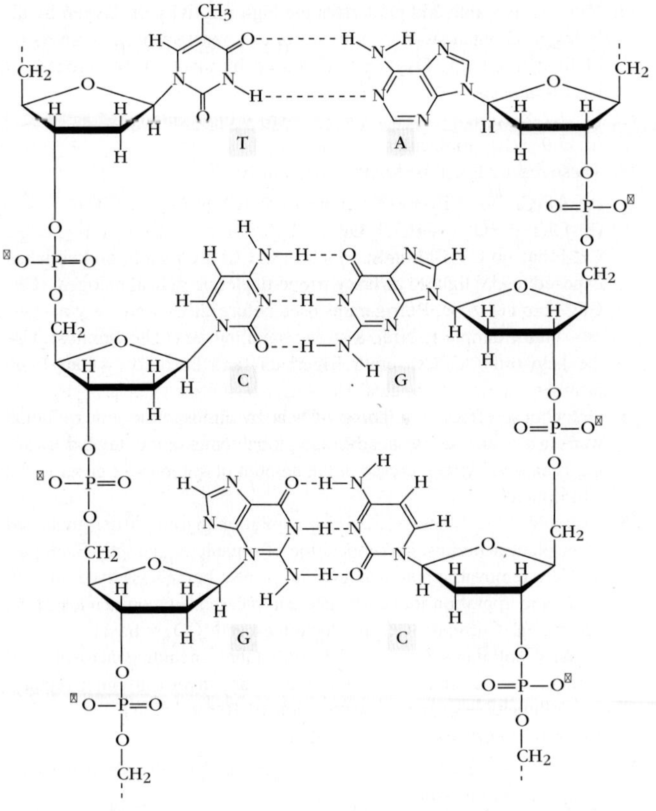

63. (a) UV > IR (b) UV < IR (c) UV > IR
65. (a) C—C stretch (b) O—H stretch
67. (a) (1) NCC angle is 180°, (2) HCH angle is 109.5°,
(3) HCO angle is 109.5°
(b) C═O (c) C═O (d) C═C
69. (a) $\ddot{O}$═C═C═C═$\ddot{O}$ (b) 180° (c) 180°
72. It has an extended conjugated pi-system, so yes, it should absorb visible light.
73. (a) 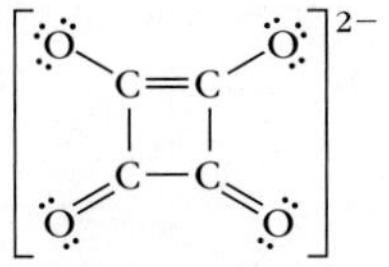 (b) Each C is sp^2 hybridized.
(c) Inconsistent, because angles are 90° not 120°
75. (a) For HCl: 2.69×10^{-20} C;
For HF: 6.95×10^{-20} C
(b) Partial negative charge on F is larger than that on Cl, so F has a higher electronegativity.
76. KF has $\delta^+ = 1.32 \times 10^{-19}$ C, which is 81.6% of the charge of the electron, so KF is not completely ionic.
78. If the two atoms are the same element, then the molecule will be nonpolar. If the two atoms have different electronegativities, the molecule will be polar.
79. (a) A, B, D, and H
(b) D and H
(c) A and G
(d) E and I
(e) B, C, D, E, F, H, and I
(f) A
(g) C, PH_3; G, F_2
(h) C, PH_3; G, F_2
(i) C, PF_3; G, HF
(j) F
81. CH_3SH has weaker intermolecular forces.
83. (a) 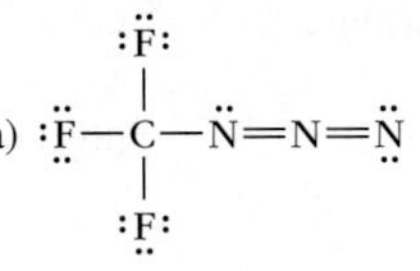
120°
(b) 180°
(c) sp^2, sp^3
(d) 
88. The molecules are polar, so they interact with dipole-dipole forces.
90. (a) Nitrogen (b) Boron (c) Phosphorus
(d) Iodine (There are other possible answers.)
92. Five

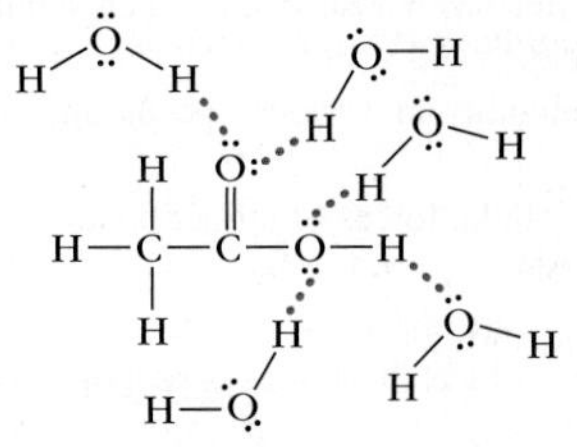

94. Diagram (d) is correct.
96. 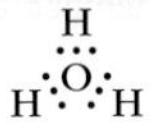 Triangular planar; Triangular planar 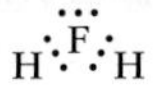 Triangular planar; bent

98. (a)

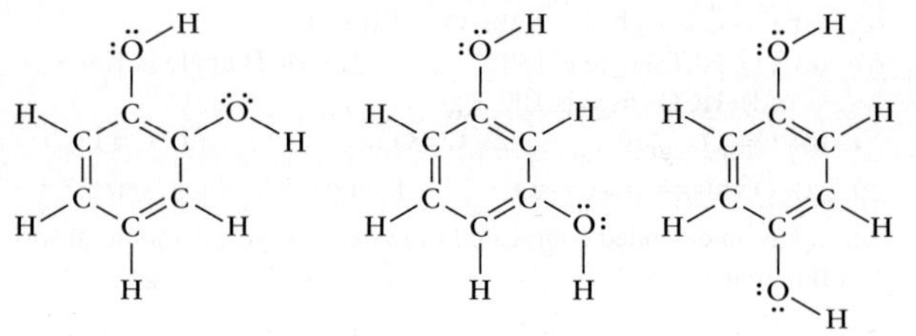

(i) (ii) (iii)

(b) (i) < (ii) < (iii)

100. (a)

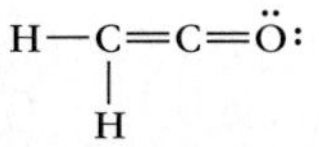

(b) Left C is triangular planar, H—C—C and H—C—H bond angles both 120°; right C is linear, H—C—C angle is 120°, and C—C—O angle is 180°.

(c) sp^2, sp, sp

(d) Polar, because of polar C=O bond

102. (a)

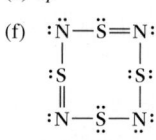

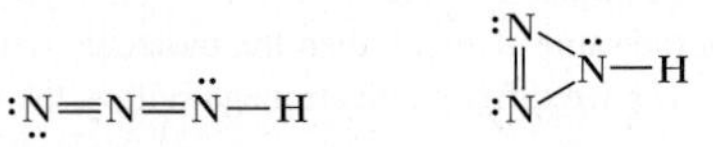

hydrogen azide cyclotriazene

(b) sp, sp, sp^2 (c) sp^2, sp^2, sp^3

(d) 3, 4 (e) 2, 1

(f) Hydrogen azide 120°, 180°; cyclotriazene 60°, 60°, 109.5°

104. (a) Angle 1: 120°, angle 2: 120°, angle 3: 109.5° (b) sp^3

(c) For O with two single bonds, sp^3; for O with one double bond, sp^2

106. (a)

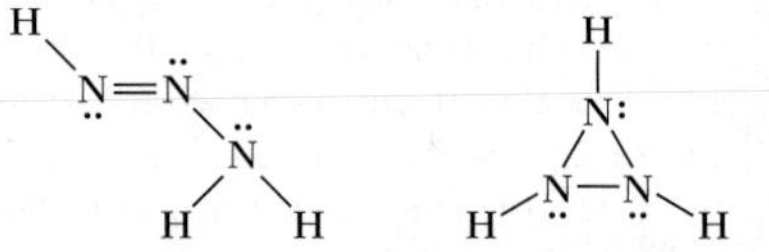

Triazene Triaziridine

(b) 120°, 60°

108. (a) sp^3 (b) sp^2 (c) sp^2 (d) 8σ, 2π

(other answers are possible)

Chapter 8

10. (a) 0.947 atm (b) 950. mmHg (c) 542 Torr (d) 98.7 kPa (e) 6.91 atm

12. 14 m

13. Atmospheric pressure can only push water up to about 33.8 feet.

15. I. A gas is composed of molecules whose size is much smaller than the distances between them.
 II. Gas molecules move randomly at various speeds and in every possible direction.
 III. Except when molecules collide, forces of attraction and repulsion between them are negligible.
 IV. When collisions occur, they are elastic.
 V. The average kinetic energy of gas molecules is proportional to the absolute temperature.

 Postulate I will become false at very high pressures. Postulates III and IV will become false at very high pressures or very low temperatures. Postulate II is most likely to always be correct.

17. 25.5 L

19. 172 mmHg

21. 26.5 mL

23. 4.00 atm

25. 501 mL

27. 0.51 atm

29. Largest number in sample (d); smallest number in sample (c)

31. 2.7×10^3 mL

33. 130. g/mol

35. 3.7×10^{-4} g/L

37. P_{He} is 7.000 times greater than P_{N_2}.

39. 8.9 L H_2

41. 10.4 L O_2, 10.4 L H_2O

43. 0.38 atm

45. 88 g/mol

47. (a) $2\ C_8H_{18}(\ell) + 25\ O_2(g) \longrightarrow 16\ CO_2(g) + 18\ H_2O(g)$
 (b) 1.4×10^3 L CO_2

50. $P_{tot} = 4.51$ atm

52. (a) 154 mmHg
 (b) $X_{N_2} = 0.777$, $X_{O_2} = 0.208$, $X_{Ar} = 0.0093$, $X_{H_2O} = 0.0053$, $X_{CO_2} = 0.0003$
 (c) 77.7% N_2, 20.8% O_2, 0.93% Ar, 0.03% CO_2, 0.54% H_2O. This sample is wet.

54. (a) $P_{tot} = 1.98$ atm
 (b) $P_{O_2} = 0.438$ atm, $P_{N_2} = 0.182$ atm, $P_{Ar} = 1.36$ atm

56. Membrane irritation: 1×10^{-4} atm; fatal narcosis: 0.02 atm

57. 0.0041; 3.1 mmHg; the mean partial pressure includes humid and dry air, summer and winter, worldwide.

58. 74.7%

61. $CH_2Cl_2 < Kr < N_2 < CH_4$

63. Ne; it has the lowest mass, so fastest average velocity, and reaches the end first

66. 18 mL $H_2O(\ell)$, 22.4 L $H_2O(g)$; No, because the vapor pressure of water at 0 °C is 4.6 mmHg; we cannot achieve 1 atm pressure for this gas at this temperature.

67. Gas molecule strikes the walls of the container with less force because of the attractive forces between it and its neighbors. Lower force leads to lower pressure.

70. N_2 is more like an ideal gas at high P than CO_2.

73. Nitrogen is stable and moderates the high reactivity of oxygen by diluting it. Oxygen sustains animal life as a reactant in the conversion of food to energy. Oxygen is produced by plants in the process of photosynthesis.

75. (a) 1.2×10^4 L (b) 2 times more
 (c) 5.9×10^{25} molecules

78. These reactions can be found in Section 10-11.
 (a) $CF_3Cl \xrightarrow{h\nu} \cdot CF_3 + \cdot Cl$ (b) $\cdot Cl + \cdot O \cdot \longrightarrow ClO\cdot$
 (c) $ClO\cdot + \cdot O \cdot \longrightarrow \cdot Cl + O_2$

80. CH_3F has no C—Cl bonds, which in CH_3Cl are readily broken when exposed to UV light to produce ozone-depleting radical halogen, ·Cl.

82. CFCs are not toxic. Refrigerants used before CFCs were very dangerous. One example is NH_3, a strong-smelling, reactive chemical. Use the keywords "CFCs" or "refrigerants" and "toxicity" in any web browser.

83. Greenhouse effect = trapping of heat by atmospheric gases. Global warming = increase of the average global temperature. Global warming is caused by an increase in the amount of greenhouse gases in the atmosphere.

85. Examples of CO_2 sources: animal respiration, burning fossil fuels and other plant materials, decomposition of organic matter, etc. Examples of CO_2 removal: photosynthesis in plants, being dissolved in rain water, incorporation into carbonate and bicarbonate compounds in the oceans, etc. Currently, CO_2 production exceeds CO_2 removal.

87. Primary pollutants (e.g., particle pollutants, including aerosols and particulates, sulfur dioxide, nitrogen oxides, hydrocarbons); secondary pollutants (e.g., ozone; see Section 10-10 for details).

89. 1.6×10^9 metric tons, 2.3×10^6 hours

91. $N_2 + O_2 \xrightarrow{heat} 2\ NO$; reaction takes elemental nitrogen and makes a compound of nitrogen

93. (a) $X_{SO_2} = 1.75 \times 10^{-7}$ (b) $P_i = 1.33 \times 10^{-4}$ mmHg
(c) 500. μg SO_2
94. (a) Before: $P_{H_2} = 3.7$ atm; $P_{Cl_2} = 4.9$ atm
(b) Before: $P_{tot} = 8.6$ atm (c) After: $P_{tot} = 8.6$ atm
(d) Cl_2; 0.5 mol remains (e) $P_{HCl} = 7.4$ atm; $P_{Cl_2} = 1.2$ atm
(f) $P = 8.9$ atm
97. Three times bigger
99. Use F = mg = PA, to derive m = PA/g and appropriate metric conversions.
102. 3.2×10^8 molecules/cm^3
105. Statements (a), (b), (c), and (d) are true.
107.

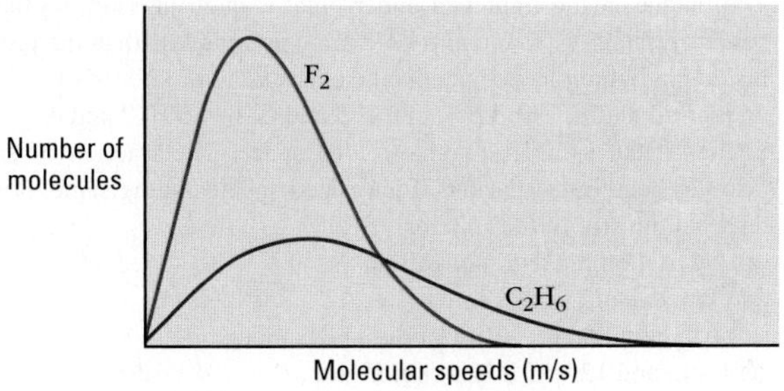
Number of molecules
F_2
C_2H_6
Molecular speeds (m/s)

110. (a) Box (iii) = Box (iv) > Box (i) = Box (ii)
(b) Box (iii) > Box (i) = Box (iv) > Box (ii)
(c) Box (i) = Box (ii) = Box (iii) = Box (iv)
(d) Average Molecular Speed: Box (ii) = Box (iv) > Box (i) = Box (iii)
112. For reference, the initial state looks like this:

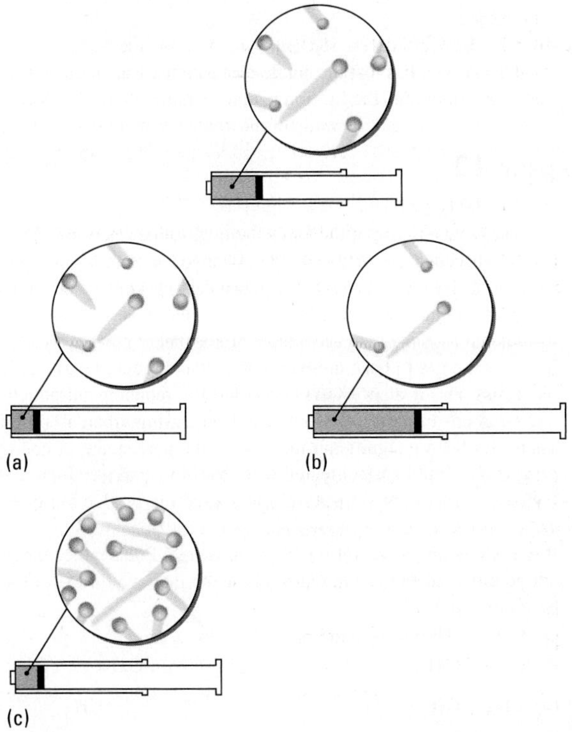
(a)
(b)
(c)

114. Box (b). The initial-to-final volume ratio is 2:1, so, for every two molecules of gas reactants, there must be one molecule of gas products. Six reactant molecules must produce 3 product molecules.
116. (a) 64.1 g/mol
(b) Empirical formula = CHF; molecular formula = $C_2H_2F_2$
(c)

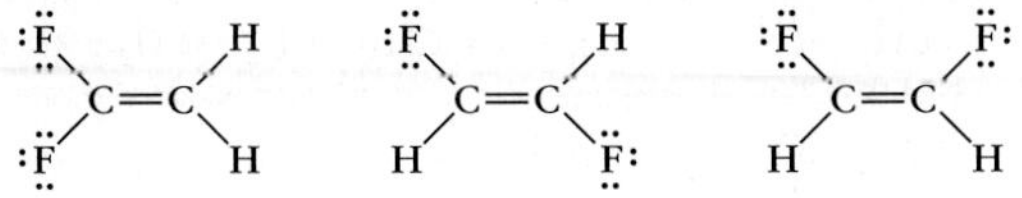

123. (a) 7.93×10^{-4} L (b) 0.000135
126. $\frac{m_{Ne}}{m_{Ar}} = 1.0$
127. 458 Torr
128. (a) More significant, because of more collisions
(b) More significant, because of more collisions
(c) Less significant, because the molecules will move faster

Chapter 9

11. (b) < (a) < (d) < (c)
(a) London forces
(b) London forces
(c) H-bonding, dipole-dipole, and London forces
(d) Dipole-dipole and London forces
13. If a surface molecule is moving in the right direction and has more kinetic energy than the potential energy it experiences with attractive intermolecular forces to other molecules, it will leave the liquid phase. A gas molecule that hits the surface of the liquid may lose some kinetic energy during the collision with other liquid molecules and new attractive intermolecular forces among them hold it in the liquid state.
15. 181 kJ
17. 73.4 kJ
19. The molecules in the liquid state must gain sufficient energy to overcome the attractive noncovalent intermolecular forces among the liquid molecules to be able to enter the gas phase; hence, vaporization is endothermic.
21. NH_3 has a relatively large boiling point because the molecules interact using relatively strong hydrogen-bonding intermolecular forces. The increase in the boiling points of the series PH_3, AsH_3, and SbH_3 is related to the increasing London dispersion intermolecular forces experienced, due to the larger central atom in the molecule. (Size: P < As < Sb)
23. Methanol molecules are capable of hydrogen bonding, whereas formaldehyde molecules use dipole-dipole forces to interact. Molecules experiencing stronger intermolecular forces (such as methanol here) will have higher boiling points and lower vapor pressures compared with molecules experiencing weaker intermolecular forces (such as formaldehyde here).
24. (a) 0.21 atm (b) approximately 57 °C
26. Substance D; it has lowest vapor pressure.
27. 5×10^2 mmHg
29. 39.4 kJ/mol
31. It is a molecular solid composed of small nonpolar molecules or atoms.
33. I_2; stronger London dispersion forces
35. 27 kJ
37. 51.9 g CCl_2F_2
38. Interionic attractions in the solid MgO (with Mg^{2+} and O^{2-}) are stronger than those in solid NaF (with Na^+ and F^-).
40. Highest melting point is (b) CaO.
Lowest melting point is (c) CO.
42. The freezer compartment of a frost-free refrigerator keeps the air so cold and dry that ice inside the freezer compartment sublimes [(s) → (g)]. The hailstones would eventually disappear.
43. (a) Region A: solid; region B: liquid; region C: gas
(b) Point 1: solid and gas; point 2: solid, liquid, and gas; point 3: liquid and gas; point 5: solid and liquid
45. Ideal gas has density of 0.13 g/mL; much smaller than the real density of CO_2 (0.47 g/mL).
46. (a) Ionic solid (b) Molecular solid
(c) Amorphous solid (vitreous silica) or network solid (quartz)
(d) Network solid
48. (a) Molecular or network solid
(b) Metallic solid (c) Network solid
(d) Molecular solid; compare properties with the various types
50. (a) Molecular solid (b) Ionic solid
(c) Metallic solid or network solid
(d) Amorphous solid

52. The crystal structure in ice maximizes the hydrogen-bonding capacity and leaves considerable open spaces compared with the liquid. See discussion in Section 9-5 for more details.
54. Surface tension is based on the ability of particles in a liquid to interact with other particles in the liquid. At higher temperatures, the molecules move more. Increased random motion disrupts the intermolecular interactions responsible for surface tension.
55. Water molecules interact using relatively strong hydrogen-bonding intermolecular forces, so greater kinetic energy is needed for molecules to escape into the gas phase.
58. See Figure 9.22 and its description.
59. 220 pm
61. Four unit cells share each of the Na^+ ions in the front face. Eight unit cells share the corner Cl^- ion in the front face. Two unit cells share the face-centered Cl^- ion in the front face.
63. No; the ratio of ions in the unit cell must reflect the empirical formula of the compound.
65. 0.533 g/cm^3
67. Carbon atoms in diamond are sp^3 hybridized and are tetrahedrally bonded to four other carbon atoms. Carbon atoms in pure graphite are sp^2 hybridized and bonded with a triangle planar shape to other carbon atoms. These bonds are partially double bonded so they are shorter than the single bonds in diamond. However, the planar sheets of sp^2 hybridized carbon atoms are only weakly attracted by intermolecular forces to adjacent layers, so these interplanar distances are much longer than the C—C single bonds. Therefore, graphite is less dense than diamond.
69. Diamond and graphite are examples of network solids. Neither is soluble in water or most other solvents. Atoms in a covalent network are held together by covalent bonds, which are much stronger than intermolecular forces among the atoms and molecules of a solvent. Therefore, dissolving would be highly endothermic and is not likely to occur.
71. $\nu = 5.30 \times 10^{17}\ s^{-1}$ (a) 3.51×10^{-16} J/photon (b) 2.11×10^{8} J/mol, X-ray
73. 361 pm
75. Metals (Fe), ceramics (brick), polymer (wool), composites (fiber-reinforced polymer)
78. In a conductor, the valence band is only partially filled, whereas in an insulator, the valence band is completely full, the conduction band is empty, and there is a wide energy gap between the two. In a semiconductor, the gap between the valence band and the conduction band is very small so that electrons are easily excited into the conduction band.
80. Substance (c), Ag, has the greatest electrical conductivity because it is a metal. Substance (d), P_4, has the smallest electrical conductivity because it is a nonmetal. (The other two are metalloids.)
82. A superconductor is a substance that is able to conduct electricity with no resistance. Two examples are $YBa_2Cu_3O_7$ and $Hg_{0.8}Tl_{0.2}Ba_2Ca_2Cu_3O_{8.23}$.
84. Doping is the intentional addition of small amounts of specific impurities into very pure silicon. Group III elements are used because they have one less electron per atom than the Group IV silicon. Group V elements are used because they have one more electron per atom.
87. The amorphous solids known as glasses are different from SiO_2 because they lack symmetry or long-range order, whereas crystalline covalent network solids such as SiO_2 are extremely symmetrical. SiO_2 must be heated above its melting point and then cooled quickly to make a glass.
89. (a) Two examples of silicate ceramics are aluminosilicates, such as kaolinite, and calcium silicate, Ca_2SiO_4.
(b) Two examples of oxide ceramics: Al_2O_3 and MgO.
(c) Two examples of nonoxide ceramics: Si_3N_2 and SiC.
90. With the lid on, the temperature and internal pressure increase more rapidly, creating higher vapor pressure and faster boiling.
93. Liquid phase; at the given pressure, the given temperature is higher than the melting point and lower than the boiling point.
95. (a) Approximately 560 mmHg (b) Benzene (c) 73 °C
(d) Methyl ethyl ether, approximately 7 °C; carbon disulfide, approximately 47 °C; benzene, approximately 81 °C
97. 2.18 kJ
99. Dipole-dipole forces and London forces
103. (a) Ionic; solubility in water and high density suggests large heavy ions.
(b) Molecular; solubility in benzene and low density suggests it is a nonpolar molecular compound.
107. Vapor-phase water condenses on contact with skin and condensation is exothermic, which imparts more energy to the skin.
110. The butane in the lighter is under great enough pressure so that the vapor pressure of butane at room temperature is less than the pressure inside the lighter. Hence the butane exists as a liquid.
112. 1 and C, 2 and E, 3 and B, 4 and F, 5 and G, 6 and H, 7 and A
114. Each has the same fraction of filled space. The fraction of spaces filled by the closest packed, equal-sized spheres is the same, no matter what the size of the spheres.
116. (a) Two: rhombic and monoclinic
(b) Three: B, C, and E
(c) At 1 atm and 80 °C it is a rhombic solid.
At 1 atm and 125 °C it is a liquid.
(d) At low pressures, it is a gas, between 5×10^{-4} atm and 10^3 atm it is a liquid, and above 10^3 atm it is a rhombic solid.
(e) $P < 10^{-4}$ atm and $T < 120$ °C
(f) Monoclinic solid
(g) Rhombic to monoclinic to liquid (briefly) to vapor
(h) Normal melting point = 120 °C
120. (a) 152 pm
(b) $r_{I^-} = 212$ pm, $r_{Li^+} = 88.0$ pm
(c) The Li^+ ion has lost the single electron that was in the $n = 2$ shell in the Li atom; thus, the Li^+ ion is much smaller than a Li atom.

Chapter 10

18. (a) 20–200 °C
(b) The octane rating of the straight-run gasoline fraction is 55.
(c) No; because the octane rating is far lower than regular gasoline we buy at the pump (86–94), which means it would cause far more pre-ignition than we expect from the gasoline. It would need to be reformulated to make it an acceptable motor fuel.
21. It allows the gas to burn more completely and reduces carbon monoxide emissions. It also lowers the volatility, reducing introduction of hydrocarbons into the atmosphere and decreasing urban air pollution.
23. There has been a significant increase in the percentage of energy derived from natural gas compared with that from coal (see Figure 10.4).
25. Hydraulic fracturing ("fracking") increased the production of natural gas from 1% in 2000 to more than 20% in 2010.
28. It will show up earlier on the chromatograph. Polar molecules would not be attracted to the stationary phase, so they would exit the chamber more quickly.
30. (a) $CH_3—CH_2—CH_2—OH$
(b) 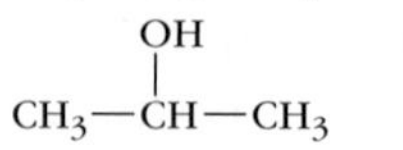
OH
CH₃—CH—CH₃
(c) 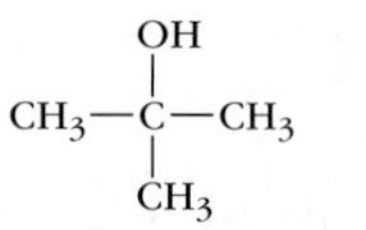
OH
CH₃—C—CH₃
CH₃
(There are other correct answers.)
32. (a) Tertiary (b) Primary (c) Secondary (d) Secondary (e) Tertiary (f) Secondary
34. (a) (b)
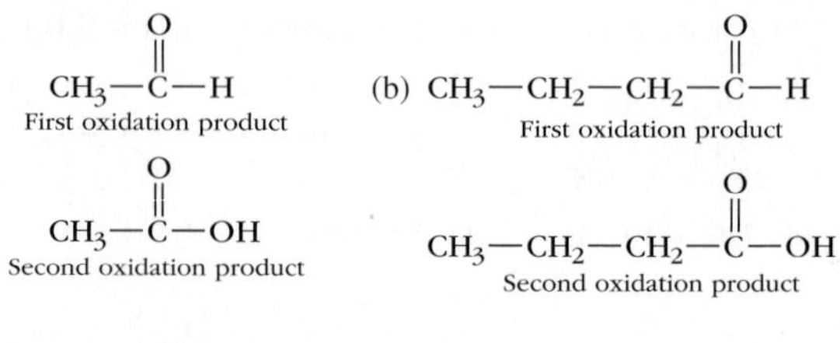
CH₃—C—H
First oxidation product
CH₃—CH₂—CH₂—C—H
First oxidation product
CH₃—C—OH
Second oxidation product
CH₃—CH₂—CH₂—C—OH
Second oxidation product

36. (a) 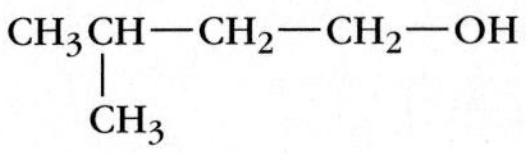

(b) 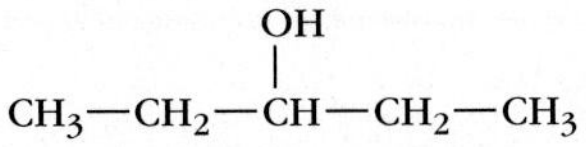

(c) 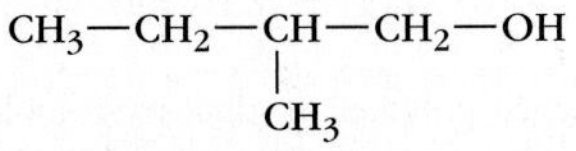

38. Wood alcohol (methanol) is made by heating hardwoods such as beech, hickory, maple, or birch. Grain alcohol (ethanol) is made from the fermentation of plant materials such as grains.

40. —OH groups are a common site of hydrogen-bonding intermolecular forces. Their presence would increase the solubility of the biological molecule in water and create specific interactions with other biological molecules.

41. Radio waves

43. 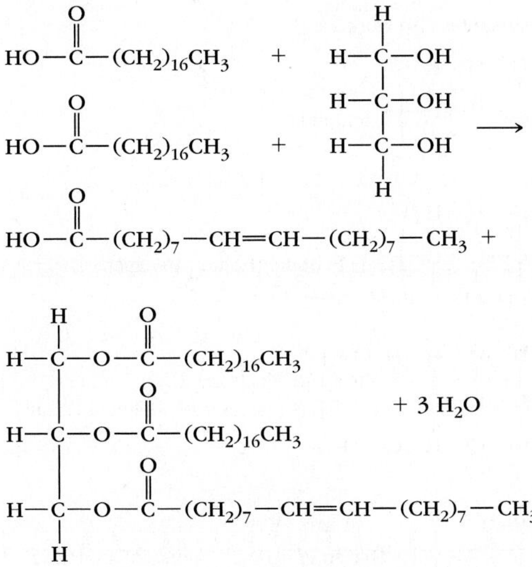

45. (a) (i) polyunsaturated (ii) monounsaturated (iii) saturated

(b)

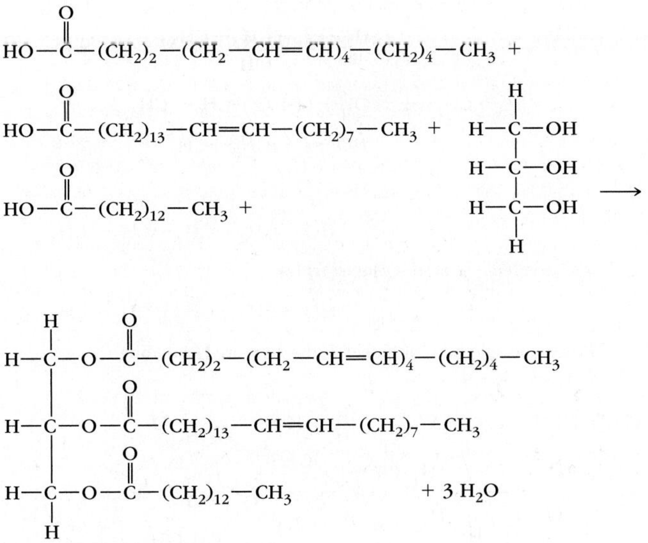

47. (a) Ester
(b) $CH_3—(CH_2)_{24}—COOH$, $HO—(CH_2)_{29}—CH_3$

50. (a) $CH_3COOCH_2CH_3$ (b) $CH_3CH_2COOCH_2CH_2CH_3$
(c) $CH_3CH_2COOCH_3$

52. (a) $CH_3CH_2COOH + CH_3OH$
(b) $HCOOH + CH_3CH_2OH$
(c) $CH_3COOH + CH_3CH_2OH$

55. Examples of thermoplastics are milk jugs (polyethylene), sunglasses and toys (polystyrene), and CD audio discs (polycarbonates). Thermoplastics soften and flow when heated.

57. (a) 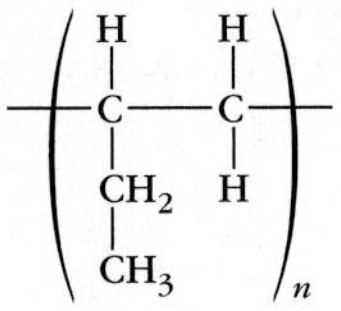 (b) 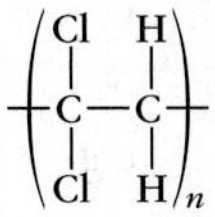

(c) 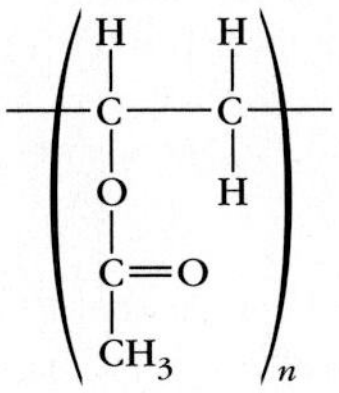

59. (a) $CH_2{=}C(CH_3)COOCH_3 + 2\ R\bullet \longrightarrow$

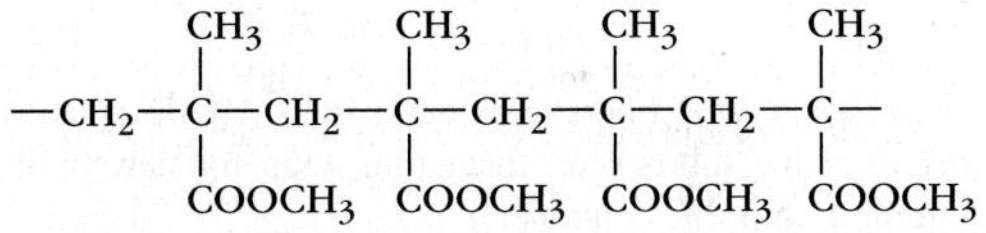

(b) 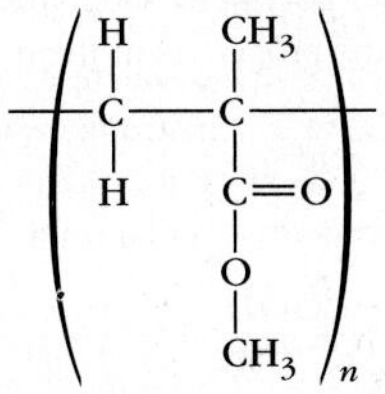

61. (a) $5\ H_2C{=}CCl_2 + 2\ \bullet OR \longrightarrow$

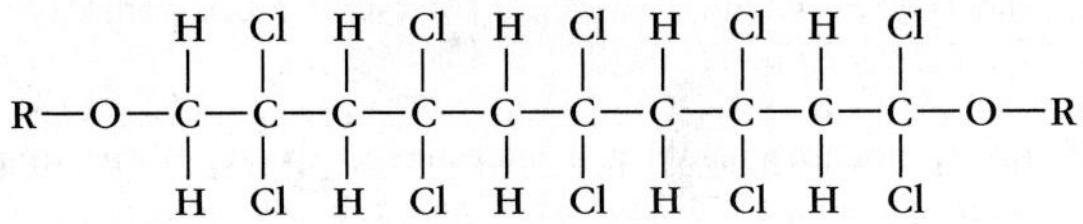

(b) $5\ F_2C{=}CF_2 + 2\ \bullet OR \longrightarrow$

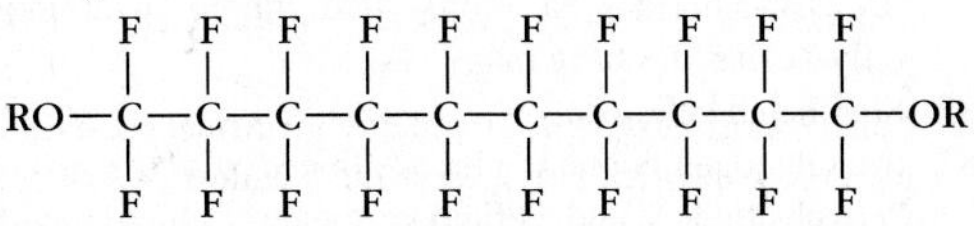

62. (a) 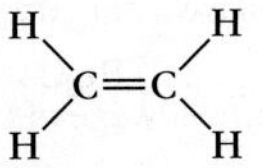 (b) 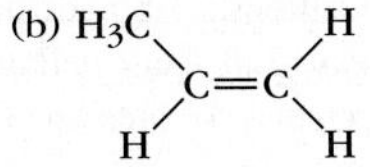

(c) 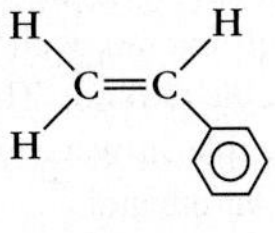 (d) 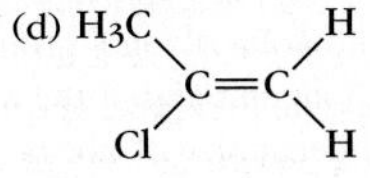 (e) 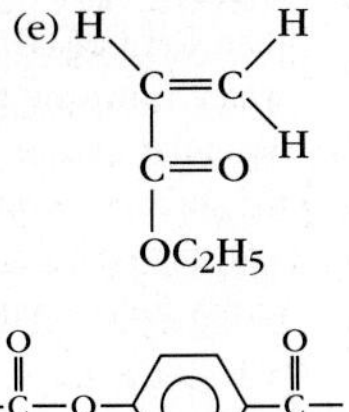

64. 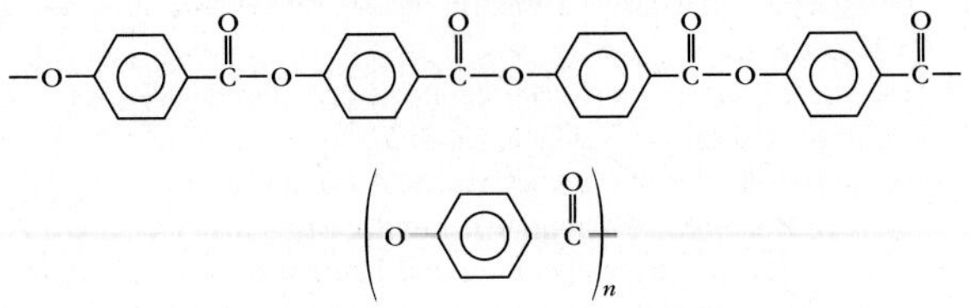

66. Carboxylic acid and alcohol

69. (a) HOOC—$(CH_2)_8$—COOH, H_2N—$(CH_2)_6$—NH_2

(b) 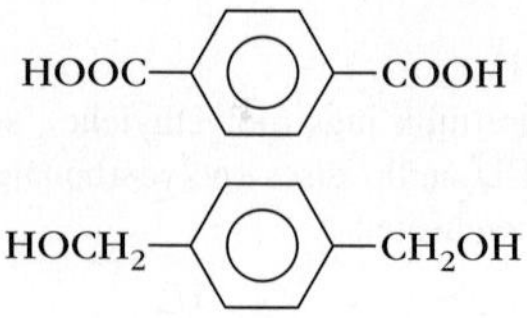

71. 1500 monomers

74. (a) 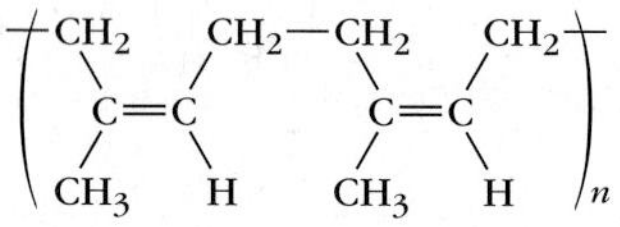

(b) 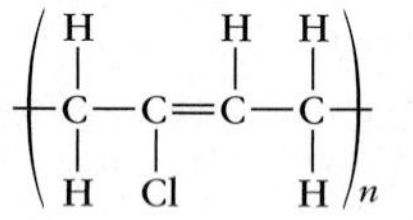

(c) 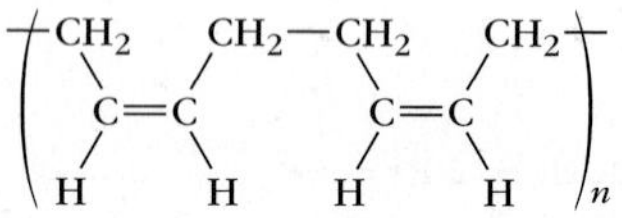

75. Major end uses for recycled PET include fiberfill for ski jackets and sleeping bags, carpet fibers, and tennis balls. HDPE is converted into a fiber used for sportswear, insulating wrap for new buildings, and very durable shipping containers.
76. One major difference is that the protein polymer's monomers are not all alike. Different side chains change the properties of the protein.
78. Proteins, DNA, RNA
80. Amine, amide, alcohol, and carboxylic acid
81. 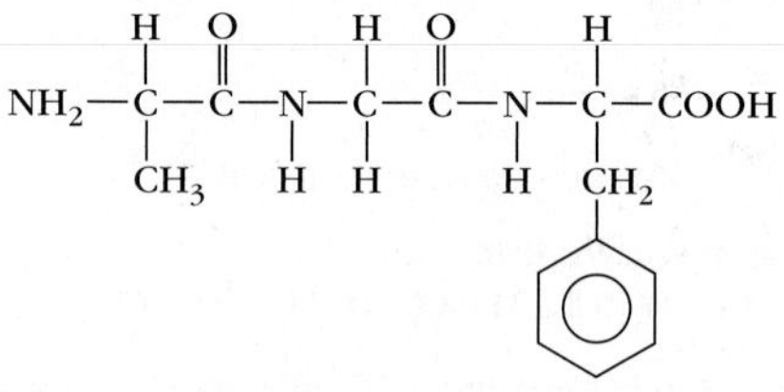

83. (a) A monosaccharide is a molecule composed of one simple sugar molecule, whereas disaccharides are molecules composed of two simple sugar molecules.
(b) Disaccharides have only two simple sugar molecules, whereas polysaccharides have many.
85. Starch and cellulose
87. (a) Glycogen contains glucose linked with the glycosidic linkages in "*cis*-positions," and cellulose contains glucose with the glycosidic linkages in "*trans*-positions." Humans do not have the enzymes required to break the *trans*-linkage in cellulose.
(b) Cows have the enzymes for breaking the *trans*-linkage of cellulose.
88. 1.0×10^{-3}%, 10. ppm
91. $CH_3CH_2CH_2CH_2CH_2CH_2CH_2CH_2CH_2CH_2OH$ is a larger molecule than CH_3CH_2OH. The polar alcohol group will interact well with the water; however, the nonpolar end of the molecule will not. The longer nonpolar end of the decanol will not be miscible in water, lowering the solubility compared with smaller, more polar ethanol.
93. Vulcanized rubber has short chains of sulfur atoms that bond together (cross-link) the polymer chains of natural rubber.

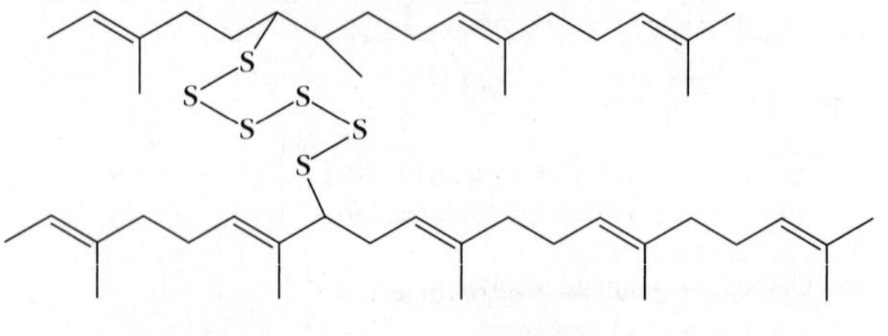

94. 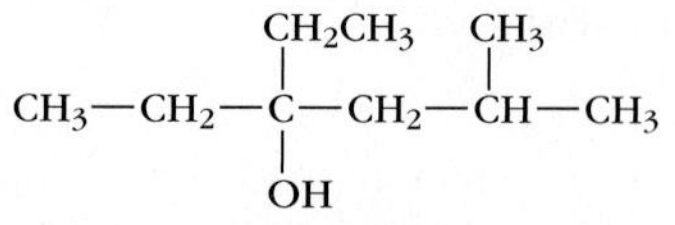

A tertiary alcohol

97. (a) $2\ C_8H_{18}(\ell) + 25\ O_2(g) \rightarrow 16\ CO_2(g) + 18\ H_2O(g)$
(b) 1.4×10^3 L CO_2
99. Glycogen is a highly branched polymer of glucose so molecules of glycogen can pack together like spheres to form granules. Cellulose is a linear polymer with many —OH groups along the sides so the molecules can easily pack side-by-side held by hydrogen bonds to form layers.
101. (a) CH_2
(b) Alkene
(c) 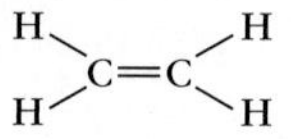

(Other structures also work.)

103. 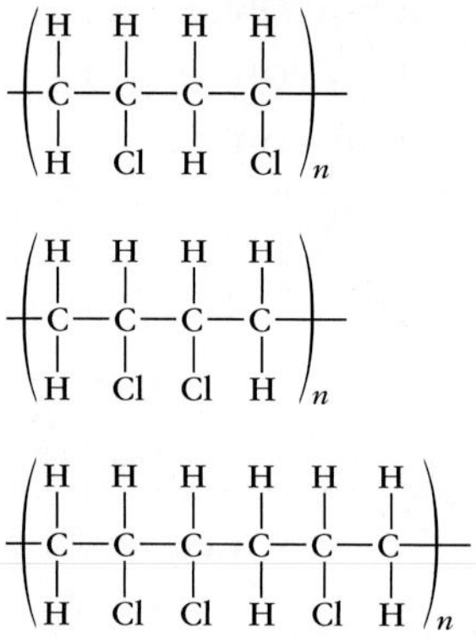

105. 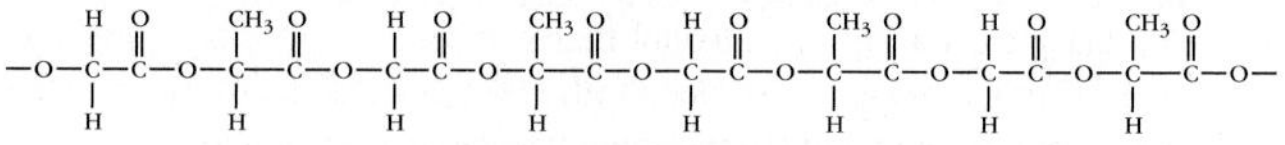

107. 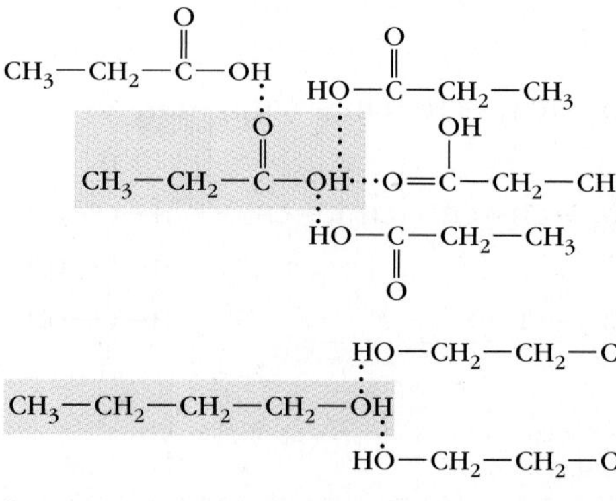

Propanoic acid boils at higher *T*, due to more H-bonding.

109. $CH_3—C\equiv C—H$

111. 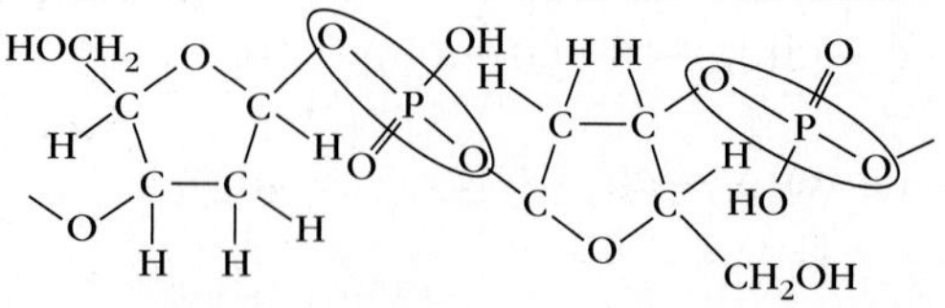

113. Some data that you would need to know are sources and amounts of CO_2 generated over time to determine additional CO_2, photosynthesis rate of depletion per tree per year, average number of trees per acre, the number of acres of land in Australia that could support trees, and the allowable tree density. Other data besides these may need to be contemplated, so consider it a challenge to think of other things you would need to know.

115. (a) Box C (b) Box G (c) Box G (d) Box F
(e) Box D (f) Box B (g) Boxes A and H
(h) Box C (i) Box I (j) Box E

117. (a) C_3H_8O
(b) $C_3H_8O(\ell) + \frac{9}{2}\,O_2(g) \rightarrow 3\,CO_2(g) + 4\,H_2O(g)$
(c) Empirical formula, because it only identifies the smallest ratio of atoms
(d)

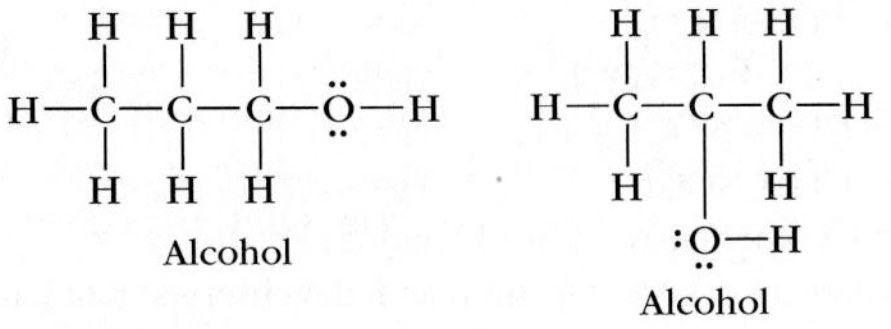
Alcohol
Alcohol
Ether

(e)

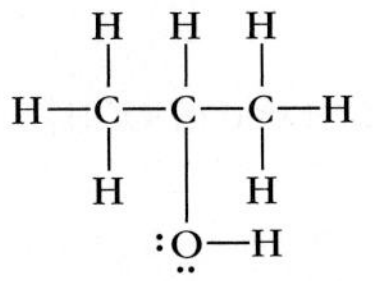

(f) −315 kJ/mol

123. (a) X is ethanol, Y is acetic acid, and Z is acetaldehyde
(b)

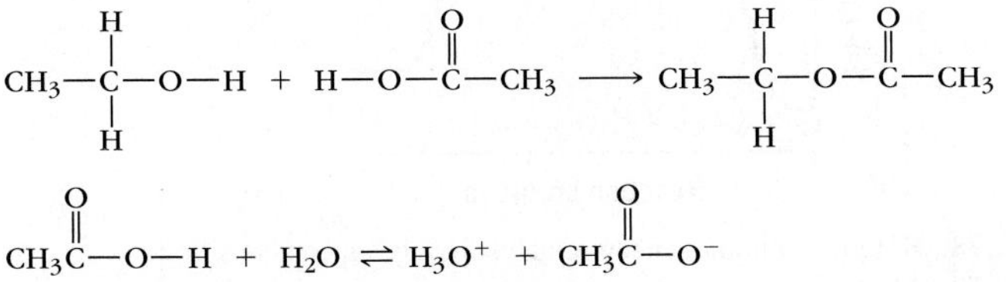

(c) Ethanol is a primary alcohol that is initially oxidized to acetaldehyde; further oxidation of acetaldehyde produced acetic acid. When acetaldehyde is the initial reactant, it is directly oxidized to acetic acid.

124.

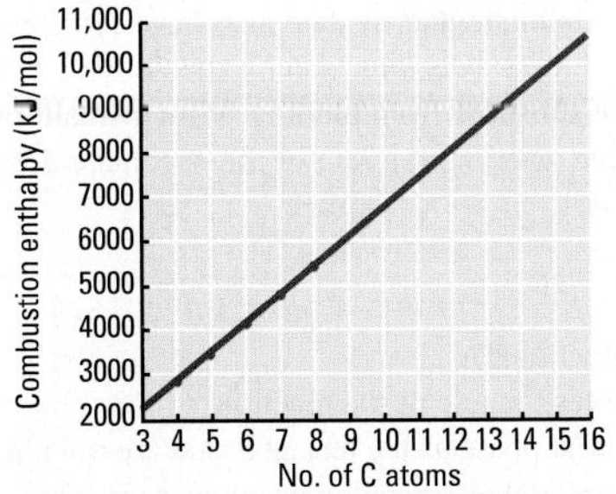
Combustion enthalpy (kJ/mol)
11,000
10,000
9000
8000
7000
6000
5000
4000
3000
2000
3 4 5 6 7 8 9 10 11 12 13 14 15 16
No. of C atoms

(a) Estimate: C_3H_8, 2200 kJ/mol; C_9H_{20}, 6100 kJ/mol; $C_{16}H_{34}$, 10,700 kJ/mol
(b) The bond enthalpies increase because breaking C—H bonds requires more energy than breaking C—C bonds.

126. (a)

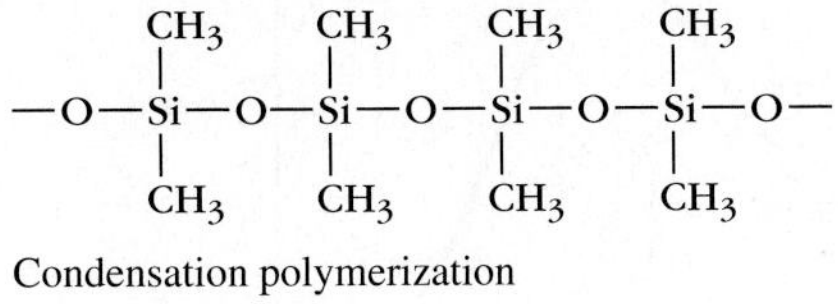

(b) Condensation polymerization

Chapter 11

9. (d) Dissolves fastest; (a) dissolves slowest. Rate of dissolving is larger when the grains of sugar are smaller, because there is more surface contact with the solvent.

11. (a)

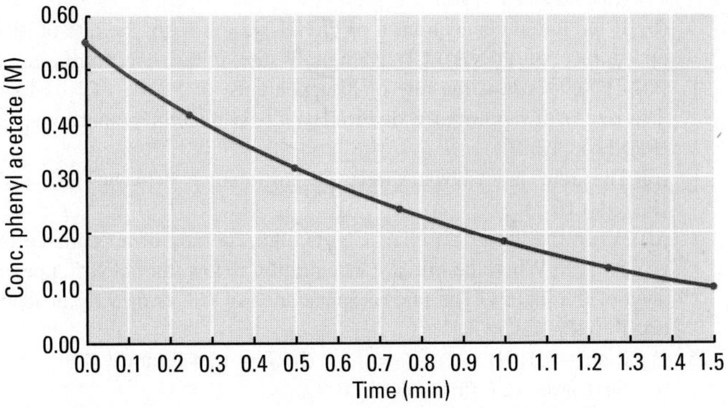
Conc. phenyl acetate (M)
0.60
0.50
0.40
0.30
0.20
0.10
0.00
0.0 0.1 0.2 0.3 0.4 0.5 0.6 0.7 0.8 0.9 1.0 1.1 1.2 1.3 1.4 1.5
Time (min)

It has a declining nonlinear shape, similar to that of Figure 11.3.
(b) 0.5 mol L^{-1} min^{-1}, 0.2 mol L^{-1} min^{-1}, concentration of reactant is decreasing
(c) 0.2 mol L^{-1} min^{-1}

13. (a) 0.23 mol L^{-1} h^{-1} (b) 0.20 mol L^{-1} h^{-1}
(c) 0.161 mol L^{-1} h^{-1} (d) 0.12 mol L^{-1} h^{-1}
(e) 0.090 mol L^{-1} h^{-1} (f) 0.066 mol L^{-1} h^{-1}

14. (a) Calculate the average concentration for each time interval and plot average concentration vs. average rate:

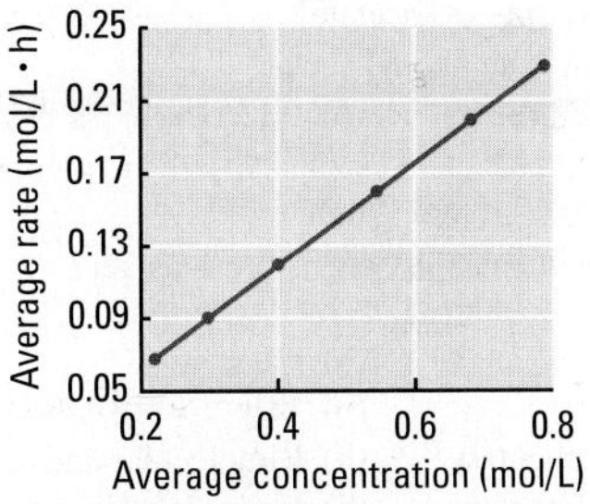
Average rate (mol/L • h)
0.25
0.21
0.17
0.13
0.09
0.05
0.2
0.4
0.6
0.8
Average concentration (mol/L)

The linear relationship shows: Rate = $k[N_2O_5]$.
(b) $k = 0.29\ h^{-1}$ (c) 0.12 mol L^{-1} h^{-1}

15.

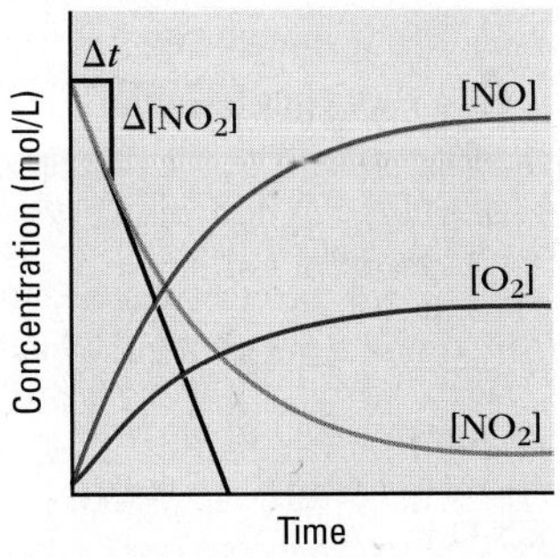
Δt
Δ[NO2]
[NO]
[O2]
[NO2]
Concentration (mol/L)
Time

(a) To calculate initial rate, obtain values for Δt and Δ(concentration) near initial time, where the curve can still be approximated by a straight line (the black line, above) then divide Δ(concentration) by Δt and change the sign to make it positive, for example,

$$\text{Initial rate} = \frac{-\Delta[NO_2]}{\Delta t}$$

(b) The curves on the graph become horizontal after a very long time, so the final rate will be zero.

17. (a) $\frac{-\Delta[N_2O_4]}{\Delta t}$ (b) $\frac{\Delta[N_2]}{\Delta t}$

19. (a) The rate increases by a factor of nine.
(b) The rate will be one fourth as fast.

21. (a) Rate = $k[NO_2]^2$
(b) The rate will be one fourth as fast.
(c) The rate is unchanged.

23. (a) (i) 9.0×10^{-4} M/h (ii) 1.8×10^{-3} M/h (iii) 3.6×10^{-3} M/h
(b) If the initial concentration of $Pt(NH_3)_2Cl_2$ is high, the rate of disappearance of $Pt(NH_3)_2Cl_2$ is high. If the initial concentration of $Pt(NH_3)_2Cl_2$ is low, the rate of disappearance of $Pt(NH_3)_2Cl_2$ is low. The rate of disappearance of $Pt(NH_3)_2Cl_2$ is directly proportional to $[Pt(NH_3)_2Cl_2]$.
(c) The rate law shows direct proportionality between rate and $[Pt(NH_3)_2Cl_2]$.
(d) When the initial $[Pt(NH_3)_2Cl_2]$ is high, the rate of appearance of Cl^- is high. When the initial concentration is low, the rate of appearance of Cl^- is low. The rate of appearance of Cl^- is directly proportional to $[Pt(NH_3)_2Cl_2]$.

25. (a) Rate $= k$[I][II] (b) $k = 1.04$ L mol^{-1} s^{-1}
(c) First-order in I, first-order in II
(d) 5.9×10^{-9} mol L^{-1} s^{-1}

28. (a) First-order in A and third-order in B, fourth-order overall
(b) First-order in A and first-order in B, second-order overall
(c) First-order in A and zero-order in B, first-order overall
(d) Third-order in A and first-order in B, fourth-order overall

30. Equation (a) cannot be right

32. (a) $\dfrac{-\Delta[NH_2]}{\Delta t} = k$
(b) 2.5×10^{-4} mol L^{-1} min^{-1}

33. (a) Rate $= k$[phenyl acetate]
(b) First-order in phenyl acetate
(c) $k = 1.259$ s^{-1} (d) 0.13 mol L^{-1} s^{-1}

35. (a) Rate $= k[A]^2$
(b) 1.6×10^{-5} L mol^{-1} min^{-1}
(c) The order for A is two.

37. 4.3 mg

39. (a) 0.16 mol/L (b) 90. s (c) 120 s

41. (a) 0.02 M (b) 14 y (c) 5.55 y

42. (a) Not elementary (b) Bimolecular and elementary
(c) Not elementary (d) Unimolecular and elementary

44. $NO + O_3$; NO is an asymmetric molecule and Cl is a symmetric atom.

46. Ratio = 1.8

48. 10.7 times faster

50. 3×10^2 kJ/mol

51. (a) $E_a = 115$ kJ/mol, $A = 1.98 \times 10^{13}$ s^{-1}
(b) $k = 1.5 \times 10^{-3}$ s^{-1}

53. (a) $E_a = 22.2$ kJ/mol, $A = 6.66 \times 10^7$ L^2 mol^{-2} s^{-1}
(b) $k = 8.39 \times 10^4$ L^2 mol^{-2} s^{-1}

55. (a) 3×10^{-20} (b) 4×10^{-16}
(c) 4×10^{-10} (d) 1.9×10^{-6}

57. (a) 1×10^{-5} mol L^{-1} s^{-1} (b) 2×10^1 mol L^{-1} s^{-1}

59. (a) Rate $= k$[Cl][ICl]
(b) Rate $= k[CH_3N{=}NCH_3]$
(c) Rate $= k[N_2O_4]$

61. (a)

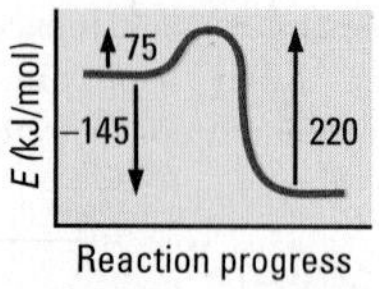
75
−145
220
E (kJ/mol)
Reaction progress

(b)

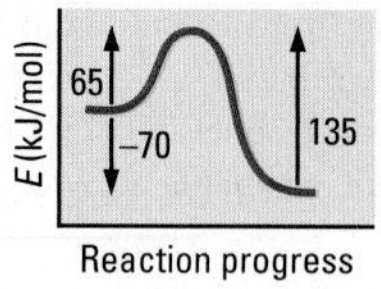
65
−70
135
E (kJ/mol)
Reaction progress

(c)

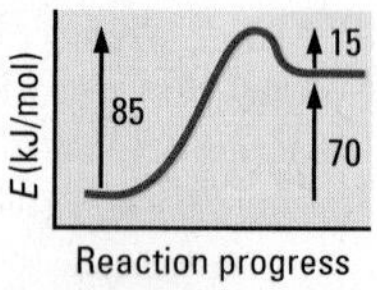
15
85
70
E (kJ/mol)
Reaction progress

63. (a) Reaction (b) (b) Reaction (c)

65. (a) Reaction (c) (b) Reaction (a)

67. (a) Rate $= k[NO][NO_3]$ (b) Rate $= k[O][O_3]$
(c) Rate $= k[(CH_3)_3CBr]$ (d) Rate $= k[HI]^2$

69. (a)
$$\begin{array}{l} NO_2 + F_2 \longrightarrow FNO_2 + \cancel{F} \\ + NO_2 + \cancel{F} \longrightarrow FNO_2 \\ \hline 2\,NO_2 + F_2 \longrightarrow 2\,FNO_2 \end{array}$$
(b) The first step is rate-determining.

71. (a) Rate $= k'[NO]^2[Cl_2]$
(b) $NO + NO \rightleftharpoons N_2O_2$ fast
$N_2O_2 + Cl_2 \rightarrow 2\,NOCl$ slow
(c) $NO + Cl_2 \rightarrow NOCl + Cl$ slow
$NO + Cl \rightarrow NOCl$ fast
There can be other answers for (b) and (c).

72. Only mechanism (a) is compatible with the observed rate law.

74. (a) is true; (b), (c), and (d) are false.
(b) A catalyst is never consumed in a reaction.
(c) A catalyst need not be the same phase as the reactants.
(d) A catalyst can change the course of a reaction, but the same products are always formed.

76. (a) $CH_3COOCH_3 + H_2O \rightleftharpoons CH_3COOH + CH_3OH$
(b) Rate $= k'[CH_3COOCH_3][H_3O^+]$
(c) Catalyst: H^+
(d) Intermediates: $H_3C(OH)OCH_3^+$, $H_3C(H_2O)(OH)OCH_3^+$, $H_3C(OH)_2OHCH_3^+$, and H_2O
(e)

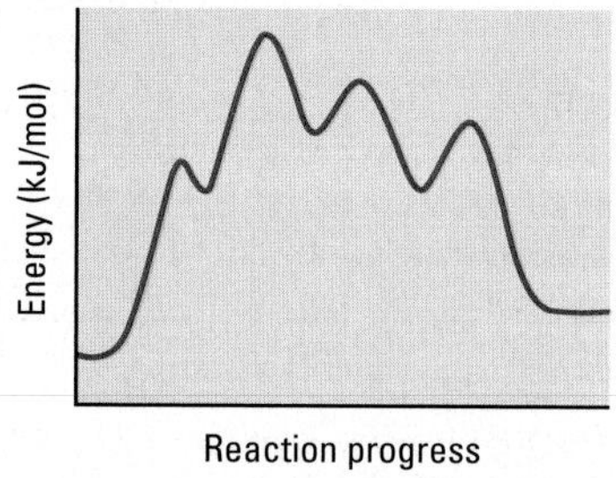
Energy (kJ/mol)
Reaction progress

78. $H^+(aq)$ is a homogeneous catalyst in this aqueous reaction.

79. 38 times

80. 30. times faster

82. The active site on the enzyme (where the substrate binds to form an intermediate complex) is designed to accommodate the substrate's four terminal O atoms. Malonate and oxalate also have four terminal O atoms. When either of these two ions occupy the active site of an enzyme, they prevent the substrate from binding. With a smaller number of active sites free for reaction, the rate of the succinate dehydrogenation reaction decreases.

84. (a) Rate $= k\dfrac{[X][HA]}{[A^-]}$
(b) First-order with respect to HA
(c) Doubling [HA] doubles the rate of the reaction

86. They have the same physical phase as the reactants and are often part of a homogeneous mixture, so simple separation techniques are ineffective.

87. (a) To maximize the surface area.
(b) Too little of the metal would be on the surface.

89.

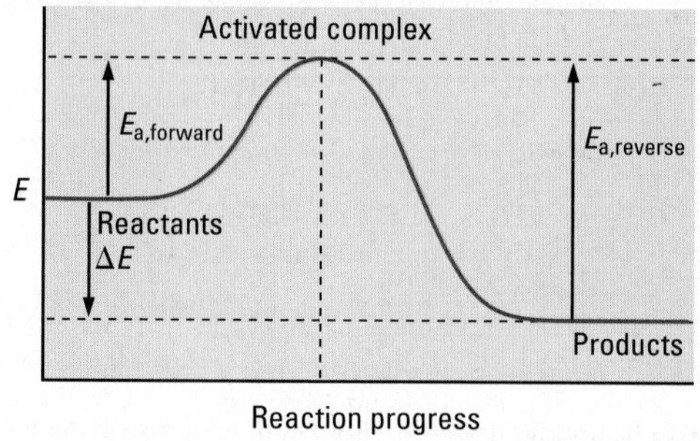
Activated complex
$E_{a,forward}$
$E_{a,reverse}$
E
Reactants
ΔE
Products
Reaction progress

$\Delta E = E_{a,\ forward} - E_{a,\ reverse}$

91. (a)

Time(s)	$[C_6H_{12}]$ (M)	$[C_{12}H_{10}]$ (M)	$[H_2]$ (M)
0.0	0.200	0.000	0.000
1.00	0.159	0.021	0.144
2.00	0.132	0.034	0.238
3.00	0.088	0.044	0.308

(b) 0.036 mol L^{-1} s^{-1}

92. (a) False. "The reaction might occur in a single step."
(b) True.
(c) False. "Raising the temperature will cause the value of k to increase."
(d) False. "The activation energy is independent of temperature."
(e) False. "If the concentration of both reactants are doubled, the rate will quadruple."
(f) True.

94. (a) NO is second-order, O_2 is first-order.
(b) Rate = $k[NO]^2[O_2]$
(c) 25 L^2 mol^{-2} s^{-1}
(d) 7.8×10^{-4} mol L^{-1} s^{-1}
(e) $-\dfrac{\Delta[NO]}{\Delta t} = 2.0 \times 10^{-4}$ mol L^{-1} s^{-1}

$+\dfrac{\Delta[NO_2]}{\Delta t} = 2.0 \times 10^{-4}$ mol L^{-1} s^{-1}

96. (a) First-order in $HCrO_4^-$, first-order in H_2O_2, and first-order in H_3O^+
(b) Cancel intermediates, H_2CrO_4 and $H_2CrO(O_2)_2$, and add the three reactions.
(c) Second step, because the rate law derived from the mechanism is the same as the observed rate law

99. (a) 2.8×10^3 s (b) 1.4×10^4 s (c) 2.0×10^4 s

101. (a) 270 kJ/mol (b) 2×10^3 s

102. Curve A represents $[H_2O(g)]$ increase with time, Curve B represents $[O_2(g)]$ increase with time, and Curve C represents $[H_2O_2(g)]$ decrease with time.

104. Snapshot (b); products form more slowly at lower temperatures, so choose the snapshot with fewer HI molecules

106. Rate = $k[A]^3[B][C]^2$

107. (a) Three, because each hill in the reaction energy diagram represents one elementary reaction
(b) Exothermic, because products have lower energy than reactants

108. E_a is very, very small—approximately zero.

110. 29.6 s, 94.7 s

113. (a) Rate = 2.4×10^{-7} mol L^{-1} s^{-1} + $k[BSC][F^-]$
(b) 0.3 L mol^{-1} s^{-1}

115. 0.127 s^{-1}, 54.6 s

117. Note: there are other correct answers to this question; the following are examples:

(a) $CH_3CO_2CH_3 + H^+ \longrightarrow CH_3COHOCH_3^+$	slow
$CH_3COHOCH_3^+ + H_2O \longrightarrow CH_3COH(OH_2)OCH_3^+$	fast
$CH_3COH(OH_2)OCH_3^+ \longrightarrow CH_3C(OH)_2^+ + CH_3OH$	fast
$CH_3C(OH)_2^+ \longrightarrow CH_3COOH + H^+$	fast
(b) $H_2 + I_2 \longrightarrow 2\ HI$	slow
(c) $H_2 + Pt(s) \longrightarrow PtH_2$	fast
$PtH_2 + I_2 \longrightarrow PtH_2I_2$	slow
$PtH_2I_2 + O_2 \longrightarrow PtI_2O + H_2O$	fast
$PtI_2O + H_2 \longrightarrow Pt(s) + I_2 + H_2O$	fast
(d) $H_2 \longrightarrow 2\ H$	fast
$H + CO \longrightarrow HCO$	slow
$HCO + H \longrightarrow H_2CO$	fast

119. The speed of the reaction is related to the energy of the production of a photon that can break a bond in iodine and make a free radical catalyst. Wavelengths less than 800 nm have photons with energy less than the bond enthalpy of I_2.

121. (i) Define the reaction rate in terms of [A]: Rate = $-\dfrac{\Delta[A]}{\Delta t}$.

(ii) Write the rate law using [A], k and t: $-\dfrac{\Delta[A]}{\Delta t} = k[A]$.

Calculus uses "d" instead of "Δ": $-\dfrac{d[A]}{dt} = k[A]$.

(iii) Separate the variables: $\dfrac{d[A]}{[A]} = -kdt$.

(iv) Do a separate definite integral on each side of the equation:

$$\int_0^t \frac{d[A]}{[A]} = -k\int_0^t dt$$

$$\ln[A]_t - \ln[A]_0 = -k(t_t - t_0)$$

(v) With $t = t_t - t_0$:

$$\ln[A]_t - \ln[A]_0 = -kt$$

$$\ln[A]_t = -kt + \ln[A]_0$$

The above equation is in the proper form, with $y = \ln[A]_t$, $m = -k$, $x = t$, and $b = \ln[A]_0$.

(vi) $t = t_{1/2}$, when $[A]_t = \dfrac{1}{2}[A]_0$

$$\ln\frac{1}{2}[A]_0 - \ln[A]_0 = -kt_{1/2}$$

$$\ln[A]_0 - \ln 2 - \ln[A]_0 = -kt_{1/2}$$

$$-\ln 2 = -kt_{1/2}$$

$$t_{1/2} = \frac{-\ln 2}{-k} = \frac{\ln 2}{k}$$

Chapter 12

8. There are many answers to this question; this is an example: Prepare a sample of N_2O_4 in which the N atoms are the heavier isotopes ^{15}N. Introduce the heavy isotope of N_2O_4 into an equilibrium mixture of N_2O_4 and NO_2. Use spectroscopic methods, such as infrared spectroscopy, to observe the distribution of the radioisotope among the reactants and products.

10. (a) 0 °C
(b) Dynamic equilibrium; Molecules at the interface between the water and the ice may detach from the ice and enter the liquid phase or may attach to the solid phase leaving the liquid phase.

12.

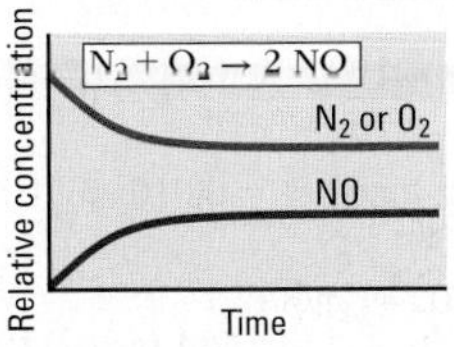
N2 + O2 → 2 NO
N2 or O2
NO
Relative concentration
Time

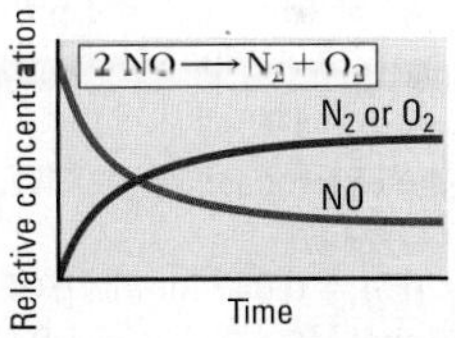
2 NO → N2 + O2
N2 or O2
NO
Relative concentration
Time

14. (a) $K = \dfrac{[O_3]^2}{[O_2]^3}$ (b) $K = \dfrac{[Fe(CO)_5]}{[CO]^5}$
(c) $K = [NH_3]^2[CO_2][H_2O]$
(d) $K = [Ag^+]^2[SO_4^{2-}]$

16. (a) $K_c = \dfrac{[H_2O]^2[O_2]}{[H_2O_2]^2}$ (b) $K_c = \dfrac{[PCl_5]}{[PCl_3][Cl_2]}$
(c) $K_c = [CO]^2$ (d) $K_c = \dfrac{[H_2S]}{[H_2]}$

18. (a) $K_c = [PCl_3][Cl_2]$ (b) $K_c = \dfrac{[CoCl_4^{2-}]}{[Co(H_2O)_6^{2+}][Cl^-]^4}$
(c) $K_c = \dfrac{[CH_3COO^-][H^+]}{[CH_3COOH]}$ (d) $K_c = \dfrac{[OF_2][HF]^2}{[F_2]^2[H_2O]}$

20. (a) $K_c = [H_2O]^2$ (b) $K_c = \dfrac{[HF]^4}{[SiF_4][H_2O]^2}$

(c) $K_c = \dfrac{[HCl]^2}{[H_2O]}$

22. (a) $2\ SO_2(g) + 2\ H_2O(g) \longrightarrow 2\ H_2S(g) + 3\ O_2(g)$
(b) $IF(g) \longrightarrow \frac{1}{2}\ F_2(g) + \frac{1}{2}\ I_2(g)$
(c) $Cl_2(g) + 2\ Br^-(aq) \longrightarrow Br_2(\ell) + 2\ Cl^-(aq)$
24. Equation (e)
26. (a) 0.87 (b) 1.3
28. (a) $K_P = \dfrac{P^2_{H_2O}P_{O_2}}{P^2_{H_2O_2}}$ (b) $K_P = \dfrac{P_{PCl_5}}{P_{PCl_3}P_{Cl_2}}$

(c) $K_P = P^2_{CO}$ (d) $K_P = \dfrac{P_{H_2S}}{P_{H_2}}$

30. 2.6×10^{11}
32. $K_c = 0.0161$
34. 2.6×10^{-9}
36. $K_P = 1.6$
38. (a) [CO] = 0.0071 mol/L, [$COCl_2$] = 0.00308 mol/L
(b) 1.4×10^2
40. $K_c = 0.075$
42. 3.9×10^{-4}
44. 1.4×10^3
45. Reactions (b) and (c) are product-favored. Most reactant-favored is (a), then (b), and then (c).
47. (a), (b), and (c)
48. (a) $Ag_2SO_4(s) \rightleftharpoons 2\ Ag^+(aq) + SO_4^{2-}(aq)$
$K = [Ag^+]^2[SO_4^{2-}] = 1.7 \times 10^{-5}$
$Ag_2S(s) \rightleftharpoons 2\ Ag^+(aq) + S^{2-}(aq)$
$K = [Ag^+]^2[S^{2-}] = 6 \times 10^{-30}$
(b) $Ag_2SO_4(s)$, larger K_c (c) $Ag_2S(s)$, larger K_c
50. (a)

	butane $\rightleftharpoons$ 2-methylpropane	
Conc. initial	0.100 mol/L	0.100 mol/L
Change conc.	$-x$	$+x$
Equilibrium conc.	$0.100 - x$	$0.100 + x$

(b) $K_c = \dfrac{[\text{2-methylpropane}]}{[\text{butane}]} = 2.5 = \dfrac{0.100 + x}{0.100 - x}$ $x = 0.043$

(c) [2-methylpropane] = 0.024 mol/L, [butane] = 0.010 mol/L
52. 3.39 g $C_6H_{12}(g)$
54. [Br_2] = [F_2] = 0.047 mol/L; [BrF] = 0.347 mol/L
55. (a) 1.94 mol HI (b) 1.92 mol HI (c) 1.98 mol HI
57. (a) [CO] = [H_2O] = 0.95 mol/L, [CO_2] = [H_2] = 0.0489 mol/L
(b) [CO] = [H_2O] = 1.90 mol/L, [CO_2] = [H_2] = 0.0977 mol/L
59. (a) No (b) Proceed toward products
61. (a) No (b) Proceed toward reactants
(c) [N_2] = [O_2] = 0.025 M; [NO] = 0.0010 M
64. (a) Left (b) Left (c) Left (d) Right
66. (a) Reverse (b) Forward (c) Forward
68.

Change	[Br_2]	[HBr]	K_c	K_P
Some H_2 is added to the container.	Decrease	Increase	No change	No change
Temperature of the gases in the container is increased.	Increase	Decrease	Decrease	Decrease
The pressure of HBr is increased.	Increase	Increase	No change	No change

70. (a) No change (b) Left (c) Left
72. (a) (i) (b) (ii) (c) (i) (d) (iii) (e) (iii)
73. (a) Decrease
(b)

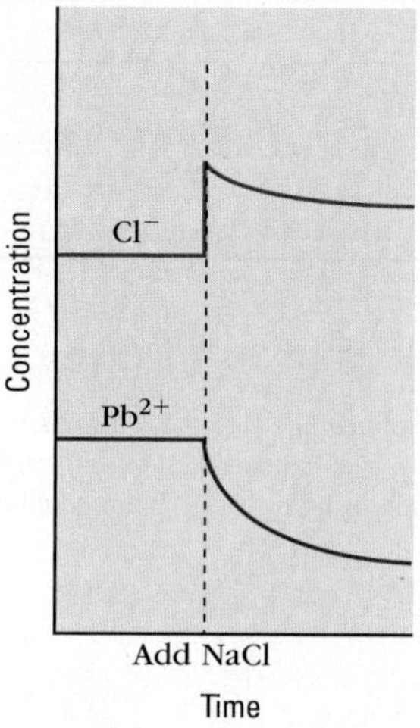

Concentration
Cl⁻
Pb2+
Add NaCl
Time

75. (a) (i) Increase (ii) Increase (iii) No change
(iv) Increase (v) Increase
(b) No change increases *K;* change (v) decreases *K.*
76. (a) Left
(b) [PCl_5] = 0.0198 M; [PCl_3] = 0.0231 M; [Cl_2] = 0.0403 M
78. (a) Energy effect (b) Entropy effect (c) Neither
80. (a) Insufficient information is available.
(b) Greater than 1, products favored
(c) Less than 1, reactants favored
82. A reaction will only go significantly toward products if it is product-favored. If a reaction is accompanied by an increase in entropy (a favorable entropy effect) and the products are lower in energy (a favorable enthalpy effect), then products are favored. However, if the reaction is favored only by an entropy effect, then it needs high temperatures to assist the endothermic process. If the reaction is favored only by an enthalpy effect, then it needs low temperatures to keep randomness at a minimum.
84. (a) First reaction: $\Delta_r H = -296.830$ kJ, second reaction: $\Delta_r H = -197.78$ kJ, third reaction: $\Delta_r H = -132.44$ kJ
(b) All three are exothermic, none are endothermic
(c) None of the reactions has an entropy increase; the second and third reactions have an entropy decrease, and the first reaction entropy is about the same.
(d) Low temperature favors more products in all three reactions.
86. $K_c = [H^+][OH^-]$, $K_c = \dfrac{[CH_3COO^-][H^+]}{[CH_3COOH]}$,

$K_c = \dfrac{[NH_3]^2}{[N_2][H_2]^3}$, $K_c = \dfrac{[CO_2][H_2]}{[CO][H_2O]}$

First reaction
(a) $(1.0 + x)(1.0 + x) = 1.0 \times 10^{-14}$
(b) Quadratic (c) N/A
Second reaction
(a) $\dfrac{(1.0 + x)(1.0 + x)}{(1.0 - x)} = 1.8 \times 10^{-5}$
(b) Quadratic (c) N/A
Third reaction
(a) $\dfrac{(1.0 + 2x)^2}{(1.0 - x)(1.0 - 3x)^3} = 3.5 \times 10^8$
(b) Not quadratic
(c) Use approximation techniques
Fourth reaction
(a) $\dfrac{(1.0 + x)(1.0 + x)}{(1.00 - x)(1.00 - x)} = 4.00$
(b) This equation is quadratic.
(c) Not applicable

88. $K_P = 0.108$
90. (a)

Species	Br_2	Cl_2	F_2	H_2	N_2	O_2
[E] (mol/L)	0.28	0.057	1.44	1.76×10^{-5}	4×10^{-14}	4.0×10^{-6}

(b) F_2; (At this temperature, the lowest bond energy is predicted from the reaction that gives the most products.) Compared with Table 6.2, the product production decreases as the bond energy increases: 156 kJ F_2, 193 kJ Br_2, 242 kJ Cl_2, 436 kJ H_2, 498 kJ O_2, 946 kJ N_2. Lewis structures of F_2, Br_2, Cl_2, H_2 have a single bond and more products are produced than O_2 with double bond and N_2 with triple bonds.
94. (a) 0.927 atm (b) 0.0420 mol
95. Cases c and d will have increased [HI] and Cases a and b will have decreased [HI] at equilibrium.
97. (a) 0.18
(b) $P_{CO} = 1.02$ atm; $P_{Br_2} = 0.52$ atm; $P_{COBr_2} = 0.10$ atm
99. 71.5 g $CaCO_3$
102. It is at equilibrium at 600. K. No more experiments are needed.
105. In the warmer sample, the molecules would be moving faster and more NO_2 molecules would be seen. In the cooler sample, the molecules would be moving somewhat slower and fewer NO_2 molecules would be seen. In both samples, the molecules are moving very fast. The average speed of gas molecules is commonly hundreds of miles per hour. In both samples, one would see a dynamic equilibrium with some N_2O_4 molecules decomposing and some NO_2 molecules reacting with each other, at equal rates.
107. Diagrams (b), (c), and (d)
109. This is an example of a diagram. (Other answers are possible.)

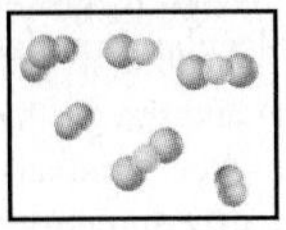

111. Diagram (b)
113. Dynamic equilibria with small values of K introduce a small amount of D^+ ions in place of H^+ ions in the place of the acidic hydrogen.
115. Dynamic equilibria representing the decomposition of the dimer $N_2O_4(g)$ produce $NO_2(g)$ and $*NO_2(g)$, which will occasionally recombine into the mixed dimer, $O_2N*{-}NO_2(g)$

$O_2N{-}NO_2(g) \rightleftharpoons 2\ NO_2(g)$

$O_2N*\ *NO_2(g) \rightleftharpoons 2\ *NO_2(g)$

$O_2N*{-}NO_2(g) \rightleftharpoons *NO_2(g) + NO_2(g)$

117. 1.34 atm
119. $K_c = \dfrac{A_f e^{(-E_{a,f}/RT)}}{A_r e^{(-E_{a,r}/RT)}} = \dfrac{A_f}{A_r} e^{[(E_{a,r}-E_{a,f})/RT]}$

When a catalyst is added, the activation energies of both the forward and reverse reactions get smaller:

$$K_{c,\,cat} = \frac{A_{f,cat}}{A_{r,cat}} e^{[(E_{a,r,cat}-E_{a,f,cat})/RT]}$$

The frequency factors did not change and, because the reaction has to conserve energy, the E_a values must each be reduced by the same amount of energy X:

$A_{f,cat} = A_f$ $\quad E_{a,f,cat} = E_{a,f} - X$

$A_{r,cat} = A_r$ $\quad E_{a,r,cat} = E_{a,r} - X$

Notice that

$E_{a,f,cat} - E_{a,r,cat} = (E_{a,f} - X) - (E_{a,r} - X)$

$= E_{a,f} - X - E_{a,r} + X = E_{a,f} - E_{a,r}$

Substituting into the $K_{c,cat}$ equation gives

$$K_{c,cat} = \frac{A_f}{A_r} e^{[(E_{a,r}-E_{a,f})/RT]} = K_c$$

Therefore, the equilibrium state, as described quantitatively by the equilibrium constant, does not change with the addition of a catalyst.

122. (a) $K_p = 0.03126$; $K_c = 0.0128$
(b) $K_P = 1$
(c) $K_c = 1/(RT_{bp})$
124. 1.15%

Chapter 13

13. If the solid interacts with the solvent using similar (or stronger) intermolecular forces, it will dissolve readily. If the solute interacts with the solvent using different intermolecular forces than those experienced in the solvent, it will be almost insoluble. For example, consider dissolving an ionic solid in water and oil. The interactions between the ions in the solid and water are very strong, because ions would be attracted to the highly polar water molecule; hence the solid would have a high solubility. However, the ions in the solid interact with each other much more strongly than the London dispersion forces experienced between the nonpolar hydrocarbons in the oil; hence the solid would have a low solubility. (Other examples exist.)
15. Beaker (c)
17. When an organic acid has a large (nonpolar) piece, it interacts primarily using London dispersion intermolecular forces. Because water interacts via hydrogen-bonding intermolecular forces, it would rather interact with itself than with the acid. Hence the solubility of the large organic acids drops, and some are completely insoluble.
18. Mix >143 g oxalic acid in 1 L water, agitate and heat until solid dissolves, then cool to 25 °C. Keep some undissolved solid in the bottom of the container when stored.
20. (a) Supersaturated (b) Unsaturated
22. Endothermic
24. 157 kJ/mol
26. The partial positive H end of the very polar water molecule interacts with the negative ions. The partial negative O end of the very polar water molecule interacts with the positive ions.

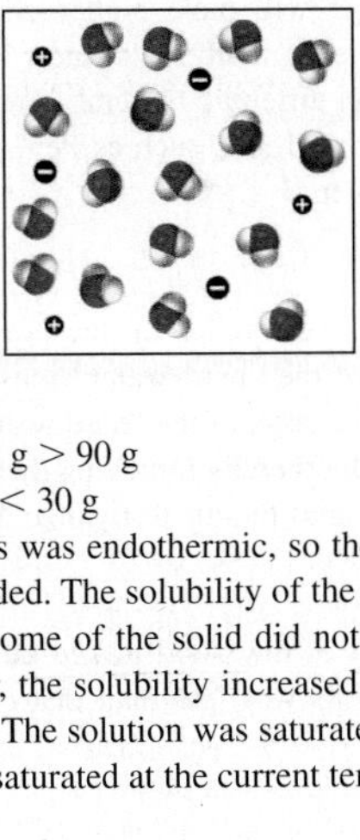

27. (a) No; 65 g < 90 g
(b) Supersaturated; 95 g > 90 g
(c) Unsaturated; 25 g < 30 g
29. The dissolving process was endothermic, so the temperature dropped as more solute was added. The solubility of the solid at the lower temperature is lower, so some of the solid did not dissolve. As the solution warmed, however, the solubility increased again. What remained of the solid dissolved. The solution was saturated at the lower temperature but is no longer saturated at the current temperature.
31. 1.2×10^{-3} M N_2
33. 0.05%
35. 0.00732% by weight
37. $1\text{ ppb} = \dfrac{1\text{ g part}}{10^9\text{ g whole}} \times \dfrac{1\ \mu\text{g part}}{10^{-6}\text{ g part}} \times \dfrac{1000\text{ g whole}}{1\text{ kg whole}} = \dfrac{1\ \mu\text{g part}}{1\text{ kg whole}}$
39. 90. g ethanol
42. 2.0×10^{-6} g Pb
44. (a) 160. g NH_4Cl (b) 83.9 g KCl (c) 7.46 g Na_2SO_4
46. (a) 0.762 M (b) 0.174 M
(c) 0.0126 M (d) 0.167 M
48. 96% H_2SO_4
50. 0.1 M NaCl
52. 59 g
54. 0.764-M sucrose

56. (a) 108 ppm
(b) 5.64×10^{-4} mol/L
(c) 5.59×10^{-4} mol/kg

58. (a) 0.106, 10.6% (b) 0.127, 12.7% (c) 0.308, 30.8%
(d) 2.49×10^{-3}, 0.249%

61. Curve (a) is for benzene and curve (b) is for the solution. The addition of solute lowers the vapor pressure of the solvent.

62. 100.26 °C

64. Freezing point of (a) > freezing point of (d) > freezing point of (b) > freezing point of (c)

66. $T_f = -1.65$ °C, $T_b = 100.46$ °C

68. $X_{H_2O} = 0.79999$, 712 g sucrose

70. 1.9×10^2 g/mol

72. 3.6×10^2 g/mol; $C_{20}H_{16}Fe_2$

75. 1.8×10^2 g/mol; $C_{14}H_{10}$

77. (a) 2500 g (b) 104.2 °C

79. 29 atm

80. 1.2×10^4 g/mol

82. (a) A foam with a gas dispersed phase and a liquid continuous phase
(b) An emulsion with a liquid dispersed phase and a liquid continuous phase
(c) An aerosol with a solid dispersed phase and a gas continuous phase
(d) A gel with a liquid dispersed phase and a solid continuous phase

84. Protein molecules (the dispersed phase) have multiple sites that interact with water (the continuous phase) through hydrogen bonding and dipole-dipole attractions.

86. 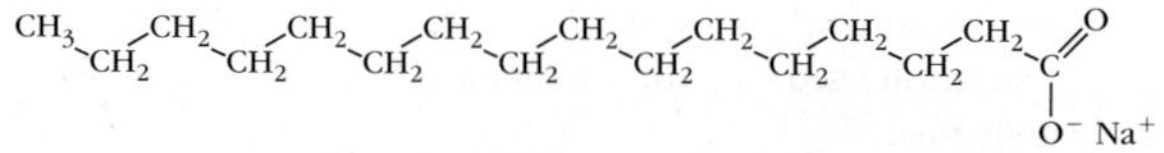
CH3 CH2 CH2 CH2 CH2 CH2 CH2 CH2 CH2 CH2 CH2 CH2 CH2 CH2 CH2 CH2 C O
O⁻ Na⁺

It has a long nonpolar hydrocarbon chain that interacts with other molecules using London dispersion forces and a carboxylate anion salt at one end that interacts with water molecules using ion-dipole forces.

88. See Figure 13.24.

90. No; surface water will have higher concentrations of atmospheric gases (O_2, CO_2, etc.) than well water. Rock and soil deeper in the ground will have different mineral content; for example, some reduced forms of metal ions such as Fe^{2+} may exist in well water and not in surface water.

92. 1×10^{-4} g As

94. No

96. The lime-soda process relies on the precipitation of insoluble compounds to remove the "hard water" ions. The ion-exchange process relies on the high charge of the "hard water" ions to attract them to an ion-exchange resin, thereby removing them from the water.

98. Water is sprayed into the air to oxidize organic substances dissolved in it.

101. 4%

103. Water in the cells of the wood leaked out, because the osmotic pressure inside the cells was less than that of the seawater in which the wood was sitting.

105. 28% NH_3

107. 0.982 *m*, 10.2%

108. Freezing point of (a) = freezing point of (c) > freezing point of (d) > freezing point of (b)

111. 1.77 g

113. (a) $Na_2SO_4(aq) + Ba(NO_3)_2(aq) \rightarrow 2\ NaNO_3(aq) + BaSO_4(s)$
(b) 15.0 mL (c) 12.0 mL

114. (a) No (b) 108.9 °C

117. (a)
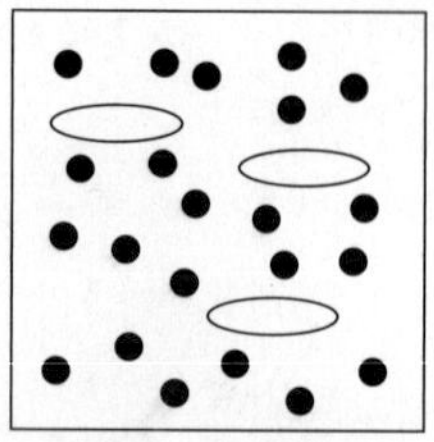
(b)
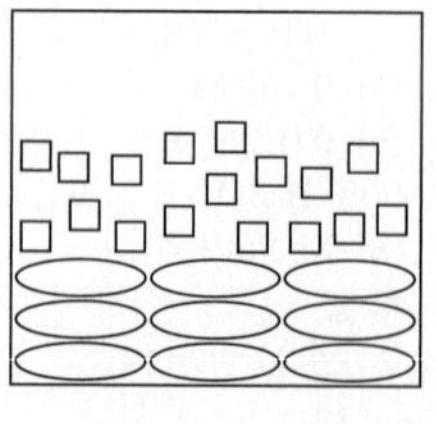

119. (a) Unsaturated (b) Supersaturated
(c) Supersaturated

121.

Compound	Mass of Compound	Mass of Water	Mass Fraction of Solute	Weight Percent of Solute	ppm of Solute
Lye	75.0 g	125 g	0.375	37.5%	3.75×10^5
Glycerol	33 g	200. g	0.14	14%	1.4×10^5
Acetylene	0.0015	2×10^2 g	0.000009	0.0009%	9

123. Molecules slow down. The reduced motion prevents them from randomly translocating as they had in the liquid state. As a result, the intermolecular forces between one molecule and the next begin to organize them into a crystal form. The presence of a nonvolatile solute disrupts the formation of the crystal. Its size and shape will be different from that of the solute. Intermolecular forces between the solute and solvent are also different from those of solvent molecules with each other. To form the crystalline solid, the solute has to be excluded. If the ice in an iceberg is in regular crystalline form, the water will be pure. Only if the ice is crushed or dirty will other particles be included. So, melting an iceberg will produce relatively pure water.

125. (a) Seawater contains more dissolved solutes than freshwater. The presence of a solute lowers the freezing point. That means a lower temperature is required to freeze the seawater than to freeze freshwater.
(b) Salt added to a mixture of ice and water will lower the freezing point of the water. If the ice cream is mixed at a lower temperature, its temperature will drop faster; hence, it will freeze faster.

128. The empirical and molecular formulas are both $C_{18}H_{24}Cr$.

130. 28 m

131. 0.30 M

133. (a)
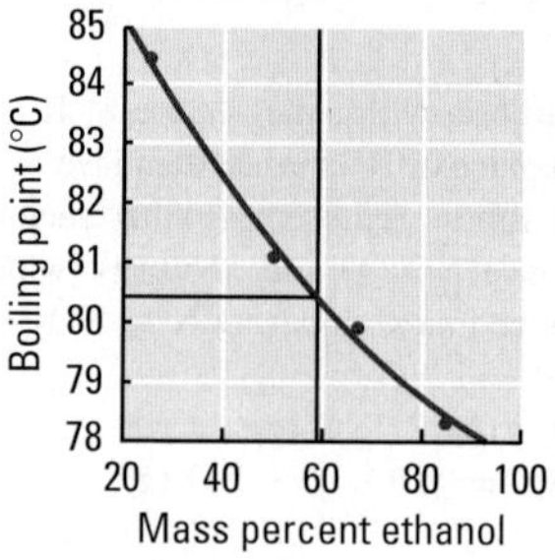
Boiling point (°C)
85
84
83
82
81
80
79
78
20
40
60
80
100
Mass percent ethanol

Approximately 80.4 °C
(b) 12.3 M; 44.1 mol/kg

134. (a)
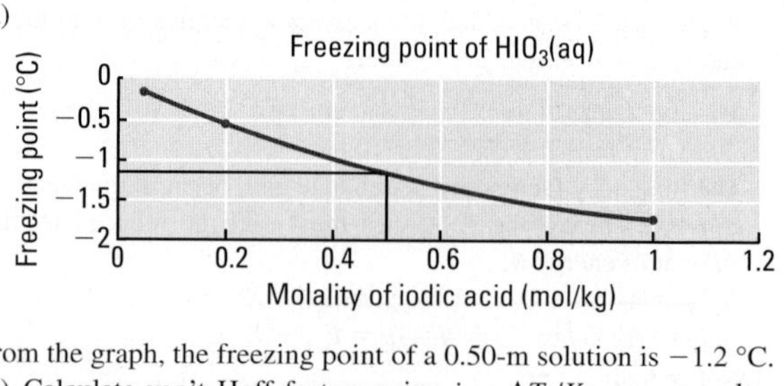
Freezing point of HIO3(aq)
Freezing point (°C)
0
−0.5
−1
−1.5
−2
0
0.2
0.4
0.6
0.8
1
1.2
Molality of iodic acid (mol/kg)

From the graph, the freezing point of a 0.50-m solution is −1.2 °C.
(b) Calculate van't Hoff factors using $i = \Delta T_f / K_f\ m_{solute}$; results are 1.7, 1.5, and 0.92.
(c) If iodic acid did not dissociate at all (0% dissociation), $i = 1$; if it completely dissociated (100% dissociation) into H^+ and IO_3^- ions, then $i = 2$. The percent dissociation can be calculated as

$$\%\ \text{dissociation} = [i_{exp} - i_{0\%}]/[i_{100\%} - i_{0\%}] \times 100\%$$

Values are 67%, 46%, and −8%. The negative value for 1.0-m HIO_4 indicates that the solution does not behave ideally.

135. (a)

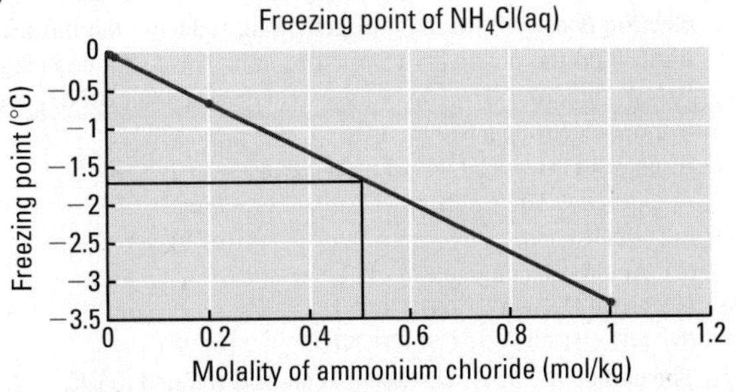
Freezing point of $NH_4Cl(aq)$
Freezing point (°C)
0
−0.5
−1
−1.5
−2
−2.5
−3
−3.5
0
0.2
0.4
0.6
0.8
1
1.2
Molality of ammonium chloride (mol/kg)

From the graph, the freezing point of a 0.50-m solution is −1.7 °C.
(b) Calculate van't Hoff factors using $i = \Delta T_f/K_f\, m_{solute}$; results are 1.9, 1.9, 1.8, and 1.8.
(c) If ammonium chloride did not dissociate at all (0% dissociation), $i = 1$; if it completely dissociated (100% dissociation) into NH_4^+ and Cl^- ions, then $i = 2$. The percent dissociation can be calculated as

$$\% \text{ dissociation} = [i_{exp} - i_{0\%}]/[i_{100\%} - i_{0\%}] \times 100\%$$

Values are 94%, 91%, 83%, and 79%.

137. (a) 6300 ppm, 6300000 ppb
(b) 0.040 M (c) 4.99×10^5 bottles

Chapter 14

9. (a) $HCO_3^- + H_2O \rightleftharpoons CO_3^{2-} + H_3O^+$
(b) $HCl + H_2O \rightleftharpoons Cl^- + H_3O^+$
(c) $CH_3COOH + H_2O \rightleftharpoons CH_3COO^- + H_3O^+$
(d) $HCN + H_2O \rightleftharpoons CN^- + H_3O^+$

11. (a) $HSO_4^- + H_2O \rightleftharpoons H_2SO_4 + OH^-$
(b) $CH_3NH_2 + H_2O \rightleftharpoons CH_3NH_3^+ + OH^-$
(c) $I^- + H_2O \rightleftharpoons HI + OH^-$
(d) $H_2PO_4^- + H_2O \rightleftharpoons H_3PO_4 + OH^-$

13. (a) I^-, iodide, conjugate base
(b) HNO_3, nitric acid, conjugate acid
(c) HCO_3^-, hydrogen carbonate ion, conjugate acid
(d) HCO_3^-, hydrogen carbonate ion, conjugate base
(e) SO_4^{2-}, sulfate ion, conjugate base; H_2SO_4 sulfuric acid, conjugate acid
(f) HSO_3^-, hydrogen sulfite ion, conjugate acid

15. Pairs (b), (c), and (d)

17. (a) Reactant acid = H_2O, reactant base = HS^-, product conjugate acid = H_2S, product conjugate base = OH^-
(b) Reactant acid = NH_4^+, reactant base = S^{2-}, product conjugate acid = HS^-, product conjugate base = NH_3
(c) Reactant acid = HSO_4^-, reactant base = HCO_3^-, product conjugate acid = H_2CO_3, product conjugate base = SO_4^{2-}
(d) Reactant acid = NH_3, reactant base = NH_2^-, product conjugate base = NH_2^-, product conjugate acid = NH_3

19. (a) D (b) E (c) E

21. $C_5H_5N(aq) + H_2O(\ell) \rightleftharpoons C_5H_5NH^+(aq) + OH^-(aq)$; basic due to OH^- produced

23. $CH_3COCOOH(aq) + H_2O(\ell) \rightleftharpoons H_3O^+(aq) + CH_3COCOO^-(aq)$; acidic due to H_3O^+ produced

25. 3×10^{-11} M, basic

27. pH = 12.40, pOH = 1.60

29. pOH = 12.51

31. 2 g HCl

33.

	pH	$[H_3O^+]$ (M)	$[OH^-]$ (M)	Acidic or Basic
(a)	—	1.0×10^{-1}	1.0×10^{-13}	Acidic
(b)	—	3×10^{-11}	3×10^{-4}	Basic
(c)	3.74	—	5.6×10^{-11}	Acidic
(d)	9.36	4.3×10^{-10}	—	Basic

35. (a) Six times more acidic
(b) Three times more basic
(c) Three times more basic
(d) 10,000,000 times more acidic

37. (a) 5.0×10^{-9} M (b) Basic

38. (a) A, because K_a is larger (b) X, because K_b is larger
(c) Y, because K_b is smaller, so conjugate K_a is larger
(d) A, because K_a is larger, so conjugate K_b is smaller

39.

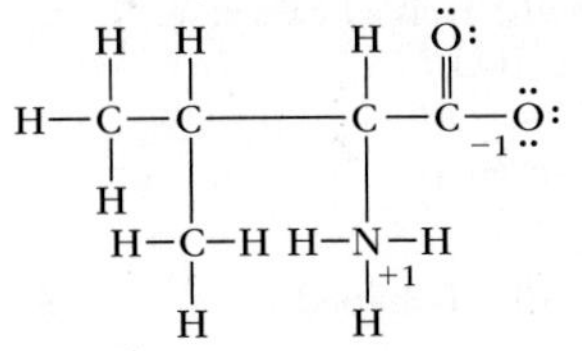

41. (a)

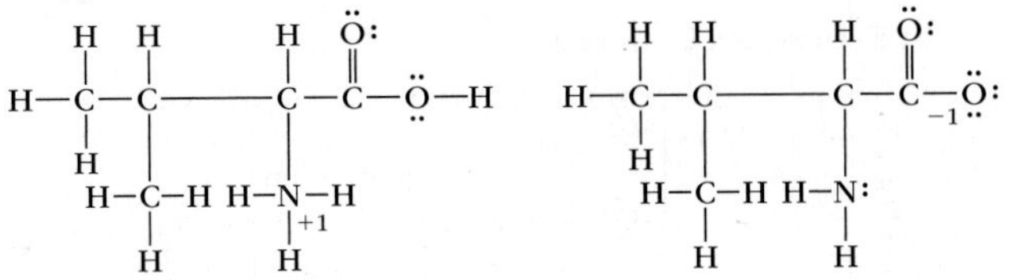

(b) At high pH, the net-negative form exists.
(c) At low pH, the net-positive form exists.

43. (a) $CH_3COOH(aq) + H_2O(\ell) \rightleftharpoons H_3O^+(aq) + CH_3COO^-(aq)$

$$K = \frac{[H_3O^+][CH_3COO^-]}{[CH_3COOH]}$$

(b) $HCN(aq) + H_2O(\ell) \rightleftharpoons H_3O^+(aq) + CN^-(aq)$

$$K = \frac{[H_3O^+][CN^-]}{[HCN]}$$

(c) $SO_3^{2-}(aq) + H_2O(\ell) \rightleftharpoons HSO_3^-(aq) + OH^-(aq)$

$$K = \frac{[HSO_3^-][OH^-]}{[SO_3^{2-}]}$$

(d) $PO_4^{3-}(aq) + H_2O(\ell) \rightleftharpoons HPO_4^{2-}(aq) + OH^-(aq)$

$$K = \frac{[HPO_4^{2-}][OH^-]}{[PO_4^{3-}]}$$

(e) $NH_4^+(aq) + H_2O(\ell) \rightleftharpoons H_3O^+(aq) + NH_3(aq)$

$$K = \frac{[H_3O^+][NH_3]}{[NH_4^+]}$$

(f) $H_2SO_4(aq) + H_2O(\ell) \rightleftharpoons H_3O^+(aq) + HSO_4^-(aq)$

$$K = \frac{[H_3O^+][HSO_4^-]}{[H_2SO_4]}$$

45. (a) 0.10 M H_2CO_3 (b) 0.10 M $KHSO_4$
(c) 0.1 M $NaHCO_3$ (d) 0.1 M H_2S

48. (a) H_2SO_4 (b) HNO_3 (c) $HClO_4$
(d) $HClO_3$ (e) H_2SO_4

51. (a) $CO_3^{2-} + H_2O \rightleftharpoons HCO_3^- + OH^-$
$HCO_3^- + H_2O \rightleftharpoons H_2CO_3 + OH^-$
(b) $H_3AsO_4 + H_2O \rightleftharpoons H_2AsO_4^- + H_3O^+$
$H_2AsO_4^- + H_2O \rightleftharpoons HAsO_4^{2-} + H_3O^+$
$HAsO_4^{2-} + H_2O \rightleftharpoons AsO_4^{3-} + H_3O^+$
(c) $NH_2CH_2COO^- + H_2O \rightleftharpoons {}^+NH_3CH_2COO^- + OH^-$
${}^+NH_3CH_2COO^- + H_2O \rightleftharpoons {}^+NH_3CH_2COOH + OH^-$

54. (a) 2.09 (b) 5.13 (c) 8.09 (d) 9.02
(e) 13.30 (f) 1.55 (g) 9.68 (h) 7.00

55. 3.6×10^{-4}

57. $K_a = 1.4 \times 10^{-5}$

58. $[H_3O^+] = 1.9 \times 10^{-3}$ M, $[CH_3COO^-] = 1.9 \times 10^{-3}$ M, $[CH_3COOH] = 0.20$ M

60. 8.84

62. 10.47

64. 3.27

66. (a) NH_4^+; reactant-favored (b) CH_3COO^-; product-favored (c) NH_2^-; product-favored
68. Reactions (a) and (c) are product-favored; reactions (b) and (d) are reactant-favored.
70. (a) pH < 7 (b) pH > 7 (c) pH $= 7$
73. (a), (b), (c), (d), and (e)
74. $[Ni(H_2O)_6]^{2+} + H_2O(\ell) \rightarrow [Ni(H_2O)_5(OH)]^+ + H_3O^+(aq)$
76. All three are Lewis bases; CO_2 is a Lewis acid.
78. Cr^{3+} and SO_3 are Lewis acids. CH_3NH_2 is a Lewis base.
80. (a) Lewis acid: SO_2; Lewis base: H_2O
(b) Lewis acid: H_3BO_3; Lewis base: OH^-
(c) Lewis acid: Cu^{2+}; Lewis base: NH_3
(d) Lewis acid: Sn^{2+}; Lewis base: Cl^-

83. 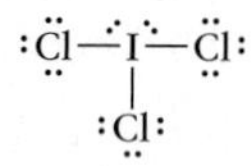 T-shaped

It functions as a Lewis acid.

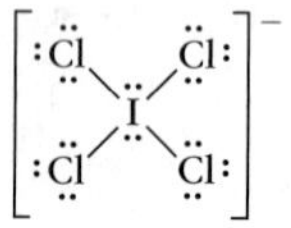 Square planar

84. $[Zn(H_2O)(OH)_3]^-$
86. $HCO_3^-(aq) + H_2PO_4^-(aq) \rightarrow HPO_4^{2-}(aq) + CO_2(g) + H_2O(\ell)$; base: HCO_3^-; acid: $H_2PO_4^-$
87. $Na_2CO_3 + 2\ CH_3(CH_2)_{16}COOH \rightarrow 2\ CH_3(CH_2)_{16}COONa + H_2O + CO_2$
89. Dishwasher detergent is very basic and should not be used to wash anything by hand, including a car. If it gets into the engine area, it can also dissolve automobile grease and oil, which could prevent the engine from running correctly.
91. Lemon juice is acidic and neutralizes the basic amines. The acid formed from the neutralized base is an ion and not volatile.
92. (a) Weak acid (b) Strong base (c) Strong acid (d) Weak base (e) Strong base (f) Amphiprotic
95. (a) Less than 7 (b) Equal to 7 (c) Greater than 7
96. 2.85
97. (a) Decreases (b) Increases (c) Stays the same
100. 9.59
101. $Al^{3+}(aq) + 6\ H_2O(\ell) \rightleftharpoons Al(H_2O)_6^{3+}(aq)$
$Al(H_2O)_6^{3+}(aq) + H_2O(\ell) \rightleftharpoons Al(OH)_5(OH)^{2+}(aq) + H_3O^+(aq)$
$H_3O^+(aq) + HCO_3^-(aq) \rightleftharpoons CO_2(g) + 2\ H_2O(\ell)$
Dissolving in water allows ions and molecules to come into contact. Water molecules hydrate the aluminum ions making the H atoms in water more acidic. The acid reacts with hydrogen carbonate ions to generate carbon dioxide gas.
106. At 10 °C, pH = 7.27; at 25 °C, pH = 6.998; and at 50 °C, pH = 6.631. At 10 °C, at 25 °C, and at 50 °C, the solutions are neutral, because $[H_3O^+] = [OH^-]$.
108. (a) $H_2O > Cl^- = H_3O^+ \gg OH^-$
(b) $H_2O > Na^+ = ClO_4^- \gg H_3O^+ = OH^-$
(c) $H_2O > HNO_2 > H_3O^+ = NO_2^- \gg OH^-$
(d) $H_2O > Na^+ \cong ClO^- > OH^- = HClO \gg H_3O^+$
(e) $H_2O > NH_4^+ \cong Cl^- > H_3O^+ = NH_3 \gg OH^-$
(f) $H_2O > Na^+ = OH^- \gg H_3O^+$
110. Conjugates must differ by just one H^+.
112. (a) 2.02 (b) 23 times
115. (a) HY; the calculated K_a is largest.
(b) Z^-; the smallest K_a acid has the strongest conjugate base.
118. (a) Boxes C, F, and G (b) Box I (c) Boxes A and H
(d) Boxes D and E (e) Boxes B and E
(f) Box I (g) Boxes D and E
121. (a) 10.46 (b) 10.45 (c) 10.45
124. Lactic acid sample
125. 0.76 L; no
126. Yes, higher pH
128. Br has a higher electronegativity than H. The bromine withdraws electron density from the nitrogen atom, reducing the nitrogen's ability to bind the positively charged proton. The K_b value is smaller for $BrNH_2$ than for NH_3. $ClNH_2$ is a weaker base than $BrNH_2$ because Cl is more electronegative than Br.
129. Strongest HM > HQ > HZ weakest; $K_{a,HZ} = 1 \times 10^{-5}$, $K_{a,HQ} = 1 \times 10^{-3}$, $K_{a,HM} = 1 \times 10^{-1}$ or larger
130. (a) Weak (b) Weak (c) Acid/base conjugates (d) 6.29
133. (a) $NH_2CH_2CH_2CH_2CH_2CH_2NH_2$
(b) $NH_3CH_2CH_2CH_2CH_2CH_2NH_3^{2+}$ (c) 9.13
135. The carboxylic acid substituent is an acid (related to pK_a = 2.18). The two NH_2 groups are bases (related to pK_a = 8.95 and 10.53).

Chapter 15

12. (b) and (c)
14. Combination (b), because the pK_a for the acid is closest to 7
16. (a) 3.38 (b) 9.25 (c) 3.74
18. (a) Lactic acid/lactate ion
(b) Dihydrogen phosphate ion/hydrogen phosphate ion
(c) Acetic acid/acetate ion
(d) Hydrogen carbonate ion/carbonate ion; in each case $[H_3O^+]$ is closest to K_a.
20. 4.2 g $NaCH_3COO$
22. 13 g C_6H_5COOH per liter
24. 7.23
26. Samples (a) and (b) are buffers; they both contain an acid-base conjugate pair. (c) is not a buffer because there is no base. (d) is not a buffer because too much base was added.
29. (a) ΔpH = 0.1 (b) ΔpH = 3.8 (c) ΔpH = 7.25
31. (a) pH = 5.02 (b) pH = 4.99 (c) pH = 2.64
32. (a) Acid 2; higher initial pH
(b) Acid 1 pH ≅ 8.0; Acid 2 pH ≅ 10.0
(c) The weaker acid produces a more basic salt.
(d) Acid 1 is more highly ionized.
(e) Bromthymol blue for Acid 1. Phenolphthalein for Acid 2.
35. (a) Bromthymol blue
(b) Phenolphthalein
(c) Methyl red
(d) Bromthymol blue; suitable pH color changes
37. 0.0253-M HCl
39. 93.6%
41. (a) 29.2 mL (b) 600. mL (c) 1.20 L (d) 2.7 mL
43. (a) pH = 3.62 (b) pH = 8.31 (c) pH = 12.15
44. (a) 13.176 (b) 12.699 (c) 10.2 (d) 7.000
(e) 3.8 (f) 1.52

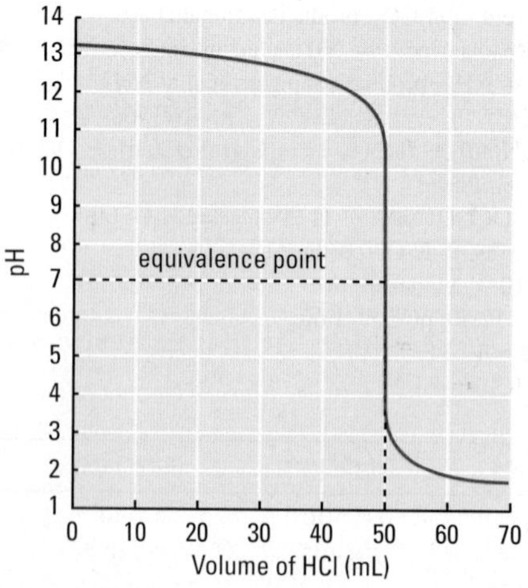
pH
equivalence point
14
13
12
11
10
9
8
7
6
5
4
3
2
1
0
10
20
30
40
50
60
70
Volume of HCl (mL)

47. NO_2 reaction: $2\,NO_2(g) + H_2O(g) \rightarrow HNO_3(g) + HNO_2(g)$
SO_3 reactions: $2\,SO_2(g) + O_2(g) \rightarrow 2\,SO_3(g)$
$SO_3(g) + H_2O(g) \rightarrow H_2SO_4(g)$
49. $CaCO_3(s) + 2\,H^+(aq) \rightarrow Ca^{2+}(aq) + CO_2(g) + H_2O(\ell)$
50. (a) $Ag_3AsO_4(s) \rightleftharpoons 3\,Ag^+(aq) + AsO_4^{3-}(aq)$ $K_{sp} = [Ag^+]^3[AsO_4^{3-}]$
(b) $Ag_2SO_4(s) \rightleftharpoons 2\,Ag^+(aq) + SO_4^{2-}(aq)$ $K_{sp} = [Ag^+]^2[SO_4^{2-}]$
(c) $Ca_3(PO_4)_2(s) \rightleftharpoons 3\,Ca^{2+}(aq) + 2\,PO_4^{3-}(aq)$ $K_{sp} = [Ca^{2+}]^3[PO_4^{3-}]^2$
(d) $Mn(OH)_3(s) \rightleftharpoons Mn^{3+}(aq) + 3\,OH^-(aq)$ $K_{sp} = [Mn^{3+}][OH^-]^3$
(e) $FeCO_3(s) \rightleftharpoons Fe^{2+}(aq) + CO_3^{2-}(aq)$ $K_{sp} = [Fe^{2+}][CO_3^{2-}]$
52. $K_{sp} = 2.8 \times 10^{-22}$
54. $K_{sp} = 2.22 \times 10^{-4}$
56. $K_{sp} = 2.2 \times 10^{-12}$
58. $K_{sp} = 1.7 \times 10^{-5}$
60. 1.2×10^{-9} M or lower
62. pH = 9.16
64. (a) More $PbCl_2$ solid forms
(b) More precipitate forms. Some $PbCl_2$ solid dissolves, but more AgCl solid forms.
(c) More $PbCl_2$ solid forms
66. 6.2×10^{-11}-M Zn^{2+}
68. (a) $K_{sp} = 1.8 \times 10^{-11}$ (b) $[OH^-]$ must be 0.0093 M or higher.
70. (a) 3.7×10^{-6} M
(b) 2.8×10^{-10} M
(c) 2.8×10^{-10} M
73. (a) $Ag^+(aq) + 2\,CN^-(aq) \rightleftharpoons [Ag(CN)_2]^-(aq)$ $K_f = \dfrac{[[Ag(CN)_2]^-]}{[Ag^+][CN^-]^2}$
(b) $Cd^{2+}(aq) + 4\,NH_3(aq) \rightleftharpoons [Cd(NH_3)_4]^{2+}(aq)$
$K_f = \dfrac{[[Cd(NH_3)_4]^{2+}]}{[Cd^{2+}][NH_3]^4}$
75. 0.0063 mol or more
77. (a) $Zn(OH)_2(s) + 2\,H_3O^+(aq) \rightarrow Zn^{2+}(aq) + 4\,H_2O(\ell)$
$Zn(OH)_2(s) + 2\,OH^-(aq) \rightarrow [Zn(OH)_4]^{2-}(aq)$
(b) $Sb(OH)_3(s) + 3\,H_3O^+(aq) \rightarrow Sb^{3+}(aq) + 6\,H_2O(\ell)$
$Sb(OH)_3(s) + OH^-(aq) \rightarrow [Sb(OH)_4]^-(aq)$
80. (a) $BaSO_4$ precipitates first (b) 2.5×10^{-4}-M SO_4^{2-}
82. (a) CaF_2 (b) 79%
85. (a) H_2O, CH_3COO^-, Na^+, CH_3COOH, H_3O^+, OH^-
(b) pH = 4.95
(c) pH = 5.05
(d) $CH_3COOH(aq) + H_2O(\ell) \rightleftharpoons CH_3COO^-(aq) + H_3O^+(aq)$
87. [*o*-ethylbenzoic acid] must be approximately 0.63 times the [potassium *o*-ethylbenzoate]
89. 0.012 mol HNO_2
91. (a) pH = 2.78 (b) pH = 5.39
93. $K_a = 3.5 \times 10^{-6}$
95. No change
96. (a) 5.004 (b) 6.84
(c) Pure water did not act like a buffer due to its very large pH change.
99. The tiny amount of base (CH_3COO^-) present is insufficient to prevent the pH from changing dramatically if a strong acid is introduced into the solution.
101. $K_a = 2.3 \times 10^{-4}$
103. Blood pH decreases; acidosis
105. (a) Adding Ca^{2+} drives the reaction more toward reactants, making more apatite.
(b) Acid reacts with OH^-, removing a product and driving the reaction toward the products, causing apatite to decompose.
107. Acidosis; increase
109. (a) Box C (b) Boxes C and D
(c) Box E (d) Box D (e) Box A
(f) Box B (g) Box E
111. (a) Box III (b) Box IV (c) Box II (d) Box I
113. 5.64
115. 3.22
117. (a) 1.1×10^{-9} M (b) 0.010 M
122. (a) 13.00 (b) 5.0×10^{-4}; T is different from 25 °C.
123. Sample A: $NaHCO_3$; Sample B: NaOH; Sample C: Mixture of NaOH and Na_2CO_3 and/or $NaHCO_3$; Sample D: Na_2CO_3

Chapter 16

13. (a) $2\,H_2O(\ell) \rightarrow 2\,H_2(g) + O_2(g)$, reactant-favored
(b) $C_8H_{18}(\ell) \rightarrow C_8H_{18}(g)$, product-favored
(c) $C_{12}H_{22}O_{11}(s) \rightarrow C_{12}H_{22}O_{11}(aq)$, product-favored
15. (a) $\frac{1}{2}$ (b) $\frac{1}{2}$ (c) 50 of each
17. (a) Probability of $\frac{1}{2}$ in flask A; probability of $\frac{1}{2}$ in flask B
(b) 50 in flask A and 50 in flask B
19. (a) Negative (b) Positive (c) Positive
21. (a) Positive (b) Positive (c) Positive
23. (a) Item 2 (b) Item 2 (c) Item 2
25. (a) NaCl(s) (b) $P_4(g)$ (c) $NH_4NO_3(aq)$
27. (a) Negative (b) Positive
(c) Negative (d) Positive
29. (a) Negative (b) Negative (c) Positive
31. 112 $J\,K^{-1}\,mol^{-1}$
33. (a) 2.63 J/K (b) 1000 J/K (c) 2.45 J/K
35. (a) −120.64 J/K (b) 156.9 J/K
(c) −198.76 J/K (d) 160.6 J/K
37. (a) −173.01 $J\,K^{-1}$ (b) −326.69 $J\,K^{-1}$ (c) 137.55 $J\,K^{-1}$
39. $\Delta_r S^\circ_{system} = -247.7$ J/K; Cannot tell without $\Delta_r H^\circ_{system}$ also, because that is needed to calculate $\Delta_r G^\circ_{system}$. $\Delta_r H^\circ_{system} = -88.13$ kJ and $\Delta_r G^\circ_{system} = -14.3$ kJ, so it is product-favored.
41. Product-favored at low temperatures; the exothermicity is sufficient to favor products if the temperature is low enough to overcome the decrease in entropy.
43. Exothermic reactions with an increase in disorder, exhibited by a larger number of gas-phase products than gas-phase reactants, never need help from the surroundings to favor products.
45. (a) Enthalpy change ($\Delta_r H^\circ$) is negative; entropy change ($\Delta_r S^\circ$) is positive.
(b) Using enthalpy of formations and Hess's law:
$\Delta_r H^\circ = -184.28$ kJ; $\Delta_r S^\circ = -7.7\ J\,K^{-1}$
The enthalpy change is negative, as predicted in (a). But the entropy change is negative, not as predicted in (a). The aqueous solute must have sufficient order to compensate for the high disorder of the gas. The value of $-7.7\ J\,K^{-1}$ is small.
49. (a) $\Delta_r S^\circ = -37.52$ J/K, $\Delta_r H^\circ = -851.5$ kJ, product-favored at low temperature
(b) $\Delta_r S^\circ = -21.77$ J/K, $\Delta_r H^\circ = 66.36$ kJ, never product-favored
51. (a) $\Delta_r S_{universe} = 4.92 \times 10^3$ J/K (b) -1.47×10^3 kJ
(c) Yes; ethane is used as a fuel; hence we might expect that its combustion reaction is product-favored.
53. $\Delta_r G^\circ = \Delta_r H^\circ - T\Delta_r S^\circ$; here $\Delta_r H^\circ$ is negative, and ΔS° is positive, so $\Delta_r G^\circ = -|\Delta_r H^\circ| - |T\Delta_r S^\circ| = -(|\Delta_r H^\circ| + |T\Delta_r S^\circ|) < 0$.
55. $\Delta_r G^\circ = 28.63$ kJ; reactant-favored
57. $\Delta_r G^\circ = 462.28$ kJ; uncatalyzed; it would not be a good way to make Si.
59. (a) 385.7 K (b) 835.1 K
61. (a) 49.7 kJ (b) −178.2 kJ (c) −1267.5 kJ
63. (a) $\Delta_r H^\circ = 178.32$ kJ, $\Delta_r S^\circ = 160.6$ J/K, $\Delta_r G^\circ = 130.5$ kJ
(b) Reactant-favored
(c) No; it is only product-favored at high temperatures.
(d) 1110. K
65. $\Delta_f G^\circ(Ca(OH)_2) = -867.8$ kJ/mol, which is very close to the value in Appendix J, −868.07 kJ/mol.
67. (a) $K = 4 \times 10^{-34}$ (b) $K = 5 \times 10^{-31}$
69. (a) $\Delta_r G^\circ = -100.97$ kJ, product-favored
(b) $K = 5 \times 10^{17}$. When $\Delta_r G^\circ$ is negative, K is larger than 1.
71. (a) $K^\circ_{800} = 8.7 \times 10^{26}$, product-favored
(b) $K^\circ_{500} = 7 \times 10^{10}$, product-favored
(c) $K^\circ_{2000} = 1.3 \times 10^{-15}$, reactant-favored

73. (a) $\Delta_rG° = -106$ kJ/mol (b) $\Delta_rG° = 8.55$ kJ/mol
(c) $\Delta_rG° = -33.8$ kJ/mol
75. (a) Can be harnessed; (b) and (c) require work to be done.
77. Reaction 75(b), 5.068 g graphite oxidized; reaction 75(c), 26.94 g graphite oxidized
79. (a) $2\,CuO(s) \rightarrow 2\,Cu(s) + O_2(g)$;
$CuO(s) + C(graphite) \rightarrow Cu(s) + CO(g)$; $\Delta G° = -7.5$ kJ
(b) $2\,Ag_2O(s) \rightarrow 4\,Ag(s) + O_2(g)$;
$Ag_2O(s) + C(graphite) \rightarrow 2\,Ag(s) + CO(g)$; $\Delta G° = -125.97$ kJ
(c) $2\,HgO(s) \rightarrow 2\,Hg(\ell) + O_2(g)$;
$HgO(s) + C(graphite) \rightarrow Hg(\ell) + CO(g)$; $\Delta G° = -125.97$ kJ
(d) $2\,MgO(s) \rightarrow 2\,Mg(s) + O_2(g)$;
$MgO(s) + C(graphite) \rightarrow Mg(s) + CO(g)$; $\Delta G° = 432.26$ kJ
(e) $2\,PbO(s) \rightarrow 2\,Pb(s) + O_2(g)$;
$PbO(s) + C(graphite) \rightarrow Pb(s) + CO(g)$; $\Delta G° = 50.72$ kJ;
Cu, Ag, and Hg can be obtained by this method.
80. CuO, Ag_2O, HgO, and PbO
82. (a) Five O—H bonds, seven C—O bonds, seven C—H bonds, five C—C bonds, and six O═O bonds are broken. Twelve C═O bonds and 12 O—H bonds are formed. $\Delta_rH° \cong -2873$ kJ.
(b) Interactive forces in condensed phases (solid glucose and liquid water) are neglected in this calculation.
84. (a) 6.46 mol ATP per mol of glucose
(b) $\Delta_rG° = -106$ kJ (c) Product-favored
86. The combustions of coal, petroleum, and natural gas are the most common sources used to supply free energy. We also use solar and nuclear energy as well as the kinetic energy of wind and water. (There may be other answers.)
88. (a) $\Delta G° = -86.5$ kJ, product-favored
(b) $\Delta G° = -873.1$ kJ (c) No (d) Yes
90. $CH_4(g) + 2\,O_2(g) \rightarrow CO_2(g) + 2\,H_2O(\ell)$; $\Delta_rG° = -817.90$ kJ, therefore product-favored;
$C_6H_6(g) + \frac{15}{2}\,O_2(g) \rightarrow 6\,CO_2(g) + 3\,H_2O(\ell)$; $\Delta_rG° = -3202.0$ kJ, therefore product-favored;
$CH_3OH(\ell) + \frac{3}{2}\,O_2(g) \rightarrow CO_2(g) + 2\,H_2O(\ell)$; $\Delta_rG° = -702.34$ kJ, therefore product-favored.
Complex molecules require significant rearrangement of atoms, indicating kinetic stability is likely.
92. (a) +113.0 J/K (b) +38.17 J/K
94. (a) 3.5×10^{18} J/yr (b) 9.5×10^{15} J/day
(c) 1.1×10^{11} J/s (d) 1.1×10^{11} W
(e) 3×10^2 W/person
96. (a) Reaction 2 (b) Reactions 1 and 5
(c) Reaction 2 (d) Reactions 2 and 3
(e) None of them
98. (a) $\Delta_rG° = 31.8$ kJ (b) $K_P = P_{Hg(g)}$
(c) $K° = 2.7 \times 10^{-6}$ (d) $P_{Hg(g)} = 2.7 \times 10^{-6}$ atm
(e) $T = 450$ K
100. Scrambled is a very disordered state for an egg. The second law of thermodynamics says that the more disordered state is the more probable state. Putting the delicate tissues and fluids back where they were before the scrambling occurred would take a great deal of energy. Humpty Dumpty is a fictional character who was also an egg. He fell off a wall. A very probable result of that fall is for an egg to become scrambled. The story goes on to tell that all the energy of the king's horses and men was not sufficient to put Humpty together again.
101. Absolute entropies can be determined because the minimum value of $S°$ is zero at $T = 0$ K. It is not possible to define conditions for a specific minimum value for internal energy, enthalpy, or Gibbs free energy of a substance, so relative quantities must be used.
103. The reactions are exothermic and product-favored. The oxide ion has a noble-gas electron configuration, the metal ions are relatively easy to form because ionization energies for metals are low, and the large charges on the ions produce strong ionic bonding.
105. (a) False (b) False (c) True
(d) True (e) True
107. The energy obtained from nutrients is stored as ATP. The source of the energy needed to synthesize the sugars was sunlight used by plants to produce the sugars and other carbohydrates.
109. $\Delta_rG < 0$ means products are favored; however, the equilibrium state will always have some reactants present, too. To convert all the reactants to products requires the removal of the products from the reactants, so the reaction continues forward.
111. (a) (c) (b) (a)
113. (a) 58.78 J/K (b) −53.29 J/K (c) −173.93 J/K
Adding more hydrogen makes the $\Delta_rS°$ more negative.
116. (a) 331.51 K (b) 371. K
118. (a) $K_P = 1.5 \times 10^7$ (b) Product-favored (c) $K_c = 3.7 \times 10^8$
120. (a) $7\,C(s) + 6\,H_2O(g) \rightarrow 2\,C_2H_6(g) + 3\,CO_2(g)$
$5\,C(s) + 4\,H_2O(g) \rightarrow C_3H_8(g) + 2\,CO_2(g)$
$3\,C(s) + 4\,H_2O(g) \rightarrow 2\,CH_3OH(\ell) + CO_2(g)$
(b) For C_2H_6, $\Delta H° = 101.02$ kJ, $\Delta G° = 122.72$ kJ, $\Delta_rS° = -72.71$ J/K
For C_3H_8, $\Delta_rH° = 76.5$ kJ, $\Delta_rG° = 102.08$ kJ, $\Delta_rS° = -86.6$ J/K
For CH_3OH, $\Delta_rH° = 96.44$ kJ, $\Delta_rG° = 187.39$ kJ, $\Delta_rS° = -305.2$ J/K
None of these is feasible. $\Delta_rG°$ is positive. In addition, $\Delta_rH°$ is positive and $\Delta_rS°$ is negative, suggesting that there is no temperature at which the products would be favored.
122. $\Delta_rG° = -RT \ln K = \Delta_rH° - T\Delta_rS°$

$$\ln K = -\frac{\Delta_rH°}{RT} + \frac{\Delta_rS°}{R}$$

If ln K is plotted against $1/T$, the slope would be $-\Delta_rH°/R$, and the y-intercept would be $\Delta_rS°/R$. The linear graph shows that $\Delta_rS°$ and $\Delta_rH°$ are independent of temperature.
124. (a) (ii) (b) (i) (c) (ii)
126. 205.4 J mol^{-1} K^{-1}

Chapter 17

6. Oxidation numbers in reactants are H: +1, O: −2, Mn: +7, Fe: +2; in products are H: +1, O: −2, Mn: +2, Fe: +3. Fe^{2+} is oxidized and Mn in MnO_4^- is reduced. MnO_4^- is the oxidizing agent. Fe^{2+} is the reducing agent.
10. (a) $Cd(s) \rightarrow Cd^{2+}(aq) + 2\,e^-$
(b) $Fe^{3+}(aq) + 3\,e^- \rightarrow Fe(s)$
(c) $Sn^{4+}(aq) + 2\,e^- \rightarrow Sn^{2+}(aq)$
(d) $Cl_2(g) + 2\,e^- \rightarrow 2\,Cl^-(aq)$
(e) $6\,H_2O(\ell) + SO_2(g) \rightarrow SO_4^{2-}(aq) + 4\,H_3O^+(aq) + 2\,e^-$
12. $Fe^{2+}(aq) \rightarrow Fe^{3+}(aq) + e^-$
$MnO_4^-(aq) + 8\,H^+(aq) + 5\,e^- \rightarrow Mn^{2+}(aq) + 4\,H_2O(\ell)$
14. $8\,H^+(aq) + MnO_4^-(aq) + 5\,Fe^{2+}(aq) \rightarrow 5\,Fe^{3+}(aq) + Mn^{2+}(aq) + 4\,H_2O(\ell)$
16. (a) $3\,CO(g) + O_3(g) \rightarrow 3\,CO_2(g)$
O_3 is the oxidizing agent; CO is the reducing agent.
(b) $H_2(g) + Cl_2(g) \rightarrow 2\,HCl(g)$
Cl_2 is the oxidizing agent; H_2 is the reducing agent.
(c) $H_2O_2(aq) + Ti^{2+}(aq) \rightarrow 2\,H^+(aq) + TiO_2(s)$
H_2O_2 is the oxidizing agent; Ti^{2+} is the reducing agent.
(d) $2\,MnO_4^-(aq) + 6\,Cl^-(aq) + 8\,H^+(aq) \rightarrow 2\,MnO_2(s) + 3\,Cl_2(g) + 4\,H_2O(\ell)$
MnO_4^- is the oxidizing agent; Cl^- is the reducing agent.
(e) $4\,FeS_2(s) + 11\,O_2(g) \rightarrow 2\,Fe_2O_3(s) + 8\,SO_2(g)$
O_2 is the oxidizing agent; FeS_2 is the reducing agent.
(f) $O_3(g) + NO(g) \rightarrow O_2(g) + NO_2(g)$
O_3 is the oxidizing agent; NO is the reducing agent.
(g) $Zn(s) + H_2O(\ell) + HgO(s) \rightarrow Zn(OH)_2(s) + Hg(\ell)$
HgO is the oxidizing agent; Zn(s) is the reducing agent.
18. The generation of electricity occurs when electrons are transmitted through a wire from the metal to the cation. Here, the transfer of electrons would occur directly from the metal to the cation and the electrons would not flow through any wire.

21. (a) $Zn(s) + Pb^{2+}(aq) \rightarrow Zn^{2+}(aq) + Pb(s)$
(b) Oxidation of zinc occurs at the anode. The reduction of lead occurs at the cathode. The anode is metallic zinc. The cathode is metallic lead.
(c)

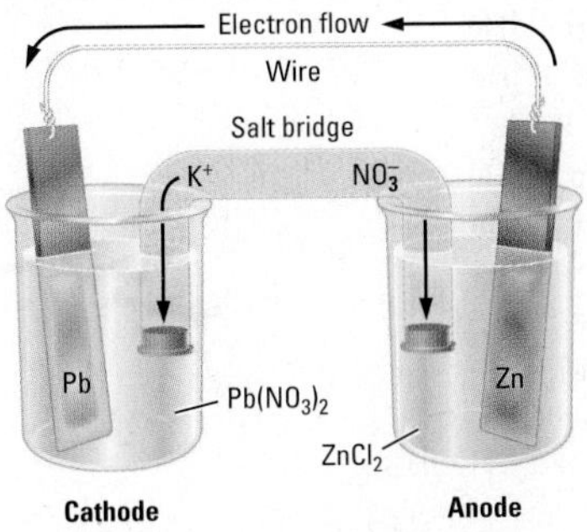
Electron flow
Wire
Salt bridge
K+
NO3−
Pb
Pb(NO3)2
Zn
ZnCl2
Cathode
Anode

23. Assume that the salt bridge contains KNO_3 in both cases.
(a)

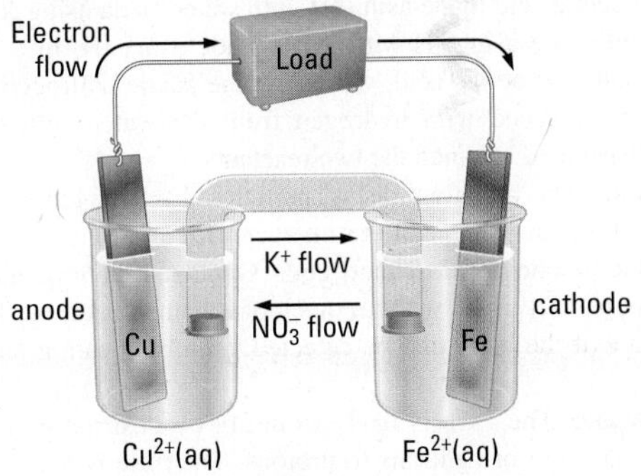
Electron flow
Load
K+ flow
NO3− flow
anode
cathode
Cu
Fe
Cu2+(aq)
Fe2+(aq)

(b)

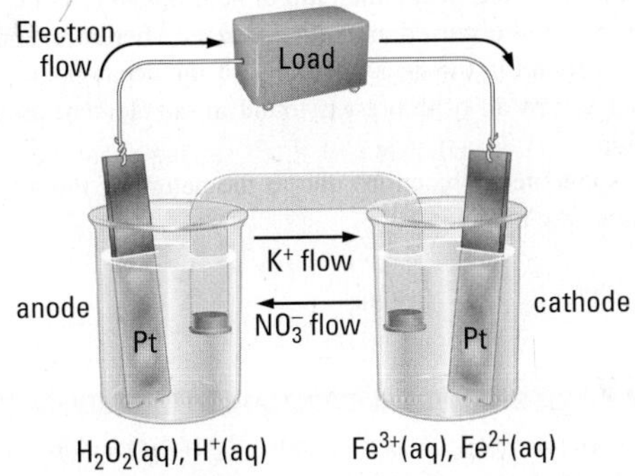
Electron flow
Load
K+ flow
NO3− flow
anode
cathode
Pt
Pt
H2O2(aq), H+(aq)
Fe3+(aq), Fe2+(aq)

24. 9.0×10^4 J; 7.5×10^3 C
25. (a) $Cu(s) \rightarrow Cu^{2+}(aq) + 2\,e^-$
$Ag^+(aq) + e^- \rightarrow Ag(s)$
(b) The copper half-reaction is oxidation and it occurs in the anode compartment. The silver half-reaction is reduction and it occurs in the cathode compartment.
27. Li is the strongest reducing agent and Li^+ is the weakest oxidizing agent. F_2 is the strongest oxidizing agent and F^- is the weakest reducing agent.
29. (a) < (c) < (e) < (b) < (d)
31. (a) 2.89 V (b) −0.028 V
(c) 0.65 V (d) 1.164 V
Reactions (a), (c), and (d) are product-favored.
32. (a) Mn^{2+} (b) Au^{3+} (c) Mn (d) Au
(e) Yes (f) No (g) Au^{3+} and Pt^{2+}
35. greater; less
36. (a) 1.55 V (b) −1196 kJ/mol, 1.55 V
38. $Zn(s) + Cl_2(g) \rightarrow Zn^{2+}(aq) + 2\,Cl^-(aq)$ $\Delta G° = -409$ kJ/mol
40. $K° = 1 \times 10^{25}$, $\Delta_r G° = -143$ kJ/mol
42. $K° = 2 \times 10^{-53}$, $\Delta_r G° = 301.5$ kJ/mol
45. $K° = 0.12$
46. (a) 0.360 V (b) 0.20 M (c) 0.33 V
49. −0.378 V
51. (conc. Pb^{2+})/(conc. Sn^{2+}) = 0.38
52. Advantages: rechargeable, durable, inexpensive, reliable, and simple; disadvantages: heavy, dangerous (can be corrosive and could explode), contains lead, a known environmental danger
54. (a) $Ni^{2+}(aq) + Cd(s) \rightarrow Ni(s) + Cd^{2+}(aq)$
(b) Cd is oxidized, Ni^{2+} is reduced, Ni^{2+} is the oxidizing agent, and Cd is the reducing agent.
(c) Metallic Cd is the anode and metallic Ni is the cathode.
(d) 0.15 V
(e) Electrons flow from the Cd electrode to the Ni electrode.
(f) Toward the anode compartment
56. A fuel cell has a continuous supply of reactants and will be usable for as long as the reactants are supplied. A battery contains all the reactants of the reaction. Once the reactants are gone, the battery is no longer usable.
58. (a) The N_2H_4 half-reaction occurs at the anode and the O_2 half-reaction occurs at the cathode.
(b) $N_2H_4(g) + O_2(g) \rightarrow N_2(g) + 2\,H_2O(\ell)$
(c) 7.5 g N_2H_4 (d) 7.5 g O_2
59. $O_2(g)$ produced at anode; $H_2(g)$ produced at cathode; 2 mol H_2 per mol O_2
61. Au^{3+}, Hg^{2+}, Ag^+, Hg_2^{2+}, Fe^{3+}, Cu^{2+}, Sn^{4+}, Sn^{2+}, Ni^{2+}, Cd^{2+}, Fe^{2+}, Zn^{2+}
63. H_2, Br_2, and OH^- are produced. After the reaction is complete, the solution contains Na^+, OH^-, a small amount of dissolved Br_2 (though it has low solubility in water), and a very small amount of H_3O^+. H_2 is formed at the cathode. Br_2 is formed at the anode.
65. 0.16 g Ag
67. 5.93 g Cu
69. 2.7×10^5 g Al
71. 1.9×10^2 g Pb
73. 0.043 g Li; 0.64 g Pb
75. 6.85 min
78. Ions increase the electrolytic capacity of the solution.
80. Chromium is highly resistant to corrosion and protects iron in steel from oxidizing.
84. 0.00689 g Cu, 0.00195 g Al
86. (a) 9.5×10^6 g HF (b) 1.7×10^3 kWh
88. 4+
89. 75 s
91. Worst reducing agent B < D < A < C best reducing agent
93. (a) B is oxidized and A^{2+} is reduced.
(b) A^{2+} is the oxidizing agent and B is the reducing agent.
(c) B is the anode and A is the cathode.
(d) $A^{2+} + 2e^- \rightarrow A$
$B \rightarrow B^{2+} + 2e^-$
(e) A gains mass.
(f) Electrons flow from B to A.
(g) K^+ ions will migrate toward the A^{2+} solution.
95. (a) The reaction with water is product-favored:
$4\,Co^{3+}(aq) + 2\,H_2O(\ell) \rightarrow 4\,Co^{2+}(aq) + O_2(aq) + 4\,H^+(aq)$
$E°_{cell} = 0.591$ V
(b) The reaction with oxygen in air is product-favored:
$4\,Fe^{2+}(aq) + O_2(aq) + 4\,H^+(aq) \rightarrow 4\,Fe^{3+}(aq) + 2\,H_2O(\ell)$
$E°_{cell} = 0.458$ V
101. (conc. Ag^+) = 1.0×10^{-2} M; $K_{sp} = 9.5 \times 10^{-17}$
103. Calculate pH from the cell potential for the test solution: pH = 3.72.
106. (a) 48.1 min (b) 1.93 L
108. $Cl_2(g) + 2\,Br^-(aq) \rightarrow Br_2(\ell) + 2\,Cl^-(aq)$
111. 4×10^{-6}-M Cu^{2+}

Chapter 18

10.

	Symbol	Mass (g/particle)	Charge
α particle	^4_2He	6.65×10^{-24}	2+
β particle	$^{0}_{-1}e$	9.11×10^{-28}	1−
γ radiation	$^0_0\gamma$	0	0

12. (a) α emission (b) β emission
(c) Electron capture or positron emission
(d) β emission
14. (a) ${}^{238}_{92}U$ (b) ${}^{32}_{15}P$ (c) ${}^{10}_{5}B$
(d) ${}^{0}_{-1}e$ (e) ${}^{15}_{7}N$
16. (a) ${}^{28}_{12}Mg \rightarrow {}^{28}_{13}Al + {}^{0}_{-1}e$
(b) ${}^{238}_{92}U + {}^{12}_{6}C \rightarrow 4\,{}^{1}_{0}n + {}^{246}_{98}Cf$
(c) ${}^{2}_{1}H + {}^{3}_{2}He \rightarrow {}^{4}_{2}He + {}^{1}_{1}H$
(d) ${}^{38}_{19}K \rightarrow {}^{38}_{18}Ar + {}^{0}_{+1}e$
(e) ${}^{175}_{78}Pt \rightarrow {}^{4}_{2}He + {}^{171}_{76}Os$
18. ${}^{231}_{90}Th$, ${}^{231}_{91}Pa$, ${}^{227}_{89}Ac$, ${}^{227}_{90}Th$, ${}^{223}_{88}Ra$
20. ${}^{222}_{86}Rn \rightarrow {}^{4}_{2}He + {}^{218}_{84}Po$
${}^{218}_{84}Po \rightarrow {}^{4}_{2}He + {}^{214}_{82}Pb$
${}^{214}_{82}Pb \rightarrow {}^{0}_{-1}e + {}^{214}_{83}Bi$
${}^{214}_{83}Bi \rightarrow {}^{0}_{-1}e + {}^{214}_{84}Po$
${}^{214}_{84}Po \rightarrow {}^{4}_{2}He + {}^{210}_{82}Pb$
${}^{210}_{82}Pb \rightarrow {}^{0}_{-1}e + {}^{210}_{83}Bi$
${}^{210}_{83}Bi \rightarrow {}^{0}_{-1}e + {}^{210}_{84}Po$
${}^{210}_{84}Po \rightarrow {}^{4}_{2}He + {}^{206}_{82}Pb$
21. (a) ${}^{19}_{10}Ne \rightarrow {}^{19}_{9}F + {}^{0}_{+1}e$ (b) ${}^{230}_{90}Th \rightarrow {}^{0}_{-1}e + {}^{230}_{91}Pa$
(c) ${}^{82}_{35}Br \rightarrow {}^{0}_{-1}e + {}^{82}_{36}Kr$ (d) ${}^{212}_{84}Po \rightarrow {}^{4}_{2}He + {}^{208}_{82}Pb$
23. The binding energy per nucleon of ${}^{10}B$ is 6.252×10^8 kJ/mol nucleon. The binding energy per nucleon of ${}^{11}B$ is 6.688×10^8 kJ/mol nucleon. ${}^{11}B$ is more stable than ${}^{10}B$ because its binding energy is larger.
25. -1.7394×10^{11} kJ/mol ${}^{238}U$; 7.3086×10^8 kJ/mol nucleon
28. 5 mg
30. (a) ${}^{131}_{53}I \rightarrow {}^{131}_{54}Xe + {}^{0}_{-1}e$ (b) 1.56 mg
32. 4.0 yr
34. 34.8 d
36. 2.58×10^3 yr; approximately 570 B.C.
38. 1.6×10^5 yr $= 5.05 \times 10^{12}$ s
40. ${}^{18}_{8}O + {}^{1}_{1}p \rightarrow {}^{18}_{9}F + {}^{1}_{0}n$
42. ${}^{209}_{83}Bi + {}^{1}_{0}n \rightarrow {}^{210}_{84}Po + {}^{0}_{-1}e$
45. ${}^{239}_{94}Pu + 2\,{}^{1}_{0}n \rightarrow {}^{0}_{-1}e + {}^{241}_{95}Am$
47. ${}^{12}_{6}C$
49. ${}^{4}_{2}He + {}^{4}_{2}He \rightarrow {}^{8}_{4}Be$
51. ${}^{12}_{6}C + {}^{16}_{8}O \rightarrow {}^{28}_{14}Si$
53. Cadmium rods (a neutron absorber to control the rate of the fission reaction), uranium rods (a source of fuel, because uranium is a reactant in the nuclear equation), and water (used for cooling by removing excess heat and used in steam/water cycle for the production of turning torque for the generator)
55. (a) ${}^{140}_{54}Xe$ (b) ${}^{101}_{41}Nb$ (c) ${}^{92}_{36}Kr$
57. 6.9×10^3 barrels
59. (a) 400 rad (b) 4 Gy (c) 7.3×10^6 mrem
61. A rad is a measure of the amount of radiation absorbed. A rem includes a quality factor that better describes the biological impact of a radiation dose. The unit rem would be more appropriate when talking about the effects of an atomic bomb on humans. The unit gray (Gy) is 100 rad.
63. Because most elements have some proportion of unstable isotopes that decay and we are composed of these elements (e.g., ${}^{14}C$), our bodies emit radiation particles.
65. The gamma ray is a high-energy photon. Its interaction with matter is most likely just to impart large quantities of energy. The alpha and beta particles are charged particles of matter, which could interact, and possibly react, with the matter composing the food.
67. 0.13 L
69. (a) ${}^{0}_{-1}e$ (b) ${}^{4}_{2}He$ (c) ${}^{87}_{35}Br$
(d) ${}^{216}_{84}Po$ (e) ${}^{68}_{31}Ga$
71. 2 mg
73. 3.6×10^9 yr
74. (a) ${}^{247}_{99}Es$ (b) ${}^{16}_{8}O$ (c) ${}^{4}_{2}He$
(d) ${}^{12}_{6}C$ (e) ${}^{10}_{5}B$
76. (a) ${}^{48}_{20}Ca + {}^{249}_{97}Bk \rightarrow {}^{294}_{117}Uus + 3\,{}^{1}_{0}n$
${}^{48}_{20}Ca + {}^{249}_{97}Bk \rightarrow {}^{293}_{117}Uus + 4\,{}^{1}_{0}n$
(b) ${}^{293}_{117}Uus \rightarrow {}^{289}_{115}Uup + {}^{4}_{2}He$
${}^{289}_{115}Uup \rightarrow {}^{285}_{113}Uut + {}^{4}_{2}He$
${}^{285}_{113}Uut \rightarrow {}^{281}_{111}Rg + {}^{4}_{2}He$
(c) ${}^{294}_{117}Uus \rightarrow {}^{290}_{115}Uup + {}^{4}_{2}He$
${}^{290}_{115}Uup \rightarrow {}^{286}_{113}Uut + {}^{4}_{2}He$
${}^{286}_{113}Uut \rightarrow {}^{282}_{111}Rg + {}^{4}_{2}He$
${}^{282}_{111}Rg \rightarrow {}^{278}_{109}Mt + {}^{4}_{2}He$
${}^{278}_{109}Mt \rightarrow {}^{274}_{107}Bh + {}^{4}_{2}He$
${}^{274}_{107}Bh \rightarrow {}^{270}_{105}Db + {}^{4}_{2}He$
(d) 6A(17)
(e) Halogens
78. (a) 1.70×10^8 kJ/g (b) 3.82×10^{28} kJ (c) 1.2×10^{-9} %
83. ${}^{3}H$; react acetic acid made using ${}^{1}H$ with water made using ${}^{3}H$ or react acetic acid made using ${}^{3}H$ with water made using ${}^{1}H$. In either case, by analyzing the acetic acid, we expect the acidic hydrogen in acetic acid to be replaced with hydrogen from the water until the ${}^{3}H$ is evenly distributed between the two reactants.
85. Alpha and beta radiation decay particles are charged (${}^{4}_{2}He^{21}$ and ${}^{0}_{-1}e^{-}$), so they are better able to interact with and ionize tissues, disrupting the function of the cancer cells. Gamma radiation, like X-rays, goes through soft tissue without much of it being absorbed. This radiation can exit the body and be detected, thereby locating the cancerous cells.
87. ${}^{20}Ne$ is stable. The ${}^{17}Ne$ is likely to decay by positron emission, increasing the ratio of neutrons to protons. The ${}^{23}Ne$ is likely to decay by beta emission, decreasing the ratio of neutrons to protons.
89. A nuclear reaction occurred, making products. Therefore, some of the lost mass is found in the decay particles, if the decay is alpha or beta decay, and almost all of the rest is found in the element produced by the reaction.
92. They are spontaneous reactions due to the nature of the nucleus. No other species are involved.
94. 75.1 yr
96. 3.92×10^3 yr

Chapter 19

13. (a) ${}^{16}_{8}O + {}^{4}_{2}He \rightarrow {}^{20}_{10}Ne$ (b) ${}^{12}_{6}C + {}^{12}_{6}C \rightarrow {}^{20}_{10}Ne + {}^{4}_{2}He$
15. "Burning" and capture of neutrons, see Section 19-1b and Figure 19.1
17. Nitrogen, oxygen, sulfur, argon, platinum, gold (other answers are also correct)
19. Toxicity is determined by the persistence in human tissue. Amphibole asbestos is composed of long, thin, straight fibers that can penetrate narrow lung passages, so it is relatively insoluble and persists in human tissue. In contrast, serpentine asbestos fibers are curly, so they ball up like yarn and are more easily rejected by the body; they tend to be soluble and disappear in human tissue.
22. Amphiboles are formed by joining two pyroxene chains (made of two SiO_3^{2-}) such that they share one oxygen atom to give the total formula $Si_4O_{11}^{6-}$. The amphibole mineral is discussed in detail in Section 19-2 and the combining of pyroxene chains into an amphibole structure is shown in Figure 19.6.
23. The boiling points of N_2, O_2, and Ar differ.
25. (a) Nitrogen (b) Sulfur
27. (a) Sodium metal and chlorine gas;
$2\,NaCl(\ell) \rightarrow 2\,Na(\ell) + Cl_2(g)$
(b) Sodium hydroxide, chlorine, and hydrogen
$2\,Cl^-(aq) + 2\,H_2O(\ell) \rightarrow Cl_2(g) + 2\,OH^-(aq) + H_2(g)$
(c) Magnesium metal;
$Mg^{2+}(\text{in melt}) + 2\,Cl^-(\text{in melt}) \rightarrow Mg(\text{in melt}) + Cl_2(g)$
(d) Aluminum metal;
$4\,Al^{3+}(\text{in melt}) + 3\,C(s) + 3\,O_2(g) \rightarrow 4\,Al(\text{in melt}) + 3\,CO_2(g)$

29. (a) $2\ Cl^-(aq) + 2\ H_2O(\ell) \rightarrow Cl_2(g) + 2\ OH^-(aq) + H_2(g)$
(b) The 2002 Cl_2/NaOH ratio (1.27) is higher than the balanced equation ratio (0.88639). Electrolysis of brine may not be the only method used to produce Cl_2 and NaOH. (Other answers may also explain the difference.)
31. (a) Cathode (b) Chlorine gas (c) 2.027×10^9 C
(d) 1.6×10^3 kJ/mol
33. 7×10^3 kWh
35. 67.1 g Al
36. Phosphine does not form strong hydrogen bonds, but ammonia does.
38. It is product-favored.
40. (a) Nitric acid;
$4\ NH_3(g) + 5\ O_2(g) \rightarrow 4\ NO(g) + 6\ H_2O(\ell)$
$2\ NO(g) + O_2(g) \rightarrow 2\ NO_2(g)$
$3\ NO_2(g) + H_2O(\ell) \rightarrow 2\ HNO_3(aq) + NO(g)$
(b) Sulfuric acid;
$S_8(s) + 8\ O_2(g) \rightarrow 8\ SO_2(g)$
$2\ SO_2(g) + O_2(g) \rightarrow 2\ SO_3(g)$
$SO_3(g) + H_2O(\ell) \rightarrow H_2SO_4(aq)$
(c) Ammonia;
$N_2(g) + 3\ H_2(g) \rightarrow 2\ NH_3(g)$
42. 0.2–0.4 g
44. Oxidizing agent is SnO_2; reducing agent is C.
46. (a) No (b) Yes (c) Yes
(d) Yes (e) No
The molecules that do not dissolve are polar and interact with H-bonding. The molecules that do dissolve are nonpolar like nonpolar CS_2.
48. Answers are in **bold**.

Formula	Name	Oxidation State of Phosphorus
P_4	Phosphorus	**0**
$(NH_4)_2HPO_4$	**Ammonium hydrogen phosphate**	**+5**
H_3PO_4	Phosphoric acid	**+5**
P_4O_{10}	Tetraphosphorus decaoxide	**+5**
$Ca_3(PO_4)_2$	**Calcium phosphate**	**+5**
$Ca(H_2PO_4)_2$	Calcium dihydrogen phosphate	**+5**

50. 1.7×10^7 tons
52.
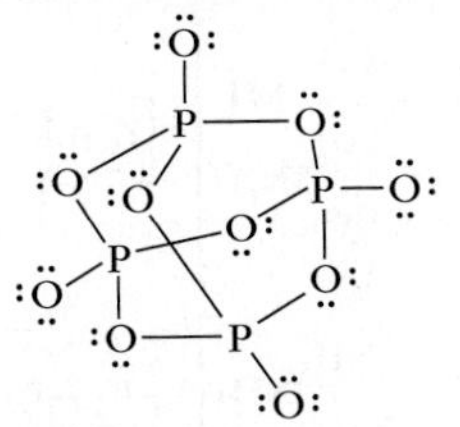
54. 962 K
56. Raw materials: sulfur, water, oxygen, and catalyst (Pt or VO_5)
$S_8(s) + 8\ O_2(g) \rightarrow 8\ SO_2(g)$
$2\ SO_2(g) + O_2(g) \rightarrow 2\ SO_3(g)$
$SO_3(g) + H_2SO_4(\ell) \rightarrow H_2S_2O_7(\ell)$
$H_2S_2O_7(\ell) + H_2O(\ell) \rightarrow 2\ H_2SO_4(aq)$
58.
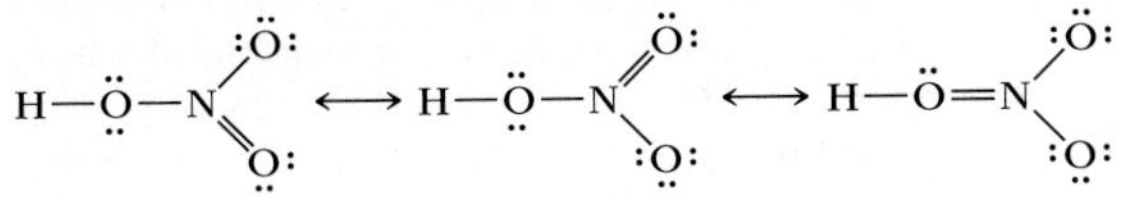
60. $K_c = 2.7$
62. 3 L
64. $K = 3.1 \times 10^5$; product-favored reaction produces $Mg(OH)_2(s)$. Putting seawater in the presence of $Ca(OH)_2$ will cause a precipitate of $Mg(OH)_2$ to form. The solid can be isolated after it settles.
66. $\ddot{N}=N=\ddot{N}-\ddot{N}=\ddot{O}$ (Other structures are possible.)
68. (a) $H-\ddot{N}=\ddot{N}=\ddot{N} \longleftrightarrow H-\ddot{N}-N\equiv N:$
(b) +140. kJ/mol with single and triple bond; +410. kJ/mol with two double bonds
70. (a) $NO_2(g) + NO(g) \rightarrow N_2O_3(g)$
(b)
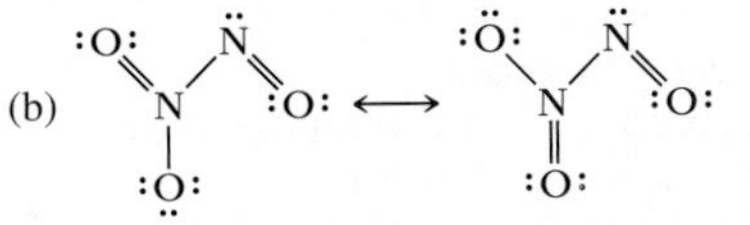
(c) The O—N—O angle is 120° and the N—N—O angle is slightly less than 120°.
73. (a) 660 kJ/mol
(b) 3.4 V must be used to overcome the cell potential.
75. (a) Three: 140 °C, 10^3 atm; 95 °C, 10^{-5} atm; and 120 °C, 1×10^{-4} atm
(b) (i) Solid rhombic (ii) Liquid (iii) Solid monoclinic (iv) Vapor
77. (a) $4\ H^+(aq) + MnO_2(s) + 2\ Br^-(aq) \rightarrow Mn^{2+}(aq) + 2\ H_2O(\ell) + Br_2(\ell)$
(b) 0.0813 mol Br^- (c) 3.54 g MnO_2
85. 6.3×10^5 J/mol
88. (a) $P_4(s) + 5\ O_2(g) \rightarrow P_4O_{10}(s)$
(b) pH = 2.25
(c) $3\ Ca(NO_3)_2(aq) + 2\ H_3PO_4(aq) \rightarrow Ca_3(PO_4)_2(s) + 6\ HNO_3(aq)$; 25.0 g $Ca_3(PO_4)_2$
(d) $H_2(g)$; 5.42 L
90. (a) Cl is reduced; Al and N are oxidized. (b) −2674 kJ
92. −2708 kJ/mol
94. $K_p = 1.99$
96. 2961 kJ, Mg^{2+} is a much smaller ion with higher charge density than Sr^{2+}, so the $Mg^{2+} + F^-$ ions are closer together and the ionic bond is much stronger.
98. B_6H_{12}

Chapter 20

13. (a) Ag [Kr] $4d^{10}5s^1$; Ag^+ [Kr] $4d^{10}$
(b) Au [Xe] $4f^{14}5d^{10}6s^1$; Au^+ [Xe]$4f^{14}\ 5d^{10}$; Au^{3+} [Xe] $4f^{14}\ 5d^8$
15. Cr^{2+} and Cr^{3+}
17. $Cr_2O_7^{2-}$ (in acid) > Cr^{3+} > Cr^{2+}; the species with a more positive oxidation state of Cr has greater tendency to be reduced.
E°_{red}, respectively, is +1.33 V > −0.74 V > −0.91 V, so the more positive the oxidation state, the better the oxidizing agent.
19. (a) $Fe_2O_3(s) + 3\ CO(g) \rightarrow 2\ Fe(s) + 3\ CO_2(g)$
(b) $Fe(s) + 2\ H^+(aq) \rightarrow Fe^{2+}(aq) + H_2(g)$ or
$2\ Fe(s) + 6\ H^+(aq) \rightarrow 2\ Fe^{3+}(aq) + 3\ H_2(g)$
21. $6\ H^+(aq) + 3\ NO_3^-(aq) + Fe(s) \rightarrow Fe^{3+}(aq) + 3\ NO_2(g) + 3\ H_2O(\ell)$
23. (a) V: +5 (b) Cr: +6 (c) Mn: +4 (d) Os: +8
25. Size depends on the largest occupied orbitals and all the transition elements have electrons filling orbitals that are not in the largest shell.
27. (a) Ag (b) Ti (c) Hg
29. Fe(s) is produced by reduction because natural sources of iron are compounds containing Fe^{2+} and Fe^{3+}. The cations must be reduced to make Fe(s). To make steel, oxidation is needed to remove the carbon impurity introduced during the reduction process.
31. The basic oxygen process has pure oxygen blown through a water-cooled tube inserted in the surface of molten, impure iron. At 1900 °C, dissolved carbon reacts with oxygen to form gaseous CO and CO_2, which are vented.
33. Bronze contains copper alloyed with tin. Brass contains copper alloyed with zinc.
35. In electrorefining, pure Cu metal is the cathode and blister copper is the anode. The electrolyte solution contains $CuSO_4$ and H_2SO_4. Copper from the blister copper is oxidized into aqueous Cu^{2+} ions at the anode. The Cu^{2+} ions migrate to the pure Cu cathode where they are reduced to pure copper metal. (See Figure 20.7)
37. Ag reacts with HNO_3 to form Ag^+. Au does not react with HNO_3, unless HCl is also present.
39. Increase hardness

41. (a) +3 (b) +5 (c) +4 (d) +2

43. (a) $[Cr(NH_3)_2(H_2O)_3(OH)]^{2+}$; 2+
(b) The counter ion would be a doubly charged anion, such as SO_4^{2-}.

45. (a) $C_2O_4^{2-}$ with 2− charge and Cl^- with 1− charge
(b) 3+ charge
(c) $[Co(NH_3)_4Cl_2]^+$, with 1+ charge

48. For $Na_3[IrCl_6]$: (a) Six Cl^- (b) Ir with 3+ charge
(c) $[IrCl_6]^{3-}$ with 3− charge (d) Na^+
For $[Mo(CO)_4]Br_2$: (a) Four CO (b) Mo with 2+ charge
(c) $[Mo(CO)_4]^{2+}$ with 2+ charge (d) Br^-

49. (a) Four (b) Four

52. (a) $[Pt(NH_3)_2Br_2]$ (b) $[Pt(en)(NO_2)_2]$ (c) $[Pt(NH_3)_2BrCl]$

54. (a) Four (b) Four
(c) Six (d) Six

56. (a) Monodentate (b) Tetradentate
(c) Tridentate (d) Monodentate

58. For example, $FeCl_3$ (Other answers are possible.)

59. (a) $K[Co(H_2O)_2(ox)_2]$
(b) $[Cr(NH_3)_2(H_2O)_3OH]NO_3$
(c) $(NH_4)_2[CuCl_4]$

61. (a) Tetrachloromanganate (II)
(b) Potassium trioxalatoferrate (III)
(c) Diamminedicyanoplatinum (II)

63. (a)
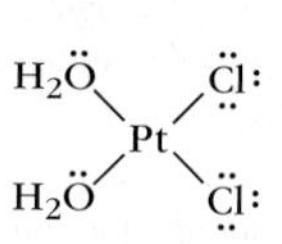

(b)
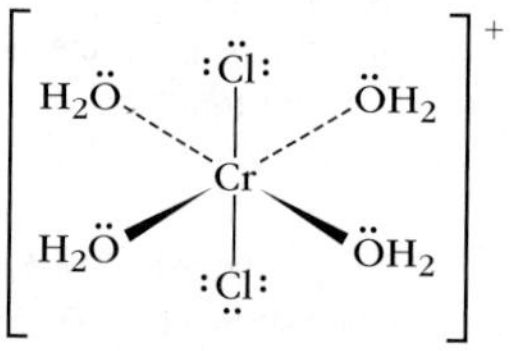

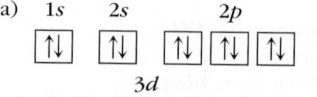

65.
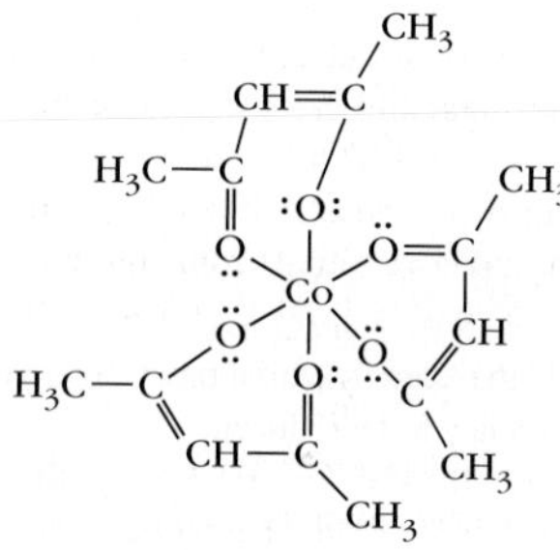

67. (a) and (b)

69. (a), (c), and (d)

71.
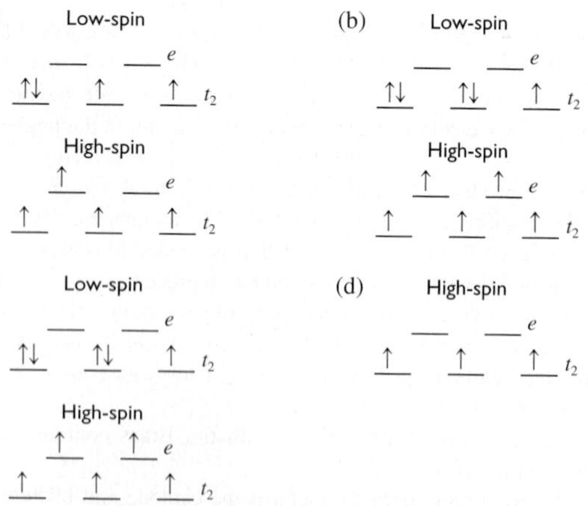
(a) Low-spin
e
t2
High-spin
(b) Low-spin
High-spin
(c) Low-spin
High-spin
(d) High-spin

74. Cr^{2+} has four electrons. Cr^{3+} has three electrons. Since there are three low-energy t_2 orbitals, the first three electrons can fill one into each t_2 orbital. The fourth electron (only found in Cr^{2+}) is required to pair up with one of the other electrons (low spin) or to span the Δ_o gap to go into an e orbital (high spin).

76. If Δ_o is large, electrons fill the t_2 orbitals and end up all paired. If Δ_o is small, all five orbitals are half-filled before any pairing begins, resulting in four unpaired electrons. So, high-spin Co^{3+} complexes are paramagnetic and low-spin Co^{3+} complexes are diamagnetic.

78. Cu^{2+} has nine d-electrons. All t_2 orbitals are filled, so high- and low-spin complexes do not form.

79. violet light; Approximately 400 nm

81. With CN^-: Δ_o increases and the observed color shifts closer to yellow-orange.
With Cl^-: Δ_o decreases and the observed color shifts closer to blue.

84. (a) $[Ar]3d^4$ (b) $[Ar]3d^{10}$ (c) $[Ar]3d^7$ (d) $[Ar]3d^3$

86. (a) 1s [↑↓] 2s [↑↓] 2p [↑↓][↑↓][↑↓] 3s [↑↓] 3p [↑↓][↑↓][↑↓]
3d [↑][↑][↑][↑][ ]
Four unpaired electrons

(b) 1s [↑↓] 2s [↑↓] 2p [↑↓][↑↓][↑↓] 3s [↑↓] 3p [↑↓][↑↓][↑↓]
3d [↑↓][↑↓][↑↓][↑↓][↑↓]
Zero unpaired electrons

(c) 1s [↑↓] 2s [↑↓] 2p [↑↓][↑↓][↑↓] 3s [↑↓] 3p [↑↓][↑↓][↑↓]
3d [↑↓][↑↓][↑][↑][↑]
Three unpaired electrons

(d) 1s [↑↓] 2s [↑↓] 2p [↑↓][↑↓][↑↓] 3s [↑↓] 3p [↑↓][↑↓][↑↓]
3d [↑][↑][↑][ ][ ]
Three unpaired electrons

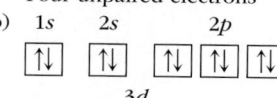

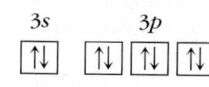

89. 14.8 g Cu

91. 0.40 ton SO_2

92. (a) Four (b) Six (c) Four (d) Six

94.
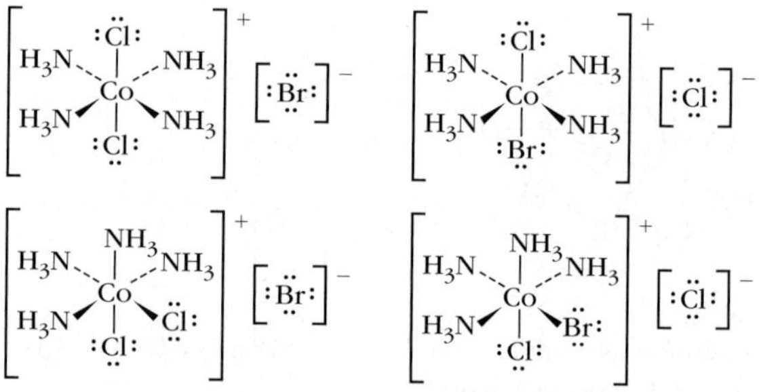

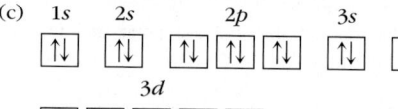

97. (a) False; the coordination number is six.
(b) False; Cu^+ has no unpaired electrons.
(c) False; the net charge is 3+.

99. (a) +2
(b)
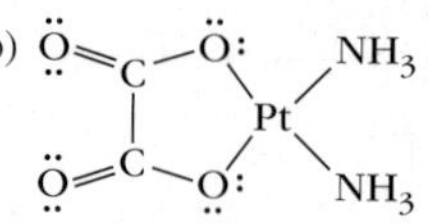

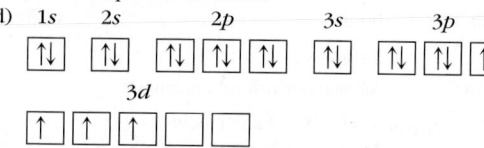

101. $Fe(s) + 2\ Fe^{3+}(aq) \rightarrow 3\ Fe^{2+}(aq)$

103. Four

105.

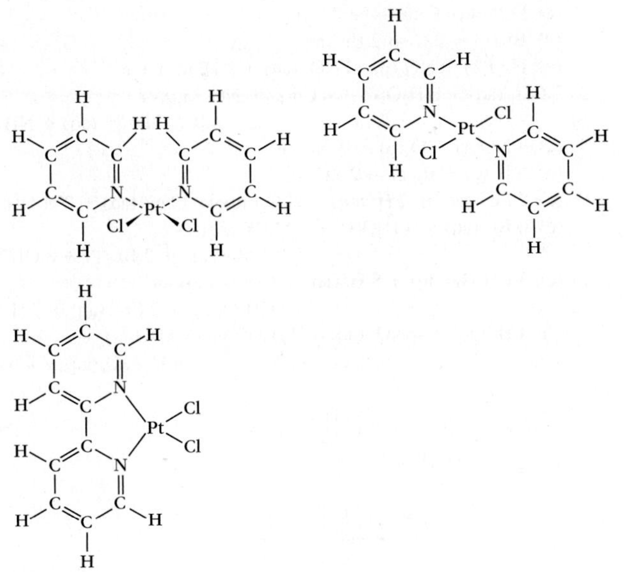

The bidentate 1,10-phenanthroline is not able to form the *trans*-isomer.

107. (a) $2\,Ag^{+}(aq) + Ni(s) \rightarrow Ni^{2+}(aq) + 2\,Ag(s)$
(b) +1.05V

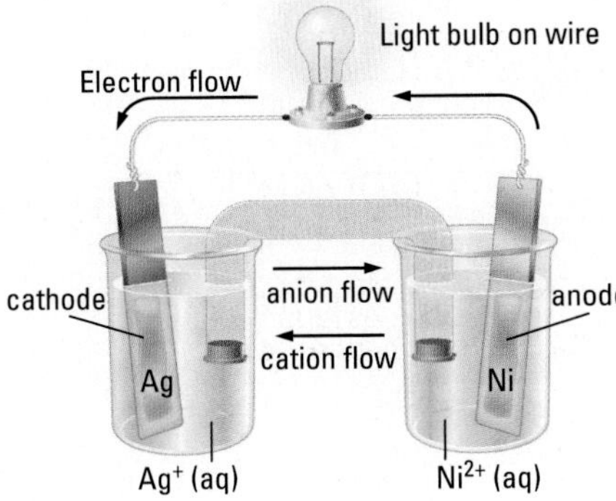
Light bulb on wire
Electron flow
cathode
anion flow
cation flow
anode
Ag
Ni
Ag⁺ (aq)
Ni²⁺ (aq)

109. 17.4% Fe
111. (a) −1.35 V (b) 3×10^{-21}-M Ag^{2+}
113. (a) No ions (b) 2.00 mol ions
(c) 4.00 mol ions (d) 3.00 mol ions

116.

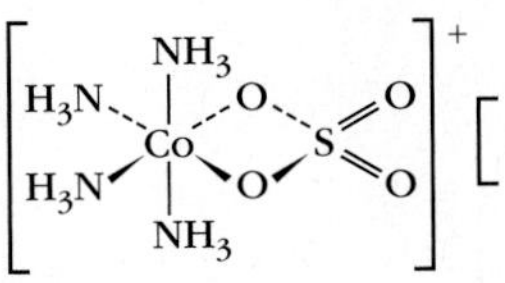

118. (a) $[Co(NH_3)_6]Cl_3$
(b) $[Co(NH_3)_6]Cl_3 \rightarrow [Co(NH_3)_6]^{3+}(aq) + 3\,Cl^{-}(aq)$

120.

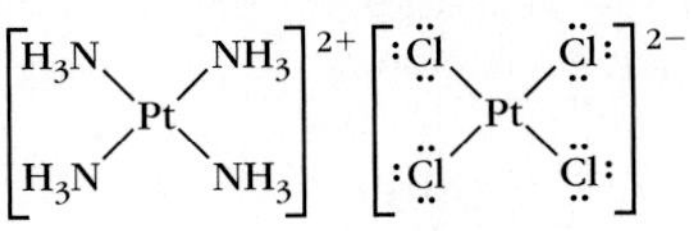

122.

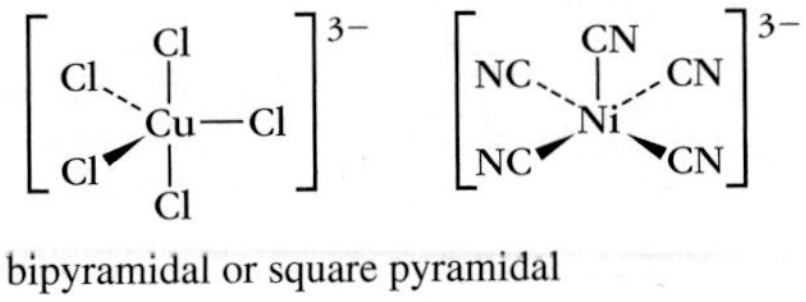

Triangular bipyramidal or square pyramidal

124. A mixture of high-spin and low-spin complexes, in which the high-spin species is favored by a 2:1 ratio, would give the observed 2.67 unpaired electrons per iron ion. High-spin iron(II) has four unpaired electrons, while low-spin iron(II) has zero unpaired electrons. This translates to 8 unpaired electrons for every three iron ions or 8/3 = 2.67 unpaired electrons per iron ion.

126.

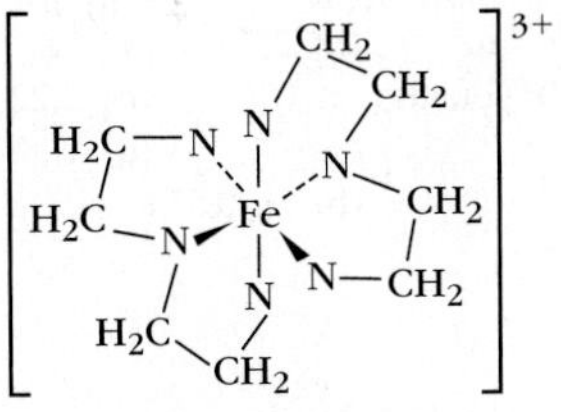

(One other structure is possible, too.)

Appendix A

4. The calculation is incomplete or incorrect. Check for incomplete unit conversions (e.g., cm^3 to L, g to kg, mmHg to atm, etc.) and check to see that the numerator and denominator of unit factors are placed such that the unwanted units can cancel (e.g., $\frac{g}{mL}$ or $\frac{mL}{g}$).
6. Answer (c) gives the properly reported observation. Answers (a) and (b) do not have the proper units. Answer (d) is incomplete, with no units. Answer (e) shows the conversion of the observed measurement to new units, making (e) the result of a calculation, not the observed measurement.
8. (a) The units should not be squared. To do this correctly, add values with common units (g) and give the answer the same units (g). The result is 9.95 g.
(b) The unit factor is not set up to cancel unwanted units (g); the units reported (g) are not the units resulting from this calculation ($\frac{g^2}{mL}$). To do this correctly, $5.23\text{ g} \times \frac{1.00\text{ mL}}{4.87\text{ g}} = 1.07$ mL.
(c) The unit factor must be cubed to cancel all unwanted units; the units reported (m^3) are not the units resulting from this calculation ($m \cdot cm^2$). To do this correctly, $3.57\text{ cm}^3 \times (\frac{1\text{ m}}{100\text{ cm}})^3 = 3.57 \times 10^{-6}\text{ m}^3$.
10. (a) 95.9 ± 0.59 in
(b) Three, because uncertainty is in the tenths place.
(c) The result is accurate (i.e., the true answer, 8.000 ft, is within the range of the uncertainty described by these values, 7.951–8.037 ft), though not very precise.
12. (a) 4 (b) 2
(c) 2 (d) 4
14. (a) 43.32 (b) 43.32 (c) 43.32
(d) 43.32 (e) 43.32 (f) 43.32
16. (a) 13.7 (b) 0.247 (c) 12.0
18. (a) 7.6003×10^4 (b) 3.7×10^{-4} (c) 3.4×10^4
20. (a) 2.415×10^{-3} (b) 2.70×10^8
(c) 3.236 (d) 116 cm^3
22. (a) −0.1351 (b) 3.541 (c) 23.7797
(d) 54.7549 (e) −7.455
24. (a) 5.404 (b) 1×10^{15} (c) 110.
(d) 1.000320 (e) 3.755
26. (a) 0.480, −1.80 (b) 3.23, 1.34

28.

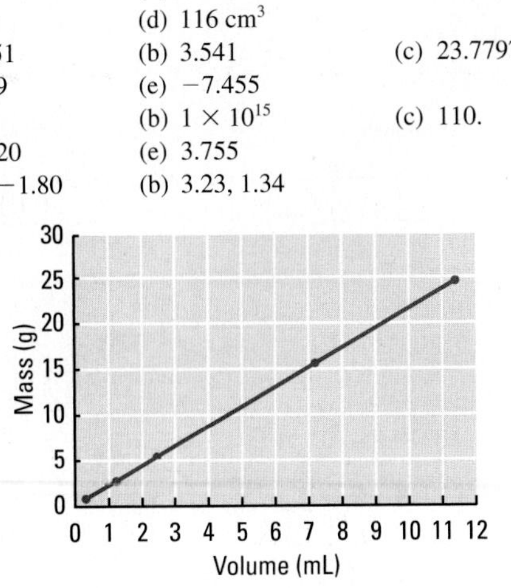
Mass (g)
Volume (mL)

Yes, *y* is proportional to *x*.

Appendix B

1. (a) Kilogram, no additional prefix
   (b) Cubic meters, nano- (c) Meters, milli-
3. SI base (fundamental) units are set by convention; derived units are based on the fundamental units.
5. (a) 4.75×10^{-10} m (b) 5.6×10^{7} kg (c) 4.28×10^{-6} A
7. (a) 8.7×10^{-18} m^2 (b) 2.73×10^{-17} J (c) 2.73×10^{-5} N
9. (a) 2.20×10^{0} pounds (b) 2.22×10^{3} kg
   (c) 60 in^3 (d) 1×10^{5} pascal, 1 bar
11. (a) 99 °C (b) −10.5 °F (c) −40.0 °F

Appendix F

1. (a) $Ni(s) \rightarrow Ni^{2+}(aq) + 2e^-$
   (b) $2\,H^+(aq) + 2e^- \rightarrow H_2(g)$
   (c) $Fe^{3+}(aq) + e^- \rightarrow Fe^{2+}(aq)$
   (d) $Br_2(\ell) + 2e^- \rightarrow 2\,Br^-(aq)$
   (e) $H_2O(\ell) + NO_2(g) \rightarrow NO_3^-(aq) + 2\,H^+(aq) + e^-$
3. $7\,OH^-(aq) + 6\,H_2O(\ell) + 4\,Cd(s) + NO_3^-(aq) \rightarrow 4\,[Cd(OH)_4]^{2-}(aq) + NH_3(aq)$
5. (a) $3\,NO(g) + O_3(g) \rightarrow 3\,NO_2(g)$
   (b) $3\,H_2(g) + P_2(g) \rightarrow 2\,PH_3(g)$
   (c) $2\,Fe^{2+}(aq) + 2\,H^+(aq) + H_2O_2(aq) \rightarrow 2\,Fe^{3+}(aq) + 2\,H_2O(\ell)$
   (d) $6\,Br^-(aq) + 4\,H_2O(\ell) + 2\,MnO_4^-(aq) \rightarrow 2\,MnO_2(s) + 3\,Br_2(\ell) + 8\,OH^-(aq)$
   (e) $3\,CH_3OH(aq) + 8\,H^+(aq) + Cr_2O_7^{2-}(aq) \rightarrow 3\,CH_2O(aq) + 2\,Cr^{3+}(aq) + 7\,H_2O(\ell)$
   (f) $4\,H^+(aq) + 4\,NO_3^-(aq) + 7\,H_2O(\ell) + 3\,As_2O_3(s) \rightarrow 6\,H_3AsO_4(aq) + 4\,NO(g)$

Glossary

A

A See **ampere**.

A See **activity**; see **frequency factor**; see **mass number**.

α See **alpha carbon**; see **alpha particle**.

absolute temperature scale (See also **Kelvin temperature scale, thermodynamic temperature scale**) A temperature scale on which the zero is the lowest possible temperature and the degree is the same size as the Celsius degree.

absolute zero Lowest possible temperature, 0 K.

absorb To draw a substance into the bulk of a liquid or a solid (compare with adsorb).

accuracy The degree to which a measurement (or series of measurements) agrees with the accepted value.

achiral Describes an object that can be superimposed on its mirror image; opposite of chiral.

acid (Arrhenius) A substance that increases the concentration of hydrogen ions, $H^+(aq)$, in aqueous solution. (See also **Brønsted-Lowry acid, Lewis acid**.)

acid-base neutralization reaction A reaction in which an acid reacts with a base and each neutralizes (negates) the properties of the other.

acid-base reaction (Brønsted-Lowry) A reaction in which one molecule or ion (an acid) transfers a proton, H^+, to another molecule or ion (a base).

acid-base reaction (Lewis) A reaction in which one molecule or ion (a base) donates a pair of electrons to another molecule or ion (an acid) forming a coordinate covalent bond.

acid ionization constant (K_a) The equilibrium constant for the reaction of a weak acid with water to produce hydronium ions and the conjugate base of the weak acid.

acid ionization constant expression Mathematical expression in which the product of the equilibrium concentrations of hydronium ion and conjugate base is divided by the equilibrium concentration of the un-ionized conjugate acid.

acid rain Rain (or other precipitation) with a pH below about 5.6 (the pH of unpolluted rain water).

acidic oxide A nonmetal oxide, such as SO_2, that reacts with bases or dissolves in water to give an acidic solution.

acidic solution An aqueous solution in which the concentration of hydrogen ions, $H^+(aq)$, exceeds the concentration of hydroxide ions.

actinides The elements after actinium in the seventh period; in actinides the 5*f* subshell is being filled.

activated complex A molecular structure corresponding to the maximum of a plot of energy versus reaction progress; also known as the transition state.

activation energy (E_a) The potential energy difference between reactants and activated complex; the minimum energy that reactant molecules must have to be converted to product molecules.

active site The part of an enzyme molecule that binds the substrate to help it to react.

activity (*A*) A measure of the rate of nuclear decay, given as disintegrations per unit time.

activity series of metals A list of metals ranked in order of decreasing ability to donate electrons during redox reactions.

actual yield The quantity of a reaction product obtained experimentally; less than the theoretical yield.

addition polymers Polymers formed when monomer molecules join directly with one another, with no other products formed in the reaction.

addition reaction A reaction in which two molecules combine to form a third, larger molecule and no other product molecule. Example: addition of molecules such as H_2 and Cl_2 to molecules containing C=C bonds.

adsorb To attract and hold a substance on a surface (compare with absorb).

aerosols Small particles (1 nm to about 10,000 nm in diameter) that remain suspended indefinitely in air.

air pollutant A substance that degrades air quality.

alcohol An organic compound containing a hydroxyl group (—OH) covalently bonded to a saturated carbon atom.

aldehyde An organic compound characterized by a carbonyl group in which the carbon atom is bonded to a hydrogen atom; a molecule containing the —CHO functional group.

alkali metals The Group 1A elements in the periodic table (except hydrogen).

alkaline-earth metals The elements in Group 2A of the periodic table.

alkaline solution See **basic solution**.

alkane Any of a class of hydrocarbons characterized by the presence of only single carbon-carbon and carbon-hydrogen bonds.

alkene Any of a class of hydrocarbons characterized by the presence of a carbon-carbon double bond.

alkyl group A fragment of an alkane structure that results from the removal of a hydrogen atom from the alkane.

alkyne Any of a class of hydrocarbons characterized by the presence of a carbon-carbon triple bond.

allotropes Different forms of the same element that exist in the same physical state under the same conditions of temperature and pressure.

alloy A solid that has metallic properties and is a mixture of a metal with one or more other elements, usually other metals.

alpha (α) carbon The carbon atom adjacent to a carbon atom of interest; for example, the carbon atom adjacent to the acid group (—COOH) in an amino acid.

alpha (α) particle A positively charged (2+) particle ejected at high speed from a radioactive nucleus; the nucleus of a helium atom.

alpha (α) radiation Radiation composed of alpha particles (helium nuclei).

amide An organic compound characterized by the presence of a carbonyl group in which the carbon atom is bonded to a nitrogen atom ($—CONH_2$, —CONHR, $—CONR_2$); the product of the reaction of an amine with a carboxylic acid.

amide group A functional group consisting of a carbonyl group bonded to a nitrogen atom. The nitrogen atom may have zero, one, or two other carbon atoms bonded to it.

amide linkage The linkage $-\overset{\text{O}}{\overset{\|}{\text{C}}}-\overset{\text{H}}{\overset{|}{\text{N}}}-$ that connects monomers in a polymer.

amine An organic compound containing an $—NH_2$, —NHR, or $—NR_2$ functional group.

amino acid An organic molecule containing a carboxylic acid group and an amine group. Alpha amino acids have an amine group and an R group on the alpha carbon atom and are the building-block monomers of proteins.

amorphous solid A solid whose constituent nanoscale particles have no long-range repeating structure.

amount of substance (*n*) A measure of the number of elementary entities (such as atoms, ions, molecules) in a sample of matter compared with the number of elementary entities in exactly 0.012 kg pure ^{12}C. Also called amount or molar amount.

ampere (A) The SI unit of electrical current; involves the flow of one coulomb of charge per second.

amphiprotic Describes a molecule or ion that can either donate or accept a proton, H^+, in acid-base reactions; an amphiprotic species can serve as either an acid or a base depending on the circumstances.

amphoteric Refers to a substance that can act as either an acid or a base.

amu See **unified atomic mass unit**.

anion An ion with a negative electrical charge.

anode The electrode of an electrochemical cell at which oxidation occurs.

anodic inhibition The prevention of oxidation of an active metal by painting it, coating it with grease or oil, or allowing a thin film of metal oxide to form.

antibonding molecular orbital A higher-energy molecular orbital that, if occupied by electrons, does not result in attraction between the atoms.

antioxidant Reducing agent that converts free radicals and other reactive oxygen species into less reactive substances.

aqueous solution A solution in which water is the solvent.

aromatic compound Any of a class of organic compounds characterized by the presence of one or more benzene rings or benzene-like rings.

Arrhenius equation Mathematical relation that gives the temperature dependence of the reaction rate constant; $k = Ae^{-E_a/RT}$.

asymmetric Describes a molecule or object that is not symmetrical. Refers to chiral atoms and structures that can have a molecular dipole.
atmosphere The blanket of air surrounding Earth to an altitude of about 50 km. The atmosphere consists principally of nitrogen and oxygen, plus argon, water vapor, and many trace gases.
atmosphere (atm) A unit of pressure; one atmosphere, 1 atm, is defined as 101.325 kPa, approximately the pressure of the atmosphere at Earth's surface.
atom The smallest particle of an element that can be involved in chemical combination with another element.
atom economy The fraction of atoms of starting materials incorporated into the desired final product in a chemical reaction.
atomic force microscope Analytical instrument that produces images of individual atoms or molecules on a surface.
atomic mass unit (u or amu) See **unified atomic mass unit**.
atomic number The number of protons in the nucleus of an atom of an element.
atomic radius One-half the distance between the nuclei centers of two like atoms in a molecule.
atomic structure The identity and arrangement of subatomic particles in an atom.
atomic theory Dalton's theory that each element consists of atoms that are the smallest particles that embody the chemical properties of the element and that are different from atoms of every other element.
atomic weight The average mass of an atom in a representative sample of atoms of an element.
autoionization The equilibrium reaction in which water molecules react with each other to form hydronium ions and hydroxide ions.
average atomic mass The weighted average mass of a representative sample of atoms of an element; average atomic mass is expressed in unified atomic mass units, u.
average reaction rate A reaction rate calculated from a change in concentration divided by a change in time.
Avogadro constant (N_A) The number of particles per mole of any substance, 6.022×10^{23}/mol.
Avogadro's law The volume of a gas, at a given temperature and pressure, is directly proportional to the amount of gas.
Avogadro's number The number of particles in a mole of any substance, 6.022×10^{23}.
axial positions Positions above and below the equatorial plane in a triangular bipyramidal structure.

B

β See **beta particles**; see **beta radiation**.
background radiation Radiation from natural and synthetic radioactive sources to which all members of a population are exposed.
balanced chemical equation A chemical equation that shows equal numbers of atoms of each kind in the products and the reactants.
bar A pressure unit equal to 100,000 Pa.
barometer A device for measuring atmospheric pressure.
basal metabolic rate (BMR) The energy required to maintain an organism that is awake, at rest, and not digesting or metabolizing food.
base (Arrhenius) A substance that increases the concentration of hydroxide ions, OH^-, in aqueous solution. (See also **Brønsted-Lowry base, Lewis base.**)
base ionization constant (K_b) The equilibrium constant for the reaction of a weak base with water to produce hydroxide ions and the conjugate acid of the weak base.
base ionization constant expression Mathematical expression in which the product of the equilibrium concentrations of hydroxide ion and conjugate acid is divided by the equilibrium concentration of the conjugate base.
basic oxide A metal oxide, such as BaO and Na_2O, that reacts with acids or dissolves in water to give basic solutions.
basic (alkaline) solution An aqueous solution in which the concentration of hydroxide ions is greater than the concentration of hydrogen ions, H^+(aq).
battery Two or more voltaic cells in which a product-favored oxidation-reduction reaction is used to produce an electric current. In everyday speech, battery also refers to a single voltaic cell.
becquerel (Bq) A unit of radioactivity equal to 1 nuclear disintegration per second.
beta (β) particles Electrons ejected from certain radioactive nuclei.
beta (β) radiation Radiation composed of electrons.
bidentate ligand A ligand that has two atoms with lone pairs that can form coordinate covalent bonds to the same metal ion.
bimolecular reaction An elementary reaction in which two particles must collide for products to be formed.
binary molecular compound A molecular compound whose molecules contain atoms of only two elements.
binding energy The energy required to separate all nucleons in an atomic nucleus.
binding energy per nucleon The energy per nucleon required to separate all nucleons in an atomic nucleus.
biochemicals Substances that make up plants and animals; with few exceptions, a major one being water, biochemicals are organic compounds.
biodegradable Capable of being decomposed by biological means, especially by bacterial action.
biodiesel A fuel made from biological sources, mainly vegetable oils.
BMR See **basal metabolic rate**.
boiling The process whereby a liquid vaporizes throughout when its vapor pressure equals atmospheric pressure.
boiling point The temperature at which the equilibrium vapor pressure of a liquid equals the external pressure on the liquid.
boiling-point elevation A colligative property; the difference between the higher boiling point of a solution containing a nonvolatile solute and the normal boiling point of the pure solvent.
bond Attractive force between two atoms holding them together, for example, as part of a molecule.
bond angle The angle between the bonds to two atoms that are bonded to the same third atom.
bond energy See **bond enthalpy**.
bond enthalpy The change in enthalpy when a mole of chemical bonds of a given type is broken, separating the bonded atoms; the atoms and molecules must be in the gas phase.
bond length The distance between the centers of the nuclei of two bonded atoms.
bond order The number of electron-pair bonds connecting two atoms; in molecular orbital (MO) theory, half the difference between the number of electrons in bonding MOs, n_B, and the number of electrons in antibonding MOs, n_A, that is, $(n_B - n_A)/2$.
bonding electrons Electron pairs shared in covalent bonds.
bonding molecular orbital A lower-energy molecular orbital that can be occupied by bonding electrons.
bonding pair A pair of valence electrons that are shared between two atoms.
Born-Haber cycle A stepwise thermochemical cycle in which the constituent elements are converted to ions and combined to form an ionic compound.
boundary surface A surface within which there is a specified probability (often 90%) that an electron will be found.
Boyle's law The volume of a confined ideal gas varies inversely with the applied pressure, at constant temperature and amount of gas.
Bq See **Becquerel**.
Brønsted-Lowry acid A hydrogen ion donor.
Brønsted-Lowry acid-base reaction A reaction in which an acid donates a hydrogen ion and a base accepts the hydrogen ion.
Brønsted-Lowry base A hydrogen ion acceptor.
buckyball Buckminsterfullerene; an allotrope of carbon consisting of molecules in which 60 carbon atoms are arranged in a cage-like structure consisting of five-membered rings sharing edges with six-membered rings.
buffer A chemical system that resists change in pH. See **buffer solution**.
buffer capacity The quantity of acid or base a buffer can accommodate without a significant pH change (more than one pH unit).
buffer solution A solution that resists changes in pH when limited amounts of acids or bases are added; it contains a weak acid and its conjugate base, or a weak base and its conjugate acid.
building-block elements Four nonmetallic elements—hydrogen, oxygen, carbon, and nitrogen—that are the major components of biomolecules.

C

cal See **calorie**.
Cal See **Calorie**.
caloric value The energy of complete combustion of a stated size sample of a food, usually reported in Calories (kilocalories).

calorie (cal) A unit of energy equal to 4.184 J. Approximately 1 cal is required to raise the temperature of 1 g of liquid water by 1 °C.
Calorie (Cal) A unit of energy equal to 4.184 kJ = 1 kcal. (See also **kilocalorie**.)
calorimeter A device for measuring the quantity of thermal energy transferred during a chemical reaction or some other process.
capillary action The process whereby a liquid rises in a small-diameter tube due to noncovalent interactions between the liquid and the tube's material.
carbohydrates Biochemical compounds with the general formula $C_x(H_2O)_y$, in which x and y are whole numbers.
carbonyl group An organic functional group consisting of carbon bonded to two other atoms and double bonded to oxygen; >C=O.
carboxylic acid An organic compound characterized by the presence of the carboxyl functional group (—COOH).
catalyst A substance that increases the rate of a reaction but is not consumed in the overall reaction.
catalytic cracking A petroleum refining process using a catalyst, high temperature, and pressure to break long-chain hydrocarbons into shorter-chain hydrocarbons, including both alkanes and alkenes suitable for gasoline.
catalytic reforming A petroleum refining process in which straight-chain hydrocarbons are converted to branched-chain hydrocarbons and aromatics for use in gasoline and the manufacture of other organic compounds.
catenation Formation of chains and rings by bonds between atoms of the same element.
cathode The electrode of an electrochemical cell at which reduction occurs.
cathodic protection A process of protecting a metal from corrosion whereby the metal is made the cathode by connecting it electrically to a more reactive metal or to the negative pole of a DC voltage supply.
cation An ion with a positive electrical charge.
cell potential The electrical potential energy difference between the cathode and anode of a voltaic cell. Cell potential can be measured with a voltmeter and is the quantity of work in joules a cell can produce per coulomb of charge that the chemical reaction transfers.
cell voltage See **cell potential**.
Celsius temperature scale A scale defined by the freezing (0 °C) and boiling (100 °C) points of pure water, at 1 atm.
cement A solid consisting of microscopic particles containing compounds of calcium, iron, aluminum, silicon, and oxygen in varying proportions and tightly bound to one another.
ceramics Materials fashioned from clay or other natural materials at room temperature and then hardened by heating to a high temperature.
CFCs See **chlorofluorocarbons**.
chain termination A step in a polymerization reaction that stops the growth of the polymer chain. For example, combination of two free-radical chains where a new bond forms by pairing their unpaired electrons.
change of state A physical process in which one state of matter is changed into another (such as melting a solid to form a liquid).
Charles's law The volume of an ideal gas at constant pressure and amount of gas varies directly with its absolute temperature.
chelating ligand A ligand that uses more than one atom to bind to the same metal ion in a coordination complex.
chemical bond Attractive force between two atoms holding them together, for example, as part of a molecule.
chemical change (chemical reaction) A process in which substances (reactants) change into other substances (products) by rearrangement, combination, or separation of atoms.
chemical compound (compound) A pure substance (for example, sucrose or water) that can be decomposed into two or more different pure substances; homogeneous, constant-composition matter that consists of two or more chemically combined elements.
chemical element (element) A substance (for example, carbon, hydrogen, or oxygen) that cannot be decomposed into two or more new substances by chemical or physical means.
chemical equilibrium A state in which the concentrations of reactants and products remain constant because the rates of forward and reverse reactions are equal.
chemical formula (formula) A notation combining element symbols and numerical subscripts that shows the relative numbers of each kind of atom in a molecule or formula unit of a substance.
chemical fuel A substance that reacts exothermically with atmospheric oxygen and is available at reasonable cost and in reasonable quantity.
chemical kinetics The study of the speeds of chemical reactions and the nanoscale pathways or rearrangements by which atoms, ions, and molecules are converted from reactants to products.
chemical periodicity, law of Law stating that the properties of the elements are periodic functions of atomic number.
chemical properties The kinds of chemical reactions that chemical elements or compounds can undergo.
chemical reaction (chemical change) A process in which substances (reactants) change into other substances (products) by rearrangements, combination, or separation of atoms.
chemical sciences Chemistry and other sciences that are closely related to chemistry; sciences that rely on chemical concepts and knowledge.
chemical symbol A one- or two-letter abbreviation for the name of an element.
chemistry The study of matter and the changes it can undergo.
chemotrophs (See also **phototrophs**.) Organisms that must depend on phototrophs to create the chemical substances from which they obtain Gibbs free energy.
chiral Describes a molecule or object that is not superimposable on its mirror image.
chlor-alkali process Electrolysis process for producing chlorine and sodium hydroxide from aqueous sodium chloride.
chlorination Addition of chlorine or a chlorine compound to kill bacteria in municipal water supplies; HOCl formed in water is the antibacterial agent.
chlorofluorocarbon (CFC) Compounds of carbon, fluorine, and chlorine. CFCs have been implicated in stratospheric ozone depletion.
Ci See **curie**.
***cis* isomer** The isomer in which two like substituents are on the same side of a carbon-carbon double bond, the same side of a ring of carbon atoms, or the same side of a coordination complex.
***cis-trans* isomerism** A form of stereoisomerism in which the isomers have the same molecular formula and the same atom-to-atom bonding sequence, but differ in the location of pairs of substituents on the same side (*cis*) or on opposite sides (*trans*) of a molecule or ion. Also called geometric isomerism.
Clausius-Clapeyron equation Equation that gives the relationship between vapor pressure and temperature.
closest packing Arranging spherical atoms so that they occupy the minimum volume.
coagulation The process in which the protective charge layer on colloidal particles is overcome, causing them to aggregate into a soft, semisolid, or solid mass.
coefficients The multiplying numbers assigned to the formulas in a chemical equation to balance the equation. Also called stoichiometric coefficients.
cofactor An inorganic or organic molecule or ion required by an enzyme to carry out its catalytic function.
colligative properties Properties of solutions that depend only on the concentration of solute particles in the solution, not on the nature of the solute particles.
colloid A state intermediate between a solution and a suspension, in which solute particles are large enough to scatter light, but too small to settle out; found in gas, liquid, and solid states.
combined gas law A form of the ideal gas law that relates the P, V, T of a given amount of gas before and after a change: $P_1V_1/T_1 = P_2V_2/T_2$.
combining volumes, law of At constant temperature and pressure, the volumes of reacting gases are always in the ratios of small whole numbers.
combustion analysis A quantitative method to obtain percent composition data for compounds that can burn in oxygen.
combustion reaction A reaction in which an element or compound burns in air or oxygen.
common ion effect Shift in equilibrium position that results from addition of an ion identical to one in the equilibrium.
complementary base pairs Pairs of bases, each in a different DNA strand, that hydrogen-bond to each other: guanine with cytosine and adenine with thymine.
complex ion An ion with several molecules or ions connected to a central metal ion by coordinate covalent bonds.
composites Materials with components that may be metals, polymers, and ceramics.
compound See **chemical compound**.

compressibility The property of a gas that allows it to be compacted into a smaller volume by application of pressure.
concentration The relative quantities of solute and solvent in a solution.
concentration cell An electrochemical cell in which the voltage is generated because of a difference in concentrations of the same chemical species.
concrete A mixture of cement, sand, and aggregate (crushed stone or pebbles) in varying proportions that reacts with water and carbon dioxide to form a rock-hard solid.
condensation Process whereby a molecule in the gas phase enters the liquid phase.
condensation enthalpy ($\Delta_{cond}H$) The enthalpy change when a gas condenses; the quantity of energy transferred by heating during condensation at constant pressure.
condensation polymer A polymer made from the condensation reaction of monomer molecules that contain two or more functional groups.
condensation reaction A chemical reaction in which two (or more) molecules combine to form a larger molecule, simultaneously producing a small molecule such as water.
condensed formula A chemical formula of an organic compound indicating how atoms are grouped together in a molecule.
conduction band In a solid, an energy band that contains electrons of higher energy than those in the valence band, the electrons are free to move about the solid.
conductor See **electrical conductor**.
conjugate acid-base pair A pair of molecules or ions related to one another by the loss and gain of a single hydrogen ion.
conjugated Refers to a system of alternating single and double bonds in a molecule.
conservation of electric charge, law of Law stating that the total electric charge must be the same on both sides of a balanced equation; arises because electrons are neither created nor destroyed in a chemical reaction.
conservation of energy, law of (first law of thermodynamics) Law stating that energy can be neither created nor destroyed—the total energy of the universe is constant.
conservation of matter, law of Law stating that there is no detectable change in mass during an ordinary chemical reaction.
constant composition, law of Law stating that a chemical compound always contains the same elements in the same proportions by mass.
constitutional isomers (structural isomers) Compounds with the same molecular formula that differ in the order in which their atoms are bonded together.
continuous phase The solvent-like dispersing medium in a colloid.
continuous spectrum A spectrum consisting of all possible wavelengths.
conversion factor (proportionality factor) A relationship between two measurement units derived from the proportionality of one quantity to another (for example, density is the conversion factor between mass and volume).
coordinate covalent bond A chemical bond in which both of the two electrons forming the bond were originally associated with the same one of the two bonded atoms.
coordination complex An ion or uncharged molecule in which several molecules or ions are connected to a central metal ion by coordinate covalent bonds.
coordination compound A compound in which complex ions are combined with oppositely charged ions to form an uncharged compound; an uncharged coordination complex.
coordination number The number of coordinate covalent bonds between ligands and a central metal ion in a coordination complex.
copolymer A polymer formed by combining two different types of monomers.
core electrons The electrons in the filled inner shells of an atom.
corrosion Oxidation of a metal exposed to the environment.
coulomb The unit of electrical charge equal to the quantity of charge that passes a fixed point in an electrical circuit when a current of one ampere flows for one second.
Coulomb's law Law that represents the force of attraction between two charged particles; $F = k(q_1q_2/d^2)$.
covalent bond Interatomic attraction resulting from the sharing of electrons between two atoms.
critical mass The minimum quantity of fissionable material needed to support a self-sustaining chain reaction.
critical point The temperature and pressure above which there is no longer any distinction between liquid and gas phases; the end of the liquid-gas equilibrium line in a phase diagram.
critical pressure (P_c) The vapor pressure of a liquid at its critical temperature.
critical temperature (T_c) The temperature above which there is no distinction between liquid and vapor phases.
cryogen A substance that can maintain very low temperatures; a liquefied gas with boiling point below -150 °C.
crystal-field splitting energy, Δ Energy difference between sets of *d* orbitals on the central metal ion in a coordination compound.
crystal-field theory Theory that predicts spectra and magnetism of coordination compounds based on electrostatic bonding between ligands and a metal ion.
crystal lattice The ordered, repeating arrangement of ions, molecules, or atoms in a crystalline solid.
crystalline solids Solids with an ordered arrangement of atoms, molecules, or ions that results in planar faces and sharp angles of the crystals.
crystallization The process in which mobile atoms, molecules, or ions in a liquid or solution convert into a crystalline solid.
crystallization enthalpy The enthalpy change when a liquid solidifies; the quantity of energy transferred by heating during crystallization at constant pressure; has the same magnitude as the molar fusion enthalpy, but the opposite sign.
cubic closest packed Describes the three-dimensional crystal structure that results when atoms or ions are closest packed in the *abcabc* arrangement; this structure is the same as face-centered cubic.
cubic unit cell A unit cell with equal-length edges that meet at 90° angles.
curie (Ci) A unit of radioactivity equal to 3.7×10^{10} disintegrations per second.

D

Δ See **crystal-field splitting energy**.
ΔH See **enthalpy change**.
$\Delta_f H°$ See **standard formation enthalpy**.
$\Delta_r H°$ See **standard reaction enthalpy**.
ΔG See **Gibbs free energy**.
$\Delta_f G°$ See **standard formation Gibbs free energy**.
$\Delta_r G°$ See **standard reaction Gibbs free energy**.
ΔS See **entropy**.
***d*-to-*d* transition** Excitation of an electron from a lower-energy *d*-orbital to a higher-energy *d*-orbital in a transition-metal ion.
Dalton's law of partial pressures The total pressure exerted by a mixture of gases is the sum of the partial pressures of the individual gases in the mixture.
decomposition reaction A reaction in which a compound breaks down chemically to form two or more simpler substances.
degree of polymerization (*n*) Number of repeating units in a polymer chain.
delocalized electrons Electrons, such as in benzene, that are spread over several atoms in a molecule or polyatomic ion.
denaturation Disruption in protein secondary and tertiary structure brought on by high temperature, heavy metals, and other substances.
density The ratio of the mass of an object to its volume.
deoxyribonucleic acid (DNA) A double-stranded polymer of nucleotides that stores genetic information.
deposition The process in which a gas converts directly to a solid.
deposition enthalpy The enthalpy change when a gas converts directly to a solid; has the same magnitude as the sublimation enthalpy, but the opposite sign.
detergent A molecule whose structure contains a long hydrocarbon portion that is hydrophobic and a polar end that is hydrophilic.
diamagnetic Describes atoms or ions in which all the electrons are paired in filled shells so their magnetic fields effectively cancel each other.
diatomic molecule A molecule that contains only two atoms.
dietary minerals Essential elements that are not carbon, hydrogen, oxygen, or nitrogen.
diffusion Spread of gas molecules of one type through those of another type.
dilution Reduction of the concentration of solute(s) in a solution by adding solvent (water in the case of an aqueous solution).
dimensional analysis A method of using units in calculations to check for correctness.
dimer A molecule made from two smaller units (monomers).

dipole-dipole attraction The noncovalent force of attraction between any two polar molecules or two polar regions within the same large molecule.
dipole moment The product of the magnitude of the partial charges ($\delta+$ and $\delta-$) of a molecule times the distance of separation between the charges.
disaccharides Carbohydrates such as sucrose consisting of two monosaccharide units.
dispersed phase The larger-than-molecule-sized particles that are distributed uniformly throughout a colloid.
disproportionation reaction An oxidation-reduction reaction in which the same reactant serves as both the oxidizing and reducing agent.
dissociation The separation of cations from anions when an ionic solid dissolves in water.
DNA See **deoxyribonucleic acid**.
doping Adding a tiny concentration of one substance (a *dopant*) to improve the semiconducting properties of another.
double bond A bond formed by sharing two pairs of electrons between the same two atoms.
dynamic equilibrium A balance between opposing reactions occurring at equal rates.

E

E_a See **activation energy**.
$E°$ See **standard half-cell potentials**.
effective nuclear charge The attraction toward the nucleus experienced by outer-shell electrons in a many-electron atom; the positive charge of the nucleus corrected for the repulsions of all other electrons for a specified electron.
effusion Escape of gas molecules from a container through a tiny hole into a vacuum.
electrical conductor A material that carries an electric current; a material with overlapping valence and conduction bands.
electrical insulator A material that does not carry an electric current; a material in which the valence band is filled and the unfilled conduction band is at much higher energy than the valence band.
electrochemical cell A combination of anode, cathode, and other materials arranged so that a product-favored oxidation-reduction reaction can cause a current to flow or an electric current can cause a reactant-favored redox reaction to occur.
electrochemistry The study of the relationship between electron flow and oxidation-reduction reactions.
electrode A device such as a metal plate or wire that conducts electrons into and out of a system.
electrolysis The use of electrical energy to produce a chemical change.
electrolytes Substances that ionize or dissociate when dissolved in water to form an electrically conducting solution.
electromagnetic radiation Radiation that consists of perpendicular, oscillating electric and magnetic fields that travel through space at the same rate (the speed of light: 186,000 miles/s or 2.998×10^8 m/s in a vacuum).
electromotive force (emf) The difference in electrical potential energy between the two electrodes in an electrochemical cell; emf is measured with a voltmeter.
electron (e^-) A negatively charged subatomic particle that occupies most of the volume of an atom and carries electric current in metals.
electron affinity The energy change when a mole of electrons is added to a mole of atoms in the gas phase.
electron capture A radioactive decay process in which one of an atom's inner-shell electrons is captured by the nucleus, which decreases the atomic number by 1.
electron configuration The complete description of the orbitals occupied by all the electrons in an atom or ion.
electron density The probability of finding an electron within a tiny volume in an atom; determined by the square of the wave function.
electron-pair geometry See **electron-region geometry**.
electron-region geometry The arrangement of regions of electron density (bonds and lone electron pairs) around a central atom.
electronegativity A measure of the ability of an atom in a molecule to attract bonding electrons to itself.
electronically excited molecule A molecule whose potential energy is greater than the minimum (ground-state) energy because of a change in its electronic structure.
element (chemical element) A substance (for example, carbon, hydrogen, and oxygen) that cannot be decomposed into two or more new substances by chemical or physical means.
elementary reaction A nanoscale reaction whose equation indicates exactly which atoms, ions, or molecules collide or change as the reaction occurs.
empirical formula A formula showing the simplest possible ratio of atoms of elements in a compound.
emulsion A colloid consisting of a liquid dispersed in a second liquid; formed by the presence of an *emulsifier* that coats and stabilizes dispersed-phase particles.
enantiomers A pair of molecules consisting of a chiral molecule and its mirror-image isomer.
end point The point at which the indicator changes color during a titration.
endergonic Refers to a reaction that requires input of Gibbs free energy; applies to biochemical reactions that are reactant-favored at body temperature.
endothermic Describes a process in which thermal energy must be transferred into a thermodynamic system to maintain constant temperature.
energy (E) The capacity to do work.
energy band In a solid, a large group of orbitals whose energies are closely spaced; in an atomic solid, the average energy of a band equals the energy of the corresponding orbital in an individual atom.
energy change The energy of the final state of a system minus the energy of the initial state.
energy conservation The conservation of useful energy, that is, of Gibbs free energy.
energy density The quantity of energy released per unit volume of a fuel.
energy level A specific value of the energy of an atom or molecule; according to the quantum theory, only certain energy levels are available to an atom or molecule. In a H atom, all orbitals having the same principal quantum number, n.
enthalpy change (ΔH) The quantity of energy transferred by heating when a process takes place at constant temperature and pressure.
enthalpy of fusion See **fusion enthalpy**.
enthalpy of solution See **solution enthalpy**.
enthalpy of sublimation See **sublimation enthalpy**.
enthalpy of vaporization See **vaporization enthalpy**.
entropy (S) A measure of the number of ways energy can be distributed in a system; a measure of the dispersal of energy in a system.
enzyme A highly efficient biochemical catalyst for one or more reactions in a living system.
enzyme-substrate complex The combination formed by the binding of an enzyme with a substrate through noncovalent forces.
equatorial positions Positions lying on the equator of an imaginary sphere around a triangular bipyramidal molecular or ionic structure.
equilibrium concentration The concentration of a substance (usually expressed as molarity) in a system that has reached the equilibrium state.
equilibrium constant (K_c) A quotient of equilibrium concentrations of product and reactant substances that has a constant value for a given reaction at a given temperature.
equilibrium constant (K_P) A quotient of equilibrium partial pressures of product and reactant substances that has a constant value for a given gas-phase reaction at a given temperature.
equilibrium constant expression The mathematical expression associated with an equilibrium constant.
equilibrium vapor pressure Pressure of the vapor of a substance in equilibrium with its liquid or solid in a closed container.
equivalence point The point in a titration at which a stoichiometrically equivalent amount of one substance has been added to another substance.
esters Organic compounds formed by the reaction of an alcohol and a carboxylic acid and containing the functional group:

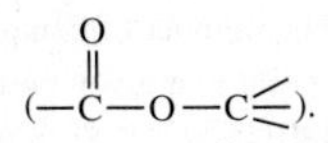

evaporation The process of conversion of a liquid to a gas.
exchange reaction A reaction in which cations and anions that were partners in the reactants are interchanged in the products.
excited state The unstable state of an atom or molecule in which at least one electron does not have its lowest possible energy.
exergonic Refers to a reaction that releases Gibbs free energy; applies to biochemical reactions that are product-favored at body temperature.
exothermic Refers to a process in which thermal energy must be transferred out of a thermodynamic system to maintain constant temperature.
extent of reaction The fraction of reactants that has been converted to products.

F

F See **Faraday constant**.
Faraday constant (*F*) The quantity of electric charge on one mole of electrons, 9.6485×10^4 C/mol.
fat A solid triester of fatty acids with glycerol.
ferromagnetic A substance that contains clusters of atoms with unpaired electrons whose magnetic spins become aligned, causing permanent magnetism.
first law of thermodynamics (law of conservation of energy) Energy can neither be created nor destroyed—the total energy of the universe is constant.
formal charge The charge a bonded atom would have if its electrons were shared equally.
formation constant (K_f) The equilibrium constant for the formation of a complex ion.
formula (chemical formula) A notation combining element symbols and numerical subscripts that shows the relative numbers of each kind of atom in a molecule or formula unit of a substance.
formula unit The simplest cation-anion grouping represented by the formula of an ionic compound; also the collection of atoms represented by any formula.
formula weight The sum of the atomic weights in u (amu) of all the atoms in a compound's formula.
fracking See **hydraulic fracturing**.
fractional distillation The process of refining petroleum (or another mixture) by distillation to separate it into groups (fractions) of compounds having distinctive boiling point ranges.
Frasch process Process for recovering sulfur from underground deposits by melting the sulfur with superheated water.
free radical An atom, ion, or molecule that contains one or more unpaired electrons; usually highly reactive.
freezing point Temperature at which a pure liquid freezes—same as the melting point.
freezing-point lowering A colligative property; the difference between the freezing point of a pure solvent and the freezing point of a solution in which a nonvolatile nonelectrolyte solute is dissolved in the solvent.
frequency (ν) The number of complete traveling waves passing a point in a given period of time (cycles per second).
frequency factor (*A*) The factor in the Arrhenius equation that depends on how often molecules collide when all concentrations are 1 mol/L and on whether the molecules are properly oriented to react when they collide.
fuel cell An electrochemical cell that converts the chemical energy of fuels directly into electricity.
fuel value The quantity of energy released when 1 g of a fuel is burned.
fullerenes Allotropic forms of carbon that consist of many five- and six-membered rings of carbon atoms sharing edges.
functional group An atom or group of atoms that imparts characteristic properties and defines a given class of organic compounds (for example, the —OH group is present in all alcohols).
fusion enthalpy ($\Delta_{fus}H$) The enthalpy change when a substance melts; the quantity of energy that must be transferred by heating when a substance melts at constant temperature and pressure.

G

γ See **gamma radiation**.
galvanized Has a thin coating of zinc metal that forms an oxide coating impervious to oxygen, thereby protecting a less active metal, such as iron, from corrosion.
gamma (γ) radiation Radiation composed of highly energetic photons.
gas A phase or state of matter in which a substance has no definite shape and has a volume determined by the volume of its container.
gas chromatography (GC) An important instrumental method used to separate, identify, and determine the quantity of chemical substances in complex mixtures.
gas-forming reaction A reaction in which a gas is generated.
gasohol A blended motor fuel consisting of 90% gasoline and 10% ethanol by mass.
gene The unique sequence of nitrogen bases in DNA that codes for the synthesis of a specific protein; carrier of a genetic trait.
geometric isomerism A form of stereoisomerism in which the isomers have the same molecular formula and the same atom-to-atom bonding sequence, but differ in the location of pairs of substituents on the same side (*cis*) or on opposite sides (*trans*) of a molecule or ion. Also called *cis-trans* isomerism.
Gibbs free energy A thermodynamic function that decreases for any product-favored system. For a process at constant temperature and pressure, $\Delta G = \Delta H - T\Delta S$.
glass Amorphous, clear solids formed from silicates and other oxides.
global warming Increase in temperature at Earth's surface as a result of the greenhouse effect amplified by increasing concentrations of carbon dioxide and other greenhouse gases.
glycogen A highly branched, high-molar-mass polymer of glucose found in animals.
glycosidic linkage The C—O—C bond that connects monosaccharides in disaccharides and polysaccharides; forms between carbons 1 and 4 or 1 and 6 of linked monosaccharides.
Graham's law Law that states that the rate of effusion of a gas is inversely proportional to the square root of its molar mass.
gram The basic unit of mass in the metric system; equal to 1×10^{-3} kg.
graphene A two-dimensional covalent network consisting of a single layer of carbon atoms in a "chicken-wire" hexagonal arrangement.
gray (Gy) The SI unit of absorbed radiation dose equal to the absorption of 1 joule per kilogram of material.
green chemistry The design, development, and implementation of chemical products and processes to reduce or eliminate the use and generation of substances hazardous to human health and the environment.
greenhouse effect Atmospheric warming caused when atmospheric carbon dioxide, water vapor, methane, ozone, and other greenhouse gases absorb infrared radiation reradiated from Earth.
ground state The state of an atom or molecule in which all of the electrons are in their lowest possible energy levels.
groups The vertical columns of the periodic table of the elements.
Gy See **gray**.

H

h See **Planck's constant**.
H See **enthalpy change**.
Haber-Bosch process The process developed by Fritz Haber and Carl Bosch for the direct synthesis of ammonia from its elements.
half-cell One half of an electrochemical cell in which only the anode or cathode is located.
half-life ($t_{1/2}$) The time required for the concentration of one reactant to reach half its original value; radioactivity—the time required for the activity of a radioactive sample to reach half of its original value.
half-reaction A reaction that represents either an oxidation or a reduction process.
halide ions All monoatomic ions (1−) of halogens.
halogens The elements in Group 7A of the periodic table.
heat (heating) The energy-transfer process between two samples of matter at different temperatures.
heat capacity The quantity of energy that must be transferred to an object to raise its temperature by 1 °C.
heat of See **enthalpy of**.
heating curve A plot of the temperature of a substance versus the quantity of energy transferred to it by heating.
helium burning The fusion of helium nuclei to form beryllium-8, as it occurs in stars.
Henderson-Hasselbalch equation The equation describing the relationships among the pH of a buffer solution, the pK_a of the acid, and the concentrations of the acid and its conjugate base.
Henry's law A mathematical expression for the relationship of gas pressure and solubility; $S_g = k_H P_g$.
Hess's law If two or more chemical equations can be combined to give another equation, the enthalpy change for that equation will be the sum of the enthalpy changes for the equations that were combined. A similar situation applies to entropy and Gibbs free energy.
heterogeneous catalyst A catalyst that is in a different phase from that of the reaction mixture.
heterogeneous mixture A mixture in which components remain separate and can be observed as individual substances or phases.
heterogeneous reaction A reaction that takes place at an interface between two phases, solid and gas, for example.
hexadentate ligand A ligand in which each of six different atoms donates an electron pair to form a coordinate covalent bond to a central metal ion.
hexagonal closest packed Describes the three-dimensional crystal structure that results when

layers of atoms in a solid are closest packed in the *ababab* arrangement.
HFCs See **hydrofluorocarbons**.
high-spin complex A complex ion that has the maximum possible number of unpaired electrons.
homogeneous catalyst A catalyst that is in the same phase as that of the reaction mixture.
homogeneous mixture A mixture of two or more substances in a single phase that is uniform throughout; also called a solution.
homogeneous reaction A reaction in which the reactants and products are all in the same phase.
Hund's rule Electrons pair only after each orbital in a subshell is occupied by a single electron.
hybrid orbitals Orbitals formed on a single atom by combining atomic orbitals of appropriate energy and orientation.
hybridized Refers to atomic orbitals of proper energy and orientation that have combined to form hybrid orbitals.
hydrate A solid compound that has a stoichiometric amount of water molecules bonded to metal ions or trapped within its crystal lattice.
hydration The binding of one or more water molecules to an ion or molecule within a solution or within a crystal lattice.
hydraulic fracturing ("fracking") The use of high pressure (up to 100 MPa) to force a fluid down a bored hole, thereby fracturing shale to release natural gas and crude oil.
hydrocarbons Organic compounds composed only of carbon and hydrogen.
hydrofluorocarbons (HFCs) Compounds of carbon, hydrogen, and fluorine used to replace chlorofluorocarbons, CFCs (which deplete stratospheric ozone); HFCs have much shorter atmospheric lifetimes than CFCs.
hydrogen bond The attraction of a hydrogen atom in a molecule or part of a molecule X—H for another atom or group of atoms.
hydrogen bonding An attraction between two molecules or two parts of the same molecule that involves one or more hydrogen bonds.
hydrogen burning The fusion of hydrogen nuclei (protons) to form helium, as it occurs in stars.
hydrogen ion A hydrogen atom that has lost its single electron; a proton. In aqueous solution, a proton surrounded by water molecules hydrogen bonded into a unit.
hydrogenation An addition reaction in which hydrogen adds to the double bond of an alkene; the catalyzed reaction of H_2 with a liquid triglyceride to produce saturated fatty acid chains, which convert the triglyceride into a semisolid or solid.
hydrolysis A reaction in which a bond is broken by reaction with a water molecule and the —H and —OH of the water add to the atoms of the broken bond.
hydronium ion H_3O^+; the simplest proton-water complex; responsible for acidity. See **hydrogen ion**.
hydrophilic "Water-loving," a term describing a polar molecule or part of a molecule that is strongly attracted to water molecules.
hydrophobic "Water-fearing," a term describing a nonpolar molecule or part of a molecule that is not attracted to water molecules.
hydroxide ion OH^- ion; bases increase the concentration of hydroxide ions in solution.
hypertonic Refers to a solution having a higher concentration of nanoscale particles and therefore a higher osmotic pressure than another solution.
hypothesis A tentative explanation for an observation and a basis for experimentation.
hypotonic Refers to a solution having a lower solute concentration of nanoscale particles and therefore a lower osmotic pressure than another solution.

I

ideal gas A gas that behaves exactly as described by the ideal gas law, and by Boyle's, Charles's, and Avogadro's laws.
ideal gas constant (*R*) The proportionality constant, *R*, in the equation $PV = nRT$; $R = 0.0821\ \text{L atm mol}^{-1}\ \text{K}^{-1} = 8.314\ \text{J K}^{-1}\ \text{mol}^{-1} = 8.314\ \text{dm}^3\ \text{Pa K}^{-1}\ \text{mol}^{-1}$.
ideal gas law A law that relates pressure, volume, amount (moles), and temperature for an ideal gas; the relationship expressed by the equation $PV = nRT$.
IE (ionization energy) The energy needed to remove an electron from an atom or ion in the gas phase.
immiscible Describes two liquids that form two separate phases when mixed because each is only slightly soluble in the other.
indicator A substance that can be added in a titration to show when the equivalence point has been reached.
induced dipole A temporary dipole created by a momentary uneven distribution of electrons in a molecule or atom.
induced fit The change in the shape of an enzyme, its substrate, or both when they bind.
inhibitor A molecule or ion other than the substrate that causes a decrease in catalytic activity of an enzyme.
initial rate The instantaneous rate of a reaction determined at the very beginning of the reaction.
initiation The breaking of a carbon-carbon double bond in a polymerization reaction to produce a molecule with highly reactive sites that react with other molecules to produce a polymer; first step in a chain reaction.
inorganic compounds Chemical compounds that are not organic compounds; usually of mineral or nonbiological origin.
insoluble Describes a solute, almost none of which dissolves in a solvent.
instantaneous reaction rate The rate at a particular time after a reaction has begun.
insulator A material that has a large energy gap between fully occupied and empty energy bands, and does not conduct electricity.
intermediate An atom, molecule, or ion that is produced in one step of a reaction mechanism and used up in a later step of the mechanism.
intermolecular forces Noncovalent attractions between separate molecules.
internal energy The sum of the individual energies (kinetic and potential) of all of the nanoscale particles (atoms, molecules, or ions) in a sample of matter.
ions Atoms, or groups of atoms, that have lost or gained one or more electrons so that they are no longer electrically neutral.
ion product (*Q*) A value found from an expression with the same mathematical form as the solubility product expression (K_{sp}) but using actual concentrations of the species involved, rather than equilibrium concentrations. (See **reaction quotient**.)
ionic bonding Forces of attraction between cations and anions in the crystal lattice of an ionic compound.
ionic compound A compound that consists of positive and negative ions (cations and anions).
ionic hydrate Ionic compounds that incorporate water molecules in the ionic crystal lattice.
ionic radius Radius of an anion or cation in an ionic compound.
ionic solid A solid in which electrostatic interactions between cations and anions hold the ions tightly in a crystal lattice. Example: table salt, NaCl.
ionization The process of forming an ion or ions. Examples are loss or gain of electrons by atoms and formation of $H^+(aq)$ and $Cl^-(aq)$ when HCl dissolves in water.
ionization constant for water (K_w) The equilibrium constant that is the mathematical product of the concentration of aqueous hydrogen ions and the concentration of hydroxide ions in any aqueous solution; $K_w = 1.0 \times 10^{-14}$ at 25 °C.
ionization energy (IE) The energy needed to remove an electron from an atom or ion in the gas phase.
ionizes See **ionization**.
isoelectronic Refers to atoms and ions that have identical electron configurations.
isomers Compounds that have the same molecular formula but different arrangements of atoms.
isotonic Refers to a solution having the same concentration of nanoscale particles and therefore the same osmotic pressure as another solution.
isotopes Forms of an element composed of atoms with the same atomic number but different mass numbers owing to a difference in the number of neutrons.

J

joule (J) A unit of energy equal to 1 kg m^2/s^2. The kinetic energy of a 2-kg object traveling at a speed of 1 m/s.

K

k See **rate constant**.
K See **kelvin**.
K See **equilibrium constant**.
$K°$ See **standard equilibrium constant**.
K_a See **acid ionization constant**.
K_b See **base ionization constant**.
K_{sp} See **solubility product constant**.
K_w See **ionization constant for water**.
kelvin (K) The SI unit of temperature. The lowest possible temperature is 0 K; one kelvin is the same size as one Celsius degree.

Kelvin temperature scale (See also **absolute temperature scale, thermodynamic temperature scale.**) A temperature scale on which the zero is the lowest possible temperature and the degree is the same size as a Celsius degree.
ketone An organic compound characterized by the presence of a carbonyl group in which the carbon atom is bonded to two other carbon atoms ($R_2C{=}O$).
kilocalorie (kcal or Cal) (See also **calorie.**) A unit of energy equal to 4.184 kJ. Approximately 1 kcal (1 Cal) is required to raise the temperature of 1 kg of liquid water by 1 °C. The food Calorie.
kilopascal (kPa) One kilopascal, 10^3 Pa, is approximately the pressure exerted by a mass of 10. g resting on a 1 cm^2 area at Earth's surface.
kinetic energy Energy that an object has because of its motion. Equal to $\frac{1}{2}mv^2$, where m is the object's mass and v is its velocity.
kinetic-molecular theory The theory that matter consists of nanoscale particles that are in constant, random motion.

L

ℓ See **orbital quantum number**.
λ See **wavelength**.
lanthanide contraction The decrease in atomic radii across the fourth-period and the fifth-period *f*-block elements (lanthanides and actinides) due to the lack of electron screening by electrons in the 4*f* and 5*f* orbitals.
lanthanides The elements after lanthanum in the sixth period in which the 4*f* subshell is being filled.
lattice energy Enthalpy of formation of 1 mol of an ionic solid from its separated gaseous ions.
law A statement that summarizes a wide range of experimental results and has not been contradicted by experiments.
law of ____ For specific laws see the name of the law; for example, the law of chemical periodicity is found under chemical periodicity, law of.
lead-acid storage battery A battery consisting of six cells and based on the oxidation of lead by lead(IV) oxide to form lead(II) sulfate. The familiar automobile battery.
Le Chatelier's principle If a system is at equilibrium and the conditions are changed so that it is no longer at equilibrium, the system will react to give a new equilibrium in a way that partially counteracts the change.
Lewis acid A molecule or ion that can accept an electron pair from another atom, molecule, or ion to form a new bond.
Lewis base A molecule or ion that can donate an electron pair to another atom, molecule, or ion to form a new bond.
Lewis dot symbol An atomic symbol with dots representing valence electrons. For example, the Lewis dot symbol for an oxygen atom is :Ö· .
Lewis structure A structural formula for a molecule that shows all valence electrons as dots or as lines that represent covalent bonds. For example, the Lewis structure for O_2 is :Ö=Ö:.
ligands Atoms, molecules, or ions bonded to a central atom, such as the central metal ion in a coordination complex.
limiting reactant The reactant present in limited supply that controls the amount of product formed in a reaction.
line drawing A molecular structure in which bonds between carbon atoms are shown, but not the carbon atoms. Also, neither hydrogen atoms nor carbon-hydrogen bonds are shown. All atoms other than C and H are shown as atomic symbols. Also called line structure.
line emission spectrum A spectrum produced by excited atoms and consisting of discrete wavelengths of light.
line structure A molecular structure in which bonds between carbon atoms are shown, but not the carbon atoms. Also, neither hydrogen atoms nor carbon-hydrogen bonds are shown. All atoms other than C and H are shown as atomic symbols. Also called line drawing.
linear Molecular geometry in which there is a 180° angle between bonded atoms.
lipid bilayer The structure of cell membranes that are composed of two aligned layers of phospholipids with their hydrophobic regions within the bilayer.
liquid A phase of matter in which a substance has no definite shape but a definite volume.
London forces Forces resulting from the attraction between positive and negative regions of momentary (induced) dipoles in neighboring molecules. Also referred to as dispersion forces.
lone-pair electrons Paired valence electrons unused in bond formation; also called nonbonding or unshared pairs.
low-spin complex A coordination complex that has the minimum possible number of unpaired electrons.

M

m_ℓ See **magnetic quantum number**.
m_s See **spin quantum number**.
macromolecule A very large polymer molecule made by chemically joining many small molecules (monomers).
macroscale Refers to samples of matter that can be observed by the unaided human senses; samples of matter large enough to be seen, measured, and handled.
macroscopic See **macroscale**.
magnetic quantum number (m_ℓ) The quantum number that designates the orientation of different atomic orbitals within the same subshell; it can have any integer value between ℓ and $-\ell$, including zero.
main-group elements Elements in the eight A groups to the left and right of the transition elements in the periodic table; the *s*- and *p*-block elements.
major minerals Dietary minerals present in humans in quantities greater than 100 mg per kg of body weight.
mass A measure of an object's resistance to acceleration.
mass fraction The ratio of the mass of one component to the total mass of a sample.
mass number (*A*) The number of protons plus neutrons in the nucleus of an atom of an element.
mass percent The mass fraction multiplied by 100%.
mass spectrometer An analytical instrument used to measure atomic and molecular masses directly.
mass spectrum A plot of ion abundance versus the mass of the ions; produced by a mass spectrometer.
materials science The science of the relationships between the structure and the chemical and physical properties of materials.
matter Anything that has mass and occupies space.
measurement A process in which the quantity to be measured, such as the volume of ethanol, is compared with a unit, such as mL.
melting The process of changing from solid to liquid.
melting point The temperature at which the structure of a pure solid collapses and the solid changes to a liquid.
meniscus A concave or convex surface that forms on a liquid as a result of the balance of noncovalent forces in a narrow container.
metabolism All of the chemical reactions that occur as an organism converts food nutrients into constituents of living cells, to stored Gibbs free energy, or to thermal energy.
metals Elements that are malleable, ductile, form alloys, and conduct electricity when solid or liquid.
metal activity series A ranking of relative reactivity of metals in order of their ability to donate electrons and form positive ions in aqueous solution.
metallic bonding In solid metals, the nondirectional attraction between positive metal ions and the surrounding sea of negatively charged electrons.
metallic solid A solid in which the positive cores of metal atoms are held together by attractions with valence electrons that are delocalized over all atoms. Example: iron.
metalloids Elements that have some typically metallic properties and other properties that are more characteristic of nonmetals.
methyl group A $-CH_3$ group.
metric system A decimalized measurement system.
micelles Colloid-sized particles built up from many surfactant molecules; micelles can transport various materials within them.
microscale Refers to samples of matter so small that they have to be viewed with a microscope.
microscopic See **microscale**.
millimeter of mercury (mmHg) A unit of pressure related to the height of a column of mercury in a mercury barometer (760 mmHg = 1 atm = 101.3 kPa).
mineral A naturally occurring inorganic compound with a characteristic composition and crystal structure.
miscible Describes two liquids that dissolve in each other in any proportion.
mmHg See **millimeters of mercury**.
MO See **molecular orbital**.
model A mechanical or mathematical way to make a theory more concrete. Example: a ball-and-stick molecular model.

molality (m) The composition of a solution expressed as the amount (mol) of solute per kilogram of solvent.
molar Describes a property that refers to one mole of a substance or process.
molar amount See **amount of substance**.
molar heat capacity The quantity of energy that must be transferred to 1 mol of a substance to increase its temperature by 1 °C.
molar mass The mass in grams of 1 mol of atoms, molecules, or formula units of one kind, numerically equal to the atomic or molecular weight in u (amu).
molar solubility The solubility of a solute in a solvent, expressed in moles per liter.
molarity Solution composition expressed as the amount (mol) of solute per unit volume (L) of solution.
mole (mol) The amount of substance that contains as many elementary entities as there are atoms in exactly 0.012 kg of carbon-12 isotope.
mole fraction (X) The ratio of amount of one component to the total amount of all components in a mixture of substances.
mole of reaction A mole of reaction means that the process specified in a balanced chemical equation occurs 6.022×10^{23} times; that is, a mole of reaction processes (specified by the equation) takes place.
mole ratio A mole-to-mole ratio relating the molar amount of a reactant or product to the molar amount of another reactant or product. Also called stoichiometric factor.
molecular compound A compound composed of atoms of two or more elements chemically combined in molecules.
molecular formula A formula that expresses the number of atoms of each type within one molecule of a substance.
molecular geometry The three-dimensional arrangement of atoms in a molecule.
molecular orbital (MO) One of several orbitals generated by combining atomic orbitals; an MO can extend over an entire molecule.
molecular orbital (MO) theory The theory that treats molecular bonding in terms of molecular orbitals that can extend over all atoms in an entire molecule.
molecular solid A solid in which molecules are held together by intermolecular interactions such as London forces, dipole-dipole forces, and hydrogen bonds. Examples: dry ice ($CO_2(s)$); water ice.
molecular weight The sum of the atomic weights of all the atoms in a substance's formula.
molecule The smallest particle of an element or compound that exists independently, and retains the chemical properties of that element or compound.
momentum The product of the mass (m) times the velocity (v) of an object in motion.
monoatomic ion An ion consisting of a single atom bearing an electrical charge.
monodentate ligand A ligand that donates one electron pair to a coordinated metal ion.
monomers Small-molecule repeating units from which a polymer is formed.
monoprotic acids Acids that can donate a single hydrogen ion per molecule.
monosaccharides The simplest carbohydrates, composed of one saccharide unit.
monounsaturated fatty acid Refers to fatty acids, such as oleic acid, that contain only one carbon-carbon double bond per molecule.
mortar A mixture of cement, sand, and lime that reacts with water and carbon dioxide to form a rock-hard solid.
multiple covalent bonds Double or triple covalent bonds.
multiple proportions, law of When two elements A and B can combine in two or more ways, the mass ratio A : B in one compound is a small-whole-number multiple of the mass ratio A : B in the other compound.

N

n See **amount of substance, degree of polymerization**, or **principal quantum number**.
N_A See **Avogadro constant**.
ν See **frequency**.
***n*-type semiconductor** A material made by doping a semiconductor with an impurity that leaves extra valence electrons.
nanoscale Refers to samples of matter (for example, atoms and molecules) whose normal dimensions are in the 1 to 100 nanometer range. Compare with macroscale, microscale.
nanotubes Members of the family of fullerenes in which graphite-like layers of carbon atoms form cylindrical shapes.
Nernst equation The equation relating the potential of an electrochemical cell to the concentrations of the chemical species involved in the oxidation-reduction reactions occurring in the cell.
net ionic equation A chemical equation in which only those molecules or ions undergoing chemical changes in the course of the reaction are represented.
network solid A solid in which atoms are held in a network of covalent bonds. The network of bonds can be one-dimensional (example: polyethylene), two-dimensional (example: graphite), or three dimensional (example: diamond).
neutral solution A solution containing equal concentrations of $H^+(aq)$ and $OH^-(aq)$; a solution that is neither acidic nor basic.
neutralization reaction A reaction of an acid with a base where each neutralizes (negates) the properties of the other.
neutron An electrically neutral subatomic particle found in the nucleus.
newton (N) The SI unit of force; the force required to accelerate one kilogram by one meter per second every second; 1 kg m s^{-2}.
NiCad and NiMH batteries Small, lightweight secondary batteries that are used in cordless appliances and other applications.
nitrogen cycle The natural cycle of chemical transformations involving nitrogen and its compounds.
nitrogen fixation The conversion of atmospheric nitrogen, N_2, to nitrogen compounds utilizable by plants or industry.
NMR See **nuclear magnetic resonance**.
NO_x Oxides of nitrogen.
noble-gas notation An abbreviated electron configuration of an element in which filled inner shells are represented by the symbol of the preceding noble gas in brackets. For Al, this would be $[Ne]3s^23p^1$.
noble gases Gaseous elements in Group 8A; the least reactive elements.
nonbiodegradable Not capable of being decomposed by microorganisms.
noncovalent interactions All forces of attraction other than covalent, ionic, or metallic bonding.
nonelectrolyte A substance that dissolves in water to form a solution that does not conduct electricity.
nonmetals Elements that do not have the chemical and physical properties of a metal.
nonpolar covalent bond A bond in which the electron pair is shared equally by the bonded atoms.
nonpolar molecule A molecule that is not polar either because it has no polar bonds or because its polar bonds are oriented symmetrically so that they cancel each other.
normal boiling point The temperature at which the vapor pressure of a liquid equals 1 atm.
nuclear burning The nuclear fusion reactions by which elements are formed in stars.
nuclear decay Spontaneous emission of radioactivity by an unstable nucleus that is converted into a more stable nucleus.
nuclear fission The highly exothermic process by which very heavy fissionable nuclei split to form lighter nuclei.
nuclear fusion The highly exothermic process by which very light nuclei combine to form heavier nuclei.
nuclear magnetic resonance (NMR) The process in which the nuclear spins of atoms align in a magnetic field and absorb radio frequency photons to become excited. These excited atoms then return to a lower energy state when they emit the absorbed radiofrequency photons.
nuclear medicine The use of radioisotopes in medical diagnosis and therapy.
nuclear reaction A process in which one or more atomic nuclei change into one or more different nuclei.
nuclear reactor A container in which a controlled nuclear reaction takes place.
nucleon A nuclear particle, either a neutron or a proton.
nucleotide Repeating unit in DNA, composed of one sugar unit, one phosphate group, and one cyclic nitrogen base.
nucleus The tiny central core of an atom; contains protons and neutrons. (There are no neutrons in hydrogen-1.)
nutrients The chemical raw materials, eaten as food, that are needed for survival of a living organism.

O

octahedral Molecular geometry of six groups around a central atom in which all groups are at angles of 90° to other groups.

octahedron A geometric solid with eight faces that are equilateral triangles and six corners.
octane number A measure of the ability of a gasoline to burn smoothly in an internal-combustion engine.
octet rule In forming bonds, many main-group elements gain, lose, or share electrons to achieve a stable electron configuration characterized by eight valence electrons.
optical fiber A fiber made of glass constructed so that light can pass through it with little loss of intensity; used for transmission of information.
optical isomers Isomeric, nonsuperimposable molecular structures that differ only because one is a mirror image of the other.
orbital A region of an atom or molecule within which there is a significant probability that an electron will be found.
orbital quantum number (ℓ) The second quantum number; it designates subshells and determines the shape of the electron-density distribution for an orbital.
orbital shape The shape of an electron density distribution determined by an orbital.
order of reaction The reaction rate dependency on the concentration of a reactant or product, expressed as an exponent of a concentration term in the rate equation.
ore A mineral containing a sufficiently high concentration of an element to make the element's extraction profitable.
organic compounds Compounds of carbon with hydrogen, possibly also oxygen, nitrogen, sulfur, phosphorus, or other elements.
osmosis The movement of a solvent (such as water) through a semipermeable membrane from a region of lower solute concentration to a region of higher solute concentration.
osmotic pressure (Π) The pressure that must be applied to a solution to stop osmosis from a sample of pure solvent.
overall reaction order The sum of the exponents for all concentration terms in the rate equation.
oxidation The loss of electrons by an atom, ion, or molecule, leading to an increase in oxidation number.
oxidation number The hypothetical charge an atom would have if all bonds to that atom were completely ionic. Also called oxidation state.
oxidation-reduction reaction A reaction involving the transfer of one or more electrons from one species to another so that oxidation numbers change. Also called redox reaction.
oxidation state See **oxidation number**.
oxides Compounds of oxygen combined with another element.
oxidized The result when an atom, molecule, or ion loses one or more electrons.
oxidizing agent The substance that accepts electron(s) and is reduced in an oxidation-reduction reaction.
oxoacids Acids in which the acidic hydrogen is bonded directly to an oxygen atom.
oxoanions Polyatomic anions that contain oxygen.
oxygenated gasolines Blends of gasoline with oxygen-containing organic compounds such as methanol, ethanol, and *tertiary*-butyl alcohol.
ozone hole Regions of ozone depletion in the stratosphere centered on Earth's poles, most significantly the South Pole.
ozone layer Region of maximum relative ozone concentration in the stratosphere.

P

Pa See **pascal**.
***p*-block elements** Main-group elements in Groups 3A through 8A whose valence electrons consist of outermost *s* and *p* electrons.
***p-n* junction** An interface between *p*-type and *n*-type semiconductors that produces a rectifier that allows current to flow in only one direction.
***p*-type semiconductor** A material made by doping a semiconductor with an impurity that leaves a deficiency of valence electrons.
paramagnetic Refers to atoms, molecules, or ions that are attracted to a magnetic field because they have unpaired electrons in incompletely filled electron subshells.
partial hydrogenation Addition of hydrogen to some of the carbon-carbon double bonds in a triglyceride (a fat or oil).
partial pressure The pressure that one gas in a mixture of gases would exert if it occupied the same volume at the same temperature as the mixture.
particulate Atmospheric solid particles, generally larger than 10,000 nm in diameter.
parts per billion (ppb) One part in one billion (10^9) parts.
parts per million (ppm) One part in one million (10^6) parts.
parts per trillion (ppt) One part in one trillion (10^{12}) parts.
pascal (Pa) The SI unit of pressure; $1\ \text{Pa} = 1\ \text{N/m}^2$.
Pauli exclusion principle An atomic principle that states that, at most, two electrons can be assigned to the same orbital in the same atom or molecule, and these two electrons must have opposite spins.
peptide bond The amide linkage $-\overset{\overset{\Large O}{\|}}{C}-\overset{\overset{\Large H}{|}}{N}-$ formed when amino acids polymerize to make a protein molecule.
peptide linkage See **peptide bond**.
percent abundance The percentage of atoms of a particular isotope in a natural sample of a pure element.
percent composition by mass The percentage of the mass of a compound represented by each of its constituent elements.
percent yield The ratio of actual yield to theoretical yield, multiplied by 100%.
periodic table A table of elements arranged in order of increasing atomic number so that those with similar chemical and physical properties fall in the same vertical groups.
periods The horizontal rows of the periodic table of the elements.
petroleum fraction A mixture of hundreds of hydrocarbons in the same boiling point range obtained from the fractional distillation of petroleum.
pH The negative logarithm of the hydronium ion concentration ($-\log\,[H_3O^+]$).
pH meter An instrument based on electrochemical principles for measuring pH of solutions.
phase Any of the three states of matter: gas, liquid, solid. Also, one of two or more solid-state structures of the same substance, such as iron in a body-centered cubic or face-centered cubic structure.
phase change A physical process in which one state or phase of matter is changed into another (such as melting a solid to form a liquid).
phase diagram A diagram showing the relationships among the phases of a substance (solid, liquid, and gas), at different temperatures and pressures.
phospholipid Glycerol derivative with two long, nonpolar fatty-acid chains and a polar phosphate group; present in cell membranes.
photochemical reactions Chemical reactions that take place as a result of absorption of photons by reactant molecules.
photodissociation Splitting of a molecule into two free radicals by absorption of an ultraviolet photon.
photoelectric effect The emission of electrons by some metals when illuminated by light of certain wavelengths.
photon A massless particle of light whose energy is given by $h\nu$, where ν is the frequency of the light and h is Planck's constant.
photosynthesis A series of reactions in a green plant that combines carbon dioxide with water to form carbohydrate and oxygen.
phototrophs (See also **chemotrophs**.) Organisms that can carry out photosynthesis and therefore can use sunlight to supply their Gibbs free energy needs.
physical changes Changes in the physical properties of a substance, such as the transformation of a solid to a liquid.
physical properties Properties (for example, melting point or density) that can be observed and measured without changing the composition of a substance.
physical quantity A number and units that report the result of a measurement.
physical state Describes whether a sample is a solid, a liquid, or a gas.
pi bond (π bond) A bond formed by the sideways overlap of parallel *p* orbitals.
pK_w The negative logarithm of the ionization constant for water. At 25 °C, $pK_w = \text{pH} + \text{pOH} = 14.00$.
Planck's constant (h) The proportionality constant that relates the energy of a photon to its frequency. The value of h is 6.626×10^{-34} Js.
plasma A state of matter consisting of unbound nuclei and electrons.
plastic A polymeric material that has a soft or liquid state in which it can be molded or otherwise shaped. (See also **thermoplastic** and **thermosetting plastic**.)
pOH The negative logarithm of the concentration of hydroxide ions. At 25 °C, $\text{pOH} = pK_w - \text{pH} = 14.00 - \text{pH}$.
polar covalent bond A covalent bond between atoms with different electronegativities; bonding electrons are shared unequally between the atoms.

polar molecule A molecule that is polar because it has polar bonds arranged so that electron density is concentrated at one end of the molecule.
polarization The induction of a temporary dipole in a molecule or atom by shifting of electron distribution.
polluted water Water that is unsuitable for an intended use, such as drinking, washing, irrigation, or industrial use.
polyamide A polymer in which the monomer units are connected by amide bonds.
polyatomic ion An ion consisting of more than one atom.
polydentate ligand A ligand that can form two or more coordinate covalent bonds to the same metal ion.
polyester A polymer in which the monomer units are connected by ester bonds.
polymer A large molecule composed of many smaller repeating units, usually arranged in a chain-like structure.
polypeptide A polymer composed of 20 to 50 amino acid residues joined by peptide linkages (amide linkages).
polyprotic acids Acids that can donate more than one hydrogen ion per molecule.
polysaccharides Carbohydrates that consist of many monosaccharide units.
polyunsaturated acid A carboxylic acid containing two or more carbon-carbon double bonds; commonly refers to a fatty acid.
positron A nuclear particle having the same mass as an electron, but a positive charge.
potential energy Energy that an object has because of its position.
pound per square inch (psi) A force of 1 lb on a 1 in^2 area.
power The energy transferred per unit time. The unit of power is the watt, W.
ppb See **parts per billion**.
ppm See **parts per million**.
ppt See **parts per trillion**.
precipitate An insoluble solid product of an exchange reaction in aqueous solution.
precipitation reaction A reaction in aqueous solutions where at least one insoluble product, a precipitate, is formed.
precision The degree to which several determinations of the same measurement give the same value.
pressure The force exerted on an object divided by the area over which the force is exerted; force per unit area.
primary battery A voltaic cell (or battery of cells) in which the oxidation and reduction half-reactions cannot easily be reversed to restore the cell to its original state.
primary structure (protein) The sequence of amino acids along the polymer chain in a protein molecule.
principal energy level An energy level containing orbitals with the same quantum number (n = 1, 2, 3 . . .).
principal quantum number (n) An integer assigned to each of the allowed main electron energy levels in an atom.
products One or more substances formed as a result of a chemical reaction.
product-favored Describes a reaction in which, when the reaction appears to be over, products predominate over reactants.
product-favored system A system in which, when a reaction appears to be over, products predominate over reactants.
propagation A step in the process of forming a polymer in which the polymer chain is lengthened.
proportionality factor (conversion factor) A relationship between two measurement units derived from the proportionality of one quantity to another (for example, density is the conversion factor between mass and volume).
proton A positively charged subatomic particle found in the nucleus. A hydrogen atom that has lost its single electron (see **hydrogen ion**).
purified Refers to a substance that has been separated from all other substances and therefore displays characteristic properties.
pyrometallurgy The extraction of a metal from its ore using chemical reactions carried out at high temperatures.

Q

Q See **reaction quotient**; see **ion product**.
qualitative In observations, nonnumerical experimental information, such as a description of color or texture.
quantitative Numerical information, such as the mass or volume of a substance, expressed in appropriate units.
quantum The smallest possible unit of a distinct quantity; for example, the smallest possible unit of energy for electromagnetic radiation of a given frequency.
quantum number A number that defines a quantum state or energy level. For electrons in atoms there are four quantum numbers: n, ℓ, m_ℓ, and m_s.
quantum theory The theory that energy comes in very small packets (quanta); this is analogous to matter occurring in very small particles—atoms.

R

R See **roentgen**.
R See **ideal gas constant**.
racemic mixture A mixture of equal amounts of enantiomers of a chiral compound.
rad A unit of radioactivity; a measure of the energy of radiation absorbed by a substance, 1.00×10^{-2} J/kg.
radial distribution plot A graph showing the probability of finding an electron as a function of distance from the nucleus of an atom.
radioactive series A series of nuclear reactions in which a radioactive isotope undergoes successive nuclear transformations resulting ultimately in a stable, nonradioactive isotope.
radioactivity The spontaneous emission of energy and/or subatomic particles by unstable atomic nuclei; the energy or particles so emitted.
Raoult's law A mathematical expression for the vapor pressure of the solvent in a solution; $P_1 = X_1P_1^0$.
rate The change in some measurable quantity per unit time.
rate constant (k) A proportionality constant relating reaction rate and concentrations of reactants and other species that affect the rate of a specific reaction.
rate law A mathematical equation that summarizes the relationship between concentrations and reaction rate. Also called rate equation.
rate-limiting step The slowest step in a reaction mechanism.
reactants One or more substances that are initially present and undergo change in a chemical reaction.
reactant-favored Describes a reaction in which, when the reaction appears to be over, reactants predominate over products.
reactant-favored system A system in which, when a reaction appears to be over, reactants predominate over products.
reaction intermediate An atom, molecule, or ion produced in one step and used in a later step in a reaction mechanism; does not appear in the equation for the overall reaction.
reaction mechanism A sequence of unimolecular and bimolecular elementary reactions by which an overall reaction may occur.
reaction quotient (Q) A value found from an expression with the same mathematical form as the equilibrium constant expression but with the actual concentrations in a mixture not at equilibrium.
reaction rate The change in concentration of a reactant or product per unit time.
redox Abbreviation of *red*uction/*ox*idation.
redox reaction See **oxidation-reduction reaction**.
reduced The result when an atom, molecule, or ion gains one or more electrons.
reducing agent The atom, molecule, or ion that donates electron(s) and is oxidized in an oxidation-reduction reaction.
reduction The gain of electrons by an atom, ion, or molecule, leading to a decrease in its oxidation number.
reformulated gasolines Oxygenated gasolines with lower volatility and containing a lower percentage of aromatic hydrocarbons than regular gasoline.
rem A unit of radioactivity; 1 rem has the physiological effect of 1 roentgen of radiation.
replication The copying of DNA during regular cell division.
representative elements Elements in which s and p subshells are being filled with electrons. In the periodic table, elements in Groups 1A and 2A on the left and Groups 3A through 8A at the right, or elements in Groups 1, 2, and 13–18.
resonance hybrid The actual structure of a molecule that can be represented by more than one Lewis structure.
resonance structures The possible structures of a molecule for which more than one Lewis structure can be written, differing by the arrangement of electrons but having the same arrangement of atomic nuclei.
reverse osmosis Application of pressure greater than the osmotic pressure to cause solvent to flow through a semipermeable membrane from a

concentrated solution to a solution of lower solute concentration.
reversible process A process for which a very small change in conditions will cause a reversal in direction.
roentgen (R) A unit of radioactivity; 1 R corresponds to deposition of 93.3×10^{-7} J per gram of tissue.

S

***s*-block elements** Main-group elements in Groups 1A and 2A whose valence electrons are *s* electrons.
salt An ionic compound whose cation comes from a base and whose anion comes from an acid.
salt bridge A device for maintaining balance of ion charges in the compartments of an electrochemical cell.
saponification The hydrolysis of a triglyceride (a fat or oil) by reaction with NaOH to give sodium salts that are soaps.
saturated fats Fats (or oils) that contain only carbon-carbon single bonds in their hydrocarbon chains.
saturated hydrocarbon Hydrocarbon in which carbon atoms are bonded to the maximum number of hydrogen atoms.
saturated solution A solution in which the concentration of solute is the concentration that would be in equilibrium with undissolved solute at a given temperature.
scanning tunneling microscope An analytical instrument that produces images of individual atoms or molecules on a surface.
screening effect Reduction of the effective attraction between nucleus and valence electrons as a result of repulsion of the outer valence electrons by electrons in inner shells.
second law of thermodynamics The total entropy of the universe (the system and surroundings) is continually increasing. In any product-favored system, the entropy of the universe is greater after a reaction than it was before.
secondary battery A voltaic cell (or battery of cells) in which the oxidation and reduction half-reactions can be reversed to restore the cell to its original state after discharge.
secondary structure (protein) Regular repeating patterns of molecular structure in proteins, such as α-helix, β-sheet.
semiconductor Material with a narrow energy gap between the valence band and the conduction band; a conductor when an electric field or a higher temperature is applied.
semipermeable membrane A thin layer of material through which only certain kinds of molecules can pass.
shell A collection of orbitals with the same value of the principal quantum number, *n*.
shifting an equilibrium Changing the conditions of an equilibrium system so that the system is no longer at equilibrium and there is a net reaction in either the forward or reverse direction until equilibrium is reestablished.
SI base units A set of units from which all other SI units are derived.
SI prefixes A letter that represents a power of 10 that is multiplied times a base unit to get a new unit that is larger or smaller than the base unit.
SI units The International System of Units (Système International), the officially recognized measurement system used by scientists throughout the world.
sievert (Sv) The SI unit of effective dose of absorbed radiation, 1 Sv = 100 rem.
sigma bond (σ bond) A bond formed by head-to-head orbital overlap along the bond axis.
significant figures (digits) All digits in an experimentally determined numerical value that correspond with the precision of the experimental measurement; if a measurement is precise to between 1 part in 100 and 1 part in 1000, then the result has three significant figures, such as 3.25×10^2.
simple sugars Monosaccharides and disaccharides.
single covalent bond A bond formed by sharing one pair of electrons between the same two atoms.
smog A mixture of smoke (particulate matter), fog (an aerosol), and other substances that degrade air quality.
solar cell A device that converts solar photons into electricity; based on doped silicon.
solid A state of matter in which a substance has a definite shape and volume.
solubility The maximum amount of solute that will dissolve in a given volume of solvent at a given temperature when pure solute is in equilibrium with the solution.
solubility product constant (K_{sp}) An equilibrium constant that is the product of concentrations of ions in a solution in equilibrium with a solid ionic compound.
solubility product constant expression A mathematical expression in which concentrations of a cation and an anion are each raised to a power equal to the ion's coefficient in the balanced chemical equation for the solubility equilibrium.
solubility rules General guidelines for predicting the water solubilities of ionic compounds based on the ions they contain. The rules are given in Table 3.1, Sec. 3-3c.
solute The material dissolved in a solution.
solution A homogeneous mixture of two or more substances in a single phase.
solution enthalpy The enthalpy change when a solute dissolves to form a solution; the quantity of energy transferred by heating when a solution forms at constant *T* and *P*.
solvation The process in which solvent molecules surround and stabilize ions in solution.
solvent The medium in which a solute is dissolved to form a solution.
***sp* hybrid orbitals** Orbitals of the same atom formed by the combination of one *s* orbital and one *p* orbital.
***sp^2* hybrid orbitals** Orbitals of the same atom formed by the combination of one *s* orbital and two *p* orbitals.
***sp^3* hybrid orbitals** Orbitals of the same atom formed by the combination of one *s* orbital and three *p* orbitals.
specific heat capacity The quantity of energy that must be transferred to 1 g of a substance to increase its temperature by 1 °C.
spectator ion An ion that is present in a solution in which a reaction takes place, but is not involved in the net process.
spectrochemical series A list of ligands in the order of their crystal-field splitting energy.
spectroscopy Use of electromagnetic radiation to study the nature of matter.
spectrum A plot of the intensity of light (photons per unit time) as a function of the wavelength or frequency of light.
spin quantum number (m_s) The quantum number that indicates direction of spin of an electron; it can be either $+\frac{1}{2}$ or $-\frac{1}{2}$.
standard atmosphere (atm) See **atmosphere (atm)**.
standard-state conditions These are 1 bar pressure for all gases, 1-M concentration for all solutes, at a specified temperature.
standard electrode potential ($E°$) See **standard half-cell potential**.
standard enthalpy change The enthalpy change when a process occurs with reactants and products all in their standard states.
standard equilibrium constant ($K°$) An equilibrium constant in which each concentration (or pressure) is divided by the standard-state concentration (or pressure); if concentrations are expressed in moles per liter (or pressures in bars) then the concentration (or pressure) equilibrium constant equals the standard equilibrium constant $\Delta G° = -RT \ln K°$.
standard formation enthalpy ($\Delta_f H°$) The standard reaction enthalpy for formation of one mole of a compound from its elements in their standard states.
standard formation Gibbs free energy ($\Delta_f G°$) The standard reaction Gibbs free energy for formation of one mole of a compound from its elements in their standard states.
standard half-cell potential ($E°$) A half-cell potential measured under standard conditions by comparison with a standard hydrogen electrode.
standard hydrogen electrode The electrode against which standard reduction potentials are measured, consisting of a platinum electrode at which 1-M hydrogen ion is reduced to hydrogen gas at 1 bar.
standard molar enthalpy of formation See **standard formation enthalpy**.
standard molar volume The volume occupied by exactly 1 mol of an ideal gas at standard temperature (0 °C) and pressure (1 atm), equal to 22.414 L.
standard potential ($E°$) Potential measured under standard-state conditions.
standard reaction enthalpy ($\Delta_r H°$) The enthalpy of pure, unmixed reaction products minus the enthalpy of pure, unmixed reactants at the standard pressure of 1 bar and a specified temperature.
standard reaction Gibbs free energy ($\Delta_r G°$) The Gibbs free energy of pure, unmixed reaction products minus the Gibbs free energy of pure, unmixed reactants at the standard pressure of 1 bar and a specified temperature.

standard reduction potential ($E°$) The potential of an electrochemical cell when a given electrode as the cathode is paired with a standard hydrogen electrode as the anode under standard conditions.
standard solution A solution whose concentration is known accurately.
standard state The most stable form of a substance in the physical state in which it exists at 1 bar and a specified temperature.
standard temperature and pressure (STP) A temperature of 0 °C and a pressure of 1 atm.
standard voltages Electrochemical cell voltages measured under standard conditions.
state function A property whose value is invariably the same if a system is in the same state.
steel A material made from iron with most P, S, and Si impurities removed, a low carbon content, and possibly other alloying metals.
steric factor A factor in the expression for rate of reaction that reflects the fact that some three-dimensional orientations of colliding molecules are more likely to result in reaction than others.
stoichiometric coefficients The multiplying numbers assigned to the formulas in a chemical equation to balance the equation. Also called just coefficients.
stoichiometric factor A factor relating the amount of a reactant or product to the amount of another reactant or product. Also called mole ratio.
stoichiometry The study of the quantitative relations between amounts of reactants and products in chemical reactions.
STP See **standard temperature and pressure**.
stratosphere The region of the atmosphere approximately 12 to 50 km above sea level.
stratospheric ozone layer Region of maximum relative concentration of ozone in the stratosphere.
strong acid An acid that ionizes completely in aqueous solution.
strong base A base that ionizes completely in aqueous solution.
strong electrolyte An electrolyte that consists solely of ions in aqueous solution.
structural formula A formula written to show how atoms in a molecule or polyatomic ion are connected to each other.
structural isomers (constitutional isomers) Compounds with the same molecular formula that differ in the order in which their atoms are bonded.
sublimation Conversion of a solid directly to a gas with no formation of liquid.
sublimation enthalpy The enthalpy change when a solid transforms directly to the gas phase; the quantity of energy, at constant T and P, that must be transferred to cause a solid to vaporize.
subshell A group of atomic orbitals with the same n and ℓ quantum numbers.
substance Matter of a particular kind; each substance, when pure, has a well-defined composition and a set of characteristic properties that differ from the properties of any other substance.
substitution reaction A reaction in which one atom or group of atoms substitutes for another. For example, $I^- + CH_3Cl \rightarrow CH_3I + Cl^-$ involves substitution of I for Cl.
substrate A molecule or molecules whose reaction is catalyzed by an enzyme.
sugars Sweet-tasting carbohydrates. Classified as mono-, di-, or polysaccharides. Example: glucose, $C_6H_{12}O_6$.
superconductor A substance that, below some temperature, offers no resistance to the flow of electric current.
supercritical fluid A substance above its critical temperature and pressure; has density characteristic of a liquid, but flow properties of a gas.
supersaturated solution A solution that temporarily contains more solute per unit volume than a saturated solution at a given temperature.
surface tension The energy required to overcome the attractive forces between molecules at the surface of a liquid.
surfactant A compound that consists of molecules that have both a hydrophobic part and a hydrophilic part.
surroundings Everything that can exchange energy with a thermodynamic system.
Sv See **sievert**.
system In thermodynamics, that part of the universe that is singled out for observation and analysis. The region of primary concern.

T

temperature The physical property of matter that determines whether one object can heat another.
tertiary structure (protein) The overall three-dimensional folding of protein molecules.
tetrahedral Molecular geometry of four atoms or groups of atoms around a central atom with bond angles of 109.5°.
theoretical yield The maximum quantity of product theoretically obtainable from a given quantity of reactant in a chemical reaction.
theory A unifying principle that explains a body of facts and the laws based on them.
thermal equilibrium The condition of equal temperatures achieved between two samples of matter that are in contact.
thermochemical expression A balanced chemical equation, including specification of the states of matter of reactants and products, together with the corresponding value of the enthalpy change.
thermodynamic temperature scale The temperature scale on which the zero is the lowest possible temperature and the degree is the same size as a Celsius degree; also called the absolute or Kelvin scale.
thermodynamics The science of heat, work, and the transformations of each into the other.
thermoplastics Plastics that can be repeatedly softened by heating and hardened by cooling.
thermosetting plastics Polymers that melt upon initial heating and form cross-links so that they cannot be melted again without decomposition.
third law of thermodynamics A perfect crystal of any substance at 0 K has the lowest possible entropy.
titrant The solution being added from a buret to another solution during a titration.
titration A procedure whereby a substance in a standard solution reacts with a known stoichiometry with a substance whose concentration is to be determined.
titration curve A plot of the progress of a titration as a function of the volume of titrant added.
torr (Torr) A unit of pressure equivalent to 1 mmHg.
trace elements (See also **major minerals.**) The dietary minerals that are present in smaller concentrations than the major minerals, sometimes far smaller concentrations.
tracer A radioisotope used to track the pathway of a chemical reaction, industrial process, or medical procedure.
***trans* isomer** The isomer in which two like substituents are on opposite sides of a carbon-carbon double bond, a ring of carbon atoms, or a coordination complex.
transition elements Elements that lie in rows 4 through 7 of the periodic table in which d or f subshells are being filled; comprising scandium through zinc, yttrium through cadmium, lanthanum through mercury, and actinium and elements of higher atomic number.
transition state A molecular structure corresponding to the maximum of a plot of energy versus reaction progress; also known as the activated complex.
triangular bipyramid Molecular geometry of five groups around a central atom that consists of two triangular pyramids with a common face; three groups are in equatorial positions and two are in axial positions.
triangular planar Molecular geometry of three groups at the corners of an equilateral triangle around a central atom at the center of the triangle.
triglycerides Esters in which a glycerol molecule is joined with three fatty acid molecules.
triple bond A bond formed by sharing three pairs of electrons between two atoms.
triple point The point on a temperature/pressure phase diagram of a substance where solid, liquid, and gas phases are all in equilibrium.
troposphere The lowest region of the atmosphere, extending from Earth's surface to an altitude of about 12 km.
Tyndall effect Scattering of visible light by a colloid.

U

u See **unified atomic mass unit**.
uncertainty principle The statement that it is impossible to determine simultaneously the exact position and the exact momentum of an electron.
unified atomic mass unit (u) One-twelfth the mass of an atom of carbon-12. Also abbreviated amu.
unimolecular reaction A reaction in which the rearrangement of the structure of a single molecule produces the product molecule or molecules.
unit cell A small portion of a crystal lattice that can be replicated in each of three directions to generate the entire lattice.
unsaturated fats Fats (or oils) that contain one or more carbon-carbon double bonds in their hydrocarbon chains.

unsaturated hydrocarbon A hydrocarbon containing double or triple carbon-carbon bonds.
unsaturated solution A solution that contains a smaller concentration of solute than the concentration of a saturated solution at a given temperature.

V

V See **volt**.
valence band In a solid, an energy band (group of closely spaced orbitals) that contains valence electrons.
valence bond model A theoretical model that describes a covalent bond as resulting from an overlap of one orbital on each of the bonded atoms.
valence bond theory The description of covalent bonding as pairs of valence electrons localized in bonds between atoms.
valence electrons Electrons in an atom's highest occupied principal shell and in partially filled subshells of lower principal shells; electrons available to participate in bonding.
valence-shell electron-pair repulsion (VSEPR) model A simple model used to predict the shapes of molecules and polyatomic ions based on repulsions between bonding pairs and lone pairs around a central atom.
van der Waals equation An equation of state for gases that takes into account the volume occupied by molecules and noncovalent attractions between molecules:

$$\left(P + a\frac{n^2}{V^2}\right)(V - bn) = nRT$$

van't Hoff factor (i_{solute}) The factor that takes account of the number of particles per formula unit of solute when calculating colligative properties.
vaporization The change of a substance from the liquid to the gas phase.
vaporization enthalpy ($\Delta_{vap}H$) The enthalpy change when a substance vaporizes: the quantity of energy that must be transferred by heating when a liquid vaporizes at constant temperature and pressure.
vapor pressure The pressure of the vapor of a substance in equilibrium with its liquid or solid in a sealed container.
vapor pressure lowering The reduction in vapor pressure of a pure liquid when a solute dissolves in the liquid.
volatility The tendency of a liquid to vaporize.
volt (V) Electrical potential energy difference defined so that 1 joule (work) is performed when 1 coulomb (charge) moves through 1 volt (potential difference).
voltaic cell An electrochemical cell in which a product-favored oxidation-reduction reaction is used to produce an electric current.
VSEPR See **valence-shell electron-pair repulsion model**.

W

water of hydration The water molecules trapped within the crystal lattice of an ionic hydrate or coordinated to a metal ion in a crystal lattice or in solution.
wave functions Solutions to the Schrödinger wave equation that describe the behavior of an electron in an atom.
wavelength (λ) The distance between adjacent crests (or troughs) in a wave.
weak acid An acid that is only partially ionized in aqueous solution.
weak base A base that is only partially ionized in aqueous solution.
weak electrolyte An electrolyte that is only partially ionized in aqueous solution.
weight percent A mass fraction expressed as a percent by multiplying by 100%; used for elemental composition of a compound, composition of a solute in solution.
work (working) A mechanical process that transfers energy to or from an object.

X

X-ray crystallography The science of determining nanoscale crystal structure by measuring the diffraction of X-rays by a crystal.

Z

zone refining A purification process in which a molten zone is moved through a solid sample, causing the impurities to concentrate in the liquefied portion.
zwitterion A structure containing both a positive charge and a negative charge, commonly due to loss and gain of a hydrogen ion within the same molecule.

Index

Page numbers in **bold face** refer to key term definitions.
Page numbers in *italics* refer to images or diagrams.
Page numbers followed by a "t" indicate a table.
ES-2 and ES-3 refer to inside front cover.
ES-6 and ES-7 refer to inside back cover.
A.1 indicates appendix page 1.

A

B

C

D

F

G

J

K

L

M

Q

R

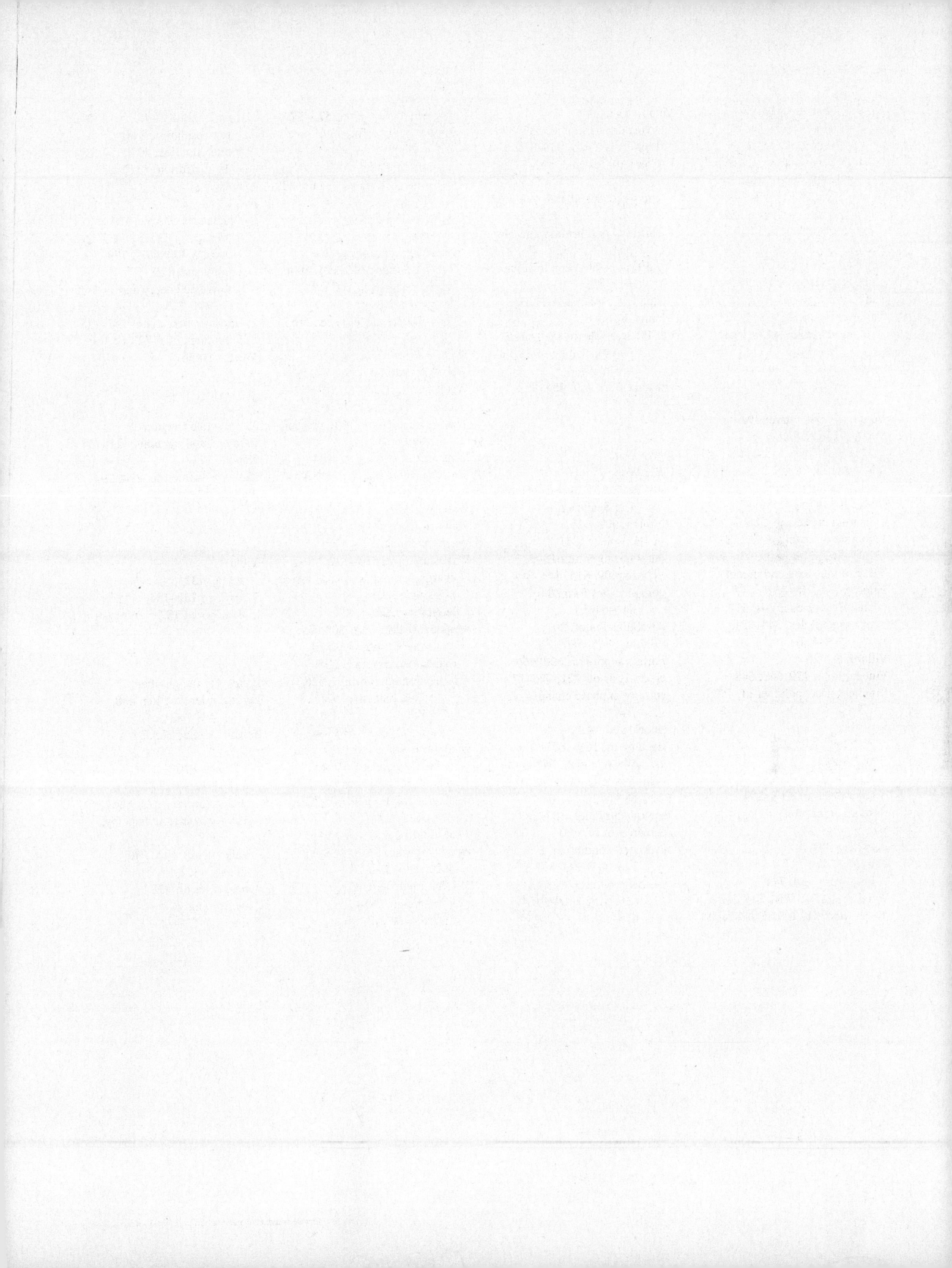

Useful Conversion Factors and Relationships

Length

SI unit: meter (m)

1 kilometer = 1000 meters
= 0.62137 mile
1 meter = 100 centimeters
1 centimeter = 10 millimeters
1 nanometer = 1×10^{-9} meter
1 picometer = 1×10^{-12} meter
1 inch = 2.54 centimeter (exactly)
1 Ångström = 1×10^{-10} meter

Mass

SI unit: kilogram (kg)

1 kilogram = 1000 grams
1 gram = 1000 milligrams
1 pound = 453.59237 grams = 16 ounces
1 ton = 2000 pounds
1 tonne = 1000 kilograms

Volume

SI unit: cubic meter (m^3)

1 liter (L) = 1×10^{-3} m^3
= 1 dm^3 = 1000 cm^3
= 1.056688 quarts
1 gallon = 4 quarts

Energy

SI unit: joule (J)

1 joule = 1 kg m^2/s^2
= 0.2390057 calorie
= 1 C × 1 V
1 calorie = 4.184 joules

Note: Numbers that reflect definitions, such as 1000 meters = 1 kilometer or 1 ton = 2000 pounds are exact, that is, they have an infinite number of significant figures

Pressure

SI unit: pascal (Pa)

1 pascal = 1 N/m^2
= 1 kg m^{-1} s^{-1}
1 atmosphere = 101.325 kilopascals
= 760. mmHg = 760. Torr
= 14.70 $lb/in.^2$
1 bar = 1×10^5 Pa

Temperature

SI unit: kelvin (K)

If T_K is the numeric value of the temperature in kelvins, t_C the numeric value of the temperature in °C, and t_F the numeric value of the temperature in °F, then

$$T_K = t_C + 273.15$$

$$t_C = \left(\frac{5}{9}\right)(t_F - 32)$$

$$t_F = \left(\frac{9}{5}\right)t_C + 32$$

Location of Useful Tables and Figures

Atomic, Ionic, and Molecular Properties

Atomic radii	Figures 5.25, 5.26
Bond energies	Table 6.2
Bond lengths	Table 6.1
Electron affinities	Table 5.7
Electron configurations	Table 5.4, Figure 5.19, App. D
Electronegativity	Figure 6.8
Ionic radii	Figure 5.27
Ionization energies	Table 5.6, Figures 5.29, 5.30
Molecular geometries	Tables 7.1, 7.2, 7.3
Polyatomic ions	Table 2.2
Properties of molecular and ionic compounds	Table 2.4
Properties of solids	Table 9.4

Thermodynamic Properties

Specific heat capacities	Table 4.1
Standard molar formation enthalpies	Table 4.2, App. J
Composition and caloric values of some foods	Table 4.3
Standard molar entropies	Table 16.1, App. J
Enthalpy, Gibbs free energy, entropy	App. J
Equilibrium constants	Table 12.1, Apps. G, H

Acids, Bases, and Salts

Acid-base properties of ions in solution	Table 14.5
Common acids and bases	Table 3.2
Common polyatomic ions	Table 2.2
Solubility guidelines for ionic compounds	Table 3.1
Common monoatomic ions	Figure 2.7
Amino acids	Table 10.8
Buffer systems	Table 15.1
Acid and base ionization constants	Table 14.2, App. G
Solubility product constants	Table 15.2, App. H

Other Useful Tables and Figures

Common oxidizing and reducing agents	Table 3.3
Common ligands	Table 20.4, Figure 20.10
Electromagnetic spectrum	Figure 5.2
Composition of dry air	Table 8.6
Standard half-cell potentials	Table 17.1, App. I
Density values	Table 1.1